W9-CIE-705

Plant Physiology

SECOND EDITION

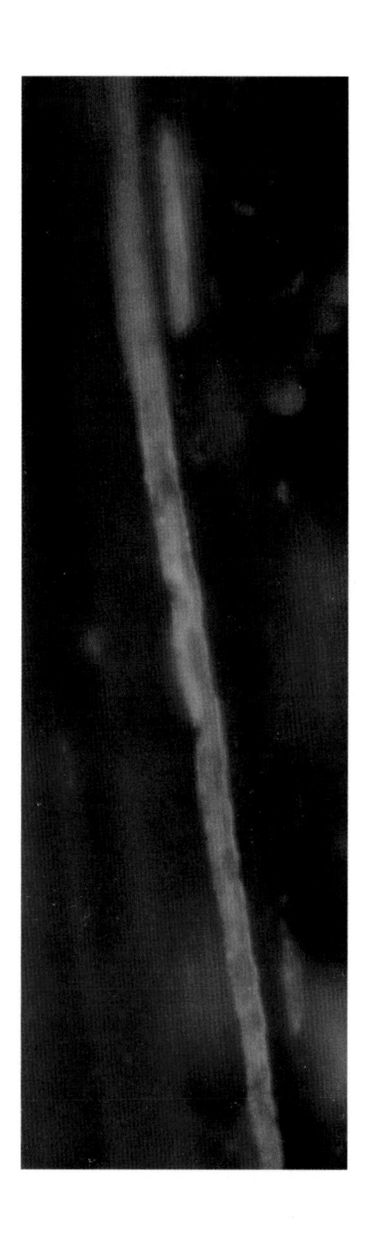

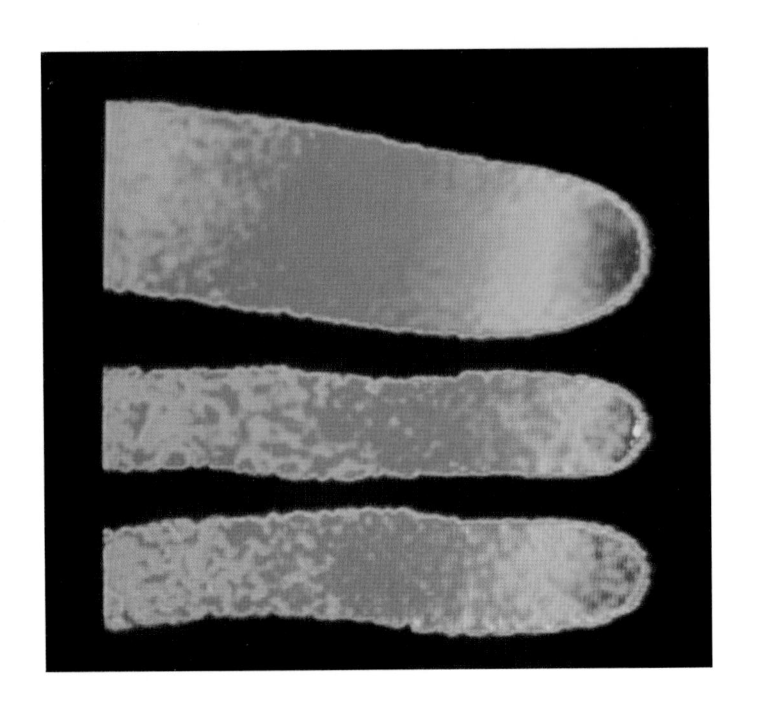

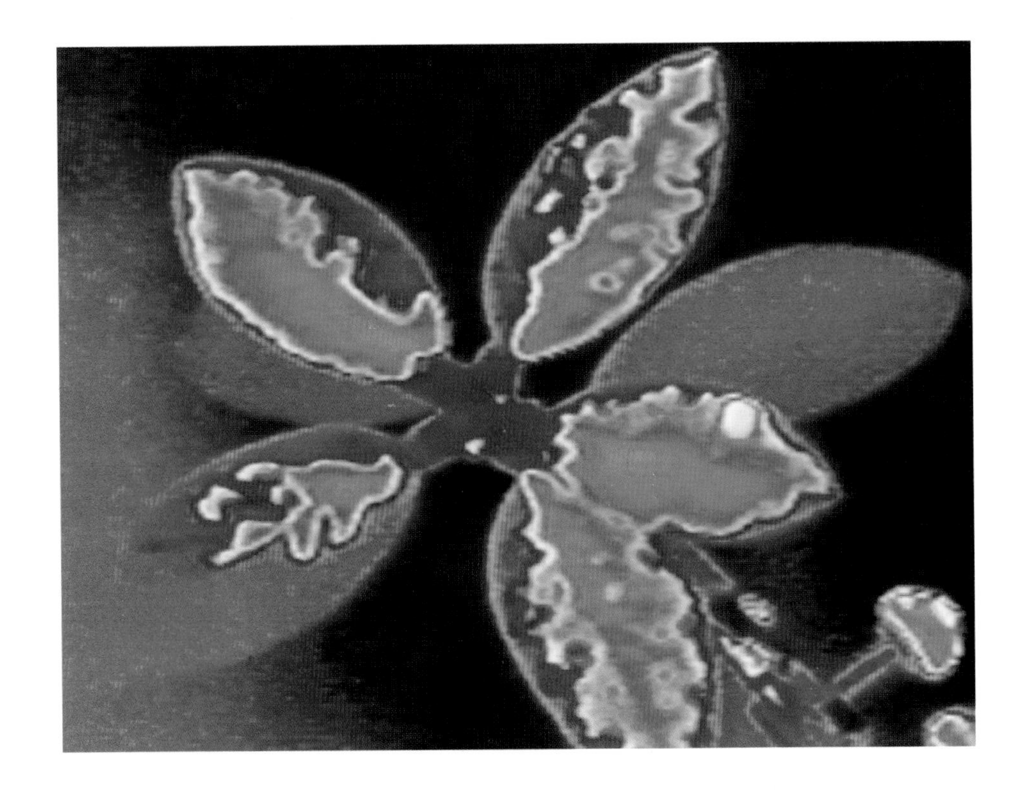

Plant Physiology

SECOND EDITION

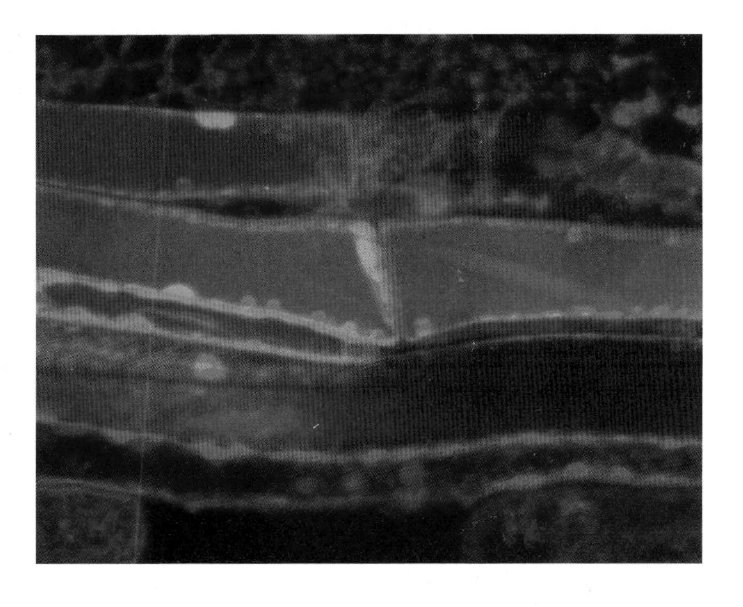

Lincoln Taiz
University of California, Santa Cruz

Eduardo Zeiger
University of California, Los Angeles

Sinauer Associates, Inc., Publishers
Sunderland, Massachusetts

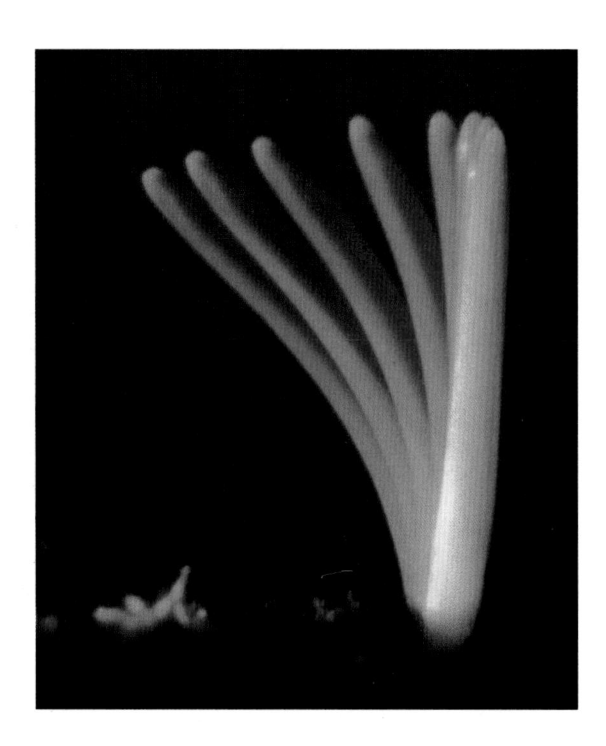

PLANT PHYSIOLOGY, *Second Edition*

For information or to order, address:
Sinauer Associates, Inc., P. O. Box 407
23 Plumtree Road, Sunderland, MA, 01375 U.S.A.
FAX: 413-549-1118
Internet: publish@sinauer.com;
http://www.sinauer.com

Library of Congress Cataloging-in-Publication Data

Taiz, Lincoln.
Plant physiology / Lincoln Taiz, Eduardo Zeiger.
—2nd ed.
p. cm.
Includes bibliographical references (p.)
and index.
ISBN 0-87893-831-1 (hardcover)
1. Plant physiology. I. Zeiger, Eduardo. II. Title.

QK711.2.T35 1998
571.2—dc21 98-34409
CIP

Printed in U. S. A. on recycled paper, 50% pre-consumer/10% post-consumer waste

5 4 3 2

FRONT COVER
Mexican goldpoppies blooming beside the vascular skeleton of a cholla cactus (*Opuntia* sp.).

BACK COVER
Close-up of the spines and flower of a strawberry hedgehog cactus. Front and back cover photographs by Willard Clay.

THE HALF-TITLE PAGE
Localization of the sucrose–H^+ symporter in the phloem. This micrograph shows a single companion cell from broad-leaved plantain (*Plantago major*) stained with two fluorescent dyes. One of the dyes (green) is (indirectly) linked to an antibody that is specific for the PmSUC2 sucrose–H^+ symporter. The second dye (blue) binds to DNA. Since the two dyes are found on a single phloem cell, which is always adjacent to a sieve element, the sucrose symporter is concluded to be located in the companion cell membrane in this species. (From Stadler et al. 1995, courtesy of N. Sauer. See page 266 in Chapter 10.)

THE FRONTISPIECE
TOP: Tip-growing pollen tubes exhibit a steep intracellular gradient of calcium ion concentration, with the highest levels at the growing tip. Pollen tubes from three species of plants were microinjected with a fluorescent calcium indicator dye to demonstrate this Ca^{2+} gradient, ranging from about 1 μM at the extreme tip to about 0.2 μM at the base. (Photo from Hepler 1997, courtesy of P. Hepler. See page 436 in Chapter 15.)

BOTTOM: The nucleation and propagation of ice in the stem, buds, and leaves of rhododendron (*Rhododendron* sp.). Emission of heat (colored blue and rust) during ice formation was detected by infrared thermography. Ice nucleation was observed to begin in the stem (lower right-hand corner), and then to spread to buds and leaves (blue and rust-colored areas). Even though a drop of a suspension of the ice-nucleating bacteria *Pseudomonas syringae* (yellow spot on the center leaf of the right-hand side) froze, initial nucleation and freezing started in the stem. (From Wisniewski et al. 1997, courtesy of M. Wisniewski, © American Society of Plant Physiologists, reprinted with permission. See page 740 in Chapter 25.)

THE TITLE PAGE
Phloem tissue of a bean doubly stained with a locally applied dye (red) and a translocated dye (green). Protein deposited against the plasma membrane and the sieve plate does not impede translocation. A crystalline P protein body is stained by the green dye. Phloem plastids are evenly distributed around the periphery of the cell. (From Knoblauch and van Bel 1998, courtesy of A. van Bel. See page 275 in Chapter 10.)

THIS PAGE
Time-lapse photograph of a corn coleoptile growing toward unilateral blue light. Unilateral blue light was given from the left. The consecutive exposures were made 30 minutes apart. Note the increasing angle of curvature as the coleoptile bends. (Photograph courtesy of M. A. Quiñones. See page 519 in Chapter 18.)

Preface

IT IS A PRIVILEGE TO PRESENT TO OUR READERS the second edition of *Plant Physiology*, which follows the first edition by seven years. The challenge of condensing, organizing, and synthesizing all the available knowledge in the field was daunting enough at the time of the first edition; the explosion of progress since 1991 makes these tasks even more demanding for the second edition.

The strength of the second edition, like that of the first, lies primarily with the outstanding group of scientists who have contributed their expertise and historical perspectives to this complex effort. They deserve all the credit for selecting the information that best represents the true conceptual advances in the field of plant physiology. Our task has been to ensure that all the topics were adequately covered and that the various topics were presented in a uniform style and level of difficulty. The editorial division of labor was as follows: E.Z. was responsible for Chapters 3, 4, 5, 7, 8, 9, 12, 18, and 25; L.T. was responsible for Chapters 1, 2, 6, 13, 14, 15, 16, 17, 19, 20, 21, 22, 23, and 24; Chapters 10, 11, and 25 have been edited by both of us. Several of the chapter authors from the first edition did not join us for this second edition, but their important contributions are still central to many of the chapters in the book. We wish to thank them for their previous efforts on behalf of the book: George W. Bates, Donald P. Briskin, Anthony L. Fink, Shimon Gepstein, Adrienne R. Hardham, Frank Harold, George H. Lorimer, John W. Radin, Stanley J. Roux, Thomas David Sharkey, Richard G. Stout, Daphne Vince-Prue, and Stephen M. Wolniak.

As in the first edition, much of the credit for integration and pedagogical style belongs to our developmental editor, James Funston. We feel fortunate to have engaged such a wise, creative, and endlessly patient advisor for both the first and second editions of *Plant Physiology*. A major improvement in the preparation of the second edition has been the convenience of using email and the internet, which enabled us to track down information much more efficiently than before, often from the comfort of our own home offices. The availability of email enabled us to rapidly check the accuracy of information with scientists spread all over the globe. Email also helped us to make pedagogical decisions, since we were able to contact large numbers of people to determine their preferences. Thus the number of colleagues around the world who have provided critical input into the preparation of this book is truly unprecedented.

Perhaps the most important innovation of the second edition is the new publisher, Sinauer Associates. We wish to extend our gratitude to Andy Sinauer for his initial faith in the book and for his continued encouragement during the ensuing months of its preparation; to our editor Nan Sinauer for her infinite patience, adroit decision making, and tireless shepherding of the manuscripts from first drafts to final copy; and to Suzette Stephens, Stephanie Hiebert, Chris Small, Janice Holabird, and Jefferson Johnson for their significant contributions.

Last but not least, we wish to thank our departmental colleagues, postdoctoral fellows, and students for their precious help and for enduring the extended periods of "total immersion" that were often required to get the job done. Finally, L.T. wishes to thank his wife, Lee Taiz, whose faith in the project sometimes exceeded his own, and whose love and support sustained him whenever the road became bumpy.

Lincoln Taiz

Eduardo Zeiger

July 1998

The Authors

Lincoln Taiz is a Professor of Biology at the University of California at Santa Cruz. He received his Ph.D. in Botany from the University of California at Berkeley in 1971. Dr. Taiz's current research interests include the structure and regulation of vacuolar H^+-ATPases, mechanisms of heavy metal tolerance, and the roles of aminopeptidases in plant development.

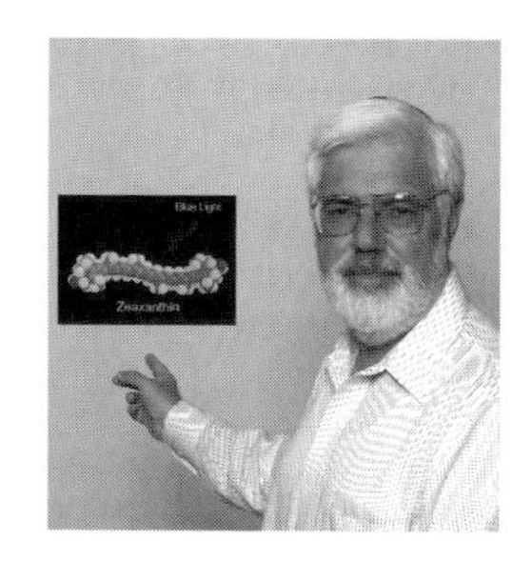

Eduardo Zeiger is a Professor of Biology at the University of California at Los Angeles. He received a Ph.D. in Plant Genetics at the University of California at Davis in 1970. His research interests include stomatal function, the sensory transduction of blue light responses, and the study of stomatal acclimations associated with increases in crop yields.

Principal Contributors

Richard Amasino is a Professor in the Department of Biochemistry at the University of Wisconsin–Madison. He received a Ph.D. in Biology from Indiana University in 1982 in the laboratory of Carlos Miller, where his interests in the induction of flowering were kindled. One of his research interests continues to be the mechanisms by which plants regulate the timing of flower initiation. (Chapter 24)

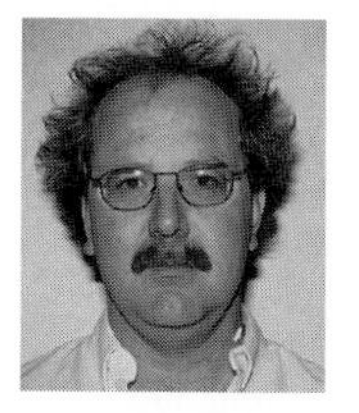

Paul Bernasconi is Director of Biochemistry at Novartis Crop Protection, Inc., in Research Triangle Park. His Ph.D. in Plant Biology and Protein Biochemistry was earned in 1987 at Lausanne University in Switzerland. His present work is focused on the biochemistry of herbicide, insecticide, and fungicide targets for crop protection. (Chapter 19)

Robert E. Blankenship is a Professor of Chemistry and Biochemistry at Arizona State University in Tempe. He received his Ph.D. in Chemistry from the University of California at Berkeley in 1975. His professional interests include mechanisms of energy and electron transfer in photosynthetic organisms, and the origin and early evolution of photosynthesis. (Chapter 7)

Arnold J. Bloom is a Professor in the Department of Vegetable Crops at the University of California at Davis. He received a Ph.D. in Biological Sciences at Stanford University in 1979. His research focuses on plant-nitrogen relationships, especially the differences in plant responses to ammonium and nitrate as nitrogen sources. (Chapters 5 and 12)

Ray A. Bressan is Professor of Plant Physiology at Purdue University. He received a Ph.D. in Plant Physiology from Colorado State University in 1976. Dr. Bressan has studied the basis of salinity and drought tolerance for several years. His recent interests have also turned toward the way plants defend themselves against insects and fungal disease. (Chapter 25)

John A. Browse is a Professor in the Institute of Biological Chemistry at Washington State University. He received his Ph.D. from the University of Aukland, New Zealand, in 1997. Dr. Browse's research interests include the biochemistry of lipid metabolism and the responses of plants to low temperatures. (Chapter 11)

Bob B. Buchanan is Professor of Plant and Microbial Biology at the University of California at Berkeley. After working on photosynthesis, Dr. Buchanan turned his attention to seed germination, where his findings have led to technologies that are currently under precommercial development. (Chapter 8)

Daniel J. Cosgrove is a Professor of Biology at the Pennsylvania State University at University Park. His Ph.D. in Biological Sciences was earned at Stanford University. Dr. Cosgrove's research interest is focused on plant growth, specifically the biochemical and molecular mechanisms governing cell enlargement and cell wall expansion. His research team discovered the cell wall loosening proteins called expansins and is currently studying the structure, function, and evolution of this gene family. (Chapters 3, 4, and 15)

Peter J. Davies is a Professor of Plant Physiology at Cornell University. He received his Ph.D. in Plant Physiology from the University of Reading in England. His present interests are using genotypes and polygene analysis to elucidate the role of hormones in potato tuberization, stem elongation, and plant senescence. He is the compiler and editor of the principal monograph on plant hormones and has also worked on the isolation of genes of gibberellin biosynthesis. (Chapter 20)

Malcolm C. Drew is Professor of Plant Physiology and Plant Biotechnology at Texas A & M University. His Ph.D. in Plant Nutrition was earned at the University of Oxford, England, in 1966. Dr. Drew's research interests in plant physiology include the physiology of plant roots and plant responses to environmental stress, particularly excess salinity, desiccation, oxygen shortage, and nutrient deficiency. His current work is on programmed cell death in roots and phytoremediation. (Chapter 25)

Susan Dunford is an Associate Professor of Biological Sciences at the University of Cincinnati. She received her Ph.D. from the University of Dayton in 1973 with a specialization in plant and cell physiology. Dr. Dunford's research interests include long-distance transport systems in plants, especially translocation in the phloem, and plant water relations. (Chapter 10)

Donald E. Fosket is Professor of Developmental and Cell Biology at the University of California at Irvine. He received his Ph.D. in Biology from the University of Idaho and subsequently did postdoctoral work at Brookhaven National Laboratory and at Harvard. Currently he is investigating the molecular mechanism controlling plant cell polarity at the Salk Institute in La Jolla, California. (Chapters 16, 21, and 24)

Jonathan Gershenzon is a Director of the newly established Max Planck Institute for Chemical Ecology, Jena, Germany. He received his Ph.D. from the University of Texas at Austin in 1984 and did postdoctoral work at Washington State University. His research focuses on the biosynthesis of plant secondary metabolites, and in establishing the roles of these compounds in plant–herbivore interactions. (Chapter 13)

Paul M. Hasegawa is Professor of Plant Physiology at Purdue University. He earned a Ph.D. in Plant Physiology at the University of California at Riverside. His research has focused on plant morphogenesis and the genetic transformation of plants. He has used his expertise in these areas to study many aspects of stress tolerance in plants, especially ion homeostasis. (Chapter 25)

Joseph Kieber is an Assistant Professor in the Department of Biological Sciences at the University of Illinois at Chicago. He earned his Ph.D. in Biology from the Massachusetts Institute of Technology in 1990. Dr. Kieber's research interests include the role of hormones in plant development, with a focus on the signaling pathways for ethylene and ctyokinin, as well as circuitry regulating ethylene biosynthesis. (Chapter 22)

Ronald J. Poole is a Professor of Biology at McGill University, Montreal. He received his Ph.D. from the University of Birmingham, England, in 1960. Dr. Poole's research interests are in ion transport in plant cells, including electrophysiology, biochemistry, and molecular biology of ion pumps and channels. (Chapter 6)

James N. Siedow is a Professor in the Department of Botany at Duke University. He earned his Ph.D. in Botany at Indiana University in 1972. Professor Siedow's long-term research interests have involved the study of redox reactions in biological systems, with a focus on the pathways of plant respiration. Included among these studies have been the characterization of the cyanide-resistant respiratory pathway and its regulation in plants. (Chapter 11)

Wendy K. Silk is Professor and Plant Scientist at the University of California at Davis. She received her Ph.D. from the University of California at Berkeley. Dr. Silk uses approaches from mathematics and physics to analyze the physiology of growth and plant responses to environmental variation. Research interests include morphogenesis, biomechanics, and stress physiology. (Chapter 16)

Jane Silverthorne is an Associate Professor in the Department of Biology at the University of California at Santa Cruz. She received her Ph.D. in Biology from the University of Warwick in the United Kingdom in 1980. Her research interests focus on the role of phytochrome in the regulation of molecular aspects of plant development. (Chapter 17)

Thomas C. Vogelmann is a Professor of Plant Physiology at the University of Wyoming. He received his Ph.D. from Syracuse University in 1980, specializing in plant development, and he subsequently did postdoctoral work in plant photobiology at the University of Lund, Sweden. His current research is focused on how plants interact with light. Specific research areas include plant tissue optics, leaf structure function related to photosynthesis and environmental stress, and plant adaptations to the environment. (Chapter 9)

Ricardo A. Wolosiuk is Professor in the Instituto de Investigaciones Bioquímicas at the University of Buenos Aires. He received his Ph.D. in Chemistry from the same university in 1974. Dr. Wolosiuk's research interests concern the modulation of chloroplast metabolism and the structure and function of plant proteins. (Chapter 8)

Other Contributors & Box Authors

Ton Bisseling (Material on nitrogen fixation, Chapter 12)
Agricultural University, Dreijenlaan

Joanne Chory (Box 17.4)
The Salk Insitute

Shimon Gepstein (Material on senescence, Chapter 16)
Technion/Israel Institute of Technology

Peter Hepler (Box 15.4)
University of Massachusetts

John Radin (Box 4.1)
USDA/Maryland

Alice Tarun (Box 22.1)
Plant Gene Expression Center

Eduardo Zeiger (Box 25.2)
University of California, Los Angeles

Reviewers

Philip N. Benfey
New York University

Wade Berry
University of California, Los Angeles

Mary A. Bisson
State University of New York, Buffalo

Anthony Bleeker
University of Wisconsin

Winslow R. Briggs
Carnegie Institute of Washington

Nicholas C. Carpita
Purdue University

Joe Chappell
University of Kentucky, Lexington

Parag R. Chitnis
Iowa State University

Robert Cleland
University of Washington

Eric E. Conn
University of California, Davis

Grant R. Cramer
University of Nevada, Reno

Thomas E. Elthon
University of Nebraska, Lincoln

Emanuel Epstein
University of California, Davis

Harold J. Evans
Oregon State University

Donald Geiger
University of Dayton

Simon Gilroy
Pennsylvania State University

Johann Peter Gogarten
University of Connecticut

Wieslaw Gruszecki
University of Lublin

Charles L. Guy
University of Florida

Candace H. Haigler
Texas Tech University

Tuan-hua David Ho
Washington University

Benjamin Horwitz
Israeli Institute of Technology

Steven C. Huber
USDA/North Carolina State University

Daniel W. Israel
USDA/North Carolina State University

Russell Jones
University of California, Berkeley

André Läuchli
Technische Hochschule, Darmstadt

Carl McDaniel
Rensselaer Polytechnic Institute

Anastasios Melis
University of California, Bekeley

Roy O. Morris
University of Missouri

Ann Oaks
University of Guelph

John Ohlrogge
Michigan State University

Thomas W. Okita
Washington State University

Neil Olszewski
University of Minnesota

Peter Quail
University of California, Berkeley

Douglas D. Randall
University of Missouri

Philip Rea
University of Pennsylvania

Karen Schumaker
University of Arizona

Thomas D. Sharkey
University of Wisconsin

Edgar Spalding
University of Wisconsin

Roger M. Spanswick
Cornell University

Ernst Steudle
Universitaet Bayreuth

Renee Sung
University of California, Berkeley

Heven Sze
University of Maryland

Gary Tallman
Willamette University

Athanasios Theologis
University of California, Berkeley

Elaine Tobin
University of California, Los Angeles

Robert Turgeon
Cornell University

Michael Venis
Horticulture Research International, Wellesbourne

Detlef Weigel
The Salk Institute for Biological Sciences

Contents in Brief

Table of Contents

6 Solute Transport 125

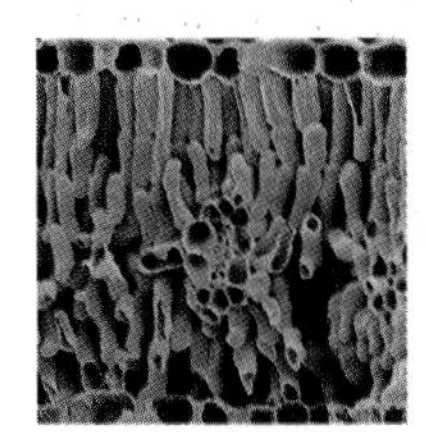

UNIT II
Biochemistry and Metabolism 153

7 Photosynthesis: The Light Reactions 155

8 Photosynthesis: Carbon Reactions 195

9 *Photosynthesis: Physiological and Ecological Considerations 227*

10 *Translocation in the Phloem 251*

11 *Respiration and Lipid Metabolism 287*

12 *Assimilation of Mineral Nutrients 323*

UNIT III

Growth and Development 377

14 Gene Expression and Signal Transduction 379

15 Cell Walls: Structure, Biogenesis, and Expansion 409

16 *Growth, Development, and Differentiation 445*

17 *Phytochrome 483*

18 *Blue-Light Responses: Stomatal Movements and Morphogenesis 517*

19 *Auxins 543*

20 *Gibberellins 591*

21 *Cytokinins 621*

24 *The Control of Flowering 691*

25 Stress Physiology 725

Overview of Essential Concepts

1 Plant and Cell Architecture

THE TERM "CELL" IS DERIVED FROM THE LATIN *CELLA*, meaning storeroom or chamber. It was first used in biology in 1665 by the English botanist Robert Hooke to describe the individual units he observed in cork tissue under a compound microscope. The cork "cells" Hooke observed were actually the cell walls surrounding the empty lumens of dead cells, but the term is an apt one because cells are the basic building blocks that define plant structure.

This book will emphasize the physiological and biochemical functions of plants, but it is important to recognize that these functions depend on structures, whether the process is gas exchange in the leaf, water conduction in the xylem, photosynthesis in the chloroplast, or ion transport across the plasma membrane. At every level, structure and function represent different frames of reference of a biological unity.

This chapter provides an overview of plant anatomy and cell structure and function, elements of which will be treated in greater detail in the chapters that follow.

Plant Life: Unifying Principles

The spectacular diversity of plant size and form is familiar to everyone. Plants range in size from less than 1 cm tall to greater than 100 m. Plant morphology, or shape, is also surprisingly diverse. At first glance, the tiny plant *Lemna* (duckweed), seems to have little in common with a giant saguaro cactus or a redwood tree. Yet regardless of their specific adaptations, all plants carry out fundamentally similar processes and are based on the same architectural plan. We can summarize the major design elements of plants as follows:

1. As Earth's primary producers, green plants are the ultimate solar collectors. They harvest the energy of sunlight by converting light energy to chemical energy, which they store in bonds formed when they synthesize carbohydrates from carbon dioxide and water.
2. Other than certain reproductive cells, plants are nonmotile. As a substitute for motility, they have evolved the ability to grow toward essential resources such as light, water, and mineral nutrients throughout their life span.
3. Terrestrial plants are structurally reinforced to support their mass as they grow toward sunlight against the pull of gravity.
4. Terrestrial plants lose water continuously by evaporation and have evolved mechanisms for avoiding desiccation.
5. Terrestrial plants have mechanisms for moving water and minerals from the soil to the sites of photosynthesis and growth, and mechanisms for moving the products of photosynthesis to nonphotosynthetic organs and tissues.

The Plant Kingdom

Since the time of Linnaeus (1707–1778), biologists have sought to classify organisms. At first the purpose was ease of identification ("artificial" classification schemes). Later, after Darwin, the goal of classification was to reflect evolutionary relationships ("natural" classification schemes). For the past 100 years or so, biologists have emphasized natural systems of classification and have attempted to define morphological criteria that best reflect evolutionary relationships. However, we now know that morphology, the form and structure of organisms, is the end product of the action of genes. All of the information needed to form a complete organism is encoded in its DNA sequences. DNA sequence analysis has thus provided evolutionary biologists with a powerful new tool for arriving at a truly natural classification system.

On the basis of phylogenetic analyses of highly conserved DNA sequences, living organisms have been divided into three major domains: Bacteria, Archaea, and Eucarya (Figure 1.1A) (Woese et al. 1990). The **Eucarya** include the **eukaryotes**, organisms whose cells contain a true nucleus. The **Bacteria**, or **eubacteria**, which include the cyanobacteria, lack a true nucleus and are therefore prokaryotic. The **Archaea**, or **archaebacteria**, are also prokaryotic, but they differ from the Bacteria: Besides their morphological and biochemical differences, they are often adapted to extreme environments, such as sulfur hot springs or saline ponds. Phylogenetic studies have suggested that the Archaea and Eucarya split after the Bacteria separated from the common ancestor (Gogarten et al. 1989). Thus Archaea and Eucarya represent sister groups. This closer relation between Archaea and Eucarya is reflected in their similar promoter structures and RNA polymerases, the presence of histones, and many other characteristics.

Before the three domains of life were recognized, the five-kingdom classification scheme of living organisms was widely accepted. According to this scheme, all living organisms were divided into five kingdoms: Monera, Protista, Fungi, Plantae, and Animalia. The **Monera** include all prokaryotic organisms: the eubacteria and the archaebacteria. Some biologists still adhere to this grouping, which is based on the dichotomy between prokaryotes and eukaryotes. However, the discovery that the archaebacteria and eukaryotes are sister groups suggests that the archaebacteria do not belong in the same kingdom with the eubacteria. Hence, biologists who prefer a natural classification scheme reject the Monera grouping, while those who prefer an artificial classification scheme (i.e., one based on criteria other than evolutionary relationships) continue to use it. Whereas in the natural system the distinction between prokaryote and eukaryote is ignored, in the artificial system this distinction becomes the primary criterion for dividing the organisms.

The "kingdom" **Protista** presents a similar dilemma. Protists are usually defined as unicellular eukaryotes, although the classification also includes multicellular algae with relatively simple structures. Proponents of a natural classification scheme divide the protists into three separate kingdoms according to their relatedness: Archezoa, Protozoa, and Chromista (Figure 1.1B). The **Archezoa** are extremely primitive eukaryotic cells, such as the intestinal parasite *Giardia*, which lacks both mitochondria and chloroplasts. *Giardia* is thought to have branched off before the endosymbiotic bacteria that gave rise to these organelles were incorporated in the eukaryotes (Figure 1.1C). The **Protozoa** include both photosynthetic unicellular cells that were formerly grouped with the algae, such as dinoflagellates and *Euglena*, and unicellular organisms that have lost their chloroplasts, such as the ciliates. The **Chromista**, which include the brown algae, water molds, and diatoms, are characterized by both multicellular and unicellular forms, many of which have chloroplasts that differ biochemically from the chloroplasts of green algae and plants.

Fungi were formerly classified with the plants. However, fungi are sufficiently different from plants and the other eukaryotes to be placed in a separate kingdom. Having lost their chloroplasts, fungi are heterotrophic; that is, they depend on other organisms for their food, and they satisfy their nutritional needs by absorbing inorganic ions and organic molecules from the external environment. Most fungal species are filamentous and possess cell walls made of chitin, the same substance that is found in insect exoskeletons.

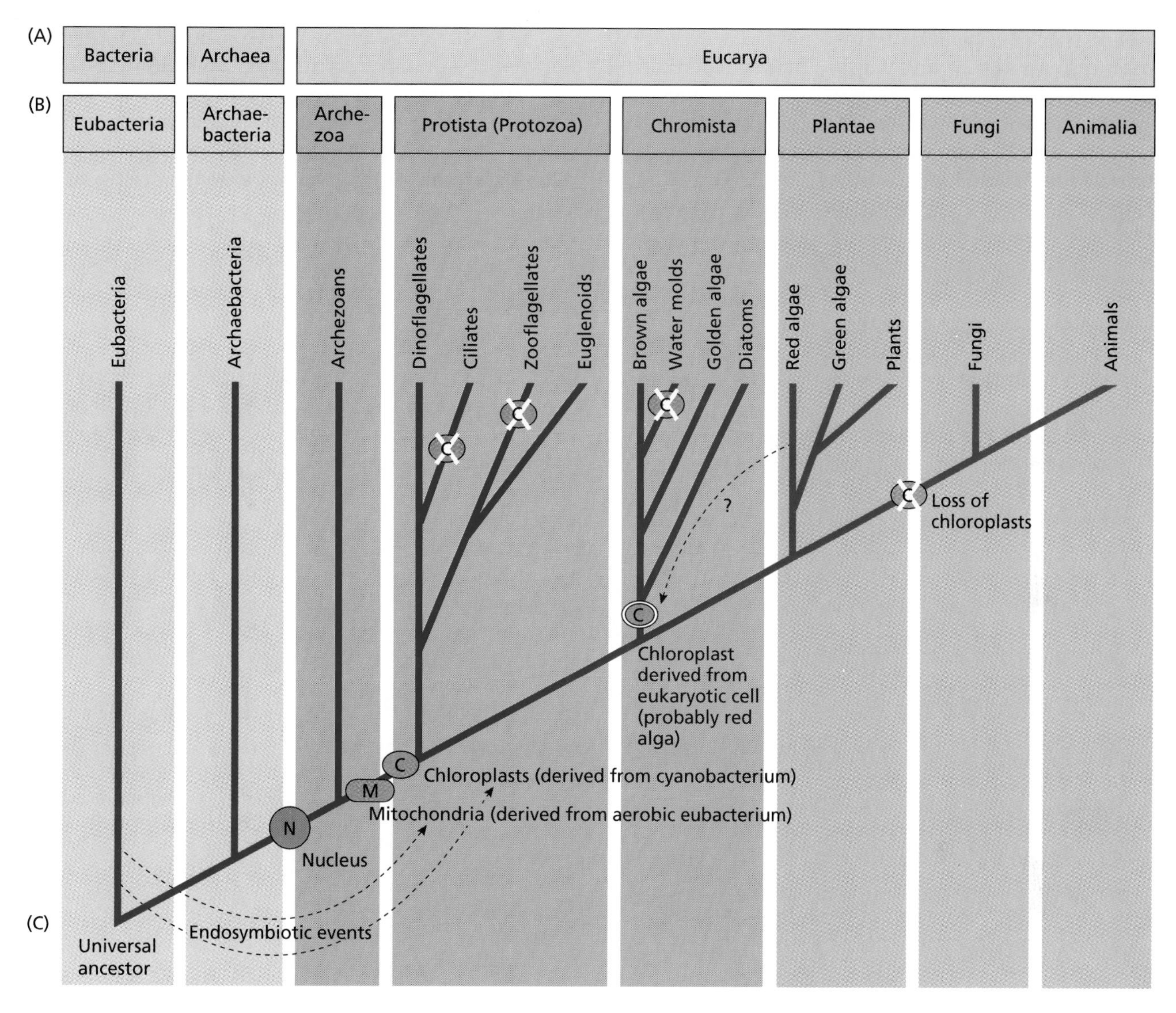

Figure 1.1 Natural classification scheme and phylogeny of living organisms. (A) The three domains of life. (B) Division of organisms into two prokaryotic kingdoms and six eukaryotic kingdoms. (C) Hypothetical phylogeny showing the origin of the major groups of eukaryotes. The appearance of the nucleus and the acquisition of mitochondria and chloroplasts as endosymbiotic organelles are indicated by the dashed lines and arrows. (After Campbell 1996.)

The kingdom **Plantae**, newly constituted on the basis of biochemical and molecular criteria, includes the red and green algae, as well as the terrestrial photosynthetic organisms traditionally regarded as "plants." From a plant physiologist's perspective, this grouping makes sense because the green algae have long been used as model systems for studying physiological processes ranging from photosynthesis to phytochrome. For example, the unicellular green alga *Chlorella* was used to solve the CO_2 fixation reactions of photosynthesis (see Chapter 8), and the filamentous green alga *Mougeotia* has been used to study light-regulated chloroplast movements (see Chapter 17). The terrestrial plants include the bryophytes (which reproduce mainly by spores), the vascular plants, and the seed plants.

Bryophytes are small (rarely more than 4 cm in height), very simple land plants, and the least abundant in terms of number of species and overall population. Bryophytes do not appear to be in the direct line of evolution leading to the vascular plants; rather they seem to constitute a separate minor branch. Bryophytes include mosses, liverworts, and hornworts. These small plants have life cycles that depend on water during the sexual phase. Water facilitates **fertilization**, the fusion of gametes to produce a diploid **zygote**, a feature also seen in the algal precursors of these plants. Bryophytes are like algae in other respects as well: They have neither true roots nor true leaves, they lack a vascular system, and they produce no hard tissues for structural support. The absence of these structures that are important for growth on land greatly restricts the potential size of bryophytes, which, unlike algae, are terrestrial rather than aquatic.

The **ferns** represent the largest group of spore-bearing **vascular plants**. In contrast to the bryophytes, ferns have true roots, leaves, and vascular tissues, and they produce hard tissues for support. These architectural features enable ferns to grow to the size of small trees. Although ferns are better adapted to the drying condi-

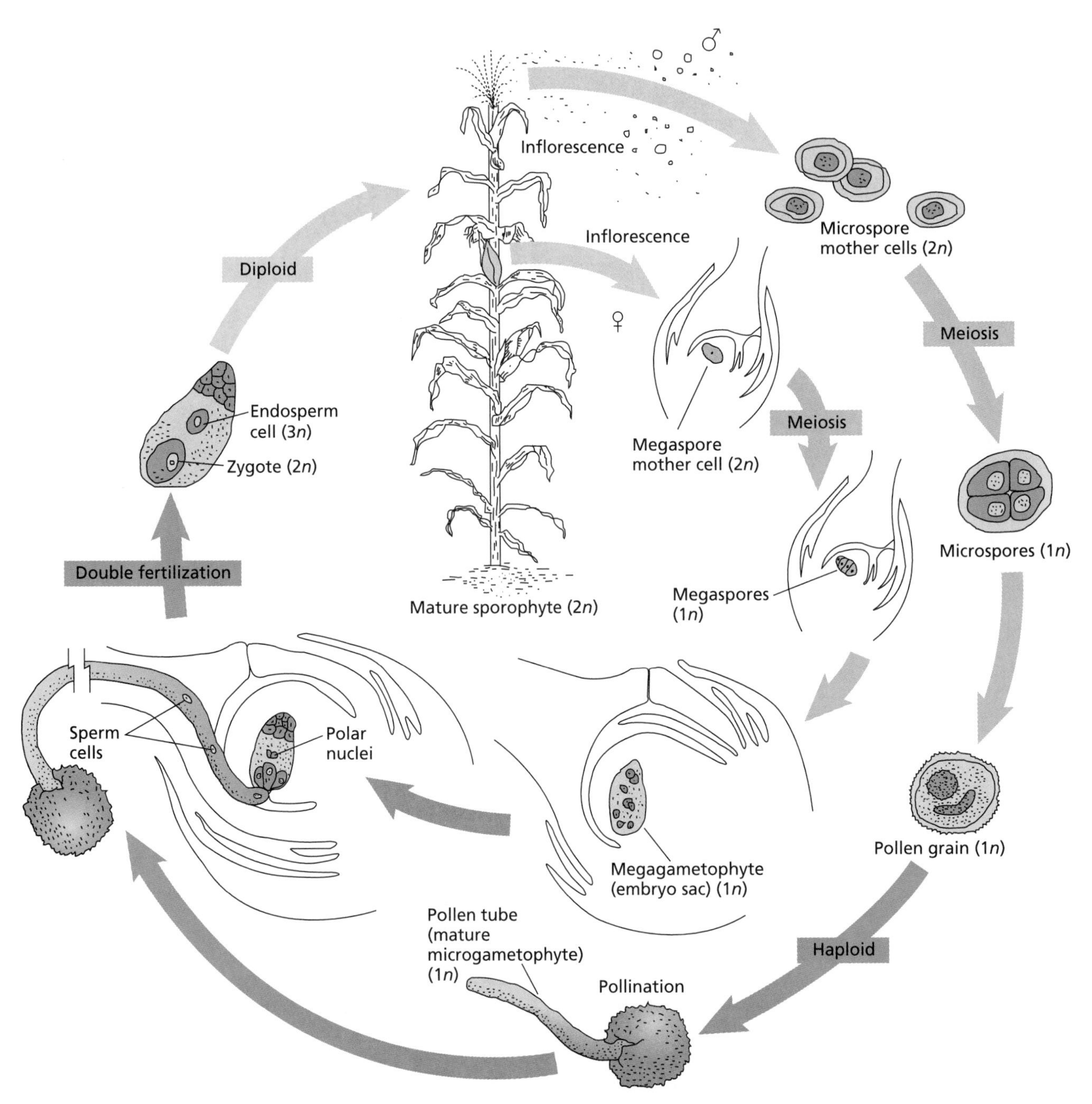

Figure 1.2 Life cycle of corn (*Zea mays*), a monocotyledonous angiosperm. The vegetative plant represents the diploid sporophyte generation. Meiosis occurs in the male and female flowers, represented by the tassels and ears, respectively. The haploid microspores (male spores) develop into pollen grains, and the single surviving haploid megaspore (female spore) divides mitotically to form the embryo sac (megagametophyte). Pollination leads to the formation of a pollen tube containing two sperm cells (the microgametophyte). Double fertilization results in the formation of the diploid zygote, the first stage of the new sporophyte generation, and the triploid endosperm cell.

tions of terrestrial life than bryophytes are, they still depend on water as a medium for the movement of sperm to the egg. This dependence on water during a critical stage of their life cycle restricts the ecological range of ferns to relatively moist habitats.

The most successful terrestrial plants are the **seed plants**. Seed plants have been able to adapt to an extraordinary range of habitats. The embryo, protected and nourished inside the seed, is able to survive in a dormant state during unfavorable growing conditions such as drought. Seed dispersal also facilitates the dissemination of the embryos away from the parent plant. Another important feature of seed plants is their mode of fertilization. Fertilization in seed plants is brought about by wind- or insect-mediated transfer of **pollen**, the gamete-producing structure of the male, to the sexual structure of the female, the **pistil**. Pollination is independent of external water, a distinct advantage in terrestrial environments. Many seed plants produce massive amounts of woody tissues, which enable them to grow to extraordinary heights. These features of seed plants have contributed to their success and account for their wide range.

There are two categories of seed plants: gymnosperms (from the Greek for "naked seed") and angiosperms (based on the Greek for "vessel seed," or seeds contained in a vessel). **Gymnosperms** are the less advanced type; about 700 species of gymnosperms are known. The largest group of gymnosperms is the conifers ("cone-bearers"), which include such commercially important forest trees as pine, fir, spruce, and redwood. Two types of cones are present: male cones, which produce pollen, and female cones, which bear ovules. The ovules are located on the surfaces of specialized structures called **cone scales**. After wind-mediated pollination, the sperm reaches the egg via a **pollen tube**, and the fertilized egg develops into an embryo. Upon maturation, the cone scales, which are appressed during early development, separate from each other, allowing the naked seeds to fall to the ground. **Angiosperms**, the more advanced type of seed plant, first became abundant during the Cretaceous period, about 100 million years ago. Today, they dominate the landscape, easily outcompeting their cousins, the gymnosperms. About 250,000 species are known, but many more remain to be characterized. A typical angiosperm life cycle is shown in Figure 1.2.

The major innovation of the angiosperms is the flower; hence they are referred to as *flowering plants.* There are other anatomical differences between angiosperms and gymnosperms, but none so crucial and far-reaching as the mode of reproduction. The flower consists of several leaflike structures attached to a specialized region of the stem called the **receptacle** (Figure 1.3). **Sepals** and **petals** are the most leaflike.

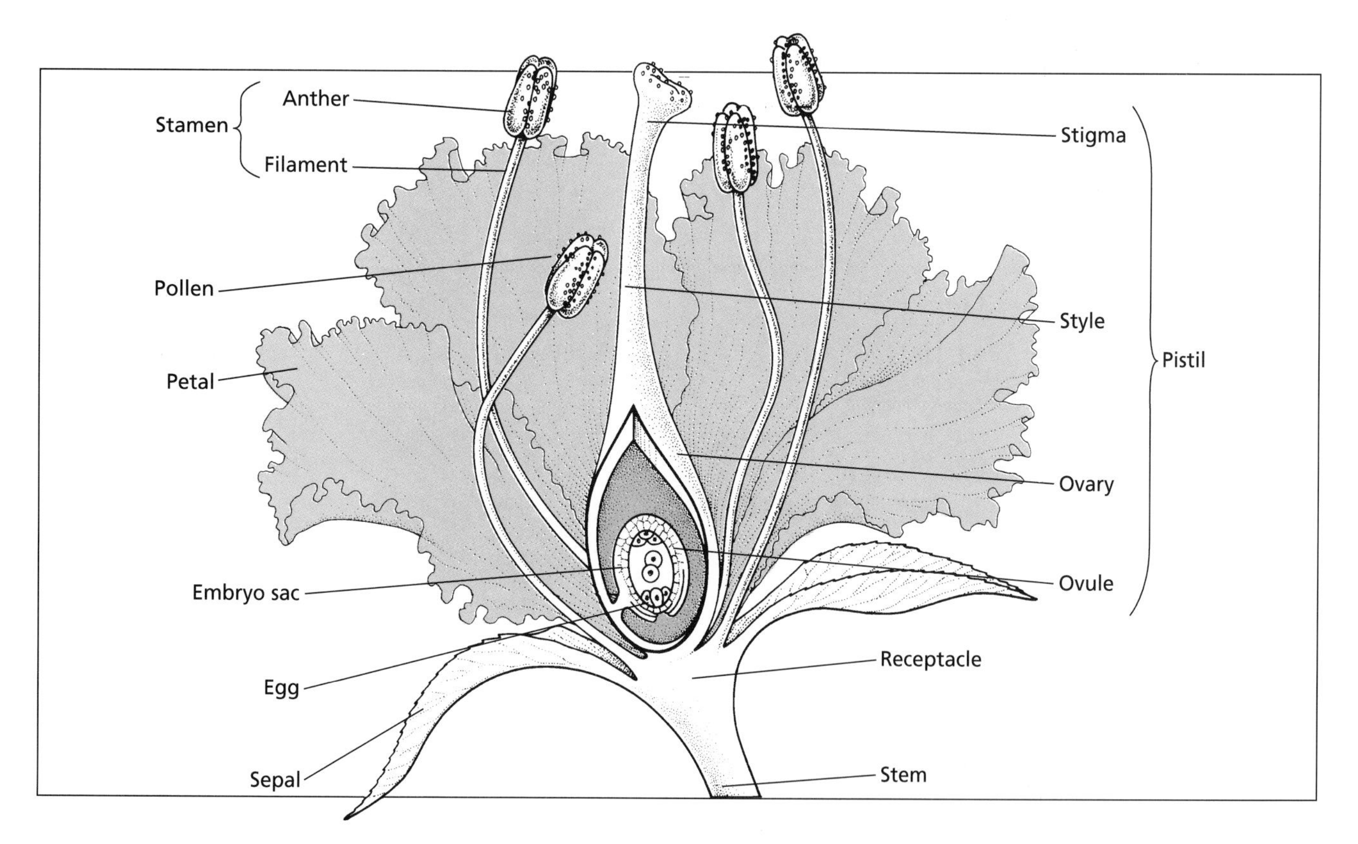

Figure 1.3 Schematic representation of an idealized flower of the angiosperms.

Petals have the primary function of attracting insects to serve as pollinators, accounting for their often showy and brightly colored appearance. The **stamen** is the male sexual structure, and the **pistil** is the female sexual structure. The pistil is composed of one or more united **carpels**; the pistil, or in some flowers a whorl of pistils, is sometimes referred to as the **gynoecium**. The stamen consists of a narrow stalk called the **filament** and a chambered structure called the **anther**. The anther contains tissue that gives rise to **pollen grains**. The pistil consists of the **stigma** (the tip where pollen lands during pollination), the **style** (an elongated structure), and the **ovary**. The ovary, the hollow basal portion of the pistil, completely encloses one or more **ovules**. Each ovule, in turn, contains an **embryo sac**, the structure that gives rise to the female gamete, the **egg**.

After landing on the stigma, the pollen grain germinates to form a long **pollen tube**, which penetrates the tissues of the style and ultimately enters the cavity of the ovary, which houses the ovule. Within the ovary, the pollen tube enters the ovule and deposits two haploid sperm cells in the embryo sac (see Figure 1.2). One

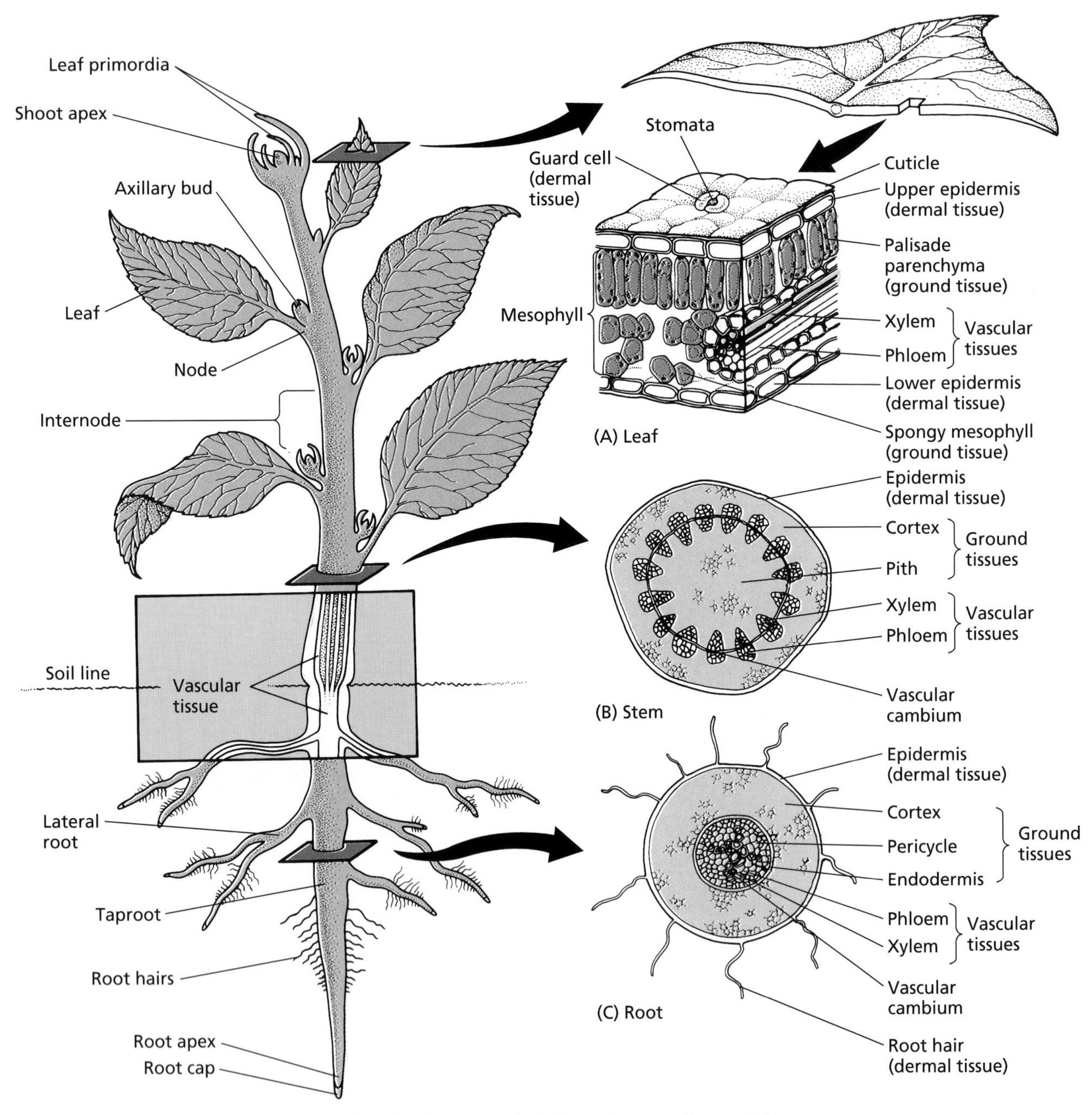

Figure 1.4 Schematic representation of the body of a typical dicot. Cross sections of (A) the leaf, (B) the stem, and (C) the root are also shown.

sperm cell fuses with the egg to produce the zygote; the other typically fuses with the two **polar nuclei** to produce a specialized storage tissue termed the **endosperm**, which provides nutrients to the growing embryo. Endosperm tissue also provides the bulk of the world's food supply in the form of cereal grains. As in conifers, in angiosperms the outer tissues of the ovule harden into a protective **seed coat**. Angiosperm seeds have a second layer of protective tissues, the fruit. The **fruit** consists of the ovary wall and, in some cases, receptacle tissue.

Angiosperms are divided into two major groups, **dicotyledons** (dicots) and **monocotyledons** (monocots). This distinction is based primarily on the number of **cotyledons**, or seed leaves, present in the embryo. In addition, the two groups differ with respect to other anatomical features, such as the arrangement of their vascular tissues, and their floral structure.

As the dominant plant group on Earth, and because of their great economic and agricultural importance, angiosperms have been studied much more intensively than other types of plants, and they are discussed extensively in this book. Plant physiologists have focused on a relatively small number of species that represent convenient experimental systems for the study of specific phenomena. Therefore, while we focus on these famous few, it is important to keep in mind the tremendous diversity of form and function that exists within the angiosperms, and the even greater diversity of form and function that is found within the plant kingdom as a whole.

The Plant: An Overview of Structure

Despite their apparent diversity, all seed plants have the same basic body plan (Figure 1.4). The vegetative body is composed of three organs: **leaf**, **stem**, and **root**. The primary function of a leaf is photosynthesis, that of the stem is support, and that of the root is anchorage and absorption of water and minerals. Leaves are attached to the stem at **nodes**, and the region of the stem between two nodes is termed the **internode**. The stem together with its leaves is commonly referred to as the **shoot**.

A fundamental difference between plants and animals is that each plant cell is surrounded by a rigid **cell wall**. In animals, embryonic cells can migrate from one location to another, resulting in the development of tissues and organs containing cells that originated in different parts of the organism. In plants, each walled cell and its neighbor are cemented together by a **middle lamella**. Plant cells are like bricks held together with mortar. This architecture prevents any cell migration, and as a consequence, plant development depends solely on the pattern of cell division and cell enlargement.

Plant cells have two types of walls: primary and secondary (Figure 1.5). **Primary cell walls** are typically thin (less than 1 μm) and are characteristic of young, growing cells. **Secondary cell walls** are thicker and stronger than primary walls and are deposited when most cell enlargement has ended. Secondary cell walls owe their strength and toughness to **lignin**, a brittle, gluelike material (see Chapters 13 and 15). The evolution of lignified secondary cell walls provided plants with the structural reinforcement necessary to grow vertically above the soil and to colonize the land. The bryophytes, which lack lignified cell walls, are unable to grow more than a few centimeters above the ground.

New Cells Are Produced by Dividing Tissues Called Meristems

Plant growth is concentrated in localized regions of cell division called **meristems**. Nearly all nuclear divisions (mitosis) and cell divisions (cytokinesis) occur in these meristematic regions. In a young plant, the most active meristems are called **apical meristems**; they are located at the tips of the stem and the root (Figure 1.6). At the nodes, **axillary buds** contain the apical meristems for branch shoots. Lateral roots arise from the **pericycle**, an internal meristematic tissue (see Figure 1.4C). Proximal to (i.e., next to) and overlapping the meristematic regions are zones of cell elongation in which cells increase dramatically in length and width. Cells usually differentiate into specialized types after they elongate.

The phase of plant development that gives rise to new organs and to the basic plant form is called **primary growth**. Primary growth results from the activity of apical meristems, in which cell division is followed by pro-

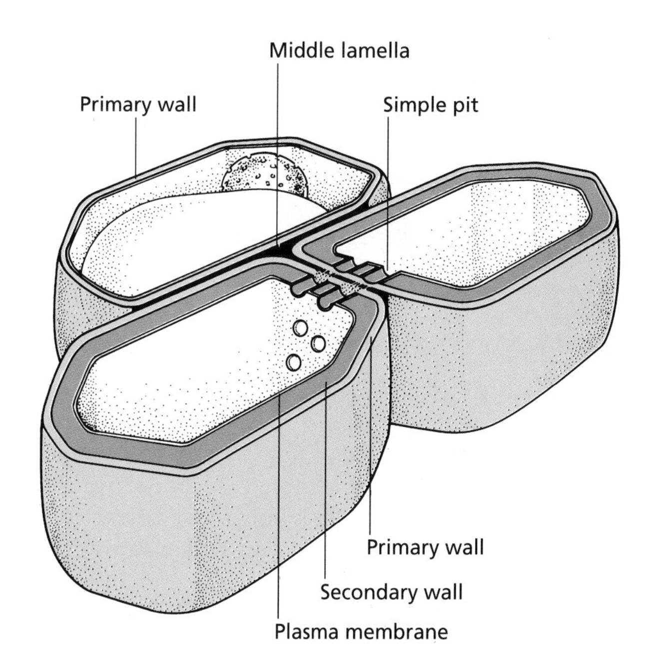

Figure 1.5 Schematic representation of primary and secondary cell walls and their relationship to the rest of the cell.

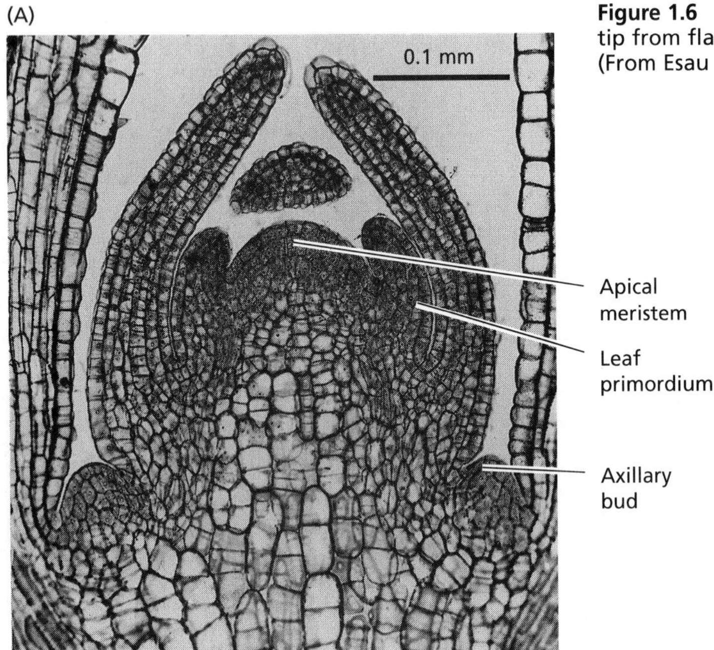

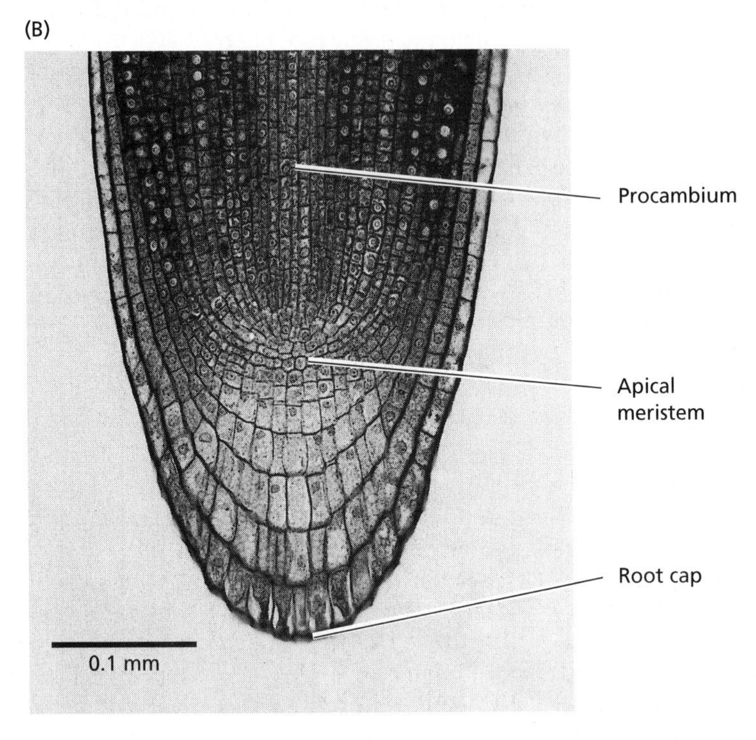

Figure 1.6 Longitudinal sections of (A) a shoot tip and (B) a root tip from flax (*Linum usitatissimum*), showing the apical meristems. (From Esau 1960, courtesy of R. Evert.)

gressive cell enlargement, typically elongation. After elongation in a given region is complete, **secondary growth** may occur. Secondary growth involves two lateral meristems, the **vascular cambium** and the **cork cambium**.

Plants Are Composed of Three Major Tissue Systems

Three major tissue systems are found in all plant organs: dermal tissue, ground tissue, and vascular tissue (see Figure 1.4).

Dermal. The **epidermis** is the **dermal tissue** of young plants undergoing primary growth (Figure 1.7A). It is generally composed of specialized, flattened polygonal cells that occur on all plant surfaces. Shoot surfaces are usually coated with a waxy **cuticle** to prevent water loss and are often covered with hairs, or trichomes, which are epidermal cell extensions. Pairs of specialized epidermal cells, the **guard cells**, are found surrounding microscopic pores in all leaves (see Figure 1.4A). The guard cells and pores are called **stomata** (singular stoma), and they permit gas exchange (water loss, CO_2 uptake, and O_2 release or uptake) between the atmosphere and the interior of the leaf. The root epidermis is adapted for absorption of water and minerals, and its outer wall surface typically does not have a waxy cuticle. Extensions from the root epidermal cells, the **root hairs**, increase the surface area over which absorption can take place (see Figure 1.4C).

Ground. Making up the bulk of the plant are cells termed the **ground tissue**. There are three types of ground tissue: parenchyma, collenchyma, and sclerenchyma. **Parenchyma**, the most abundant ground tissue, consists of thin-walled, metabolically active cells that carry out a variety of functions in the plant, including photosynthesis and storage (Figure 1.7B). **Collenchyma** tissue is composed of narrow, elongated cells with thick primary walls (Figure 1.7C). Collenchyma cells provide structural support to the growing plant body, particularly shoots, and their thickened walls are nonlignified, so they can stretch as the organ elongates. Collenchyma cells are typically arranged in bundles or layers near the periphery of stems or leaf petioles. **Sclerenchyma** consists of two types of cells, sclereids and fibers (Figure 1.7D). Both have thick secondary walls and are frequently dead at maturity. **Sclereids** occur in a variety of shapes, ranging from roughly spherical to branched, and are widely distributed throughout the plant. In contrast, **fibers** are narrow, elongated cells that are commonly associated with vascular tissues. The main function of sclerenchyma is to provide mechanical support, particularly to parts of the plant that are no longer elongating.

In the stem, the pith and the cortex make up the ground tissue (see Figure 1.4B). The **pith** is located

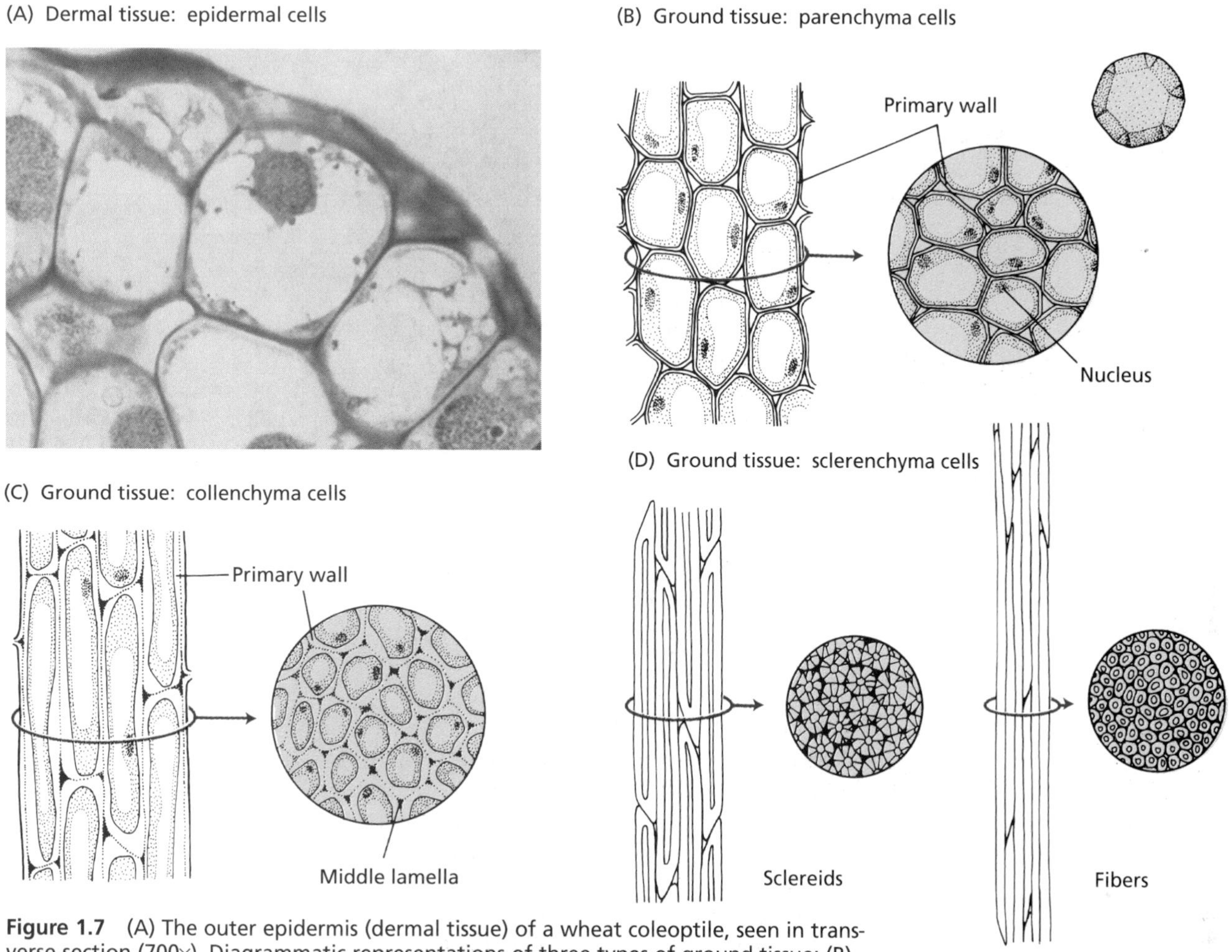

(A) Dermal tissue: epidermal cells

(B) Ground tissue: parenchyma cells

Primary wall

Nucleus

(C) Ground tissue: collenchyma cells

Primary wall

Middle lamella

(D) Ground tissue: sclerenchyma cells

Sclereids

Fibers

Figure 1.7 (A) The outer epidermis (dermal tissue) of a wheat coleoptile, seen in transverse section (700×). Diagrammatic representations of three types of ground tissue: (B) parenchyma, (C) collenchyma, and (D) sclerenchyma cells. (A from O'Brien and McCully 1969; B–D from Esau 1977.)

within the cylinder of vascular tissue, where it often exhibits a spongy texture because of the presence of large intercellular air spaces. If the growth of the pith fails to keep up with that of the surrounding tissues, the pith may degenerate, producing a hollow stem. In general, roots lack piths, although there are exceptions to this rule. In contrast, the **cortex**, which is located between the epidermis and the vascular cylinder, is present in both stems and roots (see Figure 1.4B and C).

At the boundary between the ground tissue and the vascular tissue in roots, and occasionally in stems, is a specialized layer of cortex known as the **endodermis** (see Figure 1.4C). This single layer of cells originates from cortical tissue at the innermost layer of the root cortex and forms a cylinder that surrounds the central vascular tissue, or **stele**. Early in root development, a narrow band composed of the waxy substance **suberin** is formed in the cell walls circumscribing each endodermal cell (see Figure 4.4). These suberin deposits, called **Casparian strips**, form a barrier in the endodermal walls to the intercellular movement of water, ions, and other water-soluble solutes to the vascular cells.

Leaves have two interior layers of ground tissue that are collectively known as the **mesophyll** (see Figure 1.4A). The **palisade parenchyma** consists of closely spaced, columnar cells located beneath the upper epidermis. There is usually one layer of palisade parenchyma in the leaf. Palisade parenchyma cells are rich in chloroplasts and are a primary site of photosynthesis in the leaf. Below the palisade parenchyma are irregularly shaped, widely spaced **spongy mesophyll** cells. The spongy mesophyll cells are also photosynthetic, and the large spaces between these cells allow diffusion of carbon dioxide. The spongy mesophyll also contributes to leaf flexibility in the wind, and this flexibility facilitates the movement of gases within the leaf.

Vascular: Xylem and phloem. The vascular tissue is composed of two major conducting systems: the xylem and the phloem. The **xylem** transports water and mineral ions from the root to the rest of the plant. The **phloem** distributes the products of photosynthe-

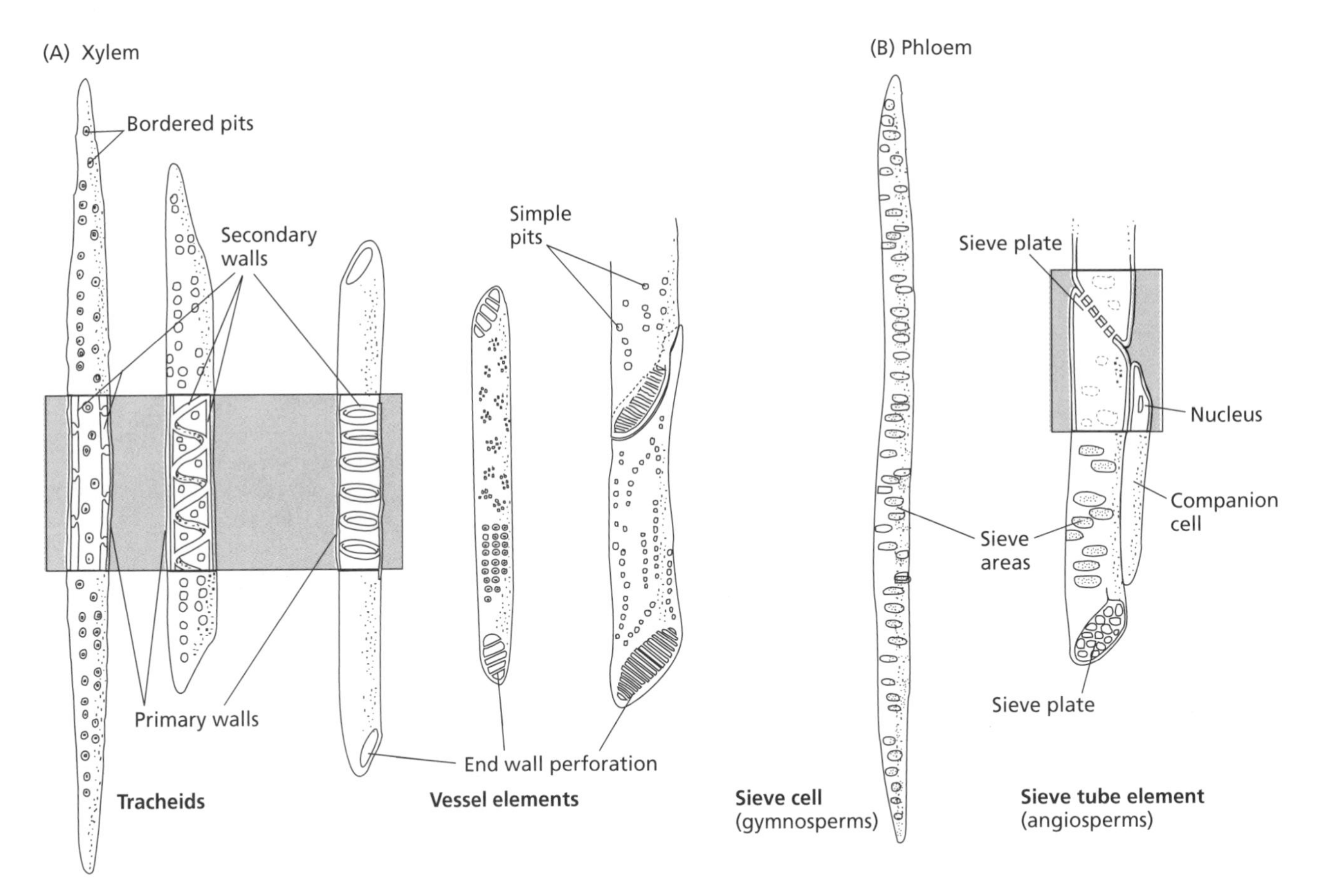

Figure 1.8 Diagrammatic representation of (A) xylem and (B) phloem conducting cells.

sis and a variety of other solutes throughout the plant (see Figure 1.4B and C).

The **tracheids** and **vessel elements** are the conducting cells of the xylem (Figure 1.8A). Both of these cell types have elaborate secondary-wall thickenings and lose their cytoplasm at maturity; that is, they are dead when functional. Tracheids overlap each other, whereas vessel elements have open end walls and are arranged end to end to form a larger unit called a **vessel**. Other cell types present in the xylem include parenchyma cells, which are important for the storage of energy-rich molecules and phenolic compounds, and sclerenchyma fibers.

The **sieve elements*** and **sieve cells** are responsible for sugar translocation in the phloem (Figure 1.8B). The former are found in angiosperms; the latter perform the same function in gymnosperms. Like vessel elements, sieve elements are often stacked in vertical rows, forming larger units called **sieve tubes**, whereas sieve cells form overlapping arrays. Both types of conducting cells are living when functional, but they lack nuclei and central vacuoles and have relatively few cytoplasmic organelles. Substances are translocated from sieve cell to sieve cell laterally through circular or oval zones containing enlarged pores, called **sieve areas**. In contrast, sieve tubes translocate substances through large pores in the end walls of the sieve elements, called **sieve plates**. Sugar movement through sieve tubes is more efficient and rapid than through sieve cells and represents a more evolutionarily advanced mechanism.

Sieve elements are associated with, and depend on, densely cytoplasmic parenchyma cells called **companion cells**. The analogous cells adjacent to the sieve cells of gymnosperms are called **albuminous cells**. Companion cells provide proteins and metabolites necessary for the functions of the sieve tube elements. In addition, the phloem frequently contains storage parenchyma and fibers that provide mechanical support.

The Plant Cell

Plants are multicellular organisms composed of millions of cells with specialized functions. At maturity, such specialized cells may differ greatly from one another in their structures. However, all plant cells have the same

* Vessel elements are sometimes referred to as "vessel members;" "sieve elements" may also be called "sieve tube members."

basic eukaryotic organization; they contain a nucleus, a cytoplasm, and subcellular organelles, and they are enclosed in a membrane that defines their boundaries. Certain structures, including the nucleus, can be lost during cell maturation, but all plant cells *begin* with a similar complement of organelles. An additional characteristic feature of plant cells is that they are surrounded by a cellulosic cell wall. The following sections provide an overview of the membranes and organelles of plant cells. The structure and function of the cell wall will be treated in detail in Chapter 15.

Biological Membranes Are Phospholipid Bilayers That Contain Proteins

All cells are enclosed in a membrane that serves as their outer boundary, separating the cytoplasm from the external environment (Figure 1.9). This **plasma membrane** (also called **plasmalemma**) allows the cell to take up and retain certain substances while excluding others. Various transport proteins embedded in the plasma membrane are responsible for this selective traffic of solutes across the membrane. The accumulation of ions or molecules in the cytosol through the action of transport proteins consumes metabolic energy. Membranes also delimit the boundaries of the specialized internal organelles of the cell and regulate the fluxes of ions and metabolites into and out of these compartments.

All biological membranes have the same basic molecular organization. They consist of a double layer (*bilayer*) of either phospholipids or, in the case of chloroplasts, glycosylglycerides, in which proteins are embedded (Figure 1.10A). Such a bilayer is sometimes referred to as a **unit membrane**. The composition of the lipid components and the properties of the proteins vary from membrane to membrane, conferring on each membrane its unique functional characteristics.

Phospholipids are a class of lipids in which two fatty acids are covalently linked to glycerol, which is covalently linked to a phosphate group. Also attached to this phosphate group is a variable component, called the *head group*, such as serine, choline, glycerol, or inositol (see Figure 1.10C). In contrast to the fatty acids, the head groups are highly polar; consequently, phospholipid molecules display both hydrophilic and hydrophobic

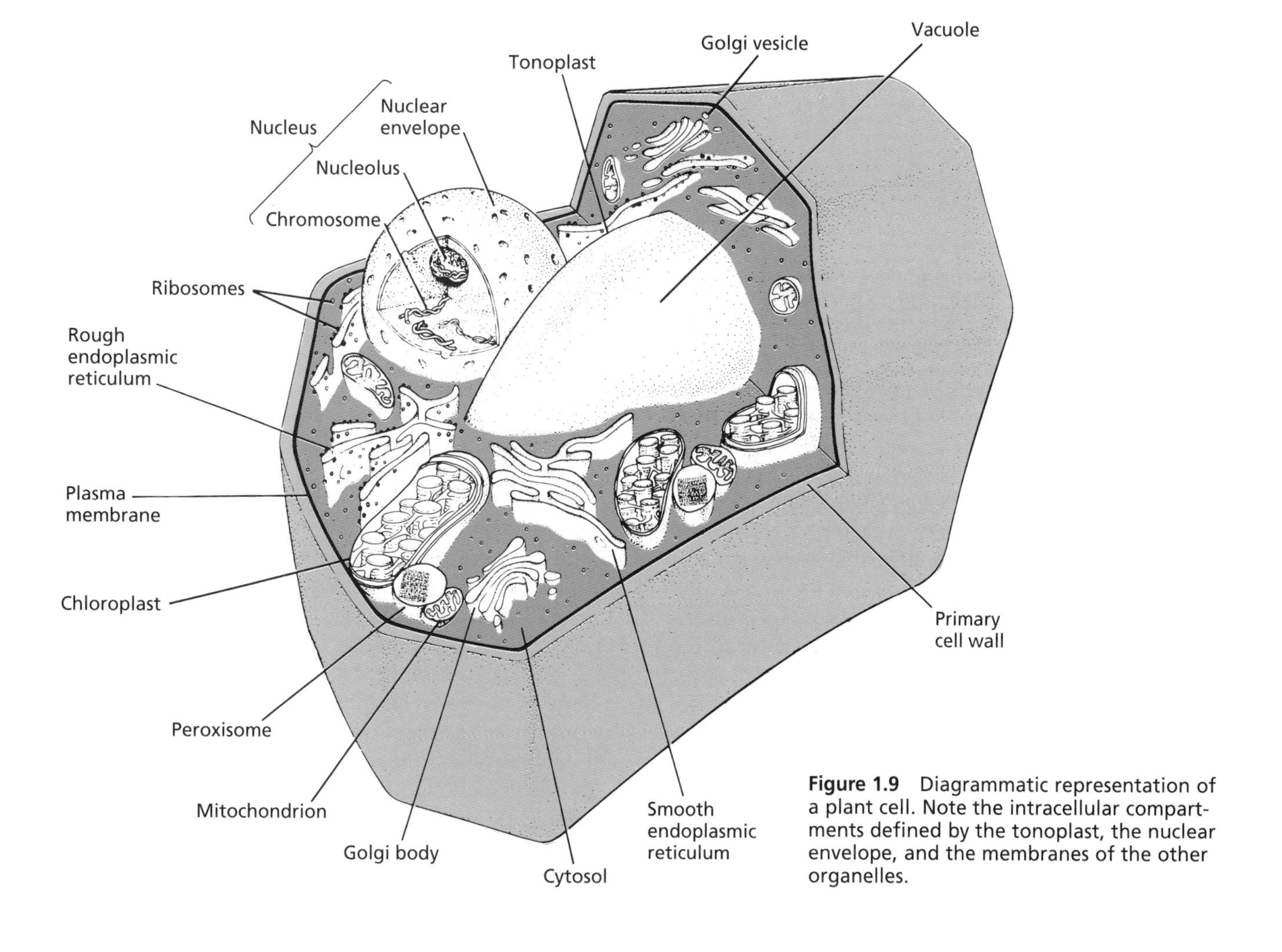

Figure 1.9 Diagrammatic representation of a plant cell. Note the intracellular compartments defined by the tonoplast, the nuclear envelope, and the membranes of the other organelles.

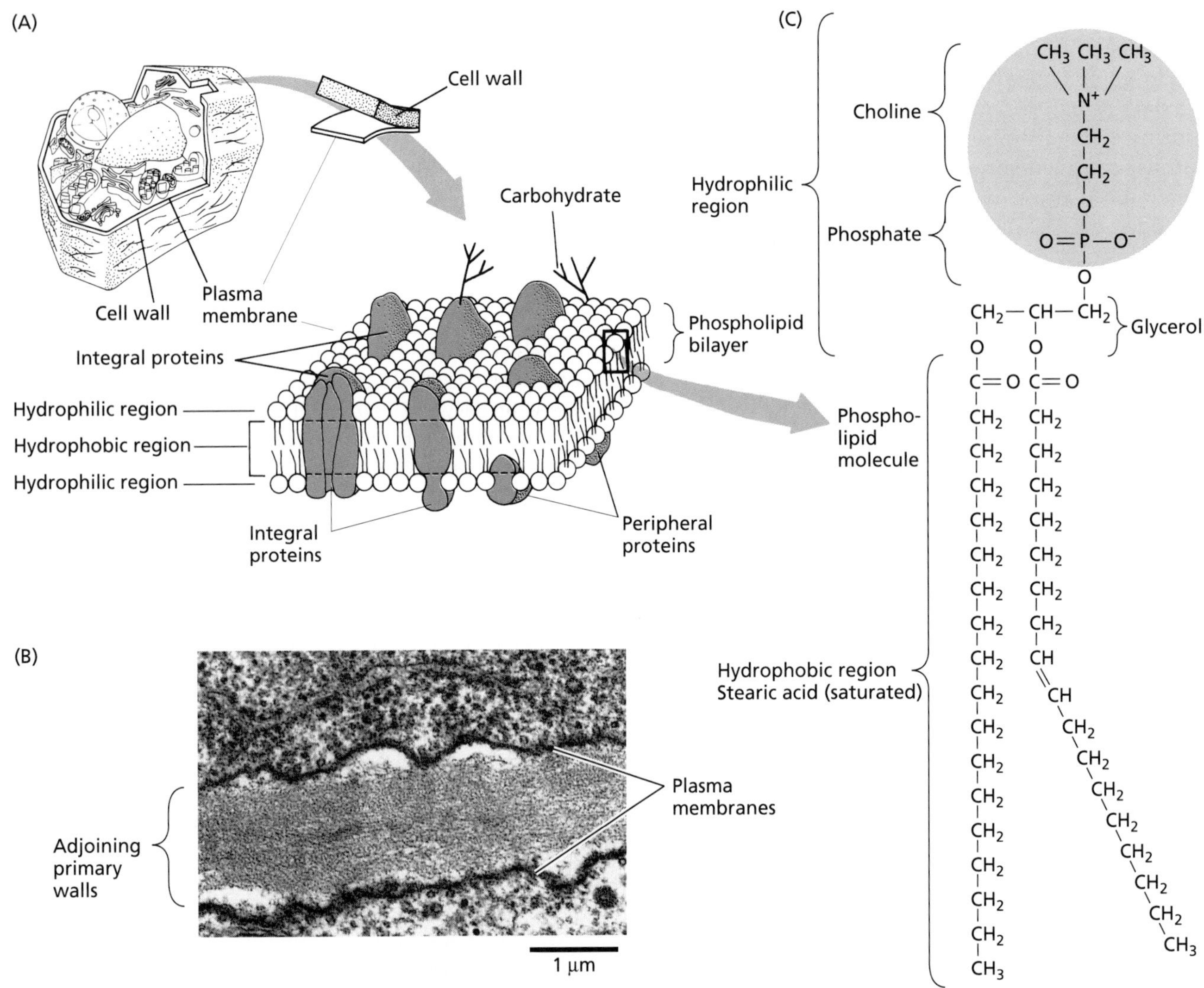

Figure 1.10 (A) The plasma membrane, endoplasmic reticulum, and other endomembranes of plant cells consist of proteins embedded in a phospholipid bilayer. (B) This transmission electron micrograph shows plasma membranes in cells from the meristematic region of a root tip of cress (*Lepidium sativum*). The overall thickness of the plasma membrane, viewed as two dense lines and an intervening space, is 8 nm. (C) Chemical structure of a typical phospholipid, phosphatidylcholine. (B from B. Gunning and M. Steer, *Plant Cell Biology: Structure and Function*, Jones and Bartlett, 1996.)

properties (that is, they are *amphipathic*). The nonpolar hydrocarbon chains of the fatty acids form a region that is exclusively hydrophobic—that is, that excludes water.

Plastid membranes are unique in that their lipid component consists almost entirely of **glycosylglycerides** rather than phospholipids. In glycosylglycerides, the polar head group consists of galactose, digalactose, or sulfated galactose, without a phosphate group. The structures of the three plant glycosylglycerides, monogalactosyldiacylglycerol (MGDG), digalactosyldiacylglycerol (DGDG) and sulfoquinovosyldiacylglycerol (SQDG), are shown in Figure 1.11. The two galactolipids, MGDG and DGDG, are uncharged but polar; the sulfolipid, SQDG, is negatively charged as well as being polar (Harwood 1997).

The fatty acid chains of phospholipids and glycosylglycerides are variable in length, but they usually consist of 14 to 24 carbons. One of the fatty acids is typically *saturated* (i.e., it contains no double bonds); the other fatty acid chain usually has one or more *cis* double bonds (i.e., it is *unsaturated*). The presence of *cis* double bonds creates a kink in the chain that prevents tight packing of the phospholipids in the bilayer. As a result, the fluidity of the membrane is increased. The fluidity of the membrane, in turn, plays a critical role in many membrane functions. Membrane fluidity is also strongly influenced by temperature. As *poikilothermic organisms*—that is, organisms that cannot regulate their body temperature—plants are often faced with the problem of maintaining membrane fluidity under conditions of low

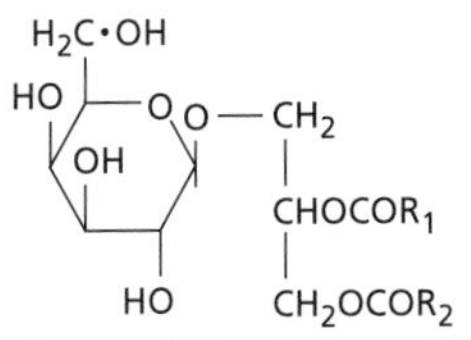

Monogalactosyldiacylglycerol (MGDG)

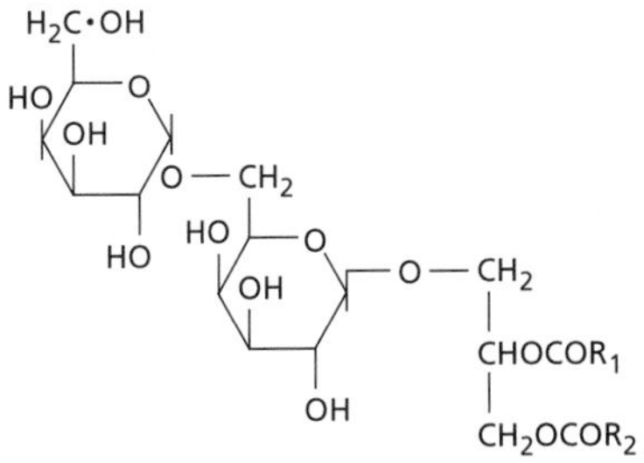

Digalactosyldiacylglycerol (DGDG)

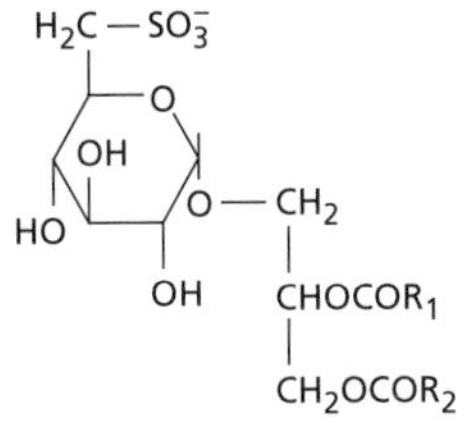

Plant sulfolipid (sulfoquinovosyldiacylglycerol; SQDG)

Figure 1.11 The structures of chloroplast glycosylglycerides. MGDG and DGDG are galactolipids; SQDG is a sulfolipid. Note that the two galactolipids are uncharged, while the sulfolipid is charged. (After Harwood 1997.)

temperature, which tends to decrease membrane fluidity. Thus, plant phospholipids have a high percentage of unsaturated fatty acids, such as oleic acid (one double bond), linoleic acid (two double bonds) and α-linolenic acid (three double bonds).

The proteins associated with the lipid bilayer are of two types, integral and peripheral (they have also been called intrinsic and extrinsic, respectively). **Integral proteins** are embedded in the lipid bilayer. Most integral proteins span the entire width of the phospholipid bilayer, so one part of the protein interacts with the outside of the cell, another part interacts with the hydrophobic core of the membrane, and a third part interacts with the interior of the cell, the cytosol. Proteins that serve as ion channels (see Chapter 6) are always integral membrane proteins. **Peripheral proteins** are attached to the membrane surface by noncovalent bonds, such as ionic bonds or hydrogen bonds. Peripheral proteins have several roles in membrane function. For example, some are involved in interactions between the plasma membrane and components of the cytoskeleton, such as microtubules and actin microfilaments, which are discussed later in this chapter.

The Nucleus Contains Most of the Genetic Material of the Cell

The **nucleus** is the organelle that contains the genetic information primarily responsible for regulating the metabolism, growth, and differentiation of the cell. Collectively, these genes and their intervening sequences are referred to as the **nuclear genome**. The size of the nuclear genome in plants is highly variable, ranging from 1.5×10^8 base pairs for the diminutive dicot *Arabidopsis thaliana* to 2×10^{11} base pairs for certain gymnosperms. The remainder of the genetic information of the cell is contained in the two semiautonomous organelles, the chloroplasts and mitochondria, which we will discuss a little later in this chapter.

The membrane system that surrounds the nucleus, called the **nuclear envelope**, consists of two discrete lipid bilayer membranes (Figure 1.12A). The space between the two membranes of the nuclear envelope is called the **perinuclear space**, and the two membranes of the nuclear envelope are fused at sites called **nuclear pores**

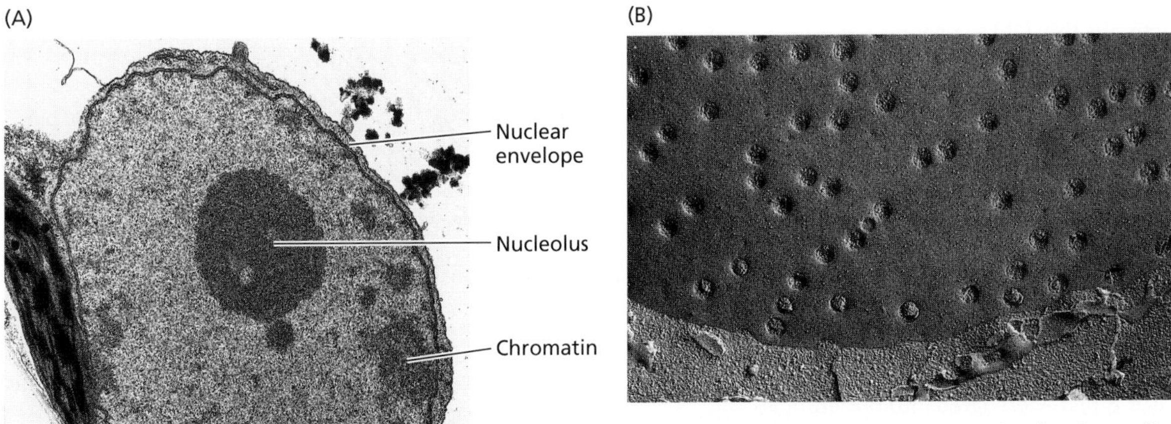

Figure 1.12 (A) Transmission electron micrograph of a plant cell, showing the nucleolus and the nuclear envelope. (B) A freeze-etched preparation of nuclear pores from a cell of an onion root. (A courtesy of R. Evert; B courtesy of D. Branton.)

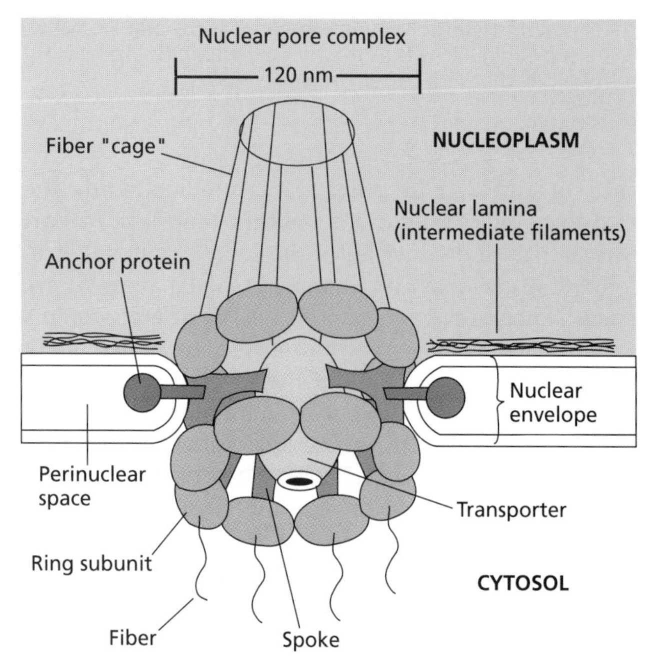

Figure 1.13 Schematic model of the structure of the nuclear pore complex. Parallel rings composed of eight subunits each are arranged octagonally near the inner and outer membranes of the nuclear envelope. Various proteins form the other structures, such as spokes, fibers, anchor proteins, and the barrel-shaped transporter that runs through the center of the pore complex. A cage-like structure composed of fibers extends into the nucleus. The precise functions of the different components are not known. (From Becker et al. 1996.)

(Figure 1.12B). The nuclear "pore" is actually an elaborate structure composed of more than a hundred different proteins arranged octagonally to form a **nuclear pore complex** (Figure 1.13). There can be very few to many thousands of nuclear pore complexes on an individual nuclear envelope. Macromolecules from the nucleus, including ribosomal subunits, are able to pass through the nuclear pores into and out of the cytosol.

The nucleus is the site of storage and replication of the **chromosomes**, composed of DNA and its associated proteins. Collectively, this DNA–protein complex is known as **chromatin**. The linear length of all the DNA within any plant genome is usually millions of times greater than the diameter of the nucleus in which it is found. To solve the problem of packaging this chromosomal DNA within the nucleus, segments of the linear double helix of DNA are coiled twice around a solid cylinder of eight **histone** protein molecules, forming a **nucleosome**. Nucleosomes are arranged like beads on a string along the length of each chromosome.

During mitosis, the chromatin condenses, first by coiling tightly into a **30 nm chromatin fiber**, with six nucleosomes per turn, followed by further folding and packing processes that depend on interactions between proteins and nucleic acids (Figure 1.14). At interphase, two types of chromatin are visible: heterochromatin and euchro-

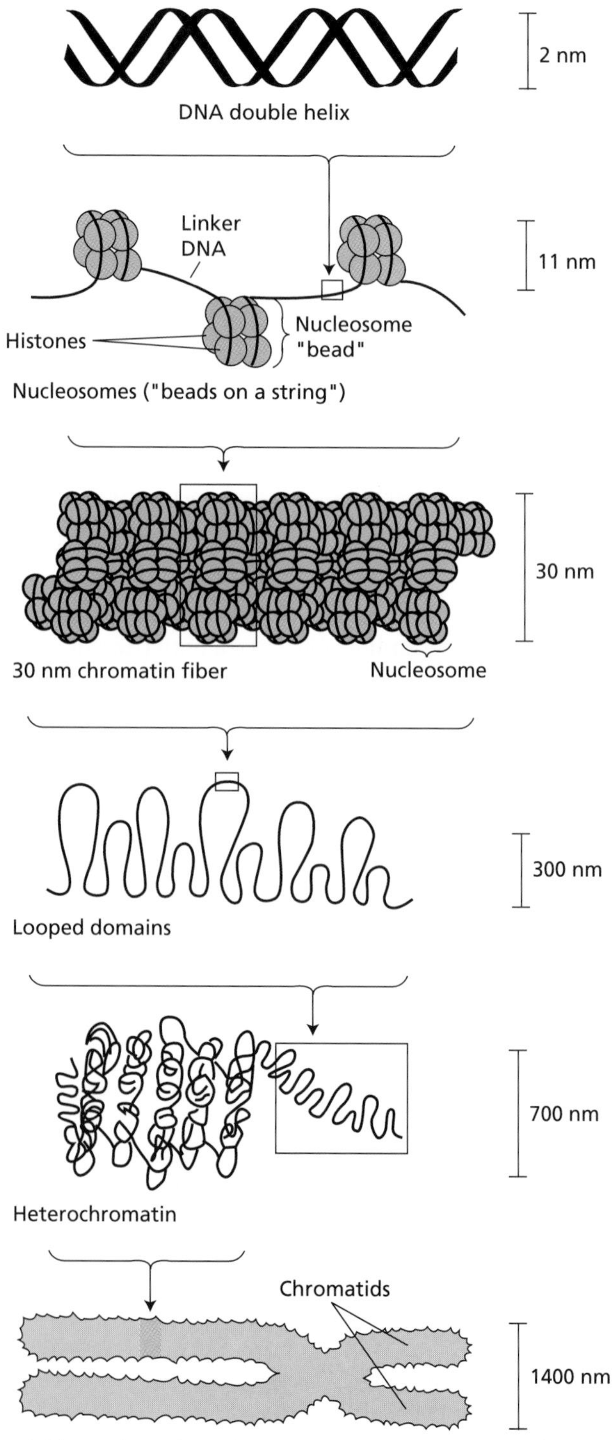

Figure 1.14 Packaging of DNA in a metaphase chromosome. The DNA is first aggregated into nucleosomes and then wound to form the 30 nm chromatin fibers. Further coiling leads to the condensed metaphase chromosome. (From Alberts et al. 1994.)

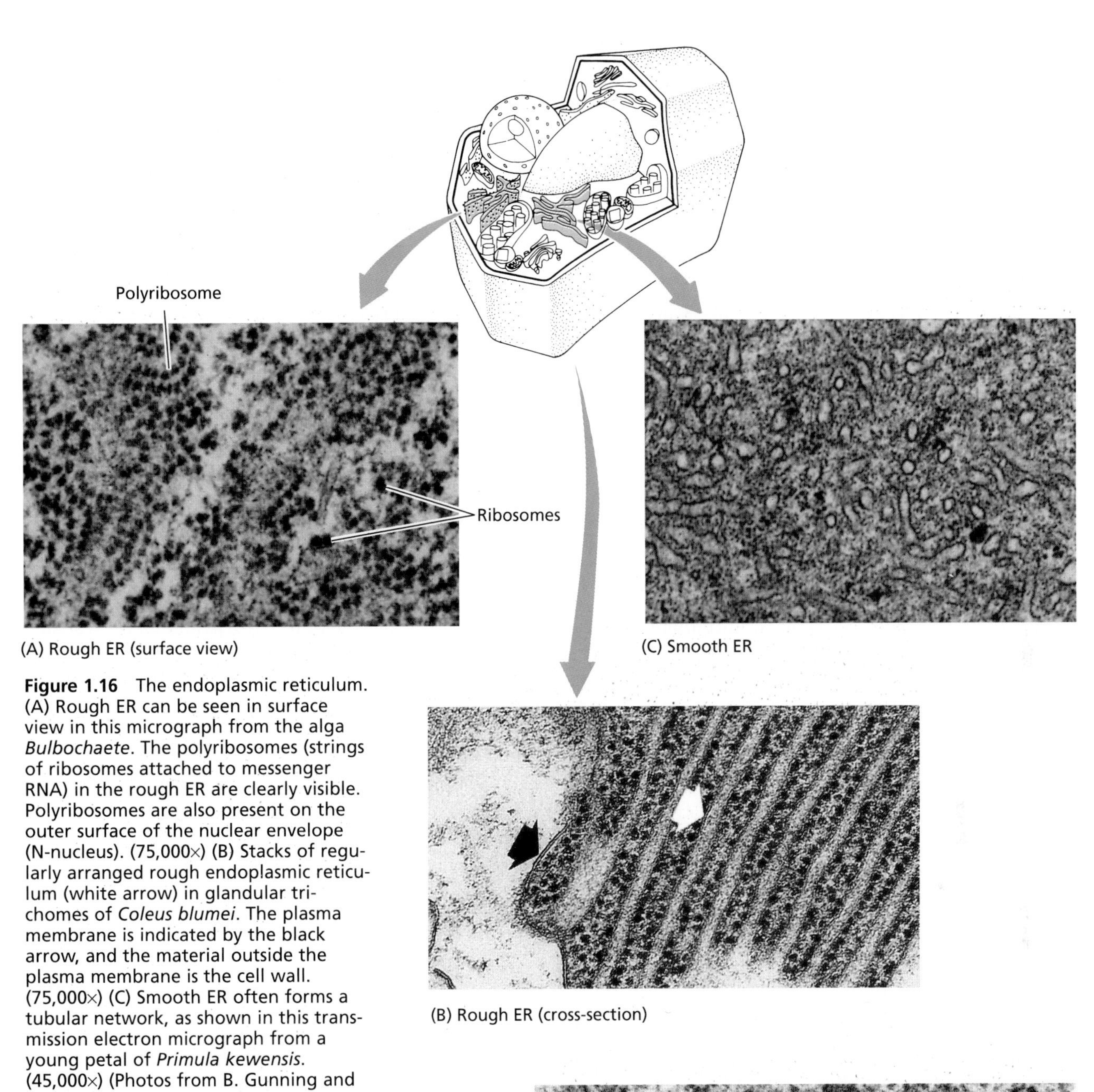

Figure 1.16 The endoplasmic reticulum. (A) Rough ER can be seen in surface view in this micrograph from the alga *Bulbochaete*. The polyribosomes (strings of ribosomes attached to messenger RNA) in the rough ER are clearly visible. Polyribosomes are also present on the outer surface of the nuclear envelope (N-nucleus). (75,000×) (B) Stacks of regularly arranged rough endoplasmic reticulum (white arrow) in glandular trichomes of *Coleus blumei*. The plasma membrane is indicated by the black arrow, and the material outside the plasma membrane is the cell wall. (75,000×) (C) Smooth ER often forms a tubular network, as shown in this transmission electron micrograph from a young petal of *Primula kewensis*. (45,000×) (Photos from B. Gunning and M. Steer, *Plant Cell Biology: Structure and Function*, Jones and Bartlett, 1996.)

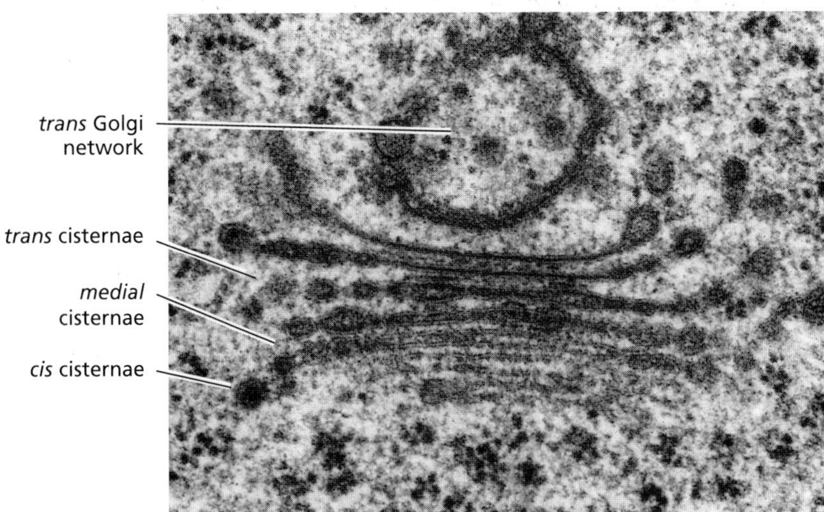

Figure 1.17 Electron micrograph of a Golgi apparatus in a tobacco (*Nicotiana tabacum*) root cap cell. The *cis*, *medial*, and *trans* cisternae are indicated. The *trans* Golgi network is associated with the *trans* cisterna. (60,000×) (From B. Gunning and M. Steer, *Plant Cell Biology: Structure and Function*, Jones and Bartlett, 1996.)

protein cross-links that hold the cisternae together. Whereas in animal cells Golgi bodies tend to be clustered in one part of the cell and are interconnected with one another via tubules, plant cells contain up to several hundred apparently separate Golgi bodies dispersed throughout the cytoplasm (Driouich et al. 1994).

The Golgi apparatus of plants has been described as a carbohydrate factory because of its role in the assembly of the oligosaccharide side chains of glycoproteins (proteins with covalently bound sugars) and in the synthesis and secretion of complex polysaccharides (polymers composed of different types of sugars) (Driouich et al. 1994). As already noted, the polypeptide chains of future glycoproteins are first synthesized on the polysomes of the rough ER, then transferred across the ER membrane and glycosylated (covalently bound to sugars) on the —NH_2 groups of asparagine residues. The so-called **N-linked glycoproteins** are then transported to the Golgi apparatus via small vesicles. The vesicles move through the cytosol and fuse with cisternae on the *cis* face of the Golgi apparatus.

The exact pathway of glycoproteins through the plant Golgi apparatus is not yet known. Since there appears to be no direct membrane continuity between successive cisternae, the contents of one cisterna may be transferred to the next cisterna via small vesicles budding off from the margins, as occurs in the Golgi apparatus of animals. However, there is increasing evidence that entire cisternae may progress through the Golgi body and emerge from the *trans* face.

Within the lumens of the Golgi cisternae, the glycoproteins are enzymatically modified. Certain sugars, such as mannose, are removed from the oligosaccharide chains, and other sugars are added. Glycosylation of the —OH groups of serine, threonine, and tyrosine residues (**O-linked oligosaccharides**) also takes place in the Golgi. After being processed within the Golgi, the glycoproteins leave the organelle in other vesicles, usually from the *trans* side of the stack. All of this processing appears to confer on each protein a specific tag or marker that specifies the ultimate destination of that protein inside or outside the cell.

In plant cells, the Golgi body is important in cell wall formation (see Chapter 15). Noncellulosic cell wall polysaccharides (hemicellulose and pectin) are synthesized, and a variety of hydroxyproline-rich glycoproteins are processed, within the Golgi. **Secretory vesicles** derived from the Golgi carry the polysaccharides and glycoproteins to the plasma membrane, where the vesicles fuse with the plasma membrane and empty their contents into the region of the cell wall. Two types of vesicles are present: **smooth secretory vesicles** and **coated vesicles**.

Unlike the smooth vesicles, coated vesicles are surrounded by the protein **clathrin**. Clathrin subunits are arranged in the form of three-pronged structures called **triskelions**, which associate with each other to form cages around the vesicles (Figure 1.18). In animal cells, coated vesicles function in endocytosis and in the transfer of materials within the endomembrane system. In plants, coated vesicles have been implicated in the transport of storage proteins to specialized protein-storing vacuoles.

The Central Vacuole Contains Water and Solutes

Mature plant cells contain large, water-filled central vacuoles that can occupy 80 to 90% of the total volume of the cell (see Figure 1.9). They are surrounded by a **vacuolar membrane**, or **tonoplast**. In meristematic tissue, vacuoles are less prominent, though they are always present as

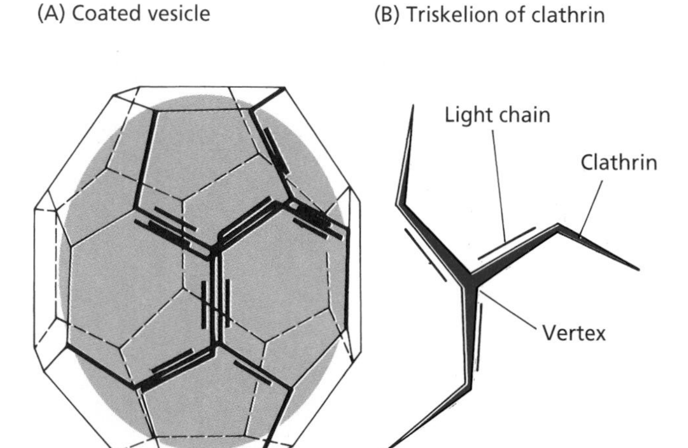

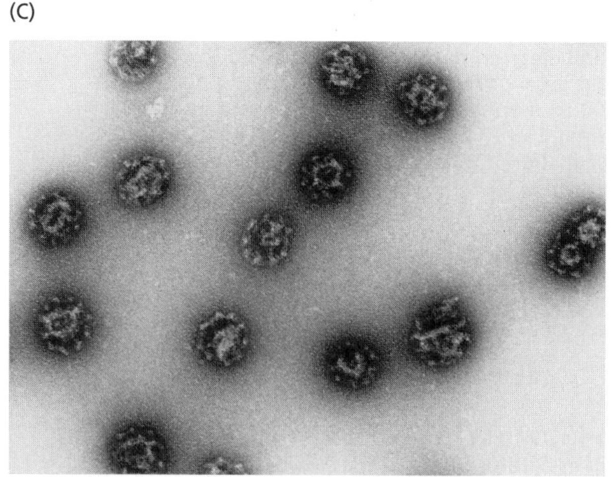

Figure 1.18 Coated vesicles. (A) The organization of triskelions into the clathrin coat. (B) Individual triskelion, composed of three clathrin heavy chains and three light chains. (C) Preparation of coated vesicles isolated from bean leaves. (102,000×) (Photo courtesy of D. G. Robinson.)

small **provacuoles**. Provacuoles are produced by the *trans* Golgi network (see Figure 1.17). As the cell begins to mature, the provacuoles fuse together to produce the large central vacuoles that are characteristic of most mature plant cells. In such cells, the cytoplasm is restricted to a thin layer surrounding the vacuole. Many cells also have cytoplasmic strands that run through the vacuole, but each transvacuolar strand is surrounded by the tonoplast.

The vacuole contains water and dissolved inorganic ions, organic acids, sugars, enzymes, and a variety of secondary metabolites, which often play roles in plant defense. Active solute accumulation provides the osmotic driving force for water uptake by the vacuole, which is required for plant cell enlargement. The turgor pressure generated by this water uptake provides the structural rigidity needed to keep herbaceous plants upright, since they lack the lignified support tissues of woody plants. Vacuoles are also rich in hydrolytic enzymes, including proteases, ribonucleases, and glycosidases, which, upon release into the cytosol during senescence, participate in the degradation of the cell. These hydrolytic enzymes may also participate in the turnover of cellular constituents throughout the life of the cell. Specialized protein-storing vacuoles, called **protein bodies**, are abundant in seeds. During germination the storage proteins in the protein bodies are hydrolyzed to amino acids and exported to the cytosol for use in protein synthesis.

Mitochondria and Chloroplasts Are Sites of Energy Conversion

A typical plant cell has two types of energy-producing organelles: mitochondria and chloroplasts. Both types of organelles are separated from the cytosol by a double membrane (an outer and an inner membrane). **Mitochondria** are the cellular sites of respiration, a process in which the energy released from sugar metabolism is used for the synthesis of ATP from ADP and inorganic phosphate (P_i).

Mitochondria can vary in shape from spherical to tubular, but they all have a smooth outer membrane and a highly convoluted inner membrane (Figure 1.19). The infoldings of the inner membrane are called **cristae**. The compartment enclosed by the inner membrane is the mitochondrial **matrix**, and it contains the enzymes of the pathway of intermediary metabolism called the Krebs cycle. In contrast to the mitochondrial outer membrane and all other membranes in the cell, the inner membrane of a mitochondrion is almost 70% protein and contains some phospholipids that are unique to the organelle (e.g., cardiolipin). The proteins present in and on the inner membrane have special enzymatic and transport capacities. The inner membrane is highly impermeable to the passage of H^+; it serves as a barrier to the movement of protons. This important feature underlies the formation of electrochemical gradients. Dissipation of such gradi-

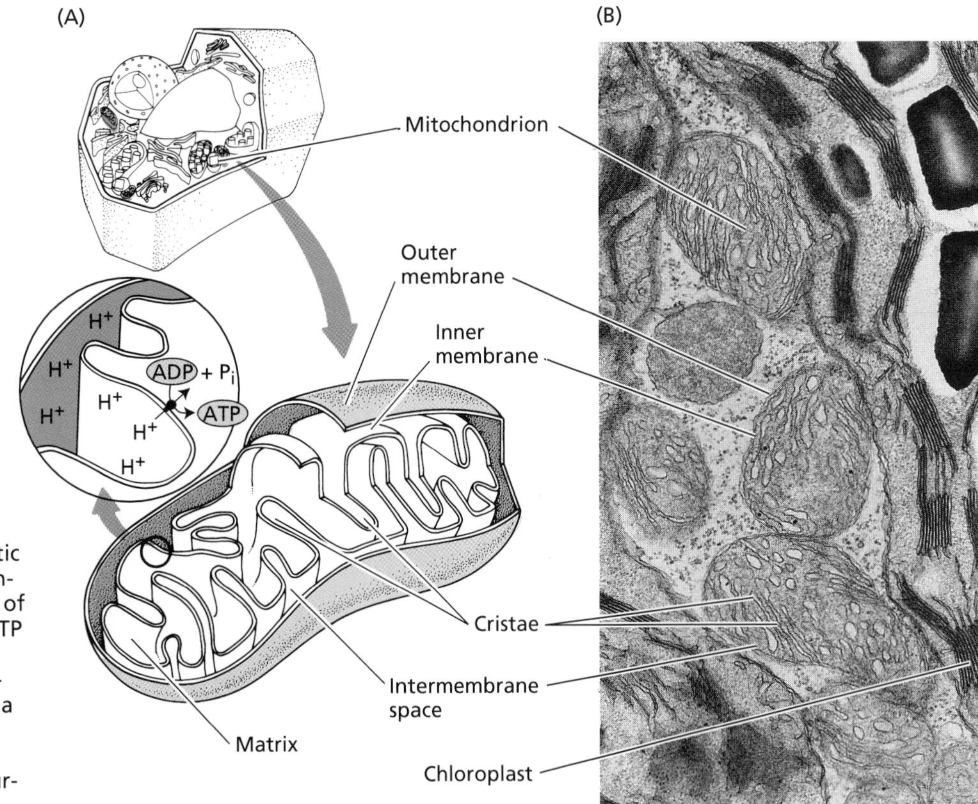

Figure 1.19 (A) Diagrammatic representation of a mitochondrion, including the location of the H^+-ATPases involved in ATP synthesis on the inner membrane. (B) An electron micrograph of mitochondria from a leaf cell of Bermuda grass, *Cynodon dactylon*. (26,000×) (Photo by S. E. Frederick, courtesy of E. H. Newcomb.)

ents by the controlled movement of H^+ ions through the transmembrane enzyme called ATP synthase or H^+-ATPase* is coupled to the phosphorylation of ADP to produce ATP. ATP can then be released to other cellular sites where energy is needed to drive specific reactions.

Chloroplasts (Figure 1.20) belong to a group of double membrane–enclosed organelles called **plastids**. Chloroplasts contain chlorophyll and its associated proteins and are the sites of photosynthesis. In addition to their inner and outer envelope membranes, chloroplasts possess a third system of membranes called **thylakoids**. A stack of thylakoids forms a **granum** (plural grana). Proteins and pigments (chlorophylls and carotenoids) that function in the photochemical events of photosyn-

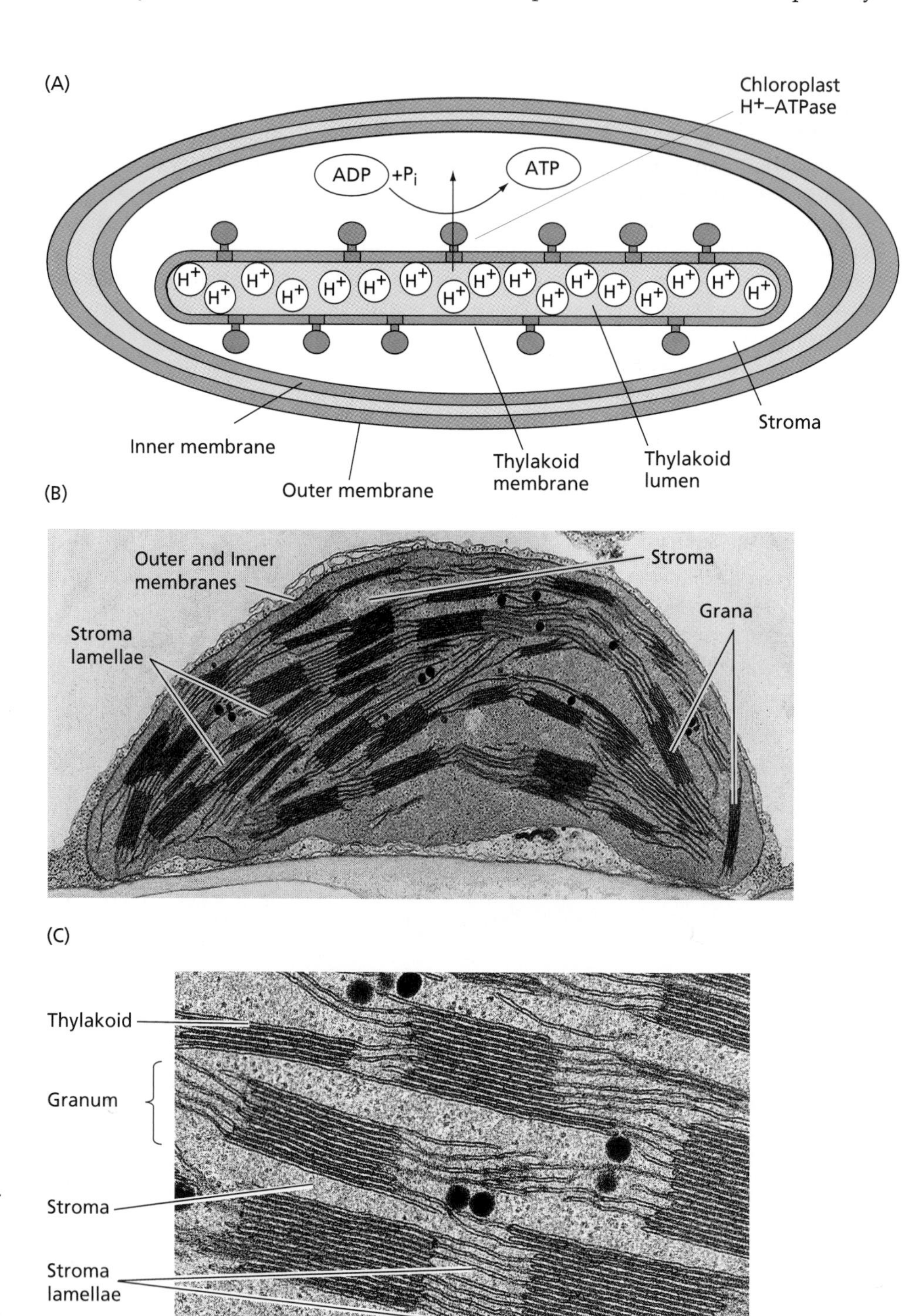

Figure 1.20 (A) Diagrammatic representation of a chloroplast, showing the location of the H^+-ATPases on the thylakoid membranes. (B) Electron micrograph of a chloroplast from a leaf of timothy grass, *Phleum pratense*. (18,000×) (C) The same preparation at higher magnification. (52,000×) (Micrographs by W. P. Wergin, courtesy of E. H. Newcomb.)

* The enzyme is called an *ATPase* when it functions in the forward reaction to hydrolyze ATP, and an *ATP synthase* when it functions in the reverse reaction to synthesize ATP.

thesis are part of the thylakoid. The fluid compartment surrounding the thylakoids is called the **stroma** and is analogous to the matrix of the mitochondrion. Adjacent grana are connected by nonstacked thylakoids called **stroma lamellae**.

The different components of the photosynthetic apparatus are localized in different areas of the grana and the stroma lamellae. The ATP synthases of the chloroplast are located on the thylakoid membranes (see Figure 1.20A). During photosynthesis, light-driven electron transfer reactions result in a proton gradient across the thylakoid membrane. As in the mitochondria, ATP is synthesized when the proton gradient is dissipated via the ATP synthase.

Plastids that contain high concentrations of carotenoid pigments rather than chlorophyll are called **chromoplasts**, and they are one of the causes of the yellow, orange, or red colors of many fruits and flowers, as well as of autumn leaves (Figure 1.21). Nonpigmented plastids are called **leucoplasts**. The most important type of leucoplast is the **amyloplast**, a starch-storing plastid. Amyloplasts are abundant in storage tissues of the shoot and root, and in seeds. Specialized amyloplasts in the root cap also serve as gravity sensors that direct root growth downward into the soil (see Chapter 19).

Mitochondria and Chloroplasts Are Semiautonomous Organelles

Both mitochondria and chloroplasts contain their own DNA and protein-synthesizing machinery (ribosomes, tRNAs, and other components) and are believed to have evolved from endosymbiotic bacteria. Both plastids and mitochondria divide by fission, and mitochondria can also undergo extensive fusion to form elongated structures or networks. The DNA of these organelles is in the form of circular chromosomes, similar to those of bacteria and very different from the linear chromosomes in the nucleus. These DNA circles are localized in specific regions of the mitochondrial matrix or plastid stroma called **nucleoids**. DNA replication in both mitochondria and chloroplasts is independent of DNA replication in the nucleus. On the other hand, the numbers of these organelles within a given cell type remain approximately constant, suggesting that some aspects of organelle replication are under cellular regulation.

The mitochondrial genome of plants is about 200 kilobase pairs (200,000 base pairs), a size considerably larger than that of most animal mitochondria. The mitochondria of meristematic cells are typically polyploid; that is, they contain multiple copies of the circular chromosome. However, the number of copies per mitochondrion gradually decreases as cells mature, because the mitochondria continue to divide in the absence of DNA synthesis. Most of the proteins encoded by the mitochondrial genome are 70S ribosomal proteins and components of the electron transfer system. The majority of mitochondrial proteins, including Krebs cycle enzymes, are encoded by nuclear genes and are imported from the cytosol.

The chloroplast genome is smaller than that of the mitochondria, about 145 kilobase pairs (145,000 base pairs). Whereas mitochondria are polyploid only in the meristems, chloroplasts become polyploid during cell maturation. Thus the average amount of DNA per chloroplast in the plant is much greater than that of the mitochondria. The total amount of DNA from the mitochondria and plastids combined is about one-third of the nuclear genome (Gunning and Steer 1996).

Chloroplast DNA encodes rRNA; transfer RNA (tRNA); the large subunit of the enzyme that fixes CO_2, ribulose-1,5-bisphosphate carboxylase/oxygenase (rubisco); and

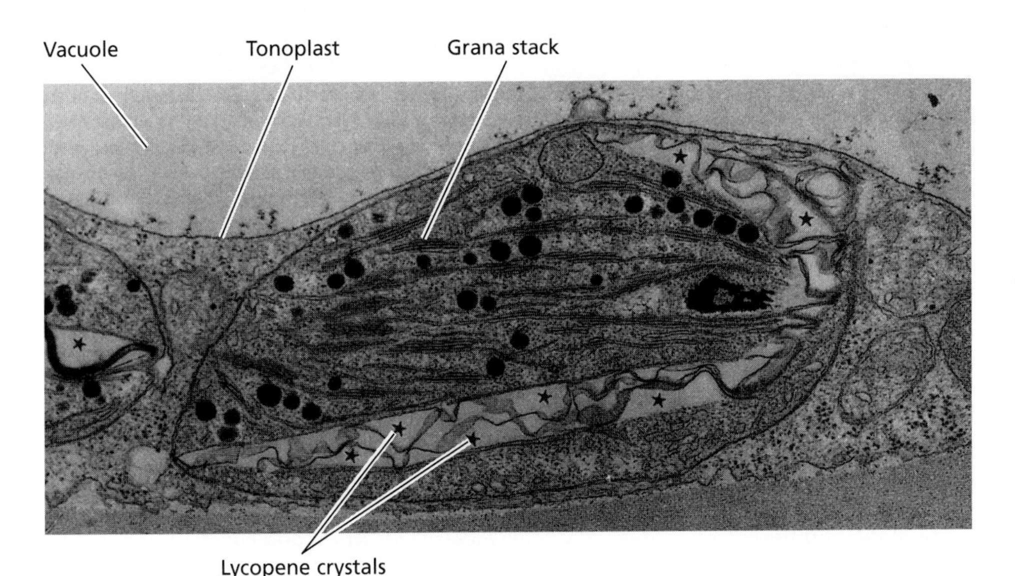

Figure 1.21 Electron micrograph of a chromoplast from tomato (*Lycopersicum esculentum*) fruit at an early stage in the transition from chloroplast to chromoplast. Small grana stacks are still visible. Crystals of the carotenoid lycopene are indicated by the stars. (27,000×) (From B. Gunning and M. Steer, *Plant Cell Biology: Structure and Function*, Jones and Bartlett, 1996.)

several of the proteins that participate in photosynthesis. Nevertheless, the majority of chloroplast proteins, like those of mitochondria, are encoded by nuclear genes, synthesized in the cytosol, and transported to the organelle. Although mitochondria and chloroplasts have their own genomes and can divide independently of the cell, they are characterized as *semiautonomous organelles* because they depend on the nucleus for the majority of their proteins.

Different Plastid Types Are Interconvertible

Meristem cells contain **proplastids**, which have few or no internal membranes, no chlorophyll, and an incomplete complement of the enzymes necessary to carry out photosynthesis (Figure 1.22A and B). In angiosperms and some gymnosperms, chloroplast development from proplastids is triggered by light. Upon illumination, enzymes are formed inside the proplastid or imported

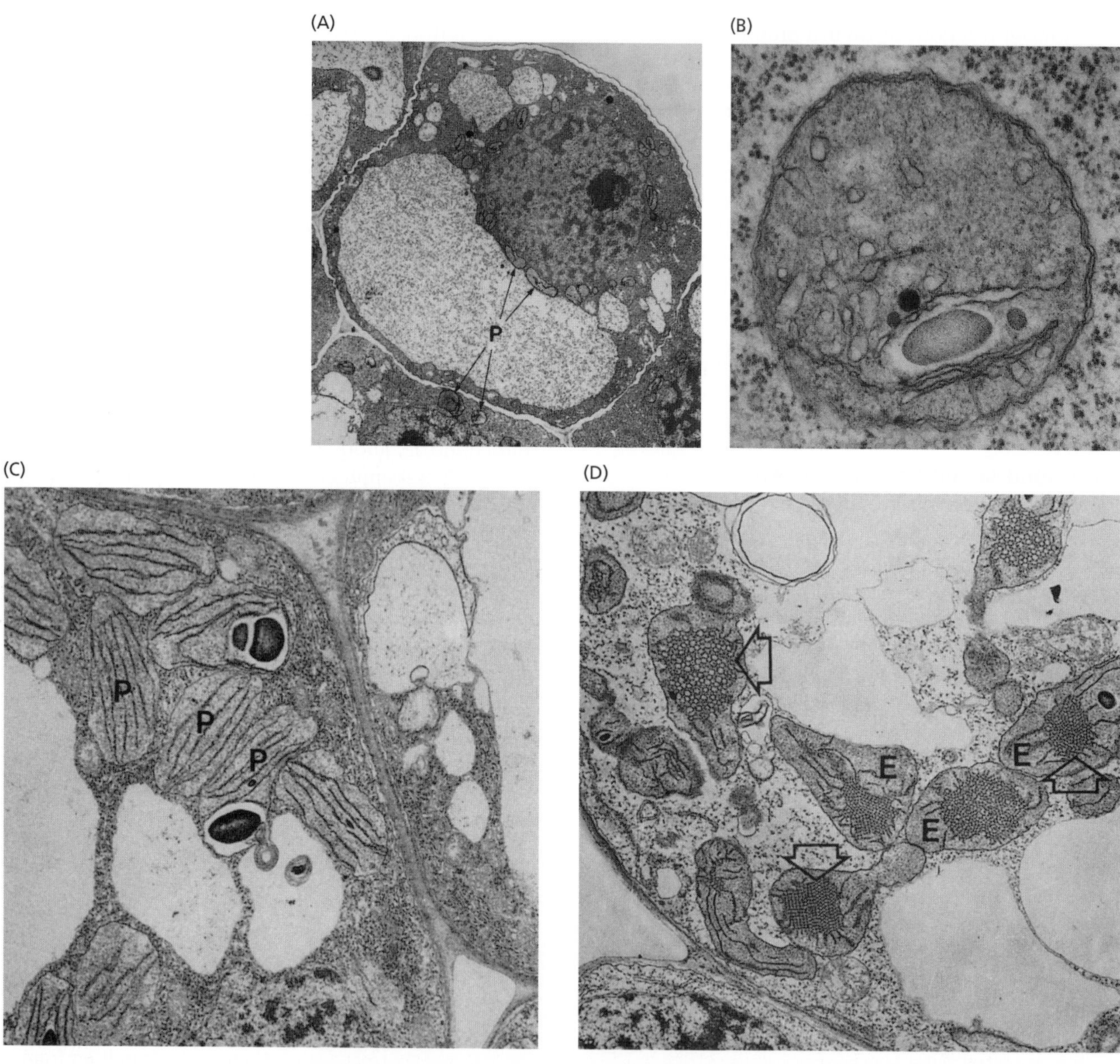

Figure 1.22 Electron micrographs illustrating several stages of plastid development. (A) Shoot apical meristem cells of the common oat (*Avena sativa*). The proplastids (P) are visible as small organelles. (3,300×) (B) A higher-magnification view of a proplastid from the root apical meristem of the broad bean (*Vicia faba*). The internal membrane system is rudimentary, and grana are absent. (47,000×) (C) A mesophyll cell of a young oat leaf at an early stage of differentiation in the light. The plastids (P) are developing grana stacks. (D) A cell from a young oat leaf from a seedling grown in the dark. The plastids have developed as etioplasts (E), with elaborate semicrystalline lattices of membrane tubules called prolamellar bodies (arrows). When exposed to light, the etioplast can convert to a chloroplast by the disassembly of the prolamellar body and the formation of grana stacks. (7,200×) (From B. Gunning and M. Steer, *Plant Cell Biology: Structure and Function*, Jones and Bartlett, 1996.)

from the cytosol, light-absorbing pigments are produced, and membranes proliferate rapidly, giving rise to stroma lamellae and grana stacks (Figure 1.22C).

Seeds usually germinate in the soil away from light, and chloroplasts develop only when the young shoot is exposed to light. If seeds are germinated in the dark, the proplastids differentiate into **etioplasts**, which contain semicrystalline tubular arrays of membrane known as **prolamellar bodies** (Figure 1.22D). Instead of chlorophyll, the etioplast contains a pale yellow green precursor pigment, protochlorophyll.

Within minutes after exposure to light, the etioplast differentiates, converting the prolamellar body into thylakoids and stroma lamellae and the protochlorophyll into chlorophyll. The maintenance of chloroplast structure depends on the presence of light, and mature chloroplasts can revert to etioplasts during extended periods of darkness. Chloroplasts can be converted to chromoplasts, as in the case of autumn leaves and ripening fruit, and in some cases this process is reversible. Amyloplasts can also be converted to chloroplasts, which explains why exposure of roots to light often results in greening of the roots.

Microbodies Play Specialized Metabolic Roles in Leaves and Seeds

Plant cells also contain **microbodies**, a class of spherical organelles surrounded by a single membrane and specialized for one of several metabolic functions. The two main types of microbodies are peroxisomes and glyoxysomes. **Peroxisomes** are found in all eukaryotic organisms, and in plants they are present in photosynthetic cells (Figure 1.23). Peroxisomes function in the removal of hydrogens from organic substrates, consuming O_2 in the process, according to two reactions:

$$RH_2 + O_2 \rightarrow R + H_2O_2 \quad (1.1)$$

$$H_2O_2 \rightarrow H_2O + \frac{1}{2}O_2 \quad (1.2)$$

where R is the organic substrate. The peroxide produced in copious amounts in these reactions gives this organelle its name.

The endogenous substrate for plant peroxisomes is glycolic acid, the two-carbon organic acid that is generated in the chloroplast as the result of the wasteful oxygenase reaction of rubisco. This salvage pathway, which also involves mitochondria, is referred to as *photorespiration* because oxygen is consumed and CO_2 is released (see Chapter 8). Peroxisomes are thus a major site of O_2 consumption in the cell. One peroxisomal enzyme, catalase, constitutes up to 40% of the total protein of the peroxisome (see Figure 1.23). Catalase breaks down the hydrogen peroxide generated as a consequence of glycolate oxidation to the less harmful products water and oxygen (Equation 1.2) (see Chapter 11).

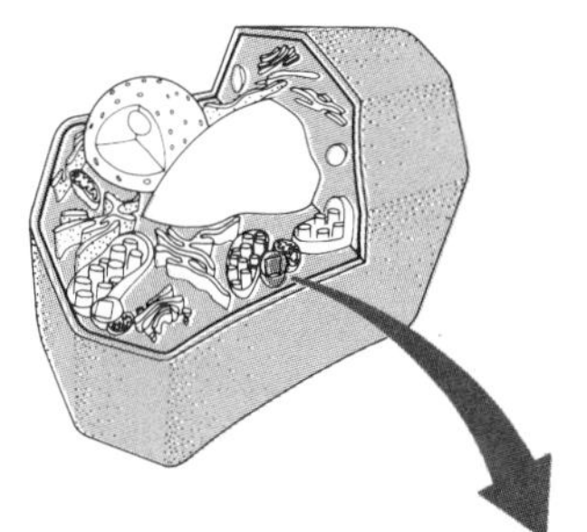

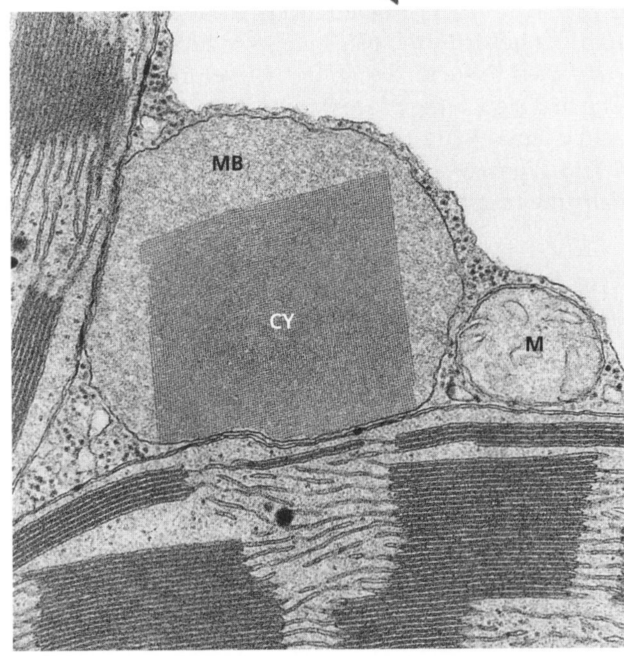

Figure 1.23 Electron micrograph of a peroxisome from a mesophyll cell, showing a crystalline core. (27,000×) This peroxisome is seen in close association with two chloroplasts and a mitochondrion, probably reflecting the cooperative role of these three organelles in photorespiration. MB, microbody; CY crystalline core; M, mitochondrion. (Photo courtesy of E. H. Newcomb.)

Another type of microbody, the **glyoxysome**, is present in oil-storing seeds. Glyoxysomes contain the **glyoxylate cycle** enzymes, which help convert stored fatty acids into sugars that can be translocated throughout the young plant to provide energy for growth (see Chapter 11).

Oleosomes Store Lipids

In addition to starch and protein, many plants synthesize and store large quantities of triacylglycerol in the form of oil during seed development. These oils accumulate in organelles called **oleosomes**, also referred to as *lipid bodies* or *spherosomes*. Oleosomes are unique among the organelles in that they are surrounded by a "half–unit membrane"—that is, a phospholipid monolayer—probably derived from the ER (Harwood 1997). The phospholipids in the half–unit membrane are oriented with their polar head groups toward the aqueous phase and their hydrophobic fatty acid tails facing the lumen, dissolved in the stored lipid. Oleosomes are thought to arise from the deposition of lipids within the bilayer itself. Proteins, called **oleosins**, are present in the half–unit membrane (Murphy 1990). One of the functions of the oleosins may be to maintain each oleosome as a discrete organelle by preventing fusion.

Oleosins may also help other proteins to bind to the organelle surface. As noted earlier, during seed germination the lipids in the oleosomes are broken down and converted to sucrose with the help of the glyoxysome. The first step in the process is the hydrolysis of the fatty acid chains from the glycerol backbone by the enzyme lipase. Lipase is tightly associated with the surface of the half–unit membrane, and may be attached to the oleosins.

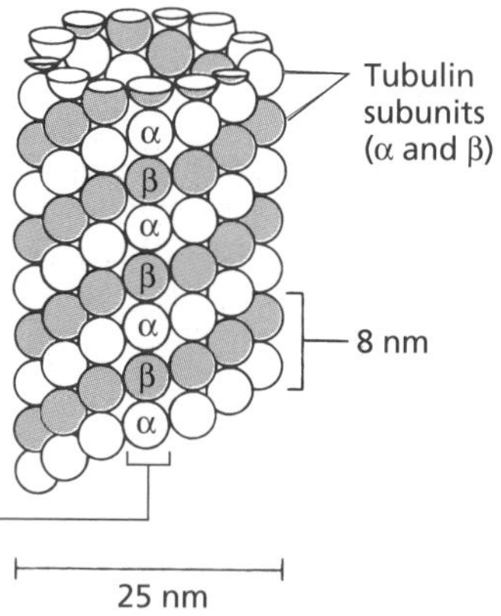

The Cytoskeleton

The cytosol is organized into a three-dimensional network of filamentous proteins called the **cytoskeleton.** This network provides the spatial organization for the organelles and serves as a scaffolding for the movements of organelles and other cytoskeletal components. It also plays fundamental roles in mitosis, meiosis, cytokinesis, wall deposition, the maintenance of cell shape, and cell differentiation.

Plant Cells Contain Microtubules, Microfilaments, and Intermediate Filaments

Three types of cytoskeletal elements have been demonstrated in plant cells: microtubules, microfilaments, and intermediate filaments. Each type is filamentous, having a fixed diameter and a variable length, up to many micrometers. **Microtubules** are hollow cylinders with an outer diameter of 25 nm. **Microfilaments** are solid, with a diameter of 7 nm. **Intermediate filaments** are a diverse group of helically wound fibrous elements, 10 nm in diameter.

Microtubules and microfilaments are macromolecular assemblies of globular proteins. Microtubules are composed of polymers of the protein **tubulin**. The tubulin monomer of microtubules is a heterodimer composed of two similar polypeptide chains (α- and β-tubulin), each having an apparent molecular mass of 55,000 daltons (Figure 1.24). A single microtubule consists of hundreds of thousands of tubulin monomers arranged in 13 columns called *protofilaments.*

Microfilaments are composed of a special form of the protein found in muscle, globular actin or **G-actin**. Each actin molecule is composed of a single polypeptide with a molecular mass of approximately 42,000 daltons. A microfilament consists of two chains of polymerized actin subunits that intertwine in a helical fashion (Figure 1.25).

Intermediate filaments are composed of linear polypeptide monomers of various types. For example, in animal cells, the **nuclear lamins** are composed of a specific polypeptide monomer, while the **keratins**, a type of intermediate filament found in the cytoplasm, are composed of a different polypeptide monomer. Pairs of parallel monomers (i.e., aligned with their $—NH_2$ groups at the same ends) are helically wound around each other in a **coiled coil.** Two coiled-coil dimers then align with each other in an antiparallel fashion (i.e., aligned with their NH_2 groups at opposite ends) to form a tetrameric unit. The tetrameric units then assemble into the final intermediate filament (Figure 1.26).

Figure 1.24 Drawing of a microtubule in longitudinal view. Each microtubule is composed of 13 protofilaments. The organization of the α and β subunits is shown.

Microtubules and Microfilaments Can Assemble and Disassemble

In the cell, actin and tubulin monomers exist as pools of free proteins that are in dynamic equilibrium with the polymerized forms. Polymerization requires energy: ATP is required for microfilament polymerization, GTP for microtubule polymerization. The attachments between subunits in the polymer are noncovalent, but they are strong enough to render the structure stable under cellular conditions. Both microtubules and microfilaments are polarized. In microtubules, the polarity arises from the polarity of the α- and β-tubulin heterodimer; in microfilaments, the polarity arises from the polarity of the actin monomer itself. The opposite ends of microtubules and microfilaments are termed **plus** and **minus**, and polymerization is more rapid at the positive end. Once formed, microtubules and microfilaments can disassemble. The overall *rate* of assembly and disas-

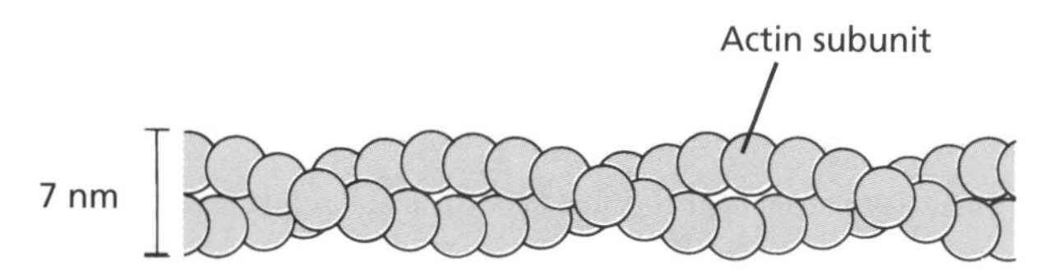

Figure 1.25 Diagrammatic representation of a microfilament, showing two strands of actin subunits.

Figure 1.26 The current model for the assembly of intermediate filaments from protein monomers. (A) Monomer with an extended α-helical region. (B) Coiled-coil dimer in parallel orientation (i.e., with amino and carboxyl termini at the same ends). (C) A tetramer of two dimers. Note that the dimers are arranged in an antiparallel fashion, and that one is slightly offset from the other. (D) Two tetramers. (E) Tetramers packed together to form the 10 nm intermediate filament. (From Alberts et al. 1994.)

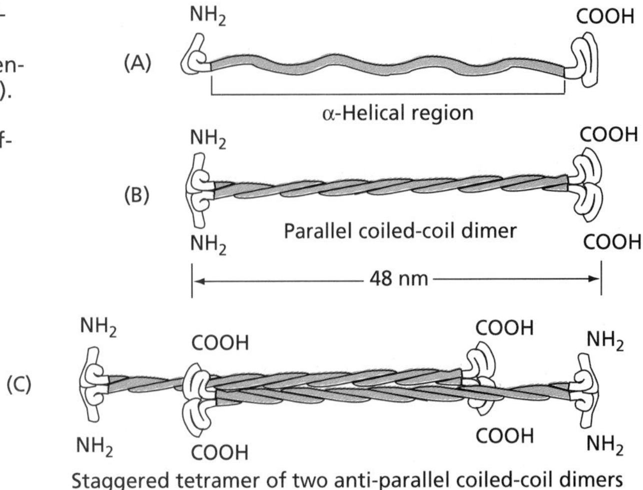

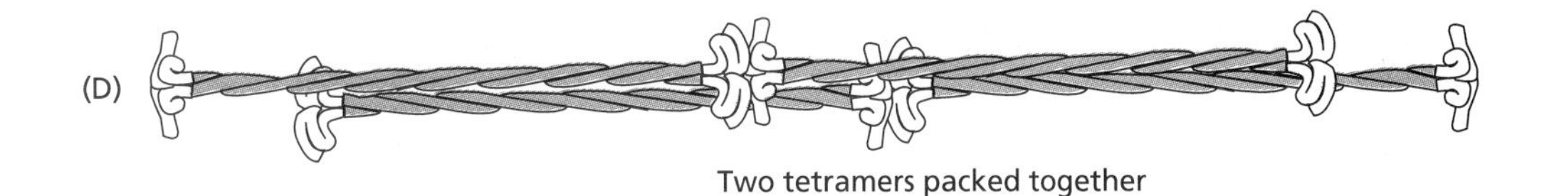

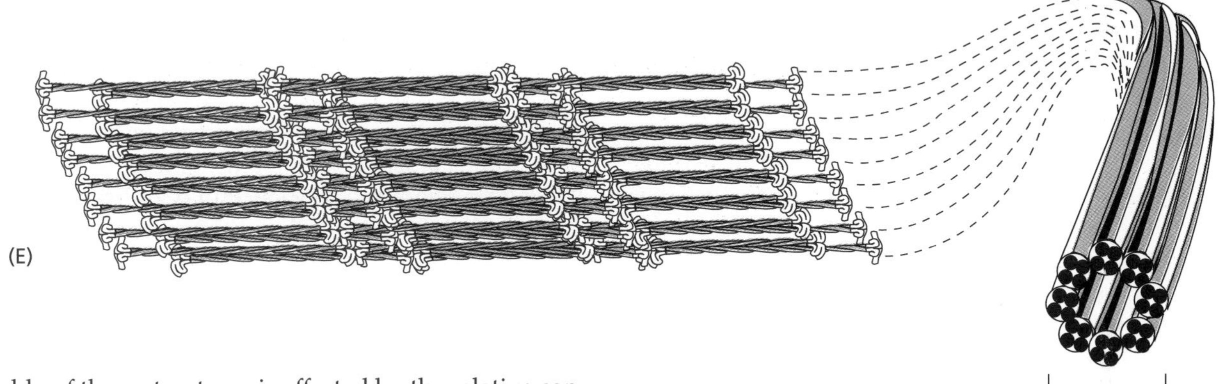

sembly of these structures is affected by the relative concentrations of free or assembled subunits. In general, microtubules are more unstable than microfilaments. In animal cells, the half-life for an individual microtubule is about 10 minutes. Thus microtubules are said to exist in a state of **dynamic instability**.

In contrast to microtubules and microfilaments, intermediate filaments lack polarity because of the antiparallel orientation of the dimers that make up the tetramers. In addition, intermediate filaments appear to be much more stable than either microtubules or microfilaments. Although relatively little is known about the intermediate filaments of plant cells, in animal cells nearly all of the intermediate-filament protein exists in the polymerized state.

Microtubules Function in Mitosis and Cytokinesis

Mitosis is the process by which previously replicated chromosomes are aligned, separated, and distributed in an orderly fashion to daughter cells (Figure 1.27). Microtubules are an integral part of mitosis. Before mitosis begins, microtubules in the cytoplasm depolymerize, breaking down into their constituent subunits. The subunits then repolymerize early in prophase to form the microtubules that are characteristic of the **spindle apparatus**. In plants, the *spindle pole region* at each end of the spindle contains a **microtubule organizing center (MTOC)**, a poorly defined region of the cytosol that is the site of microtubule formation. (In animals, structures called *centrioles* serve as MTOCs during mitosis.) Some of the microtubules of the spindle apparatus become attached to the chromosomes at their **kinetochores** (kinetochore microtubules), while others remain unattached (nonkinetochore microtubules). The kinetochores are located in the **centromeric** regions of the chromosomes. Some of the unattached microtubules overlap with microtubules from the opposite polar region in the spindle midzone.

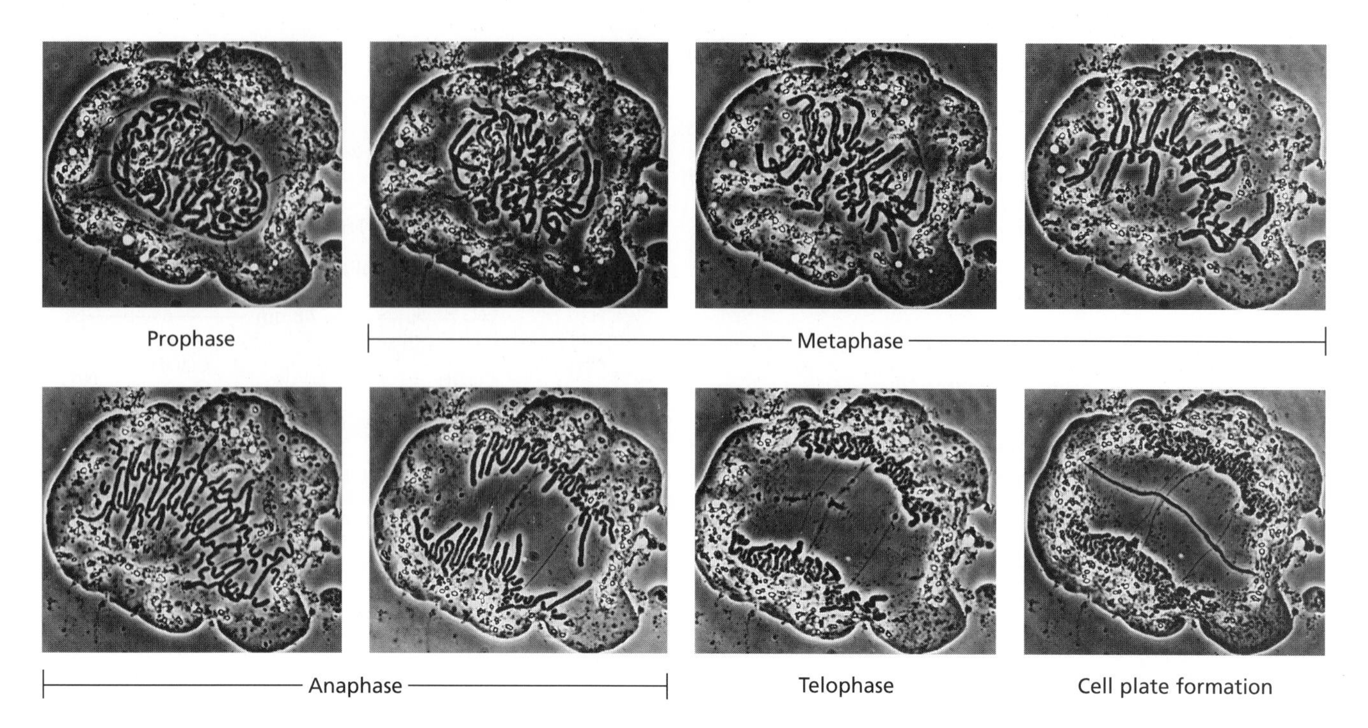

Figure 1.27 Mitosis in a living endosperm cell of the African blood lily, *Haemanthus katharinae*, recorded with time-lapse microscopy. (Photos courtesy of W. T. Jackson.)

Cytokinesis is the process whereby a cell is partitioned into two progeny cells. Cytokinesis usually begins late in mitosis. The precursor of the new wall that forms between incipient daughter cells is called the **cell plate**, and it is rich in pectins. The process of vesicle aggregation in the spindle midzone is organized by the **phragmoplast**, a complex of microtubules and ER that forms during late anaphase or early telophase from dissociated spindle subunits.

Cell plate formation in higher plants is a multi-step process. In the first step, Golgi vesicles, some of which are interconnected via *fusion tubes*, aggregate in the spindle midzone area. This structure is called the fusion tube network (FTN) (Figure 1.28A and F). The contents of the vesicles, mainly pectins, represent the precursors from which the new middle lamella is assembled outside the cell. In the next stage, vesicle fusion increases, forming a tubulo-vesicular network (TVN), and the membranes become coated with either clathrin or other proteins (Figure 1.28B). The transition from the first to the second stage of cell plate formation can be inhibited by caffeine. In the third stage, the central region of the growing cell plate forms a tubular network (TN), with vesicle fusion occurring at the growing edges where the remaining microtubules are located (Figure 1.28C). In the final stage, the cell plate contacts and adheres to the plasma membrane of the parent cell. At the same time the tubular network expands to form a fenestrated sheet (Figure 1.28D). At the end of mitosis, the phragmoplast disappears, the cell enters interphase, and microtubules reappear in the cytosol near the plasma membrane, where they play a role in the deposition of cellulose microfibrils during cell wall growth (Figure 1.28E) (see Chapter 15).

Cytoskeletal Components Determine the Plane of Cell Division

The plane in which the cell plate forms establishes the relationship of the progeny cells, and microtubules play a role in this process. After the cytoplasmic microtubules break down, but before the spindle forms and mitosis begins, a circular array of microtubules forms just below the plasma membrane, forming a conspicuous ring around the nucleus. This microtubule array is called the **preprophase band** (**PPB**), and it appears in the region where the cell plate forms after the completion of mitosis (see Chapter 16). In many cells, a band of actin microfilaments co-localizes with the microtubular PPB. This actin band is usually wider than the ring of microtubules, in part because cortical actin microfilaments are not restricted to this band during prophase (Cleary et al. 1992).

At the end of prophase, both the microtubules and the actin microfilaments of the PPB disappear from the cell cortex, but cortical actin remains elsewhere, resulting in the formation of an **actin-depleted zone** (**ADZ**) at the site of the former PPB (Cleary et al. 1992). It has been suggested that the ADZ, which persists throughout mitosis and cytokinesis, may provide the "memory" that directs the phragmoplast to the site of the PPB. In addi-

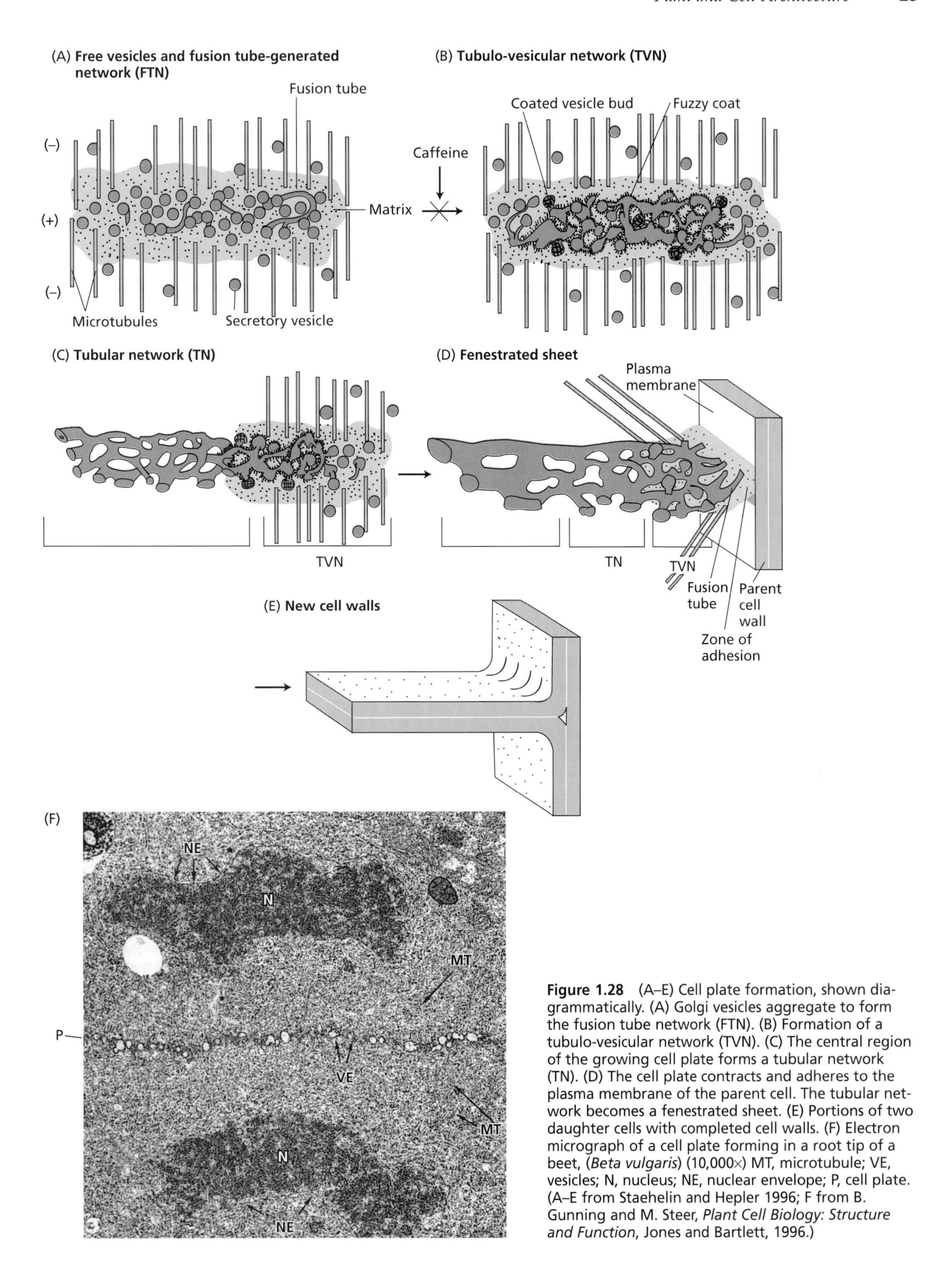

Figure 1.28 (A–E) Cell plate formation, shown diagrammatically. (A) Golgi vesicles aggregate to form the fusion tube network (FTN). (B) Formation of a tubulo-vesicular network (TVN). (C) The central region of the growing cell plate forms a tubular network (TN). (D) The cell plate contracts and adheres to the plasma membrane of the parent cell. The tubular network becomes a fenestrated sheet. (E) Portions of two daughter cells with completed cell walls. (F) Electron micrograph of a cell plate forming in a root tip of a beet, (*Beta vulgaris*) (10,000×) MT, microtubule; VE, vesicles; N, nucleus; NE, nuclear envelope; P, cell plate. (A–E from Staehelin and Hepler 1996; F from B. Gunning and M. Steer, *Plant Cell Biology: Structure and Function*, Jones and Bartlett, 1996.)

tion, actin filaments connecting the edges of the phragmoplast to the cortical cytoplasm have been observed—for example, by Valster and Hepler (1997). These actin microfilaments may guide the growing phragmoplast to the correct site of fusion with the plasma membrane, thus establishing the correct plane of cell division.

Microfilaments Are Involved in Cytoplasmic Streaming and in Tip Growth

Cytoplasmic streaming is the coordinated movement of particles and organelles through the cytosol in a helical path down one side of a cell and up the other side. Cytoplasmic streaming occurs in most plant cells and has been studied extensively in the large cells of the green algae *Chara* and *Nitella*, in which speeds up to 75 $\mu m\ s^{-1}$ have been measured.

The mechanism of cytoplasmic streaming involves bundles of microfilaments that are arranged parallel to the longitudinal direction of particle movement. The shear forces necessary for movement may be generated by an interaction of the microfilament protein actin with the protein myosin in a fashion comparable to that which occurs during muscle contraction in animals. **Myosins** are proteins that have the ability to hydrolyze ATP to ADP and P_i when activated by binding to an actin microfilament. The energy released by ATP hydrolysis propels myosin molecules along the actin microfilament from the minus end to the plus end. Thus, myosins belong to the general class of **motor proteins** that drive cytoplasmic streaming and the movements of organelles within the cell. Examples of other motor proteins include the **kinesins** and **dyneins**, which drive movements of organelles and other cytoskeletal components along the surfaces of microtubules.

Microfilaments also participate in the growth of the pollen tube. Upon germination, a pollen grain forms a tubular extension that grows down the style toward the embryo sac. As the tip of the pollen tube extends, new cell wall material is continually deposited to maintain the integrity of the cell wall. A network of microfilaments appears to guide vesicles containing wall precursors from their site of formation in the Golgi through the cytosol to the site of new wall formation at the tip. Fusion of these vesicles with the plasma membrane deposits wall precursors outside the cell, where they are assembled into wall material. The regulation of pollen tube growth and the important role of calcium ions in this process will be discussed in Chapter 15.

Intermediate Filaments Occur in the Cytosol and the Nucleus of Plant Cells

Relatively little is known about plant intermediate filaments, but the information available suggests that they are as diverse as animal intermediate filaments, in terms of both their protein monomers and their locations within the cell. Intermediate filaments composed of keratin, also found in animal epithelial cells and their derivatives (hair and nails), have been identified in the leaves and cotyledons of Chinese cabbage (*Brassica pekinensis*) (Yang et al. 1995). Intermediate filament–like structures have been observed connecting the surface of the nucleus to the cell periphery of interphase tobacco cells, and around the spindle poles (Mizuno 1995). The functions of these intermediate filaments are unknown, but they may, as in animals, provide structural rigidity to the cell.

In animal cells, the inner surface of the inner nuclear envelope is coated with a dense network of specialized intermediate filaments called **lamins**. This mesh of intermediate filaments is called the **nuclear lamina** (see Figure 1.13). Lamin-like intermediate filaments have been detected in plant nuclei (Frederick et al. 1992), but it is not yet known whether the lamin-like intermediate filaments of plants are composed of the same type of protein monomer as the animal lamins are.

The Cell Cycle Consists of Four Phases

The cell division cycle, or cell cycle, is the process by which cells reproduce themselves and their genetic material, the nuclear DNA. The four phases of the cell cycle are designated G_1, S, G_2, and M (Figure 1.29A). Each cell cycle phase is characterized by a specific set of biochemical and cellular activities. Nuclear DNA is prepared for replication in G_1 by the assembly of a prereplication complex at the origins of replication along the chromatin. DNA is replicated during the S phase, and G_2 cells prepare for mitosis. The whole architecture of the cell is altered as cells enter mitosis; the nuclear envelope breaks down, chromatin condenses to form recognizable chromosomes, the mitotic spindle forms, and the replicated chromosomes attach to the spindle fibers. The transition from metaphase to anaphase of mitosis marks a major transition point when the two chromatids of each replicated chromosome, which were held together at their kinetochores, are separated and the daughter chromosomes are pulled to opposite poles by spindle fibers.

At a key regulatory point early in G_1 of the cell cycle, the cell becomes committed to the initiation of DNA synthesis. In yeast, this point is called START. Once a cell has passed START, it is irreversibly committed to initiating DNA synthesis and completing the cell cycle through mitosis and cytokinesis. After the cell has completed mitosis, it may initiate another complete cycle (G_1 through mitosis) or it may leave the cell cycle and differentiate. This choice is made at the critical G_1 point, before the cell begins to replicate its DNA.

DNA replication and mitosis are linked in mammalian cells. Often mammalian cells that have stopped dividing can be stimulated to reenter the cell cycle by a variety of hormones and growth factors. When they do so, they reenter the cell cycle at the critical point in early G_1. In contrast, plant cells can leave the cell division cycle either before or after replicating their DNA (i.e., during

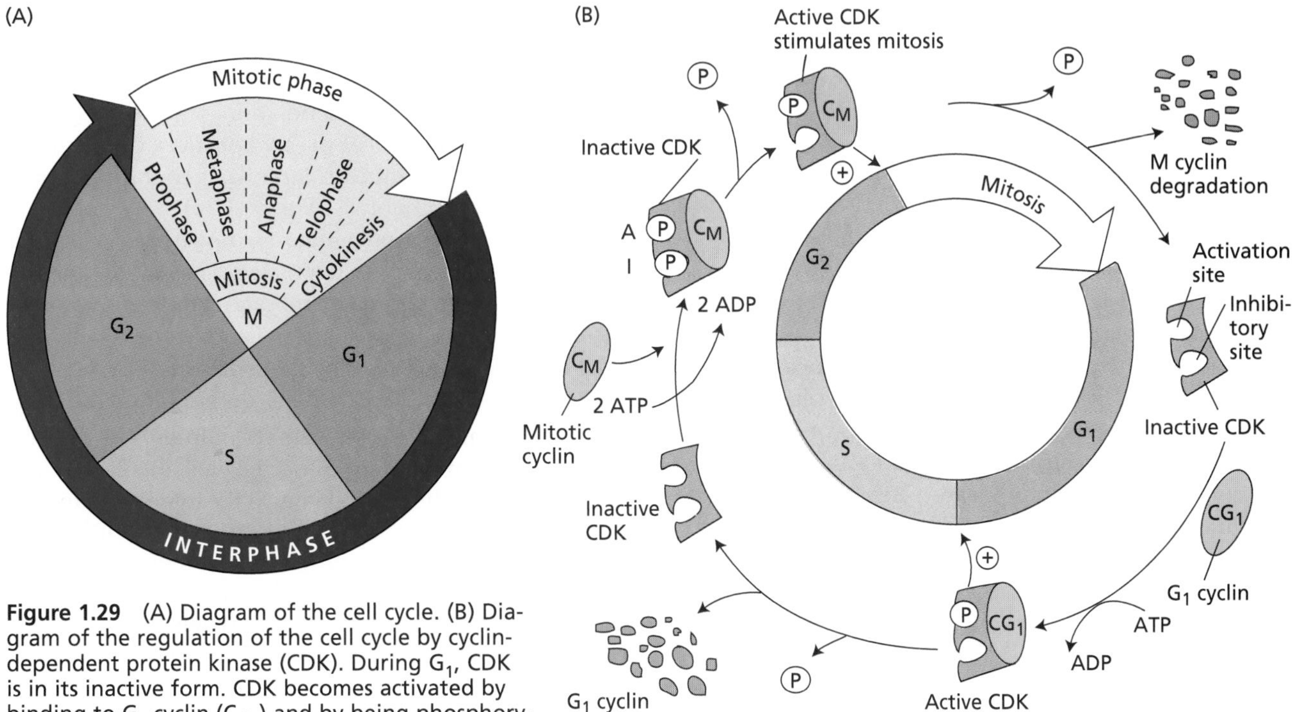

Figure 1.29 (A) Diagram of the cell cycle. (B) Diagram of the regulation of the cell cycle by cyclin-dependent protein kinase (CDK). During G_1, CDK is in its inactive form. CDK becomes activated by binding to G_1 cyclin (C_{G_1}) and by being phosphorylated (P) at the activation site (A). The activated CDK–cyclin complex allows the transition to the S phase. At the end of the S phase, the G_1 cyclin is degraded and the CDK is dephosphorylated, resulting in an inactive CDK. The cell enters G_2. During G_2, the inactive CDK binds to the mitotic cyclin (C_M), or M cyclin. At the same time, the CDK–cyclin complex becomes phosphorylated at both its activation (A) and its inhibitory (I) sites. The CDK–cyclin complex is still inactive because the inhibitory site is phosphorylated. The inactive complex becomes activated when the phosphate is removed from the inhibitory site by a protein phosphatase. The activated CDK then stimulates the transition from G_2 to mitosis. At the end of mitosis, the mitotic cyclin is degraded and the remaining phosphate at the activation site is removed by the phosphatase, and the cell enters G_1 again. (After Campbell 1996, and Alberts et al. 1994.)

G_1 or G_2). As a consequence, whereas most animal cells are diploid (having two sets of chromosomes), plant cells frequently are tetraploid (having four sets of chromosomes), or even polyploid (having more than two sets of chromosomes), after going through additional cycles of nuclear DNA replication without mitosis.

The Cell Cycle Is Regulated by Protein Kinases

The mechanism regulating the progression of cells through their division cycle is highly conserved in evolution, and plants have retained the basic components of this mechanism (Renaudin et al. 1996). The key enzymes that control the transitions between the different states of the cell cycle, and the entry of nondividing cells into the cell cycle, are the cyclin-dependent protein kinases, or CDKs (see Figure 1.29B). Protein kinases are enzymes that phosphorylate proteins using ATP. Most multicellular eukaryotes use several protein kinases that are active in different phases of the cell cycle. All depend on regulatory subunits called cyclins for their activities. The regulated activity of CDKs is essential for the transitions from G_1 to S and from G_2 to M, and for the entry of nondividing cells into the cell cycle.

CDK activity can be regulated in various ways, but two of the most important mechanisms are (1) cyclin synthesis and destruction and (2) the phosphorylation and dephosphorylation of key amino acid residues within the CDK protein. CDKs are inactive unless they are associated with a cyclin. Most cyclins turn over rapidly. They are synthesized and then actively degraded (using ATP) at specific points in the cell cycle. Cyclins are degraded in the cytosol, a process mediated by a large proteolytic complex called the *26S proteasome*. Before being degraded by the proteasome, the cyclins are marked for destruction by the attachment of a small protein called *ubiquitin*, a process that requires ATP (see Chapter 14).

The transition from G_1 to S requires a set of cyclins (known as **G_1 cyclins**) different from those required in the transition from G_2 to mitosis, where **mitotic cyclins** activate the CDK (see Figure 1.29B). The activity of many proteins is controlled by phosphorylation, including that of the cyclin-dependent protein kinases. While the phosphorylation of certain tyrosine residues within CDKs is necessary for their activity, another tyrosine residue, when phosphorylated, inhibits its activity. Specific kinases carry out both the stimulatory and the inhibitory phosphorylations. Similarly, some phosphatase enzymes can remove phosphate from CDKs, either stimulating or inhibiting their activity, depending on the position of the phosphate. The addition or removal of phosphate groups from CDKs is highly regulated and an important mechanism for the control of cell cycle progression (Jacobs 1995). Cyclin inhibitors play an important role in regulating the cell cycle in animals, and probably in plants as well, although little is known about plant cyclin inhibitors.

Plasmodesmata Interconnect Living Plant Cells

Plasmodesmata (singular, plasmodesma) are tubular extensions of the plasma membrane, 40 to 50 nm in diameter, that traverse the cell wall and connect the cytoplasms of adjacent cells. Because most plant cells are interconnected in this way, their cytoplasms form a continuum referred to as the **symplast**. Intercellular transport of solutes through plasmodesmata is thus called **symplastic transport** (see Chapters 4 and 6).

Developmentally, plasmodesmata can be classified into two types: primary and secondary. **Primary plasmodesmata** form during cytokinesis when Golgi-derived vesicles containing cell wall precursors fuse to form the cell plate (the future middle lamella). Rather than forming a continuous uninterrupted sheet, the newly deposited cell plate is penetrated by numerous pores (see Figure 1.28) where remnants of the spindle apparatus, consisting of ER and microtubules, disrupt vesicle fusion. Further deposition of wall polymers increases the thickness of the two primary cell walls on either side of the middle lamella, generating linear membrane-lined channels (Figure 1.30).

Secondary plasmodesmata form between cells after their cell walls have been deposited. They arise either by evagination of the plasma membrane at the cell surface, or by branching from a primary plasmodesma (Lucas and Wolf 1993). Besides increasing the communication between cells that are clonally related (i.e., derived from the same mother cell), secondary plasmodesmata allow symplastic continuity between cells that are not clonally related.

Plasmodesmata can also be classified according to their morphology, as simple or branched. **Simple plasmodesmata** are single unbranched channels, which may be either primary or secondary in origin. **Branched plasmodesmata** consist of multiple channels, usually interconnected at a median cavity, and can arise during cell plate formation (primary) or de novo (secondary) (McLean et al. 1997).

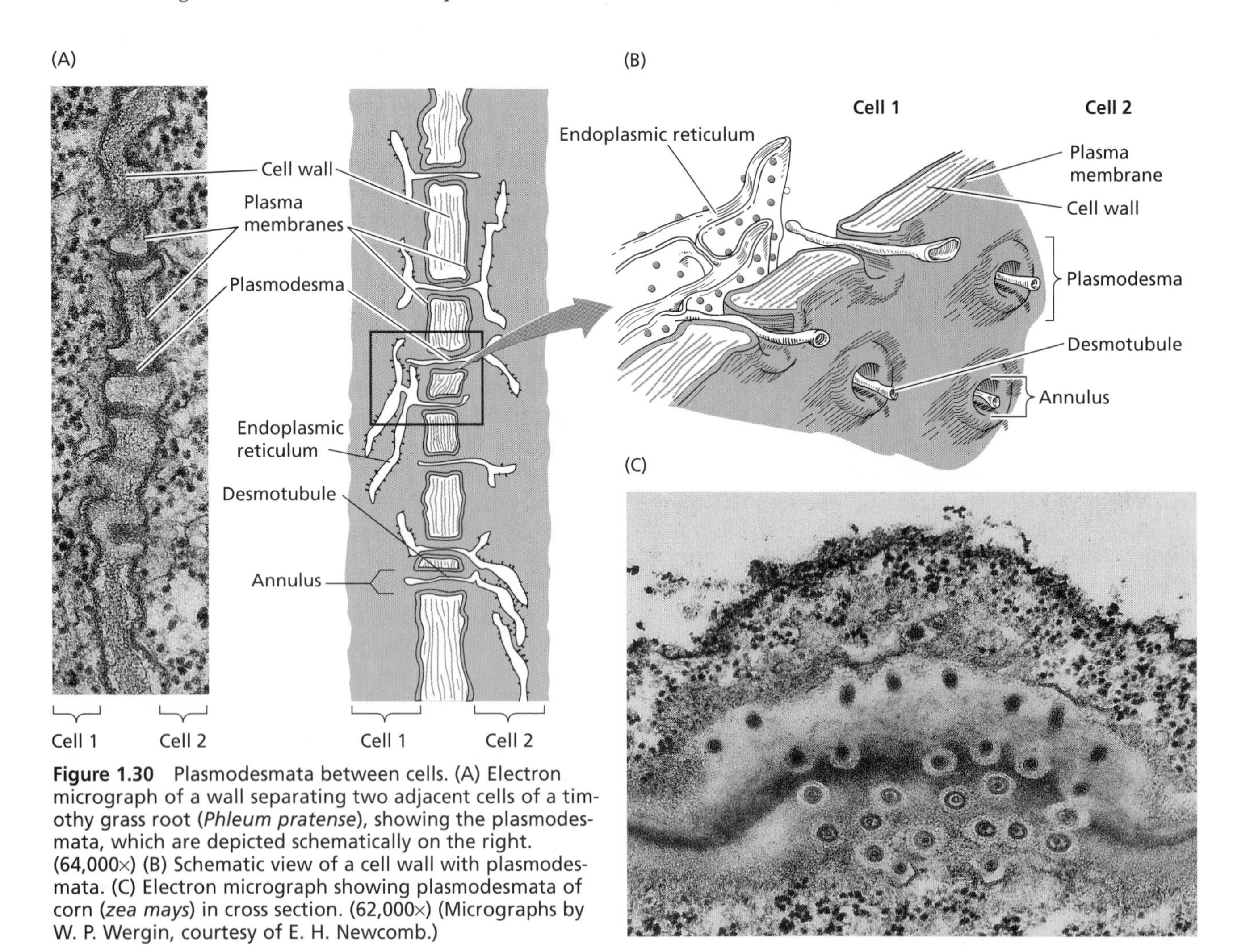

Figure 1.30 Plasmodesmata between cells. (A) Electron micrograph of a wall separating two adjacent cells of a timothy grass root (*Phleum pratense*), showing the plasmodesmata, which are depicted schematically on the right. (64,000×) (B) Schematic view of a cell wall with plasmodesmata. (C) Electron micrograph showing plasmodesmata of corn (*zea mays*) in cross section. (62,000×) (Micrographs by W. P. Wergin, courtesy of E. H. Newcomb.)

Each plasmodesma contains a narrow tubule of ER called a **desmotubule** (Figure 1.31). The desmotubule is continuous with the ER of the adjacent cells. Thus, the symplast joins not only the cytosol of neighboring cells, but the contents of the ER lumens as well. However, it is not clear that the desmotubule actually represents a passage since there does not appear to be a space between the membranes, which are tightly appressed.

Globular proteins are associated with both the desmotubule membrane and the plasma membrane within the pore (see Figure 1.31). These globular proteins appear to be interconnected, dividing the pore into eight to ten microchannels (Ding et al. 1992). Some molecules can pass from cell to cell through plasmodesmata, probably by flowing through the microchannels, although the exact pathway of communication has not been established.

By following the movement of fluorescent dye molecules of different sizes through plasmodesmata connecting leaf epidermal cells, the limiting molecular mass for transport was determined to be about 700 to 1000 daltons, equivalent to a molecular size of about 1.5 to 2.0 nm (Robards and Lucas 1990). This is the **size exclusion limit**, or **SEL**, of plasmodesmata. Since the width of the cytoplasmic sleeve is approximately 5 to 6 nm, how are molecules larger that 1.6 nm excluded? The proteins attached to the plasma membrane and the ER within the plasmodesmata appear to act to restrict the size of molecules that can pass through the pore. As we'll see in Chapter 16, the SELs of plasmodesmata can be regulated. The mechanism for regulating the SEL is poorly understood, but the localization of both actin and myosin within plasmodesmata suggests that they may participate in the process (White et al. 1994, Radford and White 1996).

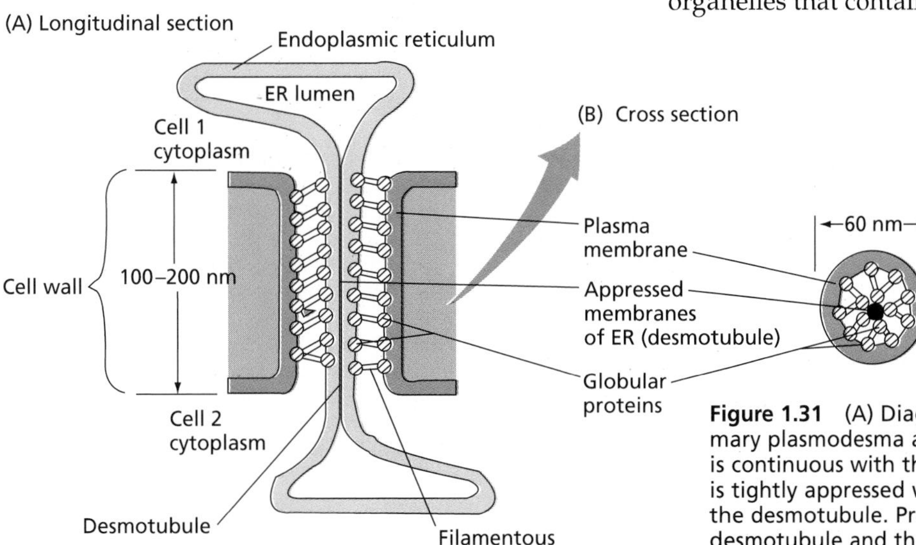

Figure 1.31 (A) Diagrammatic representation of a primary plasmodesma and desmotubule. The desmotubule is continuous with the ER of the adjoining cells. The ER is tightly appressed within the plasmodesma to form the desmotubule. Proteins line the outer surface of the desmotubule and the inner surface of the plasma membrane; they are thought to be connected by filamentous proteins. (B) Diagram of a cross section through the primary plasmodesma, showing the same structures. (After Lucas and Wolf 1993.)

Summary

Despite their great diversity in form and size, all plants carry out similar physiological processes. As primary producers, plants convert solar energy to chemical energy. Being nonmotile, plants must grow toward light, and they must have efficient vascular systems for movement of water, mineral nutrients, and photosynthetic products throughout the plant body. Green land plants must also have mechanisms for avoiding desiccation.

The major vegetative organ systems of seed plants are the shoot and the root. The shoot consists of two types of organs: stems and leaves. Unlike animal development, plant growth is indeterminate because of the presence of permanent meristem tissue at the shoot and root apices, which gives rise to new tissues and organs during the entire vegetative phase of the life cycle. Lateral meristems (the vascular cambium and the cork cambium) produce growth in girth, or secondary growth.

Three major tissue systems are recognized: dermal, ground, and vascular. Each of these tissues contains a variety of cell types specialized for different functions, such as support (sclerenchyma, collenchyma), photosynthesis (mesophyll parenchyma), storage (parenchyma), gas exchange (guard cells), water conduction (xylem tracheids and vessel elements), and sugar translocation (phloem sieve tube elements).

Plants are eukaryotes and have the typical eukaryotic cell organization consisting of nucleus and cytoplasm. The nuclear genome directs the growth and development of the organism. The cytoplasm is enclosed by a plasma membrane and contains numerous membrane-enclosed organelles, including plastids, mitochondria, microbodies, oleosomes, and a large central vacuole. Chloroplasts and mitochondria are semiautonomous organelles that contain their own DNA. Nevertheless, most of their proteins are encoded by nuclear DNA and are imported from the cytosol.

The cytoskeletal components—microtubules, microfilaments, and intermediate filaments—participate in a variety of processes involving intracellular movements, such as mitosis, cytoplasmic streaming, secretory vesicle transport, cell plate formation, and cellulose microfibril deposition. The process by which cells reproduce is called the cell cycle. The cell cycle consists of the G_1, S, G_2, and M phases. The transition from one phase to another is regulated by cyclin-dependent protein kinases. The activity of the CDKs is regulated by cyclins and by protein phosphorylation.

During cytokinesis, the phragmoplast gives rise to the cell plate in a multistep process that involves vesicle fusion. After cytokinesis, primary cell walls are deposited. The cytosol of adjacent cells is continuous through the cell walls because of the presence of membrane-lined channels called plasmodesmata, which play a role in cell–cell communication.

General Reading

*Alberts, B., Bray, D., Lewis, J., Raff, M., Roberts, K., and Watson, J. D. (1994) *Molecular Biology of the Cell*, 3rd ed. Garland, New York.

Becker, W. M., Reece, J. B., and Poenie, M. F. (1996) *The World of the Cell*, 3rd ed. Benjamin/Cummings, Menlo Park, CA.

Burgess, I. (1985) *An Introduction to Plant Cell Development.* Cambridge University Press, Cambridge.

Cutter, E. G. (1971) *Plant Anatomy: Experiment and Interpretation*, Vols. 1–2. Addison-Wesley, Reading, Mass.

Dey, P. M., and Harborne, J. B., eds. (1997) *Plant Biochemistry.* Academic Press, San Diego.

*Esau, K. (1960) *Anatomy of Seed Plants.* Wiley, New York.

*Esau, K. (1977) *Anatomy of Seed Plants*, 2nd ed. Wiley, New York.

*Gunning, B. E. S., and Steer, M. W. (1996) *Plant Cell Biology: Structure and Function of Plant Cells.* Jones and Bartlett, Boston.

Karp, G. (1996) *Cell and Molecular Biology. Concepts and Experiments.* Wiley, New York.

*Ledbetter, M. C., and Porter, K. R. (1970) *Introduction to the Fine Structure of Plant Cells.* Springer, Berlin.

Lodish, H., Baltimore, D., Berk, A., Zipursky, S., Matsidaira, P., and Darnell, J. (1995) *Molecular Cell Biology.* Scientific American Books, New York.

Lucas, W. J., Ding, B., and Van Der Schoot, C. (1993) Plasmodesmata and the supracellular nature of plants. Tansley Review No. 58. *New Phytol.* 125: 435–476.

Meylan, B. A., and Burterfield, B. G. (1972) *Three Dimensional Structure of Wood: A Scanning Electron Microscope Study.* Syracuse University Press, Syracuse, NY.

*O'Brien, T. P., and McCully, M. E. (1969) *Plant Structure and Development.* Macmillan, London.

Tobin, A. J., and Morel, R. E. (1997) *Asking about Cells.* Harcourt Brace, Fort Worth, TX.

* Indicates a reference that is general reading in the field and is also cited in this chapter.

Chapter References

Campbell, N.A. (1996) *Biology*, 4th ed. Benjamin/Cummings. Menlo Park, CA.

Cleary, A. L., Gunning, B. E. S., Wasteneys, G. O. and Hepler, P. K. (1992) Microtubule and F-actin dynamics at the division site in living *Tradescantia* stamen hair cells. *J. Cell Sci.* 103: 977–988.

Depta, H., Freundt, H., Hartmann, D., and Robinson, D. G. (1987) Preparation of a homogeneous coated vesicle fraction from bean leaves. *Protoplasma* 136: 154–160.

Ding, B., Turgeon, R., and Parthasarathy, M. V. (1992) Substructure of freeze substituted plasmodesmata. *Protoplasma* 169: 28–41.

Driouich, A., Levy, S., Staehelin, L. A., and Faye, F. (1994) Structural and functional organization of the Golgi apparatus in plant cells. *Plant Physiol. Biochem.* 32: 731–749.

Faye, L., Fitchette-Lainé, A. C., Gomord, V., Chekkafi, A., Delaunay, A. M., and Driouich, A. (1992) Detection, biosynthesis and some functions of glycans N-linked to plant secreted proteins. In *Post-translational Modifications in Plants*, N. H. Battey, H. G. Dickinson, and A. M. Heatherington, eds., SEB Seminar series, no. 53, Cambridge University Press, Cambridge, pp. 213–242.

Frederick, S. E., Mangan, M. E., Carey, J. B., and Gruber, P. J. (1992) Intermediate filament antigens of 60 and 65 kDa in the nuclear matrix of plants: Their detection and localization. *Exp. Cell Res.* 199: 213–222.

Gogarten, J. P., Kibak, H., Dittrich, P., Taiz, L., Bowman, E. J., Bowman, B. J., Manolson, M. F., Poole, R. J., Date, R., Oshima, T., Denda, K., and Yoshida, M. (1989) Evolution of the vacuolar proton ATPase: Implications for the origin of eukaryotes. *Proc. Natl. Acad. Sci. USA* 86: 6661–6665.

Harwood, J. L. (1997) Plant lipid metabolism. In *Plant Biochemistry*, P. M. Dey and J. B. Harborne, eds., Academic Press, San Diego, pp. 237–272.

Jacobs, T. W. (1995) Cell cycle control. *Annu. Rev. Plant Physiol. Plant Mol. Biol.* 46: 317–339.

Lucas, W. J., and Wolf, S. (1993) Plasmodesmata: The intercellular organelles of green plants. *Trends Cell Biol.* 3: 308–315.

McLean, B. G., Hempel, F. D., and Zambryski, P. C. (1997) Plant intercellular communication via plasmodesmata. *Plant Cell* 9: 1043–1054.

Mizuno, K. (1995) A cytoskeletal 50 kDa protein in higher plants that forms intermediate-sized filaments and stabilizes microtubules. *Protoplasma* 186: 99–112.

Murphy, D. J. (1990) Storage lipid bodies in plants and other organisms. *Prog. Lipid Res.* 29: 299–324.

Radford, J., and White, R. G. (1996) Preliminary localization of myosin to plasmodesmata. Third Intnl. Workshop Basic Appl. Res. Plasmodesma Biol. Zichron-Takov, Israel. pp. 37–38.

Renaudin, J-P., Doonan, J. H., Freeman, D., Hashimoto, J., Hirt, H., Inze, D., Jacobs, T., Kouchi, H., Rouze, P., Sauter, M., et al. (1996) Plant cyclins: A unified nomenclature for plant A-, B- and D-type cyclins based on sequence organization. *Plant Mol. Biol.* 32: 1003–1018.

Robards, A. W. and Lucas, W. J. (1990) Plasmodesmata. *Annu. Rev. Plant Physiol. Plant Mol. Biol.* 41: 369–419.

Staehelin, L. A., and Hepler, P. K. (1996) Cytokinesis in higher plants. *Cell* 84: 821–824.

Valster, A. H. and Hepler, P. K. (1997) Caffeine inhibition of cytokinesis: Effect on the phragmoplast cytoskeleton in living *Tradescantia* stamen hair cells. *Protoplasma* 196: 155–166.

White, R. G., Badelt, K., Overall, R. L., and Wesk, M. (1994) Actin associated with plasmodesmata. *Protoplasma* 180: 169–184.

Woese, C. R., Kandler, O., and Wheelis, M. L. (1990) Towards a natural system of organisms: Proposal for the domains Archaea, Bacteria, and Eucarya. *Proc. Natl. Acad. Sci. USA* 87: 4576–4579.

Yang, C., Min, G. W., Tong, X. J., Luo, Z., Liu, Z. F., and Zhai, Z. H. (1995) The assembly of keratins from higher plant cells. *Protoplasma* 188: 128–132.

2 Energy and Enzymes

The force that through the green fuse drives the flower
Drives my green age; that blasts the roots of trees
Is my destroyer.
And I am dumb to tell the crooked rose
My youth is bent by the same wintry fever.

The force that drives the water through the rocks
Drives my red blood; that dries the mouthing streams
Turns mine to wax.
And I am dumb to mouth unto my veins
How at the mountain spring the same mouth sucks.

Dylan Thomas, *Collected Poems* (1952)

In these opening stanzas from Dylan Thomas's famous poem, the poet proclaims the essential unity of the forces that propel animate and inanimate objects alike, from their beginnings to their ultimate decay. Scientists call this force energy. Energy transformations play a key role in all the physical and chemical processes that occur in living systems. But energy alone is insufficient to drive the growth and development of organisms. Protein catalysts called enzymes are required to ensure that the rates of biochemical reactions are rapid enough to support life. In this chapter we will examine basic concepts about energy, the way in which cells transform energy to perform useful work (bioenergetics), and the structure and function of enzymes.

Energy Flow through Living Systems

The flow of matter through individual organisms and biological communities is part of everyday experience; the flow of energy is not, even though it is central to the very existence of living things.

What makes concepts such as energy, work, and order so elusive is their insubstantial nature: We find it far easier to visualize the dance of atoms and molecules than the forces and fluxes that determine the direction and extent of natural processes. The branch of physical science that deals with such matters is thermodynamics, an abstract and demanding discipline that most biologists are content to skim over lightly. Yet bioenergetics is so shot through with concepts and quantitative relationships derived from thermodynamics that it is scarcely possible to discuss the subject without frequent reference to free energy, potential, entropy, and the second law.

The purpose of this chapter is to collect and explain, as simply as possible, the fundamental thermodynamic concepts and relationships that recur throughout this book. Readers who prefer a more extensive treatment of the subject should consult either the introductory texts by Klotz (1967) and by Nicholls and Ferguson (1992) or the advanced texts by Morowitz (1978) and by Edsall and Gutfreund (1983).

Thermodynamics evolved during the nineteenth century out of efforts to understand how a steam engine works and why heat is produced when one bores a cannon. The very name "thermodynamics," and much of the language of this science, recall these historical roots, but it would be more appropriate to speak of energetics, for the principles involved are universal. Living plants, like all other natural phenomena, are constrained by the laws of thermodynamics. By the same token, thermodynamics supplies an indispensable framework for the quantitative description of biological vitality.

Energy and Work

Let us begin with the meanings of "energy" and "work." **Energy** is defined in elementary physics, as in daily life, as the capacity to do work. The meaning of work is harder to come by and more narrow. **Work**, in the mechanical sense, is the displacement of any body against an opposing force. The work done is the product of the force and the distance displaced, as expressed in the following equation:*

$$W = f\,\Delta l \tag{2.1}$$

Mechanical work appears in chemistry because whenever the final volume of a reaction mixture exceeds the initial volume, work must be done against the pressure of the atmosphere; conversely, the atmosphere performs work when a system contracts. This work is calculated by the expression $P\Delta V$ (where P stands for pressure and V for volume), a term that appears frequently in thermodynamic formulas. *In biology, work is employed in a broader sense to describe displacement against any of the forces that living things encounter or generate: mechanical, electric, osmotic, or even chemical potential.*

A familiar mechanical illustration may help clarify the relationship of energy to work. The spring in Figure 2.1 can be extended if force is applied to it over a particular distance—that is, if work is done on the spring. This work can be recovered by an appropriate arrangement of pulleys and used to lift a weight onto the table. The extended spring can thus be said to possess energy that is numerically equal to the work it can do on the weight (neglecting friction). The weight on the table, in turn, can be said to possess energy by virtue of its position in Earth's gravitational field, which can be utilized to do other work, such as turning a crank. The weight thus illustrates the concept of **potential energy**, a capacity to do work that arises from the position of an object in a field of force, and the sequence as a whole illustrates the conversion of one kind of energy into another, or **energy transduction**.

The First Law: The Total Energy Is Always Conserved

It is common experience that mechanical devices involve both the performance of work and the produc-

* We may note in passing that the dimensions of work are complex—ml^2t^{-2}—where m denotes mass, l distance, and t time, and that work is a scalar quantity, that is, the product of two vectorial terms.

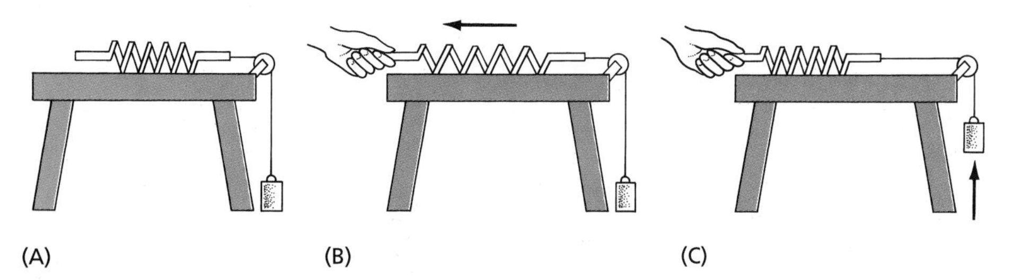

Figure 2.1 Energy and work in a mechanical system. (A) A weight resting on the floor is attached to a spring via a string. (B) Pulling on the spring places the spring under tension. (C) The potential energy stored in the extended spring performs the work of raising the weight when the spring contracts.

tion or absorption of heat. We are at liberty to vary the amount of work done by the spring, up to a particular maximum, by using different weights, and the amount of heat produced will also vary. But much experimental work has shown that, under ideal circumstances, the sum of the work done and of the heat evolved is constant and depends only on the initial and final extensions of the spring. We can thus envisage a property, the internal energy of the spring, with the characteristic described by the following equation:

$$\Delta U = \Delta Q + \Delta W \quad (2.2)$$

Here Q is the amount of heat absorbed by the system, and W is the amount of work done on the system.* In Figure 2.1 the work is mechanical, but it could just as well be electrical, chemical, or any other kind of work. Thus ΔU is the net amount of energy put into the system, either as heat or as work; conversely, both the performance of work and the evolution of heat entail a decrease in the internal energy. We cannot specify an absolute value for the energy content; only changes in internal energy can be measured. Note that Equation 2.2 assumes that heat and work are equivalent; its purpose is to stress that, under ideal circumstances, ΔU depends only on the initial and final states of the system, not on how heat and work are partitioned.

Equation 2.2 is a statement of the first law of thermodynamics, which is the principle of energy conservation. If a particular system exchanges no energy with its surroundings, its energy content remains constant; if energy is exchanged, the change in internal energy will be given by the difference between the energy gained from the surroundings and that lost to the surroundings. The change in internal energy depends only on the initial and final states of the system, not on the pathway or mechanism of energy exchange. Energy and work are interconvertible; even heat is a measure of the kinetic energy of the molecular constituents of the system. To put it as simply as possible, Equation 2.2 states that no machine, including the chemical machines that we recognize as living, can do work without an energy source.

An example of the application of the first law to a biological phenomenon is the energy budget of a leaf. Leaves absorb energy from their surroundings in two ways: as direct incident irradiation from the sun and as infrared irradiation from the surroundings. Some of the energy absorbed by the leaf is radiated back to the surroundings as infrared irradiation and heat, while a fraction of the absorbed energy is stored, as either photosynthetic products or leaf temperature changes. Thus we can write the following equation:

Total energy absorbed by leaf = energy emitted from leaf + energy stored by leaf

Note that although the energy absorbed by the leaf has been transformed, the total energy remains the same, in accordance with the first law.

The Change in the Internal Energy of a System Represents the Maximum Work It Can Do

We must qualify the equivalence of energy and work by invoking "ideal conditions"—that is, by requiring that the process be carried out reversibly. The meaning of "reversible" in thermodynamics is a special one: The term describes conditions under which the opposing forces are so nearly balanced that an infinitesimal change in one or the other would reverse the direction of the process.† Under these circumstances the process yields the maximum possible amount of work. Reversibility in this sense does not often hold in nature, as in the example of the leaf. Ideal conditions differ so little from a state of equilibrium that any process or reaction would require infinite time and would therefore not take place at all. Nonetheless, the concept of thermodynamic reversibility is useful: If we measure the change in internal energy that a process entails, we have an upper limit to the work that it can do; for any real process the maximum work will be less.

In the study of plant biology we encounter several sources of energy—notably light and chemical transformations—as well as a variety of work functions, including mechanical, osmotic, electrical, and chemical work. The meaning of the first law in biology stems from the certainty, painstakingly achieved by nineteenth-century physicists, that the various kinds of energy and work are measurable, equivalent, and, within limits, interconvertible. Energy is to biology what money is to economics: the means by which living things purchase useful goods and services.

Each Type of Energy Is Characterized by a Capacity Factor and a Potential Factor

The amount of work that can be done by a system, whether mechanical or chemical, is a function of the size of the system. Work can always be defined as the product of two factors—force and distance, for example. One is a potential or intensity factor, which is independent of the size of the system; the other is a capacity factor and is directly proportional to the size (Table 2.1).

* Equation 2.2 is more commonly encountered in the form $\Delta U = \Delta Q - \Delta W$, which results from the convention that Q is the amount of heat absorbed by the system from the surroundings and W is the amount of work done by the system on the surroundings. This convention affects the sign of W but does not alter the meaning of the equation.

† In biochemistry, reversibility has a different meaning: Usually the term refers to a reaction whose pathway can be reversed, often with an input of energy.

Table 2.1
Potential and capacity factors in energetics

Type of energy	Potential factor	Capacity factor
Mechanical	Pressure	Volume
Electrical	Electric potential	Charge
Chemical	Chemical potential	Mass
Osmotic	Concentration	Mass
Thermal	Temperature	Entropy

In biochemistry, energy and work have traditionally been expressed in calories; 1 calorie is the amount of heat required to raise the temperature of 1 g of water by 1°C, specifically, from 15.0 to 16.0°C . In principle, one can carry out the same process by doing the work mechanically with a paddle; such experiments led to the establishment of the mechanical equivalent of heat as 4.186 joules per calorie (J cal^{-1}).* We will also have occasion to use the equivalent electrical units, based on the volt: A volt is the potential difference between two points when 1 J of work is involved in the transfer of a coulomb of charge from one point to another. (A coulomb is the amount of charge carried by a current of 1 ampere [A] flowing for 1 s. Transfer of 1 mole [mol] of charge across a potential of 1 volt [V] involves 96,500 J of energy or work.) The difference between energy and work is often a matter of the sign. Work must be done to bring a positive charge closer to another positive charge, but the charges thereby acquire potential energy, which in turn can do work.

The Direction of Spontaneous Processes

Left to themselves, events in the real world take a predictable course. The apple falls from the branch. A mixture of hydrogen and oxygen gases is converted into water. The fly trapped in a bottle is doomed to perish, the pyramids to crumble into sand; things fall apart. But there is nothing in the principle of energy conservation that forbids the apple to return to its branch with absorption of heat from the surroundings or that prevents water from dissociating into its constituent elements in a like manner. The search for the reason that neither of these things ever happens led to profound philosophical insights and generated useful quantitative statements about the energetics of chemical reactions and the amount of work that can be done by them. Since living things are in many respects chemical machines, we must examine these matters in some detail.

* In current standard usage based on the meter, kilogram, and second, the fundamental unit of energy is the joule (1 J = 0.24 cal) or the kilojoule (1 kJ = 1000 J).

The Second Law: The Total Entropy Always Increases

From daily experience with weights falling and warm bodies growing cold, one might expect spontaneous processes to proceed in the direction that lowers the internal energy—that is, the direction in which ΔU is negative. But there are too many exceptions for this to be a general rule. The melting of ice is one exception: An ice cube placed in water at 1°C will melt, yet measurements show that liquid water (at any temperature above 0°C) is in a state of higher energy than ice; evidently, some spontaneous processes are accompanied by an increase in internal energy. Our melting ice cube does not violate the first law, for heat is absorbed as it melts. This suggests that there is a relationship between the capacity for spontaneous heat absorption and the criterion determining the direction of spontaneous processes, and that is the case. The thermodynamic function we seek is called **entropy**, the amount of energy in a system not available for doing work, corresponding to the degree of randomness of a system. Mathematically, entropy is the capacity factor corresponding to temperature, Q/T. We may state the answer to our question, as well as the second law of thermodynamics, thus: The direction of all spontaneous processes is to increase the entropy of a system plus its surroundings.

Few concepts are so basic to a comprehension of the world we live in, yet so opaque, as entropy—presumably because entropy is not intuitively related to our sense perceptions, as mass and temperature are. The explanation given here follows the particularly lucid exposition by Atkinson (1977), who states the second law in a form bearing, at first sight, little resemblance to that given above:

> We shall take [the second law] as the concept that any system not at absolute zero has an irreducible minimum amount of energy that is an inevitable property of that system at that temperature. That is, a system requires a certain amount of energy just to be at any specified temperature.

The molecular constitution of matter supplies a ready explanation: Some energy is stored in the thermal motions of the molecules and in the vibrations and oscillations of their constituent atoms. We can speak of it as isothermally unavailable energy, since the system cannot give up any of it without a drop in temperature (assuming that there is no physical or chemical change). The isothermally unavailable energy of any system increases with temperature, since the energy of molecular and atomic motions increases with temperature. Quantitatively, the isothermally unavailable energy for a particular system is given by ST, where T is the absolute temperature and S is the entropy.

But what is this thing, entropy? Reflection on the nature of the isothermally unavailable energy suggests that, for any particular temperature, the amount of such energy will be greater the more atoms and molecules are free to move and to vibrate—that is, the more chaotic is the system. By contrast, the orderly array of atoms in a crystal, with a place for each and each in its place, corresponds to a state of low entropy. At absolute zero, when all motion ceases, the entropy of a pure substance is likewise zero; this statement is sometimes called the third law of thermodynamics.

A large molecule, a protein for example, within which many kinds of motion can take place, will have considerable amounts of energy stored in this fashion—more than would, say, an amino acid molecule. But the entropy of the protein molecule will be less than that of the constituent amino acids into which it can dissociate, because of the constraints placed on the motions of those amino acids as long as they are part of the larger structure. Any process leading to the release of these constraints increases freedom of movement, and hence entropy.

This is the universal tendency of spontaneous processes as expressed in the second law; it is why the costly enzymes stored in the refrigerator tend to decay and why ice melts into water. The increase in entropy as ice melts into water is "paid for" by the absorption of heat from the surroundings. As long as the net change in entropy of the system plus its surroundings is positive, the process can take place spontaneously. That does not necessarily mean that the process will take place: The rate is usually determined by kinetic factors separate from the entropy change. All the second law mandates is that the fate of the pyramids is to crumble into sand, while the sand will never reassemble itself into a pyramid; the law does not tell how quickly this must come about.

A Process Is Spontaneous If ΔS for the System and Its Surroundings Is Positive

There is nothing mystical about entropy; it is a thermodynamic quantity like any other, measurable by experiment and expressed in entropy units. One method of quantifying it is through the heat capacity of a system, the amount of energy required to raise the temperature by 1°C. In some cases the entropy can even be calculated from theoretical principles, though only for simple molecules. For our purposes, what matters is the sign of the entropy change, ΔS: A process can take place spontaneously when ΔS for the system and its surroundings is positive; a process for which ΔS is negative cannot take place spontaneously, but the opposite process can; and for a system at equilibrium, the entropy of the system plus its surroundings is maximal and ΔS is zero.

"Equilibrium" is another of those familiar words that is easier to use than to define. Its everyday meaning implies that the forces acting on a system are equally balanced, such that there is no net tendency to change; this is the sense in which the term "equilibrium" will be used here. A mixture of chemicals may be in the midst of rapid interconversion, but if the rates of the forward reaction and the backward reaction are equal, there will be no net change in composition, and equilibrium will prevail.

The second law has been stated in many versions. One version forbids perpetual-motion machines: Because energy is, by the second law, perpetually degraded into heat and rendered isothermally unavailable ($\Delta S > 0$), continued motion requires an input of energy from the outside. The most celebrated yet perplexing version of the second law was provided by R. J. Clausius (1879): "The energy of the universe is constant; the entropy of the universe tends towards a maximum."

How can entropy increase forever, created out of nothing? The root of the difficulty is verbal, as Klotz (1967) neatly explains. Had Clausius defined entropy with the opposite sign (corresponding to order rather than to chaos), its universal tendency would be to diminish; it would then be obvious that spontaneous changes proceed in the direction that decreases the capacity for further spontaneous change. Solutes diffuse from a region of higher concentration to one of lower concentration; heat flows from a warm body to a cold one. Sometimes these changes can be reversed by an outside agent to reduce the entropy of the system under consideration, but then that external agent must change in such a way as to reduce its own capacity for further change. In sum, "entropy is an index of exhaustion; the more a system has lost its capacity for spontaneous change, the more this capacity has been exhausted, the greater is the entropy" (Klotz 1967). Conversely, the farther a system is from equilibrium, the greater is its capacity for change and the less its entropy. Living things fall into the latter category: *A cell is the epitome of a state that is remote from equilibrium.*

Free Energy and Chemical Potential

Many energy transactions that take place in living organisms are chemical; we therefore need a quantitative expression for the amount of work a chemical reaction can do. For this purpose, relationships that involve the entropy change in the system plus its surroundings are unsuitable. We need a function that does not depend on the surroundings but that, like ΔS, attains a minimum under conditions of equilibrium and so can serve both as a criterion of the feasibility of a reaction and as a measure of the energy available from it for the performance of work. The function universally employed for this purpose is free energy, abbreviated G in honor of the nineteenth-century physical chemist J. Willard Gibbs, who first introduced it.

ΔG Is Negative for a Spontaneous Process at Constant Temperature and Pressure

Earlier we spoke of the isothermally unavailable energy, ST. **Free energy** is defined as the energy that is available under isothermal conditions, and by the following relationship:

$$\Delta H = \Delta G + T\Delta S \tag{2.3}$$

The term H, **enthalpy** or heat content, is not quite equivalent to U, the internal energy (see Equation 2.2). To be exact, ΔH is a measure of the total energy change, including work that may result from changes in volume during the reaction, whereas ΔU excludes this work. (We will return to the concept of enthalpy a little later.) However, in the biological context we are usually concerned with reactions in solution, for which volume changes are negligible. For most purposes, then,

$$\Delta U \cong \Delta G + T\Delta S \tag{2.4}$$

and

$$\Delta G \cong \Delta U - T\Delta S \tag{2.5}$$

What makes this a useful relationship is the demonstration that *for all spontaneous processes at constant temperature and pressure, ΔG is negative.* The change in free energy is thus a criterion of feasibility. Any chemical reaction that proceeds with a negative ΔG can take place spontaneously; a process for which ΔG is positive cannot take place, but the reaction can go in the opposite direction; and a reaction for which ΔG is zero is at equilibrium, and no net change will occur. For a given temperature and pressure, ΔG depends only on the composition of the reaction mixture; hence the alternative term "chemical potential" is particularly apt. Again, nothing is said about rate, only about direction. Whether a reaction having a given ΔG will proceed, and at what rate, is determined by kinetic rather than thermodynamic factors.

There is a close and simple relationship between the change in free energy of a chemical reaction and the work that the reaction can do. Provided the reaction is carried out reversibly,

$$\Delta G = \Delta W_{max} \tag{2.6}$$

That is, for a reaction taking place at constant temperature and pressure, $-\Delta G$ is a measure of the maximum work the process can perform. More precisely, $-\Delta G$ is the maximum work possible, exclusive of pressure–volume work, and thus is a quantity of great importance in bioenergetics. Any process going toward equilibrium can, in principle, do work. We can therefore describe processes for which ΔG is negative as "energy-releasing," or **exergonic**. Conversely, for any process moving away from equilibrium, ΔG is positive, and we speak of an "energy-consuming," or **endergonic**, reaction. Of course, an endergonic reaction cannot occur: All real processes go toward equilibrium, with a negative ΔG. The concept of endergonic reactions is nevertheless a useful abstraction, for many biological reactions appear to move away from equilibrium. A prime example is the synthesis of ATP during oxidative phosphorylation, whose apparent ΔG is as high as 67 kJ mol^{-1} (16 kcal mol^{-1}). Clearly, the cell must do work to render the reaction exergonic overall. The occurrence of an endergonic process in nature thus implies that it is coupled to a second, exergonic process. Much of cellular and molecular bioenergetics is concerned with the mechanisms by which energy coupling is effected.

The Standard Free-Energy Change, ΔG^0, Is the Change in Free Energy When the Concentration of Reactants and Products Is 1 *M*

Changes in free energy can be measured experimentally by calorimetric methods. They have been tabulated in two forms: as the free energy of formation of a compound from its elements, and as ΔG for a particular reaction. It is of the utmost importance to remember that, by convention, the numerical values refer to a particular set of conditions. *The standard free-energy change, ΔG^0, refers to conditions such that all reactants and products are present at a concentration of 1* M; in biochemistry it is more convenient to employ $\Delta G^{0\prime}$, which is defined in the same way except that the pH is taken to be 7. The conditions obtained in the real world are likely to be very different from these, particularly with respect to the concentrations of the participants. To take a familiar example, $\Delta G^{0\prime}$ for the hydrolysis of ATP is about –33 kJ mol^{-1} (–8 kcal mol^{-1}). In the cytoplasm, however, the actual nucleotide concentrations are approximately 3 m*M* ATP, 1 m*M* ADP, and 10 m*M* P_i. As we will see, changes in free energy depend strongly on concentrations, and ΔG for ATP hydrolysis under physiological conditions thus is much more negative than $\Delta G^{0\prime}$, about –50 to –65 kJ mol^{-1} (–12 to –15 kcal mol^{-1}). *Thus, whereas values of $\Delta G^{0\prime}$ for many reactions are easily accessible, they must not be used uncritically as guides to what happens in cells.*

The Value of ΔG Is a Function of the Displacement of the Reaction from Equilibrium

The preceding discussion of free energy shows that there must be a relationship between ΔG and the equilibrium constant of a reaction: At equilibrium, ΔG is zero, and the farther a reaction is from equilibrium, the larger ΔG is and the more work the reaction can do. The quantitative statement of this relationship is

$$\Delta G^0 = -RT \ln K = -2.3RT \log K \tag{2.7}$$

where R is the gas constant, T the absolute temperature, and K the equilibrium constant of the reaction. This equation is one of the most useful links between thermodynamics and biochemistry and has a host of applications. For example, the equation is easily modified to allow computation of the change in free energy for concentrations other than the standard ones. For the reactions shown in the equation

$$A + B \Leftrightarrow C + D \tag{2.8}$$

the actual change in free energy, ΔG, is given by the equation

$$\Delta G = \Delta G^0 + RT \ln \frac{[C][D]}{[A][B]} \tag{2.9}$$

where the terms in brackets refer to the concentrations at the time of the reaction. Strictly speaking, one should use activities, but these are usually not known for cellular conditions, so concentrations must do.

Equation 2.9 can be rewritten to make its import a little plainer. Let q stand for the mass:action ratio, [C][D]/[A][B]. Substitution of Equation 2.7 into Equation 2.9, followed by rearrangement, then yields the following equation:

$$\Delta G = -2.3\ RT \log \frac{K}{q} \tag{2.10}$$

In other words, the value of ΔG is a function of the displacement of the reaction from equilibrium. In order to displace a system from equilibrium, work must be done on it and ΔG must be positive. Conversely, a system displaced from equilibrium can do work on another system, provided that the kinetic parameters allow the reaction to proceed and a mechanism exists that couples the two systems. Quantitatively, a reaction mixture at 25°C whose composition is one order of magnitude away from equilibrium ($\log K/q = 1$) corresponds to a free-energy change of 5.7 kJ mol^{-1} (1.36 kcal mol^{-1}). The value of ΔG is negative if the actual mass:action ratio is less than the equilibrium ratio and positive if the mass:action ratio is greater.

The point that ΔG is a function of the displacement of a reaction (indeed, of any thermodynamic system) from equilibrium is central to an understanding of bioenergetics. Figure 2.2 illustrates this relationship diagrammatically for the chemical interconversion of substances A and B, and the relationship will reappear shortly in other guises.

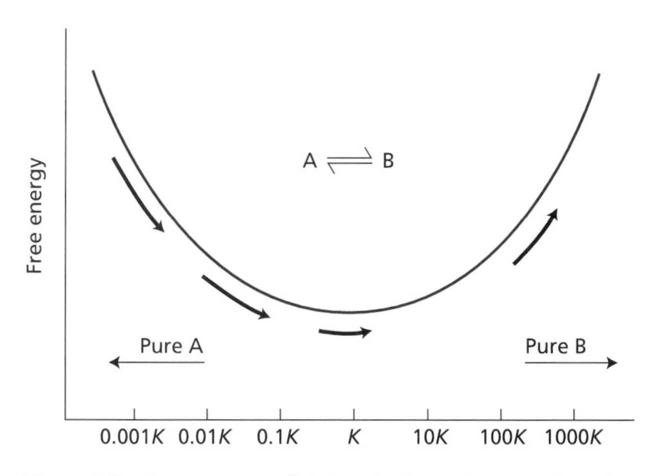

Figure 2.2 Free energy of a chemical reaction as a function of displacement from equilibrium. Imagine a closed system containing components A and B at concentrations [A] and [B]. The two components can be interconverted by the reaction A ↔ B, which is at equilibrium when the mass:action ratio, [B]/[A], equals unity. The curve shows qualitatively how the free energy, G, of the system varies when the total [A] + [B] is held constant but the mass:action ratio is displaced from equilibrium. The arrows represent schematically the change in free energy, ΔG, for a small conversion of [A] into [B] occurring at different mass:action ratios. (After Nicholls and Ferguson 1992.)

The Enthalpy Change Measures the Energy Transferred as Heat

Chemical and physical processes are almost invariably accompanied by the generation or absorption of heat, which reflects the change in the internal energy of the system. The amount of heat transferred and the sign of the reaction are related to the change in free energy, as set out in Equation 2.3. The energy absorbed or evolved as heat under conditions of constant pressure is designated as the change in heat content or enthalpy, ΔH. Processes that generate heat, such as combustion, are said to be **exothermic**; those in which heat is absorbed, such as melting or evaporation, are referred to as **endothermic**. The oxidation of glucose to CO_2 and water is an exergonic reaction ($\Delta G^0 = -2858$ kJ mol^{-1} [–686 kcal mol^{-1}]); when this reaction takes place during respiration, part of the free energy is conserved through coupled reactions that generate ATP. The combustion of glucose dissipates the free energy of reaction, releasing most of it as heat ($\Delta H = -2804$ kJ mol^{-1} [–673 kcal mol^{-1}]).

Bioenergetics is preoccupied with energy transduction and therefore gives pride of place to free-energy transactions, but at times heat transfer may also carry biological significance. For example, water has a high heat of vaporization, 44 kJ mol^{-1} (10.5 kcal mol^{-1}) at 25°C, which plays an important role in the regulation of leaf temperature. During the day, the evaporation of water from the leaf surface (transpiration) dissipates heat to the surroundings and helps cool the leaf. Conversely, the condensation of water vapor as dew heats the leaf, since water condensation is the reverse of evaporation, is exothermic. The abstract enthalpy function is a direct measure of the energy exchanged in the form of heat.

Redox Reactions

Oxidation and reduction refer to the transfer of one or more electrons from a donor to an acceptor, usually to another chemical species; an example is the oxidation of ferrous iron by oxygen, which forms ferric iron and water. Reactions of this kind require special consideration, for they play a central role in both respiration and photosynthesis.

The Free-Energy Change of an Oxidation–Reduction Reaction Is Expressed as the Standard Redox Potential in Electrochemical Units

Redox reactions can be quite properly described in terms of their change in free energy. However, the participation of electrons makes it convenient to follow the course of the reaction with electrical instrumentation and encourages the use of an electrochemical notation. It also permits dissection of the chemical process into separate oxidative and reductive half-reactions. For the oxidation of iron, we can write

$$2Fe^{2+} \Leftrightarrow 2Fe^{3+} + 2e^- \quad (2.11)$$

$$\tfrac{1}{2}O_2 + 2H^+ + 2E^- \Leftrightarrow H_2O \quad (2.12)$$

$$2Fe^{2+} + \tfrac{1}{2}O_2 + 2H^+ \Leftrightarrow 2Fe^{3+} + H_2O \quad (2.13)$$

The tendency of a substance to donate electrons, its "electron pressure," is measured by its standard reduction (or redox) potential, E_0, with all components present at a concentration of 1 *M*. In biochemistry, it is more convenient to employ E'_0, which is defined in the same way except that the pH is 7. By definition, then, E'_0 is the electromotive force given by a half cell in which the reduced and oxidized species are both present at 1 *M*, 25°C, and pH 7, in equilibrium with an electrode that can reversibly accept electrons from the reduced species. By convention, the reaction is written as a reduction. The standard reduction potential of the hydrogen electrode* serves as reference: at pH 7, it equals –0.42 V. The standard redox potential as defined here is often referred to in the bioenergetics literature as the **midpoint** potential, E_m. A negative midpoint potential marks a good reducing agent; oxidants have positive midpoint potentials.

The redox potential for the reduction of oxygen to water is +0.82 V; for the reduction of Fe^{3+} to Fe^{2+} (the direction opposite to that of Equation 2.11), +0.77 V. We can therefore predict that, under standard conditions, the Fe^{2+}–Fe^{3+} couple will tend to reduce oxygen to water rather than the reverse. A mixture containing Fe^{2+}, Fe^{3+}, and oxygen will probably not be at equilibrium, and the extent of its displacement from equilibrium can be expressed in terms of either the change in free energy for Equation 2.13 or the difference in redox potential, $\Delta E'_0$, between the oxidant and the reductant couples (+0.05 V in the case of iron oxidation). In general,

$$\Delta G^{0\prime} = -nF\,\Delta E'_0 \quad (2.14)$$

where *n* is the number of electrons transferred and *F* is Faraday's constant (23.06 kcal V^{-1} mol^{-1}). In other words, the standard redox potential is a measure, in electrochemical units, of the change in free energy of an oxidation–reduction process.

As with free-energy changes, the redox potential measured under conditions other than the standard ones depends on the concentrations of the oxidized and reduced species, according to the following equation (note the similarity in form to Equation 2.9):

$$E_h = E'_0 + \frac{2.3RT}{nF} \log \frac{[\text{oxidant}]}{[\text{reductant}]} \quad (2.15)$$

Here E_h is the measured potential in volts, and the other symbols have their usual meanings. It follows that the redox potential under biological conditions may differ substantially from the standard reduction potential.

The Electrochemical Potential

In the preceding section we introduced the concept that a mixture of substances whose composition diverges from the equilibrium state represents a potential source of free energy (see Figure 2.2). Conversely, a similar amount of work must be done on an equilibrium mixture in order to displace its composition from equilibrium. In this section, we will examine the free-energy changes associated with another kind of displacement from equilibrium—namely, gradients of concentration and of electric potential.

Transport of an Uncharged Solute against Its Concentration Gradient Decreases the Entropy of the System

Consider a vessel divided by a membrane into two compartments that contain solutions of an uncharged solute at concentrations C_1 and C_2, respectively. The work required to transfer 1 mol of solute from the first compartment to the second is given by the following equation:

$$\Delta G = 2.3RT \log \frac{C_2}{C_1} \quad (2.16)$$

This expression is analogous to the expression for a chemical reaction (Equation 2.10) and has the same meaning. If C_2 is greater than C_1, ΔG is positive, and work must be done to transfer the solute. Again, the free-energy change for the transport of 1 mol of solute against a tenfold gradient of concentration is 5.7 kJ, or 1.36 kcal.

The reason that work must be done to move a substance from a region of lower concentration to one of

* The standard hydrogen electrode consists of platinum, over which hydrogen gas is bubbled at a pressure of 1 atm. The electrode is immersed in a solution containing hydrogen ions. When the activity of hydrogen ions is 1, approximately 1 *M* H^+, the potential of the electrode is taken to be 0.

higher concentration is that the process entails a change to a less probable state and therefore a decrease in the entropy of the system. Conversely, diffusion of the solute from the region of higher concentration to that of lower concentration takes place in the direction of greater probability; it results in an increase in the entropy of the system and can proceed spontaneously. The sign of ΔG becomes negative, and the process can do the amount of work specified by Equation 2.16, provided a mechanism exists that couples the exergonic diffusion process to the work function.

The Membrane Potential Is the Work That Must Be Done to Move an Ion from One Side of the Membrane to the Other

Matters become a little more complex if the solute in question bears an electric charge. Transfer of positively charged solute from compartment 1 to compartment 2 will then cause a difference in charge to develop across the membrane, the second compartment becoming electropositive relative to the first. Since like charges repel one another, the work done by the agent that moves the solute from compartment 1 to compartment 2 is a function of the charge difference; more precisely, it depends on the difference in electric potential across the membrane. This difference, called membrane potential for short, will appear again in later pages.

The **membrane potential**, ΔE,* is defined as the work that must be done by an agent to move a test charge from one side of the membrane to the other. When 1 J of work must be done to move 1 coulomb of charge, the potential difference is said to be 1 V. The absolute electric potential of any single phase cannot be measured, but the potential difference between two phases can be. By convention, the membrane potential is always given in reference to the movement of a positive charge. It states the intracellular potential relative to the extracellular one, which is defined as zero.

* Many texts use the term $\Delta\Psi$ for the membrane potential difference. However, to avoid confusion with the use of $\Delta\Psi$ to indicate water potential (see Chapter 3), the term ΔE will be used here and throughout the text.

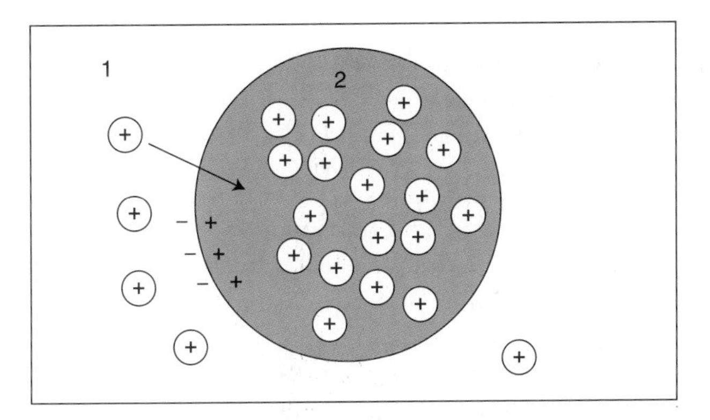

Figure 2.3 Transport against an electrochemical-potential gradient. The agent that moves the charged solute (from compartment 1 to compartment 2) must do work to overcome both the electrochemical-potential gradient and the concentration gradient. As a result, cations in compartment 2 have been raised to a higher electrochemical potential than those in compartment 1. Neutralizing anions have been omitted.

The work that must be done to move 1 mol of an ion against a membrane potential of ΔE volts is given by the following equation:

$$\Delta G = zF\,\Delta E \qquad (2.17)$$

where z is the valence of the ion and F is Faraday's constant. The value of ΔG for the transfer of cations into a positive compartment is positive and so calls for work. Conversely, the value of ΔG is negative when cations move into the negative compartment, so work can be done. *The electric potential is negative across the plasma membrane of the great majority of cells; therefore cations tend to leak in but have to be "pumped" out.*

The Electrochemical-Potential Difference, $\Delta\tilde{\mu}$, Includes Both Concentration and Electric Potentials

In general, ions moving across a membrane are subject to gradients of both concentration and electric potential. Consider, for example, the situation depicted in Figure 2.3, which corresponds to a major event in energy transduction during photosynthesis. A cation of valence z moves from compartment 1 to compartment 2, against both a concentration gradient ($C_2 > C_1$) and a gradient of membrane electric potential (compartment 2 is electropositive relative to compartment 1). The free-energy change involved in this transfer is given by the following equation:

$$\Delta G = zF\Delta E + 2.3RT\,\log\frac{C_2}{C_1} \qquad (2.18)$$

ΔG is positive, and the transfer can proceed only if coupled to a source of energy, in this instance the absorption of light. As a result of this transfer, cations in compartment 2 can be said to be at a higher electrochemical potential than the same ions in compartment 1.

The electrochemical potential for a particular ion is designated $\tilde{\mu}_{ion}$. Ions tend to flow from a region of high electrochemical potential to one of low potential and in so doing can in principle do work. The maximum amount of this work, neglecting friction, is given by the change in free energy of the ions that flow from compartment 2 to compartment 1 (see Equation 2.6) and is numerically equal to the electrochemical-potential difference, $\Delta\tilde{\mu}_{ion}$. This principle underlies much of biological energy transduction.

The electrochemical-potential difference, $\Delta\tilde{\mu}_{ion}$, is properly expressed in kilojoules per mole or kilocalories per mole. However, it is frequently convenient to express the driving force for ion movement in electrical terms, with the dimensions of volts or millivolts. To convert $\Delta\tilde{\mu}_{ion}$ into millivolts (mV), divide all the terms in Equation 2.18 by F:

$$\frac{\Delta\tilde{\mu}_{ion}}{F} = z\Delta E + \frac{2.3RT}{F} \log \frac{C_2}{C_1} \qquad (2.19)$$

An important case in point is the proton motive force, which will be considered at length in Chapter 6.

Equations 2.18 and 2.19 have proved to be of central importance in bioenergetics. First, they measure the amount of energy that must be expended on the active transport of ions and metabolites, a major function of biological membranes. Second, since the free energy of chemical reactions is often transduced into other forms via the intermediate generation of electrochemical-potential gradients, these gradients play a major role in descriptions of biological energy coupling. It should be emphasized that the electrical and concentration terms may be either added, as in Equation 2.18, or subtracted, and that the application of the equations to particular cases requires careful attention to the sign of the gradients. We should also note that free-energy changes in chemical reactions (see Equation 2.10) are scalar, whereas transport reactions have direction; this is a subtle but critical aspect of the biological role of ion gradients.

Ion distribution at equilibrium is an important special case of the general electrochemical equation (Equation 2.18). Figure 2.4 shows a membrane-bound vesicle (compartment 2) that contains a high concentration of the salt K_2SO_4, surrounded by a medium (compartment 1) containing a lower concentration of the same salt; the membrane is impermeable to anions but allows the free passage of cations. Potassium ions will therefore tend to diffuse out of the vesicle into the solution, whereas the sulfate anions are retained. Diffusion of the cations generates a membrane potential, with the vesicle interior negative, which restrains further diffusion. At equilibrium, ΔG and $\Delta\tilde{\mu}_{K^+}$ equal zero (by definition). Equation 2.18 can then be arranged to give the following equation:

$$\Delta E = \frac{-2.3RT}{zF} \log \frac{C_2}{C_1} \qquad (2.20)$$

where C_2 and C_1 are the concentrations of K^+ ions in the two compartments; z, the valence, is unity; and ΔE is the membrane potential in equilibrium with the potassium concentration gradient.

This is one form of the celebrated **Nernst equation**. It states that at equilibrium, a permeant ion will be so distributed across the membrane that the chemical driving force (outward in this instance) will be balanced by the electric driving force (inward). For a univalent cation at 25°C, each tenfold increase in concentration factor corresponds to a membrane potential of 59 mV; for a divalent ion the value is 29.5 mV.

The preceding discussion of the energetic and electrical consequences of ion translocation illustrates a point that must be clearly understood—namely, that an electric potential across a membrane may arise by two distinct mechanisms. The first mechanism, illustrated in Figure 2.4, is the diffusion of charged particles down a preexisting concentration gradient, an exergonic process. A potential generated by such a process is described as a **diffusion potential** or as a Donnan potential. (**Donnan potential** is defined as the diffusion potential that occurs in the limiting case where the counterion is completely impermeant or fixed, as in Figure 2.4.) Many ions are unequally distributed across biological membranes and differ widely in their rates of diffusion across the barrier; therefore diffusion potentials always contribute to the observed membrane potential. But in most biological systems the measured electric potential differs from the value that would be expected on the basis of passive ion diffusion. In these cases one must invoke electrogenic ion pumps, transport systems that carry out the exergonic process indicated in Figure 2.3 at the expense of an external energy source. Transport systems of this kind transduce the free energy of a chemical reaction into the electrochemical potential of an ion gradient and play a leading role in biological energy coupling.

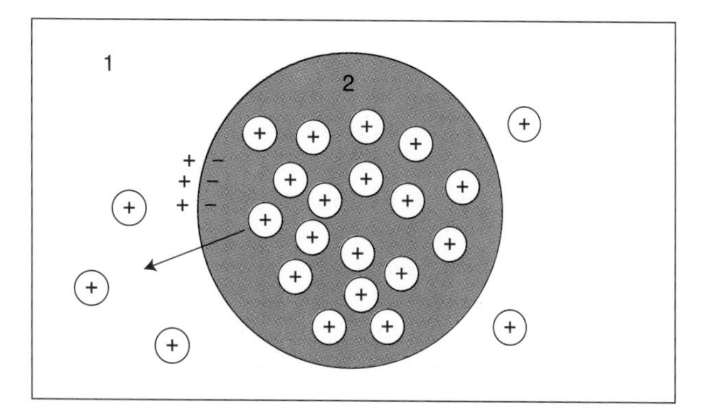

Figure 2.4 Generation of an electric potential by ion diffusion. Compartment 2 has a higher salt concentration than compartment 1 (anions are not shown). If the membrane is permeable to the cations but not to the anions, the cations will tend to diffuse out of compartment 2 into compartment 1, generating a membrane potential in which compartment 2 is negative.

Enzymes: The Catalysts of Life

Proteins constitute about 30% of the total dry weight of typical plant cells. If we exclude inert materials, such as the cell wall and starch, which can account for up to 90% of the dry weight of some cells, proteins and amino acids represent about 60 to 70% of the dry weight of the living cell. As we saw in Chapter 1, cytoskeletal structures such as microtubules and microfilaments are composed of protein. Proteins can also occur as storage forms, particularly in seeds. But the major function of proteins in metabolism is to serve as enzymes, biological catalysts that greatly increase the rates of biochemical reactions, making life possible. Enzymes participate in these reactions but are not themselves fundamentally changed in the process (Mathews and Van Holde 1996).

Enzymes have been called the "agents of life"—a very apt term, since they control almost all life processes. A typical cell has several thousand different enzymes, which carry out a wide variety of actions. The most important features of enzymes are their *specificity*, which permits them to distinguish among very similar molecules, and their *catalytic efficiency*, which is far greater than that of ordinary catalysts. The stereospecificity of enzymes is remarkable, allowing them to distinguish not only between enantiomers (mirror-image stereoisomers), for example, but between apparently identical atoms or groups of atoms (Creighton 1983).

This ability to discriminate between similar molecules results from the fact that the first step in enzyme catalysis is the formation of a tightly bound, noncovalent complex between the enzyme and the substrate(s): the **enzyme–substrate complex**. Enzyme-catalyzed reactions exhibit unusual kinetic properties that are also related to the formation of these very specific complexes. Another distinguishing feature of enzymes is that they are subject to various kinds of regulatory control, ranging from subtle effects on the catalytic activity by effector molecules (inhibitors or activators) to regulation of enzyme synthesis and destruction by the control of gene expression and protein turnover.

Enzymes are unique in the large rate enhancements they bring about, orders of magnitude greater than those effected by other catalysts. Typical orders of rate enhancements of enzyme-catalyzed reactions over the corresponding uncatalyzed reactions are 10^8 to 10^{12}. Many enzymes will convert about a thousand molecules of substrate to product in 1 s. Some will convert as many as a million!

Unlike most other catalysts, enzymes function at ambient temperature and atmospheric pressure and usually in a narrow pH range near neutrality (there are exceptions; for instance, vacuolar proteases and ribonucleases are most active at pH 4 to 5). A few enzymes are able to function under extremely harsh conditions; examples are pepsin, the protein-degrading enzyme of the stomach, which has a pH optimum around 2.0, and the hydrogenase of the hyperthermophilic ("extreme heat–loving") archaebacterium *Pyrococcus furiosus*, which oxidizes H_2 at a temperature optimum greater than 95°C (Bryant and Adams 1989). The presence of such remarkably heat-stable enzymes enables *Pyrococcus* to grow optimally at 100°C.

Enzymes are usually named after their substrates by the addition of the suffix "-ase"—for example, α-amylase, malate dehydrogenase, β-glucosidase, phosphoenolpyruvate carboxylase, horseradish peroxidase. Many thousands of enzymes have already been discovered, and new ones are being found all the time. Each enzyme has been named in a systematic fashion, on the basis of the reaction it catalyzes, by the International Union of Biochemistry. In addition, many enzymes have common, or trivial, names. Thus the common name *rubisco* refers to D-ribulose-1,5-bisphosphate carboxylase/oxygenase (EC 4.1.1.39*).

The versatility of enzymes reflects their properties as proteins. The nature of proteins permits both the exquisite recognition by an enzyme of its substrate and the catalytic apparatus necessary to carry out diverse and rapid chemical reactions (Stryer 1995).

Proteins Are Chains of Amino Acids Joined by Peptide Bonds

Proteins are composed of long chains of amino acids (Figure 2.5) linked by amide bonds, known as **peptide bonds** (Figure 2.6). The 20 different amino acid side chains endow proteins with a large variety of groups that have different chemical and physical properties, including hydrophilic (polar, water-loving) and hydrophobic (nonpolar, water-avoiding) groups, charged and neutral polar groups, and acidic and basic groups. This diversity, in conjunction with the relative flexibility of the peptide bond, allows for the tremendous variation in protein properties, ranging from the rigidity and inertness of structural proteins to the reactivity of hormones, catalysts, and receptors. The three-dimensional aspect of protein structure provides for precise discrimination in the recognition of **ligands**, the molecules that interact with proteins, as shown by the ability of enzymes to recognize their substrates and of antibodies to recognize antigens, for example.

All molecules of a particular protein have the same sequence of amino acid residues, determined by the sequence of nucleotides in the gene that codes for that protein. Although the protein is synthesized as a linear chain on the ribosome, upon release it folds spontaneously into a specific three-dimensional shape, the **native** state. The chain of amino acids is called a polypeptide. The three-dimensional arrangement of the atoms in the molecule is referred to as the **conformation**.

* The Enzyme Commission (EC) number indicates the class (4 = lyase) and subclasses (4.1 = carbon–carbon cleavage; 4.1.1 = cleavage of C—COO^- bond).

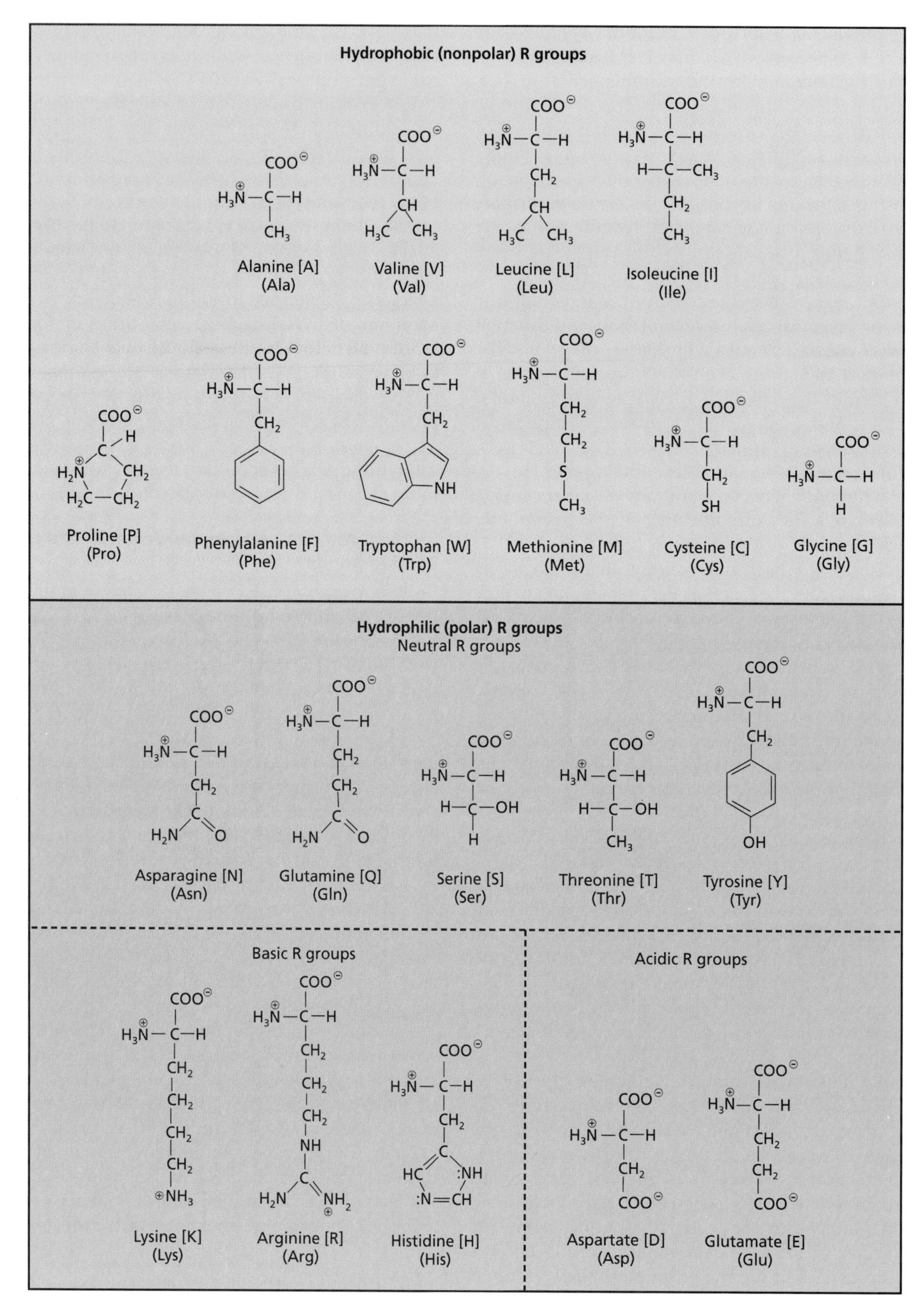

Figure 2.5 The structures, names, single-letter codes (in square brackets), three-letter abbreviations, and classification of the amino acids.

Figure 2.6 (A) The peptide (amide) bond links two amino acids. (B) Sites of free rotation, within the limits of steric hindrance, about the N—C_α and C_α—C bonds (ψ and ϕ); there is no rotation about the peptide bond, because of its double-bond character.

Changes in conformation do not involve breaking of covalent bonds. *Denaturation* involves the loss of this unique three-dimensional shape and results in the loss of catalytic activity.

The forces that are responsible for the shape of a protein molecule are noncovalent (Figure 2.7). These noncovalent interactions include hydrogen bonds; electrostatic interactions (also known as ionic bonds or salt bridges); van der Waals interactions (dispersion forces), which are transient dipoles between spatially close atoms; and hydrophobic "bonds"—the tendency of nonpolar groups to avoid contact with water and thus to associate with themselves. In addition, covalent disulfide bonds are found in many proteins. Although each of these types of noncovalent interaction is weak, there are so many noncovalent interactions in proteins that in total they contribute a large amount of free energy to stabilizing the native structure.

Protein Structure Is Hierarchical

Proteins are built up with increasingly complex organizational units. The **primary structure** of a protein refers to the sequence of amino acid residues. The **secondary structure** refers to regular, local structural units, usually held together by hydrogen bonding. The most common of these units are the α helix and β strands forming parallel and antiparallel β pleated sheets and turns (Figure 2.8). The **tertiary structure**—the final three-dimensional structure of the polypeptide—results from the packing together of the secondary structure units and the exclusion of solvent. The **quaternary structure** refers to the association of two or more separate three-dimensional polypeptides to form complexes. When associated in this manner, the individual polypeptides are called **subunits**.

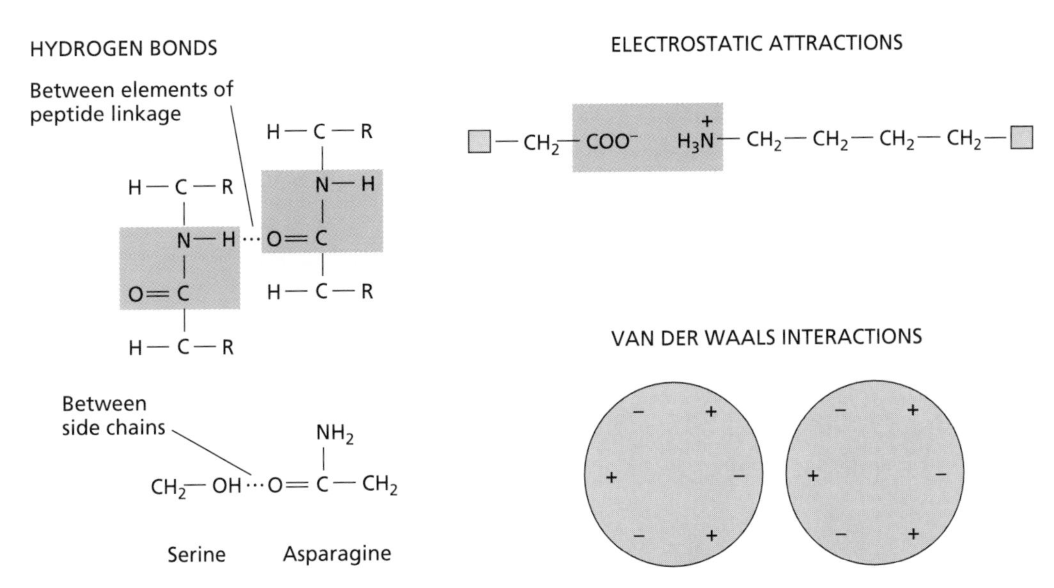

Figure 2.7 Examples of noncovalent interactions in proteins. Hydrogen bonds are weak electrostatic interactions involving a hydrogen atom between two electronegative atoms. In proteins the most important hydrogen bonds are those between the peptide bonds. Electrostatic interactions are ionic bonds between positively and negatively charged groups. The van der Waals interactions are short-range transient dipole interactions. Hydrophobic interactions (not shown) involve restructuring of the solvent water around nonpolar groups, minimizing the exposure of nonpolar surface area to polar solvent; these interactions are driven by entropy.

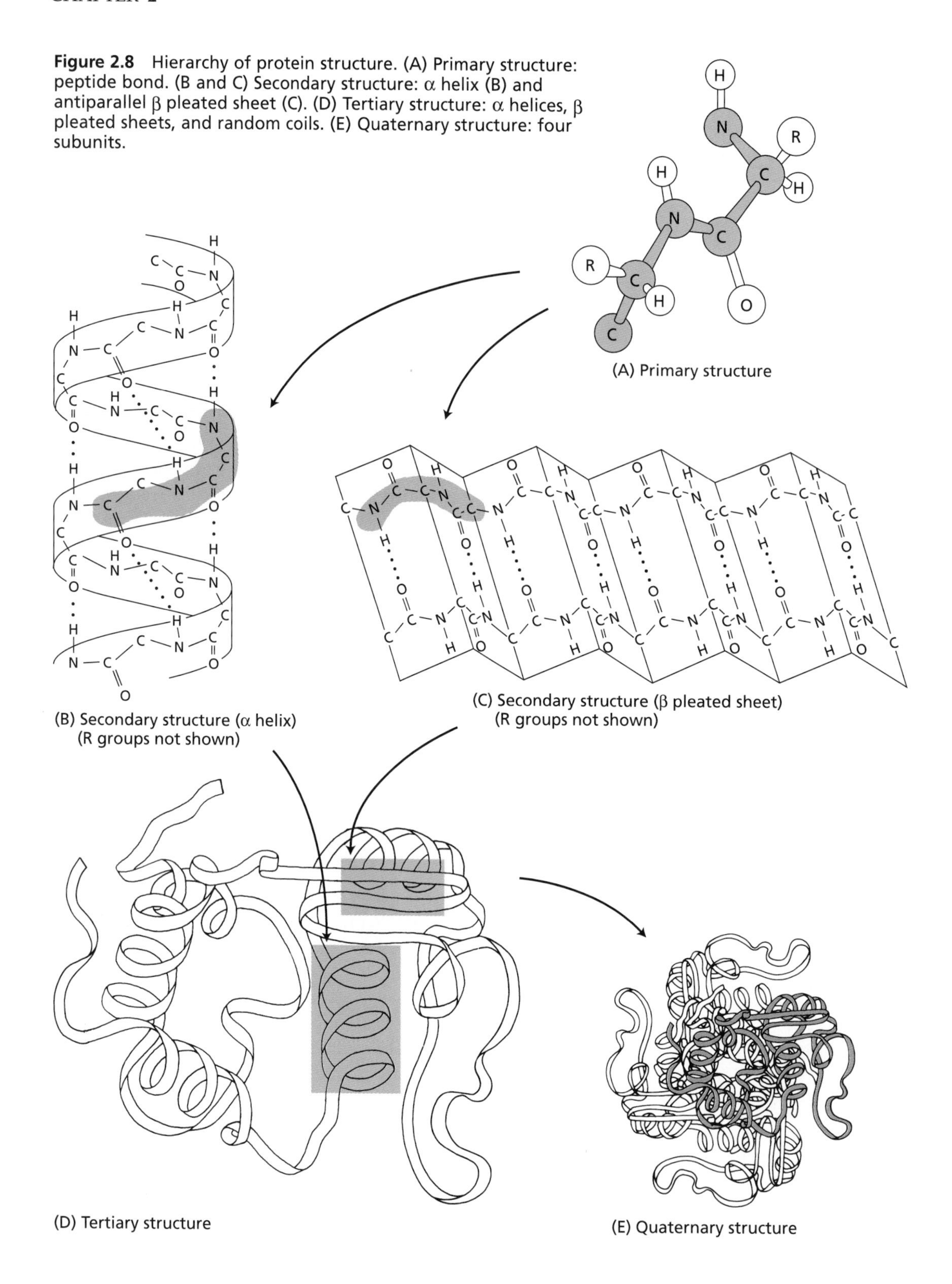

Figure 2.8 Hierarchy of protein structure. (A) Primary structure: peptide bond. (B and C) Secondary structure: α helix (B) and antiparallel β pleated sheet (C). (D) Tertiary structure: α helices, β pleated sheets, and random coils. (E) Quaternary structure: four subunits.

A protein molecule consisting of a large single polypeptide chain is composed of several independently folding units known as **domains**. Typically, domains have a molecular mass of about 10^4 daltons. The active site of an enzyme—that is, the region where the substrate binds and the catalytic reaction occurs—is often located at the interface between two domains. For example, in the enzyme papain (a vacuolar protease that is found in papaya and is representative of a large class of plant thiol proteases), the active site lies at the junction of two domains (Figure 2.9). Helices, turns, and β sheets contribute to the unique three-dimensional shape of this enzyme.

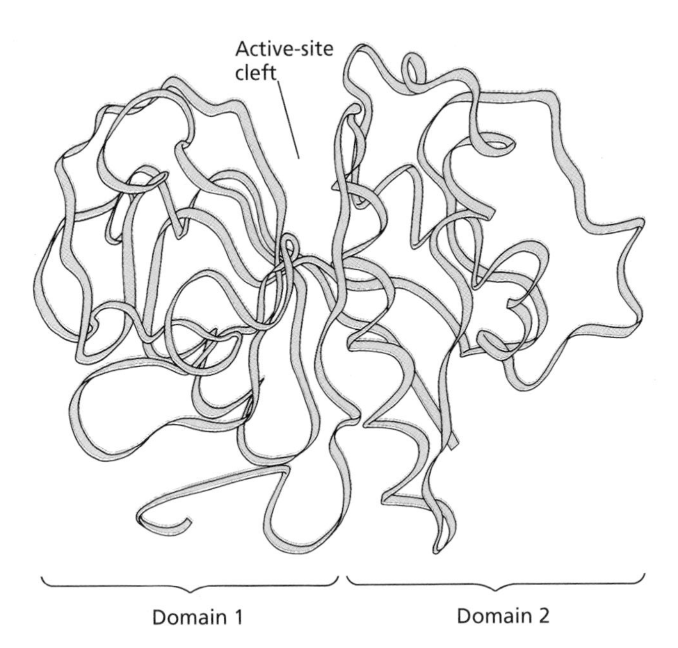

Figure 2.9 The backbone structure of papain, showing the two domains and the active-site cleft between them.

Determinations of the conformation of proteins have revealed that there are families of proteins that have common three-dimensional folds, as well as common patterns of supersecondary structure, such as β-α-β.

Enzymes Are Highly Specific Protein Catalysts

All enzymes are proteins, although recently some small ribonucleic acids and protein–RNA complexes have been found to exhibit enzymelike behavior in the processing of RNA. Proteins have molecular masses ranging from 10^4 to 10^6 daltons, and they may be a single folded polypeptide chain (subunit, or protomer) or oligomers of several subunits (oligomers are usually dimers or tetramers). Normally, enzymes have only one type of catalytic activity associated with the same protein; **isoenzymes**, or **isozymes**, are enzymes with similar catalytic function that have different structures and catalytic parameters and are encoded by different genes. For example, various different isozymes have been found for peroxidase, an enzyme in plant cell walls that is involved in the synthesis of lignin. An isozyme of peroxidase has also been localized in vacuoles. Isozymes may exhibit tissue specificity and show developmental regulation.

Enzymes frequently contain a nonprotein **prosthetic group** or **cofactor** that is necessary for biological activity. The association of a cofactor with an enzyme depends on the three-dimensional structure of the protein. Once bound to the enzyme, the cofactor contributes to the specificity of catalysis. Typical examples of cofactors are metal ions (e.g., zinc, iron, molybdenum), heme groups or iron–sulfur clusters (especially in oxidation–reduction enzymes), and coenzymes (e.g., nicotinamide adenine dinucleotide [NAD^+/NADH], flavin adenine dinucleotide [FAD/$FADH_2$], flavin mononucleotide [FMN], and pyridoxal phosphate [PLP]). Coenzymes are usually vitamins or are derived from vitamins and act as carriers. For example, NAD^+ and FAD carry hydrogens and electrons in redox reactions, biotin carries CO_2, and tetrahydrofolate carries one-carbon fragments. Peroxidase has both heme and Ca^{2+} prosthetic groups and is glycosylated; that is, it contains carbohydrates covalently added to asparagine, serine, or threonine side chains. Such proteins are called **glycoproteins**.

A particular enzyme will catalyze only one type of chemical reaction for only one class of molecule—in some cases, for only one particular compound. Enzymes are also very stereospecific and produce no by-products. For example, β-glucosidase catalyzes the hydrolysis of β-glucosides, compounds formed by a glycosidic bond to D-glucose. The substrate must have the correct anomeric configuration: it must be β-, not α-. Furthermore, it must have the glucose structure; no other carbohydrates, such as xylose or mannose, can act as substrates for β-glucosidase. Finally, the substrate must have the correct stereochemistry, in this case the D absolute configuration. Rubisco (D-ribulose-1,5-bisphosphate carboxylase/oxygenase) catalyzes the addition of carbon dioxide to D-ribulose-1,5-bisphosphate to form two molecules of 3-phospho-D-glycerate, the initial step in the C_3 photosynthetic carbon reduction cycle, and is the world's most abundant enzyme. Rubisco has very strict specificity for the carbohydrate substrate, but it also catalyzes an oxygenase reaction in which O_2 replaces CO_2, as will be discussed further in Chapter 8.

Enzymes Lower the Free-Energy Barrier between Substrates and Products

Catalysts speed the rate of a reaction by lowering the energy barrier between substrates (reactants) and products and are not themselves used up in the reaction, but are regenerated. Thus a catalyst increases the rate of a reaction but does not affect the equilibrium ratio of reactants and products, because the rates of the reaction in both directions are increased to the same extent. It is important to realize that enzymes cannot make a nonspontaneous (energetically uphill) reaction occur. However, many energetically unfavorable reactions in cells proceed because they are coupled to an energetically more favorable reaction usually involving ATP hydrolysis (Figure 2.10).

Enzymes act as catalysts because they lower the free energy of activation for a reaction. They do this by a combination of raising the **ground state** ΔG of the substrate and lowering the ΔG of the **transition state** of the reaction, thereby decreasing the barrier against the reaction (Figure 2.11). The presence of the enzyme leads to

$$
\begin{array}{lll}
A + B \rightarrow C & & \Delta G = +4.0 \text{ kcal mol}^{-1} \\
ATP + H_2O \rightarrow & ADP + P_i + H^+ & \Delta G = -7.3 \text{ kcal mol}^{-1} \\
\hline
A + B + ATP + H_2O \rightarrow & C + ADP + P_i + H^+ & \Delta G = -3.3 \text{ kcal mol}^{-1}
\end{array}
$$

$$
\begin{array}{ll}
A + ATP \rightarrow & A-P + ADP \\
A-P + B + H_2O \rightarrow & C + H^+ + P_i \\
\hline
A + B + ATP + H_2O \rightarrow & C + ADP + P_i + H^+
\end{array}
$$

Figure 2.10 Coupling of the hydrolysis of ATP to drive an energetically unfavorable reaction. The reaction A + B → C is thermodynamically unfavorable, whereas the hydrolysis of ATP to form ADP and inorganic phosphate (P_i) is thermodynamically very favorable (it has a large negative ΔG). Through appropriate intermediates, such as A–P, the two reactions are coupled, yielding an overall reaction that is the sum of the individual reactions and has a favorable free-energy change.

a new reaction pathway that is different from that of the uncatalyzed reaction.

Catalysis Occurs at the Active Site

The **active site** of an enzyme molecule is usually a cleft or pocket on or near the surface of the enzyme that takes up only a small fraction of the enzyme surface. It is convenient to consider the active site as consisting of two components: the **binding site** for the substrate (which attracts and positions the substrate) and the **catalytic groups** (the reactive side chains of amino acids or cofactors, which carry out the bond-breaking and bond-forming reactions involved).

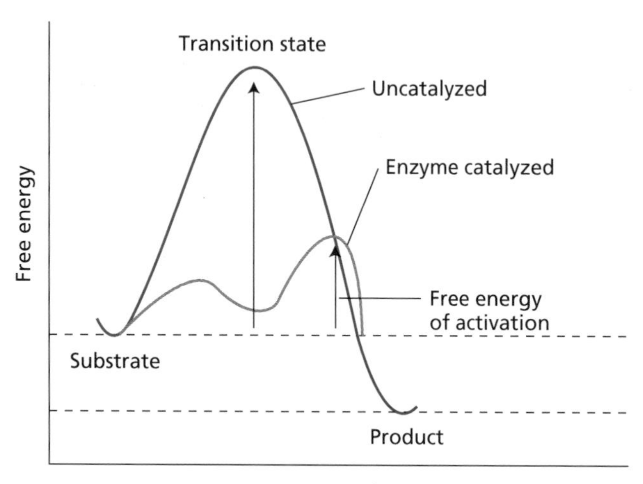

Figure 2.11 Free-energy curves for the same reaction, either uncatalyzed or enzyme catalyzed. As a catalyst, an enzyme lowers the free energy of activation of the transition state between substrates and products compared with the uncatalyzed reaction. It does this by forming various complexes and intermediates, such as enzyme–substrate and enzyme–product complexes. The ground state free energy of the enzyme–substrate complex in the enzyme-catalyzed reaction may be higher than that of the substrate in the uncatalyzed reaction, and the transition state free energy of the enzyme-bound substrate will be signficantly less than that in the corresponding uncatalyzed reaction.

Binding of substrate at the active site initially involves noncovalent interactions between the substrate and either side chains or peptide bonds of the protein. The rest of the protein structure provides a means of positioning the substrate and catalytic groups, flexibility for conformational changes, and regulatory control. The shape and polarity of the binding site account for much of the specificity of enzymes, and there is complementarity between the shape and the polarity of the substrate and those of the active site. In some cases, binding of the substrate induces a conformational change in the active site of the enzyme. Conformational change is particularly common where there are two substrates. Binding of the first substrate sets up a conformational change of the enzyme that results in formation of the binding site for the second substrate. Hexokinase is a good example of an enzyme that exhibits this type of conformational change (Figure 2.12).

The catalytic groups are usually the amino acid side chains and/or cofactors that can function as catalysts. Common examples of catalytic groups are acids (—COOH from the side chains of aspartic acid or glutamic acid, imidazole from the side chain of histidine), bases (—NH_2 from lysine, imidazole from histidine, —S^- from cysteine), nucleophiles (imidazole from histidine, —S^- from cysteine, —OH from serine), and electrophiles (often metal ions, such as Zn^{2+}). The acidic catalytic groups function by donating a proton, the basic ones by accepting a proton. Nucleophilic catalytic groups form a transient covalent bond to the substrate.

The decisive factor in catalysis is the direct interaction between the enzyme and the substrate. In many cases, there is an intermediate that contains a covalent bond between the enzyme and the substrate. Although the details of the catalytic mechanism differ from one type of enzyme to another, a limited number of features are involved in all enzyme catalysis. These features include acid–base catalysis, electrophilic or nucleophilic catalysis, and ground state distortion through electrostatic or mechanical strains on the substrate.

A Simple Kinetic Equation Describes an Enzyme-Catalyzed Reaction

Enzyme-catalyzed systems often exhibit a special form of kinetics, called Michaelis–Menten kinetics, which are characterized by a hyperbolic relationship between reaction velocity, v, and substrate concentration, [S] (Figure 2.13). This type of plot is known as a saturation plot because when the enzyme becomes saturated with

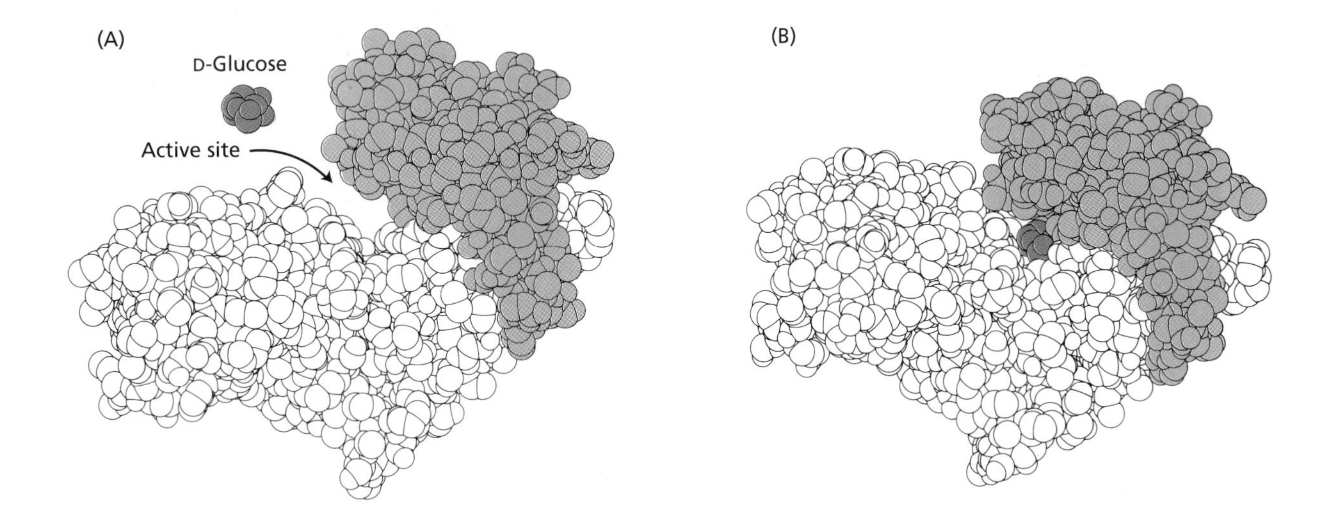

Figure 2.12 Conformational change in hexokinase, induced by the first substrate of the enzyme, D-glucose. (A) Before glucose binding. (B) After glucose binding. The binding of glucose to hexokinase induces a conformational change in which the two major domains come together to close the cleft that contains the active site. This change sets up the binding site for the second substrate, ATP. In this manner the enzyme prevents the unproductive hydrolysis of ATP by shielding the substrates from the aqueous solvent. The overall reaction is the phosphorylation of glucose and the formation of ADP.

substrate (i.e., each enzyme molecule has a substrate molecule associated with it), the rate becomes independent of substrate concentration. Saturation kinetics implies that an equilibrium process precedes the rate-limiting step:

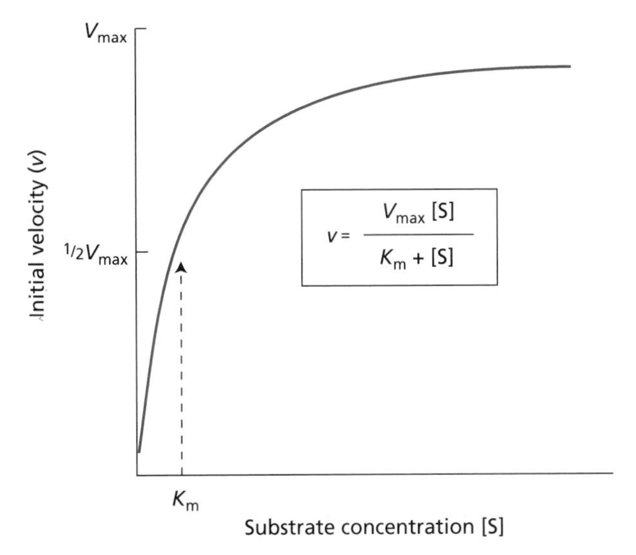

Figure 2.13 Plot of initial velocity, v, versus substrate concentration, [S], for an enzyme-catalyzed reaction. The curve is hyperbolic. The maximal rate, V_{max}, occurs when all the enzyme molecules are fully occupied by substrate. The value of K_m, defined as the substrate concentration at $\frac{1}{2}V_{max}$, is a reflection of the affinity of the enzyme for the substrate. The smaller the value of K_m, the tighter the binding.

$$E + S \xleftrightarrow{\text{fast}} ES \xrightarrow{\text{slow}} E + P$$

where E represents the enzyme, S the substrate, P the product, and ES the enzyme–substrate complex. Thus, as the substrate concentration is increased, a point will be reached at which all the enzyme molecules are in the form of the ES complex, and the enzyme is saturated with substrate. Since the rate of the reaction depends on the concentration of ES, the rate will not increase further, because there can be no higher concentration of ES.

When an enzyme is mixed with a large excess of substrate, there will be an initial very short time period (usually milliseconds) during which the concentrations of enzyme–substrate complexes and intermediates build up to certain levels; this is known as the pre–steady-state period. Once the intermediate levels have been built up, they remain relatively constant until the substrate is depleted; this period is known as the **steady state**.

Normally enzyme kinetic values are measured under steady-state conditions, and such conditions usually prevail in the cell. For many enzyme-catalyzed reactions the kinetics under steady-state conditions can be described by a simple expression known as the Michaelis–Menten equation:

$$v = \frac{V_{max}[S]}{K_m + [S]} \qquad (2.21)$$

where v is the observed rate or velocity (in units such as moles per liter per second), V_{max} is the maximum velocity (at infinite substrate concentration), and K_m (usually measured in units of molarity) is a constant that is characteristic of the particular enzyme–substrate system and is related to the association constant of the enzyme for the substrate (see Figure 2.13). K_m represents the concentration of substrate required to half-saturate the enzyme and thus is the substrate concentration at $V_{max}/2$. In many cellular systems the usual substrate concentration is in the vicinity of K_m. The smaller the value of K_m, the more strongly the enzyme binds the substrate. Typical values for K_m are in the range of 10^{-6} to 10^{-3} *M*.

We can readily obtain the parameters V_{max} and K_m by fitting experimental data to the Michaelis–Menten equation, either by computerized curve fitting or by a linearized form of the equation. An example of a linearized form of the equation is the Lineweaver–Burk double-reciprocal plot shown in Figure 2.14A. When divided by the concentration of enzyme, the value of V_{max} gives the **turnover number**, the number of molecules of substrate converted to product per unit of time per molecule of enzyme. Typical turnover number values range from 10^2 to 10^3 s^{-1}.

Enzymes Are Subject to Various Kinds of Inhibition

Any agent that decreases the velocity of an enzyme-catalyzed reaction is called an inhibitor. Inhibitors may exert their effects in many different ways. Generally, if inhibition is irreversible the compound is called an **inactivator**. Other agents can increase the efficiency of an enzyme; they are called **activators**. Inhibitors and activators are very important in the cellular regulation of enzymes. Many agriculturally important insecticides and herbicides are enzyme inhibitors. The study of enzyme inhibition can provide useful information about kinetic mechanisms, the nature of enzyme–substrate intermediates and complexes, the chemical mechanism of catalytic action, and the regulation and control of metabolic enzymes. In addition, the study of inhibitors of potential target enzymes is essential to the rational design of herbicides.

Inhibitors can be classified as reversible or irreversible. **Irreversible inhibitors** form covalent bonds with an enzyme or they denature it. For example, iodoacetate (ICH_2COOH) irreversibly inhibits thiol proteases such as papain by alkylating the active-site —SH group. One class of irreversible inhibitors is called affinity labels, or active site–directed modifying agents, because their structure directs them to the active site. An example is tosyl-lysine chloromethyl ketone (TLCK), which irreversibly inactivates papain. The tosyl-lysine part of the inhibitor resembles the substrate structure and so binds in the active site. The chloromethyl ketone part of the bound inhibitor reacts with the active-site histidine side chain. Such compounds are very useful in mechanistic studies of enzymes, but they have limited practical use as herbicides because of their chemical reactivity, which can be harmful to the plant.

Reversible inhibitors form weak, noncovalent bonds with the enzyme, and their effects may be competitive, noncompetitive, or mixed. For example, the widely used broad-spectrum herbicide glyphosate (Roundup®) works by competitively inhibiting a key enzyme in the biosynthesis of aromatic amino acids, 5-**e**nol**p**yruvyl**s**hikimate-3-**p**hosphate (EPSP) synthase (see Chapter 13). Resistance to glyphosate has recently been achieved by genetic engineering of plants so that they are capable of overproducing EPSP synthase (Donahue et al. 1995).

Competitive inhibition. Competitive inhibition is the simplest and most common form of reversible inhibition. It usually arises from binding of the inhibitor to the active site with an affinity similar to or stronger

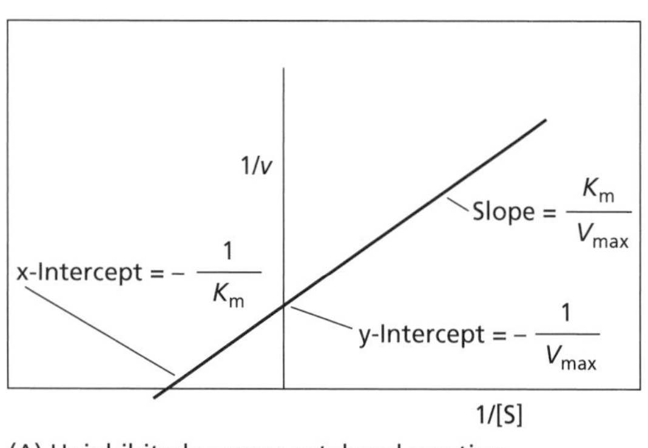

(A) Uninhibited enzyme-catalyzed reaction

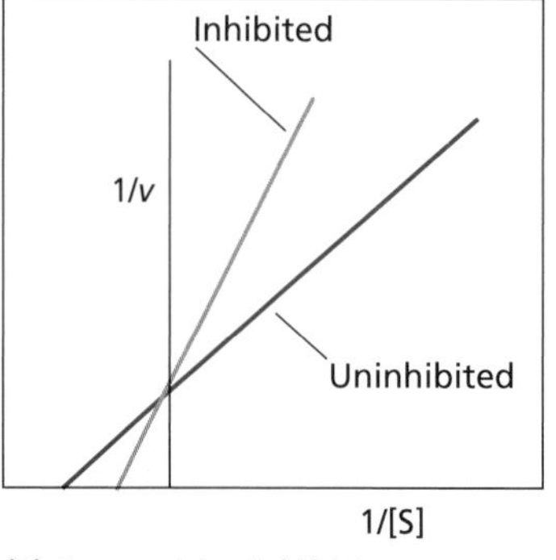

(B) Competitive inhibition

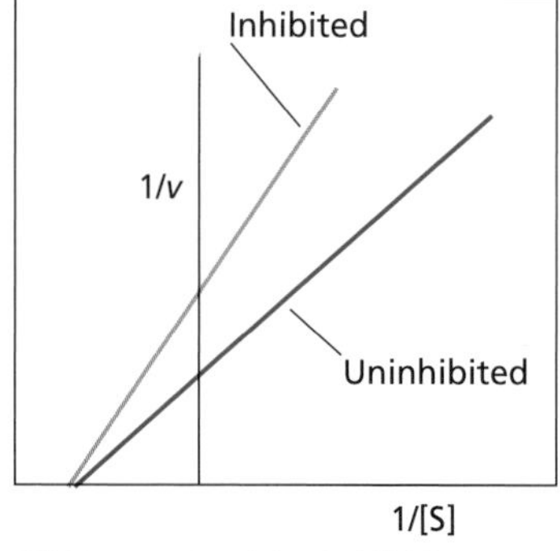

(C) Noncompetitive inhibition

Figure 2.14 Lineweaver–Burk double-reciprocal plots. A plot of 1/*v* versus 1/[S] yields a straight line. (A) Uninhibited enzyme-catalyzed reaction showing the calculation of K_m from the x-intercept and of V_{max} from the y-intercept. (B) The effect of a competitive inhibitor on the parameters K_m and V_{max}. The apparent K_m is increased, but the V_{max} is unchanged. (C) A noncompetitive inhibitor reduces V_{max} but has no effect on K_m.

than that of the substrate. Thus the effective concentration of the enzyme is decreased by the presence of the inhibitor, and the catalytic reaction will be slower than if the inhibitor were absent. Competitive inhibition is usually based on the fact that the structure of the inhibitor resembles that of the substrate; hence the strong affinity of the inhibitor for the active site. Competitive inhibition may also occur in **allosteric enzymes**, where the inhibitor binds to a distant site on the enzyme, causing a conformational change that alters the active site and prevents normal substrate binding. Such a binding site is called an **allosteric site**. In this case, the competition between substrate and inhibitor is indirect.

Competitive inhibition results in an apparent increase in K_m and has no effect on V_{max} (see Figure 2.14B). By measuring the apparent K_m as a function of inhibitor concentration, one can calculate K_i, the inhibitor constant, which reflects the affinity of the enzyme for the inhibitor.

Noncompetitive inhibition. In noncompetitive inhibition, the inhibitor does not compete with the substrate for binding to the active site. Instead, it may bind to another site on the protein and obstruct the substrate's access to the active site, thereby changing the catalytic properties of the enzyme, or it may bind to the enzyme–substrate complex and thus alter catalysis. Noncompetitive inhibition is frequently observed in the regulation of metabolic enzymes. The diagnostic property of this type of inhibition is that K_m is unaffected, whereas V_{max} decreases in the presence of increasing amounts of inhibitor (see Figure 2.14C).

Mixed inhibition. Mixed inhibition is characterized by effects on both V_{max} (which decreases) and K_m (which increases). Mixed inhibition is very common and results from the formation of a complex consisting of the enzyme, the substrate, and the inhibitor that does not break down to products.

pH and Temperature Affect the Rate of Enzyme-Catalyzed Reactions

Enzyme catalysis is very sensitive to pH. This sensitivity is easily understood when one considers that the essential catalytic groups are usually ionizable ones (imidazole, carboxyl, amino) and that they are catalytically active in only one of their ionization states. For example, imidazole acting as a base will be functional only at pH values above 7. Plots of the rates of enzyme-catalyzed reactions versus pH are usually bell-shaped, corresponding to two sigmoidal curves, one for an ionizable group acting as an acid and the other for the group acting as a base (Figure 2.15A). Although the effects of pH on enzyme catalysis usually reflect the ionization of the catalytic group, they may also reflect a pH-dependent conformational change in the protein that leads to loss of activity as a result of disruption of the active site.

The temperature dependence of most chemical reactions also applies to enzyme-catalyzed reactions. Thus, most enzyme-catalyzed reactions show an exponential increase in rate with increasing temperature. However, because the enzymes are proteins, another major factor comes in to play—namely, denaturation. After a certain temperature is reached, enzymes show a very rapid decrease in activity as a result of the onset of denaturation (Figure 2.15B). The temperature at which denaturation begins, and hence at which catalytic activity is lost, varies with the particular protein as well as the environmental conditions, such as pH. Frequently, denaturation begins at about 40 to 50°C and is complete over a range of about 10°C.

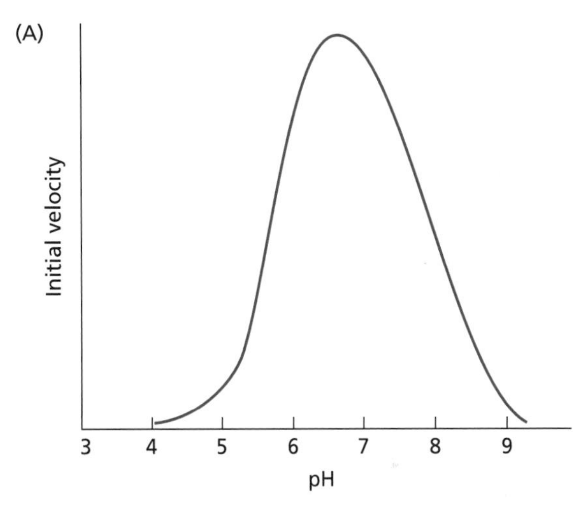

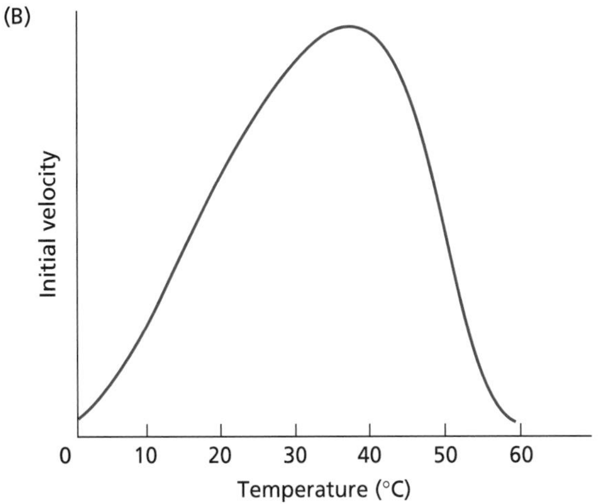

Figure 2.15 pH and temperature curves for typical enzyme reactions. (A) Many enzyme-catalyzed reactions show bell-shaped profiles of rate versus pH. The inflection point on each shoulder corresponds to the pK_a of an ionizing group (that is, the pH at which the ionizing group is 50% dissociated) in the active site. (B) Temperature causes an exponential increase in the reaction rate until the optimum is reached. Beyond the optimum, thermal denaturation dramatically decreases the rate.

Cooperative Systems Increase the Sensitivity to Substrates and Are Usually Allosteric

Cells control the concentrations of most metabolites very closely. To keep such tight control, the enzymes that control metabolite interconversion must be very sensitive. From the plot of velocity versus substrate concentration (see Figure 2.13), we can see that the velocity of an enzyme-catalyzed reaction increases with increasing substrate concentration up to V_{max}. However, we can calculate from the Michaelis–Menten equation (Equation 2.21) that raising the velocity of an enzyme-catalyzed reaction from 0.1 V_{max} to 0.9 V_{max} requires an enormous (81-fold) increase in the substrate concentration:

$$0.1V_{max} = \frac{V_{max}[S]}{K_m + [S]}, \quad 0.9V_{max} = \frac{V_{max}[S]'}{K_m + [S]'}$$

$$0.1K_m = 0.9[S], \qquad 0.9K_m = 0.1[S]'$$

$$\frac{0.1}{0.9} = \frac{0.9}{0.1} \times \frac{[S]}{[S]'}$$

$$\frac{[S]}{[S]'} = \left(\frac{0.1}{0.9}\right)^2 = \frac{0.01}{0.81}$$

This calculation shows that reaction velocity is insensitive to small changes in substrate concentration. The same factor applies in the case of inhibitors and inhibition. In **cooperative systems**, on the other hand, a small change in one parameter, such as inhibitor concentration, brings about a *large change* in velocity. A consequence of a cooperative system is that the plot of v versus [S] is no longer hyperbolic, but becomes *sigmoidal* (Figure 2.16). The advantage of cooperative systems is that a small change in the concentration of the critical effector (substrate, inhibitor, or activator) will bring about a large change in the rate. In other words, the system behaves like a switch.

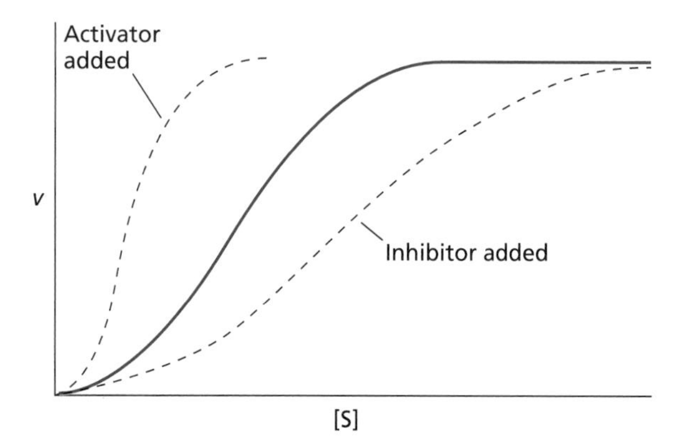

Figure 2.16 Allosteric systems exhibit sigmoidal plots of rate versus substrate concentration. The addition of an activator shifts the curve to the left; the addition of an inhibitor shifts it to the right.

Cooperativity is typically observed in allosteric enzymes that contain multiple active sites located on multiple subunits. Such oligomeric enzymes usually exist in two major conformational states, one active and one inactive (or relatively inactive). Binding of ligands (substrates, activators, or inhibitors) to the enzyme perturbs the position of the equilibrium between the two conformations. For example, an inhibitor will favor the inactive form; an activator will favor the active form. The cooperative aspect comes in as follows: A positive cooperative event is one in which binding of the first ligand makes binding of the next one easier. Similarly, negative cooperativity means that the second ligand will bind less readily than the first.

Cooperativity in substrate binding (homoallostery) occurs when the binding of substrate to a catalytic site on one subunit increases the substrate affinity of an identical catalytic site located on a different subunit. Effector ligands (inhibitors or activators), in contrast, bind to sites other than the catalytic site (heteroallostery). This relationship fits nicely with the fact that the end products of metabolic pathways, which frequently serve as feedback inhibitors, usually bear no structural resemblance to the substrates of the first step.

The Kinetics of Some Membrane Transport Processes Can Be Described by the Michaelis–Menten Equation

Membranes contain proteins that speed up the movement of specific ions or organic molecules across the lipid bilayer. Some membrane transport proteins are enzymes, such as ATPases, that use the energy from the hydrolysis of ATP to pump ions across the membrane. When these reactions run in the reverse direction, the ATPases of mitochondria and chloroplasts can synthesize ATP. Other types of membrane proteins function as carriers, binding their substrate on one side of the membrane and releasing it on the other side.

The kinetics of carrier-mediated transport can be described by the Michaelis–Menten equation in the same manner as the kinetics of enzyme-catalyzed reactions are (see Chapter 6). Instead of a biochemical reaction with a substrate and product, however, the carrier binds to the solute and transfers it from one side of a membrane to the other. Letting X be the solute, we can write the following equation:

$$X_{out} + \text{carrier} \rightarrow [\text{X-carrier}] \rightarrow X_{in} + \text{carrier}$$

Since the carrier can bind to the solute more rapidly than it can transport the solute to the other side of the membrane, solute transport exhibits saturation kinetics. That is, a concentration is reached beyond which adding more solute does not result in a more rapid rate of transport (Figure 2.17). V_{max} is the maximum rate of transport of X across the membrane; K_m is equivalent to the bind-

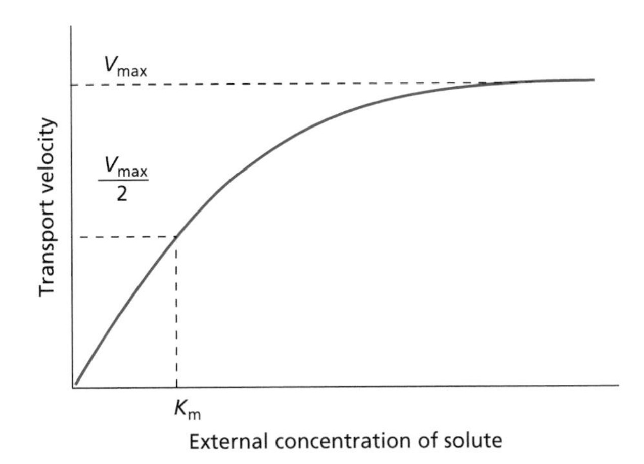

Figure 2.17 The kinetics of carrier-mediated transport of a solute across a membrane are analogous to those of enzyme-catalyzed reactions. Thus, plots of transport velocity versus solute concentration are hyperbolic, becoming asymptotic to the maximal velocity at high solute concentration.

ing constant of the solute for the carrier. Like enzyme-catalyzed reactions, carrier-mediated transport requires a high degree of structural specificity of the protein. The actual transport of the solute across the membrane apparently involves conformational changes, also similar to those in enzyme-catalyzed reactions.

Enzyme Activity Is Often Regulated

Cells can control the flux of metabolites by regulating the concentration of enzymes and their catalytic activity. By using allosteric activators or inhibitors, cells can modulate enzymatic activity and obtain very carefully controlled expression of catalysis.

Control of enzyme concentration. The amount of enzyme in a cell is determined by the relative rates of synthesis and degradation of the enzyme. The rate of synthesis is regulated at the genetic level by a variety of mechanisms, which are discussed in greater detail in the last section of this chapter.

Compartmentalization. Different enzymes or isozymes with different catalytic properties (e.g., substrate affinity) may be localized in different regions of the cell, such as mitochondria and cytosol. Similarly, enzymes associated with special tasks are often compartmentalized; for example, the enzymes involved in photosynthesis are found in chloroplasts. Vacuoles contain many hydrolytic enzymes, such as proteases, ribonucleases, glycosidases, and phosphatases, as well as peroxidases. The cell walls contain glycosidases and peroxidases. The mitochondria are the main location of the enzymes involved in oxidative phosphorylation and energy metabolism, including the enzymes of the tricarboxylic acid (TCA) cycle.

Covalent modification. Control by covalent modification of enzymes is common and usually involves their phosphorylation or adenylylation*, such that the phosphorylated form, for example, is active and the nonphosphorylated form is inactive. These control mechanisms are normally energy dependent and usually involve ATP.

Proteases are normally synthesized as inactive precursors known as zymogens or proenzymes. For example, papain is synthesized as an inactive precursor called propapain and becomes activated later by cleavage (hydrolysis) of a peptide bond. This type of covalent modification avoids premature proteolytic degradation of cellular constituents by the newly synthesized enzyme.

Feedback inhibition. Consider a typical metabolic pathway with two or more end products such as that shown in Figure 2.18. Control of the system requires that if the end products build up too much, their rate of formation is decreased. Similarly, if too much reactant A builds up, the rate of conversion of A to products should be increased. The process is usually regulated by control of the flux at the first step of the pathway and at each branch point. The final products, G and J, which might bear no resemblance to the substrate A, inhibit the enzymes at A → B and at the branch point.

By having two enzymes at A → B, each inhibited by one of the end metabolites but not by the other, it is possible to exert finer control than with just one enzyme. The first step in a metabolic pathway is usually called

* Although some texts refer to the conjugation of a compound with adenylic acid (AMP) as "adenylation," the chemically correct term is *"adenylylation."*

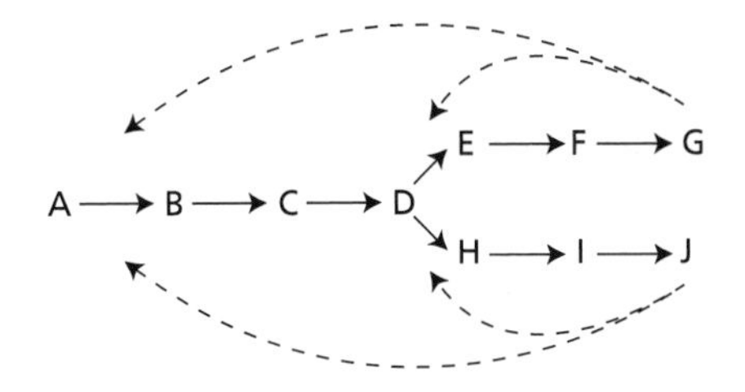

Figure 2.18 Feedback inhibition in a hypothetical metabolic pathway. The letters (A–J) represent metabolites, and each arrow represents an enzyme-catalyzed reaction. The boldface arrow for the first reaction indicates that two different enzymes with different inhibitor susceptibilities are involved. Broken lines indicate metabolites that inhibit particular enzymes. The first step in the metabolic pathway and the branch points are particularly important sites for feedback control.

the *committed step*. At this step enzymes are subject to major control.

Fructose-2,6-bisphosphate plays a central role in the regulation of carbon metabolism in plants. It functions as an activator in glycolysis (the breakdown of sugars to generate energy) and an inhibitor in gluconeogenesis (the synthesis of sugars). Fructose-2,6-bisphosphate is synthesized from fructose-6-phosphate in a reaction requiring ATP and catalyzed by the enzyme fructose-6-phosphate 2-kinase. It is degraded in the reverse reaction catalyzed by fructose-2,6-bisphosphatase, which releases inorganic phosphate (P_i). Both of these enzymes are subject to metabolic control by fructose-2,6-bisphosphate, as well as ATP, P_i, fructose-6-phosphate, dihydroxyacetone phosphate, and 3-phosphoglycerate. The role of fructose-2,6-bisphosphate in plant metabolism will be discussed further in Chapters 8 and 11.

Summary

Living organisms, including green plants, are governed by the same physical laws of energy flow that apply everywhere in the universe. These laws of energy flow have been encapsulated in the laws of thermodynamics.

Energy is defined as the capacity to do work, which may be mechanical, electrical, osmotic, or chemical work. The first law of thermodynamics states the principle of energy conservation: Energy can be converted from one form to another, but the total energy of the universe remains the same. The second law of thermodynamics describes the direction of spontaneous processes: A spontaneous process is one that results in a net increase in the total entropy (ΔS), or randomness, of the system plus its surroundings. Processes involving heat transfer, such as the cooling due to water evaporation from leaves, are best described in terms of the change in heat content, or enthalpy (ΔH), defined as the amount of energy absorbed or evolved as heat under constant pressure.

The free-energy change, ΔG, is a convenient parameter for determining the direction of spontaneous processes in chemical or biological systems without reference to their surroundings. The value of ΔG is negative for all spontaneous processes at constant temperature and pressure. The ΔG of a reaction is a function of its displacement from equilibrium. The greater the displacement from equilibrium, the more work the reaction can do. Living systems have evolved to maintain their biochemical reactions as far from equilibrium as possible.

The redox potential represents the free-energy change of an oxidation–reduction reaction expressed in electrochemical units. As with changes in free energy, the redox potential of a system depends on the concentrations of the oxidized and reduced species.

The establishment of ion gradients across membranes is an important aspect of the work carried out by living systems. The membrane potential is a measure of the work required to transport an ion across a membrane. The electrochemical-potential difference includes both concentration and electric potentials.

The laws of thermodynamics predict whether and in which direction a reaction can occur, but they say nothing about the speed of a reaction. Life depends on highly specific protein catalysts called enzymes to speed up the rates of reactions. All proteins are composed of amino acids linked together by peptide bonds. Protein structure is hierarchical; it can be classified into primary, secondary, tertiary, and quaternary levels. The forces responsible for the shape of a protein molecule are noncovalent and are easily disrupted by heat, chemicals, or pH, leading to loss of conformation, or denaturation.

Enzymes function by lowering the free-energy barrier between the substrates and products of a reaction. Catalysis occurs at the active site of the enzyme. Enzyme-mediated reactions exhibit saturation kinetics and can be described by the Michaelis–Menten equation, which relates the velocity of an enzyme-catalyzed reaction to the substrate concentration. The substrate concentration is inversely related to the affinity of an enzyme for its substrate. Since reaction velocity is relatively insensitive to small changes in substrate concentration, many enzymes exhibit cooperativity. Typically, such enzymes are allosteric, containing two or more active sites that interact with each other and that may be located on different subunits.

Enzymes are subject to reversible and irreversible inhibition. Irreversible inhibitors typically form covalent bonds with the enzyme; reversible inhibitors form noncovalent bonds with the enzyme and may have competitive, noncompetitive, or mixed effects.

Enzyme activity is often regulated in cells. Regulation may be accomplished by compartmentalization of enzymes and/or substrates; covalent modification; feedback inhibition, in which the end products of metabolic pathways inhibit the enzymes involved in earlier steps; and control of the enzyme concentration in the cell by gene expression and protein degradation.

General Reading

Alberts, B., Bray, D., Lewis, J., Raff, M., Roberts, K., and Watson, J. D. (1994) *Molecular Biology of the Cell*, 3rd ed. Garland, New York.

Atchison, M. L. (1988) Enhancers: Mechanisms of action and cell specificity. *Annu. Rev. Cell Biol.* 4: 127–153.

*Atkinson, D. E. (1977) *Cellular Energy Metabolism and Its Regulation.* Academic Press, New York.

*Creighton, T. E. (1983) *Proteins: Structures and Molecular Principles.* W. H. Freeman, New York.

Darnell, J., Lodish, H., and Baltimore, D. (1995) *Molecular Cell Biology*, 3rd ed. Scientific American Books, W. H. Freeman, New York.

*Edsall, J. T., and Gutfreund, H. (1983) *Biothermodynamics: The Study of Biochemical Processes at Equilibrium.* Wiley, New York.

Fersht, A. (1985) *Enzyme Structure and Mechanism*, 2nd ed. W. H. Freeman, New York.
*Klotz, I. M. (1967) *Energy Changes in Biochemical Reactions.* Academic Press, New York.
*Morowitz, H. J. (1978) *Foundations of Bioenergetics.* Academic Press, New York.
Walsh, C. T. (1979) *Enzymatic Reaction Mechanisms.* W. H. Freeman, New York.
Webb, E. (1984) *Enzyme Nomenclature.* Academic Press, Orlando, Fla.

* Indicates a reference that is general reading in the field and is also cited in this chapter.

Chapter References

Bryant, F. O., and Adams, M. W. W. (1989) Characterization of hydrogenase from the hyperthermophilic archaebacterium? *Pyrococcus furiosus. J. Biol. Chem.* 264: 5070–5079.
Clausius, R. (1879) *The Mechanical Theory of Heat.* Tr. by Walter R. Browne. Macmillan, London.
Donahue, R. A., Davis, T. D., Michler, C. H., Riemenschneider, D. E., Carter, D. R., Marquardt, P. E., Sankhla, N., Sahkhla, D. Haissig, B. E., and Isebrands, J. G. (1995) Growth, photosynthesis, and herbicide tolerance of genetically modified hybrid poplar. *Can. J. Forest Res.* 24: 2377–2383.
Mathews, C. K., and Van Holde, K. E. (1996) *Biochemistry*, 2nd ed. Benjamin/Cummings, Menlo Park, CA.
Nicholls, D. G., and Ferguson, S. J. (1992) *Bioenergetics 2.* Academic Press, San Diego.
Stryer, L. (1995) *Biochemistry*, 4th ed. W. H. Freeman, New York.

Transport and Translocation of Water and Solutes

UNIT

I

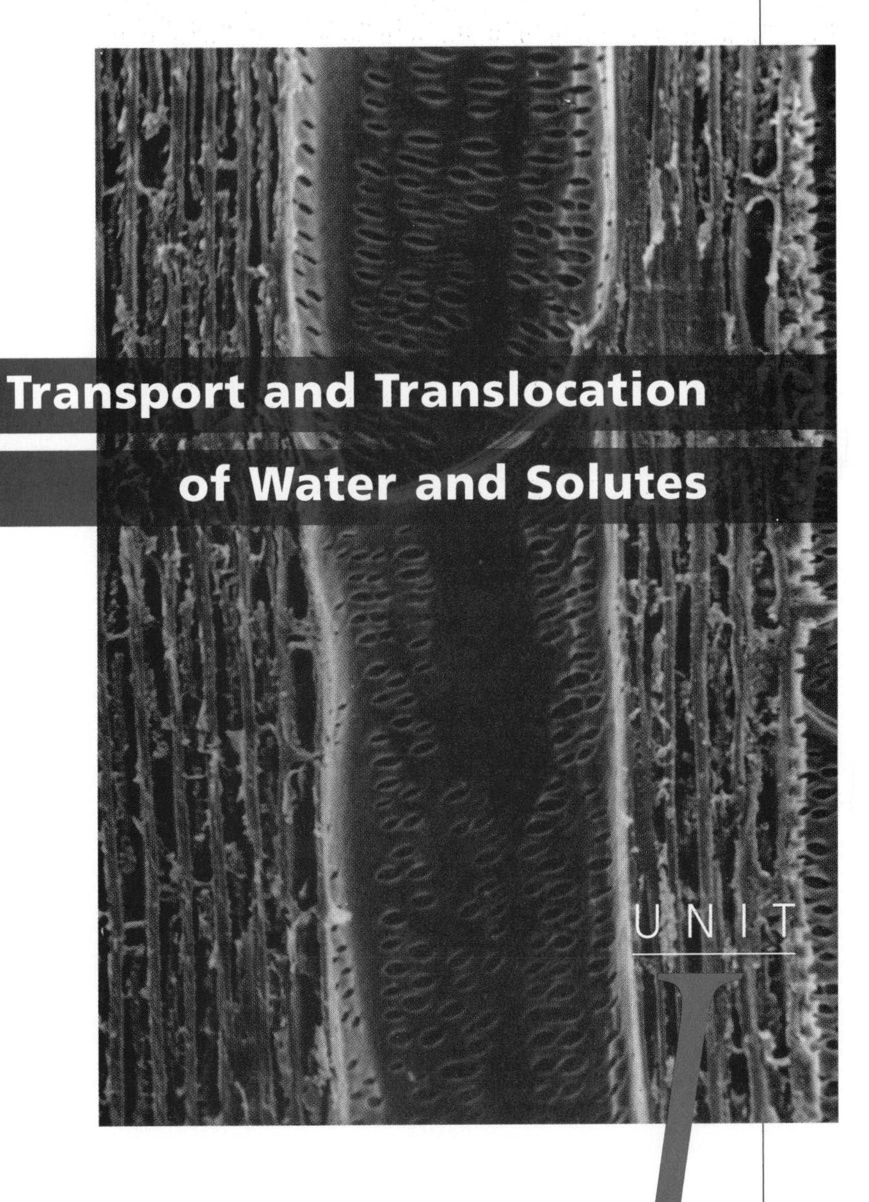

3 Water and Plant Cells

THIS CHAPTER AND THE FOLLOWING ONE deal with plant water relations. This is an important topic because water plays a crucial role in the life of the plant. For every gram of organic matter made by the plant, approximately 500 g of water is absorbed by the roots, transported through the plant body and lost to the atmosphere. Even slight imbalances in this flow of water can cause water deficits and severe malfunctioning of many cellular processes. Thus, every plant must delicately balance its uptake and loss of water. This balancing is a serious challenge for land plants because their need to draw carbon dioxide from the atmosphere (for photosynthesis) inevitably exposes them to the threat of dehydration by the atmosphere.

Unlike animal cells, plant cells build up a large intracellular pressure, called **turgor pressure**, as a consequence of their normal water balance. Turgor pressure is essential for many physiological processes, including cell enlargement, gas exchange in the leaves, transport in the phloem, and various transport processes in membranes. Turgor pressure also contributes to the rigidity and mechanical stability of nonlignified plant tissues. In this chapter we will consider how water moves into and out of plant cells, emphasizing the physical forces that influence water movement at the cell level. In the next chapter we will examine water transport at the whole-plant level and how land plants cope with the inevitable loss of water to the atmosphere.

But first we will describe the major functions of water in plant life. Water makes up most of the mass of plant cells, as we can readily appreciate if we look at microscopic sections of mature plant cells: Each cell is packed with a large water-filled vacuole. In such cells, the cytoplasm makes up only 5 to 10% of the cell volume; the

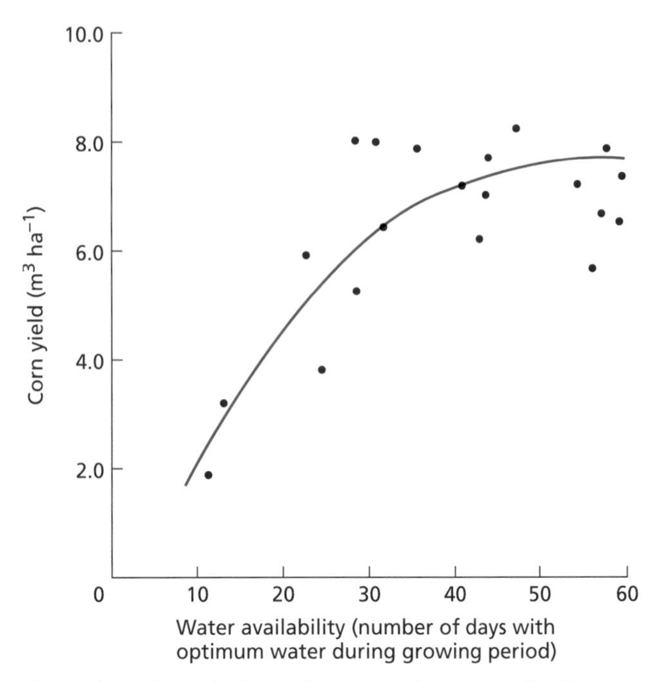

Figure 3.1 Corn yield as a function of water availablity. The data plotted here were gathered at an Iowa farm over a 4-year period. Water availability was assessed as the number of days without water stress during a 9-week growing period. (Data from CAED Report 20.)

remainder is a vacuole. Water typically constitutes 80 to 95% of the mass of growing plant tissues. Common vegetables such as carrots and lettuce may contain 85 to 95% water. Wood, which is composed mostly of dead cells, has a lower water content; sapwood, which functions in transport in the xylem, contains 35 to 75% water; and heartwood has a slightly lower water content. Seeds, with a water content of 5 to 15%, are among the driest of plant tissues, yet before germinating they must absorb a considerable amount of water.

Water is the most abundant and arguably the best solvent known. As a solvent, it makes up the medium for the movement of molecules within and between cells and greatly influences the structure of proteins, nucleic acids, polysaccharides, and other cell constituents. Water forms the environment in which most of the biochemical reactions of the cell occur, and it directly participates in many essential chemical reactions, such as those involving hydrolysis and dehydration.

Plants continuously absorb and lose water. On a warm, dry, sunny day a leaf will exchange up to 100% of its water in a single hour. During the plant's lifetime, water equivalent to 100 times the fresh weight of the plant may be lost through the leaf surfaces. Such water loss, called **transpiration**, is an important means of dissipating the heat input from sunlight. Heat dissipates because the water molecules that escape into the atmosphere have higher-than-average energy, which breaks the bonds holding them in the liquid. When these molecules escape, they leave behind a mass of molecules with lower-than-average energy and thus a cooler body of water. For a typical leaf, nearly half of the net heat input from sunlight is dissipated by transpiration. The stream of water taken up by the roots is also an important means of bringing dissolved soil minerals to the root surface for absorption.

Of all the resources that plants need to grow and function, water is the most abundant and at the same time the most limiting for agricultural productivity (Figure 3.1). The fact that water is limiting is the reason for the practice of crop irrigation. Water availability likewise limits the productivity of natural ecosystems (Figure 3.2). Thus an understanding of the uptake and loss of water by plants is very important.

We will begin our study of water by considering how its structure gives rise to some of its unique physical properties. We will then examine the physical basis for water movement, the concept of water potential, and the application of this concept to cell water relations.

The Structure and Properties of Water

Water has special properties that enable it to act as a solvent and to be readily transported through the body of the plant. These properties derive primarily from the polar structure of the water molecule. Here we will examine water's hydrogen bonds and how they contribute to the properties of water that are so necessary for life.

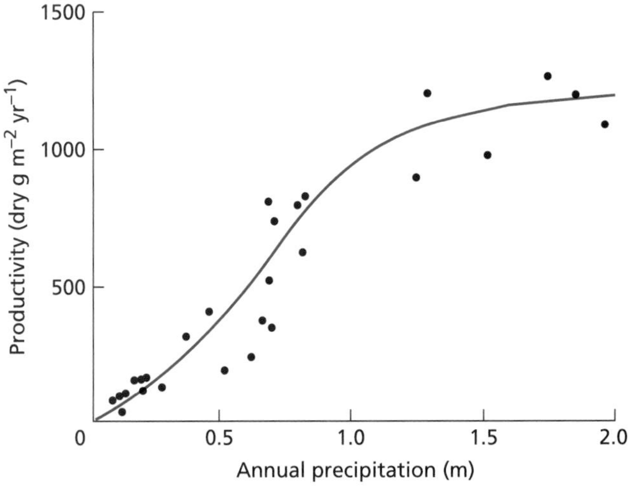

Figure 3.2 Productivity of various ecosystems as a function of annual precipitation. Productivity was estimated as net aboveground accumulation of organic matter through growth and reproduction. (After Whittaker 1970.)

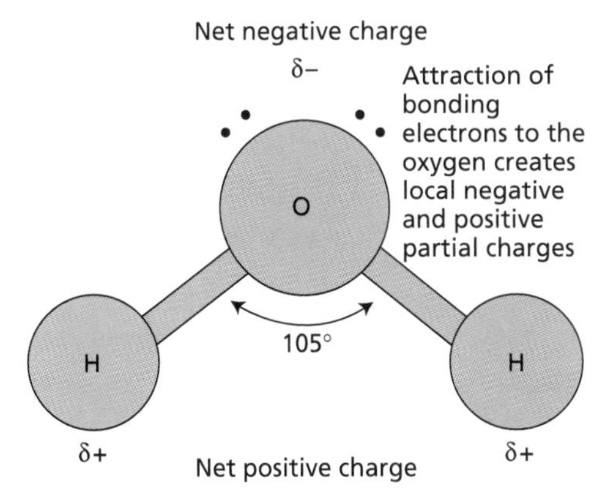

Figure 3.3 Diagram of the water molecule. The two intramolecular hydrogen–oxygen bonds form an angle of 105°. The opposite partial charges (δ– and δ+) on the water molecule result in the formation of intermolecular hydrogen bonds with other water molecules.

The Polarity of Water Molecules Gives Rise to Hydrogen Bonds

The water molecule consists of an oxygen atom covalently bonded to two hydrogen atoms. The two O–H bonds form an angle of 105° (Figure 3.3). Because the oxygen atom is more **electronegative** than hydrogen, it tends to attract the electrons of the covalent bond. This attraction results in a partial negative charge at the oxygen end of the molecule and a partial positive charge at each hydrogen. These partial charges are equal, so the water molecule carries no *net* charge. Nevertheless, this separation of partial charges, together with the shape of the water molecule, makes water a *polar molecule*, and the opposite partial charges between neighboring water molecules tend to attract each other. The weak electrostatic attraction between water molecules, known as a **hydrogen bond**, is responsible for many of the unusual physical properties of water.

Hydrogen bonds can also form between water and other molecules that contain electronegative atoms (O or N). The structure of proteins, polysaccharides, nucleic acids, and other molecules in the cell is strongly influenced by hydrogen bonds. Hydrogen bonding is responsible for the stable base pairing between complementary strands of DNA. In aqueous solutions, hydrogen bonding between water molecules leads to local, ordered clusters of water that, because of the continuous thermal agitation of the water molecules, continually form, break up, and re-form (Figure 3.4).

The Polarity of Water Makes It an Excellent Solvent

The physical properties of water make it uniquely suitable as a medium for life. First, water is an excellent solvent: It dissolves greater amounts of a wider variety of substances than do other related solvents. This versatility as a solvent is due in part to the small size of the water molecule and in part to its polar nature, which makes water a particularly good solvent for ionic substances and for molecules such as sugars and proteins that contain polar —OH or —NH_2 residues.

The water molecules orient themselves around ions and polar solutes in solution and effectively shield their electric charges. This shielding decreases the electrostatic interaction between the charged substances and thereby increases their solubility. Furthermore, the polar ends of water molecules can orient themselves next to charged or partially charged groups in macromolecules, forming **shells of hydration**. Hydrogen bonding between macromolecules and water reduces the interaction between the macromolecules and helps draw them into solution.

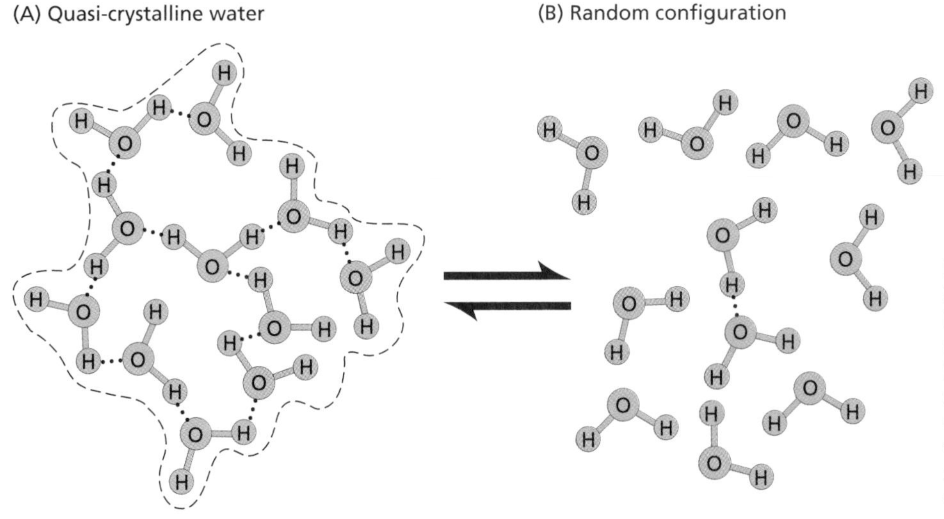

Figure 3.4 Hydrogen bonding between water molecules results in local aggregations of ordered, quasi-crystalline water (A). Because of the continuous thermal agitation of the water molecules, these aggregations are very short-lived; they break up rapidly to form much more random configurations (B).

The Thermal, Cohesive, and Adhesive Properties of Water Result from Hydrogen Bonding

The extensive hydrogen bonding between water molecules results in unusual thermal properties, such as high specific heat and high latent heat of vaporization. **Specific heat** refers to the heat energy required to raise the temperature of a substance by a specific amount. When the temperature of water is raised, the molecules must vibrate faster, so a great deal of energy must be put into the system to break the hydrogen bonds between water molecules. Thus, compared with other liquids, water requires a relatively large energy input to raise its temperature. This large energy input requirement is important for plants because it helps slow potentially harmful temperature fluctuations.

Latent heat of vaporization is the energy needed to separate molecules from the liquid phase and move them into the gas phase at constant temperature—a process that occurs during transpiration. For water at 25°C, the heat of vaporization is 44 kJ mol^{-1}—the highest value known for any liquid. Most of this energy is used to break hydrogen bonds between water molecules. The high latent heat of vaporization of water enables plants to cool themselves by evaporating water from leaf surfaces, which are prone to heat up because of the radiant input from the sun. Transpiration is an important component of temperature regulation in many plants.

Water molecules at an air–water interface are more strongly attracted to neighboring water molecules than to the gas phase on the other side of the surface. As a consequence of this unequal attraction, the air–water interface tends to minimize its surface area. In effect, the water molecules exert a force at the air–water interface. This force not only influences the shape of the surface but also may create a pressure in the rest of the liquid. The condition that exists at the interface is known as **surface tension**. As we will see later, surface tension at the evaporative surfaces of leaves generates the physical forces that pull a stream of water through the plant's vascular system.

In addition, the extensive hydrogen bonding in water gives rise to the property known as **cohesion**, the mutual attraction between molecules. A related property, called **adhesion**, is the attraction of water to a solid phase such as a cell wall or glass surface. Cohesion, adhesion, and surface tension give rise to a phenomenon known as **capillarity**, the movement of water a small distance up a glass capillary tube. The upward movement of the water is due to attraction of water at the periphery to the polar surface of a clean glass tube (adhesion) and to the surface tension of water, which tends to minimize the surface area. Together, adhesion and surface tension exert a tension on the water molecules at and just below the surface, causing them to move up the tube until the force of adhesion is balanced by the weight of the water column. The smaller the tube, the higher the capillary rise, which may be calculated using the following formula:

$$\text{Capillary rise} = \frac{14.9 \times 10^{-6}\,\text{m}^2}{\text{radius}} \tag{3.1}$$

where both capillary rise and radius are expressed in meters.

For a xylem vessel with 25 μm radius, the capillary rise is about 0.6 m. This distance is much too small to be significant for water transport up tall trees. However, fibrous materials such as cell walls can act like wicks to draw water by capillarity from nearby xylem. This action ensures that cell wall surfaces that are directly exposed to the air, such as those in leaf mesophyll, remain wetted and do not dry out. Because the cell wall capillaries have a tiny radius, about 10 m^{-8}, very large physical forces can be generated in the water just below the evaporative surfaces of cell walls.

Water Has a High Tensile Strength

Cohesion gives water a high **tensile strength**, defined as the ability to resist a pulling force. We do not usually think of liquids as having tensile strength; however, such a property must exist for a water column to be pulled up a capillary tube without breaking.

We can demonstrate the tensile strength of water by placing it in a capped syringe (Figure 3.5). When we *push* on the plunger, the water is compressed and a positive **hydrostatic pressure** builds up. The pressure is measured in units called *pascals* (Pa) or, more conve-

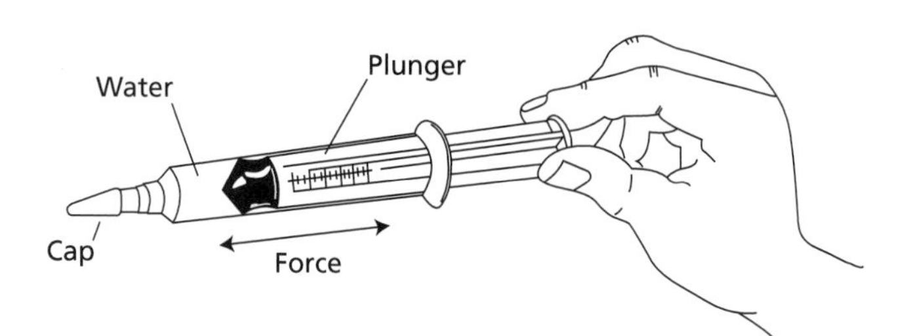

Figure 3.5 A sealed syringe can be used to create positive and negative pressures in a fluid like water. Pushing on the plunger compresses the fluid, and a positive pressure builds up. If a small air bubble is trapped within the syringe, it shrinks as the pressure increases. Pulling on the plunger causes the fluid to develop a tension, or negative pressure. Any air bubbles in the syringe will expand as the pressure is reduced from atmospheric pressure. As a gas bubble in the syringe expands, the pressure never goes below a pure vacuum, because the gas phase can expand indefinitely. If the syringe is filled with a degassed solution and lacks bubbles, pressures below vacuum (i.e., negative pressures) can develop because the hydrogen bonds holding the fluid together can withstand considerable tension before breaking.

TABLE 3.1
Comparison of units of pressure

1 atmosphere = 14.7 pounds per square inch
= 760 mm Hg (at sea level, 45° latitude)
= 1.013 bar
= 0.1013 MPa
= 1.013×10^5 Pa
A car tire is typically inflated to about 0.2 MPa.
The water pressure in home plumbing is typically 0.2–0.3 MPa.
The water pressure under 15 feet (5 m) of water is about 0.05 MPa.

niently, *megapascals* (MPa).* Pressure is equivalent to a force per unit area (1 Pa = 1 N m^{-2}) and to an energy per unit volume (1 Pa = 1 J m^{-3}). Table 3.1 compares pascals with some other units of pressure. If instead of pushing on the plunger we *pull* on it, a tension, or negative hydrostatic pressure, develops in the water to resist the pull. How hard must we pull on the plunger before the water molecules are torn away from each other and the water column breaks? Breaking the water column requires sufficient energy to overcome the forces that attract water molecules to one another.

In the syringe shown in Figure 3.5, small bubbles often interfere with this measurement by expanding as the pressure is reduced. However, careful studies have demonstrated that water in small capillaries can resist tensions more negative than –30 MPa (the negative sign indicates tension, as opposed to compression). This value is only a fraction of the theoretical strength of water computed on the basis of the strength of hydrogen bonds. Nevertheless, it is about 10% of the tensile strength of copper or aluminum wire and is thus quite substantial.

The presence of gas bubbles reduces the tensile strength of a water column. The lowest absolute pressure possible in a gas phase is 0 MPa (pure vacuum) because the intermolecular forces of attraction needed to resist a negative pressure (or tension) do not exist in ideal gases. In contrast, in solids and liquids the intermolecular attractions can resist tensile forces. Therefore, if even a tiny gas bubble forms in a column of water under tension, the gas bubble will expand indefinitely, with the result that the tension in the liquid phase collapses, a phenomenon known as **cavitation**. As we will see in Chapter 4, cavitation can have a devastating effect on water transport through the xylem of trees.

Water Transport Processes

When water moves from the soil through the plant to the atmosphere, it travels through a widely variable medium, and the mechanisms of water transport also vary with the type of medium (cell wall, cytoplasm, membrane, air spaces). We will now consider the two major processes in water transport: molecular diffusion and bulk flow.

Diffusion Is the Movement of Molecules by Random Thermal Agitation

Water molecules in a solution are not static; they are in continuous motion, colliding with one another and exchanging kinetic energy. **Diffusion** is the process by which molecules intermingle as a result of their random thermal agitation. Such agitation gives rise to the random but progressive movement of substances from regions of high free energy to regions of low free energy. As long as other forces are not acting on the molecules, diffusion causes molecules to move from regions of high concentration to regions of low concentration—that is, down a concentration gradient (Figure 3.6). Fick discovered that the rate of diffusional movement is directly proportional to the concentration gradient ($\partial c/\partial x$). In symbols, we write this relation as Fick's first law:

$$J_s = -D_s \frac{\partial c_s}{\partial x} \tag{3.2}$$

The rate of transport, or the **flux density** (J_s), is the amount of substance *s* crossing a unit area per unit time (e.g., J_s may have units of moles per square meter per second [mol m^{-2} s^{-1}]). The **diffusion coefficient** (D_s) is a proportionality constant that measures how easily substance *s* moves through a particular medium. The diffusion coefficient is a characteristic of the substance (larger molecules have smaller diffusion coefficients) and depends on the medium (diffusion in air is much faster than diffusion in a liquid, for example). The concentration gradient ($\partial c_s/\partial x$)is usually approximated as $\Delta c_s/\Delta x$—that is, as the difference in concentration of substance *s* (Δc_s) between two points separated by the distance Δx. The negative sign in the equation indicates that the flux moves down a concentration gradient. Fick's first law says that a substance will diffuse faster when the concentration gradient becomes steeper or when the diffusion coefficient is increased. This equation accounts only for movement in response to a concentration gradient, and not for movement in response to other forces (e.g., pressure, electric fields, and so on).

Diffusion Is Rapid over Short Distances but Extremely Slow over Long Distances

From Fick's law, one can derive (with difficulty) an expression for the time it takes for a substance to diffuse a particular distance. If the initial conditions are such that all the solute molecules are concentrated at the starting position (Figure 3.7A), then the concentration front moves away from the starting position, as shown

* 1 MPa = approximately 9.9 atmospheres.

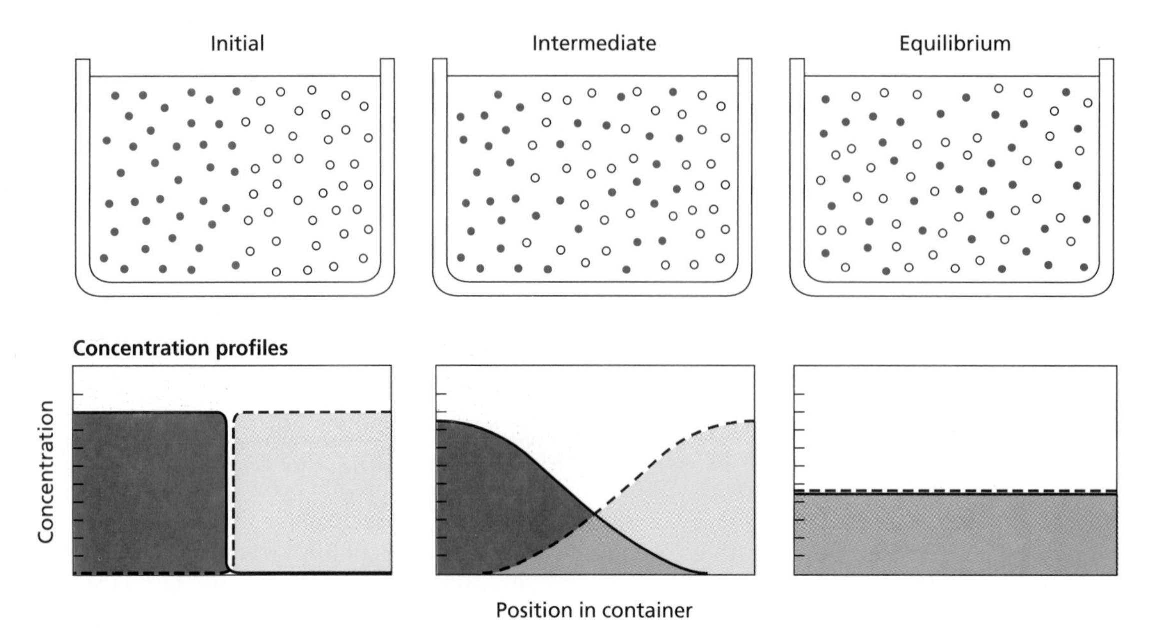

Figure 3.6 Thermal motion of molecules leads to diffusion—the gradual mixing of molecules and eventual dissipation of concentration differences. Initially, two materials containing different molecules are brought into contact. The materials may be gas, liquid, or solid. Diffusion is fastest in gases, intermediate in liquids, and slowest in solids. The initial separation of the molecules is depicted graphically in the upper panels, and the corresponding concentration profiles are shown in the lower panels as a function of position. With time, the mixing and randomization of the molecules diminishes net movement. At equilibrium the two types of molecules are randomly (evenly) distributed.

for a later time point in Figure 3.7B. As the substance diffuses away from the starting point, the concentration gradient becomes less steep, and thus net movement becomes slower. The time it takes for the substance at any given distance from the starting point to reach one-half of the concentration at the starting point ($t_c = 1/2$) is given by the following equation:

$$t_{c=1/2} = \frac{(\text{distance})^2}{D_s} K \qquad (3.3)$$

where K is a constant that depends on the shape of the system (for convenience, we will use a K value of 1 in the following calculations) and D_s is the diffusion coefficient. Equation 3.3 shows that the time required for a substance to diffuse a given distance increases in proportion to the *square* of that distance.

We can see what this means by considering two numerical examples. First, how long it would take a small molecule to diffuse across a typical cell? The diffusion coefficient for a small molecule like glucose is about 10^{-9} m^2 s^{-1}, and the cell size may be 50 μm. Thus, for this example:

$$t_{c=1/2} = \frac{(50 \times 10^{-6}\,\text{m})^2}{10^{-9}\,\text{m}^2\text{s}^{-1}} = 2.5\ \text{s}$$

This calculation shows that small molecules diffuse over cellular dimensions rapidly. What about diffusion over longer distances? Calculating the time needed for the same substance to move a distance of 1 m (e.g., the length of a corn leaf), we find:

$$t_{c=1/2} = \frac{(1\ \text{m})^2}{10^{-9}\,\text{m}^2\,\text{s}^{-1}} = 10^9\ \text{s} \approx 32\ \text{years}$$

a value that exceeds by orders of magnitude the life span of a corn plant, which lives only a few months.

From these numerical examples we see that diffusion in solutions can be effective within cellular dimensions but is far too slow for mass transport over long distances. As we'll see in Chapter 4, diffusion is of great importance during loss of water vapor from leaves because the diffusion coefficient in air is much greater than in aqueous solutions.

Pressure-Driven Bulk Flow Drives Long-Distance Water Transport

A second process by which water moves is known as **bulk flow** or **mass flow**. Bulk flow is the concerted movement of groups of molecules en masse, most often in response to a pressure gradient. Among many common examples of bulk flow are convection currents, water moving through a garden hose, a river flowing, and rain falling.

If we consider bulk flow through a pipe, the rate of volume flow depends on the radius (r) of the pipe, the

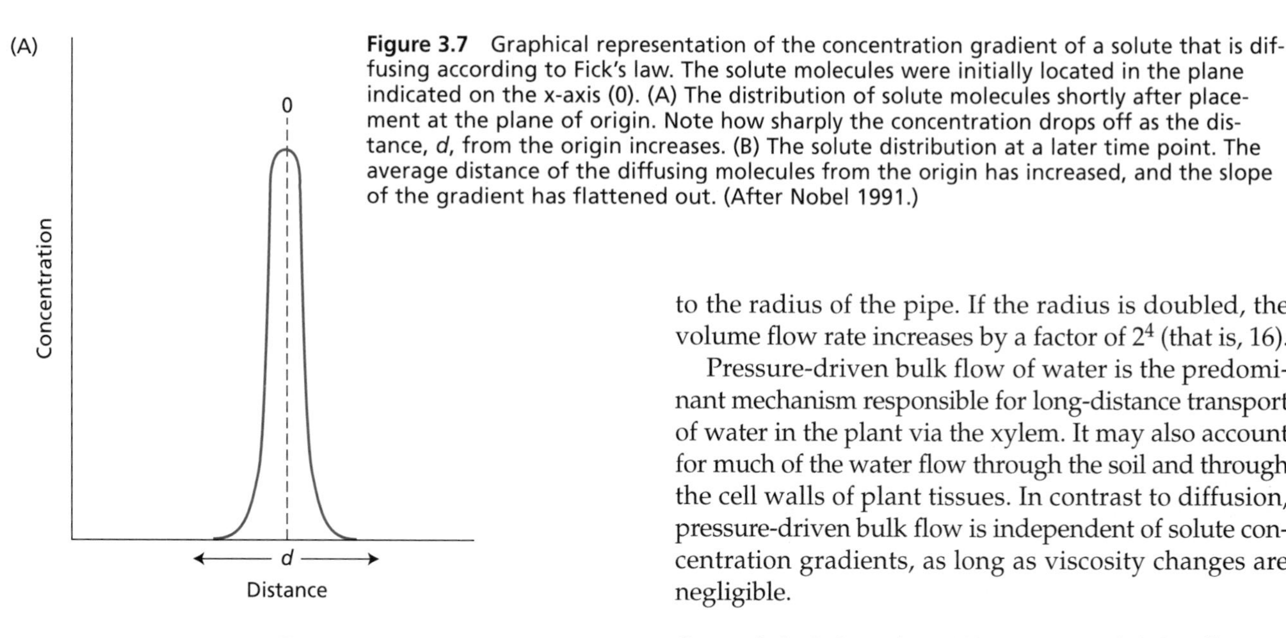

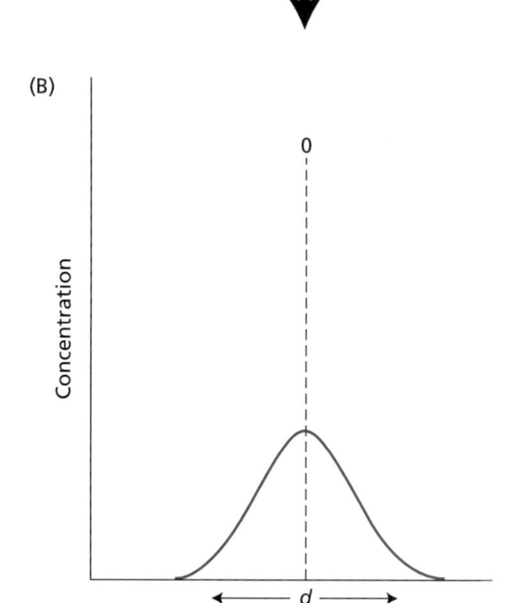

Figure 3.7 Graphical representation of the concentration gradient of a solute that is diffusing according to Fick's law. The solute molecules were initially located in the plane indicated on the x-axis (0). (A) The distribution of solute molecules shortly after placement at the plane of origin. Note how sharply the concentration drops off as the distance, *d*, from the origin increases. (B) The solute distribution at a later time point. The average distance of the diffusing molecules from the origin has increased, and the slope of the gradient has flattened out. (After Nobel 1991.)

viscosity (η) of the liquid, and the pressure gradient ($\partial\Psi_z/\partial x$) that drives the flow. This relation is given by one form of **Poiseuille's equation**:*

$$\text{Volume flow rate} = \left(\frac{\pi r^4}{8\eta}\right)\left(\frac{\partial\Psi_p}{\partial x}\right) \tag{3.4}$$

expressed in cubic meters per second ($m^3\ s^{-1}$). This equation demonstrates that bulk flow is very sensitive to the radius of the pipe. If the radius is doubled, the volume flow rate increases by a factor of 2^4 (that is, 16).

Pressure-driven bulk flow of water is the predominant mechanism responsible for long-distance transport of water in the plant via the xylem. It may also account for much of the water flow through the soil and through the cell walls of plant tissues. In contrast to diffusion, pressure-driven bulk flow is independent of solute concentration gradients, as long as viscosity changes are negligible.

Osmosis Is Driven by a Water Potential Gradient

Movement of a solvent such as water through a membrane is called **osmosis**. Although water can be taken up and lost by plant cells relatively quickly, uptake and loss are significantly limited by the plasma membrane, which acts as a barrier to the movement of most substances. Membranes of plant cells are **selectively permeable**; that is, they allow the movement of water and other small uncharged substances across them more readily than the movement of larger solutes and charged substances. To facilitate the transport of inorganic ions, sugars, amino acids, and other metabolites across the various cell membranes, special transport proteins are required. This aspect of membrane transport will be discussed in Chapter 6. Here we restrict our discussion to the movement of water across membranes.

Like molecular diffusion and pressure-driven bulk flow, osmosis occurs spontaneously in response to a driving force. In simple diffusion, substances move down a concentration gradient; in pressure-driven bulk flow, substances move down a pressure gradient; in osmosis, both types of gradients influence transport. The direction and rate of water flow across a membrane are determined not solely by the concentration gradient of water or by the pressure gradient, but by the sum of these two driving forces. This observation leads to the concept of a composite or total driving force, representing the free-energy gradient of water. In practice, this driving force is expressed as a **chemical-potential gradient** or, more commonly by plant physiologists, as a **water potential gradient**.

In the case of water movement into plant cells, the mechanism of osmosis involves a combination of (1) diffusion of single water molecules across the membrane bilayer and (2) bulk flow through tiny water-filled pores of molecular dimensions (Figure 3.8). Whether water

* Jean-Léonard-Marie Poiseuille (1797–1869) was a French physician and physiologist.

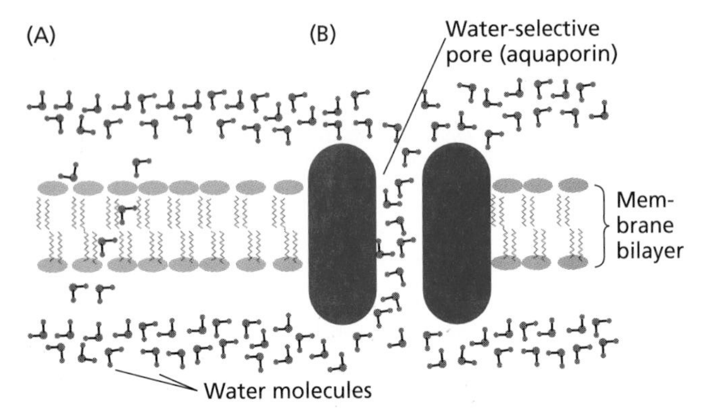

Figure 3.8 Water can cross plant membranes (A) by diffusion of individual water molecules through the membrane bilayer and (B) by bulk flow of files of water molecules through a water-selective pore formed by integral membrane proteins such as aquaporins.

moves by diffusion or by microscopic bulk flow, the relevant driving force is the water potential gradient, for reasons that were worked out in a theoretical analysis of osmosis by Peter Ray (1960). According to Ray's analysis, the exclusion of solutes from the water-filled pore in the membrane creates a pressure gradient within the pore. Thus, gradients in concentration and pressure give rise to equivalent transport across the membrane. This equivalence of pressure and concentration is strictly true only for ideal membranes—that is, membranes with much greater permeability for water than for dissolved solutes. For nonideal membranes, more advanced analyses are needed (Stein 1986; Finkelstein 1987).

For many years there was much uncertainty about whether water actually moved through microscopic pores within plant membranes. Some studies indicated that simple diffusion was not sufficient to account for osmosis, but the evidence in support of microscopic pores was not compelling. This uncertainty was put to rest with the recent discovery of **aquaporins** (see Figure 3.8B). These are integral membrane proteins that form water-selective channels across the membrane (see also Chapter 6). Such channels facilitate water movement across the membrane (Weig et al. 1997; Schäffner 1998). The ability of aquaporins to transport water may be regulated by their phosphorylation state (Maurel et al. 1995). In other words, the cell can regulate its membrane permeability to water by adding or removing phosphate groups to specific amino acid residues on the aquaporin protein. Such a modulation of aquaporin activity may alter the *rate* of water movement across the membrane; however, it does not change the direction of transport or the driving force for water movement.

To begin our detailed discussion of cell water relations, consider the following "thought experiment": Place a flaccid cell—that is, one with negligible turgor pressure—in water. The cell takes up water, initially quickly and then more slowly, and eventually reaches an equilibrium where net water uptake ceases. At this point, the free energy of water is the same inside and outside the cell. If we measured cell volume and turgor pressure, we would find that they increase and then stabilize with a similar time course. Now we may ask the following questions: What is the nature of this equilibrium and how is it defined quantitatively? What are the values for the changes in cell volume and cell turgor? What is the time course for reaching equilibrium and on what does it depend? These issues are examined in the sections that follow.

The Chemical Potential of Water Represents the Free-Energy Status of Water

The **chemical potential** of water (or *water potential*) is a quantitative expression of the free energy associated with water. In thermodynamics, free energy represents a potential for performing work. All living things, including plants, require a continuous input of free energy in order to maintain and repair their structures and their organized states, as well as to grow and reproduce. Processes such as biochemical reactions, solute accumulation, and long-distance transport are all driven by an input of free energy into the plant.

The concepts of free energy and chemical potential derive from thermodynamics, the study of the transformations of energy. As we discussed in Chapter 2, thermodynamics provides a useful framework for studying the energetics of many processes essential to life. Here we will restrict ourselves to a discussion of the thermodynamic basis for transport of water in plants.

Osmosis is an energetically spontaneous process. That is, water moves down a chemical-potential gradient, from a region of high chemical potential to a region of low chemical potential. In the thought experiment described in the previous section, this means that water moves from outside to inside the cell. No net work is involved in osmosis; rather, free energy is released. To understand water movement by osmosis, we need to examine more closely what influences the chemical potential of water.

In general, the chemical potential of water may be influenced by three major factors: *concentration*, *pressure*, and *gravity*. Note that chemical potential is a relative quantity: It is usually expressed as the difference between the potential of a substance in a given state and the potential of the same substance in a standard state. The unit of chemical potential is energy per mole of substance (J mol^{-1}).

For historical reasons, plant physiologists have most often used a related quantity called **water potential**, defined as the chemical potential of water divided by the partial molal volume of water (the volume of 1 mol of water): 18×10^{-6} m^3 mol^{-1}. Water potential is a mea-

TABLE 3.2
Values of *RT* and osmotic potential of solutions at various temperatures

Temperature (°C)	RT^a (L MPa mol^{-1})	Osmotic potential (MPa) of solution with solute concentration in mol L^{-1} water			Osmotic potential of seawater (MPa)
		0.01	0.10	1.00	
0	2.271	−0.0227	−0.227	−2.27	−2.6
20	2.436	−0.0244	−0.244	−2.44	−2.8
25	2.478	−0.0248	−0.248	−2.48	−2.8
30	2.519	−0.0252	−0.252	−2.52	−2.9

[a] R = 0.0083143 L MPa mol^{-1} K^{-1}.

sure of the free energy of water per unit volume (J m^{-3}). These units are equivalent to pressure units such as pascals, which is the common measurement unit for water potential. We will now consider more fully the important concept of water potential.

Three Major Factors Contribute to Cell Water Potential

Water potential is symbolized by Ψ_w (Greek letter psi), and the water potential of solutions may be dissected into individual components, usually written as the following sum:

$$\Psi_w = \Psi_s + \Psi_p + \Psi_g \tag{3.5}$$

The terms Ψ_s, Ψ_p, and Ψ_g denote the effects of solutes, pressure, and gravity, respectively, on the free energy of water. The reference state used to define water potential is liquid water at ambient pressure and temperature. This means that Ψ_w is proportional to the work required to move 1 mol of pure water at ambient pressure and temperature to another state at the same temperature. In most cases Ψ_w inside plant cells is negative, because pure water has a higher potential than the water inside the cell. Let's consider each of the terms on the right side of Equation 3.5.

Solutes. The term Ψ_s, called the **solute potential** or the **osmotic potential**, represents the effect of dissolved solutes on water potential. Solutes reduce the free energy of water by diluting the water. This effect is primarily an entropy effect; that is, the mixing of solutes and water increases the disorder of the system and thereby lowers free energy. The entropy effect of dissolved solutes is revealed in other physical effects known as the **colligative properties** (called "colligative" because they all occur together, or collectively; colligative properties depend on the number of dissolved particles and not on the nature of the solute) of solutions: Solutes reduce the vapor pressure of a solution, raise its boiling point, and lower its freezing point. The specific nature of the solute does not matter. For dilute solutions of nondissociating substances, the osmotic potential may be estimated by the **van't Hoff equation**:

$$\Psi_s = -RTc_s \tag{3.6}$$

where R is the gas constant (8.32 J mol^{-1} K^{-1}), T is the absolute temperature (in degrees Kelvin, or K), and c_s is the solute concentration of the solution, expressed as **osmolality** (moles of total dissolved solutes per liter of water [mol L^{-1}]). The minus sign indicates that dissolved solutes reduce the water potential of a solution. Table 3.2 shows the values of RT at various temperatures and the Ψ_s values of solutions of different solute concentrations.*

For ionic solutes that dissociate into two or more particles, c_s must be multiplied by the number of dissociated particles to account for the increased number of dissolved particles. Thus, if we dissolve 0.1 mol of sucrose in 1 L of water, we obtain a solution with an osmolality of 0.1 mol L^{-1}. If instead we dissolve 0.1 mol of salt (NaCl) in 1 L of water, we obtain an osmolality of 0.2 mol L^{-1} because salt dissociates into two particles. At 20°C (293 K), these values translate into the following osmotic potential (Ψ_s) values: –0.244 MPa for the sucrose solution and –0.488 MPa for the salt solution.

Equation 3.6 is valid for "ideal" solutions at dilute concentration. Real solutions frequently deviate from the ideal, especially at high concentrations—for example, greater than 0.1 mol L^{-1}. In our treatment of water

* The great German plant physiologist Wilhelm Pfeffer (1845–1920) made the fundamental discoveries about solutes, osmotic potentials, and the need for a membrane for osmosis to work. Pfeffer's results led to the Dutch chemist Jacobus Hendricus van't Hoff's (1852–1911) theory of solutions, for which van't Hoff received a Nobel prize in 1901. See Bünning 1989 for a fascinating biography of Pfeffer.

BOX 3.1

Alternative Conventions for Components of Water Potential

STUDENTS PLANNING FURTHER STUDY of plant water relations should note that the components of water potential defined in the text are sometimes given different names and symbols. In particular, the equation

$$Y_w = Y_s + Y_p$$

(Equation 3.7 in the text) is often replaced by the following equivalent equation:

$$Y_w = -p + P$$

In this alternative convention, P is the same as Y_p. It is the hydrostatic pressure of the solution, and may be positive, as in turgid cells, or negative, as in xylem water. The symbol p is called **osmotic pressure** and is the negative of Y_s. That is, p has positive values, and Y_s has negative values. "Osmotic pressure" is the term that physical chemists, zoologists, and many others use to denote the effect of dissolved solutes on the free energy of water. Most handbooks of physics and chemistry use the term "osmotic pressure" and the symbol p. The negative sign in front of p in the equation above accounts for the reduction in water potential (Y_w) by dissolved solutes. Thus $Y_s = -p$. A very interesting, if somewhat unconventional, account of the history and physical meaning of osmotic pressure is given by Hammel and Scholander (1976).

Unfortunately, some authors have mixed the conventions for Y_s and p, leading to unnecessary confusion about what is meant by the symbol p. Thus, p is sometimes incorrectly called osmotic potential instead of osmotic pressure, and it may be used either as a positive quantity or as a negative quantity. Let the reader beware!

potential, we will assume that we are dealing with ideal solutions.* Readers are referred to more advanced treatments for further discussion of nonideal behavior (see Friedman 1986; Nobel 1991).

Pressure. The term Y_p is the **hydrostatic pressure** of the solution. Positive pressures raise the water potential; negative pressures reduce it. Sometimes Y_p is called *pressure potential*. When referring to the positive hydrostatic pressure within cells, Y_p is usually called **turgor pressure**. The value of Y_p may be negative—for example, in the xylem and in the walls between cells, where a *tension*, or a *negative hydrostatic pressure*, can develop. As we will see, negative pressures outside cells are very important in moving water long distances through the plant.

Hydrostatic pressure is measured as the deviation from ambient pressure (Box 3.2). Remember that water in the reference state is at ambient pressure, so by this definition $Y_p = 0$ MPa for water in the standard state. Thus, the value of Y_p for pure water in an open beaker is 0 MPa, even though its absolute pressure is approximately 0.1 MPa (1 atmosphere). Water under a perfect vacuum has a Y_p value of –0.1 MPa; its absolute pressure is 0 MPa. It is important to keep in mind the distinction between absolute pressure and Y_p, which will be relevant to later discussions.

Gravity. Gravity causes water to move downward, unless the force of gravity is opposed by an equal and opposite force. The potential for water movement thus depends on height. The effect of gravity on water potential, Y_g, depends on the height (h) of the water above the reference-state water, the density of water (r_w), and the acceleration due to gravity (g). In symbols, we write these relationships as follows: $Y_g = r_w gh$, where $r_w g$ has a value of 0.01 MPa m^{-1}. Thus, a vertical distance of 10 m translates into a 0.1 MPa change in water potential.

Matric potential. In discussions of dry soils, seeds, and cell walls, one often finds reference to yet another component of Y_w: the **matric potential** (Y_m).† The matric potential is used to account for the reduction in free energy of water when it exists as a thin surface layer, one or two molecules thick, adsorbed onto the surface of relatively dry soil particles, cell walls, and other materials. The matric potential does not represent a new force acting on water, because the effect of the surface interaction can theoretically be accounted for by its effects on Y_s and Y_p (see Passioura 1980; Nobel 1991). As a practical matter, however, this surface interaction effect often cannot be easily separated into Y_p and Y_s components in dry materials, so frequently they are bulked together and designated as the matric potential.

It is generally not valid to add Y_m to independent measurements of Y_s and Y_p to arrive at a total water potential. This is particularly true for water inside

* Reference books of chemistry and physics often give tables listing exact osmotic potentials of actual solutions. These are usually given as osmotic pressures (usually the symbol p). Osmotic pressure and osmotic potential differ only in sign: $Y_s = -p$. In other words, osmotic potentials are negative, and osmotic pressures are positive (see Box 3.1).

† In the alternative naming system discussed in Box 3.1, matric potential is usually designated by the symbol t.

hydrated cells and cell walls, where matric effects are either negligible or they are accounted for by a reduction in Ψ_p. For instance, what we describe in this chapter as the negative pressure in water held by cell wall microcapillaries at the evaporative surfaces of leaves is sometimes described as a wall matric potential. Care is needed to avoid inconsistencies when accounting for this physical effect in definitions of Ψ_p, Ψ_s, and Ψ_m (Passioura 1980).

Simplifications for cell water relations. When dealing with water transport at the cell level, Equation 3.5 is usually simplified as follows:

$$\Psi_w = \Psi_s + \Psi_p \tag{3.7}$$

or sometimes $\Psi = \Psi_s + \Psi_p$. The gravitational component (Ψ_g) is ignored because it is negligible when vertical distances are small (say, less than 5 m). Thus the only significant components of cell Ψ_w are due to dissolved solutes and hydrostatic pressure.

Water Enters the Cell along a Water Potential Gradient

Now let's extend our thought experiment by illustrating the osmotic behavior of plant cells with some numerical examples. First, imagine an open beaker full of pure water at 20°C (Figure 3.9A). Since the water is open to

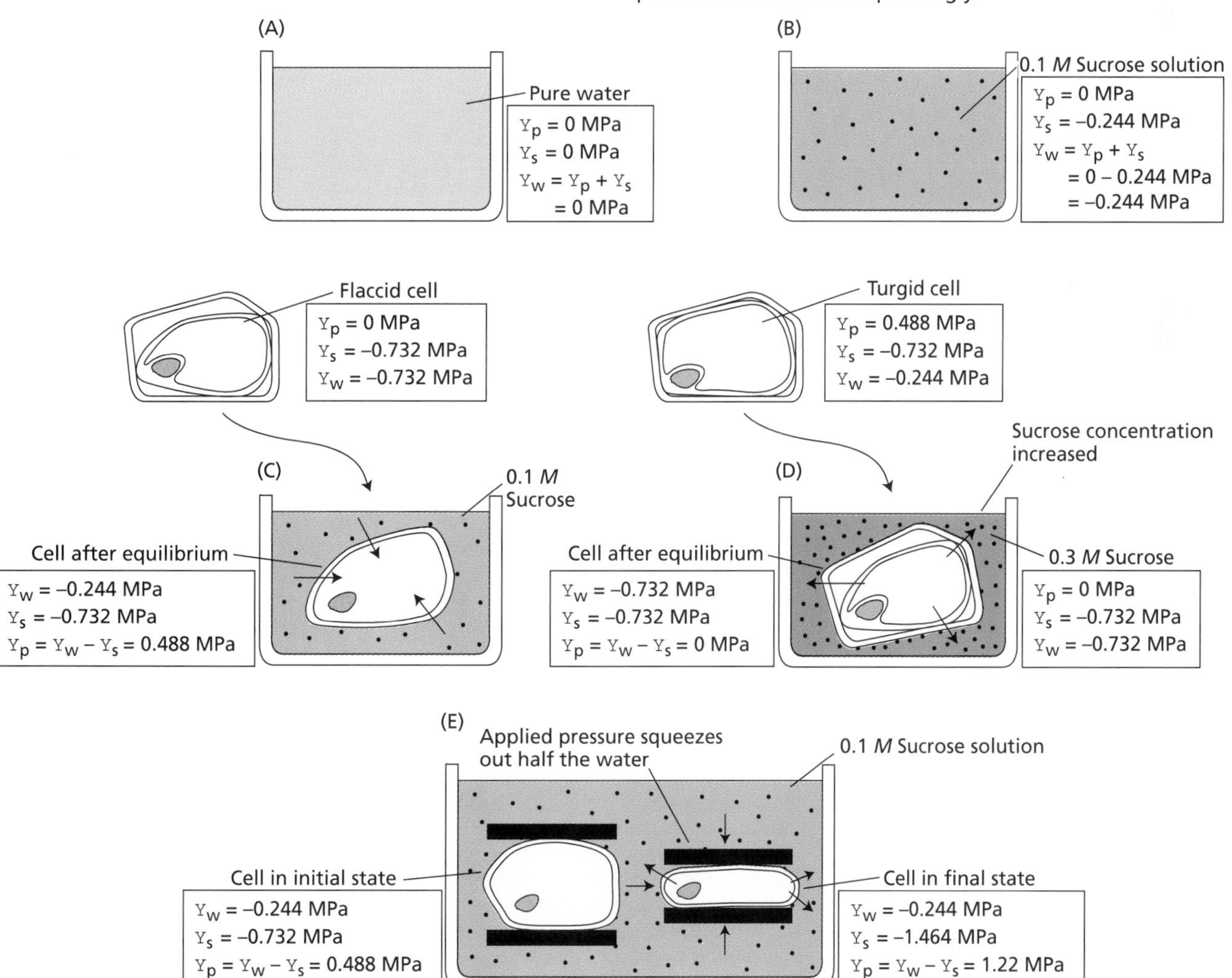

Figure 3.9 Five examples illustrating the concept of water potential and its components. (A) Pure water. (B) A solution containing 0.1 *M* sucrose. (C) After a flaccid cell is dropped in the 0.1 *M* sucrose solution, because the starting water potential of the cell is less than the water potential of the solution, the cell takes up water. After equilibration, the water potential of the cell rises to equal the water potential of the solution, and the result is a cell with a positive turgor pressure. (D) Increasing the concentration of sucrose in the solution makes the cell lose water. The increased sucrose concentration lowers the solution water potential, draws water out from the cell, and thereby reduces the cell's turgor pressure. (E) Another way to make the cell lose water is by slowly pressing it between two plates. In this case, half of the cell water is removed, so cell osmotic potential decreases (becomes more negative) and turgor pressure increases correspondingly.

BOX 3.2

Measuring Water Potential

CELL GROWTH, photosynthesis, and crop productivity are all strongly influenced by water potential and its components. Like the body temperature of humans, water potential is a good overall indicator of plant health. Plant scientists have thus expended considerable effort in devising accurate and reliable methods for evaluating the water status of a plant. Four instruments that have been used extensively to measure Ψ_w, Ψ_s, and Ψ_p are described here: psychrometer, pressure chamber, cryoscopic osmometer, and pressure probe.

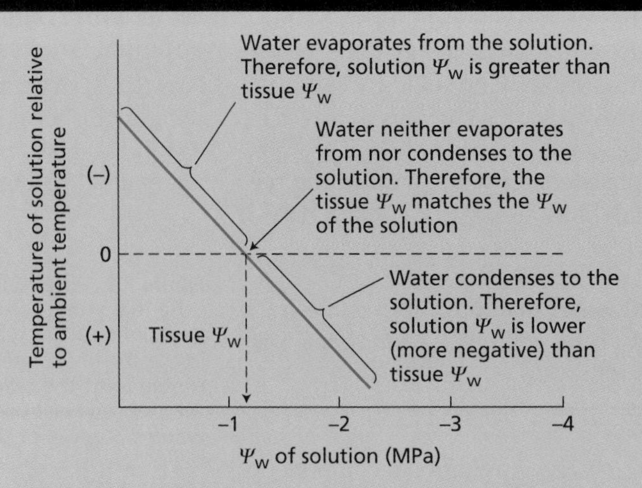

Figure 2 The temperature of the sensor in the psychrometer shown in Figure 1 depends on the water potential of the solution relative to the water potential of the tissue sample.

Psychrometer (Ψ_w measurement)

Psychrometry (the prefix "psychro-" comes from the Greek word *psychein*, "to cool") is based on the fact that the vapor pressure of water is lowered as its water potential is reduced. This is one of the *colligative properties* of solutions. Psychrometers measure the water vapor pressure of a solution or plant sample, on the basis of the principle that evaporation of water from a surface cools the surface.

One psychrometric technique, known as *isopiestic psychrometry*, has been used extensively by John Boyer and coworkers (Boyer and Knipling 1965) and is illustrated in Figure 1. Investigators make a measurement by placing a piece of tissue sealed inside a small chamber that contains a temperature sensor (in this case, a thermocouple) in contact with a small droplet of water. Initially, water evaporates from both the tissue and the water droplet, raising the humidity of the air inside the sealed chamber.

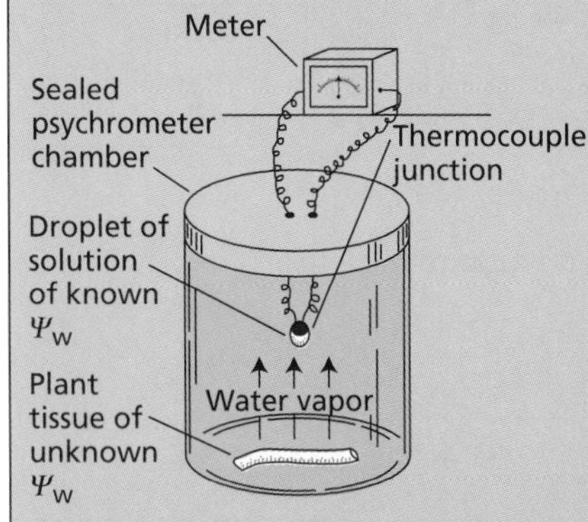

Figure 1 Diagram illustrating the use of isopiestic psychrometry to measure the water potential of a plant tissue.

Evaporation continues until the air becomes saturated or nearly saturated with water vapor. At this point, if the plant tissue and the water droplet have the same water potential, the net movement of water from the droplet stops, and the temperature of the droplet, measured with the temperature sensor, is the same as the ambient temperature. But if the tissue has a lower water potential than that of the droplet, water evaporates from the droplet, diffuses through the air, and is absorbed by the tissue. This slight evaporation of water cools the drop. The larger the difference in water potential between the tissue and the droplet, the higher the rate of water transfer and hence the cooler the droplet.

Rather than placing pure water on the temperature sensor, one may place a standard solution of known solute concentration (known Ψ_s and thus known Ψ_w). If the standard solution has a lower water potential than that of the sample to be measured, water will diffuse from the tissue to the droplet, causing warming of the droplet. Measuring the change in temperature of the droplet for several solutions of known Ψ_w makes it possible to match exactly the water potential of the solution with that of the sample (Figure 2). When the match is perfect, the change in temperature of the droplet is zero.

Psychrometers have been used to measure the water potentials of excised and intact plant tissue. Moreover, the method is applicable to solutions in which Ψ_w equals Ψ_s. Thus psychrometry can measure both the water potential of living tissue and the osmotic potential of a solution. Frequently, the Ψ_w of a tissue is measured with a psychrometer, and then the tissue is crushed and the Ψ_s value of the expressed cell sap is measured with the same instrument. By combining the two measurements, researchers can estimate the turgor pressure that existed in the cells before the tissue was crushed ($\Psi_p = \Psi_w - \Psi_s$).

This method is very useful, but it is very sensitive to temperature fluctuations. For example, a change in temperature of 0.01°C may correspond to a change in water potential of about 0.1 MPa (the value varies with the type of temperature sensor). Since it is often desirable to have a resolution of 0.01 MPa, the instrument must be kept under stringent conditions of constant temperature. For this reason, the method is used primarily in laboratory settings and has found only limited use for fieldwork, where temperature is not easily controlled. There are many variations in psychrometric technique; interested readers should consult Brown and Van Haveren 1972 and Slavik 1974.

Pressure chamber (Ψ_w measurement)

A relatively quick method for estimating the water potential of large pieces of tissues, such as whole leaves and shoots, is by use of the **pressure chamber**. This method was pioneered by Henry Dixon at Trinity College, Dublin, at the beginning of the twentieth century, but it did not come into widespread use until P. Scholander

BOX 3.2 (continued)

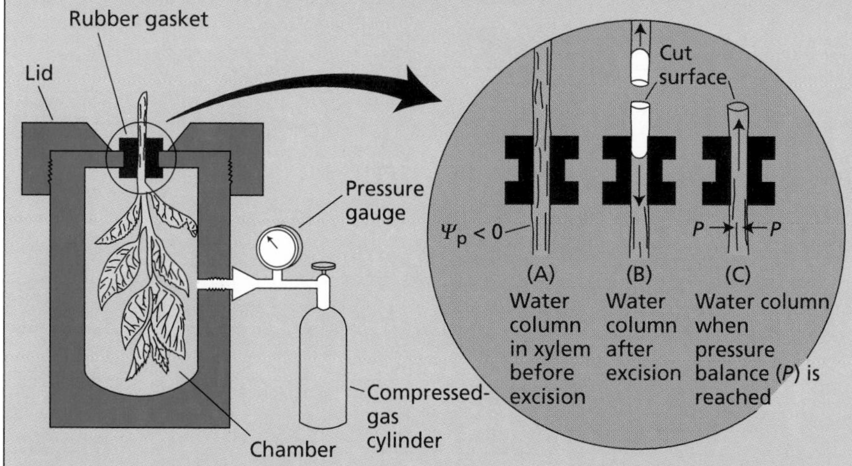

Figure 3 The pressure chamber method for measuring plant water potential. The diagram at left shows a shoot sealed into a chamber, which may be pressurized with compressed gas. The diagrams at right show the state of the water columns within the xylem at three points in time: (A) The xylem is uncut and under a negative pressure, or tension. (B) The shoot is cut, causing the water to pull back into the tissue, away from the cut surface, in response to the tension in the xylem. (C) The chamber is pressurized, bringing the xylem sap back to the cut surface.

and coworkers at the Scripps Institution of Oceanography improved the instrument design and showed its practical use (Scholander et al. 1965).

The pressure chamber measures the negative hydrostatic pressure (tension) that exists in the xylem of most plants. The water potential of the xylem is assumed to be fairly close to the average water potential of the whole organ—an assumption that is probably valid, because (1) in many cases the osmotic potential of the xylem solution is negligible, so the major component of the water potential in the xylem is the (negative) hydrostatic pressure in the xylem column, and (2) the xylem is in intimate contact with most cells in the plant.

In this technique, the organ to be measured is excised from the plant and is partly sealed in a pressure chamber (Figure 3). Before excision, the water column in the xylem is under some tension. When the water column is broken by excision of the organ, the water is pulled into the xylem capillary by the now unopposed tension. The cut surface consequently appears dull and dry. To make a measurement, the investigator pressurizes the chamber with compressed gas until the water in the xylem is brought back to the cut surface. The pressure needed to bring the water back to the surface is called the *balance pressure* and is readily detected by the change in the appearance of the cut surface, which becomes wet and shiny when this pressure is attained.

The balance pressure is equal in magnitude (but opposite in sign) to the negative pressure that existed in the xylem column before the plant material was excised. For example, if a balance pressure of 0.5 MPa is found, then Ψ_p in the xylem before excision was –0.5 MPa. If we know Ψ_s for the xylem sap from other measurements, we may calculate the water potential of the xylem, which, as already stated, is assumed to be close to the water potential of the whole organ.

In many outdoor plants, Ψ_p in the xylem may be –1 to –2 MPa, whereas Ψ_s may be only –0.05 to –0.2 MPa. Therefore, in many situations pressure chamber measurements by themselves provide an adequate estimate of the water potential of the plant. Because the pressure chamber method is rapid and does not require delicate instrumentation or elaborate temperature control, it has been used extensively under field conditions to estimate water potential (Tyree and Hammel 1972).

Cryoscopic osmometer (Ψ_s measurement)

The **cryoscopic osmometer** measures the osmotic potential of a solution by measuring its freezing point. One of the *colligative properties* of solutions is the decrease in the freezing point as the solute concentration increases. For example, a solution containing 1 mol of solutes per kilogram of water has a freezing point of –1.86°C, compared with 0°C for pure water.

Various instruments can be used to measure the freezing-point depression of solutions (for two examples, see Prager and Bowman 1963 and Bearce and Kohl 1970). With a cryoscopic osmometer, solution samples as small as 1 nanoliter (10^{-9} L) are placed in an oil medium located on a temperature-controlled stage (Figure 4). The very small sample size allows sap from single cells to be measured and permits rapid thermal equilibration with the stage. To prevent evaporation, the investigator suspends the samples in oil-filled wells in a silver plate (silver has high thermal conductivity). The temperature of the stage is rapidly decreased to about –30°C, which causes the sample to freeze. The temperature is then raised very slowly, and the melting process in the sample is observed through a microscope. When the last ice crystal in the sample melts, the temperature of the stage is recorded (note that the melting and freezing points are the same). It is a straightforward job to calculate the solute concentration from the freezing-point depression; and from the solute concentration (c_s), Ψ_s is calculated as $-RTc_s$. This technique has been used to measure droplets extracted from single cells (Malone and Tomos 1992).

Pressure probe (Ψ_p measurement)

If a cell were as large as a watermelon or even a grape, measuring its hydrostatic pressure would be a relatively easy task. Because of the small size of plant cells, however, the development of methods for direct measurement of turgor pressure has been slow. Paul Green at the University of Pennsylvania developed one of the first direct methods, using a micromanometer, for measuring turgor pressure in plant cells (Green and Stanton 1967). In this technique, an air-filled glass tube sealed at one end is inserted into a cell (Figure 5). The high pressure in the cell compresses the trapped gas, and from the change in volume one can readily calculate the pressure of the cell from the ideal gas law (pressure × volume = constant). This method works only for cells of relatively large volume, such as the giant cell of the filamentous

BOX 3.2 (*continued*)

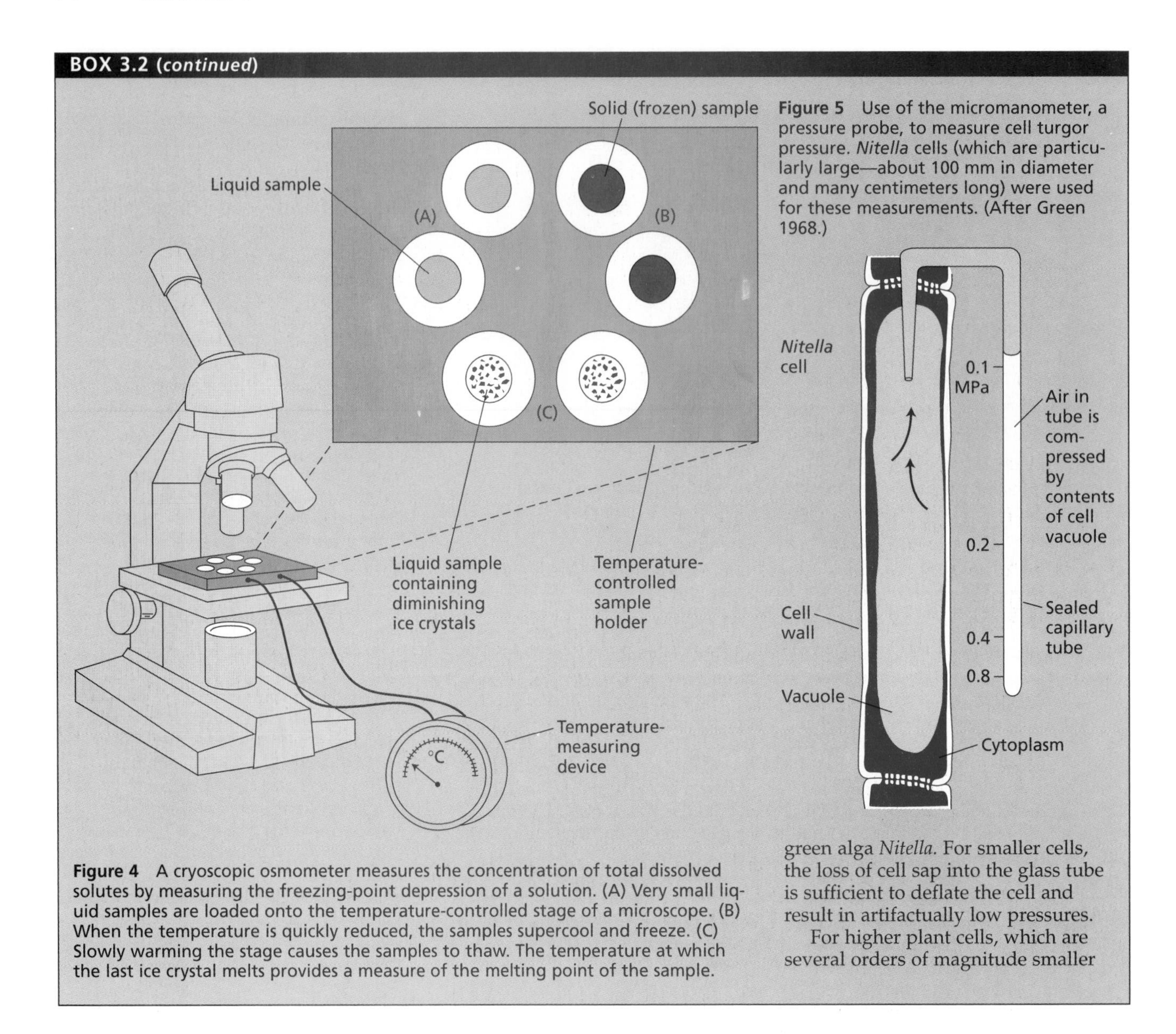

Figure 5 Use of the micromanometer, a pressure probe, to measure cell turgor pressure. *Nitella* cells (which are particularly large—about 100 mm in diameter and many centimeters long) were used for these measurements. (After Green 1968.)

Figure 4 A cryoscopic osmometer measures the concentration of total dissolved solutes by measuring the freezing-point depression of a solution. (A) Very small liquid samples are loaded onto the temperature-controlled stage of a microscope. (B) When the temperature is quickly reduced, the samples supercool and freeze. (C) Slowly warming the stage causes the samples to thaw. The temperature at which the last ice crystal melts provides a measure of the melting point of the sample.

green alga *Nitella*. For smaller cells, the loss of cell sap into the glass tube is sufficient to deflate the cell and result in artifactually low pressures.

For higher plant cells, which are several orders of magnitude smaller

the atmosphere, the hydrostatic pressure of the water is the same as atmospheric pressure (Ψ_p = 0 MPa). There are no solutes in the water, so Ψ_s = 0 MPa; therefore the water potential is 0 MPa ($\Psi_w = \Psi_s + \Psi_p$). Now imagine dissolving sucrose in the water to a concentration of 0.1 *M* (Figure 3.9B). This addition lowers the osmotic potential (Ψ_s) to –0.244 MPa (see Table 3.2) and decreases the water potential (Ψ_w) to –0.244 MPa.

Next, consider a flaccid plant cell (i.e., a cell with no turgor pressure) that has a total internal solute concentration of 0.3 *M* (Figure 3.9C). This solute concentration gives an osmotic potential (Ψ_s) of –0.732 MPa. Because the cell is flaccid, the internal pressure is the same as ambient pressure, so the hydrostatic pressure (Ψ_p) is 0 MPa and the water potential of the cell is –0.732 MPa. What happens if this cell is placed in the beaker containing 0.1 *M* sucrose (see Figure 3.9C)? Because the water potential of the sucrose solution (Ψ_w = –0.244 MPa; see Figure 3.9B) is greater than the water potential of the cell (Ψ_w = –0.732 MPa), water will move from the sucrose solution to the cell (from high to low water potential).

Even a slight increase in cell volume causes a large increase in the hydrostatic pressure within the cell, because plant cells are surrounded by relatively rigid cell walls. As water enters the cell, the cell wall is stretched by the contents of the enlarging protoplast. The wall resists such stretching by pushing back on the cell. This phenomenon is analogous to inflating a basketball with air, except that air is compressible, whereas water is nearly incompressible. Thus, as water moves into the cell, the hydrostatic pressure or turgor pressure (Ψ_p) of the cell increases. Consequently, the cell water potential (Ψ_w) increases and the difference between inside and outside water potentials ($\Delta\Psi_w$) is reduced.

BOX 3.2 (continued)

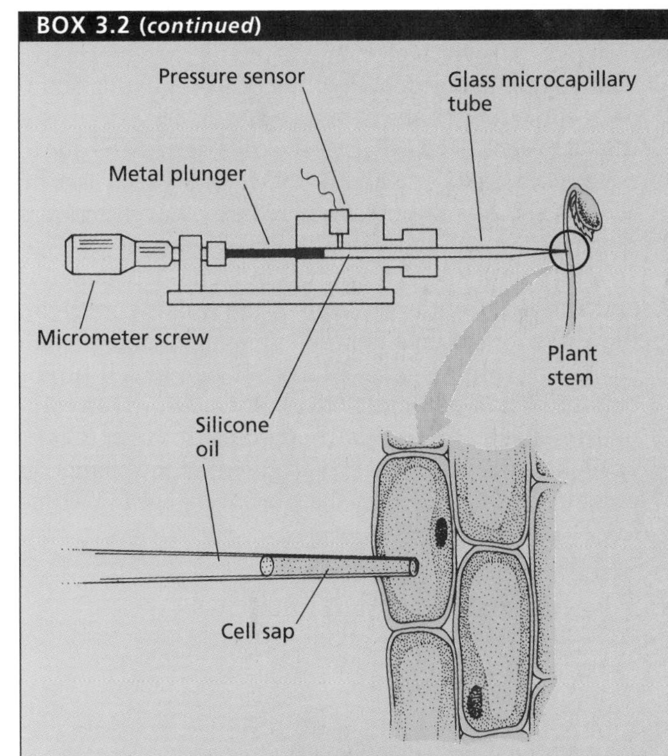

Figure 6 Diagram of the simplest pressure probe (not to scale). The primary advantage of this method over the one shown in Figure 5 is that cell volume is minimally disturbed. Minimal disturbance is of great importance for the tiny cells that are typical of higher plants, in which loss of even a few picoliters (10^{-12} L) of fluid can substantially reduce turgor pressure.

in volume than *Nitella*, a more sophisticated device, the **pressure probe**, was developed by Ernest Steudle, Ulrich Zimmermann, and their colleagues in Germany (Husken et al. 1978). This instrument is similar to a miniature syringe (Figure 6). A glass microcapillary tube is pulled to a fine point and is inserted into a cell. The microcapillary is filled with silicone oil, a relatively incompressible fluid that can be readily distinguished from cell sap under a microscope. When the tip of the microcapillary is first inserted into the cell, cell sap begins to flow into the capillary because of the initial low pressure of that region. Investigators can observe such movement of sap under the microscope and counteract it by pushing on the plunger of the device, thus building up a pressure. In such fashion the boundary between the oil and the cell sap can be pushed back to the tip of the microcapillary. When the boundary is returned to the tip and is held in a constant position, the initial volume of the cell is restored and the pressure inside the cell is exactly balanced by the pressure in the capillary. This pressure is measured by a pressure sensor in the device. Thus the hydrostatic pressure of individual cells may be measured directly.

This method has been used to measure Ψ_p and other parameters of water relations in cells of both excised and intact tissues of a variety of plant species (Steudle 1993). The primary limitation of this method is that some cells are too small for current instruments to measure. Furthermore, some cells tend to leak after being stabbed with the capillary, and others plug up the tip of the capillary, thereby preventing valid measurements. The pressure probe has also been adapted to measure positive and negative values of Ψ_p in the xylem (Heydt and Steudle 1991). However, technical problems with cavitation (see Chapter 4) limit the measurement of negative Ψ_p by this technique.

Eventually, cell Ψ_p increases enough to raise the cell Ψ_w to the same value as the Ψ_w of the sucrose solution. At this point, equilibrium is reached ($\Delta\Psi_w = 0$ MPa), and net water transport ceases. The tiny amount of water taken up by the cell does not significantly affect the solute concentration of the sucrose solution, the volume of which is very much larger than that of the cell. Hence, Ψ_s, Ψ_p, and Ψ_w of the sucrose solution are not altered. Therefore, at equilibrium $\Psi_{w(cell)} = \Psi_{w(solution)} = -0.244$ Mpa.

The calculation of cell Ψ_p and Ψ_s is straightforward. If we assume that the cell has a very rigid cell wall, then very little water will enter. Thus we can assume to a first approximation that $\Psi_{s(cell)}$ is unchanged during the equilibration process and that $\Psi_{s(solution)}$ remains at –0.732 MPa. We can obtain cell hydrostatic pressure by rearranging Equation 3.7: $\Psi_p = \Psi_w - \Psi_s = (-0.244) - (-0.732) = 0.488$ MPa.

Water Can Also Leave the Cell in Response to a Water Potential Gradient

Water can also leave the cell by osmosis. If, in the previous example, we remove our plant cell from the 0.1 *M* sucrose solution and place it in a 0.3 *M* sucrose solution (Figure 3.9D), $\Psi_{w(solution)}$ (–0.732 MPa) is more negative than $\Psi_{w(cell)}$ (–0.244 MPa), and water will move from the cell to the solution. As water leaves the cell, the cell volume decreases. As the cell volume decreases, cell Ψ_p and Ψ_w decrease also until $\Psi_{w(cell)} = \Psi_{w(solution)} = -0.732$ MPa. From the water potential equation (Equation 3.7) we can calculate that at equilibrium $\Psi_p = 0$ MPa. As before, we assumed that the change in cell volume is small, so we can ignore the change in Ψ_s.

If we slowly squeeze the cell by pressing it between two plates (Figure 3.9E), we effectively raise the cell Ψ_p, consequently raising the cell Ψ_w and creating a $\Delta\Psi_w$ such that water now flows *out* of the cell. If we continue squeezing until half the cell water is removed and then hold the cell in this condition, the cell will reach a new equilibrium. As in the previous example, at equilibrium $\Delta\Psi_w = 0$ MPa, and the amount of water added to the external solution is so small that it can be ignored. The cell will thus return to the Ψ_w value that it had before the squeezing procedure. However, the components of the cell Ψ_w will be quite different.

Because half of the water was squeezed out of the cell while the solutes remained inside the cell (the membrane is selectively permeable), the cell solution is concentrated twofold, and thus Ψ_s is lower $(-0.732 \times 2 = -1.464$ MPa). Knowing the final values for Ψ_w and Ψ_s, we can calculate the turgor pressure from Equation 3.7 as $\Psi_p = \Psi_w - \Psi_s = (-0.244) - (-1.464) = 1.22$ MPa. This final example is unusual in that we used an external force to change cell volume without a change in water potential. In most cases, the water potential of the cell's environment changes and the cell gains or loses water until its Ψ_w matches that of its environment.

One point common to all these examples deserves emphasis: *Water flow is a passive process. That is, water moves in response to physical forces, toward regions of low water potential or free energy.* There are no metabolic "pumps" (reactions driven by ATP hydrolysis) that push water from one place to another. This rule is valid as long as water is the only substance being transported. When solutes are transported, however, as occurs for short distances across membranes (see Chapter 6) and for long distances in the phloem (see Chapter 7), then water transport may be coupled to solute transport and this coupling may move water against a water potential gradient. For example, when phloem transports phloem sap from the leaves to the roots, it moves water against a water potential gradient (see Nobel 1991: 516–520).

On another scale, the movement of sugars, amino acids, or other small molecules by various membrane transport proteins was found to be coupled directly to the movement of up to 260 water molecules across the membrane per molecule of solute (Loo et al. 1996). Such exceptions defy no laws of thermodynamics, because the loss of free energy by the solute more than compensates for the gain of free energy by the water. Thus, the net change in free energy remains negative. *These exceptions notwithstanding, the vast majority of water transport in plants is energetically downhill, toward lower water potential.*

Small Changes in Plant Cell Volume Cause Large Changes in Turgor Pressure

Because plants cells have rigid walls, a change in cell (protoplast) volume and Ψ_w is usually accomplished by a change in Ψ_p, with little change in cell Ψ_s. This phenomenon is illustrated in greater detail in the **Höfler diagram** of Figure 3.10, which plots Ψ_w, Ψ_p, and Ψ_s as a function of relative cell volume for a hypothetical cell. In this example, as the cell volume is reduced by 5%, Ψ_w decreases from 0 to about –1.8 MPa. Most of this decrease is due to a reduction in Ψ_p (by about 1.5 MPa); Ψ_s decreases by about 0.3 MPa as a result of water loss by the cell and consequent increased concentration of cell solutes.

The exact shape of the curves in Figure 3.10 depends on the rigidity of the cell wall. If the wall is very rigid, the pressure–volume curve is very steep and a small change in volume gives a large change in cell turgor pressure. The wall rigidity is usually measured as **volumetric elastic modulus**, symbolized by ε (Greek epsilon). This wall property is given by the change in pressure ($\Delta\Psi_p$) divided by the relative change in volume ($\Delta V/V$); that is,

$$\varepsilon = \frac{\Delta\Psi_p}{\Delta V/V} \tag{3.8}$$

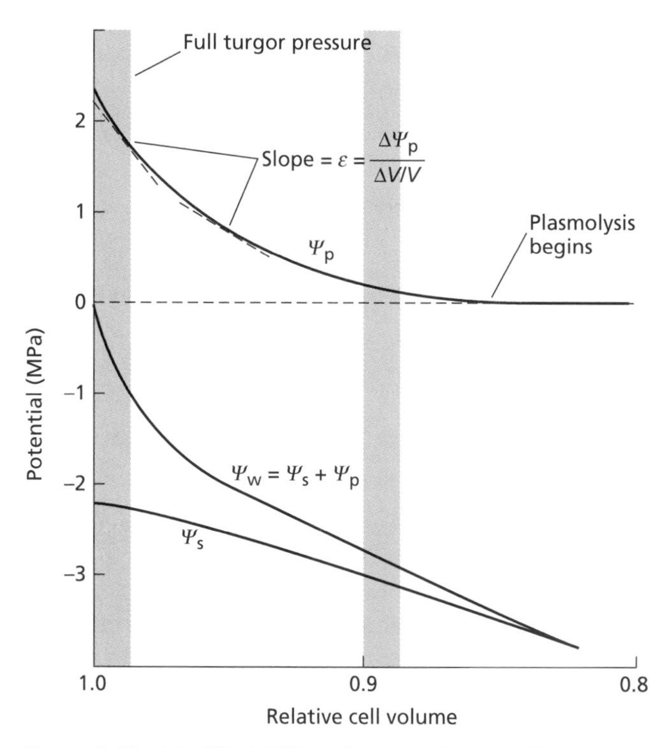

Figure 3.10 Modified Höfler diagram, showing that turgor pressure (Ψ_p) decreases steeply with the initial decrease in cell volume. In comparison, osmotic potential (Ψ_s) changes very little (see screened area at left). As cell volume decreases below 0.9 in this example, the situation reverses: Most of the change in water potential is due to a drop in cell Ψ_s accompanied by relatively little change in turgor pressure (see screened area at right). The slope of the curve that illustrates Ψ_p versus volume relationship is a measure of the cell's volumetric elastic modulus (ε). Note that ε is not constant but decreases as the cell loses turgor. (After Tyree and Jarvis 1982, based on a shoot of Sitka spruce.)

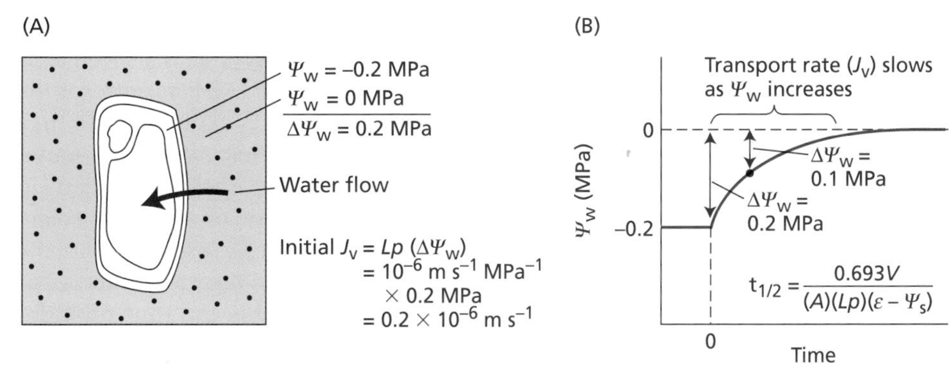

Figure 3.11 The rate of water transport into a cell depends on the water potential difference ($\Delta\Psi_w$) and the hydraulic conductivity of the cell membranes (Lp). In this example, (A) the initial water potential difference is 0.2 MPa and Lp is 10^{-6} m s^{-1} MPa^{-1}. These values give an initial transport rate (J_v) of 0.2×10^{-6} m s^{-1}. (B) As water is taken up by the cell, the water potential difference decreases with time, leading to a slowing in the rate of water uptake. This effect follows an exponentially decaying time course with a half-time ($t_{1/2}$) that depends on the following cell parameters: volume (V), surface area (A), Lp, volumetric elastic modulus (ε), and cell osmotic potential (Ψ_s).

The volumetric elastic modulus, ε, is the slope of the Ψ_p curve in Figure 3.10 and has units of pressure, with typical values on the order of 10 MPa.

Equation 3.8 shows that when ε = 10 MPa, a 1% increase in volume increases turgor pressure (Ψ_p) by 0.1 MPa. This increase is equivalent to a 10% increase in turgor pressure for a typical cell (in which Ψ_p = 1.0 MPa). In contrast, a 1% increase in volume would result in only a 1% increase in cell Ψ_s. Thus, when the water potential of turgid cells increases as a result of water uptake, most of this change is effected through an increase in Ψ_p rather than an increase in Ψ_s.

Figure 3.10 illustrates additional aspects that are characteristic of plant cell wall relations. First, turgor pressure (Ψ_p) approaches zero as cells lose a mere 10 to 15% of their cell volume. (For cells with very stretchy walls, such as guard cells, this volume fraction may be substantially larger.) Second, note that ε is not constant but decreases as turgor pressure is lowered. This effect is most evident in the flatness of the Ψ_p curve in the region where cell volume is 85 to 90% (see Figure 3.10). Third, when ε and Ψ_p are low, changes in water potential are due mostly to changes in Ψ_s (look at the Ψ_p and Ψ_s curves where volume = 85%). In fact, this relationship is characteristic of animal cells that lack cell walls and occurs because nonlignified plant cell walls usually are rigid only when turgor pressure puts them in tension. Such cells act like a basketball: The wall is stiff (has high ε) when the ball is inflated, but becomes soft and collapsible (ε = 0) when the ball loses pressure.

The Rate of Water Transport Depends on Driving Force and Hydraulic Conductivity

So far, we have seen that water moves across a membrane in response to a water potential gradient. The direction of flow is determined by the direction of the Ψ_w gradient, which is the **driving force** for transport. But what determines the *rate* at which the water moves?

Consider a cell with an initial water potential of –0.2 MPa, submerged in pure water. From this information we know that water will flow into the cell and that the driving force is $\Delta\Psi_w$ = 0.2 MPa, but what is the initial rate of movement? The rate depends on permeability of the membrane to water, a property usually called the **hydraulic conductivity** (***Lp***) of the membrane (Figure 3.11).

Driving force, membrane permeability, and flow rate are related by the following equation:

$$\text{Flow rate} = \text{driving force} \times \text{hydraulic conductivity} \qquad (3.9)$$

Hydraulic conductivity expresses how readily water can move across a membrane and has units of volume of water per unit area of membrane per unit time per unit driving force (for instance, m^3 m^{-2} s^{-1} MPa^{-1} or m s^{-1} MPa^{-1}). The larger the hydraulic conductivity, the larger the flow rate. In our example in Figure 3.11, the hydraulic conductivity of the membrane is 10^{-6} m s^{-1} MPa^{-1}. The transport (flow) rate (J_v) can then be calculated from the following equation:*

$$J_v = Lp(\Delta\Psi_w) \qquad (3.10)$$

where J_v is the volume of water crossing the membrane per unit area of membrane and per unit time (m^3 m^{-2} s^{-1} or, equivalently, m s^{-1}). In this example, J_v has a value of 0.2×10^{-6} m s^{-1}. Note that J_v has the physical meaning of a velocity. We can calculate the flow rate in volumetric terms (m^3 s^{-1}) by multiplying J_v by the surface area of the cell.

* This equation assumes that the membrane is ideal—that is, that solute transport is negligible and water transport is equally sensitive to $\Delta\Psi_s$ and $\Delta\Psi_p$ across the membrane. Nonideal membranes require a more complicated equation that separately accounts for water flow induced by $\Delta\Psi_s$ and by $\Delta\Psi_p$ (see Nobel 1991).

The resulting value is the *initial* rate of water transport. As water is taken up, cell Ψ_w increases and the driving force ($\Delta\Psi_w$) decreases. As a result, water transport slows with time. The rate approaches zero in an exponential manner (see Dainty 1976), with a half-time given by:

$$t_{1/2} = \left(\frac{0.693}{(A)(Lp)}\right)\left(\frac{V}{\varepsilon - \Psi_s}\right) \qquad (3.11)$$

where V and A are the volume and surface area of the cell, respectively. A small value for t means fast equilibration. This equation shows that the rate of water potential equilibration is affected by cell geometry (V and A) and by hydraulic conductivity (Lp) of the membrane. The quantity $1/(A)(Lp)$ is sometimes called the cell **hydraulic resistance** and $V/(\varepsilon - \Psi_s)$ is its **hydraulic capacitance**. Hydraulic capacitance determines the total volume of water that must move for any given change in water potential. For instance, if the cell walls are very rigid (meaning ε is large), very little water transport is required to effect a change in water potential, and thus the cell half-time is small. Cell half-times typically range from 1 to 10 s, although some are much shorter (Steudle 1989). These low half-time values mean that single cells come to water potential equilibrium with their surroundings in less than 1 minute. For multicellular tissues, the half-time values may be much larger.

The Water Potential Concept Helps Us Evaluate the Water Status of a Plant

The concept of water potential has two principal uses. First, water potential is the quantity that governs transport across cell membranes, as we have described. Second, water potential is often used as a measure of the *water status* of a plant. Because of transpirational water loss to the atmosphere, plants are seldom fully hydrated. They suffer from water deficits that lead to inhibition of plant growth and photosynthesis, as well as to other detrimental effects. Figure 3.12 lists some of the physiological changes that plants experience as they become dry. The process that is most affected by water deficit is cell growth. More severe water stress leads to inhibition of cell division, inhibition of wall and protein synthesis, accumulation of solutes, closing of stomata, and inhibition of photosynthesis. Water potential is one measure of how hydrated a plant is and thus provides a relative index of the *water stress* the plant is experiencing (see Chapter 25).

Figure 3.12 also shows representative values for Ψ_w at various stages of water stress. In leaves of well-watered plants, Ψ_w ranges from –0.2 to about –0.6 MPa, but the leaves of plants in arid climates can have much lower values, perhaps –2 to –5 MPa under extreme conditions. Because water transport is a passive process, plants can take up water only when the plant Ψ_w is less than the soil Ψ_w. As the soil becomes drier, the plant similarly becomes less hydrated (attains a lower Ψ_w). If this were not the case, the soil would begin to extract water from the plant.

The Components of Water Potential Vary with Growth Conditions and Location within the Plant

Just as Ψ_w values depend on the growing conditions and the type of plant, so too the values of Ψ_s can vary considerably. Within cells of well-watered garden plants (examples include lettuce, cucumber seedlings, and bean leaves), Ψ_s may be as high as –0.5 MPa, although values of –0.8 to –1.2 MPa are more typical. The upper limit for cell Ψ_s is set probably by the minimum concentration of dissolved ions, metabolites, and proteins in the cytoplasm of living cells. At the other extreme, plants sometimes attain a much lower Ψ_s. For instance, water stress typically leads to an accumulation of solutes in the cytoplasm and vacuole, thus maintaining turgor pressure despite low water potentials.

Plant tissues that store high concentrations of sucrose or other sugars, such as sugar beet roots, sugarcane stems, or grape berries, also attain low values of Ψ_s. Val-

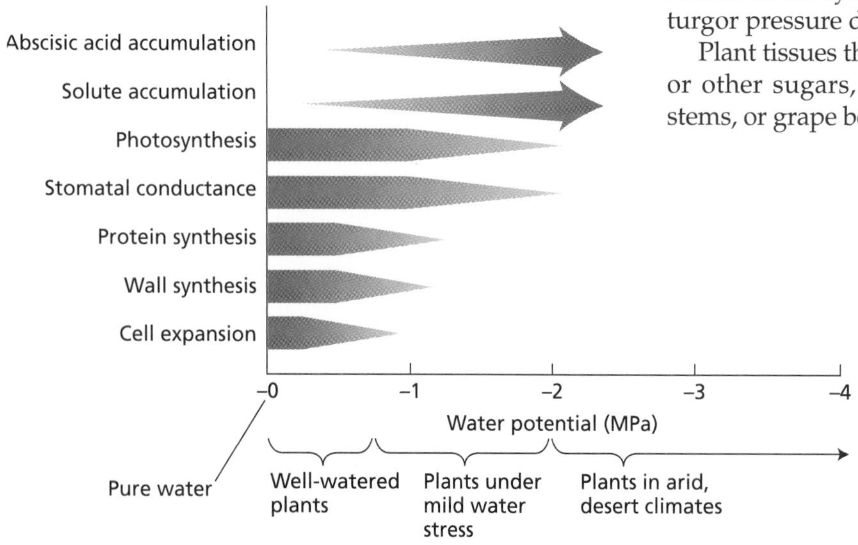

Figure 3.12 Water potential of plants under various growing conditions, and sensitivity of various physiological processes to water potential. The intensity of the bar color corresponds to the magnitude of the process. For example, cell expansion decreases as water potential falls (becomes more negative). Abscisic acid is a hormone that induces stomatal closure during water stress (see Chapter 23). (After Hsiao 1979.)

ues as low as –2.5 MPa are not unusual. **Halophytes**, plants that grow in saline environments, typically have very low values of Ψ_s; by this means they reduce cell Ψ_w enough to extract water from salt water. Most crop plants cannot survive in seawater which, because of the dissolved salts, has a lower water potential than the tissues of the plant can attain while maintaining their functional competence. Although Ψ_s *within* cells may be quite negative, the apoplastic solution surrounding the cells—that is, in the cell walls and in the xylem—may contain only low concentrations of solutes. Thus Ψ_s of this phase of the plant is typically much higher—for example, –0.1 to 0 MPa. Negative water potentials in the xylem and cell walls are usually due to negative Ψ_p (in the wall, this is sometimes attributed to a low matric potential, Ψ_m).

Values for Ψ_p within cells of well-watered garden plants may range from 0.1 to perhaps 1 MPa, depending on the value of Ψ_s inside the cell. A positive turgor pressure (Ψ_p) is important for two principal reasons. First, growth of plant cells requires turgor pressure to stretch the cell walls. The loss of Ψ_p under water deficits can explain in part why cell growth is so sensitive to water stress (see Chapter 25). Second, turgor pressure increases the mechanical rigidity of cells and tissues. This function of cell turgor pressure is particularly important for young nonlignified tissues, which cannot support themselves mechanically without a high internal pressure. A plant **wilts** when the turgor pressure inside the cells of such tissues falls toward zero. In cells submerged in a solution, **plasmolysis** may result when additional water is removed. Plasmolysis is the condition in which the membrane pulls away from the wall.

Whereas the solution inside cells may have a positive and large Ψ_p, the water outside the cell may have negative values for Ψ_p. In the xylem of rapidly transpiring plants, Ψ_p is negative and may attain values of –1 MPa or lower. The magnitude of Ψ_p in the cell walls and xylem varies considerably, depending on the rate of transpiration and the height of the plant. During the middle of the day, when transpiration is maximal, xylem Ψ_p reaches its lowest, most negative values, whereas it tends to increase at night, when transpiration is low, and the plant rehydrates.

Summary

Water is important in the life of plants because it makes up the matrix and medium in which most biochemical processes essential for life take place. The structure and properties of water strongly influence the structure and properties of proteins, membranes, nucleic acids, and other cell constituents.

In most land plants, water is continually lost to the atmosphere and taken up from the soil. The movement of water is driven by a reduction in free energy, and water may move by diffusion, by bulk flow, or by a combination of these fundamental transport mechanisms. Water diffuses because molecules are in constant thermal agitation, which tends to even out concentration differences. Water moves by bulk flow in response to a pressure difference, whenever there is a suitable pathway for bulk movement of water. Osmosis, the movement of water across membranes, depends on a gradient in free energy of water across the membrane—a gradient commonly measured as a difference in water potential.

Solute concentration and hydrostatic pressure are the two major factors that affect water potential, although gravity is also important when large vertical distances are involved. These components of the water potential may be summed: $\Psi_w = \Psi_s + \Psi_p + \Psi_g$. In soils, seeds, and dry cell walls, the reduction in water potential is usually attributed to a matric potential (Ψ_m). Plant cells quickly come into water potential equilibrium with their local environment by absorbing or losing water. Usually this change in cell volume results in a change in cell Ψ_p, accompanied by minor changes in cell Ψ_s. The rate of water transport across a membrane depends on the water potential difference across the membrane and the hydraulic conductivity of the membrane.

In addition to its importance in transport, water potential is a useful measure of the water status of plants. As we will see in Chapter 4, diffusion, bulk flow, and osmosis all help move water from the soil through the plant to the atmosphere.

General Reading

*Finkelstein, A. (1987) *Water Movement through Lipid Bilayers, Pores, and Plasma Membranes: Theory and Reality*. Wiley, New York.

*Friedman, M. H. (1986) *Principles and Models of Biological Transport*. Springer, Berlin.

Kramer, P. J., and Boyer, J. S. (1995) *Water Relations of Plants and Soils*. Academic Press, San Diego.

Milburn, J. A. (1979) *Water Flow in Plants*. Longman, London.

*Nobel, P. S. (1991) *Physicochemical and Environmental Plant Physiology*. Academic Press, San Diego.

* Indicates a reference that is general reading in the field and is also cited in this chapter.

Chapter References

Bearce, B. C., and Kohl, H. C., Jr. (1970) Measuring osmotic pressure of sap within live cells by means of a visual melting point apparatus. *Plant Physiol.* 46: 515–519.

Brown, R. W., and Van Haveren, B. P., eds. (1972) *Psychrometry in Water Relations Research*. Proceedings of the Symposium on Thermocouple Psychrometers. Utah Agricultural Experiment Station, Utah State University, Provo.

Boyer, J. S., and Knipling, E. B. (1965) Isopiestic technique for measuring leaf water potentials with thermocouple psychrometer. *Proc. Natl. Acad. Sci. USA* 54: 1044–1051.

Bünning, E. (1989) *Ahead of His Time: Wilhelm Pfeffer. Early Advances in Plant Biology* (translated by Helmut William Pfeffer). Carleton University Press, Ottawa, Ontario.

CAED Report 20: Center for Agricultural and Economic Development (CAED). 1964. *Weather and our Food Supply*. Report 20. Iowa State University of Science and Technology, Ames, IA.

Dainty, J. (1976) Water relations of plant cells. In *Transport in Plants II, Part A, Cells,* U. Lüttge and M. G. Pitman, eds. *Encyclopedia of Plant Physiology*, New Series, Vol. 2. Springer-Verlag, Berlin, pp. 12–35.

Green, P. B. (1968) Growth physics in *Nitella*: A method for continuous in vivo analysis of extensibility based on a micro-manometer technique for turgor pressure. *Plant Physiol.* 43: 1169–1184.

Green, P. B., and Stanton, F. W. (1967) Turgor pressure: Direct manometric measurement in single cells of *Nitella. Science* 155: 1675–1676.

Hammel, H. T., and Scholander, P. F. (1976) Osmosis and tensile solvent. Springer, Berlin.

Heydt, H., and Steudle, E. (1991) Measurement of negative pressure in the xylem of excised roots. Effects on water and solute relations. *Planta* 184: 389–396.

Hsiao, T. C. (1979) Plant responses to water deficits, efficiency, and drought resistance. *Agricult. Meteorol.* 14: 59–84.

Hüsken, D., Steudle, E., and Zimmermann, U. (1978) Pressure probe technique for measuring water relations of cells in higher plants. *Plant Physiol.* 61: 158–163.

Loo, D. D. F., Zeuthen, T., Chandy, G., and Wright, E. M. (1996) Cotransport of water by the Na+/glucose cotransporter. *Proc. Natl. Acad. Sci. USA* 93: 13367–13370.

Malone, M., and Tomos, A. D. (1992) Measurement of gradients of water potential in elongating pea stem by pressure probe and picolitre osmometry. *J. Exp. Bot.* 43: 1325–1331.

Maurel, C., Kado, R. T., Guern, J., and Chrispeels, M. J. (1995) Phosphorylation regulates the water channel activity of the seed-specific aquaporin a-TIP. *EMBO J.* 14: 3028–3035.

Passioura, J. B. (1980) The meaning of matric potential. *J. Exp. Bot.* 31: 1161–1169.

Prager, D. J., and Bowman, R. L. (1963) Freezing-point depression: New method for measuring ultramicro quantities of fluids. *Science* 142: 237–239.

Ray, P. M. (1960) On the theory of osmotic water movement. *Plant Physiol.* 35: 783–795.

Schäffner, A. R. (1998) Aquaporin function, structure, and expression: Are there more surprises to surface in water relations? *Planta* 204: 131–139.

Scholander, P. F., Hammel, H. T., Bradstreer, E. D., and Hemmingsen, E. A. (1965) Sap pressure in vascular plants. *Science* 148: 339–346.

Slavik, B. (1974) *Methods of Studying Plant Water Relations.* Academia, Prague.

*Stein, W. D. (1986) *Transport and Diffusion across Cell Membranes*. Academic Press, Orlando, FL.

Steudle, E. (1989) Water flow in plants and its coupling to other processes: An overview. *Methods Enzymol.* 174: 183–225.

Steudle, E. (1993) Pressure probe techniques: Basic principles and application to studies of water and solute relations at the cell, tissue and organ level. In *Water Deficits: Plant Responses from Cell to Community*, J. A. C. Smith and H. Griffiths, eds., BIOS Scientific, Oxford, pp. 5–36.

Tyree, M. T., and Hammel, H. T. (1972) The measurement of the turgor pressure and the water relations of plants by the pressure-bomb technique. *J. Exp. Bot.* 23: 266–282.

Tyree, M. T., and Jarvis, P. G. (1982) Water in tissues and cells. In *Physiological Plant Ecology.* II. *Water Relations and Carbon Assimilation* (Encyclopedia of Plant Physiology, new series, vol. 12B), O. L. Lange, P. S. Nobel, C. B. Osmond, and H. Ziegler, eds., Springer, Berlin, pp. 35–77.

Weig, A., Deswarte, C., and Chrispeels, M. J. (1997) The major intrinsic protein family of *Arabidopsis* has 23 members that form three distinct groups with functional aquaporins in each group. *Plant Physiol.* 114: 1347–1357.

Whittaker R. H. (1970) *Communities and Ecosystems.* Macmillan, London.

4 Water Balance of the Plant

LIFE IN EARTH'S ATMOSPHERE presents a formidable challenge to land plants. On the one hand, the atmosphere is the source of carbon dioxide, which is needed for carbon fixation during photosynthesis. Plants therefore need ready access to the atmosphere so that they can photosynthesize. On the other hand, the atmosphere is relatively dry and can dehydrate the plant. To meet the contradictory demands of maximizing carbon dioxide uptake while limiting water loss, plants have evolved adaptations that minimize water loss from the leaf, allow some degree of control over water loss, and permit rapid transport of water from the soil to replace water lost to the atmosphere.

In this chapter we will examine how water moves from the soil through the plant to the atmosphere. This pathway is continuous, but it is not homogeneous. In traveling along this pathway, water in the liquid phase moves through the soil, across membranes, through cells, along hollow conduits, and through cell walls; and then it escapes as a vapor into the air spaces inside the leaf and diffuses to the outside atmosphere. Different driving forces and mechanisms of transport are called into play along the various parts of this pathway (Figure 4.1).

Transpirational water loss from the leaf is driven by a gradient in water vapor concentration. Long-distance transport in the xylem is driven by pressure gradients, as is water movement in the soil. Water transport through cell layers such as the root is complex, but it responds to water potential gradients across the tissue. Throughout this journey water transport is passive, in the sense that the free energy of water decreases as it moves. Despite its passive nature, water transport is finely regulated by the plant to minimize dehydration, largely by regulating transpiration to the atmosphere. We will examine the nature of water transport and its control in these various regions, beginning with the soil.

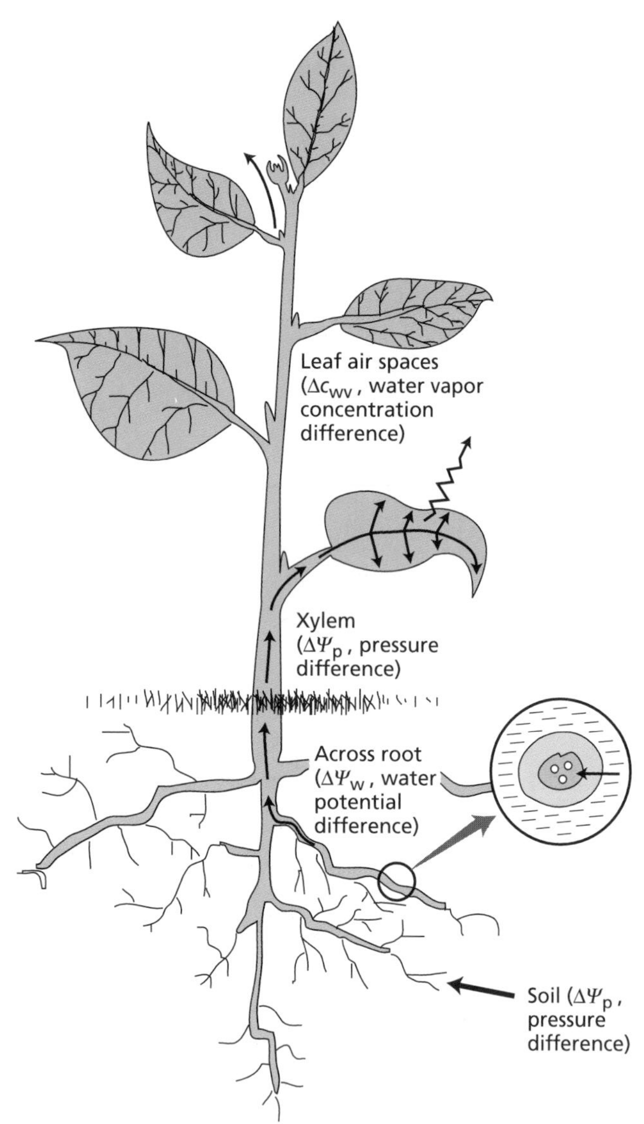

Figure 4.1 Main driving forces for water flow from the soil through the plant to the atmosphere: differences in water vapor concentration (Δc_{wv}), hydrostatic pressure ($\Delta \Psi_p$), and water potential ($\Delta \Psi_w$).

Water in the Soil

The water content and the rate of water movement in soils depend to a large extent on soil type and soil structure. Table 4.1 shows that the physical characteristics of different soils can vary greatly. At one extreme is sand, in which the soil particles may be 1 mm or more in diameter. Sandy soils have a relatively low surface area per gram of soil and have large spaces or channels between particles. At the other extreme is clay, in which particles are smaller than 2 μm in diameter. Clay soils have much greater surface areas and smaller channels between particles. With the aid of organic substances such as humus (decomposing organic matter), clay particles may aggregate into "crumbs" that help improve soil aeration and infiltration of water.

Table 4.1
Physical characteristics of different soils

Soil	Particle diameter (μm)	Surface area per gram (m²)
Coarse sand	2000–200	<1–10
Fine sand	200–20	
Silt	20–2	10–100
Clay	<2	100–1000

When a soil is heavily watered by rain or by irrigation, the water percolates downward by gravity through the spaces between soil particles, partly displacing, and in some cases trapping, air in these channels. Water in the soil may exist as a film adhering to the surface of soil particles, or it may fill the entire channel between particles. In sandy soils, the spaces between particles are so large that water tends to drain from them and remain only on the particle surfaces and at interstices between particles. In clay soils, the channels are small enough that water does not freely drain from them; it is held more tightly (see Box 4.1). This phenomenon is reflected in the moisture-holding capacity, or field capacity, of soils. **Field capacity** is the water content of a soil after it has been saturated with water and excess water has been allowed to drain away. Field capacity is large for clay soils and soils that have a high humus content and much lower for sandy soils.

In the following sections we will examine how the negative pressure in soil water alters soil water potentials, how water moves in the soil, and how roots absorb the water needed by the plant.

A Negative Hydrostatic Pressure in Soil Water Lowers Soil Water Potential

Like the water potential of cell solutions, the water potential of wet soils may be dissected into two components, the osmotic potential and the hydrostatic pressure. The osmotic potential (Ψ_s) (see Chapter 3) of soil water is generally negligible because solute concentrations are low; a typical value might be –0.02 MPa. For saline soils that contain a substantial concentration of salts, however, Ψ_s is significant, perhaps –0.2 MPa or lower. The second component of soil water potential is hydrostatic pressure (Ψ_p). For wet soils, Ψ_p is very close to zero. As a soil dries out, Ψ_p decreases. Some texts attribute this lowering of water potential when a soil dries to what they refer to as a matric potential.*

* As described in Chapter 3, dry soils contain so little free water that often it is not practical to measure soil $\Psi_s + \Psi_p$. In this case the soil water potential is defined by the soil matric potential, Ψ_m. The soil matric potential is due primarily to a local negative pressure, caused by capillarity and interaction of water with solid surfaces in the soil. See Passioura 1980.

BOX 4.1

Irrigation

PEOPLE HAVE USED irrigation to provide water for crops almost as long as they have practiced agriculture. In the harsh, hot deserts of southern Arizona, the prehistoric Hohokam Native American communities dug extensive networks of irrigation canals to divert water from the nearby Salt River to their crops. Thus their society thrived where it would otherwise have perished. Aerial photography reveals the location of these ancient canals, which follow many of the same routes as today's concrete-lined waterways that carry water throughout the area of Phoenix, Arizona. In the Tigris and Euphrates Valleys of Mesopotamia, which have been called the "cradle of civilization," the irrigation of crops yielded similar results. Civilization may have declined when the irrigated fields were "salted out"—made useless by the continuing addition of salts to the soil from poor-quality water. The Incas of ancient Peru also practiced extensive irrigation, including some innovative methods involving raised beds and provision of water from below.

Irrigation as practiced today in world agriculture is governed by the same general principles as in ancient times. Water is expensive and therefore must be used as efficiently as possible. If surface water is used, dams and waterways must be constructed and maintained. If water is pumped from the ground, energy must be expended to raise it to the surface. Perhaps surprisingly, most of the inefficiency in the use of water is encountered before the water ever reaches the plant in the field. Evaporation from lakes, seepage from canals, transpiration from aquatic weeds that grow in the canals, and uneven application in the fields are just some of the problems faced by irrigation specialists. A system is considered efficient if 50% of the water available is delivered to the root zone of the crop it is supporting.

Various methods have been developed for delivering water to fields. If the land can be graded and sloped appropriately, furrows with a very slight downhill slope can be placed between rows of plants. Water is then applied at the uphill end and allowed to flow through the furrows. Some water is wasted, because the soil at the uphill end becomes saturated before enough has been delivered in the middle of the field. Water may also "puddle" at the downhill end.

Level-basin irrigation is being used increasingly to overcome these problems. Large basins are leveled to within 10 to 20 mm by laser-directed machinery. If the soil is uniformly permeable throughout the basin, this leveling allows water to be applied evenly. If the land cannot be leveled or if the amounts of water applied are small, sprinklers are often used. However, sprinklers not only require energy to pump and pressurize the water, but they also produce small droplets from which evaporation is excessive. In addition, if the water is of poor quality, sprinkling will deposit salts directly on the leaves, where they can injure the plants. In some cases, sprinkling at night can prevent much of the injury because the salt-laden droplets do not evaporate as fast.

Some soils with very high clay content expand as they are wetted and shrink and crack as they dry. When they are wet they tend to become sealed, and water enters very slowly. For these cracking clay soils, a technique known as *surge irrigation* has been developed. Water is supplied to the field extremely rapidly so that it can flow down the cracks and enter the root zone before the soil swells and the cracks disappear.

In recent years a technique known as *drip irrigation* (also called *trickle irrigation* or *microirrigation*) has come into use in some areas of the world. Water is pumped directly to the base of a plant by plastic tubing and bled through an emitter at a slow rate that just meets the plant's needs. This approach is very efficient, but it is also very expensive and requires diligent maintenance of the hardware to keep the system working. For instance, the emitters tend to become plugged by both mineral deposits and slime produced by microorganisms. Periodically, the system must be flushed out with acid or with disinfectant. So far, drip irrigation has been used mostly for high-value crops for which quality (and price) depends strongly on a reliable supply of water. Under these conditions the profit from drip irrigation will pay for the extra costs. Fresh fruits such as blueberries and strawberries are irrigated extensively by this method in the United States.

With the availability of extremely efficient irrigation systems such as drip irrigation, more attention can be shifted to inefficient use of water by the crops themselves and to possible improvements in plant water use efficiency. The physiology of water in plants is an exciting area of research because many problems are waiting to be solved. For example, grain crops are most sensitive to drought when they are flowering, when abortion of the very young embryos can result in a barren plant. The solving of these problems will allow agronomists to make grain yields reliable even without resorting to irrigation.

Where does the negative pressure in soil water come from? Recall from our discussion of capillarity in Chapter 3 that water has a high surface tension that tends to minimize air–water interfaces. As a soil dries out, water is first removed from the middle of the largest spaces between particles. Because of adhesive forces, water tends to cling to the surfaces of soil particles, so a large surface area between soil water and soil air develops (Figure 4.2). As the water content of the soil decreases, the water recedes into the interstices between soil particles, and the air–water surface stretches and develops curved menisci. Water under these curved surfaces develops a negative pressure that may be estimated by the following formula:

$$\Psi_p = \frac{-2T}{r} \quad (4.1)$$

where T is the surface tension of water (7.28×10^{-8} MPa m) and r is the radius of curvature of the meniscus.

The value of Ψ_p in soil water can become quite negative because the radius of curvature of air–water sur-

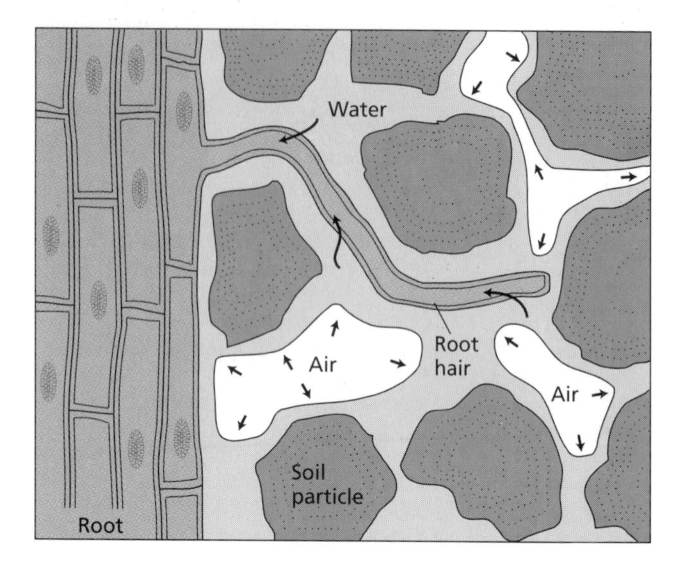

Figure 4.2 Root hairs make intimate contact with soil particles and amplify the surface area that can be used for water absorption by the plant (large arrows). The soil is a mixture of particles (both mineral and organic), water, dissolved solutes, and air. As water is absorbed by the plant, the air spaces expand (small arrows) and the soil solution recedes into smaller pockets, channels, and crevices between the soil particles. This recession causes the surface of the soil solution to develop concave menisci (curved interfaces between air and water), thus bringing the solution into tension, or negative pressure, by surface tension. As more water is removed from the soil, more acute menisci are formed, resulting in greater tensions (more negative pressures).

faces may become very small in drying soils. For instance, a curvature $r = 1$ µm (about the size of the largest clay particles) corresponds to a Ψ_p value of –0.15 MPa. The value of Ψ_p may easily reach –1 to –2 MPa as the air–water interface recedes into the smaller cracks between clay particles. At this point the availability of water is very low, and soil scientists thus often prefer to describe soil water potential in terms of a matric potential.*

Water Moves through the Soil by Bulk Flow

Water moves through soils predominantly by bulk flow driven by a pressure gradient, although diffusion also accounts for some water movement. As a plant absorbs water from the soil, it depletes the soil of water near the surface of the roots. This depletion reduces Ψ_p in the water near the root surface and establishes a pressure gradient with respect to neighboring regions of soil that have higher Ψ_p values. Because the water-filled pore spaces in the soil are interconnected, water moves to the root surface by bulk flow through these channels down the pressure gradient.

The rate of water flow in soils depends on the size of the pressure gradient through the soil and on the hydraulic conductivity of the soil. **Soil hydraulic conductivity** is a measure of the ease with which water moves through the soil, and it varies with the type of soil. Sandy soils, with their large spaces between particles, have large hydraulic conductivities, whereas clay soils, with the minute spaces between their particles, have appreciably smaller hydraulic conductivities. Besides soil type, water content also affects the hydraulic conductivity of soil. Figure 4.3 shows that as the water content (and hence the water potential) of a soil decreases, the hydraulic conductivity decreases drastically (note the log-

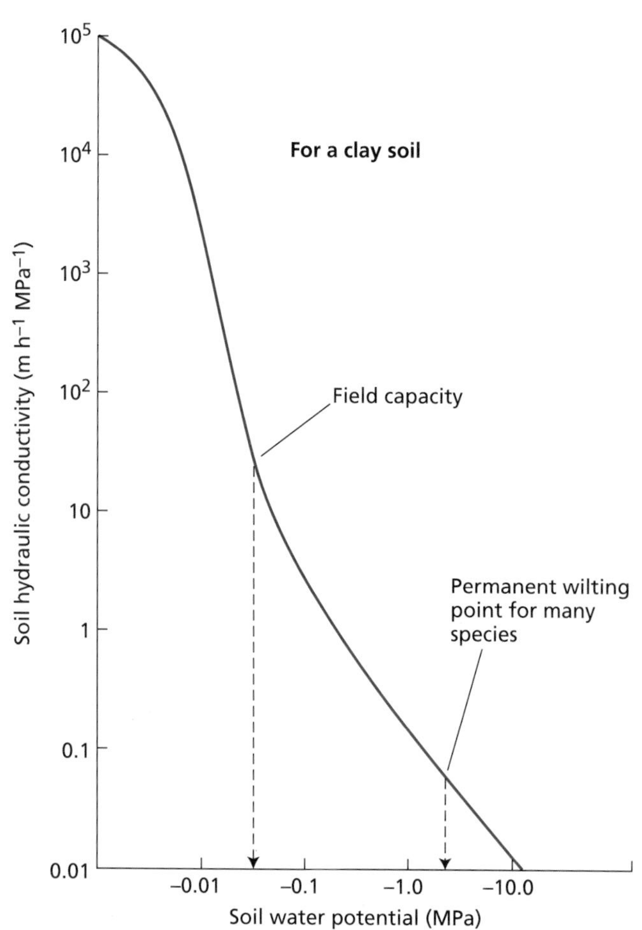

Figure 4.3 Soil hydraulic conductivity as a function of the water potential of the soil. Conductivity measures the ease with which water moves through the soil. The decrease in conductivity as the soil dries is due primarily to the movement of air into the soil to replace the water. As air moves in, the pathways for water flow between soil particles become smaller and more tortuous, and flow becomes more difficult. The overall shape of this curve is representative of many soils, but the shape for a particular soil may be influenced by the size distribution of its particles and by its organic matter content. The field capacity is the amount of water the soil is able to retain against gravitational forces. The permanent wilting point is the soil water potential value at which plants cannot regain turgor pressure even at night, in the absence of transpiration.

* See the footnote on page 82.

arithmic scales). This decrease in soil hydraulic conductivity is due primarily to the replacement of water in the soil spaces by air. When air moves into a soil channel previously filled with water, water movement through that channel is restricted to the periphery of the channel. As more of the soil spaces become filled with air, water can flow through fewer and narrower channels, and the hydraulic conductivity falls.

In very dry soils, the water potential (Ψ_w) may fall below what is called the **permanent wilting point**. At this point the water potential of the soil is so low that plants cannot regain turgor pressure even if all water loss through transpiration stops. The plants remain wilted (that is, cell turgor pressure is null) even at night, when transpiration ceases almost entirely. This means that the water potential of the soil (Ψ_w) is less than or equal to the osmotic potential (Ψ_s) of the plant. Since cell Ψ_s varies with plant species, the permanent wilting point is clearly not a unique property of the soil; it depends on the plant species as well.

Water Absorption by the Root

Intimate contact between the surface of the root and the soil is essential for effective water absorption by the root. This contact provides the surface area for water uptake by the root and is maximized by the growth of the root and of root hairs into the soil. **Root hairs** are microscopic extensions of root epidermal cells that greatly increase the surface area of the root, thus providing greater capacity for absorption of soil ions and, to a lesser extent, soil water (see Figure 4.2). When 4-month-old rye (*Secale*) plants were examined, their root hairs were found to constitute more than 60% of the surface area of the roots. Water enters the root most readily in the apical part of the root that includes the root hair zone. More mature regions of the root often have an outer layer of protective tissue, called an *exodermis* or *hypodermis*, that contains hydrophobic materials in its walls and is relatively impermeable to water.

The intimate contact between the soil and the root surface is easily ruptured when the soil is disturbed. It is for this reason that newly transplanted seedlings and plants need to be protected from water loss for the first few days after transplantation. Thereafter, new root growth into the soil reestablishes soil–root contact, and the plant can better withstand water stress.

Let's look at how the water moves within the root, and the different factors that determine the rates of water uptake into the root.

Water Moves in the Root via the Apoplast, Transmembrane, and Symplast Pathways

In the soil, water is transported predominantly by bulk flow. However, when water comes in contact with the root surface, the nature of water transport becomes more complex to analyze. Figure 4.4 shows that from the

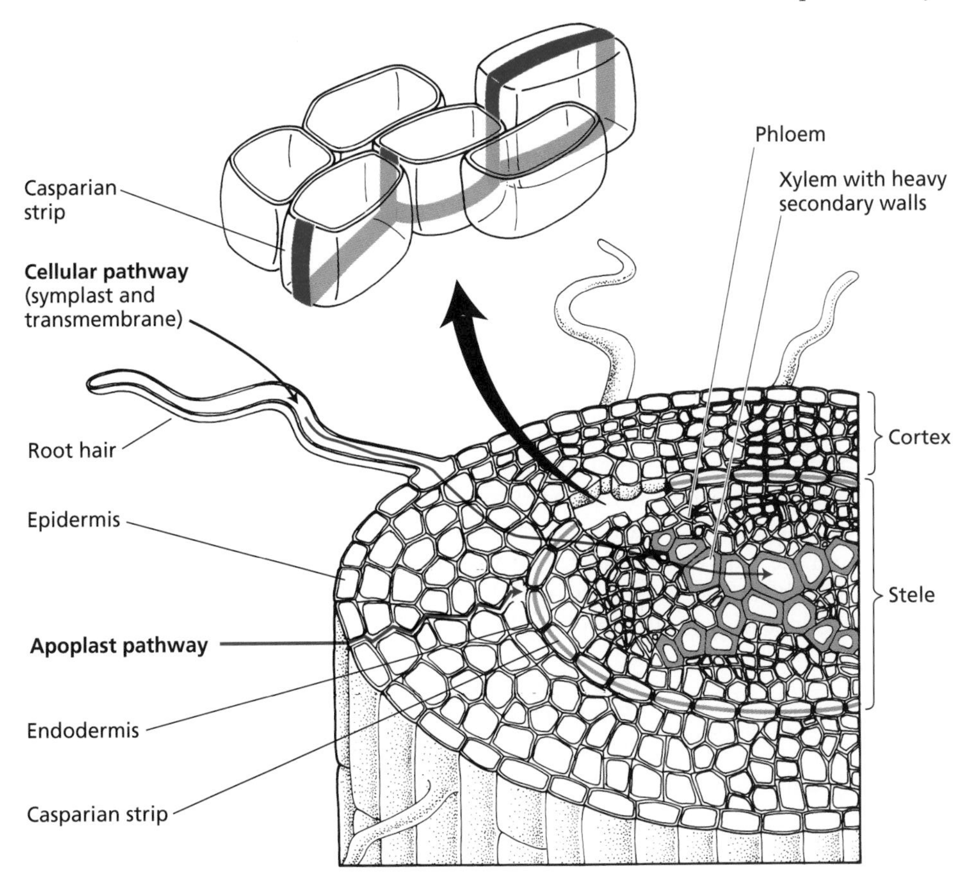

Figure 4.4 Pathways for water uptake by the root. Through the cortex, water may travel via the apoplast pathway or the cellular pathway, which includes the transmembrane and symplast pathways. In the symplast, water flows between cells through the plasmodesmata without crossing the plasma membrane. In the transmembrane pathway, water moves across the plasma membranes, with a short visit to the cell wall space. At the endodermis, the apoplast pathway is blocked by the Casparian strip.

epidermis to the endodermis of the root, there are three pathways through which water can flow: the apoplast, the transmembrane pathway, and the symplast. In the **apoplast pathway**, water moves exclusively through the cell wall without crossing any membranes. The **apoplast** is the continuous system of cell walls and intercellular air spaces in plant tissues. Water may also move via the **cellular pathway**, which has two components. The first component, the **transmembrane pathway**, is the route followed by water that sequentially enters a cell on one side, exits the cell on the other side, enters the next in the series, and so on. In this pathway, water crosses at least two membranes for each cell in its path (the plasma membrane on entering and on exiting). Transport across the tonoplast may also be involved. In the second component of the cellular pathway, water moves through the symplast, traveling from one cell to the next via the plasmodesmata (see Chapter 1). The **symplast** consists of the entire network of cell cytoplasm interconnected by plasmodesmata.

Although the relative importance of the apoplast, transmembrane, and symplast pathways has not yet been clearly established, it is always a combination of these three pathways that transports water across the root. Experiments with the pressure probe (see Figure 6 in Box 3.2) indicate that the apoplast pathway is particularly important for water uptake by young corn roots (Frensch et al. 1996; Steudle and Frensch 1996).

At the endodermis, water movement through the apoplast pathway may be obstructed by the Casparian strip (see Figure 4.4). The **Casparian strip** is a band of radial cell walls in the endodermis that is impregnated with the waxlike, hydrophobic substance **suberin.** Suberin acts as a barrier to water and solute movement. The endodermis is suberized in the nongrowing part of the root, some distance behind the tip. The apical, unsuberized region of the root is most permeable to water (Figure 4.5). Even in the suberized region, however, water may be transported across the endodermis via the symplast pathway or via breaks in the Casparian strip that form during the outgrowth of secondary roots.

The foregoing discussion has made clear that radial movement of water across the root is a complex process. One way to simplify the process conceptually is to treat the whole multicellular pathway—from the root hairs to the root xylem—as if it were a single membrane with a single hydraulic conductance. With this simplification one can quantify the **root hydraulic conductance** (L_{root}) as follows:

$$L_{root} = \frac{J_V}{\Delta\Psi_w} \tag{4.2}$$

where J_V is the rate of water flow and $\Delta\Psi_w$ is the radial water potential difference across the root.

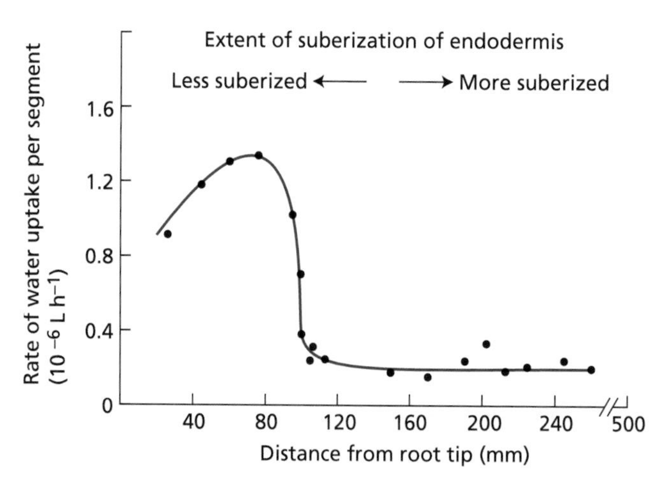

Figure 4.5 The rate of water uptake at various positions along pumpkin root. The suberized region of the root is indicated at the top of the graph. (After Kramer and Boyer 1995).

Researchers have measured the hydraulic conductance of roots under various conditions and found that it is not constant, but varies with flow rate and according to the way in which the radial water potential gradient is established. The explanation for this complex behavior is that roots behave not as simple membranes, but as complex barriers in which pressure differences are more effective for water transport than are the differences in osmotic potential (Steudle and Frensch 1996). In part, this is so because the transport mechanism and the pathway across the root radius differ when water fluxes are driven by pressure or solute differences.

Moreover, root hydraulic conductance decreases when roots are subjected to low temperature or anaerobic conditions, or treated with respiratory inhibitors (such as cyanide). These treatments inhibit root respiration, and the roots consequently transport less water. The exact explanation for this effect is not yet clear, but it may be due to inactivation of aquaporins (see Chapter 3) or inactivation of another process required for water transport across the root. This decrease in water transport in the roots provides an explanation for the wilting of plants in waterlogged soils: Submerged roots soon run out of oxygen, which is normally provided by diffusion through the air spaces in the soil (diffusion through gas is 10^4 times faster than diffusion through water). The anaerobic roots transport less water to the shoots, which consequently suffer net water loss and begin to wilt.

Solute Accumulation in the Xylem Can Generate "Root Pressure"

Plants sometimes exhibit a phenomenon referred to as **root pressure** or positive hydrostatic pressure. For example, if the stem of a young seedling is cut off just

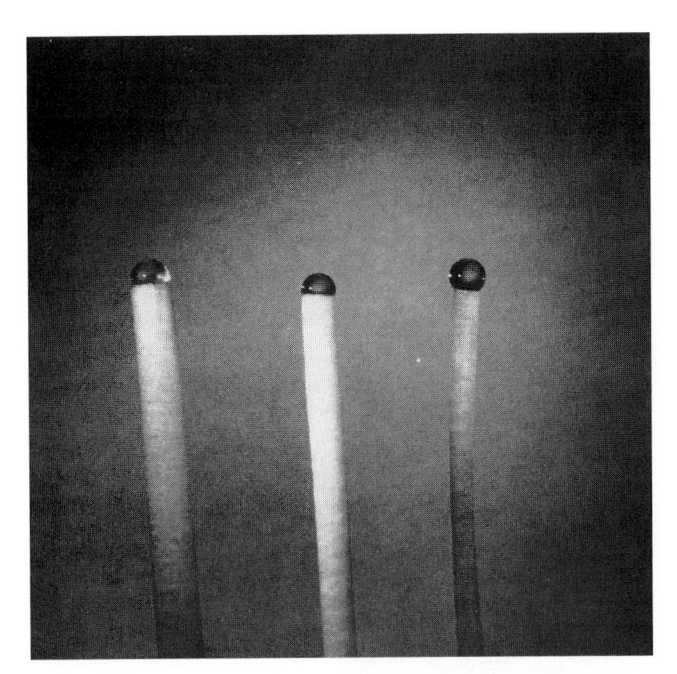

Figure 4.6 Exudation from cut stems of zucchini (left), soybean (center), and cucumber (right). The photograph was taken about 30 minutes after excision of the zucchini and soybean stems and about 15 minutes after excision of the cucumber stem. The exudation is a manifestation of the positive pressure (referred to as root pressure) in the xylem of such plants. (From Cosgrove 1987, © Springer-Verlag.)

above the soil line, the stump will often exude sap from the cut xylem for many hours (Figure 4.6). If a manometer is sealed over the stump, positive pressures can be measured. These pressures can be as high as 0.05 to 0.5 MPa.

Root pressure can be understood as a positive hydrostatic pressure in the xylem. The root absorbs ions from the dilute soil solution and transports them into the xylem. The buildup of solutes in the xylem sap leads to a decrease in the xylem osmotic potential (Ψ_s) and thus a decrease in the xylem water potential (Ψ_w). This lowering of the xylem Ψ_w provides a driving force for water absorption, which in turn leads to a positive hydrostatic pressure in the xylem. In effect, the whole root acts like an osmotic cell; the multicellular root tissue behaves as an osmotic membrane does, building up a positive hydrostatic pressure in the xylem in response to the accumulation of solutes. Root pressure is most prominent in well-hydrated plants under high humidity where there is little transpiration. Under drier conditions, when transpiration rates are high, water is taken up so rapidly into the leaves and lost to the atmosphere that a positive pressure never develops in the xylem.

Plants that develop root pressure frequently exhibit exudation of liquid from the leaves, a phenomenon known as **guttation**. Positive xylem pressure causes exudation of xylem sap through **hydathodes**, structures that are located near terminal tracheids of the bundle ends around the margins of leaves. The "dewdrops" that can be seen on the tips of grass leaves in the morning are actually guttation droplets exuded from such specialized pores. Guttation is most noticeable when transpiration is suppressed and the relative humidity is high, such as during the night.

Water Is Transported through Tracheids and Vessels

In most plants, the xylem constitutes the longest part of the pathway of water transport. In a plant 1 m tall, more than 99.5% of the water transport pathway through the plant is within the xylem, and in tall trees the xylem represents an even greater fraction of the pathway. Compared with the complex pathway across the root tissue, the xylem is a simple pathway of low resistance. In the following sections we will examine how water movement through the xylem is optimally suited to carry water from the roots to the leaves, and how negative hydrostatic pressure generated by leaf transpiration pulls water up the water column in the xylem.

The conducting cells in the xylem have a specialized anatomy that enables them to transport large quantities of water with great efficiency. There are two important types of **tracheary elements** in the xylem: tracheids and vessel elements (Figure 4.7A). Vessel elements are found only in angiosperms and a small group of gymnosperms called the Gnetales. Tracheids are present in both angiosperms and gymnosperms.

Both of these tracheary elements are dead when functional: they have no membranes or organelles. They are like hollow tubes reinforced by lignified secondary walls. **Tracheids** are elongated, spindle-shaped cells that communicate with adjoining tracheids by means of the numerous pits in their lateral walls. These **pits** are microscopic regions where the secondary wall is absent and the primary wall is thin and porous (Figure 4.7C). Pits of one tracheid are typically located opposite pits of an adjoining tracheid, forming **pit pairs**. Pit pairs constitute a low-resistance path for water movement between tracheids. The porous layer between pit pairs, consisting of two primary walls and a middle lamella, is called the **pit membrane**.

The pit membranes in tracheids of gymnosperms usually have a central thickening, called a torus (Figure 4.7C). The torus acts like a valve to close the pit by lodging itself in the circular or oval wall thickenings bordering these pits. Such lodging of the torus is an effective way of preventing dangerous gas bubbles from invading neighboring tracheids (we will discuss this formation of bubbles, called cavitation, shortly).

Vessel elements tend to be shorter and wider than tracheids and have perforated end walls that form a **perforation plate** at each end of the cell. Like tracheids, ves-

(A)

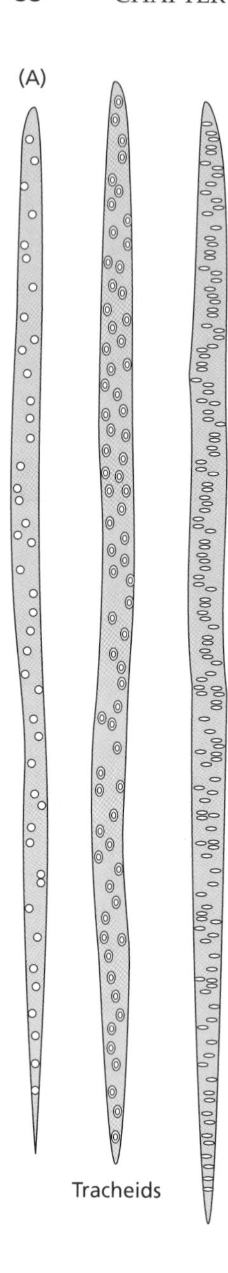

(B)

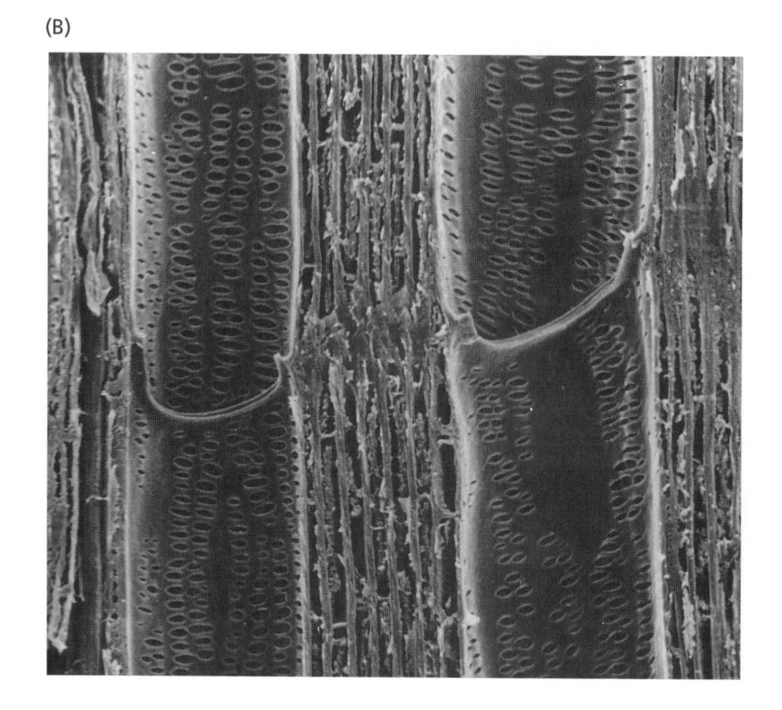

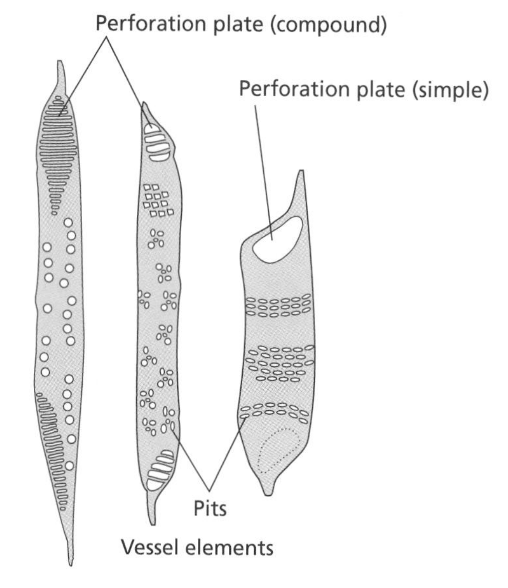

(C)

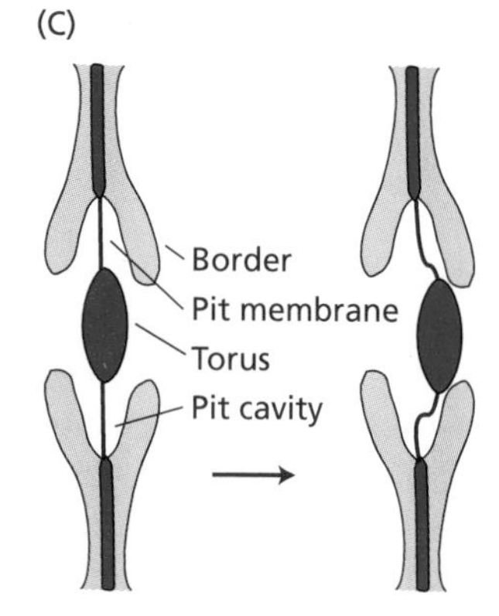

Figure 4.7 Tracheary elements and their interconnections. (A) Structural comparison of tracheids and vessel elements, two classes of tracheary elements involved in xylem water transport. Tracheids are elongate, hollow, dead cells with highly lignified walls. The walls contain numerous pits—regions where secondary wall is absent but primary wall remains. The shape and pattern of wall pitting vary with species and organ type. Tracheids are present in all vascular plants. Vessels consist of a stack of two or more vessel elements. Like tracheids, vessel elements are dead cells and are connected to one another through perforation plates—regions of the wall where pores or holes have developed. Vessels are connected to other vessels and to tracheids through pits. Vessels are found in most angiosperms and are lacking in most gymnosperms. (B) Scanning electron micrograph of oak wood showing two vessel elements that make up a portion of a vessel. Large pits are visible on the side walls and the end walls are open at the perforation plate. (420×) (C) Diagram of a bordered pit with a torus either centered in the pit cavity or lodged to one side of the cavity, thereby blocking flow. (B © G. Shih-R. Kessel/Visuals Unlimited; C after Zimmerman 1983.)

sel elements also have pits on their lateral walls (see Figure 4.7A). Unlike tracheids, which are arranged in overlapping vertical files, vessel members are stacked end to end to form a larger unit called a **vessel** (Figure 4.7B). The exact length of vessels is difficult to ascertain, but they may extend 1 to many meters. Because of their open cross-walls, vessels provide a very efficient low-resistance pathway for water movement. The vessel members at the extreme ends of a vessel lack perforations at the end walls and communicate with neighboring vessels via pit pairs (see Figure 4.7C).

Water Movement through the Xylem Requires Less Pressure than Movement through Living Cells

The lack of membranes in the tracheary elements and the perforations in the walls of vessel elements allow water to move freely through these water-filled capillaries in response to a pressure gradient. A simple calculation can help us appreciate the extraordinary efficiency of a vessel. We will calculate the driving force required to move water through the xylem at a typical velocity and compare it with the driving force that would be needed to move water through a comparable cell-to-cell pathway.

Let's begin by noting that the velocity with which water travels up the trunk of a tree depends on both the type of tree and the transpirational demand placed on the xylem. For trees with wide vessels (radii of 100 to 200 μm), peak velocities of 16 to 45 m h^{-1} (4 to 13 mm s^{-1}) have been measured. Trees with smaller vessels (radii of 25 to 75 μm) have lower peak velocities, from 1 to 6 m h^{-1} (0.3 to 1.7 mm s^{-1}). For our calculation we will use a figure of 4 mm s^{-1} for the xylem transport velocity and 40 μm as the vessel radius. This is a high velocity for such a narrow vessel, so it will tend to exaggerate the pressure gradient required to support water flow in the xylem.

A version of Poiseuille's equation (see Equation 3.4) can be used to estimate the pressure gradient ($\partial\Psi_p/\partial x$) needed to move water at this velocity (4×10^{-3} m s^{-1}) through a pipe of radius (r) 40 μm. By dividing Equation 3.4 by the cross-sectional area (πr^2) of the xylem vessel, we find that the rate of transport (J_v, in m s^{-1}) is given by the following equation:

$$J_v = \left(\frac{(\text{radius})^2}{8\ (\text{viscosity})}\right)\left(\frac{\partial\Psi_p}{\partial x}\right) \qquad (4.3)$$

Taking the viscosity of xylem sap to be that of water (10^{-3} Pa s), we find that the pressure gradient required is 2×10^4 Pa m^{-1} (or 0.02 MPa m^{-1}). This is the pressure gradient needed to overcome the viscous drag that arises as water moves through an *ideal* vessel at a rate of 4 mm s^{-1}. *Real* vessels have irregular inner wall surfaces and constrictions, such as perforation plates, at the points where vessel elements meet. Tracheids, with their smaller diameters and pitted walls, offer even greater resistance to water flow. Such deviations from an ideal pipe will increase the frictional drag above that calculated from Poiseuille's equation, but since we selected a low value for vessel radius, our estimate of 0.02 MPa m^{-1} should be in the correct range for pressure gradients found in real trees.

Let's now compare this value (0.02 MPa m^{-1}) with the driving force that would be necessary to move water at the same velocity through a layer of *living cells.* We will ignore water movement in the apoplast pathway in this example and focus on water moving from cell to cell, crossing the plasma membrane each time. As we described in Chapter 3, the velocity (J_v) of water flow across a membrane depends on the membrane hydraulic conductivity (Lp) and on the difference in water potential ($\Delta\Psi_w$) across the membrane:

$$J_v = Lp(\Delta\Psi_w) \qquad (4.4)$$

A high value for the Lp of higher plant cells is about 4×10^{-7} m s^{-1} MPa^{-1}. Thus, to move water across a membrane at 4×10^{-3} m s^{-1} would require a driving force ($\Delta\Psi_w$) of 10^4 MPa (4×10^{-3} m s^{-1} divided by 4×10^{-7} m s^{-1} MPa^{-1}). This is the driving force needed to move the water across a *single membrane.* To move through a cell, water must cross at least two membranes, so the total driving force across one cell would be 2×10^4 MPa. If we estimate the cell length as 100 μm (10^{-4} m, a generous estimate), then the water potential gradient needed for water to move at a velocity of 4 mm s^{-1} through a layer of cells would be 2×10^4 MPa divided by 10^{-4} m, or 2×10^8 MPa m^{-1}. This is an enormous driving force, and it illustrates that water flow through the xylem is exceedingly more efficient than water flow across the membranes of cells. Comparing the two driving forces (for open vessel and cell transport), we see that the two pathways show a difference of a factor of 10^{10}. This is a huge difference indeed.

The Pressure Difference Needed to Lift Water to the Top of a 100-Meter Tree Is About 3 MPa

With the foregoing example in mind, let's see how large a pressure gradient is needed to move water up the tallest tree. Redwoods are among the tallest trees in the world, and the tallest redwood is approximately 100 m high. If we think of the stem of a tree as a long pipe, we can estimate that the pressure difference, from the ground to the top of the tree, that is needed to overcome the frictional drag of water moving through the xylem is about 2 MPa (0.02 MPa $m^{-1} \times 100$ m). Most gymnosperms, such as redwoods, have tracheids rather than vessels in their xylem. Tracheids have end walls that increase the resistance to water flow through the xylem. Thus, our estimate of 2 MPa is probably low, but it will serve as a reasonable order-of-magnitude estimate. In addition to frictional resistance, we must consider gravity. The weight of a standing column of water 100 m tall creates a pressure of 1 MPa at the bottom of the water column (100 m × 0.01 MPa m^{-1}). This pressure gradient due to gravity must be added to that required to cause water movement through the xylem. Thus we calculate that a pressure difference of about 2 + 1 = 3 MPa, from the base to the top branches, is needed to carry water up the tallest trees.

The Conducting Cells of the Xylem Are Adapted for the Transport of Water under Tension

Does this pressure difference arise because the tree builds up a positive pressure at its base, or because the tree develops a tension (negative pressure) at its top? We mentioned previously that some roots develop a positive hydrostatic pressure in their xylem—a so-called root pressure. But the root pressure is typically less than 0.1 MPa and disappears when the transpiration rate is high, so it is clearly inadequate to move water up a tall tree. Instead, the water at the top of a tree develops a large tension (a negative hydrostatic pressure), and this tension *pulls* water up the long water columns in the xylem. This mechanism is usually called the **cohesion–tension theory of sap ascent**, because it requires the cohesive properties of water to be able to support large tensions in the xylem water columns.

The cohesion–tension theory has been a controversial subject for more than a century and continues to generate lively debate. The main controversy surrounds the question of whether water columns in the xylem can sustain the large tensions (negative pressures) necessary to pull water up tall trees. The most recent argument began when Ulrich Zimmermann and coworkers punctured xylem with a fine glass capillary connected to a pressure probe (see Figure 6 in Box 3.2). They failed to find the expected large negative pressures in the xylem. Soon thereafter Martin Canny challenged the conventional view that negative pressures in the xylem are accurately measured by the pressure chamber (or pressure bomb) technique (see Canny 1998). (For a description of the pressure chamber method, see Figure 3 in Box 3.2.)

Prompted by these controversial results and conclusions, other researchers have reexamined the cohesion–tension theory. New results have shown that water in the xylem can indeed sustain large negative tensions (Holbrook et al. 1995; Pockman et al. 1995). Most researchers have thus concluded that the basic theory is sound and that the pressure probe measurements are in error, probably because of artifactual cavitation when the xylem walls are punctured with the glass capillary of the pressure probe (Tyree 1997). Researchers can readily demonstrate xylem tensions by puncturing intact xylem through a drop of ink on the surface of a stem from a transpiring plant. When this is done correctly, the ink is drawn instantly into the xylem, resulting in long ink streaks as the xylem tension is relieved upon puncture.

The large tensions that develop in the xylem of trees and other plants create special problems. First, the water under tension transmits an inward force to the walls of the xylem. If the cell walls were weak or pliant, they would collapse under the influence of this tension. The secondary wall thickenings and lignification of tracheids and vessels are adaptations that offset this tendency to collapse.

A second, and more serious, problem is that water under such tensions is in a *physically metastable state.* We mentioned in Chapter 3 that the experimentally determined breaking strength of degassed water (water that has been boiled to remove gases) is greater than 30 MPa. This value is much larger than the estimated tension of 3 MPa needed to pull water up the tallest trees. As the tension in water increases, however, there is an increased tendency for air to be pulled through microscopic pores in the xylem cell walls. This phenomenon is sometimes called "air seeding." Once a gas bubble has formed within the water column under tension, it will expand, since gases cannot resist tensile forces. This phenomenon of bubble formation is known as **cavitation** or **embolism**. It is similar to vapor lock in the fuel line of an automobile or embolism in a blood vessel. Cavitation of the xylem breaks the continuity of the water column and stops water transport (Tyree and Sperry 1989).

Such breaks in the water columns in plants are not unusual. With the proper equipment, one can "hear" the water columns break. John Milburn, Melvin Tyree, and others have attached ultrasonic detectors to the stems of plants to detect cavitation. When the plants are deprived of water, clicks are detected. The clicks correspond to the formation and rapid expansion of air bubbles in the xylem, resulting in high-frequency acoustic shock waves through the rest of the plant. These breaks in xylem water continuity, if not repaired, would be disastrous to the plant. By blocking the main transport pathway of water, such embolisms would cause the dehydration and death of the leaves.

The impact of xylem cavitation on the plant is minimized by several means. Because the tracheary elements in the xylem are interconnected, one gas bubble might, in principle, expand to fill the whole network. In practice, gas bubbles do not spread far, because they are stopped at the pitted walls between overlapping tracheids and vessels . Their expansion is stopped because the gas cannot easily squeeze through the small pores of the pit membranes, an effect also due to the surface tension of water. Since the capillaries in the xylem are interconnected, one gas bubble does not completely stop water flow through a vessel or column of tracheids. Instead, water can detour around the blocked point by traveling through neighboring, connected conduits (Fig-

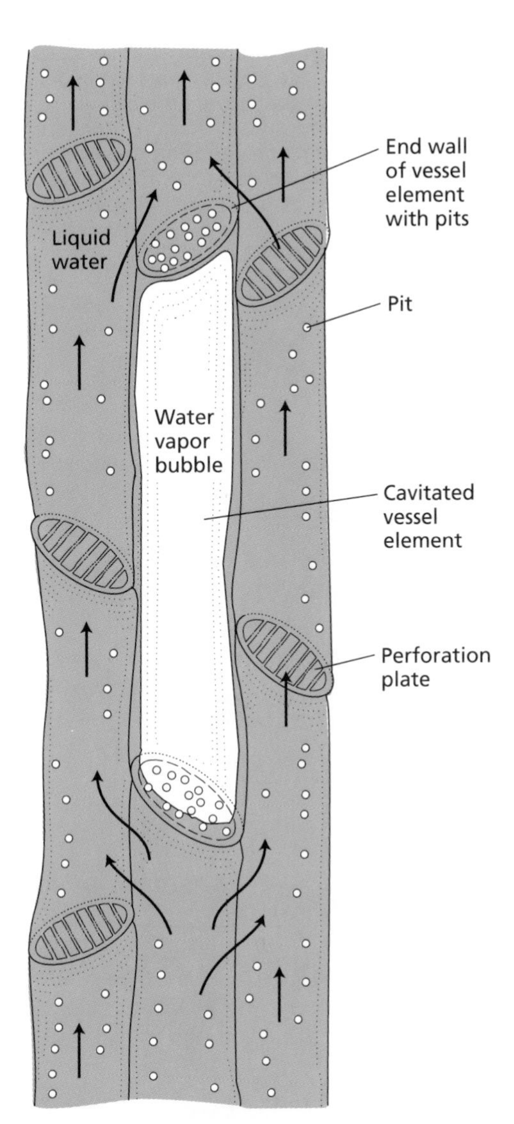

Figure 4.8 Detours around a vapor-locked vessel element. Tracheids and vessels constitute multiple, parallel, interconnected pathways for water movement. Cavitation in this example blocks water movement within the cavitated vessel element. Because these water conduits are interconnected through wall pits, however, cavitation of a vessel element or tracheid does not completely stop water movement in the cell file. Water can detour around the blocked vessel by moving through adjacent tracheary elements. The spread of the vapor bubble throughout the xylem is eventually stopped by an end wall that lacks a perforation plate.

ure 4.8). Thus while the presence of pitted walls in the xylem increases the resistance to water flow, it also provides a way to restrict cavitation.

Gas bubbles can also be eliminated from the xylem. At night, when transpiration is low, xylem Ψ_p increases and the water vapor and gases may simply dissolve back into the solution of the xylem. Moreover, as we have seen, some plants develop positive pressures (root pressures) in the xylem. Such pressures shrink the gas bubble and cause the gases to dissolve. Finally, many plants have secondary growth in which new xylem forms each year. The new xylem becomes functional before the old xylem ceases to function because of occlusion by gas bubbles or by substances secreted by the plant.

Water Evaporation in the Leaf Generates a Negative Pressure in the Xylem

We mentioned earlier (Chapter 3) that an actively transpiring leaf may exchange all of its water in the course of 1 hour. Such a leaf would very quickly wilt and die if the water lost by evaporation were not replaced. In the intact plant, water is brought to the leaves via the **xylem** of the leaf vascular bundle, which branches into a very fine and sometimes intricate network of **veins** throughout the leaf (Figure 4.9). This **venation pattern** becomes so finely divided that most cells in a typical leaf are within 0.5 mm of a minor vein. From the xylem, water is drawn into the cells of the leaf and along the cell walls.

It is at the surface of the cell walls in the leaf that the negative pressure that causes water to move up through the xylem develops. The situation is analogous to that in

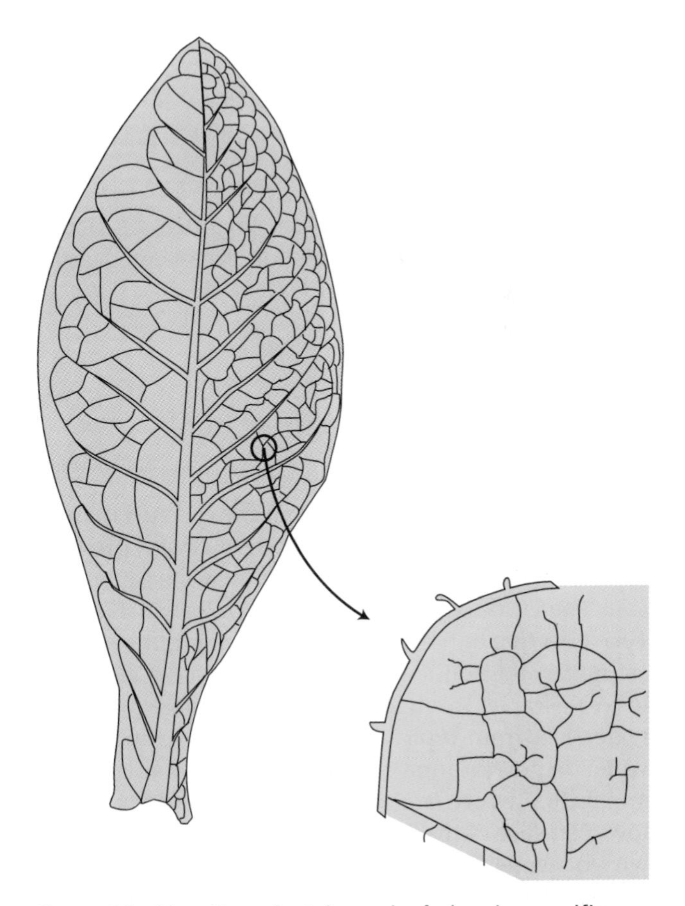

Figure 4.9 Venation of a tobacco leaf, showing ramification of the midrib vein into finer lateral veins. This venation pattern brings xylem water close to every cell in the leaf. (After Kramer and Boyer 1995.)

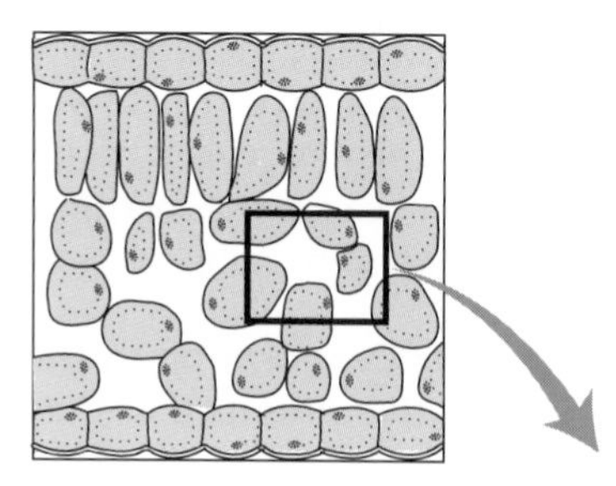

Figure 4.10 The origin of tensions or negative pressures in cell wall water of the leaf. As water evaporates from the surface film that covers the cell walls of the mesophyll, water withdraws farther into the interstices of the cell wall, and surface tension causes a negative pressure in the liquid phase. As the radius of curvature decreases, the pressure decreases (becomes more negative), as calculated from the equation $P = -2T/r$, where T is the surface tension of water and r is the radius.

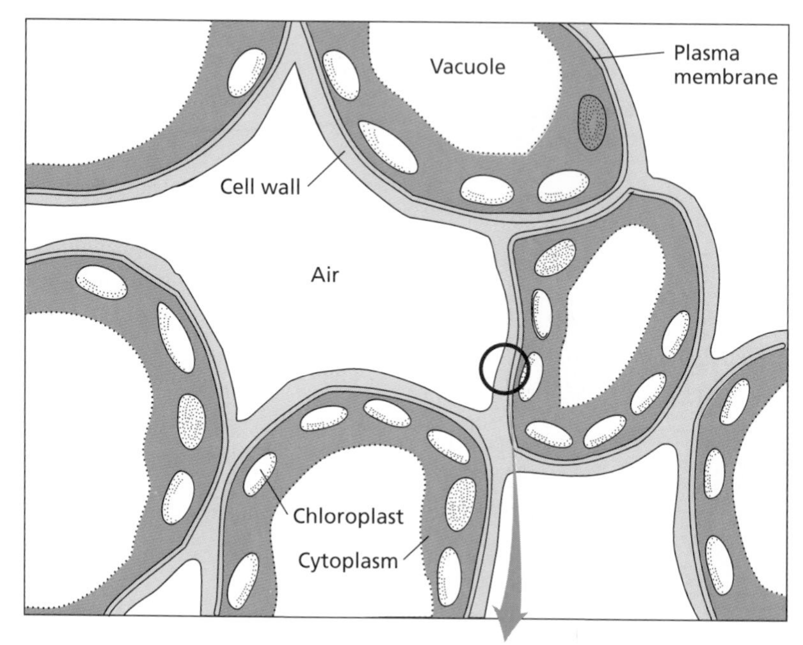

the soil. The cell wall acts like a very fine capillary wick soaked with water. Water adheres to the cellulose microfibrils and other hydrophilic components of the wall. The mesophyll cells within the leaf are in direct contact with the atmosphere through an extensive system of intercellular air spaces. Initially, water evaporates from a thin film lining these air spaces. As water is lost to the air, the surface of the remaining water is drawn into the interstices of the cell wall (Figure 4.10). The air–water surface becomes curved, forming microscopic menisci, and the surface tension induces a tension, or negative pressure, in the water. As more water is removed from the wall, the water surface develops more acute menisci, and the pressure of the water becomes more negative (see Equation 4.1). Thus, the motive force for xylem transport is generated at the air–water interfaces within the leaf.

Water Vapor Moves from the Leaf to the Atmosphere by Diffusion through Stomata

When water has evaporated from the cell surface into the intercellular air space, diffusion is the primary means of any further movement of the water out of the leaf. The waxy cuticle that covers the leaf surface is a very effective barrier to water movement. It has been estimated that only about 5% of the water lost from leaves escapes through the cuticle. Almost all of the water lost from typical leaves is lost by diffusion of water vapor through the tiny pores of the stomatal apparatus, which are usually most abundant on the lower surface of the leaf. Figure 4.11A shows the pathway of water vapor loss from inside the leaf to the atmosphere. Water moves along this pathway predominantly by diffusion, so this water movement is controlled by the *concentration gradient* of water vapor. In principle, water vapor loss could also be described in terms of water potential gradients, but these equations

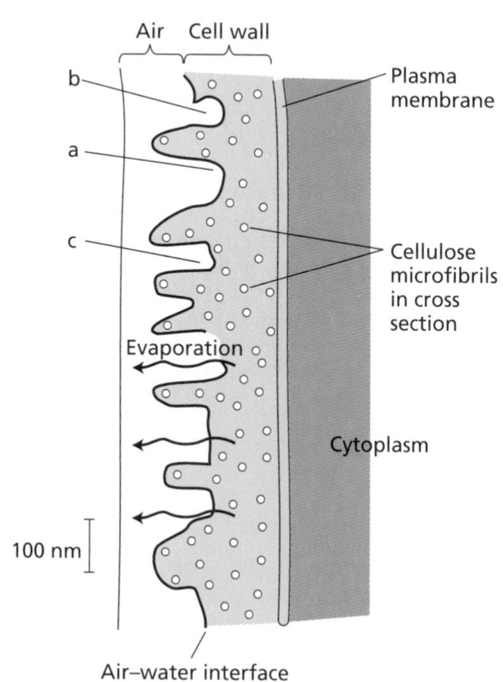

	Radius of curvature (μm)	Hydrostatic pressure (MPa)
(a)	0.5	–0.3
(b)	0.05	–3
(c)	0.01	–15

become nonlinear because of the logarithmic relationship between water vapor concentration and water potential (see Equation 4.6).

We will now examine the driving force for leaf transpiration, the main resistances in the diffusion pathway from the leaf to the atmosphere, and the anatomical features of the leaf that regulate transpiration.

Diffusion of Water Vapor in Air Is Fast

We saw in Chapter 3 that diffusion in liquids is slow and is effective only at cellular dimensions. How long would it take for water vapor to diffuse from the cell wall surfaces inside the leaf to the outside atmosphere? To answer this question, we will again make use of Equation 3.3, which relates distance to the time required for the concentration of a diffusing substance to reach a certain value—for example, half of the starting concentration. The distance through which water vapor must diffuse from the surface inside the leaf to the turbulent air outside is approximately 1 mm (10^{-3} m). The diffusion

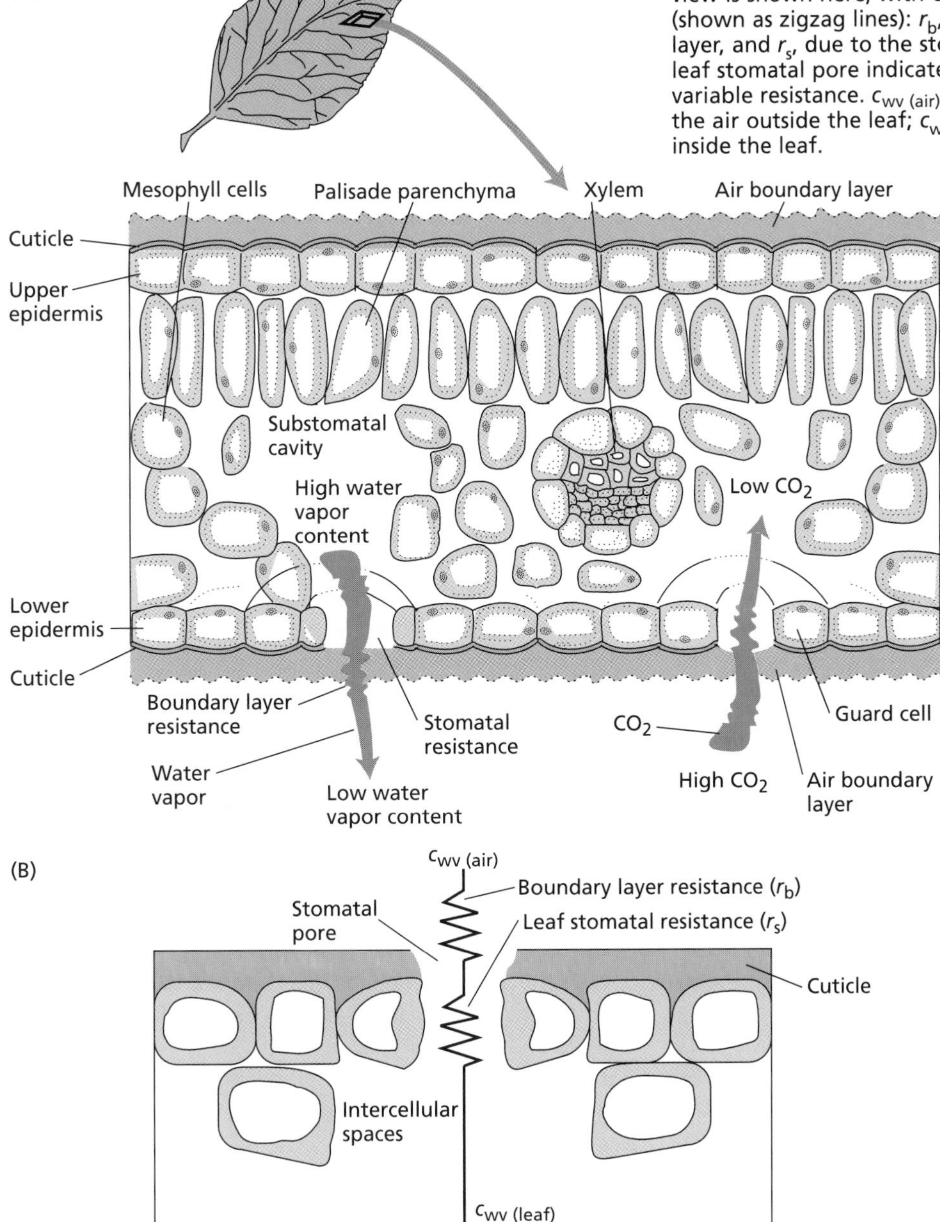

Figure 4.11 The water pathway through the leaf. (A) Water is pulled from the xylem into the cell walls of the mesophyll, where it evaporates into the air spaces within the leaf. By diffusion, water vapor then moves through the leaf air space, through the stomatal pore, and across the boundary layer of still air that adheres to the leaf surface. CO_2 diffuses in the opposite direction along its concentration gradient (low inside, higher outside). (B) In the electrical analog of transpiration, water vapor flows through a set of resistances in series and in parallel. A very simplified view is shown here, with only two resistances in series (shown as zigzag lines): r_b, due to the unstirred boundary layer, and r_s, due to the stomatal pore. The arrow across the leaf stomatal pore indicates that stomatal resistance is a variable resistance. $c_{wv\,(air)}$, water vapor concentration in the air outside the leaf; $c_{wv\,(leaf)}$, water vapor concentration inside the leaf.

coefficient of water vapor in air is $2.4 \times 10^{5}\ m^2\ s^{-1}$. Therefore the diffusion time is

$$t_{c=1/2} = \frac{(\text{distance})^2}{\text{diffusion coefficient}} = \frac{(10^{-3}\ m)^2}{2.4 \times 10^{-5}\ m^2\ s^{-1}}$$
$$= 0.042\ s$$

The reason that this time is so short is that diffusion is much more rapid in a gas than in a liquid. Thus we see that diffusion is adequate to move water vapor through the gas phase of the leaf.

Transpiration from the leaf depends on two major factors: (1) the **difference in water vapor concentration** between the leaf air spaces and the external air and (2) the **diffusional resistance** (*r*) of this pathway. This concept of transpiration is analogous to the flow of electrons in an electric circuit. Indeed, an electrical analog is commonly used as a model for water vapor loss from the leaf (Figure 4.11B). In this analog, resistances are associated with each part of the pathway; the major ones are the resistance at the stomatal pore (r_s) and the resistance due to the layer of unstirred air at the surface of the leaf (r_b) (the so-called boundary layer). Transpiration rate (*E*, in mol $m^{-2}\ s^{-1}$) may be related to diffusional resistances (*r*, in s m^{-1}) by the following equation:

$$E = \frac{c_{wv(leaf)} - c_{wv\ (air)}}{r_s + r_b} \qquad (4.5)$$

where $c_{wv\ (leaf)}$ is the water vapor concentration inside the leaf and $c_{wv\ (air)}$ is the water vapor concentration of the air outside the leaf, both expressed in moles per cubic meter (mol m^{-3}). Resistance (*r*) is the inverse of conductance; that is, resistance = 1/conductance. Thus, a high resistance is the same as a low conductance. Expression of this value in terms of resistances is preferred over expression in terms of conductances in some instances because resistances in series may be summed to calculate a total resistance, as in Equation 4.5, whereas a similar calculation of conductances in series is more complicated. In the leaf, the total resistance is due mostly to the diffusion limitation imposed by the stomatal pore, but other parts of the pathway for water vapor loss, such as the boundary layer (which we will discuss shortly), may contribute significantly to *r*. Let's examine the factors in Equation 4.5 in greater detail.

Table 4.2
Relationships among water vapor concentration (c_{wv}), water vapor pressure (P_{wv}), relative humidity (*RH*), and water potential (Ψ_w)

c_{wv} (mol m^{-3})	P_{wv} (kPa)	RH	Ψ_w (MPa)[a]
0.961	2.34	1	0.00
0.957	2.33	0.996	–0.54
0.951	2.32	0.990	–1.36
0.923	2.25	0.960	–5.51
0.865	2.11	0.900	–14.2
0.480	1.17	0.500	–93.6
0	0	0	–infinity

Note: These data are for 20°C.
[a] Calculated using Equation 4.6, with a value of 135 MPa for $RT/\bar{V}_w$.

The Driving Force for Water Loss Is the Difference in Water Vapor Concentration

The difference in water vapor concentration is expressed as $c_{wv\ (leaf)} - c_{wv\ (air)}$. Sometimes vapor pressures are used instead of concentrations, and the difference is called the **water vapor pressure deficit**. Water vapor pressure (p_{wv}) is measured in kilopascals (kPa) and is proportional to water vapor concentration (Table 4.2). Vapor pressures and concentrations are equivalent; in our analysis we will use the latter.

The water vapor concentration of bulk air ($c_{wv\ (air)}$) can be readily measured, but that of the leaf ($c_{wv\ (leaf)}$) is more difficult to assess. We can estimate it by assuming that the air space in the leaf is close to water potential equilibrium with the cell wall surfaces. This approximation is not strictly true, because water is diffusing away from these surfaces. However, it introduces little error because the major resistance to vapor loss is at the stomatal pore. Moreover, the volume of air space inside the leaf is small, whereas the wet surface from which water evaporates is comparatively large. Air space volume is about 5% of the total leaf volume for pine needles, 10% for corn leaves, 30% for barley, and 40% for tobacco leaves. The internal surface area from which water evaporates may be from 7 to 30 times the external leaf area. This high ratio of surface area to volume makes for rapid vapor equilibration inside the leaf.

Making the assumption of equilibrium, we can calculate the water vapor concentration in the leaf air spaces if we know (1) the water potential of the leaf (which is the same as the water potential of the wall surfaces from which water is evaporating) and (2) the leaf temperature. Let's take as an example a leaf with a water potential of –1.0 MPa. To reach vapor equilibrium, water evaporates from the cell wall surfaces until the water potential of the air inside the leaf equals the water potential of the leaf. The water potential of the air is given by the following equation:

$$\Psi_w = \frac{RT}{\bar{V}_w} \ln (RH) \qquad (4.6)$$

where *R* is the gas constant, *T* is temperature (in degrees Kelvin), $\bar{V}_w$ is the partial molar volume of liquid water, and *RH* is the relative humidity of the air. **Relative humidity** is the water vapor concentration expressed as a fraction of the **saturation water vapor concentration**, $c_{wv(sat.)}$:

$$RH = \frac{c_{wv}}{c_{wv(sat.)}} \qquad (4.7)$$

RH varies between 0 and 1; RH multiplied by 100 is the percentage relative humidity.

Table 4.2 shows RH values as a function of water potential, calculated from Equation 4.6. This table shows that the air spaces of living leaves must have a high RH, a value of nearly 1 (100%), when water potentials are in the physiological range. Moreover, outside air, with RH of, for example, 0.5 (50%), has a remarkably low water potential.

To convert from RH to c_{wv}, we need to know $c_{wv(sat.)}$. The saturation water vapor concentration is *strongly dependent on temperature.* As the air temperature rises, the water-holding capacity of air increases sharply, (Figure 4.12). In the range of 10 to 35°C, an increase in air temperature of 12°C doubles the water vapor concentration of saturated air. This is an important observation. If our leaf with a water potential of –1.0 MPa warms up abruptly from 20 to 32°C, the relative humidity in the leaf air space drops abruptly from 99.3 to almost 50%. This drop in RH results because the water-holding capacity of the air, $c_{wv(sat.)}$, doubles. As a result of the drop in RH, water will evaporate in the air space until RH returns to a value of 99.3% and the air is again in water potential equilibrium with the leaf. As a consequence of this change in temperature, $c_{wv\,(leaf)}$ (the water vapor concentration of the leaf air space) increases from 0.95 to 1.87 mol m^{-3}, which makes for a steeper concentration difference driving the diffusional loss of water from the leaf. For this reason, leaf temperature is an important determinant of the transpiration rate.

Table 4.3 illustrates how RH, c_{wv}, and water potential change at various points in the transpiration pathway. We see that c_{wv} decreases along each step of the pathway from the cell wall surface to the bulk air outside the leaf. Keep in mind that RH can increase along part of this pathway because the external air temperature may be lower than the temperature of the leaf. The important points to remember are (1) that the driving force for water loss from the leaf is the *absolute* concentration difference (difference in c_{wv}, not in RH), and (2) that this difference depends on leaf temperature.

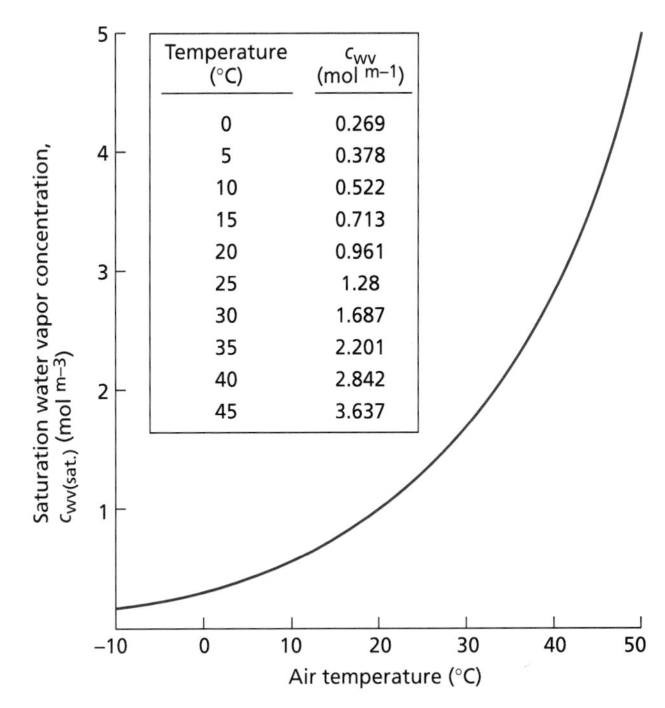

Figure 4.12 Concentration of water vapor in saturated air as a function of air temperature. As its temperature increases, air holds more water.

Water Loss Is Also Regulated by the Pathway Resistances

The second important factor governing water loss from the leaf is the diffusional resistance of the transpiration pathway ($r_s + r_b$ in Equation 4.5). Two major resistances limit water loss by the leaf; both are variable. The first, and more important, is the resistance associated with diffusion through the stomatal pore, the **leaf stomatal resistance** (r_s). The second is the resistance due to the layer of unstirred air next to the leaf surface through which water vapor must diffuse to reach the turbulent

Table 4.3
Representative values for relative humidity, absolute water vapor concentration, and water potential for four points in the pathway of water loss from a leaf

Location	Relative humidity	Water vapor Concentration (mol m^{-3})	Water vapor Potential (MPa)[a]
Inner air spaces (25°C)	0.99	1.27	−1.38
Just inside stomatal pore (25°C)	0.95	1.21	−7.04
Just outside stomatal pore (25°C)	0.47	0.60	−103.7
Bulk air (20°C)	0.50	0.50	−93.6

Source: Adapted from Nobel 1991.
Note: See Figure 4.11.
[a] Calculated using Equation 4.6; with values for $RT/\bar{V}_w$ of 135 MPa at 20°C and 137.3 MPa at 25°C.

air of the atmosphere (see Figure 4.11). This second resistance, r_b, is called the leaf **boundary layer resistance**.

Because the cuticle covering the leaf is nearly impermeable to water, most of the leaf transpiration is due to loss of water vapor by diffusion through the stomatal pore. Water loss through the stomatal pore will be discussed in detail in the next section. For now, keep in mind that the microscopic stomatal pores provide a *low-resistance pathway* for diffusional movement of gases across the epidermis and cuticle. That is, the stomatal pores lower the leaf stomatal resistance (or equivalently, raise leaf stomatal conductance). Changes in stomatal resistance are important in regulating water loss by the plant and controlling the rate of carbon dioxide uptake necessary for sustained CO_2 fixation during photosynthesis.

Once water vapor has diffused through the stomatal pore, the last step is diffusion across the boundary layer on the surface of the leaf. The **boundary layer** is a thin film of still air that hugs the surface of the leaf, and its resistance to water vapor diffusion is proportional to its thickness. The thickness of the boundary layer is determined primarily by wind speed. When the air surrounding the leaf is very still, the layer of unstirred air on the surface of the leaf may be so thick that it is the primary deterrent to water vapor loss from the leaf. As Figure 4.13 shows, increases in stomatal apertures under such conditions have little effect on transpiration rate (although closing the stomata completely will still reduce transpiration).

When wind velocity is high, the drag of the moving air reduces the thickness of the boundary layer at the leaf surface, so it effectively reduces the resistance of this layer. Under such conditions, the stomatal resistance has the largest amount of control over water loss by the leaf. This situation is more representative of conditions in the field, where even on windless days small local convection currents reduce boundary layer resistances. Wind may sometimes influence transpiration in other ways. For example, when leaf temperature is higher than air temperature, greater heat conduction to the air by the wind may reduce leaf temperature, thereby lowering $c_{wv\,(leaf)}$ and E (see Equation 4.5).

Various anatomical and morphological aspects of the leaf can influence the thickness of the boundary layer. Hairs on the surface of leaves can serve as microscopic windbreaks. Some plants have sunken stomata that provide a dead-volume region, outside the stomatal pore, that is protected from wind. The size and shape of leaves also influence the way the wind sweeps across the leaf surface. Although these and other factors may influence the boundary layer, they are not characteristics that can be altered on an hour-to-hour or even day-to-day basis. For short-term regulation, control of stomatal apertures by the guard cells plays the most crucial role in the regulation of leaf transpiration.

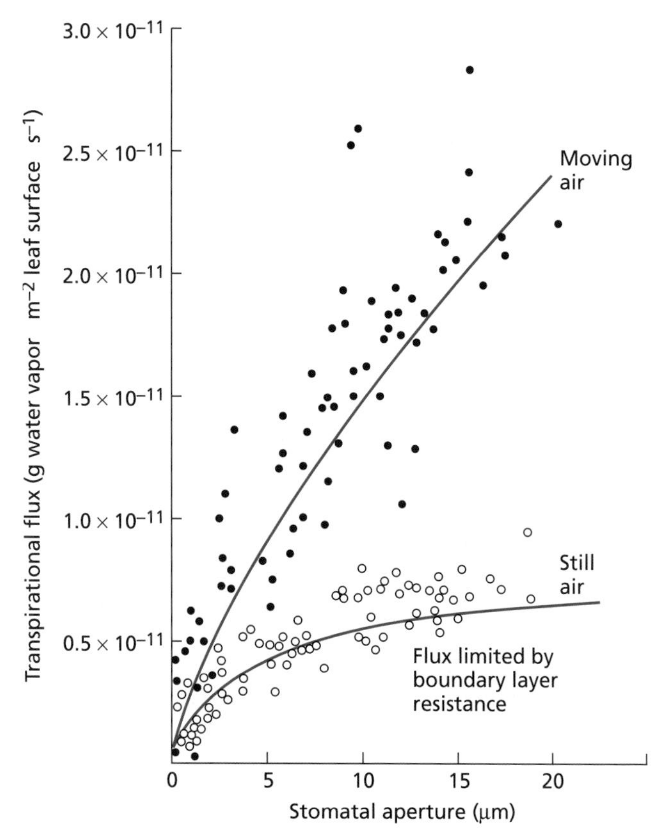

Figure 4.13 Dependence of transpiration flux on the stomatal aperture of zebra plant (*Zebrina pendula*) in still air and in moving air. The boundary layer is larger and more rate limiting in still air than in moving air. As a result, the stomatal aperture has less control over transpiration in still air. (From Bange 1953.)

Stomatal Control Couples Leaf Transpiration to Leaf Photosynthesis

As already noted, land plants are faced with competing demands to take up CO_2 from the atmosphere while limiting water loss. The cuticle that covers exposed plant surfaces serves as an effective barrier to water loss and thus protects the plant from desiccation. However, plants cannot prevent outward diffusion of water without simultaneously excluding CO_2 from the leaf. This problem is compounded because the concentration gradient for CO_2 uptake is much, much smaller than the concentration gradient driving water loss.

The functional solution to this dilemma is the *temporal* regulation of stomatal apertures. At night, when there is no photosynthesis and thus no demand for CO_2 inside the leaf, stomatal apertures are kept small, preventing unnecessary loss of water. On a sunny morning when the supply of water is abundant and the solar radiation incident on the leaf favors high photosynthetic activity, the demand for CO_2 inside the leaf is large, and the stomatal pores are wide open, decreasing the stomatal resistance to CO_2 diffusion. Water loss by transpiration is also substantial under these conditions, but since the water supply is plentiful, it is advantageous for the plant to trade water for the products of photosynthesis, which are essential for growth and reproduction.

On the other hand, on another sunny morning later in the season, when the water in the soil has been depleted, the stomata will open less or even remain closed. By keeping its stomata closed in dry conditions, the plant avoids potentially lethal dehydration. The values in Equation 4.5 for $(c_{wv\,(leaf)} - c_{wv\,(air)})$ and for r_b are largely a matter of physics and are not readily amenable to biological control. Stomatal resistance (r_s) is the remaining factor that controls leaf transpiration, and it is regulated by opening and closing of the stomatal pore. Stomatal resistance is the plant's primary means of regulating the exchange of gases across the leaf surfaces. This biological control is exerted by a pair of specialized epidermal cells, the **guard cells**, which surround the stomatal pore (Figure 4.14).

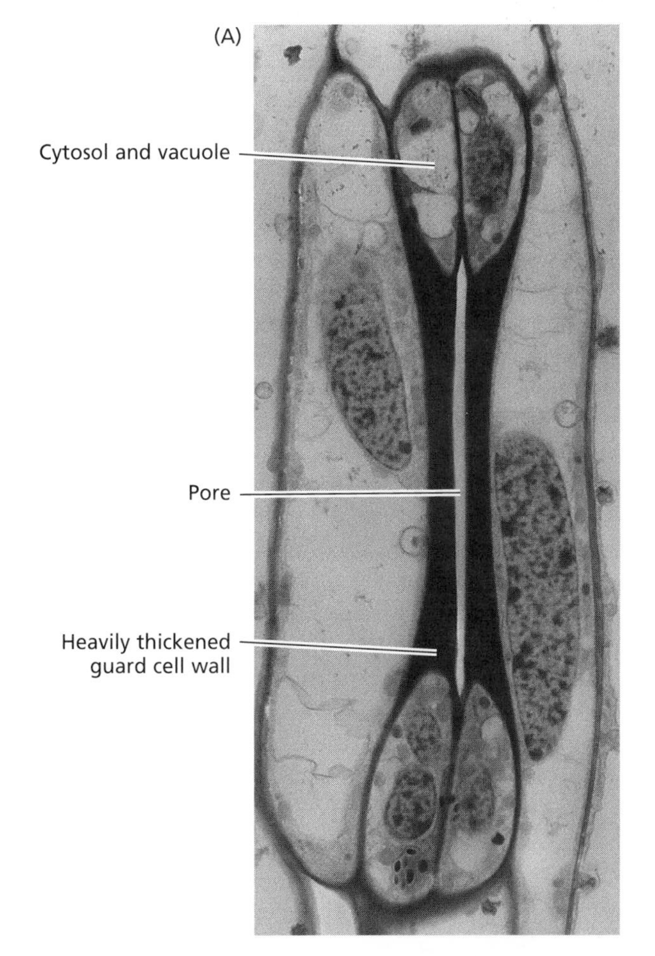

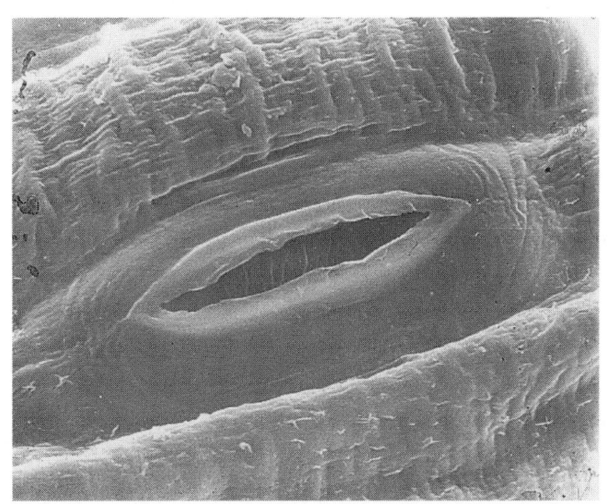

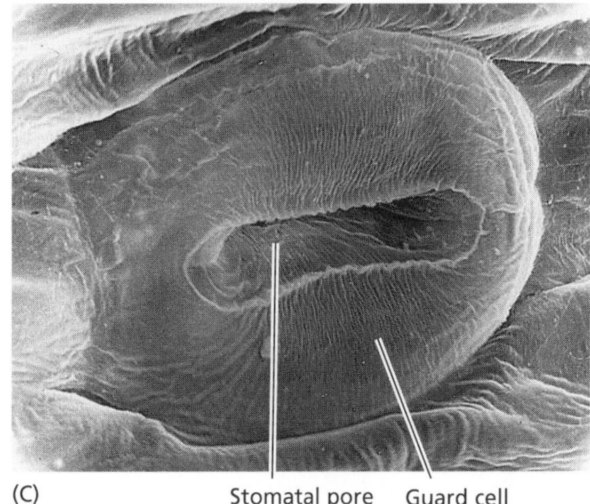

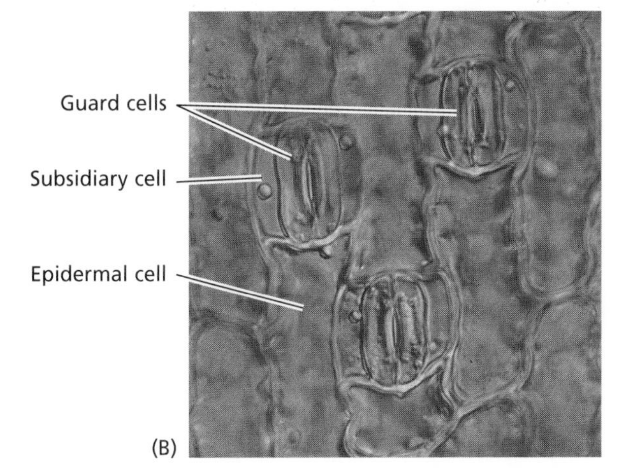

Figure 4.14 Electron micrographs of stomata. (A) A stoma from a grass. The bulbous ends of each guard cell show their cytosolic content and are joined by the heavily thickened walls. The stomatal pore separates the two midportions of the guard cells. (2560×) (B) Stomatal complexes of the sedge, *Carex*, viewed with differential interference contrast light microscopy. Each complex consists of two guard cells surrounding a pore and two flanking subsidiary cells. (550×) (C) Scanning electron micrographs of onion epidermis. The top panel shows the outside surface of the leaf, with a stomatal pore inserted in the cuticle. The bottom panel shows a pair of guard cells facing the stomatal cavity, toward the inside of the leaf. (1640×) (A from Palevitz 1981, B courtesy of B. Palevitz; micrographs in C from Zeiger and Hepler 1976 [top] and E. Zeiger and N. Burnstein [bottom].)

Cell Walls of Guard Cells Have Unique Structures

Guard cells can be found in leaves of all vascular plants, and they are present in organs from more primitive plants, such as the liverworts and the mosses (Ziegler 1987). Guard cells show considerable morphological diversity, but we can distinguish two main types: One is typical of grasses and a few other monocots, such as palms; the other is found in all dicots, in many monocots, and in mosses, ferns, and gymnosperms. In grasses (see Figure 4.14A), guard cells have a characteristic dumbbell shape, with bulbous ends. The pore proper is a long slit located between the two "handles" of the dumbbells. These guard cells are always flanked by a pair of differentiated epidermal cells called **subsidiary cells**, which help the guard cells control the stomatal pores (see Figure 4.14B). The guard cells, subsidiary cells, and pore are collectively called the **stomatal complex**. In dicot plants and nongrass monocots, kidney-shaped guard cells have an elliptical contour with the pore at its center (see Figure 4.14C). Although subsidiary cells are not uncommon in species with kidney-shaped stomata, they are often absent, in which case the guard cells are surrounded by ordinary epidermal cells.

A distinctive feature of the specialized organization of the guard cells is the structure of their walls. Portions of these walls are substantially thickened (Figure 4.15) and may be up to 5 μm across, in contrast to the 1 to 2 μm typical of epidermal cells. In kidney-shaped guard cells, a differential thickening pattern results in very thick inner and outer (lateral) walls, a thin dorsal wall (the wall in contact with epidermal cells), and a somewhat thickened ventral (pore) wall (see Figure 4.15). The portions of the wall that face the atmosphere extend into well-developed ledges, which form the pore proper. This thickening pattern is an essential element of guard cell mechanics, associated with the alignment of their **cellulose microfibrils**, which reinforce the walls and are an important determinant of cell shape (see Chapter 15).

In ordinary cells with a cylindrical shape, cellulose microfibrils are oriented transverse to the long axis of the cell. As a result, the cell expands in the direction of its long axis, since the cellulose reinforcement offers the least resistance at right angles to its orientation. In guard cells the microfibril organization is different. Kidney-shaped guard cells have cellulose microfibrils fanning out radially from the pore (Figure 4.16A). Thus the cell girth is reinforced like a steel-belted radial tire, and the guard cells curve outward during stomatal opening (Sharpe et al. 1987). In grasses, the dumbbell-shaped guard cells function like beams with inflatable ends. As the bulbous ends of the cells increase in volume and swell, the beams are separated from each other and the slit between them widens (Figure 4.16B).

An Increase in Guard Cell Turgor Pressure Opens the Stomata

Guard cells function as multisensory hydraulic valves. Environmental factors such as light intensity and quality, temperature, relative humidity, and intracellular CO_2 concentrations are sensed by guard cells, and these signals are integrated into well-defined stomatal responses. If leaves kept in the dark are suddenly illuminated, the light stimulus is perceived by the guard cells as an open-

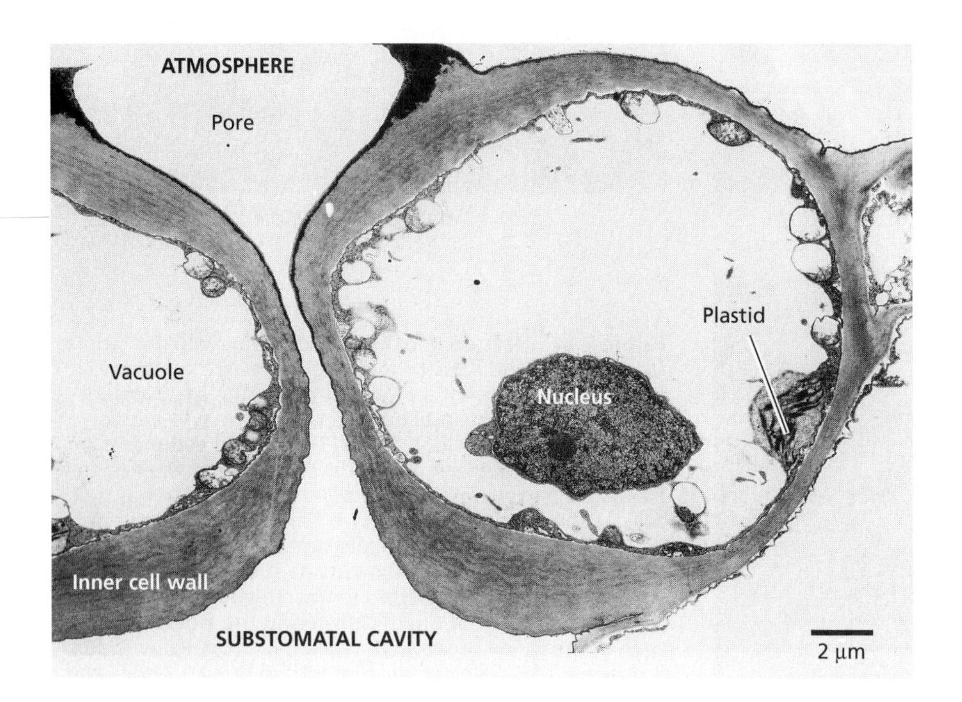

Figure 4.15 Electron micrograph showing a pair of guard cells from the dicot *Nicotiana tabacum* (tobacco). The section was made perpendicular to the main surface of the leaf. The pore faces the atmosphere; the bottom faces the substomatal cavity inside the leaf. Note the uneven thickening pattern of the walls, which determines the asymmetric deformation of the guard cells when their volume increases during stomatal opening. (From Sack 1987, courtesy of F. Sack.)

(A)

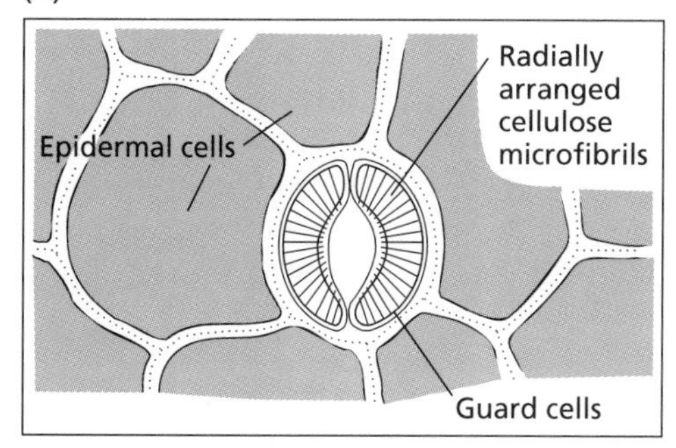

(B)

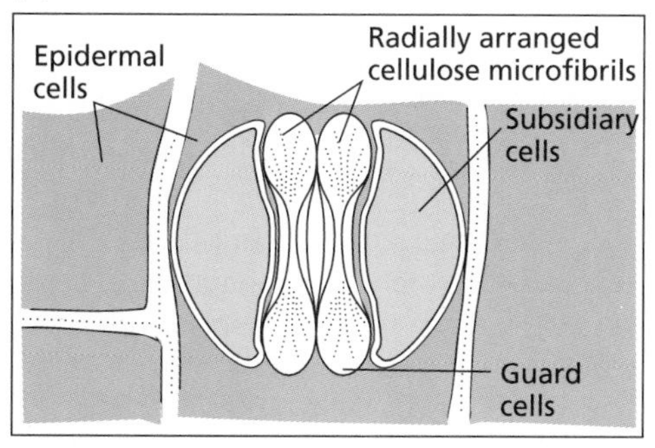

Figure 4.16 The radial alignment of the cellulose microfibrils in guard cells and epidermal cells of (A) a kidney-shaped stoma and (B) a grasslike stoma. (From Meidner and Mansfield 1968.)

ing signal, triggering a series of responses that result in opening of the stomatal pore.

The early aspects of this process are ion uptake and other metabolic changes in the guard cells, which will be discussed in detail in Chapter 18. Here we will note the effect of decreases in osmotic potential (Ψ_s) resulting from ion uptake and from biosynthesis of organic molecules in the guard cells. Water relations in guard cells follow the same rules as in other cells. As Ψ_s decreases, the water potential decreases and water consequently moves into the guard cells. As water enters the cell, turgor pressure increases and compensates for the lower Ψ_s. Because of the elastic properties of their walls, guard cells can reversibly increase their volume by 40 to 100%, depending on the species. The deformation of the guard cell walls imposed by this increase in volume is a central aspect of stomatal movements.

The Transpiration Ratio Is a Measure of the Relationship between Water Loss and Carbon Gain

The effectiveness of plants in moderating water loss while allowing sufficient CO_2 uptake for photosynthesis can be assessed by a parameter called the **transpiration ratio**. This value is defined as the amount of water transpired by the plant divided by the amount of carbon dioxide assimilated by photosynthesis:

$$\text{Transpiration ratio} = \frac{\text{moles of } H_2O \text{ transpired}}{\text{moles of } CO_2 \text{ fixed}} \quad (4.8)$$

For typical plants in which the first stable product of carbon fixation is a three-carbon compound (called C_3 plants; see Chapter 8), about 500 molecules of water are lost for every molecule of CO_2 fixed by photosynthesis, giving a transpiration ratio of 500.*

The large ratio of H_2O efflux to CO_2 influx depends on two factors. First, the concentration gradient driving water loss is about 50 times larger than that driving CO_2 influx. In large part, this difference is due to the low concentration of CO_2 in air (about 0.03%) and the relatively high concentration of water vapor within the leaf air spaces. Second, CO_2 diffuses about 1.6 times more slowly through air than water does (the CO_2 molecule is larger than H_2O and has a smaller diffusion coefficient). Furthermore, CO_2 has a longer diffusion path, because it must cross the plasma membrane, the cytoplasm, and the chloroplast envelope before it is assimilated in the chloroplast. These membranes add to the resistance of the CO_2 diffusion pathway.

Plants with C_4 photosynthesis (in which a four-carbon compound is the first stable product of photosynthesis; see Chapter 8) generally transpire less water per molecule of CO_2 fixed; a typical transpiration ratio for C_4 plants is about 250. Desert-adapted plants with CAM (crassulacean acid metabolism) photosynthesis, in which CO_2 is initially fixed into four-carbon organic acids at night, have even lower transpiration ratios; values of about 50 are not unusual.

Overview: The Soil–Plant–Atmosphere Continuum

We have seen that movement of water from the soil through the plant to the atmosphere involves different mechanisms of transport. In the soil and the xylem, water moves by bulk flow in response to a pressure gradient. In the vapor phase, water moves primarily by diffusion, at least until it reaches the outside air, where convection (a form of bulk flow) becomes dominant. When water is transported across membranes, the driving force is the water potential difference across the membrane. Such osmotic flow occurs when cells absorb water and when roots transport water from the soil to the xylem. In all of these situations, *water moves toward regions of low water potential or free energy.*

* Sometimes the reciprocal of the transpiration ratio, called the *water use efficiency*, is cited. Plants with a transpiration ratio of 500 have a water use efficiency of 1/500, or 0.002.

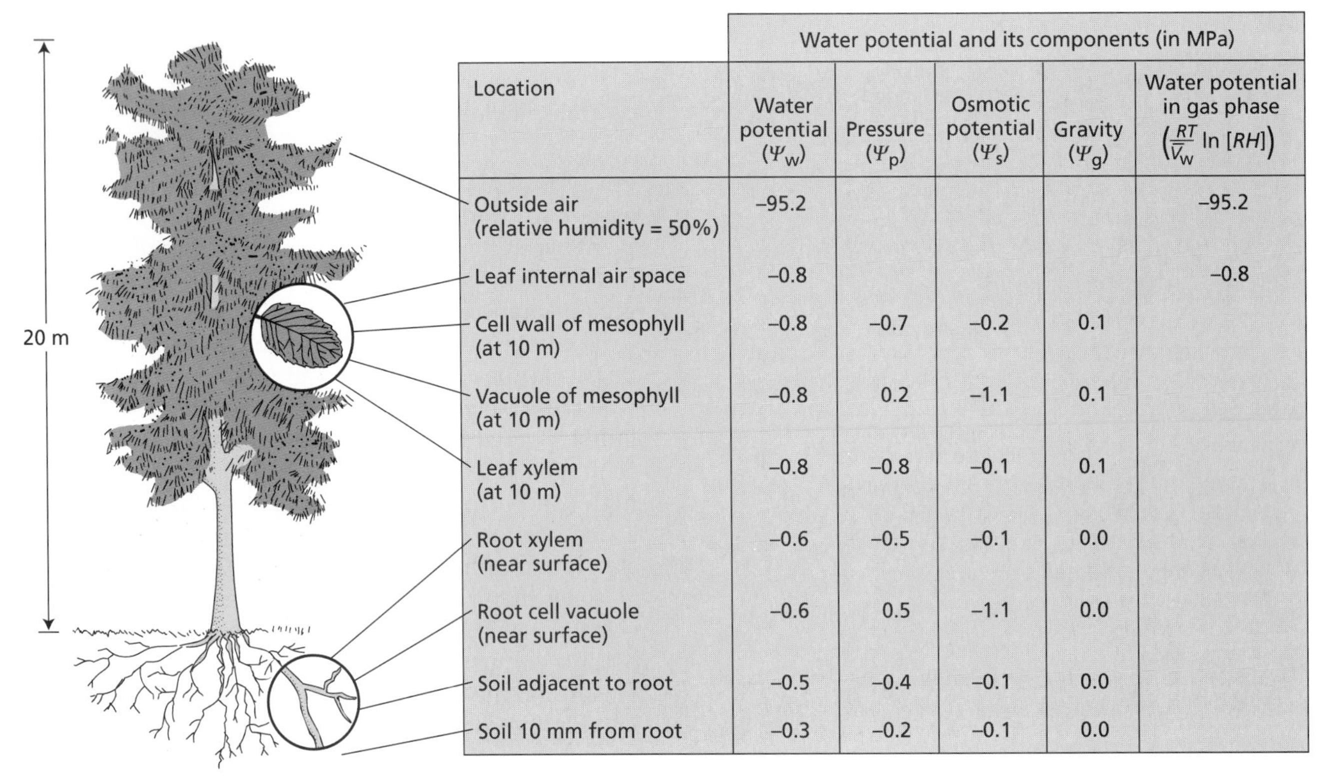

Location	Water potential and its components (in MPa)				
	Water potential (Ψ_w)	Pressure (Ψ_p)	Osmotic potential (Ψ_s)	Gravity (Ψ_g)	Water potential in gas phase $\left(\frac{RT}{\bar{V}_w} \ln [RH]\right)$
Outside air (relative humidity = 50%)	–95.2				–95.2
Leaf internal air space	–0.8				–0.8
Cell wall of mesophyll (at 10 m)	–0.8	–0.7	–0.2	0.1	
Vacuole of mesophyll (at 10 m)	–0.8	0.2	–1.1	0.1	
Leaf xylem (at 10 m)	–0.8	–0.8	–0.1	0.1	
Root xylem (near surface)	–0.6	–0.5	–0.1	0.0	
Root cell vacuole (near surface)	–0.6	0.5	–1.1	0.0	
Soil adjacent to root	–0.5	–0.4	–0.1	0.0	
Soil 10 mm from root	–0.3	–0.2	–0.1	0.0	

Figure 4.17 Representative overview of water potential and its components at various points in the transport pathway from the soil through the plant to the atmosphere. Water potential Ψ_w can be measured through this continuum, but the components vary. In the liquid part of the pathway, pressure (Ψ_p), osmotic potential (Ψ_s), and gravity (Ψ_g), determine Ψ_w. In the air, only the relative humidity ($RT/\bar{V}_w \times \ln[RH]$) is important. Note that although the water potential is the same in the vacuole of the mesophyll cell and in the surrounding cell wall, the components of Ψ_w can differ greatly (e.g., in this example Ψ_p is 0.2 MPa inside the mesophyll cell and –0.7 MPa outside). (After Nobel 1991.)

This phenomenon is illustrated in Figure 4.17, which shows representative values for water potential and its components at various points along the water transport pathway. From the soil to the leaves water potential decreases consistently. In addition, the components of water potential can be quite different at different parts of the pathway. For example, inside the leaf cells, such as in the mesophyll, the water potential is approximately the same as that in the neighboring xylem, yet the components of Ψ_w are quite different. The dominant component of Ψ_w in the xylem is the negative pressure (Ψ_p), whereas in the leaf cell Ψ_p is generally positive. This large difference in Ψ_p occurs across the plasma membrane of the leaf cells. Within the leaf cells, water potential is reduced by a high concentration of dissolved solutes (low Ψ_s). Finally, in the vapor phase we cannot dissect Ψ_w into Ψ_s and Ψ_p; instead, the relative humidity is the significant factor determining Ψ_w.

Summary

Water is the essential medium of life. Land plants are faced with potentially lethal desiccation by water loss to the atmosphere. This problem is aggravated by the large surface area of leaves, their high radiant-energy gain, and their need to have an open pathway for CO_2 uptake. Thus there is a conflict between the need for water conservation and the need for CO_2 assimilation.

The need to resolve this vital conflict determines much of the structure of land plants: (1) an extensive root system to extract water from the soil; (2) a low-resistance pathway through the xylem vessels and tracheids to bring water to the leaves; (3) a hydrophobic cuticle covering the surfaces of the plant to reduce evaporation; (4) microscopic stomata on the leaf surface to allow gas exchange; and (5) guard cells to regulate the diameter (and diffusive resistance) of the stomatal aperture.

The result is an organism that transports water from the soil to the atmosphere purely in response to physical forces; no energy is expended directly by the plant to translocate water, although development and maintenance of the structures needed for efficient and controlled water transport require considerable energy input.

The mechanisms of water transport from the soil through the plant body to the atmosphere include diffusion, bulk flow, and osmosis. Each of these processes is coupled to different driving forces.

Water in the plant can be considered a continuous hydraulic system, connecting the water in the soil with the water vapor in the atmosphere. Transpiration is regulated principally by the guard cells, which regulate the stomatal pore size to meet the photosynthetic demand for CO_2 uptake while minimizing water loss to the atmosphere. Water evaporation from the cell walls of the leaf mesophyll cells generates large negative pressures (or tensions) in the apoplastic water. These negative pressures are transmitted to the xylem and pull water through the long xylem conduits.

Although aspects of the cohesion–tension theory of sap ascent are intermittently debated, an overwhelming body of evidence supports the idea that water transport in the xylem is driven by pressure gradients. When transpiration is high, negative pressures in the xylem water may cause cavitation (embolisms) in the xylem. Such embolisms can block water transport and lead to severe water deficits in the leaf. Water deficits are commonplace in plants, necessitating a host of adaptive responses that modify the physiology and development of plants.

General Reading

Dainty, J. (1976) Water relations of plant cells. In *Transport in Plants* (*Encyclopedia of Plant Physiology*, new series, vol. 2A), U. Luttge and M. Pitman, eds., Springer, Berlin, pp. 12–35.

Milburn, J. A. (1979) *Water Flow in Plants*. Longman, London.

*Nobel, P. S. (1991) *Physicochemical and Environmental Plant Physiology*. Academic Press, San Diego.

Smith, J. A. C., and Griffiths, H. (1993) *Water Deficits. Plant Responses from Cell to Community*. BIOS Scientific, Oxford.

Tyree, M. T., and Jarvis, P. G. (1982) Water in tissues and cells. In *Physiological Plant Ecology II: Water Relations and Carbon Assimilation* (*Encyclopedia of Plant Physiology*, new series, vol. 12B), O. L. Lange, P. S. Nobel, C. B. Osmond, and H. Ziegler, eds., Springer, Berlin, pp. 35–77.

Weatherly, P. E. (1982) Water uptake and flow in roots. In *Physiological Plant Ecology II: Water Relations and Carbon Assimilation* (*Encyclopedia of Plant Physiology*, new series, vol. 12B), O. L. Lange, P. S. Nobel, C. B. Osmond, and H. Ziegler, eds., Springer, Berlin, pp. 79–109.

Zeiger, E. (1983) The biology of stomatal guard cells. *Annu. Rev. Plant Physiol.* 34: 441–475.

Zeiger, E., Farquhar, G., and Cowan, I., eds. (1987) *Stomatal Function*. Stanford University Press, Stanford, CA.

Zimmermann, M. H. (1983) *Xylem Structure and the Ascent of Sap*. Springer, Berlin.

*Ziegler, H. (1987) The evolution of stomata. In *Stomatal Function*, E. Zeiger, G. Farquhar, and I. Cowan, eds., Stanford University Press, Stanford, CA, pp. 29–57.

* Indicates a reference that is general reading in the field and is also cited in this chapter.

Chapter References

Bange, G. G. J. (1953) On the quantitative explanation of stomatal transpiration. *Acta Botanica Neerlandica* 2: 255–296.

Canny, M. J. (1998) Transporting water in plants. *Am. Sci.* 86: 152–159.

Frensch J., Hsiao T. C., and Steudle E. (1996) Water and solute transport along developing maize roots. *Planta* 198: 348–355.

Holbrook, N. M., Burns, M. J., and Field, C. B. (1995) Negative xylem pressures in plants: A test of the balancing pressure technique. *Science* 270: 1193–1194.

Jarvis, P., and Mansfield, T. (1981) *Stomatal Physiology*. Cambridge University Press, Cambridge.

Kramer, P. J., and Boyer, J. S. (1995) *Water Relations of Plants and Soils*. Academic Press, San Diego.

Meidner, H., and Mansfield, D. (1968) *Stomatal Physiology*. McGraw-Hill, London.

Palevitz, B. A. (1981) The structure and development of guard cells. In *Stomatal Physiology*, P. G. Jarvis and T. A. Mansfield, eds., Cambridge University Press, Cambridge, pp. 1–23.

Passioura, J. B. (1980) The meaning of matric potential. *J. Exp. Bot.* 31: 1161–1169.

Pockman, W. T., Sperry, J. S., and O'Leary, J. W. (1995) Sustained and significant negative pressure in xylem. *Nature* 378: 715–716.

Sack, F. D. (1987) The development and structure of stomata. In *Stomatal Function*, E. Zeiger, G. Farquhar, and I. Cowan, eds., Stanford University Press, Stanford, CA, pp. 59–89.

Sharpe, P. J. H., Wu, H., and Spence, R. D. (1987) Stomatal mechanics. In *Stomatal Function*, E. Zeiger, G. Farquhar, and I. Cowan, eds., Stanford University Press, Stanford, CA, pp. 91–114.

Steudle, E., and Frensch, J. (1996) Water transport in plants: Role of the apoplast. *Plant Soil* 187: 67–79.

Tyree, M. T. (1997) The cohesion-tension theory of sap ascent: Current controversies. *J. Exp. Bot.* 48: 1753–1765.

Tyree, M. T., and Sperry, J. S. (1989) Vulnerability of xylem to cavitation and embolism. *Annu. Rev. Plant Physiol.* 40: 19–38.

Zeiger, E., and Hepler, P. K. (1976) Production of guard cell protoplasts from onion and tobacco. *Plant Physiol.* 58: 492–498.

5 Mineral Nutrition

MINERAL ELEMENTS, those acquired primarily in the form of inorganic ions, continually cycle through all organisms and their environment. Mineral elements enter the biosphere predominantly through the root systems of plants, so in a sense plants act as the "miners" of Earth's crust (Epstein 1972, 1994). The large surface area of roots and their ability to absorb inorganic ions at low concentrations in the soil solution make mineral absorption by plants a very effective process. After being absorbed at the roots, the mineral elements are translocated to the various parts of the plant where they are utilized in important biological functions. Other organisms, such as mycorrhizal fungi and nitrogen-fixing bacteria, often participate with roots in nutrient acquisition.

The study of how plants absorb and assimilate inorganic ions is called **mineral nutrition**. This area of research is central to modern agriculture and environmental protection. High agricultural yields depend strongly on fertilization with mineral elements. In fact, yields of most crop plants increase linearly with the amount of fertilizer that they absorb (Loomis and Conner 1992). To meet increased demand for food, world consumption of the primary fertilizer mineral elements—nitrogen, phosphorus, and potassium—rose steadily from 112 million metric tons in 1980 to 143 million metric tons in 1990 (Lauriente 1995). Crop plants, however, typically use less than half of the fertilizer applied (Loomis and Conner 1992). The remaining minerals may leach into groundwater, become fixed in the soil, or contribute to air pollution.

As a consequence of fertilizer leaching, many water wells in the United States no longer meet federal standards for nitrate concentrations in drinking water (U.S. Congress. Office of Technology Assessment 1990). Similarly, increased atmospheric deposition of nitrogen, derived in part from fertilizer volatilization, has been linked to the rapid decline of European forests (Schulze 1989). On a brighter note, plants are the traditional means for recycling animal wastes and are proving useful for removing deleterious minerals from toxic-waste dumps (Erickson et al. 1994). Because of the complex nature of plant–soil–atmosphere relationships, studies in the area of mineral nutrition involve atmospheric chemists, soil scientists, hydrologists, microbiologists, and ecologists, as well as plant physiologists.

In this chapter we will discuss first the nutritional needs of plants, the symptoms of specific nutritional deficiencies, and the use of fertilizers to ensure proper plant nutrition. Then we will examine how soil and root structure influence the transfer of inorganic nutrients from the environment into a plant. Finally, we will introduce the topic of mycorrhizal associations. Chapters 6 and 12 address additional aspects of solute transport and nutrient assimilation, respectively.

Essential Nutrients, Deficiencies, and Plant Disorders

Only certain elements have been determined to be essential for plant growth. An **essential element** is defined as one that has a clear physiological role and whose absence prevents a plant from completing its life cycle (Arnon and Stout 1939). If plants are given these essential elements, as well as energy from sunlight, they can synthesize all the compounds they need for normal growth. Table 5.1 lists the elements that are considered to be essential for most, if not all, higher plants. The first three elements—hydrogen, carbon, and oxygen—are not considered mineral elements, because they are obtained primarily from water or carbon dioxide.

Mineral essential elements are usually classified as macronutrients or micronutrients, according to their relative concentration in plant tissue. In some cases, the differences in tissue content of macronutrients and micronutrients are not as great as those indicated in Table 5.1. For example, some plant tissues, such as the leaf mesophyll, have almost as much iron or manganese as they do sulfur or magnesium. Many elements may be present in concentrations greater than the plant's minimum requirements.

Some researchers have argued that a classification into macronutrients and micronutrients is difficult to justify physiologically. Mengel and Kirkby (1987) have proposed that the essential elements be classified instead according to their biochemical role and physi-

Table 5.1
Adequate tissue levels of elements that may be required by plants

Element	Chemical symbol	Atomic weight	Concentration in dry matter		Relative number of atoms with respect to molybdenum
			$\mu mol\ g^{-1}$	ppm or %[a]	
OBTAINED FROM WATER OR CARBON DIOXIDE					
Hydrogen	H	1.01	60,000	6	60,000,000
Carbon	C	12.01	40,000	45	40,000,000
Oxygen	O	16.00	30,000	45	30,000,000
OBTAINED FROM THE SOIL					
Macronutrients					
Nitrogen	N	14.01	1,000	1.5	1,000,000
Potassium	K	39.10	250	1.0	250,000
Calcium	Ca	40.08	125	0.5	125,000
Magnesium	Mg	24.32	80	0.2	80,000
Phosphorus	P	30.98	60	0.2	60,000
Sulfur	S	32.07	30	0.1	30,000
Silicon	Si	28.09	30	0.1	30,000
Micronutrients					
Chlorine	Cl	35.46	3.0	100	3,000
Iron	Fe	55.85	2.0	100	2,000
Boron	B	10.82	2.0	20	2,000
Manganese	Mn	54.94	1.0	50	1,000
Sodium	Na	22.91	0.40	10	400
Zinc	Zn	65.38	0.30	20	300
Copper	Cu	63.54	0.10	6	100
Nickel	Ni	58.69	0.002	0.1	2
Molybdenum	Mo	95.95	0.001	0.1	1

Source: Epstein 1972, 1994.

[a] The values for the nonmineral elements (H, C, O) and the macronutrients are percentages. The values for micronutrients are expressed in parts per million.

ological function. Table 5.2 shows such a classification, in which plant nutrients have been divided into four basic groups. The first group includes the elements that form the organic compounds of the plant. Plants assimilate these nutrients via biochemical reactions involving oxidation and reduction. Elements of the second group are important in energy transfer reactions or in maintaining structural integrity. They are often present in plant tissues as phosphate, borate, and silicate esters in which the elemental group is bound to the hydroxyl group of an organic molecule (i.e., sugar–phosphate). Elements of the third group are present in plant tissue as either free ions or ions bound to substances such as the pectic acids present in the plant cell wall. Of particular importance are their roles as enzyme cofactors and in the regulation of osmotic potentials. Members of the fourth group of essential elements have important roles in reactions involving electron transfer.

Elements other than those listed in Table 5.1 can also accumulate in plant tissues. Aluminum is not considered to be an essential element, but plants commonly contain from 0.1 to 500 ppm aluminum, and addition of low levels of aluminum to a nutrient solution may stimulate plant growth (Marschner 1995). Many species in the genera *Astragalus*, *Xylorrhiza*, and *Stanleya* accumulate selenium, although it is not known whether they have a specific requirement for this element (Läuchli 1993). Cobalt is part of cobalamin (vitamin B_{12} and its derivatives), a component of several enzymes in nitrogen-fixing microorganisms. Thus cobalt deficiency

Table 5.2
Classification of plant mineral nutrients according to biochemical function

Nutrient element	Functions
Group 1	**Nutrients that form the organic compounds of plants**
N	Constituent of amino acids, amides, proteins, nucleic acids, nucleotides, coenzymes, hexoamines, etc.
S	Component of cysteine, cystine, and methionine, and proteins. Constituent of lipoic acid, coenzyme A, thiamine pyrophosphate, glutathione, biotin, adenosine-5′-phosphosulfate, and 3-phosphoadenosine.
Group 2	**Nutrients that are important in energy storage or structural integrity**
P	Component of sugar phosphates, nucleic acids, nucleotides, coenzymes, phospholipids, phytic acid, etc. Has a key role in reactions in which ATP is involved.
B	Complexes with mannitol, mannan, polymannuronic acid, and other constituents of cell walls. Involved in cell elongation and nucleic acid metabolism.
Si	Deposited as amorphous silica in cell walls. Contributes to cell wall mechanical properties, including rigidity and elasticity.
Group 3	**Nutrients that remain in ionic form**
K	Required as a cofactor for more than 40 enzymes. Principal cation in establishing cell turgor and maintaining cell electroneutrality.
Na	Involved with the regeneration of phosphoenolpyruvate in C_4 and CAM plants. Substitutes for potassium in some functions.
Mg	Required by many enzymes involved in phosphate transfer. Constituent of the chlorophyll molecule.
Ca	Constituent of the middle lamella of cell walls. Required as a cofactor by some enzymes involved in the hydrolysis of ATP and phospholipids. Acts as a second messenger in metabolic regulation.
Mn	Required for activity of some dehydrogenases, decarboxylases, kinases, oxidases, peroxidases. Involved with other cation-activated enzymes and photosynthetic O_2 evolution.
Cl	Required for the photosynthetic reactions involved in O_2 evolution.
Group 4	**Nutrients that are involved in electron transfers**
Fe	Constituent of cytochromes and nonheme iron proteins involved in photosynthesis, N_2 fixation, and respiration.
Cu	Component of ascorbic acid oxidase, tyrosinase, monoamine oxidase, uricase, cytochrome oxidase, phenolase, laccase, and plastocyanin.
Zn	Constituent of alcohol dehydrogenase, glutamic dehydrogenase, carbonic anhydrase, etc.
Mo	Constituent of nitrogenase, nitrate reductase, and xanthine dehydrogenase.
Ni	Constituent of urease. In N_2-fixing bacteria, constituent of hydrogenases.

Source: After Evans and Sorget 1966 and Mengel and Kirkby 1987.

blocks the development and function of nitrogen-fixing nodules. Nonetheless, plants that do not fix nitrogen, as well as nitrogen-fixing plants that are supplied with ammonium or nitrate, do not require cobalt. Plants normally contain only small amounts of other elements.

Special Techniques Are Used in Nutritional Studies

To demonstrate that an element is essential requires that plants be grown under experimental conditions in which only the element under investigation is absent. Such conditions are extremely difficult to achieve with plants grown in a complex medium such as soil. In the nineteenth century, several researchers, including De Saussure, Sachs, Boussingault, and Knop, approached this problem by growing plants with their roots immersed in a **nutrient solution** containing only inorganic salts. Their demonstration that plants could grow normally with no soil or organic matter proved unequivocally that plants can fulfill all their needs from only inorganic elements and sunlight.

The technique of growing plants with their roots immersed in nutrient solution without soil is called solution culture or **hydroponics** (Figure 5.1A). Successful hydroponic culture requires large solution volumes or frequent adjustment of nutrient solutions to prevent nutrient uptake by roots from producing radical changes in nutrient concentrations and pH of the medium. A sufficient supply of oxygen to the root sys-

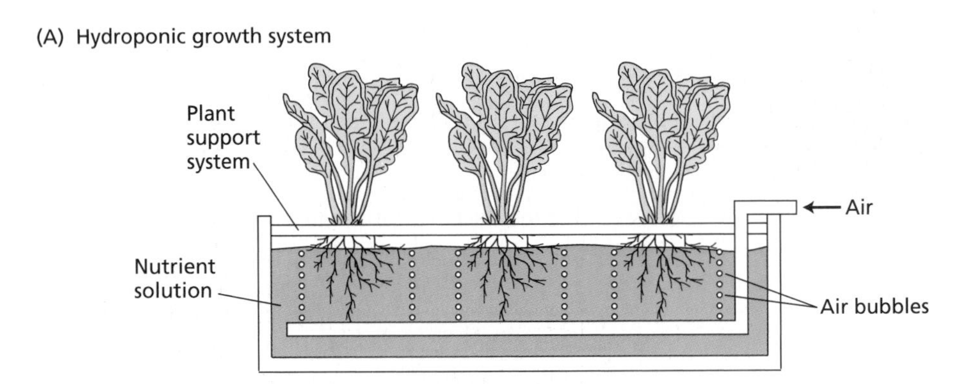

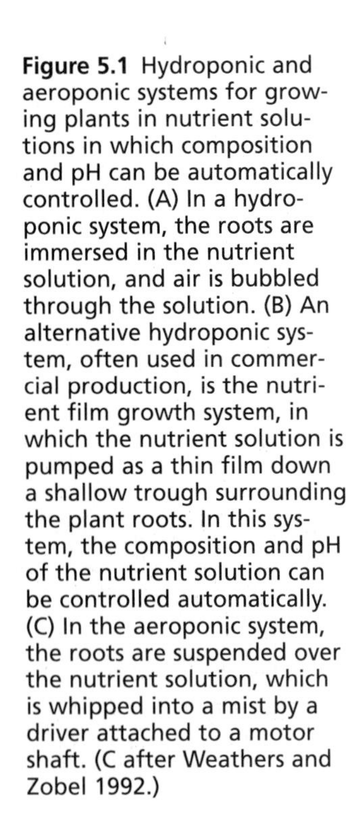

Figure 5.1 Hydroponic and aeroponic systems for growing plants in nutrient solutions in which composition and pH can be automatically controlled. (A) In a hydroponic system, the roots are immersed in the nutrient solution, and air is bubbled through the solution. (B) An alternative hydroponic system, often used in commercial production, is the nutrient film growth system, in which the nutrient solution is pumped as a thin film down a shallow trough surrounding the plant roots. In this system, the composition and pH of the nutrient solution can be controlled automatically. (C) In the aeroponic system, the roots are suspended over the nutrient solution, which is whipped into a mist by a driver attached to a motor shaft. (C after Weathers and Zobel 1992.)

tem—also critical—may be achieved by vigorous bubbling of air through the medium. Hydroponics is used in the commercial production of many greenhouse crops. In one form of commercial hydroponic culture, plants are grown in a supporting material such as sand, gravel, vermiculite, or expanded clay (i.e., kitty litter). Nutrient solutions are then flushed through the supporting material, and old solutions are removed by leaching. In another form of hydroponic culture, plant roots lie on the surface of a trough, and nutrient solutions flow in a thin layer along the trough over the roots (Asher and Edwards 1983). This nutrient film growth system ensures that the roots receive an ample supply of oxygen (Figure 5.1B).

Another alternative is to grow the plants **aeroponically** (Weathers and Zobel 1992). In this technique, plants are grown with their roots suspended in air while being sprayed continuously with a nutrient solution (Figure 5.1C). This approach provides easy manipulation of the gaseous environment around the root, but it requires higher levels of nutrients than hydroponic culture does to sustain rapid plant growth. For this reason and other technical difficulties, the use of aeroponics is not widespread.

Nutrient Solutions Can Sustain Rapid Plant Growth

Over the years, many formulations have been used for nutrient solutions. Early formulations developed by Wilhelm Knop in Germany included only KNO_3, $Ca(NO_3)_2$, KH_2PO_4, $MgSO_4$, and an iron salt. At the time, this nutrient solution was believed to contain all the minerals required by the plant, but these experiments were carried out with chemicals that were contaminated with other elements that are now known to be essential (such as boron or molybdenum). Table 5.3 shows a more modern formulation for a nutrient solution. This formulation is called a modified **Hoagland**

Table 5.3
Composition of a modified Hoagland nutrient solution for growing plants

Compound	Molecular weight	Concentration of stock solution	Concentration of stock solution	Volume of stock solution per liter of final solution	Element	Final concentration of element	
	$g\ mol^{-1}$	m*M*	$g\ L^{-1}$	mL		μ*M*	ppm
Macronutrients							
KNO_3	101.10	1,000	101.10	6.0	N	16,000	224
$Ca(NO_3)_2 \cdot 4H_2O$	236.16	1,000	236.16	4.0	K	6,000	235
$NH_4H_2PO_4$	115.08	1,000	115.08	2.0	Ca	4,000	160
$MgSO_4 \cdot 7H_2O$	246.48	1,000	246.49	1.0	P	2,000	62
					S	1,000	32
					Mg	1,000	24
Micronutrients							
KCl	74.55	25	1.864	2.0 (combined, KCl through H_2MoO_4)	Cl	50	1.77
H_3BO_3	61.83	12.5	0.773		B	25	0.27
$MnSO_4 \cdot H_2O$	169.01	1.0	0.169		Mn	2.0	0.11
$ZnSO_4 \cdot 7H_2O$	287.54	1.0	0.288		Zn	2.0	0.13
$CuSO_4 \cdot 5H_2O$	249.68	0.25	0.062		Cu	0.5	0.03
H_2MoO_4 (85% MoO_3)	161.97	0.25	0.040		Mo	0.5	0.05
NaFeDTPA (10% Fe)	558.50	53.7	30.0	0.3–1.0	Fe	16.1–53.7	1.00–3.00
Optional[a]							
$NiSO_4 \cdot 6H_2O$	262.86	0.25	0.066	2.0	Ni	0.5	0.03
$Na_2SiO_3 \cdot 9H_2O$	284.20	1,000	284.20	1.0	Si	1,000	28

Source: After Epstein 1972.

Note: The macronutrients are added separately from stock solutions to prevent precipitation during the preparation of the nutrient solution. A combined stock solution is made up containing all micronutrients except iron. Iron is added as sodium ferric diethylenetriaminepentaacetate (NaFeDTPA, trade name Ciba-Geigy Sequestrene 330 Fe; see Figure 5.2); some plants, such as maize, require the higher level of iron shown in the table.

[a] Nickel is usually present as a contaminant of the other chemicals, so it may not need to be added explicitly. Silicon, if included, should be added first and the pH adjusted with HCl to prevent precipitation of the other nutrients.

solution, named after Dennis R. Hoagland, a researcher who was prominent in the development of modern mineral nutrition research in the United States.

A modified Hoagland solution contains all the mineral elements needed for rapid plant growth. The concentrations of these elements are set at the highest possible levels without producing toxicity symptoms or salinity stress and thus may be several orders of magnitude higher than those found in the soil around plant roots. For example, phosphorus is present in the soil solution at concentrations normally less than 0.06 ppm, whereas here it is offered at 62 ppm (Epstein 1972). Such high initial levels permit plants to be grown in a medium for extended periods without replenishment of the nutrients. Many researchers, however, dilute their nutrient solutions severalfold and replenish them frequently in order to minimize fluctuations of nutrient concentration in the medium and in plant tissue.

Another important property of the Hoagland formulation is that nitrogen is supplied as both ammonium (NH_4^+) and nitrate (NO_3^-). Supplying nitrogen in a balanced mixture of cations and anions tends to reduce the rapid rise in medium pH that is commonly observed when the nitrogen is supplied solely as nitrate anion (Asher and Edwards 1983). Even when the pH of the medium is kept neutral, most plants grow better if they have access to both NH_4^+ and NO_3^- because absorption and assimilation of the two nitrogen forms promotes cation–anion balance within the plant (Raven and Smith 1976; Bloom 1994).

A significant problem with nutrient solutions is maintaining the availability of iron. When supplied as an inorganic salt such as $FeSO_4$ or $Fe(NO_3)_2$, iron can precipitate out of solution as iron hydroxide. If phosphate salts are present, insoluble iron phosphate will also form. Precipitation of the iron out of solution makes it physically unavailable to the plant, unless iron salts are added at frequent intervals. Earlier researchers approached this problem by adding iron together with citric acid or tartaric acid. Compounds such as citric acid or tartaric acid are called **chelating agents** because they form soluble complexes with cations such as iron and calcium. More modern nutrient solutions use the chemicals ethylenediaminetetraacetic acid (EDTA) or diethylenetriaminepentaacetic acid (DTPA, or pentetic acid) as chelating agents (Sievers and Bailar 1962). Figure 5.2 shows the structure of DTPA. The fate of the chelation complex during iron uptake by the root cells is not clear; iron may be released from the chelator when it is reduced from Fe^{3+} to Fe^{2+} at the root surface. The chelator may then diffuse back into the nutrient (or soil) solution and react with another Fe^{3+} ion. After uptake, iron is kept soluble by chelation with organic compounds present in plant cells. Citric acid may play a major role in iron chelation and its long-distance transport in the xylem (Olsen et al. 1981).

Mineral Deficiencies Disrupt Plant Metabolism and Function

Inadequate supply of an essential element results in a nutritional disorder manifested by characteristic deficiency symptoms. In hydroponic culture, withholding of an essential element can be readily correlated with a given symptom. Diagnosis of soil-grown plants may be more complex, because several elements may be deficient at once, excessive amounts of one element may induce deficiencies of another, and some virus-induced plant diseases may produce symptoms similar to those of nutrient deficiencies.

Nutrient deficiency symptoms in a plant are the expression of metabolic disorders resulting from the insufficient supply of an essential element. These disorders are related to the roles played by essential elements in normal plant metabolism and function. Table 5.2 lists some of the roles of essential elements. Even though each essential element participates in many different metabolic reactions, some general statements about the functions of essential elements in plant metabolism are possible. In general, the essential elements function (1) as constituents of compounds, (2) in the activation of enzymes, and (3) in the osmoregulation of plant cells. Other roles may be related to the ability of divalent cations such as calcium or magnesium to modify the permeability of plant membranes. In addition, research continues to reveal specific

(A)

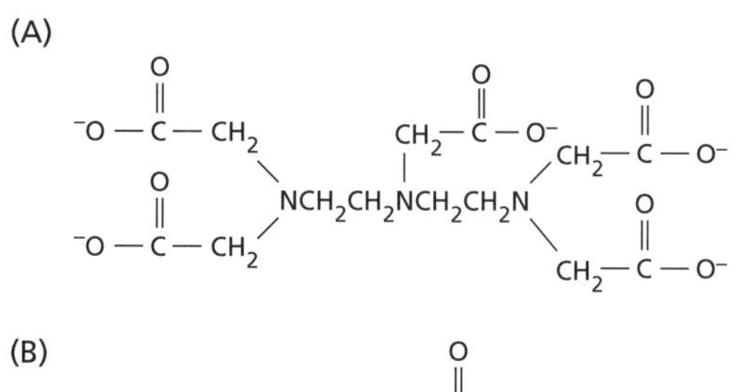

(B)

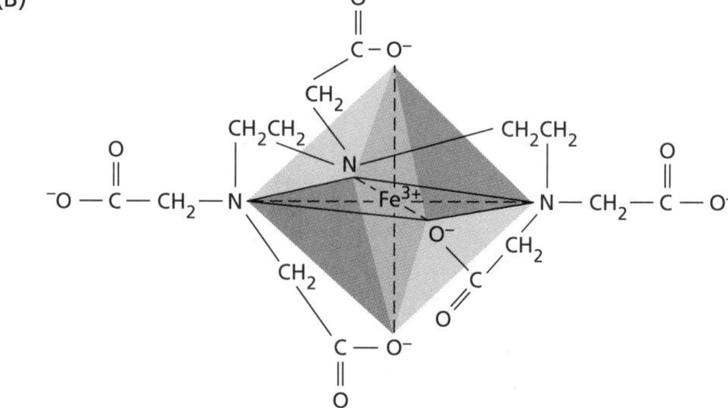

Figure 5.2 Chemical structure of the chelator DTPA by itself (A) and chelated to an Fe^{3+} ion (B). Iron binds to DTPA through interaction with three nitrogen atoms and the three ionized oxygen atoms of the carboxylate groups (Sievers and Bailar 1962). The resulting ring structure clamps the metallic ion and effectively neutralizes its reactivity in solution. During the uptake of iron at the root surface, Fe^{3+} appears to be reduced to Fe^{2+}, which is released from the DTPA–iron complex. The chelator can then bind to other available Fe^{3+} ions.

roles of these elements in plant metabolism; for example, calcium regulates key enzymes in the cytosol (Hepler and Wayne 1985). Thus, most essential elements have multiple roles in plant metabolism.

When relating deficiency symptoms to the role of an essential element, it is important to consider the extent to which an element can be recycled from older to younger leaves. Some elements, such as nitrogen, phosphorus, and potassium, can readily move from leaf to leaf; others, such as boron, iron, and calcium, are relatively immobile in most plant species (Table 5.4). If an essential element is mobile, deficiency symptoms will appear first in older leaves. Deficiency of an immobile essential element will become evident first in younger leaves. Although the precise mechanisms of nutrient mobilization are not well understood, plant hormones such as cytokinins appear to be involved (see Chapter 21). In the discussion that follows, we will describe the specific deficiency symptoms and functional roles for the mineral essential elements in the order of their general prevalence in plant tissues (see Table 5.1).

Nitrogen. Nitrogen is the mineral element that plants require in greatest amounts. It serves as a constituent of many plant cell components, including amino acids and nucleic acids. Therefore, nitrogen deficiency rapidly inhibits plant growth. If such a deficiency persists, most species show **chlorosis** (yellowing of the leaves), especially in the older leaves near the base of the plant. Under severe nitrogen deficiency, these leaves become completely yellow (or tan) and fall off the plant. Younger leaves may not show these symptoms initially, because nitrogen can be mobilized from older leaves. Thus, a nitrogen-deficient plant may have light green upper leaves and yellow or tan lower leaves. When nitrogen deficiency develops slowly, plants may have markedly slender and often woody stems. This woodiness may be due to a buildup of excess carbohydrates that cannot be used in the synthesis of amino acids or other nitrogen compounds. Carbohydrates not used in nitrogen metabolism may also be used in anthocyanin synthesis, leading to accumulation of the pigment. This condition is revealed as a purple coloration in leaves, petioles, and stems of some nitrogen-deficient plants, such as tomato and certain varieties of corn.

Table 5.4
Mineral elements classified on the basis of their mobility within a plant and their tendency to retranslocate during deficiencies

Mobile	Immobile
Nitrogen	Calcium
Potassium	Sulfur
Magnesium	Iron
Phosphorus	Boron
Chlorine	Copper
Sodium	
Zinc	
Molybdenum	

Note: Elements are listed in the order of their abundance in the plant.

Potassium. Potassium, present within plants as the cation K^+, plays an important role in regulation of the osmotic potential of plant cells (see Chapters 3 and 6). It also activates many enzymes involved in respiration and photosynthesis. The first observable symptom of potassium deficiency is mottled or marginal chlorosis, which then develops into necrosis primarily at the leaf tips, at the margins, and between veins. In many monocots, these necrotic lesions may initially form at the leaf tips and margins and then extend toward the leaf base. Because potassium can be mobilized to the younger leaves, these symptoms appear initially on the more mature leaves toward the base of the plant. The leaves may also curl and crinkle. The stems of potassium-deficient plants may be slender and weak, with abnormally short internodal regions. In potassium-deficient corn, the roots may have an increased susceptibility to root-rotting fungi present in the soil, and this susceptibility, together with effects on the stem, results in an increased tendency for the plant to be easily bent to the ground (lodging).

Calcium. Calcium ions (Ca^{2+}) are used in the synthesis of new cell walls, particularly the middle lamellae that separate newly divided cells. Calcium is also used in the mitotic spindle during cell division. It is required for the normal functioning of plant membranes and has been implicated as a second messenger for various plant responses to both environmental and hormonal signals (Hepler and Wayne 1985). In its function as a second messenger, calcium may bind to **calmodulin**, a protein found in the cytosol of plant cells. The calmodulin–calcium complex regulates many metabolic processes. Characteristic symptoms of calcium deficiency include necrosis of young meristematic regions, such as the tips of roots or young leaves, where cell division and wall formation are most rapid. Necrosis in slowly growing plants may be preceded by a general chlorosis and downward hooking of the young leaves. Young leaves may also appear deformed. The root system of a calcium-deficient plant may appear brownish, short, and highly branched. Severe stunting may result if the meristematic regions of the plant die prematurely.

Magnesium. In plant cells, magnesium ions (Mg^{2+}) have a specific role in the activation of enzymes involved in respiration, photosynthesis, and the synthesis of DNA and RNA. Magnesium is also a part of the ring structure of the chlorophyll molecule (see Figure 7.5A). A characteristic symptom of magnesium deficiency is chlorosis between the leaf veins, occurring first in the older leaves because of the mobility of this element. This pattern of chlorosis results because chlorophyll in the vascular bundles remains unaffected for longer periods than the chlorophyll in the cells between the bundles does. If the deficiency is extensive, the leaves may become yellow or white. An additional symptom of magnesium deficiency may be premature leaf abscission.

Phosphorus. Phosphorus (as phosphate, PO_4^{3-}) is an integral component of important compounds of plant cells, including the sugar–phosphate intermediates of respiration and photosynthesis, and the phospholipids that make up plant membranes. It is also a component of nucleotides used in plant energy metabolism and in DNA and RNA. Characteristic symptoms of phosphorus deficiency include stunted growth in young plants and a dark green coloration of the leaves, which may be malformed and contain small spots of dead tissue called **necrotic spots**. As in nitrogen deficiency, some species may produce excess anthocyanins, giving the leaves a slight purple coloration. In contrast to nitrogen deficiency, this purple coloration is not associated with chlorosis. In fact, the leaves may be a dark greenish purple. Additional symptoms of phosphorus deficiency include the production of slender (but not woody) stems and the death of older leaves. Maturation of the plant may also be delayed.

Sulfur. Many of the symptoms of sulfur deficiency are similar to those of nitrogen deficiency, including chlorosis, stunting of growth, and anthocyanin accumulation. This similarity is not surprising, since sulfur and nitrogen are both constituents of proteins. However, the chlorosis caused by sulfur deficiency generally arises initially in young leaves, rather than on the more mature leaves as in nitrogen deficiency, because unlike nitrogen, sulfur is not easily remobilized to the younger leaves in most species. Nonetheless, in many plant species chlorosis may occur simultaneously in all leaves or even initially in the older leaves.

Silicon. Only members of the family Equisetaceae—called "scouring rushes" because at one time their ash, rich in gritty silica, was used to scour pots—require silicon to complete their life cycle. Nonetheless, many other species accumulate substantial amounts of silicon within their tissues and show enhanced growth and fertility when supplied with adequate amounts of silicon. Plants deficient in silicon are more susceptible to lodging (falling over) and fungal infection. Silicon is deposited primarily in the endoplasmic reticulum, cell walls, and intercellular spaces as hydrated, amorphous silica ($SiO_2 \bullet nH_2O$). It also forms complexes with polyphenols and thus serves as an alternative to lignin in the reinforcement of cell walls.

Chlorine. The element chlorine is found in plants as the chloride ion (Cl^-). It is required for the water-splitting reaction of photosynthesis (see Chapter 7) in which oxygen is produced (Haehnel 1984). In addition, chlorine may be required for cell division in both leaves and roots (Terry 1977). Plants deficient in chlorine develop wilting of the leaf tips followed by general leaf chlorosis and necrosis. The leaves may also exhibit reduced growth. Eventually, the leaves may take on a bronzelike color ("bronzing"). Roots of chlorine-deficient plants may appear stunted and thickened near the root tips. Chloride ions are very soluble and generally available in soils because seawater is swept into the air by wind and is delivered to soil when it rains. Therefore, chlorine deficiency is unknown in plants grown in native or agricultural habitats. Most plants generally absorb chlorine at levels much higher than those required for normal functioning.

Iron. Iron has an important role as a component of enzymes involved in the transfer of electrons (redox reactions), such as cytochromes. In this role, it is reversibly oxidized from Fe^{2+} to Fe^{3+} during electron transfer. As in magnesium deficiency, characteristic symptoms of iron deficiency involve intervenous chlorosis. In contrast to magnesium deficiency symptoms, these symptoms appear initially on the younger leaves because iron cannot be readily mobilized from older leaves. Under conditions of extreme or prolonged deficiency, the veins may also become chlorotic, causing the whole leaf to become white. The leaves become chlorotic because iron is required for the synthesis of some of the chlorophyll-protein complexes in the chloroplast. The low mobility of iron is probably due to the precipitation in the older leaves as insoluble oxides or phosphates or to the formation of complexes with phytoferritin, an iron-binding protein found in the leaf (Bienfait and Van der Mark 1983). The precipitation of iron reduces subsequent mobilization of the metal to the phloem for long-distance translocation.

Boron. Although the precise function of boron in plant metabolism is unclear, evidence suggests that it plays roles in cell elongation, nucleic acid synthesis, hormone responses, and membrane function (Shelp 1993). Boron-deficient plants may exhibit a wide variety of symptoms, depending on the species and the age of the plant. A characteristic symptom is black necrosis of the young leaves and terminal buds. The necrosis of the young leaves occurs primarily at the base of the leaf blade. Stems may be unusually stiff and brittle. Apical dominance may also be lost, causing the plant to become highly branched; however, the terminal apices of the branches soon become necrotic because of inhibition of cell division. Structures such as the fruit, fleshy roots, and tubers may exhibit necrosis or abnormalities related to the breakdown of internal tissues.

Manganese. Manganese ions (Mn^{2+}) activate several enzymes in plant cells. In particular, decarboxylases and dehydrogenases involved in the tricarboxylic acid (Krebs) cycle are specifically activated by manganese. The best-defined function of manganese is in the photosynthetic reaction in which oxygen is produced from water (Marschner 1995). The major symptom of manganese deficiency is intervenous chlorosis associated with the development of small necrotic spots. This chlorosis may occur on younger or older leaves, depending on plant species and growth rate.

Sodium. Most species utilizing the C_4 and CAM pathways of carbon fixation (see Chapter 8) require sodium ions (Na^+). In these plants, sodium appears vital for regenerating phosphoenolpyruvate, the substrate for the first carboxylation in the C_4 and CAM pathways (Johnstone et al. 1988). Under sodium deficiency, these plants exhibit chlorosis and necrosis, or even fail to form flowers. In addition, many C_3 species can benefit from exposure to sodium ions. Sodium stimulates growth through enhanced cell expansion, and it can partly substitute for potassium as an osmotically active solute.

Zinc. Many enzymes require zinc ions (Zn^{2+}) for their activity, and zinc may be required for chlorophyll biosynthesis in some plants. Zinc deficiency is characterized by a reduction in internodal growth, and as a result plants display a rosette habit of growth in which the leaves form a circular cluster radiating at or close to the ground. The leaves may also be small and distorted, with leaf margins having a puckered appearance. These symptoms may result from loss of the capacity to produce sufficient amounts of the auxin, indoleacetic acid. In some species (corn, sorghum, beans), the older leaves may become intervenously chlorotic and then develop white necrotic spots. This chlorosis may be an expression of a zinc requirement for chlorophyll biosynthesis.

Copper. Like iron, copper is associated with enzymes involved in redox reactions ($Cu^{2+}e^- \longleftrightarrow Cu^+$). An example of such an enzyme is plastocyanin, which is involved in electron transfer during the light reactions of photosynthesis (Haehnel 1984). The initial symptom of copper deficiency is the production of dark green leaves, which may contain necrotic spots. The necrotic spots appear first at the tips of the young leaves and then extend toward the leaf base along the margins. The leaves may also be twisted or malformed. Under extreme copper deficiency, leaves may abscise prematurely.

Nickel. Urease is the only known nickel-containing enzyme in higher plants, although nitrogen-fixing microorganisms require nickel for the enzyme that reprocesses some of the hydrogen gas generated during fixation (hydrogen uptake hydrogenase) (see Chapter 12). Nickel-deficient plants accumulate urea in their leaves and, consequently, show leaf tip necrosis. Plants grown in soil seldom, if ever, show signs of nickel deficiency, because the amounts of nickel required are so minuscule.

Molybdenum. Molybdenum ions (Mo^{4+} through Mo^{6+}) are components of several enzymes, including nitrate reductase and nitrogenase. Nitrate reductase catalyzes the reduction of nitrate to nitrite during its assimilation by the plant cell; nitrogenase converts nitrogen gas to ammonia in nitrogen-fixing microorganisms (see Chapter 12). The first indication of a molybdenum deficiency is general chlorosis between veins and necrosis of the older leaves. In some plants, such as cauliflower or broccoli, the leaves may not become necrotic but instead may appear twisted and subsequently die (whiptail disease). Flower formation may be prevented, or the flowers may abscise prematurely. Because molybdenum is involved with both nitrate assimilation and nitrogen fixation, a molybdenum deficiency may bring about a nitrogen deficiency if the nitrogen source is primarily nitrate or if the plant depends on symbiotic nitrogen fixation. Although plants require only small amounts of molybdenum, some soils supply inadequate levels. Small additions of molybdenum to such soils can greatly enhance crop or forage growth at negligible cost.

Analysis of Plant Tissues Reveals Mineral Deficiencies

Requirements for mineral elements change during the growth and development of a plant. In crop plants, nutrient levels at certain stages of growth influence the yield of the economically important tissues (tuber, grain, etc.). To optimize yields, farmers use analyses of nutrient levels in soil and in plant tissue to determine fertilizer schedules.

Soil analysis is the chemical determination of the nutrient content in a soil sample from the root zone. As discussed later in the chapter, both the chemistry and the biology of soils are complex, and the results of soil analyses vary with sampling methods, storage conditions for the samples, and nutrient extraction techniques. Moreover, soil analysis reflects the levels of nutrients *potentially* available to the plant roots, but it fails to evaluate the uptake conditions and the amounts of nutrients actually absorbed by plants. This additional information is made available by plant tissue analysis.

Proper use of **plant tissue analysis** requires an understanding of the relationship between plant growth (or yield) and the mineral content of plant tissue samples (Bouma 1983). When the nutrient content in a tissue sample is low, growth is reduced (Figure 5.3). In this **deficiency zone** of the curve, an increase in the tissue mineral content is directly related to an increase in growth or yield. As the nutrient level in the tissue sample increases further, a point is reached at which additional increases in tissue mineral content are no longer related to increases in growth or yield. This region of the curve is often called the **adequate zone**. The transition between the deficiency and adequate zones of the curve reveals the **critical concentration** of the nutrient, which may be defined as the minimum tissue content of the nutrient that is correlated with maximal growth or yield. As the nutrient content of the tissue increases beyond the adequate zone, growth or yield declines because of toxicity.

To evaluate the relationship between growth and the nutrient content of a tissue, researchers grow plants in soil or nutrient solution in which all the nutrients are present in adequate amounts except the nutrient under consideration. This nutrient is then added in increasing concentrations to different sets of plants, and the content of the nutrient in specific tissues is correlated with a given yield parameter. Several curves are established for each element, one for each tissue and tissue age. Because the elements nitrogen, phosphorus, and potassium limit agricultural productivity to the greatest extent, most farms routinely use curves for at least these elements.

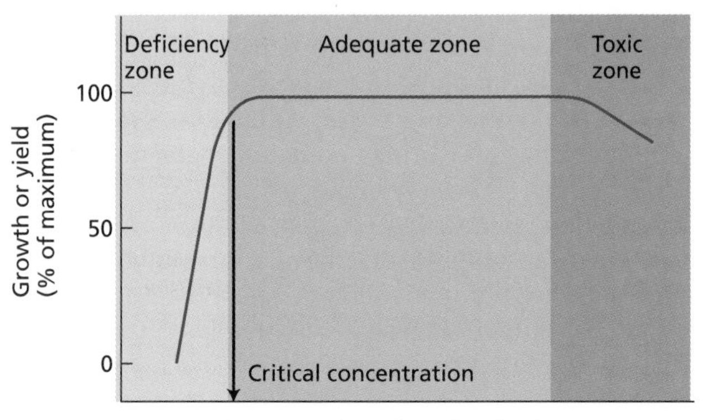

Figure 5.3 Relationship between yield (or growth) and the nutrient content of the plant tissue. The yield parameter may be expressed in terms of shoot dry weight or height. The deficiency, adequate, and toxic zones are indicated on the graph. To yield data of this type, plants are grown under conditions in which the concentration of one essential nutrient is varied while all others are in adequate supply. The effect of varying the concentration of this nutrient during plant growth is reflected in the growth or yield. The critical concentration for that nutrient is the concentration below which yield or growth is reduced.

Plant tissue analysis has been applied to many different plant tissues and organs, including leaves of different ages and positions, stems, roots, seeds, fruit, grain, or sap from petioles. When a particular nutrient is in the deficiency zone, changes in nutrient content of a tissue or organ may be highly correlated with its nutritional status (Bouma 1983). Plant analysis provides a more accurate indicator of nutrient deficiency than the morphological deficiency symptoms discussed in the previous section do. The critical concentration level for a mineral nutrient (see Figure 5.3) represents the tissue content below which nutrient disorders will occur. If a nutrient deficiency is suspected, plant tissue analysis should be conducted immediately and steps taken to correct the deficiency before it reduces growth or yield. Plant analysis has proven useful in establishing fertilizer schedules that sustain yields and ensure the food quality of many crops.

Treating Nutritional Deficiencies

Many traditional and subsistence farming practices promote the recycling of mineral elements. Crop plants absorb the nutrients from the soil, humans and animals consume locally grown crops, and crop residues and manure from humans and animals return the nutrients to the soil. The main losses of nutrients from such agricultural systems ensue from leaching that carries dissolved ions away with drainage water. Leaching may be decreased by the addition of lime—a mix of CaO, $CaCO_3$, and $Ca(OH)_2$—to make the soil more alkaline, because many mineral elements form less soluble compounds when the pH is higher than 6 (Figure 5.4). In the high-production agricultural systems of industrial countries, the removal of nutrients from the soil becomes unidirectional because a large portion of crop biomass leaves the area of cultivation. Because plants synthesize all their components from basic inorganic substances and sunlight, the use of chemical fertilizers is essential to restore mineral nutrients in high-production agricultural practices.

Crop Yields Can Be Improved by Addition of Fertilizers

Most chemical fertilizers contain inorganic salts of the macronutrients nitrogen, phosphorus, and potassium

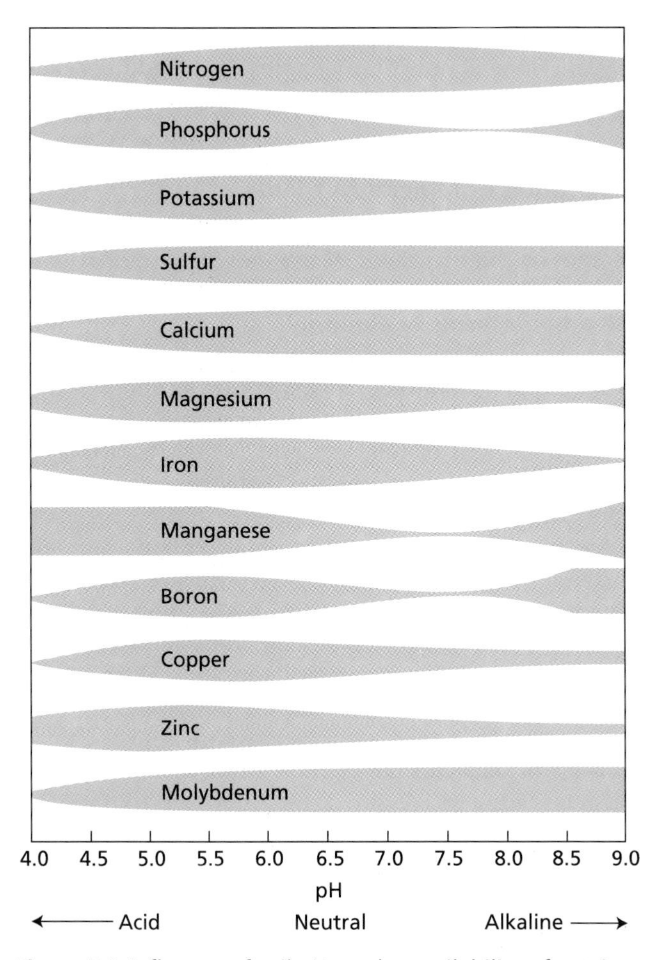

Figure 5.4 Influence of soil pH on the availability of nutrient elements in organic soils. The width of the shaded areas indicates the degree of nutrient availability to the plant root. All of these nutrients are available in the pH range of 5.5 to 6.5. (From Lucas and Davis 1961.)

(see Table 5.1). Fertilizers that contain only one of these three nutrients are termed **straight fertilizers**. Some examples of straight fertilizers are superphosphate, ammonium nitrate, and muriate of potash (a source of potassium). Fertilizers that contain two or more mineral nutrients are called **compound** or **mixed fertilizers**, and the numbers on the package label, such as 10-14-10, refer to the effective percentages of nitrogen, phosphorus, and potassium, respectively, in the fertilizer. With increased agricultural production, consumption of micronutrients can increase to the point at which they, too, must be added to the soil as fertilizers. Micronutrients may also be added to the soil to correct a preexisting deficiency. For example, some soils in the United States are deficient in boron, copper, zinc, manganese, or iron (Mengel and Kirkby 1987) and can benefit from nutrient supplementation.

Chemicals may also be applied to the soil to modify soil pH. As Figure 5.4 shows, soil pH affects the availability of several mineral nutrients. Addition of lime, as mentioned previously, can raise the pH of acidic soils; addition of elemental sulfur can lower the pH of alkaline soils. In the latter case, microorganisms absorb the sulfur and subsequently release sulfate and hydrogen ions that acidify the soil.

Organic fertilizers, in contrast to chemical fertilizers, originate from the residues of plant or animal life or from natural rock deposits. Plant and animal residues contain nutrient elements in the form of organic compounds. Before crop plants can acquire the nutrient elements from the residues, the organic compounds must be broken down, usually by the action of soil microorganisms through a process called **mineralization.** Mineralization depends on many factors, including temperature, water and oxygen availability, and the type and number of microorganisms present in the soil. As a consequence, the rate of mineralization is highly variable, and nutrients from organic residues become available to plants over periods that range from days to months. This unpredictability hinders efficient fertilizer use so that farms that rely solely on organic fertilizers may suffer even higher nutrient losses than farms that use chemical fertilizers. However, organic residues do improve the physical structure of most soils, enhancing water retention during drought and increasing drainage in wet weather.

Some Mineral Nutrients Can Be Absorbed by Leaves

In addition to being added as fertilizers to the soil, some mineral nutrients can be applied to the leaves as sprays, a process known as **foliar application**, and the leaves can absorb the applied nutrients. This method may have some agronomic advantages over the application of nutrients to the soil. Foliar nutrition can reduce the lag time between application and uptake by the plant, which could be important during a phase of rapid growth. It can also circumvent the problem of restricted uptake of a nutrient from the soil. For example, foliar application of mineral nutrients such as iron, manganese, and copper may be more efficient than application through the soil, where they are adsorbed on soil particles and hence are less available to the root system.

Nutrient uptake by plant leaves is most effective when the nutrient solution remains on the leaf as a thin film (Mengel and Kirkby 1987). Production of a thin film often requires that the nutrient solutions be supplemented with surfactant chemicals, such as the detergent Tween 80, that reduce surface tension. Nutrient movement into the plant seems to involve diffusion through the cuticle and uptake by leaf cells. Although uptake through the stomatal pore could provide a pathway into the leaf, the architecture of the pore (see Figure 4.15) largely prevents liquid penetration (Ziegler 1987).

For foliar nutrient application to be successful, damage to the leaves must be minimized. If foliar sprays are applied on a hot day, when evaporation is high, salts may accumulate on the leaf surface and cause burning or scorching. Spraying on cool days or in the evening alleviates this problem. Addition of lime to the spray diminishes the solubility of many nutrients and limits toxicity. Foliar application has proved successful mainly with tree crops and vines such as grapes, and it is also used with cereals. Nutrients applied to the leaves could save an orchard or vineyard when soil-applied nutrients would be too slow to correct a deficiency. In wheat, nitrogen applied to the leaves during the later stages of growth enhances the protein content of seeds.

Soil, Roots, and Microbes

The soil is a complex physical, chemical, and biological substrate. It is a heterogeneous material containing solid, liquid, and gaseous phases (see Chapter 4). All of these phases interact with mineral elements. The inorganic particles of the solid phase provide a reservoir of nutrients such as potassium, calcium, magnesium, and iron. Also associated with this solid phase are particles containing organic compounds that contain nitrogen, phosphorus, and sulfur, among other elements. The liquid phase of the soil constitutes the soil solution, which contains dissolved mineral ions and serves as the medium for ion movement to the root surface. Gases such as oxygen, carbon dioxide, and nitrogen are dissolved in the soil solution, but their uptake by roots is predominantly from the air gaps between soil particles.

From a biological perspective, soil constitutes a diverse ecosystem in which plant roots and microorganisms compete strongly for mineral nutrients. In spite of this competition, roots and microorganisms can form alliances for their mutual benefit (**symbiosis**). In this section we will discuss the importance of soil properties, root structure, and mycorrhizal symbiotic relationships to plant mineral nutrition. Chapter 12 will address symbiotic relationships with nitrogen-fixing bacteria.

Negatively Charged Soil Particles Affect the Adsorption of Mineral Nutrients

Soil particles, both inorganic and organic, have negative charges on their surfaces. Many inorganic soil particles are crystal lattices that are tetrahedral arrangements of the cationic forms of aluminum and silicon (Al^{3+} and Si^{4+}). These tetrahedral arrangements are covalently bound to oxygen atoms, thus forming aluminates and silicates. These particles become negatively charged by the replacement of the Al^{3+} and Si^{4+} with cations of lesser charge, a process termed **isomorphous substitution**. Organic soil particles originate from the products of the microbial decomposition of dead plants, animals, and microorganisms. The negative surface charges of organic particles result from the dissociation of hydrogen ions from the carboxylic acid and phenolic groups present in this component of the soil. Most of the world's soil particles, however, are inorganic.

Inorganic soils are categorized by particle size: Gravel consists of particles larger than 2 mm, coarse sand of particles between 0.2 and 2 mm, fine sand of particles between 0.02 and 0.2 mm, silt of particles between 0.002 and 0.02, and clay of particles smaller than 0.002 mm (see Table 4.1). The silicate-containing clay materials are further divided into three major groups—kaolinite, smectite, and illite—on the basis of differences in their structure and physical properties (Table 5.5). The kaolinite group is generally found in well-weathered soils; the smectite and illite groups are found in younger soils.

Mineral cations such as ammonium (NH_4^+) and potassium (K^+) adsorb to the negative surface charges of inorganic and organic soil particles. This adsorption is an important factor in soil fertility. Mineral cations adsorbed on the surface of soil particles are not easily lost when the soil is leached by water, and they provide a nutrient

Table 5.5
Comparative properties of three major types of silicate clays found in the soil

	Type of clay		
Property	**Smectite**	**Illite**	**Kaolinite**
Size (μm)	0.01–1.0	0.1–2.0	0.1–5.0
Shape	Irregular flakes	Irregular flakes	Hexagonal crystals
Cohesion	High	Medium	Low
Water-swelling capacity	High	Medium	Low
Cation exchange capacity (milliequivalents 100 g^{-1})	80–100	15–40	3–15

Source: After Brady 1974.

reserve available to plant roots. Mineral nutrients adsorbed in this way can be replaced by other cations in a process known as **cation exchange** (Figure 5.5). The degree to which a soil can adsorb and exchange ions is termed its **cation exchange capacity** and is highly dependent on the soil type. Soils with small particles, such as clays, have a high ratio of surface area to volume. Keep in mind that a sphere has a surface area of $4\pi r^2$ (where r is the radius of the sphere) and a volume of $4/3\ \pi r^3$, so the ratio of surface area to volume equals $3r^{-1}$. The high ratio of surface area to volume of soils with small particles exposes more surface charges per volume of soil and thus generates a higher cation exchange capacity. A soil with higher cation exchange capacity generally has a larger reserve of mineral nutrients.

Mineral anions such as nitrate (NO_3^-) and chloride (Cl^-) tend to be repelled by the negative charge on the surface of soil particles and remain dissolved in the soil solution. Thus, the anion exchange capacity of most agricultural soils is small compared to their cation exchange capacity. Among the required anions, nitrate remains mobile in the soil solution, where it is susceptible to leaching by water moving through the soil. Phosphate ions (PO_4^{3-}) may bind to soil particles containing aluminum or iron because the positively charged iron and aluminum ions (Fe^{2+}, Fe^{3+}, and Al^{3+}) have hydroxyl (OH^-) groups that exchange with phosphate; as a result, phosphate can be tightly bound, and its mobility and availability in soil can be limited. Sulfate (SO_4^{2-}) in the presence of calcium (Ca^{2+}) forms gypsum ($CaSO_4$).Gypsum is only slightly soluble, but it releases sufficient sulfate to support plant growth. Most soils contain substantial amounts of calcium, and consequently sulfate mobility in these soils is low and not highly susceptible to leaching.

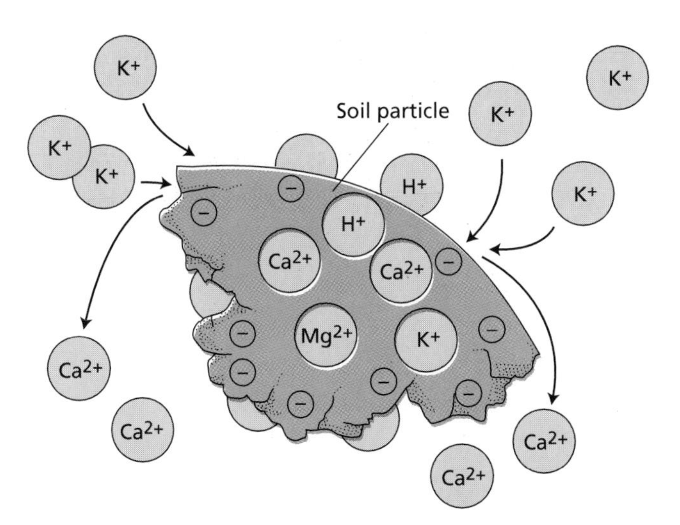

Figure 5.5 The principle of cation exchange on the surface of a soil particle. Cations are bound to the surface of soil particles because the surface is negatively charged. Addition of a cation such as potassium (K^+) can displace another cation such as calcium (Ca^{2+}) from its binding on the surface of the soil particle and make it available for uptake by the root.

Soil pH Affects Nutrient Availability, Soil Microbes, and Root Growth

Hydrogen ion concentration (pH) is an important property of soils because it affects the growth of plant roots and soil microorganisms. Root growth is generally favored in slightly acidic soils, at pH values between 5.5 and 6.5. Fungi generally predominate in acidic soils; bacteria become more prevalent in alkaline soils. Soil pH also determines the availability of soil nutrients (see Figure 5.4). Acidity promotes the weathering of rocks that releases K^+, Mg^{2+}, Ca^{2+}, and Mn^{2+} and increases the solubility of carbonates, sulfates, and phosphates. Increasing solubility facilitates absorption by the root. The amount of rainfall and decomposition of organic matter in soils are major factors in lowering the soil pH. Carbon dioxide is produced as a result of the decomposition of organic material and equilibrates with soil water in the following reaction:

$$CO_2 + H_2O \rightleftharpoons H^+ + HCO_3^-$$

This reaction releases hydrogen ions, lowering the pH of the soil. Microbial decomposition of organic material also produces ammonia and hydrogen sulfide that can be oxidized in the soil to form the strong acids nitric acid (HNO_3) and sulfuric acid (H_2SO_4), respectively. Hydrogen ions also displace K^+, Mg^{2+}, Ca^{2+}, and Mn^{2+} from the cation exchange complex in a soil. Leaching then may remove these ions from the upper soil layers, leaving a more acid soil. By contrast, the weathering of rock in arid regions releases K^+, Mg^{2+}, Ca^{2+}, and Mn^{2+} to the soil, but because of the low rainfall, these ions do not leach from the upper soil layers, and the soil remains alkaline.

Excessive Minerals in the Soil Limit Plant Growth

When excess minerals are present in the soil, the soil is said to be saline, and plant growth may be restricted if these mineral ions reach levels that limit water availability or exceed the adequate zone for a particular nutrient (see Chapter 25). Sodium chloride and sodium sulfate are the most common salts in saline soils. Excess minerals in soils can be a major problem in arid and semiarid regions because rainfall is insufficient to leach the mineral ions from the soil. Moreover, irrigated agriculture may foster soil salinization. Irrigation water can contain 100 to 1000 g of minerals per cubic meter. An average crop requires about 4000 m^3 of water per acre. Consequently, 400 to 4000 kg of minerals can be added to the soil per crop (Marschner 1995). Another important problem with excess minerals is the accumulation of heavy metals in the soil, which can cause severe toxicity (Box 5.1).

BOX 5.1

Heavy-Metal Stress and Homeostasis

HEAVY METALS, such as copper, cobalt, zinc, and nickel, are among the essential micronutrients required in trace amounts by plants to complete their life cycles. When their concentrations in the soil exceed trace levels, however, heavy metals can be extremely toxic. During the last 100 years, unrestricted mining, manufacturing, and municipal waste disposal practices have resulted in the addition of large amounts of heavy metals to the biosphere. Such metals can persist indefinitely, posing an ever-increasing threat to agriculture and human health. Of particular concern are the highly toxic nonnutrient metals: mercury, lead, cadmium, silver, and chromium. Indeed, the total toxicity of heavy-metal pollutants added to the environment each year now exceeds the toxicity of all organic and radioactive wastes combined (Nriagu and Pacyna 1988)!

In response to these concerns, plant physiologists have recently increased their efforts to understand the nature of heavy-metal homeostasis and toxicity in plants. The main goals of this research are to develop metal-tolerant plants that can grow in metal-contaminated soils, and to develop new varieties of plants that can mine the metals from the soil so that the soil can be reclaimed for agriculture—a practice called **phytoremediation**.

The toxicity of many heavy metals is due to their ability to cause oxidative damage to tissues (Stohs and Bagchi 1995). Damage includes enhanced lipid peroxidation, DNA damage, and the oxidation of protein sulfhydryl groups. Yet despite the extreme toxicity of heavy metals, plants have the ability to adapt to metal-rich soils through natural selection (Antonovics et al. 1971). Metal-adapted species have been studied in serpentine soils (soils rich in nickel, chromium, manganese, magnesium, and cobalt), uraniferous soils (uranium-rich soils), calamine soils (soils rich in zinc and cadmium), and soils rich in chromium and cobalt. Metal-tolerant plants that have the ability to accumulate high concentrations of metals in their vacuoles are called **hyperaccumulators** (Baker and Brooks 1989). For example, the serpentine endemic *Thlaspi goesingense* can accumulate nickel to 9.5 mg/g, or about 1% of its dry weight, and *Sebertia acuminata*, a tree native to New Caledonia, accumulates a remarkable 20% of its dry weight in nickel. Hyperaccumulators are currently being studied for their potential in phytoremediation.

Plants cope with excess heavy-metal ions by relying on two main mechanisms for detoxifying metals taken up into the cell: complexation with organic compounds and compartmentation within the vacuole. Plants produce a variety of compounds that are capable of forming complexes with metal ions. For example, citrate and malate have been shown to bind cadmium and zinc, respectively. The vacuoles of metal-treated plants often contain high concentrations of metals complexed with organic acids. Amino acids can also form complexes with metal ions. Recently, Krämer and colleagues (1996) found histidine to be complexed with nickel in the nickel hyperaccumulator *Alyssum lesbiacum*, and they have proposed that histidine synthesis is an important mechanism for nickel hyperaccumulation.

The **phytochelatins** are a group of metal-binding polypeptides that have the general formula $[\text{Glu(-Cys)}]_n$-Gly, where n is a number from 2 to 11. Phytochelatins are synthesized from the tripeptide glutathione (Glu-Cys-Gly) (Grill et al. 1989). The enzyme phytochelatin synthase is constitutively expressed in plants and is strongly activated by metal ions. Phytochelatins accumulate in the vacuoles of plants exposed to excess heavy metals, and they are considered to be indicators of metal stress. Phytochelatin–metal complexes are actively transported into plant vacuoles by a group of organic solute transporters that are directly energized by Mg-ATP (see Chapter 6) (Salt and Rauser 1995; Rea et al. in press). These transporters recognize and transport a variety of solutes that are conjugated to glutathione, including anthocyanins and herbicides (see Chapter 13).

Plants also have genes that encode **metallothioneins**, small metal-binding proteins originally identified in animals. In *Arabidopsis*, three main types of metallothionein genes have been identified in vegetative tissues: *MT1*, *MT2*, and *MT3* (Murphy et al. 1997), and the expression of one of these genes (*MT2*) was related to copper tolerance in seedlings (Murphy and Taiz 1995, 1997). Thus, the metallothioneins appear to protect cellular constituents from oxidative damage due to metals, although their precise roles in metal tolerance and homeostasis remain to be elucidated.

In saline soil, plants encounter **salt stress**. Some plants are affected adversely by the presence of excess minerals; other plants can survive (**salt-tolerant plants**) or even thrive (**halophytes**). The mechanisms by which plants tolerate salinity are complex (see Chapter 25), involving molecular synthesis, enzyme induction, and membrane transport. In some species, excess minerals are not taken up; in others, minerals are taken up but excreted from the plant by salt glands associated with the leaves. To prevent toxic buildup of mineral ions in the cytosol, many plants may sequester them in the vacuole (Stewart and Ahmad 1983). Efforts are under way to bestow salt tolerance on salt-sensitive crop species using both classic plant-breeding and biotechnology techniques (Bohnert and Jensen 1996).

Plants Develop Extensive Root Systems

The ability of plants to obtain both water and mineral nutrients from the soil is related to their capacity to develop an extensive root system. In the late 1930s, H. J. Dittmer examined the root system of a single winter rye plant after 16 weeks of growth and estimated that the plant had 13×10^6 primary and lateral root axes, extending more than 500 km in length and providing 200 m^2 of surface area (Dittmer 1937). This plant also had more than 10^{10} root hairs, providing another 300 m^2 of surface area.

In the desert, the roots of mesquite (genus *Prosopis*) may extend down more than 50 m to reach groundwater. In annual crop plants, roots usually grow between 0.1 and 2.0 m in depth and extend laterally to distances of 0.3 to 1.0 m, and in trees planted 0.5 m apart the major root systems reach a total length of 12 to 18 km. In natural ecosystems, the annual production of roots may easily surpass that of shoots, so in many respects, the aboveground portions of a plant represent only a minor fraction of the plant.

Plant roots may grow continuously throughout the year. Their proliferation, however, depends on the availability of water and minerals in the immediate microenvironment surrounding the root, the so-called **rhizosphere**. If the rhizosphere is poor in nutrients or too dry, root growth is slow. As rhizosphere conditions improve, root growth increases. If fertilization and irrigation provide abundant nutrients and water, root growth may not keep pace with shoot growth. Plant growth under such conditions becomes carbohydrate limited, and a relatively small root system meets the nutrient needs of the whole plant (Bloom et al. 1993). Roots growing below the soil surface are studied by special techniques (Box 5.2).

Root Systems Differ in Form but Are Based on Common Structures

The *form* of the root system differs greatly among plant species. In monocots, root development starts with the emergence of three to six **primary** (or seminal) root axes from the germinating seed. With further growth, new adventitious roots, called **nodal** or brace roots, form.

BOX 5.2

Observing Roots below Ground

STUDYING ROOT GROWTH below the soil surface requires a means of observing the root system directly. As early as 1873, the German botanist Julius von Sachs studied root systems by using simple soil-filled boxes with one glass wall. Since that time, facilities for studying root growth in the soil have become much more complex. Large laboratories with subterranean chambers for the observation of root growth have been constructed and allow the analysis of root growth while the aerial parts of the plant are exposed to natural field conditions (Klepper and Kaspar 1994). These laboratories are called **rhizotrons** (from the Greek *rhizos*, meaning "root," and *tron*, meaning "a device for studying"). In a rhizotron, roots grow in glass-walled chambers that line underground passageways. Grid lines on the glass walls indicate soil depth. Details of root morphology (root size and distribution) under natural growing conditions can be observed with specially designed microscopes, mounted adjacent to the glass walls of the root chambers. In addition, the growth of the roots over a period of time can be measured with time-lapse photography.

Because rhizotrons are expensive to construct and maintain, they have been largely supplanted by minirhizotrons or root periscopes. Minirhizotrons are transparent plastic tubes previously buried at an angle in the soil near the plants to be observed. An optical device such as a tilted mirror with a magnifying lens or a miniature videocamera is inserted into a tube to monitor roots growing along its surface (see the figure). This device provides information on rooting density in the bulk soil, as well as on root growth and phenology. Plant physiologists use such information in conjunction with changes in nutrient and water levels in the soil to assess root activity through the soil profile.

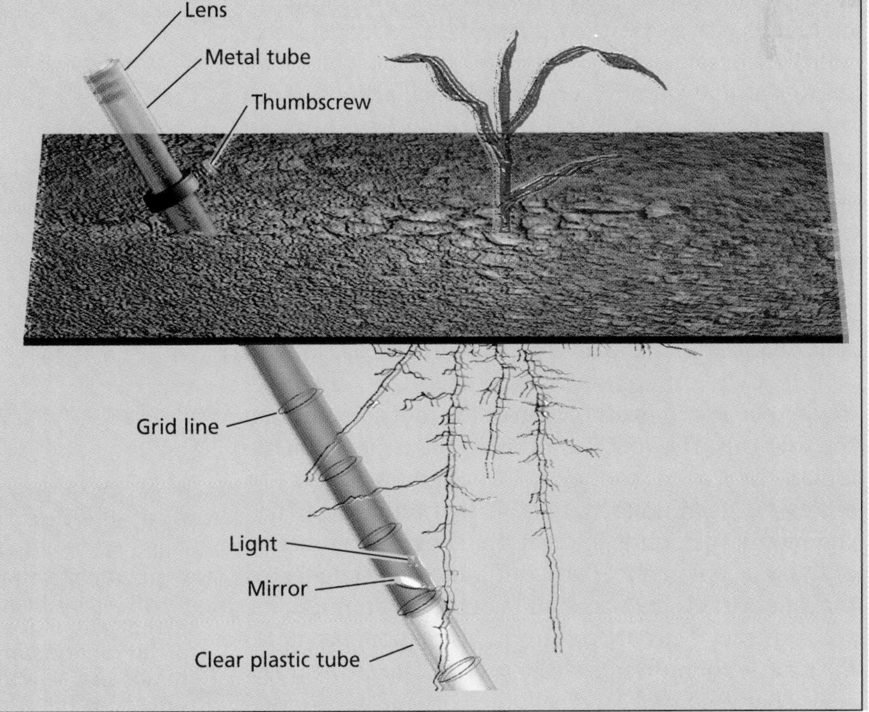

Drawing of a minirhizotron installed near an oat seedling. A metal tube that slides within a clear plastic tube is held in place by a thumbscrew. The metal tube contains the optics for viewing roots growing along the surface of the plastic tube. A dark grid scribed onto the surface of the plastic tube assists in positioning the metal tube. The optics include lenses, mirror, and lights. Many modern configurations use a miniature television camera instead of lenses to obtain the image.

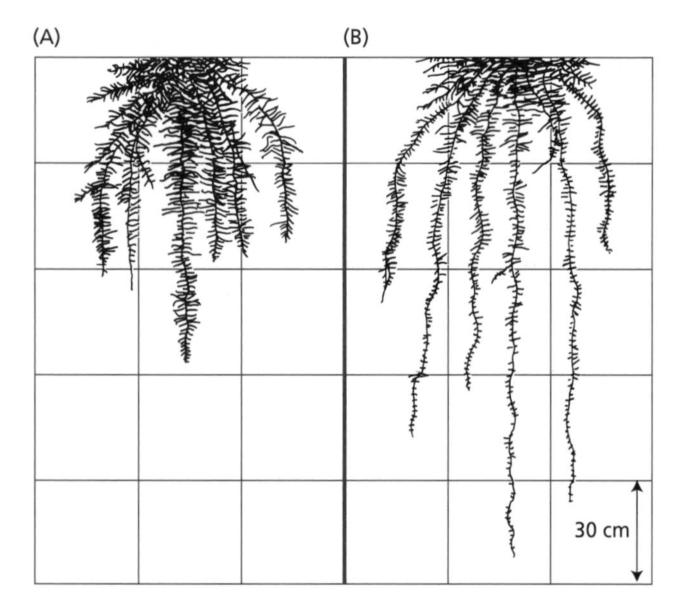

Figure 5.6 Fibrous root systems of wheat (monocot). (A) The root system of a mature (3-month-old) wheat plant growing in dry soil. (B) The root system of a wheat plant growing in irrigated soil. In the fibrous root system, the primary root axes are no longer distinguishable. It is also apparent that the morphology of the root system is affected by the amount of water present in the soil. (After Weaver 1926.)

Over time, the primary and nodal root axes grow and branch extensively to form a complex fibrous root system (Figure 5.6). In the fibrous root system, all the roots generally have the same diameter (except where environmental conditions or pathogenic interactions modify the root structure), so it is difficult to distinguish a main root axis. By contrast, dicots develop root systems with a main single root axis, called a **taproot**, which may thicken as a result of secondary cambial activity. From this main root axis, lateral roots develop to form an extensively branched root system (Figure 5.7).

The development of the root system in both monocots and dicots depends on the activity of the root apical meristem and the production of lateral root meristems. Figure 5.8 shows a generalized diagram of the apical region of a plant root. In the meristematic zone, cells divide both in the direction of the root base to form cells that will differentiate into the tissues of the functional root and in the direction of the root apex to form the **root cap**. The root cap protects the delicate meristematic cells as the root moves through the soil. It also secretes a gelatinous material called **mucigel**, which commonly surrounds the root tip. The precise function of the mucigel is uncertain, but it has been suggested that it lubricates the penetration of the root through the soil, protects the root apex from desiccation, promotes the transfer of nutrients to the root, or affects the interaction between roots and soil microorganisms (Russell 1977). The root cap is central to the perception of gravity, the signal that directs the growth of roots downward. This process is termed the **gravitropic response** (see Chapter 19).

Cell division at the root apex proper is relatively slow, and thus this region is called the **quiescent center**. After a few generations of slow cell divisions, root cells displaced from the apex by about 0.1 mm begin to divide more rapidly. Cell division again tapers off at about 0.4 mm from the apex, and the cells expand equally in all directions.

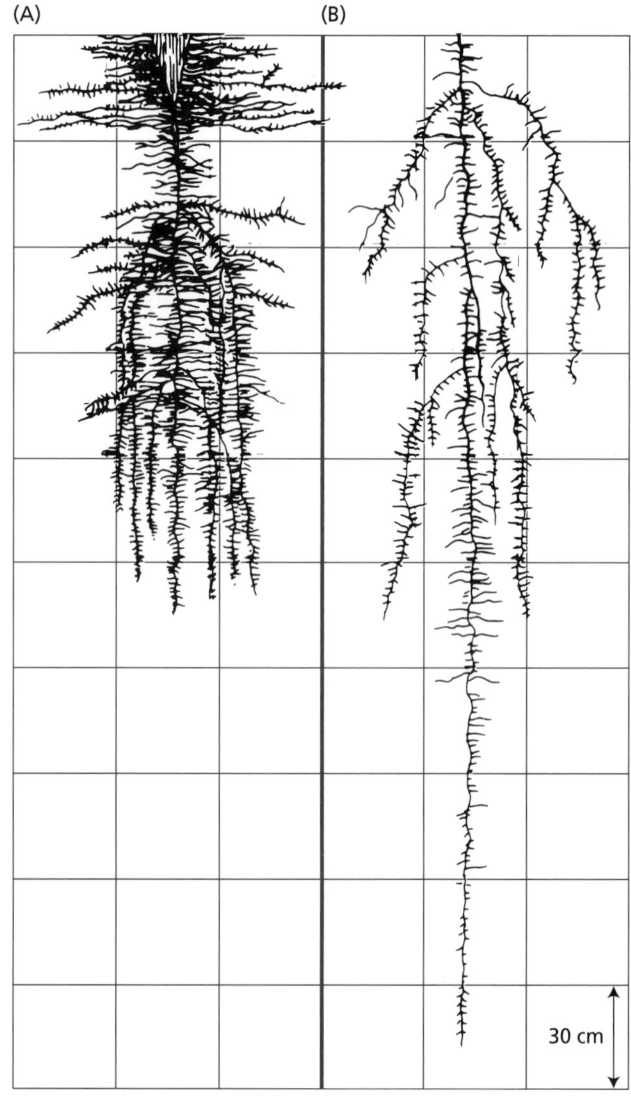

Figure 5.7 Taproot system of two dicots: sugar beet (A) and alfalfa (B). The sugar beet root system is typical of 5 months of growth; the alfalfa root system is typical of 2 years of growth. In both dicots, the root system shows a major vertical root axis. In the case of sugar beet, the upper portion of the taproot system is thickened because of its function as storage tissue. Both plants were grown under adequate water conditions. (After Weaver 1926.)

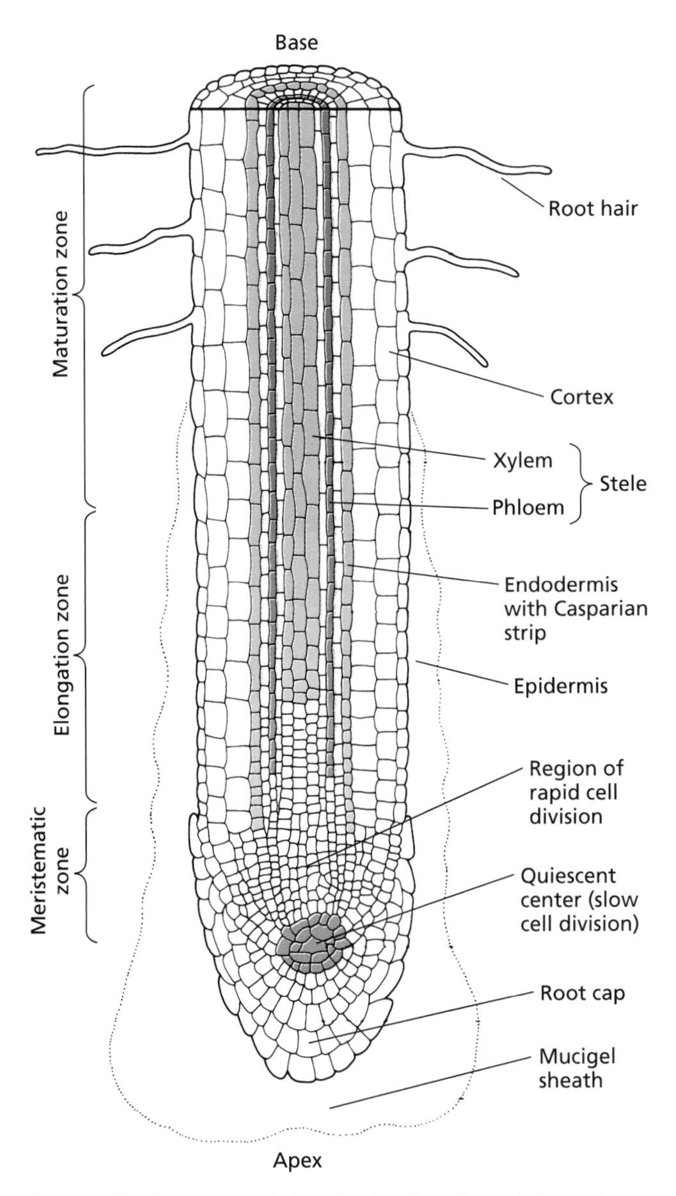

Figure 5.8 Diagrammatic longitudinal section of the apical region of the root. The meristematic cells are located near the tip of the root. These cells generate the root cap and the upper tissues of the root. In the elongation zone, cells differentiate to produce xylem, phloem, and cortex. Root hairs, formed in epidermal cells, first appear in the maturation zone.

The elongation zone begins 0.7 to 1.5 mm from the apex. In this zone, cells elongate rapidly and undergo a final round of divisions to produce a central ring of cells called the **endodermis**. The walls of this endodermal cell layer become thickened, and suberin (see Chapter 13) deposited on the radial walls forms the **Casparian strip**, a hydrophobic structure that prevents the apoplastic movement of water or solutes across the root (see Figure 4.4). The endodermis divides the root into two regions: the **cortex** toward the outside and the **stele** toward the inside. The stele contains the vascular elements of the root: the **phloem**, which transports metabolites from the shoot to the root, and the **xylem**, which transports water and solutes to the shoot.

Phloem develops more rapidly than xylem, attesting to the fact that phloem function is critical near the root apex. Large quantities of carbohydrates must flow through the phloem to the growing apical zones in order to support cell division and elongation. Carbohydrates provide rapidly growing cells with an energy source and with carbon skeletons required to synthesize organic compounds. Six-carbon sugars (hexoses) also function as osmotically active solutes in the root tissue. At the root apex, where the phloem is not yet developed, carbohydrate movement depends on symplastic diffusion and is relatively slow (Bret-Harte and Silk 1994). The low rates of cell division in the quiescent center may result from the fact that insufficient carbohydrates reach this centrally located region. Root hairs, with their large surface area for absorption of water and solutes, first appear in the maturation zone, and it is here that the xylem develops the capacity to translocate substantial quantities of water and solutes to the shoot.

Different Areas of the Root Absorb Different Mineral Ions

The precise point of entry of minerals into the root system has been a topic of considerable interest. Some researchers have claimed that nutrients are absorbed only at the apical regions of the root axes or branches (Bar-Yosef et al. 1972); others claim that nutrients are absorbed over the entire root surface (Nye and Tinker 1977; Greenwood 1982). Experimental evidence supports both possibilities, depending on the nutrient being investigated.

Root absorption of calcium in barley appears to be restricted to the apical region. Iron may be taken up either at the apical region, as in barley (Clarkson and Sanderson 1978), or over the entire root surface, as in corn (Kashirad et al. 1973). Potassium, nitrate, ammonium, and phosphate can be absorbed freely at all locations of the root surface (Clarkson and Hanson 1980), but in corn the elongation zone has the maximum rates of potassium accumulation (Sharp et al. 1990) and nitrate absorption (Colmer and Bloom 1998). In corn and rice, the root apex absorbs ammonium more rapidly than the elongation zone does (Colmer and Bloom 1998). And for several species, root hairs are the most active in phosphate absorption (Fohse et al. 1991).

The high rates of nutrient absorption in the apical root zones result from the strong demand for nutrients in these tissues and the relatively high nutrient availability in the soil surrounding them. For example, cell elongation depends on the accumulation of solutes such as potassium and nitrate to increase the osmotic pressure within the cell (see Chapter 16). Ammonium is the preferred nitrogen source to support cell division in the meristem because meristematic tissues are often carbohydrate limited and the assimilation of ammonium consumes less energy than that of nitrate (see Chapter 12). The root apex and root hairs grow into fresh soil, where nutrients have not yet been depleted.

Within the soil, nutrients can move to the root surface both by bulk flow and by diffusion (see Chapter 3). In **bulk flow**, nutrients are carried by water moving through the soil toward the root. The amount of nutrient provided to the root by bulk flow depends on the rate of water flow through the soil toward the plant, which depends on transpiration rates and on nutrient levels in the soil solution. When both the rate of water flow and the concentrations of nutrients in the soil solution are high, bulk flow can play an important role in nutrient supply. In **diffusion**, mineral nutrients move from a region of higher concentration to a region of lower concentration. Nutrient uptake by the roots lowers the concentration of nutrients at the root surface, generating concentration gradients in the soil solution surrounding the root. Diffusion of nutrients down their concentration gradient and bulk flow resulting from transpiration can increase nutrient availability at the root surface.

When absorption of nutrients by the roots is high and the nutrient concentration in the soil is low, bulk flow can supply only a small fraction of the total nutrient requirement (Mengel and Kirkby 1987). Under these conditions, diffusion rates limit the movement of nutrients to the root surface. When diffusion is too slow to maintain high nutrient concentrations near the root, a **nutrient depletion zone** forms adjacent to the root surface (Figure 5.9). Nutrient depletion decreases as the distance to the root surface increases. Nonetheless, nutrient uptake depends on the relationship between the affinity of the root uptake mechanism for a given nutrient and the prevailing concentration of that nutrient at the root surface. This aspect of nutrient uptake is discussed in Chapter 6.

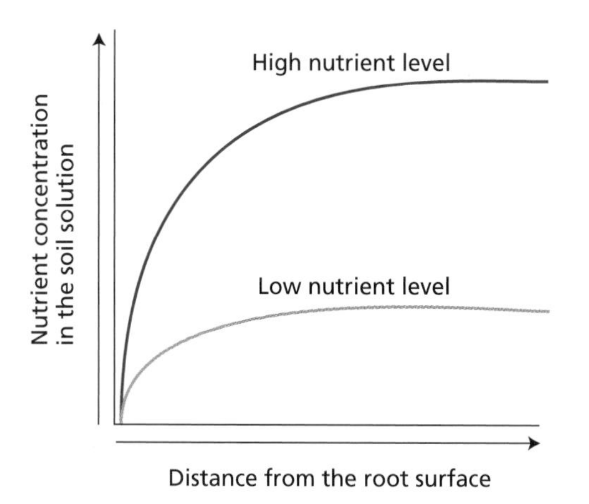

Figure 5.9 Formation of a nutrient depletion zone in the region of the soil adjacent to the plant root. A nutrient depletion zone forms when the rate of nutrient uptake by the cells of the root exceeds the rate of replacement of the nutrient by diffusion in the soil solution. This depletion causes a localized decrease in the nutrient concentration in the area adjacent to the root surface. (After Mengel and Kirkby 1987.)

The formation of a depletion zone tells us something important about mineral nutrition. Because roots deplete the mineral supply in the rhizosphere, their effectiveness in mining minerals from the soil is determined not only by the rate at which they can remove nutrients from the soil solution, but by their continuous growth. Without growth, roots would rapidly deplete the soil adjacent to their surface. Optimal nutrient acquisition therefore depends both on the capacity for nutrient uptake and on the ability of the root system to grow into fresh soil.

Mycorrhizal Fungi Facilitate Nutrient Uptake by Roots

Our discussion thus far has centered on the direct acquisition of mineral elements by the root, but this process may be modified by the association of mycorrhizal fungi with the root system. **Mycorrhizae** (singular mycorrhiza, from the Greek words for "fungus" and "root") are not unusual; in fact, they are widespread under natural conditions. Much of the world's vegetation appears to have roots associated with mycorrhizal fungi: Eighty-three percent of dicots, 79% of monocots, and all gymnosperms regularly form mycorrhizal associations (Wilcox 1991). On the other hand, plants from the families Cruciferae (cabbage), Chenopodiaceae (spinach), and Proteaceae (macadamia nuts), as well as aquatic plants, rarely, if ever, have mycorrhizae. Mycorrhizae are absent from roots in very dry, saline, or flooded soils, or where soil fertility is extreme, either high or low. In particular, plants grown under hydroponics and young, rapidly growing crop plants seldom have mycorrhizae.

Mycorrhizal fungi are composed of fine, tubular filaments called *hyphae* (singular hypha). The *mycelium* is the mass of hyphae that forms the body of a fungus. There are two major classes of mycorrhizal fungi: ectotrophic mycorrhizae and vesicular-arbuscular mycorrhizae (Rovira et al. 1983; Smith et al. 1997). Minor classes of mycorrhizal fungi include the ericaceous and orchidaceous mycorrhizae, which may have limited importance in terms of mineral nutrient uptake (Tinker and Gildon 1983).

Ectotrophic mycorrhizal fungi typically show a thick sheath, or "mantle," of fungal mycelium around the roots, and some of the mycelium penetrates between the cortical cells (Figure 5.10). The cortical cells themselves are not penetrated by the fungal hyphae but instead are surrounded by a network of hyphae called the **Hartig**

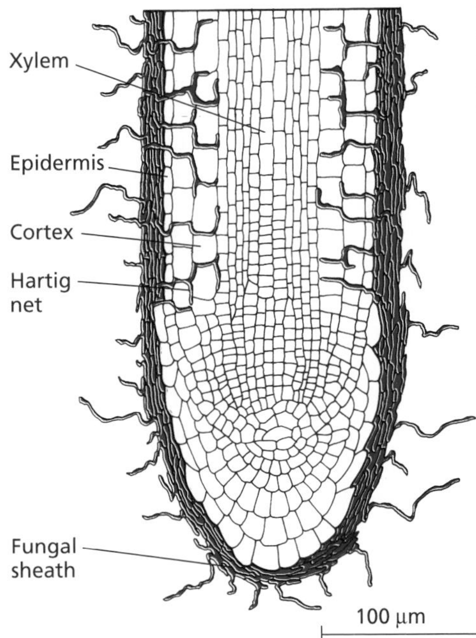

Figure 5.10 A root infected with ectotrophic mycorrhizal fungi. In the infected root, the fungal hyphae surround the root to produce a dense fungal sheath and penetrate the intercellular spaces of the cortex to form the Hartig net. The total mass of fungal hyphae may be comparable to the root mass itself. (From Rovira et al. 1983.)

net. Often, the amount of fungal mycelium is so extensive that its total mass is comparable to that of the roots themselves. The fungal mycelium also extends into the soil, away from this compact mantle, where it forms individual hyphae or strands containing fruiting bodies (Tinker and Gildon 1983). The capacity of the root system to absorb nutrients is improved by the presence of external fungal hyphae that are much finer than plant roots and can reach beyond the areas of nutrient-depleted soil near the roots (Clarkson 1985). Ectotrophic mycorrhizal fungi exclusively infect tree species, including gymnosperms and woody angiosperms.

Unlike the ectotrophic mycorrhizal fungi, **vesicular-arbuscular mycorrhizal fungi** do not produce a compact mantle of fungal mycelium around the root. Instead, the hyphae grow in a less dense arrangement, both within the root itself and extending outward from the root into the surrounding soil (Figure 5.11). After entering the root through either the epidermis or a root hair, the hyphae not only extend through the regions between cells but also penetrate individual cells of the cortex. Within the cells, the hyphae can form oval structures called **vesicles** and branched structures called **arbuscules**. The arbuscules appear to be sites of nutrient transfer between the fungus and the host plant. Outside the root, the external mycelium can extend several centimeters away from the root and may contain spore-bearing structures. Unlike the ectotrophic mycorrhizae, vesicular-arbuscular mycorrhizae make up only a small mass of fungal material, which is unlikely to exceed 10% of the root weight. Vesicular-arbuscular mycorrhizae are found in association with the roots of most species of herbaceous angiosperms (Smith et al. 1997).

The association of vesicular-arbuscular mycorrhizae with plant roots facilitates the uptake of phosphorus and trace metals such as zinc and copper. By extending beyond the depletion zone for phosphorus around the root, the external mycelium improves phosphorus absorption. Calculations by Sanders and Tinker (1971)

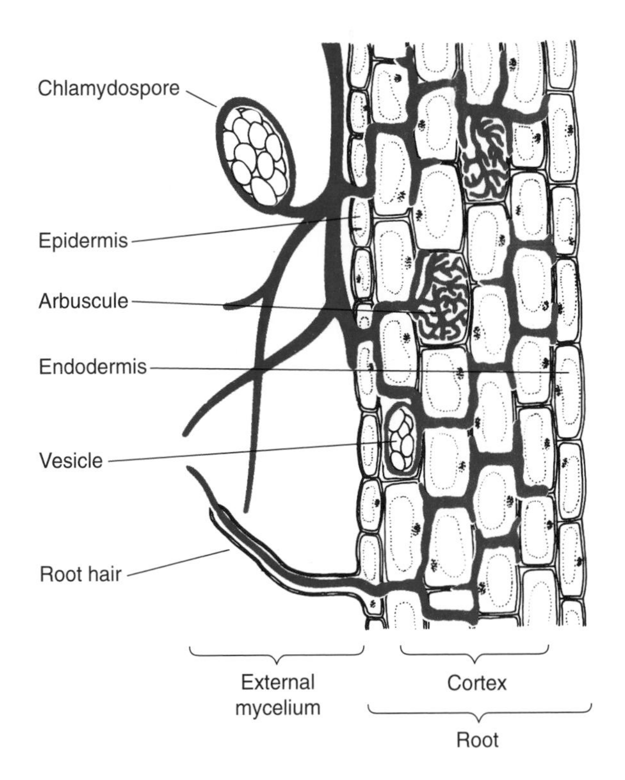

Figure 5.11 The association of vesicular-arbuscular mycorrhizal fungi with a section of a plant root. The external mycelium can bear reproductive chlamydospores and extend out from the root into the surrounding soil. The fungal hyphae grow into the intercellular wall spaces of the cortex and penetrate individual cortical cells. As they extend into the cell, they do not break the plasma membrane or the tonoplast of the host cell. Instead, the hypha is surrounded by these membranes as it occupies intracellular space. In this process, the fungal hyphae may form ovoid structures known as vesicles or branched structures known as arbuscules. The arbuscules participate in nutrient ion exchange between the host plant and the fungus. Arbuscules develop and proliferate following the penetration of the hyphae into the cortical cells of the root. In later stages, the arbuscules separate from the hyphae and degenerate. (From Mauseth 1988.)

show that a root associated with mycorrhizal fungi can transport phosphate at a rate more than four times higher than that of a root not associated with mycorrhizae. The external mycelium of the ectotrophic mycorrhizae can also absorb phosphate and make it available to the plant. In addition, it has been suggested that ectotrophic mycorrhizae proliferate in the organic litter of the soil and hydrolyze organic phosphorus for transfer to the root (Smith et al. 1997).

Nutrients Move from the Mycorrhizal Fungi to the Root Cells

Little is known about the mechanism by which the mineral nutrients absorbed by mycorrhizal fungi are transferred to the cells of plant roots. In ectotrophic mycorrhizae, inorganic phosphate may simply diffuse from the hyphae in the Hartig net and be absorbed by the root cortical cells. In vesicular-arbuscular mycorrhizae, the situation may be more complex. Nutrients may diffuse from intact arbuscules to root cortical cells. Alternatively, because some root arbuscules are continually degenerating while new ones are forming, degenerating arbuscules may release their internal contents to the host root cells.

A key factor in the extent of mycorrhizal association with the plant root is the nutritional status of the host plant. Deficiency of a nutrient such as phosphorus tends to promote infection, whereas plants under a surfeit of nutrients tend to suppress mycorrhizal infection. Mycorrhizal association in well-fertilized soils may shift from a symbiotic relationship to a parasitic one in that the fungus still obtains carbohydrates from the host plant, but the host plant no longer benefits from improved nutrient uptake efficiency. Under such conditions, the host plant may treat mycorrhizal fungi as it does other pathogens (Brundrett 1991; Marschner 1995).

Summary

Plants are autotrophic organisms capable of using the energy from sunlight to synthesize all their components from carbon dioxide, water, and mineral elements dissolved in the soil solution. Studies of plant nutrition have shown that specific mineral elements are essential for plant life. These elements are classified as macronutrients or micronutrients, depending on the relative amounts required.

A plant that is deficient in a macronutrient or micronutrient exhibits a nutritional disorder with characteristic symptoms. Nutritional disorders occur because nutrients have key roles in plant metabolism. They serve as components of organic compounds, in energy storage, to maintain plant structures, as enzyme cofactors, and in electron transfer reactions. Mineral nutrition can be studied through the use of hydroponics or aeroponics, which allow the characterization of specific nutrient requirements. Soil and plant tissue analysis can provide information on the nutritional status of plants and can suggest corrective actions to avoid deficiencies.

When crop plants are grown under modern high-production conditions, substantial amounts of nutrients are removed from the soil. To prevent the development of deficiencies, nutrients can be added back to the soil in the form of fertilizers. Fertilizers that provide nutrients in inorganic forms are called chemical fertilizers; those that derive from plant or animal residues are considered organic fertilizers. In both cases, plants absorb the nutrients primarily as inorganic ions. Most fertilizers are applied to the soil, but some are sprayed on leaves.

The soil is a complex substrate—physically, chemically, and biologically. The size of soil particles and the cation exchange capacity of the soil determine the extent to which a soil provides a reservoir for water and nutrients. Soil pH also has a large influence on the availability of mineral elements.

If mineral elements, especially sodium or heavy metals, are present in excess in the soil, plant growth may be adversely affected. Certain plants are able to tolerate excess mineral elements, and a few species—known as halophytes in the case of sodium or hyperaccumulators in the case of heavy metals—even thrive under these conditions.

To obtain nutrients from the soil, plants develop extensive root systems. Plant root systems can be directly studied with rhizotrons, which allow the examination of roots under field conditions. Roots have a relatively simple structure with radial symmetry and few differentiated cell types. Roots continually deplete the nutrients from the immediate soil around them, and such a simple structure may permit rapid growth into fresh soil.

Plant roots often form associations with mycorrhizal fungi. The fine hyphae of mycorrhizae extend the reach of roots into the surrounding soil and facilitate the acquisition of mineral elements, particularly those like phosphorus that are relatively immobile in the soil. In return, plants provide mycorrhizae with carbohydrates. Plants tend to suppress mycorrhizal associations under conditions of high nutrient availability.

General Reading

*Baker, A. J. M., and Brooks, R. R. (1989) Terrestrial higher plants which hyperaccumulate metallic elements—A review of their distribution, ecology and phytochemistry. *Biorecovery* 1: 81–126.

*Epstein, E. (1972) *Mineral Nutrition of Plants: Principles and Perspectives.* Wiley, New York.

*Loomis, R. S., and Conner, D. J. (1992) *Crop Ecology: Productivity and Management in Agricultural Systems.* Cambridge University Press, Cambridge.

*Mengel, K., and Kirkby, E. A. (1987) *Principles of Plant Nutrition*. International Potash Institute, Bern.

*Smith, S. E., Read, D. J., and Harley, J. L. (1997) *Mycorrhizal symbiosis*. Academic Press, San Diego.

* Indicates a reference that is general reading in the field and is also cited in this chapter.

Chapter References

Antonovics, J., Bradshaw, A. D., and Turner, R. G. (1971) Heavy metal tolerance in plants. *Adv. Ecol. Res.* 7: 1–85.

Arnon, D. I., and Stout, P. R. (1939) The essentiality of certain elements in minute quantity for plants with special reference to copper. *Plant Physiol.* 14: 371–375.

Asher, C. J., and Edwards, D. G. (1983) Modern solution culture techniques. In *Inorganic Plant Nutrition* (*Encyclopedia of Plant Physiology*, new series, vol. 15B), A. Läuchli and R. L. Bieleski, eds., Springer, Berlin, pp. 94–119.

Bar-Yosef, B., Kafkafi, U., and Bresler, E. (1972) Uptake of phosphorus by plants growing under field conditions. I. Theoretical model and experimental determination of its parameters. *Soil Sci.* 36: 783–800.

Bienfait, H. F., and Van der Mark, F. (1983) Phytoferritin and its role in iron metabolism. In *Metals and Micronutrients: Uptake and Utilization by Plants*, D. A. Robb and W. S. Pierpoint, eds., Academic Press, New York, pp. 111–123.

Bloom, A. J. (1994) Crop acquisition of ammonium and nitrate. In *Physiology and Determination of Crop Yield*, K. J. Boote, J. M. Bennett, T. R. Sinclair, and G. M. Paulsen, eds., Agronomy Society of America, Crop Science Society of America, Soil Science Society of America, Madison, WI, pp. 303–309.

Bloom, A. J., Jackson, L. E., and Smart, D. R. (1993) Root growth as a function of ammonium and nitrate in the root zone. *Plant Cell Environ.* 16: 199–206.

Bohnert, H. J., and Jensen, R. G. (1996) Metabolic engineering for increased salt tolerance: The next step. *Aust. J. Plant Physiol.* 23: 661–667.

Bouma, D. (1983) Diagnosis of mineral deficiencies using plant tests. In *Inorganic Plant Nutrition* (*Encyclopedia of Plant Physiology*, new series, vol. 15B), A. Läuchli and R. L. Bieleski, eds., Springer, Berlin, pp. 120–146.

Brady, N. C. (1974) *The Nature and Properties of Soils*, 8th ed. Macmillan Publishing, New York.

Bret-Harte, M. S., and Silk, W. K. (1994) Nonvascular, symplasmic diffusion of sucrose cannot satisfy the carbon demands of growth in the primary root tip of *Zea mays* L. *Plant Physiol.* 105: 19–33.

Brundrett, M. C. (1991) Mycorrhizas in natural ecosystems. *Adv. Ecol. Res.* 21: 171–313.

Clarkson, D. T. (1985) Factors affecting mineral nutrient acquisition by plants. *Annu. Rev. Plant Physiol.* 36: 77–115.

Clarkson, D. T., and Hanson, J. B. (1980) The mineral nutrition of higher plants. *Annu. Rev. Plant Physiol.* 31: 239–298.

Clarkson, D. T., and Sanderson, J. (1978) Sites of absorption and translocation of iron in barley roots. Tracer and microautoradiographic studies. *Plant Physiol.* 61: 731–736.

Colmer, T. D., and Bloom, A. J. (1998) A comparison of net NH_4^+ and NO_3^- fluxes along roots of rice and maize. *Plant Cell Environ.* 21: 240–246.

Dittmer, H. J. (1937) A quantitative study of the roots and root hairs of a winter rye plant (*Secale cereale*). *Am. J. Bot.* 24: 417–420.

Epstein, E. (1994) The anomaly of silicon in plant biology. *Proc. Natl. Acad. Sci. USA* 91: 11–17.

Erickson, L. E., Banks, M. K., Davis, L. C., Schwab, A. P., Muralidharan, N., Reilley, K., and Tracy, J. C. (1994) Using vegetation to enhance in situ bioremediation. *Environ. Prog.* 13: 226–231.

Evans, H. J., and Sorger, G. J. (1966) Role of mineral elements with emphasis on the univalent cations. *Annu. Rev. Plant Physiol.* 17: 47–76.

Fohse, D., Claassen, N., and Jungk, A. (1991) Phosphorus efficiency of plants. 2. Significance of root radius, root hairs and cation-anion balance for phosphorus influx in 7 plant species. *Plant Soil* 132: 261–272.

Greenwood, D. J. (1982) Modelling of crop response of nitrogen fertilizer. *Philos. Trans. R. Soc. Lond. [Biol.]* 296: 351–362.

Grill, E., Loffler, S., Winnacker, E-L. and Zenk, M. H. (1989) Phytochelatins, the heavy metal-binding peptides of plants, are synthesized from glutathione by a specific γ-glutamylcysteine dipeptidyl transpeptidase (phytochelatin synthase). *Proc. Natl. Acad. Sci. USA* 86: 6838–6842.

Haehnel, W. (1984) Photosynthetic electron transport in higher plants. *Annu. Rev. Plant Physiol.* 35: 659–693.

Hepler, P. K., and Wayne, R. O. (1985) Calcium and plant development. *Annu. Rev. Plant Physiol.* 36: 397–439.

Johnstone, M., Grof, C. P. L., and Brownell, P. F. (1988) The effect of sodium nutrition on the pool sizes of intermediates of the C4 photosynthetic pathway. *Aust. J. Plant Physiol.* 15: 749–760.

Kashirad, A., Marschner, H., and Richter, C. H. (1973) Absorption and translocation of 59Fe from various parts of the corn plant. *Z. Pflanzenernähr. Bodenk.* 134: 136–147.

Klepper, B., and Kaspar, T. C. (1994) Rhizotrons—Their development and use in agricultural research. *Agron. J.* 86: 745–753.

Krämer, U., Cotter-Howells, J. D., Charnock, J. M., Baker, A. J. M., and Smith, J. A. C. (1996) Free histidine as a metal chelator in plants that accumulate nickel. *Nature* 379: 635–638.

Läuchli , A. (1993) Selenium in plants: Uptake, functions, and environmental toxicity. *Bot. Acta* 106: 455–468.

Lauriente, D. H. (1995) World fertilizer overview. In *Chemical Economics Handbook*, Stanford Research Institute, ed., SRI, Menlo Park, CA, pp. 166.

Lucas, R. E., and Davis, J. F. (1961) Relationships between pH values of organic soils and availabilities of 12 plant nutrients. *Soil Sci.* 92: 177–182.

Marschner, H. (1995) *Mineral Nutrition of Higher Plants*, 2nd ed. Academic Press, London.

Mauseth, J. D. (1988) *Plant Anatomy*. Benjamin/Cummings, Redwood City, CA.

Murphy, A., and Taiz, L. (1995) Comparison of metallothionein gene expression and nonprotein thiols in ten *Arabidopsis* ecotypes: Correlation with copper tolerance. *Plant Physiol.* 109: 945–954.

Murphy, A., and Taiz, L. (1997) Correlation between long term K^+ leakage and copper tolerance in ten *Arabidopsis* ecotypes. *New Phytol.* 136: 211–222.

Murphy, A., Zhou, J., Goldsbrough, P. B., and Taiz, L. (1997) Purification and immunological identification of metallothioneins 1 and 2 from *Arabidopsis thaliana*. *Plant Physiol.* 113: 1293–1301.

Nriagu, J. O., and Pacyna, J. M. (1988) Quantitative assessment of worldwide contamination of air, water and soils by trace metals. *Nature* 333: 134–139.

Nye, P. H., and Tinker, P. B. (1977) *Solute Movement in the Soil-Root System*. University of California Press, Berkeley, CA.

Olsen, R. A., Clark, R. B., and Bennet, I. H. (1981) The enhancement of soil fertility by plant roots. *Am. Sci.* 69: 378–384.

Raven, J. A., and Smith, F. A. (1976) Nitrogen assimilation and transport in vascular land plants in relation to intracellular pH regulation. *New Phytol.* 76: 415–431.

Rea, P. A., Li, Z-S., Lu, Y-P., Drozdowics, Y. M., and Martinoia, E. (In press) From vacuolar GS-X pumps to multispecific ABC transporters. *Annu. Rev. Plant Physiol. Plant Mol. Biol.*

Rovira, A. D., Bowen, C. D., and Foster, R. C. (1983) The significance of rhizosphere microflora and mycorrhizas in plant nutrition. In *Inorganic Plant Nutrition* (*Encyclopedia of Plant Physiology*, new series, vol. 15B A. Läuchli and R. L. Bieleski, eds., Springer, Berlin, pp. 61–93.

Russell, R. S. (1977) *Plant Root Systems: Their Functions and Interaction with the Soil.* McGraw-Hill, London.

Salt, D. E., and Rauser, W. E. (1995) MgATP-dependent transport of phytochelatins across the tonoplast of oat roots. *Plant Physiol.* 107: 1293–1301.

Sanders, F. E., and Tinker, P. B. (1971) Mechanism of absorption of phosphate from soil by endomycorrhizas. *Nature* 233: 278–279.

Schulze, E. D. (1989) Air pollution and forest decline in a spruce (*Picea abies*) forest. *Science* 244: 776–783.

Sharp, R. E., Hsiao, T. C., and Wilk, W. K. (1990) Growth of the maize primary root at low water potentials. 2. Role of growth and deposition of hexose and potassium in osmotic adjustment. *Plant Physiol.* 93: 1337–1346.

Shelp, B. J. (1993) Physiology and biochemistry of boron in plants. In *Boron and Its Role in Crop Production*, U. C. Gupta, ed., CRC Press, Boca Raton, FL, pp. 53–85.

Sievers, R. E., and Bailar, J. C., Jr. (1962) Some metal chelates of ethylenediaminetetraacetic acid, diethylenetriaminepentaacetic acid, and triethylenetriaminehexaacetic acid. *Inorganic Chem.* 1: 174–182.

Stewart, G. R., and Ahmad, I. (1983) Adaptation to salinity in angiosperm halophytes. In *Metals and Micronutrients: Uptake and Utilization by Plants*, D. A. Robb and W. S. Pierpoint, eds., Academic Press, New York, pp. 33–50.

Stohs, S. J., and Bagchi, D. (1995) Oxidative mechanisms in the toxicity of metal ions. *Free Radic. Biol. Med.* 18: 321–336.

Terry, N. (1977) Photosynthesis, growth and the role of chloride. *Plant Physiol.* 60: 69–75.

Tinker, P. B., and Gildon, A. (1983) Mycorrhizal fungi and ion uptake. In *Metals and Micronutrients: Uptake and Utilization by Plants*, D. A. Robb and W. S. Pierpoint, eds., Academic Press, New York, pp. 21–32.

U.S. Congress. Office of Technology Assessment. (1990) *Beneath the Bottom Line: Agricultural Approaches to Reduce Agrichemical Contamination of Groundwater*. U.S. Government Printing Office, Washington, DC.

Weathers, P. J., and Zobel, R. W. (1992) Aeroponics for the culture of organisms, tissues, and cells. *Biotech. Adv.* 10: 93–115.

Weaver, J. E. (1926) *Root Development of Field Crops*. McGraw-Hill, New York.

Wilcox, H. E. (1991) Mycorrhizae. In *The Plant Root, the Hidden Half*, Y. Waisel and U. Kafkafi, eds., Marcel Dekker, New York, pp. 731–765.

Ziegler, H. (1987) The evolution of stomata. In *Stomatal Function*, E. Zeiger, G. Farquhar, and I. Cowan, eds., Stanford University Press, Stanford, CA, pp. 29–57.

6 Solute Transport

PLANT CELLS ARE SEPARATED from their environment by a plasma membrane that is only two lipid molecules thick. This thin sheet separates a relatively constant internal milieu from a highly variable external environment. It must accommodate and promote continual inward and outward traffic of molecules and ions, as the cell takes up nutrients and exports wastes. As the cell's only contact with the environment, the plasma membrane must also relay information about its physical environment, about molecular signals from other cells, and about the presence of invading pathogens. To accomplish these tasks, the plasma membrane must provide second-by-second control of the rate of movement of each of a multitude of solutes. The same is true of the internal membranes that separate the various compartments within each cell.

Molecular and ionic movement between different compartments in biological systems is known as **transport**. Transport can be considered at the level of individual cellular membranes, or between a plant and its environment, or from one part of the plant to another. However, such higher-order transport is generally driven and controlled by transport at the cellular level. For example, the transport of sucrose from leaf to root through the phloem, referred to as *translocation*, may be driven and regulated by membrane transport into the phloem cells of the leaf, and from the phloem to the storage cells of the root (see Chapter 10).

In this chapter, we will consider first the physical and chemical principles that govern the movements of molecules in solution. Next we will show how these principles apply to biological systems; we will also discuss the molecular mechanisms of transport in living cells and the great variety of membrane transport proteins that are responsible for the particular transport properties of plant cells. Finally, we will examine the radial pathway that ions take when they enter the root, and the mechanism of *xylem loading*, the process whereby ions are released into the vessel elements and tracheids of the stele.

Passive and Active Transport

According to Fick's law (see Equation 3.2), the movement of molecules by diffusion will always proceed spontaneously, down a concentration or chemical gradient, until equilibrium is reached. The movement of molecules by diffusion is termed **passive transport**. At equilibrium, no further net movements of solute can occur without the application of a driving force. The movement of substances against a concentration or chemical gradient, termed **active transport**, is not spontaneous and requires that work be done on the system by the application of cellular energy. One way of accomplishing this task is to couple transport to the hydrolysis of ATP.

We can calculate the driving force for diffusion, or the energy input necessary to move substances against a gradient, by measuring the potential-energy gradient, which is often a simple function of the difference in concentration. As we stated in Chapter 3, biological transport can also be driven by other forces: hydrostatic pressure, gravity, and electric fields. (In biological systems, however, gravity seldom contributes substantially to the force that drives transport.) Along with concentration gradients, these sources of potential energy define the **chemical potential** of any solute (see Chapter 2) according to the following relation:

$$\tilde{\mu}_j = \mu_j^* + RT \ln C_j + z_j FE + \bar{V}_j P \qquad (6.1)$$

$\tilde{\mu}_j$: Chemical potential for a given solute, *j*
μ_j^*: Chemical potential of *j* under standard conditions
$RT \ln C_j$: Concentration (activity) component
$z_j FE$: Electric-potential component
$\bar{V}_j P$: Hydrostatic-pressure component

Here $\tilde{\mu}_j$ is the chemical potential of the solute species *j* in joules per mole (J mol^{-1}), μ_j^* is its chemical potential under standard conditions (a correction factor that will cancel out in future equations and so can be ignored), *R* is the universal gas constant, *T* is the absolute temperature, and C_j is the concentration (more accurately the activity) of *j*. The electrical term, $z_j FE$, applies only to dissolved ions; *z* is the electrostatic charge of the ion (+1 for monovalent cations, –1 for monovalent anions, +2 for divalent cations, and so on), *F* is Faraday's constant (equivalent to the electric charge on 1 mol of protons), and *E* is the overall electric potential of the solution (with respect to ground). The final term, $\bar{V}_j P$, expresses the contribution of the partial molal volume of *j* (V_j) and pressure (*P*) to the chemical potential of *j*. (The partial molal volume of *j* is the change in volume per mole of substance *j* added to the system, for an infinitesimal addition.) This final term makes a much smaller contribution to $\tilde{\mu}$ than do the concentration and electrical terms, except in the very important case of osmotic water movements. As discussed in Chapter 3, the chemical potential of water (i.e., the water potential) depends on the concentration of dissolved solutes and the hydrostatic pressure on the system.

As indicated in Chapter 2, writing an equation in terms of the chemical potential, $\tilde{\mu}_j$, allows one to consider the contribution of an individual solute (in this case *j*) to the free energy of the system. The term $\tilde{\mu}$ can be described as the partial molal free energy for the solute *j*: It represents the increase in free energy per mole of *j* added, for an infinitesimal addition under constant conditions, and $\Delta\tilde{\mu}_j$ is the contribution of the solute *j* to the total free-energy change (ΔG) of the system. *The importance of the concept of chemical potential is that it sums all the forces that may act on a molecule to drive net transport* (Nobel 1991). In general, diffusion (or passive transport) always moves molecules from areas of higher chemical potential downhill to areas of lower chemical potential. Movement against a chemical-potential gradient is indicative of active transport (Figure 6.1).

Taking as an example the diffusion of sucrose across a permeable membrane, the chemical potential of sucrose in any compartment is accurately approximated by the concentration term alone (unless a solution is very concentrated, causing hydrostatic pressure to build up). The chemical potential of sucrose inside a cell can be described as follows (in the next three equations, the subscript "s" stands for sucrose and the superscripts "i" and "o" stand for inside and outside, respectively):

$$\tilde{\mu}_s^i = \mu_s^* + RT \ln C_s^i$$

$\tilde{\mu}_s^i$: Chemical potential of sucrose solution inside the cell
μ_s^*: Chemical potential of sucrose solution under standard conditions
$RT \ln C_s^i$: Concentration component

The chemical potential of sucrose outside the cell is

$$\tilde{\mu}_s^o = \mu_s^* + RT \ln C_s^o$$

We can calculate the difference in the chemical potential of sucrose between the solutions inside and outside the cell, regardless of the mechanism of transport. For inward transport, sucrose is being removed from outside the cell and added to the inside, so the change in free energy due to sucrose transport will be as follows:

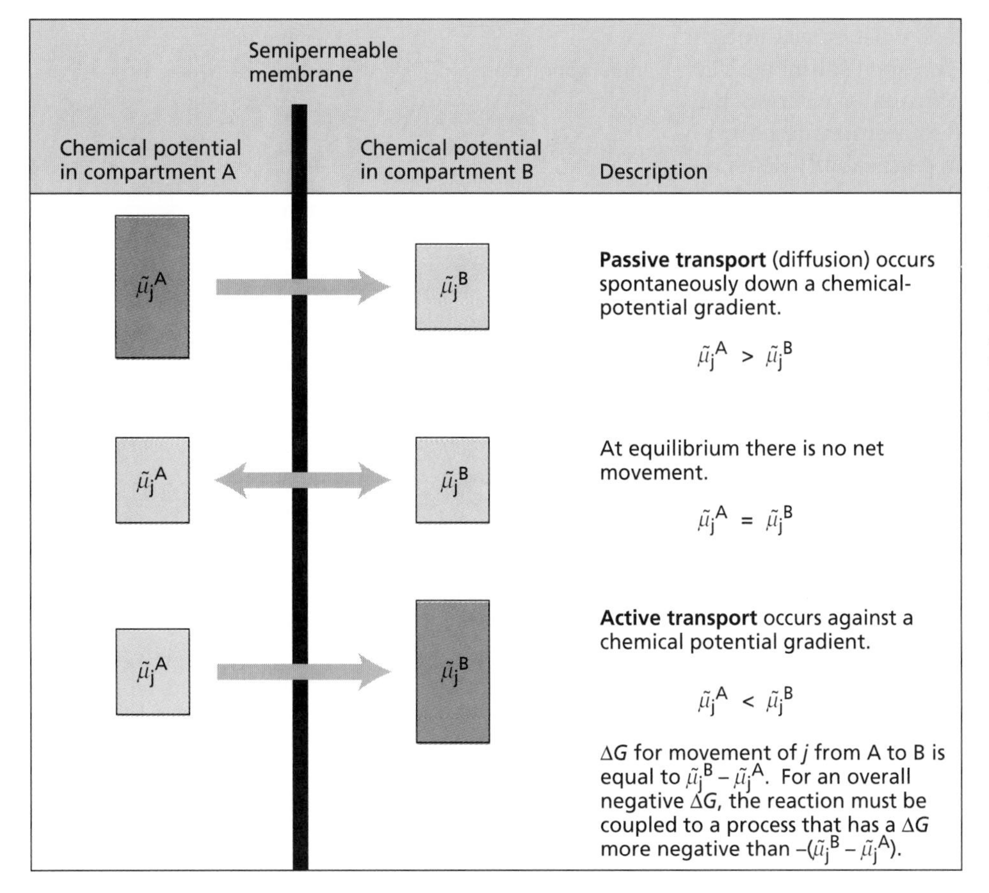

Figure 6.1 Relationship between the chemical potential, $\tilde{\mu}$, and the transport of molecules across a permeability barrier. The net movement of molecular species j between compartments A and B depends on the relative magnitude of the chemical potential of j in each compartment, represented here by the height of the bars. Movement down a chemical gradient occurs spontaneously; movement against a gradient requires energy and is called active transport.

$$\begin{aligned}\Delta\tilde{\mu}_s &= \tilde{\mu}_s^{\,i} - \tilde{\mu}_s^{\,o} \\ &= \left(\tilde{\mu}_s^{\,*} + RT\ln C_s^{\,i}\right) - \frac{RT}{z_jF}\left(\tilde{\mu}_s^{\,*} + RT\ln C_s^{\,o}\right) \\ &= RT\left(\ln C_s^{\,i} - \ln C_s^{\,o}\right) \\ &= RT\ln\frac{C_s^{\,i}}{C_s^{\,o}}\end{aligned} \tag{6.2}$$

If this difference in chemical potential is negative, sucrose could diffuse inward spontaneously (provided the membrane had a finite permeability to sucrose; see the next section). In other words, the driving force ($\Delta\tilde{\mu}_s$) for solute diffusion is related to the magnitude of the concentration gradient ($C_s^{\,i}/C_s^{\,o}$).

If the solute carries an electric charge (as does the potassium ion), the electrical component of the chemical potential must also be considered. Suppose the membrane is permeable to K^+ and Cl^- rather than to sucrose. Because the ionic species (K^+ and Cl^-) diffuse independently, each has its own chemical potential. Thus for inward K^+ diffusion

$$\begin{aligned}\Delta\tilde{\mu}_{K^+} &= \tilde{\mu}^{i}_{K^+} - \tilde{\mu}^{o}_{K^+} \\ &= RT\ln\frac{[K^+]^i}{[K^+]^o} + zF\left(E^i - E^o\right)\end{aligned}$$

and because the electrostatic charge of K^+ is +1, $z = +1$ and

$$\Delta\tilde{\mu}_{K^+} = RT\ln\frac{[K^+]^i}{[K^+]^o} + F(E^i - E^o) \tag{6.3}$$

The magnitude and sign of this expression will indicate the driving force for K^+ diffusion across the membrane, and its direction. A similar expression can be written for Cl^- (but remember that $z_{Cl^-} = -1$).

Equation 6.3 shows that ions, such as K^+, diffuse in response to both their concentration gradients ($[K^+]^i/[K^+]^o$) *and* any electric-potential difference between the two compartments ($E^i - E^o$). One very important implication of this equation is that ions can be driven passively *against* their concentration gradients if an appropriate voltage (electric field) is applied between the two compartments. Because of the importance of electric fields in biological transport, $\tilde{\mu}$ is often called the **electrochemical potential** and $\Delta\tilde{\mu}$ is the difference in electrochemical potential between two compartments.

Transport of Solutes across a Membrane Barrier

If the two KCl solutions in the previous example are separated by a real biological membrane, diffusion is complicated by the fact that the ions must move through the membrane as well as across the open solutions. The extent to which a membrane permits or restricts the movement of a substance is called **membrane permeability**. As will be discussed later, permeability depends on the composition of the membrane as well as on the chemical nature of the solute. In a loose sense, permeability can be expressed in terms of a diffusion coefficient for the solute in the membrane. However, permeabilities are influenced by several additional factors, such as membrane thickness, that are difficult to measure. Despite their theoretical complexity, permeabilities can readily be measured by a determination of the rate at which a solute passes through a membrane under a specific set of conditions. Generally, the membrane will slow diffusion and thus reduce the speed with which equilibrium is reached. The membrane itself, however, cannot alter the final equilibrium conditions.

In this section we will discuss the factors that influence the passive distribution of ions across a membrane. These parameters can be used to predict the relationship between the electric gradient and the concentration gradient of an ion.

Initial conditions: $[KCl]_A > [KCl]_B$

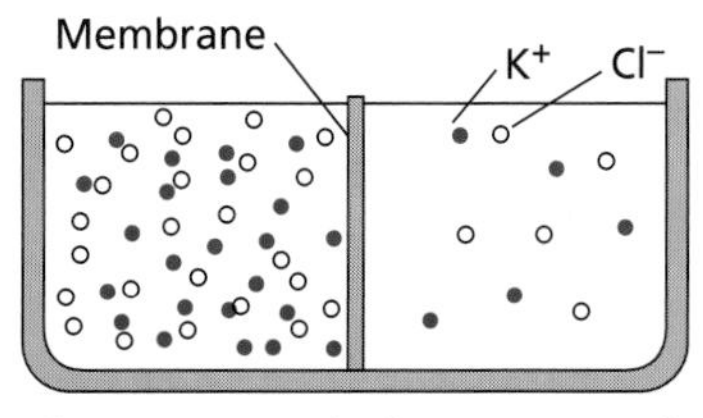

Diffusion potential exists until chemical equilibrium is reached

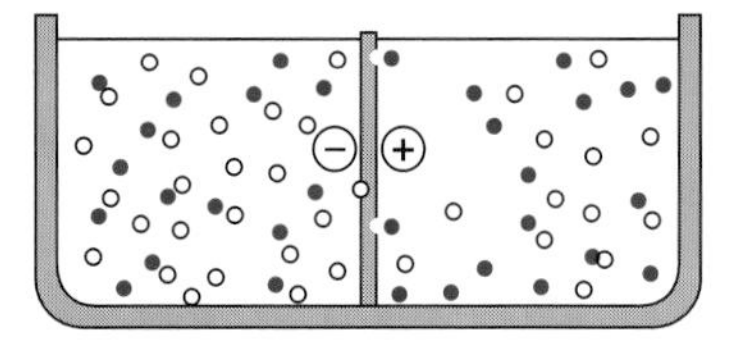

Equilibruim conditions: $[KCl]_A = [KCl]_B$

At chemical equilibrium, diffusion potential equals zero

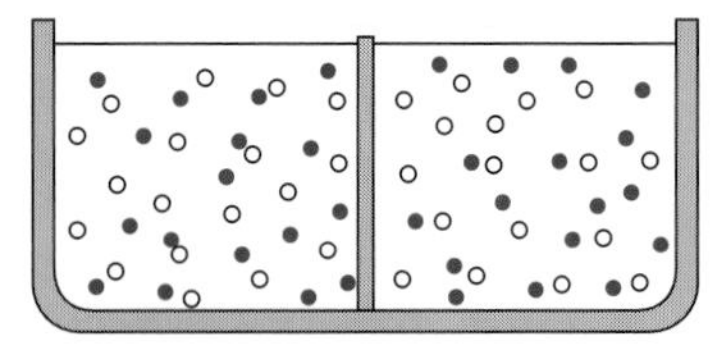

Figure 6.2 Development of a diffusion potential and a charge separation between two compartments separated by a membrane that is preferentially permeable to cations. If the concentration of potassium chloride is higher in compartment A ($[KCl]_A > [KCl]_B$), potassium and chloride ions will diffuse at a higher rate into compartment B, and a diffusion potential will be established. When membranes are more permeable to potassium than to chloride, potassium ions will diffuse faster than chloride ions, and charge separation (+ and –) will develop.

Diffusion Potentials Develop When Oppositely Charged Ions Move across a Membrane at Different Rates

When salts diffuse across a membrane, an electric membrane potential (voltage) can develop. Consider the two KCl solutions separated by a membrane in Figure 6.2. The K^+ and Cl^- ions will permeate the membrane independently as they diffuse down their respective gradients of electrochemical potential. Unless the membrane is very porous, its permeabilities for the two ions will differ. Consequently, K^+ and Cl^- initially will diffuse across the membrane at different rates. The result will be a slight separation of charge, which instantly creates an electric potential across the membrane. In biological systems, membranes are usually more permeable to K^+ than to Cl^-. Therefore, K^+ will diffuse out of the cell (compartment A in Figure 6.2) faster than Cl^-, causing the cell to develop a negative electric charge with respect to the medium. The membrane potential that develops as a result of diffusion is called the **diffusion potential**.

An important principle that must always be kept in mind when the movement of ions across membranes is considered is the **principle of electric neutrality**. Bulk solutions always contain equal numbers of anions and cations. The existence of a membrane potential implies that the distribution of charges across the membrane is uneven; however, *the actual number of unbalanced ions is negligible in chemical terms*. For example, a membrane potential of –100 mV (millivolts), like that found across the plasma membranes of many plant cells, results from the presence of only one extra anion out of every 100,000 within the cell!

As Figure 6.2 shows, all of these extra anions are found immediately adjacent to the surface of the membrane; there is no charge imbalance throughout the bulk of the cell. In our example of KCl diffusion across a membrane, electric neutrality is preserved because, as K^+ moves ahead of Cl^- in the membrane, the resulting diffusion potential retards the movement of K^+ and speeds that of Cl^-. Ultimately, both ions diffuse at the same rate, but the diffusion potential persists. As the system moves toward equilibrium and the concentration gradient collapses, the diffusion potential also collapses.

The Nernst Equation Relates the Membrane Potential and the Distribution of an Ion at Equilibrium

In the preceding example, because the membrane is permeable to both ions (K^+ and Cl^-), equilibrium will not be reached for either ion until the concentration gradients decrease to zero. However, if the membrane were permeable only to K^+, diffusion of K^+ would carry charges across the membrane until the membrane potential balanced the concentration gradient. Since a change in potential requires very few ions, this balance would be reached instantly. Transport would then be at equilibrium even though the concentration gradients were unchanged.

When the distribution of any solute across a membrane reaches **equilibrium**, the passive flux (J) (that is, the amount of solute crossing a unit area of membrane per unit time) is the same in the two directions, outside to inside and inside to outside:

$$J_{o \to i} = J_{i \to o}$$

We have already seen that fluxes are proportional to the driving forces; thus, at equilibrium, the electrochemical potentials will be the same:

$$\tilde{\mu}_j^{o} = \tilde{\mu}_j^{i}$$

and for any given ion (the ion is symbolized here by the subscript j):

$$\mu_j^* + RT \ln C_j^{o} + z_j F E^{o} = \mu_j^* + RT \ln C_j^{i} + z_j F E^{i}$$

By rearranging this equation, we can obtain the difference in electric potential between the two compartments at equilibrium ($E^i - E^o$):

$$E^{i} - E^{o} = \frac{RT}{z_j F}\left(\ln \frac{C_j^{o}}{C_j^{i}}\right)$$

This electric-potential difference is known as the **Nernst potential** (ΔE_n) for that ion:

$$\Delta E_n = E^{i} - E^{o}$$

and

$$\Delta E_n = \frac{RT}{z_j F}\left(\ln \frac{C_j^{o}}{C_j^{i}}\right)$$

or

$$\Delta E_n = \frac{2.3\,RT}{z_j F}\left(\log \frac{C_j^{o}}{C_j^{i}}\right) \qquad (6.4)$$

This relationship, known as the **Nernst equation**, states that at equilibrium the difference in concentration of an ion between two compartments is balanced by the voltage difference between the compartments. The Nernst equation can be further simplified for a univalent cation at 25°C:

$$\Delta E_n = 59 \log \frac{C_j^{o}}{C_j^{i}} \qquad (6.5)$$

where ΔE_n is expressed in millivolts. Note that a tenfold difference in concentration corresponds to a Nernst potential of 59 mV ($C^o/C^i = 10/1$; log 10 = 1). That is, a membrane potential of 59 mV would maintain a tenfold concentration gradient of an ion that is transported by passive diffusion. Similarly, if a tenfold concentration gradient of an ion existed across the membrane, passive diffusion of that ion down its concentration gradient would result in a difference of 59 mV across the membrane.

All living cells exhibit a membrane potential that is due to the asymmetric ion distribution between the inside and outside of the cell. We can readily determine these membrane potentials by inserting a microelectrode into the cell and measuring the voltage difference between the inside of the cell and the external bathing medium (Figure 6.3).

The Nernst equation can be used at any time to determine whether a given ion is at equilibrium across a membrane. However, a distinction must be made between equilibrium and steady state. **Steady state** is the condition in which influx and efflux of a given solute are equal and therefore the ion concentrations are constant with respect to time. Steady state is *not* the same as equilibrium (see Figure 6.1); in steady state, the existence of active transport across the membrane prevents many diffusive fluxes from ever reaching equilibrium. For an ion that is in steady state in a biological system, a simple experimental test for active transport is to determine the concentrations of the ion inside and outside of the cell, use these to calculate the Nernst potential for the ion in question, and compare this value with the experimentally measured value of the membrane potential. Any significant disparity between ΔE_n and the measured membrane potential (ΔE) suggests that the ion is being transported actively.

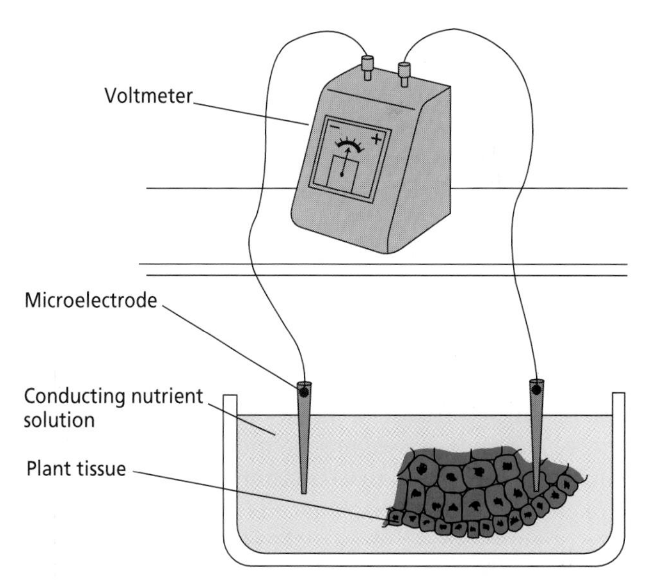

Figure 6.3 Diagram of a pair of microelectrodes used to measure membrane potentials across cell membranes. One of the glass micropipette electrodes is inserted into the cell compartment under study (usually the vacuole or the cytoplasm), while the other is kept in an electrolytic solution that serves as a reference. The microelectrodes are connected to a voltmeter, which records the electric-potential difference between the cell compartment and the solution. Typical membrane potentials across plant cell membranes range from –60 to –240 mV.

TABLE 6.1
Comparison of observed and predicted ion concentrations in pea root tissue

Ion	Concentration in external medium (mmol L^{-1})	Internal Concentration (mmol L^{-})[1]	
		Predicted	Observed
K^+	1	74	75
Na^+	1	74	8
Mg^{2+}	0.25	1,340	3
Ca^{2+}	1	5,360	2
NO_3^-	2	0.0272	28
Cl^-	1	0.0136	7
$H_2PO_4^-$	1	0.0136	21
SO_4^{2-}	0.25	0.00005	19

Source: Data from Higinbotham et al. 1967.
Note: The membrane potential was measured as –110 mV.

The Nernst Equation Can Be Used to Distinguish between Active and Passive Transport

Table 6.1 shows how the experimentally measured ion gradients at steady state for pea root cells compare with predicted values calculated from the Nernst equation (Higinbotham et al. 1967). In this example, the external concentration of each ion in the solution bathing the tissue and the measured membrane potential were substituted into the Nernst equation, and a predicted internal concentration was calculated for that ion. Notice that, of all the ions, *only K^+ is at or near equilibrium*. The anions—NO_3^-, Cl^-, $H_2PO_4^-$, and SO_4^{2-}—all have higher internal concentrations than predicted, indicating that their uptake is active. The cations Na^+, Ca^{2+}, and Mg^{2+} have lower internal concentrations than predicted; therefore, these ions enter the cell by diffusion down their electrochemical-potential gradients and then are actively extruded.

The example shown in Table 6.1 is an oversimplification, for plant cells have several internal compartments, each of which can differ in their ionic composition. The cytosol and the vacuole are the most important intracellular compartments that determine the ionic relations of plant cells. In mature plant cells, the central vacuole often occupies 90% or more of the cell's volume, and the cytosol is restricted to a thin layer around the periphery of the cell. Because of its small volume, the cytosol of most angiosperm cells is difficult to assay chemically. For this reason, much of the early work on the ionic relations of plants focused on certain green algae, such as *Chara* and *Nitella*, whose cells are several inches long and can contain an appreciable volume of cytosol. Figure 6.4 diagrams the conclusions from these studies and from related work with higher plants.

Potassium is accumulated passively by both the cytosol and the vacuole except when extracellular K^+ concentrations are very low, in which case it is taken up actively. Sodium is pumped actively out of the cytosol into the extracellular spaces and vacuole. Excess protons, generated by intermediary metabolism, are also actively extruded from the cytosol. This process helps maintain the cytosolic pH near neutrality, while the vacuole and the extracellular medium are generally more acidic by 1 or 2 pH units. All the anions are taken up actively into the cytosol. Calcium is actively transported out of the cytosol at both the cell membrane and the vacuolar membrane, which is called the tonoplast (see Figure 6.4).

The Goldman Equation Relates Diffusion Potential and Prevailing Ion Gradients across a Membrane

Many different ions permeate the membranes of living cells simultaneously; therefore, the Nernst potential for any single ionic species seldom describes the membrane's diffusion potential. Instead, the contributions of all the ions to the diffusion potential across the membrane are described by the **Goldman equation**:

$$\Delta E = \frac{RT}{F}\left(\ln \frac{P_{K^+}C_{K^+}{}^{o} + P_{Na^+}C_{Na^+}{}^{o} + P_{Cl^-}C_{Cl^-}{}^{i}}{P_{K^+}C_{K^+}{}^{i} + P_{Na^+}C_{Na^+}{}^{i} + P_{Cl^-}C_{Cl^-}{}^{o}}\right) \tag{6.6}$$

This equation relates the standing ion gradients across a membrane to the diffusion potential that develops. P_{K^+}, P_{Na^+}, and P_{Cl^-} represent the membrane permeabilities for K^+, Na^+, and Cl^-, respectively. Although the equation should include terms for all ions passing through the membrane, K^+, Na^+, and Cl^- have the largest membrane permeabilities and highest concen-

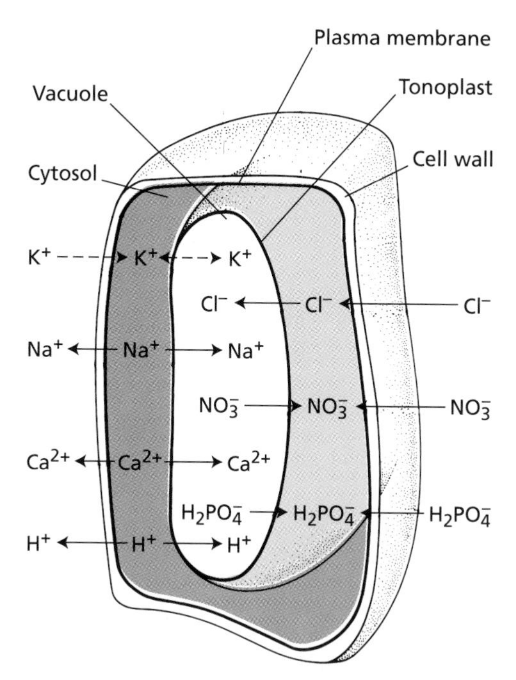

Figure 6.4 Ion concentrations in the cytosol and the vacuole are controlled by passive (dashed arrows) and active (solid arrows) transport processes. In most plant cells the vacuole occupies up to 90% of the cell's volume and contains the bulk of the cell solutes. Control of the ion concentrations in the cytosol is important for the regulation of metabolic enzymes. The cell wall surrounding the plasma membrane does not represent a permeability barrier and hence is not a factor in solute transport.

trations in plant cells and therefore dominate the equation.

The relationship between the Goldman equation (Equation 6.6) and the Nernst equation (Equation 6.4) can be seen if one imagines a membrane that is permeable to only one ion, say K^+, so that both P_{Cl^-} and P_{Na^+} equal zero. Under these conditions, the Goldman equation reduces to the Nernst equation for K^+. Although biological membranes are never permeable to only a single ionic species, artificial membranes can approximate this situation and are used in the manufacture of pH electrodes and other ion-selective electrodes.

Proton Transport Is the Major Determinant of the Membrane Potential

When permeabilities and ion gradients are known, it is possible to calculate a diffusion potential for the membrane from the Goldman equation (Equation 6.6). In most plant cells, K^+ has both the greatest internal concentration and the highest membrane permeability, so the diffusion potential may approach the Nernst potential for K^+. Under some conditions, this phenomenon is observed. In plants and fungi, however, the experimentally measured membrane potentials are often much more negative than those calculated from the Goldman equation. For example, cells in stems and roots of young seedlings generally have membrane potentials of –130 to –110 mV, whereas their calculated diffusion potentials are usually only –80 to –50 mV. Thus, in addition to the diffusion potential, the membrane potential has a second component. The excess voltage is provided by the plasma membrane electrogenic H^+-ATPase.

Whenever an ion moves into or out of a cell without being balanced by countermovement of an ion of opposite charge, a voltage is created across the membrane. Any active transport mechanism that results in the movement of a net electric charge will tend to move the membrane potential away from the value predicted by the Goldman equation. Such a transport mechanism is called an electrogenic pump and is common in living cells.

The energy required for active transport is often provided by the hydrolysis of ATP. In plants we can demonstrate the dependence of the membrane potential on ATP by observing the effect of cyanide (CN^-) on the membrane potential (Figure 6.5). Cyanide rapidly poisons the mitochondria, and the cell's ATP consequently becomes depleted. As ATP synthesis is inhibited, the membrane potential falls to the level of the Goldman diffusion potential, which is due to the passive movements of K^+, Cl^-, and Na^+ (see Equation 6.6). Thus, the membrane potentials of plant cells have two components: a diffusion potential and a component resulting

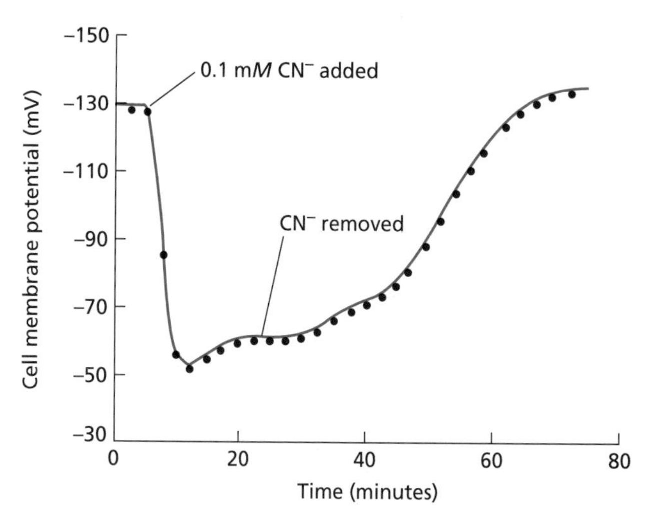

Figure 6.5 The membrane potential of a pea cell collapses when cyanide (CN^-) is added to the bathing solution. Cyanide blocks ATP production in the cells by poisoning the mitochondria. The collapse of the membrane potential upon addition of cyanide indicates that an ATP supply is necessary for maintenance of the potential. Washing the cyanide out of the tissue results in a slow recovery of ATP production and restoration of the membrane potential. (From Higinbotham et al. 1970.)

from electrogenic ion transport (transport that results in the generation of a membrane potential) (Spanswick 1981). When cyanide inhibits electrogenic ion transport, the pH of the external medium increases while the cytosol becomes acidic because H^+ remains inside the cell. This is one piece of evidence that the active transport of H^+ out of the cell is electrogenic.

As discussed earlier, a change in the membrane potential caused by an electrogenic pump will change the driving forces for diffusion of all ions that cross the membrane. For example, the outward transport of H^+ can create a driving force for the passive diffusion of K^+ into the cell. H^+ is transported electrogenically across the cell membrane not only in plants but also in bacteria, algae, fungi, and some animal cells, such as those of the kidney epithelia. ATP synthesis in mitochondria and chloroplasts also depends on an H^+-ATPase. In mitochondria and chloroplasts, this transport protein is sometimes called an ATP synthase because it forms ATP rather than hydrolyzing it (see Chapter 11). The mitochondrial ATP synthase is structurally different from the H^+-ATPases located in plasma membranes. The structure and function of membrane proteins involved in active and passive transport in plant cells will be discussed in the next section.

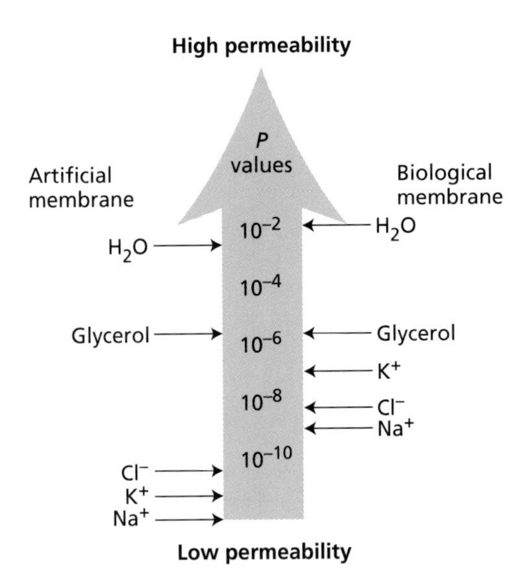

Figure 6.6 Permeability values, *P*, in centimeters per second, for some substances diffusing across an artificial phospholipid bilayer and across a biological membrane. For nonpolar molecules, and some small uncharged substances such as water, *P* values are similar in both systems, but for most polar molecules, and for ions such as potassium or sodium, *P* values are much higher in biological membranes, reflecting the role of transport proteins. Note the logarithmic scale.

Membrane Transport Proteins

Artificial membranes have been used extensively to study the permeability of pure phospholipid membranes. Because of the nonpolar nature of the membrane's interior, pure phospholipid bilayers are highly impermeable to ions or polar molecules. Figure 6.6 shows the permeabilities of artificial bilayers for a variety of substances. Nonpolar molecules such as O_2 pass through these membranes very rapidly, and the membrane permeabilities for small polar molecules (for example, water, CO_2 and glycerol) are slightly lower but still significant. Among polar molecules, water has an unusually high membrane permeability, probably because of its strong interactions with the polar head groups of the lipids and because of its small size. In general, as molecules increase in size and polarity, their membrane permeabilities decrease. Ions penetrate pure phospholipid bilayers so slowly that these artificial membranes may be considered impermeable to them.

When the permeabilities of artificial bilayers for ions and molecules are compared with those of biological membranes, important similarities and differences emerge (see Figure 6.6). For nonpolar molecules and many small polar molecules, both types of membranes have similar permeabilities. On the other hand, for ions and large polar molecules such as sugars, biological membranes are much more permeable than artificial bilayers. The reason is that, unlike artificial bilayers, biological membranes contain **transport proteins** that facilitate the passage of ions and other polar molecules. The molecules for which biological and artificial membranes exhibit similar permeabilities apparently diffuse directly through the lipid phase of the membrane. However, most of the substances that are important to cell nutrition and metabolism cannot diffuse across the lipid bilayer directly and must be transported by membrane proteins. These transport proteins can be grouped into three main categories: *channels, carriers, and pumps.*

Transport proteins exhibit specificity in the solutes they transport, hence their great diversity in cells. The simple prokaryote *Haemophilus influenzae,* the first organism for which the complete genome has been sequenced, has only 1743 genes, yet more than 200 of these genes (greater than 10% of the genome) encode various proteins involved in membrane transport. Although a particular transport protein is usually highly specific for the kinds of substances it will transport, it generally also transports a family of related substances. For example, in plants a K^+ transporter on the plasma membrane may transport Rb^+ and Na^+ in addition to K^+, but K^+ is usually preferred. At the same time, the K^+ transporter is completely ineffective in transporting anions such as Cl^- or uncharged solutes such as sucrose. Similarly, a protein involved in the transport of neutral amino acids may move glycine, alanine, and valine with equal ease but not accept aspartic acid or lysine.

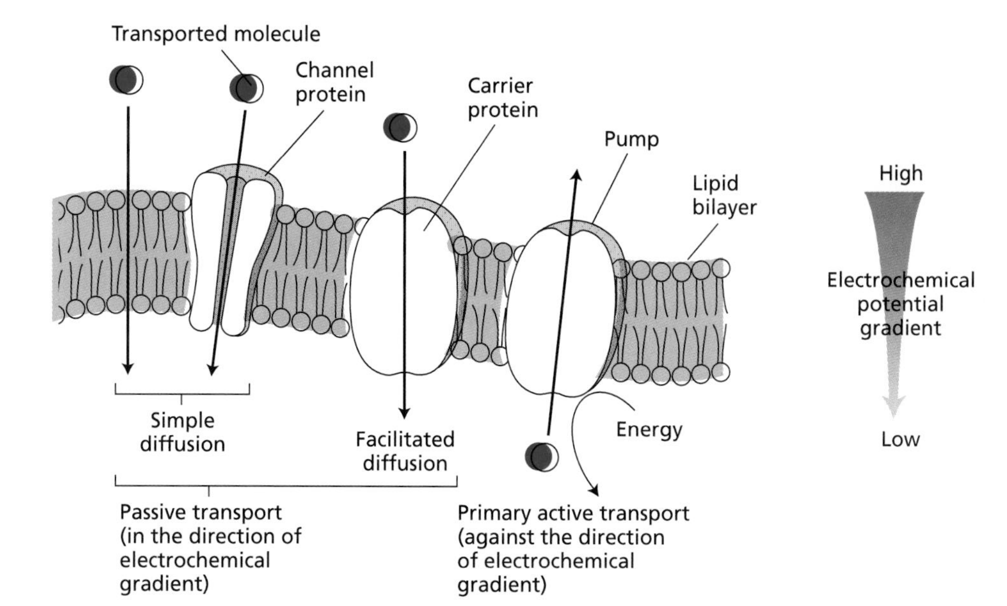

Figure 6.7 The three classes of membrane transport proteins: channels, carriers, and pumps. Channels and carriers can mediate the passive transport of solutes across membranes (by simple diffusion or facilitated diffusion), down the solute's gradient of electrochemical potential. Channel proteins act as membrane pores, and their specificity is determined primarily by the biophysical properties of the channel. Carrier proteins bind the transported molecule on one side of the membrane and release it on the other side. Primary active transport is carried out by pumps and uses energy directly, usually from ATP hydrolysis, to pump solutes against their gradient of electrochemical potential.

In this section we will consider the structures, functions, and physiological roles of the various channels, carriers, and pumps found in plant cells, especially on the plasma membrane and tonoplast. We begin with a discussion of the role of channels and carriers in promoting the diffusion of solutes across membranes. We then distinguish between primary and secondary active transport, and discuss the roles of the electrogenic H^+-ATPase and various symporters (proteins that transport two substances in the same direction simultaneously) in driving proton-coupled secondary active transport. Finally, we examine the structure and functional domains of various transport proteins present on the plasma membrane and tonoplast.

Two Types of Transporters Enhance Solute Diffusion across Membranes

Two types of membrane transporters enhance the movement of solutes across membranes: *channels* and *carriers* (Figure 6.7). In general, **channels** are transmembrane proteins that function as selective pores in the membrane. The size of a pore and the density of surface charges on its interior lining determine its transport specificity. Molecules that can move through channels move through the pore. Transport through a channel may or may not involve transient binding of the solute to the channel protein. In any case, as long as the channel pore is open, solutes that can penetrate the pore diffuse through it extremely rapidly: about 10^8 ions s^{-1} through each channel protein. However, channels do not usually stay open for long periods: They have "gates" that open and close the pore in response to external signals (Figure 6.8). Transport through channels is always passive, and because the specificity of transport depends on pore size and electric charge more than on selective binding, channel transport is limited mainly to ions or water. The region of the channel that determines specificity is called the **selectivity filter** (see Figure 6.8).

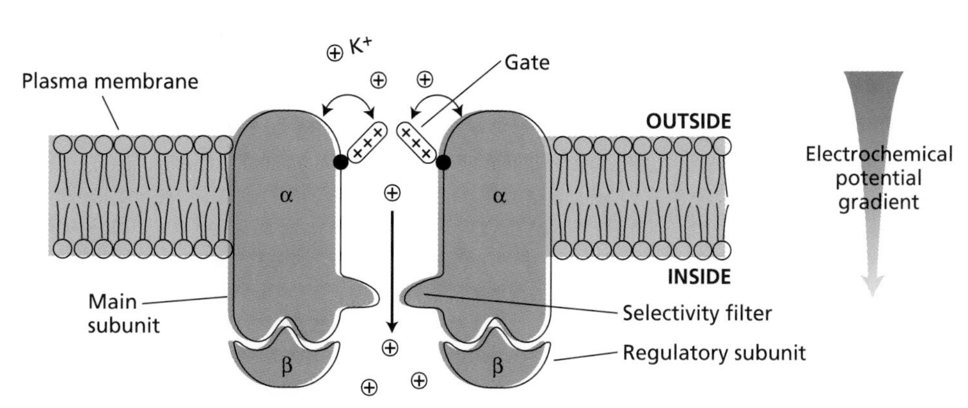

Figure 6.8 Model of a voltage-gated K^+ channel in a plant. The channel itself is composed of a tetramer of the main subunit (α), which contains the selectivity filter and the voltage gate. The voltage gate sequence consists of a cluster of basic amino acids that provide the positive charge. In response to the membrane potential, the voltage gate opens or closes the channel. Regulatory β subunits may also be present.

BOX 6.1

Patch Clamp Studies in Plant Cells

THE USE OF THE PATCH CLAMP METHOD in plant physiology is generating information about the properties of plant membranes that is beyond the scope of other techniques. Conventional electrophysiological studies with intracellular electrodes (see Figure 6.3) are valuable for the measurement of membrane potentials and other electrical properties of plant cells, but these methods have limitations for the characterization of ion pumps and single ion channels. Patch clamping, on the other hand, is very well suited for that purpose (Hedrich and Schroeder 1989).

During conventional intracellular recording of cellular electric potentials, cells are impaled with a glass micropipette electrode, and the electrode tip is located inside the cell, usually in the vacuole. For patch clamp experiments, the cell walls are first digested away by treatment with the appropriate cell wall enzymes, resulting in naked protoplasts. The tip of a glass micropipette is then brought into contact with the plasma membrane of the protoplast, and gentle suction is applied to facilitate the formation of a tight seal between the micropipette and the membrane (Figures 1 and 2). The unusual tightness of the seal that forms between the patch pipette and the membrane reduces the background electrical noise (due to leakage) sufficiently to allow high-resolution recordings of the currents through single ion channels.

When a tight seal is attained, several options become available to the investigator. Further suction can remove the portion of the membrane delimited by the opening in the pipette tip, exposing the interior of the cell to the micropipette solution (thus different solutions can be introduced into the cell through the micropipette). In this configuration, measured electric currents reflect the sum total of all electric charges carried by active and passive ion fluxes across the plasma membrane of the **whole cell** (see Figure 2).

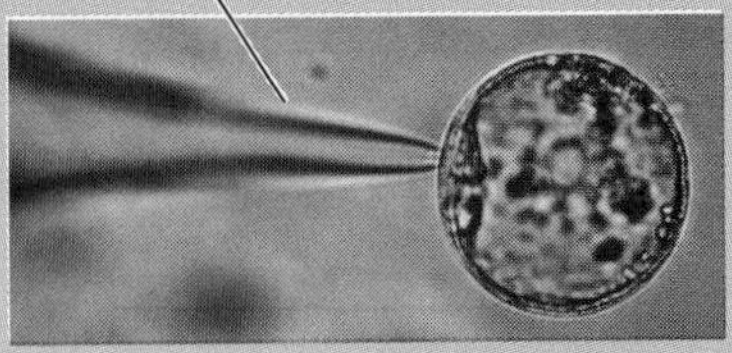

Patch pipette

Figure 1 Micrograph of a patch pipette attached to the surface of a barley aleurone cell protoplast. (Photo courtesy of J. Schroeder and D. Bush.)

Pulling the electrode away from the cell produces a **membrane patch** that becomes exposed to both the pipette solution and the bathing medium (see Figure 2). Because the membrane patches contain only a few channels, this configuration makes it possible to study the abrupt changes in current caused by the opening and closing of single ion channels (see Figure 3C). Remarkably, this technique is sensitive enough to detect the change in conformation of a single protein molecule.

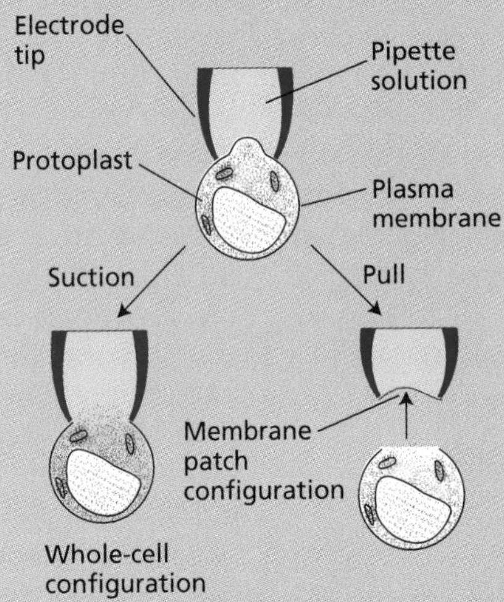

Figure 2 Diagram of the whole-cell and membrane patch configuration.

Because protoplasts are spherical, their volume is easy to measure, and their lack of electrical connections with other cells facilitates calculations of charge fluxes per unit area of membrane. Other advantages of patch clamping are the ability to distinguish between electrical events at the plasma membrane and the tonoplast and the possibility of controlling the composition of both the cytosol and the bathing medium.

Many of the patch clamp studies on plant membranes have been performed with guard cells. Figure 3 shows some of the results obtained by measuring electric current at controlled membrane potentials (voltage clamping). Whole-cell currents are shown in Figure 3A, using solutions in which K^+ is the principal ion that can permeate the membrane. Initially, the membrane potential is held at a value close to the K^+ equilibrium potential. Then voltage pulses (in this case about 3 s long) are applied to the protoplast (as indicated at right in the figure), and the electric current (in picoamperes) is followed with time (an upward deflection of the recording instrument indicates positive charges moving out of the cell). In this case, the current does not change instantaneously to a new stable value after a voltage change.

The gradual increase in current after a change in voltage reflects the

As will be discussed in more detail in Chapter 23, patch clamp studies (see Box 6.1) have revealed that gated channels can also be distinguished on the basis of how long the gates remain open in response to a prolonged stimulus. In guard cell plasma membranes, two types of anion channels have been characterized: rapidly activated (*R-type*) and slowly activated (*S-type*). **R-type channels** open and close very rapidly in response to a voltage stimulus; **S-type channels** remain open for the duration of the stimulus. Rapid and slow channels have also been identified in vacuolar membranes, and they are referred to as **fast vacuolar** (**FV**) or **slow vacuolar** (**SV**) **channels** (see Figure 6.12).

In transport mediated by a **carrier**, the substance being transported is initially bound to a specific site on the carrier protein. Binding causes a conformational change in the protein, which exposes the substance to the solution on the other side of the membrane. Transport is complete when the substance dissociates from the carrier's binding site. Because a conformational

BOX 6.1 (continued)

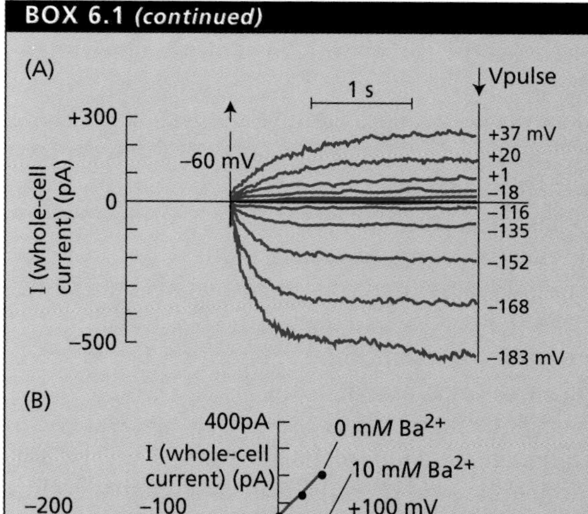

Figure 3 Recordings of K^+ currents in whole cells and in single K^+-selective channels of guard cell protoplasts. (A) K^+ currents recorded in the whole-cell configuration (*see* Figure 2) when the membrane potential is clamped at different values. Upward deflections show outward currents; downward deflections show inward currents. K^+ concentrations were 105 m*M* in the pipette and cytoplasm, and 11 m*M* in the bathing solution outside the cell. (B) Data from the same experiment plotted as an I/V (current/voltage) curve, before and after the addition of Ba^{2+}, an ion that blocks K^+ channels. (C) Inward K^+ current through a single channel in a membrane patch. (From Schroeder et al. 1987.)

(B)
400pA
I (whole-cell current) (pA)
0 m*M* Ba^{2+}
10 m*M* Ba^{2+}
−200
−100
+100 mV
V, membrane potential (mV)
−500

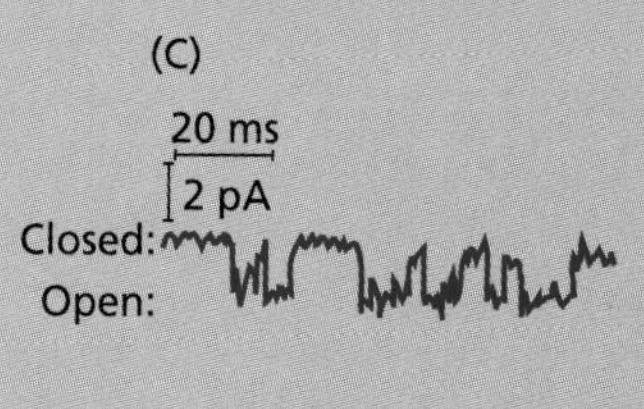

opening of increasing numbers of channels in response to the membrane potential. Channels responding in this way are termed **voltage-gated channels**. In general, the gating (opening and closing) of a channel is controlled independently of its ion selectivity, which is generally constant for a particular kind of channel protein. The same data are plotted in Figure 3B as an **I/V (current/voltage) curve**. Again the current does not depend in a simple way on the driving force for K^+ diffusion (that is, the deviation of the potential from the K^+ equilibrium potential). An outward current is activated only at potentials more positive than –40 mV, and an inward current is activated only at potentials more negative than about –100 mV. This activation is a reflection of the gating of individual channel proteins with different properties.

The inward and outward currents can be shown to have different single-channel *conductances* (the ability of current to cross the membrane) and varying response times, and they are differentially affected by various activators and inhibitors. It is now known that the inward and outward currents are carried by two different kinds of K^+ channels, as demonstrated by the cloning and expression of the respective genes (Maathuis et al. 1997). The **outward-rectifying channels** open only when the membrane potential favors outward diffusion of K^+, and the **inward-rectifying channels** open only when the membrane potential favors the inward diffusion of K^+. Since each type of channel has its own regulatory mechanisms, the cell can independently control the rates of inward and outward diffusion.

Additional information about K^+ channels can be obtained from studies of currents in isolated patches of membrane in which single channels can be visualized (see Figure 3C). When held at a constant voltage, these patches show abrupt jumps in the level of current, as single K^+ channels open and close. This behavior is seen only in membrane channels, never in ion pumps or carriers, because a single conformational change of a channel can allow hundreds of thousands of ions to cross the membrane, whereas a conformational change of a pump or a carrier can transport only one, or a few, molecules or ions.

Single-channel recordings (see Figure 3C) allow one to distinguish the effects of channel conductance (ion permeability) and of channel gating. They can show how the probability of opening of an individual channel depends on the membrane potential, or on intracellular signals initiated by light, hormones, toxins, or other factors. In many cases, these channel responses play an essential role in signal transduction in the cell. The patch clamp technique is clearly able to give unique information about ion channels, but it remains a delicate and technically difficult procedure.

change in the protein is required to transport individual molecules or ions, the rate of transport by a carrier is many orders of magnitude slower than through a channel. Typically, carriers may transport 100 to 1000 ions or molecules per second, which is about 10^6 times slower than transport through a channel. The binding and release of a molecule at a specific site on a protein that occur in carrier-mediated transport are similar to the binding and release of molecules from an enzyme in an enzyme-catalyzed reaction. In fact, as will be discussed later in the chapter, enzyme kinetics (see Chapter 2) are often used to characterize transport carrier proteins.

Carrier-mediated transport (unlike transport through channels) can be either passive or active, and it can transport a much wider range of possible substrates. Passive transport on a carrier is sometimes called **facilitated diffusion**, although it resembles diffusion only in that it transports substances *down* their gradient of electrochemical potential, without an additional input of energy.

Primary Active Transport Is Directly Coupled to Metabolic or Light Energy

To carry out active transport, a carrier must couple the uphill transport of the solute with another, energy-releasing, event so that the overall free-energy change is negative. **Primary active transport** is coupled directly to a metabolic source of energy, such as ATP hydrolysis, an oxidation–reduction reaction (the electron transport chain of mitochondria and chloroplasts), or the absorption of light by the carrier protein (in halobacteria, bacteriorhodopsin). The membrane proteins that carry out primary active transport are called **pumps** (see Figure 6.7). Most pumps transport ions, such as H^+ or Ca^{2+}. However, as we will see later in the chapter, pumps belonging to the "ATP-binding cassette" family of transporters can carry large organic molecules into the vacuole.

Ion pumps can be further characterized as either electrogenic or electroneutral. In general, **electrogenic transport** refers to ion transport involving the *net movement of charge* across the membrane. In contrast, **electroneutral transport**, as the name implies, involves no net movement of charge. For example, the Na^+/K^+-ATPase of animal cells pumps three Na^+ ions out for every two K^+ ions in, resulting in a net outward movement of one positive charge. The Na^+/K^+-ATPase is therefore an electrogenic ion pump. In contrast, the H^+/K^+-ATPase of the animal gastric mucosa pumps one H^+ out of the cell for every one K^+ in, so there is no net movement of charge across the membrane. Therefore, the H^+/K^+-ATPase is an electroneutral pump.

In the plasma membranes of plants, fungi, and bacteria, as well as in plant tonoplasts and other plant and animal endomembranes, H^+ is the principal ion that is electrogenically pumped across the membrane. The **plasma membrane H^+-ATPase** creates the gradient of electrochemical potentials of H^+ in the plasma membranes, while

BOX 6.2

Chemiosmosis in Action

THE BRITISH NOBEL LAUREATE PETER MITCHELL coined the term **chemiosmosis** to explain the coupling between gradient of electrochemical potentials of protons across selectively permeable membranes and the performance of cellular work (Harold 1986). In plants, proton gradients play a role in transport across the plasma membrane and the tonoplast and drive both transport and ATP synthesis on the inner membranes of chloroplasts and mitochondria. The passage of protons along their electric and chemical gradients can be coupled to cellular work because the change in free energy is negative. This process can be represented by the following equation:

$$\Delta G = \Delta\tilde{\mu}_{H^+} = F\Delta E + 2.3\ RT\left(\log\frac{[H^+]^i}{[H^+]^o}\right)$$

where $\Delta\mu_{H+}$ is the proton gradient of electrochemical potential, F is the Faraday constant, ΔE is the membrane potential, R is the gas constant, T is the absolute temperature, and the superscripts "i" and "o" refer to the inside and outside, respectively, of the cell or other membrane-enclosed compartment.

Mitchell introduced the term "proton motive force" (Δp) for the difference in electrochemical potential between protons inside and outside a cellular compartment, μ_{H+}. It is convenient to express Δp in units of electric potential, which we can do by dividing both sides of the foregoing equation by the Faraday constant, F, which changes the units to millivolts:

$$\Delta p = \frac{\Delta\tilde{\mu}_{H^+}}{F} = \Delta E + \frac{2.3\ RT}{F}\left(\log\frac{[H^+]^i}{[H^+]^o}\right)$$

Since pH = –log[H^+], the term $\log([H^+]^i/[H^+]^o)$ simplifies to $-(pH^i - pH^o)$. If ΔpH is defined as $pH^i - pH^o$, the general expression for proton motive force results:

$$\frac{\Delta\tilde{\mu}_{H^+}}{F} = \Delta p = \Delta E - \frac{2.3\ RT}{F}\Delta pH$$

At 25°C, $2.3RT/F$ = 59 mV, and substituting this value into the foregoing expression results in the most commonly used equation for proton motive force:

$$\Delta p = \Delta E - 59\Delta pH \quad (1)$$

where Δp is expressed in millivolts.

Let's consider the example of a cell bathed in a solution of 1 m*M* KCl and 1 m*M* sucrose. Proton pumping by an H^+-ATPase results in a membrane potential of –120 mV and a pH difference between the inside and the outside of the cell of 2 pH units. Thus, from Equation 1,

$$\Delta p = -120\text{ mV} - 59(2)\text{ mV} = -238\text{ mV}$$

How much K^+ can the cell take up by using this Δp? Because potassium is positively charged and the membrane potential of the inside of the cell is negative, potassium will be taken up through ion channels by the electrical component of Δp, which equals –120 mV. Using the Nernst equation (Equation 6.5), we can calculate that an external K^+ concentration of 1 m*M* and ΔE value of –120 mV will equilibrate with an internal K^+ concentration of 100 m*M*. Thus, if the electrical component of Δp is used, the cell can generate a 100-fold K^+ gradient across the membrane.

The proton motive force can also be used to take up sucrose against a concentration gradient, usually via a proton–sucrose symporter. Because sucrose is cotransported with a proton, both components of the gradient of electrochemical potential (–238 mV in our example) can be used for sucrose uptake (Harold 1986). At equilibrium, $\Delta\tilde{\mu}_s$ (see Equation 6.2) will be equal to $F\Delta p$; thus we can calculate that an external sucrose concentration of 1 m*M* and a Δp of –238 mV will equilibrate with an internal sucrose concentration of 10 *M*. In real life, however, such concentration gradients would not exist: Sucrose would diffuse back out of the cell, and regulatory mechanisms at the membrane would repress the function of the symporter after certain critical concentrations were attained.

the **vacuolar H^+-ATPase (V-ATPase)** and the **H^+-pyrophosphatase (H^+-PPase)** electrogenically pump protons into the lumen of the vacuole and the Golgi cisternae.

In plant plasma membranes, only H^+ and Ca^{2+} appear to be transported by pumps, and the direction of pumping is outward, not inward. Therefore another mechanism is needed to drive the active uptake of most mineral nutrients. The other important way that solutes can be actively transported across a membrane against their gradient of electrochemical potential is by coupling of the uphill transport of one solute to the downhill transport of another. This type of carrier-mediated cotransport is termed *secondary active transport*, and it is driven indirectly by pumps.

Secondary Active Transport Processes Use the Energy Stored in the Proton Motive Force

When protons are extruded from the cytosol by electrogenic H^+-ATPases, both at the plasma membrane and at the vacuole membrane, a membrane potential and a pH gradient are created at the expense of ATP hydrolysis. This gradient of electrochemical potential for H^+, called the **proton motive force (PMF)**, or Δp, represents stored free energy in the form of the H^+ gradient (see Box 6.2).

The proton motive force generated by electrogenic H^+ transport is used to drive the transport of many other substances against their gradient of electrochemical potentials in **secondary active transport**. Figure 6.9 shows how secondary transport may operate. The carrier is a transmembrane protein with a site on the outside of the membrane that can bind a proton. Proton binding causes a second site to be exposed. This site binds the ion or solute that is being actively transported. With both molecules bound, the transporter undergoes a conformational change that exposes the binding sites to the opposite side of the membrane. The cycle is completed by diffusion of the proton and the substrate molecule away from their binding sites, causing the transporter to regain its original, or "relaxed," conformation.

The example shown in Figure 6.9 is called a **symport** (and the protein involved is called a **symporter**) because the two substances are moving in the same direction through the membrane. **Antiport** (facilitated by a protein called an **antiporter**) refers to coupled transport in

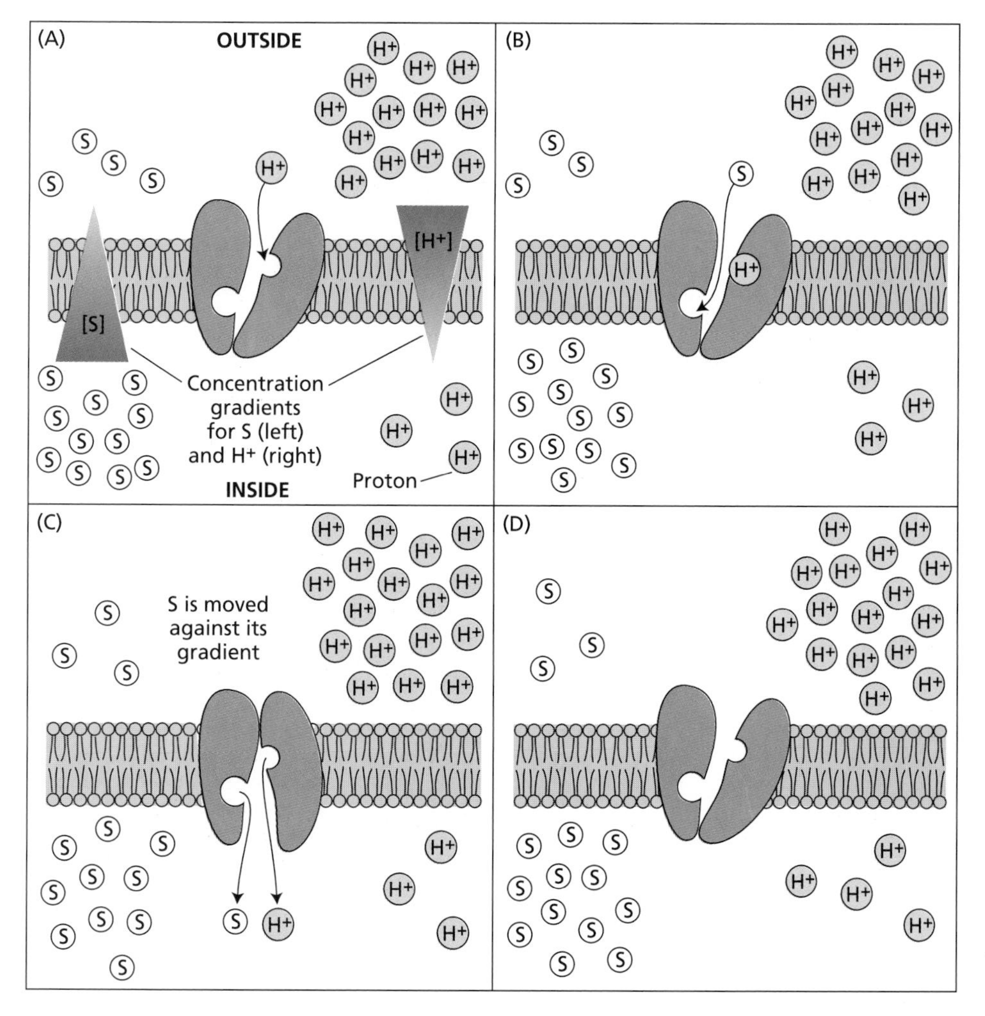

Figure 6.9 Hypothetical model for secondary active transport. The energy that drives the process has been stored in a proton gradient (symbolized by the triangle on the right in A) and is being used to take up a substrate (S) against its concentration gradient (left-hand triangle). (A) In the initial conformation, the binding sites on the protein are exposed to the outside environment and can bind a proton. (B) This binding results in a conformational change that permits a molecule of S to be bound. (C) The binding of S causes another conformational change that exposes the binding sites and their substrates to the inside of the cell. (D) Release of a proton and a molecule of S to the cell's interior restores the original conformation of the carrier and allows a new pumping cycle to begin.

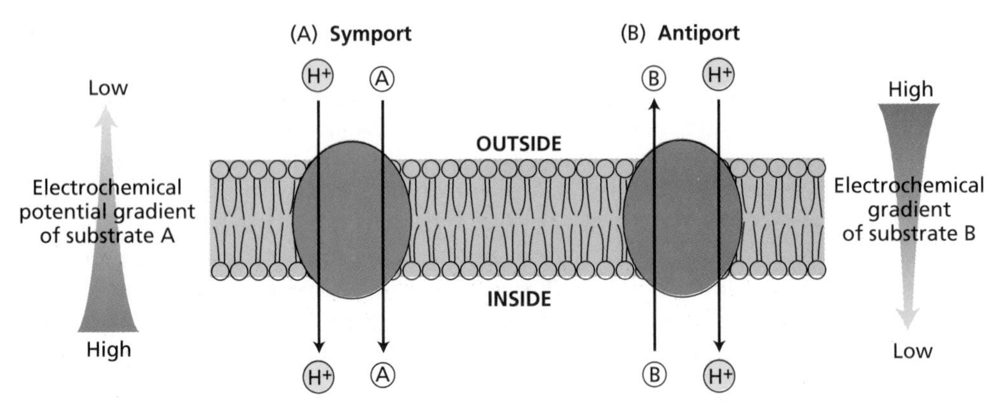

Figure 6.10 Two examples of secondary active transport coupled to a primary proton gradient. (A) In symport, the energy dissipated by a proton moving back into the cell is coupled to the uptake of one molecule of a substrate (e.g., a sugar) into the cell. (B) In antiport, the energy dissipated by a proton moving back into the cell is coupled to the active transport of a substrate (for example, a sodium ion) out of the cell. In both cases, the substrate under consideration is moving against its gradient of electrochemical potential. Both neutral and charged substrates can be transported by such secondary active transport processes.

which the downhill movement of protons drives the active (uphill) transport of a solute in the opposite direction (Figure 6.10). In both types of secondary transport, the ion or solute being transported simultaneously with the protons is moving against its gradient of electrochemical potential, so its transport is active. However, the energy driving this transport is provided by the proton motive force rather than by ATP hydrolysis, and it is captured by the cell by the inward movement of H^+ through the transporter, which dissipates the proton motive force.

Typically, a cell depends on one primary active transport system coupled to ATP hydrolysis to set up a gradient of electrochemical potential of one ion—for example, H^+. Many other ions or organic substrates can then be transported by a variety of secondary active transport proteins, which energize the transport of their respective substrates by simultaneously carrying one or two H^+ ions down their energy gradient. Thus H^+ ions circulate across the membrane, outward through the primary active transport proteins, and back into the cell through the secondary transport proteins.

In plants and fungi, sugars and amino acids are taken up by symport with protons. Figure 6.11 shows some of the experimental evidence for this conclusion. When glucose is supplied to a plant cell bathed in a simple solution of mineral salts, a reduction in the membrane potential, an increase in external pH, and uptake of glucose occur simultaneously (Novacky et al. 1980). The decrease in membrane potential is due to the positive charges (H^+) that move into the cell along with glucose. This membrane depolarization is transitory, however, because the reduced membrane voltage allows the H^+ pump to work faster and thereby restore the membrane voltage and pH gradient in the presence of continuing glucose uptake.

Research has led to the view that most of the ionic gradients across membranes of higher plants are generated and maintained by electrochemical-potential gradients of H^+ (Tazawa et al. 1987). In turn, these H^+ gradients are generated by the electrogenic proton pumps.

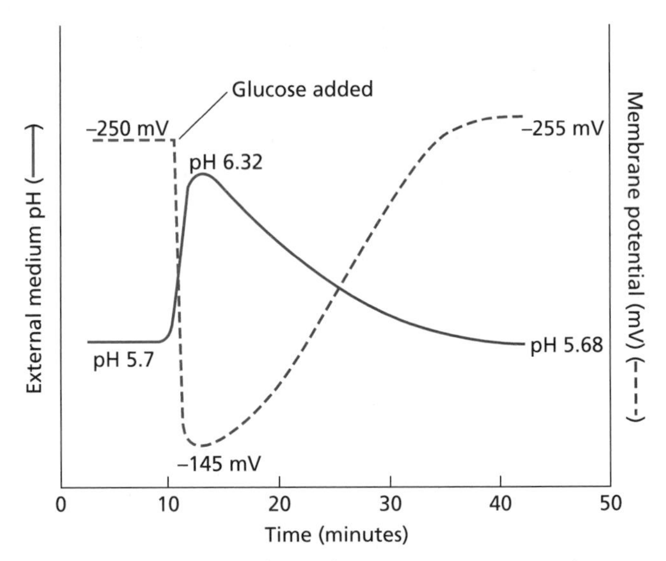

Figure 6.11 Evidence for a glucose–proton symport is shown by simultaneous measurements of the pH of the medium that bathes the surface of the aquatic plant duckweed (*Lemna gibba*) and the membrane potential of one cell. The early portions of the curves show steady values of pH and membrane potential, conditions that change when 50 m*M* glucose is added to the solution. The observed increase in pH indicates that protons are disappearing from the medium at the same time that the membrane potential of the cell is decreasing. These observations are predicted by a glucose–proton symport if glucose is cotransported with a proton into the cell. With time, an increase in pump activity restores the initial pH and membrane potential values. (After Novacky et al. 1980.)

Evidence suggests that Na^+ is transported out of the cell by a Na^+–H^+ antiporter and that Cl^-, NO_3^-, $H_2PO_4^-$, sucrose, amino acids, and other substances enter the cell via specific proton symporters. Potassium at very low external concentrations can be taken up by active symport proteins, but at higher concentrations K^+ can enter the cell by diffusion through specific K^+ channels. However, even transport through channels is driven by the H^+-ATPase, in the sense that K^+ diffusion is driven by the membrane potential, which is maintained at a value more negative than the K^+ equilibrium potential by the action of the electrogenic H^+ pump. Several representative transport processes located on the plasma membrane and the tonoplast are illustrated in Figure 6.12.

Kinetic Analyses Can Elucidate Transport Mechanisms

Thus far, we have described cellular transport in terms of its energetics. However, cellular transport can also be studied by use of enzyme kinetics, because transport involves the binding and dissociation of molecules at active sites on transport proteins (see Chapter 2). One advantage of the kinetic approach is that it gives new insights into the regulation of transport.

Kinetic experiments involve measurements of the effects of external ion (or other solute) concentrations on transport rates, and they lead to a view of transport different from that yielded by other methods. In simple diffusion, the rate of inward transport is proportional to the external concentration of the transported molecule. Carrier-mediated transport, on the other hand, tends toward a maximum rate (V_{max}) that cannot be exceeded regardless of the concentration of substrate. V_{max} is approached when the substrate-binding site on the carrier is always occupied. The concentration of carrier, not the concentration of solute, becomes rate limiting. The constant K_m, which is numerically equal to the solute concentration that yields half the maximal rate of transport, tends to reflect the properties of the particular binding site.

Usually, transport displays both active and passive properties when a wide range of solute concentrations is studied. Figure 6.13 shows sucrose uptake by soybean cotyledon protoplasts as a function of the external sucrose concentration (Lin et al. 1984). Uptake increases sharply with concentration and begins to saturate at about 10 m*M*. At concentrations above 10 m*M*, uptake becomes linear and nonsaturable. Inhibition of ATP synthesis with metabolic poisons blocks the saturable component but not the linear one. The interpretation is that sucrose uptake at low concentrations is an active carrier-mediated process (sucrose–H^+ symport). At higher concentrations, sucrose enters the cells down its concentration gradient and is therefore not sensitive to metabolic poisons. However, additional information is needed to determine whether the nonsaturating component represents uptake by a carrier with very low affinity, or by a channel. (Transport by a carrier is more likely in the case of an organic solute.)

Sometimes kinetic analysis of transport appears to show multiple saturable components for the same solute (Figure 6.14). This result has often been interpreted as indicating multiple carriers in the same tissue, each of which has a different K_M value for the transported solute (Epstein 1972). In the case of K^+, now that inward-rectifying K^+ channels have been characterized, and genes for both K^+ channels and K^+ active symport have been cloned and expressed (see the next section), the high-affinity transport system responsible for K^+ uptake at low concentrations of K^+ can be attributed to secondary active transport by a symporter carrying K^+ along with H^+ (or Na^+, discussed in the next section). In contrast, detailed kinetic analyses of the low-affinity transport system (K^+ uptake at high K^+ concentrations) have indicated that the rate of uptake is not saturable, as was initially believed. Low-affinity K^+ transport is more readily explained by uptake through K^+ channels.

The Genes for Many Protein Transporters Have Been Cloned

Transport mutants can be readily isolated whenever a suitable selection strategy can be devised. Nitrate transport is of interest not only because of its nutritional importance, but also because of its complexity. Nitrate transport not only shows high-affinity and low-affinity components, but it is also strongly regulated according to nitrate availability: The enzymes required for nitrate transport as well as nitrate assimilation are induced in the presence of nitrate in the environment, and uptake can also be repressed if nitrate accumulates in the cells. Mutants in nitrate transport or nitrate reduction can be selected by growth in the presence of chlorate (ClO_3^-). Chlorate is a nitrate analog that is taken up and reduced in wild-type plants to the toxic product chlorite. If plants resistant to chlorate are selected, they are likely to show mutations that block nitrate transport or reduction.

Several such mutations have been identified in *Arabidopsis*, a small crucifer that is ideal for genetic studies. The first transport gene identified in this way encodes a low-affinity inducible nitrate–proton symporter. Selection of additional mutants at different chlorate concentrations in the presence and absence of nitrate indicates that at least four separate genes are involved in nitrate uptake, including two constitutively expressed genes for low- and high-affinity uptake, respectively, and two nitrate-inducible genes for low- and high-affinity uptake, respectively (Crawford 1995).

Researchers have cloned some plant transport genes by identifying regions of sequence similarity with transport genes of other organisms, and occasionally they

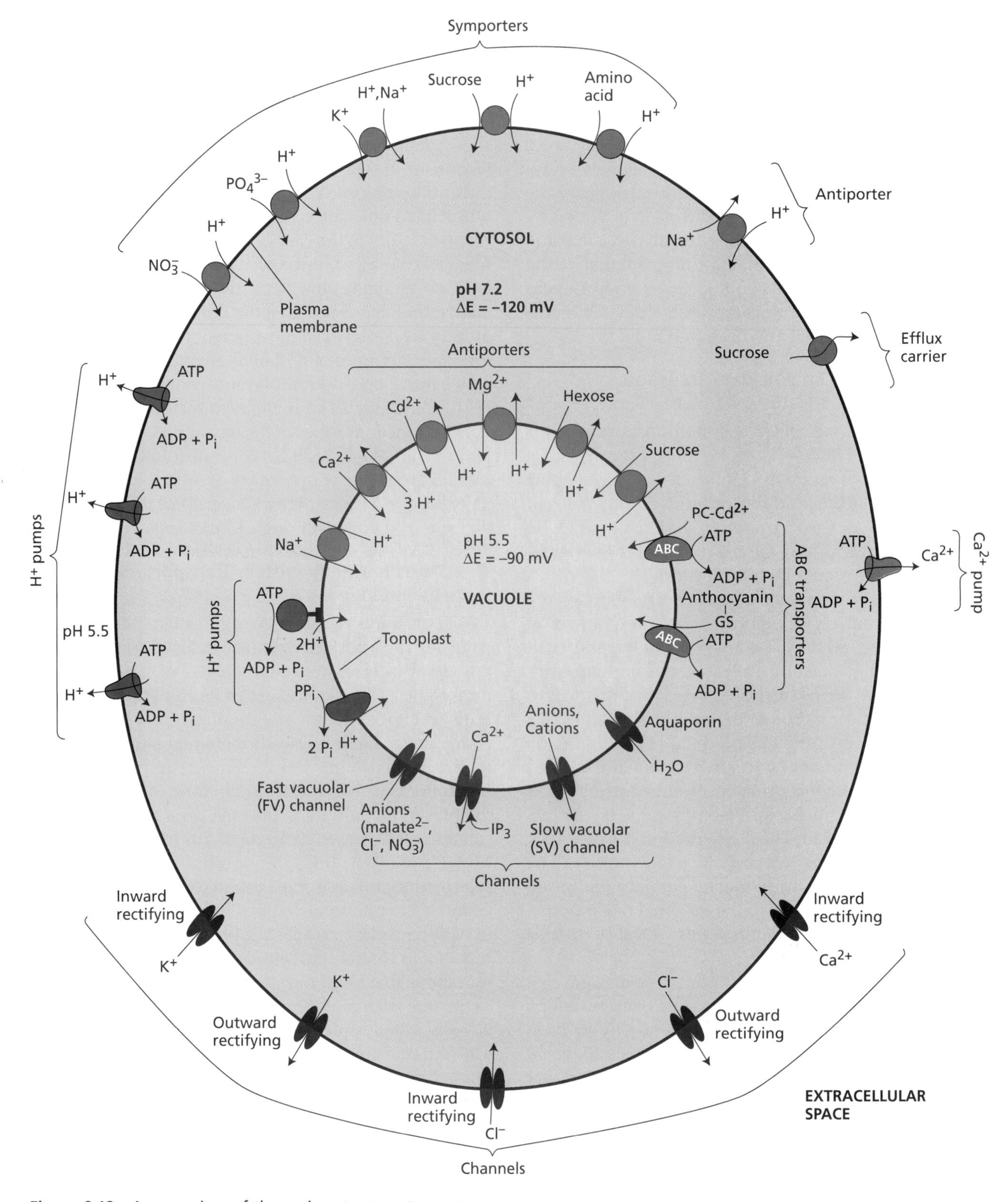

Figure 6.12 An overview of the various transport processes on the plasma membrane and tonoplast of plant cells.

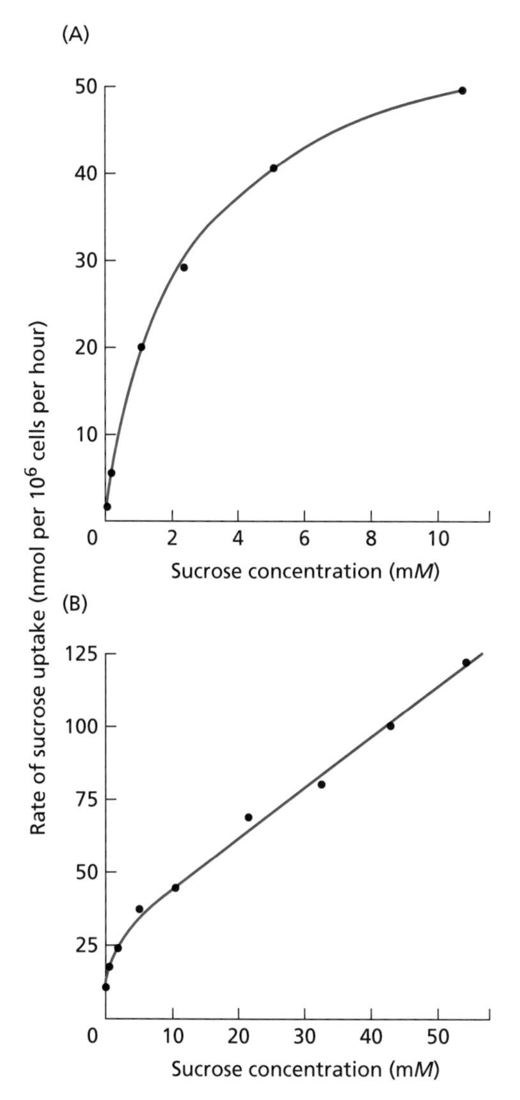

Figure 6.13 The transport properties of a solute can change at different solute concentrations. For example, at low concentrations (1 to 10 m*M*) (A), the rate of uptake of sucrose by soybean cells shows saturation kinetics, typical of carriers. At higher sucrose concentrations (B), the uptake rate increases linearly over a broad range of concentrations, suggesting the existence of other sucrose transporters, which might be carriers with very low affinity for the substrate. (From Lin et al. 1984.)

have been able to identify the gene after purifying the transport protein. More commonly, sequence similarity is limited, and individual transport proteins represent too small a fraction of total protein for these strategies to be effective. Another way to identify transport genes is to screen plant cDNA (complementary DNA) libraries for genes that complement transport deficiencies in yeast. Many yeast transport mutants are known and have been used to identify corresponding plant genes by complementation. Several inward-rectifying K^+ channels have been identified in this way.

After identifying channels, researchers have studied the behavior of the channel proteins by expressing the genes in oocytes of the toad *Xenopus*, which, because of their large size, are convenient for electrophysiological studies. Of the inward-rectifying K^+ channel genes identified so far, one is expressed strongly in stomatal guard cells, another in roots, and a third in leaves. In addition, complementation of a yeast strain that is defective in high-affinity K^+ uptake permitted the cloning and characterization of a wheat gene for high-affinity active transport of K^+. This gene was expected to encode a K^+–proton symporter, but surprisingly, the gene product carries Na^+, instead of H^+, with the K^+ ions. Physiological studies indicate that K^+–proton symporters are probably also present in plants. Genes for plant vacuolar H^+–Ca^{2+} antiporters have been isolated by functional cloning in yeasts (Hirshi et al. 1996). Genes for the proton symport of several amino acids and sugars have also been identified by complementation of yeast mutants and/or characterized by expression in *Xenopus* oocytes (Frommer and Ninnemann 1995; Tanner and Caspari 1996).

One class of proteins that is relatively abundant in plant membranes revealed no ion currents when expressed in oocytes, but when the osmolarity of the external medium is reduced, these proteins cause swelling and bursting of the oocytes. This result is due to rapid influx of water across the oocyte plasma mem-

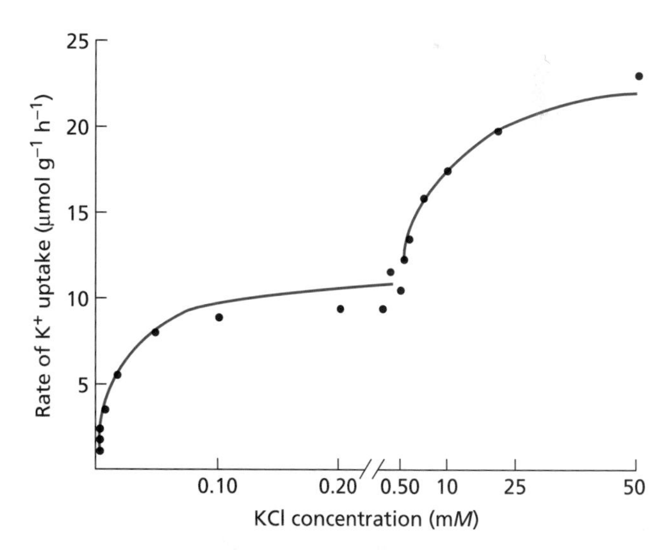

Figure 6.14 The transport of potassium into barley roots shows two different phases. The biphasic kinetics of potassium uptake, accentuated in this figure by the change of scale at around 1 m*M*, suggests the presence of different types of transport systems for potassium. The high-affinity transport system, having a K_m value of 0.02 to 0.03 m*M*, is attributed to active transport by symporters; the low-affinity system (which may or may not show saturation) is attributed to diffusion through K^+ channels. (After Epstein 1972.)

brane, which normally has a very low water permeability. These proteins that appear to form water channels in membranes have been named **aquaporins** (see Figure 6.12). Their existence was a surprise at first, since it was thought that the lipid bilayer is itself sufficiently permeable to water. Nevertheless, aquaporins are common in plant and animal membranes, and their expression and activity appear to be regulated, possibly by protein phosphorylation, in response to water availability (Maurel 1997).

It appears to be the rule that a family of genes, rather than an individual gene, exists in the plant genome for each transport function. Variations within the gene family in transport characteristics such as K_m, in mode of regulation, and in differential tissue expression give plants enormous versatility in responding to physiological needs and environmental conditions. Studies of K_m values have shown a feedback inhibition (see Chapter 2) of carrier-mediated transport that allows the cell to maintain stable cytoplasmic concentrations over a wide range of external concentrations. For example, the apparent K_m value of the high-affinity K^+ carrier in barley roots (see Figure 6.13) changes with the plant's demand for K^+. Barley plants starved for K^+ have high-affinity K_m values in the range of 0.02 to 0.03 m*M*, whereas well-fertilized plants have K_m values that are four- to fivefold greater (remember, an increase in K_m reflects a decrease in the carrier's affinity for K^+; see Chapter 2). The change in apparent K_m probably reflects the expression of different members of a gene family. In this way, plant cells can match nutrient uptake to cellular demand. A change in the number of protein molecules of a particular carrier would appear kinetically as a change in V_{max} without a change in K_m.

Plants that have been starved and then resupplied with sulfate or phosphate often respond in this way. As the internal sulfate and phosphate levels recover, the V_{max} values for their uptake drop to the usual values as excess carriers are removed from the membrane (Glass 1983).

The Plasma Membrane H^+-ATPase Has Several Functional Domains

The active transport of H^+ outward across the plasma membrane creates gradients of pH and electric potential that drive the transport of many other substances through the various secondary active transport proteins, and it contributes to the driving forces for diffusion through ion channels. Figure 6.15 shows how a plasma membrane H^+-ATPase might work. At rest, the protein forms an occluded pore in the membrane with sites on its cytoplasmic side for binding both the transportable ion and ATP. When these substances are bound, a phosphate group is transferred from ATP to phosphorylate an aspartic acid residue on the protein. The protein then undergoes a conformational change, opening the transport pathway to the outside and simultaneously closing it on the cytoplasmic side. The ion then leaves its binding site, and the protein reverts to its original conformation. Finally, the phosphate group leaves and the cycle can start again.

ATPases that are phosphorylated as part of the catalytic cycle are known as P-type ATPases. Plant and fungal plasma membrane H^+-ATPases and Ca^{2+}-ATPases

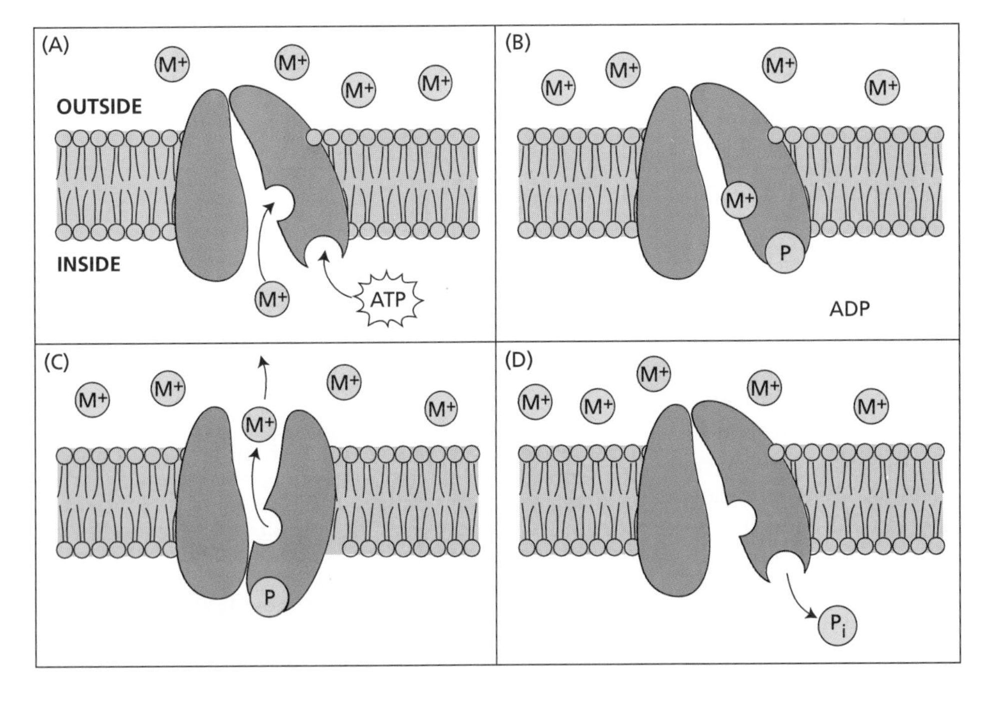

Figure 6.15 Hypothetical steps in the transport of a cation (the hypothetical M^+) against its chemical gradient by an electrogenic pump. The protein, embedded in the membrane, binds the cation on the inside of the cell (A) and is phosphorylated by ATP (B). This phosphorylation leads to a conformational change that exposes the cation to the outside of the cell and makes it possible for the cation to diffuse away (C). Release of the phosphate ion (P) from the protein into the cytosol (D) restores the initial configuration of the membrane protein and allows a new pumping cycle to begin.

are P-type ATPases. Because of this feature, the plasma membrane ATPases are strongly inhibited by orthovanadate, a phosphate analog that competes with the phosphate from ATP for the aspartic acid phosphorylation site on the enzyme.

Molecular studies have revealed that the plasma membrane H^+-ATPase consists of a single polypeptide chain with a molecular mass of about 100,000 Da (100 kDa); kinetic studies suggest that the protein normally functions as a dimer. Plasma membrane H^+-ATPases are encoded by a family of about ten genes that encode various isoforms of the enzyme (Sussman 1994). Different members of the family are expressed in different tissues of the plant. Thus one form is found predominantly at the surface of the root, another in the seed, yet another in the phloem, and so on. This variety in isoforms may allow transport to be regulated in different ways for each tissue.

Figure 6.16 shows a model of the functional domains of the plasma membrane H^+-ATPase of yeast, which is similar to that of plants. The protein has ten membrane-spanning domains that cause it to loop back and forth across the membrane. Some of the membrane-spanning domains make up the cation channel through which protons are pumped. The catalytic domain, including the aspartic acid residue that becomes phosphorylated during the catalytic cycle, is on the cytosolic face of the membrane.

Like other enzymes, the plasma membrane ATPase is regulated by variables such as the concentration of substrate (ATP), pH, and temperature. In addition, individual H^+-ATPase molecules can be reversibly activated or deactivated in response to a variety of signals, such as light, hormones, pathogen attack, and the like. This type of regulation is mediated by a specialized **autoinhibitory domain** at the C-terminal end of the polypeptide chain, which acts as a gate regulating the activity of the proton pump (see Figure 6.16) (Palmgren 1991). This domain is more extensive in plants than in yeast. If the autoinhibitory domain is removed through the action of a protease, the enzyme becomes irreversibly activated.

The autoinhibitory effect of the C-terminal domain can also be regulated through the action of protein kinases and phosphatases that add or remove phosphate groups to serine or threonine residues on the enzyme. For example, one mechanism of response to pathogens in tomato involves the activation of protein phosphatases that dephosphorylate residues on the plasma membrane H^+-ATPase, thereby activating it (Vera-Estrella et al. 1994). This is one step in a cascade of responses that activate plant defenses. Other mechanisms by which the H^+-ATPase can be activated include the binding of specific lipids to the ATPase, which has been suggested to be involved in the stimulation of proton extrusion by the plant growth hormone auxin, or the binding of a protein that serves as a receptor for fusicoccin, a fungal toxin that has a remarkable activating effect on the enzyme (see Chapter 19).

The Vacuolar H^+-ATPase Drives Solute Accumulation into Vacuoles

Since plant cells enlarge primarily by the uptake of water into large, central vacuoles, the osmotic pressure of the vacuole must be maintained sufficiently high for water to enter from the cytoplasm. The tonoplast regulates the traffic of ions and metabolites between the cytosol and the vacuole, just as the plasma membrane regulates uptake into the cell. However, tonoplast transport remained relatively inaccessible for study until methods were developed for isolating intact vacuoles and tonoplast vesicles (see Box 6.3) (Sze 1985).

Studies with isolated vacuoles and purified tonoplast vesicles led to the discovery of a new type of proton-

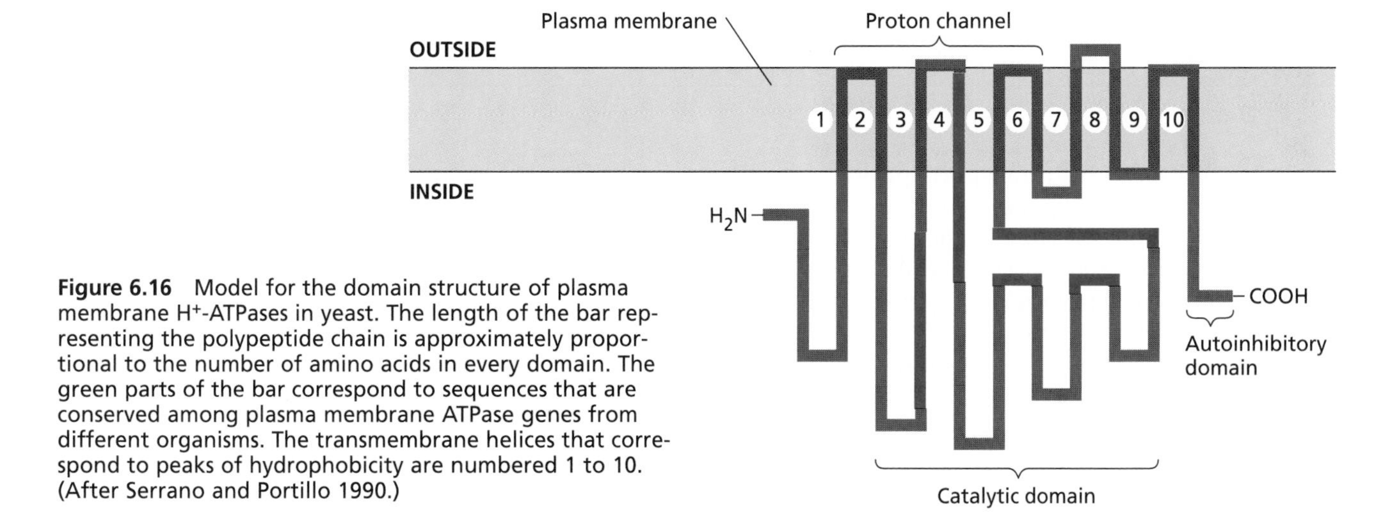

Figure 6.16 Model for the domain structure of plasma membrane H^+-ATPases in yeast. The length of the bar representing the polypeptide chain is approximately proportional to the number of amino acids in every domain. The green parts of the bar correspond to sequences that are conserved among plasma membrane ATPase genes from different organisms. The transmembrane helices that correspond to peaks of hydrophobicity are numbered 1 to 10. (After Serrano and Portillo 1990.)

BOX 6.3

Transport Studies with Isolated Vacuoles and Membrane Vesicles

JUST AS STUDIES with isolated thylakoid membranes led to the elucidation of the electron transfer reactions of photosynthesis (see Chapter 7), the development of techniques for isolating functionally competent tonoplasts and plasma membranes has enabled scientists to study the transport reactions of these membranes in vitro. Two types of preparations have been used: intact vacuoles, and isolated plasma membrane or tonoplast vesicles.

When protoplasts prepared by enzymatic digestion of the cell walls are maintained in the plasmolyzed state by incubation in medium containing a suitable osmotic solute, such as mannitol, the plasma membranes can be carefully lysed without breaking the tonoplast by a gentle osmotic shock. The intact vacuoles that are released by this treatment can then be purified by centrifugation through a sucrose density gradient. Intact vacuoles (along with contaminating cellular debris) are centrifuged in a tube containing a concentration gradient of sucrose (or another suitable solute). Because intact vacuoles tend to exclude the solute, they sediment to a lighter density on the density gradient than the other organelles and membranes do. In this way preparations of highly purified intact vacuoles can be obtained.

Alternatively, tonoplasts and plasma membranes can be purified as small membrane vesicles. When tissues are ground in a mortar and pestle or homogenized in a blender, broken pieces of various membranes tend to seal up to form small vesicles less than 1 μm in diameter. Because of a low ratio of protein to lipid in the tonoplast membrane, tonoplast vesicles usually have a relatively low density, and they can be separated from other cellular membranes by density-gradient centrifugation (Sze 1985). On the other hand, plasma membrane vesicles can be purified by phase partition in a mixture of dextran and polyethylene glycol solutions (Larsson et al. 1987). The use of isolated membrane vesicles to study transport avoids interference from metabolism, and since the contents of the vesicles can be fairly readily exchanged, the concentrations of solutes on each side of the membrane can be better defined.

Tissue homogenization often produces a mixture of vesicles of different orientation (right-side-out and inside-out vesicles). In the case of plasma membrane vesicles, phase partition selects for right-side-out vesicles, which therefore show little reactivity to ATP because the catalytic sites of the enzymes are facing inside the vesicle. If a detergent is added to such a preparation, ATP penetrates the vesicles and is hydrolyzed, but since the vesicles are no longer sealed, no transport is observed.

One kind of detergent (Brij 58) in low doses does not make the membranes leaky but, remarkably, causes them to reverse their orientation (Johansson et al. 1995). Apparently, this detergent accumulates in the inner half of the lipid bilayer, causing it to expand and flip the vesicles inside out. Since the catalytic site of the enzyme is now facing the external solution, in the presence of this detergent the vesicles respond to ATP by creating a gradient of pH and membrane potential.

Radioisotopes can be used to monitor the transport of various solutes into membrane vesicles. The vesicles are incubated for various times with the labeled solute, then separated by rapid filtration, and their radioactivity is measured in a scintillation counter. Other methods have been developed to measure the gradients of pH and membrane potential created by proton pumps. One such method uses fluorescent weak bases such as 9-aminoacridine (9-AA) or quinacrine, which are lipid-soluble in their uncharged form and readily diffuse across the membrane and enter the vesicle. Within the acidified vesicle, the amino group becomes protonated, trapping the charged form of the molecule inside (see part A of the figure). As the dye molecules accumulate inside the vesicles, they form dimers and other aggregates that result in a reduction in their fluorescence, a phenomenon termed **fluorescence quenching**. One can thus study ATP-driven proton transport across the vesicle membrane by measuring the rate of fluorescence quenching in an instrument called a fluorescence spectrophotometer.

As part B of the figure shows, addition of ATP to membrane vesicles isolated from the plasma membrane or tonoplast leads to a rapid decrease in the fluorescence (quenching). A steady state is reached when the rate of leakage of protons out of the vesicle equals the rate of transport of protons into the vesicle. Gramicidin, a compound that induces nonspecific holes in the membrane, rapidly dissipates the pH gradient, as indicated by the recovery of fluorescence. Other molecular probes, such as oxonol dyes, which associate with lipid membranes and show changes in fluores-

pumping ATPase, which transports protons into the vacuole (see Figure 6.12). The vacuolar H^+-ATPase (V-ATPase) differs both structurally and functionally from the plasma membrane H^+-ATPase (P-ATPase), and the V-ATPase is more closely related to the F-ATPases of mitochondria and chloroplasts (see Chapter 11). Because catalysis does not involve the formation of a phosphorylated intermediate, V-ATPases are insensitive to vanadate, an inhibitor of P-ATPases. However, V-ATPases are specifically inhibited by the antibiotic bafilomycin, as well as by high concentrations of nitrate, neither of which inhibit P-ATPases.

V-ATPases belong to a general class of vacuolar ATPases that are present on the endomembrane systems of all eukaryotes. They are large enzyme complexes, about 750 kDa, composed of at least ten different subunits (Lüttge and Ratajczak 1997). As in the F-ATPases, these subunits are organized into a peripheral catalytic complex, V_1, and an integral membrane channel complex, V_o when viewed in the electron microscope, except that the "ball" (representing the catalytic complex) has a characteristic lobed appearance (Figure 6.17). Because of their similarities to F-ATPases, V-ATPases are assumed to operate like tiny rotary motors (for a

BOX 6.3 (*continued*)

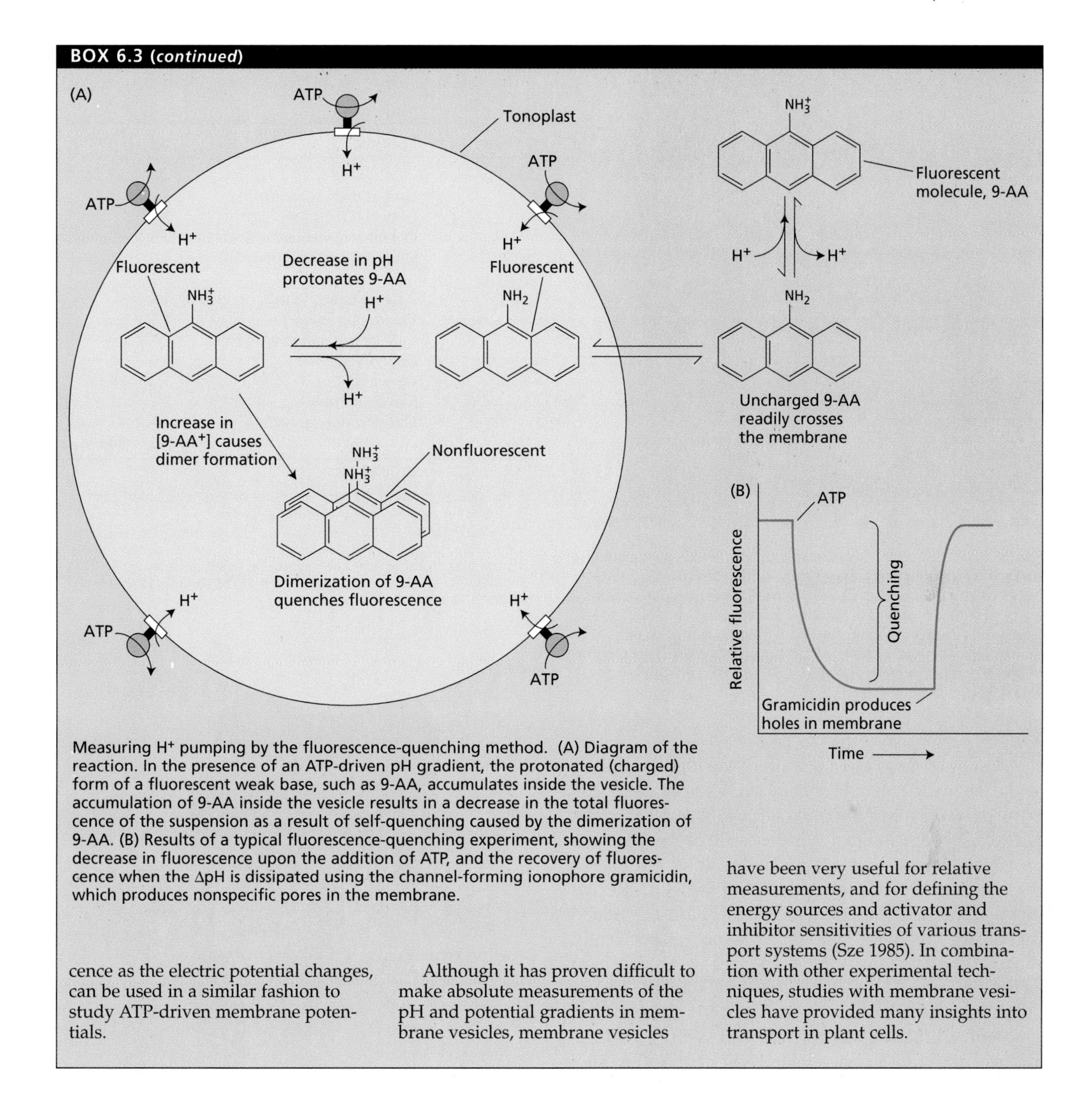

Measuring H^+ pumping by the fluorescence-quenching method. (A) Diagram of the reaction. In the presence of an ATP-driven pH gradient, the protonated (charged) form of a fluorescent weak base, such as 9-AA, accumulates inside the vesicle. The accumulation of 9-AA inside the vesicle results in a decrease in the total fluorescence of the suspension as a result of self-quenching caused by the dimerization of 9-AA. (B) Results of a typical fluorescence-quenching experiment, showing the decrease in fluorescence upon the addition of ATP, and the recovery of fluorescence when the ΔpH is dissipated using the channel-forming ionophore gramicidin, which produces nonspecific pores in the membrane.

cence as the electric potential changes, can be used in a similar fashion to study ATP-driven membrane potentials.

Although it has proven difficult to make absolute measurements of the pH and potential gradients in membrane vesicles, membrane vesicles have been very useful for relative measurements, and for defining the energy sources and activator and inhibitor sensitivities of various transport systems (Sze 1985). In combination with other experimental techniques, studies with membrane vesicles have provided many insights into transport in plant cells.

description, see Box 11.1). Figure 6.18 illustrates how V-ATPases might work.

V-ATPases are electrogenic proton pumps that are capable of generating a proton motive force across the tonoplast. This ability accounts for the fact that the vacuole is typically 20 to 30 mV more positive than the cytoplasm, although it is still negative relative to the external medium. To maintain bulk electric neutrality, anions such as Cl^- or malate^{2-} are transported from the cytoplasm into the vacuole through channels in the membrane (Barkla and Pantoja 1996). Without the simultaneous movement of anions along with the pumped protons, the charge buildup across the tonoplast would make the pumping of additional protons energetically unfavorable. In the presence of ongoing anion transport, the tonoplast H^+-ATPase can generate a concentration (pH) gradient of protons across the tonoplast. This gradient accounts for the fact that the pH of the vacuolar sap is typically about 5.5, while the cytoplasmic pH is always 7.0 to 7.5. Whereas the electrical component of the proton motive force drives the uptake of anions into the vacuole, the pH gradient is harnessed to drive the uptake of cations and sugars into the vacuole via secondary transport (antiporter) systems.

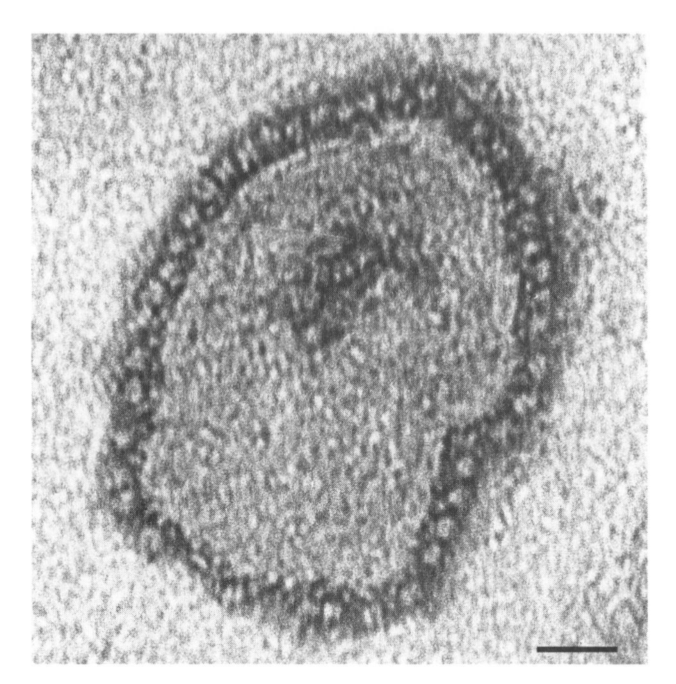

Figure 6.17 Electron micrograph of V-ATPases studding the surface of a tonoplast vesicle obtained from purified vacuoles of carrot (*Daucus carota*) root. The tonoplast vesicles were first dried onto an electron microscope grid and then treated by a procedure called negative staining, which makes the enzyme appear white against a dark background in the electron microscope. Bar ≅ 60 nm. (From Taiz and Taiz 1991.)

Although the pH of most plant vacuoles is mildly acidic (about 5.5), the pH of the vacuoles of some species is much lower, a phenomenon termed *hyperacidification.* Vacuolar hyperacidification is the cause of the sour taste of certain fruits (lemons) and vegetables (rhubarb). Some extreme examples are listed in Table 6.2. Recent biochemical studies with lemon fruits have suggested that the low pH of the lemon fruit vacuoles (specifically, those of the juice sac cells) is due to a combination of two main factors: (1) the low permeability of the vacuolar membrane to protons, which permits a steeper pH gradient to build up; (2) a specialized V-ATPase that is able to pump protons more efficiently (with less wasted energy) than normal V-ATPases can (Müller et al. 1996, 1997). Another important factor contributing to the sour taste of fruits and vegetables is the accumulation of organic acids, such as citric, malic, and oxalic acids, which help maintain the low pH of the vacuole by acting as buffers.

Sequence analyses of the two major subunits, A and B, of the catalytic complex of V-ATPases, suggest that they arose through an ancient gene duplication, a finding that has unexpectedly contributed to our understanding of the early evolution of life (Gogarten et al. 1989; Taiz and Nelson 1996). As previously noted, eukaryotic V-ATPases are related to the F-ATPases of

TABLE 6.2
The vacuolar pH of some hyperacidifying plant species

Tissue	Species	pH[a]
Fruits		
	Lime (*Citrus aurantifolia*)	1.7
	Lemon (*Citrus limonia*)	2.5
	Cherry (*Prunus cerasus*)	2.5
	Grapefruit (*Citrus paradisi*)	3.0
Leaves		
	Rosette oxalis (*Oxalis deppei*)	1.3
	Wax begonia (*Begonia semperflorens*)	1.5
	Begonia lucerna	0.9 – 1.4
	Oxalis sp.	1.9 – 2.6
	Sorrel (*Rumex*)	2.6
	Opuntia phaeacanth[b]	1.4 (6:45 A.M.) 5.5 (4:00 P.M.)

Source: Data from Small 1946.

[a] The values represent the pH of the juice or expressed sap of each tissue, usually a close indicator of vacuolar pH.

[b] The vacuolar pH of the cactus *Opuntia phaeacantha* varies with the time of day. As will be discussed in Chapter 8, many desert succulents have a specialized type of photosynthesis, called crassulacean acid metabolism (CAM), which causes the pH of the vacuoles to decrease during the night.

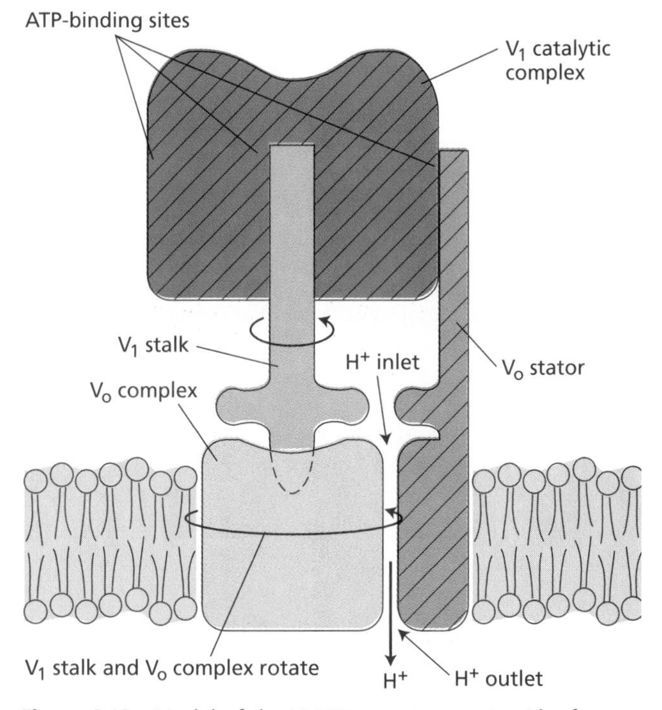

Figure 6.18 Model of the V-ATPase rotary motor. The four structural components are the V_1 catalytic complex; the V_1 stalk; the V_o complex, which acts like a rotor; and the V_o stator, which holds the V_1 catalytic complex in a stationary position. Hydrolysis of ATP by the V_1 catalytic complex drives the rotation of the V_1 stalk, which acts like a shaft. The rotation of the V_1 stalk, in turn, drives the rotation of the V_o complex. When the V_o complex turns, protons are transported from one side of the membrane to the other. (From Boekema et al. 1997, © National Academy of Sciences, U.S.A.)

mitochondria and chloroplasts, which are thought to have evolved from eubacterial endosymbionts. Remarkably, vacuolar H^+-ATPases are even more closely related to the H^+-ATPases of the archaebacteria. The discovery of the close relationship between eukaryotic V-ATPases and archaebacterial H^+-ATPases, along with data from the analysis of other duplicated genes, ultimately led to the recognition that archaebacteria and eukaryotes are sister groups (see Chapter 1) (Gogarten et al. 1996).

Plant Vacuoles Are Energized by a Second Proton Pump, the H^+-Pyrophosphatase

Another type of proton pump, an H^+-pyrophosphatase (H^+-PPase) (Rea and Poole 1993), appears to work in parallel with the V-ATPase to create a proton gradient across the tonoplast (see Figure 6.12). This enzyme consists of a single polypeptide that has a molecular mass of 80 kDa. The H^+-PPase gains its energy from the hydrolysis of inorganic pyrophosphate (PP_i). Plants seem to have a more versatile metabolism than animals, and they can make use of the energy of pyrophosphate hydrolysis in some glycolytic reactions as well as in H^+ transport at the tonoplast. Since the vacuolar H^+-PPase and other enzymes utilizing PP_i are induced in some plants by low O_2 levels (hypoxia) or by chilling, when ATP levels are depleted they seem to represent a backup system to maintain essential cell activities under conditions of stress.

Although the free energy released by PP_i hydrolysis is less than that from ATP hydrolysis, this lack is compensated for by the fact that the vacuolar H^+-PPase carries only one H^+ ion per PP_i molecule hydrolyzed, whereas the V-ATPase appears to transport two H^+ per ATP hydrolyzed. Thus the energy available per H^+ ion transported is about the same, and each enzyme can therefore generate about the same gradient of electrochemical potential of H^+.

Vacuolar ABC Transporters Pump Large Organic Molecules

Members of another large family of active transport proteins energized directly by ATP hydrolysis are known as the **ATP-binding cassette (ABC) transporters** (see Figure 6.12). ABC transporters are able to use the energy of ATP hydrolysis directly to pump organic molecules across a membrane, entirely independently of the transmembrane H^+ gradient of electrochemical potential. Like the P-type H^+-ATPases, ABC transporters form a phosphorylated intermediate during catalysis, and they are therefore inhibited by vanadate. Originally identified in microbial and animal cells, the ABC transporters form a large protein family that is divided into two main subclasses: the *multidrug resistance proteins* (MDRs) and the *multidrug resistance–associated proteins* (MRPs). Genes encoding both types of ABC transporters have been identified in plants, but only the MRPs have been defined functionally.

Plant MRPs are also referred to as **glutathione conjugate pumps**, or **GS-X pumps**, since they typically transport molecules that have been covalently attached to glutathione.* However, plant GS-X pumps are also capable of transporting certain nonglutathionated compounds as well (reviewed by Rea et al. in press). The GS-X pumps of plant cells are specifically localized on the vacuolar membrane, where they function in herbicide detoxification, protection against oxidative damage, pigment accumulation, and the storage of antimicrobial compounds (see Chapter 13). A family of enzymes called glutathione transferases (GSTs) are responsible for attaching glutathione to the molecule to be transported. The common chemical determinant for GS conjugation is a carbon–carbon double bond adjacent to an electron-withdrawing group (CH_2=CH—Z) (Talalay et al. 1998). Compounds that contain this determinant include the anthocyanins (see Figure 6.12), IAA (indoleacetic acid, or auxin), and various phenolic compounds. Phytochelatins (see Box 5.1), which are synthesized from glutathione, are also transported into plant vacuoles along with their chelated heavy metals via GS-X pumps (see Figure 6.12). So far, four genes encoding vacuolar ABC transporters have been cloned from *Arabidopsis*. However, since the yeast genome contains 29 ABC protein genes, probably more will be discovered in plants as well. Each of the four *Arabidopsis* genes thus far characterized encodes a polypeptide of 170 to 180 kDa, containing four to six transmembrane helices and a cytoplasmically oriented catalytic domain (Lu et al. 1997; Rea et al. in press). The transmembrane domain provides the pathway for solute transport and determines the molecular specificity of the transporter.

By expressing the plant genes in yeast, researchers have been able to demonstrate that the proteins differ in their affinities for various substrates and in their transport efficiencies. For example, the *Arabidopsis* ABC transporter AtMRP2 has been shown to transport glutathione-conjugated chlorophyll catabolites, which are generated during leaf senescence, into the vacuole with high efficiency (low K_m, high V_{max}) (Lu et al. 1998).

Calcium Pumps, Antiports, and Channels Regulate Intracellular Calcium

Calcium is another important ion whose concentration is strongly regulated. The wall and apoplastic spaces are rich in calcium, but cytosolic Ca^{2+} concentrations are maintained at very low levels (in the micromolar range) despite a strong electrochemical-potential gradient

* Glutathione (GS) is a tripeptide (Glu-Cys-Gly) that functions as an important cellular antioxidant molecule.

driving Ca^{2+} diffusion into the cell. Small fluctuations in cytosolic Ca^{2+} concentration drastically alter the activities of many enzymes. Some crucial metabolic reactions are regulated through changes in cytosolic Ca^{2+} levels (see Chapter 14). Plant cells achieve this regulation by controlling the opening of Ca^{2+} channels that allow calcium to diffuse in, as well as by modulating the activity of pumps that drive Ca^{2+} out of the cytoplasm back into the extracellular spaces. Most of the calcium in the cell is stored in the central vacuole, where it is taken up via Ca^{2+}–H^+ antiporters, which use the proton gradient of electrochemical potential to energize the accumulation of calcium into the vacuole (Bush 1995).

Calcium efflux from the vacuole into the cytosol may in some cells be triggered by inositol trisphosphate (IP_3). IP_3, which appears to act as a "second messenger" in certain signal transduction pathways (see Chapter 14), induces the opening of IP_3-gated calcium channels on the tonoplast and endoplasmic reticulum. P-type (phosphorylated) Ca^{2+}-ATPases are found at the plasma membrane and in some endomembranes of plant cells (see Figure 6.12). The plasma membrane calcium pumps move calcium out of the cell, whereas the calcium pumps on the ER transport calcium into the ER lumen, another reservoir of intracellular calcium.

Ion Transport in Roots

Mineral nutrients absorbed by the root are carried to the shoot by the transpiration stream moving through the xylem. Just as the initial uptake of nutrients is a highly specific, well-regulated process, so too is the subsequent movement of mineral ions from the root surface across the cortex and into the xylem. Ion transport across the root obeys the same biophysical laws that govern cellular transport. However, as we have seen in the case of water movement (see Chapter 4), the anatomy of roots imposes some special constraints on the pathway of ion movement. In this section we will discuss the pathways and mechanisms involved in the radial movement of ions from the root surface to the tracheary elements of the xylem.

Solutes Move through Both Apoplast and Symplast

Thus far, our discussion of cellular ion transport has not included the cell wall. In terms of the transport of small molecules, the cell wall is an open lattice of polysaccharides through which mineral nutrients diffuse readily. Because all plant cells are separated by cell walls, ions can diffuse across a tissue (or be carried passively by water flow) entirely through the cell wall space without ever entering a living cell. This continuum of cell walls is called the free space or *apoplast* (see Chapter 4). We can determine the apoplastic volume of a piece of plant tissue by comparing the uptake of ^{3}H-labeled water and ^{14}C-labeled mannitol. Mannitol is a nonpermeating sugar alcohol that equilibrates with the free space but cannot enter the cells. Water, on the other hand, freely penetrates both the cells and the cell walls. Measurements of this type usually show that 5 to 20% of the plant tissue volume is occupied by cell walls.

Just as the cell walls form a continuous phase, so too do the cytoplasms of neighboring cells, collectively referred to as the *symplast*. Plant cells are interconnected by cytoplasmic bridges called *plasmodesmata* (see Chapter 1), cylindrical pores 20 to 60 nm in diameter (see Figures 1.30 and 1.31). Each plasmodesma is lined with a plasma membrane and contains a narrow tubule, the desmotubule, that is a continuation of the endoplasmic reticulum. In tissues in which significant amounts of intercellular transport occur, neighboring cells contain large numbers of plasmodesmata, up to 15 per square micrometer of cell surface (Figure 6.19). Specialized secretory cells, such as floral nectaries and leaf salt glands, appear to have high densities of plasmodesmata; so do the cells near root tips, where most nutrient absorption occurs.

By injecting dyes or by making electric-resistance measurements on cells containing large numbers of plasmodesmata, investigators have shown that ions and small solutes can move from cell to cell through these pores. As discussed in Chapter 4, water also moves through plasmodesmata. Because each plasmodesma is partly occluded by the desmotubule and associated proteins (see Chapter 1), movement of large molecules such as proteins through the plasmodesmata requires special mechanisms (Ghoshroy et al. 1997). Ions, on the other hand, appear to move from cell to cell through the entire plant by simple diffusion through the symplast (see Chapter 4).

Ions Moving through the Root Cross Both Symplastic and Apoplastic Spaces

Ion absorption by the roots (see Chapter 5) is more pronounced in the root hair zone than in the meristem and elongation zones. Cells in the root hair zone have completed their elongation but have not yet begun secondary growth. The root hairs are simply extensions of specific epidermal cells that greatly increase the surface area available for ion absorption.

The apoplast forms a continuous phase from the root surface through the cortex. At the boundary between the vascular cylinder (the stele) and the cortex is a layer of specialized cells, the endodermis. As discussed in Chapters 4 and 5, a suberized cell layer in the endodermis, known as the Casparian strip, effectively blocks the entry of water and mineral ions into the stele via the apoplast. An ion that enters a root may enter the symplast immediately by crossing an epidermal cell plasma

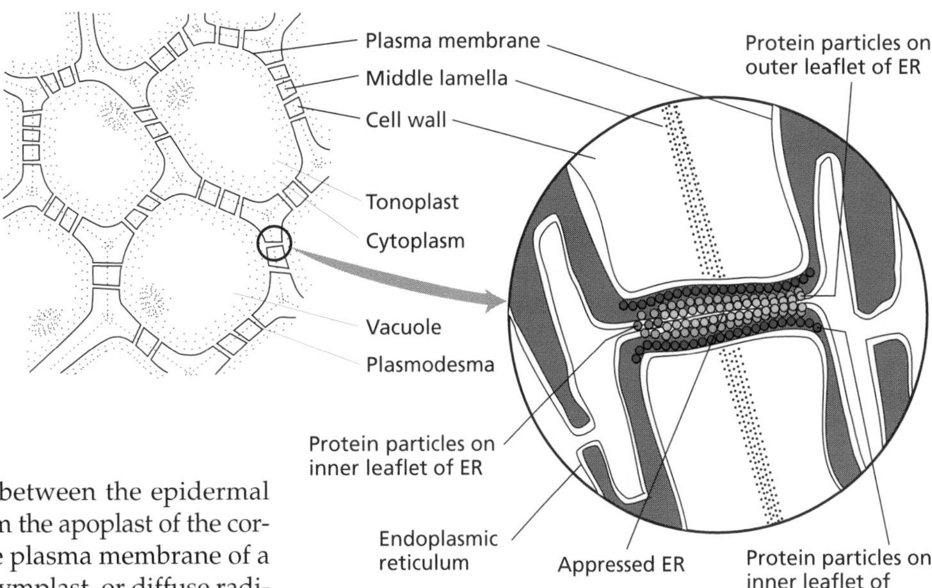

Figure 6.19 Diagram illustrating how plasmodesmata connect the cytoplasms of neighboring cells. Plasmodesmata are about 40 nm in diameter and allow diffusion of water and small molecules from one cell to the next. In addition, the size of the opening can be regulated by rearrangements of the internal proteins to allow the passage of larger molecules.

membrane, or it may diffuse between the epidermal cells through the cell walls. From the apoplast of the cortex, an ion may either cross the plasma membrane of a cortical cell, thus entering the symplast, or diffuse radially all the way to the endodermis via the apoplast. In all cases, ions must enter the symplast before they can enter the stele, because of the presence of the Casparian strip.

Once an ion has entered the stele through the symplastic connections across the endodermis, it continues to diffuse from cell to cell into the xylem. Finally, the ion reenters the apoplast as it diffuses into a xylem tracheid or vessel element. Again, the Casparian strip prevents the ion from diffusing back out of the root through the apoplast. Because of the presence of the Casparian strip, the plant can maintain a higher ionic concentration in the xylem than exists in the soil water surrounding the roots.

Xylem Parenchyma Cells Participate in Xylem Loading

Once ions have been taken up into the symplast of the root at the epidermis or cortex, they must be loaded into the tracheids or vessel elements of the stele to be translocated to the shoot. Since the xylem tracheary elements are dead cells, the ions must exit the symplast by crossing a plasma membrane a second time.

The process whereby ions exit the symplast and enter the conducting cells of the xylem is called **xylem loading**. The mechanism of xylem loading has long baffled scientists. Ions could enter the tracheids and vessel elements of the xylem by simple passive diffusion. In this case, the movement of ions from the root surface to the xylem would take only a single step requiring metabolic energy. The site of this single energy-dependent uptake would be the plasma membrane surfaces of the root epidermal, cortical, and endodermal cells. According to the passive-diffusion model, ions move passively into the stele via the symplast down a concentration gradient, and then leak out of the living cells of the stele (possibly because of lower oxygen availability in the interior of the root) into the nonliving conducting cells of the xylem.

Support for the passive-diffusion model is shown in Figure 6.20. Bowling and coworkers used ion-specific microelectrodes to measure the electrochemical potentials of various ions across maize roots (Dunlop and Bowling 1971). Data from this and other studies indicate that K^+, Cl^-, Na^+, SO_4^{2-}, and NO_3^- are all taken up actively by the epidermal and cortical cells and are maintained in the xylem against a gradient of electrochemical potential when compared with the external medium (Lüttge and Higinbotham 1979). However, none of these ions is at a higher electrochemical potential in the xylem than in the cortex or living portions of the stele. Therefore, the final movement of ions into the xylem could be due to passive diffusion.

However, other observations have led to the view that this final step of xylem loading may also involve active processes within the stele (Lüttge and Higinbotham 1979). With the type of apparatus shown in Figure 6.21, it is possible to make simultaneous measurements of ion uptake into the epidermal or cortical cytoplasm and of ion loading into the xylem. By using treatments with inhibitors and plant hormones, investigators have shown that ion uptake by the cortex and ion loading into the xylem operate independently. For example, treatment with the protein synthesis inhibitor cycloheximide or with the cytokinin benzyladenine inhibits xylem loading without affecting uptake by the cortex. This result indicates that efflux from the stelar cells is regulated independently from uptake by the cortical cells.

Recent biochemical studies have supported a role for the xylem parenchyma cells in xylem loading. The

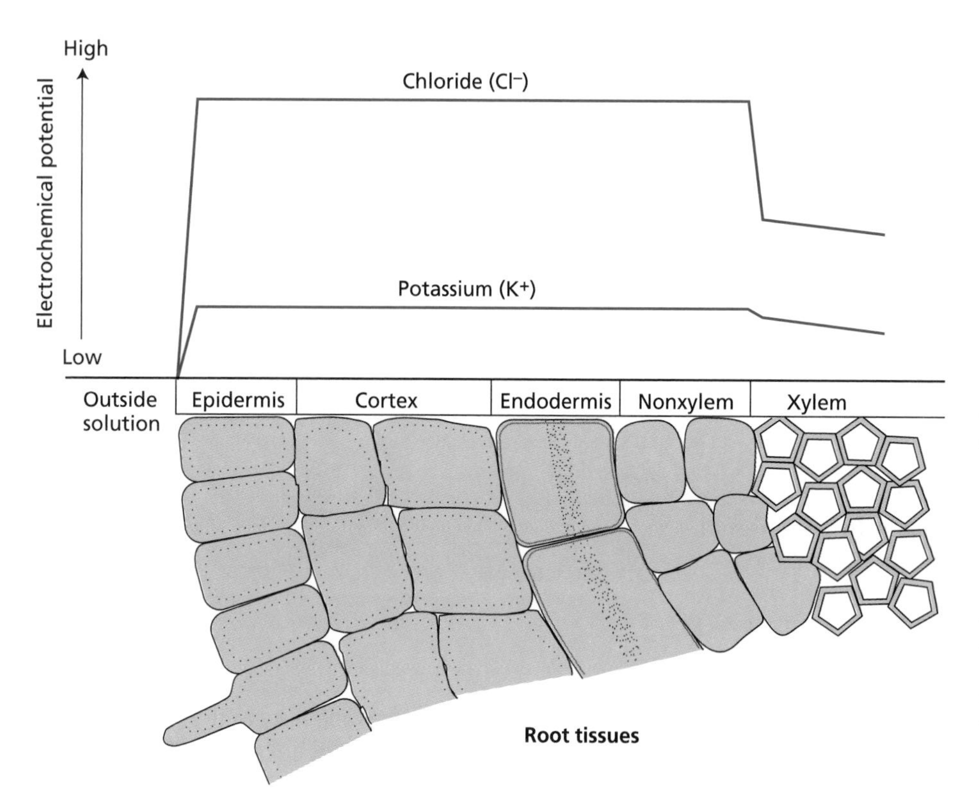

Figure 6.20 Diagram showing electrochemical potentials of K^+ and Cl^- across a maize root. To determine the electrochemical potentials, the root was bathed in a solution containing 1 m*M* KCl and 0.1 m*M* $CaCl_2$. A reference electrode was positioned in the bathing solution, and an ion-sensitive measuring electrode was inserted in different cells of the root. The horizontal axis shows the different tissues found in a root cross section. The substantial increase in electrochemical potential for both K^+ and Cl^- between the bathing medium and the epidermis indicates that ions are taken up into the root by an active transport process. In contrast, the potentials decrease at the xylem vessels, suggesting that ions are transported into the xylem by passive diffusion down the gradient of electrochemical potential. (After Dunlop and Bowling 1971.)

plasma membranes of xylem parenchyma cells contain proton pumps, water channels, and a variety of ion channels specialized for influx or efflux (Maathuis et al. 1997). In barley xylem parenchyma, two types of cation efflux channels have been identified: K^+-specific efflux channels and nonselective cation efflux channels. These channels are regulated by both the membrane potential and the cytosolic calcium concentration (De Boer and Wegner 1997). This finding suggests that the flux of ions from the xylem parenchyma cells into the xylem tracheary elements, rather than being due to simple leakage, is under tight metabolic control through regulation of the plasma membrane H^+-ATPase and ion efflux channels.

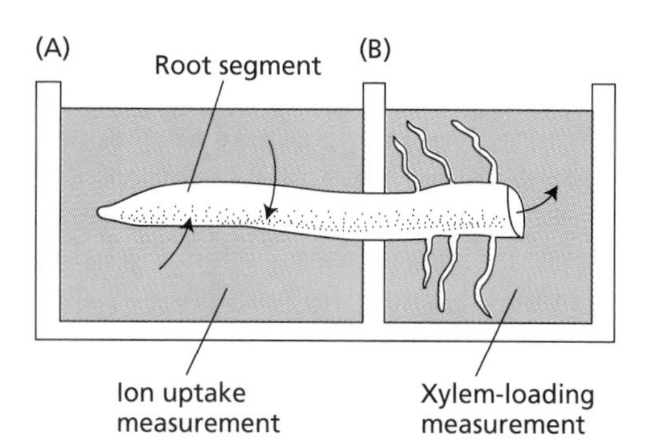

Figure 6.21 We can measure the relationship between ion uptake into the root and xylem loading by placing a root segment across two compartments and adding a radioactive tracer to one of them (in this case compartment A). The rate of disappearance of the tracer from compartment A gives a measure of ion uptake, and the rate of appearance in compartment B provides a measurement of xylem loading. (From Lüttge and Higinbotham 1979.)

Summary

Molecular movement between different compartments of biological systems is known as transport. Plants exchange solutes with their environment and among their tissues and organs. Transport at the cellular level is responsible for all higher-order activities. Transport between cells is specifically controlled by their plasma membranes.

Forces that drive biological transport, which include concentration gradients, electric-potential gradients, and hydrostatic pressures, are integrated by an expression called the electrochemical potential. Transport of solutes down chemical gradients (e.g., by diffusion) is known as passive transport. Movement of solutes against a chemical-potential gradient is known as active transport and requires energy input.

The extent to which a membrane permits or restricts the movement of a substance is called membrane permeability. The composition of the membrane and the chemical properties of the solute are major determinants of membrane permeability. When cations and anions move across a membrane at different rates, the electric potential that develops is called the diffusion potential.

For each ion, the relationship between the voltage difference across the membrane and the distribution of the ion at equilibrium is described by the Nernst equation. The Nernst equation shows that the difference in concentration of an ion between two compartments is balanced by the voltage difference between the compartments. That voltage difference, or membrane potential, is seen in all living cells because of the asymmetric ion distributions between the inside and outside of the cells. All diffusion potentials across a cell membrane are summed by the Goldman equation. Electrogenic pumps, which carry out active transport and carry a net charge, change the membrane potential from the value created by diffusion.

Membranes contain specialized proteins—channels, carriers, and pumps—that facilitate solute transport. Channels are transport proteins that span the membrane, forming pores through which solutes diffuse down their gradient of electrochemical potentials. Carriers bind a solute on one side of the membrane and release it on the other side. Transport specificity is determined largely by the properties of channels and carriers.

In plants, a family of H^+-pumping ATPases provides the primary driving force for transport across the plasma membrane. Two other kinds of electrogenic proton pumps serve this purpose at the tonoplast. Plant cells also have calcium-pumping ATPases that participate in the regulation of intracellular calcium concentrations, as well as ATP-binding cassette transporters that use the energy of ATP to transport large anionic molecules. The gradient of electrochemical potential generated by H^+ pumping is used to drive the transport of other substances in a process called secondary transport. Genetic studies have revealed the multitude of genes, and their corresponding transport proteins, that account for the versatility of plant transport. Patch clamp electrophysiology provides unique information on ion channels, and enables the permeability and gating of individual channel proteins to be measured.

Solutes move between cells either through the extracellular spaces (the apoplast) or from cytoplasm to cytoplasm (via the symplast). Cytoplasms of neighboring cells are connected by plasmodesmata, which facilitate symplastic transport. When an ion enters the root, it may be taken up into the cytoplasm of an epidermal cell or it may diffuse through the apoplast into the root cortex and enter the symplast through a cortical cell. From the symplast, the ion is loaded into the xylem and transported to the shoot.

General Reading

Alberts, B., Bray, D., Lewis, J., Raff, M., Roberts, K., and Watson, J. D. (1994) *Molecular Biology of the Cell*, 3rd ed. Garland, New York.

*Bush, D. S. (1995) Calcium regulation in plant cells and its role in signaling. *Annu. Rev. Plant Physiol. Plant Mol. Biol.* 46: 95–122.

*Epstein, E. (1972) *Mineral Nutrition of Plants: Principles and Perspectives*. Wiley, New York.

*Frommer, W. B., and Ninnemann, O. (1995) Heterologous expression of genes in bacterial, fungal, animal, and plant cells. *Annu. Rev. Plant Physiol. Plant Mol. Biol.* 46: 419–444.

*Ghoshroy, S., Lartey, R., Sheng, J., and Citovsky, V. (1997) Transport of proteins and nucleic acids through plasmodesmata. *Annu. Rev. Plant Physiol. Plant Mol. Biol.* 48: 27–50.

*Glass, A. D. M. (1983) Regulation of ion transport. *Annu. Rev. Plant Physiol.* 34: 311–326.

*Harold, F. M. (1986) *The Vital Force: A Study of Bioenergetics*. W. H. Freeman, New York.

*Hedrich, R., and Schroeder, J. I. (1989) The physiology of ion channels and electrogenic pumps in higher plants. *Annu. Rev. Plant Physiol.* 40: 539–569.

*Lüttge, U., and Higinbotham, N. (1979) *Transport in Plants*. Springer, Berlin.

*Maurel, C. (1997) Aquaporins and water permeability of plant membranes. *Annu. Rev. Plant Physiol. Plant Mol. Biol.* 48: 399–429.

Sze, H., Ward, J. M., and Lai, S. (1992) Vacuolar H^+-translocating ATPases from plants: Structure, function, and isoforms. *J. Bioenerg. Biomembr.* 24: 371–381.

*Tanner, W., and Caspari, T. (1996) Membrane transport carriers. *Annu. Rev. Plant Physiol. Plant Mol. Biol.* 47: 595–626.

*Tazawa, M., Shimmen, T., and Mimura, T. (1987) Membrane control in the Characeae. *Annu. Rev. Plant Physiol.* 38: 95–117.

Ward, J. (1997) Patch-clamping and other molecular approaches for the study of plasma membrane transporters demystified. *Plant Physiol.* 114: 1151–1159.

Zeiger, E. (1983) The biology of stomatal guard cells. *Annu. Rev. Plant Physiol.* 34: 441–475.

* Indicates a reference that is general reading in the field and is also cited in this chapter.

Chapter References

Barkla, B. J., and Pantoja, O. (1996) Physiology of ion transport across the tonoplast of higher plants. *Annu. Rev. Plant Physiol.* 47: 159–184.

Boekema, E. J., Ubbink-Kok, T., Lolkema, J. S., Brisson, A., and Konings, W. N. (1997) Visualization of a peripheral stalk in V-type ATPase: Evidence for a stator structure essential to rotational catalysis. *Proc. Natl. Acad. Sci. USA* 94: 14291–14293.

Crawford, N. M. (1995) Nitrate: Nutrient and signal for plant growth. *Plant Cell* 7: 859–868.

De Boer, A. H., and Wegner, L. H. (1997) Regulatory mechanisms of ion channels in xylem parenchyma cells. *J. Exp. Bot.* 48: 441–449.

Dunlop, J., and Bowling, D. J. F. (1971) The movement of ions to the xylem exudate of maize roots. *J. Exp. Bot.* 22: 453–464.

Gogarten, J. P., Hilario, E., and Olendzenski, L. (1996) Gene duplications and horizontal gene transfer during early evolution. In *Evolution of Microbial Life*, D. M. Roberts, P. Sharp, G. Alderson, and M. Collins, eds., Cambridge University Press, Cambridge, pp. 267–292.

Gogarten, J. P., Kibak, H., Dittrich, P., Taiz, L., Bowman, E. J., Bowman, B. J., Manolson, M. F., Poole, R. J., Date, T., Oshima, T., Konishi, J., Denda, K., and Yoshida, M. (1989) Evolution of the vacuolar H^+-ATPase: Implications for the origin of eukaryotes. *Proc. Natl. Acad. Sci. USA* 86: 6661–6665.

Higinbotham, N., Etherton, B., and Foster, R. J. (1967) Mineral ion contents and cell transmembrane electropotentials of pea and oat seedling tissue. *Plant Physiol.* 42: 37–46.

Higinbotham, N., Graves, J. S., and Davis, R. F. (1970) Evidence for an electrogenic ion transport pump in cells of higher plants. *J. Membr. Biol.* 3: 210–222.

Hirshi, K. D., Zhen, R-G., Rea, P. A., and Fink, G. R. (1996) *CAX1*, an H^+/Ca^{2+} antiporter from *Arabidopsis*. *Proc. Natl Acad. Sci. USA* 93: 8782–8786.

Johansson, F., Olbe, M., Sommarin, M., and Larsson, C. (1995) Brij 58, a polyoxyethylene acyl ether, creates membrane vesicles of uniform sidedness. A new tool to obtain inside-out (cytoplasmic side-out) plasma membrane vesicles. *Plant J.* 7: 165–173.

Larsson, C., Widell, S., and Kjellbom, P. (1987) Preparation of high-purity plasma membranes. *Methods Enzymol.* 148: 558–568.

Lin, W., Schmitt, M. R., Hitz, W. D., and Giaquinta, R. T. (1984) Sugar transport into protoplasts isolated from developing soybean cotyledons. *Plant Physiol.* 75: 936–940.

Lu, Y-P., Li, Z-S., Drozdowicz, Y. M., Hortensteiner, S., Martinoia, E., and Rea, P. A. (1998) AtMRP2, an *Arabidopsis* ATP binding cassette transporter able to transport glutathione S-conjugates and chlorophyll catabolites: Functional comparisons with AtMRP1. *Plant Cell* 10: 267–282.

Lu, Y. P., Li, Z. S., and Rea, P. A. (1997) *AtMRP1* gene of *Arabidopsis* encodes a glutathione S-conjugate pump: Isolation and functional definition of a plant ATP-binding cassette transporter gene. *Proc. Natl. Acad. Sci. USA* 94: 8243–8248.

Lüttge, U., and Ratajczak, R. (1997) The Physiology, Biochemistry and Molecular Biology of the Plant Vacuolar ATPase. *Adv. Bot. Res.* 25: 253–296.

Maathuis, F. J. M., Ichida, A. M., Sanders, D., and Schroeder, J. I. (1997) Roles of higher plant K^+ channels. *Plant Physiol.* 114: 1141–1149.

Müller, M., Irkens-Kiesecker, U., Kramer, D., and Taiz, L. (1997) Purification and reconstitution of the vacuolar H^+-ATPases from lemon fruits and epicotyls. *J. Biol. Chem.* 272: 12762–12770.

Müller, M. L., Irkens-Kiesecker, U., Rubinstein, B., and Taiz, L. (1996) On the mechanism of hyperacidification in lemon: Comparison of the vacuolar H^+-ATPase activities of fruits and epicotyls. *J. Biol. Chem.* 271: 1916–1924.

Nobel, P. (1991) *Physicochemical and Environmental Plant Physiology.* Academic Press, San Diego.

Novacky, A., Ullrich-Eberius, C. I., and Lüttge, U. (1980) pH and membrane potential changes during glucose uptake in *Lemna gibba* G1 and their response to light. *Planta* 149: 321–326.

Palmgren, M. G. (1991) Regulation of plant plasma membrane H^+-ATPase activity. *Physiol. Plantarum* 83: 314–323.

Rea, P. A., Li, Z-S., Lu, Y-P., Drozdowicz, Y., and Martinoia, E. (In press) From vacuolar GS-X pumps to multispecific ABC transporters. *Annu. Rev. Plant Physiol. Plant Mol. Biol.*

Rea, P. A., and Poole, R. J. (1993) Vacuolar H^+-translocating pyrophosphatase. *Annu. Rev. Plant Physiol. Plant Mol. Biol.* 44: 157–180.

Schroeder, J. L., Raschke, K., and Neher, E. (1987) Voltage dependence of K^+ channels in guard-cell protoplasts. *Proc. Natl Acad. Sci. USA* 84: 4108–4112.

Serrano, R., and Portillo, F. (1990) Catalytic and regulatory sites of yeast plasma membrane H^+-ATPase studied by directed mutagenesis. *Biochim. Biophys. Acta* 1018: 195–199.

Small, J. (1946) *pH and Plants.* D. Van Nostrand, New York.

Spanswick, R. M. (1981) Electrogenic ion pumps. *Annu. Rev. Plant Physiol.* 32: 267–289.

Sussman, M. R. (1994) Molecular analysis of proteins in the plant plasma membrane. *Annu. Rev. Plant Physiol. Plant Mol. Biol.* 45: 211–234.

Sze, H. (1985) H^+-translocating ATPase: Advances using membrane vesicles. *Annu. Rev. Plant Physiol.* 36: 175–208.

Taiz, L., and Nelson, N. (1996) Evolution of V- and F-ATPases. In *Origin and Evolution of Biological Energy Conversion*, H. Baltscheffsky, ed., VCH, New York, pp. 291–306.

Taiz, S. L., and Taiz, L. (1991) Ultrastructural comparison of the vacuolar and mitochondrial H^+-ATPases of *Daucus carota*. *Bot. Acta* 104: 117–121.

Talalay, P., De Long, M., and Prochaska, H. J. (1998) Identification of a common chemical signal regulating the induction of enzymes that protect against chemical carcinogenesis. *Proc. Natl. Acad. Sci. USA* 85: 8261–8265.

Vera-Estrella, R., Barkla, B. J., Higgins, V. J., and Blumwald, E. (1994) Plant defense response to fungal pathogens. Activation of host-plasma-membrane H^+-ATPase by elicitor-induced enzyme dephosphorylation. *Plant Physiol.* 104: 209–215.

Biochemistry and Metabolism

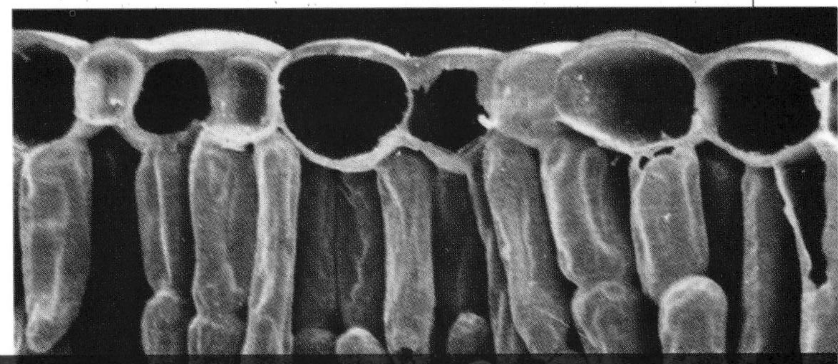

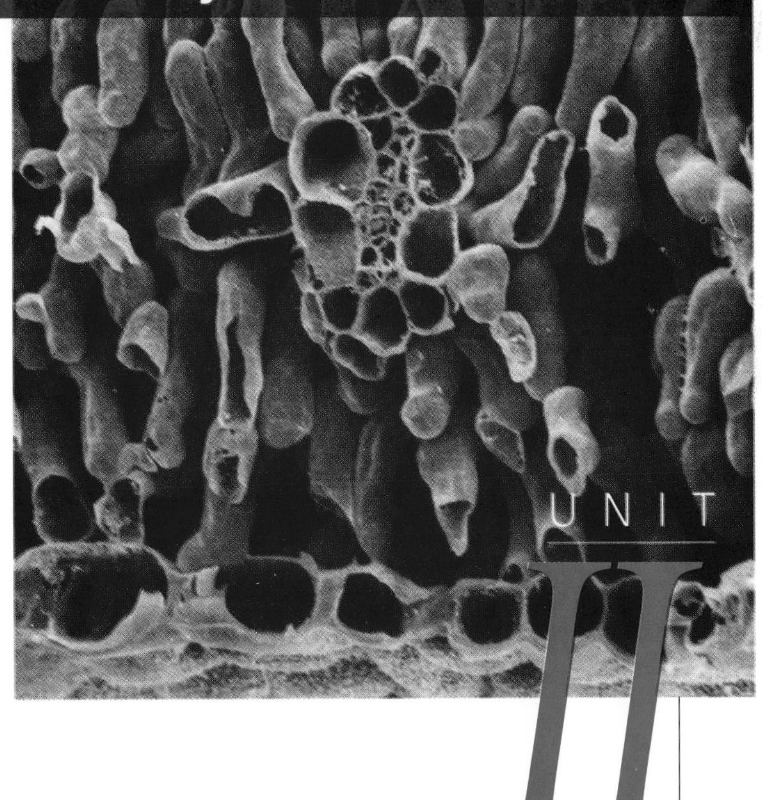

UNIT

II

7 Photosynthesis: The Light Reactions

LIFE ON EARTH ULTIMATELY DEPENDS on energy derived from the sun. Photosynthesis is the only process of biological importance that can harvest this energy. In addition, a large fraction of the planet's energy resources results from photosynthetic activity in either recent (biomass) or ancient (fossil fuel) times. This chapter introduces the basic physical principles that underlie photosynthetic energy storage and the current understanding of the structure and function of the photosynthetic apparatus.

The term "photosynthesis" means literally "synthesis using light." As we will see in this chapter, photosynthetic organisms use solar energy to synthesize organic compounds that cannot be formed without the input of energy. Energy stored in these molecules can be used later to power cellular processes in the plant and can serve as the energy source for all forms of life. This chapter deals with the role of light in photosynthesis, the structure of the photosynthetic apparatus, and the processes that begin with the excitation of chlorophyll by light and culminate in the synthesis of ATP and NADPH.

Photosynthesis in Higher Plants

The most active photosynthetic tissue in higher plants is the mesophyll of leaves (see Chapter 1). Mesophyll cells have many chloroplasts, which contain the specialized light-absorbing green pigments, the **chlorophylls**. In photosynthesis, the plant uses solar energy to oxidize water, thereby releasing oxygen, and to reduce carbon dioxide into organic compounds, primarily sugars. The complex series of reactions that culminate in the reduction of CO_2 include the **thylakoid reactions** and the **carbon fixation reactions**. The thylakoid reactions of photosynthesis take place in the specialized internal membranes of the chloroplast called thylakoids (see Chapter 1). The end products of these thylakoid reactions are the high-energy compounds ATP and NADPH, which are used for the synthesis of sugars in the carbon fixation reactions. These synthetic processes take place in the stroma of the chloroplasts, the aqueous region that surrounds the thylakoids. The thylakoid reactions of photosynthesis are the subject of this chapter; the carbon fixation reactions are discussed in Chapter 8.

In the chloroplast, light energy is harvested by two different functional units called *photosystems*. The absorbed light energy is used to power the transfer of electrons through a series of compounds that act as electron donors and electron acceptors. The majority of electrons ultimately reduce $NADP^+$ to NADPH. Light energy is also used to generate a proton motive force (see Chapter 6) across the thylakoid membrane, which is used to synthesize ATP.

General Concepts and Historical Background

In this section we will explore the essential concepts that provide a foundation for an understanding of photosynthesis. These concepts include the nature of light, the properties of pigments, the various roles of pigments, and the overall chemical equations for photosynthesis. In addition, we will discuss some of the critical experiments that have made the conceptual understanding of photosynthesis possible.

Light Has Characteristics of Both a Particle and a Wave

A triumph of physics in the early twentieth century was the realization that light has properties of both particles and waves. A wave (Figure 7.1) is characterized by a **wavelength**, denoted by the Greek letter lambda (λ), which is the distance between successive wave crests. The **frequency**, represented by the Greek letter nu (ν), is the number of wave crests that pass an observer in a given time. A simple equation relates the wavelength, the frequency, and the speed of any wave:

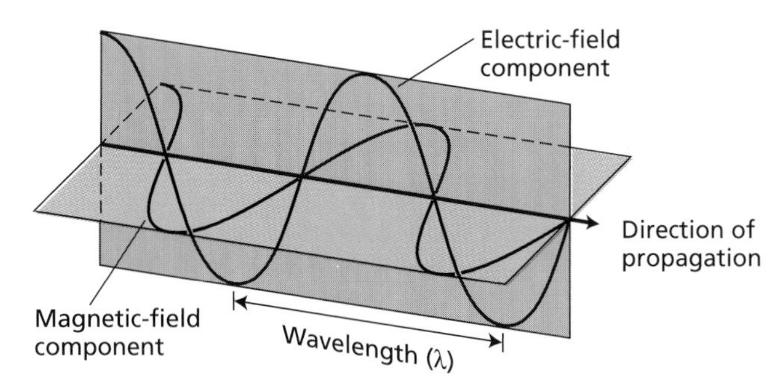

Figure 7.1 Light is a transverse electromagnetic wave, consisting of oscillating electric and magnetic fields that are perpendicular to each other and perpendicular to the direction of propagation of the light. Light moves at a speed of 3×10^8 m s^{-1}. The wavelength (λ) is the distance between successive crests of the wave.

$$c = \lambda\nu \qquad (7.1)$$

where c is the speed of the wave—in the present case, the speed of light (3.0×10^8 m s^{-1}). The light wave is a transverse (side-to-side) electromagnetic wave, in which both electric and magnetic fields oscillate perpendicular to the direction of propagation of the wave and at 90° with respect to each other.

Light is also a particle, which we call a **photon**. Each photon contains an amount of energy that is called a **quantum** (plural quanta). The energy content of light is not continuous but rather is delivered in these discrete packets, the quanta. The energy (E) of a photon depends on the frequency of the light according to a relation known as Planck's law:

$$E = h\nu \qquad (7.2)$$

where h is Planck's constant (6.626×10^{-34} J s).

Sunlight is like a rain of photons of different frequencies. Our eyes are sensitive to only a small range of frequencies—the visible-light region of the electromagnetic spectrum (Figure 7.2). Light of slightly higher fre-

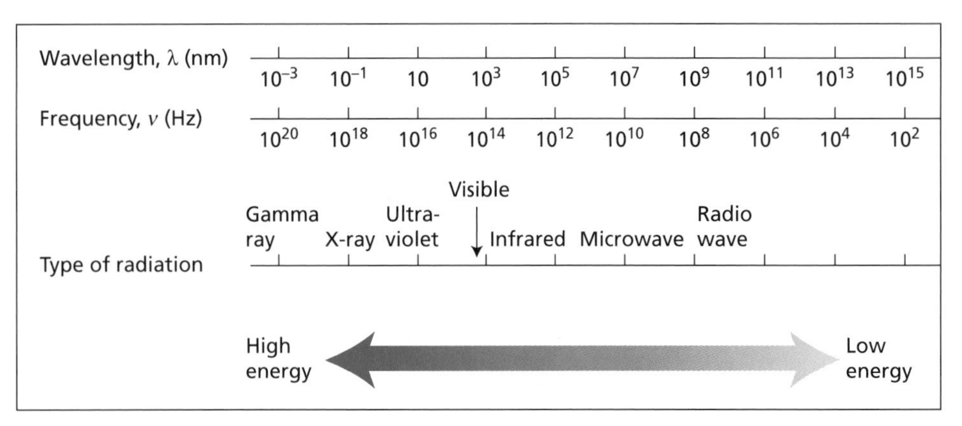

Figure 7.2 The electromagnetic spectrum. Wavelength (λ) and frequency (ν) are inversely related. Our eyes are sensitive to only a narrow range of wavelengths of radiation, the visible region, which extends from about 400 nm (violet) to about 700 nm (red). Short-wavelength (high-frequency) light has a high energy content; long-wavelength (low-frequency) light has a low energy content.

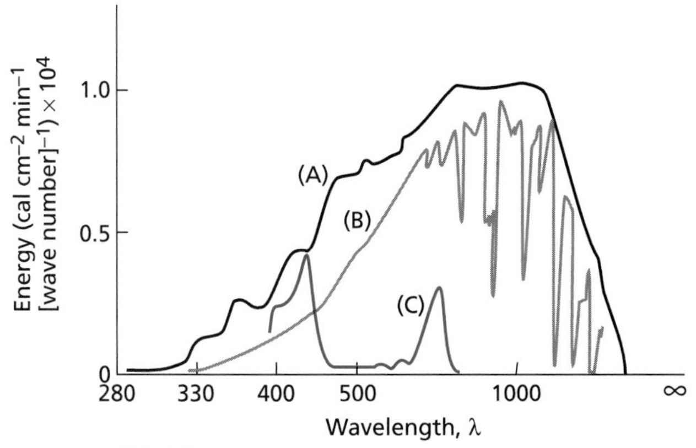

Figure 7.3 The solar spectrum and its relation to the absorption spectrum of chlorophyll. Curve A is the energy output of the sun as a function of wavelength. Curve B is the energy that strikes the surface of Earth. The sharp valleys in the infrared region beyond 700 nm represent the absorption of solar energy by molecules in the atmosphere, chiefly water vapor. Curve C is the absorption spectrum of chlorophyll *a*, which absorbs strongly in the blue (about 430 nm) and the red (about 660 nm) portions of the spectrum. Since the green light in the middle of the visible region is absorbed only weakly, much of it is therefore reflected into our eyes and gives plants their characteristic green color. Note that curve C has absorbance rather than energy units. (Solar spectra from Calvin 1976.)

quencies (or shorter wavelengths) is in the ultraviolet region of the spectrum, and light of slightly lower frequencies (or longer wavelengths) is in the infrared region. The output of the sun is shown in Figure 7.3, along with the energy density that strikes the surface of Earth. The absorption spectrum of chlorophyll *a* (curve C in Figure 7.3) indicates approximately the portion of the solar output that is utilized by plants. Spectrophotometry, the technique used to measure the absorption of light by a sample, is discussed in Box 7.1.

When Molecules Absorb or Emit Light, They Change Their Electronic State

Chlorophyll appears green to our eyes because it absorbs light in the red and blue parts of the spectrum, so only some of the light enriched in green wavelengths (about 550 nm) is reflected into our eyes (see Figure 7.3). The absorption of light is represented by Equation 7.3, in which chlorophyll (Chl) in its lowest-energy, or ground, state absorbs a photon (represented by $h\nu$) and makes a transition to a higher-energy, or excited, state (Chl*):

$$\text{Chl} + h\nu \rightarrow \text{Chl}^* \quad (7.3)$$

The distribution of electrons in the excited molecule is somewhat different from the distribution in the ground-state molecule. Figure 7.4 describes the absorption and emission of light by chlorophyll molecules. Absorption of blue light excites the chlorophyll to a higher energy state than absorption of red light, because the energy of

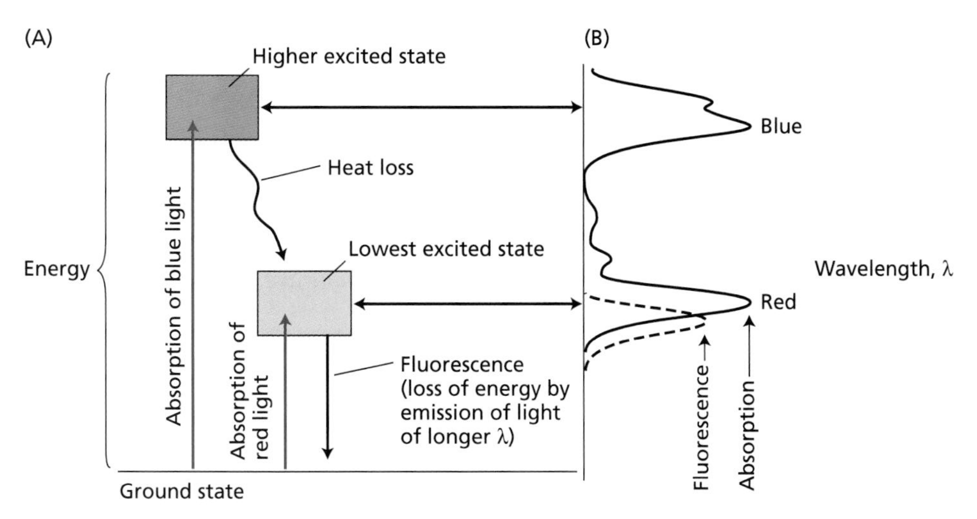

Figure 7.4 Light absorption and emission by chlorophyll. (A) Energy level diagram. Absorption or emission of light is indicated by vertical lines that connect the ground state with excited electron states. The blue and red absorption bands of chlorophyll (which absorb blue and red photons, respectively) correspond to the upward vertical arrows, signifying that energy absorbed from light causes the molecule to change from the ground state to an excited state. The downward-pointing arrow indicates fluorescence, in which the molecule goes from the lowest excited state to the ground state while re-emitting energy as a photon. (B) The spectra of absorption and fluorescence. The long-wavelength absorption band of chlorophyll corresponds to light that has the energy required to cause the transition from the ground state to the first excited state. The short-wavelength absorption band corresponds to a transition to a higher excited state. (After Sauer 1975.)

BOX 7.1

Principles of Spectrophotometry

MUCH OF WHAT WE KNOW about the photosynthetic apparatus was learned through **spectroscopy**—that is, measurements of the interaction of light and molecules. Spectrophotometry is an important branch of spectroscopy that focuses on the technique of measurement. Here we will examine four topics: Beer's law, the measurement of absorbance, action spectra, and difference spectra.

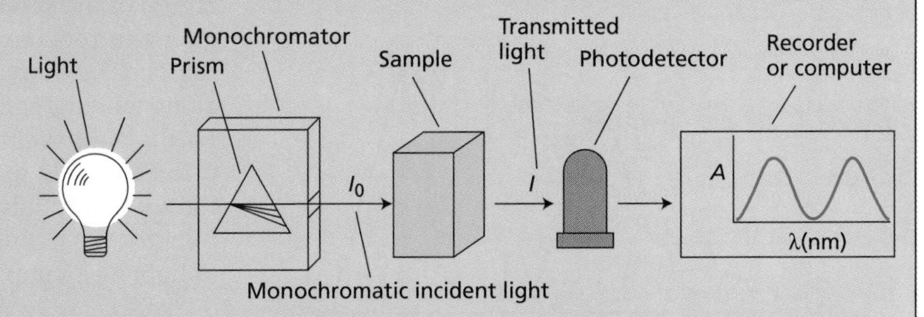

Figure 2 Schematic diagram of a spectrophotometer. The instrument consists of a light source, a monochromator that contains a wavelength selection device such as a prism, a sample holder, a photodetector, and a recorder or computer. The output wavelength of the monochromator can be changed by rotation of the prism; the graph of absorbance versus wavelength is called a spectrum.

Beer's Law

An essential piece of information about any molecular species is how much of it is present. Quantitative measures of concentration are the cornerstone of much biological science. Of all the methods that have been devised for measuring concentration, by far the most widely applied is absorption spectrophotometry. In this technique, the amount of light that a sample absorbs at a particular wavelength is measured and used to determine the concentration of the sample by comparison with appropriate standards or reference data. The most useful measure of light absorption is the **absorbance** (A), also commonly called the optical density (OD) (Figure 1). The absorbance is defined as

$$A = \log I_0/I$$

where I_0 is the intensity of light that is incident on the sample and I is the intensity of light that is transmitted by the sample.

The absorbance of a sample can be related to the concentration of the absorbing species through **Beer's law**:

$$A = \varepsilon cl$$

where c is concentration, usually measured in moles per liter; l is the length of the light path, usually 1 cm; and ε is a proportionality constant known as the **molar extinction coefficient**, with the units of liters per mole per centimeter. The value of ε is a function of both the particular compound being measured and the wavelength. Chlorophylls typically have an ε value of about 100,000 L mol^{-1} cm^{-1}. When more than one component of a complex mixture absorbs at a given wavelength, the absorbances due to the individual components are generally additive.

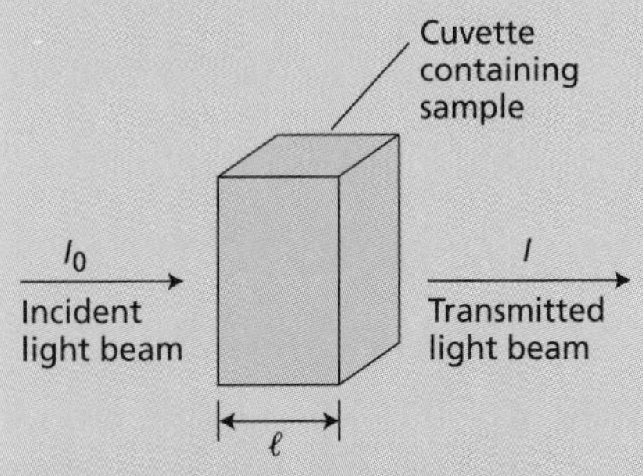

Figure 1 Definition of absorbance. A monochromatic incident light beam of intensity I_0 traverses a sample contained in a cuvette of length ℓ. Some of the light is absorbed by the chromophores in the sample, and the intensity of light that emerges is I.

The Spectrophotometer

The absorbance is measured by an instrument called a **spectrophotometer** (Figure 2). The essential parts of a spectrophotometer include a light source, a wavelength selection device such as a monochromator or filter, a sample chamber, a light detector, and a readout device. Modern instruments usually also include a computer, which is used for storage and analysis of the spectra. The most useful machines scan the wavelength of the light that is incident on the sample and produce, as output, spectra of absorbance versus wavelength, such as those shown in Figure 7.6.

Action Spectra

The use of action spectra has been central to the development of our current understanding of photosynthesis. An **action spectrum** is a graph of the magnitude of the biological effect observed as a function of wavelength. Examples of effects measured by action spectra are oxygen evolution (Figure 3) and hormonal growth responses due to the action of phytochrome (see Chapter 17). Often an action spectrum can identify the chromophore responsible for a particular light-induced phenomenon. Action spectra were instrumental in the discovery of the existence of the two photosystems in O_2-evolving photosynthetic organisms.

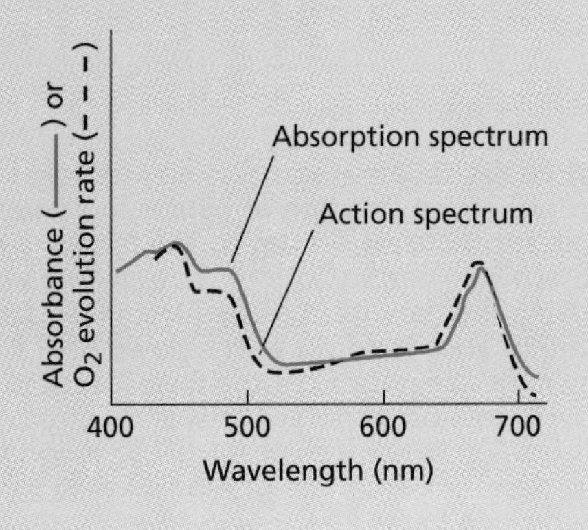

Figure 3 An action spectrum compared to an absorption spectrum. The absorption spectrum is measured as shown in Figure 2. An action spectrum is measured by plotting a response to light such as oxygen evolution, as a function of wavelength. If the pigments used to obtain the absorption spectrum are the same as those that cause the response, the absorption and action spectra will match. In the example shown here, the action spectrum for oxygen evolution matches the absorption spectrum of intact chloroplasts quite well, indicating that light absorption by the chlorophylls mediates oxygen evolution. Discrepancies are found in the region of carotenoid absorption, from 450 to 550 nm, indicating that energy transfer from carotenoids to chlorophylls is not as effective as energy transfer between chlorophylls.

▶

BOX 7.1 *(continued)*

Some of the first action spectra were measured by T. W. Engelmann in the late 1800s (Figure 4). Engelmann used a prism to disperse sunlight into a rainbow that was allowed to fall on an aquatic algal filament. A population of O_2-seeking bacteria was introduced into the system. The bacteria congregated in the regions of the filaments that evolved the most O_2. These were the regions illuminated by blue light and red light, which are strongly absorbed by chlorophyll. Today, action spectra can be measured in room-sized spectrographs in which the scientist enters a huge monochromator and places samples for irradiation in a large area of the room bathed by monochromatic light. But the principle of the experiment is the same as that of Engelmann's experiments.

Difference Spectra

An important technique in studies of photosynthesis is light-induced **difference spectroscopy**, which measures changes in absorbance (Figure 5). In this technique, bright light, often called **actinic** light, is used to illuminate a sample, while a dim beam of light is used to measure the absorbance of the sample at wavelengths other than that of the actinic beam. In this way a difference spectrum is obtained, which represents the changes in the absorption spectrum of the sample induced by illumination with the actinic light. Absorption bands that disappear upon illumination appear as negative peaks; new bands that appear upon illumination appear as positive peaks. Difference spectra give important clues to the identity of molecular species participating in the photoreactions of photosynthesis. The difference spectrum of the photooxidation of P700 (a chlorophyll that absorbs light of wavelength 700 nm, as we'll discuss later in the chapter) is shown in Figure 6 (Ke 1973).

By the use of special flash techniques, it is possible to record the difference spectrum at a given time after flash excitation. Multiple difference spectra recorded at different times after flash excitation can be used to measure the kinetics of the chemical reactions that follow photon excitation of a reaction center. These techniques can have extraordinary time resolution, in some cases less than a picosecond (10^{-12} s), and have provided great insights into the earliest events in the photosynthetic energy storage process.

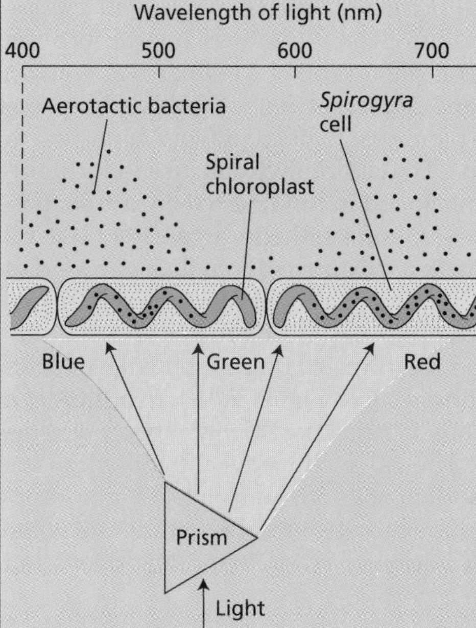

Figure 4 Schematic diagram of the action spectrum measurements by T. W. Engelmann. Engelmann projected a spectrum of light onto the spiral chloroplast of the filamentous green alga *Spirogyra* and observed that oxygen-seeking bacteria introduced into the system collected in the region of the spectrum where chlorophyll pigments absorb. This was the first action spectrum ever measured and gave the first indication of the effectiveness of light absorbed by accessory pigments in driving photosynthesis.

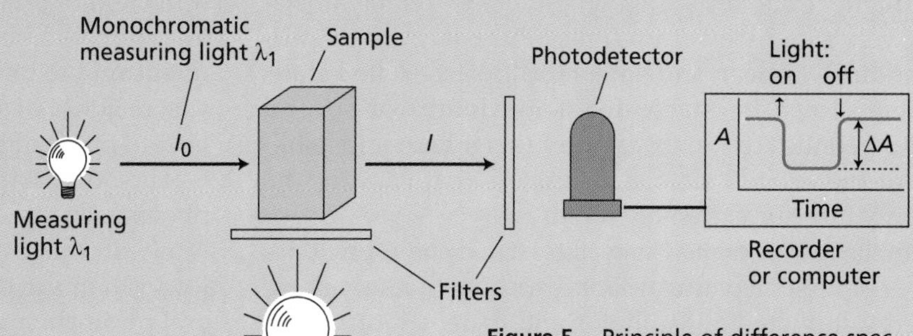

Figure 5 Principle of difference spectroscopy. We measure difference spectra by observing as a function of time the change in the absorbance of a measuring light λ_1 in a sample when an actinic light is turned on. The actinic light λ_2 causes chemical changes in the sample that change its absorption spectrum. Blocking filters are necessary to prevent scattered actinic light from entering the detector. The up and down arrows signify the times when the actinic light was turned on and off, respectively. The change in absorbance induced by the actinic light, ΔA, can be either positive or negative. Usually, measurements are made at one wavelength at a time. A difference spectrum is built up by repetition of the measurement at many different wavelengths.

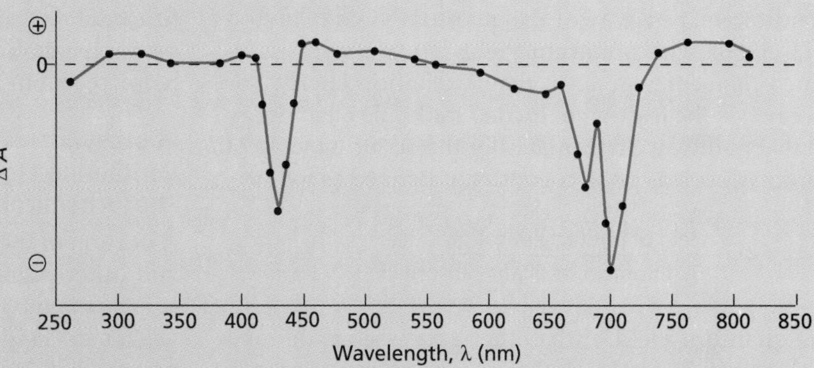

Figure 6 Light-minus-dark difference spectrum for photooxidation of P700, measured as shown in Figure 5. The decreases of absorption (bleaching) at 430 and 700 nm are due to loss of the absorbance of P700. Increases observed around 450 nm and beyond 730 nm are due to absorption by $P700^+$ (oxidized P700). (After Ke 1973.)

photons is higher when their wavelength is shorter. In the higher excited state, chlorophyll is extremely unstable, very rapidly gives up some of its energy to the surroundings as heat, and enters the lowest excited state, where it can be stable for a maximum of several nanoseconds (10^{-9} s). Because of this inherent instability of the excited state, any process that captures its energy must be extremely rapid.

In the lowest excited state, the excited chlorophyll has several possible pathways for disposing of its available energy. It can re-emit a photon and thereby return to its ground state—a process known as **fluorescence**. When it does so, the wavelength of fluorescence is almost always slightly longer than the wavelength of absorption of the same electron state, because a portion of the excitation energy is converted into heat before the fluorescent photon is emitted. Conservation of energy therefore requires that the energy of the fluorescent photon be lower than that of the excitation photon—hence the shift to longer wavelength. Chlorophylls fluoresce in the red region of the spectrum.

Alternatively, the excited chlorophyll can return to its ground state by directly converting its excitation energy into heat, with no emission of a photon. A third process that deactivates excited chlorophyll is **energy transfer**, in which an excited chlorophyll transfers its energy to another molecule. A fourth process is **photochemistry**, in which the energy of the excited state causes chemical reactions to occur. The rate of the earliest steps in the photosynthetic energy storage process are among the fastest known chemical reactions. This extreme speed is necessary for photochemistry to compete with the other possible reactions of the excited state.

The Quantum Yield Gives Information about the Fate of the Excited State

The process with the highest rate will be the one most likely to deactivate excited chlorophyll; slower processes only occasionally win out in the race to dispose of the energy of the excited state. This concept is expressed quantitatively by way of the quantum yield (Clayton 1971, 1980). The **quantum yield** (Φ) of a process in which molecules give up their excitation energy (or "decay") is the fraction of excited molecules that decay via that pathway. Mathematically, the quantum yield of a process, such as photochemistry, is defined as follows:

$$\Phi = \frac{\text{yield of photochemical products}}{\text{total number of quanta absorbed}} \tag{7.4}$$

The quantum yield of the other processes is defined analogously. The value of Φ for a particular process can range from 0 (if that process is never involved in the decay of the excited state) to 1.0 (if that process always deactivates the excited state). The sum of the quantum yields of all possible processes is 1.0.

In functional chloroplasts kept in dim light, the quantum yield of photochemistry is approximately 0.95, the quantum yield of fluorescence is 0.05 or lower, and the quantum yields of other processes are negligible. The vast majority of excited chlorophyll molecules therefore lead to photochemistry.

The quantum yield of formation of the products of photosynthesis, such as O_2, can be measured quite accurately. In this case the quantum yield is substantially lower than the value for photochemistry, because several photochemical events must take place before any O_2 molecules form. For O_2 production the measured maximum quantum yield is approximately 0.1, meaning that 10 quanta are absorbed for each O_2 molecule released. The reciprocal of the quantum yield is called the **quantum requirement**. The minimum quantum requirement for O_2 evolution is therefore about 10. Quantitative measurements of the absorption of light and the fate of the energy contained in the light are essential to an understanding of photosynthesis.

Photosynthetic Pigments Absorb the Light That Powers Photosynthesis

The energy of sunlight is first absorbed by the pigments of the plant. All pigments active in photosynthesis are found in the chloroplast. Structures and absorption spectra of several photosynthetic pigments are shown in Figures 7.5 and 7.6, respectively. Table 7.1 shows the distribution of pigments in different types of photosynthetic organisms. The **chlorophylls** and **bacteriochlorophylls** (pigments found in certain bacteria) are the typical pigments of photosynthetic organisms, but all organisms contain a mixture of more than one kind of pigment, each serving a specific function. All chlorophylls have a complex ring structure that is chemically related to the porphyrin-like groups found in hemoglobin and cytochromes (see Figure 7.5A). In addition, a long hydrocarbon tail is almost always attached to the ring structure. The tail anchors the chlorophyll to the hydrophobic portion of its environment. The ring structure contains some loosely bound electrons and is the part of the molecule involved in electron transitions and redox reactions.

Photosynthesis Takes Place in Complexes Containing Light-Gathering Antennas and Photochemical Reaction Centers

A portion of the light energy absorbed by chlorophylls and other pigments is eventually stored as chemical energy via the formation of chemical bonds. This conversion of energy from one form to another is a complex process that depends on cooperation between many pigment molecules and a group of electron transfer proteins. The majority of the pigments serve as an **antenna**, collecting light and transferring the energy to the **reaction center**, where the chemical reactions leading to

(A) Chlorophylls

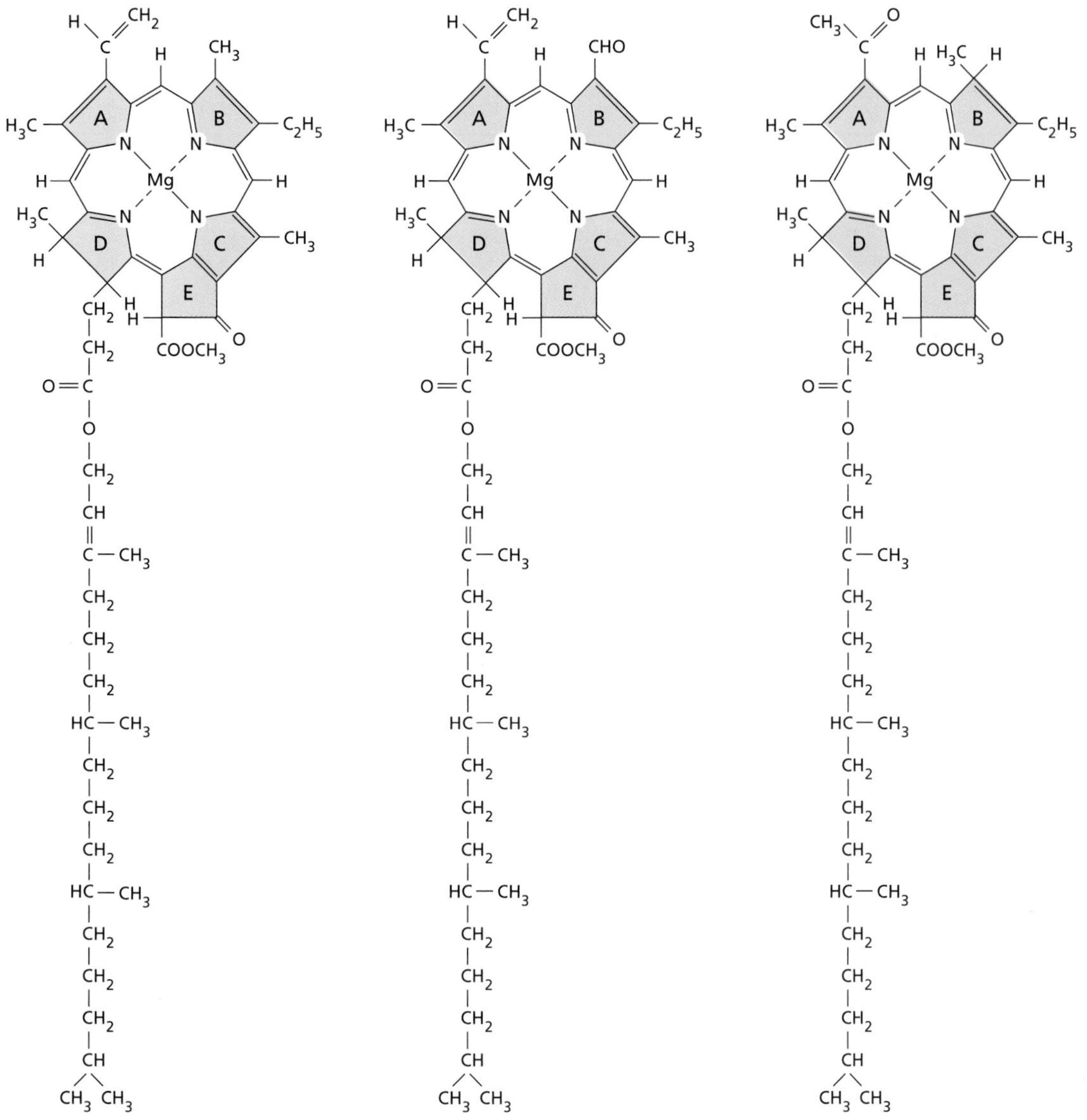

Chlorophyll *a*

Chlorophyll *b*

Bacteriochlorophyll *a*

(B) Carotenoids

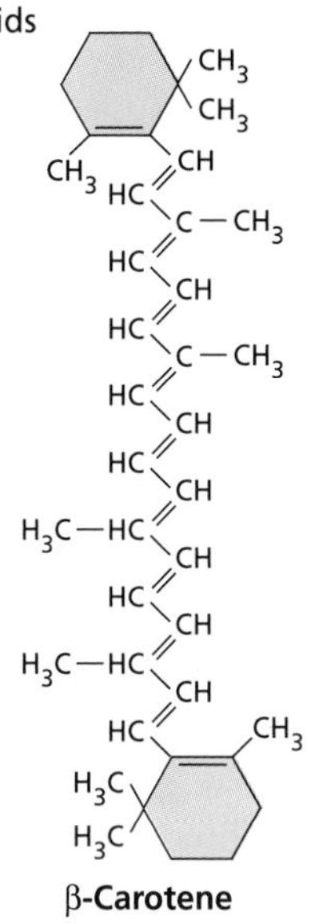

β-Carotene

(C) Bilin pigments

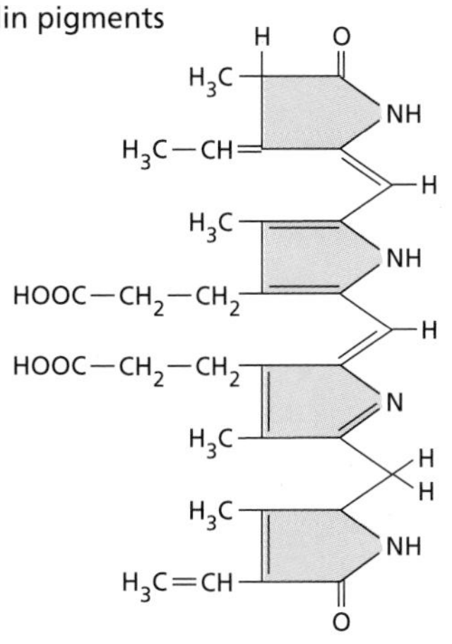

Phycoerythrobilin

Figure 7.5 Molecular structure of some photosynthetic pigments. (A) The chlorophylls have a porphyrin-like ring structure with a magnesium atom (Mg) coordinated in the center and a long hydrophobic hydrocarbon tail that anchors them in the photosynthetic membrane. The porphyrin-like ring is the site of the electron rearrangements that occur when the chlorophyll is excited and of the unpaired electrons when it is either oxidized or reduced. Various chlorophylls differ chiefly in the substituents around the rings and the pattern of double bonds. (B) Carotenoids are linear polyenes that serve as both antenna pigments and photoprotective agents. (C) Bilin pigments are open-chain tetrapyrroles found in antenna structures known as phycobilisomes that occur in cyanobacteria and red algae.

Table 7.1
Distribution of chlorophylls and other photosynthetic pigments

	Chlorophylls				Bacteriochlorophylls							
Organism	a	b	c	d	a	b	c	d	e	g	**Carotenoids**	**Phycobiliproteins**
Eukaryotes												
Mosses, ferns, seed plants	+	+	–	–							+	–
Green algae	+	+	–	–							+	–
Euglenoids	+	+	–	–							+	–
Diatoms	+	–	+	–							+	–
Dinoflagellates	+	–	+	–							+	–
Brown algae	+	–	+	–							+	–
Red algae	+	–	–	+							+	+
Prokaryotes												
Cyanobacteria	+	–	–	+							+	+
Prochlorophytes	+	+	–	–							+	–
Sulfur purple bacteria					+ or +		–	–	–	–	+	–
Nonsulfur purple bacteria					+ or +		–	–	–	–	+	–
Green bacteria					+	–	+ or + or +			–	+	–
Heliobacteria					–	–	–	–	–	+	+	–

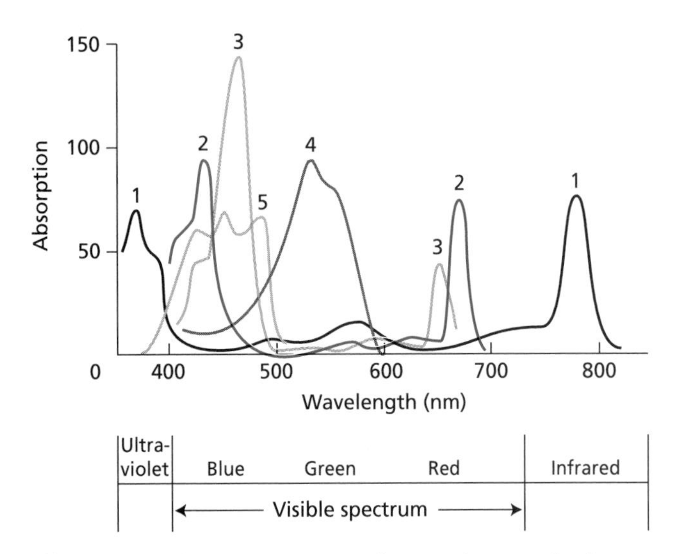

Figure 7.6 Absorption spectra of some photosynthetic pigments. Curve 1, bacteriochlorophyll *a*; curve 2, chlorophyll *a*; curve 3, chlorophyll *b*; curve 4, phycoerythrobilin; curve 5, β-carotene. The absorption spectra shown are for pure pigments dissolved in nonpolar solvents, except for curve 4, which represents an aqueous buffer of phycoerythrin, a protein from cyanobacteria that contains a phycoerythrobilin chromophore covalently attached to the peptide chain. In many cases, the spectra of photosynthetic pigments in vivo are substantially affected by the environment of the pigments in the photosynthetic membrane. (From Avers 1985.)

long-term energy storage take place. Figure 7.7 diagrams the basic concept of light absorption and energy transfer during photosynthesis. Molecular structures of some of the antenna and reaction center complexes are discussed later in the chapter.

How does the plant benefit from this division of labor between antenna and reaction center pigments? Even in bright sunlight, a chlorophyll molecule absorbs only a few photons each second. If every chlorophyll had a complete reaction center associated with it, the enzymes that make up this system would be idle most of the time, only occasionally being activated by photon absorption. However, if many pigments can send energy into a common reaction center, the system is kept active a large fraction of the time.

In 1932, Robert Emerson and William Arnold performed a classic experiment that provided the first evidence for the cooperation of many chlorophyll molecules in energy conversion during photosynthesis (Emerson and Arnold 1932). They gave very brief (10^{-5} s) flashes of light to a suspension of the green alga *Chlorella pyrenoidosa* and measured the amount of oxygen produced. The flashes were spaced about 0.1 s apart, a time that they had determined in earlier work was long enough for the enzymatic steps of the process to be completed before the arrival of the next flash. The investigators varied the energy of the flashes and found that at high energies the oxygen production did not increase when a more intense flash was given: The photosynthetic system was saturated with light.

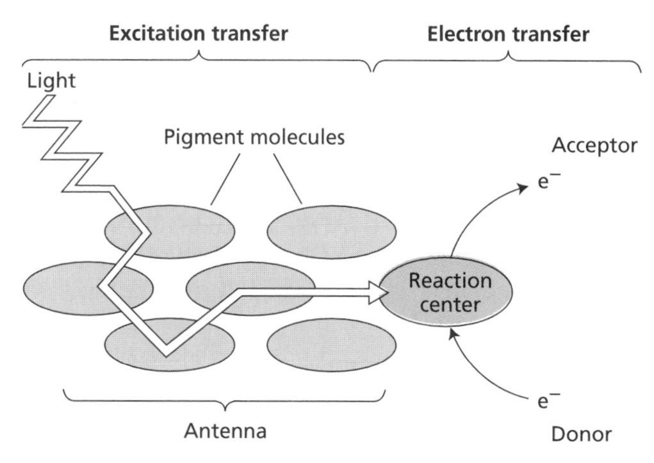

Figure 7.7 The basic concept of energy transfer during photosynthesis. Many pigments together serve as an antenna, collecting light and transferring its energy to the reaction center, where chemical reactions store some of the energy by transferring electrons from a chlorophyll pigment to an electron acceptor molecule. An electron donor then reduces the chlorophyll again. The transfer of excitation in the antenna is a purely physical phenomenon and involves no chemical changes. The first chemical reactions after photon absorption take place in the reaction center. Subsequent reactions stabilize the unstable products of the initial chemical reaction.

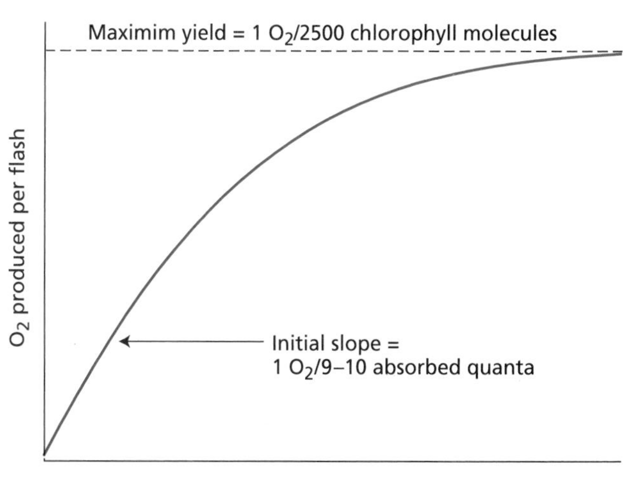

Figure 7.8 The relationship of oxygen production to flash energy, the first evidence for the interaction between the antenna pigments and the reaction center. This light saturation curve, obtained in 1932 by Robert Emerson and William Arnold, showed that at saturating energies, the maximum amount of O_2 produced is 1 molecule per 2500 chlorophyll molecules. The slope of the low-intensity part of the curve gives the quantum requirement of oxygen production.

Figure 7.8 shows the relationship of oxygen production to flash energy measured by Emerson and Arnold. They were surprised to find that under saturating conditions, only one molecule of oxygen was produced for each 2500 chlorophyll molecules in the sample. We know now that several hundred pigments are associated with each reaction center and that each reaction center must operate four times to produce one molecule of oxygen; hence the value of 2500 chlorophylls per O_2.

The reaction centers and most of the antenna complexes are integral components of the photosynthetic membrane. In eukaryotic photosynthetic organisms, these membranes are found within the chloroplast; in photosynthetic prokaryotes, the site of photosynthesis is the plasma membrane or membranes derived from it.

The Chemical Reaction of Photosynthesis Is Driven by Light

Establishing the overall chemical equation of photosynthesis required several hundred years and contributions by many scientists (reviewed by Rabinowitch and Govindjee 1969). In 1771, Joseph Priestley observed that a sprig of mint growing in air in which a candle had burned out improved the air so that another candle could burn. He had discovered oxygen evolution by plants. A Dutchman, Jan Ingenhousz, documented the essential role of light in photosynthesis in 1779. Other scientists established the roles of CO_2 and H_2O and showed that organic matter, specifically carbohydrate, is a product of photosynthesis along with oxygen. By the end of the nineteenth century, the balanced overall chemical reaction for photosynthesis could be written as follows:

$$6\,CO_2 + 6\,H_2O \xrightarrow{\text{light, plant}} C_6H_{12}O_6 + 6\,O_2 \qquad (7.5)$$

where $C_6H_{12}O_6$ represents a simple sugar such as glucose. As discussed in Chapter 8, glucose is not the actual product of the carbon fixation reactions. However, the energetics for the actual products are approximately the same, so the appearance of glucose in Equation 7.5 should be regarded as a convenience but not taken literally.

It is important to realize that equilibrium for the chemical reaction shown in Equation 7.5 lies very far in the direction of the reactants (see Equation 2.9 in Chapter 2). The equilibrium constant for Equation 7.5, calculated from tabulated free energies of formation for each of the compounds involved, is about 10^{-500}. This number is so close to zero that one can be quite confident that in the entire history of the universe no molecule of glucose has formed spontaneously from H_2O and CO_2 without external energy being provided. The energy needed to drive the photosynthetic reaction comes from light. A simpler form of Equation 7.5 is

$$CO_2 + H_2O \xrightarrow{\text{light, plant}} (CH_2O) + O_2 \qquad (7.6)$$

where (CH_2O) is one-sixth of a glucose molecule. About nine or ten photons of light are required to drive the reaction of Equation 7.6.

If red light of wavelength 680 nm is absorbed, the total energy input (see Equation 7.2) is 1760 kJ per mole of oxygen formed. This amount of energy is more than enough to drive the reaction in Equation 7.6, which has a standard-state free-energy change of +467 kJ mol^{-1}. The efficiency of conversion of light energy at the optimal wavelength into chemical energy is therefore about 27%, which is remarkably high for an energy conversion system. Most of this stored energy is used for cellular maintenance processes, so the amount diverted to the formation of biomass is much less.

There is no conflict between the fact that the photochemical quantum efficiency (quantum yield) is nearly 100% and the energy conversion efficiency is only 27%. The quantum efficiency is a measure of the percentage of absorbed photons that give rise to photochemistry, while the energy efficiency is a measure of how much energy in the absorbed photons is stored as chemical products. The numbers indicate that almost all the absorbed photons engage in photochemistry but that only about a fourth of the energy in each photon is stored, the remainder being converted to heat.

Photosynthesis Is a Light-Driven Redox Process

The reaction shown in Equation 7.6 is an overall reaction for photosynthesis. It tells us what goes into the chloroplast and what comes out, but it tells us nothing about how the transformation takes place. The chemical mechanism for the reaction is complex; at least 50 intermediate steps have been identified, and undoubtedly some have not yet been discovered. The first clue to the nature of the mechanism came from studies by C. B. van Niel in the 1920s that utilized non-oxygen-evolving photosynthetic bacteria such as *Chromatium vinosum*. *C. vinosum* can grow by using organic acids or reduced sulfur compounds as reductants, as follows:

$$CO_2 + 2H_2A \xrightarrow{\text{light, bacterium}} (CH_2O) + 2A + H_2O \quad (7.7)$$

where H_2A represents any one of these reduced compounds—for example, H_2S. In this equation the reduced compound is oxidized, and CO_2 is reduced. Van Niel drew the analogy to the process shown in Equation 7.6 and correctly concluded that photosynthesis is a redox (reduction–oxidation) process. In the special case of oxygen-evolving photosynthetic organisms, H_2O serves as the reductant. Although some aspects of van Niel's formulation of photosynthesis were incorrect, his conclusion about the redox nature of the process stands as the central tenet of our understanding of the subject.

In 1937, Robert Hill found that isolated chloroplast thylakoids photoreduce a variety of compounds, such as iron salts (Hill 1939). These compounds serve as oxidants in place of CO_2, as the following equation shows:

$$4\,Fe^{3+} + 2\,H_2O \rightarrow 4\,Fe^{2+} + O_2 + 4\,H^+ \quad (7.8)$$

Many compounds have since been shown to act as artificial electron acceptors in what has come to be known as the Hill reaction. Their use has been invaluable in elucidating the reactions that precede carbon reduction. We now know that during the normal functioning of the photosynthetic system, light reduces nicotinamide adenine dinucleotide phosphate (NADP), which in turn serves as the reducing agent for carbon fixation in the Calvin cycle (see Chapter 8). ATP is also formed during the electron flow from water to NADP, and it too is used in carbon reduction. The chemical reactions in which water is oxidized to oxygen, NADP is reduced, and ATP is formed are known as the *thylakoid reactions*; the carbon reduction reactions are called the *stroma reactions*. This division, while somewhat arbitrary, is reasonable in that almost all the reactions up to NADP reduction take place within the thylakoids, while the carbon reduction reactions take place in the aqueous region of the chloroplast, the stroma.

Oxygen-Evolving Organisms Have Two Photosystems That Operate in Series

By the late 1950s, several experiments were puzzling the scientists who studied photosynthesis. One of these experiments measured the quantum yield of photosynthesis as a function of wavelength and revealed an effect known as the red drop (Figure 7.9). If the quantum yield is measured for the wavelengths at which chlorophyll absorbs light, the values found throughout most of the range are remarkably constant, indicating that any photon absorbed by chlorophyll or other pigments is as effective as any other photon in driving photosynthesis. At the extreme red edge of chlorophyll absorption (greater than 680 nm), however, the yield drops dramatically. This drop does not simply reflect the fact that

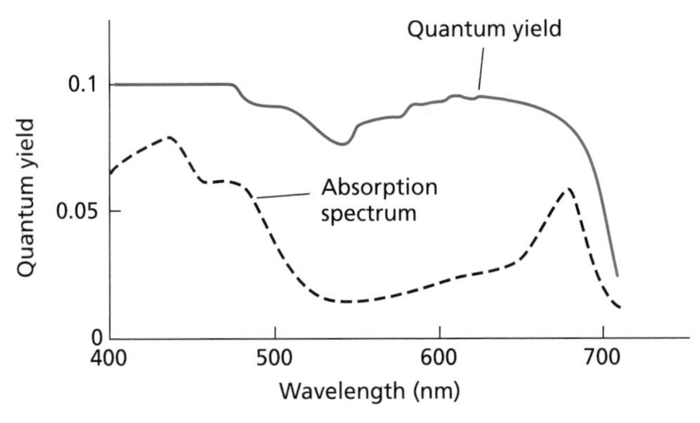

Figure 7.9 The red drop effect. The quantum yield of photosynthesis (solid curve) falls off drastically for far-red light of wavelengths greater than 680 nm, indicating that far-red light alone is inefficient in driving photosynthesis. The slight dip near 500 nm reflects the somewhat lower efficiency of photosynthesis using light absorbed by accessory pigments, carotenoids. The absorption spectrum of the chloroplast is shown by the dashed curve.

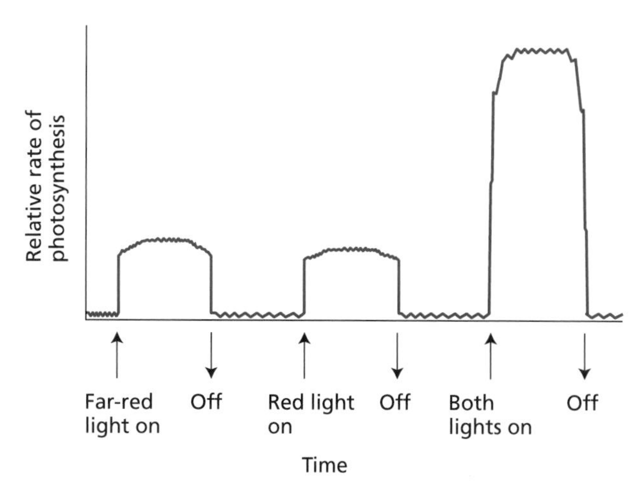

Figure 7.10 The enhancement effect. The rate of photosynthesis when red and far-red light are given together is greater than the sum of the rates when they are given apart. The results of this experiment baffled scientists in the 1950s, but it provided essential evidence in favor of the concept that photosynthesis is carried out by two photochemical systems working in tandem but with slightly different wavelength optima.

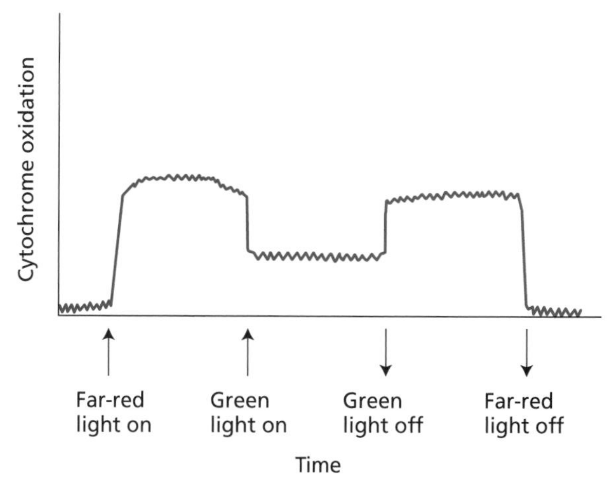

Figure 7.11 Antagonistic effects of light on cytochrome oxidation. Far-red light is very effective in oxidizing the cytochrome *f* in the chloroplast. If green light is also present, some of the cytochrome becomes reduced. The two wavelengths have opposite effects—hence the term "antagonistic." This experiment is one of the clearest demonstrations of the existence of two photochemical systems in photosynthesis: one that reduces cytochrome and one that oxidizes it. This particular experiment was done with a red alga, in which photosystem II (see Figure 7.12) is driven best by green light and photosystem I is driven best by far-red light. (After Duysens et al. 1961.)

the absorption is falling off, because the quantum yield measures only light that has actually been absorbed. Thus, light with a wavelength greater than 680 nm is much less efficient than light of shorter wavelengths.

Another puzzling experimental result was the **enhancement effect**, discovered by Emerson (Emerson et al. 1957). He measured the rate of photosynthesis separately with light of two different wavelengths and then used the two beams simultaneously (Figure 7.10). Certain conditions, especially when one of the wavelengths was in the far-red region (greater than 680 nm), yielded the surprising result that the rate of photosynthesis obtained with both wavelengths was greater than the sum of the individual rates.

These observations were explained by experiments performed by Louis Duysens of the Netherlands (Duysens et al. 1961). Chloroplasts contain cytochromes, iron-containing proteins that function as intermediate electron carriers in photosynthesis. Duysens found that when a sample of a red alga was illuminated with long-wavelength light, the cytochrome became mostly oxidized. If light of a shorter wavelength was also present, the effect was partly reversed (Figure 7.11). These **antagonistic effects** can be explained by a mechanism involving two photochemical events: one that tended to oxidize the cytochrome and one that tended to reduce it (Hill and Bendall 1960).

We know now that in the red region of the spectrum, one of the photoreactions, known as **photosystem I (PS-I)**, absorbs preferentially far-red light of wavelengths greater than 680 nm, while the second, known as **photosystem II (PS-II)**, absorbs red light of 680 nm well and is driven very poorly by far-red light. This wavelength dependence explains the enhancement effect and the red drop effect. Another difference between the photosystems is that photosystem I produces a strong reductant, capable of reducing $NADP^+$, and a weak oxidant. Photosystem II produces a very strong oxidant, capable of oxidizing water, and a weaker reductant than the one produced by photosystem I. This reductant rereduces the oxidant produced by photosystem I, which explains the antagonistic effect. These properties are shown schematically in Figure 7.12.

The scheme of photosynthesis depicted, called the Z (for "zigzag") scheme, has become the basis for understanding O_2-evolving (oxygenic) photosynthetic organisms. It accounts for the operation of two physically and chemically distinct photosystems (I and II), each with its own antenna pigments and photochemical reaction center. The two photosystems are linked by an electron transport chain. Non-O_2-evolving (anoxygenic) organisms, such as the purple photosynthetic bacteria of the genera *Rhodobacter* and *Rhodopseudomonas*, contain only a single photosystem and therefore do not display the effects described here. However, these simpler organisms have been very useful for detailed structural and functional studies that contribute to a better understanding of oxygenic photosynthesis.

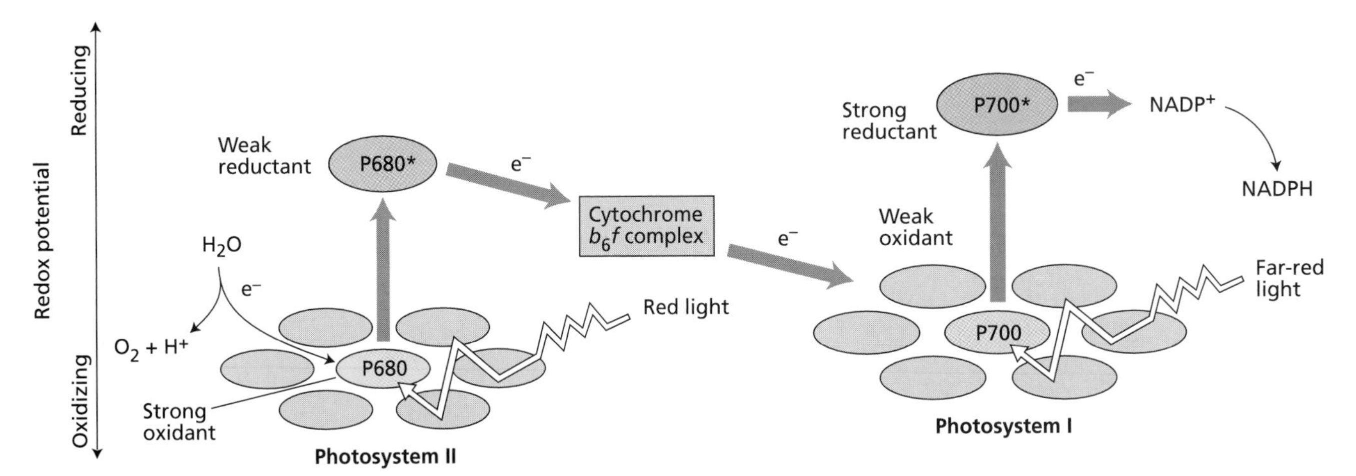

Figure 7.12 The Z scheme of photosynthesis. Red light absorbed by photosystem II (PS-II) produces a strong oxidant and a weak reductant. Far-red light absorbed by photosystem I (PS-I) produces a weak oxidant and a strong reductant. The strong oxidant generated by PS-II oxidizes water, while the strong reductant produced by PS-I reduces $NADP^+$. PS-II produces electrons that reduce the cytochrome b_6f complex, while PS-I produces an oxidant that oxidizes the cytochrome b_6f complex. This scheme is basic to an understanding of photosynthetic electron transport. P680 and P700 refer to the wavelengths of maximum absorption of the reaction center chlorophylls in PS-II and PS-I, respectively. These designations are discussed in more detail later in the chapter.

Structure of the Photosynthetic Apparatus

The previous section provided some of the physical principles underlying photosynthesis, some aspects of the functional roles of various pigments, and some of the chemical reactions carried out by photosynthetic organisms. We now turn to the architecture of the photosynthetic apparatus and the structure of its components, with the goal of describing the structure of the organelles and complexes that carry out photosynthesis.

The Chloroplast Is the Site of Photosynthesis

In photosynthetic eukaryotes, photosynthesis takes place in the subcellular organelle known as the chloroplast. Figure 7.13 shows a transmission electron micrograph of a thin section taken from a pea chloroplast. The most striking aspect of the structure of the chloroplast is the extensive system of internal membranes known as **thylakoids**. All the chlorophyll is contained within this membrane system, which is the site of the light reactions of photosynthesis. The carbon reduction reactions, which are catalyzed by water-soluble enzymes, take place in the **stroma**, the region of the chloroplast outside the thylakoids. Most of the thylakoids appear to be very closely associated with each other. These stacked membranes are known as **grana lamellae** (each stack is called *granum*), while the exposed membranes in which stacking is absent are known as **stroma lamellae.**

Two separate membranes, each composed of a lipid bilayer and together known as the **envelope**, surround most types of chloroplasts (Figure 7.14). This double-membrane system contains a variety of metabolite transport systems (see Chapter 6). The chloroplast also con-

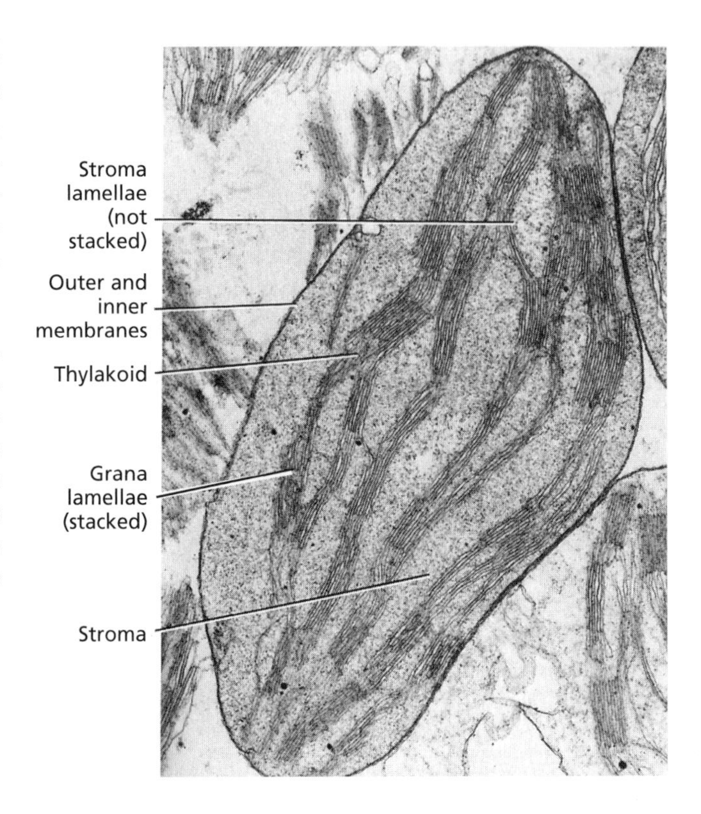

Figure 7.13 Transmission electron micrograph of a chloroplast from pea (*Pisum sativum*), fixed in glutaraldehyde and OsO_4, embedded in plastic resin, and thin-sectioned with an ultramicrotome. (14,500×) (Courtesy of J. Swafford.)

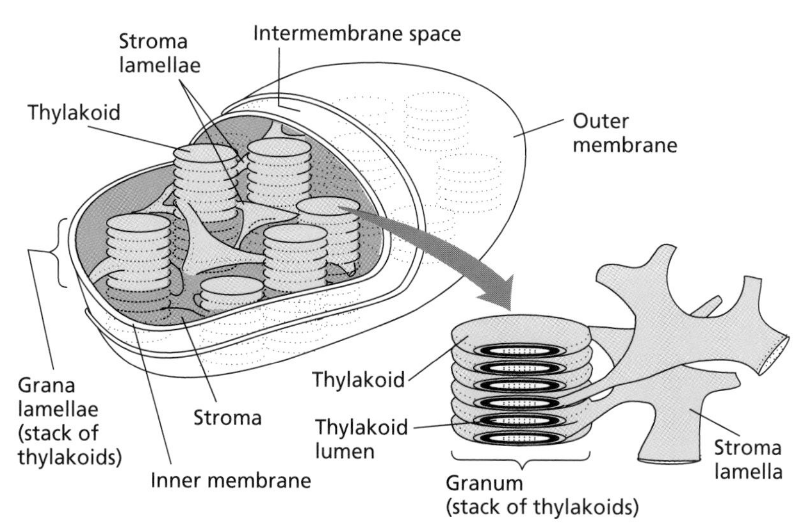

Figure 7.14 Schematic picture of the overall organization of the membranes in the chloroplast. The chloroplast of higher plants is surrounded by the inner and outer membranes. The region of the chloroplast that is inside the inner membrane but outside the thylakoid membranes is known as the stroma. It contains the enzymes that catalyze carbon fixation and other biosynthetic pathways. The thylakoid membranes are highly folded and appear in many pictures to be stacked like coins, although in reality they form one or a few large interconnected membrane systems, with a well-defined interior and exterior with respect to the stroma. The inner space within a thylakoid is known as the lumen. (From Becker 1986.)

tains its own DNA, RNA, and ribosomes. Many of the chloroplast proteins are products of transcription and translation within the chloroplast itself, whereas others are encoded by nuclear DNA, synthesized on cytoplasmic ribosomes, and then imported into the chloroplast. This remarkable division of labor, extending in many cases to different subunits of the same enzyme complex, will be discussed in more detail in a later section.

Thylakoids Contain Integral Membrane Proteins

A wide variety of proteins essential to photosynthesis are embedded in the thylakoid membrane. In many cases, portions of these proteins extend into the aqueous regions on both sides of the thylakoid. These **integral membrane proteins** contain a large proportion of hydrophobic amino acids and are therefore much more stable in a nonaqueous medium such as the hydrocarbon portion of the membrane (see Figure 1.10A). The reaction centers, the antenna pigment–protein complexes, and most of the electron transport enzymes are all integral membrane proteins and are inserted in the lipid bilayer of the membrane. The three-dimensional structures of these complexes have been difficult to elucidate because purified membrane proteins have been resistant to crystallization, the essential first step in an X-ray structure determination. However, much progress has been made in this area in recent years, and structures of several membrane proteins are now available. This information is invaluable in improving our understanding of photosynthesis.

In all known cases, integral membrane proteins of the chloroplast have a unique orientation within the membrane. Many thylakoid membrane proteins have one region pointing toward the stromal side of the membrane and the other oriented toward the interior portion of the thylakoid, known as the **lumen** (Figure 7.15). This vectorial arrangement of proteins is a universal feature of energy-conserving membranes.

The chlorophylls and accessory light-gathering pigments in the thylakoid membrane are always associated in a noncovalent but highly specific way with proteins, forming **chlorophyll proteins**. Both antenna and reaction center chlorophylls are associated with chlorophyll proteins that are inserted into the membrane. In many cases histidine residues in the protein serve as ligands to the magnesium in a chlorophyll. Often several pigments are associated with each peptide. The distances between pigments and their orientations with respect to each other appear to be relatively fixed. This precisely determined geometric arrangement optimizes energy transfer in antenna complexes and electron transfer in reaction centers, while at the same time minimizing wasteful processes.

Photosystems I and II Are Spatially Separated in the Thylakoid Membrane

The PS-II reaction center, along with its antenna chlorophylls and associated electron transport proteins, is located predominantly in the grana lamellae (Figure 7.16) (Anderson and Andersson 1988). The PS-I reaction center and its associated antenna pigments and electron transfer proteins, as well as the coupling-factor enzyme that catalyzes the formation of ATP, are found almost exclusively in the stroma lamellae and at the edges of the grana lamellae. The cytochrome b_6f complex that connects the photosystems is evenly distributed. Thus, the two photochemical events that take place in O_2-evolving photosynthesis are spatially separated. This separation implies that one or more of the electron carriers that function between the photosystems diffuses from the grana region of the membrane to the stroma region, where electrons are delivered to photosystem I.

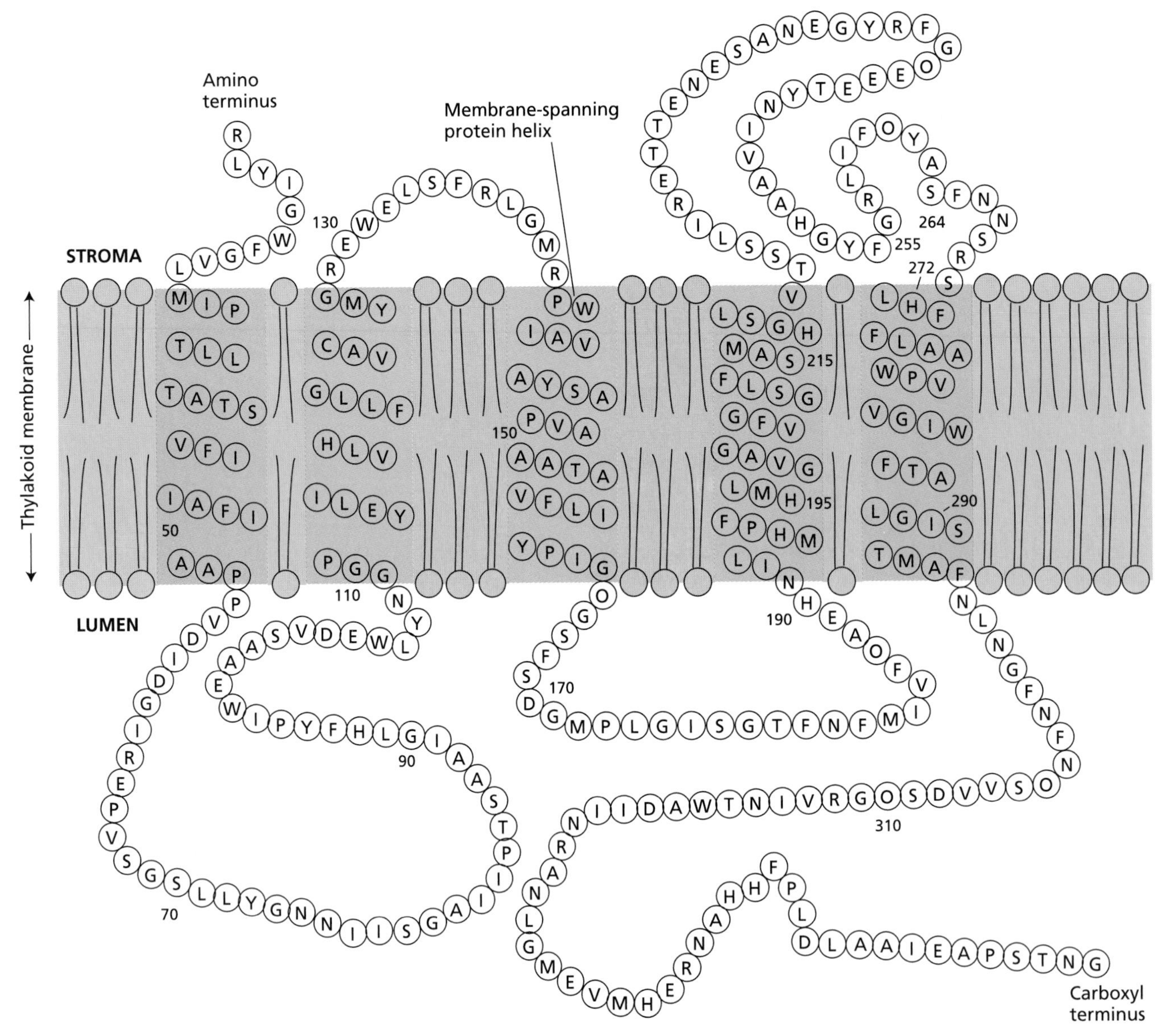

Figure 7.15 Predicted folding pattern of the D1 protein of the PS-II reaction center. The hydrophobic portion of the membrane is traversed five times by the peptide chain rich in hydrophobic amino acid residues. The protein is asymmetrically arranged in the thylakoid membrane, with the amino (NH_2) terminus on the stromal side of the membrane and the carboxyl (COOH) terminus on the lumen side. Specific amino acids are identified by their single-letter abbreviations. The numbers indicate the positions of amino acids within the polypeptide chain. (After Trebst 1986.)

The protons produced by the oxidation of water must also be able to diffuse to the stroma region, where ATP is synthesized. The possible functional role of this large separation (many tens of nanometers) between photosystems I and II is not entirely clear, although some researchers believe that the separation is involved in the regulation of the input of light energy to the two photosystems (Trissl and Wilhelm 1993).

The spatial separation between photosystems I and II indicates that a strict one-to-one stoichiometry between the two photosystems is not required. Instead, PS-II reaction centers feed reducing equivalents into a common intermediate pool of one of the electron carriers, plastoquinone, and PS-I reaction centers remove the reducing equivalents from the common pool, rather than from any specific PS-II reaction center complex.

Most measurements of the relative quantities of photosystems I and II have shown that there is an excess of photosystem II in chloroplasts. Under many conditions the ratio of PS-II to PS-I is about 1.5:1, but it can change when plants are grown in different conditions of light. Cyanobacteria contain a large excess of photosystem I, as do the bundle sheath chloroplasts of C_4 plants (see Chapter 8).

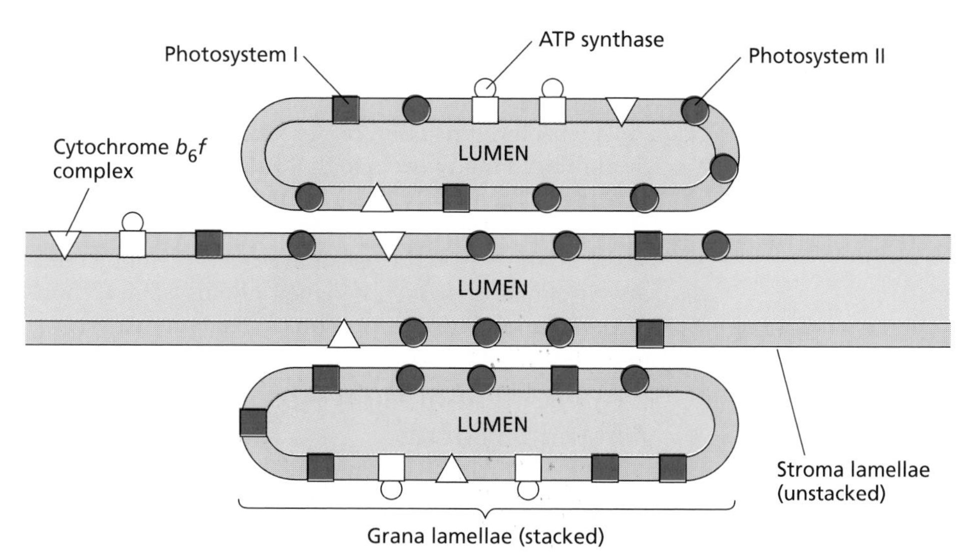

Figure 7.16 Organization of the protein complexes of the thylakoid membrane. Photosystem II is located predominantly in the stacked regions of the thylakoid membrane; photosystem I and ATP synthase are found in the unstacked regions protruding into the stroma. Cytochrome b_6f complexes are evenly distributed. This lateral separation of the two photosystems requires that electrons and protons produced by photosystem II be transported a considerable distance before they can be acted on by photosystem I and the ATP-coupling enzyme.

The Structures of Two Bacterial Reaction Centers Have Been Determined at Very High Resolution

In 1984, Hartmut Michel, Johann Deisenhofer, Robert Huber, and coworkers in Munich solved the three-dimensional structure of the reaction center from the purple photosynthetic bacterium *Rhodopseudomonas viridis* (Deisenhofer and Michel 1989). This landmark achievement, for which a Nobel prize was awarded in 1988, was the first high-resolution X-ray structure determination for an integral membrane protein and the first structure determination for a reaction center complex. The structure is shown in Figure 7.17.

The protein part of the complex consists of four separate polypeptides. Two of them, called L and M (for *l*ight and *m*edium mass) bind all of the bacteriochlorophyll, quinone, and carotenoid cofactors of the complex. The structure has a twofold symmetry about an axis perpendicular to the plane of the membrane, hinting at a dimeric nature of the reaction center. The ten transmembrane portions of the L and M peptides (five from each) are arranged in α helices, and there are almost no charged amino acid residues in the interior of the membrane. The H (*h*eavy) protein has a single transmembrane helix and is localized mostly on the cytoplasmic

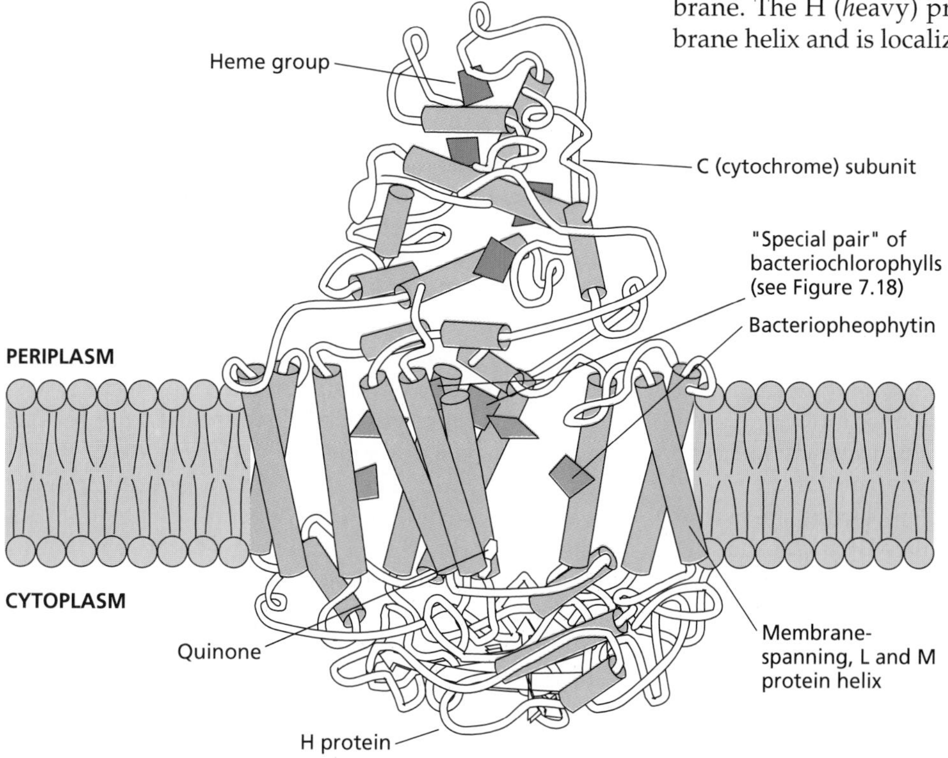

Figure 7.17 Structure of the reaction center of the purple bacterium *Rhodopseudomonas viridis*, resolved by X-ray crystallography. The 11 tubes within the membrane represent transmembrane protein helices. The chlorophyll-type pigments are shown in light green; the heme groups of the tightly bound iron-containing cytochrome are shown in dark gray at the top of the diagram. The C (cytochrome subunit) points into the periplasm or outside the cell; the H subunit is in contact with the cytoplasm of the cell. The quinones and the cytochrome participate in electron transfer reactions with the bacteriochlorophylls. (Courtesy of J. Richardson.)

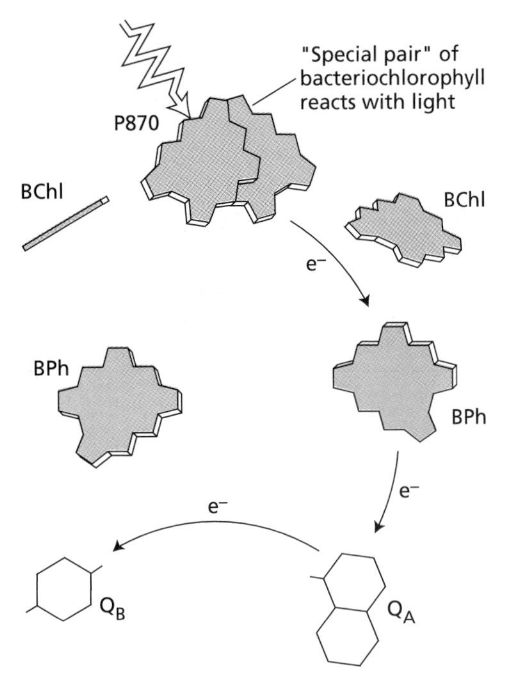

Figure 7.18 Geometric arrangement of pigments and other prosthetic groups in the bacterial reaction center: bacteriochlorophyll (BChl); bacteriopheophytin (BPh), a chlorophyll molecule in which the magnesium atom has been replaced by two hydrogen atoms; and Q_A and Q_B, the first and second quinone acceptors. The cytochrome heme groups have been deleted from this diagram; if present, they would be at the top. The two bacteriochlorophylls at the top (P870) are the "special pair" that react with light within the reaction center. The electron transfer sequence begins at the special pair and proceeds down the right side of the diagram to bacteriopheophytin. Next, the electrons are transferred to Q_A and then to Q_B. The distances between the depicted molecules and their angles within the planes of symmetry of the reaction center have been determined with precision, making it possible to analyze the path of photons and electrons in the reaction center in remarkable detail.

side of the membrane. The C (cytochrome) subunit is located in the periplasmic region (the region between the bacterial plasma membrane and the outer membrane). The geometric arrangement of the pigments and the quinones (electron acceptors) is shown in Figure 7.18, with the protein removed. A similar arrangement is found in the reaction center of another purple photosynthetic bacterium, *Rhodobacter sphaeroides*, except that the C subunit is not present.

Two of the bacteriochlorophyll molecules are in intimate contact with each other and are known as the **special pair**. This dimer, whose existence was predicted from magnetic-resonance studies, is the photoactive portion of the complex. An electron is transferred from this dimer along the sequence of electron carriers on the right side of the complex. Detailed analysis of these structures, along with analysis of numerous mutants, has revealed many of the principles involved in the energy storage processes that are carried out by all reaction centers.

The bacterial reaction center structure is thought to be similar in many ways to that found in photosystem II from oxygen-evolving organisms, especially in the electron acceptor portion of the chain. The proteins that make up the core of the bacterial reaction center are relatively similar in sequence to their photosystem II counterparts, implying an evolutionary relatedness.

Organization of Light-Absorbing Antenna Systems

The antenna systems of different classes of photosynthetic organisms are remarkably varied, in contrast to the reaction centers, which appear to be similar in even distantly related organisms. The variety of antenna complexes reflects evolutionary adaptation to the diverse environments in which different organisms live, as well as the need in some organisms to balance energy input to the two photosystems (Grossman et al. 1995).

Antenna systems function to deliver energy efficiently to the reaction centers with which they are associated (van Grondelle et al. 1994; Pullerits and Sundström 1996). The size of the antenna system varies considerably in different organisms: a low of 20 to 30 bacteriochlorophylls per reaction center in some photosynthetic bacteria, generally 200 to 300 chlorophylls per reaction center in higher plants, and a few thousand pigments per reaction center in some types of algae and bacteria. The molecular structures that serve as antennas are also quite diverse, although all of them are associated in some way with the photosynthetic membrane.

The mechanism by which excitation energy is conveyed from the chlorophyll that absorbs the light to the reaction center is thought to be **resonance transfer** (also known as **Förster transfer**, after the scientist who first described the phenomenon). In this process, photons are not simply emitted by one molecule and reabsorbed by another; rather the excitation energy is transferred from one molecule to another by a nonradiative process. A useful analogy for resonance transfer is the transfer of energy between two tuning forks. If one tuning fork is struck and properly placed near another, the second tuning fork receives some energy from the first and begins to vibrate. The efficiency of energy transfer between the two tuning forks depends on their distance from each other and their relative orientation, as well as their pitches or vibrational frequencies, just as in resonance energy transfer in antenna complexes.

The end result of this process is that approximately 95 to 99% of the photons absorbed by the antenna pigments have their energy transferred to the reaction center, where it can be used for photochemistry. It is impor-

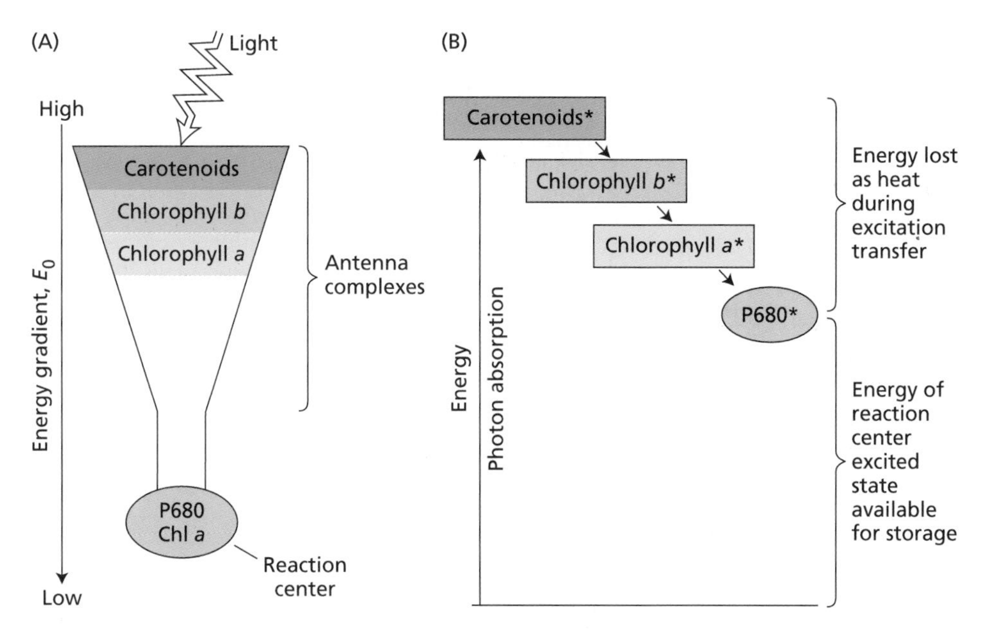

Figure 7.19 Funneling of excitation from the antenna system toward the reaction center. (A) The excited-state energy of pigments increases with distance from the reaction center; that is, pigments closer to the reaction center are lower in energy than those farther from the reaction center. This energy gradient ensures that excitation transfer toward the reaction center is energetically favorable and that excitation transfer back out to the peripheral portions of the antenna is energetically unfavorable. (B) Some energy is lost as heat to the environment by this process, but under optimal conditions almost all the excitations absorbed in the antenna complexes can be delivered to the reaction center.

tant to distinguish the energy transfer among pigments in the antenna from the electron transfer that occurs in the reaction center. Whereas energy transfer is a purely physical phenomenon, electron transfer involves chemical changes in molecules.

The Antenna Funnels Energy to the Reaction Center

The energy absorbed in antenna pigments is funneled toward the reaction center by a sequence of pigments with absorption maxima that are progressively shifted toward longer red wavelengths (Figure 7.19). This red shift in absorption maximum means that the energy of the excited state is somewhat lower nearer the reaction center than in the more peripheral portions of the antenna system. For example, when excitation is transferred from a chlorophyll *b* molecule absorbing maximally at 650 nm to a chlorophyll *a* molecule absorbing maximally at 670 nm, the difference in energy between these two excited chlorophylls is lost to the environment as heat.

For the excitation to be transferred back to the chlorophyll *b*, the lost energy would have to be resupplied. The probability of reverse transfer is therefore smaller simply because thermal energy is not sufficient to make up the deficit between the lower-energy and higher-energy pigments. This effect gives the energy-trapping process a degree of directionality or irreversibility and makes the delivery of excitation to the reaction center more efficient. In essence, the system sacrifices some energy from each quantum so that nearly all of the quanta can be trapped by the reaction center.

Many Antenna Complexes Have a Common Structural Motif

In all eukaryotic photosynthetic organisms that contain both chlorophyll *a* and chlorophyll *b*, the most abundant antenna proteins are members of a large family of structurally related proteins. Some of these proteins are associated primarily with photosystem II and are called **light-harvesting complex II (LHCII)** proteins; others are associated with photosystem I and are called *LHCI proteins*. These antenna complexes are also known as **chlorophyll *a*/*b* antenna proteins** (Paulsen 1995; Green and Durnford 1996).

The structure of one of the LHCII proteins has been determined by a combination of electron microscopy and electron crystallography* (Figure 7.20) (Kühlbrandt et al. 1994). The protein contains three α-helical regions, about 15 chlorophyll pigments, and carotenoids, only some of which are visible in the resolved structure. The structure of the LHCI proteins has not yet been determined but is probably similar to that of the LHCII proteins. All of these proteins have significant sequence similarity and

* Like X-ray crystallography, electron crystallography uses the diffraction patterns of soft-energy electrons to resolve macromolecule structures.

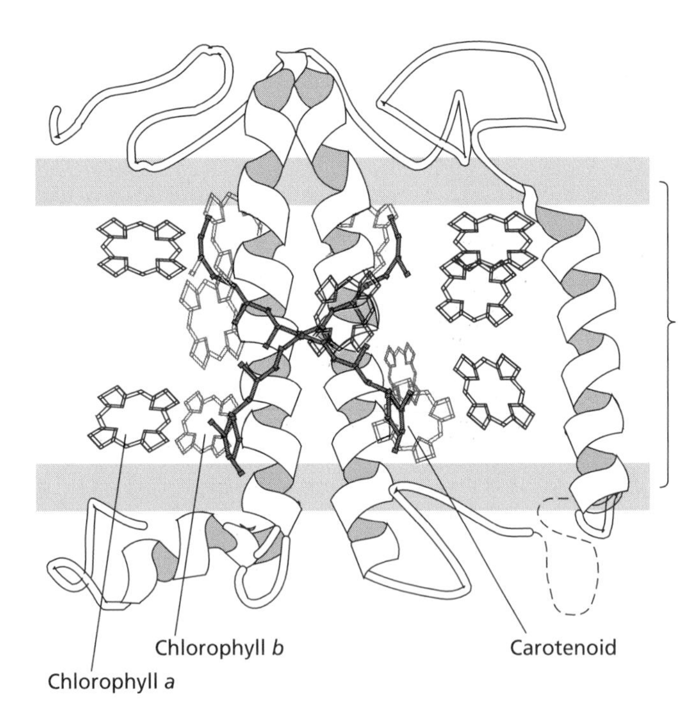

Figure 7.20 Structure of the LHCII antenna complex from higher plants. The complex is a transmembrane pigment protein, with three helical regions that cross the nonpolar part of the membrane. Approximately 15 chlorophyll *a* and *b* molecules are associated with the complex, as well as several carotenoids. The positions of several of the chlorophylls are shown, and two of the carotenoids form an X in the middle of the complex. In the membrane, the complex is trimeric and aggregates around the periphery of the PS-II reaction center complex. (After Kühlbrandt et al. 1994.)

are almost certainly descendants of a common ancestral protein (Grossman et al. 1995; Green and Durnford 1996).

Light absorbed by carotenoids or chlorophyll *b* in the LHC proteins is rapidly transferred to chlorophyll *a* and then to other antenna pigments that are intimately associated with the reaction center. The LHCII complex is also involved in regulatory processes, which are discussed later in the chapter.

Figure 7.21 Detailed Z scheme for O_2-evolving photosynthetic organisms. The redox carriers are placed at their accepted midpoint redox potentials (at pH 7). The vertical arrows signify photon absorption by the reaction center chlorophylls: P680 for photosystem II and P700 for photosystem I. The excited PS-II reaction center chlorophyll, P680*, transfers an electron to pheophytin (Pheo). On the oxidizing side of PS-II (to the left of the arrow joining P680 with P680*), P680 oxidized by light is rereduced by Y_Z, the immediate donor to PS-II. Y_Z is a tyrosine on the reaction center protein D1. (The electron donor, Z, has been shown to be a tyrosine in the reaction center protein D1. The tyrosine amino acid is designated by Y. Hence what has been designated Z in chemical schemes of electron transport, is now called Y_Z.) Electrons are extracted from water by the oxygen-evolving complex and rereduce Y_Z^+ (oxidized Y_Z). On the reducing side of PS-II (to the right of the arrow joining P680 with P680*), pheophytin transfers electrons to the plastoquinone (PQ) acceptors Q_A and Q_B. The cytochrome b_6f complex transfers electrons to plastocyanin (PC), a soluble protein, which in turn reduces P700+ (oxidized P700). The b_6f complex contains a Rieske iron–sulfur protein (FeS_R), two *b*-type cytochromes (Cyt *b*) and cytochrome *f* (Cyt *f*). Electron flow within this complex is shown in Figure 7.28. The acceptor of electrons from P700* (A_0) is thought to be a chlorophyll, and the next acceptor (A_1) is a quinone. A series of membrane-bound iron–sulfur proteins (FeS_X, FeS_B, and FeS_A) transfers electrons to soluble ferredoxin (Fd). The soluble flavoprotein ferredoxin–NADP reductase (Fp) reduces $NADP^+$ to NADPH, which is used in the Calvin cycle to reduce CO_2 (see Chapter 8). The dashed line indicates cyclic electron flow around PS-I. (After Blankenship and Prince 1985.)

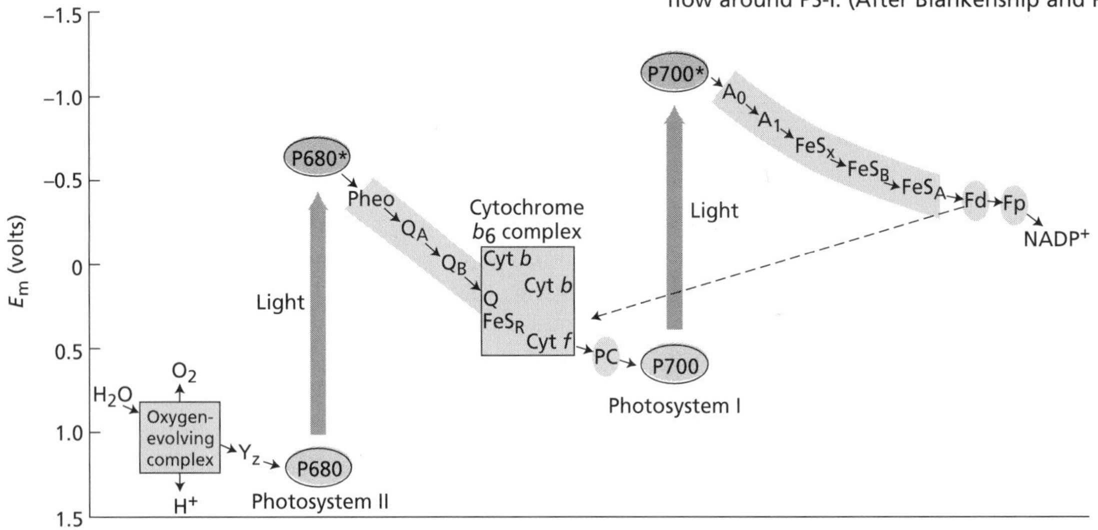

Mechanisms of Electron and Proton Transport

Some of the evidence that led to the idea of two photochemical reactions operating in series was discussed earlier in this chapter. Here we will consider in detail the chemical reactions involved in electron transfer during photosynthesis. We will discuss the excitation of chlorophyll by light and the reduction of the first electron acceptor, the flow of electrons through photosystems II and I, the oxidation of water as the primary source of electrons, the reduction of the final electron acceptor ($NADP^+$), and the chemiosmotic mechanism that mediates ATP synthesis.

Figure 7.21 shows a current version of the Z scheme, in which all the electron carriers known to function in electron flow from H_2O to $NADP^+$ are arranged vertically at their midpoint redox potentials (see Box 7.2). Components known to react with each other are connected by arrows, so the Z scheme is really a synthesis of both kinetic and thermodynamic information. The large vertical arrows represent the input of light energy into the system.

Four Thylakoid Protein Complexes Carry Out Electron and Proton Transport

Almost all the chemical processes that make up the light reactions of photosynthesis are carried out by four major protein complexes: photosystem II, the cytochrome b_6f complex, photosystem I, and the ATP synthase. These integral membrane complexes are vectorially oriented in the thylakoid membrane so that water is oxidized to O_2 in the thylakoid lumen, $NADP^+$ is reduced to NADPH on the stromal side of the membrane, and ATP is released into the stroma by H^+ moving from the lumen to the stroma (Figure 7.22).

Energy Is Captured When an Excited Chlorophyll Reduces an Electron Acceptor Molecule

As discussed earlier in the chapter, the function of light is to excite a specialized chlorophyll that is present in the reaction center from the ground electronic state to an excited electronic state, either by direct absorption or, more likely, via energy transfer from an antenna pigment. This excitation process can be envisioned as the promotion of an electron from the highest-energy filled orbital of the chlorophyll to the lowest-energy unfilled orbital (Figure 7.23). The electron in the upper orbital is only loosely bound to the chlorophyll and is easily lost if a molecule that can accept the electron is nearby. The excited state of the reaction center chlorophyll is therefore an extremely strong reducing agent, strong enough to reduce molecules that do not readily accept an electron. It is not possible to carry out a standard redox titration (see Box 7.2) on the excited chlorophyll, because of its short lifetime. However, values for the redox potential of the excited state can be estimated (see Figure 7.21) (Blankenship and Prince 1985).

The first reaction that converts electron energy into chemical energy—that is, the primary photochemical event—is the transfer of an electron from the excited state of a chlorophyll in the reaction center to an acceptor molecule. The energy of light is used to convert the chlorophyll from a weak reducing agent to a very strong one. An equivalent way to view this process is that the absorbed photon causes an electron rearrangement in the reaction center chlorophyll, followed by an electron transfer process in which part of the energy in the photon is captured in the form of redox energy.

Immediately after the photochemical event, the reaction center chlorophyll is in an oxidized state and the nearby electron acceptor molecule is reduced. The sys-

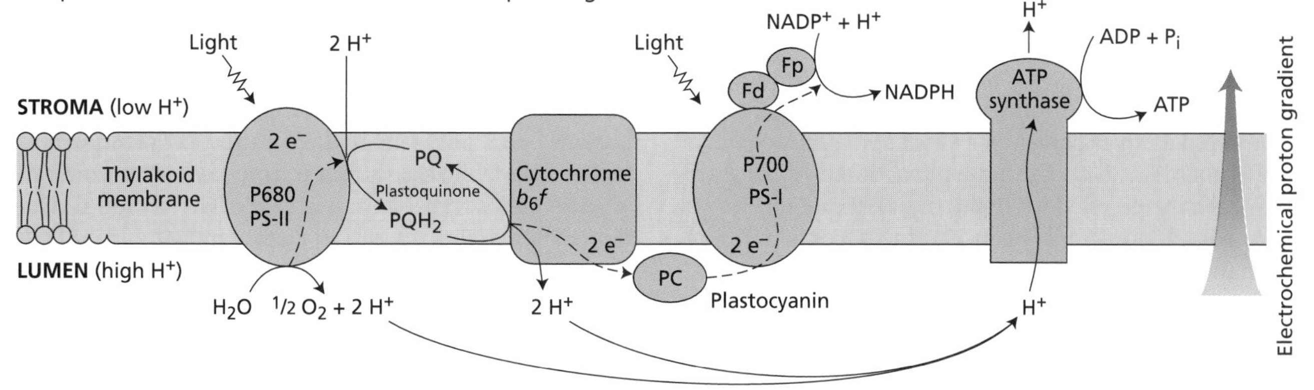

Figure 7.22 The transfer of electrons and protons in the thylakoid membrane is carried out vectorially by four protein complexes. Water is oxidized and protons are released in the lumen by PS-II. PS-I reduces $NADP^+$ to NADPH in the stroma, via the action of ferredoxin (Fd) and the flavoprotein ferredoxin–NADP reductase (Fp). Protons are also transported into the lumen by the action of the cytochrome b_6f complex and contribute to the electrochemical proton gradient. These protons must then diffuse to the ATP synthase enzyme, where their diffusion down the electrochemical energy gradient is used to synthesize ATP in the stroma. Reduced plastoquinone (PQH_2) and plastocyanin transfer electrons to cyytochrome b_6f and to PS-I, respectively. Dashed lines represent electron transfer; solid lines represent proton movement.

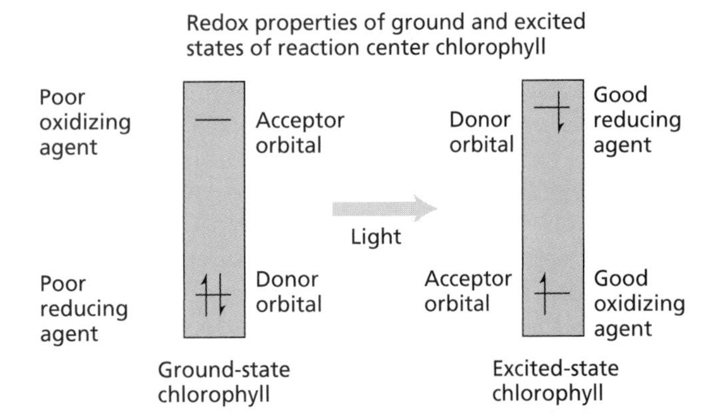

Figure 7.23 Orbital occupation diagram for the ground and excited states of reaction center chlorophyll. In the ground state the molecule is a poor reducing agent (loses electrons from a low-energy orbital) and a poor oxidizing agent (accepts electrons only into a high-energy orbital). In the excited state the situation is reversed, and an electron can be lost from the high-energy orbital, making the molecule an extremely powerful reducing agent. This is the reason for the extremely negative excited-state redox potential shown by P680* and P700* in Figure 7.21. The excited state can also act as a strong oxidant by accepting an electron into the lower-energy orbital, although this pathway is not significant in reaction centers. (After Blankenship and Prince 1985.)

tem is now at a critical juncture. The lower-energy orbital of the positively charged oxidized reaction center chlorophyll shown in Figure 7.23 has a vacancy and can accept an electron. If the acceptor molecule donates its electron back to the reaction center chlorophyll, the system will be returned to the same state that existed before the light excitation, and all the absorbed energy will be converted into heat. However, this wasteful *recombination* process does not appear to occur to any substantial degree in functioning reaction centers. Instead, the acceptor transfers its extra electron to a secondary acceptor and so on down the electron transport chain. The oxidized reaction center of the chlorophyll that had donated an electron is rereduced by a secondary donor, which in turn is reduced by a tertiary donor. In plants, the ultimate electron donor is H_2O, and the ultimate electron acceptor is $NADP^+$ (see Figures 7.12 and 7.21).

The essence of photosynthetic energy storage is thus the transfer of an electron from an excited chlorophyll to an acceptor molecule, followed by separation of the positive and negative charges by a very rapid series of chemical reactions. These secondary reactions separate the charges to opposite sides of the thylakoid membrane in approximately 200 picoseconds (1 picosecond = 10^{-12} s). With the charges thus separated, the reversal reaction is many orders of magnitude slower, and the energy has been captured. Each of the secondary electron transfers is accompanied by a loss of some energy, thus making the process effectively irreversible.

The quantum yield for the production of stable products in purified reaction centers from photosynthetic bacteria has been measured as 1.0; that is, every photon produces stable products and no reversal reactions occur. Although these types of measurements have not been made on purified reaction centers from higher plants, the measured quantum requirements for O_2 production under optimal conditions (low-intensity light) indicate that the values for the primary photochemical events are very close to 1.0. This conclusion follows from the fact that four electrons must be transferred to form one oxygen molecule, as we'll see shortly, and that the two photosystems must be activated for the overall process to be completed. The theoretical minimum quantum requirement is therefore 8, compared to the best measured values of 9 to 10 for oxygen evolution (see Figure 7.8). Of course, this represents the maximum attainable efficiency. Under less than ideal conditions, such as high light intensity, the quantum requirement is much higher.

The structure of the reaction center appears to be extremely fine-tuned for maximal rates of productive reactions and minimal rates of energy-wasting reactions. How this task is accomplished is not well understood and is a very active research area. The determination of the three-dimensional structure of the bacterial reaction centers described earlier is an important advance toward understanding photosynthetic energy conversion. Knowledge of the structure of a biological system is an essential step toward understanding its function.

The Reaction Center Chlorophylls of the Two Photosystems Absorb at Different Wavelengths

The reaction center chlorophyll is transiently in an oxidized state after losing an electron and before being rereduced by its electron donor. In the oxidized state, the strong light absorbance in the red region of the spectrum that is characteristic of chlorophylls is lost, or **bleached**. It is therefore possible to monitor the redox state of these chlorophylls by time-resolved optical absorbance measurements in which this bleaching is monitored directly (see Box 7.1). Using such techniques, Bessel Kok found that the reaction center chlorophyll of photosystem I absorbs maximally at 700 nm in its reduced state. Accordingly, this chlorophyll is named **P700** (the P stands for "pigment"). H. T. Witt and coworkers found the analogous optical transient of photosystem II at 680 nm, so its reaction center chlorophyll is known as **P680**. Earlier, Louis Duysens had identified the reaction center bacteriochlorophyll from purple photosynthetic bacteria as **P870**. The X-ray structure of the bacterial reaction center described earlier clearly indicates that P870 is a closely coupled pair or dimer of bacteriochlorophylls, rather than a single molecule. The primary donor of photosystem I, P700, is a dimer of chlorophyll *a* molecules. The aggregation state of P680 is a matter of debate and is yet to be fully characterized.

BOX 7.2

Midpoint Potentials and Redox Reactions

REDOX REACTIONS, midpoint potentials, and their relationship to the laws of thermodynamics were discussed in Chapter 2. These concepts are useful for our discussion of electron flow from H_2O to $NADP^+$ and the interactions between the different electron carriers.

Recall that the midpoint potential (E_m) is a measure of the tendency of a compound to take electrons from other compounds. A large positive midpoint potential means that the compound is a strong oxidant; a large negative value means that the compound is a strong reductant (in both cases relative to the standard hydrogen electrode).

Equilibrium constants can easily be predicted from midpoint potentials, in the same way that free energies were related to equilibrium constants (see Equation 2.15). Midpoint potentials for many chemical and biochemical reactions have been measured and tabulated. The y-axis on the Z scheme in Figure 7.21 shows midpoint potentials of the electron carriers, with negative values higher than positive ones. This choice makes reactions that are spontaneous (releasing free energy) appear "downhill" on the graph.

Knowledge of the midpoint potentials of the various electron carriers is important in establishing the pathway of electron flow in any biochemical electron transport system, such as those found in chloroplasts or mitochondria. Researchers make this measurement usually by carrying out a **redox titration** (Dutton 1978). They adjust, or poise, the sample at a particular redox potential, usually by adding small amounts of oxidants or reductants. **Redox mediators**, small molecules that permit rapid equilibration between the sample and the electrodes of the measurement system, must be included to ensure that the system is at equilibrium when the measurement is made. Several measurements are made at a variety of redox potentials. The sample is stirred in a special cell that contains platinum and reference electrodes, and chemical oxidants and reductants are added to adjust the redox potential, which is read with a voltmeter (part A of the figure). We can measure the extent of the redox reaction by following a particular property of the sample, usually absorbance, at each potential. In the example illustrated in part B of the figure, the reduced form of the compound has an absorbance that decreases as the compound is oxidized. The fraction of reduced form at each potential is plotted against redox potential, and the midpoint potential (E_m) is determined as the potential at which the compound is half oxidized and half reduced (part C of the figure).

(A) The setup for measuring midpoint potential

Voltmeter
Platinum electrode
Syringe containing chemical titrant
+153
Reference electrode
Reaction vessel
I_0
I
Absorbance measurement
Stir bar

(B) Changes in absorbance as a function of redox potential

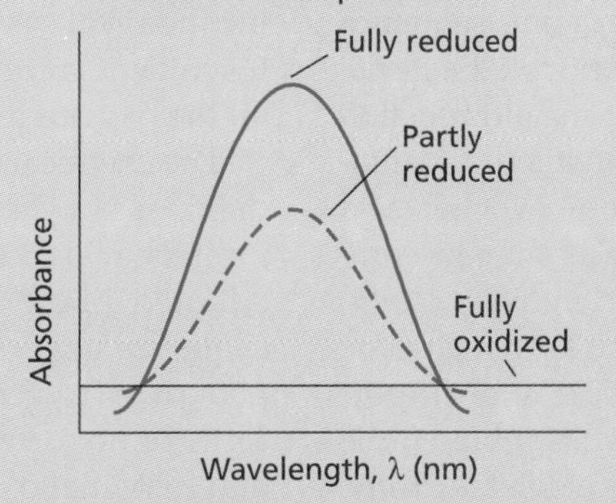

(C) Plot of redox state versus redox potential

1.0
0.5
0.0
Fraction reduced
Midpoint potential, E_m
Redox potential (mV)

In the oxidized state, reaction center chlorophylls contain an unpaired electron. Molecules with unpaired electrons often can be detected by a magnetic-resonance technique known as **electron spin resonance (ESR)**. ESR studies, along with the spectroscopic measurements already described, have led to the discovery of many intermediate electron carriers in the photosynthetic electron transport system.

The Photosystem II Reaction Center Is a Multisubunit Pigment–Protein Complex

Photosystem II is contained in a multisubunit protein complex (Figure 7.24). The core of the reaction center consists of two membrane proteins with molecular masses of 32 and 34 kilodaltons (kDa), known as D1 and D2, respectively. P680, additional chlorophylls, pheophytins, carotenoids, and plastoquinones are bound to

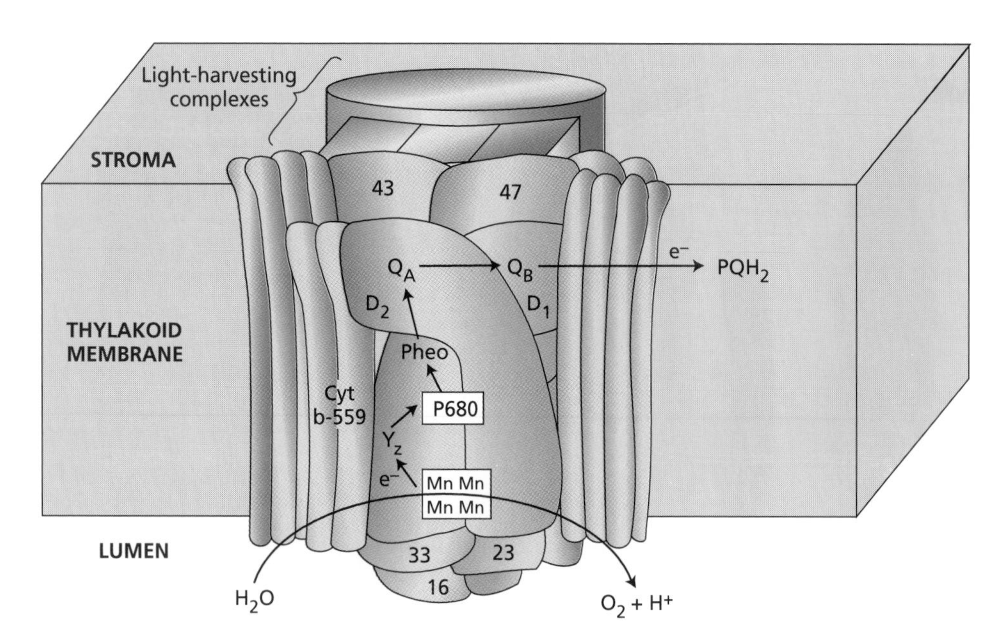

Figure 7.24 Schematic model of the PS-II reaction center and the oxygen-evolving complex. P680 is associated with the D1 and D2 proteins, which are analogous to the L and M proteins of the bacterial reaction center (see Figure 7.17). Electrons are transferred from P680 to pheophytin (Pheo) and then to Q_A and Q_B (PQH_2). $P680^+$ (oxidized P680) is rereduced by Y_Z, a tyrosine residue in the D1 protein. The oxygen evolution complex involves Mn, Ca^{2+}, and Cl^- as cofactors. (From Hankamer et al. 1997.)

the membrane proteins. D1 and D2 have some sequence similarity to the L and M peptides of purple bacteria. The current understanding of the structure and function of photosystem II owes much to the detailed information available on the bacterial system (Michel and Deisenhofer 1988). Other proteins serve as antenna complexes (43 and 47 kDa) or are involved in oxygen evolution (33, 23, and 16 kDa). Some, such as cytochrome b_{559}, have no known function but appear to be important because no stable PS-II complex is assembled if the genes that code for the polypeptides are deleted. Some other, smaller proteins, some of unknown function, are also found in photosystem II.

The structure of photosystem II is not yet known in atomic detail, but considerable information is available from a variety of techniques, in particular electron microscopy (Rögner et al. 1996). In most cases PS-II appears to be dimeric, with two complete reaction center complexes and some antenna complexes associated into a large supercomplex.

Water Is Oxidized to Oxygen by Photosystem II

Water is oxidized according to the following chemical reaction (Ghanotakis and Yocum 1990; Govindjee and Coleman 1990; Debus 1992; Yachandra et al. 1996):

$$2\ H_2O \rightarrow O_2 + 4\ H^+ + 4\ e^- \qquad (7.9)$$

This equation indicates that four electrons are removed from two water molecules, generating an oxygen molecule and four hydrogen ions. (For an introduction to oxidation–reduction reactions, see Chapter 2 and Box 7.2.) Water is a very stable molecule. Oxidation of water to form molecular oxygen is very difficult, and the photosynthetic oxygen-evolving complex is the only known biochemical system that carries out this reaction. Photosynthetic oxygen evolution is also the source of almost all the oxygen in Earth's atmosphere.

The chemical mechanism of photosynthetic water oxidation is not yet known, although there is a great deal of indirect evidence about the process. If a sample of dark-adapted photosynthetic membranes is exposed to a sequence of very brief, intense flashes, a characteristic pattern of oxygen production is observed (Figure 7.25A): Little or no oxygen is produced on the first two flashes, and maximal oxygen is released on the third flash and every fourth flash thereafter, until eventually the yield per flash damps to a constant value. This remarkable result was first observed by Pierre Joliot in the 1960s.

A schematic model explaining these observations, proposed by Kok and coworkers, has been widely accepted (Kok et al. 1970). This model for the photooxidation of water, called the **S state mechanism**, consists of a series of five states, known as S_0 to S_4, which represent successively more oxidized forms of the water-oxidizing enzyme system, or **oxygen-evolving complex** (Figure 7.25B). The light flashes advance the system from one S state to the next, until state S_4 is reached. State S_4 produces O_2 without further light input and returns the system to S_0. Occasionally, a center does not advance to the next S state upon flash excitation, and less frequently, a center is activated twice by a single flash. These misses and double hits cause the synchrony achieved by dark adaptation to be lost and the oxygen yield eventually to damp to a constant value. After this steady state has been reached, a complex has the same probability of being in any of the states S_0 to S_3 (S_4 is unstable and occurs only transiently), and the yield of O_2 becomes constant. States S_2 and S_3 decay in the dark,

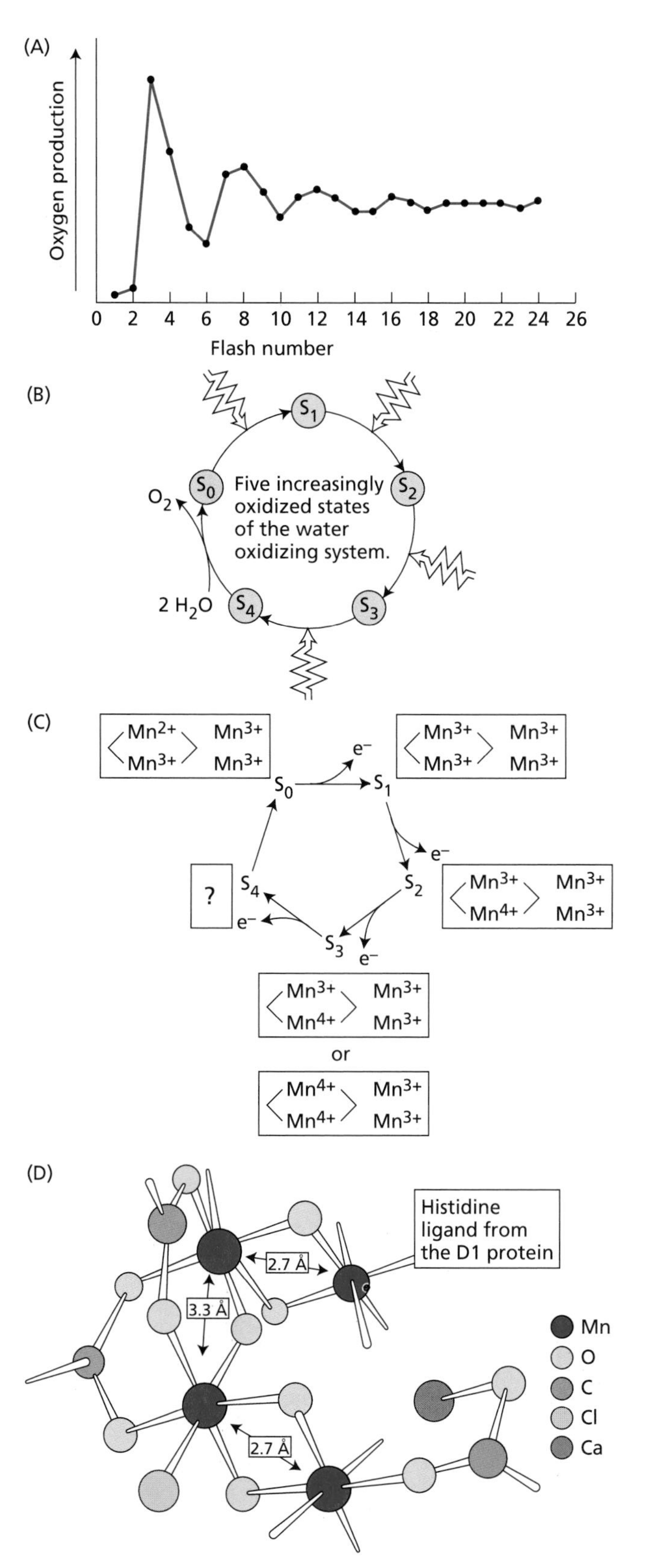

Figure 7.25 The pattern of oxygen evolution in flashing light and the S state mechanism for oxygen evolution. (A) A sample of dark-adapted chloroplasts is exposed to a series of short light flashes. The maximum production of oxygen occurs on the third flash, secondary maxima occur on every fourth flash, and the yield damps to a steady-state value after about 20 flashes. (B) The S state model developed by Kok and coworkers was advanced to explain the observations of part A. The oxygen-evolving system can exist in five states: S_0 to S_4. Successive photons (squiggly arrows) trapped by PS-II advance the system from S_0 to S_1, S_1 to S_2, and so on until state S_4 is reached. S_4 is unstable and reacts with two water molecules to produce O_2. (C) A structural model for the oxygen-evolving complex based on spectroscopic studies. The four Mn ions are linked to amino acids in the D1 protein, and to oxygen, chloride, and calcium. (D) One possible scheme for the redox states (S_0 to S_4) of the water-oxidizing Mn complex. Upon the release of oxygen, S_4 converts into S_0. No data are yet available for S_4. (C from Rutherford et al. 1992; D from Yachandra et al. 1996, © American Chemical Society, reprinted with permission).

but only as far back as S_1, which is stable in the dark. Therefore, after adaptation to the dark, approximately three-fourths of the oxygen-evolving complexes appear to be in state S_1 and one-fourth in state S_0. This distribution of states explains why the maximum yield of O_2 is observed after the third of a series of flashes given to dark-adapted chloroplasts.

This S state mechanism explains the observed pattern of O_2 release, but not the chemical nature of the S states or whether any partly oxidized intermediates, such as H_2O_2, are formed. Additional information has been obtained by measurement of the pattern of proton release (see Equation 7.9) as the S states are advanced with flashes. In addition to oxygen, hydrogen ions are a product of water oxidation. The release of protons is not strictly coupled to O_2 release, which occurs only at the last step of the cycle. Evidence suggests that the protons released before O_2 is produced come not directly from partly oxidized water but from ionizable protein groups such as certain amino acids in the region of the oxygen-evolving complex.

The protons produced by water oxidation are released into the lumen of the thylakoid, not directly into the stromal compartment (see Figures 7.14 and 7.22). They are released into the lumen because of the vectorial nature of the membrane and the fact that the oxygen-evolving complex is localized on the interior surface of the thylakoid. These protons are eventually released from the lumen to the stroma through ATP synthesis. In this way, the electrochemical potential formed by the release of protons during water oxidation contributes to ATP formation.

It has been known for many years that manganese (Mn) is an essential cofactor in the water-oxidizing process (see Chapter 5), and a classic hypothesis in photosynthesis research postulates that the S states represent successively oxidized states of an Mn-containing enzyme (Figure 7.25C) (Rutherford 1989, Rutherford et al. 1992; Debus 1992). This hypothesis has received strong support from a variety of experiments, most notably X-ray absorption and ESR studies, both of which detect the manganese directly (Yachandra et al. 1993, 1996). Analytical experiments indicate that four Mn ions are associated with each oxygen-evolving complex. Other experiments have shown that Cl^- and Ca^{2+} ions are essential for O_2 evolution, although their precise mechanistic roles have not yet been determined. A proposed model for the Mn cluster is shown in Figure 7.25D.

Considerable progress has been made in understanding the structural aspects of the oxygen-evolving system. Several amino acids from the D1 protein that may coordinate Mn have been identified by the use of site-directed mutagenesis (Vermaas 1993). Other proteins, including three peripheral membrane polypeptides of molecular mass 16, 23, and 33 kDa, respectively, have been implicated in the water-oxidizing process and its control. A tentative model of these proteins and their relation to the thylakoid membrane and the PS-II reaction center is shown in Figure 7.24.

One electron carrier, generally identified as Y_Z, functions between the oxygen-evolving complex and P680. To function in this region, Y_Z must have an extremely positive midpoint redox potential—that is, a very strong tendency to retain its electrons. This species has been identified as a radical formed from a tyrosine residue in the D1 protein of the PS-II reaction center (Barry 1993).

Pheophytin and Two Quinones Are Early Electron Acceptors of Photosystem II

Evidence from spectral and ESR studies indicates that pheophytin acts as an early acceptor in photosystem II, followed by a complex of two plastoquinones in close proximity to an iron atom. **Pheophytin** is a chlorophyll in which the central magnesium atom has been replaced by two hydrogen atoms. This chemical change gives the pheophytin chemical and spectral properties that are slightly different from those of chlorophyll. The precise arrangement of the carriers in the electron acceptor complex is not known, but it is probably very similar to that of the reaction center of purple bacteria (see Figure 7.18).

The plastoquinones bound to the reaction center operate as a **two-electron gate** (Okamura and Feher 1992). An electron is transferred from pheophytin to the first quinone of the complex and then transferred rapidly to the second quinone, where it remains. During this time, oxidized P680 regains an electron from Y_Z (see Figure 7.21), thereby returning to a photochemically competent state. A second electron is transferred from pheophytin to the first quinone and then to the second quinone. Two protons are picked up from the surrounding medium, producing a fully reduced hydroquinone (QH_2) (Figure 7.26). The hydroquinone then dissociates from the complex and enters the hydrocarbon portion of the membrane, where it in turn transfers its electrons to the cytochrome b_6f complex.

Electron Flow through the Cytochrome b_6f Complex Transports Protons to the Thylakoid Lumen

The **cytochrome b_6f complex** is a large multisubunit protein with several prosthetic groups (Cramer et al. 1996). It contains two *b*-type hemes and one *c*-type heme (**cytochrome *f***). In *c*-type cytochromes the heme is covalently attached to the peptide; in *b*-type cytochromes the chemically similar protoheme group is not covalently attached (Figure 7.27). In addition, the complex contains a **Rieske iron–sulfur protein** (named for the scientist who discovered it), in which two iron atoms are bridged by two sulfur atoms. The structures of cytochrome *f* and the related cytochrome bc_1 complex (discussed later in this section) have been determined.

The mechanism of electron and proton flow through the cytochrome b_6f complex is not yet fully understood, but a mechanism known as the **Q cycle** accounts for

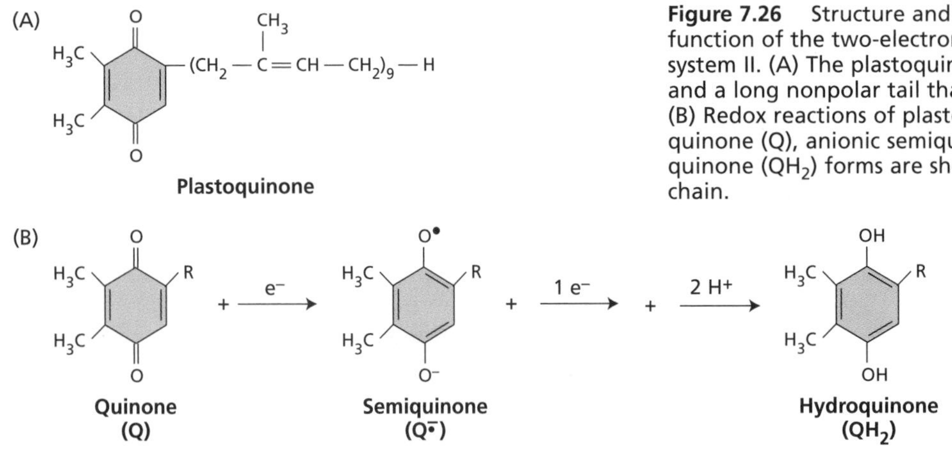

Figure 7.26 Structure and reactions of plastoquinone and function of the two-electron gate that operates in photosystem II. (A) The plastoquinone consists of a quinoid head and a long nonpolar tail that anchors it in the membrane. (B) Redox reactions of plastoquinone. The fully oxidized quinone (Q), anionic semiquinone ($Q^{\bullet-}$), and reduced hydroquinone (QH_2) forms are shown; R represents the side chain.

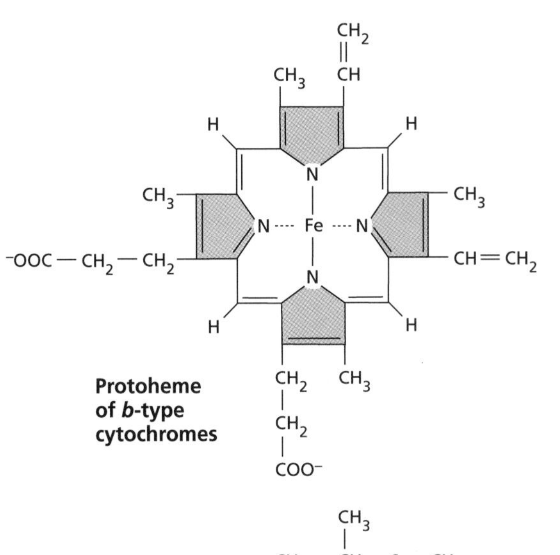

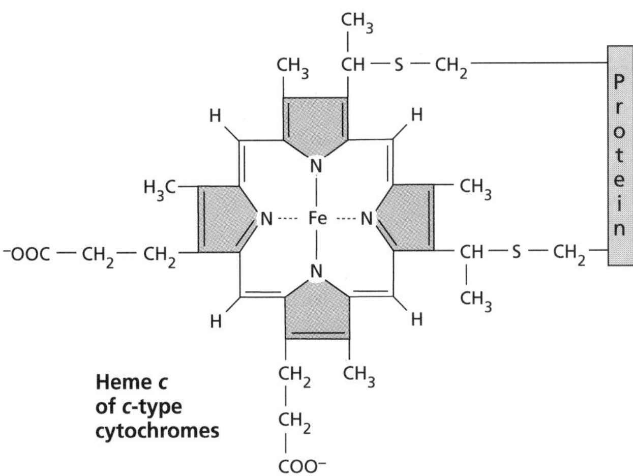

Figure 7.27 Structure of prosthetic groups of *b*- and *c*-type cytochromes. The protoheme group (also called protoporphyrin IX) is found in *b*-type cytochromes, the heme *c* group in *c*-type cytochromes. The heme *c* group is covalently attached to the protein by thioether linkages with two cysteine residues in the protein; the protoheme group is not covalently attached to the protein. The Fe ion is in the 2+ oxidation state in reduced cytochromes and in the 3+ oxidation state in oxidized cytochromes.

most of the observations. In this mechanism, plastohydroquinone is oxidized, with one of the two electrons passed along a linear electron transport chain toward photosystem I, while the other electron goes through a cyclic process that increases the number of protons pumped across the membrane.

In the linear electron transport chain, the oxidized Rieske protein (FeS_R) accepts an electron from **plastohydroquinone** (**QH_2**) and transfers it to cytochrome *f* (Figure 7.28). Cytochrome *f* then transfers an electron to the blue-colored copper protein **plastocyanin (PC)**, which in turn reduces oxidized P700 of PS-I. In the cyclic part of the process, the plastosemiquinone transfers its other electron to one of the *b*-type hemes, releasing both its protons to the lumenal side of the membrane. The *b*-type heme transfers its electron through the second *b*-type heme to an oxidized quinone molecule, reducing it to the semiquinone form near the stromal surface of the complex. Another similar sequence of electron flow fully

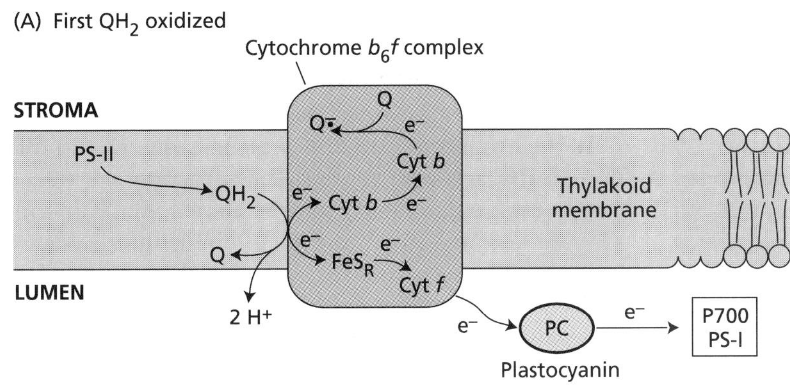

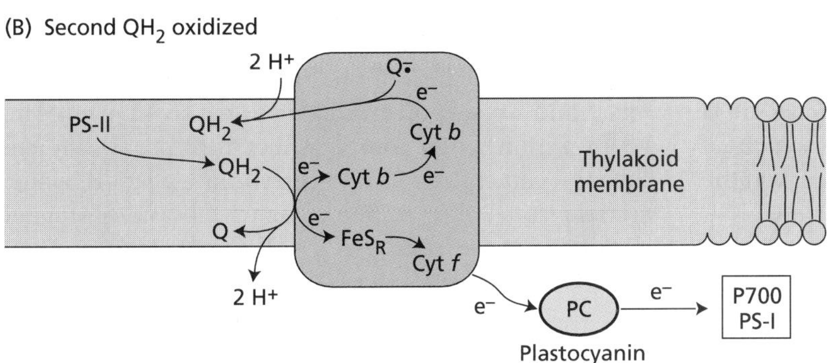

Figure 7.28 Mechanism of electron and proton transfer in the cytochrome b_6f complex. This complex contains two *b*-type cytochromes (Cyt *b*), a *c*-type cytochrome (Cyt *c*, historically called cytochrome *f*) a Rieske Fe–S protein (FeS_R), and two quinone oxidation–reduction sites. (A) A plastohydroquinone (QH_2) molecule produced by the action of PS-II (see Figure 7.26) is oxidized near the lumenal side of the complex, transferring its two electrons to the Rieske Fe–S protein and one of the *b*-type cytochromes and simultaneously expelling two protons to the lumen. The electron transferred to FeS_R is passed to cytochrome *f* (Cyt *f*) and then to plastocyanin (PC), which reduces P700 of PS-I. The reduced *b*-type cytochrome transfers an electron to the other *b*-type cytochrome, which reduces a quinone (Q) to the semiquinone ($Q^{\bullet-}$). state (see Figure 7.26). (B) A second QH_2 is oxidized, with one electron going from FeS_R to PC and finally to P700. The second electron goes through the two *b*-type cytochromes and reduces the semiquinone to the plastohydroquinone, at the same time picking up two protons from the stroma. Overall, four protons are transported across the membrane for every two electrons delivered to P700.

reduces the plastoquinone, which picks up protons from the stromal side of the membrane and is released from the b_6f complex as plastohydroquinone.

The net result of two turnovers of the complex is that two electrons are transferred to P700, two plastohydroquinones are oxidized to the quinone form, and one oxidized plastoquinone is reduced to the hydroquinone form. In addition, four protons are transferred from the stromal to the lumenal side of the membrane. In this way, electron flow connecting the acceptor side of the PS-II reaction center to the donor side of the PS-I reaction center also gives rise to an electrochemical potential across the membrane, due in part to H^+ concentration differences on the two sides of the membrane. This electrochemical potential is used to power the synthesis of ATP. The cyclic electron flow through the cytochrome *b* and plastoquinone increases the number of protons pumped per electron beyond what could be achieved in a strictly linear sequence.

Many types of bacteria, as well as mitochondria, contain a **cytochrome bc_1 complex** similar to the cytochrome b_6f complex (Knaff 1993; Trumpower and Gennis 1994). The reaction sequence just described is fairly well established in these bc_1 complexes, but the precise sequence in the cytochrome b_6f complex in chloroplasts is not yet certain. The structure of the bc_1 complex from mitochondria has recently been determined by X-ray diffraction (Xia et al. 1997) revealing the localization of the iron centers and inhibitor-binding sites within the complex, and the membrane-spanning portions of the polypeptides.

Plastoquinone and Plastocyanin Are Putative Electron Carriers between Photosystems II and I

The lateral heterogeneity of the photosynthetic membrane (see Figure 7.16) requires that at least one component be capable of moving along or within the membrane in order to deliver electrons produced by photosystem II to photosystem I. The cytochrome b_6f complex is distributed equally between the grana and the stroma regions of the membrane, but its large size makes it unlikely that it is the mobile carrier. Instead, plastoquinone or plastocyanin or possibly both are thought to serve as mobile carriers to connect the two photosystems. Plastocyanin is a small (10.5 kDa), water-soluble, copper-containing protein that transfers electrons between the cytochrome b_6f complex and P700. This protein is found in the lumenal space. In certain green algae and cyanobacteria, a *c*-type cytochrome is sometimes found instead of plastocyanin; which of these two proteins is synthesized depends on the amount of copper available to the organism.

Some of the cytochrome b_6f complexes are found in the stroma region of the membrane, where photosystem I is located. Under certain conditions **cyclic electron flow** from the reducing side of photosystem I, through the b_6f complex and back to P700, is known to occur. This cyclic electron flow is coupled to proton pumping into the lumen, which can be utilized for ATP synthesis but does not oxidize water or reduce $NADP^+$. Cyclic electron flow is especially important as an ATP source in the bundle sheath chloroplasts of some plants that carry out C_4 carbon fixation (see Chapter 8).

The Photosystem I Reaction Center Reduces $NADP^+$

The PS-I reaction center is composed of a multiprotein complex. P700 and about 100 core antenna chlorophylls are bound to two peptides with molecular masses in the range of 66 to 70 kDa (Golbeck 1992; Krauss et al. 1993, 1996; Chitnis 1996). PS-I reaction center complexes have been isolated from several organisms and found to contain the 66 to 70 kDa peptides, along with a variable number of smaller peptides in the range of 4 to 25 kDa. Some of these proteins serve as binding sites for the soluble electron carriers plastocyanin and ferredoxin. The functions of some of the other peptides are not well understood. An 8-kDa protein contains some of the bound iron–sulfur centers that serve as early electron acceptors in photosystem I. The structure of the PS-I complex from a cyanobacterium has been determined to a resolution of 4 Å, and the positions of many of the chlorophylls and electron transfer components have been located (Krauss et al. 1996; Schubert et al. 1997; see Figure 7.29). The PS-I reaction center is somewhat unusual in that the core antenna consisting of about 100 chlorophylls is part of the minimal reaction center unit. The antenna pigments form a bowl surrounding the electron transfer cofactors, which are in the center of the complex. Thus the distinction between the antenna and the reaction center is blurred in this instance because they are part of the same molecular complex.

In their reduced form, the electron carriers that function in the acceptor region of photosystem I are all extremely strong reducing agents. These reduced species are very unstable and thus difficult to identify. Evidence indicates that one of these early acceptors is a chlorophyll molecule, and another is a quinone species, phylloquinone, also known as vitamin K_1 (Nugent 1996).

Additional electron acceptors include a series of three membrane-associated iron–sulfur proteins, or bound ferredoxins, also known as **Fe–S centers X**, **A**, and **B**. Fe–S center X is part of the P700-binding protein; centers A and B reside on an 8 kDa protein that is part of the PS-I reaction center complex. Electrons are transferred through centers A and B to **ferredoxin**, a small, water-soluble iron–sulfur protein. The membrane-associated flavoprotein **ferredoxin–NADP reductase** reduces $NADP^+$ to NADPH, thus completing the sequence of noncyclic electron transport that begins with the oxidation of water (Karplus et al. 1991).

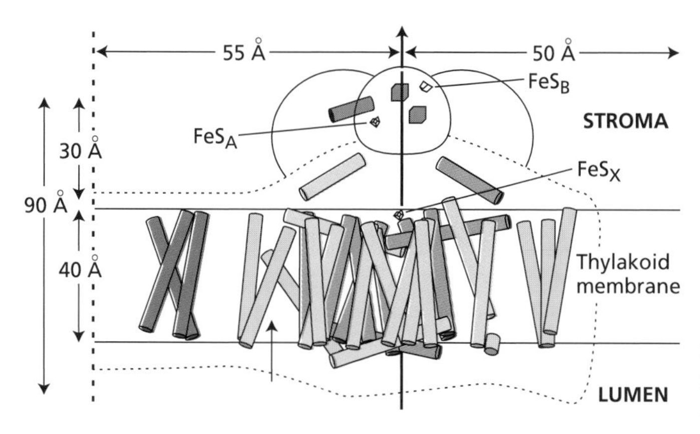

Figure 7.29 Structure of Photosystem I from cyanobacteria at 4.0 Å resolution. The figure shows an arrangement of α helices (shown as tubes) viewed parallel to the plane of the membrane. The stromal side of the membrane is at the top, and the lumenal side is at the bottom of the figure. The three FeS centers that are electron acceptors in Photosystem I, FeS_A, FeS_B, and FeS_X are visible on the stromal side as cubes. Pigments are removed for clarity, molecular dimension in Å are shown. (After Schubert et al. 1997.)

In addition to the reduction of $NADP^+$, reduced ferredoxin produced by photosystem I serves several other functions in the chloroplast, including providing reductants to reduce nitrate and regulating some of the carbon fixation enzymes (Buchanan 1991; Chapter 8).

The PS-I reaction center appears to have some functional similarity to the reaction center found in the anaerobic green sulfur bacteria and the heliobacteria. These bacteria contain low-potential Fe–S centers as early electron acceptors and are probably capable of ferredoxin-mediated NAD^+ reduction similar to the $NADP^+$ reduction function of photosystem I. There is almost certainly an evolutionary relationship between these complexes and photosystem I of oxygen-evolving organisms.

A Chemiosmotic Mechanism Converts the Energy Stored in Chemical and Electric Potentials to ATP

In addition to the energy stored as redox equivalents (NADPH) by the light reactions, a portion of the photon's energy is captured in the formation of ATP. This process was discovered in the 1950s by Daniel Arnon and coworkers (Arnon 1984). In normal cellular conditions, this **photophosphorylation** requires electron flow, although under some conditions electron flow and photophosphorylation can take place independently of each other. Electron flow without accompanying phosphorylation is said to be **uncoupled**.

It is now widely accepted that photophosphorylation works via the **chemiosmotic mechanism**, first proposed in the 1960s by Peter Mitchell (Mitchell 1979). The same general mechanism is applicable to phosphorylation during aerobic respiration in bacteria and mitochondria (see Chapter 11) and underlies the transfer of many ions and metabolites across membranes (see Chapter 6). The chemiosmotic mechanism appears to be a unifying aspect of membrane processes in all forms of life.

In Chapter 6 we discussed ATPases in the plasma membrane and their role in chemiosmosis and ion transport at the cellular level. In that context, the ATPase utilizes ATP made available by photophosphorylation in the chloroplast and oxidative phosphorylation in the mitochondrion. Here we are concerned with chemiosmosis and transmembrane proton concentration differences used to make ATP in the chloroplast.

The basic principle of chemiosmosis is that ion concentration differences and electric-potential differences across membranes are a source of free energy that can be utilized by the cell. As described by the second law of thermodynamics (see Chapter 2), any nonuniform distribution of matter or energy represents a source of energy. Differences in **chemical potential** of any molecular species whose concentrations are not the same on opposite sides of a membrane provide such a source of energy.

The asymmetric nature of the photosynthetic membrane and the fact that proton flow from one side of the membrane to the other accompanies electron flow were discussed earlier. The direction of proton translocation is such that the stroma becomes more alkaline and the lumen becomes more acidic as a result of electron transport (see Figures 7.22 and 7.28).

Some of the early evidence in support of a chemiosmotic mechanism of photosynthetic ATP formation was provided by an elegant experiment carried out by André Jagendorf and coworkers (Figure 7.30) (Jagendorf 1967). They suspended chloroplasts in a pH 4 buffer, which permeated the membrane and caused the interior, as well as the exterior, of the thylakoid to equilibrate at this acidic pH. They then rapidly injected the suspension into a pH 8 buffer solution, thereby creating a pH difference of 4 units across the thylakoid membrane, with the inside acidic relative to the outside. They found that large amounts of ATP were formed from ADP and P_i by this process, with no light input or electron transport. This result supports the predictions of the chemiosmotic hypothesis.

Mitchell proposed that the total energy available for ATP synthesis, which he called the **proton motive force** (Δp), is the sum of a proton chemical potential and a transmembrane electric potential (Lowe and Jones 1984). These two components of the proton motive force from the outside of the membrane to the inside are given by the following equation:

$$\Delta p = \Delta E - 59(pH_i - pH_o) \qquad (7.10)$$

In Equation 7.10, ΔE is the transmembrane electric potential and $pH_i - pH_o$ (or ΔpH) is the pH difference

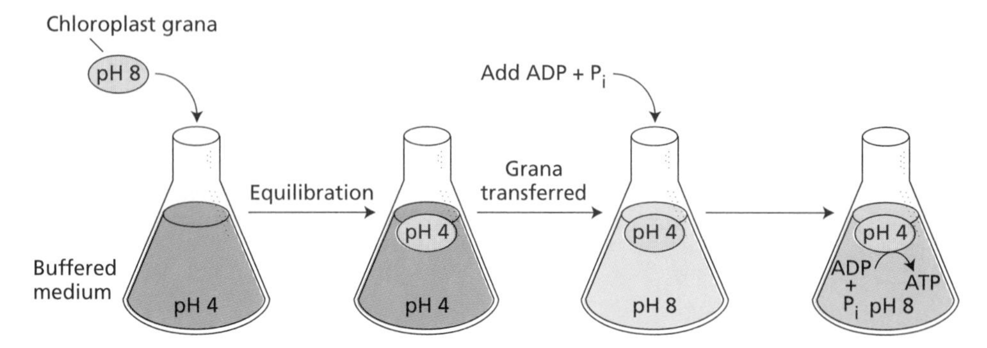

Figure 7.30 Summary of the experiment carried out by Jagendorf and coworkers. Isolated chloroplast grana previously at pH 8 were equilibrated in an acid medium at pH 4. The grana were then transferred to a medium of pH 8 that contained ADP and P_i. This manipulation generates a pH difference (pH 4 inside the grana and pH 8 outside) that in the intact chloroplast is produced by electron flow and proton translocation. In these conditions, ATP was synthesized without light. This experiment provided evidence in favor of Peter Mitchell's chemiosmotic theory, which stated that pH differences and electric potentials across a membrane are a source of energy that can be used to synthesize ATP.

across the membrane. The constant of proportionality (at 25°C) is 59 mV per pH unit, so a transmembrane pH difference of 1 unit is equivalent to a membrane potential of 59 mV.

Many experiments have established that the two components of the proton motive force are interchangeable, so a large ΔE, a large ΔpH, or intermediate amounts of both are all equally effective in forming ATP (Hangarter and Good 1988). Under conditions of steady-state electron transport in chloroplasts, the membrane electric potential is quite small because of ion movement across the membrane, so Δp is built almost entirely by ΔpH. The stoichiometry of protons translocated to ATP synthesized has recently been found to be four H^+ per ATP.

In addition to the need for mobile electron carriers discussed earlier, the uneven distribution of photosystems II and I, and of ATP synthase on the thylakoid membrane (see Figure 7.16) poses some challenges for the formation of ATP. ATP synthase is found only in the stroma lamellae and at the edges of the grana stacks. Protons pumped across the membrane by the cytochrome b_6f complex or protons produced by water oxidation in the middle of the grana must move laterally up to several tens of nanometers to reach ATP synthase.

The ATP is synthesized by a large (400 kDa) enzyme complex known by several names: **ATP synthase**, the **coupling factor**, **ATPase** (after the reverse reaction of ATP hydrolysis), and **CF_o–CF_1** (Boyer 1997). This enzyme consists of two parts: a hydrophobic membrane-bound portion called CF_o and a portion that sticks out into the stroma called CF_1 (Figure 7.31). Three copies of each of the α and β peptides are present in the CF_1 portions, arranged alternately much like the sections of an orange. The catalytic sites are located largely on the β polypeptide, while many of the other peptides are thought to have primarily regulatory functions. CF_o appears to form a channel across the membrane through which protons can pass; CF_1 is the portion of the complex that synthesizes ATP. The molecular structure of a portion of the mitochondrial ATP synthase has been determined by X-ray crystallography (Abrahams et al. 1994). Although there are significant differences between the chloroplast and mitochondrial enzymes, they have the same overall architecture and probably nearly identical catalytic sites (Figure 7.32).

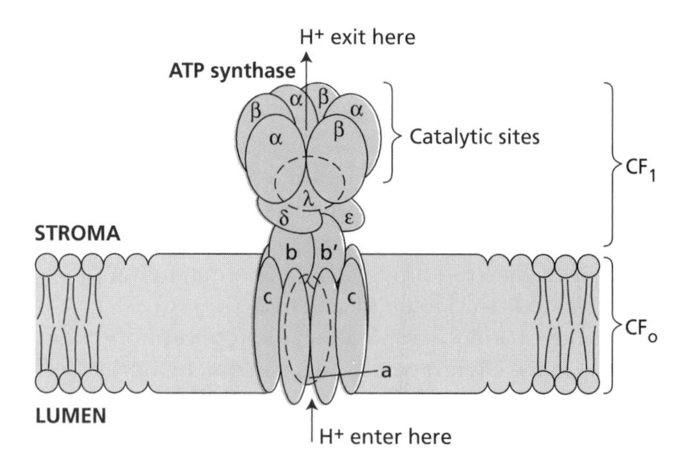

Figure 7.31 Structure of ATP synthase, or coupling factor. This enzyme consists of a large multisubunit complex, CF_1, attached on the stromal side of the membrane to an integral membrane portion, known as CF_o. Protons enter the channel through the membrane formed by CF_o and are expelled by CF_1. The catalytic sites for ATP synthesis are found on CF_1. CF_1 consists of five different polypeptides, with a stoichiometry of α_3, β_3, γ, δ, ε. CF_o contains probably four different polypeptides, with a stoichiometry of a, b, b′, c_{12}. ATP synthase has a fundamental role in the harnessing of light energy and its storage in the high-energy chemical bond of ATP.

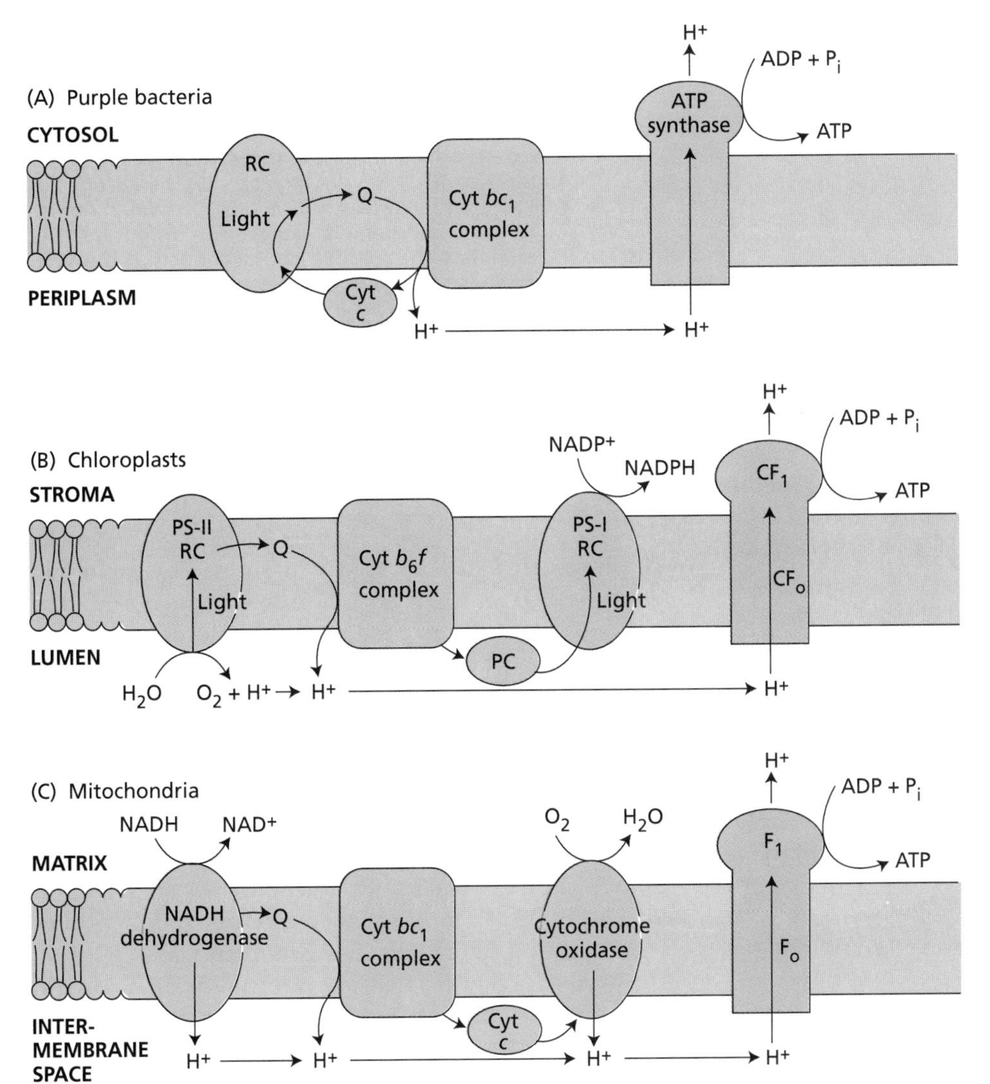

Figure 7.32 Similarities of photosynthetic and respiratory electron flow. In all three types of energy-conserving systems shown here, electron flow is coupled to proton translocation, creating a transmembrane proton motive force (Δp). The energy in the proton motive force is then used for the synthesis of ATP by the coupling-factor enzyme complex (CF_o–CF_1), also called ATP synthase. (A) A reaction center (RC) in purple photosynthetic bacteria carries out cyclic electron flow, generating a proton potential by the action of the cytochrome bc_1 complex. (B) Chloroplasts carry out noncyclic electron flow, oxidizing water and reducing $NADP^+$. Protons are produced by the oxidation of water and by the oxidation of PQH_2 (Q) by the cytochrome b_6f complex. (C) Mitochondria oxidize NADH to NAD^+ and reduce oxygen to water. Protons are pumped by the enzyme NADH dehydrogenase, the cytochrome bc_1 complex, and cytochrome oxidase. The ATP synthases in the three systems are very similar in structure.

The chemical mechanism by which ATP is synthesized is not yet understood in detail. However, considerable evidence now supports a mechanism, first proposed by Paul Boyer, in which the principal energy-requiring step is the release of bound ATP from the enzyme (Boyer 1993, 1997). It has also been proposed that during catalysis a large portion of the CF_1 complex rotates about a bearing consisting of the γ subunit (Noji et al. 1997). The γ subunit may act as a camshaft does, rotating alternately against the α and β subunits (see Box 11.1). The energy of the conformational movements is then translated into phosphoanhydride bond energy (Junge et al. 1997).

Regulation and Repair of the Photosynthetic Apparatus

Photosynthetic systems face a special challenge. They are designed to absorb large amounts of light energy and process it into chemical energy. At the molecular level, the energy in a photon is a huge perturbation, which the photosynthetic systems elegantly and efficiently process under normal conditions. Under some conditions, however, they may not be able to deal with all the incoming energy. The excess energy can lead to production of toxic species and damage of the system if it is not dissipated safely (Horton et al. 1996). Photosynthetic organisms therefore contain a complex set of regulatory and repair mechanisms.

Some of these mechanisms regulate energy flow in the antenna system, to avoid excess excitation of the reaction centers and ensure that the two photosystems are equally driven. Although very effective, these processes are not entirely foolproof, and sometimes toxic species are produced. Additional mechanisms are needed to dissipate these compounds—in particular, toxic oxygen species. Even with all these protective and scavenging mechanisms, the photosynthetic apparatus sometimes still becomes damaged, and additional

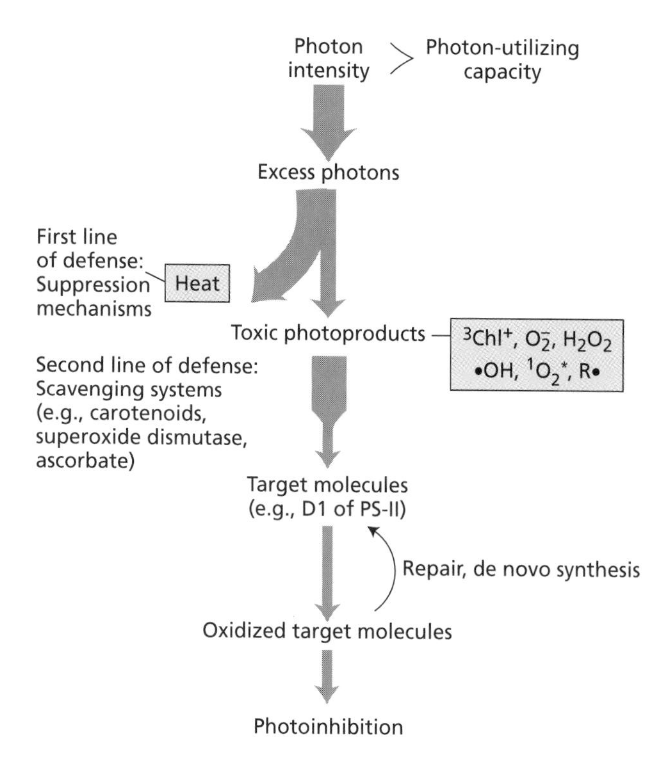

Figure 7.33 Overall picture of the regulation of photon capture and the protection and repair of photodamage. Protection against photodamage is a multilevel process. The first line of defense is suppression of damage by quenching of excess excitation as heat. If this defense is not sufficient and toxic photoproducts form (the triplet state of Chl [$^3Chl^*$], superoxide [O_2^-], singlet oxygen [$^1O_2^*$], hydrogen peroxide [H_2O_2], and hydroxyl radical [•OH]), a variety of scavenging systems (carotenoids, superoxide dismutase, ascorbate) eliminate the reactive photoproducts. If this second line of defense also fails, the photoproducts can damage certain target molecules that are susceptible to photodamage, especially the D1 protein of photosystem II. This damage leads to photoinhibition. The D1 protein is then excised from the PS-II reaction center and degraded. A newly synthesized D1 is reinserted into the PS-II reaction center to form a functional unit. (After Asada 1996.)

mechanisms are present to repair the system. Figure 7.33 shows a model for the overall organization of the regulation and repair systems.

Carotenoids Serve as Both Accessory Pigments and Photoprotective Agents

Carotenoids are found in all photosynthetic organisms, except for mutants incapable of living outside the laboratory (Frank and Cogdell 1996). The different types of carotenoids found in photosynthetic organisms are all linear molecules with multiple conjugated double bonds (see Figure 7.5B). Absorption bands in the 400 to 500 nm region give carotenoids their characteristic orange color. The color of carrots, for example, is due to the carotenoid β-carotene, whose structure and absorption spectrum are shown in Figures 7.5 and 7.6, respectively.

Carotenoids are usually associated intimately with both antenna and reaction center pigment proteins and are integral constituents of the membrane. The energy of light absorbed by carotenoids is rapidly transferred to chlorophylls, so carotenoids are termed **accessory pigments**.

Carotenoids also play an essential role in **photoprotection**. The photosynthetic membrane can easily be damaged by the large amounts of energy absorbed by the pigments if this energy cannot be stored by photochemistry; hence the need for a protection mechanism. The photoprotection mechanism can be thought of as a safety valve, venting excess energy before it can damage the organism. When the energy stored in chlorophylls in the excited state is rapidly dissipated by excitation transfer or photochemistry, the excited state is said to be **quenched**. If the excited state of chlorophyll is not rapidly quenched by excitation transfer or photochemistry, it can react with molecular oxygen to form an excited state of oxygen known as **singlet oxygen** ($^1O_2^*$). The extremely reactive singlet oxygen species goes on to react with and damage many cellular components, especially lipids. Carotenoids exert their photoprotective action by rapidly quenching the excited state of chlorophyll. The excited state of carotenoid does not have sufficient energy to form singlet oxygen, so it decays back to its ground state while losing its energy as heat.

Mutant organisms that lack carotenoids cannot live in the presence of both light and molecular oxygen—a rather difficult situation for an O_2-evolving photosynthetic organism. For non-O_2-evolving photosynthetic bacteria, mutants that lack carotenoids can be maintained under laboratory conditions if oxygen is excluded from the growth medium.

Recently carotenoids were found to play a role in nonphotochemical quenching, which is a second protective and regulatory mechanism. We will describe this process in detail after we consider the partitioning of energy between the two photosystems.

Thylakoid Stacking Permits Energy Partitioning between the Photosystems

The fact that photosynthesis in higher plants is driven by two photochemical systems with different light-absorbing properties poses a special problem. If the rate of delivery of energy to photosystems I and II is not precisely matched and conditions are such that the rate of photosynthesis is limited by the available light (low light intensity), the rate of electron flow will be limited by the photosystem that is receiving less energy. The most efficient situation would be one in which the input of energy is the same to both photosystems. However, no single arrangement of pigments would satisfy this requirement, because at different times of day the light intensity and spectral distribution tend to favor one photosystem or the other. The solution to this problem would be a mechanism that shifts energy from one photosystem to the other in response to different conditions, and such a regulating mechanism has been shown to operate in different experimental conditions. The observation that the overall quantum yield of photosynthesis is nearly independent of wavelength (see Figure 7.9) strongly suggests that such a mechanism exists.

Considerable progress has been made in understanding the molecular mechanism that is responsible for this energy redistribution (Bennett 1991; Allen 1992). Thylakoid membranes contain a protein kinase that can phosphorylate a specific threonine residue on the surface of one of the membrane-bound antenna pigment proteins. This pigment–protein complex is the LHCII described earlier (see Figure 7.20). When LHCII is not phosphorylated, it delivers more energy to photosystem II, and when it is phosphorylated it delivers more energy to photosystem I. The kinase is activated when plastoquinone, one of the electron carriers between the photosystems, accumulates in the reduced state, which occurs when photosystem II is being activated more frequently than photosystem I. The phosphorylated LHCII then migrates out of the stacked regions of the membrane into the unstacked regions (see Figure 7.16), probably because of repulsive interactions with negative charges on adjacent membranes.

The lateral migration of LHCII shifts the energy balance toward photosystem I, which is located in the stroma lamellae, and away from photosystem II, which is located in the stacked membranes of the grana. This situation is called **state 2**. If the plastoquinone becomes more oxidized because of excess excitation of photosystem I, the kinase is deactivated and the level of phosphorylation of LHCII is decreased by the action of a membrane-bound phosphatase. LHCII then moves back to the grana, and the system is in **state 1**. The net result is very precise control of the energy distribution between the photosystems, allowing the most efficient use of the available energy.

Some Xanthophylls Also Participate in Energy Dissipation

A major regulatory process involved in the delivery of excitation energy to the reaction center has been described only recently. The process, known as **nonphotochemical quenching**, appears to be an essential part of the regulation of antenna systems in most algae and plants. Nonphotochemical quenching is the quenching of chlorophyll fluorescence (see Figure 7.4) by processes other than photochemistry. It was first discovered in studies of chlorophyll fluorescence, which revealed that intense illumination produces a state in which a large fraction of the excitations in the antenna system are quenched by conversion into heat (Krause and Weiss 1991). This process, which seems wasteful at first, is now thought to be involved in protecting the organism against overexcitation and subsequent damage. The process can be thought of as a "volume knob" that adjusts the flow of excitations to the PS-II reaction center to a manageable level, depending on the light intensity and other conditions.

The molecular mechanism of nonphotochemical quenching is not yet understood in detail. Several factors seem to be involved, including the pH of the thylakoid lumen and the state of aggregation of the antenna complexes in the membrane (Horton et al. 1996). In addition, certain carotenoids known as **xanthophylls** are involved in this regulatory mechanism. Figure 7.34 shows the structure of two of these xanthophylls, violaxanthin and zeaxanthin, and an intermediate, antheraxanthin. These carotenoids can be interconverted by epoxidase and de-epoxidase enzymes that are present in the chloroplast. The same high-light regime that induces nonphotochemical quenching tends to activate the de-epoxidase enzyme that converts the xanthophylls into the zeaxanthin form; low-light conditions activate the epoxidase, resulting in violaxanthin accumulation. So zeaxanthin is associated with the quenched state and violaxanthin is found when the system is in the unquenched state. It is not yet clear if the carotenoid itself is the quenching agent, although many researchers think it is. This question is a very active area of research (Demmig-Adams and Adams 1992; Pfündel and Bilger 1994; Horton et al. 1996).

The Photosystem II Reaction Center Is Easily Damaged

Another effect that appears to be a major factor in the stability of the photosynthetic apparatus is photoinhibition, which occurs when excess excitation arriving at the PS-II reaction center leads to its inactivation and damage (Barber and Andersson 1992; Long et al. 1994).

Photoinhibition is a complex set of molecular processes. It is defined as the inhibition of photosynthesis by excess light. Much of the effects of excess light appear to be localized in photosystem II, and inhibitory targets on both the donor and acceptor sides have been identified. As discussed in detail in Chapter 9, this inhibition is reversible in early stages. Later stages of inhibition, however, result in damage to the system such that the PS-II reaction center must be disassembled and repaired. The main target of this damage is the D1 protein that makes up part of the PS-II reaction center complex. This protein is easily damaged by excess light and then must be removed from the membrane and replaced with a newly synthesized copy. The other parts of the PS-II reaction center are thought to be recycled, so the D1 protein is the only component that needs to be synthesized.

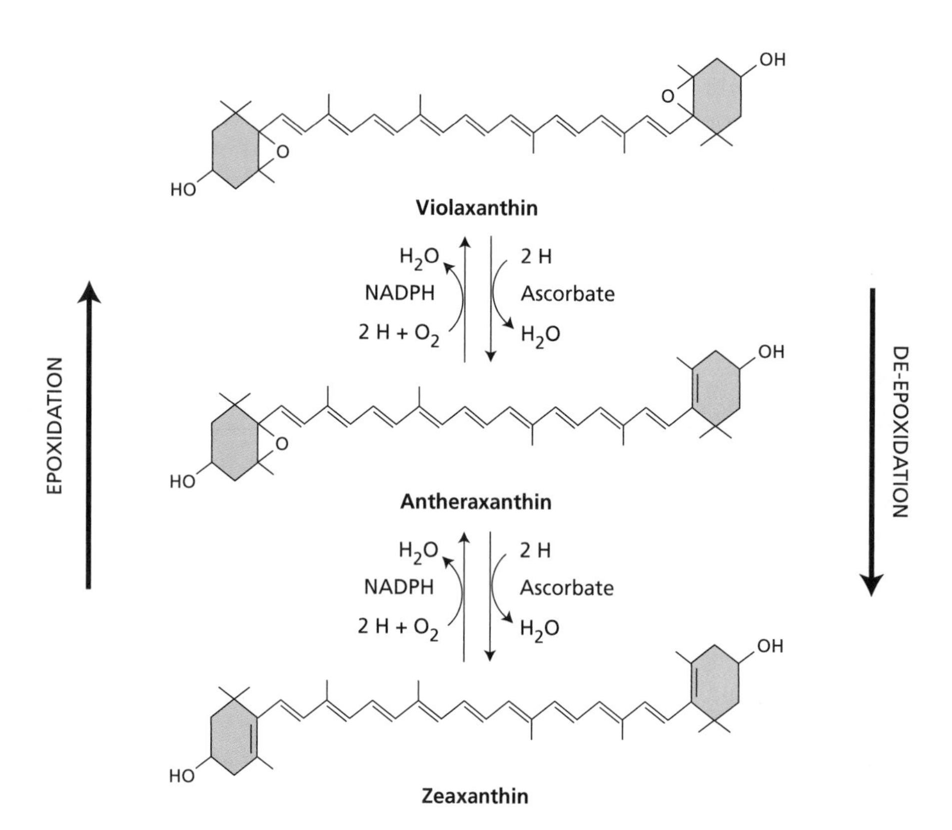

Figure 7.34 Chemical structure of violaxanthin, antheraxanthin, and zeaxanthin. The highly quenched state of photosystem II is associated with zeaxanthin, the unquenched state with violaxanthin. Enzymes interconvert these two carotenoids, with antheraxanthin as the intermediate, in response to changing conditions. Zeaxanthin formation uses ascorbate as a cofactor, and violaxanthin formation requires NADPH. (After Pfündel and Bilger 1994.)

Photosystem I Is Protected from Active Oxygen Species

Photosystem I is vulnerable to damage from active oxygen species. The ferredoxin acceptor of photosystem I is a very strongly reducing species that can easily reduce molecular oxygen to form superoxide (O_2^-). This reduction competes with the normal channeling of electrons to the reduction of carbon dioxide and other processes. Superoxide is one of a series of active oxygen species that are potentially very damaging to biological membranes. Superoxide formed in this way can be detoxified by the action of a series of enzymes, including superoxide dismutase and ascorbate peroxidase (Asada 1994, 1996; Polle 1996).

Genetics, Assembly, and Evolution of Photosynthetic Systems

The Entire Chloroplast Genome Has Been Sequenced

The complete chloroplast genomes of several organisms have been sequenced (Ohyama et al. 1988). Chloroplast DNA is circular, with a size that ranges from 120 to 160 kilobases. A remarkable feature of chloroplast DNA is a duplicated region known as an **inverted repeat**, which is a repeated region of DNA with opposite orientation. The inverted repeat is flanked by a large and a small single-copy region. The chloroplast genome contains coding sequences for approximately 120 proteins. Many of these DNA sequences code for proteins that are yet to be characterized. It is uncertain whether all these genes are transcribed into mRNA and translated into protein, but it seems likely that protein constituents of the chloroplast remain to be identified.

Chloroplast Genes Exhibit Non-Mendelian Patterns of Inheritance

Chloroplasts and mitochondria reproduce by division rather than by **de novo synthesis**. This mode of reproduction is not surprising, since these organelles contain genetic information that is not present in the nucleus. During cell division, chloroplasts are divided between the two daughter cells. In most sexual plants, however, only the maternal plant contributes chloroplasts to the zygote. In these plants the normal Mendelian pattern of inheritance does not apply to chloroplast-encoded genes, because the offspring receive chloroplasts from only one parent. The result is **non-Mendelian**, or **maternal, inheritance**. Numerous traits are inherited in this way; one example is the herbicide resistance trait discussed in Box 7.3.

Many Chloroplast Proteins Are Imported from the Cytoplasm

Proteins found in chloroplasts are encoded by either chloroplast DNA or nuclear DNA. The chloroplast-encoded proteins are synthesized on chloroplast ribo-

BOX 7.3

Some Herbicides Kill Plants by Blocking Photosynthetic Electron Flow

HERBICIDES OF ONE MAJOR CLASS—about half of the commercially important compounds—act by interrupting photosynthetic electron flow (Ashton and Crafts 1981). Part A of the figure shows the chemical structure of two of these compounds. The precise sites of action of many of these agents have been found to lie either at the reducing side of photosystem I (for example, in paraquat) or in the quinone acceptor complex in the electron transport chain between the two photosystems (for example, in diuron) (part B of the figure).

Paraquat acts by intercepting electrons between the bound ferredoxin acceptors and NADP and then reducing oxygen to **superoxide** (O_2^-). Superoxide is a free radical that reacts nonspecifically with a wide range of molecules in the chloroplast, leading to the rapid loss of chloroplast activity. Lipid molecules in cell membranes are especially sensitive.

The herbicides that act on the quinone acceptor complex compete with plastoquinone for the Q_B binding site. If herbicide is present, it displaces the oxidized form of plastoquinone and occupies the specific binding site for the quinone acceptor, which is thought to lie on the D1 herbicide-binding protein. The herbicide is not able to accept electrons, so the electron is unable to leave Q_A, the first quinone acceptor. Thus, the binding of herbicide effectively blocks electron flow and inhibits photosynthesis. Many herbicides that act in this manner also inhibit electron flow in photosynthetic bacteria that have quinone-type electron acceptor complexes (Trebst 1986).

In recent years, herbicide-resistant biotypes of common weeds have appeared in areas where a single type of herbicide has been used continuously for several years. These biotypes can be orders of magnitude more resistant to certain classes of herbicides than the nonresistant plant is. In several cases the resistance factor has been traced to a single amino acid substitution in the D1 protein. This change, presumably in the quinone (and herbicide) binding region of the peptide, lowers the binding affinity of the herbicide, making it much less effective.

The possibility of fine-tuning the herbicide sensitivity of crop plants by making subtle changes in the proteins of the PS-II reaction center has created a great deal of interest in the agricultural chemical industry. Through biotechnology, it is now possible to make a crop plant that is resistant to a particular herbicide, which can then be applied to control undesirable plants that are not resistant. The success of this approach will depend on whether undesirable side effects of the herbicide-resistant mutations can be controlled and on how rapidly the weeds acquire resistance through natural selection or gene transfer. Thus, it is a race between science and evolutionary change (Dover and Croft 1986).

A complementary approach is to insert resistance factors to certain pests, such as insects, into the plant via genetic engineering. The plants then produce the toxic species themselves, eliminating the need for insecticides.

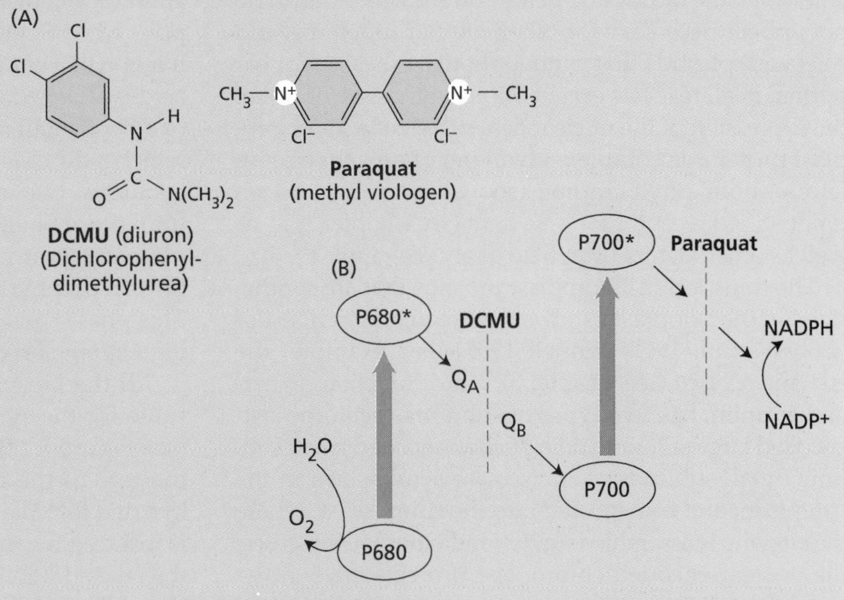

The use of herbicides to kill unwanted plants is widespread in modern agriculture. Many different classes of herbicides have been developed, and they act by blocking amino acid, carotenoid, or lipid biosynthesis or by disrupting cell division. Understanding the mode of action of herbicides has been an important tool in research on plant metabolism and has facilitated their application under different agricultural practices (Ashton and Crafts 1981). (A) Chemical structure of two herbicides that block photosynthetic electron flow. DCMU is also known as diuron. Paraquat has acquired public notoriety because of its use on marijuana crops. (B) Sites of action of herbicides. Many herbicides, such as DCMU, act by blocking electron flow at the quinone acceptors of photosystem II, by competing for the binding site of plastoquinone that is normally occupied by Q_B. Other herbicides, such as paraquat, act by accepting electrons from the early acceptors of photosystem I and then reacting with oxygen to form superoxide, O_2^-, a species that is very damaging to chloroplast components, especially lipids.

somes; the nuclear-encoded proteins are synthesized on cytoplasmic ribosomes and then transported into the chloroplast. Many nuclear genes contain introns—that is, base sequences that do not code for protein. The mRNA is processed to remove the introns, and the proteins are then synthesized in the cytoplasm. The genes needed for chloroplast function are distributed in the nucleus and in the chloroplast genome with no evident pattern, but both sets are essential for the viability of the chloroplast. Chloroplast genes, on the other hand, do not appear necessary for other cellular functions, and cells with plastids that cannot photosynthesize appear normal in all respects except photosynthesis. Control of the expression of the nuclear genes that code for chloroplast proteins is complex, involving photoactive regulators—both phytochrome (see Chapter 17) and an unidentified receptor for blue light (see Chapter 18)—as well as other factors (Mayfield et al. 1995; Link 1996).

The transport of chloroplast proteins that are synthesized in the cytoplasm shows well-regulated properties (Cohen et al. 1995; Schnell 1995). For example, the enzyme rubisco (see Chapter 8), which functions in carbon fixation, has two types of subunits, a chloroplast-encoded large subunit and a nuclear-encoded small subunit. Small subunits of rubisco are synthesized in the cytoplasm and transported into the chloroplast, where the enzyme is assembled. In this and other known cases, the nuclear-encoded chloroplast proteins are synthesized as precursor proteins containing an N-terminal amino acid sequence known as a **transit peptide**. This terminal sequence directs the precursor protein to the chloroplast, facilitates its passage through both the outer and the inner envelope membranes, and is then clipped off. Plastocyanin is a water-soluble protein that is encoded in the nucleus but functions in the lumen of the chloroplast. It therefore must cross three membranes and yet end up as a hydrophilic protein. In the case of plastocyanin, the transit peptide is very large and is processed in more than one step.

The Biosynthesis and Breakdown of Chlorophyll Are Complex Pathways

Chlorophylls are complex molecules that are exquisitely well suited to the light absorption, energy transfer, and electron transfer functions they carry out in photosynthesis (see Figure 7.5). Like all other biomolecules, chlorophylls are made by a biosynthetic pathway in which simple molecules are used as building blocks to assemble the more complex structures (von Wettstein et al. 1995; Porra 1997). Each of the steps in the biosynthetic pathway is enzymatically catalyzed.

The chlorophyll biosynthetic pathway consists of more than a dozen steps. The process can be divided into several phases, each of which can be considered separately, but which in the cell are highly coordinated and regulated. This regulation is essential because free chlorophyll and many of the biosynthetic intermediates are damaging to cellular components. The damage results largely because the molecules are highly colored and therefore absorb light efficiently but do not have efficient pathways for disposing of the energy, with the result that toxic singlet oxygen is formed.

In the first phase of chlorophyll biosynthesis, the amino acid glutamic acid is converted to 5-aminolevulinic acid (ALA) (Figure 7.35). This reaction is unusual in that it involves a covalent intermediate in which the glutamic acid is attached to a transfer RNA molecule. This is one of a very small number of examples in biochemistry in which a tRNA is utilized in a process other than protein synthesis. Two molecules of ALA are then condensed to form porphobilinogen (PBG), which ultimately form the pyrrole rings in chlorophyll. The next phase is the assembly of a porphyrin structure from four molecules of PBG. This phase consists of six distinct enzymatic steps, ending with the product protoporphyrin IX.

All the biosynthesis steps up to this point are the same for the synthesis of both chlorophyll and heme (see Figure 7.27). But here the pathway branches, and the fate of the molecule depends on which metal is inserted into the center of the porphyrin. If magnesium is inserted by an enzyme called magnesium chelatase, then the additional steps needed to convert the molecule into chlorophyll take place; if iron is inserted, the species ultimately becomes heme.

The next phase of the chlorophyll biosynthetic pathway is the formation of the fifth ring (ring E) by cyclization of one of the propionic acid side chains to form protochlorophyllide. The pathway involves the reduction of one of the double bonds in ring D, using NADPH. This process is driven by light in angiosperms and is carried out by an enzyme called protochlorophyllide oxidoreductase (POR). Non-oxygen-evolving photosynthetic bacteria carry out this reaction without light, using a completely different set of enzymes. Cyanobacteria, algae, lower plants, and gymnosperms contain both the light-dependent POR pathway and the light-independent pathway. Seedlings of angiosperms grown in complete darkness lack chlorophyll, because the POR enzyme requires light. These *etiolated* plants very rapidly turn green when exposed to light. The final step in the chlorophyll biosynthetic pathway is the attachment of the phytol tail, which is catalyzed by an enzyme called chlorophyll synthetase.

The elucidation of the biosynthetic pathways of chlorophylls and related pigments is a difficult task, in part because many of the enzymes are present in low abundance. Recently, genetic analysis has been used to clarify many aspects of these processes (Suzuki et al. in press).

The breakdown pathway of chlorophyll in senescent leaves is quite different from the biosynthetic pathway (Matile et al. 1996). The first step is removal of the phytol tail by an enzyme known as chlorophyllase, followed by removal of the Mg by magnesium dechelatase. Next, the porphyrin structure is opened by an oxygen-dependent oxygenase enzyme to form an open-chain tetrapyrrole. The tetrapyrrole is further modified to form water-

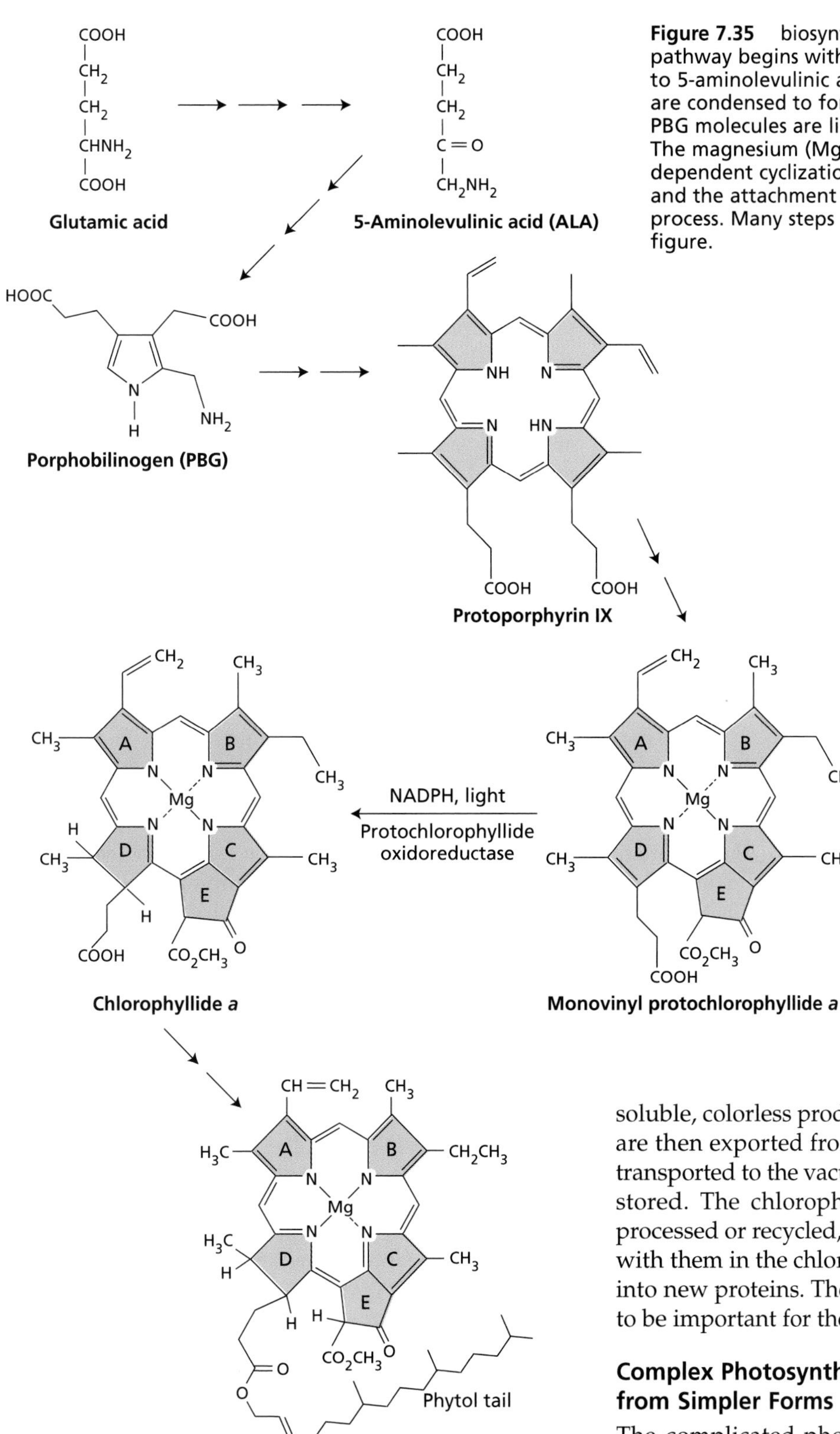

Figure 7.35 biosynthetic pathway of chlorophyll. The pathway begins with glutamic acid, which is converted to 5-aminolevulinic acid (ALA). Two molecules of ALA are condensed to form porphobilinogen (PBG). Four PBG molecules are linked to form protoporphyrin IX. The magnesium (Mg) is then inserted, and the light-dependent cyclization of ring E, the reduction of ring D, and the attachment of the phytol tail complete the process. Many steps in the process are omitted in this figure.

soluble, colorless products. These colorless metabolites are then exported from the senescent chloroplast and transported to the vacuole, where they are permanently stored. The chlorophyll metabolites are not further processed or recycled, although the proteins associated with them in the chloroplast are subsequently recycled into new proteins. The recycling of proteins is thought to be important for the nitrogen economy of the plant.

Complex Photosynthetic Organisms Have Evolved from Simpler Forms

The complicated photosynthetic apparatus found in plants and algae is the end product of a long evolutionary sequence. Much can be learned about this evolutionary process from analysis of simpler prokaryotic photosynthetic organisms, including the anoxygenic photosynthetic bacteria and the cyanobacteria.

The chloroplast is a semiautonomous cell organelle, with its own DNA and a complete protein synthesis apparatus. Many of the proteins that make up the photosynthetic apparatus and all the chlorophyll and lipids are synthesized in the chloroplast. Other proteins are imported from the cytoplasm and are encoded by nuclear genes. How did this curious division of labor come about? Most experts now agree that the chloroplast is the descendant of a symbiotic relationship between a cyanobacterium and a simple nonphotosynthetic eukaryotic cell. This process is called **endosymbiosis** (Gray 1992; Margulis 1993; Whatley 1993).

Originally, the cyanobacterium was capable of independent life, but over time much of its genetic information needed for normal cellular functions was lost and a substantial amount of information needed to synthesize the photosynthetic apparatus was transferred to the nucleus. So the chloroplast was no longer capable of life outside its host and eventually became an integral part of the cell. In some types of algae, chloroplasts are thought to have arisen by endosymbiosis of eukaryotic photosynthetic organisms (Palmer and Delwiche 1996). In these organisms, the chloroplast is surrounded by three and in some cases four membranes, which are thought to be remnants of the cell walls of the earlier organisms. Mitochondria are also thought to have originated by endosymbiosis in a separate event much earlier than chloroplast formation.

The answers to other questions related to the evolution of photosynthesis are less clear. These include the nature of the earliest photosynthetic systems, how the two photosystems became linked, and the evolutionary origin of the oxygen evolution complex (Olson and Pierson 1987; Blankenship 1992; Blankenship and Hartman in press).

Summary

Photosynthesis is the storage of solar energy carried out by plants, algae, and photosynthetic bacteria. Absorbed photons excite chlorophyll molecules, and these excited chlorophylls can dispose of this energy as heat, fluorescence, energy transfer, or photochemistry. Light is absorbed mainly in the antenna complexes, which comprise chlorophylls, accessory pigments, and proteins and are located at the thylakoid membranes of the chloroplast.

Photosynthetic antenna pigments transfer the energy to a specialized chlorophyll–protein complex known as a reaction center. The reaction center contains multiple-subunit protein complexes and hundreds or, in some organisms, thousands of chlorophylls. The antenna complexes and the reaction centers are integral components of the thylakoid membrane. The reaction center initiates a complex series of chemical reactions that capture energy in the form of chemical bonds. The relationship between the amount of absorbed quanta and the yield of a photochemical product made in a light-dependent reaction is given by the quantum yield. The quantum yield of the early steps of photosynthesis is 0.95, indicating that nearly every photon that is absorbed yields a charge separation at the reaction center.

Plants and some photosynthetic prokaryotes have two reaction centers, photosystem I and photosystem II, that function in series. The two photosystems are spatially separated: Photosystem I is found exclusively in the nonstacked stroma membranes, photosystem II largely in the stacked grana membranes. The reaction center chlorophylls of photosystem I absorb maximally at 700 nm, those from photosystem II at 680 nm. Photosystems II and I carry out noncyclic electron transport, oxidize water to molecular oxygen, and reduce $NADP^+$ to NADPH. It is energetically very difficult to oxidize water to form molecular oxygen and the photosynthetic oxygen-evolving system is the only known biochemical system that can oxidize water, thus providing almost all the oxygen in Earth's atmosphere. The photooxidation of water is modeled by the five-step S state mechanism. Manganese is an essential cofactor in the water-oxidizing process, and the five S states appear to represent successive oxidized states of a manganese-containing enzyme.

A tyrosine residue of the D1 protein of the PS-II reaction center functions as an electron carrier between the oxygen-evolving complex and P680. Pheophytin and two plastoquinones are electron carriers between P680 and the large cytochrome $b_6 f$ complex. Plastocyanin is the electron carrier between cytochrome $b_6 f$ and P700. The electron carriers that accept electrons from P700 are very strong reducing agents, and they include a quinone and three membrane-bound iron–sulfur proteins known as bound ferredoxins. The electron flow ends with the reduction of $NADP^+$ to NADPH by a membrane-bound, ferrodoxin–NADP reductase.

A portion of the energy of photons is also initially stored as chemical potential energy, largely in the form of a pH difference across the thylakoid membrane. This energy is quickly converted into chemical energy during ATP formation by action of an enzyme complex known as the ATP synthase. The photophosphorylation of ADP by the ATP synthase is driven by a chemiosmotic mechanism. Photosynthetic electron flow is coupled to proton translocation across the thylakoid membrane, and the stroma becomes more alkaline and the lumen more acidic. This proton gradient drives ATP synthesis with a stoichiometry of four H^+ per ATP. NADPH and ATP formed by the light reactions provide the energy for carbon reduction.

Excess light energy can damage photosynthetic systems, and several mechanisms minimize such damage. Carotenoids work as photoprotective agents by rapidly quenching the excited state of chlorophyll. Changes in the phosphorylated state of antenna pigment proteins can change the energy distribution between photosystems I and II when there is an imbalance between the energy absorbed by each photosystem. The xanthophyll cycle also contributes to the dissipation of excess energy by photochemical quenching.

Chloroplasts contain DNA and encode and synthesize some of the proteins that are essential for photosynthesis. Additional proteins are encoded by nuclear DNA, synthesized in the cytosol, and imported into the chloroplast. Chlorophylls are synthesized in a biosynthetic pathway involving more than a dozen steps, each of which is very carefully regulated. Once synthesized, proteins and pigments are assembled into the thylakoid membrane.

General Reading

Amesz, J., and Hoff, A. J., eds. (1996) *Biophysical Techniques in Photosynthesis.* Kluwer, Dordrecht, Netherlands.

*Avers, C. G. (1985) *Molecular Cell Biology.* Addison-Wesley, Reading, MA.

Baker, N. R., ed. (1996) *Photosynthesis and the Environment*. Kluwer, Dordrecht, Netherlands.

Barber, J., ed. (1987) *The Light Reactions* (Topics in Photosynthesis, vol. 8). Elsevier, Amsterdam.

Barber, J., ed. (1992) *The Photosystems: Structure, Function and Molecular Biology* (Topics in Photosynthesis, vol. 2). Elsevier, Amsterdam.

*Becker, W. M. (1986) *The World of the Cell.* Benjamin/Cummings, Menlo Park, CA.

Blankenship, R. E., Madigan, M. T., and Bauer, C. E., eds. (1995) *Anoxygenic Photosynthetic Bacteria* (Advances in Photosynthesis, vol. 2). Kluwer, Dordrecht, Netherlands.

Bryant, D. A., ed. (1994) *The Molecular Biology of Cyanobacteria* (Advances in Photosynthesis, vol. 1). Kluwer, Dordrecht, Netherlands.

*Clayton, R. K. (1971) *Light and Living Matter: A Guide to the Study of Photobiology.* McGraw-Hill, New York.

*Clayton, R. K. (1980) *Photosynthesis: Physical Mechanisms and Chemical Patterns.* Cambridge University Press, Cambridge.

Cramer, W. A., and Knaff, D. B. (1990) *Energy Transduction in Biological Membranes: A Textbook of Bioenergetics.* Springer, New York.

Hall, D. O., and Rao, K. K. (1987) *Photosynthesis*, 4th ed. Edward Arnold, London.

Nicholls, D. G., and Ferguson, S. J. (1992) *Bioenergetics*, 2nd ed. Academic Press, San Diego.

Ort, D. R., and Yocum, C. F., eds. (1996) *Oxygenic Photosynthesis: The Light Reactions* (Advances in Photosynthesis, vol. 4). Kluwer, Dordrecht, Netherlands.

*Rabinowitch, E., and Govindjee. (1969) *Photosynthesis.* Wiley, New York.

Staehelin, L. A., and Arntzen, C. J., eds. (1986) *Photosynthesis.* III. *Photosynthetic Membranes and Light Harvesting Systems* (*Encyclopedia of Plant Physiology*, new series, vol. 19). Springer, Berlin.

Stevens, E., Bryant, D., and Shannon, J., eds. (1988) *Light-Energy Transduction in Photosynthesis: Higher Plants and Bacterial Models.* American Society of Plant Physiologists, Rockville, Md.

Walker, D. (1992) *Energy, Plants and Man*, 2nd ed. Oxygraphics, Brighton, East Sussex, England.

* Indicates a reference that is general reading in the field and is also cited in this chapter.

Chapter References

Abrahams, J. P., Leslie, A. G. W., Lutter, R., and Walker, J. E. (1994) Structure at 2.8 Å resolution of F_1-ATPase from bovine heart mitochondria. *Nature* 370: 621–628.

Allen, J. F. (1992) How does protein phosphorylation regulate photosynthesis? *Trends Biochem. Sci.* 17: 12–17.

Anderson, J. M., and Andersson, B. (1988) The dynamic photosynthetic membrane and regulation of solar energy conversion. *Trends Biochem. Sci.* 13: 351–355.

Arnon, D. I. (1984) The discovery of photosynthetic phosphorylation. *Trends Biochem. Sci.* 9: 258–262.

Asada, K. (1994) Mechanisms for scavenging reactive molecules generated in chloroplasts under light stress. In *Photoinhibition of Photosynthesis*, N.R. Baker and J. R. Bower, eds. Bios Scientific Publishers, Oxford, pp. 131–145.

Asada, K. (1996) Radical production and scavenging in the chloroplasts. In *Photosynthesis and the Environment*, N. R. Baker, ed., Kluwer, Dordrecht, Netherlands, pp. 123–150.

Ashton, F. M., and Crafts, A. S. (1981) *Mode of Action of Herbicides*, 2nd ed. Wiley, New York.

Barber, J., and Andersson, B. (1992) Too much of a good thing: Light can be bad for photosynthesis. *Trends Biochem. Sci.* 17: 61–66.

Barry, B. (1993) The role of redox-active amino acids in the photosynthetic water oxidizing complex. *Photochem. Photobiol.* 57: 179–188.

Bennett, J. (1991) Protein phosphorylation in green plant chloroplasts. *Annu. Rev. Plant Physiol. Plant Mol. Biol.* 42: 281–311.

Blankenship, R. E. (1992) Origin and early evolution of photosynthesis. *Photosynth. Res.* 33: 91–111.

Blankenship, R. E., and Hartman, H. (In press) The origin and evolution of oxygenic photosynthesis. *Trends Biochem. Sci.*

Blankenship, R. E., and Prince, R. C. (1985) Excited-state redox potentials and the Z scheme of photosynthesis. *Trends Biochem. Sci.* 10: 382–383.

Boyer, P. D. (1993) The binding change mechanism for ATP synthase—Some probabilities and possibilities. *Biochim. Biophys. Acta* 1140: 215–250.

Boyer, P. D. (1997) The ATP synthase—A splendid molecular machine. *Annu. Rev. Biochem.* 66: 717–749.

Buchanan, B. B. (1991) Regulation of CO_2 assimilation in oxygenic photosynthesis: The ferredoxin/thioredoxin system. *Arch. Biochem. Biophys.* 288: 1–9.

Calvin, M. (1976) Photosynthesis as a resource for energy and materials. *Photochem. Photobiol.* 23: 425–444.

Chitnis, P. R. (1996) Photosystem I. *Plant Physiol.* 111: 661–669.

Cohen, Y., Yalovsky, S., and Nechushtai, R. (1995) Integration and assembly of photosynthetic protein complexes in chloroplast thylakoid membranes. *Biochim. Biophys. Acta* 1241: 1–30.

Cramer, W. A., Soriano, G. M., Ponomarev, M., Huang, D., Zhang, H., Martinez, S. E., and Smith, J. L. (1996) Some new structural aspects and old controversies concerning the cytochrome b_6f complex of oxygenic photosynthesis. *Annu. Rev. Plant Physiol. Plant Mol. Biol.* 47: 477–508.

Debus, R. J. (1992) The manganese and calcium ions of photosynthetic oxygen evolution. *Biochim. Biophys. Acta* 1102: 269–352.

Deisenhofer, J., and Michel, H. (1989) The photosynthetic reaction center from the purple bacterium *Rhodopseudomonas viridis. Science* 245: 1463–1473.

Demmig-Adams, B., and Adams, W. W., III. (1992) Photoprotection and other responses of plants to high light stress. *Annu. Rev. Plant Physiol. Plant Mol. Biol.* 43: 599–626.

Dover, M. J., and Croft, B. A. (1986) Pesticide resistance and public policy. *BioScience* 36: 78–85.

Dutton, P. L. (1978) Redox potentiometry: Determination of midpoint potentials of oxidation-reduction components of biological electron-transfer systems. *Methods Enzymol.* 54: 411–435.

Duysens, L. N. M., Amesz, J., and Kamp, B. M. (1961) Two photochemical systems in photosynthesis. *Nature* 190: 510–511.

Emerson, R., and Arnold, W. (1932) The photochemical reaction in photosynthesis. *J. Gen. Physiol.* 16: 191–205.

Emerson, R., Chalmers, R., and Cederstrand, C. (1957) Some factors influencing the long-wave limit of photosynthesis. *Proc. Natl. Acad. Sci. USA* 43: 133–143.

Frank, H. A., and Cogdell, R. J. (1996) Carotenoids in photosynthesis. *Photochem. Photobiol.* 63: 257–264.

Ghanotakis, D., and Yocum, C. F. (1990) Photosystem II and the oxygen-evolving complex. *Annu. Rev. Plant Physiol. Plant Mol. Biol.* 41: 255–276.

Golbeck, J. H. (1992) Structure and function of photosystem I. *Annu. Rev. Plant Physiol. Plant Mol. Biol.* 43: 293–324.

Govindjee, and Coleman, W. J. (1990) How plants make oxygen. *Sci. Am.* 262 (2): 50–58.

Gray, M. W. (1992) The endosymbiont hypothesis revisited. *Int. Rev. Cytol.* 141: 233–357.

Green, B. R., and Durnford, D. G. (1996) The chlorophyll-carotenoid proteins of oxygenic photosynthesis. *Annu. Rev. Plant Physiol. Plant Mol. Biol.* 47: 685–714.

Grossman, A. R., Bhaya, D., Apt, K. E., and Kehoe, D. M. (1995) Light-harvesting complexes in oxygenic photosynthesis: Diversity, control, and evolution. *Annu. Rev. Genet.* 29: 231–288.

Hangarter, R. P., and Good, N. E. (1988) Active transport ion movements and pH changes. II. Changes of pH and ATP synthesis. *Photosynth. Res.* 19: 237–250.

Hankamer, B., Barber, J., and Boekema, E. J. (1997) Structure and membrane organization of photosystem II in green plants. *Annu. Rev. Plant Physiol. Plant Mol. Biol.* 48: 641–671.

Hill, R. (1939) Oxygen produced by isolated chloroplasts. *Proc. R. Soc. London Ser. B* 127: 192–210.

Hill, R., and Bendall, F. (1960) Function of the two cytochrome components in chloroplasts: A working hypothesis. *Nature* 186: 136–137.

Horton, P., Ruban, A. V., and Walters, R. G. (1996) Regulation of light harvesting in green plants. *Annu. Rev. Plant Physiol. Plant Mol. Biol.* 47: 655–684.

Jagendorf, A. T. (1967) Acid-based transitions and phosphorylation by chloroplasts. *Fed. Proc. Am. Soc. Exp. Biol.* 26: 1361–1369.

Junge, W., Lill, H., and Engelbrecht, S. (1997) ATP synthase: An electrochemical transducer with rotatory mechanics. *Trends Biochem. Sci.* 22: 420–423.

Karplus, P. A., Daniels, M. J., and Herriott, J. R. (1991) Atomic structure of ferredoxin-$NADP^+$ reductase: Prototype for a structurally novel flavoenzyme family. *Science* 251: 60–66.

Ke, B. (1973) The primary electron acceptor of photosystem I. *Biochim. Biophys. Acta* 301: 1–33.

Knaff, D. B. (1993) The cytochrome b_c1 complexes of photosynthetic purple bacteria. *Photosynth. Res.* 35: 117–133.

Kok, B., Forbush, B., and McGloin, M. (1970) Cooperation of charges in photosynthetic O_2 evolution. I. A linear four step mechanism. *Photochem. Photobiol.* 11: 457–475.

Krause, G. H., and Weis, E. (1991) Chlorophyll fluorescence and photosynthesis: The basics. *Annu. Rev. Plant Physiol. Plant Mol. Biol.* 42: 313–349.

Krauss, N., Hinrichs, W., Witt, I., Fromme, P., Pritzkow, W., Dauter, Z., Betzel, C., Wilson, K. S., Witt, H. T., and Saenger, W. (1993) Three-dimensional structure of system I of photosynthesis at 6 Å resolution. *Nature* 361: 326–331.

Krauss, N., Schubert, W-D., Klukas, O., Fromme, P., Witt, H. T., and Saenger, W. (1996) Photosystem I at 4 Å resolution represents the first structural model of a joint photosynthetic reaction centre and core antenna system. *Nature Struct. Biol.* 3: 965–973.

Kühlbrandt, W., Wang, D. N., and Fujiyoshi, Y. (1994) Atomic model of plant light-harvesting complex by electron crystallography. *Nature* 367: 614–621.

Link, G. (1996) Green life: Control of chloroplast gene transcription. *Bioessays* 18: 465–471.

Long, S. P., Humphries, S., and Falkowski, P. G. (1994) Photoinhibition of photosynthesis in nature. *Annu. Rev. Plant Physiol. Plant Mol. Biol.* 45: 633–662.

Lowe, A. G., and Jones, M. N. (1984) Proton motive force—What price Δp? *Trends Biochem. Sci.* 9: 11–12.

Margulis, L. (1993) *Symbiosis in Cell Evolution*, 2nd ed. W. H. Freeman, San Francisco.

Matile, P., Hörtensteiner, S., Thomas, H., and Kräutler, B. (1996) Chlorophyll breakdown in senescent leaves. *Plant Physiol.* 112: 1403–1409.

Mayfield, S. P., Yohn, C. B., Cohen, A., and Danon, A. (1995) Regulation of chloroplast gene expression. *Annu. Rev. Plant Physiol. Plant Mol. Biol.* 46: 147–166.

Michel, H., and Deisenhofer, J. (1988) Relevance of the photosynthetic reaction center from purple bacteria to the structure of photosystem II. *Biochemistry* 27: 1–7.

Mitchell, P. (1979) Keilin's respiratory chain concept and its chemiosmotic consequences. *Science* 206: 1148–1159.

Noji, H., Yasuda, R., Yoshida, M., and Kinosita, K., Jr. (1997) Direct observation of the rotation of the F_1-ATPase. *Nature* 386: 299–302.

Nugent, J. H. A. (1996) Oxygenic photosynthesis: Electron transfer in photosystem I and photosystem II. *Eur. J. Biochem.* 237: 519–531.

Ohyama, K., Kohchi, T., Sano, T., and Yamada, Y. (1988) Newly identified groups of genes in chloroplasts. *Trends Biochem. Sci.* 13: 19–22.

Okamura, M. Y., and Feher, G. (1992) Proton transfer in reaction centers from photosynthetic bacteria. *Annu. Rev. Biochem.* 61: 861–896.

Olson, J. M., and Pierson, B. K. (1987) Evolution of reaction centers in photosynthetic prokaryotes. *Int. Rev. Cytol.* 108: 209–248.

Palmer, J. D., and Delwiche, C. F. (1996) Second-hand chloroplasts and the case of the disappearing nucleus. *Proc. Natl. Acad. Sci. USA* 93: 7432–7435.

Paulsen, H. (1995) Chlorophyll *a/b*-binding proteins. *Photochem. Photobiol.* 62: 367–382.

Pfündel, E., and Bilger, W. (1994) Regulation and possible function of the violaxanthin cycle. *Photosynth. Res.* 42: 89–109.

Polle, A. (1996) Mehler reaction: Friend or foe in photosynthesis? *Bot. Acta* 109: 84–89.

Porra, R. J. (1997) Recent progress in porphyrin and chlorophyll biosynthesis. *Photochem. Photobiol.* 65: 492–516.

Pullerits, T., and Sundström, V. (1996) Photosynthetic light-harvesting pigment-protein complexes: Toward understanding how and why. *Acc. Chem. Res.* 29: 381–389.

Rögner, M., Boekema, E. J., and Barber, J. (1996) How does photosystem 2 split water? The structural basis of efficient energy conversion. *Trends Biochem. Sci.* 21: 44–49.

Rutherford, A. W. (1989) Photosystem II, the water-splitting enzyme. *Trends Biochem. Sci.* 14: 227–232.

Rutherford, A. W., Zimmerman, J-L., and Boussac, A. (1992) Oxygen evolution. In *The Photosystems: Structure, Function and Molecular Biology* (Topics in Photosynthesis, vol. 11), J. Barber, ed., Elsevier, Amsterdam, pp. 179–229.

Sauer, K. (1975) Primary events and the trapping of energy. In *Bioenergetics of Photosynthesis*, Govindjee, ed., Academic Press, New York, pp. 116–181.

Schnell, D. J. (1995) Shedding light on the chloroplast protein import machinery. *Cell* 83: 521–524.

Schubert, W. D., Klukas, O., Krauss, N., Saenger, W., Fromme, P., and Witt, H. T. (1997) Photosystem I of *Synechococcus elongatus* at 4 angstrom resolution: Comprehensive structure analysis. *J. Molec. Biol.* 272: 741–769.

Suzuki, J. Y., Bollivar, D. W., and Bauer, C. E. (In press) Genetic analysis of chlorophyll biosynthesis. *Annu. Rev. Genet.*

Trebst, A. (1986) The topology of the plastoquinone and herbicide binding peptides of photosystem II in the thylakoid membrane. *Z. Naturforsch. Teil C* 41: 240–245.

Trissl, H-W., and Wilhelm, C. (1993) Why do thylakoid membranes from higher plants form grana stacks? *Trends Biochem. Sci.* 18: 415–419.

Trumpower, B. L., and Gennis, R. B. (1994) Energy transduction by cytochrome complexes in mitochondrial and bacterial respiration: The enzymology of coupling electron transfer reactions to transmembrane proton translocation. *Annu. Rev. Biochem.* 63: 675–716.

van Grondelle, R., Dekker, J. P., Gillbro, T., and Sundström, V. (1994) Energy transfer and trapping in photosynthesis. *Biochim. Biophys. Acta* 1187: 1–65.

Vermaas, W. (1993) Molecular-biological approaches to analyze photosystem II structure and function. *Annu. Rev. Plant Physiol. Plant Mol. Biol.* 44: 457–482.

von Wettstein, D., Gough, S., and Kannagara, G. (1995) Chlorophyll biosynthesis. *Plant Cell* 7: 1039–1057.

Whatley, J. M. (1993) The endosymbiotic origin of chloroplasts. *Int. Rev. Cytol.* 144: 259–299.

Xia, D., Yu, C. A., Kim, H., Xian, J. Z., Kachurin, A. M., Zhang, L., Yu, L., and Deisenhofer, J. (1997) Crystal structure of the cytochrome bc_1 complex from bovine heart mitochondria. *Science* 277: 60–66.

Yachandra, V. K., Derose, V. J., Latimer, M. J. Mukerji, I., Sauer, K., and Klein, M. P. (1993) Where plants make oxygen: A structural model for the photosynthetic oxygen-evolving manganese cluster. *Science* 260: 675–679.

Yachandra, V. K., Sauer, K., and Klein, M. P. (1996) Manganese cluster in photosynthesis: Where plants oxidize water to dioxygen. *Chem. Rev.* 96: 2927–2950.

8 Photosynthesis: Carbon Reactions

IN CHAPTER 5 WE DISCUSSED plants' requirements for mineral nutrients and light in order to grow and complete their life cycle. Because living organisms interact with one another and their environment, mineral nutrients cycle through the biosphere. These cycles involve complex interactions, and each cycle is critical in its own right. Given that the amount of matter in the biosphere remains constant, energy must be supplied to keep the cycles operational. Otherwise, increasing entropy dictates that the flow of matter ultimately stops. Members of a diverse group of living organisms—autotrophs—have the ability to convert physical and chemical sources of energy from the environment into a biologically useful energy in the absence of organic substrates. Most of the external energy is consumed in transforming CO_2 to a reduced state that is compatible with the needs of the cell (—CHOH—). Recent estimates indicate that about 200 billion tons of CO_2 is converted to biomass each year. Some 40% of this mass comes from the activities of marine phytoplankton. The bulk of the carbon is incorporated into organic compounds by the carbon reduction reactions associated with photosynthesis.

In Chapter 7 we saw how the photochemical oxidation of water to molecular oxygen is coupled to the generation of ATP and reduced pyridine nucleotide (NADPH) by reactions taking place in the chloroplast thylakoid membrane. The reactions catalyzing the reduction of CO_2 to carbohydrate are coupled to the consumption of NADPH and ATP by enzymes found in the stroma, the soluble phase of chloroplasts. These reactions were long thought to be independent of light and, as a consequence, were referred to as the "dark reactions." In the past three decades, however, it has become clear that these reactions require light. The important role of light in photosynthetic carbon fixation will be discussed in this chapter.

Because of this requirement for light, the reactions effecting the conversion of CO_2 to carbohydrate are more suitably referred to as the "carbon reactions of photosynthesis" (Figure 8.1).

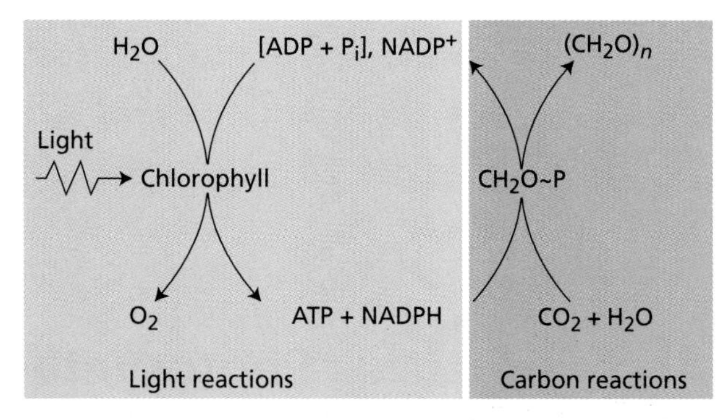

Figure 8.1 The light and carbon reactions of photosynthesis. Light is required for the generation of ATP and NADPH. The ATP and NADPH are consumed by the carbon reactions, which reduce CO_2 to carbohydrate ($[CH_2O]_n$).

The Calvin Cycle

All photosynthetic eukaryotes, from the most primitive alga to the most advanced angiosperm, reduce CO_2 to carbohydrate via the same basic mechanism, the photosynthetic carbon reduction cycle originally described for C_3 species (the **Calvin cycle**, or **reductive pentose phosphate [RPP] cycle**). Other metabolic pathways associated with the photosynthetic fixation of CO_2, such as the C_4 photosynthetic carbon assimilation cycle and the photorespiratory carbon oxidation cycle, are either auxiliary to or dependent on the basic Calvin cycle. As we will see, however, some anaerobic photosynthetic prokaryotes lack the Calvin cycle and assimilate CO_2 by other pathways.

In this section we will examine how CO_2 is fixed by the Calvin cycle using ATP and NADPH generated by the light reactions, and how the Calvin cycle is regulated.

The Calvin Cycle Includes Carboxylation, Reduction, and Regeneration

In the Calvin cycle, CO_2 and water from the environment are enzymatically combined with a five-carbon acceptor molecule to generate two molecules of a three-carbon intermediate. This intermediate (3-phosphoglycerate) is reduced to carbohydrate by use of the ATP and NADPH generated photochemically. The cycle is completed by regeneration of the five-carbon acceptor (ribulose-1,5-bisphosphate). Since the input of CO_2 is continuous, the synthesis of carbohydrates (starch, sucrose) provides the output for maintaining an adequate flow of carbon atoms in the cycle. The Calvin cycle proceeds in three stages (Figure 8.2):

1. *Carboxylation* of the CO_2 acceptor ribulose-1,5-bisphosphate, forming two molecules of 3-phosphoglycerate, the first stable intermediate of the Calvin cycle.
2. *Reduction* of 3-phosphoglycerate, forming gyceraldehyde-3-phosphate, a carbohydrate.
3. *Regeneration* of the CO_2 acceptor ribulose-1,5-bisphosphate from glyceraldehyde-3-phosphate.

We can track the reduction of the most oxidized forms of carbon in the Calvin cycle by following the oxidation number of the carbon atom being fixed. As the cycle proceeds, the oxidation state falls from +4 in the incoming CO_2 to +3 in the 3-phosphoglycerate intermediate, and subsequently to +1 in the glyceraldehyde-3-phosphate product. Overall, the early reactions of the Calvin cycle complete the reduction of atmospheric carbon and, in so doing, facilitate its incorporation into organic compounds.

The Carboxylation of Ribulose Bisphosphate Is Catalyzed by the Enzyme Rubisco

CO_2 enters the Calvin cycle by reacting with ribulose-1,5-bisphosphate to yield two molecules of 3-phosphoglycerate (Table 8.1, reaction 1), a reaction catalyzed by the chloroplast enzyme ribulose bisphosphate carboxylase/oxygenase, referred to by the acronym **rubisco**. As indicated by the full name, the enzyme also has an oxygenase activity in which O_2 competes with CO_2 for the

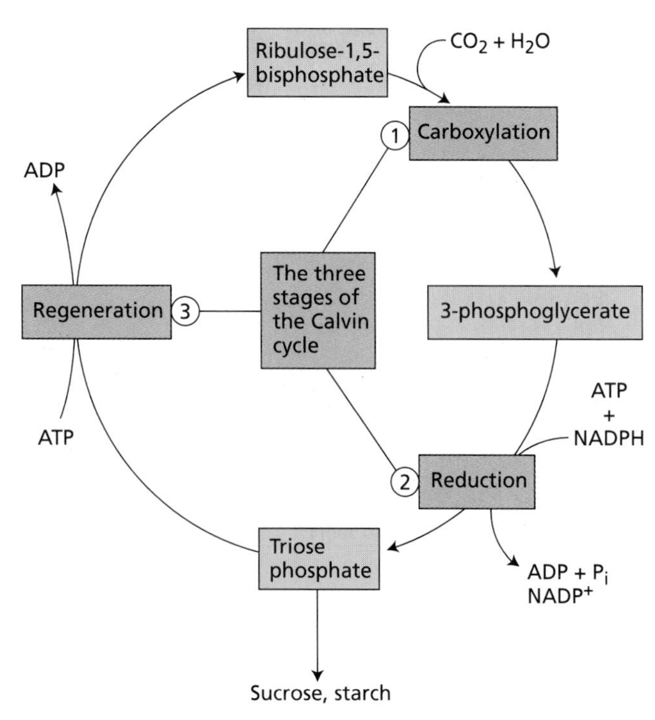

Figure 8.2 The Calvin cycle proceeds in three stages: (1) carboxylation, during which CO_2 is covalently linked to a carbon skeleton; (2) reduction, during which carbohydrate is formed at the expense of the photochemically derived ATP and reducing equivalents in the form of NADPH; and (3) regeneration, during which the CO_2 acceptor ribulose-1,5-bisphosphate re-forms.

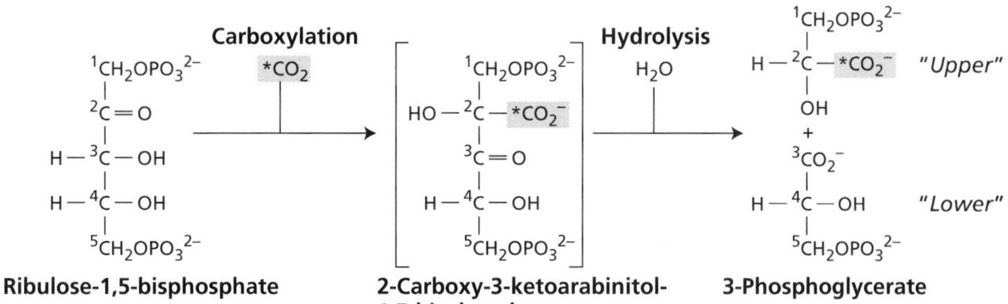

Carboxylation $*CO_2$ → Hydrolysis H_2O →

$^{1}CH_2OPO_3^{2-}$ / $^{2}C{=}O$ / $H{-}^{3}C{-}OH$ / $H{-}^{4}C{-}OH$ / $^{5}CH_2OPO_3^{2-}$

Ribulose-1,5-bisphosphate

$^{1}CH_2OPO_3^{2-}$ / $HO{-}^{2}C{-}*CO_2^-$ / $^{3}C{=}O$ / $H{-}^{4}C{-}OH$ / $^{5}CH_2OPO_3^{2-}$

2-Carboxy-3-ketoarabinitol-1,5-bisphosphate
(a transient, unstable, enzyme-bound intermediate)

$^{1}CH_2OPO_3^{2-}$ / $H{-}^{2}C{-}*CO_2^-$ / OH *"Upper"*
+
$^{3}CO_2^-$ / $H{-}^{4}C{-}OH$ / $^{5}CH_2OPO_3^{2-}$ *"Lower"*

3-Phosphoglycerate

Figure 8.3 The carboxylation of ribulose-1,5-bisphosphate, catalyzed by rubisco, proceeds in two stages: carboxylation and hydrolysis. In carboxylation, CO_2 is added to carbon 2 of ribulose-1,5–bisphosphate to form the unstable, enzyme-bound intermediate 2-carboxy-3-ketoarabinitol-1,5-bisphosphate, which is hydrolyzed to yield two molecules of the stable product 3-phosphoglycerate. The two molecules of 3-phosphoglycerate—"upper" and "lower"—are distinguished by the fact that the upper molecule contains the newly incorporated carbon dioxide, designated here as $*CO_2$.

common substrate ribulose-1,5-bisphosphate (Andrews and Lorimer 1987). As we will discuss later, this property limits net CO_2 fixation.

The carboxylase reaction brings about the incorporation of CO_2 into the carboxyl group of the "upper" molecule of the newly synthesized 3-phosphoglycerate (Figure 8.3). Two properties of the carboxylase reaction are especially important. First, the negative change in free energy (see Chapter 2) associated with the carboxylation of ribulose-1,5-bisphosphate is large (ΔG = –51 kJ [12.3 kcal] mol^{-1}); thus the forward reaction is strongly favored. Second, the affinity of rubisco for CO_2 is sufficiently high to ensure rapid carboxylation at the low concentrations of CO_2 found in photosynthetic cells. Rubisco is very abundant, representing 40% of the total soluble protein of most leaves. The concentration of rubisco active sites within the chloroplast stroma is calculated to be about 4 m*M*, or about 500 times greater than the concentration of its CO_2 substrate (see Box 8.1).

BOX 8.1

Carbon Dioxide: Some Important Physicochemical Properties

AN UNDERSTANDING of the mechanism of CO_2 fixation requires knowledge of the physical and chemical properties of CO_2, particularly those related to its interaction with water. The amount of any gas dissolved in water is proportional to its partial pressure (P_{gas}) above the solution (Henry's law) and its Bunsen absorption coefficient (α). The Bunsen absorption coefficient is the volume of gas absorbed by one volume of water at a pressure of 1 atmosphere and is temperature dependent, decreasing as the temperature rises. The solubility of a gas therefore decreases with increasing temperature. Thus, for a given temperature,

$$[\text{gas}]\,\mu M = P_{gas} \times \alpha \times \frac{10^6}{V_0}$$

where V_0 is the normal volume of an ideal gas at standard temperature and pressure (V_0 = 22.4 L mol^{-1}).

We can calculate the partial pressure of a gas by multiplying the mole fraction of the gas by the total pressure. The mole fraction of a gas is its partial volume divided by the total volume of all the gases present. Thus, the mole fractions of CO_2 and O_2 in air are 0.0345% and 20.95%, respectively. At sea level, atmospheric pressure is about 0.1 MPa, so the partial pressures of CO_2 and O_2 are 3.4×10^{-5} and 2.1×10^{-2} MPa, respectively. From these values and those of α, the corresponding solution concentrations of CO_2 and O_2 can be computed by the equation given above. The table presents values for these concentrations at different temperatures.

These values place considerable constraints on carboxylation. As a carboxylase, rubisco must be capable of operating efficiently even at the rather low concentrations of CO_2 available to it. Rubisco also functions as an oxygenase in photorespiration (as we'll see later in this chapter), so the solution concentration of O_2 is also important. Because of the different temperature dependences of the Bunsen absorption coefficients $\alpha(CO_2)$ and $\alpha(O_2)$, the concentrations of these two gases vary with temperature such that the ratio of $[CO_2]$ to $[O_2]$ decreases as the temperature increases. This effect is important biologically, because as the temperature increases, the ratio of carboxylation to oxygenation catalyzed by rubisco decreases and the ratio of photorespiration to photosynthesis thus increases.

Temperature (°C)	α (CO_2)	$[CO_2]$ (μM in solution)	α (O_2)	$[O_2]$ (μM in solution)	$[CO_2]/[O_2]$
5	1.424	21.93	0.0429	401.2	0.0515
15	1.019	15.69	0.0342	319.8	0.0462
25	0.759	11.68	0.0283	264.6	0.0416
35	0.592	9.11	0.0244	228.2	0.0376

Table 8.1
Reactions of the Calvin cycle

1. *Ribulose-1,5-bisphosphate carboxylase/oxygenase*
   Ribulose-1,5-bisphosphate + CO_2 + H_2O → 2(3-phosphoglycerate) + 2 H^+

2. *3-Phosphoglycerate kinase*
   3-Phosphoglycerate + ATP → 1,3 bisphosphoglycerate + ADP

3. *NADP:glyceraldehyde-3-phosphate dehydrogenase*
   1,3-Bisphosphoglycerate + NADPH + H^+ → glyceraldehyde-3-phosphate $NADP^+$ + $HOPO_3^{2-}$

4. *Triose phosphate isomerase*
   Glyceraldehyde-3-phosphate → dihydroxyacetone-3-phosphate

5. *Aldolase*
   Glyceraldehyde-3-phosphate + dihydroxyacetone-3-phosphate → fructose-1,6-bisphosphate

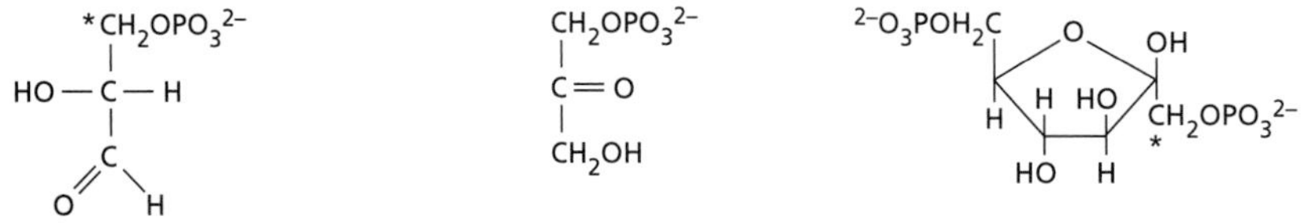

6. *Fructose-1,6-bisphosphate phosphatase*
   Fructose-1,6-bisphosphate + H_2O → fructose-1,6-phosphate + $HOPO_3^{2-}$

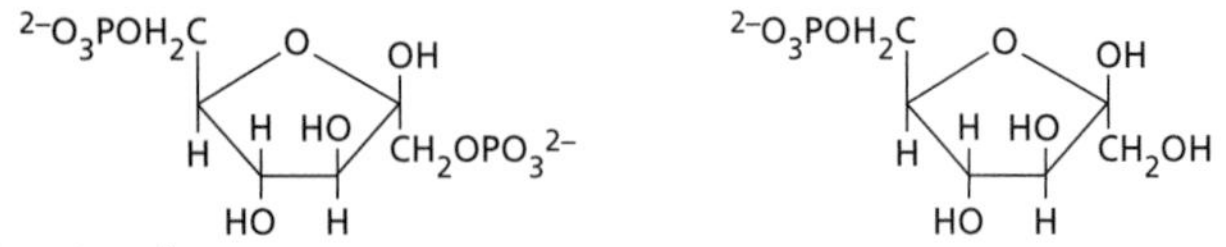

7. *Transketolase*
   Fructose-6-phosphate + glyceraldehye-3-phosphate → erythrose-4-phosphate + xylulose-5-phosphate

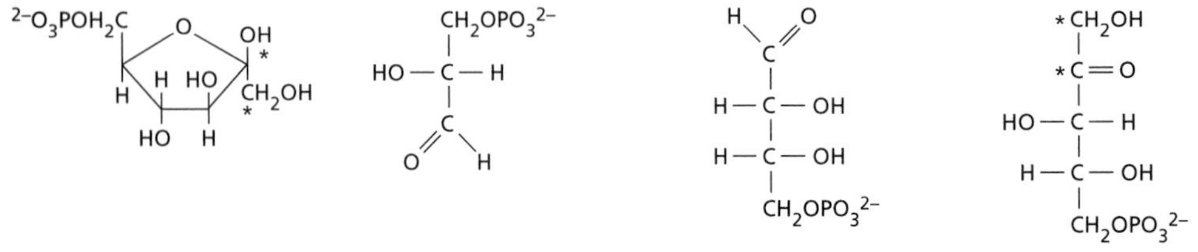

Table 8.1 (*continued*)
Reactions of the Calvin cycle

8. *Aldolase*
Erythrose-4-phosphate + dihydoxyacetone-3-phosphate → sedoheptulose-1,7-bisphosphate

H, O (aldehyde C)
H—C—OH
H—C—OH
* $CH_2OPO_3^{2-}$

CH_2OH
C=O
$CH_2OPO_3^{2-}$

$CH_2OPO_3^{2-}$, H, O, OH, H, H, HO, HO, $CH_2OPO_3^{2-}$ *, HO, H

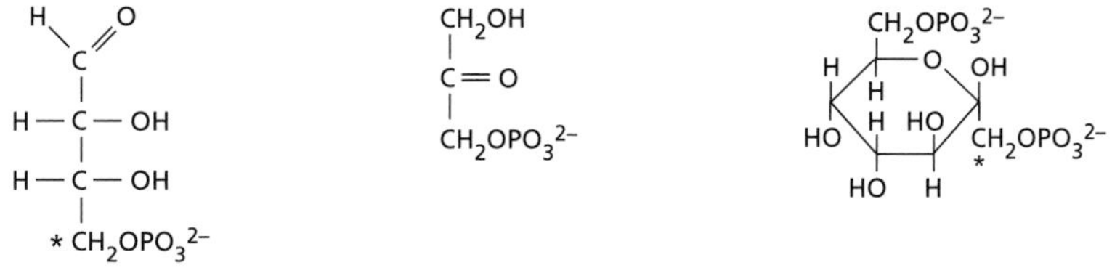

9. *Sedoheptulose-1,7,bisphosphate phosphatase*
Sedoheptulose-1,7-bisphosphate + H_2O → sedoheptulose-7-phosphate + $HOPO_3^{2-}$

$CH_2OPO_3^{2-}$, H, O, OH, H, H, HO, HO, $CH_2OPO_3^{2-}$, HO, H

$CH_2OPO_3^{2-}$, H, O, OH, H, H, HO, HO, CH_2OH, HO, H

10. *Transketolase*
Sedoheptulose-7-phosphate + glyceraldehyde-3-phosphate → ribose-5-phosphate + xylulose-5-phosphate

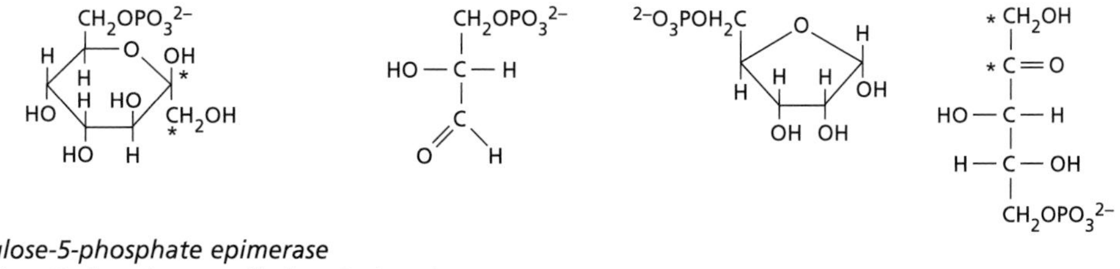

$CH_2OPO_3^{2-}$, H, O, OH, *, H, H, HO, HO, CH_2OH *, HO, H

$CH_2OPO_3^{2-}$
HO—C—H
C (O, H)

$^{2-}O_3POH_2C$, O, H, H, H, H, OH, OH OH

* CH_2OH
* C=O
HO—C—H
H—C—OH
$CH_2OPO_3^{2-}$

11. *Ribulose-5-phosphate epimerase*
Xylulose-5-phosphate → ribulose-5-phosphate

CH_2OH
C=O
HO—C—H
H—C—OH
$CH_2OPO_3^{2-}$

CH_2OH
C=O
H—C—OH
H—C—OH
$CH_2OPO_3^{2-}$

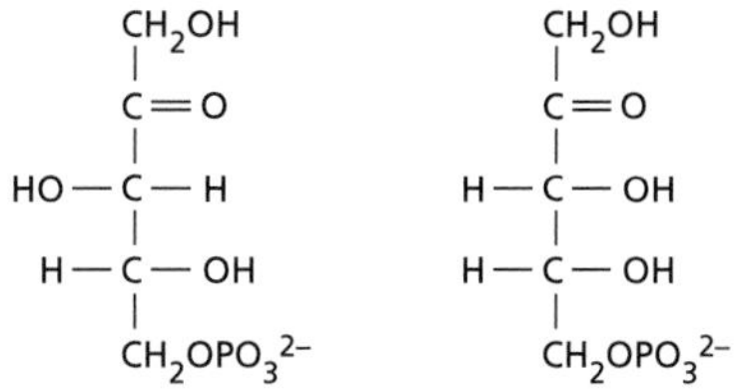

12. *Ribose-5-phosphate isomerase*
Ribose-5-phosphate → ribulose-5-phosphate

$^{2-}O_3POH_2C$, O, H, H, H, H, OH, OH OH

CH_2OH
C=O
H—C—OH
H—C—OH
$CH_2OPO_3^{2-}$

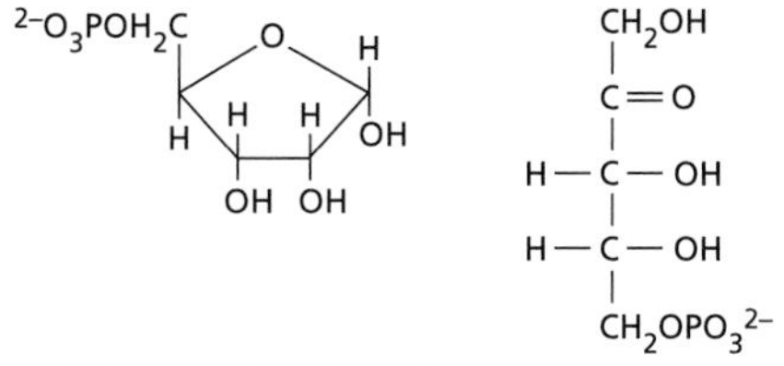

13. *Ribulose-5-phosphate kinase*
Ribulose-5-phosphate + ATP → ribulose-5-bisphosphate + ATP + H^+

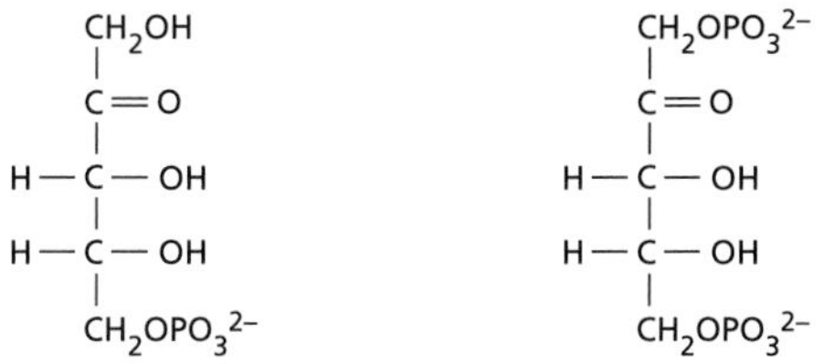

CH_2OH
C=O
H—C—OH
H—C—OH
$CH_2OPO_3^{2-}$

$CH_2OPO_3^{2-}$
C=O
H—C—OH
H—C—OH
$CH_2OPO_3^{2-}$

Notes: The name in italics following the number of each reaction is the enzyme that catalyzes the reaction.
Stars identify the same carbon atom before and after a reaction.

Rubisco occurs in two functionally analogous forms whose structure, distribution, and O_2 sensitivity are different. Oxygenic phototrophs (cyanobacteria, chloroplasts) and many photosynthetic bacteria contain a form of the enzyme made up of eight large (L) catalytic subunits (about 55 kDa each) and eight small (S) subunits (about 14 kDa each), giving a molecular mass of about 560 kDa for the complete protein (L_8S_8). In some photosynthetic purple nonsulfur bacteria, rubisco is composed of two large subunits, each 50 kDa. It was recently demonstrated that the eukaryotic algal dinoflagellates also use a dimeric enzyme.

In contrast to prokaryotes, in which gene expression and protein biosynthesis take place in the same cellular compartment (cytosol), the rubisco of most eukaryotic organisms (plants, green algae) is assembled by coordinated biochemical processes occurring in several cellular compartments. In the nuclear genome, light modulates the expression of *rbcS*, a gene that produces a precursor polypeptide of the small subunit that is subsequently translated by cytosolic ribosomes. The translocation of the precursor polypeptide of the small subunit across the chloroplast envelope removes the N-terminal plastid-targeting peptide, and the mature form appears in the stroma of chloroplasts. The chloroplast genome contains *rbcL*, the gene that codes for the mature form of the large subunit, which is synthesized entirely on chloroplast ribosomes. Finally, the assembly of rubisco from its constituent subunits is a chloroplast event. Not all photosynthetic organisms have nuclear and chloroplastic genes for rubisco. Genes encoding both the large and the small rubisco subunits occur in chloroplasts of brown and red algae.

Special proteins that modulate protein folding assist the formation of the L_8S_8 type of rubisco in a process that uses ATP. These proteins, called *chaperonins*, modify the noncovalent interactions of target proteins and bring about the active conformation of the enzyme. Several different chaperonins promote the in vitro assembly of both subunits.

Triose Phosphates are Formed in the Reduction Step of the Calvin Cycle

In the reduction stage of the Calvin cycle (Figure 8.4), the 3-phosphoglycerate formed as a result of the carboxylation of ribulose bisphosphate (see Table 8.1, reaction 1) is first phosphorylated via 3-phosphoglycerate kinase to 1,3-bisphosphoglycerate using the ATP generated in the light reactions (Table 8.1, reaction 2), and is then reduced to glyceraldehyde-3-phosphate using the NADPH generated by the light reactions (Table 8.1, reaction 3). The chloroplast enzyme NADP:glyceraldehyde-3-phosphate dehydrogenase catalyzes this step. Note that the enzyme is similar to that of glycolysis (to be discussed in Chapter 11), except that NADP rather than NAD is the coenzyme. An NADP-linked form of the enzyme is synthesized during chloroplast development (greening), and the NADP-requiring enzyme is preferentially used in biosynthesis reactions.

Activity of the Calvin Cycle Requires the Regeneration of Ribulose-1,5-Bisphosphate

The continued uptake of CO_2 requires that the CO_2 acceptor, ribulose-1,5-bisphosphate, be constantly regenerated. The cycle of intermediates is prevented from depletion by the formation of three molecules of ribulose-1,5-bisphosphate by reshuffling of the carbons from five molecules of triose phosphate to three molecules of pentose phosphate (see Figure 8.4).

One molecule of glyceraldehyde-3-phosphate is converted via triose phosphate isomerase to dihydroxyacetone-3-phosphate in an isomerization reaction (Table 8.1, reaction 4). Dihydroxyacetone-3-phosphate then undergoes aldol condensation with a second molecule of glyceraldehyde-3-phosphate (Table 8.1, reaction 5), using aldolase to give fructose-1,6-bisphosphate. This sugar bisphosphate product, which occupies a key position in the cycle, is hydrolyzed to fructose-6-phosphate (Table 8.1, reaction 6), which then reacts with the enzyme transketolase.

A two-carbon unit (C-1 and C-2 of fructose-6-phosphate) is transferred via transketolase to a third molecule of glyceraldehyde-3-phosphate to give erythrose-4-phosphate (from C-3 to C-6 of the fructose) and xylulose-5-phosphate (from C-2 of the fructose and the glyceraldehyde-3-phosphate) (Table 8.1, reaction 7). Erythrose-4-phosphate then combines via aldolase with a fourth molecule of triose phosphate (dihydroxyacetone-3-phosphate) to yield the seven-carbon sugar sedoheptulose-1,7-bisphosphate (Table 8.1, reaction 8), which is further hydrolyzed by way of a specific phosphatase to give sedoheptulose-7-phosphate (Table 8.1, reaction 9).

Sedoheptulose-7-phosphate donates a two-carbon unit to the fifth (and last) molecule of glyceraldehyde-3-phosphate via transketolase and produces ribose-5-phosphate (from C-3 to C-7 of sedoheptulose) and xylulose-5-phosphate (from C-2 of the sedoheptulose and the glyceraldehyde-3-phosphate) (Table 8.1, reaction 10).

The two molecules of xylulose-5-phosphate (see Table 8.1, reactions 7 and 10) are converted to two ribulose-5-phosphate sugars by an epimerase (Table 8.1, reaction 11). The third molecule of ribulose-5-phosphate is formed from ribose-5-phosphate by an isomerase (Table 8.1, reaction 12). Finally, using ribulose-5-phosphate

Figure 8.4 The Calvin cycle. The carboxylation of three molecules of ribulose-1,5-bisphosphate leads to the *net* synthesis of one molecule of glyceraldehyde-3-phosphate and the regeneration of the three molecules of starting material. This process starts and ends with three molecules of ribulose-1,5-bisphosphate, reflecting the cyclic nature of the pathway. ▶

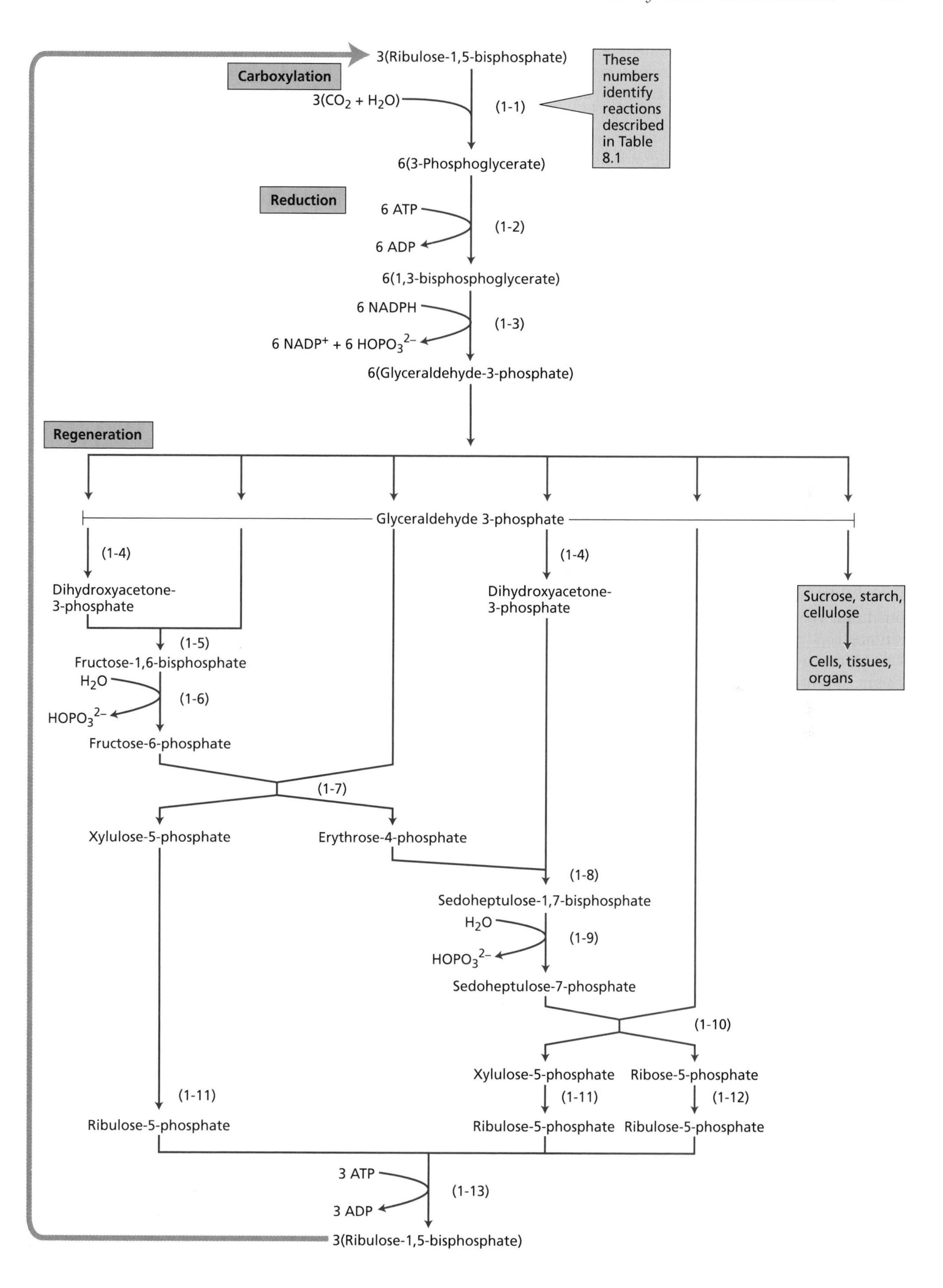

3(Ribulose-1,5-bisphosphate)
Carboxylation
$3(CO_2 + H_2O)$
(1-1)
These numbers identify reactions described in Table 8.1
6(3-Phosphoglycerate)
Reduction
6 ATP
6 ADP
(1-2)
6(1,3-bisphosphoglycerate)
6 NADPH
$6 NADP^+ + 6 HOPO_3^{2-}$
(1-3)
6(Glyceraldehyde-3-phosphate)
Regeneration
Glyceraldehyde 3-phosphate
(1-4)
Dihydroxyacetone-3-phosphate
(1-4)
Dihydroxyacetone-3-phosphate
Sucrose, starch, cellulose
Cells, tissues, organs
(1-5)
Fructose-1,6-bisphosphate
H_2O
$HOPO_3^{2-}$
(1-6)
Fructose-6-phosphate
(1-7)
Xylulose-5-phosphate
Erythrose-4-phosphate
(1-8)
Sedoheptulose-1,7-bisphosphate
H_2O
$HOPO_3^{2-}$
(1-9)
Sedoheptulose-7-phosphate
(1-10)
Xylulose-5-phosphate
Ribose-5-phosphate
(1-11)
(1-11)
(1-12)
Ribulose-5-phosphate
Ribulose-5-phosphate
Ribulose-5-phosphate
3 ATP
3 ADP
(1-13)
3(Ribulose-1,5-bisphosphate)

kinase, ribulose-5-phosphate is phosphorylated with ATP, thus regenerating the three needed molecules of the initial CO_2 acceptor, ribulose-1,5-bisphosphate (Table 8.1, reaction 13).

The Calvin Cycle Was Elucidated by the Use of Radioactive Isotopes

The elucidation of the Calvin cycle was the result of a series of elegant experiments by Melvin Calvin and his colleagues in the 1950s, for which a Nobel prize was awarded in 1961. They used suspensions of the unicellular eukaryotic green alga *Chlorella* to trace the path of carbon. The Calvin cycle is also referred to as the reductive pentose phosphate (RPP) cycle to distinguish the photosynthetic cycle from the oxidative pentose phosphate pathway (see Chapter 11), with which it shares several enzymes. Yet another designation of the Calvin cycle is photosynthetic carbon reduction (PCR) cycle.

To elucidate the cycle, the investigators first exposed the algal cells to constant conditions of light and CO_2 to establish steady-state photosynthesis. Then they added radioactive $^{14}CO_2$ for a brief period to label the intermediates of the cycle. They then killed the cells and inactivated their enzymes by plunging the suspension into boiling alcohol. They separated the ^{14}C-labeled compounds from one another and identified them by their positions on two-dimensional paper chromatograms (Figure 8.5). In this manner, 3-phosphoglycerate and the various sugar phosphates were identified as intermediates in carbon fixation. By exposing the cells to $^{14}CO_2$ for progressively shorter periods of time, Calvin's group was able to identify 3-phosphoglycerate as the first stable intermediate. Therefore, the other labeled sugar phosphates must be derived from a subsequent reduction of 3-phosphoglycerate (Bassham 1965).

To deduce the path of carbon, it was necessary to determine the distribution of ^{14}C in each of the labeled sugars. After a brief exposure to $^{14}CO_2$, individual intermediates were isolated and chemically degraded so that the amount of ^{14}C in each carbon atom could be determined. The results showed that 3-phosphoglycerate was initially labeled predominantly in the carboxyl group. This finding suggested that the initial CO_2 acceptor was a two-carbon compound, and it prompted a long and futile search for such a compound. The subsequent discovery that pentose monophosphates and a pentose bisphosphate (ribulose) participate in the cycle raised the possibility that the initial CO_2 acceptor was a five-carbon compound. This conceptual breakthrough rapidly led to the identification of ribulose-1,5-bisphosphate as the CO_2 acceptor and to formulation of the complete cycle. The operation of the cycle explained the ^{14}C-labeling pattern of the other intermediates.

To demonstrate conclusively that a metabolic pathway exists, it is necessary to prove that the postulated enzymes catalyze the proposed reactions in the test tube. Ideally, the rates of these enzyme reactions (in vitro) should be equal to or in excess of those observed in the intact cell (in vivo). However, this evidence can be used only to support a proposed pathway. Failure to demonstrate a reaction in vitro does not prove that the reaction does not occur. All of the reactions of the Calvin cycle have been demonstrated in vitro. With one exception, the in vitro rates of enzymatic activity are well in excess of the maximal observed rates of photosynthesis. That exception is rubisco, which, when assayed under the levels of CO_2 and O_2 in air, shows barely enough in vitro activity to account for the observed rate of photosynthesis in air. Although great progress has been made in increasing activity by providing proper conditions for activation,

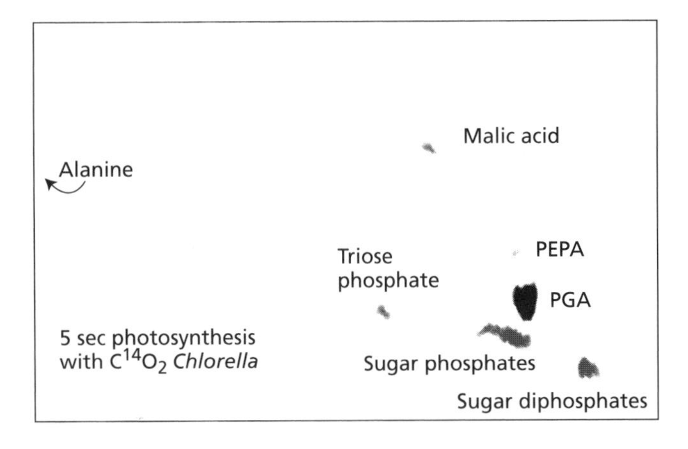

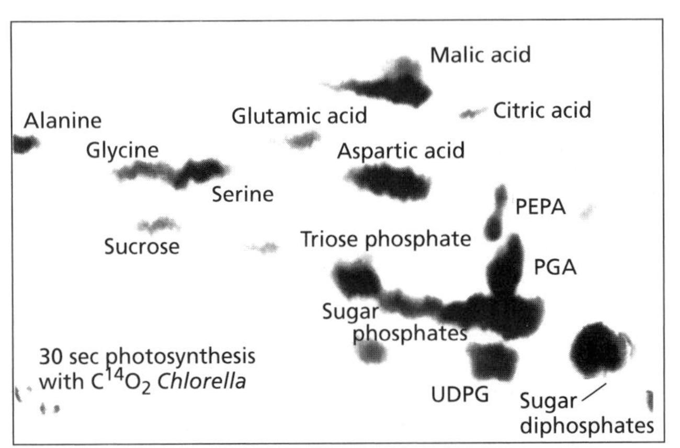

Figure 8.5 Autoradiograms showing the labeling of carbon compounds in the alga *Chlorella* after exposure to $^{14}CO_2$. The time intervals shown in the figures indicate the length of exposure to the radiolabel. At the indicated time intervals, the reaction was terminated, by plunging of the contents into boiling alcohol. The labeled compounds in the cell homogenates were then separated by paper chromatography. The heavy labeling of 3-phosphoglycerate (PGA) after the shorter exposure indicates that it is the first stable intermediate of the Calvin (reductive pentose phosphate) cycle. (After Bassham 1965.)

researchers continue to debate whether rubisco is a rate-limiting step in photosynthesis (Lorimer 1981).

The Calvin Cycle Regenerates Its Own Biochemical Components

The rate of operation of the Calvin cycle can be enhanced by increases in the concentration of its intermediates; that is, the cycle is **autocatalytic**. In addition, the Calvin cycle has the metabolically desirable feature of being able to produce more substrate (ribulose-1,5-bisphosphate, abbreviated RuBP) than is consumed as long as triose phosphate is not being diverted elsewhere:

$$5\ RuBP^{4-} + 5\ CO_2 + 9\ H_2O + 16\ ATP^{4-} + 10\ NADPH \rightarrow 6\ RuBP^{4-} + 14\ HOPO_3^{2-} + 6\ H^+ + 16\ ADP^{3-} + 10\ NADP^+$$

The importance of this autocatalytic property is shown by experiments in which previously darkened leaves or isolated chloroplasts are illuminated. In such experiments, CO_2 fixation starts only after a lag, called the induction period, and the rate of photosynthesis increases with time in the first few minutes after the onset of illumination. The increase in the rate of photosynthesis during the induction period is due in part to an increase in the concentration of intermediates of the Calvin cycle and in part to the activation of enzymes by light (discussed later).

Another important feature of the Calvin cycle is its overall stoichiometry. At the onset of illumination, most of the triose phosphates are drawn back into the cycle to facilitate the buildup of an adequate concentration of metabolites. However, when photosynthesis reaches a steady state, five-sixths of the triose phosphate contributes to regeneration of the ribulose-1,5-bisphosphate and one-sixth is exported to the cytosol for the synthesis of sucrose or other metabolites that are converted to starch in the chloroplast. Let's analyze the cycle by following the energy required for the fixation of six molecules of CO_2. (In the equations that follow, P stands for "phosphate.")

$$6\ RuBP + 6\ CO_2 + 6\ H_2O \rightarrow 12(\text{3-phosphoglycerate}) + 12\ H^+$$

$$12(\text{3-phosphoglycerate}) + 12\ ATP \rightarrow 12(\text{1,3-bisphosphoglycerate}) + 12\ ADP$$

$$12(\text{1,3-bisphosphoglycerate}) + 12(NADPH + H^+) \rightarrow 12(\text{triose P}) + 12\ NADP^+ + 12\ P_i$$

$$6(\text{triose P}) \rightarrow 3(\text{fructose-1,6-bisP})$$

$$3(\text{fructose-1,6-bisP}) + 3\ H_2O \rightarrow 3(\text{fructose-6-P}) + 3\ P_i$$

$$2(\text{fructose-6-P}) + 2(\text{triose P}) \rightarrow 2(\text{xylulose-5-P}) + 2(\text{erythrose-4-P})$$

$$2(\text{erythrose-4-P}) + 2(\text{triose P}) \rightarrow 2(\text{sedoheptulose-1,7-bisP})$$

$$2(\text{sedoheptulose-1,7-bisP}) + 2\ H_2O \rightarrow 2(\text{sedoheptulose-7-P}) + 2\ P_i$$

$$2(\text{sedoheptulose-7-P}) + 2(\text{triose P}) \rightarrow 2(\text{xylulose-5-P}) + 2(\text{ribose-5-P})$$

$$4(\text{xylulose-5-P}) \rightarrow 4(\text{ribulose-5-P})$$

$$2(\text{ribose-5-P}) \rightarrow 2(\text{ribulose-5-P})$$

$$6(\text{ribulose-5-P}) + 6\ ATP \rightarrow 6(\text{ribulose-1,5-bisP}) + 6\ ADP$$

$$\text{Net: } 6\ CO_2 + 11\ H_2O + 12\ NADPH + 18\ ATP \rightarrow \text{fructose-6-P} + 12\ NADP^+ + 6\ H^+ + 18\ ADP + 17\ P_i$$

This calculation shows that in order to synthesize the equivalent of 1 molecule of hexose, 6 molecules of CO_2 are fixed at the expense of 18 ATP and 12 NADPH. In other words, the Calvin cycle consumes two molecules of NADPH and three molecules of ATP for every molecule of CO_2 fixed into carbohydrate. We can compute the maximal overall thermodynamic efficiency of photosynthesis if we know the energy content of the light, the minimum quantum requirement (moles of quanta absorbed per mole of CO_2 fixed; see Chapter 7), and the energy stored in a mole of carbohydrate (hexose).

Red light at 680 nm contains 175 kJ (42 kcal) per quantum mole of photons. The minimum quantum requirement is usually calculated to be 8 photons per molecule of CO_2 fixed, although the number obtained experimentally is 9 to 10 (see Chapter 7). Therefore, the minimum light energy needed to reduce 6 moles of CO_2 to a mole of hexose is approximately $6 \times 8 \times 175$ kJ = 8400 kJ (2016 kcal). However, a mole of hexose such as fructose yields only 2804 kJ when totally oxidized. Comparing 8400 and 2804 kJ (673 kcal), we see that the maximum overall thermodynamic efficiency of photosynthesis is about 33%. However, most of the light energy is lost in the generation of ATP and NADPH by the light reactions (see Chapter 7) rather than during operation of the Calvin cycle.

We can calculate the efficiency of the Calvin cycle more directly by computing the changes in free energy associated with the hydrolysis of ATP and the oxidation of NADPH, which are 29 and 217 kJ (7 and 52 kcal) per mol, respectively. We saw in the list of reactions above that the synthesis of one molecule of fructose-6-phosphate from 6 molecules of CO_2 uses 12 NADPH and 18 ATP molecules. Therefore the Calvin cycle consumes (12 $\times$ 217) + (18 $\times$ 29) = 3126 kJ (750 kcal) in the form of NADPH and ATP, respectively, resulting in a thermodynamic efficiency close to 90%. An examination of these calculations shows that the bulk of the energy required for the conversion of CO_2 to carbohydrate comes from NADPH. That is, 2 mol NADPH $\times$ 52 kcal

mol^{-1} = 104 kcal, but 3 mol ATP × 7 kcal mol^{-1} = 21 kcal. Thus, 83% (104 of 125 kcal) of the energy stored comes from the reductant NADPH.

The Calvin cycle cannot be found in all autotrophic cells. Some anaerobic bacteria use other pathways for autotrophic growth: the ferrodoxin-mediated synthesis of organic acids from acetyl CoA derivatives via a reversal of the citric acid cycle (the reductive carboxylic acid cycle of green sulfur bacteria); the glyoxylate-producing cycle (the hydroxypropionate pathway of green nonsulfur bacteria); and the linear route (acetyl CoA pathway) of acetogenic, methanogenic bacteria. Thus, while the Calvin cycle is quantitatively the most important pathway of autotrophic CO_2 fixation, other pathways have been described.

The Calvin Cycle Is Regulated by Several Mechanisms

The high energy efficiency of the Calvin cycle indicates that some form of regulation must ensure that all intermediates in the cycle are present at adequate concentrations and that the cycle is turned off when it is not needed in the dark. The amount of each enzyme present in the chloroplast stroma is regulated at the genetic level by mechanisms that control expression of the nuclear and chloroplast genomes (Maier et al. 1995; Purton 1995). Short-term regulation of the Calvin cycle is achieved by several mechanisms that act together to minimize futile cycling—a phenomenon characterized by unregulated reactions that operate in opposing directions (Wolosiuk et al. 1993). As seen below, individual enzymes of the cycle are regulated by the transformation of covalent bonds—that is, the reduction of disulfide bonds, or the carbamylation of amino groups.

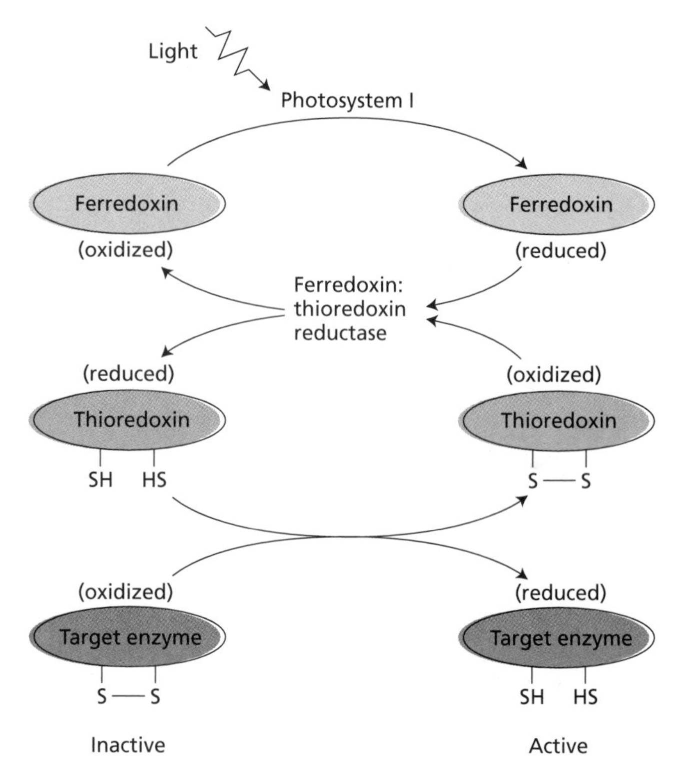

Figure 8.6 The ferredoxin–thioredoxin system reduces specific enzymes in the light: fructose-1,6-bisphosphate phosphatase, ribulose-5-phosphate kinase, sedoheptulose-1,7-bisphosphate phosphatase, and NADP:glyceraldehyde-3-phosphate dehydrogenase. On reduction, the enzymes are converted from an inactive to an active state. Ferredoxin: thioredoxin reductase has a catalytically active disulfide group (—S—S—) that acts to generate the unique —SH groups of thioredoxin. In the dark, the reduced thioredoxin is first oxidized to the disulfide form by molecular oxygen. The oxidized form of thioredoxin, in turn, oxidizes the reduced form of the target enzymes, thereby converting them to the original oxidized, inactive state.

Light-Dependent Enzyme Activation Regulates the Calvin Cycle

Five light-regulated enzymes operate in the Calvin cycle: rubisco, NADP:glyceraldehyde-3-phosphate dehydrogenase, fructose-1,6-bisphosphate phosphatase, sedoheptulose-1,7-bisphosphate phosphatase, and ribulose-5-phosphate kinase. Of these five, light controls the activity of the last four via the **ferredoxin–thioredoxin system**, a covalent thiol-based oxidation–reduction mechanism identified by B. Buchanan and colleagues (Besse and Buchanan 1997). The four enzymes targeted contain one or more disulfide (—S—S—) groups. In the dark, these residues exist in the oxidized (—S—S—) state, which renders the enzyme inactive or subactive. In the light, the —S—S— group is reduced to the sulfhydryl (—SH HS—) state. This redox change leads to activation of the enzyme (Figure 8.6).

The activation process starts in the light by a reduction of ferredoxin by photosystem I (see Chapter 7). The reduced ferredoxin plus two protons are used to reduce a catalytically active disulfide (—S—S—) group of the iron–sulfur enzyme ferredoxin:thioredoxin reductase, which in turn reduces the disulfide (—S—S—) bond of the small regulatory protein thioredoxin (see Box 8.2). The reduced form (—SH HS—) of thioredoxin then reduces the critical disulfide bond (converts —S—S— to —SH HS—) of a target enzyme and thereby leads to activation of that enzyme (Buchanan 1980). The light signal is thus converted to a sulfhydryl, or —SH, signal via ferredoxin and the enzyme ferredoxin:thioredoxin reductase.

This sulfhydryl, or dithiol (—SH HS—), signal of thioredoxin is transmitted to specific target enzymes that are converted to a fully active state. In some cases (such as fructose-1,6-bisphosphate phosphatase), the thioredoxin-linked activation is enhanced by an effector (for example, fructose-1,6-bisphosphate substrate). Inactivation of the target enzymes that is observed upon darkening appears to take place by reversal of the

BOX 8.2

Thioredoxins

THIOREDOXINS ARE 12 kDa PROTEINS that have a catalytically active disulfide group and occur in most if not all organisms. Thioredoxins have an active site with two redox-active cysteine residues (-Trp-Cys-Gly[Ala]-Pro-Cys-). The active site is located on the first α helix of a motif consisting of three flanking α helices and a four-stranded β sheet that has been identified in several redox-active proteins. The oxidized form of each thioredoxin contains a disulfide (—S—S—) bridge that is reduced to the sulfhydryl (—SH) level by either reduced ferredoxin or NADPH. The reduced form of thioredoxin is an excellent catalyst for the reduction of intramolecular (intrachain) disulfide bonds of proteins that are only slowly reduced by glutathione, the other major cellular sulfhydryl reductant. Upon reduction, the disulfide bond of enzymes targeted specifically by a thioredoxin causes a dramatic shift (usually increase) in catalytic activity. First observed to activate enzymes of photosynthesis, thioredoxins are now known to play a broad regulatory role in many cell types.

Although only one type of thioredoxin has been detected in bacteria and animals, three well-characterized variants exist in photosynthetic cells. Two of the three (*m* and *f*) are located in chloroplasts and can be distinguished from one another on the basis of both primary structure and specificity for target enzymes, the feature that led to their discovery. These two chloroplast thioredoxins are reduced by electrons from excited chlorophyll via ferredoxin and ferredoxin:thioredoxin reductase, an iron–sulfur enzyme with a catalytically active disulfide group.

The enzymes regulated by thioredoxin show differing degrees of specificity. The Calvin cycle enzymes fructose-1,6-bisphosphate phosphatase, sedoheptulose-1,7-bisphosphate phosphatase, ribulose-5-phosphate kinase, and NADP:glyceraldehyde-3-phosphate dehydrogenase are preferentially activated by thioredoxin *f*. One enzyme, ATP synthase, which catalyzes the synthesis of ATP in photophosphorylation, is activated by both chloroplast thioredoxins. Thioredoxin *m* effectively activates NADP:malate dehydrogenase , an enzyme of the C_4 photosynthetic carbon cycle and C_3 redox shuttle, and deactivates glucose-6-phosphate dehydrogenase, the main regulatory enzyme of the oxidative pentose phosphate pathway (see Chapter 11). Thioredoxin of the *m*, or bacterial, type has been found to regulate other light-dependent chloroplast processes, such as the translation of mRNA and the assimilation of sulfate.

Plants have a third class of thioredoxin, the *h* type—located in the cytosol, endoplasmic reticulum, and mitochondrion that is reduced by NADPH in a reaction catalyzed by a flavoprotein enzyme, NADP:thioredoxin reductase (NTR):

$$\text{NADPH} + \text{H}^+ + \text{thioredoxin}_{\text{ox}} \xrightarrow{\text{NTR}} \text{NADP}^+ + \text{thioredoxin}_{\text{red}}$$

The function of thioredoxin *h* is currently being investigated in several plant systems. It has been found to promote the mobilization of carbon and nitrogen of the endosperm early in the germination of cereals. Thioredoxin *h* appears to act (1) by reducing the major seed storage proteins, thereby enhancing their susceptibility to proteases, (2) by activating a newly discovered type of calcium-linked protease (thiocalsin), and (3) by neutralizing (via reduction of their intrachain disulfide bonds) low-molecular-weight proteins that inhibit enzymes of starch degradation.

Through studies of gene inactivation, thioredoxin *h* has been identified as a requirement for the development of pollen self-incompatibility and for cell division in *Arabidopsis*. Thioredoxin, again linked to NADP, has also been found to function widely in animal cells in processes ranging from transcription to development of the immune response to embryogenesis.

Extensions of the biochemical and molecular studies have led to possible applications of thioredoxin in technology and medicine. Thioredoxin may be used to improve foods by lowering allergenicity and increasing digestibility. Ongoing studies have uncovered the potential use of thioredoxin as a marker in diagnosing and monitoring certain human diseases, in neutralizing venom neurotoxins and factors involved in the inflammatory response, and in improving the success of organ transplants. Thioredoxin is an example in which principles learned in basic research are being applied to contemporary societal problems.

reduction (activation) pathway. That is, oxygen converts the thioredoxin and target enzyme from the reduced (—SH HS—) to the oxidized (—S—S—) state.

In addition to activating enzymes of the Calvin cycle, the ferredoxin–thioredoxin system activates enzymes of other chloroplast processes. Thus, the C_4 cycle, photophosphorylation, and mRNA translation are activated by thioredoxin via NADP:malate dehydrogenase, thylakoid-bound ATP synthase (coupling factor), and activator proteins of the 5′ untranslated region of chloroplast mRNAs, respectively. Recent work has shown that, in addition to its role in C_4 plants, thioredoxin-linked NADP:malate dehydrogenase functions in C_3 plants in a cyclic process that transports malate synthesized in the chloroplast to the cytosol, where it generates NADH. The oxaloacetate formed in this oxidation is transported back to the chloroplast in exchange for malate by a specific antiporter.

The chloroplast also contains an enzyme, glucose-6-phosphate dehydrogenase, that functions in the opposite way than the enzymes of photosynthesis: It is deactivated upon reduction by thioredoxin. By lowering the activity of this enzyme, light *inhibits* operation of the oxidative pentose phosphate pathway for the breakdown of carbohydrate (see Chapter 11). Chloroplasts thus use thioredoxin to link light to the control of carbon flow through two major opposing carbon pathways (Scheibe 1990; Buchanan 1992).

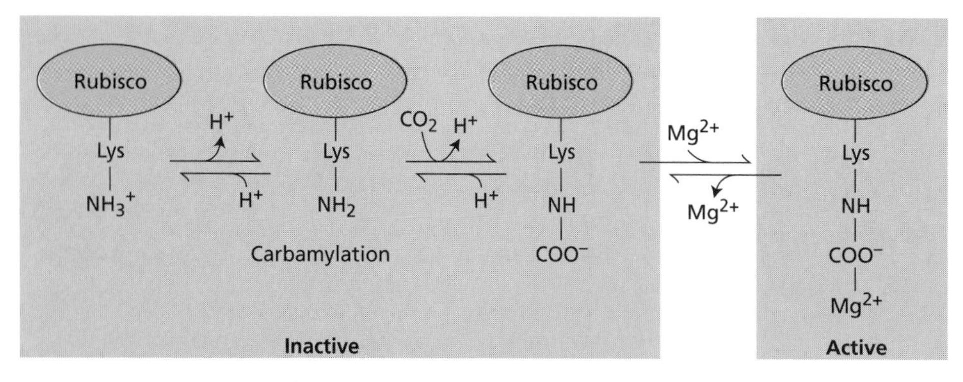

Figure 8.7 One mechanism for the activation of rubisco involves the formation of a carbamate–Mg^{2+} complex on the ε-amino group of a lysine within the active site of the enzyme. Two protons are released, so activation is enhanced by the increase in pH and Mg^{2+} concentration that result from illumination. Note that the CO_2 involved in the carbamate–Mg^{2+} reaction is not the same as the CO_2 involved in the carboxylation of ribulose-1,5-bisphosphate.

The activity of rubisco increases in the light, but this enzyme cannot be shown to respond to thioredoxin. George Lorimer and colleagues found that rubisco is activated when activator CO_2 (a different molecule from the substrate CO_2 that becomes fixed) reacts with an uncharged ε-NH_2 group of lysine within the active site of the enzyme. The resulting carbamate derivative (a new anionic site) then binds Mg^{2+} to yield the activated complex (Figure 8.7). Since two protons are released, activation is promoted by an increase both in pH and in Mg^{2+} concentration. Thus the observed activation of rubisco in the light appears to be mediated by light-dependent stromal changes in Mg^{2+} and pH.

Sugar phosphate compounds such as ribulose-1,5-bisphosphate bound to rubisco impede carbamylation until they are removed by a recently identified enzyme, **rubisco activase** (Salvucci and Ogren 1996). The isolation of *Arabidopsis* mutants that are capable of growth only at elevated (nonphysiological) levels of CO_2 made it possible to identify two structurally similar forms of rubisco activase that enhance the in vitro activity of rubisco in the presence of ATP and increase its affinity for CO_2 (Figure 8.8). The primary structures of the two forms of the enzyme were found to differ only in the C-terminal region.

The primary role of rubisco activase is to reactivate rubisco that has been deactivated by sugar bisphosphates (see Figure 8.8). Rubisco activase accelerates the release of bound sugar bisphosphates that inhibit the decarbamylated form of rubisco. Rubisco activase also removes carboxyarabinitol-1-phosphate, a rubisco inhibitor that closely resembles the six-carbon transition intermediate of the carboxylation reaction. Carboxyarabinitol-1-phosphate is present at low concentrations in leaves of many species and at high concentrations in leaves from legumes such as soybean and bean. The inhibitor binds to rubisco at night, and it is removed in the morning, when photon flux density increases.

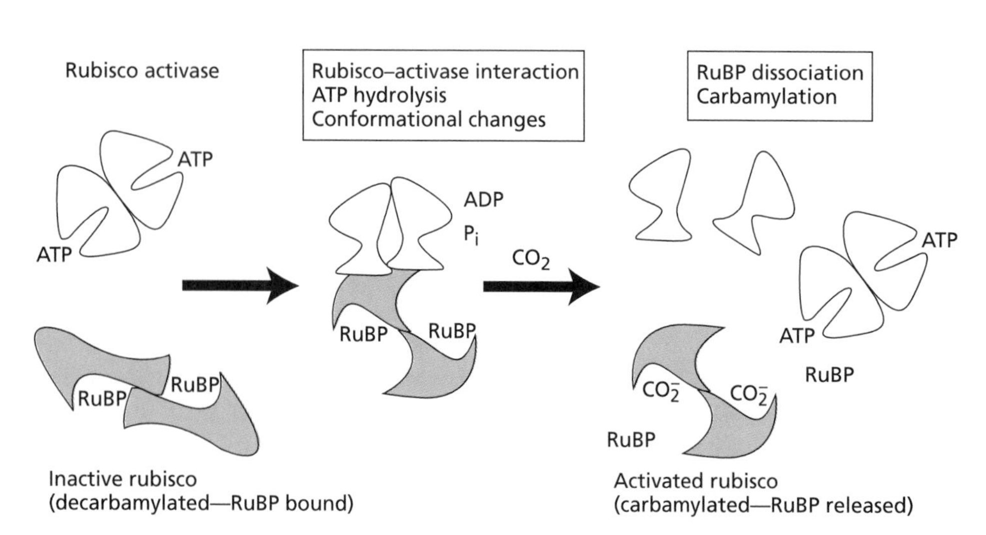

Figure 8.8 Rubisco activase functions by removing bound RuBP from the decarbamylated inactive form of rubisco. The activase enzyme appears to couple the hydrolysis of ATP to the release of ribulose or other inhibitory pentose bisphosphates. Rubisco is then activated (carbamylated) by binding CO_2 and Mg^{2+} (see Figure 8.7). (From Gutteridge 1990.)

Light-Dependent Ion Movements Regulate Calvin Cycle Enzymes

As already discussed, light causes reversible ion changes in the chloroplast stroma that influence the activity of rubisco and other chloroplast enzymes. H. W. Heldt and his colleagues have shown that upon illumination, protons are pumped from the stroma into the lumen of the thylakoids (see Chapter 7). This proton efflux, which is coupled to Mg^{2+} uptake, results in a stromal increase in pH from approximately 7 to 8. These pH changes are reversed upon darkening. Several Calvin cycle enzymes, such as rubisco, fructose-1,6-bisphosphate phosphatase, and ribulose-5-phosphate kinase are more active at pH 8 than at pH 7, so these ion fluxes enhance enzyme activities. The light-dependent increase in stromal Mg^{2+} also enhances the activity of some enzymes of the Calvin cycle (Heldt 1979).

Two distinct modes of regulation of Calvin cycle enzymes are thus apparent. Covalent changes such as the reduction of disulfide bonds and the carbamylation of amino groups result in the formation of a chemically modified enzyme. Noncovalent changes, such as the binding of metabolites, result in an isomerized enzyme. In addition, the efficiency of the Calvin cycle is enhanced by binding of the enzymes to the thylakoid membranes, thereby achieving a higher level of organization that favors the channeling and protection of substrates.

Light-Dependent Membrane Transport Regulates the Calvin Cycle

The rate at which carbon is exported from the chloroplast plays a role in regulation of the Calvin cycle. Carbon is exported as triose phosphates in exchange for orthophosphate via the phosphate translocator in the inner membrane of the chloroplast envelope (Flügge and Heldt 1991). To ensure continued operation of the Calvin cycle, at least five-sixths of the triose phosphate must be recycled (see Figure 8.4). Thus, at most one-sixth can be exported for sucrose synthesis in the cytosol or diverted to starch synthesis within the chloroplast. The regulation of this aspect of photosynthetic carbon metabolism will be discussed further when the synthesis of sucrose and starch is considered in detail later in this chapter.

The Photorespiratory Carbon Oxidation Cycle

An important property of rubisco, discovered by W. L. Ogren and colleagues, is its ability to catalyze both the carboxylation and the oxygenation (Table 8.2, reaction 1) of RuBP. Oxygenation is the primary reaction in a process known as **photorespiration**. Because photosynthesis and photorespiration work in diametrically opposite directions, photorespiration results in loss of CO_2 from cells that are simultaneously fixing CO_2 by the Calvin cycle (Ogren 1984; Leegood et al. 1995).

In this section we will describe the photorespiratory carbon oxidation cycle—the reactions that result in the partial recovery of fixed carbon, and the intracellular location of the participating reactions. This cycle has been characterized by N. E. Tolbert and a number of other investigators.

Photosynthetic CO_2 Fixation and Photorespiratory Oxygenation Are Competing Reactions

The oxygenation of ribulose-1,5-bisphosphate results in the incorporation of one atom of molecular O_2 into one of the products of the reaction, 2-phosphoglycolate (see Table 8.2, reaction 1). The ability to catalyze the oxygenation of ribulose-1,5-bisphosphate is a property of all rubiscos, regardless of taxonomic origin. Even the rubisco from anaerobic, autotrophic bacteria catalyzes the oxygenase reaction when exposed to oxygen. As alternative substrates for rubisco, CO_2 and O_2 compete for reaction with ribulose-1,5-bisphosphate, since carboxylation and oxygenation occur within the same active site of the enzyme. Offered equal concentrations of CO_2 and O_2 in a test tube, angiosperm rubiscos fix CO_2 about 80 times faster than they oxygenate. However, an aqueous solution in equilibrium with air at 25°C has a CO_2-to-O_2 ratio of 0.0416 (see the table in Box 8.1). Thus, carboxylation in air outruns oxygenation by a scant three to one.

The photorespiratory carbon oxidation (PCO) cycle acts as a scavenger operation to recover the fixed CO_2 lost by the oxygenase reaction of rubisco. The 2-phosphoglycolate formed in the chloroplast by oxygenation of ribulose-1,5-bisphosphate is rapidly hydrolyzed to glycolate by a specific chloroplast phosphatase (Table 8.2, reaction 2). Subsequent metabolism of the glycolate (Figure 8.9) involves the cooperation of two other organelles: peroxisomes and mitochondria (see Chapter 1) (Tolbert 1981). Glycolate leaves the chloroplast via a specific transporter protein in the envelope membrane and diffuses to a peroxisome. There it is oxidized to glyoxylate and hydrogen peroxide by an oxidase (Table 8.2, reaction 3). The peroxide is destroyed by the action of catalase (Table 8.2, reaction 4), and the glyoxylate undergoes transamination (Table 8.2, reaction 5). The amino donor for this transamination is probably glutamate, and the product is the amino acid glycine.

The glycine leaves the peroxisome and enters a mitochondrion. There two molecules of glycine are converted to a serine and CO_2 (Table 8.2, reactions 6 and 7). The amino acid glycine thus serves as the immediate source of the photorespiratory CO_2. The newly formed serine leaves the mitochondrion and enters a peroxisome, where it is converted first by transamination (Table 8.2, reaction 8) to hydroxypyruvate and then by reduction to glycerate (Table 8.2, reaction 9). Finally, glycerate reen-

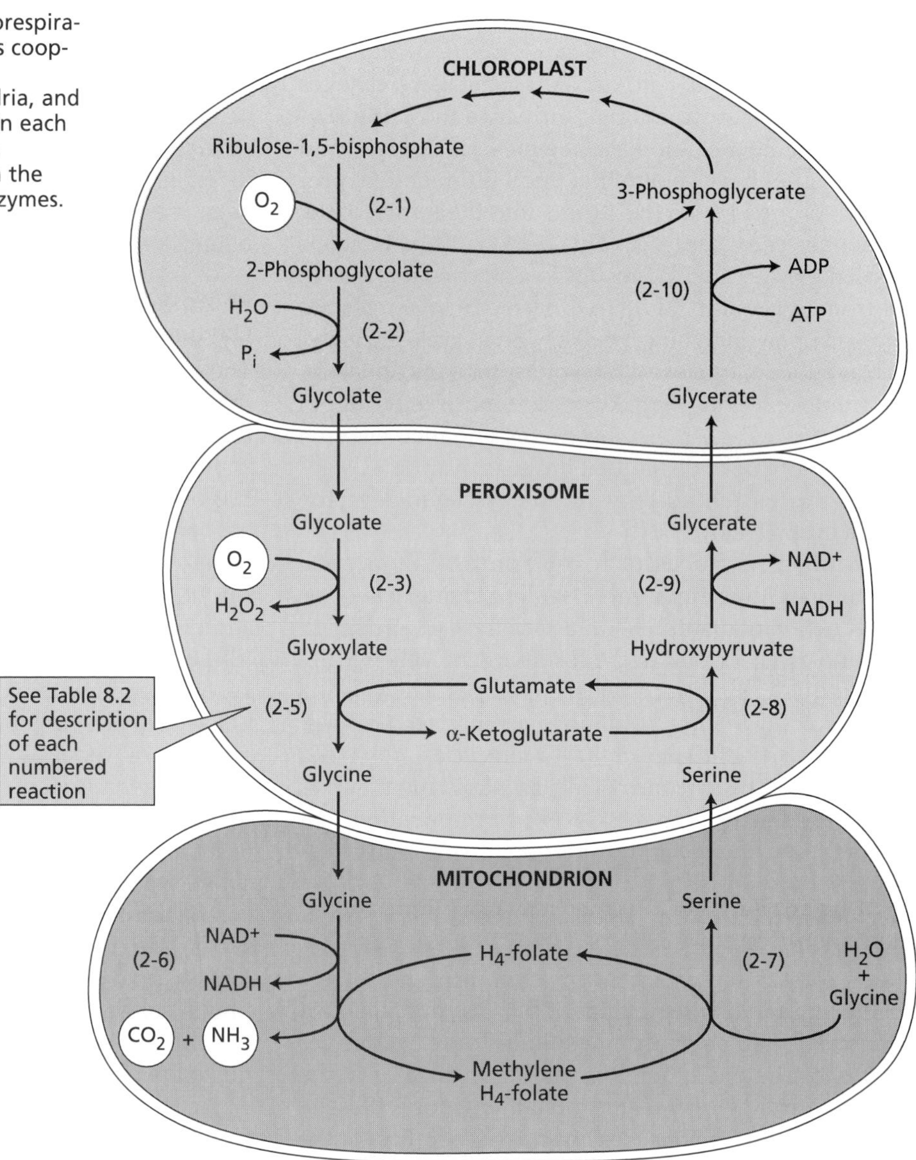

Figure 8.9 Operation of the photorespiratory carbon oxidation cycle involves cooperative interaction among three organelles: chloroplasts, mitochondria, and peroxisomes. Oxygen is consumed in each organelle. Shuttling of metabolites between organelles is evident from the localization of the participating enzymes.

ters the chloroplast, where it is phosphorylated to yield 3-phosphoglycerate (Table 8.2, reaction 10).

Thus overall, two molecules of phosphoglycolate (four carbon atoms), lost from the Calvin cycle by the oxygenation of RuBP, are converted into one molecule of 3-phosphoglycerate (three carbon atoms) and one CO_2. In other words, 75% of the carbon lost by the oxygenation of ribulose-1,5-bisphosphate is recovered by the PCO cycle and returned to the Calvin cycle (Lorimer 1981).

Evidence in support of the PCO cycle stems from several types of experiments. Using mutant *Arabidopsis* plants lacking phosphoglycolate phosphatase (the enzyme that catalyzes reaction 2 in Table 8.2), C. Somerville and W. L. Ogren demonstrated that radioactivity from $^{14}CO_2$ accumulates in 2-phosphoglycolate under conditions favoring photorespiration (high concentrations of O_2, low concentrations of CO_2), but not under nonphotorespiratory conditions (low O_2, high CO_2). In addition, glycolate synthesis in vivo is dependent on O_2 and is inhibited by CO_2 in a competitive manner, as expected from the primary role of rubisco in the cycle. Several studies have shown that, as with the Calvin cycle, the enzymes of the PCO cycle catalyze all the proposed reactions in vitro at rates higher than the in vivo rate of photorespiration. In addition, the participating enzymes are localized in the choroplast (rubisco, phosphoglycolate phosphatase, glycerate kinase), the peroxisome (glycolate oxidase, catalase, hydroxypyruvate reductase, and the two aminotransferases), and the mitochondrion (glycine decarboxylase and serine hydroxymethyltransferase) (see Table 8.2) (Hatch and Osmond 1976; Tolbert 1981).

Table 8.2
Reactions of the photorespiratory carbon oxidation cycle

1. *Ribulose-1,5-bisphosphate carboxylase/oxygenase (chloroplast)*
Ribulose-1,5-bisphosphate + O_2 → 2-phosphoglycolate + 3-phosphoglycerate + 2 H^+

$CH_2OPO_3^{2-}$ | C=O | H—C—OH | H—C—OH | $CH_2OPO_3^{2-}$ $CH_2OPO_3^{2-}$ | COO^- $CH_2OPO_3^{2-}$ | HO—C—H | COO^-

2. *Phosphoglycolate phosphatase (chloroplast)*
Phosphoglycolate + H_2O → glycolate + $HOPO_3^{2-}$

$CH_2OPO_3^{2-}$ | COO^- CH_2OH | COO^-

3. *Glycolate oxidase (peroxisome)*
Glycolate + O_2 → glyoxylate + H_2O_2

CH_2OH | COO^- H—C=O | COO^-

4. *Catalase (peroxisome)*
2 H_2O_2 → 2 H_2O + O_2

5. *Glyoxylate:glutamate aminotransferase (peroxisome)*
Glyoxylate + glutamate → glycine + α-ketoglutarate

H—C=O | COO^- COO^- | CH_2 | CH_2 | H—C—NH_2 | COO^- NH_2 | H_2C | COO^- COO^- | CH_2 | CH_2 | C=O | COO^-

6. *Glycine decarboxylase (mitochondrion)*
Glycine + NAD^+ + H_4-folate→ NADH + H^+ + CO_2 + NH_3 + methylene-H_4-folate

NH_2 | H_2C | COO^- —CH_2— H_4-folate

7. *Serine hydroxymethyltransferase (mitochondrion)*
Methylene-H_4-folate + H_2O + glycine→ serine + H_4-folate

—CH_2— H_4-folate NH_2 | H_2C | COO^- CH_2OH | H—C—NH_2 | COO^-

8. *Serine aminotransferase (peroxisome)*
Serine + α-ketoglutarate → hydroxypyruvate + glutamate

CH_2OH | H—C—NH_2 | COO^- COO^- | CH_2 | CH_2 | C=O | COO^- CH_2OH | C=O | COO^- COO^- | CH_2 | CH_2 | H—C—NH_2 | COO^-

9. *Hydroxypyruvate reductase (peroxisome)*
Hydroxypyruvate + NADH + H^+ → glycerate + NAD^+

CH_2OH | C=O | COO^- CH_2OH | H—C—H | COO^-

10. *Glycerate kinase (chloroplast)*
Glycerate + ATP → 3-phosphoglycerate + ADP + H^+

CH_2OH | HO—C—H | COO^- $CH_2OPO_3^{2-}$ | HO—C—H | COO^-

Notes: The name in italics following the number of each reaction is the enzyme that catalyzes the reaction. The 3-phosphoglycerate formed in reaction 10 is converted to ribulose-1,5-bisphosphate via the reductive and regenerative reactions of the Calvin cylce.

Competition between Carboxylation and Oxygenation Decreases the Efficiency of Photosynthesis

Since photorespiration is concurrent with photosynthesis, it is difficult to measure the rate of photorespiration in intact cells. Two molecules of 2-phosphoglycolate (4 carbon atoms) are needed to make one molecule of 3-phosphoglycerate, with the release of one molecule of CO_2; so theoretically one-fourth of the carbon entering the PCO cycle is released as CO_2. Measurements of CO_2 release by sunflower leaves support this value.

This result indicates that the rate of "true" photosynthesis is approximately 120 to 125% of the rate of net photosynthesis. The ratio of carboxylation to oxygenation in air at 25°C is computed to be between 2.5 and 3. Further calculations indicate that photorespiration lowers the efficiency of photosynthetic carbon fixation from 90% to approximately 50%. This decreased efficiency can be measured as an increase in the quantum requirement for CO_2 fixation under photorespiratory conditions (air with high O_2 and low CO_2) as opposed to nonphotorespiratory conditions (low O_2 and high CO_2).

Carboxylation and Oxygenation Are Closely Interlocked in the Intact Leaf

Photosynthetic carbon metabolism in the intact leaf reflects the integrated balance between two mutually opposing and interlocking cycles (Figure 8.10). The Calvin cycle is capable of independent operation, but the PCO cycle depends on the Calvin cycle for a supply of ribulose-1,5-bisphosphate. The balance between the two cycles is determined by three factors: the kinetic properties of rubisco, the concentrations of the substrates CO_2 and O_2, and temperature.

As the temperature increases, the concentration of CO_2 in a solution in equilibrium with air decreases more than the concentration of O_2 does (see Box 8.1). Consequently, the concentration ratio of $[CO_2]$ to $[O_2]$ decreases as the temperature rises. As a result of this property, photorespiration (oxygenation) increases relative to photosynthesis (carboxylation) as the temperature rises. This effect is enhanced by the kinetic properties of rubisco, which also result in a relative increase in oxygenation at higher temperatures (Ku and Edwards 1978). Overall, then, increasing temperatures progressively tilt the balance away from the Calvin cycle and toward the photorespiratory carbon oxidation cycle (see Chapter 9).

The Biological Function of Photorespiration Is Unknown

Although the PCO cycle recovers 75% of the carbon originally lost from the Calvin cycle as 2-phosphoglycolate, why does 2-phosphoglycolate form at all? One possible explanation is that the formation of 2-phosphoglycolate is a consequence of the chemistry of the carboxylation reaction, which requires an intermediate that can react with both CO_2 and O_2. Such a reaction would have had little consequence in early evolutionary times if the ratio of CO_2 to O_2 in air were higher than it is today. However, the low CO_2-to-O_2 ratios prevalent in modern times are conducive to photorespiration, with no other function than the recovery of some of the carbon present in 2-phosphoglycolate.

Another possible explanation is that photorespiration is important, especially under conditions of high light intensity and low intercellular CO_2 concentration (for example, when stomata are closed because of water stress), to dissipate excess ATP and reducing power from the light reactions, thus preventing damage to the photosynthetic apparatus. *Arabidopsis* mutants that are unable to photorespire grow normally under 2% CO_2, but they die rapidly if transferred to normal air. Furthermore, recent evidence with transgenic plants suggests that photorespiration protects C_3 plants from photooxidation and photoinhibition (Kozaki and Takeba 1996). Further work is needed to improve our understanding of the function of photorespiration.

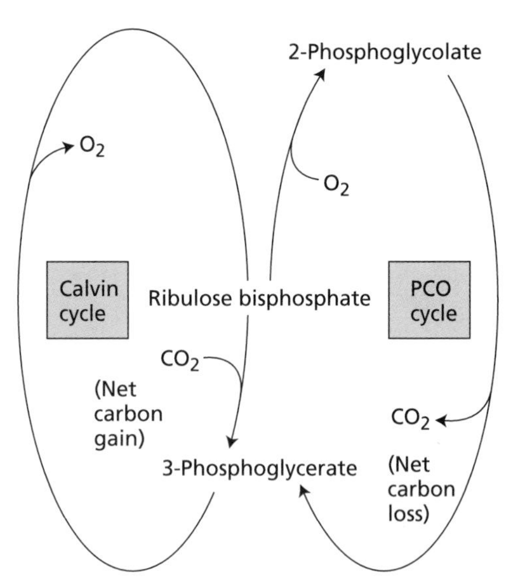

Figure 8.10 The flow of carbon in the leaf is determined by the balance between two mutually opposing cycles. Operation of the Calvin cycle results in the consumption of CO_2, the release of O_2, and a gain in dry matter (carbon gain). In contrast, operation of the photorespiratory carbon oxidation (PCO) cycle results in the release of CO_2, the consumption of O_2, and a loss of dry matter (carbon loss). In normal air (see Box 8.1) the two cycles operate simultaneously, carboxylating and oxygenating ribulose bisphosphate in a ratio of about 3:1. Whereas the Calvin cycle is capable of independent operation, the PCO cycle requires continued operation of the Calvin cycle to regenerate its starting material, ribulose-1,5-bisphosphate.

CO_2 Concentrating Mechanisms I: Algal and Cyanobacterial Pumps

Many plants either do not photorespire at all, or they may do so only to a limited extent. These plants have normal rubiscos, and their lack of photorespiration is a consequence of mechanisms that concentrate CO_2 in the rubisco environment and thereby suppress the oxygenation reaction.

In this and the two following sections we will discuss three mechanisms for concentrating CO_2 at the site of carboxylation. Two systems of CO_2 concentration are found in some angiosperms and involve "add-ons" to the Calvin cycle: the C_4 photosynthetic carbon fixation (C_4) and the crassulacean acid metabolism (CAM) pathways. Plants with C_4 metabolism are often found in hot environments; CAM plants are typical of desert environments. We will examine each of these two systems after we consider the third mechanism: a CO_2 pump found in aquatic plants that has been studied extensively in unicellular cyanobacteria and algae.

When algal and cyanobacterial cells are grown in air enriched with 5% CO_2 and then transferred to a low-CO_2 medium, they display the symptoms that are typical of photorespiration (O_2 inhibition of photosynthesis at low concentration of CO_2). But if the cells are grown in air containing 0.03% CO_2, they rapidly develop the ability to concentrate inorganic carbon (CO_2 plus HCO_3^-) internally. Under these conditions, the cells no longer photorespire.

The accumulation of inorganic carbon is accomplished by CO_2–HCO_3^- pumps in the plasma membrane that are driven by ATP derived from the light reactions. It is not always clear which species of inorganic carbon (CO_2 and/or HCO_3^-) is transported, but substantial differences in the concentration of inorganic carbon inside and outside the cells have been measured. Total inorganic carbon inside some cyanobacterial cells can reach concentrations of 50 m*M* (Ogawa and Kaplan 1987). The proteins that function as CO_2–HCO_3^- pumps are not present in cells grown in high concentrations of CO_2 but are induced upon exposure to low concentrations of CO_2. The accumulated HCO_3^- is converted to CO_2 by the enzyme carbonic anhydrase, and the CO_2 enters the Calvin cycle.

The metabolic consequence of this CO_2 enrichment is suppression of the oxygenation of ribulose bisphosphate and hence also suppression of photorespiration. The energetic cost of this adaptation is the additional ATP needed for concentrating the CO_2. Mutants impaired in their ability to grow under low CO_2 concentrations are being used to identify the genes involved in genetic regulation of the carbon concentrating mechanisms (Kaplan et al. 1994).

CO_2 Concentrating Mechanisms II: The C_4 Carbon Cycle

There are differences in leaf anatomy between plants that have a C_4 carbon cycle (called C_4 plants) and those that photosynthesize solely via the Calvin photosynthetic cycle (C_3 plants). A cross section of a typical C_3 leaf reveals one major cell type that has chloroplasts, the **mesophyll**. In contrast, a typical C_4 leaf has two distinct chloroplast-containing cell types: mesophyll and **bundle sheath** (or Kranz, German for "wreath") cells (Figure 8.11). There is considerable anatomic variation in the arrangement of the bundle sheath cells with respect to the mesophyll and vascular tissue. However, operation of the C_4 cycle requires the cooperative effort of both cell types. No mesophyll cell of a C_4 plant is more than two or three cells distant from the nearest bundle sheath cell (see Figure 8.11A). In addition, an extensive network of plasmodesmata (see Figure 1.30) connects mesophyll and bundle sheath cells, thus providing a pathway for the flow of metabolites between the cell types (see Figure 8.11E).

Malate and Aspartate Are Carboxylation Products of the C_4 Cycle

The discovery of the C_4 cycle can be traced to studies on $^{14}CO_2$ labeling of sugarcane by H. P. Kortschack and colleagues and of maize by Y. Karpilov and coworkers. When leaves were exposed for a few seconds to $^{14}CO_2$ in the light, 70 to 80% of the label was found in the C_4 acids malate and aspartate—a pattern very different from the one observed in leaves that photosynthesize solely via the Calvin cycle. In pursuing these initial observations, M. D. Hatch and C. R. Slack elucidated what is now known as the C_4 cycle. They established that the C_4 acids malate and aspartate are the first stable, detectable intermediates of photosynthesis in leaves of sugarcane and that carbon atom 4 of malate subsequently becomes carbon atom 1 of 3-phosphoglycerate (Hatch and Slack 1966). The primary carboxylation in these leaves is catalyzed not by rubisco, but by PEP (phosphoenylpyruvate) carboxylase (O'Leary 1982).

The manner in which carbon was transferred from carbon atom 4 of malate to carbon atom 1 of 3-phosphoglycerate became clear when the involvement of mesophyll and bundle sheath cells was elucidated. The participating enzymes occur in one of the two cell types: PEP carboxylase and pyruvate–orthophosphate dikinase are restricted to mesophyll cells; the decarboxylases and the enzymes of the complete Calvin cycle are confined to the bundle sheath cells. With this knowledge, Hatch and Slack were able to formulate the basic model of the cycle (Figure 8.12).

(A)

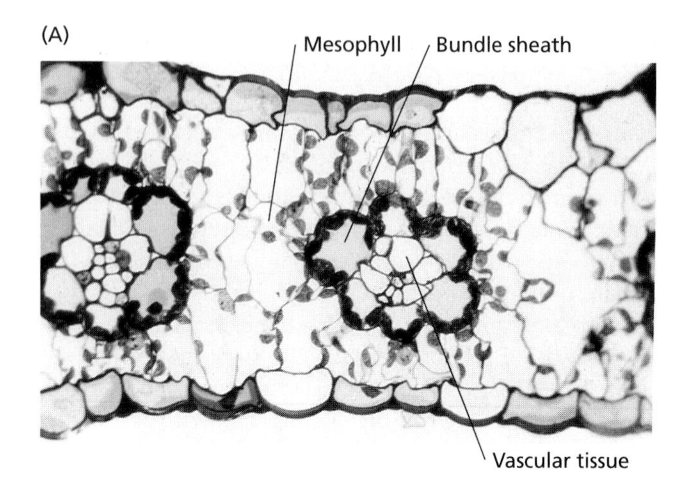

(B)

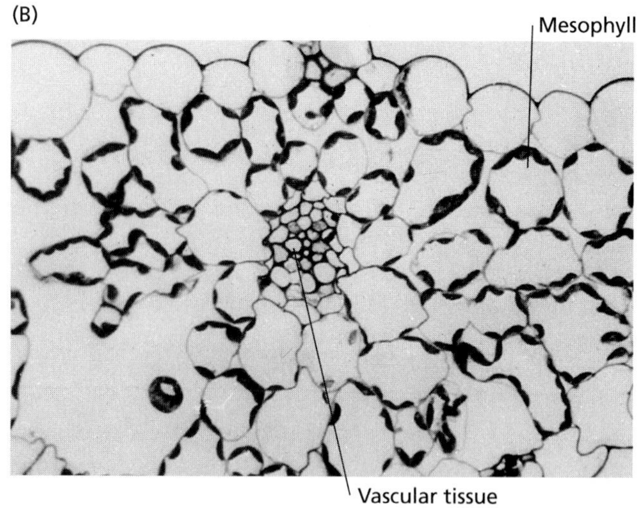

(C)

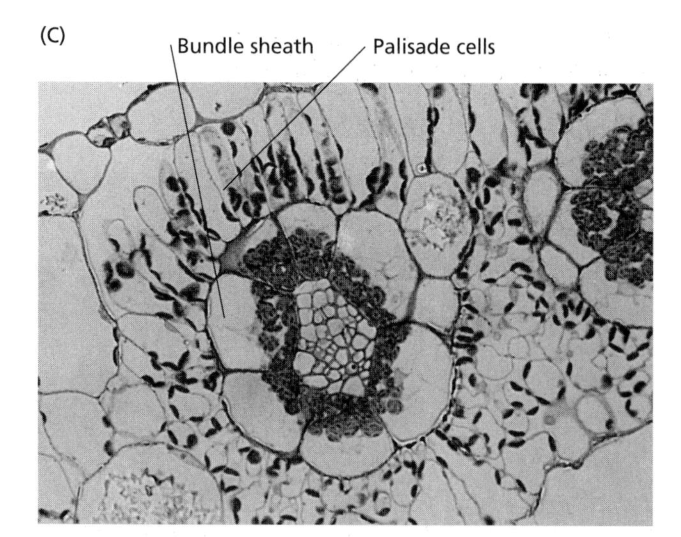

(D)

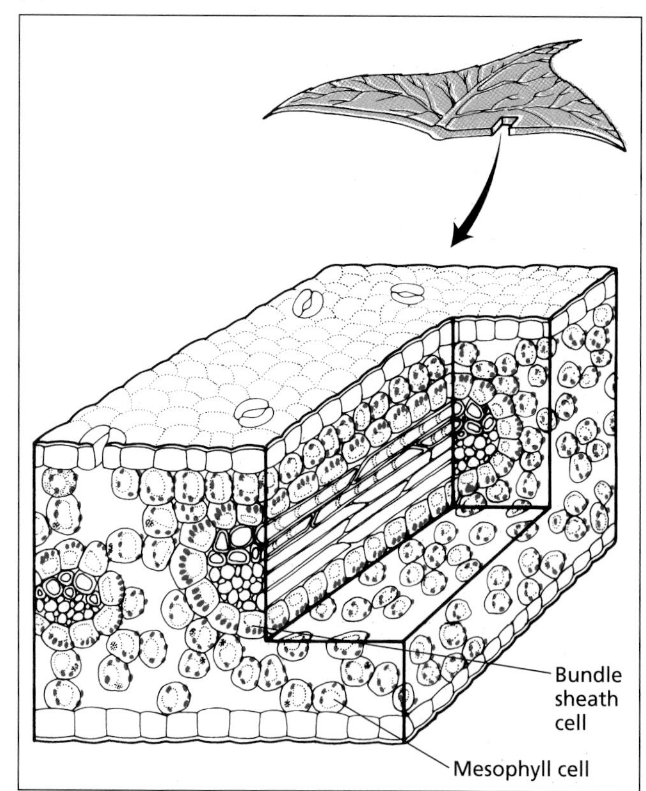

(E)

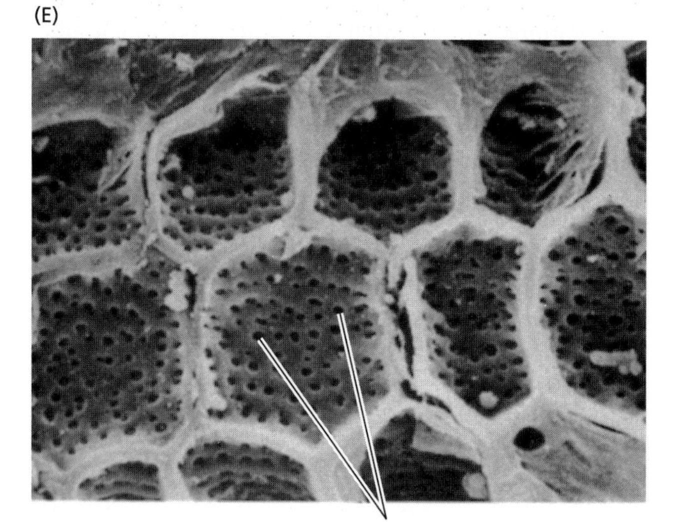

Figure 8.11 Cross sections of leaves, showing the anatomic differences between C_3 and C_4 plants. (A) A C_4 monocot, *Zea mays* (corn). (350×) (B) A C_3 monocot, *Avena sativa* (oat). (380×) (C) A C_4 dicot, *Gomphrena* (amaranth). (740×) The bundle sheath cells are large in C_4 leaves (A and C), and no mesophyll cell is more than two or three cells away from the nearest bundle sheath cell. These anatomic features are absent in the C_3 leaf (B). (D) Three-dimensional model of a C_4 leaf. (E) Scanning electron micrograph of a C_4 leaf from *Triodia irritans*, showing the plasmodesmata pits in the bundle sheath cell walls through which metabolites of the C_4 photosynthetic carbon cycle are thought to be transported. (1450×) (A, B, and C from Edwards and Walker 1983; photographs by S. E. Fredrick and E. H. Newcomb; D from Lüttge and Higinbotham 1979 ; E from Craig and Goodchild 1977.)

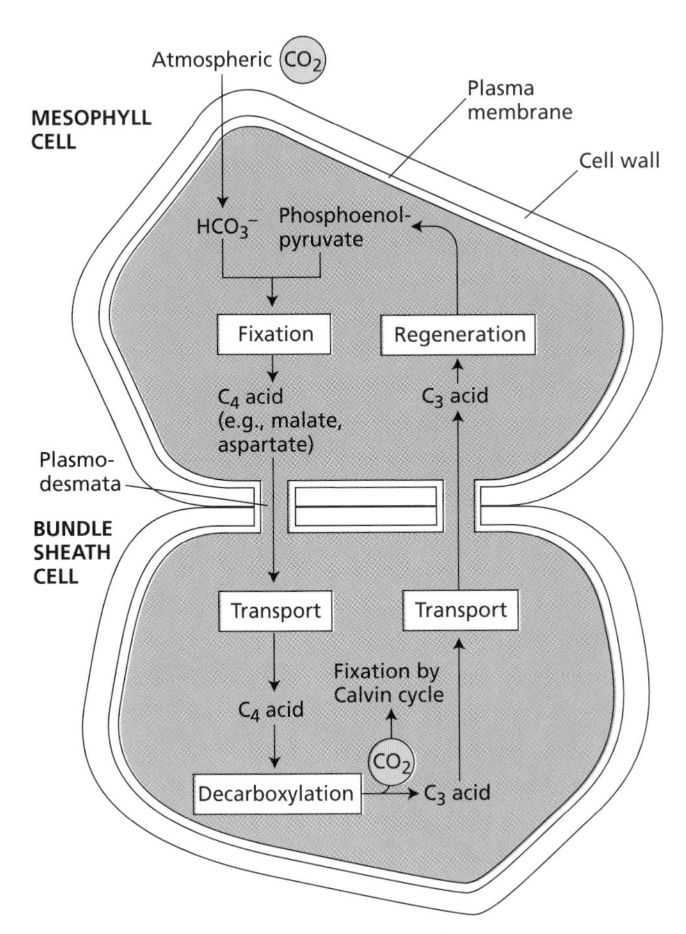

Figure 8.12 The basic C_4 photosynthetic carbon cycle involves two cell types and proceeds in four stages: (1) *fixation* of CO_2 into a four-carbon acid in the mesophyll cells; (2) *transport* of the four-carbon acid from the mesophyll cells to the bundle sheath cells; (3) *decarboxylation* of the four-carbon acid, generating a high concentration of CO_2 in the bundle sheath cells; and (4) *transport* of the residual three-carbon acid back to the mesophyll cells, where the original CO_2 acceptor, phosphoenolpyruvate, is regenerated. The C_4 cycle concentrates CO_2 in the bundle sheath cells, to which rubisco and the other enzymes of the Calvin cycle are restricted. This high concentration of CO_2 in the bundle sheath cells suppresses the oxygenation of ribulose-1,5-bisphosphate by rubisco.

The C_4 Cycle Concentrates CO_2 in Bundle Sheath Cells

The basic C_4 cycle (see Figure 8.12) consists of four stages: (1) fixation of CO_2 by the carboxylation of phosphoenolpyruvate in the mesophyll cells to form a C_4 acid (malate and/or aspartate); (2) transport of the C_4 acids to the bundle sheath cells; (3) decarboxylation of the C_4 acids within the bundle sheath cells and generation of CO_2, which is then reduced to carbohydrate via the Calvin cycle; and (4) transport of the C_3 acid (pyruvate or alanine) that is formed by the decarboxylation step back to the mesophyll cell and regeneration of the CO_2 acceptor phosphoenolpyruvate. One interesting feature of the cycle is that the enzyme that catalyzes this regeneration, pyruvate–orthophosphate dikinase, consumes two "high-energy" phosphate bonds through the conversion of ATP to AMP (reaction 7 in Table 8.3). The regeneration of ATP from AMP requires the incorporation of two phosphate bonds, one to form ADP and the second to form ATP, thereby explaining the high energy requirement of the cycle about to be described.

The cycle thus effectively shuttles CO_2 from the atmosphere into the bundle sheath cells. This transport process generates a much higher concentration of CO_2 in the bundle sheath cells than would occur in equilibrium with the external atmosphere. This elevated concentration of CO_2 at the site of carboxylation of ribulose-1,5-bisphosphate results in suppression of the oxygenation of ribulose-1,5-bisphosphate and hence of photorespiration.

Discovered in tropical grasses (e.g., sugarcane and maize), the C_4 cycle is now known to occur in 16 families of both monocotyledons and dicotyledons, and it is particularly prominent in Gramineae (sugarcane, corn, sorghum), Chenopodiaceae (*Atriplex*), and Cyperaceae (sedges). About 1% of the characterized species have C_4 metabolism (Edwards and Walker 1983).

There are three variations of the basic C_4 pathway (Figure 8.13). The variations differ principally in the C_4 acid transported into the bundle sheath cells (malate or aspartate) and in the manner of decarboxylation. They are named after the enzymes that catalyze their decarboxylation reactions: NADP-dependent malic enzyme (NADP-ME) found in chloroplasts; NAD-dependent malic enzyme (NAD-ME) in mitochondria; and phosphoenolpyruvate carboxykinase (PEP-CK) in the cytosol. The following discussion traces the route of carbon for the NADP-ME type, such as sugarcane or corn. Details of the enzyme reactions of the other C_4 pathway variants are given in Table 8.3.

The Carbon Path The primary carboxylation reaction, catalyzed by PEP carboxylase, which is common to all three variants, occurs in the cytosol of the mesophyll cells and uses HCO_3^- rather than CO_2 as a substrate (Table 8.3, reaction 1). In the NADP-ME variant, oxaloacetate is rapidly reduced to malate in the mesophyll chloroplasts by NADPH using NADP:malate dehydrogenase (Table 8.3, reaction 2). The malate formed enters the chloroplast of the bundle sheath cell and there undergoes oxidative decarboxylation (Table 8.3, reaction 4), yielding pyruvate. The CO_2 released within the bundle sheath cells is converted to carbohydrate by the Calvin cycle.

The residual C_3 acid is transported back to the mesophyll as pyruvate and converted to phosphoenolpyruvate in the mesophyll chloroplast (Table 8.3, reaction 7). This reaction takes place in the mesophyll chloroplast of

Table 8.3
Reactions of the C_4 photosynthetic carbon cycle

1. *Phosphoenolpyruvate carboxylase*
   Phosphoenolpyruvate + HCO_3^- → oxaloacetate + $HOPO_3^{2-}$

```
CH2                     COO−
||                      ||
C—O—PO3 2−              CH2
|                       |
COO−                    C=O
                        |
                        COO−
```

2. *NADP:malate dehydrogenase*
   Oxaloacetate + NADPH + H^+ → malate + $NADP^+$

```
COO−              COO−
|                 |
CH2               CH2
|                 |
C=O           H—C—OH
|                 |
COO−              COO−
```

3. *Aspartate aminotransferase*
   Oxaloacetate + glutamate ⇔ aspartate + α-ketoglutarate

```
COO−        COO−        COO−        COO−
|           |           |           |
CH2         CH2         CH2         CH2
|           |           |           |
C=O         CH2     H—C—NH2         CH2
|           |           |           |
COO−    H—C—NH2         COO−        C=O
            |                       |
            COO−                    COO−
```

4. *NAD(P) malic enzyme*
   Malate + $NADP^+$ → pyruvate + CO_2 + NADPH + H^+

```
    COO−        CH2
    |           |
    CH2         C=O
    |           |
H—C—OH          COO−
    |
    COO−
```

5. *Phosphoenolpyruvate carboxykinase*
   Oxaloacetate + ATP → phosphoenolpyruvate + CO_2 + ADP

```
COO−            CH2
|               ||
CH2             C—O—PO3 2−
|               |
C=O             COO−
|
COO−
```

6. *Alanine aminotransferase*
   Pyruvate + glutamate ⇔ alanine + α-ketoglutarate

```
CH3         COO−        CH3         COO−
|           |           |           |
C=O         CH2     H—C—NH2         CH2
|           |           |           |
COO−        CH2         COO−        CH2
            |                       |
        H—C—NH2                     C=O
            |                       |
            COO−                    COO−
```

7. *Pyruvate–orthophosphate dikinase*
   Pyruvate + $HOPO_3^{2-}$ + ATP → phosphoenolpyruvate + AMP + $H_2P_2O_7^{2-}$

```
CH3             CH2
|               ||
C=O             C—O—PO3 2−
|               |
COO−            COO−
```

Note: The name in italics following the number of each reaction is the enzyme that catalyzes the reaction.

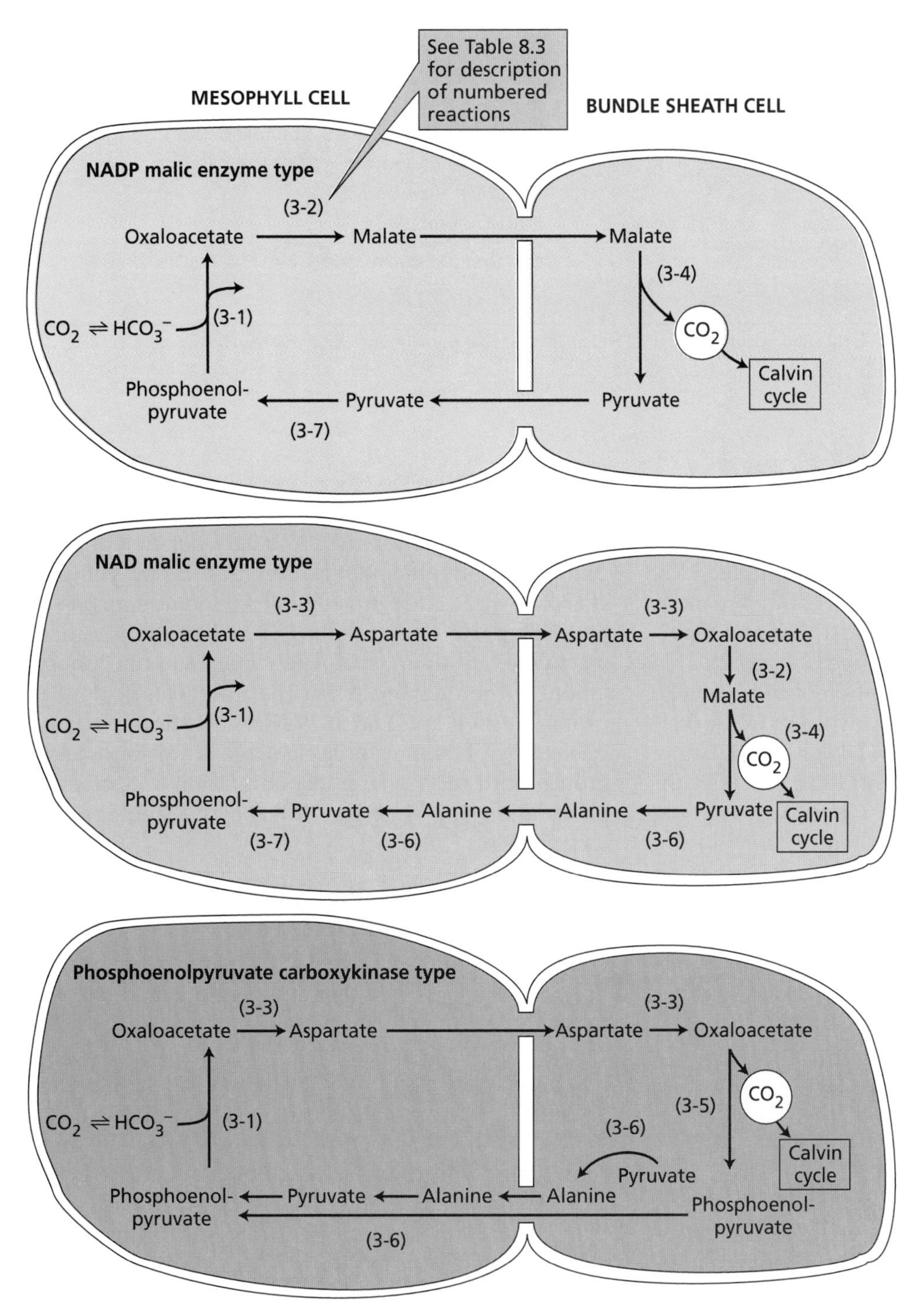

Figure 8.13 The three variants of the C_4 photosynthetic carbon cycle. The variants differ principally in (1) the nature of the four-carbon acid (malate or aspartate) transported into the bundle sheath cell and of the three-carbon acid (pyruvate or alanine) returned to the mesophyll cell and (2) the nature of the enzyme that catalyzes the decarboxylation step in the bundle sheath cell. The three variants are named after the enzymes that catalyze the decarboxylation reactions. Representatives of each variant include maize, crabgrass, sugarcane, sorghum (NADP malic enzyme); pigweed, millet (NAD malic enzyme); guinea grass (phosphoenolpyruvate carboxykinase).

all three C_4 variants. The cycle then starts again with carboxylation of the phosphoenolpyruvate to produce oxaloacetate (Gutierrez et al. 1974).

Another process localized in the mesophyll chloroplast of all three C_4 variants is the reduction of 3-phosphoglycerate to triose phosphates, which occurs within the Calvin cycle proper in C_3 chloroplasts. In contrast, all C_4 variants export 3-phosphoglycerate from the bundle sheath for reduction in the mesophyll chloroplast. After reduction, the triose phosphates are transported to the bundle sheath chloroplasts, for use in the operation of the Calvin cycle there. One possible reason for this remarkable split of the Calvin cycle between the bundle sheath and the mesophyll chloroplast is to limit electron transport activity and oxygen evolution in the bundle sheath, and thus avoid oxygen levels that would favor photorespiration (Bowyer and Leegood 1997). The transport of metabolites of the C_4 pathway is accompanied by movements of protons and other ionic species in order to maintain pH and charge balance.

Table 8.4
Energetics of the C_4 photosynthetic carbon cycle

Phosphoenolpyruvate + H_2O + NADPH + CO_2 (mesophyll)	→	malate + $NADP^+$ + $HOPO_3^{2-}$ (mesophyll)
Malate + $NADP^+$	→	pyruvate + NADPH + CO_2 (bundle sheath)
Pyruvate + $HOPO_3^{2-}$ + ATP	→	phosphoenolpyruvate + AMP + $H_2P_2O_7^{2-}$
$P_2O_7^{4-}$ + H_2O	→	2 $HOPO_3^{2-}$ (mesophyll)
AMP + ATP	→	2 ADP (mesophyll)
Net: CO_2 (mesophyll) + 2 ATP + 2 H_2O	→	CO_2 (bundle sheath cell) + 2 ADP + 2 $HOPO_3^{2-}$ + 2 H^+
Cost of concentrating CO_2 within the bundle sheath cell = 2 ATP per CO_2		

Note: As shown in reaction 1 of Table 8.3, the H_2O and CO_2 shown in the first line of this table actually react with phosphoenolpyruvate as HCO_3^-.

The Concentration of CO_2 in Bundle Sheath Cells Has an Energy Cost

The net effect of the C_4 cycle is to convert a dilute solution of CO_2 in mesophyll cells into a concentrated CO_2 solution in cells of the bundle sheath. Thermodynamics tells us that work must be done to establish and maintain such a concentration gradient (see Chapter 2). This principle also applies to the operation of the C_4 cycle. From a summation of the reactions involved, we can calculate the energy cost to the plant (Table 8.4). The calculation shows that the CO_2 concentrating process consumes two extra molecules of ATP per CO_2 molecule transported. Thus, the total energy requirement for fixing CO_2 by the combined C_4 and Calvin cycles is five ATP plus two NADPH per CO_2 fixed.

Because of this higher energy demand, C_4 plants photosynthesizing under nonphotorespiratory conditions (high CO_2 and low O_2) require more quanta of light per CO_2 than C_3 leaves do. In normal air, the quantum requirement of C_3 plants changes with factors that affect the balance between photosynthesis and photorespiration, such as temperature, while the quantum requirement of C_4 plants remains relatively constant.

Light Regulates the Activity of Key C_4 Enzymes

Operation of the C_4 cycle requires several levels of metabolic regulation. Shuttling of metabolites between mesophyll and bundle sheath cells is driven by diffusion gradients along numerous plasmodesmata, and transport is regulated by concentration gradients and the operation of specialized translocators at the chloroplast envelope. Another level of regulation is enzyme activation by light. For example, the activities of PEP carboxylase (see Table 8.3, reaction 1), NADP:malate dehydrogenase (see Table 8.3, reaction 2), and pyruvate– orthophosphate dikinase (see Table 8.3, reaction 7) are regulated in response to variations in photon flux density by two different processes, reduction–oxidation of thiol groups and phosphorylation–dephosphorylation.

NADP:malate dehydrogenase is regulated via the thioredoxin system of the chloroplast (see Figure 8.6). The enzyme is reduced (activated) upon illumination of leaves and is oxidized (inactivated) upon darkening. PEP carboxylase is activated by a light-dependent phosphorylation–dephosphorylation mechanism yet to be characterized. The third regulatory member of the C_4 pathway, pyruvate–orthophosphate dikinase, is rapidly inactivated by an unusual ADP-dependent phosphorylation of the enzyme when the photon flux density drops (Burnell and Hatch 1985). Activation is accomplished by phosphorolytic cleavage of this phosphate group. Both reactions, phosphorylation and dephosphorylation, appear to be catalyzed by a single regulatory protein.

In Hot, Dry Climates, the C_4 Cycle Reduces Photorespiration and Water Loss

Two features of the C_4 cycle in C_4 plants overcome the deleterious effects of higher temperature on photosynthesis that were noted earlier. First, the affinity of PEP carboxylase for its substrate, HCO_3^-, is sufficiently high that the enzyme is saturated by the equivalent of air levels of CO_2. Furthermore, since the substrate is HCO_3^-, oxygen is not a competitor in the reaction. This high activity of PEP carboxylase enables C_4 plants to reduce the stomatal aperture and thereby conserve water while fixing CO_2 at rates equal to or greater than those of C_3 plants. Second, concentrating CO_2 in the bundle sheath cells suppresses photorespiration. These features enable C_4 plants to photosynthesize more efficiently at high temperatures than C_3 plants, and they are probably the reason for the relative abundance of C_4 plants in drier, hotter climates. Depending on their natural environment, some plants show properties intermediate between strictly C_3 and C_4 species.

CO_2 Concentrating Mechanisms III: Crassulacean Acid Metabolism

A third mechanism for concentrating CO_2 at the site of rubisco is found in crassulacean acid metabolism (CAM). Despite its name, CAM is not restricted to the family

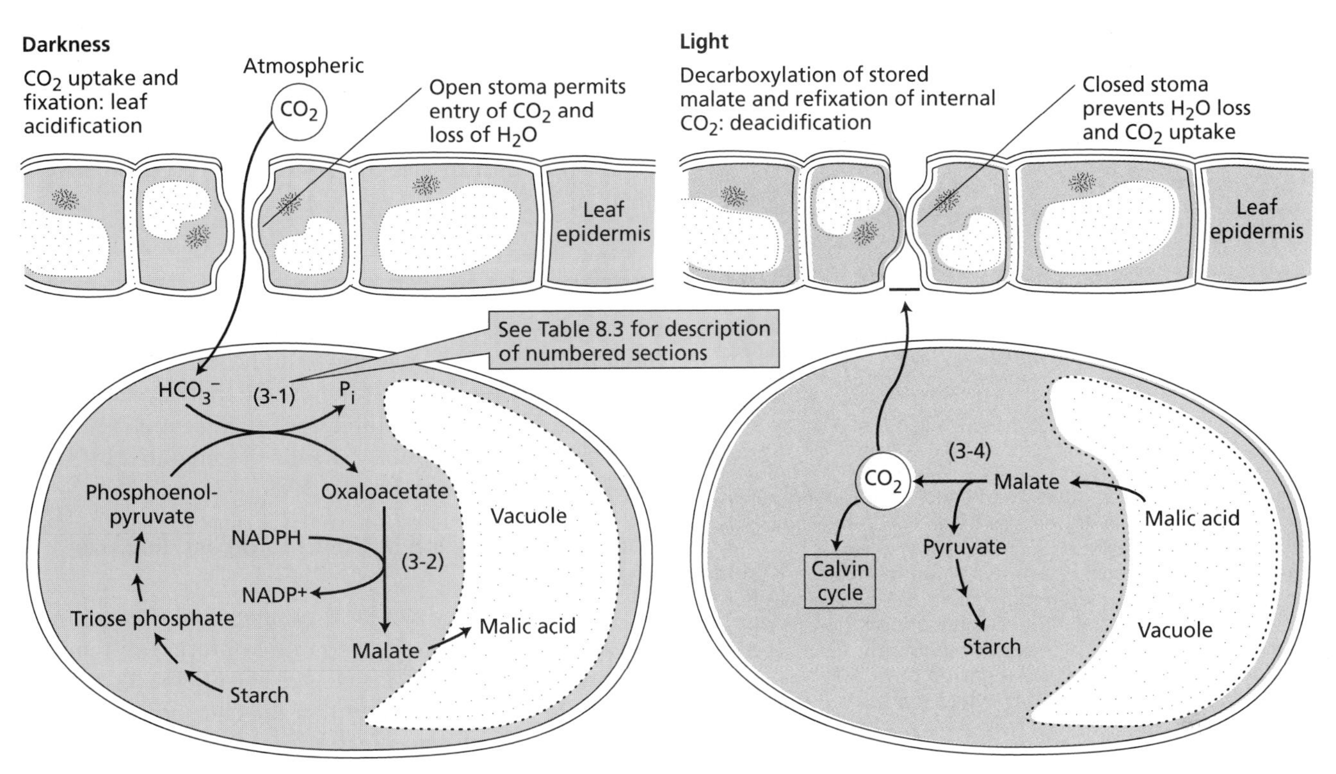

Figure 8.14 In crassulacean acid metabolism (CAM), the reactions of photosynthesis and CO_2 uptake are temporally separated: CO_2 uptake and fixation take place at night, and decarboxylation and refixation of the internally released CO_2 occur during the day. CAM is an adaptation primarily to minimize the quantity of water that is lost when stomata are opened to permit the entry of CO_2. In CAM plants, the stomata are opened in the cool of the night. CO_2 is fixed as malic acid, which is stored in the vacuole. As malic acid accumulates, the leaf vacuoles acidify in the dark. Upon illumination, the stomata close, and the leaf deacidifies. The malic acid is recovered from the vacuole and undergoes decarboxylation. The CO_2 that is released is prevented from escaping by stomatal closure and is assimilated via the Calvin cycle using photochemically generated ATP and NADPH.

Crassulaceae (*Crassula, Kalanchoe, Sedum*); it is found in numerous angiosperm families. Cacti and euphorbias are CAM plants, as well as pineapple, vanilla, and agave. The CAM mechanism enables plants to improve water use efficiency. Typically, a CAM plant loses 50 to 100 g of water for every gram of CO_2 gained, compared with values of 250 to 300 and 400 to 500 g for C_4 and C_3 plants, respectively (see Chapter 4). Thus, CAM plants have a competitive advantage in dry environments.

The CAM mechanism is similar in many respects to the C_4 cycle, but it differs in two important features. In C_4 plants, formation of the C_4 acids is spatially (not temporally) separated from decarboxylation of the C_4 acids and from refixation of the resulting CO_2 by the Calvin cycle. A specialized anatomy is needed to effect this spatial separation. In CAM plants, formation of the C_4 acids is both temporally and spatially separated. At night, CO_2 is captured by PEP carboxylase in the cytosol, and the malate that forms from the oxaloacetate product is stored in the vacuole (Figure 8.14). During the day, the stored malate is transported to the chloroplast and decarboxylated, and the released CO_2 is fixed by the Calvin cycle.

The Stomata of CAM Plants Open at Night and Close during the Day

CAM plants such as cacti achieve their high water use efficiency by opening their stomata during the cool, desert nights and by closing them during the hot, dry days. Closing the stomata during the day minimizes water loss, but since H_2O and CO_2 share the same diffusion pathway, CO_2 must then be taken up at night. CO_2 is metabolized via carboxylation of phosphoenolpyruvate to oxaloacetate, which is then reduced to malate. The malate accumulates and is stored in the large vacuoles that are a typical, but not obligatory anatomic feature of the leaf cells of CAM plants (see Figure 8.14). The accumulation of substantial amounts of malic acid, equivalent to the amount of CO_2 assimilated at night, has long been recognized as a nocturnal acidification of the leaf (Bonner and Bonner 1948). The phosphoenolpyruvate shown in Figure 8.14 originates from the breakdown of starch and other sugars by the glycolytic pathway (see Chapter 11), which operates during the night.

With the onset of day, the stomata close, preventing loss of water and further uptake of CO_2. The leaf cells deacidify as the reserves of vacuolar malic acid are con-

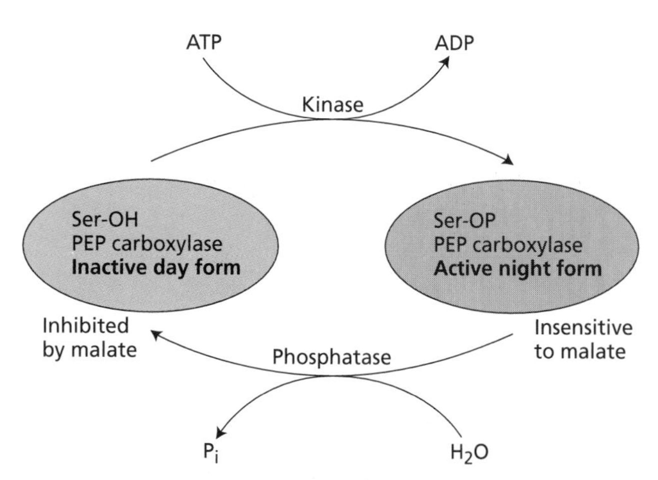

Figure 8.15 Diurnal regulation of CAM phosphoenolpyruvate (PEP) carboxylase appears to be achieved by phosphorylation of a serine residue (Ser) of the day form of the enzyme (Ser-OH), which is sensitive to inhibition by malate, to yield the phosphorylated night form (Ser-OP), which is relatively insensitive to malate. Dephosphorylation of the night form reverses the process. Consequently, PEP carboxylase is active at night during the period of malate formation but inactive during the day, when malate is decarboxylated.

sumed. Decarboxylation is achieved by the action of NADP malic enzyme on malate or PEP carboxykinase on oxaloacetate. Because the stomata are closed, the internally released CO_2 cannot escape from the leaf and instead is fixed and converted to carbohydrate by the Calvin cycle. The elevated internal concentration of CO_2 effectively suppresses the photorespiratory oxygenation of ribulose bisphosphate and favors carboxylation. The C_3 acid resulting from the decarboxylation is thought to be converted first to triose phosphate and then to starch or sucrose, thus regenerating the original carbon acceptor.

CAM Is Regulated by Phosphorylation of PEP Carboxylase

The mechanism of CAM outlined in the previous section requires that PEP carboxylase and the decarboxylases, both of which are located in the cytosol, function at different times. One way to avoid a futile cycle of carboxylation and decarboxylation is to "switch on" the carboxylase at night and "switch off" the enzyme during the day. CAM phosphoenolpyruvate carboxylase exists in two forms: The molecular species that operates at night is insensitive to malate; the one that operates during the day is inhibited by low concentrations of malate. The two forms are interconverted via phosphorylation (Figure 8.15) (Jiao and Chollet 1991).

In addition to short-term (day and night) regulation, some CAM plants show longer-term regulation and are able to adjust their pattern of CO_2 uptake to environmental conditions. Facultative CAM plants such as the ice plant (*Mesembryanthemum crystallinum*) carry on C_3 metabolism under unstressed conditions, and they shift to CAM in response to heat, water, or salt stress. This form of regulation requires the expression of numerous CAM genes in response to stress signals (Vernon et al. 1993).

Synthesis of Starch and Sucrose

Sucrose is the principal form of carbohydrate translocated throughout the plant by the phloem. Starch is an insoluble stable carbohydrate reserve that is present in almost all plants. Both starch and sucrose are synthesized from the triose phosphate that is generated by the Calvin cycle (Beck and Ziegler 1989).

Starch Is Synthesized in the Chloroplast, Sucrose in the Cytosol

The pathways for the synthesis of starch and sucrose are shown in Figure 8.16. Electron micrographs showing prominent starch deposits, as well as enzyme localization studies, leave no doubt that the chloroplast is the site of starch synthesis in leaves (Figure 8.17). The reactions for the synthesis of starch are shown in Table 8.5. Starch is synthesized from triose phosphate via fructose-1,6-bisphosphate (Table 8.5, reactions 1 through 4). The glucose-1-phosphate intermediate is converted to ADP-glucose via ADP-glucose pyrophosphorylase in a reaction that requires ATP and generates pyrophosphate (Table 8.5, reaction 5). As in many biosynthesis reactions, the pyrophosphate is hydrolyzed via a specific inorganic pyrophosphatase to two orthophosphate ($HOPO_3^{2-}$) molecules, thereby driving reaction 5 toward ADP-glucose synthesis (Table 8.5, reaction 6). Finally, the glucose moiety of ADP-glucose is transferred to the nonreducing end (carbon 4) of the terminal glucose of a growing starch chain (Table 8.5, reaction 7), thus completing the reaction sequence.

The site of sucrose synthesis has been studied by cell fractionation, in which the organelles are isolated and separated from one another. Enzyme analyses have shown that sucrose is synthesized in the cytosol from triose phosphates by a pathway similar to that of starch—that is, by way of fructose-1,6-bisphosphate and glucose-1-phosphate (Table 8.6, reactions 2 through 6). The glucose-1-phosphate is converted to UDP-glucose via a specific UDP-glucose pyrophosphorylase (Table 8.6, reaction 7) that is analogous to the ADP-glucose pyrophosphorylase of chloroplasts. The UDP-glucose that forms is combined with fructose-6-phosphate to yield sucrose-6-phosphate and UDP (Table 8.6, reaction 8).

As in starch synthesis, the pyrophosphate is hydrolyzed, but not immediately as in the chloroplasts. Because of the absence of an inorganic pyrophosphatase, the pyrophosphate is used by other enzymes, such as fructose-6-phosphate:pyrophosphate phosphotrans-

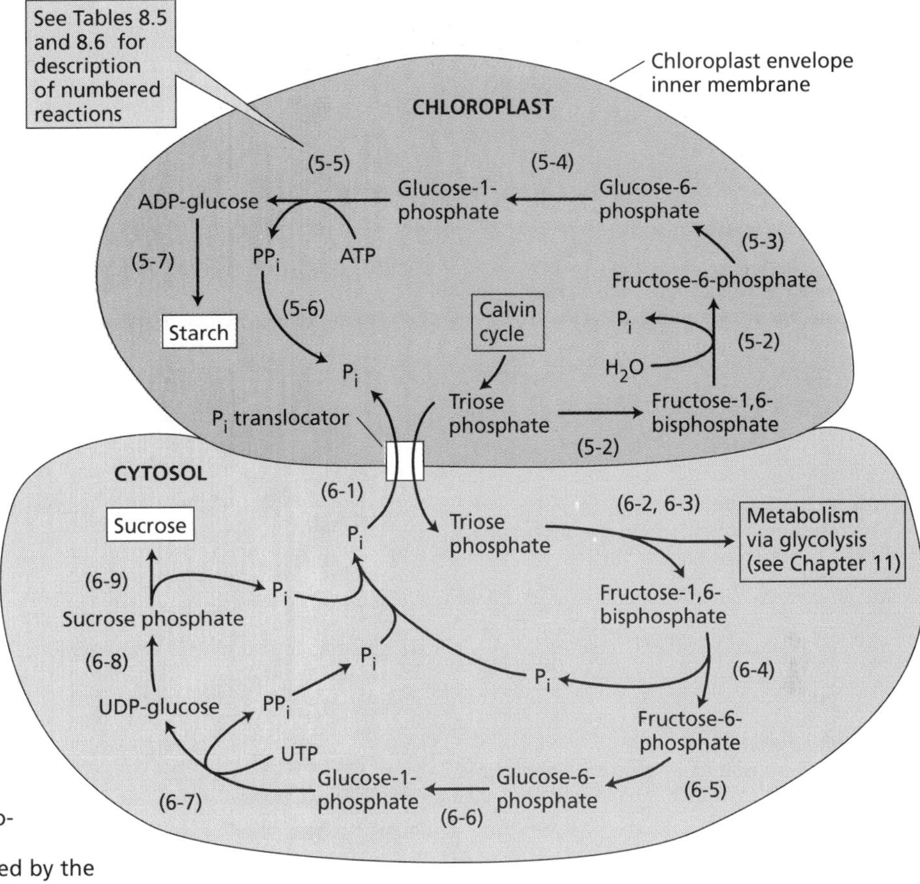

Figure 8.16 The syntheses of starch and sucrose are competing processes that occur in different cellular compartments. Starch synthesis and deposition is localized in the chloroplast. Sucrose, the principal carbohydrate transported throughout most plants, is synthesized in the cytosol. The partitioning of triose phosphate between starch synthesis in the chloroplast and sucrose synthesis in the cytosol is determined in part by the orthophosphate (P_i) concentration in these cellular compartments. When the cytosolic P_i concentration is high, chloroplast triose phosphate is exported to the cytosol in exchange for P_i, and sucrose is synthesized. When the cytosolic P_i concentration is low, triose phosphate is retained within the chloroplast, and starch is synthesized. Reversibility of the reaction effected by the P_i translocator is not shown. Recent work shows that the transport of hexoses from the chloroplast to the cytosol is also important, especially at night, when the starch is broken down by amylases. This hexose transport mechanism bypasses the cytosolic fructose-1,6-bisphosphate phosphatase, which would be largely inactive because of the high level of fructose-2,6-bisphosphate ocurring at night.

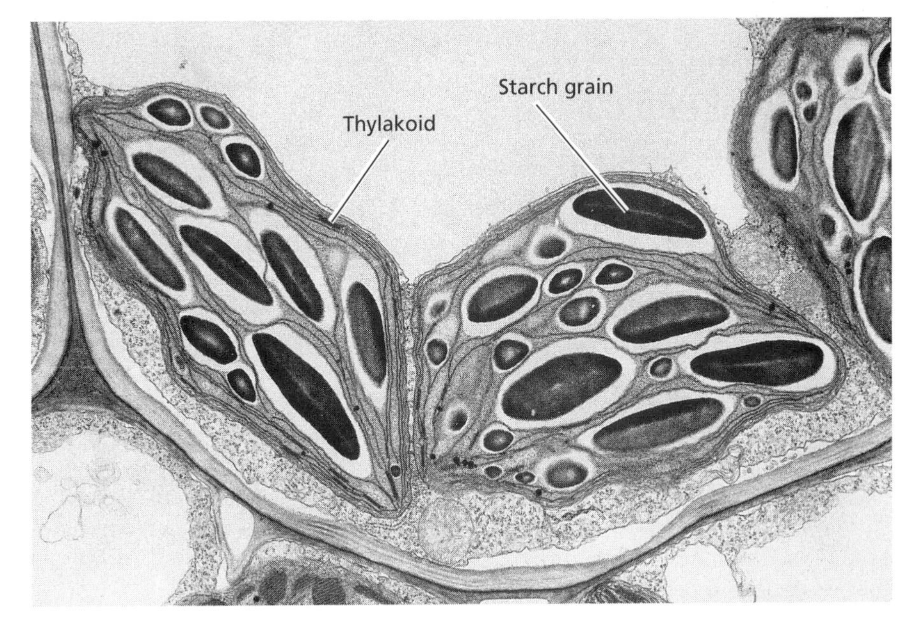

Figure 8.17 Electron micrograph of a bundle sheath cell from maize, showing the starch grains in the chloroplasts. (15,800×) (Photo by S. E. Frederick, courtesy of E. H. Newcomb.)

Table 8.5
Reactions of starch synthesis from triose phosphate in chloroplasts

1. *Fructose-1,6,bisphosphate aldolase*
   Dihydroxyacetone-3-phosphate + glyceraldehyde-3-phosphate→ fructose-1,6-bisphosphate

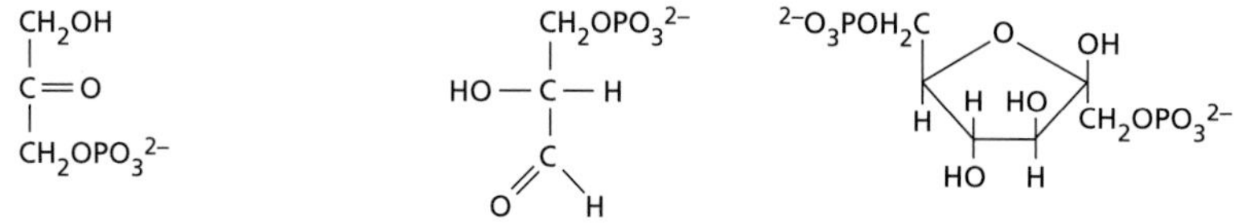

2. *Fuctose-1,6-bisphosphate phosphatase*
   Fructose-1,6-bisphosphate + H_2O → fructose-6-phosphate + $HOPO_3^{2-}$

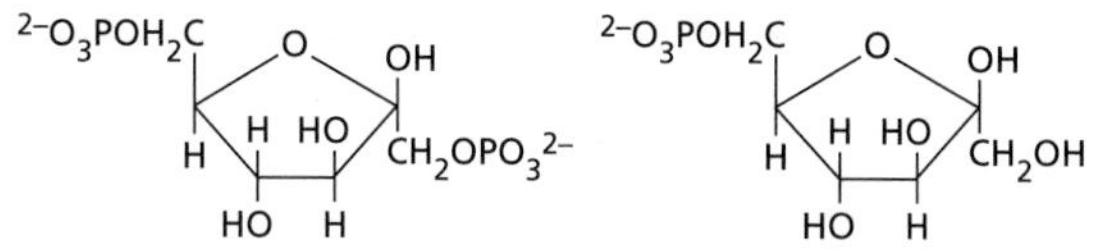

3. *Hexose phosphate isomerase*
   Fructose-6-phosphate → glucose-6-phosphate

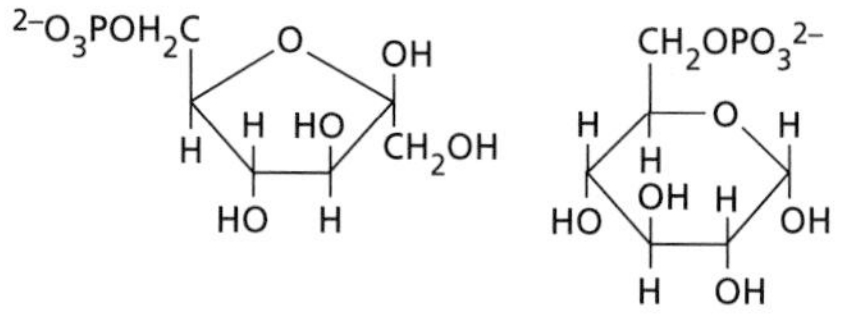

4. *Phosphoglucomutase*
   Glucose-6-phosphate → glucose-1-phosphate

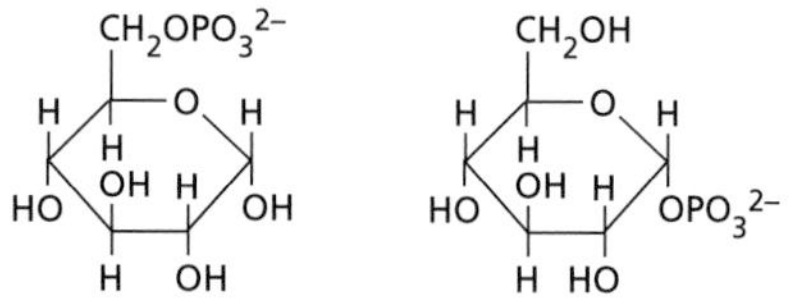

5. *ADP-glucose pyrophosphorylase*
   Glucose-1-phosphate + ATP → ADP-glucose + $H_2P_2O_7^{2-}$

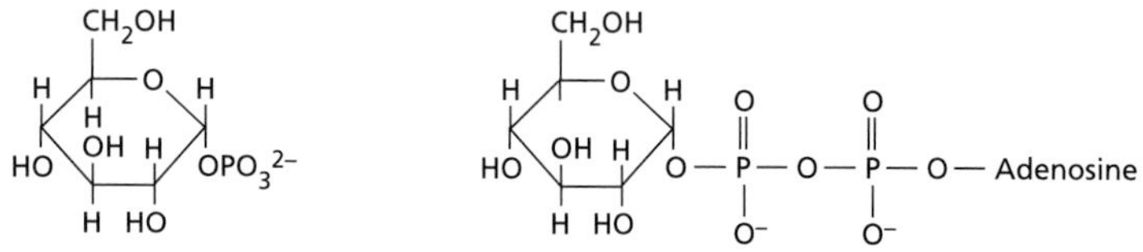

6. *Pyrophosphatase*
   $H_2P_2O_7^{2-} + H_2O \rightarrow 2\ HOPO_3^{2-} + 2\ H^+$

7. *Starch synthase*
   ADP-glucose (1,4-α-D-glucosyl)$_n$ → ADP + (1,4-α-D-glucosyl)$_{n+1}$

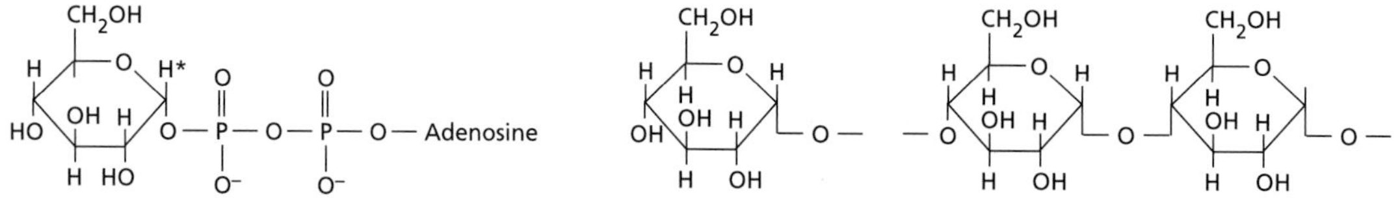

   Nonreducing end of a starch chain with *n* residues

   Elongated starch with *n* + 1 residues

Note: Reaction 4 is irreversible and "pulls" the preceding reaction to the right.

Table 8.6
Reactions of sucrose synthesis from triose phosphate in the cytosol

1. *Phosphate/triose phosphate translocator*
   Triose phosphate (chloroplast) + $HOPO_3^{2-}$ (cytosol) → triose phosphate (cytosol) + $HOPO_3^{2-}$ (chloroplast)

2. *Triose phosphate isomerase*
   Dihydroxyacetone-3-phosphate → glyceraldehye-3-phosphate

3. *Fructose-1,6-bisphosphate aldolase*
   Dihydroxyacetone-3-phosphate + glyceraldehye-3-phosphate → fructose-1,6-bisphosphate

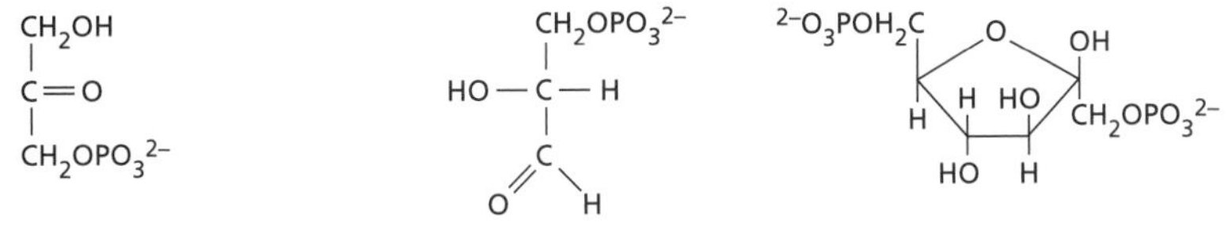

4. *Fructose-1,6-bisphosphate phosphatase*
   Fructose-1,6-bisphosphate + H_2O → fructose-6-phosphate + $HOPO_3^{2-}$

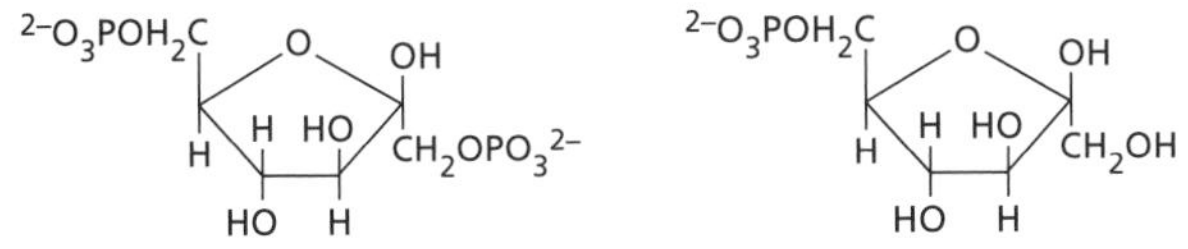

4a. *PP_i-linked phosphofructokinase*
   Fructose-6-phosphate + $H_2P_2O_7^{2-}$ → fructose-1,6-bisphosphate + $HOPO_3^{2-}$

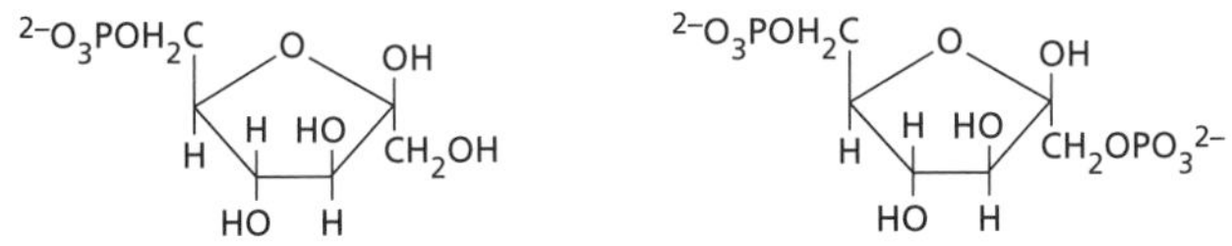

5. *Hexose-phosphate isomerase*
   Fructose-6-phosphate → glucose-6-phosphate

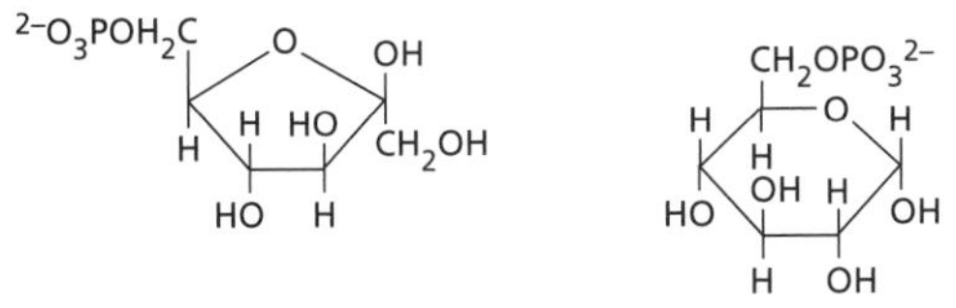

6. *Phosphoglucomutase*
   Glucose-6-phosphate → glucose-1-phosphate

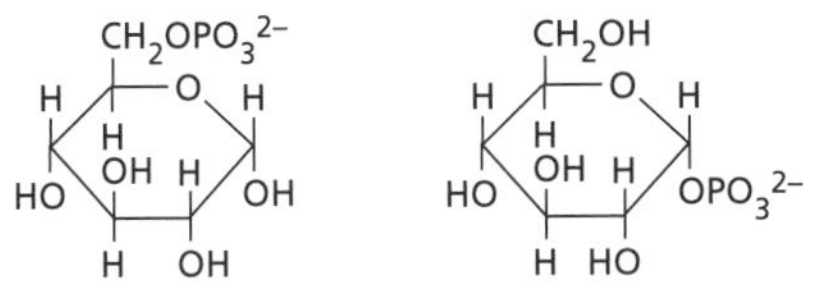

7. *UDP-glucose pyrophosphorylase*
   Glucose-1-phosphate + UTP → UDP-glucose + $H_2P_2O_7^{2-}$

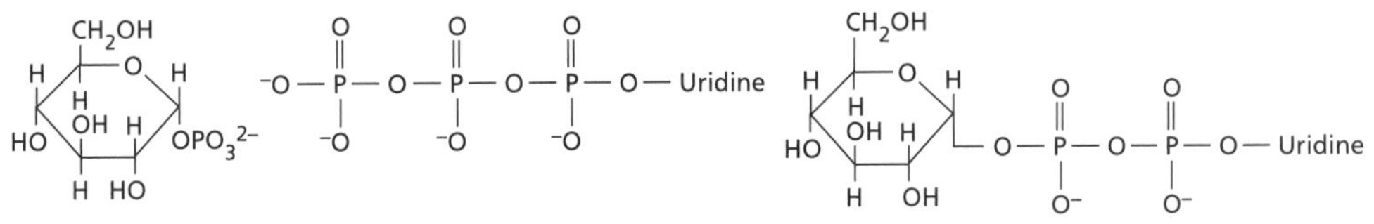

Table 8.6 (continued)
Reactions of sucrose synthesis from triose phosphate in the cytosol

8. *Sucrose phosphate synthase*
UDP-glucose + fructose-6-phosphate → UDP + sucrose-6-phosphate

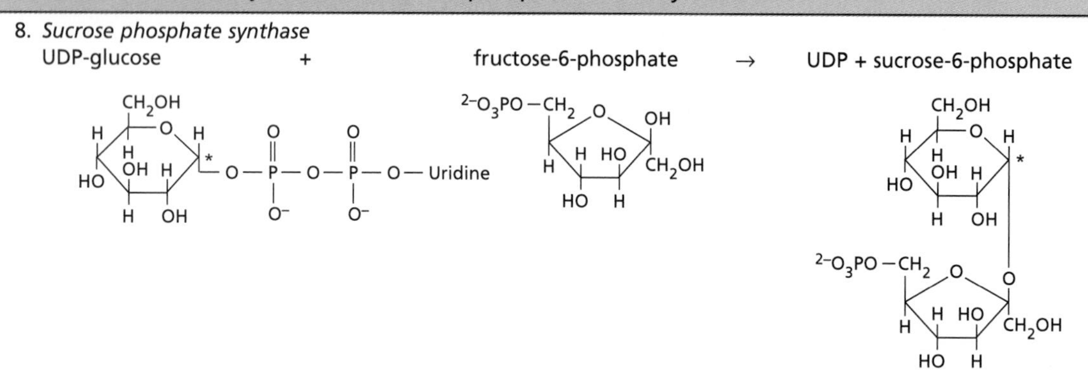

9. *Sucrose phosphate phosphatase*
Sucrose-6-phosphate + H_2O → sucrose + $HOPO_3^{2-}$

Note: Reaction 1 takes place on the chloroplast inner envelope membrane. Reactions 2 through 9 take place in the cytosol.

ferase, in transphosphorylation reactions. The sucrose-6-phosphate is hydrolyzed to yield sucrose and orthophosphate (Table 8.6, reaction 9), which drives reaction 8 in the direction of sucrose-6-phosphate synthesis.

A comparison of Tables 8.5 and 8.6 reveals that the conversion of triose phosphates to glucose-1-phosphate in the pathways leading to the synthesis of starch and sucrose have several steps in common. However, these pathways utilize isozymes (different forms of the enzymes that catalyze the same reaction) that are unique to the chloroplast or cytosol. The isozymes show markedly different properties. For example, the chloroplastic fructose-1,6-bisphosphate phosphatase is regulated by the thioredoxin system but not by fructose-2,6-bisphosphate and AMP. Conversely, the cytosolic form of the enzyme is regulated by fructose-2,6-bisphosphate (see below), is sensitive to AMP especially in the presence of fructose-2,6-bisphoshate, and is unaffected by thioredoxin.

Aside from the cytosolic fructose-1,6-bisphosphatase, sucrose synthesis is regulated at the level of sucrose phosphate synthase, an allosteric enzyme that is activated by glucose-6-phosphate and inhibited by orthophosphate. The enzyme is inactivated (in the dark) by phosphorylation of a specific serine residue and activated (in the light) via dephosphorylation. Certain metabolites that influence the phosphorylation status of sucrose phosphate synthase have recently been described, and the exact signal prompting these changes is under active investigation.

The Syntheses of Sucrose and Starch Are Competing Reactions

The relative concentrations of orthophosphate and triose phosphate are major factors that control whether photosynthetically fixed carbon is partitioned as starch in the chloroplast or as sucrose in the cytosol. The two com-

Fructose-2,6-bisphosphate
(a regulatory metabolite)

Fructose-1,6-bisphosphate
(an intermediary metabolite)

Figure 8.18 Structure of fructose-2,6-bisphosphate. This regulatory metabolite differs from the well-known fructose-1,6-bisphosphate intermediary metabolite by having a phosphate group on carbon 2 rather than on carbon 1. The carbon atom numbers are shown on the left structure.

partments communicate with one another via the phosphate/triose phosphate translocator, also called the phosphate translocator (see Table 8.6, reaction 1), a strict stoichiometric antiporter described by H. W. Heldt and his colleagues.

The phosphate translocator catalyzes the movement of orthophosphate and triose phosphate in opposite directions between chloroplast and cytosol. A low concentration of orthophosphate in the cytosol limits the export of triose phosphate from the chloroplast through the translocator, thereby promoting the synthesis of starch. Conversely, an abundance of orthophosphate in the cytosol inhibits starch synthesis within the chloroplast and promotes the export of triose phosphate into the cytosol, where it is converted to sucrose.

Orthophosphate and triose phosphate control the activity of several regulatory enzymes in the sucrose and starch biosynthetic pathways. The chloroplast enzyme ADP-glucose pyrophosphorylase (see Table 8.5, reaction 5) is the key enzyme that regulates the synthesis of starch from glucose-1-phosphate. J. Preiss and his colleagues have demonstrated that this enzyme is stimulated by 3-phosphoglycerate and inhibited by orthophosphate. A high concentration ratio of 3-phosphoglycerate to orthophosphate is typically found in illuminated chloroplasts that are actively synthesizing starch. Reciprocal conditions prevail in the dark.

Studies by B. Buchanan, H. W. Heldt, M. Stitt, and their colleagues have established the role of fructose-2,6-bisphosphate in the regulation of cytosolic sucrose synthesis (Figure 8.18). Fructose-2,6-bisphosphate is found in the cytosol in minute concentrations, and it exerts a regulatory effect on the cytosolic interconversion of fructose-1,6-bisphosphate and fructose-6-phosphate (Huber 1986). Increased cytosolic fructose-2,6-bisphosphate is associated with decreased rates of sucrose synthesis because fructose-2,6-bisphosphate is a powerful inhibitor of cytosolic fructose-1,6-bisphosphate phosphatase (see Table 8.6, reaction 4) and an activator of the pryophosphate-dependent phosphofructokinase (see Table 8.6, reaction 4a). But what, in turn, controls the cytosolic concentration of fructose-2,6-bisphosphate?

Fructose-2,6-bisphosphate is synthesized from fructose-6-phosphate by a special fructose-6-phosphate 2-kinase (not to be confused with the fructose-6-phosphate 1-kinase of glycolysis) and is degraded specifically by fructose-2,6-bisphosphate phosphatase (not to be confused with fructose-1,6-bisphosphate phosphatase of the Calvin cycle). These activities are controlled by orthophosphate and triose phosphate. Orthophosphate stimulates fructose-6-phosphate 2-kinase and inhibits fructose-2,6-bisphosphate phosphatase, whereas triose phosphate inhibits the 2-kinase (Figure 8.19). Consequently, a low cytosolic ratio of triose phosphate to orthophosphate promotes the formation of fructose-2,6-bisphosphate, which in turn inhibits the hydrolysis of cytosolic fructose-1,6-bisphosphate and slows the rate

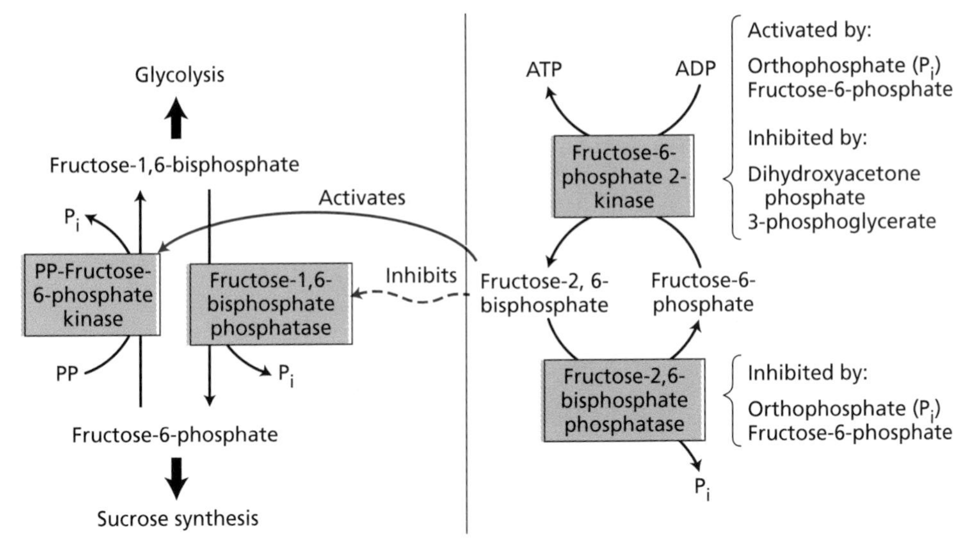

Figure 8.19 Regulation of the cytosolic interconversion of fructose-6-phosphate and fructose-1,6-bisphosphate. At left, interconversion between these key intermediates of sucrose synthesis and glycolysis is catalyzed by the enzymes pyrophosphate (PP) fructose-6-phosphate kinase and fructose-1,6-bisphosphatase. The regulatory metabolite fructose 2,6-bisphosphate regulates the interconversion by inhibiting fructose-1,6-bisphosphate phosphatase and activating PP fructose-6-phosphate kinase. At right, the synthesis of fructose-2,6-bisphosphate itself is under strict regulation by the activators and inhibitors shown in the figure, which either activate or inhibit the enzymes that catalyze fructose-2,6-bisphosphate synthesis and degradation—that is, fructose-6-phosphate 2-kinase and fructose-2,6-bisphosphate phosphatase, respectively. Light regulates the concentration of these activators and inhibitors through the reactions associated with photosynthesis and thereby controls the concentration of fructose-2,6-bisphosphate in the cytosol. Control of the fructose-2,6-bisphosphate concentration is a central mechanism that allows increased synthesis of sucrose in the light and decreased synthesis in the dark. The glycolytic enzyme phosphofructokinase (not shown) also functions in the conversion of fructose-6-phosphate to fructose-1,6-bisphosphate, but in plants it is not appreciably affected by fructose-2,6-bisphosphate. The activity of phosphofructokinase in plants appears to be regulated by the relative concentrations of ATP, ADP, and AMP. The remarkable plasticity of plants was once again illustrated by recent gene deletion experiments showing that plants can grow without a functional PP fructose-6-phosphate kinase enzyme. In this case the conversion of fructose-6-phosphate to fructose-1,6-bisphosphate is apparently catalyzed exclusively by phosphofructokinase.

of sucrose synthesis. A high cytosolic ratio of triose phosphate to orthophosphate has the opposite effect.

Summary

The reduction of CO_2 to carbohydrate via the carbon-linked reactions of photosynthesis is coupled to the consumption of NADPH and ATP synthesized by the light reactions of thylakoid membranes. Photosynthetic eukaryotes reduce CO_2 via the Calvin cycle that takes place in the stroma, or soluble phase, of chloroplasts. Here, CO_2 and water are combined with ribulose-1,5-bisphosphate to form two molecules of 3-phosphoglycerate, which are reduced and converted to carbohydrate. The continued operation of the cycle is ensured by the regeneration of ribulose-1,5-bisphosphate. The Calvin cycle consumes two molecules of NADPH and three molecules of ATP for every CO_2 fixed and, provided these substrates, has a thermodynamic efficiency close to 90%.

Several light-dependent systems act jointly to regulate the Calvin cycle—changes in ions (Mg^{2+} and H^+), effector metabolites (enzyme substrates), and protein-mediated systems (rubisco activase, ferredoxin–thioredoxin system). The ferredoxin–thioredoxin system plays a versatile role by linking light to the regulation other chloroplast processes, such as carbohydrate breakdown via the oxidative pentose phosphate pathway, photophosphorylation, sulfur assimilation, and mRNA translation.

Rubisco, the enzyme that catalyzes the carboxylation of ribulose-1,5-bisphosphate, also acts as an oxygenase. In both cases the enzyme must be carbamylated to be fully active. The carboxylation and oxygenation reactions take place at the same active site of rubisco. When reacting with oxygen, rubisco produces 2-phosphoglycolate and 3-phosphoglycerate from ribulose-1,5-bisphosphate and thereby decreases the efficiency of photosynthesis. The photorespiratory carbon oxidation cycle rescues the carbon lost as 2-phosphoglycolate by rubisco oxidase activity. The dissipative effects of photorespiration are avoided in some plants by mechanisms that concentrate CO_2 at the carboxylation sites in the chloroplast. These mechanisms include a C_4 photosynthetic carbon cycle, CAM metabolism, and "CO_2 pumps" of algae and cyanobacteria.

The carbohydrates synthesized by the Calvin cycle are converted into storage forms of energy and carbon: sucrose and starch. Sucrose, the transportable form of carbon and energy in most plants, is synthesized in the cytosol, and its synthesis is regulated by phosphorylation of sucrose phosphate synthase. Starch is synthesized in the chloroplast. The balance between the biosynthetic pathways for sucrose and starch is determined by the relative concentrations of metabolite effectors (orthophosphate, fructose-6-phosphate, 3-phosphoglycerate, and dihydroxyacetone phosphate).

These metabolite effectors function in the cytosol by way of the enzymes synthesizing and degrading fructose-2,6-bisphosphate—the regulatory metabolite that plays a primary role in controlling the partitioning of photosynthetically fixed carbon between sucrose and starch. Two of these effectors, 3-phosphoglycerate and orthophosphate, also act on starch synthesis in the chloroplast by allosterically regulating the activity of ADP-glucose pyrophosphorylase. In this way, the synthesis of starch during the day can be separated from its breakdown, which is required to provide energy to the plant at night.

General Reading

*Andrews, T. J., and Lorimer, G. H. (1987) Rubisco: Structure, mechanisms, and prospects for improvement. In *The Biochemistry of Plants*, Vol. 10: *Photosynthesis*, M. D. Hatch and N. K. Boardman, eds., Academic Press, San Diego, pp. 131–218.

*Bassham, T. A. (1965) Photosynthesis: The path of carbon. In *Plant Biochemistry*, 2nd ed., J. Bonner and E. Varner, eds., Academic Press, New York, pp. 875–902.

*Beck, E., and Ziegler, P. P. (1989) Biosynthesis and degradation of starch in higher plants. *Annu. Rev. Plant Physiol. Plant Mol. Biol.* 40: 95–117.

*Buchanan, B. B. (1980) Role of light in the regulation of chloroplast enzymes. *Annu. Rev. Plant Physiol.* 31: 341–374.

*Burnell, J. N., and Hatch, M. D. (1985) Light-dark modulation of leaf pyruvate, P_i dikinase. *Trends Biochem. Sci.* 10: 288–291.

*Edwards, G., and Walker, D. (1983) *C_3, C_4: Mechanisms, and Cellular and Environmental Regulation of Photosynthesis.* University of California Press, Berkeley.

*Flügge, U. I., and Heldt, H. W. (1991) Metabolite translocators of the chloroplast envelope. *Annu. Rev. Plant Physiol. Plant Mol. Biol.* 42: 129–144.

Hatch, M. D., and Boardman, N. K., eds. (1987) *The Biochemistry of Plants*, Vol. 10: *Photosynthesis.* Academic Press, San Diego.

*Hatch, M. D., and Osmond, C. B. (1976) Compartmentation and transport in C_4 photosynthesis. In *Transport in Plants* (*Encyclopedia of Plant Physiology*, New Series, vol. 3), C. R. Stocking and U. Heber, eds., Springer, Berlin, pp. 144–184.

*Heldt, H. W. (1979) Light-dependent changes of stromal H^+ and Mg^{+2} concentrations controlling CO_2 fixation. In *Photosynthesis II* (*Encyclopedia of Plant Physiology*, New Series, vol. 6), M. Gibbs and E. Latzko, eds., Springer, Berlin, pp. 202–207.

Huber, S. C., Huber, J. L., and McMichael, R. W., Jr. (1994) Control of plant enzyme activity by reversible protein phosphorylation. *Int. Rev. Cytol.* 149: 47–98.

*Leegood, R. C., Lea, P. J., Adcock, M. D., and Haeusler, R. E. (1995) The regulation and control of photorespiration. *J. Exp. Bot.* 46: 1397–1414.

*Lorimer, G. H. (1981) The carboxylation and oxygenation of ribulose 1,5-bisphosphate: The primary events in photosynthesis and photorespiration. *Annu Rev. Plant Physiol.* 32: 349–383.

*Ogren, W. L. (1984) Photorespiration: Pathways, regulation and modification. *Annu Rev. Plant Physiol.* 35: 415–442.

*O'Leary, M. H. (1982) Phosphoenolpyruvate carboxylase: An enzymologist's view. *Annu Rev. Plant Physiol.* 33: 297–315.

*Purton, S. (1995) The chloroplast genome of *Chlamydomonas. Sci. Prog.* 78: 205–216.

Stitt, M. (1990) Fructose-2,6-bisphosphate as a regulatory molecule in plants. *Annu. Rev. Plant Physiol. Plant Mol. Biol.* 41: 153–185.

*Tolbert, N. E. (1981) Metabolic pathways in peroxisomes and glyoxysomes. *Annu. Rev. Biochem.* 50: 133–157.

* Indicates a reference that is general reading in the field and is also cited in this chapter.

Chapter References

Besse, I., and Buchanan, B. B. (1997) Thioredoxin-linked plant and animal processes: The new generation. *Bot. Bull. Acad. Sinica* 38: 1–11.

Bonner, W., and Bonner, J. (1948) The role of carbon dioxide in acid formation by succulent plants. *Am. J. Bot.* 35: 113–117.

Bowyer, J. R., and Leegood, R. C. (1997) Photosynthesis. In *Plant Biochemistry*, P. M. Dey and J. B. Harbone, eds., Academic Press, San Diego, pp. 49–110.

Buchanan, B. B. (1992) Carbon dioxide assimilation in oxygenic and anoxygenic photosynthesis. *Photosynth. Res.* 33: 147–162.

Craig, S., and Goodchild, D. J. (1977) Leaf ultrastructure of *Triodia irritans*: A C_4 grass possessing an unusual arrangement of photosynthetic tissues. *Aust. J. Bot.* 25: 277–290.

Frederick, S. E., and Newcomb, E. H. (1969) Cytochemical localization of catalase in leaf microbodies (peroxisomes). *J. Cell Biol.* 43: 343–353.

Gutierrez, M., Gracen, V. E., and Edwards, G. E. (1974) Biochemical and cytological relationships in C4 plants. *Planta* 119: 279–300.

Gutteridge, S. (1990) *Biochim. Biophys. Acta* 1015: 1–14.

Hatch, M. D., and Slack, C. R. (1966) Photosynthesis by sugarcane leaves. A new carboxylation reaction and the pathway of sugar formation. *Biochem. J.* 101: 103–111.

Huber, S. C. (1986) Fructose-2,6-bisphosphate as a regulatory metabolite in plants. *Annu. Rev. Plant Physiol.* 37:233–246.

Jiao, J. A., and Chollet, R. (1991) Posttranslational regulation of phosphoenolpyruvate carboxylase in C_4 and CAM plants. *Plant Physiol.* 95: 981–985.

Kaplan, A., Schwarz, R., Lieman-Hurwitz, J., and Reinhold, L. (1994) Physiological and molecular studies on the response of cyanobacteria to changes in the ambient inorganic carbon concentration. In *The Molecular Biology of Cyanobacteria*, D. Bryant, ed., Kluwer, Dordrecht, Netherlands, pp. 469–485.

Kozaki, A., and Takeba, G. (1996) Photorespiration protects C_3 plants from photooxidation. *Nature* 384: 557–560.

Ku, S. B., and Edwards, G. E. (1978) Oxygen inhibition of photosynthesis. III. Temperature dependence of quantum yield and its relation to O_2/CO_2 solubility ratio. *Planta* 140: 1–6.

Lüttge, V., and Higinbotham, N. (1979) *Transport in Plants.* Springer-Verlag, New York.

Maier, R. M., Neckermann, K., Igloi, G. L., and Koessel, H. (1995) Complete sequence of the maize chloroplast genome: Gene content, hotspots of divergence and fine tuning of genetic information by transcript editing. *J. Mol. Biol.* 251: 614–628.

Ogawa, T., and Kaplan, A. (1987) The stoichiometry between CO_2 and H^+ fluxes involved in the transport of inorganic carbon in cyanobacteria. *Plant Physiol.* 83: 888–891.

Scheibe, R. (1990) Light/dark modulation: Regulation of chloroplast metabolism in a new light. *Bot. Acta* 103: 327–334.

Salvucci, M. E., and Ogran, W. L. (1996) The mechanism of Rubisco activase: Insights from stdies of the properties and structure of the enzyme. *Photosynth. Res.* 47: 1–11.

Vernon, D. M., Ostrem, J. A., and Bohnert, H. J. (1993) Stress perception and response in a facultative halophyte: The regulation of salinity-induced genes in *Mesembryanthemum crystallinum. Plant Cell Environ.* 16: 437–444.

Wolosiuk, R. A., Ballicora, M. A., and Hagelin, K. (1993) The reductive pentose phosphate cycle for photosynthetic carbon dioxide assimilation: Enzyme modulation. *FASEB J.* 7: 622–637.

9 Photosynthesis: Physiological and Ecological Considerations

THE CONVERSION OF SOLAR ENERGY to bond energy of organic compounds is a complex process that includes electron transport (see Chapter 7) and photosynthetic carbon metabolism (see Chapter 8). Previous detailed discussions of the photochemical and biochemical reactions should not overshadow the fact that, under natural conditions, photosynthesis is an integrated process taking place in intact organisms that are continuously responding to internal and external changes. This chapter addresses some of the photosynthetic responses of the intact leaf to its environment. Additional photosynthetic responses to different types of stress are included in Chapter 25.

The impact of the environment on photosynthesis is of interest to both plant physiologists and agronomists. From a physiological standpoint, we wish to understand how photosynthesis responds to environmental factors such as light, ambient CO_2 concentrations, and temperature. The dependence of photosynthetic processes on environment is important to agronomists because plant productivity, and hence crop yield, depends strongly on prevailing photosynthetic rates in a dynamic environment.

In studying the environmental dependence of photosynthesis, a central question arises: How many environmental factors can limit photosynthesis at one time? The British plant physiologist F. F. Blackman postulated in 1905 that, under any particular conditions, the rate of photosynthesis is limited by the slowest step, the so-called **limiting factor**. The implication of this hypothesis is that at any given time, photosynthesis can be limited either by light or by CO_2 concentration, but not by both factors. This hypothesis has had a marked influence on the approach used by plant physiologists to study photosynthesis—that is, varying one factor and keeping all

Table 9.1
Some characteristics of limitations to the rate of photosynthesis

	Conditions that lead to this limitation		Response of photosynthesis under this limitation to		
Limiting factor	CO_2	**Light**	CO_2	O_2	**Light**
Rubisco activity	Low	High	Strong	Strong	Absent
RuBP regeneration	High	Low	Moderate	Moderate	Strong

other environmental conditions constant (see, for example, Figure 9.8).

In the intact leaf, three major metabolic steps have been identified as important for optimal photosynthetic performance: rubisco activity, the regeneration of ribulose bisphosphate (RuBP), and the metabolism of the triose phosphates. The first two steps are the most prevalent under natural conditions. Table 9.1 provides some examples of how light and CO_2 can affect these key metabolic steps. In the following sections, biophysical, biochemical, and environmental aspects of photosynthesis in leaves are discussed in detail.

Light, Leaves, and Photosynthesis

This chapter will address some of the photosynthetic responses of the plant and the intact leaf. Scaling photosynthesis up from the chloroplast (the focus of Chapters 7 and 8) to the leaf adds new levels of complexity to photosynthesis. At the same time, it makes possible other levels of regulation to control the photosynthetic performance of the leaf. We will start by examining how leaf anatomy, and movements by chloroplasts and leaves, control the absorption of light for photosynthesis. Then we will describe how chloroplasts and leaves adapt to their light environment and how the photosynthetic response of leaves grown under low light reflects their adaptation to a low-light environment. The same is true for leaves grown under high light, illustrating that plants are physiologically flexible and that they adapt to their immediate environment.

Both the amount of light and the amount of CO_2 determine the photosynthetic response of leaves, and in some situations photosynthesis is limited by an inadequate supply of light or CO_2. On the other hand, absorption of too much light can cause severe problems, and special mechanisms protect the photosynthetic system from excess light. Multiple levels of control over photosynthesis allow plants to grow successfully in a constantly changing environment and different habitats.

Leaf Anatomy Maximizes Light Absorption

Roughly 1.3 kW m^{-2} of radiant energy from the sun reaches Earth, but only about 5% of this energy can be converted into carbohydrates by a photosynthesizing leaf (Figure 9.1). The reason that this percentage is so low is that a major fraction of the incident light is of a wavelength either too short or too long to be absorbed by the photosynthetic pigments (see Figure 7.3). In addition, much of the absorbed light energy is lost as heat, and a much smaller amount is lost as fluorescence (see Chapter 7). Radiant energy from the sun consists of many different wavelengths of light, from short (less than 400 nm—UV light) to long (greater than 700 nm—far red and infrared). The wavelengths from 400 to 700 nm are utilized in photosynthesis, and this light is called

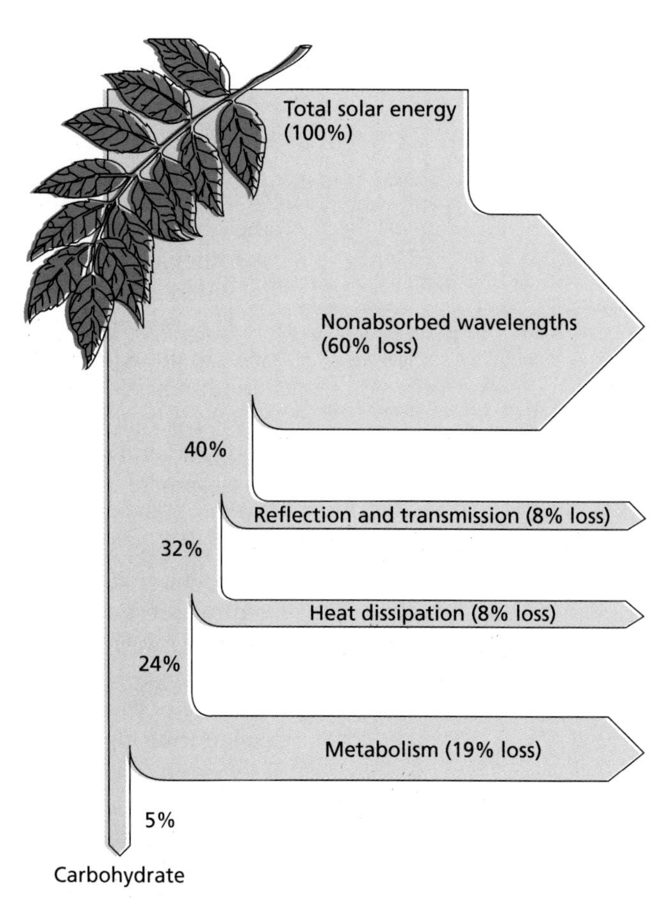

Figure 9.1 Conversion of solar energy into carbohydrates by a leaf. Of the total incident energy, only 5% is converted into carbohydrates.

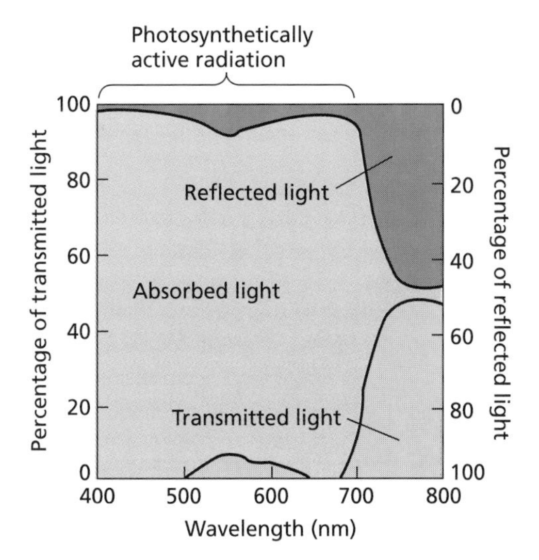

Figure 9.2 Optical properties of a bean leaf. Shown here are the percentages of light absorbed, reflected, and transmitted, as a function of wavelength. The transmitted and reflected green light in the wave band at 500 to 600 nm gives leaves their green color. Note that most of the light above 700 nm is not absorbed by the leaf. (From Smith 1986.)

photosynthetically active radiation (**PAR**) (see Box 9.1). About 85 to 90% of the PAR is absorbed by the leaf; the remainder is either reflected at the leaf surface or is transmitted through the leaf (Figure 9.2). Because chlorophyll absorbs very strongly in the blue and the red regions of the spectrum (see Figure 7.3), the transmitted and reflected light are enriched in green, yielding the green color of vegetation.

The anatomy of the leaf is highly specialized for light absorption (Terashima and Hikosaka 1995). The outermost cell layer, the epidermis, is usually transparent to visible light, and the individual cells are often convex. Epidermal cells often act as lenses and can focus light so that the amount reaching the chloroplasts is many times greater than the amount of ambient light (Vogelmann et al. 1996). Epidermal focusing is common among herbaceous plants and is especially prominent among tropical plants that grow in the forest understory, where light levels are very low.

Below the epidermis, the top layers of photosynthetic cells are called **palisade cells**; they are shaped like pillars that stand in parallel columns one to three layers deep (Figure 9.3). Some leaves have several layers of columnar palisade cells, and we may wonder how efficient it is for a plant to invest energy in the development of multiple cell layers when the high chlorophyll content of the first layer would appear to allow little transmission of the incident light to the leaf interior. In fact, more light than might be expected penetrates the first layer of palisade cells, because of the sieve effect and light channeling (Vogelmann 1993).

The **sieve effect** is due to the fact that chlorophyll is not uniformly distributed throughout cells but instead is confined to the chloroplasts. This packaging of chlorophyll results in shading between the chlorophyll molecules and creates gaps between the chloroplasts, where light is not absorbed; hence the reference to a sieve. The total absorption of light by a given amount of chlorophyll in a palisade cell is less than that by the same amount of chlorophyll in a solution. **Light channeling** occurs when some of the incident light is propagated through the central vacuole of the palisade cells and through the air spaces between the cells, an arrangement that facilitates the transmission of light into the leaf interior (Vogelmann and Martin 1993).

Below the palisade layers is the **spongy mesophyll**, where the cells are very irregular in shape and are surrounded by large air spaces (see Figure 9.3). The many

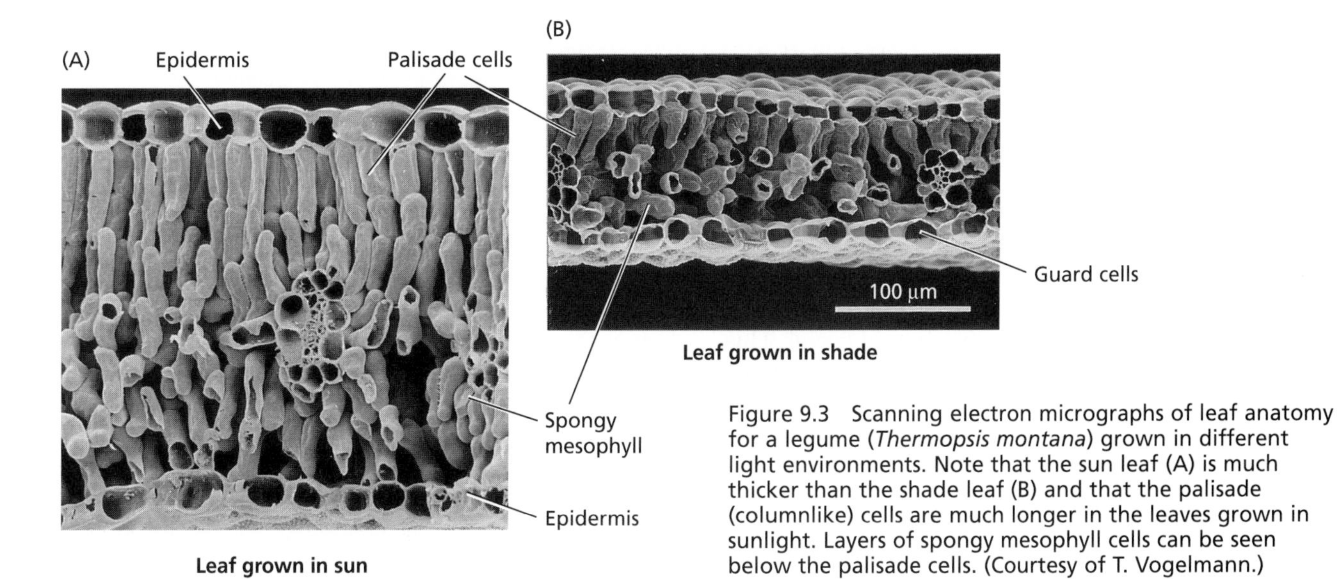

Figure 9.3 Scanning electron micrographs of leaf anatomy for a legume (*Thermopsis montana*) grown in different light environments. Note that the sun leaf (A) is much thicker than the shade leaf (B) and that the palisade (columnlike) cells are much longer in the leaves grown in sunlight. Layers of spongy mesophyll cells can be seen below the palisade cells. (Courtesy of T. Vogelmann.)

BOX 9.1

Working with Light

THREE LIGHT PARAMETERS are especially important when working with light: (1) the amount, (2) the direction, and (3) the spectral quality. The first two parameters are important with respect to the geometry of the part of the plant that intercepts the light. For a flat leaf, a flat or planar light sensor makes sense, and the amount of energy that falls on a flat sensor of known area per unit time is quantified as **irradiance** (see the table). Units can be expressed in terms of energy, such as watts per square meter (W m^{-2}). Time (seconds) is contained within the term watt: 1 W = 1 joule (J) s^{-1}. Light can also be thought of as a stream of particles, or **quanta**. In this case, units can be expressed in moles per square meter per second (mol m^{-2} s^{-1}), where "moles" refers to the number of photons (1 mol of light = 6.02×10^{23} photons, Avogadro's number). This measure is called **photon irradiance.** Quanta and energy units can be interconverted relatively easily, provided the wavelength of the light is known. The energy of a photon is related to its wavelength as follows:

$$E = \frac{hc}{\lambda} \quad (1)$$

where c is the speed of light (3×10^8 m s^{-1}), h is Planck's constant (6.63×10^{-34} J s), and λ is the wavelength of light, usually expressed in nm (1 nm = 10^{-9} m). From this equation it can be shown that a photon at 400 nm has twice the energy of a photon at 800 nm. For simplicity, Equation 1 can be reduced to the following:

$$E = \frac{1988 \times 10^{-19} J}{\lambda} \quad (2)$$

where λ is expressed in nanometers. From this calculation, we can determine that a photon of 400 nm light contains 4.97×10^{-19} J.

Concepts and units for the quantification of light

	Energy system (W m^{-2})	Photon system (mol $m^{-2}s^{-1}$)
Flat light sensor	Irradiance	Photon irradiance
	Photosynthetically active radiation (PAR, 400-700 nm, energy units)	PAR (quantum units)
	—	Photosynthetic photon flux density (PPFD)
Spherical light sensor	Fluence rate (energy units) Scalar irradiance	Fluence rate (quantum units) Quantum scalar irradiance

Now, suppose we have 3 µmol of 400 nm light falling on 1 m^2 every second, or a photon irradiance of 3 µmol m^{-2} s^{-1}. This quantity is approximately the amount of blue light at 400 nm that strikes the surface of Earth on a sunny day. If we want to convert photon irradiance to energy irradiance, we must first convert micromoles to moles (1 µmol = 10^{-6} mol): (3 µmol m^{-2} s^{-1}) × (10^{-6} mol µmol^{-1}) = 3×10^{-6} mol m^{-2} s^{-1} of photons, or quanta. Calculating the number of quanta from Avogadro's number gives: (3×10^{-6} mol quanta m^{-2} s^{-1}) × (6.02×10^{23} quanta mol^{-1}) = 1.8×10^{18} quanta m^{-2} s^{-1}. As we already calculated, each photon, or quantum, of 400 nm light contains 4.97×10^{-19} J. Thus (4.97×10^{-19} J $quantum^{-1}$) × (1.8×10^{18} quanta m^{-2} s^{-1}) = 0.9 J s^{-1} m^{-2}. Since 1 J s^{-1} = 1 W, we have an irradiance of 0.9 W m^{-2}.

Turning our attention to the direction of light, light can strike a flat surface directly from above or it can strike the surface obliquely. When light deviates from perpendicular, irradiance is proportional to the cosine of the angle at which the light rays hit the sensor (see the figure on the next page). Thus, irradiance is maximal when light strikes a surface directly from above, and it decreases as light becomes more oblique—similar to the situation with a typical leaf. Sensors that correct for the angle of incidence of light are said to be *cosine corrected*. But there are many examples in nature in which the light-intercepting object is not flat (e.g., complex shoots, whole plants, chloroplasts). In addition, in some situations light can come from many directions simultaneously (e.g., direct light from the sun plus the light that is reflected upward from sand, soil, or snow). In these situations it makes more sense to measure light with a spherical sensor that measures light omnidirectionally (from all directions). The term for this omnidirectional measurement is **fluence rate** (Rupert and Letarjet 1978), and this quantity can be expressed in watts per square meter (W m^{-2}) or moles per square meter per second (mol m^{-2} s^{-1}). It is obvious from the units whether light is being measured as energy (W) or as photons (mol). In contrast to a flat sensor, a spherical sensor is equally sensitive to light, independent of direction (see the figure). Depending on whether the light is collimated (rays are parallel) or diffuse (rays travel in random direc-

interfaces between air and water reflect and refract the light, thereby randomizing its direction of travel. This phenomenon is called *light scattering*, and it is especially important in leaves because the multiple reflections between cell–air interfaces greatly increase the length of the path over which photons travel, thereby increasing the probability for absorption. In fact, photon pathlengths within leaves are commonly four times or more longer than the leaf is thick (Richter and Fukshansky 1996). The contrasting properties of the entire mesophyll tissues—the palisade which allows light to pass through, and the spongy mesophyll which scatters as much light as possible—result in more uniform light absorption throughout the leaf.

BOX 9.1 *(continued)*

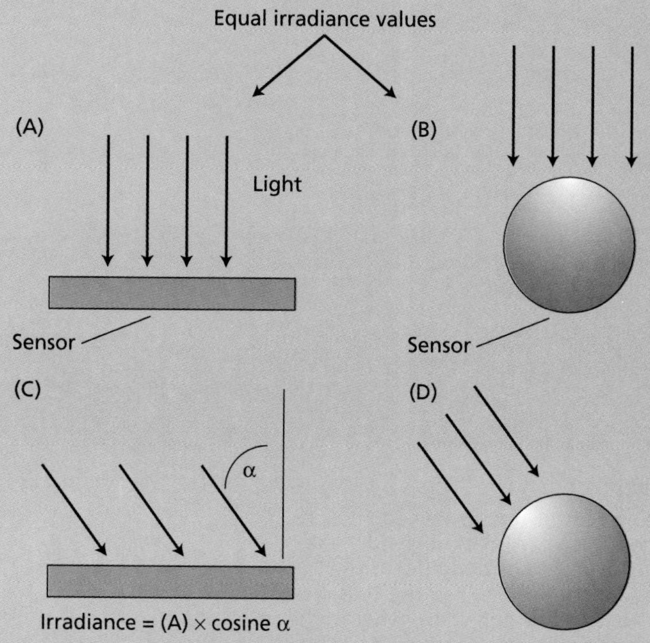

Irradiance and fluence rate. Equivalent amounts of collimated light strike a flat irradiance-type sensor (A) and a spherical sensor (B) that measure fluence rate. With collimated light, A and B will give the same light readings. When the light direction is changed 45°, the spherical sensor (D) will measure the same quantity as in B. In contrast, the flat irradiance sensor (C) will measure an amount equivalent to the irradiance in A multiplied by cosine of the angle α in C. (After Björn and Vogelmann 1994.)

tions), values for fluence rate versus irradiance can be quite different from one another. They are equivalent only under special conditions (for a detailed discussion, see Björn and Vogelmann 1994 and Kirk 1994).

Photosynthetically active radiation (*PAR*, 400 to 700 nm) may also be expressed on the basis of energy (W m^{-2}) or quanta (mol m^{-2} s^{-1}) (McCree 1981). It is important to note that PAR is an irradiance-type measurement. In research on photosynthesis, when PAR is expressed on a quantum basis, it is given the special term **photosynthetic photon flux density (PPFD)**. However, it has been suggested that the term "density" be discontinued (Holmes et al. 1985) because within the International System of Units (Système Internationale d'Unités, or SI units) "density" can mean area or volume. Moreover, area is contained within the term flux. PPFD has in some cases been shortened to PPF, but it is not clear whether this abbreviation represents an irradiance-type or a spherical measurement.

When light moves through water such as lakes or oceans, it scatters in such a way that its direction of travel is randomized. Thus, aquatic photosynthetic organisms can receive light simultaneously from above, below, and the sides. In oceanography and limnology, the light that travels from the water surface toward the bottom is called *downwelling irradiance*, and the light that travels in the reverse direction is called *upwelling irradiance*.

Before the advent of specialized light meters suitable for plant physiology, light was measured in **lux** or **foot-candles**. These measurements are based on the perception of light by the human eye, which is maximally sensitive to light within the green at 555 nm. Sensitivity falls off on both sides of this wavelength and approaches zero within the blue and red, wavelength regions that are important for photosynthesis. Thus, lux and foot-candles are of little use in plant physiology. These units can be converted to the units of irradiance and fluence rate, but only if detailed knowledge about the spectral quality of the light is available, something that can be obtained only by use of an instrument called a *spectroradiometer*, which measures the amount of light at each wavelength.

In summary, when choosing how to quantify light, it is important to match sensor geometry and spectral response with that of the plant. Flat, cosine-corrected sensors are ideally suited to measure the amount of light that strikes the surface of a leaf; spherical sensors are more appropriate in other situations, such as when studying a chloroplast suspension or a branch from a tree.

How much light is there on a sunny day and what is the relationship between PAR irradiance and PAR fluence rate? Under direct sunlight, PAR irradiance and fluence rate are both about 2000 µmol m^{-2} s^{-1}, though higher values can be measured at high altitudes. The corresponding value in energy units is about 400 W m^{-2}. When light is completely diffuse, irradiance is only 0.25 times the fluence rate.

Some environments, primarily deserts, have so much light that it is potentially harmful. In these environments leaves often have special anatomical features, such as hairs, salt glands, and epicuticular wax that increase the reflection of light from the leaf surface, thereby reducing light absorption (Ehleringer et al. 1976). Such adaptations can decrease light absorption by as much as 40%, minimizing heating and other problems associated with the absorption of too much light.

Chloroplast Movement and Leaf Movement Can Control Light Absorption

Chloroplast movement is widespread among algae, mosses, and leaves of higher plants (Haupt and

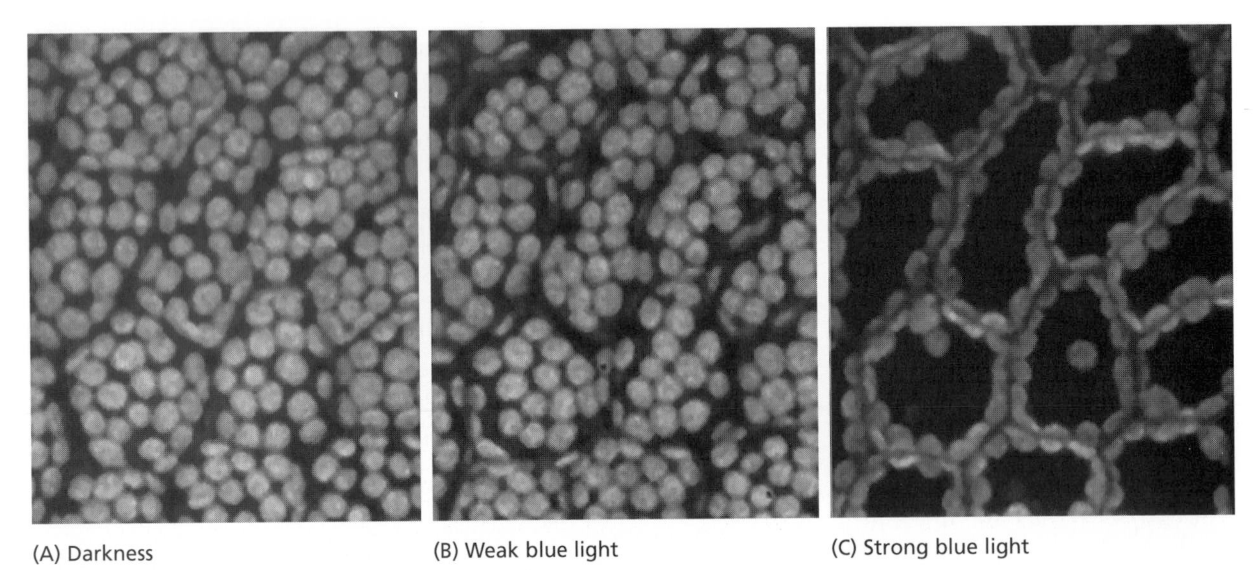

Figure 9.4 Chloroplast distribution in photosynthesizing cells of the duckweed *Lemna*. These surface views show the same cells under three conditions: (A) darkness, (B) weak blue light, and (C) strong blue light. In A and B, chloroplasts are positioned near the surface of the cells, where they can absorb maximum amounts of light. When the cells were irradiated with strong blue light (C), the chloroplasts move to the side walls, where they shade each other, thus minimizing the absorption of excess light. (Micrographs courtesy of M. Tlalka and M. D. Fricker.)

Scheuerlein 1990). By controlling chloroplast position, leaves can control how much light is absorbed. Under low light (Figure 9.4A and B), chloroplasts gather at the cell surfaces parallel to the plane of the leaf so that they are aligned perpendicularly to the incident light, a position that maximizes absorption of light. Under high light (Figure 9.4C), the chloroplasts move to the cell surfaces that are parallel to the incident light, thus avoiding excess absorption of light. Usually such movement decreases the amount of light absorbed by the leaf by about 15% (Brugnoli and Björkman 1992). Chloroplast movement has an action spectrum typical of responses to blue light (see Chapter 18) (Wada et al. 1993). Blue light also controls chloroplast orientation in many of the lower plants, but in some algae chloroplast rmovement is controlled by phytochrome (Haupt and Scheuerlein 1990). In leaves, chloroplasts move along cables of actin microfilaments in the cytoplasm, and calcium regulates their movement (Tlalka and Gabryś 1993).

Leaves absorb the most light when the leaf lamina is perpendicular to the incident light. Some plants control light absorption by **solar tracking**; that is, their leaves continuously adjust the orientation of their laminae such that they are perpendicular to the sun's rays (Figure 9.5). Alfalfa, cotton, soybean, bean, lupine, and some wild species of the mallow family (Malvaceae) are examples

Figure 9.5 Leaf movement in sun-tracking plants. (A) Initial leaf orientation in the lupine *Lupinus succulentus*. (B) Leaf orientation 4 hours after exposure to oblique light. The direction of the light beam is indicated by the arrows. Movement is generated by asymmetric swelling of a pulvinus, found at the junction between the lamina and the petiole. In natural conditions, the leaves track the sun's trajectory in the sky. (From Vogelmann and Björn 1983, courtesy of T. Vogelmann.)

of the numerous plant species that are capable of solar tracking. Solar-tracking leaves keep a nearly vertical position at sunrise, facing the eastern horizon, where the sun will rise. The leaf laminae then lock on to the rising sun and follow its movement across the sky with an accuracy of ±15° until sunset, when the laminae are nearly vertical, facing the west where the sun sets. During the night the leaf takes a horizontal position and reorients just before dawn so that it faces the eastern horizon in anticipation of another sunrise. Leaves track the sun only on clear days, and they stop when a cloud obscures the sun. In the case of intermittent cloud cover, some leaves can reorient as rapidly as 90° per hour, thus allowing them to catch up to the new solar position when the sun emerges from behind a cloud (Koller 1990).

In many cases, leaf orientation is controlled by a specialized organ called the **pulvinus** (plural pulvini), found at the junction between the blade and the petiole. The pulvinus contains motor cells that change their osmotic potential and generate mechanical forces that determine laminar orientation. In other plants, leaf orientation is controlled by small mechanical changes along the length of the petiole and by movements of the younger parts of the stem. Solar tracking is another blue-light response, and solar tracking leaves have special light-sensitive regions. In species of *Lavatera* (Malvaceae), the photosensitive region is located in or near the major leaf veins (Koller et al. 1990). In lupines, *Lupinus* (Fabaceae), leaves consist of five or more leaflets, and the photosensitive region is located in the basal part of each leaflet lamina.

Since full sunlight vastly exceeds the amount of light that can be utilized for photosynthesis, what advantage is gained by solar tracking? By keeping leaves perpendicular to the sun, solar-tracking plants maintain maximum photosynthetic rates throughout the day, including early morning and late afternoon. Moreover, these are times of day when water stress is lower, so solar tracking gives plants that grow in arid regions an advantage (Ehleringer and Forseth 1980). Some solar-tracking plants can also move their leaves such that they avoid full exposure to sunlight, thus minimizing heat absorption and water loss. Building on a term often used for sun-induced leaf movements , "**heliotropism**" (bending toward the sun), these leaves are called *paraheliotropic,* and leaves that maximize light interception by solar tracking are called *diaheliotropic*. In some species, a plant can display diaheliotropic movements when it is well watered and paraheliotropic movements when it experiences water stress.

Plants, Leaves, and Cells Adapt to Their Light Environment

Some plants are plastic enough to adapt to a range of light regimes, growing as sun plants in sunny areas and as shade plants in shady habitats (Evans et al. 1988). Some shady habitats receive less than 1% of the PAR available in an exposed habitat. Leaves that are adapted to very sunny or very shady environments are often unable to survive in the other type of habitat. Figure 9.9 illustrates some changes in the photosynthetic properties of leaves grown at high and low photon flux (number of photons per unit area and unit time).

Sun and shade leaves have some contrasting characteristics. For example, shade leaves have more total chlorophyll per reaction center, have a higher ratio of chlorophyll *b* to chlorophyll *a*, and are usually thinner than sun leaves. Sun leaves have higher concentrations of soluble protein, rubisco, and xanthophyll cycle components. Contrasting anatomical characteristics can also be found in leaves of the same plant that are exposed to different light regimes. Figure 9.3 shows some differences between a leaf grown in the sun and a leaf grown in the shade. Even different parts of a single leaf show adaptations to the immediate light environment. Cells in the upper surface of the leaf, which are exposed to the highest prevailing photon flux, have characteristics of cells from leaves grown in full sunlight; cells in the lower surface of the leaf have characteristics of cells found in shade leaves (Terashima 1989).

These morphological and biochemical modifications are associated with physiological specializations. Shade plants show adaptations to light intensity and quality, such as a 3:1 ratio of photosystem II to photosystem I reaction centers, compared with the 2:1 ratio found in sun plants (Anderson 1986). Other shade plants, rather than altering the ratio of PS-I to PS-II, add more antennae chlorophyll to PS-II. These adaptations appear to enhance light absorption and energy transfer in shady environments, where far-red light is more abundant. Far-red light is absorbed primarily by PS-I, and altering the ratio of PS-I to PS-II or changing the light-harvesting antennae associated with the photosystems, makes it possible to maintain a better balance of energy flow through the two photosystems (Melis 1996). Sun and shade plants also differ in their respiration rates, and these differences alter the relationship between respiration and photosynthesis, as we'll see a little later in this chapter.

Plants Compete for Sunlight

Sunlight is a resource for which plants normally compete. Held upright by stems and trunks, leaves configure a canopy that absorbs light and influences photosynthetic rates and growth beneath them. Leaves that are shaded by other leaves have much lower photosynthetic rates. Some plants have very thick leaves that transmit little, if any, light. Other plants, such as the dandelion (*Taraxacum* sp.), have a rosette growth habit in which leaves grow radially very close to each other and

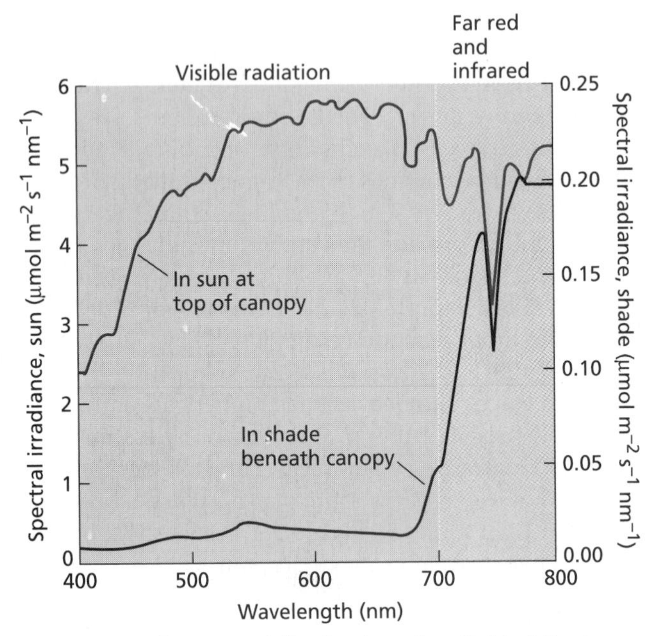

Figure 9.6 The spectral distribution of sunlight at the top of a canopy and under the canopy. For unfiltered sunlight, the total irradiance was 1900 µmol m^{-2} s^{-1}; for shade, 17.7 µmol m^{-2} s^{-1}. Most of the photosynthetically active radiation was absorbed by leaves in the canopy. (From Smith 1986.)

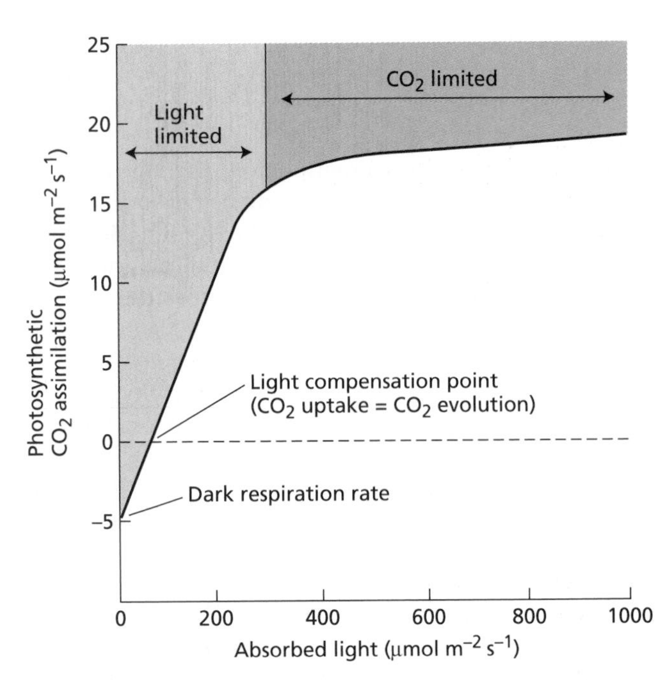

Figure 9.7 Response of photosynthesis to irradiance in a C_3 plant. In darkness, respiration causes a net efflux of CO_2 from the plant. At low irradiance the light compensation point, the point at which the uptake of CO_2 by photosynthesis balances the evolution of CO_2 by respiration, is reached. Increasing irradiance above the light compensation point results in a proportional increase in photosynthesis, indicating that photosynthesis is limited by the rate of electron transport, which in turn is limited by the amount of available light. This portion of the curve is light limited. Further increases in irradiance eventually saturate CO_2 uptake, and in this part of the curve, referred to as CO_2 limited, photosynthesis is limited by the carboxylation capacity of rubisco or the metabolism of triose phosphates.

to the stem, thus precluding the growth of any leaves below them. Trees represent an outstanding adaptation for light interception. The elaborate branching structure of trees results in a vast increase in the interception of sunlight. Very little PAR penetrates the canopy of many forests; almost all of it is absorbed by leaves (Figure 9.6).

Another feature of the shady habitat is **sunflecks,** patches of sunlight that pass through discontinuities of the canopy and move across shaded leaves as the sun moves. In a dense forest, sunflecks can change the photon flux impinging on a leaf in the forest floor more than tenfold within seconds. For some of these leaves, a sunfleck contains nearly 50% of the total light energy available during the day, but this critical energy is available for only a few minutes in a very high dose. Sunflecks also play a role in the carbon metabolism of lower leaves in dense crops that are shaded by the upper leaves of the plant. Rapid responses of both the photosynthetic apparatus and the stomata to sunflecks have been of substantial interest to plant physiologists and ecologists (Pearcy 1994).

Photosynthetic Responses to Light by the Intact Leaf

Measuring CO_2 fixation in intact leaves at increasing photon flux allows us to construct light–response curves (Figure 9.7) that provide useful information about the photosynthetic properties of leaves. In the dark, CO_2 is given off by the plant because of respiration (see Chapter 11), and by convention, CO_2 assimilation is negative in this part of the light–response curve. As the photon flux increases, photosynthetic CO_2 uptake increases until it equals CO_2 release by mitochondrial respiration. The point at which CO_2 uptake exactly balances CO_2 release, is called the **light compensation point**.

The photon flux at which different leaves reach the light compensation point varies with species and developmental conditions. One of the more interesting differences is found between plants grown in full sunlight and those grown in the shade (Figure 9.8). Light compensation points of sun plants range from 10 to 20 mmol m^{-2} s^{-1}; corresponding values for shade plants are 1 to 5 mmol m^{-2} s^{-1}. The values for shade plants are lower because respiration rates in shade plants are very low, so little net photosynthesis suffices to bring the rates of CO_2 evolution to zero. Low respiratory rates seem to represent a basic adaptation that allows shade plants to survive in light-limited environments.

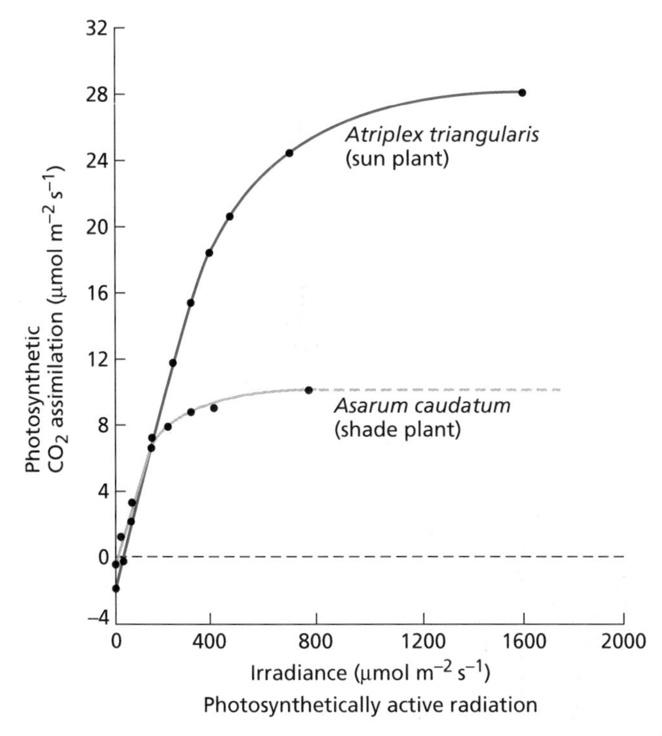

Figure 9.8 Light–response curves of photosynthetic carbon fixation as a function of irradiance. *Atriplex triangularis* (triangle orache) is a sun plant, and *Asarum caudatum* (a wild ginger) is a shade plant. Typically, shade plants have a low light compensation point and have lower maximal photosynthetic rates than sun plants. (From Harvey 1979.)

Increasing photon flux above the light compensation point results in a proportional increase in photosynthetic rate (see Figure 9.7), yielding a linear relationship between photon flux and photosynthetic rate. In other words, photosynthesis is *light limited* in this range, and providing more light allows more photosynthesis. In this linear portion of the curve, the slope of the line gives the **maximum quantum yield** of photosynthesis for the leaf. Recall from Chapter 7 that quantum yield can be expressed as the number of absorbed photons required to fix one molecule of CO_2, to evolve one molecule of O_2, or to initiate a photochemical event. Quantum yields vary from 0, where none of the light energy is used in photosynthesis, to 1, where all the absorbed light is used. Recall from Chapter 7 that the quantum yield of photochemistry is about 0.95 and the quantum yield of oxygen evolution by isolated chloroplasts is about 0.1 (10 photons per O_2).

In the intact leaf, measured quantum yields for CO_2 fixation vary between 0.04 and 0.1. Healthy leaves from many species of C_3 plants, kept under low O_2 concentrations that inhibit photorespiration, usually show a quantum yield of 0.1. In normal air, the quantum yield of C_3 plants is lower, typically 0.05. Quantum yield varies with temperature and CO_2 concentration, because of their effect on the ratio of the carboxylase and oxygenase reactions of rubisco (see Chapter 8). Below 30°C, quantum yields of C_3 plants are generally higher than those of C_4 plants; above 30°C, the situation is usually reversed (see Figure 9.23). Despite their different growth habitats, sun and shade plants show similar quantum yields.

At higher photon flux, the photosynthetic response to light starts to level off (see Figure 9.7) and reaches *saturation*. Once the saturation point is reached, further increases in photon flux no longer affect photosynthetic rates, indicating that factors such as electron transport reactions, rubisco activity, or the metabolism of triose phosphates, have become limiting. Here, photosynthesis is commonly referred to as *CO_2 limited*, reflecting the inability of the carbon metabolism enzymes to keep pace with the absorbed light energy. Light saturation levels for shade plants are substantially lower than those for sun plants (see Figure 9.8). These levels usually reflect the maximal photon flux to which the leaf was exposed during growth (Figure 9.9).

The light–response curve of most leaves saturates between 500 and 1000 mmol m^{-2} s^{-1}, photon fluxes well below full sunlight (which is about 2000 mmol m^{-2} s^{-1}).

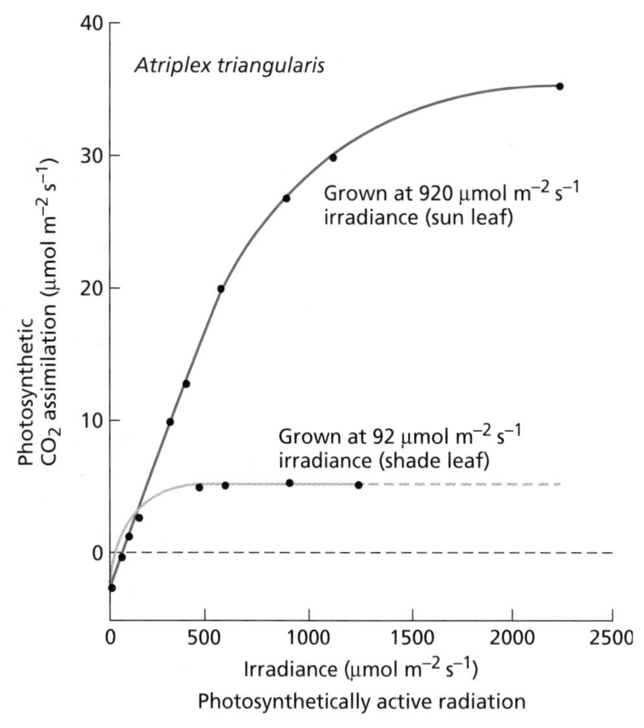

Figure 9.9 Changes in photosynthesis in leaves of *Atriplex triangularis* as a function of irradiance. The upper curve represents a leaf grown at an irradiance ten times higher than that of the lower curve. In the leaf grown at the lower light levels, photosynthesis saturates at a substantially lower irradiance, indicating that the photosynthetic properties of a leaf depend on its growing conditions. (From Björkman 1981.)

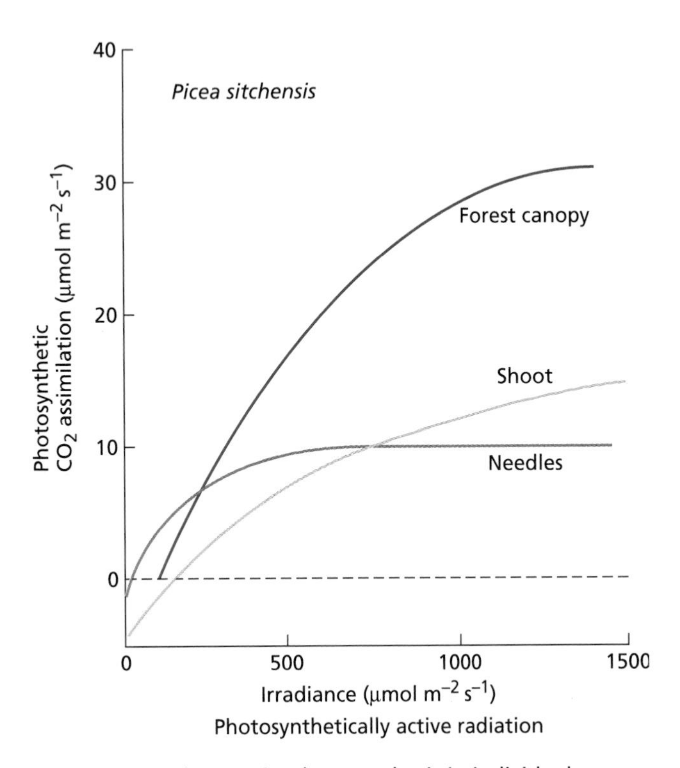

Figure 9.10 Changes in photosynthesis in individual needles, a complex shoot, and a forest canopy of Sitka spruce (*Picea sitchensis*) as a function of irradiance. Needles shade each other in complex shoots and the canopy. As a result, photosynthesis does not saturate, even at irradiance values approaching full sunlight. (From Jarvis and Leverentz 1983.)

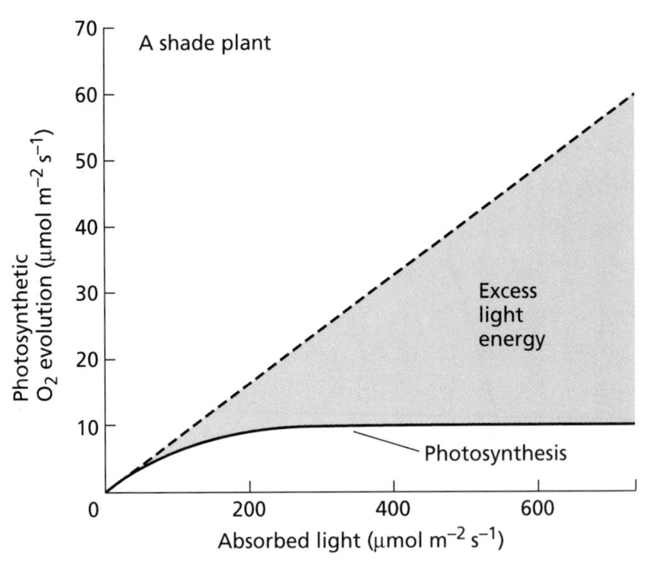

Figure 9.11 Excess light energy in relation to a photosynthetic light–response curve. The broken line shows theoretical oxygen evolution in the absence of any rate limitation to photosynthesis. At levels of photon flux up to 150 μmol m^{-2} s^{-1}, a shade plant is able to utilize the absorbed light for photosynthesis. Above 150 μmol m^{-2} s^{-1}, however, photosynthesis saturates, and an increasingly larger amount of the absorbed light energy must be dissipated. Note the large disparity at higher irradiances between the fraction of light used versus that which must be dissipated. Sun plants have higher photosynthetic rates and are able to utilize a correspondingly larger proportion of absorbed light energy. (After Osmond 1994.)

Although individual leaves are rarely able to utilize full sunlight, whole plants usually consist of many leaves that shade each other. For example, only a small fraction of a tree's leaves are exposed to full sun at any given time of the day. The rest of the leaves receive subsaturating photon fluxes in the form of small patches of light that pass through gaps in the leaf canopy or in the form of light transmitted through other leaves. Since the photosynthetic response of the intact plant is the sum of the photosynthetic activity of all the leaves, only rarely is photosynthesis saturated at the level of the whole plant. Light–response curves of individual trees and of the forest canopy show that photosynthetic rate increases with photon flux and usually does not saturate, even at full sunlight (Figure 9.10). Along these lines, crop productivity is related to the total amount of light received during the growing season, and generally, the more light a crop receives, the higher the biomass (Ort and Baker 1988).

Leaves Must Dissipate Excess Light Energy

When exposed to excess light, leaves must dissipate the surplus absorbed light energy so that it does not harm the photosynthetic apparatus (Figure 9.11). There are several routes for energy dissipation involving nonphotochemical quenching (see Chapter 7). Although the molecular mechanisms are not yet fully understood, the xanthophyll cycle appears to be an important avenue for dissipation of excess light energy. Under high light, violaxanthin (V) is converted to antheraxanthin (A) and then to zeaxanthin (Z) (see Figure 7.34). This cycle may divert a substantial amount, 60% or more, of the excitation energy to heat. The fraction of light energy that is dissipated depends on irradiance, species, growth conditions, nutrient status, and ambient temperature (Demmig-Adams et al. 1996). In leaves that are exposed directly to the sun, as irradiance increases toward midday, V decreases and A and Z together increase by 60% or more (Figure 9.12). When irradiance returns to low values, A and Z are converted enzymatically back to V. This conversion to V can be slow when the leaf is nutritionally deprived of nitrogen or when the leaf is stressed by high-light conditions.

Leaves that grow in full sunlight contain substantially higher amounts of V, Z and A than shade leaves do, and so they can dissipate greater photon fluxes. However, the xanthophyll cycle also operates in plants that grow in the low light of the forest understory,

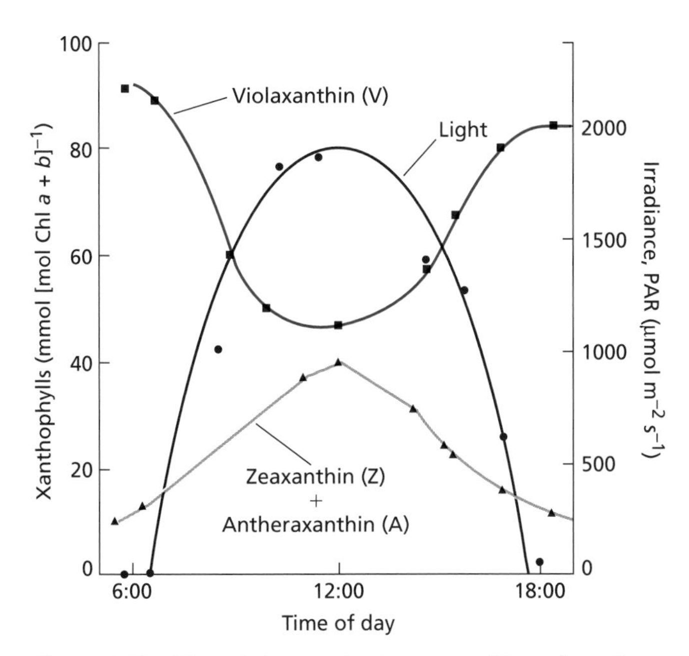

Figure 9.12 Diurnal changes in the composition of xanthophylls in response to changes in light in sunflower (*Helianthus annuus*). As the amount of light that falls on a leaf increases, a greater proportion of the xanthophylls are converted from violaxanthin (V) to antheraxanthin (A) and zeaxanthin (Z), thereby dissipating excess excitation energy and protecting the photosynthetic apparatus. (After Demmig-Adams et al. 1996.)

where they are only occasionally exposed to high light when sunlight passes through gaps in the overlying leaf canopy, forming sunflecks. Exposure to one sunfleck results in the conversion of much of the V to A and Z. Between sunflecks, the formed A and Z are retained and protect the leaf against exposure to subsequent sunflecks. The xanthophyll cycle is also found in species such as conifers, the leaves of which remain green during winter, when photosynthetic rates are very low yet light absorption remains high. Contrary to the diurnal formation of A and Z observed within leaves in the summer, the level of A and Z remains high all day during the winter. Presumably this mechanism maximizes dissipation of light energy, thereby protecting the leaves against photooxidation during winter (Demmig-Adams and Adams 1996).

Leaves Must Dissipate Vast Quantities of Heat

The heat load on a leaf exposed to full sunlight is very high. In fact, a leaf with an effective thickness of water of 300 µm would warm up by 100°C every minute if all available solar energy were absorbed and no heat was lost. However, this enormous heat load is dissipated by the emission of long-wave radiation, by sensible (or perceptible) heat loss, and by evaporative (or latent) heat loss (Figure 9.13). Air circulation around the leaf removes heat from the leaf surfaces if the temperature of the leaf is warmer than that of the air; this phenomenon is called **sensible heat loss**. On the other hand, **evaporative heat loss** occurs because the evaporation of water requires energy (see Chapter 4). Thus, as water evaporates from a leaf it withdraws heat from the leaf and cools it. The human body is cooled by the same principle, through perspiration.

Sensible heat loss and evaporative heat loss are the most important processes in the regulation of leaf temperature, and the ratio of the two is called the **Bowen ratio** (Campbell 1977):

$$\text{Bowen ratio} = \frac{\text{sensible heat loss}}{\text{evaporative heat loss}}$$

In well-watered crops, transpiration (see Chapter 4), and hence water evaporation from the leaf, is high, so the Bowen ratio is low. One can calculate the water loss from the crop by computing the amount of water that will dissipate the net radiant energy being absorbed by the crop when vaporized. Very low Bowen ratios can be measured in a lawn on a relatively still day. In these conditions there is no sensible heat loss, because the air around the leaf is at the same temperature as the leaf; the Bowen ratio therefore approaches zero, and the water loss of the lawn is determined primarily by the solar energy input.

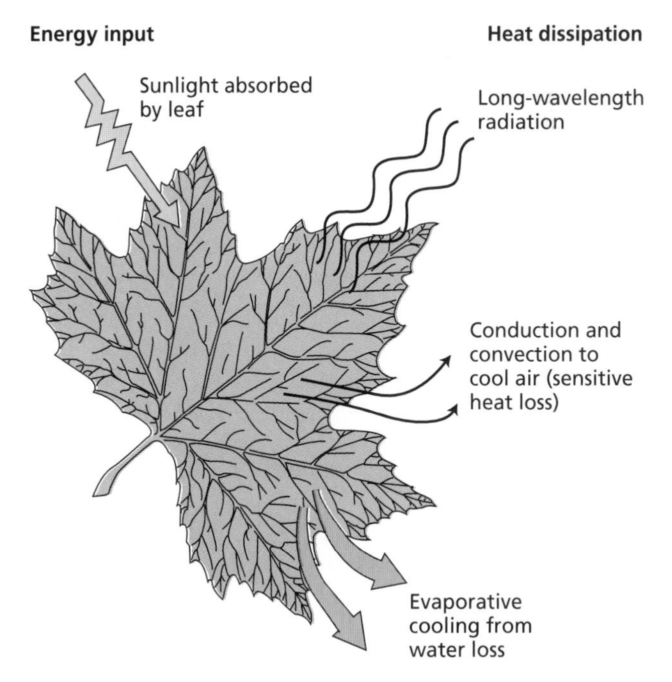

Figure 9.13 The absorption of energy from sunlight by the leaf. The imposed heat load must be dissipated in order to avoid damage to the leaf. The heat load is dissipated by emission of long-wavelength radiation, by sensible heat loss to the air surrounding the leaf, and by the evaporative cooling caused by transpiration.

In other cases, such as cotton leaves in the afternoon, water loss from stomatal transpiration cools the leaf below the air temperature by evaporative heat loss (see Box 25.2). In that case, there is sensible heat gain rather than heat loss, and the Bowen ratio becomes negative. In contrast, the Bowen ratio can be very high in leaves from desert plants. In some cacti, stomata are tightly closed, precluding any evaporative cooling; all the heat is dissipated by sensible heat loss, and the Bowen ratio is infinite. Plants with very high Bowen ratios conserve water but have to endure very high leaf temperatures in order to maintain a sufficient temperature gradient between the leaf and the air. Slow growth is usually correlated with these adaptations.

Isoprene Synthesis Is a Mechanism to Cope with Heat

We have seen how the xanthophyll cycle can protect against high light, but how do chloroplasts cope with the high leaf temperatures that usually accompany high light? Isoprene synthesis might confer stability to photosynthetic membranes at high temperatures. Many plants emit monoterpenes or isoprene (2-methyl-1,3-butadiene) (Figure 9.14) amounting, on a global scale, to 3×10^{14} g released to the atmosphere each year (Brasseur and Chatfield 1991). These gaseous hydrocarbons are responsible for the pine scent (α- and β-pinene) in conferous forests and can form a blue haze above forests on hot days. Because isoprene and related hydrocarbons are a natural form of air pollution and because they play an important role in atmospheric chemistry, they have attracted much attention from atmospheric scientists.

Isoprene emission from leaves can constitute a significant fraction of the carbon assimilated in photosynthesis. For example, up to 2% of the carbon fixed by photosynthesis in aspen and oak leaves at 30°C is released as isoprene (Sharkey 1996). Sun leaves synthesize more isoprene than shade leaves, and synthesis is proportional to leaf temperature and water stress. On the basis of observations that photosynthesis in transgenic plants, unable to emit isoprene, is more easily damaged at high temperatures than in wild-type plants, it has been proposed that isoprene emission stabilizes the photosynthetic membranes at high temperatures (Sharkey and Singsaas 1995). More research is needed to confirm this interesting hypothesis, which ascribes a functional role to the massive isoprene emission from vegetation.

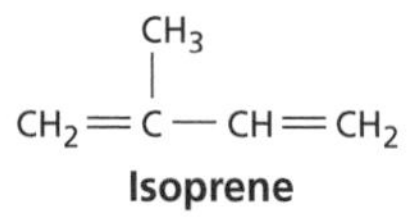

$$CH_2{=}C(CH_3){-}CH{=}CH_2$$

Isoprene

Figure 9.14 Chemical structure of isoprene. This gaseous hydrocarbon is emitted by plants and appears to make the photosynthetic system more tolerant to heat.

Interestingly, in addition to protecting the photosynthetic system against high light, the xanthophyll cycle may help protect against high temperatures. Chloroplasts are more tolerant of heat when they accumulate zeaxanthin (Havaux et al. 1996). Thus, plants may employ more than one biochemical mechanism to guard against the deleterious effect of excess heat.

Absorption of Too Much Light Can Lead to Photoinhibition

When leaves are exposed to more light than they can utilize (see Figure 9.11), photosynthesis is inhibited (see Chapter 7) and quantum efficiency decreases. This phenomenon is called **photoinhibition** (Kok 1956). The characteristics of photoinhibition depend on the amount of light to which the plant is exposed. When excess light is moderate, the quantum efficiency decreases but the maximum photosynthetic rate remains unchanged (Figure 9.15). This decrease in quantum efficiency is often temporary, and quantum efficiency can return to its initial higher value when photon flux decreases below saturation levels. This type of photoinhibitory response has been termed *dynamic photoinhibition* (Osmond 1994), and it is caused by the diversion of absorbed light energy toward heat dissipation.

On the other hand, exposure to high levels of excess light can damage the photosynthetic system, and this damage leads to decreases in both quantum efficiency and maximum photosynthetic rate (see Figure 9.15). In contrast to dynamic photoinhibition, these effects are relatively long lasting, persisting for weeks or months. This type of photoinhibition has been termed *chronic photoinhibition* (Osmond 1994). Initially, any decrease in quantum efficiency was thought to be the result of damage to the photosynthetic apparatus, but it is now recognized that short-term decreases in quantum efficiency probably reflect protective mechanisms (see Chapter 7), whereas chronic photoinhibition represents overload or failure of these mechanisms.

How significant is photoinhibition in nature? Dynamic photoinhibition appears to occur normally at midday, when leaves are exposed to maximum amounts of light and there is a corresponding reduction in carbon fixation. This effect becomes larger at low temperatures, and photoinhibition can shift to the more severe chronic form under more extreme climatic conditions. Studies of willows (Ögren and Sjöström 1990) and crops of *Brassica napus* (oilseed rape) and *Zea mays* (maize) have shown that the daily depression in photosynthetic rates caused by photoinhibition decreases biomass by 10% at the end of the growing season (Long et al. 1994). This may seem to be not a particularly large effect, but it could be significant in natural plant populations, where competition is high and where any reduction in carbon allocated to reproduction can adversely affect the survival success of the species.

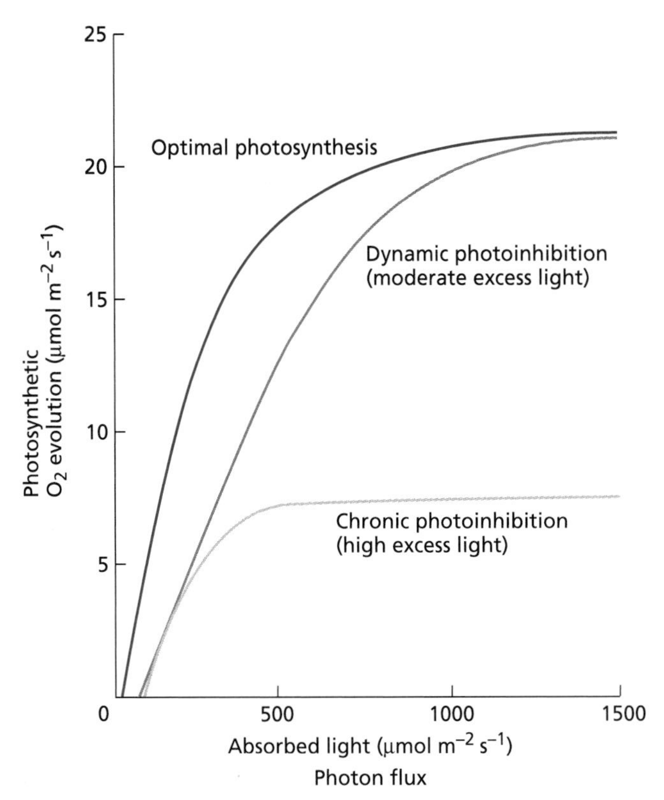

Figure 9.15 Photoinhibition-induced changes in light–response curves for photosynthesis. Exposure to moderate levels of excess light can decrease quantum efficiency without reducing maximum photosynthetic rate, a condition called dynamic photoinhibition. Dynamic photoinhibition may be caused by mechanisms that protect the chloroplasts against damaging effects of excess light. Under extreme conditions, exposure to high levels of excess light leads to chronic photoinhibition, where there is damage to the photosynthetic system leading to decreases in quantum efficiency and maximum photosynthetic rate. (After Osmond 1994.)

Photosynthetic Responses to Carbon Dioxide

We have discussed how plant growth and leaf anatomy are influenced by light. Now we turn our attention to how CO_2 concentration affects photosynthesis. CO_2 diffuses from the atmosphere into leaves, first through stomata, then through the intercellular air spaces, and ultimately into cells and chloroplasts. In the presence of adequate amounts of light, higher CO_2 concentrations support higher photosynthetic rates. The reverse is also true, and low CO_2 concentration can limit the amount of photosynthesis. Atmospheric CO_2 concentration has varied through the geological ages, and from the standpoint of most plants the current CO_2 concentration is relatively low, although it is rising because of the combustion of fossil fuels. The photosynthetic performance of C_3 plants can be limited by atmospheric CO_2 concentration; C_4 plants are able to concentrate CO_2 and thus are affected less by low CO_2 concentrations.

Atmospheric CO_2 Concentration Keeps Rising

Carbon dioxide is a trace gas in the atmosphere, accounting for about 0.036%, or 360 parts per million (ppm), of air. The partial pressure of ambient CO_2 (C_a) varies with atmospheric pressure and is approximately 36 Pa at sea level (see Box 9.2). Water vapor usually accounts for up to 2% of the atmosphere and O_2 for about 20%. The bulk of the atmosphere, nearly 80%, is nitrogen.

The current atmospheric concentration of CO_2 is almost twice the concentration prevailing during most of the last 160,000 years, as measured from air bubbles trapped in glacial ice in Antarctica (Figure 9.16A). Except for the last 200 years, CO_2 concentrations during the recent geological past have been low, fluctuating between 180 and 260 ppm. These low concentrations were typical of times extending back to the Cretaceous, when Earth was much warmer and the CO_2 concentration may have been as high as 1200 to 2800 ppm (Ehleringer et al. 1991). The current CO_2 concentration of the atmosphere is increasing by about 1 ppm each year, primarily because of the burning of fossil fuels (see Figure 9.16C). Since 1958, when systematic measurements of CO_2 began at Mauna Loa, Hawaii, atmospheric CO_2 concentrations have increased by more than 14% (Keeling et al. 1995), and by 2020 the atmospheric CO_2 concentration could reach 600 ppm. The consequences of this increase are the subject of intense scrutiny by scientists and government agencies, particularly because of predictions that the **greenhouse effect** could alter the world's climate.

A greenhouse roof transmits visible light, which is absorbed by plants and other surfaces inside the greenhouse. The absorbed light energy is converted to heat, and part of it is re-emitted as long-wavelength radiation. Because glass transmits long-wavelength radiation very poorly, this radiation cannot leave the greenhouse through the glass roof, and the greenhouse heats up. Certain gases in the atmosphere, particularly CO_2 and methane, play the same role as the glass roof in a greenhouse. The term "greenhouse effect" refers to the resulting warming of Earth's climate, caused by the trapping of long-wavelength radiation by the atmosphere.

The increased CO_2 concentration and temperature associated with the greenhouse effect can influence photosynthesis. At current atmospheric CO_2 concentrations, photosynthesis in C_3 plants is CO_2 limited (see Figure 9.18 and the discussion following it), but this situation could change as atmospheric CO_2 concentrations rise. Most C_3 plants grow 30 to 60% faster when CO_2 concentration is doubled (600 to 700 ppm) under laboratory conditions, but this result can be affected by nutrient status (Bowes 1993). In some plants the enhanced growth is only temporary. Carbon dioxide enrichment of greenhouses, where nutrient conditions are optimized, is used

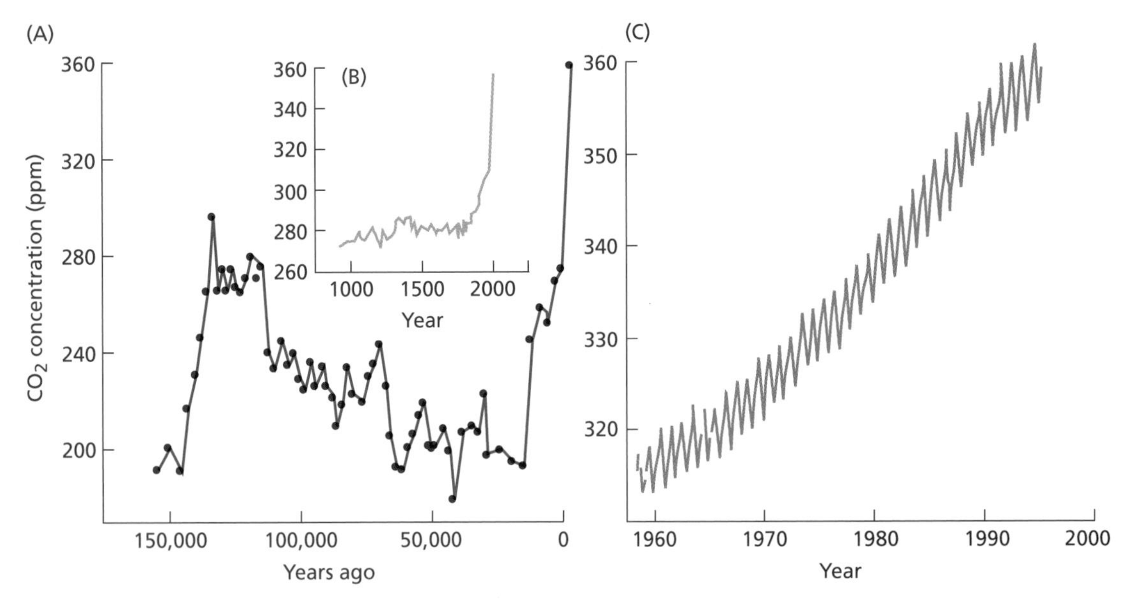

Figure 9.16 Concentration of atmospheric CO_2 from the present to 160,000 years ago. (A) Past atmospheric CO_2 concentrations, determined from bubbles trapped in glacial ice in Antarctica, were much lower than current levels. (B) In the last 1000 years, the rise in CO_2 concentration has coincided with the Industrial Revolution and the increased burning of fossil fuels. (C) Current atmospheric concentrations of CO_2 measured at Mauna Loa, Hawaii, continue to rise. Each year, the highest concentration is observed in May, just before the Northern Hemisphere growing season, and the lowest concentration is observed in October. (After Barnola et al. 1994, Keeling and Whorf 1994, Neftel et al. 1994, and Keeling et al. 1995.)

to increase the productivity of many crops, such as tomatoes, lettuce, cucumbers, and roses. The photosynthetic performance of C_3 plants under elevated CO_2 is enhanced because photorespiration decreases (see Chapter 8). In fact, the relatively low CO_2 concentrations present during the Miocene may have provided the driving force for the evolution and spread of C_4 plants found in the fossil record at this time (see Figure 9.21) (Cerling et al. 1997).

For Photosynthesis to Occur, CO_2 Must Diffuse into the Leaf

For photosynthesis to occur, carbon dioxide must diffuse from the atmosphere into the leaf and into the carboxylation site of rubisco. Since diffusion rates depend on concentration gradients (see Chapters 3 and 6), appropriate gradients are needed to ensure adequate diffusion of CO_2 from the leaf surface to the chloroplast. The cuticle that covers the leaf is nearly impermeable to CO_2, so the main port of entry of CO_2 into the leaf is the stomatal pore. CO_2 diffuses through the pore into the substomatal cavity and into the intercellular air spaces

BOX 9.2

Working with Gases

THE CONCENTRATION OF OXYGEN in the atmosphere is independent of altitude. Why, then, is it more difficult to breathe at high altitudes? The answer is that the binding of oxygen to hemoglobin in the blood depends on the partial pressure of oxygen, not on its concentration, and the partial pressure of a gas does vary with atmospheric pressure. Many biological reactions involving gases, such as the responses of rubisco to CO_2, depend on the partial pressure of the gas—hence the importance of measuring partial pressures.

The most familiar measure of concentration of a gas is mole fraction, not partial pressure. The **mole fraction** of a gas in a mixture is the number of moles of the gas of interest in a mole of the mixture. For example, the atmosphere has 21% oxygen and 360 ppm CO_2. Both of these measures are mole fractions: They tell us the number of moles of CO_2 present in a mole of air. The partial volume of a gas divided by the volume of the mixture, the number of moles divided by the total number of moles, and the partial pressure divided by the total pressure are all expressions of mole fractions and are unitless and interchangeable.

Partial pressure is the mole fraction multiplied by the total pressure. For example, at sea level, where the atmospheric pressure is about 10^5 Pa, the partial pressure of oxygen is 2.1×10^4 Pa ($21\% \times 10^5$), and the partial pressure of CO_2 is 36 Pa (360 ppm = 0.036%; $0.036\% \times 10^5 = 36$). At the top of a 1500 m mountain, where the atmospheric pressure is about 8.5×10^4 Pa, the partial pressure of oxygen is 1.8×10^4 Pa and that of CO_2 is 30.6 Pa. This is the reason that it is harder for people to breathe and for leaves to photosynthesize at higher altitudes.

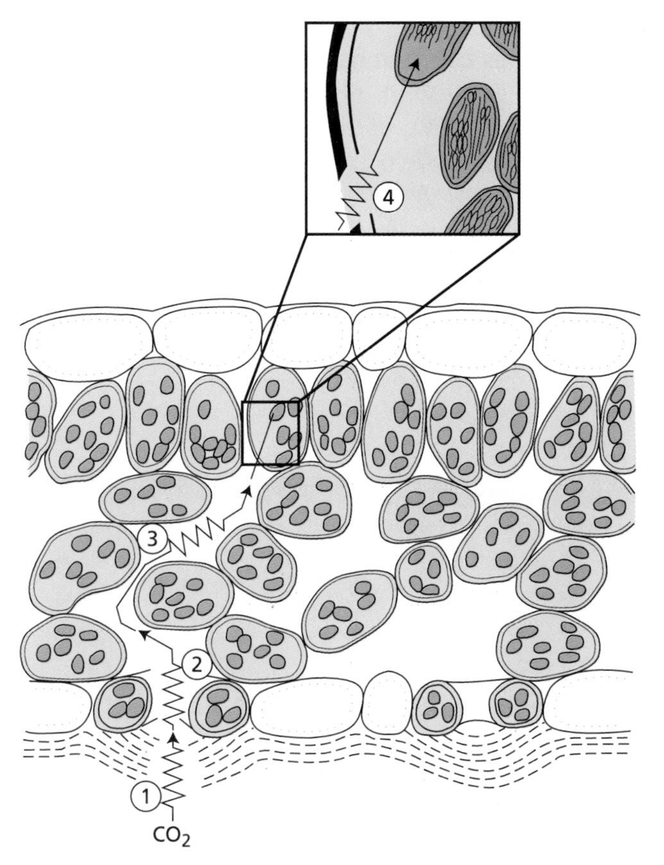

Figure 9.17 Points of resistance to the diffusion of CO_2 from outside the leaf to the chloroplasts. The boundary layer of water vapor that surrounds the leaf (boundary layer resistance) (1) and the stomatal pore (stomatal resistance) (2), are initial points of resistance to CO_2 entry into a leaf. There is some resistance to the passage of CO_2 through the intercellular air space (intercellular air space resistance) (3), and there is a resistance originating at the interface between air and the cell wall (mesophyll resistance) (4), where CO_2 is dissolved into solution and moves through liquid into the chloroplast. The stomatal pore is the major point of resistance to CO_2 diffusion to the chloroplasts.

between the mesophyll cells. This portion of the diffusion path of CO_2 into the chloroplast is a gaseous phase. The remainder of the diffusion path to the chloroplast is a liquid phase, which begins at the water layer that wets the walls of the mesophyll cells and continues through the plasma membrane, the cytosol, and the chloroplast. (For the properties of CO_2 in solution, see Box 8.1). Each portion of this diffusion pathway imposes a resistance to CO_2 diffusion, so the supply of CO_2 for photosynthesis meets a series of different points of resistance (Figure 9.17). An evaluation of the magnitude of each point of resistance is helpful for understanding CO_2 limitations to photosynthesis.

Water transpired by the leaf evaporates from the wet cell walls into the intercellular air spaces and exits the leaf through the stomatal pores (see Chapter 4). The same path is traveled in the reverse direction by CO_2 on its way to the chloroplast. The sharing of this pathway by CO_2 and water presents the plant with a functional dilemma. In air of high relative humidity, the diffusion gradient that drives water loss is about 50 times larger than the gradient that drives CO_2 uptake. In drier air, this gradient can be even larger. Therefore, a decrease in stomatal resistance, facilitating higher CO_2 uptake, unavoidably entails substantial water loss.

Recall from Chapter 4 that the gas phase of CO_2 diffusion into the leaf can be divided into three components—the boundary layer, the stomata, and the intercellular spaces of the leaf—each of which imposes a resistance to CO_2 diffusion (see Figure 9.17). The boundary layer consists of relatively unstirred air at the leaf surface, and its resistance to diffusion is called the **boundary layer resistance**. The magnitude of the boundary layer resistance decreases with leaf size and wind speed. The boundary layer resistance to water and CO_2 diffusion is physically related to the boundary layer resistance for sensible heat loss. Smaller leaves have a lower boundary layer resistance to CO_2 and water diffusion, and to sensible heat loss. Leaves of desert plants are usually small, facilitating sensible heat loss. The large leaves often found in the humid Tropics can have large boundary layer resistances, but these leaves can dissipate the radiation heat load by evaporative cooling because of the high transpiration rates made possible by the abundant water supply in these habitats.

After diffusing through the boundary layer, CO_2 enters the leaf through the stomatal pores, which impose the next type of resistance in the diffusion pathway, the **stomatal resistance**. Under most conditions in nature, in which the air around a leaf is seldom completely still, the boundary layer resistance is much smaller than the stomatal resistance, and the main limitation to CO_2 diffusion is imposed by the stomatal resistance. There is also a resistance to CO_2 diffusion in the air spaces that separate the substomatal cavity from the walls of the mesophyll cells, called the **intercellular air space resistance**. This resistance is also usually small—causing a drop of 0.5 Pa or less in partial pressure of CO_2, compared with the 36 Pa outside the leaf.

Since the stomatal pores usually impose the largest resistance to CO_2 uptake and water loss in the diffusion pathway, this regulation provides the plant with an effective way to control gas exchange between the leaf and the atmosphere. In experimental measurements of gas exchange from leaves, the boundary layer resistance and the intercellular air space resistance are usually ignored, and the stomatal resistance is used as the single parameter describing the gas phase resistance to CO_2.

The resistance to CO_2 diffusion of the liquid phase—the **liquid phase resistance**, also called **mesophyll resis-**

tance—encompasses diffusion from the intercellular leaf spaces to the carboxylation sites in the chloroplast. This point of resistance to CO_2 diffusion has been calculated as approximately one-tenth of the combined boundary layer resistance and stomatal resistance when the stomata are fully open. This low resistance can be attributed in part to the large surface area of mesophyll cells exposed to the intercellular air spaces, which can be as much as 10 to 30 times the projected leaf area (Syvertsen et al. 1995). The liquid phase resistance, however, can be larger in thick leaves.

CO_2 Imposes Limitations on Photosynthesis

Expressing photosynthetic rate as a function of the partial pressure of CO_2 in the intercellular air space (C_i) within the leaf (see Box 9.3), makes it possible to evaluate limitations imposed by CO_2 supply. At very low CO_2 concentrations, there is a negative balance between CO_2 fixed by photosynthesis and CO_2 produced by respiration, so there is a net efflux of CO_2 from the plant. Increasing CO_2 to the concentration at which these two processes balance each other defines the **CO_2 compen-**

BOX 9.3

Calculating Important Parameters in Leaf Gas Exchange

EVAPORATION (*E*) CAN BE DEFINED using Fick's law (see Chapter 3):

$$E = \frac{(e_i - e_a)}{Pr} \quad (1)$$

where E is the evaporation rate; e_i and e_a are the partial pressures of water inside the leaf and in the ambient air, respectively; P is the atmospheric pressure; and r is the stomatal resistance. The ratio $(e_i - e_a)/P$ is the difference of water vapor between the inside and the outside of the leaf (in mole fraction; see Box 9.2).

The concept of a resistance to gas diffusion is intuitively obvious but has the disadvantage that evaporation is inversely related to r. However, one can define conductance to water vapor (g) as the inverse of r. In contrast to r, g is directly related to E, and it is the parameter usually used in gas exchange analysis. Equation 1 can thus be rearranged as follows:

$$E = \frac{(e_i - e_a)g}{P} \quad (2)$$

and Equation 2 can be used to solve for g:

$$g = \frac{EP}{(e_i - e_a)} \quad (3)$$

The rate of evaporation from a leaf can be determined with devices that measure the amount of water leaving the leaf, and the vapor pressure of water in the ambient air can also be measured. The air in the intercellular spaces of the leaf is assumed to be at 100% humidity, and the vapor pressure of water at 100% relative humidity is a function of temperature. Therefore, we determine e_i by measuring the leaf temperature.

Since CO_2 diffuses along the same pathway as water, we can write an equation defining photosynthetic CO_2 assimilation (A) based on Equation 2:

$$A = \frac{(C_a - C_i)}{1.6Pr} = \frac{(C_a - C_i)g}{1.6P} \quad (4)$$

where C_a and C_i are the partial pressures of CO_2 in ambient air and in the intercellular spaces of the leaf, respectively. The CO_2 molecule is larger than H_2O and therefore has a smaller diffusion coefficient; the difference between the two diffusion coefficients has been empirically determined to be 1.6 (von Caemmerer and Farquhar 1981). Use of this correction factor in Equation 4 allows us to convert the measured resistance to water vapor diffusion into a resistance to CO_2 diffusion.

From Equation 4 we can calculate the partial pressure of CO_2 in the intercellular spaces of the leaf:

$$C_i = C_a - \frac{1.6AP}{g} \quad (5)$$

Since all the variables on the right-hand side of Equation 5 can be either measured or calculated, this equation is the basis for the calculation of C_i in experiments in which A is measured as a function of C_i (von Caemmerer and Farquhar 1981).

Water use efficiency can be quantified by a combination of Equations 2 and 4:

$$\frac{A}{E} = \frac{(C_a - C_i)}{1.6(e_i - e_a)} \quad (6)$$

Let's assume a leaf temperature of 30°C and an atmosphere with relative humidity of 50%. For a C_3 plant, $C_a - C_i$ is typically 10 Pa. At a leaf temperature of 30°C and a relative humidity of 50%, $e_i - e_a = 2.1 \times 10^3$ Pa. From Equation 6, we compute that A/E = 0.003 mol CO_2 (mol $H_2O)^{-1}$, or a transpiration ratio (E/A) of 336. If water availability is reduced, stomata close and evaporation decreases, leading to improved water use efficiency. Total CO_2 assimilation also decreases, but the plant conserves water and increases its chances of survival.

These equations have been used very successfully to advance our understanding of photosynthesis and the response of plants to their environment. However, every technique has limitations, and these equations are valid only when stomata behave uniformly across the leaf. Problems arise when some stomata are closed and others are open, especially in leaves in which discrete regions of mesophyll are sealed off from each other by leaf veins. Under some conditions, in one mesophyll region the stomata are open and there is a correspondingly high photosynthetic rate, whereas in an immediately adjacent region the stomata are partially closed and the mesophyll has a low photosynthetic rate. This phenomenon is termed *patchy stomata* and often arises when leaves are stressed (Terashima 1992). Although patchy stomata may reflect normal interactions between stomata and leaf mesophyll (Siebke and Weis 1995), patchiness can lead to inaccurate determinations of C_i.

sation point, the point at which the net efflux of CO_2 from the plant is zero. This concept is analogous to that of the light compensation point discussed earlier in the chapter: The CO_2 compensation point reflects the balance between photosynthesis and respiration as a function of CO_2 concentration, and the light compensation point reflects that balance as a function of photon flux.

In C_3 plants, increasing CO_2 above the compensation point stimulates photosynthesis over a wide concentration range. At low to intermediate CO_2 concentrations, photosynthesis is limited by the carboxylation capacity of rubisco. At high CO_2 concentrations, photosynthesis is limited by the ability of Calvin cycle enzymes to regenerate the acceptor molecule ribulose-1,5-bisphosphate, which depends on electron transport rates. Most leaves appear to regulate their C_i such that it is intermediate between limitations imposed by carboxylation capacity and the capacity to regenerate ribulose-1,5-bisphosphate.

Inherent differences between C_3 and C_4 plants are revealed by a plot of their photosynthetic performance as a function of intercellular CO_2 concentration, C_i (Figure 9.18). In C_4 plants, photosynthetic rates saturate at C_i values of about 20 Pa, reflecting the effective CO_2 concentrating mechanisms operating in these plants (see Chapter 8). On the other hand, in C_3 plants, increasing C_i levels continue to stimulate photosynthesis over a much broader range. These results indicate that C_3 plants may benefit from ongoing increases in atmospheric CO_2 concentrations (see Figure 9.16), whereas most C_4 plants are saturated or nearly so by current atmospheric CO_2 levels. In addition, plants with C_4 metabolism have a CO_2 compensation point of zero or nearly zero (see Figure 9.18), reflecting their very low levels of photorespiration (see Chapter 8). This difference between C_3 and C_4 plants is not apparent when the experiments are conducted at low oxygen concentrations, a condition that shifts rubisco activity toward carboxylation.

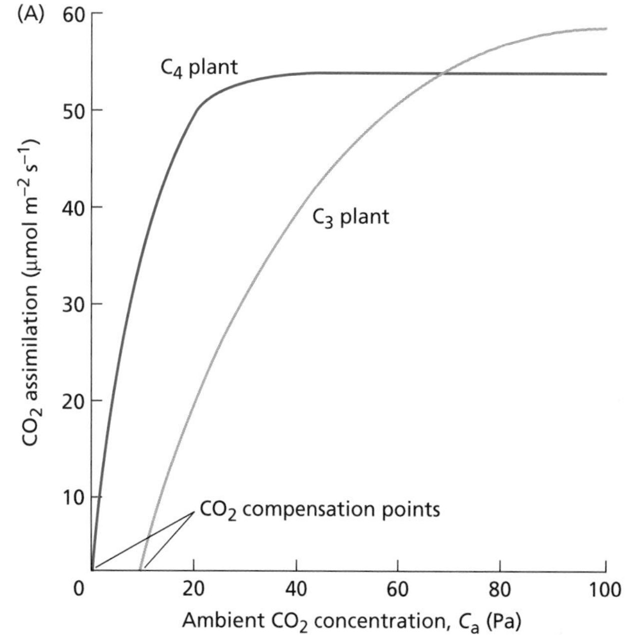

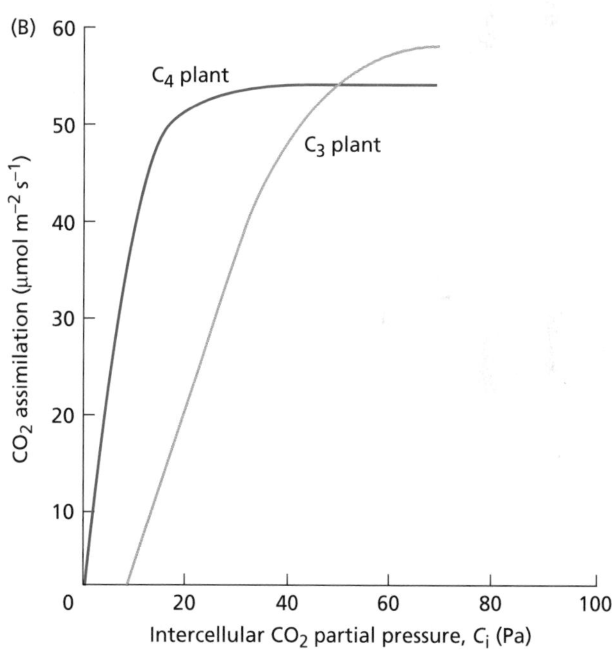

Figure 9.18 Changes in photosynthesis as a function of ambient intercellular CO_2 concentrations in *Tidestromia oblongifolia* (Arizona honeysweet), a C_4 plant, and *Larrea divaricata* (creosote bush), a C_3 plant. Photosynthetic rate is plotted against (A) ambient CO_2 concentration and (B) calculated intercellular partial pressure of CO_2 inside the leaf (see Equation 5 in Box 9.3). The concentration at which CO_2 assimilation is zero defines the CO_2 compensation point. (From Berry and Downton 1982.)

CO_2 Concentrating Mechanisms Affect Photosynthetic Responses of Leaves

In plants with CO_2 concentrating mechanisms, which include C_4 and CAM plants, the CO_2 concentrations at the carboxylation sites are often saturating. This physiological feature has several implications. Plants with C_4 metabolism need less rubisco to achieve a given rate of photosynthesis; therefore they require less nitrogen to grow. In addition, the CO_2 concentrating mechanism allows the leaf to maintain high photosynthetic rates at lower C_i values, which require lower rates of stomatal conductance for a given rate of photosynthesis. Thus, C_4 plants can use water and nitrogen more efficiently than C_3 plants can. On the other hand, the additional energy cost of the concentrating mechanism (see Chapter 8) makes C_4 plants less efficient in their utilization of light. This is probably one of the reasons that most shade-adapted plants are C_3 plants.

Many cacti and other succulent plants with CAM metabolism open their stomata at night and close them

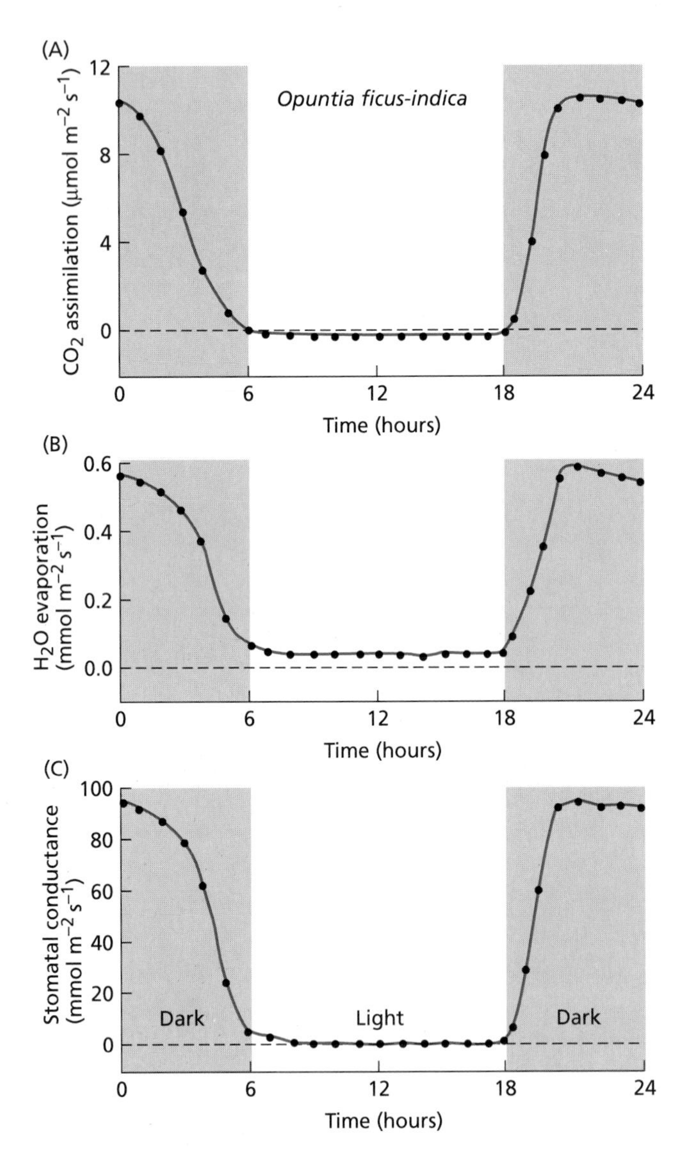

Figure 9.19 Photosynthetic carbon assimilation, evaporation, and stomatal conductance of a CAM plant, the cactus *Opuntia ficus-indica*, during a 24-hour period. The whole plant was kept in a gas exchange chamber in the laboratory. The dark period is indicated by shaded areas. In contrast to plants with C_3 or C_4 metabolism, CAM plants open their stomata and fix CO_2 at night. (From Gibson and Nobel 1986.)

during the day (Figure 9.19). The CO_2 taken up during the night is fixed into malate (see Chapter 8). Since air temperatures are much lower at night than during the day, water loss is low and a significant amount of water is saved relative to the amount of CO_2 fixed. The main constraint on CAM metabolism is that the capacity to store malic acid is limited, and this limitation restricts the amount of CO_2 uptake. However, many CAM plants can fix CO_2 via the C_3 photosynthetic reduction cycle at the end of the day, when temperature gradients are less extreme (Osmond 1978).

Cladodes (flattened stems) of the cactus *Opuntia* sp. can survive after detachment from the plant for several months without water. Their stomata are closed all the time, and the CO_2 released by respiration is refixed into malate. This process, which has been called CAM idling, allows the plant to survive for prolonged periods of time while losing remarkably little water.

Discrimination of Carbon Isotopes Unravels Different Photosynthetic Pathways

Atmospheric CO_2 contains the naturally occurring carbon isotopes ^{12}C, ^{13}C, and ^{14}C in the proportions 98.9%, 1.1%, and 10^{-10}%, respectively. $^{14}CO_2$ is present in such small quantities that it has no physiological relevance, but $^{13}CO_2$ is different. The chemical properties of $^{13}CO_2$ are identical to those of $^{12}CO_2$, but because of the slight difference in mass (2.3%), plants use less $^{13}CO_2$ than $^{12}CO_2$. In other words, plants discriminate against the heavier isotope of carbon, and they have smaller ratios of ^{13}C to ^{12}C than atmospheric CO_2 has. How effective are plants at distinguishing between the two carbon isotopes? Although discrimination against ^{13}C is subtle, the isotope composition of plants reveals a wealth of information.

Carbon isotope composition is measured by use of a mass spectrometer, which yields the following ratio:

$$R = \frac{^{13}CO_2}{^{12}CO_2} \tag{9.1}$$

The **isotope composition** of plants, $\delta^{13}C$, is quantified on a per mil (‰) basis:

$$\delta^{13}C\ ‰ = \left(\frac{R_{sample}}{R_{standard}} - 1\right) \times 1000 \tag{9.2}$$

where the standard represents the carbon isotopes contained in the Pee Dee limestone formation from South Carolina. Isotope discrimination is calculated with respect to the carbon isotopic ratio of air:

$$\Delta^{13}C\ ‰ = \left(\frac{R_{air}}{R_{plant}} - 1\right) \times 1000 \tag{9.3}$$

What are some typical values for isotope composition? C_3 plants have a $\delta^{13}C$ of about –28‰; C_4 plants have an average value of –14‰ (Figure 9.20) (O'Leary 1988). The negative sign indicates that both C_3 and C_4 plants have less ^{13}C than the isotope standard has. Since the per mil calculation involves multiplying by 1000, the actual isotope discrimination is small. Nonetheless, isotope discrimination values reveal many things. For example, measuring the $\Delta^{13}C$ of table sugar (sucrose) makes it possible to determine if the sucrose came from sugar beet (a C_3 plant) or sugarcane (a C_4 plant). Fossil fuels have a $\delta^{13}C$ of about –30‰ because the carbon in these deposits came from organisms that had a C_3 car-

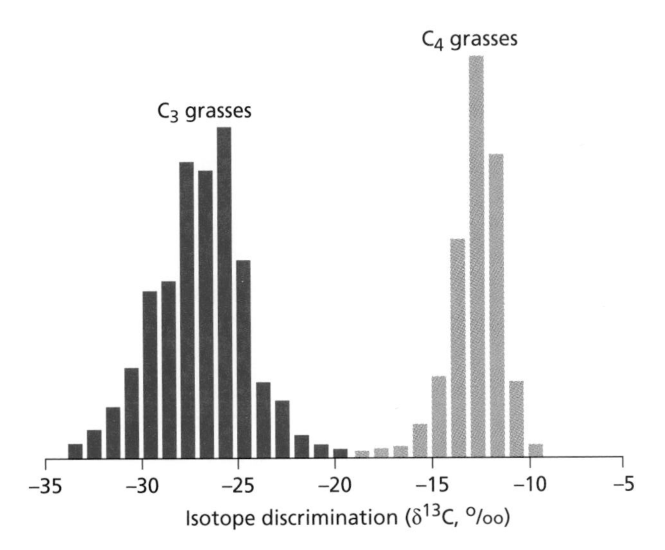

Figure 9.20 Carbon isotope discrimination in plants. C_3 plants discriminate against and take up less ^{13}C than C_4 plants do. Consequently, C_3 and C_4 grasses have distinct isotope compositions. (From Čerling et al. 1997.)

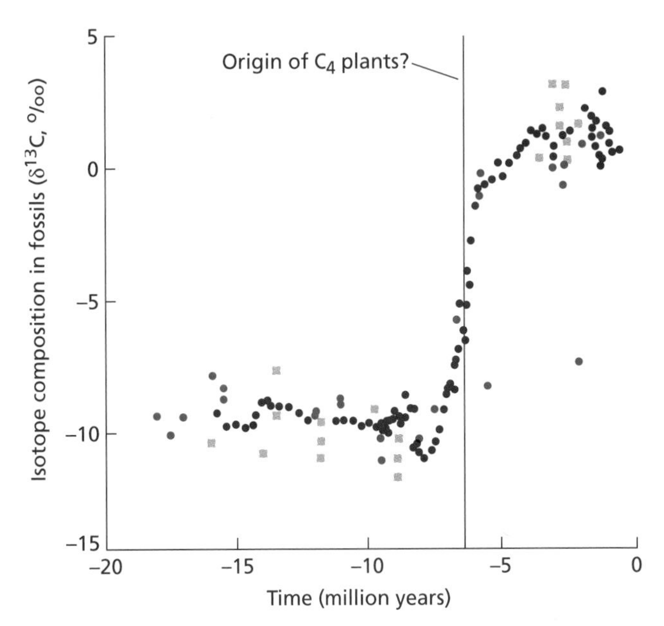

Figure 9.21 Transition in isotope composition in fossil soil and fossil animal teeth, indicating an expansion of C_4 plants in the late Miocene. Isotope values are shown for carbonates (black circles) extracted from fossil soil from Pakistan, for fossil mammalian tooth enamel from Pakistan (light green squares), and for fossil horse tooth enamel from North America (green circles). The change in isotope composition 5 to 7 million years ago suggests a shift from a flora dominated by C_3 plants to a flora dominated by C_4 plants. (After Cerling et al. 1993.)

bon fixation pathway. And measuring $\delta^{13}C$ in fossil carbonate-containing soils and fossil teeth of extinct herbivores makes it possible to determine that C_4 photosynthesis developed and became prevalent relatively recently, only 7 million years ago (Figure 9.21) (Cerling et al. 1997).

What is the physiological basis for isotope enrichment in plants? CO_2 moves from air outside of the leaf to the carboxylation site on an enzyme within the chloroplast in a series of steps involving diffusion. Since $^{12}CO_2$ is lighter than $^{13}CO_2$, it diffuses slightly faster toward the carboxylation site. However, the largest isotope discrimination step is the carboxylation reaction catalyzed by rubisco (Farquhar et al. 1989). For reasons not yet understood, this enzyme has an intrinsic discrimination value ($\Delta^{13}C$) of –30‰. By contrast, PEP carboxylase, the primary CO_2 fixation enzyme of C_4 plants, has a much smaller isotope discrimination effect ($\Delta^{13}C$ = –2 to 6‰). Thus, the inherent difference between the discrimination effects of the two carboxylating enzymes causes the different isotope compositions observed in C_3 and C_4 plants (Farquhar et al. 1989).

CAM plants can have $\delta^{13}C$ values that are intermediate between those of C_3 and C_4 plants. In CAM plants that fix CO_2 at night via PEP carboxylase, $\delta^{13}C$ is similar to that of C_4 plants. However, when some CAM plants are well watered, they switch to C_3 mode by opening their stomata and fixing CO_2 during the day via rubisco. Under these conditions the isotope composition shifts more toward that of C_3 plants. Thus, the $^{13}C/^{12}C$ values for CAM plants reflect how much carbon is fixed via the C_3 pathway versus the C_4 pathway.

Do other physiological characteristics of plants affect isotope composition? In C_3 plants, isotope composition is proportional to the partial pressure of CO_2 in the intercellular air space of leaves (C_i). More discrimination occurs when C_i is high, as when stomata are open. Open stomata also facilitate water loss. Thus, lower water use efficiency results in greater discrimination against ^{13}C (Farquhar et al. 1989). Plants also fractionate other isotopes, such as $^{18}O/^{16}O$ and $^{15}N/^{14}N$, and the various patterns of isotope enrichment can be used as indicators of particular metabolic pathways or features.

Photosynthetic Responses to Temperature

Temperature has an important effect on photosynthesis, and plants can photosynthesize in habitats that have a surprisingly broad range of temperatures. In the lower temperature range, plants growing in alpine areas are capable of net CO_2 uptake at temperatures close to 0°C; at the other extreme, plants living in Death Valley, California, have optimal rates of photosynthesis at temperatures approaching 50°C. Similar to light, temperature varies throughout the day, and plants may spend a significant amount of time below or above their optimum temperature for photosynthesis. The result may be a limit to plant productivity.

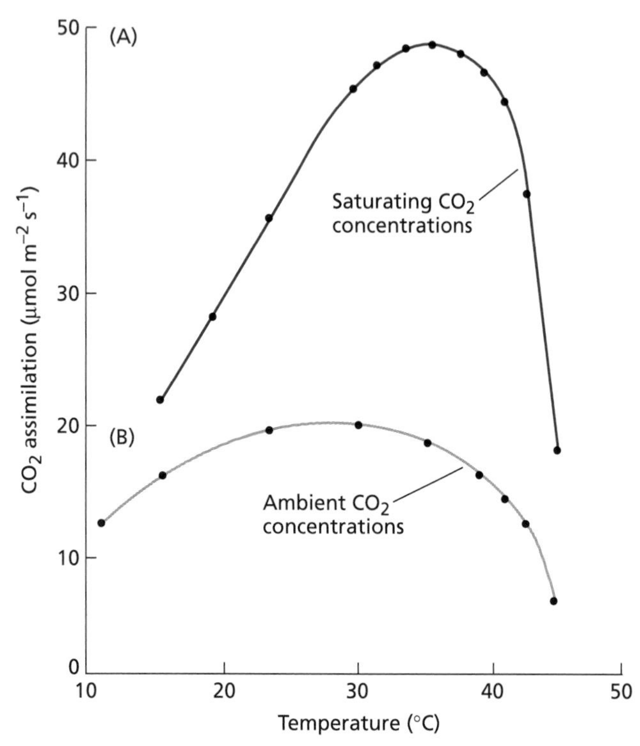

Figure 9.22 Changes in photosynthesis as a function of temperature at concentrations that saturate photosynthetic CO_2 assimilation (A) and at normal atmospheric CO_2 concentrations (B). Photosynthesis depends strongly on temperature at saturating CO_2 concentrations. (Redrawn from Berry and Björkman 1980.)

When photosynthetic rate is plotted as a function of temperature, the curve has a characteristic bell shape (Figure 9.22). The ascending arm of the curve represents a temperature-dependent stimulation of photosynthesis up to an optimum; the descending arm is associated with deleterious effects, some of which are reversible while others are not.

Temperature affects all biochemical reactions of photosynthesis, so it is not surprising that the responses to temperature are complex. We can gain insight into the underlying mechanisms by comparing photosynthetic rates in air at normal and at high CO_2 concentrations. At high CO_2 (see Figure 9.22A), there is an ample supply of CO_2 at the carboxylation sites, and the rate of photosynthesis is limited primarily by biochemical reactions connected with electron transfer (see Chapter 7). In these conditions, temperature changes have large effects on fixation rates. At ambient CO_2 concentrations (see Figure 9.22B), photosynthesis is limited by the activity of rubisco, and the response reflects two conflicting processes: an increase in carboxylation rate with temperature and a decrease in the affinity of rubisco for CO_2 as the temperature rises (see Chapter 8). These opposing effects damp the temperature response of photosynthesis at ambient CO_2 concentrations.

Respiration rates also increase as a function of temperature, and the interaction between photorespiration and photosynthesis becomes apparent in temperature responses. Figure 9.23 shows changes in quantum yield as a function of temperature in a C_3 plant and in a C_4 plant. In the C_4 plant, the quantum yield remains constant with temperature, reflecting typical low rates of photorespiration. In the C_3 plant, the quantum yield decreases with temperature, reflecting a stimulation of photorespiration by temperature and an ensuing higher energy demand per net CO_2 fixed.

At low temperatures, photosynthesis is often limited by phosphate availability at the chloroplast (Sage and Sharkey 1987). When triose phosphates are exported from the chloroplast to the cytosol, an equimolar amount of inorganic phosphate is taken up via translocators in the chloroplast membrane. If the rate of triose phosphate utilization in the cytosol decreases, phosphate uptake into the chloroplast is inhibited and pho-

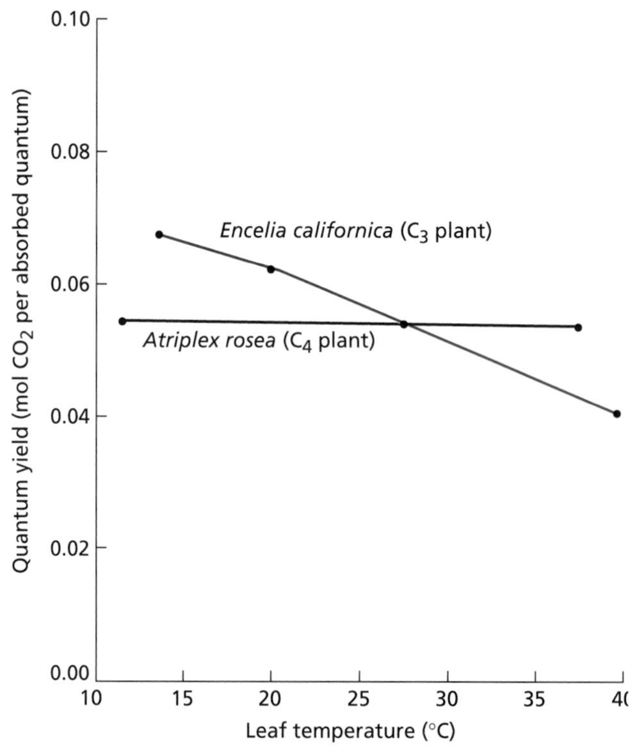

Figure 9.23 The quantum yield of photosynthetic carbon fixation in a C_3 plant and in a C_4 plant as a function of leaf temperature. In normal air, photorespiration increases with temperature in C_3 plants, and the energy cost of net CO_2 fixation increases accordingly. This higher energy cost is expressed in lower quantum yields at higher temperatures. Because of the CO_2 concentrating mechanisms of C_4 plants, photorespiration is low in these plants, and the quantum yield does not show a temperature dependence. However, at lower temperatures the quantum yield of C_3 plants is higher than that of C_4 plants, indicating that C_3 plants are more efficient at lower temperatures. (From Ehleringer and Björkman 1977.)

tosynthesis becomes phosphate limited (Geiger and Servaites 1994). Starch synthesis and sucrose synthesis decrease rapidly with temperature, reducing the demand for triose phosphates and causing the phosphate limitation observed at low temperatures.

The highest photosynthetic rates seen in temperature responses represent the so-called optimal temperature response. When these temperatures are exceeded, photosynthetic rates decrease again. It has been argued that this optimal temperature is the point at which the capacities of the various steps of photosynthesis are optimally balanced, with some of the steps becoming limiting as the temperature decreases or increases. Optimal temperatures have strong genetic and physiological components. Plants of different species growing in habitats with different temperatures have different optimal temperatures for photosynthesis, and plants of the same species, grown at different temperatures and then tested for their photosynthetic responses, show temperature optima that correlate with the temperature at which they were grown. Plants growing at low temperatures maintain higher photosynthetic rates at low temperatures than plants grown at high temperatures. These changes in photosynthetic properties in response to temperature play an important role in plant adaptations to different environments.

Summary

Photosynthetic activity in the intact leaf is an integral process that depends on many biochemical reactions. Different environmental factors can limit photosynthetic rates. Leaf anatomy is highly specialized for light absorption, and the properties of palisade and mesophyll cells ensure uniform light absorption throughout the leaf. In addition to the anatomical features of the leaf, chloroplast movements within cells and solar tracking by the leaf blade help maximize light absorption. Light transmitted through upper leaves is absorbed by leaves growing beneath them. Many properties of the photosynthetic apparatus change as a function of the available light, including the light compensation point, which is higher in sun leaves than in shade leaves. The linear portion of the light–response curve for photosynthesis provides a measure of the quantum yield of photosynthesis in the intact leaf. In temperate areas, quantum yields of C_3 plants are generally higher than those of C_4 plants.

Sunlight imposes a substantial heat load on the leaf, which is dissipated back into the air by long-wavelength radiation, by sensible heat loss, or by evaporative heat loss. Increasing CO_2 concentrations in the atmosphere are increasing the heat load on the biosphere. This process could cause damaging changes in the world's climate, but it could also reduce the CO_2 limitations on photosynthesis. At high photon flux, photosynthesis in most plants is CO_2 limited, but the limitation is substantially lower in C_4 and CAM plants because of their CO_2 concentrating mechanisms. Diffusion of CO_2 into the leaf is constrained by a series of different points of resistance. The largest resistance is usually that imposed by the stomata, so modulation of stomatal apertures provides the plant with an effective means of controlling water loss and CO_2 uptake. Both stomatal and nonstomatal factors affect CO_2 limitations on photosynthesis.

Temperature responses of photosynthesis reflect the temperature sensitivity of the biochemical reactions of photosynthesis and are most pronounced at high CO_2 concentrations. Because of the role of photorespiration, the quantum yield is strongly dependent on temperature in C_3 plants but is nearly independent of temperature in C_4 plants. Leaves growing in cold climates can maintain higher photosynthetic rates at low temperatures than leaves growing in warmer climates. Leaves grown at high temperatures perform better at high temperatures then leaves grown at low temperatures do. Functional changes in the photosynthetic apparatus in response to prevailing temperatures in their environment have an important effect on the capacity of plants to live in diverse habitats.

General Reading

*Berry, J. A., and Downton, J. S. (1982) Environmental regulation of photosynthesis. In *Photosynthesis, Development, Carbon Metabolism and Plant Productivity*, Vol. II, Govindjee, ed., Academic Press, New York, pp. 263–343.

*Björn, L. O., and Vogelmann, T. C. (1994) Quantification of light. In *Photomorphogenesis in Plants*, 2nd ed., R. E. Kendrick and G. H. M. Kronenberg, eds., Kluwer, Dordrecht, Netherlands, pp. 17–25.

*Bowes, G. (1993) Facing the inevitable: Plants and increasing atmospheric CO_2. *Annu. Rev. Plant Physiol. Plant Mol. Biol.* 44: 309–332.

*Demmig-Adams, B., and Adams, W. (1996) The role of xanthophyll cycle carotenoids in the protection of photosynthesis. *Trends Plant Sci.* 1: 21–26.

*Ehleringer, J. R., Sage, R. F., Flanagan, L. B., and Pearcy, R. W. (1991) Climate change and the evolution of C_4 photosynthesis. *Trends Ecol. Evol.* 6: 95–99.

*Evans, J. R., von Caemmerer, S., and Adams, W. W. (1988) *Ecology of Photosynthesis in Sun and Shade*. CSIRO, Melbourne.

*Farquhar, G. D., Ehleringer, J. R., and Hubick, K. T. (1989) Carbon isotope discrimination and photosynthesis. *Annu. Rev. Plant Physiol. Plant Mol. Biol.* 40: 503–537.

*Gibson, A. C., and Nobel, P. S. (1986) *The Cactus Primer*. Harvard University Press, Cambridge, MA.

*Haupt, W., and Scheuerlein, R. (1990) Chloroplast movement. *Plant Cell Environ.* 13: 595–614.

*Kirk, J. T. (1994) *Light and Photosynthesis in Aquatic Ecosystems*. Cambridge University Press, Cambridge.

*Koller, D. (1990) Light-driven leaf movements. *Plant Cell Environ.* 13: 615–632.

*Long, S. P., Humphries, S., and Falkowski, P. G. (1994) Photoinhibition of photosynthesis in nature. *Annu. Rev. Plant Physiol. Plant Mol. Biol.* 45: 633–662.

Ort, D. R., and Yocum, C. F. (1996) *Oxygenic Photosynthesis: The Light Reactions*. Kluwer, Dordrecht, Netherlands.

*Osmond, C. B. (1978) Crassulacean acid metabolism: A curiosity in context. *Annu. Rev. Plant Physiol.* 29: 379–414.

*Osmond, C. B. (1994) What is photoinhibition? Some insights from comparisons of shade and sun plants. In *Photoinhibition of Photosynthesis*, N. R. Baker and J. R. Bowyer, eds., BIOS Scientific, Oxford, pp. 1–24.

*Pearcy, R. W. (1994) Photosynthetic response to sunflecks and light gaps: Mechanisms and constraints. In *Photoinhibition of Photosynthesis*, N. R. Baker and J. R. Bowyer, eds., BIOS Scientific, Oxford, pp. 255–271.

*Sharkey, T. D. (1996) Emission of low molecular mass hydrocarbons from plants. *Trends Plant Sci.* 1: 78–82.

*Terashima, I. (1992) Anatomy of non-uniform leaf photosynthesis. *Photosynth. Res.* 31: 195–212.

*Terashima, I., and Hikosaka, K. (1995) Comparative ecophysiology of leaf and canopy photosynthesis. *Plant Cell Environ.* 18: 1111–1128.

*Vogelmann, T. C. (1993) Plant tissue optics. *Annu. Rev. Plant Physiol. Plant Mol. Biol.* 44: 231–251.

* Indicates a reference that is general reading in the field and is also cited in this chapter.

Chapter References

Anderson, J. M. (1986) Photoregulation of the composition, function, and structure of thylakoid membranes. *Annu. Rev. Plant Physiol.* 37: 93–136.

Barnola, J. M., Raynaud, D., Lorius, C., and Korotkevich, Y. S. (1994) Historical CO_2 record from the Vostok ice core. In *Trends '93: A Compendium of Data on Global Change*, T.A. Boden, D.P. Kaiser, R.J. Sepanski, and F.W. Stoss, eds., ORNL/CDIAC-65, Carbon Dioxide Information Center, Oak Ridge National Laboratory, Oak Ridge, TN, pp 7–10.

Berry, J., and Björkman, O. (1980) Photosynthetic response and adaptation to temperature in higher plants. *Annu. Rev. Plant Physiol.* 31: 491–543.

Björkman, O. (1981) Responses to different quantum flux densities. In *Encyclopedia of Plant Physiology*, New Series, Vol. 12A, O. L. Lange, P. S. Nobel, C. B. Osmond, and H. Zeigler, eds., Springer, Berlin, pp. 57–107.

Brasseur, G. P., and Chatfield, R. B. (1991) The fate of biogenic trace gases in the atmosphere. In *Trace Gas Emission from Plants*, T. D. Sharkey, E. A. Holland, and H. A. Mooney, eds., Academic Press, San Diego, pp. 1–27.

Brugnoli, E., and Björkman, O. (1992) Chloroplast movements in leaves: Influence on chlorophyll fluorescence and measurements of light-induced absorbance changes related to delta-pH and zeaxanthin formation. *Photosynth. Res.* 32: 23–35.

Campbell, G. S. (1977) *An Introduction to Environmental Biophysics*. Springer, New York.

Cerling, T. E., Harris, J. M., MacFadden, B. J., Leakey, M. G., Quade, J., Eisenmann, V., and Ehleringer, J. R. (1997) Global vegetation change through the Miocene/Pliocene boundary. *Nature* 389: 153–158.

Cerling, T. E., Wang, Y., and Quade, J. (1993) Expansion of C4 ecosystems as an indicator of global ecological change in the late Miocene. *Nature* 361: 344–345.

Demmig-Adams, B., Gilmore, A. M., and Adams, W. (1996) In vivo functions of carotenoids in higher plants. *FASEB J.* 10: 403–412.

Ehleringer, J., Björkman, O., and Mooney, H. A. (1976) Leaf pubescence: Effects on absorptance and photosynthesis in a desert shrub. *Science* 192: 376–377.

Ehleringer, J., and Forseth, I. (1980) Solar tracking by plants. *Science* 210: 1094–1098.

Ehleringer, J. A., and Björkman, O. (1977) Quantum yields for CO_2 uptake in C_3 and C_4 plants. *Plant Physiol.* 59: 86–90.

Geiger, D. R., and Servaites, J. C. (1994) Diurnal regulation of photosynthetic carbon metabolism in C_3 plants. *Annu. Rev. Plant Physiol. Plant Mol. Biol.* 45: 235–256.

Harvey, G. W. (1979) Photosynthetic performance of isolated leaf cells from sun and shade plants. *Carnegie Inst. Washington Yearbook* 79: 161–164.

Havaux, M., Tardy, F., Ravenel, J., and Parot, P. (1996) Thylakoid membrane stability to heat stress studied by flash spectroscopic measurements of the electrochromic shift in intact potato leaves: Influence of the xanthophyll content. *Plant Cell Environ.* 19: 1359–1368.

Holmes, M. G., Klein, W. H., and Sager, J. C. (1985) Photons, flux, and some light on philology. *Hort. Sci.* 20: 29–31.

Jarvis, P. G., and Leverenz, J. W. (1983) Productivity of temperate, deciduous and evergreen forests. In *Encyclopedia of Plant Physiology*, New Series, Vol. 12D, O. L. Lange, P. S. Nobel, C. B. Osmond, and H. Zeigler, eds., Springer, Berlin, pp. 233–280.

Keeling, C. D., and Whorf, T. P. (1994) Atmospheric CO_2 records from sites in the SIO air sampling network. In *Trends '93: A Compendium of Data on Global Change*, T.A. Boden, D.P. Kaiser, R.J. Sepanski, and F.W. Stoss, eds., ORNL/CDIAC-65, Carbon Dioxide Information Center, Oak Ridge National Laboratory, Oak Ridge, TN, pp 16–26.

Keeling, C. D., Whorf, T. P., Wahlen, M., and Van der Plicht, J. (1995) Interannual extremes in the rate of rise of atmospheric carbon dioxide since 1980. *Nature* 375: 666–670.

Kok, B. (1956) On the inhibition of photosynthesis by intense light. *Biochim. Biophys. Acta* 21: 234–244.

Koller, D., Ritter, S., Briggs, W. R., and Schäfer, E. (1990) Action dichroism in perception of vectorial photo-excitation in the solar-tracking leaf of *Lavatera cretica* L. *Planta* 181: 184–190.

McCree, K. J. (1981) Photosynthetically active radiation. In *Encyclopedia of Plant Physiology*, New Series, Vol. 12A, O. L. Lange, P. S. Nobel, C. B. Osmond, and H. Zeigler, eds., Springer, Berlin, pp. 41–55.

Melis, A. (1996) Excitation energy transfer: Functional and dynamic aspects of Lhc (cab) proteins. In *Oxygenic Photosynthesis: The Light Reactions*, D. R. Ort and C. F. Yocum, eds., Kluwer, Dordrecht, Netherlands, pp. 523–538.

Neftel, A., Friedle, H., Moor, E., Lötscher, H., Oeschger, H., Siegenthaler, U., and Stauffer, B. (1994) Historical CO_2 record from the Siple Station ice core. In *Trends '93: A Compendium of Data on Global Change*, T.A. Boden, D.P. Kaiser, R.J. Sepanski, and F.W. Stoss, eds., ORNL/CDIAC-65, Carbon Dioxide Information Center, Oak Ridge National Laboratory, Oak Ridge, TN, pp 11–15.

Ögren, E., and Sjöström, M. (1990) Estimation of the effect of photoinhibition on the carbon gain in leaves of a willow canopy. *Planta* 181: 560–567.

O'Leary, M. H. (1988) Carbon isotopes in photosynthesis. *BioScience* 38: 328–333.

Ort, D. R., and Baker, N. R. (1988) Consideration of photosynthetic efficiency at low light as a major determinant of crop photosynthetic performance. *Plant Physiol. Biochem.* 26: 555–565.

Richter, T., and Fukshansky, L. (1996) Optics of a bifacial leaf: 2. Light regime as affected by leaf structure and the light source. *Photochem. Photobiol.* 63: 517–527.

Rupert, C. S., and Letarjet, R. (1978) Toward a nomenclature and dosimetric scheme applicable to all radiations. *Photochem. Photobiol.* 28: 3–5.

Sage, R. F., and Sharkey, T. D. (1987) The effect of temperature on the occurrence of O_2 and CO_2 insensitive photosynthesis in field grown plants. *Plant Physiol.* 84: 658–664.

Sharkey, T. D., and Singsaas, E. L. (1995) Why plants emit isoprene. *Nature* 374: 769.

Siebke, K., and Weis, E. (1995) Assimilation images of leaves of *Glechoma hederacea*: Analysis of non-synchronous stomata related oscillations. *Planta* 196: 155–165.

Smith, H. (1994). Sensing the light environment: The functions of the phytochrome family. In *Photomorphogenesis in Plants*, R. E. Kendrick and G. H. M. Kronenberg, eds., Nijhoff, Dordrecht, Netherlands, pp 377-416.

Syvertsen, J. P., Lloyd, J., McConchie, C., Kriedemann, P. E., and Farquhar, G. D. (1995) On the relationship between leaf anatomy and CO_2 diffusion through the mesophyll of hypostomatous leaves. *Plant Cell Environ.* 18: 149–157.

Terashima, I. (1989) Productive structure of a leaf. In *Photosynthesis, Proceedings of the C. S. French Symposium* (Plant Biology, vol. 8), W. Briggs, ed., Liss, New York, pp. 207–213.

Tlalka, M., and Gabryś, H. (1993) Influence of calcium on blue-light-induced chloroplast movement in *Lemna trisulca* L. *Planta* 189: 491–498.

Vogelmann, T. C., and Björn, L. O. (1983) Response to directional light by leaves of a sun-tracking lupine (*Lupinus succulentus*). *Physiol. Plantarum* 59: 533–538.

Vogelmann, T. C., Bornman, J. F., and Yates, D. J. (1996) Focusing of light by leaf epidermal cells. *Physiol. Plantarum* 98: 43–56.

Vogelmann, T. C., and Martin, G. (1993) The functional significance of palisade tissue: Penetration of directional vs diffuse light. *Plant Cell Environ.* 16: 65–72.

von Caemmerer, S., and Farquhar, G. D. (1981) Some relationships between the biochemistry of photosynthesis and the gas exchange of leaves. *Planta* 153: 376–387.

Wada, M., Grolig, F., and Haupt, W. (1993) Light-oriented chloroplast positioning. Contribution to progress in photobiology. *J. Photochem. Photobiol.* 17: 3–25.

10 Translocation in the Phloem

SURVIVAL ON LAND POSED a new set of problems for plants newly emerged from an aquatic environment—foremost of which was the need to acquire and retain water. In response to these environmental pressures, plants evolved roots—which anchor the plant and absorb water and nutrients, and leaves—which absorb light and exchange gases. As plants increased in size, the roots and leaves became increasingly separated from each other in space. Thus, systems evolved for long-distance transport that allowed the shoot and the root efficient exchange of the products of absorption and assimilation.

The **xylem** is the tissue that transports water and minerals from the root system to the aerial portions of the plant (see Chapters 1, 4, and 6). The **phloem** is the tissue that translocates the products of photosynthesis from mature leaves to areas of growth and storage, including the roots. As we will see, the phloem also redistributes water and various compounds throughout the plant body. These compounds, some of which initially arrive in the mature leaves via the xylem, can be either transferred out of the leaves without modification or metabolized before redistribution.

The discussion that follows emphasizes translocation in the phloem of angiosperms, since most of the research has been conducted on that group of plants. Gymnosperms will be compared briefly to angiosperms in terms of the anatomy of their conducting cells and possible differences in their mechanism of translocation. First, we will examine some aspects of translocation in the phloem that have been researched extensively and are thought to be well understood. These include the pathway and patterns of translocation, materials translocated in the phloem, and rates of movement. In the second part of the chapter we will explore aspects of translocation in the phloem that need further investigation. Some of these

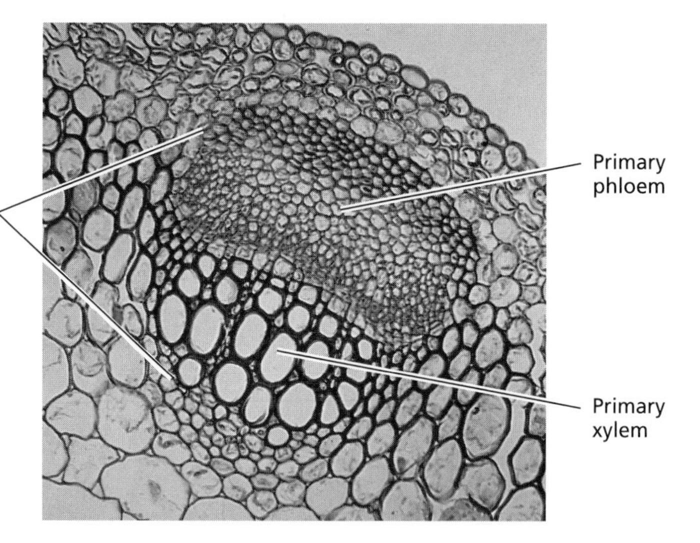

Figure 10.1 Transverse section of a vascular bundle of trefoil, a clover (*Trifolium*). (130×) The primary phloem is toward the outside of the stem. Both the primary phloem and the primary xylem are surrounded by a bundle sheath of thick-walled sclerenchyma cells, which isolates the vascular tissue from the ground tissue. (J. N. A. Lott/Biological Photo Service.)

areas, such as phloem loading and unloading and assimilate allocation and partitioning, are being studied intensively at present. The mechanism of translocation will also be discussed in this section, since it has been thoroughly investigated only in the angiosperms.

Pathways of Translocation

As already mentioned, two long-distance transport pathways, the phloem and the xylem, extend throughout the plant body. The phloem is generally found on the outer side of both primary and secondary vascular tissues (Figures 10.1 and 10.2); in plants with secondary growth the phloem constitutes the inner bark. The cells of the phloem that conduct sugars and other organic materials throughout the plant are called **sieve elements**. "Sieve element" is a comprehensive term that includes both the highly differentiated **sieve tube elements** typical of the angiosperms and the relatively unspecialized **sieve cells** of gymnosperms. In addition to sieve elements, the phloem tissue contains companion cells, parenchyma cells, and in some cases, fibers, sclereids, and laticifers (latex-containing cells). However, only the sieve elements are directly involved in translocation.

The small veins of leaves and the primary vascular bundles of stems are often surrounded by a **bundle sheath**, which consists of one or more layers of compactly arranged cells (see Figure 10.1). In the vascular tissue of leaves, the bundle sheath surrounds the small veins all the way to their ends, isolating the veins from the intercellular spaces of the leaf. In some leaves, cells similar to the bundle sheath cells extend to the upper and lower epidermal layers and are thought to aid in water conduction throughout the leaf.

We will begin our discussion of translocation pathways with the experimental evidence demonstrating that the sieve elements are the conducting cells in the phloem. Then we will examine the structure and physiology of these unusual plant cells.

Labeling Studies Have Shown That Sugar Is Translocated in Phloem Sieve Elements

In classic experiments on the translocation of organic solutes performed by the Italian anatomist Marcello Malpighi in 1686, the bark of a tree was removed in a ring around the trunk (Figure 10.3). In 1928, T. G. Mason and E. J. Maskell observed that this treatment, called **girdling**, has no immediate effect on transpiration, since water moves in the xylem, interior to the bark (Mason and Maskell 1928). However, sugar transport in the trunk is blocked at the site where the bark has been removed. Sugars accumulate above the girdle—that is, on the side toward the leaves—and are depleted below the treated region. Eventually the bark below the girdle dies, while the bark above swells and remains healthy.

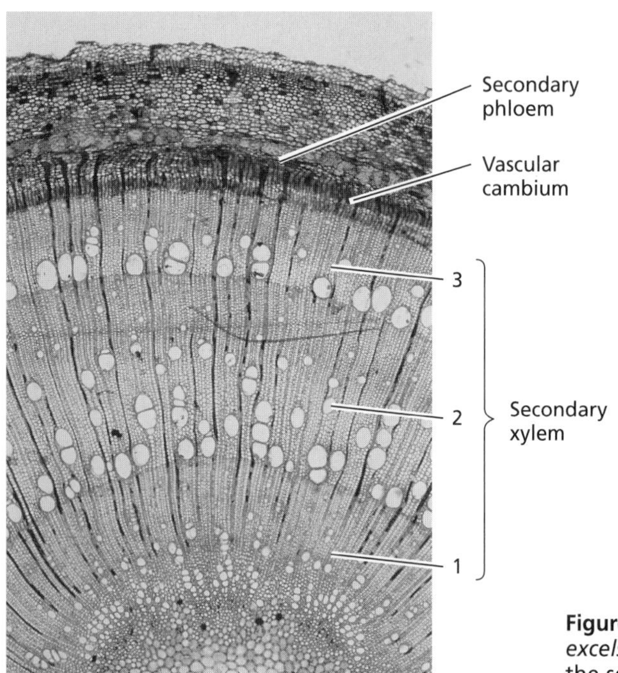

Figure 10.2 Transverse section of a 3-year-old stem of ash (*Fraxinus excelsior*) tree. (27×) The numbers 1, 2, and 3 indicate growth rings in the secondary xylem. The old secondary phloem has been crushed by expansion of the xylem. Only the most recent (innermost) layer of secondary phloem is functional. (P. Gates/Biological Photo Service.)

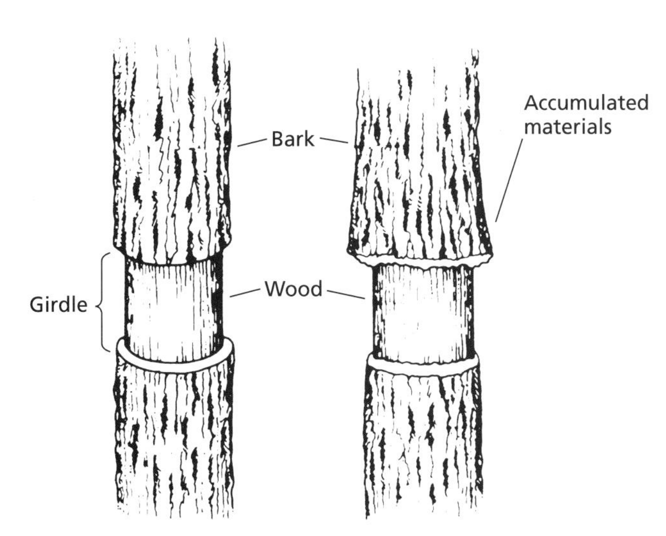

Figure 10.3 Tree trunk immediately after girdling (left) and later (right). Girdling is the removal of the bark of a tree in a ring around the trunk. At right, materials translocated from the leaves have accumulated in the region above the girdle and caused it to swell.

Mason and Maskell concluded that sugar is transported in the bark of the tree and, further, that the sieve elements are the cellular channels of sugar transport. The latter conclusion was based on an observed high correlation between leaf and bark sucrose contents and on the high sucrose concentration calculated to be present in the sieve elements.

More sophisticated experiments on phloem translocation became possible in the 1940s, when radioactive isotopes became available for scientific research. Labeled organic compounds can be introduced into the plant in various ways. For example, carbon dioxide can be labeled with ^{14}C or ^{11}C* and supplied to an intact mature leaf enclosed in a sealed chamber. The labeled CO_2 is incorporated into organic compounds via the CO_2-fixing reactions of photosynthesis (see Chapter 9). Some of the sugar phosphates formed in photosynthesis react further to form the translocated sugars, such as sucrose and stachyose. In another type of experiment, researchers bypass the photosynthetic reactions by supplying labeled transport sugar, usually sucrose, directly to the leaf. In this case, a solution containing the radioactive sugar must be applied in a way that allows it to reach cells in the leaf interior. The solution can be applied to the leaf surface after the epidermis has been peeled off or the cuticle has been abraded in a small area. Alternatively, the solution can be supplied through an isolated vein. Compounds labeled with other isotopes, such as ^{32}P, can be introduced in the same way.

To identify the pathway of sugar translocation, we must determine the tissue or cellular location of the isotope, usually by means of a technique called **autoradiography**. In tissue autoradiography, parts of labeled plants are rapidly frozen, freeze-dried, embedded in paraffin or resin, and sliced into thin sections. The sections are then coated with a film of photographic emulsion. During a period of exposure, radiation from the label exposes the film, and when the film is developed, silver grains appear wherever the label was present in the tissue. A comparison of the tissue section and the pattern of silver grains reveals the location of the label in the tissue. In terms of the transport pathway for sugars, the label initially appears in the sieve elements of the phloem, confirming the results of the earlier experiments (Figure 10.4).

(A)

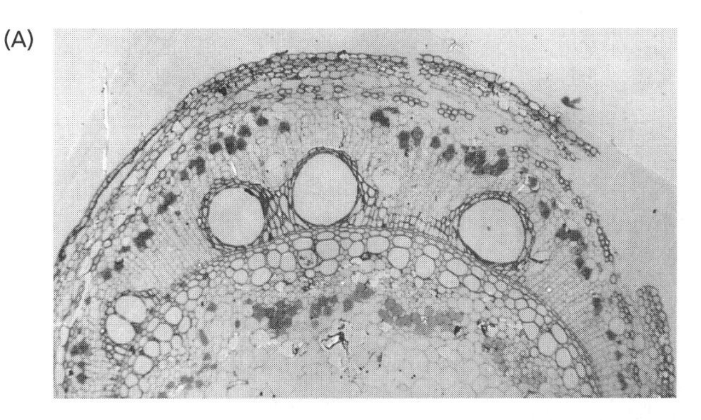

(B)

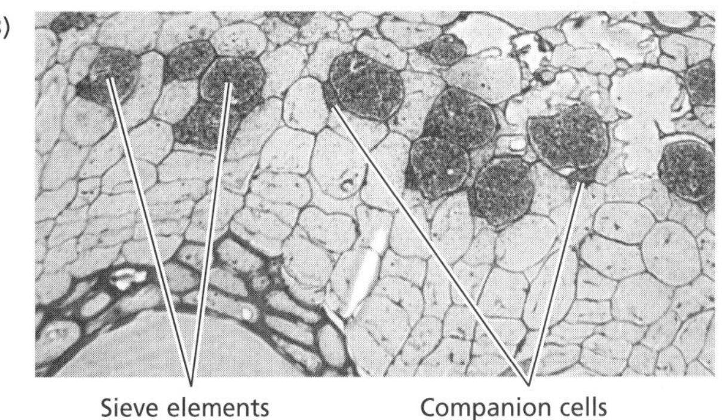

Figure 10.4 Autoradiographs of cross sections of morning glory (*Ipomea nil*) stem tissue. $^{14}CO_2$ was supplied to the source leaf. During photosynthesis, the ^{14}C was incorporated into labeled sugars, which were then transported to other parts of the plant. The location of the label is revealed by the presence of dark grains on the film. Comparison of the film with the underlying tissue section reveals that the label is confined almost entirely to the sieve elements of the phloem. (A: 50×; B: 325×) (Courtesy of D. Fisher.)

* These two isotopes of carbon, ^{11}C and ^{14}C, have quite different half-lives (the half-life is the time required for the radioactivity to decrease by one-half). The half-life for ^{11}C is 20 minutes; that for ^{14}C is 5,600 years. For this reason, the isotopes are used in different types of experiments. Carbon-14 is useful when the researcher wishes to analyze extensively the compounds in which the radioactivity is present. Carbon-11 would disappear before such an analysis could be made. Carbon-11 is most useful when the researcher wishes to repeat measurements on the same plant, since only a few hours are needed for the isotope in the plant to disappear.

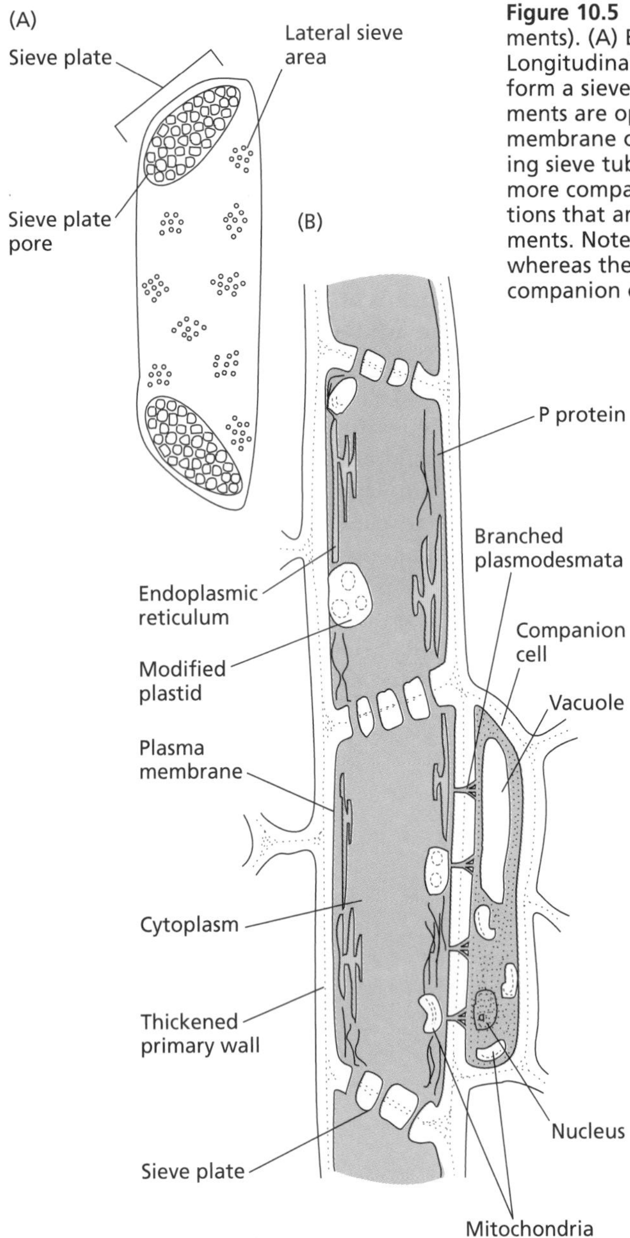

Figure 10.5 Schematic drawings of mature sieve elements (sieve tube elements). (A) External view, showing sieve plates and lateral sieve areas. (B) Longitudinal section, showing two sieve tube elements joined together to form a sieve tube. The pores in the sieve plates between the sieve tube elements are open channels for transport through the sieve tube. The plasma membrane of a sieve tube element is continuous with that of its neighboring sieve tube element. Each sieve tube element is associated with one or more companion cells, which assume some of the essential metabolic functions that are reduced or lost during differentiation of the sieve tube elements. Note that the companion cell is enriched in cytoplasmic organelles, whereas the sieve tube element has relatively few organelles. An ordinary companion cell (which will be described shortly) is depicted here.

Mature Sieve Elements Are Living Cells Highly Specialized for Translocation

Detailed knowledge of the ultrastructure of sieve elements is critical to any discussion of the mechanism of translocation in the phloem. Mature sieve elements are unique among living plant cells (Figures 10.5 and 10.6). They lack many structures normally found in living cells, including the undifferentiated cells from which mature sieve elements are formed. For example, sieve elements lose their nuclei and tonoplasts (vacuolar membrane) during development. Microfilaments, microtubules, Golgi bodies, and ribosomes are also absent from the mature cells. In addition to the plasma membrane, organelles that are retained include somewhat modified mitochondria, plastids, and smooth endoplasmic reticulum. The walls are nonlignified, though they are secondarily thickened in some cases. Thus, the sieve elements are unlike the tracheary elements of the xylem, which are dead at maturity, lack a plasma membrane, and have lignified secondary walls. As we will see, this difference is critical to the mechanism of translocation in the phloem.

Sieve Areas Are the Prominent Feature of Sieve Elements

Sieve elements are characterized by sieve areas, portions of the cell wall where pores interconnect the conducting cells. The sieve area pores range in diameter from less

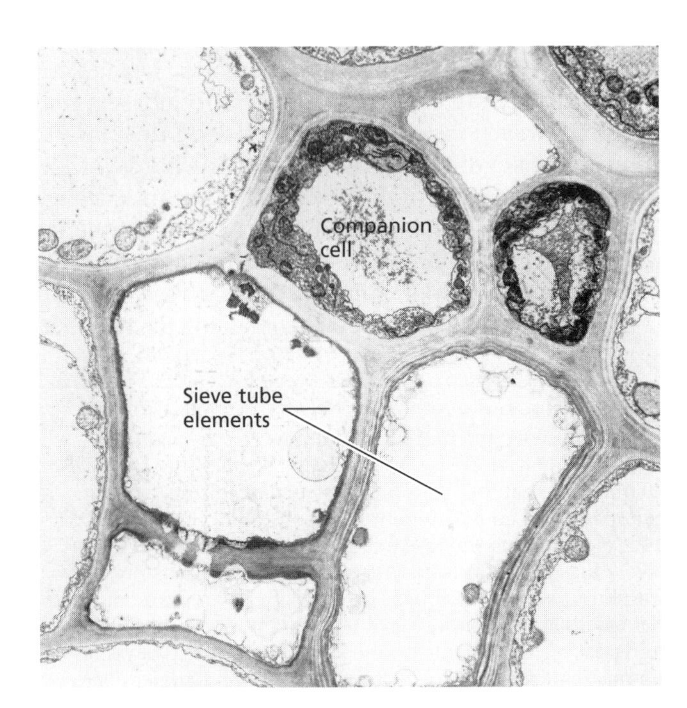

Figure 10.6 Electron micrograph of a transverse section of ordinary companion cells and mature sieve tube elements. (3600×) The cellular components are distributed along the walls of the sieve tube elements. (From Warmbrodt 1985.)

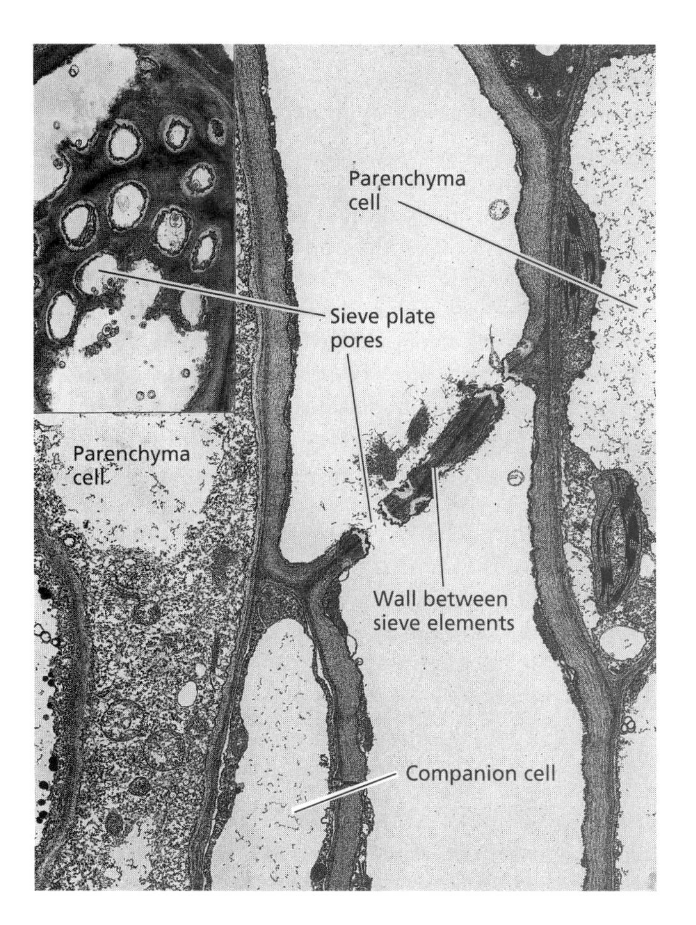

Figure 10.7 Electron micrograph of a portion of two mature sieve elements (sieve tube elements), sectioned longitudinally, in the hypocotyl of winter squash (*Cucurbita maxima*). (3685×) The wall between the sieve elements is a sieve plate. The inset shows sieve plate pores in face view. In both cases, the sieve plate pores are open—that is, unobstructed by P protein. (4280×) (From Evert 1982, courtesy of R. Evert.)

than 1 μm to approximately 15 μm. The sieve areas of sieve tube elements (angiosperms) are more specialized than those of sieve cells (gymnosperms). For instance, some of the sieve areas of sieve tube elements are differentiated into **sieve plates** (Figure 10.7 and Table 10.1).

Table 10.1
Characteristics of the two types of sieve elements in seed plants

Sieve tube elements

1. Found in angiosperms
2. Some sieve areas are differentiated into sieve plates; individual sieve tube elements are joined together into a sieve tube
3. Sieve plate pores are open channels
4. P protein is present in all dicots and many monocots
5. Companion cells are sources of ATP and perhaps other compounds and, in some species, are transfer cells or intermediary cells

Sieve cells

1. Found in gymnosperms
2. No sieve plates; all sieve areas are similar
3. Pores in sieve areas appear blocked with membranes
4. No P protein
5. Albuminous cells sometimes function as companion cells

Sieve plates have larger pores than the other sieve areas in the cell and are generally found on the end walls of sieve tube elements, where the individual cells are joined together to form a longitudinal series called a **sieve tube** (see Figure 10.5). Furthermore, the sieve plate pores of sieve tube elements are open channels that allow transport between cells (see Figure 10.7).

In sieve cells (gymnosperms), on the other hand, all of the sieve areas are more or less the same. The pores of sieve cells are relatively unspecialized and appear to be filled with numerous membranes, which meet in large median cavities in the middle of the wall and are continuous with the smooth endoplasmic reticulum opposite the sieve areas (Figure 10.8).

Deposition of P Protein and Callose Seals Off Damaged Sieve Elements

The sieve tube elements of most angiosperms are rich in a phloem protein called **P protein** (see Figure 10.5B)

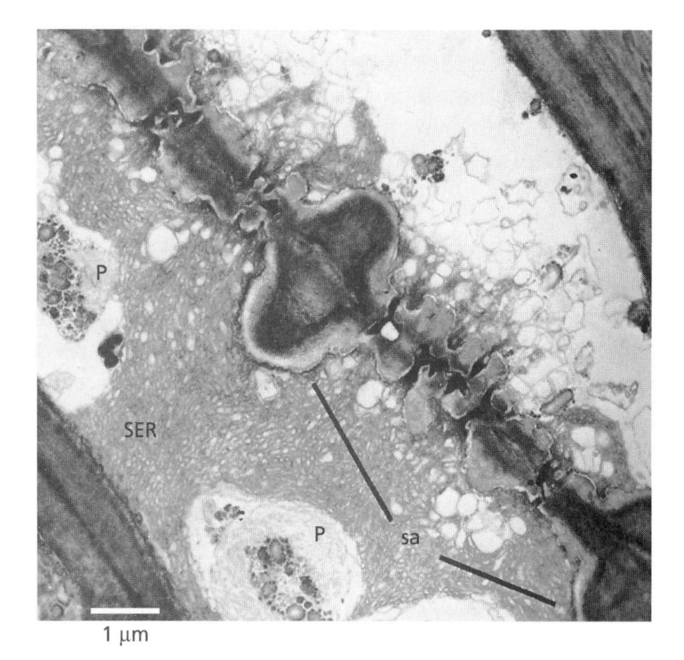

Figure 10.8 Electron micrograph showing sieve areas (sa) linking two sieve cells of Norway spruce (*Picea abies*). The pores are filled with membranes, which meet in large median cavities in the middle of the wall and are continuous with smooth endoplasmic reticulum (SER) on either side of the sieve area pores. How this arrangement of the pores affects translocation is unclear. The SER encloses sieve element plastids (P) in this section. (From Schulz et al. 1989.)

(Clark et al. 1997). (In the past, P protein was called "slime".) P protein is found in all dicots and in many monocots, and it is absent in gymnosperms. It occurs in several different forms (tubular, fibrillar, granular, and crystalline) depending on the species and maturity of the cell. In immature cells, P protein is most evident as discrete bodies in the cytosol known as **P protein bodies**. P protein bodies may be spheroidal, spindle-shaped, or twisted and coiled. They generally disperse into tubular or fibrillar forms during cell maturation.

Recently, progress has been made in characterizing P-proteins at the molecular level. In the genus *Cucurbita*, for example, P-proteins consist of two major proteins: PP1, the phloem filament protein, and PP2, the phloem lectin. The gene for PP1 in pumpkin (*Cucurbita maxima*) has been shown to encode a protein with a highly repetitive amino acid sequence, with four domains containing structural motifs similar to cysteine proteinase inhibitors. Both PP1 and PP2 are thought to be synthesized in companion cells (discussed below) and transported via the plasmodesmata to the sieve elements, where they associate to form P-protein filaments and P-protein bodies (Clark et al. 1997).

P protein appears to function in sealing off damaged sieve elements by plugging up the sieve plate pores. Sieve tubes are under very high internal turgor pressure, and the sieve elements in a sieve tube are connected to each other through open sieve plate pores. When a sieve tube is cut or punctured, the release of pressure causes the contents of the sieve elements to surge toward the cut end, from which the plant could lose much sugar-rich phloem sap if there were no sealing mechanism. ("Sap" is a general term used to refer to the fluid contents of plant cells.) When surging occurs, however, P protein and other cellular inclusions are trapped on the sieve plate pores, helping to seal the sieve element and to prevent further loss of sap.

A longer-term solution to the damage problem in most species is the production of **callose** in the sieve pores. Callose, a β-1,3-glucan, is localized by its reaction with aniline blue stain. Callose is synthesized by an enzyme in the plasma membrane and is deposited between the plasma membrane and the cell wall. The enzyme callose synthase appears to be arranged vectorially in the plasma membrane, with substrate being supplied from the cytoplasmic side and product being deposited on the wall surface. Callose is synthesized in functioning sieve elements in response to damage and other stresses, such as mechanical stimulation and high temperatures, or in preparation for normal developmental events, such as dormancy. The deposition of **wound callose** in the sieve pores efficiently seals off damaged sieve elements from surrounding intact tissue. As the sieve elements recover from damage, the callose disappears from these pores.

Companion Cells Aid the Highly Specialized Sieve Elements

Each sieve tube element is associated with one or more **companion cells** (see Figures 10.5, 10.6, and 10.7). The sieve tube element and the companion cell are formed by the division of a single mother cell. Numerous intercellular connections, the **plasmodesmata** (see Chapter 1), penetrate the walls between sieve tube elements and their companion cells, suggesting a close functional relationship and ease of transport between the two cells. The plasmodesmata are often complex and branched on the companion cell side. In some gymnosperms the companion cell function is performed by **albuminous cells**, which do not arise from the same mother cells as the sieve cells.

Companion cells have long been thought to take over some of the critical metabolic functions, such as protein synthesis, that are reduced or lost during differentiation of the sieve elements. Recent advances in experimental techniques have permitted verification of this concept (Bostwick et al. 1992). In addition, the numerous mitochondria in companion cells may supply energy as ATP to the sieve elements. Finally, companion cells of various types play a role in the transport of photosynthetic products from producing cells in mature leaves to the sieve elements in the minor (small) veins of the leaf.

There are at least three different types of companion cells in the minor veins of mature, exporting leaves: "ordinary" companion cells, transfer cells, and intermediary cells. All three cell types have dense cytoplasm and abundant mitochondria. **Ordinary companion cells** (Figure 10.9A) are further characterized by chloroplasts with well-developed thylakoids and a cell wall with a smooth inner surface. Of most significance, relatively few plasmodesmata connect this type of companion cell to any of the surrounding cells except its own sieve element.

A **transfer cell** is like an ordinary companion cell, but with one striking and characteristic difference: the development of fingerlike wall ingrowths, particularly on the cell walls that face away from the sieve element (Figure 10.9B). These wall ingrowths greatly increase the surface area of the plasma membrane, thus increasing the potential for solute transfer across the membrane. Because of their structural features (the scarcity of cytoplasmic connections to surrounding cells and the wall ingrowths in transfer cells), the ordinary companion cell and the transfer cell are thought to be specialized for taking up solutes from the apoplast or cell wall space. Xylem parenchyma cells can also be modified as transfer cells, probably serving to retrieve and reroute solutes moving in the xylem, which is also part of the apoplast.

The third type of companion cell, the **intermediary cell**, appears well suited for taking up solutes via cytoplasmic connections (Figure 10.9C). Intermediary cells have numerous plasmodesmata connecting them to sur-

(A)

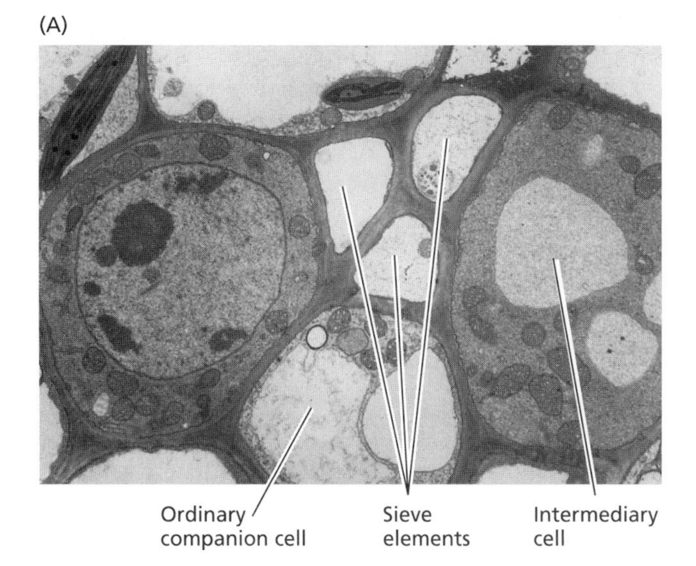

(B)

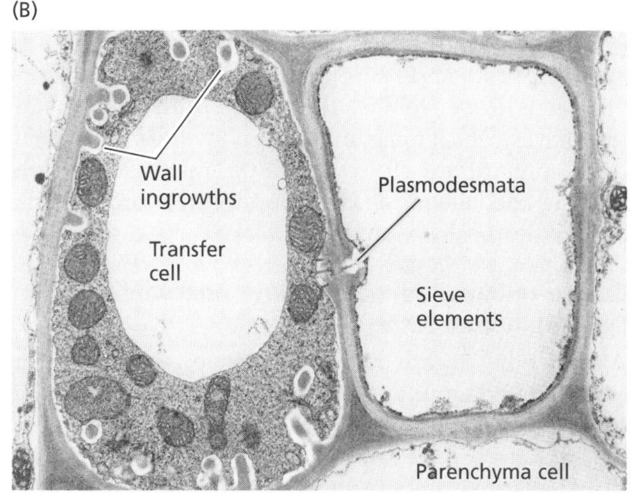

(C)

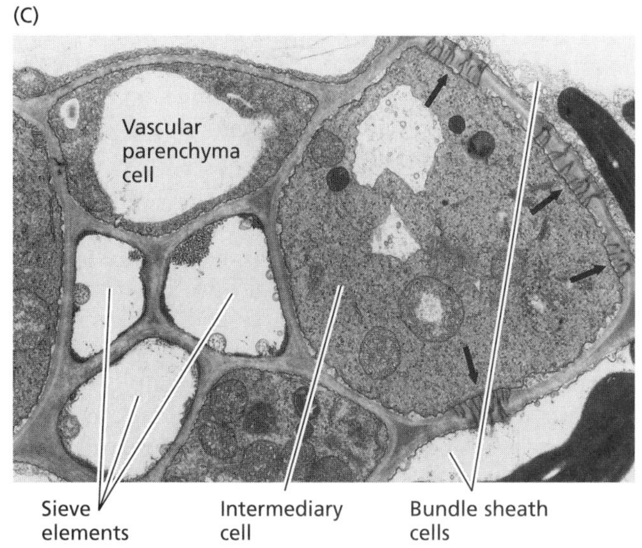

Figure 10.9 Electron micrographs of companion cells in minor veins of mature leaves. (A) A minor vein from *Mimulus cardinalis* (6585×) Three sieve elements abut two intermediary cells and a more lightly stained ordinary companion cell. (B) A transfer cell and a sieve element from pea (*Pisum sativum*). (8020×) Note the numerous wall ingrowths on the walls of the transfer cell that are adjacent to the parenchyma cells but not on the wall opposite the sieve element. Such ingrowths greatly increase the surface area of the transfer cell's plasma membrane, thus increasing the potential for solute transport and ultimately facilitating the transfer of materials from the mesophyll to the sieve elements. Note the presence of plasmodesmata between the transfer cell and the sieve element. (C) Minor-vein phloem from heartleaf maskflower (*Alonsoa warscewiczii*). (4700×) Note the typical intermediary cell with numerous fields of plasmodesmata (arrows) connecting it to neighboring bundle sheath cells. These plasmodesmata are branched on both sides, but the branches are longer and more narrow on the intermediary cell. The intermediary cells also have numerous small vacuoles and chloroplasts that lack internal membranes. (A and C from Turgeon et al. 1993, courtesy of R. Turgeon; B from Brentwood and Cronshaw 1978.)

rounding cells, particularly to the bundle sheath cells. Although the presence of many plasmodesmatal connections to surrounding cells is their most characteristic feature, intermediary cells are also distinctive in having numerous small vacuoles, as well as poorly developed thylakoids and a lack of starch grains in the chloroplasts.

The types of plants that have intermediary cells and the functional consequences of the structural differences between ordinary companion cells and intermediary cells are discussed later in the chapter.

Patterns of Translocation: Source to Sink

Materials in the phloem are not translocated exclusively in either an upward or a downward direction, and translocation in the phloem is not defined with respect to gravity. Materials are translocated from areas of supply, called sources, to areas of metabolism or storage, called sinks.

Sources include any exporting organ, typically a mature leaf, that is capable of producing photosynthate in excess of its own needs. Another type of source is a storage organ during the exporting phase of its development. For example, the storage root of the biennial wild beet (*Beta maritima*) is a sink during the growing season of the first year, when it accumulates sugars received from the source leaves. During the second growing season the same root becomes a source; the sugars are remobilized and utilized to produce a new shoot, which ultimately becomes reproductive.*

* In contrast, roots of the cultivated sugar beet (*Beta vulgaris*) can increase in dry mass during both the first and the second growing seasons, so the leaves serve as sources during both flowering and fruiting stages. This characteristic is true of cultivated varieties of beets because they have been selected for the capacity of their roots to act as sinks during all phases of development.

Sinks include any nonphotosynthetic organs of the plant, and organs that do not produce enough photosynthetic products to support their own growth or storage needs. Roots, tubers, developing fruits, and immature leaves, which must import carbohydrate for normal development, are all examples of sink tissues. Both girdling and labeling studies support the source-to-sink pattern of translocation in the phloem.

Source-to-Sink Pathways Follow Anatomical and Developmental Patterns

Although the overall pattern of transport in the phloem can be stated simply as source-to-sink movement, the specific pathways involved are often more complex. Not all sinks are equally supplied by all source leaves on a plant; rather, certain sources preferentially supply specific sinks. In the case of herbaceous plants, such as sugar beet and soybean, the following generalizations can be made:

Proximity. The proximity of the source to the sink is a significant factor. Thus, the upper mature leaves on a plant usually transport assimilates to the growing shoot tip and young, immature leaves, while the lower leaves supply predominantly the root system. (An **assimilate** is a molecule, such as sucrose, that is produced endogenously by the metabolism of the plant—in this case, photosynthesis.) Intermediate leaves export in both directions, bypassing the intervening mature leaves.

Development. The importance of various sinks may shift during plant development. Whereas the root and shoot apices are usually the major sinks during vegetative growth, fruits generally become the dominant sinks during reproductive development, particularly for adjacent and other nearby leaves.

Vascular connections. Source leaves preferentially supply sinks with which they have direct vascular connections. In the shoot system, for example, a given leaf is generally connected via the vascular system to other leaves directly above or below it on the stem. Such a vertical row of leaves is called an **orthostichy**. The number of internodes between leaves on the same orthostichy varies with the species. Figure 10.10A illustrates this pattern in sugar beet.

Altering translocation pathways. Interference with the translocation pathway by wounding or pruning can alter the patterns outlined thus far. In the absence of direct connections between source and sink, vascular interconnections, called **anastomoses**, can provide an alternative pathway. In sugar beet, for example, removing source leaves from one side of the plant can bring about cross-transfer of assimilates to young leaves (sink leaves) on the pruned side (Figure 10.10B). Upper source leaves on a plant can be forced to translocate materials to the roots by removing the lower source leaves, and vice versa. However, the flexibility of the translocation pathway depends on the extent of the interconnections between vascular bundles and thus on the species and organs studied. In some species the leaves on a branch with no fruits cannot transport photosynthate to the fruits on an adjacent defoliated branch. But in other plants, such as soybean (*Glycine max*), photosynthate is cross-transferred readily from a partly defruited side to a partly defoliated side.

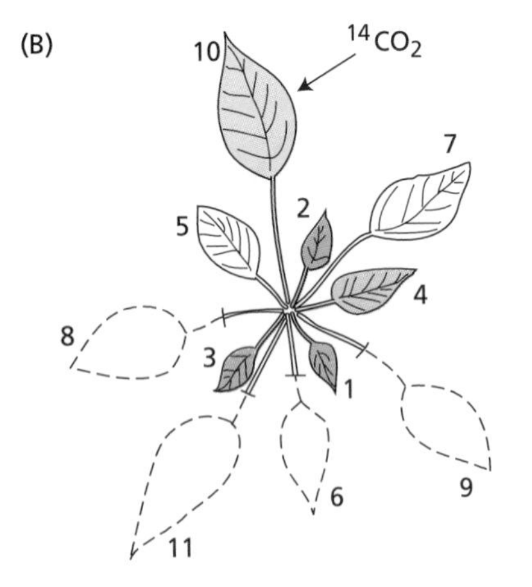

Figure 10.10 Diagrams showing the distribution of radioactivity from a single labeled source leaf in an intact plant and in a pruned plant. The degree of radioactive labeling is indicated by the intensity of shading of the leaves. (A) The distribution of radioactivity in leaves of an intact sugar beet plant (*Beta vulgaris*) 1 week after $^{14}CO_2$ was supplied for 4 hours to a single source leaf (arrow). Leaves are numbered according to their age; the youngest, newly emerged leaf is designated 1. The ^{14}C label was translocated mainly to the sink leaves directly above the source leaf (that is, sink leaves on the same orthostichy as the source). (B) Same as A, except all source leaves on the side of the plant opposite the labeled leaf were removed 24 hours before labeling. Sink leaves on both sides of the plant now receive ^{14}C-labeled assimilates from the source. (Based on data from Joy 1964.)

Materials Translocated in the Phloem: Sucrose, Amino Acids, Hormones, and Some Inorganic Ions

Water is quantitatively the most abundant substance transported in the phloem. Dissolved in the water are the translocated solutes, which consist mainly of carbohydrates (Table 10.2). Sucrose is the sugar most commonly transported in sieve elements. It is always found at some level in sieve element sap and can reach concentrations of 0.3 to 0.9 *M*. Some other organic solutes also move in the phloem. Nitrogen occurs in the phloem largely in the form of amino acids and amides, especially glutamate and aspartate and their respective amides, glutamine and asparagine. Reported levels of amino acids and organic acids vary widely, even for the same species, but they are usually low compared with carbohydrates.

Almost all the endogenous plant hormones, including auxin, gibberellins, cytokinins, and abscisic acid (see Chapters 19, 20, 21, and 23), have been found in sieve elements. The long-distance transport of hormones is thought to occur at least partly in the sieve elements, though auxin is also transported in the xylem and in the polar transport pathway. Nucleotide phosphates and proteins have also been found in phloem sap. Sap proteins associated with basic cellular functions include protein kinases (protein phosphorylation), thioredoxin (disulfide reduction), ubiquitin (protein turnover), and chaperones (protein folding) (Schobert et al. 1995). Inorganic solutes that move in the phloem include potassium, magnesium, phosphate, and chloride. In contrast, nitrate, calcium, sulfur, and iron are almost completely excluded from the phloem.

We begin the discussion that follows with a look at the methods used to identify materials translocated in the phloem. We will then examine the translocated sugars and the complexities of nitrogen transport in the plant.

Phloem Sap Can Be Collected and Analyzed

Since phloem tissue, like xylem tissue, consists of a mixture of conducting and nonconducting cells, identifying the material translocated in the sieve elements is challenging. Methods based on labeling with radioactive isotopes and direct extraction of the tissue do not distinguish between compounds in transit and those outside the translocation pathway.

The tendency of some species to exude phloem sap from wounds that sever sieve elements has been exploited to obtain relatively pure samples of the translocated material. The driving force for exudation is the high positive pressure in the sieve elements. Most species do not exude detectable amounts of phloem sap, because of the sealing mechanisms mentioned earlier. However, exudation rates can be increased by treating the cut surface with the chelating agent EDTA (ethylenediaminetetraacetic acid) (King and Zeevaart 1974). Callose synthase requires the presence of calcium, and since chelating agents bind divalent cations such as calcium very effectively, they lower the calcium concentration available for callose synthase.

Table 10.2
The composition of phloem sap from castor bean (*Ricinus communis*), collected as an exudate from cuts in the phloem

Component	Concentration (mg mL^{-1})
Sugars	80.0–106.0
Amino acids	5.2
Organic acids	2.0–3.2
Protein	1.45–2.20
Chloride	0.355–0.675
Phosphate	0.350–0.550
Potassium	2.3–4.4
Magnesium	0.109–0.122

Source: Hall and Baker 1972.

The major disadvantage of the wound exudation method is that the fluid collected may not represent the true composition of the translocated material. Contaminants can originate from damaged parenchyma cells or even from the sieve elements themselves. Furthermore, the abrupt lowering of turgor pressure in the sieve elements, as discussed in Chapter 4, causes a decrease in the water potential of these cells. As a result, water from the surroundings enters the sieve elements along a water potential gradient, causing the phloem sap to become diluted.

The ideal way to collect phloem sap would be to tap into a single sieve element by using a tiny syringe. Fortunately, nature has provided just such a probe: the aphid stylet. Aphids are small insects that feed by inserting their mouthparts, consisting of four tubular stylets, into a single sieve element of a leaf or stem (Figure 10.11). The high turgor pressure in the sieve element forces the cell contents through the insect's food canal and into its gut, where amino acids are selectively removed and some sugars are metabolized. The excess sap, still rich in carbohydrates, is then excreted as "honeydew," which can be collected for chemical analysis. Sap can also be collected from aphid stylets cut from the body of the insect, usually with a laser, after the aphid has been anesthetized with CO_2. Exudate from severed stylets provides a more accurate picture of the substances present in the sieve elements than does honeydew, the composition of which has been altered by the insect.

Experimental techniques using aphids have substantial advantages in the study of phloem physiology. Of

(A)

(B)

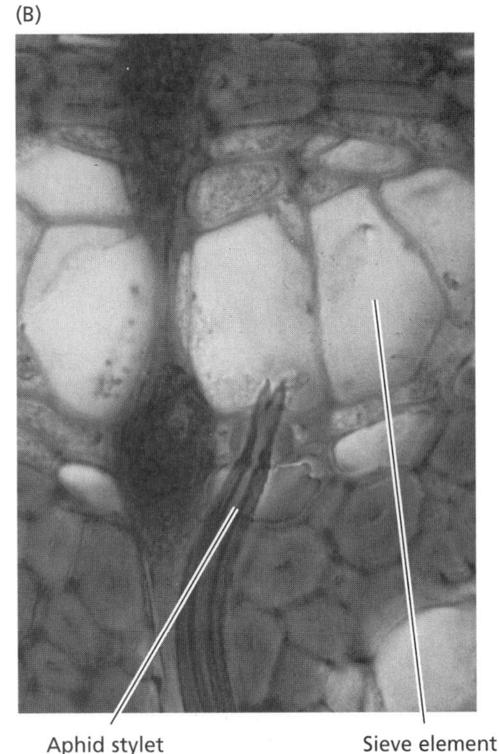

Figure 10.11 Collection of phloem sap using aphids, which tap single sieve elements. (A) An aphid (*Longistigma caryae*) feeding on a branch of linden (*Tilia americana*). The insect is just releasing a droplet of honeydew, which it does about once every 30 minutes. The honeydew excreted by the aphid consists of sieve element sap from which selected solutes have been removed in the gut of the insect. The high turgor pressure in the sieve element forces the cell contents through the food canal of the aphid and into its gut. (B) Transverse section through the bark of linden, showing the tips of aphid stylets that had been exuding before they were sectioned. The tips of the stylets are inside a single sieve element; the sheath of saliva secreted by the aphid ends outside the cell. (From Zimmermann and Brown 1971.)

particular importance is the fact that aphids tap a single sieve element, so there is no problem of contamination from other cell types. Exudation from severed stylets can continue for hours, suggesting that the aphid prevents normal sealing mechanisms from operating or that the drop in pressure that occurs when the sieve element is penetrated is not sufficient to trigger sealing mechanisms.

The disadvantages of using aphids include difficulties in placing the insects at a desired location and severing the stylets without disrupting them. In addition, aphids may induce reactions in the host plant by secreting saliva into the plant tissues. However, the magnitude of these reactions over the period of time necessary to collect sap for analysis is probably negligible.

Sugars Are Translocated in Nonreducing Form

Results from analyses of sap collected by the techniques outlined in the previous section indicate that the translocated carbohydrates are all nonreducing sugars. Reducing sugars, such as glucose and fructose, contain an exposed aldehyde or ketone group (Figure 10.12A) and can be assayed colorimetrically because of their ability to reduce Cu^{2+} to Cu^{+} in solution. In a nonreducing sugar, such as sucrose, the ketone or aldehyde group is reduced to an alcohol or combined with a similar group on another sugar (Figure 10.12B). Most researchers believe that the nonreducing sugars are the major compounds translocated in the phloem because they are less reactive than their reducing counterparts.

Sucrose is the most commonly translocated sugar; many of the other mobile carbohydrates contain sucrose bound to varying numbers of galactose molecules. Raffinose consists of sucrose and one galactose molecule, stachyose consists of sucrose and two galactose molecules, and verbascose consists of sucrose and three galactose molecules (see Figure 10.12B). Translocated sugar alcohols include mannitol and sorbitol.

Phloem and Xylem Interact to Transport Nitrogenous Compounds

Nitrogen is transported throughout the plant in either inorganic or organic form; the form that predominates depends on several factors, including the transport pathway. Whereas nitrogen is transported in the phloem almost entirely in organic form, it can be transported in the xylem either as nitrate or as part of an organic molecule (see Chapter 12). Usually, the same group of organic molecules carries nitrogen in both the xylem and the phloem.

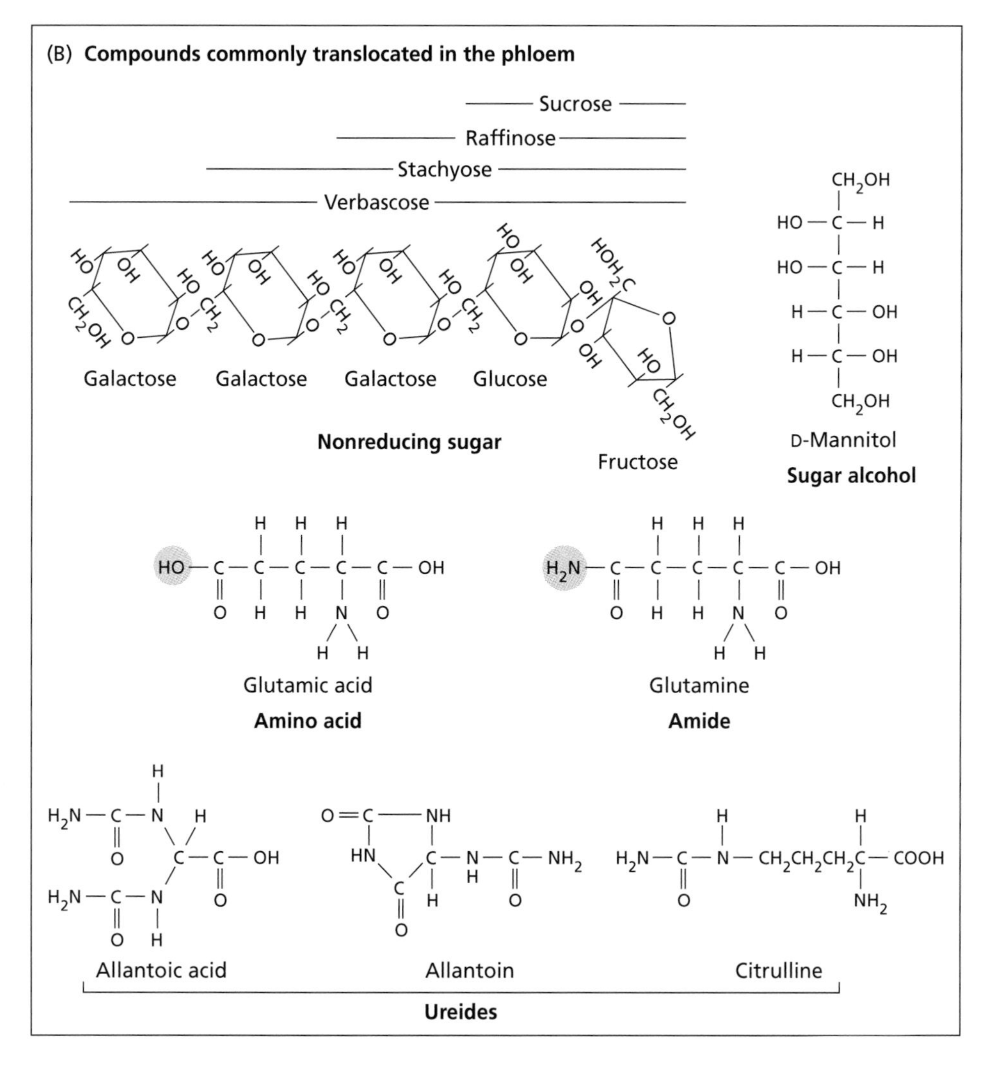

Figure 10.12 Structures of compounds not normally translocated in the phloem and of compounds commonly translocated in the phloem. (A) The reducing sugars glucose, fructose, and mannose are generally not found in phloem sap. The reducing groups are aldehyde (glucose and mannose) and ketone (fructose) groups. (B) Compounds commonly translocated in the phloem. Sucrose is a disaccharide made up of one glucose and one fructose molecule. Raffinose, stachyose, and verbascose contain sucrose bound to one, two, or three galactose molecules, respectively. Mannitol is a sugar alcohol formed by the reduction of the aldehyde group of mannose. All of these compounds are nonreducing. Glutamic acid, an amino acid, and glutamine, its amide, are important nitrogenous compounds in the phloem, in addition to aspartate and asparagine. Species with nitrogen-fixing nodules also utilize ureides as transport forms of nitrogen.

The form in which nitrogen is transported in the xylem depends on the species studied. Species that do not form a symbiotic association with nitrogen-fixing microorganisms depend on soil nitrate as their major nitrogen source (see Chapter 12). In the xylem of these species, nitrogen is usually present in the form of both nitrate and nitrogen-rich organic molecules, particularly the amides asparagine and glutamine (see Figure 10.12B). Species with nitrogen-fixing nodules on their roots (see Chapter 12) depend on atmospheric nitrogen, rather than on soil nitrate, as their major nitrogen source. After being converted to an organic form, this nitrogen is transported in the xylem to the shoot, usually in the form of amides or ureides such as allantoin, allantoic acid, or citrulline (see Figure 10.12).

In all cases in which nitrogen is assimilated into organic compounds in the roots, both the energy and the carbon skeletons required for assimilation are derived from photosynthates transported to the roots via the phloem. Nitrogen levels in mature leaves are quite stable, indicating that at least some of the excess nitrogen continuously arriving via the xylem is redistributed via the phloem to fruits or younger leaves.

One of the most widely studied species in terms of nitrogen transport is the economically important soybean, a legume that has nitrogen-fixing root nodules. Research has shown that 12 to 48% of the nitrogen needs of the developing seeds of this species are supplied from compounds newly synthesized in the leaves, and 35 to 52% from compounds synthesized in the roots and transferred from xylem to phloem (Layzell and LaRue 1982). The remaining nitrogen probably comes from the breakdown of specialized storage proteins in the soybean leaf.

Finally, levels of nitrogenous compounds in the phloem are quite high during leaf senescence. In woody species, senescing leaves mobilize and export nitrogenous compounds to the woody tissues for storage, whereas in herbaceous plants nitrogen is exported generally to the seeds. Other solutes, such as mineral ions, are redistributed from senescing leaves in the same manner.

Rates of Movement

The rate of movement of material in the sieve elements can be expressed in two ways: as **velocity**, the linear distance traveled per unit time, or as **mass transfer rate**, the quantity of material passing through a given cross section of phloem or sieve elements per unit time (Box 10.1). Mass transfer rates based on the cross-sectional area of the sieve elements are preferred, since the sieve elements are the conducting cells of the phloem.

In the original publications reporting on rates of transport in the phloem, the units of velocity were centimeters per hour (cm h^{-1}), and the units of mass transfer were grams per hour per square centimeter (g h^{-1} cm^{-2}) of phloem or sieve elements. The currently preferred units (SI units) are meters (m) or millimeters (mm) for length, seconds (s) for time, and kilograms (kg) for mass.

Velocities of Phloem Transport Exceed the Rate of Diffusion

Both velocities and mass transfer rates can be measured with radioactive tracers. In the simplest type of experiment, ^{11}C- or ^{14}C-labeled CO_2 is applied for a brief period of time to a source leaf (pulse labeling), and the arrival of label at a sink tissue or at a particular point

BOX 10.1

Monitoring Traffic on the Sugar Freeway

SUGAR MOLECULES can be thought of as cars speeding along a freeway on their way to various destinations, called *sinks*. To monitor this traffic we need to count the number of cars per unit time or, in physiological terms, the *mass transfer rate*. Measurements of mass transfer rates were among the earliest quantitative determinations made in phloem physiology. The early studies measured either gain in weight by developing fruits or storage organs (sinks) or loss of weight by mature leaves (sources) over a known time interval; the sink experiments are generally the more accurate.

For example, in 1926 Mason and Lewin determined weights of individual tubers of the greater yam (*Dioscorea alata*) over a growing season. The increase in dry weight during the growing season was divided by the time interval to obtain an average mass transfer rate. This rate, divided by the average sieve element area per stem, yields a specific mass transfer rate of 5.7 g h^{-1} cm^{-2} of sieve elements. Several similar studies obtained mass transfer rates of 0.2 to 4.8 g h^{-1} cm^{-2} of *phloem*. If sieve elements are assumed to occupy approximately one-third of the phloem cross-sectional area (a reasonable estimate), these results yield specific mass transfer rates of 0.6 to 14.5 g h^{-1} cm^{-2} for *sieve elements*.

A more recent technique for determining mass transfer rates by use of radioactive tracers gives rates in excellent agreement with the earlier measurements (Geiger and Swanson 1965). In this method the plant is usually pruned to a single mature source leaf and a single immature sink leaf. Carbon dioxide labeled with ^{14}C is supplied to the source leaf in such a way that the rate of supply is equal to the rate of fixation. The rate of arrival of label in the sink tissue is monitored for the duration of the experiment. The supplied radioactive CO_2 contains a predetermined amount of radioactivity per unit weight of carbon (specific activity). After a certain period of time, the sugars in transit from the source to the sink reach the same specific activity as the supplied CO_2. When this state, called *isotopic saturation*, has been reached, the mass transfer rate can be calculated from the rate of arrival of label in the sink tissue, as follows:

Arrival rate in sink		Specific activity of supplied label		Mass transfer rate
$\frac{Bq}{s}$	$\div$	$\frac{Bq}{\mu g \text{ of carbon}}$	$=$	$\frac{\mu g \text{ of carbon}}{s}$

where Bq stands for becquerel, a unit of radioactivity.* The rate determined in this way, corrected for the cross-sectional area of sieve elements in the transit pathway, provides an estimate of specific mass transfer. For sugar beet, which transports sucrose, the value obtained is equivalent to a rate of 4.8 g h^{-1} cm^{-2} of sieve elements, in excellent agreement with the values determined in earlier investigations.

*A becquerel (Bq) is equivalent to 1 disintegration per second. Another unit of radioactivity that has been used quite often is the curie (Ci), which is equivalent to 3.7×10^{10} disintegrations per second.

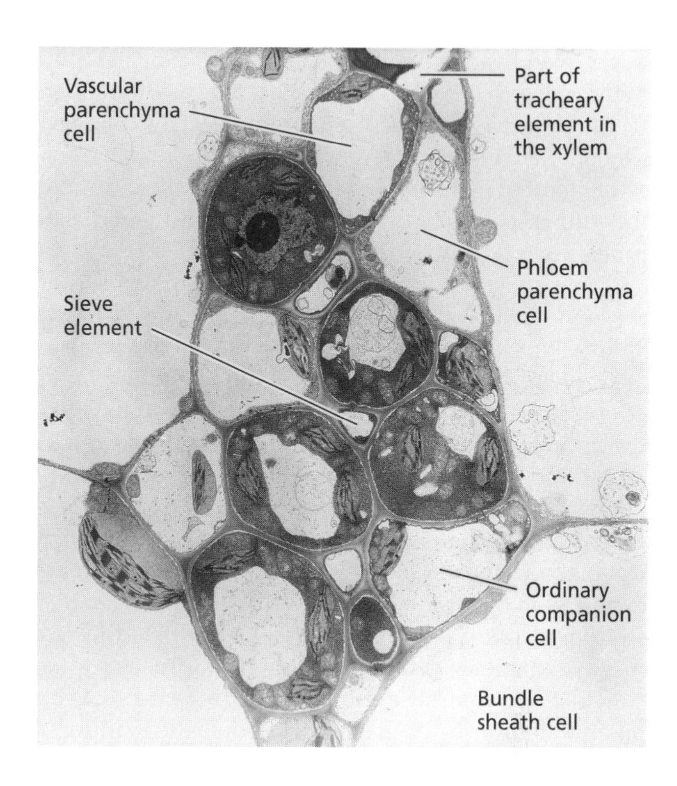

Figure 10.13 Electron micrograph of a transverse section of a small vein in a source leaf of sugar beet (*Beta vulgaris*), showing the spatial relationships between the various cell types of the vein. Surrounding the compactly arranged cells of the bundle sheath layer are the photosynthetic cells (mesophyll cells) with numerous intercellular spaces (not shown in this micrograph). Photosynthetic products moving from the mesophyll to the sieve elements of the minor vein must traverse a distance equivalent to several cell diameters before being loaded. Note that the sieve elements of the small veins of sugar beet are smaller than the companion cells, the reverse of the size relationship in the large veins and main vascular system. This relationship is typical of dicots and probably reflects the importance of the companion cells in the uptake of sugars into the sieve elements of small veins. (From Evert and Mierzwa 1985, courtesy of R. Evert.)

along the pathway is monitored with an appropriate detector. The length of the translocation pathway divided by the time interval required for label to be first detected at the sink yields a measure of velocity, at least for the fastest-moving labeled component. A more accurate measurement of velocity is obtained by monitoring the arrival of label at two points along the pathway, since this method excludes from the time measurement the time required for fixation of labeled carbon by photosynthesis, for its incorporation into transport sugar, and for accumulation of sugar in the sieve elements of the source leaf.

In general, velocities measured by a variety of techniques average about 1 m h^{-1} and range from 0.3 to 1.5 m h^{-1} (30 to 150 cm h^{-1}), with a few studies reporting speeds an order of magnitude greater than these. The variability in the measured velocities is due to species differences, differences in experimental methods, and problems inherent in the experimental methods. Despite this variability, transport velocities in the phloem are clearly quite high—well in excess of the rate of diffusion over long distances. Any proposed mechanism of phloem translocation must account for these high velocities.

Phloem Loading: From Chloroplasts to Sieve Elements

Several transport steps are involved in **phloem loading**, the movement of photosynthetic products (photosynthate) from the mesophyll chloroplasts to the sieve elements of mature leaves (Oparka and van Bel 1992):

In step one, triose phosphate formed from photosynthesis (see Chapter 8) during the day must first be transported from the chloroplast to the cytosol, where it is converted to sucrose. During the night, carbon from stored starch exits the chloroplast probably in the form of glucose and is converted to sucrose. (Other transport sugars are later synthesized from sucrose in some species.)

In step two, sucrose then moves from the mesophyll cell to the vicinity of the sieve elements in the smallest veins of the leaf (Figure 10.13). This **short-distance transport** pathway usually covers a distance of only two or three cell diameters.

In the third step, **sieve element loading**, sugars are transported into the sieve elements and companion cells, where they become more concentrated than in the mesophyll. Note that, with respect to loading, the sieve elements and companion cells are often considered a functional unit, called the sieve element–companion cell complex. Once inside the sieve elements, sucrose and other solutes are translocated away from the source, a process known as **export**. Translocation through the vascular system to the sink is referred to as **long-distance transport**.

The processes of loading at the source, and unloading at the sink, are believed to produce the driving force for translocation (discussed later in the chapter) and thus are of considerable basic as well as agricultural importance. Once the mechanisms are understood, it may be possible to increase crop productivity by increasing the accumulation of photosynthate by edible sink tissues, such as cereal grains.

Photosynthate Moves from Mesophyll Cells to the Sieve Elements via the Apoplast or the Symplast

We have seen that solutes (mainly sugars) in source leaves must move from the photosynthesizing cells in the mesophyll to the veins. Sugars might move entirely through the symplast (cytoplasm) via the plasmodesmata, or they might enter the apoplast (cell wall) at some point en route to the phloem (Figure 10.14). In the latter case, the sugars would be actively loaded from the apoplast into the sieve elements and companion cells by an energy-driven transporter located in the plasma membranes of these cells. As we will discover, both the apoplastic and symplastic routes are used, but in different species.

The earliest research on phloem loading focused on the apoplastic pathway. If phloem loading is partly apoplastic, we can make three basic predictions (Grusak et al. 1996): (1) transport sugars should be found in the apoplast; (2) transport sugars, supplied exogenously to the apoplast, should be accumulated by the sieve elements and companion cells ("exogenous" means originating from an external source; its opposite is "endogenous," an internal source); and (3) inhibition of sugar uptake from the apoplast should result in inhibition of export from the leaf.

Considerable research has been conducted to test these predictions. Although no single approach proves conclusively that an apoplastic loading route exists, taken together the data support apoplastic loading in some species. Results from tests of these predictions include the following:

1. Transport sugars are found in the apoplast. Sucrose is the predominant apoplastic sugar in species that transport mainly sucrose in the phloem—for example, sugar beet (*Beta vulgaris*) and broad bean (*Vicia faba*). Treatments and events that alter the rate of translocation from the source leaf also change the flux of sucrose through the apoplast (Geiger et al. 1974) or the apoplastic sucrose level.
2. Sucrose supplied exogenously to a source leaf of sugar beet accumulates in the sieve elements and companion cells of the minor veins, as does sucrose derived from photosynthetic CO_2 fixation (Figure 10.15). Similar observations have been made in broad bean, pea (*Pisum sativum*), castor bean (*Ricinus communis*), and other species.
3. PCMBS (*p*-chloromercuribenzenesulfonic acid) is a reagent that inhibits the transport of sucrose across plasma membranes but does not enter the symplast. PCMBS inhibits the uptake of sucrose from the apoplast when the sugar is supplied exogenously to sugar beet. Of greater importance, PCMBS also inhibits the export of sucrose synthesized from CO_2 in the mesophyll, implying that short-distance transport in sugar beet normally includes an apoplastic step (Giaquinta 1976). Assimilate loading is inhibited by PCMBS also in broad bean, pea, and other species.
4. Experiments with transgenic plants also support the interpretation that sucrose moves through the apoplast. (Transgenic organisms carry cloned genes integrated into their DNA by a variety of recombi-

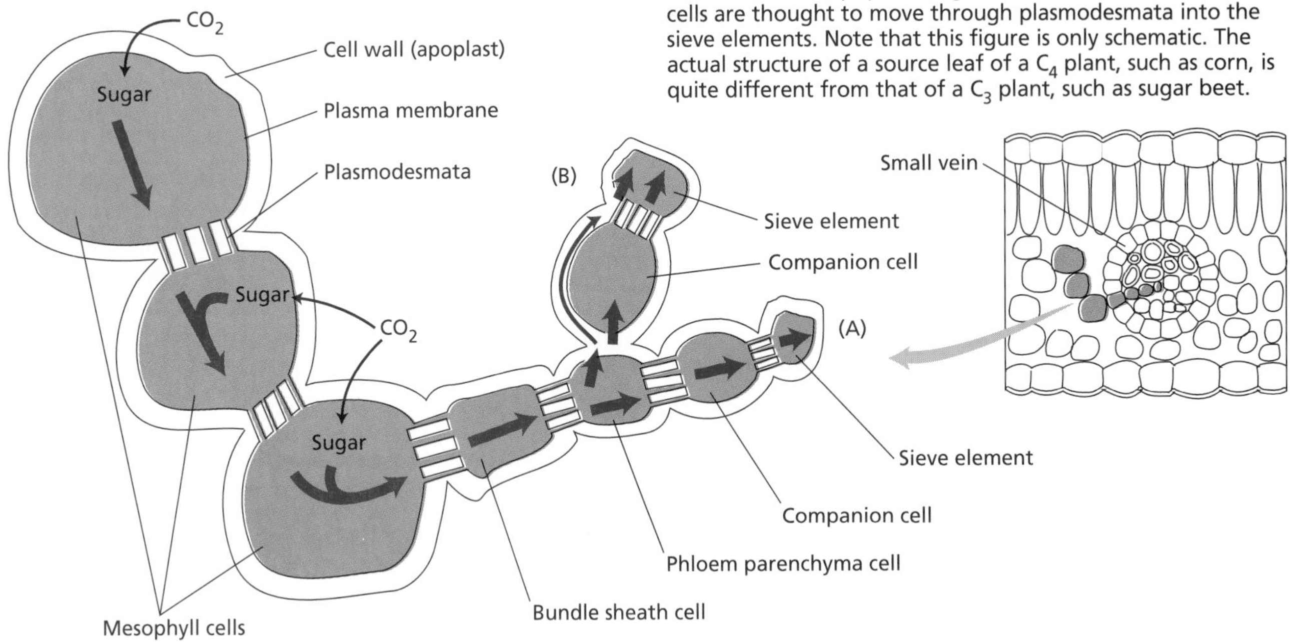

Figure 10.14 Diagram of possible pathways of phloem loading in source leaves. In the totally symplastic pathway (A), sugars move from one cell to another in the plasmodesmata, all the way from the mesophyll to the sieve elements. In the partly apoplastic pathway (B), sugars enter the apoplast at some point. (Membrane transport steps are shown in color.) For simplicity, sugars are shown here entering the apoplast near the sieve element–companion cell complex, but it is also possible that they enter the apoplast over the entire mesophyll and then move to the small veins. In any case, the sugars are actively loaded into the companion cells and sieve elements from the apoplast. Sugars loaded into the companion cells are thought to move through plasmodesmata into the sieve elements. Note that this figure is only schematic. The actual structure of a source leaf of a C_4 plant, such as corn, is quite different from that of a C_3 plant, such as sugar beet.

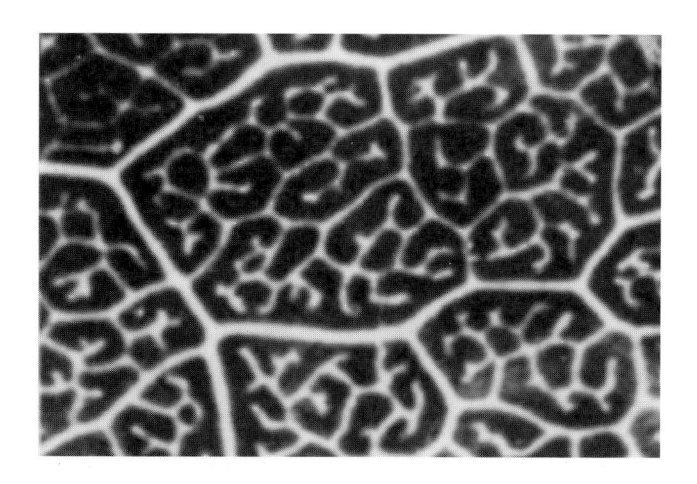

Figure 10.15 Autoradiograph of a sugar beet (*Beta vulgaris*) source leaf that was supplied with ^{14}C-labeled sucrose. (22×) The sugar solution was applied for 30 minutes to the upper surface of a leaf that had previously been kept in darkness for 3 hours. The surface of the leaf was gently rubbed with an abrasive paste to remove the cuticle and allow penetration of the solution to the interior of the leaf. The label is accumulated in the small veins, sieve elements, and companion cells of the source leaf, indicating the ability of these cells to transport sucrose against its concentration gradient. (From Fondy 1975, courtesy of D. Geiger.)

nant-DNA techniques.) Transgenic tomato plants that have a sucrose-cleaving enzyme (invertase) in their apoplast show a very slow growth rate and fail to mobilize the starch in their source leaves during a prolonged dark period (Dickinson et al. 1991). Apparently, hydrolyzing sucrose in the apoplast inhibits phloem loading in source leaves of this species. This result is expected only if the phloem-loading pathway includes an apoplastic step.

In the Apoplastic Pathway, Sucrose Uptake Requires Metabolic Energy

In source leaves, sugars become more concentrated in the sieve elements and companion cells than in the mesophyll. This difference in solute concentration can be demonstrated by measuring the solute potential (Ψ_s) of the various cell types in the leaf, using a combination of plasmolysis and light or electron microscopy (see Chapter 3). In sugar beet, for example, the solute potential of the mesophyll is approximately –1.3 MPa, and the solute potential of the sieve elements and companion cells is on the order of –3.0 MPa (Geiger et al. 1973). Most of this difference in solute potential is thought to be due to sugar, and specifically to sucrose, since sucrose is the major transport sugar in this species. Furthermore, both endogenous and exogenous sucrose accumulate in the sieve elements and companion cells of the minor veins of a sugar beet source leaf, as noted in the previous section (see Figure 10.15).

The fact that sucrose, an uncharged solute, is at a higher concentration in the sieve element–companion cell complex than in surrounding cells indicates that sucrose is transported against its chemical-potential gradient and is transported actively. Other data support the concept that sieve element loading from the apoplast requires metabolic energy. For example, treating source tissues with respiratory inhibitors decreases the ATP concentration in the tissue and inhibits the loading of exogenous sugar.

In the Apoplastic Pathway, Sieve Element Loading Uses a Sucrose–H^+ Symport

Most active-transport mechanisms use metabolic energy supplied in the form of ATP. As we saw in Chapter 6, ATP hydrolysis can be coupled to the movement of solutes across membranes in at least two ways: In **primary active transport** the same membrane protein is thought to break down ATP and use the energy released to move the solute across the membrane barrier against a chemical-potential gradient. In plant cells, the primary-active-transport ATPase is a proton pump. In **secondary transport** (see Chapter 6), solute movement against a chemical-potential gradient is driven not directly by ATP hydrolysis, but indirectly by the proton gradient established by the primary-active-transport ATPase.

The high proton concentration in the apoplast and the spontaneous tendency toward equilibrium (equal proton concentrations and electric neutrality in the apoplast and symplast) cause protons to diffuse back into the symplast. Specific transporters couple this movement to the transport of solutes such as sucrose. This type of transport is known as *sucrose–proton symport*, or **cotransport**.

Transport from the apoplast into the sieve elements and companion cells is thought to be a type of secondary transport mediated by a sucrose–H^+ symporter (Figure 10.16A). In keeping with this hypothesis, high pH (low H^+ concentration) in the apoplast reduces the uptake of exogenous sucrose into the sieve elements and companion cells of broad bean. This effect occurs because high pH in the apoplast means a low proton concentration in the apoplast, which reduces the driving force for proton diffusion into the symplast and for the sucrose–H^+ symporter. Results from other types of physiological experiments support this finding, but interpretation of the data is sometimes confused by the presence of many cell types in the tissues being studied.

Data from molecular techniques complement the physiological experiments and support the hypothesis. Proton pumps (H^+-ATPases), localized by immunological techniques, are found in the plasma membranes of companion cells (*Arabidopsis thaliana*) and transfer cells (broad bean). In the case of transfer cells, the pump is

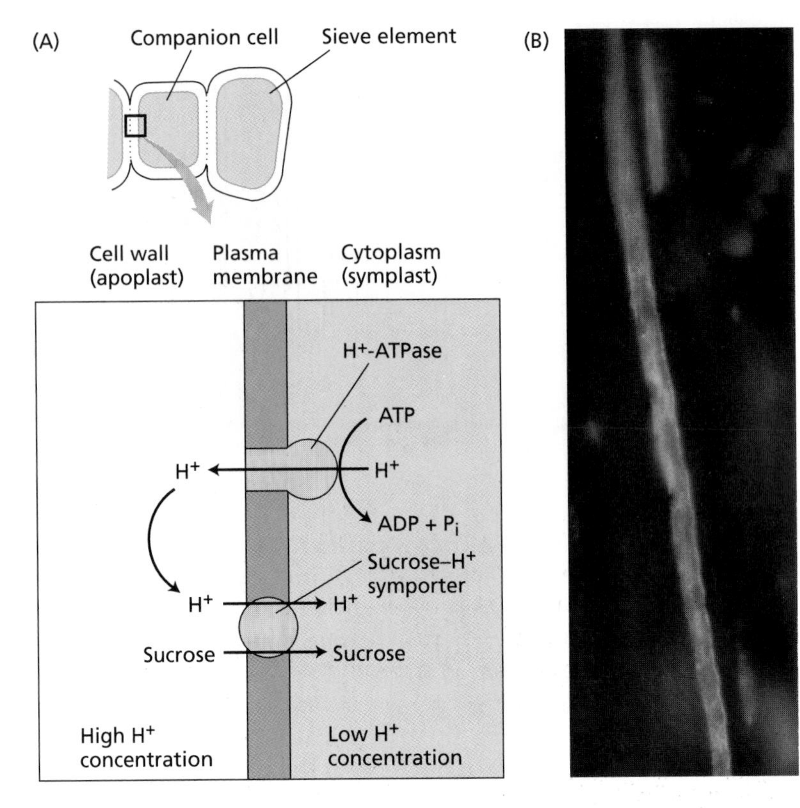

Figure 10.16 Energy utilization in sieve element loading. (A) In the cotransport model of sucrose loading into the symplast of the sieve element–companion cell complex, the plasma membrane ATPase pumps protons out of the cell into the apoplast, establishing a high proton concentration there. The energy in this proton gradient is then used to drive the transport of sucrose into the symplast of the sieve element–companion cell complex through a sucrose–H^+ symporter. (B) Localization of the sucrose–H^+ symporter in the phloem. This micrograph shows a single companion cell from broad-leaved plantain (*Plantago major*) stained with two fluorescent dyes. One of the dyes (green) is (indirectly) linked to an antibody that is specific for the PmSUC2 sucrose–H^+ symporter. The second dye (blue) binds to DNA. Since the two dyes are found on a single phloem cell, which is always adjacent to a sieve element, the sucrose symporter is concluded to be located in the companion cell membrane in this species. (B from Stadler et al. 1995, courtesy of N. Sauer. See the half-title page for a full-color version of this photograph.)

most concentrated in the plasma membrane infoldings that face the bundle sheath and phloem parenchyma cells (see Figure 10.9B), where it can energize an efficient transport of photosynthate from the apoplast (Bouche-Pillon et al. 1994). Furthermore, the distribution of the *Arabidopsis* proton pump appears to be correlated with the presence in the phloem of a sucrose–H^+ symporter called SUC2 (DeWitt and Sussman 1995). Thus, in *Arabidopsis* the proton pump that generates the proton gradient and the sucrose symporter that utilizes the gradient are found in the same location (the companion cell). The SUC2 transporter is also found in the companion cells of broad-leaved plantain (*Plantago major*) (Figure 10.16B). Interestingly, expression of the SUC2 transporter begins in the tip and proceeds to the base in developing leaves during a sink-to-source transition, the same pattern shown by assimilate export capacity, which is discussed later in the chapter.

SUC2 is only one of several sucrose–H^+ symporters that have been cloned and localized in the phloem of various species. Another sucrose transporter, SUT1, is found in tobacco, potato, and tomato (Kuhn et al. 1997). In these plants the sucrose symporter is located specifically in the plasma membrane of the sieve elements, rather than in the companion cells. The mRNA for SUT1, however, is synthesized in the companion cells. (The mRNA cannot be synthesized in the sieve elements, because they lack nuclei.)

Many fascinating questions are raised by these experiments. Does the difference in cellular location of the sucrose transporters (SUC2 and SUT1) depend on the plant species or on the function of the transporter? Is the SUT1 protein made in the companion cells or in the sieve elements, which are generally thought to lack ribosomes? What is the role of the transporters in sink tissues, where they are found at lower levels than in source leaves? The levels of *SUT1* mRNA and protein are lower after 15 hours of darkness than after being in the light. What is the mechanism of such rapid turnover? The answers to these questions await further experimentation.

Loading of sucrose from the apoplast to the sieve elements by the sucrose–H^+ symporter is thought to be regulated by the solute potential or, more likely, the turgor pressure of the sieve elements. According to this model, a decrease in sieve element turgor below a certain point would lead to a compensatory increase in loading. The entry of sucrose into the sieve elements is also affected by sucrose concentration: Higher concentrations in the apoplast increase phloem loading by the sucrose transporter.

Although considerable research has been conducted on the sieve element–loading step in the apoplastic pathway, the mechanism of sucrose efflux from the mesophyll cells into the apoplast is unknown. However, efflux is enhanced by the presence of certain substances, such as potassium, in the apoplast, thus coordinating a favorable nutrient supply, increased translocation to sinks, and enhanced sink growth.

Phloem Loading Appears to be Symplastic in Plants with Intermediary Cells

As already noted, the earliest studies of phloem loading focused on the apoplastic loading model. Many results support that model, particularly in species that transport only sucrose in the phloem and that possess either ordinary companion cells or transfer cells in the minor veins. Results from subsequent experiments, however, led to the suggestion that a symplastic pathway operates in species that transport raffinose and/or stachyose in the phloem in addition to sucrose, and that have intermediary cells in the minor veins. Some examples of such species are common coleus (*Coleus blumei*), squash (*Cucurbita pepo*), and melon (*Cucumis melo*).

The operation of a symplastic pathway would predict the presence of open plasmodesmata between the different cells in the pathway. Many species possess the potential for symplastic sieve element loading since they have numerous plasmodesmata at the interface between the sieve element–companion cell complex and the surrounding cells (see Figure 10.9C) (Gamalei 1989). Evidence from physiological experiments also indicates that symplastic continuity exists in source leaves of some species. For example, fluorescent dyes, which are mobile in the symplast but cannot cross membranes, can be microinjected into plant cells (Madore et al. 1986). Dyes injected into leaf cells can move from the mesophyll to the minor veins and from intermediary cells to bundle sheath and mesophyll cells. However, movement into sieve elements has not been demonstrated, and the availability of the pathway does not necessarily mean that it is operational.

Perhaps the most compelling evidence for a symplastic pathway is that loading of endogenous sugars is insensitive to PCMBS in plants such as *Coleus*, indicating that sugars derived from photosynthetic CO_2 fixation do not pass through the apoplast on their way to the sieve elements. Uptake of exogenous sugars is at least partly sensitive to PCMBS in these plants, as it is in other species (Turgeon and Gowan 1990). Finally, it has been suggested that some of these species load both apoplastically and symplastically, in the same sieve element, in different sieve elements in the same vein, or in sieve elements in veins of different sizes.

Phloem Loading Is Specific and Selective

The composition of sieve element sap does not generally correspond to the composition of solutes in tissues surrounding the phloem. Therefore, specific transport sugars must be selected in the source leaf. Both release from the mesophyll and loading from the apoplast of the source could be selective for the transport sugars, given the fact that membrane-bound transporters are specific for the transported molecule. However, it is more difficult to envision how diffusion through plasmodesmata during symplastic loading could be selective for certain sugars.

Species with symplastic loading appear to have a higher osmotic content in the sieve elements and companion cells than in the mesophyll (as do species with apoplastic loading); the difference in solute concentration depends on the movement of assimilate from the mesophyll. How can diffusion, which moves materials from an area of high concentration to an area of low concentration, concentrate sugars?

In the symplastic pathway, a process other than transporter-mediated membrane passage must control the availability of specific sugars for translocation in the phloem. One candidate for this selective process is the preferential conversion of nontransport sugars into transport sugars within the sieve elements and companion cells.

Robert Turgeon and his coworkers at Cornell University have suggested the following **polymer-trapping model** (Figure 10.17). Sucrose, synthesized in the mesophyll, diffuses from the bundle sheath cell into the inter-

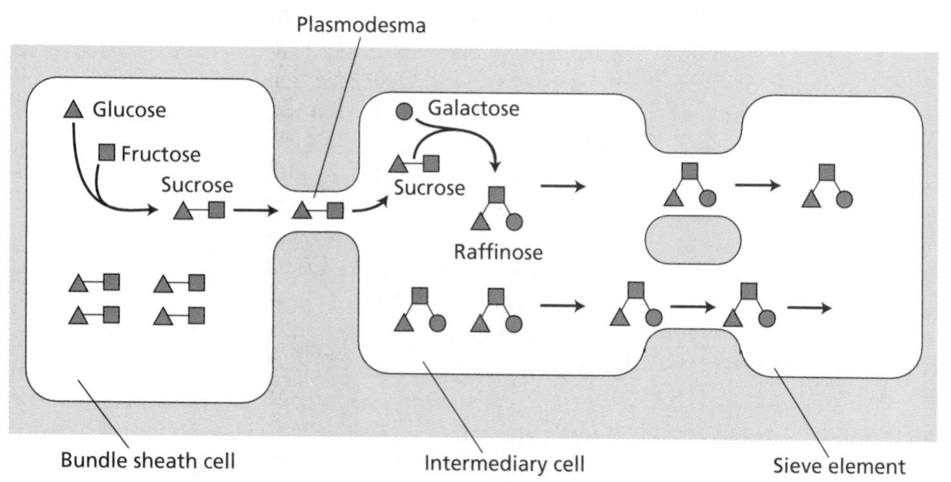

Figure 10.17 The polymer-trapping model of phloem loading. Sucrose, synthesized in the mesophyll, diffuses from the bundle sheath cells into the intermediary cells through the abundant plasmodesmata that connect the two cell types. Raffinose (and stachyose) are synthesized from sucrose and galactose in the intermediary cells and are not able to diffuse back into the mesophyll, because they are too large; they do, however, diffuse into the sieve elements. As a result, the concentration of transport sugar rises in the intermediary cells and the sieve elements. (After van Bel 1992.)

Table 10.3
Patterns in apoplastic and symplastic loading

	Apoplastic loading	Symplastic loading
Transport sugar	Sucrose	Oligosaccharides in addition to sucrose
Type of companion cell in the minor veins	Ordinary companion cells or transfer cells	Intermediary cells
Number of plasmodesmata connecting the sieve elements and companion cells to surrounding cells	Few[a] XV CC CC SE PP PP SE CC	Abundant[a] XV IC SE SE IC SE IC

Source: Drawings after Van Bel et al. 1992.
Note: Some species may load both apoplastically and symplastically, since different types of companion cells can be found within the veins of a single species.
[a] CC, companion cell; IC, intermediary cell; PP, phloem parenchyma; SE, sieve element; XV, xylem vessel. Plasmodesmata are indicated by xxxx.

mediary cell through the abundant plasmodesmata that connect the two cell types. Raffinose and stachyose are synthesized from sucrose in the intermediary cells, thus maintaining the diffusion gradient for sucrose. Raffinose and stachyose are not able to diffuse back into the mesophyll, because they are too large. As a result, the concentration of transport sugar rises in the intermediary cells and the sieve elements.

The polymer-trapping model makes the following three predictions: (1) Sucrose should be more concentrated in the mesophyll than in the intermediary cells, (2) the enzymes for raffinose and stachyose synthesis should be preferentially located in the intermediary cells, and (3) the plasmodesmata linking the bundle sheath cells and the intermediary cells should exclude molecules larger than sucrose. In support of this model, numerous biochemical and immunological studies have localized all of the enzymes required to synthesize stachyose from sucrose in the intermediary cells of cucurbits, though the mesophyll cells probably also express these enzymes.

An alternative model, based on the work of Yuri Gamalei and Aart van Bel at the Justus-Liebig University in Giessen, Germany, suggests that most stachyose is synthesized in the mesophyll cells and that the stachyose is loaded into an endoplasmic reticulum labyrinth that extends from the mesophyll to the intermediary cells across the plasmodesmata. Clearly, aspects of both models require further study and experimentation.

The Type of Phloem Loading Is Correlated with Plant Family and with Climate

As already mentioned, the use of apoplastic and symplastic phloem-loading pathways is correlated with the transport sugar, the type of companion cell in the minor veins, and the number of plasmodesmata connecting the sieve elements and companion cells to the surrounding photosynthetic cells (Table 10.3). These correlations have been verified in some species (van Bel et al. 1992).

Species in which phloem loading has been shown to be apoplastic translocate sucrose almost exclusively, have either ordinary companion cells or transfer cells in the minor veins, and possess few connections between the sieve element–companion cell complex and the surrounding cells.

Species in which phloem loading has been shown to be symplastic translocate oligosaccharides such as raffinose in addition to sucrose, have intermediary-type companion cells in the minor veins (also known as open minor veins), and possess abundant connections between the sieve element–companion cell complex and the surrounding cells. Note, however, that these correlations are not necessarily inflexible categories, but may represent specific cases of a continuum of plant types. Furthermore, new loading pathways may be discovered in the future (Flora and Madore 1996).

The degree of connection between the phloem cells and the surrounding cells is a strong family feature. Only about 10% of dicot families have more than one type of minor-vein connection. Plants that have abundant plasmodesmata between the phloem and surrounding cells are often trees, shrubs, or vines; plants with few plasmodesmata at this interface are more typically herbaceous plants. Taking global family distributions into account leads to the additional observation that plants with abundant plasmodesmata between the phloem and surrounding cells tend to be found in tropical and subtropical regions, while plants with few plasmodesmata at this interface tend to be found in temperate and arid climates.

These observations have led workers in the field to hypothesize that apoplastic loading is an adaptive response to the inhibition of symplastic transport by cold temperatures and water stress (van Bel and Gamalei 1992). However, many of the species that have abundant plasmodesmata between the phloem and surrounding cells translocate very little raffinose and stachyose, their companion cells are not intermediary cells, and little is known about their mode of loading. The questions posed by these plants offer challenging avenues for future research on loading mechanisms.

Some Substances Enter the Phloem by Diffusion

Many substances, such as organic acids and plant hormones, are found in the phloem sap at lower concentrations than the carbohydrates described in the preceding sections. These substances are probably not actively loaded into the sieve element–companion cell complex, but enter the sieve elements via other pathways and mechanisms. They may be taken up directly by diffusion across the phospholipid bilayer of the plasma membrane of the sieve element–companion cell complex or by a passive transporter in the plasma membrane of those cells, or they may diffuse into the sieve elements via the symplast.

Once in the sieve elements, these substances are swept along in the translocation stream by *bulk flow*, the motive force being generated by the active loading of only certain sugars or amino acids. Many substances not normally found in plants, such as herbicides and fungicides, can be transported in the phloem because of their ability to diffuse through membranes at an intermediate rate. In other words, they diffuse through membranes rapidly enough to allow considerable accumulation in the sieve elements, but slowly enough that they are not lost from the sieve elements completely before reaching a sink tissue. In addition, compounds that are weakly acidic tend to be "trapped" within the sieve elements, becoming negatively charged in the basic environment (low H^+ concentration) of the sieve elements and thus less likely to diffuse out of the cells across the hydrophobic membrane (Kleier 1988). Substances that are not transported in the phloem (such as calcium) apparently cannot enter the sieve elements.

Phloem Unloading and Sink-to-Source Transition

The detailed preceding discussion illustrates the events leading up to the export of sugars from sources. In many ways, the events in sink tissues are simply the reverse of the steps in sources. Transport into sink organs, such as developing roots, tubers, and reproductive structures, is termed **import**. After import, the following steps occur:

1. *Sieve-element unloading*. This is the process by which imported sugars leave the sieve elements of sink tissues.
2. *Short-distance transport*. After sieve element unloading, the sugars are transported to cells in the sink by means of a short-distance transport pathway. This pathway has also been called post–sieve element transport.
3. *Storage and metabolism*. In the final step, sugars are stored or metabolized in the cells of the sink.

Taken together, these three transport steps constitute **phloem unloading**, the movement of photosynthetic products (photosynthate) from the sieve elements to the receiver cells of the sink (Oparka and van Bel 1992).

In this section we will discuss the following questions: Is phloem unloading symplastic or apoplastic? Is sucrose hydrolyzed during the process? Does phloem unloading require energy? Finally, we will examine the process by which a young, importing leaf becomes a source (exporting) leaf.

Phloem Unloading Can Be Symplastic or Apoplastic

In sinks, sugars move from the sieve elements to the **receiver cells**. As in sources, the sugars might move entirely through the symplast via the plasmodesmata, or they might enter the apoplast at some point. Sinks vary widely from vegetative organs that are growing (root tips and young leaves) to storage tissues (roots and stems) to units of reproduction and dispersal (fruits and seeds). Because sinks vary so greatly in structure and function, there is no single scheme of phloem unloading.

Figure 10.18 diagrams several possible phloem-unloading pathways. The unloading pathway appears to be completely symplastic in some young dicot leaves, such as sugar beet and tobacco (Figure 10.18A). The evidence includes insensitivity of unloading to inhibitors of sucrose uptake from the apoplast, such as PCMBS and anoxia, and sufficient plasmodesmatal continuity. In developing leaves of monocots like corn, however, few plasmodesmata connect sieve elements and their companion cells to surrounding cells, apparently ruling out symplastic unloading (Evert and Russin 1993 and references therein). Meristematic and elongating regions of primary root tips also appear to unload symplastically (Oparka et al. 1995).

In other sink organs, part of the phloem-unloading pathway is apoplastic (Figure 10.18B). The apoplastic step could be located at the site of the sieve element–companion cell complex (type 1 in Figure 10.18B), or it could be farther removed (type 2). In developing

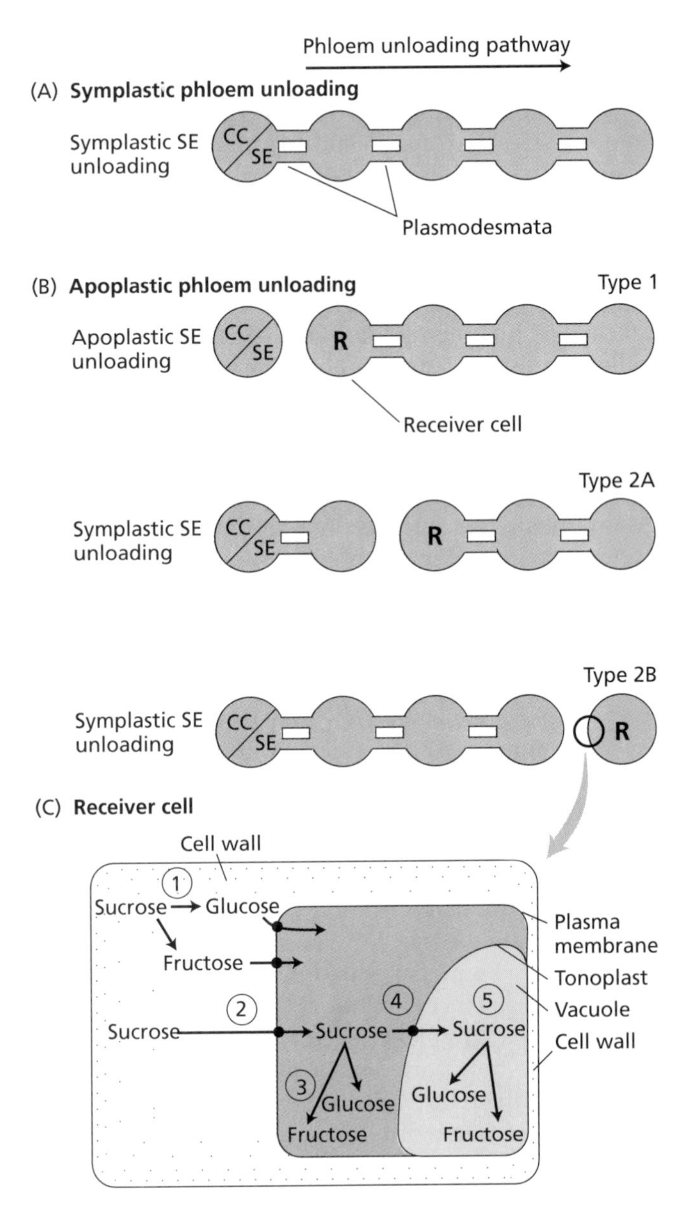

Figure 10.18 Possible pathways for phloem unloading. The sieve element–companion cell complex (CC/SE) is considered a single functional unit. The presence of plasmodesmata implies functional symplastic continuity. An absence of plasmodesmata between cells indicates an apoplastic transport step. R, receiver cell. (A) Symplastic phloem unloading. The entire pathway is symplastic. There is no receiver cell, which takes up sugars apoplastically. (B) Apoplastic phloem unloading. *Type 1*: This phloem-unloading pathway is designated apoplastic because one step, transport from the sieve element–companion cell complex to the receiver cell, occurs in the apoplast. In other words, sieve element unloading is apoplastic. Once the sugars are taken back up into the symplast of adjoining cells, transport is symplastic. *Type 2*: This type of pathway is also apoplastic phloem unloading because of an apoplastic step. However, the exit from the sieve element–companion cell complex—that is, sieve element unloading—is symplastic. The apoplastic step occurs later in the pathway. The upper figure (2A) shows an apoplastic step close to the sieve element–companion cell complex; the lower (2B), an apoplastic step that is further removed. (C) Diagram of a receiver cell, showing schematically the possible fates of sucrose unloaded apoplastically in sink tissues. The receiver cell could be one of the cells marked R in the pathway diagrams in B or a cell farther removed from the phloem. The circles represent possible sites of transporter-mediated movement of sugars, either passive or active, across membranes. Sucrose that enters the apoplast can be split into glucose and fructose by a wall invertase before entering the receiver cell (1), or it can be taken up into the receiver cell unaltered (2). Once in the symplast of the receiver cell, sucrose can be split into glucose and fructose by a cytoplasmic invertase (3), or it can enter the vacuole unaltered (4). Once in the vacuole, sucrose can be split into glucose and fructose by a vacuolar invertase (5), or it can remain unaltered. Of course, once in the cytosol or vacuole of the receiver cell, the sugars can enter pathways other than the ones shown here. For example, glucose and fructose that enter the cytoplasm by route 1 may be used to resynthesize sucrose. (A and B after Oparka and van Bel 1992; C after Giaquinta et al. 1983.)

seeds, for example, an apoplastic step is necessary because there are no symplastic connections between the maternal tissues and the tissues of the embryo. However, the step in which sugars exit the sieve elements (sieve element unloading) is itself symplastic; sugars are transferred from the symplast to the apoplast at some point removed from the sieve element–companion cell complex (type 2 in Figure 10.18B).

The phloem-unloading pathway in storage sinks may also include an apoplastic step. Some storage sinks accumulate osmotically active solutes to high concentrations; the taproot of sugar beet and the stem of sugarcane (*Saccharum officinarum*) are examples. An apoplastic step in these sinks would help prevent osmotic or turgor gradients in the symplast that would otherwise drive transport in the direction opposite that of the unloading pathway. However, such gradients could also be prevented either by the active uptake of sugar into a separate compartment such as the vacuole or by the control of turgor by changes in water relations parameters (Moore and Cosgrove 1991). In sugar beet, labeled sugars translocated from a source leaf are found in the taproot apoplast at levels thought to be too high for simple leakage, lending experimental support to the existence of an apoplastic step in phloem unloading there.

Transport Sugar May Be Hydrolyzed in the Apoplast

When phloem unloading is symplastic, transport sugars such as sucrose move through the plasmodesmata to the receiver cells. Once in the sink cells, sucrose can be metabolized in the cytosol or the vacuole before being stored or entering metabolic pathways associated with growth of the tissue. When phloem unloading is apoplastic, however, there is an additional opportunity for metabolic change. The transport sugar can be partly metabolized in the apoplast, or it can cross the apoplast unchanged (Figure 10.18C). For example, sucrose can be hydrolyzed into glucose and fructose in the apoplast by invertase, and glucose and/or fructose would then enter the sink cells. The fact that many monosaccharide transporters have been localized mainly in sink tissues supports the possible existence of this pathway.

As we will discuss later, sucrose-cleaving enzymes are thought to play a role in the control of phloem transport by sinks in general.

Transport into Sink Tissues Depends on Metabolic Activity

Numerous studies with inhibitors have shown that import into sink tissues is energy dependent. The site or sites at which energy is required vary with the species or organ studied. Phloem unloading is symplastic in growing leaves and roots and in storage sinks in which carbon is ultimately stored in polymers like starch or protein, rather than sucrose. A low sucrose concentration in the sink cells of these organs is maintained by respiration or by conversion of the transport sugars either to compounds needed in growth or to storage polymers. No membranes are crossed during the uptake into the sink cells, and unloading through the plasmodesmata is passive, since transport sugars move from a high concentration in the sieve elements to a low concentration in the sink cells. Thus, metabolic energy is required directly for respiration and biosynthesis reactions in these sinks and only indirectly for nutrient uptake. Metabolic conversions also help maintain the concentration gradient for uptake by the sink in some organs with apoplastic unloading.

In the case of apoplastic phloem unloading, sugars must cross at least two membranes: the plasma membrane of the cell that is exporting the sugar, and the plasma membrane of the receiver cell. When sugars are transported into the vacuole of the sink cell, they must also traverse the tonoplast (see Figure 10.18C). Transporters (see Chapter 6) are thought to function in movement across these membranes, and at least one of the membrane transport steps is thought to be active (dependent on metabolic energy) in sinks where phloem unloading is apoplastic.

Developing seeds have proven to be a most interesting system in which to study unloading processes. In legumes such as soybean, the embryo can be removed from the seed coat. Thus, unloading from the seed coat into the apoplast can be studied without the influence of the embryo, and uptake into the embryo can also be investigated separately. Such studies have shown that in legumes both entry of sucrose into the apoplast and uptake into the embryo are mediated by transporters and are active. In cereals like wheat, only uptake into the embryo is active; the loss of sucrose (sucrose efflux) from the maternal tissues can be passive (down the concentration gradient), because the subsequent active step keeps the sucrose concentration in the apoplast low. In corn, the cell wall invertase helps maintain a low apoplastic sucrose concentration by splitting the disaccharide into monosaccharides. In general, sugar–proton symport mechanisms appear to function in retrieval from the apoplast—for instance, in sucrose uptake into the soybean embryo.

Storage organs often accumulate sugars to high concentrations, as we have noted for sugar beet taproot and sugarcane stem. This fact is consistent with active membrane transport, since energy is generally required to move sugars into storage compartments against a concentration gradient. Sugar transport into the vacuoles of storage cells such as those of sugar beet is thought to be accomplished by a **sucrose–proton antiport** (see Chapter 6). In this case, a vacuolar H^+-ATPase pumps protons into the vacuole; the antiport carrier then moves sucrose into the vacuole in exchange for protons, which exit the vacuole down their electrochemical-potential gradient.

The Transition of a Leaf from Sink to Source Is Gradual

Dicot leaves begin their development as sink organs. A transition from sink to source status begins later in development, generally when the leaf is approximately 25% expanded, and is usually complete when the leaf is 40 to 50% expanded. Export from the leaf begins at the tip or apex of the blade and progresses toward the base until the whole leaf is exporting. During the transition period, the tip exports sugar while the base imports it from the other source leaves, as autoradiography shows (Figure 10.19).

What causes import to cease and export to begin? The maturation of leaves is accompanied by a large number of functional and anatomical changes, many of which prepare for export. In addition, the sink-to-source transition will clearly be different for species with apoplastic versus symplastic loading. In leaves with apoplastic phloem loading, a drastic switch from a symplastic unloading pathway to an apoplastic loading pathway must be made.

In the development of a leaf destined to load apoplastically, the cessation of import and the initiation of export are separate events (Turgeon 1984). In albino leaves of tobacco, which have no chlorophyll and therefore are incapable of photosynthesis, import stops at the same developmental stage as in green leaves, even though export is not possible. Therefore some other change must occur in developing leaves of tobacco that causes them to cease importing sugars from other leaves. Such a change could involve blockage of the unloading pathway at some point in the development of mature leaves. Since unloading in dicot sink leaves appears to be symplastic, plasmodesmatal closure, a

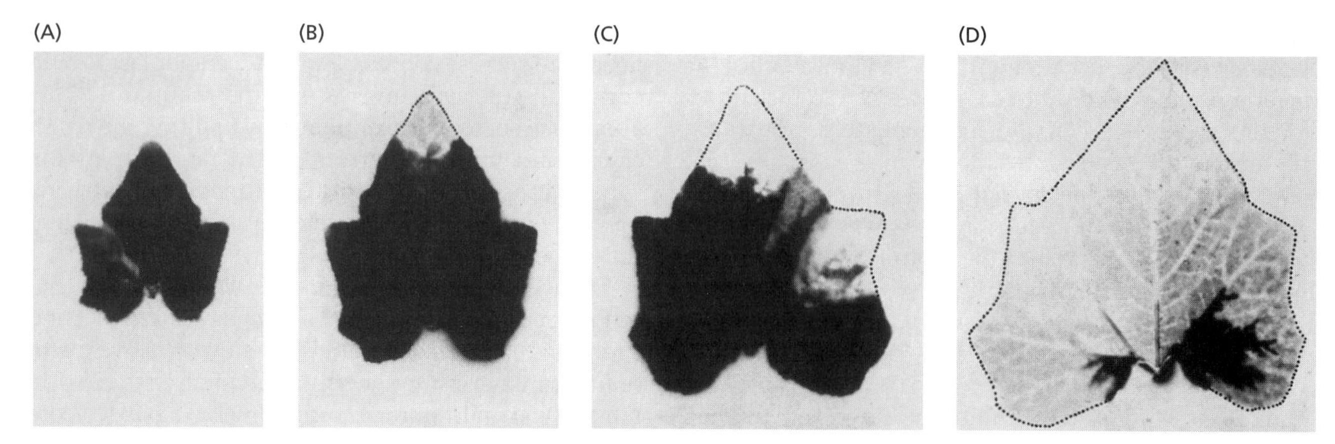

Figure 10.19 Autoradiographs of a leaf of summer squash (*Cucurbita pepo*), showing the transition of the leaf from sink to source status. In each case, the leaf imported ^{14}C from the source leaf on the plant for 2 hours. Label is visible as black accumulations. (A) The entire leaf is a sink, importing sugar from the source leaf. (B–D) The base is still a sink. As the tip of the leaf loses the ability to unload and stops importing sugar, as shown by the loss of black accumulations in B through D, it gains the ability to load and to export sugar. (From Turgeon and Webb 1973, courtesy of R. Turgeon.)

decrease in plasmodesmatal frequency, or another change in symplastic continuity could account for the cessation of unloading and import into transition leaves.

Additional data support the hypothesis that the unloading pathway is blocked in mature leaves of apoplastic loaders. Some translocation and import of labeled sugars into a mature leaf (from other mature leaves) can be induced by extreme conditions, such as darkening of the treated leaf and elimination of any alternative sinks. However, the imported sugars are not unloaded. They remain in the veins, indicating that the symplastic pathway to the mesophyll has been blocked during maturation of the leaf.

Export begins when phloem loading has accumulated sufficient photosynthate in the sieve elements to drive translocation out of the leaf. In normal leaves with apoplastic loading, export is initiated when (1) the symplastic unloading pathway is closed, (2) the leaf is synthesizing assimilates in sufficient quantity that some is available for export, (3) the sucrose-synthesizing genes are being expressed, and (4) the sucrose–H^+ symporter is in place in the plasmalemma of the sieve element–companion cell complex.

In leaves of plants like sugar beet and tobacco (including albino leaves), the ability to accumulate exogenous [^{14}C]sucrose in the sieve element–companion cell complex is gained as the leaves undergo the sink-to-source transition, presumably because the symporter responsible for loading is active. In developing leaves of *Arabidopsis*, expression of the symporter thought to be responsible for loading begins in the tip and proceeds to the base during a sink-to-source transition. The same basipetal pattern is seen in the development of export capacity. Finally, the minor veins in tobacco that are eventually responsible for most of the loading do not mature until about the time import ceases, so materials are unloaded and loaded almost entirely via different veins.

Less is known about the sink-to-source transition in leaves with symplastic loading. As might be expected, though, the transition is somewhat reversible in leaves in which the symplastic route for unloading is maintained for loading. Indeed, in variegated leaves of coleus that have both green and albino regions, the albino portions of mature leaves retain many sinklike characteristics. The green regions of the leaves can transport assimilate to the albino regions; if the green regions are removed, the albino regions can import and unload sugars from other mature leaves. The albino regions cannot, however, load exogenous sucrose (Weisberg et al. 1988). Little more is known about the sink-to-source transition in symplastic loaders. Even more intriguing questions will be raised if it proves true that both unloading and loading are *apoplastic* in some monocot leaves. For example, are the cessation of import and the initiation of export results of the same event in these leaves—perhaps expression of the symporter responsible for loading?

The Mechanism of Translocation in the Phloem: The Pressure-Flow Model

The mechanism of phloem translocation was a subject of active research from the 1930s to the mid-1970s. Now, one hypothesis is generally accepted as the correct explanation for translocation. This hypothesis, called the **pressure-flow model**, accounts for most of the experimental and structural data currently available for angiosperms.

The discussion here will begin with a description of the various types of models proposed in the past to account for phloem translocation. Then we will review the considerable evidence in favor of the pressure-flow model in angiosperms. Our discussion will end with a note of caution: Pressure-flow may not be a complete description of translocation in all species.

Active and Passive Mechanisms Have Been Proposed to Explain Phloem Translocation

All theories of phloem translocation assume that energy is required for loading in the source and for unloading in the sink. Some theories, called *active* theories, further assume that an additional expenditure of energy is required by sieve elements in the path between sources and sinks in order to drive translocation. (Descriptions of these theories are included in the review by Zimmermann and Milburn, 1975.)

A second group of theories, the so-called *passive* mechanisms, postulate that energy expended by the path sieve elements is not a major driving force for translocation, but maintains the functional integrity of the transporting cells and reloads any sugars lost to the apoplast by leakage. Keep in mind that even these passive theories acknowledge the need for energy in sources and sinks to drive translocation. The pressure-flow model belongs to this group of theories.

According to the Pressure-Flow Model, a Pressure Gradient Drives Translocation

Models in which the driving force for translocation depends solely on activities in sources and sinks include diffusion and the pressure-flow hypothesis. Diffusion is far too slow to account for the velocities of solute movement observed in the phloem. Translocation velocities average 1 m h^{-1}; the rate of diffusion is 1 m per *32 years.* (See Chapter 3 for a discussion of diffusion velocities and the distances over which diffusion is an effective transport mechanism.)

The **pressure-flow model**, on the other hand, is widely accepted as the most probable mechanism of phloem translocation. First proposed by Ernst Münch in 1930, the pressure-flow model states that a flow of solution in the sieve elements is driven by an osmotically generated pressure gradient between source and sink. The pressure gradient is established as a consequence of phloem loading at the source and phloem unloading at the sink. That is, energy-driven phloem loading generates a low solute potential in the sieve elements of the source tissue, causing a steep drop in the water potential. In response to the water potential gradient, water enters the sieve elements and causes the turgor pressure to increase.

At the receiving end of the translocation pathway, phloem unloading leads to a higher solute potential in the sieve elements of sink tissues. As the water potential of the phloem rises above that of the xylem, water tends to leave the phloem in response to the water potential gradient, causing a decrease in the turgor pressure in the sieve elements of the sink. Figure 10.20 illustrates the pressure-flow hypothesis.

If no cross-walls were present in the translocation pathway—that is, if the entire pathway were a single membrane-enclosed compartment—the different pressures at the source and sink would rapidly approach equilibrium. The presence of sieve plates greatly increases the resistance along the pathway and results in the generation and maintenance of a substantial pressure gradient in the sieve elements between source and sink. The sieve element contents are physically pushed along the translocation pathway by bulk flow, much like water flowing through a garden hose.

Close inspection of the water potential values shown in Figure 10.20 shows that water in the phloem is moving against a water potential gradient from source to sink. Such water movement does not transgress the laws of thermodynamics, however, since the water is moving by bulk flow rather than by osmosis. That is, no membranes are crossed during transport from one sieve tube to another, and solutes are moving at the same rate as the water molecules. Under these conditions, the solute potential, Ψ_s, cannot contribute to the driving force for water movement, although it still influences the water potential. Water movement in the translocation pathway is therefore driven by the pressure gradient rather than by the water potential gradient. Of course, the passive, pressure-driven, long-distance translocation in the sieve tubes ultimately depends on the active, short-distance transport mechanisms involved in phloem loading and unloading. These active mechanisms are responsible for setting up the pressure gradient.

The Predictions of the Pressure-Flow Model Have Been Confirmed

Since the pressure-flow hypothesis has gained such wide acceptance, the predictions based on the model demand special attention.

First, the sieve plate pores must be unobstructed. If P protein or other materials blocked the pores, the resistance to flow of the sieve element sap would be too great.

Second, true *bidirectional transport* (i.e., simultaneous transport in both directions) in a single sieve element cannot occur. A mass flow of solution precludes such bidirectional movement, since a solution can flow in only one direction in a pipe at any one time. Solutes within the phloem can move bidirectionally, but in different vascular bundles or in different sieve elements.

Third, great expenditures of energy are not required in order to drive translocation in the tissues along the path, although energy is required in the path to maintain the structure of the sieve elements and the integrity of the cell membranes and to reload any sugars lost to the apoplast by leakage. Therefore, treatments that

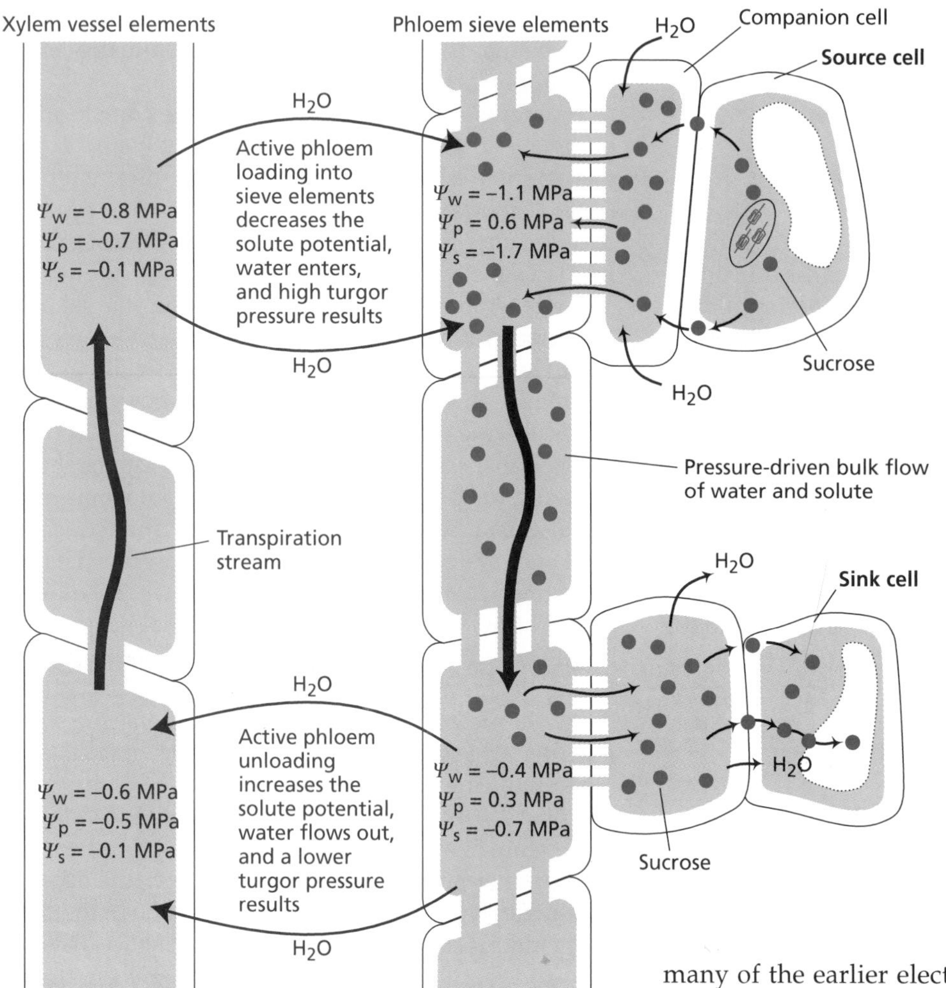

Figure 10.20 Schematic diagram showing the pressure-flow model of translocation in the phloem. At the source, sugar is actively loaded into the sieve element–companion cell complex. Water enters the phloem cells osmotically, building up a high turgor pressure. At the sink, as sugars are unloaded, water leaves the phloem cells and a lower pressure results. Water and its dissolved solutes move by bulk flow from the area of high pressure (source) to the area of low pressure (sink). Possible values for Ψ_w, Ψ_p, and Ψ_s in the xylem and phloem are illustrated. Note that this illustration shows sieve element loading as apoplastic and sieve element unloading as symplastic with a later apoplastic step; this scenario is certainly not universal. See the sections in the text on loading and unloading. (After P. S. Nobel 1991.)

restrict the supply of ATP, such as low temperature, anoxia, and metabolic inhibitors, should not stop translocation.

Fourth, the pressure-flow hypothesis demands the presence of a positive pressure gradient. Turgor pressure must be higher in sieve elements of sources than in sieve elements of sinks, and the pressure difference must be large enough to overcome the resistance of the pathway and to maintain flow at the observed velocities.

The following discussion summarizes the evidence supporting this hypothesis.

Sieve Plate Pores Are Open Channels

The ultrastructure of sieve elements is not easy to investigate, because of their high internal pressure. When the phloem is excised or fixed (killed) slowly with chemical fixatives, the turgor pressure in the sieve elements is released. The contents of the cell, including P protein, surge toward the point of pressure release, and in the case of sieve tube elements, accumulate on the sieve plates. This accumulation is probably the reason that many of the earlier electron micrographs show sieve plates that are obstructed. As mentioned earlier, if the pores of the sieve plates normally were blocked, translocation by pressure-flow would not be possible.

Now, rapid freezing and fixation techniques are available that afford a fairly reliable picture of undisturbed sieve elements. Electron micrographs of sieve tube elements prepared by such techniques indicate that P protein normally is found along the periphery of the sieve tube elements (see Figure 10.5, 10.6, and 10.7), or it is evenly distributed throughout the lumen of the cell. Furthermore, the pores contain P protein in similar positions, lining the pore or in a loose network. The open condition of the pores has been observed in a number of angiosperm species, including cucurbits, sugar beet, and bean (e.g., see Figure 10.7), consistent with the pressure-flow model.

However, the electron micrographs of rapidly frozen and fixed material do not establish unequivocally that sieve plate pores are open in vivo. To address this question, Michael Knoblauch and Aart van Bel recently devised a method for the direct observation of translocation through living sieve elements of fava bean (*Vicia faba*), using a confocal laser scanning microscope (Knoblauch and van Bel 1998). First they mounted an intact leaf upside down on the stage of a confocal microscope. Then they made two shallow cuts, several centimeters apart and parallel to the surface of the major vein, exposing the phloem tissue (Figure 10.21A). By placing the objective of the microscope over the "window" created by the basal cut (that is, near the base of the leaf), Knoblauch and van Bel were able to visualize the living sieve elements of the phloem. They then used the apical window to apply a green phloem-mobile fluorescent dye (called HPTSA), which could be translocated along with the sugars in the translocation stream. If translocation occurred, the dye would become visible in the microscope at the basal window of the leaf. In this way it could be demonstrated that the sieve elements being observed were alive and functional. Figure 10.21B shows such a sieve element.

The tissue in Figure 10.21B was also stained with another fluorescent dye; this red dye (called RH-160) stains primarily membranes and was applied locally—that is, directly to the tissue at the basal window. Figure 10.21C also shows tissue that was stained with this red dye. Note that the sieve plate shows staining within the pores that are lined with plasma membrane and that the pores are open and not occluded. Taken together, these micrographs offer graphic evidence that the sieve plate pores of living, translocating sieve elements are open.

Simultaneous Bidirectional Transport in a Single Sieve Element Has Not Been Demonstrated

Researchers have investigated bidirectional transport by applying two different tracers to two source leaves, one above the other (Eschrich 1975). Each source receives one of the tracers, and a point between the two sources is monitored for the presence of both tracers. Alternatively, a single tracer is applied to an internode and detected at points above and below the site of application.

Transport in two directions has often been detected in sieve elements of different vascular bundles in stems.

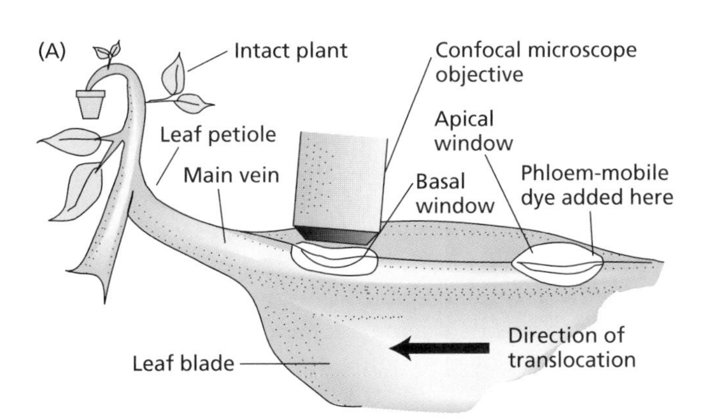

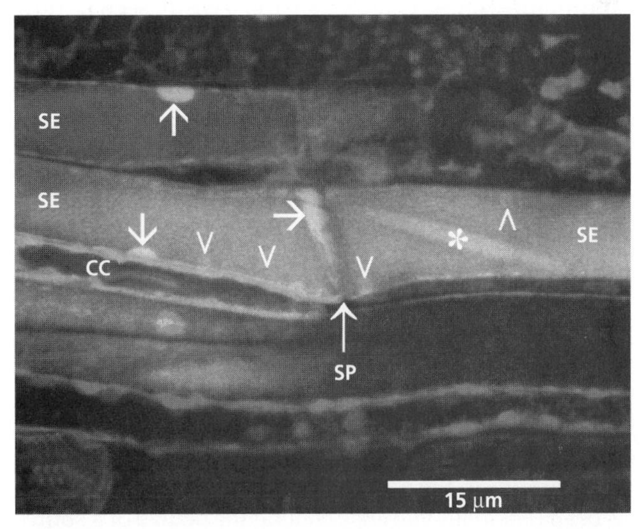

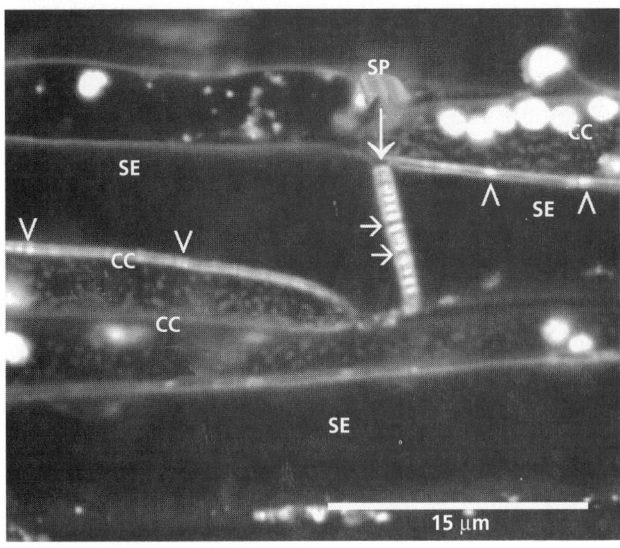

Figure 10.21 Observation of translocation in living, functional sieve elements in an intact bean (*Vicia faba*) plant. (A) Two windows were sliced paradermally (parallel to the epidermis) on the lower side of the main vein of a mature leaf, and the objective of the laser confocal microscope was positioned over the basal window. Phloem-mobile fluorescent dyes were added at the apical window. (B) Phloem tissue of bean doubly stained with a locally applied dye (red) and a translocated dye (green). Protein (arrows) deposited against the plasma membrane and the sieve plate does not impede translocation. A crystalline P protein body (asterisk) is stained by the green dye. Phloem plastids (arrowheads) are evenly distributed around the periphery of the cell. (C) Intact phloem of bean stained with a locally applied fluorescent dye that associates mainly with membranous structures. No fluorescence-labeled substances are evident inside the sieve elements. Sieve-plate pores are not blocked. Plasmodesmata between the sieve elements and the adjoining companion cells are stained clearly. Chloroplasts and probably mitochondria are visible in the companion cells. CC, companion cell; SE, sieve element; SP, sieve plate. (From Knoblauch and van Bel 1998, courtesy of A. van Bel. See the title page for a full-color version of C.)

Transport in two directions has also been seen in adjacent sieve elements of the same bundle in petioles. Bidirectional transport in adjacent sieve elements can occur in the petiole of a leaf that is undergoing the transition from sink to source and simultaneously importing and exporting assimilates through its petiole. *However, simultaneous bidirectional transport in a single sieve element has never been convincingly demonstrated.* Experiments that purport to demonstrate such bidirectional transport can be interpreted in other ways.

Translocation Rate Is Relatively Insensitive to the Energy Supply of the Path Tissues

In plants, such as sugar beet, that are able to survive periods of low temperatures, rapidly chilling a short segment of the petiole of a source leaf to approximately 1°C does not cause sustained inhibition of mass transport out of the leaf (Figure 10.22). Rather, there is a brief period of inhibition, after which transport slowly returns to the control rate. These changes are most easily seen in time courses obtained using steady-state labeling, as in Figure 10.22. Chilling reduces the respiration rate (as well as ATP production and utilization) by about 90% in the petioles at a time when translocation has recovered and is proceeding normally. When squash (*Cucurbita pepo*) petioles are treated with 100% nitrogen in the dark, transport persists even though anaerobic metabolism is the only source of ATP throughout a significant portion of the transport pathway (Sij and Swanson 1973). The main point is that the energy requirement for transport through the pathway of these plants is small, consistent with the pressure-flow hypothesis.

Extreme treatments that inhibit all energy metabolism inhibit translocation, but do not invalidate the hypothesis. For example, in the case of chilling-sensitive plants, such as bean (*Phaseolus vulgaris*), chilling the petiole of a source leaf to 10°C does inhibit translocation out of the leaf. Treating the petiole with a metabolic inhibitor (cyanide) also inhibits translocation in this plant. However, examination of the treated tissue by electron microscopy reveals blockage of the sieve plate pores by cellular debris in both cases (Giaquinta and Geiger 1973, 1977). Clearly, these results do not bear on the question of whether energy is required for translocation along the pathway.

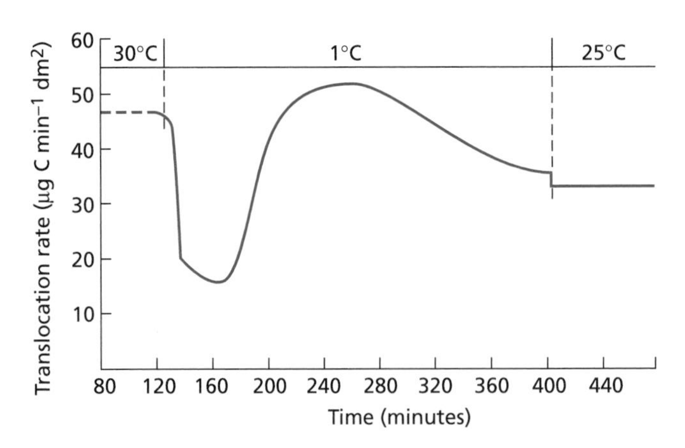

Figure 10.22 The response of translocation in sugar beet (*Beta vulgaris*) to loss of metabolic energy caused by chilling of the tissue. $^{14}CO_2$ was supplied to a source leaf, and a 2 cm portion of its petiole was chilled to 1°C. The investigators measured translocation by monitoring the rate of arrival of ^{14}C at a sink leaf. The chilling causes a temporary inhibition and subsequent recovery of translocation through the petiole. The fact that translocation can continue when ATP production and utilization are largely inhibited by chilling indicates that the energy requirement for translocation through the pathway of this plant is small. (dm [decimeter] = 0.1 meter) (Data from Geiger and Sovonick 1975.)

Pressure Gradients Are Sufficient to Drive a Mass Flow of Solution

Turgor pressures in sieve elements can be either calculated from the water potential and solute potential ($\Psi_p = \Psi_w - \Psi_s$) or directly measured. The earliest direct measurements were made with a hypodermic needle attached to a simple manometer (a thin glass capillary partly filled with fluid and sealed at the end opposite the connection to the needle). Although this technique is useful in the field, it has the disadvantage that the hypodermic needle ruptures many sieve elements when it is inserted. Because of the leakage of sieve element sap around the needle tip and the consequent pressure loss as cells lose sap into the needle, these measurements probably underestimate the true turgor (Sovonick-Dunford et al. 1981).

A newer, quite elegant technique eliminates these problems by using micromanometers or pressure transducers sealed over exuding aphid stylets (see Figure 10.11A) (Wright and Fisher 1980). The data obtained are much more accurate because aphids pierce only a single sieve element and the plasma membrane apparently seals well around the aphid stylet. However, the technique is also technically difficult.

When sieve element turgor is measured by the techniques described here, the pressure at the source is generally found to be higher than at the sink. For example, in soybean, the observed pressure difference between source and sink has been shown to be sufficient to drive a mass flow of solution through the pathway, taking into account the path resistance (mainly caused by the sieve plate pores), the path length, and the velocity of translocation (Fisher 1978). The actual pressure difference between source and sink was calculated from the water potential and solute potential to be 0.41 MPa, and the pressure difference required for translocation by pressure-flow was calculated to be 0.12 to 0.46 MPa. Thus, the observed pressure difference appears to be sufficient to drive mass flow through the phloem.

All the experiments and data that we have described in this and the preceding sections support the operation of pressure-flow in angiosperm phloem. The lack of an energy requirement in the pathway and the presence of open sieve plate pores are definitive evidence for a mechanism in which the path phloem is relatively passive. The failure to detect bidirectional transport or motility proteins, as well as the positive data on pressure gradients, are all in accord with the pressure-flow hypothesis. The method of entry of assimilates into the sieve elements, whether symplastic or apoplastic, does not affect the bulk flow mechanism.

The Mechanism of Phloem Transport in Gymnosperms May Be Different from That in Angiosperms

Although pressure-flow is adequate to explain translocation in angiosperms, it may not be sufficient in gymnosperms. Very little physiological information on gymnosperm phloem is available, and speculation about translocation in these species is based almost entirely on ultrastructural data. As discussed previously, the sieve cells of gymnosperms are similar in many respects to sieve tube elements of angiosperms. However, the sieve areas of sieve cells are relatively unspecialized and do not appear to consist of open pores (see Figure 10.8).

Instead of open pores, in gymnosperms the pores are filled with numerous membranes that are continuous with the smooth endoplasmic reticulum adjacent to the sieve areas. Such pores are clearly inconsistent with the requirements of the pressure-flow hypothesis. Either the picture presented by these electron micrographs is an artifact of the fixation process, or translocation in gymnosperms involves a different mechanism, in which perhaps the endoplasmic reticulum plays a significant role.

Assimilate Allocation and Partitioning

The photosynthetic rate determines the total amount of fixed carbon available to the leaf. However, the amount of fixed carbon available for translocation depends on subsequent metabolic events. The regulation of the diversion of fixed carbon into the various metabolic pathways is termed **allocation**.

The vascular bundles in a plant form a system of pipes that can direct the flow of photoassimilates to various sinks: young leaves, stems, roots, fruits, or seeds. However, the vascular system is often highly interconnected, forming an open network that allows source leaves to communicate with multiple sinks. Under these conditions, what determines the volume of flow to any given sink? The differential distribution of photoassimilates within the plant is termed **partitioning**.

After giving an overview of allocation and partitioning, we will examine the coordination of starch and sucrose synthesis in some detail. We will conclude by discussing how sinks compete, how sink demand might regulate photosynthetic rate in the source leaf, and how sources and sinks communicate with each other.

Allocation Includes the Storage, Utilization, and Transport of Fixed Carbon

The fate of fixed carbon in a source cell can be classified into three principal categories: storage, utilization, and transport:

1. *Synthesis of storage compounds.* Starch is synthesized and stored within chloroplasts and, in most species, is the primary storage form that is mobilized for translocation during the night. Such plants are called starch storers. In some organs of certain grasses, fructans (polymerized fructose molecules) are the storage compounds, rather than starch. (See also item 3 below.)
2. *Metabolic utilization.* Fixed carbon can be utilized within various compartments of the photosynthesizing cell to meet the energy needs of the cell or to provide carbon skeletons for the synthesis of other compounds required by the cell.
3. *Synthesis of transport compounds.* Fixed carbon can be incorporated into transport sugars for export to various sink tissues. A portion of the transport sugar can also be stored temporarily in the vacuole. In most species studied, this stored sucrose is highly transitory, providing a buffer against short-term changes in sucrose synthesis. But in some species, such as barley (*Hordeum vulgare*), fixed carbon is stored primarily as sucrose for use during the night, and very little is stored as starch. Such plants are called sucrose storers.

Allocation is a key process in sink tissues as well. Once the transport sugars have been unloaded and enter the receiver cells, they can remain as such or can be transformed into various other compounds. In storage sinks, fixed carbon can be accumulated as sucrose or hexose in vacuoles or as starch in amyloplasts. In growing sinks, sugars can be utilized for respiration and for the synthesis of other molecules required for growth.

Transport Sugars Are Partitioned among the Various Sink Tissues

The greater the ability of a sink to store or metabolize imported sugars (the process of allocation), the greater is its ability to compete for assimilates being exported by the sources. Such competition determines the distribution of transport sugars among the various sink tissues of the plant (assimilate partitioning), at least in the short term.

Of course, events in sources and sinks must be synchronized. Partitioning determines the patterns of growth, and that growth must be balanced between shoot growth (photosynthetic productivity) and root growth (water and mineral uptake) (Geiger et al. 1996). So an additional level of control lies in the interaction between areas of supply and demand. Turgor pressure in the sieve elements could be an important means of communication between sources and sinks, acting to coordinate rates of loading and unloading. Chemical messengers are also important in signaling to one organ the status of the other. Such chemical messengers include plant growth regulators (hormones) and nutrients, such as sucrose, potassium, and phosphate.

There has been a surge in research on assimilate allocation and partitioning, with the goal of improving yields of crop plants. In general, efforts by plant breeders to increase yield by increasing net photosynthetic rates have been unsuccessful (Gifford et al. 1984). However, significant improvements in yield have resulted from increases in shoot **harvest index**, the ratio of commercial or edible yield such as grains to total shoot yield (the latter including inedible portions of the shoot). An understanding of partitioning should enable plant breeders to select and develop varieties with improved transport to edible portions of the plant.

However, allocation and partitioning in the whole plant must be coordinated such that increased transport to edible tissues does not occur at the expense of other essential processes and structures. Crop yield will also be improved if photoassimilates that are normally "lost" by the plant are retained. For example, losses due to nonessential respiration or exudation from roots could be reduced. In the latter case, care must be taken not to disrupt essential processes outside the plant—for example, the growth of beneficial microbial species in the vicinity of the root.

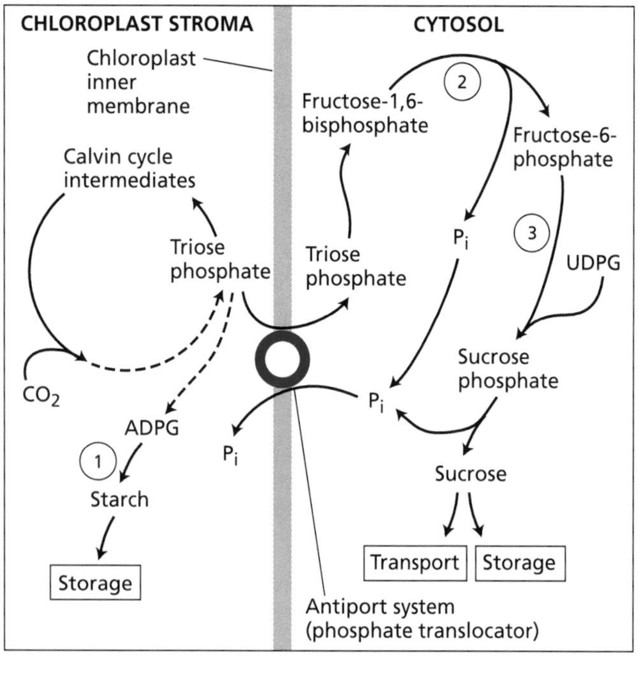

Allocation in Source Leaves Is Regulated by Key Enzymes

As Figure 10.23 shows, the quantity of sucrose available for export during the day is influenced by various biochemical reactions and carrier-mediated events, including the rate of CO_2 fixation. In fact, increasing the rate of photosynthesis in a source leaf generally results in an increase in the rate of translocation from the source. The fraction of fixed carbon that is channeled to the so-called transport pool is determined by other processes. Control points include the allocation of triose phosphates: (1) to the regeneration of intermediates in the C_3 photosynthetic carbon reduction (PCR) cycle; (2) to starch synthesis; or (3) to sucrose synthesis, and the distribution of sucrose between transport and temporary storage pools. Various enzymes operate in the pathways that process the triose phosphates, and the control of these steps is complex (Geiger and Servaites 1994).

One area of importance is the coordination of starch and sucrose synthesis during the day. The rate of starch synthesis in the chloroplast must be coordinated with sucrose synthesis in the cytosol. Triose phosphates (glyceraldehyde-3-phosphate and dihydroxyacetone phosphate) produced in the chloroplast by the C_3 PCR cycle (see Chapter 9) can be used for either starch or sucrose synthesis (disregarding for the moment the utilization of fixed carbon for regeneration of ribulose bisphosphate in the PCR cycle and for processes serving the photosynthesizing cell itself).

One mechanism for coordinating starch and sucrose synthesis involves a membrane carrier called the **phosphate translocator**, located in the inner chloroplast membrane. Since sucrose is synthesized in the cytosol, the triose phosphate destined for sucrose synthesis must leave the chloroplast. At the same time, ATP synthesis

Figure 10.23 A simplified scheme for starch and sucrose synthesis during the day. Triose phosphate, formed during photosynthesis in the Calvin (C_3 photosynthetic carbon reduction) cycle, can either be utilized in starch formation in the chloroplast or transported into the cytosol in exchange for inorganic phosphate (P_i) via the phosphate translocator in the inner chloroplast membrane. The outer chloroplast membrane is porous to small molecules and is omitted here for clarity. In the cytosol, triose phosphate can be converted to sucrose for either storage in the vacuole or transport. Key enzymes involved are starch synthetase (1), fructose-1,6-bisphosphate phosphatase (2), and sucrose phosphate synthase (3). The second and third enzymes, along with ADP-glucose pyrophosphorylase, which forms adenosine diphosphate glucose (ADPG), are regulated enzymes in sucrose and starch synthesis (see Chapter 9). UDPG, uridine diphosphate glucose. (After Preiss 1982.)

in the chloroplast requires a supply of phosphate from the cytosol. One carrier, the phosphate translocator, accomplishes the exchange of triose phosphate from the chloroplast for phosphate from the cytosol. A balance between sucrose and starch synthesis can be achieved because phosphate is released into the cytosol whenever sucrose is synthesized (see Figure 10.23).

The regulatory events occur in the following sequence: (1) sucrose synthesis in the cytosol, (2) release of phosphate, (3) exchange of phosphate from the cytosol for triose phosphate from the chloroplast, and (4) reduction in the availability of triose phosphate for starch synthesis in the chloroplast. Thus, sucrose synthesis diverts triose phosphate away from starch synthesis and storage. For example, it has been shown that when the demand for sucrose by other parts of a soybean plant is high, less carbon is stored as starch by the source leaves.

The key enzymes in the cytoplasm that are involved in the regulation of sucrose synthesis are sucrose phosphate synthase (SPS, enzyme 3 in Figure 10.23) and fructose-1,6-bisphosphatase (enzyme 2 in Figure 10.23); the key chloroplast enzyme involved in the regulation of starch synthesis is ADP-glucose pyrophosphorylase. As discussed in Chapter 8, the activities of these enzymes are controlled by the concentrations of key metabolites. One example of the control by sink demand should illustrate.

There is often a correlation between SPS activity and sucrose export in the light. SPS catalyzes the formation of sucrose phosphate, which is subsequently hydrolyzed to sucrose and inorganic phosphate. High sink demand removes sucrose, the end product of these reactions, from the source, "pulling" the reaction in the forward direction. The resulting increase in phosphate release in the source cytoplasm leads to increased efflux of triose phosphate from the chloroplast via the phosphate translocator (see Figure 10.23). Glucose-6-phosphate (which is formed from triose phosphate) stimulates the activity of SPS, thus providing the sucrose to meet the increased sink demand.

There is, however, a limit to the amount of carbon that normally can be diverted from starch synthesis in species that store carbon primarily as starch. Studies of allocation between starch and sucrose under different conditions suggest that a fairly steady rate of translocation throughout the 24-hour period is a priority for most plants. Such control is evident in studies of starch synthesis in soybean (Chatterton and Silvius 1979). In soybean, short day lengths (7 hours light, 17 hours dark) led to increased rates of starch synthesis at the expense of translocation from the source. On long days (14 hours light, 10 hours dark), rates of starch accumulation were lower. By synthesizing more starch and translocating less sucrose during the short day period, the plant has more stored materials available for translocation during the long dark period. The biochemical basis for such "anticipatory" controls is not known.

The examples discussed so far have shown a delicate balance between the synthesis of sucrose for translocation and the synthesis of starch, at least in species that store carbon in the form of starch. The controls we have discussed operate during daylight hours. Starch must then be degraded during periods of low to zero photosynthesis to provide fixed carbon for export. There is evidence that glucose, rather than triose phosphate, exits from the chloroplast under these conditions (Trethewey and ap Rees 1994). Thus, the form in which fixed carbon is supplied by the chloroplast for sucrose synthesis and export differs between the day and the night, allowing differential regulation of the two pathways.

The use of mutants and transgenic plants enables us to ask a whole new set of questions; for example, what happens when one of the competing processes, such as starch synthesis, is inhibited or even eliminated? The results have revealed the amazing flexibility of plants. For example, starch-deficient tobacco mutants synthesize only trace amounts of starch but are able to compensate for a lack of stored carbon by doubling the rate of sucrose synthesis and export during the day and by switching most of their growth to the day (Geiger et al. 1995). On the other hand, plants with enhanced starch synthesis during the day often export more of their fixed carbon during the night.

Sink Tissues Compete for Available Translocated Assimilate

As discussed earlier, translocation to sink tissues depends on the position of the sink in relation to the source and on the vascular connections between source and sink. Another factor determining the pattern of transport in whole plants is competition between sinks. For example, reproductive tissues (seeds) can compete with growing vegetative tissues (young leaves and roots) for assimilates in the translocation stream. Competition is indicated by numerous experiments in which removal of a sink tissue from a plant generally results in increased translocation to alternative, presumably competing sinks.

In the reverse type of experiment, the source supply is altered while the sink tissues are left intact. When the supply of assimilates from sources to competing sinks is suddenly and drastically reduced by shading of all the source leaves but one, the sink tissues now depend on a single source. In sugar beet and bean plants, the rates of photosynthesis and export from the single remaining source leaf usually do not change over the short term (approximately 8 hours; Fondy and Geiger 1980). However, the roots receive less sugar from the single source, while the young leaves receive relatively more. Presumably, in these plants the young leaves are stronger sinks than the roots. A stronger sink can deplete the sugar content of the sieve elements more readily and thus steepen the pressure gradient, increasing translocation toward itself.

An effect on the pressure gradient is also indicated indirectly by experiments in which investigators enhance transport to a sink over the short term by making the sink water potential more negative. Treatment of the root tips of pea seedlings with mannitol solutions increased the import of [^{14}C]sucrose, presumably because of a turgor decrease in the sink cells; 350 m*M* mannitol increased import by more than 300% (Schulz 1994).

Sink Strength Is a Function of Sink Size and Sink Activity

Various experiments indicate that the ability of a sink to mobilize assimilates toward itself, the **sink strength**, depends on two factors: **sink size** and **sink activity**.

$$\text{Sink strength} = \text{sink size} \times \text{sink activity}$$

Sink activity is the rate of uptake of assimilates per unit weight of sink tissue, and sink size is the total weight of the sink tissue. Altering either the size or the activity of the sink results in changes in transport patterns. For example, the ability of a pea pod to import ^{14}C depends on the dry weight of that pod as a proportion of the total reproductive dry weight (Jeuffroy and Warembourg 1991).

Changes in sink activity can be more complex, since various activities in sink tissues can potentially limit the rate of uptake by the sink. These activities include unloading from the sieve elements, metabolism in the cell wall, retrieval from the apoplast in some cases, and metabolic processes that use the assimilate in either growth or storage. Cooling a sink tissue inhibits activities that require metabolic energy and results in a decrease in the speed of transport toward the sink. Because cooling is so general, however, it does not permit discrimination among processes.

Single-gene mutations can be helpful here, as in so many other instances. In corn, a mutant that has a defective enzyme for starch synthesis in the kernels transports less material to the kernels than does its normal counterpart (Koch et al. 1982). In this mutant, a deficiency in assimilate storage leads to an inhibition of transport.

Sink activity and thus sink strength are also thought to be related to the presence and activity of the sucrose-splitting enzymes acid invertase and sucrose synthase, since they catalyze the first step in sucrose utilization. Acid invertase displays optimum activity at a low pH (3.5 to 5.0), catalyzes the irreversible breakdown of sucrose to glucose and fructose, and can be bound to the cell wall or found in a soluble form, probably in the vacuole. Sucrose synthase catalyzes a reversible reaction, but it is usually involved in sucrose breakdown, and generates UDP glucose and fructose in the presence of UDP.

All three enzymes (sucrose synthase and bound and soluble acid invertase) are thought to be involved in the partitioning, utilization, and storage of sucrose—their specific roles depending on the plant organ and its developmental stage. Invertase in sink cells is thought to be involved in early sink growth and expansion; sucrose synthase is associated with polysaccharide synthesis during the storage phase of sink development (Quick and Schaffer 1996).

How might these enzymes be involved in partitioning between sinks? In sinks that have an apoplastic step in phloem unloading, cell wall–bound invertase could enhance transport by maintaining a large sucrose concentration gradient between the cell releasing the sucrose and the apoplast. In sinks that have entirely symplastic unloading, invertase or sucrose synthase in receiver cells could maintain a low sucrose concentration there, with the resulting steep gradient driving diffusion from the sieve elements to the sink cells. As already noted, a stronger sink can draw phloem transport toward itself by affecting the sieve element turgor gradient.

Whether these enzymes control sink strength or are simply correlated with sink metabolism and growth are currently active topics of research and debate. Interestingly, the genes for sucrose synthase and invertase are among those regulated by carbohydrate supplies. In general "carbohydrate depletion enhances expression of genes for photosynthesis, reserve mobilization, and export processes, while abundant carbon resources favor genes for storage and utilization" (Koch 1996: 511). Thus, it was an unexpected finding that different forms of sucrose synthase and invertase, encoded by different genes, respond in opposite ways to carbohydrate supply.

For example, the mRNA for one gene for sucrose synthase in corn roots is widely distributed in root tissues and is maximally expressed when sugars are abundant. The mRNA of a second gene for sucrose synthase is most abundant in the epidermis and outer tissues of the root and is maximally expressed under conditions of sugar depletion. Thus, utilization of imported sugars is broadly maximized when sugars are abundant, but when sugar supply is low, utilization is increasingly restricted to sites that are crucial for uptake of water and minerals (Koch et al. 1996). In addition, genes for invertase and sucrose synthase are often expressed at different times during sink development, as noted earlier. In bean pods and corn kernels, changes in invertase activity are found to precede changes in assimilate import. These results are thought to indicate a key role of invertase and sucrose synthase in controlling import patterns, both during the genetic program of sink development and during responses to environmental stresses (Geiger et al. 1996).

One final note on sucrose synthase: This enzyme, which is often associated with vascular tissues, can be more specifically localized by use of antibodies to the protein. Sucrose synthase is found in the companion cells of the phloem in both loading and unloading zones of maize and citrus. In this location it is postulated to play a role in generating ATP for sieve element loading, as well as UDP glucose for callose synthesis (Nolte and Koch 1993).

Changes in the Source-to-Sink Ratio Cause Long-Term Alterations in the Source

When all but one of the source leaves on a soybean plant are shaded for an extended period (e.g., 8 days), many changes occur in the single remaining source leaf, including a decrease in starch concentration and increases in photosynthetic rate, activity of ribulose bisphosphate carboxylase (rubisco, the CO_2 fixation enzyme of photosynthesis), sucrose concentration, transport from the source, and orthophosphate concentration (Thorne and Koller 1974). Thus, while only the distribution of assimilates among different sinks may change over the short term, over longer periods the metabolism of the source adjusts to the altered conditions.

Photosynthetic rate (the net amount of carbon fixed per unit leaf area per unit time) often increases over several days when sink demand increases and decreases when sink demand decreases. Photosynthesis is most strongly inhibited under conditions of reduced sink demand in plants that normally store starch, rather than sucrose, during the day. Perhaps an accumulation of assimilates (starch, sucrose, or hexoses) in the source leaf could account for the linkage between sink demand and photosynthetic rate in these plants. Possible mechanisms include the following:

1. *Inhibition by starch.* When sink demand is low, high starch levels in the source could physically disrupt the chloroplasts, interfere with CO_2 diffusion, or block light absorption. Little evidence supports this hypothesis.
2. *Phosphate availability.* When sink demand is low, photosynthesis could be restricted by a lack of free orthophosphate in the chloroplast. Under conditions of low demand by the sink, sucrose synthesis is usually reduced, and less phosphate is thus available for exchange with triose phosphate from the chloroplast (via the phosphate translocator). If starch synthesis, which releases orthophosphate in the chloroplast, could not recycle phosphate fast enough, a deficiency in phosphate would result. ATP synthesis and thus CO_2 fixation would decline. Recent evidence provides some support for this hypothesis in potato plants transformed with antisense DNA to the phosphate translocator (Riesmeier et al. 1993). The transformed plants, which displayed reduced phosphate translocator activity, allocated proportionately more carbon into starch and less into sucrose. These effects were accompanied by a reduction in the light- and CO_2-saturated rates of photosynthesis in young plants.
3. *Regulation by sugars.* High sugar levels decrease the transcription rate and expression of genes for many photosynthetic enzymes (Koch 1996). The changes in gene expression occur over the same time frame as the source adjustments already described. For example, in source leaves of spinach (*Spinacia oleracea*), mRNA for several photosynthetic enzymes decreased when soluble carbohydrates accumulated as a result of inhibition of export from the leaf (Krapp and Stitt 1995). Although transcript levels began to decline almost immediately, changes in photosynthetic enzyme activity were apparent only after several days. In this species at least, photosynthesis appeared to be inhibited because of changes in gene expression, not because of phosphate limitation (see item 2 above).

Long-Distance Signals May Coordinate the Activities of Sources and Sinks

Signals between sources and sinks might be physical, such as turgor pressure, or chemical, such as plant hormones and carbohydrates. Changes in turgor could act as a signal from sinks that is rapidly transmitted to source tissues via the interconnecting system of sieve elements. For example, if phloem unloading were rapid, turgor pressures in the sieve elements of sinks would be reduced, and this reduction would be transmitted to the sources. If loading were controlled in part by turgor in the sieve elements of the source, it would increase in response to this signal from the sinks. The opposite response would be seen when unloading was slow in the sinks.

Evidence suggests that cell turgor affects transport across the plasma membranes of plant cells by modifying the activity of the proton-pumping ATPase in the membranes. In experiments with taproot tissue of sugar beet, for example, incubation in a mannitol solution, which lowers the turgor pressure, stimulated proton extrusion (Wyse et al. 1986).

Shoots produce growth regulators, such as auxin (indoleacetic acid, IAA; see Chapter 19), which can be rapidly transported to the roots via the phloem, and roots produce cytokinins, which move to the shoots through the xylem. Gibberellins (GA) and abscisic acid (ABA) (see Chapters 20 and 23) are also transported throughout the plant in the vascular system. Plant hormones play at least an indirect role in regulating source–sink relationships. They affect assimilate partitioning by controlling sink growth, leaf senescence, and other developmental processes.

Active transporters in plasma membranes are obvious potential sites of regulation of apoplastic loading and unloading by hormones, and experiments with exogenous hormones support this hypothesis. For example, loading of sucrose in castor bean is stimulated by exogenous IAA, but inhibited by ABA, while exogenous ABA enhances, and IAA inhibits, sucrose uptake by sugar beet taproot tissue. Other potential sites of hormone regulation of unloading include tonoplast transporters, enzymes for metabolism of incoming sucrose, wall extensibility, and plasmosdesmatal permeability in the case of symplastic unloading (see the next section). However, it remains to be established whether endogenous hormones play such roles in intact plants (Morris 1996).

As indicated earlier, carbohydrate levels can influence the expression of photosynthetic genes, as well as genes involved in sucrose hydrolysis. Many genes have been shown to be responsive to sugar depletion and abundance (Koch 1996). Thus, not only is sucrose transported in the phloem, but sucrose or its metabolites can act as signals that modify the activities of sources and sinks. In some source–sink systems, sugars and other metabolites have been shown to interact with hormonal signals to control gene expression (Thomas and Rodriguez 1994).

Plasmodesmata May Control Sites for Whole-Plant Translocation

Plasmodesmata are clearly important in the short-distance pathways of phloem loading and unloading. Although originally considered to be simple pores connecting the cytoplasms of neighboring cells, plasmodesmata have a complex structure (see Chapter 1) and can exercise dynamic control of the intercellular diffusion of small molecules (Lucas et al. 1993 and references therein.) For example, large pressure differences between cells close plasmodesmata, and the degree of closure depends on the pressure difference. Callose deposition, regulated by cytoplasmic calcium levels, is one established mechanism for sealing plasmodesmata, and some plant hormones, such as IAA and ABA, alter intracellular calcium levels via the inositol trisphosphate–diacylglycerol second-messenger system (see Chapter 14).

Information macromolecules, such as RNA and protein, also move from cell to cell in plants via plasmodesmata. Virally encoded "movement proteins" interact directly with plasmodesmata to allow the passage of viral nucleic acids. Tobacco and potato plants transformed with the tobacco-mosaic-virus movement protein show altered allocation patterns in source leaves and, at least in tobacco, modified whole-plant partitioning patterns (Olesinski et al. 1996). The source leaves of the transformed potato plants retained less sugar and starch than normal plants retain. It remains to be seen whether the changes in allocation and partitioning observed in these transformed plants were due to interactions of their altered phenotypes with controls described in earlier sections of this chapter or to a direct effect on the coordination of growth and development throughout the plant.

Plasmodesmata have been implicated in nearly every aspect of phloem translocation, from loading to long-distance transport (remember that pores in sieve areas and sieve plates are modified plasmodesmata) to allocation and partitioning. They will certainly constitute an important avenue of research in the exciting days ahead for transport in the phloem.

Summary

Translocation in the phloem is the movement of the products of photosynthesis from mature leaves to areas of growth and storage. The phloem also redistributes water and various compounds throughout the plant body.

Some aspects of phloem translocation have been well established by extensive research over a number of years. These include the following:

1. *The pathway of translocation.* Sugars and other organic materials are conducted throughout the plant in the phloem, specifically in cells called sieve elements. Sieve elements display a variety of structural adaptations that make these cells well suited for transport.
2. *Patterns of translocation.* Materials are translocated in the phloem from sources (areas of photosynthate supply) to sinks (areas of metabolism or storage of photosynthate). Sources are usually mature leaves. Sinks include organs such as roots and immature leaves and fruits.
3. *Materials translocated in the phloem.* The translocated solutes are mainly carbohydrates, and sucrose is the most commonly translocated sugar. Phloem sap also contains other organic molecules, such as amino acids and plant hormones, as well as inorganic ions.
4. *Rates of movement.* Rates of movement in the phloem are quite rapid, well in excess of rates of diffusion. Velocities average 1 m h^{-1}, and mass transfer rates range from 1 to 15 g h^{-1} cm^{-2} of sieve elements.

Other aspects of phloem translocation require further investigation, and most of these are being studied intensively at the present time. These aspects include the following:

1. *Phloem loading and unloading.* Transport of sugars into and out of the sieve elements, respectively, is called sieve element loading and unloading. In some species, sugars must enter the apoplast of the source leaf before loading. In these plants, loading into the sieve elements requires metabolic energy, provided in the form of a proton gradient. In other species, the whole pathway from the photosynthesizing cells to the sieve elements occurs in the symplast of the source leaf. In either case, phloem loading is specific for the transported sugar. Phloem unloading requires metabolic energy, but the transport pathway, site of metabolism of transport sugars, and site where energy is expended vary with the organ and species.
2. *Mechanism of translocation.* Pressure-flow is well accepted as the most probable mechanism of phloem translocation. In this model, the bulk flow of phloem sap occurs in response to an osmotically generated pressure gradient. A variety of structural and physiological data indicate that materials are translocated in the phloem of angiosperms by pressure-flow. The mechanism of translocation in gymnosperms requires further investigation.
3. *Assimilate allocation and partitioning.* Allocation is the regulation of the quantities of fixed carbon that are channeled into various metabolic pathways. In sources, these regulatory mechanisms determine the quantities of fixed carbon that will be stored, usually as starch; metabolized within the cells of the source; or immediately transported to sink tissues. In sinks, transport sugars are allocated to growth processes or to storage. Partitioning is the differential distribution of photoassimilates within the whole plant. Partitioning mechanisms determine the quantities of fixed carbon delivered to specific sink tissues. Phloem loading and unloading and assimilate allocation and partitioning are of great research interest because of their roles in crop productivity.

General Reading

Beebe, D. U., and Turgeon, R. (1991) Current perspectives on plasmodesmata: structure and function. *Physiol. Plantarum* 83: 194–199.

Frommer, W. B., and Sonnewald, U. (1995) Molecular analysis of carbon partitioning in solanaceous species. *J. Exper. Bot.* 46: 587–607.

Hall, J. L., Baker, D. A., and Oparka, K. L., eds. (1996) Transport of photoassimilates. *J. of Exper. Bot.* 47 (Special Issue): 1119–1333.

Madore, M. M., and Lucas, W. J., eds. (1995) *Carbon Partitioning and Source Sink Interactions in Plants.* American Society of Plant Physiologists, Rockville, MD.

Oparka, K. J. (1990) What is phloem unloading? *Plant Physiol.* 94: 393–396.

Patrick, J. W. (1997) Phloem unloading: Sieve element unloading and post-sieve element transport. *Annu. Rev. Plant Physiol. Plant Mol. Biol.* 48: 191–222.

Pollock, C. J., Farrar, J. F., and Gordon, A. J., eds. (1992) *Carbon Partitioning Within and Between Organisms.* BIOS Scientific Publishers Limited, Oxford.

Pontis, H. G., Salerno, G. L., and Echeverria, E. J. (1996) *Sucrose Metabolism, Biochemistry, Physiology and Molecular Biology.* American Society of Plant Physiologists, Rockville, MD.

van Bel, A. J. E. (1993) Strategies of phloem loading. *Annu. Rev. Plant Physiol. Plant Mol. Biol.* 44: 253–281.

Zamski, E., and Schaffer, A., eds. (1996) *Photoassimilate Distribution in Plants and Crops: Source–Sink Relationships.* M. Dekker, New York.

* Indicates a reference that is general reading in the field and is also cited in this chapter.

Chapter References

Bostwick, D. E., Dannenhoffer, J. M., Skaggs, M. I., Lister, R. M., Larkins, B. A., and Thompson, G. A. (1992) Pumpkin phloem lectin genes are specifically expressed in companion cells. *Plant Cell* 4: 1539–1548.

Bouche-Pillon, S., Fleurat-Lessard, P., Fromont, J-C., Serrano, R., and Bonnemain, J-L. (1994) Immunolocalization of the plasma membrane H^+-ATPase in minor veins of *Vicia faba* in relation to phloem loading. *Plant Physiol.* 105: 691–697.

Brentwood, B., and Cronshaw, J. (1978) Cytochemical localization of adenosine triphosphatase in the phloem of *Pisum sativum* and its relation to the function of transfer cells. *Planta* 140:111–120.

Chatterton, N. J. and Silvius, J. E. (1979) Photosynthate partitioning into starch in soybean leaves. I. Effects of photoperiod vs. photosynthetic period duration. *Plant Physiol.* 64: 749–753.

Christy, A. L., and Fisher, D. B. (1978) Kinetics of ^{14}C -photosynthate translocation in morning glory vines. *Plant Physiol.* 61: 283–290.

Clark, A. M., Jacobsen, K. R., Bostwick, D. E., Dannenhoffer, J. M., Skaggs, M. I., and Thompson, G. A. (1997) Molecular characterization of a phloem-specific gene encoding the filament protein, phloem protein 1 (PP1), from *Cucurbita maxima. Plant J.* 12: 49–61.

DeWitt, N. D., and Sussman, M. R. (1995) Immunocytological localization of an epitope-tagged plasma membrane proton pump (H^+-ATPase) in phloem companion cells. *Plant Cell* 7: 2053–2067.

Dickinson, C. D., Altabella, T., and Chrispeels, M. J. (1991) Slow-growth phenotype of transgenic tomato expressing apoplastic invertase. *Plant Physiol.* 95: 420–425.

Eschrich, W. (1975) Bidirectional transport. In *Transport in Plants. 1. Phloem Transport (Encyclopedia of Plant Physiology,* New Series, Vol. 1), M. H. Zimmermann and J. A. Milburn, eds., Springer-Verlag, New York, pp. 245–255.

Evert, R. F. (1982) Sieve-tube structure in relation to function. *BioScience.* 32: 789–795.

Evert, R. F., and Mierzwa, R. J. (1985) Pathway(s) of assimilate movement from mesophyll cells to sieve tubes in the *Beta vulgaris* leaf. In *Phloem Transport. Proceedings of an International Conference on Phloem Transport,* Asilomar, CA. Cronshaw, J., Lucas, W. J., and Giaquinta, R. T., eds.,. Alan R. Liss, New York, pp. 419–432.

Evert, R. F., and Russin, W. A. (1993) Structurally, phloem unloading in the maize leaf cannot be symplastic. *Am. J. Bot.* 90: 1310–1317.

Fisher, D. B. (1978) An evaluation of the Munch hypothesis for phloem transport in soybean. *Planta* 139: 25–28.

Flora, L. L., and Madore, M. A. (1996) Significance of minor-vein anatomy to carbohydrate transport. *Planta* 198: 171–178.

Fondy, B. R. (1975) *Sugar Selectivity of Phloem Loading in* Beta Vulgaris, L. *and* Fraxinus Americanus, L. Thesis, University of Dayton, Dayton, OH.

Fondy, B. R., and Geiger, D. R. (1977) Sugar selectivity and other characteristics of phloem loading in *Beta vulgaris* L. *Plant Physiol.* 59: 953–960.

Fondy, B. R., and Geiger, D. R. (1980) Effect of rapid changes in sink-source ratio, on export and distribution of products of photosynthesis in leaves of *Beta vulgaris* L. and *Phaseolus vulgaris* L. *Plant Physiol.* 66: 945–949.

Gamalei, Y. (1989) Structure and function of leaf minor veins in trees and herbs. A taxonomic review. *Trees* 3: 96–110.

Geiger, D. R., and Servaites, J. C. (1994) Diurnal regulation of photosynthetic carbon metabolism in C_3 plants. *Annu. Rev. Plant Physiol. Plant Mol. Biol.* 45: 235–256.

Geiger, D. R., Koch, K. E., and Shieh, W.-J. (1996) Effect of environmental factors on whole plant assimilate partitioning and associated gene expression. *J. of Exper. Bot.* 47 (special issue): 1229–1238.

Geiger, D. R., Shieh, W.-J., and Yu, X.-M. (1995) Photosynthetic carbon metabolism and translocation in wild-type and starch-deficient mutant *Nicotiana sylvestris* L. *Plant Physiol.* 107: 507–514.

Geiger, D. R., and Sovonick, S. A. (1975) Effects of temperature, anoxia and other metabolic inhibitors on translocation. In *Transport in Plants. 1. Phloem Transport (Encyclopedia of Plant Physiology,* New Series, Vol. 1), M. H. Zimmerman and J. A. Milburn, eds., Springer-Verlag, New York, pp. 256–286.

Geiger, D. R., and Swanson, C. A. (1965) Evaluation of selected parameters in a sugar beet translocation system. *Plant Physiol.* 40: 942–947.

Geiger, D. R., Giaquinta, R. T., Sovonick, S. A., and Fellows, R. J. (1973) Solute distribution in sugar beet leaves in relation to phloem loading and translocation. *Plant Physiol.* 52: 585–589.

Geiger, D. R., Sovonick, S. A., Shock, T. L., and Fellows, R. J. (1974) Role of free space in translocation in sugar beet. *Plant Physiol.* 54: 892–898.

Giaquinta, R. (1976) Evidence for phloem loading from the apoplast. Chemical modification of membrane sulfhydryl groups. *Plant Physiol.* 57: 872–875.

Giaquinta, R. T., and Geiger, D. R. (1973) Mechanism of inhibition of translocation by localized chilling. *Plant Physiol.* 51: 372–377.

Giaquinta, R. T., and Geiger, D. R. (1977) Mechanism of cyanide inhibition of phloem translocation. *Plant Physiol.* 59: 178–180.

Giaquinta, R. T., Lin, W., Sadler, N. L., and Franceschi, V. R. (1983) Pathway of phloem unloading of sucrose in corn roots. *Plant Physiol.* 72: 362–367.

Gifford, R. M., Thorne, J. H., Hitz, W. D., and Giaquinta, R. T. (1984) Crop productivity and photoassimilate partitioning. *Science* 225: 801–808.

Grusak, M. A., Beebe, D. U., and Turgeon, R. (1996) Phloem loading. In *Photoassimilate Distribution in Plants and Crops: Source—Sink Relationships*, E. Zamski, A. and Schaffer, eds., M. Dekker, New York, pp. 209–227.

Hall, S. M., and Baker, D. A. (1972) The chemical composition of *Ricinus* phloem exudate. *Planta* 106: 131–140.

Jeuffroy, M.-H. and Warembourg, F. R. (1991) Carbon transfer and partitioning between vegetative and reproductive organs in *Pisum sativum* L. *Plant Physiol.* 97: 440–448.

Joy, K. W. (1964) Translocation in sugar beet. I. Assimilation of $^{14}CO_2$ and distribution of materials from leaves. *J. of Exper. Bot.* 15: 485–494.

King, R. W., and Zeevaart, J. A. D. (1974) Enhancement of phloem exudation from cut petioles by chelating agents. *Plant Physiol.* 53: 96–103.

Kleier, D. A. (1988) Phloem mobility of xenobiotics. I. Mathematical model unifying the weak acid and intermediate permeability theories. *Plant Physiol.* 86: 803–810.

Knoblauch, M., and van Bel, A. J. E. (1998) Sieve tubes in action. *Plant Cell* 10: 35–50.

Koch, K. E. (1996) Carbohydrate-modulated gene expression in plants. *Annu. Rev. Plant Physiol. Plant Mol. Biol.* 47: 509–540.

Koch, K. E., Tsui, C.-L., Schrader, L. E., and Nelson, O. E. (1982) Source-sink relations in maize mutants with starch deficient endosperms. *Plant Physiol.* 70: 322–325.

Koch, K. E., Wu, Y., and Xu, J. (1996) Sugar and metabolic regulation of genes for sucrose metabolism: potential influence of maize sucrose synthase and soluble invertase responses on carbon partitioning and sugar sensing. *J. Exper. Bot.* 47 (special issue): 1179–1185.

Krapp, A., and Stitt, M. (1995) An evaluation of direct and indirect mechanisms for the "sink-regulation" of photosynthesis in spinach: Changes in gas exchange, carbohydrates, metabolites, enzyme activities and steady-state transcript levels after cold-girdling source leaves. *Planta* 195: 313–323.

Kuhn, C., Franceschi, V. R., Schulz, A., Lemoine, R., and Frommer, W. B. (1997) Macromolecular trafficking indicated by localization and turnover of sucrose transporters in enucleate sieve elements. *Science* 275: 1298–1300.

Layzell, D. B., and LaRue, T. A. (1982) Modeling C and N transport to developing soybean fruits. *Plant Physiol.* 70: 1290–1298.

Lucas, W. J., Ding, B., and Van der Schoot, C. (1993) Plasmodesmata and the supracellular nature of plants. *New Phytol.* 125: 435–476.

Madore, M. A., Oross, J. W., and Lucas, W. J. (1986) Symplastic transport in *Ipomea tricolor* source leaves: Demonstration of functional symplastic connections from mesophyll to minor veins by a novel dye-tracer method. *Plant Physiol.* 82: 432–442.

Mason, T. G., and Lewin, C. J. (1926) On the rate of carbohydrate transport in the greater yam, *Dioscorea alata. Sci. Proc. R. Dublin Soc.* 18: 203–205.

Mason, T. G., and Maskell, E. J. (1928) Studies in the transport of carbohydrates in the cotton plant. II The factors determining the rate and the direction of the movement of sugars. *Ann. Bot.* 42: 571–636.

Moore, P. H., and Cosgrove, D. J. (1991) Developmental changes in cell and tissue water relations parameters in storage parenchyma of sugarcane. *Plant Physiol.* 96: 794–801.

Morris, D. A. (1996) Hormonal regulation of source-sink relationships: an overview of potential control mechanisms. In *Photoassimilate Distribution in Plants and Crops: Source—Sink Relationships*, E. Zamski, and A. Schaffer, eds., M. Dekker, New York, pp. 441–465.

Münch, E. (1930) *Die Stoffbewegungen in der Pflanze.* Gustav Fischer, Jena.

Nobel, P. S. (1991) *Physicochemical and Environmental Plant Physiology.* Academic Press, San Diego.

Nolte, K. D., and Koch, K. E. (1993) Companion-cell specific localization of sucrose synthase in zones of phloem loading and unloading. *Plant Physiol.* 101: 899–905.

Olesinski, A. A., Almon, E., Navot, N., Perl, A., Galun, E., Lucas, W. J., and Wolf, S. (1996) Tissue-specific expression of the tobacco mosaic virus movement protein in transgenic potato plants alters plasmodesmatal function and carbohydrate partitioning. *Plant Physiol.* 111: 541–550.

Oparka, K. J., and van Bel, A. J. E. (1992) Pathways of phloem loading and unloading: a plea for a uniform terminology. In *Carbon Partitioning Within and Between Organisms,* C. J. Pollock, J. F. Farrar, and A. J. Gordon, eds., BIOS Scientific Publishers Limited, Oxford, pp. 249–254.

Oparka, K. J., Prior, D. A. M., and Wright, K. M. (1995) Symplastic communication between primary and developing lateral roots of *Arabidopsis thaliana. J. Exper. Bot.* 46: 187–197.

Preiss, J. (1982) Regulation of the biosynthesis and degradation of starch. *Annu. Rev. Plant Physiol.* 33: 431–454.

Quick, W. P. and Schaffer, A. A. (1996) Sucrose metabolism in sources and sinks. In *Photoassimilate Distribution in Plants and Crops: Source—Sink Relationships.* E. Zamski and A. Schaffer, eds.,. M. Dekker, New York, pp.115–156.

Riesmeier, J. W., Flugge, U.-I., Schulz, B., Heineke, D., Heldt, H.-W., Willmitzer, L., and Frommer, W. B. (1993) Antisense repression of the chloroplast triose phosphate translocator affects carbon partitioning in transgenic potato plants. *Proc. Natl. Acad. Sci. USA* 90: 6160–6164.

Schobert, C., Groβmann, P., Gottschalk, M., Komor, E., Pecsvaradi, A., and zur Nieden, U. (1995) Sieve-tube exudate from *Ricinus communis* L. seedlings contains ubiquitin and chaperones. *Planta* 196: 205–210.

Schulz, A. (1994) Phloem transport and differential unloading in pea seedlings after source and sink manipulations. *Planta* 192: 239–248.

Schulz, A., Alosi, M. C., Sabnis, D. D. and Park, R. B. (1989) A phloem-specific, lectin-like protein is located in pine sieve-element plastids by immunocytochemistry. *Planta* 179: 506–515.

Sij, J. W., and Swanson, C. A. (1973) Effect of petiole anoxia on phloem transport in squash. *Plant Physiol.* 51: 368–371.

Sovonick-Dunford, S., Lee, D. R., and Zimmermann, M. H. (1981) Direct and indirect measurements of phloem turgor pressure in white ash. *Plant Physiol.* 68: 121–126.

Stadler, R., Brandner, J., Schulz, A., Gahrtz, M., and Sauer, N. (1995) Phloem loading by the PmSUC2 sucrose carrier from *Plantago major* occurs into companion cells. *Plant Cell* 7: 1545–1554.

Thomas, B. R. and Rodriguez, R. L. (1994) Metabolite signals regulate gene expression and source/sink relations in cereal seedlings. *Plant Physiol.* 106: 1235–1239.

Thorne, J. H., and Koller, H. R. (1974) Influence of assimilate demand on photosynthesis, diffusive resistances, translocation, and carbohydrate levels of soybean leaves. *Plant Physiol.* 54: 201–207.

Trethewey, R. N., and ap Rees, T. (1994) A mutant of *Arabidopsis thaliana* lacking the ability to transport glucose across the chloroplast envelope. *Biochem. J.* 301: 449–454.

Turgeon, R. (1984) Termination of nutrient import and development of vein loading capacity in albino tobacco leaves. *Plant Physiol.* 76: 45–48.

Turgeon, R., Beebe, D. U., and Gowan, E. (1993) The intermediary cell: Minor-vein anatomy and raffinose oligosaccharide synthesis in the Scrophulariaceae. *Planta* 191: 446–456.

Turgeon, R., and Gowan, E. (1990) Phloem loading in *Coleus blumei* in the absence of carrier-mediated uptake of export sugar from the apoplast. *Plant Physiol.* 94: 1244–1249.

Turgeon, R., and Webb, J. A. (1973) Leaf development and phloem transport in *Cucurbita pepo:* Transition from import to export. *Planta* 113: 179–191.

van Bel, A. J. E. (1992) Different phloem-loading machineries correlated with the climate. *Acta Bot. Neerl.* 41: 121–141.

van Bel, A. J. E., and Gamalei, Y. V. (1992) Ecophysiology of phloem loading in source leaves. *Plant Cell Environ.* 15: 265–270.

van Bel, A. J. E., Gamalei, Y. V., Ammerlaan, A., and Bik, L. P. M. (1992) Dissimilar phloem loading in leaves with symplasmic or apoplasmic minor-vein configurations. *Planta* 186: 518–525.

Warmbrodt, R. D. (1985) Studies on the root of *Hordeum vulgare* L.—ultrastructure of the seminal root with special reference to the phloem. *Amer. J. Bot.* 72: 414–432.

Weisberg, L. A., Wimmers, L. E., and Turgeon, R. (1988) Photoassimilate-transport characteristics of nonchlorophyllous and green tissue in variegated leaves of *Coleus blumei* Benth. *Planta* 175: 1–8.

Wright, J. P., and Fisher, D. B. (1980) Direct measurement of sieve tube turgor pressure using severed aphid stylets. *Plant Physiol.* 65: 1133–1135.

Wyse, R. E., Zamski, E., and Tomos, A. D. (1986) Turgor regulation of sucrose transport in sugar beet taproot tissue. *Plant Physiol.* 81: 478–481.

Zimmermann, M. H. and Brown, C. L. (1971) *Trees: Structure and Function.* Springer-Verlag, Berlin.

Zimmermann, M. H. and Milburn, J. A., eds. (1975) *Transport in Plants. I. Phloem Transport* (*Encyclopedia of Plant Physiology*, new series, vol. 1). Springer, New York.

11 Respiration and Lipid Metabolism

WHEREAS PHOTOSYNTHESIS PROVIDES the carbohydrate substrate upon which plants (and nearly all life) depend, glycolysis and respiration are the processes whereby the energy stored in carbohydrates is released in a controlled manner. In the first part of this chapter we will review the basic features of these two vital interconnecting processes, taking care to note the special features that are peculiar to plants. We will also examine recent advances in our understanding of the molecular biology of mitochondria. In addition, many plants store their photosynthate in a more highly reduced form as lipids, a phenomenon common in seeds. In the second part of the chapter we will describe the pathways of lipid biosynthesis that lead to the accumulation of fats and oils. We will also examine lipid synthesis and the influence of lipids on membrane properties. Finally, we will discuss the catabolic pathways involved in the breakdown of lipids and the conversion of the degradation products to sugars that occur during seed germination.

Overview of Plant Respiration

Aerobic respiration is common to all eukaryotic organisms, and in its broad outlines, the respiratory process in plants is similar to that found in animals and lower eukaryotes. However, some specific aspects of plant respiration distinguish it from its animal counterpart. **Aerobic respiration** is the biological process by which reduced organic compounds are mobilized and subsequently oxidized in a controlled manner. During respiration, free energy is released and incorporated into a form, ATP, that can be readily utilized for the maintenance and development of the plant. From a chemical standpoint, respiration is most commonly expressed in terms of the oxidation of the six-carbon sugar glucose:

$$C_6H_{12}O_6 + 6\ O_2 + 6\ H_2O \rightarrow 6\ CO_2 + 12\ H_2O$$

This equation is the reverse of that used to describe the photosynthetic process (see Equation 7.5), and it represents a coupled redox reaction in which glucose is completely oxidized to CO_2 while oxygen serves as the ultimate electron acceptor, being reduced to water. The standard free-energy change for the reaction as written derives from the release of roughly 2880 kJ (686 kcal) per mole (180 g) of glucose oxidized. The controlled release of this free energy, together with its coupling to the synthesis of ATP, is the primary, though by no means the only, role of respiratory metabolism. Although glucose is most commonly cited as the substrate for respiration, in a functioning plant cell the reduced carbon is derived from sources such as the glucose polymer starch, the di-

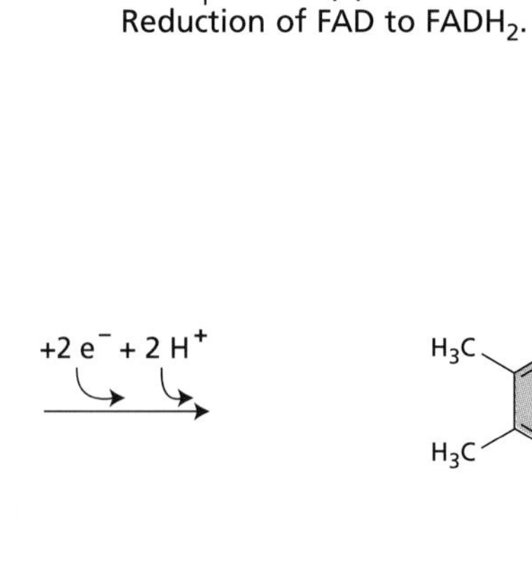

Figure 11.1 Structures and reactions of the major cofactors involved in respiratory bioenergetics. (A) Conversion of ADP and P_i to ATP. (B) Reduction of $NAD(P)^+$ to NAD(P)H. (C) Reduction of FAD to $FADH_2$.

saccharide sucrose, fructose-containing polymers (fructans), and other sugars, as well as lipids (primarily triacylglycerols), organic acids, and on occasion, proteins.

To prevent damage (incineration) of cellular structures, the cell liberates the large amount of free energy released in the oxidation of glucose by a multistep process in which glucose is oxidized through a series of reactions. These reactions can be subdivided into three stages: (1) glycolysis, (2) the tricarboxylic acid cycle, and (3) the electron transport chain.

Glycolysis is a series of reactions carried out by a group of soluble enzymes located in the cytosol. Chemically, glucose is partly oxidized to produce two molecules of pyruvate (a three-carbon compound), a little ATP, and stored reducing power in the form of a reduced pyridine nucleotide, NADH.

The tricarboxylic acid cycle and the electron transport chain both take place within the confines of the membrane-enclosed organelle known as the mitochondrion. In the **tricarboxylic acid (TCA) cycle**, pyruvate is oxidized completely to CO_2, and a considerable amount of reducing power (about 10 NADH equivalents per glucose) is generated in the process. With one exception (succinate dehydrogenase), these reactions involve a series of soluble enzymes located in the internal aqueous compartment, or matrix, of the mitochondrion (see Figure 11.4). As we will discuss later, succinate dehydrogenase is localized on the inner mitochondrial membrane.

The **electron transport chain** consists of a collection of electron transport proteins bound to the inner of the two mitochondrial membranes. This system transfers electrons from NADH (and related species), produced during glycolysis and the TCA cycle, to oxygen. This electron transfer releases a large amount of free energy, much of which is conserved through the conversion of ADP and P_i (inorganic phosphate) to ATP (Figure 11.1). This final stage completes the oxidation of glucose. We can now formulate a more complete picture of respiration as related to its role in cellular energy metabolism:

$$C_6H_{12}O_6 + 6\ O_2 + 6\ H_2O + 32\ ADP + 32\ P_i \rightarrow 6\ CO_2 + 12\ H_2O + 32\ ATP$$

As will become apparent, not all the carbon that enters the respiratory pathway ends up as CO_2. Many important metabolic intermediates appear in the reactions of the glycolytic and TCA cycle pathways, allowing these pathways to serve as starting points for many other cellular pathways in addition to their primary role in energy metabolism (see Figure 11.10). Examples of metabolites derived from respiratory intermediates include several amino acids, the pentoses used in cell wall and nucleotide biosynthesis, precursors of porphyrin biosynthesis, and the glycerol needed to synthesize phospholipids.

Glycolysis: A Cytosolic Process

In glycolysis (from the Greek words *glykos*, "sugar," and *lysis*, "splitting")—the first stage of respiration—glucose, a six-carbon sugar, is split into two three-carbon sugars. These three-carbon sugars are then oxidized and rearranged to yield two molecules of pyruvate.

Besides preparing the substrate for oxidation in the TCA cycle, glycolysis yields a small amount of chemical energy in the form of ATP and NADH. Before the evolution of photosynthesis and the appearance of oxygen in the atmosphere, glycolysis was probably the main source of energy for early cells via the fermentation pathway. The ancient origin of this pathway is evident from its localization in the cytosol; the subsequent stages of respiration, which require molecular oxygen, are located in the mitochondria. In this section we will describe the basic glycolytic pathway, as well as some features that are specific for plant cells.

Glycolysis Converts Glucose into Pyruvate, Capturing Released NADH and ATP

Glycolysis occurs in all living organisms (prokaryotes and eukaryotes). From an evolutionary standpoint, it is the oldest of the three stages of respiration. No oxygen is required to convert glucose to pyruvate, and glycolytic metabolism can become the primary mode of energy production in plant tissues when oxygen levels are low—for example, in roots in a flooded soil. As already noted, glycolysis consists of reactions catalyzed by a series of soluble enzymes located in the cytosol, and until recently this process was not associated with a particular organelle or subcellular organizational network. However, evidence now suggests that in animal cells the enzymes of the glycolytic pathway do not exist independently in the cytosol but are associated in a supramolecular complex that may be loosely bound to the surface of the outer mitochondrial membrane. This complex is thought to facilitate the conversion of substrates to products in the multistep glycolytic process.

The principal reactions associated with the classic glycolytic and fermentative pathways are shown in Figures 11.2 and 11.3. Because sucrose is the major translocated sugar in plants and is therefore the form of carbon that most nonphotosynthetic tissues import, sucrose (*not* glucose) should be thought of as the true "sugar substrate" for plant respiration. Sucrose is broken down to the two monosaccharides glucose and fructose, which can readily enter the glycolytic pathway. Two pathways for the degradation of sucrose are known in plants. In most plant tissues sucrose synthase, which is located in the cytosol and combines sucrose with UTP to produce fructose, UDP-glucose, and inorganic phosphate is used to degrade sucrose. But in some tissues, invertase, which is localized in the cell

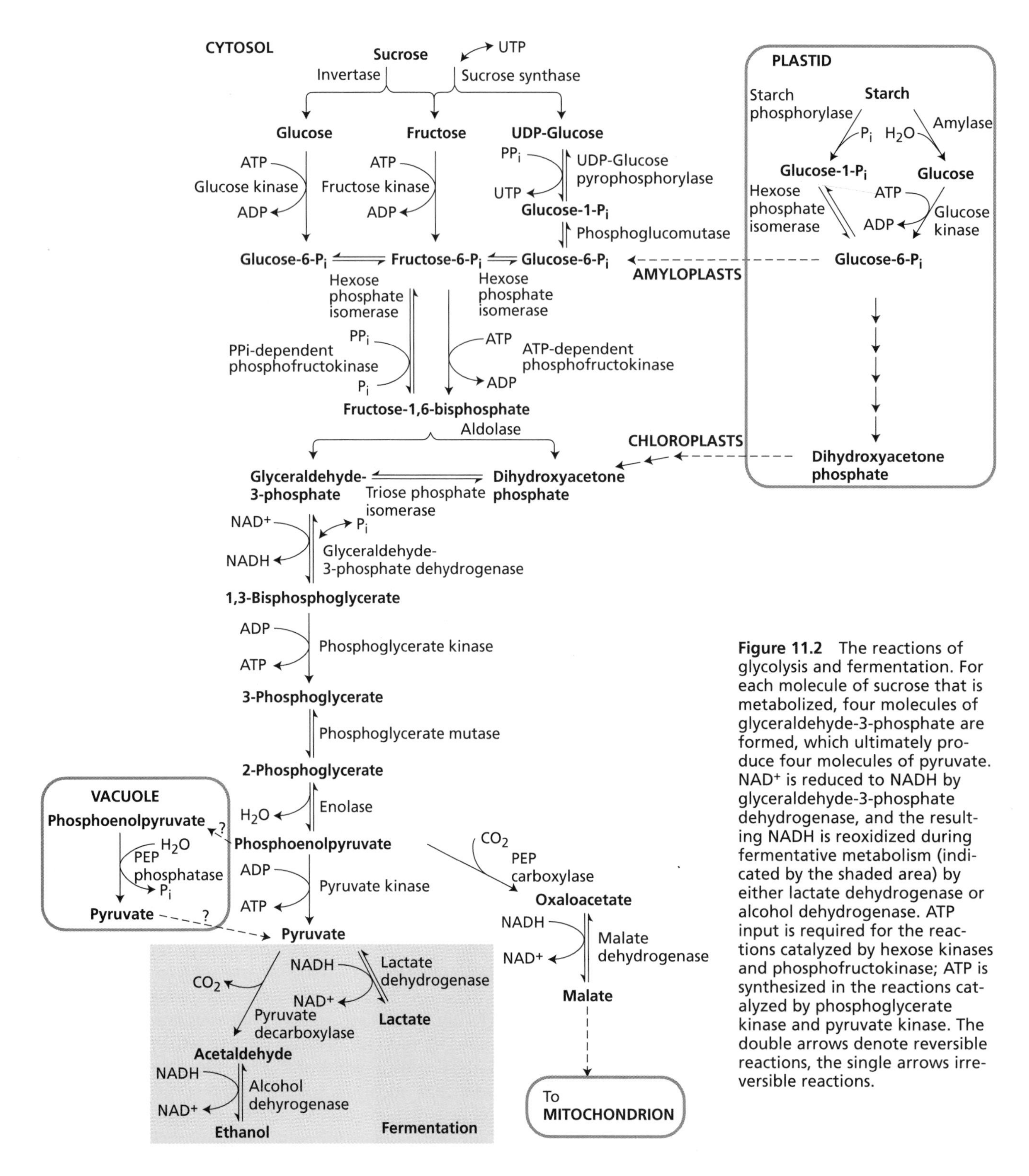

Figure 11.2 The reactions of glycolysis and fermentation. For each molecule of sucrose that is metabolized, four molecules of glyceraldehyde-3-phosphate are formed, which ultimately produce four molecules of pyruvate. NAD^+ is reduced to NADH by glyceraldehyde-3-phosphate dehydrogenase, and the resulting NADH is reoxidized during fermentative metabolism (indicated by the shaded area) by either lactate dehydrogenase or alcohol dehydrogenase. ATP input is required for the reactions catalyzed by hexose kinases and phosphofructokinase; ATP is synthesized in the reactions catalyzed by phosphoglycerate kinase and pyruvate kinase. The double arrows denote reversible reactions, the single arrows irreversible reactions.

wall, simply hydrolyzes sucrose to its two component hexoses (glucose and fructose).

Because in plants starch is synthesized and catabolized in plastids, carbon obtained from starch degradation enters the glycolytic sequence in the cytosol primarily at the level of either the three-carbon sugars glyceraldehyde-3-phosphate and dihydroxyacetone phosphate, in the case of chloroplasts, or the hexose glucose-1-phosphate, which is translocated out of amyloplasts (see Chapter 1). In the case of the three-carbon sugars, a separate set of isozymes catalyzing the same series of reactions that convert six-carbon sugars to three-carbon sugars, as occurs in glycolysis, is localized in the plastid. All the enzymes shown in Figure 11.2 have been measured at levels sufficient to support the respiration rates observed in intact plant tissues.

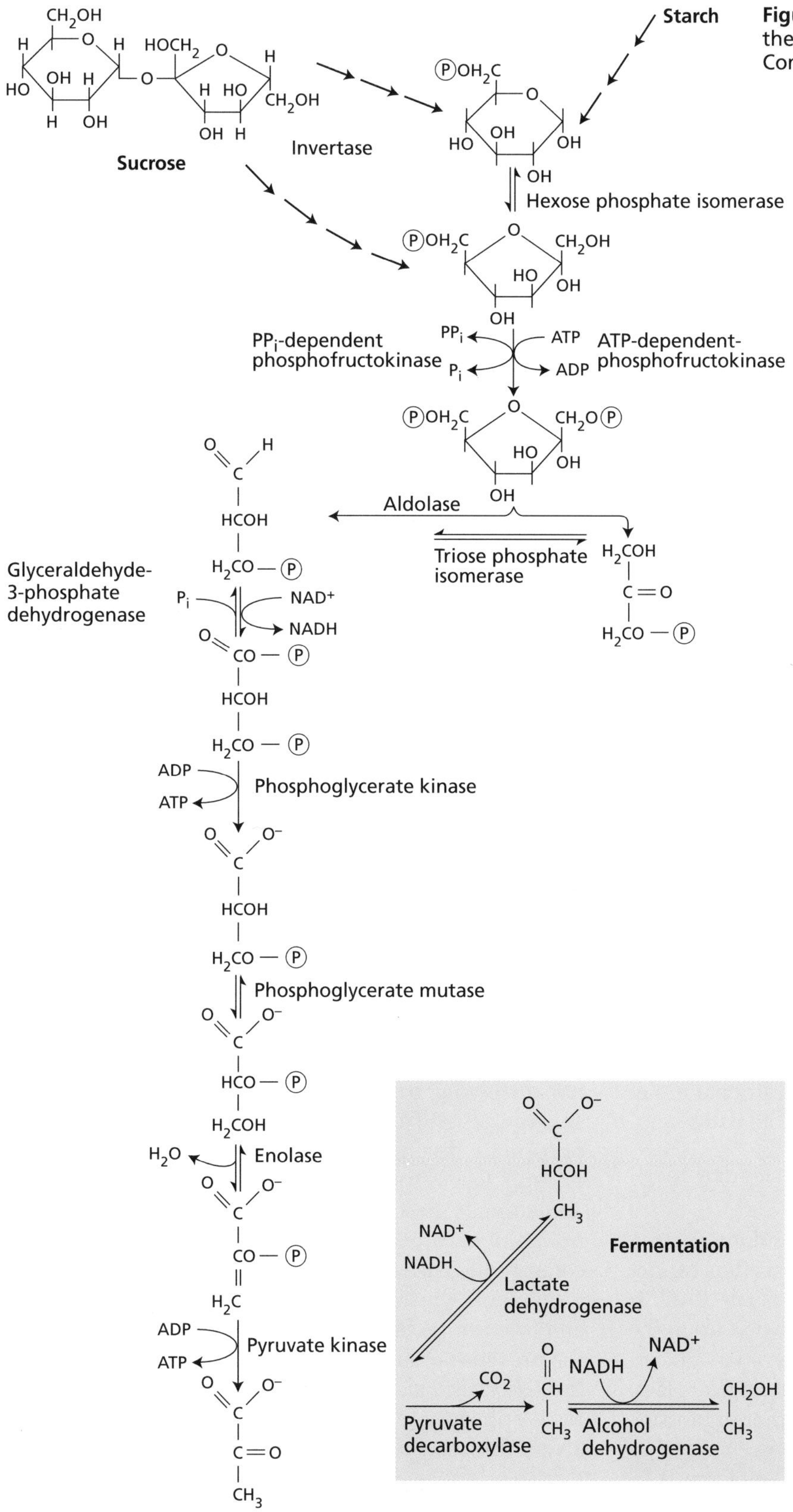

Figure 11.3 Structure of the intermediates of the glycolytic and fermentative pathways. Compare with Figure 11.2.

In the initial series of glycolytic reactions, the entering hexose (glucose or fructose) is phosphorylated twice and then split, producing two molecules of the three-carbon sugar glyceraldehyde-3-phosphate. This series of reactions requires the input of two molecules of ATP per glucose and includes two of the three irreversible reactions (i.e., reactions for which the value of K_{eq} is greater than 500; see Chapter 2) of the glycolytic pathway that are catalyzed by hexose kinases (including glucose kinase, and fructose kinase) and phosphofructokinase. The phosphofructokinase reaction is one of the control points of glycolysis in both plants and animals. Recent work on plants indicates that this part of the pathway and its regulation are complex; they will be discussed later in the chapter.

Once glyceraldehyde-3-phosphate is formed, the glycolytic pathway can begin to extract usable energy. The enzyme glyceraldehyde-3-phosphate dehydrogenase catalyzes the oxidation of the aldehyde to a carboxylic acid, releasing sufficent free energy to allow the concomitant reduction of NAD^+ to NADH and the phosphorylation (using inorganic phosphate) of glyceraldehyde-3-phosphate to produce 1,3-bisphosphoglycerate.

Nicotinamide adenine dinucleotide (NAD^+/NADH) is an organic cofactor associated with many en-zymes that catalyze cellular redox reactions. NAD^+ is the oxidized form of the cofactor, and it undergoes a reversible two-electron reaction that yields NADH ($NAD^+ + 2\ e^- + H^+$; see Figure 11.1B). The standard reduc-tion potential for this redox couple is about –320 mV, which, biologically speaking, makes it a relatively strong reductant (i.e., electron donor). NADH is thus a good species in which to store free energy released during the stepwise oxidations of glycolysis and the TCA cycle. The subsequent oxidation of NADH by oxygen via the electron transport chain then releases sufficient free energy (220 kJ mol^{-1}, or 52 kcal mol^{-1}) to drive the synthesis of ATP. A related compound, NADP (nicotinamide adenine dinucleotide phosphate), performs a redox function in photosynthesis (see Chapter 8) and in the oxidative pentose phosphate pathway, which will be discussed later in the chapter.

Now let's return to the intermediates and reactions of glycolysis, starting with 1,3-bisphosphoglycerate, which we already mentioned. The phosphorylated carboxylic acid on carbon 1 of 1,3-bisphosphoglycerate (see Figure 11.3) represents a mixed-acid anhydride that has a large free energy of hydrolysis (–18.9 kJ mol^{-1}, or –4.5 kcal mol^{-1}). Thus 1,3-bisphosphoglycerate is a strong donor of phosphate groups. In the next step of glycolysis, catalyzed by phosphoglycerate kinase, the phosphate on carbon 1 is transferred to a molecule of ADP, yielding ATP and 3-phosphoglycerate. For each glucose entering the pathway, two ATPs are generated by this reaction—one for each molecule of 1,3-bisphosphoglycerate.

This type of ATP synthesis is referred to as **substrate-level phosphorylation** because it involves the direct transfer of a phosphate moiety from a substrate molecule to ADP to form ATP. Substrate-level phosphorylation is distinct from the mechanism of ATP synthesis during oxidative phosphorylation that is utilized by the electron transport chain in the final stage of respiration and from photophosphorylation that is used to synthesize ATP during photosynthetic electron transfer (see Chapter 7).

In the last series of glycolytic reactions, the phosphate on 3-phosphoglycerate is transferred to carbon 2 and a molecule of water is removed, yielding the compound phosphoenolpyruvate (PEP). The phosphate group on PEP also has a high free energy of hydrolysis (–30.5 kJ mol^{-1}, or –7.3 kcal mol^{-1}), by virtue of its locking the parent compound, pyruvate, into an energetically unfavored enol (—C═C—OH) configuration, and this high level of free energy makes PEP an extremely good phosphate donor for ATP formation as well. In the final step of glycolysis, the enzyme pyruvate kinase catalyzes a second substrate-level phosphorylation to yield ATP and pyruvate. This final step, which is the third irreversible step in glycolysis, yields two additional molecules of ATP for each hexose that enters the pathway.

Fermentation Regenerates the NAD^+ Needed for Glycolysis in the Absence of O_2

In the absence of oxygen, the TCA cycle and the electron transport chain cannot function. Glycolysis thus cannot continue to operate, because the cell's supply of NAD^+ is limited, and once all the NAD^+ becomes tied up in the reduced state (NADH), the reaction catalyzed by glyceraldehyde-3-phosphate dehydrogenase cannot take place. To overcome this problem, plants and other organisms can further metabolize pyruvate by carrying out one or more forms of **fermentative metabolism** (see Figure 11.2).

In alcoholic fermentation (common in plants, but more widely recognized for its role in brewer's yeast), the two enzymes pyruvate decarboxylase and alcohol dehydrogenase act on pyruvate, ultimately producing ethanol and CO_2 and oxidizing NADH in the process. In lactic acid fermentation (common to mammalian muscle but also found in plants), the enzyme lactate dehydrogenase uses NADH to reduce pyruvate to lactate, thus regenerating NAD^+.

Under some circumstances, plant tissues may be subjected to low (**hypoxic**) or zero (**anoxic**) concentrations of ambient oxygen, forcing them to carry out fermentative metabolism. The best-known example involves flooded or waterlogged soils in which the diffusion of oxygen to the roots is sufficiently reduced to cause root tissues to become hypoxic. In corn, the initial response is lactic acid fermentation, but the subsequent response is alcoholic fermentation (Roberts et al. 1984; Xia and Roberts 1994)). Ethanol is thought to be a less deleterious end product of fermentation than lactate, because accumulation of lactate promotes acidification of the cytosol. There are numerous other examples of plants functioning under near-anaerobic conditions by carrying out some form of fermentation.

Plants Have Alternative Glycolytic Reactions

The reactions described thus far in this chapter occur in all organisms that carry out glycolysis. In addition, organisms can utilize this pathway operating in the opposite direction to synthesize glucose from other organic molecules. The synthesis of glucose through reversal of the glycolytic pathway is known as **gluconeogenesis**. Gluconeogenesis is not common in plants, but it does operate in the seeds of some plants, such as castor bean and sunflower, that store a significant quantity of their carbon reserves in the form of oils (triacylglycerols). After the seed germinates, much of the oil is converted via gluconeogenesis to sucrose, which is then used to support the growing seedling.

Because the glycolytic reaction catalyzed by ATP-dependent phosphofructokinase is irreversible, an additional enzyme, fructose-1,6-bisphosphatase, acts at this step to convert fructose-1,6-bisphosphate to fructose-6-phosphate during gluconeogenesis. Fructose-1,6-bisphosphatase is found in plants. Further, in both plants and animals these two enzymes represent a major control point of carbon flux through the glycolytic and gluconeogenic pathways.

In plants, the interconversion of fructose-6-phosphate and fructose-1,6-bisphosphate is made more complex by the presence of an additional enzyme, a *pyrophosphate* *(PP_i)-dependent phosphofructokinase* (pyrophosphate:fructose-6-phosphate 1-phosphotransferase), which catalyzes this reaction reversibly.

$$\text{Fructose-6-P} + PP_i \leftrightarrow \text{fructose-1,6-}P_2 + P_i$$

where "P" represents phosphate and "P_2" bisphosphate. Pyrophosphate:fructose-6-phosphate 1-phosphotransferase is found in the cytosol of many plant tissues at levels that can be considerably higher than those of the

ATP-dependent phosphofructokinase. Because the reaction catalyzed by the PP_i-dependent phosphofructokinase is readily reversible, this enzyme can operate during either glycolytic or gluconeogenic carbon fluxes, although its operation in glycolysis requires a source of pyrophosphate (Black et al. 1985). Of greater importance, this enzyme, like the ATP-dependent phosphofructokinase and the fructose bisphosphatase, appears to be under metabolic control (discussed later in the chapter), suggesting that under some circumstances, operation of the glycolytic (or gluconeogenic) pathway in plants differs from that in other organisms.

At the end of the glycolytic sequence, plants have alternative pathways for metabolizing PEP. In one pathway PEP is carboxylated by the ubiquitous cytosolic enzyme PEP carboxylase to form oxaloacetate (OAA). The OAA is then reduced to malate by the action of malate dehydrogenase, which uses NADH as a source of electrons and thus performs a role similar to that of the dehydrogenases used during fermentative metabolism (see Figure 11.2). The resulting malate is then transported to the mitochondrion, where it is metabolized to form pyruvate by NAD^+ malic enzyme, as discussed more fully at the end of the section that focuses on the TCA cycle.

In the second pathway for metabolizing PEP in plant glycolysis, the enzyme PEP phosphatase catalyzes the hydrolysis of PEP to pyruvate and P_i (see Figure 11.2). This enzyme is localized in the vacuole and is found to increase in activity in plants grown under phosphate-limited conditions (Plaxton 1996). Because the transport of PEP into the vacuole and pyruvate out of the vacuole has not yet been demonstrated, this pathway remains hypothetical, as indicated by the question marks in Figure 11.2.

Fermentation Does Not Liberate All the Energy Available in Each Sugar Molecule

Before we leave the topic of glycolysis, we need to consider the efficiency of fermentation. The standard free-energy change ($\Delta G^{0\prime}$) for the complete oxidation of glucose is –2880 kJ mol^{-1} (–686 kcal mol^{-1}). The value of $\Delta G^{0\prime}$ for the synthesis of ATP is –31.8 kJ mol^{-1} (–7.6 kcal mol^{-1}). However, under the nonstandard conditions that normally exist in plant cells, the synthesis of ATP requires an input of free energy of approximately 50.2 kJ mol^{-1} (12 kcal mol^{-1}). Given the net synthesis of two molecules of ATP for each glucose that is converted to ethanol (or lactate), the efficiency of anaerobic fermentation—that is, the energy stored as ATP relative to the energy potentially available in a molecule of glucose—is only about 4%. Most of the energy available in glucose remains in the reduced by-product of fermentation: lactate or ethanol. During aerobic respiration, the pyruvate produced by glycolysis is transported into mitochondria, where it is further oxidized, resulting in much more efficient conversion of the free energy originally available in the glucose .

The Tricarboxylic Acid Cycle: A Mitochondrial Matrix Process

During the nineteenth century, biologists discovered that in the absence of air, cells produce ethanol or lactic acid, while in the presence of air, cells consume O_2 and produce CO_2 and H_2O. In 1937, the German-born British biochemist Hans A. Krebs reported the discovery of the tricarboxylic acid (TCA) cycle—also called the *citric acid cycle* or *Krebs cycle*. The elucidation of the TCA cycle not only explained how pyruvate was broken down to CO_2 and H_2O; it also introduced the key concept of cycles in metabolic pathways. For his discovery Hans Krebs was awarded the Nobel prize in medicine and physiology in 1953.

Since the TCA cycle is localized in the matrix of mitochondria, we will begin with a description of mitochondria as semiautonomous organelles, including the mitochondrial genome and the alterations that can result through genomic rearrangements. We will then review the steps of the TCA cycle and discuss features that are specific for plants.

Mitochondria Are Semiautonomous Organelles

The breakdown of glucose to pyruvate releases less than 25% of the total energy in glucose; the remaining energy is stored in the two molecules of pyruvate. The next two stages of respiration (the TCA cycle, and electron transport coupled with ATP synthesis) take place within an organelle enclosed by a double membrane, the **mitochondrion**.

Plant mitochondria were originally identified by light microscopy as particles that stained with the dye Janus green B. With the advent of electron microscopy, scientists were able to clearly define mitochondrial morphology (Figure 11.4). Isolated plant mitochondria generally appear as spherical or rodlike entities ranging from 0.5 to 1.0 μm in diameter and up to 3 μm in length (Douce 1985). In intact cells, highly reticulate networks representing single complex mitochondria are commonly observed in animal cells, and such networks have been reported in unicellular algae and plants. The number of mitochondria observed per plant cell, with some exceptions, is far below that found in a typical animal cell. However, the number of mitochondria per plant cell can vary and is usually directly related to the metabolic activity of the tissue, reflecting the mitochondrial role in energy metabolism. Guard cells, for example, are unusually rich in mitochondria.

The ultrastructural features of plant mitochondria are similar to those of mitochondria in nonplant tissues (see

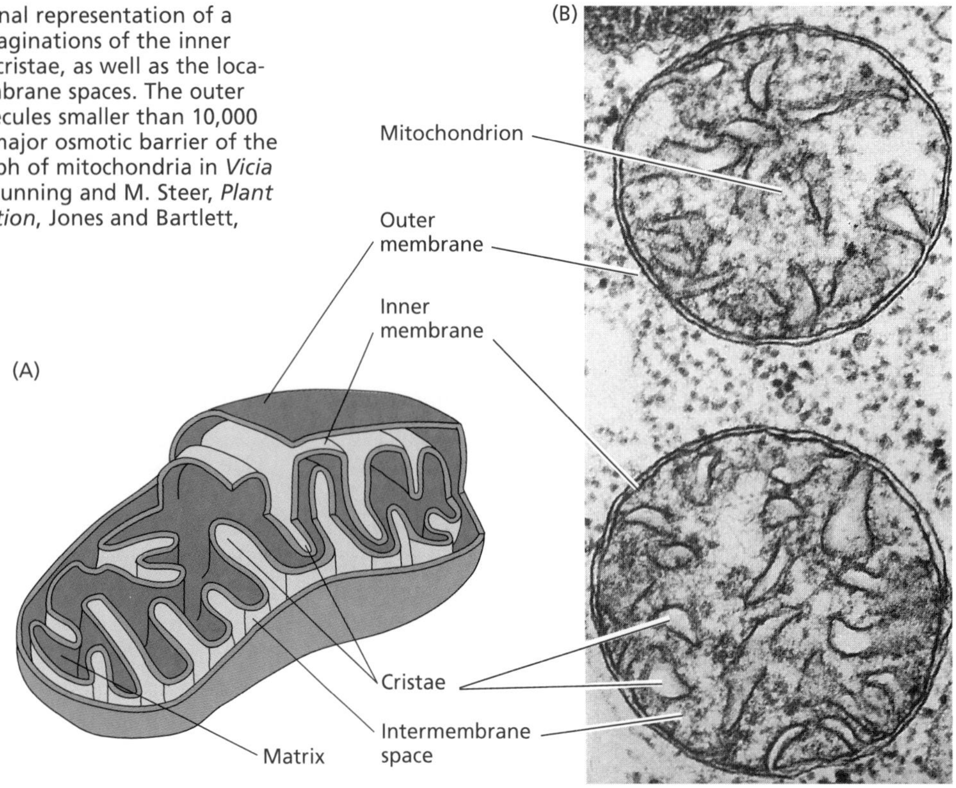

Figure 11.4 (A) Three-dimensional representation of a mitochondrion, showing the invaginations of the inner membrane that give rise to the cristae, as well as the location of the matrix and intermembrane spaces. The outer membrane is permeable to molecules smaller than 10,000 Da; the inner membrane is the major osmotic barrier of the organelle. (B) Electron micrograph of mitochondria in *Vicia faba*. (43,550×) (Photo from B. Gunning and M. Steer, *Plant Cell Biology: Structure and Function*, Jones and Bartlett, 1996.)

Figure 11.4). Plant mitochondria have two membranes: a smooth **outer membrane** that completely surrounds a highly invaginated **inner membrane**. The aqueous phase contained within the inner membrane is referred to as the mitochondrial **matrix**, and the region between the two mitochondrial membranes is known as the **intermembrane space**. Invaginations of the inner membrane give rise to structures known as **cristae**.

Intact mitochondria are osmotically active; that is, they take up water and swell when placed in a hypoosmotic medium. Most inorganic ions and charged organic molecules are not able to diffuse freely into the matrix space. The site of the osmotic barrier is the inner membrane; the outer membrane is permeable to solutes that have a molecular mass less than approximately 10,000 Da (i.e., most cellular metabolites and ions). The lipid fraction of both membranes is made up primarily of phospholipids, 80% of which are either phosphatidylcholine or phosphatidylethanolamine.

Like chloroplasts, mitochondria are semiautonomous organelles, because they contain ribosomes, RNA, and DNA, which encodes a limited number of mitochondrial proteins. Plant mitochondria are thus able to carry out the various steps of protein synthesis and to transmit their genetic information. Mitochondria proliferate through the division by fission of preexisting mitochondria and not through de novo biogenesis of the organelle.

Plant Mitochondrial DNA Has Special Characteristics

One vestige of the endosymbiotic origin of mitochondria and chloroplasts is the presence of two genetic systems in plants in addition to the primary genetic system associated with the nucleus. Many features of the molecular genetics of mitochondria in plants are similar to those found in other eukaryotic organisms (Schuster and Brennicke 1994). The mitochondrial genome (**mtDNA**) of plants encodes most of the tRNAs required for mitochondrial protein synthesis, all three ribosomal proteins, and various components of the electron transport system. The vast majority of mitochondrial proteins, including the TCA cycle enzymes, are encoded by nuclear genes and are imported from the cytosol (see Chapter 1). As with other eukaryotic organisms, the 70S ribosomes that catalyze plant mitochondrial protein synthesis are distinctly prokaryotic, on the basis of their component subunit sizes, protein composition, and antibiotic sensitivities. However, some features of the plant mitochondrial genetic system are not generally found in the mitochondria of animals, protozoans, or even fungi.

Plant mitochondrial genomes can be more than 100 kilobases (kb) in size, but they also vary considerably among different, and even closely related, plants. This size compares with the relatively compact and uniform 16 kb genome found in mammalian mitochondria. In

1992, the complete 186 kb nucleotide sequence of the mitochondrial genome from the liverwort *Marchantia polymorpha*, was reported. The delineation of the liverwort mtDNA sequence verified directly that much of the "extra" DNA found in plant mitochondria consists of noncoding regions, including numerous introns. Mammalian mtDNA is known to encode only 13 proteins. Protein synthesis assays using isolated mitochondria from angiosperms indicate a number somewhat larger, 20 to 30 proteins, but this number is considerably less than the 94 open reading frames identified in *Marchantia*. Whatever the exact number of mitochondrial proteins encoded by plant mtDNA, it still represents only a fraction of the total number of proteins found in plant mitochondria. Another feature unique to the plant mitochondrial genetic system is that it strictly observes the universal genetic code, showing none of the limited number of deviations found in mtDNA in all other kingdoms.

One of the more surprising features of plant mtDNA is the extent to which the nucleotide sequence within its coding regions is not strictly complementary to the mRNA that is translated into the encoded protein product. This lack of complementarity, which under normal circumstances might present a problem, is possible because a considerable amount of **RNA editing** takes place after transcription but before translation of the mRNA. If this editing did not take place, the resulting RNA would not encode the correct amino acid sequence for its particular protein, so the editing is a very important process. RNA editing is found in other mitochondria, but plant mitochondria are noteworthy in the high frequency with which genes that are encoded by the mitochondria are edited. It is not uncommon for 20 to 40 nucleotides to be modified within a single mRNA. Most of the editing involves the conversion of cytosine (C) to uracil (U) on the mRNA as a result of the deamination of the amino group at carbon 6 on the cytosine. Neither the general role of RNA editing nor the reason that plants do it to such a great extent are known at present.

A third feature directly linked to the plant mitochondrial genome is a phenomenon known as **cytoplasmic male sterility**, or ***cms***. Plant lines that display *cms* do not form viable pollen; hence the connotation "male sterility." The term "cytoplasmic" here refers to the fact that this trait is transmitted in a non-Mendelian fashion; the *cms* genotype is always maternally inherited. *cms* is potentially very important from an agricultural perspective because a stable male sterile line can facilitate the production of hybrid seed stock. In most plants, both the mitochondrial and the chloroplastic genomes are maternally inherited. However, all plants carrying the *cms* trait that have been characterized at the molecular level show the presence of distinct rearrangements in their mtDNA, relative to wild-type plants. Such rearrangements in the mitochondrial genome create novel open reading frames and have been strongly correlated with *cms* phenotypes in various systems. The morphological aberrations that indicate *cms*-associated degeneration are usually restricted to tissues of the anther (Conley and Hanson 1995).

The consequences of these changes in mtDNA associated with *cms* manifest themselves only at the specific stage of pollen formation; plants carrying *cms* generally show no major effects of having the trait throughout all other stages of their life cycle. Why pollen development is specifically affected, and why such a broad range of mtDNA rearrangements can result in a common phenotype in so many different plants are questions yet to be answered. Equally interesting is the presence of nuclear restorer genes that are capable of overcoming the effects of the mtDNA rearrangements and restoring to fertility plants that have the *cms* genotype. Such restorer genes are essential for the commercial utilization of *cms* because once the hybrid seed has been produced, plants that develop from the resulting seed usually need to be capable of self-pollinating and forming viable seeds, if seeds are the harvested product.

An interesting consequence of the use of the *cms* gene appeared in the late 1960s, when more than 80% of the hybrid feed corn grown in the United States was derived from the use of a *cms* line of maize called *cms*-T (Texas). In *cms*-T maize, the mtDNA rearrangements give rise to a unique 13 kDa protein, URF13 (Levings and Siedow 1992). How the URF13 protein acts to bring about male sterility is not known, but in the late 1960s a specific race of the fungus *Bipolaris maydis* (formerly *Helminthosporium maydis*) appeared that synthesizes a compound (HmT-toxin) that specifically interacts with the URF13 protein to produce pores in the inner mitochondrial membrane, making it more generally permeable. This interaction between HmT-toxin and URF13 made *Bipolaris maydis* race T a particularly virulent pathogen on *cms*-T maize and led to an epidemic in the corn-growing regions of the United States, especially in the Southeast, that was known as southern corn leaf blight. As a result of this epidemic, the use of *cms*-T in the production of hybrid maize was discontinued. No other *cms* maize has been found to be a suitable replacement, so current production of hybrid corn seed has reverted to relying on manual detasseling to prevent self-pollination.

Pyruvate Enters the Mitochondrion and Is Oxidized via the TCA Cycle

Although mitochondria possess their own DNA, their functions are very much dependent on the nuclear-encoded proteins from the cytosol, which include the all-important TCA cycle enzymes. As already noted, the TCA cycle is also known as the *citric acid cycle,* because of the importance of citrate as an early intermediate (Figure 11.5). This cycle represents the second stage in respiration and takes place in the mitochondrial matrix. Its operation necessitates transport of the pyruvate generated in the cytosol during glycolysis through the impermeable inner mitochondrial membrane. This transport involves a pyruvate (monocarboxylate) translocator that catalyzes an electroneutral exchange of pyruvate and OH^- across the inner membrane (Figure 11.6; see also Figure 11.8).

Once inside the mitochondrial matrix, pyruvate is oxidatively decarboxylated by the enzyme pyruvate dehydrogenase to produce NADH (from NAD^+), CO_2, and acetic acid. The acetic acid is linked via a thioester bond to a sulfur-containing cofactor, coenzyme A (CoA), to form acetyl CoA (see Figure 11.5). Pyruvate dehydrogenase exists as a large complex of several enzymes that catalyze the overall reaction in a three-step process: decarboxylation, oxidation, and conjugation to CoA. After this reaction, the enzyme citrate synthase combines acetyl CoA with a four-carbon dicarboxylic acid, oxaloacetate (OAA), to give a six-carbon tricarboxylic acid, citrate. Citrate is then isomerized to isocitrate by the enzyme aconitase, and the next two reactions are successive oxidative decarboxylations, each of which produces one NADH and releases one molecule of CO_2, ultimately producing a four-carbon molecule, succinyl CoA. At this point, three molecules of CO_2 have been produced for each pyruvate that entered the mitochondrion, so glucose has been completely oxidized.

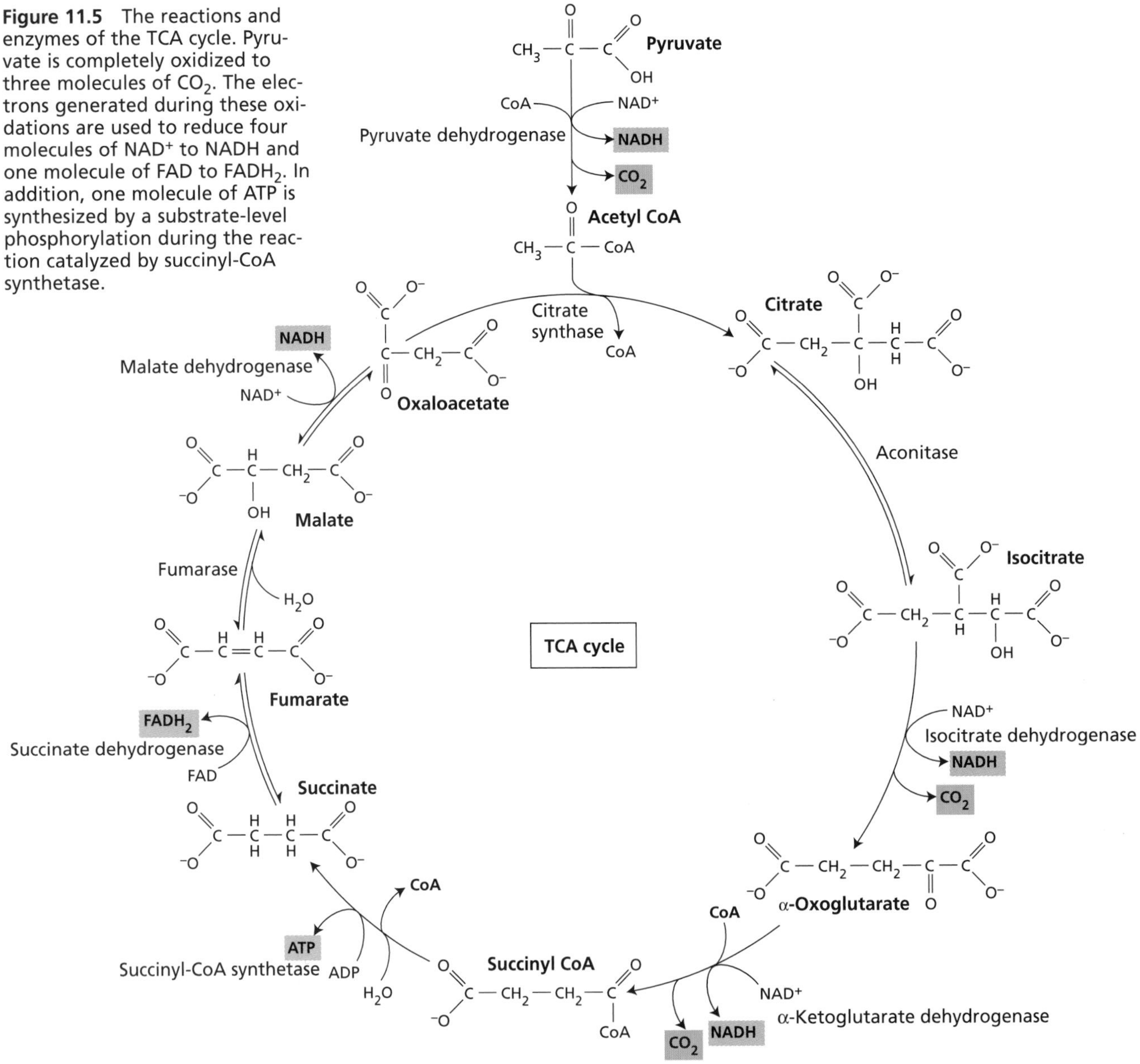

Figure 11.5 The reactions and enzymes of the TCA cycle. Pyruvate is completely oxidized to three molecules of CO_2. The electrons generated during these oxidations are used to reduce four molecules of NAD^+ to NADH and one molecule of FAD to $FADH_2$. In addition, one molecule of ATP is synthesized by a substrate-level phosphorylation during the reaction catalyzed by succinyl-CoA synthetase.

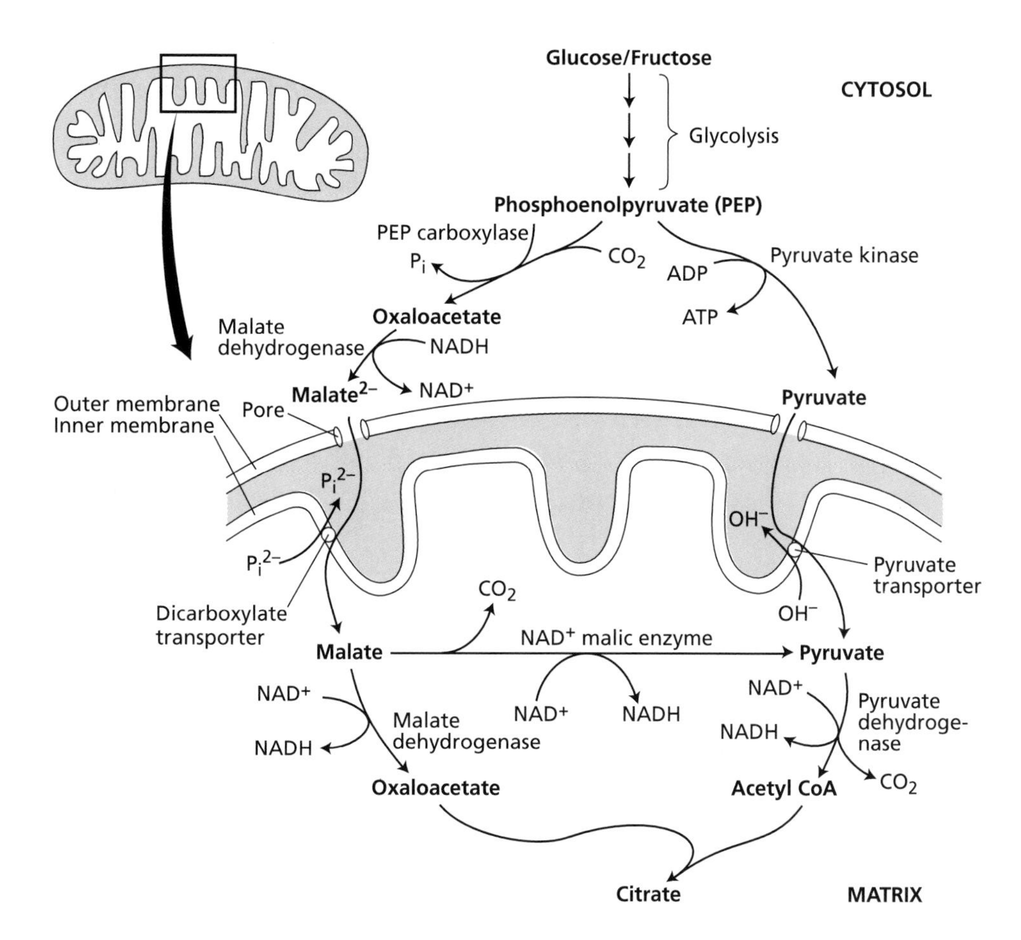

Figure 11.6 Pathways in higher plants by which malate and pyruvate formed in glycolysis move into the mitochondrion and enter the TCA cycle. Pyruvate generated in the cytosol during glycolysis is transported across the inner mitochondrial membrane in an electroneutral exchange with OH^- that involves a pyruvate-specific transporter. Because plant mitochondria contain NAD^+ malic enzyme, they are also capable of completely oxidizing malate (in the absence of added pyruvate). In this pathway, PEP generated in the cytosol during glycolysis is converted to malate, which moves into the mitochondria through the dicarboxylate transporter. This transporter catalyzes the electroneutral exchange of malate (and other dicarboxylates) and P_i^{2-} across the inner mitochondrial membrane. See Figures 11.1, 11.3, and 11.5 for structures of compounds.

During the remainder of the TCA cycle, succinyl CoA is converted to OAA, allowing the continued operation of the cycle. Initially, the large amount of free energy available in the thioester bond of succinyl CoA is conserved through the synthesis of ATP from ADP and P_i via a substrate-level phosphorylation catalyzed by succinyl-CoA synthetase. (Recall that the free energy available in acetyl CoA was used to form a carbon–carbon bond in the step catalyzed by citrate synthase.) The resulting succinate is oxidized to fumarate by succinate dehydrogenase, which is the only membrane-associated enzyme of the TCA cycle and is also considered to be a component of complex II (to be discussed shortly) of the electron transport chain. The electrons removed from succinate end up not on NAD^+ but on another electron-transferring cofactor, FAD (flavin adenine dinucleotide). FAD is covalently bound to the active site of succinate dehydrogenase and undergoes a reversible two-electron reduction to produce $FADH_2$ (FAD + 2 e^- + 2 H^+; see Figure 11.1C). In the final two reactions of the TCA cycle, fumarate is hydrated to produce malate, which is subsequently oxidized by malate dehydrogenase to regenerate OAA and produce another molecule of NADH. The OAA produced is now able to react with another acetyl CoA and continue the cycle.

The stepwise oxidation of pyruvate in the mitochondrion gives rise to three molecules of CO_2, and much of the free energy released during these oxidations is stored in the form of reduced NAD^+ (four NADH) and FAD (one $FADH_2$). In addition, one molecule of ATP is produced by a substrate-level phosphorylation during the TCA cycle. All the enzymes associated with the TCA cycle have been measured in plant mitochondria, and it is likely some of them are associated in a multienzyme complex.

In Plants, the TCA Cycle Has Unique Features

The TCA cycle reactions outlined in Figure 11.5 are not all identical to those carried out by animal mitochondria. For example, the step catalyzed by succinyl-CoA synthetase produces ATP in plants and GTP in animals. In animal mitochondria, ATP is formed in a second enzyme reaction that involves the transfer of a phosphate from GTP to ADP.

Another feature of the TCA cycle that is generally unique to plants is the significant activity of NAD^+ malic enzyme, which has been observed in all plant mitochondria analyzed to date. This enzyme catalyzes the oxidative decarboxylation of malate:

$$\text{Malate} + NAD^+ \rightarrow \text{pyruvate} + CO_2 + \text{NADH}$$

The presence of NAD^+ malic enzyme enables plant mitochondria to operate an alternative pathway for the metabolism of PEP derived from glycolysis. As already described, malate can be synthesized from PEP in the cytosol via the enzymes PEP carboxylase and malate dehydrogenase (see Figures 11.2 and 11.6). The malate is then readily transported into the mitochondrial matrix by a dicarboxylate transporter on the inner membrane that catalyzes the electroneutral exchange of malate and P_i (see Figures 11.6 and 11.8). In the matrix, NAD^+ malic enzyme promotes the metabolism of malate by oxidizing it to pyruvate, which is then oxidized via the TCA cycle as already outlined.

The presence of NAD^+ malic enzyme thus allows the complete oxidation of organic acids (e.g., malate, citrate, 2-oxoglutarate) in the absence of the normal TCA cycle substrate, pyruvate (Wiskich and Dry 1985). The presence of an alternative pathway for the oxidation of malate is consistent with the observation that many plants, in addition to those that carry out Crassulacean acid metabolism (see Chapter 9), store significant levels of malate in their central vacuole.

Electron Transport and ATP Synthesis: Mitochondrial Membrane Processes

Since phosphorylation energy is the form of energy used by cells to drive living processes, the high-energy electrons captured during the TCA cycle (in the form of NADH and $FADH_2$) must be converted to ATP to perform useful work in the cell. This O_2-dependent process occurs on the inner mitochondrial membrane and involves a series of electron carriers known as the *electron transport chain.*

In this section we will describe the process by which the energy of the electrons is lowered in a stepwise fashion, leading to the formation of a proton gradient across the inner mitochondrial membrane. Although fundamentally similar in all aerobic cells, the electron transport chain of plants has certain unique features, such as *cyanide-resistant respiration.* Next we will examine recent exciting advances in our understanding of the enzyme that uses the energy of the proton gradient to synthesize ATP, the F_oF_1-ATP synthase. After examining the various stages in the production of ATP, we will summarize the energy conservation steps at each stage, as well as the regulatory mechanisms that coordinate the different pathways.

Finally, we will briefly describe an alternative pathway for oxidizing glucose and for synthesizing five-carbon sugars (pentoses): the *oxidative pentose phosphate pathway.*

The Electron Transport Chain Catalyzes an Electron Flow from NADH to O_2

For each molecule of glucose oxidized through glycolysis and the TCA cycle pathways, two molecules of NADH are generated in the cytosol, and eight molecules of NADH plus two molecules of $FADH_2$ (associated with succinate dehydrogenase) appear in the mitochondrial matrix. As discussed previously, these reduced compounds must be reoxidized or the entire respiratory process will come to a halt. This is especially true in mitochondria, where the NAD^+ pool size is limited. The electron transport chain catalyzes an electron flow from NADH (and $FADH_2$) to oxygen, the final electron acceptor of the respiratory process. For the oxidation of NADH, the overall two-electron transfer is represented as follows:

$$NADH + H^+ + \frac{1}{2} O_2 \rightarrow NAD^+ + H_2O$$

From the midpoint reduction potentials for the NADH–NAD^+ couple (–320 mV) and the H_2O–$\frac{1}{2}$ O_2 couple (+810 mV), the standard free energy released during this overall reaction ($-nF\Delta E^{0\prime}$) is about 220 kJ mol^{-1} (52 kcal mol^{-1}) per two electrons (see Chapter 2). Because the $FADH_2$–FAD reduction potential (–45 mV) is somewhat higher than that of NADH–NAD^+, only 167.5 kJ mol^{-1} (40 kcal mol^{-1}) is released for each two electrons generated during the oxidation of succinate. The role of the electron transport chain is to bring about the oxidation of NADH (and $FADH_2$) and, in the process, utilize some of the free energy released to generate a $\Delta\tilde{\mu}_{H^+}$ across the inner mitochondrial membrane.

The electron transport chain of plants contains approximately the same general set of electron carriers found in nonplant mitochondria (Figure 11.7). The individual electron transport proteins are organized into a series of four multiprotein complexes (identified by Roman numerals I through IV), each localized in the inner mitochondrial membrane. Electrons from NADH generated in the mitochondrial matrix during the TCA cycle are oxidized by complex I (an NADH dehydrogenase), which in turn transfers these electrons to **ubiquinone**, a *p*-benzoquinone molecule that is within the inner membrane and is not tightly associated with any protein. Chemically and functionally, ubiquinone is similar to plastoquinone in the photosynthetic electron transport chain (see Chapter 7). The electron carriers in complex I include a tightly bound cofactor (flavin mononucleotide, FMN, which is chemically similar to FAD) and several iron–sulfur proteins. The iron–sulfur proteins can be distinguished on the basis of their redox potentials and their electron spin resonance (ESR; see Chapter 7) spectra (Siedow 1995).

The TCA cycle enzyme succinate dehydrogenase is a component of complex II, so electrons derived from the oxidation of succinate are transferred via the $FADH_2$ and a group of three iron–sulfur proteins into the ubiquinone pool. Complex III acts as a ubiquinol:cytochrome *c* oxidoreductase, oxidizing reduced ubiquinone (ubiquinol) and transferring the electrons via an

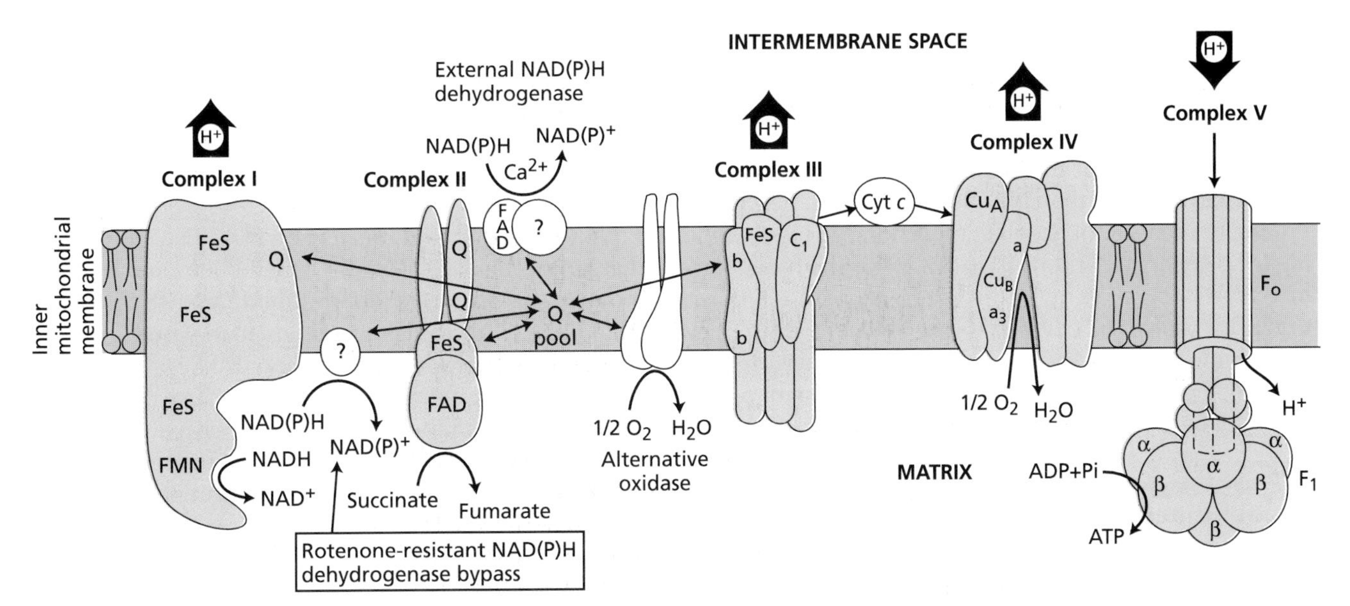

Figure 11.7 Organization of the plant mitochondrial electron transport chain in the inner mitochondrial membrane. In addition to the five standard protein complexes found in all other mitochondria, plant mitochondria contain an alternative oxidase, an alternative rotenone-resistant NAD(P)H dehydrogenase on the matrix side of the membrane, as well as an external NAD(P)H dehydrogenase that can accept electrons directly from NAD(P)H produced in the cytosol. "Q pool" refers to the large pool of ubiquinone that freely diffuses within the inner membrane and serves as the electron carrier that transfers electrons derived from the four dehydrogenases to either complex III or the alternative oxidase. "FMN" is the cofactor flavin mononucleotide. "Cyt *c*" is cytochrome c, a peripheral protein that transfers electrons from complex III to complex IV. Bold, upward-facing arrows designate the three complexes at which protons are translocated, leading to generation of the proton electrochemical gradient that drives ATP synthesis. The downward-facing arrow shows the path of proton flow through the ATP synthase complex that is coupled to the conversion of ADP and P_i to ATP .

iron–sulfur center, two *b*-type cytochromes (b_{565} and b_{560}), and a membrane-bound cytochrome c_1 to cytochrome *c* (Siedow 1995). Cytochrome *c* is the only protein in the electron transport chain that is not an integral membrane protein, and it serves as a mobile carrier to transfer electrons between complexes III and IV. Complex IV represents the cytochrome *c* oxidase, which contains two copper centers, Cu_A and Cu_B) and cytochromes *a* and a_3. Complex IV brings about the four-electron reduction of O_2 to two molecules of H_2O (see Figure 11.7).

The organization of these four complexes on the inner membrane is quite specific (see Figure 11.7). NADH and succinate are oxidized on the matrix side of the membrane, where oxygen is reduced. Cytochrome *c* is loosely bound to the outside of the inner mitochondrial membrane; that is, it faces the intermembrane space. In complex III, the iron–sulfur center, cytochrome c_1, and cytochrome b_{565} all appear to face the outside of the inner membrane, while cytochrome b_{560} of complex III is localized more toward the matrix side of the inner membrane. This organization necessitates transmembrane movement of electrons as they traverse the electron transport chain, and its functional significance will become clear when we consider the mechanism of ATP synthesis later in this chapter.

These four individual electron transfer complexes are not associated into larger "supercomplexes"; rather, electrons are transferred during random collisions between reacting units of the chain (e.g., between complex I and ubiquinone) as they diffuse laterally within the fluid phospholipid bilayer.

Some Plant Electron Carriers Are Absent from Animal Mitochondria

In addition to the collection of electron carriers described in the previous section, plant mitochondria contain some components not commonly found in other mitochondria (see Figure 11.7). One example is the one or more NAD(P)H dehydrogenase complexes that face the intermembrane space of the inner membrane and apparently facilitate the oxidation of cytoplasmic NADH and, possibly, NADPH (Møller et al. 1993). Electrons from these *external* NAD(P)H dehydrogenases enter the main electron transport chain at the level of the ubiquinone pool.

A second unusual feature of plant mitochondria is the presence of two pathways for oxidizing matrix NADH. Electron flow through the classic complex I, described in the previous section, is sensitive to inhibition by several compounds, including rotenone and piericidin. Isolated plant mitochondria consistently display the presence of a rotenone-resistant pathway for the oxidation of NADH derived from TCA cycle substrates. The role of this rotenone-resistant bypass in plant metabolism is still a matter of speculation, but the catalytic properties of this dehydrogenase suggest that one of its primary roles is the oxidation of NADPH formed in the matrix rather than NADH (Møller et al. 1993).

Finally, most, if not all, plant mitochondria have varying levels of an "alternative" pathway for the reduction of oxygen. This pathway involves an oxidase that, unlike cytochrome *c* oxidase, is insensitive to inhibition by cyanide, azide, or carbon monoxide. The nature and physiological significance of this cyanide-resistant pathway will be considered more fully later in the chapter.

ATP Synthesis in the Mitochondrion Is Coupled to Electron Transport

The transfer of electrons to oxygen via complexes I through IV is coupled to the synthesis of ATP, and the number of ATPs synthesized depends on the nature of the electron donor. With isolated mitochondria, electrons derived from internal (matrix) NADH give **ADP:O ratios** (the number of ATPs synthesized per two electrons transferred to oxygen) of 2.4 to 2.7 (Table 11.1). Succinate and externally added NADH each give values in the range of 1.6 to 1.8, while ascorbate, which serves as an artificial electron donor to cytochrome *c*, gives values of 0.8 to 0.9. Results such as these (for both plant and animal mitochondria) have led to the general concept that there are three sites of energy conservation along the electron transport chain, at complexes I, III, and IV. Because of the observed connection between mitochondrial electron flow and ATP synthesis, this conversion of ADP and P_i to ATP became known as **oxidative phosphorylation** long before any of its mechanistic features were understood. In recent years a remarkable amount of progress has been achieved in improving our understanding of the mechanism of this energy transformation.

The currently accepted mechanism of mitochondrial ATP synthesis is based on the chemiosmotic hypothesis, first proposed in 1961 by Nobel laureate Peter Mitchell as a general mechanism of energy conservation across biological membranes (Nicholls and Ferguson 1992). According to this theory, the asymmetric orientation of electron carriers within the mitochondrial inner membrane allows for the transfer of protons (H^+) across the inner membrane during electron flow (Whitehouse and Moore 1995). Numerous studies have confirmed that mitochondrial electron transport is associated with a net transfer of protons from the mitochondrial matrix to the intermembrane space (Figure 11.8).

Because the inner mitochondrial membrane is impermeable to H^+, a proton electrochemical gradient can build up. As discussed in Chapters 6 and 7, the free energy associated with the formation of this *proton electrochemical gradient* ($\Delta\tilde{\mu}_{H^+}$, also referred to as a **proton motive force,** Δp, when expressed in units of volts) is made up of an electric-potential (membrane potential) component (ΔE) and a chemical-potential component (ΔpH) according to the following equation:

$$\Delta p = \Delta E - 59\Delta \text{pH}$$

The ΔE results from the asymmetric distribution of a charged species (H^+) across the membrane, and the ΔpH is due to the proton concentration difference across the membrane. Both values contribute to the proton motive force in plant mitochondria, although ΔE is consistently found to be of greater magnitude. Because protons are translocated from the mitochondrial matrix to the intermembrane space, the orientation of the resulting ΔE across the inner mitochondrial membrane is positive outside and negative inside.

The input of free energy required to generate $\Delta\tilde{\mu}_{H^+}$ comes from the free energy released during electron transport. How electron transport is coupled to proton translocation is not well understood in all cases. Mitchell originally proposed a series of alternating hydrogen (H^+ + e^-) and electron (e^-) carriers traversing the membrane and facilitating the uptake of a proton from one side of the membrane and its release on the opposite side of the membrane. Such a "redox loop" is found, in modified form, for proton translocation at complex III but does not explain the coupling of proton and electron transfer at complexes I and IV, where the linkage between electron transfer and proton translocation is not well understood.

Once generated, the $\Delta\tilde{\mu}_{H^+}$ is reasonably stable, because of the low permeability (conductance) of the inner membrane to protons, and the free energy stored in the proton electrochemical gradient can be utilized to carry out chemical work (ATP synthesis). The $\Delta\tilde{\mu}_{H^+}$ is coupled to the synthesis of ATP by an additional protein complex associated with the inner membrane, the **F_0F_1-ATP synthase** (also called complex V) (see Chapter 7). This complex consists of two major components, F_1 and F_o (see Figure 11.7). F_1 is a peripheral membrane protein complex that is composed of at least five different subunits and contains the catalytic site for converting ADP and P_i to ATP (or hydrolyzing ATP to ADP and P_i when ATPase activity is being measured). This complex is attached to the matrix side of the inner membrane. F_o is an integral

Table 11.1
ADP:O ratios in isolated plant mitochondria

Substrate	ADP:O ratio
Malate	2.4–2.7
Succinate	1.6–1.8
NADH (external)	1.6–1.8
Ascorbate	0.8–0.9

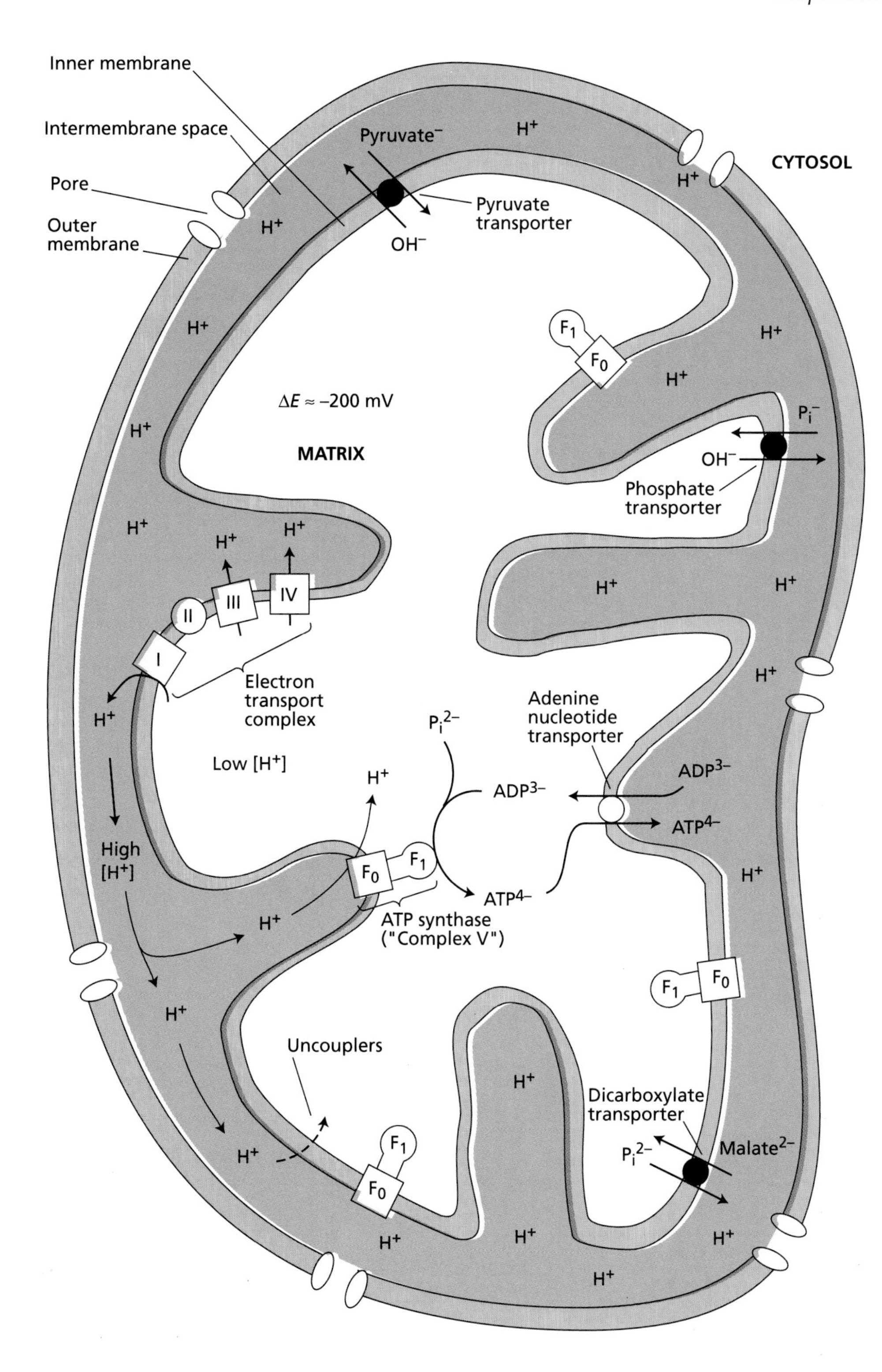

Figure 11.8 A proton electrochemical gradient ($\Delta \tilde{\mu}_{H^+}$) is established across the inner mitochondrial membrane (acidic outside) during electron transport as outlined in the text. The membrane potential component (ΔE) of the gradient drives the electrogenic exchange of ADP in the cytosol for ATP in the mitochondrion via the adenine nucleotide transporter, and the ΔpH drives the electroneutral uptake of P_i through the phosphate transporter. The free energy stored in the proton gradient is then coupled to the synthesis of ATP from ADP and P_i via the many F_0F_1-ATP synthase complexes that line the inner membrane. The F_0 component forms a channel for the movement of protons across the inner membrane; the F_1 complex provides the catalytic site for the condensation of ADP and P_i to ATP. Uncouplers provide a pathway for the movement of protons across the inner membrane, preventing buildup of the proton electrochemical gradient and inhibiting ATP synthesis but not electron transfer. (After Douce 1985.)

membrane protein complex that consists of at least three different polypeptides that form the channel through which protons are able to cross the inner membrane.

The passage of protons through the channel is coupled to the catalytic cycle of the F_1 component of the ATP synthase, allowing the ongoing synthesis of ATP and simultaneously dissipating the $\Delta\tilde{\mu}_{H^+}$. Recently a high-resolution X-ray structure of most of the F_1 complex of the mammalian mitochondrial ATP synthase has supported a "rotational model" for the catalytic mechanism of ATP synthesis (see Box 11.1) (Abrahams et al. 1994). The structure and function of the mitochondrial ATP synthase is similar to that of the CF_o–CF_1 ATPase in photosynthetic photophosphorylation (see Chapter 7).

BOX 11.1

F_0F_1-ATPases: The World's Smallest Rotary Motors

F_0F_1-ATP SYNTHASES (also called F-ATPases) are present in the inner membranes of mitochondria, chloroplasts, and bacteria. As discussed in Chapter 7, these large, multisubunit enzymes consist of a water-soluble catalytic complex (F_1) attached to an integral membrane protein complex (F_o) that transports protons across the membrane (see Figure 7.32). The F_1 complex is composed of at least five different types of subunits (three α, three β, one γ, one δ, and one ε). When the F_1 complex is dissociated from the membrane, it is active as an ATPase. In fact, under the appropriate conditions the intact F_0F_1-ATP synthase can run in reverse and act as a proton pump, using the energy of ATP hydrolysis to move H^+ across the membrane.

Our understanding of how ATP is synthesized was advanced by Paul Boyer at the University of California, Los Angeles, who proposed the *binding-change mechanism* for catalysis by F-ATPases (Figure 1) (Boyer 1989, 1993, 1997). The binding-change mechanism for ATP synthesis contains three important components:

1. The major energy-requiring step is not the synthesis of ATP from ADP and P_i, but the release of ATP from the enzyme.
2. Substrate is bound and products are released at three separate but interacting catalytic sites, corresponding to the three catalytic subunits (β subunits). Each catalytic site can exist in one of three conformations: tight, loose, or open.
3. The binding changes are coupled to proton transport by rotation of the γ subunit. That is, the flow of protons down their electrochemical gradient through the F_0 complex causes the γ subunit to rotate. Rotation of the γ subunit then brings about the conformational changes in the catalytic complex that allow the release of ATP from the enzyme, and the reaction is driven forward. The reverse occurs when the enzyme functions as an ATP-driven proton pump.

The first two predictions of the binding-change model are supported by many lines of evidence, mainly kinetic studies, and are now generally accepted. However, the prediction of a rotary mechanism for coupling proton flow to ATP synthesis has been more difficult to demonstrate. Recently, two major breakthroughs have led to confirmation of the third prediction as well.

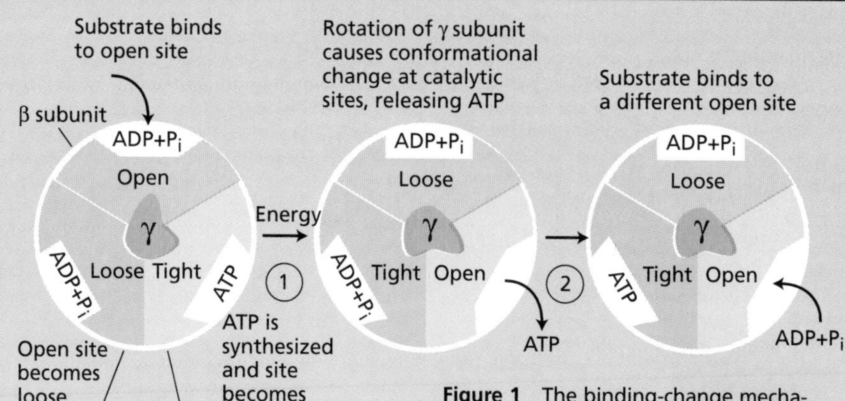

Figure 1 The binding-change mechanism as seen from the top of the F_1 complex. There are three catalytic sites in three different conformations: loose, open, and tight. (For clarity, only the three β subunits are shown.) Substrate (ADP +P_i) initially binds to the open site and is converted to ATP at the tight site. In step 1, rotation of the γ subunit causes a conformational change resulting in a change in the formation of the sites. As a result, ATP is released from the enzyme. In step 2, substrate again binds to the open site, and another ATP is synthesized at the tight site. (After Duncan et al. 1995.)

The first breakthrough was the determination of the crystal structure of the F_1-ATPase of bovine mitochondria by the laboratory of John Walker in Cambridge, England (Abrahams et al. 1994). The crystal structure showed that the three catalytic β subunits differ in their conformations and in the nucleotide bound to them, consistent with the binding-change mechanism. Even more exciting was the discovery that the γ subunit is inserted like a shaft through the center of the catalytic complex, which consists of three α subunits and three β subunits arranged alternately in a doughnut-like structure. Moreover, the interface between the γ subunit and the α and β subunits is highly hydrophobic. The hydrophobicity of the interface minimizes the interactions between the subunits, consistent with the rotation of the γ subunit within the hole formed by the catalytic complex. In other words, the γ subunit looks like a molecular bearing lubricated by a hydrophobic interface.

Although many questions were answered by elucidation of the crystal structure of the F_1-ATPase, the rotational model cannot be tested by a static "snapshot" of the enzyme. Definitive demonstration of rotation requires a video recording of the spinning of the enzyme in real time. But although they are large for proteins, F_1-ATPases are still far too small to be visualized in a light microscope.

To visualize the rotation of the enzyme, Masasuke Yoshida and his colleagues at the Tokyo Institute of Technology came up with an inge-

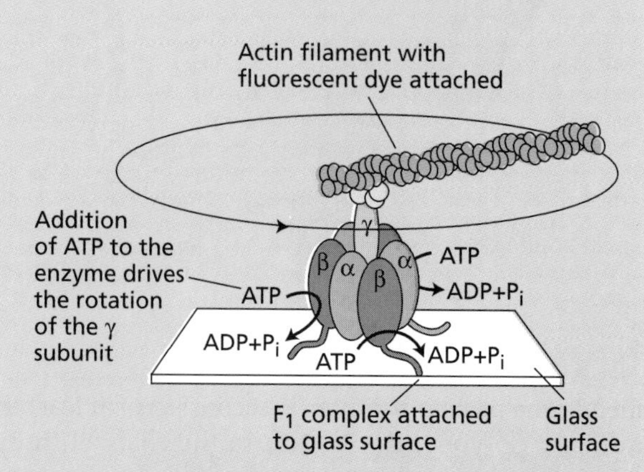

Figure 2 A method for visualizing rotation of the γ subunit. A fluorescently labeled actin filament was attached to one protruding end of the γ subunit. The F_1 complex was then attached upside down to a coverslip. When ATP was added to the coverslip, the actin filament rotated. (After Noji et al. 1997.)

BOX 11.1 *(continued)*

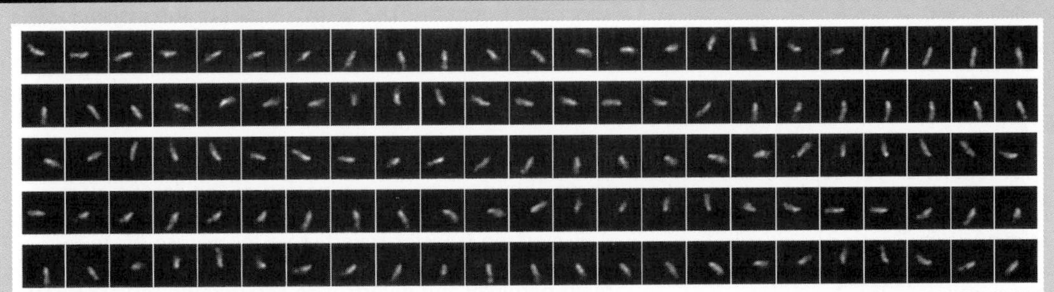

Figure 3 Sequential images of the rotating actin filament attached to the γ subunit, as viewed in a fluorescence microscope. (From Noji et al. 1997, courtesy of S. Noji.)

nious method to make the enzyme much larger than it is (Noji et al. 1997). As Figure 2 shows, they attached an actin filament labeled with a fluorescent dye to the base of the γ subunit using another protein as a "glue." They then attached the F_1 complex upside down to a glass surface. If the γ subunit rotates with respect to the catalytic complex, the actin filament should swing around with it. Since the filament is very long compared to the ATPase (about 1 μm), its rotation should be visible in a fluorescence microscope. In other words, the fluorescently tagged actin filament, which is large enough to visualize in a light microscope, reports the rotation of the γ subunit.

The results were spectacular! When ATP was added to the modified enzyme, the actin filaments were seen to swing around in a circle at about 4 revolutions per second in a fluorescence microscope (Figure 3). To give some idea of scale, if you were a γ subunit, this rotation rate would be equivalent to swinging a rope 60 m long around your head at 4 revolutions per second. However, the measured velocity is undoubtedly an underestimate of the actual velocity in vivo, because of the enormous torque required to swing such a large mass.

Demonstration of the rotary motion of the γ subunit made it possible to put together a model of how the ATP synthase works (Figure 4). For their contributions to elucidation of the mechanism of ATP synthesis, Paul Boyer and John Walker shared the Nobel prize in medicine in 1997.

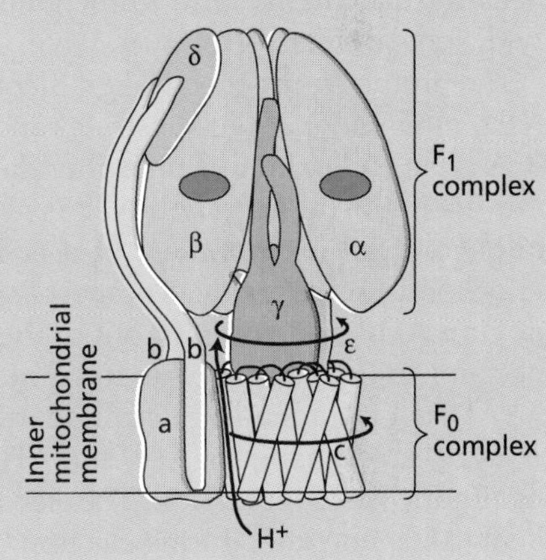

Figure 4 Model of the F_0F_1-ATPase, showing the attachment of the catalytic complex to the membrane via the b subunit and the δ subunit. When the reaction runs in reverse (ATP synthesis), protons diffuse through the F_0 complex down their electrochemical gradient. The movement of protons through the channel drives the rotation of the entire F_0 complex within the membrane. The γ subunit, which is attached to the F_0 complex, then turns within the catalytic complex, causing the conformational changes that are required for ATP synthesis. It is assumed that the catalytic complex itself does not rotate, but is anchored to the membrane. The δ subunit is located on the outside of the β subunit and serves as the site of attachment of the b subunit, which anchors the catalytic complex to the membrane and prevents it from spinning. In mechanical terms, the F_1 complex and its membrane anchor act as a stationary housing, or "stator," while γ subunit (and possibly the F_0 complex) serves as the "rotor." (From Junge et al. 1997.)

The general features of a chemiosmotic mechanism of ATP synthesis have several implications. First, the true site of ATP formation on the mitochondrial inner membrane is the ATP synthase, not complex I, III, or IV. These complexes serve as sites of *energy conservation* whereby electron transport is coupled to the generation of a $\Delta\tilde{\mu}_{H^+}$. In fact, ATP synthesis is not obligatorily linked to electron transport if an alternative method of generating a proton gradient across the inner membrane exists. ATP synthesis in the absence of electron transport can be demonstrated experimentally by inhibiting electron transport and subjecting the mitochondria to an artificially generated pH gradient (Nicholls and Ferguson 1992).

The chemiosmotic theory also explains the mechanism of action of **uncouplers**, a wide range of chemically unrelated compounds (including dinitrophenol, FCCP [*p*-trifluoromethoxycarbonylcyanide phenylhydrazone], and many detergents) that inhibit mitochondrial ATP synthesis but are often observed to stimulate the rate of electron transport. All of these compounds make the inner membrane leaky to protons, which prevents the buildup of a sufficiently large $\Delta\tilde{\mu}_{H^+}$ to drive ATP synthesis. The stimulation of electron transport upon addition of uncouplers is related to an effect seen with isolated mitochondria. In experiments on isolated mitochondria, higher rates of electron flow (measured as the rate of oxygen uptake in the presence of a TCA cycle substrate such as succinate) are observed upon addition of ADP (referred to as *state 3*) than in its absence (*state 4*) (Figure 11.9). ADP provides a substrate that stimulates dissipation of the $\Delta\tilde{\mu}_{H^+}$ through the F_oF_1-ATP synthase during ATP synthesis. Once all the ADP has been converted to ATP, the $\Delta\tilde{\mu}_{H^+}$ builds up again and reduces the rate of electron flow. The ratio of the rates with and without ADP (state 3:state 4) is

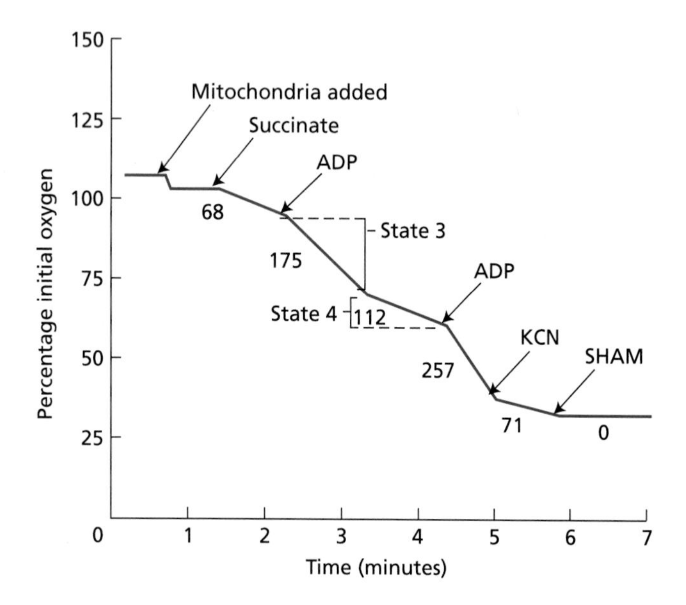

Figure 11.9 Regulation of respiratory rate by ADP in isolated mitochondria from mung bean (*Vigna radiata*) during succinate oxidation. The arrows indicate successive additions of purified mitochondria, 10 m*M* succinate, 150 μ*M* ADP, 450 μ*M* ADP, 0.5 m*M* KCN (potassium cyanide), and 1.0 m*M* SHAM (salicylhydroxamic acid). Addition of succinate initiates mitochondrial electron transfer, which is measured with an oxygen electrode as the rate of oxygen reduction (to H_2O). Addition of ADP stimulates electron transfer (state 3) by facilitating dissipation of the proton electrochemical gradient. Once all the ADP has been converted to ATP, electron transfer reverts to a lower rate (state 4). Addition of cyanide inhibits electron flow through the main pathway and diverts it to the cyanide-resistant pathway, which is subsequently inhibited by the addition of SHAM. The numbers below the traces refer to the rate of oxygen uptake expressed as O_2 consumed (nmol min^{-1} mg $protein^{-1}$). (Courtesy of Steven J. Stegink.)

referred to as the respiratory control ratio (RCR). The RCR and the ADP:O (see Table 11.1) ratio can be used to measure the quality of a mitochondrial preparation (with some caveats; see Douce 1985).

The proton electrochemical gradient also plays a role in the movement of the substrates and products of the TCA cycle and oxidative phosphorylation into and out of mitochondria. Although ATP is synthesized in the mitochondrial matrix, most of it is used outside the mitochondrion, so an efficient mechanism is needed for moving ADP into and ATP out of the organelle. This mechanism involves another inner-membrane protein, the ADP/ATP (adenine nucleotide) transporter, which catalyzes an exchange of ADP and ATP across the inner membrane (see Figure 11.8). The electric-potential gradient (ΔE) generated during electron transfer (negative inside) is such that there will be a net movement of the more negatively charged ATP^{4-} out of the mitochondria in exchange for ADP^{3-}.

Like the export of ATP, the uptake of P_i involves an active phosphate transporter protein that uses the ΔpH component of the proton motive force to catalyze the electroneutral exchange of P_i^{2-} (in) for OH^- (out). As long as a ΔpH is maintained across the inner membrane, the P_i content within the matrix will remain high. Similar reasoning applies to the uptake of pyruvate, which is driven by the electroneutral exchange of pyruvate for OH^-, leading to continued uptake of pyruvate from the cytoplasm (see Figure 11.8).

Aerobic Respiration Yields 32 to 36 Molecules of ATP per Molecule of Hexose

The complete oxidation of a hexose leads to the net formation of four molecules of ATP by substrate-level phosphorylation (two during glycolysis and two in the TCA cycle), two molecules of NADH in the cytosol, and eight molecules of NADH plus two molecules of $FADH_2$ (via succinate dehydrogenase) in the mitochondrial matrix. On the basis of measured ADP:O values (see Table 11.1), a total of approximately 28 molecules of ATP will be generated per hexose by oxidative phosphorylation. The result is a total of 32 ATPs synthesized per hexose. However, the exact number of ATPs synthesized per two electrons remains controversial. The ADP:O ratio is a function of the number of protons translocated per two electrons, multiplied by the number of ATPs synthesized per proton translocated, which need not be an integral value.

The uncertainty of the exact number of protons translocated by the mitochondrial electron transfer chain per two electrons has led some to suggest that the ADP:O ratios in vivo are the rounded-up integral values (i.e., 3.0 and 2.0 for the oxidation of NADH and succinate, respectively), rather than the measured values given in Table 11.1. This assumption accounts for the difference between the value of 32 ATPs per hexose cited here and the value of 36 that is commonly seen in textbooks.

Using 50.2 kJ mol^{-1} (12 kcal mol^{-1}) as the actual free energy of formation of ATP in vivo, we find that about 1606 kJ mol^{-1} (384 kcal mol^{-1}) of free energy is conserved in the form of ATP per mole of glucose oxidized during aerobic respiration. This amount represents about 56% of the total free energy available from the complete oxidation of glucose; the rest is lost as heat. This percentage is a marked improvement over the conversion of only 3 to 5% of the energy available in glucose as ATP that is associated with fermentative metabolism.

Plants Have Cyanide-Resistant Respiration

If cyanide (1 m*M*) is added to actively respiring animal tissues, cytochrome *c* oxidase is inhibited and the respiration rate quickly drops to less than 1% of its initial level. However, most plant tissues display a level of

cyanide-resistant respiration that can represent 10 to 25%, and, in some tissues, up to 100%, of the uninhibited control rate. The enzyme responsible for this oxygen uptake has been identified as a cyanide-resistant oxidase component of the plant mitochondrial electron transport chain called the **alternative oxidase** (Siedow and Umbach 1995).

Current evidence indicates that electrons feed off the main electron transport chain onto the **alternative pathway** at the level of the ubiquinone pool (see Figure 11.7). The alternative oxidase catalyzes a four-electron reduction of oxygen to water and is specifically inhibited by several compounds, most notably salicylhydroxamic acid (SHAM). When electrons pass to the alternative pathway from the ubiquinone pool, two sites of energy conservation (at complexes III and IV) are bypassed, and no ATP is formed. Since there is no energy conservation site on the alternative pathway between ubiquinone and oxygen, the free energy that would normally be stored as ATP is lost as heat when electrons are shunted through the alternative pathway.

Recent discoveries have helped scientists understand how the alternative pathway is regulated (Siedow and Umbach 1995). The alternative oxidase is now known to be associated with a 32 kDa integral membrane protein that, in plants, exists as a dimer of two identical polypeptides and is subject to the action of two regulatory systems. One regulatory system is the reversible oxidation–reduction of a sulfhydryl–disulfide system. In the reduced state (—SH HS—), the oxidase is considerably more active than when the component sulfhydryls are oxidized, giving rise to an intermolecular disulfide bond (—S—S—) that covalently links the two protein monomers. The second regulatory feature is the stimulation of alternative oxidase activity by the action of α-keto acids such as pyruvate or 2-oxoglutarate. This stimulation markedly increases the activity of the oxidase at low concentrations of reduced ubiquinone, allowing the "activated" form of the alternative oxidase to compete effectively with the main pathway for electrons.

How can a process as seemingly energetically wasteful as the cyanide-resistant pathway contribute to plant metabolism? One specific example of its functioning appears during floral development in certain members of the Araceae (the arum family)—for example, the voodoo lily (*Sauromatum guttatum*). Just before pollination, tissues of the clublike inflorescence, called the *appendix*, which bears male and female flowers, exhibit a dramatic increase in the rate of respiration via the alternative pathway, and the temperature of the upper appendix increases by as much as 25°C. This condition lasts for about 7 hours. This extraordinary production of heat volatilizes certain amines and indoles, thereby giving off a putrid odor that attracts insect pollinators. Salicylic acid, a phenolic compound related to aspirin (see Chapter 13), has been identified as the chemical signal responsible for initiating this thermogenic event in *S. guttatum* (Raskin et al. 1989).

In most plants, however, both the respiratory rates and the level of cyanide-resistant respiration are too low to generate significant levels of heat, so what role does the alternative pathway play? It has been suggested that the alternative pathway serves in an "energy overflow" capacity, oxidizing respiratory substrates that accumulate in excess of those needed for growth, storage, or ATP synthesis (Lambers 1985). This view suggests that electrons flow through the alternative pathway only when the activity of the main pathway is saturated. Such saturation is reached in vitro when cyanide is added; in vivo, it could take place if the respiration rate exceeds the cell's demand for ATP (i.e., when ADP is very low, as in state 4; see Figure 11.9). However, studies have shown that the alternative oxidase is able to compete for electrons with an unsaturated main pathway. If this is true, the function of the alternative oxidase in plant tissues needs to be reevaluated.

Although further supporting evidence is required, there are reasons to believe that the cyanide-resistant pathway plays a role in the response of plants to a variety of stresses (chilling, drought, osmotic stress), many of which can inhibit mitochondrial respiration (see Chapter 25) (Wagner and Krab 1995). By draining off electrons from the electron transport chain, the alternative pathway prevents the potential overreduction of the ubiquinone pool (see Figure 11.7), which, if left unchecked, could lead to the generation of such deleterious reactive oxygen species as superoxide anions and hydroxyl radicals. In this way, the alternative pathway may attenuate the inhibitory effects of stress on respiration.

Respiration Is Regulated by Key Metabolites

The control of carbohydrate metabolism in plants is poorly understood, but clearly it operates at numerous levels, starting with the rate at which reduced carbon is synthesized (by photosynthesis) and/or imported into a plant tissue. In addition, because respiration generates ATP, the rate of ATP utilization by a cell or tissue will determine the level of ADP present. The level of ADP appears to be the ultimate influence on the rates of glycolysis in the cytosol, as well as the TCA cycle and oxidative phosphorylation in the mitochondria, as will be described shortly.

Control points exist at all three stages of respiration; here we will give just a brief overview of some major features (see Douce 1985 and Plaxton 1996 for a more detailed discussion). In vivo, glycolysis appears to be regulated at the level of the metabolic conversions catalyzed by the enzymes phosphofructokinase and pyruvate kinase, as it is in animal respiration. ATP, however, which plays a prominent role as an allosteric inhibitor of both the ATP-dependent phosphofructokinase and pyruvate kinase in animals, is not a major effector of these two enzymes in most plant tissues. The cytosolic concentration of PEP, which is a potent inhibitor of the plant ATP-dependent phosphofructokinase, is a more important regulator of plant glycolysis than ATP is.

This inhibitory effect of PEP on phosphofructokinase is strongly reduced by inorganic phosphate, making the cytosolic ratio of PEP to P_i a critical factor regulating plant glycolytic activity. Pyruvate kinase, the enzyme that converts PEP to pyruvate (with the synthesis of ATP) during glycolysis (see Figure 11.2), is in turn sensitive to feedback inhibition by various TCA cycle intermediates, including citrate, 2-oxoglutarate, and malate. In plants, the regulation of glycolysis therefore comes from the "bottom up," with primary regulation at the level of the pyruvate kinase reaction and secondary regulation being exerted at the conversion of fructose-6-phosphate to fructose-1,6-bisphosphate. In animals, primary control exists at phosphofructokinase, secondary control at pyruvate kinase.

However, evidence suggests that the reaction catalyzed by pyruvate kinase can be bypassed in plant cells. For example, the ubiquitous presence of the enzyme PEP carboxylase in the cytosol of plants, coupled with the ability of plant mitochondria to metabolize malate and the presence of a vacuolar PEP phosphatase in plants, makes it possible, in theory, for plant cells to bypass the regulatory features of the pyruvate kinase reaction. Experimental support for multiple pathways of PEP metabolism has come from the study of transgenic tobacco plants lacking cytosolic pyruvate kinase in their leaves (Plaxton 1996). In these transgenic plants, rates of both leaf respiration and photosynthesis were unaffected relative to controls having wild-type levels of pyruvate kinase, indicating that the pyruvate kinase reaction could be circumvented.

Additional complications appear with regulation of the conversion of fructose-6-phosphate to fructose-1,6-bisphosphate. Fructose-2,6-bisphosphate (fructose-2,6-P_2) strongly stimulates PP_i-dependent phosphofructokinase activity. In animals, fructose-2,6-P_2 plays an important regulatory role in glycolysis, stimulating phosphofructokinase activity and inhibiting fructose-1,6-bisphosphatase (see Figure 11.2). Fructose-2,6-P_2 is found in the cytosol of plant tissues, as are the enzymes that act to synthesize and break down fructose-2,6-P_2: fructose-6-phosphate 2-kinase and fructose-2,6-bisphosphatase. Although the role of fructose-2,6-P_2 in the regulation of plant carbohydrate metabolism is not fully understood, several points have been established (Stitt 1990). Fructose-2,6-P_2 is present at varying levels in the cytoplasm. It markedly inhibits the activity of fructose-1,6-bisphosphatase, stimulates the activity of PP_i-dependent phosphofructokinase, and has little or no effect on the activity of the ATP-dependent phosphofructokinase. These observations suggest that fructose-2,6-P_2 plays a central role in the regulation of glycolysis and gluconeogenesis, although further information is needed before we will fully understand regulation at this point of the pathway.

The best-characterized site of regulation of the TCA cycle is at the pyruvate dehydrogenase complex (see Figure 11.5), which is reversibly phosphorylated in response to the actions of a regulatory kinase and a phosphatase. Pyruvate dehydrogenase is inactive in the phosphorylated state, and the regulatory kinase is inhibited by pyruvate, allowing the enzyme to be active when substrate is available. Activation of the phosphorylated enzyme involves a phosphatase activity whose regulatory features are less understood. In addition, several TCA cycle enzymes, including pyruvate dehydrogenase and citrate synthase, are directly inhibited by NADH and ATP.

The primary control of TCA cycle oxidations, and subsequently respiration, seems to be at the level of cellular adenine nucleotides. Which feature is most important is less certain. Both the **adenylate energy charge**—([ATP] + 1/2[ADP])/([ATP] + [ADP] + [AMP])—and the **[ATP]:[ADP] ratio** have been cited as regulatory indicators of the cell's energy status. However, from the limited number of studies of the mechanisms of regulatory control in plant mitochondria, the absolute concentrations of ADP and P_i available in the cytosol appear to be the most important factors in the regulation of mitochondrial respiratory rates.

If we focus on absolute levels of ADP, as the cell's demand for ATP in the cytosol decreases relative to the rate of synthesis of ATP in the mitochondria, less ADP will be available, and the electron transport chain will operate at a reduced rate (see Figure 11.9). This slowdown could be communicated to TCA cycle enzymes through increases in (1) matrix NADH, inhibiting several TCA cycle dehydrogenase activities, and (2) the matrix concentration of ATP, which can directly inhibit both the pyruvate dehydrogenase and citrate synthase reactions, causing a decrease in TCA cycle activity and a buildup of TCA cycle intermediates (Wiskich and Dry 1985). The buildup of TCA cycle intermediates, such as malate, in the cytosol will inhibit the action of pyruvate kinase, increasing the cytosolic PEP concentration, which in turn reduces the rate of conversion of fructose-6-phosphate to fructose-1,6-bisphosphate, thus inhibiting glycolysis.

In summary, plant respiratory rates are controlled from the "bottom up" by the cellular level of ADP. ADP initially regulates the rate of electron transfer and oxidative phosphorylation that in turn regulates TCA cycle activity, which, finally, regulates the rate of the glycolytic reactions.

The decrease in demand for ATP could result from various causes, including exposure of the plant to transient environmental stresses, such as drought or cold. A stress-induced decrease in the electron flow through the main electron transport chain would increase the reduction state of the mitochondrial matrix and increase levels of metabolites such as pyruvate. These changes activate the regulatory systems that modulate alternative oxidase activity, allowing it to compete effectively with the main electron transfer chain for electrons and maintain respiratory rates during the course of the stress. Thus, the alternative oxidase appears to provide a homeostatic mechanism for maintaining the respiratory rate in times of stress. In spite of this general understanding of the regulation of plant respiration, much more needs to be learned about the various controls that modulate plant respiration rates and about the internal and external features that affect them.

Respiration Is Tightly Coupled to Other Pathways

Although much of this chapter has focused on the role of respiration in energy metabolism, glycolysis and the TCA cycle are both linked to some other important metabolic pathways, which will be covered in greater detail in Chapter 13. Glycolysis and the TCA cycle are central to the production of a wide variety of plant metabolites, including amino acids, lipids and related compounds, isoprenoids, and porphyrins (Figure 11.10). Indeed, much of the reduced carbon that is metabolized by glycolysis and the TCA cycle is diverted to biosynthetic purposes and not oxidized to CO_2.

Two additional reactions associated with plant mitochondria are not shown in Figure 11.10. First, two enzymes found in the matrix of mitochondria from chlorophyll-containing tissues—glycine decarboxylase and serine hydroxymethyltransferase—catalyze the CO_2-

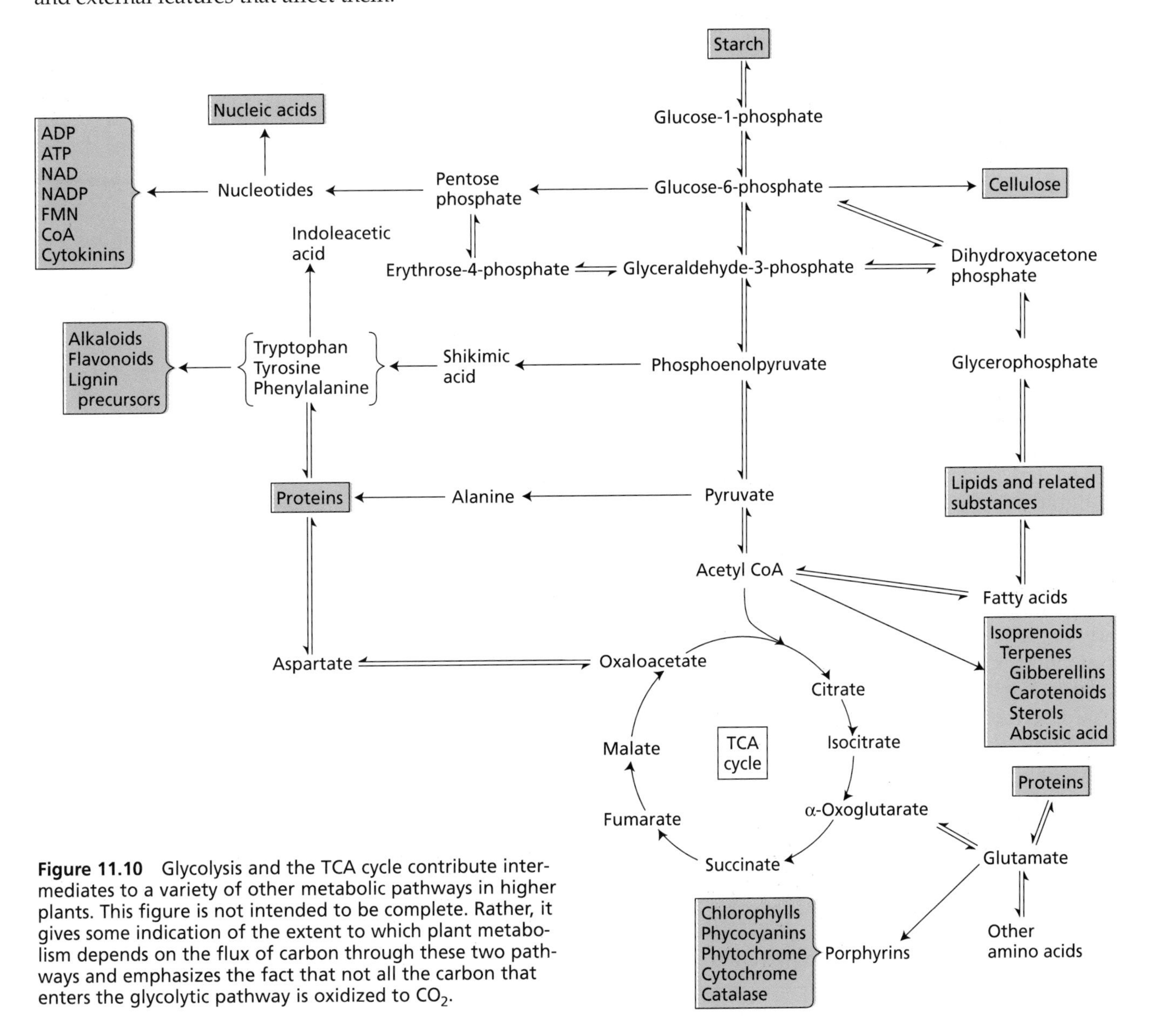

Figure 11.10 Glycolysis and the TCA cycle contribute intermediates to a variety of other metabolic pathways in higher plants. This figure is not intended to be complete. Rather, it gives some indication of the extent to which plant metabolism depends on the flux of carbon through these two pathways and emphasizes the fact that not all the carbon that enters the glycolytic pathway is oxidized to CO_2.

releasing step in photorespiration (see Chapter 9). Second, plant mitochondria provide some of the reactions associated with the utilization of stored fats in developing seeds, which is considered more extensively later in this chapter.

The Pentose Phosphate Pathway Oxidizes Glucose-6-Phosphate and Produces NADPH

The glycolytic pathway is not the only route available for the oxidation of glucose in the cytosol of plant cells. The **oxidative pentose phosphate pathway** (also known as the hexose monophosphate shunt) can also accomplish this task, using enzymes that are soluble in the cytosol (Figure 11.11). The first two reactions of this series represent the oxidative events of the pathway, converting the six-carbon glucose-6-phosphate to a five-carbon sugar, ribulose-5-phosphate, with loss of a CO_2 and generation of two molecules of NADPH (not NADH). The remaining reactions of the pathway convert ribulose-5-phosphate to the glycolytic intermediates glyceraldehyde-3-phosphate and fructose-6-phosphate. Studies of the release of $^{14}CO_2$ from isotopically labeled glucose indicate that glycolysis is the more dominant pathway, accounting for 80 to 95% of the total carbon flux in most plant tissues. However, the pentose phosphate pathway does contribute, and developmental studies indicate that its contribution increases as plant cells go from a meristematic to a more differentiated state (ap Rees 1980).

The oxidative pentose phosphate pathway plays several roles in plant metabolism. (1) The products of the two oxidative steps are NADPH, and this NADPH is thought to drive reductive steps associated with various biosynthetic reactions that occur in the cytosol. Such a role has commonly been accepted for the operation of this pathway in animal tissues. (2) However, because the NADH dehydrogenase facing the cytosol on the mitochondrial inner membrane is also capable of oxidizing NADPH, in plant cells some of the reducing power generated by this pathway may contribute to cellular energy metabolism; that is, electrons from NADPH may end up reducing O_2 and generating ATP. (3) During the early stages of greening, before leaf tissues become fully photoautotrophic, the oxidative pentose phosphate pathway is thought to be involved in generating Calvin cycle intermediates. (4) The pathway also produces ribose-5-phosphate, a precursor of the ribose and deoxyribose needed in the synthesis of RNA and DNA, respectively. (5) Another intermediate in this pathway, the four-carbon erythrose-4-phosphate, combines with PEP in the initial reaction in the production of plant phenolic compounds, including the aromatic amino acids and the precursors of lignin, flavonoids, and phytoalexins (see Chapter 13).

The oxidative pentose phosphate pathway is controlled by the initial reaction of the pathway catalyzed by glucose-6-phosphate dehydrogenase, the activity of which is markedly inhibited as the ratio of NADPH to NADP increases. However, measurements of the enzyme activities of the oxidative pentose phosphate pathway in green tissues are complicated by the fact that many of the same catalytic activities are also associated with chloroplast enzymes that catalyze the reactions of the reductive pentose phosphate pathway, or Calvin cycle (see Chapter 9). Both of the dehydrogenase activities of the oxidative pentose phosphate pathway appear in chloroplasts and other plastids, leading to speculation that this pathway functions in chloroplasts under some conditions, notably in the dark. However, little operation of the oxidative pathway is likely to occur in the chloroplast in the light, because the end products of the pathway, fructose-6-phosphate and glyceraldehyde-3-phosphate, are being synthesized by the Calvin cycle.

Thus, mass action will drive the nonoxidative interconversions of the pathway in the direction of pentose synthesis. Moreover, glucose-6-phosphate dehydrogenase will be inhibited in the light by the high ratio of NADPH to NADP in the chloroplast, as well as by a reductive inactivation involving the ferredoxin–thioredoxin system (see Chapter 9).

Whole-Plant Respiration

Detailed and revealing studies of plant respiratory biochemistry and its regulation have been carried out on isolated organelles and on cell-free extracts of plant tissues. But how does this knowledge relate to the functioning of the whole plant in a natural or agricultural setting? In this section we'll examine respiration in the context of the whole plant under a variety of conditions. First, when green tissues are exposed to light, respiration and photosynthesis operate simultaneously and interact in ways that are not fully understood. Next we will discuss the fact that different tissues have different rates of respiration, which may be under developmental control. Finally, we will look at the influence of various environmental factors (such as O_2, H_2O, and temperature) on respiration rates.

Respiration Operates Simultaneously with Photosynthesis

Many factors can affect the respiration rates of a plant, as well as of the individual organs of the plant. Such factors include the species and growth habit of the plant, the type and age of the specific organ, and environmental variables such as the external oxygen concentration, temperature, and plant water status (see Chapter 25). Whole-plant respiration rates, particularly when considered on a fresh-weight basis, are generally lower than respiration rates reported for animal tissues. This difference is due in large part to the extensive central vacuole and cell wall compartments that contain no mitochondria. Nonetheless, respiration rates in some plant tissues equal the rates observed in actively respiring animal tissues, so the plant respiratory process is not inherently slower than respiration in animals (Seymour et al. 1983).

Relative to the maximum rate of photosynthesis, respiration rates measured in green tissues are far slower, generally by a factor ranging from 6- to 20-fold. Given that rates of photorespiration (see Chapter 8) can often reach 20 to 40 percent of the gross photosynthetic rate, mitochondria-linked respiration operates at rates well below those associated with photorespiration as well. A

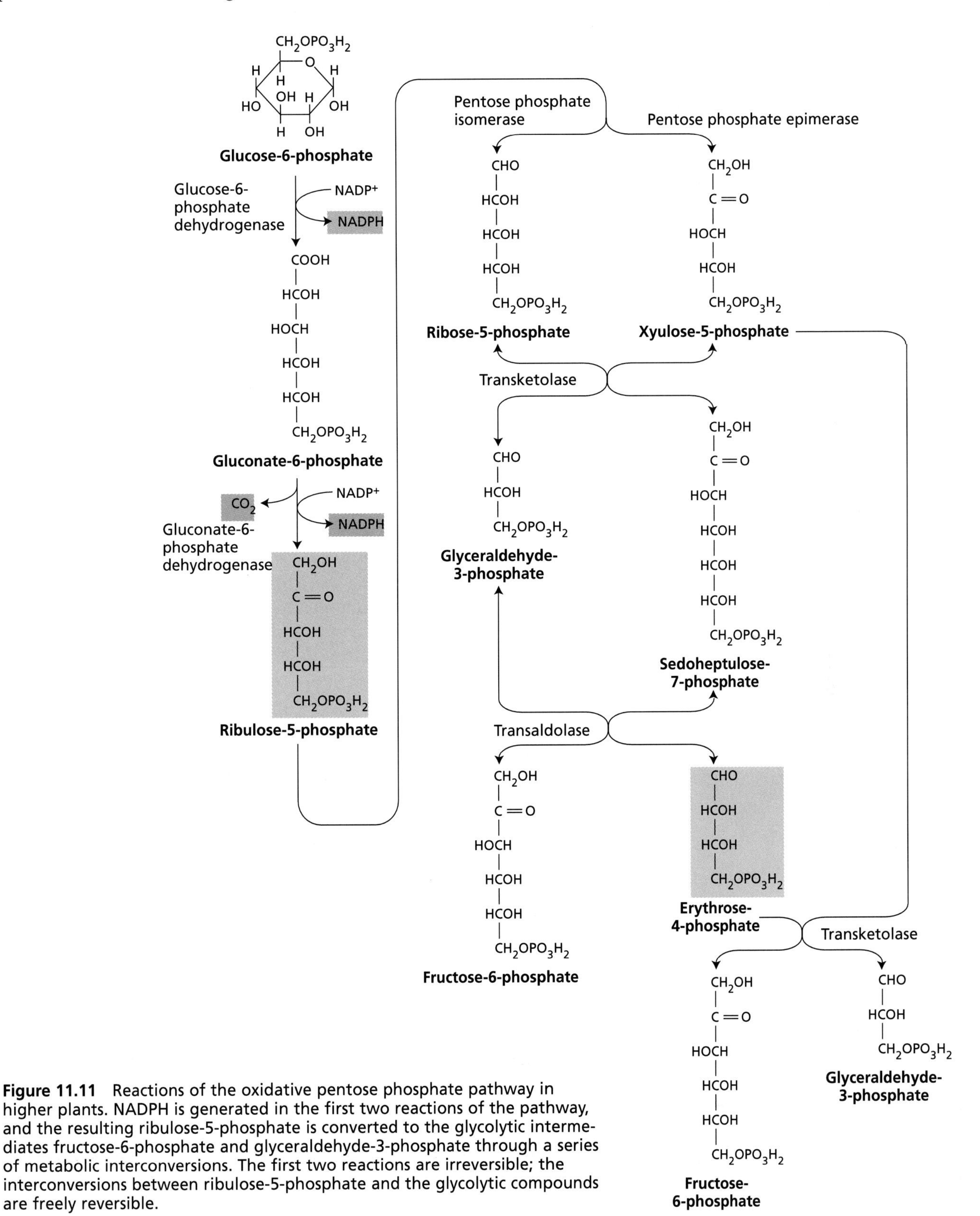

Figure 11.11 Reactions of the oxidative pentose phosphate pathway in higher plants. NADPH is generated in the first two reactions of the pathway, and the resulting ribulose-5-phosphate is converted to the glycolytic intermediates fructose-6-phosphate and glyceraldehyde-3-phosphate through a series of metabolic interconversions. The first two reactions are irreversible; the interconversions between ribulose-5-phosphate and the glycolytic compounds are freely reversible.

question that has not been adequately answered is how much mitochondria-linked respiration (independent of its involvement in the photorespiratory carbon oxidation cycle) operates simultaneously with photosynthesis in illuminated green tissues. It seems clear that the rate of respiration decreases in the light, but to what extent is less certain at present.

Even though plants generally have low respiration rates, the contribution of respiration to the overall carbon economy of the plant can be substantial (see Box 11.2). Nongreen tissues only respire, and they do so 24 hours a day. Even in photosynthetically active tissues, respiration, if integrated over the entire day, can represent a substantial fraction of gross photosynthesis. A survey of several herbaceous species indicated that 30 to 60% of the daily gain in photosynthetic carbon was lost to respiration, although these values tended to decrease with age of the plants (Lambers 1985). Young trees lose roughly a third of their daily photosynthate to respiration, and this loss can double in older trees as the ratio of photosynthetic to nonphotosynthetic tissue decreases. In tropical areas, 70 to 80% of the daily photosynthetic gain can be lost to respiration, because of the high nighttime respiration rates associated with elevated night temperatures.

Different Tissues and Organs Respire at Different Rates

A useful rule of thumb is that the greater the overall metabolic activity of a given tissue, the higher its respiration rate. Developing buds usually show very high rates of respiration (on a dry-weight basis), and respiration rates of vegetative tissues usually decrease from the growing tip to more differentiated regions. In mature vegetative tissues, stems generally have the lowest respiration rates, and leaf and root respiration varies with the plant species and the conditions under which the plants are growing.

When a plant tissue has reached maturity, its respiration rate will either remain roughly constant or decrease slowly as the tissue ages and ultimately senesces. An exception to this behavior is the marked rise in respiration, known as the **climacteric**, that accompanies the onset of ripening in many fruits (avocado, apple, banana) and senescence in detached leaves and flowers. Both ripening and the climacteric respiratory rise are triggered by the endogenous production of ethylene, and both processes can be stimulated to occur prematurely by exogenous application of the gaseous plant hormone ethylene (see Chapter 22). Although ripening in these fruits is associated with increased rates of protein synthesis, which require cellular energy, the exact role of the respiratory rise is not clear. Some fruits (citrus, grape) do not show a respiratory climacteric, and these fruits are also not generally stimulated to ripen by addition of ethylene. Many studies have shown that fruits that exhibit a climacteric respiratory rise display a high level of the cyanide-resistant pathway before ripening, but the exact role of this pathway in ripening is not clear (see Box 11.2).

Environmental Factors Alter Respiration Rates

Oxygen. The environmental factors that can affect plant respiration include oxygen because of its role as a substrate in the overall process. At 25°C, the equilibrium concentration of O_2 in an air-saturated (21% O_2)

BOX 11.2

Does Respiration Reduce Crop Yields?

PLANT RESPIRATION can consume an appreciable amount of the carbon fixed each day during photosynthesis over and above the losses due to photorespiration (see Chapter 9). To what extent can changes in a plant's respiratory metabolism affect crop yields? Attempts to establish a quantitative relation between respiratory energy metabolism and the various processes going on in the cell have led to a splitting of respiration into two components (Lambers 1985). In **growth respiration**, reduced carbon is processed to bring about the addition of new biomass. **Maintenance respiration** is the component of respiration needed to keep existing, mature cells in a viable state. Utilization of energy by the latter is not well understood, but estimates indicate that it can represent more than 50% of the total respiratory flux. In addition, there is the specter of the cyanide-resistant pathway, with its potential for utilizing considerable amounts of the cell's reduced carbon to no apparent useful end.

Although numerous questions remain regarding these issues, there are several examples of empirical relations between plant respiration rates and crop yield. In the forage crop perennial ryegrass (*Lolium perenne*), yield increases of 10 to 20% were correlated with a 20% decrease in the leaf respiration rate (Wilson and Jones 1982). Similar correlations have been found for other plants, including corn (*Zea mays*)and tall fescue (*Festaca arundinacea*) (Lambers 1985). Although there is a potential for increasing crop yield through reduction of respiration rates, a better understanding of the sites and mechanisms that control plant respiration is needed before such changes can be exploited commercially in a logical fashion by plant physiologists, geneticists, and molecular biologists. Furthermore, much remains to be established regarding the general applicability of such observations and the conditions under which slower-respiring lines might actually be at a disadvantage, causing a reduction in crop yields rather than an increase.

aqueous solution is about 265 μ*M*. The value of K_m for oxygen in the reaction catalyzed by cytochrome *c* oxidase is difficult to measure exactly, but it is well below 1 μ*M*, so there should be no apparent dependence of the respiration rate on external O_2 concentrations (see Chapter 2 for discussion of K_m). In fact, little effect on measured respiration rates is observed until the atmospheric oxygen concentration drops below 5% for whole tissues or below 2 to 3% for tissue slices.

Isolated mitochondria are unaffected by oxygen depletion approaching 0%. That there is any effect of oxygen concentration is due to limitations on the diffusion of oxygen through the aqueous phase between the atmosphere and the mitochondria. The diffusion coefficient of oxygen dissolved in an aqueous medium is 10^4 less than that of oxygen diffusing through air. The fact that atmospheric oxygen tensions as high as 5% can reduce the observed respiration rate in whole tissues indicates that the diffusive movement of oxygen represents a limitation on plant respiration. Further, the diffusion limitations of the aqueous pathway for the supply of oxygen imply that the intercellular air spaces present in plant tissues are important in facilitating oxygen movement within plant tissues. If there were no gaseous diffusion pathway throughout the plant, the cellular respiration rates of many plants would be limited by an insufficient supply of oxygen.

This diffusion limitation is even more significant when the medium in which the tissue is growing is liquid. When plants are grown hydroponically, the solutions must be aerated vigorously to keep oxygen levels high in the vicinity of the roots. The problem of oxygen supply also arises with plants growing in predominantly wet or flooded soils (see Chapter 25). Some plants, particularly trees, are restricted in geographic distribution by the need to maintain a supply of oxygen to their roots. For instance, dogwood and tulip tree poplar can survive only in well-drained, aerated soils because their roots are unable to tolerate more than a limited exposure to a flooded condition. However, many other plants can adapt to growth in flooded soils. Herbaceous species such as rice and sunflower often rely on a network of intercellular air spaces running from the leaves to the roots to provide a continuous, gaseous pathway for the movement of oxygen to the flooded roots.

The problem can be even more acute for trees that have very deep roots growing in wet soils. Such roots must survive on anaerobic (fermentative) metabolism and/or develop structures that facilitate the movement of oxygen to the roots. Examples of the latter are outgrowths of the roots, called **pneumatophores**, that protrude out of the water and provide a gaseous pathway for oxygen diffusion into the roots. Pneumatophores are found in *Avicennia* and *Rhizophora*, trees that grow in mangrove swamps under continuously flooded conditions. While superficially similar in appearance to mangrove pneumatophores, the "knees" of cypress (*Taxodium*) do not perform such a function.

Although lowering the external oxygen concentration ultimately reduces the rate of aerobic respiration, it does not necessarily decrease the rate of carbohydrate utilization. In 1861, Louis Pasteur discovered that yeast cells consume less glucose in the presence of air than under anaerobic conditions. This response, known as the **Pasteur effect**, occurs in plant tissues and manifests itself as an increased rate of fermentative metabolism under anaerobic conditions. In aerobic respiration, cytosolic levels of PEP and ATP are high, thus inhibiting the regulatory enzymes of the glycolytic pathway. In tissues maintained under anaerobic conditions, the cytosolic PEP and ATP levels and their attendant inhibitory effects are decreased, leading to higher rates of hexose utilization.

Temperature. In the physiological temperature range, respiration is temperature dependent. The increase in respiration rate for every 10°C increase in ambient temperature, commonly referred to as the Q_{10}, is slightly greater than 2, although in some plants an unexplained decrease in the Q_{10} has been observed at temperatures above 30°C. High night temperatures are thought to account for the large respiratory component in tropical plants, and lowered temperatures are utilized to retard postharvest respiration rates during the storage of fruits and vegetables. However, complications may arise from this practice. For instance, when potato tubers are stored at temperatures above 10°C, respiration and ancillary metabolic activities are sufficient to allow sprouting. Below 5°C, respiration rates and sprouting are reduced in most tissues, but the breakdown of stored starch and its conversion to sucrose impart an unwanted sweetness to the tubers. As a compromise, potatoes are stored at 7 to 9°C, which prevents the breakdown of starch while minimizing respiration and germination.

CO_2 concentration. One process that takes advantage of the effects of atmospheric oxygen and temperature on respiration is fruit storage, in which it is common to maintain fruits in the cold under conditions of 2 to 3% oxygen and 3 to 5% CO_2. The reduced temperature lowers the respiration rate, as does the reduced oxygen. Low levels of oxygen are used instead of anoxic conditions to avoid lowering tissue oxygen tensions so far as to stimulate fermentative metabolism. Carbon dioxide has a limited direct inhibitory effect on the respiration rate at a concentration of 3 to 5%, which is well in excess of the 0.036% normally found in the atmosphere. More specifically, high concentrations of CO_2 are used in fruit storage to block the effect of the hormone ethylene on fruit ripening (see Chapter 22).

Ions. Addition of ions to plants previously grown in distilled water stimulates respiration—a phenomenon called **salt respiration**. The simplest explanation for this effect is that respiration-linked ATP is required to support enhanced ion uptake, but other factors also seem to be involved.

Injury. Physical injury to plant tissues often stimulates oxygen uptake because of increases in respiration (mitochondria-linked oxygen uptake) and in nonmitochondrial oxygen-consuming enzyme activities (such as reactions catalyzed by lipoxygenase, polyphenol oxidase, and peroxidase; see Chapter 13). For example, the rapid browning observed when apple fruits or potato tubers are cut results from the cellular damage that leads to the mixing of several of these oxidative enzymes with their substrates. It is important to recognize that many enzymes in plant tissues use molecular oxygen as a substrate and can therefore contribute to the "apparent" respiration rate when the latter is measured as oxygen consumption, but not when respiration is considered in terms of either utilization of reduced carbon or release of CO_2.

Lipid Metabolism

Starch and sucrose are not the only substrates available for energy production in the plant. Fats and oils are important storage forms of reduced carbon in many seeds, including those of agriculturally important species such as soybean, sunflower, peanut, and cotton. Oils often serve a major storage function in nondomesticated plants that produce small seeds. Some fruits, such as olives and avocados, also store fats and oils. In this final part of the chapter we describe the biosynthesis of two types of glycerolipids: the *triacylglycerols* (the fats and oils stored in seeds) and the *polar glycerolipids* (which form the lipid bilayers of cellular membranes) (Figure 11.12). We will see that the biosynthesis of triacylglycerols and polar glycerolipids requires the cooperation of two organelles: the plastids and the endoplasmic reticulum. Finally, we will examine the complex process by which germinating seeds obtain metabolic energy from the oxidation of fats and oils.

Fats and Oils Store Large Amounts of Energy

Fats and oils belong to the general class lipids, a structurally diverse group of hydrophobic compounds that are soluble in organic solvents and highly insoluble in water. They represent a more reduced form of carbon than carbohydrates do, so the complete oxidation of 1 g of fat or oil (which contains about 40 kJ or 9.3 kcal of energy) can produce considerably more ATP than the oxidation of 1 g of starch (about 15.9 kJ or 3.8 kcal). Conversely, the biosynthesis of fats, oils, and related molecules, such as the phospholipids of membranes, requires a correspondingly large investment of metabolic energy. Other lipids are important for plant structure and function but are not used for energy storage. Included are waxes, which make up the protective cuticle that reduces water loss from exposed plant tissues, and terpenoids (also known as isoprenoids), which include carotenoids involved in photosynthesis and sterols present in many plant membranes (see Chapter 13).

Triacylglycerols Are Stored in Oleosomes

Fats and oils exist mainly in the form of **triacylglycerols** ("acyl" refers to the fatty acid portion), or *triglycerides*, in which fatty acid molecules are linked by ester bonds to the three hydroxyl groups of glycerol (see Figure 11.12). The fatty acids in plants are usually straight-chain carboxylic acids having an even number of carbon atoms. The carbon chains can be as short as 12 units and as long as 20, but more commonly they are 16 or 18 carbons long. Oils are liquid at room temperature, primarily because of the presence of unsaturated bonds in their component fatty acids; fats, which have a higher proportion of saturated fatty acids, are solid at room temperature. The major fatty acids in plant lipids are shown

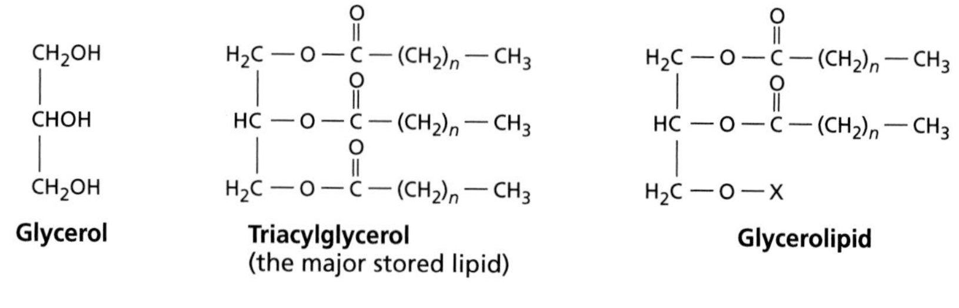

CH_2OH — $CHOH$ — CH_2OH
Glycerol

$H_2C—O—C(=O)—(CH_2)_n—CH_3$
$HC—O—C(=O)—(CH_2)_n—CH_3$
$H_2C—O—C(=O)—(CH_2)_n—CH_3$
Triacylglycerol
(the major stored lipid)

$H_2C—O—C(=O)—(CH_2)_n—CH_3$
$HC—O—C(=O)—(CH_2)_n—CH_3$
$H_2C—O—X$
Glycerolipid

X = H	Diacylglycerol (DAG)
$X = HPO_3^{2-}$	Phosphatidic acid
$X = PO_3^{2-}—CH_2—CH_2—\overset{+}{N}(CH_3)_3$	Phosphatidylcholine
$X = PO_3^{2-}—CH_2—CH_2—NH_2$	Phosphatidylethanolamine
X = galactose	Galactolipids

Figure 11.12 The structural features of triacylglycerols and polar glycerolipids in higher plants. The carbon chain lengths of the fatty acids, which always have an even number of carbons, range from 12 to 20 but are typically 16 or 18. Thus, the value of *n* is usually 14 or 16.

in Table 11.2. The composition of fatty acids in plant lipids varies with the species. For example, peanut oil is about 9% palmitic acid, 59% oleic acid, and 21% linoleic acid, while cottonseed oil is 20% palmitic acid, 30% oleic acid, and 45% linoleic acid. The biosynthesis of these fatty acids will be discussed shortly.

Triacylglycerols in most seeds are stored in the cytoplasm of either cotyledon or endosperm cells in organelles known as **oleosomes** (also called *spherosomes* or *oil bodies*) (see Chapter 1). Oleosomes have an unusual membrane barrier that separates the triglycerides from the aqueous cytoplasm. A single layer of phospholipids (i.e., a half-bilayer) surrounds the oil body with the hydrophilic ends of the phospholipids exposed to the cytosol and the hydrophobic acyl hydrocarbon chains facing the triacylglycerol interior (see Chapter 1). The oleosome is stabilized by the presence of specific proteins, called **oleosins**, that coat the surface and prevent the phospholipids of adjacent oil bodies from coming in contact and fusing (Huang 1992). This unique structure appears to result from the pattern of triacylglycerol biosynthesis. Triacylglycerol synthesis is completed by enzymes located in the membranes of the endoplasmic reticulum (ER), and the resulting fats accumulate between the two monolayers of the ER membrane bilayer. The bilayer swells apart as more fats are added to the growing structure, and ultimately a mature oil body buds off from the ER (Napier et al. 1996).

Polar Glycerolipids Are the Main Structural Lipids in Membranes

As outlined in Chapter 1, each membrane in the cell is a bilayer of amphipathic (i.e., having both hydrophilic and hydrophobic regions) lipid molecules in which a polar head group interacts with the aqueous phase while hydrophobic fatty acid chains form the center of the membrane. This hydrophobic core prevents random diffusion of solutes between cell compartments and thereby allows the biochemistry of the cell to be organized.

The main structural lipids in membranes are the *polar glycerolipids* (see Figure 11.12), in which the hydrophobic portion consists of two 16-carbon or 18-carbon fatty acid chains esterified to positions 1 and 2 of a glycerol backbone. The polar head group is attached to position 3 of the glycerol. There are two categories of polar glycerolipids: **glyceroglycolipids**, in which sugars form the head group; and **glycerophospholipids**, in which the head group contains phosphate (Figure 11.13).

There are additional structural lipids in plant membranes, including sphingolipids and sterols (see Chapter 13), but these are minor components. Other lipids perform specific roles in photosynthesis and other processes. Included among these lipids are chlorophylls, plastoquinone, carotenoids, and tocopherols, which together account for about one-third of the lipids in plant leaves. There are eight major glycerolipid classes in plants, each of which can be associated with many different fatty acid combinations. The structures shown in Figure 11.13 illustrate some of the more common molecular species.

Chloroplast membranes, which account for 70% of the membrane lipids in photosynthetic tissues, are dominated by glyceroglycolipids, while other membranes of the cell contain glycerophospholipids (see Table 11.3). In nonphotosynthetic tissues, phospholipids are the major membrane glycerolipids.

Fatty Acid Biosynthesis Consists of Cycles of Two-Carbon Addition

Fatty acid biosynthesis involves the cyclic condensation of two-carbon units in which acetyl CoA is the precursor. In plants, fatty acids are synthesized exclusively in the plastids; in animals, fatty acids are synthesized pri-

Table 11.2
Common fatty acids in higher plant tissues

Name	Structure
SATURATED FATTY ACIDS	
Lauric acid (12:0)[a]	$CH_3(CH_2)_{10}CO_2H$
Myristic acid (14:0)	$CH_3(CH_2)_{12}CO_2H$
Palmitic acid (16:0)	$CH_3(CH_2)_{14}CO_2H$
Stearic acid (18:0)	$CH_3(CH_2)_{16}CO_2H$
UNSATURATED FATTY ACIDS	
Oleic acid (18:1)	$CH_3(CH_2)_7CH{=}CH(CH_2)_7CO_2H$
Linoleic acid (18:2)	$CH_3(CH_2)_4CH{=}CH{-}CH_2{-}CH{=}CH(CH_2)_7CO_2H$
Linolenic acid (18:3)	$CH_3CH_2CH{=}CH{-}CH_2{-}CH{=}CH{-}CH_2{-}CH{=}CH{-}(CH_2)_7CO_2H$

[a] Each fatty acid has a numerical abbreviation. The number before the colon represents the total number of carbons, while the number after the colon is the number of double bonds.

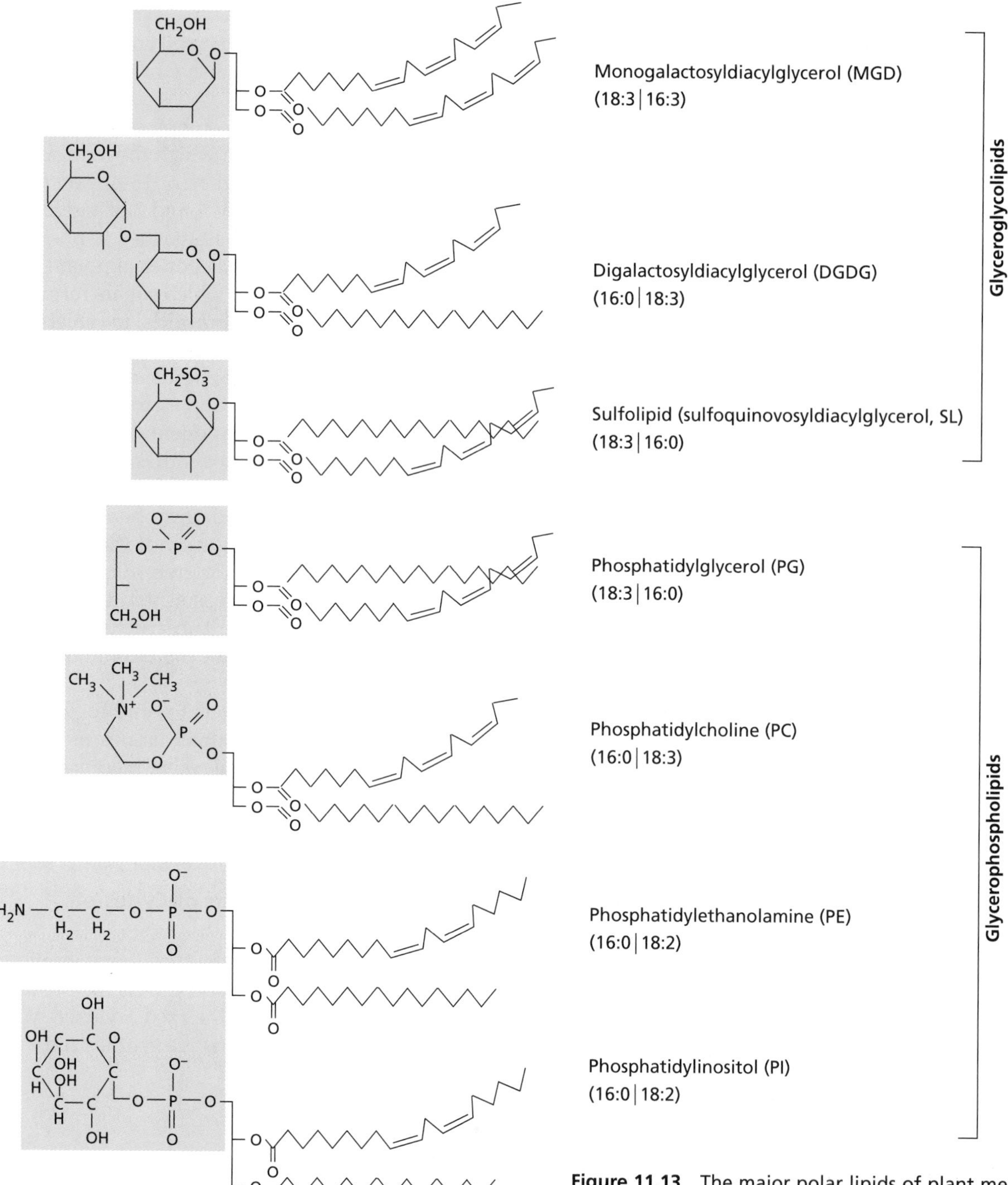

Figure 11.13 The major polar lipids of plant membranes: glyceroglycolipids and glycerophospholipids. At least six different fatty acids may be attached to the glycerol backbone. One of the more common molecular species is shown for each lipid. The numbers given below each name refer to the number of carbons (number before the colon) and the number of double bonds (number after the colon).

marily in the cytosol. The enzymes of the pathway are thought to be held together in a complex, collectively referred to as **fatty acid synthase**. The complex probably allows the series of reactions to occur more efficiently than it would if the enzymes were physically separated from each other. In addition, the growing acyl chains are tethered to a low-molecular-weight, acidic protein called **acyl carrier protein** (**ACP**). When conjugated to the acyl carrier protein, the fatty acid chain is referred to as **acyl-ACP**.

The first committed step in the pathway (i.e., the first step unique to the synthesis of fatty acids) is the synthesis of **malonyl CoA** from acetyl CoA and CO_2 by the enzyme *acetyl-CoA carboxylase* (Figure 11.14) (Sasaki et al. 1995). The activity of acetyl-CoA carboxylase is probably tightly regulated to determine the overall rate of

Table 11.3
Glycerolipid components of cellular membranes

	Lipid composition (percentage of total)		
	Chloroplast	Endoplasmic reticulum	Mitochondrion
Phosphatidylcholine	4	47	43
Phosphatidylethanolamine	—	34	35
Phosphatidylinositol	1	17	6
Phosphatidylglycerol	7	2	3
Diphosphatidylglycerol	—	—	13
Monogalactosyldiacylglycerol	55	—	—
Digalactosyldiacylglycerol	24	—	—
Sulfolipid	8	—	—

fatty acid synthesis (Ohlrogge and Jaworski 1997). In the first cycle of fatty acid synthesis, the acetate group from acetyl CoA is transferred to a specific cysteine of **condensing enzyme** (3-ketoacyl-ACP synthase) and then combined with **malonyl-ACP** to form **acetoacetyl-ACP**. Next, the keto group at carbon 3 is removed (reduced) by the action of three enzymes to form a new acyl chain (**butyryl-ACP**), which is now four carbons long (see Figure 11.14). The four-carbon acid and another molecule of malonyl-ACP then become the new substrates for

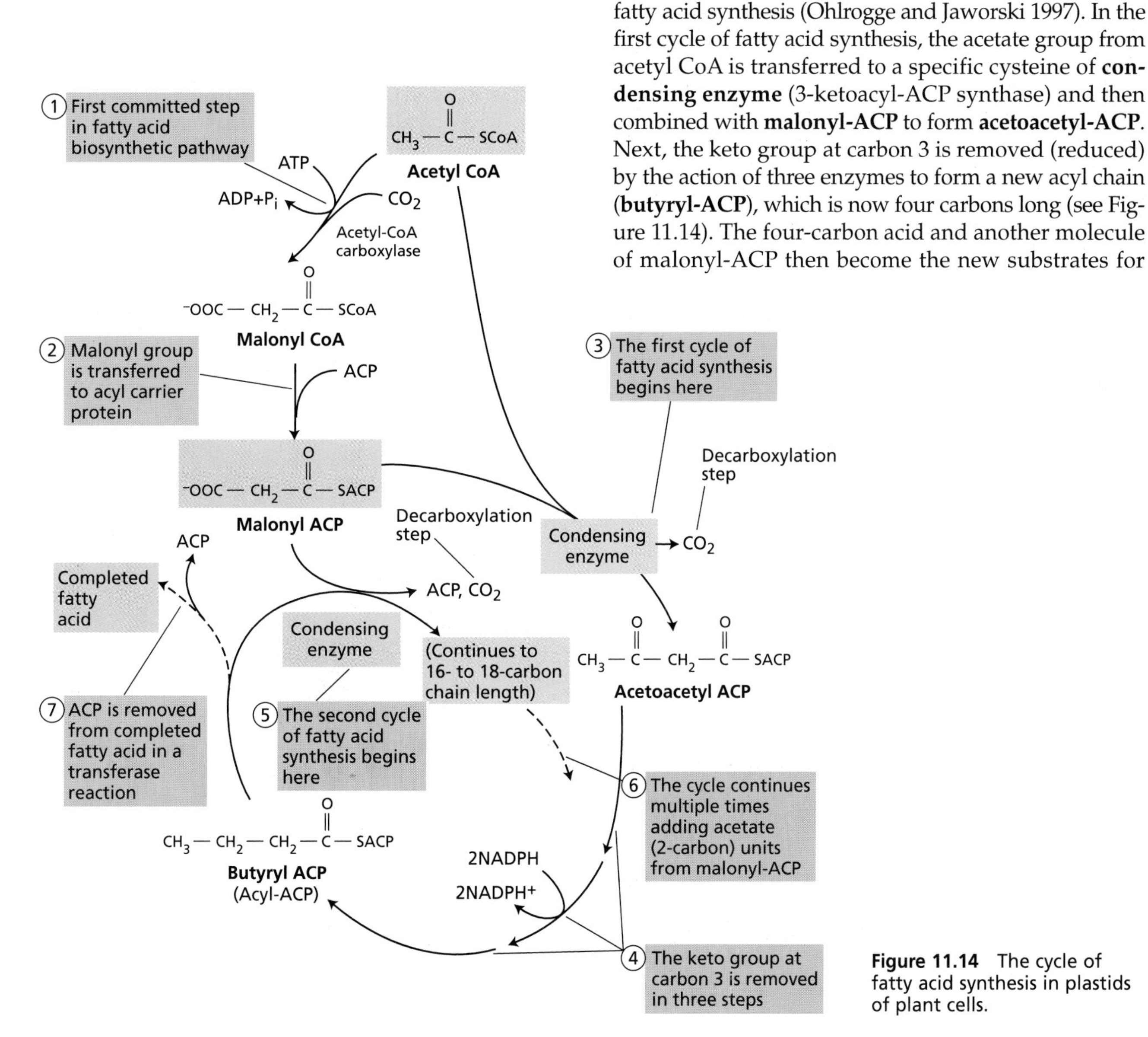

Figure 11.14 The cycle of fatty acid synthesis in plastids of plant cells.

condensing enzyme, resulting in the addition of another two-carbon unit to the growing chain, and the cycle continues until 16 or 18 carbons have been added.

Some 16:0-ACP is released from the fatty acid synthase machinery, but most molecules that are elongated to 18:0-ACP are efficiently converted to 18:1-ACP by a stromal stearoyl-ACP desaturase (Lindqvist et al. 1996). Thus, 16:0-ACP and 18:1-ACP are the major products of fatty acid synthesis in plastids (see Figure 11.16).

Glycerolipids Are Synthesized in the Plastids and the ER

The fatty acids synthesized in the plastid are next used to make the glycerolipids of membranes and oleosomes. The first steps of glycerolipid synthesis are two acylation reactions that transfer fatty acids from acyl-ACP or acyl CoA to glycerol-3-phosphate to form phosphatidic acid. Diacylglycerol (DAG) is produced from phosphatidic acid by a specific phosphatase. Phosphatidic acid can be converted to phosphatidylinositol or phosphatidylglycerol, while DAG can give rise to phosphatidylethanolamine or phosphatidylcholine (see Figure 11.15).

In the last 15 years, information on localization of the enzymes of glycerolipid synthesis has revealed an intriguingly complex and highly regulated interaction between the chloroplast, where fatty acids are synthesized, and other membrane systems of the cell. In simple terms, the biochemistry involves two pathways. The 16:0- and 18:1-ACP products of chloroplast fatty acid synthesis either may be used directly for production of chloroplast lipids by a pathway in the plastids (the "prokaryotic" pathway) or may be exported to the cytoplasm as CoA esters, which then are incorporated into lipids in the endoplasmic reticulum by an independent set of acyltransferases (the "eukaryotic" pathway) (Browse and Somerville 1991). A simplified version of this model is depicted in Figure 11.15.

In some higher plants, including *Arabidopsis* and spinach, the two pathways contribute almost equally to chloroplast lipid synthesis. In many other angiosperms, however, phosphatidylglycerol is the only product of the prokaryotic pathway, and the remaining chloroplast lipids are synthesized entirely by the eukaryotic pathway.

The glycerolipids are first synthesized using 16:0 and 18:1 acyl groups. Further double bonds are placed in the fatty acids by a series of desaturase enzymes that are integral membrane proteins (Ohlrogge and Browse 1995). Desaturase isozymes exist in the chloroplast and the endoplasmic reticulum. Each desaturase inserts a double bond at a specific position in the fatty acid chain, and the enzymes act sequentially to produce the final 18:3 and 16:3 products.

The biochemistry of triacylglycerol synthesis in oilseeds is generally the same as described for the glycerolipids. 16:0 and 18:1 are synthesized in the plastids of the cell and exported as CoA thioesters for incorporation into DAG in the endoplasmic reticulum (see Figure 11.15). The key enzyme in oilseed metabolism, not shown in Figure 11.15, is the acyl-CoA:DAG acyltransferase, which catalyzes the final step in triacylglycerol synthesis. As noted earlier, triacylglycerol molecules accumulate in specialized subcellular structures—the oleosomes—from which they can be mobilized during germination and converted to sugar.

Lipid Composition Influences Membrane Function

A central dilemma in membrane biology is the need to explain lipid diversity. Each membrane system of the cell has a characteristic and distinct complement of lipid types, and within a single membrane each class of lipids has a distinct fatty acid composition. Nevertheless, the predominant view of a membrane is one in which lipids simply compose the fluid, semipermeable bilayer that is the matrix for the functional membrane proteins. Since this

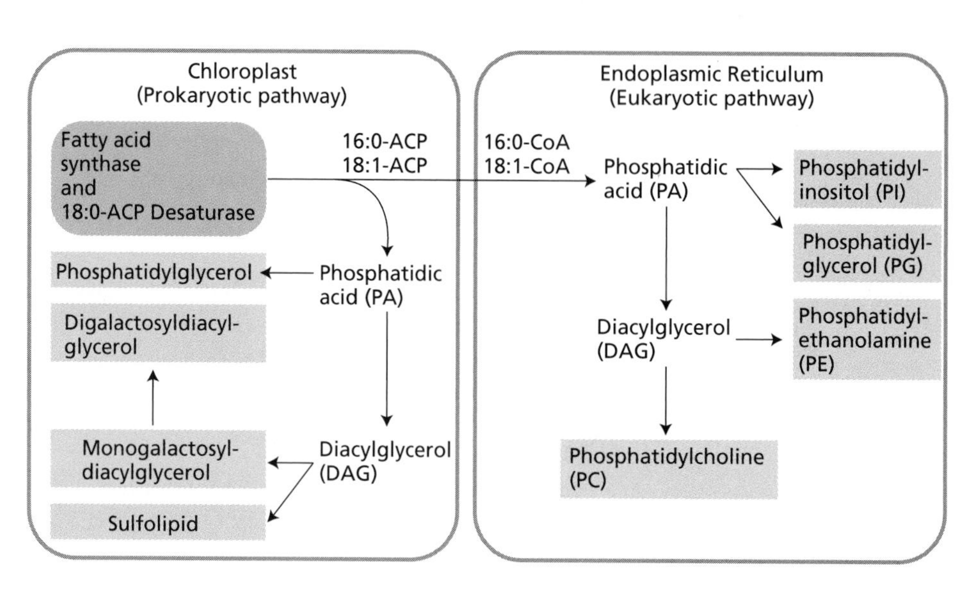

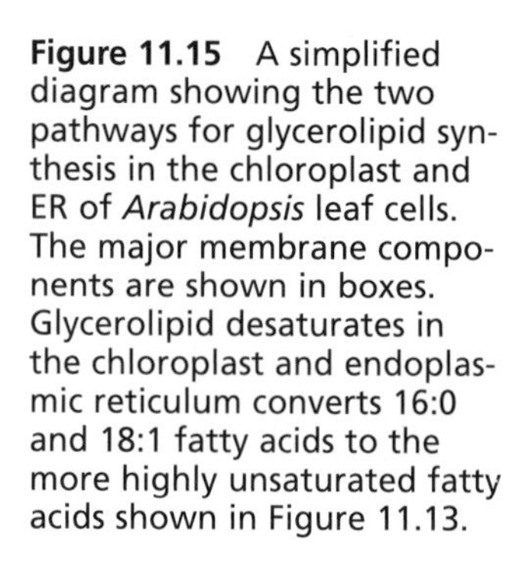
Figure 11.15 A simplified diagram showing the two pathways for glycerolipid synthesis in the chloroplast and ER of *Arabidopsis* leaf cells. The major membrane components are shown in boxes. Glycerolipid desaturates in the chloroplast and endoplasmic reticulum converts 16:0 and 18:1 fatty acids to the more highly unsaturated fatty acids shown in Figure 11.13.

bulk lipid role could be satisfied by a single unsaturated species of phosphatidylcholine, obviously such a simple model is unsatisfactory. Why is lipid diversity needed?

The characterization of lipid mutants and the experimental modification of lipid composition by molecular-genetic means have now revealed some broad generalizations about the roles of membrane lipids in the cell biology and physiology of plants. However, the biochemical and physical bases for the phenotypes that are observed remain uncertain. Changes in fatty acid composition have been shown to alter chloroplast size and architecture. Mutations that eliminate polyunsaturated fatty acids block photosynthesis and prevent the plant from growing autotrophically (McConn and Browse 1998). Surprisingly, such mutant plants can grow on sucrose, indicating that photosynthesis is the only plant function that absolutely requires a polyunsaturated membrane. As described in Chapter 13, however, polyunsaturated fatty acids are also required as precursors of signaling compounds, such as *jasmonic acid*, that are necessary for such diverse functions as pollen development and plant defense (Howe et al. 1996; McConn and Browse 1998; Vijayan et al. 1998).

One of the most extensively studied issues in membrane biology is the relationship between lipid composition and the ability of organisms to adjust to temperature changes (Wolter et al. 1992). For example, chill-sensitive plants experience sharp reductions in growth rate and development at temperatures between 0 and 12°C. The physical and physiological changes in chill-sensitive plants that are induced by exposure to low temperature, together with the subsequent expression of stress symptoms, are termed chilling injury (see Chapter 25). Many economically important crops, such as cotton, soybean, maize, rice, and many tropical and subtropical fruits, are classified as chill sensitive. In contrast, most plants of temperate origin are able to grow and develop at chilling temperatures and are classified as chill-resistant plants.

In attempts to link the biochemical and physiological changes associated with chilling injury with a single "trigger" or site of damage, it has been suggested that the primary event of chilling injury is a transition from a liquid-crystalline phase to a gel phase in the cellular membranes. According to this proposal, the transition from liquid-crystalline phase to gel phase would result in alterations in the metabolism of chilled cells and lead to injury and death of the chill-sensitive plants. The degree of unsaturation of the fatty acids would determine the temperature at which such damage occurs. A similar hypothesis has been proposed for chloroplast membranes in which the levels of disaturated phosphatidylglycerol (molecules containing no *cis* double bonds in the fatty acid chains) would determine the chilling sensitivity of plant species (Murata et al. 1992).

However, the most recent research results indicate that the relationship of membrane unsaturation to plant temperature responses is more subtle and complex than is suggested in these earlier hypotheses. On the one hand, studies of five different *Arabidopsis* mutants have demonstrated that reduced unsaturation can result in plants that grow well at 22°C but that are less robust than wild-type plants when grown at 2 to 5°C. These results were observed even though the lipid changes in most of the mutants are insufficient to cause a lipid phase transition. In addition, the responses of these mutants to low temperature appear quite distinct from classic chilling sensitivity, suggesting that normal chilling injury may not be related to the level of unsaturation of membrane lipids (Hugly and Somerville 1992; Miquel et al. 1993).

In one particular mutant, *fab1*, saturated forms of phosphatidylglycerol account for 43% of the total leaf phosphatidylglycerol—a higher percentage than is found in many chill-sensitive plants. Nevertheless, the mutant was completely unaffected (when compared with wild-type controls) by a range of low-temperature treatments that quickly led to the death of cucumber and other chill-sensitive plants (Wu and Browse 1995).

A complementary series of experiments has been carried out in tobacco, which is a chill-sensitive plant. The transgenic expression of exogenous genes in tobacco has been used specifically to decrease the level of saturated phosphatidylglycerol or to bring about a general increase in membrane unsaturation (Murata et al. 1992; Kodama et al. 1994; Ishizaki-Nishizawa et al. 1996). In each case, damage caused by chilling was alleviated to some extent. These new findings make it clear that the extent of membrane unsaturation or the presence of particular lipids, such as disaturated phosphatidylglycerol, can affect the responses of plants to low temperature. However, membrane lipid composition is not the major determinant of chilling sensitivity in plants.

In Germinating Seeds, Storage Lipids Are Converted into Carbohydrates

After germinating, oil-containing seeds metabolize stored triacylglycerols by converting lipids to carbohydrates. Plants are not able to transport fats from the endosperm to the root and shoot tissues of the germinating seedling, so they must convert stored lipids to a more mobile form of carbon, generally sucrose. This process involves several steps that are located in different cellular compartments: oleosomes, glyoxysomes, mitochondria, and cytosol (Huang et al. 1983).

The conversion of lipid to sugar in oilseeds is triggered by germination and begins with the hydrolysis of triacylglycerols stored in the oil bodies to free fatty acids, followed by oxidation of the fatty acids to produce acetyl CoA (Figure 11.16A). The fatty acids are oxidized in a type of peroxisome called a **glyoxysome**, an organelle

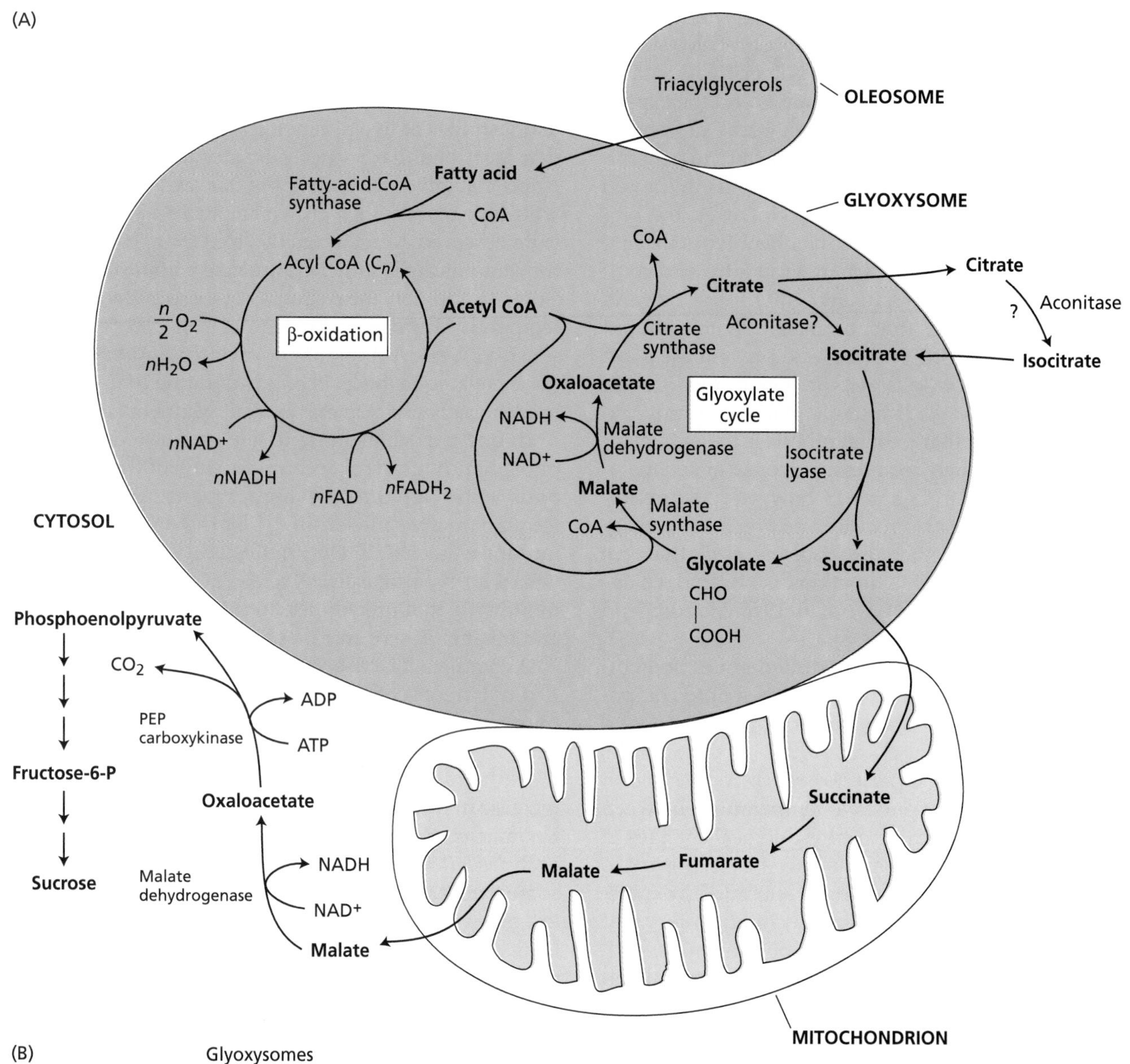

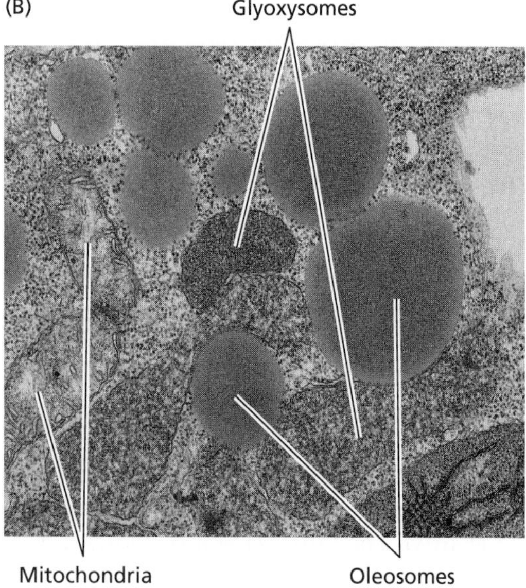

Figure 11.16 The conversion of fats to sugars during germination in oil-storing seeds. (A) Carbon flow during fatty acid breakdown and gluconeogenesis. Triacylglycerols in the oleosome are hydrolyzed to yield fatty acids, which (with the possible exception of the step involving aconitase) are metabolized to acetyl CoA in the glyoxysome. The latter conversion consists of a series of reactions known as the β-oxidation pathway. In this pathway, fatty-acyl-CoA synthase first converts the free fatty acid to its fatty acyl CoA adduct, and the fatty acyl CoA is then degraded in two-carbon units through a reaction sequence involving a series of four enzymes to release one molecule of acetyl CoA. Every two molecules of acetyl CoA produced are further metabolized through the glyoxylate cycle to generate one succinate, which moves into the mitochondrion and is converted to malate. The resulting malate is transported into the cytosol and oxidized to oxaloacetate, which is converted to phosphoenolpyruvate by the enzyme PEP carboxykinase. The resulting PEP is then metabolized to produce glucose via the gluconeogenic pathway. (Refer to Figures 11.1 and 11.3 for structures.) (B) Electron micrograph of a cell from the oil-storing cotyledon of a cucumber seedling, showing glyoxysomes, mitochondria, and oleosomes. (Photo courtesy of R. N. Trelease.)

enclosed by a single membrane that is found in the oil-rich storage tissues of seeds. The discovery of glyoxysomes in castor bean (*Ricinus communis*) endosperm, and much of our understanding of glyoxysomal function, are due to the work of Harry Beevers and his colleagues at Purdue University and the University of California, Santa Cruz. Acetyl CoA is metabolized in the glyoxysome (see Figure 11.16A) to produce succinate, which is transported from the glyoxysome to the mitochondrion, where it is converted to oxaloacetate. The process ends in the cytosol with the conversion of oxaloacetate to glucose via gluconeogenesis, and then to sucrose. Although some of this fatty acid–derived carbon is diverted to other metabolic reactions in certain oilseeds, in castor bean the process is so efficient that each gram of lipid metabolized results in the formation of 1 g of carbohydrate, equivalent to a 40% recovery of free energy in the form of carbon bonds ([15.9 kJ/40 kJ] × 100 = 40%).

Lipase hydrolysis. The initial step in the conversion of lipids to carbohydrate is the breakdown of triglycerides stored in the oil bodies by the enzyme lipase, which, at least in castor bean endosperm, is located on the half-membrane that serves as the outer boundary of the oil body. The lipase hydrolyzes triacylglycerols to three molecules of fatty acid and glycerol. Corn and cotton also contain a lipase activity in the oil body, but peanut, soybean, and cucumber show lipase activity in the glyoxysome instead. During the breakdown of lipids, oil bodies and glyoxysomes are generally in close physical association.

β-Oxidation of fatty acids. After hydrolysis of the triacylglycerols, the resulting fatty acids enter the glyoxysome, where they undergo activation by conversion to fatty acyl CoA by the enzyme fatty-acyl-CoA synthase. Fatty acyl CoA is the initial substrate for the **β-oxidation** series of reactions, in which C_n fatty acids (fatty acids composed of n number of carbons) are sequentially broken down to $n/2$ molecules of acetyl CoA (see Figure 11.16A). This reaction sequence is similar to that associated with the breakdown of fatty acids in animal tissues and involves the reduction of $\frac{1}{2}$ O_2 to H_2O and the formation of one NADH and one $FADH_2$ for each acetyl CoA produced.

In mammalian tissues, the four enzymes associated with β-oxidation are present in the mitochondrion; in plant seed storage tissues, they are localized exclusively in the glyoxysome. Interestingly, in plant vegetative tissues (e.g., mung bean hypocotyl and potato tuber), the β-oxidation reactions are localized in a related organelle, the peroxisome, rather than in the mitochondria (Gerhardt 1983), suggesting that the limited breakdown of lipids that does occur in plant nonstorage tissues takes place in peroxisomes and not mitochondria, as in mammalian tissues.

The glyoxylate cycle. The acetyl CoA produced by β-oxidation is further metabolized in the glyoxysome through a series of reactions that make up the **glyoxylate cycle** (see Figure 11.16A). Initially, the acetyl CoA reacts with oxaloacetate to give citrate, which in turn is isomerized to isocitrate. Both of the enzymes involved here (citrate synthase and aconitase) catalyze the same reactions as in the TCA cycle in the mitochondria (see Figure 11.5). However, aconitase is present in very low abundance in glyoxysomes, and recently some researchers have claimed that it is absent altogether from the glyoxysomes of castor bean endosperm (Courtois-Verniquet and Douce 1993). Since there is a highly active aconitase in the cytosol, citrate may be converted to isocitrate in the cytosol, but the issue is still unresolved, as the question marks on Figure 11.16A indicate.

Although the source of the isocitrate is unclear, the next two reactions are known to occur within the glyoxysome and are unique to the glyoxylate pathway. First, isocitrate (C_6) is cleaved by the enzyme isocitrate lyase to give succinate (C_4) and glyoxylate (C_2). Next, malate synthase combines a second molecule of acetyl CoA with glyoxylate to produce malate. According to the classic view of the operation of the glyoxylate cycle, the malate is then oxidized by malate dehydrogenase to oxaloacetate, which can combine with another acetyl CoA to continue the cycle (see Figure 11.16A).

The mitochondrial role. The function of the glyoxylate cycle is to convert two molecules of acetyl CoA to succinate. The succinate then moves to the mitochondrion, where it is converted to malate by the normal TCA cycle reactions. The resulting malate can be moved out of the mitochondria in exchange for the incoming succinate via the dicarboxylate transporter on the inner mitochondrial membrane. The malate is then oxidized to oxaloacetate by malate dehydrogenase in the cytosol, and the resulting oxaloacetate is converted to carbohydrate.

This conversion requires circumventing the irreversibility of the pyruvate kinase reaction (see Figure 11.2) and is facilitated by the enzyme PEP carboxykinase, which utilizes the phosphorylating ability of ATP to convert oxaloacetate to PEP and CO_2 (see Figure 11.16A). From PEP, gluconeogenesis can proceed to the production of glucose, as described earlier. Sucrose is the ultimate product of this process, and the primary form in which reduced carbon is translocated from the cotyledons to the growing seedling tissues.

Although the pathway for the mobilization of triacylglycerols has been best characterized in castor bean, it seems to be similar in the storage tissues of other oilseeds. However, not all seeds quantitatively convert fat to sugar. In castor bean, the endosperm degenerates after the fat and protein reserves are fully utilized. In many oilseeds—such as sunflower (*Helianthus annuus*), cotton (*Gossipium hirsutum*), and members of the squash family (Cucurbitaceae)—the cotyledons become green and photosynthesize after the food reserves are used up.

In these tissues only part of the stored lipid is converted to exported carbohydrate. Much of the lipid-derived carbon remains in the cotyledons, where it contributes to the synthesis of chloroplasts and other cellular structures. As the greening process takes place, there is a transition in the peroxisome population of these cells: Some peroxisomes have fewer of the characteristics of glyoxysomes and more of leaf-type peroxisomes. Such a transition is in keeping with the decreased requirement for the breakdown of stored lipids and the increased need to metabolize the products of photorespiration as the tissue goes from a heterotrophic to a more autotrophic mode of metabolism.

Summary

Plant respiration couples the complete oxidation of reduced cellular carbon that is generated during photosynthesis to the synthesis of ATP. Respiration takes place in three stages: glycolysis, the tricarboxylic acid cycle, and the electron transport chain. In glycolysis, carbohydrate is converted in the cytosol to pyruvate while a small amount of ATP is synthesized via substrate-level phosphorylation. Pyruvate is subsequently oxidized within the mitochondrial matrix through the TCA cycle, generating a large number of reducing equivalents in the form of NADH and $FADH_2$. In the third stage, NADH and $FADH_2$ are oxidized by the electron transport chain, which is associated with the inner mitochondrial membrane.

The free energy released during electron transport is used to synthesize a large amount of ATP in a process known as oxidative phosphorylation. ATP synthesis is accomplished by the generation of a proton gradient across the inner mitochondrial membrane during electron transfer. The F_oF_1-ATP synthase complex then couples breakdown of the proton gradient to conversion of ADP and P_i to ATP.

Substrate oxidation during respiration is regulated at control points in glycolysis, the TCA cycle, and the electron transport chain, but ultimately substrate oxidation is controlled by the level of cellular ADP. Aerobic respiration in plants has several unique features, including the presence of a cyanide-resistant oxidative pathway. Carbohydrates can also be oxidized via the oxidative pentose phosphate pathway, in which the reducing power generated produces NADPH for biosynthetic purposes. Numerous glycolytic and TCA cycle reactions also provide the starting points for different biosynthetic pathways.

Many factors can affect the respiration rate observed at the whole-plant level. Included are the nature and age of the plant tissue, as well as environmental factors such as the external oxygen concentration and temperature. Because respiration rates contribute to the overall net carbon balance of a plant, variations in whole-plant respiration rates can affect final agronomic yields.

Lipids play a major role in plants; amphipathic lipids serve as the primary nonprotein components of plant membranes, while fats and oils are an efficient storage form of reduced carbon, particularly in seeds. Glycerolipids play important roles as structural components of membranes. Fatty acids are synthesized in plastids using acetyl CoA. Fatty acids from the plastid can be transported to the ER, where they are further modified. Membrane function may be influenced by the lipid composition. The degree of unsaturation of the fatty acids influences the sensitivity of plants to cold, but does not seem to be involved in normal chilling injury. On the other hand, certain membrane lipid breakdown products, such as jasmonic acid, can act as signaling agents in plant cells.

Triacylglycerol is synthesized in the ER and accumulates within the phospholipid bilayer, forming oil bodies. During germination in oil-storing seeds, the stored lipids are metabolized to carbohydrate in a series of reactions that involve a metabolic sequence known as the glyoxylate cycle. This cycle takes place in glyoxysomes, and subsequent steps occur in the mitochondria. The reduced carbon generated during lipid breakdown in the glyoxysomes is ultimately converted to carbohydrate in the cytosol by gluconeogenesis.

General Reading

*ap Rees, T. (1980) Assessment of the contributions of metabolic pathways to plant respiration. In *The Biochemistry of Plants*, Vol. 2, D. D. Davies, ed., Academic Press, New York, pp. 1–29.

Coté, G. G., and Crain, R. C. (1993) Biochemistry and phosphoinositides. *Annu. Rev. Plant Physiol. Plant Mol. Biol.* 44: 333–356.

*Douce, R. (1985) *Mitochondria in Higher Plants: Structure, Function, and Biogenesis*. Academic Press, Orlando, Fla.

Harwood, J. L. (1996) Recent advances in the biosynthesis of plant fatty acids. *Biochim. Biophys. Acta* 1301: 7–56.

*Huang, A. H. C., Trelease, R. N., and Moore, T. S., Jr. (1983) *Plant Peroxisomes*. Academic Press, New York.

Joyard, J., Block, M., and Douce, R. (1991) Molecular aspects of plastid envelope biochemistry. *Eur. J. Biochem.* 199: 489–509.

Moore, T. S., Jr., ed. (1993) *Lipid Metabolism in Plants*. CRC Press, Boca Raton, FL.

*Nicholls, D. G., and Ferguson, S. J. (1992) *Bioenergetics 2*. Academic Press, London.

*Ohlrogge, J. B., and Browse, J. A. (1995) Lipid biosynthesis. *Plant Cell* 7: 957–970.

*Ohlrogge, J. B., and Jaworski, J. G. (1997) Regulation of fatty acid synthesis. *Annu. Rev. Plant Physiol. Plant Mol. Biol.* 48: 109–136.

*Plaxton, W. C. (1996) The organization and regulation of plant glycolysis. *Annu. Rev. Plant Physiol. Plant Mol. Biol.* 47: 185–214.

*Schuster, W., and Brennicke, A. (1994) The plant mitochondrial genome: Physical structure, information content, RNA editing, and gene migration to the nucleus. *Annu. Rev. Plant Physiol. Plant Mol. Biol.* 45: 61–78.

*Siedow, J. N. (1995) Bioenergetics: The plant mitochondrial electron transfer chain. In *The Molecular Biology of Plant Mitochondria*, C. S. Levings III and I. Vasil, eds., Kluwer, Dordrecht, Netherlands, pp. 281–312.

*Siedow, J. N., and Umbach, A. L. (1995) Plant mitochondrial electron transfer and molecular biology. *Plant Cell* 7: 821–831.

Stumpf, P. K., ed. (1980) *Lipids: Structure and Function* (*The Biochemistry of Plants*, vol. 4). Academic Press, New York.

Stumpf, P. K., ed. (1985) *Lipids: Structure and Function* (*The Biochemistry of Plants*, vol. 9). Academic Press, New York.

*Whitehouse, D. G., and Moore, A. L. (1995) Regulation of oxidative phosphorylation in plant mitochondria. In *The Molecular Biology of Plant Mitochondria*, C. S. Levings III and I. Vasil, eds., Kluwer, Dordrecht, Netherlands, pp. 313–344.

*Wiskich, J. T., and Dry, I. B. (1985) The tricarboxylic acid cycle in plant mitochondria: Its operation and regulation. In *Higher Plant Cell Respiration* (Encyclopedia of Plant Physiology, new series, vol. 18), R. Douce and D. A. Day, eds., Springer, Berlin, pp. 281–313.

* Indicates a reference that is general reading in the field and is also cited in this chapter.

Chapter References

Abrahams, J. P., Leslie, A. G. W., Lutter, R., and Walker, J. E. (1994) Structure at 2.8 Å resolution of F_1-ATPase from bovine heart mitochondria. *Nature* 370: 621–628.

Black, C. C., Smyth, D. A., and Wu, M. Y. (1985) Pyrophosphate dependent glycolysis and regulation by fructose 2,6-bisphosphate in plants. In *Nitrogen Fixation and CO_2 Metabolism*, P. W. Ludden and I. E. Burris, eds., Elsevier, New York, pp. 361–370.

Boyer, P. D. (1989) A perspective of the binding change mechanism of ATP synthesis. *FASEB J.* 3: 2164–2178.

Boyer, P. D. (1993) The binding change mechanism for ATP synthase: Some probabilities and possibilities. *Biochim. Biophys. Acta* 1140: 215–250.

Boyer, P. D. (1997) The ATP synthase: A splendid molecular machine. *Annu. Rev. Biochem.* 66: 717–749.

Browse, J., and Somerville, C. (1991) Glycerolipid metabolism: Biochemistry and regulation. *Annu. Rev. Plant Physiol. Plant Mol. Biol.* 42: 467–506.

Conley, A. C., and Hanson, M. R. (1995) How do alterations in plant mitochondrial genomes disrupt pollen development? *J. Bioenerg. Biomembr.* 27: 447–457.

Courtois-Verniquet, F., and Douce, R. (1993) Lack of aconitase in glyoxysomes and peroxisomes. *Biochem. J.* 294: 103–107.

Duncan, T. M., Bulygin, V. V., Zhou, Y., Hutcheon, M. L., and Cross, R. L. (1995) Rotation of subunits during catalysis by *Escherichia coli* F1-ATPase. *Proc. Natl. Acad. Sci. USA* 92: 10964–10968.

Gerhardt, B. (1983) Localization of β-oxidation enzymes in peroxisomes isolated from nonfatty plant tissues. *Planta* 159: 238–246.

Gunning, B. E. S., and Steer, M. W. (1996) *Plant Cell Biology: Structure and Function of Plant Cells*. Jones and Bartlett, Boston.

Howe, G. A., Lightner, J., Browse, J., and Ryan, C. A. (1996) An octadecanoid pathway mutant (JL5) of tomato is compromised in signaling for defense against insect attack. *Plant Cell* 8: 2067–2077.

Huang, A. H. C. (1992) Oil bodies and oleosins in seeds. *Annu. Rev. Plant Physiol. Plant Mol. Biol.* 43: 177–200.

Hugly, S., and Somerville, C. (1992) A role for membrane lipid polyunsaturation in chloroplast biogenesis at low temperature. *Plant Physiol.* 99: 197–202.

Ishizaki-Nishizawa, O., Fujii, T., Azuma, M., Sekiguchi, K., Murata, N., Ohtani, T., and Toguri, T. (1996) Low-temperature resistance of higher plants is significantly enhanced by a nonspecific cyanobacterial desaturase. *Nature Biotechnol.* 14: 1003–1006.

Junge, W., Lill, H., and Engelbrecht, S. (1997) ATP synthase: An electrochemical transducer with rotary mechanics. *TIBS* 22: 420–423.

Kodama, H., Hamada, T., Horiguchi, G., Nishimura, M., and Iba, K. (1994) Genetic enhancement of cold tolerance by expression of a gene for chloroplast omega-3 fatty acid desaturase in transgenic tobacco. *Plant Physiol.* 105: 601–605.

Lambers, H. (1985) Respiration in intact plants and tissues. Its regulation and dependence on environmental factors, metabolism and invaded organisms. In *Higher Plant Cell Respiration* (*Encyclopedia of Plant Physiology*, new series, vol. 18), R. Douce and D. A. Day, eds., Springer, Berlin, pp. 418–473.

Levings, C. S., III, and Siedow, J. N. (1992) Molecular basis of disease susceptibility in the Texas cytoplasm of maize. *Plant Mol. Biol.* 19: 135–147.

Lindqvist, Y., Huang, W., Schneider, G., and Shanklin, J. (1996) Crystal structure of δ^{-9} stearoyl-acyl carrier protein desaturase from castor seed and its relationship to other di-iron proteins. *EMBO J.* 15: 4081–4092.

McConn, M., and Browse, J. (1996) The critical requirement for linolenic acid is for pollen development, not photosynthesis, in an *Arabidopsis* mutant. *Plant Cell* 8: 403–416.

McConn, M., and Browse, J. (1998) Polyunsaturated membranes are required for photosynthetic competence in a mutant of *Arabidopsis*. *Plant J.* (in press).

McConn, M., Creelman, R. A., Bell, E., Mullet, J. E., and Browse, J. (1997) Jasmonate is essential for insect defense in *Arabidopsis*. *Proc. Natl. Acad. Sci. USA* 94: 5473–5477.

Miquel, M., James, D., Dooner, H., and Browse, J. (1993) *Arabidopsis* requires polyunsaturated lipids for low temperature survival. *Proc. Natl. Acad. Sci. USA* 90: 6208–6212.

Møller, I. M., Rasmusson, A. G., and Fredlund, K. M. (1993) NAD(P)H-ubiquinone oxidoreductases in plant mitochondria. *J. Bioenerg. Biomembr.* 25: 377–384.

Murata, N., Ishizaki-Nishizawa, O., Higashi, S., Hayashi, H., Tasaka, Y., and Nishida, I. (1992) Genetically engineered alteration in the chilling sensitivity of plants. *Nature* 356: 710–713.

Napier, J. A., Stobart, A. K., and Shewry, P. R. (1996) The structure and biogenesis of plant oil bodies: The role of the ER membrane and the oleosin class of proteins. *Plant Mol. Biol.* 31: 945–956.

Noji, H., Yasuda, R., Yoshida, M., and Kinosita, K. (1997) Direct observation of the rotation of F1-ATPase. *Nature* 386: 299–302.

Raskin, I., Turner, I. M., and Melander, W. R. (1989) Regulation of heat production in the inflorescences of an *Arum* lily by endogenous salicylic acid. *Proc. Natl. Acad. Sci. USA* 86: 2214–2218.

Roberts, J. K. M., Callis, J., Wemmer, D., Walbot, V., and Jardetzky, O. (1984) Mechanism of cytoplasmic pH regulation in hypoxic maize root tips and its role in survival under hypoxia. *Proc. Natl. Acad. Sci. USA* 81: 3379–3383.

Sasaki, Y., Konishi, T., and Nagano, Y. (1995) The compartmentation of acetyl-coenzyme A carboxylase in plants. *Plant Physiol.* 108: 445–449.

Seymour, R. S., Bartholomew, G. A., and Barnhart, M. C. (1983) Respiration and heat production by the inflorescence of *Philodendron selloum* Koch. *Planta* 157: 336–343.

Stitt, M. (1990) Fructose-2,6-bisphosphate as a regulatory molecule in plants. *Annu. Rev. Plant Physiol. Plant Mol. Biol.* 41: 153–185.

Trelease, R. N., Gruber, P. J., Becker, W. M., and Newcomb, F. H. (1971) *Plant Physiol.* 48: 461.

Vijayan, P., Shockey, J., Levesque, C. A., Cook, R. J., and Browse, J. (1998) A role for jasmonate in pathogen defense of *Arabidopsis*. *Proc. Natl. Acad. Sci. USA* 95: 7209–7214.

Wagner, A. M., and Krab, K. (1995) The alternative respiration pathway in plants: Role and regulation. *Physiol. Plantarum* 95: 318–325.

Wilson, D., and Jones, J. G. (1982) Effect of selection for dark respiration rate of mature leaves on crop yields of *Lolium perenne cv.* S23. *Ann. Bot.* 49: 313–320.

Wolter, F. P., Schmidt, R., and Heinz, E. (1992) Chilling sensitivity in *Arabidopsis thaliana* with genetically engineered membrane lipids. *EMBO J.* 11: 4685–4692.

Wu, J., and Browse, J. (1995) Elevated levels of high-melting-point phosphatidylglycerols do not induce chilling sensitivity in a mutant of *Arabidopsis*. *Plant Cell* 7: 17–27.

Xia, J.-H., and Roberts, J. K. M. (1994) Improved cytoplasmic pH regulation, increased lactate efflux, and reduced cytoplasmic lactate levels are biochemical traits expressed in root tips of whole maize seedlings acclimated to a low-oxygen environment. *Plant Physiol.* 105: 651–657.

12 Assimilation of Mineral Nutrients

HIGHER PLANTS ARE AUTOTROPHIC ORGANISMS that can synthesize their organic molecular components out of inorganic nutrients obtained from their surroundings. For many mineral nutrients, this process involves root absorption from the soil (see Chapter 5) and incorporation into the organic compounds that are essential for growth and development. This incorporation of mineral nutrients into organic substances such as pigments, enzyme cofactors, lipids, nucleic acids, or amino acids is termed **nutrient assimilation**.

Assimilation of some nutrients—particularly nitrogen and sulfur—requires a complex series of biochemical reactions that are among the most energy-requiring reactions in living organisms. In nitrate (NO_3^-) assimilation, the nitrogen in NO_3^- is converted to a higher energy form in nitrite (NO_2^-), then to a yet higher energy form in ammonium (NH_4^+), and finally into the amide nitrogen of glutamine. This process consumes the equivalent of 12 ATPs per nitrogen (Bloom et al. 1992). Plants such as legumes form symbiotic relationships with nitrogen-fixing bacteria to convert molecular nitrogen (N_2) into ammonia (NH_3); this process of biological nitrogen fixation, together with the subsequent assimilation of NH_3 into an amino acid, consumes about 16 ATPs per nitrogen (Pate and Layzell 1990; Vande Broek and Vanderleyden 1995).

The primary pathway for sulfate (SO_4^{2-}) assimilation in plants is still uncertain, but in all postulated pathways sulfate assimilation consumes about 14 ATPs per SO_4^{2-} converted to the amino acid cysteine (Hell 1997). For some perspective on the enormous energies involved, consider that if these reactions run rapidly in reverse, say from NH_4NO_3 (ammonium nitrate) to N_2, they become explosive. In fact, nearly all explosives are based on the rapid oxidation of nitrogen or sulfur compounds.

Table 12.1
The major processes of the biogeochemical nitrogen cycle

Process	Definition	Rate (10^{12} g yr^{-1})[a]
Industrial fixation	Industrial conversion of molecular nitrogen to ammonia	80
Atmospheric fixation	Lightning and photochemical conversion of molecular nitrogen to nitrate	19
Biological fixation	Prokaryotic conversion of molecular nitrogen to ammonia	170
Plant acquisition	Plant absorption and assimilation of ammonium or nitrate	1200
Immobilization	Microbial absorption and assimilation of ammonium or nitrate	N/C
Ammonification	Bacterial and fungal catabolism of soil organic matter to ammonium	N/C
Nitrification	Bacterial (*Nitrosomonas* sp.) oxidation of ammonium to nitrite and subsequent bacterial (*Nitrobacter* sp.) oxidation of nitrite to nitrate	N/C
Mineralization	Bacterial and fungal catabolism of soil organic matter to mineral nitrogen through ammonification or nitrification	N/C
Volatilization	Physical loss of gaseous ammonia to the atmosphere	100
Ammonium fixation	Physical imbedding of ammonium into soil particles	10
Denitrification	Bacterial conversion of nitrate to nitrous oxide and molecular nitrogen	210
Nitrate leaching	Physical flow of nitrate dissolved in groundwater out of the topsoil and eventually into the oceans	36

Note: Terrestrial organisms, the soil, and the oceans contain about 5.2×10^{15} g, 95×10^{15} g, and 6.5×10^{15} g, respectively, of organic nitrogen that is active in the cycle. Assuming that the amount of atmospheric N_2 remains constant (inputs = outputs), "the mean residence time," the average time that a nitrogen molecule remains in organic forms is about 370 years [(pool size)/(fixation input) = (5.2×10^{15} g + 95×10^{15} g)/(80×10^{12} g yr^{-1} + 19×10^{12} g yr–1+ 170×10^{12} g yr^{-1})] (Schlesinger 1991).

[a] N/C, not calculated.

Assimilation of other nutrients, especially the macronutrient and micronutrient cations (see Chapter 5), involves the formation of complexes with organic compounds. For example, Mg^{2+} associates with chlorophyll pigments, Ca^{2+} associates with pectates within the cell wall, and Mo^{6+} associates with enzymes such as nitrate reductase and nitrogenase. These complexes are highly stable, and removal of the nutrient from the complex may result in total loss of function.

This chapter outlines the primary reactions through which the predominant nutrients (nitrogen, sulfur, phosphate, cations, and oxygen) are assimilated. We emphasize the physiological implications of the required energy expenditures and introduce the topic of symbiotic nitrogen fixation.

Nitrogen in the Environment

Many biochemical compounds present in plant cells contain nitrogen (see Chapter 5). For example, nitrogen is found in the nucleoside phosphates and amino acids that form the building blocks of nucleic acids and proteins, respectively. Only the elements oxygen, carbon, and hydrogen are more abundant in plants than nitrogen. Most natural and agricultural ecosystems show dramatic gains in productivity after fertilization with inorganic nitrogen, attesting to the importance of this element. In this section we will discuss the biogeochemical cycle of nitrogen, the crucial role of nitrogen fixation in the conversion of molecular nitrogen into ammonium and nitrate, and the fate of nitrate and ammonium in plant tissues.

Nitrogen Passes through Several Forms in a Biogeochemical Cycle

Nitrogen is present in many forms in the biosphere. The atmosphere contains vast quantities (about 78% by volume) of molecular nitrogen (N_2) (see Chapter 9). For the most part, this large reservoir of nitrogen is not directly available to living organisms. Acquisition of nitrogen from the atmosphere requires the breaking of an exceptionally stable triple covalent bond between two nitrogen atoms (N≡N) to produce ammonia (NH_3) or nitrate (NO_3^-). These reactions, known as **nitrogen fixation**, result from both industrial and natural processes.

Under elevated temperature (about 200°C) and high pressure (about 200 atmospheres), N_2 combines with hydrogen to form ammonia. The extreme conditions are required to overcome the high activation energy of the reaction. This nitrogen fixation reaction, called the Haber process, is a starting point for the manufacture of many industrial and agricultural products. Worldwide industrial production of nitrogen fertilizers amounts to more than 80×10^{12} g yr^{-1} (FAO/UNIDO/World Bank Working Group on Fertilizers 1996).

Natural processes fix about 190×10^{12} g yr^{-1} of nitrogen (Table 12.1). Of this total, lightning is responsible for about 8%. Lightning converts water vapor and oxygen into highly reactive hydroxyl free radicals, free hydrogen atoms, and free oxygen atoms that attack molecular nitrogen to form nitric acid (HNO_3). This nitric acid subsequently falls to Earth with rain. Approximately 2% of the nitrogen fixed derives from photochemical reactions between gaseous nitric oxide and ozone that produce nitric acid. The remaining 90% results from **biological**

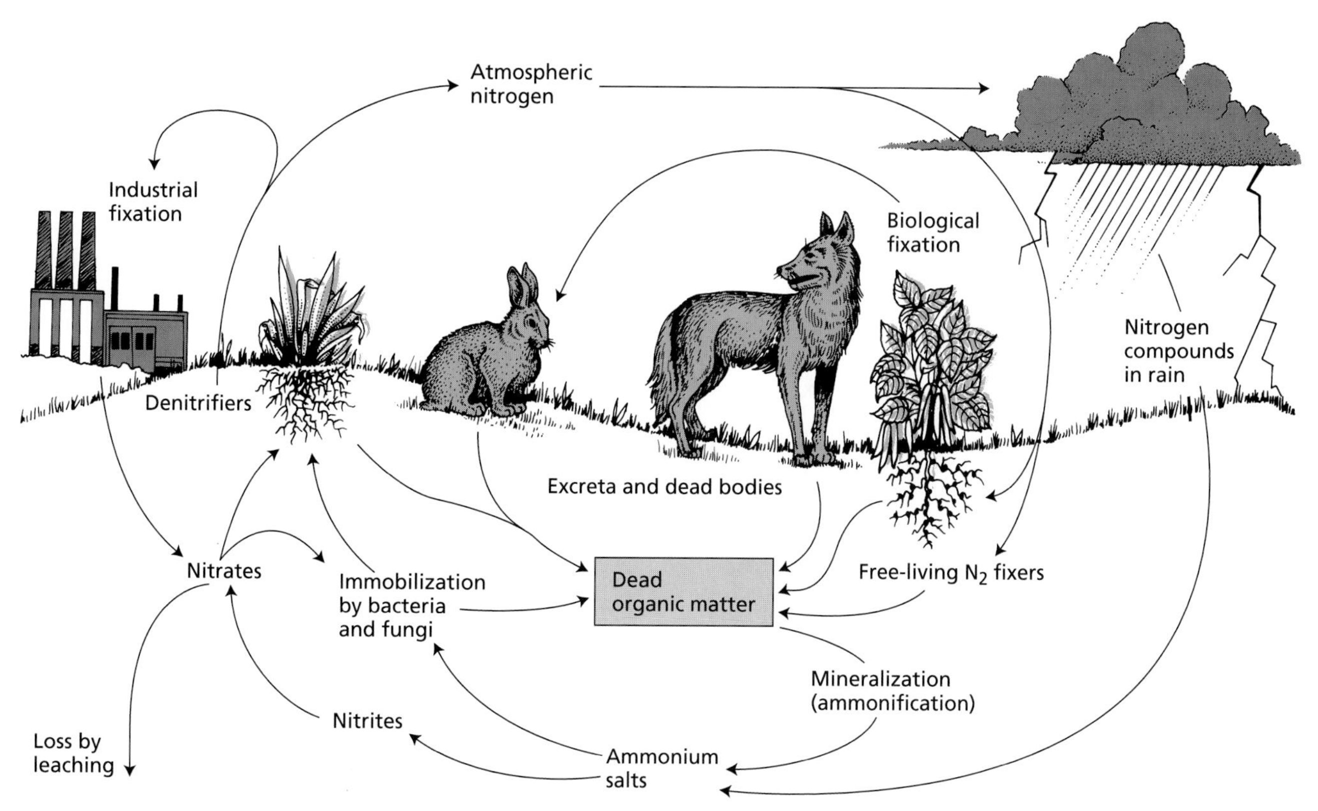

Figure 12.1 Nitrogen cycles through the atmosphere as it changes from a gaseous form to reduced ions before being incorporated into organic compounds in living organisms. Some of the steps involved in the nitrogen cycle are shown.

nitrogen fixation, in which bacteria or blue-green algae (cyanobacteria) fix N_2 into ammonium. From an agricultural standpoint, biological nitrogen fixation is critical because industrial production of nitrogen fertilizers seldom meets agricultural demand (FAO/UNIDO/World Bank Working Group on Fertilizers 1996).

Once fixed into ammonium or nitrate, nitrogen enters a biogeochemical cycle and passes through several organic or inorganic forms before it eventually returns to molecular nitrogen (Figure 12.1; see also Table 12.1). The ammonium and nitrate that are generated through fixation or released through decomposition of soil organic matter become the object of intense competition among plants and microorganisms. To remain competitive, plants have developed mechanisms for scavenging these ions from the soil solution as quickly as possible (see Chapters 5 and 6). Under the elevated soil concentrations that occur after fertilization, the absorption of ammonium and nitrate by the roots may exceed the capacity of a plant to assimilate these ions, leading to their accumulation within the plant's tissues.

Stored Ammonium or Nitrate Can Be Toxic

Plants can store high levels of nitrate, or they can translocate it from tissue to tissue without deleterious effect. However, if livestock and humans consume plant material that is high in nitrate, they may suffer methemoglobinemia, a disease in which the liver reduces nitrate to nitrite, which combines with hemoglobin and renders the hemoglobin unable to bind oxygen. Humans and other animals may also convert nitrate into nitrosamines, which are potent carcinogens. Some countries limit the nitrate content in plant materials sold for human consumption.

In contrast to nitrate, high levels of ammonium are toxic to both plants and animals. Ammonium dissipates transmembrane proton gradients (Figure 12.2) that are required for both photosynthetic (see Chapter 7) and respiratory (see Chapter 11) electron transport and for sequestering metabolites in the vacuole (see Chapter 6). Because high levels of ammonium are dangerous, animals have developed a strong aversion to its smell. The active ingredient in smelling salts is ammonium carbonate. Plants assimilate ammonium near the site of absorption or generation and rapidly store any excess in their vacuoles, thus avoiding toxic effects on membranes and the cytosol.

We will now discuss how the nitrate absorbed by the roots is assimilated into organic compounds, and the enzymatic processes mediating the reduction of nitrate first into nitrite and then into ammonium.

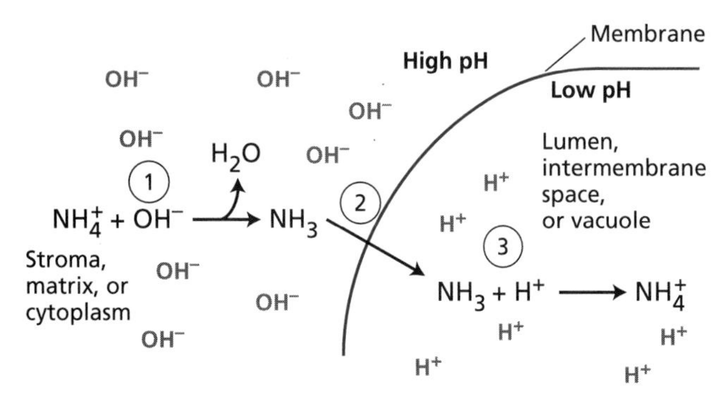

Figure 12.2 A schematic showing how NH_4^+ toxicity may result from the dissipation of pH gradients. The left side represents the stroma, matrix, or cytoplasm (higher pH); the right side represents the lumen, intermembrane space, or vacuole (lower pH). The membrane represents the thylakoid, inner mitochondrial membrane, or tonoplast membrane for a chloroplast, a mitochondrion, or a root cell, respectively. (1) If NH_4^+ is present at sufficient concentrations, it will react with OH^- on the left side of the membrane to produce NH_3 and H_2O. (2) This NH_3 is membrane permeable and will diffuse across the membrane along its concentration gradient. (3) On the right side of the membrane, the NH_3 will react with H^+ to form NH_4^+. The net result is that both the OH^- concentration on the left side and the H^+ concentration on the right side have been diminished; that is, the pH gradient has been dissipated. (After Bloom 1997.)

Nitrate Assimilation

Plants assimilate most of the nitrate absorbed by their roots into organic nitrogen compounds. The first step of this process is the reduction of nitrate to nitrite in the cytosol (Oaks 1994). The enzyme **nitrate reductase** catalyzes this reaction:

$$NO_3^- + NAD(P)H + H^+ + 2\,e^- \rightarrow NO_2^- + NAD(P)^+ + H_2O \qquad (12.1)$$

where NAD(P)H indicates NADH or NADPH. The most common form of nitrate reductase uses only NADH as an electron donor; another form of the enzyme that is found predominantly in nongreen tissues such as roots can use either NADH or NADPH (Warner and Kleinhofs 1992).

The nitrate reductases of higher plants are homodimers; that is, they are composed of two identical subunits, each with a molecular mass of 100 kDa (Campbell 1996). Each subunit contains three prosthetic groups: FAD (flavin adenine dinucleotide), heme, and a molybdenum complex. Molybdenum is bound to the enzyme via a complex with an organic molecule called a *pterin* (see Figure 18.19), which acts as a chelator of the metal (Mendel and Stallmeyer 1995). Nitrate reductase is the principal molybdenum-containing protein in vegetative tissues, and one symptom of molybdenum deficiency is the accumulation of nitrate that results from diminished nitrate reductase activity.

The genes for nitrate reductase from several higher plants have been cloned (Lam et al. 1996). Comparison between the deduced amino acid sequences and those of other well-characterized proteins that bind FAD, heme, or molybdenum has led to the structural model for nitrate reductase shown in Figure 12.3. The FAD-binding domain accepts two electrons from NAD or NADPH. The electrons then pass through the heme domain to the molybdenum complex, where they are transferred to nitrate.

Nitrate, Light, and Carbohydrates Regulate Nitrate Reductase

Expression of the genes coding for nitrate reductase and the activity of the enzyme vary with nitrate concentration and with light or carbohydrate levels. These factors induce the de novo synthesis of the enzyme (Sivasankar and Oaks 1996). In barley seedlings, nitrate reductase mRNA was detected approximately 40 minutes after addition of nitrate, and maximum levels were attained within 3 hours (Figure 12.4). The pattern of rapid mRNA accumulation contrasts with the gradual linear increase in nitrate reductase activity because the nitrate reductase protein accumulates slowly as the accumulated mRNA is translated. In addition, the protein is subject to post-translational modulation (involving a reversible phosphorylation) that is analogous to the regulation of sucrose phosphate synthase (see Chapters 8 and 10). Light, carbohydrate levels, and other environmental factors stimulate a protein phosphatase that dephosphorylates several serine residues on the nitrate reductase protein and thereby activates the enzyme. In the reverse direction, darkness and Mg^{2+} stimulate a protein kinase that phosphorylates the same serine residues and thereby inactivates nitrate reductase (Kaiser and Huber 1994). Regulation of nitrate reductase activity through phosphorylation and dephosphorylation provides more rapid control than can be achieved through synthesis or degradation of the enzyme (minutes versus hours).

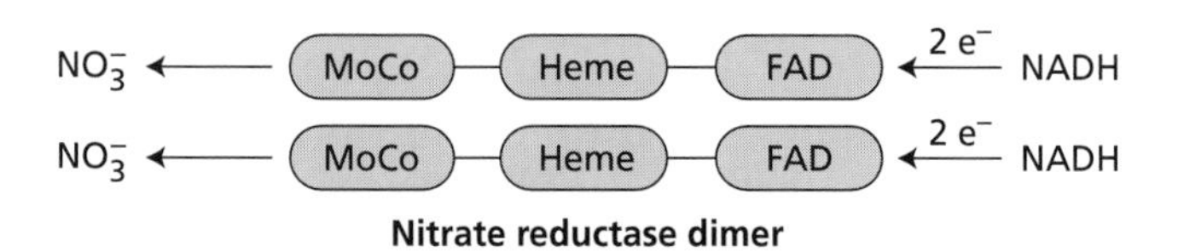

Figure 12.3 A model of the nitrate reductase dimer, illustrating the three binding domains for which the polypeptide sequences are similar in eukaryotes: molybdenum complex (MoCo), heme, and FAD. NADH binds at the FAD-binding region of each subunit and initiates a two-electron transfer through each of the electron transfer components. Nitrate is reduced at the molybdenum complex.

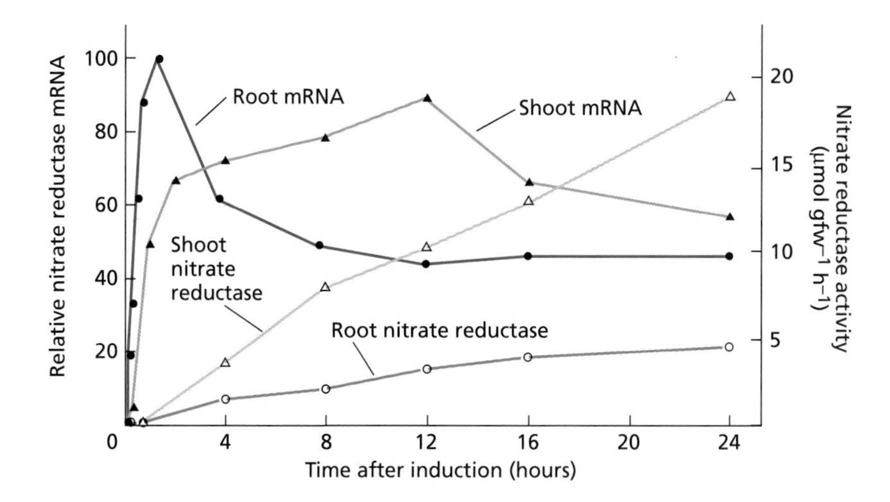

Figure 12.4 The pattern of nitrate reductase activity and nitrate reductase mRNA induction in shoots and roots of barley. The induction of mRNA precedes the appearance of the enzyme in both cases, although the precise timing is slightly different in the two tissues. gfw, grams fresh weight. (From Kleinhofs et al. 1989.)

Nitrite Reductase Converts Nitrite to Ammonium

Nitrite (NO_2^-) is a highly reactive, potentially toxic ion. Plant cells immediately transport the nitrite generated during nitrate reduction (see Equation 12.1) from the cytosol into chloroplasts in leaves and plastids in roots. In these organelles, the enzyme **nitrite reductase** reduces nitrite to ammonium. Chloroplasts and root plastids contain different forms of the enzyme, but both forms transfer electrons from ferredoxin to nitrite according to the following overall reaction:

$$NO_2^- + 6\ Fd_{red} + 8\ H^+ + 6\ e^- \rightarrow NH_4^+ + 6\ Fd_{ox} + 2\ H_2O \quad (12.2)$$

where Fd is ferredoxin, and the subscripts "red" and "ox" stand for "reduced" and "oxidized," respectively. Reduced ferredoxin derives from photosynthetic electron transport in the chloroplasts (see Chapter 7) and from NADPH generated by the oxidative pentose phosphate pathway in nongreen tissues (see Chapter 11).

Both the chloroplast and the root plastid forms of nitrite reductase consist of a single 63 kDA polypeptide and contain two prosthetic groups: an iron–sulfur cluster (Fe_4S_4) and a specialized heme (Siegel and Wilkerson 1989). Kinetic experiments suggest that the coupled Fe_4S_4–heme pair on the enzyme binds nitrite and reduces it directly to ammonium, without accumulation of nitrogen compounds of intermediate redox states. Thus, the pathway of electron flow through ferredoxin can be represented as shown in Figure 12.5.

Nitrite reductase is encoded in the nucleus and synthesized as a precursor carrying an N-terminus transit peptide that targets it to the plastids (Wray 1993). NO_3^- and light induce the transcription of nitrite reductase mRNA. Asparagine and glutamine repress this induction; sucrose enhances it.

Plants Can Assimilate Nitrate in Both Roots and Shoots

In most plant species, both roots and shoots have the capacity to assimilate nitrate first as nitrite and then as ammonium. The relative extent to which nitrate is reduced in the roots or in the leaves depends on several factors, including the level of nitrate supplied to the roots and the plant species. In many plants, when the roots receive small amounts of nitrate, nitrate is reduced primarily in the roots. As the supply of nitrate increases, a greater proportion of the absorbed nitrate is translocated to the shoot and assimilated there (Marschner 1995).

Even under similar conditions of nitrate supply, the balance between root and shoot nitrate metabolism—measured by the proportion of nitrate reductase activity in each of the two tissues or by the relative concentrations of nitrate and reduced nitrogen in the xylem

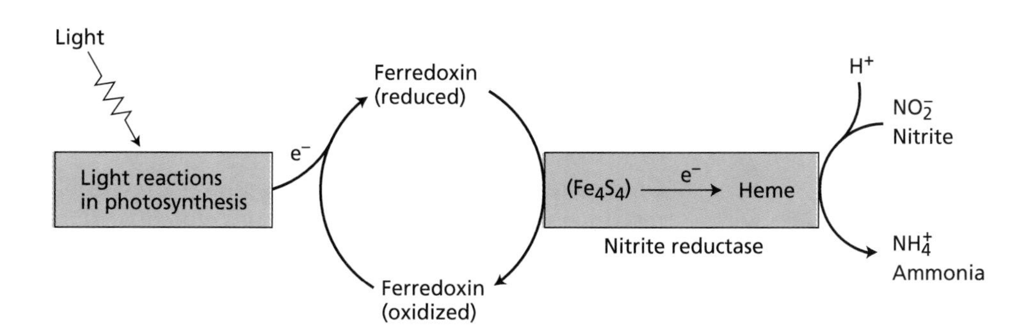

Figure 12.5 Model for coupling photosynthetic electron flow, via ferredoxin, to the reduction of nitrite by nitrite reductase. The enzyme contains two prosthetic groups, Fe_4S_4 and heme, which participate in the reduction of nitrite to ammonium.

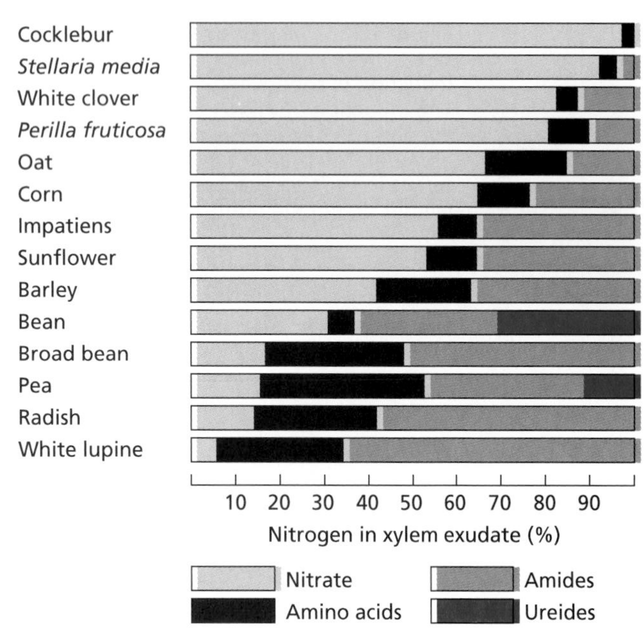

Figure 12.6 Relative amounts of nitrate and other nitrogen compounds in the xylem exudate of various plant species. The investigators grew the plants with their roots exposed to nitrate solutions and collected xylem sap by severing the stem. Note the presence of ureides, specialized nitrogen compounds, in bean and pea (which will be discussed later in the text). (After Pate 1973.)

sap—vary from species to species (Andrews 1986). In plants such as the cocklebur (*Xanthium strumarium*), nitrate metabolism is restricted to the shoot; in other plants, such as white lupine (*Lupinus albus*), most nitrate is metabolized in the roots (Figure 12.6). Generally, species native to temperate regions rely more heavily on nitrate assimilation by the roots than do species of tropical or subtropical origins.

Ammonium Assimilation

Plant cells avoid ammonium toxicity by rapidly converting the ammonium generated from nitrate assimilation or photorespiration (see Chapter 8) into amino acids. The primary pathway for this conversion involves the sequential actions of glutamine synthetase and glutamate synthase (Lea et al. 1992). In this section we will discuss the enzymatic processes that mediate the assimilation of ammonium into essential amino acids, and the role of amides in the regulation of nitrogen and carbon metabolism.

Multiple Forms of Two Enzymes Convert Ammonium to Amino Acids

Glutamine synthetase (**GS**) combines ammonium with glutamate to form glutamine (Figure 12.7A).

$$\text{Glutamate} + NH_4^+ + \text{ATP} \rightarrow \text{glutamine} + \text{ADP} + P_i \qquad (12.3)$$

This reaction requires the hydrolysis of one ATP and involves a divalent cation such as Mg^{2+}, Mn^{2+}, or Co^{2+} as a cofactor. GS has a molecular mass of 350 kDa and is composed of eight nearly identical subunits (Sivasankar and Oaks 1996). Plants contain two classes of GS, one in the cytosol and the other in root plastids or shoot chloroplasts. The cytosolic forms are expressed in germinating seeds or in the vascular bundles of roots and shoots and produce glutamine for intracellular nitrogen transport. The GS in root plastids generates amide nitrogen for local consumption; the GS in shoot chloroplasts reassimilates photorespiratory NH_4^+ (Lam et al. 1996). Light and carbohydrate levels alter the expression of the plastid forms of the enzyme, but they have little effect on the cytosolic forms.

Elevated plastid levels of glutamine stimulate the activity of **glutamate synthase** (also known as glutamine:2-oxoglutarate aminotransferase, or **GOGAT**). This enzyme transfers the amide group of glutamine to 2-oxoglutarate, yielding two molecules of glutamate (see Figure 12.7A). Plants contain two types of GOGAT: One accepts electrons from NADH, the other accepts electrons from ferredoxin (Fd).

$$\text{Glutamine} + \text{2-oxoglutarate} + \text{NADH} + H^+ \rightarrow 2\ \text{glutamate} + \text{NAD}^+ \qquad (12.4)$$

$$\text{Glutamine} + \text{2-oxoglutarate} + \text{Fd}_{red} \rightarrow 2\ \text{glutamate} + \text{Fd}_{ox} \qquad (12.5)$$

The NADH type of the enzyme (NADH-GOGAT) is located in plastids of nonphotosynthetic tissues such as roots or vascular bundles of developing leaves. In roots, NADH-GOGAT is involved in the assimilation of NH_4^+ absorbed from the rhizosphere (the soil near the surface of the roots); in vascular bundles of developing leaves, NADH-GOGAT assimilates glutamine translocated from roots or senescing leaves.

The ferredoxin-dependent type of glutamate synthase (Fd-GOGAT) is found in chloroplasts and has a molecular mass of 165 kDa, including a transit peptide for transport into the chloroplast. Fd-GOGAT in the shoot serves in photorespiratory nitrogen metabolism. Both the amount of protein and its activity increase with light levels. Roots, particularly those under nitrate nutrition, have Fd-GOGAT in plastids. Fd-GOGAT in the roots presumably functions to incorporate the glutamine generated during nitrate assimilation.

Ammonium Can Be Assimilated via an Alternative Pathway

Glutamate dehydrogenase (**GDH**) catalyzes a reversible reaction that synthesizes or deaminates glutamate (Figure 12.7B):

$$\text{2-Oxoglutarate} + NH_4^+ + \text{NAD(P)H} \leftrightarrow \text{glutamate} + H_2O + \text{NAD(P)}^+ \qquad (12.6)$$

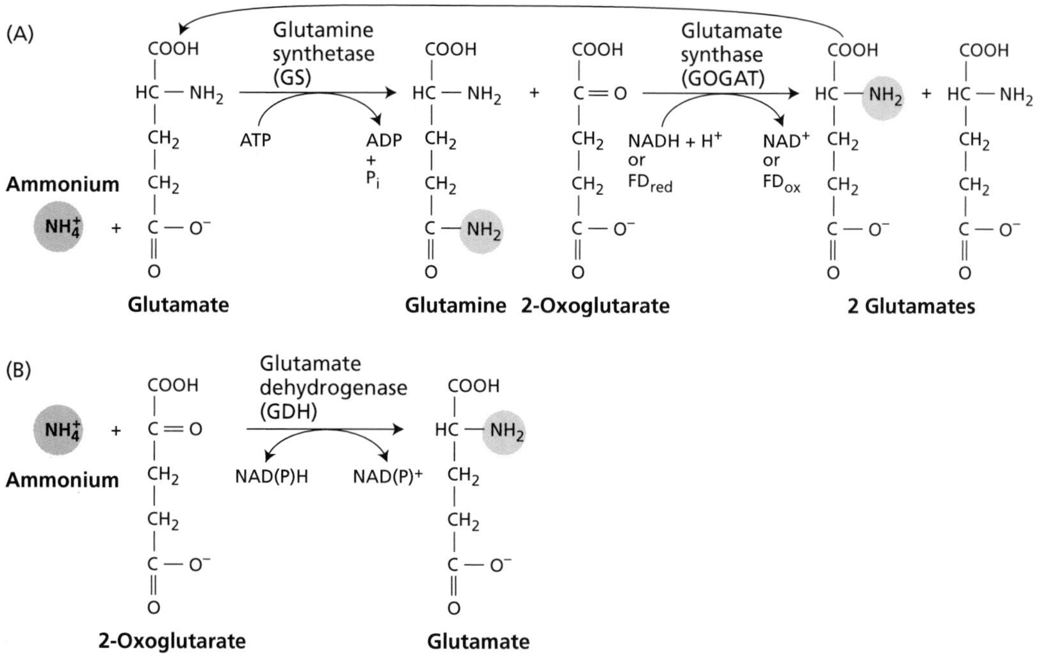

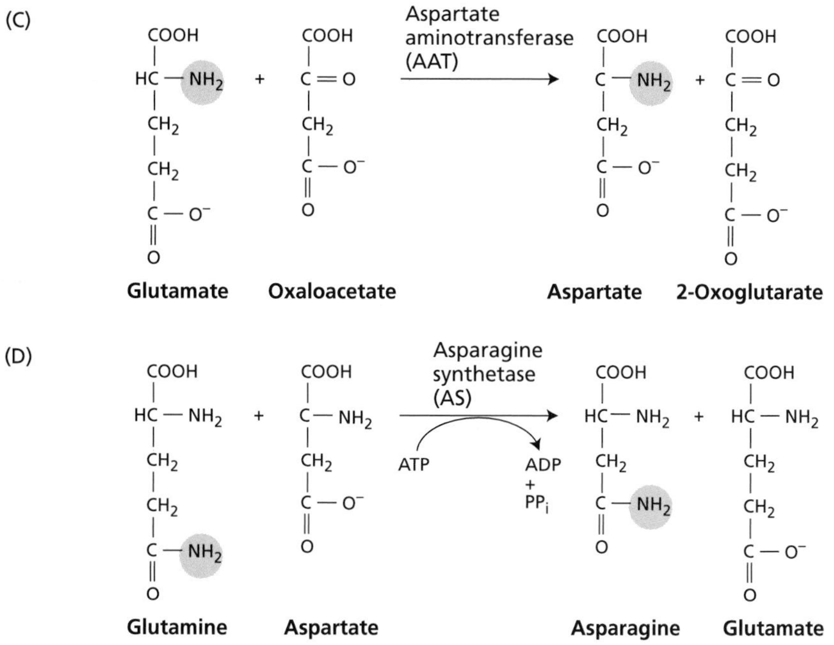

(A) Ammonium NH_4^+ + Glutamate —Glutamine synthetase (GS); ATP → ADP + P_i→ Glutamine + 2-Oxoglutarate —Glutamate synthase (GOGAT); NADH + H^+ or FD_{red} → NAD^+ or FD_{ox}→ 2 Glutamates

(B) Ammonium NH_4^+ + 2-Oxoglutarate —Glutamate dehydrogenase (GDH); NAD(P)H → NAD(P)$^+$→ Glutamate

(C) Glutamate + Oxaloacetate —Aspartate aminotransferase (AAT)→ Aspartate + 2-Oxoglutarate

(D) Glutamine + Aspartate —Asparagine synthetase (AS); ATP → ADP + PP_i→ Asparagine + Glutamate

Figure 12.7 Structure and pathways of compounds involved in ammonium metabolism. Ammonium can be assimilated by one of several processes. (A) Combination with glutamate to form the amide glutamine (catalyzed by glutamine synthetase, GS). This glutamine then combines with 2-oxoglutarate to form two molecules of glutamate (catalyzed by glutamate synthase, GOGAT). (B) Deamination of 2-oxoglutarate to produce glutamate (catalyzed by glutamate dehydrogenase). A reduced cofactor is required for the reaction: ferredoxin in green leaves and NADH in nonphotosynthetic tissue. (C) Transfer of the amino group from glutamate to aspartate (catalyzed by aspartate aminotransferase). (D) Synthesis of asparagine by transfer of an amino group from glutamine to aspartate (catalyzed by asparagine synthetase).

An NADH-dependent form of GDH is found in mitochondria, and an NADPH-dependent form is localized in the chloroplasts of photosynthetic organs. Although both forms are relatively abundant, their specific roles remain unresolved. At one time, the mitochondrial form was thought to be involved in photorespiratory nitrogen metabolism because photorespiration generates ammonium in the mitochondrion (see Chapter 8). However, GS or Fd-GOGAT are required for reassimilation of photorespiratory ammonium because mutant plants that are deficient in either enzyme can survive only under conditions that minimize photorespiration, such as an atmosphere enriched in carbon dioxide (Lam et al. 1996). Moreover, methionine sulfoximine, a specific inhibitor of GS, blocks all incorporation of ammonium into glutamate or glutamine. These results indicate that GDH cannot substitute for the GS–GOGAT pathway for assimilation of ammonium and that the primary function of GDH is to deaminate glutamate (see Figure 12.7B).

Transamination Reactions Transfer Nitrogen

Once assimilated into glutamine and glutamate, nitrogen is incorporated into other amino acids via transamination reactions. The enzymes that catalyze these reactions are known as aminotransferases. An example is **aspartate aminotransferase** (**AAT**), which catalyzes the reaction

$$\text{Glutamate + oxaloacetate} \rightarrow \text{aspartate + 2-oxoglutarate} \qquad (12.7)$$

in which the amino group of glutamate is transferred to the carboxyl atom of aspartate (Figure 12.7C). Aspartate is an amino acid that participates in the malate–aspartate shuttle to transfer reducing equivalents from the mitochondrion and chloroplast into the cytosol (see Chapter 11) and in the transport of carbon from mesophyll to bundle sheath for C_4 carbon fixation (see Chapter 8). All transamination reactions require pyridoxal phosphate (vitamin B_6) as a cofactor. Aminotransferases are found in the cytoplasm, chloroplasts, mitochondria, glyoxysomes, and peroxisomes. The aminotransferases localized in the chloroplasts may have a significant role in amino acid biosynthesis because plant leaves or isolated chloroplasts exposed to radioactively labeled carbon dioxide rapidly incorporate the label into glutamate, aspartate, alanine, serine, and glycine (Roberts et al. 1970).

Asparagine and Glutamine Link Carbon and Nitrogen Metabolism

Asparagine, isolated from asparagus as early as 1806, was the first amide to be identified (Lam et al. 1996). It serves not only as a protein precursor, but as a key compound for nitrogen transport and storage because of its stability and high nitrogen-to-carbon ratio (2 N to 4 C for asparagine versus 2 N to 5 C for glutamine or 1 N to 5 C for glutamate). The major pathway for asparagine synthesis involves the transfer of the amide nitrogen from glutamine to aspartate (Figure 12.7D):

$$\text{Glutamine + aspartate + ATP} \rightarrow \text{asparagine + glutamate + AMP + PP}_i \qquad (12.8)$$

Asparagine synthetase (**AS**), the enzyme that catalyzes this reaction, is found in the cytosol of leaves and roots and in nitrogen-fixing nodules (see the discussion of biological nitrogen fixation in the next section). It is a dimer with two identical subunits of 160 kDa. In maize roots, particularly those under potentially toxic levels of ammonia, ammonium may replace glutamine as the source of the amide group (Sivasankar and Oaks 1996).

High levels of light and carbohydrate—conditions that stimulate plastid GS and Fd-GOGAT—inhibit gene expression and activity of AS. The differential regulation of these competing pathways helps balance the metabolism of carbon and nitrogen in plants (Lam et al. 1996). Conditions of ample energy (i.e., high levels of light and carbohydrates) stimulate GS and GOGAT, inhibit AS, and thus favor nitrogen assimilation into glutamine and glutamate, compounds that are rich in carbon and participate in the synthesis of new plant materials.

By contrast, energy-limited conditions inhibit GS and GOGAT, stimulate AS, and thus favor nitrogen assimilation into asparagine, a compound that is rich in nitrogen and sufficiently stable for long-distance transport or long-term storage.

Biological Nitrogen Fixation

Biological nitrogen fixation accounts for most of the fixation of atmospheric N_2 into ammonium, thus representing the key entry point of molecular nitrogen into the biogeochemical cycle of nitrogen. The following sections describe the properties of the nitrogenase enzymes that fix nitrogen, the symbiotic relations between nitrogen-fixing organisms and higher plants, the specialized structures that form in roots when infected by nitrogen-fixing bacteria, and the genetic and signaling interactions that regulate nitrogen fixation by symbiotic prokaryotes and their hosts.

Nitrogen Fixation Requires Anaerobic Conditions

Some bacteria, as stated earlier, can convert atmospheric nitrogen into ammonium (Table 12.2). Most of these nitrogen-fixing prokaryotes are free-living microorganisms found in the soil. A few form symbiotic associations with higher plants in which the prokaryote directly provides the host plant with fixed nitrogen in exchange for other nutrients and carbohydrates (top portion of Table 12.2). Such symbioses occur in nodules that form on the roots of the plant and contain the nitrogen-fixing bacteria.

The most common type of symbiosis occurs between members of the plant family Leguminosae and soil bacteria of the genera *Rhizobium*, *Bradyrhizobium*, *Azorhizobium*, *Sinorhizobium*, and *Photorhizobium* (collectively called **rhizobia**; Table 12.3 and Figure 12.8). Another common type of symbiosis occurs between several woody plant species, such as alder trees, and soil bacteria of the genus *Frankia*. The South American herb *Gunnera* and the tiny water fern *Azolla* form associations with the cyanobacteria *Nostoc* and *Anabaena*, respectively (see Table 12.2 and Figure 12.9).

Because oxygen irreversibly inactivates the **nitrogenase** enzymes involved in nitrogen fixation, nitrogen must be fixed under anaerobic conditions. Thus, each of the nitrogen-fixing organisms listed in Table 12.2 either functions under natural anaerobic conditions or can create an internal anaerobic environment in the presence of oxygen. In cyanobacteria, anaerobic conditions are created in specialized cells called **heterocysts** (Figure 12.9). Heterocysts are thick-walled cells that differentiate

Table 12.2
Some examples of organisms that can carry out nitrogen fixation

Symbiotic nitrogen fixation	
Host plants	**N-fixing prokaryotic genera**
Leguminous plants: Legumes, *Parasponia*	*Azorhizobium, Bradyrhizobium, Photorhizobium, Rhizobium, Sinorhizobium*
Actinorhizal plants: alder (tree), *Ceanothus* (shrub), *Casuarina* (tree), *Datisca* (shrub)	*Frankia*
Gunnera	*Nostoc*
Azolla (water fern)	*Anabaena*
Free-living nitrogen fixation	
Type of bacterium	**N-fixing prokaryotic genera**
Cyanobacteria (blue-green algae)	*Anabaena, Calothrix, Gloeotheca, Nostoc*
Other bacteria	
Aerobic	*Azotobacter, Azospirillum, Beijerinckia, Derxia*
Facultative	*Bacillus, Klebsiella*
Anaerobic	
Nonphotosynthetic	*Clostridium, Methanococcus* (archaebacterium)
Photosynthetic	*Chromatium, Rhodospirillum*

when filamentous cyanobacteria are deprived of NH_4^+. These cells lack photosystem II, the oxygen-producing photosystem of chloroplasts (see Chapter 7), so they do not generate oxygen (Burris 1976). Heterocysts appear to represent an adaptation for nitrogen fixation, in that they are widespread among aerobic cyanobacteria that fix nitrogen (Beevers 1976).

Cyanobacteria that lack heterocysts can fix nitrogen only under anaerobic conditions such as those that occur in flooded fields. In Asian countries, nitrogen-fixing cyanobacteria of both the heterocyst and nonheterocyst types are a major means for maintaining an adequate nitrogen supply to the soil in rice fields (Brady 1979). These microorganisms fix nitrogen when the fields are flooded and die as the fields dry, releasing the fixed nitrogen to the soil. Another important source of available nitrogen in flooded rice fields is the water fern, *Azolla*, which associates with the cyanobacterium *Anabaena*. The *Azolla–Anabaena* association can fix as much as 0.5 kg of atmospheric nitrogen per hectare per day.

Free-living bacteria that are capable of fixing nitrogen are aerobic, facultative, or anaerobic (see Table 12.2). Aerobic nitrogen-fixing bacteria such as *Azotobacter* are thought to maintain reduced oxygen conditions (**microaerobic** conditions) through their high levels of respiration (Burris 1976). Others, such as *Gloeotheca*, evolve O_2 photosynthetically during the day and fix nitrogen during the night. Facultative organisms, which are able to grow under both aerobic and anaerobic conditions, generally fix nitrogen only under anaerobic conditions. For anaerobic nitrogen-fixing bacteria, oxygen does not pose a problem, because it is absent in their

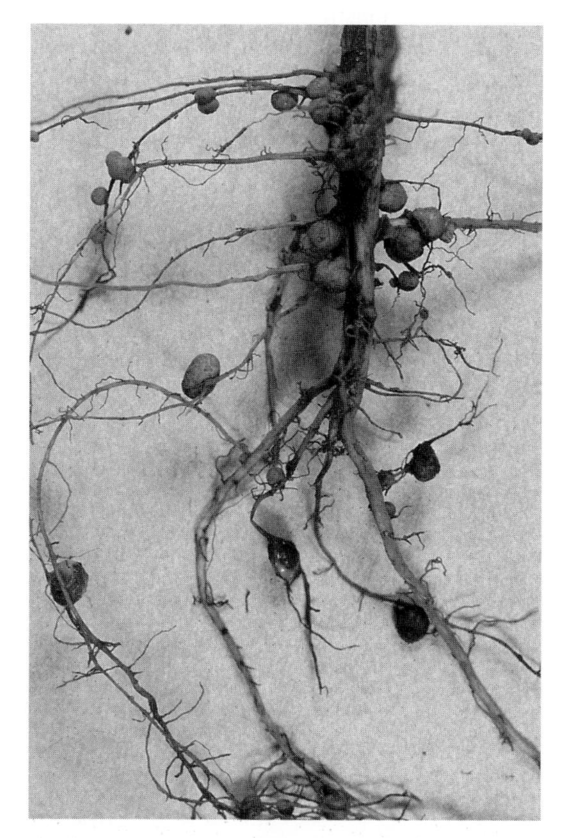

Figure 12.8 Root nodules on soybean. The nodules are a result of infection by *Rhizobium japonicum*. (© Robert Calentine/Visuals Unlimited.)

Table 12.3
Associations between host plants and rhizobia

Plant host	Rhizobial symbiont
Parasponia (a nonlegume, formerly called *Trema*)	*Bradyrhizobium* spp.
Soybean (*Glycine max*)	*Bradyrhizobium japonicum* (slow-growing); *Sinorhizobium fredii* (fast-growing type)
Alfalfa (*Medicago sativa*)	*Sinorhizobium meliloti*
Sesbania (aquatic)	*Azorhizobium* (forms both root and stem nodules; the stems have adventitious roots)
Bean (*Phaseolus*)	*Rhizobium leguminosarum* bv. *phaseoli*; *Rhizobium tropicii*; *Rhizobium etli*
Clovers (*Trifolium*)	*Rhizobium leguminosarum* bv. *trifolii*
Pea (*Pisum sativum*)	*Rhizobium leguminosarum* bv. *viciae*
Aeschenomene (aquatic)	*Photorhizobium* (photosynthetically active rhizobia that form stem nodules, probably associated with adventitious roots)

habitat. These anaerobic organisms can be either photosynthetic (e.g., *Rhodospirillum*), or nonphotosynthetic (e.g., *Clostridium*).

Symbiotic Nitrogen Fixation Occurs in Specialized Structures

Symbiotic nitrogen-fixing prokaryotes dwell within **nodules**, the special organs of the plant host that are separated from the plant cytoplasm by membranes derived from the plant plasma membrane. In the case of *Gunnera*, these organs are existing stem glands that develop independently of the symbiont. In the case of legumes and actinorhizal plants, the nitrogen-fixing bacteria induce the plant to form root nodules.

Grasses can also develop pseudosymbiotic relationships with nitrogen-fixing organisms, but in these associations root nodules are not produced (Smith et al. 1976). Instead, the nitrogen-fixing bacteria seem to be anchored in the root surfaces, mainly around the elongation zone and the root hairs. Bacteria such as *Azospirillum* have been studied intensively for possible application in the cultivation of corn and other grasses. However, recent evidence indicates that *Azospirillum* fixes little nitrogen when associated with plants and that any stimulation of plant development probably derives from the bacterial production of phytohormones such as indoleacetic acid (an auxin discussed in Chapter 19) (Vande Broek and Vanderleyden 1995).

Legumes and actinorhizal plants regulate gas permeability in their nodules, maintaining a level of oxygen within the nodule that can support respiration but is sufficiently low to avoid inactivation of the nitrogenase (Kuzma et al. 1993). Gas permeability increases in the light and decreases under drought or upon exposure

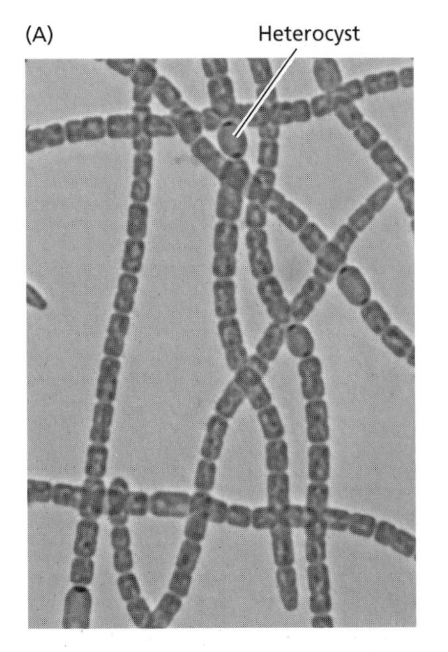

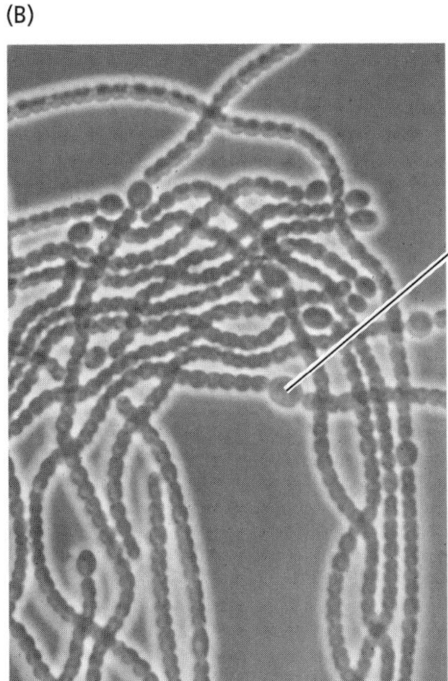

Figure 12.9 Heterocysts in filaments of the nitrogen-fixing cyanobacteria *Anabaena* (A) and *Nostoc* (B) The thick-walled heterocysts, interspaced amongst vegetative cells, have an anaerobic inner environment which allows cyanobacteria to fix nitrogen in aerobic conditions. (From Meeks 1998, courtesy of J. Meeks, © 1998 American Institute of Biological Sciences.)

to nitrate. The mechanism for regulating gas permeability is not yet known.

Nodules contain an oxygen-binding heme protein called **leghemoglobin**. Leghemoglobin is present in the cytoplasm of infected nodule cells at high concentrations (700 μ*M* in soybean nodules) and gives the nodules a pink color. The host plant produces the globin portion of leghemoglobin in response to infection by the bacteria (Marschner 1995); the bacterial symbiont produces the heme portion (Appleby 1984). Leghemoglobin has a high affinity for oxygen; half-saturation occurs at 10 to 20 n*M* (10^{-9} *M*) oxygen, compared with 126 n*M* for the β chain of human hemoglobin. Although leghemoglobin was once thought to provide a buffer for nodule oxygen, recent studies indicate that it stores only enough oxygen to support nodule respiration for a few seconds (Denison and Harter 1995). Its function is to facilitate the diffusion of oxygen to the respiring symbiotic bacterial cells in a manner analogous to hemoglobin in animal circulatory systems (Ludwig and de Vries 1986).

Establishing Symbiosis Requires an Exchange of Signals

The symbiosis between legumes and rhizobia is not obligatory. Legume seedlings germinate without any association with rhizobia, and they may remain unassociated throughout their life cycle. Rhizobia also occur as free-living organisms in the soil. Under nitrogen-limited conditions, however, the symbionts seek out one another through an elaborate exchange of signals. This signaling, the subsequent infection process, and the development of nitrogen-fixing nodules have been studied extensively. Let's look at the specific genes in both the host and the symbionts that are involved in this process.

Plant genes specific to nodules are called **nodulin** (*Nod*) genes; rhizobial genes that participate in nodule formation are called **nodulation** (*nod*) genes (Heidstra and Bisseling 1996). The *nod* genes are classified as common *nod* genes or host-specific *nod* genes. The common *nod* genes—*nodA*, *nodB*, and *nodC*—are found in all rhizobial strains; the host-specific *nod* genes, such as *nodP*, *nodQ* and *NodH*, or *nodF*, *nodE*, and *nodL* differ among rhizobial species and determine the host range. Only one of the *nod* genes, the regulatory *nodD*, is constitutively expressed, and as we will explain in detail, its protein product (NodD) regulates the transcription of the other *nod* genes.

The first stage in the formation of the symbiotic relationship between the nitrogen-fixing bacteria and its host is migration of the bacteria toward the roots of the host plant. This migration is a chemotactic response mediated by chemical attractants, especially (iso)-flavonoids and betaines, secreted by the roots. These attractants activate the rhizobial NodD protein, which then induces transcription of the other *nod* genes

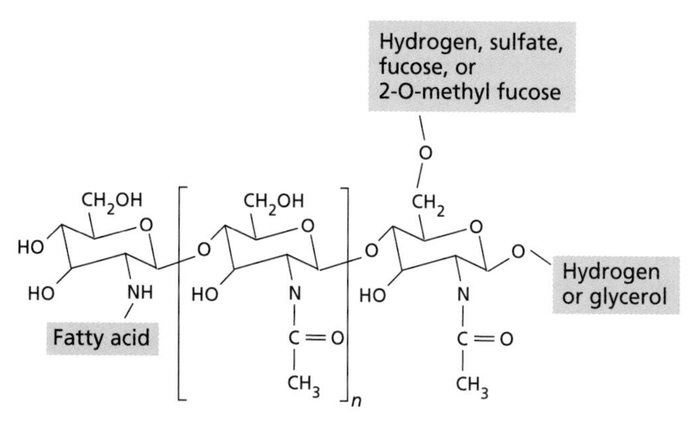

Figure 12.10 Structure of the Nod factors that are lipochitin oligosaccharides. The fatty acid chain typically has 16 to 18 carbons. The number of repeated middle sections (*n*) is usually 2 to 3. (After Stokkermans et al. 1995.)

(Phillips and Kapulnik 1995). The promoter region of all *nod* operons, except that of *nodD*, contains a highly conserved sequence called the *nod* box. Most likely, binding of the activated NodD to the *nod* box induces transcription of the other *nod* genes.

Nod Factors Produced by Bacteria Act as Signals for Symbiosis

The *nod* genes activated by NodD code for nodulation proteins, most of which are involved in the biosynthesis of Nod factors. **Nod factors** are lipochitin oligosaccharide signal molecules, all of which have a chitin β-1→4-linked *N*-acetyl-D-glucosamine backbone, varying in length between three to six sugar units, and a fatty acyl chain on the C-2 position of the nonreducing sugar (Figure 12.10). Three of the *nod* genes—*nodA*, *nodB*, and *nodC*—are required for synthesizing this basic structure. NodA is an *N*-acyltransferase that catalyzes the addition of a fatty acyl chain. NodB is a chitin-oligosaccharide deacetylase that removes the acetyl group from the terminal nonreducing sugar, and NodC is a chitin-oligosaccharide synthase that links *N*-acetyl-D-glucosamine monomers (Stokkermans et al. 1995).

Host-specific *nod* genes, varying in different rhizobial species, are involved in the modification of the fatty acyl chain or the addition of strain-specific substitutions that are important in determining host specificity (Carlson et al. 1995). The host-specific *nodE* and *nodF* genes determine the length and degree of saturation of the fatty acyl chain; those of *Rhizobium leguminosarum* bv. *viciae* and *R. meliloti* determine the synthesis of an 18:4 and a 16:2 fatty acyl group*, respectively. Other genes for host specificity control the addition of specific substitutions

* The number before the colon gives the total number of carbons in the fatty acyl chain; the number after the colon gives the number of double bonds.

at the reducing or nonreducing sugar moieties of the chitin backbone. An example is the *O*-sulfate substitution at the C-6 position of the sugar moiety at the reducing end of the *R. meliloti* Nod factor.

Legume root hairs produce specific lectins (sugar-binding proteins). Nod factors activate these lectins, and this lectin activation may direct particular rhizobia to appropriate hosts and facilitate attachment of the rhizobia to the cell walls of a root hair (van Rhijn et al. in press).

Nodule Formation Involves Several Phytohormones

Two processes, infection and nodule organogenesis, occur simultaneously during root nodule formation. During the infection process, rhizobia attached to the root hairs release Nod factors that induce a pronounced curling of the root hair cells (Figure 12.11A and B). The rhizobia become enclosed in the small compartment formed by the curling. The cell wall of the root hair degrades in these regions, also in response to Nod factors, allowing the bacterial cells direct access to the outer surface of the plant plasma membrane (Lazarowitz and Bisseling 1997).

The next step is formation of the **infection thread** (Figure 12.11C), an internal tubular extension of the plasma membrane that is produced by the fusion of Golgi-derived membrane vesicles at the site of infection. The thread grows at its tip by the fusion of secretory vesicles to the end of the tube. The infected root cortical cells dedifferentiate and start dividing, forming a distinct area within the cortex, called a *nodule primordium*, from which the nodule will develop. The nodule primordia form opposite the protoxylem poles of the root vascular bundle. Different signaling compounds, acting either positively or negatively, control the position of nodule primordia. Uridine diffuses from the stele into the cortex in the protoxylem zones of the root and stimulates cell division (Lazarowitz and Bisseling 1997). Ethylene is synthesized in the region of the pericycle, diffuses into the cortex, and blocks cell division opposite phloem poles.

The infection thread filled with proliferating rhizobia elongates through the root hair and cortical cell layers, in the direction of the nodule primordium. When the infection thread reaches specialized cells within the nodule, the tip of the infection thread fuses with the plasma membrane of the host cell, releasing bacterial cells that are packaged in a membrane derived from the host cell plasma membrane (Figure 12.11D). Branching of the infection thread enables the bacteria to infect many cells (Figure 12.11E and F) (Mylona et al. 1995).

At first, the bacteria continue to divide, and the surrounding membrane increases in surface area to accommodate this growth by fusing with smaller vesicles. Soon thereafter, upon an undetermined signal from the plant, the bacteria stop dividing and begin to enlarge and to differentiate into nitrogen-fixing endosymbiotic organelles called **bacteroids**. The membrane surrounding the bacteroids is called the *peribacteroid membrane*.

The nodule as a whole develops such features as a vascular system (which facilitates the exchange of fixed nitrogen produced by the bacteroids for nutrients contributed by the plant) and a layer of cells to exclude O_2 from the root nodule interior. In some temperate legumes (e.g., peas), the nodules are elongated and cylindrical because of the presence of a nodule meristem. The nodules of tropical legumes, such as soybeans and peanuts, lack a persistent meristem and are spherical (Rolfe and Gresshoff 1988).

The Nitrogenase Enzyme Complex Fixes N_2

Biological nitrogen fixation, like industrial nitrogen fixation, produces ammonia from molecular nitrogen. The overall reaction is

$$N_2 + 8\ e^- + 8\ H^+ + 16\ ATP \rightarrow 2\ NH_3 + H_2 + 16\ ADP + 16\ P_i \tag{12.9}$$

Note that the reduction of N_2 to 2 NH_3, a six-electron transfer, appears to be coupled with the reduction of two protons to evolve H_2. The **nitrogenase enzyme complex** catalyzes this reaction.

The nitrogenase enzyme complex can be separated into two components, the Fe protein and the MoFe protein, neither of which has catalytic activity by itself. The Fe protein is the smaller of the two components and has two identical subunits of 30 to 72 kDa each, depending on the organism. It contains one 4Fe and $4S^{2-}$ per dimer; the iron–sulfur cluster participates in the redox reactions involved in the conversion of N_2 to NH_3. The Fe protein is extremely sensitive to oxygen; it is irreversibly inactivated by O_2 with typical half decay times of 30 to 45 seconds (Dixon and Wheeler 1986). The MoFe protein has four subunits, with a total molecular mass of 180 to 235 kDa, depending on the species. It has two molybdenum atoms per molecule in two Mo–Fe–S clusters, which contain a variable number of Fe–S clusters. The MoFe protein is also inactivated by oxygen, and the half decay time in air is 10 minutes. In the overall nitrogen reduction reaction, ferredoxin (see Chapter 7) serves as an electron donor to the Fe protein, which in turn hydrolyzes ATP and reduces the MoFe protein; the MoFe protein then reduces the N_2 substrate (Figure 12.12).

Nitrogenase can reduce numerous substrates (Table 12.4), although under natural conditions it reacts only with N_2 and H^+. One of the reactions catalyzed by nitrogenase, the reduction of acetylene to ethylene, is used in estimating nitrogenase activity. Direct measurement of

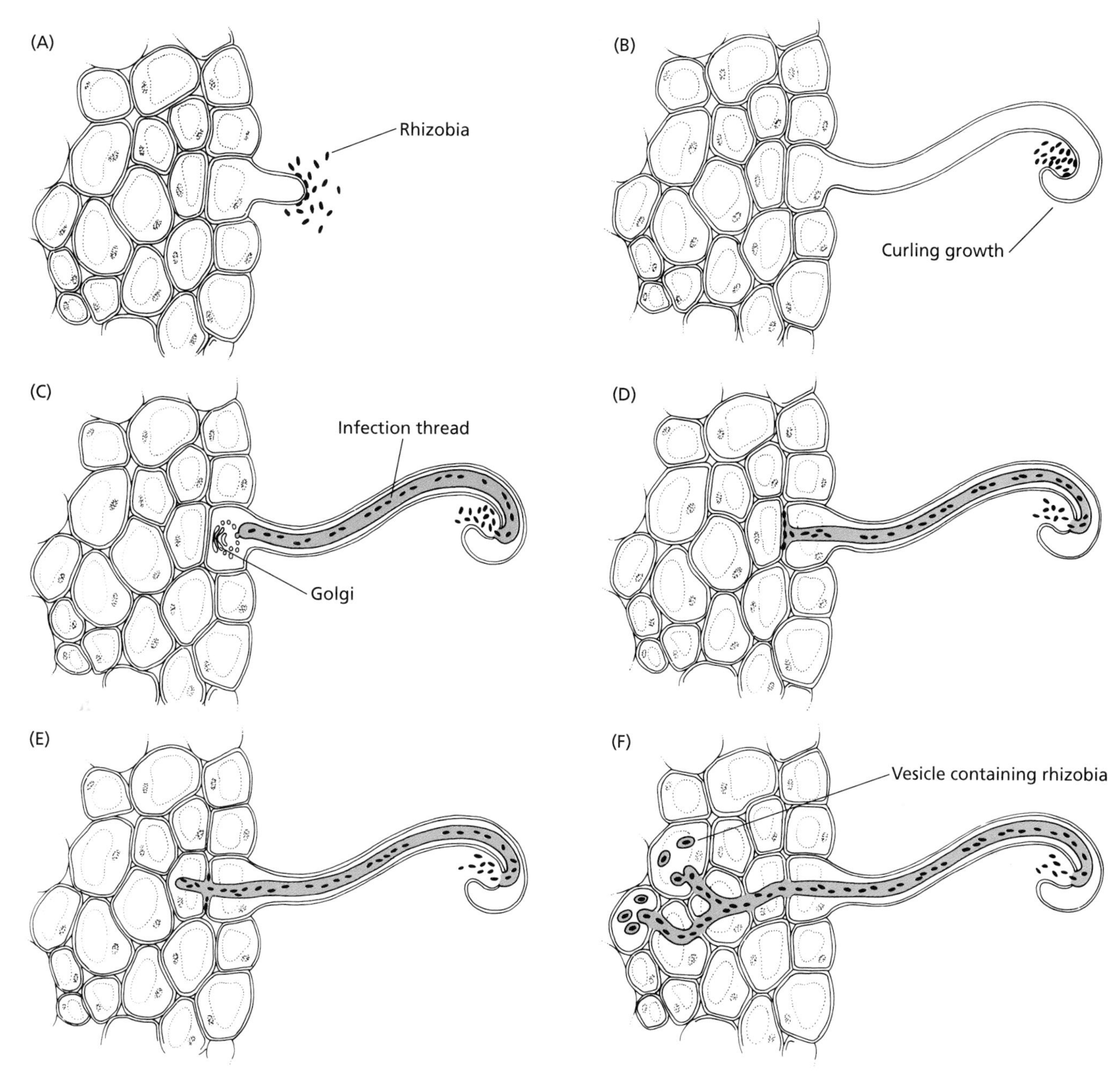

Figure 12.11 The infection process. (A) Rhizobia bind to an emerging root hair in response to chemical attractants sent by the plant. (B) In response to factors produced by the bacteria, the root hair exhibits abnormal curling growth, and rhizobia cells proliferate within the coils. (C) Localized degradation of the root hair wall leads to infection and formation of the infection thread from Golgi secretory vesicles. (D) The infection thread reaches the end of the cell, and its membrane fuses with the plasma membrane of the root hair cell. (E) Rhizobia are released into the apoplast and penetrate the compound middle lamella to the subepidermal cell plasma membrane, leading to the initiation of a new infection thread, which forms an open channel with the first. The two infection threads are discontinuous, however, because the infection thread membrane is absent from the apoplast. (F) The infection thread extends and branches until it reaches target cells, where vesicles containing bacterial cells are released into the cytosol.

TABLE 12.4

Reactions catalyzed by nitrogenase

$N_2 \rightarrow NH_3$	Molecular nitrogen fixation
$N_2O \rightarrow N_2 + H_2O$	Nitrous oxide reduction
$N_3^- \rightarrow N_2 + NH_3$	Azide reduction
$C_2H_2 \rightarrow C_2H_4$	Acetylene reduction
$2\ H^+ \rightarrow H_2$	H_2 production
$ATP \rightarrow ADP + P_i$	ATP hydrolytic activity

Source: After Burris 1976.

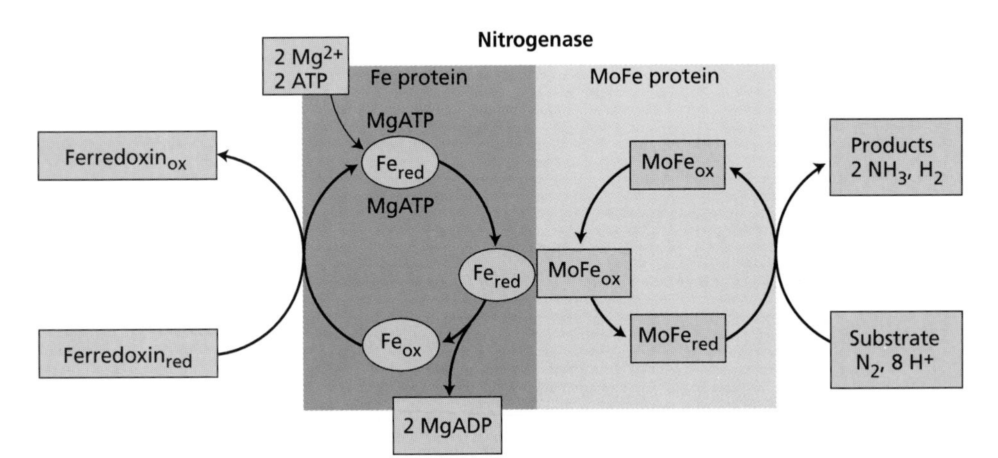

Figure 12.12 The reaction catalyzed by nitrogenase. Ferredoxin reduces the Fe protein. Binding and hydrolysis of ATP to the Fe protein is thought to cause a conformational change of the Fe protein which facilitates the redox reactions. The Fe protein reduces the MoFe protein, and the MoFe protein reduces the N_2. The darker arrows trace the flow of electrons. (From Dixon and Wheeler 1986.)

N_2 fixation requires a mass spectrometer, instrumentation that is not readily available, but the reduction of acetylene to ethylene can easily be measured by gas chromatography (Dilworth 1966). Acetylene reduction involves two electrons, as opposed to the eight electrons required for the reduction of N_2 and 2 H^+ (see Equation 12.9). Therefore, four ethylene molecules correspond to the reduction of one N_2 molecule. The acetylene method has limitations, however: Acetylene blocks gas diffusion into nodules, and rhizosphere microorganisms produce ethylene. For these reasons, the acetylene method is suitable for comparative studies, while mass spectroscopy is used for direct measurements of N_2 reduction required for precise quantification of nitrogen fixation.

The energetics of nitrogen fixation is complex. The production of NH_3 from N_2 and H_2 is an exergonic reaction (see Chapter 2), with a $\Delta G^{0'}$ (change in free energy) of –27 kJ mol^{-1} (Dixon and Wheeler 1986). However, industrial production of NH_3 from N_2 and H_2 requires a very large energy input because of the activation energy needed to break the triple bond in N_2. For the same reason, the enzymatic reduction of N_2 by nitrogenase also requires a large investment of energy (see Equation 12.9), although the exact changes in free energy remain to be established.

The energy cost of biological nitrogen fixation can be estimated from carbohydrate utilization by plants (Heytler et al. 1984). Calculations show that a plant consumes 12 g of organic carbon per gram of N_2 fixed. On the basis of Equation 12.9, the $\Delta G^{0'}$ for the overall reaction of biological nitrogen fixation is about –200 kJ mol^{-1}. Since the overall reaction is highly exergonic, ammonium production is limited by the rather slow turnover rate (number of N_2 molecules reduced per unit time) of the nitrogenase complex (Ludwig and de Vries 1986).

Under natural conditions, the reduction of H^+ to H_2 gas can be substantial and can compete with N_2 for electrons from nitrogenase. In rhizobia, 30 to 60% of the energy supplied to nitrogenase may be lost as H_2, diminishing the efficiency of nitrogen fixation (Schubert et al. 1978). Some rhizobia, however, contain hydrogenase, an enzyme that can split the H_2 formed and generate electrons for N_2 reduction, thus improving the efficiency of nitrogen fixation (Marschner 1995).

Nitrogen-Fixing Plants Export Amides and Ureides

The symbiotic nitrogen-fixing prokaryotes release ammonia that, to avoid toxicity, must be rapidly converted into organic forms in the root nodules before being transported to the shoot via the xylem. On the basis of the composition of the xylem sap, nitrogen-fixing legumes can be divided into amide exporters or ureide exporters. Temperate-region legumes, such as pea (*Pisum*), clover (*Trifolium*), broad bean (*Vicia*), and lentil (*Lens*), tend to export amides, principally the amino acids asparagine or glutamine. Legumes of tropical origin, such as soybean (*Glycine*), kidney bean (*Phaseolus*), peanut (*Arachis*), and southern pea (*Vigna*), export nitrogen in the form of ureides.

The three major ureides are allantoin, allantoic acid, and citrulline (Figure 12.13). Allantoin is synthesized in peroxisomes from uric acid, and allantoic acid is synthesized from allantoin in the endoplasmic reticulum. The site of citrulline synthesis from the amino acid ornithine has not yet been determined. All three compounds are ultimately released into the xylem and transported to the shoot, where they are rapidly catabolized to ammonium. This ammonium enters the assimilation pathway described earlier.

Sulfur Assimilation

Sulfur is among the most versatile elements in living organisms (Hell 1997). Disulfide bridges in proteins play structural and regulatory roles (see Chapter 8). Sulfur participates in electron transport through iron–sulfur clusters (see Chapters 7 and 11). The catalytic sites for several enzymes and coenzymes, such as urease and coenzyme A, contain sulfur. Secondary sulfur com-

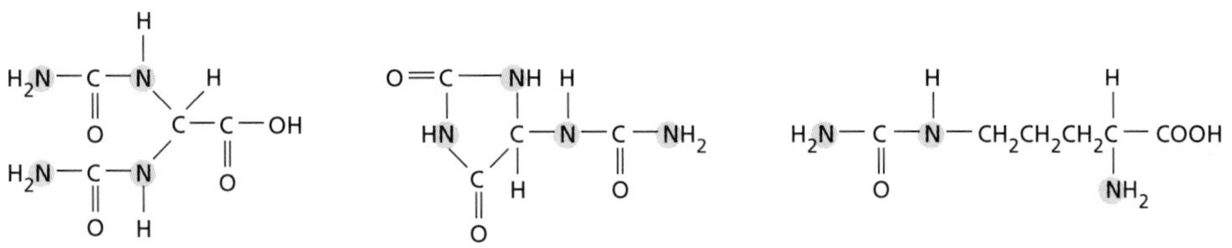

Figure 12.13 Structure of the major ureide compounds found in nitrogen-fixing plants.

pounds range from the rhizobial Nod factors discussed in the previous section to antiseptic alliin in garlic and anticarcinogen sulforaphane in broccoli.

The versatility of sulfur derives in part from the property that it shares with nitrogen: multiple stable oxidation states. In this section, we discuss the enzymatic steps that mediate sulfur assimilation, and the biochemical reactions that catalyze the reduction of sulfate into the two sulfur-containing amino acids, cysteine and methionine.

Sulfate Is the Absorbed Form of Sulfur in Plants

Most of the sulfur in higher-plant cells derives from sulfate (SO_4^{2-}) absorbed from the soil solution. Sulfate in the soil comes predominantly from the weathering of parent rock material. Industrialization, however, adds an additional source of sulfate: atmospheric pollution. The burning of fossil fuels releases several gaseous forms of sulfur, including sulfur dioxide (SO_2) and hydrogen sulfide (H_2S), which find their way to the soil in rain. When dissolved in water, SO_2 is hydrolyzed to become sulfuric acid (H_2SO_4), a strong acid, which is the major source of acid rain. Plants can also metabolize sulfur dioxide taken up in the gaseous form through their stomata. Nonetheless, prolonged exposure (more than 8 hours) to high atmospheric concentrations (greater than 0.3 parts per million) of H_2S causes extensive tissue damage because of the formation of sulfuric acid.

Sulfate Assimilation Requires the Reduction of Sulfate to Cysteine

The first step in the synthesis of sulfur-containing organic compounds is the reduction of sulfate to the amino acid cysteine (Figure 12.14). Sulfate is very stable and thus needs to be activated before any subsequent reactions may proceed. Activation begins with the reaction between sulfate and ATP to form adenosine-5′-phosphosulfate (APS) and pyrophosphate (PP_i) (see Figure 12.14B):

$$\text{Sulfate} + \text{ATP} \rightarrow \text{APS} + \text{PP}_i \quad \Delta G^{0\prime} = 45 \text{ kJ mol}^{-1} \quad (12.10)$$

The enzyme that catalyzes this reaction, ATP sulfurylase, has cytosolic and plastid forms in spinach and potato, but it seems restricted to plastids in *Arabidopsis thaliana* (Leustek 1996). The activation reaction is energetically unfavorable, as indicated by a positive $\Delta G^{0\prime}$ and a small equilibrium constant ($K_{sp} \cong 10^{-7}$).* To drive this reaction forward, the products APS and PP_i must be converted rapidly to other compounds.

PP_i is hydrolyzed to inorganic phosphate (P_i) according to the following reaction:

$$\text{PP}_i + \text{H}_2\text{O} \rightarrow 2\ \text{P}_i \quad \Delta G^{0\prime} = -33.5 \text{ kJ mol}^{-1} \quad (12.11)$$

This reaction, which is catalyzed by inorganic pyrophosphatase, provides additional driving force for sulfate activation. The other product, APS, is short-lived, and its fate is unknown (Leustek 1996; Hell 1997). It may react with ATP to form 3′-phosphoadenosine-5′-phosphosulfate (PAPS) (see Figure 12.14A):

$$\text{APS} + \text{ATP} \rightarrow \text{PAPS} + \text{ADP} \quad \Delta G^{0\prime} = -25 \text{ kJ mol}^{-1} \quad (12.12)$$

PAPS is then reduced to sulfite (SO_3^{2-}) and then to sulfide (S^{2-}).This is the major pathway for sulfate assimilation in bacteria and fungi. Alternatively, the sulfur in APS may be converted to an enzyme-bound thiosulfonate ($R–SO_3^-$), which in turn is reduced to a thiosulfide ($R–S^-$) (see Figure 12.14B). Yet a third possibility is that APS is directly reduced to sulfite and then to sulfide (Figure 12.14C) (Hell 1997). In all three pathways, the resultant thiosulfide or sulfide reacts with *O*-acetylserine to form cysteine and acetate (Figure 12.14D). The enzymes involved in cysteine synthesis have been found in the cytosol, plastids, and mitochondria of various plants, probably reflecting the inability of organelles to transport cysteine across their membranes (Hell 1997).

Sulfate Assimilation Occurs Mostly in Leaves

The reduction of sulfate to cysteine changes the oxidation number of sulfur from +6 to –4, thus entailing the

* K_{sp} stands for *solubility product constant* and replaces K_{eq} (equilibrium constant, see Chapter 2) when one of the substances in a chemical reaction has low solubility and readily precipitates into a solid. A low K_{sp} value indicates that the substances in the reaction that remain in solution can only be present in low concentration without precipitating.

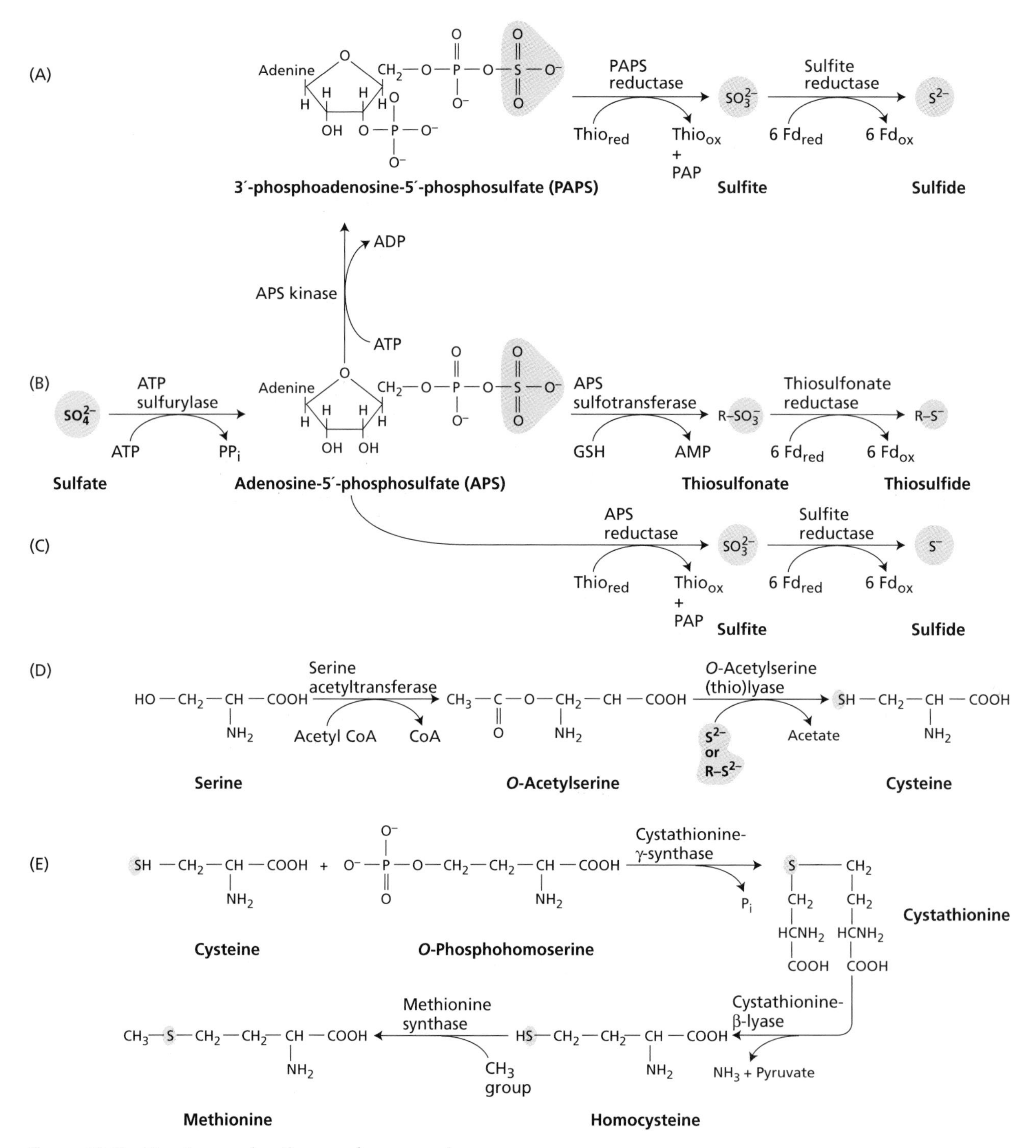

Figure 12.14 Structure and pathways of compounds involved in sulfur assimilation. (A–C) Sulfide or thiosulfide is produced from adenosine-5′-phosphosulfate (APS) through one of three pathways. The pathway in A operates mainly in bacteria and fungi. The enzyme ATP sulfurylase cleaves pyrophosphate from ATP and replaces it with sulfate in the pathway in B. (D) Serine reacts with acetyl-CoA to produce *O*-acetylserine and the sulfide or thiosulfide produced reacts with *O*-acetylserine to form cysteine. (E) Cysteine may be further modified to produce methionine. Fd, ferredoxin; GSH, glutathione; Thio, thioredoxin.

transfer of 10 electrons. Glutathione, thioredoxin, ferredoxin, NAD(P)H, or *O*-acetylserine may serve as electron donors at various steps of the pathway (see Figure 12.14). Leaves are generally much more active than roots in sulfur assimilation, presumably because photosynthesis provides reduced thioredoxin and ferredoxin and photorespiration generates serine that may stimulate the production of *O*-acetylserine (see Chapter 8). Sulfur assimilated in leaves is exported via the phloem to sites of protein synthesis (shoot and root apices, and fruits) mainly as glutathione (Bergmann and Rennenberg 1993). Glutathione also acts as a signal that coordinates the absorption of sulfate by the roots and the assimilation of sulfate by the shoot.

Methionine Is Synthesized from Cysteine

Methionine, the other sulfur-containing amino acid found in proteins, is synthesized in plastids from cysteine. In the first step of the pathway, cysteine and *O*-phosphohomoserine react to form cystathionine via the enzyme cystathionine-γ-synthase (see Figure 12.14E). Cystathionine is cleaved into homocysteine, pyruvate, and ammonia by cystathionine-β-lyase. Finally, methionine synthase methylates homocysteine to form methionine. After cysteine and methionine are synthesized, sulfur can be incorporated into proteins and a variety of other compounds, such as acetyl coenzyme A and *S*-adenosylmethionine. The latter compound is important in the synthesis of ethylene (see Chapter 22) and in reactions involving the transfer of methyl groups, as in lignin synthesis (see Chapter 13).

Phosphate Assimilation

Phosphate (HPO_4^{2-}) in the soil solution is readily absorbed by plant roots and incorporated into a variety of organic compounds, including sugar phosphates, phospholipids, and nucleotides. The main entry point of phosphate into assimilatory pathways occurs during the formation of ATP, the energy "currency" of the cell. In the overall reaction for this process, inorganic phosphate is added to the second phosphate group in adenosine diphosphate to form a phosphate ester bond.

In mitochondria, the energy for ATP synthesis derives from the oxidation of NADH by oxidative phosphorylation (see Chapter 11). ATP synthesis is also driven by light-dependent photophosphorylation in the chloroplasts (see Chapter 7). In addition to these reactions in mitochondria and chloroplasts, reactions in the cytosol assimilate phosphate. Glycolysis incorporates inorganic phosphate into 1,3-bisphosphoglyceric acid, forming a high-energy acyl phosphate group. This phosphate can be donated to ADP to form ATP in a **substrate-level phosphorylation** reaction (Figure 12.15). Once incorporated into ATP, the phosphate group may be transferred via many different reactions to form the various phosphorylated compounds found in higher-plant cells.

Cation Assimilation

Cations taken up by plant cells form complexes with organic compounds in which the cation becomes bound to a carbon compound without covalent bonds being formed. Macronutrient cations such as potassium, magnesium, and calcium, as well as micronutrient cations such as copper, iron, manganese, cobalt, sodium, and zinc, are taken up in this manner. In this section we will describe coordination bonds and electrostatic bonds, which mediate the absorption of several cations that plants require as nutrients, and the special requirements for the assimilation of iron by roots.

Cations Form Coordination Bonds or Electrostatic Bonds with Carbon Compounds

In the formation of a coordination complex, oxygen or nitrogen atoms of a carbon compound donate unshared electrons to form a bond with the cation nutrient. As a result, the positive charge on the cation is neutralized. Coordination complexes typically form between polyvalent cations and organic molecules, such as copper and tartaric acid (Figure 12.16A) or magnesium and chlorophyll *a* (Figure 12.16B), in which the magnesium is bound by coordination bonds with the nitrogen atoms in the porphyrin ring. The nutrients that are assimilated as coordination complexes include copper, zinc, iron, and magnesium. Calcium can also form coordination complexes with the polygalacturonic acid of cell walls (Figure 12.16C).

Electrostatic bonds form because of the attraction of a positively charged cation for a negatively charged group on an organic compound. Unlike the situation in coordination bonds, the cation in an electrostatic bond retains its positive charge. An important negatively charged group is the ionized form of a carboxylic acid

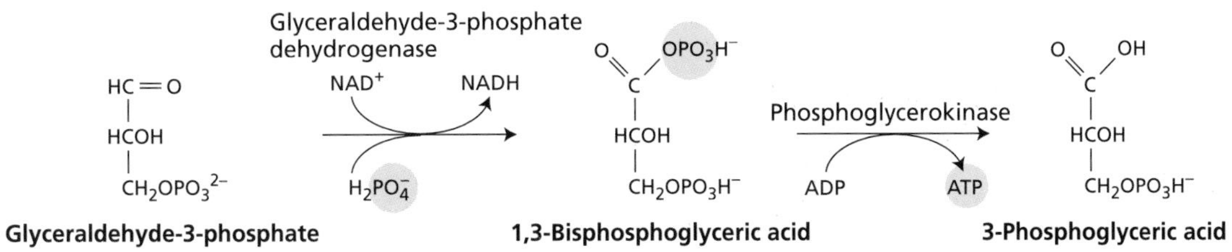

Figure 12.15 Substrate-level phosphorylation. In this reaction, which occurs in glycolysis, inorganic phosphate is incorporated into 1,3-bisphosphoglyceric acid (first reaction), and this compound is broken down to produce ATP from ADP (second reaction).

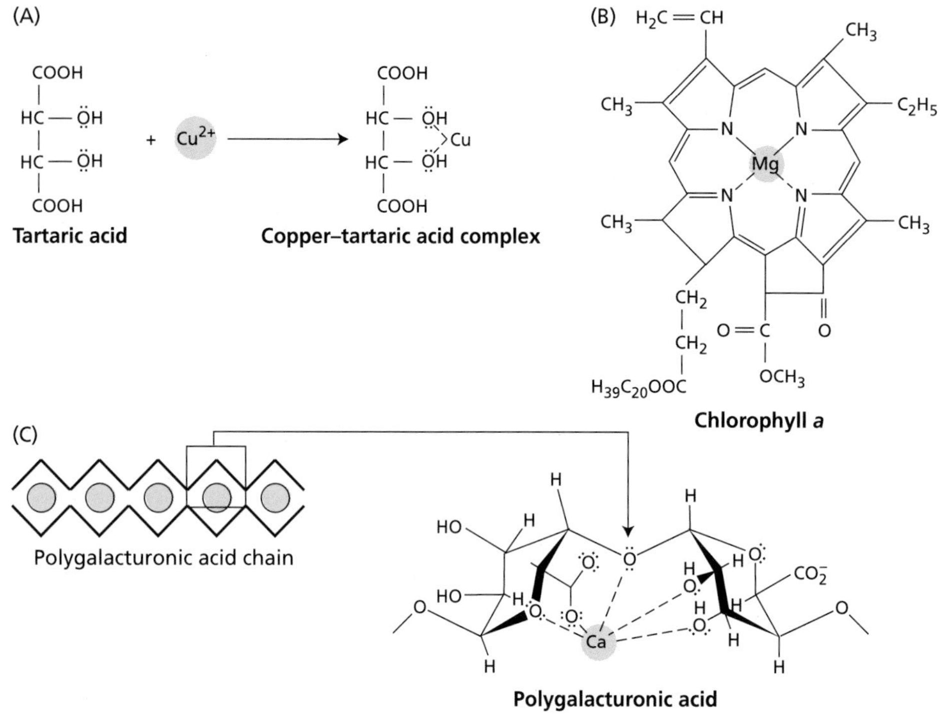

Figure 12.16 Examples of coordination complexes. Coordination complexes form when oxygen or nitrogen atoms of a carbon compound donate unshared electrons (represented by pairs of dots) to form a bond with a cation. (A) Copper ions share electrons with the hydroxyl oxygens of tartaric acid. (B) Magnesium ions share electrons with nitrogen atoms in chlorophyll *a*. Dashed lines represent a coordination bond between unshared electrons from the nitrogen atoms and the magnesium cation. (C) The "egg-box" model of the interaction of polygalacturonic acid and calcium ions. On the left, calcium ions are held in the spaces between two polygalacturonic acid chains, indicated by the kinked, horizontal lines. At right is an enlargement of a single calcium ion forming a coordination complex with the hydroxyl oxygens of the galacturonic acid residues. Much of the calcium in the cell wall is thought to be bound in this fashion. (After Rees 1977.)

(–COO^-). Monovalent cations such as potassium (K^+) can form electrostatic bonds with the carboxylic groups of many organic acids (Figure 12.17A). However, much of the potassium that is accumulated by plant cells and functions in osmotic regulation and enzyme activation remains in the cytosol and the vacuole as the free ion. Divalent ions such as calcium form electrostatic bonds with pectates (Figure 12.17B) and the carboxylic groups of polygalacturonic acid (see Chapter 15).

In general, cations such as magnesium (Mg^{2+}) and calcium (Ca^{2+}) are assimilated by the formation of both coordination complexes and the formation of electrostatic bonds with amino acids, phospholipids, and other negatively charged molecules.

Roots Modify the Rhizosphere to Acquire Iron

Iron is important in iron–sulfur proteins (see Chapter 7) and as a catalyst in enzyme-mediated redox reactions

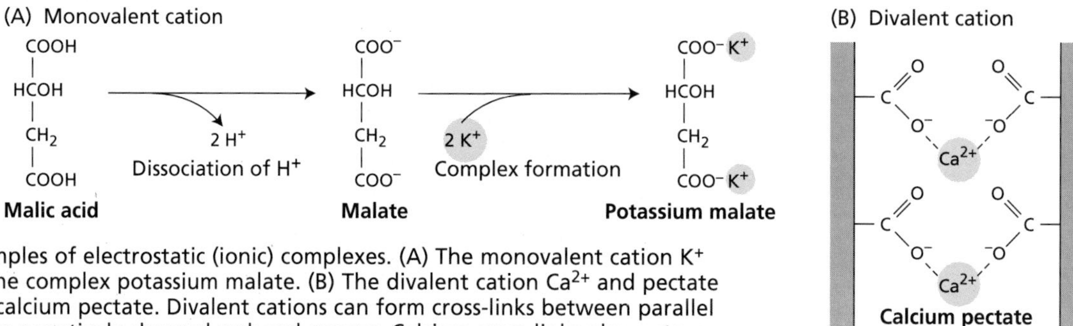

Figure 12.17 Examples of electrostatic (ionic) complexes. (A) The monovalent cation K^+ and malate form the complex potassium malate. (B) The divalent cation Ca^{2+} and pectate form the complex calcium pectate. Divalent cations can form cross-links between parallel strands that contain negatively charged carboxyl groups. Calcium cross-links play a structural role in the cell walls.

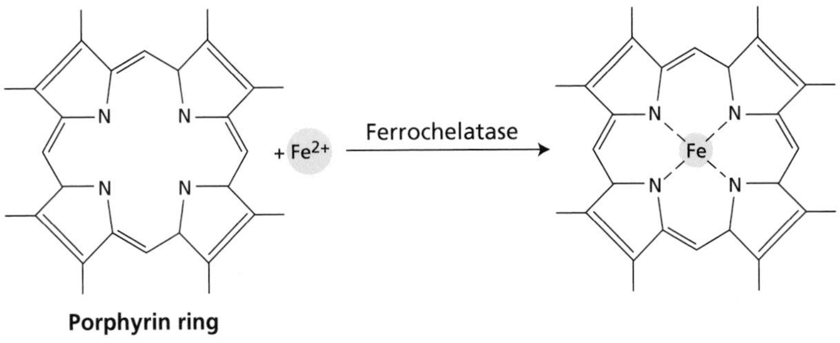

Figure 12.18 The ferrochelatase reaction. The enzyme ferrochelatase catalyzes the insertion of iron into the porphyrin ring to form a coordination complex. See Figure 7.35 for illustration of the biosynthesis of the porphyrin ring.

(see Chapter 5). Plants obtain iron from the soil, where it is present primarily as ferric iron (Fe^{3+}) in oxides such as $Fe(OH)^{2+}$, $Fe(OH)_3$, and $Fe(OH)_4^-$. At neutral pH, ferric iron is highly insoluble, as indicated by a K_{sp} value of 2×10^{-39}. To absorb sufficient amounts of iron from the soil solution, roots have developed several mechanisms that increase iron solubility and thus its availability. These mechanisms involve (1) soil acidification that increases the solubility of ferric iron, (2) reduction of ferric iron to the more soluble ferrous form (Fe^{2+}), or (3) release of chelators (see Figure 5.2) that form stable, soluble complexes with iron (Marschner 1995).

Roots generally acidify the soil around them. They extrude protons during the absorption and assimilation of cations, particularly ammonium, and release organic acids such as malic acid and citric acid that enhance iron and phosphate availability (see Figure 5.4). Iron deficiencies stimulate the extrusion of protons by roots. In addition, plasma membranes in roots contain an enzyme, called iron-chelating reductase, that reduces ferric iron to the ferrous form, with NADH or NADPH serving as the electron donor. The activity of this enzyme increases under iron deprivation.

Several root exudates form stable chelates with iron. Examples include malic acid, citric acid, phenolics, and piscidic acid. Grasses produce a special class of iron chelators called **phytosiderophores**. Phytosiderophores are made of amino acids that are not found in proteins, such as mugineic acid, and form highly stable complexes with Fe^{3+} ($K_{sp} \cong 10^{33}$). Root cells of grasses have Fe^{3+}–phytosiderophore transport systems in their plasma membrane that bring the chelate into the cytoplasm. Under iron deficiency, grass roots release more phytosiderophores into the soil and increase the capacity of their Fe^{3+}–phytosiderophore transport system.

Iron Forms Complexes with Carbon and Phosphate

Once the roots absorb iron or an iron chelate, they oxidize it to a ferric form and translocate much of it to the leaves as an electrostatic complex with citrate (Olsen et al. 1981). Most of the iron in the plant is found in the heme molecule of cytochromes within the chloroplasts and mitochondria (see Chapters 7 and 11). An important assimilatory reaction for iron is its insertion into the porphyrin precursor of heme (see Chapter 7). This reaction, called the ferrochelatase reaction (Figure 12.18), is catalyzed by the enzyme ferrochelatase (Jones 1983). In addition, iron–sulfur proteins of the electron transport chain (see Chapter 7) contain nonheme Fe covalently bound to the sulfur atoms of cysteine residues in the apoprotein. Iron is also found in Fe_2S_2 centers, which contain two irons, each complexed with the sulfur atoms of cysteine residues, and two inorganic sulfides.

Free iron (iron that is not complexed with organic compounds) may interact with oxygen to form superoxide anions (O_2^-), which can damage membranes by degrading unsaturated lipid components (Trelstad et al. 1981). Plant cells, however, store surplus iron in an iron–protein complex, called **phytoferritin** (Bienfait and Van der Mark 1983). Phytoferritin consists of a protein shell with 24 identical subunits forming a hollow sphere that has a molecular mass of about 480 kDa. Within this sphere is a core of 5400 to 6200 iron atoms present as a ferric oxide–phosphate complex. How iron is released from phytoferritin is uncertain, but breakdown of the protein shell appears to be involved. The level of free iron in plant cells regulates the de novo biosynthesis of phytoferritin (Lobreaux et al. 1992).

Oxygen Assimilation

Respiration accounts for the bulk (about 90%) of the oxygen (O_2) utilized by plant cells (see Chapter 11). Another major pathway for the assimilation of O_2 into organic compounds involves the incorporation of O_2 from water (see reaction 1 in Table 8.1). A small proportion of oxygen can be directly assimilated into organic compounds in the process of **oxygen fixation**. In oxygen fixation, molecular oxygen is added directly to an organic compound in reactions carried out by enzymes known as oxygenases (Metzler 1977). Recall from Chaper 8 that oxygen is directly incorporated into an organic compound during photorespiration in a reaction that involves the oxygenase activity of ribulose-1,5-bisphosphate carboxylase/oxygenase (rubisco), the enzyme of CO_2 fixation (Ogren 1984). The first stable product that contains oxygen originating from molecular oxygen is 2-phosphoglycolate.

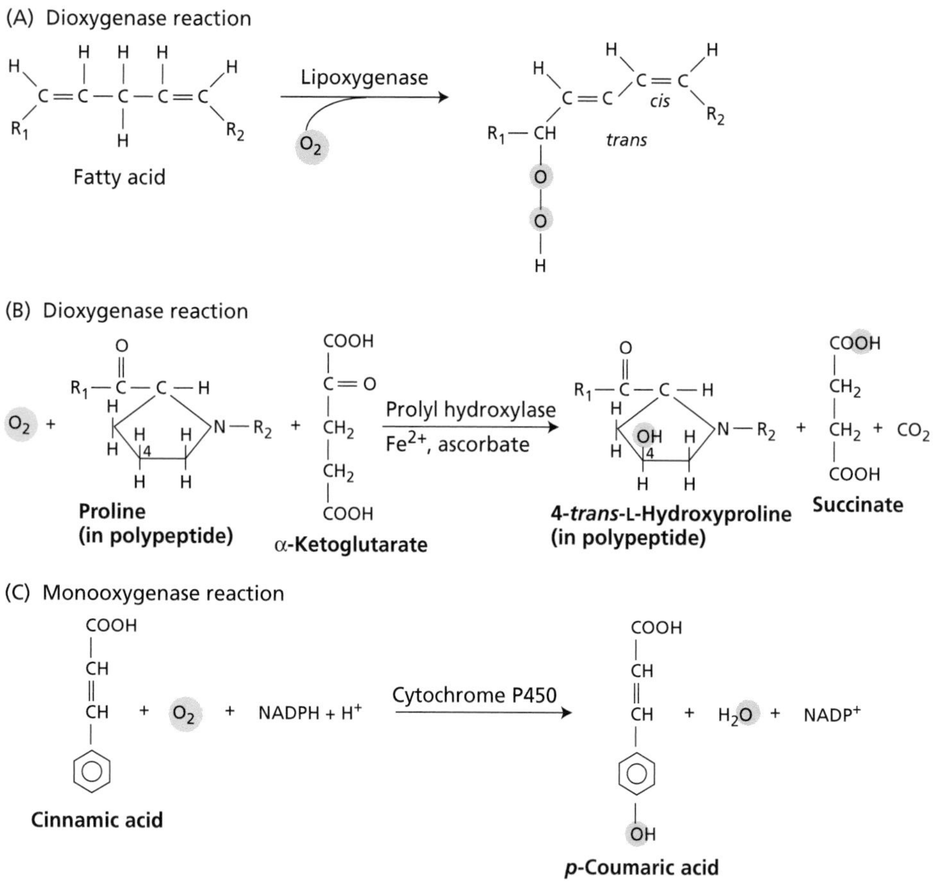

Figure 12.19 Examples of the two types of oxygenase reactions in cells of higher plants. (A) The dioxygenase lipoxygenase catalyzes the addition of two atoms of oxygen to the conjugated portion of an unsaturated fatty acid to form a hydroperoxide with a pair of *cis–trans* conjugated double bonds. The hydroxy peroxy fatty acid may then be enzymatically converted to hydroxy fatty acids and other metabolites. (B) The dioxygenase prolyl hydroxylase catalyzes the addition of one oxygen from O_2 to proline in a polypeptide chain to produce hydroxyproline, and one oxygen to α-ketoglutarate to produce succinate and CO_2. (C) The monooxygenase cytochrome P450 uses one oxygen from O_2 to hydroxylate cinnamic acid (and other substrates) and the other oxygen to produce water. NAD(P)H serves as the electron donor for monooxygenase reactions.

In general, oxygenases are classified as **dioxygenases** or **monooxygenases**, according to the number of atoms of oxygen that are transferred to an organic compound in the catalyzed reaction. Examples of dioxygenases in plant cells are lipoxygenase, which catalyzes the addition of two atoms of oxygen to unsaturated fatty acids (Figure 12.19A), and prolyl hydroxylase, the enzyme that converts proline to the less common amino acid hydroxyproline (Figure 12.19B). Hydroxyproline is an important component of the cell wall protein **extensin** (see Chapter 15). Prolyl hydroxylase is considered a dioxygenase because one oxygen atom is incorporated into proline to form hydroxyproline, while a second oxygen is used to convert α-ketoglutarate to succinate. Ferrous iron and ascorbate are also needed as cofactors, but they do not participate directly in the redox reaction (see Figure 12.19B). The synthesis of hydroxyproline from proline differs from the synthesis of all other amino acids in that the reaction occurs *after* the proline has been incorporated into protein and is therefore a posttranslational modification reaction. Prolyl hydroxylase is localized in the endoplasmic reticulum, suggesting that most proteins containing hydroxyproline are found in the secretory pathway.

Monooxygenases add one of the atoms in molecular oxygen to an organic compound; the other oxygen atom is converted into water. Monooxygenases are sometimes referred to as *mixed-function oxidases* because of their ability to catalyze simultaneously both the oxygenation reaction and the oxidase reaction (reduction of oxygen to water). The monooxygenase reaction also requires a reduced substrate (NADH or NADPH) as an electron donor, according to the following equation:

$$A + O_2 + BH_2 \rightarrow AO + H_2O + B$$

where A represents an organic compound and B represents the electron donor.

An important monooxygenase in plants is the family of heme proteins collectively called cytochrome P450, which catalyzes the hydroxylation of cinnamic acid to *p*-coumaric acid (Figure 12.19C). In monooxygenases, the oxygen is first activated by being combined with the iron atom of the heme group; NADPH serves as the electron donor. The mixed-function oxidase system is localized on the endoplasmic reticulum and is capable of oxidizing a variety of substrates, including mono- and diterpenes and fatty acids.

The Energetics of Nutrient Assimilation

Nutrient assimilation generally requires large amounts of energy to convert stable, low-energy inorganic compounds into high-energy organic compounds. For example, the reduction of nitrate to nitrite and then to ammonium requires the transfer of about ten electrons and accounts for about 25% of the total energy expenditures in both roots and shoots (Bloom 1997). Consequently, a plant may use one-fourth of its energy to assimilate nitrogen, a constituent that accounts for less than 2% of the total dry weight of the plant.

Many of these assimilatory reactions occur in the stroma of the chloroplast, where they have ready access to powerful reducing agents such as NADPH, thioredoxin, and ferredoxin generated during photosynthetic electron transport. This process—coupling nutrient assimilation to photosynthetic electron transport—is called **photoassimilation**. Photoassimilation and the Calvin cycle occur in the same compartment, but they do not seem to compete for reductant. That is, only when photosynthetic electron transport generates reductant in excess of the needs of the Calvin cycle (for instance, under conditions of high light and low CO_2), does photoassimilation proceed (Robinson 1988), and high levels of CO_2 inhibit photoassimilation (Bloom 1997). As a result, C_4 plants (see Chapter 8) conduct the majority of their photoassimilation in mesophyll cells, where the CO_2 concentrations are lower (Becker et al. 1993). The mechanisms that regulate the partitioning of reductant between the Calvin cycle and photoassimilation warrant investigation because atmospheric levels of CO_2 are expected to double during the next century (see Chapter 9), so this phenomenon may affect the future of plant–nutrient relations.

Summary

Nutrient assimilation is the process by which nutrients acquired by plants are incorporated into the carbon constituents necessary for growth and development. These processes often involve chemical reactions that are highly energy intensive and thus may depend directly on reductant generated through photosynthesis.

For nitrogen, assimilation is but one in a series of steps that constitute the nitrogen cycle. The nitrogen cycle encompasses the various states of nitrogen in the biosphere and their interconversions. The principal sources of nitrogen available to plants are nitrate (NO_3^-) and ammonium (NH_4^+). The nitrate absorbed by roots is assimilated in either roots or shoots, depending on nitrate availability and plant species. In nitrate assimilation, nitrate is reduced to nitrite (NO_2^-) in the cytosol via the enzyme nitrate reductase; then nitrite is reduced to ammonium in root plastids or chloroplasts via the enzyme nitrite reductase.

Ammonium, derived either from root absorption or generated through nitrate assimilation or photorespiration, is converted to glutamine and glutamate through the sequential actions of glutamine synthetase and glutamate synthase, which are located in the cytosol and root plastids or chloroplasts. Once assimilated into glutamine or glutamate, nitrogen may be transferred to many other organic compounds through various reactions, including the transamination reactions. Interconversion between glutamine and asparagine via asparagine synthetase balances carbon metabolism and nitrogen metabolism within a plant.

Many plants form a symbiotic relationship with nitrogen-fixing bacteria that contain an enzyme complex, nitrogenase, that can reduce atmospheric nitrogen to ammonia. Legumes and actinorhizal plants form associations with rhizobia and *Frankia*, respectively. These associations result from a finely tuned interaction between the symbiont and host plant that involves the recognition of specific signals, the induction of a specialized developmental program within the plant, the uptake of the bacteria by the plant, and the development of nodules, unique organs that house the bacteria within plant cells. Some nitrogen-fixing prokaryotic microorganisms do not form symbiotic relationships with higher plants but benefit plants by enriching the nitrogen content of the soil.

Like nitrate, sulfate (SO_4^{2-}) must be reduced by assimilation. In sulfate reduction, an activated form of sulfate called adenosine-5′-phosphosulfate (APS) forms. Sulfide (H_2S), the end product of sulfate reduction, does not accumulate in plant cells, but is instead rapidly incorporated into the amino acids cysteine and methionine.

Phosphate (HPO_4^{2-}) is present in a variety of compounds found in plant cells, including sugar phosphates, lipids, nucleic acids, and free nucleotides. The initial product of its assimilation is ATP, which is produced by substrate-level phosphorylations in the cytosol, oxidative phosphorylation in the mitochondria, and photophosphorylation in the chloroplasts.

Whereas the assimilation of nitrogen, sulfur, and phosphorus requires the formation of covalent bonds with carbon compounds, many macro- and micronutrient cations (e.g., K^+, Mg^{2+}, Ca^{2+}, Cu^{2+}, Fe^{3+}, Mn^{2+}, Co^{2+}, Na^+, Zn^{2+}) simply form complexes. These complexes may be held together by electrostatic bonds or coordination bonds. Iron assimilation may involve chelation, oxidation–reduction reactions, and the formation of complexes. In order to store large amounts of iron, plant cells synthesize phytoferritin, an iron storage protein. An important function of iron in plant cells is to act as a redox component in the active site of enzymes, often as an iron–porphyrin complex. Iron is inserted into a porphyrin group in the ferrochelatase reaction.

In addition to being utilized in respiration, molecular oxygen can be assimilated in the process of oxygen fixation, the direct addition of oxygen to organic compounds. This process is catalyzed by enzymes known as oxygenases, which are classified as monooxygenases or dioxygenases.

General Reading

*Appleby, C. A. (1984) Leghemoglobin and rhizobium respiration. *Annu. Rev. Plant Physiol.* 35: 443–478.

*Beevers, L. (1976) *Nitrogen Metabolism in Plants.* Elsevier, London.

Crawford, N. M., and Arst, H. N. (1993) The molecular genetics of nitrate assimilation in fungi and plants. *Annu. Rev. Genet.* 27: 115–146.

*Dixon, R. O. D., and Wheeler, C. T. (1986) *Nitrogen Fixation in Plants.* Chapman and Hall, New York.

George, E., Marschner, H., and Jakobsen, I. (1995) Role of arbuscular mycorrhizal fungi in uptake of phosphorus and nitrogen from soil. *Crit. Rev. Biotechnol.* 15(3-4): 257–270.

*Lam, H-M., Coschigano, K. T., Oliveira, I. C., Melo-Oliveira, R., and Coruzzi, G. M. (1996) The molecular-genetics of nitrogen assimilation into amino acids in higher plants. *Annu. Rev. Plant Physiol. Plant Mol. Biol.* 47: 569–593.

Long, S. R. (1996) Rhizobium symbiosis: Nod factors in perspective. *Plant Cell* 8: 1885–1898.

*Marschner, H. (1995) *Mineral Nutrition of Higher Plants*, 2nd ed. Academic Press, London.

Marzluf, G. A. (1997) Molecular genetics of sulfur assimilation in filamentous fungi and yeast. *Annu. Rev. Microbiol.* 51: 73–96.

Mathews, C. K., and Van Holde, K. E. (1996) *Biochemistry*, 2nd ed. Benjamin/Cummings, Menlo Park, CA.

*Rolfe, B. G., and Gresshoff, P. M. (1988) Genetic analysis of legume nodule initiation. *Annu. Rev. Plant Physiol. Plant Mol. Biol.* 39: 297–319.

*Schlesinger, W. H. (1991) *Biogeochemistry: An Analysis of Global Change.* Academic Press, San Diego.

*Vande Broek, A., and Vanderleyden, J. (1995) Review: Genetics of the *Azospirillum*-plant root association. *Crit. Rev. Plant Sci.* 14: 445–466.

* Indicates a reference that is general reading in the field and is also cited in this chapter.

Chapter References

Andrews, M. (1986) The partitioning of nitrate assimilation between root and shoot of higher plants. *Plant Cell Environ.* 9: 511–519.

Becker, T. W., Perrot-Rechenmann, C., Suzuki, A., and Hirel, B. (1993) Subcellular and immunocytochemical localization of the enzymes involved in ammonia assimilation in mesophyll and bundle-sheath cells of maize leaves. *Planta* 191: 129–136.

Bergmann, L., and Rennenberg, H. (1993) Glutathione metabolism in plants. In *Sulfur Nutrition and Assimilation in Higher Plants. Regulatory, Agricultural and Environmental Aspects*, L. J. De Kok, I. Stulen, H. Rennenberg, C. Brunold, and W. E. Rauser, eds., SPB Academic Publishers, The Hague, Netherlands, pp. 102–123.

Bienfait, H. F., and Van der Mark, F. (1983) Phytoferritin and its role in iron metabolism. In *Metals and Micronutrients: Uptake and Utilization by Plants*, D. A. Robb and W. S. Pierpoint, eds., Academic Press, New York, pp. 111–123.

Bloom, A. J. (1997) Nitrogen as a limiting factor: Crop acquisition of ammonium and nitrate. In *Ecology in Agriculture*, L. E. Jackson, ed., Academic Press, San Diego, pp. 145–172.

Bloom, A. J., Sukrapanna, S. S., and Warner, R. L. (1992) Root respiration associated with ammonium and nitrate absorption and assimilation by barley. *Plant Physiol.* 99: 1294–1301.

Brady, N. C. (1979) *Nitrogen and Rice.* International Rice Research Institute, Manila, Philippines.

Burris, R. H. (1976) Nitrogen fixation. In *Plant Biochemistry*, J. Bonner and J. E. Varner, eds., Academic Press, New York, pp. 887–908.

Campbell, W. H. (1996) Nitrate reductase biochemistry comes of age. *Plant Physiol.* 111: 355–361.

Carlson, R. W., Forsberg, L. S., Price, N. P. J., Bhat, U. R., Kelly, T. M., and Raetz, C. R. H. (1995) The structure and biosynthesis of *Rhizobium leguminosarum* lipid A. In *Progress in Clinical and Biological Research*, Vol. 392: *Bacterial Endotoxins: Lipopolysaccharides from Genes to Therapy*, J. Levin, et al., eds., Wiley, New York, pp. 25–31.

Denison, R. F., and Harter, B. L. (1995) Nitrate effects on nodule oxygen permeability and leghemoglobin. *Plant Physiol.* 107: 1355–1364.

Dilworth, M. J. (1966) Acetylene reduction by nitrogen-fixing preparations from *Clostridium pasteurianium. Biochim. Biophys. Acta* 127: 285–294.

Dixon, R. O. D., and Wheeler, C. T. (1986) *Nitrogen Fixation in Plants.* Chapman and Hall, New York.

FAO/UNIDO/World Bank Working Group on Fertilizers. (1996) *Current World Fertilizer Situation and Outlook 1994/95–2000/2001.* Food and Agriculture Organization of the United Nations, Rome.

Heidstra, R., and Bisseling, T. (1996) Nod factor-induced host responses and mechanisms of Nod factor perception. *New Phytol.* 133: 25–43.

Hell, R. (1997) Molecular physiology of plant sulfur metabolism. *Planta* 202: 138–148.

Heytler, P. G., Reddy, G. S., and Hardy, R. W. F. (1984) *In vivo* energetics of symbiotic nitrogen fixation in soybeans. In *Nitrogen Fixation and CO_2 Metabolism*, P. W. Ludden and J. E. Burris, eds., Elsevier, New York, pp. 283–292.

Jones, O. T. G. (1983) Ferrochelatase. In *Metals and Micronutrients: Uptake and Utilization by Plants*, D. A. Robb and W. S. Pierpoint, eds., Academic Press, New York, pp. 125–144.

Kaiser, W. M., and Huber, S. C. (1994) Posttranslational regulation of nitrate reductase in higher plants. *Plant Physiol.* 106: 817–821.

Kleinhofs, A., Warner, R. L., and Melzer, J. M. (1989) Genetics and molecular biology of higher plant nitrate reductases. In *Recent Advances in Phytochemistry*, Vol. 23: *Plant Nitrogen Metabolism*, J. E. Poulton, J. T. Romeo, and E. Conn, eds., Plenum, New York, pp. 117–155.

Kuzma, M. M., Hunt, S., and Layzell, D. B. (1993) Role of oxygen in the limitation and inhibition of nitrogenase activity and respiration rate in individual soybean nodules. *Plant Physiol.* 101: 161–169.

Lazarowitz, S. G., and Bisseling, T. (1997) Plant development from the cellular perspective: Integrating the signals. *Plant Cell* 9: 1884–1900.

Lea, P. J., Blackwell, R. D., and Joy, K. W. (1992) Ammonia assimilation in higher plants. In *Nitrogen Metabolism of Plants* (Proceedings of the Phytochemical Society of Europe, 33), K. Mengel and D. J. Pilbean, eds., Clarendon, Oxford, pp. 153–186.

Leustek, T. (1996) Molecular genetics of sulfate assimilation in plants. *Physiol. Plantarum* 97: 411–419.

Lobreaux, S., Massenet, O., and Briat, J. F. (1992) Iron induces ferritin synthesis in maize plantlets. *Plant Mol. Biol.* 19: 563–575.

Ludwig, R. A., and de Vries, G. E. (1986) Biochemical physiology of *Rhizobium* dinitrogen fixation. In *Nitrogen Fixation*, Vol. 4: *Molecular Biology*, W. J. Broughton and S. Puhler, eds., Clarendon, Oxford, pp. 50–69.

Meeks, J. C. (1998) Symbiosis between nitrogen fixing cyanobacteria and plants. *Biosci.* 48: 266–276.

Mendel, R. R., and Stallmeyer, B. (1995) Molybdenum cofactor (nitrate reductase) biosynthesis in plants: First molecular analysis. In *Current Plant Science and Biotechnology in Agriculture*, Vol. 22: *Current Issues in Plant Molecular and Cellular Biology*, M. R. C. Terzi and A. Falavigna, eds., Kluwer, Dordrecht, Netherlands, pp. 577–582.

Metzler, D. E. (1977) *Biochemistry.* Academic Press, New York.

Mylona, P., Pawlowski, K., and Bisseling, T. (1995) Symbiotic nitrogen fixation. *Plant Cell* 7: 869–885.

Oaks, A. (1994) Primary nitrogen assimilation in higher plants and its regulation. *Can. J. Bot.* 72: 739–750.

Ogren, W. L. (1984) Photorespiration: Pathways, regulation, and modification. *Annu. Rev. Plant Physiol.* 35: 415–442.

Olsen, R. A., Clark, R. B., and Bennet, I. H. (1981) The enhancement of soil fertility by plant roots. *Am. Sci.* 69: 378–384.

Pate, J. S. (1973) Uptake, assimilation and transport of nitrogen compounds by plants. *Soil Biol. Biochem.* 5: 109–119.

Pate, J. S., and Layzell, D. B. (1990) Energetics and biological costs of nitrogen assimilation. In *The Biochemistry of Plants*, Vol. 16: *Intermediary Nitrogen Metabolism*, B. J. Miflin and P. J. Lea, eds., Academic Press, San Diego, pp. 1–42.

Phillips, D. A., and Kapulnik, Y. (1995) Plant isoflavonoids, pathogens and symbionts. *Trends Microbiol.* 3: 58–64.

Rees, D. A. (1977) *Polysaccharide Shapes*. Chapman and Hall, London.

Roberts, G. R., Keys, A. I., and Whittingham, C. P. (1970) The transport of photosynthetic products from the chloroplast of tobacco leaves. *J. Exp. Bot.* 21: 683–692.

Robinson, J. M. (1988) Spinach leaf chloroplast CO_2 and NO_2^- photoassimilations do not compete for photogenerated reductant. Manipulation of reductant levels by quantum flux density titrations. *Plant Physiol.* 88: 1373–1380.Schubert, K. R., Jennings, N. T., and Evans, H. J. (1978) Hydrogen reactions of nodulated leguminous plants. *Plant Physiol.* 61: 398–401.

Schubert, K. R., Jennings, N. T., and Evans, H. J. (1978) Hydrogen reactions of nodulated leguminous plants. *Plant Physiol.* 61: 398–401

Siegel, L. M., and Wilkerson, J. Q. (1989) Structure and function of spinach ferredoxin-nitrite reductase. In *Molecular and Genetic Aspects of Nitrate Assimilation*, J. L. Wray and J. R. Kinghorn, eds., Oxford Science, Oxford, pp. 263–283.

Sivasankar, S., and Oaks, A. (1996) Nitrate assimilation in higher plants—The effect of metabolites and light. *Plant Physiol. Biochem.* 34: 609–620.

Smith, R. I., Bouton, J. H., Schank, S. C., Quesenberry, K. H., Tyler, M. E., Milam, J. R., Gaskins, M. H., and Littell, R. C. (1976) Nitrogen fixation in grasses inoculated with *Spirillum lipoferum*. *Science* 193: 1003–1005.

Stewart, W. D. P. (1977) Blue-green algae. In *A Treatise on Dinitrogen Fixation*, Vol. 3, R. W. F. Hardy and W. S. Silver, eds., Wiley, New York, pp. 63–123.

Stokkermans, T. J. W., Ikeshita, S., Cohn, J., Carlson, R. W., Stacey, G., Ogawa, T., and Peters, N. K. (1995) Structural requirements of synthetic and natural product lipo-chitin oligosaccharides for induction of nodule primordia on *Glycine soja*. *Plant Physiol.* 108: 1587–1595.

Trelstad, R. L., Lawley, K. R., and Holmes, L. B. (1981) Nonenzymatic hydroxylations of proline and lysine by reduced oxygen derivatives. *Nature* 289: 310–312.

van Rhijn, P., Goldberg, R. B., and Hirsch, A. M. (In press) *Lotus corniculatus* nodulation specificity is changed by the presence of a soybean lectin gene. *Plant Cell.*

Warner, R. L., and Kleinhofs, A. (1992) Genetics and molecular biology of nitrate metabolism in higher plants. *Physiol. Plantarum* 85: 245–252.

Wray, J. L. (1993) Molecular biology, genetics and regulation of nitrite reduction in higher plants. *Physiol. Plantarum* 89: 607–612.

13 Plant Defenses: Surface Protectants and Secondary Metabolites

IN NATURAL HABITATS, plants are surrounded by an enormous number of potential enemies. Nearly all ecosystems contain a wide variety of bacteria, viruses, fungi, nematodes, mites, insects, mammals, and other herbivorous animals. By their nature, plants cannot avoid these herbivores and pathogens simply by moving away; they must protect themselves in other ways. The cuticle (a waxy outer layer) and the periderm (secondary protective tissue), in addition to retarding water loss (see Chapter 4), provide barriers to bacterial and fungal entry. In addition, plant compounds belonging to a group known as secondary metabolites defend plants against a variety of herbivores and pathogenic microbes.

In this chapter we will discuss some of the mechanisms by which plants protect themselves against both herbivory and pathogenic organisms. We begin with a discussion of the three compounds that provide surface protection to the plant: cutin, suberin, and waxes. Next we describe the structures and biosynthetic pathways for the three major classes of secondary metabolites: terpenes, phenolics, and nitrogen-containing compounds. In addition to their roles in plant defense, secondary compounds may serve other important functions, such as structural support, as in the case of lignin, or as pigments, as in the case of the anthocyanins. Finally, we will examine specific plant responses to pathogen attack, the genetic control of host–pathogen interactions, and cell signaling processes associated with infection.

Cutin, Suberin, and Waxes

All plant parts exposed to the atmosphere are coated with layers of lipid material that reduce water loss and help block the entry of pathogenic fungi and bacteria. The principal types of coatings are cutin, suberin, and waxes. Cutin is found on most aboveground parts, while suberin is present on underground parts, woody stems, and healed wounds. Waxes are associated with both cutin and suberin.

Cutin, Suberin, and Waxes Are Made Up of Hydrophobic Compounds

Cutin is a macromolecule, a polymer consisting of many long-chain fatty acids that are attached to each other by ester linkages, creating a rigid three-dimensional network. Cutin is formed from 16:0 and 18:1 fatty acids* with hydroxyl or epoxide groups situated either in the middle of the chain or at the end opposite the carboxylic acid function (Figure 13.1A).

Cutin is a principal constituent of the **cuticle**, a multilayered secreted structure that coats the outer cell walls of the epidermis on the aerial parts of all herbaceous plants (Figure 13.2). The cuticle is composed of a top coating of wax, a thick middle layer containing cutin embedded in wax (the cuticle proper), and a lower layer formed of cutin and wax blended with the cell wall substances pectin, cellulose, and other carbohydrates (the cuticular layer). Recent research suggests that, in addition to cutin, the cuticle may contain a second lipid polymer, made up of long-chain hydrocarbons, that has been named **cutan** (Jeffree 1996).

Waxes are not macromolecules, but complex mixtures of long-chain acyl lipids that are extremely hydrophobic. The most common components of wax are straight-chain alkanes and alcohols of 25 to 35 carbon atoms (see Figure 13.1C). Long-chain aldehydes, ketones, esters, and free fatty acids are also found. The waxes of the cuticle are synthesized by epidermal cells. They leave the epi-

* Recall from Chapter 11 that the nomenclature for fatty acids is X:Y, where X is the number of carbon atoms and Y is the number of *cis* double bonds.

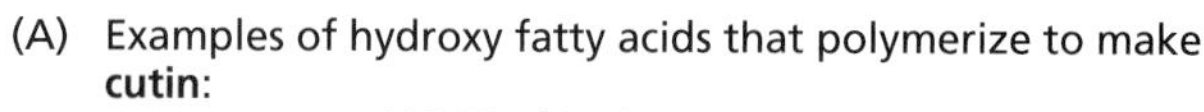

(A) Examples of hydroxy fatty acids that polymerize to make **cutin**:

$HOCH_2(CH_2)_{14}COOH$

$CH_3(CH_2)_8CH(OH)(CH_2)_5COOH$

(B) Examples of hydroxy fatty acids that polymerize along with other constituents to make **suberin**:

$HOCH_2(CH_2)_{14}COOH$

$HOOC(CH_2)_{14}COOH$ (a dicarboxylic acid)

(C) Examples of common **wax** components:

Straight-chain alkanes	$CH_3(CH_2)_{27}CH_3$
	$CH_3(CH_2)_{29}CH_3$
Fatty acid ester	$CH_3(CH_2)_{22}C(=O)-O(CH_2)_{25}CH_3$
Long-chain fatty acid	$CH_3(CH_2)_{22}COOH$
Long-chain alcohol	$CH_3(CH_2)_{24}CH_2OH$

Figure 13.1 Constituents of (A) cutin, (B) suberin, and (C) waxes.

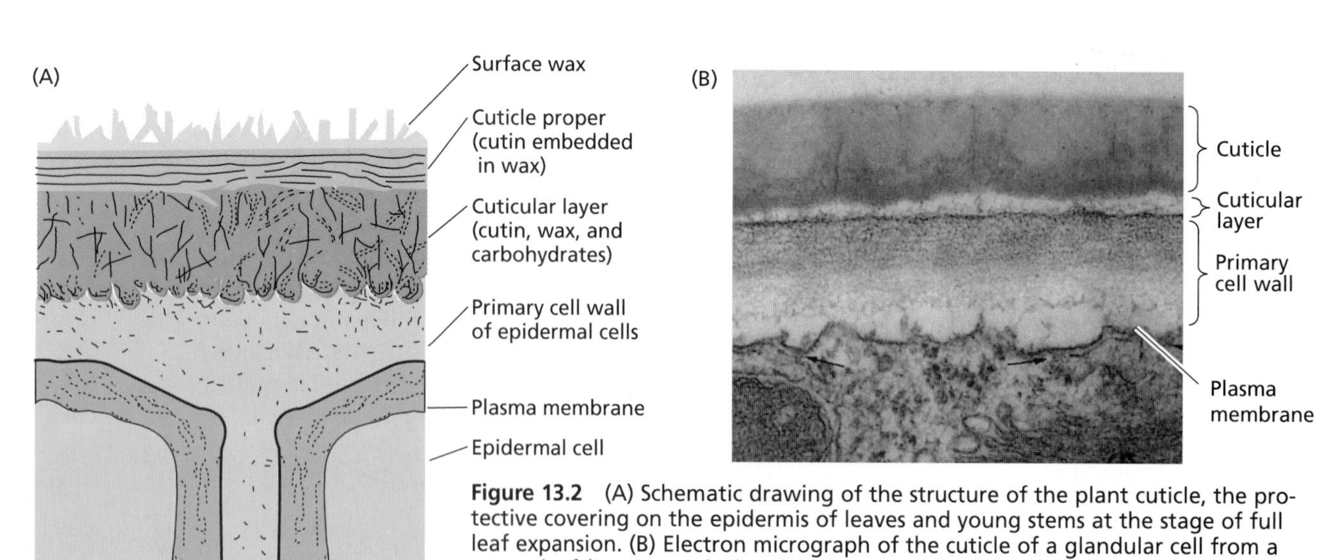

Figure 13.2 (A) Schematic drawing of the structure of the plant cuticle, the protective covering on the epidermis of leaves and young stems at the stage of full leaf expansion. (B) Electron micrograph of the cuticle of a glandular cell from a young leaf (*Lamium* sp.), showing the presence of the cuticle layers indicated in A, except for surface waxes, which are not visible. (51,000×) (A after Jeffree 1996; B from B. Gunning and M. Steer, *Plant Cell Biology: Structure and Function*, Jones and Bartlett, 1996.)

Figure 13.3
Surface wax deposits, which form the top layer of the cuticle, adopt a variety of different forms. These scanning electron micrographs show the leaf surfaces of different lines of Brassica oleracea, which differ in wax abundance and crystal structure. Prominent wax deposits may help protect plants against insect herbivores by hindering insect attachment. However, recent research suggests that plants without surface wax may also enjoy protection from insect herbivores because predatory insects are more mobile and more effective in the absence of wax deposits. (From Eigenbrode et al. 1991, courtesy of S. D. Eigenbrode, with permission from the Entomological Society of America.)

dermal cells as droplets that pass through pores in the cell wall by an unknown mechanism. The top coating of cuticle wax often crystallizes in an intricate pattern of rods, tubes, or plates (Figure 13.3).

Suberin is a polymer whose structure is very poorly understood. Like cutin, it is formed from hydroxy or epoxy fatty acids joined by ester linkages. However, suberin differs from cutin in that it has dicarboxylic acids (see Figure 13.1B), more long-chain components, and a significant proportion of phenolic compounds as part of its structure. Suberin is a cell wall constituent found in many locations throughout the plant. We have already noted its presence in the Casparian strip of the root endodermis, which forms a barrier between the apoplast of the cortex and the stele (see Chapter 4). Suberin is a principal component of the outer cell walls of all underground organs and is associated with the cork cells of the **periderm**, the tissue that forms the outer bark of stems and roots during secondary growth of woody plants. Suberin also forms at sites of leaf abscission and in areas damaged by disease or wounding.

Cutin, Waxes, and Suberin Help Reduce Transpiration and Pathogen Invasion

Cutin, suberin, and their associated waxes form barriers between the plant and its environment that function to keep water in and pathogens out. The cuticle is very effective at limiting water loss from aerial parts of the plant but does not block transpiration completely, since even with the stomata closed some water is lost. The thickness of the cuticle varies with environmental conditions. Plant species native to arid areas typically have thicker cuticles than plants from moist habitats have, but plants from moist habitats often develop thick cuticles when grown under dry conditions.

The cuticle and suberized tissue are both important in excluding fungi and bacteria, although they do not appear to be as important in pathogen resistance as some of the other defenses we will discuss in this chapter. Because of its waxy nature, the cuticle repels water and may therefore slow fungal germination on plant surfaces, since most fungal spores require moist conditions for germination. After germination, some fungi produce cutinase, an enzyme that hydrolyzes cutin and thus facilitates the entry of fungal hyphae into the plant.

Secondary Metabolites

Plants produce a large, diverse array of organic compounds that appear to have no direct function in growth and development. These substances are known as **secondary metabolites**, secondary products, or natural products. Unlike primary metabolites, such as chlorophyll, amino acids, nucleotides, simple carbohydrates or membrane lipids, secondary metabolites have no generally recognized roles in the processes of photosynthesis, respiration, solute transport, translocation, nutrient assimilation, and differentiation discussed elsewhere in this book.

Secondary metabolites also differ from primary metabolites in having a restricted distribution in the plant kingdom. That is, particular secondary metabolites are often found in only one plant species or a taxonomically related group of species, whereas the basic primary metabolites are found throughout the plant kingdom.

This section provides a brief overview of the functions of secondary metabolites before we begin detailed discussions of their major classes.

Secondary Metabolites Defend Plants against Herbivores and Pathogens

For many years, the adaptive significance of most plant secondary metabolites was unknown. These compounds were thought to be simply functionless end products of metabolism, or metabolic wastes. Study of these substances was pioneered by organic chemists of the nineteenth and early twentieth centuries who were interested in these substances because of their importance as medicinal drugs, poisons, flavors, and industrial materials (Agosta 1996).

More recently, many secondary metabolites were suggested to have important ecological functions in plants. Chief among these functions is protection against being eaten by herbivores (herbivory) and infection by microbial pathogens. Secondary products have also been shown to serve as attractants for pollinators and seed-dispersing animals and as agents of plant–plant competition. In the remainder of this chapter we will discuss the major types of plant secondary metabolites, their biosynthesis, and what is known about their functions in the plant, particularly their roles in defense.

Plant Defenses Evolved to Maintain Reproductive Fitness

We can begin by asking how plants came to have defenses. According to evolutionary biologists, plant defenses must have arisen through heritable mutations, natural selection, and evolutionary change. Random mutations in basic metabolic pathways led to the appearance of new compounds that happened to be toxic or deterrent to herbivores and pathogenic microbes. As long as these compounds were not unduly toxic to the plants themselves and the metabolic cost of producing them was not excessive, they gave the plants that possessed them greater reproductive fitness than undefended plants had. Thus, the defended plants left more descendants than undefended ones and passed their defensive traits on to the next generation.

Interestingly, the very defense compounds that increase the reproductive fitness of plants by warding off fungi, bacteria, and herbivores may also make them undesirable as food for humans. Many important crop plants have been artificially selected for producing relatively low levels of these compounds, which, of course, can make them more susceptible to insects and disease.

There Are Three Principal Groups of Secondary Metabolites

Plant secondary metabolites can be divided into three chemically distinct groups: terpenes, phenolics, and nitrogen-containing compounds. Figure 13.4 shows in simplified form the pathways involved in the biosynthesis of secondary metabolites and their interconnections with primary metabolism. **Terpenes** are lipids synthesized from acetyl CoA or from basic intermediates of glycolysis. **Phenolic compounds** are aromatic substances formed via the shikimic acid pathway or the malonic acid pathway. The **nitrogen-containing secondary products**, such as alkaloids, are biosynthesized primarily from amino acids.

Terpenes

The terpenes, or terpenoids, constitute the largest class of secondary products. The diverse substances of this class are generally insoluble in water and are united by their common biosynthetic origin from acetyl CoA or glycolytic intermediates. After discussing the biosynthesis of terpenes, we'll examine their antiherbivore activities and how some herbivores circumvent the toxic effects of terpenes.

Terpenes Are Formed by the Fusion of Five-Carbon Isoprene Units

All terpenes are derived from the union of 5-carbon elements that have the branched carbon skeleton of isopentane:

$$(H_3C)_2CH—CH_2—CH_3$$

The basic structural elements of terpenes are sometimes called **isoprene units** because terpenes can decompose at high temperatures to give isoprene:

$$H_3C(H_2C{=})C—CH{=}CH_2$$

Thus, all terpenes are occasionally referred to as isoprenoids.

Terpenes are classified by the number of five-carbon units they contain, although because of extensive metabolic modifications it is sometimes difficult to pick out the original five-carbon residues. Ten-carbon terpenes, which contain two C_5 units, are called **monoterpenes**; 15-carbon terpenes (three C_5 units) are **sesquiterpenes**; and 20-carbon terpenes (four C_5 units) are **diterpenes**. Larger terpenes include **triterpenes** (30 carbons), **tetraterpenes** (40 carbons), and **polyterpenoids** ($[C_5]_n$ carbons where $n > 10$).

There Are Two Pathways for Terpene Biosynthesis

Terpenes are biosynthesized from primary metabolites in at least two different ways. In the well-studied **meval-**

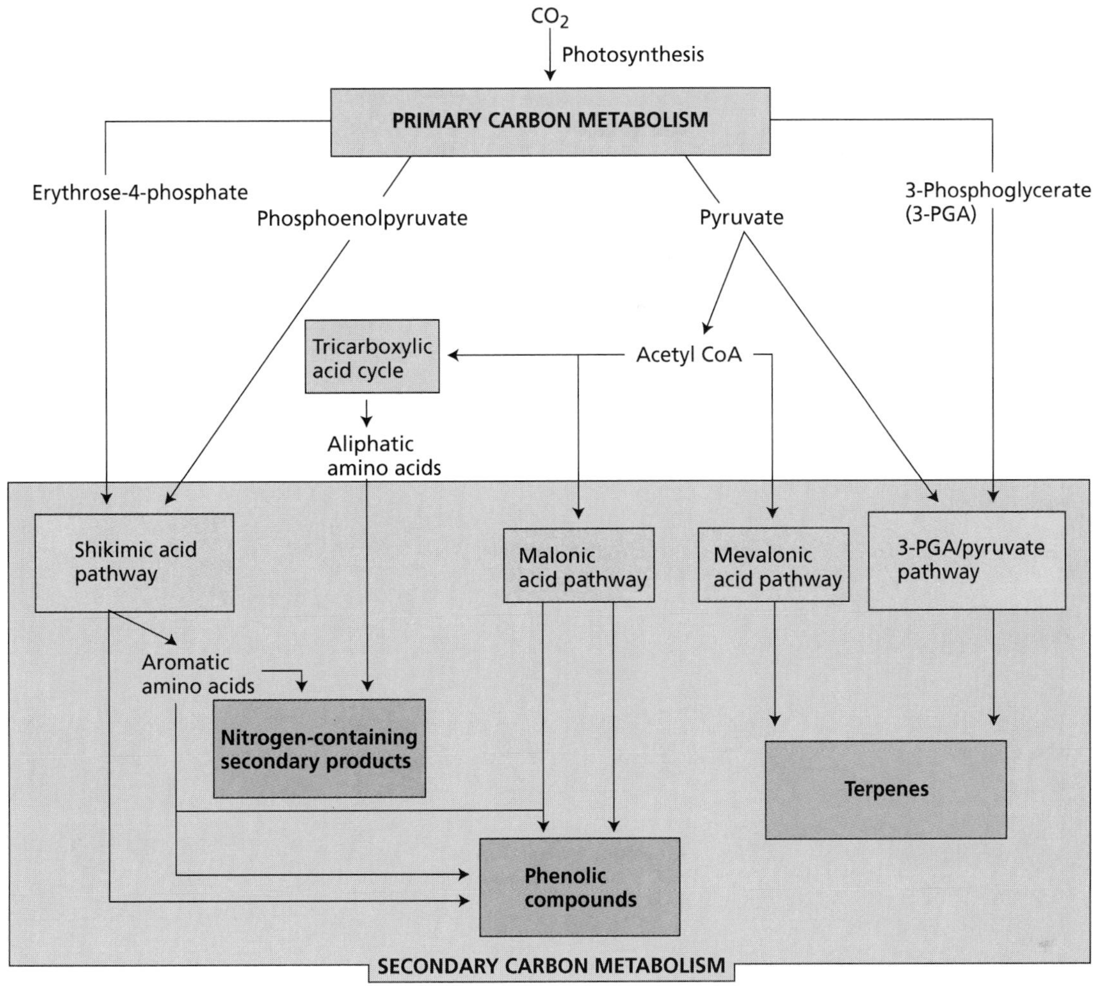

Figure 13.4 A simplified view of the major pathways of secondary-metabolite biosynthesis and their interrelationships with primary metabolism.

onic acid pathway, three molecules of acetyl CoA are joined together stepwise to form mevalonic acid (Figure 13.5). This key six-carbon intermediate is then pyrophosphorylated, decarboxylated, and dehydrated to yield **isopentenyl pyrophosphate (IPP)**. IPP is the activated five-carbon building block of terpenes. Recently, it was discovered that IPP can be formed from intermediates of glycolysis or the photosynthetic carbon reduction cycle via a separate pathway that operates in chloroplasts and other plastids (Lichtenthaler et al. 1997). Although all the details have not yet been elucidated, 3-phosphoglycerate and two carbon atoms derived from pyruvate appear to combine to generate an intermediate that is eventually converted to IPP.

Isopentenyl Pyrophosphate and Its Isomer Combine to Form Larger Terpenes

Isopentenyl pyrophosphate and its isomer, dimethylallyl pyrophosphate (DPP), are the activated five-carbon building blocks of terpene biosynthesis that join together to form larger molecules. First, IPP and DPP react to give geranyl pyrophosphate (GPP), the ten-carbon precursor of nearly all the monoterpenes (Figure 13.6). GPP can then link to another molecule of IPP to give the 15-carbon compound farnesyl pyrophosphate (FPP), the precursor of nearly all the sesquiterpenes. Addition of yet another molecule of IPP gives the 20-carbon compound geranylgeranyl pyrophosphate (GGPP), the precursor of the diterpenes. Finally, FPP and GGPP can dimerize to give the triterpenes (C_{30}) and the tetraterpenes (C_{40}), respectively.

Certain terpenes have a well-characterized function in plant growth or development and so can be considered primary rather than secondary metabolites. For example, the gibberellins, an important group of plant hormones (see Chapter 20), are diterpenes. Sterols are triterpene derivatives that are essential components of cell membranes, which they stabilize by interacting with

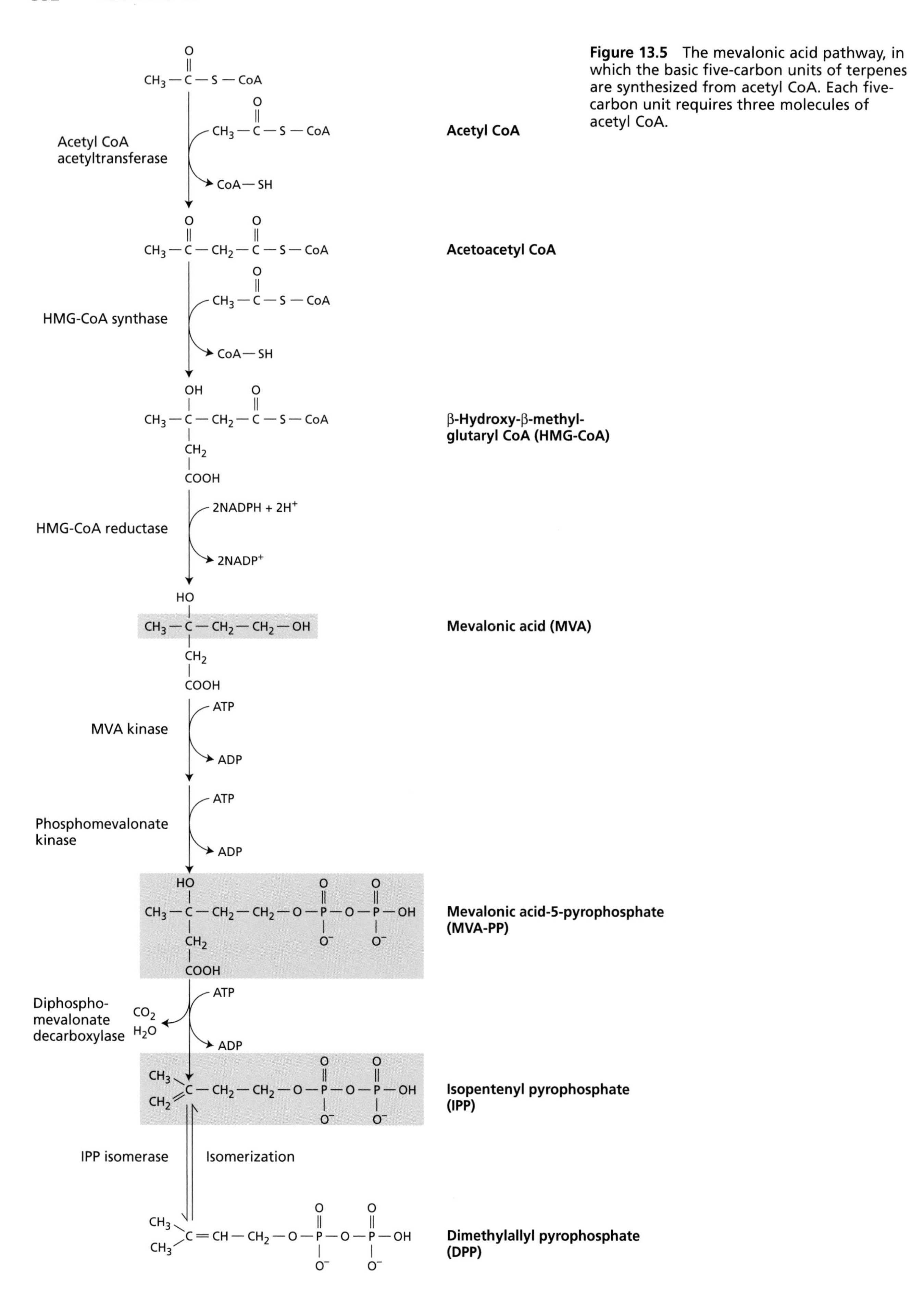

Figure 13.5 The mevalonic acid pathway, in which the basic five-carbon units of terpenes are synthesized from acetyl CoA. Each five-carbon unit requires three molecules of acetyl CoA.

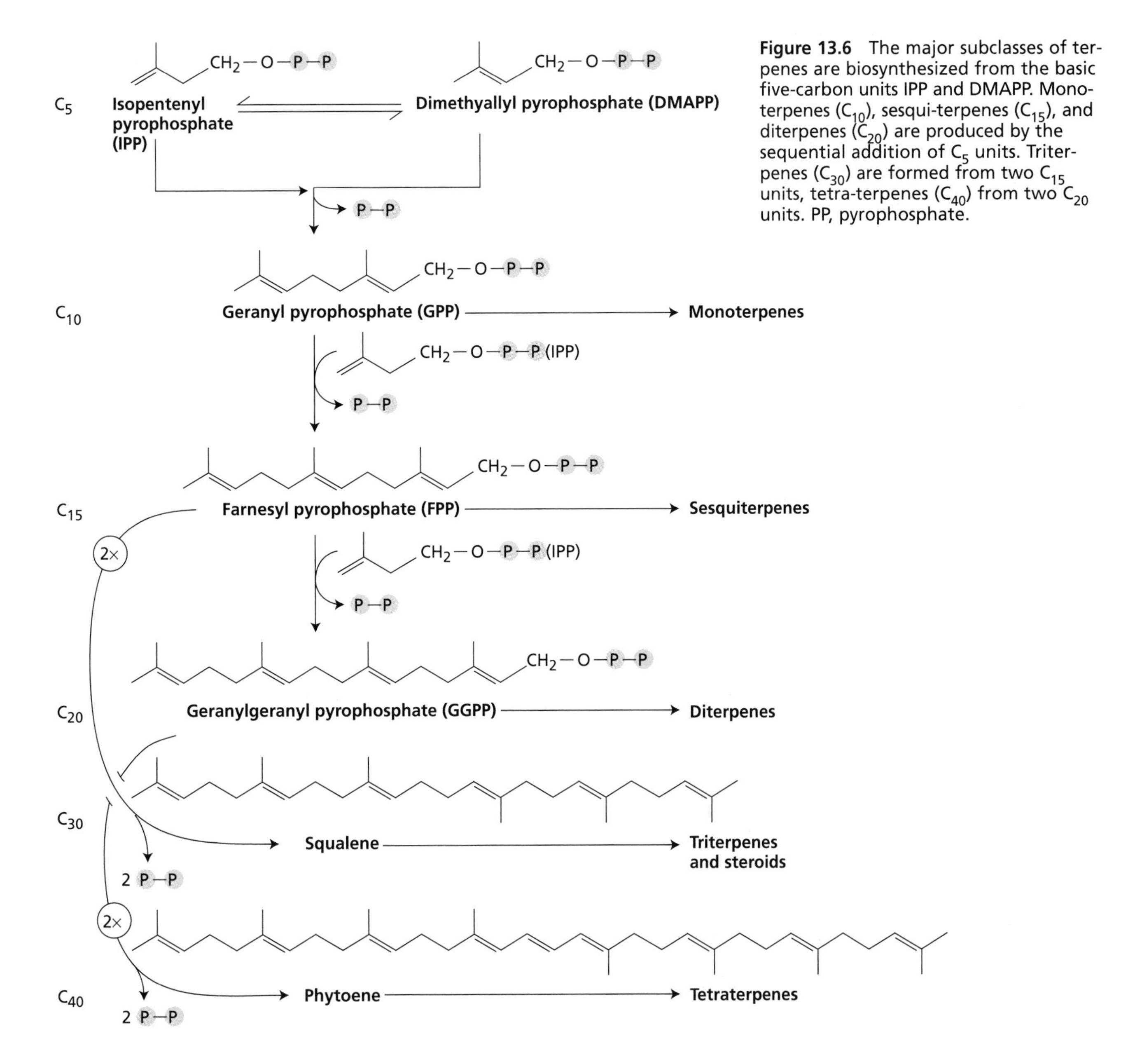

Figure 13.6 The major subclasses of terpenes are biosynthesized from the basic five-carbon units IPP and DMAPP. Monoterpenes (C_{10}), sesqui-terpenes (C_{15}), and diterpenes (C_{20}) are produced by the sequential addition of C_5 units. Triterpenes (C_{30}) are formed from two C_{15} units, tetra-terpenes (C_{40}) from two C_{20} units. PP, pyrophosphate.

phospholipids (see Chapter 11). The red, orange, and yellow carotenoids are tetraterpenes that function as accessory pigments in photosynthesis and protect photosynthetic tissues from photooxidation (see Chapter 7). The hormone abscisic acid (see Chapter 23) is a C_{15} terpene produced by degradation of a carotenoid precursor. Long-chain polyterpene alcohols known as dolichols function as carriers of sugars in cell wall and glycoprotein synthesis (see Chapters 1 and 15). Terpene-derived side chains, such as the phytol side chain of chlorophyll (see Chapter 7), help anchor certain molecules in membranes. Thus, various terpenes have important primary roles in plants. However, the vast majority of the different terpene structures produced by plants are secondary metabolites that are presumed to be involved in defense.

Terpenes Serve as Antiherbivore Defense Compounds in Many Plants

Terpenes are toxins and feeding deterrents to a large number of plant-feeding insects and mammals; thus they appear to play important defensive roles in the plant kingdom (Gershenzon and Croteau 1991). To illustrate various aspects of these roles, we will discuss examples drawn from each of the five major subclasses of terpenes: monoterpenes, sesquiterpenes, diterpenes, triterpenes, and polyterpenes.

Monoterpenes (C_{10}). Many monoterpenes and their derivatives are important agents of insect toxicity. For example, the monoterpene esters called **pyrethroids** that occur in the leaves and flowers of *Chrysanthemum* species show very striking insecticidal activity. Both

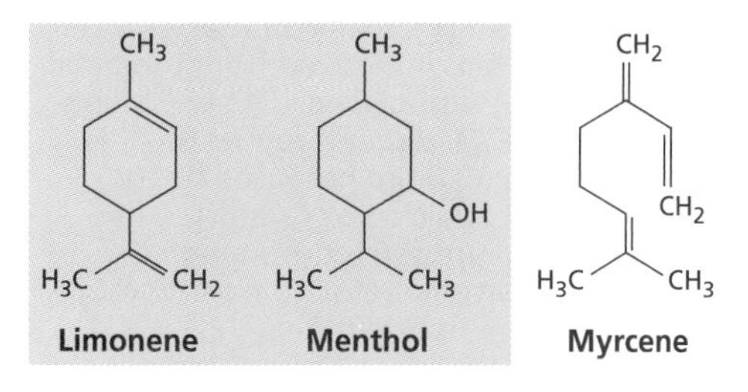

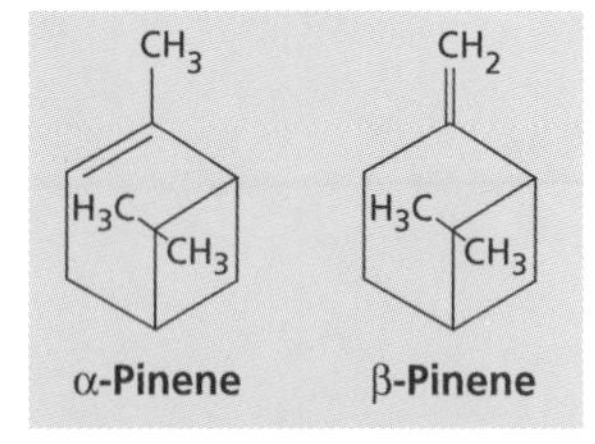

Figure 13.7 Structure of some well-known monoterpenes. These are common constituents of the resin of conifers and the essential oils of certain herbs, and they serve as defenses against insects and other organisms that feed on these plants.

natural and synthetic pyrethroids are popular ingredients in commercial insecticides because of their low persistence in the environment and their negligible toxicity to mammals.

In conifers such as pine and fir, monoterpenes accumulate in resin ducts found in the needles, twigs, and trunk. The principal monoterpenes of conifer resin are α-pinene, β-pinene, limonene, and myrcene (Figure 13.7). These compounds are toxic to numerous insects, including bark beetles, which are serious pests of conifer species throughout the world. Many conifers respond to bark beetle infestation by producing additional quantities of monoterpenes.

Many plants contain mixtures of volatile monoterpenes and sesquiterpenes, called **essential oils**, that lend a characteristic odor to their foliage. Peppermint, lemon, basil, and sage are examples of plants that contain essential oils. The chief monoterpene constituent of peppermint oil is menthol; that of lemon oil is limonene (see Figure 13.7). Essential oils have well-known insect repellent properties. They are frequently found in glandular hairs that project outward from the epidermis and serve to "advertise" the toxicity of the plant, repelling potential herbivores even before they take a trial bite. In the glandular hairs, the terpenes are stored in a modified extracellular space in the cell wall (Figure 13.8). Essential oils can be extracted from plants by steam distillation and are important commercially in flavoring foods and making perfumes.

Recent research in the United States and the Netherlands has revealed an interesting twist on the role of volatile terpenes in plant protection. In corn, cotton, and other species, certain monoterpenes and sesquiterpenes are produced and emitted only after insect feeding has already begun. These substances attract natural enemies, including predatory and parasitic insects, that kill plant-feeding insects and so help minimize further damage (Turlings et al. 1995). Thus, volatile terpenes are not only defenses in their own right, but also provide a way for plants to enlist defensive help from other organisms.

Sesquiterpenes (C_{15}). Among the many sesquiterpenes known to be antiherbivore agents are the **sesquiterpene lactones** found in the glandular hairs of members of the composite family, such as sunflower and sagebrush (Figure 13.9B). These compounds are characterized by a five-membered lactone ring, a cyclic ester (Figure 13.9A). Experiments with sesquiterpene lactones have shown that they are strong feeding repellents to many herbivorous insects and mammals (Picman 1986). Like many other mammalian feeding deterrents, sesquiterpene lactones taste bitter to humans. Ecologists believe that bitterness is not necessarily an intrinsic property of defensive plant compounds, but that through evolution, herbivores have learned to associate unpleasant tastes with certain defensive substances and thus avoid feeding on plants containing those substances.

Another defensive compound of sesquiterpene origin is **gossypol**, an aromatic sesquiterpene dimer from cotton (see Figure 13.9B). Found in subepidermal pigment glands, gossypol is responsible for significant resistance to insects in certain varieties of cotton. Gossypol and other sesquiterpenes are also important defenses against fungal and bacterial pathogens.

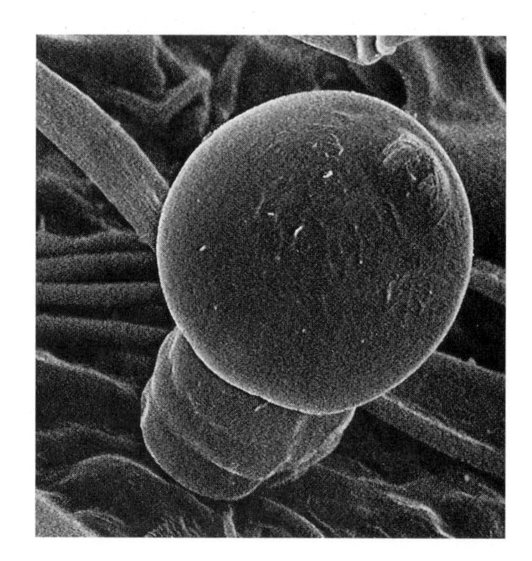

Figure 13.8 Monoterpenes and sesquiterpenes are commonly found in glandular hairs on the plant surface. This scanning electron micrograph shows a glandular hair on a young leaf of spring sunflower (*Balsamorhiza sagittata*). Terpenes are thought to be synthesized in the cells of the hair and are stored in the rounded cap at the top. This "cap" is an extracellular space that forms when the cuticle and a portion of the cell wall pull away from the remainder of the cell. (× 1105) (J. N. A. Lott/Biological Photo Service.)

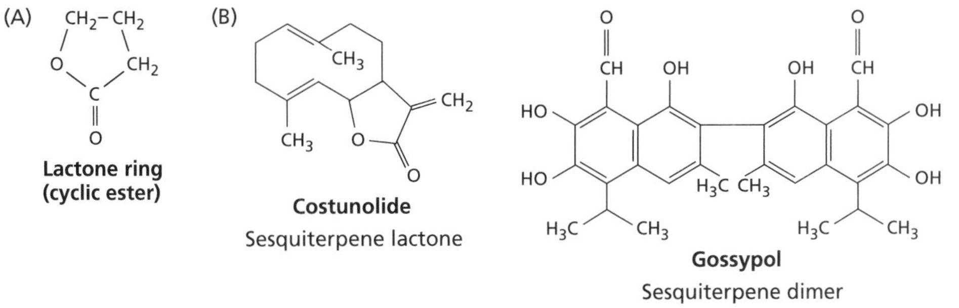

(A) CH_2-CH_2, O, CH_2, C, O — **Lactone ring (cyclic ester)**

(B) **Costunolide** Sesquiterpene lactone; **Gossypol** Sesquiterpene dimer

Figure 13.9 (A) A lactone ring. (B) Structures of two sesquiterpenes. Costunolide is a sesquiterpene lactone. Gossypol is a sesquiterpene dimer, a compound made up of two identical sesquiterpene units.

Diterpenes (C_{20}). Many diterpenes have been shown to be toxins and feeding deterrents to herbivores. Plant resins, including those of pines and certain tropical leguminous trees, often contain significant amounts of diterpenes such as abietic acid (Figure 13.10). When resin canals in the trunk are pierced by insect feeding, the outflow of resin may physically block feeding and serve as a chemical deterrent to continued predation. On exposure to air, the resin polymerizes and seals the wound. For example, upon wounding, the tropical leguminous tree *Hymenaea courbaril* exudes large quantities of resin from resin-secreting pockets located in the cambium (Figure 13.11).

Plants in the spurge family (the Euphorbiaceae) produce diterpene esters of phorbol (see Figure 13.10) and other compounds that are severe skin irritants and internal toxins to mammals. Currently, phorbol-type diterpenes are of great interest as model tumor promoters in studies of carcinogenesis in animals. Another diterpene, taxol from the Pacific yew (*Taxus brevifolia*), is a powerful new anticancer drug.

Triterpenes (C_{30}). The triterpenes comprise a variety of structurally diverse compounds, including steroids, many of which have been modified to have fewer than 30 carbon atoms. As mentioned previously, several steroid alcohols (sterols) are important components of plant cell membranes. Found especially in the plasma membrane, sterols decrease the permeability of the membrane to small molecules by decreasing the motion of the fatty acid chains. Other steroids are defensive secondary products. For example, the **phytoecdysones** are a group of plant steroids that have the same basic structure as insect molting hormones

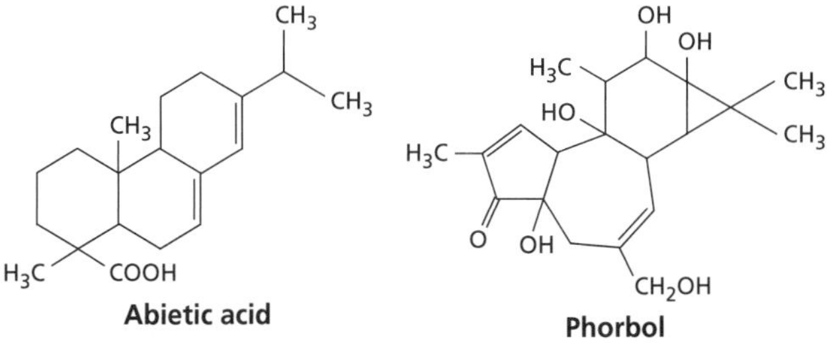

Abietic acid **Phorbol**

Figure 13.10 Structure of two diterpenes. Abietic acid is found in the resins of pines. Phorbol esters are protective compounds produced by species of the Euphorbiaceae, the spurge family.

(A)

(B)

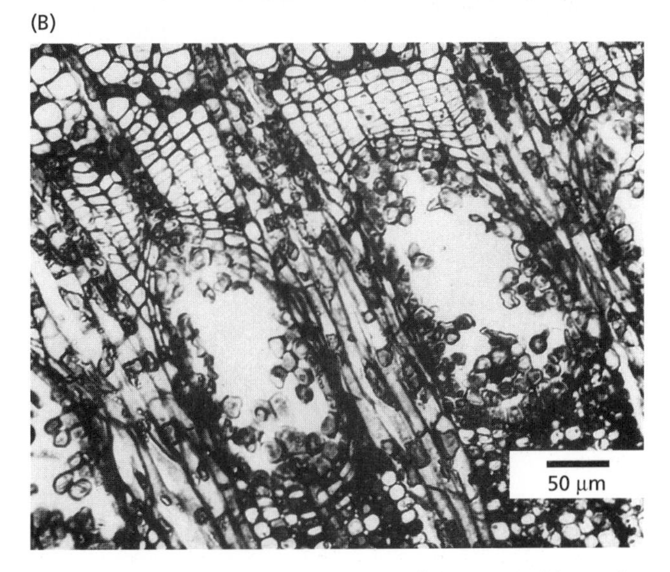

Figure 13.11 (A) A cut in the trunk of the tropical leguminous tree *Hymenaea courbaril*, showing the exudation of resin from the wound site at the cambium. (B) Structure of the resin-secreting glandular pockets in the cambium layer of *H. courbaril*. (Photos courtesy of J. Langenheim.)

(Figure 13.12). Ingestion of phytoecdysones by insects disrupts molting and other developmental processes, often with lethal consequences.

Another group of triterpene antiherbivore compounds is the **limonoids**, which are well known as bitter substances in citrus fruit. Perhaps the most powerful deterrent to insect feeding known is *azadirachtin* (see Figure 13.12), a complex limonoid from the neem tree (*Azadirachta indica*) of Africa and Asia. Azadirachtin is a feeding deterrent to some insects at doses as low as 50 parts per billion, and it exerts a variety of toxic effects (Mordue and Blackwell 1993). It has considerable potential as a commercial insect control agent because of its low toxicity to mammals, and several preparations containing azadirachtin are now being marketed in North America and India.

Triterpenes that are active against vertebrate herbivores include cardenolides and saponins. **Cardenolides** are glycosides (compounds containing an attached sugar or sugars) that taste bitter and are extremely toxic to higher animals. In humans, they have dramatic effects on the heart muscle through their influence on Na^+/K^+-activated ATPases. In carefully regulated doses, they slow and strengthen the heartbeat. Cardenolides extracted from species of foxglove (*Digitalis*) are prescribed to millions of patients for the treatment of heart disease. Digitoxigenin (see Figure 13.12) is the aglycone (sugarless) triterpene portion of the naturally occurring digitanides, which contain one molecule of the rare sugar D-digitoxose and one molecule of acetyldigitoxose.

Saponins are steroid and triterpene glycosides, so named because of their soaplike properties. The presence of both lipid-soluble (the triterpene) and water-soluble (the sugar) elements in one molecule gives saponins detergent properties, and they form a soapy lather when shaken with water. The toxicity of saponins is thought to be a result of their ability to form complexes with sterols. Saponins may interfere with sterol uptake from the digestive system or disrupt cell membranes after absorption into the bloodstream. The yam *Dioscorea* contains saponins—e.g., yamogenin (see Figure 13.12)—that are widely used as starting materials in the synthesis of progesterone-like compounds for birth control pills.

Polyterpenes ($[C_5]_n$). Several high-molecular-weight polyterpenes occur in plants. Rubber, the best known of these, is a polymer containing 1500 to 15,000 isopentenyl units in which nearly all of the carbon–carbon double bonds have a *cis* (Z) configuration (Figure 13.13).

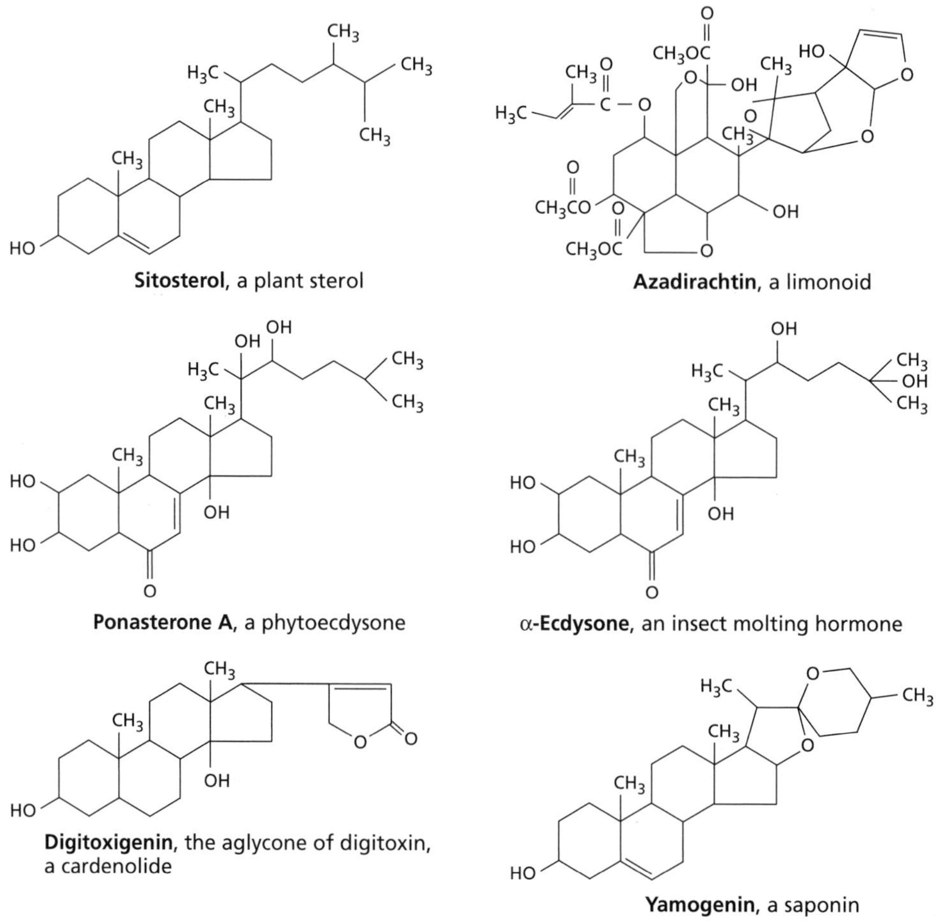

Sitosterol, a plant sterol

Azadirachtin, a limonoid

Ponasterone A, a phytoecdysone

α-Ecdysone, an insect molting hormone

Digitoxigenin, the aglycone of digitoxin, a cardenolide

Yamogenin, a saponin

Figure 13.12 Structure of various triterpenes.

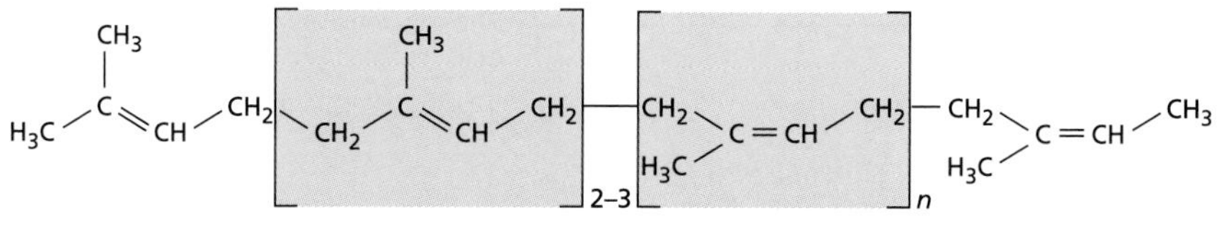

Rubber

Figure 13.13 Structure of rubber, a terpene polymer containing thousands of C_5 units. All of the double bonds in rubber have a *cis* (Z) configuration, except for those in the first two or three units. Other plant polyterpenes include gutta, in which the double bonds are all *trans* (E).

Rubber is found in numerous plants, but the most commercially important one is the rubber tree, *Hevea brasiliensis.* Gutta rubber has its double bonds in the *trans* (E) configuration and is produced by sapodilla (*Manilkara zapota*) and other plants. In the rubber tree, rubber occurs as small particles suspended in a milky fluid called latex, which is found in long vessels known as **laticifers.** Laticifers are found in many plant species and may contain other secondary products, such as triterpenes, in addition to or instead of rubber. The function of laticifers in the plant has been debated since their discovery by H. A. de Bary in 1877. Since rubber is not readily mobilized, its role as a storage compound is doubtful. The most probable function of laticifers appears to be protection, both as a mechanism for wound healing and as a defense against herbivores.

Some Herbivores Can Circumvent the Toxic Effects of Secondary Metabolites

Scientists have sometimes hesitated to accept the defensive functions of terpenes and other secondary metabolites because herbivores are often seen to eat plant parts containing high levels of these compounds without ill effect, and they may even be attracted by certain secondary metabolites rather than repelled. These observations can be explained by the evolution of counteradaptations to secondary metabolites in certain herbivore species that permit them to ingest a considerable amount of "defended" plant material without being poisoned. Some herbivores, for instance, efficiently detoxify lipophilic secondary metabolites by converting them to water-soluble derivatives that can be excreted. Other herbivores have physiological modifications that make them no longer susceptible to the toxic effects of particular products.

Once an herbivore species has developed the ability to ingest toxic compounds with impunity, it may come to specialize on plants containing those toxins, since such plants should be relatively unavailable to other herbivores. In these cases, it is often advantageous for adapted herbivores to use the distinctive secondary chemistry of their host plant species as an aid to locating that species in nature. For example, bark beetles have evolved the ability to metabolize the major monoterpenes of certain conifer species and have become specialist feeders on these trees. They are also attracted to various volatile monoterpenes of conifers (and to some metabolites of their own terpene detoxification systems), and they use these compounds as cues for finding their host trees (Byers 1989).

The ability of specialist feeders such as bark beetles to detoxify, at least partly, host secondary metabolites means that the secondary compounds of a plant may provide it with only partial protection against certain herbivores. Other types of defense, such as tough leaves or low nutrient concentrations, could be important deterrents to specialist herbivores. These findings do not invalidate the generalization that secondary metabolites are protective. Secondary products still protect the plant against nonspecialist herbivores, and sufficient quantities may deter specialists, since toxicity often depends on the quantity ingested.

Some herbivores not only use secondary metabolites to find their hosts but also are adapted to store ingested toxic secondary metabolites for protection against their own predators. The classic example of this phenomenon is the monarch butterfly, which is a specialist feeder on milkweeds (*Asclepias* spp.). Milkweeds contain cardenolides, which are toxic to most herbivores. Investigations by Lincoln Brower at the University of Florida have shown that, while feeding on milkweeds, monarch caterpillars accumulate cardenolides in their bodies with no observable ill effects. Both the caterpillars and the resulting brightly colored adult monarch butterflies are toxic to predators such as birds.

Phenolic Compounds

Plants produce a large variety of secondary products that contain a phenol group, a hydroxyl functional group on an aromatic ring:

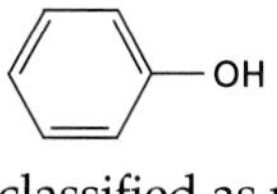

These substances are classified as phenolic compounds. Plant phenolics are a chemically heterogeneous group: Some are soluble only in organic solvents, some are water-soluble carboxylic acids and glycosides, and others are large, insoluble polymers.

In keeping with their chemical diversity, phenolics play a variety of roles in the plant. After a brief account of phenolic biosynthesis, we will discuss several princi-

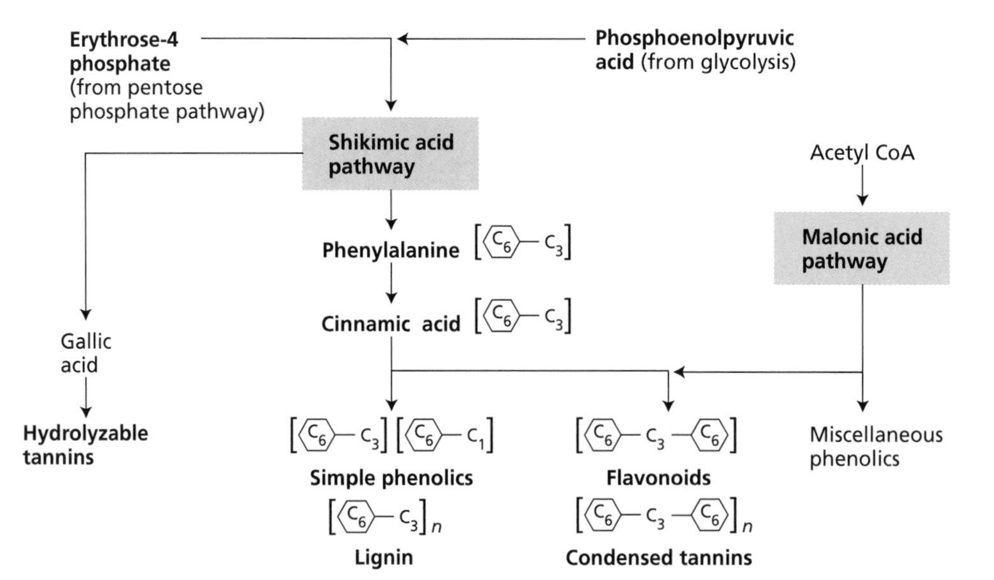

Figure 13.14 Plant phenolics are biosynthesized in several different ways. In higher plants, most phenolics are derived at least in part from phenylalanine, a product of the shikimic acid pathway. Formulas in brackets indicate the basic arrangement of carbon skeletons: C_6 indicates a benzene ring, and C_3 is a three-carbon chain. More detail on the pathway from phenylalanine onward is given in Figure 13.16.

pal groups of phenolic compounds and what is known about their roles in the plant. Many serve as defense compounds against herbivores and pathogens. Others function in mechanical support, in attracting pollinators and fruit dispersers, in absorbing harmful ultraviolet radiation, or in reducing the growth of nearby competing plants.

Phenylalanine Is an Intermediate in the Biosynthesis of Most Plant Phenolics

Plant phenolics are biosynthesized by several different routes and thus constitute a heterogeneous group from a metabolic point of view. Two basic pathways are involved: the shikimic acid pathway and the malonic acid pathway (Figure 13.14). The shikimic acid pathway participates in the biosynthesis of most plant phenolics. The malonic acid pathway, although an important source of phenolic secondary products in fungi and bacteria, is of less significance in higher plants.

The **shikimic acid pathway** converts simple carbohydrate precursors derived from glycolysis and the pentose phosphate pathway to the aromatic amino acids (Figure 13.15). One of the pathway intermediates is shikimic acid, which has given its name to this whole sequence of reactions. The well-known, broad-spectrum herbicide glyphosate (available commercially as Roundup) kills plants by blocking a step in this pathway (see Chapter 2). The shikimic acid pathway is present in plants, fungi, and bacteria but is not found in animals. Animals have no way to synthesize the three aromatic amino acids—phenylalanine, tyrosine, and tryptophan—which are therefore essential nutrients in animal diets.

The most abundant classes of secondary phenolic compounds in plants are derived from phenylalanine via the elimination of an ammonia molecule to form cinnamic acid (Figure 13.16). This reaction is catalyzed by **phenylalanine ammonia lyase** (**PAL**), perhaps the most studied enzyme in plant secondary metabolism. PAL is situated at a branch point between primary and secondary metabolism and so is an important regulatory step in the formation of many phenolic compounds. Studies with several different species of plants have shown that the activity of PAL is increased by environmental factors, such as low nutrient levels, light (through its effect on phytochrome), and fungal infection (Hahlbrock and Scheel 1989). The point of control appears to be the initiation of transcription. Fungal invasion, for example, triggers the transcription of messenger RNA that codes for PAL, thus increasing the amount of PAL in the plant, which then stimulates the synthesis of phenolic compounds. The regulation of PAL activity in plants is made more complex by the existence in many species of multiple PAL-encoding genes, some of which are expressed only in specific tissues or only under certain environmental conditions (Logemann et al. 1995).

The product of PAL, *trans*-cinnamic acid, is converted to *para*-coumaric acid by the addition of a hydroxyl group on the aromatic ring in a *para* position (see Figure 13.16). Subsequent reactions lead to the addition of more hydroxyl groups and other substituents. *Trans*-cinnamic acid, *para*-coumaric acid, and their derivatives are simple phenolic compounds called **phenylpropanoids** because they contain a benzene ring

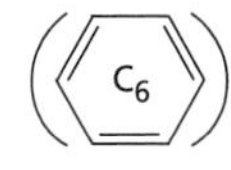

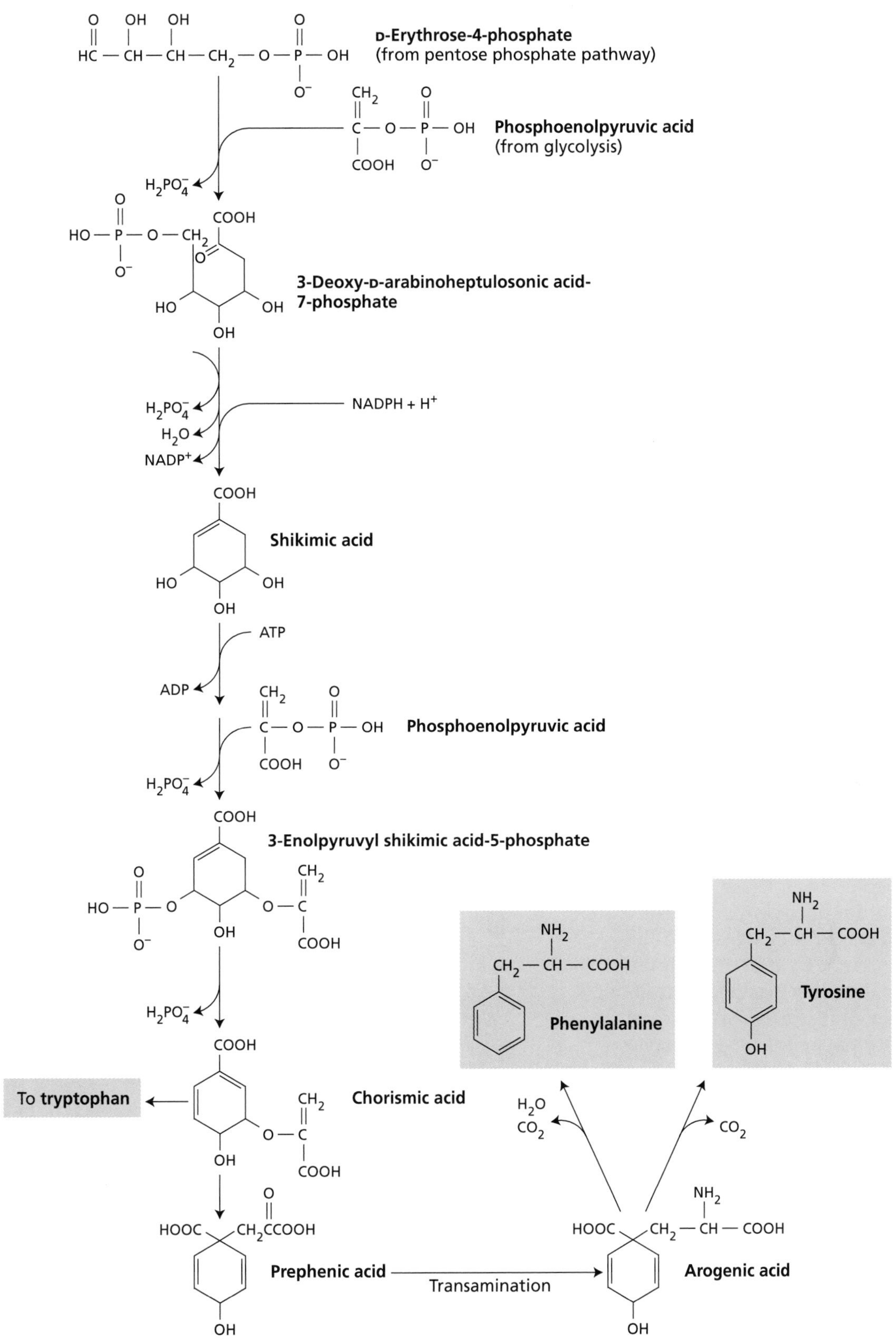

D-Erythrose-4-phosphate (from pentose phosphate pathway)

Phosphoenolpyruvic acid (from glycolysis)

$H_2PO_4^-$

3-Deoxy-D-arabinoheptulosonic acid-7-phosphate

NADPH + H^+

$H_2PO_4^-$

H_2O

$NADP^+$

Shikimic acid

ATP

ADP

Phosphoenolpyruvic acid

$H_2PO_4^-$

3-Enolpyruvyl shikimic acid-5-phosphate

$H_2PO_4^-$

Chorismic acid

To **tryptophan**

Prephenic acid

Transamination

Arogenic acid

H_2O

CO_2

CO_2

Phenylalanine

Tyrosine

Figure 13.15 In the shikimic acid pathway, the aromatic amino acids are synthesized from carbohydrate precursors derived from the pentose phosphate pathway (D-erythrose-4-phosphate) and glycolysis (phosphoenolpyruvic acid).

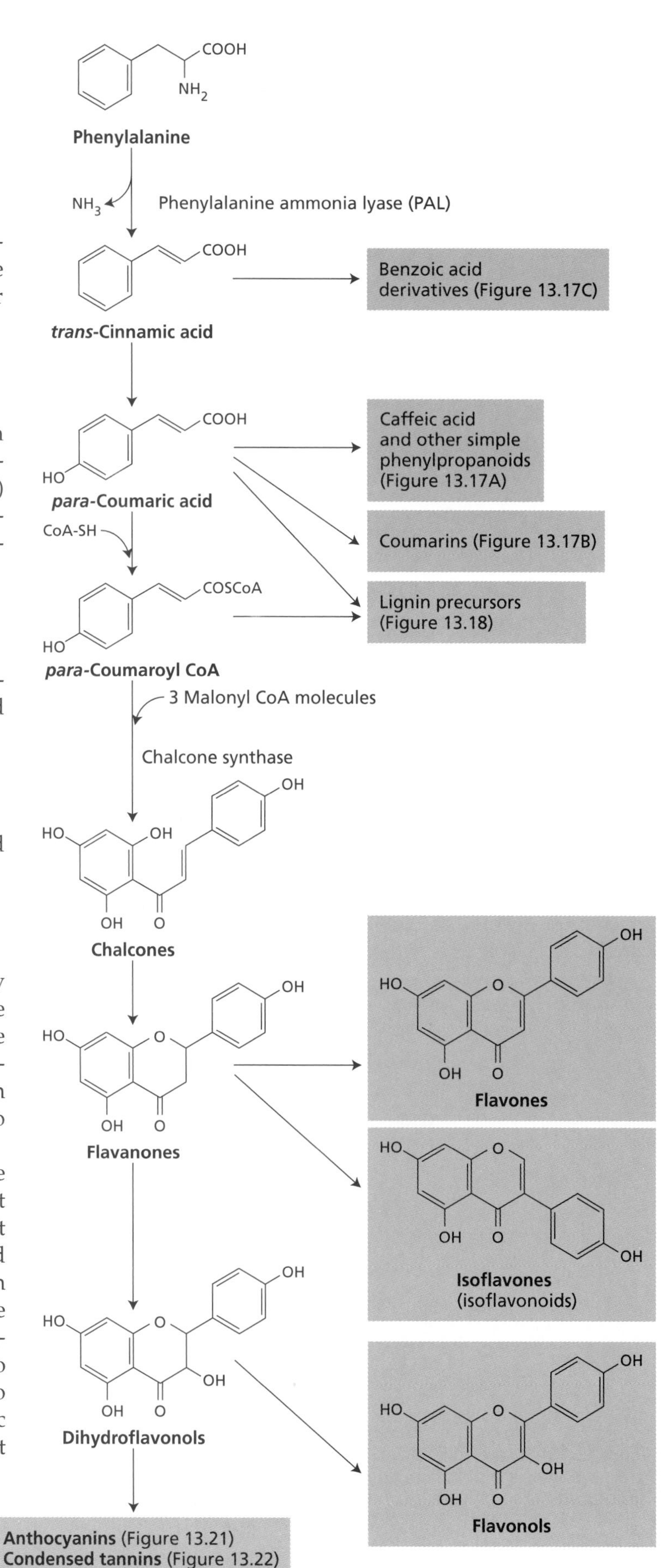

Figure 13.16 Outline of phenolic biosynthesis from phenylalanine. The formation of many plant phenolics, including simple phenylpropanoids, coumarins, benzoic acid derivatives, lignin, anthocyanins, isoflavones , condensed tannins, and other flavonoids, begins with phenylalanine.

and a three-carbon side chain. Phenylpropanoids are important building blocks of the more complex phenolic compounds discussed later in this chapter.

Some Simple Phenolics Are Activated by Ultraviolet Light

Simple phenolic compounds are widespread in vascular plants and appear to function in different capacities. Their structures include: (1) simple phenylpropanoids, such as *trans*-cinnamic acid, *para*-coumaric acid, and their derivatives, which have a basic

C_6—C_3

carbon skeleton (Figure 13.17A); (2) phenylpropanoid lactones (cyclic esters) called coumarins, also with a

C_6—C_3

skeleton (Figure 13.17B); and (3) benzoic acid derivatives that have a

C_6—C_1

skeleton, formed from phenylpropanoids by cleavage of a two-carbon fragment from the side chain (Figure 13.17C) (see also Figure 13.16). As with many other secondary products, plants can elaborate on the basic carbon skeleton of simple phenolic compounds to make more complex products.

Many simple phenolic compounds have important roles in plants as defenses against insect herbivores and fungi. Of special interest is the phototoxicity of certain coumarins called **furanocoumarins**, which have an attached furan ring (see Figure 13.17B). These compounds are not toxic until they are activated by light. Sunlight in the ultraviolet A (UV-A) region (320 to 400 nm) causes some furanocoumarins to become activated to a high-energy electronic state. Activated furanocoumarins can insert

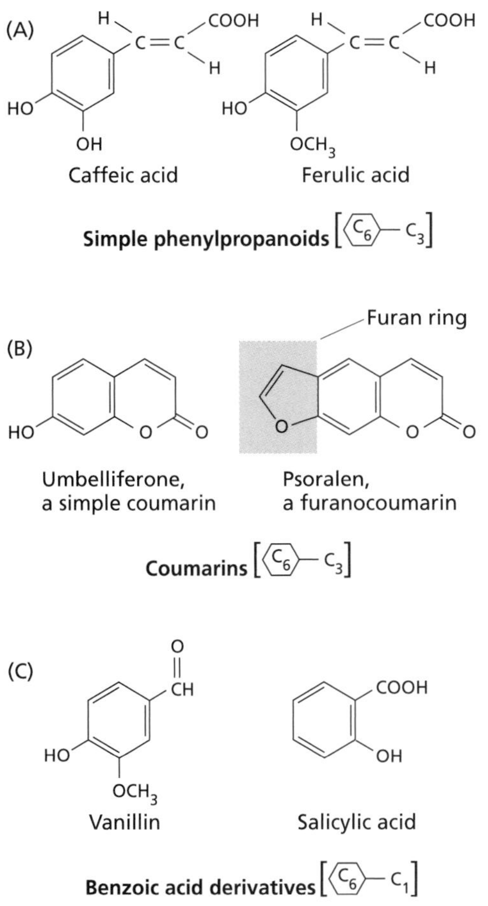

Figure 13.17 Simple phenolic compounds play a great diversity of roles in plants. Caffeic acid and ferulic acid may be released into the soil and inhibit the growth of neighboring plants. Psoralen is a furanocoumarin that exhibits phototoxicity to insect herbivores. Salicylic acid is a plant growth regulator that is involved in systemic resistance to plant pathogens.

themselves into the double helix of DNA and bind to the pyrimidine bases cytosine and thymine, thus blocking transcription and repair and leading eventually to cell death.

Phototoxic furanocoumarins are especially abundant in members of the Umbelliferae family, including celery, parsnip, and parsley. In celery, the level of these compounds can increase about 100-fold if the plant is stressed or diseased. Celery pickers, and even some grocery shoppers, have been known to develop skin rashes from handling stressed or diseased celery. Some insects have adapted to survive on plants that contain furanocoumarins and other phototoxic compounds by living in silken webs or rolled-up leaves, which screen out the activating wavelengths (Sandberg and Berenbaum 1989).

The Release of Simple Phenolics May Affect the Growth of Other Plants

From leaves, roots, and decaying litter, plants release a variety of primary and secondary metabolites into the environment. Investigation of the effects of these compounds on neighboring plants is the study of **allelopathy**. If a plant can reduce the growth of nearby plants by releasing chemicals into the environment, it may increase its access to light, water, and nutrients and thus its evolutionary fitness. Generally speaking, the term "allelopathy" has come to be applied to the harmful effects of plants on their neighbors, although a precise definition also includes beneficial effects.

Simple C_6–C_1 and C_6–C_3 phenolic compounds are frequently cited as having allelopathic activity. Compounds such as caffeic acid and ferulic acid (see Figure 13.17A) occur in soil in appreciable amounts and have been shown in laboratory experiments to inhibit the germination and growth of many plants (Rice 1987).

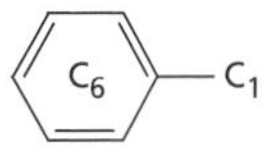

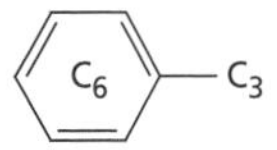

In spite of results such as these, the importance of allelopathy in natural ecosystems is still controversial. Many scientists doubt that allelopathy is a significant factor in plant–plant interactions because good evidence for this phenomenon has been hard to obtain. It is easy to show that extracts or purified compounds from one plant can inhibit the growth of other plants in laboratory experiments, but it has been very difficult to demonstrate that these compounds are present in the soil in sufficient concentration to inhibit growth. Furthermore, organic substances in the soil are often bound to soil particles and may be rapidly degraded by microbes (Dao 1987).

In spite of the lack of supporting evidence, allelopathy is currently of great interest because of its potential agricultural applications. Reductions in crop yields caused by weeds or residues from the previous crop may in some cases be a result of allelopathy. An exciting future prospect is the development of crop plants genetically engineered to be allelopathic to weeds.

Lignin Is a Highly Complex Phenolic Macromolecule

After cellulose, the most abundant organic substance in plants is **lignin**, a highly branched polymer of C_6–C_3 phenylpropanoid groups that plays both primary and secondary roles. The precise structure of lignin is not known, because it is difficult to extract lignin from plants where it is covalently bound to cellulose and other polysaccharides of the cell wall.

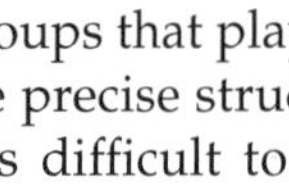

Lignin is generally formed from three different phenylpropanoid alcohols: coniferyl, coumaryl, and sinapyl alcohols (Figure 13.18), which are synthesized

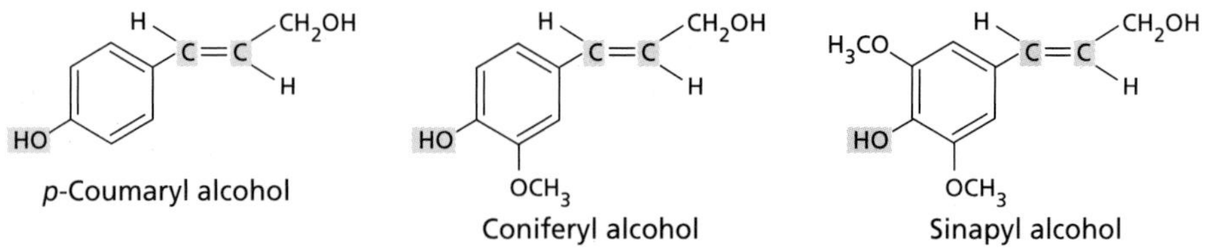

Figure 13.18 The three phenylpropanoid alcohols that join to form lignin. The shaded portions represent the sites at which the monomers are most frequently linked (see Figure 13.19). Less frequently, the monomers are linked via the CH_2OH groups.

p-Coumaryl alcohol

Coniferyl alcohol

Sinapyl alcohol

from phenylalanine via various cinnamic acid derivatives. The phenylpropanoid alcohols are joined into a polymer through the action of enzymes that generate free-radical intermediates. There are often multiple C—C and C—O—C bonds in each phenylpropanoid alcohol unit in lignin, resulting in a complex structure that branches in three dimensions. Therefore, unlike polymers such as starch, rubber, or cellulose, the units of lignin are not linked in a simple, repeating way. To add to the complexity, the proportions of the three monomeric units in lignin vary among species, plant organs, and even layers of a single cell wall (Lewis and Yamamoto 1990). Figure 13.19 shows a part of the structure of a hypothetical lignin molecule.

Lignin is found in the cell walls of various types of supporting and conducting tissue, notably the tracheids and vessel elements of the xylem. It is deposited chiefly in the thickened secondary wall but can also occur in the

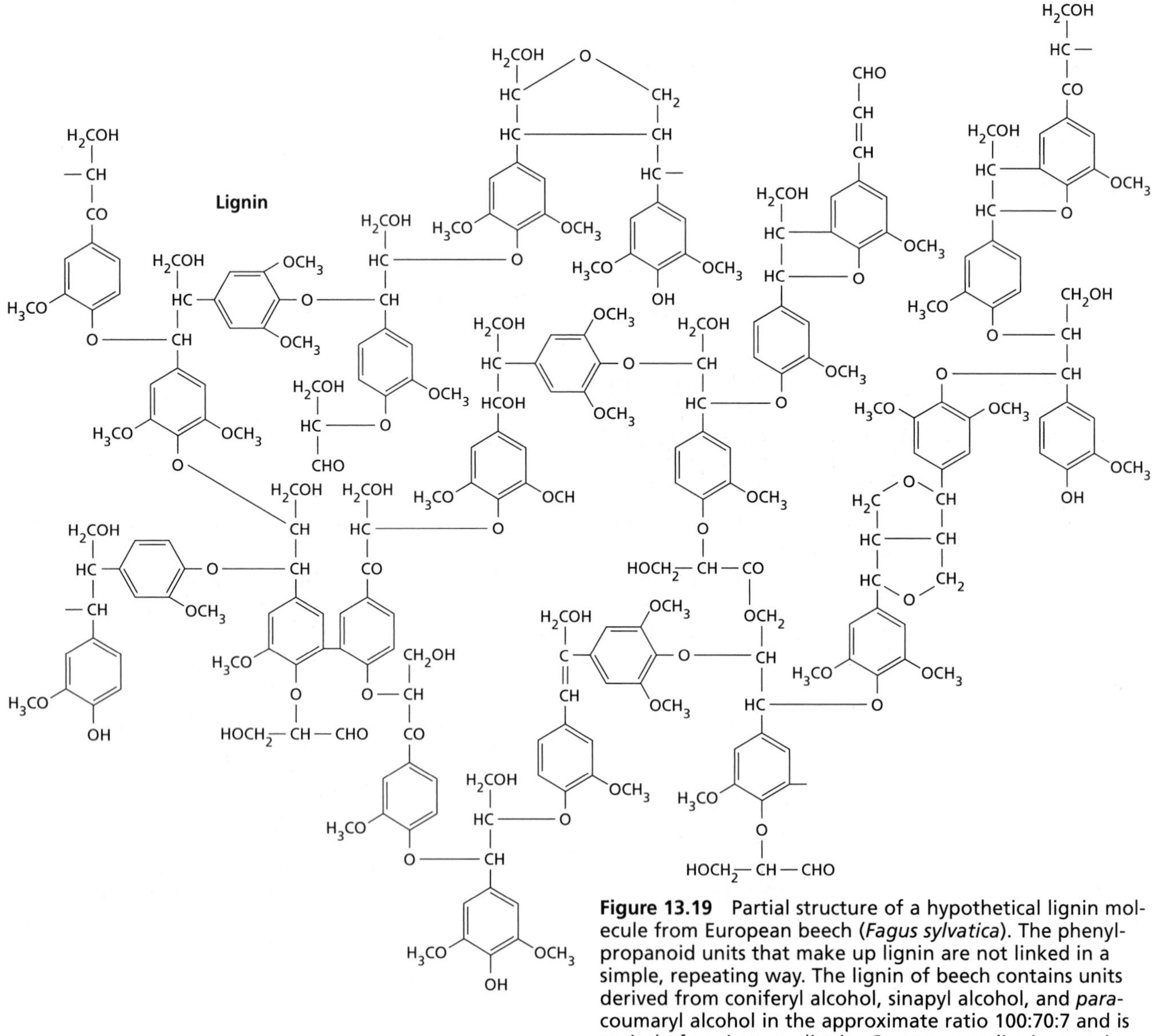

Lignin

Figure 13.19 Partial structure of a hypothetical lignin molecule from European beech (*Fagus sylvatica*). The phenylpropanoid units that make up lignin are not linked in a simple, repeating way. The lignin of beech contains units derived from coniferyl alcohol, sinapyl alcohol, and *para*-coumaryl alcohol in the approximate ratio 100:70:7 and is typical of angiosperm lignin. Gymnosperm lignin contains relatively fewer sinapyl alcohol units. (After Nimz 1974.)

primary wall and middle lamella in close contact with the celluloses and hemicelluloses already present. The mechanical rigidity of lignin strengthens stems and vascular tissue, allowing upward growth and permitting water and minerals to be conducted through the xylem under negative pressure without collapse of the tissue. Because lignin is such a key component of water transport tissue, the ability to make lignin must have been one of the most important adaptations permitting primitive plants to colonize dry land.

Besides providing mechanical support, lignin has significant protective functions in plants. Its physical toughness deters feeding by animals, and its chemical durability makes it relatively indigestible to herbivores. By bonding to cellulose and protein, lignin also reduces the digestibility of these substances. Lignification blocks the growth of pathogens and is a frequent response to infection or wounding.

Lignin biosynthesis is presently a very active area of research. Investigators are seeking to understand how the composition of individual phenylpropanoid units in lignin is controlled and what regulates the types of bonds formed during polymerization. The modification of plant lignin composition by genetic engineering may make it easier to remove lignin from wood pulp used in paper manufacture and may increase the digestibility of certain forage species.

Flavonoids Are Formed by Two Different Biosynthetic Pathways

One of the largest classes of plant phenolics is the **flavonoids**. The basic carbon skeleton of a flavonoid contains 15 carbons in a

C_6—C_3—C_6

arrangement, with two aromatic rings connected by a three-carbon bridge (Figure 13.20). This structure results from two separate biosynthetic pathways. The bridge and one aromatic ring (ring B) constitute a phenylpropanoid unit biosynthesized from phenylalanine, itself a product of the shikimic acid pathway. The six carbons of the other aromatic ring (ring A) originate from the condensation of three acetate units via the malonic acid pathway. The fusion of these two parts involves the stepwise condensation of a phenylpropanoid, *para*-coumaroyl CoA, with three malonyl CoA residues (each of which donates two carbon atoms) in a reaction catalyzed by **chalcone synthase** (see Figure 13.16).

This series of condensations is analogous to the cycle of two-carbon addition during fatty acid biosynthesis that is described in Chapter 11, except that construction of the flavonoid skeleton starts with *para*-coumaroyl CoA, rather than acetyl CoA, and the growing acyl chain is not reduced or dehydrated. The chalcones formed in this process give rise to all the other types of flavonoids. Chalcone synthase, like PAL, is an important regulatory enzyme in plant secondary metabolism. Its activity is modified by various environmental conditions, and control is exercised at the level of gene transcription (Hahlbrock and Scheel 1989).

Flavonoids are classified into different groups based primarily on the degree of oxidation of the three-carbon bridge (see Figure 13.20). We will discuss four of the groups shown in Figure 13.16: the anthocyanins, the flavones, the flavonols, and the isoflavones. The basic flavonoid carbon skeleton may have numerous substituents. Hydroxyl groups are usually present at positions 4, 5, and 7, but they may also be found at other positions. Sugars are very common as well; in fact, the majority of flavonoids exist naturally as glycosides. Whereas both hydroxyl groups and sugars increase the water solubility of flavonoids, other substituents, such as methyl ethers or modified isopentyl units, make flavonoids lipophilic (hydrophobic). Different types of flavonoids perform very different functions in the plant, including pigmentation and defense.

Anthocyanins Are Colored Flavonoids That Attract Animals for Pollination and Seed Dispersal

Interactions between plants and animals are not limited to antagonistic relationships, such as predator–prey interactions. There are also mutualistic associations. In return for the reward of ingesting nectar or fruit pulp, animals perform extremely important services for plants as carriers of pollen and seeds. Secondary metabolites are involved in these plant–animal interactions, helping to attract animals to flowers and fruit by providing visual and olfactory signals.

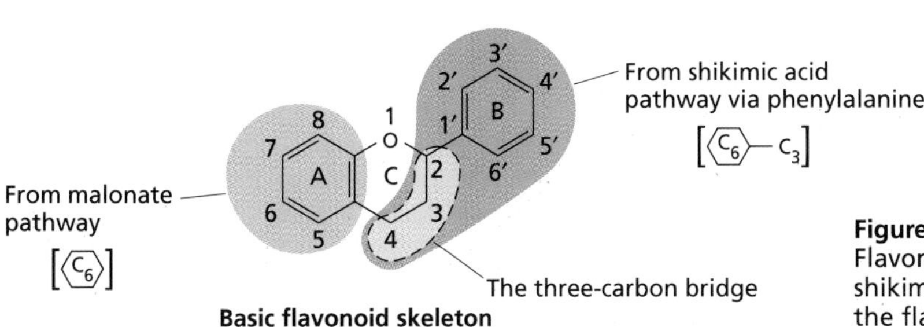

Figure 13.20 Basic flavonoid carbon skeleton. Flavonoids are biosynthesized from products of the shikimic acid and malonate pathways. Positions on the flavonoid ring system are numbered as shown.

The colored pigments of plants are of two principal types: carotenoids and flavonoids. Carotenoids, as we have already seen, are yellow, orange, and red terpenoid compounds that also serve as accessory pigments in photosynthesis (see Chapter 7). Flavonoids are phenolic compounds that include a wide range of colored substances. The most widespread group of pigmented flavonoids is the **anthocyanins**, which are responsible for most of the red, pink, purple, and blue colors observed in plant parts. By coloring flowers and fruits, the anthocyanins are vitally important in attracting animals for pollination and seed dispersal.

Anthocyanins are glycosides that have sugars at position 3 (see Figure 13.21) and sometimes elsewhere. Without their sugars, anthocyanins are known as **anthocyanidins**. The most common anthocyanidin structures and their colors are shown in Figure 13.21. Anthocyanin color is influenced by many factors, including the number of hydroxyl and methoxyl groups in ring B of the anthocyanidin, the presence of aromatic acids esterified to the main skeleton, and the pH of the cell vacuole in which these compounds are stored. Recent research in Japan has shown that anthocyanins may exist in supramolecular complexes along with chelated metal ions and flavone copigments. The blue pigment of *Commelina communis* (dayflower) was found to consist of a large complex of six anthocyanin molecules, six flavones, and two associated magnesium ions (Kondo et al. 1992).

Considering the variety of factors affecting anthocyanin coloration and the possible presence of carotenoids as well, it is not surprising that so many different shades of flower and fruit color are found in nature. The evolution of flower color may have been governed by selection pressures for different sorts of pollinators, which often have different color preferences (see Box 13.1). Color, of course, is just one type of signal used to attract pollinators to flowers. Volatile chemicals, particularly monoterpenes, frequently provide attractive scents.

(A) Anthocyanidin (B) Anthocyanin

Anthocyanidin	Substituents	Color
Pelargonidin	4′—OH	Orange red
Cyanidin	3′—OH, 4′—OH	Purplish red
Delphinidin	3′—OH, 4′—OH, 5′—OH	Bluish purple
Peonidin	3′—OCH_3, 4′—OH	Rosy red
Petunidin	3′—OCH_3, 4′—OH, 5′—OCH_3	Purple

Figure 13.21 The structures of anthocyanidins (A) and anthocyanin (B). The colors of anthocyanidins depend in part on the substituents attached to ring B. An increase in the number of hydroxyl groups shifts absorption to a longer wavelength and gives a bluer color. Replacement of a hydroxyl group with a methoxyl group (OCH_3) shifts absorption to a slightly shorter wavelength, resulting in a redder color.

Flavonoids May Protect against Damage by Ultraviolet Light

Two other major groups of flavonoids found in flowers are **flavones** and **flavonols** (see Figure 13.16). These flavonoids generally absorb light at shorter wavelengths than anthocyanins do, so they are not visible to the human eye. However, insects such as bees, which see farther into the ultraviolet range of the spectrum than humans do, may respond to flavones and flavonols as attractant cues. By using ultraviolet photography, investigators such as Morris Levy and his coworkers at Purdue University have shown that flavonols in a flower often form symmetric patterns of stripes, spots, or concentric circles called nectar guides (see the figure in Box 13.1) (McCrea and Levy 1983). These patterns may be conspicuous to insects and are thought to help indicate the location of pollen and nectar.

Flavones and flavonols are not restricted to flowers; they are also present in the leaves of all green plants. These two classes of flavonoids function to protect cells from excessive UV-B radiation (280 to 320 nm) because they accumulate in the epidermal layers of leaves and stems and absorb light strongly in the UV-B region while letting the visible (photosynthetically active) wavelengths pass through uninterrupted. In addition, exposure of plants to increased UV-B light has been demonstrated to increase the synthesis of flavones and flavonols.

Recently, the importance of flavonoids as UV protectants was confirmed in Rob Last's laboratory at Cornell University. Last and coworkers studied *Arabidopsis thaliana* mutants that lack chalcone synthase activity and thus produce no flavonoids. These plants are much more sensitive to UV-B radiation than wild-type individuals are, and they grow very poorly under normal conditions. When shielded from UV light, however, they grow normally (Li et al. 1993). In addition to flavonoids, certain phenylpropanoid esters also serve as effective UV protectants in *Arabidopsis*.

Other functions of flavonoids have recently been discovered. For example, flavones and flavonols secreted into the soil by legume roots mediate the interaction of legumes and nitrogen-fixing symbionts, a phenomenon described in Chapter 12.

BOX 13.1

A Bee's-Eye View of a Golden-Eye's "Bull's-Eye"

ULTRAVIOLET-ABSORBING PIGMENTS are invisible to humans, since the human eye cannot perceive UV light. Hence, patterns formed by such pigments in flowers cannot be detected by the human eye. In contrast, honeybees, which can perceive UV light, are able to detect such pigments. Plants have evolved to exploit the visual range of pollinating insects in order to attract them to their flowers. The figure shows photographs of a golden-eye (*Viguiera*) as seen by humans (A) and as it might appear to honeybees (B). Special lighting was used to simulate the spectral sensitivity of the honeybee visual system.

To humans, the golden-eye has yellow rays and a brown central disc; to bees, the tips of the rays appear "light yellow," the inner portion of the rays "dark yellow," and the central disc "black." Ultraviolet-absorbing flavonols are found in the inner parts of the rays but not in the tips. The distribution of flavonols in the rays and the sensitivity of insects to part of the UV spectrum contribute to the "bull's-eye" pattern seen by honeybees, which presumably helps them locate pollen and nectar. Bees and other insects perceive not only the presence of UV-absorbing pigments but also the whole floral color pattern, which is a result of carotenoids and other flavonoid pigments as well as flavonols (McCrea and Levy 1983).

(A)

(B)

The golden-eye as seen by (A) humans and (B) honeybees. (Courtesy of M. Levy.)

Isoflavonoid Defense Compounds Are Synthesized Immediately Following Infection by Fungi or Bacteria

The **isoflavonoids** (isoflavones) are a group of flavonoids in which the position of one aromatic ring (ring B) is shifted (see Figure 13.16). Isoflavonoids are found mostly in legumes and have several different biological activities. Some, such as the rotenoids, have strong insecticidal actions; others have antiestrogenic effects and so cause infertility in mammals.

In the past few years, isoflavonoids have become best known for their role as *phytoalexins*, antimicrobial compounds synthesized in response to bacterial or fungal infection that help limit the spread of the invading pathogen. Phytoalexins are discussed in more detail later in this chapter.

Tannins Deter Feeding by Herbivores

Besides lignin, a second category of plant phenolic polymers with defensive properties is the **tannins**. The term "tannin" was first used to describe compounds that could convert raw animal hides into leather in the process known as tanning. Tannins bind the collagen proteins of animal hides, increasing their resistance to heat, water, and microbes.

There are two categories of tannins, condensed and hydrolyzable. **Condensed tannins** are

$$(C_6 - C_3 - C_6)_n$$

compounds formed by the polymerization of flavonoid units (Figure 13.22A). They are frequent constituents of woody plants. Because condensed tannins can often be hydrolyzed to anthocyanidins by treatment with strong acids, they are sometimes called proanthocyanidins. **Hydrolyzable tannins** are heterogeneous polymers containing phenolic acids, especially gallic acid, and simple sugars (Figure 13.22B). They are smaller than condensed tannins and may be hydrolyzed more easily; only dilute acid is needed. Most tannins have molecular masses between 600 and 3000.

Tannins are general toxins that significantly reduce the growth and survivorship of many herbivores when added to their diets. In addition, tannins act as feeding repellents to a great diversity of animals. In humans, tannins cause a sharp, astringent sensation in the mouth as a result of their binding of salivary proteins. Mammals, such as cattle, deer, and apes, characteristically avoid plants or parts of plants with high tannin contents (Oates et al. 1980). Unripe fruits, for instance, frequently have very high tannin levels, which may be concentrated in the outer cell layers. Interestingly, humans often prefer a certain level of astringency in tannin-containing foods such as apples, blackberries, tea, and red wine.

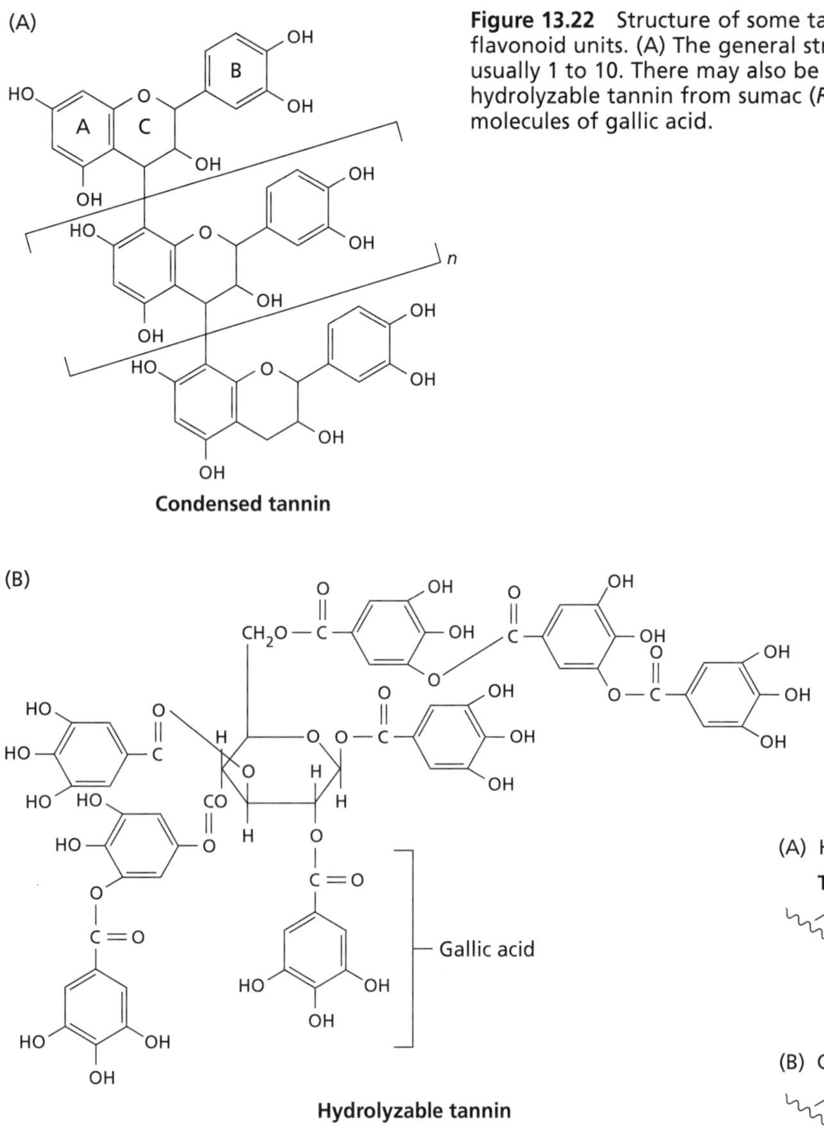

Figure 13.22 Structure of some tannins formed from phenolic acids or flavonoid units. (A) The general structure of a condensed tannin, where *n* is usually 1 to 10. There may also be a third —OH group on ring B. (B) The hydrolyzable tannin from sumac (*Rhus semialata*) consists of glucose and eight molecules of gallic acid.

The defensive properties of tannins are generally attributed to their ability to bind proteins. It has long been thought that plant tannins complex proteins in the guts of herbivores by forming hydrogen bonds between their hydroxyl groups and electronegative sites on the protein (Figure 13.23A). More recent evidence indicates that tannins and other phenolics can also bind to dietary protein in a covalent fashion (Figure 13.23B). The foliage of many plants contains enzymes that oxidize phenolics to their corresponding quinone forms in the guts of herbivores (Felton et al. 1989). Quinones are highly reactive electrophilic molecules that readily react with the nucleophilic —SH and —NH_2 groups of proteins (see Figure 13.23B). By whatever mechanism protein–tannin binding occurs, this process has a negative impact on herbivore nutrition. Tannins can inactivate herbivore digestive enzymes and create complex aggregates of tannins and plant proteins that are difficult to digest.

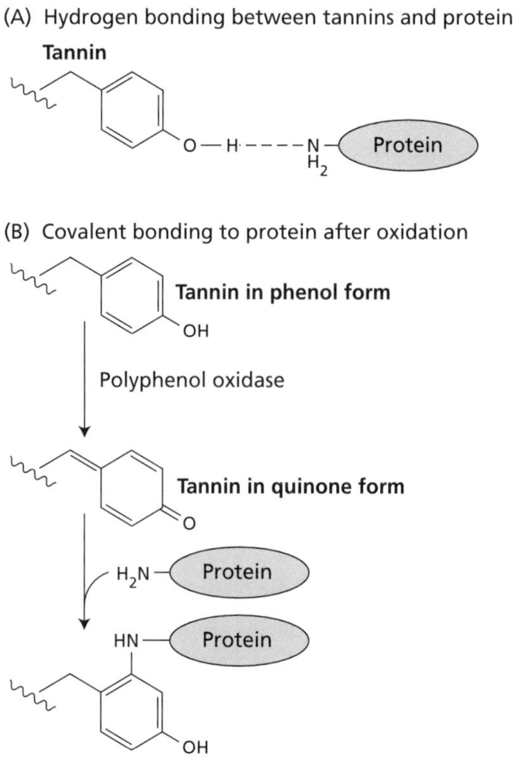

Figure 13.23 Proposed mechanisms for the interaction of tannins with proteins. (A) Hydrogen bonds may form between the phenolic hydroxyl groups of tannins and electronegative sites on the protein. (B) Phenolic hydroxyl groups may bind covalently to proteins following activation by oxidative enzymes, such as polyphenol oxidase.

Herbivores that habitually feed on tannin-rich plant material appear to possess some interesting adaptations to remove tannins from their digestive systems. For example, some mammals, such as rodents and rabbits, produce salivary proteins with a very high proline content (25 to 45 percent) that have a high affinity for tannins. Secretion of these proteins is induced by ingestion of food with a high tannin content and greatly diminishes the toxic effects of tannins (Butler 1989). The large number of proline residues gives these proteins a very flexible, open conformation and a high degree of hydrophobicity that facilitates binding to tannins.

Plant tannins also serve as defenses against microorganisms. For example, the nonliving heartwood of many trees contains high concentrations of tannins that help prevent fungal and bacterial decay.

Nitrogen-Containing Compounds

A large variety of plant secondary metabolites have nitrogen in their structure. Included in this category are such well-known antiherbivore defenses as alkaloids and cyanogenic glycosides, which are of considerable interest because of their toxicity to humans and their medicinal properties. Most nitrogenous secondary metabolites are biosynthesized from common amino acids. In this section we will examine the structure and biological properties of various nitrogen-containing secondary metabolites, including alkaloids, cyanogenic glycosides, glucosinolates, and nonprotein amino acids. In addition, we will discuss the ability of *systemin*, a protein released from damaged cells, to serve as a wound signal to the rest of the plant.

Alkaloids Have Marked Physiological Effects on Animals

The **alkaloids** are a large family of nitrogen-containing secondary metabolites found in approximately 20% of the species of vascular plants. As a group, they are best known for their striking pharmacological effects on vertebrate animals. The nitrogen atom in these substances is usually part of a **heterocyclic ring**, a ring that contains both nitrogen and carbon atoms. As their name would suggest, most alkaloids are alkaline. At pH values commonly found in the cytosol (pH 7.2) or the vacuole (pH 5 to 6), the nitrogen atom is protonated; hence, alkaloids are positively charged and are generally water soluble.

Alkaloids are usually synthesized from one of a few common amino acids—in particular, aspartic acid, lysine, tyrosine, and tryptophan. However, the carbon skeleton of some alkaloids contains a component derived from the terpene pathway. Table 13.1 lists the major alkaloid types and their amino acid precursors. Several different

Table 13.1
Major types of alkaloids, their amino acid precursors, and well-known examples of each type

Alkaloid class	Structure	Biosynthetic precursor	Examples
Pyrrolidine	N	Ornithine	Nicotine
Tropane	N	Ornithine	Atropine, cocaine
Piperidine	N	Lysine (or acetate)	Coniine
Pyrrolizidine	N	Ornithine	Retrorsine
Quinolizidine	N	Lysine	Lupinine
Isoquinoline	N	Tyrosine	Codeine, morphine
Indole	N	Tryptophan	Psilocybin, reserpine, strychnine

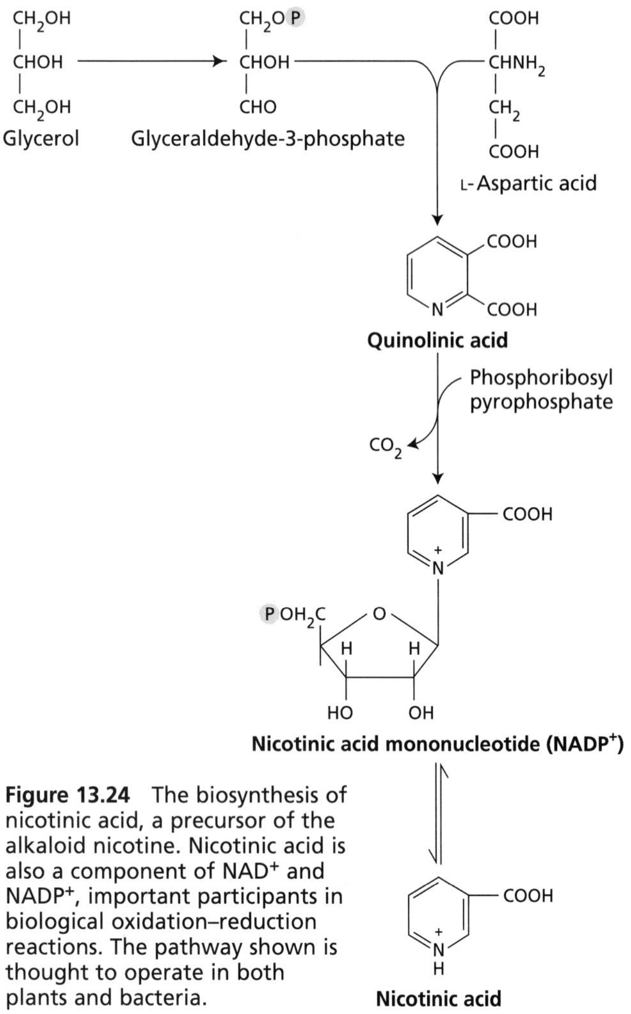

Glycerol → Glyceraldehyde-3-phosphate + L-Aspartic acid → Quinolinic acid → (Phosphoribosyl pyrophosphate; CO_2) → Nicotinic acid mononucleotide (NADP⁺) ⇌ Nicotinic acid

Figure 13.24 The biosynthesis of nicotinic acid, a precursor of the alkaloid nicotine. Nicotinic acid is also a component of NAD^+ and $NADP^+$, important participants in biological oxidation–reduction reactions. The pathway shown is thought to operate in both plants and bacteria.

types, including nicotine and its relatives, are derived from ornithine, an intermediate in arginine biosynthesis. The biosynthetic pathway leading to nicotine is of special interest not only because nicotine is a bioactive component of tobacco but also because the B vitamin nicotinic acid (niacin) is a precursor of one of the two rings of this alkaloid. The pyrrolidine (five-membered) ring of nicotine arises from ornithine; the pyridine (six-membered) ring is derived from nicotinic acid (see Figure 13.25). Nicotinic acid is also a constituent of NAD^+ and $NADP^+$, which serve as electron carriers in metabolism. In plants and bacteria, nicotinic acid is probably synthesized from aspartic acid and glyceraldehyde-3-phosphate (Figure 13.24), whereas animals and higher fungi utilize tryptophan as a precursor.

The role of alkaloids in plants has been a subject of speculation for at least 100 years. Alkaloids were once thought to be nitrogenous wastes (analogous to urea and uric acid in animals), nitrogen storage compounds, or growth regulators, but there is little evidence to support any of these functions. Most alkaloids are now believed to function as defenses against predators, especially mammals, because of their general toxicity and deterrence capability (Hartmann 1991).

Large numbers of livestock deaths are caused by the ingestion of alkaloid-containing plants. In the United States, a significant percentage of all grazing livestock are poisoned each year by consumption of large quantities of alkaloid-containing plants such as lupines (*Lupinus*), larkspur (*Delphinium*), and groundsel (*Senecio*). This phenomenon may be due to the fact that domestic animals, unlike wild animals, have not been subjected to natural selection for the avoidance of toxic plants. Indeed, some livestock seem to prefer alkaloid-containing plants to less harmful forage.

Nearly all alkaloids are also toxic to humans when taken in sufficient quantity. For example, strychnine, atropine, and coniine (from poison hemlock) are classic alkaloid poisoning agents. At lower doses, however, many are useful pharmacologically. Morphine, codeine, atropine, and ephedrine are just a few of the plant alkaloids currently used in medicine. Other alkaloids, including cocaine, nicotine, and caffeine (Figure 13.25), enjoy widespread nonmedical use as stimulants or sedatives.

On a cellular level, the mode of action of alkaloids in animals is quite variable. Many interfere with components of the nervous system, especially the chemical transmitters; others affect membrane transport, protein synthesis, or miscellaneous enzyme activities.

Cyanogenic Glycosides Release the Poison Hydrogen Cyanide

Various nitrogenous protective compounds other than alkaloids are found in plants. Two groups of these substances, cyanogenic glycosides and glucosinolates, are not in themselves toxic but are readily broken down to

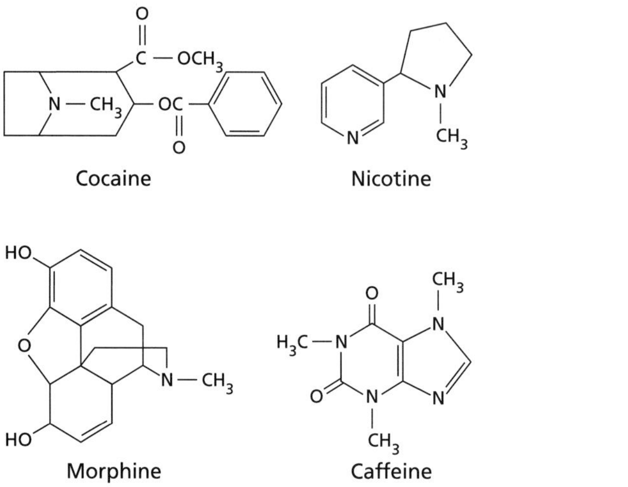

Cocaine · Nicotine · Morphine · Caffeine

Representative alkaloids

Figure 13.25 Examples of alkaloids, a diverse group of secondary metabolites that contain nitrogen, usually as part of a heterocyclic ring. Note that caffeine is a purine-type alkaloid similar to the nucleic acid bases adenine and guanine.

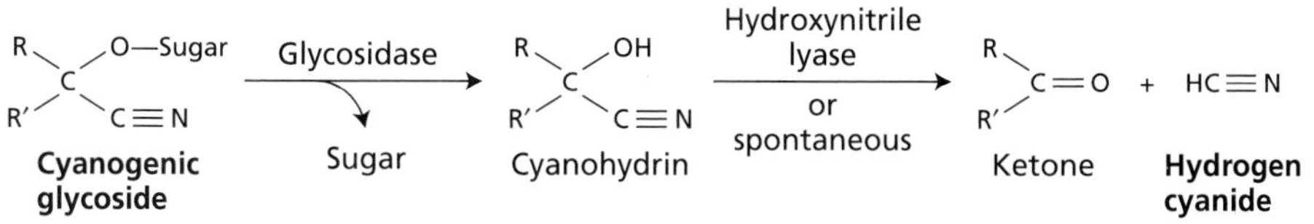

Figure 13.26 The enzyme-catalyzed hydrolysis of cyanogenic glycosides to release hydrogen cyanide. R and R′ represent various alkyl or aryl substituents. For example, if R is phenyl, R′ is hydrogen, and the sugar is the disaccharide β-gentiobiose, the compound is amygdalin, the common cyanogenic glycoside found in the seeds of almonds, apricots, cherries, and peaches.

give off volatile poisons when the plant is crushed. Cyanogenic glycosides release the well-known poisonous gas hydrogen cyanide (HCN).

The breakdown of cyanogenic glycosides in plants is a two-step enzymatic process. Species that make cyanogenic glycosides also make the enzymes necessary to hydrolyze the sugar and liberate HCN. In the first step, the sugar is cleaved by a **glycosidase**, an enzyme that separates sugars from other molecules to which they are linked (Figure 13.26). The resulting hydrolysis product, called an α-hydroxynitrile or cyanohydrin, can decompose spontaneously at a low rate to liberate HCN. This second step can be accelerated by the enzyme **hydroxynitrile lyase**.

Cyanogenic glycosides are not normally broken down in the intact plant, because the glycoside and the degradative enzymes are spatially separated, in different cellular compartments or in different tissues. For example, in sorghum, the cyanogenic glycoside dhurrin is present in the vacuoles of epidermal cells, whereas the hydrolytic and lytic enzymes are found in the mesophyll (Poulton 1990). Under ordinary conditions, this compartmentation prevents decomposition of the glycoside. However, when the leaf is damaged, as during herbivore feeding, the cell contents of different tissues mix, and HCN forms. Cyanogenic glycosides are widely distributed in the plant kingdom and are frequently encountered in legumes, grasses, and species of the rose family.

Considerable evidence indicates that cyanogenic glycosides have a protective function in certain plants. HCN is a fast-acting toxin that blocks cellular respiration by binding to the iron-containing heme group of cytochrome oxidase and other respiratory enzymes. The presence of cyanogenic glycosides deters feeding by insects and other herbivores, such as snails and slugs. As with other classes of secondary metabolites, however, some herbivores have adapted to feed on cyanogenic plants and can tolerate large doses of HCN.

The tubers of cassava (*Manihot esculenta*), a high-carbohydrate, staple food in many tropical countries, contain high levels of cyanogenic glycosides. Traditional processing methods, such as grating, grinding, soaking, and drying, lead to the removal or degradation of a large fraction of the cyanogenic glycosides present in cassava tubers. However, chronic cyanide poisoning is still widespread in regions where cassava is a major food source, suggesting that the traditional detoxification methods employed are not completely effective. Efforts are currently under way to reduce the cyanogenic glycoside content of cassava using both conventional breeding and genetic engineering approaches. The complete elimination of cyanogenic glycosides may not be desirable, since these substances are probably responsible for the fact that cassava can be stored for very long periods of time without being attacked by pests.

Glucosinolates Also Release Volatile Toxins

A second class of plant glycosides that break down to release volatile defensive substances consists of the **glucosinolates**, or mustard oil glycosides. Found principally in the Cruciferae and related plant families, glucosinolates give off the compounds responsible for the smell and taste of vegetables such as cabbage, broccoli, and radishes. The release of these mustard-smelling volatiles from glucosinolates is catalyzed by a hydrolytic enzyme, called a **thioglucosidase** or myrosinase, that cleaves glucose from its bond with the sulfur atom (Figure 13.27). The resulting **aglycone**, the nonsugar portion of the molecule, rearranges with loss of the sulfate to give pungent and chemically reactive products, includ-

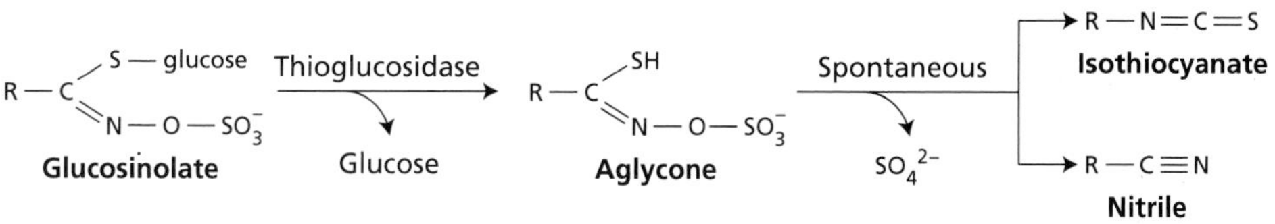

Figure 13.27 The hydrolysis of glucosinolates to mustard-smelling volatiles. R represents various alkyl or aryl substituents. For example, if R is $CH_2=CH—CH_2^-$, the compound is sinigrin, a major glucosinolate of black mustard seeds and horseradish roots.

ing isothiocyanates and nitriles, depending on the conditions of hydrolysis. These products function in defense as herbivore toxins and feeding repellents. Like cyanogenic glycosides, glucosinolates are stored in the intact plant separately from the enzymes that hydrolyze them, and they are brought into contact with these enzymes only when the plant is crushed.

As with other secondary metabolites, certain animals are adapted to feed on glucosinolate-containing plants without ill effects. For adapted herbivores, such as the cabbage butterfly, glucosinolates often serve as stimulants for feeding and egg laying, and the isothiocyanates produced after glucosinolate hydrolysis act as volatile attractants (Renwick et al. 1992). Most of the recent research on glucosinolates in plant defense has concentrated on rape, or canola (*Brassica napus*), a major oil crop in both North America and Europe. Plant breeders have tried to lower the glucosinolate levels of rapeseed so that the high-protein seed meal remaining after oil extraction can be used as animal food. The first low-glucosinolate varieties tested in the field were unable to survive because of severe pest problems. However, more recently developed varieties with low glucosinolate levels in seeds but high glucosinolate levels in leaves are able to hold their own against pests and still provide a protein-rich seed residue for animal feeding.

Nonprotein Amino Acids Function as Antiherbivore Defenses

Plants and animals incorporate the same 20 amino acids into their proteins. However, many plants also contain unusual amino acids, called nonprotein amino acids, that are not incorporated into proteins but are present instead in the free form and act as protective substances. Nonprotein amino acids are often very similar to common protein amino acids. Canavanine, for example, is a close analog of arginine, and azetidine-2-carboxylic acid has a structure very much like that of proline (Figure 13.28).

Nonprotein amino acids exert their toxicity in various ways. Some block the synthesis or uptake of protein amino acids; others can be mistakenly incorporated into proteins. The basis of canavanine's toxicity to herbivores has been well studied in Gerald Rosenthal's laboratory at the University of Kentucky. After ingestion, canavanine is recognized by the herbivore enzyme that normally binds arginine to the arginine transfer RNA molecule, and so becomes incorporated into proteins in place of arginine. The usual result is a nonfunctional protein because either its tertiary structure or its catalytic site is disrupted. Canavanine is less basic than arginine and may alter the ability of an enzyme to bind substrates or catalyze chemical reactions (Rosenthal 1991).

Plants that synthesize nonprotein amino acids are not susceptible to the toxicity of these compounds. The jack bean (*Canavalia ensiformis*), which synthesizes large amounts of canavanine in its seeds, has protein-synthesizing machinery that can discriminate between canavanine and arginine, and it does not incorporate canavanine into its own proteins. Some insects that specialize on plants containing nonprotein amino acids have similar biochemical adaptations.

Some Plant Proteins Inhibit Herbivore Digestive Processes

Among the diverse components of plant defense arsenals are proteins that interfere with herbivore digestion. For example, some legumes synthesize **α-amylase inhibitors** that block action of the starch-digesting enzyme α-amylase. Other plant species produce **lectins**, defensive proteins that bind to carbohydrates or carbohydrate-containing proteins. After being ingested by an herbivore, lectins bind to the epithelial cells lining the digestive tract and interfere with nutrient absorption (Peumans and Van Damme 1995).

The best-known antidigestive proteins in plants are the **proteinase inhibitors**. Found in legumes, tomatoes, and other plants, these substances block the action of herbivore proteolytic enzymes. After entering the herbivore's digestive tract, they hinder protein digestion by binding tightly and specifically to the active site of protein-hydrolyzing enzymes such as trypsin and chymotrypsin. Insects that feed on plants containing proteinase inhibitors suffer reduced rates of growth and development that can be offset by supplemental amino acids in their diet. The defensive role of proteinase

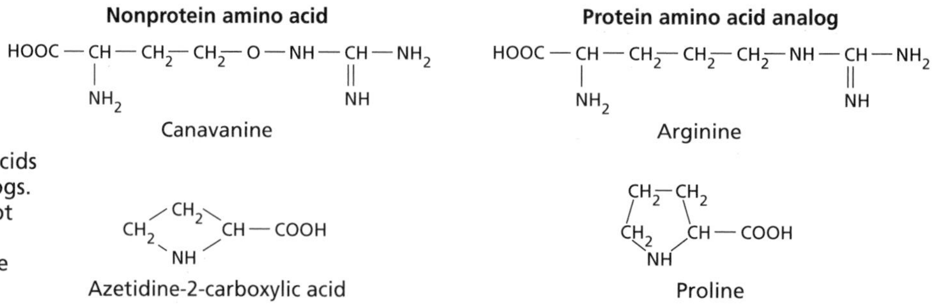

Nonprotein amino acid

$HOOC-CH(NH_2)-CH_2-CH_2-O-NH-C(=NH)-NH_2$

Canavanine

CH_2–CH_2–CH–$COOH$ ring with NH

Azetidine-2-carboxylic acid

Protein amino acid analog

$HOOC-CH(NH_2)-CH_2-CH_2-CH_2-NH-C(=NH)-NH_2$

Arginine

CH_2-CH_2, CH_2, $CH-COOH$, NH (ring)

Proline

Figure 13.28 Nonprotein amino acids and their protein amino acid analogs. The nonprotein amino acids are not incorporated into proteins but are defensive compounds found in free form in plant cells.

inhibitors has been confirmed by experiments with transgenic tobacco. Plants that had been transformed to accumulate increased levels of proteinase inhibitors suffered less damage from insect herbivores than untransformed control plants did (Johnson et al. 1989).

Some Damaged Cells Release a Protein That Serves as a Wound Signal

Proteinase inhibitors and certain other defenses are not continuously present in plants, but are synthesized only after initial herbivore or pathogen attack. In tomatoes, insect feeding leads to the rapid accumulation of proteinase inhibitors throughout the plant, even in undamaged areas far from the initial feeding site. Research in Clarence A. Ryan's laboratory at Washington State University has shown that the systemic production of proteinase inhibitors in young tomato plants is triggered by a complex sequence of events. First, wounded tomato leaves produce an 18–amino acid polypeptide called **systemin**, the first (and so far only) polypeptide hormone discovered in plants (Pearce et al. 1991). Systemin is released from damaged cells into the apoplast and is transported out of the wounded leaf via the phloem (Figure 13.29).

In target cells, systemin is believed to bind to a site on the plasma membrane and initiate the biosynthesis of **jasmonic acid**, a plant growth regulator that has wide-ranging effects (Creelman and Mullet 1997). The structure and biosynthesis of jasmonic acid have intrigued plant biologists because of the parallels to some eicosanoids that are central to inflammatory responses and other physiological processes in mammals (see Chapter 14). In plants, jasmonic acid is synthesized from linolenic acid (18:3), which is released from membrane lipids and then converted to jasmonic acid by the series of enzymatic steps shown in Figure 13.30. Jasmonic acid eventually activates the expression of genes that encode proteinase inhibitors (see Figure 13.29). Other signals, such as ABA (abscisic acid), salicylic acid, and pectin fragments from damaged plant cell walls also appear to participate in this wound-signaling cascade, but their specific roles are still unclear.

Some other plant defenses are induced by initial predator or pathogen damage. Earlier we mentioned the formation of resin monoterpenes in conifers infested by bark beetles. In the last section of this chapter, we will discuss the induction of defenses by pathogenic microorganisms. Defenses that are produced only after initial herbivore damage theoretically require a smaller investment of plant resources than defenses that are always present, but they must be activated quickly to be effective. Like proteinase inhibitors, other induced defenses appear to be triggered by complex signal transduction networks, which often involve jasmonic acid.

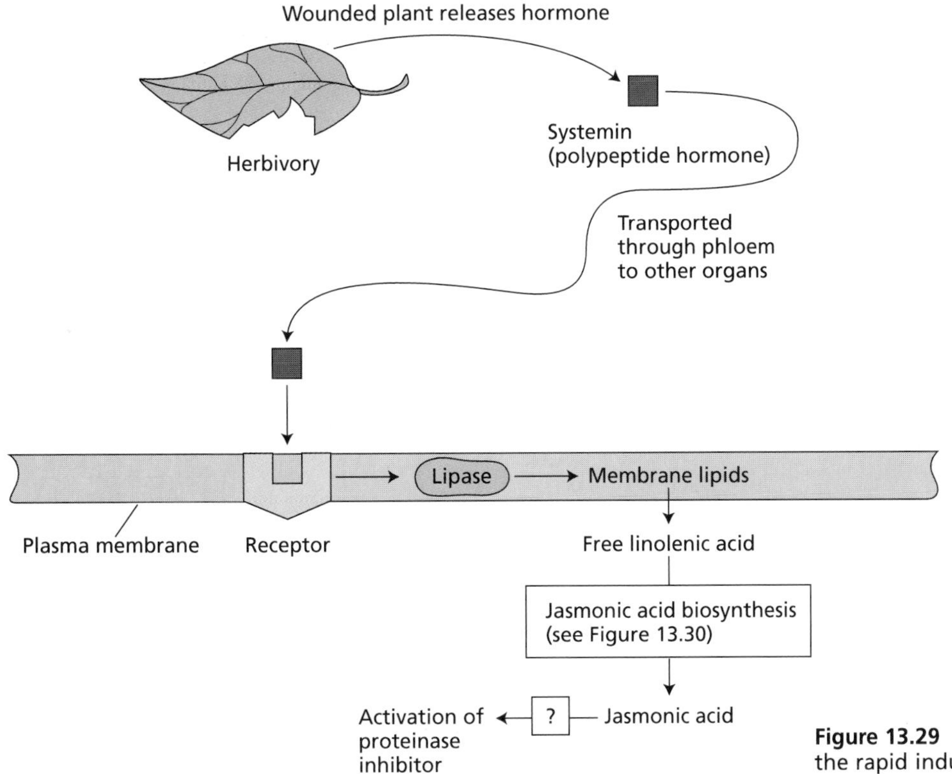

Figure 13.29 Proposed signaling pathway for the rapid induction of proteinase inhibitor biosynthesis in wounded tomato plants.

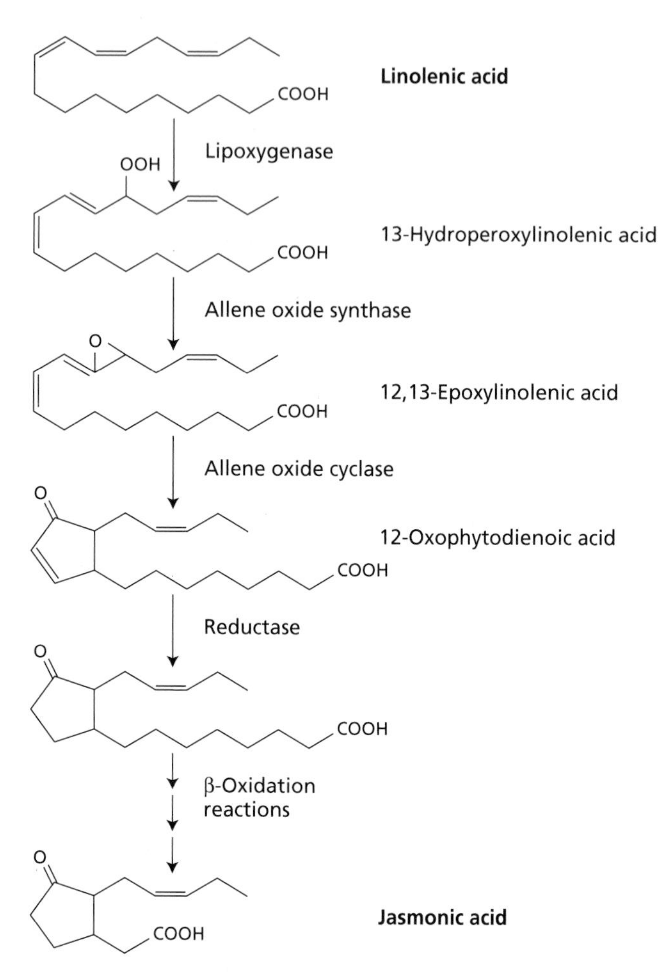

Figure 13.30 The pathway for conversion of linolenic acid (18:3) to jasmonic acid.

Plant Defense against Pathogens

Even though they lack an immune system, plants are surprisingly resistant to diseases caused by the fungi, bacteria, viruses, and nematodes that are ever present in the environment. In this section we will examine the diverse array of mechanisms that plants have evolved to resist infection, including the production of antimicrobial agents and a type of programmed cell death (see Chapter 16) called the *hypersensitive response*. Finally, we will discuss a special type of plant immunity called *systemic acquired resistance*.

Some Antimicrobial Compounds Are Synthesized before Pathogen Attack

Several classes of secondary metabolites that we have already discussed have strong antimicrobial activity when tested in vitro; thus they have been proposed to function as defenses against pathogens in the intact plant. Among these are saponins and simple phenylpropanoids. Saponins, a group of triterpenes, are thought to disrupt fungal membranes by binding to sterols. Simple phenylpropanoids have been suggested to serve as antipathogen defenses by recent experiments performed in the laboratories of C. Lamb at the Salk Institute in California and R. Dixon at the Noble Foundation in Oklahoma (Maher et al. 1994).

Genetically transformed tobacco plants were created with suppressed levels of PAL activity. Plants with low PAL activity did not exhibit substantial chemical differences from nontransformed control plants except that they accumulated much less of the simple phenylpropanoid conjugate chlorogenic acid. However, these low-PAL plants were significantly less resistant to a tobacco fungal pathogen than untransformed plants were, implying that chlorogenic acid functions as a preformed antimicrobial defense compound.

Other Antipathogen Defenses Are Induced by Infection

After being infected by a pathogen, plants deploy a broad spectrum of defenses against invading microbes. A common defense is the **hypersensitive response**, in which cells immediately surrounding the infection site die rapidly, depriving the pathogen of nutrients and preventing its spread. After a successful hypersensitive response, a small region of dead tissue is left at the site of the attempted invasion, but the rest of the plant is unaffected.

The hypersensitive response is often preceded by the production of **reactive oxygen species**. Cells in the vicinity of infection synthesize a burst of toxic compounds formed by the reduction of molecular oxygen, including the superoxide anion (O_2^-), hydrogen peroxide (H_2O_2) and the hydroxyl radical (•OH). An NADPH-dependent oxidase located on the plasma membrane (Figure 13.31) is thought to produce O_2^-, which is in turn converted to •OH and H_2O_2. The hydroxyl radical is the strongest oxidant of these active oxygen species and can initiate radical chain reactions with a range of organic molecules, leading to lipid peroxidation, enzyme inactivation, and nucleic acid degradation (Lamb and Dixon 1997). Active oxygen species may contribute to cell death as part of the hypersensitive response or act to kill the pathogen directly.

Many species react to fungal or bacterial invasion by synthesizing lignin or callose (see Chapter 10). These polymers are thought to serve as barriers, walling off the pathogen from the rest of the plant and physically blocking its spread. A related response is the modification of cell wall proteins. Certain proline-rich proteins of the wall become oxidatively cross-linked after pathogen attack in an H_2O_2-mediated reaction (see Figure 13.31) (Bradley et al. 1992). This process strengthens the walls of the cells in the vicinity of the infection site, increasing their resistance to microbial digestion.

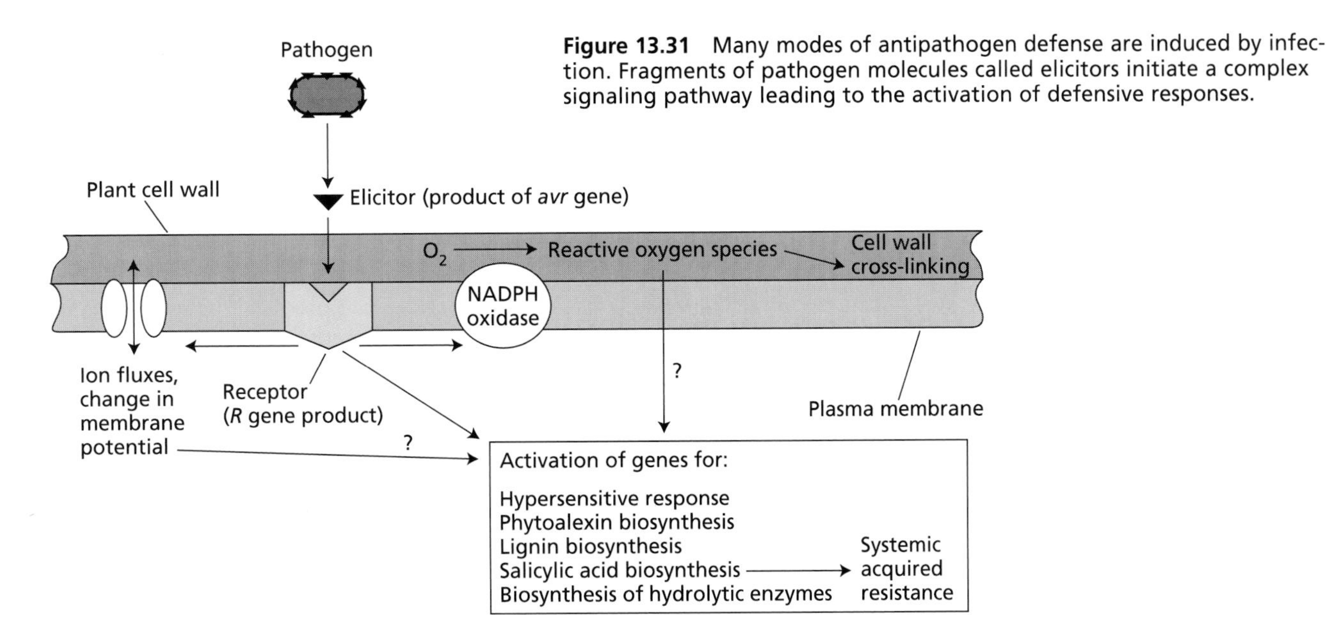

Figure 13.31 Many modes of antipathogen defense are induced by infection. Fragments of pathogen molecules called elicitors initiate a complex signaling pathway leading to the activation of defensive responses.

Another defensive response to infection is the formation of hydrolytic enzymes that attack the cell wall of the pathogen. An assortment of glucanases, chitinases, and other hydrolases are induced by fungal invasion. Chitin, a polymer of *N*-acetylglucosamine residues, is a principal component of fungal cell walls. These hydrolytic enzymes belong to a group of proteins that are closely associated with pathogen infection and so are known as **PR (pathogenesis-related) proteins**.

Phytoalexins. Perhaps the best-studied response of plants to bacterial or fungal invasion is the synthesis of **phytoalexins**. Phytoalexins are a chemically diverse group of secondary metabolites with strong antimicrobial activity that accumulate around the site of infection. Phytoalexin production appears to be a common mechanism of resistance to pathogenic microbes in a wide range of plants. However, different plant families employ different types of secondary products as phytoalexins. For example, isoflavonoids are common phytoalexins in the legume family, whereas in plants of the potato family, (Solanaceae) such as potato, tobacco, and tomato, various sesquiterpenes are produced as phytoalexins (Figure 13.32).

Phytoalexins are generally undetectable in the plant before infection, but they are synthesized very rapidly after microbial attack because of the activation of new biosynthetic pathways. The point of control is usually the initiation of gene transcription. Thus, plants do not appear to store any of the enzymatic machinery required for phytoalexin synthesis. Instead, soon after microbial invasion they begin transcribing and translating the appropriate mRNAs and synthesizing the enzymes de novo.

Although phytoalexins accumulate in concentrations that have been shown to be toxic to pathogens in bioassays, the defensive significance of these compounds in the intact plant is not fully known. Recent experiments using plants and pathogens that have been modified by the introduction of foreign genes have provided the first direct proof of phytoalexin function in vivo. For example, when tobacco was transformed with a gene catalyzing biosynthesis of the phenylpropanoid phytoalexin resveratrol, it became much more resistant to a fungal pathogen than nontransformed control plants were (Hain et al. 1993). In other experiments, pathogens that had been transformed with genes encoding phytoalexin-degrading enzymes were then able to infect plants that were normally resistant to them (Kombrink and Somssich 1995).

To Initiate Defenses Rapidly, Some Plants Recognize Specific Substances from Pathogens

Within a species, individual plants often differ greatly in their resistance to microbial pathogens. These differences often lie in the speed and intensity of a plant's reactions. Resistant plants respond more rapidly and more vigorously to pathogens than susceptible plants. Hence, it is important to learn how plants sense the presence of pathogens and initiate defense.

In the last few years, researchers have isolated numerous plant resistance genes, known as ***R* genes**, that function in defense against fungi, bacteria, and nematodes. Most of the *R* genes are thought to encode receptors that recognize and bind specific molecules originating from pathogens and alert the plant to the pathogen's presence (see Figure 13.31). The specific pathogen molecules recognized are referred to as **elici-**

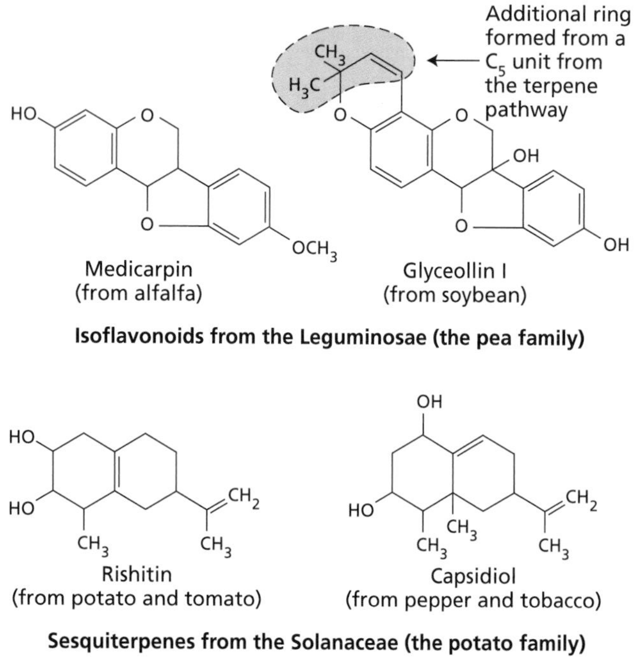

Additional ring formed from a C_5 unit from the terpene pathway

Medicarpin (from alfalfa)

Glyceollin I (from soybean)

Isoflavonoids from the Leguminosae (the pea family)

Rishitin (from potato and tomato)

Capsidiol (from pepper and tobacco)

Sesquiterpenes from the Solanaceae (the potato family)

Figure 13.32 Structure of some phytoalexins—secondary metabolites with antimicrobial properties that are rapidly synthesized after microbial infection.

tors, and include proteins, peptides, lipids, or polysaccharide fragments arising from the pathogen wall, the outer membrane, or a secretion process (Boller 1995). The *R* gene products themselves are nearly all proteins, with a leucine-rich domain that is repeated inexactly several times in the amino acid sequence (Hammond-Kosack and Jones 1997) (see Chapter 14).

Such domains mediate protein–protein interactions in other eukaryotic systems. Hence, *R* gene products may initiate signaling pathways that activate the various modes of antipathogen defense. Within a few minutes after exposure to elicitors, plant cells experience changes in membrane potential, ion flux, protein phosphorylation, and the level of active oxygen species, all of which may be important in signal transduction.

Studies of plant disease have revealed complex patterns of host relationships between plants and pathogen strains. Plant species are generally susceptible to the attack of certain pathogen strains, but resistant to others. This specificity is thought to be determined by interaction between the products of host *R* genes and pathogen *avr* (*avir*ulence) genes believed to encode specific elicitors. According to current thinking, successful resistance requires the elicitor, a product of the pathogen *avr* gene, to be rapidly recognized by a host plant receptor, the product of an *R* gene.

A Single Episode of Infection May Increase Resistance to Future Pathogen Attack

When a plant survives the infection of a pathogen at one site, it often develops increased resistance to subsequent attack at sites throughout the plant and enjoys protection against a wide range of pathogen species. This phenomenon, called **systemic acquired resistance (SAR)**, develops over a period of several days following initial infection (Ryals et al. 1996). Systemic acquired resistance appears to result from increased levels of certain defense compounds that we have already mentioned, including chitinases and other hydrolytic enzymes.

Although the mechanism of SAR induction is still unknown, one of the endogenous signals is likely to be **salicylic acid**. The level of this benzoic acid derivative, a

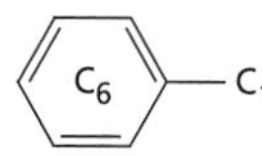

C_6—C_1

compound, rises dramatically in the zone of infection after initial attack, and it is thought to establish SAR in other parts of the plant (Figure 13.33). In addition to salicylic acid, recent studies suggest that its methyl ester, methyl salicylate, acts as a volatile SAR-inducing signal transmitted to distant parts of the plant and even to neighboring plants (Shulaev et al. 1997). Thus, even though plants lack immune systems like those present in many animals, they have developed elaborate mechanisms to protect themselves from disease-causing microbes.

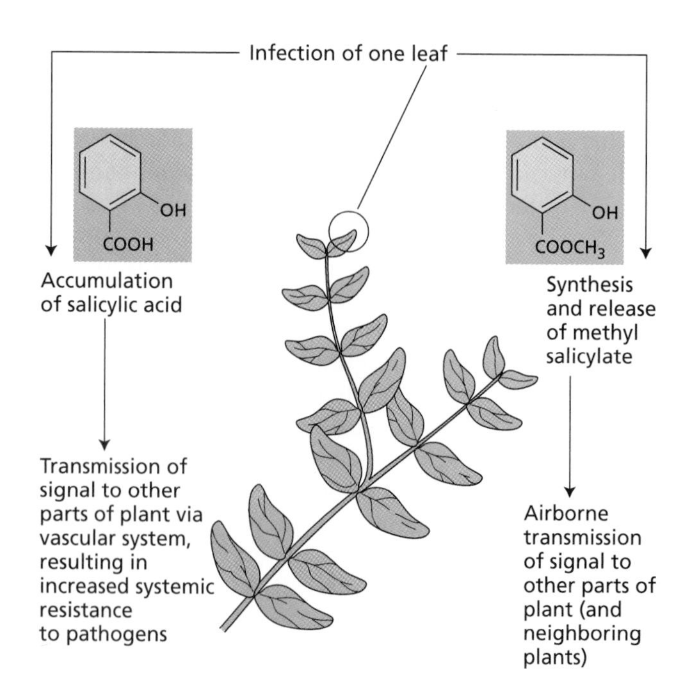

Figure 13.33 Initial pathogen infection may increase resistance to future pathogen attack through development of systemic acquired resistance.

Summary

Plants produce an enormous diversity of substances that have no apparent roles in growth and development processes and so are classified under the heading of secondary metabolites. Scientists have long speculated that these compounds protect plants from predators and pathogens on the basis of their toxicity and repellency to herbivores and microbes when tested in vitro. Recent experiments using plants whose secondary-metabolite expression has been altered by modern molecular methods have begun to confirm these defensive roles.

There are three major groups of secondary metabolites: terpenes, phenolics, and nitrogen-containing compounds. Terpenes, composed of five-carbon isopentanoid units, are toxins and feeding deterrents to many herbivores. Phenolics, which are synthesized primarily from products of the shikimic acid pathway, have several important roles in plants. Lignin mechanically strengthens cell walls. Flavonoid pigments function as shields against harmful ultraviolet radiation and as attractants for pollinators and fruit dispersers. Finally, lignin, flavonoids, and other phenolic compounds serve as defenses against herbivores and pathogens. Members of the third major group, nitrogen-containing secondary metabolites, are synthesized principally from common amino acids. Compounds such as alkaloids, cyanogenic glycosides, glucosinolates, nonprotein amino acids, and proteinase inhibitors protect plants from a variety of herbivorous animals.

Plants have evolved multiple defense mechanisms against microbial pathogens. Besides antimicrobial secondary metabolites, some of which are preformed and some of which are induced by infection, other modes of defense include the construction of polymeric barriers to pathogen penetration and the synthesis of enzymes that degrade pathogen cell walls. In addition, plants employ specific recognition and signaling systems enabling the rapid detection of pathogen invasion and initiation of a vigorous defensive response. Once infected, some plants also develop an immunity to subsequent microbial attacks.

For millions of years, plants have produced defenses against herbivory and microbial attack. Well-defended plants have tended to leave more survivors than poorly defended plants, so the capacity to produce effective defensive products has become widely established in the plant kingdom. In response, many species of herbivores and microbes have evolved the ability to feed on or infect plants containing secondary products without being adversely affected, and this herbivore and pathogen pressure has in turn selected for new defensive products in plants.

The study of plant secondary metabolites has many practical applications. By virtue of their biological activities against herbivorous animals and microbes, many of these substances are employed commercially as insecticides, fungicides, and pharmaceuticals, while others find uses as fragrances, flavorings, medicinal drugs, and industrial materials. The breeding of increased levels of secondary metabolites into crop plants has made it possible to reduce the need for certain costly and potentially harmful pesticides. In some cases, however, it has been necessary to reduce the levels of naturally occurring secondary metabolites to minimize toxicity to humans and domestic animals.

General Reading

*Agosta, W. (1996) *Bombardier Beetles and Fever Trees: A Close-up Look at Chemical Warfare and Signals in Animals and Plants*. Addison-Wesley, Reading, MA.

Eisner, T., and Meinwald, J., eds. (1995) *Chemical Ecology: The Chemistry of Biotic Interaction*. National Academy Press, Washington, DC.

Ellis, B. E., Kuroki, G. W., and Stafford, H. A., eds. (1994) *Genetic Engineering of Plant Secondary Metabolism* (Recent Advances in Phytochemistry, vol. 28). Plenum, New York.

Hammond-Kosack, K. E., and Jones, J. D. G. (1996) Resistance gene-dependent plant defense responses. *Plant Cell* 8: 1773–1791.

Harborne, J. B. (1993) *Introduction to Ecological Biochemistry*, 4th ed. Academic Press, New York.

Hashimoto, T., and Yamada, Y. (1994) Alkaloid biogenesis: Molecular aspects. *Annu. Rev. Plant Physiol. Plant Mol. Biol.* 45: 257–285.

Kerstiens, G., ed. (1996) *Plant Cuticles: An Integrated Functional Approach*. BIOS Scientific, Oxford.

Kinghorn, A. D., and Balandrin, M. F., eds. (1993) *Human Medicinal Agents from Plants* (ACS Symposium Series, no. 534). American Chemical Society, Washington, DC.

Rosenthal, G. A., and Berenbaum, M. R., eds. (1991) *Herbivores: Their Interactions with Secondary Plant Metabolites*, Vol. I: *The Chemical Participants*, 2nd ed. Academic Press, San Diego.

Rosenthal, G. A., and Berenbaum, M. R., eds. (1992) *Herbivores: Their Interactions with Secondary Plant Metabolites*, Vol. II: *Ecological and Evolutionary Processes*, 2nd ed. Academic Press, San Diego.

Schaller, A., and Ryan, C. A. (1996) Systemin—A polypeptide signal in plants. *Bioessays* 18: 27–33.

von Wettstein-Knowles, P. M. (1993) Waxes, cutin and suberin. In *Lipid Metabolism in Plants*, T. S. Moore, Jr., ed., CRC Press, Boca Raton, FL, pp. 127–166.

* Indicates a reference that is general reading in the field and is also cited in this chapter.

Chapter References

Boller, T. (1995) Chemoperception of microbial signals in plant cells. *Annu. Rev. Plant Physiol. Plant Mol. Biol.* 46: 189–214.

Bradley, D. J., Kjellbom, P., and Lamb, C. J. (1992) Elicitor- and wound-induced oxidative cross-linking of a proline-rich plant cell wall protein: A novel, rapid defense response. *Cell* 70: 21–30.

Butler, L. G. (1989) Effects of condensed tannin on animal nutrition. In *Chemistry and Significance of Condensed Tannins*, R. W. Hemingway and J. J. Karchesy, eds., Plenum, New York, pp. 391–402.

Byers, J. A. (1989) Chemical ecology of bark beetles. *Experientia* 45: 271–283.

Creelman, R. A., and Mullet, J. E. (1997) Biosynthesis and action of jasmonates in plants. *Annu. Rev. Plant Physiol. Plant Mol. Biol.* 48: 355–381.

Dao, T. H. (1987) Sorption and mineralization of plant phenolic acids in soil. In *Allelochemicals: Role in Agriculture and Forestry* (ACS

Symposium Series, no. 330), G. R. Waller, ed., American Chemical Society, Washington, DC., pp. 358–370.

Eigenbrode, S. D. (1996) Plant surface waxes and insect behavior. In *Plant Cuticles: An Integrated Functional Approach*, G. Kerstiens, ed., BIOS Scientific, Oxford, pp. 201–221.

Eigenbrode, S. D., Stoner, K. A., Shelton, A. M., and Kain, W. C. (1991) Characteristics of leaf waxes of *Brassica oleracea* associated with resistance to diamondback moth. *J. Econ. Entomol.* 83: 1609–1618.

Felton, G. W., Donato, K., Del Vecchio, R. J., and Duffey, S. S. (1989) Activation of plant foliar oxidases by insect feeding reduces nutritive quality of foliage for noctuid herbivores. *J. Chem. Ecol.* 15: 2667–2694.

Gershenzon, J., and Croteau, R. (1991) Terpenoids. In *Herbivores: Their Interactions with Secondary Plant Metabolites*, Vol. I: *The Chemical Participants*, 2nd ed., G. A. Rosenthal and M. R. Berenbaum, eds, Academic Press, San Diego, pp. 165–219.

Gunning, B. E. S., and Steer, M. W. (1996) *Plant Cell Biology: Structure and Function of Plant Cells.* Jones and Bartlett, Boston.

Hahlbrock, K., and Scheel, D. (1989) Physiology and molecular biology of phenylpropanoid metabolism. *Annu. Rev. Plant Physiol. Plant Mol. Biol.* 40: 347–369.

Hain, R., Reif, H-J., Krause, E., Langebartels, R., Kindl, H., Vornam, B., Wiese, W., Schmelzer, E., Schreier, P. H., Stoecker, R. H., and Stenzel, K. (1993) Disease resistance results from foreign phytoalexin expression in a novel plant. *Nature* 361: 153–156.

Hammond-Kosack, K. E., and Jones, J. D. G. (1997) Plant disease resistance genes. *Annu. Rev. Plant Physiol. Plant Mol. Biol.* 48: 575–607.

Hartmann, T. (1991) Alkaloids. In *Herbivores: Their Interactions with Secondary Plant Metabolites*, Vol. I: *The Chemical Participants*, 2nd ed., G. A. Rosenthal and M. R. Berenbaum, eds., Academic Press, San Diego, pp. 79–121.

Jeffree, C. E. (1996) Structure and ontogeny of plant cuticles. In *Plant Cuticles: An Integrated Functional Approach*, G. Kerstiens, ed., BIOS Scientific, Oxford, pp. 33–85.

Johnson, R., Narvaez, J., An, G., and Ryan, C. (1989) Expression of proteinase inhibitors I and II in transgenic tobacco plants: Effects on natural defense against *Manduca sexta* larvae. *Proc. Natl. Acad. Sci. USA* 86: 9871–9875.

Kombrink, E., and Somssich, I. E. (1995) Defense responses of plants to pathogens. *Adv. Bot. Res.* 21: 1–34.

Kondo, T., Yoshida, K., Nakagawa, A., Kawai, T., Tamura, H., and Goto, T. (1992) Structural basis of blue-color development in flower petals from *Commelina communis*. *Nature* 358: 515–518.

Lamb, C., and Dixon, R. A. (1997) The oxidative burst in plant disease resistance. *Annu. Rev. Plant Physiol. Plant Mol. Biol.* 48: 251–275.

Lewis, N. G., and Yamamoto, E. (1990) Lignin: Occurrence, biogenesis and biodegradation. *Annu. Rev. Plant Physiol. Plant Mol. Biol.* 41: 455–496.

Li, J., Ou-Lee, T-M., Raba, R., Amundson, R. G., and Last, R. L. (1993) *Arabidopsis* flavonoid mutants are hypersensitive to UV-B irradiation. *Plant Cell* 5: 171–179.

Lichtenthaler, H. K., Schwender, J., Disch, A., and Rohmer, M. (1997) Biosynthesis of isoprenoids in higher plant chloroplasts proceeds via a mevalonate-independent pathway. *FEBS Lett.* 400: 271–274.

Logemann, E., Parniske, M., and Hahlbrock, K. (1995) Modes of expression and common structural features of the complete phenylalanine ammonia-lyase gene family in parsley. *Proc. Natl. Acad. Sci. USA* 92: 5905–5909.

Maher, E. A., Bate, N. J., Ni, W., Elkind, Y., Dixon, R. A., and Lamb, C. J. (1994) Increased disease susceptibility of transgenic tobacco plants with suppressed levels of preformed phenylpropanoid products. *Proc. Natl. Acad. Sci. USA* 91: 7802–7806.

McCrea, K. D., and Levy, M. (1983) Photographic visualization of floral colors as perceived by honeybee pollinators. *Am. J. Bot.* 70: 369–375.

Mordue, A. J., and Blackwell, A. (1993) Azadirachtin: An update. *J. Insect Physiol.* 39: 903–924.

Nimz, H. (1974) Beech lignin—Proposal of a constitutional scheme. *Angew. Chem. Int. Ed.* 13: 313–321.

Oates, J. F., Waterman, P. G., and Choo, G. M. (1980) Food selection by the South Indian leaf-monkey, *Presbytis johnii*, in relation to leaf chemistry. *Oecologia* 45: 45–56.

Pearce, G., Strydom, D., Johnson, S., and Ryan, C. A. (1991) A polypeptide from tomato leaves induces wound-inducible proteinase inhibitor proteins. *Science* 253: 895–898.

Peumans, W. J., and Van Damme, E. J. M. (1995) Lectins as plant defense proteins. *Plant Physiol.* 109: 347–352.

Picman, A. K. (1986) Biological activities of sesquiterpene lactones. *Biochem. Sys. Ecol.* 14: 255–281.

Poulton, J. E. (1990) Cyanogenesis in plants. *Plant Physiol.* 94: 401–405.

Renwick, J. A. A., Radke, C. D., Sachdev-Gupta, K., and Staedler, E. (1992) Leaf surface chemicals stimulating oviposition by *Pieris rapae* (Lepidoptera: Pieridae) on cabbage. *Chemoecology* 3: 33–38.

Rice, E. L. (1987) Allelopathy: An overview. In *Allelochemicals: Role in Agriculture and Forestry* (ACS Symposium Series, no. 330), G. R. Waller, ed., American Chemical Society, Washington, DC., pp. 8–22.

Rosenthal, G. A. (1991) The biochemical basis for the deleterious effects of L-canavanine. *Phytochemistry* 30: 1055–1058.

Ryals, J. A., Neuenschwander, U. H., Willits, M. G., Molina, A., Steiner, H-Y., and Hunt, M. D. (1996) Systemic acquired resistance. *Plant Cell* 8: 1809–1819.

Sandberg, S. L., and Berenbaum, M. R. (1989) Leaf-tying by tortricid larvae as an adaptation for feeding on phototoxic *Hypericum perforatum*. *J. Chem. Ecol.* 15: 875–885.

Shulaev, V., Silverman, P., and Raskin, I. (1997) Airborne signalling by methyl salicylate in plant pathogen resistance. *Nature* 385: 718–721.

Turlings, T. C. J., Loughrin, J. H., McCall, P. J., Roese, U. S. R., Lewis, W. J., and Tumlinson, J. H. (1995) How caterpillar-damaged plants protect themselves by attracting parasitic wasps. *Proc. Natl. Acad. Sci. USA* 92: 4169–4174.

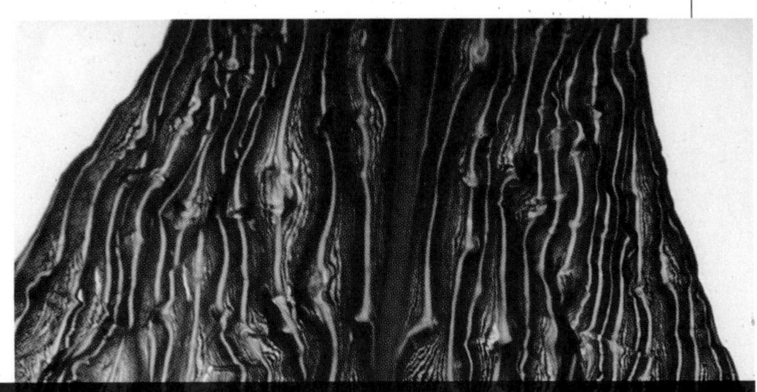

Growth and Development

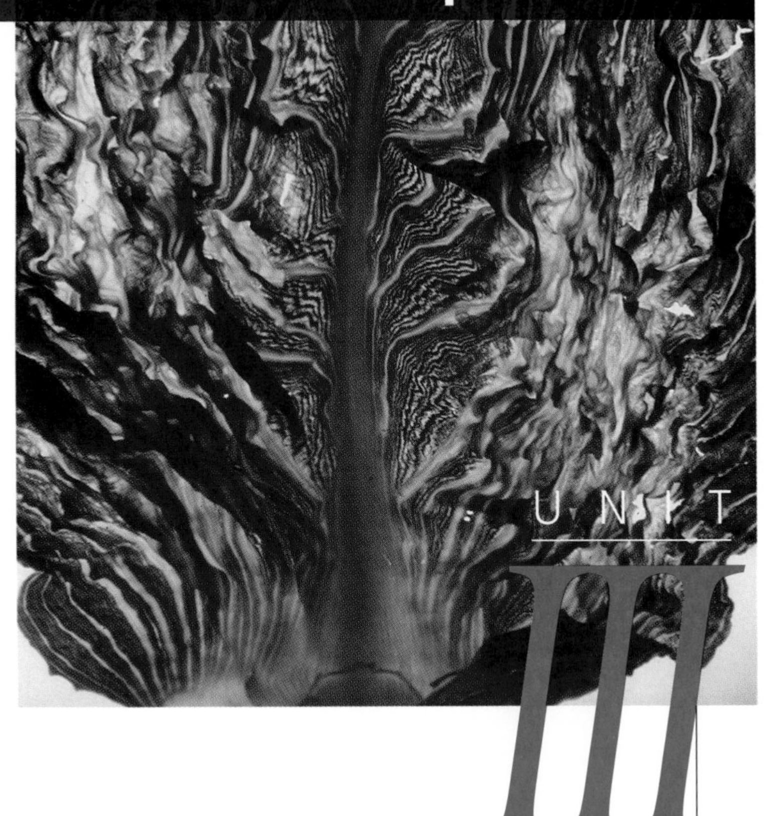

UNIT

III

14 Gene Expression and Signal Transduction

PLANT BIOLOGISTS MAY BE FORGIVEN for taking abiding satisfaction in the fact that Mendel's classic studies on the role of heritable factors in development were carried out on a flowering plant: the garden pea. The heritable factors that Mendel discovered, which control such characters as flower color, flower position, pod shape, stem length, seed color, and seed shape, came to be called **genes**. Genes are the DNA sequences that encode the RNA molecules directly involved in making the enzymes and structural proteins of the cell. Genes are arranged linearly on **chromosomes**, which form linkage groups—that is, genes that are inherited together. The total amount of DNA or genetic information contained in a cell, nucleus, or organelle is termed its **genome**.

Since Mendel's pioneering discoveries in his garden, the principle has become firmly established that the growth, development, and environmental responses of even the simplest microorganism are determined by the programmed expression of its genes. Among multicellular organisms, turning genes on (**gene expression**) or off alters a cell's complement of enzymes and structural proteins, allowing cells to differentiate. In the chapters that follow, we will discuss various aspects of plant development in relation to the regulation of gene expression.

Various internal signals are required for coordinating the expression of genes during development and for enabling the plant to respond to environmental signals. Such internal (as well as external) signaling agents typically bring about their effects by means of sequences of biochemical reactions, called **signal transduction pathways**, that greatly amplify the original signal and ultimately result in the activation or repression of genes.

Much progress has been made in the study of signal transduction pathways in plants in recent years. However, before describing what is known about these pathways in plants, we will provide background information on gene expression and signal transduction in other organisms, such as bacteria, yeasts, and animals, making reference to plant systems wherever appropriate. These models will provide the framework for the recent advances in the study of plant development that are discussed in subsequent chapters.

Genome Size, Organization, and Complexity

As might be expected, the size of the genome bears some relation to the complexity of the organism. For example, the genome size of *E. coli* is 4.7×10^6 bp (base pairs), that of the fruit fly is 2×10^8 bp per haploid cell, and that of a human is 3×10^9 bp per haploid cell. However, genome size in eukaryotes is an unreliable indicator of complexity because not all of the DNA encodes genes.

In prokaryotes, nearly all of the DNA consists of **unique sequences** that encode proteins or functional RNA molecules. In addition to unique sequences, however, eukaryotic chromosomes contain large amounts of noncoding DNA whose main functions appear to be chromosome organization and structure. Much of this noncoding DNA consists of multicopy sequences, called **repetitive DNA**. The remainder of the noncoding DNA is made up of single-copy sequences called **spacer DNA**. Together, repetitive and spacer DNA can make up the majority of the total genome in some eukaryotes. For example, in humans only about 5% of the total DNA consists of genes, the unique sequences that encode for RNA and protein synthesis.

The genome size in plants is more variable than in any other group of eukaryotes. In angiosperms, the haploid genome ranges from about 1.5×10^8 bp for *Arabidopsis thaliana* (smaller than that of the fruit fly) to 1×10^{11} bp for the monocot *Trillium*, which is considerably larger than the human genome. Even closely related beans of the genus *Vicia* exhibit genomic DNA contents that vary over a 20-fold range. Why are plant genomes so variable in size?

Studies of plant molecular biology have shown that most of the DNA in plants with large genomes is repetitive DNA. *Arabidopsis* has the smallest genome of any plant because only 10% of its nuclear DNA is repetitive DNA. The genome size of rice is estimated to be about five times that of *Arabidopsis*, yet the total amount of unique sequence DNA in the rice genome is about the same as in *Arabidopsis*. Thus the difference in genome size between *Arabidopsis* and rice is due mainly to repetitive and spacer DNA.

Most Plant Haploid Genomes Contain 20,000 to 30,000 Genes

Until recently, the total number of genes in an organism's genome was difficult to assess. Thanks to recent advances in many genomic sequencing projects, such numbers are now becoming available, although precise values are still lacking. According to Miklos and Rubin (1996), the number of genes in bacteria varies from 500 to 8,000 and overlaps with the number of genes in many simple unicellular eukaryotes. For example, the yeast genome appears to contain about 6,000 genes. More complex eukaryotes, such as protozoans, worms, and flies, all seem to have gene numbers in the range of 12,000 to 14,000. The *Drosophila* (fruit fly) genome contains about 12,000 genes. Thus, the current view is that it takes roughly 12,000 basic types of genes to form a eukaryotic organism, although values as high as 43,000 genes are common, as a result of multiple copies of certain genes, or **multigene families**.

The best-studied plant genome is that of *Arabidopsis thaliana*. Chris Somerville and his colleagues at Stanford University have estimated that the *Arabidopsis* genome contains roughly 20,000 genes (Rounsley et al. 1996). This estimate is based on more than one approach. For example, since large regions of the genome have been sequenced, we know there is one gene for every 5 kb (kilobases) of DNA. Since the entire genome contains about 100,000 kb, there must be about 20,000 genes. However, 6% of the genome encodes ribosomal RNA, and another 2% consists of highly repetitive sequences, so the number could be lower. Similar values likely will be found for the genomes of other plants as well. The current consensus is that the genomes of most plants will be found to contain from 20,000 to 30,000 genes.

Some of these genes encode proteins that perform housekeeping functions, basic cellular processes that go on in all the different kinds of cells. Such genes are permanently turned on; that is, they are **constitutively expressed**. Other genes are highly **regulated**, being turned on or off at specific stages of development or in response to specific environmental stimuli.

Prokaryotic Gene Expression

The first step in gene expression is **transcription**, the synthesis of an mRNA copy of the DNA template that encodes a protein (Alberts et al. 1994; Lodish et al. 1995). Transcription is followed by **translation**, the synthesis of the protein on the ribosome. Developmental studies have shown that each plant organ contains large numbers of organ-specific mRNAs. Transcription is controlled by proteins that bind DNA, and these DNA-binding proteins are themselves subject to various types of regulation.

Much of our understanding of the basic elements of transcription is derived from early work on bacterial systems; hence we precede our discussion of eukaryotic gene expression with a brief overview of transcriptional regulation in prokaryotes. However, it is now clear that gene regulation in eukaryotes is far more complex than in prokaryotes. The added complexity of gene expression in eukaryotes is what allows cells and tissues to differentiate and makes possible the diverse life cycles of plants and animals.

DNA-Binding Proteins Regulate Transcription in Prokaryotes

In prokaryotes, genes are arranged in **operons**, sets of contiguous genes that include **structural genes** and **regulatory sequences**. A famous example is the *E. coli* lactose (*lac*) operon, which was first described in 1961 by François Jacob and Jacques Monod of the Pasteur Institute in Paris. The *lac* operon is an example of an **inducible** operon—that is, one in which a key metabolic intermediate induces the transcription of the genes.

The *lac* operon is responsible for the production of three proteins involved in utilization of the disaccharide lactose. This operon consists of three structural genes and three regulatory sequences. The structural genes (*z*, *y*, and *a*) code for the sequence of amino acids in three proteins: β-galactosidase, the enzyme that catalyzes the hydrolysis of lactose to glucose and galactose; permease, a carrier protein for the membrane transport of lactose into the cell; and transacetylase, the significance of which is unknown.

The three regulatory sequences (*i*, *p*, and *o*) control the transcription of mRNA for the synthesis of these proteins (Figure 14.1). Gene *i* is responsible for the synthesis of a **repressor protein** that recognizes and binds to a specific nucleotide sequence, the **operator**. The operator, *o*, is located downstream (i.e., on the 3′ side) of

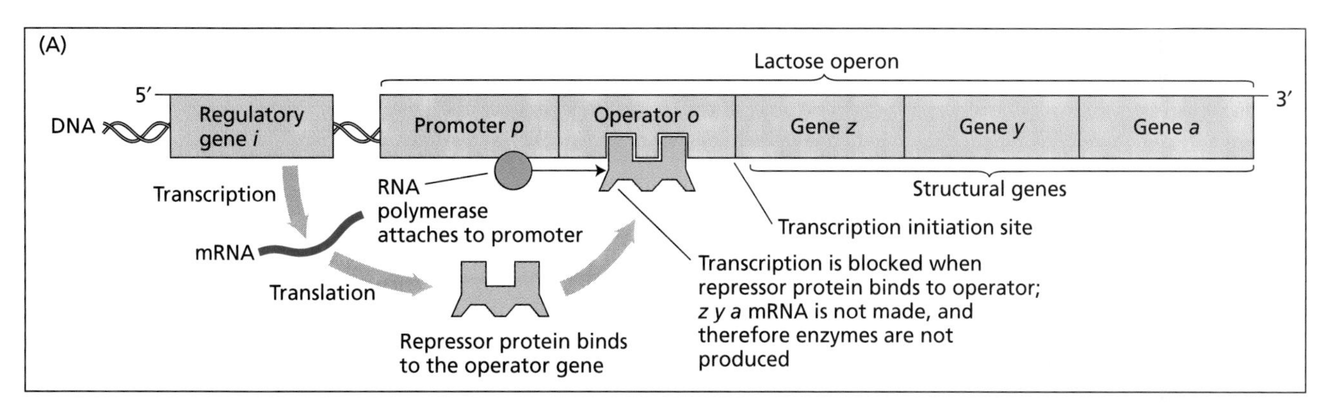

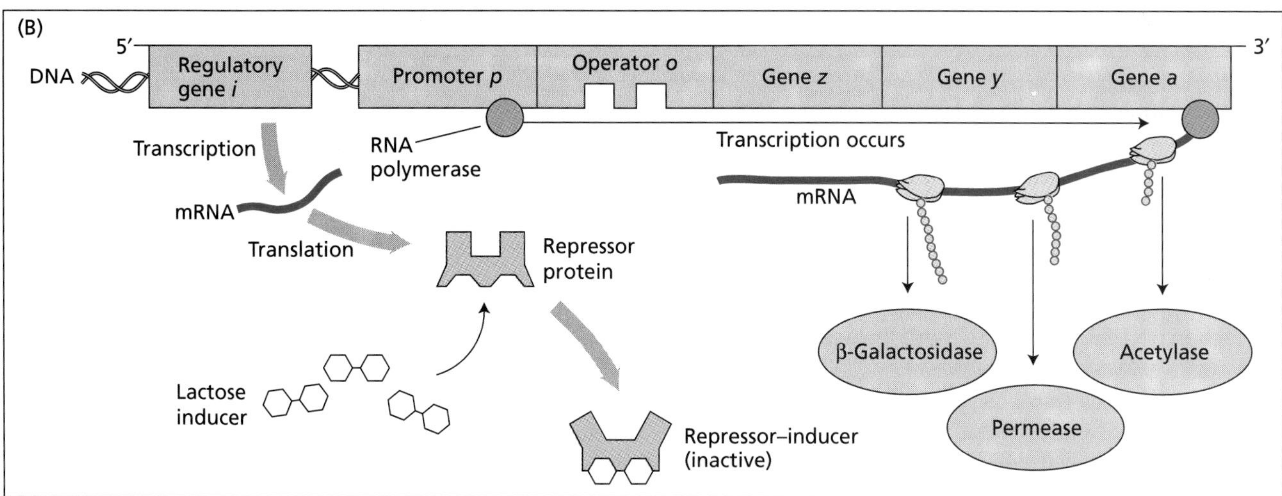

Figure 14.1 The *lac* operon of *E. coli* uses negative control. (A) The regulatory gene *i*, located upstream of the operon, is transcribed to produce an mRNA that encodes a repressor protein. The repressor protein binds to the operator gene *o*. The operator is a short stretch of DNA located between the promoter sequence *p* (the site of RNA polymerase attachment to the DNA) and the three structural genes, *z*, *y*, and *a*. Upon binding to the operator, the repressor prevents RNA polymerase from binding to the transcription initiation site. (B) When lactose (inducer) is added to the medium and is taken up by the cell, it binds to the repressor and inactivates it. The inactivated repressor is unable to bind to *o*, and transcription and translation can proceed. The mRNA produced is termed "polycistronic" because it encodes multiple genes. Note that translation begins while transcription is still in progress.

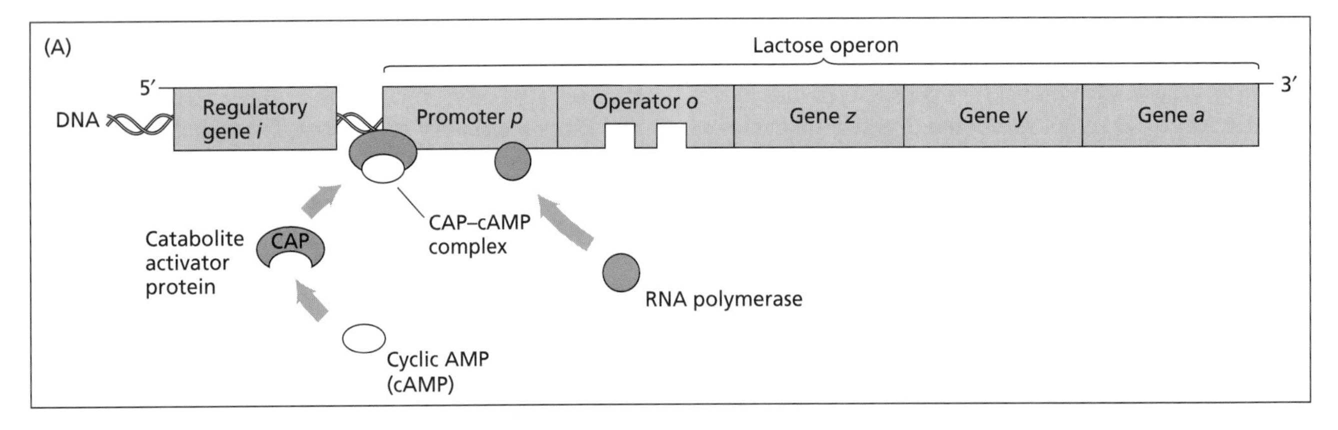

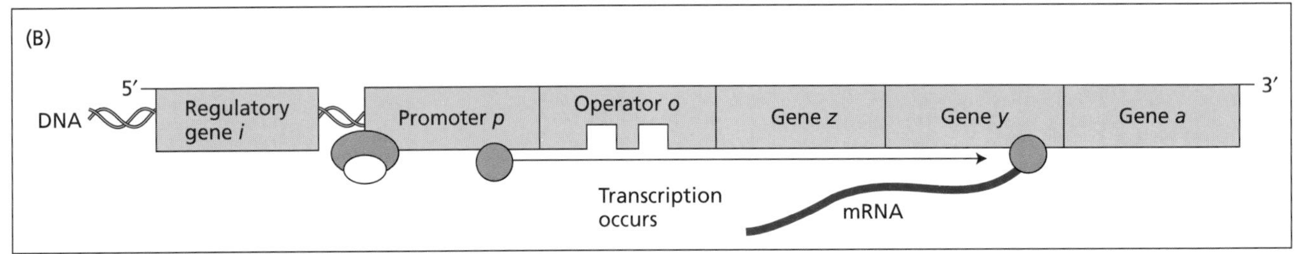

Figure 14.2 Stimulation of transcription by the catabolite activator protein (CAP) and cyclic AMP (cAMP). CAP has no effect on transcription until cAMP binds to it. (A) The CAP–cAMP complex binds to a specific DNA sequence near the promoter region of the *lac* operon. (B) Binding of the CAP–cAMP complex makes the promoter region more accessible to RNA polymerase, and transcription rates are enhanced.

the **promoter** sequence, *p*, where RNA polymerase attaches to the operon to initiate transcription, and immediately upstream (i.e., on the 5′ side) of the transcription start site, where transcription begins. (The initiation site is considered to be at the 5′ end of the gene, even though the RNA polymerase transcribes from the 3′ end to the 5′ end along the opposite strand. This convention was adopted so that the sequence of the mRNA would match the DNA sequence of the gene.)

In the absence of lactose, the lactose repressor forms a tight complex with the operator sequence and blocks the interaction of RNA polymerase with the transcription start site, effectively preventing transcription (see Figure 14.1A). When present, lactose binds to the repressor, causing it to undergo a conformational change (see Figure 14.1B). The *lac* repressor is thus an allosteric protein whose conformation is determined by the presence or absence of an **effector** molecule, in this case lactose. As a result of the conformational change due to binding lactose, the *lac* repressor detaches from the operator. When the operator sequence is unobstructed, the RNA polymerase can move along the DNA, synthesizing a continuous mRNA. The translation of this mRNA yields the three proteins, and lactose is said to induce their synthesis.

The *lac* repressor is an example of **negative control**, since the repressor blocks transcription upon binding to the operator region of the operon. The *lac* operon is also regulated by **positive control**, which was discovered in connection with a phenomenon called the *glucose effect*. If glucose is added to a nutrient medium that includes lactose, the *E. coli* cells metabolize the glucose and ignore the lactose. Glucose suppresses expression of the *lac* operon and prevents synthesis of the enzymes needed to degrade lactose. Glucose exerts this effect by lowering the cellular concentration of cyclic AMP (cAMP). When glucose levels are low, cAMP levels are high. Cyclic AMP binds to an **activator protein**, the *catabolite activator protein* (CAP), which recognizes and binds to a specific nucleotide sequence immediately upstream of the *lac* operator and promoter sites (Figure 14.2).

In contrast to the behavior of the lactose repressor protein, when the CAP is complexed with its effector, cAMP, its affinity for its DNA-binding site is dramatically *increased* (hence the reference to positive control). The ternary complex formed by CAP, cAMP, and the lactose operon DNA sequences induces bending of the DNA, which activates transcription of the lactose operon structural genes by increasing the affinity of RNA polymerase for the neighboring promoter site. Bacteria synthesize cyclic AMP when they exhaust the glucose in their growth medium. The lactose operon genes are thus under opposing regulation by the absence of glucose (high levels of cyclic AMP) and the presence of lactose, since glucose is a catabolite of lactose.

In bacteria, metabolites can also serve as *corepressors*, activating a repressor protein that blocks transcription. Repression of enzyme synthesis is often involved in the regulation of biosynthetic pathways in which one or

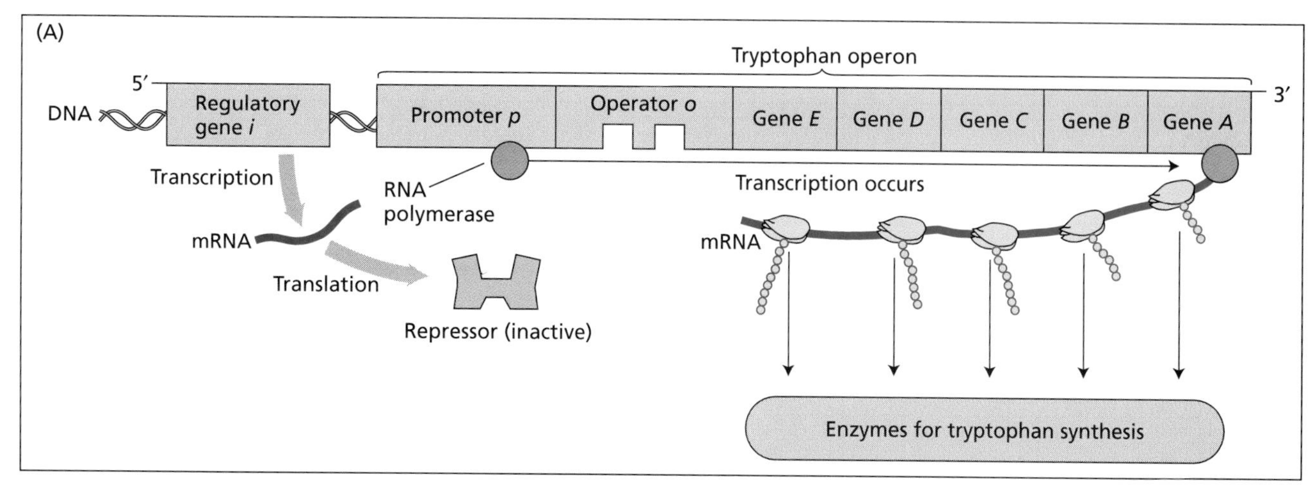

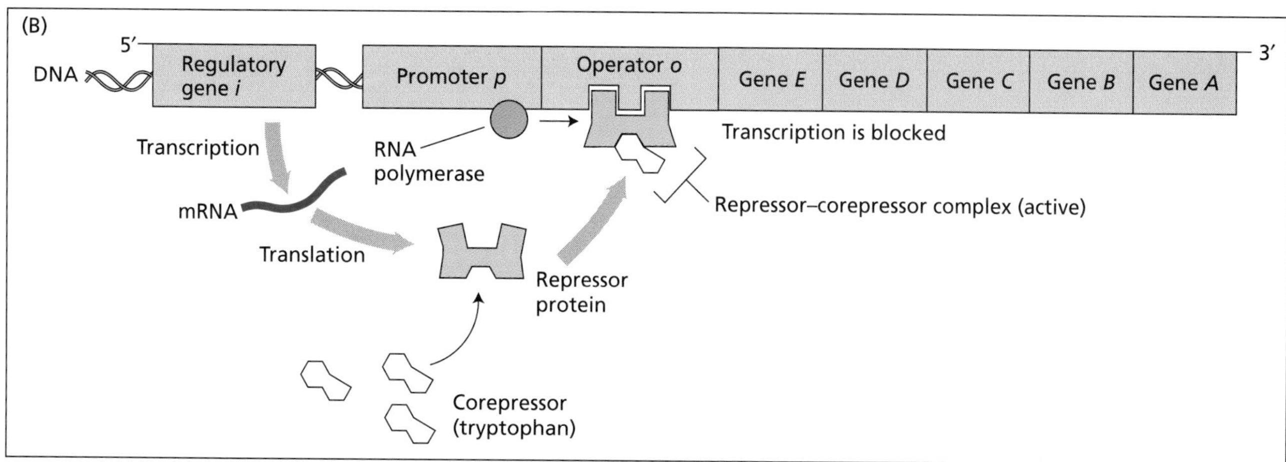

Figure 14.3 The tryptophan (*trp*) operon of *E. coli.* Tryptophan (Trp) is the end product of the pathway catalyzed by tryptophan synthetase and other enzymes. Transcription of the repressor genes results in the production of a repressor protein. However, the repressor is inactive until it forms a complex with its corepressor, Trp. (A) In the absence of Trp, transcription and translation proceed. (B) In the presence of Trp, the activated repressor–corepressor complex blocks transcription by binding to the operator sequence.

more enzymes are synthesized only if the end product of the pathway—an amino acid, for example—is not available. In such a case the amino acid acts as a corepressor: It complexes with the repressor protein, and this complex attaches to the operator DNA, preventing transcription. The tryptophan (*trp*) operon in *E. coli* is an example of an operon that works by corepression (Figure 14.3).

Eukaryotic Gene Expression

The study of bacterial gene expression has provided models that can be tested in eukaryotes. However, the details of the process are quite different and more complex in eukaryotes. In prokaryotes, translation is coupled to transcription: As the mRNA transcripts elongate, they bind to ribosomes and begin synthesizing proteins (translation). In eukaryotes, however, the nuclear envelope separates the genome from the translational machinery. The transcripts must first be transported to the cytoplasm, adding another level of control.

Eukaryotic Nuclear Transcripts Require Extensive Processing

Eukaryotes differ from prokaryotes also in the organization of their genomes. In most eukaryotic organisms, each gene encodes a single polypeptide. The eukaryotic nuclear genome contains no operons, with one notable exception.* Furthermore, eukaryotic genes are divided into coding regions called **exons** and noncoding regions

* About 25% of the genes in the nematode *Caenorhabditis elegans* are in operons. The operon pre-mRNAs are processed into individual mRNAs that encode single polypeptides (monocistronic mRNAs) by a combination of cleavage, polyadenylation, and splicing (Kuersten et al. 1997).

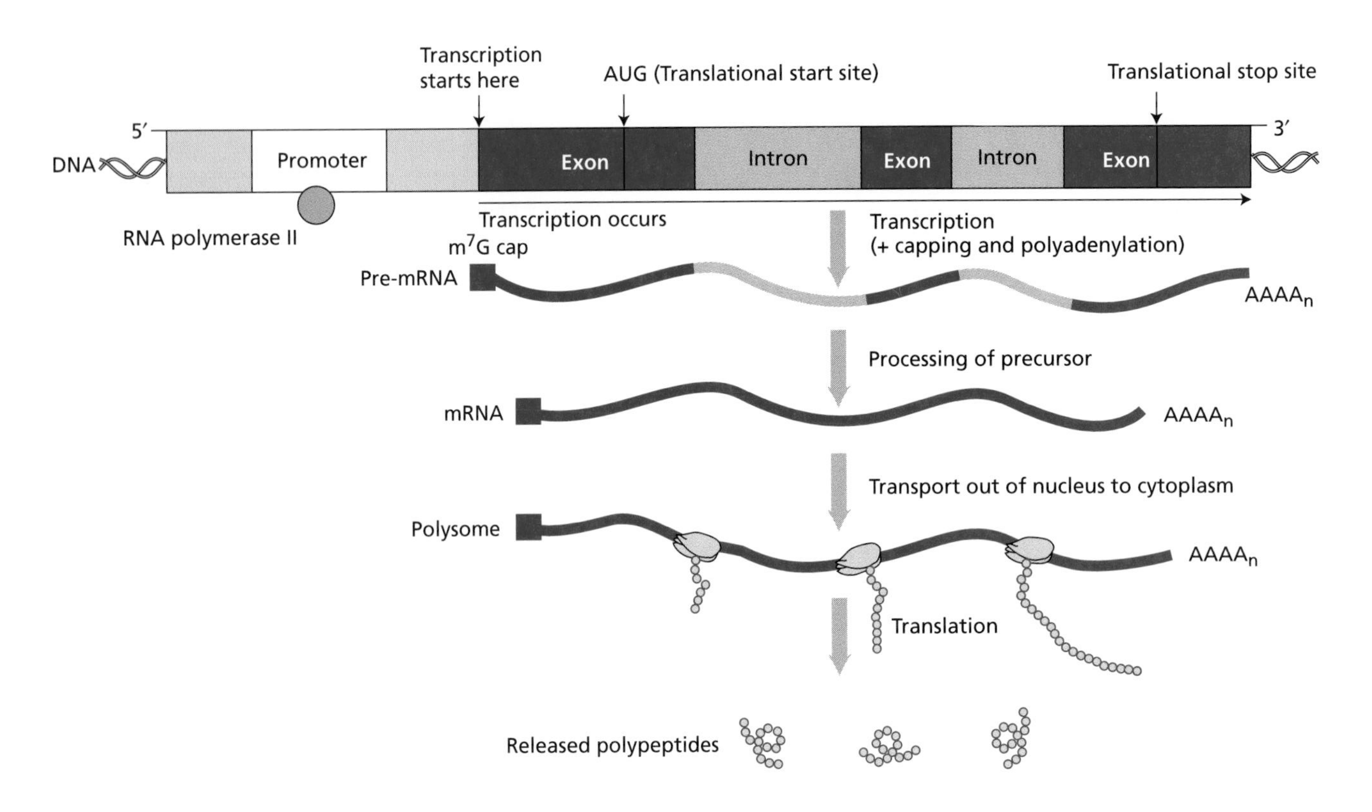

Figure 14.4 Gene expression in eukaryotes. RNA polymerase II binds to the promoter of genes that encode proteins. Unlike prokaryotic genes, eukaryotic genes are not clustered in operons, and each is divided into introns and exons. Transcription from the template strand proceeds in the 3′-to-5′ direction at the transcription start site, and the growing RNA chain extends one nucleotide at a time in the 5′-to-3′ direction. Translation begins with the first AUG encoding methionine, as in prokaryotes, and ends with the stop codon. The pre-mRNA transcript is first "capped" by the addition of 7-methylguanylate (m^7G) to the 5′ end. The 3′ end is shortened slightly by cleavage at a specific site, and a poly-A tail is added. The capped and polyadenylated pre-mRNA is then spliced by a spliceosome complex, and the introns are removed. The mature mRNA exits the nucleus through the pores and initiates translation on ribosomes in the cytosol. As each ribosome progresses toward the 3′ end of the mRNA, new ribosomes attach at the 5′ end and begin translating, leading to the formation of polysomes.

called **introns** (Figure 14.4). Since the primary transcript, or pre-mRNA, contains both exon and intron sequences, the pre-mRNA must be processed to remove the introns.

RNA processing involves multiple steps. The newly synthesized pre-mRNA is immediately packaged into a string of small protein-containing particles, called **heteronuclear ribonucleoprotein particles**, or **hnRNP particles**. Some of these particles are composed of proteins and small nuclear RNAs, and are called **small nuclear ribonucleoproteins**, or **snRNPs** (pronounced "snurps"). Various snRNPs assemble into **spliceosome complexes** at exon–intron boundaries of the pre-mRNA and carry out the splicing reaction.

In some cases, the primary transcript can be spliced in different ways, a process called **alternative RNA splicing**. For example, an exon that is present in one version of a processed transcript may be spliced out of another version. In this way, the same gene can give rise to different polypeptide chains. Approximately 15% of human genes are processed by alternative splicing. Although alternative splicing is rare in plants, it is involved in the synthesis of rubisco activase, RNA polymerase II, and the gene product of a rice homeobox gene (discussed later in the chapter), as well as other proteins (Golovkin and Reddy 1996).

Before splicing, the pre-mRNA is modified in two important ways. First it is **capped** by the addition of 7-methylguanylate to the 5′ end of the transcript via a 5′-to-5′ linkage. The pre-mRNA is capped almost immediately after the initiation of mRNA synthesis. One of the functions of the 5′ cap is to protect the growing RNA transcript from degradation by RNases. At a later stage in the synthesis of the primary transcript, the 3′ end is

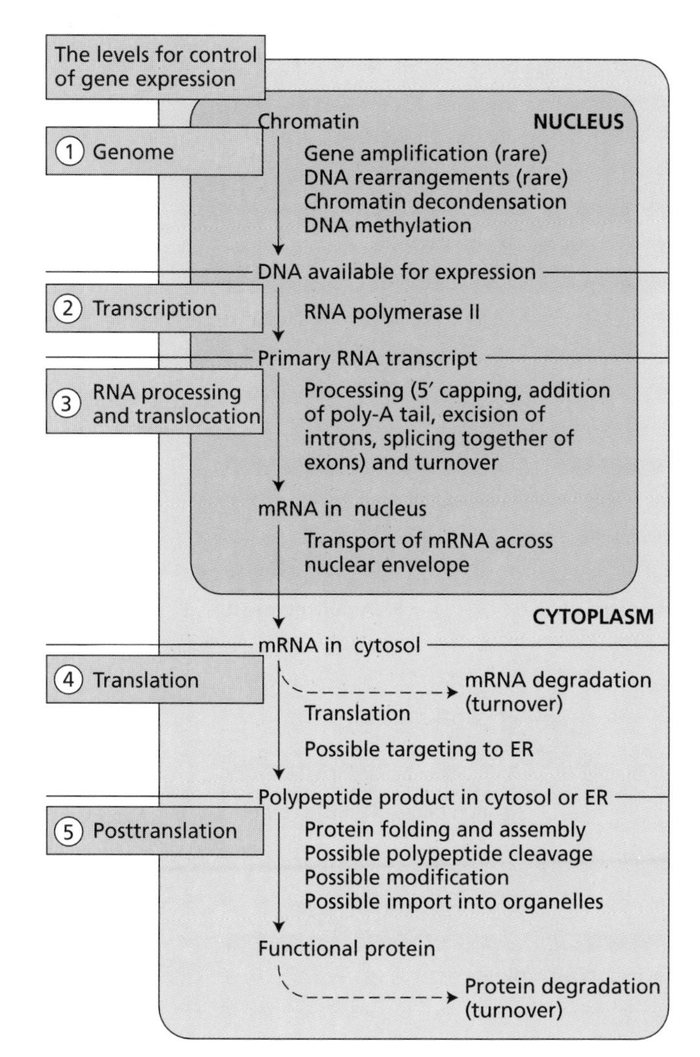

Figure 14.5 Eukaryotic gene expression can be regulated at multiple levels. (1) genomic regulation, by gene amplification, DNA rearrangements, chromatin decondensation or condensation, or DNA methylation; (2) transcriptional regulation; (3) RNA processing, and RNA turnover in the nucleus and translocation out of the nucleus; (4) translational control (including binding to ER in some cases); (5) posttranslational control, including mRNA turnover in the cytosol, and the folding, assembly, modification, and import of proteins into organelles. (After Becker et al. 1996.)

cleaved at a specific site, and a **poly-A tail**, usually consisting of about 100 to 200 adenylic acid residues, is added by the enzyme poly-A polymerase (see Figure 14.4).

The poly-A tail has several functions: (1) It protects against RNases and therefore increases the stability of mRNA molecules in the cytoplasm, (2) both it and the 5′ cap are required for transit through the nuclear pore, and (3) it increases the efficiency of translation on the ribosomes. The requirement of eukaryotic mRNAs to have both a 5′ cap and a poly-A tail ensures that only properly processed transcripts will reach the ribosome and be translated.

Each step in eukaryotic gene expression can potentially regulate the amount of gene product in the cell at any given time (Figure 14.5). Like transcription initiation, splicing may be regulated. Export from the nucleus is also regulated. For example, to exit the nucleus an mRNA must possess a 5′ cap and a poly-A tail, and it must be properly spliced. Incompletely processed transcripts remain in the nucleus and are degraded.

Various Posttranscriptional Regulatory Mechanisms Have Been Identified

The stabilities or **turnover rates** of mRNA molecules differ from one another, and may vary from tissue to tissue, depending on the physiological conditions. For example, in bean (*Vicia faba*), fungal infection causes the rapid degradation of the mRNA that encodes the proline-rich protein PvPRP1 of the bean cell wall. Another example of the regulation of gene expression by RNA degradation is the regulation of expression of one of the genes for the small subunit of rubisco in roots of the aquatic duckweed *Lemna gibba*. *Lemna* roots are photosynthetic and therefore express genes for the small subunit of rubisco, but the expression of one of the genes (*SSU5B*) is much lower in roots than in the fronds (leaves). Jane Silverthorne and her colleagues at the University of California, Santa Cruz, showed that the low level of SSU5B in the roots is due to a high rate of turnover of the *SSU5B* pre-mRNA in the nucleus (Peters and Silverthorne 1995).

In addition to RNA turnover, the **translatability** of mRNA molecules is variable. For example, RNAs fold into molecules with varying secondary and tertiary structures that can influence the accessibility of the translation initiation codon (the first AUG sequence) to the ribosome. Another factor that can influence translatability of an mRNA is codon usage. There is redundancy in the triplet codons that specify a given amino acid during translation, and each cell has a characteristic ratio of the different aminoacylated tRNAs available, known as **codon bias**. If a message contains a large number of triplet codons that are rare for that cell, the small number of charged tRNAs available for those codons will slow translation. Finally, the **cellular location** at which translation occurs seems to affect the rate of gene expression. Free polysomes may translate mRNAs at very different rates from those at which polysomes bound to the endoplasmic reticulum do; even within the endoplasmic reticulum, there may be differential translation rates.

Although examples of posttranscriptional regulation have been demonstrated for each of the steps described above and summarized in Figure 14.5, *the expression of most eukaryotic genes, like their prokaryotic counterparts, appears to be regulated at the level of transcription.*

Transcription in Eukaryotes Is Modulated by *cis*-Acting Regulatory Sequences

The synthesis of most eukaryotic proteins is regulated at the level of transcription. However, transcription in eukaryotes is much more complex than in prokaryotes. First, there are three different RNA polymerases in eukaryotes: I, II, and III. RNA polymerase I is located in the nucleolus and functions in the synthesis of most ribosomal RNAs. RNA polymerase II, located in the nucleoplasm, is responsible for pre-mRNA synthesis. RNA polymerase III, also located in the nucleoplasm, synthesizes small RNAs, such as tRNA and 5S rRNA.

A second important difference between transcription in prokaryotes and in eukaryotes is that the RNA polymerases of eukaryotes require additional proteins called **general transcription factors** to position them at the correct start site. While prokaryotic RNA polymerases also require accessory polypeptides called sigma factors (σ), these polypeptides are considered to be subunits of the RNA polymerase. In contrast, eukaryotic general transcription factors make up a large, multisubunit **transcription initiation complex**. For example, seven general transcription factors constitute the initiation complex of RNA polymerase II, each of which must be added in a specific order during assembly (Figure 14.6).

According to one current model, transcription is initiated when the final transcription factor, TFIIH (*tran*scription *f*actor for RNA polymerase *II* protein *H*), joins the complex and causes phosphorylation of the RNA polymerase. RNA polymerase II then separates from the initiation complex and proceeds along the antisense strand in the 3′-to-5′ direction. While some of the general transcription factors dissociate from the complex at this point, others remain to bind another RNA polymerase molecule and initiate another round of transcription.

A third difference between transcription in prokaryotes and in eukaryotes is in the complexity of the promoters, the sequences upstream (5′) of the initiation site that regulate transcription. We can divide the structure of the eukaryotic promoter into two parts, the **core** or **minimum promoter**, consisting of the minimum upstream sequence required for gene expression, and the additional **regulatory sequences**, which control the activity of the core promoter.

Each of the three RNA polymerases has a different type of promoter. An example of a typical RNA polymerase II promoter is shown schematically in Figure 14.7A. The minimum promoter for genes transcribed by RNA polymerase II typically extends about 100 bp upstream of the transcription initiation site and includes several sequence elements referred to as **proximal promoter sequences**. About 25 to 35 bp upstream of the transcriptional start site is a short sequence called the **TATA box**, consisting of the sequence TATAAA(A). The

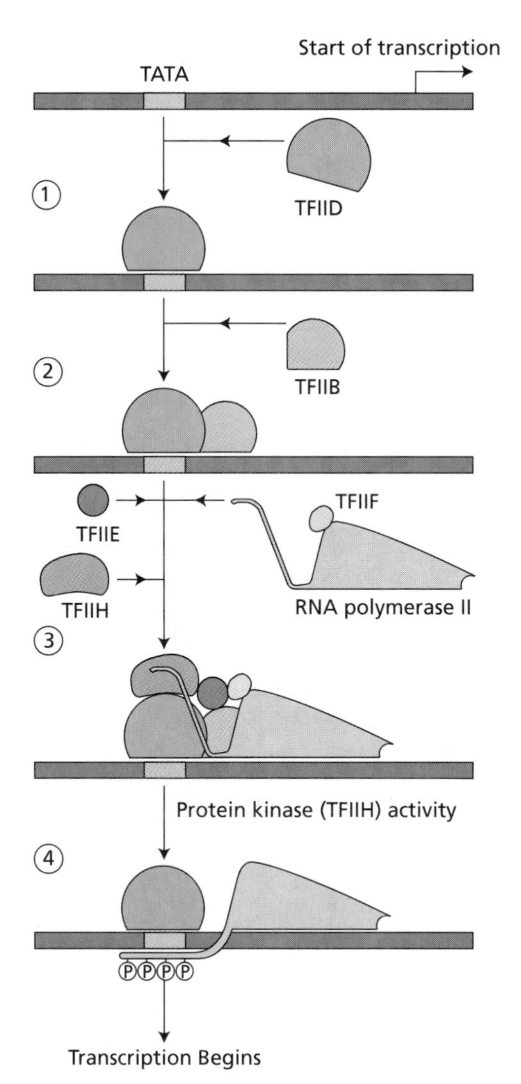

Figure 14.6 Ordered assembly of the general transcription factors required for transcription by RNA polymerase II. (1) TFIID, a multisubunit complex, binds to the TATA box via the TATA-binding protein. (2) TFIIB joins the complex. (3) TFIIF bound to RNA polymerase II associates with the complex, along with TFIIE and TFIIH. The assembly of proteins is referred to as the transcription initiation complex. (4) TFIIH, a protein kinase, phosphorylates the RNA polymerase, some of the general transcription factors are released, and transcription begins. (From Alberts et al. 1994.)

TATA box plays a crucial role in transcription because it serves as the site of assembly of the transcription initiation complex. Approximately 85% of the plant genes sequenced thus far contain TATA boxes.

In addition to the TATA box, the minimum promoters of eukaryotes also contain two additional regulatory sequences: the **CAAT box** and the **GC box** (see Figure 14.7A). These two sequences are the sites of binding of **transcription factors**, proteins that enhance the rate of transcription by facilitating the assembly of the initiation complex. The DNA sequences themselves are

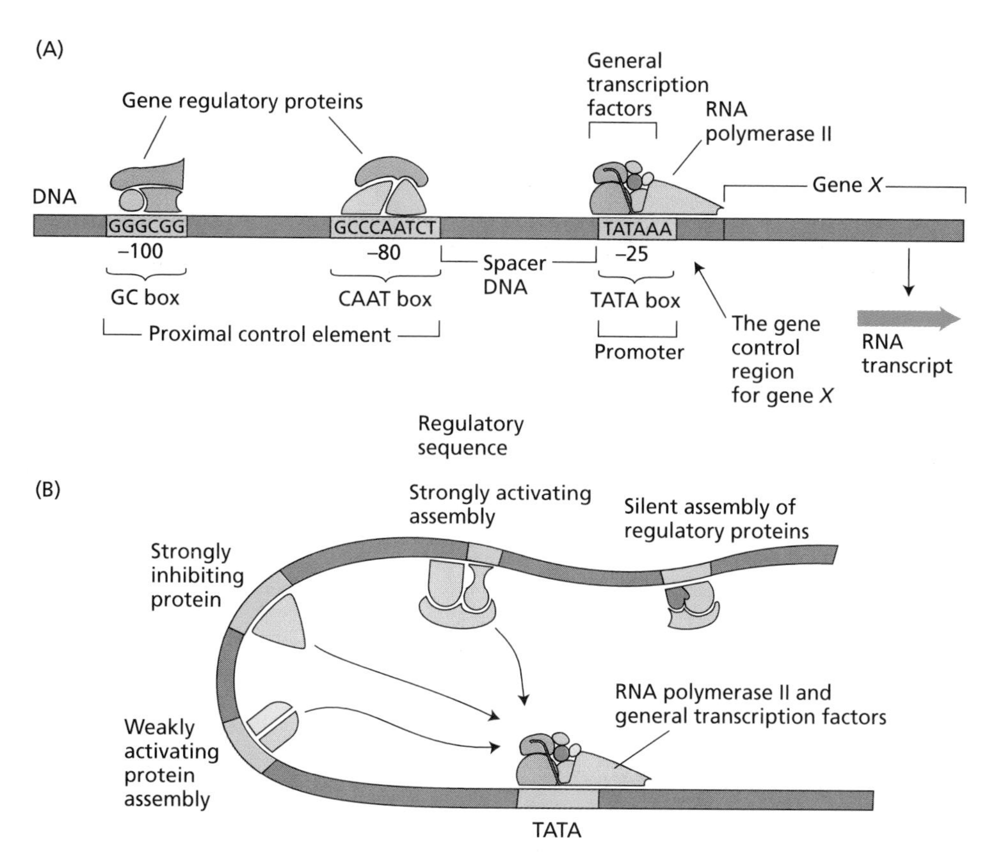

Figure 14.7 Organization and regulation of a typical eukaryotic gene. (A) Features of a typical eukaryotic RNA polymerase II minimum promoter and proteins that regulate gene expression. RNA polymerase II is situated at the TATA box in association with the general transcription factors about 25 bp upstream of the transcription start site. Two *cis*-acting regulatory sequences that enhance the activity of RNA polymerase II are the CAAT box and the GC box, located at about 80 and 100 bp upstream, respectively, of the transcription start site. The DNA proteins that bind to these elements are indicated. (B) Regulation of transcription by distal regulatory sequences and *trans*-acting factors. *trans*-acting factors bound to distal regulatory sequences can act in concert to activate transcription by making direct physical contact with the transcription initiation complex. The details of this process are not well understood. (A after Alberts et al. 1994; B from Alberts et al 1994.)

termed ***cis*-acting sequences**, since they are adjacent to the transcription units they are regulating. The transcription factors that bind to the *cis*-acting sequences are called ***trans*-acting factors**, since the genes that encode them are located elsewhere in the genome.

Numerous other *cis*-acting sequences located farther upstream of the proximal promoter sequences can exert either positive or negative control over eukaryotic promoters. These sequences are termed the **distal regulatory sequences** and they are usually located within 1000 bp of the transcription initiation site. As with prokaryotes, the positively acting transcription factors that bind to these sites are called **activators**, while those that inhibit transcription are called **repressors**.

As we will see in Chapters 19 and 20, the regulation of gene expression by the plant hormones and by phytochrome is thought to involve the deactivation of repressor proteins. *Cis*-acting sequences involved in gene regulation by hormones and other signaling agents are called **response elements**. As will be discussed in Chapters 17 and 19 through 23 (on phytochrome and the plant hormones), numerous response elements that regulate gene expression have been identified in plants.

In addition to having regulatory sequences within the promoter itself, eukaryotic genes can be regulated by control elements located tens of thousands of base pairs away from the start site. Distantly located positive regulatory sequences are called **enhancers**. Enhancers may be located either upstream or downstream from the promoter. In plants, many developmentally important plant genes have been shown to be regulated by enhancers (Sundaresan et al. 1995).

How do all the DNA-binding proteins on the *cis*-acting sequences regulate transcription? During formation

Table 14.1
DNA-Binding Motifs

Name	Examples of proteins	Key structural features	Illustration
Helix-turn-helix	Transcription factors that regulate genes in anthocyanin biosynthesis pathway	Two α helices separated by a turn in the polypeptide chain; function as dimers	COOH, NH_2
Zinc finger	COP1 in *Arabidopsis*	Various structures in which zinc plays an important structural role; bind to DNA either as monomers or as dimers	His, His, Zn, Cys, Cys; Cys, Cys, Zn, Cys, Cys
Helix-loop-helix	GT element–binding protein of phytochrome-regulated genes	A short α helix connected by a loop to a longer α helix; function as dimers	
Leucine zipper	Fos and Jun	An α helix of about 35 amino acids containing leucine at every seventh position; dimerization occurs along the hydrophobic surface	H_3^+N, NH_3^+, Leu, Leu, Leu, Leu, Leu, Leu, Leu, COO^-, COO^-
Basic zipper (bZip)	Opaque 2 protein in maize, G box factors of phytochrome-regulated genes, transcription factors that bind ABA response elements	Variation of the leucine zipper motif in which other hydrophobic amino acids substitute for leucine and the DNA-binding domain contains amino acids	H_3^+N, NH_3^+, Leu, Ala, Val, Ile, Ala, Val, COO^-, COO^-

of the initiation complex, the DNA between the core promoter and the most distally located control elements loops out in such a way as to allow all of the transcription factors bound to that segment of DNA to make physical contact with the initiation complex (see Figure 14.7B). Through this physical contact the transcription factor exerts its control, either positive or negative, over transcription. Given the large number of control elements that can modify the activity of a single promoter, the possibilities for differential gene regulation in eukaryotes are nearly infinite.

Transcription Factors Contain Specific Structural Motifs

Transcription factors generally have three structural features: a DNA-binding domain, a transcription-activating domain, and a ligand-binding domain. To bind to a specific sequence of DNA, the DNA-binding domain must have extensive interactions with the double helix through the formation of hydrogen, ionic, and hydrophobic bonds. Although the particular combination and spatial distribution of such interactions are unique for each sequence, analyses of many DNA-binding proteins have led to the identification of a small number of highly conserved DNA-binding structural motifs, which are summarized in Table 14.1.

Most of the transcription factors characterized thus far in plants belong to the basic zipper (bZIP) class of DNA-binding proteins. DNA-binding proteins containing the zinc finger domain are relatively rare in plants.

Homeodomain Proteins Are a Special Class of Helix-Turn-Helix Proteins

The term "homeodomain protein" is derived from a group of *Drosophila* (fruit fly) genes called **selector genes** or **homeotic genes**. *Drosophila* homeotic genes encode transcription factors that determine which structures develop at specific locations on the fly's body; that is, they act as major developmental switches that activate a large number of genes that constitute the entire genetic program for a particular structure. Mutations in homeotic genes cause **homeosis**, the transformation of one body part into another. For example, a homeotic mutation in the *ANTENNAPEDIA* gene causes a leg to form in place of an antenna. When the sequences of various homeotic genes in *Drosophila* were compared, the proteins were all found to contain a highly conserved stretch of 60 amino acids called the **homeobox**.

Homologous homeobox sequences have now been identified in developmentally important genes of vertebrates and plants. As will be discussed in Chapter 16, the *KN1* (*KNOTTED*) gene of maize encodes a homeodomain protein that can affect cell fate during development. Maize plants with the *kn1* mutation exhibit abnormal cell divisions in the vascular tissues, giving rise to the "knotted" appearance of the leaf surface. However, the *kn1* mutation is not a homeotic mutation, since it does not involve the substitution of one entire structure for another. Rather, the plant homeodomain protein, KN1, is involved in the regulation of cell division. Thus, not all genes that encode homeodomain proteins are homeotic genes, and vice versa. As will be discussed in Chapter 24, four of the floral homeotic genes in plants encode proteins with the DNA-binding helix-turn-helix motif called the **MADS domain**.

Eukaryotic Genes Can Be Coordinately Regulated

Although eukaryotic nuclear genes are not arranged into operons, they are often coordinately regulated in the cell. For example, in yeast, many of the enzymes involved in galactose metabolism and transport are inducible and coregulated, even though the genes are located on different chromosomes. Incubation of wild-type yeast cells in galactose-containing media results in more than a thousandfold increase in the mRNA levels for all of these enzymes.

The six yeast genes that encode the enzymes in the galactose metabolism pathway are under both positive and negative control (Figure 14.8). Most yeast genes are regulated by a single proximal control element called an *upstream activating sequence* (UAS). The *GAL4* gene encodes a transcription factor that binds to UAS elements located about 200 bp upstream of the transcription start sites of all six genes. The UAS of each of the six genes, while not identical, consists of one or more copies of a similar 17 bp repeated sequence. The GAL4 protein can bind to each of them and activate transcription. *In this way a single transcription factor can control the expression of many genes.*

Protein–protein interactions can modify the effects of DNA-binding transcription factors. Another gene on a different yeast chromosome, *GAL80*, encodes a negative transcription regulator that forms a complex with the GAL4 protein when it is bound to the UAS. When the GAL80 protein is complexed with GAL4, transcription is blocked. In the presence of galactose, however, the metabolite formed by the enzyme that is encoded by the *GAL3* gene acts as an inducer by causing the dissociation of GAL80 from GAL4 (Johnston 1987; Mortimer et al. 1989).

There are many other examples of coordinate regulation of genes in eukaryotes. In plants, the developmental effects induced by light and hormones (see Chapters 17 through 23), as well as the adaptive responses caused by various types of stress (see Chapter 25), involve the coordinate regulation of groups of genes that share a common response element upstream of the promoter. In addition, genes that act as major developmental switches, such as the homeotic genes, encode transcription factors that bind to a common regulatory sequence that is present on dozens, or even hundreds, of genes scattered throughout the genome (see Chapters 16 and 24).

The Ubiquitin Pathway Regulates Protein Turnover

An enzyme molecule, once synthesized, has a finite lifetime in the cell, ranging from a few minutes to several hours. Hence, steady-state levels of cellular enzymes are attained as the result of an equilibrium between enzyme synthesis and enzyme degradation, or turnover. Protein turnover plays an important role in development. In etiolated seedlings, for example, the red-light photoreceptor, phytochrome, is regulated by proteolysis. The phytochrome synthesized in the dark is highly stable and accumulates in the cells to high concentrations. Upon exposure to red light, however, the phytochrome is converted to its active form and simultaneously becomes highly susceptible to degradation by proteases (see Chapter 17).

In animal cells there are two distinct pathways of protein turnover, one in specialized digestive vacuoles called lysosomes and the other in the cytosol. Proteins destined to be digested in lysosomes appear to be specifically targeted to these organelles. Upon entering the lysosomes, the proteins are rapidly degraded by lysosomal proteases. Lysosomes are also capable of engulfing and digesting entire organelles by an autophagic process. The central vacuole of plant cells is rich in proteases and is the plant equivalent of lysosomes, but as yet there is no clear evidence that plant vacuoles either engulf organelles or participate in the turnover of cytosolic proteins, except during senescence.

The nonlysosomal pathway of protein turnover involves the ATP-dependent formation of a covalent bond to a small, 76-amino-acid polypeptide called **ubiquitin**. Ubiquitination of an enzyme molecule apparently marks it for destruction by a large ATP-dependent proteolytic complex (**26S proteasome**) that specifically recognizes the "tagged" molecule (Coux et al. 1996). More than 90% of the short-lived proteins in eukaryotic cells are degraded via the ubiquitin pathway (Lam 1997). The ubiquitin pathway regulates cytosolic protein turnover in plant cells as well (Shanklin et al. 1987).

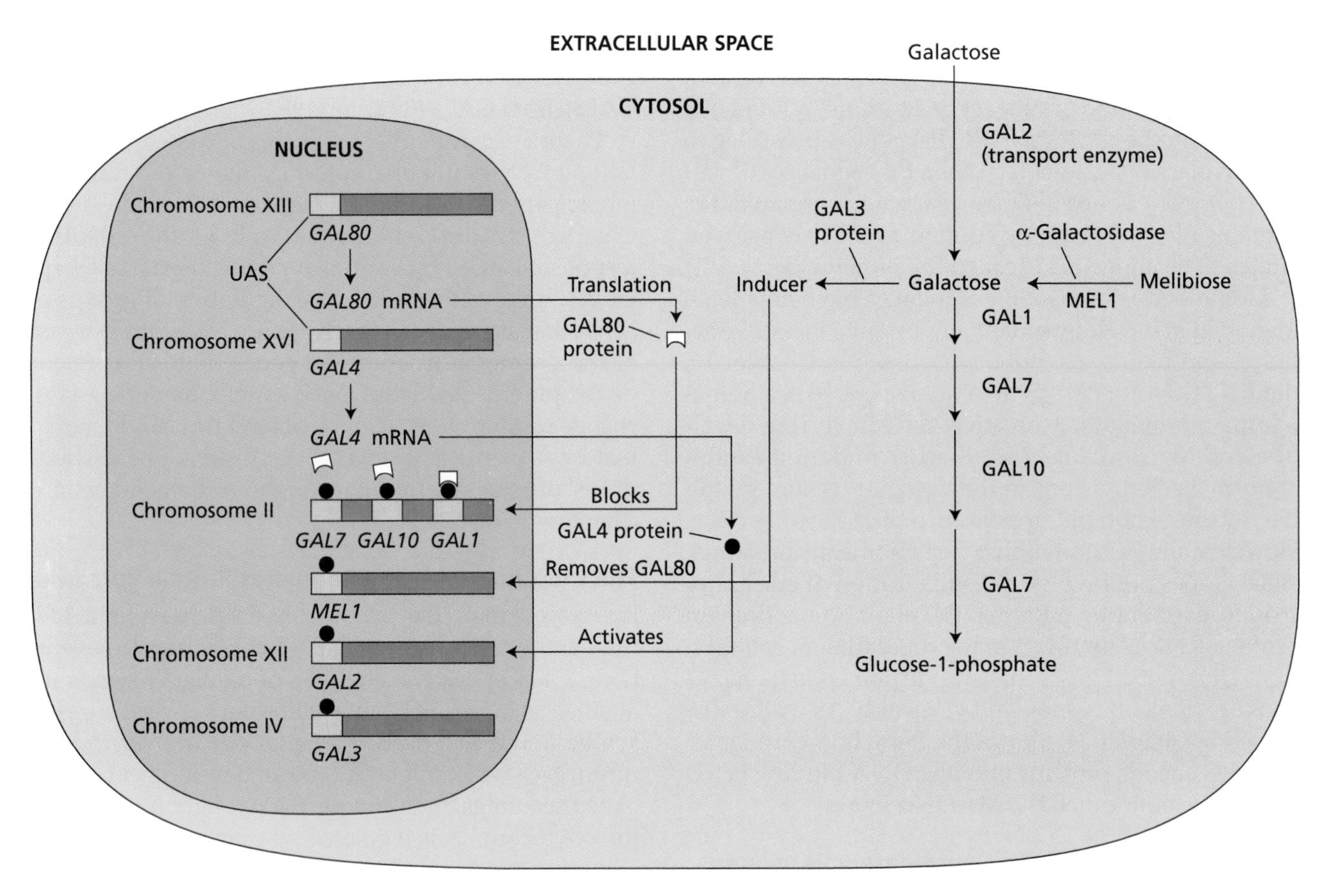

Figure 14.8 Model for eukaryotic gene induction: the galactose metabolism pathway of the yeast *Saccharomyces cerevisiae*. Several enzymes involved in galactose transport and metabolism are induced by a metabolite of galactose. The genes *GAL7*, *GAL10*, *GAL1*, and *MEL1* are located on chromosome II; *GAL2* is on chromosome XII; *GAL3* is on chromosome IV. *GAL4* and *GAL80*, located on two other chromosomes, encode positive and negative *trans*-acting regulatory proteins, respectively. The GAL4 protein binds to an upstream activating sequence located upstream of each of the genes in the pathway, indicated by the hatched lines. The GAL80 protein forms an inhibitory complex with the GAL4 protein. In the presence of galactose, the metabolite formed by the GAL3 gene product diffuses to the nucleus and stimulates transcription by causing dissociation of the GAL80 protein from the complex. (After Darnell et al. 1990.)

Before it can take part in protein tagging, free ubiquitin must be activated (Figure 14.9). The enzyme **E1** catalyzes the ATP-dependent adenylylation of the C terminus of ubiquitin. The adenylylated ubiquitin is then transferred to a second enzyme, called **E2**. Proteins destined for ubiquitination form complexes with a third protein, **E3**. Finally, the E2–ubiquitin conjugate is used to transfer ubiquitin to the lysine residues of proteins bound to E3. This process can occur multiple times to form a polymer of ubiquitin. The ubiquitinated protein is then targeted to the proteasome for degradation. As we shall see in Chapter 19, recent evidence suggests that

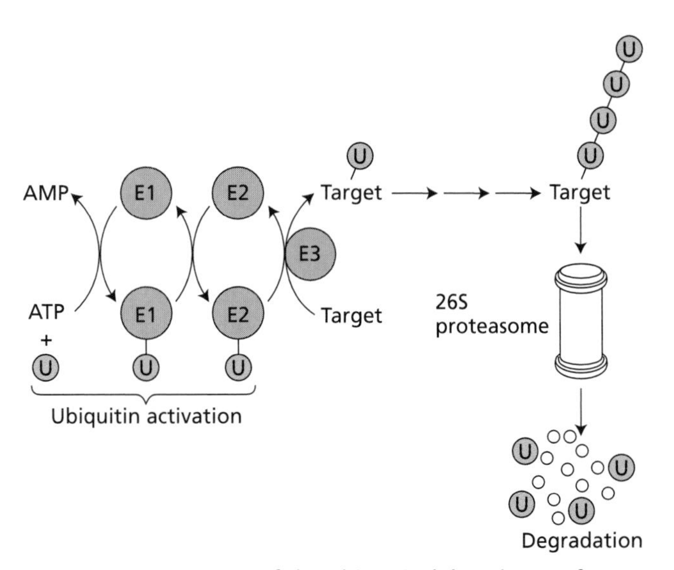

Figure 14.9 Diagram of the ubiquitin (U) pathway of protein degradation in the cytosol. ATP is required for the initial activation of E1. E1 tranfers ubiquitin to E2. E3 mediates the final transfer of ubiquitin to a target protein, which may be ubiquinated multiple times. The ubiquinated target protein is then degraded by the 26S proteasome.

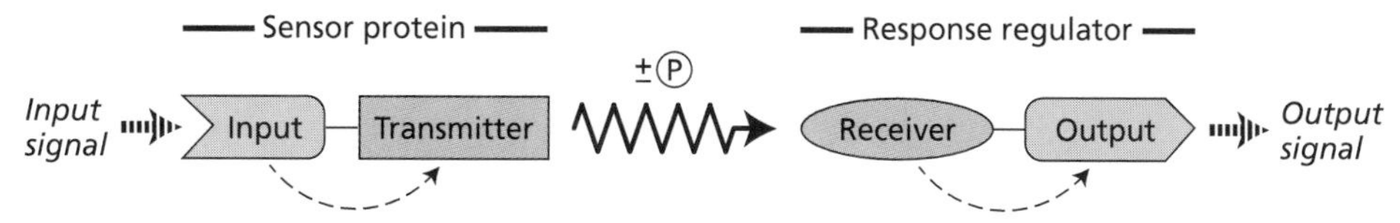

Figure 14.10 Signaling via bacterial two-component systems. The sensor protein detects the stimulus via the input domain and transfers the signal to the transmitter domain by means of a conformational change (indicated by the first dashed arrow). The transmitter domain of the sensor then communicates with the response regulator by protein phosphorylation of the receiver domain. Phosphorylation of the receiver domain induces a conformational change (second dashed arrow) that activates the output domain and brings about the cellular response. (After Parkinson 1993.)

the regulation of gene expression by the phytohormone, auxin, may be mediated in part by the activation of the ubiquitin pathway.

Signal Transduction in Prokaryotes

Prokaryotic cells could not have survived billions of years of evolution without an exquisitely developed ability to sense their environment. As we have seen, bacteria respond to the presence of a nutrient by synthesizing the proteins involved in the uptake and metabolism of that nutrient. Bacteria can also respond to nonnutrient signals, both physical and chemical. Motile bacteria can adjust their movements according to the prevailing gradients of light, oxygen, osmolarity, temperature, and toxic chemicals in the medium.

The basic mechanisms that enable bacteria to sense and to respond to their environment are common to all cell sensory systems, and include stimulus *detection*, signal *amplification*, and the appropriate *output responses*. Many bacterial signaling pathways have been shown *to* consist of modular units called *transmitters* and *receivers*. These modules form the basis of the so-called two-component regulatory systems.

Bacteria Employ Two-Component Regulatory Systems to Sense Extracellular Signals

Bacteria sense chemicals in the environment by means of a small family of cell surface receptors, each involved in the response to a defined group of chemicals (hereafter referred to as ligands). A protein in the plasma membrane of bacteria binds directly to a ligand, or binds to a soluble protein that has already attached to the ligand, in the **periplasmic space** between the plasma membrane and the cell wall. Upon binding, the membrane protein undergoes a conformational change that is propagated across the membrane to the cytosolic domain of the receptor protein. This conformational change initiates the signaling pathway that leads to the response.

A broad spectrum of responses in bacteria, including osmoregulation, chemotaxis, and sporulation, are regulated by two-component systems. **Two-component regulatory systems** are composed of a **sensor protein** and a **response regulator protein** (Figure 14.10) (Parkinson 1993). The function of the sensor is to receive the signal and to pass the signal on to the response regulator, which brings about the cellular response, typically gene expression. Sensor proteins have two domains, an **input domain**, which receives the environmental signal, and a **transmitter domain**, which transmits the signal to the response regulator. The response regulator also has two domains, a **receiver domain**, which receives the signal from the transmitter domain of the sensor protein, and an **output domain**, such as a DNA-binding domain, which brings about the response.

The signal is passed from transmitter domain to receiver domain via protein phosphorylation. Transmitter domains have the ability to phosphorylate themselves, using ATP, on a specific histidine residue near the amino terminus (Figure 14.11A). For this reason, sensor proteins containing transmitter domains are called **autophosphorylating histidine kinases**. These proteins normally

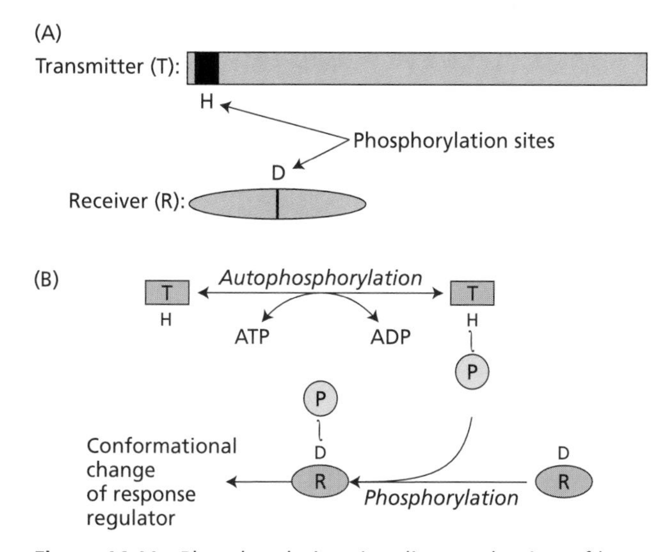

Figure 14.11 Phosphorylation signaling mechanism of bacterial two-component systems. (A) The transmitter domain of the sensor protein contains a conserved histidine (H) at its N-terminal end, while the receiver domain of the response regulator contains a conserved aspartate (D). (B) The transmitter phosphorylates itself at its conserved histidine and transfers the phosphate to the aspartate of the response regulator. The response regulator then undergoes a conformational change leading to the response. (After Parkinson 1993.)

function as dimers in which the catalytic site of one subunit phosphorylates the acceptor site on the other.

Immediately after the transmitter domain becomes autophosphorylated on a histidine residue, the phosphate is transferred to a specific aspartate residue near the middle of the receiver domain of the response regulator protein (see Figure 14.11A). As a result, a specific aspartate residue of the response regulator becomes phosphorylated (Figure 14.11B). Phosphorylation of the aspartate residue causes the response regulator to undergo a conformational change that results in its activation.

Osmolarity Is Detected by a Two-Component System

An example of a relatively simple bacterial two-component system is the signaling system involved in sensing osmolarity in *E. coli*. *E. coli* is a Gram-negative bacterium and thus has two cell membranes, an inner membrane and an outer membrane, separated by a cell wall. The inner membrane is the primary permeability barrier of the cell. The outer membrane contains large pores composed of two types of porin proteins, **OmpF** and **OmpC**. Pores made with OmpF are larger than those made with OmpC.

When *E. coli* is subjected to high osmolarity in the medium, it synthesizes more OmpC than OmpF, resulting in smaller pores on the outer membrane. These smaller pores filter out the solutes from the periplasmic space, shielding the inner membrane from the effects of the high solute concentration in the external medium. When the bacterium is placed in a medium with low osmolarity, more OmpF is synthesized, and the average pore size increases.

As Figure 14.12 shows, expression of the genes that encode the two porin proteins is regulated by a two-component system. The sensor protein, **EnvZ**, is located on the inner membrane. It consists of an N-terminal periplasmic input domain that detects the osmolarity changes in the medium, flanked by two membrane-spanning segments, and a C-terminal cytoplasmic transmitter domain.

When the osmolarity of the medium increases, the input domain undergoes a conformational change that is transduced across the membrane to the transmitter domain. The transmitter then autophosphorylates its histidine residue. The phosphate is rapidly transferred to an aspartate residue of the receiver domain of the response regulator, **OmpR**. The N terminus of OmpR consists of a DNA-binding domain. When activated by phosphorylation, this domain interacts with RNA polymerase at the promoters of the porin genes, enhancing the expression of *ompC* and repressing the expression of *ompF*. Under conditions of low osmolarity in the medium, the nonphosphorylated form of OmpR stimulates *ompF* expression and represses *ompC* expression. In this way the osmolarity stimulus is communicated to the genes.

Related Two-Component Systems Have Been Identified in Eukaryotes

Recently, combination sensor–response regulator proteins related to the bacterial two-component systems have been discovered in yeast and in plants. For example, The *SLN1* gene of the yeast *Saccharomyces cerevisiae* encodes a 134-kilodalton protein that has sequence similarities to both the transmitter and the receiver domains of bacteria and appears to function in osmoregulation (Ota and Varshavsky 1993).

There is increasing evidence that several plant signaling systems evolved from bacterial two-component systems. For example, the red/far-red–absorbing pigment, phytochrome, has now been demonstrated in

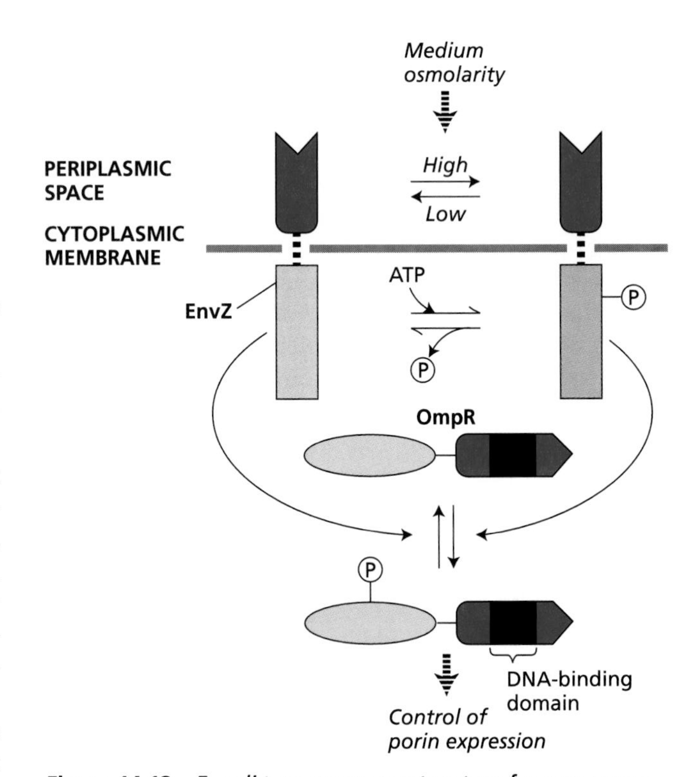

Figure 14.12 *E. coli* two-component system for osmoregulation. When the osmolarity of the medium is high, the membrane sensor protein, EnvZ (in the form of a dimer), acts as an autophosphorylating histidine kinase. The phosphorylated EnvZ then phosphorylates the response regulator, OmpR, which has a DNA-binding domain. Phosphorylated OmpR binds to the promoters of the two porin genes, *ompC* and *ompF*, enhancing expression of the former and repressing expression of the latter. When the osmolarity of the medium is low, EnvZ acts as a protein phosphatase instead of a kinase and dephosphorylates OmpR. When the nonphosphorylated form of OmpR binds to the promoters of the two porin genes, *ompC* expression is repressed and *ompF* expression is stimulated. (From Parkinson 1993.)

cyanobacteria, and it appears to be related to bacterial sensor proteins (see Chapter 17). In addition, the genes that encode putative receptors for two plant hormones, cytokinin and ethylene, both contain autophosphorylating histidine kinase domains, as well as contiguous response regulator motifs. These proteins will be discussed further in Chapters 21 and 22.

Signal Transduction in Eukaryotes

Many eukaryotic microorganisms use chemical signals in cell–cell communication. For example, in the slime mold *Dictyostelium*, starvation induces certain cells to secrete cyclic AMP (cAMP). The secreted cAMP diffuses across the substrate and induces nearby cells to aggregate into a sluglike colony. Yeast mating-type factors are another example of chemical communication between the cells of simple microorganisms. Around a billion years ago, however, cell signaling took a great leap in complexity when eukaryotic cells began to associate together as multicellular organisms. After the evolution of multicellularity came a trend toward ever-increasing cell specialization, as well as the development of tissues and organs to perform specific functions.

Coordination of the development and environmental responses of complex multicellular organisms required an array of signaling mechanisms. Two main systems evolved in animals: the nervous system and the endocrine system. Plants, lacking motility, never developed a nervous system, but they did evolve hormones as chemical messengers. As photosynthesizing organisms, plants also evolved mechanisms for adapting their growth and development to the amount and quality of light.

In the sections that follow we will explore some of the basic mechanisms of signal transduction in animals, emphasizing pathways that may have some parallel in plants. However, keep in mind that plant signal transduction pathways may differ in significant ways from those of animals. To illustrate this point, we end the chapter with an overview of some of the known plant-specific transmembrane receptors.

Two Classes of Signals Define Two Classes of Receptors

Hormones fall into two classes based on their ability to move across the plasma membrane: *lipophilic hormones*, which diffuse readily across the hydrophobic bilayer of the plasma membrane; and *water-soluble hormones*, which are unable to enter the cell. Lipophilic hormones bind mainly to receptors in the cytoplasm or nucleus; water-soluble hormones bind to receptors located on the cell surface. In either case, ligand binding alters the receptor, typically by causing a conformational change.

Some receptors, such as the steroid hormone receptors (see the next section), can regulate gene expression directly. In the vast majority of cases, however, the receptor initiates one or more sequences of biochemical reactions that connect the stimulus to a cellular response. Such a sequence of reactions is called a **signal transduction pathway**. Typically, the end result of signal transduction pathways is to regulate transcription factors, which in turn regulate gene expression.

Signal transduction pathways often involve the generation of **second messengers**, transient secondary signals inside the cell that greatly amplify the original signal. For example, a single hormone molecule might lead to the activation of an enzyme that produces hundreds of molecules of a second messenger. Among the most common second messengers are 3′,5′-cyclic AMP (cAMP); 3′,5′-cyclic GMP (cGMP); 1,2-diacylglycerol (DAG); inositol 1,4,5-trisphosphate (IP_3); and Ca^{2+} (Figure 14.13). Hormone binding normally causes elevated levels of one or more of these second messengers, resulting in the activation or inactivation of enzymes or regulatory proteins. Protein kinases and phosphatases are nearly always involved.

Most Steroid Receptors Act as Transcription Factors

The steroid hormones, thyroid hormones, retinoids, and vitamin D all pass freely across the plasma membrane because of their hydrophobic nature and they bind to

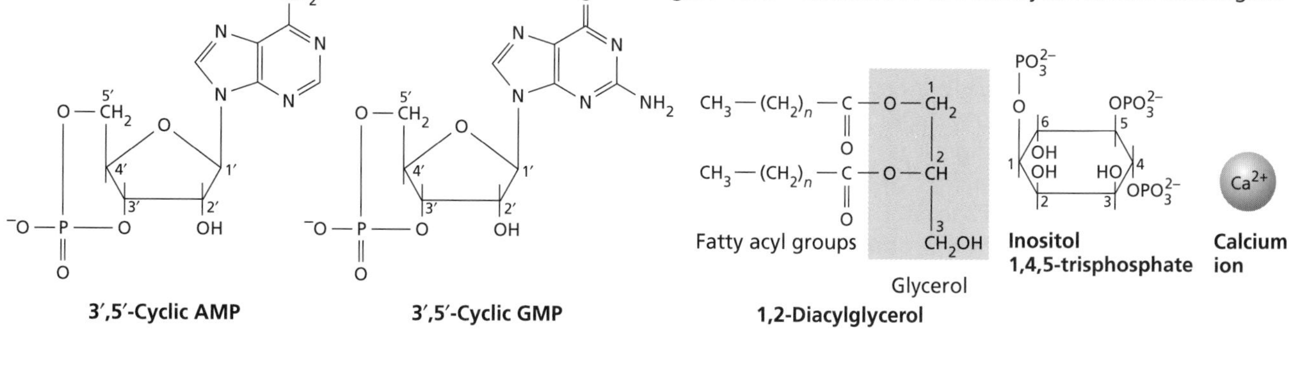

Figure 14.13 Structure of five eukaryotic second messengers.

intracellular receptor proteins. When activated by binding to their ligand, these proteins function as transcription factors. All such steroid receptor proteins have similar DNA-binding domains. Steroid response elements are typically located in enhancer regions of steroid-stimulated genes. Most steroid receptors are localized in the nucleus, where they are anchored to nuclear proteins in an inactive form.

When the receptor binds to the steroid, it is released from the anchor protein and becomes activated as a transcription factor. The activated transcription factor then binds to the enhancer and stimulates transcription. The receptor for thyroid hormone deviates from this pattern in that it is already bound to the DNA but is unable to stimulate transcription in the absence of the hormone. Binding to the hormone converts the receptor to an active transcription factor.

Not all intracellular steroid receptors are localized in the nucleus. The receptor for glucocorticoid hormone (cortisol) differs from the others in that it is located in the cytosol, anchored in an inactive state to a cytosolic protein. Binding of the hormone causes the release of the receptor from its cytosolic anchor, and the receptor–hormone complex then migrates into the nucleus, where it binds to the enhancer and stimulates transcription (Figure 14.14).

Although most studies on animal steroid hormones have focused on their roles in regulating gene expression via receptors that act as transcription factors, increasing evidence suggests that steroids can also interact with proteins on the cell surface (McEwen 1991). As will be discussed in Chapter 17, **brassinosteroid** has recently been demonstrated to be an authentic steroid hormone in plants, and the gene for a **brassinosteroid receptor** has recently been cloned and sequenced. It encodes a type of transmembrane receptor called a *leucine-rich repeat receptor*, which is described at the end of this chapter.

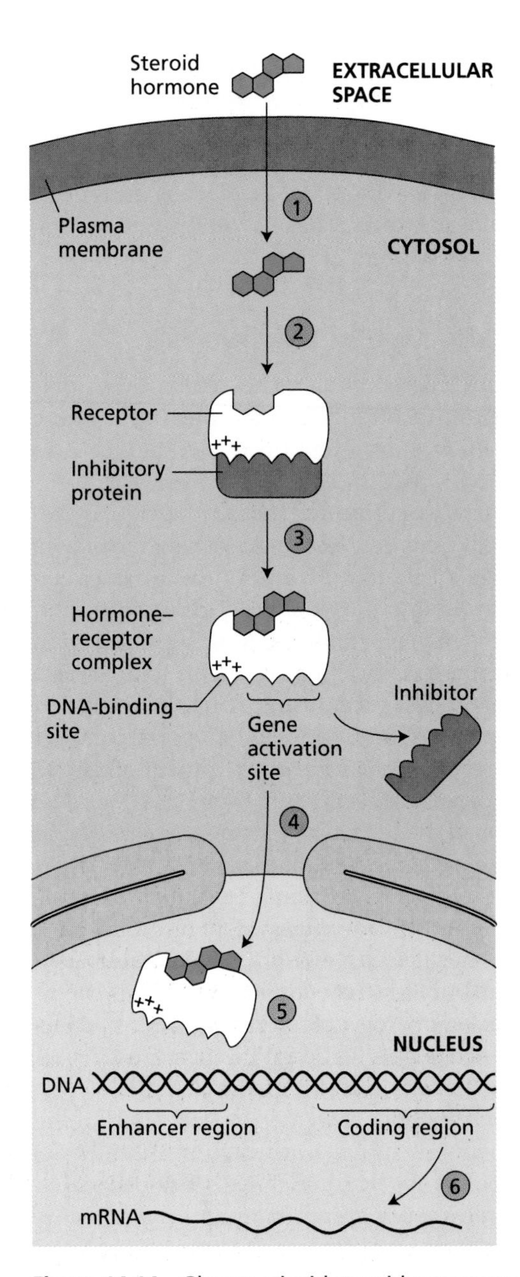

Figure 14.14 Glucocorticoid steroid receptors are transcription factors. (1) Glucocorticoid hormone is lipophilic and diffuses readily through the membrane to the cytosol. (2) Once in the cytosol, the hormone binds to its cytosolic receptor, (3) causing the release of an inhibitory protein from the receptor. (4) The activated receptor then diffuses into the nucleus. (5) In the nucleus, the receptor–hormone complex binds to the enhancer regions of steroid-regulated genes. (6) Transcription of the genes is stimulated. (From Becker et al. 1996.)

Cell Surface Receptors Can Interact with G Proteins

All water-soluble mammalian hormones bind to cell surface receptors. Members of the largest class of mammalian cell surface receptors interact with signal-transducing, GTP-binding regulatory proteins called **heterotrimeric G proteins**. The activated G proteins, in turn, activate an **effector enzyme**. The activated effector enzyme generates an intracellular second messenger, which stimulates a variety of cellular processes.

Receptors using heterotrimeric G proteins are structurally similar and functionally diverse. Their overall structure is similar to that of bacteriorhodopsin, the purple pigment involved in photosynthesis in bacteria of the genus *Halobacterium*, and to that of rhodopsin, the visual pigment of the vertebrate eye. The recently characterized olfactory receptors of the vertebrate nose also belong to this group. The receptor proteins consist of **seven transmembrane α helices** (Figure 14.15). These receptors are sometimes referred to as *seven-spanning, seven-pass*, or *serpentine* receptors.

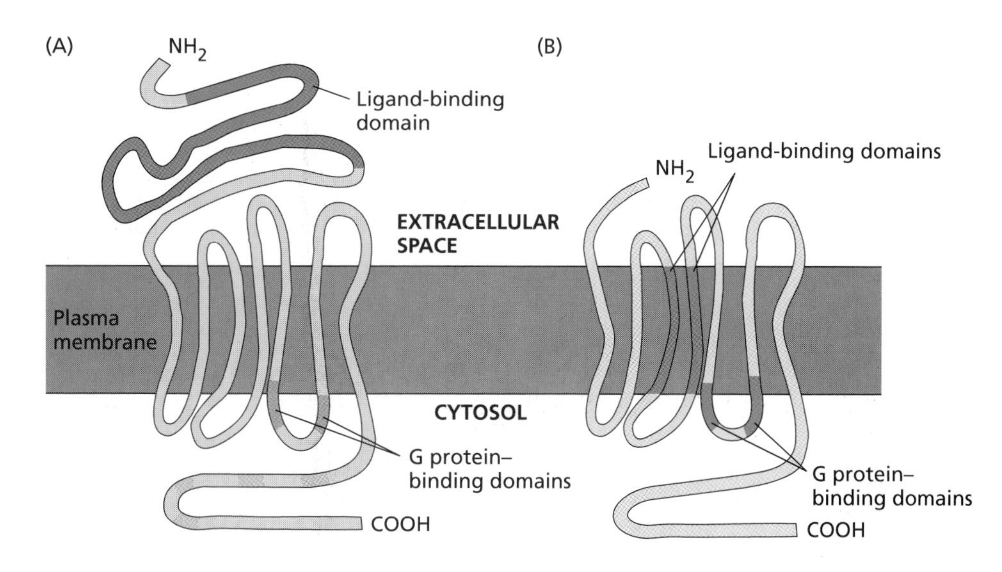

Figure 14.15 Schematic drawing of two types of seven-spanning receptors. (A) Large extracellular ligand-binding domains are characteristic of seven-spanning receptors that bind proteins. The region of the intracellular domain that interacts with the heterotrimeric G protein is indicated. (B) Small extracellular domains are characteristic of seven-spanning receptors that bind to small ligands such as epinephrine. The ligand-binding site is usually formed by several of the transmembrane helices within the bilayer. (After Alberts et al. 1994.)

Heterotrimeric G Proteins Cycle between Active and Inactive Forms

The G proteins that transduce the signals from the seven-spanning receptors are called *heterotrimeric* G proteins because they are composed of three different subunits: α, β, and γ (gamma). They are distinct from the monomeric G proteins, which will be discussed later. Heterotrimeric G proteins cycle between active and inactive forms, thus acting as molecular switches. The β and γ subunits form a tight complex that anchors the trimeric G protein to the membrane on the cytoplasmic side (Figure 14.16). The G protein becomes activated upon binding to the ligand-activated seven-spanning receptor. In its inactive form, G exists as a trimer with GDP bound to the α subunit. Binding to the receptor–ligand complex induces the α subunit to exchange GDP for GTP. This exchange causes the α subunit to dissociate from β and γ, allowing α to associate instead with an effector enzyme.

The α subunit has a GTPase activity that is activated when it binds to the effector enzyme, in this case **adenylyl cyclase** (also called *adenylate cyclase*) (see Figure 14.16). GTP is hydrolyzed to GDP, thereby inactivating the α subunit, which in turn inactivates adenylyl cyclase. The α subunit bound to GDP reassociates with the β and γ subunits and can then be reactivated by associating with the hormone–receptor complex.

Activation of Adenylyl Cyclase Increases the Level of Cyclic AMP

Cyclic AMP is an important signaling molecule in both prokaryotes and animal cells, and increasing evidence suggests that cAMP plays a similar role in plant cells. In vertebrates, adenylyl cyclase is an integral membrane protein that contains two clusters of six membrane-spanning domains separating two catalytic domains that extend into the cytoplasm. Activation of adenylyl cyclase by heterotrimeric G proteins raises the concentration of cAMP in the cell, which is normally maintained at a low level by the action of **cyclic AMP phosphodiesterase**, which hydrolyzes cAMP to 5′-AMP.

Nearly all the effects of cAMP in animal cells are mediated by the enzyme **protein kinase A (PKA)**. In unstimulated cells, PKA is in the inactive state because of the presence of a pair of inhibitory subunits. Cyclic AMP binds to these inhibitory subunits, causing them to dissociate from the two catalytic subunits, thereby activating the catalytic subunits. The activated catalytic subunits then are able to phosphorylate specific serine or threonine residues of selected proteins, which may also be protein kinases. An example of an enzyme that is phosphorylated by PKA is *glycogen phosphorylase kinase*. When phosphorylated by PKA, glycogen phosphorylase kinase phosphorylates (activates) *glycogen phosphorylase*, the enzyme that breaks down glycogen in muscle cells to glucose-1-phosphate.

In cells in which cAMP regulates gene expression, PKA phosphorylates a transcription factor called **CREB** (*c*yclic AMP *r*esponse *e*lement–*b*inding protein). Upon activation by PKA, CREB binds to the cAMP response element (CRE), which is located in the promoter regions of genes that are regulated by cAMP.

In addition to activating PKA, cAMP can interact with specific cAMP-gated cation channels. For example, in olfactory receptor neurons, cAMP binds to and opens Na^+ channels on the plasma membrane, resulting in Na^+ influx and membrane depolarization.

Because of the extremely low levels of cyclic AMP that have been detected in plant tissue extracts, the role of cAMP in plant signal transduction has been highly controversial (Assmann 1995). Nevertheless, various

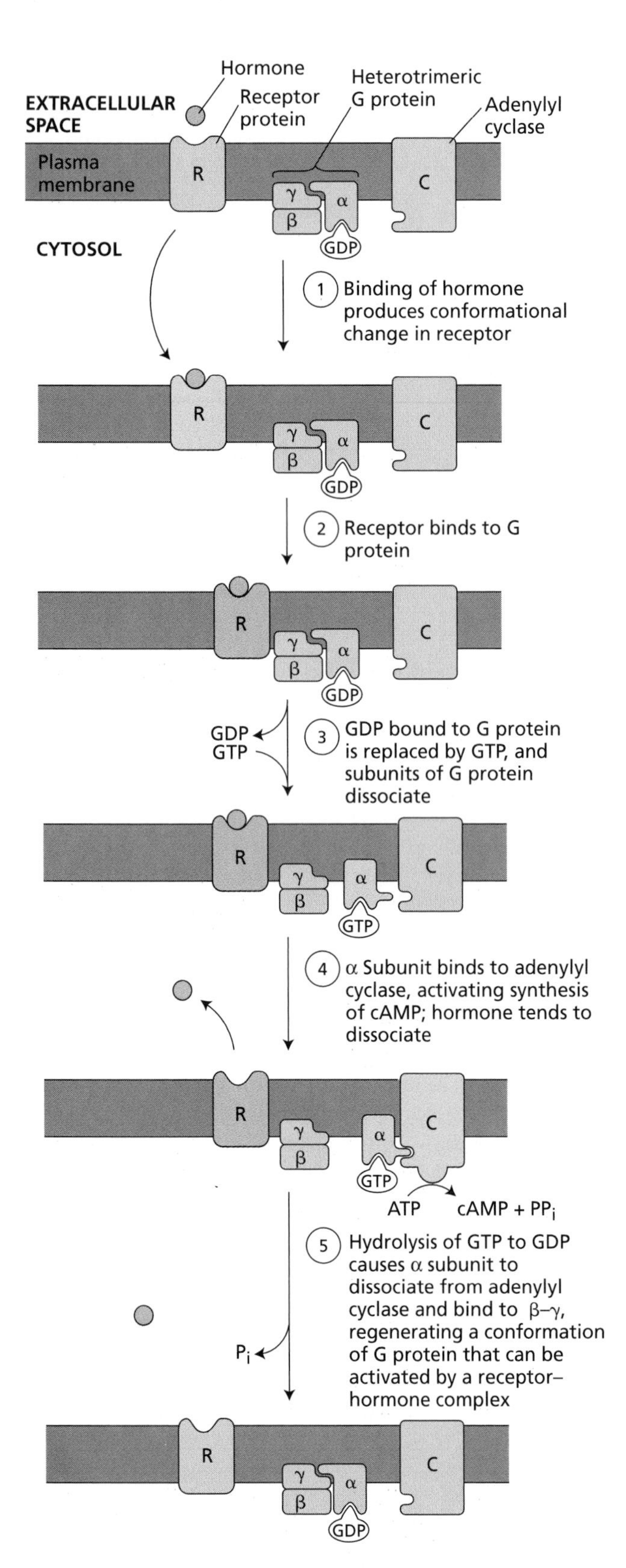

Figure 14.16 Hormone-induced activation of an effector enzyme is mediated by the α subunit of a heterotrimeric G protein. (1) Upon binding to its hormonal ligand, the seven-spanning receptor undergoes a conformational change. (2) The receptor binds to the heterotrimeric G protein. (3) Contact with the receptor induces the α subunit of the heterotrimeric G protein to exchange GDP for GTP, and the α subunit then dissociates from the complex. (4) The G protein α subunit associates with the effector protein (adenylyl cyclase) in the membrane, causing its activation. At the same time the hormone is released from its receptor. (5) The effector enzyme becomes inactivated when GTP is hydrolyzed to GDP. The α subunit then reassociates with the heterotrimeric G protein and is ready to be reactivated by a second hormonal stimulus. (From Lodish et al. 1995.)

lines of evidence supporting a role of cAMP in plant cells have accumulated. For example, genes that encode homologs of CREB have been identified in plants (Kategiri et al. 1989). Pollen tube growth in lily has been shown to be stimulated by concentrations of cAMP as low as 10 n*M* (Tezuka et al. 1993). Li and colleagues (1994) showed that cAMP activates K^+ channels in the plasma membrane of fava bean (*Vicia faba*) mesophyll cells. And Ichikawa and coworkers (1997) recently identified possible genes for adenylyl cyclase in tobacco (*Nicotiana tabacum*) and *Arabidopsis*. Thus, despite years of doubt, the role of cAMP as a universal signaling agent in living organisms, including plants, seems likely.

Activation of Phospholipase C Initiates the IP_3 Pathway

Calcium serves as a second messenger for a wide variety of cell signaling events. This role of calcium is well established in animal cells, and as we will see in later chapters, circumstantial evidence suggests a role for calcium in signal transduction in plants as well. The concentration of free Ca^{2+} in the cytosol normally is maintained at extremely low levels (1×10^{-7} *M*). Ca^{2+}-ATPases on the plasma membrane and on the endoplasmic reticulum pump calcium ions out of the cell and into the lumen of the ER, respectively. In plant cells, most of the calcium of the cell accumulates in the vacuole. The proton electrochemical gradient across the vacuolar membrane that is generated by tonoplast proton pumps drives calcium uptake via Ca^{2+}–H^+ antiporters (see Chapter 6).

In animal cells, certain hormones can induce a transient rise in the cytosolic Ca^{2+} concentration to about 5×10^{-6} *M*. This increase may occur even in the absence of extracellular calcium, indicating that the Ca^{2+} is being released from intracellular compartments by the opening of intracellular calcium channels. However, the coupling of hormone binding to the opening of intracellular calcium channels is mediated by yet another second messenger, inositol trisphosphate (IP_3).

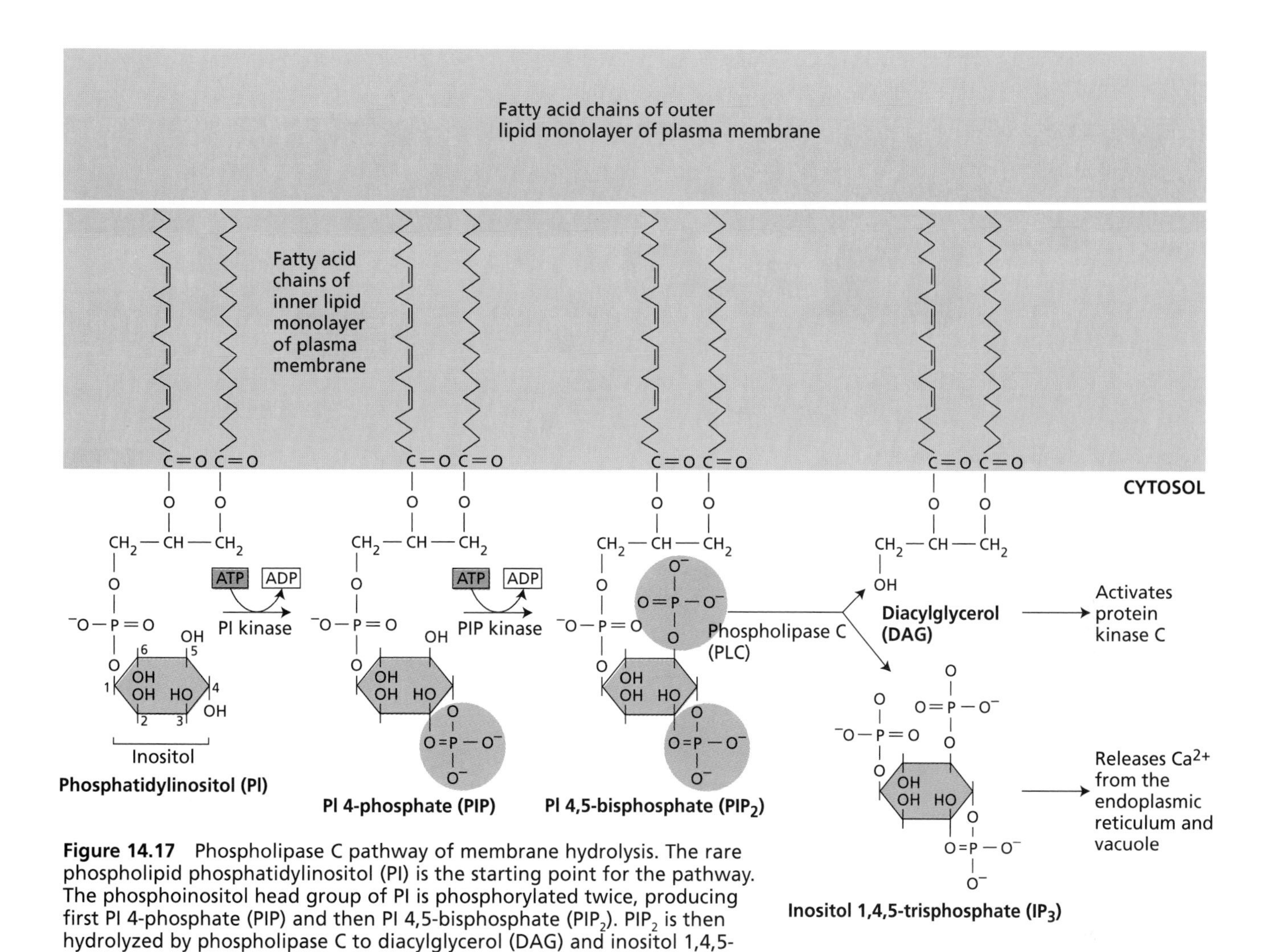

Figure 14.17 Phospholipase C pathway of membrane hydrolysis. The rare phospholipid phosphatidylinositol (PI) is the starting point for the pathway. The phosphoinositol head group of PI is phosphorylated twice, producing first PI 4-phosphate (PIP) and then PI 4,5-bisphosphate (PIP_2). PIP_2 is then hydrolyzed by phospholipase C to diacylglycerol (DAG) and inositol 1,4,5-trisphosphate (IP_3). (After Alberts et al. 1994.)

Phosphatidylinositol (PI) is a minor phospholipid component of cell membranes (see Chapter 11). PI can be converted to the polyphosphoinositides PI phosphate (PIP) and **PI bisphosphate (PIP_2)** by kinases (Figure 14.17). Although PIP_2 is even less abundant in the membrane than PI is, it plays a central role in signal transduction. In animal cells, binding of a hormone, such as vasopressin, to its receptor leads to the activation of heterotrimeric G proteins. The α subunit then dissociates from G and activates a phosphoinositide-specific phospholipase, **phospholipase C (PLC)**. The activated PLC rapidly hydrolyzes PIP_2, generating **inositol trisphosphate (IP_3)** and **diacylglycerol (DAG)** as products. Each of these two molecules plays an important role in cell signaling.

IP_3 Opens Calcium Channels on the ER and on the Tonoplast

The IP_3 generated by the activated phospholipase C is water soluble and diffuses through the cytosol until it encounters IP_3-binding sites on the ER and (in plants) on the tonoplast. These binding sites are IP_3-gated Ca^{2+} channels that open when they bind IP_3 (Figure 14.18). Since these organelles maintain internal Ca^{2+} concentrations in the millimolar range, calcium diffuses rapidly into the cytosol down a steep concentration gradient. The response is terminated when IP_3 is broken down by specific phosphatases or when the released calcium is pumped out of the cytoplasm by Ca^{2+}-ATPases.

Studies with Ca^{2+}-sensitive fluorescent indicators, such as fura-2 and aequorin, have shown that the calcium signal often originates in a localized region of the cell and propagates as a wave throughout the cytosol. Repeated waves called *calcium oscillations* can follow the original signal, each lasting from a few seconds to several minutes. The biological significance of calcium oscillation is still unclear, although it has been suggested that it is a mechanism for avoiding the toxicity that might result from a sustained elevation in cytosolic levels of free calcium. Such wavelike oscillations have

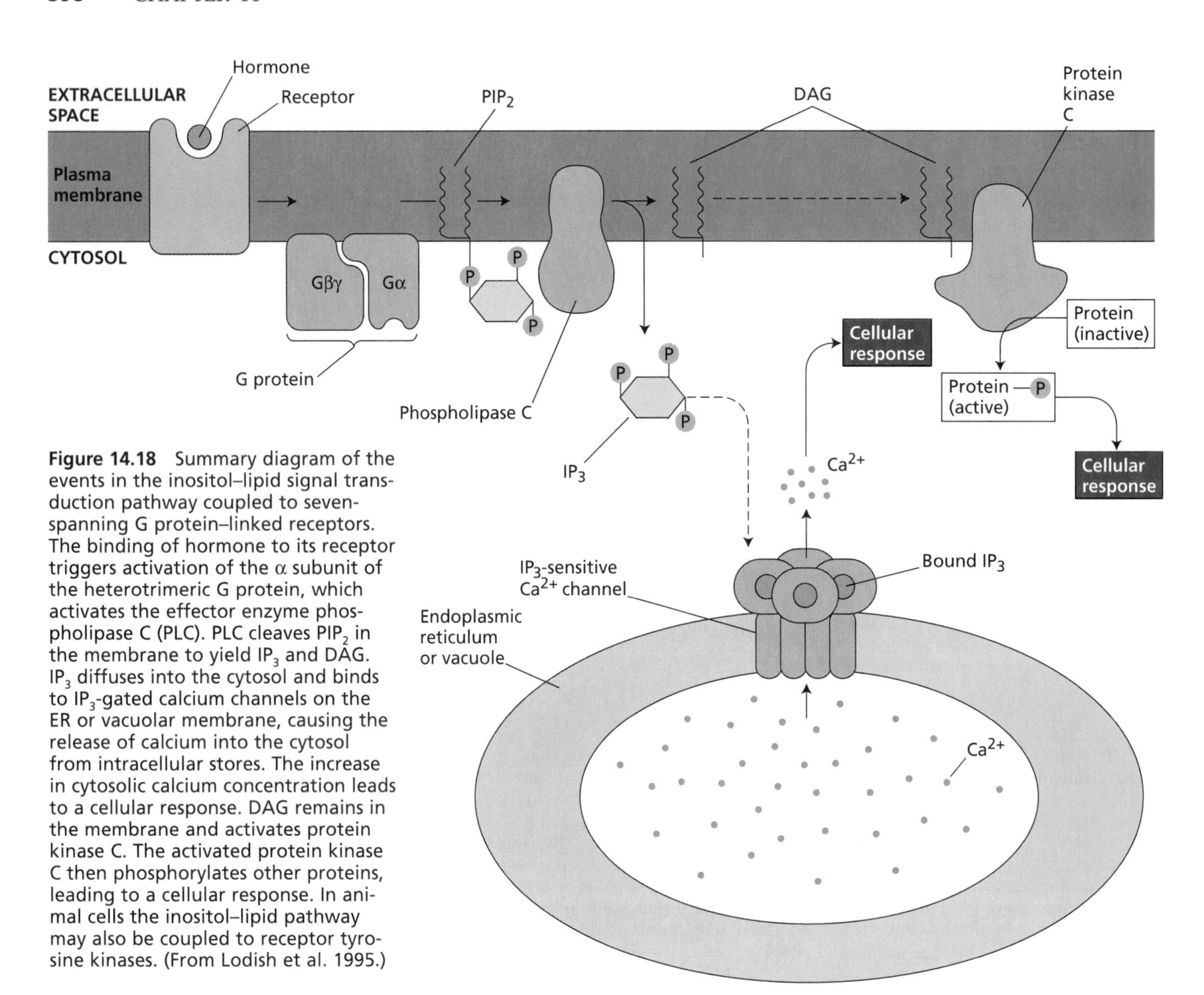

Figure 14.18 Summary diagram of the events in the inositol–lipid signal transduction pathway coupled to seven-spanning G protein–linked receptors. The binding of hormone to its receptor triggers activation of the α subunit of the heterotrimeric G protein, which activates the effector enzyme phospholipase C (PLC). PLC cleaves PIP_2 in the membrane to yield IP_3 and DAG. IP_3 diffuses into the cytosol and binds to IP_3-gated calcium channels on the ER or vacuolar membrane, causing the release of calcium into the cytosol from intracellular stores. The increase in cytosolic calcium concentration leads to a cellular response. DAG remains in the membrane and activates protein kinase C. The activated protein kinase C then phosphorylates other proteins, leading to a cellular response. In animal cells the inositol–lipid pathway may also be coupled to receptor tyrosine kinases. (From Lodish et al. 1995.)

recently been detected in plant stomatal guard cells (McAinsh et al. 1995).

Some Protein Kinases Are Activated by Calcium–Calmodulin Complexes

As we have seen with IP_3-gated channels, calcium can activate some proteins, such as channels, by binding directly to them. However, most of the effects of calcium result from the binding of calcium to the regulatory protein **calmodulin** (Figure 14.19). Calmodulin is a highly conserved protein that is abundant in all eukaryotic cells, but it appears to be absent from prokaryotic cells. The same calcium-binding site is found in a wide variety of calcium-binding proteins and is called an EF hand. The name is derived from the two α helices, E and F, that are part of the calcium-binding domain of the protein parvalbumin (Kretsinger 1980).

Each calmodulin molecule binds four Ca^{2+} ions and changes conformation, enabling it to bind to and activate other proteins. The Ca^{2+}–calmodulin complex can stimulate some enzymes directly, such as the plasma membrane Ca^{2+}-ATPase, which pumps calcium out of the cell. Most of the effects of calcium, however, are brought about by activation of **Ca^{2+}–calmodulin-dependent protein kinases (CaM kinases)**. CaM kinases phosphorylate serine or threonine residues of their target enzymes, causing enzyme activation. Thus, the effect that calcium has on a particular cell depends to a large extent on which CaM kinases are expressed in that cell.

Calcium signaling has been strongly implicated in many developmental processes in plants, ranging from the regulation of development by phytochrome (see Chapter 17) to the regulation of stomatal guard cells by abscisic acid (see Chapter 23). Thus far, however, there have been few reports of CaM kinase activity in plants. Recently, however, a gene that codes for a CaM kinase has been cloned from lily and shown to be specifically expressed in anthers. The lily CaM kinase is a

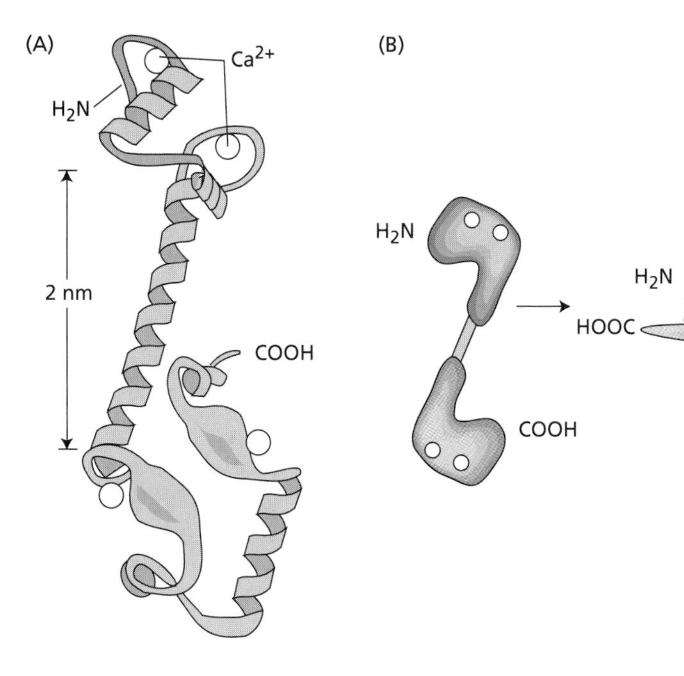

Figure 14.19 Structure of calmodulin. (A) Calmodulin consists of two globular ends separated by a flexible α helix. Each globular end has two calcium-binding sites. (B) When the calcium–calmodulin complex associates with a protein, it literally wraps around it. (From Alberts et al. 1994.)

serine/threonine kinase that phosphorylates various protein substrates in vitro in a Ca^{2+}–calmodulin-dependent manner (Takezawa et al. 1996). The occurrence and regulatory roles of such plant CaM kinases remain to be determined.

Plants Contain Calcium-Dependent Protein Kinases

The most abundant calcium-regulated protein kinases in plants appear to be the **calcium-dependent protein kinases (CDPKs)** (Harper et al. 1991; Roberts and Harmon 1992). CDPKs are strongly activated by calcium, but are insensitive to calmodulin. The proteins are characterized by two domains: a catalytic domain that is similar to those of the animal CaM kinases, and a calmodulin-like domain. The presence of a calmodulin-like domain may explain why the enzyme does not require calmodulin for activity.

CDPKs are widespread in plants and are encoded by multigene families. A CDPK has also been identified in *Chara*, the giant freshwater green alga thought to be a precursor of land plants (McCurdy and Harmon 1992).

In *Chara* the enzyme was shown to be associated with the actin microfilaments that line the outer cortex of the cytoplasm along the inner surface of the plasma membrane. The function of these microfilaments is to drive cytoplasmic streaming around the cell. The rate of cytoplasmic streaming is inhibited by increases in cytosolic calcium, and it has been proposed that CDPKs mediate the effects of calcium by phosphorylating the heavy chain of myosin, a component of the microfilaments (McCurdy and Harmon 1992).

CDPKs may also mediate the effects of calcium in guard cells. Abscisic acid–induced stomatal closure involves calcium as a second messenger (see Chapter 23). Recent studies using isolated vacuoles from guard cells of *Vicia faba* (fava bean) suggest that CDPKs can regulate anion channels on the tonoplast (Pei et al. 1996). Thus, CDPKs may be a component of the abscisic acid signaling pathway.

Diacylglycerol Activates Protein Kinase C

Cleavage of PIP_2 by phospholipase C produces diacylgycerol (DAG) in addition to IP_3 (see Figure 14.17). Whereas IP_3 is hydrophilic and diffuses rapidly into the cytoplasm, DAG is a lipid and remains in the membrane. In animal cells, DAG can associate with and activate the serine/threonine kinase **protein kinase C (PKC)**. The inactive form of PKC is a soluble enzyme that is located in the cytosol. Upon binding to calcium, the soluble, inactive PKC undergoes a conformational change and associates with a PKC receptor protein that transports it to the inner surface of the plasma membrane, where it encounters DAG.

PKCs have been shown to phosphorylate ion channels, transcription factors, and enzymes in animal cells. One of the enzymes phosphorylated by PKC is another protein kinase that regulates cell proliferation and differentiation, *MAP kinase kinase kinase* (discussed later in the chapter). G proteins, phospholipase C, and various protein kinases have been identified in plant membranes (Millner and Causier 1996). PKC activity has also been detected in plants (Elliott and Kokke 1987; Chen et al. 1996), and a plant gene encoding the PKC receptor protein that transports the soluble enzyme to the membrane has recently been cloned (Kwak et al. 1997). However, there is as yet no evidence that activation of PKC by DAG plays a role in plant signal transduction.

Phospholipase A_2 Generates Other Membrane-Derived Signaling Agents

In animals, the **endocrine system** is involved in signaling between hormone-producing cells at one location of the body and hormone-responding cells at another location; in contrast, the **autocrine system** involves cells sending signals to themselves and their immediate

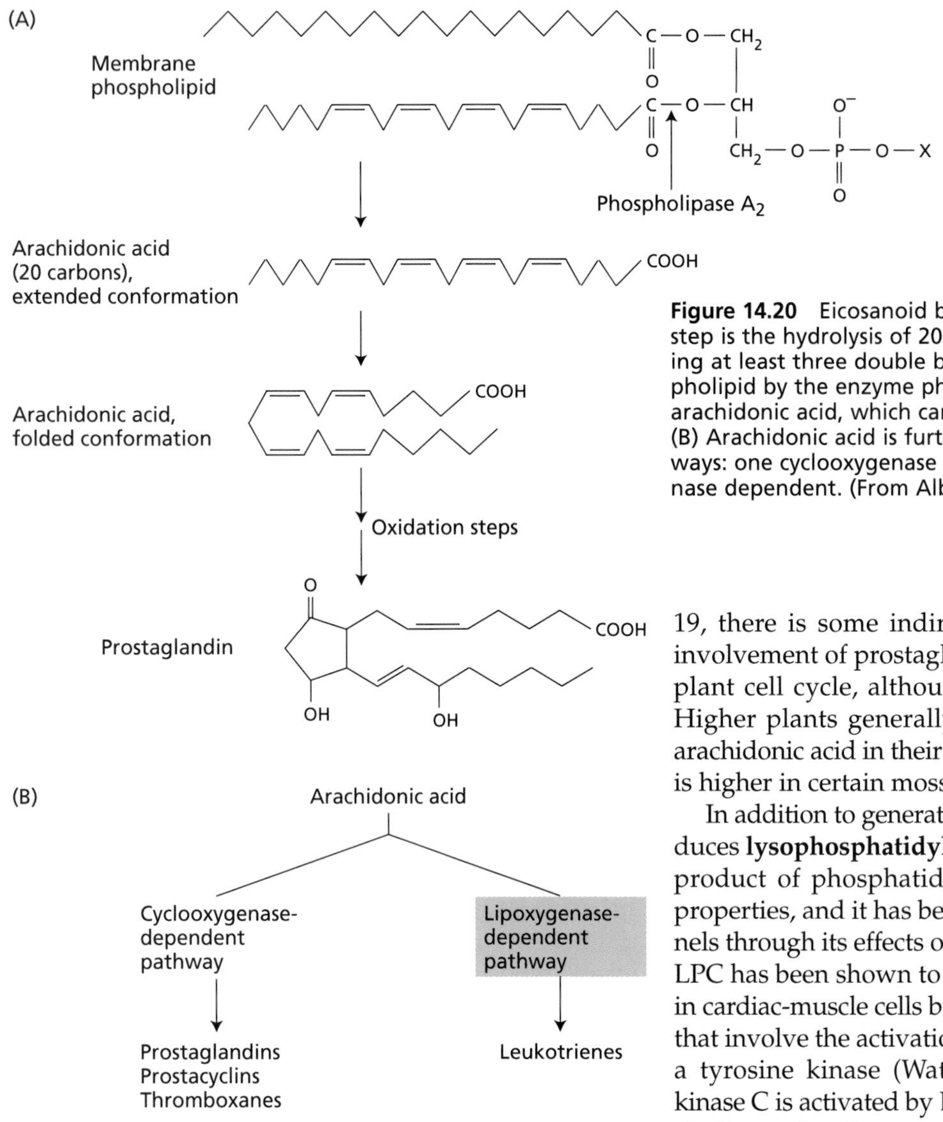

(A)

Membrane phospholipid

C—O—CH$_2$

O

C—O—CH O$^-$

O CH$_2$—O—P—O—X

Phospholipase A_2 O

Arachidonic acid (20 carbons), extended conformation

COOH

Arachidonic acid, folded conformation

COOH

Oxidation steps

O

Prostaglandin

COOH

OH OH

(B)

Arachidonic acid

Cyclooxygenase-dependent pathway

Lipoxygenase-dependent pathway

Prostaglandins
Prostacyclins
Thromboxanes

Leukotrienes

Figure 14.20 Eicosanoid biosynthetic pathway. (A) The first step is the hydrolysis of 20-carbon fatty acid chains containing at least three double bonds from a membrane phospholipid by the enzyme phospholipase A_2, producing arachidonic acid, which can be oxidized by prostaglandin. (B) Arachidonic acid is further metabolized by two pathways: one cyclooxygenase dependent, the other lipoxygenase dependent. (From Alberts et al. 1994.)

neighbors. One type of autocrine signaling system that plays important roles in pain and inflammatory responses, as well as platelet aggregation and smooth-muscle contraction, is called the **eicosanoid pathway**.

There are four major classes of eicosanoids: prostaglandins, prostacyclins, thromboxanes, and leukotrienes. All are derived from the breakdown of membrane phospholipids, and in this respect the eicosanoid pathway resembles the IP_3 pathway. There the resemblance ends, however. For whereas the IP_3 pathway begins with the cleavage of IP_3 from PIP_2 by phospholipase C, the eicosanoid pathway is initiated by the cleavage of the 20-carbon fatty acid **arachidonic acid** from the intact phospholipid by the enzyme **phospholipase A_2** (**PLA_2**) (Figure 14.20A). Two oxidative pathways—one cyclooxygenase dependent, the other lipoxygenase dependent—then convert arachidonic acid to the four eicosanoids (Figure 14.20B). As we will see in Chapter 19, there is some indirect evidence for the possible involvement of prostaglandins in the regulation of the plant cell cycle, although direct evidence is lacking. Higher plants generally have negligible amounts of arachidonic acid in their membranes, although the level is higher in certain mosses.

In addition to generating arachidonic acid, PLA_2 produces **lysophosphatidylcholine** (**LPC**) as a breakdown product of phosphatidylcholine. LPC has detergent properties, and it has been shown to regulate ion channels through its effects on protein kinases. For example, LPC has been shown to modulate the sodium currents in cardiac-muscle cells by signal transduction pathways that involve the activation of both protein kinase C and a tyrosine kinase (Watson and Gold 1997). Protein kinase C is activated by LPC independently of the phospholipase C pathway.

In recent years plant biologists have become increasingly interested in the eicosanoid pathway because it now appears that an important signaling agent in plant defense responses, **jasmonic acid**, is produced by a similar pathway, which was described in Chapter 13. In addition, LPC has been shown to activate plant protein kinases in vitro. As we will see in Chapter 19, LPC is one of many candidates for a second messenger in the rapid responses of plant cells to auxin.

In Vertebrate Vision, a Heterotrimeric G Protein Activates Cyclic GMP Phosphodiesterase

The human eye contains two types of photoreceptor cells: rods and cones. Rods are responsible for monochromatic vision in dim light; cones are involved in color vision in bright light. Signal transduction in response to light has been studied more intensively in rods. The rod is a highly specialized tubular cell that contains an elongated stack of densely packed membrane sacs called **discs** at the tip, or **outer segment**, reminiscent of the grana stacks of chloroplasts. The disc membranes of rod cells contain the photosensitive protein pigment rhodopsin, a member of the seven-spanning transmembrane family of receptors. **Rhodopsin** consists of the protein opsin covalently bound to the light-absorbing molecule **11-*cis*-retinal**. When 11-*cis*-retinal absorbs a single photon of light (400 to 600 nm) it immediately isomerizes to **all-*trans*-retinal** (Figure 14.21). This change causes a slower conformational change in the protein, converting it to *meta*-rhodopsin II, or **activated opsin**.

Activated opsin, in turn, lowers the concentration of the cyclic nucleotide **3′5′-cGMP**. Cyclic GMP is synthesized from GTP by the enzyme guanylate cyclase. In the dark, guanylate cyclase activity results in the buildup of a high concentration of cGMP in the rod cells. Because the plasma membrane contains cGMP-gated Na^+ channels, the high cGMP concentration in the cytosol maintains the Na^+ channels in the open position in the absence of light. When the Na^+ channels are open, Na^+ can enter the cell freely, and this passage of Na^+ tends to depolarize the membrane potential.

When opsin becomes activated by light, however, it binds to the heterotrimeric G protein **transducin**. This binding causes the α subunit of transducin to exchange GDP for GTP and dissociate from the complex. The α subunit of transducin then activates the enzyme cyclic GMP phosphodiesterase, which breaks down 3′5′-cGMP to 5′-GMP (Figure 14.22). Light therefore has the effect of decreasing the concentration of cGMP in the rod cell. A lower concentration of cGMP has the effect of closing the cGMP-gated Na^+ channels on the plasma membrane, which are kept open in the dark by a high cGMP concentration. To give some idea of the signal amplification provided, a single photon may cause the closure of hundreds of Na^+ channels, blocking the uptake of about 10 million Na^+ ions.

By preventing the influx of Na^+, which tends to depolarize the membrane, the membrane polarity increases—that is, becomes hyperpolarized. In this way a light signal is converted into an electric signal. Membrane hyperpolarization, in turn, inhibits neurotransmitter release from the synaptic body of the rod cell. Paradoxically, the nervous system detects light as an *inhibition* rather than a stimulation of neurotransmitter release.

Cyclic GMP, which regulates ion channels and protein kinases in animal cells, appears to be an important regulatory molecule in plant cells as well. Cyclic GMP has been definitively identified in plant extracts by gas chromatography combined with mass spectrometry (Janistyn 1983; Newton and Brown 1992). Moreover, cGMP has been implicated as a second messenger in the responses of phytochrome (see Chapter 17) and gibberellin (see Chapter 20).

Cell Surface Receptors May Have Catalytic Activity

Some cell surface receptors are enzymes themselves or are directly associated with enzymes. Unlike the seven-

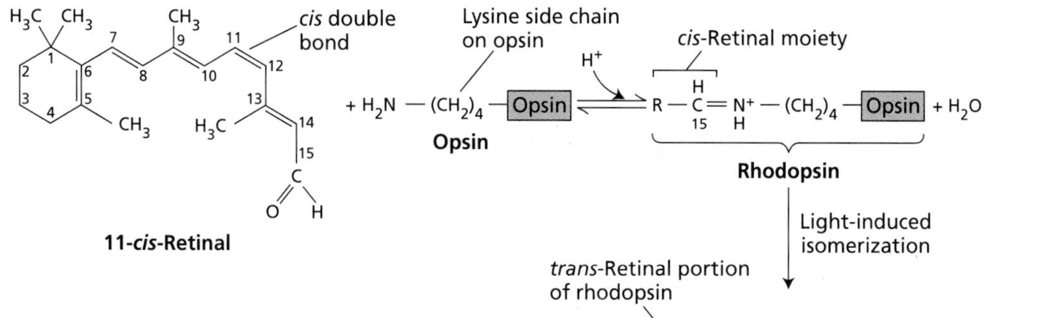

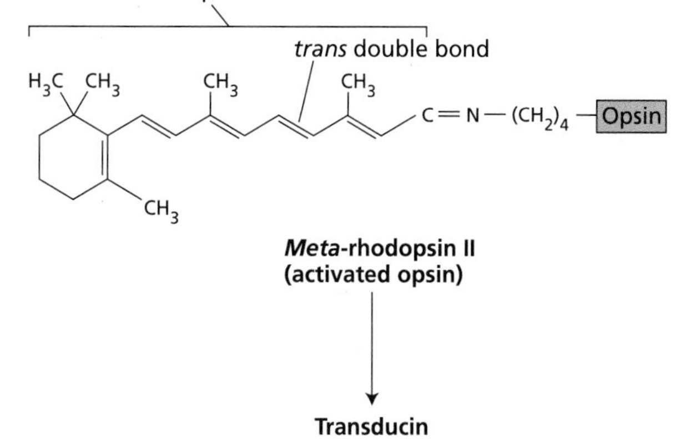

Figure 14.21 Transduction of the light signal in vertebrate vision. The photoreceptor pigment is rhodopsin, a transmembrane protein composed of the protein opsin and the chromophore 11-*cis*-retinal. Light absorption causes the rapid isomerization of *cis*-retinal to *trans*-retinal. The formation of *trans*-retinal then causes a conformational change in the protein opsin, forming *meta*-rhodopsin II, the activated form of opsin. The activated opsin then interacts with the heterotrimeric G protein transducin. (After Lodish et al. 1995.)

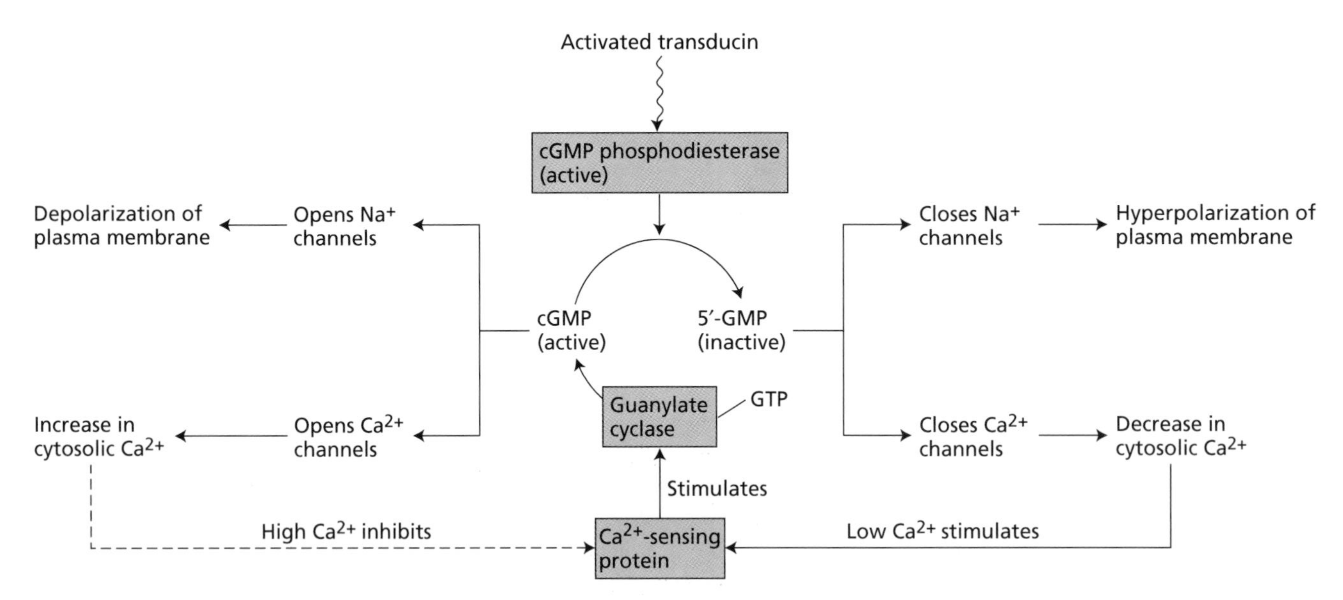

Figure 14.22 The role of cyclic GMP (cGMP) and calcium as second messengers in vertebrate vision. Activation of the heterotrimeric G protein transducin by activated opsin causes the activation of cGMP phosphodiesterase, which lowers the concentration of cGMP in the cell. The reduction in cGMP closes cGMP-activated Na^+ channels. Closure of the Na^+ channels blocks the influx of Na^+, causing membrane hyperpolarization. Cyclic GMP also regulates calcium channels. When the cGMP concentration in the cell is high, the calcium channels open, raising the cytosolic calcium concentration. Guanylate cyclase, the enzyme that synthesizes cGMP from GTP, is inhibited by high levels of calcium. Conversely, when cGMP levels are low, closure of calcium channels lowers the cytosolic calcium concentration. This lowering of the calcium concentration stimulates guanylate cyclase. Calcium thus provides a feedback system for regulating cGMP levels in the cell.

spanning receptors, the **catalytic receptors**, as these enzyme or enzyme-associated receptors are called, are typically attached to the membrane via a single transmembrane helix and do not interact with heterotrimeric G proteins. The six main categories of catalytic receptors in animals include: (1) receptor tyrosine kinases, (2) receptor tyrosine phosphatases, (3) receptor serine/threonine kinases, (4) tyrosine kinase–linked receptors, (5) receptor guanylate cyclases, and (6) cell surface proteases. Of these, the receptor tyrosine kinases are probably the most abundant in animal cells.

Thus far, no receptor tyrosine kinases (RTKs) have been identified in plants. However, plant cells do contain a class of receptors called receptorlike kinases (RLKs) that are structurally similar to the animal RTKs. In addition, some of the components of the RTK signaling pathway of animals have been identified in plants. After first reviewing the animal RTK pathway, we will examine the RLK receptors of plants.

Ligand Binding to Receptor Tyrosine Kinases Induces Autophosphorylation

The **receptor tyrosine kinases** (**RTKs**) make up the most important class of enzyme-linked cell surface receptors in animal cells, although so far they have not been found in either plants or fungi. Their ligands are soluble or membrane-bound peptide or protein hormones, including insulin, epidermal growth factor (EGF), platelet-derived growth factor (PDGF), and several other protein growth factors.

Since the transmembrane domain that separates the hormone-binding site on the outer surface of the membrane from the catalytic site on the cytoplasmic surface consists of only a single α helix, the hormone cannot transmit a signal directly to the cytosolic side of the membrane via a conformational change. Rather, binding of the ligand to its receptor induces dimerization of adjacent receptors, which allows the two catalytic domains to come into contact and phosphorylate each other on multiple tyrosine residues (autophosphorylation) (Figure 14.23). Dimerization may be a general mechanism for activating cell surface receptors that contain single transmembrane domains.

Intracellular Signaling Proteins That Bind to RTKs Are Activated by Phosphorylation

Once autophosphorylated, the catalytic site of the RTKs binds to a variety of cytosolic signaling proteins. After binding to the RTK, the inactive signaling protein is itself phosphorylated on specific tyrosine residues. Some transcription factors are activated in this way, after which they migrate to the nucleus and stimulate gene expression directly. Other signaling molecules take part in a signaling cascade that ultimately results in the activation of

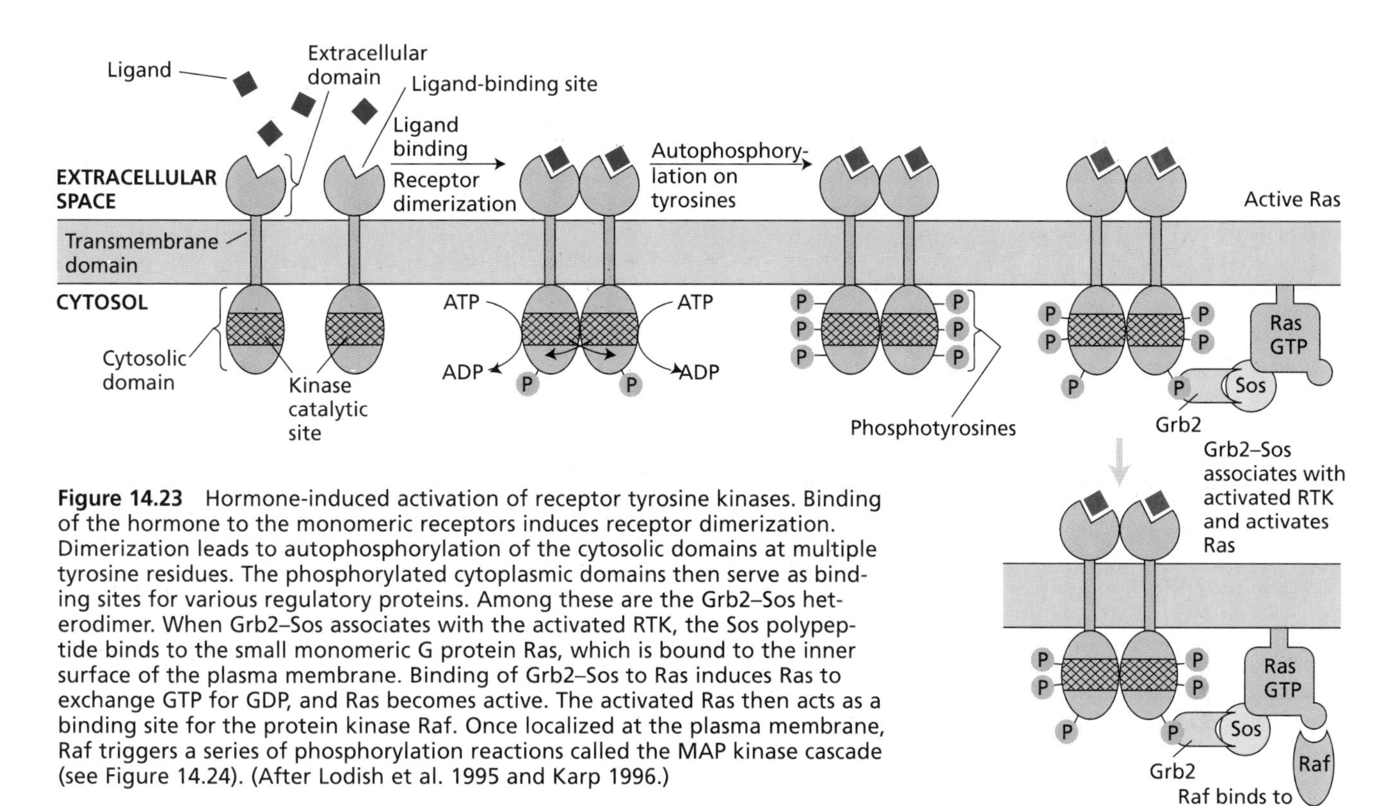

Figure 14.23 Hormone-induced activation of receptor tyrosine kinases. Binding of the hormone to the monomeric receptors induces receptor dimerization. Dimerization leads to autophosphorylation of the cytosolic domains at multiple tyrosine residues. The phosphorylated cytoplasmic domains then serve as binding sites for various regulatory proteins. Among these are the Grb2–Sos heterodimer. When Grb2–Sos associates with the activated RTK, the Sos polypeptide binds to the small monomeric G protein Ras, which is bound to the inner surface of the plasma membrane. Binding of Grb2–Sos to Ras induces Ras to exchange GTP for GDP, and Ras becomes active. The activated Ras then acts as a binding site for the protein kinase Raf. Once localized at the plasma membrane, Raf triggers a series of phosphorylation reactions called the MAP kinase cascade (see Figure 14.24). (After Lodish et al. 1995 and Karp 1996.)

transcription factors. The signaling cascade initiated by RTKs begins with the small, monomeric G protein Ras.

The Ras superfamily. In addition to possessing heterotrimeric G proteins, eukaryotic cells contain small **monomeric G proteins** that are related to the α subunits of the heterotrimeric G proteins. The three families, Ras, Rab, and Rho/Rac, all belong to the **Ras superfamily of monomeric GTPases**. Rho and Rac relay signals from surface receptors to the actin cytoskeleton; members of the Rab family of GTPases are involved in regulating intracellular membrane vesicle traffic; the Ras proteins, which are located on the inner surface of the membrane, play a crucial role in initiating the kinase cascade that relays signals from RTKs to the nucleus.

The *RAS* gene was originally discovered as a viral oncogene (cancer-causing gene) and was later shown to be present as a normal gene in animal cells. Ras is a G protein that cycles between an inactive GDP-binding form and an active GTP-binding form. Ras also possesses GTPase activity that hydrolyzes bound GTP to GDP, thus terminating the response. The *RAS* oncogene is a mutant form of the protein that is unable to hydrolyze GTP. As a result, the molecular switch remains in the on position, triggering uncontrolled cell division.

The study of small GTP-binding proteins in plants is still in its infancy. Thus far, about 30 genes encoding members of monomeric G protein families have been cloned, including homologs of *RAB* and *RHO*. Surprisingly, *RAS* itself has so far not yet been identified in plants (Terryn et al. 1996).

Ras Recruits Raf to the Plasma Membrane

The initial steps in the Ras signaling pathway are illustrated in Figure 14.23. First, binding of the hormone to the RTK induces dimerization followed by autophosphorylation of the catalytic domain. Autophosphorylation of the receptor causes binding to the **Grb2** protein, which is tightly associated with another protein, called **Sos**. As a result, the Grb2–Sos complex attaches to the RTK at the phosphorylation site. The Sos protein then binds to the inactive form of Ras, which is associated with the inner surface of the plasma membrane. Upon binding to Sos, Ras releases GDP and binds GTP instead, which converts Ras to the active form. The activated Ras, in turn, provides a binding site for the soluble serine/threonine kinase **Raf**. The primary function of the activated Ras is thus to recruit Raf to the plasma membrane. Binding to Ras activates Raf and initiates a chain of phosphorylation reactions called the MAPK cascade (see the next section).

As we will see in later chapters, increasing evidence suggests that plant signaling pathways also employ the

MAPK cascade. For example, the ethylene receptor, ETR1, probably passes its signal to CTR1, a protein kinase of the Raf family (see Chapter 22).

The Activated MAP Kinase Enters the Nucleus

The **MAPK** (*m*itogen-*a*ctivated *p*rotein *k*inase) **cascade** owes its name to a series of protein kinases that phosphorylate each other in a specific sequence, much like runners in a relay race passing a baton (Figure 14.24). The first kinase in the sequence is Raf, referred to in this context as **MAP kinase kinase kinase** (**MAPKKK**). MAPKKK passes the phosphate baton to **MAP kinase kinase** (**MAPKK**), which hands it off to **MAP kinase** (**MAPK**). MAPK, the "anchor" of the relay team, enters the nucleus, where it activates still other protein kinases, specific transcription factors, and regulatory proteins.

The transcription factors that are activated by MAPK are called **serum response factors** (**SRFs**) because all of the growth factors that bind to RTKs are transported in the serum. Serum response factors bind to specific nucleotide sequences on the genes they regulate called **serum response elements** (**SREs**). The entire process from binding of the growth factor to the receptor to transcriptional activation of gene expression can be very rapid, taking place in a few minutes.

Some of the genes that are activated encode other transcription factors that regulate the expression of other genes. Because these genes are important for cell proliferation and growth, many of them are proto-oncogenes. For example, one of the genes whose expression is stimulated by MAPK is the proto-oncogene *FOS*. A **proto-oncogene** is a normal gene that potentially can cause malignant tumors when mutated. When the **Fos** protein combines with the phosphorylated **Jun** protein (one of the nuclear proteins that is phosphorylated by MAPK), it forms a heterodimeric transcription factor called **AP-1**, which turns on other genes. Other important proto-oncogenes that encode nuclear transcription factors include *MYC* and *MYB*. Both phytochrome (see Chapter 17) and gibberellin (see Chapter 20) are believed to regulate gene expression via the up-regulation of MYB-like transcription factors.

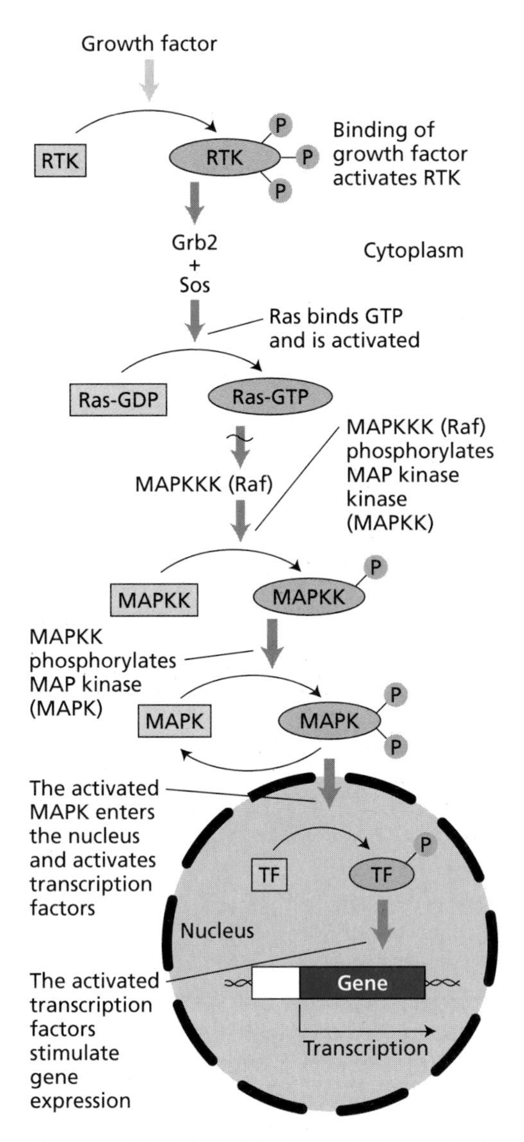

Figure 14.24 The MAPK cascade. Hormonal stimulation of the receptor tyrosine kinase leads to the activation of Raf (see Figure 14.23), also known as MAP kinase kinase kinase (MAPKKK). (1) MAPKKK phosphorylates MAP kinase kinase (MAPKK). (2) MAPKK phosphorylates MAP kinase (MAPK). (3) The activated MAPK enters the nucleus. and activates transcription factors (TF). (4) The activated transcription factors stimulate gene expression. (After Karp 1996.)

Plant Receptorlike Kinases Are Structurally Similar to Animal Receptor Tyrosine Kinases

Because of the recent progress in sequencing the genomes of plants such as *Arabidopsis* and rice, it is possible to use computers to search for DNA nucleotide sequences that correspond to the amino acid sequences of proteins identified in other organisms. Such database searches of plant DNA sequences have successfully identified a large family of **receptorlike protein kinases** (**RLKs**) by homology to animal receptor tyrosine kinases. These plant RLKs are structurally similar to the animal RTKs. They have a large extracellular domain, span the membrane only once, and contain a catalytic domain on the cytoplasmic side. Although they resemble the RTKs in their general structure, they differ in catalytic activity. Whereas RTKs of animals are autophosphorylating tyrosine kinases, the plant RLKs are autophosphorylating serine/threonine kinases (Walker 1994).

Three types of RLKs have been identified in plants, primarily on the basis of their extracellular domains. The first class is characterized by an extracellular S domain and is called **S receptor kinase** or **SRK**. The S domain was first identified in a group of secreted glycoproteins, called S locus glycoproteins (SLGs), which regulate self-incompatibility in *Brassica* species. Self-incompatibility is characterized by the failure of pollen tubes to grow when placed on pistils from the same plant, and self-incompatibility loci are genes that regulate this phenotype.

The S domain consists of ten cysteines in a particular arrangement with other amino acids. The high degree of homology between the S domains of SRKs and those of SLGs suggests that they are functionally related and are involved in the recognition pathways involved in pollen tube growth. Consistent with this idea, SRK genes are expressed predominantly in pistils. Several other S domain RLKs with highly divergent sequences have been identified in other species, and each of these may play unique roles in plant cell signaling.

The **leucine-rich repeat** (**LRR**) family of receptors constitute the second group of RLKs. They were first identified as disease resistance genes that may play key roles in the cell surface recognition of ligands produced by pathogens and the subsequent activation of the intracellular defense response (Bent 1996; Song et al. 1995). However, plant LRR receptors have been implicated in normal developmental functions as well. For example, a pollen-specific LRR receptor has been identified in sunflower that may be involved in cell–cell recognition during pollination (Reddy et al. 1995), and the *Arabidopsis ERECTA* gene, which regulates the shape and size of organs originating from the shoot apical meristem, encodes an LRR receptor (Torii et al. 1996). More recently, the receptor for the plant steroid hormone brassinosteroid has been identified as an LRR receptor (see Chapter 17).

The LRR receptors are members of a larger family of LRR proteins that includes soluble forms with lower molecular mass that are widespread in plants and animals. The most conserved element of the LRR domain forms a β sheet with an exposed face that participates in protein–protein interactions (Buchanan and Gay 1996). The small soluble LRR proteins may participate in cell signaling by hydrophobic binding to LRR receptors. For example, in tomato a protein that contains four tandem repeats of a canonical 24-amino-acid leucine-rich repeat motif is up-regulated during virus infection. This protein is apparently secreted into the apoplast along with a protease that digests it to lower molecular weight peptides (Tornero et al. 1996). These peptides could form part of a signaling pathway by interacting with cell surface LRR receptors.

Finally, a third type of RLK that contains an epidermal growth factor–like repeat has been identified in *Arabidopsis*. Interestingly, the receptor, called **PRO25**, is localized in the chloroplast and interacts with a light-harvesting chlorophyll *a*/*b*–binding protein (LHCP) (Walker 1994). Little or nothing is known about signaling within plastids, which undoubtedly will be an important area for future research.

Summary

The size of the genome (the total amount of DNA in a cell, a nucleus, or an organelle) is related to the complexity of the organism. However, not all of the DNA in a genome codes for genes. Prokaryotic genomes consist mainly of unique sequences (genes). Much of the genome in eukaryotes, however, consists of repetitive DNA and spacer DNA. The genome size in plants is highly variable, ranging from 1.5×10^8 bp in *Arabidopsis* to 1×10^{11} bp in *Trillium*. Plant genomes contain about 25,000 genes; by comparison, the *Drosophila* genome contains about 12,000 genes.

In prokaryotes, structural genes involved in related functions are organized into operons, such as the *lac* operon. Regulatory genes encode DNA-binding proteins that may repress or activate transcription. In inducible systems, the regulatory proteins are themselves activated or inactivated by binding to small effector molecules.

Similar control systems are present in eukaryotic genomes. However, related genes are not clustered in operons, and genes are subdivided into exons and introns. Pre-mRNA transcripts must be processed by splicing, capping, and addition of poly-A tails to produce the mature mRNA, and the mature mRNA must then exit the nucleus to initiate translation in the cytosol. Despite these differences, most eukaryotic genes are regulated at the level of transcription, as in prokaryotes.

Transcription in eukaryotes is characterized by three different RNA polymerases whose activities are modulated by a diverse group of *cis*-acting regulatory sequences. RNA polymerase II is responsible for the synthesis of pre-mRNA. General transcription factors assemble into a transcription initiation complex at the TATA box of the minimum promoter, which lies within 100 bp of the transcription start site of the gene. Additional *cis*-acting regulatory sequences, such as the CAAT box and GC box, bind transcription factors that enhance expression of the gene. Distal regulatory sequences located farther upstream bind to other transcription factors called activators or repressors. Many plant genes are also regulated by enhancers, distantly located positive regulatory sequences.

Despite being scattered throughout the genome, many eukaryotic genes are both inducible and coregulated. Genes that are coordinately regulated have common *cis*-acting regulatory sequences in their promoters. Most transcription factors in plants contain the basic zipper (bZIP) motif. An important group of transcription factors in plants, the floral homeotic genes, contain the MADS domain.

Enzyme concentration is also regulated by protein degradation, or turnover. As yet there is no evidence that plant vacuoles function like animal lysosomes in protein turnover, except during senescence, when the contents of the vacuole are released. However, protein turnover via the covalent attachment of the short polypeptide ubiquitin and subsequent proteolysis is an important mechanism for regulating the cytosolic protein concentration in plants.

Signal transduction pathways coordinate gene expression with environmental conditions and with development. Prokaryotes employ two-component regulatory systems that include a sensor protein and a response regulator protein that facilitates the response, typically gene expression. The sensor and the response regulator communicate via protein phosphorylation. Receptor proteins related to the bacterial two-component systems have recently been identified in yeast and plants.

In multicellular eukaryotes, lipophilic hormones usually bind to intracellular receptors, while water-soluble hormones bind to cell surface receptors. Binding to a receptor initiates a signal transduction pathway, often involving the generation of second messengers, such as cyclic nucleotides, inositol trisphosphate, and calcium, which greatly amplify the original signal and bring about the cellular response. Such pathways normally lead to changes in gene expression. In plants, the receptor for the phytohormone brassinosteroid is a cell surface receptor.

The seven-spanning receptors of animal cells interact with heterotrimeric G proteins, which act as molecular switches by cycling between active (GTP-binding) forms and inactive (GDP-binding) forms. Dissociation of the α subunit from the complex allows it to activate the effector enzyme. Activation of adenylyl cyclase increases cAMP levels, resulting in the activation of protein kinase A. Cyclic AMP can also regulate cation channels directly.

When heterotrimeric G proteins activate phospholipase C, it initiates the IP_3 pathway. IP_3 released from the membrane opens intracellular calcium channels, releasing calcium from the ER and vacuole into the cytosol. The increase in calcium concentration, in turn, activates protein kinases and other enzymes. In plants, calcium dependent protein kinases, which have a calmodulin domain, are activated by calcium directly. The other byproduct of phospholipase C, diacylglycerol, can also act as a second messenger by activating protein kinase C.

There is increasing evidence that cyclic GMP operates as a second messenger in plant cells as it does in animal cells. In animal cells, cyclic GMP has been shown to regulate ion channels and protein kinases.

The most common family of cell surface catalytic receptors in animals consists of the receptor tyrosine kinases. RTKs dimerize upon binding to the hormone; then their multiple tyrosine residues are autophosphorylated. The phosphorylated receptor then acts as an assembly site for various protein complexes, including the Ras superfamily of monomeric GTPases. Binding of Ras leads to recruitment of the protein kinase Raf to the membrane. Raf initiates the MAPK cascade. The last kinase to be phosphorylated (activated) is MAP kinase, which enters the nucleus and activates various transcription factors (serum response factors), which bind to *cis*-acting regulatory sequences called serum response elements.

Plants appear to lack RTKs, but they have structurally similar receptors called receptorlike kinases, which are serine/threonine kinases. The three main categories of plant RLKs are the S receptor kinases, the leucine-rich repeat receptors, and a receptor on the chloroplast called PRO25. Little is known about the signaling pathways used by these receptors, although enzymes of the MAPK cascade have been identified in plants.

General Reading

*Alberts, B., Bray, D., Lewis, J., Raff, M., Roberts, K., and Watson, J. D. (1994) *Molecular Biology of the Cell*, 3rd ed. Garland, New York.

*Becker, W. M., Reece, J. B., and Poenie, M. F. (1996) *The World of the Cell*, 3rd ed. Benjamin/Cummings, Menlo Park, Calif.

Dey, P. M., and Harborne, J. B., eds. (1997) *Plant Biochemistry*. Academic Press, San Diego.

Fosket, D. E. (1994) *Plant Growth and Development: A Molecular Approach*. Academic Press, San Diego.

*Karp, G. (1996) *Cell and Molecular Biology: Concepts and Experiments*. Wiley, New York.

*Lodish, H., Baltimore, D., Berk, A., Zipursky, S., Matsidaira, P., and Darnell, J. (1995) *Molecular Cell Biology*. Scientific American Books, New York.

Tobin, A. J., and Morel, R. E. (1997) *Asking about Cells*. Harcourt Brace, Fort Worth, Tex.

* Indicates a reference that is general reading in the field and is also cited in this chapter.

Chapter References

Assmann, A. M. (1995) Cyclic AMP as a second messenger in higher plants: Status and future prospects. *Plant Physiol.* 91: 624–628.

Bent, A. F. (1996) Plant disease resistance genes: Function meets structure. *Plant Cell* 8: 1757–1771.

Buchanan, S. G., and Gay, N. J. (1996) Structural and functional diversity in the leucine-rich repeat family of proteins. *Prog. Biophys. Mol. Biol.* 65: 1–44.

Chen, X., Xiao Z-A., and Zhang, C-H. (1996) A preliminary study on plant protein kinase C. *Acta Phytophysiol. Sinica* 22: 437–440.

Coux, O., Tanaka, K., and Goldberg, A. L. (1996) Structure and functions of the 20S and 26S proteasomes. *Annu. Rev. Biochem.* 65: 2069–2076.

Darnell, J., Lodish, H., and Baltimore, D. (1990) *Molecular Cell Biology*, 2nd ed. Scientific American Books, W. H. Freeman, New York.

Elliott, D. C., and Kokke, Y. S. (1987) Partial purification and properties of a protein kinase C type enzyme from plants. *Phytochemistry* 26: 2929–2936.

Golovkin, M., and Reddy, A. S. N. (1996) Structure and expression of a plant U1 snRNP 70K gene—Alternative splicing of U1 snRNP pre-mRNA produces two different transcripts. *Plant Cell* 8: 1421–1435.

Harper, J. J., Sussman, M. R., Schaller, G. E., Putnam-Evans, C., Charbonneau, H., and Harmon, A. C. (1991) A calcium-dependent protein kinase with a regulatory domain similar to calmodulin. *Science* 252: 951–954.

Ichikawa, T., Suzuki, Y., Czaja, I., Schommer, C., Lesnick, A., Schell, J., and Walden, R. (1997) Identification and role of adenylyl cyclase in auxin signalling in higher plants. *Nature* 390: 698–701.

Janistyn, B. (1983) Gas chromatographic-mass spectroscopic identification and quantification of cyclic guanosine-3′:5′-monophosphate in maize seedlings (*Zea mays*). *Planta* 159: 382–385.

Johnston, M. (1987) A model fungal gene regulatory mechanism: The GAL genes of *Saccharomyces cerevisiae*. *Microbiol. Rev.* 51: 458–476.

Kategiri, F., Lam, E., and Chua, N-H. (1989) Two tobacco DNA binding proteins with homology to the nuclear factor CREB. *Nature* 340: 727–730.

Kretsinger, R. (1980) Structure and evolution of calcium-modulated proteins. *CRC Crit. Rev. Biochem.* 8: 119–174.

Kuersten, S., Lea, K., Macmorris, M., Spieth, J., and Blumenthal, T. (1997) Relationship between 3′ end formation and SL2-specific trans-splicing in polycistronic *Caenorhabditis elegans* pre-mRNA processing. *RNA* 3: 269–278.

Kwak, J. M., Kim, S. A., Lee, S. K., Oh, S-A., Byoun, C-H., Han, J-K., and Nam, H. G. (1997) Insulin-induced maturation of *Xenopus* oocytes is inhibited by microinjection of a *Brassica napus* cDNA clone with high similarity to a mammalian receptor for activated protein kinase C. *Planta* 201: 245–251.

Lam, E. (1997) Nucleic acids and proteins. In *Plant Biochemistry*, P. M. Dey and J. B. Harborne, eds., Academic Press, San Diego, pp. 316–352.

Li, W., Luan, S., Schreiber, S. L., and Assmann, S. M. (1994) Cyclic AMP stimulates K^+ channel activity in mesophyll cells of *Vicia faba* L. *Plant Physiol.* 106: 957–961.

McAinsh, M. R., Webb, A. A. R., Taylor, J. E., and Hetherington, A. M. (1995) Stimulus-induced oscillations in guard cell cytosolic free calcium. *Plant Cell* 7: 1207–1219.

McCurdy, D. W., and Harmon, A. C. (1992) Calcium-dependent protein kinase in the green alga *Chara*. *Planta* 188: 54–61.

McEwen, B. S. (1991) Non-genomic and genomic effects of steroids on neural activity. *TIPS* 12: 141–147.

Miklos, L. G., and Rubin, G. M. (1996) The role of the genome project in determining gene function: Insights from model organisms. *Cell* 86: 521–529.

Millner, P. A., and Causier, B. E. (1996) G-protein coupled receptors in plant cells. *J. Exp. Bot.* 47: 983–992.

Mortimer, R. K., Schild, D., Contopoulou, C. R., and Kans, J. A. (1989) Genetic map of *Saccharomyces cerevisiae*, Edition 10. *Yeast* 5: 321–403.

Newton, R. P., and Brown, E. G. (1992) Analytical procedures for cyclic nucleotides and their associated enzymes in plant tissues. *Phytochem. Anal.* 3: 1–13.

Ota, I. M., and Varshavsky, A. (1993) A yeast protein similar to bacterial two-component regulators. *Science* 262: 566–569.

Parkinson, J. S. (1993) Signal transduction schemes of bacteria. *Cell* 73: 857–871.

Pei, Z-M., Ward, J. M., Harper, J. F., and Schroeder, J. I. (1996) A novel chloride channel in *Vicia faba* guard cell vacuoles activated by the serine-threonine kinase, CDPK. *EMBO J.* 15: 6564–6574.

Peters, J. L., and Silverthorne, J. (1995) Organ-specific stability of two *Lemna* rbcS mRNAs is determined primarily in the nuclear compartment. *Plant Cell* 7: 131–140.

Reddy, J. T., Dudareva, N., Evrard, J-L., Krauter, R., Steinmetz, A., and Pillay, D. T. N. (1995) A pollen-specific gene from sunflower encodes a member of the leucine-rich-repeat protein superfamily. *Plant Sci.* 111: 81–93.

Roberts, D. M., and Harmon, A. C. (1992) Calcium-modulated protein targets of intracellular calcium signals in higher plants. *Annu. Rev. Plant Physiol. Plant Mol. Biol.* 43: 375–414.

Rounsley, S. D., Glodek, A., Sutton, G., Adams, M. D., Somerville, C. R., Venter, J. C., and Kerlavage, A. R. (1996) The construction of *Arabidopsis* EST assemblies: A new resource to facilitate gene identification. *Plant Physiol.* 112: 1177–1183.

Shanklin, J., Jabben, M., and Vierstra, R.D. (1987) Red light induced formation of ubiquitin-phytochrome conjugates: Identification of possible intermediates of phytochrome degradation. *Proc. Natl. Acad. Sci. USA* 84:359–363.

Song, W-Y., Wang, G-L., Chen, L-L., Kim, H-S., Pi, L-Y., Holsten, T., Gardner, J., Wang, B., Zhai, W-X., Zhu, L-H., et al. (1995) A receptor kinase-like protein encoded by the rice disease resistance gene, Xa21. *Science* 270: 1804–1806.

Sundaresan, V., Springer, P., Volpe, T., Haward, S., Jones, J. D. G., Dean, C., Ma, H., and Martienssen, R. (1995) Patterns of gene action in plant development revealed by enhancer trap and gene trap transposable elements. *Genes Dev.* 9: 1797–1810.

Takezawa, D., Ramachandiran, S., Paranjape, V., and Poovaiah, B. W. (1996) Dual regulation of a chimeric plant serine-threonine kinase by calcium and calcium-calmodulin. *J. Biol. Chem.* 271: 8126–8132.

Terryn, N., Inze, D., and Van Montagu, M. (1996) Small GTP-binding proteins in plants. In *Proceedings of the Phytochemical Society of Europe. Plant Membrane Biology*, I. M. Møller and P. Brodelius, eds., Clarendon, Oxford, pp. 19–27.

Terzaghi, W. B., and Cashmore, A. R. (1995) Light-regulated transcription. *Annu. Rev. Plant Physiol. Plant Mol. Biol.* 46: 445–474.

Tezuka, T., Hiratsuka, S., and Takahasi, S. Y. (1993) Promotion of the growth of self-incompatible pollen tubes in lily by cAMP. *Plant Cell Physiol.* 27: 193–197.

Torii, K. U., Mitsukawa, N., Oosumi, T., Matsuura, Y., Yokayama, R., Whittier, R. F., and Komeda, Y. (1996) The *Arabidopsis* ERECTA gene encodes a putative receptor protein kinase with extracellular leucine-rich repeats. *Plant Cell* 8: 735–746.

Tornero, P., Mayda, E., Gomez, M. D., Canas, L., Conejero, V., and Vera, P. (1996) Characterization of LRP, a leucine-rich repeat (LRR) protein from tomato plants that is processed during pathogenesis. *Plant J.* 10: 315–330.

Walker, J. C. (1994) Structure and function of the receptor-like protein kinases of higher plants. *Plant Mol. Biol.* 26: 1599–1609.

Watson, C. L., and Gold, M. R. (1997) Lysophosphatidylcholine modulates cardiac I-Na via multiple protein kinase pathways. *Circ. Res.* 81: 387–395.

15 Cell Walls: Structure, Biogenesis, and Expansion

PLANT CELLS, UNLIKE ANIMAL CELLS, are surrounded by a relatively thin but mechanically strong cell wall. However, cell walls are not unique to plants. The prokaryotic cells of eubacteria and archaebacteria are also enclosed by cell walls, as are the eukaryotic cells of fungi and algae. Thus, the loss of cell walls by animal cells is the exception rather than the rule. Plant cell walls consist of a complex mixture of polysaccharides and other polymers that are secreted by the cell and are assembled into an organized network linked together by a mixture of covalent and noncovalent bonds. Plant cell walls also contain structural proteins, enzymes, phenolic polymers, and other materials that modify the wall's physical and chemical characteristics. The cell walls of prokaryotes, fungi, algae, and plants all serve two primary functions: regulating cell volume and determining cell shape. As we will see, however, plant cell walls have acquired additional functions that are not apparent in the walls of other organisms. Because of these diverse functions, the structure and composition of plant cell walls are complex and variable.

In addition to having biological functions, the plant cell wall is important in human economics. As a natural product, the plant cell wall is used commercially in the form of paper, textiles, fibers (cotton, flax, hemp, and others), charcoal, lumber, and other wood products. Another major use of cell walls is in the form of extracted polysaccharides that have been modified to make plastics, films, coatings, adhesives, gels, and thickeners in a huge variety of products (Lapasin and Pricl 1995). As the most abundant reservoir of organic carbon in nature, the plant cell wall also takes part in the processes of carbon flow through ecosystems. The organic substances that make up humus in the soil and that enhance soil structure and fertility are derived from cell walls. Finally, as an important source of roughage in our diet, the plant cell wall is a significant factor in human health and nutrition.

We begin this chapter with a description of the general structure and composition of cell walls and the mechanisms of the biosynthesis and secretion of cell wall materials. We then turn to the role of the primary cell wall in cell expansion. The mechanisms of *tip growth* will be contrasted with those of *diffuse growth*, particularly with respect to the establishment of cell polarity and the control of the rate of cell expansion. Finally, we will describe the dynamic changes in the cell wall that often accompany cell differentiation, along with the role of cell wall fragments as signaling molecules.

The Structure of Plant Cell Walls: A Functional Overview

The cell wall is essential for many processes in plant physiology and development. A tough outer coating enclosing the cell, the wall acts as a cellular exoskeleton that controls cell shape and allows high turgor pressures to develop. Because of its influence on turgor pressure and on the relationship between the pressure and volume of cells, the cell wall is required for the normal water relations of plants (see Chapter 3). Because the expansion growth of plant cells is limited principally by the ability of the cell wall to expand, plant morphogenesis depends largely on the control of cell wall properties. Cell walls also glue cells together, preventing them from sliding past one another. This constraint on cellular movement contrasts markedly to the situation in animal cells, and it dictates the way in which plants develop (see Chapter 16). Plant cell walls also determine the mechanical strength of plant structures, allowing them to grow to great heights. In addition, the pressure-driven flow of water in the xylem requires a mechanically tough wall that resists collapse by the negative pressure in the xylem.

Much of the carbon that is assimilated in photosynthesis ends up as polysaccharides in the wall. During specific phases of development, these polymers may be hydrolyzed into their constituent sugars, which may be scavenged by the cell and used to make new polymers. This phenomenon is most notable in many seeds, in which wall polysaccharides of the endosperm or cotyledons function primarily as food reserves. Furthermore, oligosaccharide components of the cell wall act as important signaling molecules during cell differentiation and during recognition of pathogens and symbionts. Although permeable to small molecules, the wall acts as a diffusion barrier that limits the size of macromolecules that can reach the plasma membrane from outside, and it is a major structural barrier to pathogen invasion.

The diversity of functions of the plant cell wall derives from the diversity and complexity of plant cell wall structure. In this section we will begin with a brief description of the morphology and basic architecture of plant cell walls. Then we will discuss the organization, composition, and synthesis of primary and secondary cell walls.

Plant Cell Walls Have Varied Architecture

Stained sections of plant tissues reveal that the cell wall is not uniform, but varies greatly in appearance and composition in different cell types (Figure 15.1). Cell walls of the cortical parenchyma are generally thin and have few distinguishing features. In contrast, the walls of some specialized cells, such as epidermal cells, collenchyma, phloem fibers, xylem tracheary elements, and other forms of sclerenchyma have thicker, multilayered walls. Often these walls are intricately sculpted and are impregnated with specific substances such as lignin, cutin, silica, or structural proteins. The individual sides of a wall surrounding a cell may also vary in thickness, embedded substances, sculpting, and frequency of pitting and plasmodesmata. For example, the outer wall of the epidermis is usually much thicker than the other walls of the cell; moreover, this wall lacks plasmodesmata and is embedded with cutin and waxes. In guard cells, the side of the wall adjacent to the stomatal pore is much thicker than the walls on the other sides of the cell.

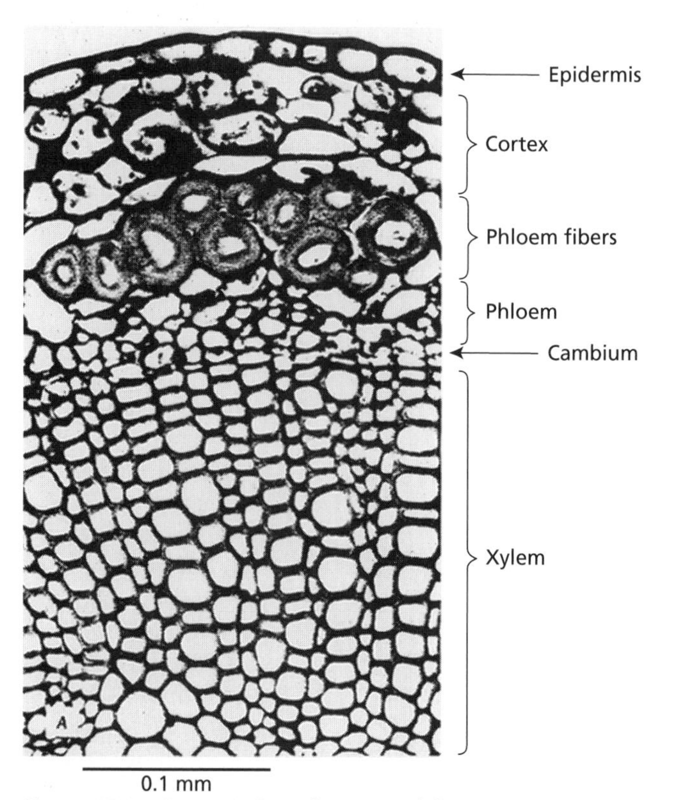

Figure 15.1 Cross section of a stem of flax, showing cells with varying wall morphology. Note the highly thickened walls of the phloem fibers. (From Esau 1977, courtesy of R. Evert.)

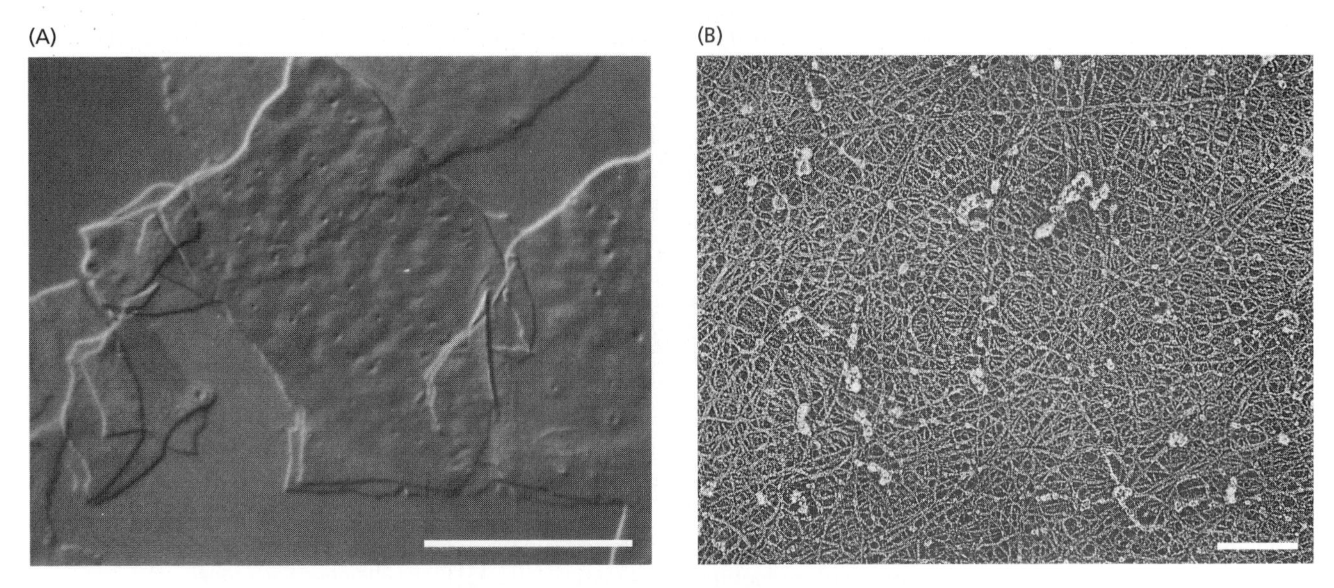

Figure 15.2 Primary cell walls from onion parenchyma. (A) This surface view of cell wall fragments was taken using Nomarski optics. Note that the wall looks like a very thin sheet with small surface depressions; these depressions may be pit fields, places where plasmodesmatal connections between cells are concentrated. Bar = 100 μm. (B) This surface view of a cell wall was prepared by a freeze-etch replica technique. It shows the fibrillar nature of the cell wall. Bar = 200 nm. (From McCann et al. 1990, courtesy of M. McCann.)

Such variations in wall architecture for a single cell reflect the polarity and differentiated functions of the cell.

Despite this diversity in cell wall morphology, cell walls commonly are classified into two major types: primary walls and secondary walls. **Primary walls** are formed by growing cells and are usually considered to be relatively unspecialized and similar in molecular architecture in all cell types. Nevertheless, the ultrastructure of primary walls also shows wide variation. Some primary walls, such as those of the onion bulb parenchyma, are very thin (100 nm) and architecturally simple (Figure 15.2). Other primary walls, such as those found in collenchyma or in the epidermis (Figure 15.3), may be much thicker and consist of multiple layers.

Secondary walls are the cell walls that form after cell growth (enlargement) has ceased. Secondary walls may

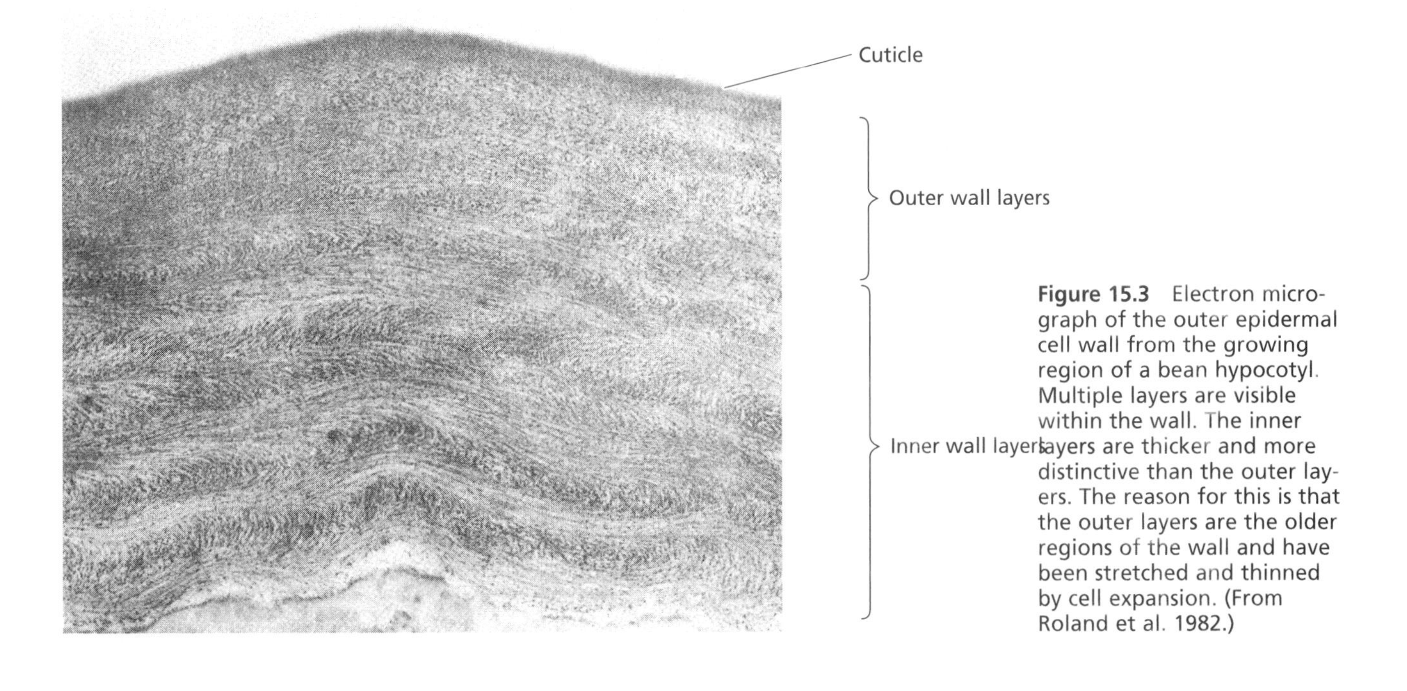

Figure 15.3 Electron micrograph of the outer epidermal cell wall from the growing region of a bean hypocotyl. Multiple layers are visible within the wall. The inner layers are thicker and more distinctive than the outer layers. The reason for this is that the outer layers are the older regions of the wall and have been stretched and thinned by cell expansion. (From Roland et al. 1982.)

become highly specialized in structure and composition, reflecting the differentiated state of the cell. Xylem cells, such as those found in wood, are notable for possessing highly thickened secondary walls that are strengthened by **lignin** (see Chapter 13). A thin layer of material, the **middle lamella**, can usually be seen at the junction where the walls of neighboring cells come into contact. The composition of the middle lamella differs from the rest of the wall in that it is high in pectin and contains a different complement of proteins than the remainder of the wall has. Its origin can be traced to the cell plate that formed during cell division (see Figure 1.28).

As we saw in Chapter 1, the cell wall is usually penetrated by tiny membrane-lined channels, called **plasmodesmata** (singular plasmodesma), which connect neighboring cells (see Figures 1.30 and 1.31). Plasmodesmata are thought to function in communication between cells, by allowing passive transport of small molecules and active transport of proteins and nucleic acids between the cytoplasms of adjacent cells (see Chapter 6).

The Primary Cell Wall Is Composed of Cellulose Microfibrils Embedded in a Polysaccharide Matrix

In primary cell walls, *cellulose microfibrils* are embedded in a highly hydrated, amorphous matrix (Figure 15.4). In some ways this structure resembles fiberglass and other *composite materials,* in which rigid crystalline fibers are used to reinforce a more flexible epoxy matrix. In the case of cell walls, the **matrix** consists of two major groups of polysaccharides, usually called *hemicelluloses* and *pectins,* plus a small amount of *structural protein.* The matrix polysaccharides consist of a variety of polymers that may vary according to cell type and plant species (Table 15.1).

Cellulose microfibrils are relatively stiff structures that contribute to the strength and structural bias of the cell wall. The individual polysaccharide chains that make up the microfibril are closely aligned and bonded

Table 15.1
Structural components of cell walls

Class	Examples
Cellulose	Microfibrils of (1→4)β-D-glucan
Pectins	Homogalacturonan Rhamnogalacturonan Arabinan Galactan
Hemicellulose	Xyloglucan Xylan Glucomannan Arabinoxylan Callose (1→3)β-D-glucan (1→3,1→4)β-D-glucan [grasses only]
Lignin	(see Chapter 13)
Structural proteins	(see Table 15.2)

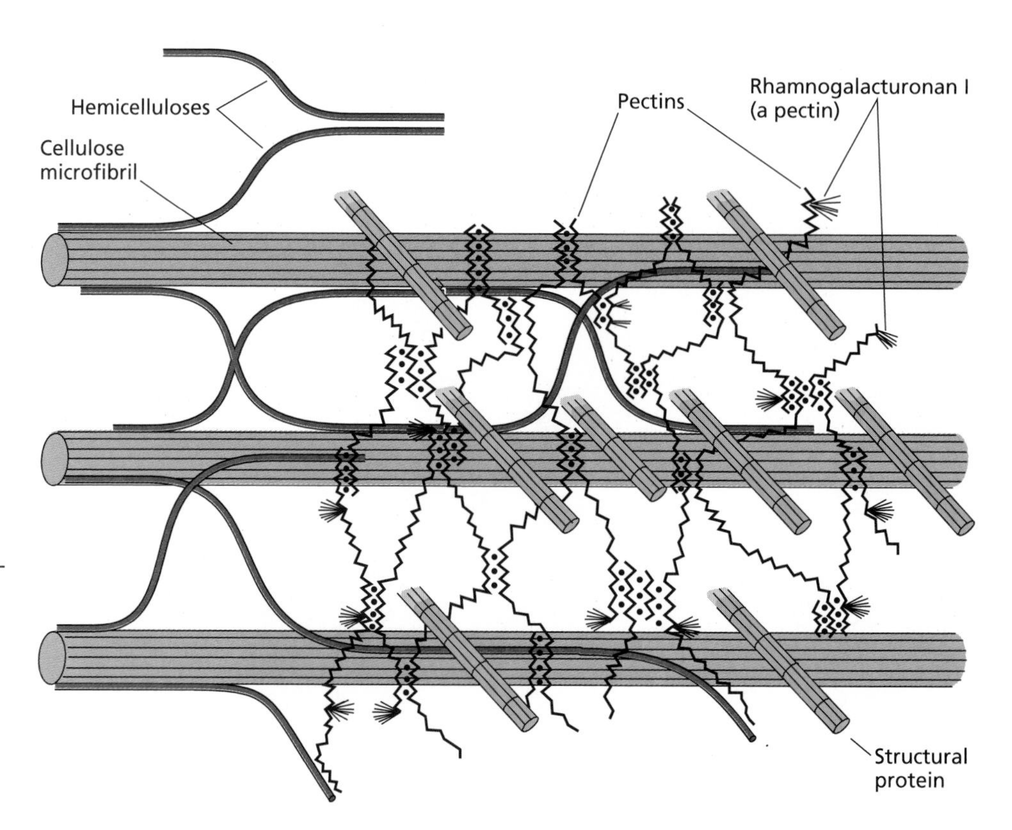

Figure 15.4 Schematic diagram of the major structural components of the primary cell wall and their likely arrangement. Cellulose microfibrils are coated with hemicelluloses (such as xyloglucan), which may also tether the microfibrils to one another. Pectins form an interlocking matrix gel, perhaps interacting with structural proteins. (From Brett and Waldron 1996.)

to each other to make a **crystalline** or **paracrystalline structure** that is relatively inaccessible to enzymatic attack. (Paracrystalline materials contain some noncrystalline regions.) Hence, cellulose is very stable and only breaks down at specific times in development, such as senescence (see Chapter 16) or abscission (see Chapter 22). Cell wall degradation will be discussed at the end of the chapter.

Hemicelluloses are flexible polysaccharides that characteristically bind to the surface of cellulose. They may form tethers that bind cellulose microfibrils together into a cohesive network (see Figure 15.4), or they may act as a slippery coating to prevent direct microfibril–microfibril contact. **Pectins** form a gel phase in which the cellulose–hemicellulose network is embedded. They act as hydrophilic filler, to prevent aggregation and collapse of the cellulose network. They also determine the porosity of the cell wall to macromolecules. The precise role of the **structural proteins** is uncertain, but they may add mechanical strength to the wall and assist in the proper assembly of other wall components.

The primary wall is composed of approximately 25% cellulose, 25% hemicelluloses, and 35% pectins, with perhaps 1 to 8% structural protein, on a dry-weight basis. However, large deviations from these values may be found. For example, the walls of grass coleoptiles consist of 60 to 70% hemicelluloses, 20 to 25% cellulose, and only about 10% pectins. Cereal endosperm walls are mostly (about 85%) hemicelluloses. Secondary walls typically contain much higher cellulose contents. In this chapter we will present a basic model of the primary wall, but note that plant cell walls are more diverse than represented in this model. The composition of matrix polysaccharides and structural proteins in walls is highly variable among different species and cell types (Bacic et al. 1988).

The primary wall also contains much water. This water is located mostly in the matrix, which is perhaps 75 to 80% water. The hydration state of the matrix is an important determinant of the physical properties of the wall; for example, removal of water makes the wall stiffer and less extensible. This stiffening effect of dehydration may play a role in growth inhibition by water deficits. We will examine the structure of each of the major polymers of the cell wall in more detail in the sections that follow.

Cellulose Microfibrils Are Synthesized at the Plasma Membrane

Cellulose is a tightly packed microfibril of linear chains of 1→4-linked β-D-glucose (see Box 15.1 for a description of polysaccharide terminology, and Figure 15.5 for the structures of sugars commonly found in plant cell walls). Because of the alternating spatial configuration of the glu-

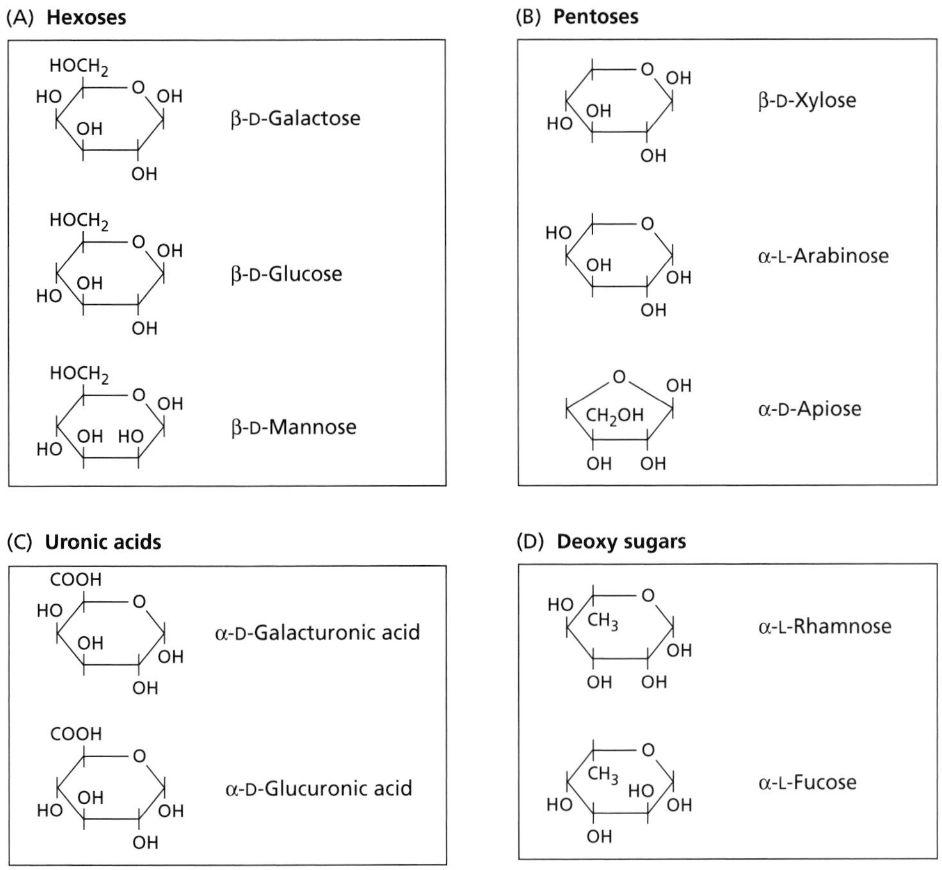

(A) **Hexoses**

β-D-Galactose

β-D-Glucose

β-D-Mannose

(B) **Pentoses**

β-D-Xylose

α-L-Arabinose

α-D-Apiose

(C) **Uronic acids**

α-D-Galacturonic acid

α-D-Glucuronic acid

(D) **Deoxy sugars**

α-L-Rhamnose

α-L-Fucose

Figure 15.5 Structure of sugars commonly found in plant cell walls. (A) Hexoses (six-carbon sugars). (B) Pentoses (five-carbon sugars). (C) Uronic acids (acidic sugars). (D) Deoxy sugars.

BOX 15.1

Terminology for Polysaccharide Chemistry

POLYSACCHARIDES ARE NAMED after the principal sugars they contain. For example, a *glucan* is a polymer made up of glucose; a *xylan* is a polymer made up of xylose; a *galactan* is a polymer made up of galactose. For branched polysaccharides, the backbone of the polysaccharide is usually given by the last part of the name. Thus, *xyloglucan* has a glucan backbone (a linear chain of glucose residues) with xylose sugars attached to it in the side chains; *arabinoxylan* has a xylan backbone (made up of xylose subunits) with arabinose side chains. However, a compound name does not necessarily imply a branched structure. For example, "glucomannan" is the name given to a polymer containing both glucose and mannose in its backbone. Because sugars may assume multiple forms and may be connected in many ways, the nomenclature for sugars is not simple and is briefly summarized here.

Monosaccharides (single-sugar residues) are polyhydroxyl aldehydes or ketones with the empirical formula $(CH_2O)_n$. They typically contain one asymmetric carbon, so they can exist as two stereoisomers, designated as the D or L enantiomers (see part A of the figure).

Monosaccharides usually form a cyclic, rather than a linear, structure, because the carbonyl oxygen can react with the hydroxyl groups, typically forming either five-member rings (**furanose** rings) or six-member rings (**pyranose** rings). These rings have a new chiral center and may take two forms, designated as α or β **anomers**. Thus, β-D-glucopyranose is D-glucose taking the form of a six-member ring in the β configuration (see part B of the figure). In solution, monosaccharides freely mutorotate (convert between α or β anomers); when linked in polymers, the anomeric configuration is stable.

The sugars in polysaccharides are linked together by **O-glycosidic bonds**. We indicate the linkage pattern by specifying the carbon atoms connected by the glycosidic bond. In cellulose, for example, carbon atom 1 is linked to carbon atom 4 in the neighboring glucose residues. Thus, cellulose is designated as a (1→4)β-D-glucan, because it is a string of β-D-glucose molecules linked through carbons 1 and 4 (see part C of the figure). It may also be called a polymer of 1→4-linked β-D-glucopyranose. Finally, an alternative naming convention allows cellulose to be called a β-1,4-D-glucan.

Terminology for the structure of polysaccharides, using cellulose as an example. (A) D- and L-glucose shown in the linear form. The carbons are numbered from the end with the aldehyde (CHO) (C = O) group. (B) Glucose may form five- or six-member rings, in the α or β configuration. The five-member rings (furanose; structures 1 and 2) are unstable and make up less than 1% of the total. The six-member rings (pyranose) are found mostly in the β configuration (64% of the total); the remainder (36%) are found in the α form. (C) A linear chain of 1→4-linked β-D-glucopyranose, showing the numbering system for the carbons and the configuration of the glycosidic bond. This is the glucan that makes up cellulose. The sugar rings are connected through an oxygen atom (the glycosidic oxygen). Because of the configuration of the glycosidic bonds, the repeating unit in this polymer is cellobiose, a disaccharide. (D) The repeating cellobiose unit, with the glucose residues drawn in the conformational ("chair") form. This way of drawing the glucose ring represents better the actual positions of the individual atoms than in the Haworth project (shown in C). Note that the carbon atoms in the ring structure are not shown in C and D; they are represented by the ring vertices.

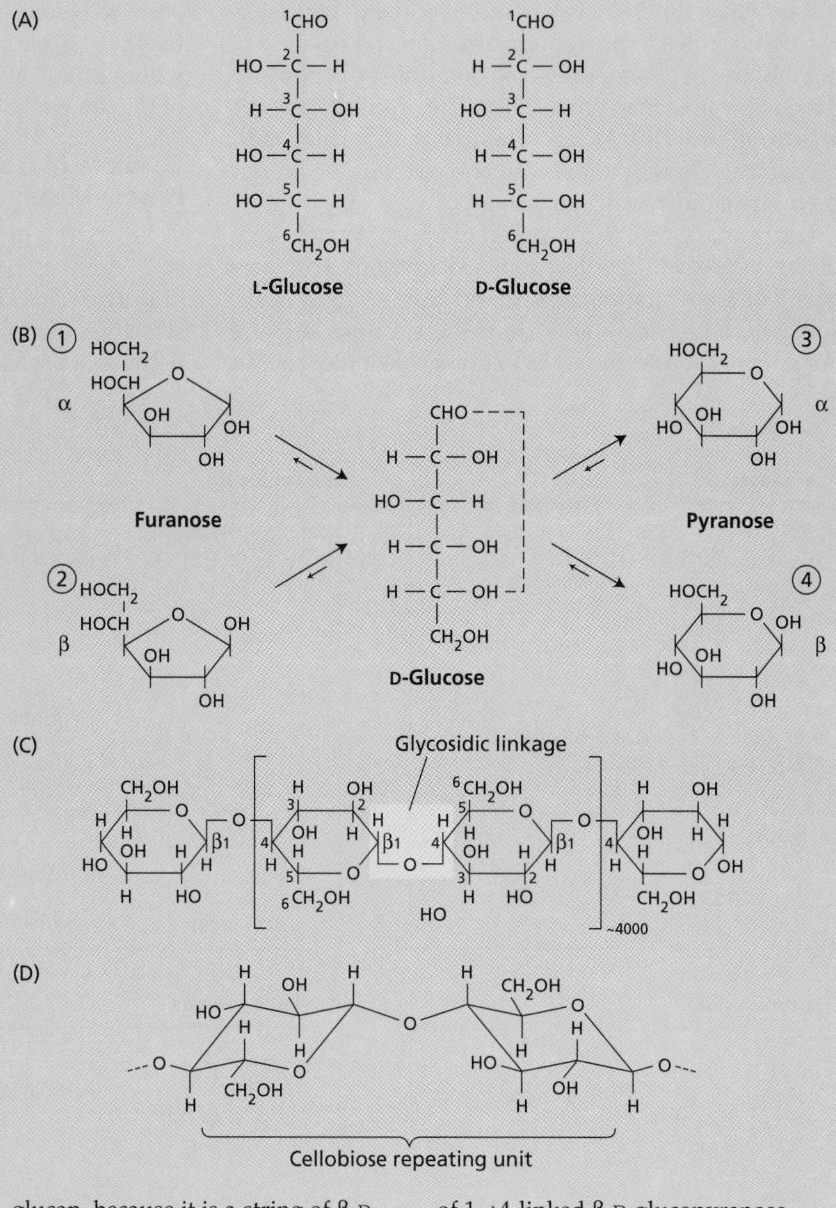

cosidic bonds linking adjacent glucose residues, the repeating unit in cellulose is considered to be cellobiose, a 1→4-linked β-D-glucose disaccharide. Cellulose microfibrils are of indeterminate length and vary considerably in width and in degree of order, depending on the source. For instance, cellulose microfibrils in land plants appear under the electron microscope to be 4 to 10 nm wide, whereas those formed by algae may be up to 30 nm wide and more crystalline. This variety in width corresponds to a variation in the number of parallel chains that make up the cross section of a microfibril (variously estimated to be about 20 to 40 in the thinner microfibrils).

The precise molecular structure of the cellulose microfibril is uncertain. Current models of microfibril organization suggest that it has a substructure consisting of highly crystalline domains linked together by less organized "amorphous" regions (Figure 15.6). Within the crystalline domains, adjacent glucans are highly ordered and bonded to each other by noncovalent bonding, such as hydrogen bonds and hydrophobic interactions. The individual glucan chains of cellulose are composed of 2000 to more than 25,000 glucose residues (Brown et al. 1996). These chains are long enough (about 1 to 5 μm long) to extend through multiple crystalline and amorphous regions within a microfibril. When cellulose is degraded—for example, by fungal cellulases—the amorphous regions are degraded first, releasing

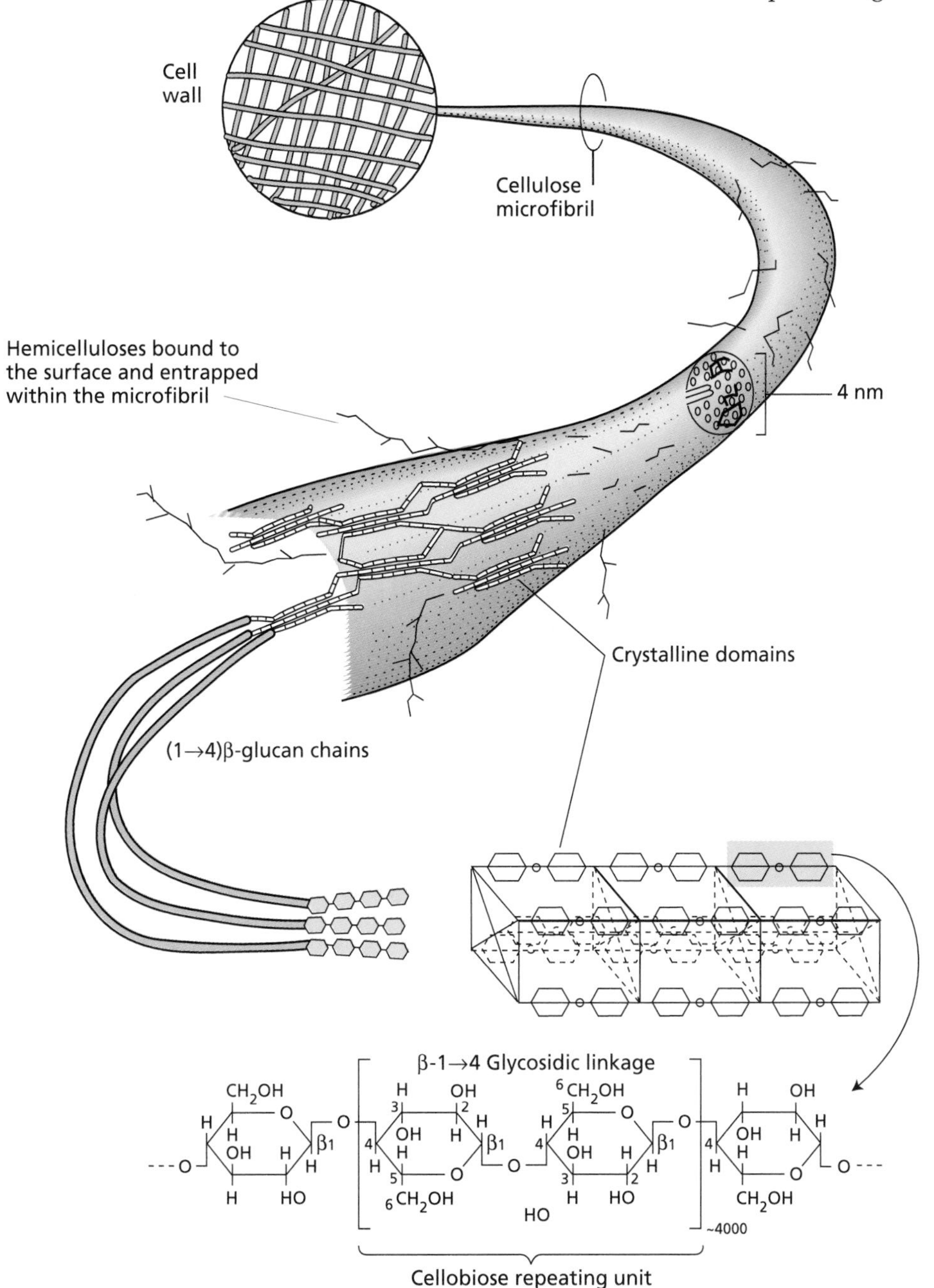

Figure 15.6 Structural model of a cellulose microfibril. The microfibril has regions of high crystallinity intermixed with less-organized glucans. Some hemicelluloses may also be trapped within the microfibril and bound to the surface.

small crystallites that are thought to correspond to the crystalline domains of the microfibril.

The extensive noncovalent bonding between adjacent glucans within a cellulose microfibril gives this structure remarkable properties. Cellulose has a high tensile strength, equivalent to that of steel. Cellulose is also insoluble, chemically stable, and relatively immune to chemical and enzymatic attack. These properties make cellulose an excellent structural material for building a strong cell wall.

Evidence from electron microscopy suggests that cellulose microfibrils are synthesized by large, ordered protein complexes, called **particle rosettes** or **terminal complexes**, that are embedded in the plasma membrane (Figure 15.7) (Brown et al. 1996). These structures are believed to contain many units of **cellulose synthase**, the enzyme that synthesizes the individual 1→4-linked β-D-glucans that make up the microfibril (Figure 15.8A). Cellulose synthase transfers a glucose residue from a sugar nucleotide donor to the growing glucan chain. The sugar nucleotide donor is probably uridine diphosphate D-glucose (UDP-glucose). Recent evidence suggests that the glucose used for the synthesis of cellulose may be obtained from sucrose (a disaccharide composed of fructose and glucose) (Amor et al. 1995). According to this still-tentative hypothesis, the enzyme **sucrose synthase** acts as a metabolic channel to transfer glucose taken from sucrose, via UDP-glucose, to the growing cellulose chain (Figure 15.8B, see page 418).

(A)

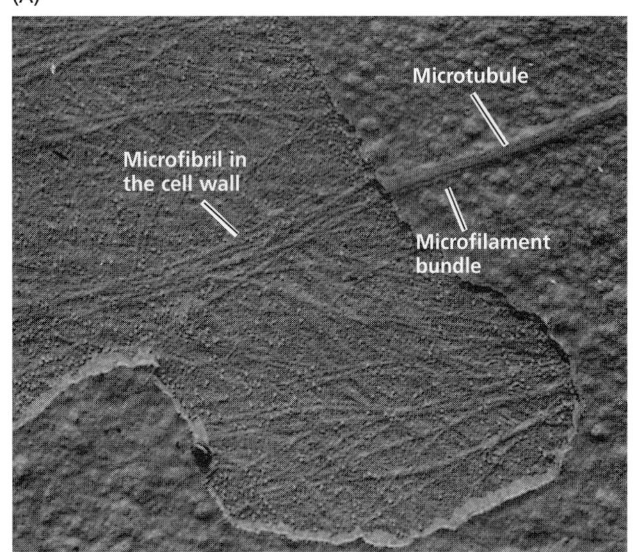

(B)

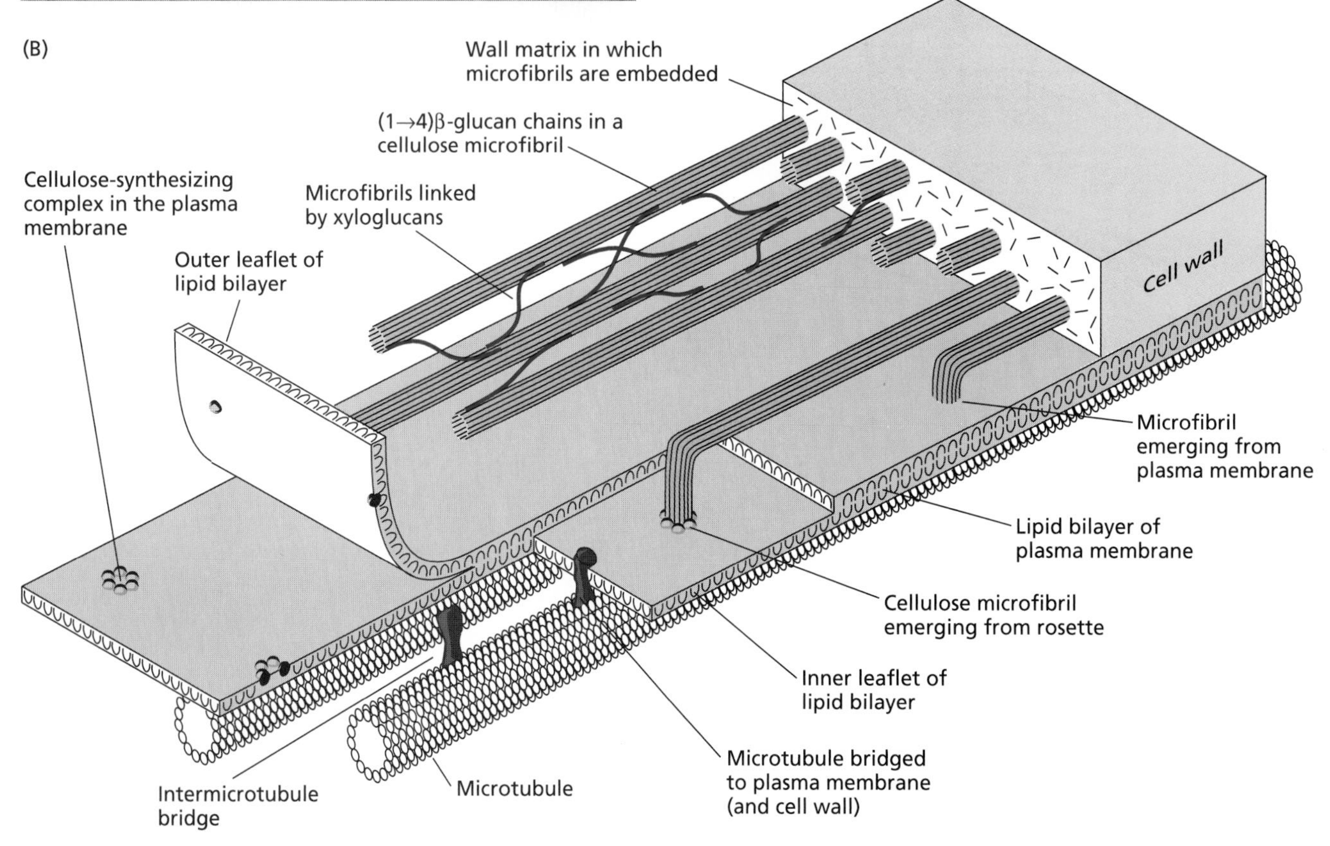

Figure 15.7 Electron micrograph of terminal complexes in the plasma membrane (A), and the hypothetical relationship between these complexes and other cell components (B). (From B. Gunning and M. Steer, *Plant Cell Biology: Structure and Function*, Jones and Bartlett, 1996.)

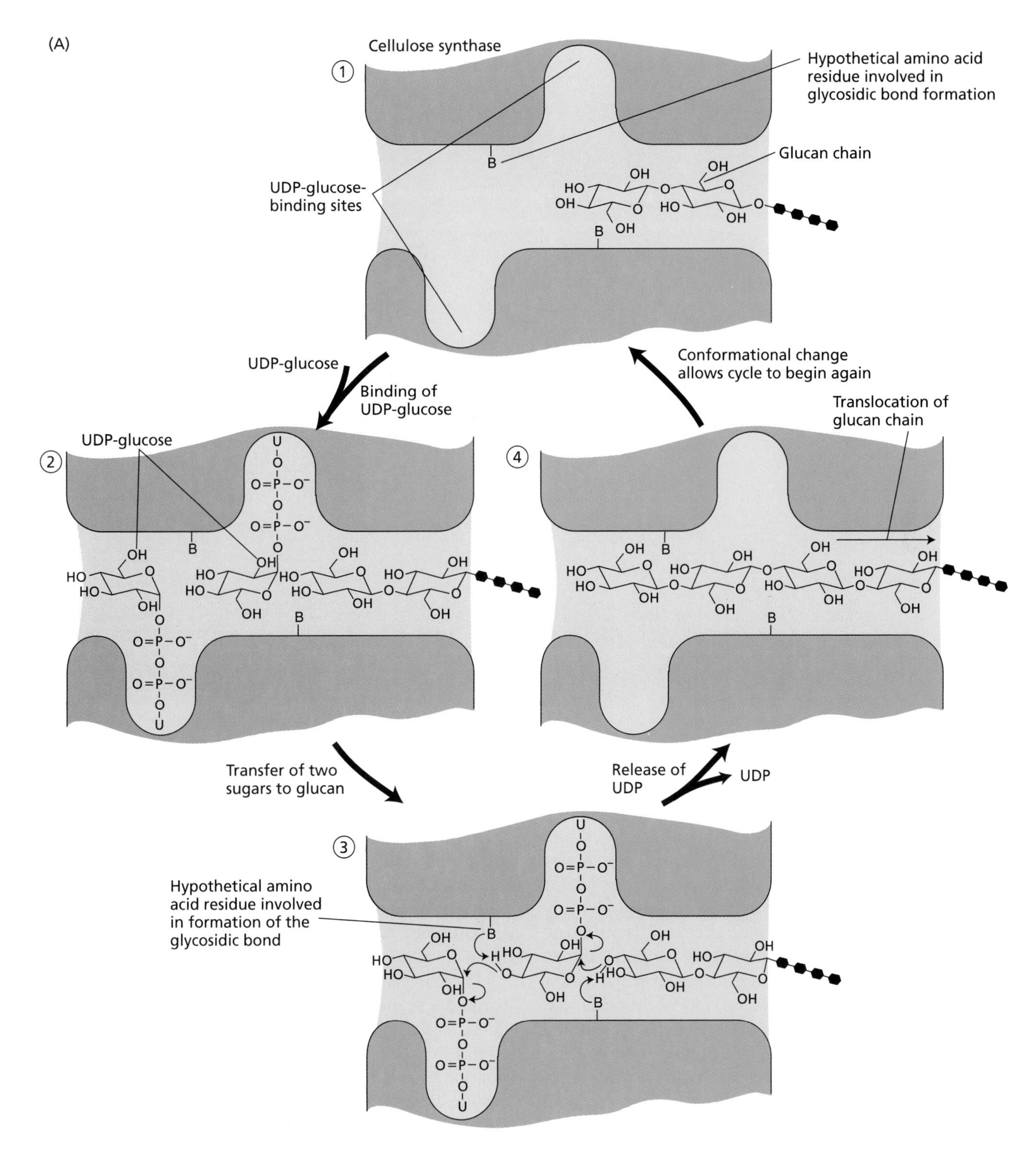

Figure 15.8 (A) Molecular model for the synthesis of cellulose and other wall polysaccharides that consist of a disaccharide repeat. Cellulose synthase contains two binding sites for UDP-glucose (or other sugar nucleotides). After binding UDP-glucose (step 2), the sugars are transferred to the nonreducing end of the glucan (step 3). At step 4, the glucan chain is translocated by two glucose residues, restoring the complex for the start of another round of synthesis (step 1). (A after Koyama et al. 1997.) (B, see next page)

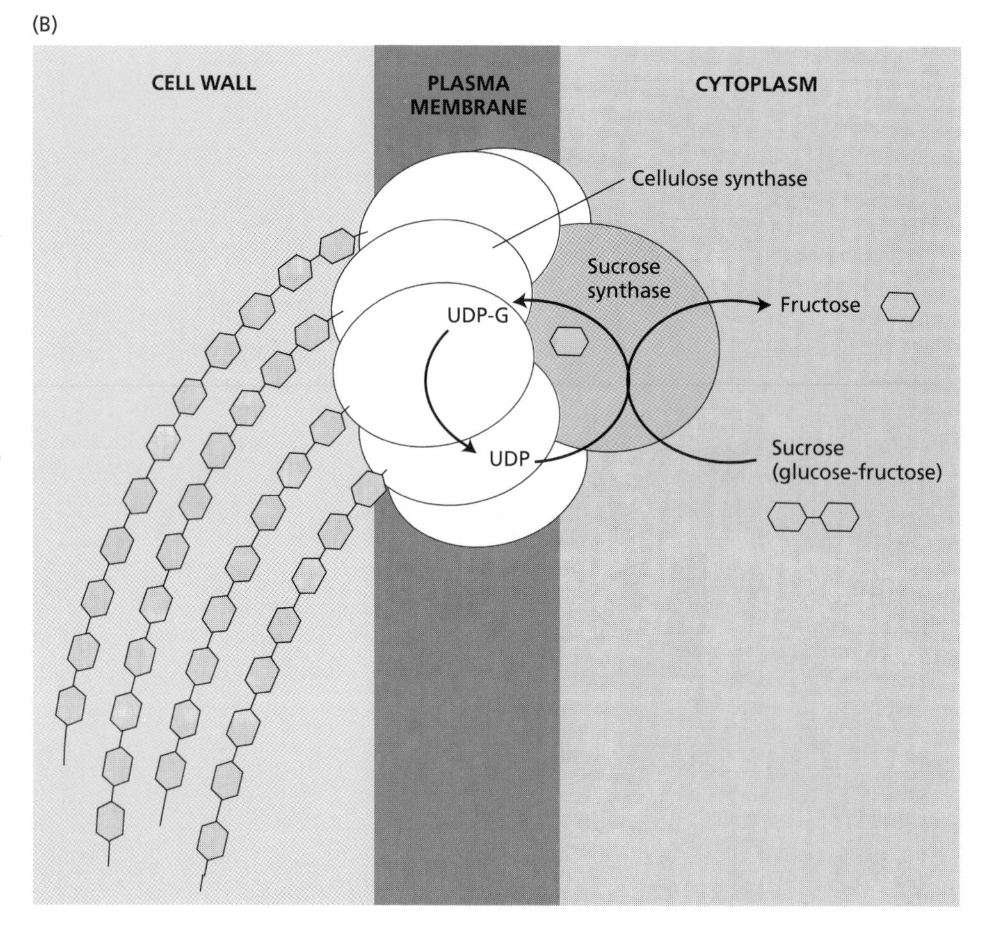

Figure 15.8 (*continued*) (B) Speculative model of cellulose synthesis by a multisubunit complex containing cellulose synthase. Glucose residues are donated to the growing glucan chains by UDP-glucose (UDP-G). Sucrose synthase may act as a metabolic channel to transfer glucose taken from sucrose to UDP-glucose, or UDP-glucose may be obtained directly from the cytoplasm. (B after Delmer and Amor 1995.)

After many years of searching, researchers recently identified the genes for cellulose synthase in higher plants (Pear et al. 1996; Ariole et al. 1998). These genes code for glycosyltransferases with two catalytic sites for transferring glycosyl residues from their nucleotide donors to the nascent polymer acceptor (see Figure 15.8A). This two-site arrangement may explain why the cellobiose disaccharide is the repeating subunit in cellulose: It has been proposed that two glucose residues are added to the glucan chain for each catalytic cycle of the enzyme. A similar arrangement in other polysaccharide synthases could explain why disaccharide repeats commonly make up the backbones of many other cell wall polysaccharides in plants.

The formation of cellulose involves not only the synthesis of the glucan, but also the crystallization of multiple glucans into a microfibril. Little is known about the control of this process, except that the direction of microfibril deposition may be guided by microtubules adjacent to the membrane.

When the cellulose microfibril is synthesized, it is secreted into a milieu (the wall) that contains a high concentration of other polysaccharides that are able to interact with and perhaps modify the growing microfibril. In vitro binding studies have shown that hemicelluloses such as xyloglucan and xylan may bind to the surface of cellulose. Some hemicelluloses may also become physically entrapped within the microfibril during its formation, thereby reducing the crystallinity and order of the microfibril (Hayashi 1989).

The bacterium *Acetobacter xylinum*, which normally forms large, highly ordered cellulose microfibrils, has been used as a model system to study cellulose microfibril formation. The microfibrils formed by *Acetobacter* are not embedded in a matrix, so their morphology is easy to observe by microscopy. When *Acetobacter* makes cellulose in the presence of materials that bind to (1→4) β-D-glucans, the resulting microfibrils are altered in shape, being more fragmented and frayed (Haigler et al. 1982). In addition, X-ray diffraction analysis shows that these modified microfibrils are less organized at the molecular level. They resemble microfibrils synthesized by land plants, which are less ordered than the bacterial cellulose. Such results suggest that the entrapment of hemicelluloses during cellulose formation in land plants significantly alters microfibril morphology.

Matrix Polymers Are Synthesized in the Golgi and Secreted in Vesicles

The matrix is a highly hydrated phase in which the cellulose microfibrils are embedded. The major polysaccharides of the matrix are synthesized by membrane-bound enzymes in the Golgi apparatus and are delivered to the cell wall via exocytosis of tiny vesicles (Figure 15.9). The enzymes responsible for synthesis are **sugar-nucleotide polysaccharide glycosyltransferases**. These enzymes transfer monosaccharides from sugar nucleotides to the growing end of the polysaccharide chain.

Unlike cellulose, which forms a crystalline microfibril, the matrix polysaccharides are much less ordered and are often described as amorphous. This noncrystalline character is a consequence of the structure of these polysaccharides—their branching and their nonlinear conformation. Nevertheless, spectroscopy studies indicate that there is partial order in the orientation of hemicelluloses and pectins in the cell wall, probably as a result of a physical tendency for these polymers to become aligned along the long axis of cellulose (Morikawa et al. 1978; Séné et al. 1994).

Hemicelluloses Are Matrix Polysaccharides That Bind to Cellulose

Hemicelluloses* are a heterogeneous group of polysaccharides (Figure 15.10) that are bound tightly in the wall. Typically they are solubilized from depectinated walls by the use of a strong alkali (0.1 to 4 M NaOH). Several kinds of hemicelluloses are found in plant cell walls, and walls from different tissues and different species vary in their hemicellulose composition.

* A more recent name for hemicelluloses is **cellulose-linking glycans**, a name that reflects current thinking about the function of these polysaccharides and avoids the false connotation that hemicelluloses are modified celluloses (Bacic et al. 1988). In this chapter we will adhere to the more traditional term, "hemicellulose."

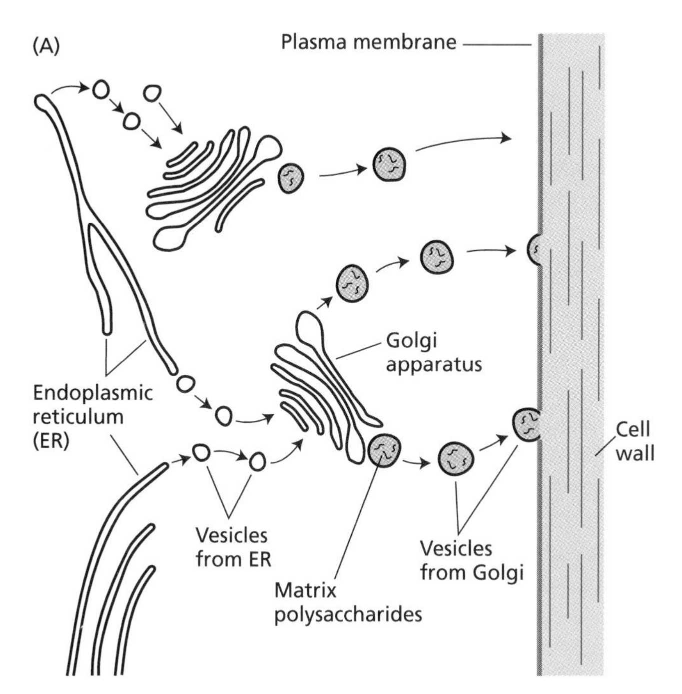

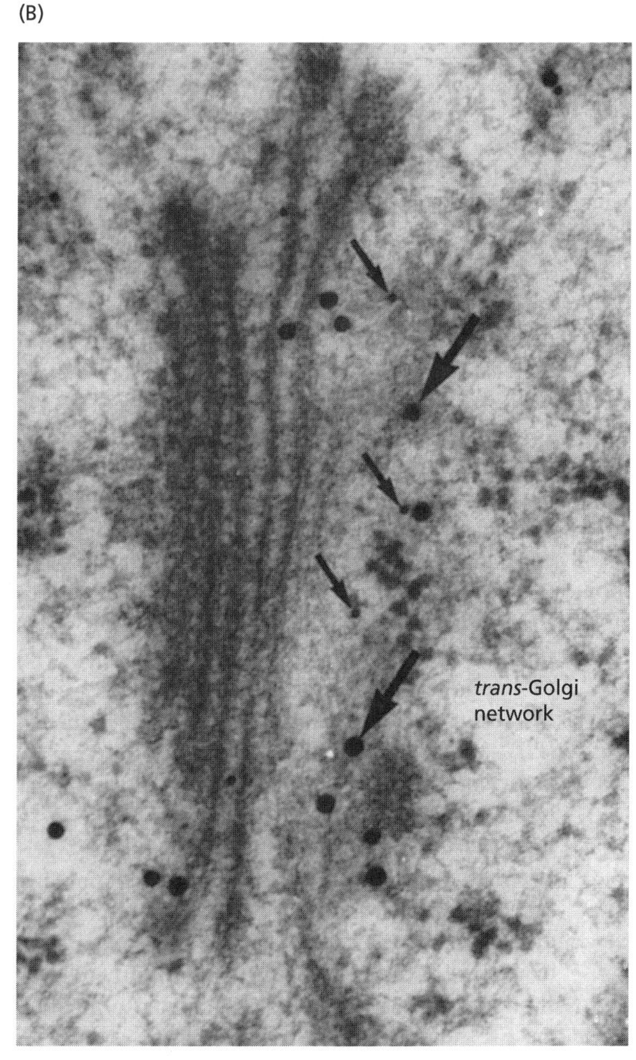

Figure 15.9 (A) Diagram showing the scheme for synthesis and delivery of matrix polysaccharides to the cell wall. Polysaccharides are synthesized by enzymes in the Golgi apparatus and then secreted to the wall by fusion of membrane vesicles to the plasma membrane. (B) Electron micrograph showing Golgi stacks and vesicles containing xyloglucan (large arrows) and glycosylated proteins (small arrows). This section, taken from a sycamore suspension-cultured cell, was labeled with two types of antibodies conjugated to colloidal gold particles (the large particles are attached to the antixyloglucan antibody; the small particles are attached to the antibody that detects glycosylated proteins). (A after Brett and Waldron 1996; B from B. Gunning and M. Steer, *Plant Cell Biology: Structure and Function*, Jones and Bartlett, 1996, photo by A. Staehelin.)

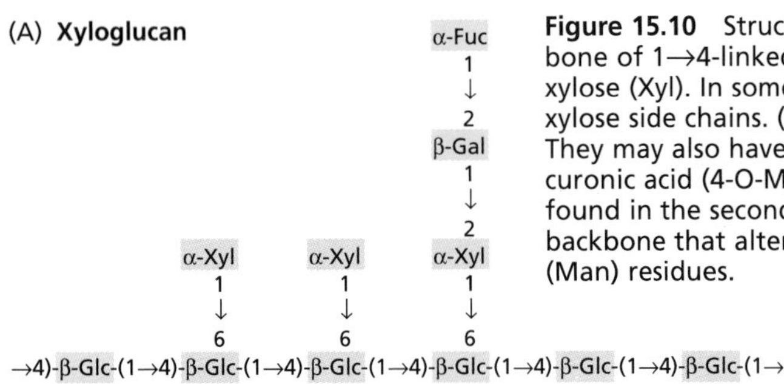

Figure 15.10 Structure of common hemicelluloses. (A) Xyloglucan has a backbone of 1→4-linked β-D-glucose (Glc), with 1→6-linked branches containing xylose (Xyl). In some cases galactose (Gal) and fucose (Fuc) are added to the xylose side chains. (B) Xylans have a 1→4-linked backbone of β-D-xylose (Xyl). They may also have side chains containing arabinose (Ara), 4-O-methyl-glucuronic acid (4-O-Me-β–GlcA), or other sugars. (C) Glucomannans are most often found in the secondary walls, especially of conifers. They have a 1→4-linked backbone that alternates one β-D-glucose (Glc) residue with two β-D-mannose (Man) residues.

(B) **Xylans**

→4)-β-Xyl-(1→4)-β-Xyl-(1→4)-β-Xyl-(1→4)-β-Xyl-(1→4)-β-Xyl-(1→4)-β-Xyl-(1→
2 ↑ 1 4-O-Me-β-GlcA
2 ↑ 1 α-Ara

(C) **Glucomannans**

→4)-β-Glc-(1→4)-β-Man-(1→4)-β-Man-(1→4)-β-Glc-(1→4)-β-Man-(1→4)-β-Man-(1→

In the primary wall of dicotyledons (the best-studied example), the most abundant hemicellulose is **xyloglucan** (see Figure 15.10A). This polysaccharide has a backbone of 1→4-linked β-D-glucose residues, just like cellulose has. Unlike cellulose, however, xyloglucan has short side chains that contain xylose, galactose, and often, though not always, a terminal fucose. By interfering with the linear alignment of the glucan backbones with one another, these side chains prevent the assembly of xyloglucan into a crystalline microfibril. The terminal L-fucose on the side chains is thought to stabilize the xyloglucan backbone in a planar configuration. According to recent molecular models, the terminal fucose residues of the xyloglucan side chains fold back and interact with the glucan chain in such a way that the nearby glucose residues (three to four) are held in a linear (planar) configuration, rather than turning or twisting. This flattened structure facilitates the binding of short segments of the xyloglucan backbone to the surface of the cellulose microfibril (Levy et al. 1997). Since xyloglucans are longer (about 50 to 500 nm) than the spacing between cellulose microfibrils (20 to 40 nm), they have the potential to link several microfibrils together.

Depending on developmental state and plant species, the hemicellulose fraction of the wall also contains other important polysaccharides—for example, **xylans** (including branched xylans such as *arabinoxylan* and *glucuronoxylan*) and **glucomannans** (see Figure 15.10B and C). Secondary walls often contain less xyloglucan and more xylans and glucomannans, which bind hemicelluloses to cellulose in a fashion similar to that of xyloglucan.

Although the cell walls of cereal grasses (monocots such as rice, oats, and maize) contain some xyloglucan, other polymers are more abundant in the hemicellulose fraction of grasses, specifically a mixed-linked (1→3,1→4)β-D-glucan and an arabinoxylan, which has a (1→4)β-D-xylan backbone with short side chains of arabinose and glucuronic acid. These hemicelluloses are believed to serve the same role that xyloglucans play in dicotyledons.

Pectins Are Gel-Forming Components of the Matrix

Like the hemicelluloses, pectins constitute a heterogeneous group of polysaccharides (Figure 15.11), characteristically containing acidic sugars such as galacturonic acid and neutral sugars such as rhamnose, galactose, and arabinose. Pectins are the most soluble of the wall polysaccharides; they can be extracted with hot water or with calcium chelators.

Some pectins have a relatively simple primary structure, such as **homogalacturonan** (see Figure 15.11B), a linear polymer of 1→4-linked α-D-galacturonic acid, with occasional rhamnosyl residues that put a kink in the chain. One of the most abundant of the pectins is **rhamnogalacturonan I (RG I)**, which has a long backbone and a variety of side chains (see Figure 15.11A). The backbone consists of a repeating disaccharide of rhamnose and galacturonic acid, whereas the side chains are arabinans, galactans, arabinogalactans, and perhaps homogalacturonans. Rhamnogalacturonan I forms a very large molecule, thought to contain highly branched ("hairy") regions interspersed with "smooth" regions of homogalacturonan (Figure 15.12A).

Pectins may be very complex. A striking example is a highly branched pectic polysaccharide called **rhamnogalacturonan II (RG II)**, which contains at least ten different sugars in a complicated pattern of linkages.* Indeed, the complexity of RG II suggests that it may have functions other than being a simple structural component of the wall, such as cell signaling.

*Although they have similar-sounding names, RG I and RG II have very different structures.

(A) **Rhamnogalacturonan I (RG I)**

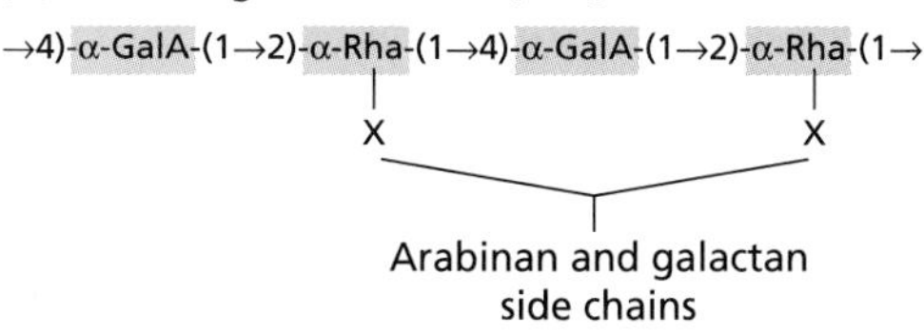

(B) **Homogalacturonan (polygalacturonic acid)**

→4)-α-GalA-(1→4)-α-GalA-(1→4)-α-GalA-(1→4)-α-GalA-(1→4)-α-GalA-(1→

(C) **Arabinans**

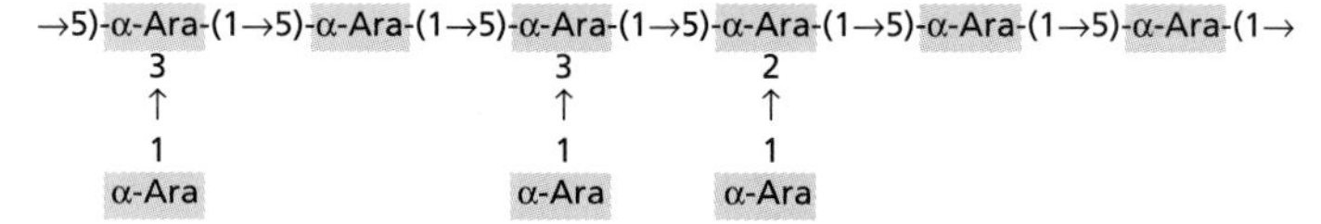

(D) **Arabinogalactan I**

→4)-β-Gal-(1→4)-β-Gal-(1→4)-β-Gal-(1→4)-β-Gal-(1→4)-β-Gal-(1→4)-β-Gal-(1→

3 ↑ 1: α-Ara-(1→5)-α-Ara-(1→5)-α-Ara

3 ↑ 1: α-Ara

Figure 15.11 Structure of the most common pectins. (A) Rhamnogalacturonan I (RG I) is a very large and heterogeneous pectin, with a backbone of 1,4-linked α-galacturonic acid (GalA) and 1,2-linked α-rhamnose (Rha). Side chains (X) are attached to rhamnose and are composed principally of arabinans and galactans, respectively. These side chains may be short or quite long. The galacturonic acid residues are often methyl esterified. (B) Homogalacturonan, also known as polygalacturonic acid or pectic acid, is made up of 1,4-linked α-galacturonic acid (GalA). The carboxyl residues are often methyl esterified. (C) Arabinans are highly branched molecules composed principally of arabinose (Ara). (D) Arabinogalactan I has a backbone of galactose (Gal) residues and side chains containing arabinose (Ara).

(A) **Rhamnogalacturonan I structure**

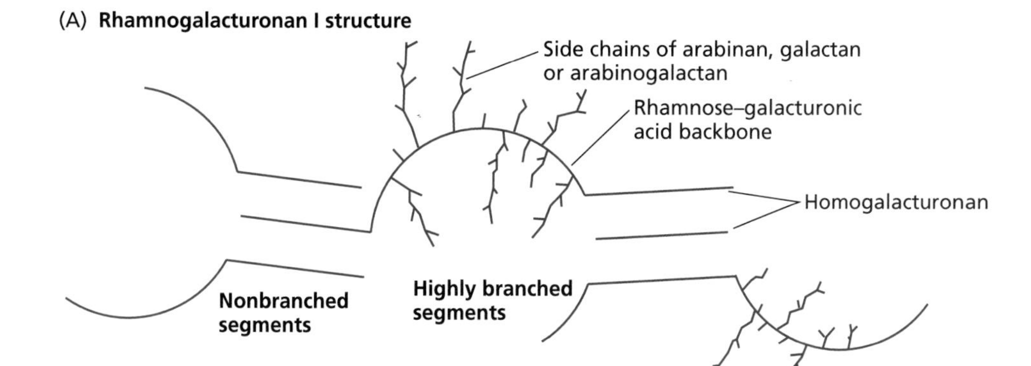

(B) **Calcium–pectin gel formation**

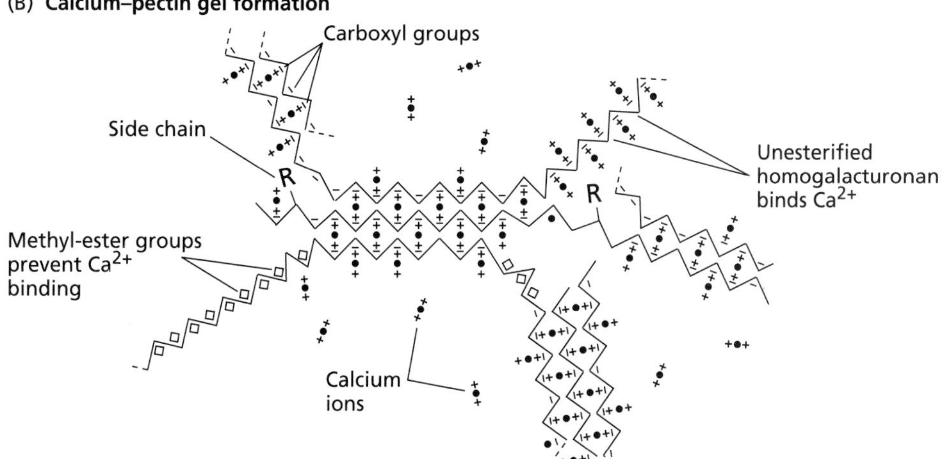

Figure 15.12 Pectin structure. (A) Proposed structure of rhamnogalacturonan I, containing highly branched segments interspersed with nonbranched segments, and a backbone of rhamnose and galacturonic acid. (B) Formation of a pectin network involves ionic bridging of the nonesterified carboxyl groups (COO^-) by calcium ions. Methyl-esterified groups cannot participate in this type of interchain network formation. Likewise, the presence of side chains on the backbone interferes with network formation.

Pectins form hydrated gels in which the charged carboxyl (COO^-) groups of neighboring pectin molecules are ionically linked together via Ca^{2+}. A large calcium-bridged network may thus form, as illustrated in Figure 15.12B. In addition to being connected by calcium bridging, pectins may be linked to each other by various covalent bonds, including ester linkages through phenolic dimers such as diferulic acid (see Chapter 13). Borate diesters of RG II have also been identified.

Pectins are subject to modifications that may alter their conformation and linkage in the wall. Many of the acidic residues are esterified with methyl, acetyl, and other unidentified groups during biosynthesis in the Golgi apparatus. Such esterification prevents calcium bridging between pectins and thus reduces the gel-forming character of the pectin. Once the pectin is secreted into the wall, the ester groups may be removed by pectin esterases found in the wall, increasing the ability of the pectin to form a rigid gel. By creating free carboxyl groups, de-esterification also increases the electric-charge density in the wall, which in turn may influence the concentration of ions in the wall and the activities of wall enzymes.

Structural Proteins Become Cross-Linked in the Wall

In addition to the major polysaccharides described in the previous section, the cell wall contains several classes of structural proteins (Showalter 1993). These proteins usually are classified according to their predominant amino acid composition—e.g., *hydroxyproline-rich glycoprotein* (*HRGP*), *glycine-rich protein* (*GRP*), *proline-rich protein* (*PRP*), and so on (Table 15.2). Some wall proteins have sequences that are characteristic of more than one class. Many structural proteins of walls have highly repetitive primary structures and are highly glycosylated. In vitro extraction studies have shown that newly secreted wall structural proteins are relatively soluble, but they become more and more insoluble during cell maturation or in response to wounding. The biochemical nature of the insolubilization process is uncertain, but intermolecular diphenylether linkages between tyrosines have been suggested.

Wall structural proteins vary greatly in their abundance, depending on cell type, maturation, and previous stimulation. Wounding, pathogen attack, and treatment with elicitors (molecules that activate plant defense responses; see Chapter 13) increase expression of the genes that code for many of these proteins. In histological studies, wall structural proteins are often localized to specific cell and tissue types. For example, HRGPs are associated mostly with cambium, phloem parenchyma , and various types of sclerenchyma. GRPs and PRPs are most often localized to xylem vessels and fibers and thus are more characteristic of a differentiated cell wall.

Hydroxyproline-rich proteins (**HRGPs**) are the best-studied class of wall structural proteins in plants; they constitute a large family of proteins (Kieliszewski and Lamport 1994). HRGPs are related to the wall structural proteins of *Chlamydomonas* and other volvocine species of green algae, which form a noncellulosic protein-based wall (Woessner and Goodenough 1994). A subfamily of plant HRGPs with the serine–hydroxyproline repeating motif, Ser-Hyp-Hyp-Hyp-Hyp, was originally named "extensin" because of a mistaken belief that it controlled the extensibility of the cell wall, but this function has not proved to be correct. The precise role of HRGPs is still unknown. However, the fact that HRGP gene expression is induced by wounding and by pathogen infections suggests that HRGPs are involved in protection against pathogens and desiccation and perhaps in strengthening of walls. HRGPs may also serve as nucleating sites for lignification during the formation of secondary walls.

In addition to the structural proteins already listed, cell walls contain **arabinogalactan proteins** (**AGPs**). These water-soluble proteins are very heavily glycosylated: More than 90% of the mass of AGPs may be sugar residues, primarily galactose and arabinose (Bacic et al. 1996). Multiple AGP forms are found in plant tissues, either in the wall or associated with the plasma membrane, and they display tissue- and cell-specific expression patterns. Less than 1% of the dry mass of the wall is made up of AGPs. AGPs may function in cell adhesion and in cell signaling during cell differentiation. As evidence for the latter idea, treatment of suspension cultures with exogenous AGPs or with agents that specifically

Table 15.2
Structural proteins of the cell wall

Class of cell wall proteins	Percentage carbohydrate	Localization typically in:
HRGP (hydroxyproline-rich glycoprotein)	~55	Phloem, cambium, schlerids
PRP (Proline-rich protein)	~0–20	Xylem, fibers, cortex
GRP (glycine-rich protein)	0	Xylem

bind AGPs is reported to influence cell proliferation and embryogenesis. AGPs are also implicated in the growth, nutrition, and guidance of pollen tubes through stylar tissues, as well as in other developmental processes (Cheung et al. 1996).

New Primary Walls Are Assembled during Cytokinesis

Primary walls originate de novo during the final stages of cell division, when the newly formed **cell plate** separates the two daughter cells and solidifies into a stable wall that is capable of bearing a physical load from turgor pressure (see Figure 1.28). The cell plate forms when Golgi vesicles and ER cisternae aggregate in the spindle midzone area of a dividing cell. This aggregation is organized by the **phragmoplast**, a complex assembly of microtubules, membranes, and vesicles that forms during late anaphase or early telophase. The membranes of the vesicles fuse with each other, and with the lateral plasma membrane, to become the new plasma membrane separating the daughter cells. The contents of the vesicles are the precursors from which the new middle lamella and the primary wall are assembled. A similar process of de novo wall formation occurs during cellularization of noncellular endosperm (in some species, endosperm nuclei initially divide freely without cytokinesis; only later do walls form to separate the endosperm into cellular compartments).

After a wall forms, it can grow and mature through a process that may be outlined as follows: synthesis → secretion → assembly → expansion (in growing cells) → cross-linking and secondary-wall formation. The synthesis and secretion of the major wall polymers were described earlier. Now we will consider the assembly and expansion of the wall.

After their secretion into the extracellular space, the wall polymers must be assembled into a cohesive structure; that is, the individual polymers must attain the physical arrangement and bonding relationships that are characteristic of the wall. Although the details of wall assembly are not understood, the prime candidates for this process are self-assembly and enzyme-mediated assembly.

Self-assembly is attractive because it is mechanistically simple. Wall polysaccharides possess a marked tendency to aggregate spontaneously into organized structures. For example, isolated cellulose may be dissolved in strong solvents and then spun into stable fibers, called rayon. Similarly, hemicelluloses may be dissolved in strong alkali; when the alkali is removed, these polysaccharides aggregate into concentric, ordered networks that resemble the native wall at the ultrastructural level (Roland et al. 1977). This tendency to aggregate can make the separation of hemicellulose into its component polymers technically difficult. In contrast, pectins are more soluble and tend to form dispersed, isotropic networks (gels). These observations indicate that the wall polymers have an inherent ability to aggregate into partly ordered structures.

In addition to self-assembly, wall enzymes may take part in putting the wall together. A prime candidate for enzyme-mediated wall assembly is **xyloglucan endotransglycosylase (XET)**. This enzyme has the ability to cut the backbone of xyloglucans and to join one end of the cut xyloglucan with the free end of an acceptor xyloglucan (Figure 15.13). Such **transglycosylation** might assist the integration of newly synthesized xyloglucans into the existing load-bearing wall. Newly synthesized xyloglucans are much smaller than the bulk xyloglucans in the wall. Evidently these polymers can be "stitched" together in some way, end to end. XET may serve in this role, or it may facilitate wall expansion (Fry 1995; Nishitani 1997).

Other wall enzymes that might aid in assembly of the wall include glycosidases, pectin methyl esterases, and various oxidases. Some glycosidases remove the side chains of hemicelluloses. This "debranching" activity

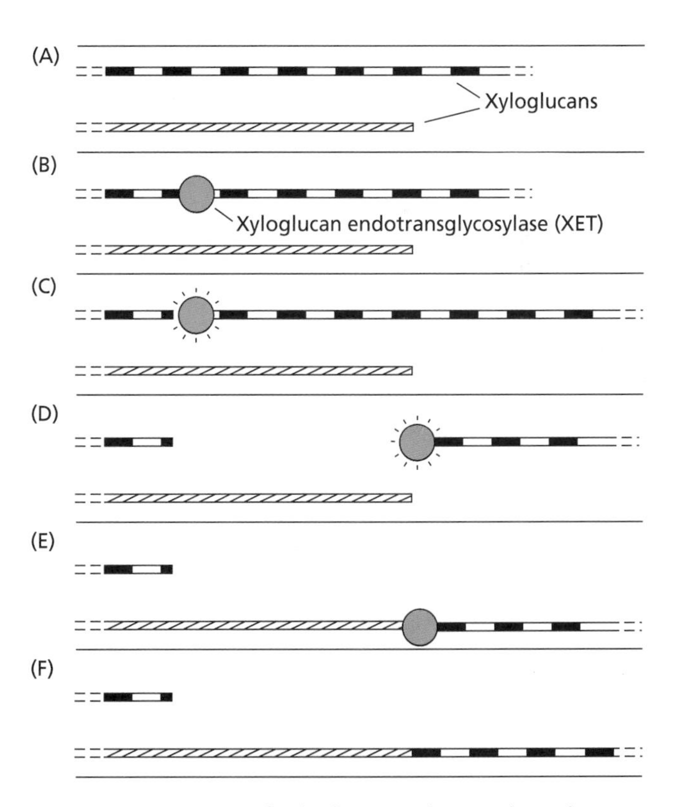

Figure 15.13 Action of xyloglucan endotransglycosylase (XET) to cut and stitch xyloglucan polymers into new configurations. Two xyloglucan chains are shown in (A) with two distinct patterns to emphasize their rearrangement. XET binds to the middle of one xyloglucan (B), cuts it (C), and transfers one end to the end of a second xyloglucan (D, E), resulting in one shorter and one longer xyloglucan (F). (After Smith and Fry 1991.)

increases the tendency of hemicelluloses to adhere to the surface of cellulose microfibrils. Pectin methyl esterases hydrolyze the methyl esters that block the carboxyl groups of pectins. By unblocking the carboxyl groups, these enzymes increase the concentration of acidic groups on the pectins and enhance the ability of pectins to form a Ca^{2+}-bridged gel network. Oxidases such as peroxidase may catalyze cross-linking between phenolic groups (tyrosine, phenylalanine, ferulic acid) in wall proteins, pectins, and other wall polymers. Such phenolic coupling is clearly important for the formation of lignin cross-links, and it may likewise link diverse components of the wall together.

Secondary Walls Form in Some Cells after Expansion Ceases

After wall expansion ceases, cells sometimes continue to synthesize a wall, known as a secondary wall. Secondary walls are often quite thick, as in tracheids, fibers, and other cells that serve in mechanical support of the plant (Figure 15.14). Often such walls are multilayered and differ in structure and composition from the primary wall. For example, the secondary walls in wood contain xylans rather than xyloglucans, as well as a higher proportion of cellulose. The orientation of the cellulose microfibrils may be more neatly aligned parallel to each other in secondary walls than in primary walls. Secondary walls are often (but not always) impregnated with lignin.

Lignin is a phenolic polymer with a complex, irregular pattern of linkages that link the aromatic alcohol subunits together (see Figures 13.18 and 13.19). These subunits (coumaryl, coniferyl, and sinapyl alcohols) are synthesized from phenylalanine and are secreted to the wall, where they are oxidized in place by the enzymes peroxidase and laccase. As lignin forms in the wall, it displaces water from the matrix and forms a hydrophobic meshwork that bonds tightly to cellulose and prevents wall enlargement (Figure 15.15). Lignin adds significant mechanical strength to cell walls and reduces the susceptibility of walls to attack by pathogens. Lignin also reduces the digestibility of plant material by animals. Genetic engineering of lignin content and structure may improve the digestibility and nutritional content of plants used as animal fodder.

The Polarity of Diffuse Growth

During plant cell enlargement, new wall polymers are continuously synthesized and secreted at the same time that the preexisting wall is expanding. Wall expansion may be localized (as in the case of **tip growth**) or evenly distributed over the wall surface (**diffuse growth**) (Figure 15.16). Even in cells with diffuse growth, however, different parts of the wall may enlarge at different rates or in different directions. These differences may be due to structural or enzymatic variations in specific regions of the wall or variations in the stresses borne by differ-

(A)

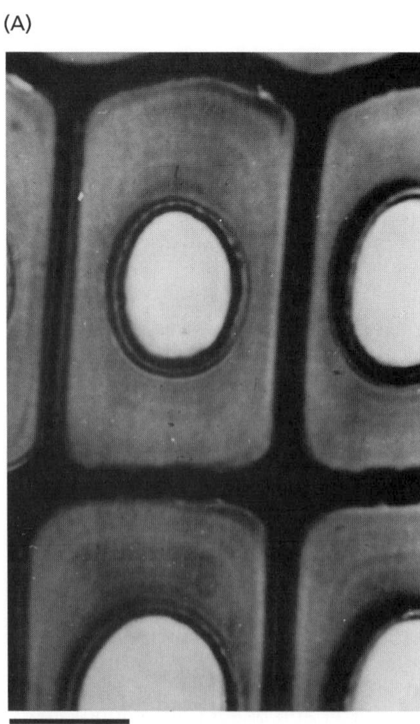

(B)

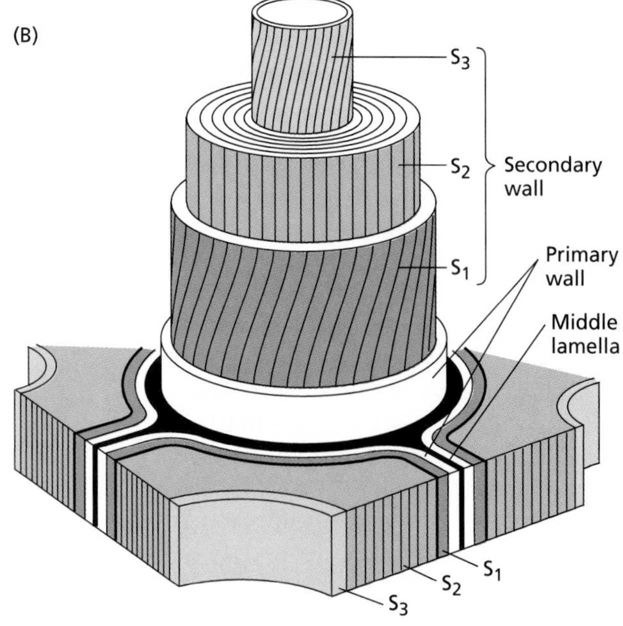

Figure 15.14 (A) Cross section of a pine tracheid, in which three distinct layers in the secondary wall are visible. (B) Diagram of the cell wall organization often found in tracheids and other cells with thick secondary walls. Three distinct layers (S_1, S_2, S_3) are formed interior to the primary wall. (A from Fahn 1990, courtesy of A. Fahn.)

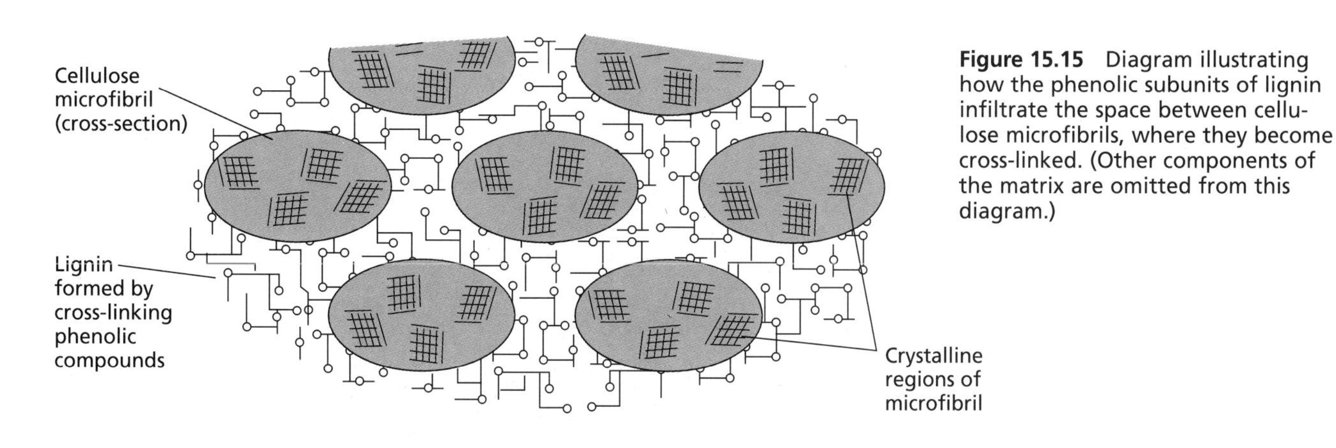

Figure 15.15 Diagram illustrating how the phenolic subunits of lignin infiltrate the space between cellulose microfibrils, where they become cross-linked. (Other components of the matrix are omitted from this diagram.)

ent regions of the wall. As a consequence of this uneven pattern of wall expansion, plant cells may assume very irregular forms.

Typically, however, plant cells have a characteristic cylindrical shape, which serves as the basis for cellular polarity. Plants, especially at the early stages of development, are extremely polar organisms, characterized by a root apical meristem at one end and a shoot apical meristem at the other. This polarity can be traced to the cellular level, and it is evident in phenomena such as polar auxin transport (see Chapter 19) and the earliest cell divisions of the zygote (see Chapter 16). In this section we will discuss how polarity is established in both diffuse- and tip-growing cells.

Microfibril Orientation Determines Polarity in Diffuse-Growing Cells

As described in Chapter 3, water is pulled into cells by osmosis, but the cell wall resists water uptake, creating a strong positive turgor pressure within the cell. This internal hydrostatic pressure is directed outward, equally in all directions. Cellulose microfibrils have great tensile strength, and the ability to resist expansion driven by water uptake is determined by reinforcement of the amorphous cell wall matrix by cellulose microfibrils. The orientation of the cellulose microfibrils within the matrix can also determine the polarity of cell expansion.

When cells first form in the meristem, they are isodiametric; that is, they have equal diameters in all directions. If the orientation of cellulose microfibrils in the primary cell wall were **isotropic** (randomly arranged), the cell would grow equally in all directions, expanding radially to generate a sphere (Figure 15.17A). In most plant cell walls, however, the arrangement of cellulose microfibrils is **anisotropic** (nonrandom) (see Chapter 1). Cellulose microfibrils are deposited mainly in the lateral walls of cylindrically shaped, expanding cells such as cortical and vascular cells of stems and roots, or the giant internode cells of the filamentous green alga *Nitella.* Moreover, the cellulose microfibrils are deposited circumferentially (transversely) in these lateral walls, at right angles to the long axis of the cell. The circumferentially arranged cellulose microfibrils have been likened to hoops in a barrel, restricting growth in girth and promoting growth in length (Figure 15.17B). Thus, circumferentially reinforced cells elongate, increasing only minimally in girth.

Cell wall deposition continues as cells enlarge. Each successive wall layer is stretched and thinned during cell expansion, so the microfibrils become passively reoriented in the longitudinal direction—that is, in the direction of growth. Successive layers of microfibrils thus show a gradation in their degree of reorientation across the thickness of the wall, and those in the outer layers are longitudinally oriented as a result of wall stretching. Because of thinning, these outer layers have

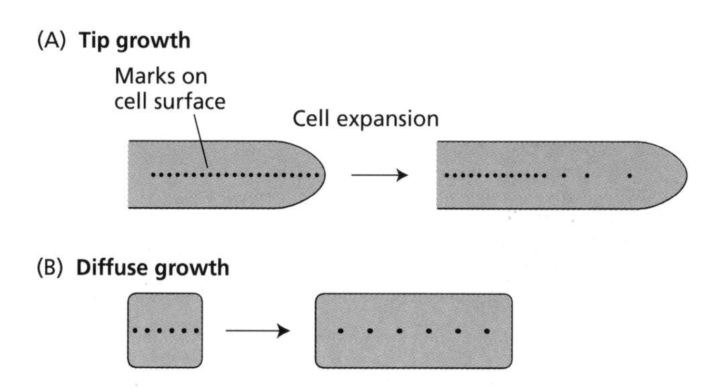

Figure 15.16 The cell surface expands differently during tip growth and diffuse growth. (A) Expansion of a tip-growing cell is confined to an apical dome at one end of the cell. If marks are placed on the cell surface and the cell is allowed to continue to grow, only the marks that were initially within the apical dome grow farther apart. Root hairs and pollen tubes are examples of plant cells that exhibit tip growth. (B) If marks are placed on the surface of a diffuse-growing cell, the distance between all the marks increases as the cell grows. Most cells in multicellular plants grow by diffuse growth.

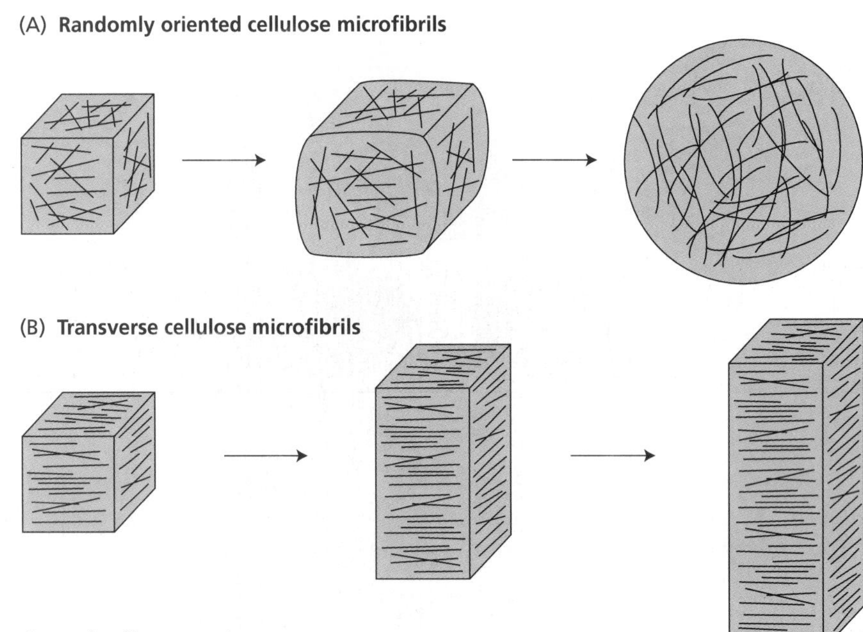

Figure 15.17 The orientation of newly deposited cellulose microfibrils determines the direction of cell expansion. (A) If the cell wall is reinforced by randomly oriented cellulose microfibrils, the cell will expand equally in all directions, forming a sphere. (B) When most of the reinforcing cellulose microfibrils have the same orientation, the cell expands at right angles to the microfibril orientation and is constrained in the direction of the reinforcement. Here the microfibril orientation is transverse, so cell expansion is longitudinal.

much less influence on the direction of cell expansion than do the newly deposited inner layers. The inner one-fourth of the wall is largely responsible for determining the polarity of cell growth (see Box 15.2).

Cortical Microtubules Determine the Orientation of Newly Deposited Microfibrils

Newly deposited cellulose microfibrils and cytoplasmic microtubules in cell walls usually are coaligned, suggesting that microtubules determine the orientation of cellulose microfibril deposition. The main evidence for the involvement of microtubules in the deposition of cellulose microfibrils is that the orientation of the microfibrils is perturbed when cells are treated with drugs that disrupt cytoplasmic microtubules. The orientation of microtubules in the cortical cytoplasm usually mirrors that of the newly deposited microfibrils in the adjacent cell wall, and both are usually coaligned in the transverse direction, at right angles to the axis of polarity (Figure 15.18). In some cell types, such as tracheids, the microfibrils in the wall alternate between transverse and longitudinal orientations, and in such cases the

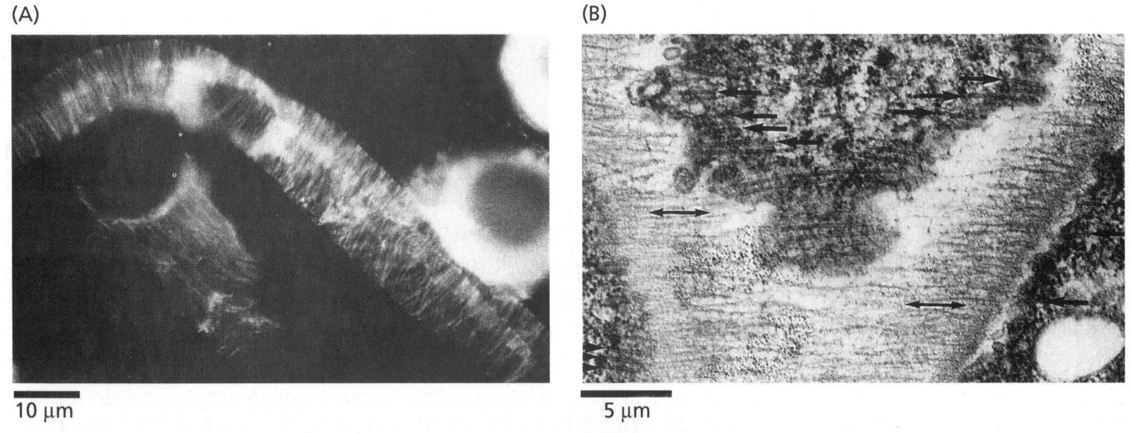

Figure 15.18 The orientation of microtubules in the cortical cytoplasm mirrors the orientation of newly deposited cellulose microfibrils in the cell wall of cells that are elongating. (A) The arrangement of microtubules can be revealed with fluorescently labeled antibodies to the microtubule protein tubulin. In this long cylindrical cell from an onion (*Allium cepa*) root, the microtubules run transversely (circumferentially) around the cell. (B) The alignment of cellulose microfibrils in the cell wall can sometimes be seen in grazing sections prepared for electron microscopy, as in this micrograph of a developing sieve tube element in a root of *Azolla*. The longitudinal axis of the root and the sieve tube element runs vertically. Both the wall microfibrils (double-headed arrows) and the cortical microtubules (single-headed arrows) are aligned transversely. (Photos courtesy of A. Hardham.)

BOX 15.2

Using *Nitella* to Study the Mechanical Properties of Cell Walls

THE SMALL SIZE of typical plant cells (20 to 100 μm) has been a serious impediment to the study of cell wall mechanical properties. To measure the extensibility of isolated cell walls from higher plants, researchers must place entire frozen and thawed organs or tissues in an *extensometer*, as illustrated in Figures 15.21 and 15.22. Although such measurements can give qualitative answers, the extensions measured cannot be related to the growth of living cells in a quantitative way, because the applied stress is **uniaxial** (unidirectional), whereas the in vivo stress due to turgor pressure is **multiaxial** (multidirectional).

The giant internodal cells of the freshwater green algae *Nitella* and *Chara* have figured prominently in classic studies on ion transport, water permeability, and cytoplasmic streaming (Hope and Walker 1975). These algae are characterized by erect, branched coenocytic (multinucleate) filaments, differentiated into nodes and internodes, each node bearing a whorl of specialized lateral internodes (Figure 1).

The internode cell begins as a shortened disc 20 mm long, which elongates to produce a slender cylinder about 6 cm long and 0.5 mm wide at maturity. The cell expands by diffuse growth in a predominantly longitudinal direction. Because individual internode cells are large, their cell walls can be isolated and the mechanical properties of the walls measured. It is even possible to study the anisotropic (unequal) mechanical properties of individual cell walls by the use of multiaxial stress to mimic the in vivo turgor pressure.

Figure 2 shows the experimental setup for studying multiaxial extensibility in *Nitella* internode cells. Empty cell wall cylinders are glued to a pipette and inflated with mercury under pressure. Liquid mercury is insoluble and therefore inert under these conditions. In response to the multiaxial stress caused by the pressurized mercury, the cell wall cylinder expands, and investigators can monitor the expansion in both length and width by viewing the process through a horizontal microscope (Figure 3) (Metraux et al. 1980).

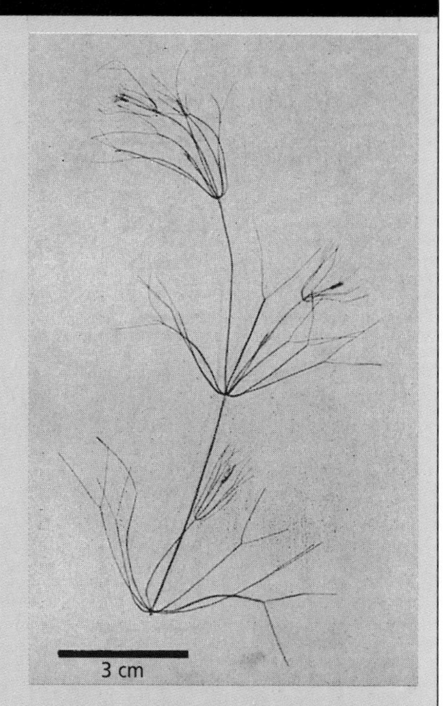

Figure 1 Filament of *Nitella axillaris* containing several internodes and lateral internodes with branches. (From Taiz et al. 1981.)

With the multiaxial stress procedure on *Nitella* internodes, one can use microtubule inhibitors to alter the orientation of the cellulose microfibrils and test the hypothesis that microfibril orientation is responsible for the mechanical anisotropy of the cell wall. As Figure 3 shows, the multiaxial extension of control cells is highly anisotropic. Not only is extension in length four times greater than extension in diameter, but much of the extension in length is irreversible, whereas the extension in diameter is almost completely reversible. These results are consistent with the anisotropic growth pattern of the cells. Figure 3 also shows the effects of pretreating the cells with the microtubule poison isopropyl *N*-phenylcarbamate (IPC), which, like colchicine, induces the random depo-

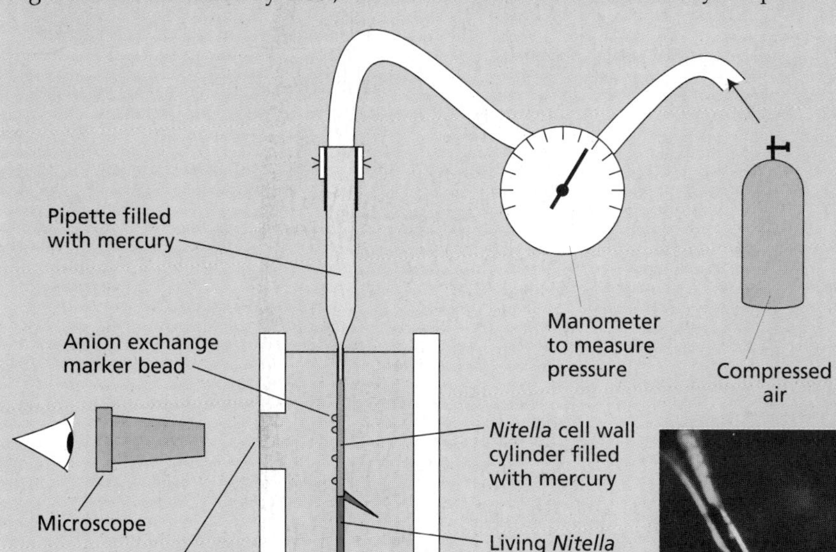

Figure 2 Apparatus for measuring the multiaxial extensibility of *Nitella* cell walls. The pressure source is a compressed air tank. Air passes through high-pressure tubing into a manometer and from there into a mercury-filled pipette glued to the open end of a *Nitella* cell wall cylinder. A living internode is maintained at the distal end. The wall is immersed in a perfusion chamber with a glass window. Displacement of resin beads attached to the surface of the wall is measured with a horizontally oriented microscope. The inset shows a view of a mercury-inflated cell wall cylinder with marker beads attached as seen through the traveling microscope. (Photo from L. Taiz.)

BOX 15.2 *(continued)*

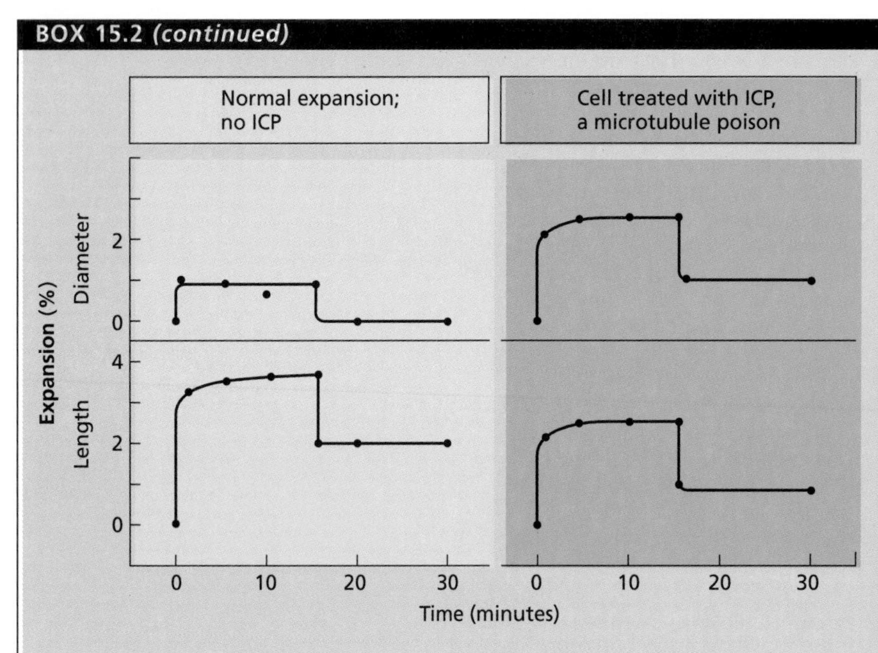

Figure 3 Data from an in vitro multiaxial cell expansion experiment showing expansion in length and diameter expressed as a percentage of the initial value. (After Taiz et al. 1981.)

sition of cellulose microfibrils in the wall. The result is that the extensions of the wall in length and diameter are about equal, consistent with the isotropic growth pattern induced by this inhibitor.

Nitella has also proven to be useful for studying cell wall properties in vivo. Many of the concepts applied to contemporary studies on the biophysical control of cell expansion are derived from the pioneering experiments on living *Nitella* cells by Paul Green and his colleagues at Stanford University (Green 1968). An important insight to come out of such studies is that the inner 25% of the cell wall determines the polarity of cell expansion. Taking advantage of the fact that cellulose microfibrils are crystalline and therefore exhibit birefringence (double refraction) when viewed with polarized light, Paul Richmond at the University of the Pacific in California devised an optical technique to measure cell expansion in length and diameter, and the orientation of the cellulose microfibril in the cell wall, simultaneously (Richmond 1983). This technique made it possible to determine the relationship between microfibril angle and the direction of cell expansion.

As Figure 4 shows, the microfibril angle began to change from transverse to random almost immediately after colchicine was added to the growth medium. However, the anisotropy of growth did not change until about 8 hours later. The transition from anisotropic growth to isotropic growth is complete after a relative increase in cell surface area of about 25%. Since wall thickness remains constant during this period because of continued wall synthesis, we can infer that the directionality of cell expansion changes when the inner 25% of the cell wall has been replaced. Thus the inner 25% of the cell wall appears to represent the load-bearing portion of the wall. By implication, the outer 75% has become too weakened by stretching to restrict cell expansion.

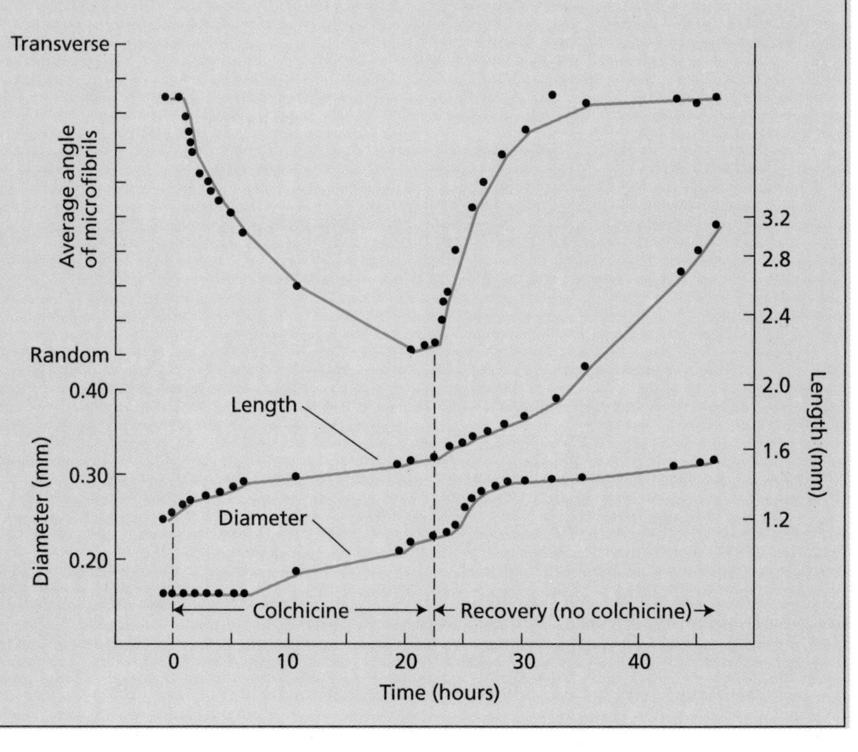

Figure 4 Effect of colchicine addition and removal on cellulose microfibril angle and anisotropic growth of a young *Nitella* internode. The growth response to colchicine treatment is an increase in the rate of growth in diameter and a decrease in the rate of growth in length. Length and diameter are plotted on log scales; the slopes of the curves express relative growth rates (see Chapter 16). Colchicine was added at time zero, and was removed at 24 hours. Note that the average angle of the cellulose microfibrils starts to become less transverse and more random almost immediately after colchicine is added, and it begins to become more transverse immediately after colchicine is removed. However, growth anisotropy does not change until 8 hours after colchicine addition and takes about 8 hours to recover after colchicine removal. (After Taiz et al. 1981.)

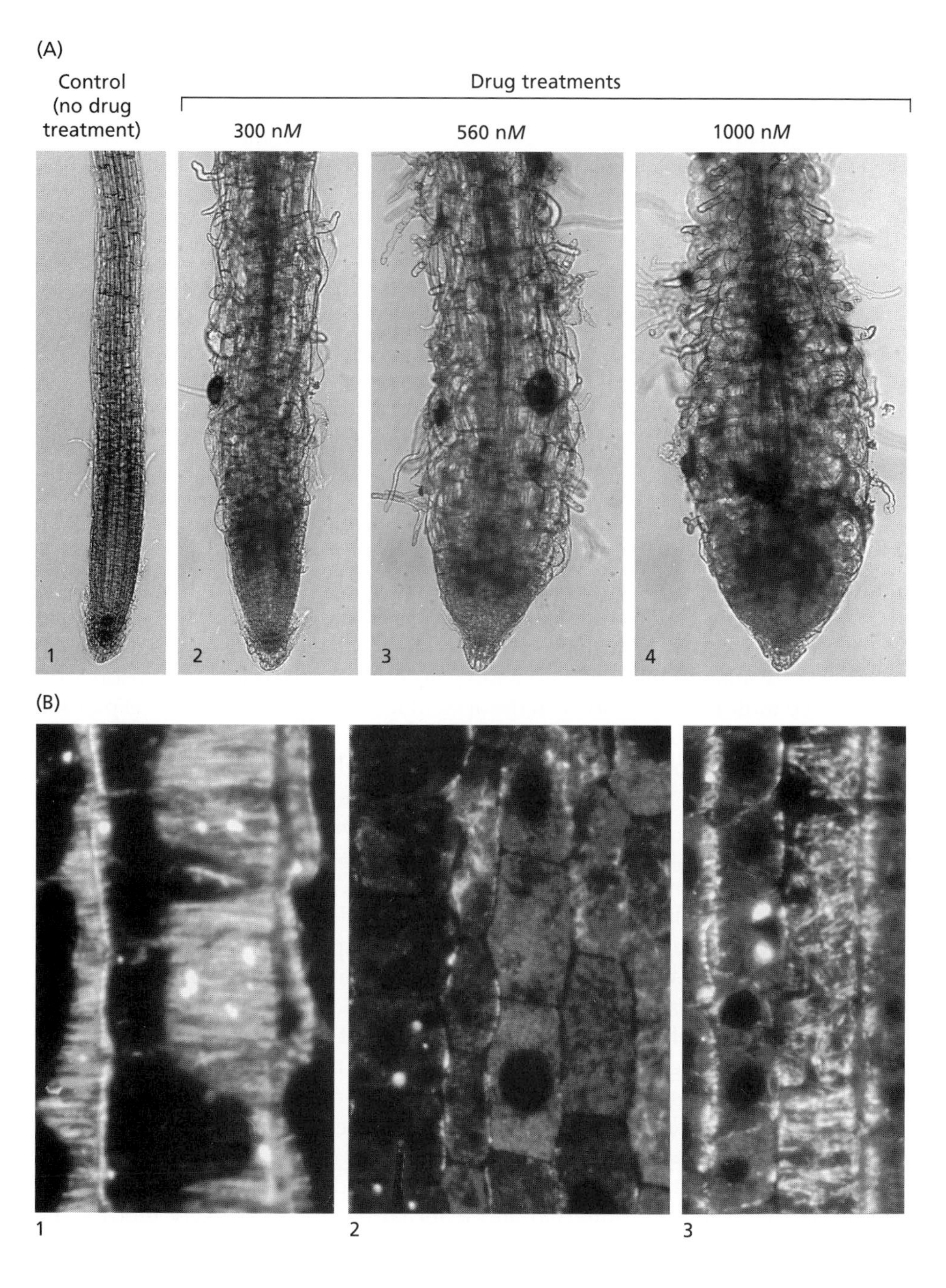

Figure 15.19 The disruption of cortical microtubules results in a dramatic increase in radial cell expansion and a concomitant decrease in elongation. (A) Roots of *Arabidopsis* seedlings were treated with different concentrations of the microtubule-depolymerizing drug oryzalin for 2 days before these photomicrographs were taken. With increasing concentration of oryzalin, there is progressively more lateral cell expansion and less cell elongation. The drug has altered the polarity of growth. (B) Microtubules were visualized by means of an indirect immunofluorescence technique and an antitubulin antibody. While cortical microtubules in the controls are oriented at right angles to the direction of cell elongation (1), those in roots treated with 300 n*M* oryzalin are more disordered (2), while very few microtubules remain in roots treated with 1000 n*M* oryzalin (3). (From Baskin et al. 1994, courtesy of T. Baskin.)

microtubules are always observed parallel to the microfibrils of the most recently deposited wall layer.

Several drugs bind to tubulin, the subunit protein of microtubules, causing them to depolymerize. When growing roots are treated with a microtubule-depolymerizing drug, such as oryzalin, the region of elongation expands laterally, becoming bulbous and tumorlike (Figure 15.19) (Baskin et al. 1994). The reason is that the enlarging cells in the growth zone expand isotropically, instead of elongating in a polar fashion. The drug-induced destruction of microtubules in the growing cells also disrupts the transverse orientation of cellulose microfibrils in the most recently deposited layers of the wall. Cell wall deposition continues in the absence of microtubules, but the cellulose microfibrils are deposited randomly and the cells expand equally in all directions. Since the antimicrotubule drugs specifically target the microtubules, these results suggest that microtubules act as guides for the orientation of cellulose microfibril deposition.

We do not know how cortical microtubules control the orientation of cellulose microfibril deposition. As discussed earlier, cellulose microfibrils are constructed outside the plasma membrane by the enzyme cellulose synthase, a complex of integral membrane proteins that form the particle rosettes. As the cell wall is deposited, the cellulose synthase complexes must move in the plane of the plasma membrane, trailing cellulose microfibrils behind them (Giddings and Staehelin 1988). Cortical microtubules are cross-linked to the cytoplasmic face of the plasma membrane and may act as guidance elements, creating oriented channels or barriers within the membrane (see Figure 15.7B). The synthase complexes might travel in these channels, propelled by forces generated by the polymerization and crystallization of the cellulose microfibrils. Microtubules, however, can generate movement as well as guidance.

Several molecular motors are known to be capable of utilizing energy from the hydrolysis of ATP to move along microtubules, transporting proteins, or other structures to which they are bound. These molecular motors include *dynein* and *kinesin*, which are known to be involved in the movement of flagella, in the transport of vesicles along microtubules, and in the movement of chromosomes to the poles at mitosis (Asada and Collings 1997). A kinesinlike microtubule motor may also be attached to the cellulose synthase complex, actively moving it in a specific direction, as determined by the microtubule.

Polarity in Diffuse-Growing Cells Is Regulated

Polarized cells have an asymmetric distribution of membrane proteins and cellular structures that is brought about by external signals. Although microtubules may be important in determining the orientation of cellulose microfibrils, we do not know how microtubule orientation is determined. Microtubule orientation may be only a manifestation of the polarity of the cell. Microtubules are transversely oriented in elongating cells because these cells have established a polar growth axis that in turn determines where microtubules are assembled within the cytoplasm. But how is plant cell polarity determined? We do not know the answer to this question either.

In budding yeast, cells become polarized in response to signals within their walls, or in response to gradients of chemicals or pheromones secreted by cells of the opposite mating type. Haploid yeast has an axial budding pattern in which the next bud will form near the bud scar that remains in the cell wall from the previous division. A signal transduction pathway transmits this information to molecules that orient the cytoskeleton, polarizing the cell (Chant 1995; Chant and Pringle 1995).

Although we do not know how polarity is established in plant cells, we do know that cell polarity can be influenced by hormones. For example, it is well established that the gaseous hormone ethylene can alter the orientation of both microtubules and cellulose microfibrils, inducing lateral cell expansion (see Chapter 22). In addition, some evidence suggests that gibberellins (see Chapter 20) can promote the deposition of transverse microfibrils (Shibaoka 1994).

Control of the Rate of Cell Elongation

Once their polarity is established, plant cells typically expand 10- to 100-fold in volume before reaching maturity. In extreme cases, cells may enlarge more than 10,000-fold in volume (e.g., xylem vessel elements). The cell wall undergoes this profound expansion without losing its mechanical integrity and generally without becoming thinner (but there are exceptions; see Roland et al. 1977). Thus, newly synthesized polymers are integrated into the wall without destabilizing it. Exactly how this integration is accomplished is uncertain. This process may be particularly critical for rapidly growing root hairs, pollen tubes, and other specialized cells that exhibit tip growth, in which the region of wall deposition and surface expansion is localized to the hemispherical dome at the apex of the tubelike cell, and cell expansion and wall deposition must be closely coordinated.

In rapidly growing cells with tip growth, the wall doubles its surface area and is displaced to the nonexpanding part of the cell within minutes. This is a much greater rate of wall expansion than is typically found in cells with diffuse growth, and it may increase the danger of wall thinning and bursting. Although diffuse growth and tip growth appear to be different growth patterns, both types of wall expansion must have analogous, if not identical, processes of polymer integration, stress relaxation, and wall polymer creep.

Many factors influence the rate of cell wall expansion. Cell type and age are important developmental factors. So too are hormones such as auxin and gibberellin. Environmental conditions such as light and water availability may likewise modulate cell expansion. These internal and external factors most likely modify cell expansion by influencing the yielding (stretching) properties of the cell wall.

In this section, we will first examine the biomechanical and biophysical parameters that control the rate of cell expansion. For cells to expand at all, the rigid cell wall must be loosened in some way. The type of wall loosening thought to be involved in plant cell expansion is termed *stress relaxation*. According to one model, the *acid-growth hypothesis*, stress relaxation is induced by cell wall acidification resulting from proton extrusion across the plasma membrane. We will explore the biochemical basis for acid-induced stress relaxation, including the role of a special class of wall-loosening proteins called *expansins*. As the cell approaches its maximum size, its growth rate diminishes and finally ceases altogether. At the end of this section we will consider the process of cell wall rigidification that leads to the cessation of growth.

Stress Relaxation of the Cell Wall Drives Water Uptake and Cell Elongation

Because the cell wall is the major mechanical restraint that limits cell expansion, much attention has been given to its physical properties. As a hydrated polymeric material, the plant cell wall has physical properties that are intermediate between those of a solid and those of a liquid. We call these **viscoelastic**, or **rheological** (flow), **properties**. Growing cell walls are generally found to be less rigid than walls from nongrowing cells, and under appropriate conditions they exhibit a long-term irreversible stretching, or **yielding**, that is lacking or nearly lacking in nonexpanding walls.

Stress relaxation is a crucial concept for understanding how cell walls enlarge (Cosgrove 1997). "Stress" is used here in the mechanical sense, as force per unit area. Wall stresses arise as an inevitable consequence of cell turgor. The turgor pressure in growing plant cells is typically between 0.3 and 1.0 MPa. Turgor pressure stretches the cell wall and generates a counterbalancing physical stress or tension in the wall. Because of cell geometry, this wall tension is equivalent to 10 to 100 MPa of tensile stress—a very large stress indeed.

This simple fact has important consequences for the mechanics of cell enlargement. Whereas animal cells can change shape in response to cytoskeleton-generated forces, such forces are negligible compared with the turgor-generated forces that are resisted by the plant cell wall. To change shape, plant cells must thus control the direction and rate of wall expansion, which they do by secreting cellulose in a biased orientation (which determines the directionality of cell wall expansion) and by selectively loosening the bonding between cell wall polymers. This biochemical loosening enables the wall polymers to slip by each other, thereby increasing the wall surface area. At the same time, such loosening reduces the physical stress in the wall.

Such wall stress relaxation is crucial because by this means growing plant cells reduce their turgor and water potential, which enables them to absorb water and to expand. Without stress relaxation, wall synthesis would only thicken the wall, not expand it. During secondary-wall deposition in nongrowing cells, for example, stress relaxation does not occur.

The water uptake that is induced by wall stress relaxation enlarges the cell and tends to restore wall stress and turgor pressure to their equilibrium values (see Box 15.3). However, if growing cells are physically prevented from taking up water, wall relaxation progressively reduces cell turgor. This situation may be detected, for example, by turgor measurements with a **pressure probe** or by water potential measurements with a **psychrometer** or a **pressure chamber**. Figure 15.20 shows such an experiment.

The time course of stress relaxation derived from these experiments illustrates two common properties of growing cells: Wall stress relaxation exhibits a *yield threshold* (see Box 15.3), and above the yield threshold the rate of relaxation depends on wall stress (or turgor pressure). These characteristics are directly related to the turgor dependence of cell wall enlargement, which also exhibits a yield threshold, or minimum turgor required for growth, and a dependence of growth rate on the turgor in excess of the yield threshold (see Box 15.3). The slope of this growth–turgor relation is usually called *wall extensibility*, often represented by the symbol m. Experimental studies indicate that wall extensibility and yield threshold may be modulated by hormones and by growth conditions (see Chapters 19 and 20). Nongrowing cells exhibit negligible wall relaxation in vivo.

Acid-Induced Growth Is Mediated by Expansins

An important characteristic of growing cell walls is that they extend much faster at acidic pH than at neutral pH (Rayle and Cleland 1992). This phenomenon is called **acid growth**. In living cells, acid growth is evident when growing cells are treated with acid buffers or with fusicoccin, which induces acidification of the cell wall solution by activating an H^+-ATPase in the plasma membrane. There is some controversy about whether the acidification caused by auxin is sufficient to account for

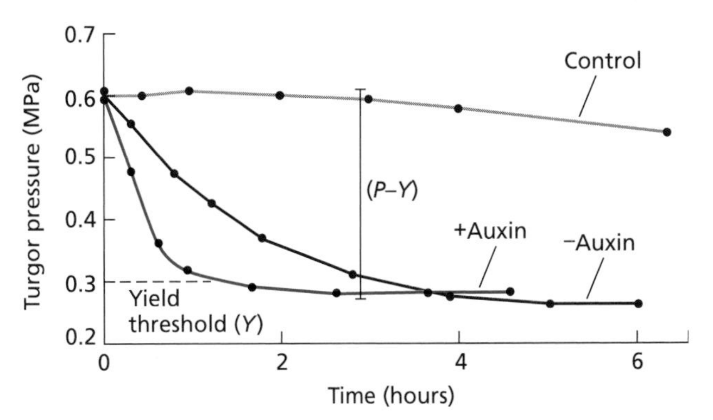

Figure 15.20 Reduction of cell turgor pressure (water potential) by stress relaxation. In this experiment, the excised stem segments from growing pea seedlings were incubated in solution with or without auxin, then blotted dry and sealed in a humid chamber. Cell turgor pressure (*P*) was measured at various time points. The segments treated with auxin rapidly reduced their turgor to the yield threshold (*Y*), as a result of rapid wall relaxation. The segments without auxin showed a slower rate of relaxation. The control segments were treated the same as the group treated with auxin, except that they remained in contact with a drop of water, which prevented wall relaxation. (After Cosgrove 1985.)

BOX 15.3

The Biophysics of Plant Cell Enlargement

WHEN PLANT CELLS ENLARGE before maturation, the increase in volume is generated mostly by uptake of water. This water ends up principally in the vacuole, which takes up an ever-larger proportion of the cell volume as the cell grows. Here we will discuss the biophysical equations that describe how growing cells regulate their water uptake and how this uptake is coordinated with wall expansion.

Water uptake by growing cells is a passive process. There are no active water pumps; instead the growing cell is able to lower the water potential inside the cell so that water is taken up spontaneously in response to a water potential difference, without direct energy expenditure. We define the water potential difference, $\Delta\Psi_w$ (expressed in megapascals), as the water potential outside the cell minus the water potential inside (see Chapters 3 and 4). The rate of uptake also depends on the surface area of the cell (*A*, in square meters) and the hydraulic conductivity of the membrane (*Lp*, in meters per second per megapascal). Thus we have the rate of water uptake in volume units (dV/dt, expressed in cubic meters per second):

$$dV/dt = A \times Lp(\Delta\Psi_w) = -A \times Lp(\Psi_s + \Psi_p) \quad (1)$$

The right side of this equation was obtained from the assumption that the outside water potential is zero (pure water), and from the dissection of the cell water potential into its components of osmotic potential (Ψ_s) and turgor (Ψ_p). This equation states that *the rate of water uptake depends only on the cell area, hydraulic conductivity, cell turgor, and osmotic potential.*

Equation 1 is valid for both growing and nongrowing cells in pure water. But how can we account for the fact that growing cells can continue to take up water for a long time, whereas nongrowing cells soon cease water uptake? In a nongrowing cell, water absorption would increase cell volume, causing the protoplast to push harder against the cell wall, thereby increasing cell turgor. This increase in Ψ_p would increase cell Ψ_w, quickly bringing $\Delta\Psi_w$ to zero. Net water uptake would then cease. In a growing cell, $\Delta\Psi_w$ is prevented from reaching zero because the cell wall is "loosened." It yields irreversibly to the forces generated by turgor and thereby reduces simultaneously the wall stress and the cell turgor. This process is called **stress relaxation**, and it is the essential biophysical difference between growing and nongrowing cells.

Stress relaxation can be understood as follows. In a turgid cell, the cell contents push against the wall, causing the wall to stretch elastically (that is, reversibly) and giving rise to a counterforce, a wall stress. In a growing cell, biochemical loosening enables the wall to yield inelastically to the wall stress. Because water is nearly incompressible, only an infinitesimal expansion of the wall is needed to reduce cell turgor pressure and, simultaneously, wall stress. Thus, *stress relaxation is a decrease in wall stress with nearly no change in wall dimensions.* As a consequence of wall stress relaxation, the cell water potential is reduced and water flows into the cell, causing a measurable extension of the cell wall and increasing cell surface area and volume. Sustained growth of plant cells entails simultaneous stress relaxation of the wall, which tends to reduce turgor pressure, and water absorption, which tends to increase turgor pressure.

Empirical evidence has shown that wall relaxation and expansion depend on turgor pressure. As turgor is reduced, wall relaxation and growth slow down. Growth usually ceases before turgor reaches zero. The turgor value at which growth ceases is called the **yield threshold** (usually represented by the symbol *Y*). This dependence of cell wall expansion on turgor pressure is embodied in the following equation:

$$GR = m(\Psi_p - Y) \quad (2)$$

where *GR* is the cell growth rate and *m* is the coefficient that relates growth rate to the turgor that is in excess of the yield threshold. The coefficient *m* is usually called **wall extensibility**.

Under conditions of steady-state growth, *GR* in Equation 2 is the same as dV/dt in Equation 1. That is, the increase in the volume of the cell equals the volume of water taken up. The two equations are plotted in the figure. The two processes of wall

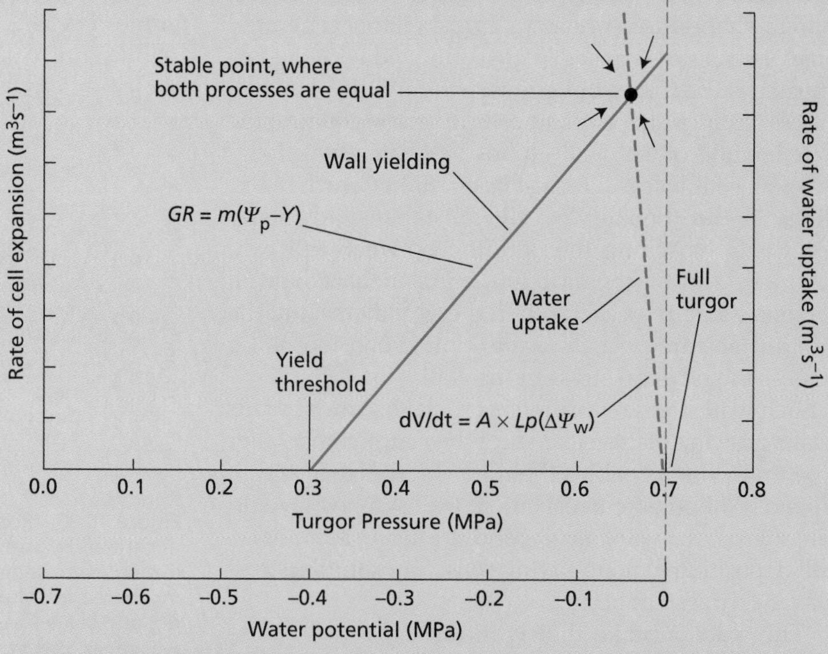

Graphic representation of the two equations that relate water uptake and cell expansion to cell turgor pressure and cell water potential. The values for the rates of cell expansion and water uptake are arbitrary. Steady-state growth is attained only at the point where the two equations intersect. Any imbalance between water uptake and wall expansion will result in changes in cell turgor and bring the cell back to this stable point of intersection between the two processes.

BOX 15.3 (continued)

expansion and water uptake show opposing reactions to a change in turgor. For example, an increase in turgor increases wall extension but reduces water uptake. Under normal conditions, the turgor is dynamically balanced in a growing cell exactly at the point where the two lines intersect. At this point both equations are satisfied, and water uptake is exactly matched by enlargement of the wall chamber. This is the steady-state condition, and any deviations from this point will cause transient imbalances between the processes of water uptake and wall expansion. The result of these imbalances is that turgor will return to the point of intersection.

The regulation of cell growth—for example, by hormones or by light—typically is accomplished by regulation of the biochemical processes that regulate wall loosening and stress relaxation. Such changes can be measured as a change in m or in Y.

the entire growth induction by this hormone (see Chapter 19). Nevertheless, this pH-dependent mechanism of wall extension appears to be an evolutionarily conserved process common to all land plants (Cosgrove 1996).

Acid growth may also be observed in isolated cell walls, which lack normal cellular, metabolic, and synthetic processes. Such observation requires the use of an extensometer to put the walls under tension and to measure the pH-dependent **wall creep** (Figure 15.21). "Creep" refers to a time-dependent irreversible extension, typically the result of slippage of wall polymers relative to one another. When growing walls are incubated in neutral buffer (pH 7) and clamped in an extensometer, the walls extend briefly when tension is applied, but extension soon ceases. When transferred to an acidic buffer (pH 5 or less), the wall begins to extend rapidly, in some instances continuing for many hours.

This acid-induced creep is characteristic of walls from growing cells, but it is not found in mature (nongrowing) walls. When walls are pretreated with heat, proteases, or other agents that denature proteins, they lose their acid-growth ability. Such results indicate that acid growth is not simply due to the physical chemistry of the wall (e.g., a weakening of the pectin gel), but is catalyzed by one or more wall proteins.

The idea that proteins are required for acid growth was confirmed in reconstitution experiments, in which heat-inactivated walls were restored to nearly full acid-growth responsiveness by addition of proteins extracted from growing walls (Figure 15.22). The active components proved to be a group of proteins that were named **expansins** (McQueen-Mason et al. 1992). These proteins catalyze the pH-dependent extension and stress relaxation of cell walls. They are effective in catalytic amounts (about 1 part protein per 5,000 parts wall, by dry weight).

The molecular basis for expansin action on wall rheology is still uncertain, but most evidence indicates that expansins cause wall creep by loosening hydrogen bonding between wall polysaccharides (Cosgrove 1997). Binding studies suggest that expansins act at the interface between cellulose and one or more hemicelluloses.

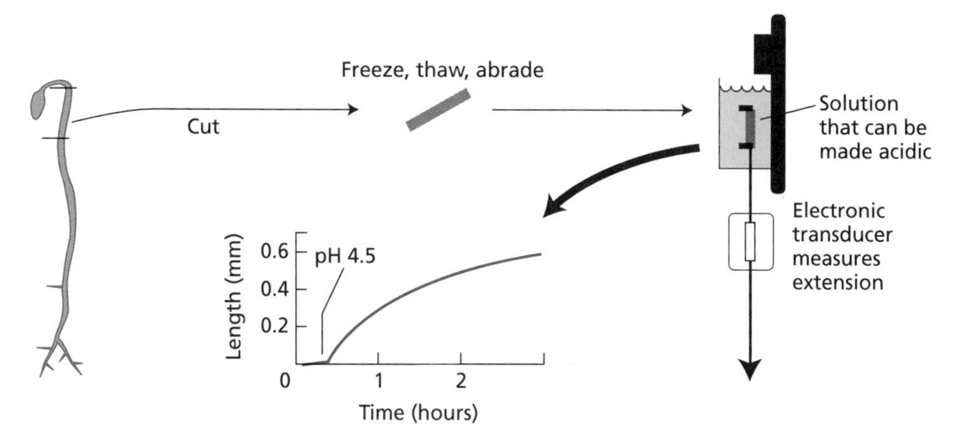

Figure 15.21 Acid-induced extension of isolated cell walls, measured in an extensometer. The walls are prepared as follows: The growing region of a seedling is excised, frozen, and thawed to kill the cells, and the surface cuticle is lightly abraded to facilitate buffer exchange. The wall sample is clamped and put under tension in an extensometer that measures the length with an electronic transducer attached to a clamp. When the solution surrounding the wall is replaced with an acidic buffer (e.g., pH 4.5), the wall extends irreversibly in a time-dependent fashion (it *creeps*).

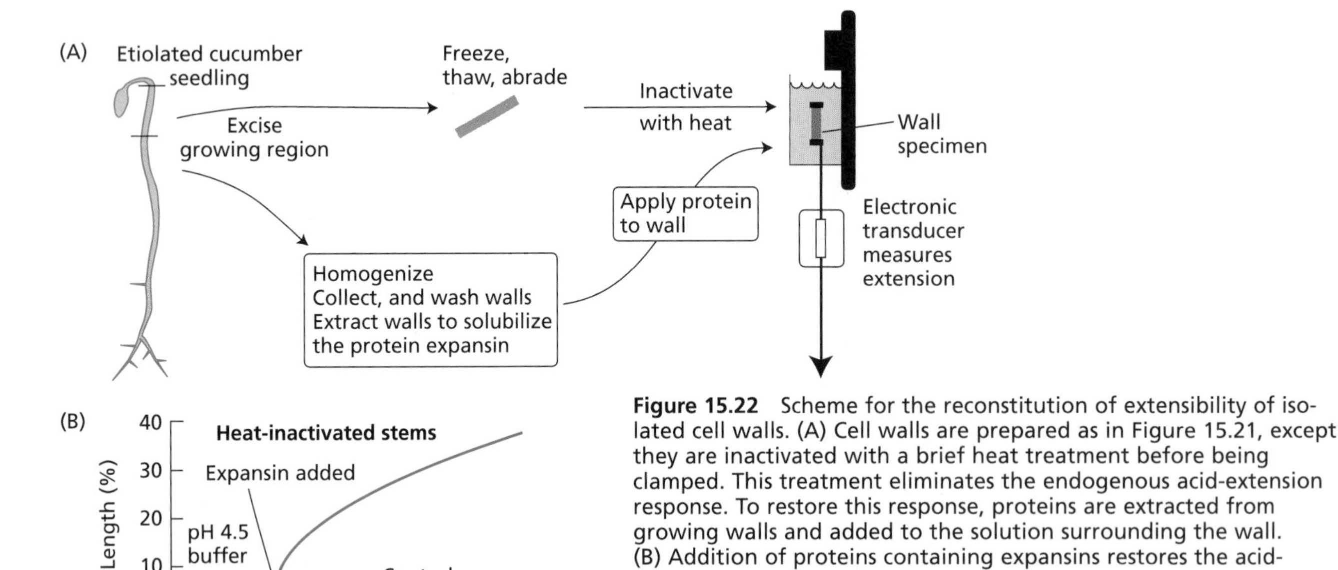

Figure 15.22 Scheme for the reconstitution of extensibility of isolated cell walls. (A) Cell walls are prepared as in Figure 15.21, except they are inactivated with a brief heat treatment before being clamped. This treatment eliminates the endogenous acid-extension response. To restore this response, proteins are extracted from growing walls and added to the solution surrounding the wall. (B) Addition of proteins containing expansins restores the acid-extension properties of the wall. (After Cosgrove 1996.)

Glucanases and Other Hydrolytic Enzymes May Modify the Matrix

Several types of experiments implicate (1→4)β-D-glucanases in cell wall loosening, especially during auxin-induced cell elongation (see Chapter 19). For example, matrix glucans such as xyloglucan show enhanced hydrolysis and turnover in excised segments when growth is stimulated by auxin. Interference with this hydrolytic activity by use of antibodies or lectins reduces growth in excised segments. Expression of (1→4)β-D-glucanases is associated with growing tissues, and application of glucanases to cells in vitro may stimulate growth. Such results support the idea that wall stress relaxation and expansion are the direct result of the activity of glucanases that digest xyloglucan in dicotyledons or (1→3,1→4)β-D-glucans in grass cell walls (Hoson 1993).

However, glucanases and related wall hydrolases do not cause walls to extend in the same way that expansins do. Instead, treatment of walls with glucanases or pectinases enhances the subsequent extension response to expansins (Cosgrove and Durachko 1994). These results suggest that wall hydrolytic enzymes such as (1→4)β-D-glucanases may not be the principal catalysts of wall expansion, but they may act indirectly by modulating expansin-mediated polymer creep.

Many Structural Changes Accompany the Cessation of Wall Expansion

The growth cessation that occurs during cell maturation is generally irreversible and is typically accompanied by a reduction in wall extensibility, as measured by various biophysical methods. These physical changes in the wall might come about by (1) a reduction in wall-loosening processes, (2) an increase in wall cross-linking, or (3) an alteration in the composition of the wall, making for a more rigid structure or one less susceptible to wall loosening. There is some evidence for each of these ideas (Cosgrove 1997).

Several modifications of the maturing wall may contribute to wall rigidification. Newly secreted matrix polysaccharides may be altered in structure, so as to form tighter complexes with cellulose or other wall polymers, or they may be resistant to wall-loosening activities. An example is the arabinoxylan of maize coleoptiles, which becomes less branched as the coleoptile matures and may form tighter complexes with cellulose (Carpita and Gibeaut 1993). Removal of mixed-link β-D-glucans is also coincident with growth cessation in these walls. De-esterification of pectins, leading to more rigid pectin gels, is similarly associated with growth cessation in both grasses and dicotyledons. Cross-linking of phenolic groups in the wall (such as tyrosine residues in HRGPs, ferulic acid residues attached to pectins, and lignin) generally coincides with wall maturation and is believed to be mediated by peroxidase, a putative wall rigidification enzyme. Many structural changes occur in the wall during and after cessation of growth, and it has not yet been possible to dissect out the significance of individual processes for cessation of wall expansion.

The Polarity of Tip Growth

Plant cells perceive their position within a tissue and assume the polarity appropriate for that tissue. In addition, cells can perceive and act on information coming from their environment to alter their polarity and redirect growth when necessary. Cellular polarity must be established to orient subsequent expansion. Cells that are initially apolar, such as zygotes, spores, and pollen grains, later become polarized and initiate tip growth and have been especially valuable in the analysis of the induction and establishment of polarity. A wide variety of external gradients, including temperature, light intensity, pH, ions, and electric-potential gradients, can polarize and orient the axis of growth of these cells. Thus, polarity can be experimentally manipulated without the complexities associated with multicellular tissues. Much of our understanding of the establishment of polarity in tip-growing cells is derived from studies of rhizoid emergence in zygotes of the brown alga *Fucus*.

Unlike higher plants, and even unlike most other algae, certain brown algae, particularly *Fucus* and *Pelvetia*, shed their eggs and sperm into the sea, where fertilization occurs. The zygote initially lacks a cell wall and is completely apolar. However, a polar axis is established within a few hours of fertilization in response to unilateral light. The axis is not stable at first and can be changed many times simply by a change in the direction of the light. The polar axis becomes fixed within 10 to 14 hours after fertilization, after which it cannot be altered by a change in the direction of the light, and a structure that will become the rhizoid begins to protrude from the darkened side.

Even though the zygote is still a single cell, it is polarized, with two structurally different poles, the thallus and rhizoid poles, which have radically different fates. The first zygotic division is asymmetric and occurs at right angles to the polar axis. It establishes two cells, each having the identity of the pole from which it emerges. The smaller *rhizoid cell* produces a filament of cells and the holdfast that attaches the plant to rocks. The larger *thallus cell* divides to produce a globular mass of cells, from which the main body of the plant, the thallus, forms (Quatrano and Shaw 1997).

In this section we will describe the sequence of events thought to be involved in the establishment of cell polarity in *Fucus* zygotes. These events include the localization of ion channels at one end of the cell; the development of a calcium ion gradient within the cell, as well as a polarized ionic current; the assembly of a polarized actin cytoskeleton; and the localized secretion of Golgi-derived vesicles.

Calcium Channel Redistribution Leads to a Transcellular Current

A dye known as fluorescent dihydropyridine (FL-DHP), which binds to calcium channels in mammalian cells and can interfere with calcium uptake, has been used as a probe to examine the polarization of the *Fucus* zygote cell membrane. During the first 6 to 8 hours after fertilization, and in response to unilateral light, FL-DHP receptors, presumably calcium channels, are redistributed and become localized on the dark side of the *Fucus* zygote at what will become the rhizoid pole. If the direction of the light is changed, the location of the FL-DHP–binding proteins also changes, again becoming concentrated on the dark side of the zygote (Shaw and Quatrano 1996).

One of the earliest events in the establishment of polarity after exposure of the zygotes to unilateral light is the induction of a positive current that flows into the cell on the shaded side and loops back through the external medium (Figure 15.23). This **transcellular current**, which is due largely to calcium ions, is on the order of 100 pA (1 picoampere = 10^{-12} ampere) and develops in advance of any morphological sign of polarity. The relevance of this phenomenon to the establishment of cell polarity is indicated by the fact that the site at which positive charge flows into the cell is also the site of subsequent rhizoid outgrowth. Transcellular ionic currents also have been observed in some tip-growing cells of higher plants, including root hairs and germinating pollen grains, but they have not yet been detected in plant cells that expand by diffuse growth (Jaffe and Nuccitelli 1977).

The ionic current in polarizing *Fucus* zygotes likely arises from the redistribution of ion channels in the plasma membrane. This current generates an electric-potential difference across the membrane that could alter other membrane properties, such as the distribution of membrane proteins. The current also alters the ionic composition of the cytosol; the magnitude of the effect depends on the ions that form the current, which vary in different cell types. Because the Ca^{2+} concentration in the cytosol is usually very low (between 10^{-6} and 10^{-8} *M*), even a small increase in intracellular Ca^{2+} concentration could have a marked effect.

The fluorescent calcium-indicator dye *fura-2* has been used to detect intracellular cytosolic calcium gradients in both plant and animal cells. Since the excitation spectrum of fura-2 shifts when it binds calcium, we can determine the intracellular concentrations of both forms of the dye (with and without bound calcium) by exciting them with the appropriate two wavelengths. The ratio of the two emissions provides a measure of the calcium concentration that is independent of the dye concentration. Using this technique to measure the cytosolic Ca^{2+} gradient in growing *Fucus* rhizoids, researchers observed a gradient from 450 n*M* at the tip to 100 n*M* at the base (Brownlee and Pulsford 1988). This calcium ion gradient may be important for the polarized assembly of actin filaments and vectoral discharge of Golgi-derived vesicles at the rhizoid pole. Calcium ion gradients also have been shown to be established in other tip-growing cells, such as germinating pollen grains, where the direction of

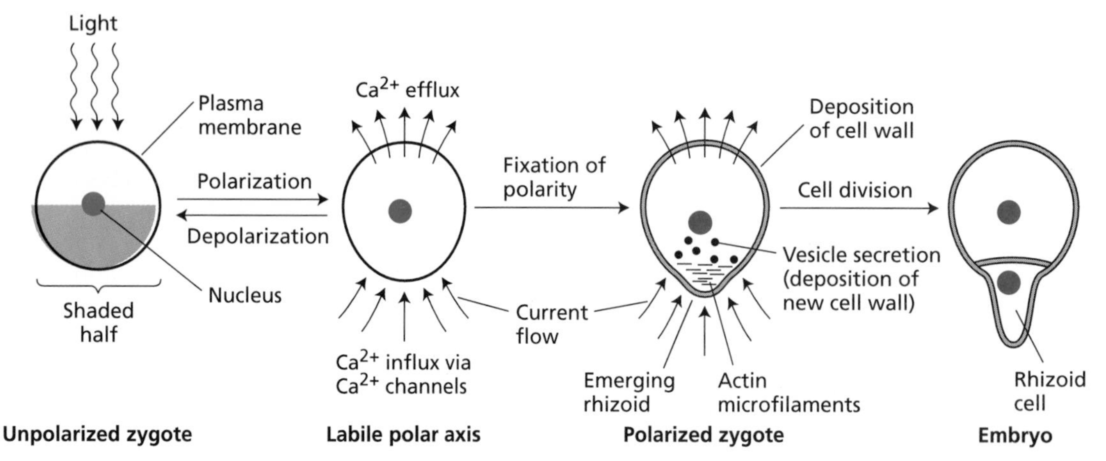

Figure 15.23 A summary of the events involved in the establishment of polarity in zygotes of the brown alga *Fucus*. The zygote is polarized by an asymmetric stimulus from the environment, such as unilateral light. A current flows (charged calcium ions move) through the polarized but still spherical zygote at the site at which the rhizoid will emerge, driven by Ca^{2+} uptake in the shaded half of the cell, from which the rhizoid will emerge. The current (Ca^{2+}) flows out on the opposite side. Cell polarity becomes fixed when actin microfilaments assemble at the site of rhizoid emergence and a cell wall is assembled around the zygote. The cell wall completely surrounds the zygote, but its composition differs in the rhizoid and thallus halves. F granules (vesicles containing sulfated polysaccharides) are transported by actin filaments to the plasma membrane and deposited in the cell wall only at the site of rhizoid emergence. Finally, the zygote divides, and the rhizoid cell grows at its tip.

growth and the region in which Golgi vesicles carrying cell wall components are deposited are predicted by the calcium gradient (Figure 15.24) (Box 15.4).

The Actin Cytoskeleton Becomes Polarized

Inhibition of protein synthesis by treatment with the drug *cycloheximide* delays rhizoid formation but does not stop the establishment of polarity. This result suggests that although rhizoid formation requires the synthesis of new proteins, the establishment of polarity does not, so polarity development must involve a redistribution of existing components. Actin filaments become preferentially localized in the rhizoid half of the zygote and extend from the vicinity of the nucleus to the plasma membrane at the site marked by localized FL-DHP–binding proteins as polarity develops (Kropf 1994; 1997).

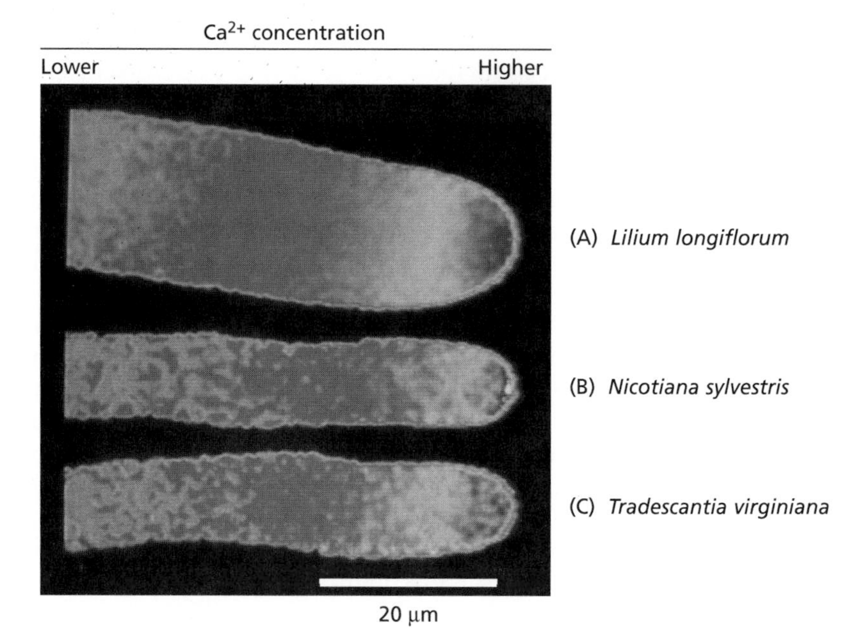

Figure 15.24 Tip-growing pollen tubes exhibit a steep intracellular gradient of calcium ion concentration, with the highest levels at the growing tip. Pollen tubes from three species of plants were microinjected with a fluorescent calcium indicator dye to demonstrate this Ca^{2+} gradient, ranging from about 1 μM at the extreme tip to about 0.2 μM at the base. (Photo from Hepler 1997, courtesy of P. Hepler. See frontispiece for a full-color version of this photograph.)

BOX 15.4

Calcium Gradients and Oscillations in Growing Pollen Tubes

POLLEN TUBE GROWTH delivers the two sperm cells to the embryo sac (see Chapter 1), and thus is essential for sexual reproduction in higher plants (Taylor and Hepler 1997). The process has important and unique features: It is extremely fast, reaching rates of up to 1 cm per hour in corn; it is highly polarized, with growth being confined to the tip (tip growth); and it possesses a guidance mechanism that determines the direction of growth. Several aspects of tip growth have been deciphered; for example, vesicles containing cell wall precursor material are produced by the Golgi apparatus, and through cytoplasmic streaming they flow to the apex of the pollen tube, where they fuse with the plasma membrane and secrete their contents into the cell wall (Taylor and Hepler 1997). It is less clear how pollen tube growth is regulated, although recent work indicates that calcium ions play an important role (Holdaway-Clarke et al. 1997; Messerli and Robinson 1997).

It has been known for many years that calcium ions are essential for growth (Taylor and Hepler 1997 for review). Calcium ions must be present in the growth medium at a concentration above 10 μ*M*, but below 10 m*M*. Further, when pollen tubes are cultured in a medium that contains a radioactive isotope of calcium (^{45}Ca), the tracer accumulates in the apical regions. More recent work has focused on the intracellular and extracellular status of the ion, using technology that has become available only within the last decade. Studies on intracellular calcium, using fluorescent indicator dyes (Holdaway-Clarke et al. 1997), or the calcium-sensitive fluorescent protein aequorin (Messerli and Robinson 1997), reveal that the growing pollen tube possesses a highly focused gradient of free calcium at its extreme apex (see Figure 15.24). Thus the intracellular calcium concentration immediately adjacent to the plasma membrane is 2,000 n*M* or higher, but it declines abruptly to a basal level 200 n*M* or less within 20 μm from the tip.

Researchers have studied extracellular calcium using the calcium-selective *vibrating electrode*, a device that rapidly measures the local calcium ion concentration in the medium at two different points, one close to the pollen tube and another 10 μm away. With knowledge of these concentrations and their difference, it is possible, using Fick's law, to calculate the flux or movement of calcium ions relative to the growing pollen tube (Kühtreiber and Jaffe 1990). Results from these investigations show that growing pollen tubes exhibit a tip-directed influx of calcium. We know that both the intracellular gradient and the extracellular flux are correlated with growth, since treatments that block growth (such as injection of a calcium ion chelator, or elevation of the osmotic value of the medium with sucrose) reduce these calcium signals to basal levels (Pierson et al. 1996). Further, when the pollen tube is permitted to recover, by removal of the inhibiting agent, both the gradient and the influx reemerge as growth resumes. Additional studies also indicate that the spatial point at which the intracellular gradient is the highest is the precise position of the cell that elongates the most (Malhó and Trewavas 1996; Pierson et al. 1996). The intracellular calcium gradient thus marks the direction of pollen tube growth.

Most recent investigations show that the magnitudes of both the intracellular calcium gradient (Holdaway-Clarke et al. 1997; Messerli and Robinson 1997) and the extracellular calcium influx oscillate (Holdaway-Clarke et al. 1997), with a period of 15 to 50 seconds. Thus the intracellular calcium concentration at the tip will periodically change (oscillate) from a 3 to 4 μ*M* to less than 1 μ*M*, while the extracellular calcium influx varies between 0 and 17 pmol cm^{-2} s^{-1}. Of particular interest has been the observation of a similar periodic oscillation in the rate of pollen tube growth. Simultaneous measurements of the intracellular calcium concentration and the growth rate have revealed that they are exactly in phase (Holdaway-Clarke et al. 1997; Messerli and Robinson 1997); thus the time of the fastest growth correlates closely with the time at which the intracellular Ca^{2+} gradient reaches its greatest magnitude. In contrast, the tip-directed influx of extracellular calcium, although exhibiting the same period as growth, does not show the same phase. Careful analysis indicates that the moment of maximum influx is delayed by 11 seconds from the time of most rapid growth (see the figure).

In effect, the magnitude of pollen tube growth rate at any point in time predicts the magnitude of the inward current 11 seconds later. The delay indicates that calcium ions do not immediately enter the cytosolic pool; if they did, they would be detected by the intracellular fluorescent indicator. Although a definitive explanation is not yet available, the influx of calcium is thought to be governed by changes in the calcium ion–binding properties within the cell wall rather than by movement across the plasma membrane. For example, mature cell walls have been shown to have greater amounts of bound calcium, possibly as a result of de-esterification of

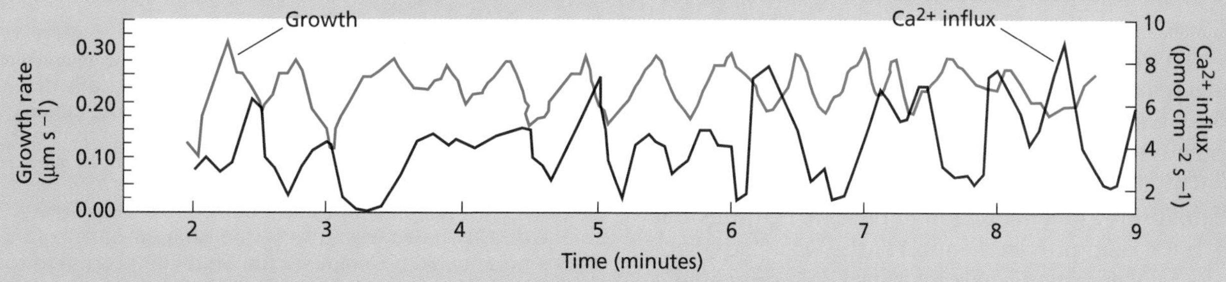

Calcium influx into the pollen tip oscillates about 11 seconds out of phase with the growth rate. (After Holdaway-Clarke et al. 1997.)

BOX 15.4 *(continued)*

pectins (which would increase the number of free carboxyl groups) (Goldberg et al. 1986).

Calcium ions clearly play a major role in pollen tube growth, but what are they doing? Comparison of different animal and plant cells suggests that the intracellular calcium ions facilitate the fusion of vesicles at the tip of the pollen tube and thus stimulate secretion and new cell wall formation. However, the role of the calcium may be more complicated than just the control of vesicle fusion; for example, if the extracellular calcium concentration exceeds 10 m*M*, growth stops, possibly because of an ionic cross-linking of acidic pectin residues in the cell wall. It seems plausible that there is a subtle interplay between cytoplasmic free calcium and calcium bound to the cell wall such that the former controls secretion, while the latter influences the yielding properties of the cell wall. A close interdependence could be imagined if there were stretch-activated calcium channels on the plasma membrane at the apex of the pollen tube. Thus, a decrease in cell wall tensile strength, together with wall yielding, would open stretch-activated channels and allow the entry of ions into the cytosol, while an increase in wall strength would close stretch channels and reduce calcium entry. These ideas need to be tested.

In summary, calcium may play a dual role in the control of pollen tube growth. Within the cell calcium could facilitate vesicle fusion and thus control the amount and location of secretion. In the extracellular space calcium could regulate the rigidity of the cell wall and thus the degree and position of cell wall yielding. These two factors may contribute fundamentally to the control of pollen tube growth.

Actin filaments are dynamic, and their assembly from a pool of monomeric actin subunits is highly regulated. Actin filaments are localized in the rhizoid half of the cell presumably in response to a change in the assembly or disassembly dynamics, and this localization involves a redistribution of the cellular actin. The fixation of polarity can be blocked by *cytochalasin B*, a drug that destroys actin filaments. Treatment with cytochalasin B also disrupts the ionic current that develops during polarization (Brawley and Robinson 1985). Actin filaments participate in cell secretion by transporting secretory vesicles to the plasma membrane in many types of cells.

Secretion of Golgi-Derived Vesicles Is Localized

Cytoplasmic organelles become asymmetrically distributed before the polarity of the cell is fixed (Figure 15.25). Golgi-derived vesicles accumulate at the site of the future outgrowth in the rhizoid half of the *Fucus* zygote as polarization is induced by light (Quatrano 1978). For several hours after the induction of polarity, the asymmetric distribution of vesicles can be changed by subsequent treatment with light from a different direction. During this phase, cytoplasmic vesicles accumulate at the prospective site of rhizoid formation on the shaded side of the cell. Subsequently, the vesicles fuse with the plasma membrane, secreting their contents into the cell wall, leading to rhizoid emergence. Evidence suggests that calcium gradients play a role in triggering vesicle discharge at the pole (Derksen et al. 1995).

The fixation of light-induced polarity in *Fucus* zygotes also requires the presence of the cell wall. Removal of the wall with cell wall–digesting enzymes does not prevent the induction of polarity by unilateral light, but it does prevent fixation of the axis. The orientation of the embryonic axis remains labile until a new cell wall is synthesized (Kropf et al. 1988).

(A)

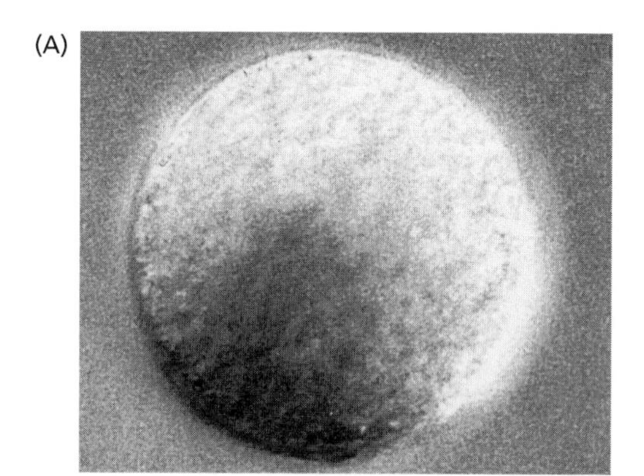

(B)

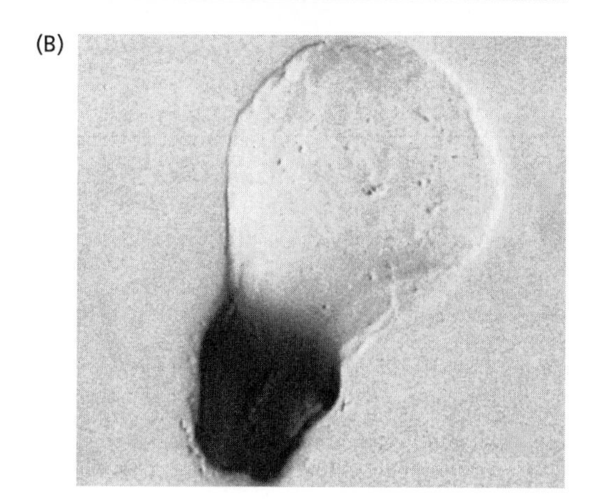

Figure 15.25 Cytoplasmic vesicles (F granules) that contain specific cell wall components become asymmetrically distributed early in the development of *Fucus* zygote polarity. The dye toluidine blue stains cytoplasmic vesicles that carry a component of the cell wall (sulfated fucoidan), as well as the wall itself if it contains this polysaccharide. (A) A toluidine blue O-stained *Fucus* zygote before fixation of the polar axis. (B) The embryo after establishment of rhizoid–thallus polarity, with toluidine blue O-staining confined to the cell wall in the rhizoid portion of the embryo. (From Quatrano and Shaw 1997, courtsey of R. S. Quatrano.)

From these observations it is possible to envisage a chain of events that might occur during the establishment of cell polarity in tip-growing cells. The ionic currents that arise early in the process may initiate the formation of an intracellular Ca^{2+} gradient. The increase in intracellular calcium at the tip may help organize and stabilize the microfilament components of the cytoskeleton, which in turn direct vesicle traffic to the growing tip. In addition, a calcium gradient, perhaps immediately adjacent to the plasma membrane, may promote vesicle fusion with the plasma membrane, although this remains to be demonstrated. Finally, cell wall deposition fixes the polarity of the cell, possibly by interacting with transmembrane proteins.

Wall Degradation and Plant Defense

The plant cell wall is not simply an inert and static exoskeleton. In addition to its role in mechanical restraint, the plant cell wall serves as an extracellular matrix which interacts with cell surface proteins, providing positional and developmental information. It contains numerous enzymes and smaller molecules that are biologically active and that can modify the physical properties of the wall, sometimes within seconds. In some cases, wall-derived molecules can also act as signals to inform the cell of environmental conditions, such as the presence of pathogens. This is an important aspect of the defense response of plants (see Chapter 13).

Walls may also be substantially modified long after growth has ceased. For instance, the cell wall may be massively degraded, such as occurs in ripening fruit or in the endosperm of germinating seeds. In cells that make up the *abscission zones* of leaves and fruits (see Chapter 22), the middle lamella may be selectively degraded, with the result that the cells become unglued and separate. Cells may also separate selectively during the formation of intercellular air spaces, during radicle emergence from germinating seeds, and during other developmental processes. Plant cells may also modify their walls during pathogen attack as a form of defense.

In this section we will consider two types of dynamic changes that can occur in mature cell walls: hydrolysis and oxidative cross-linking. We will also discuss how fragments of the cell wall released during pathogen attack, or even during normal cell wall turnover, may act as cellular signals that influence metabolism and development.

Enzymes Mediate Wall Hydrolysis and Degradation

Hemicelluloses and pectins may be modified and broken down by a variety of enzymes that are found naturally in the cell wall. This process has been studied in greatest detail in ripening fruit, in which softening is thought to be the result of disassembly of the wall (Fischer and Bennett 1991). Glucanases and related enzymes may hydrolyze the backbone of hemicelluloses. Xylosidases and related enzymes may remove the side branches from the backbone of xyloglucan. Transglycosylases may cut and join hemicelluloses together. Such enzymatic changes may alter the physical properties of the wall, for example, by changing the viscosity of the matrix or by altering the tendency of the hemicelluloses to stick to cellulose.

Messenger RNAs for expansin have recently been found to be highly and selectively expressed in ripening tomato fruit, suggesting that they play a role in wall disassembly (Rose et al. 1997). Similarly, softening fruits express high levels of pectin methyl esterase, which hydrolyzes the methyl esters from pectins. This hydrolysis makes the pectin more susceptible to subsequent hydrolysis by pectinases and related enzymes. The presence of these and related enzymes in the cell wall indicates that walls are capable of significant modification during development.

Oxidative Bursts Accompany Pathogen Attack

When plant cells are wounded or treated with certain low-molecular-weight elicitors (see Chapter 13), they activate a defense response that results in the production of high concentrations of hydrogen peroxide, superoxide radicals, and other active oxygen species in the cell wall. This "oxidative burst" appears to be part of a defense response against invading pathogens (see Chapter 13) (Brisson et al. 1994; Otte and Barz 1996). Active oxygen species may directly attack the pathogenic organisms, and they may indirectly deter subsequent invasion by the pathogenic organisms by causing a rapid cross-linking of phenolic components of the cell wall. In tobacco stems, for example, proline-rich structural proteins of the wall become rapidly insolubilized upon wounding or elicitor treatment, and this cross-linking is associated with an oxidative burst and with a mechanical stiffening of the cell walls.

Wall Fragments Can Act as Signaling Molecules

Degradation of cell walls can result in the production of biologically active fragments, called **oligosaccharins**, that may be involved in natural developmental responses and in defense responses (Figure 15.26). Some of the reported physiological and developmental effects of oligosaccharins include stimulation of phytoalexin synthesis, oxidative bursts, ethylene synthesis, membrane depolarization, changes in cytoplasmic calcium, induced synthesis of pathogen-related proteins such as chitinase and glucanase, other systemic and local "wound" signals, and alterations in the growth and morphogenesis of isolated tissue samples (John et al. 1997).

(A) **Galacturonic acid oligomers**

$(1\rightarrow4)$-α-D-GalA $-\left[(1\rightarrow4)\text{-}\alpha\text{-D-GalA}\right]_n-$ $(1\rightarrow4)$-α-D-GalA

(B) **Three oligosaccharins from xyloglucan**

```
                                     Fuc
                                      ↓
                                Gal  Gal
                                 ↓    ↓
Xyl  Xyl  Xyl              Xyl  Xyl  Xyl                Xyl
 ↓    ↓    ↓                ↓    ↓    ↓                  ↓
Glc→Glc→Glc→Glc            Glc→Glc→Glc→Glc          Glc→Glc→Glc
```

(C) **Oligosaccharin from fungal cell wall**

```
β-D-Glc-(1→6)-β-D-Glc-(1→6)-β-D-Glc-(1→6)-β-D-Glc-(1→6)-Glucitol
               3                                 3
               ↑                                 ↑
               1                                 1
            β-D-Glc                           β-D-Glc
```

Figure 15.26 Structure of biologically active oligosaccharins. (A) Oligomers of galacturonic acid are released during pectin degradation. The most active fragments are 10 to 15 residues long. (B) Three oligosaccharins derived from xyloglucan. (C) An oligosaccharin derived from the degradation of fungal (*Phytophthora*) cell walls. Such oligosaccharides stimulate defense responses. (For explanation of abbreviations, see Figures 15.10 and 15.11.)

The best-studied examples are oligosaccharide elicitors produced during pathogen invasion (see Chapter 13). For example, the fungus *Phytophthora* secretes an endopolygalacturonase (a type of pectinase) during its attack on plant tissues. As this enzyme degrades the pectin component of the plant cell wall, it produces pectin fragments—**oligogalacturonans**—that elicit multiple defense responses by the plant cell (Figure 15.27). The oligogalacturonans that are 10 to 13 residues long are most active in these responses. Plant cell walls also contain a β-D-glucanase that attacks the β-D-glucan that is specific to the fungal cell wall. When this enzyme attacks the fungal wall, it releases glucan oligomers with potent elicitor activity. The wall components serve in this case as part of a sensitive system for the detection of pathogen invasion.

Oligosaccharins may also function during the normal control of cell growth and differentiation. For example, a specific *nonasaccharide* (an oligosaccharide containing nine sugar residues) derived from xyloglucan has been found to inhibit growth promotion by the auxin 2,4-D. The nonasaccharide acts at an optimal concentration of 10^{-9} *M*. This xyloglucan oligosaccharin may act as a feedback

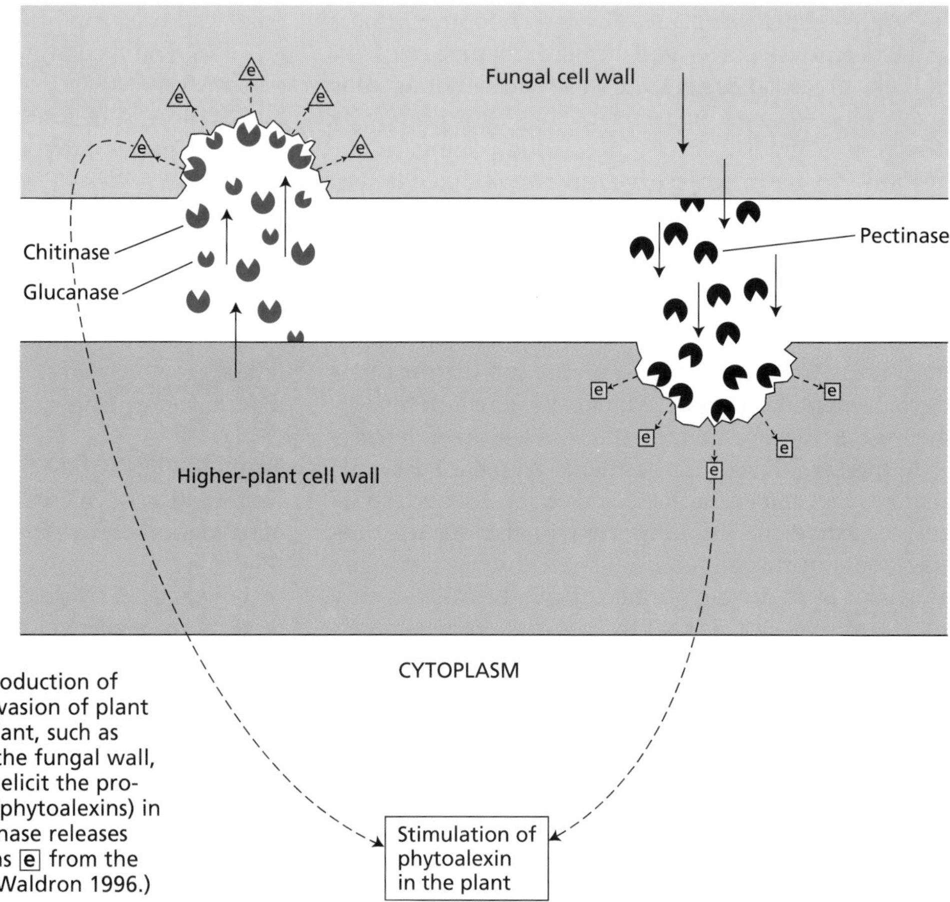

Figure 15.27 Scheme for the production of oligosaccharins during fungal invasion of plant cells. Enzymes secreted by the plant, such as chitinase and glucanase, attack the fungal wall, releasing oligosaccharins (e) that elicit the production of defense compounds (phytoalexins) in the plant. Similarly, fungal pectinase releases biologically active oligosaccharins (e) from the plant cell wall. (After Brett and Waldron 1996.)

inhibitor of growth; for example, when auxin-induced breakdown of xyloglucan is maximal, it may prevent excessive weakening of the cell wall. Related xyloglucan oligomers have also been reported to influence organogenesis in tissue cultures and may play a wider role in cell differentiation (Creelman and Mullet 1997).

Summary

The architecture, mechanics, and function of plants depend crucially on the structure of the cell wall. The wall is secreted and assembled as a complex structure that varies in form and composition as the cell differentiates. Primary cell walls are synthesized in actively growing cells, and secondary cell walls are deposited in certain cells, such as xylem vessel elements and sclerenchyma, after cell expansion ceases.

The basic model of the primary wall is a network of cellulose microfibrils embedded in a matrix of hemicelluloses, pectins, and structural proteins. Cellulose microfibrils are crystalline or paracrystalline arrays of glucan chains synthesized on the membrane by protein complexes called particle rosettes. The genes for cellulose synthase in plants have recently been identified. The matrix is secreted into the wall via the Golgi. Hemicelluloses and proteins cross-link microfibrils, while pectins form hydrophilic gels that can become cross-linked by calcium ions. Wall assembly may be mediated by enzymes. For example, xyloglucan endotransglycosylase has the ability to carry out transglycosylation reactions that might integrate newly synthesized xyloglucans into the wall.

Secondary walls differ from primary walls in that they contain a higher percentage of cellulose, have different hemicelluloses, and lignin replaces pectins in the matrix. Secondary walls can also become highly thickened, sculpted, and embedded with specialized structural proteins.

In diffuse-growing cells, polarity is determined by the orientation of the cellulose microfibrils, which is determined by the orientation of microtubules in the cytoplasm. Once polarity is established, plant cells typically elongate greatly. Cell enlargement is limited by the ability of this network to undergo polymer creep, which in turn is controlled in a complex way by the adhesion of wall polymers to one another and by the influence of pH on wall-loosening proteins such as expansins, glucanases, and other enzymes. According to the acid-growth hypothesis, proton extrusion by the plasma membrane H^+-ATPase acidifies the wall, activating the protein expansin. Expansins induce stress relaxation of the wall by loosening hydrogen bonds between the hemicelluloses and microfibrils. The cessation of cell elongation appears to be due to cell wall rigidification caused by an increase in the number of cross-links.

Polarity in tip-growing cells is determined by the localized secretion of cell wall polymers. In *Fucus* zygotes, the polarity of rhizoid growth develops in several steps: the localization of calcium channels at one end of the cell; the development of a calcium ion gradient within the cell, as well as a polarized ionic current; the assembly of an actin cytoskeleton at the site of growth; and the localized secretion of Golgi-derived vesicles.

Mature cell walls may be degraded completely or selectively by hydrolytic enzymes during fruit ripening, seed germination, and the formation of abscission layers. Cell walls can also undergo oxidative cross-linking in response to pathogen attack. In addition, pathogen attack may release cell wall fragments, and certain wall fragments have been shown to be capable of acting as cell signaling agents.

General Reading

Aldington, S., and Fry, S. C. (1993) Oligosaccharins. *Adv. Bot. Res.* 19: 1–101.

*Bacic, A., Harris, P. J., and Stone, B. A. (1988) Structure and function of plant cell walls. In *The Biochemistry of Plants*, Vol. 14, J. Preiss, ed., Academic Press, New York, pp. 297–371.

*Carpita, N. C., and Gibeaut, D. M. (1993) Structural models of primary cell walls in flowering plants: Consistency of molecular structure with the physical properties of the walls during growth. *Plant J.* 3: 1–30.

Cosgrove, D. J. (1993) Wall extensibility: Its nature, measurement, and relationship to plant cell growth. *New Phytol.* 124: 1–23.

Cosgrove, D. J. (1997) Assembly and enlargement of the primary cell wall in plants. *Annu. Rev. Cell Dev. Biol.* 13: 171–201.

*Delmer, D. P., and Amor, Y. (1995) Cellulose biosynthesis. *Plant Cell* 7: 987–1000.

Fry, S. C. (1988) *The Growing Plant Cell Wall: Chemical and Metabolic Analysis.* Longman, London.

*Fry, S. C. (1995) Polysaccharide-modifying enzymes in the plant cell wall. *Annu. Rev. Plant Physiol. Plant Mol. Biol.* 46: 497–520.

*McCann, M. C., Wells, B., and Roberts, K. (1990) Direct visualization of cross-links in the primary plant cell wall. *J. Cell Sci.* 96: 323–334.

McNeil, M., Darvill, A. G., Fry, S. C., and Albersheim, P. (1984) Structure and function of the primary cell walls of plants. *Annu. Rev. Biochem.* 53: 625–663.

* Indicates a reference that is general reading in the field and is also cited in this chapter.

Chapter References

Amor, Y., Haigler, C. H., Johnson, S., Wainscott, M., and Delmer, D. P. (1995) A membrane-associated form of sucrose synthase and its potential role in synthesis of cellulose and callose in plants. *Proc. Natl. Acad. Sci. USA* 92: 9353–9357.

Ariole, T., Peng, L. C., Betzner, A. S., Burn, J., Wittke, W., Herth, W., Camilleri, C., Hπfte, H., Plazinski, J., Birch, R., Cork, A., Glover, J., Redmond, J., and Williamson, R. E. (1998) Molecular analysis of cellulose biosynthesis in *Arabidopsis*. *Science* 279: 717–720.

Asada, T., and Collings, D. (1997) Molecular motors in higher plants. *Trends Plant Sci.* 2: 29–37.

Bacic, A., Du, H., Stone, B. A., and Clarke, A. E. (1996) Arabinogalactan proteins: A family of cell-surface and extracellular matrix plant proteoglycans. *Essays Biochem.* 31: 91–101.

Baskin, T. I., Wilson, J. E., Cork, A., and Williamson, R. E. (1994) Morphology and microtubule organization in *Arabidopsis* roots exposed to oryzalin or taxol. *Plant Cell Physiol.* 35: 935–942.

Brawley, S. H., and Robinson, K. R. (1985) Cytochalasin treatment disrupts the endogenous currents associated with cell polarization in fucoid zygotes: Studies of the role of F-actin in embryogenesis. *J. Cell Biol.* 100: 1173–1184.

Brett, C., and Waldron, K. (1996) *Physiology and Biochemistry of Plant Cell Walls*, 2nd ed. Chapman and Hall, London.

Brisson, L. F., Tenhaken, R., and Lamb, C. (1994) Function of oxidative cross-linking of cell wall structural proteins in plant disease resistance. *Plant Cell* 6: 1703–1712.

Brown, R. M., Jr., Saxena, I. M., and Kudlicka, K. (1996) Cellulose biosynthesis in higher plants. *Trends Plant Sci.* 1: 149–155.

Brownlee, C., and Pulsford, A. L. (1988) Visualization of the cytoplasmic Ca^{2+} gradient in *Fucus serratus* rhizoids: Correlation with cell ultrastructure and polarity. *J. Cell Sci.* 91: 249–256.

Chant, J. (1995) Control of cell polarity by internal programs and external signals in yeast. *Sem. Dev. Biol.* 6: 13–23.

Chant, J., and Pringle, J. R. (1995) Patterns of bud site selection in the yeast *Saccharomyces cerevisiae*. *J. Cell Biol.* 129: 751–765.

Cheung, A. Y., Zhan, X. Y., Wang, H., and Wu, H. M. (1996) Organ-specific and Agamous-regulated expression and glycosylation of a pollentube growth-promoting protein. *Proc. Natl. Acad. Sci. USA* 93: 3853–3858.

Cosgrove, D. J. (1985) Cell wall yield properties of growing tissues. Evaluation by in vivo stress relaxation. *Plant Phys.* 78: 347–356.

Cosgrove, D. J. (1996) Plant cell enlargement and the action of expansins. *BioEssays* 18: 533–540.

Cosgrove, D. J. (1997) Relaxation in a high-stress environment: The molecular bases of extensible cell walls and cell enlargement. *Plant Cell* 9: 1031–1041.

Cosgrove, D. J., and Durachko, D. M. (1994) Autolysis and extension of isolated walls from growing cucumber hypocotyls. *J. Exp. Bot.* 45: 1711–1719.

Creelman, R. A., and Mullet, J. E. (1997) Oligosaccharins, brassinolides, and jasmonates: Nontraditional regulators of plant growth, development, and gene expression. *Plant Cell* 9: 1211–1223.

Derksen, J., Rutten, T. L. M., Van Amstel, T., De Win, A., Doris, F., and Steer, M. (1995) Regulation of pollen tube growth. *Acta Bot. Neerl.* 44: 93–119.

Esau, K. (1977) *Anatomy of Seed Plants*, 2nd ed. Wiley, New York.

Fahn, A. (1990) *Plant Anatomy*, 4th ed., Pergamon, Tarrytown, NY.

Fischer, R. L., and Bennett, A. B. (1991) Role of cell wall hydrolases in fruit ripening. *Annu. Rev. Plant Physiol. Plant Mol. Biol.* 42: 675–703.

Giddings, T. H., Jr., and Staehelin, L. A. (1988) Spatial relationship between microtubules and plasma-membrane rosettes during the deposition of primary wall microfibrils in *Closterium* sp. *Planta* 173: 22–30.

Goldberg, R., Morvan, C., and Roland, J. C. (1986) Composition, properties and localisation of pectins in young and mature cells of the mung bean hypocotyl. *Plant Cell Physiol.* 27: 417–429.

Green, P. B. (1968) Cell morphogenesis. *Annu. Rev. Plant Physiol.* 20: 365–394.

Gunning, B. S., and Steer, M. W. (1996) *Plant Cell Biology*. Jones and Bartlett, Boston.

Haigler, C. H., White, A. R., Brown, R. M., Jr., and Cooper, K. M. (1982) Alteration of in vivo cellulose ribbon assembly by carboxylmethylcellulose and other cellulose derivatives. *J. Cell Biol.* 94: 64–69.

Hayashi, T. (1989) Xyloglucans in the primary cell wall. *Annu. Rev. Plant Physiol. Plant Mol. Biol.* 40: 139–168.

Hepler, P. K. (1997) Tip growth in pollen tubes: Calcium leads the way. *Trends Plant Sci.* 2: 79–80.

Holdaway-Clarke, T. L., Feijó, J. A., Hackett, G. R., Kunkel, J. G., and Hepler, P. K. (1997) Pollen tube growth and the intracellular cytosolic calcium gradient oscillate in phase while extracellular calcium influx is delayed. *Plant Cell* 9: 1999–2010.

Hope, A. B., and Walker, N. A. (1975) *The Physiology of Giant Algal Cells*. Cambridge University Press, London.

Hoson, T. (1993) Regulation of polysaccharide breakdown during auxin-induced cell wall loosening. *J. Plant Res.* 103: 369–381.

Jaffe, L. F., and Nuccitelli, R. (1977) Electrical controls of development. *Annu. Rev. Biophys. Bioeng.* 6: 445–476.

John, M., Röhrig, H., Schmidt, J., Walden, R., and Schell, J. (1997) Cell signalling by oligosaccharides. *Trends Plant Sci.* 2: 111–115.

Kieliszewski, M. J., and Lamport, D. T. A. (1994) Extensin: Repetitive motifs, functional sites, post-translational codes, and phylogeny. *Plant J.* 5: 157–172.

Koyama, M., Helbert, W., Imai, T., Sugiyama, J., and Henrissat, B. (1997) Parallel-up structure evidences the molecular directionality during biosynthesis of bacterial cellulose. *Proc. Natl. Acad. Sci. USA* 94: 9091–9095.

Kropf, D. L., Kloareg, B., and Quatrano, R. S. (1988) Cell wall is required for fixation of the embryonic axis in *Fucus* zygotes. *Science* 239: 187–190.

Kropf, D. L. (1994) Cytoskeletal control of cell polarity in a plant zygote. *Dev. Biol.* 165: 361–371.

Kropf, D. L. (1997) Induction of polarity in fucoid zygotes. *Plant Cell* 9: 1011–1020.

Kühtreiber, W. M., and Jaffe, L. F. (1990) Detection of extracellular calcium gradients with a calcium-specific vibrating electrode. *J. Cell Biol.* 110: 1565–1573.

Lapasin, R., and Pricl, S. (1995) *The Rheology of Industrial Polysaccharides. Theory and Applications*. Blackie, London.

Levy, S., Maclachlan, G., and Staehelin, L. A. (1997) Xyloglucan sidechains modulate binding to cellulose during in vitro binding assays as predicted by conformational dynamics simulations. *Plant J.* 11: 373–386.

Malhó, R., and Trewavas, A. J. (1996) Localized apical increases of cytosolic free calcium control pollen tube orientation. *Plant Cell* 8: 1935–1949.

McQueen-Mason, S., Durachko, D. M., and Cosgrove, D. J. (1992) Two endogenous proteins that induce cell wall expansion in plants. *Plant Cell* 4: 1425–1433.

Messerli, M., and Robinson, K. R. (1997) Tip localized Ca^{2+} pulses are coincident with peak pulsatile growth rates in pollen tubes of *Lilium longiflorum*. *J. Cell Sci.* 110: 1269–1278.

Metraux, J-P., Richmond, P., and Taiz, L. (1980) Control of cell elongation in *Nitella* by endogenous cell wall pH gradients. Multiaxial extensibility and growth studies. *Plant Physiol.* 65: 204–210.

Morikawa, H., Hayashi, R., and Senda, M. (1978) Infrared analysis of pea stem cell walls and oriented structure of matrix polysaccharides in them. *Plant Cell Physiol.* 19: 1151–1159.

Nishitani, K. (1997) The role of endoxyloglucan transferase in the organization of plant cell walls. *Int. Rev. Cytol.* 173: 157–206.

Otte, O., and Barz, W. (1996) The elicitor-induced oxidative burst in cultured chickpea cells drives the rapid insolubilization of two cell wall structural proteins. *Planta* 200: 238–246.

Pear, J. R., Kawagoe, Y., Schreckengost, W. E., Delmer, D. P., and Stalker, D. M. (1996) Higher plants contain homologs of the bacterial *celA* genes encoding the catalytic subunit of cellulose synthase. *Proc. Natl. Acad. Sci. USA* 93: 12637–12642.

Pierson, E. S., Miller, D. D., Callaham, D. A., Shipley, A. M., Rivers, B. A., Cresti, M., and Hepler, P. K. (1996) Pollen tube growth is coupled to the extracellular calcium ion flux and the intracellular calcium gradient: Effect of BAPTA-type buffers and hypertonic media. *Plant Cell* 6: 1815–1828.

Quatrano, R. S. (1978) Development of cell polarity. *Annu. Rev. Plant Physiol.* 29: 487–510.

Quatrano, R. S., and Shaw, S. L. (1997) Role of the cell wall in the determination of cell polarity and the plane of cell division in *Fucus* embryos. *Trends Plant Sci.* 2: 15–21.

Rayle, D. L., and Cleland, R. E. (1992) The acid growth theory of auxin-induced cell elongation is alive and well. *Plant Physiol.* 99: 1271–1274.

Richmond, P. A. (1983) Patterns of cellulose deposition and rearrangement in *Nitella*: In vivo analysis by birefringence index. *J. Appl. Polym. Sci.* 37: 107–122.

Roland, J. C., Vian, B., and Reis, D. (1977) Further observations on cell wall morphogenesis and polysaccharide arrangement during plant growth. *Protoplasma* 91: 125–141.

Roland, J. C., Reis, D., Mosiniak, M., and Vian, B. (1982) Cell wall texture along the growth gradient of the mung bean hypocotyl: Ordered assembly and dissipative processes. *J. Cell Sci.* 56: 303–318.

Rose, J. K. C., Lee, H. H., and Bennett, A. B. (1997) Expression of a divergent expansin gene is fruit-specific and ripening-regulated. *Proc. Natl. Acad. Sci. USA* 94: 5955–5960.

Séné, C. F. B., McCann, M. C., Wilson, R. H., and Grinter, R. (1994) Fourier-transform Raman and Fourier-transform infrared spectroscopy. An investigation of five higher plant cell walls and their components. *Plant Physiol.* 106: 1623–1631.

Shaw, S. L., and Quatrano, R. (1996) Polar localization of a dihydropyridine receptor on living *Fucus* zygotes. *J. Cell Sci.* 109: 335–342.

Shibaoka, H. (1994) Plant hormone-induced changes in the orientation of cortical microtubules: Alterations in the cross-linking between microtubules and the plasma membrane. *Annu. Rev. Plant Physiol. Plant Mol. Biol.* 45: 527–544.

Showalter, A. M. (1993) Structure and function of plant cell wall proteins. *Plant Cell* 5: 9–23.

Smith, R. C. and Fry, S. C. (1991) Endotransglycosylation of xyloglucans in plant cell suspension cultures. *Biochem. J.* 279: 529–535.

Taiz, L., Metraux, J. P., and Richmond, P. A. (1981) Control of cell expansion in the *Nitella* internode. In *Cytomorphogenesis in Plants* (Cell Biology Monographs, vol. 8), O. Kiermayer, ed., Springer, Wien, pp. 231–264.

Taylor, L. P., and Hepler, P. K. (1997) Pollen germination and tube growth. *Annu. Rev. Plant Physiol. Plant Mol. Biol.* 48: 461–491.

Woessner, J. P., and Goodenough, U. W. (1994) Volvocine cell walls and their constituent glycoproteins: An evolutionary perspective. *Protoplasma* 181: 245–258.

16 Growth, Development, and Differentiation

THE VEGETATIVE PHASE OF PLANT DEVELOPMENT begins with a single cell, the **zygote**. The zygote undergoes three basic developmental processes: cell division, cell expansion, and cell differentiation. If the zygote were to carry out these functions in a random fashion, the result would be a clump of disorganized cells with no definable form or function. Such disorganized development can result when cells initiate growth in tissue culture, where they may form a tissue known as **callus**. The fact that a zygote gives rise to an organized embryo with a predictable and species-specific structure tells us that the zygote is programmed to develop in a particular way, and that cell division, cell expansion, and cell differentiation are tightly controlled during embryogenesis.

In plants, unlike animals, organogenesis can continue indefinitely at the growing tips, or **meristems**, of the mature root and shoot. This ability gives rise to the phenomenon referred to as **indeterminate growth**. Although the potential for indeterminate growth is rarely realized in nature, some long-lived trees, such as bristlecone pines and the California redwoods, continue to grow for thousands of years. Eventually, the adult plant undergoes a transition from vegetative to reproductive development, culminating in the production of a zygote, and the process begins again.

The fundamental importance of the problem of development in plant biology was first recognized by the German botanist Matthias Schleiden (1804–1881):

> *The mode in which one cell forms many, and how these, dependent on the influence of the former, assume their proper figure and arrangement, is exactly the point upon which the whole knowledge of plants turns; and whosoever does not propose this*

question . . . or does not reply to it, can never connect a clear scientific idea with plants and their life.

(Schleiden 1842)

Whereas Schleiden and his contemporaries were concerned chiefly with the descriptive aspects of plant growth and development, today's plant scientists seek to go beyond description to the underlying molecular causes of these patterns. In Chapter 15, we examined the polarity and expansion of individual cells in relation to the mechanical properties of their cell walls. We now turn our attention to the patterns of cell expansion and cell division during the development of the whole plant, and to the differentiation of specialized cells and tissues.

What are the basic principles that govern the size increase (growth) that occurs throughout plant development? How does a zygote give rise to an embryo, an embryo to a seedling, and so forth? How do new plant structures arise from preexisting structures? Organs are generated by cell division and expansion, but they are also composed of tissues in which groups of cells have acquired specialized function, and these tissues are arranged in specific patterns. How do these tissues come to be formed in a particular pattern, and how do cells differentiate? Understanding how growth, cell differentiation, and pattern formation are regulated at the cellular, biochemical, and molecular levels is the ultimate goal of developmental biologists. Such an understanding must include the genetic basis of development. Which genes are involved, what is their hierarchical order, and how do they bring about developmental change?

In this chapter, we will explore what is known about these questions, beginning with growth. Derivatives of the apical meristems exhibit specific patterns of cell expansion, and these expansion patterns contribute to the overall shape and size of the plant. Analysis of the spatial and temporal aspects of these growth patterns has made it possible to study the process at the cellular level.

Embryogenesis initiates plant development, but unlike animal development, plant development is an ongoing process. Embryogenesis establishes the basic plant body plan and forms the meristems that generate additional organs in the adult. After discussing the formation of the embryo, we will examine the root and shoot meristems. Root and shoot meristems exhibit specific cell division patterns, and we will discuss how the planes of cell division are controlled. Finally, despite their indeterminate growth habit, plants, like all other multicellular organisms, senesce and die. At the end of the chapter we will consider death as a developmental phenomenon, at both the cellular and the organismal levels. The transition to reproductive development will be discussed in Chapter 24.

The Analysis of Plant Growth

How do plants grow? This deceptively simple question has challenged plant scientists for more than 150 years. New cells form continually at the apical meristems. Cells enlarge slowly in the apical meristem and more rapidly in the subapical regions. The resulting increase in cell size can range from severalfold to 100-fold, depending on the species and environmental conditions. Classically, plant growth has been analyzed in terms of cell number or overall size (or mass). However, this is only part of the story.

Tissue growth is neither uniform nor random. The derivatives of the apical meristems expand in predictable and site-specific ways, and the expansion patterns in these subapical regions largely determine the size and shape of the primary plant body. The total growth of the plant can be thought of as the sum of the local patterns of cell expansion, while **morphogenesis** (the acquisition of shape or form) is the sum of the patterns of both cell division and cell expansion.

The peculiar way in which plants grow from their tips sets up a dynamic, morphogenic pattern analogous to the flow of water in a fountain. The analysis of the motions of cells or "tissue elements" (and the related problem of cell expansion) is called *kinematics*. In this section we will discuss both the classical definitions of growth and the more modern, kinematic approach. As we will see, the advantage of the kinematic approach is that it allows one to describe the growth patterns of organs mathematically in terms of the expansion patterns of their component cells.

Plant Growth Can Be Measured in Different Ways

Growth in plants is defined as *an irreversible increase in volume*. The largest component of plant growth is cell expansion driven by turgor pressure. During this process, cells increase in volume manyfold and become highly vacuolate. However, size is only one criterion that may be used to measure growth. For example, growth can be measured in terms of change in **fresh weight**—that is, the weight of the living tissue—over a particular period of time. However, the fresh weight of plants growing in soil fluctuates in response to changes in the water status, so this may be a poor indicator of actual growth. In these situations, measurements of **dry weight** are often more appropriate.

Cell number is a common and convenient parameter by which to measure the growth of unicellular organisms, such as the green alga *Chlamydomonas* (Figure 16.1), but in multicellular plants cells can divide in the absence of an increase in volume. For example, during the early stages of embryogenesis the zygote subdivides into progressively smaller cells *with no net increase in the size of the embryo*. Only after it reaches the eight-cell stage

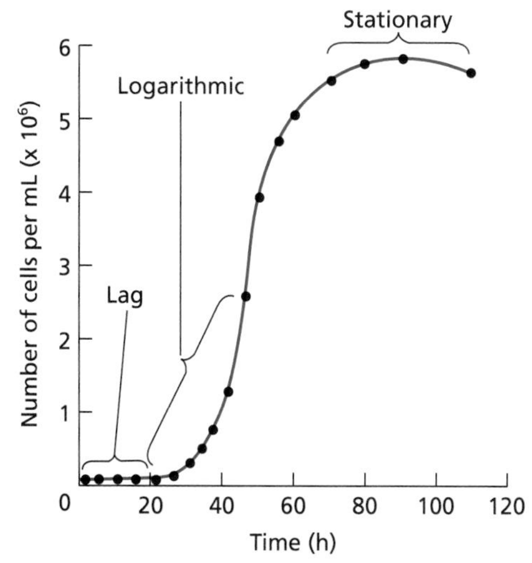

Figure 16.1 Growth of the unicellular green alga *Chlamydomonas*. Growth is assessed by a count of the number of cells per milliliter at increasing times after the cells are placed in fresh growth medium. Temperature, light, and nutrients provided are optimal for growth. An initial lag period during which cells may synthesize enzymes required for rapid growth is followed by a period in which cell number increases exponentially. This period of rapid growth is followed by a period of slowing growth in which the cell number increases linearly. Then comes the stationary phase, in which the cell number remains constant or even declines as nutrients are exhausted from the medium.

does the increase in cell number begin to mirror the increase in volume. Since the zygote is an especially large cell, this lack of correspondence between an increase in cell number and growth may be unusual, but it points out the potential problem in equating an increase in cell number with growth.

Although cell number may not be a reliable measure of plant growth, under most circumstances dividing cells, particularly in meristems, double in volume during their cell cycle. Cell expansion in the apical meristems therefore contributes to plant growth, but not as much as the rapid cell expansion that occurs in the subapical region after cell division ceases. This phenomenon is illustrated by the behavior of the *Arabidopsis* mutant *stunted plant 1* (*stp1*), in which the rapid cell elongation in the subapical region is blocked while the cell expansion that accompanies cell division in the apical meristem proceeds more or less normally. Because the rapid phase of cell elongation is blocked, plants homozygous for the *stp1* mutation are remarkably reduced in size (Baskin et al. 1995).

Since all the cells of the plant axis elongate under normal conditions, the greater the number of cells produced by the apical meristem, the longer the axis will be. For example, when *Arabidopsis* plants are transformed with a gene that encodes cyclin, a key component of the cell cycle regulatory machinery (see Chapter 1), the cells of the apical meristem progress through their cell cycles more rapidly, so more cells form per unit time. As a result, the roots of these transgenic plants have more cells and are substantially longer than the roots of wild-type plants grown under similar conditions (Doerner et al. 1996).

New cells form continually at the apical meristems. With each new round of cell division and associated cell expansion, the older derivatives are displaced a small distance from the apex. As the cells recede farther from the apex, the rate of displacement is greatly accelerated. By viewing plant growth as a process of *cell displacement from the apex*, one can apply the principles of kinematics.

Kinematics Can Be Used to Analyze Plant Growth

Moving fluids such as waterfalls, fountains, and the wakes of boats can generate specific forms. The study of the motion of fluid particles and the shape changes that the fluids undergo is called **kinematics**. The ideas and numerical methods used to study these fluid structures are useful for characterizing meristematic growth. In both cases, an unchanging structure is produced, even though it is composed of moving and changing elements. For example, as seedlings emerge from the seed coat, the apical end of the dicot hypocotyl bends back on itself to form a hook (Figure 16.2). The hook is thought to protect the seedling apex from damage during growth through the soil. During seedling growth (in soil or dim light) the hook migrates up the stem, from the hypocotyl into the epicotyl and then to the first and second internodes, but the hook structure remains constant. That is, if we look from the perspective of the plant apex, the hook summit remains at a fixed distance, perhaps 5 mm behind the apical dome. The stem on the apical and basal sides of the hook summit remains straight. Thus the hook can be seen to have a form that is "steady" or constant in time, although the specific tissues that make up the hook are constantly changing, like the flow of a fountain.

The "fluid flow" aspect of the hook structure can be verified with a simple observation of a developing seedling over time. If we mark a specific epidermal cell on the seedling stem located close to the seedling apex, we can watch it as it flows into the hook summit, then down into the straight region below the hook (see Figure 16.2). The mark is not crawling over the plant surface, of course; plant cells are cemented together and do not experience much relative motion during development. The change in position of the mark relative to the hook implies that the hook is composed of a procession of tissue elements, each of which first curves and then straightens as it is displaced from the plant apex during growth. The steady form is produced by a parade of changing cells. And at any one time the seedling stem has elements that are curving (on the apical side of the hook summit) and elements that are straightening (on the basal side of the hook summit).

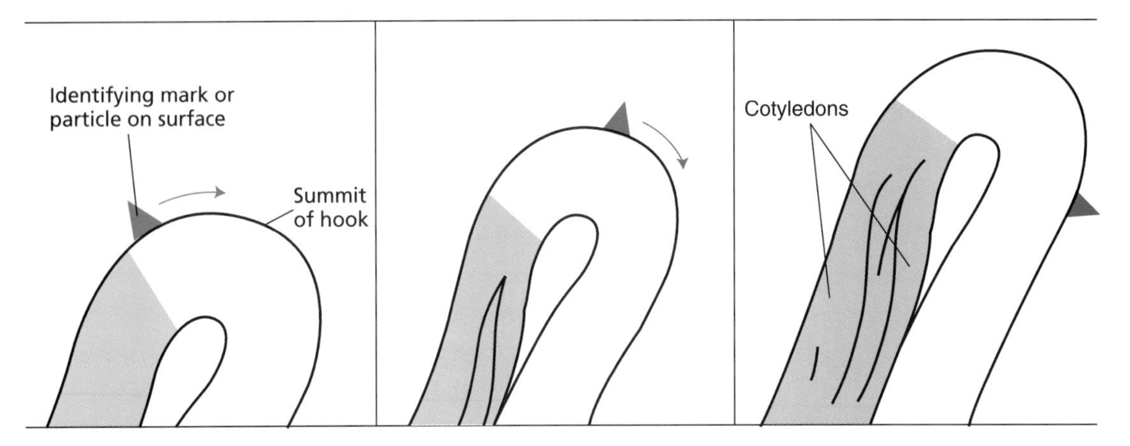

Figure 16.2 The dicot hypocotyl hook is a steady structure composed of changing elements. The form (curvature distribution) is maintained over time, while different tissues first curve and then straighten as they are displaced from the seedling apex during growth. If a mark is placed at a fixed point on the surface, it will be displaced (indicated by the arrow), appearing to flow through the hook over time. (After Silk 1984.)

A root tip is another example of a steady form composed of changing tissue elements. Here, too, the form is observed to be steady only when distance is measured from the root tip. In general, longitudinal sections of roots show that the cell length pattern is similar in young roots as well as in roots a day older, if cell lengths are compared at the same distance from the root tip. In physical terms, we say that the steady patterns are apparent in the moving coordinate frame attached to the plant apex. A region of cell division occupies perhaps 2 mm of the root tip (see Figure 16.10). The elongation zone extends for about 10 mm behind the root tip. Phloem differentiation is first observed beginning at 3 mm from the tip, while functional xylem elements may be seen at about 12 mm from the tip. A marked cell near the tip will seem to flow first through the region of cell division, then through the remainder of the elongation zone and into the region of xylem differentiation, and so on. This shifting implies that developing tissue elements first divide and elongate, and then differentiate.

In an analogous fashion, the shoot bears a succession of leaves of different developmental stages. During a period of 24 hours, a leaf may grow to have the same size, shape, and biochemical composition that its neighbor had a day earlier. Thus, shoot form is also produced by a parade of changing elements that can be analyzed using kinematics. Such an analysis is not merely descriptive; it permits calculations of the growth and biosynthetic rates of individual tissue elements (cells) within a dynamic structure.

Plant Growth Can Be Described in Both Spatial and Material Terms

There are two different ways to think about the growing plant structures. One is to visualize the *spatial pattern*, such as the hypocotyl hook or the shoot that bears leaves of different ages, at any given time. These spatial aspects of growth can be described with a drawing or a photograph. The other way to think about a growing organ is to consider the fate of the individual tissue elements that produce the pattern over time, such as the moving elements of stem tissue that first curve and then straighten. This is called a *material description*, because it follows the real or material elements as they grow. Marked sites on a tissue must be followed over time to appreciate the material aspects of growth.

The spatial description of a fountain represents a collection of water droplets at an instant in time. To understand the fountain, it is important to think also about the individual droplets as they move through the spatial pattern—the material description. An understanding of both spatial and material aspects of growth is required for unraveling the physiology of plant growth.

Tissue Elements Are Displaced during Expansion

As we have seen, growth in shoots and roots is localized in regions at the tips of these organs. Regions with expanding tissue are called *growth zones*. With time, meristems are displaced from the plant base by the growth of the cells in the growth zone. If successive marks are placed on the stem or root, the distance between these will change, depending on where they are within the growth zone. In addition, all of these marks will move away from the tip of the root or shoot, but their rate of movement will differ depending on their distance from the tip. From another perspective, if one could sit on the tip and observe the marks below, one would see that all marks on the plant axis recede with time. Thus we know that discrete **tissue elements**, minute regions ultimately consisting of individual cells, experience displacement as well as expansion during growth and development.

(A) Growth trajectories: Growth trajectories plot the positions of points spaced along the root axis as a function of time. Note that distance is measured from the root tip. This is termed a material specification of growth because real or material particles are followed.

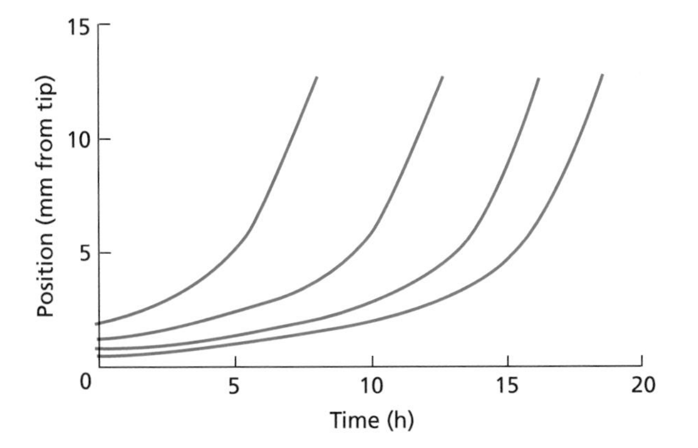

(B) Growth velocity profile: The growth velocity profile plots the velocity of movement away from the tip of points at different distances from the tip. This tells us that growth velocity increases with distance from the tip until it reaches a uniform velocity equal to the rate of elongation of the root.

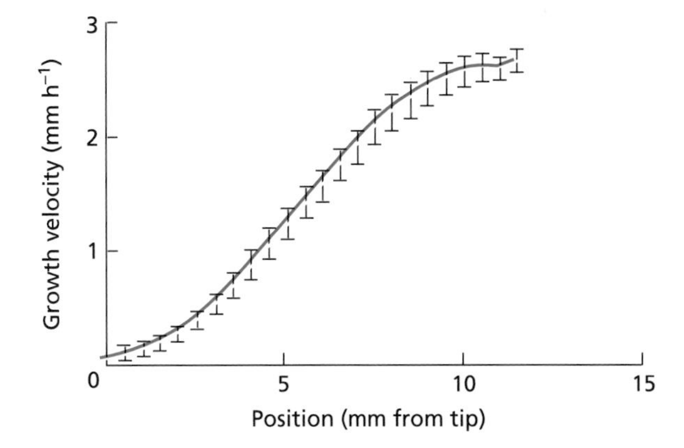

Figure 16.3 The growth of the primary root of *Zea mays* (maize) can be represented kinematically by three related growth curves. (From Silk 1994.)

(C) Relative elemental growth rate: The relative elemental growth rate tells us the rate of expansion of any particular point on the root. It is the most useful for the physiologist because it tells us where the most rapidly expanding regions are located.

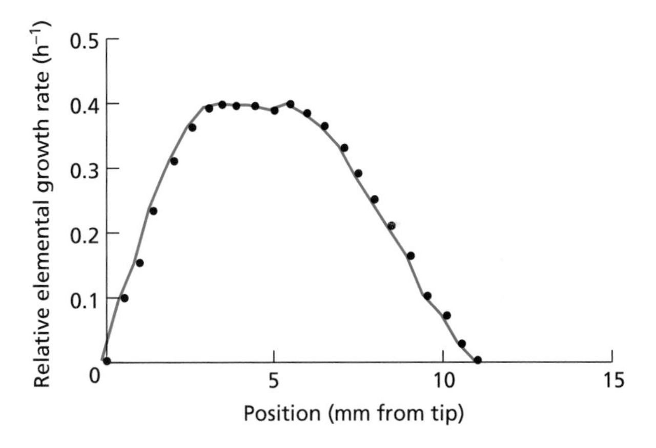

A Growth Trajectory Shows the Displacement of a Tissue Element from the Apex versus Time

One way to characterize the growth pattern along an axis mathematically is to plot the distance of a tissue element from the apex versus time. The resulting curve is called the **growth trajectory**. The slope of the curve at any given point is equivalent to the growth velocity at that region of the axis (see the next section). With the apex as the reference point, the growth trajectory is concave. Marked tissue elements at different locations may be followed simultaneously to generate a family of growth trajectories (Figure 16.3A).

As the tissue elements move away from the apex, their growth velocity increases (the cells accelerate) until a constant limiting velocity is reached equal to the overall organ extension rate. The reason for this increase in growth velocity is that with time, progressively more tissue is located between the moving particle and the apex, and progressively more cells are expanding, so the particle is displaced more and more rapidly. In a rapidly growing maize root, a tissue element takes about 8 hours to move from 2 mm (the end of the meristem) to 12 mm (the end of the elongation zone). The root tissue element accelerates through this region (see Figure 16.3A). Beyond the growth zone, elements do not separate; neighboring elements have the same velocity, and the rate at which particles are displaced from the tip is the same as the rate at which the tip is pushed away from the soil surface. The root tip of maize is pushed through the soil at 3 mm h^{-1}. This is also the rate at which the nongrowing region recedes from the apex, and it is equal to the final slope of the growth trajectory.

The growth trajectory is a material description of growth because an individual element is followed through time. The growth trajectory provides a way to convert spatial information into a developmental time course. For example, if the amount of suberin in the root increases with position, the growth trajectory will give information on the time required for the observed accumulation of suberin in cells moving between any two spatial points.

The Growth Velocity Profile Is a Spatial Description of Growth

The velocities of different tissue elements are plotted against their distance from the apex to give the spatial pattern of growth velocity, or **velocity profile** (Figure 16.3B). Growth velocity increases with position in the growth zone. A constant value is obtained at the base of the growth zone. The final growth velocity is the final, constant slope of the growth trajectory equal to the elongation rate of the organ, as discussed in the previous section. In the rapidly growing maize root, the growth velocity is 1 mm h^{-1} at 4 mm, and reaches its final value of nearly 3 mm h^{-1} at 12 mm (see Figure 16.3B).

The growth velocity profile is a spatial description of growth, because the velocities at all locations are characterized at a single instant in time. (Recall that in a material description of growth [see Figure 16.3A], individual tissue elements are monitored through time or space.) The growth velocity profile is important for calculating rates of biosynthesis and net import in the growing tissue.

The Relative Elemental Growth Rate Characterizes the Expansion Pattern in the Growth Zone

To characterize the local expansion rate, we must consider the growth velocities at the apical and basal ends of a small segment of tissue. If the apical end of the segment is moving faster than the basal end, the segment is elongating. The velocity difference, divided by the segment length, gives the relative growth rate of the segment. Now imagine that the segment shrinks to a point x. To find the relative growth rate, r, at point x, we can use calculus: The value of r is given by the velocity gradient, dv/dx, where v is the growth velocity. This measure of the local growth rate is called the **relative elemental growth rate** (Erickson and Sax 1956). It represents the fractional change in length per unit time and has units of h^{-1}. If the growth velocity is known, the relative elemental growth rate can be evaluated by differentiation of the growth velocity with respect to position (Figure 16.3C). Or we can approximate r by measuring the relative growth rates of closely spaced segments of initial length L. For each segment, $r = (1/L)(dL/dt)$.

The relative elemental growth rate profile, $r(x)$, is a spatial description of the expansion pattern in the growth zone. In the maize root, the relative elemental growth rate has a maximum of 0.40 h^{-1} at 4 mm and drops to zero by 12 mm. That is, the tissue element at 4 mm is expanding 40% h^{-1}, while tissue beyond 12 mm has ceased expansion. Again, to understand the profile in physiological terms, we must consider that individual cells expand more and more rapidly as they are displaced to 4 mm and then less and less rapidly as they are displaced through the basal part of the growth zone.

The relative elemental growth rate profile shows the location and magnitude of the extension rate and can be used to quantify the effects of environmental variation on the growth pattern. It is also the basis for studying the physiology of growth. The graph of $r(x)$ can be compared to the distribution pattern of any substance thought to be regulating the growth rate. A close correlation between the two curves would provide circumstantial evidence for the role of the substance as a limiting factor for growth.

Embryogenesis

We now turn from growth, which is governed mainly by patterns of cell expansion, to other aspects of development, which depend on patterns of cell division and differentiation. The most dramatic developmental changes occur during embryogenesis, the transformation of a single-celled zygote into a multicellular embryonic plant.

Embryogenesis can be defined as the part of plant development that takes place in the embryo sac of the ovule or immature seed (see Chapter 1). During embryogenesis, several basic features of the primary plant body are established in rudimentary form. This process occurs in well-defined stages resulting from specific patterns of cell division. In this section we will examine these morphological changes in greater detail. We will also discuss recent molecular genetic studies that use *Arabidopsis* as a model and that have led to the identification of genes whose expression is required for the normal pattern of embryo development (see Box 16.1).

Three Essential Features of the Mature Plant Are Established during Embryogenesis

Plants differ from most animals in that embryogenesis does not generate the tissues and organs of the adult. Instead, angiosperm embryogenesis establishes only a rudimentary plant body, typically consisting of an embryonic axis and two cotyledons (if it is a dicot). Most of the structures that make up the adult plant are generated after embryogenesis through the activity of meristems. However, embryogenesis in plants establishes three basic developmental patterns that persist in the adult plant (Jürgens 1992, 1995):

1. The radial pattern of tissues found in plant organs.
2. The apical–basal axial developmental pattern.
3. The primary meristems that will generate the majority of the plant body during development after germination.

Both the radial and the axial patterns established during embryogenesis can be observed readily in the mature plant. Almost all plants exhibit an axial pattern in which the root and the shoot are at opposite ends of a linear axis. Any individual segment of either the root or the shoot also will exhibit apical and basal ends with different, distinct, physiological and structural properties. For example, adventitious roots form from the basal ends of stem cuttings, while buds develop from the apical ends. The radial pattern of tissues can be observed when we examine the arrangement of tissues extending from the outside of a stem or a root into its center. If we

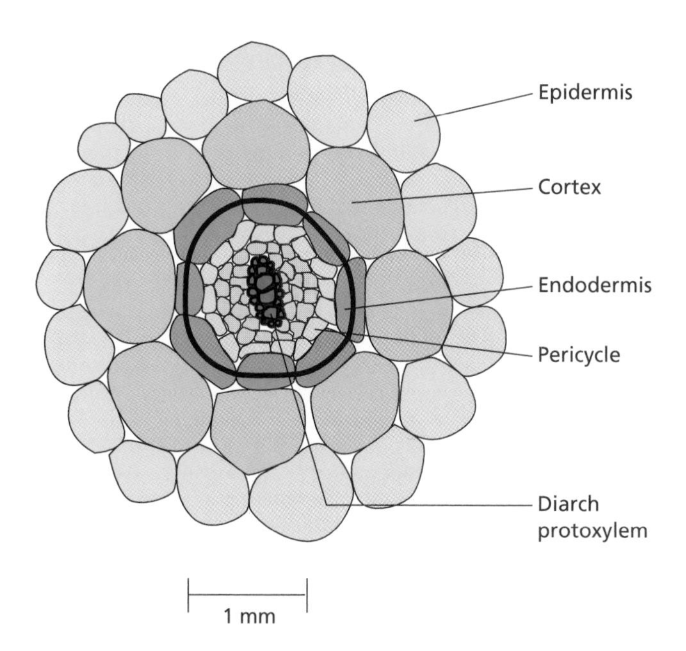

Figure 16.4 The radial pattern of tissues found in plant organs can be observed in a cross section of a root. This cross section of an *Arabidopsis* root was taken approximately 1 mm back from the root tip, a region in which the different tissues have formed. (After Dolan et al. 1993.)

examine a root in cross section, for example, we see three concentric rings of tissues arrayed along a radial axis: an outermost layer of epidermal cells covering a cylinder of cortical tissue, which in turn overlies the vascular cylinder (Figure 16.4).

The *epidermis* is a single cell layer, consisting of two or three kinds of cells, depending on the organ it surrounds. Root epidermal layers typically consist of only two kinds of cells: *trichoblasts*, which can give rise to root hairs, and cells that do not form hairs. The *cortex* underlies the epidermis and is several cells thick in both the root and the stem in most plants. The innermost layer of

BOX 16.1

Arabidopsis: Identifying Genes That Control Development

ARABIDOPSIS THALIANA, a member of the Cruciferae (Brassicaceae), or mustard family, has been called the *Drosophila* of plant biology because of its widespread use as a laboratory organism for studying developmental genetics in plants. The advantages of *Arabidopsis* for genetic studies were first recognized by several investigators, particularly George P. Redei, before the advent of molecular biology (Redei 1975). *Arabidopsis* is a small plant, well suited for laboratory culture, and it can complete its life cycle in approximately 6 weeks. However, the features that have made this plant so attractive for molecular genetics are its remarkably small genome and its low content of repetitive-sequence DNA (Leutwiler et al. 1984; Meyerowitz 1987). These traits greatly facilitate gene identification, cloning, and sequencing.

Although not the smallest of angiosperm genomes, the *Arabidopsis* genome is among the smallest. Estimates of the haploid genome size range from 0.7×10^8 to 1.5×10^8 base pairs, and this DNA is contained in five chromosomes. This size is comparable to the genome size of the nematode *Caenorhabditis elegans* and smaller than that of the fruit fly *Drosophila melanogaster*. Only about 10% of the *Arabidopsis* nuclear DNA is highly repetitive, and another 10% consists of moderately repetitive, chiefly ribosomal DNA. Each haploid genome in *Arabidopsis* has been estimated to contain approximately 20,000 genes (see Chapter 14). This number means that the gene density is high, with one gene per 5 kilobase pairs of nonrepetitive DNA. Unlike most crop plants, *Arabidopsis* allows relatively easy cloning of genes that have been identified by mutation.

Many plant genes can be identified by mutations that impair or delete the function of that gene. Although mutations can be induced by a variety of agents, including X rays and chemical mutagens such as ethylmethane sulfonate, in many respects the most useful mutagens are genetic elements that insert into genes, such as transposons and T-DNA of the *Agrobacterium* Ti plasmid. Insertional mutagenesis is useful because the mutated gene often can be cloned subsequently by screening of genomic libraries, using the mutagenic element as a probe. Two types of insertional mutagenesis, or gene tagging, have been particularly fruitful in plants: T-DNA insertions and transposon insertions (Bancroft et al. 1992; Gibson and Somerville 1993; Osborne and Baker 1995).

For T-DNA insertional mutagenesis, a kanamycin resistance gene (kanamycin is an antibiotic toxic to plants) was cloned into the T-DNA of a disarmed *Agrobacterium* vector and subsequently inserted into *Arabidopsis* by a seed transformation technique that did not involve tissue culture (Feldmann 1992). T-DNA insertion in *Arabidopsis* generates a high frequency of tagged, mutant genes because of the high gene density of the *Arabidopsis* genome. More than 1000 transformants were obtained initially, with an average of 1.4 T-DNA inserts per transformant. Many interesting mutant phenotypes were identified and characterized, from which a number of important genes have been isolated by screening of a genomic library of the DNA from the transformed mutant plants with a probe constructed from the T-DNA. These have included genes regulating trichome (hair) development and genes responsible for floral organ identity (Rerie et al. 1994; Yanofsky et al. 1990).

Some *Arabidopsis* genes that have been identified by mutation have been mapped and subsequently cloned by a procedure called "chromosome walking," with cosmid, bacterial or yeast artificial chromosome (BAC or YAC) libraries. These include the *ABI3* gene, which is necessary for seed maturation (see Chapter 23) (Giraudat et al. 1992), a gene that controls omega-3 fatty acid desaturase (Arondel et al. 1992), and the *DET1* gene, which encodes a component of the phytochrome signal transduction pathway (see Chapter 17) (Pepper et al. 1994). Before carrying out chromosome walking to a given genetic

BOX 16.1 *(continued)*

locus, the position of the locus must be accurately mapped. This position can be mapped by a comparison of the segregation of the mutation at that locus with known markers whose map position has been determined. High-resolution restriction fragment length polymorphism (RFLP) maps of the *Arabidopsis* genome have been constructed, as well as libraries of cloned *Arabidopsis* genomic DNA (Grill and Somerville 1991; Schmidt et al. 1992; Liu et al. 1995). These libraries can be screened with probes that map close to the mutant gene of interest to generate a series of overlapping clones containing the gene.

Despite the success of map-based cloning for the isolation of *Arabidopsis* genes, this technique probably will not be successful with all important genes. It has been estimated that only 25% of functional genes can be identified by mutation, and without a mutant allele it is much more difficult to map a gene. There are at least two reasons for this difficulty. First, many important activities are controlled by redundant genes. Elimination of a gene in one pathway may not eliminate the biological activity and yield a new phenotype, because another pathway may provide the necessary function. For example, nine genes in the *Arabidopsis* genome encode β-tubulin (Snustad et al. 1992). Although some of these genes may have unique functions, probably many of them are functionally interchangeable. No known *Arabidopsis* mutations affect the expression of a β-tubulin gene and these genes might not have been identified or cloned, had it not been for the significant homology they exhibit to the β-tubulin genes of other organisms.

A second reason that important genes may not be able to be identified by mutation is that they are lethal to gametes and therefore are never seen. This problem is particularly pronounced in plants, since they exhibit alternation of haploid and diploid generations in their reproductive development. The haploid male and female generations of angiosperms exhibit considerable developmental complexity. A mutation in a gene with an essential function during mega- or microsporogenesis would be lost. Sperm and eggs would not form in individuals receiving the mutant gene, while the normal allele would be transmitted.

For these reasons, many important genes can be expected to be identified as a result of the sequencing of the entire *Arabidopsis* genome. In June 1994, the North American *Arabidopsis* Steering Committee proposed a large-scale, federally funded *Arabidopsis* genome project, which would coordinate with a similar project already under way in Europe. The initial goals are to complete the *Arabidopsis* physical and genetic maps and establish a collection of sequence-ready clones representing the entire genome. Sequencing would then proceed, with a projected completion date of 2004.

the cortex in the root is a specialized layer known as the *endodermis*, in which the radial walls are impregnated with suberin. The *stele* forms a central cylinder and contains the vascular tissues. The outermost layer of the stele in the root is a specialized cell layer, the *pericycle*, from which lateral root meristems can develop. *Arabidopsis* has a diarch stele, meaning it has two strands of each of the conducting tissues, xylem and phloem.

Arabidopsis Embryos Pass through Three Distinct Stages of Development

Embryogenesis is a developmental process that typically is initiated from a zygote, and in angiosperms it usually occurs within an ovule as part of the process of seed formation. However, some somatic cells also can undergo embryogenesis, either as a result of experimental manipulations in the laboratory or as part of the normal course of their development (Zimmerman 1993). Angiosperms exhibit many different patterns of embryonic development (Raghavan 1986). However, the pattern of embryogenesis found in crucifers, including *Arabidopsis*, has been studied extensively and is the one we will present here (Mansfield and Briarty 1991; West and Harada 1993; Goldberg et al. 1994).

In *Arabidopsis*, embryogenesis has three distinct morphological stages. First, a precise pattern of initially synchronous cell divisions establishes a radially symmetrical sphere of cells, known as the **globular stage** embryo (Figure 16.5D). Next, rapid cell divisions in two regions on either side of the future shoot apex produce the cotyledonary primordia, giving rise to the bilaterally symmetrical **heart stage** embryo (Figure 16.5E and F). Finally, elongation of the axis and further development of the cotyledons produces the **torpedo stage** embryo (Figure 16.5G).

The cotyledons in many species grow extensively after the torpedo stage, increasing in cell number and size until they make up almost 90% of the total mass of the embryo. As the cotyledons of many angiosperms grow, they bend back on the axis, forming what is known as the *walking-stick embryo*, but this does not occur in *Arabidopsis*. Cotyledons are food storage organs for many species, and during the cotyledon growth phase they synthesize and deposit storage proteins, starch, and lipids that will be utilized by the seeding during the heterotrophic growth that occurs after germination. Although food reserves are stored in the *Arabidopsis* cotyledons, the growth of the cotyledons is not as extensive as it is in many other dicots.

The Axial Pattern of the Embryo Is Established during the First Cell Division

The apical–basal axial pattern is imposed on the embryo during the first division of the zygote, which is asymmetric (see Figure 16.5A). The first division of the zygote functionally polarizes the embryo into an apical and a

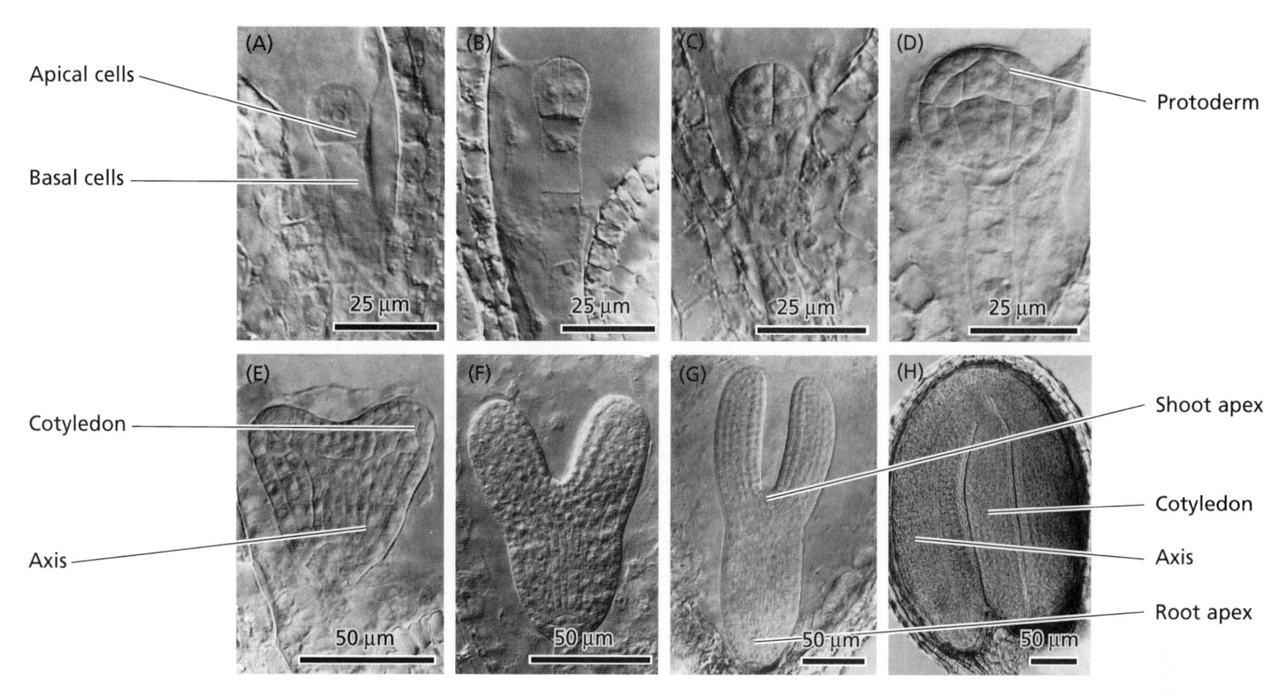

Figure 16.5 *Arabidopsis* embryogenesis is characterized by a precise pattern of cell division. Successive stages of embryogenesis are depicted. (A) One-cell embryo after the first division of the zygote, which forms the apical and basal cells. (B) Two-cell embryo. (C) Eight-cell embryo. (D) Early globular stage, which has developed a distinct protoderm (surface layer). (E) Early heart stage.(F) Late heart stage. (G) Torpedo stage. (H) Mature embryo. (From West and Harada 1993, photographs taken by K. Matsudaira Yee, courtesy of John Harada, © American Society of Plant Physiologists, reprinted with permission.)

basal cell that have very different fates. The smaller, apical daughter cell receives more cytoplasm than the larger, basal cell, which inherits the large zygotic vacuole. The apical cell subsequently divides repeatedly to generate most of the embryo, while the larger, vacuolate **basal cell** divides only a few more times before its derivatives differentiate to become the **suspensor**. The suspensor connects the embryo to the basal cell. Only the **hypophysis**, the apicalmost progeny of the basal cell, contributes to the embryo. The hypophysis will form part of the root apical meristem.

Even though the embryo is spherical throughout the globular stage of embryogenesis, the cells within the apical and basal halves of the sphere have different identities and functions. As the embryo continues to grow and reaches the heart stage, three axial regions can be recognized: (1) the apical region, which gives rise to the cotyledons and shoot apical meristem; (2) a middle region, which gives rise to the hypocotyl; and (3) the basal region, which gives rise to the root (Figure 16.6).

The Apical–Basal Pattern of the Axis Requires Specific Gene Expression

Isolation of *Arabidopsis* mutants that are missing major parts of the embryonic axis has led to the identification of genes that could be involved in establishing the axial pattern (Jürgens et al. 1991; Meinke 1986). We will consider four of these mutants: *gurke*, *fackel*, *monopteros*, and *gnom*.

Seedlings homozygous for mutations in the *GURKE* gene usually lack cotyledons (Torres-Ruiz et al. 1996). Lacking cotyledons, seedlings of *gurke* mutants look like tiny cucumbers, thus giving this mutation its name* (Figure 16.7A). Strong *gurke* mutant alleles eliminate the entire apical region, leaving the root, with a normal radial tissue pattern, intact. Mutations in the *FACKEL* gene result in deletion of the middle part of the axis, such that the seedling consists of cotyledons attached directly to the root. Mutations in the *MONOPTEROS* gene result in seedlings lacking both a hypocotyl and a root, but with largely normal cotyledons and an apical region, including shoot apical meristem (Figure 16.7B) (Berleth and Jürgens 1993). Seedlings homozygous for mutations in the *GNOM* gene lack both roots and cotyledons (Figure 16.7C) (Mayer et al. 1993). In the most extreme form, *gnom* mutant seedlings are spherical and lack axial polarity entirely.

The wild-type *GURKE*, *MONOPTEROS*, *FACKEL*, and *GNOM* genes are required for the determination of the apical–basal axial pattern of the embryo, but these genes are not involved in establishing the basic tissue

* The English word "gherkin," for a type of cucumber, is derived from the Dutch word *gurk*.

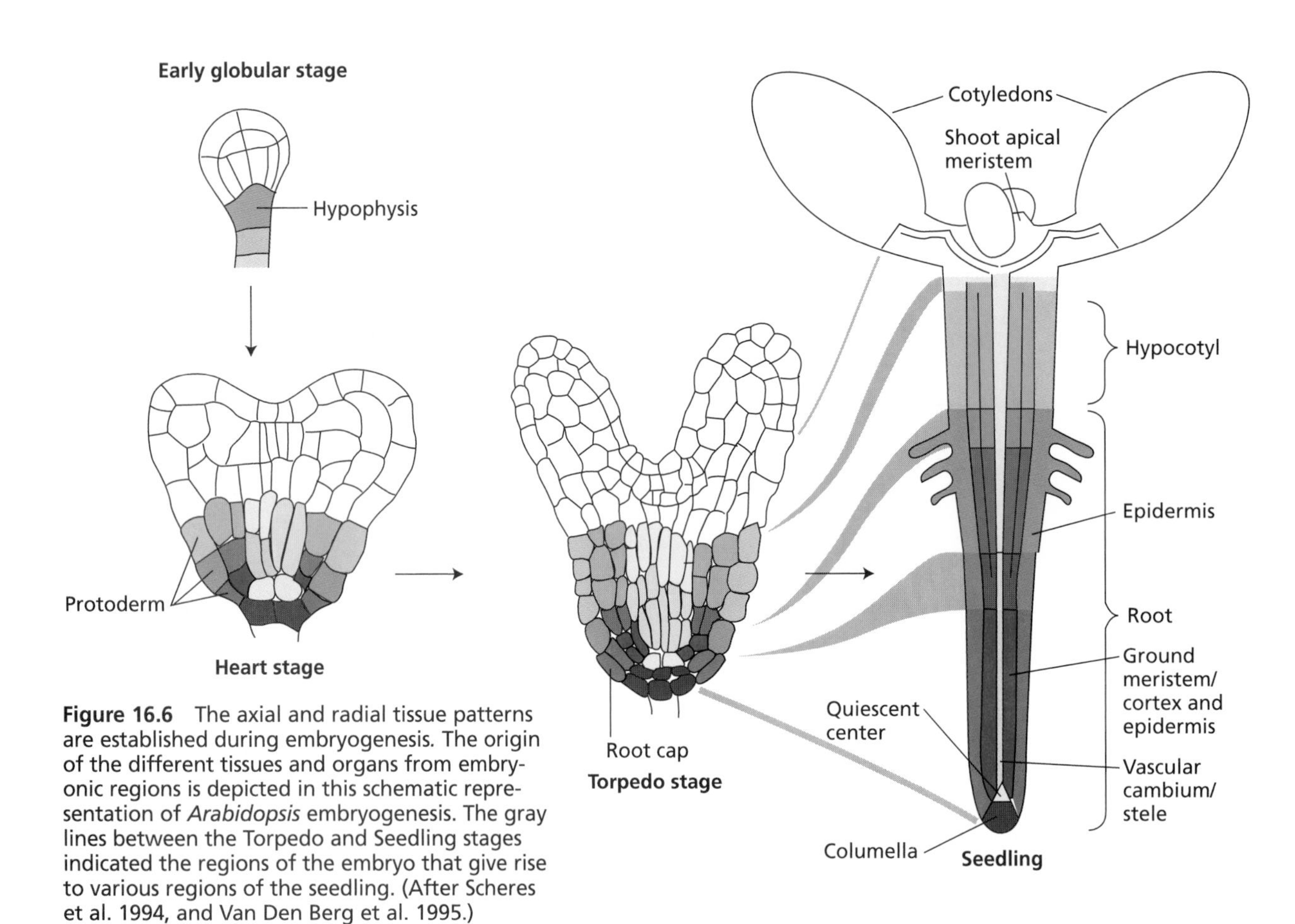

Figure 16.6 The axial and radial tissue patterns are established during embryogenesis. The origin of the different tissues and organs from embryonic regions is depicted in this schematic representation of *Arabidopsis* embryogenesis. The gray lines between the Torpedo and Seedling stages indicated the regions of the embryo that give rise to various regions of the seedling. (After Scheres et al. 1994, and Van Den Berg et al. 1995.)

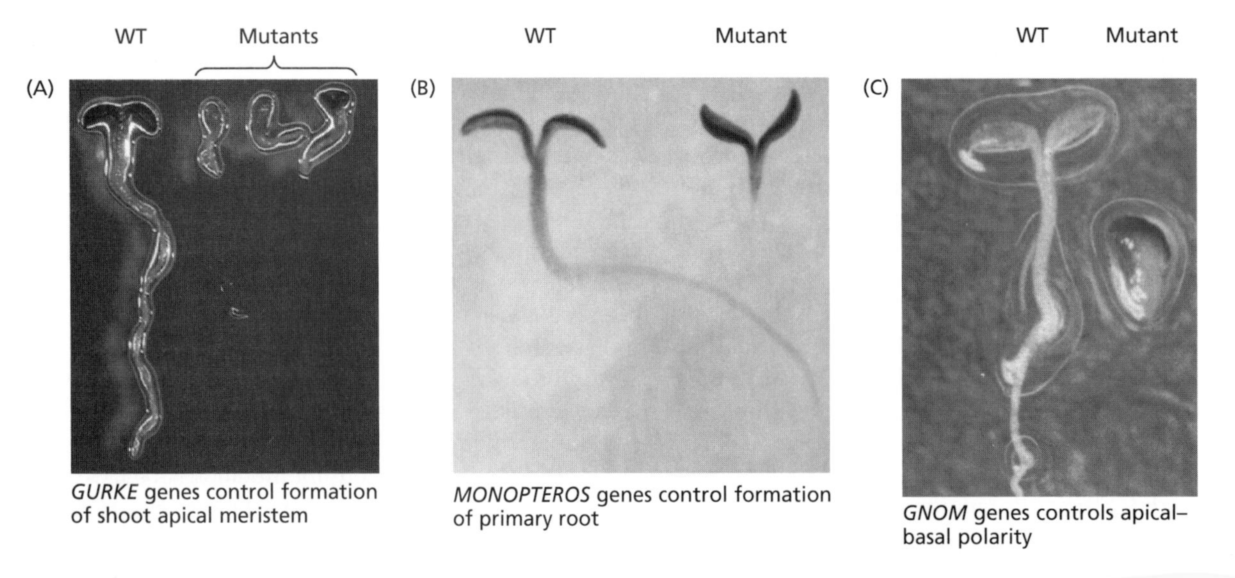

Figure 16.7 Genes that control major events in *Arabidopsis* embryogenesis have been identified by selection of mutants in which a stage of embryogenesis is blocked. (A) The development of three *gurke* mutant seedlings (right) is compared with that of the wild type (left) at the same stage of development. The *GURKE* gene is required for the formation of the apical region; *gurke* mutants lack cotyledons and, with the most severe alleles, they also lack the shoot apical meristem. (B) Mutants of the *MONOPTEROS* gene (right) are compared with the wild type at the same stage of development (left). The *MONOPTEROS* gene is required for basal patterning. Seedlings homozygous for *monopteros* mutants have a hypocotyl, a normal shoot apical meristem, and cotyledons, but they lack the primary root. (C) An *Arabidop-sis* seedling that is homozygous for a *gnom* mutation (right) is compared with that of the wild type (left). The *GNOM* gene is required for normal apical–basal polarity in the embryo. (A from Torres Ruiz et al. 1996, courtesy of R. Torres Ruiz; B from Berleth and Jürgens 1993; C from Mayer et al. 1993.)

types or in controlling cell differentiation. For example, *gnom* mutants contain epidermal, ground, and vascular tissue in the proper order, along the radial axis, although the organization of these tissues may be somewhat disrupted. For example, the vessel elements of *gnom* mutants are not connected with each other and do not form a vascular system. At present we do not know how these axis-specifying genes define the different regions of the embryo, although some of them have been cloned. In fact, it has yet to be demonstrated that these genes are primarily involved in establishing the initial patterns of the developing embryo.

The Radial Pattern of Tissue Differentiation Is First Visible at the Globular Stage

After the first zygotic division, the next several divisions of the *Arabidopsis* embryo are spatially precise and synchronous. The apical cell undergoes a series of highly ordered, stereotypic divisions, generating an eight-cell globular embryo by 30 hours after fertilization (see Figure 16.5C). At this eight-cell stage the radial pattern of tissue differentiation first becomes evident. The upper and lower tiers of cells of the early globular stage differ and divide the embryo into apical and basal halves, reflecting the axial pattern imposed on the embryo at the first zygotic division.

Additional transverse divisions divide the lower tier of cells into three regions representing the major tissue types that are radially arranged in the root and stem axes. The outermost cells form a surface layer one cell thick, known as the **protoderm** (see Figures 16.5D and 16.6). The protoderm covers both halves of the embryo and will generate the epidermal tissues of the plant. Underlying the protoderm are the cells of the **ground meristem**, which will give rise to the cortical tissues and, in the root and hypocotyl, will produce the endodermis. The **procambium** is the inner core of elongated cells that will generate the vascular cylinder and, in the root, the pericycle.

Cells Acquire Tissue Identities as the Result of Specific Gene Expression

Genes important for the establishment of the radial pattern and the specification of the different tissues within the embryo have been identified, mostly in *Arabidopsis*, but also in maize and other plants. Some of these regulatory genes have been identified by their homology to homeobox genes (see Chapter 14). **Homeobox genes** encode proteins that contain a special DNA-binding domain, known as the **homeodomain**, which enables these proteins to regulate the transcription of other genes. In *Drosophila*, the transcription factors encoded by homeobox genes determine the locations where entire organs form. Mutations in these genes can cause body segments and other structures to form at inappropriate locations. Such mutations are called **homeotic mutations**. Although "homeotic mutations" have been identified in plants in relation to flower development, they do not seem to involve homeobox genes (see Chapter 24). Thus, the homeobox genes of plants play a role in development that is different from the role played by homeobox genes in *Drosophila*.

Additional genes have been identified that function in the establishment of the radial tissue pattern in the root and hypocotyl during embryogenesis and are important for the maintenance of that pattern during postembryonic development (Scheres et al. 1995). To identify these genes, investigators isolated *Arabidopsis* mutants in which roots grew more slowly than in the wild type. Subsequent analysis of these mutants identified several that have defects in the radial tissue pattern. These mutations affect tissue organization and cell differentiation not only in the embryo, but also in both primary and secondary roots and in the hypocotyl. For example, the *shortroot* (*shr*) and *scarecrow* (*scr*) mutants both lack cell layers derived from the ground meristem (Figure 16.8). Roots of plants that are homozygous for the *shr* mutation lack the endodermal layer, although the tissue pattern otherwise is normal, and they have only a single layer of ground tissue in which the cells have the characteristics of cortex. Plants with the *scr* mutation also have roots with a single cell layer of ground tissue, but in this case the cells of this layer are of mixed identity and have characteristics of both endodermal and cortical cells. As we will see in Chapter 19, *scr* mutants also lack the cell layer called the starch sheath , a structure in the hypocotyl and stem that has been implicated in the growth response to gravity.

The *SCR* gene has been cloned and sequenced. It encodes a novel protein that likely is required for the division of the cell that gives rise to the endodermal and cortical cell lineages. Although considerably more work is necessary before we can understand how the radial pattern of tissues is established, it probably comes about through the position-dependent expression of cell or tissue identity genes, which, because they are transcription factors, control the expression of other genes whose products are required for the development and function of these tissue or cell types.

A Specific Gene Is Required for the Formation of the Shoot Protomeristem

The primary root and shoot meristems are established during embryogenesis. Since they often are not active at this time, the term **protomeristem** may be more appropriate to describe these structures. A protomeristem may be defined as a structure that will become the root or shoot meristem upon germination. The shoot protomeristem can be recognized by the torpedo stage of embryogenesis in *Arabidopsis*, when cells between the cotyledons assume a characteristic layered appearance as a result of oriented cell divisions (see the next section). Probably certain cells acquired the identity of shoot apical meristem cells earlier, during the globular stage.

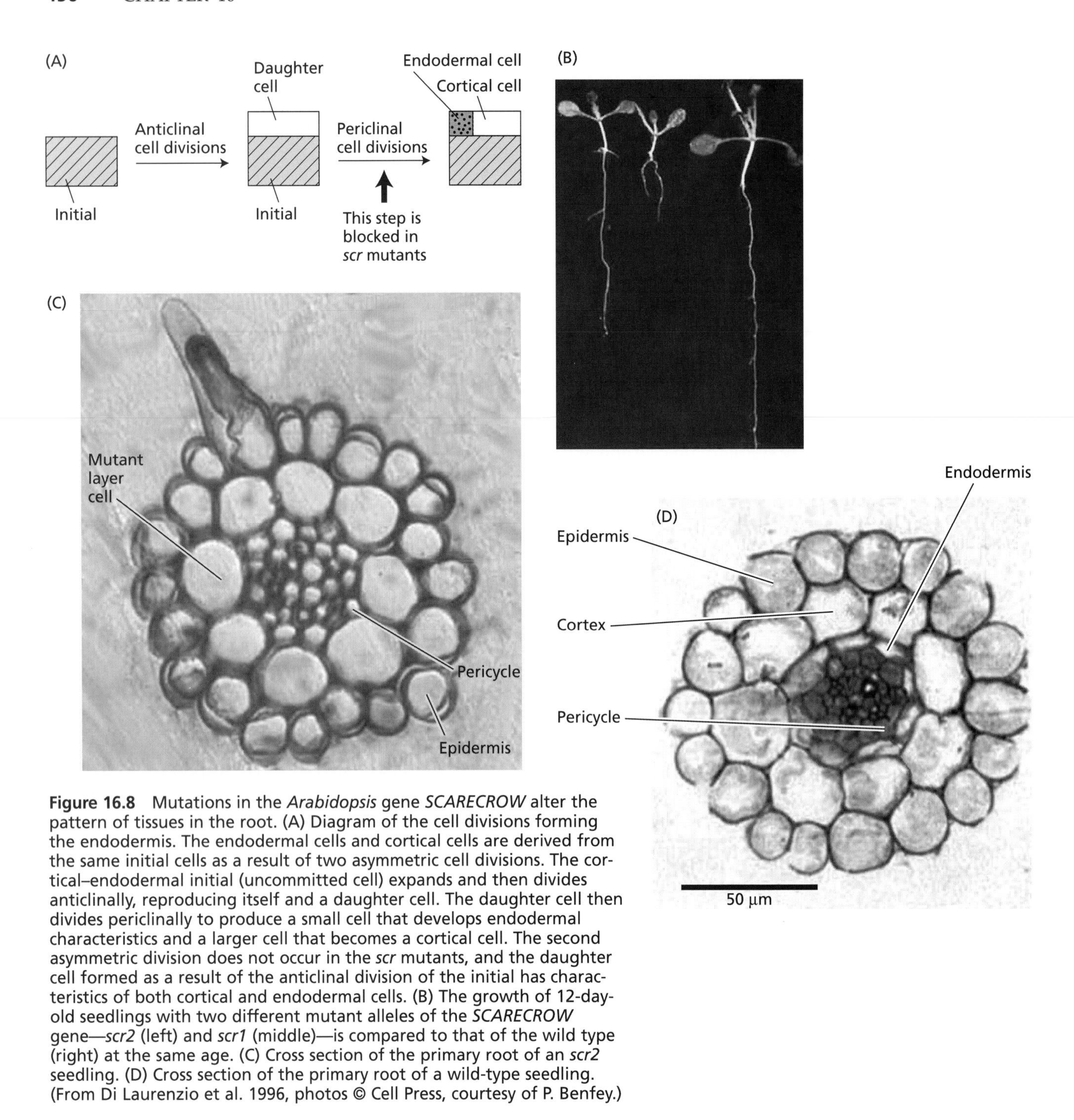

Figure 16.8 Mutations in the *Arabidopsis* gene *SCARECROW* alter the pattern of tissues in the root. (A) Diagram of the cell divisions forming the endodermis. The endodermal cells and cortical cells are derived from the same initial cells as a result of two asymmetric cell divisions. The cortical–endodermal initial (uncommitted cell) expands and then divides anticlinally, reproducing itself and a daughter cell. The daughter cell then divides periclinally to produce a small cell that develops endodermal characteristics and a larger cell that becomes a cortical cell. The second asymmetric division does not occur in the *scr* mutants, and the daughter cell formed as a result of the anticlinal division of the initial has characteristics of both cortical and endodermal cells. (B) The growth of 12-day-old seedlings with two different mutant alleles of the *SCARECROW* gene—*scr2* (left) and *scr1* (middle)—is compared to that of the wild type (right) at the same age. (C) Cross section of the primary root of an *scr2* seedling. (D) Cross section of the primary root of a wild-type seedling. (From Di Laurenzio et al. 1996, photos © Cell Press, courtesy of P. Benfey.)

The *SHOOTMERISTEMLESS* (*STM*) gene is expressed specifically in the cells that will become the shoot apical meristem, and its expression in these cells is required for the formation of the shoot promeristem. *Arabidopsis stm* mutants, homozygous for a mutated, loss-of-function *STM* gene, do not form a shoot apical meristem (Lincoln et al. 1994). The product of the wild-type *STM* gene appears to suppress cell differentiation, ensuring that the meristem cells remain undifferentiated. The expression of *STM* is first observed in one or two cells at the apical end of the midglobular embryo. By the heart stage, *STM* expression is confined to a few cells between the cotyledons (Long et al. 1996). Since *STM* acts as a marker for these cells, the shoot apical meristem must be specified long before it can be recognized morphologically. The *STM* gene is necessary not only for the formation of the embryonic shoot apical meristem, but also for the maintenance of shoot apical meristem identity in the adult plant.

A molecular marker for the root protomeristem has not yet been identified, but it also appears to be determined early in embryogenesis. Two genes called *ROOT-MERISTEMLESS* (*RML1* and *RML2*) have been identified in *Arabidopsis*. These genes differ from *STM* in that mutations of the *RML* genes do not prevent the formation of the embryonic root protomeristem, or of lateral root meristems in the seedlings. Mutations in the *RML* genes reduce the number of cell divisions root meristems undergo, thus resulting in short roots rather than absent roots (Cheng et al. 1995). Root cap initials (the cells that divide to produce the root cap) are formed from the hypophysis at the heart stage of embryogenesis, indicating that the root protomeristem is established at least by this stage of embryogenesis.

Meristems in Plant Development

Meristems are populations of small, isodiametric (having equal dimensions on all sides) cells with "embryonic" characteristics (Steeves and Sussex 1989). A meristematic cell population typically contains a few hundred to a thousand cells, although the *Arabidopsis* shoot apical meristem has only about 60 cells. Vegetative meristems are self-perpetuating. Not only do they produce the tissues that will form the body of the root or stem, but they also continuously regenerate themselves. A meristem can retain its embryonic character indefinitely, possibly even thousands of years in the case of trees. The reason for this ability is that some meristematic cells do not become committed to a differentiation pathway, and they retain the capacity for cell division, at least as long as the meristem remains vegetative.

Uncommitted cells that divide to produce the cells of the plant body are called **initials**, which are comparable to the stem cells of animals (Barlow 1978, 1994). When stem cells in animals divide, one of the daughter cells retains the identity of the stem cell while the other is committed to a particular developmental pathway that may include a period of rapid cell division before its descendants can be recognized as a specific cell type. The initials represent the ultimate source of all the cells in the meristem and the entire stem or root, but they themselves usually divide slowly.

Different Meristems Produce Different Kinds of Tissues or Organs

The different plant meristems are distinguished by their position, their origin, and the structures they produce (Steeves and Sussex 1989). **Primary meristems** generate the primary plant body. They form during embryogenesis and are found at the tip of the root and the shoot, where their activity generates the primary tissues and organs that make up the primary axis of the plant (Figure 16.9). **Secondary meristems** form later in development. The **axillary meristems** that form in the axils of leaves are secondary meristems, as are the meristems of lateral roots. Secondary meristems have a structure similar to that of primary meristems, and they have the potential to form a secondary root or shoot, although their activity may be suppressed or retarded while the primary meristem is active.

Lateral meristems are another kind of secondary meristem that may arise later in the development of woody stems and roots as cylinders of meristematic cells. The activity of lateral meristems increases the girth of the stem or the root. The vascular cambium and the cork cambium are two types of lateral meristems. The **vascular cambium** contains two types of meristematic cells: fusiform initials and ray initials. **Fusiform initials** are highly elongated, vacuolate cells that divide longitudinally to regenerate themselves, and whose derivatives differentiate into the conducting cells of the secondary xylem and phloem. **Ray initials** are small cells whose derivatives include the radially oriented files of parenchyma cells known as rays. The **cork cambium** is a meristematic layer that develops within mature cells of the cortex and the secondary phloem. Derivatives of the cork cambium differentiate as **cork cells** that make up the secondary protective layer called the **periderm**. The periderm forms the protective outer surface of the secondary plant body, replacing the epidermis.

Root Development

Roots are adapted for growing through soil and absorbing the water and mineral nutrients in the capillary spaces between soil particles. These functions have placed constraints on the evolution of root structure, resulting in a streamlined axis with no lateral appendages emerging from the apical meristem that would interfere with penetration of the soil. Branches arise internally in roots and form well behind the growing

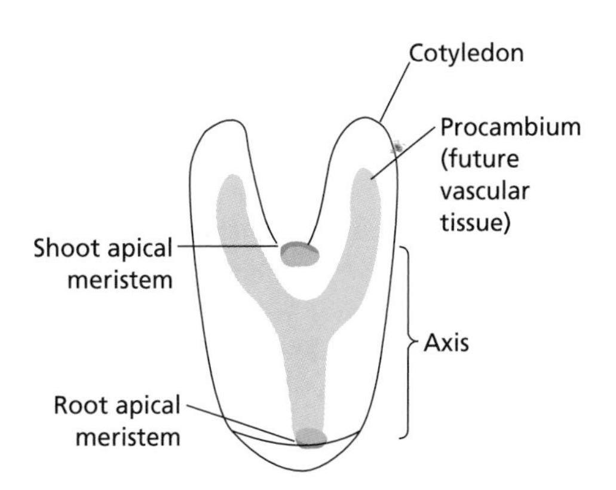

Figure 16.9 Primary meristems are at the ends of the plant axis and generate the primary plant body. (After Sussex 1989.)

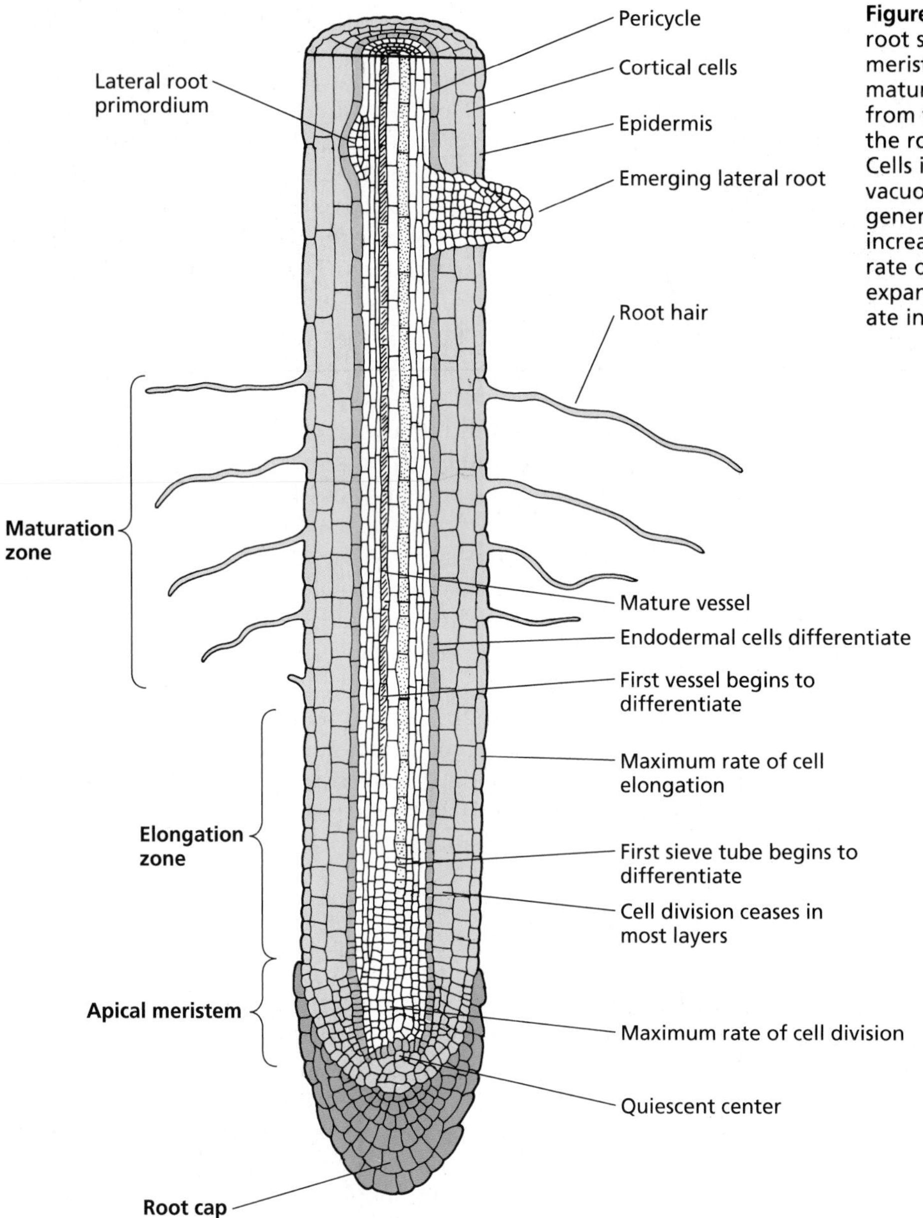

Figure 16.10 Diagram of a primary root showing the root cap, the apical meristem, the elongation zone, and the maturation zone at increasing distances from the tip. The cellular structure of the root has been simplified for clarity. Cells in the apical meristem have small vacuoles and expand and divide rapidly, generating many files of cells. At increasing distances from the tip, as the rate of cell division decreases, the cells expand greatly and begin to differentiate into specialized cell types.

tip. Absorption is enhanced by fragile root hairs, which also form behind the growth zone.

In this section we will discuss root morphogenesis, beginning with a description of the four developmental zones of the root tip. We will then turn to the apical meristem. The absence of lateral appendages, such as leaves or buds, makes cell lineages easier to follow in roots, thus facilitating molecular genetic studies on the role of patterns of cell division in root development.

The Root Tip Has Four Developmental Zones

Roots grow and develop from their distal ends. Although the boundaries are not sharp, four developmental zones can be distinguished in a root tip: the root cap, the meristematic zone, the elongation zone, and the maturation zone (Figure 16.10). These four developmental zones occupy only a little more than a millimeter of the tip of the *Arabidopsis* root (Dolan et al. 1993). The developing region is larger in other species, but growth is still confined to the tip. With the exception of the root cap, the boundaries of these zones overlap considerably.

The **root cap** protects the apical meristem from mechanical injury as the root pushes its way through the soil. Root cap cells form by specialized **root cap initials**. As new cells are produced by the root cap initials, older cells are progressively displaced toward the tip, where they are eventually sloughed off. As root cap cells differentiate, they acquire the ability to perceive gravitational stimuli and secrete mucopolysaccharides (slime) that help the root penetrate the soil.

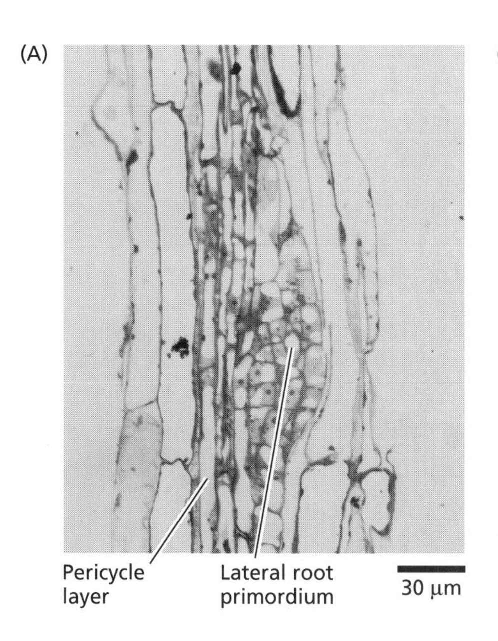

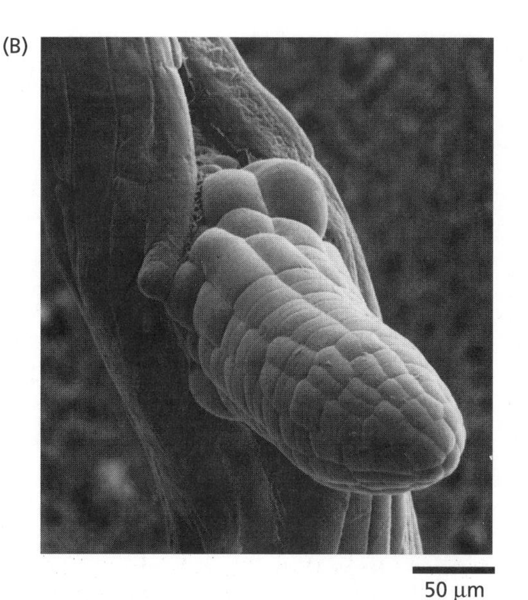

Figure 16.11 Lateral roots originate from the pericycle in the mature region of the root (shown here in *Arabidopsis*). (A) Cell divisions in the pericycle layer lead to the formation of a lateral root primordium. (B) The growing lateral root primordium pushes its way through endodermis, cortex, and epidermis to emerge as a lateral or branch root. (From Dolan et al. 1993.)

The **meristematic zone** lies just under the root cap, and in *Arabidopsis* it is about a quarter of a millimeter long. The root meristem generates only one organ, the primary root. It produces no lateral appendages. Lateral or branch roots arise from the pericycle in mature regions of the root. Cell divisions in the pericycle establish secondary meristems that grow out through the cortex and epidermis, establishing a new growth axis (Figure 16.11). Both the primary and the secondary root meristems behave similarly in that divisions of the cells in the meristem give rise to progenitors of all the cells of the root.

The **elongation zone**, as its name implies, is the site of rapid and extensive cell elongation. Although some cells may continue to divide while they elongate within this zone, the rate of division decreases progressively to zero with increasing distance from the meristem. After division and elongation have ceased, cells enter the **maturation zone**, in which they acquire their differentiated characteristics. Differentiation may begin much earlier, but the cells do not achieve the mature state until they reach this zone. The radial pattern of differentiated tissues becomes obvious in the maturation zone. Later in the chapter we will examine the differentiation and maturation of one of these cell types, the vessel element.

Root Initials Generate Longitudinal Files of Cells

Meristems are populations of dividing cells, but all cells in the meristematic region do not divide at the same rate or with the same frequency. Typically, the central cells divide much more slowly than the surrounding cells, if at all. These rarely dividing cells are called the **quiescent center** of the root meristem (see Figure 16.10). Cells are more sensitive to ionizing radiation when they are dividing. As a result, the rapidly dividing cells of the meristem can be killed by doses of radiation that nondividing and slowly dividing cells, such as those of the quiescent center, can survive. If the rapidly dividing cells of the root are killed by ionizing radiation, the root can in many cases regenerate from the cells of the quiescent center (Clowes 1964). This ability suggests that quiescent-center cells are important for the patterning involved in forming a root. This interpretation is supported by work of Feldman and Torrey (1976), who surgically isolated the quiescent center from maize roots and demonstrated that it could grow in culture to regenerate a normal root with a typical radial tissue pattern.

The most striking structural feature of the root tip, when viewed in longitudinal section, is the long files of clonally related cells. Most cell divisions in the root tip are transverse, or **anticlinal** (see Figure 16.8A), with the plane of cytokinesis oriented at right angles to the axis of the root. There are relatively few **periclinal** divisions (the plane of division is parallel to the root axis). Most of these periclinal divisions occur near the root tip and establish new files of cells. As a result, the ultimate origin of any particular mature cell can be traced back to one or a few cells in the meristem, the initial cells of a particular file (see Figure 16.10). In *Arabidopsis*, the initial cells surround the quiescent center, but they are not part of the quiescent center. The initials may be derived from quiescent-center cells, but this origin must occur during embryogenesis, since the quiescent-center cells do not divide after germination (Scheres et al. 1994).

Fern Meristems Have a Single Apical Initial Cell

Our undersanding of the mechanics of meristem function has been enhanced by investigations of the apical meristems of ferns and other primitive vascular plants. The root and shoot apical meristems of these organisms

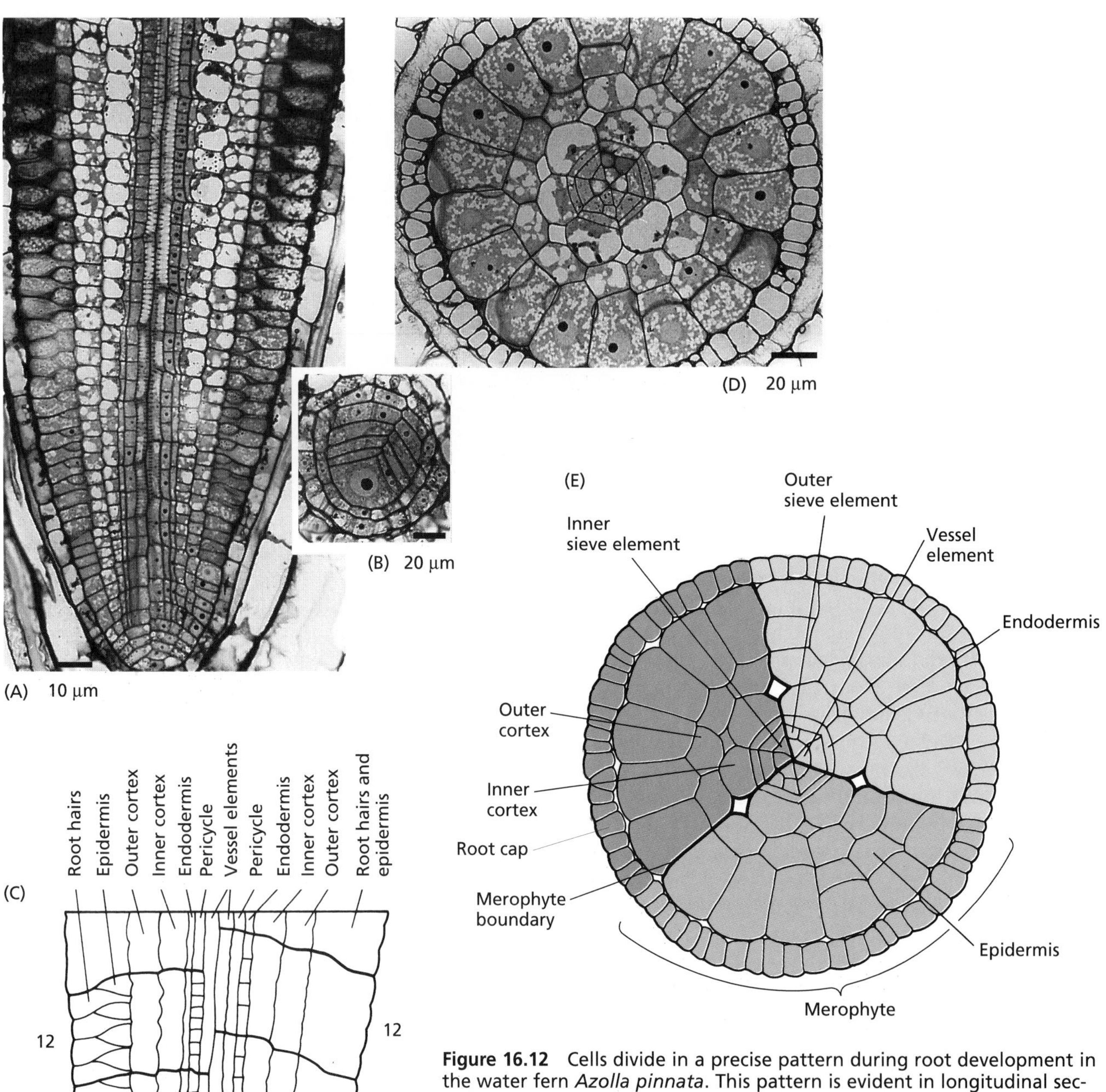

Figure 16.12 Cells divide in a precise pattern during root development in the water fern *Azolla pinnata*. This pattern is evident in longitudinal sections of the root (A). (B) Shown here at higher magnification, the single apical cell divides at the three triangular faces to form wedge-shaped cells called merophytes. (C) This diagram of the root identifies the 12 merophytes that form on each side of the root. Subsequent divisions of each merophyte are also in a precise pattern to generate a specific number of cells that will differentiate according to their position within the merophyte, as shown in the cross section (D) and the diagram (E). (A from Hardham and Gunning 1979 ; A, B, and D courtesy of A. Hardham.)

have a single large apical cell, which resides at the center of the meristem. The apical cell is the ultimate source of all the cells in the meristem, as well as the root or shoot body. Brian Gunning at the Australian National University studied in detail the patterns of cell divisions in roots of the water fern *Azolla* (Gunning 1982). *Azolla* roots are unusual in that they do not grow throughout the life of the plant. The *Azolla* root is a **determinate** organ, and the root meristem has a genetically determined limit to its growth. This condition is in contrast to most roots, in which the activity of the meristem is **indeterminate**—growth persisting as long as environmental conditions permit it.

The *Azolla* apical cell, which is shaped something like a pyramid with four more or less triangular sides, divides only 50 to 55 times, cutting off derivative cells in each of its four planes, before it stops dividing entirely. With the exception of the root cap, all the cells in the *Azolla* root can be traced to the apical cell, whose division planes are determined with great precision (Figure 16.12). The cells that will become the root body are generated by divisions parallel to the three internal triangular faces of the apical cell, forming a wedge-shaped cell known as a **merophyte**. Each of the three derivative cells subsequently divides a set number of times to form a sector that makes up one-third of the root axis. The planes and numbers of divisions of the apical cell and each of its derivatives are very precise. We can recognize two different kinds of divisions—formative and proliferative—which differ in the fate of the daughter cells produced. Both daughter cells resulting from a **proliferative division** have the same fate.

Formative divisions may be either periclinal or anticlinal, but they differ from proliferative divisions in that the two daughter cells have different fates. That is, one of the daughters may remain undifferentiated while the other daughter cell will differentiate as a particular cell type. Formative divisions establish the files of cells that make up the various tissues of the root; proliferative divisions increase the number of cells in each file. In the root, most formative divisions are periclinal (longitudinal), while proliferative divisions tend to be anticlinal (transverse). Longitudinal, formative divisions within the merophytes give rise to progenitor cells for the files of cells in the root. For example, the division that generates the endodermis and inner cortical cells occurs in the fourth merophyte on both sides of the root, as do the divisions that delineate the vessel element, pericycle, outer cortex, and root hair–epidermis cell files. Once formed, the progenitor cells usually complete several transverse, proliferative divisions before the progeny differentiate.

Root Apical Meristems of Seed Plants Have Multiple Initial Cells

The patterns of cellular organization found in the root meristems of seed plants are substantially different from those observed in *Azolla*. All seed plants have several initial cells instead of the single initial found in *Azolla* and other ferns. However, they are similar to *Azolla* in that it is possible to follow files of cells from the region of maturation into the meristem and, in some cases, to identify the initial cell from which the file was produced. The *Arabidopsis* root meristem is typical of many angiosperms and will be described here because of the dramatic progress that has been made recently in the identification of genes whose activity may control key events in root development.

At the center of the *Arabidopsis* root meristem is a group of four cells known as the **central cells** (Figure 16.13) (Dolan et al. 1993). The central cells of the *Arabidopsis* root normally do not divide after embryogenesis and are the quiescent center of the *Arabidopsis* root meristem. Surrounding the quiescent center is a ring of

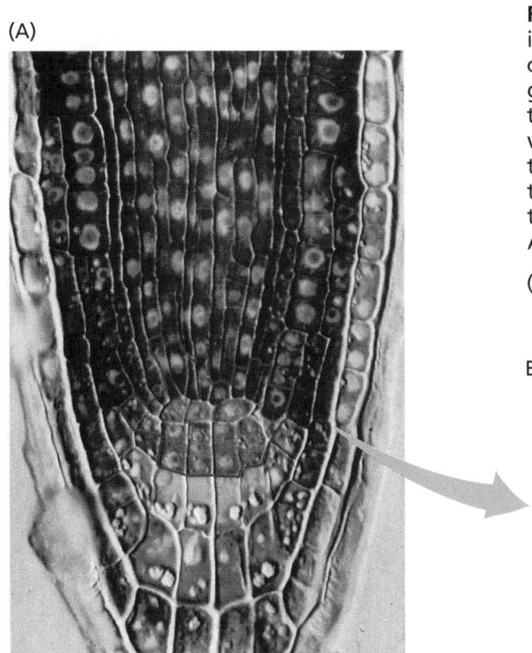

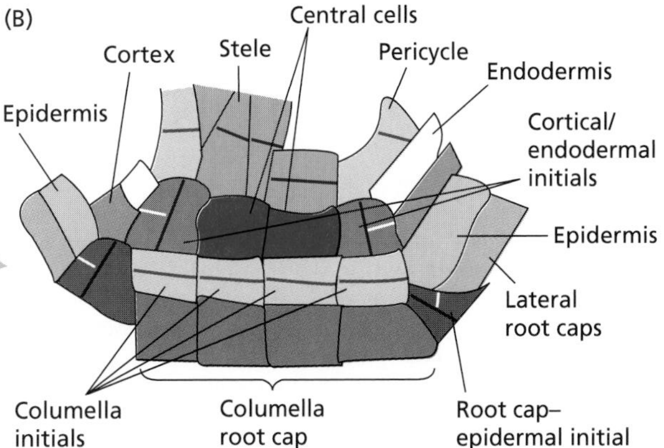

Figure 16.13 All the tissues in the root are derived from a small number of initial cells in the root apical meristem. (A) Longitudinal section through the center of an *Arabidopsis* root. The promeristem containing the initials that give rise to all the tissues of the root is outlined in green. (B) Diagram of the promeristem region outlined in A. Only two of the four central cells are visible in this section. The dark green lines indicate the cell division planes that occur in the initials. White lines indicate the secondary cell divisions that occur in the cortical–endodermal and lateral root cap–epidermal initials. (From Schiefelbein et al. 1997, courtesy of J. Schiefelbein, © the American Society of Plant Physiologists, reprinted with permission.)

initial cells that generate the cortical and endodermal layers. The **cortical–endodermal initials** first divide once anticlinally (i.e., perpendicular to the longitudinal axis); then these daughters divide periclinally (i.e., parallel to the longitudinal axis) to establish the files that become the cortex and the endodermis, each of which constitutes only one cell layer in the *Arabidopsis* root (see Figure 16.8A).

The cells immediately above (apical to) the central cells are the **columella initials**, which divide anticlinally and periclinally to generate a sector of the root cap known as the columella. The **root cap–epidermal** initials are in the same tier as the columella initials but form a ring surrounding them. Anticlinal divisions of the root cap–epidermal initials generate the epidermal cell layer, while periclinal divisions of the same initials, followed by subsequent anticlinal divisions of the derivatives, produce the lateral root cap. The pericycle and vascular tissue arise from a tier of cells just behind the quiescent-center cells. These cells represent the **stele initials**. The initial cells, together with their immediate derivatives in the apical meristem, are called the **promeristem**.

The Fate of a Cell Is Determined by Its Position Rather Than by Its Clonal Ancestry

Roots can be cultured in vitro, where they will grow indefinitely if provided with an adequate supply of nutrients. The radial tissue pattern is maintained during this growth, although some specific details of the pattern can be altered. These characteristics suggest that the tissue pattern is determined by the root meristem, whose continued activity generates the tissues of the root. One hypothesis is that the initial cells in the meristem are already committed for a particular type of cell differentiation. For example, the stele initials might be committed to form vascular tissue and pericycle, while the cortical initials are committed to differentiate as endodermal and cortical cells. The proliferative divisions that occur after the establishment of a particular cell file simply multiply the number of endodermal or cortical cells, without changing their character, before these cells acquire their final differentiated characteristics.

To some degree this hypothesis is supported by the fact that genes whose expression marks specific tissues also are expressed in the initials for that tissue. However, the hypothesis overstates the importance of the clonal origin of tissues. While cell fate may be determined early, perhaps even in the initial cells, this determination is not irreversible. The most important determinant of the fate of a plant cell is not clonal ancestry, but rather cell position within the organ. This fact has been elegantly demonstrated by the work of Ben Scheres's group at Utrecht University in the Netherlands, who followed the consequences of ablating

(A)

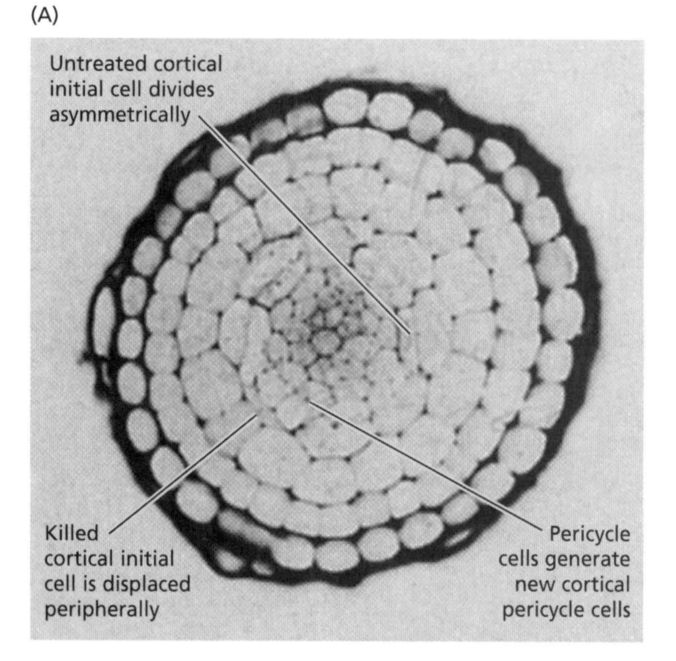

(B)

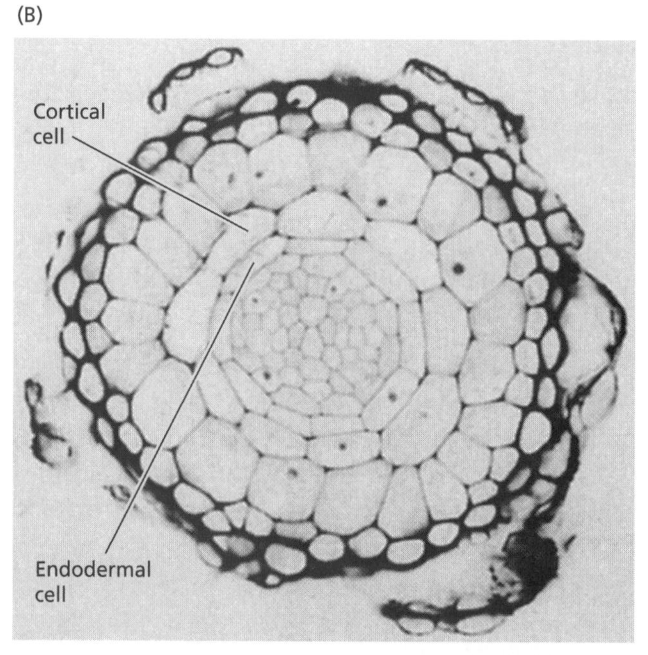

Figure 16.14 Plant cell identity is determined by position rather than by clonal origin. This fact is demonstrated by an experiment in which a cortical initial cell is ablated with a laser (A). The dead cell is displaced peripherally, and two pericycle cells have pushed into the position previously occupied by the dead initial and have assumed the function of the initial. These cells have divided periclinally to generate cells in both the cortical and the pericycle layers. (B) Three days after ablation of the cortical initial, the invading pericycle cells have generated a larger cortical cell and a smaller endodermal cell. (From Van Den Berg et al. 1995, courtesy of B. Scheres.)

(killing) specific initial cells within the root meristem with a laser (Van Den Berg et al. 1995). For example, it was possible to ablate the cortical initials that would normally give rise to the cortex and endodermal cells. However, this ablation did not eliminate the cortical and endodermal cell files.

Instead, when cortical initials were ablated, pericycle cells adjacent to the deleted cortical initials divided periclinally to regenerate the cortical and endodermal cell files (Figure 16.14). Since pericycle cells are smaller than cortical cells, more than one pericycle cell replaced the ablated cortical initial. The pericycle cells that invaded the cortex switched their fate to that of the cortical initials when they crossed into the cortical region. Thus, their fate is determined by the position they occupy within the radial tissue pattern. These experiments demonstrate that determination of plant cell fate is relatively plastic and can be changed when the positional signals necessary for its maintenance are altered.

Control of the Plane of Cell Division

One of the most striking features of tissue organization in the root in *Azolla* and *Arabidopsis* is the remarkably precise pattern of oriented cell divisions that generate the cell files radiating basipetally along the axis from the meristem. Although not as precise in many other species, the basic pattern of tissue formation is similar. Two important question arise from the precise pattern of root development: (1) How is the plane of cell division determined in plants? (2) How does control of the plane of cell division affect plant development?

Nuclear division in plant cells typically is followed by cytokinesis, in which the daughter nuclei are separated by the formation of a new cell wall (see Chapter 1). To understand how the plane of cell division is determined, we first need to review the mechanism of plant cytokinesis and the role of the cytoskeleton in this process. We begin with a discussion of the roles of cytoskeletal elements in determining the plane of cell division.

A Band of Microtubules May Determine the Orientation of the Mitotic Spindle

As we saw in Chapter 15, microtubules in the cortical cytoplasm of growing plant cells are important in determining the orientation of cellulose microfibrils in the cell wall. When dividing cells enter the G_2 phase of the cell cycle, the interphase cortical microtubule array is replaced by a narrow band of interconnected microtubules and microfilaments encircling the nucleus. This circular array of microtubules and microfilaments is called the **preprophase band** (Figure 16.15A). The preprophase band was first observed in 1966 by Jeremy Pickett-Heaps and Donald Northcote at Cambridge University (Pickett-Heaps and Northcote 1966). It appears as a transient ring of cytoskeletal elements in the cortical cytoplasm, near the plasma membrane. Generally

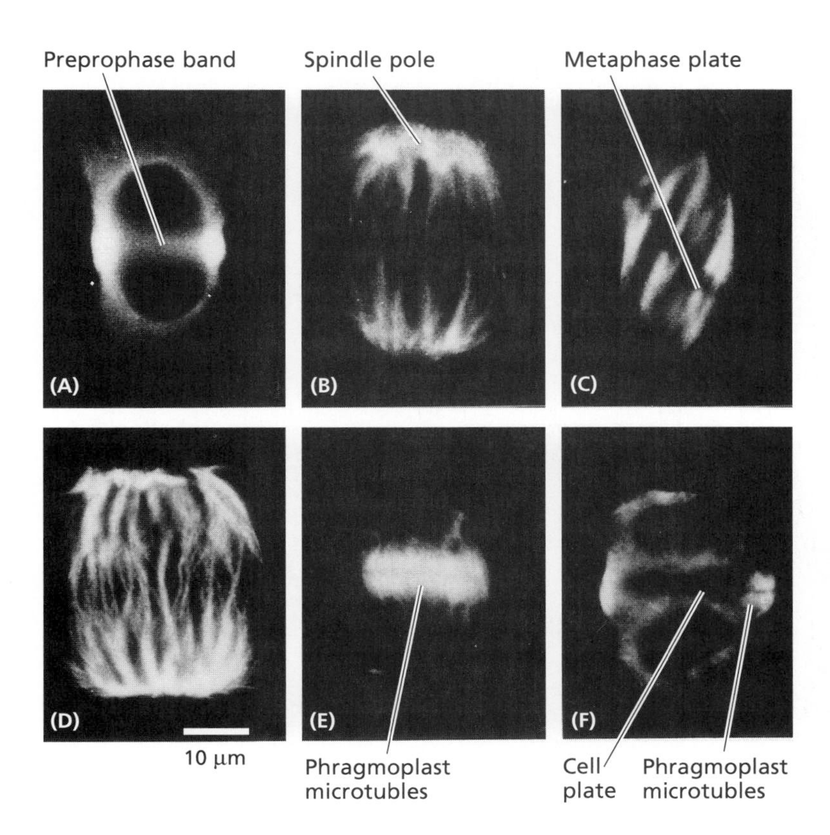

Figure 16.15 Immunofluorescent staining of the cytoskeleton in onion (*Allium cepa*) root tip cells with an antibody to the microtubule protein, tubulin. As cells progress through their cell cycle, different cytoskeletal arrays that contain microtubules are formed. (A) A preprophase band appears after DNA replication is complete, but before the cell enters mitosis. It can be seen as two bright spots on either side of the unstained nucleus; the image of the band is fainter as it passes across the nucleus. (B) The mitotic spindle forms in late prophase, with spindle microtubules emanating from the polar regions to form the spindle fibers. (C) The spindle fibers hold the chromosomes (dark line) in the metaphase plate. In this cell the alignment of the plate is oblique, which is relatively unusual. (D) The chromosomes move toward their poles within the array of spindle microtubules. (E) The phragmoplast forms in telophase and is seen here as a bright band in the center of the cell that consists of microtubules, microfilaments, and vesicles in the early stage of cell plate formation. (F) Phragmoplast microtubules form an annular ring around the perimeter of the cell plate and appear only at the edges of the cell. Bar = 10 μm. (Courtesy of A. Hardham.)

the preprophase band is 2 to 3 μm wide and contains from 10 to more than 100 microtubules, interspersed and interconnected with actin microfilaments.

As the cell enters mitosis, microtubules begin to assemble at the surface of the nuclear envelope, and as the nuclear envelope breaks down, they invade the nuclear area to form the **mitotic spindle**. The preprophase band disappears as the spindle forms, leaving behind an actin-depleted zone (Figure 16.15B) (Chapter 1). The plane in which cytokinesis occurs is determined before the cell initiates mitosis and is accurately predicted by the position of the preprophase band. It has been suggested that the preprophase band determines the orientation of the mitotic spindle and that it marks the site at which the new cell wall will form during cytokinesis, but this has not been established. The position of the preprophase band and the plane of cytokinesis both may be determined by another mechanism, possibly sensitive positional information.

Arabidopsis Mutants Lacking Preprophase Bands and Ordered Cell Divisions Have Been Isolated

Two groups of investigators have isolated *Arabidopsis* mutants lacking both preprophase bands and ordered patterns of cell division in the apical meristems. Surprisingly, the normal radial pattern of tissue differentiation is maintained in the embryo and the root despite the aberrant patterns of cell division in these mutants. The mutations, called *ton* (from *tonneau*, "barrel" in French [Traas et al. 1995]) or *fass* (German for "barrel" [Torres-Ruiz and Jürgens 1994]), by the groups that identified them, have dramatic effects on development.

The effects of *ton* or *fass* mutations are seen from the earliest stages of embryogenesis and persist throughout development. The plants are tiny, never reaching more than 2 to 3 cm in height; have misshapen leaves, roots, and stems; and are sterile (Figure 16.16). Nevertheless, the mutant plants not only establish an axial pattern, but have all the cell types and organs of the wild-type plant, and these occur nearly in their correct positions. The *fass* mutant has recently been shown to contain twofold higher levels of the growth hormone indole-3-acetic acid (IAA), which might cause the abnormal growth. The increase in free IAA may be due to a defect in the enzyme that normally conjugates free IAA to amino acids, inactivating the hormone (Fisher et al. 1996) (see Chapter 19).

A mutation similar to *fass*, named *trapu* (French for "dumpy"), has been identified in *Petunia*, although it is not known if this mutation affects a homologous gene (Dubois et al. 1996). The fact that these mutations do not completely prevent the establishment of the radial tissue

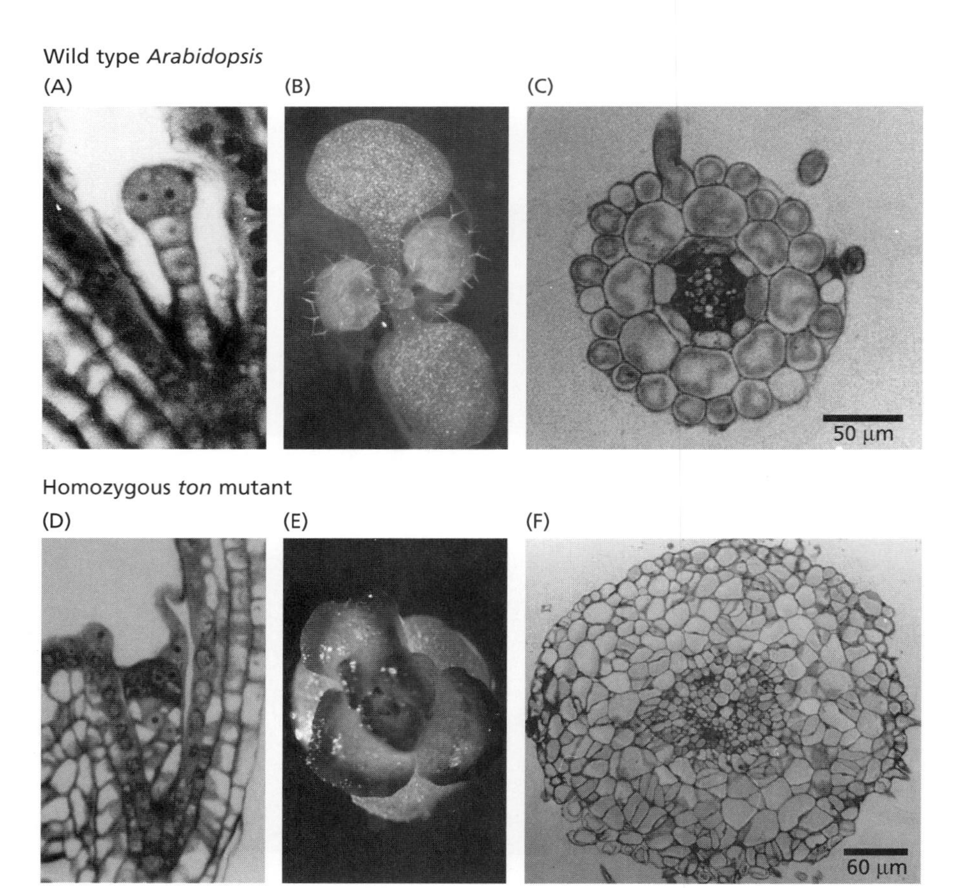

Figure 16.16 Although abnormal, the radial tissue pattern is still established in *Arabidopsis* plants with the *ton* (*fass*) mutation, which prevents the formation of preprophase bands in cells at any stage. Plants carrying this mutation are highly irregular in their cell division and expansion planes, and as a result they are severely deformed. However, they continue to produce tissues and organs in their correct positions. (A–C) Wild-type *Arabidopsis*: (A) early globular stage embryo; (B) seedling seen from the top; (C) cross section of a root. (D–F) Comparable stages of *Arabidopsis* plants homozygous for the *ton* mutant: (D) early embryogenesis; (E) mutant seedling; (*f*) cross section of the mutant root, showing the random orientation of the cells, but a near wild-type tissue order; an outer epidermal layer covers a multicellular cortex, which in turn surrounds the vascular cylinder. (From Traas et al. 1995, courtesy of J. Traas.)

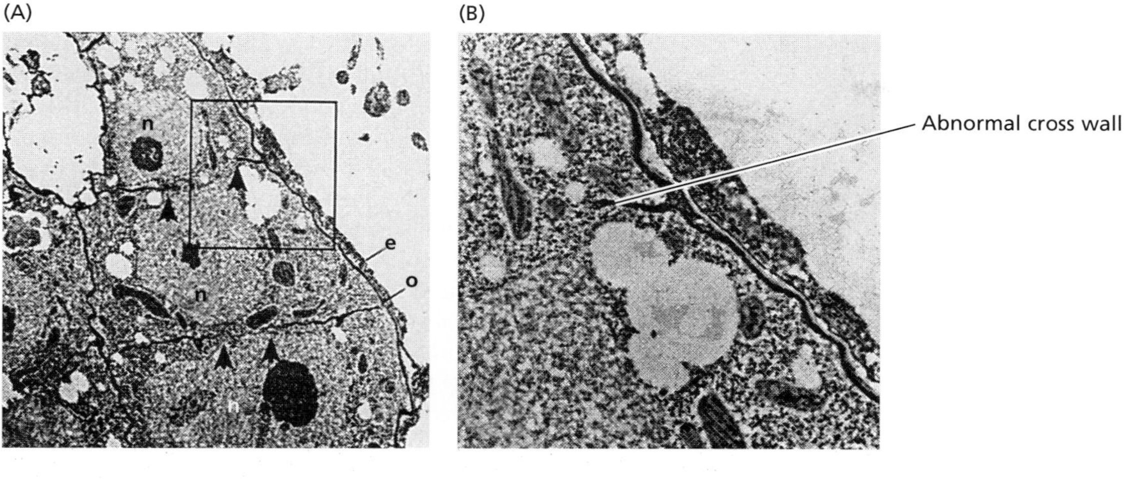

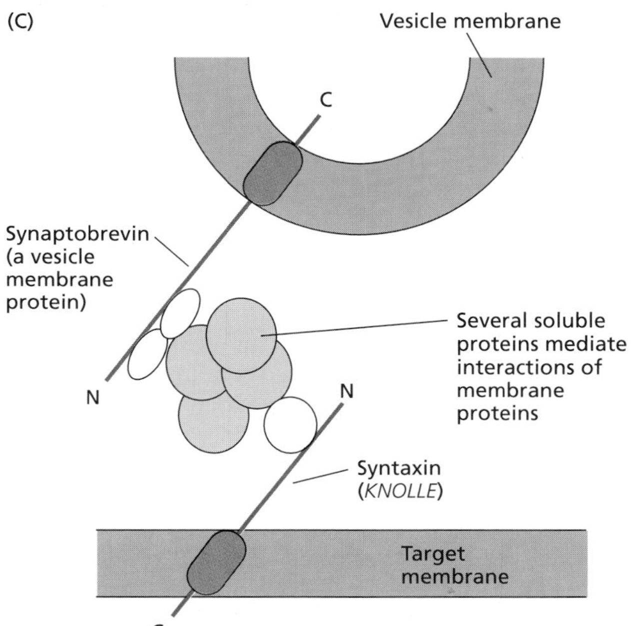

Figure 16.17 Syntaxin proteins play a critical role in the fusion of Golgi-derived membranes in many organisms. In *Arabidopsis*, the syntaxin-encoding *KNOLLE* gene is essential for normal cytokinesis. (A) Electron micrograph of a region of an *Arabidopsis* embryo with the *knolle* mutation. e, endosperm; o, parent cell wall; n, nucleus. (B) Magnification of the region boxed in (A) showing an incomplete and abnormal cross wall attached to the parent cell wall. The box width in (A) is 5 μm. (C) A model for the fusion of vesicles during cell plate formation. The interaction of synaptobrevin with syntaxin on the target membrane is mediated by a complex of soluble proteins. (A and B from Lukowitz et al. 1996, courtesy of G. Jürgens; C after Assaad et al. 1997.)

pattern calls into question the importance of both the preprophase band and the stereotypic cell division patterns found in the *Arabidopsis* embryo for the radial pattern of tissue differentiation. Since the tissue pattern is imposed on the cells regardless of their orientation, we can conclude that the ordered, precise control of the plane of cell division is not essential for its establishment.

An *Arabidopsis* Mutant with Defective Cytokinesis Cannot Establish the Radial Tissue Pattern

The *Arabidopsis* mutant *knolle* is defective in cytokinesis because of the inactivation of a gene that encodes a syntaxin-like protein that is important for vesicle fusion (Lukowitz et al. 1996). **Syntaxins** are proteins that integrate into membranes, permitting the membranes to fuse (Figure 16.17). Although cell division is not blocked by the *knolle* mutation, the cell plate is irregular and often incomplete. As a result, many cells are binucleate, while other cells are only partly separated or are connected by large cytoplasmic bridges. The division planes also are irregular. These irregularities have severe effects on development. Plants homozygous for the mutation go through embryogenesis, but the radial tissue pattern is severely disrupted and an epidermal layer does not form in early embryogenesis. The *knolle* mutation does not prevent formation of the apical–basal axis and embryogenesis is completed, although the seedlings are very short-lived and die soon after germination. The plants also lack functional meristems.

The conclusion drawn from studies of the *knolle* mutation appears to contradict what we learned from the *ton* (*fass*) mutations. Both the *knolle* and the *ton* mutations disrupt the normal pattern of cell division in embryonic and postembryonic development. The *knolle* mutations, however, block the establishment of the radial pattern of tissues, while the pattern is established in the *ton* mutants. One difference between the *ton* and the *knolle* mutations is that the latter usually prevents the effective separation of daughter cells during cytokinesis, since the cell plate is incomplete. Since cell–cell communication increasingly has been shown to be important for pattern formation, it may be necessary for cells to be isolated effectively so that the exchange of information can be regulated. Thus, the *ton* mutants are able to perceive positional information correctly, while the *knolle* mutants cannot.

Formative Cell Divisions Are Necessary for the Differentiation of Some Cell Types

Since the plane in which cells divide does not appear to be important for pattern formation, can we conclude that formative divisions also play no essential role in the establishment of cell differentiation and tissue patterns? Formative divisions are asymmetric in that the daughter cells, which may or may not differ in size, have different fates. Can cells acquire different fates without going through a formative division?

Only a few experiments have tested the hypothesis that formative divisions are important in plant development. In an often cited study, Foard, Haber, and Fishman (1965) treated roots with the antimicrotubule drug colchicine to block cell division to see if the asymmetric division of pericycle cells that precedes the initiation of lateral roots is essential for the formation of lateral root primordia. They concluded it was not, because pericycle cells altered their growth axis in the absence of cell division, growing outward by means of cell expansion, forming a lateral bulge that protruded into the cortex and resembled the initial stages of lateral root primordium formation. Since cell division was blocked, a lateral root primordium was not established, but the swollen pericycle cells clearly had different growth characteristics.

Although this study demonstrated that the growth axis of the pericycle cell is altered in the absence of cell division, it did not demonstrate that the pericycle outgrowth acquires the identity of the lateral root meristem. As a result, it does not rule out the possibility that formative divisions are required for a pericycle cell to become a root primordium cell.

There are many other instances in plant development in which asymmetric divisions precede a change in cell commitment. One such system that is particularly amenable to experimental manipulation is pollen development. Here the first mitotic division of the haploid microspore is asymmetric, forming two daughter cells with dramatically different developmental fates. In this case, the division is also morphologically asymmetric, producing a large cell that receives most of the microspore cytoplasm, and a much smaller daughter cell. The nucleus of the small daughter cell condenses, and the cell differentiates as the **generative cell**. The larger daughter differentiates as the **vegetative cell**, which can germinate to form the pollen tube that carries the sperm cells to the embryo sac.

Genes have been identified whose expression is characteristic of the vegetative cell, including one known as *LAT52*. The transcription of these vegetative cell–specific genes, including *LAT52*, can be detected only after the asymmetric division and is confined to the vegetative cell (Eady et al. 1994). David Twell and his associates blocked the first pollen division of tobacco microspores with the antimicrotubule drug colchicine and demonstrated that, although the generative cell did not differentiate, the now diploid uninucleate microspore developed as a vegetative cell (Figure 16.18B) (Eady et al. 1995). Not only did it express the vegetative cell–specific *LAT52* gene, but it also acquired the ability to germinate and form a pollen tube. The asymmetric division obviously is not required for formation of the vegetative cell. However, in the absence of the asymmetric division the generative cell did not form.

These investigators also found that a much lower concentration of colchicine did not block pollen mitosis I, although it prevented the asymmetry of the division. Instead of two unequal daughter cells, the microspores divided symmetrically in the presence of a low concentration of colchicine (Figure 16.18C). Both daughter cells expressed the *LAT52* gene, and neither cell exhibited the chromatin condensation that is characteristic of the generative cell. Thus, differentiation of the generative cell appears to depend on the asymmetric division of mitosis I in pollen.

Differentiation of the Endodermis Requires Specific Gene Expression

The importance of formative divisions for the development of the radial pattern of tissues in the root also has been investigated. As already noted, Ben Scheres and his group in the Netherlands, in collaboration with Philip Benfey in New York, isolated several *Arabidopsis* mutants with specific defects in the tissue pattern (Scheres et al. 1995). The mutations *shr* and *scr* result in defects in the formation of the ground tissues; both mutants form only a single layer of ground tissue in the root instead of the two layers, cortex and endodermis, found in the wild type. No endodermal markers are present in the ground tissue layer of the *shr* mutants, but *scr* mutants form a single layer with characteristics of both the cortex and the endodermis. These two cell layers are initiated by an asymmetric division of the cortical cell initials in which one daughter cell gives rise to the cortical cells while the other gives rise to the endodermal cells. This asymmetric division does not occur in the *scr* mutants.

To determine whether the failure of these mutants to differentiate normal ground tissue layers was due to a defect in cell division or to the inability of the cells to establish the proper identity, Scheres crossed these mutants with plants carrying the *fass* mutation. These mutants are able to establish a normal pattern of tissues despite a disruption of the normal division patterns found in *Arabidopsis* root development. An additional consequence of the *fass* mutation is that the cells divide more often than in the wild type, so all tissue layers contain additional cells. As a result, the plants carrying both

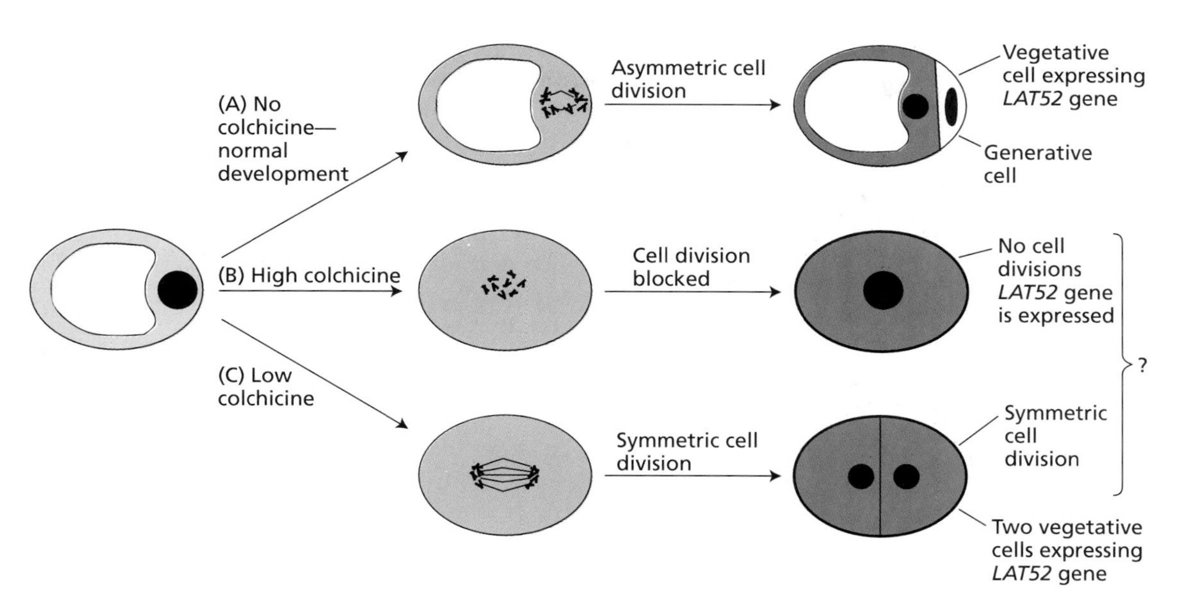

Figure 16.18 Experiment showing that an asymmetric formative division is required for differentiation of the generative cell during pollen development. Uninucleate microspores can differentiate in culture in one of three ways, depending on whether they are exposed to the antimicrotubule drug colchicine and depending on the concentration of colchicine they receive. (A) Without colchicine, development and expression of the *LAT52* gene are similar to what is observed in the plant. (B) With a high concentration of colchicine, the asymmetric division is blocked because the cells are unable to form a mitotic spindle, and no generative cell is produced. Differentiation of vegetative cells is not affected, however, and the *LAT52* gene is expressed. (C) With a low concentration of colchicine, the division of the uninucleate microspore is not blocked, but the division is symmetric instead of asymmetric. Both daughter cells differentiate as vegetative cells. (From Eady et al. 1995.)

the *fass* and the *shr* mutations have multiple layers of ground tissue instead of only one, as in the *shr* single mutant, but they still do not form an endodermis.

We can conclude that expression of the *SHR* gene is essential for the establishment of endodermal cell identity. In contrast, an endodermis is produced in the *scr/fass* double-mutant plants, showing that the *SCR* gene does not determine cell identity, but rather controls the asymmetric division of the cortical initial that is essential for the differentiation of cortical and endodermal cell layers (Di Laurenzio et al. 1996).

Shoot Development

Shoots are adapted for growth upward through the atmosphere in the direction of sunlight. Their primary function is to photosynthesize; hence the proliferation of leaves and buds at the apical meristem. As the plant progresses through different stages of maturity, its morphology may change, ultimately giving rise to reproductive structures (see Chapter 24). The continual production of lateral organs, and the variable nature of these organs, make the apical meristem of the shoot much more dynamic (and more difficult to analyze) than that of the root. Nevertheless, as described in this section, much progress has been made in recent years using genetic approaches to the problem.

The Shoot Apical Meristem Is a Highly Dynamic Structure

The vegetative shoot apical meristem generates the stem, as well as the lateral organs attached to the stem (leaves and lateral buds). Although the shoot meristem itself is at the extreme tip of the shoot, it is surrounded by immature leaves that fold over and cover it. These leaves are produced by the activity of the meristem. Therefore, it is useful to distinguish the **shoot apex**, the apical meristem plus the most recently formed leaf primordia, from the meristem proper, which refers to the undifferentiated cell population only and not any of its derivative organs.

The shoot apical meristem is a flat or slightly mounded region, 100 to 300 μm in diameter, composed mostly of small, thin-walled cells, with a dense cytoplasm, lacking large central vacuoles. It is a dynamic structure that changes during its cycle of leaf and stem formation. In addition, many plants exhibit seasonal growth. They may grow rapidly in the spring, enter a period of slower growth during the summer, and become dormant in the fall. The size and structure of the apex change with this seasonal activity.

Shoots develop and grow at their tips, as is the case with roots, but the developing regions are not as stratified and precisely ordered as they are in the root. Moreover, growth occurs over a broader region of the shoot than is the case for roots. At any given time, several of the terminal internodes, up to 10 to 15 cm long, may be undergoing primary growth. In grasses, cells divide and elongate at the base of each internode, a region known as the **intercalary meristem**.

The Shoot Apical Meristem Can Be Divided into Cell Layers and Zones

The vegetative shoot apical meristems of angiosperms usually have a highly stratified appearance in which the cells nearest the surface are organized into two or more distinct layers. These constitute the **tunica** layers of the meristem and are designated L1, L2, and L3, of which L1 is the outermost layer (Figure 16.19A). Cell divisions are *anticlinal* in the tunica layers; that is, the new cell wall separating the daughter cells is oriented at right angles to the meristem surface. The tunica layers overlie more internal cells, known collectively as the **corpus**, in which the plane of cell divisions is not highly ordered. Divisions by a small number of cells in the tunica and corpus layers give rise to progenitor cells for the leaf primordia, lateral buds, and, in floral apices, the floral organs. For this reason, the cell layers of the tunica and the corpus are sometimes called *histogenic* (tissue-producing) layers.

Active apical meristems have an additional organizational pattern called **cytohistological zonation** superimposed on the tunica–corpus organization. Each zone is composed of cells that may be distinguished not only on the basis of their division planes, but also by differences in size and by degrees of vacuolation (Figure 16.19B). The center of active meristems contains a cluster of relatively large, highly vacuolate cells known as the **central zone**, somewhat comparable to the quiescent center of root meristems. This zone is flanked by a doughnut-shaped region of smaller cells with inconspicuous vacuoles called the **peripheral zone**. The peripheral zone is the region in which the first cell divisions leading to the formation of leaf primordia occur (Kerstetter and Hake 1997). A **rib meristem** underneath the central cell zone gives rise to the internal tissues of the stem. Cytohistological zonation is a reflection of the differing rates of mitotic activity in the different regions of the meristem. Cells in the peripheral zone divide more rapidly than those in the central zone. Consequently, central-zone cells tend to be larger than peripheral-zone cells. As cell division ceases in the apical meristem with the onset of dormancy, the cytohistological zones may disappear, leaving only the tunica–corpus organization.

Position Is More Important than Clonal Ancestry in Shoot Histogenesis

The organization of apical meristems is less regular and uniform than that of the root meristem. In both cases, a small number of initial cells are the ultimate source of any particular tissue, and most of the cells in a given tissue are clonal and have arisen from the same initial. However, as in the root, cell fate does not depend on cell lineage, but instead is determined by positional information. In the vast majority of cases, shoot epidermal cells are derived from a small number of initial cells in

(A) Leaf primordia
Apical meristem
Corpus with randomly oriented cell divisions
Tunica with anticlinal cell divisions

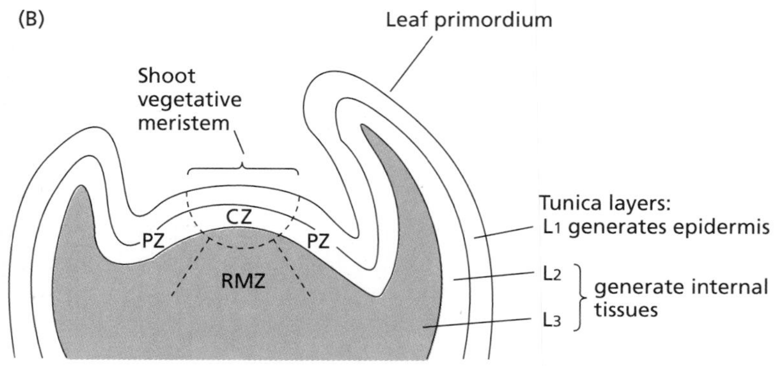

Figure 16.19 The shoot apical meristem generates the aerial organs of the plant. (A) This longitudinal section through the center of the shoot apex of *Coleus blumei* shows that the meristem consists of three tunica layers, with anticlinal cell divisions, which surround the corpus. (B) The L1 layer generates the epidermis of the shoot; the L2 and L3 layers generate internal tissues. An active shoot apical meristem also is seen to have cytohistological zonation, in which a peripheral zone (PZ) of rapidly dividing cells surrounds a region of more slowly dividing cells, the central zone (CZ). The central zone contains the apical initials. Below the central zone is the rib meristem zone (RMZ), which generates the internal tissues of the stem. (A from J. N. A. Lott/Biological Photo Service; B after Huala and Sussex 1993.)

the L1 layer. However, the derivatives of the L1 layer are committed to become epidermal cells because they occupy the outermost layer and lie on top of the cortical cell layer, not because they were clonally derived from the initial cells in the L1 layer.

The plane in which a cell divides will determine the position of its daughter cells within a tissue, and this positioning in turn plays the most significant role in deciding the fate of the daughters. The strongest evidence for the importance of position in determining a cell's ultimate fate comes from an examination of the fate of cells that are displaced from their usual position such that they come to occupy a different layer. The vast majority of the divisions in the tunica layers are anticlinal, and anticlinal division is responsible for creating the layers in the first place. Nevertheless, occasional *periclinal* divisions occur, causing one derivative to occupy the next tunica layer. This periclinal division does not alter the composition of the tissue derived from this layer. Instead, the derivatives assume a function that is appropriate for a cell occupying that location.

Further support for the role of cell location is shown by observations of leaves from a periclinal chimera (a mixture of mutant and wild-type cells) of English ivy that had a mutation in the apical initial cells of the L2 tunica layer such that none of its derivatives could form chloroplasts (Figure 16.20). However, chloroplasts were able to develop in cells derived from the L1 and L3 layers because these cells lacked the mutation. The majority of cells of photosynthetic mesophyll tissue are derived from the L2 layer, so the leaves of this chimeric ivy plant contain a great deal of white tissue. However, they are not totally white; rather they are variegated, with sectors of green tissue that must be derived from L1 or L3 layers. Occasional periclinal divisions in the L1 and L3 layers early in leaf development establish clones of normal cells that differentiate as green mesophyll cells. In this case, then, cell differentiation is not dependent on cell lineage. The fate of a cell during development is determined by the position it occupies in the plant body.

Vegetative, Floral, and Inflorescence Meristems Are Variants of the Shoot Meristem

Several different types of shoot meristems can be distinguished on the basis of their developmental origin, the types of lateral organs they generate, and whether they are determinate or indeterminate. The vegetative shoot apical meristem usually is indeterminate in its development. It repetitively forms phytomeres as long as environmental conditions favor growth but do not generate a flowering stimulus. A **phytomere** is a developmental unit consisting of one or more leaves, the node to which the leaves are attached, the internode below the node, and one or more axillary buds (see Figure 16.21. **Axillary buds** are secondary meristems; if they are also vegetative meristems, they will have a structure and developmental potential similar to that of the apical meristem.

Vegetative meristems may be converted directly into floral meristems when the plant is induced to flower. **Floral meristems** differ from vegetative meristems in that, instead of leaves, they produce floral organs:

Figure 16.20 Periclinal chimeras demonstrate that the mesophyll tissue has more than a single clonal origin in English ivy (*Hedera helix*). When the initial cells in the meristem mutate, all the cells in the plant that are derived from those initials carry the mutation. When this happens, the plant will be a chimera, a mixture of mutant and wild-type cells. The analysis of chimeras is useful for studies on the clonal origin of different tissues. When the mutation affects the ability of chloroplasts to differentiate, the presence of albino sectors shows that these sectors were derived from the initials carrying the mutation. In the ivy plant shown here, the L2 layer carried a mutation causing albinism, while the L3 layer had an unmutated copy of the same gene. The green tissue in these leaves was derived from the L3 layer, the colorless regions from the L2 layer. The L1 layer gives rise to the leaf and stem epidermis, but it is colorless because chloroplasts do not differentiate in most epidermal cells. (Photograph courtesy of S. Poethig.)

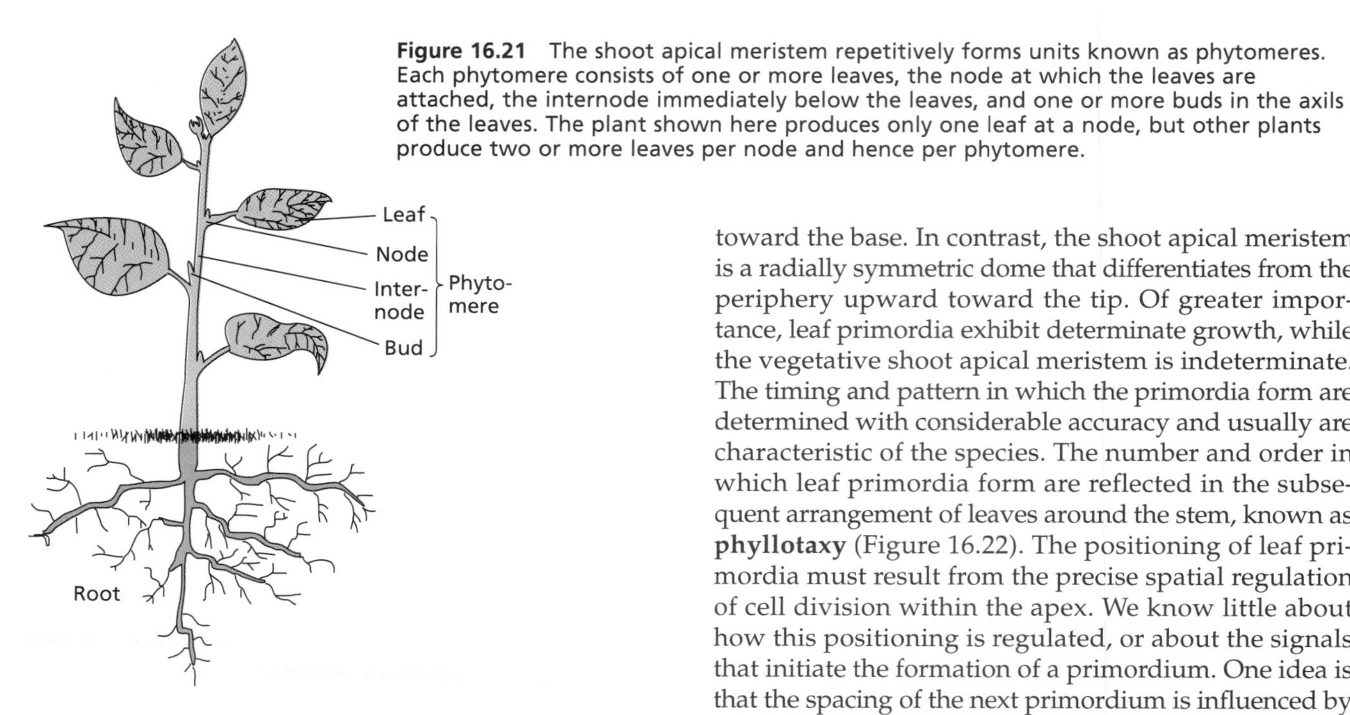

Figure 16.21 The shoot apical meristem repetitively forms units known as phytomeres. Each phytomere consists of one or more leaves, the node at which the leaves are attached, the internode immediately below the leaves, and one or more buds in the axils of the leaves. The plant shown here produces only one leaf at a node, but other plants produce two or more leaves per node and hence per phytomere.

sepals, petals, stamens, and carpels. In addition, floral meristems are determinate: All meristematic activity stops after the last floral organs are produced. In many cases, vegetative meristems are not directly converted to floral meristems. Instead, the vegetative meristem is first transformed into an **inflorescence meristem**, which produces bracts and floral meristems in the axils of the bracts, rather than leaves and axillary buds. Inflorescence meristems may be determinate or indeterminate, depending on the species.

The Arrangement of Leaf Primordia Is Regulated

Leaves or other lateral appendages are initiated as a result of localized cell divisions in the L1 and L2 layers in specific regions in the peripheral zone on the flanks of the apical dome (see Figure 16.19B). These localized regions of higher cell division activity lead to the formation of primordia that will grow and develop to become leaves or other lateral organs. Leaves usually have dorsiventral polarity, and they mature from the tip toward the base. In contrast, the shoot apical meristem is a radially symmetric dome that differentiates from the periphery upward toward the tip. Of greater importance, leaf primordia exhibit determinate growth, while the vegetative shoot apical meristem is indeterminate. The timing and pattern in which the primordia form are determined with considerable accuracy and usually are characteristic of the species. The number and order in which leaf primordia form are reflected in the subsequent arrangement of leaves around the stem, known as **phyllotaxy** (Figure 16.22). The positioning of leaf primordia must result from the precise spatial regulation of cell division within the apex. We know little about how this positioning is regulated, or about the signals that initiate the formation of a primordium. One idea is that the spacing of the next primordium is influenced by inhibitory fields generated by existing primordia.

Genes That Affect Phyllotactic Patterns Have Been Identified

Although we do not know how phyllotactic patterns are established, genes that modify them have been identified. In most plants, the different kinds of shoot meristems have different phyllotactic patterns. In snapdragon (*Antirrhinum*), leaves are initiated in a **decussate** pattern, with two opposite leaves per node and with successive leaf pairs oriented at right angles to each other during most of their vegetative development (see Figure 16.22C). The inflorescence of this plant has **spiral phyllotaxy** (see Figure 16.22E). A single leaflike bract is produced at each node, with subsequent bracts produced at a 137° angle, and a floral bud forms in the axil of each bract. The flowers, however, have a third type of phyllotaxy, called **whorled**, in which all floral organs of a given type are attached to the stem at the same node (see Figure 16.22D).

The whorled arrangement of floral organs in snapdragon depends on two genes: *FLORICAULA* (*FLO*) and

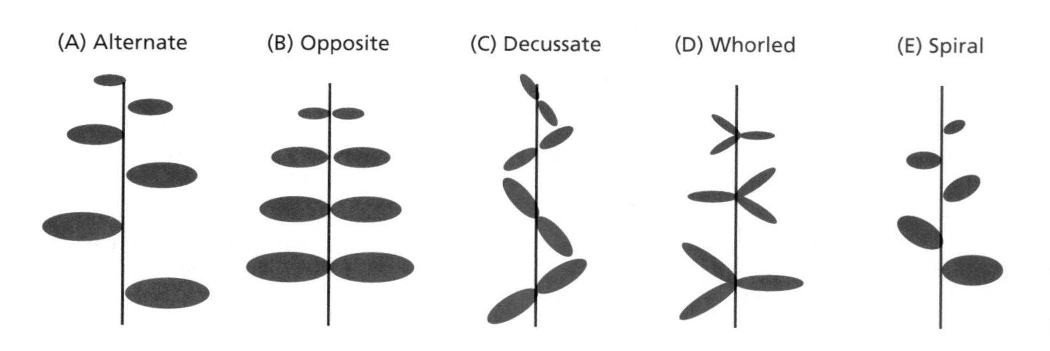

Figure 16.22 Five patterns of leaf arrangement (phyllotactic patterns). (From Sinha et al. 1993b.)

SQUAMOSA (*SQUA*) (Carpenter et al. 1995). Plants with *flo* mutations form indeterminate shoots in the axils of the bracts instead of flowers because the *FLO* gene also is necessary for the formation of floral meristems. These indeterminate shoots produce a succession of bracts in a spiral pattern, indicating that *FLO*, which is required for floral meristem identity, also controls the whorled phyllotactic pattern of floral organs (Figure 16.23). How these genes control the phyllotactic pattern remains to be determined.

Homeobox Genes Are Required for Meristem Cell Identity

How does the indeterminate shoot apical meristem give rise to determinate structures such as leaves? During the formation of primordia, the cells of the meristem must lose their identity as meristematic cells and assume the identities of the appropriate lateral appendage. Considerable evidence suggests that certain homeodomain proteins belonging to the KN1 (*kn*otted *1*) class are involved in maintaining the indeterminacy of the shoot apical meristem. Homeodomain proteins are encoded by homeobox genes and act as transcription factors, regulating the expression of other genes (see Chapter 14). The original *knotted* (*kn1*) mutation is a gain-of-function mutation. A gain-of-function mutation gives the plant characteristics that are not found in the wild type. Plants with the *kn1* mutation have small, irregular, tumorlike knots along the leaf veins. These knots result from abnormal cell divisions within the vascular tissues that distort the veins and form the knots, which protrude from the leaf surface (Figure 16.24).

Cell differentiation is relatively normal in the leaves of the *kn1* mutant plants, except in the vicinity of the knots. The knots are similar to meristems in that they contain undifferentiated cells and continue to divide after cells around them have matured and ceased dividing. This behavior suggests that the *KN1* gene controls meristem function. The mutant phenotype results from the ectopic (abnormal) expression of the gene in new tissues, rather than the loss of the normal developmental expression pattern.

Sarah Hake's group at the USDA Plant Gene Expression Center in Albany, California cloned the *KN1* gene (Hake et al. 1989). They also demonstrated that its ectopic expression caused gain-of-function mutations. Tobacco plants that have been transformed with the maize *KN1* gene, driven by a promoter that expresses the gene throughout the plant, form abnormally small leaves and develop adventitious shoot

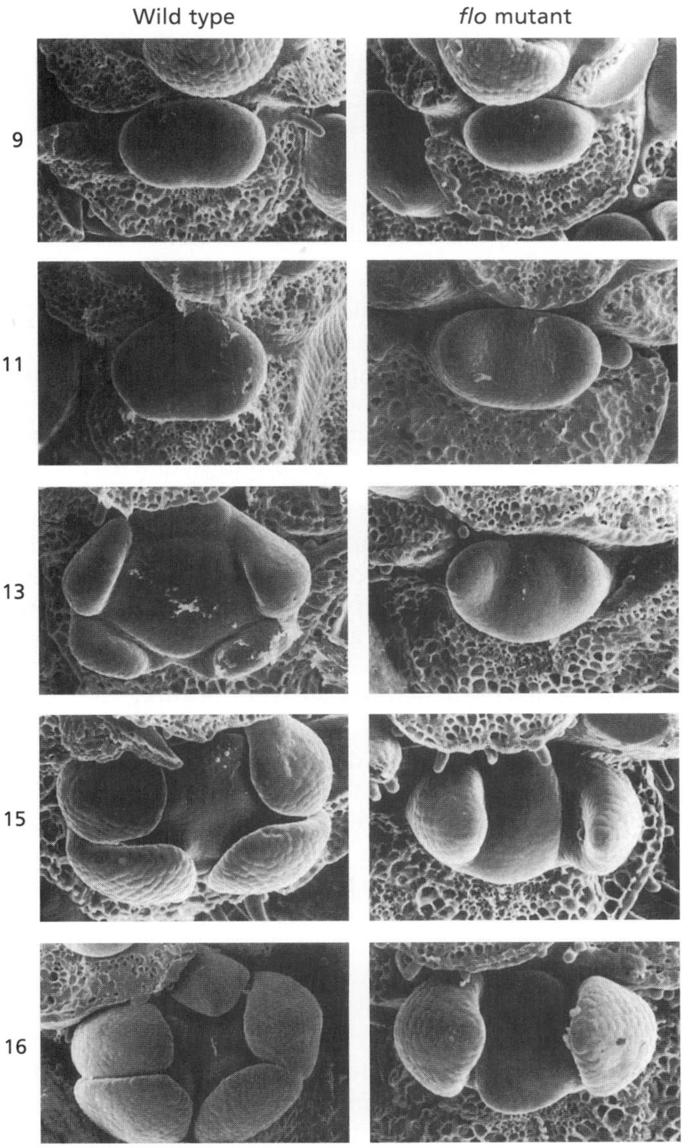

Figure 16.23 The *FLORICAULA* (*FLO*) gene of *Antirrhinum* participates in regulating the phyllotactic pattern of organ initiation by the meristem. Lateral meristems are shown at successive developmental stages in wild-type and *flo* mutant plants. Node number is shown to the left of each set of photomicrographs, numbering from the primordium most recently formed by the inflorescence meristem. In both the wild-type *Antirrhinum* and the *flo* mutant, the inflorescence meristem initiates a succession of lateral meristems, each subtended by a bract, with spiral phyllotaxy. In the wild-type inflorescence, the lateral meristems are floral meristems, each of which then initiates floral organs with whorled phyllotaxy. However, the lateral meristems produced by the *flo* mutant inflorescence give rise to branch shoots instead of flowers, each of which develops with spiral phyllotaxy. Early in their development, at node 9, there is little difference in meristem structure between the wild type and the *flo* mutant. However, the wild-type meristem begins to develop a pentagonal shape at node 11, and by node 13 sepal primordia are formed with a whorled arrangement. The *flo* mutant lateral meristem is producing a series of bracts in a spiral pattern. (After Carpenter et al. 1995, courtesy of E. Coen, © American Society of Plant Physiologists, reprinted with permission.)

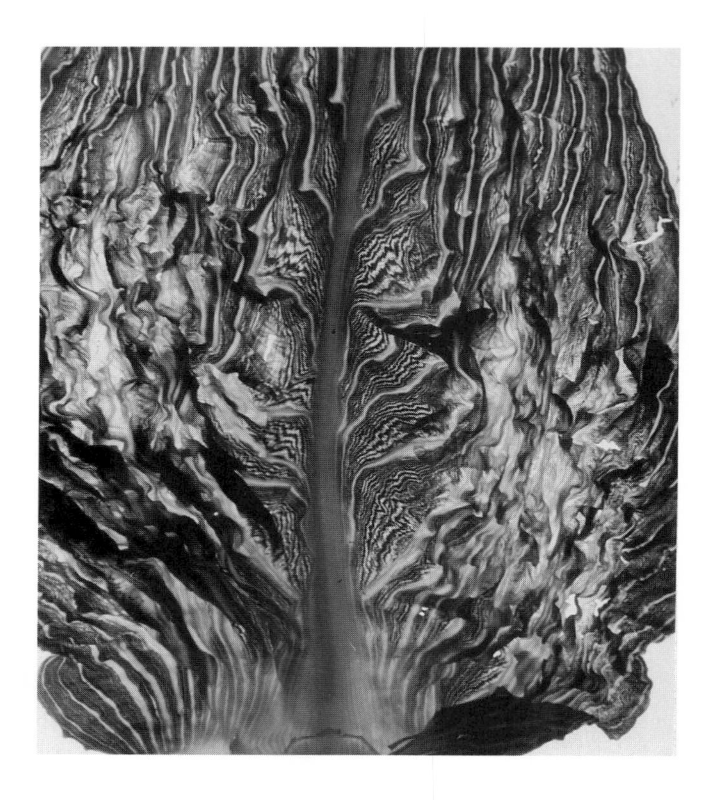

Figure 16.24 The *knotted1* (*kn1*) mutation of maize causes the gene to be expressed at the wrong time and place in development. Expression of the *KN1* gene during leaf development causes severe abnormalities around the leaf veins, especially near the base of the leaf. Extra cell proliferation occurs after normal cell division ceases, and the division planes are abnormal, causing gross distortion of the blade surface. (From Sinha et al. 1993b, courtesy of S. Hake.)

meristems along the leaf surface (Sinha et al. 1993a,b). *Arabidopsis* has several *KN1* homologs, including *KNAT1*, *KNAT2*, and *STM*.

As we have seen, expression of the *STM* (*SHOOTMERISTEMLESS*) gene is essential for the formation of the shoot apical meristem in the embryo. The *STM* gene is also expressed in the shoot meristem after germination. Expression of *STM* is confined to the apical dome of the vegetative meristem and is lost when the leaf primordia are initiated; *STM* is not normally expressed in the developing leaf primordia. Similarly, *STM* is expressed in the dome of the floral meristem, but it is silenced as floral organs appear. Since cells actively divide in the early stages of leaf and floral organ primordia development, *STM* is not necessary for cell division. Rather *KN1*, *STM*, and their functional homologs are necessary to maintain the identity of meristem cells, keeping them in an indeterminate state.

Control of Cell Differentiation

Differentiation is the process by which a cell acquires metabolic, structural, and functional properties that are distinct from those of its progenitor cell. In plants, unlike animals, *cell differentiation* is frequently reversible, particularly when differentiated cells are removed from the plant and placed in tissue culture. Under these conditions, cells lose their differentiated characteristics, reinitiate cell division, and in some cases, when provided with the appropriate nutrients and hormones, even regenerate whole plants. This ability to *de*differentiate demonstrates that differentiated plant cells retain all the genetic information required for the development of a complete plant, a property termed **totipotency**. The only exceptions to this rule are cells that lose their nuclei, such as sieve tube elements of phloem, and cells that are dead at maturity, such as vessel elements and tracheids in xylem.

As an example of the process of cell differentiation, we will discuss the formation of vessel elements. The development of these cells from the meristematic to the fully differentiated state illustrates the types of control that plants exercise over cell specialization and provides an example of the cellular changes that are brought about by differentiation (Fukuda 1996). We can recognize four important events in the formation of any differentiated cell:

1. Inductive signaling and signal perception.
2. Expression of genes that specify cell identity.
3. Expression of genes required for the specialized activities or structures of the differentiated cell.
4. Activity of gene products and changes in cell structure that are necessary for the differentiated functions of the cell.

Although much remains to be learned about the differentiation of any plant cell type, vessel element differentiation illustrates some of what we have learned about how some of these steps are accomplished.

Auxin Is an Inductive Signal That Initiates the Formation of Tracheary Elements

A substantial body of evidence suggests that auxin represents an *inductive signal* that initiates tracheary element differentiation (Aloni 1995). Much of this evidence has come from studies of the formation of vessel elements and tracheids in response to stem wounds. Maintenance of intact, functional vascular tissue is crucial to the life of the plant. The shoot would wilt and die after the vascular strands in the stem were severed if it were not possible to reestablish vascular continuity by forming new conducting elements that bypass the damaged area. To accomplish this task, differentiated cells, usually cortical parenchyma, are reprogrammed to become conducting tracheary elements and phloem sieve elements.

Wounding first initiates several rounds of cell division in the cortical cells around the damaged area. These divisions precede the formation of files of tracheary and sieve elements that connect one end of the severed vascular strand to the other and that often lie parallel to the contours of the wound.

Auxin is produced in young developing leaves and is actively transported into and down the stem, with a strong basipetal gradient.* When the leaves above a wound are removed before wounding, the tracheary elements do not regenerate. However, if auxin is applied to the cut petiole after the leaf blade is removed, tracheary element differentiation is restored in the parenchyma near the wound. Auxin may be transported primarily in vascular tissue, so a wound that severs a vascular bundle may release auxin into the cortical cells, inducing them to divide and differentiate into vascular elements. The wound is not essential for the response, and the direct application of lanolin containing auxin to the stem also induces some of the cortical cells to differentiate as tracheary elements (Sachs 1969). The applied auxin is transported basipetally through the tissues, inducing strands of vascular tissue to form.

Expression of Homeobox Genes May Identify Cells as Tracheary Elements

Auxin that forms in young expanding leaves and is transported into the stem may be the signal for the development of vascular tissue during normal stem growth, but this has not been established with certainty. Also remaining to be determined is how the auxin signal is perceived by the target cells. Some of what we know about the auxin signal transduction pathway will be presented in Chapter 19, but at present we do not have sufficient information to explain the mechanisms that underlie this pathway.

Cellular perception of the inductive signal for tracheary element differentiation leads to the expression of the homeobox gene *ATHB-8* in *Arabidopsis* (Baima et al. 1996). In other cases, homeobox genes confer specific cell identity on the cells in which they are expressed. The expression of these transcription factor–encoding genes controls the expression of specific downstream genes whose products transform and characterize the differentiated cell. Some of these downstream genes have been identified in the case of tracheary elements and include genes for proteases and ribonucleases that are involved in the autolytic final step in the formation of functioning water-conducting xylem elements (Ye and Droste 1996; Ye and Varner 1993). However, the target genes of *ATHB-8* are still unknown.

* "Basipetal" means "toward the base of the plant," which is where the shoot and the root systems are joined. Basipetal transport is movement of materials away from the apical meristems, in both root and shoot.

During Tracheary Element Differentiation a Secondary Cell Wall Forms

As we saw in Chapter 1, tracheary elements are the conducting cells in which water and solutes move through the plant. They are dead at maturity, but before their death they are highly active and construct a secondary wall, often with an elaborate pattern, and they may grow extensively. Cell death (discussed later in this chapter) is the genetically programmed finale to tracheary element differentiation. After completing its growth phase and the construction of the secondary wall, the cell dies; the nucleus and cytoplasm are completely degraded, and only the thickened cell wall remains as a monument to the cell's efforts.

The formation of secondary walls during tracheary element differentiation involves the deposition of cellulose microfibrils and other, noncellulosic polysaccharides at specific sites on the primary wall, resulting in characteristically patterned wall thickenings (see Chapter 15). The secondary walls of tracheary elements have a higher content of cellulose than primary walls, and they are impregnated with lignin, which is not usually present in primary walls. In rapidly growing regions, the secondary-wall material is deposited as discrete annular rings, or in a spiral pattern, with the thickenings separated by bands of primary wall (Figure 16.25). As the cell grows, the primary wall extends and the rings

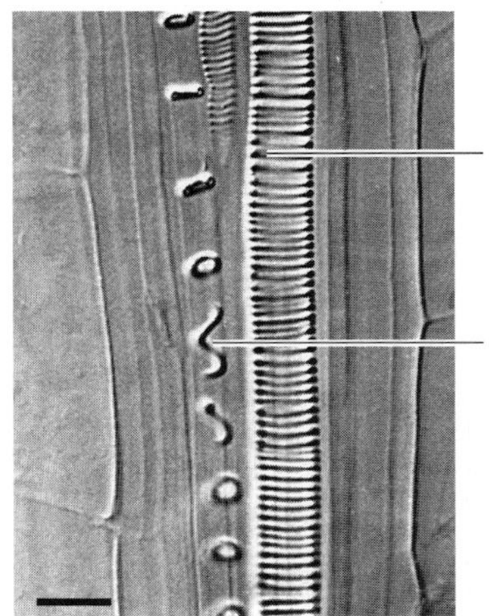

Figure 16.25 The pattern of secondary-wall deposition during vessel element development varies according to the rate of cell elongation. In this root of water hyacinth (*Salvinia auriculata*), the first file of vessel elements to differentiate—the protoxylem—is seen on the left side of the vascular bundle. During differentiation, short spiral thickenings of secondary-wall material, the "annular rings," were deposited in the protoxylem elements. Subsequent elongation pulled adjacent rings apart and stretched the short spiral thickenings. The metaxylem elements differentiated after the protoxylem, and the closely apposed rings or tight spirals have not been extended by cell elongation. (Courtesy of A. Hardham.)

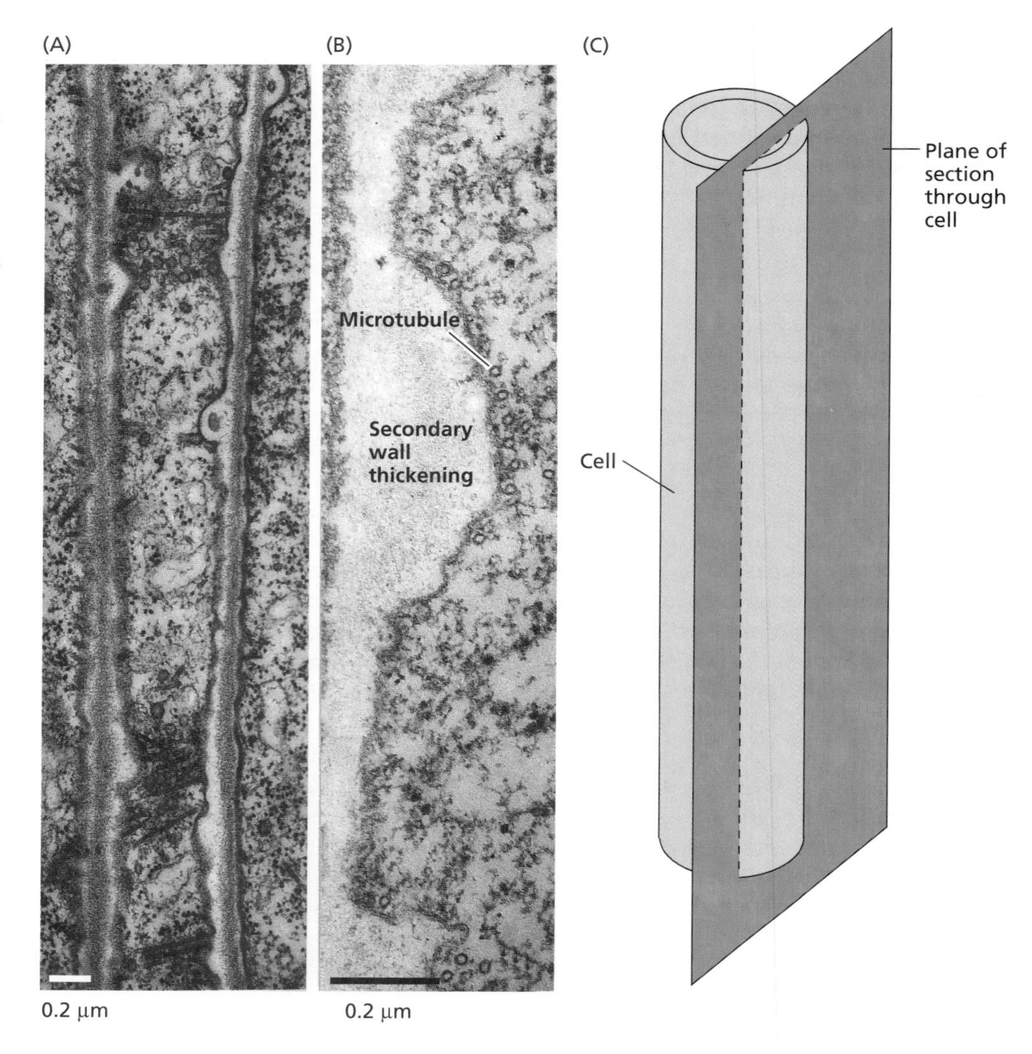

Figure 16.26 Development of secondary-wall thickenings in vessel elements in roots of the water fern *Azolla*, as seen in the electron microscope. (A) In this grazing section (illustrated in part C) through a differentiating cell, groups of microtubules are seen in the cell cortex, forming bands at the site of wall thickening before the secondary wall begins to form. Many small vesicles lie among the microtubules. (B) Annular thickenings develop beneath the bands of microtubules and are hemispheric in profile. (C) Drawing showing the plane of section for part A. (From Hardham and Gunning 1979, courtesy of A. Hardham.)

or spirals are pulled apart. The tracheary elements that form after elongation stops usually have walls that are thickened either uniformly or in a reticulate pattern. These cells cannot be stretched by growth.

Microtubules participate in determining the pattern of secondary-wall deposition. Before any alteration in the pattern of wall deposition is evident, cortical microtubules change from being more or less evenly distributed along the longitudinal walls of the cell to being clustered into bands (Figure 16.26*A*). Secondary wall is then deposited beneath the microtubule clusters (Figure 16.26B). The orientation of the cellulose microfibrils within the secondary-wall thickening is reflected in the alignment of microtubules in the cortical cytoplasm (Hepler 1981). As the thickenings develop, the number of microtubules adjacent to the longitudinal walls increases, reaching a maximum when the thickenings are about half formed. If the microtubules are destroyed with an antimicrotubule agent such as colchicine, cell wall deposition can continue, but the cellulose microfibrils are no longer precisely ordered within the thickening, and the pattern of the secondary wall is disrupted (Figure 16.27).

Cell–Cell Interactions and Cell Differentiation

Cells in multicellular plants usually are in close contact with others around them, and the behavior of each cell is carefully coordinated with that of its neighbors throughout the life of the plant. Furthermore, each cell occupies a specific position within the tissue and organ of which it is a part. Coordination of cellular activity requires cell–cell communication. Cells can communicate either symplastically via plasmodesmata or apoplastically through the cell wall. In this section, we will examine how cells communicate with each other to coordinate and integrate their differentiation with that of their neighbors during growth and development.

Cells Communicate Symplastically via Plasmodesmata

Most living cells in a plant are connected symplastically to their neighbors by **plasmodesmata** that pass through the adjoining cell walls and provide some degree of cytosolic continuity between them (Lucas 1995). Primary

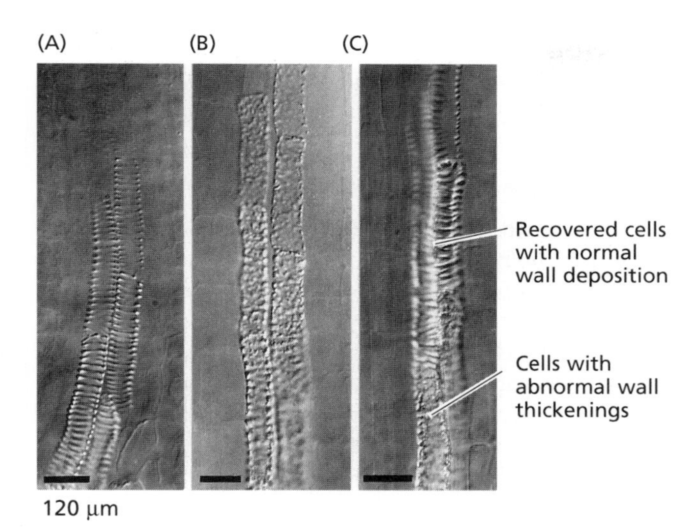

Figure 16.27 Colchicine treatments that destroy microtubules also disrupt the normal formation of secondary-wall thickenings in differentiating vessel elements. (A) During normal root growth in *Azolla* the wall thickenings are spaced evenly along the side walls. (B) In the presence of colchicine, secondary-wall materials are deposited in irregular patterns. (C) Normal growth resumes when the roots are transferred to fresh medium that lacks colchicine, and the newly differentiated vessel elements form with normal annular thickenings. (A from Hardham and Gunning 1979; B and C from Hardham and Gunning 1980; courtesy of A. Hardham.)

plasmodesmata form during cytokinesis, but secondary plasmodesmata also can be inserted in cell walls to connect adjacent cells that are not clonally related. Primary plasmodesmata are plasma membrane–lined pores through the cell wall that join the cytoplasm of neighboring cells (see Figure 1.30). The endoplasmic reticula (ERs) of adjacent cells also are connected by a tubule of ER that extends through the plasmodesmata called the **desmotubule**. Although called a tubule, it may not function as a passage since there appears to be no space between the membranes. Globular proteins are associated with both the desmotubule membrane and the plasma membrane within the pore. These globular proteins appear to be interconnected, dividing the pore into eight to ten microchannels (Ding et al. 1992).

Some molecules can pass from cell to cell through plasmodesmata, probably by flowing through the microchannels, although the exact pathway of communication has not been established. By following the movement of fluorescent dye molecules of different sizes through plasmodesmata that connect leaf epidermal cells, the limiting molecular mass for transport was determined to be about 700 to 1000 daltons, which is equivalent to a molecular size of about 1.6 nm (Robards and Lucas 1990). This is the **size exclusion limit** (**SEL**) of the plasmodesmata. Since the width of the cytoplasmic sleeve is approximately 5 to 6 nm, it is somewhat surprising that molecules larger than only 1.6 nm are excluded. However, the protein subunits attached to the plasma membrane and to the ER within the plasmodesmata may act to restrict the size of molecules that can pass through the pore.

The classic view is that plasmodesmata are passive cytoplasmic channels that connect the cytoplasm of all living cells throughout the plant. However, evidence is accumulating that plasmodesmata are, to some extent, gated so that entry through them can be regulated. Experiments with plant viruses show that the SEL of the plasmodesmata can be modified to permit entry to much larger molecules. Viruses spread from cell to cell by passing through plasmodesmata after invading the plant. A 30-kilodalton **movement protein** (**MP**) produced by tobacco mosaic virus increases the SEL so that the MP with an attached viral genome can pass through plasmodesmata (Wolf et al. 1991). Most, if not all, plant viruses have genes that encode one or more MPs. The synthesis of MPs during virus infection appears to be essential for the spread of the virus through the plant, since if the MP gene is mutated or deleted, the virus can still infect plant cells, even though it cannot move. MPs are localized on the cytoskeleton in infected cells, and they likely utilize the cytoskeleton for transport to the plasmodesmata (Heinlein et al. 1995). Although the mechanism by which MPs modify the SEL is not known, viral MPs probably mimic plant proteins that are important for development or regulating certain physiological processes.

Sieve tubes are files of connected cells, known as sieve elements, that have lost their nucleus during differentiation and that function in the transport of photosynthate (see Chapter 10). Because they lack a nucleus, they are unable to produce mRNAs or ribosomes and are therefore unable to synthesize proteins. Yet they contain a substantial number of proteins (Fisher et al. 1992). Companion cells and sieve tube elements are daughter cells and products of the same formative division, and companion cells have long been suspected to be important for the function of sieve elements. There is evidence that the functional proteins of the sieve elements are synthesized by companion cells and move into the sieve element through the plasmodesmata connecting them. Since these proteins are larger than the SEL, they also must be able to modify this limit, although direct evidence for such an ability has not yet been obtained.

Additional evidence that plasmodesmata are important for cell–cell communication during development arose from the discovery that the mRNA of the maize meristem identity gene *KN1* cannot be detected in the L1 layer of the maize vegetative shoot apical meristem. The KN1 protein, however, is detected in all regions of the apical meristem, including the L1 layer. Since it is unlikely that the KN1 protein is synthesized in the L1 layer, it may be transported into the L1 layer from the

L2 layer, possibly by modification of the SEL of the plasmodesmata, which enables the protein to utilize these channels to pass from cell to cell, just as viral MPs do (Lucas et al. 1995). At present it isn't clear why information exchange would be organized in this manner, but this type of organization may be a fairly general phenomenon in plant development. In *Antirrhinum*, expression of the *FLO* gene in the L1 layer activates expression of the floral organ identity genes in all cell layers of the meristem (Carpenter and Coen 1995). Although many explanations for this relationship are possible, one is that the FLO protein moves into these other layers from the cells in which it is synthesized by passing through the plasmodesmata.

The number of plasmodesmata per unit of wall area is specifically controlled during development. In axial structures such as shoots and roots, the frequency of plasmodesmata is often higher in transverse walls than in longitudinal walls, suggesting that communication between cells along a cell file is greater than that between cells in adjacent files. The number of plasmodesmata in a wall is determined largely by the number inserted into the developing cell plate at cytokinesis, but it can be increased by the de novo formation of secondary plasmodesmata or decreased by their closure.

Just as the cell can open the plasmodesmata for some proteins that exceed the SEL, it is also able to close these channels, even to molecules much smaller than the SEL. The development of root epidermal cells provides an example. As noted earlier, epidermal cells are derived from a ring of initials located just below the quiescent center in the *Arabidopsis* root. These files of cells elongate and progressively differentiate as either hairless cells or trichoblasts (cells that form hairs) with increasing distance from the meristem. Keith Robert's group examined the ability of fluorescent dyes to pass between epidermal cells as differentiation proceeded. They found that, although cells near the meristem are symplastically connected, as differentiation progresses symplastic connections are lost as the plasmodesmata close. By the time the root hairs grow out, the epidermal cells are symplastically isolated (Duckett et al. 1994).

Cells Also Communicate via the Apoplast

Good evidence suggests that some developmentally important molecules move in the apoplast, which consists mostly of the cell wall space. The plant hormone auxin is transported by movement in both the symplast and the apoplast. Auxin transport is highly polar, occurring preferentially from apex to base, and does not involve plasmodesmata (see Chapter 19).The apparent pore size of cell walls is 3.5 to 5.0 nm, which means that only molecules with a mass less than approximately 15 kilodaltons diffuse freely across and within the wall. If the wall is impregnated with substances such as lignin or suberin (see Chapter 13), the pore size can be so reduced that almost all movement is forced to occur through the symplast.

Senescence and Programmed Cell Death

Every autumn, people who live in temperate regions can enjoy the beautiful color changes that precede the loss of leaves from deciduous trees. The leaves change color because changing day length and cooling temperatures trigger developmental processes that lead to **leaf senescence** and death. Senescence is not simply death, or **necrosis**, which could be caused by poisons or other external factors. Rather, senescence refers to an active developmental process that is controlled by the plant's own genetic program. Leaves are genetically programmed to die, and their senescence can be initiated by environmental cues.

As new leaves are initiated from the shoot apical meristem, older leaves often are shaded and lose the ability to function efficiently in photosynthesis. Since the plant invested valuable resources in leaf formation, senescence recovers a portion of those resources. Hydrolytic enzymes, produced as part of the senescence program, will break down many cellular proteins, carbohydrates, and nucleic acids. The component sugars, nucleosides, and amino acids are then transported back into the plant via the phloem, where they will be reused for synthetic processes. Many minerals also are transported out of senescing organs, back into the main body of the plant.

Senescence of plant organs is frequently associated with **abscission**, a process whereby specific cells in the petiole differentiate to form an abscission layer, allowing the senescent organ to separate from the plant. In Chapter 22 we will have more to say about the control of abscission by ethylene. In this section we will examine the roles that senescence and programmed cell death play in plant development. We will see that there are many types of senescence, each with its own genetic program. In Chapters 21 and 22 we will describe how hormones can act as signaling agents that regulate plant senescence.

Plants Exhibit Various Types of Senescence

Senescence occurs in a variety of organs and in response to many different cues. Many annual plants, including major crop plants such as wheat, maize, and soybeans, abruptly yellow and die following fruit production, even under optimal growing conditions. Senescence of the entire plant after a single reproductive cycle is called **monocarpic senescence**. Other types of senescence include the senescence of aerial shoots in herbaceous perennials, seasonal leaf senescence (as in deciduous trees), sequential leaf senescence (in which the leaves die when they reach a certain age), senescence of dry fruits, senescence (ripening) of fleshy fruits, senescence of storage cotyledons and floral organs, and senescence of specialized cell types (e.g., trichomes, tracheids, and vessel elements).

The triggers for the various types of senescence are different and can be internal, as in monocarpic senescence, or external, such as day length and temperature in the autumnal leaf senescence of deciduous trees. Regardless of the initial stimulus, the different senescence patterns may share common internal programs in which a regulatory senescence gene initiates a cascade of secondary gene expression that eventually brings about senescence and death.

Senescence Is a Genetically Programmed Process

Senescence is an ordered series of cytological and biochemical events. On the cytological level, some organelles are destroyed while others remain active. The chloroplast is the first organelle to deteriorate during the onset of leaf senescence, with the destruction of thylakoid protein components and stromal enzymes (Hensel et al. 1993). In contrast, nuclei remain structurally and functionally intact until the late stages of senescence. Senescing tissues carry out catabolic processes that require the de novo synthesis of various hydrolytic enzymes, such as proteases, nucleases, lipases, and chlorophyll-degrading enzymes. The synthesis of these senescence-specific enzymes involves the activation of specific genes.

Not surprisingly, the levels of most leaf mRNAs decline significantly during the senescence phase, while the abundance of certain other transcripts increases. Genes whose expression decreases during senescence are called **senescence down-regulated genes (SDGs)**. SDGs include genes that encode proteins involved in photosynthesis (Hensel et al. 1993, John et al. 1995), in accordance with previous studies that demonstrated deterioration in photosynthetic activity and the sequential disappearance of chloroplast components (Bate et al. 1991; Hensel et al. 1993). However, senescence involves much more than the simple switching off of photosynthesis genes. Many genes whose expression is induced during leaf and flower senescence have been identified through differential screening of cDNA (complementary DNA) libraries (Grbić and Bleecker 1995; Drake et al. 1996).

Genes whose expression is induced during senescence are called **senescence-associated genes (SAGs)**. The functions of most SAGs have been deduced from sequence homology to known genes, and confirmed by their enzyme activities. Among them are genes that encode hydrolytic enzymes, such as proteases, ribonucleases, and lipases (Hensel et al. 1993; Taylor et al. 1993; Ryu and Wang 1995; Oh et al. 1996), as well as enzymes involved in the biosynthesis of ethylene, such as ACC (l-aminocyclopropane-l-carboxylic acid) synthase and ACC oxidase (O'Neill et al. 1993). SAGs of another class have secondary functions in senescence. These genes encode enzymes involved in the conversion or remobilization of breakdown products, such as glutamine synthetase, which catalyzes the conversion of ammonium to glutamine (see Chapter 12) and is responsible for nitrogen recycling from senescing tissues (Watanabe et al. 1994).

Stay-Green Mutants That Are Not Senescence Mutants Have Been Isolated

The regulatory mechanisms underlying tissue senescence are largely unknown. Various genetic strategies have been used to identify and characterize regulatory genes involved in senescence. Two of the most promising approaches have been (1) the identification and selection of "stay-green" mutants in natural populations and (2) the production of chemically or genetically engineered delayed-senescence mutants. Several stay-green mutants have been described, but unfortunately, most of them have been found not to be true delayed-senescence mutants, but rather nonyellowing mutants. During senescence these mutants retain their chlorophyll at high levels, but they degrade the photosynthetic apparatus and other subcellular components as in the wild type.

Other potential senescence mutants are mutants with altered hormone levels or mutants defective in their response to plant hormones. Although all the plant hormones contribute to the regulation of senescence, three of them—ethylene, cytokinin, and abscisic acid—appear to function as major regulators that govern, in opposite ways, the initiation, timing, and rate of senescence. In general, ethylene and abscisic acid accelerate senescence, whereas cytokinins delay it. The roles of cytokinins, ethylene, and abscisic acid in senescence are discussed in Chapters 21, 22 and 23.

Programmed Cell Death Is a Specialized Type of Senescence

Senescence can occur at the level of the whole plant, as in monocarpic senescence; at the organ level, as in leaf senescence; and at the cellular level, as in tracheary element differentiation. The process whereby individual cells activate an intrinsic senescence program is called **programmed cell death (PCD)**. PCD plays an important part in animal development, in which the molecular mechanism has been studied extensively. PCD can be initiated by specific signals, such as errors in DNA replication during division, and involves the expression of a characteristic set of genes. The expression of these genes results in cell death. Much less is known about PCD in plants (Pennell and Lamb 1997).

PCD in animals is usually accompanied by a distinct set of morphological and biochemical changes called **apoptosis** (from a Greek word meaning "falling off," as in autumn leaves). During apoptosis, the cell nucleus condenses and the nuclear DNA fragments in a specific pattern caused by degradation of the DNA between nucleosomes (see Chapter 2). Some plant cells, particularly in senescing tissues, exhibit similar cytological changes (Havel and Durzan 1996). PCD also appears to occur during the differentiation of xylem tracheary elements, during which the nuclei and chromatin degrade and the cytoplasm disappears. These changes result from the activation of genes that encode nucleases and proteases.

One of the important functions of PCD in plants is protection against pathogenic organisms. When a plant is infected by a pathogenic organism, the cells at the site of the infection quickly accumulate high concentrations of toxic phenolic compounds and die, forming a small circular island of cell death called a **necrotic lesion**. The function of the necrotic lesion is to quickly surround the pathogen with a toxic and nutritionally depleted environment, effectively isolating it and preventing it from spreading to surrounding healthy tissues. This rapid, localized cell death due to pathogen attack is called the **hypersensitive response**.

Originally considered to be a simple case of necrosis caused by toxins from the infecting pathogen, the hypersensitive response is now recognized to be another example of PCD, in which signals released by the pathogen initiate biochemical pathways that cause the cells to commit suicide. In tomato, the hypersensitive response induced by fungal toxins has been shown to exhibit the stereotypic hallmarks of apoptosis in animals, including the specific fragmentation of the chromatin to form "DNA ladders" that are visible when analyzed by electrophoresis (Wang et al. 1996). This evidence suggests that the basic mechanism of apoptosis is conserved in plants.

That the hypersensitive response is a genetically programmed process rather than simple necrosis is amply demonstrated by the existence of *Arabidopsis* mutants that can mimic the effect of infection and trigger the entire cascade of events leading to the formation of necrotic lesions even in the absence of the pathogen! For example, *Isd1* mutants exhibit impaired control of cell death in the absence of the pathogen, and they are unable to control the spread of cell death once it is initiated, a phenomenon referred to as **runaway cell death** (Dietrich et al. 1994). The immediate cause of cell death in the mutant appears to be the accumulation of superoxide (O_2^-), and the lack of responsiveness to signals derived from it (Jabs et al. 1996). Superoxide also appears to be a signal initiating the hypersensitive response to avirulent pathogens in wild type plants (Keller et al. 1998).

Summary

Growth in plants is defined as *an irreversible increase in volume*. Growth in the apical meristem is brought about by cell expansion accompanied by cell division, while growth in the subapical regions, representing the largest component of growth, is by cell expansion alone. Plant growth can be quantitatively analyzed using kinematics, the study of particle movement and shape change. Plant growth can be described in both spatial and material terms. Spatial descriptions focus on the patterns generated by all the cells located at different positions in the growth zones. Material analyses focus on the fate of the individual cells or tissue elements at various stages of development. A growth trajectory shows the distance of a tissue element from the apex over time, and is therefore a material description of growth. The growth velocity is the speed at which the tissue elements are being displaced from the apex. The relative elemental growth rate is a measure of the fractional increase in length of the axis per unit time and represents the magnitude of growth at a particular location.

The basic body plan of the mature plant is established during embryogenesis; in this process, tissues are arranged radially, an outer epidermal layer surrounding a cylinder of vascular tissue that is embedded within cortical or ground tissues. The apical–basal axial pattern of the mature plant, with root and shoot polar axes, also is established during embryogenesis, as are the primary meristems that will generate the adult plant.

One common type of angiosperm embryonic development, exemplified by *Arabidopsis thaliana*, is characterized by precise patterns of cell divisions, forming successive stages: the globular, heart, and torpedo stages. The axial body pattern is established during the first division of the zygote, and mutant genes eliminate part of the embryo. The radial tissue pattern is established during the globular stage, apparently as a result of the expression of genes that control cell identity. The *SHOOTMERISTEMLESS* (*STM*) gene is expressed in the region that gives rise to the shoot apical meristem during the heart stage of embryogenesis, and its continued expression suppresses the differentiation of the cells of the shoot apical meristem.

Meristems are populations of small, isodiametric cells that have "embryonic" characteristics. Vegetative meristems generate a specific portion of the plant body, and they regenerate themselves. In many plants, the root and shoot apical meristems are capable of indefinite growth. The root and shoot apical meristems are primary meristems formed during embryogenesis. Secondary meristems are initiated during postembryonic development and include the vascular cambium, cork cambium, axillary meristems, and secondary root meristems.

Roots grow from their distal ends. The root apical meristem is subterminal and covered by a root cap. Cell divisions in the root apex generate files of cells that subsequently elongate and differentiate to acquire specialized function. Four developmental zones are recognized in the root: root cap, meristem, elongation zone, and maturation zone. In *Arabidopsis*, files of mature cells can be traced to initial cells within the meristem cell population. Despite this precise pattern of division, a cell's fate is determined by its position within the root and not its lineage. Plant cell fate is relatively plastic and can be changed when the positional signals necessary for its maintenance are altered.

A complete explanation of the mechanisms responsible for establishing and maintaining these patterns is not possible at present, but there is evidence that an association of microtubules and microfilaments known as the preprophase band is important in determining the plane of cell division. Cell differentiation does not depend on cell lineage; however, the division of the initial cell is essential for this process. Expression of the *SCR* (*SCARECROW*) gene, which has been cloned and encodes a novel protein, is necessary for the division of the initial cell, and the *SHR* (*SHORTROOT*) gene must be expressed for the establishment of endodermal cell identity.

The vegetative shoot apical meristem repetitively generates lateral organs (leaves and lateral buds), as well as segments of the stem. Shoot apical meristems in angiosperms typically are organized into two or more distinct tunica layers, designated L1, L2, and L3. The epidermal layer usually is derived from the L1 layer, while leaf mesophyll cells are derived from the L2 layer. However, as in the root, cell position is more important in determining cell fate than is lineage. Below the tunica layers are internal cells, known collectively as the corpus, in which the plane of cell divisions is not highly ordered. The repetitive activity of the vegetative shoot apical meristem generates a succession of developmental units, each consisting of one or more leaves, the node, the internode, and one or more axillary buds. Divisions of a small number of cells in the tunica and corpus layers give rise to progenitor cells of the leaf primordia, lateral buds, and, in floral apices, the floral organs. The vegetative shoot apical meristem is indeterminate in its activity in that it may function indefinitely, but it gives rise to leaf primordia that are determinate in their growth.

The expression of homeobox genes similar to the maize genes *KNOTTED1* and *SHOOTMERISTEMLESS* is necessary for the continued indeterminate character of the shoot apical meristem, and the loss of expression of these genes in the leaf primordia appears to be important in the shift to determinate growth in these structures. Leaves form in a characteristic pattern. The leaf primordia must be positioned as a result of the precise spatial regulation of cell division within the apex, but the factors controlling this activity are not known.

Differentiation is the process by which cells acquire metabolic, structural, and functional properties distinct from those of their progenitors. Four events in the formation of any differentiated cell are (1) inductive signaling and signal perception, (2) expression of cell identity genes in response to inductive signals, (3) expression of genes for specialized activities or structures, and (4) changes in cell structure and function.

Tracheary element differentiation is an example of plant cell differentiation. Microtubules participate in determining the pattern in which the cellulose microfibrils are deposited in the secondary walls of tracheary elements. Wounding can reprogram mesophyll cells to differentiate into tracheary elements without cell division. Auxin appears to be a signal that initiates tracheary element differentiation, and involves expression of the homeobox gene *ATHB-8* in *Arabidopsis*. Homeobox genes encode transcription factors that control the expression of other genes whose products transform and characterize the differentiated cell. The genes expressed during tracheary element differentiation encode proteases and ribonucleases that are necessary for the final formation of water-conducting xylem elements.

Plant cells can communicate either symplastically via plasmodesmata or apoplastically through the cell wall by means of hormones or other signaling molecules. Auxin moves through tissues in both the symplast and the apoplast. However, symplastic cell–cell communication involving larger signaling molecules occurs via plasmodesmata, which connect most living cells to their neighbors. Molecules ranging in size up to about 1.6 nm (700 to 1000 daltons) can pass from cell to cell through plasmodesmata connecting leaf epidermal cells. Plasmodesmata are, to some extent, gated so that passage through them can be regulated, and their size exclusion limit can be modified to permit the passage of much larger molecules, such as viruses.

Senescence and programmed cell death are essential aspects of plant development. Plants exhibit a variety of different senescence phenomena. Leaves are genetically programmed to senesce and die. Senescence is an active developmental process that is controlled by the plant's genetic program and initiated by specific environmental or developmental cues. Senescence is an ordered series of cytological and molecular events. The expression of most genes is reduced during senescence, but the expression of some genes is initiated. The newly active genes encode various hydrolytic enzymes, such as proteases, nucleases, lipases, and chlorophyll-degrading enzymes, which will carry out the degradative processes as the tissues die.

General Reading

*Barlow, P. W. (1978) The concept of the stem cell in the context of plant growth and development. In *Stem Cells and Tissue Homeostasis*, B. I. Lord, C. S. Potten, and R. J. Cole, eds., Cambridge University Press, Cambridge, pp. 87–113.

*Barlow, P. W. (1994) Evolution of structural initial cells in apical meristems of plants. *J. Theor. Biol.* 169: 163–177.

Bleecker, A. B., and Patterson, S. E. (1997) Last exit: Senescence, abscission, and meristem arrest in *Arabidopsis*. *Plant Cell* 9: 1169–1179.

Bowman, J., ed. (1994) *Arabidopsis*: An atlas of morphology and development. Springer, Berlin.

Clark, S. E. (1997) Organ formation at the vegetative shoot meristem. *Plant Cell* 9: 1067–1076.

Esau, K. (1965) *Plant Anatomy*, 2nd ed. Wiley, New York.

Fosket, D. E. (1994) *Plant Growth and Development: A Molecular Approach.* Academic Press, San Diego.

*Fukuda, H. (1996) Xylogenesis: Initiation, progression and cell death. *Annu. Rev. Plant Physiol. Plant Mol. Biol.* 47: 299–325.

*Fukuda, H. (1992) Tracheary element formation as a model system of cell differentiation. *Int. Rev. Cytol.* 136: 289–332.

Fukuda, H. (1997) Tracheary element differentiation. *Plant Cell* 9: 1147–1156.

Gasser, C. S., and Robinson-Beers, K. (1993) Pistil development. *Plant Cell* 5: 1231–1239.

*Gibson, S., and Somerville, C. (1993) Isolating plant genes. *Trends Biotechnol.* 11: 306–313.

*Goldberg, R. B., de Paiva, G., and Yadegari, R. (1994) Plant embryogenesis: Zygote to seed. *Science* 266: 605–614.

Green, P. B. (1976) Growth and cell pattern formation on an axis: Critique of concepts, terminology, and modes of study. *Bot. Gaz.* 137: 187–202.

*Jürgens, G. (1992) Genes to greens: Embryonic pattern formation in plants. *Science* 256: 487–488.

*Jürgens, G. (1995) Axis formation in plant embryogenesis: Cues and clues. *Cell* 81: 467–470.

Kerstetter, R. A., and Hake, S. (1997) Shoot meristem formation in vegetative development. *Plant Cell* 9: 1001–1010.

Kropf, D. L. (1997) Induction of polarity in fucoid zygotes. *Plant Cell* 9: 1011–1020.

Laux, T., and Jürgens, G. (1997) Embryogenesis: A new start in life. *Plant Cell* 9: 989–1000.

*Leutwiler, L. S., Hough-Evans, B. R., and Meyerowitz, E. M. (1984) The DNA of *Arabidopsis thaliana*. *Mol. Gen. Genet.* 194: 15–23.

*Lucas, W. J. (1995) Plasmodesmata: Intercellular channels for macromolecular transport in plants. *Curr. Opin. Cell Biol.* 7: 673–680.

McLean, B. G., Hempel, F. D., and Zambryski, P. C. (1997) Plant intercellular communication via plasmodesmata. *Plant Cell* 9: 1043–1054.

Meyerowitz, E. M. (1996) Plant development: Local control, global patterning. *Curr. Opin. Genet. Dev.* 6: 475–479.

*Osborne, B. I., and Baker, B. (1995) Movers and shakers: Maize transposons as tools for analyzing other plant genomes. *Curr. Opin. Cell Biol.* 7: 406–413.

*Pennell, R. I., and Lamb, C. (1997) Programmed cell death in plants. *Plant Cell* 9: 1157–1168.

Poethig, R. S. (1997) Leaf morphogenesis in flowering plants. *Plant Cell* 9: 1077–1087.

Quatrano, R. (1978) Development of cell polarity. *Annu. Rev. Plant Physiol.* 29: 487–510.

*Raghavan, V. (1986) *Embryogenesis in Angiosperms.* Cambridge University Press, Cambridge.

*Robards, A. W., and Lucas, W. J. (1990) Plasmodesmata. *Annu. Rev. Plant Physiol. Plant Mol. Biol.* 41: 369–419.

*Sachs, T. (1969) Polarity and the induction of organized vascular tissues. *Ann. Bot.* 33: 263–275.

*Steeves, T. A., and Sussex, I. M. (1989) *Patterns in Plant Development.* Cambridge University Press, Cambridge.

*West, M. A. L., and Harada, J. J. (1993) Embryogenesis in higher plants: An overview. *Plant Cell* 5: 1361–1369.

* Indicates a reference that is general reading in the field and is also cited in this chapter.

Chapter References

Aloni, R. (1995) The induction of vascular tissue by auxin and cytokinin. In *Plant Hormones and their Role in Plant Growth Development*, 2nd ed., Davies P. J., ed., Kluwer, Dordrecht, Netherlands, pp. 531–546.

Arondel, V., Lemieux, B., Hwang, I., Gibson, S., Goodman, H. M., and Somerville, C. R. (1992) Map-based cloning of a gene controlling omega-3 fatty acid desaturation in *Arabidopsis*. *Science* 258: 1353–1355.

Assaad, F. F., Mayer, U., Lukowitz, W., and Jürgens, G. (1997) Cytokinesis in somatic plant cells. *Plant Physiol. Biochem.* 35: 177–184.

Baima, S., Nobili, F., Sessa, G., Lucchetti, S., Ruberti, I., and Morelli, G. (1996) The expression of the *Athb-8* homeobox gene is restricted to provascular cells in *Arabidopsis thaliana*. *Development* 121: 4171–4182.

Bancroft, I., Bhatt, A. M., Sjodin, C., Scofield, S., Jones, J. D. G., and Dean, C. (1992) Development of an efficient two-element transposon tagging system in *Arabidopsis thaliana*. *Mol. Gen. Genet.* 233: 449–461.

Baskin, T. I., Cork, A., Williamson, R. E., and Gorst, J. (1995) *STUNTED PLANT* 1, a gene required for expansion in rapidly elongating but not in dividing cells and mediating root growth responses to applied cytokinin. *Plant Physiol.* 107: 233–243.

Bate, N. J., Rothstein, S. J., and Thompson, J. E. (1991) Expression of nuclear and chloroplast photosynthesis-specific genes during leaf senescence. *J. Exp. Bot.* 42: 801–812.

Berleth, T., and Jürgens, G. (1993) The role of the monopteros gene in organising the basal body region of the *Arabidopsis* embryo. *Development* 118: 575–587.

Carpenter, R., and Coen, E. S. (1995) Transposon induced chimeras show that *floricaula*, a meristem identity gene, acts nonautonomously between cell layers. *Development* 121: 19–26.

Carpenter, R., Copsey, L., Vincent, C., Doyle, S., Magrath, R., and Coen, E. (1995) Control of flower development and phyllotaxy by meristem identity genes in *Antirrhinum*. *Plant Cell* 7: 2001–2011.

Cheng, J. C., Seeley, K. A., and Sung, Z. R. (1995) *RML1* and *RML2*, *Arabidopsis* genes required for cell proliferation at the root tip. *Plant Physiol.* 107: 365–376.

Clowes, F. A. L. (1964) The quiescent center in meristems and its behavior after irradiation. *Brookhaven Symp. Biol.* 16: 46–57.

Dietrich, R. A., Delaney, T. P., Uknes, S. J., Ward, E. R., Ryals, J. A., and Dangl, J. L. (1994) *Arabidopsis* mutants simulating disease resistance response. *Cell* 77: 565–577.

Di Laurenzio, L., Wysocka-Diller, J., Malamy, J. E., Pysh, L., Helariutta, Y., Freshour, G., Hahn, M. G., Feldmann, K. A., and Benfey, P. N. (1996) The *SCARECROW* gene regulates an asymmetric cell division that is essential for generating the radial organization of the *Arabidopsis* root. *Cell* 86: 423–433.

Ding, B., Turgeon, R., and Parthasarathy, M. V. (1992) Substructure of freeze substituted plasmodesmata. *Protoplasma* 169: 28–41.

Doerner, P., Jorgensen, J-E., You, R., Steppuhn, J., and Lamb, C. (1996) Control of root growth and development by cyclin expression. *Nature* 380: 520–523.

Dolan, L., Janmaat, K., Willemsen, V., Linstead, P., Poethig, S., Roberts, K., and Scheres, B. (1993) Cellular organisation of the *Arabidopsis thaliana* root. *Development* 119: 71–84.

Drake, R., John, I., Farrell, J. A., Cooper, W., Schuch, W., and Grierson, D. (1996) Isolation and analysis of cDNAs encoding tomato cysteine proteases expressed during leaf senescence. *Plant Mol. Biol.* 30: 755–767.

Dubois, F., Bui-Dang-Ha, D., Sangwan, R. S., and Durand, J. (1996) The petunia *tra1* gene controls cell elongation and plant development, and mediates responses to cytokinins. *Plant J.* 10: 47–59.

Duckett, C. M., Oparka, K. J., Prior, D. A. M., Dolan, L., and Roberts, K. (1994) Dye-coupling in the root epidermis of *Arabidopsis* is progressively reduced during development. *Development* 120: 3247–3255.

Eady, C., Lindsey, K., and Twell, D. (1994) Differential activation and conserved vegetative cell-specific activity of a late pollen promoter in species with bicellular and tricellular pollen. *Plant J.* 5: 543–550.

Eady, C., Lindsey, K., and Twell, D. (1995) The significance of microspore division and division symmetry for vegetative cell-specific transcription and generative cell differentiation. *Plant Cell* 7: 65–74.

Erickson, R. O., and Sax , K. B. (1956) Elementary growth rate of the primary root of *Zea mays*. *Proc. Am. Philos. Soc.* 100: 487–498.

Feldman, L. J., and Torrey, J. G. (1976) The isolation and culture in vitro of the quiescent center of *Zea mays*. *Am. J. Bot.* 63: 345–355.

Feldmann, K. A. (1992) T-DNA insertion mutagenesis in *Arabidopsis*: Mutational spectrum. *Plant J.* 1: 71–82.

Fisher, D. B., Wu, Y., and Ku, M. S. B. (1992) Turnover of soluble proteins in the wheat sieve tube. *Plant Physiol.* 100: 1433–1441.

Fisher, R. H., Barton, M. K., Cohen, J. D., and Cooke, T. J. (1996) Hormonal studies of *fass*, an *Arabidopsis* mutant that is altered in organ elongation. *Plant Physiol.* 59: 91–93.

Fleming, A. J., McQueen-Mason, S., Mandel, T., and Kuhlemeir, C. (1997) Induction of leaf primordia by the cell wall protein expansin. *Science* 276: 1415–1418.

Foard, D. E., Haber, A. N., and Fishman, T. N. (1965) Initiation of lateral root primordia without completion of mitosis and without cytokinesis in uniseriate pericycle. *Am. J. Bot.* 52: 580–590.

Giraudat, J., Hauge, B. M., Valon, C., Smalle, J., Parcy, F., and Goodman, H. M. (1992) Isolation of the *Arabidopsis AB13* gene by positional cloning. *Plant Cell* 4: 1251–1261.

Grbić, V., and Bleecker, A. B. (1995) Ethylene regulates the timing of leaf senescence in *Arabidopsis*. *Plant J.* 8: 595–602.

Green, P. B., Steele, C. S., and Rennich, S. C. (1996) Phyllotactic patterns: A biophysical mechanism for their origin. *Ann. Botany* 77: 515–527.

Grill, E., and Somerville, C. (1991) Construction and characterization of a yeast artificial chromosome library which is suitable for chromosome walking. *Mol. Gen. Genet.* 226: 484–490.

Gunning, B. E. S. (1982) The root of the water fern *Azolla*: Cellular basis of development and multiple roles for cortical microtubules. In *Developmental Order: Its Origin and Regulation*, S. Subtelny and P. B. Green, eds., Alan R. Liss, New York, pp. 379–421.

Hake, S., Vollbrecht, E., and Freeling, M. (1989) Cloning *Knotted*, the dominant morphological mutant in maize using *Ds2* as a transposon tag. *EMBO J.* 8: 15–22.

Hardham, A. R., and Gunning, B. E. S. (1979) Interpolation of microtubules into cortical arrays during cell elongation and differentiation in roots of *Azolla pinnata*. *J. Cell Sci.* 37: 411–442.

Hardham, A. R., and Gunning, B. E. S. (1980) Some effects of colchicine on microtubules and cell division of *Azolla pinnata*. *Protoplasma* 102: 31–51.

Havel, L., and Durzan, D. J. (1996). Apoptosis during diploid parthenogenesis and early somatic embryogenesis of Norway spruce. *Int. J. Plant Sci.* 157: 8–16.

Heinlein, M., Epel, B. L., Padgett, J. S., and Beachy, R. N. (1995) Interaction of tobamovirus movement proteins with the plant cytoskeleton. *Science* 270: 1983–1985.

Hensel, L. L., Grbić, V., Baumgarten, D. A., and Bleecker, A. B. (1993) Developmental and age-related processes that influence the longevity and senescence of photosynthetic tissues in *Arabidopsis*. *Plant Cell* 5: 553–564.

Hepler, P. K. (1981) Morphogenesis of tracheary elements and guard cells. In *Cytomorphogenesis in Plants*. Kiermayer, O., ed., Springer, Berlin, pp. 327–347.

Huala, E., and Sussex, I. M. (1993) Determination and cell interaction in reproductive meristems. *Plant Cell* 5: 1157–1165.

Jabs, T., Dietrich, R. A., and Dangl, J. L. (1996) Initiation of runaway cell death in an *Arabidopsis* mutant by extracellular superoxide. *Science* 273: 1853–1856.

John, I., Drake, R., Farrell, A., Cooper, W., Lee, P., Horton, P., and Grierson, D. (1995) Delayed leaf senescence in ethylene-deficient ACC-oxidase antisense tomato plants: Molecular and physiological analysis. *Plant J.* 7: 483–490.

Jürgens, G., Mayer, U., Torres Ruiz, R., Berleth, T., and Misera, S. (1991) Genetic analysis of pattern formation in the *Arabidopsis* embryo. *Development* Suppl. 1: 27–38.

Keller, T., Damude, H. G., Werner, D., Doerner, P., Dixon, R. A., and Lamb, C. (1998) A plant homolog of the neutrophil NADPH oxidase gp91-phox subunit encodes a plasma membrane protein with Ca^{2+} binding motifs. *Plant Cell* 10: 255–266.

Kerstetter, R. A., and Hake., S. (1997) Shoot meristem formation in vegetative development. *Plant Cell* 9: 1001–1010.

Lincoln, C., Long, J., Yamaguchi, J., Serikawa, K., and Hake, S. (1994) A *knotted1*-like homeobox gene in *Arabidopsis* is expressed in the vegetative meristem and dramatically alters leaf morphology when overexpressed in transgenic plants. *Plant Cell* 6: 1859–1876.

Liu, Y-G., Mitsukawa, N., Vazquez-Tello, A., and Whittier, R. F. (1995) Generation of a high-quality P1 library of *Arabidopsis* suitable for chromosome walking. *Plant J.* 7: 351–358.

Long, J., Moan, E. I., Medford, J. I., and Barton, M. K. (1996) A member of the KNOTTED class of homeodomain proteins encoded by the *STM* gene of *Arabidopsis*. *Nature* 379: 66–69.

Lucas, W. J., Bouche-Pillon, S., Jackson, D. P., Nguyen, L., Baker, L., Ding, B., and Hake, S. (1995) Selective trafficking of KNOTTED1 homeodomain protein and its mRNA through plasmodesmata. *Science* 270: 1980–1983.

Lukowitz, W., Mayer, U., and Jürgens, G. (1996) Cytokinesis in the *Arabidopsis* embryo involves the syntaxin-related KNOLLE gene product. *Cell* 84: 61–71.

Mansfield, S. G., and Briarty, L. G. (1991) Early embryogenesis in *Arabidopsis thaliana*. II. The developing embryo. *Can. J. Bot.* 69: 461–476.

Mayer, U., Guttner, G., and Jürgens, G. (1993) Apical-basal pattern formation in the *Arabidopsis* embryo: Studies on the role of the gnom gene. *Development* 117: 149–162.

Meinke, D. W. (1986) Embryo-lethal mutants and the study of plant embryo development. *Oxford Surv. Plant Mol. Cell. Biol.* 3: 122–165.

Meyerowitz, E. M. (1987) *Arabidopsis thaliana*. *Annu. Rev. Genet.* 21: 93–111.

Oh, S. A., Lee, S. Y., Chung, I. K., Lee, C-H., and Nam, J. G. (1996) A senescence-associated gene of *Arabidopsis thaliana* is distinctively regulated during natural and artificially induced leaf senescence. *Plant Mol. Biol.* 30: 739–754.

O'Neill, S. D., Nadeau, J. A., Zhang, X. S., Bui, A. Q., and Halevy, A. H. (1993) Interorgan regulation of ethylene biosynthetic genes by pollination. *Plant Cell* 5: 419–432.

Pepper, A., Delaney, T., Washburn, T., Poole, D., and Chory, J. (1994) *DET1*, a negative regulator of light-mediated development and gene expression in *Arabidopsis*, encodes a novel nuclear-localized protein. *Cell* 78: 109–116.

Pickett-Heaps, J. D., and Northcote, D. H. (1966) Organization of microtubules and endoplasmic reticulum during mitosis and cytokinesis in wheat meristems. *J. Cell Sci.* 1: 109–120.

Redei, G. P. (1975) *Arabidopsis* as a genetic tool. *Annu. Rev. Genet.* 9: 111–127.

Rerie, W. G., Feldmann, K. A., and Marks, M. D. (1994) The *GLABRA2* gene encodes a homeo domain protein required for normal trichome development in *Arabidopsis*. *Genes Dev.* 8: 1388–1399.

Ryu, S. B., and Wang, X. (1995) Expression of phospholipase D during castor bean leaf senescence. *Plant Physiol.* 108: 713–719.

Scheres, B., Di Laurenzio, L., Willemsen, V., Hauser, M-T., Janmaat, K., Weisbeek, P., and Benfey, P. N. (1995) Mutations affecting the radial organisation of the *Arabidopsis* root display specific defects throughout the embryonic axis. *Development* 121: 53–62.

Scheres, B., Wolkenfelt, H., Willemsen, V., Terlouw, M., Lawson, E., Dean, C., and Weisbeek, P. (1994) Embryonic origin of the *Arabidopsis* primary root and root meristem initials. *Development* 120: 2475–2487.

Schiefelbein, J. W., Masucci, J. D., and Wang, H. (1997) Building a root: The control of patterning and morphogenesis during root development. *Plant Cell* 9: 1089–1098.

Schleiden, M. (1842). *Principles of Scientific Botany*. English trans. E. Lankester. (New York: Johnson Reprint Corp., 1969). p. 102.

Schmidt, R., Cnops, G., Bancroft, I., and Dean, C. (1992) Construction of an overlapping YAC library of the *Arabidopsis thaliana* genome. *Aust. J. Plant Physiol.* 19: 341–351.

*Silk, W. K. (1984) Quantitative descriptions of development. *Ann. Rev. Plant Physiol.* 35: 479–518.

Silk, W. K. (1994) Kinematics and dynamics of primary growth. *Biomimetics* 2: 199–214.

Sinha, N. R., Williams, R. E., and Hake, S. (1993a) Overexpression of the maize homeo box gene, *KNOTTED-1*, causes a switch from determinate to indeterminate cell fates. *Genes Dev.* 7: 787–795.

Sinha, N., Hake, S., and Freeling, M. (1993b) Genetic and molecular analysis of leaf development. *Curr. Topics Dev. Biol.* 28: 47–80.

Snustad, D. P., Haas, N. A., Kopczak, S. D., and Silflow, C. D. (1992) The small genome of *Arabidopsis* contains at least nine expressed β-tubulin genes. *Plant Cell* 4: 549–556.

Sussex, I. (1989) Developmental programming of the shoot meristem. *Cell* 56: 225–229.

Taylor, C. B., Bariola, P. A., Delcardayre, S. B., Raines, R. T., and Green, P. J. (1993) Rns2, A senescence-associated RNase of *Arabidopsis* that diverged from the S RNases before speciation. *Proc. Natl. Acad. Sci. USA* 90: 5118–5122.

Thomas, H., and Smart, C. M. (1993) Crops that stay green. *Ann. Appl. Biol.* 123: 193–219.

Torres-Ruiz, R. A., and Jürgens, G. (1994) Mutations in the *FASS* gene uncouple pattern formation and morphogenesis in *Arabidopsis* development. *Development* 120: 2967–2978.

Torres-Ruiz, R. A., Lohner, A., and Jürgens, G. (1996) The GURKE gene is required for normal organization of the apical region in the *Arabidopsis* embryo. *Plant J.* 10: 1005–1016.

Traas , J., Bellini, C., Nacry, P., Kronenberger, J., Bouchez, D., and Caboche, M. (1995) Normal differentiation patterns in plants lacking microtubular preprophase bands. *Nature* 375: 676–677.

Van Den Berg, C., Willemsen, V., Hage, W., Weisbeek, P., and Scheres, B. (1995) Cell fate in the *Arabidopsis* root meristem determined by directional signaling. *Nature* 378: 62–65.

Wang, H., Li, J., Bostock, R. M., and Gilchrist, D. G. (1996). Apoptosis: A functional paradigm for programmed plant cell death induced by a host-selective phytotoxin and invoked during development. *Plant Cell* 8: 375–391.

Watanabe, A., Hamada, K., Yokoi, H., and Watanabe, A. (1994) Biophysical and differential expression of cytosolic glutamine synthetase genes of radish during seed germination and senescence of cotyledons. *Plant Mol. Biol.* 26: 1807–1817.

Wolf, S., Deom, C. M., Beachy, R., and Lucas, W. J. (1991) Plasmodesmatal function is probed using transgenic tobacco plants that express a virus movement protein. *Plant Cell* 3: 593–604.

Yanofsky, M. F., Ma, H., Bowman, J. L., Drews, G. N., Feldmann, K. A., and Meyerowitz, E. M. (1990) The protein encoded by the *Arabidopsis* homeotic gene *agamous* resembles transcription factors. *Nature* 346: 35–39.

Ye, Z-H., and Droste, D. L. (1996) Isolation and characterization of cDNAs encoding xylogenesis-associated and wounding-induced ribonucleases in *Zinnia elegans*. *Plant Mol. Biol.* 30: 697–709.

Ye, Z-H., and Varner, J. E. (1993) Gene expression patterns associated with in vitro tracheary element formation in isolated single mesophyll cells of *Zinnia elegans*. *Plant Physiol.* 103: 805–813.

Zimmerman, J. L. (1993) Somatic embryogenesis: A model for early development in higher plants. *Plant Cell* 5: 1411–1423.

17 Phytochrome

SEEDLINGS GROWN IN DARKNESS have a pale, almost ethereal appearance. This spindly, "etiolated" form of growth is dramatically different from the stockier, green appearance of seedlings grown in the light and is the result of a developmental program known as **skotomorphogenesis** (from the Greek word *skotos*, meaning "darkness") (Figure 17.1). Given the key role of photosynthesis in plant metabolism, one would be tempted to attribute much of this contrast to differences in the availability of light-derived metabolic energy. However, it takes very little light or time to initiate the transformation from the etiolated to the green appearance.

Within hours of applying a single flash of relatively dim light to a dark-grown bean seedling, one can measure a decrease in the rate of stem extension, the beginning of apical-hook straightening, and initiation of the synthesis of pigments that are characteristic of green plants. Light has acted as a signal to induce a change in the form of the seedling from one that facilitates growth beneath the soil to one that is more adaptive to growth above ground in the light. This change in form requires an alteration in the pattern of gene expression. Photosynthesis cannot be the driving force of this transformation, because chlorophyll is not present during this time. Full de-etiolation does require some photosynthesis, but the initial rapid changes are induced by a distinctly different light response, called **photomorphogenesis**.

Among the different pigments that can promote photomorphogenic responses in plants, the most important are those that absorb blue and red light. The blue-light photoreceptors will be discussed in relation to guard cells and phototropism in Chapter 18. The

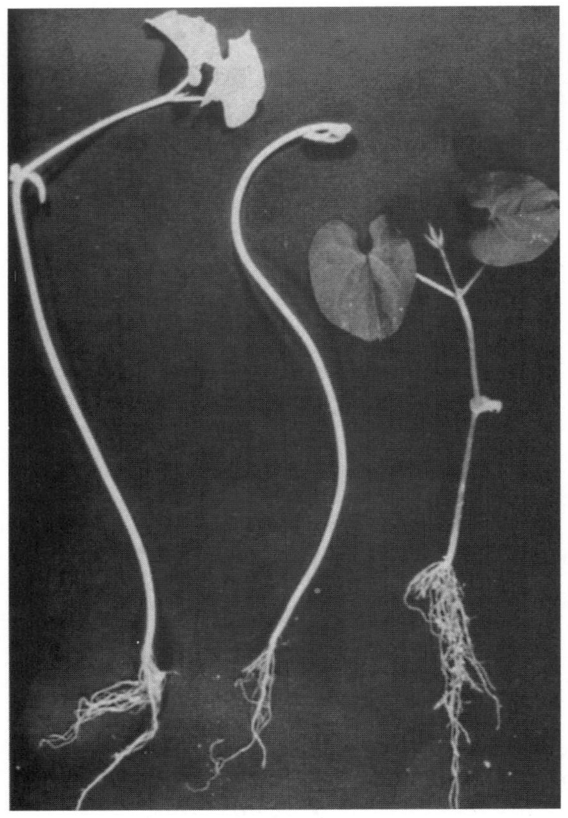

Figure 17.1 Bean (*Phaseolus vulgaris*) seedlings grown under different light conditions for 6 days. Five minutes of dim red light per day is sufficient to prevent some of the symptoms of etiolation that appear under conditions of total darkness, such as reduced leaf size and maintenance of the apical hook. (Photo courtesy of H. Smith.)

focus of this chapter is **phytochrome**, a protein pigment that absorbs red and far-red light most strongly, but that also absorbs blue light. As we will see in this chapter and in Chapter 24, phytochrome plays a key role in light-regulated vegetative and reproductive development.

We begin with the discovery of phytochrome and the phenomenon of red/far-red photoreversibility. Next we will discuss the biochemical and photochemical properties of phytochrome, and the conformational changes induced by light. Different types of phytochromes are encoded by different members of a multigene family, and different phytochromes regulate distinct processes in the plant. These different phytochrome responses can be classified according to the amount of light required to produce the effect. Finally, we will examine what is known about the mechanism of phytochrome action at the cellular and molecular levels, including signal transduction pathways and gene regulation.

The Photochemical and Biochemical Properties of Phytochrome

Phytochrome, a blue protein pigment with a molecular mass of about 125 kDa (kilodaltons), was not identified as a unique chemical species and named until 1959, mainly because of technical difficulties in isolating and purifying the protein. However, many of the biological properties of phytochrome had been established earlier in studies of whole plants. The first clues regarding the role of phytochrome in plant development came from studies begun in the 1930s on red light–induced morphogenic responses, especially seed germination. The list of such responses is now enormous and includes one or more responses at almost every stage in the life history of a wide range of different green plants (Table 17.1).

A key breakthrough in the history of phytochrome was the discovery that the effects of red light (650 to 680 nm) on morphogenesis could be reversed by a subsequent irradiation with light of longer wavelengths (710 to 740 nm), called far-red light. This phenomenon was first demonstrated in germinating seeds, but was also observed in relation to stem and leaf growth, as well as floral induction (see Chapter 24). The initial observations that led to the discovery were made by Lewis Flint, a plant physiologist at the U.S. Department of Agriculture's Seed Testing Laboratory. Flint demonstrated that the germination of lettuce seeds is stimulated by red light and inhibited by far-red light (Flint 1936). But the real breakthrough was made many years later by Harry Borthwick, Sterling Hendricks, and their colleagues at the USDA's laboratory in Beltsville, Maryland. These workers achieved spectacular results by exposing lettuce seeds to alternating treatments of red and far-red light. Nearly 100% of the seeds that received red light as the final treatment germinated; in the seeds that received far-red light as the final treatment, however, germination was strongly inhibited. A demonstration of this effect is shown in Figure 17.2.

There are two possible ways in which the Beltsville group could have interpreted these results. They could have proposed that there were two pigments, one that absorbs red light and one that absorbs far-red light, and that the two pigments act antagonistically in the regulation of seed germination. But Borthwick and colleagues (1952) opted for the more radical interpretation: that there was a single pigment that could exist in two interconvertible forms, a red light–absorbing form and a far-red light–absorbing form. This model was more radical than the two-pigment model because there was no precedent for such a photoreversible pigment.

Several years later phytochrome was demonstrated in plant extracts for the first time by Butler and coworkers (1959), and its unique photoreversible properties

Table 17.1
Typical photoreversible responses induced by phytochrome in a variety of higher and lower plants

Group	Genus	Stage of development	Effect of red light
Angiosperms	*Lactuca* (lettuce)	Seed	Promotes germination
	Avena (oat)	Seedling (etiolated)	Promotes de-etiolation (e.g., leaf unrolling)
	Sinapis (mustard)	Seedling	Promotes formation of leaf primordia, development of primary leaves and anthocyanin production
	Pisum (pea)	Adult	Inhibits internode elongation
	Xanthium (cocklebur)	Adult	Inhibits flowering (photoperiodic response)
Gymnosperms	*Pinus* (pine)	Seedling	Enhances rate of chlorophyll accumulation
Pteridophytes	*Onoclea* (sensitive fern)	Young gametophyte	Promotes growth
Bryophytes	*Polytrichum* (moss)	Germling	Promotes plastid replication
Chlorophytes	*Mougeotia* (alga)	Mature gametophyte	Promotes chloroplast orientation to directional dim light

(A) Treatments ending in red light

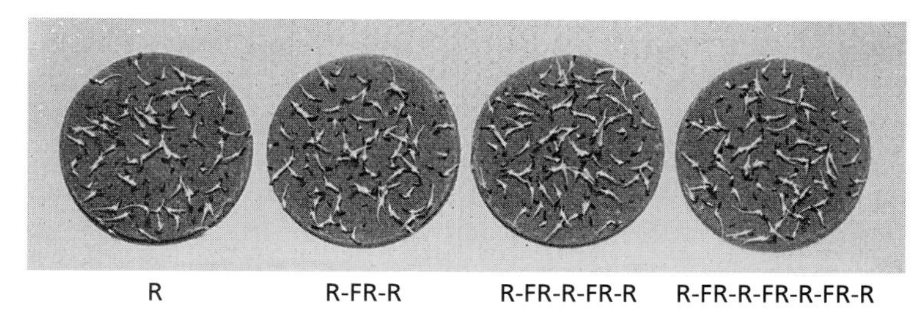

(B) Treatments ending in far-red light

Figure 17.2 Red light (R) promotes lettuce seed germination, but this effect is reversed by far-red light (FR). Imbibed (water-moistened) seeds were given alternating treatments of red followed by far-red light (R-FR). The effect of the light treatment depended on the last treatment given. Germination was stimulated when the last treatment was red light (A), but was inhibited when the last treatment was far-red light (B). Lettuce seed germination is a typical photoreversible response controlled by phytochrome. (Photos from Toole et al. 1953, courtesy of W. Briggs.)

were confirmed in vitro. Thus, the model of Borthwick and colleagues was vindicated, and it remains to this day one of biology's most prescient predictions.

In this section we will describe photoreversibility and its relationship to phytochrome responses. We will then examine the structure of phytochrome, its synthesis and assembly, and the conformational changes associated with the interconversions of the two main spectral forms of phytochrome, Pr and Pfr. Finally, we will discuss the phytochrome gene family, the members of which have different functions in photomorphogenesis.

Phytochrome Can Interconvert between Pr and Pfr Forms

In etiolated plants, phytochrome is present in a red light–absorbing form, referred to as **Pr**, because phytochrome is synthesized as the Pr form in dark-grown plants. The Pr form, which is blue, is converted by red light to a far-red light–absorbing form called **Pfr**, which is blue-green. The Pfr form, in turn, can be converted back to Pr by far-red light. This quality of photoreversibility is the most distinctive property of phytochrome, and it may be expressed in abbreviated form as follows:

$$\text{Pr} \underset{\text{Far-red light}}{\overset{\text{Red light}}{\rightleftharpoons}} \text{Pfr}$$

The interconversion of the Pr and Pfr forms can be measured in vivo or in vitro. In fact, most of the spectral properties of carefully purified phytochrome measured in vitro are the same as those observed in vivo.

Although the two forms of phytochrome are referred to by their maximum absorbance peaks (red and far red), the absorbance spectra of Pr and Pfr overlap significantly in the red region of the spectrum, and the Pr form of phytochrome absorbs a small amount of light in the far-red region (Figure 17.3). As a consequence, a dynamic equilibrium exists between the two forms. When Pr molecules are exposed to red light, most of them absorb it and are converted to Pfr, but some of the Pfr also absorbs the red light and is converted back to Pr, because both Pr and Pfr absorb red light. Thus, the proportion of phytochrome in the Pfr form after saturating irradiation by red light is only about 85%. Similarly, the very small amount of far-red light absorbed by Pr makes it impossible to convert Pfr entirely to Pr by broad-spectrum far-red light. Instead, an equilibrium of 97% Pr and 3% Pfr is achieved. This equilibrium is termed the **photostationary state**.

In addition to absorbing in the red region, both forms of phytochrome absorb in the blue region of the spectrum (see Figure 17.3). Therefore, phytochrome effects can be elicited also by blue light, which can convert Pr to Pfr. As will be discussed in Chapter 18, blue-light responses can also result from the action of a specific blue-light photoreceptor. Whether phytochrome is

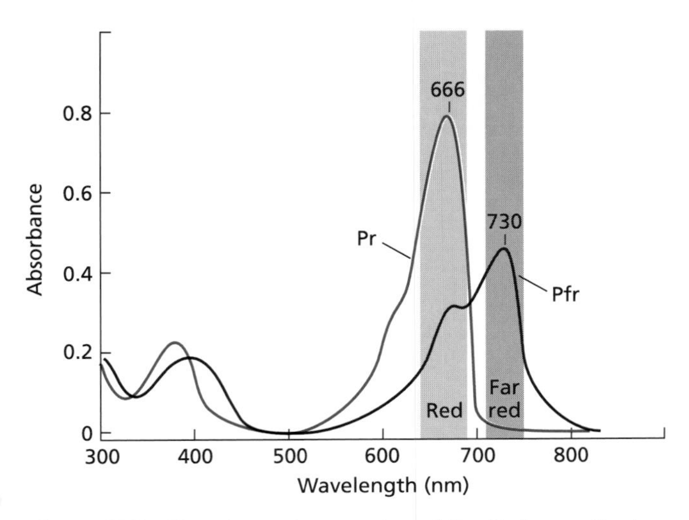

Figure 17.3 The absorption spectra of the Pr (green line) and Pfr (solid line) forms of phytochrome overlap. (After Vierstra and Quail 1983.)

involved in a response to blue light is often determined by a test of the ability of far-red light to reverse the response, since only phytochrome-induced responses are reversed by far-red light. Another way to discriminate between photoreceptors is to study mutants that are deficient in one of the photoreceptors.

Short-lived phytochrome intermediates. The photoconversions of Pr to Pfr, and of Pfr to Pr, are not one-step processes. By irradiating phytochrome with very brief flashes of light, we can observe absorption changes that occur in less than a millisecond. This technique shows that after Pr absorbs a flash of red light, several short-lived spectral forms are generated in a reproducible sequence before Pfr is formed. A different set of intermediate spectral forms occurs in the photoconversion of Pfr to Pr (Rüdiger and Thümmler 1994; Andel et al. 1997). Of course, sunlight includes a mixture of all visible wavelengths. Under such white-light conditions, both Pr and Pfr are excited, and phytochrome cycles continuously between Pr and Pfr. In this situation, the intermediate forms of phytochrome accumulate and make up a significant fraction of the total phytochrome. Such intermediates could even play a role in initiating or amplifying phytochrome responses under natural sunlight, but this question has yet to be resolved.

Pfr Is the Physiologically Active Form of Phytochrome

Because phytochrome responses are induced by red light, they could, in theory, result either from the appearance of Pfr or from the disappearance of Pr. In most cases studied, a quantitative relationship holds between the magnitude of the physiological response and the amount of Pfr generated by light, but no such relationship holds between the physiological response and the loss of Pr. Evidence such as this has led to the conclusion that Pfr is the physiologically active form of phytochrome. In cases in which it has been shown that a phytochrome response is not quantitatively related to the absolute amount of Pfr, it has been proposed that the ratio between Pfr and Pr, or between Pfr and the total amount of phytochrome, determines the magnitude of the response.

The conclusion that Pfr is the physiologically active form of phytochrome is supported by studies with null mutants of *Arabidopsis* that are unable to synthesize phytochrome. Investigators have selected such mutants by growing mutagenized *Arabidopsis* seedlings under continuous white light. In wild-type seedlings, hypocotyl elongation is strongly inhibited by white light, and phytochrome is one of the photoreceptors involved in this response. Mutant seedlings with long hypocotyls when grown under continuous white light are called *hy* mutants, and the different *hy* mutants are designated by numbers: *hy1*, *hy2*, and so on. Some, but not all, of the *hy* mutants have been shown to be blocked in phytochrome synthesis.

The phenotypes of phytochrome-deficient mutants have been useful in identifying the physiologically active form of phytochrome. If the phytochrome-induced response to white light (hypocotyl inhibition) is caused by the absence of Pr, such phytochrome-deficient mutants (which have neither Pr nor Pfr) should have short hypocotyls in both darkness and white light. Instead, the opposite occurs; that is, they have long hypocotyls in both darkness and white light. In other words, the absence of Pfr is what prevents the seedlings from responding to white light. This statement is the same as saying that Pfr brings about the physiological response.

Phytochrome Is a Dimer Composed of Two Polypeptides

Methods for phytochrome purification were developed through research pioneered by the agricultural biochemist H. William Siegelman, working at the USDA lab at Beltsville, Maryland. After surveying many different plants and plant tissues, Siegelman found that the young etiolated seedlings of cereals, such as oat and rye, were particularly rich sources of phytochrome. Some initial success in purifying phytochrome from etiolated oats (*Avena sativa*) was reported in 1964 (Siegelman and Firer 1964), and many workers subsequently used dark-grown oat seedlings for phytochrome extractions. As a result, much of the biochemical information on phytochrome is derived from studies on oat seedlings, although there is a considerable amount of data on phytochrome isolated from dark-grown rye (*Secale* spp.) and pea (*Pisum sativum*) seedlings as well.

While on a fresh-weight basis there is 50 times more phytochrome in etiolated tissue than in equivalent green tissue of the same age, the overall content of phytochrome in etiolated plants is low—about 0.2% of the total extractable protein. Thus, to purify even milligram quantities usually requires more than a kilogram fresh weight of starting material. Etiolated seedlings are also rich in proteases, and much of the phytochrome is susceptible to proteolysis. To minimize the loss of phytochrome, researchers carry out these extractions rapidly, at a low temperature, and in the presence of protease inhibitors. The stability of phytochrome during purification is also favored by the use of green-light illumination, which is poorly absorbed by phytochrome, instead of white light. Despite these precautions, successful purification of fully intact, native phytochrome from etiolated oat seedlings was not reported until 1983 (Vierstra and Quail 1983).

Native phytochrome is a soluble protein with a molecular mass of about 250 kDa. It occurs as a dimer made up of two equivalent subunits. Each subunit consists of two components: a light-absorbing pigment molecule called the **chromophore**, and a polypeptide chain called the **apoprotein**. The apoprotein monomer has a molecular mass of about 125 kDa. Together, the apoprotein and its chromophore make up the **holoprotein**. The chromophore of phytochrome is a linear tetrapyrrole termed **phytochromobilin**. There is only one chromophore per monomer of apoprotein, and it is attached to the protein through a thioether linkage to a cysteine residue (Figure 17.4).

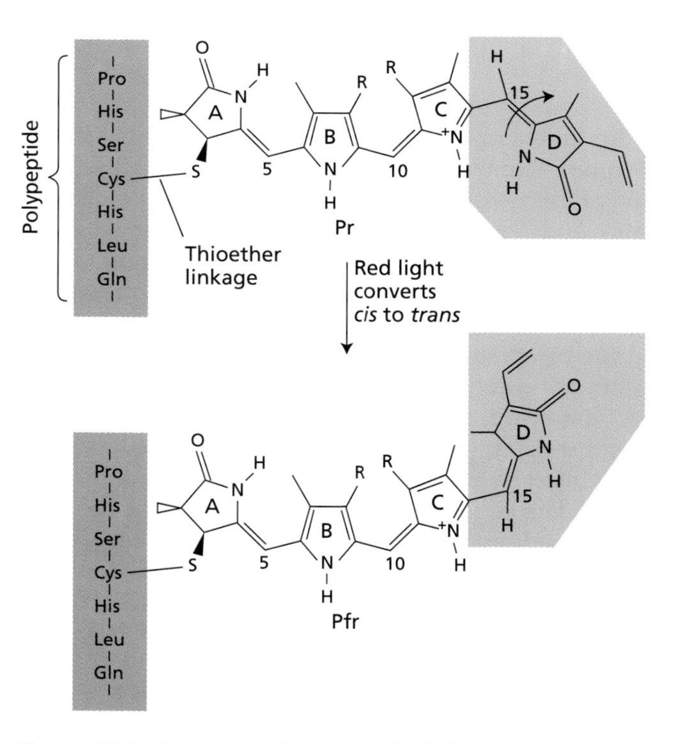

Figure 17.4 Structure of the Pr and Pfr forms of the chromophore (phytochromobilin) and the peptide region bound to the chromophore through a thioether linkage. The chromophore undergoes a *cis-trans* isomerization at carbon 15 in response to red and far-red light. (After Andel et al. 1997.)

(A)

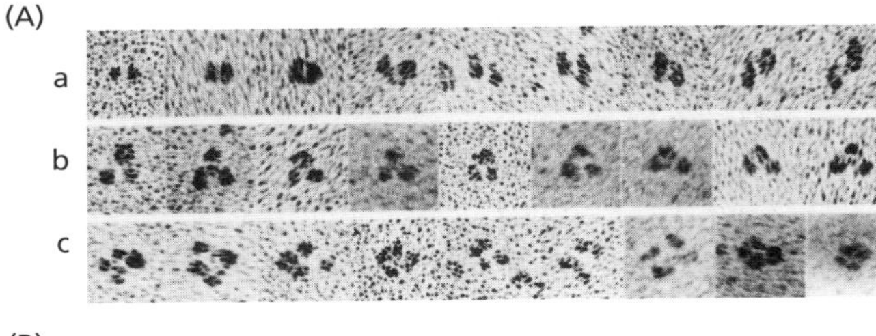

(B)

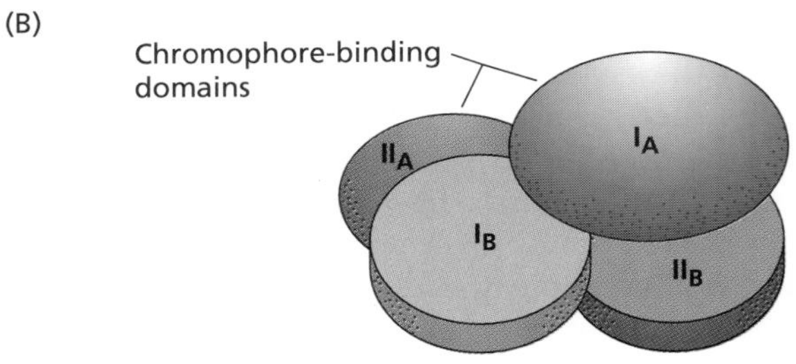

Figure 17.5 Visualization of phytochrome. (A) Phytochrome from pea, purified and then sprayed onto a mica flake. After being dried, the specimen was shadowed with platinum to enhance the three-dimensional structure. A carbon replica was prepared and viewed in the electron microscope. Various images were produced, including pairs of spots (a), three spots (b), and clusters of spots (c). The variation in the pattern in A is caused by different orientations of the protein. (B) Structure of the phytochrome dimer, based on a type of X-ray scattering that does not require crystallization. The monomers are labeled I and II. Each monomer consists of a chromophore-binding domain (A) and a smaller non-chromophore domain (B). The molecule as a whole has an ellipsoidal rather than globular shape. (After Tokutomi et al. 1989, courtesy of M. Wada.)

Researchers have visualized the Pr form of phytochrome using electron microscopy by spraying a solution of the protein onto mica flakes, drying it under vacuum, and "shadowing" the dried molecules by using platinum atoms to enhance its three-dimensional structure. As Figure 17.5A shows, phytochrome molecules appear as a collection of spots, the number depending on the orientation of the protein. On the basis of this information and more accurate measurements using X-ray scattering, the model shown in Figure 17.5B has been proposed (Nakasako et al. 1990). The polypeptide folds into two major domains separated by a "hinge" region. The larger N-terminal domain is approximately 70 kDa and bears the chromophore; the smaller C-terminal domain is approximately 55 kDa and contains the site where the two monomers associate with each other to form the dimer.

Phytochromobilin Is Synthesized in Plastids

The phytochrome apoprotein alone cannot absorb red or far-red light. Light can be absorbed only when the polypeptide is covalently linked with phytochromobilin to form the holoprotein. Phytochromobilin is synthesized inside plastids and is derived from 5-aminolevulinic acid via a pathway that branches from the chlorophyll biosynthetic pathway (see Chapter 8). It is thought to leak out of the plastid into the cytosol by a passive process (Figure 17.6).

Assembly of the phytochrome apoprotein with its chromophore is **autocatalytic**; that is, it occurs spontaneously when purified phytochrome polypeptide is mixed with purified chromophore in the test tube, with no additional proteins or cofactors. It is now possible to synthesize large quantities of native phytochrome apoprotein by expressing the gene in *Saccharomyces* yeast cells, which lack phytochromobilin, and to use the purified apoprotein in assembly studies with the purified chromophore (Li and Lagarias 1992). The resultant holoprotein assembled in vitro has spectral properties similar to those observed for the protein purified from plants, and it exhibits red/far-red reversibility. As might be expected, if an altered form of phytochrome polypeptide that lacks the cysteine required for chromophore attachment is used, the holoprotein does not assemble and no red or far-red light is absorbed. Assembly in vivo is also thought to be autocatalytic.

Mutant plants that lack the ability to synthesize the chromophore are defective in processes that require the action of photoreversible phytochrome, even though the apoprotein polypeptides are present. For example, several of the *hy* mutants noted earlier, in which white light fails to suppress hypocotyl elongation, have defects in chromophore biosynthesis. In *hy1*, *hy2*, and *hy6* mutant plants, phytochrome apoprotein levels are normal, but there is little or no spectrally active holoprotein. When a chromophore precursor is supplied to these seedlings, normal growth is restored. The same type of

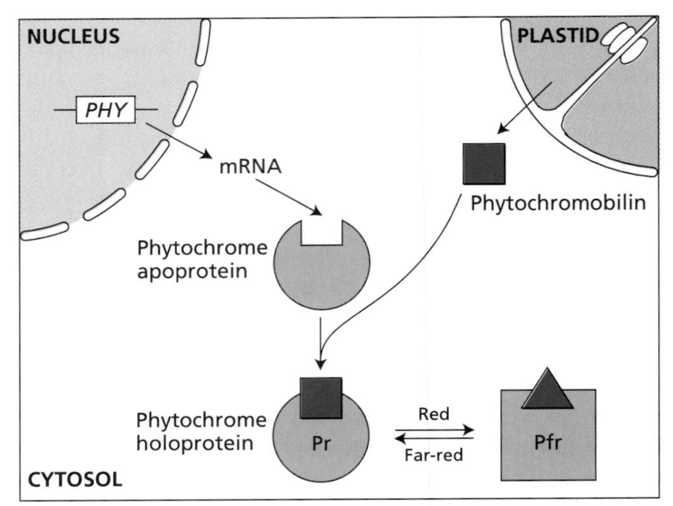

Figure 17.6 Phytochromobilin is synthesized in plastids and released into the cytosol, where it assembles with the phytochrome apoprotein. (After Kendrick et al. 1997.)

mutation has been observed in other species. For example, the *yellow-green* mutant of tomato has properties similar to those of *hy* mutants, suggesting that it is also a chromophore mutant.

Both the Chromophore and the Protein Undergo Conformational Changes

The photoconversion of Pr to Pfr involves subtle conformational changes in both the chromophore and the polypeptide components of the molecule. Since the chromophore is what absorbs the light, conformational changes in the protein are initiated by changes in the chromophore. Upon absorption of light, the Pr chromophore undergoes a *cis–trans* isomerization by rotation around the double bond between carbons 15 and 16 (see Figure 17.4) (Rüdiger and Thümmler 1994; Andel et al. 1997). This change results in a more extended conformation of the tetrapyrrole, consistent with the observation that the chromophore is more accessible to chemical probes when phytochrome is in the Pfr form (see Figure 17.4).

During the conversion of Pr to Pfr, the protein moiety of phytochrome (the apoprotein) also undergoes a subtle conformational change. This change has been inferred from biochemical studies showing that Pr and Pfr differ in their susceptibilities to proteases and in their phosphorylation by exogenous protein kinases. The Pr and Pfr forms of phytochrome can also be distinguished immunologically. However, some of the differential protein reactivity of Pr and Pfr may be due to a rearrangement of the chromophore relative to the protein rather than to a conformational change in the protein itself.

Several lines of evidence suggest that the predominant light-induced change in the conformation of the polypeptide occurs in the N terminus. The N terminus of phytochrome is more exposed in the Pr form than in the Pfr form. Predictions of secondary structure that are based on the amino acid sequence suggest that the N terminus would assume β-turn and random coil conformations. Circular dichroism measurements, a spectral assay that provides information about the secondary structures of proteins, have shown that the α helicity of phytochrome increases by just a few percentage points during the conversion of Pr to Pfr, and that the effect is photoreversible (Furuya and Song 1994). The slight shift from random coil to α helix occurs in the N terminus, which is the light-absorbing domain of the molecule.

Two Types of Phytochrome Have Been Identified

Phytochrome is most abundant in etiolated seedlings; thus most biochemical studies have been carried out on phytochrome purified from nongreen tissues. Very little phytochrome is extractable from green tissues, and a portion of the phytochrome that can be extracted differs in molecular mass from the abundant form of phytochrome found in etiolated plants. In the last decade, research has shown that there are two different types of phytochrome, with distinct properties. These have been termed Type I and Type II phytochrome (Konomi et al. 1987; Furuya 1993). Type I is about nine times more abundant than Type II in dark-grown pea seedlings; in light-grown pea seedlings the amounts of the two types are about equal. The two types seem to be distinct proteins, because they react differently with monoclonal antibodies and differ in their susceptibility to proteolysis.

The existence of two types of phytochrome in green plants had also been suggested by physiological experiments. The successful cloning of genes that encode phytochrome polypeptides has clarified the nature of the proteins present in etiolated and green seedlings: Even in etiolated seedlings, phytochrome is a mixture of related proteins encoded by different genes.

Phytochrome Is Encoded by a Multigene Family

The cloning of phytochrome genes made it possible to carry out a detailed comparison of the amino acid sequences of the related proteins. It also allowed the study of their expression patterns, at both the mRNA and protein levels. In addition, analyses of mutant forms of individual phytochrome genes has helped elucidate the functions of the different types of phytochrome.

The first phytochrome sequences cloned were cDNA (complementary DNA) copies of mRNAs isolated from etiolated oat seedlings (Hershey et al. 1985). The deduced amino acid composition indicated that phytochrome is a soluble protein, consistent with the purification studies. A similar cDNA clone that encodes phytochrome was isolated from the dicot zucchini (*Cucurbita pepo*). Using the zucchini clone as a probe, researchers identified five structurally related phytochrome genes in *Arabidopsis* (Sharrock and Quail 1989). This **gene family** is named *PHY*, and its five individual members are *PHYA*, *PHYB*, *PHYC*, *PHYD*, and *PHYE*. The apoprotein by itself (without the chromophore) is called PHY; the holoprotein (with the chromophore) is called phy. By convention, phytochrome sequences from other higher plants are named according to their homology with the *Arabidopsis PHY* genes (Quail et al. 1994).

Some of the *hy* mutants have turned out to be selectively deficient in specific phytochromes. For example, *hy8* is deficient in phyA, and *hy3* is deficient in phyB. Such mutants have been useful in determining the physiological functions of the different phytochromes.

PHY Genes Encode Type I and Type II Phytochromes

On the basis of their expression patterns, the products of members of the *PHY* gene family can be classified as either Type I or Type II phytochromes. *PHYA* is the only gene that encodes a Type I phytochrome. This conclusion is based on the expression pattern of the *PHYA* promoter, as well as on the accumulation of the mRNA and the polypeptide in response to light. Additional studies of plants that contain mutated forms of the *PHYA* gene (termed *phyA* alleles) have confirmed this conclusion, and have given some clues as to the role of this phytochrome in whole plants.

The gene is transcriptionally active in dark-grown monocots, like oat, but its expression is strongly inhibited in the light. In dark-grown oat, treatment with red light reduces phytochrome synthesis. This effect is produced because the Pfr form of phytochrome inhibits the expression of its own gene (*PHYA*) in oat. In addition, *PHYA* mRNA is unstable, so once etiolated oat seedlings are transferred to the light, *PHYA* mRNA rapidly disappears. However, the inhibitory effect of light on *PHYA* transcription is less dramatic in dicots, and in *Arabidopsis* red light has no measurable effect.

The amount of phyA in the cell is also regulated by proteolysis. The Pfr form of the protein encoded by the *PHYA* gene, called **PfrA**, is unstable. There is evidence that PfrA may become marked or tagged for proteolytic destruction by the ubiquitin system (Vierstra 1994). As discussed in Chapter 14, ubiquitin is a small polypeptide that binds covalently to proteins and serves as a recognition site for a large proteolytic complex, the proteasome. Therefore, oats and other monocots rapidly lose most of their Type I phytochrome (phyA) in the light as a result of a combination of factors: inhibition of transcription, mRNA degradation, and proteolysis. In dicots, phyA levels also decline in the light as a result of proteolysis, but not as dramatically. Figure 17.7 summarizes the various processes that can regulate the steady-state levels of PfrA.

The remaining *PHY* genes (*PHYB* through *PHYE*) encode the Type II phytochromes. Although detected in green plants, these phytochromes are also present in etiolated plants. The reason is that the expression of their mRNAs is not significantly changed by light, and the encoded phyB through phyE proteins are more stable in the Pfr form than phyA is.

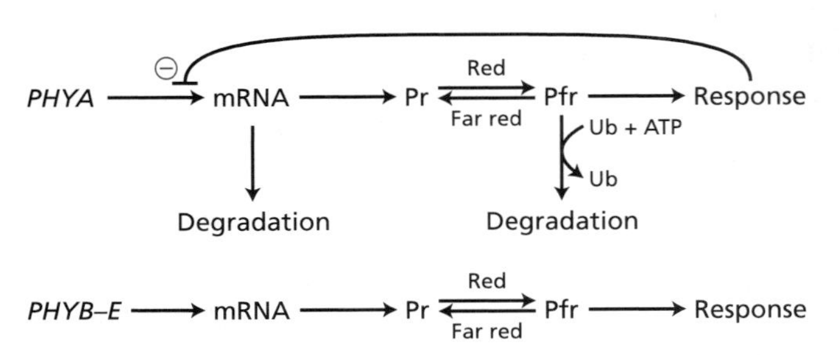

Figure 17.7 Regulation of the expression of phytochrome genes. Phytochrome A (Type I) is synthesized in the Pr form. The level of phyA is regulated by three factors: mRNA degradation, proteolysis of Pfr via the ubiquitin (Ub) pathway, and Pfr-induced repression of gene transcription. Phytochromes B through E (phyB–phyE) are synthesized at much lower rates, but the mRNAs and proteins are more stable, and transcription is not inhibited by Pfr. (After Clough and Vierstra 1997.)

Localization of Phytochrome in Tissues and Cells

Valuable insights into the function of a protein can be gained from determining where it is located. It is not surprising, therefore, that much effort has been devoted to the localization of phytochrome at the organ and tissue levels and within individual cells. Our current understanding of phytochrome localization is based on spectrophotometric analyses and on techniques involving various types of molecular probes. In this section we will discuss the locations of phytochrome at the organ, tissue, cellular, and subcellular levels.

Phytochrome Can Be Detected in Tissues Spectrophotometrically

The photoreversible spectral changes that phytochrome undergoes after irradiations with red and far-red light uniquely identify the presence of phytochrome and permit its quantitation. However, the presence of chlorophyll makes it very difficult to detect phytochrome spectrally in green tissue, although, as discussed earlier, phytochrome can easily be detected and measured by in vivo spectroscopy of etiolated plants. Using this assay method, scientists have confirmed the presence of phytochrome in many different tissues of monocots and dicots and in gymnosperms, ferns, mosses, and algae.

In etiolated seedlings, the highest phytochrome levels are usually found in meristematic regions or in regions that were recently meristematic, such as the bud and first node of pea and the tip and node regions of the coleoptile in oat (Figure 17.8). However, differences in expression patterns between monocots and dicots and between Type I and Type II phytochromes are apparent when other, more sensitive methods are used.

Antibodies and Reporter Genes Localize Phytochrome in Tissues and Cells

Physical and chemical fixation methods are available that can immobilize phytochrome in cells while preserving its reactivity with antibodies. There are also many techniques for visualizing in which antibodies have bound to phytochrome in fixed tissue. Such immunocytochemical methods have been used to localize phytochrome in plant cells and tissues.

The early studies revealed a tissue-level distribution similar to that shown by spectrophotometric methods, but immunocytochemistry added information on the location of phytochrome in roots (it is concentrated in

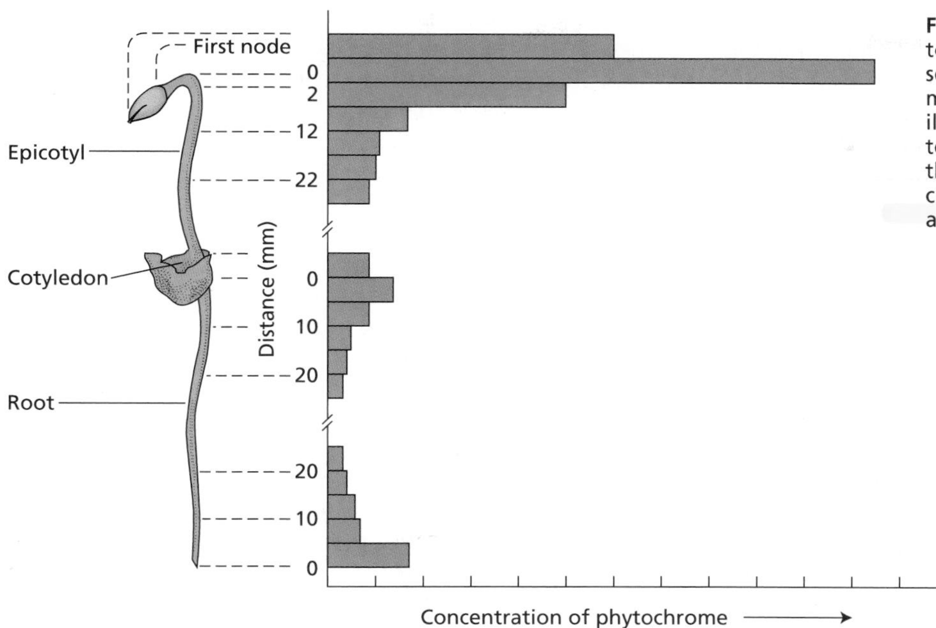

Figure 17.8 Distribution of phytochrome in an etiolated pea seedling, as measured spectrophotometrically. Phytochrome is most heavily concentrated in the apical meristems of the epicotyl and root, where the most dramatic developmental changes are occurring. (From Kendrick and Frankland 1983.)

the cap region). Moreover, immunocytochemical methods showed for the first time what spectrophotometric methods could not resolve: the subcellular distribution. Phytochrome was found both in the cytosol and in association with organelles, including plastids and nuclei. However, these studies measured the total phytochrome pool and did not distinguish between phytochrome Types I and II. In etiolated tissues, such studies probably detected mostly phyA.

Early immunocytochemical studies also revealed that the photoactivation of phytochrome induces a dramatic change in its subcellular distribution. When phytochrome is converted to Pfr, its distribution in coleoptile parenchyma cells of oat and rice seedlings changes within minutes from mainly diffuse throughout the cytosol to mainly sequestered in discrete regions. When Pfr is photoconverted back to Pr by far-red light, phytochrome slowly returns to the original diffuse distribution. In pea seedlings, phytochrome seems to become inaccessible to antibodies immediately after the red-light treatment, and the sequestered distribution appears only after a subsequent 10-minute incubation in the dark (Saunders et al. 1983). Ultrastructural studies have revealed that most of the sequestered Pfr in oat cells is associated with discrete structures that are globular to oval in shape, are about 300 nm in size, and have no morphologically identifiable membranes surrounding them. The function or significance of these structures is not yet known, though according to one hypothesis they represent intracellular sites where the Pfr-dependent dark destruction of phytochrome (phyA) occurs.

The cloning of individual *PHY* genes has allowed the development of methods to detect the transcription of phytochrome Types I and II in specific tissues. The sequences can be used directly to probe mRNAs isolated from different tissues or to analyze transcriptional activity by means of a reporter gene, which visually reveals sites of gene expression. In the latter approach, the promoter of a *PHYA* or *PHYB* gene is joined to the coding portion of a reporter gene, such as the gene for the enzyme β-glucuronidase, or *GUS* (recall that the promoter is the sequence upstream of the gene that is required for transcription).

The advantage of using the *GUS* sequence is that it encodes an enzyme that, even in very small amounts, converts a colorless substrate to a colored precipitate when the substrate is supplied to the plant. Thus, cells in which the *PHYA* promoter is active will be stained blue while other cells will be colorless. This hybrid gene, or fused gene, is then placed back into the plant using the Ti plasmid of *Agrobacterium tumefaciens* as a vector (see Box 21.1).

When this method was used to examine the transcription of two different *PHYA* genes in tobacco, dark-grown seedlings were found to contain the highest amount of stain in the apical hook and the root tips, in keeping with earlier immunological studies (Adam et al. 1994). The pattern of staining in light-grown seedlings was similar but, as might be expected, was of much lower intensity. Similar studies with *Arabidopsis PHYA–GUS* and *PHYB–GUS* fusions placed back in *Arabidopsis* confirmed the *PHYA* results for tobacco, and indicated that *PHYB–GUS* is expressed at much lower levels than *PHYA–GUS* in all tissues, as would be expected for a Type II phytochrome (Somers and Quail 1995).

A recent study comparing the expression patterns of *PHYB–GUS*, *PHYD–GUS*, and *PHYE–GUS* fusions in *Arabidopsis* has revealed that, while these Type II promoters are less active than the Type I promoters, they do show distinct expression patterns (Goosey et al. 1997). Thus the general picture emerging from these studies is that the phytochromes are expressed in distinct but overlapping patterns.

In summary, phytochromes are most abundant in young, undifferentiated tissues, in the cells where the mRNAs are most abundant and the promoters are most active. The strong correlation between phytochrome abundance and cells that have the potential for dynamic developmental changes is consistent with the important role of phytochromes in controlling such developmental changes. Since the expression patterns of individual phytochromes overlap, they may function cooperatively, although they may also use distinct signal transduction pathways. Support for this idea also comes from the study of phytochrome mutants, which we will discuss later in this chapter.

Phytochrome-Induced Whole-Plant Responses

The variety of different phytochrome responses in intact plants is great, in terms of both the kinds of responses (see Table 17.1) and the quantity of light needed to induce the responses. A survey of this variety will show how diversely the effects of a single photoevent—the absorption of light by Pr—are manifested throughout the plant.

Phytochrome-induced responses may be logically grouped, for ease of discussion, into two types: rapid biochemical events and slower morphological changes, including movements and growth. Some of the early biochemical reactions affect later developmental responses. The nature of these early biochemical events, which comprise signal transduction pathways, will be treated in detail later in the chapter. Here we will focus on the effects of phytochrome on whole-plant responses. As we will see, such responses can be classified into various types depending on the amount and duration of light required, and on their action spectra. Two important ecological roles of phytochrome are shade avoidance and circadian rhythms. We will learn how plants sense and respond to shading by other plants, and how phytochrome is involved in regulating various daily rhythms. Finally, we will examine the specialized functions of the different phytochrome gene family members in all of these processes.

Phytochrome Responses Vary in Lag Time and Escape Time

Morphological responses to the photoactivation of phytochrome may be observed visually after a lag time as brief as a few minutes (for example, the initiation of chloroplast rotation in the filamentous green alga *Mougeotia*; see Box 17.1) to as long as a few weeks (as in floral induction in many photoperiodic species; see Chapter 24). The more rapid of these responses are usually reversible movements of organelles or reversible dimensional changes (swelling, shrinking) in cells, but even some growth responses occur remarkably fast. Red-light inhibition of the stem elongation rate of light-grown *Chenopodium album* (pigweed) is observed within 8 minutes after its relative level of Pfr is increased (Morgan and Smith 1978). Information about the lag time for a phytochrome response helps scientists evaluate the kinds of biochemical events that could precede and cause the induction of that response. The shorter the lag time, the more limited the range of biochemical events that could have been involved.

Variety in phytochrome responses can also be seen in the phenomenon called **escape from photoreversibility**. Red light–induced events are reversible by far-red light for only a limited period of time, after which the response is said to have "escaped" from reversal control by light. A currently favored model to explain this phenomenon assumes that phytochrome-controlled morphological responses are the result of a step-by-step sequence of linked biochemical reactions in the responding cells. Each of these sequences has a point of no return beyond which it proceeds irrevocably to the response. The escape time for different responses ranges from less than a minute to, remarkably, hours.

Phytochrome Responses Can Be Distinguished by the Amount of Light Required

In addition to being distinguished by lag times and escape times, phytochrome responses can be distinguished by the amount of light required to induce them. The amount of light is referred to as the **fluence**,* which is defined as the number of photons absorbed per unit surface area. The most commonly used units for fluence are moles of quanta per square meter (mol m^{-2}) over a particular time period. In addition to the fluence, some phytochrome responses are sensitive to the **irradiance**† of light. The units of irradiance in terms of photons are

* For definitions of "fluence," "irradiance," and other terms involved in light measurement, see Box 9.1.

† Irradiance is sometimes loosely equated with light intensity. The term "intensity," however, refers to light emitted by the source, whereas "irradiance" refers to light that is incident on the object.

BOX 17.1

Mougeotia: A Chloroplast with a Twist

MOUGEOTIA IS A FILAMENTOUS GREEN ALGA whose cells contain a single ribbonlike chloroplast nestled between two large vacuoles and surrounded by a layer of cytoplasm (part A of the figure on the next page). The chloroplast can rotate about its long axis (part B of the figure) and can respond to light by orienting *perpendicularly* to the direction of light. The perpendicular orientation is referred to as "face position." Phytochrome is implicated in the *Mougeotia* chloroplast rotation response because the system is most sensitive to red light and because the red-light effect can be reversed by far-red light. When illuminated with far-red light, the chloroplast rotates so that it is oriented parallel to the direction of the light This parallel orientation is termed the "profile position" (see part A of the figure).

Using microbeams of red and far-red light, Wolfgang Haupt and his coworkers in Germany showed that in *Mougeotia*, phytochrome is localized near the periphery of the cytoplasm in the vicinity of the plasma membrane. When a microbeam of red light was directed at the cell surface under a microscope, the edge of the chloroplast adjacent to the microbeam rotated 90°, even though the chloroplast itself was not illuminated.

The orientation of individual phytochrome molecules within the cell was probed by the use of plane-polarized light. Each phytochrome molecule has a preferred orientation for the absorption of plane-polarized light. The greatest absorption occurs when the plane-polarized light is parallel to the absorbing surface of the pigment. If we think of the pigment as a rod and the polarized light as a narrow slit of light, more light will be absorbed when the light slit is parallel to the long axis of the pigment than in any other orientation (part C of the figure).

Haupt and his coworkers discovered that the ability of plane-polarized red light to cause chloroplast rotation depends on the plane of polarization relative to the long axis of the cell. This phenomenon is depicted in the experiment shown in part D of the figure. Half of a cell was illuminated over the entire surface with polarized red light (R) vibrating *perpendicular* to the cell axis; the other half was illuminated with light vibrating *parallel* to the cell axis. The chloroplast rotated from the profile to the face position only in response to red light vibrating in the perpendicular direction, producing a twist in the chloroplast. On the basis of these experiments, it was concluded that phytochrome molecules have a defined orientation (Haupt 1982).

Two additional experiments demonstrated that the phytochrome orientation changes during the conversion from Pr to Pfr (Kraml 1994). In the first experiment, shown on the left-hand side of part E of the figure, the cytoplasmic layer was irradiated with a microbeam of plane-polarized red light vibrating either *parallel* (upper light treatment) or *perpendicular* (lower light treatment) to the cell's long axis. In apparent contradiction to the results in part D, the chloroplast rotated only when the red light was vibrating *parallel* to the cell's long axis. The apparent contradiction is due to the cylindrical shape of the cell. Thus, the phytochrome molecules near the two edges will be oriented 90° relative to the phytochrome molecules at the center. The phytochrome molecules that absorb red light when the polarized microbeam is aimed at the edge of the cell in the experiments of part E of the figure, will be oriented 90° from the phytochrome molecules that absorb light in the *whole-cell* irradiation experiments of part D.

In the second experiment, shown on the right-hand side of part E of the figure, the cytoplasmic layer on the right side of the cell was irradiated with a microbeam of polarized red light vibrating *parallel* to the cell's long axis. On the basis of the previous experiment, we can predict that red light vibrating parallel to the cell's long axis will induce chloroplast rotation. However, the red-light treatment was immediately followed by a beam of far-red light. One half of the cell (top) was given a second irradiation of far-red light vibrating *perpendicular* to the cell axis; the other half of the cell (bottom) was given a second irradiation of far-red light vibrating *parallel* to the cell axis. As shown in the right-hand side of part E of the figure, far-red light prevented red light–induced chloroplast rotation only when it was vibrating perpendicular to the cell axis. This result was explained by the proposal that phytochrome undergoes a 90° change in orientation upon photoconversion of Pr to Pfr and vice versa (part F of the figure).

Finally, what is the mechanism of chloroplast rotation in *Mougeotia*? The effects already described are best accounted for by a model in which phytochrome is located either on the plasma membrane of *Mougeotia* or on the microtubules that form a cylindrical scaffold directly beneath the membrane. It is difficult to imagine how Pr and Pfr could maintain their respective orientation without being associated either with the membrane or with the cylinder of microtubules. The nucleotide sequence that codes for *Mougeotia* phytochrome, consistent with those of other known phytochromes, suggests that the phytochrome is not an integral membrane protein, because it lacks hydrophobic transmembrane domains. On the other hand, the amino acid sequence at the carboxy-terminal end is similar to that of proteins that bind microtubules (Winands and Wagner 1996). So the evidence suggests that phytochrome is bound to the cytoskeleton adjacent to the plasma membrane.

During rotation, the chloroplast moves away from a localized region of Pfr and toward a region of Pr (see part B of the figure). *Mougeotia* chloroplasts move in response to unilateral light as a result of gradients of the active form of phytochrome, Pfr, that form an alternating pigment pattern around the periphery of the cell cylinder, as part B of the figure shows. The gradients of Pfr are rapidly translated into gradients of actin–myosin microfilament interactions, which generate the motive force for chloroplast rotation (Wagner 1996). The primary mechanism of Pfr action remains obscure. Calcium ions from internal stores and calcium-activated calmodulin are known to participate in the response by interacting with the scaffold of microtubules, but the precise relationship between phytochrome and calcium is not yet undertstood.

BOX 17.1 *(continued)*

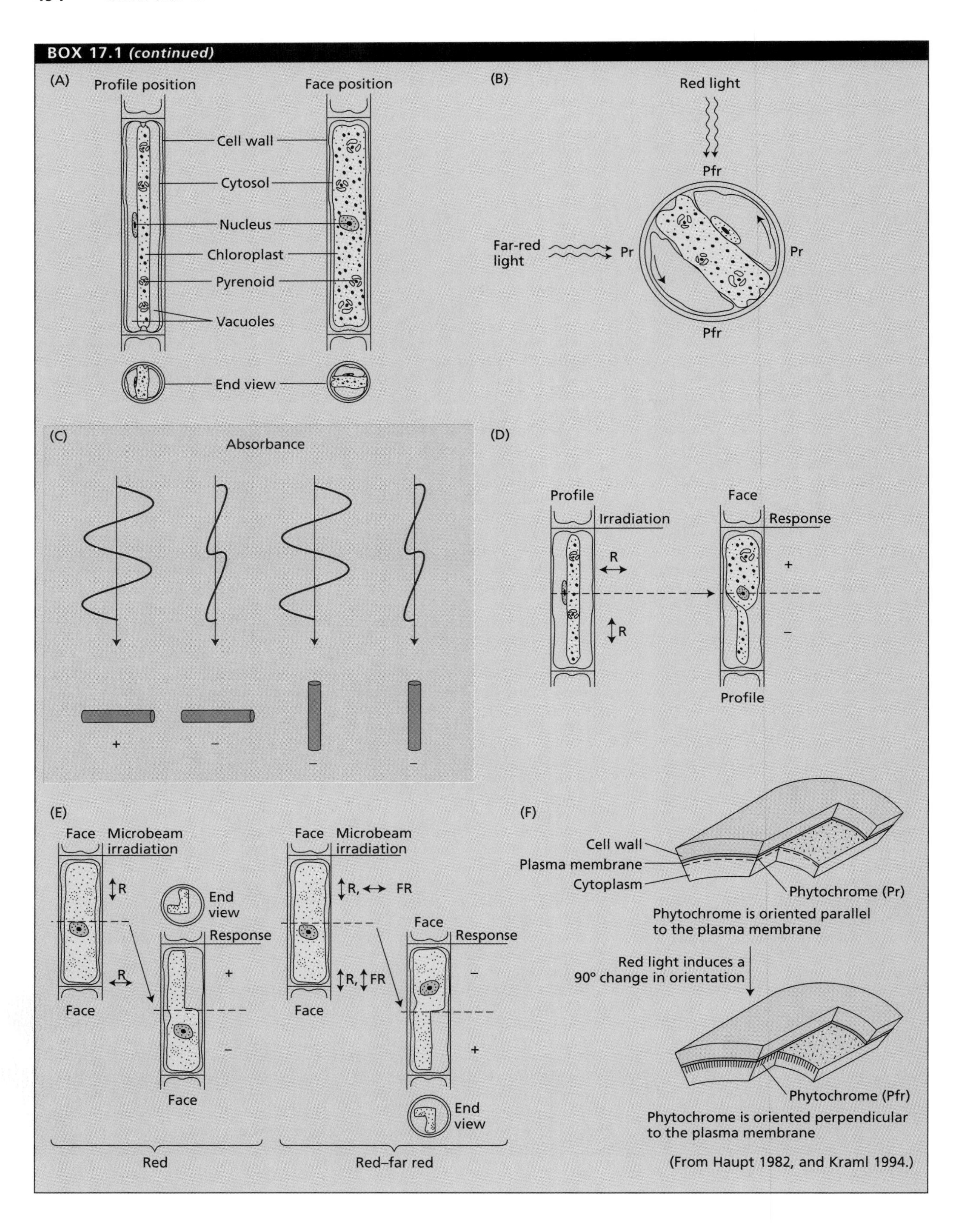

(From Haupt 1982, and Kraml 1994.)

moles of quanta per square meter per second (mol m^{-2} s^{-1}), which is also referred to as *fluence rate*.

Each phytochrome response has a characteristic range of light fluences over which the magnitude of the response is proportional to the fluence. As Figure 17.9 shows, these responses fall into three major categories based on the amount of light required: very low fluence response (VLFR), low fluence response (LFR), and high irradiance response (HIR).

Very Low Fluence Responses Are Nonphotoreversible

Some phytochrome responses can be initiated by fluences as low as 0.1 nmol m^{-2} (one-tenth of the amount of light emitted from a firefly in a single flash), and saturate (that is, reach a maximum) at about 50 nmol m^{-2}. For example, in dark-grown oat seedlings, red light can stimulate the growth of the coleoptile and inhibit the growth of the mesocotyl (the elongated axis between the coleoptile and the root) at fluences this low. *Arabidopsis* seeds can be induced to germinate with red light in the range of 1 to 100 nmol m^{-2}. These remarkable effects of vanishingly low levels of illumination are called **very low fluence responses (VLFRs)**.

The minute amount of light needed to induce VLFRs converts less than 0.02% of the total phytochrome to Pfr. Since the far-red light that would normally reverse a red-light effect converts 97% of the Pfr to Pr (as discussed earlier), about 3% of the phytochrome remains as Pfr, significantly more than is needed to induce VLFRs (Mandoli and Briggs 1984). Thus, far-red cannot reverse VLFRs. The VLFR action spectrum exhibits broad peaks in the red and blue regions (Shinomura et al. 1996). The peak in the red suggests that phytochrome is the photoreceptor for VLFRs. The pigment responsible for the blue peak may be either phytochrome or cryptochrome.

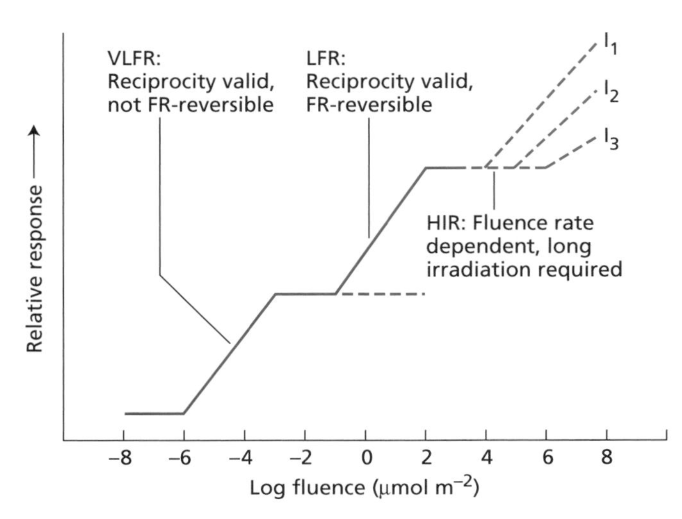

Figure 17.9 The three types of phytochrome responses, based on their sensitivities to fluence. The relative magnitudes of representative responses are plotted against increasing fluences of red light. Short light pulses activate VLFRs and LFRs. Since HIRs are also proportional to the irradiance, the effects of three different irradiances given continuously are illustrated ($I_1 > I_2 > I_3$). (From Briggs et al. 1984.)

Low Fluence Responses Are Photoreversible

Another set of phytochrome responses cannot be initiated until the fluences reach 1.0 μmol m^{-2}, and they are saturated at 1000 μmol m^{-2}. These responses are referred to as **low fluence responses (LFRs)**, and they include most of the classic red/far-red photoreversible responses, such as the promotion of lettuce seed germination and the regulation of leaf movements, that are mentioned in Table 17.1. The LFR action spectrum for *Arabidopsis* seed germination is shown in Figure 17.10. LFR spectra consist of a main peak for stimulation in the red, and a major peak for inhibition in the far red.

Both VLFRs and LFRs can be induced by brief pulses of light, provided that the total amount of light energy adds up to the required fluence. The total fluence is a function of two factors: the fluence rate (mol m^{-2} s–1) and the irradiation time. Thus a brief pulse of red light will induce a response, provided that the light is sufficiently bright, and conversely, very dim light will work if the irradiation time is long enough. This reciprocal relationship between fluence rate and time is known as the **law of reciprocity**, first formulated by Bunsen and Roscoe in 1850. VLFRs and LFRs obey the law of reciprocity.

High Irradiance Responses Are Proportional to the Irradiance

Phytochrome responses of the third type are termed **high irradiance responses (HIRs)**, several of which are listed in Table 17.2. HIRs require prolonged or continuous exposure to light of relatively high irradiance, and the response is proportional to the irradiance within a certain range. The reason that these responses are called high irradiance responses rather than high fluence responses is that they are proportional to irradiance

Table 17.2
Some plant photomorphogenic responses induced by high irradiances

Anthocyanin synthesis in various dicot seedings and in apple skin segments
Inhibition of hypocotyl elongation in mustard, lettuce, and petunia seedlings
Induction of flowering in *Hyoscyamus* (henbane)
Plumular hook opening in lettuce
Enlargement of cotyledons in mustard
Ethylene production in sorghum

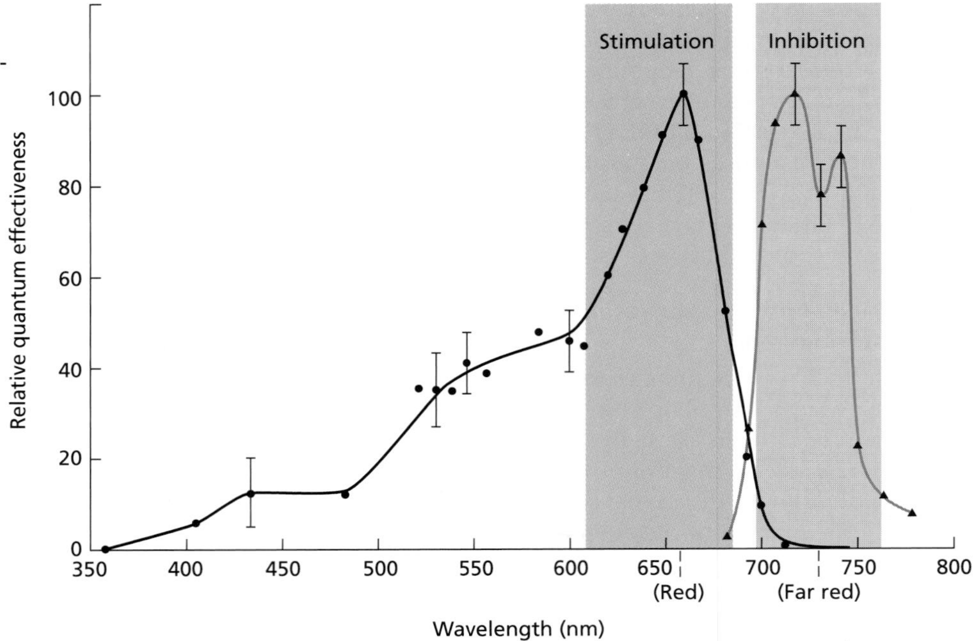

Figure 17.10 LFR action spectra for the photoreversible stimulation and inhibition of seed germination in *Arabidopsis*. (After Shropshire et al. 1961.)

(loosely speaking, the brightness of the source) rather than to fluence. HIRs saturate at much higher fluences than LFRs—at least 100 times higher—and are not photoreversible. Since neither continuous exposure to dim light nor transient exposure to bright light are effective, HIRs do not obey the law of reciprocity.

Many of the photoreversible LFRs listed in Table 17.1, particularly those involved in de-etiolation, also qualify as HIRs. For example, at low fluences, the action spectrum for anthocyanin production in seedlings of white mustard (*Sinapis alba*) shows a single peak in the red region of the spectrum, the effect is reversible with far-red light, and the response obeys the law of reciprocity. However, if the dark-grown seedlings are instead exposed to high-irradiance light for several hours, the action spectrum now includes peaks in the far-red and blue regions (see the next section), the effect is no longer photoreversible, and the response becomes proportional to the irradiance. Thus, the same effect can be either an LFR or an HIR.

The High Irradiance Responses of Etiolated Seedlings Have Peaks in the Far-Red, Blue, and UV-A Regions of the Spectrum

HIRs, such as the inhibition of stem or hypocotyl growth, have usually been studied in dark-grown, etiolated seedlings. The HIR action spectrum for the inhibition of hypocotyl elongation in dark-grown lettuce seedlings is shown in Figure 17.11. Note that the main peak of activity is in the far-red rather than the red region of the spectrum, and there are peaks in the blue and UV-A regions as well. Because the absence of a peak in the red is unusual for a phytochrome-mediated response, at first researchers believed that another pigment might be involved. In the late 1960s, however, Karl M. Hartmann at Freiburg proposed a model to explain the apparent discrepancy. Taking a cue from the enhancement effect of photosynthesis that was discovered by R. Emerson and coworkers (see Chapter 7), Hartmann speculated that the HIR might be the result of an interaction between the Pr and Pfr forms of phytochrome.

To probe such an interaction, Hartmann irradiated plants with not one but two wavelengths of light simultaneously, using two wavelengths that were inactive in the HIR for lettuce hypocotyl inhibition: 658 nm and 768 nm (see Figure 17.11). Although physiologically inactive, the two wavelengths are still *photochemically* active in driving phytochrome conversions, each in the opposite direction. By carefully varying the irradiances of each wavelength, Hartmann was able to induce a range of ratios of Pfr to total phytochrome in the plant from near zero to higher values. He was able to show that the optimum ratio of Pfr to total phytochrome for the inhibition of lettuce hypocotyl growth is approximately 0.03. Remarkably, this is precisely the ratio that results when the plants are irradiated with a single wavelength at 720 nm, the far-red peak of the HIR.

Although a large body of evidence now supports the view that phytochrome is one of the photoreceptors involved in HIRs, it has long been suspected that the peaks in the UV-A and blue regions are due to a separate photoreceptor that absorbs UV-A and blue light. To test this hypothesis, Goto and colleagues (1993) determined the HIR action spectrum for the inhibition of hypocotyl elongation of the dark-grown *hy2* mutant of *Arabidopsis*, which has little or no phytochrome. Whereas the wild-

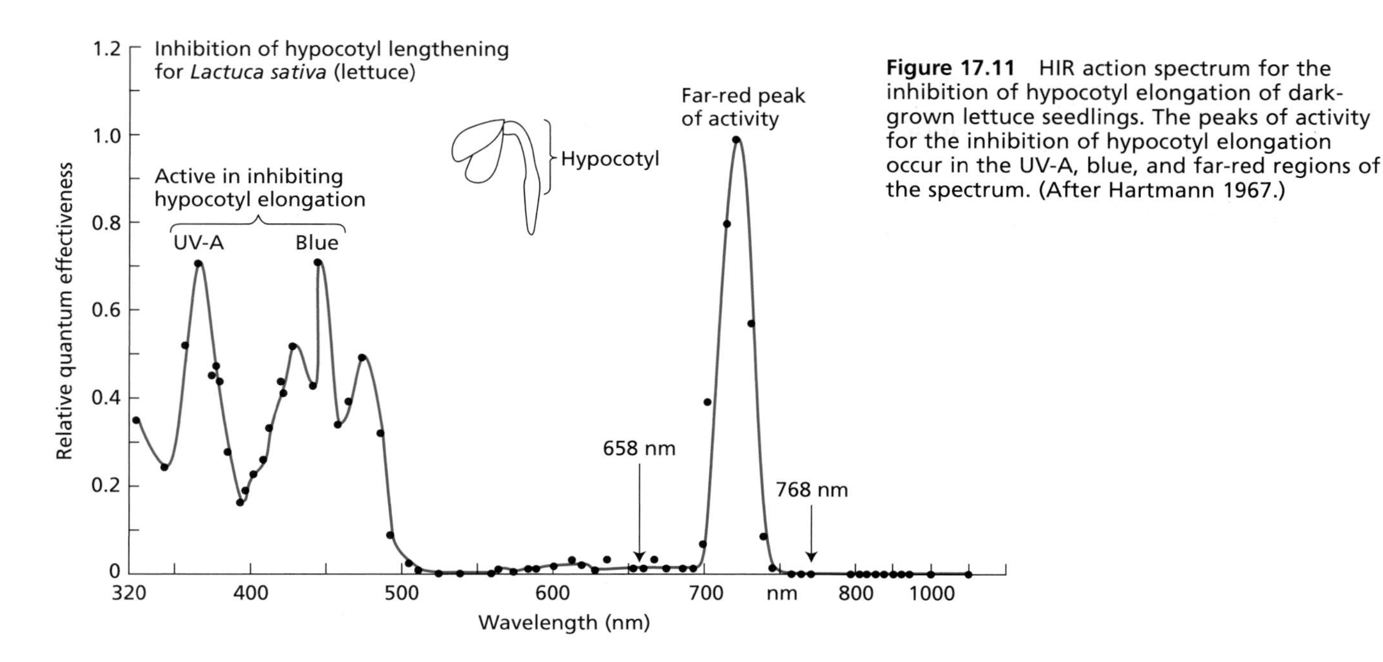

Figure 17.11 HIR action spectrum for the inhibition of hypocotyl elongation of dark-grown lettuce seedlings. The peaks of activity for the inhibition of hypocotyl elongation occur in the UV-A, blue, and far-red regions of the spectrum. (After Hartmann 1967.)

type seedlings exhibit peaks in the UV-A, blue, and far-red regions of the spectrum, the *hy2* mutant fails to respond to either far-red or red light. Although the phytochrome-deficient mutant exhibits no peak in the far red, it shows a normal response to UV-A and blue light. These results demonstrate that phytochrome is not involved in the HIR to UV-A or blue light.

The High Irradiance Responses of Green Plants Exhibit a Major Peak in the Red Region of the Spectrum

During studies of the HIR of etiolated seedlings, it was observed that the response to continuous far-red light declines rapidly as the seedlings begin to green. For example, the action spectrum for the inhibition of hypocotyl growth of light-grown green *Sinapis alba* (white mustard) seedlings is shown in Figure 17.12. Note the absence of peaks in the far-red and blue regions of the spectrum. In general, HIR action spectra for light-grown plants exhibit a single major peak in the red, similar to the action spectra of LFRs (see Figure 17.10), except that the effect is nonphotoreversible.

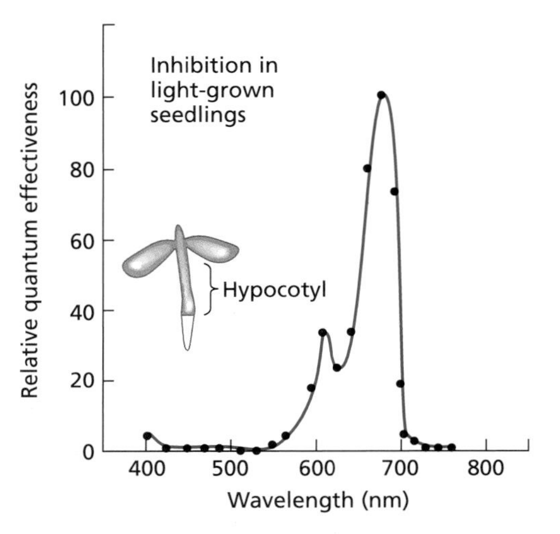

Figure 17.12 HIR action spectra for the inhibition of hypocotyl elongation of light-grown white mustard (*Sinapis alba*) seedlings. (After Beggs et al. 1980.)

The loss of responsiveness to continuous far-red light is strongly correlated with the depletion of the light-labile pool of Type I phytochrome, which we now know consists mostly of phyA. This finding suggests that the HIR of etiolated seedlings to far-red light is mediated by phyA, whereas the HIR of green seedlings to red light is mediated by the Type II phytochrome phyB.

Phytochrome Enables Plants to Adapt to Changes in Light Conditions

The presence of a red/far-red reversible pigment in all green plants, from algae to dicots, suggests that these wavelengths of light provide information that helps plants adjust to their environment. What environmental conditions change the relative levels of these two wavelengths of light in natural radiation?

The ratio of red light (R) to far-red light (FR) varies remarkably in different environments. This ratio can be defined as follows:

$$R/FR = \frac{\text{photon fluence rate in 10-nm band centered on 660 nm}}{\text{photon fluence rate in 10-nm band centered on 730 nm}}$$

Table 17.3
Ecologically important light parameters

	Photon flux density (μmol m^{-2} s^{-1})	R/FR[a]
Daylight	1900	1.19
Sunset	26.5	0.96
Moonlight	0.005	0.94
Ivy canopy	17.7	0.13
Lakes, at a depth of 1 m		
Black Loch	680	17.2
Loch Leven	300	3.1
Loch Borralie	1200	1.2
Soil, at a depth of 5 mm	8.6	0.88

Source: Smith 1982, p. 493.

Note: The light intensity factor (400–800 nm) is given as the photon flux density and phytochrome-active light is given as the R/FR ratio.

[a]Absolute values taken from spectroradiometer scans; the values should be taken to indicate the relationships between the various natural conditions and not as actual environmental means.

(Smith 1982). Table 17.3 summarizes both the total photons (400 to 800 nm) and the R/FR values in eight natural environments. Both parameters vary greatly in different environments. Compared with direct daylight, there is relatively more far-red light in sunset light, under 5 mm of soil, and especially under the canopy of other plants (as on the floor of a forest). The canopy phenomenon results from the fact that green leaves absorb red light because of their high chlorophyll content but are relatively transparent to far-red light.

Phytochrome, then, can serve as an indicator of the degree of shading of a plant by other plants. Plants that increase stem extension in response to shading are said to exhibit a **shade avoidance response**. As shading increases, the R/FR value decreases. The greater proportion of far-red light converts more Pfr to Pr, and the ratio of Pfr to total phytochrome (Pfr/P_{total}) decreases. When simulated natural radiation was used to vary the far-red content, it was found that for so-called sun plants (plants that normally grow in an open-field habitat), the higher the far-red content (i.e., the lower the Pfr/P_{total}), the higher the rate of stem extension (Figure 17.13A) (Morgan and Smith 1979).

In other words, simulated canopy shading (high levels of far-red light) induced these plants to allocate more of their resources to growing taller. This correlation did not hold for "shade plants," which normally grow in a shaded environment. Shade plants showed little or no reduction in their stem extension rate as they were exposed to higher R/FR values (Figure 17.13B). Thus, there appears to be a systematic relationship between phytochrome-controlled growth and species habitat. Such results are taken as an indication of the involvement of phytochrome in shade perception.

For a "shade-avoiding" plant there is a clear adaptive value in allocating its resources toward more rapid extension growth when it is shaded by another plant. In this way it can enhance its chances of growing above the canopy and acquiring a greater share of unfiltered, photosynthetically active light. The price for favoring internode elongation is usually reduced leaf area and reduced branching, but at least in the short run this adaptation to canopy shade seems to work.

Light quality also plays a role in regulating the germination of some seeds. Large-seeded species, with ample food reserves to sustain prolonged seedling growth in darkness (e.g., underground), generally do not require light for germination. However, such a requirement is often observed in the small seeds of herbaceous and grassland species, many of which remain dormant, even while hydrated, if they are buried below the depth to which light penetrates. Even when such seeds are on or near the soil surface, their level of shading from the vegetation canopy (i.e., the R/FR ratio they receive) is likely to affect their germination. For example, it is well-documented that far-red enrichment imparted by a leaf canopy inhibits germination in a range of small-seeded species.

For seeds of the tropical species *Cecropia obtusifolia* (trumpet tree) and *Piper auritum* (Veracruz pepper) planted on the floor of a deeply shaded forest, this inhi-

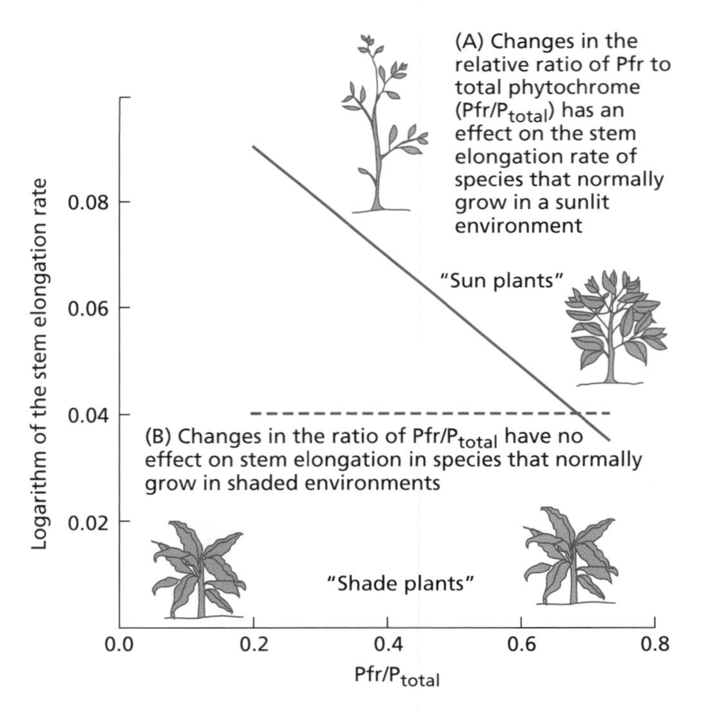

Figure 17.13 The role of phytochrome in shade perception in sun plants (A, solid line) versus shade plants (B, dashed line). (After Morgan and Smith 1979.)

bition can be reversed if a light filter is placed immediately above the seeds that permits the red component of the canopy-shaded light to pass through while blocking the far-red component. Although the canopy transmits very little red light, it is enough to stimulate the seeds to germinate, probably because most of the inhibitory far-red light is excluded by the filter and the R/FR ratio is very high. These seeds would also be more likely to germinate in spaces receiving sunlight through gaps in the canopy than in densely shaded spaces. The sunlight would help ensure that the seedlings became photosynthetically self-sustaining before their seed food reserves were exhausted.

Phytochrome Regulates Certain Daily Rhythms

Various metabolic processes in plants, such as oxygen evolution and respiration, cycle alternately through high-activity and low-activity phases with a regular periodicity of about 24 hours. These rhythmic changes are referred to as **circadian rhythms**, from the Latin *circa diem*, meaning "approximately a day." The **period** of a rhythm is the time that elapses between successive peaks or troughs in the cycle, and because the rhythm persists in the absence of external controlling factors, it is considered to be **endogenous**.

Light is a strong modulator of rhythms in both plants and animals. Although circadian rhythms that persist under controlled laboratory conditions usually have periods one or more hours longer or shorter than 24 hours, in nature their periods tend to be uniformly closer to 24 hours because of the synchronizing effects of light at daybreak, referred to as **entrainment**. Both red and blue light are effective in entrainment. The red-light effect is photoreversible by far-red light, indicative of phytochrome, whereas the blue-light effect is mediated by the blue-light photoreceptor.

The sleep movements of leaves, referred to as nyctinasty, are a well-described example of a plant circadian rhythm that is regulated by light. In **nyctinasty**, leaves and/or leaflets extend horizontally (open) to face the light during the day and fold together vertically (close) at night (Figure 17.14). Nyctinastic leaf movements are exhibited by many legumes, such as *Mimosa*, *Albizia*, and *Samanea*, as well as members of the oxalis family.

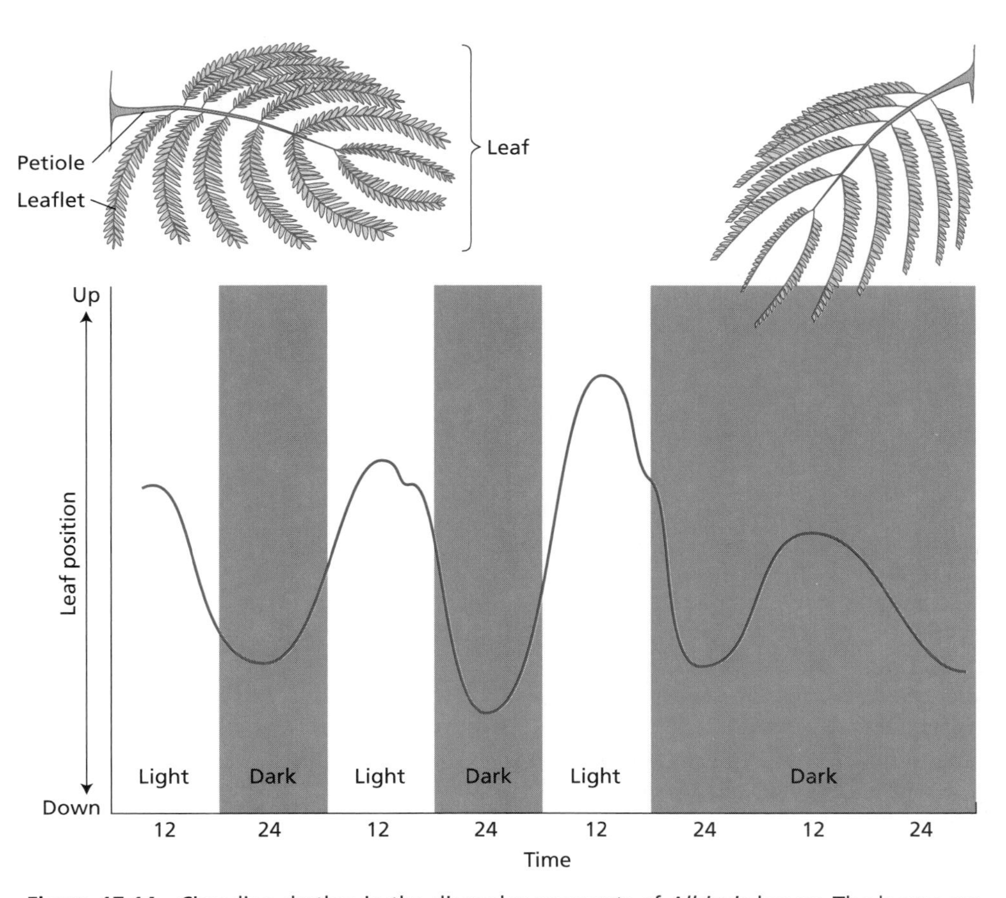

Figure 17.14 Circadian rhythm in the diurnal movements of *Albizzia* leaves. The leaves are elevated in the morning and lowered in the evening. In parallel with the raising and lowering of the leaves, the leaflets open and close. The rhythm persists at a lower amplitude for a limited time in total darkness.

The change in leaf or leaflet angle is caused by rhythmic turgor changes in the cells of the **pulvinus** (plural pulvini), a specialized structure at the base of the petiole.

Once initiated, the rhythm of opening and closing persists even in constant darkness, both in whole plants and in isolated leaflets. The phase of the rhythm (see Chapter 24), however, can be shifted by various exogenous signals, including red or blue light. Light also directly affects movement: Blue light stimulates closed leaflets to open (Satter and Galston 1981), and red light followed by darkness causes open leaflets to close. The leaflets begin to close within 5 minutes after being transferred to darkness, and closure is complete in 30 minutes. Since the effect of red light can be canceled by far-red light, phytochrome regulates leaflet closure.

The basic physiological mechanism of leaf movement is well understood. It results from turgor changes in cells located on opposite sides of the pulvinus, called **extensors** and **flexors**. As with stomatal guard cells (see Chapter 18), these changes in turgor depend on K^+ and Cl^- fluxes across the plasma membranes of the extensor and flexor cells. Leaflets open when the extensor cells accumulate K^+ and Cl^-, causing them to swell, while the flexor cells release K^+ and Cl^-, causing them to shrink. Reversal of this process results in leaflet closure. Leaflet closure is therefore an example of a rapid response to phytochrome involving ion fluxes across membranes.

Gene expression and circadian rhythms. Phytochrome can also interact with circadian rhythms at the level of gene expression. The expression of the *LHCB* gene family, encoding the light-harvesting chlorophyll *a/b*–binding proteins of photosystem II, is regulated at the transcriptional level by both circadian rhythms and phytochrome. In leaves of pea and wheat, the level of *LHCB* mRNA has recently been found to oscillate during daily light–dark cycles, rising in the morning and falling in the evening. Since the rhythm persists even in continuous darkness, it appears to be a circadian rhythm. But phytochrome can perturb this cyclical pattern of expression.

When wheat plants are transferred from a cycle of 12 hours light and 12 hours dark to continuous darkness, the rhythm persists for a while, but it slowly damps out. If, however, the plants are given a pulse of red light before being transferred to continuous darkness, no damping occurs (that is, the levels of *LHCB* mRNA are the same as during the light–dark cycles). In contrast, a far-red flash at the end of the day prevents the expression of *LHCB* in continuous darkness, and the effect of far-red is reversed by red light (Nagy et al. 1988).

How can light and the circadian rhythm both regulate *LHCB* expression? According to one model, the circadian rhythm acts as a negative regulator of gene expression, while phytochrome has two effects: (1) It regulates the phase and period of the circadian rhythm; (2) it acts as a positive regulator of gene expression (Anderson and Kay 1995).

Phytochrome B Determines the Response to Continuous Red or White Light

The first identified *phy* mutation was isolated from the collection of long-hypocotyl *hy* mutants of *Arabidopsis*, which also yielded the chromophore mutants discussed earlier. No phytochrome gene mutations other than *phyB* were found in the original *hy* collection, suggesting that a mutation in any other single phytochrome gene is not sufficient to prevent hypocotyl shortening in response to white light. The mutant *hy3* was found to have an altered *PHYB* gene. In these mutants, *PHYB* mRNA was reduced in amount or was absent, and little or no phyB could be detected. In contrast, the levels of *PHYA* mRNA and phyA protein were normal. Phytochrome B mutants of this type have also been obtained for cucumber and tomato, and as expected, they all share common features. Analysis of these mutants and comparison of their phenotypes with those of normal plants overexpressing *PHYB* have allowed partial elucidation of phyB function.

Phytochrome B is responsible for regulating hypocotyl length in response to red light. Both LFRs and HIRs are included in this category. One aspect of this function is the regulation of shade avoidance. The *hy3* mutant is unable to respond to shading by increasing hypocotyl extension. In addition, these plants do not extend their hypocotyl in response to far-red light given at the end of each photoperiod (called the *end-of-day far-red response*). Both of these responses are likely to involve perception of the Pfr/P_{total} ratio and occur in the low-fluence region of the spectrum. Although phyB is centrally involved in the shade avoidance response, evidence suggests that other phytochromes play important roles as well (Smith and Whitelam 1997).

The most dramatic effect of the *hy3* mutation is on the ability of seedlings to respond to continuous red or white light. In addition to having increased hypocotyl length (absence of inhibition) in constant red or white light, *hy3* is deficient in chlorophyll and in some mRNAs that encode chloroplast proteins, and it is impaired in its ability to respond to plant hormones. Since a mutation in *PHYB* results in impaired perception of continuous red light, the presence of the other phytochromes must not be sufficient to confer responsiveness to continuous red or white light.

Finally, phyB appears to regulate photoreversible seed germination, the phenomenon that originally led to the discovery of phytochrome. Wild-type *Arabidopsis* seeds require light for germination, and the response shows red/far-red reversibility in the low-fluence range. Mutants that lack phyA respond normally to red light, whereas mutants deficient in phyB are unable to respond to low-fluence red light (Shinomura et al. 1996). This experiment provides strong experimental evidence that phyB mediates photoreversible seed germination.

Phytochrome A Is Required for the Response to Continuous Far-Red Light

Since no phytochrome gene mutations other than *phyB* were found in the original *hy* collection, the identification of *phyA* mutants depended on development of an alternative method to select for the lack of phyA function. Since, as discussed previously, the far-red HIRs were known to require light-labile (Type I) phytochrome, it was postulated that phyA is the photoreceptor involved in the perception of continuous far-red light. Mutants were therefore screened for the ability to respond to continuous far-red light. Such mutants could fail to respond because of a deficiency of either phyA or the chromophore.

To distinguish between the two types of mutants (phyA-deficient versus chromophore-deficient), the mutants were then grown under continuous red light. The phyA-deficient mutants can grow normally under this regime, but a chromophore-deficient mutant, which also lacks phyB, does not respond. Seedlings selected in this screen had no obvious phenotype when grown in normal white light, confirming that phyA has no discernible role in sensing white light. Thus, phyA appears to have a limited role in photomorphogenesis, restricted primarily to de-etiolation and far-red responses. For example, phyA would be important when seeds germinate under a canopy, which filters out much of the red light. It is also clear from this constant far-red light phenotype that none of the other phytochromes is sufficient for the perception of constant far-red light, and despite the ability of all phytochromes to absorb red and far-red light, at least phyA and phyB have distinct roles in this regard.

Phytochrome A also appears to be involved in the seed germination VLFR of *Arabidopsis* seeds. Thus, mutants lacking phyA cannot germinate in response to red light in the very-low-fluence range, but show a normal response to red light in the low-fluence range (Shinomura et al. 1996). This result demonstrates that phyA functions as the photoreceptor for this VLFR.

Table 17.4 summarizes the different roles of phyA and phyB in the various phytochrome responses.

Phytochrome Interactions Are Important Early in Germination

Although the isolation of *Arabidopsis* plants having mutations in *PHYA* or *PHYB* genes that give rise to distinct, nonoverlapping phenotypes means that these phytochromes have distinct roles in development, it does not exclude the possibility that these phytochromes also have overlapping, redundant, or even antagonistic functions (Figure 17.15A).

The differential responses of phyA and phyB to constant far-red light and constant red light, respectively, have implications for the initiation of photomorphogenesis in etiolated seedlings upon exposure to white light in the natural environment. White light contains a mixture of wavelengths, including red and far-red light, the ratio of which varies depending on whether the seedling emerges in full sunlight (red light–enriched) or in canopy shade (far-red light–enriched) (see Table 17.3).

The absorption of continuous red light by phyB inhibits hypocotyl elongation, while absorption of continuous far-red light by phyB partly suppresses this response, acting in a way that is analogous to its function in shade avoidance. Conversely, the absorption of continuous far-red light by phyA also induces de-etiolation, while the absorption of continuous red light by phyA suppresses this response (Figure 17.15B). Thus, early in de-etiolation, photomorphogenesis depends on the relative amounts of phyA and phyB, as well as on the red and far-red light available. However, the instability of phyA and the stability of phyB in the Pfr form means that phyB action becomes dominant with time.

As Figure 17.15B shows, seedlings that emerge in open sunlight (enriched in red light) are depleted in phyA, and de-etiolation thus is regulated primarily by phyB. Seedlings emerging under vegetational shade receive a higher proportion of far-red light, and de-etiolation is thus regulated at first by phyA. With time, however, phyA is depleted, and phyB takes over. In fully

Table 17.4
Comparison of the VLFR, LFR, and HIR responses

Type of Response	Photoreversibility	Reciprocity	Peaks of action spectra[a]	Photoreceptor
VLFR	No	Yes	R, B	phyA
LFR	Yes	Yes	R, FR	phyB
HIR	No	No	Dark-grown: FR, B, UV-A	Dark-grown: phyA, cryptochrome
			Light-grown: R	Light-grown: phyB

[a] B, blue; FR, far red; R, red.

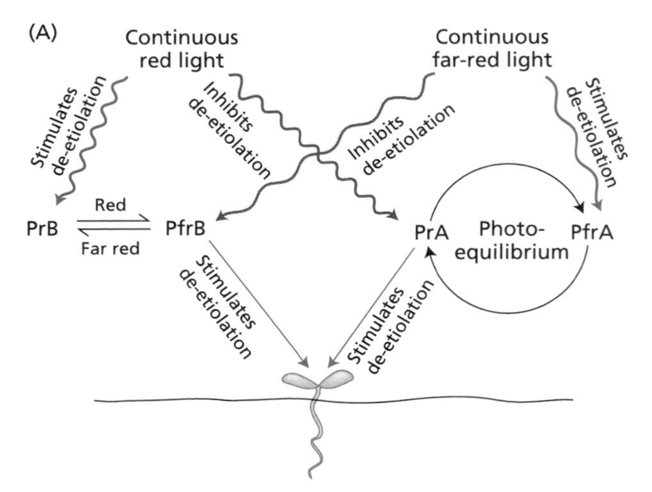

Figure 17.15 Mutually antagonistic roles of phyA and phyB. (A) Diagram showing the action of constant red and far-red light absorbed separately by the phyA and phyB systems. Continuous red light absorbed by phyB stimulates de-etiolation by maintaining high levels of PfrB. Continuous far-red light absorbed by PfrB prevents this stimulation by reducing the amount of PfrB. The stimulation of de-etiolation by phyA depends on the photostationary state of phytochrome (indicated by the circular arrows). Continuous far-red light stimulates de-etiolation when absorbed by the phyA system; continuous red light inhibits the response. (B) The effects of phyA and phyB on seedling development in sunlight versus canopy shade (enriched in far-red light). In open sunlight, which is enriched in red light compared with canopy shade, de-etiolation is mediated primarily by the phyB system (left). A seedling emerging under canopy shade, enriched in far-red light, initiates de-etiolation primarily through the phyA system (center). However, because phyA is labile, the response is taken over by phyB (right). Note that switching over to phyB releases the stem from growth inhibition (see part A), allowing for the accelerated rate of stem elongation that is part of the shade avoidance response. (After Quail et al. 1995.)

green plants, the dominance of phyB leads to a shade avoidance response by relieving stem growth inhibition in constant far-red light, which converts PfrB back to PrB (see Figure 17.15A). In canopy shade, phyA is therefore critical for early seedling development, while phyB is more important in the mature green plant.

Phytochrome Structure and Function

Because phytochrome was first believed to be a single protein, it was difficult to imagine how it could regulate such a diverse set of processes in the cell. The discovery that phytochrome is encoded by a multigene family, and the elucidation of the different expression patterns of the *PHY* genes, provided a plausible explanation: Each phytochrome-mediated response is regulated by a specific phytochrome, or by an interaction between specific phytochromes. As discussed earlier, this hypothesis was supported by the phenotypes of mutants deficient in either phyA or phyB.

As a corollary to this hypothesis, it was further postulated that specific regions of the PHY proteins must be specialized to allow them to perform their distinct functions. However, it was not possible to test these two hypotheses by conventional biochemical approaches. The strength of molecular biology is that it provides the tools to answer such difficult questions. In this section we will describe what is known about the functional domains of the phytochrome holoprotein.

Overexpression of Phytochrome Causes a Dwarfed, Dark Green Phenotype

Just as mutations *reducing* the amount of a particular phytochrome have yielded information about its role, plants genetically engineered to *overexpress* a specific phytochrome are also useful. First, they allow an extension of the range of phytochrome levels testable in regard to function. Second, as we will see, a particular phytochrome sequence can be changed and reintroduced into a normal plant to test its phenotypic effects.

Usually, plants overexpressing an introduced *PHYA* or *PHYB* gene have a dramatically altered phenotype. The plants are often dwarfed, are dark green because of elevated chlorophyll levels, and show reduced apical dominance. This phenotype requires elevated levels of an intact, photoactive holoprotein, because overexpression of a mutated form of phytochrome that is unable to combine with its chromophore has a normal phenotype. Similarly, plants expressing only the N-terminal domain of each phytochrome have a normal phenotype, even though elevated levels of photoactive fragment accumulate.

Together, these studies support the idea that, while the light-detecting part of each phytochrome resides in the N-terminal end with the chromophore, the C-terminal end is required to transmit the signal detected by the light. However, overexpression of phyA activates responses not normally under its control, since these plants are dwarf while *phyA* mutants are normal in white light. Thus, although these studies are useful in dissecting phytochrome structure in relationship to function, the effects of perturbing the composition of the total phytochrome protein pool or expressing phytochrome in cells where it is not normally present cannot be ignored.

The Functional Domains of Phytochrome Have Been Mapped

Plants that overexpress altered phytochromes have proved useful in helping reveal the function of specific portions of the phyA sequence. For example, such experiments can be used to determine which parts of phyA and phyB determine their action in relation to continuous red or far-red light. Researchers have addressed this question by making synthetic versions of phytochrome polypeptides comprising the N-terminal half of oat phyA fused to the C-terminal half of rice phyB (**phyAB**), as well as the reverse construct (**phyBA**), and overexpressing the resultant fusion proteins in *Arabidopsis*. These studies have made clear that the ability to respond to constant red light (a phyB function) or to far-red light (a phyA function) resides in the amino termini of the proteins (Figure 17.16). Thus, plants that contain phyAB respond to constant far-red light as if additional phyA were present. In contrast, plants that contain phyBA show an enhanced sensitivity to constant red light, as if they contained additional phyB.

Since the Pfr form of the phyAB fusion protein is light-labile just like phyA, the N-terminal domain plays a role in the photodegradation of phyA. A candidate for the sequence motif in the N terminus involved in photodegradation of phyA is the so-called PEST sequence, which occurs close to the site of chromophore binding (see Figure 17.16). The PEST sequence has been named according to the single-letter amino acid codes for proline (P), glutamate (E), serine (S), and threonine (T) because it is a loosely defined hydrophilic sequence rich in these amino acids. This sequence is not present in phyB. The mechanism of proteolysis regulated by the PEST sequence has not yet been determined.

Although the phyAB fusion protein is unstable, the overexpressed N-terminal half of phyA alone is stable, suggesting that some C-terminal sequences present in both phyA and phyB must also be required for degradation. A candidate for the sequence motif in the C terminus that is involved in photodegradation is the ubiquitination site (see Figure 17.16). This site is present in both phyA and phyB. As discussed earlier, ubiquitin is a small protein that is often found attached to proteins targeted for degradation by the 26S proteasome (see Chapter 14).

Sequences required for dimerization have been localized near the N-terminal end of the C-terminal domain. This area also overlaps with a region that has been identified by deletion analysis as being involved in the regulation of phytochrome activity. In addition, deletion of 35 amino acids from the C terminus of oat phyA does not affect dimerization or spectral properties, but the dwarf overexpression phenotype seen in tobacco with normal oat phyA is lost. Thus, the C-terminal end must also be important in signaling, since its removal results

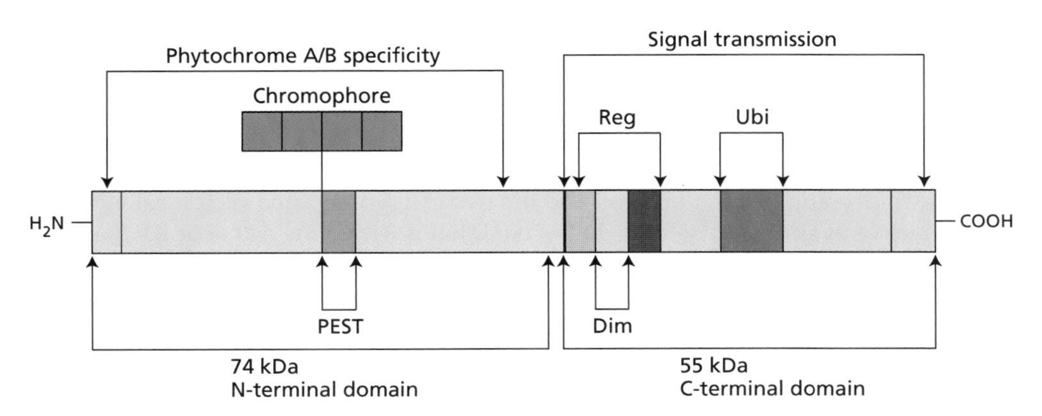

Figure 17.16 Schematic diagram of the phytochrome holoprotein, showing the various functional domains. The chromophore-binding site and PEST sequence are located in the N-terminal domain, which confers photosensory specificity to the molecule—that is, whether it responds to continuous red (phyB) or far-red (phyA) light. The C-terminal domain contains a dimerization site (Dim), a ubiquitination site (Ubi), and a regulatory region (Reg). The C-terminal domain transmits signals to proteins that act downstream of phytochrome.

in loss of function (see Figure 17.16). Consistent with the importance of the C terminus in signal transmission, this region has recently been shown to be related to the histidine kinase of the bacterial two-component signaling systems (see Box 17.2).

The structural and functional studies yield a picture of phytochrome as a molecule having two domains linked by a hinge: an N-terminal light-sensing domain in which the light specificity and stability reside, and a C-terminal domain that contains the signal-transmitting sequences, in addition to the dimerization site and ubiquitination sites (see Figure 17.16).

Cellular and Molecular Modes of Action

All phytochrome-regulated changes in plants begin with absorption of light by the pigment. After light absorption, the molecular properties of phytochrome are altered, causing the signal-transmitting sequences in the C terminus to interact with one or more components of a signal transduction pathway that ultimately bring about changes in the growth, development, or position of an organ (see Table 17.1). Some of the signal-transmitting motifs may interact with multiple signal transduction pathways; others may be unique to a specific pathway. Furthermore, it is reasonable to assume that the different phytochrome proteins utilize different sets of signal transduction pathways.

Although it may be relatively easy to document an early biochemical change induced by a stimulus, it is far more difficult to show that this change is causally related to a specific visible alteration in the plant. Molecular and biochemical techniques are helping unravel the early steps in phytochrome action and the signal transduction pathways that lead to a physiological or developmental response. These responses fall into two general categories: relatively rapid responses involving ion fluxes, and slower, long-term processes involved in photomorphogenesis. The rapid responses involve changes in membrane permeability; the slower responses require alterations in gene expression.

In this section we will examine the effects of phytochrome on both membrane permeability and gene expression, as well as the possible chain of events that constitute the signal transduction pathways that brings about these effects.

Phytochrome Regulates Membrane Potentials and Ion Fluxes

Phytochrome can rapidly alter the properties of membranes. We have already seen that low-fluence red light is required before the dark period to induce rapid leaflet closure during nyctinasty, and that fluxes of K^+ and Cl^- into and out of extensor and flexor cells mediate the response. However, the rapidity of leaf closure in the dark (lag time about 5 minutes) would seem to rule out mechanisms based on gene expression. Instead, rapid phytochrome-induced changes in membrane permeability and transport appear to be involved.

The role of the plasma membrane proton pump in *Samanea* pulvini (the structures at the base of the petiole that include the extensor and flexor cells) was studied by use of a proton-sensitive liquid membrane electrode (Lee and Satter 1988, 1989). During phytochrome-mediated leaflet closure, the apoplastic pH of the flexor cells (the cells that swell during leaflet closure) decreased, while the apoplastic pH of the extensor cells (the cells that shrink during leaflet closure) increased. Thus, the plasma membrane H^+ pump of the flexor cells appears to be activated by darkness (provided that phytochrome is in the Pfr form), whereas the H^+ pump of the extensor cells appears to be deactivated under the same conditions.

Increased H^+ pump activity on the plasma membranes of the flexor cells should increase the membrane potential, thereby allowing more K^+ to enter the cells (see Chapter 6). Conversely, a reduction in the plasma membrane H^+ pump activity in the extensor cells would promote the loss of K^+ ions. Consistent with this model, the reverse pattern of apoplastic pH change was seen during leaflet opening.

Studies have also been carried out on phytochrome regulation of K^+ channels. Since it is difficult to study membrane channels in the intact pulvinus, Richard Crain and his colleagues at the University of Connecticut isolated protoplasts (cells without their cell walls) of both extensor and flexor cells from *Samanea* leaves (Kim et al. 1993). They determined the open or closed state of the K^+ channels indirectly by measuring the membrane potential changes that occurred when the extracellular K^+ concentration was raised from 20 to 200 m*M*. When the extracellular K^+ concentration was raised, K^+ entered the protoplasts and depolarized the membrane potential only if the K^+ channels were open. When the extensor and flexor protoplasts were transferred to constant darkness, the state of the K^+ channels exhibited a circadian rhythmicity during a 21-hour incubation period and the two cell types varied reciprocally, just as they do in vivo. That is, when the extensor K^+ channels were open, the flexor K^+ channels were closed, and vice versa. Thus, the circadian rhythm of leaf movements has its origins in the circadian rhythm of K^+ channel opening.

To test the effects of phytochrome on K^+ channels, researchers first treated protoplasts with white light, then exposed them to a brief pulse of either red light alone or red light followed by far-red light and transferred them to continuous darkness. Red light caused the K^+ channels of the flexor protoplasts to open, and far-red light reversed the effect of red light. The extensor protoplasts were closed by both treatments. In con-

BOX 17.2

The Origins of Phytochrome in Bacterial Sensor Proteins

AS WE SAW IN CHAPTER 14, bacteria often use **two-component signaling systems** to sense and respond to environmental signals such as nutrients or light. The first component is a **sensor protein**, consisting of an *input domain* that receives the input signal, linked to a *transmitter domain* that transmits the signal to the second component. The second component is the **response regulator protein**, composed of a *receiver domain* that accepts the signal from the transmitter of the sensor protein, linked to an *output domain* that initiates the cellular response (see Figure 14.10).

These two components interact by a series of phosphorylation and dephosphorylation reactions. When it receives a signal, the sensor autophosphorylates a specific histidine residue in the transmitter domain. This phosphate is then transferred to a specific aspartate residue on the receiver domain of the response regulator (see Chapter 14). But what has all this to do with phytochrome?

Although they lack the specific histidine residue that becomes autophosphorylated, the C termini of phytochromes have some sequence similarity to the transmitter domains of bacterial sensor proteins (Figure 1) (Schneider-Poetsch 1992). This sharing of sequences raises the interesting possibility that phytochrome evolved from the sensor protein of a bacterial two-component system.

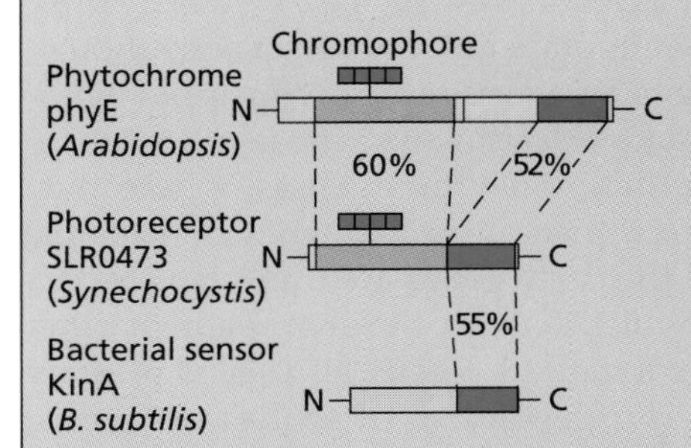

Figure 1 Homologous regions of phytochrome E, cyanobacterial photoreceptor, and sensor protein from *Bacillus subtilis*. The percentages indicate the proportion of identical amino acids between pairs of sequences. (After Yeh et al. 1997.)

Provocative new evidence for the idea that phytochrome evolved from the histidine kinase of a bacterial two-component system has recently emerged from studies of cyanobacteria. The cyanobacterium *Fremyella* uses light to regulate the composition of its photosynthetic pigments contained within the light-harvesting phycobilisomes (see Figure 7.5C), a response termed **chromatic adaptation**. Like phytochromes, phycobiliproteins bind linear tetrapyrrole chromophores that are closely related in structure to the phytochrome chromophore, but there is no evidence for red/far-red reversible responses of the type seen in higher plants. Instead, chromatic adaptation occurs in response to growth in red light and green light.

When the gene for the photoreceptor for this response (*RcaE*) was cloned, it was found to encode a 74 kDa polypeptide, the N-terminal end of which is related to the N terminus of higher-plant phytochromes, while the C-terminal end is related to histidine kinases (Kehoe and Grossman 1996). However, the lack of the conserved chromophore-binding site and of red/far-red reversibility suggests that this protein is a distant relative of higher-plant phytochromes.

Publication of the complete sequence of the genome of the cyanobacterium *Synechocystis* in 1996 led to great excitement because this organism contains a gene that has between 50 and 60% amino acid identity with *Arabidopsis* phyE along its length, including a bona fide chromophore attachment site. In addition, the amino acid sequence of the C terminus of the *Synechocystis* pigment protein is related to the KinA sensor protein of *Bacillus subtilis* (see Figure 1). This *Synechocystis* gene has been named *cph1* (cyanobacterial *ph*ytochrome *1*) (Yeh et al. 1997).

Recently it was shown that the purified Cph1 apoprotein can assemble with phycocyanobilin, a cyanobacterial linear tetrapyrrole, in vitro, and the resultant holoprotein exhibits red/far-red reversibility and absorption spectra similar (but not identical) to higher-plant phytochrome assembled in vitro with the same chromophore (Figure 2) (Hughes et al. 1997).

Perhaps the most exciting finding of all was the identification, 10 base pairs downstream from the cyanobacterial phytochrome gene, of another gene having the features of a bacterial response regulator (see Figure 1). Since the C-terminal end of the cyanobacterial phytochrome contains all the conserved features of a histidine kinase transmitter module, and the downstream response regulator resembles aspartate kinase receiver modules (see Chapter 14), it was now possible to test whether this phytochrome acts as part of a light-regulated transmitter–receiver pair, as had been proposed earlier (Schneider-Poetsch 1992).

Using the purified proteins, Clark Lagarias and his colleagues at the University of California, Davis were able to show that the Cph1 protein can phosphorylate itself in a manner consistent with its having a histidine kinase activity, and further, that the phosphate could be transferred to the response regulator (Yeh et al. 1997). Surprisingly, Cph1 is far more active as a kinase in the Pr form than in the Pfr form, suggesting that the light signal is transduced through regulation of Pr abundance, rather than through Pfr abundance as in higher plants.

Taken together, the results from these studies suggest that plant phytochromes originally evolved from a class of photoreversible transmitter histidine kinases in cyanobacteria. Are plant phytochromes also kinases? In support of this idea, a phytochrome gene isolated from the moss *Ceratodon* contains an additional C-terminal sequence that has a kinase homology (Thümmler et al. 1992). In addition, highly purified preparations

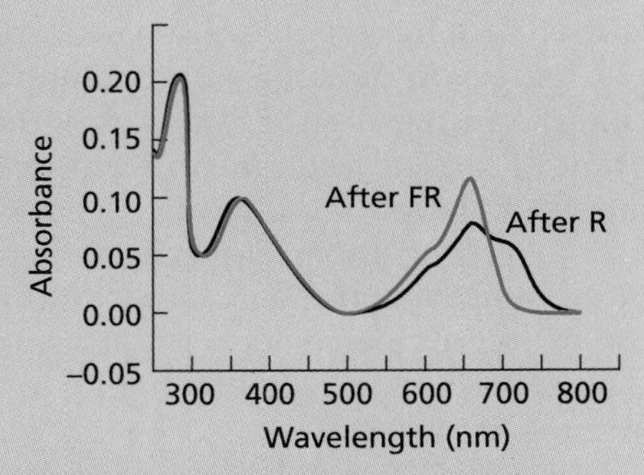

Figure 2 Absorption spectra of cyanobacterial phytochrome after irradiation with red (R) or far-red (FR) light. Note the similarities to and differences from the plant phytochrome spectra shown in Figure 17.3. (After Hughes et al. 1997.)

BOX 17.2 ***(continued)***

of higher-plant phytochrome have been found to contain a kinase activity that is also greater in the Pr form than in the Pfr form (Wong et al. 1989). How-ever, the kinase activity associated with these phytochromes is of the serine–threonine type, and consistent with this observation, these proteins lack the conserved histidines of the bacterial sensor proteins. Thus, while higher-plant phytochromes share a homology with bacterial two-component signaling systems, the reaction mechanism and components may have diverged considerably. Nevertheless, the discovery of cyanobacterial phytochromes has opened up many exciting new areas of research that have already contributed much to our understanding of phytochrome evolution and function.

trast to red light, blue light caused the K^+ channels of the extensors to open, and those of the flexors to close. Thus, phytochrome seems to regulate specifically the K^+ channels of the flexor cells. Blue light regulates primarily the extensor cells, although it can also affect the flexors at specific times of day (Kim et al. 1993).

On the basis of the evidence thus far, we can conclude that phytochrome brings about leaflet closure by regulating the activities of the primary proton pumps and/or the K^+ channels of the flexor and extensor cells. Although the effect is rapid, it is not instantaneous, and it is therefore unlikely to be due to a direct effect of phytochrome on the membrane. Instead, phytochrome acts indirectly via one or more signal transduction pathways, as in the case of the regulation of gene expression by phytochrome (see the next section).

However, some effects of red and far-red light on the membrane potential are so rapid that phytochrome may also interact directly with the membrane. Such rapid modulation has been measured in individual cells and has been inferred from the effects of red and far-red light on the surface potential of roots and oat (*Avena*) coleoptiles. The lag between the production of Pfr and the onset of measurable potential changes varies from organism to organism; for example, it is about 1.7 s for depolarization (inside becomes less negative relative to outside the plasma membrane) in the giant alga *Nitella* and 4.5 s for hyperpolarization in *Avena.*

Changes in the bioelectric potential of cells imply changes in the flux of ions across the plasma membrane. The *Mougeotia* microbeam irradiation experiments of Haupt (see Box 17.1) seemed to support the idea that phytochrome is localized on the plasma membrane, and membrane isolation studies carried out in the 1970s provided evidence that a small portion of the total phytochrome is tightly bound to various organellar membranes.

These findings led some workers to suggest that membrane-bound phytochrome represents the physiologically active fraction, and that all the effects of phytochrome on gene expression are initiated by changes in membrane permeability. On the basis of sequence analysis, however, it is now clear that phytochrome is a hydrophilic protein without membrane-spanning domains. The current view is that phytochrome in *Mougeotia* is associated with microtubules located directly beneath the plasma membrane.

If phytochrome exerts its effects on membranes from some distance, no matter how small, involvement of a "second messenger" would be implied. For example, rapid changes in cytosolic free calcium have been implicated as second messengers in several signal transduction pathways, and there is evidence that calcium plays a role in chloroplast rotation in *Mougeotia* (see Box 17.1). More recently, however, the role of calcium as a second messenger in phytochrome action has been questioned, so calcium-independent mechanisms may also be involved.

Phytochrome Regulates Gene Expression

As the term "photomorphogenesis" implies, plant development is profoundly influenced by light. Etiolation symptoms include spindly stems, small leaves (in dicots), and the absence of chlorophyll. Complete reversal of these symptoms by light involves major long-term alterations in metabolism that can be brought about only by changes in gene expression.

We have seen that phytochrome-mediated light responses at the cellular level can involve rapid effects on membrane properties. The stimulation and repression of transcription by light can also be very rapid, with lag times as short as 5 minutes. Such early-gene expression is likely to be regulated by the direct activation of transcription factors by one or more phytochrome-initiated signal transduction pathways. The activated transcription factors then enter the nucleus, where they stimulate the transcription of specific genes. Some of the gene products of the early genes are transcription factors themselves, which activate the expression of other genes. Expression of the early genes, also called **primary response genes**, is independent of protein synthesis; expression of the late genes, or **secondary response genes**, requires the synthesis of new proteins.

As with cytokinin-regulated gene expression (see Chapter 21), the photoregulation of gene expression has focused on the nuclear genes that encode messages for chloroplast proteins: the small subunit (SSU) of ribulose-1,6-bisphosphate carboxylase/oxygenase (rubisco) and the major light-harvesting chlorophyll *a/b*–binding proteins associated with the light-harvesting complex of photosystem II (LHCIIb proteins). These proteins play important roles in chloroplast development and greening; hence their regulation by phytochrome has been studied in detail. The genes for both of these proteins—*RBCS* and *LHCB* (also called *CAB* in some studies)—are present in multiple copies in the genome.

The abundance of these mRNAs has been shown to increase in response to light and phytochrome action. The abundance of other mRNAs, however, can be decreased by phytochrome action; this latter class includes mRNAs that encode oat phyA and the chlorophyll biosynthesis enzyme NADPH–protochlorophyllide oxidoreductase (see Chapter 7). Although the abundance of most of these mRNAs is regulated by light given in the low-fluence range, some *LHCB* mRNAs also respond to light in the very-low-fluence range.

We can demonstrate phytochrome-regulated induction of mRNA abundance (for example *RBCS* mRNAs) experimentally by giving etiolated plants a brief pulse of low-fluence red or far-red light, returning them to darkness to allow the signal transduction pathway to operate, and then measuring the abundance of specific mRNAs in total RNA prepared from each set of plants. If its abundance is regulated by phytochrome, the mRNA is absent or present at low levels in etiolated plants, but is increased by red light. The red light–induced increase in expression can be reversed by immediate treatment with far-red light, but far-red light alone has little effect on mRNA abundance.

In the case of mRNAs (e.g., *LHCB*) that respond to very-low-fluence red light, expression increases in response to both red and far-red light, and additional experiments are needed to confirm the involvement of phytochrome. In the case of mRNAs whose expression is negatively regulated by phytochrome, for example oat *PHYA* mRNA, levels are *high* in darkness, *reduced* by a pulse of red light, and *restored* to dark levels by red/far-red light treatment. However, none of these experiments prove the action of phytochrome on transcription; instead they show the effect of phytochrome only on overall mRNA abundance in total RNA, which reflects both transcription and degradation processes. As mentioned earlier, the abundance of oat *PHYA* mRNA in total RNA is the result of both transcription and degradation events.

The involvement of phytochrome in transcriptional regulation of mRNA abundance has been demonstrated by two approaches. In the first approach, etiolated plants are treated as was described for the mRNA studies, and then transcriptionally active nuclei are isolated and allowed to elongate transcripts initiated under the different in vivo light treatments in the presence of radioactive precursors (Figure 17.17). If the relative incorporation into an mRNA is similar to that already described, then the transcription of the mRNA is under phytochrome control.

The second approach involves fusing the promoter of the gene to be tested with the coding sequence of a reporter gene such as *GUS*, placing the fused gene back into the plant and measuring the abundance of the reporter mRNA or its encoded enzyme activity. Both of these approaches have been used extensively, and the latter has allowed the definition of the minimum amount of the upstream promoter region required to confer phytochrome-regulated transcription of the linked reporter sequence.

Recently, Elaine Tobin and her colleagues at the University of California, Los Angeles, characterized a MYB-related transcription factor (see Chapter 14) whose mRNA level increases rapidly when *Arabidopsis* is transferred from the dark to the light (Figure 17.18). This DNA-binding protein appears to bind to the promoter of the *LHCB* gene and regulate its transcription, which, as Figure 17.18 shows, occurs later than the increase in the MYB-related protein (Wang et al. 1997). The gene that encodes the MYB-related protein is therefore probably a primary response gene, while the *LHCB* gene is probably a secondary response gene.

Combinations of *cis*-Acting Elements Control Light-Regulated Transcription

The *cis*-acting regulatory sequences required to confer light regulation of gene expression have been studied extensively. Most eukaryotic promoters for genes that encode proteins comprise two functionally distinct regions: a short sequence that determines the transcription start site (the TATA box, named for its most abundant nucleotides) and upstream sequences, called *cis*-acting regulatory elements, that regulate the amount and pattern of transcription (see Chapter 14). These regulatory sequences bind specific proteins, called *trans*-acting factors, that modulate the activity of the general transcription factors that assemble around the transcription start site with RNA polymerase II.

The precise number and type of *cis*-acting sequences involved in regulating transcription of a given gene has been determined by various approaches:

Conserved sequences. The sequences of the promoter regions of a given gene (for example, *RBCS*) are first compared within members of the gene family in a particular species, and between gene family members of different species. Sequences acting as *cis*-acting regulatory elements are often represented in many species (that is, they are **conserved**) because they are essential for function, while the DNA sequences around them may be less essential, and therefore more variable. This approach gives an idea of the locations of sequences that may be involved in regulating the expression of a

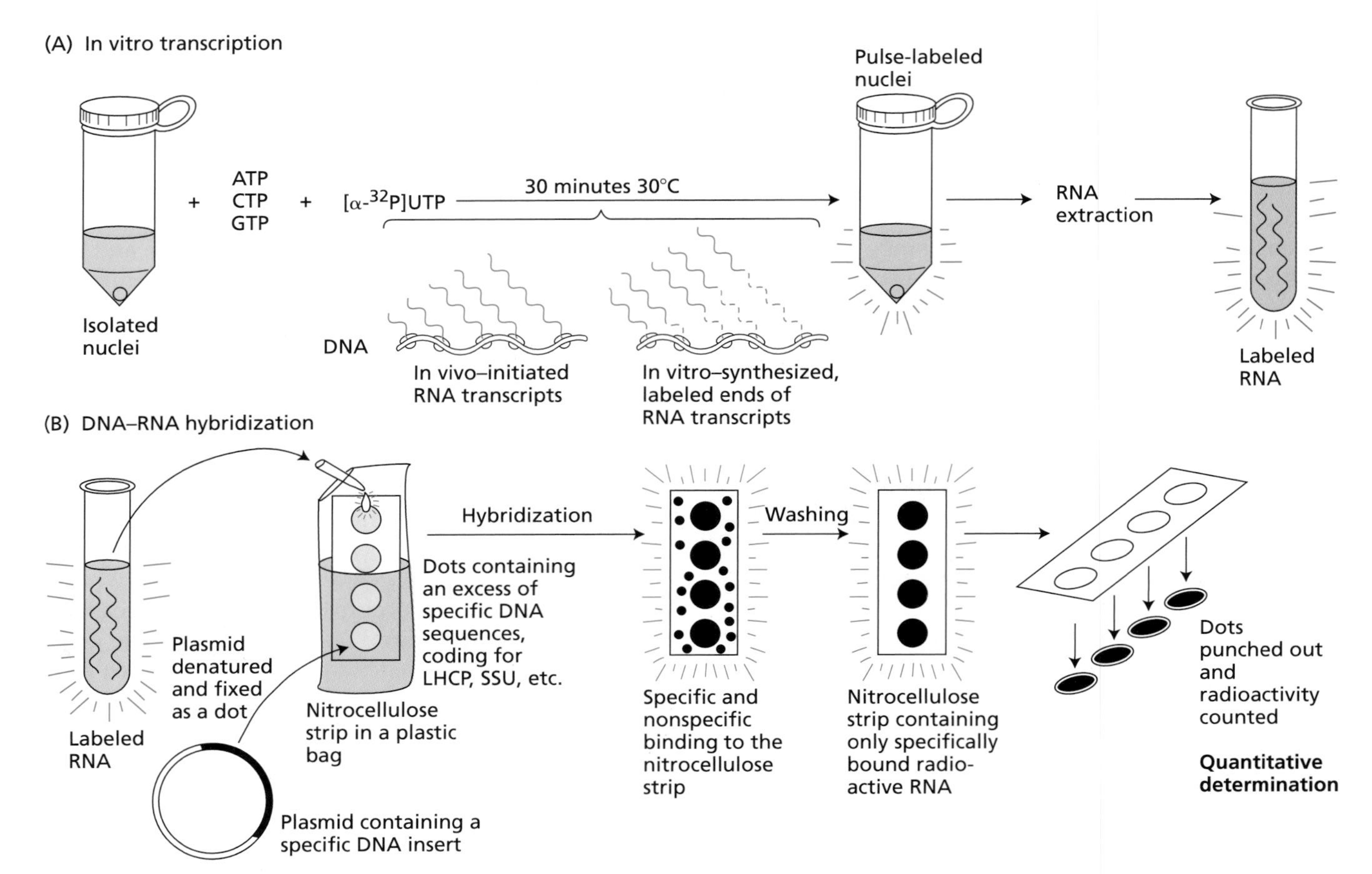

Figure 17.17 Nuclear runoff assay. (A) In vitro transcription. Isolated nuclei are incubated in a small tube with unlabeled ATP, CTP, and GTP, plus labeled UTP. During a 30-minute incubation, transcripts already initiated before nuclear isolation incorporate labeled UTP. The end-labeled RNA transcripts are purified. (B) DNA–RNA hybridization. Plasmids containing cDNA inserts for the gene of interest are spotted onto a strip of nitrocellulose paper. The strip is incubated in a plastic bag containing buffer and a drop of the labeled RNA. The mRNA that is complementary to the cDNA hybridizes to the spot, and the remainder of unbound RNA is removed by washing. The spots can then be punched out and the radioactivity quantified by counting. Alternatively, the radioactivity in each spot can be estimated by exposure to X-ray film. The size of the developed spot is roughly proportional to the radioactivity. (After Schäfer et al. 1986.)

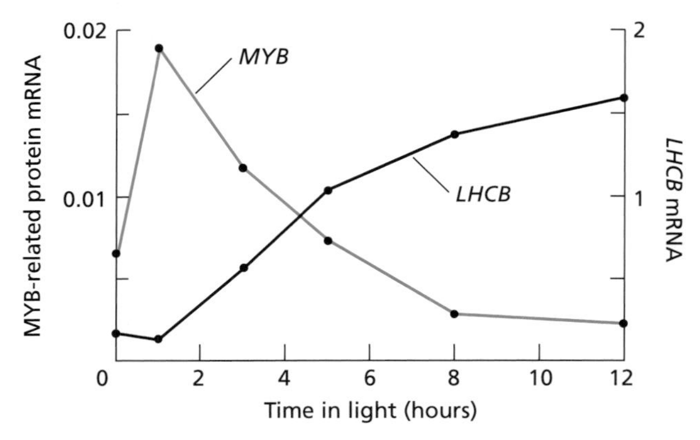

Figure 17.18 Time course of the induction of transcripts for a Myb-related transcription factor (Myb) and the light-harvesting chlorophyll *a*/*b*–binding protein (Lhcb) in *Arabidopsis* after transfer of the seedlings from darkness to continuous white light. (After Wang et al. 1997.)

gene. For *RBCS* genes, the regulatory element typically encompasses a region about 1 kilobase of DNA upstream from the transcriptional start site.

Identification of response elements. Once general locations of the conserved sequences have been identified, reporter constructs of the type already described can be tested to determine whether this fragment is necessary and sufficient to confer light regulation on the expression of the reporter sequence. Then increasing amounts of the 5′ end of the promoter region can be deleted to determine the minimum piece of DNA required for expression. Typically, 250 base pairs of DNA upstream of the transcription start site is required for light-regulated transcription. Internal deletions or the insertion of small, unrelated sequences, called *linkers,* can be used to define *cis*-acting regulatory elements.

Conferring light regulation. Synthetic copies of putative elements that contain one or more copies of a sequence can be inserted into non-light-regulated promoters to see if they can confer light regulation.

Identification of *trans*-acting factors. In vitro experiments in which nuclear protein extracts are allowed to bind to these different DNA fragments can be used to characterize the protein factors that bind to the *cis*-acting elements (see Chapter 20).

Together, these four approaches have yielded detailed promoter maps for several genes, including *RBCS* and oat *PHYA*. Some of the important *cis*-acting elements present in many light-regulated genes include **GT-1 sites** (GGTTAA), **I boxes** (GATAA) and **G boxes** (CACGTC). The *trans*-acting factors that bind to these three elements have also been characterized (Terzaghi and Cashmore 1995). However, each of these three *cis*-acting regulatory elements has also been found in the promoters of some non-light-regulated genes. Moreover, none of these elements alone is able to confer light regulation onto a non-light-regulated gene. Thus, light regulation arises from various combinations of different general elements rather than from a single regulatory element. Recent evidence suggests that the minimal sequence required for light-inducible gene expression is a pairwise combination of two elements such as GT-1 sites, G boxes, or I boxes (Puente et al. 1996).

Overall, the picture emerging for light-regulated plant promoters is similar to that for other eukaryotic genes: a collection of modular elements, the number, position, flanking sequences, and binding activities of which can lead to a wide range of transcriptional patterns. No single DNA sequence or binding protein is common to all phytochrome-regulated genes. At first it may appear paradoxical that light-regulated genes have such a range of elements, any combination of which can confer light-regulated expression. However, this array of sequences allows for the differential light- and tissue-specific regulation of many genes.

Phytochrome Acts through Multiple Signal Transduction Pathways

Two major approaches have been used to investigate the components of the signal transduction pathways between phytochrome and the activation of primary response genes. The first, discussed earlier, involves the isolation of mutants impaired in the ability to carry out light or phytochrome-regulated responses. The second approach involves the biochemical assay of signal transduction events in whole-plant and cellular systems. Using biochemical approaches, researchers have shown that signaling involves several different mechanisms, including G proteins, Ca^{2+}, and phosphorylation. We will consider the evidence for each of these in turn.

Well-characterized signaling pathways in other systems (for example, mating in yeasts) often include **G proteins** (reviewed in Chapter 14). These protein complexes are normally membrane associated, have three different subunits, and bind GTP or GDP on one subunit. The hydrolysis of GTP to GDP is required for regulation of G protein function. Sequences that encode G protein subunits have been cloned from plants, indicating that this type of system is present. One way that the function of G proteins can be tested is to treat cells with chemicals that activate or inhibit the ability of the complex to bind or break down GTP.

Early experiments with intact seedlings and cultured cells indicated that the control of gene expression by phytochrome involves the action of a heterotrimeric G protein and the calcium-binding regulatory protein **calmodulin**. Subsequently, a single-cell system has been developed that can be microinjected with various proposed components of the pathway and monitored for changes in gene expression (Neuhaus et al. 1993; Bowler et al. 1994). Although these experiments have not yet led to the identification of specific pathway components, they have been informative about the involvement of G proteins and calmodulin.

Microinjection of phytochrome (mainly phyA) purified from etiolated oat seedlings into single cells prepared from the tomato *aurea* mutant (a mutant that has reduced levels of phyA) results in stimulation of chloroplast development, accumulation of anthocyanin pigment, and expression of a GUS-coding sequence fused to a phytochrome-regulated promoter sequence. Etiolated seedlings of the *aurea* mutant show poor induction of these responses when transferred to light, probably because of the reduction in the levels of functional phyA. Inhibitors of G proteins block the phytochrome-stimulated events in the single-cell system, while activators or intermediates stimulate the response. Increased levels of Ca^{2+} in the medium also stimulate *GUS* expression and chloroplast development when calmodulin is present, but they do not affect anthocyanin biosynthesis.

Together, these studies indicate that phytochrome signaling can occur in single cells and does not require light after activation of phytochrome. At least one G protein may function downstream of phytochrome. After the G protein step, there are at least two branching pathways. One of these pathways, chloroplast development and gene expression, requires Ca^{2+} and calmodulin; the other, anthocyanin synthesis, does not.

The branching pathways can be distinguished further by the *cis*-acting regulatory elements targeted and the signaling intermediate employed. For many years, it has been known that both cyclic AMP (cAMP) and cyclic GMP (cGMP) are important intermediates in hormone- and light-induced signaling pathways in animals (see Chapter 14). Although the presence of cAMP has been

BOX 17.3

Genes That Suppress Photomorphogenesis

AS WE HAVE SEEN, mutant plants such as the *hy* mutants, which are impaired in the ability to respond to light, have been instrumental in the identification of genes that code for photoreceptors such as phytochrome and cryptochrome (see also Chapter 18). An additional class of mutants that have proved useful in studying genes involved in light regulation block the plant's response to darkness, referred to as *skotomorphogenesis.* These mutants include the *cop* (*co*nstitute *p*hotomorphogenesis) and *det* (*de*-*e*tiolated) mutants of *Arabidopsis.* As of this writing, ten different genes have been identified. Each of these is highly **pleiotropic**—that is, having many phenotypic effects (Kwok et al. 1996).

When grown in the dark, *cop* and *det* mutants exhibit short hypocotyls, open and expanded cotyledons, chloroplast differentiation, and derepression of light-inducible genes (Torii and Deng 1997). As a result, dark-grown *cop* and *det* seedlings resemble normal light-grown *Arabidopsis* seedlings, except that they lack chlorophyll. (Recall that one of the final steps in chlorophyll biosynthesis—involving protochlorophyllide oxidoreductase —requires direct activation by light; see Chapter 7.)

The *cop* and *det* mutations are recessive, indicating that they probably involve the loss of function of proteins whose normal function is to suppress photomorphogenesis in the dark. When a negative regulator is eliminated by a mutation, the pathway that it blocks becomes activated, suggesting that the photomorphogenesis pathway is *repressed* by darkness. Light causes *derepression* of the photomorphogenesis pathway, while repressing the skotomorphogenesis pathway. Thus, normal seedlings cannot initiate photomorphogenesis in the dark, because factors encoded by the *COP* and *DET* genes block these events.

The *DET1, DET2, COP1, COP9,* and *COP11* genes have been cloned and sequenced. These genes encode polypeptides of distinct size and amino acid sequence. Except for *DET2* (see Box 17.4), they all contain protein sequences for nuclear localization, and therefore presumably act in the nucleus. The *COP1* gene encodes a novel protein that has a DNA-binding sequence on the N-terminal end and sequences involved in protein–protein interaction on the C-terminal end. The COP1 protein accumulates in the nucleus in the dark but moves to the cytoplasm in the light (von Arnim and Deng 1994). It is thus likely to function in conjunction with other proteins as a transcriptional repressor. According to this model (see the figure), COP1 is normally transported to the nucleus in the dark, where it suppresses the expression of genes involved in photomorphogenesis. Light causes COP1 to exit the nucleus, allowing photomorphogenesis to take place. In the *cop1* mutant, the absence of functional COP1 allows photomorphogenesis to take place in the dark as well.

Where does COP1 act in relation to phytochrome? We can determine the order of molecular events by analyzing the results of genetic crosses with mutants blocked at other steps in the pathway. Such studies have indicated that COP1, and many other COP and DET proteins, act downstream of phytochrome, of the signal transduction pathways involved in phytochrome action, and of the *HY5* gene product, which has been identified as a transcription factor. How COP represses gene expression and which genes it represses remain to be elucidated.

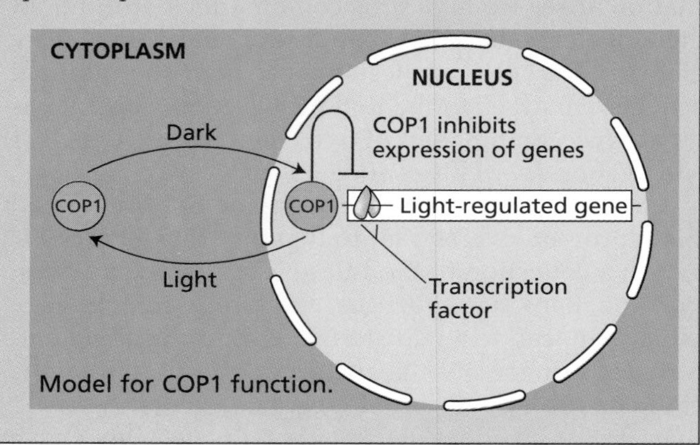

Model for COP1 function.

difficult to demonstrate in plants, the presence of cGMP in plant tissues is well established. Indeed, recent studies have shown that cGMP may serve as a second messenger in phytochrome action.

In the *aurea* cell system cGMP has been shown to act as a signaling intermediate in at least two signal transduction pathways (Wu et al. 1996). Specifically, the pathway that activates anthocyanin biosynthesis requires cGMP. Inhibitors of cGMP action inhibit this pathway; addition of cGMP activates it. In contrast, the pathway that requires calcium to activate chloroplast development and gene expression has a variable requirement for cGMP. Activation of genes that encode several chloroplast proteins, including RBCS and LHCB, does not require cGMP, while activation of genes that encode some photosystem I polypeptides and the cytochrome b_6f complex does.

The difference in signaling requirements is due in part to the different *cis*-acting regulatory elements found upstream of these genes. Specific DNA elements upstream of *RBCS* genes appear to be the targets for the calcium pathway, while other distinct elements located upstream of anthocyanin biosynthesis genes are the targets for cGMP. The differential regulation of the expression of phytochrome-regulated genes can now be explained in terms of their *cis*-acting regulatory elements, different combinations of which afford both tissue-specific and light-specific expression.

The evidence for a potential role of phosphorylation in phytochrome action comes from red-light regulation of protein phosphorylation and phosphorylation-dependent binding of transcription factors to the promoters of phytochrome-regulated genes. Some highly purified preparations of phytochrome have been reported to have

BOX 17.4

Brassinosteroids: A New Class of Plant Steroid Hormones

BY SCREENING FOR *ARABIDOPSIS* SEEDLINGS that have a light-grown morphology after 7 days growth in total darkness, Joanne Chory and her colleagues at the Salk Institute in San Diego isolated various *det* mutants, altered in their photomorphogenic response (see Box 17.3) (Chory and Li 1997). These mutants had short, thick hypocotyls, expanded cotyledons, primary leaf buds, and anthocyanins when grown in darkness, in marked contrast to dark-grown wild-type seedlings, which had long hypocotyls and folded cotyledons, and which did not accumulate anthocyanins under the same conditions. One of these mutants, *det2*, is shown in Figure 1.

In addition to their unusual dark-grown phenotype, *det2* homozygotes have an abnormal phenotype when grown in the light (see Figure 1B). They are small and dark green, because of a reduction in cell size, and they have reduced apical dominance and reduced male fertility. In addition, *det2* mutants have a prolonged vegetative phase and delayed leaf and chloroplast senescence, even after flowering. The leaves of *det2* mutants remain green even after seeds have formed, and *det2* mutants show delayed senescence in a detached leaf assay. These phenotypes indicate that the protein DET2 plays an important role throughout *Arabidopsis* development, including senescence. The phenotype of another mutant, called *cpd* (constitutive photomorphogenesis and dwarfism), is almost identical to that of *det2*.

When the *DET2* locus was cloned, it was shown to encode a protein that has an amino acid sequence similar to that of the mammalian steroid 5α-reductases (Li et al. 1996). Mammalian steroid 5α-reductases catalyze an NADPH-dependent conversion of testosterone to dihydrotesterone—a key step in steroid metabolism that is essential for the normal embryonic development of male external genitalia and the prostate. The cloning of *CPD* also suggested that it plays a role in steroid synthesis. *CPD* encodes a protein that exhibits some homology to mammalian cytochrome P450 proteins, including steroid hydroxylases. Could DET2 and CPD be involved in the synthesis of a plant steroid hormone?

At the time *det2* was first analyzed, many steroids had been identified in plants, but only the **brassinosteroids** had been shown to cause marked biological effects on plant growth at very low concentrations, and to be widely distributed throughout the plant kingdom. Brassinolide (see Figure 2), the first brassinosteroid to be isolated, was purified from rape pollen (*Brassica napus*) in 1979 (reviewed in Mandava 1988). Physiological studies

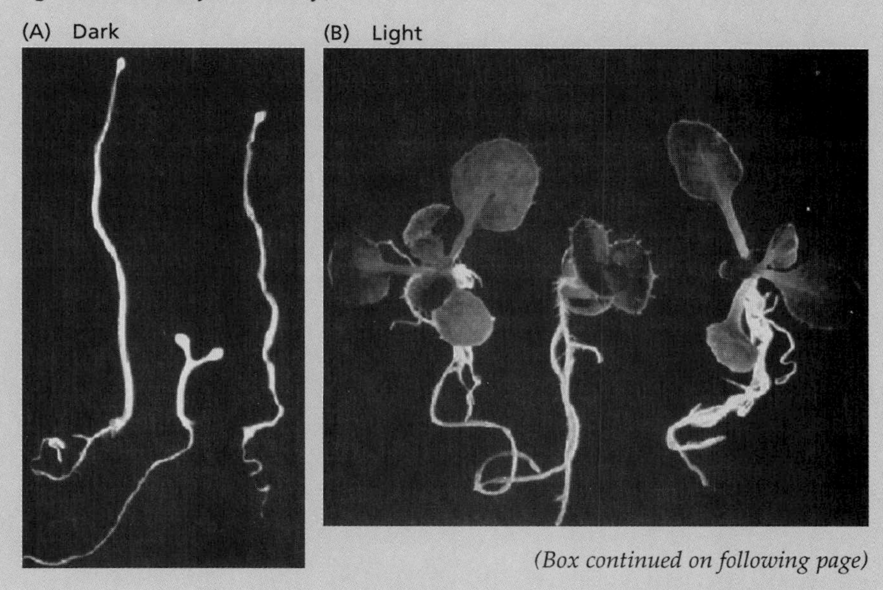

Figure 1 Phenotypes of the *Arabidopsis* wild-type (left), *det2* mutant (center), and *det2* mutant treated with brassinolide (right), grown in either the dark (A) or the light (B). (Courtesy of J. Chory.)

(Box continued on following page)

kinase activity. Kinases are enzymes that have the capacity to transfer phosphate groups from ATP to amino acids such as serine or tyrosine, either on themselves or on other proteins. Kinases are often found in signal transduction pathways in which the addition or removal of phosphate groups regulates enzyme activity.

It is not clear whether the kinase activity associated with phytochrome is due to phytochrome itself or to a tightly associated protein. However, at least some lower-plant phytochromes may have a kinase activity. The moss *Ceratodon purpureus* has an unusual chimeric phytochrome resembling phyB plus an extra nonphytochrome sequence on the C-terminal end that is homologous to known kinases, as well as kinase activity. Since cyanobacteria have been shown to contain a prokaryotic phytochrome homologous to bacterial kinases (see Box 17.2), the ancestral form of phytochrome was probably a kinase. Some phytochromes may still have this kinase activity, or the kinase activity may have been taken over by a tightly associated protein.

Overall, kinases, G proteins, Ca^{2+}, calmodulin, and other regulatory proteins such as COP or DET proteins (see Boxes 17.3 and 17.4) are probably involved in transmitting the signal from phytochrome to the primary response genes, such as those that encode MYB-related transcription factors. Phytochrome is distributed throughout cells, and although there is evidence that phyB can move to the nucleus in response to light (Sakamoto and

BOX 17.4 *(continued)*

had established that exogenous brassinosteroid causes cell elongation and cell division in excised stem sections. Brassinosteroids also inhibit root growth, enhance gravitropism, promote xylem differentiation, and delay leaf abscission.

Despite the extensive research on brassinosteroid chemistry and potential agricultural applications during the 1970s and 1980s, the status of brassinosteroid as an endogenous hormone remained uncertain. Recognition of brassinosteroids as authentic plant hormones was delayed because the developmental effects of brassinosteroids qualitatively resemble those of auxins, and exogenously applied brassinosteroids result in phenotypes that show complex interactions with the better-known phytohormones: gibberellic acid, abscisic acid, ethylene, and cytokinin.

Nevertheless, the brassinosteroids can affect elongation and gene expression independently of auxin in soybean hypocotyls. Further, root elongation in an auxin-insensitive mutant of *Arabidopsis* is inhibited by brassinosteroids, but not by auxin. But despite the accumulating evidence that brassinosteroids have properties characteristic of endogenous plant growth regulators, widespread acceptance of brassinosteroids as plant hormones awaited the discovery and analysis of the two photomorphogenesis mutants of *Arabidopsis*: *det2* and *cpd*.

Supporting evidence that the genes *DET2* and *CPD* encode enzymes involved in the synthesis of brassinosteroids came from feeding brassinosteroid biosynthesis intermediates to the mutant plants. The dark- and light-grown phenotypes of *det2* and *cpd* mutants can be completely reversed by addition of a brassinosteroid to the growth medium (see Figure 1).

Brassinosteroids are derived from the plant sterol *campesterol* by a reduction step followed by many oxidation steps (Figure 2) (Fujioka and Sakurai 1997). More than 60 brassinosteroids have now been identified. Using various biosynthesis intermediates, researchers showed that *det2* mutants are blocked in the first step of the biosynthetic pathway, in the reduction of campesterol to campestanol (Figure 2). Likewise, feeding experiments suggested that the *cpd* mutation blocks the C_{21} hydroxylation step between cathasterone and teasterone in the proposed biosynthetic pathway.

The pleiotropic effects of *det2* and *cpd* mutations on *Arabidopsis* development suggest that brassinosteroids play important roles in several light- and hormone-regulated processes, including the expression of light-regulated genes, the promotion of cell elongation, leaf and chloroplast senescence, and floral induction.

What is the nature of the interaction between light and endogenous brassinosteroids? Light may act by modulating the brassinosteroid signal transduction pathway, perhaps by regulating the biosynthesis of brassinosteroids or by altering the cell's responsiveness to brassinosteroids. To identify the components of the brassinosteroid signal transduction pathway, numerous laboratories carried out genetic screens and isolated a total of 20 brassinolide-insensitive mutants of *Arabidopsis*. Surprisingly, all 20 mutations were in the same gene. This gene has been cloned and shown to encode a protein that has sequence homology to transmembrane leucine-rich repeat (LRR) receptor kinases (see Chapter 14) (Li and Chory 1997). Thus the plant steroid hormone receptor may be an LRR receptor located on the plasma membrane.

Although the receptors of animal steroid hormones are generally considered to be transcription factors, as noted in Chapter 14 increasing evidence suggests that animal steroid hormones bind to cell surface receptors as well. Perhaps the apparent LRR steroid receptor of plants will one day lead to the identification of a similar receptor in animal cells.

Figure 2 Brassinosteroid biosynthetic pathway. The reactions catalyzed by the DET2 and CPD enzymes are indicated.

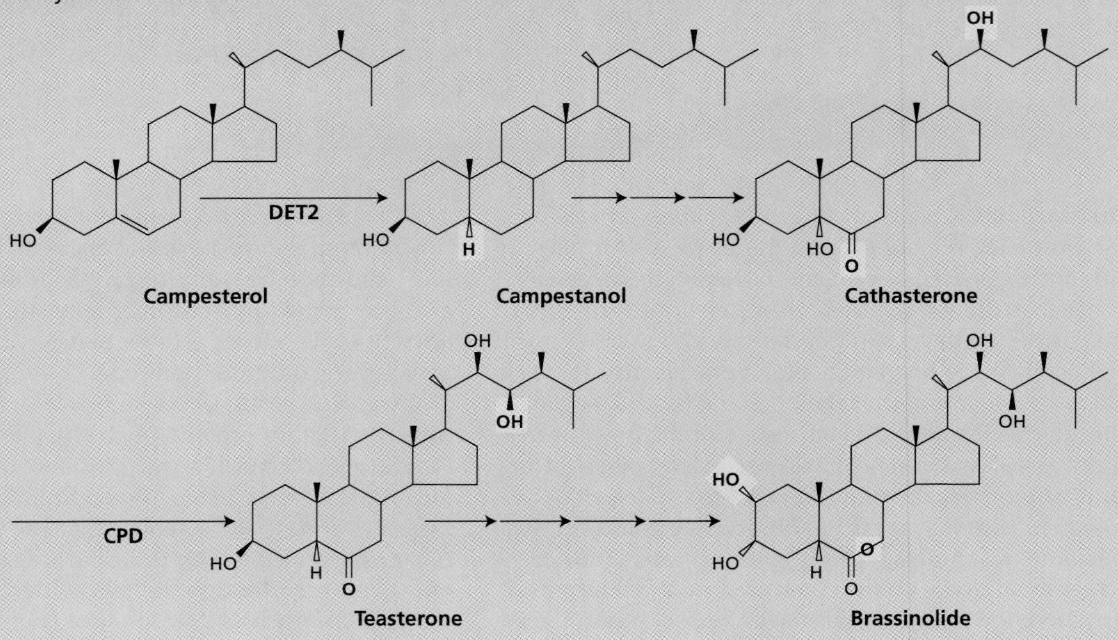

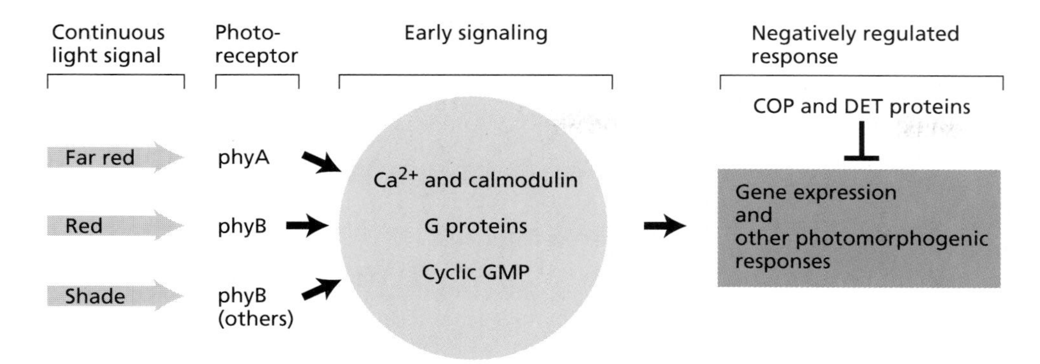

Figure 17.19 Summary of known signaling events involved in photomorphogenesis. First, the light signal is perceived by the appropriate phytochrome photoreceptor, which then activates a signal transduction pathway, possibly involving Ca^{2+} and calmodulin, G proteins, and cyclic GMP. The early signaling events lead to the activation of transcription factors, which overcome the negative regulation of the phytochrome-responsive genes by the COP and DET proteins.

Nagatani 1996), it is not clear whether this is true of other phytochromes, or whether this is the sole functional phytochrome. Another possibility is that the Pfr form of phytochrome transmits a signal to the nucleus indirectly through downstream reaction partners, perhaps involving one or more kinases. The involvement of phosphorylation in the regulation of the activity of at least one transcription factor of light-induced genes would support this view. Since COP1, an inhibitor of photomorphogenesis that may act as a transcriptional repressor, is located in the nucleus in the dark and in the cytoplasm in the light, its removal must also involve light-activated components, perhaps including a trimeric G protein (Figure 17.19).

Summary

"Photomorphogenesis" refers to the dramatic effects of light on plant development and cellular metabolism. Red light exerts the strongest influence, and the effects of red light are often reversible by far-red light. Phytochrome is the pigment involved in most photomorphogenic phenomena. Phytochrome exists in two forms, a red light–absorbing form (Pr) and a far-red light–absorbing form (Pfr). Phytochrome is synthesized in the dark in the Pr form. Absorption of red light by Pr converts it to Pfr, and absorption of far-red light by Pfr converts it to Pr. However, the absorption spectra of the two forms overlap in the red region of the spectrum, leading to an equilibrium between the two forms called a photostationary state. Pfr is considered to be the active form that gives rise to the physiological response. Other factors in addition to light regulate the steady-state level of Pfr, including the expression level of the protein and its stability in the Pfr form.

Phytochrome is a large dimeric protein made up of two equivalent subunits. The monomer has a molecular mass of about 125 kDa and is covalently bound to a chromophore molecule, an open-chain tetrapyrrole called phytochromobilin.

Phytochrome is encoded by a family of divergent genes that give rise to two types of protein, Type I and Type II. Type I, which is encoded by the *PHYA* gene, is abundant in etiolated tissue. However, Type I phytochrome is present at low levels in light-grown plants because of its instability in the Pfr form, the phyA-mediated suppression of transcription of its own gene, and the instability of its mRNA. Type II phytochrome (encoded by the *PHYB*, *PHYC*, *PHYD*, and *PHYE* genes) is present at low levels in both light-grown and dark-grown plants because its genes are constitutively expressed at low levels and the protein is stable in the Pfr form.

Spectrophotometric and immunological studies indicate that the phytochromes are concentrated in meristematic regions and that phyA is more abundant in etiolated tissues. During photoconversion of phyA to the Pfr form in higher plants, the distribution of the phytochrome within the cell changes from a diffuse to a sequestered pattern, which may be connected with its breakdown. Thus far, no organelle has been specifically associated with Pfr in higher-plant cells. In the alga *Mougeotia*, phytochrome appears to be located on the microtubular scaffold directly underneath the plasma membrane, where it mediates chloroplast rotation.

Phytochrome responses have been classified into very low fluence, low fluence, and high irradiance responses (VLFRs, LFRs, and HIRs). These three types of responses differ not only in their fluence requirements but also in other parameters, such as their escape times, action spectra, and photoreversibility. Phytochrome B plays an important role in the detection of shade in plants adapted to high levels of sunlight; phytochrome A has a more limited role, mediating the far-red HIR during early greening.

Phytochrome is known to regulate the transcription of numerous genes. Many of the genes involved in greening, such as the nuclear-encoded genes for the small subunit of rubisco and the chlorophyll *a/b*–binding protein of the light-harvesting complex, are transcriptionally regulated by phytochrome (both phyA and phyB). Phytochrome also represses the transcription of various genes, including *PHYA*. Activation or repression of these genes is thought to be mediated by general transcription factors that bind to *cis*-acting regulatory elements within the promoter regions of these genes in a combinatorial fashion. These transcription factors, in turn, are linked to phytochrome action by a complex series of signal transduction pathways involving COP and DET proteins, kinases, cyclic GMP, trimeric G proteins, Ca^{2+}, and calmodulin.

The recent discovery and characterization of bacterial phytochrome suggest that phytochrome evolved from a bacterial histidine kinase that participates in two-component signaling pathways.

In addition to the long-term effects involving changes in gene expression, phytochrome induces a variety of rapid responses, including chloroplast rotation in the alga *Mougeotia*, leaf closure during nyctinasty, and alterations in membrane potential. These responses involve rapid changes in membrane properties. The current view is that even these rapid effects of phytochrome involve signal transduction pathways.

General Reading

Barnes, S. A., Quaggio, R. B., and Chua, N-H. (1995) Phytochrome signal-transduction: Characterization of pathways and isolation of mutants. *Philos. Trans. R. Soc. London [Biol.]* 350: 67–74.

Chory, J., Cook, R. K., Dixon, R., Elich, T., Li, H. M., Lopez, E., Mochizuki, N., Nagpal, P., Pepper, A., Poole, D., and Reed, J. (1995) Signal transduction pathways controlling light-regulated development in *Arabidopsis*. *Philos. Trans. R. Soc. London [Biol.]* 350: 59–65.

Clouse, S. D. (1996) Molecular genetic studies confirm the role of brassinosteroids in plant growth and development. *Plant J.* 10: 1–8.

Elich, T. D., and Chory, J. (1994) Initial events in phytochrome signaling: Still in the dark. *Plant Mol. Biol.* 26: 1315–1327.

Kendrick, R. E., and Kronenberg, G. H. M., eds. (1994) *Photomorphogenesis in Plants*, 2nd ed. Kluwer, Dordrecht, Netherlands.

McNellis, T. W., and Deng, X-W. (1995) Light control of seedling morphogenetic pattern. *Plant Cell* 7: 1749–1761.

Millar, A. J., McGrath, R. B., and Chua, N-H. (1994) Phytochrome phototransduction pathways. *Annu. Rev. Genet.* 28: 325–349.

*Quail, P. H., Boylan, M. T., Parks, B. M., Short, T. W., Xu, Y., and Wagner, D. (1995) Phytochrome: Photosensory perception and signal transduction. *Science* 268: 675–680.

Sage, L. C. (1992) *Pigment of the Imagination. A History of Phytochrome Research*. Academic Press, San Diego.

Smith, H. (1995) Physiological and ecological function within the phytochrome family. *Annu. Rev. Plant Physiol. Plant Mol. Biol.* 46: 289–315.

Terry, M. J., Wahleithner, J. A., and Lagarias, J. C. (1993) Biosynthesis of the plant photoreceptor phytochrome. *Arch. Biochem. Biophys.* 306: 1–15.

*Terzaghi, W. B., and Cashmore, A. R. (1995) Light-regulated transcription. *Annu. Rev. Plant Physiol. Plant Mol. Biol.* 46: 445–474.

Tobin, E. M., and Kehoe, D. M. (1994) Phytochrome-regulated gene expression. *Sem. Cell Bio.* 5: 335–346.

* Indicates a reference that is general reading in the field and is also cited in this chapter.

Chapter References

Adam, E., Szell, M., Szekeres, M., Schaefer, E., and Nagy, F. (1994) The developmental and tissue-specific expression of tobacco phytochrome-A genes. *Plant J.* 6: 283–293.

Andel, F., Hasson, K. C., Gai, F., Anfinrud, P. A., and Mathies, R. A. (1997) Femtosecond time-resolved spectroscopy of the primary photochemistry of phytochrome. *Biospectroscopy* 3: 421–433.

Anderson, S. L., and Kay, S. A. (1995) Functional dissection of circadian clock- and phytochrome-regulated transcription of the *Arabidopsis CAB2* gene. *Proc. Natl. Acad. Sci. USA* 92: 1500–1504.

Beggs, C. J., Holmes, M. G., Jabben, M., and Schaefer, E. (1980) Action spectra for the inhibition of hypocotyl growth by continuous irradiation in light- and dark-grown *Sinapis alba* L. seedlings. *Plant Physiol.* 66: 615–618.

Borthwick, H. A., Hendricks, S. B., Parker, M. W., Toole, E. H., and Toole, V. K. (1952) A reversible photoreaction controlling seed germination. *Proc. Natl. Acad. Sci. USA* 38: 662–666.

Bowler, C., Yamagata, H., Neuhaus, G., and Chua, N-H. (1994) Phytochrome signal transduction pathways are regulated by reciprocal control mechanisms. *Genes Dev.* 8: 2188–2202.

Briggs, W. R., Mandoli, D. F., Shinkle, J. R., Kaufman, L. S., Watson, J. C., and Thompson, W. F. (1984) Phytochrome regulation of plant development at the whole plant, physiological, and molecular levels. In *Sensory Perception and Transduction in Aneural Organisms*. G. Colombetti, F. Lenci, and P.-S. Song, eds., Plenum, New York, pp.265–280.

Butler, W. L., Norris, K. H., Siegelman, H. W., and Hendricks, S. B. (1959) Detection, assay, and preliminary purification of the pigment controlling photosensitive development of plants. *Proc. Natl. Acad. Sci. USA* 45: 1703–1708.

Chory, J., and Li, J. (1997) Gibberellins, brassinosteroids and light-regulated development. *Plant Cell Environ.* 20: 801–806.

Clough, R. C., and Vierstra, R. D. (1997) Phytochrome degradation. *Plant Cell Environ.* 20: 713–721.

Flint, L. H. (1936) The action of radiation of specific wave-lengths in relation to the germination of light-sensitive lettuce seed. *Proc. Int. Seed Test. Assoc.* 8: 1–4.

Fujioka, S., and Sakurai, A. (1997) Biosynthesis and metabolism of brassinosteroids. *Physiol. Plantarum* 100: 710–715.

Furuya, M. (1993) Phytochromes: Their molecular species, gene families and functions. *Annu. Rev. Plant Physiol. Mol. Biol.* 44: 617–645.

Furuya, M., and Song, P-S. (1994) Assembly and properties of holophytochrome. In *Photomorphogenesis in Plants*, 2nd ed., R. E. Kendrick and G. H. M. Kronenberg, eds., Kluwer, Dordrecht, Netherlands, pp. 105–140.

Goosey, L., Palecanda, L. and Sharrock, R. A. (1997) Differential patterns of expression of the *Arabidopsis* PHYB, PHYD and PHYE phytochrome genes. *Plant Physiol.* 115: 959–969.

Goto, N., Tamamoto, K. T., and Watanabe, M. (1993) Action spectra for inhibition of hypocotyl growth of wild-type plants and of the hy2 long-hypocotyl mutants of *Arabidopsis thaliana* L. (1993) *Photochem. Photobiol.* 57: 867–871.

Hartmann, K. M. (1967) Ein Wirkungsspecktrum der Photomorphogenese unter Hochenergiebedingungen und seine Interpretation auf der Basis des Phytochroms (Kypokotylwachstumshemmung bei *Lactuca sativa* L.). *Z. Naturforsch.* 22b: 1172–1175.

Haupt, W. (1982) Light-mediated movement of chloroplasts. *Annu. Rev. Plant Physiol.* 33: 205–233.

Hershey, H. P., Barker, R. F., Idler, K. B., Lissemore, J. L., and Quail, P. H. (1985) Analysis of cloned cDNA and genomic sequences for phytochrome: Complete amino acid sequences for two gene products expressed in etiolated *Avena*. *Nucleic Acids Res.* 13: 8543–8558.

Hughes, J., Lamparter, T., Mittmann, F., Hartmann, E., Gaertner, W., Wilde, A., and Boerner, T. (1997) A prokaryotic phytochrome. *Nature* 386: 663.

Kehoe, D. M., and Grossman, A. R. (1996) Similarity of a chromatic adaptation sensor to phytochrome and ethylene receptors. *Science* 273: 1409–1412.

Kendrick R. E., and Frankland, B. (1983) *Phytochrome and Plant Growth,* 2nd ed. Edward Arnold, London.

Kendrick, R. E., Kerckhoffs, L. H. J., Van Tuinen, A., and Koornneef, M. (1997) Photomorphogenic mutants of tomato. *Plant Cell and Env.* 20: 746–751.

Kim, H. Y., Cote, G. G., and Crain, R. C. (1993) Potassium channels in *Samanea*-Saman protoplasts controlled by phytochrome and the biological clock. *Science* 260: 960-962.

Konomi, K., Abe, H., and Furuya, M. (1987) Changes in the content of phytochrome I and II apoproteins in embryonic axes of pea seeds during imbibition. *Plant Cell Physiol.* 28: 1443–1449.

Kraml, M. (1994) Light direction and polarization. In *Photomorphogenesis in Plants*, 2nd ed., R. E. Kendrick and G. H. M. Kronenberg, eds., Kluwer, Dordrecht, Netherlands, pp. 417–445.

Kwok, S. F., Piekos, B., Miséra, S., and Deng, X-W. (1996) A complement of ten essential and pleiotropic *Arabidopsis COP/DET/FUS* genes is necessary for repression of photomorphogenesis in darkness. *Plant Physiol.* 110: 731–742.

Lee, Y., and Satter, R. L. (1988) Effects of temperature on proton uptake and release during circadian rhythmic movements of excised *Samanea* motor organs. *Plant Physiol.* 86: 352–354.

Lee, Y., and Satter, R. L. (1989) Effects of white, blue, red light and darkness on pH of the apoplast in the *Samanea* pulvinus. *Planta* 178: 31–40.

Li, J., and Chory, J. (1997) A putative leucine-rich repeat receptor kinase involved in brassinosteroid signal transduction. *Cell* 90: 929–938.

Li, J., Nagpal, P., Vitart, V., McMorris, T., and Chory, J. (1996) A role for brassinosteroids in light-dependent development of *Arabidopsis*. *Science* 272: 398–401.

Li, L. M., and Lagarias, J. C. (1992) Phytochrome assembly—Defining chromophore structural requirements for covalent attachment and photoreversibility. *J. Biol. Chem.* 267: 19204–19210.

Mandava, N. B. (1988) Plant growth-promoting brassinosteroids. *Annu. Rev. Plant Physiol. Plant Mol. Biol.* 39: 23–52.

Mandoli, D. F., and Briggs, W. R. (1984) Fiber optics in plants. *Sci. Am.* 251: 90–98.

Morgan, D. C., and Smith, H. (1978) Simulated sunflecks have large, rapid effects on plant stem extension. *Nature* 273: 534-536.

Morgan, D. C., and Smith, H. (1979) A systematic relationship between phytochrome-controlled development and species habitat, for plants grown in simulated natural irradiation. *Planta* 145: 253–258.

Nagy, F., Kay, S. A., and Chua, N.-H. (1988) A circadian clock regulates transcription of the *cab-1* gene. *Genes Dev.* 2: 376–382.

Nakasako, M., Wada, M., Tokutomi, S., Yamamoto, K. T., Sakai, J., Kataoka, M., Tokunaga, F., and Furuya, M. (1990) Quaternary structure of pea phytochrome I dimer studied with small angle X-ray scattering and rotary-shadowing electron microscopy. *Photochem. Photobiol.* 52: 3–12.

Neuhaus, G., Bowler, C., Kern, R., and Chua, N-H. (1993) Calcium/calmodulin-dependent and -independent phytochrome signal transduction pathways. *Cell* 73: 937–952.

Puente, P., Wei, N., and Deng, X-W. (1996) Combinatorial interplay of promoter elements constitutes the minimal determinants for light and developmental control of gene expression in *Arabidopsis*. *EMBO J.* 15: 3732–3743.

Quail, P. H., Briggs, W. R., Chory, J., Hangarter, R. P., Harberd, N. P., Kendrick, R. E., Koorneef, M., Parks, B., Sharrock, R. A., Schäfer, E., Thompson, W. F., and Whitelam, G. C. (1994) Spotlight on phytochrome nomenclature. *Plant Cell* 6: 468–471.

Rüdiger, W., and Thümmler, F. (1994) The phytochrome chromophore. In *Photomorphogenesis in Plants*, 2nd ed., R. E. Kendrick and G. H. M. Kronenberg, eds., Martinus Nijhoff, Dordrecht, Netherlands, pp. 51–69.

Sakamoto, K., and Nagatani, A. (1996) Nuclear localization activity of phytochrome B. *Plant J.* 10: 859–868.

Satter, R. L., and Galston, A. W. (1981) Mechanisms of control of leaf movements. *Annu. Rev. Plant Physiol.* 32: 83–110.

Saunders, M. J., Cordonnier, M-M., Palevitz, B. A., and Pratt, L. H. (1983) Immunofluorescence visualizaton of phytochrome in *Pisum sativum* L. epicotyls using monoclonal antibodies. *Planta* 159: 545–553.

Schäfer, E., Apel, K., Batschauer, A., and Mösinger, E. (1986) The molecular biology of phytochrome action. In *Photomorphogenesis in Plants*, R. E. Kendrick, and G. H. M. Kronenberg, eds., Martinus Nijhoff, Dordrecht, The Netherlands, pp. 83–98.

Schneider-Poetsch, H. A. W. (1992) Signal transduction by phytochrome: Phytochromes have a module related to the transmitter modules of bacterial sensor protein. *Photochem. Photobiol.* 56: 839–846.

Sharrock, R. A., and Quail, P. H. (1989) Novel phytochrome sequences in *Arabidopsis thaliana*: Structure, evolution, and differential expression of a plant regulatory photoreceptor family. *Genes Dev.* 3: 1745–1757.

Shinomura, T., Nagatani, A., Hanzawa, H., Kubota, M., Watanabe, M., and Furuya, M. (1996) Action spectra for phytochrome A- and B-specific photoinduction of seed germination in *Arabidopsis thaliana*. *Proc. Natl. Acad. Sci. USA* 93: 8129–8133.

Shropshire, W., Jr., Klein, W. H., and Elstad, V. B. (1961) Action spectra of photomorphogenic induction and photoinactivation of germination in *Arabidopsis thaliana*. *Plant Cell Physiol.* 2: 63–69.

Siegelman, H. W., and Firer, E. M. (1964) Purification of phytochrome from oat seedlings. *Biochemistry* 3: 418–423.

Smith, H. (1982) Light quality photoperception and plant strategy. *Annu. Rev. Plant Physiol.* 33: 481–518.

Smith, H., and Whitelam, G. C. (1997) The shade avoidance syndrome: Multiple responses mediated by multiple phytochromes. *Plant Cell Environ.* 20: 840–844.

Somers, D. E., and Quail, P. H. (1995) Temporal and spatial expression patterns of *PHYA* and *PHYB* genes in *Arabidopsis*. *Plant J.* 7: 413–427.

Thümmler, F., Dufner, M., Kreisl, P., and Dittrich, P. (1992) Molecular cloning of a novel phytochrome gene of the moss *Ceratodon purpureus* which encodes a putative light-regulated protein kinase. *Plant Mol. Biol.* 20: 1003–1017.

Tokutomi, S., Nakasako, M., Sakai, J., Kataoka, M., Yamamoto, K. T., Wada, M., Tokunaga, F., and Furuya, M. (1989) A model for the dimeric molecular structure of phytochrome based on small angle x-ray scattering. *FEBS Lett.* 247: 139–142.

Toole, E. H., Borthwick, H. A., Hendricks, S. B., and Toole, V. K. (1953) Physiological studies of the effects of light and temperature on seed germination. *Proc. Int. Seed Test. Assoc.* 18: 267–276.

Torii, K. U., and Deng, X. W. (1997) The role of COP1 in light control of *Arabidopsis* seedling development. *Plant Cell Environ.* 20: 728–733.

Vierstra, R. D. (1994) Phytochrome degradation. In *Photomorphogenesis in Plants*, 2nd ed., R. E. Kendrick and G. H. M. Kronenberg, eds., Martinus Nijhoff, Dordrecht, Netherlands, pp. 141–162.

Vierstra, R. D., and Quail, P. H. (1983) Purification and initial characterization of 124-kilodalton phytochrome from *Avena*. *Biochemistry* 22: 2498–2505.

von Arnim, A. G. and Deng, X-W. (1994) Light inactivation of *Arabidopsis* photomorphogenic repressor COP1 involves a cell-specific regulation of its nucleocytoplasmic partitioning. *Cell* 79: 1035–1045.

Wagner, G. (1996) Intracellular movement. *Prog. Bot.* 57: 68–80.

Wang, Z-Y., Kenigsbuch, D., Sun, L., Harel, E., Ong., M. S., and Tobin, E. M. (1997) A myb-related transcription factor is involved in the phytochrome regulation of an *Arabidopsis Lhcb* gene. *Plant Cell* 9: 491–507.

Winands, A., and Wagner, G. (1996) Phytochrome of the green alga *Mougeotia*: cDNA sequence, autoregulation and phylogenetic position. *Plant Mol. Biol.* 32: 589–597.

Wong, Y-S., McMichael, R. W., Jr., and Lagarias, J. C. (1989) Properties of a polycation-stimulated protein kinase associated with purified *Avena* phytochrome. *Plant Physiol.* 91: 709–718.

Wu, Y., Hiratsuka, K., Neuhaus, G., and Chua, N-H. (1996) Calcium and cGMP target distinct phytochrome-responsive elements. *Plant J.* 10: 1149–1154.

Yeh, K-H., Wu, S-H., Murphy, J. T., and Lagarias, J. C. (1997) A cyanobacterial phytochrome two-component light sensory system. *Science* 277: 1505–1508.

18 Blue-Light Responses: Stomatal Movements and Morphogenesis

MOST OF US are familiar with the observation that house plants placed near a window have branches that grow toward the incoming light. This response, called **phototropism**, is an example of how plants alter their growth patterns in response to the direction of incident radiation. This response to light is intrinsically different from light trapping by photosynthesis. In photosynthesis, plants harness light and convert it into chemical energy (see Chapters 7 and 8). In contrast, phototropism is an example of the use of light as an *environmental signal*. There are two major families of plant responses to light signals: the phytochrome responses, which were covered in Chapter 17, and the **blue-light responses**.

As you might recall, Chapter 9 described some blue-light responses: for example, chloroplast movement within cells in response to incident photon fluxes, and sun tracking by leaves. As with the family of the phytochrome responses, there are numerous plant responses to blue light. Besides phototropism, they include inhibition of hypocotyl elongation, stimulation of chlorophyll and carotenoid synthesis, activation of gene expression, stomatal movements, phototaxis (the movement of motile microorganisms toward or away from light), enhancement of respiration, and anion uptake in algae (Senger 1984). Blue-light responses have been reported in higher plants, algae, ferns, and fungi.

Some responses, such as electrical events at the plasma membrane, can be detected within seconds of irradiation by blue light. More complex metabolic or morphogenetic responses, such as blue light–stimulated pigment biosynthesis in the fungus *Neurospora* or branching in the alga *Vaucheria*, might require minutes, hours, or even days (Horwitz 1994).

Alert readers might have noticed an important difference in how this text has alluded to phytochrome and blue-light responses. The

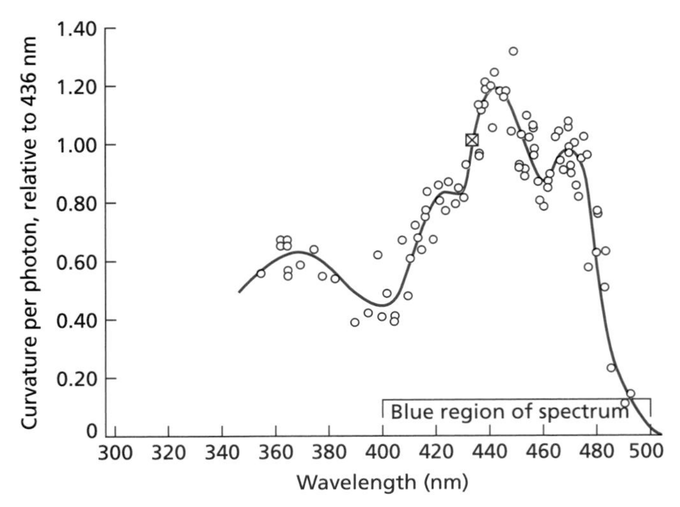

Figure 18.1 Action spectrum for blue light–stimulated phototropism in oat coleoptiles. Recall from Figure 3 in Box 7.1 that an action spectrum shows the relationship between a response to light and the effectiveness of a region of the electromagnetic spectrum, expressed as a function of wavelength. Note the "three-finger" structure in the 400 to 500 nm region, which is typical of specific blue-light responses. (After Thimann and Curry 1960.)

former are identified by a specific photoreceptor (phytochrome), the latter by the blue-light region of the visible spectrum. Several straightforward spectroscopic and biochemical properties of phytochrome, particularly its red/far-red reversibility, made possible its early identification (see Chapter 17). In contrast, the spectroscopy of blue-light responses is complex. Both chlorophylls and phytochrome absorb blue light in the visible (400 to 500 nm) range of the spectrum, and other chromophores and some amino acids, such as tryptophan, absorb light in the ultraviolet (250 to 400 nm) region. How can we then distinguish specific responses to blue light?

One important distinction is that blue light cannot be replaced by any red-light treatment in specific blue-light responses, and there is no red/far-red reversibility. One would expect red or far-red light to be effective if photosynthesis or phytochrome were involved. Another key distinction is that many blue-light responses of higher plants, such as phototropism, stomatal movements, and inhibition of hypocotyl elongation, share a characteristic *action spectrum* (see Figure 3 in Box 7.1). Action spectra can be compared with the *absorption spectra* of putative photoreceptors. A close correspondence between action and absorption spectra provides evidence that a given pigment is the photoreceptor mediating the light response under study.

Action spectra for blue light–stimulated phototropism, stomatal movements, inhibition of hypocotyl elongation and other key blue-light responses share a characteristic "three-finger" fine structure in the 400 to 500 nm region (Figure 18.1) that is not observed in spectra for responses to light that are mediated by photosynthesis, phytochrome, or other photoreceptors (Cosgrove 1994).

In this chapter we describe representative blue-light responses in plants: phototropism, inhibition of stem elongation, activation of gene expression, and stomatal movements. The stomatal responses to blue light are discussed in detail, because of the importance of stomata in leaf gas exchange (see Chapter 9) and because of the importance of stomata in plant acclimations and adaptations to their environment. We also discuss putative blue-light photoreceptors and the signal transduction cascade that links light perception with the final expression of blue-light sensing in the organism.

The Photophysiology of Blue-Light Responses

Blue-light signals are utilized by the plant in many responses that provide means to sense the presence of light and its direction. This section describes the major morphological, physiological, and biochemical changes associated with typical blue-light responses.

Blue Light Stimulates Asymmetric Growth and Bending

Directional responses toward the light can be observed in fungi, ferns, and higher plants. Bending toward the light is a *photomorphogenetic* response that is particularly dramatic in dark-grown seedlings of both monocots and dicots. Detailed studies of phototropic responses as a function of photon fluxes have shown that the response is complex and under certain experimental conditions can include bending toward and away from a unilateral light source (*positive* and *negative* phototropism, respectively; see Box 19.1). However, bending toward the light is the major, if not the sole, functional role of phototropism. Unilateral light is commonly used in experimental studies, but phototropism can also be observed when a seedling is exposed to two unequally bright light sources (Figure 18.2), a situation that is more typical of natural conditions.

Credit for one of the earliest reports on phototropism goes to Charles Darwin, who was one of the first scientists to note that grass seedlings grow toward the light. As they protrude through the soil, the shoots of grasses are protected by a modified leaf that covers the shoot, called a **coleoptile** (Figure 18.3; see also Figure 19.1). As discussed in detail in Chapter 19, unequal light perception in the coleoptile results in unequal concentrations of auxin in the lighted and shaded sides of the coleoptile, unequal growth, and bending.

It is important to keep in mind that phototropic bending occurs only in growing organs, and that coleoptiles and shoots that have stopped elongating will not bend when exposed to unilateral light. In grass seedlings

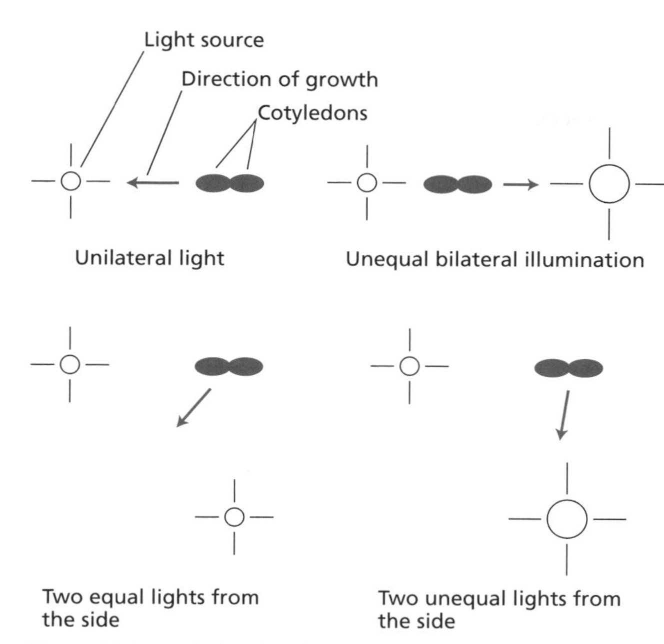

Figure 18.2 Relationship between direction of growth and unequal incident light. Cotyledons from a young seedling are shown as viewed from the top. The diagrams illustrate how the direction of growth varies with the location and the intensity of the light source. The arrows indicate the direction of phototropic curvature. (After Firn 1994.)

growing in soil under sunlight, coleoptiles stop growing as soon as the shoot has emerged from the soil and the first true leaf has pierced the tip of the coleoptile. On the other hand, dark-grown, *etiolated* coleoptiles continue to elongate at high rates for several days and, depending on the species, can attain several centimeters in length. The large phototropic response of these etiolated coleoptiles (see Figure 18.3) has made them a classic model for studies of phototropism (Firn 1994).

The action spectrum shown in Figure 18.1 was obtained through measurements of angles of curvature from oat coleoptiles that were irradiated with light of different wavelengths. The spectrum shows a peak at around 370 nm, and the "three-finger" pattern in the 400 to 500 nm region discussed earlier. An action spectrum for phototropism in the dicot alfalfa (*Medicago sativa*) was found to be very similar to that of oat coleoptiles, suggesting that a common photoreceptor mediates phototropism in the two species (Baskin and Iino 1987).

Phototropism in sporangiophores of the mold *Phycomyces* has been studied to identify genes involved in phototropic responses. The sporangiophore consists of a sporangium (spore-bearing spherical structure) that develops on a stalk consisting of a long, single cell. Growth in the sporangiophore is restricted to a growing zone just below the sporangium. When irradiated with unilateral blue light, the sporangiophore bends toward the light with an action spectrum similar to that of coleoptile phototropism (Cerda-Olmedo and Lipson 1987). These studies of *Phycomyces* have led to the isolation of many mutants with altered phototropic responses and the identification of several genes that are required for normal phototropism. In recent years, phototropism in the small dicot *Arabidopsis* (Steinitz and Poff 1986) has attracted much attention because of the short life cycle of this plant and because of the ease with which mutants of it can be obtained. Recent studies on the genetics and the molecular biology of phototropism in *Arabidopsis* are discussed later in this chapter.

How do plants sense the direction of the light signal? *Light gradients* between lighted and shaded sides have been measured in coleoptiles and hypocotyls from dicot seedlings irradiated with unilateral blue light. When a coleoptile is illuminated with 450 nm blue light, the ratio between the light that is incident to the surface of the illuminated side and the light that reaches the shaded side is 4:1 at the tip and the midregion of the coleoptile, and 8:1 at the base (Figure 18.4). On the other hand, there is a *lens* effect in the sporangiophore of the mold *Phycomyces* irradiated with unilateral blue light, and as a result, the light measured at the distal cell surface of the sporangiophore is about twice the light that is incident at the surface of the illuminated side. Light gradi-

Figure 18.3 Time-lapse photograph of a corn coleoptile growing toward unilateral blue light. Unilateral blue light was given from the left. The consecutive exposures were made 30 minutes apart. Note the increasing angle of curvature as the coleoptile bends. (Photograph courtesy of M. A. Quiñones. See the copyright page for a full-color version of this photograph.)

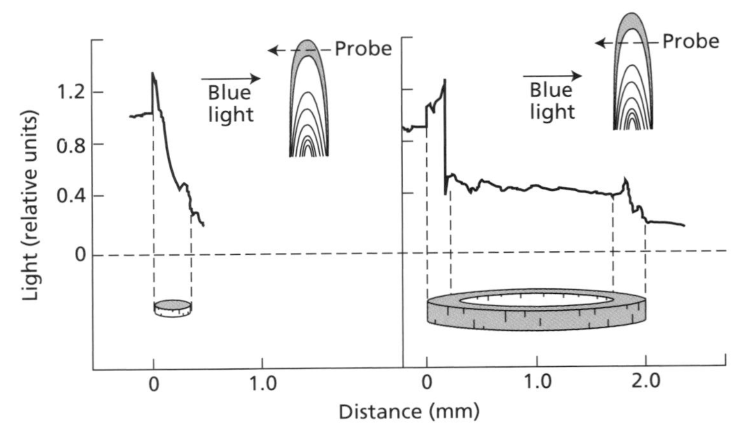

Figure 18.4 Distribution of transmitted, 450 nm blue light in an etiolated corn coleoptile. The diagram in the upper right of each frame shows the area of the coleoptile being measured by a fiber-optic probe. A cross section of the tissue appears at the bottom of each frame. The trace above it shows the amount of light sensed by the probe at each point. A sensing mechanism that depended on light gradients would sense the difference in the amount of light between the lighted and shaded sides of the coleoptile, and this information would be transduced into an unequal auxin concentration and bending. (After Vogelmann and Haupt 1985.)

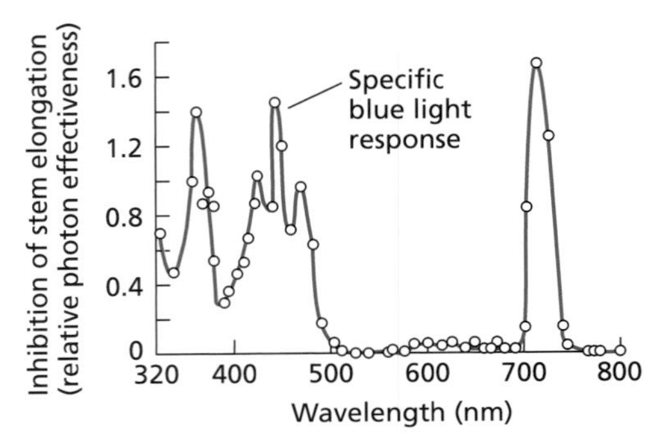

Figure 18.5 Action spectrum for the inhibition of stem elongation in dark-grown lettuce and mustard. The three-finger structure in the 400 to 500 nm region is typical of specific blue-light responses. The responses in the 600 to 750 nm region are indicative of a phytochrome-mediated response. (From Hartmann 1967.)

ents and lens effects could play a role in how the bending organ senses the direction of the unilateral light (Vogelmann 1994).

Blue Light Inhibits Stem Elongation

As discussed in Chapter 17, the stem of germinating seedlings elongates very rapidly, and the inhibition of stem elongation by light is a key morphogenetic response of the seedling emerging from the soil surface. The conversion of Pr to Pfr (the red and far-red absorbing forms of phytochrome, respectively) in etiolated seedlings causes a phytochrome-dependent, sharp decrease in elongation rates (see Figure 17.1). However, action spectra for that response show strong activity in the blue region, which cannot be explained by the absorption properties of phytochrome (Figure 18.5). In fact, the 400 to 500 nm blue region of the action spectrum for the inhibition of stem elongation closely resembles that of phototropism (see Figure 18.1). Many experiments have shown that the inhibition of hypocotyl elongation elicited by blue light is primarily a specific blue-light response, independent from phytochrome activity.

Some of the genetic evidence separating the responses to the two photoreceptor systems was discussed in Chapter 17. Photophysiological evidence is provided by results showing that low fluence rates of blue light applied under strong background yellow light can inhibit the rate of hypocotyl elongation of lettuce seedlings by more than 50% (Figure 18.6). These doses of blue light are too small to significantly change the ratio of Pfr to Pr established by the background yellow light, thus ruling out a phytochrome effect on the observed response.

Another important difference between phytochrome and blue light–mediated hypocotyl responses is the swiftness of the response. Whereas phytochrome-mediated changes in elongation rates can be detected within 15 to 90 minutes, depending on the species, blue-light responses can be measured within 15 seconds. One of the fastest responses elicited by blue light is a depolarization of the membrane of hypocotyl cells that precedes the inhibition of growth rate (Figure 18.7). The membrane depolarization is caused by the activation of anion channels (see Chapter 6), which facilitates the efflux of

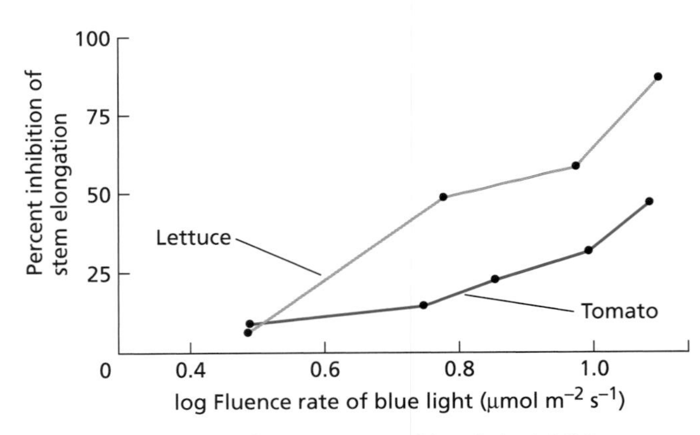

Figure 18.6 A specific response to blue light inhibits stem elongation of lettuce and tomato seedlings under a background of a strong yellow (589 nm) light that maintained a constant photostationary state of phytochrome. Inhibition of more than 50% of the rate of stem elongation caused by the addition of blue light that was less than 0.5% of the background yellow light demonstrates a specific blue-light response, distinct from that mediated by phytochrome. (After Thomas and Dickinson 1979.)

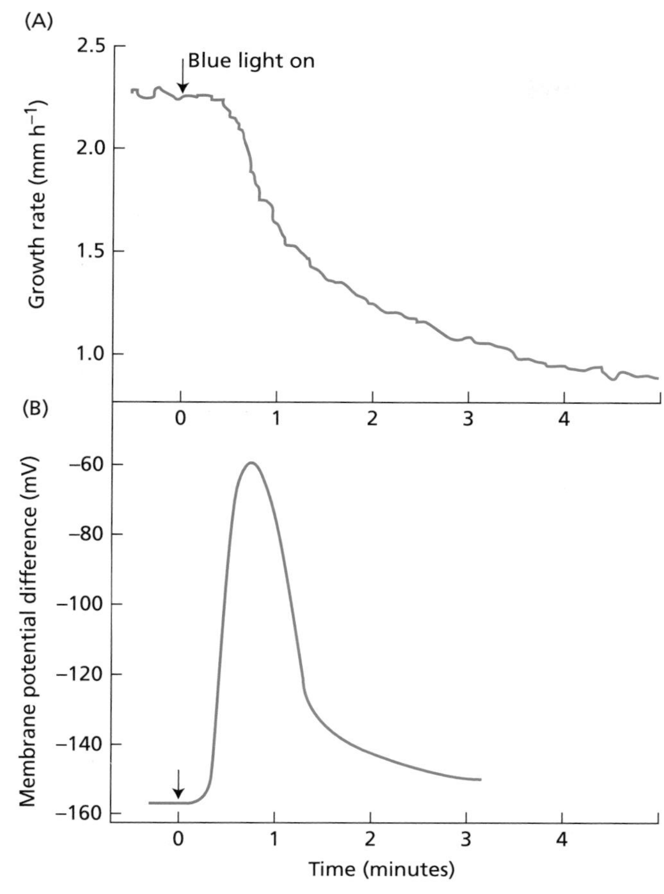

Figure 18.7 Blue light–induced changes in elongation rates of etiolated cucumber seedlings (A) and transient membrane depolarization of hypocotyl cells (B). As the membrane depolarization (measured with intracellular electrodes) reaches its maximum, growth rate (measured with position transducers) declines sharply. Note that the membrane starts to depolarize before the growth rate begins to decline, suggesting a cause–effect relation between the two phenomena. (After Spalding and Cosgrove 1989.)

anions such as chloride. Use of an anion channel blocker prevented the blue light–dependent membrane depolarization and decreased the inhibitory effect of blue light on hypocotyl elongation (Cho and Spalding 1996).

Blue Light Regulates Gene Expression

Blue light can also regulate the expression of genes involved in several important morphogenetic processes (Tilghman et al. 1997). Some of these light-activated genes have been studied in detail—for example, the genes that code for the enzyme chalcone synthase, which catalyzes the first committed step in flavonoid biosynthesis (see Chapter 13), for the small subunit of rubisco (see Chapter 8), and for the proteins that bind chlorophylls *a* and *b* (see Chapter 7). The regulatory elements of the light-activated genes are under investigation (see Chapters 14 and 17). Most of the light-activated genes studied to date show sensitivity to both blue and red light, as well as red/far-red reversibility, implicating both phytochrome (see Chapter 17) and specific blue-light responses.

One well-documented instance of gene expression that is mediated solely by a blue light–sensing system involves the *GSA* gene in the photosynthetic unicellular alga *Chlamydomonas reinhardtii* (Matters and Beale 1995). This gene encodes the enzyme glutamate-1-semialdehyde aminotransferase, a key enzyme in the chlorophyll biosynthesis pathway (see Chapter 7). The absence of phytochrome in *C. reinhardtii* simplifies the analysis of blue-light responses in this experimental system. In synchronized cultures of *C. reinhardtii*, levels of *GSA* mRNA are strictly regulated by blue light, and 2 hours after the onset of illumination, *GSA* mRNA levels are 26-fold higher than they are in the dark (Figure 18.8). These blue light–mediated mRNA increases precede increases in chlorophyll content, indicating that chlorophyll biosynthesis is being regulated by activation of the *GSA* gene.

Blue Light Stimulates Stomatal Opening

We now turn our attention to the stomatal response to blue light. Several characteristics of blue light–dependent stomatal movements make guard cells an attractive experimental system for the study of blue-light responses. The stomatal response to blue light is rapid and reversible, and it is localized in a single cell type, the guard cell. Unlike phototropism or hypocotyl elonga-

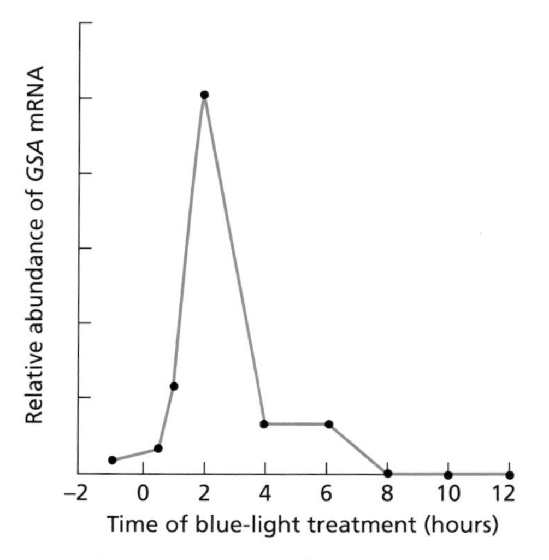

Figure 18.8 Time course of blue light–dependent expression of the *GSA* gene in *Chlamydomonas reinhardtii*. The *GSA* gene encodes the enzyme glutamate-1-semialdehyde aminotransferase, which regulates an early step in chlorophyll biosynthesis. The levels of *GSA* mRNA were probed in Northern blots with *GSA* cDNA (complementary DNA) and were expressed relative to levels in the dark. (After Matters and Beale 1995.)

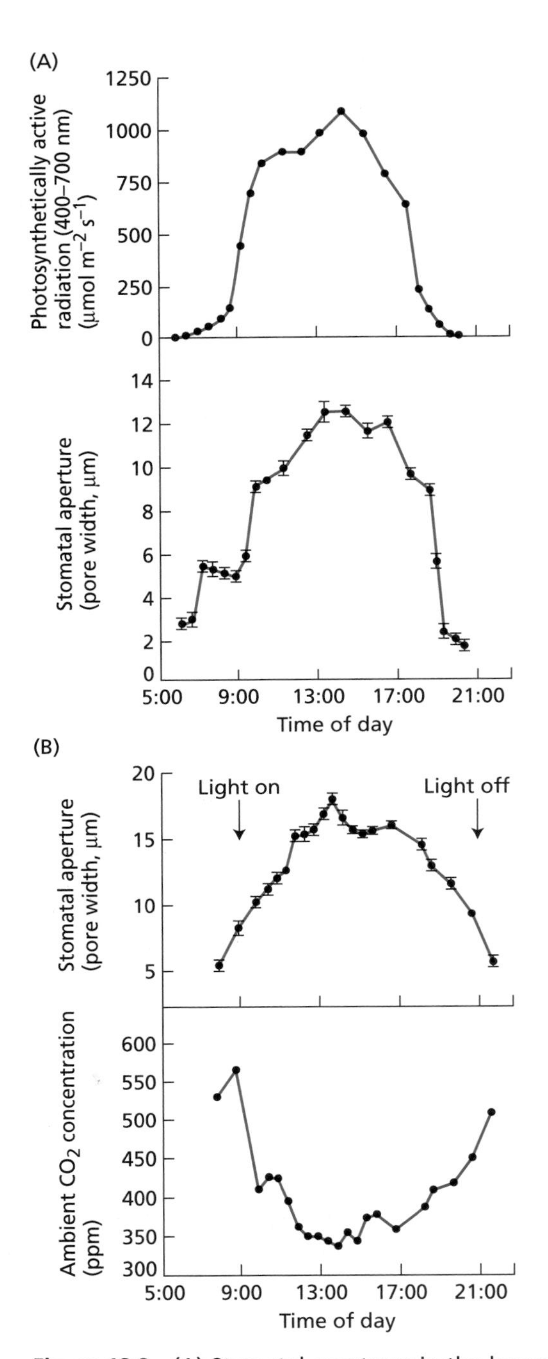

Figure 18.9 (A) Stomatal apertures in the lower surface of leaves of *Vicia faba* grown in a greenhouse, measured as the width of the stomatal pore, closely follow the levels of photosynthetically active radiation (400 to 700 nm) incident to the leaf, indicating that the response to light was the dominant response regulating stomatal apertures. (B) Stomatal apertures in the lower surface of leaves of *Vicia faba* grown in a growth chamber, which were kept at constant light and temperature, closely follow ambient CO_2 concentrations, indicating that in these growing conditions, stomata have a high sensitivity to CO_2. (A after Srivastava and Zeiger 1995a; B after Talbott et al. 1996.)

tion, which are functionally important at early stages of development, the stomatal response to blue light regulates stomatal movements throughout the life of the plant. Stomata have a major regulatory role in gas exchange in leaves (see Chapter 9), and under certain conditions they can affect yields of agricultural crops (see Chapter 25). A putative chromophore for the photoreception of blue light in guard cells has been identified, and the signal transduction cascade that links the perception of blue light with the opening of stomata is understood in considerable detail.

In the sections that follow we will discuss the two components of the stomatal response to light, the osmoregulatory mechanisms that drive stomatal movements, and the role of a blue light–activated H^+-ATPase in ion uptake by guard cells.

Light is the dominant environmental signal controlling stomatal movements in leaves of well-watered plants growing in natural environments. In greenhouse-grown leaves of broad bean (*Vicia faba*), stomatal movements closely track photon fluxes at the leaf surface: The stomata open as incident photon fluxes increase and close as they decrease (Figure 18.9A). Recall from Chapter 9 that intercellular CO_2 concentrations change as a function of photosynthetic rates. Because stomata can be very sensitive to ambient CO_2 concentrations (Figure 18.9B), it is difficult to separate the stomatal responses to light and to CO_2 in the intact leaf. However, work with stomata from detached epidermis (Figure 18.10), kept at a constant ambient CO_2 concentration, has resolved a specific stomatal response to light that can be separated from other environmental signals.

Photobiological and metabolic studies have shown that this response to light is the integrated expression of two distinct photoreceptor systems, one depending on guard cell photosynthesis (Figure 18.11), and the other driven by a specific response to blue light. Early studies of the stomatal response to white light showed that DCMU (dichlorophenyldimethylurea), an inhibitor of photosynthetic electron transport (see the figure in Box 7.3), causes a partial inhibition of light-stimulated stomatal opening. These results indicated that photosynthesis in the guard cell chloroplast is involved in light-dependent stomatal opening, but the observation that the inhibition is not complete suggested a nonphotosynthetic component of the stomatal response to light. More definitive evidence showing that the stomatal response to light is driven by two distinct photoreceptor systems was obtained in dual-beam experiments (Figure 18.12).

In these experiments, red light was used to *saturate* the photosynthetic response, and low photon fluxes of blue light were added after the response to the saturating red light was completed. The addition of blue light caused substantial further stomatal opening that cannot

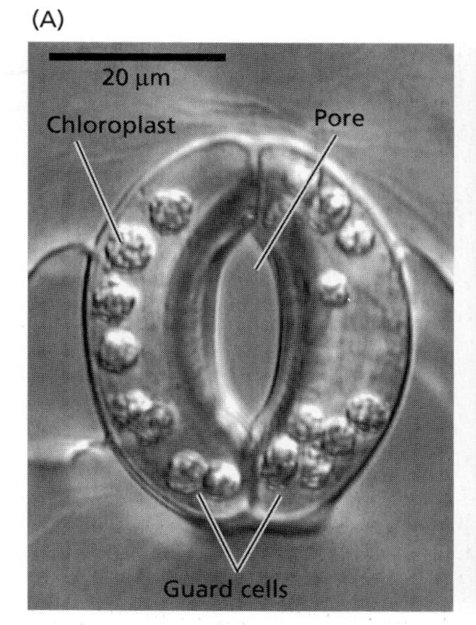

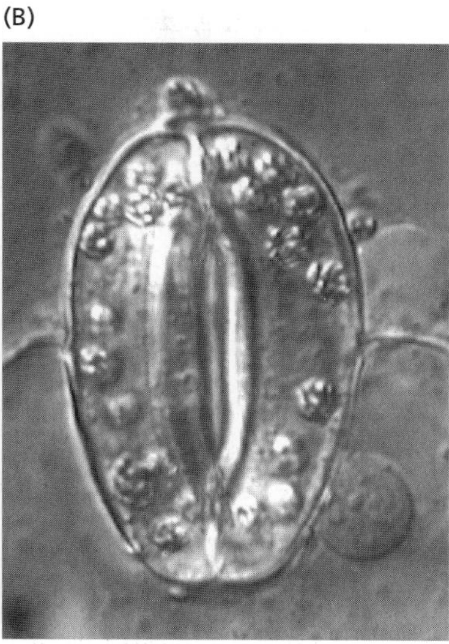

Figure 18.10 Stomata in detached epidermis of *Vicia faba*, showing the open (A) and closed (B) states. Stomatal apertures are measured under a microscope as aperture widths. Measurements of the change in aperture as a function of experimental conditions allow the characterization of stomatal responses to different environmental stimuli. (Photos courtesy of E. Raveh.)

be explained on the basis of guard cell photosynthesis, because additional red light did not cause further opening. An action spectrum for the stomatal response to blue light under background red illumination clearly shows the typical three-finger pattern (Figure 18.13) (Karlsson 1986) that was discussed earlier in the chapter for other blue-light responses. This three-finger pattern, typical of blue-light responses and distinctly different from the action spectrum for photosynthesis, provides further evidence that in addition to photosynthesis, guard cells respond specifically to blue light.

Studies of guard cell protoplasts have shown that they *swell* in response to blue light (Figure 18.14), indicating that blue light is perceived within the guard cells proper. The swelling of guard cell protoplasts also illustrates the functioning of guard cells as turgor valves. The uptake of ions and the accumulation of organic solutes in guard cells cause an increase in osmotic

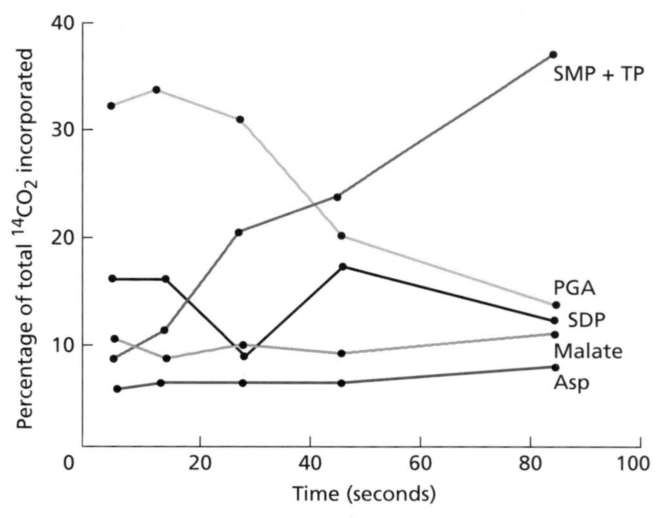

Figure 18.11 Photosynthetic carbon fixation in guard cell protoplasts of *Vicia faba.* The guard cell protoplasts were exposed to radiolabeled CO_2 at time zero. Radioactivity was incorporated into organic acids and into the key Calvin cycle metabolite 3-phosphoglycerate (PGA). Photosynthetic carbon fixation by guard cells is one of the osmoregulatory pathways that provides a source of sucrose. Asp, aspartate; SDP, sugar diphosphates; SMP, sugar monophosphates; TP, triose phosphates. (From Gotow et al. 1988, © American Society of Plant Physiologists, reprinted with permission.)

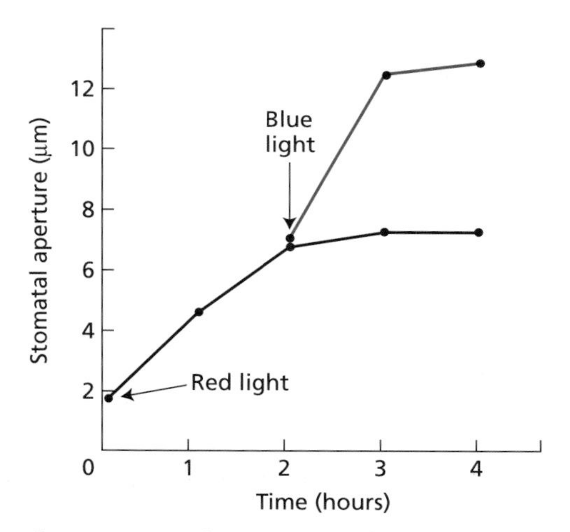

Figure 18.12 The response of stomata to blue light under a red-light background. Stomata from detached epidermis of *Commelina communis* (common dayflower) were treated with saturating photon fluxes of red light (dark trace). In a parallel treatment, stomata illuminated with red light were also illuminated with blue light, as indicated by the arrow (green trace). The increase in stomatal apertures above the level reached in the presence of saturating red light indicates that a different photoreceptor system, stimulated by blue light, is mediating the additional increases in aperture. (From Schwartz and Zeiger 1984.)

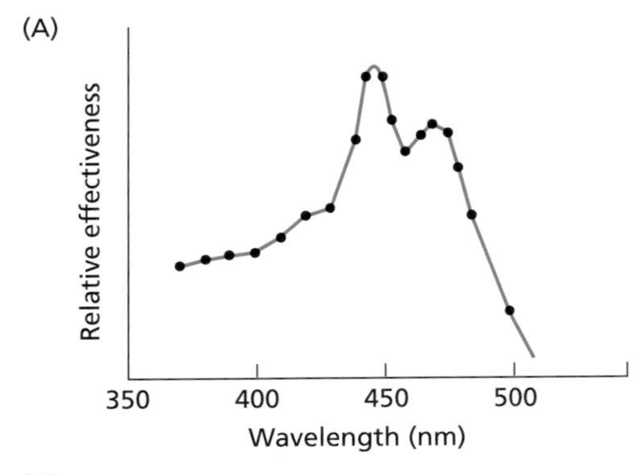

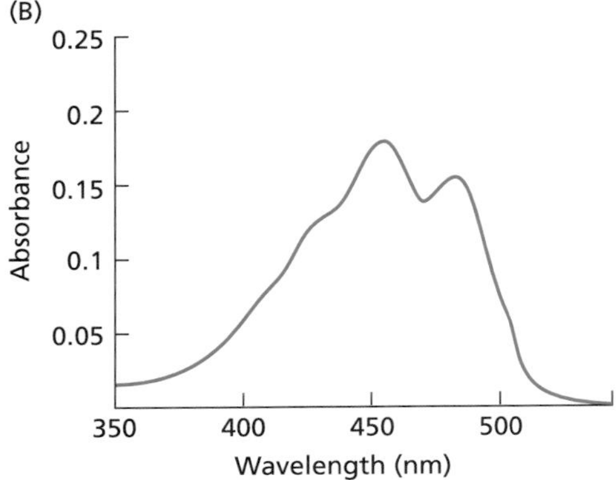

Figure 18.13 The matching between the action spectrum for blue light–stimulated stomatal opening (under a red-light background) (A) and the absorption spectrum of zeaxanthin in ethanol (B). (A after Karlsson 1986.)

potential and turgor that is mechanically transduced into an increase in stomatal apertures (see Chapter 4). In the absence of a cell wall, the blue light–mediated increase in osmotic pressure causes the guard cell protoplast to swell, and thus to increase in volume.

Blue Light Activates a Proton Pump in the Guard Cell Plasma Membrane

When guard cell protoplasts from broad bean (*Vicia faba*) kept under saturating fluence rates of red light are irradiated with low photon fluxes of blue light, the pH of the suspension medium becomes more acidic (Figure 18.15). This blue light–induced acidification is sensitive to inhibitors that are known to dissipate pH gradients, and to blockers of the proton-pumping H^+-ATPase, such as vanadate (see Figure 18.14C; see also Chapter 6). Such sensitivity indicates that the acidification results from the activation by blue light of a proton-pumping ATPase in the guard cell plasma membrane (Shimazaki et al. 1986). The extrusion of protons into the suspension medium lowers its pH.

The activation of electrogenic pumps such as the proton-pumping ATPase can be measured in patch-clamp-

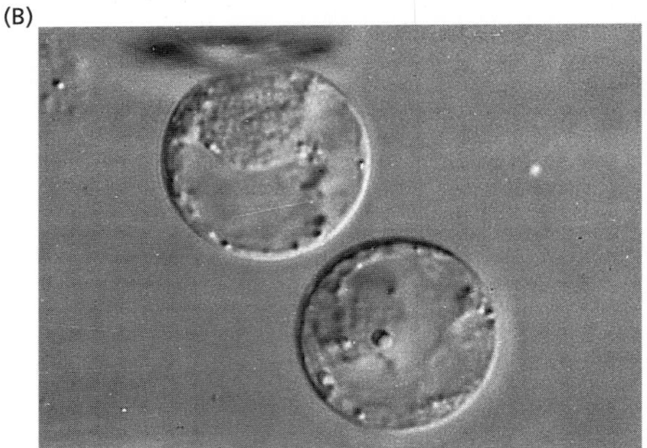

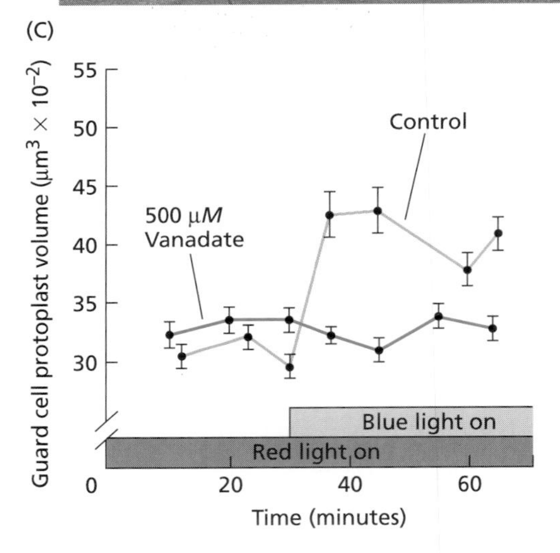

Figure 18.14 Blue light–stimulated swelling of guard cell protoplasts. Dark-adapted (A) and blue light–irradiated (B) guard cell protoplasts from onion (*Allium cepa*). (C) Blue light stimulates the swelling of guard cell protoplasts of broad bean (*Vicia faba*), and vanadate, an inhibitor of the H^+-ATPase, inhibits this swelling. Blue light stimulates ion and water uptake in the guard cell protoplasts, which in the intact guard cells provides a mechanical force that drives increases in stomatal apertures. In the absence of a rigid cell wall, the guard cell protoplasts swell. (A and B from Zeiger and Hepler 1977; C after Amodeo et al. 1992.)

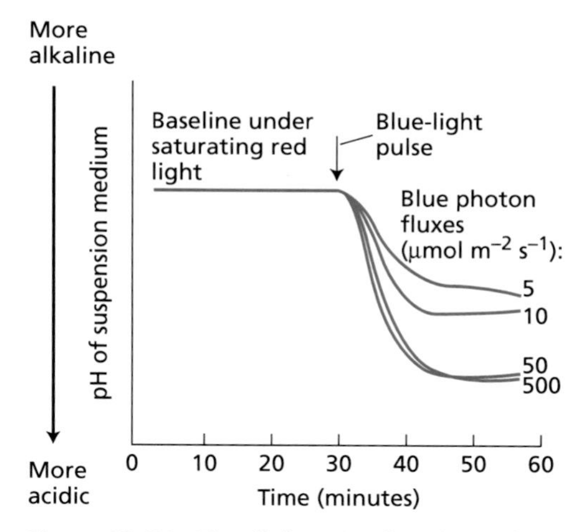

Figure 18.15 Blue light–stimulated acidification of a suspension medium of guard cell protoplasts of *Vicia faba*. Guard cell protoplasts acidify their suspension medium when exposed to pulses of blue light given under saturating photon fluxes of red light. A blue-light pulse 30 s long and of different photon fluxes (5, 10, 50, and 500 μmol m^{-2} s^{-1}) was administered at the point indicated by the arrow. The acidification results from the stimulation of an H^+-ATPase at the plasma membrane by blue light, and it is associated with the swelling shown in Figure 18.14. (After Shimazaki et al. 1986.)

ing experiments as an outward electric current at the plasma membrane (see Chapter 6). Figure 18.16A shows a patch clamp recording of a guard cell protoplast treated with the fungal toxin fusicoccin (see Chapter 19), a well-characterized activator of plasma membrane ATPases. Exposure to fusicoccin stimulates an outward electric current, which is abolished by the proton ionophore carbonyl cyanide *m*-chlorophenylhydrazone (CCCP). This proton ionophore makes the plasma membrane highly permeable to protons, thus precluding the formation of a proton gradient across the membrane and abolishing net proton efflux. The relationship between proton pumping at the guard cell plasma membrane and stomatal opening is evident from the observation that fusicoccin stimulates stomatal opening, and that CCCP inhibits the fusicoccin-stimulated opening.

Pulses of blue light given under a saturating red-light background also stimulate an outward electric current from guard cell protoplasts (Figure 18.16B) (Assmann et al. 1985). The acidification measurements shown in Figure 18.15 indicate that the outward electric current is carried by protons. Pulses of red light given under a red-light background do not elicit net proton efflux, indicating that the stimulation of proton pumping by blue-light pulses is a specific blue-light response.

Some of the characteristics of the responses to blue-light pulses shown in Figures 18.15 and 18.16B underscore very interesting properties of blue-light responses. In contrast to typical photosynthetic responses, which are activated very quickly after a "light on" signal, and cease when the light goes off (see, for instance, Figure 7.10), responses to blue-light pulses proceed at maximal rates for several minutes after the pulse is applied. This property, also shown for blue light–stimulated coleoptile bending (Iino 1988), can be explained on the basis of a physiologically inactive form of the blue-light photoreceptor that is converted to an active form by blue light, with the active form reverting slowly to the physiologically inactive form in the absence of blue light (Iino et al. 1985). The rate of the response to a blue-light pulse would thus depend on the time course of the reversion of the active form to the inactive one.

Another characteristic of responses to blue-light pulses is a lag time, which lasts about 25 s in both the

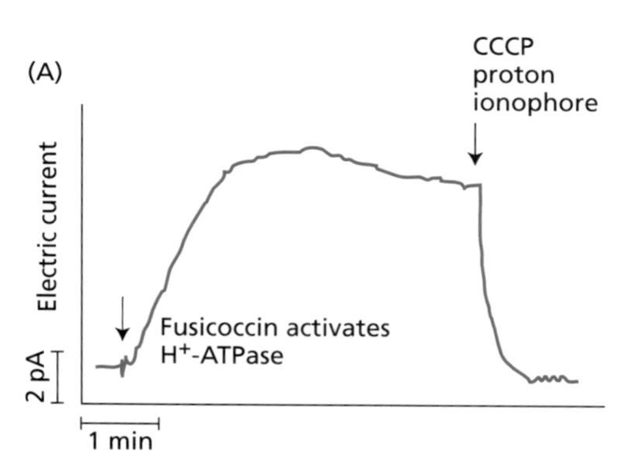

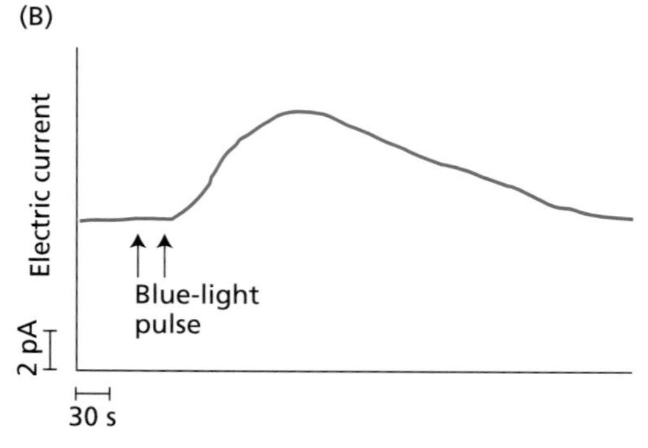

Figure 18.16 Electric currents at the plasma membrane of guard cell protoplasts, measured in patch clamp experiments. (A) Outward electric current at the plasma membrane of a guard cell protoplast stimulated by fusicoccin, an activator of the H^+-ATPase. The current is abolished by the proton ionophore CCCP (carbonyl cyanide *m*-chlorophenylhydrazone). (B) Outward electric current at the plasma membrane of a guard cell protoplast stimulated by a blue-light pulse. Stimulation of an outward electric current at the guard cell plasma membrane in response to a blue-light pulse indicates that blue light–stimulated stomatal opening is mediated by the stimulation by blue light of the H^+-ATPase. (A after Serrano et al. 1988; B after Assmann et al. 1985.)

acidification and the outward electric currents stimulated by blue light (see Figures 18.15 and 18.16). This amount of time is probably required for the signal transduction cascade to proceed from the photoreceptor site to the proton-pumping ATPase and for the proton gradient to form. Similar lag times have been measured for blue light–dependent inhibition of hypocotyl elongation, which was discussed earlier.

Figure 18.15 also illustrates the fact that pulses of the same duration but of increasing photon flux elicit correspondingly larger responses, until saturation is reached. The blue-light response of stomata obeys the *reciprocity law*. Recall from Chapter 17 that a photobiological response obeys the reciprocity law when it depends on the total dose of light, and not on the photon flux or the time of exposure used. This is the case with the net increases in aperture elicited by blue light (Karlsson 1986). This close relationship among the number of incident blue-light photons, proton pumping at the guard cell plasma membrane, and stomatal opening (Iino et al. 1985) suggests that the blue-light response of stomata can function as the component of the stomatal photoresponse that senses how much light is reaching the leaf surface.

Blue light–stimulated proton extrusion is nearly 15-fold more efficient, on a photon flux basis, than ATP synthesis in the guard cell chloroplast (Shimazaki et al. 1986). Such a high efficiency is not typical of energy conversion processes such as photosynthesis (see the discussion on quantum yields in Chapter 7) and suggests that the blue-light signal is *amplified*, as is typical of many signal transduction cascades, such as vision in vertebrates (see Chapter 14).

The magnitude of the stomatal response to blue light increases with the fluence rate of background red light. The regulation of blue-light responses by red light can be shown in experiments in which saturating blue-light pulses are applied under increasing fluence rates of *subsaturating* red-light background. In these conditions, the response to a saturating blue-light pulse increases with the fluence rates of background red light (see Figure 18.22).

A similar dependence of blue light–stimulated phototropic bending on the fluence rates of red-light pretreatment has been interpreted as an interaction between the blue-light photoreceptor and phytochrome (see Chapter 17). However, phytochrome has no known effects on blue light–stimulated stomatal apertures. Rather, the effect of background red light on blue light–stimulated stomatal opening suggests an interaction between the specific blue-light response and guard cell photosynthesis. A mechanism for that interaction, resulting from a photosynthesis-dependent accumulation of a guard cell carotenoid implicated in blue-light photoreception, is discussed later in this chapter.

Blue Light Regulates Osmotic Relations of Guard Cells

Blue light affects guard cell osmoregulation through its effects on proton pumping (described in the preceding section) and on the synthesis of organic solutes. Before discussing these blue-light responses, we will briefly describe the major osmotically active solutes in guard cells.

Potassium and its counterions malate and chloride increase in guard cells when stomata open and decrease when stomata close (Outlaw 1983). Potassium concentrations can increase severalfold in open stomata, from 100 m*M* in the closed state to 400 to 800 m*M* in the open state, depending on the plant species and the experimental conditions. These vast changes in the positively charged potassium ions are electrically balanced by the anions Cl^- and malate^{2-}. In species of the genus *Allium*, such as onion (*Allium cepa*), K^+ ions are balanced by Cl^-. In most species, however, potassium fluxes are balanced by varying amounts of Cl^- and the organic anion malate^{2-}.

Like potassium, chloride is taken up into the guard cells during stomatal opening and extruded during stomatal closing. Malate, on the other hand, is synthesized in the guard cell cytosol, in a metabolic pathway that uses carbon skeletons generated by starch hydrolysis (see Figure 2A in Box 18.1). The malate content of guard cells decreases during stomatal closing, but it is unclear whether malate is utilized as a substrate for respiration in the mitochondria or is extruded into the apoplast.

The proportion of malate and chloride used to balance the potassium ions in guard cells from intact *Vicia* leaves varies with environmental conditions. In guard cells from leaves grown in a greenhouse, most of the potassium ions were balanced with chloride; in guard cells from leaves grown in a growth chamber, a larger proportion of potassium was balanced by malate (Talbott et al. 1996). Because greenhouse-grown guard cells used little malate and the greenhouse environment is quite similar to a natural environment, these findings suggest that, at least in *Vicia*, Cl^- is the dominant physiological counterion for K^+ under natural conditions.

Potassium and chloride fluxes depend on secondary transport mechanisms driven by the H^+-ATPase–generated proton motive force discussed earlier in the chapter. Proton extrusion makes the difference in electric potential across the guard cell plasma membrane more negative; light-dependent hyperpolarizations as high as 50 mV have been measured. In addition, proton pumping generates a pH gradient of about 0.5 to 1 pH unit. The electrical component of the proton motive force provides a driving force for the passive uptake of potassium ions via voltage-regulated potassium channels (Schroeder et al. 1994; see Chapter 6). Chloride is thought to be taken up through a Cl^-–H^+ symporter. Thus, blue light–dependent stimulation of proton pumping is one of the mechanisms that mediates the regulation of osmotic relations in guard cells by blue light.

Plant physiologists working on stomatal function at the beginning of the twentieth century were intrigued about osmoregulation in guard cells (Meidner 1987). The botanist H. von Mohl had proposed in 1856 that turgor changes in guard cells provide the driving force for stomatal movements, and the plant physiologist F. E. Lloyd hypothesized in 1908 that these turgor changes depend on starch–sugar interconversions, a concept that led to a starch–sugar hypothesis of stomatal movements. Guard cell chloroplasts contain large, prominent starch grains and their starch content decreases during stomatal opening and increases during closing. Starch, an insoluble, high-molecular-weight polymer of glucose, does not contribute to the cell's osmotic potential, but the hydrolysis of starch into soluble sugars causes an increase in the osmotic potential of guard cells. In the reverse process, starch synthesis decreases the sugar concentration, resulting in a lowering of the cell's osmotic potential, which the starch–sugar hypothesis predicted to be associated with stomatal closing.

This hypothesis was widely accepted until the discovery of potassium fluxes in guard cells by S. Imamura in 1943, which were later confirmed by M. Fujino and R. A. Fischer. In the latter twentieth century, many studies have characterized a major role of ions, primarily potassium, in guard cell osmoregulation, and these developments have overshadowed the starch–sugar hypothesis (Outlaw 1983). As the next section describes, however, recent studies have found a major osmoregulatory phase in guard cells in which sucrose is the dominant osmotically active solute.

Sucrose Is an Osmotically Active Solute in Guard Cells

Studies of daily courses of stomatal movements in leaves grown in a greenhouse or in a growth chamber have shown that the potassium content in guard cells increases in parallel with early-morning opening, but it decreases in the early afternoon under conditions in which apertures continue to increase. The sucrose content of guard cells increases slowly in the morning, but upon potassium efflux, sucrose becomes the dominant osmotically active solute, and stomatal closing at the end of the day parallels a decrease in the sucrose content of guard cells (Figure 18.17). These osmoregulatory features imply that stomatal opening is associated primarily with K^+ uptake and closing is associated with a decrease in sucrose content. The need for distinct potassium- and sucrose-dominated osmoregulatory phases is unclear, but it might underlie distinct regulatory aspects

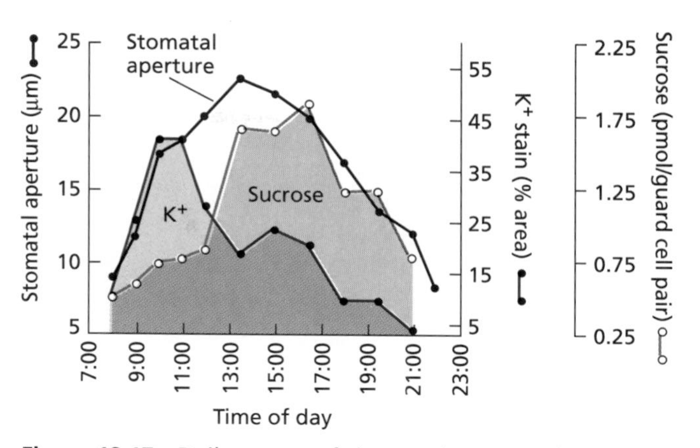

Figure 18.17 Daily course of changes in stomatal aperture, and in potassium and sucrose content in guard cells from intact leaves of broad bean (*Vicia faba*). A potassium-dependent osmoregulatory phase is observed in the morning, and a sucrose-dependent phase in the afternoon. (After Talbott and Zeiger 1998.)

of stomatal function. Potassium might be the osmotically active solute of choice for the consistent daily opening that occurs at sunrise. The sucrose phase might be associated with the functional coupling of stomatal conductance in the epidermis to rates of photosynthesis in the mesophyll (Talbott and Zeiger 1998).

Where do osmotically active solutes originate? Experimental studies have characterized four distinct metabolic pathways that can supply osmotically active solutes to guard cells: (1) the uptake of K^+ and Cl^- coupled to the biosynthesis of malate^{2-}, (2) the production of sucrose from starch hydrolysis, (3) the production of sucrose by photosynthetic carbon fixation in the guard cell chloroplast, and (4) the uptake of apoplastic sucrose generated by mesophyll photosynthesis (see Figure 2 in Box 18.1). For instance, red light–stimulated stomatal opening in detached epidermis depends solely on sucrose generated by guard cell photosynthesis, with no detectable K^+ uptake. Other experimental conditions can activate one of the other pathways, or several pathways simultaneously. It is unclear, however, how these different pathways are regulated during stomatal movements in the intact leaf.

Photoreceptors

Early hypotheses on blue-light photoreceptors were formulated in the 1930s and 1940s, following the characterization of action spectra showing that light-stimulated coleoptile phototropism is a blue-light response. Dark-grown coleoptiles look yellow because of their high carotenoid content. On the basis of the similarity between the action spectrum for phototropism and the absorption spectra of carotenoids (Figure 18.18), E. Bun-

BOX 18.1

A Blue Light–Activated Metabolic Switch

STUDIES ON PHOTOSYNTHETIC ALGAE, such as *Chlorella*, have shown that irradiation by blue light increases respiratory rates of dark-grown cells, with an action spectrum that is typical of specific blue-light responses (Kamiya and Miyachi 1974). The changes in respiratory rates are associated with a drastic shift in metabolism, which can also be observed in experiments using blue and red light. Red light stimulates photosynthetic carbon fixation in these algae, and the fixed carbon is incorporated into glucose and sucrose. The metabolic shift is observed when low photon fluxes of blue light are added to the red-light treatment; glucose and sucrose are no longer synthesized, and the fixed carbon is used in the biosynthesis of organic acids, amino acids, and proteins.

Blue light is also a powerful regulator of guard cell metabolism. Guard cells irradiated with red light accumulate sucrose (Talbott and Zeiger 1998); if low photon fluxes of blue light are added to the red-light irradiation, sucrose stops accumulating and guard cells accumulate K^+ and malate^{2-} (Figure 1). Thus, the blue light–stimulated shift to the synthesis of the organic acid malate in guard cells parallels the shift to the synthesis of organic acids and amino acids in photosynthetic algae.

The experimental use of blue and red light has been useful for investigations of osmoregulatory pathways associated with stomatal movements (Figure 2). The rapid activation of proton pumping by blue light discussed earlier in the chapter generates a proton at the guard cell plasma membrane that drives potassium and chloride uptake. Blue light also stimulates starch hydrolysis and malate biosynthesis. Prolonged incubations of guard cells in detached epidermis under low fluence rates of blue light result in large stomatal apertures accompanied by extensive starch hydrolysis. In these conditions, an initial osmoregulatory phase depends on potassium uptake and malate biosynthesis, and it is followed by a second phase that depends on sucrose accumulation (Talbott and Zeiger 1998).

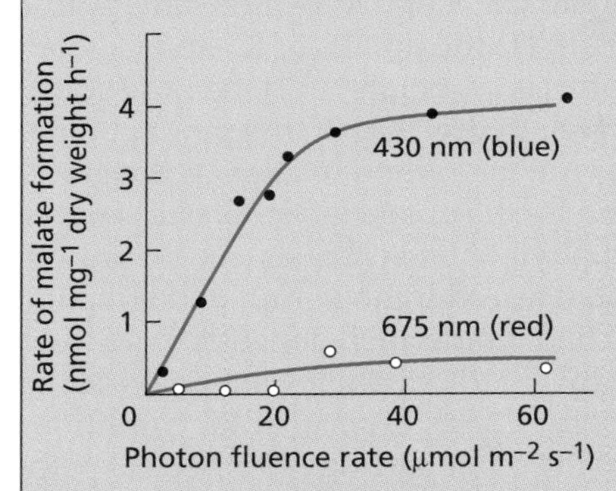

Figure 1 Blue light stimulates malate synthesis in guard cells. Under red light, guard cells synthesize mostly glucose and sucrose; this metabolic pathway is deactivated by blue light, and metabolism switches to starch degradation and malate biosynthesis. (From Ogawa et al. 1978.)

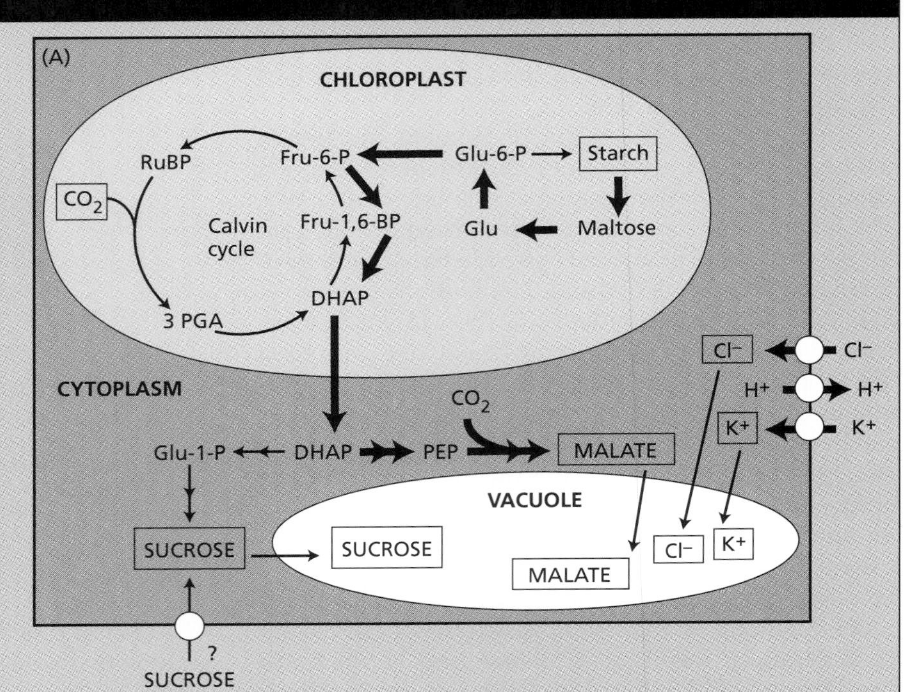

Figure 2 Three distinct osmoregulatory pathways in guard cells. The dark arrows identify the major metabolic steps of each pathway that lead to the accumulation of osmotically active solutes in the guard cells. (A) Potassium and its counterions. Potassium and chloride are taken up in secondary transport processes driven by a proton gradient; malate is formed from the hydrolysis of starch. (B) Accumulation of sucrose from starch hydrolysis. (C) Accumulation of sucrose from photosynthetic carbon fixation. The possible uptake of apoplastic sucrose is also indicated. DHAP, dihydroxyacetone 3-phosphate ; Fru-1,6-BP, fructose-1,6-bisphosphate; Fru-6-P, fructose-6-phosphate; Glu, glucose; Glu-1-P, glucose-1-phosphate; Glu-6-P, glucose-6-phosphate; PEP, phosphoenolpyruvate; PGA, phosphoglycerate; RuBP, ribulose bisphosphate. (From Talbott and Zeiger 1998.)

ning proposed in 1937 that β-carotene (see Figure 7.5B) is the photoreceptor for phototropism.

Coleoptiles are also enriched in riboflavin (for its chemical structure, see Figure 18.23A). Like all oxidized (but not reduced) flavins, oxidized riboflavin absorbs blue light with an absorption spectrum showing a maximum that matches the maximum of the action spectrum for phototropism (see Figure 18.18A). Arthur W. Galston and his colleagues found in the late 1940s that, when reduced by blue light, riboflavin could mediate the photodestruction of an auxin in vitro, and they proposed the destruction of auxin by blue light as a possi-

BOX 18.1 (*continued*)

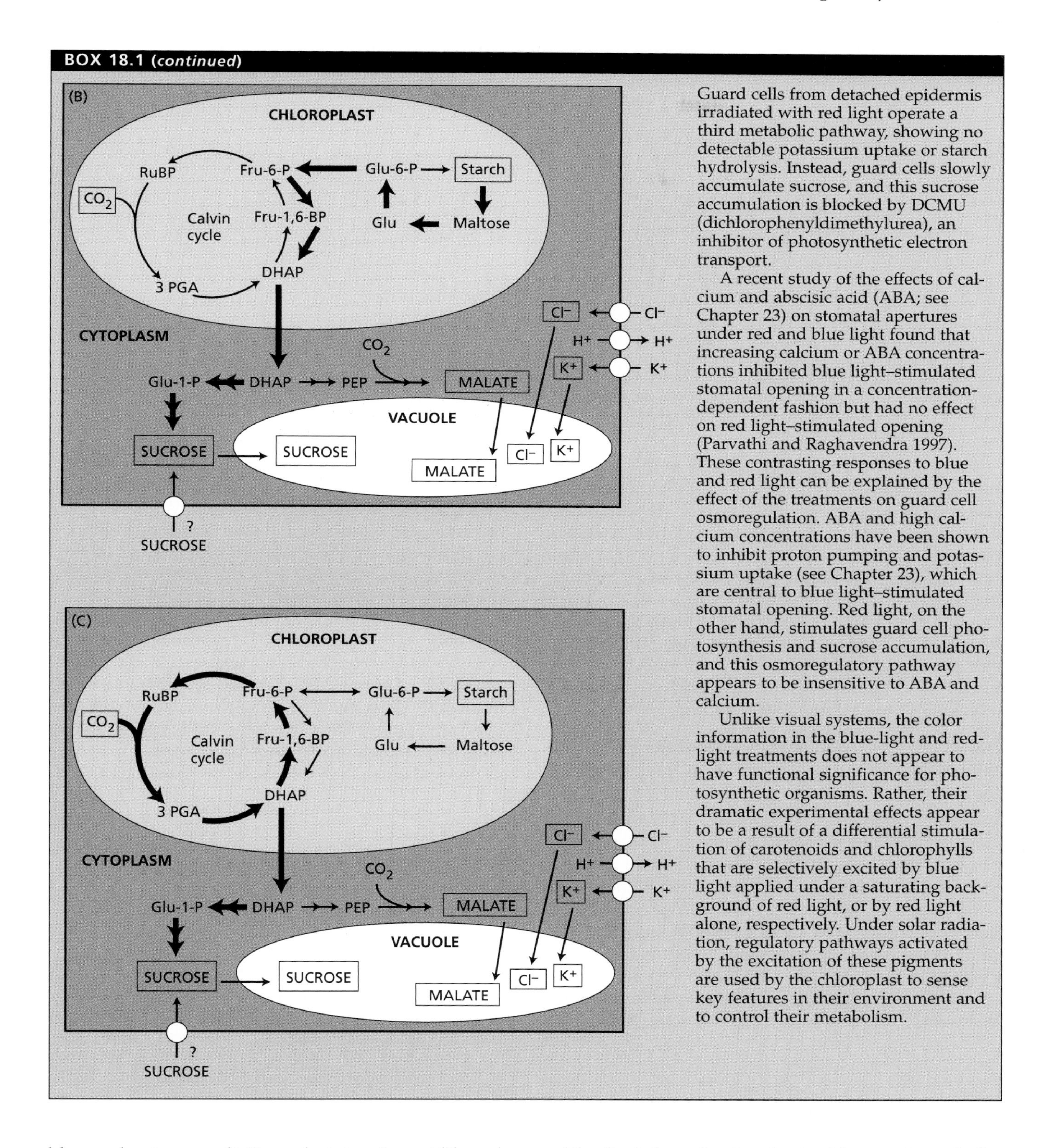

Guard cells from detached epidermis irradiated with red light operate a third metabolic pathway, showing no detectable potassium uptake or starch hydrolysis. Instead, guard cells slowly accumulate sucrose, and this sucrose accumulation is blocked by DCMU (dichlorophenyldimethylurea), an inhibitor of photosynthetic electron transport.

A recent study of the effects of calcium and abscisic acid (ABA; see Chapter 23) on stomatal apertures under red and blue light found that increasing calcium or ABA concentrations inhibited blue light–stimulated stomatal opening in a concentration-dependent fashion but had no effect on red light–stimulated opening (Parvathi and Raghavendra 1997). These contrasting responses to blue and red light can be explained by the effect of the treatments on guard cell osmoregulation. ABA and high calcium concentrations have been shown to inhibit proton pumping and potassium uptake (see Chapter 23), which are central to blue light–stimulated stomatal opening. Red light, on the other hand, stimulates guard cell photosynthesis and sucrose accumulation, and this osmoregulatory pathway appears to be insensitive to ABA and calcium.

Unlike visual systems, the color information in the blue-light and red-light treatments does not appear to have functional significance for photosynthetic organisms. Rather, their dramatic experimental effects appear to be a result of a differential stimulation of carotenoids and chlorophylls that are selectively excited by blue light applied under a saturating background of red light, or by red light alone, respectively. Under solar radiation, regulatory pathways activated by the excitation of these pigments are used by the chloroplast to sense key features in their environment and to control their metabolism.

ble mechanism mediating phototropism. Although auxin photodestruction was later proven not to play a role in vivo (see Chapter 19), this hypothesis set the stage for a lively controversy between proponents of flavins and proponents of carotenoids as candidates for blue-light photoreception in plants.

The flavin hypothesis gained wide acceptance in the late 1960s and early 1970s, when Warren L. Butler and his coworkers showed that, in a test tube, blue light can reduce a flavin such as riboflavin, and the reduced flavin can in turn reduce cytochrome *c* (see Figures 18.23 and 7.27).

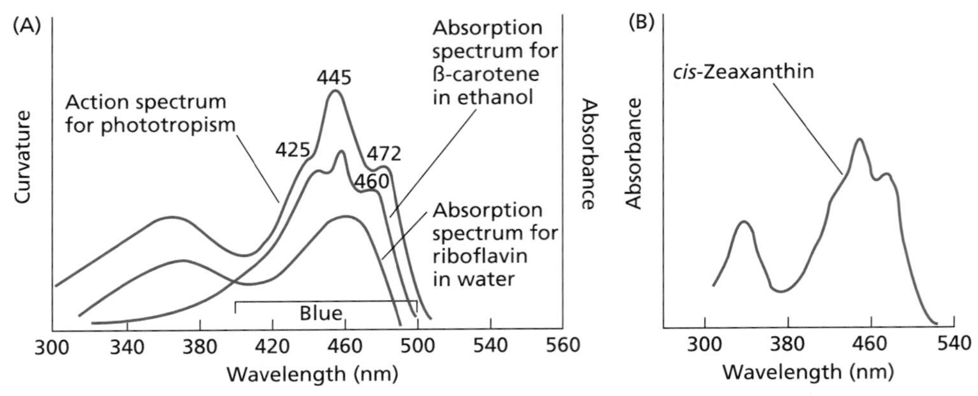

Figure 18.18 The relationship between the action spectrum for phototropism and the absorption spectra of flavins and carotenoids. (A) Absorption spectra of β-carotene in ethanol and riboflavin in water, compared with an action spectrum for phototropism (curvature), as measured by Thimann and Curry (1960). (B) Absorption spectrum of the carotenoid zeaxanthin in a *cis* configuration. In contrast to *trans*-zeaxanthin, the *cis* configuration has an absorption peak at around 350 nm. (After Molnar and Szabolcs 1993.)

These in vitro findings suggested an in vivo signal transduction cascade in which a membrane-bound flavin or flavoprotein excited by blue light reduces a cytochrome and starts a series of oxidation–reduction reactions in a membrane-bound electron transport chain that mediates blue-light responses. Redox reactions between flavins and cytochromes can be measured spectrophotometrically as absorption changes, and they have been studied extensively in cell-free solutions, cell extracts, and intact tissues. It has not been possible, however, to demonstrate that these photoreactions have a biological role in blue-light responses.

Genes Involved in Blue Light–Dependent Inhibition of Hypocotyl Elongation Have Been Identified

In 1993, Anthony R. Cashmore and his coworkers isolated the gene that is defective in the *hy4* mutant of *Arabidopsis,* which lacks the blue light–stimulated inhibition of hypocotyl elongation described earlier in the chapter (Ahmad and Cashmore 1993). The DNA sequence of the isolated gene showed significant homology to that of **photolyase,** a blue light–activated enzyme that repairs pyrimidine dimers in DNA that have been damaged by ultraviolet radiation. Photolyases are pigment proteins that contain a flavin adenine dinucleotide (FAD; see Figure 18.23) and a pterin. **Pterins** are colored pteridine derivates that often function as pigments in insects, fishes, and birds (Figure 18.19). In most photolyases, pterins absorb ultraviolet and blue light in the 350 to 450 nm region (maximum at around 406 nm) and transfer excitation energy to FAD, which is part of the catalytic site engaged in DNA repair.

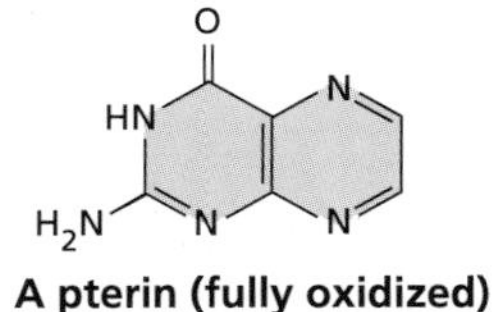

Figure 18.19 Structural formula of an oxidized pterin. Pterins are photoreceptors in photolyase, a blue light–activated enzyme that repairs photodamaged DNA, and they might be involved in specific blue-light responses.

Despite its sequence similarity with photolyase, the *HY4* gene, recently renamed *CRY1,* has no photolyase activity. On the other hand, overexpression of the CRY1 gene product in transgenic tobacco or *Arabidopsis* plants resulted in a stronger blue light–stimulated inhibition of hypocotyl elongation than in the wild type, as well as increased production of anthocyanin, another blue-light response (Figure 18.20). Thus, overexpression of CRY1

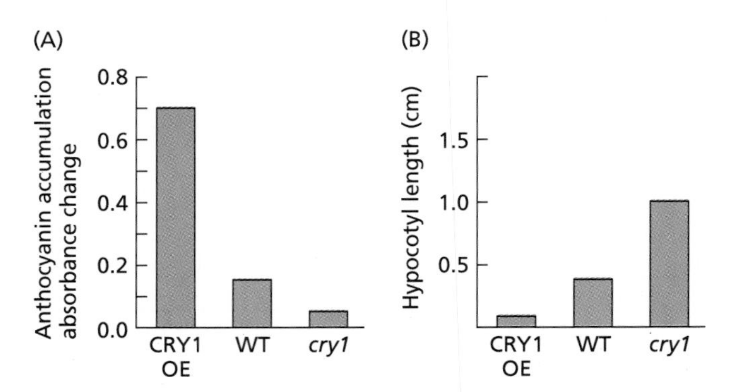

Figure 18.20 Blue light–stimulated accumulation of anthocyanin (A) and inhibition of stem elongation (B) in transgenic and mutant seedlings of *Arabidopsis.* Shown are the values for a transgenic phenotype overexpressing CRY1 (CRY1 OE), the wild type (WT), and *cry1* mutants. The enhanced blue-light response of the transgenic plant overexpressing CRY1 demonstrates the important role of this gene product in stimulating anthocyanin biosynthesis and inhibiting stem elongation. (After Ahmad et al. 1998.)

caused an enhanced sensitivity to blue light in the transgenic plants. Other blue-light responses, such as phototropism and blue light–dependent stomatal movements, appear to be normal in the *hy4* (*cry1*) mutant phenotype.

Genes with sequence homology to *CRY1* have been isolated from white mustard (*Sinapis alba*), the alga *Chlamydomonas*, pea (*Pisum sativum*), tomato (*Lycopersicum esculentum*), and rice (*Oryza sativa*) (Cashmore 1997; Horwitz and Berrocal 1997). Another gene product that is homologous to CRY1, named CRY2, has been isolated from *Arabidopsis*. Transgenic plants overexpressing CRY2 show a small enhancement of the inhibition of hypocotyl elongation, indicating that unlike CRY1, CRY2 does not play a primary role in inhibiting stem elongation. On the other hand, the transgenic plants overexpressing CRY2 show a large increase in blue light–stimulated cotyledon expansion, yet another blue-light response. It is therefore clear that the *CRY1* and *CRY2* genes play a central role in blue-light responses.

The *CRY1* gene might mediate a light-dependent redox reaction, triggering a signal transduction cascade that is yet to be characterized (Cashmore 1997). The CRY1 protein expressed in virus-infected insect cells was found to bind a flavin; however, flavin binding by CRY1 in plant cells that are involved in blue-light responses is yet to be demonstrated. This demonstration will be important to determine whether the postulated redox reactions of the bound flavin occur in vivo and whether they are triggered by blue light, and to establish the identity of the bound chromophore in plant cells.

The Carotenoid Zeaxanthin Mediates the Photoreception of Blue Light in Guard Cells

The carotenoid zeaxanthin has recently been implicated as a blue-light photoreceptor. Recall from Chapters 7 and 9 that zeaxanthin is one of the three members of the xanthophyll cycle of chloroplasts, which protects photosynthetic pigments from excess excitation energy. Guard cell chloroplasts have a typical carotenoid content and a functional xanthophyll cycle. The precise photoprotective role of the xanthophyll cycle of guard cells is yet to be characterized, but the cycle clearly plays a role in signal transduction.

One example of this signal transduction role of the xanthophyll cycle of guard cells is provided by a comparison of the zeaxanthin content of guard cells and mesophyll cells from *Vicia* (broad bean) leaves growing in a greenhouse (Figure 18.21). There is no detectable zeaxanthin in mesophyll cells in early morning or late afternoon. This lack of zeaxanthin under low levels of solar radiation is explained by the fact that no photoprotection is needed at the low photon fluxes of solar radiation that prevail at these times (see Figure 18.9A).

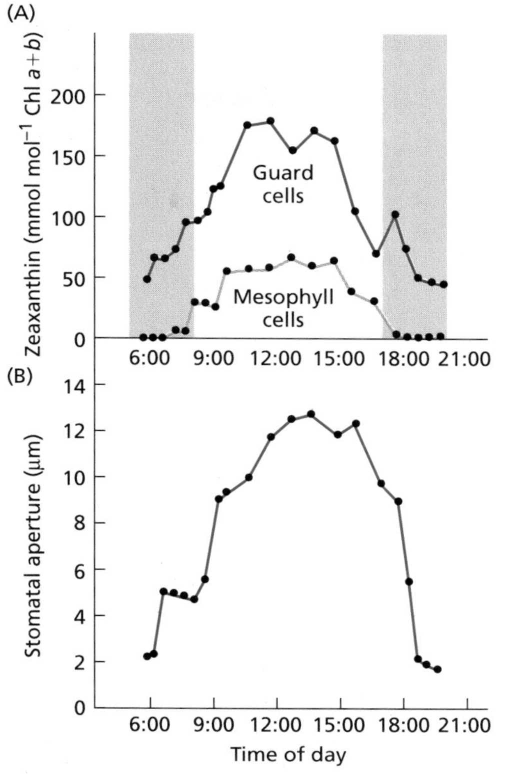

Figure 18.21 The zeaxanthin content of guard cells closely follows stomatal apertures. (A) Daily course of zeaxanthin content of guard cells and mesophyll cells in greenhouse-grown leaves of *Vicia faba*. The shaded areas illustrate the contrasting sensitivity of the xanthophyll cycle in mesophyll and guard cell chloroplasts under the low irradiances prevailing early and late in the day (for actual irradiance values, see Figure 18.9A). (B) Stomatal apertures in the same leaves used to measure guard cell zeaxanthin content. The close relationships among the zeaxanthin content of guard cells, the incident radiation, and the stomatal apertures support a role of zeaxanthin in the sensing of blue light in guard cells, contrasting with its photoprotective role in mesophyll cells. (After Srivastava and Zeiger 1995a.)

In contrast, the zeaxanthin content in guard cells closely follows incident solar radiation at the leaf surface throughout the day, and it is nearly linearly proportional to incident photon fluxes in early morning and late afternoon. This close relationship between incident photon fluxes and zeaxanthin content in guard cells throughout the day is the expected response of a pigment involved in the sensing of blue light.

The absorption spectrum of zeaxanthin closely matches the action spectrum for blue light–stimulated stomatal opening (see Figure 18.13). In addition, biochemical and metabolic studies indicate that zeaxanthin is functionally related to the blue-light component of the stomatal response to light. For example, zeaxanthin formation is blocked by dithiothreitol (DTT), a reducing

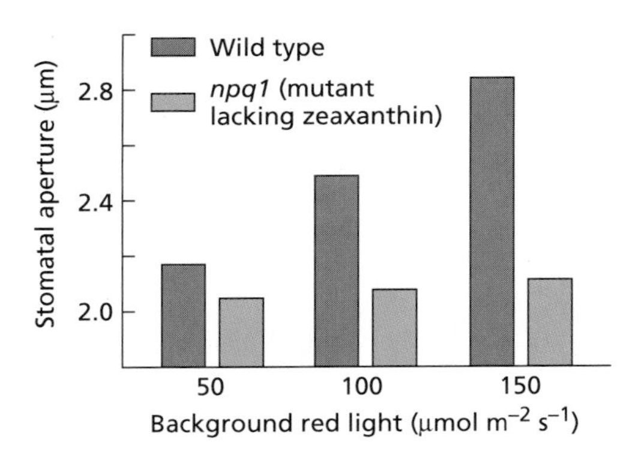

Figure 18.22 Stomatal responses to blue light in the wild type and the *npq1* mutant, a mutant that lacks zeaxanthin, of *Arabidopsis*. Stomata in detached epidermis were irradiated for 2 hours with 50, 100, and 150 μmol m^{-2} s^{-1} of red light. All treatments were then given 20 μmol m^{-2} s^{-1} of blue light for 1 hour. The wild-type stomata open further upon the addition of blue light, and the increases in aperture were proportional to the photon fluxes of background red light. In contrast, the *npq1* stomata did not respond to the blue-light treatment beyond their response to red light alone, indicating that these aperture changes were stimulated by guard cell photosynthesis. The table below the figure provides initial and final aperture values for the red light treatment of 100 μmol m^{-2} s^{-1}, showing that the mutant lacking zeaxanthin responded equally to blue and red light. (From Zeiger and Zhu 1998.)

Stomatal aperture (μm)	Wild type		*npq1*	
Initial apertures	1.47 ± 0.05		1.80 ± 0.06	
Background red light (100 μmol $m^{-2}s^{-1}$) for 2 hours	1.60 ± 0.10		1.78 ± 0.07	
Additional 1 hour light (20 μmol $m^{-2}s^{-1}$)	Blue light 2.44 ± 0.10	Red light 1.93 ± 0.06	Blue light 2.14 ± 0.07	Red light 2.11 ± 0.08

agent that reduces S—S bonds and effectively inhibits the enzyme that de-epoxidizes violaxanthin into zeaxanthin. Blue light–stimulated stomatal opening decreases as a function of DTT concentration, and the blue-light response is completely inhibited by 3 m*M* DTT. In contrast, red light–stimulated opening is insensitive to DTT, indicating that the inhibitor does not alter the component of the stomatal response to light that is mediated by guard cell photosynthesis.

Other inhibitors tested in blue-light responses, such as potassium iodine and phenylacetate, have been less useful because of poor specificity or concentration dependence. On the other hand, the specificity of the inhibition of blue light–stimulated stomatal opening by DTT, and its concentration dependence, indicate that zeaxanthin in the guard cells is required for the stomatal response to blue light (Srivastava and Zeiger 1995b).

The zeaxanthin hypothesis was also tested with *npq1* (*n*on*p*hotochemical *q*uenching), an *Arabidopsis* mutant defective in the enzyme that converts violaxanthin into zeaxanthin (see Figure 18.24) (Niyogi et al. in press). Because of this mutation, neither mesophyll nor guard cell chloroplasts of *npq1* can accumulate zeaxanthin in the light or in the dark (Zeiger and Zhu 1998). The blue-light responses of stomata from wild-type and *npq1* leaves were tested under increasing photon fluxes of subsaturating red light. As shown for stomata from other species, the blue-light responses of stomata from wild-type *Arabidopsis* increased as a function of the fluence rate of background red light (Figure 18.22). In contrast, *npq1* stomata showed baseline apertures in response to blue or red light, driven by guard cell photosynthesis, and failed to show any specific blue-light response. Thus, in both *Vicia* and *Arabidopsis*, a lack of guard cell zeaxanthin elicited by either metabolic or genetic means eliminated the stomatal response to blue light.

The zeaxanthin hypothesis provides an explanation for the enhancement of the blue-light sensitivity of guard cells by increasing photon fluxes of background red light that was described earlier in the chapter. As shown for mesophyll chloroplasts, zeaxanthin accumulation is stimulated by high photosynthetic rates, which can be elicited with both blue or red light. Thus, guard cells treated with increasing photon fluxes of red light have correspondingly higher zeaxanthin content, and the accumulation of zeaxanthin increases the stomatal response to blue light.

The zeaxanthin hypothesis might also provide an answer to the puzzling question of a need for two photoreceptor systems in guard cells. The stomatal response to blue light has been implicated in the function of stomata under special conditions, such as stomatal opening in shade leaves exposed to sun flecks (see Chapter 9) and stomatal opening at dawn, a time of day when solar radiation is relatively enriched in blue photons. However, stomatal opening under saturating red light is always lower than under white light, indicating that the photosynthetic component of the light response of stomata cannot support optimal light-dependent stomatal opening without the blue-light component. The close relationship between incident solar radiation and zeaxanthin content in guard cells, and the role of zeaxanthin in blue-light photoreception suggest that the blue-light component of the stomatal response to light functions as a light sensor that couples stomatal apertures to incident photon fluxes at the leaf surface. The photosynthetic component, on the other hand, could function in the coupling of the stomatal responses with photosynthetic rates in the mesophyll (see Chapter 9).

Blue-Light Photoreceptors Might Differ in Their Apoproteins or in Their Chromophores

Most, if not all, known photoreceptor systems are pigment proteins (see Chapters 7 and 17), consisting of an apoprotein and a chromophore. Several chromophores (flavins, zeaxanthin, and pterins) and two apoproteins (CRY1 and CRY2) have been implicated in blue-light photoreception.

How many blue-light photoreceptors can we anticipate? Large apoprotein variation among photoreceptor families is well known. In animals, bacteria, and the alga *Chlamydomonas*, the chromophore retinal (see Figure 14.21) mediates light-dependent proton pumping, phototaxis, and vision. In all cases, retinal is the invariant chromophore, and the specificity is provided by the apoprotein, which varies broadly among species and different functions. In addition, in Chapter 17 we saw that several phytochrome genes encode different apoproteins that bind the same chromophore. Thus, photoreceptor families built around a single chromophore and several different apoproteins are well known in bacteria, animals, and plants. The remarkable similarity between the action spectra for several key blue-light responses (see Figures 18.1, 18.5, and 18.13) suggests that at least a group of blue-light responses share a single chromophore and derive tissue or function specificity from variation in the apoprotein.

Should we also expect chromophore variation among different blue-light responses? As mentioned earlier, both flavins and carotenoids are leading candidates for blue-light photoreceptors. In the sections that follow we discuss some important, contrasting properties to be expected in blue-light photoreceptor systems based in flavins or carotenoids.

Spectroscopy. The absorption spectra of flavins and carotenoids have important differences in the 400 to 500 nm blue region of the spectrum. Absorption spectra of oxidized flavins or flavoproteins at room temperature show a single major peak around 450 nm (see Figure 18.18A). Absorption spectra of flavins or flavoproteins do show distinct shoulders, or minor peaks, in the 400 to 500 nm region under special conditions, such as liquid nitrogen temperature (77 K). Absorption spectra of carotenoids and carotenoproteins at room temperature consistently show this "three-finger" structure (see Figures 18.13B and 18.18A).

Carotenoids are linear polyenes arranged as a planar zigzag chain, with the repeating unit —CH=CH—CCH_3=CH—. The strong absorption bands in the visible region of the spectrum of carotenoids are a consequence of their long, conjugated double-bond system (Britton 1995). The main absorption band is due to the electronic transitions of delocalized electrons between molecular orbitals that are shared by the conjugated double bonds. The rigid structure of the conjugated polyene backbone strongly favors some of the transitions; hence the distinct shape and the sharpness of the peaks. On the other hand, electronic configurations of the blue light–absorbing isoalloxazine ring of flavins (Figure 18.23A) are less rigid and thus less homogeneous at room temperature, resulting in broader absorption peaks (see Figure 18.18A). At liquid nitrogen temperature, the vibrational and rotational freedom of the flavin molecule is limited, thus resulting in a sharper absorption spectrum, revealing shoulders in addition to the main absorption band.

The action spectra for the blue-light responses shown earlier (see Figures 18.1 and 18.5) match the 400 to 500 nm region of the absorption spectra from carotenoids remarkably well, as Figure 18.13 illustrates. On the other hand, the 360 nm peak of the action spectrum for phototropism is matched by an absorption peak of flavins in that region, which is absent in typical absorption spectra from carotenoids. However, carotenoids in the *cis* configuration have a sharp absorption peak in the 300 to 400 nm region (see Figure 18.18B).

Binding of a chromophore to its apoprotein can affect the allowed transition states of the pigment and alter its absorption spectrum typical of a solvent solution (see Figure 18.18A) in a significant way. On the other hand, an absorption spectrum of the protein CRY1 containing bound FAD (Lin et al. 1995) is substantially different from an action spectrum for the inhibition of hypocotyl elongation (see Figure 18.5). Thus more research is needed to reconcile the flavin hypothesis of blue-light photoreception with the spectroscopic features just mentioned.

Localization. The intracellular localization of a photoreceptor pigment protein determines the early steps in the signal transduction cascade that links the sensing of blue light with the functional response to the blue-light signal. Flavins are ubiquitous in intracellular compartments, but flavoprotein models hypothesize a membrane-bound flavoprotein that is a component of an electron transport chain within the membrane. The sensing of blue light within the plasma membrane would make transduction of the blue-light signal to membrane-bound targets such as the H^+-ATPase functionally expeditious. On the other hand, CRY1 and CRY2 are soluble proteins that lack the hydrophobic regions typical of transmembrane proteins (see Chapter 7). More information is therefore needed to understand how the localization of these important apoproteins shapes their function.

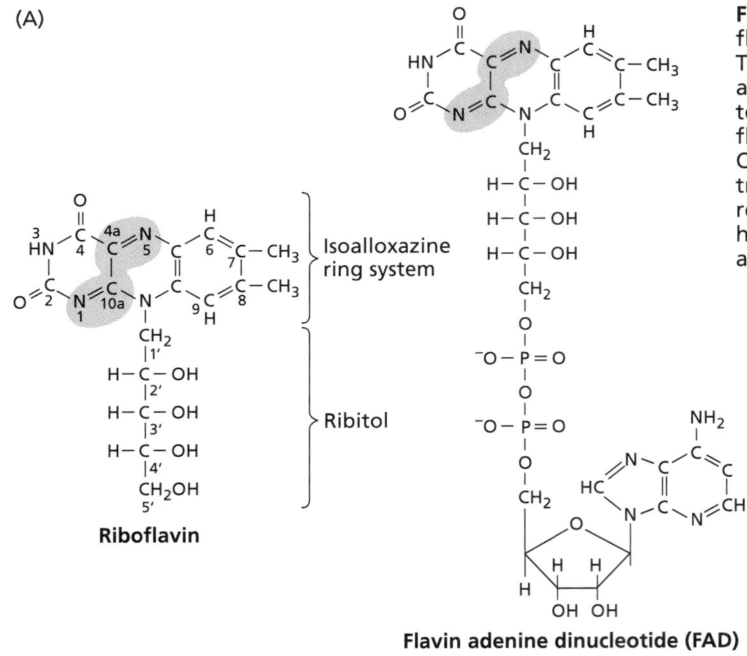

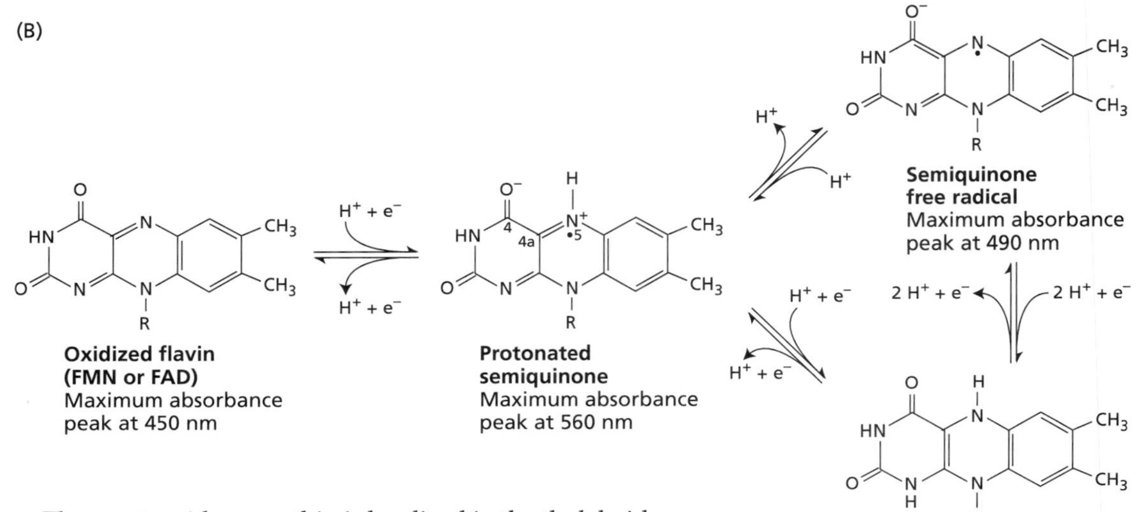

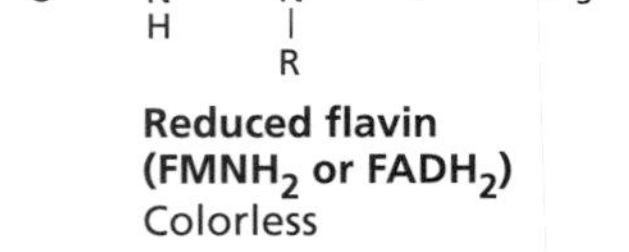

Figure 18.23 Flavins. (A) Structural formulas of the flavins riboflavin and flavin adenine dinucleotide (FAD). The coordinated double bonds between N-1 and C-10 and between N-5 and C-4 of the isoalloxazine ring system participate in the oxidation–reduction reactions of flavins catalyzed by electron donors or by blue light. (B) Oxidized flavins absorb in the blue region of the spectrum at around 450 nm (see Figure 18.18), and can be reduced by blue light in a two electron reaction, which has a semiquinone free radical that absorbs at 490 nm as a stable intermediate.

The carotenoid zeaxanthin is localized in the thylakoid membrane of the chloroplast. The localization of the *beginning* of the signal transduction cascade for a blue-light response at the chloroplast is a major departure from the plasma membrane–based models mentioned earlier. But such a conceptual shift is relatively straightforward in the case of the stomatal response to blue light, because of the well-known role of the guard cell chloroplast in the light responses of stomata. The implications of the localization of zeaxanthin in the guard cell chloroplast for blue-light signals that need to reach targets outside the chloroplast will be discussed shortly.

Photochemistry. Recall from Chapter 11 that the flavin-containing electron carriers FAD and FMN (flavin mononucleotide) are important components of the electron transport chain of mitochondria. Oxidized flavins can be reduced by blue light in a two-electron reaction, which has a semiquinone free radical as a stable intermediate (see Figure 18.23B). Thus, the operation of a flavin-based blue-light photoreceptor implies a sequence of redox reactions at the beginning of the signal transduction cascade. Redox reactions are critical for electron transport at the thylakoid membrane (see Chapter 7), but they are yet to be shown to participate in signal transduction systems in plants or animals (see Chapter 14).

The photochemistry of carotenoids, on the other hand, would be expected to involve light-induced isomerizations such as those that are well characterized for the polyene retinal (see Figure 14.21). Conformational changes resulting from isomerization of the polyene backbone would be transduced to the apoprotein of the pigment protein, and these conformational changes would start a signal transduction cascade. The contrasting differences between early photochemical events based on redox reactions as opposed to conformational changes should be helpful in further research of blue-light responses. For example, in vivo characterization of blue light–induced isomerization of zeaxanthin associated with early events of a blue-light response, and identification of the electron donors and acceptors in the postulated flavin-mediated redox cascade would add critical information to our knowledge of blue-light responses.

Signal Transduction

Available knowledge on the signal transduction of blue-light responses in plants is incomplete. Vertebrate vision provides an excellent example of a well-characterized photosensory transduction cascade. Recall from Chapter 14 that light perception in vertebrate vision starts by the light-induced isomerization of *cis*- to *trans*-retinal, followed by a conformational change of the apoprotein of rhodopsin (opsin), the activation of a G protein (transducin), the activation of a cGMP phosphodiesterase, the lowering of cGMP concentrations, the closing of sodium channels, and the generation of electric stimuli that convert the light signals into electric currents that reach the brain (see Figure 14.21). Much more research is needed to achieve a comparable understanding of photosensory transduction processes in plants. In this section, we discuss available information on the signal transduction of blue-light responses, focusing first on the stomatal response to blue light, and then describing the more limited information on the cascades for other responses.

Several Steps Link Zeaxanthin Excitation and Blue Light–Dependent Stomatal Opening

The zeaxanthin hypothesis of blue-light photoreception implies that the signal transduction cascade for the stomatal response to blue light starts with the excitation of zeaxanthin by blue photons. The magnitude of the stomatal response to blue light depends on the amount of available zeaxanthin and on the flux of incident blue light.

Zeaxanthin is localized in the light-harvesting complexes of photosystems I and II of the antenna bed of the chloroplast. The concentration of zeaxanthin in the chloroplast depends on the size of the carotenoid pool and on regulation of the xanthophyll cycle. Etiolated plastids contain low amounts of violaxanthin and antheraxanthin, but no zeaxanthin. Zeaxanthin is synthesized in the carotenoid biosynthetic pathway (Figure 18.24A), and the total pool of carotenoids, including the xanthophylls, increases upon greening.

The pool size of carotenoids changes with growing conditions. For example, guard cells from the upper surface of leaves, which grow in an environment of high irradiance, have a higher pigment content than do guard cells from the shaded, lower surface of the leaf. Thus changes in the blue-light response of stomata that are involved with adaptations and acclimations of guard cells to different growth environments will be regulated partly by changes in the size of the carotenoid pool.

Zeaxanthin concentrations also vary with the activity of the xanthophyll cycle (Figure 18.24B). The enzyme that converts violaxanthin to zeaxanthin is an integral protein of the thylakoid membrane that faces the chloroplast lumen (see Figure 7.22), and has a pH optimum at pH 5.2 (Yamamoto 1979). Photosynthetic electron transport pumps protons into the lumen, and these protons are the driving force for ATP synthesis (see Chapter 7). If ATP utilization slows down, proton accumulation in the lumen lowers the lumen pH, and the formation of zeaxanthin accelerates. Conversely, dissipation of the pH gradient at the lumen by ATP synthesis, coupled to a high rate of ATP utilization, favors the conversion of zeaxanthin to violaxanthin.

Some of the properties of the guard cell chloroplast have important implications for the xanthophyll cycle. When compared with their mesophyll counterparts, guard cell chloroplasts are enriched in photosystem II, have unusually high rates of photosynthetic electron transport, and low rates of photosynthetic carbon fixation. These properties, which have long puzzled plant physiologists (see Box 18.2), favor lumen acidification at low photon fluxes, and can explain the difference in the daily courses of zeaxanthin formation in mesophyll and guard cell chloroplasts that were discussed earlier in the chapter (see Figure 18.21).

The dependence of the stomatal response to blue light on the zeaxanthin concentrations of guard cells, and on the regulation of zeaxanthin concentrations by the xanthophyll cycle, provides remarkable flexibility to this blue light–sensing system. Any change in the guard cell environment that causes changes in electron transport or carbon fixation rates at the guard cell chloroplast would alter lumen pH and result in a change in zeaxanthin content and thus in the guard cell sensitivity to light. This is particularly the case with red-light irradiation (see Figure 18.22), and it appears to apply also to the stomatal response to CO_2 and to light–CO_2 interactions (Figure 18.25).

The precise sequence of events following zeaxanthin excitation by blue light needs characterization. As discussed earlier, the most likely photochemical reaction

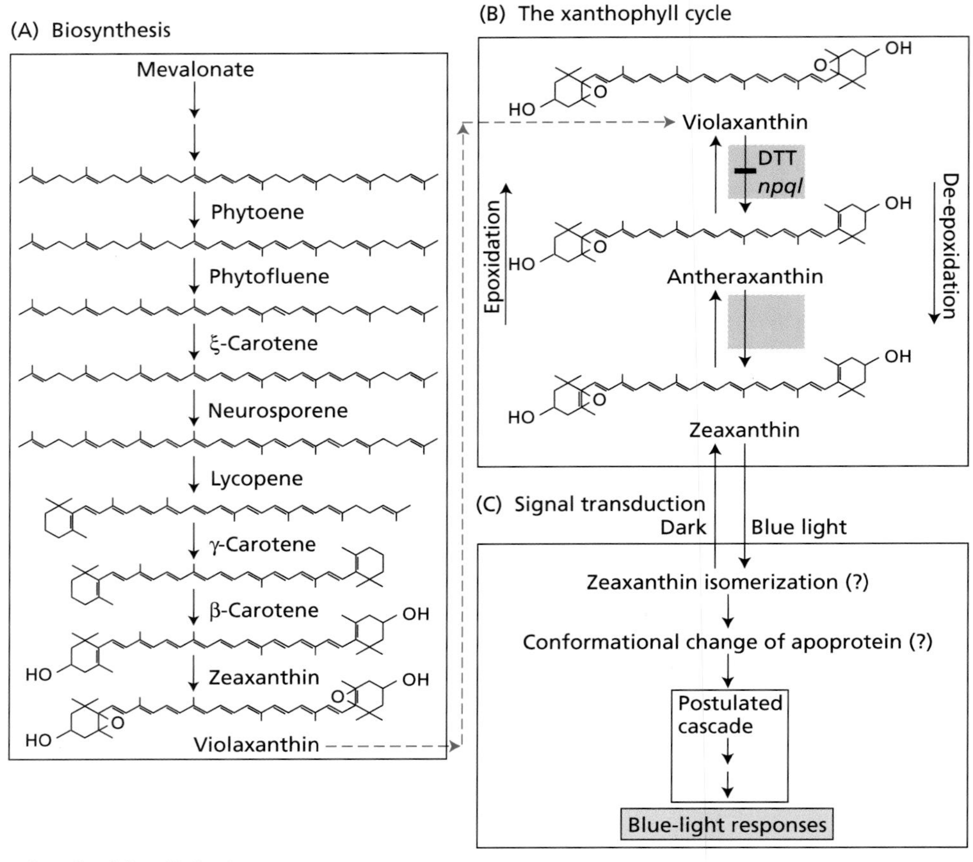

Figure 18.24 Regulation of zeaxanthin content in guard cell chloroplasts by the carotenoid biosynthetic pathway (A) and the xanthophyll cycle (B). The site of action of DTT (an inhibitor of zeaxanthin accumulation) and the *npq1* mutation are shown. Signal transduction of the stomatal response to blue light (C) is hypothesized to start with the excitation of zeaxanthin by blue light. (After Zeiger and Zhu 1998.)

ensuing from excitation of zeaxanthin by blue light is an isomerization, which would in turn cause a conformational change of an apoprotein (see Figure 18.24C). The blue-light signal would then be transmitted by a *second messenger* across the chloroplast envelope and eventually lead to activation of the H^+-ATPase at the plasma membrane (Figure 18.26). Calcium is one of the possible second messengers. The chloroplast envelope has a Ca^{2+}-ATPase, and mesophyll chloroplasts are known to take up calcium when illuminated. A lowering of cytosolic calcium activates the H^+-ATPase in guard cells (Kinoshita et al. 1995). Thus in its simplest scheme, excitation of zeaxanthin by blue light would lead to the activation of a Ca^{2+}-ATPase at the chloroplast envelope, which would take up calcium into guard cell chloroplasts. The lowering of cytosolic calcium would activate

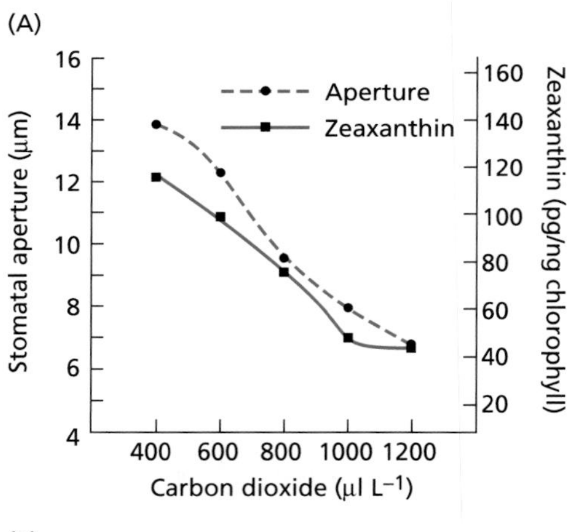

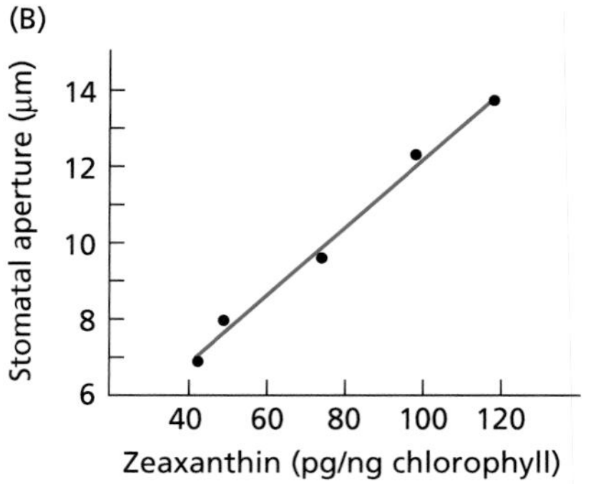

Figure 18.25 The relationships among stomatal aperture, guard cell zeaxanthin concentration, and ambient CO_2 concentration in intact leaves of *Vicia faba* (broad bean) kept at constant photon fluxes and temperature in a growth chamber. The high CO_2 sensitivity of stomata from leaves grown in a growth chamber is paralleled by a high sensitivity of the xanthophyll cycle to CO_2 under constant light conditions (A). The close correlation between stomatal apertures and guard cell zeaxanthin content (B) suggests that the xanthophyll cycle of guard cells also functions in signal transduction of the stomatal response to CO_2 in illuminated stomata. (After Zhu et al. in press.)

BOX 18.2

The Coleoptile Chloroplast

GUARD CELL CHLOROPLASTS are a conspicuous structural feature of guard cells (see Figure 18.10), and their physiology and biochemistry have been studied for several decades. In contrast, the chloroplasts of phototropically sensitive coleoptiles have only recently been analyzed in detail. The coleoptile chloroplast had been suggested as a possible site for photoreception in oat coleoptiles by Kenneth V. Thimann and his coworkers in 1960, but interest in the flavin hypothesis later in that decade shifted attention away from coleoptile chloroplasts in subsequent phototropism studies.

The coleoptile chloroplast evolves oxygen, fixes CO_2 photosynthetically, and operates the xanthophyll cycle. Dark-grown, etiolated coleoptile tips have violaxanthin and antheraxanthin but no zeaxanthin, and they accumulate zeaxanthin when irradiated (Quiñones and Zeiger 1994). When dark-grown coleoptiles are pretreated with red light for different lengths of time and then exposed to a brief pulse of blue light, their zeaxanthin content is proportional to the length of the red-light treatment, and the phototropic bending elicited by the blue-light pulse is linearly related to the zeaxanthin content of the coleoptiles (part A of the figure).

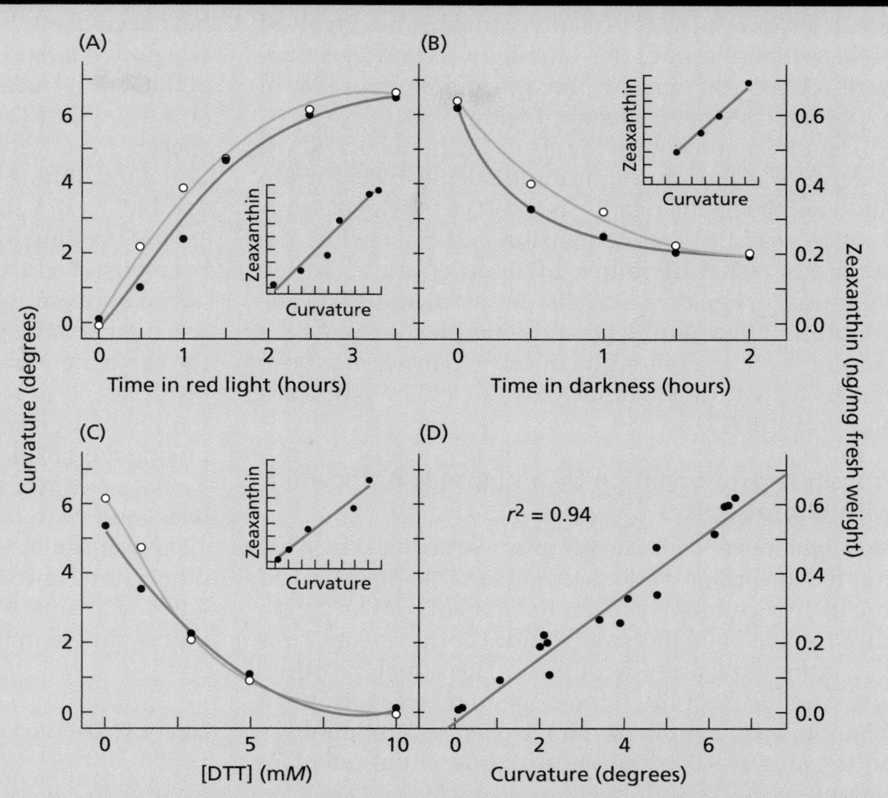

The relationship between the zeaxanthin content of corn coleoptile tips and bending in response to a pulse of blue light. Zeaxanthin content was altered by preillumination time under red light (A), transfer of zeaxanthin-containing coleoptiles to the dark (B), and a saturating red-light pretreatment in the presence of increasing concentrations of DTT, an inhibitor of zeaxanthin formation (C). Curvature was linearly related to zeaxanthin content in all treatments (D), and zero zeaxanthin content extrapolated to zero bending. (After Quiñones and Zeiger 1994.)

Red light–treated coleoptiles with a high zeaxanthin content convert their zeaxanthin to violaxanthin when transferred to the dark. Phototropic bending in response to a pulse of blue light given at different time intervals after the transfer to the dark is linearly related to zeaxanthin content (part B of the figure). When corn coleoptiles are given a saturating red-light pretreatment and a blue-light pulse in the presence of increasing concentrations of DTT (an inhibitor of zeaxanthin formation), bending and zeaxanthin content decrease linearly as a function of DTT concentration (part C of the figure). Thus the relationship between the zeaxanthin content of coleoptile chloroplasts and blue light–dependent phototropic sensitivity is very similar to the relationship between the zeaxanthin content of the guard cell chloroplast and the blue-light sensitivity of stomatal opening.

Coleoptile and guard cell chloroplasts share some properties that are clearly distinct from mesophyll chloroplasts, including high rates of oxygen evolution and low rates of photosynthetic carbon fixation, a high sensitivity of their xanthophyll cycle to low levels of incident photon fluxes, and a high starch content (Zhu et al. 1995). The two chloroplast types also show a specific blue-light response with an action spectrum that closely matches the action spectra for blue light–stimulated stomatal opening and coleoptile phototropism. These remarkable similarities suggest that these distinct characteristics are typical of chloroplasts specialized in signal transduction, as opposed to the classic role of the mesophyll chloroplast, specialized in photosynthetic carbon fixation.

the H^+-ATPase. On the other hand, other signal transduction schemes can be envisioned and await further research. Furthermore, kinases, phosphatases, inositol trisphosphate (IP_3), and calmodulin have been associated with blue light–activated stomatal opening, so the cytosolic portion of the cascade could be complex.

Electric Signaling Links the Perception of Blue Light with the Inhibition of Stem Elongation

Cell expansion and eventually cell division are slowed during the inhibition of hypocotyl elongation by blue light. As discussed earlier, blue light inhibits hypocotyl elongation within 20 to 30 s of illumination. The shortness of this lag time indicates that changes in cell expansion rates, as opposed to cell division rates, are involved in the initial phases of this blue-light response. A transient depolarization of the plasma membrane of hypocotyl cells can be detected within 15 s.

The fact that the growth rate of the hypocotyl starts to decrease at the time of maximum depolarization indicates that the depolarization is a part of the signal transduction cascade. This conclusion is supported by the observation that inhibition of anion efflux by an ion channel blocker prevents the depolarization and the inhibition of growth (Cho and Spalding 1996). At the cellular level, the inhibition of cell expansion has been associated with an inhibition of wall-yielding properties (Cosgrove 1994).

Protein Phosphorylation by a Kinase Is Associated with Phototropism

Blue light–dependent phototropism is mediated by a lateral redistribution of the hormone indole-3-acetic acid (auxin), which causes differential rates of cell elongation on the shaded and illuminated sides of the coleoptile, as well as bending (see Chapter 19). The signal transduction cascade for blue light–dependent phototropism therefore links light perception by a blue-light photoreceptor, auxin redistribution, and differential cell elongation.

Some newly isolated *Arabidopsis* mutants impaired in blue light–dependent phototropism of the hypocotyl have provided valuable information on cellular events preceding bending. One of these mutants, the *nph1* (*non*phototropic *h*ypocotyl) allele, has been found to be genetically independent of the *hy4* (*cry1*) mutant discussed earlier: The *nph1* mutant lacks the phototropic response in the hypocotyl but has normal blue light–stimulated inhibition of hypocotyl elongation, while *hy4* has the converse phenotype.

The *nph1* mutant, but not *hy4*, has also been found to be defective in a blue light–stimulated phosphorylation reaction that has been studied extensively both in vivo and in vitro (Huala et al. 1997). The phosphorylated substrate is a 120 kDa membrane protein found in several species. In oat (*Avena sativa*), the protein is tightly bound to the membrane in young coleoptiles, and it becomes soluble in later stages of development (Salomon et al. 1996). The blue light–dependent phosphorylation has been observed in membrane fractions, indicating that a chromophore, a kinase mediating the phosphorylation reaction, and the phosphorylated protein are all part of the same membrane-bound complex. The fact that the *nph1* phenotype lacks both the phototropic response and the 120 kDa protein has substantiated earlier evidence linking the 120 kDa protein to the phototropic response.

Oat coleoptiles display both a longitudinal (tip to base) and a lateral gradient in phosphorylation of the 120 kDa protein (see Chapter 19). Thus, blue-light gradients (see Figure 18.4) are associated with phosphorylation gradients, which could in turn generate auxin gradients, differential elongation rates, and bending.

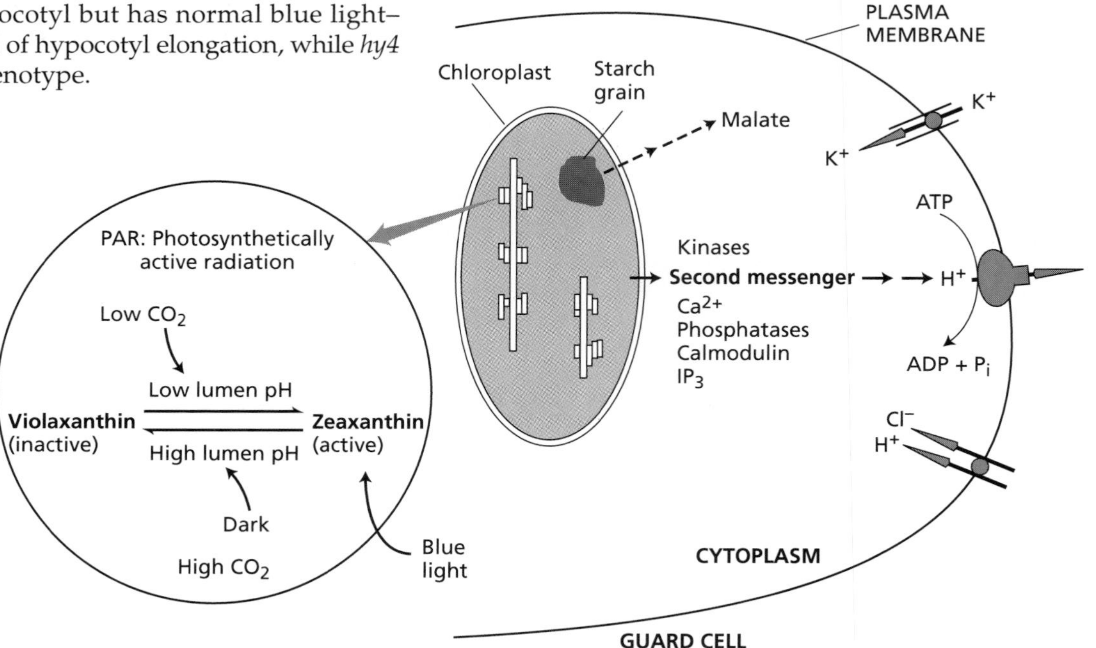

Figure 18.26 Signal transduction of the stomatal response to blue light. Zeaxanthin concentration in the thylakoid of the guard cell chloroplast is regulated by lumen pH and primarily depends on photosynthetic rates driven by photosynthetically active radiation (PAR, maximum sensitivity in the blue and red regions of the visible spectrum). Excitation of zeaxanthin by blue light starts the signal transduction process. Putative transduction steps are zeaxanthin isomerization, conformational change of an apoprotein, and transduction of the signal from the chloroplast to the H^+-ATPase at the plasma membrane. CO_2 concentrations at the thylakoid can also change zeaxanthin concentration.

The *NPH1* gene was recently isolated and sequenced. The determined DNA sequence showed a kinase domain and similarity to bacterial genes that are known to bind flavins (Huala et al. 1997). This study concluded that the NPH1 gene product regulates a very early event of the signal transduction of blue light–stimulated phototropism. Surprisingly for two blue-light responses that share a very similar action spectrum, *NPH1* has no sequence similarity to *CRY1* or *CRY2*. A recent study of a *cry1/cry2* double mutant has shown that, in contrast with *cry1* or the *cry2* phenotypes, the double mutant is deficient in phototropism, suggesting that the *CRY1* and *CRY2* genes are also involved in the phototropic response (Ahmad et al. 1998).

Of further interest is the finding that the NPH1 protein can be found in normal amounts in the *cry1/cry2* double mutant but light-dependent phosphorylation is absent. These results suggest that the NPH1 gene product acts downstream from the CRY1 and CRY2 gene products in the signal transduction cascade, and that functional *CRY1* and *CRY2* genes are needed for the NPH1-dependent phosphorylation to take place.

Much has been learned in the past decade about the mechanism that mediates blue-light responses. The isolation of the *CRY1*, *CRY2*, and *NHP1* genes is making it possible to study blue-light photoreception from the gene to the physiological response. Because of our rudimentary understanding of the different elements of the signal transduction cascade, these molecular approaches pose demanding challenges, including identification of the functional chromophore(s) and their precise role in the signal transduction cascade.

On the other hand, decades of physiological and biochemical studies on the blue-light response of guard cells have led to the identification of zeaxanthin as a blue-light photoreceptor in guard cells, and perhaps also in coleoptiles. Challenges in this research area include a need to characterize zeaxanthin photochemistry upon excitation by blue light, and to identify the apoprotein and the second messengers that link photoreception in the chloroplast with extrachloroplastic targets. Finally, integration of the role of different photoreceptors and regulatory proteins in each blue-light response is required for an understanding of the complex use of blue-light signals by plants.

Summary

Plants utilize light as a source of energy and as a signal providing information about their environment. A large family of blue-light responses is used to sense light quantity and direction. These blue-light signals are transduced into electrical, metabolic, and genetic processes that allow plants to alter growth, development, and function in order to acclimate to changing environmental conditions. Blue-light responses include phototropism, stomatal movements, inhibition of stem elongation, gene activation, pigment biosynthesis, tracking of the sun by leaves, and chloroplast movements within cells.

Specific blue-light responses can be distinguished from other responses that have some sensitivity to blue light, by a characteristic "three-finger" action spectrum in the 400 to 500 nm region, and by their insensitivity to red light.

The physiology of blue-light responses varies broadly. In phototropism, stems grow toward unilateral light sources by asymmetric growth on their shaded side. In the inhibition of stem elongation, perception of blue light depolarizes the membrane potential of elongating cells, and the rate of elongation rapidly decreases. In gene activation, blue light stimulates transcription and translation, leading to the accumulation of gene products that are required for the morphogenetic response to light.

Blue light–stimulated stomatal movements are driven by blue light–dependent changes in the osmoregulation of guard cells. Blue light stimulates an H^+-ATPase at the guard cell plasma membrane, and the resulting pumping of protons across the membrane generates an electrochemical potential gradient that provides a driving force for ion uptake. Blue light also stimulates starch degradation and malate biosynthesis. Solute accumulation within the guard cells leads to stomatal opening. Guard cells also utilize sucrose as a major osmotically active solute, and light quality can change the activity of different osmoregulatory pathways that modulate stomatal movements.

Flavins and carotenoids are leading candidates for blue-light photoreceptors. CRY1 and CRY2 are two *Arabidopsis* genes involved in blue light–dependent inhibition of stem elongation, and perhaps also phototropism. It has been proposed that CRY1 and CRY2 are apoproteins of flavin-containing pigment proteins that mediate blue-light photoreception, and several research groups are actively seeking definitive proof of this hypothesis.

The chloroplastic carotenoid zeaxanthin has been implicated in blue-light photoreception in guard cells. Blue light–stimulated stomatal opening is blocked if zeaxanthin accumulation in guard cells is prevented by genetic or biochemical manipulations. The regulation of zeaxanthin content in guard cells makes it possible to regulate their response to blue light. The signal transduction cascade for the blue-light response of guard cells comprises blue-light perception in the guard cell chloroplast, transduction of the blue-light signal across the chloroplast envelope, activation of the H^+-ATPase, turgor buildup, and stomatal opening.

In *Arabidopsis* hypocotyls, the signal transduction cascade includes the CRY1, CRY2, and NPH1 gene products. The *CRY1* and *CRY2* genes have sequence similarity to the blue light–activated enzyme photolyase, and the CRY1 gene product binds the flavin FAD in an insect gene expression system. NPH1 has been shown to mediate a light-dependent phosphorylation by a kinase, also characterized in coleoptiles. The kinase appears to function downstream from the CRY1 and CRY2 gene products. Later steps of the signal transduction cascade in hypocotyls include membrane depolarization, anion efflux, and slowing of cell elongation. In the coleoptile, later steps include asymmetrical distribution of auxin and bending.

General Reading

Assmann, S. M. (1993) Signal transduction in guard cells. *Annu. Rev. Cell Biol.* 9: 345–375.

Batschauer, A., Gilmartin, P. M., Nagy, F., and Schafer, E. (1994) A molecular and genetic approach to photomorphogenesis. In *Photomorphogenesis in Plants*, 2nd ed., R. E. Kendrick and G. H. M. Kronenberg, eds., Kluwer, Dordrecht, Netherlands, pp. 559–599.

*Cosgrove, D. J. (1994) Photomodulation of growth. In *Photomorphogenesis in Plants*, 2nd ed., R. E. Kendrick and G. H. M. Kronenberg, eds., Kluwer, Dordrecht, Netherlands, pp. 631–658.

Firn, R. D. (1994) Phototropism. In *Photomorphogenesis in Plants*, 2nd ed., R. E. Kendrick and G. H. M. Kronenberg, eds., Kluwer, Dordrecht, Netherlands, pp. 659–681.

Gilmore, A. M. (1997) Mechanistic aspects of the xanthophyll cycle-dependent photoprotection in higher plant chloroplasts and leaves. *Physiol. Plantarum* 99: 197–209.

Grabov, A., and Blatt, M. R. (1998) Co-ordination of signalling elements in guard cell ion channel control. *J. Exp. Bot.* 49: 351–360.

*Horwitz, B. A. (1994) Properties and transduction chains of the UV and blue light photoreceptors. In *Photomorphogenesis in Plants*, 2nd ed., R. E. Kendrick and G. H. M. Kronenberg, eds., Kluwer, Dordrecht, Netherlands, pp. 327–350.

*Horwitz, B. A., and Berrocal, T. G. M. (1997) A spectroscopic view of some recent advances in the study of blue light photoreception. *Bot. Acta* 110: 360–368.

Hoth, S., Dreyer, I., Dietrich, P., Becker, D., Mueller-Roeber, B., and Hedrich, R. (1997) Molecular basis of plant specific acid activation of K^+ uptake channels. *Proc. Natl. Acad. Sci. USA* 94: 4806–4810.

*Meidner, H. (1987) Three hundred years of research into stomata. In *Stomatal Function*, E. Zeiger, G. D. Farquhar, and I. R. Cowan, eds., Stanford University Press, Stanford, CA, pp. 7–27.

Moran, N., Yueh, Y. G., and Crain, R. C. (1996) Signal transduction and cell volume regulation in plant leaflet movement. *News Physiol. Sci.* 11: 108–114.

*Outlaw, W. H., Jr. (1983) Current concepts on the role of potassium in stomatal movements. *Physiol. Plantarum* 59: 302–311.

*Schroeder, J. I., Ward, J. M., and Gassmann, W. (1994) Perspectives on the physiology and structure of inward-rectifying K^+ channels in higher plants: Biophysical implications for K^+ uptake. *Annu. Rev. Biophys. Biomol. Struct.* 23: 441–471.

*Senger, H. (1984) *Blue Light Effects in Biological Systems*. Springer, Berlin.

Short, T. W., and Briggs, W. R. (1994) The transduction of blue light signals in higher plants. *Annu. Rev. Plant Physiol. Plant Mol. Biol.* 45: 143–171.

Tallman, G. (1992) The chemiosmotic model of stomatal opening revisited. *Crit. Rev. Plant Sci.* 11: 35–57.

*Vogelmann, T. C. (1994) Light within the plant. In *Photomorphogenesis in Plants*, 2nd ed., R. E. Kendrick and G. H. M. Kronenberg, eds., Kluwer, Dordrecht, Netherlands, pp. 491–533.

* Indicates a reference that is general reading in the field and is also cited in this chapter.

Chapter References

Ahmad, M., and Cashmore, A. R. (1993) *HY4* gene of *A. thaliana* encodes a protein with characteristics of a blue light photoreceptor. *Nature* 366: 162–166.

Ahmad, M., Jarillo, J. A., Smirnova, O., and Cashmore, A. R. (1998) Cryptochrome blue light photoreceptors of *Arabidopsis* implicated in phototropism. *Nature* 392: 720–723.

Amodeo, G., Srivastava, A., and Zeiger, E. (1992) Vanadate inhibits blue light–stimulated swelling of *Vicia* guard cell protoplasts. *Plant Physiol.* 100: 1567–1570.

Assmann, S. M., Simoncini, L., and Schroeder, J. I. (1985) Blue light activates electrogenic ion pumping in guard cell protoplasts of *Vicia faba*. *Nature* 318: 285–287.

Baskin, T. I., and Iino, M. (1987) An action spectrum in the blue and ultraviolet for phototropism in alfalfa. *Photochem. Photobiol.* 46: 127–136.

Britton, G. (1995) UV/visible spectroscopy. In *Carotenoids*, G. Britton, S. Liaaen-Jensen, and H. Pfander, eds., Birkhauser, Basel, Switzerland, pp. 13–62.

Cashmore, A. R. (1997) The cryptochrome family of photoreceptors. *Plant Cell Environ.* 20: 764–767.

Cerda-Olmedo, E., and Lipson, E. D. (1987) *Phycomyces*. Cold Spring Harbor Laboratory, Cold Spring Harbor, NY.

Cho, M. H., and Spalding, E. P. (1996) An anion channel in *Arabidopsis* hypocotyls activated by blue light. *Proc. Natl. Acad. Sci. USA* 93: 8134–8138.

*Firn, R. D. (1994) Phototropism. In *Photomorphogenesis in Plants*, 2nd ed., R. E. Kendrick and G. H. M. Kronenberg, eds., Kluwer, Dordrecht, Netherlands, pp. 659–681.

Gotow, K., Taylor, S., and Zeiger, E. (1988) Photosynthetic carbon fixation in guard cell protoplasts of *Vicia faba* L: Evidence from radiolabel experiments. *Plant Physiol.* 86: 700–705.

Hartmann, K. M. (1967) Ein Wirkungsspectrum der Photomorphogenese unter Hochenergiebedingungen und seine Interpretation auf der Basis des Phytochrome. *Z. Naturforsch. Teil B* 22: 1172–1175.

Huala, E., Oeller, P. W., Liscum, E., Han, I-S., Larsen, E., and Briggs, W. R. (1997) *Arabidopsis* NPH1: A protein kinase with a putative redox-sensing domain. *Science* 278: 2120–2123.

Iino, M. (1988) Pulse-induced phototropisms in oat and maize coleoptiles. *Plant Physiol.* 88: 823–828.

Iino, M., Ogawa, T., and Zeiger, E. (1985) Kinetic properties of the blue light response of stomata. *Proc. Natl. Acad. Sci. USA* 82: 8019–8023.

Kamiya, A., and Miyachi, S. (1974) Effects of blue light on respiration and carbon dioxide fixation in colorless *Chlorella* mutant cells. *Plant Cell Physiol.* 15: 927–937.

Karlsson, P. E. (1986) Blue light regulation of stomata in wheat seedlings. II. Action spectrum and search for action dichroism. *Physiol. Plantarum* 66: 207–210.

Kinoshita, T., Nishimura, M., and Shimazaki, K-I. (1995) Cytosolic concentration of Ca^{2+} regulates the plasma membrane H^+ ATPase in guard cells of fava bean. *Plant Cell* 7: 1333–1342.

Lin, C., Robertson, D. E., Ahmad, M., Raibekas, A. A., Jorns, M. S., Dutton, P. L., and Cashmore, A. R. (1995) Association of flavin adenine dinucleotide with the *Arabidopsis* blue light receptor CRY1. *Science* 269: 968–970.

Matters, G. L., and Beale, S. I. (1995) Blue-light-regulated expression of genes for two early steps of chlorophyll biosynthesis in *Chlamydomonas reinhardtii*. *Plant Physiol.* 109: 471–479.

Molnar, P., and Szabolcs, J. (1993) (Z/E)-Photoisomerization of C_{40}-carotenoids by iodine. *J. Chem. Soc. Perkin. Trans.* 2: 261–266.

Niyogi, K. K., Grossman, A.R., and Björkman, O. (1998) *Arabidopsis* mutants define a central role for the xanthophyll cycle in the regulation of photosynthetic energy conversion. *Plant Cell* 10: 1121–1134.

Ogawa, T., Ishikawa, H., Shimada, K., and Shibata, K. (1978) Synergistic action of red and blue light and action spectrum for malate formation in guard cells of *Vicia faba* L. *Planta* 142: 61–65.

Parvathi, K., and Raghavendra, A. S. (1997) Blue light–promoted stomatal opening in abaxial epidermis of *Commelina benghalensis* is maximal at low calcium. *Physiol. Plantarum* 101: 861–864.

Quiñones, M. A., and Zeiger, E. (1994) A putative role of the xanthophyll, zeaxanthin, in blue light photoreception of corn coleoptiles. *Science* 264: 558–561.

Salomon, M., Zacherl, M., and Rüdiger, W. (1996) Changes in blue-light-dependent protein phosphorylation during the early development of etiolated oat seedlings. *Planta* 199: 336–342.

Schwartz, A., and Zeiger, E. (1984) Metabolic energy for stomatal opening. Roles of photophosphorylation and oxidative phosphorylation. *Planta* 161: 129–136.

Serrano, E. E., Zeiger, E., and Hagiwara, S. (1988) Red light stimulates an electrogenic proton pump in *Vicia* guard cell protoplasts. *Proc. Natl. Acad. Sci. USA* 85: 436–440.

Shimazaki, K., Iino, M., and Zeiger, E. (1986) Blue light–dependent proton extrusion by guard cell protoplasts of *Vicia faba*. *Nature* 319: 324–326.

Spalding, E. P., and Cosgrove, D. J. (1989) Large membrane depolarization precedes rapid blue-light induced growth inhibition in cucumber. *Planta* 178: 407–410.

Srivastava, A., and Zeiger, E. (1995a) Guard cell zeaxanthin tracks photosynthetic active radiation and stomatal apertures in *Vicia faba* leaves. *Plant Cell Environ.* 18: 813–817.

Srivastava, A., and Zeiger, E. (1995b) The inhibitor of zeaxanthin formation, dithiothreitol, inhibits blue-light-stimulated stomatal opening in *Vicia faba*. *Planta* 196: 445–449.

Steinitz, B., and Poff, K. L. (1986) A single positive phototropic response induced with pulsed light in hypocotyls of *Arabidopsis thaliana* seedlings. *Planta* 168: 305–315.

Talbott, L. D., Srivastava, A., and Zeiger, E. (1996) Stomata from growth-chamber-grown *Vicia faba* have an enhanced sensitivity to CO_2. *Plant Cell Environ.* 19: 1188–1194.

Talbott, L. D., and Zeiger, E. (1998) The role of sucrose in guard cell osmoregulation. *J. Exp. Bot.* 49: 329–337.

Thimann, K. V., and Curry, G. M. (1960) Phototropism and phototaxis. In *Comparative Biochemistry*, Vol. 1, M. Florkin and H. S. Mason, eds., Academic Press, New York, pp. 243–306.

Thomas, B., and Dickinson, H. G. (1979) Evidence for two photoreceptors controlling growth in de-etiolated seedlings. *Planta* 146: 545–550.

Tilghman, J. A., Gao, J., Anderson, M. B., and Kaufman, L. S. (1997) Correct blue-light regulation of pea Lhcb genes in an *Arabidopsis* background. *Plant Mol. Biol.* 35: 293–302.

Vogelmann, T. C., and Haupt, W. (1985) The blue light gradient in unilaterally irradiated maize coleoptiles: Measurements with a fiber optic probe. *Photochem. Photobiol.* 41: 569–576.

Yamamoto, H. Y. (1979) Biochemistry of the violaxanthin cycle in higher plants. *Pure Appl. Chem.* 51: 639–648.

Zeiger, E., and Hepler, P. K. (1977) Light and stomatal function: Blue light stimulates swelling of guard cell protoplasts. *Science* 196: 887–889.

Zeiger, E., and Zhu, J. (1998) Role of zeaxanthin in blue light photoreception and the modulation of light-CO_2 interactions in guard cells. *J. Exp. Bot.* 49: 433–442.

Zhu, J., Talbott, L. D., and Zeiger, E. (1998) Guard cell zeaxanthin tracks stomatal apertures and ambient CO_2 concentrations in a constant-light environment. *Plant Cell Environ.* 21: 813–820.

Zhu, J., Zeiger, R., and Zeiger, E. (1995) Structural and functional properties of the coleoptile chloroplast: Photosynthesis and photosensory transduction. *Photosynth. Res.* 44: 207–219.

19 Auxins

THE FORM AND FUNCTION of a multicellular organism depends to a large extent on efficient communication among the vast numbers of its constituent cells. In higher plants, regulation and coordination of metabolism, growth, and morphogenesis often depend on chemical signals from one part of the plant to another. This idea originated in the nineteenth century with the pioneering work of the German botanist Julius von Sachs (1832–1897). Sachs proposed that chemical messengers are responsible for the formation and growth of different plant organs. He also suggested that external factors such as gravity could affect the distribution of these substances within a plant. Although Sachs did not know the identity of these chemical messengers, his ideas led to their eventual discovery.

As discussed in Chapter 14, many of our current concepts about intercellular communication in plants have been derived from similar studies in animals. In animals the chemical messengers that mediate intercellular communication are called **hormones**. Hormones interact with specific cellular proteins called **receptors**. Most animal hormones are synthesized and secreted in one part of the body and are transferred, typically via the bloodstream and the endocrine system, to specific target sites in another part of the body. Animal hormones fall into four general categories: proteins, small peptides, amino acid derivatives, and steroids.

Until quite recently, plant development was thought to be regulated by only five types of hormones: auxins, gibberellins, cytokinins, ethylene, and abscisic acid. However, there is now compelling evidence for a steroid family of plant hormones involved in morphological changes induced by light (see Chapter 20), and a variety of other signaling molecules have been discovered, such as jasmonic acid, salicylic acid, and the protein systemin, which play roles in resistance to pathogens and defense against herbivores (see Chapter 13). Thus the number and types of hormone-like signaling agents in plants keep expanding. Specific molecular receptors corresponding to each of these signaling agents are assumed to be present in plant cells.

The first plant hormone we will consider is auxin. Auxin deserves pride of place in any discussion of plant hormones, since it was the first growth hormone to be discovered in plants, and much of the early physiological work on the mechanism of plant cell expansion was carried out in relation to auxin action. Moreover, both auxin and cytokinin differ from the other plant hormones and signaling agents in one important respect: They are required for viability. Thus far, no true mutants lacking either of these hormones has been found, suggesting that mutations that eliminate them are lethal. Whereas all the other hormones of plants, as well as those of animals, seem to act as on–off switches that regulate specific developmental processes as needed, auxin and cytokinin appear to be required at some level more or less continuously.

We begin our discussion of auxins with a brief history of their discovery, followed by a description of their chemical structures and the methods used to detect auxins in plant tissues. A look at the pathways of auxin biosynthesis and the polar nature of auxin transport follows. We will then review the various developmental processes controlled by auxin, such as stem elongation, apical dominance, root initiation, and fruit development, focusing on auxin-regulated growth and the type of oriented growth known as *tropisms*. Finally, we will examine what is currently known about the mechanism of auxin-induced growth at the cellular and molecular levels.

The Emergence of the Concept of Auxins

During the latter part of the nineteenth century, Charles Darwin and his son Francis studied plant growth phenomena involving tropisms. One of their interests was the bending of plants toward light. This phenomenon, which is caused by differential growth, is called *phototropism*. In some experiments the Darwins used seedlings of canary grass (*Phalaris canariensis*), in which, as in many other grasses, the youngest leaves are sheathed in a protective organ called the **coleoptile** (Figure 19.1).

Coleoptiles are very sensitive to light, especially to blue light (see Chapter 18). If illuminated on one side with a short pulse of dim blue light, they will bend (grow) toward the source of the light pulse within an hour. The Darwins found that the tip of the coleoptile perceived the light, for if they covered the tip with foil, the coleoptile would not bend. But the growth zone of the coleoptile (the region that is responsible for the bending toward the light) is several millimeters below the tip. Thus, they concluded that some sort of signal is produced in the tip, travels to the growth zone, and causes the shaded side to grow faster than the illuminated side. The results of their experiments were published in 1881 in a wonderful book entitled *The Power of Movement in Plants*.

A long period of experimentation on the nature of the growth stimulus in coleoptiles followed the insightful conclusions of the Darwins. This research culminated in the demonstration in 1926 by Frits Went of the presence of a growth-promoting chemical in the tip of oat (*Avena sativa*) coleoptiles. It was known that if the tip of a coleoptile was removed, coleoptile growth ceased. Previous workers had attempted to identify the growth-promoting chemical by grinding up coleoptile tips and testing the activity of the extracts. This approach failed completely because in grinding up the tissue, inhibitory substances that normally were compartmentalized in the cell were released into the extract.

Went's major breakthrough was to avoid grinding by allowing the material to diffuse out of excised coleoptile tips directly into gelatin blocks. These blocks could then be tested for their ability to restore growth in decapitated coleoptiles. Moreover, if the gelatin block was placed asymmetrically on a coleoptile stump, the coleoptile bent away from the side containing the block (see Figure 19.1). Because the substance promoted the elongation of the coleoptile tissue, F. Kögl and other researchers named Went's diffusible compound **auxin** from the Greek *auxein*, "to increase" or "to grow."

Biosynthesis, Transport, and Metabolism of Auxin

Went's studies with agar blocks demonstrated unequivocally that the growth-promoting "influence" from the coleoptile tip was a chemical substance. The fact that it was produced at one location and transported in minute amounts to its site of action qualified it as an authentic plant hormone. In the years that followed, the chemical identity of the "growth substance" was determined, and because of its potential agricultural uses, many related chemical analogs were tested. This testing led to generalizations about the chemical requirements for auxin activity. In parallel with these studies, the agar block diffusion technique was being applied to the problem of auxin transport. Technological advances, especially the use of isotopes as tracers, enabled plant biochemists to unravel the pathways of auxin biosynthesis and breakdown.

Our discussion begins with the chemical nature of auxin, continuing with a description of its biosynthesis, transport, and metabolism. Increasingly powerful ana-

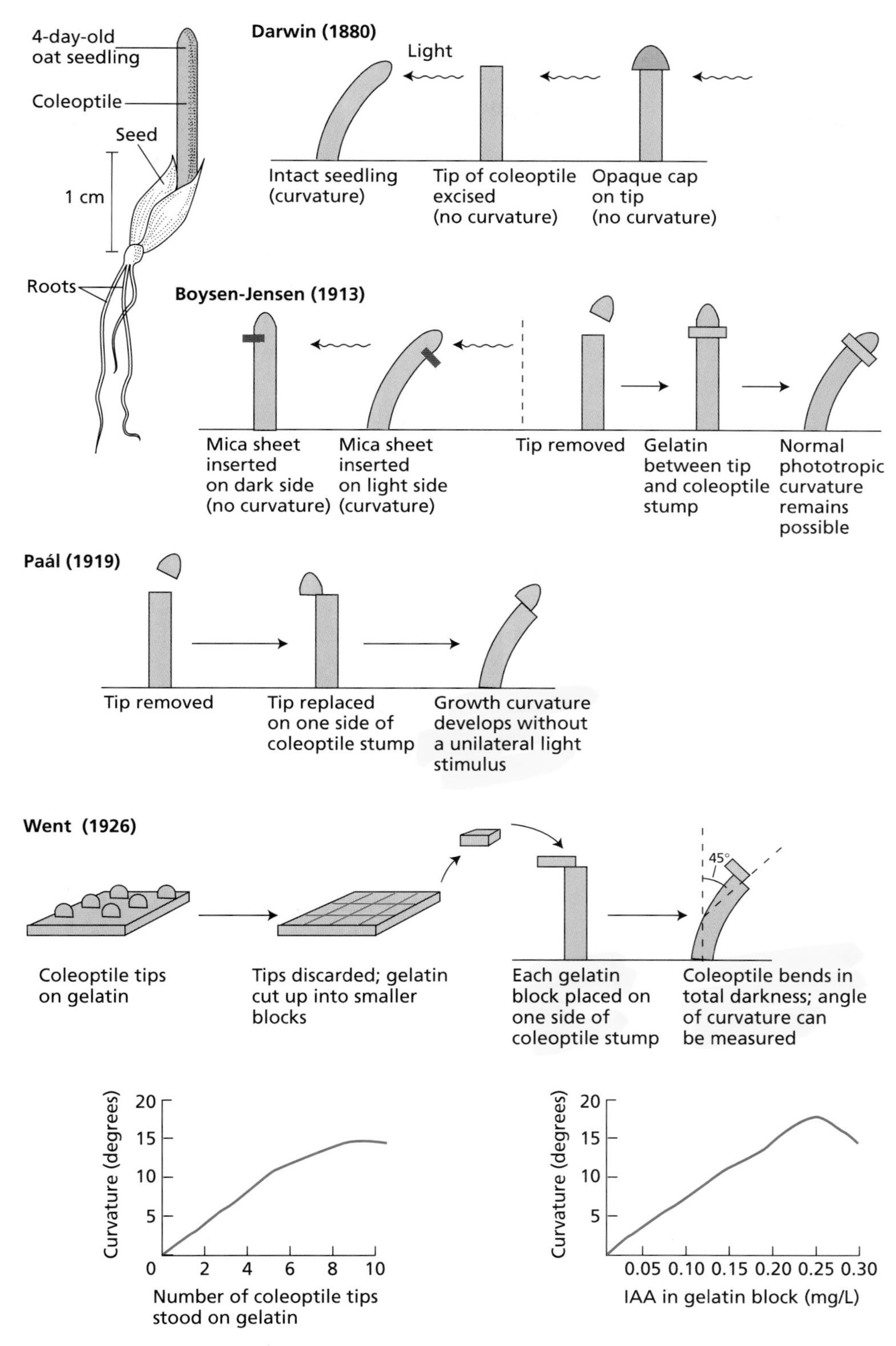

Figure 19.1 Summary of early experiments in auxin research. From experiments on coleoptile phototropism, Darwin concluded that a growth stimulus is produced in the coleoptile tip and is transmitted to the growth zone. In 1913, P. Boysen-Jensen discovered that the growth stimulus passes through material such as gelatin but not through water-impermeable barriers such as mica. In 1919, A. Paál provided evidence that the growth-promoting stimulus produced in the tip was chemical in nature. In 1926, F. W. Went showed that the active growth-promoting substance can diffuse into a gelatin block. He also devised a coleoptile-bending assay for quantitative auxin analysis.

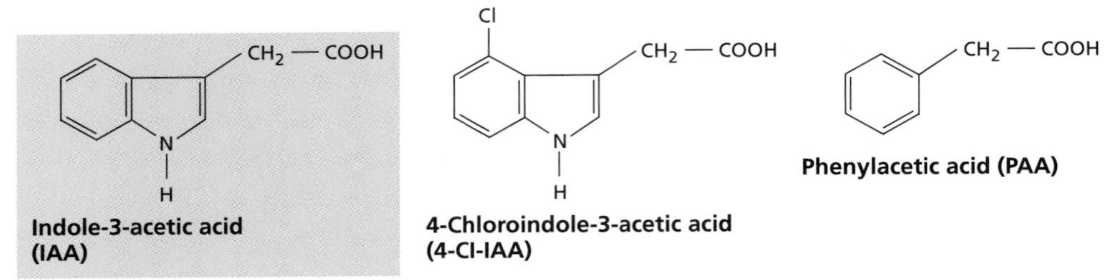

Figure 19.2 Structure of some natural auxins. Although indole-3-acetic acid (IAA) occurs in all plants, other related compounds in plants have auxin activity. Peas, for example, contain 4-chloroindole-3-acetic acid. Other compounds that are not indoles, such as phenylacetic acid, also possess auxin activity. It is not clear what roles these other natural auxins play in plant development.

lytical methods and the application of molecular biological approaches have recently allowed scientists to identify auxin precursors and to study auxin turnover and distribution within the plant.

The Principal Auxin in Higher Plants Is Indole-3-Acetic Acid

In the mid-1930s two groups of investigators, one in the Netherlands and the other in the United States, determined the chemical structure of the growth-promoting auxin. Work done primarily by F. Kögl and A. J. Haagen-Smit in Holland and K. V. Thimann in the United States led to the discovery that auxin is indole-3-acetic acid (IAA). Several other auxins in higher plants were discovered later (Figure 19.2), but IAA is by far the most abundant and physiologically relevant. Since the structure of IAA is relatively simple, academic and industrial laboratories were quickly able to synthesize a wide array of molecules with auxin activity; Figure 19.3A gives some examples.

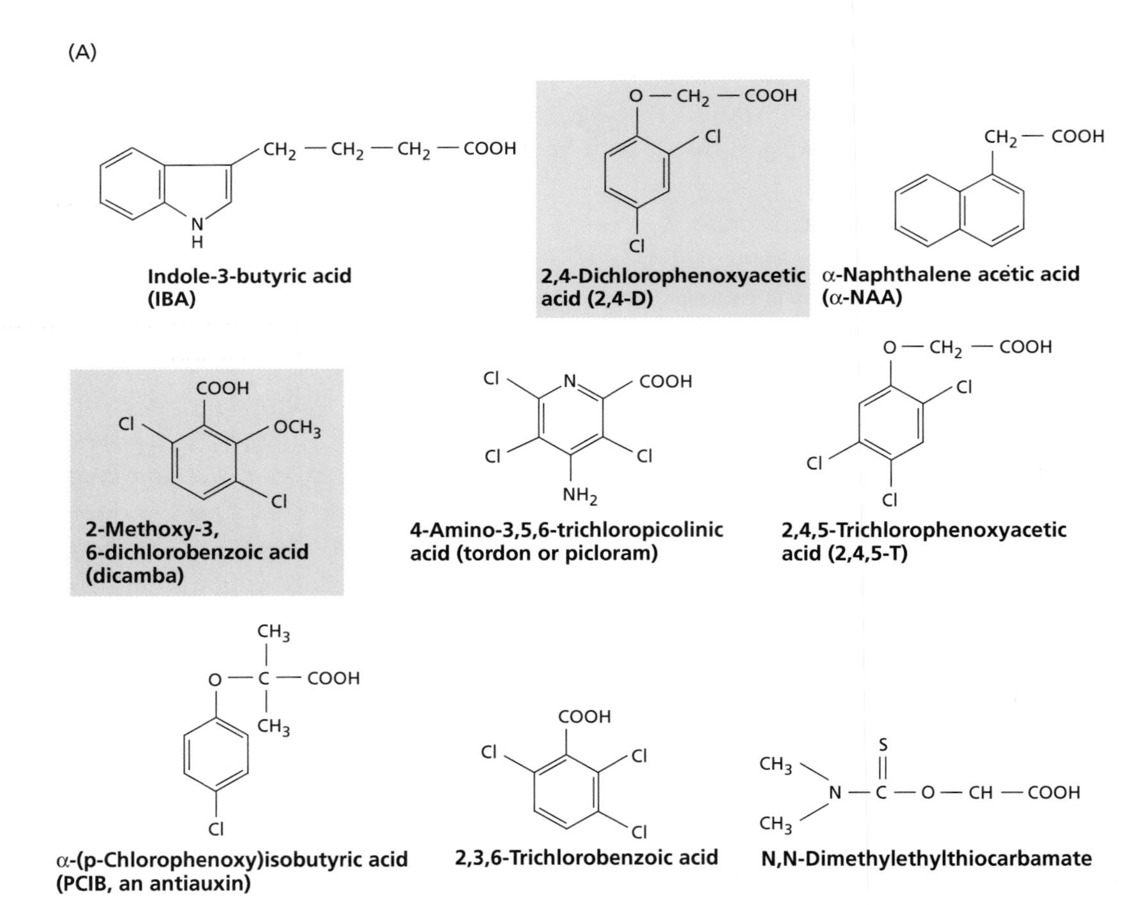

An early definition of auxins included all natural and synthetic chemical substances that stimulate elongation in coleoptiles and stem sections. However, auxins affect many developmental processes besides cell elongation. Cleland (1996: 126) recently recommended the following definition of an auxin: "*A compound that has a spectrum of biological activities similar to, but not necessarily identical with those of IAA. This includes the ability to: 1) induce cell elongation in isolated coleoptile or stem sections, 2) induce cell division in callus tissues in the presence of a cytokinin, 3) promote lateral root formation at the cut surfaces of stems, 4) induce parthenocarpic tomato fruit growth, and 5) induce ethylene formation.*" Although the new definition is more comprehensive than the original, the list of auxin effects is by no means exhaustive. The more we learn about how auxin works at the molecular level, the more precise our definitions of the hormone will become.

Figure 19.3 Structure of some synthetic auxins. (A) Most of these synthetic auxins are used as herbicides in horticulture and agriculture. The most widely used are probably dicamba and 2,4-D, which are not subject to breakdown by the plant and are very stable. (B) The dissociated forms of IAA, phenylacetic acid, α-NAA, and 2,4-D (in descending order), showing the negative charge on the carboxyl group and the fractional positive charge on the ring, separated by a distance of about 0.5 nm. (B after Farrimond et al. 1978.)

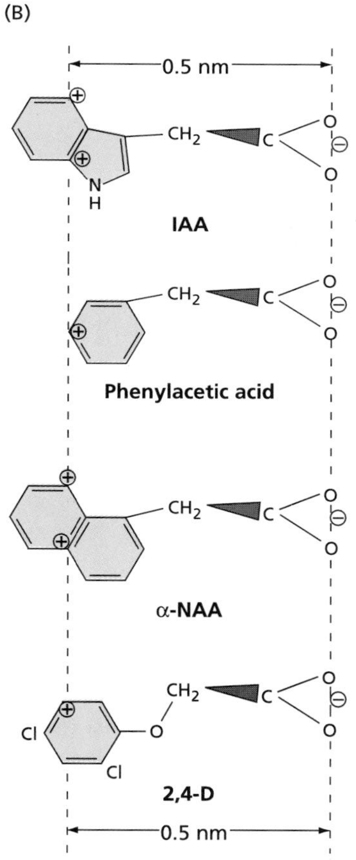

(B)

0.5 nm

IAA

Phenylacetic acid

α-NAA

2,4-D

0.5 nm

The Structures of Active Auxins Are Chemically Diverse

How can such a diverse group of chemicals all be active auxins? No other plant hormone, with the possible exception of the cytokinins, has such a wide variety of active analogs. For example, few people would have predicted that either 2,4-dichlorophenoxyacetic acid (2,4-D) or α-naphthalene acetic acid (α-NAA) (see Figure 19.3A), both of which bear little resemblance to IAA, would be able to stimulate coleoptile elongation or have activity in other auxin bioassays. A comparison of the compounds that possess auxin activity reveals that at neutral pH they all have a strong negative charge on the carboxyl group of the side chain that is separated from a weaker positive charge on the ring structure by a distance of about 0.5 nm (see Figure 19.3B) (Porter and Thimann 1965; Farrimond et al. 1978). This charge separation may be an essential structural requirement for auxin activity.

The main part of the molecule, the indole ring, is not essential for activity, although an aromatic or fused aromatic ring of a certain size range is required (Katekar 1979). Recently Edgerton and colleagues (1994) proposed a set of molecular requirements for auxin activity based on studies of the binding of various auxin analogs to a protein, auxin-binding protein 1 (ABP1), which may be the auxin receptor. Their model defines three essential regions of the binding site: a planar aromatic ring–binding platform, a carboxylic acid–binding site, and a hydrophobic transition region that separates the two binding sites.

Antiauxins are another class of synthetic auxin analogs. These compounds, such as α-(*p*-chlorophenoxy)isobutyric acid, or PCIB (see Figure 19.3A), have little or no auxin activity but specifically inhibit the effects of auxin. When applied to plants, antiauxins may compete with IAA for specific receptors without triggering an auxin response, thus inhibiting normal auxin action. It is possible to overcome the inhibition of an antiauxin by adding excess IAA.

Auxins in Biological Samples Can Be Quantified

Depending on the information a researcher needs, the amounts and/or identity of auxins in biological samples can be determined by bioassay, mass spectrometry, or radioimmunoassay. A **bioassay** is a measurement of the effect of a known or suspected biologically active substance on living material. In his pioneering work more than 60 years ago, Went used *Avena sativa* (oat) coleoptiles in a technique called the ***Avena* coleoptile curvature test** (see Figure 19.1). The coleoptile curved because the increase in auxin on one side stimulated cell elongation, and the decrease in auxin on the other side (due to the absence of the coleoptile tip) caused a decrease in the growth rate. Went found that he could estimate the

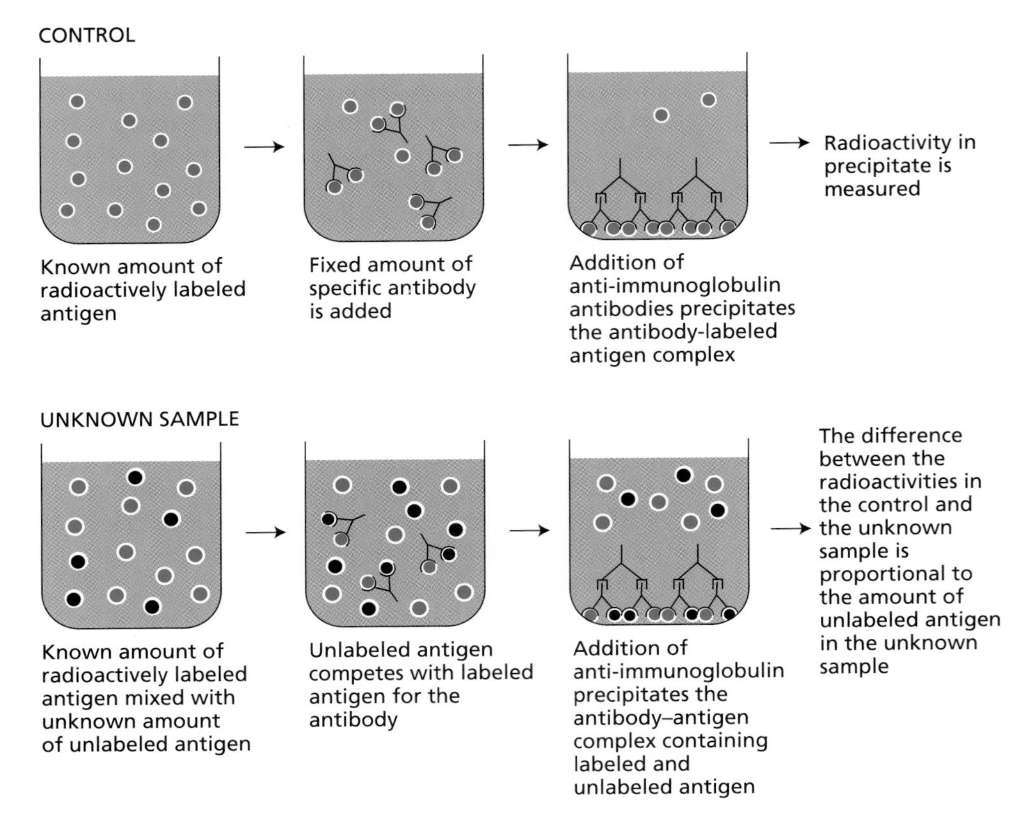

Figure 19.4 Radioimmunoassay for the amount of auxin in a plant tissue extract. In the preparation of auxin antibodies, IAA is first conjugated to a large protein and injected into a mouse or rabbit. The animal produces antibodies to all of the antigenic determinants on the protein, including the IAA. The anti-IAA antibodies can then be purified for use in the RIA. Unlabeled antigen (IAA) competes with a known amount of radioactively labeled antigen—e.g., [^{14}C]IAA—for binding to antibodies. The more unlabeled antigen present, the less radioactivity from the radioactive antigen will be found in the antibody precipitate.

amount of auxin in a sample by measuring the resulting coleoptile curvature.

Auxin bioassays are still used today to detect the presence of auxin activity in a sample. Some bioassays can also provide a rough quantitative estimate of auxin. The *Avena* coleoptile curvature assay is a sensitive measure of auxin activity (it is effective for IAA concentrations of about 0.02 to 0.2 mg L^{-1}). Another method used for auxin bioassays and auxin research is the **coleoptile straight-growth test**. This assay is based on the ability of auxin to stimulate the elongation of coleoptile sections floated in a buffered solution. Both of these bioassays can establish the presence of an auxin in a sample, but they cannot be used for precise quantitation or identification of the specific compound.

A **radioimmunoassay (RIA)** allows the measurement of physiological levels (10^{-9} g = 1 ng) of IAA in plant tissues. An RIA (Figure 19.4), requires two chemicals: (1) specific antibodies that recognize IAA; and (2) a radioactively labeled IAA in which several hydrogen atoms have been replaced by radioactive tritium. Because of the sensitivity and selectivity of the antibodies used in this technique, RIA is as sensitive as some of the best physicochemical methods (Caruso et al. 1995). Radioimmunoassays also allow one to quantify a specific auxin. In a modified RIA, the radiolabeled IAA is replaced by an auxin conjugated to an enzyme, generally an alkaline phosphatase. This *enzyme-linked immunosorbent assay (ELISA)* affords sensitivities that are similar or better than the RIA.

Mass spectrometry is the method of choice when information about both the chemical structure and the amount of IAA is needed. This method is used in conjunction with separation protocols that may involve *thin-layer chromatography (TLC)*, *high-performance liquid chromatography (HPLC)*, and *gas chromatography (GC)*. It allows the precise quantification and identification of auxins, and can detect as little as 10^{-12} g (1 picogram, or pg) of IAA, which is well within the range of auxin found in a single pea stem section or a corn kernel.

These sophisticated techniques have enabled researchers to accurately analyze auxin precursors, auxin turnover, and auxin distribution within the plant (Normanly et al. 1995).

Multiple Pathways Exist for the Biosynthesis of IAA

IAA biosynthesis is associated with rapidly dividing tissues, especially in shoots. Shoot apical meristems, young leaves, and developing fruits are the primary sites of IAA synthesis. We do not know exactly where in these tissues IAA is synthesized. Although IAA may also be produced in mature leaves and in root tips, the level of IAA production in these tissues is usually lower.

IAA is structurally related to the amino acid tryptophan, and early studies on auxin biosynthesis focused on tryptophan as the probable precursor. However, the incorporation of exogenous labeled tryptophan (e.g., [^{3}H]tryptophan) into IAA by plant tissues has proved difficult to demonstrate. Nevertheless, an enormous body of evidence has now accumulated showing that plants convert tryptophan to IAA by several pathways. This evidence includes both in vivo and in vitro labeling studies using radiolabeled tryptophan, as well as the demonstration of the enzymes for each of the steps. In addition to the tryptophan-dependent pathways (which are shown in Figure 19.5), recent genetic studies have provided evidence that plants can synthesize IAA via one or more tryptophan-*independent* pathways. The existence of multiple pathways for IAA biosynthesis makes it nearly impossible for plants to run out of auxin and is probably a reflection of the essential role of this hormone in plant development.

The **indole-3-pyruvic acid (IPA) pathway** (center of Figure 19.5), is probably the most common of the tryptophan-dependent pathways and involves a *deamination*

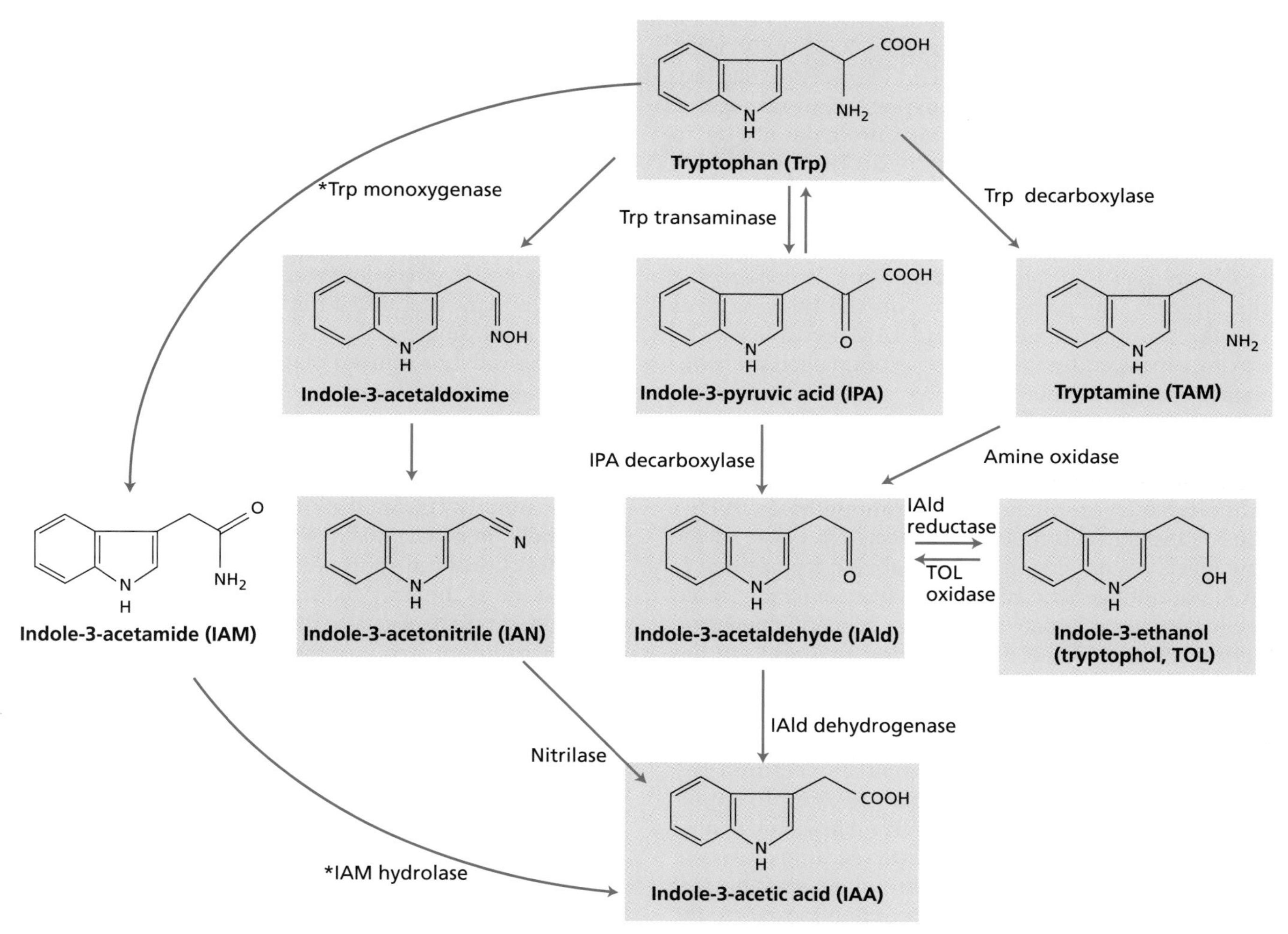

Figure 19.5 Tryptophan-dependent pathways of IAA biosynthesis in plants and bacteria. The enzymes that are present only in bacteria are marked with an asterisk. (After Bartel 1997.)

reaction to form IPA, followed by a *decarboxylation* reaction to form indole-3-acetaldehyde (IAld). Indole-3-acetaldehyde is then oxidized to IAA by a specific dehydrogenase. The **tryptamine (TAM) pathway** (at right in Figure 19.5) is similar to the IPA pathway, except that the order of the deamination and decarboxylation reactions is reversed, and different enzymes are involved. Species that do not utilize the IPA pathway possess the TAM pathway. In at least one case (tomato), there is evidence for both the IPA and the TAM pathways (DeLuca et al. 1989; Nonhebel et al. 1993). However, the simultaneous presence of both pathways seems to be rare, and may be limited to tomato (Normanly et al. 1995).

In the **indole-3-acetonitrile (IAN) pathway** (left of center in Figure 19.5), tryptophan is first converted to indole-3-acetaldoxime and then to indole-3-acetonitrile. The enzyme that converts IAN to IAA is called *nitrilase*. In a survey of 21 plant families, nitrilase and, by implication, the IAN pathway, were found to be present in only three: Cruciferae, Graminae, and Musaceae (Thimann and Mahadevan 1964). Four genes (*NIT1* through *NIT4*) encoding nitrilase enzymes have now been cloned from *Arabidopsis* (Bartel and Fink 1994; Bartling et al. 1994). When *NIT2* was expressed in transgenic tobacco, the resultant plants acquired the ability to respond to IAN as an auxin by hydrolyzing it to IAA (Schmidt et al. 1996).

Finally, the **indole-3-acetamide (IAM) pathway** (far left in Figure 19.5) is used by various pathogenic bacteria, such as *Pseudomonas savastanoi* and *Agrobacterium tumefaciens*. This pathway involves the two enzymes tryptophan monoxygenase and IAM hydrolase. The auxins produced by these bacteria often elicit morphological changes in their hosts.

IAA Is Also Synthesized from Indole or from Indole-3-Glycerol Phosphate

Although a tryptophan-independent pathway of IAA biosynthesis had long been suspected because of the low levels of conversion of radiolabeled tryptophan to IAA, not until genetic approaches were available could the existence of such pathways be confirmed and defined. Perhaps the most striking of these studies involves the *orange pericarp* (*orp*) mutant of maize. This plant has mutations in both of the genes encoding subunits of the enzyme that catalyzes the final step in tryptophan biosynthesis, *tryptophan synthase* (Figure 19.6). The *orp* mutant is a true tryptophan auxotroph, requiring exogenous tryptophan to survive. However, neither the *orp* seedlings nor the wild-type seedlings can convert tryptophan to IAA, even when the mutant seedlings are given enough tryptophan to reverse the lethal effects of the mutation.

Despite the block in tryptophan biosynthesis, the *orp* mutant contains amounts of IAA 50-fold higher than those of a wild-type plant (Wright et al. 1991). Signficantly, when *orp* seedlings were fed [^{15}N]anthranilate (see Figure 19.6), the label subsequently appeared in IAA, but not in tryptophan. These results provided the best experimental evidence for a tryptophan-independent pathway of IAA biosynthesis. Further evidence was provided by labeling studies with tryptophan auxotrophic mutants of *Arabidopsis* (Normanly et al. 1993). Together, these studies established that the branch point for IAA biosynthesis is either indole or its precursor, indole-3-glycerol phosphate (see Figure 19.6).

Studies with the *Arabidopsis* mutants *trp2* and *trp3* , which are blocked in the last two steps of tryptophan biosynthesis, have shown that IAN accumulates up to 11-fold in the mutants compared to the wild type (Radwanski et al. 1996). In tomato, IPA has been shown to be synthesized independently of tryptophan (Nonhebel et al. 1993). Depending on the species, then, either IAN or IPA may serve as the intermediate between either indole-3-glycerol phosphate or indole, respectively, and IAA.

The discovery of the tryptophan-independent pathway has drastically altered our view of IAA biosynthesis, but the relative importance of the two pathways (tryptophan-dependent versus tryptophan-independent) is poorly understood. Some plants, such as bean seedlings, synthesize IAA primarily by the tryptophan-dependent pathway. The same is true of embryogenic carrot cell suspension cultures. However, upon removal from the medium containing 2,4-D, which initiates embryogenesis in these cultures, the cells switch over to the tryptophan-*independent* pathway of IAA biosynthesis. This result suggests that the specific pathway of IAA biosynthesis that is utilized may be under developmental control.

Auxin Is Transported Polarly in Shoots and Roots

More than 50 years ago it was discovered that IAA moves mainly from the apex to the base (i.e., basipetally) in isolated oat coleoptile sections. This type of unidirectional transport has been termed **polar transport**. Auxin is the only plant growth hormone that is transported polarly. Because the shoot apex serves as the primary source of auxin for the entire plant, polar transport contributes to the formation of an auxin gradient from the shoot to the root. The longitudinal gradient of auxin is thought to affect various developmental processes, including stem elongation, apical dominance, wound healing, and leaf senescence. Each of these topics will be discussed later in the chapter.

To study polar transport, researchers have employed the *donor–receiver agar block method* (Figure 19.7). An agar block containing radioisotope-labeled auxin (donor block) is placed on one end of a tissue segment, and a receiver block is placed on the other end. The movement of auxin through the tissue into the receiver block can be determined over time by measuring the

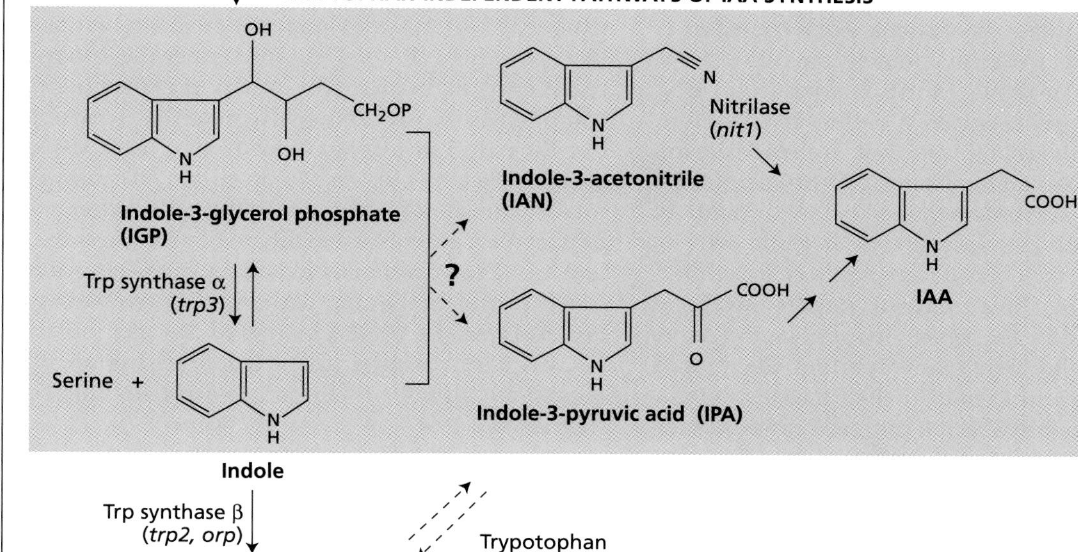

Figure 19.6 Tryptophan-independent pathways of IAA biosynthesis in plants. The tryptophan biosynthetic pathway is shown on the left, and mutants discussed in the text are indicated in parentheses. The suggested branch points from the tryptophan (Trp) biosynthetic pathway are at indole-3-glycerol phosphate or at indole. IAN and IPA are two possible intermediates. The conversion of tryptophan to IPA is hypothetical. (After Bartel 1997.)

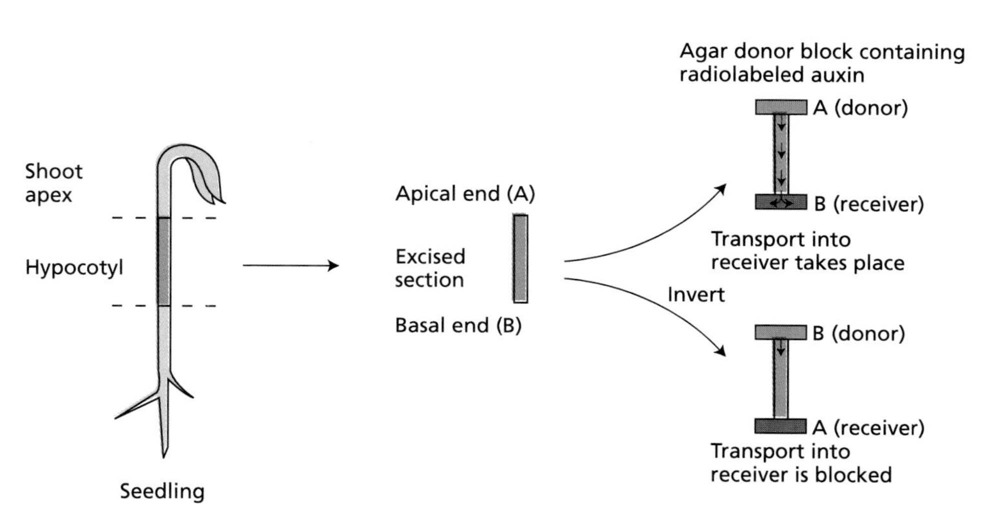

Figure 19.7 The standard method for measuring polar auxin transport. The polarity of transport is independent of orientation with respect to gravity. (From Lomax et al. 1995.)

radioactivity in the receiver block. To confirm that the radioactivity represents IAA and not a breakdown product, we can analyze the material in the receiver block using a sensitive detection method, such as mass spectrometry.

From a multitude of such studies, the general properties of polar IAA transport have emerged. In coleoptiles and vegetative shoots, basipetal transport predominates. It is independent of the orientation of the tissue (at least over short periods of time), so it is not affected by gravity. The exact sites of polar transport in shoot tissues have not been definitively established, and they may vary in different tissues. In coleoptiles, for example, polar transport appears to occur in the nonvascular tissues, whereas in intact dicot stems, polar transport is thought to be restricted to the parenchyma cells associated with the vascular tissue (Morris and Johnson 1985).

Polar transport proceeds in a cell-to-cell fashion, rather than through the symplast. That is, auxin exits the cell through the plasma membrane, diffuses across the two primary walls, and enters the cell below through its plasma membrane. The loss of auxin from cells is termed *auxin efflux*; the entry of auxin into cells is called *auxin uptake* or *influx*. These processes require metabolic energy, as evidenced by the sensitivity of polar transport to O_2 deprivation and metabolic inhibitors. The rate of polar transport is about 1 cm h^{-1}—ten times faster than the rate of diffusion, but about a hundred times slower than the rate of transport in the phloem (see Chapter 10). Polar transport is also specific for active auxins, both natural and synthetic. Neither inactive auxin analogs nor auxin metabolites are transported polarly, suggesting that polar transport involves protein carriers on the plasma membrane that can recognize the hormone and its active analogs.

Most of the auxin in the root is transported acropetally (from root base to root tip) and transport occurs mainly in the stele. The auxin in the root vascular cylinder is thought to play an important role in stimulating the pericycle to form branch roots (Lomax et al. 1995). However, a small amount of basipetal auxin transport from the root tip has also been demonstrated. In maize roots, radiolabeled IAA applied to the root tip is transported basipetally about 2 to 8 mm. As will be discussed later in the chapter in relation to gravitropism, basipetal auxin transport in the root appears to occur in the cortical tissues and possibly in the epidermis (Yang et al. 1990; Young and Evans 1996).

Inhibitors of Auxin Transport Block Auxin Efflux

At about the same time that synthetic auxins became available, several compounds that inhibit the polar transport of auxin—including NPA (1-*N*-naphthylphthalamic acid) and TIBA (2,3,5-triiodobenzoic acid)—were synthesized. These compounds are called **auxin transport inhibitors** (**ATIs**). When applied to intact plants, members of a subset of these ATIs, called **phytotropins**, also cause stunting, severe inhibition of root growth, and loss of gravitropic and phototropic responses. Phytotropins have the common structural theme of benzoic acid ortho-linked to a second aromatic ring system. NPA is the best-studied phytotropin; 1-pyrenoylbenzoic acid (PBA) is the most active phytotropin (Figure 19.8). TIBA is an example of an ATI that is not a phytotropin, since it has neither the phytotropin chemical structure nor the ability to interfere with gravitropism and phototropism.

In general, ATIs prevent polar transport by blocking auxin efflux rather than by inhibiting auxin uptake (Rubery 1990). We can demonstrate this relationship by incorporating NPA or TIBA into either the donor or the receiver block in an auxin transport experiment. Both compounds inhibit auxin efflux into the receiver block, but they do not affect auxin uptake from the donor block. In fact, by blocking auxin efflux, ATIs increase the total accumulation of auxin in the cells. When stem or hypocotyl segments are incubated in solutions that contain an ATI and radiolabeled auxin, increasing amounts of ATI stimulate the net uptake of IAA into the segment. For example, 1 μM NPA increased the accumulation of [^{14}C]IAA by zucchini (*Cucurbita pepo*) hypocotyl segments by eight- to tenfold relative to the control (Bernasconi 1996).

Although all ATIs inhibit polar transport by preventing auxin efflux, the mechanism of action of phytotropins appears to be different from that of nonphytotropin ATIs, such as TIBA. Since TIBA itself has weak auxin activity and is transported polarly (Thomson et al. 1973), it may inhibit polar transport directly by binding to the auxin-binding site on the efflux carrier (discussed below). In contrast, phytotropins, which are not transported polarly, are believed to bind to an associated protein that regulates the efflux carrier.

Phytotropins Bind to Specific Sites on the Plasma Membrane

The availability of radiolabeled NPA has allowed the biochemical characterization of the phytotropin-binding site. [^{3}H]NPA binds specifically to plasma membranes from a wide range of species (Lomax et al. 1995). Analysis of the binding kinetics suggests that NPA binds to a single site, presumably on a protein, and that other phytotropins can displace NPA by competing for this site. Neither IAA nor TIBA compete with NPA for its binding site (Sussman and Goldsmith 1981). In some studies the binding activity seems to be associated with a peripheral membrane protein (Cox and Muday 1994), whereas in other studies NPA appears to bind to an integral membrane protein on the plasma membrane (Bernasconi 1996).

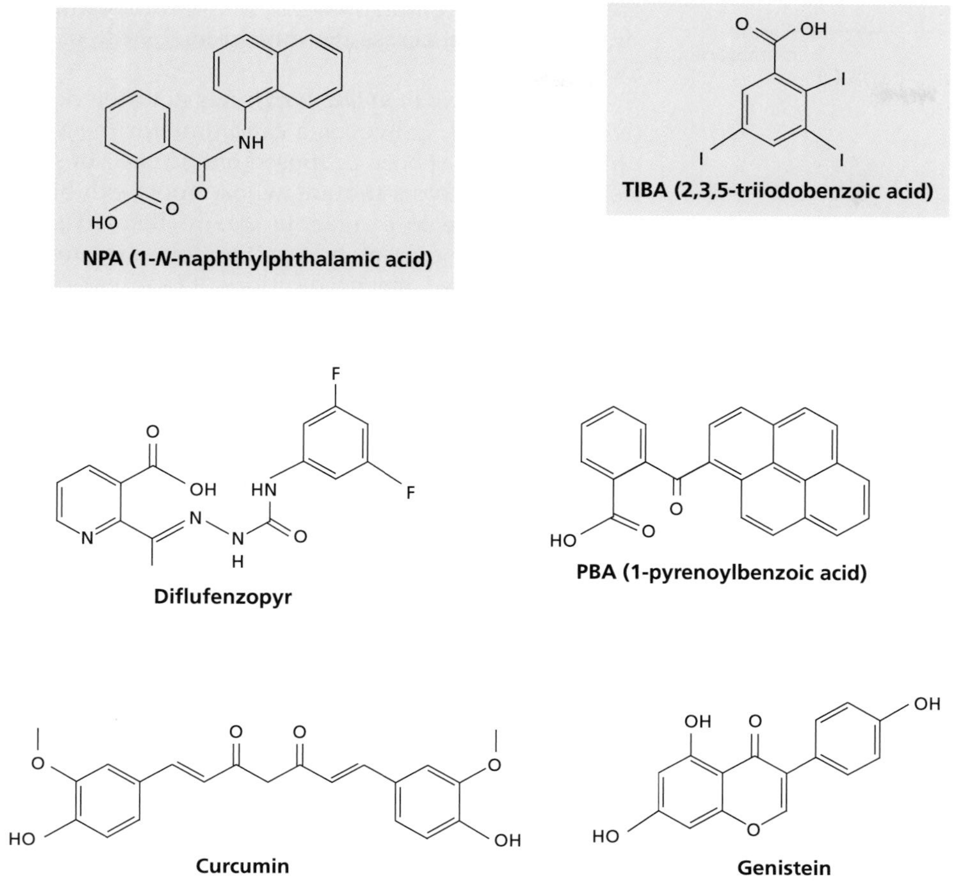

NPA (1-*N*-naphthylphthalamic acid)

TIBA (2,3,5-triiodobenzoic acid)

Diflufenzopyr

PBA (1-pyrenoylbenzoic acid)

Curcumin

Genistein

Figure 19.8 Structure of auxin transport inhibitors. The most active ATI is PBA. Curcumin (a phenolic) and genistein (a flavonoid) are naturally occurring plant compounds.

Inasmuch as NPA is not present in plants, scientists have sought endogenous molecules that might bind to the same site as NPA. Such compounds would be candidates for "natural phytotropins" that might regulate polar auxin transport during normal development. Indeed, naturally occurring flavonoid compounds, such as *quercitin*, are able to compete with [^{3}H]NPA for its binding site on membranes (Jacobs and Rubery 1988). Since quercitin also blocks auxin efflux from hypocotyl segments, it has the properties of a naturally occurring phytotropin, but its role in regulating polar auxin transport has not yet been established.

A Chemiosmotic Model Explains Polar Auxin Transport

The elucidation of the chemiosmotic mechanism of solute transport in the late 1960s (see Chapter 6) led to the application of this model to polar auxin transport (Goldsmith 1977). According to the widely accepted **chemiosmotic model** for polar auxin transport, the total proton motive force ($\Delta E + \Delta pH$) across the plasma membrane drives the uptake of auxin, whereas transport of auxin from the cell (efflux) is driven solely by the membrane potential, ΔE. A crucial feature of the polar transport model is that the auxin efflux carriers are localized predominantly at the basal ends of the conducting cells (Figure 19.9). The evidence for each step in this model is considered separately in the discussion that follows.

The first step in polar transport is auxin influx. Auxin enters plant cells by either of two mechanisms: passive diffusion across the phospholipid bilayer or secondary active transport via a proton symporter. The dual pathway of auxin uptake arises because the passive permeability of the membrane to auxin depends strongly on the apoplastic pH. The undissociated form of indole-3-acetic acid, in which the carboxyl group is protonated, is lipophilic and readily diffuses across lipid bilayer membranes. In contrast, the dissociated form of IAA is negatively charged and therefore does not cross membranes unaided. Since the plasma membrane H^+-ATPase normally maintains the cell wall solution at about pH 5, about half of the auxin ($pK_a = 4.75$) in the apoplast will be in the undissociated form, and will diffuse passively across the plasma membrane down a

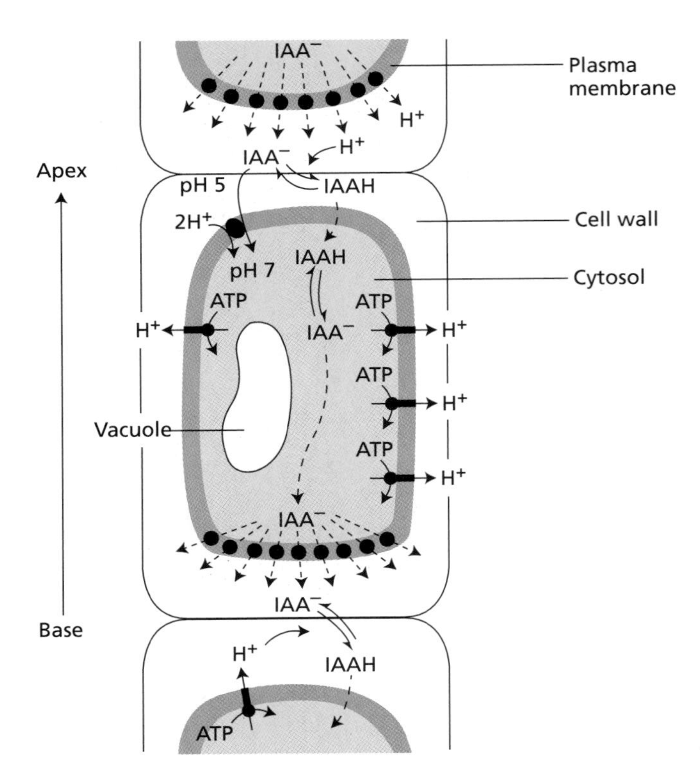

Figure 19.9 The chemiosmotic model for polar auxin transport. Shown here is one cell in a column of auxin-transporting cells. The cell wall is maintained at an acidic pH by the activity of the plasma membrane H^+-ATPase. IAA enters the cell either passively in the undissociated form (IAAH) or in the anionic form (IAA^-) by secondary active cotransport with protons via a permease. In the cytosol, which has a neutral pH, the anionic form predominates. The anions exit the cell via auxin anion efflux carriers that are concentrated at the basal ends of each cell in the longitudinal pathway. (From Jacobs 1983.)

concentration gradient. Experimental support for pH-dependent, passive auxin uptake was first provided by Rubery and Sheldrake (1973), who showed that IAA uptake by plant cells increases as the extracellular pH is lowered from a neutral to a more acidic value.

Evidence for a carrier-mediated, secondary active uptake mechanism has been provided by Lomax and colleagues (reviewed by Lomax et al. 1995). In studies using isolated membrane vesicles from zucchini (*Cucurbita pepo*) hypocotyls and radiolabeled IAA, auxin uptake was shown to be saturable and specific for active auxins (Lomax 1986). In experiments in which the ΔpH and ΔE of isolated membrane vesicles were manipulated artificially, the uptake of radiolabeled auxin was shown to be stimulated in the presence of a pH gradient, as in passive uptake, but also when the inside of the vesicle was negatively charged relative to the outside. On the basis of these and other experiments, Lomax and her colleagues concluded that an H^+–IAA^- symporter cotransports two protons along with the auxin anion. This secondary active transport of auxin allows for greater auxin accumulation than simple diffusion does, since it is driven across the membrane by the proton motive force.

Thus far the *auxin uptake carrier* has not been definitively identified. However, a candidate for the auxin uptake carrier has been proposed on the basis of studies of an *Arabidopsis* mutant whose roots exhibit an altered response to exogenous auxin (Bennett et al. 1996). Roots are extremely sensitive to auxin and are normally inhibited at concentrations of IAA that stimulate the growth of the stem. One *Arabidopsis* mutant, *aux1*, exhibits a dramatic dwarfed phenotype. The roots of the *aux1* mutants are both insensitive to auxin and agravitropic (that is, they do not grow downward in response to gravity). The *AUX1* gene was cloned and shown to encode a polypeptide similar in sequence to amino acid permeases, suggesting that AUX1 may mediate the transport of an amino acid–like signaling molecule. Since IAA is chemically similar to the amino acid tryptophan, it is a possible ligand for the AUX1 carrier protein.

Once IAA enters the cytosol, which has a pH of approximately 7.2, nearly all of it will be in the anionic form. Because the membrane is less permeable to IAA^- than to IAAH , auxin will tend to accumulate in the cytosol. However, much of the auxin that enters the cell escapes via an *auxin efflux carrier*. Transport of IAA^- out of the cell is driven by the inside negative membrane potential. As noted earlier, a crucial feature of the chemiosmotic model for polar transport is that IAA^- efflux takes place preferentially at the basal end of each cell. According to the model, the repetition of auxin uptake at the apical end of the cell and preferential release from the base of each cell in the pathway gives rise to the total polar transport effect.

Is the auxin efflux carrier actually localized at the basal ends of cells? Using mouse monoclonal antibodies that had been selected for their ability to compete with [^{3}H]NPA for binding to a membrane fraction from pea epicotyls, Jacobs and Gilbert (1983) obtained immunocytochemical evidence that the NPA-binding protein is indeed localized at the basal ends of bundle sheath parenchyma cells in pea stems. If these results can be confirmed, they would provide compelling evidence in support of the chemiosmotic model of polar auxin transport.

Auxin Transport Mutants Have Been Isolated in *Arabidopsis*

Some *Arabidopsis* mutants are impaired in auxin transport or in ATI perception. Identification of the genes for such mutations will help elucidate the polar transport mechanism. For example, the *tir3* (*t*ransport *i*nhibitor *r*esponse *3*) mutant is much less sensitive to NPA than is the wild type, and it exhibits a reduction in polar auxin transport and NPA-binding activity (Ruegger et al. 1997). Although it responds normally to exogenous auxin, *tir3* has a dwarf phenotype and is deficient in the production of lateral roots, a process known to depend on polar auxin transport from the shoot to the root. The *TIR3* gene may thus encode either the NPA-binding protein itself, or a gene required for the expression, localization, or stability of the NPA-binding protein.

Another mutant recently isolated from *Arabidopsis* exhibits a peculiar response to NPA: The roots curl. Designated *rcn1* (*r*oots *c*url in *N*PA), the mutant exhibits an enhanced sensitivity to NPA in other assays as well, including inhibition of hypocotyl elongation and auxin efflux. The *RCN1* gene was cloned and found to be closely related to the regulatory subunit of *protein phosphatase 2A*, a serine/threonine phosphatase (Garbers et al. 1996). Protein phosphatases are known to play important roles in enzyme regulation, gene expression, and signal transduction by removing regulatory phosphate groups from proteins (see Chapter 14). Although the full implications of this finding are unclear, it suggests that a signal transduction pathway involving protein kinases and protein phosphatases may be involved in signaling between the NPA-binding protein and the auxin efflux carrier. Consistent with this idea, many of the flavonoids that displace NPA from its binding site on membranes are also protein kinase inhibitors (Bernasconi 1996).

Auxin Is Transported Nonpolarly in the Phloem

Most of the IAA that is synthesized in mature leaves appears to be transported to the rest of the plant nonpolarly via the phloem. Auxin, along with the other components of the phloem, can move from these leaves up or down the plant at velocities much higher than those of polar transport (see Chapter 10). Phloem transport is largely passive, not requiring energy directly. As discussed in the next section, most of the auxin found in the plant is conjugated (bound) to other molecules. In corn seedlings, conjugated IAA appears to be exported and distributed from the germinating seed kernel to the coleoptile tip via the phloem. The overall importance of the phloem pathway versus the polar transport system for the long-distance movement of IAA in plants remains unclear. However, the evidence suggests that long-distance auxin transport in the phloem is important for controlling processes such as cambium division and branch root formation (Aloni 1995).

Polar transport and phloem transport are not independent of each other. Recent studies with radiolabeled IAA suggest that in pea, auxin can be transferred from the nonpolar phloem pathway to the polar transport pathway (Cambridge and Morris 1996). [^{14}C]IAA applied to the upper surface of a mature leaf was transported nonpolarly, and could be detected in aphids feeding on the stem, indicative of phloem transport. Four hours after application of the label, internode segments 3 cm long were excised and tested for auxin efflux. [^{14}C]IAA diffused from the segments basipetally, but not acropetally, and basipetal efflux was strongly inhibited by NPA. These results suggest that auxin was transported from the phloem to the polar transport system. This transfer seems to take place mainly in the immature tissues of the shoot apex. As yet there is no evidence that auxin can be transferred from the polar transport system to the phloem pathway.

Most IAA in the Plant Is in a Covalently Bound Form

Although free IAA is the biologically active form of the hormone, the vast majority of auxin in plants is found in a covalently bound state. These **conjugated**, or "bound," auxins have been identified in all higher plants (Bandurski et al. 1995) and are generally regarded as hormonally inactive (Hangarter and Good 1981). IAA has been found to be conjugated to both high- and low-molecular-weight compounds. Examples of low-molecular-weight conjugated auxins include esters of IAA with glucose or *myo*-inositol and amide conjugates such as IAA-*N*-aspartate (Figure 19.10). High-molecular-weight IAA conjugates include IAA-glucan (7 to 50 glucose units per IAA) and IAA-glycoproteins found in cereal seeds. The compound to which IAA is conjugated and the extent of the conjugation depend on the specific conjugating enzymes. The best-studied reaction is the conjugation of IAA to glucose in *Zea mays*. The gene that encodes the enzyme uridine 5′-diphosphate-glucose:indole-3-acetyl—D-glycosyl transferase, which catalyzes this reaction, has been cloned (Szerszen et al. 1994).

The highest concentrations of free auxin in the living plant are in the apical meristems of shoots and in young leaves, since these are the primary sites of auxin synthesis. However, auxins are widely distributed in the plant. Metabolism of conjugated auxin may be a major contributing factor in the regulation of the levels of free auxin. For example, during the germination of seeds of *Zea mays*, IAA-*myo*-inositol is translocated from the endosperm to the coleoptile via the phloem. At least a portion of the free IAA produced in coleoptile tips of *Zea mays* is believed to be derived from the hydrolysis of IAA-*myo*-inositol. In addition, environmental stimuli such as light and gravity have been shown to influence both the rate of auxin conjugation (removal of free auxin) and the release of free auxin (hydrolysis of conjugated auxin). The formation of conjugated auxins may serve other functions as well, including storage and protection against oxidative degradation.

IAA Is Degraded by Multiple Pathways

Like IAA biosynthesis, the enzymatic breakdown (oxidation) of IAA may involve more than one pathway. For some time it has been thought that peroxidative enzymes are chiefly responsible for IAA oxidation, pri-

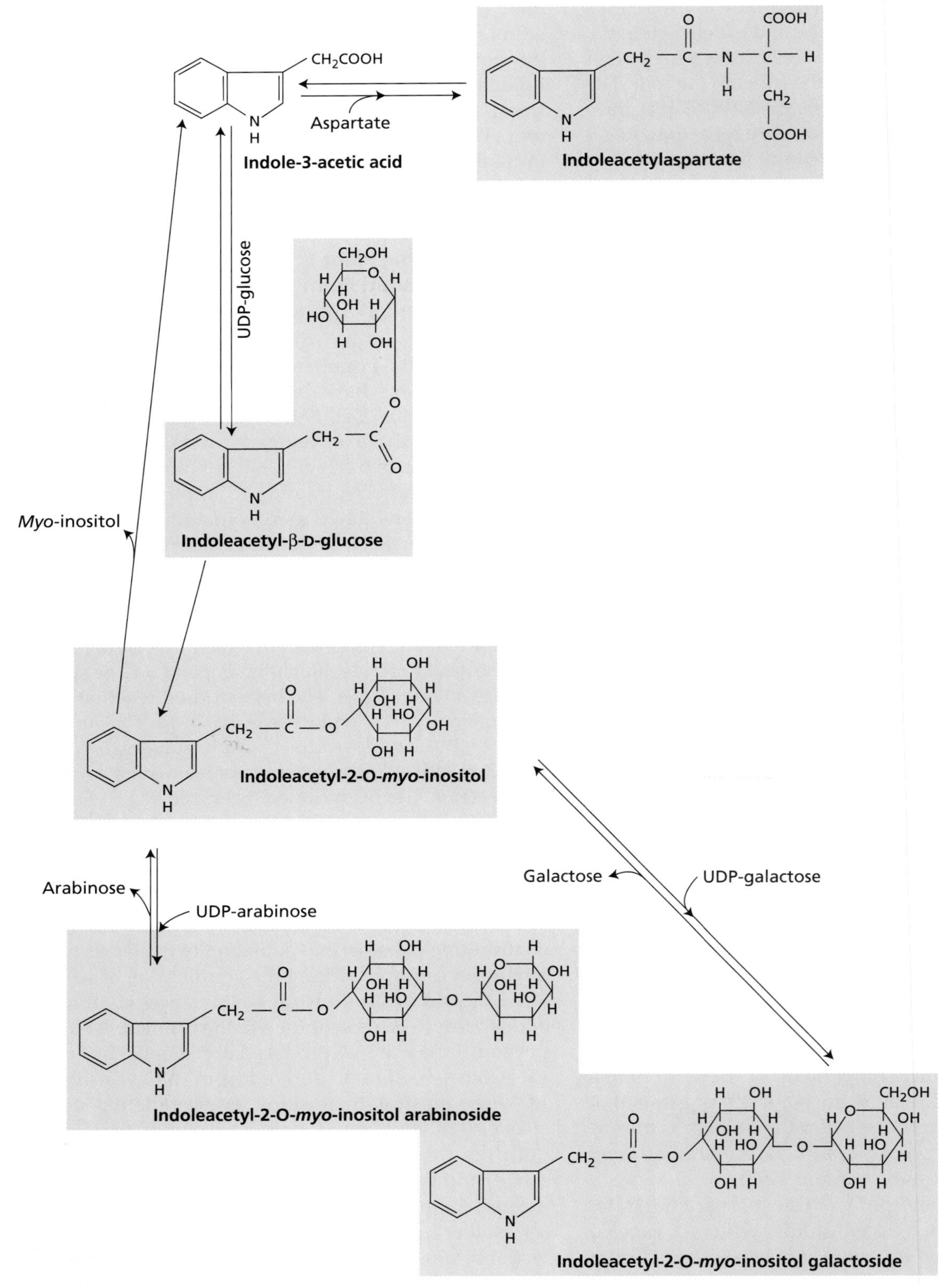

Figure 19.10 Structure and proposed metabolic pathways of bound auxins. The diagram shows structures of various IAA conjugates and proposed metabolic pathways involved in their synthesis and breakdown. Single arrows indicate irreversible pathways; double arrows indicate reversible ones.

marily because these enzymes are ubiquitous in higher plants and their ability to degrade IAA can be demonstrated in vitro. The peroxidase pathway leads to an oxidized product, 3-methyleneoxindole, in which the side chain (the acetic acid residue) is shortened by one carbon (decarboxylated) (Figure 19.11A). However, the physiological significance of the peroxidase pathway is unclear. For example, no change in the IAA levels of transgenic plants was observed with either a tenfold increase in peroxidase expression or a tenfold repression of peroxidase activity (Normanly et al. 1995).

On the basis of isotopic labeling and metabolite identification, two other oxidative pathways have been proposed that are more likely to be involved in the controlled degradation of IAA (Figure 19.11B), with oxindole-3-acetic acid (OxIAA) as the final product. OxIAA has been shown to be a naturally occurring compound in the endosperm and shoot tissues of *Zea mays*. In one pathway, IAA is oxidized without decarboxylation to OxIAA. In another pathway, the IAA-aspartate conjugate is oxidized first to the intermediate dioxindole-3-acetylaspartate, and then to OxIAA. These pathways differ from the peroxidase pathway in that oxidation is carried out without decarboxylation and the acetic acid side chain remains intact (Bandurski et al. 1995).

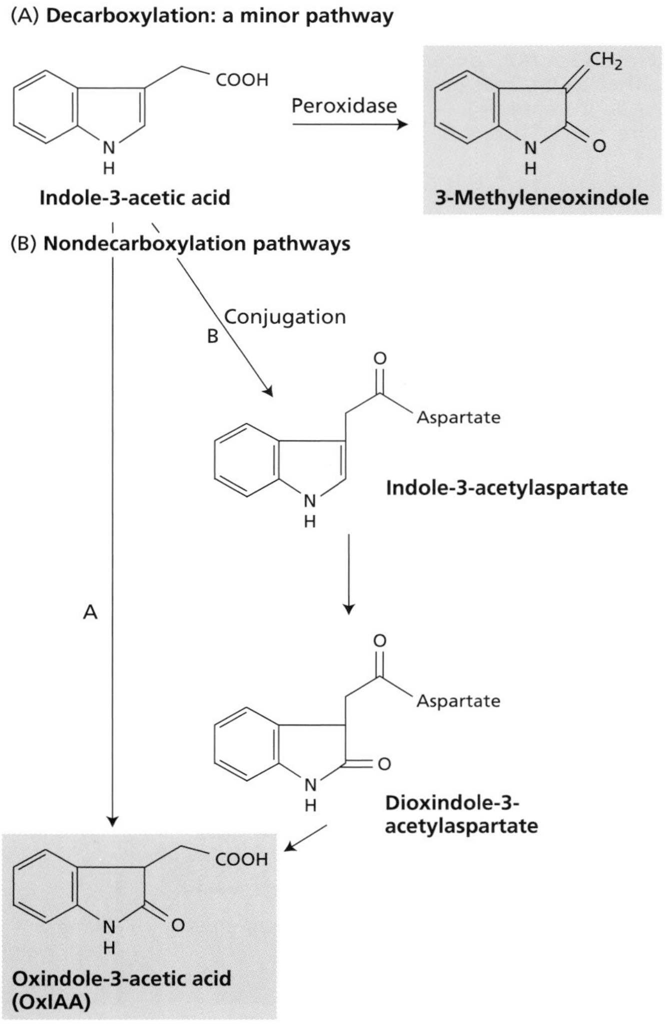

In vitro, IAA can be oxidized nonenzymatically when exposed to light, and its photodestruction may be promoted by plant pigments such as riboflavin. The products of such photooxidation have been isolated from plants, but the pathway of IAA photooxidation and its role, if any, in plant growth regulation are not known.

(A) **Decarboxylation: a minor pathway**

COOH

Peroxidase

N H

Indole-3-acetic acid

CH_2

N H O

3-Methyleneoxindole

(B) **Nondecarboxylation pathways**

Conjugation

B

O

Aspartate

N H

Indole-3-acetylaspartate

A

O

Aspartate

N H O

Dioxindole-3-acetylaspartate

COOH

N H O

Oxindole-3-acetic acid (OxIAA)

Figure 19.11 Biodegradation of IAA. (A) The peroxidase route (decarboxylation pathway) plays a relatively minor role. (B) The two nondecarboxylation routes of IAA oxidative degradation, A and B, are the most common metabolic pathways. (After Tuominen et al. 1994.)

There Are Two Subcellular Pools of IAA: The Cytosol and the Chloroplasts

The distribution of IAA in the cell appears to be regulated largely by pH. Because the dissociated form does not cross membranes unaided, whereas the protonated form readily diffuses across membranes, auxin tends to accumulate in the more alkaline compartments of the cell. Göran Sandberg and his colleagues in Sweden have studied the distribution of IAA and its metabolites in wild-type and transgenic tobacco cells. In the wild-type cells, about one-third of the IAA is found in the chloroplast, and the remainder is located in the cytosol. IAA conjugates are located exclusively in the cytosol (Sitbon et al. 1993).

In the transgenic tobacco cells that express the *Agrobacterium tumefaciens* genes for auxin biosynthesis, the intermediate indole-3-acetamide is located exclusively in the cytosol. Thus there are two main pools of IAA in the cell: the cytosol and the chloroplasts. The two compartments differ with respect to IAA metabolism. IAA in the cytosol is metabolized either by conjugation or by nondecarboxylative catabolism (see Figure 19.11). The IAA in the chloroplast is protected from these processes, but it is regulated by the amount of IAA in the cytosol, with which it is in equilibrium (Sitbon et al. 1993).

Multiple Factors Regulate Steady-State IAA Levels

The steady-state level of free IAA in the cytosol is determined by several interconnected processes (Figure 19.12). Auxin metabolism includes the synthesis, oxidation, and conjugation of IAA. In addition to IAA metabolism, its compartmentation in the choroplast and transport also play important roles in regulating the level of free IAA (Normanly et al. 1995). The sum total of these processes at any given location in the plant determines the amount of free IAA available to a particular cell.

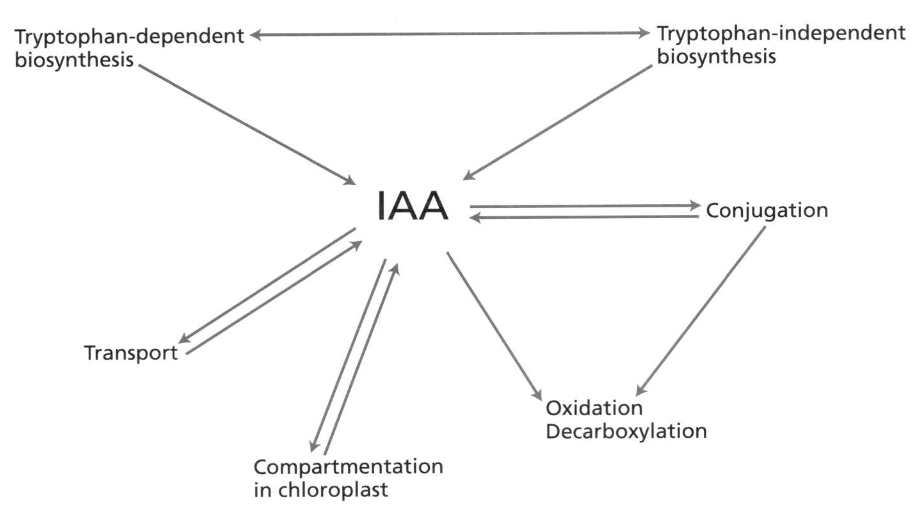

Figure 19.12 Factors that influence the steady-state levels of free IAA in plant cells. Biosynthesis by the tryptophan-dependent and tryptophan-independent pathways can lead only to an increase in the concentration of free IAA. Degradation (either by nondecarboxylative oxidation or by decarboxylation) leads only to a decrease in IAA concentration, while conjugation is reversible and can therefore lead to either an increase or a decrease. Both transport and compartmentation can cause either an increase or a decrease in the cytosolic IAA concentration, depending on the direction of hormone movement. (After Normanly et al. 1995.)

The amounts of IAA in different regions of a wild-type tobacco plant are shown in Figure 19.13A. The highest concentration of free IAA is in the apical bud. The level of free IAA in the mature regions of the stem is nearly constant. Since there is little IAA synthesis in the mature regions of the shoot, we can infer that there must be little catabolism or conjugation in these regions; otherwise the free IAA concentrations would continue to decline.

What happens to the IAA distribution if we perturb the system by transforming the tobacco plant with the two IAA biosynthesis genes from *Agrobacterium tumefaciens*? The results of such an experiment are shown in Figure 19.13B. The level of free IAA in the upper part of the shoot increased considerably, while the free IAA concentrations in the mature parts of the shoot hardly changed. Because of the particular promoter used, the auxin biosynthesis genes are expressed in all the cells of the plant, yet the level of free IAA in the mature shoot is unchanged. We can infer that processes that decrease the amount of free IAA must have been activated in the mature shoot. Sitbon and coworkers (1991) showed, in fact, that the level of conjugated IAA is several times higher in the transformed tobacco plant. Thus, the system is balanced so that the concentration of free IAA is maintained close to its normal level.

Physiological Effects of Auxin: Cell Elongation

Auxin was discovered as the hormone involved in the bending of coleoptiles toward light. The coleoptile bends because of the unequal rates of cell elongation on its shaded versus its illuminated side (see Figure 19.1). The ability of auxin to regulate the rate of cell elongation has long fascinated plant scientists, and more is known about this aspect of auxin activity than any other. In this section we will review the physiology of auxin-induced

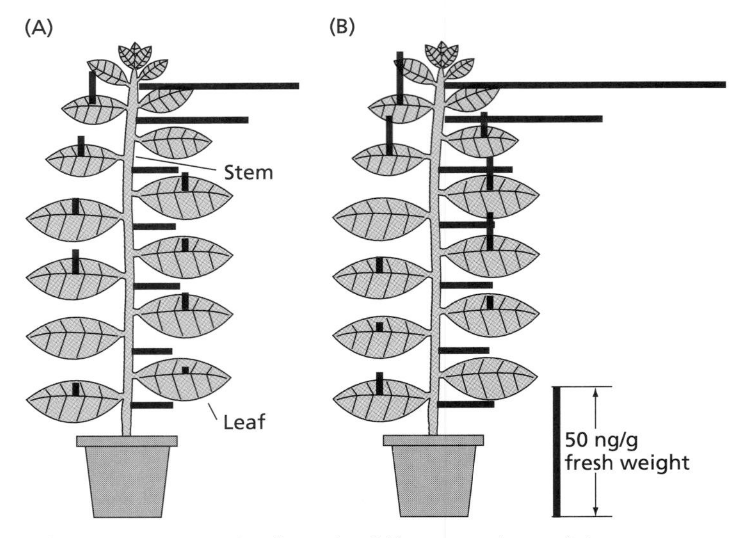

Figure 19.13 Levels of IAA in different regions of the shoot of (A) a wild-type tobacco plant and (B) a transformed tobacco plant expressing the bacterial genes that encode the two enzymes involved in the IAM pathway of IAA biosynthesis: Trp monoxygenase and IAM hydrolase (see Figure 19.5). The bars at the different regions indicate the auxin levels. (After Sitbon et al. 1991.)

cell elongation, some aspects of which were discussed in Chapter 15.

Auxins Promote Growth in Stems and Coleoptiles, but Inhibit Growth in Roots

As we have seen, auxin is synthesized in the shoot apex and transported basipetally to the tissues below. The steady supply of auxin arriving at the subapical region of the stem or coleoptile is required for the continued elongation of these cells. Because the level of endogenous auxin in the elongation region of a normal healthy plant is nearly optimal for growth, spraying the plant with exogenous auxin causes only a modest and short-lived stimulation in growth, and may even be inhibitory in the case of dark-grown seedlings, which are more sensitive to supraoptimal auxin concentrations than light-grown plants are. However, when the endogenous source of auxin is removed by excision of sections containing the elongation zones, the growth rate rapidly decreases to a low basal rate. Such excised sections will often respond dramatically to exogenous auxin by rapidly increasing their growth rate back to the level in the intact plant.

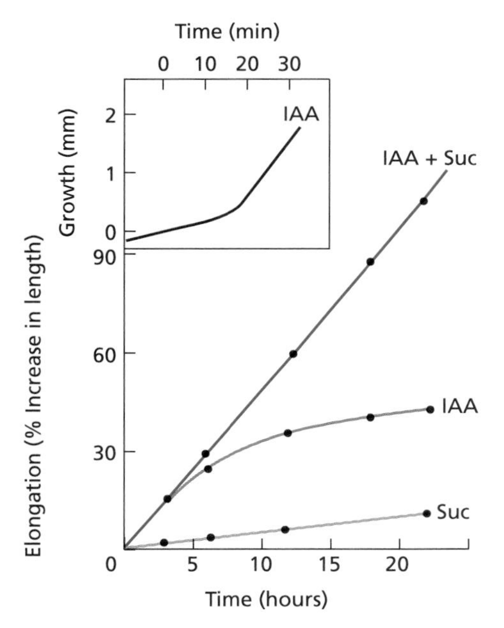

Figure 19.14 Time course for auxin-induced growth of *Avena* (oat) coleoptile sections. Growth is plotted as the percent increase in length. When sucrose (Suc) is included in the medium, the response can continue for as long as 20 hours. Sucrose prolongs the growth response to auxin mainly by providing osmotically active solute that can be taken up for the maintenance of turgor pressure during cell elongation. KCl can substitute for sucrose. The inset shows a short-term time course using an electronic position-sensing transducer. In this graph, growth is plotted as the absolute length in millimeters versus time. The curve shows a lag time of about 15 minutes for auxin-stimulated growth to begin. (From Cleland 1995.)

In long-term experiments, treatment of excised sections of coleoptiles or dicot stems with auxin stimulates the rate of elongation of the section for up to 20 hours (Figure 19.14). The optimal auxin concentration for elongation growth is typically 10^{-6} to 10^{-5} *M* (Figure 19.15). The inhibition beyond the optimal concentration is generally attributed to auxin-induced ethylene biosynthesis. As we will see in Chapter 22, the gaseous hormone ethylene inhibits stem elongation in many species.

The control of root elongation growth is less well understood. Originally it was hypothesized that the auxin responses in roots and shoots are similar, except that the optimal auxin concentration is much lower in roots. However, auxin-induced root growth has been difficult to demonstrate. A possible source of confusion is the fact that auxin induces the synthesis of the plant hormone ethylene, a root growth inhibitor. If ethylene biosynthesis is specifically blocked, low concentrations (10^{-10} to 10^{-9} *M*) of auxin *promote* the growth of intact roots, but higher concentrations (10^{-6} *M*) still inhibit growth. Thus, roots may require a minimum concentration of auxin to grow, but root growth is strongly inhibited by auxin at concentrations that promote elongation in stems and coleoptiles.

Young expanding leaves are a rich source of free auxin for the growing plant. Although auxin may regulate the growth of leaf veins, the role of auxin and other plant hormones in regulating leaf growth is poorly understood. Apart from auxin, light has been shown to be an important factor stimulating leaf expansion (Van Volkenburgh and Cleland 1979), while cytokinins stimulate the expansion of some cotyledons (see Chapter 21).

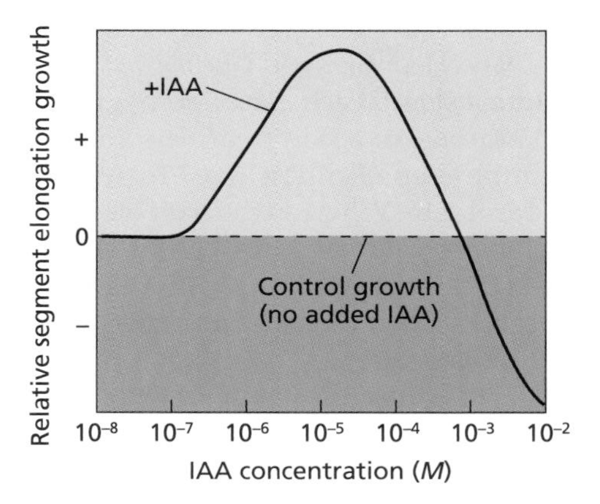

Figure 19.15 Typical dose–response curve for IAA-induced growth in stem or coleoptile sections. Elongation growth of excised sections of coleoptiles or young stems is plotted versus increasing concentrations of exogenous IAA. At higher concentrations (above 10^{-5} *M*), IAA becomes less and less effective; above about 10^{-4} *M* it becomes inhibitory, as shown by the fact that the curve falls below the dashed line, which represents growth in the absence of added IAA.

The Outer Tissues of Dicot Stems Are the Targets of Auxin Action

Dicot stems are composed of many types of tissues and cells, only some of which may limit the growth rate. This point is illustrated by a simple experiment. When stem sections from growing regions of a dicot stem are split lengthwise and incubated in buffer, the two halves bend outward. This result indicates that, in the absence of auxin, the central tissues, including the pith, vascular tissues, and inner cortex, elongate at a faster rate than the outer tissues, consisting of the outer cortex and epidermis. By implication, the outer tissues must be limiting the extension rate of the stem in the absence of auxin. When the split sections are incubated in buffer plus auxin, the two halves now curve inward, demonstrating that the outer tissues of dicot stems are the primary targets of auxin action during cell elongation.

The observation that the outer cell layers are the targets of auxin seems to conflict with the localization of polar transport in the parenchyma cells of the vascular bundles (discussed earlier). However, there is evidence that auxin moves laterally from the vascular tissues of dicot stems to the outer tissues of the elongation zone (Sánchez-Bravo et al. 1991). In coleoptiles, on the other hand, all of the nonvascular tissues (epidermis plus mesophyll) seem to be capable of transporting auxin as well as responding to it.

The Minimum Lag Time for Auxin-Induced Growth Is 10 Minutes

When a stem or coleoptile section is excised and inserted into a sensitive growth-measuring device, the growth response to auxin can be monitored with extremely high precision (Evans 1974). Without auxin in the medium, the growth rate declines rapidly. Addition of auxin markedly stimulates the growth rate after a lag period of only 10 to 12 minutes (see inset in Figure 19.14). Both *Avena* (oat) coleoptiles and *Glycine max* (soybean) hypocotyls (dicot stem) reach a maximum growth rate after 30 to 60 minutes of auxin treatment (Figure 19.16). This maximum represents a five- to tenfold increase over the basal rate. Whereas the oat coleoptile section maintains the maximum rate for up to 18 hours (in the presence of sucrose or KCl), the growth rate of the soybean hypocotyl section oscillates and then gradually declines to the basal rate.

As might be expected, the stimulation of growth by auxin requires energy, and metabolic inhibitors rapidly inhibit the response within minutes. Auxin-induced growth is also extremely sensitive to inhibitors of protein synthesis, such as cycloheximide, suggesting that proteins with high turnover rates are involved. Inhibitors of RNA synthesis also inhibit auxin-induced growth, after a slightly longer delay (Cleland 1995). Although the length of the lag time for auxin-stimulated

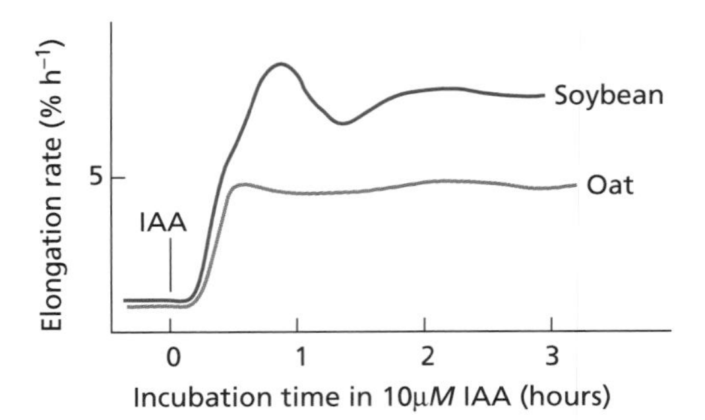

Figure 19.16 Comparison of the growth kinetics of oat coleoptile and soybean hypocotyl sections, incubated with 10 μ*M* IAA and 2% sucrose. Growth is plotted as the rate at each time point, rather than of the absolute length. The growth rate of the soybean hypocotyl oscillates after 1 hour, whereas that of the oat coleoptile is constant. In longer-term studies, the rate of growth of the soybean hypocotyl declines gradually over an 18-hour period, whereas oat coleoptiles maintain a nearly constant growth rate. (After Cleland 1995.)

growth can be increased by lowering the temperature or by using suboptimal auxin concentrations, the lag time cannot be shortened by raising the temperature, by using supraoptimal auxin concentrations, or by removing the waxy cuticle to allow more rapid penetration of the tissue. Thus, the minimum lag time of 10 minutes is not determined by the time required for auxin to reach its site of action. Rather, it reflects the time needed for the biochemical machinery of the cell to bring about the increase in the growth rate.

Auxin Rapidly Increases the Extensibility of the Cell Wall

How does auxin cause a five- to tenfold increase in the growth rate with such a short lag period? To understand the mechanism, we must first review the process of cell enlargement in plants (see Chapter 15). Plant cells expand in two steps : (1) osmotic uptake of water across the plasma membrane, driven by the gradient in water potential ($\Delta\Psi$), and (2) extension of the cell wall in response to the internal turgor pressure. The effects of these parameters on the growth rate are encapsulated in the growth rate equation:

$$\text{Growth rate} = m(P - Y)$$

where m is the irreversible extensibility of the wall, P is the turgor pressure, and Y is the yield threshold. In principle, auxin could increase the growth rate by increasing m, increasing P, or decreasing Y. As discussed in Chapter 15, experiments have shown that auxin does *not* increase turgor pressure when it stimulates growth (Cosgrove 1985). Conflicting results have been obtained regarding Y. Depending on the method of measurement, auxin has been shown either to have no effect on Y or to cause a large decrease (Cleland 1995).

In contrast to the uncertainty about Y, there is general agreement that auxin causes an increase in m, the irreversible wall extensibility. The increase in m has been measured in various ways. Early measurements of wall extensibility focused on the mechanical properties of isolated cell walls. Coleoptile or stem sections that had been incubated in the presence or absence of auxin were first killed by boiling in methanol or by repeated freeze-thawing, and then were clamped into an extensometer (see Chapter 15).

Such studies indicated that auxin increases the plastic (irreversible) extensibility of the cell walls after the living sections are treated with auxin for 1 hour. However, the results with isolated walls must be interpreted with caution. Wall extension in living cells requires a continuous input of metabolic energy. Thus, if growing stem sections are treated with metabolic inhibitors, growth ceases almost immediately. Growth would not cease if auxin acted by causing major modifications of cell wall structure that could be detected in boiled walls using an extensometer. Instead, inhibitor experiments suggest that auxin acts through more subtle cell wall–loosening events requiring a continuous supply of metabolic energy. At the biochemical level, wall loosening involves rearrangements of the load-bearing bonds of the cell wall. As discussed in Chapter 15, the effect of these molecular rearrangements at the biophysical level is to reduce wall stress, a phenomenon called *stress relaxation*.

Auxin-Induced Proton Extrusion Acidifies the Cell Wall, Increasing Its Extensibility

The idea that hydrogen ions can act as the intermediate between auxin and cell wall loosening was first proposed independently by Rayle and Cleland (1970) at the University of Washington (Seattle) and Hager, Menzel, and Krauss (1971) in Germany. According to the original model, auxin causes responsive cells to extrude protons actively into the cell wall, and the resulting decrease in apoplastic pH activates wall-loosening enzymes that promote the breakage of load-bearing bonds, thus increasing wall extensibility (see Chapter 15). The plasma membrane H^+-ATPase is responsible for the extrusion of protons.

The value of any model is that it allows specific predictions that can be tested experimentally. In the case of the **acid growth hypothesis**, the following three predictions could be made: (1) Auxin should increase the rate of proton extrusion, and the kinetics of proton extrusion should closely match those of auxin-induced growth; (2) neutral buffers should inhibit auxin-induced growth; (3) compounds (other than auxin) that promote proton extrusion should stimulate growth.

All three of these predictions have been confirmed. Auxin stimulates proton extrusion into the cell wall after 10 to 15 minutes of lag time, consistent with the growth kinetics (see Figure 19.19B). Auxin-induced growth has also been shown to be inhibited by neutral buffers, which prevent acidification of the cell wall. And *fusicoccin*, a fungal phytotoxin (Figure 19.17), stimulates both proton extrusion and transient growth in stem and coleoptile sections.

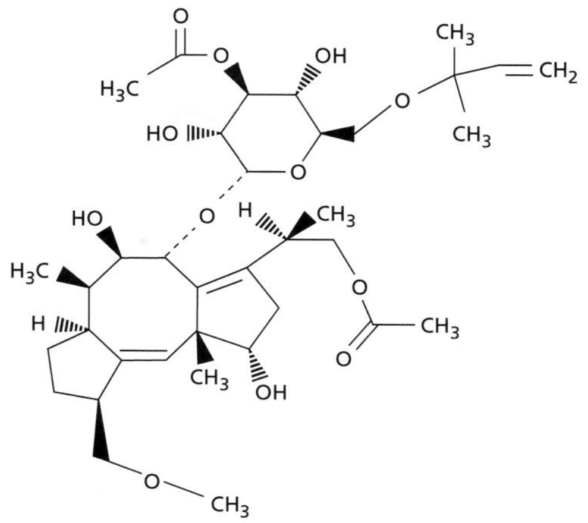

Figure 19.17 The structure of fusicoccin A, a phytotoxin produced by the fungus *Fusicoccum amygdali*.

Specific Proteins Mediate Acid-Induced Loosening of the Cell Wall

Cell wall hydrolases, such as cellulases, hemicellulases, or pectinases, were among the early candidates for cell wall–loosening enzymes that were thought to be activated by acidic pH during auxin-stimulated growth. For example, in monocots long-term growth is associated with a decrease in molecular weight of some hemicelluloses, especially β-glucans (see Figure 19.19C). However, hydrolytic enzymes weaken cell walls irreversibly, whereas auxin-induced growth is rapidly reversible with metabolic inhibitors. Hence, hydrolases are no longer regarded as candidates for the cell wall–loosening enzymes involved in cell elongation.

On the other hand, a *transglycosylase* that transfers sugar groups from one polysaccharide to another might allow cell wall polymers to slip past one another while maintaining the structural integrity of the wall. As discussed in Chapter 15, the enzyme *xyloglucan endotransglycosylase* (*XET*) is a candidate for a wall-loosening enzyme in dicots, because xyloglucan is a major hemicellulosic component of dicot cell walls, and experiments have shown that xyloglucans turn over during auxin-induced growth. However, no direct evidence links XET action to acid-induced wall loosening.

There is now compelling evidence that shows that a family of cell wall proteins called **expansins** causes cell wall loosening in response to acid pH. As described in Chapter 15, expansins restore cell wall loosening in response to acidic buffers when added to boiled cell wall preparations. Expansins can even mechanically weaken filter paper composed only of purified cellulose! Thus, expansins seem to loosen cell walls by weakening the hydrogen bonds between the polysaccharide components of the wall.

The Mechanism of Auxin-Induced Proton Extrusion Is Unresolved

Auxin could increase the rate of proton extrusion by stimulating one of two processes: (1) activation of preexisting plasma membrane H^+-ATPases; and (2) synthesis of new H^+-ATPases on the plasma membrane. Evidence for both mechanisms has been obtained (Cleland 1995). For example, Peltier and Rossignol (1996) have been able to demonstrate an in vitro auxin stimulation (about 20%) of the ATP-driven proton-pumping activity of isolated plasma membrane vesicles from tobacco cells. A greater in vitro stimulation (about 40%) was observed if the living cells were treated with IAA before the membranes were isolated. Although the in vitro stimulation is modest, it suggests that auxin can activate preexisting plasma membrane ATPases.

The ability of protein synthesis inhibitors, such as cycloheximide, to rapidly inhibit auxin-induced proton extrusion and growth first suggested that auxin might stimulate proton pumping by increasing the synthesis of the H^+-ATPase. Indeed, fusicoccin-induced acidification and growth are both *insensitive* to protein synthesis inhibitors, indicating that fusicoccin activates preexisting H^+-ATPases. Immunological evidence that auxin increases the amount of plasma membrane H^+-ATPase in cells has been presented (Hager et al. 1991). Using immunoblotting techniques, an increase in the amount

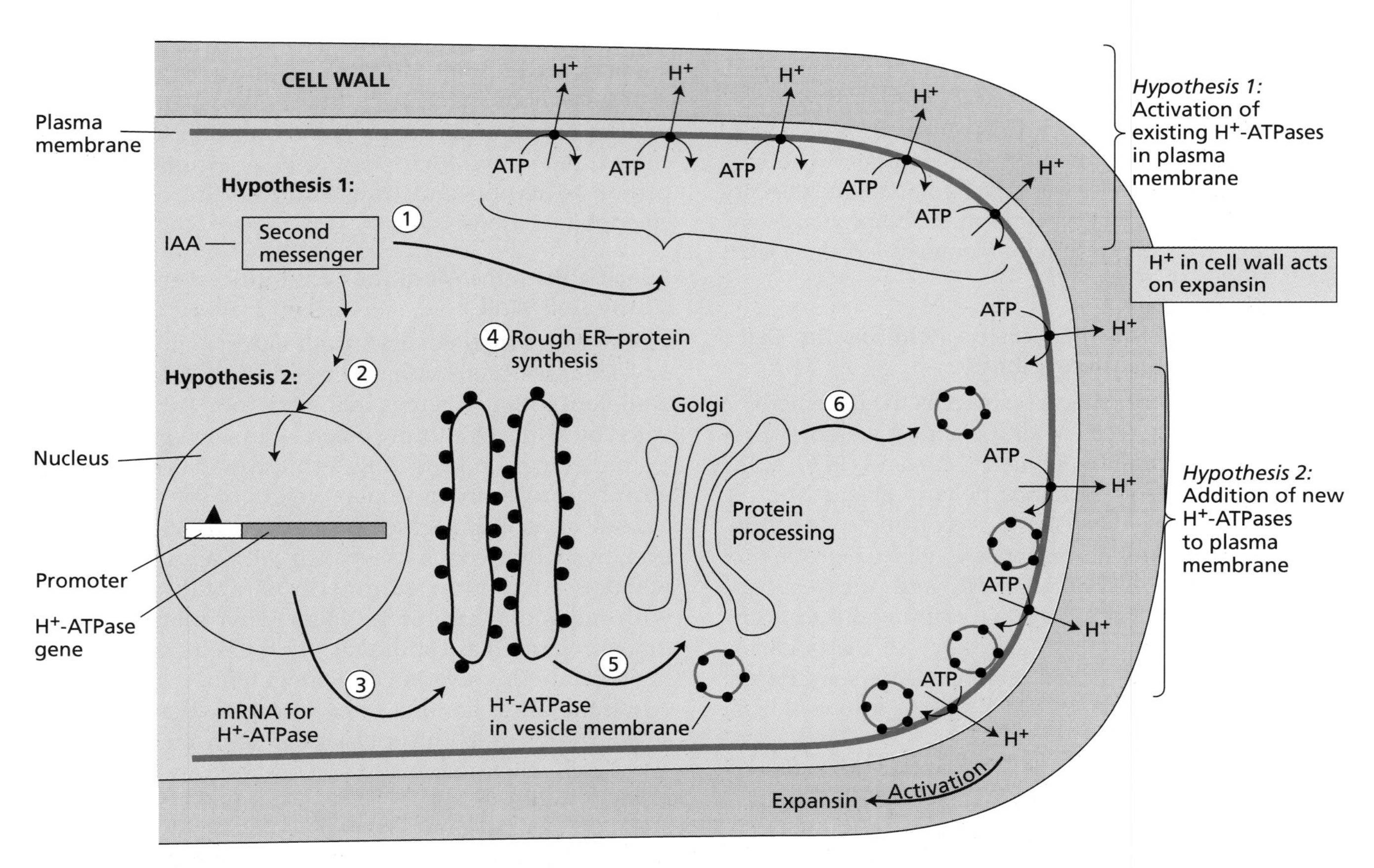

Figure 19.18 Current models for IAA-induced H^+ extrusion. According to Hypothesis 1, auxin initiates a signal transduction pathway that results in the production of second messengers that directly activate preexisting H^+-ATPases in the plasma membrane (step 1). Hypothesis 2 suggests that the IAA-induced second messengers activate the expression of genes (step 2) that encode the plasma membrane H^+-ATPase (step 3). The protein is synthesized on the rough endoplasmic reticulum (step 4) and targeted via the secretory pathway to the plasma membrane (steps 5 and 6). The increase in proton extrusion results from an increase in the number of proton pumps on the membrane. Regardless of how the H^+ pumping is increased, acid-induced wall relaxation is thought to be mediated by expansins.

of plasma membrane ATPase was detected after only 5 minutes of auxin treatment, and a doubling of the H^+-ATPase was observed by 40 minutes of treatment. Addition of cycloheximide resulted in a decline to the control level within an hour. These results suggest that the H^+-ATPase itself may be the short-lived protein required for auxin-induced acidification and growth. Unfortunately, these results could not be repeated by another laboratory (Jahn et al. 1996).

A possible resolution of the discrepancy has now been proposed. Using the technique of **in situ hybridization**, in which RNA probes that are complementary to a specific mRNA sequence are used to localize specific mRNAs in tissues, Frias and colleagues (1996) demonstrated that auxin causes a threefold stimulation of expression of the gene for the plasma membrane ATPase specifically in the *nonvascular tissues* of the coleoptile. Detection of the auxin-induced increase in the level of H^+-ATPase in homogenates of maize coleoptiles is, therefore, complicated by the presence of a large amount of non-auxin-stimulated background activity associated with the vascular tissues. The question of activation versus synthesis is still unresolved, and it is possible that auxin may stimulate proton extrusion by both activation *and* stimulation of synthesis of the H^+-ATPase.

Figure 19.18 summarizes the proposed mechanisms of auxin-induced cell wall loosening via proton extrusion.

Auxin Maintains the Capacity for Acid-Induced Loosening of the Cell Wall

Cell wall acidification is not the only way in which auxin induces plant cell elongation. Acid-stimulated growth is transient, lasting less than an hour, but auxin-enhanced growth can last much longer, especially in coleoptiles. Auxin must therefore affect other processes important to plant cell growth, such as the uptake or generation of osmotic solutes, the maintenance of cell wall structure, and the hydraulic conductivity of the plasma membrane. The precise role of auxin in regulating these factors is not clear. Auxin increases the uptake of certain osmotic solutes, at least partly through its stimulation of the plasma membrane H^+-ATPase, which provides the driving force for solute uptake. In this way, solute accumulation keeps pace with auxin-induced cell elongation. Consistent with this idea, auxin-stimulated proton pumping is accompanied by plasma membrane hyperpolarization, indicative of a buildup of an electric gradient, as well as a pH gradient.

Continued growth also depends on the synthesis of new wall material. Auxin is known to increase the activity of certain enzymes involved in wall polysaccharide biosynthesis. A steady supply of new wall polymers maintains the *capacity for acid-induced wall loosening*. This can be demonstrated by incubating sections in the presence or absence of auxin for a period of time and then testing the frozen and thawed cell walls for the ability to extend in response to acid pH. The cell walls of auxin-treated sections have a greater capacity for acid-stimulated growth than do either control sections or sections treated with fusicoccin.

A composite time course for auxin-induced growth and several biochemical events in *Avena* coleoptile sections are shown in Figure 19.19: the rapid kinetics of auxin-induced growth, expressed both as elongation

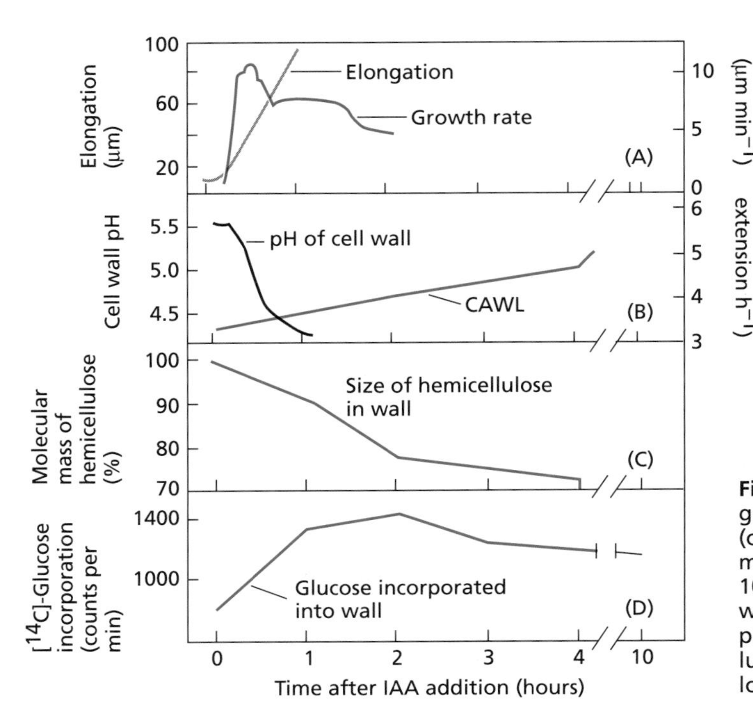

Figure 19.19 Composite time course for auxin-induced growth, proton extrusion (as reflected by pH), CAWL (capacity for acid-induced wall loosening), and cell wall metabolism of *Avena* coleoptile sections during the first 10 hours. The only parameter that correlates strongly with the rapid growth response to auxin (A) is cell wall pH (B). The other parameters: CAWL (B), size of hemicellulose (C), and cell wall synthesis (D), are part of the long-term growth response. (From Taiz 1984.)

and growth rate (Figure 19.19A); the kinetics of proton extrusion and the capacity for acid-induced wall loosening (CAWL) (Figure 19.19B); the degradation of hemicelluloses (Figure 19.19C); and the synthesis of new cell wall material (Figure 19.17D). Proton extrusion correlates well with the rapid increase in the growth rate, whereas CAWL is a long-term effect of auxin. Neither the degradation of hemicelluloses nor the synthesis of cell wall material correlates well with the rapid kinetics of growth; thus both are part of the long-term growth response to auxin.

Auxin-Stimulated Growth Has Been Demonstrated in Intact Plants

Much has been learned about auxin action using isolated stem and coleoptile sections. However, until recently it was not known whether the results obtained with isolated sections could be applied to intact plants. A close correlation has been observed between the level of free auxin and the rate of stem growth in a range of genetic lines of peas differing in height. This correlation suggests that the level of free auxin is one of the important factors regulating growth in plants (Law and Davies 1990). However, the inability to obtain an appreciable stimulatory response to auxin in intact plants has been a barrier to full acceptance of the role of auxin in normal growth.

Recently, Peter Davies and his colleagues have been able to demonstrate a typical auxin growth response in intact light-grown dwarf peas (Yang et al. 1993, 1996). The growth responses of the intact dwarf pea plants to both IAA and gibberellin, another plant growth hormone (see Chapter 20), are shown in Figure 19.20. Growth is strongly stimulated by IAA, and the kinetics are similar to those obtained with isolated sections. Gibberellin induces a slight growth stimulation, but after a lag time of about 3 hours. Gibberellin also prevents the steady decline in the growth rate that occurs with IAA alone. This experiment demonstrates that the level of free auxin can be a limiting factor for growth in intact plants as well as in sections.

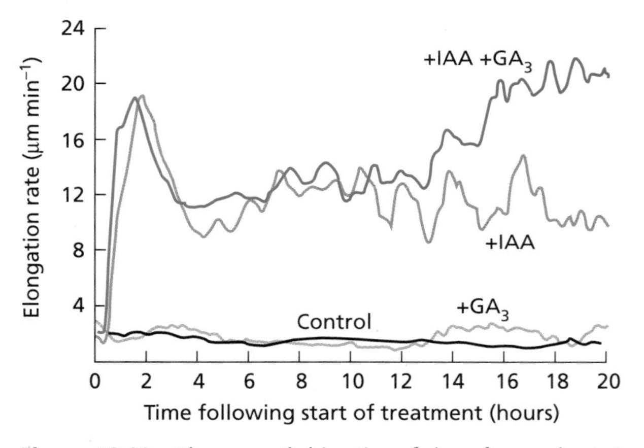

Figure 19.20 The growth kinetics of dwarf pea plants in response to applied IAA, GA_3 (gibberellic acid), or IAA and GA_3 combined, as compared to a control treatment. Treatments were started at time zero, and stem growth was measured minute by minute by position transducers attached to the top of the stem. IAA was continuously applied to the stem by a wick with flowing IAA solution. When the dwarf is treated with IAA and GA_3, its growth rate becomes similar to that of a tall pea plant. (From Yang et al. 1996, © American Society of Plant Physiologists, reprinted with permission.)

Physiological Effects of Auxin: Phototropism and Gravitropism

Two main guidance systems control the orientation of plant growth: phototropism and gravitropism. **Phototropism**, or growth with respect to light, is expressed in all shoots and some roots; it ensures that leaves will receive optimal sunlight for photosynthesis. **Gravitropism**, growth in response to gravity, enables roots to grow downward into the soil and shoots to grow upward away from the soil, which is especially critical during the early stages of germination. *Thigmotropism*, or growth with respect to touch, enables roots to grow around rocks and is responsible for the ability of the shoots of climbing plants to wrap around other structures for support. In this section we will examine the evidence that bending in response to tropic stimuli results from the lateral redistribution of auxin. We will also consider the cellular mechanisms involved in generating lateral auxin gradients during bending growth.

Phototropism May Be Mediated by the Lateral Redistribution of Auxin

As we saw earlier, Charles and Francis Darwin provided the first clue concerning the mechanism of phototropism by demonstrating that the sites of perception and differential growth (bending) are separate: Light is perceived at the tip, but bending occurs below the tip. The Darwins proposed that some "influence" that was transported from the tip to the growing region brought about the observed asymmetric growth response. This "influence" was later shown to be indole-3-acetic acid—auxin. When a shoot is growing vertically, auxin is transported polarly from the growing tip to the elongation zone. The polarity of auxin transport from tip to base is developmental and is independent of orientation with respect to gravity. However, auxin can also be transported laterally, and this lateral movement of auxin lies at the heart of the Cholodny–Went model for tropisms.

According to the classical **Cholodny–Went model** for phototropism, the tips of grass coleoptiles have three specialized functions: (1) the production of free IAA, (2) the perception of a unilateral light stimulus, and (3) the lateral transport of IAA in response to the phototropic stimulus. Thus, in response to a directional light stimulus, the auxin produced at the tip, instead of being transported basipetally, is transported laterally toward the shaded side.

The precise sites of auxin production, light perception, and lateral transport have been difficult to define. In maize coleoptiles, auxin appears to be produced mainly in the upper 1 to 2 mm of the tip. The zones of photosensing and lateral transport extend farther, within the upper 5 mm of the tip (Iino 1995; Kaldenhoff and Iino 1997). The response also depends on the light fluence (see Box 19.1).

As described in Chapter 18, the level of zeaxanthin appears to correlate with the phototropic bending response of coleoptiles of maize (*Zea mays*), leading to the suggestion that zeaxanthin could be a photoreceptor for phototropism (see Box 18.2). However, Palmer et al. (1996) studied the phototropic bending response of maize seedlings which were either carotenoid deficient through a genetic lesion or which had been chemically treated to block carotenoid biosynthesis. The phototropic bending of the carotenoid deficient seedlings was the same as that of the wild type, making it unlikely that zeaxanthin or any other carotenoid is the chromophore of the photoreceptor for phototropism.

On the other hand, recent molecular/genetic studies in the laboratory of Winslow Briggs have led to the identification of a 116 kDa (kilodalton) flavoprotein, NPH1 (see Chapter 18), which has all the characteristics of the blue light photoreceptor for phototropism (Liscum and Briggs 1995). This flavoprotein has been shown to be an autophosphorylating protein kinase whose activity is stimulated by blue light. The action spectrum for blue light activation of the kinase activity closely matches the action spectrum for phototropism, including the multiple peaks in the blue region (Briggs 1998). Consistent with the role of NPH1 in phototropism, a lateral gradient in phosphorylation of the 116 kDa protein has been demonstrated during exposure to unilateral blue light. According to the current hypothesis, the phosphorylation gradient induces the movement of auxin to the shaded side of the coleoptile (Box 19.1).

Once the auxin reaches the shaded side, it is transported basipetally to the elongation zone where it stimulates cell elongation. The acceleration of growth on the shaded side, and the slowing of growth on the illuminated side, called *differential growth*, give rise to the curvature toward light (Figure 19.21).

Direct tests of the Cholodny–Went model's prediction that auxin is transported laterally in response to unilateral light were carried out by Frits Went and others. The amount of auxin diffusing into an agar block from a single coleoptile tip in darkness was shown to cause a curvature of 25.8° in the *Avena* coleoptile curvature test (Figure 19.22A). The total amount of auxin diffusing into the block is the same in the presence of unilateral light as in darkness (Figure 19.22B). This result indicates that light does not cause the photodestruction of auxin on the illuminated side, as had been proposed by some investigators. Splitting the coleoptile and agar block in half and separating them with an impermeable barrier resulted in a 1:1 distribution of auxin diffusing into the two half-blocks (Figure 19.22C). This result confirmed that light does not cause the destruction of auxin on the illuminated side. However, if the upper portion of the tip is allowed to remain intact, twice as much auxin diffuses into the half-block on the shaded side as into the half-block on the illuminated side (Figure 19.22D). Since the total amount of auxin was unaffected, this experiment demonstrated that auxin produced at the tip is transported laterally from the illuminated side to the shaded side. Similar results were obtained when [^{14}C]IAA was applied to the coleoptile tip and the radioactivity diffusing into divided agar half-blocks was measured (Briggs 1963).

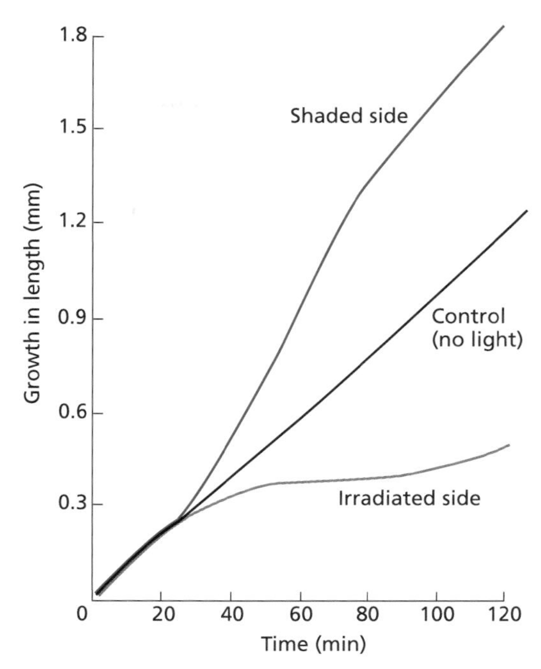

Figure 19.21 Time course of growth on the illuminated and shaded sides of a coleoptile responding to a 30-second pulse of unidirectional blue light. Control coleoptiles were not given a light treatment. (After Iino and Briggs 1984.)

Over the years, the Cholodny–Went model for phototropism has been tested and challenged repeatedly (Firn 1994). The bulk of the evidence continues to support the model, although certain discrepancies have emerged. For example, it has long been known that a fully decapitated coleoptile can regain the ability to bend in response to unilateral light after a recovery period of 3 hours. This observation seems to contradict the crucial role of the tip in photosensing and auxin production. However, we now know that this phenomenon is due to the stimulation of auxin production in the cells near the cut surface, as well as to the activation of the photosensing and auxin lateral-transport mechanisms in this region (for a review of the history, see Kaldenhoff and Iino 1997).

BOX 19.1

The Fluence Response of Phototropism

PHOTOTROPISM INVOLVES TWO FUNDAMENTAL PROCESSES: blue-light perception and differential cell elongation. The identity of the blue-light photoreceptor involved in phototropism was discussed in Chapter 18. There is now compelling evidence that a 116-kilodalton (kDa) flavoprotein associated with the plasma membrane (NPH1) is the photoreceptor for phototropism (Huala et al., 1997). This protein becomes phosphorylated in response to blue light. As discussed in this chapter, differential cell elongation in response to unilateral light is thought to be caused by the lateral redistribution of auxin to the shaded side.

If phototropism were a simple photochemical reaction, one would expect the fluence response curve to rise from a threshold level of light to a plateau at which the photoreceptor becomes saturated. However, this is not what happens. The figure shows a diagram of a typical fluence response curve, which is similar to what has been observed in a wide range of species, including *Avena* coleoptiles and *Arabidopsis* stems (Poff et al. 1994). Starting at the light threshold, there is a bell-shaped curve that has been termed the **first positive curvature**. The first positive curvature is followed by a **neutral zone** in which little or no measurable response is observed. Under certain conditions, *Avena* coleoptiles may actually bend *away from* the light in this range, a phenomenon referred to as the **first negative curvature**. Both the first positive and the first negative curvatures (neutral zone) are restricted to the *coleoptile tip*. At higher light fluences the curve ascends again, and this second bending response is termed the **second positive curvature**. The second positive curvature differs from the first positive curvature in that the bending occurs at the base rather than at the tip of the coleoptile.

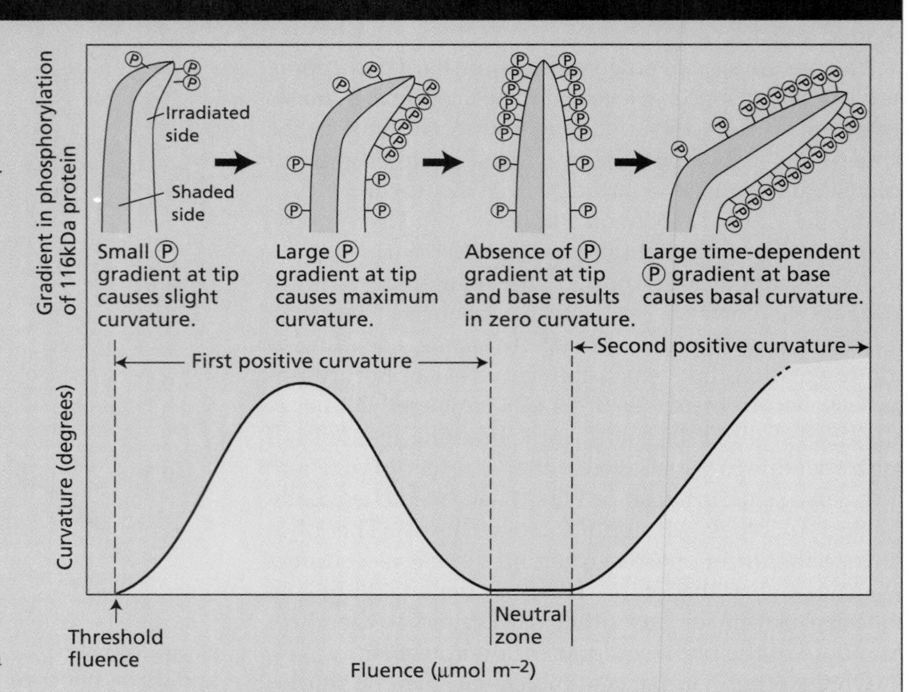

A typical fluence response curve for phototropism. The phosphorylation model for the first positive curvature, the neutral zone, and the second positive curvature of coleoptiles is shown above the curve.

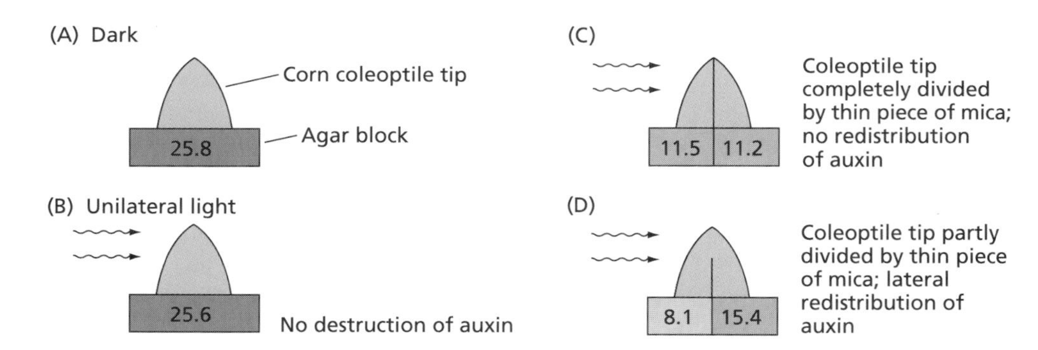

Figure 19.22 Evidence that the lateral redistribution of auxin is stimulated by unidirectional light in corn coleoptiles. (A) When an agar block is placed under an excised corn coleoptile tip in the dark, auxin diffuses into the block. (B) If the coleoptile is exposed to unidirectional light and auxin is allowed to diffuse into the block, approximately the same amount of auxin is found as in the dark. This result demonstrates that unilateral light does not cause the photodestruction of auxin on the illuminated side. (C) If the coleoptile and the agar block are divided completely in half, no lateral redistribution of auxin is observed, nor does the biosynthesis of auxin increase on the shaded side. However, if the tip is only partly divided (D), lateral redistribution of auxin in response to unidirectional light is observed at the tip.

BOX 19.1 (*continued*)

Plant physiologists have long puzzled over the meaning of this complex relationship between light fluence and phototropism. Most studies of phototropism have been carried out in the range of the first positive curvature. The first positive curvature and the neutral zone obey the Bunsen–Roscoe reciprocity law, which states that the response will be the same for a brief exposure to bright light as for a longer exposure to dim light, as long as the product of time and the photon fluence rate are the same (see Chapter 17). Thus the first positive curvature is a response to short light pulses.

In contrast to the first positive curvature, the second positive curvature is a response to irradiations of long duration. The second positive curvature does not obey the reciprocity law; instead, it is proportional to the *duration* and the photon fluence rate. For the second positive response to occur, both the time threshold and the fluence threshold must be exceeded (Poff et al. 1994). Since these are the conditions that prevail normally in nature, phototropism in nature usually represents the second positive curvature.

What causes the gradient in auxin to form at the three fluences? It has long been assumed that the light gradient that is established across the coleoptile during unilateral irradiation with blue light must generate a biochemical gradient between the cells of the shaded and the irradiated sides, but the nature of this biochemical gradient has, until recently, eluded detection. With the discovery of blue light–induced phosphorylation of the 116 kDa plasma membrane protein, much work has focused on this protein, NPH1, which is widely considered to be the photoreceptor for phototropism (Palmer et al. 1993; Liscum and Briggs 1995).

Recently, Michael Salomon and his colleagues at the University of Munich demonstrated that unilateral blue light causes a lateral gradient in the phosphorylation of NPH1 from the irradiated to the shaded side (Salomon et al. 1997a, 1997b). Moreover, the concentration of NPH1 declines exponentially from the tip of the coleoptile to the base. Studies on the rates of phosphorylation of the 116 kDa protein at the tip versus at the base of oat coleoptiles have led to a new model for the fluence response curve of phototropism.

According to this new model proposed by Salomon and colleagues, the gradient in concentration of NPH1 from tip to base makes the tip more sensitive to light than the base is (see the figure). Thus, at threshold fluences, a gradient in phosphorylation occurs only at the tip, leading to a small bending response. Increasing the fluence increases the gradient of phosphorylation at the tip, until phosphorylation on the irradiated side saturates and a maximum phosphorylation gradient is obtained. This maximum gradient represents the peak of the first positive curvature. As the fluence is increased beyond this point, phosphorylation increases on the shaded side due to the penetration of light, reducing the gradient. When phosphorylation of NPH1 is equal on both sides of the tip, no bending occurs (the neutral zone).

However, with continued exposure to light of sufficient fluence, phosphorylation increases at the base on the irradiated side. Phosphorylation at the base is dependent on both fluence and time, unlike phosphorylation at the tip which is strictly dependent on fluence. Thus, phosphorylation at the base appears to involve some type of time-dependent biochemical process. Since recent immunological studies have shown that the total amount of the 116 kD protein remains constant during phototropism, the time-dependent increase in phosphorylation at the base does not involve synthesis of new photoreceptor.

The model proposed by Salomon and colleagues requires further testing, and other interpretations are certainly possible. Nevertheless, the new model provides the first plausible explanation for the fluence response curve of phototropism. However, even if this model is confirmed, the question remains as to how a gradient in phosphorylation can bring about a gradient in auxin.

Phototropism in dicot stems is similar to that of coleoptiles, but clearly more complex. The same requirements for auxin biosynthesis, photosensing, and lateral transport apply to phototropism in stem and hypocotyl tissues. Complexities arise because the sites of auxin biosynthesis and photosensing may occur in both the apical bud and the young leaves. In addition, other hormones, such as gibberellins, as well as the production or transport of growth inhibitors, may contribute to the overall response. Nevertheless, increasing evidence suggests that lateral auxin gradients also play a role in regulating the phototropism of dicot stems and hypocotyls.

The Cholodny–Went Model Also Applies to Gravitropism

When dark-grown *Avena* seedlings are oriented horizontally, the coleoptiles bend upward in response to gravity. The prediction of the Cholodny–Went model that auxin from the tip is redirected to the lower side of the coleoptile, causing the lower side to grow faster than the upper side, has been confirmed by the use of radioisotope-labeled IAA, physicochemical methods, and radioimmunoassay (Mertens and Weiler 1983; Migliaccio and Rayle 1984). Early experimental evidence indicated that the tip of the coleoptile can perceive gravity and redistribute auxin to the lower side. For example, if coleoptile tips are oriented horizontally, a greater amount of auxin diffuses from the lower half than the upper half. However, auxin redistribution can occur to some extent below the tip as well.

Tissues below the tip are able to respond to gravity. For example, when vertically oriented maize coleoptiles are decapitated by removal of the upper 2 mm of the tip and immediately reoriented 30° from the vertical, gravitropic bending continues at a slower rate for several hours after decapitation. Application of IAA to the cut surface restores the rate of bending to normal levels. This finding indicates that both the perception of the gravitational stimulus and the lateral redistribution of auxin can occur in the tissues below the tip, although the tip is still required for auxin production (Iino 1995).

Some critics of the Cholodny–Went model have questioned whether the rate and magnitude of the lateral auxin redistribution are sufficient to account for the kinetics and magnitude of differential growth during gravitropism (Firn and Digby 1980). When maize seedlings are oriented horizontally, the growth rate of the upper side of the coleoptiles decreases to zero within 1 hour. Such a sharp decrease in growth on the upper side seems inconsistent with the modest amount of lateral auxin redistribution that is detectable at this time. Hence, other factors in addition to lateral auxin gradients may contribute to gravitropic curvatures.

Indirect support for the Cholodny–Went model has been provided by studies on the effect of tropic stimuli on the pattern of proton extrusion. When hypocotyls bend in response either to unilateral light or to gravity, the side that is growing more rapidly (the convex side) becomes more acidic than the concave side (Mulkey et al. 1981). Similarly, roots responding to gravity are more acidic on their upper side than on their lower side (Versel and Pilet 1986). These results provide indirect support for lateral auxin gradients and are thus consistent with the Cholodny–Went model

Studies on the pattern of gene expression in soybean hypocotyls responding gravitropically have provided additional evidence in favor of the Cholodny–Went model. As will be discussed later in the chapter, auxin has been shown to stimulate the expression of specific genes. In soybean hypocotyls, gravitropism leads to a rapid asymmetry in the accumulation of a group of auxin-stimulated mRNAs called **small auxin up-regulated RNAs** (**SAURs**) (McClure and Guilfoyle 1989). In vertical seedlings, SAUR gene expression is symmetrically distributed. Within 20 minutes after the seedling is oriented horizontally, SAURs begin to accumulate on the lower half of the hypocotyl. Under these conditions, gravitropic bending first becomes evident after 45 minutes, well after the induction of the SAURs (Figure 19.23). The existence of a lateral gradient in SAUR gene expression is indirect evidence for the existence of a lateral gradient in auxin.

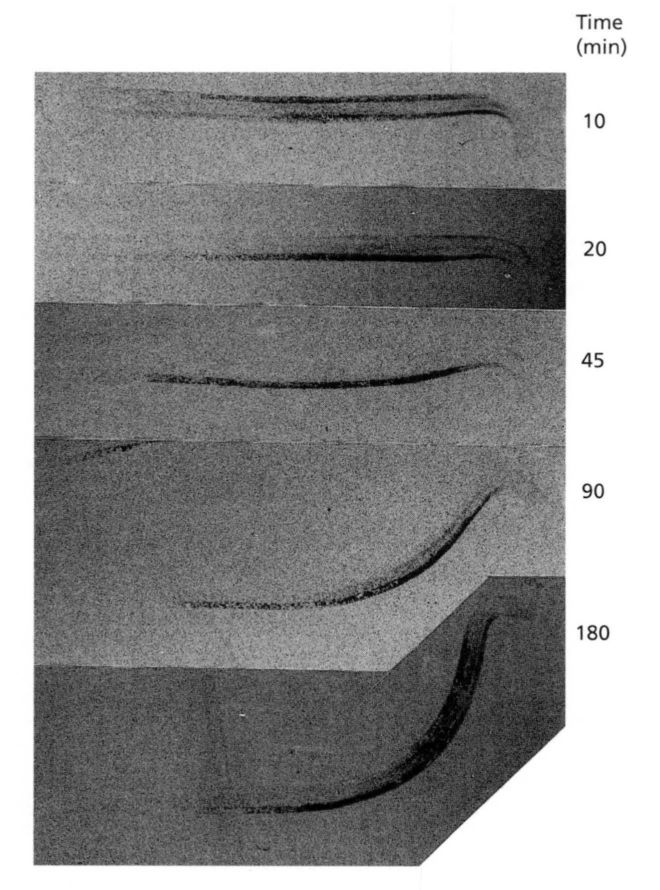

Figure 19.23 Pattern of expression of the SAUR genes in a soybean hypocotyl responding gravitropically. Etiolated soybean hypocotyls were oriented longitudinally for varying lengths of time. At the time intervals indicated, the hypocotyl was sliced in half longitudinally and the cut surface was blotted onto a nylon membrane, a process called tissue printing. The membrane was then incubated with a mixture of radioactively labeled antisense RNA probes for three SAUR genes, and autoradiograms were produced. At 10 minutes, SAUR genes are still expressed equally on the upper and lower sides. A shift in SAUR gene expression to the lower side begins at 20 minutes, preceding the gravitropic bending response, which begins at 45 minutes. (From Guilfoyle et al. 1990, courtesy of T. Guilfoyle.)

Agravitropic Auxin Mutants Support the Cholodny–Went Model

Mutants that do not respond to gravity, termed *agravitropic mutants*, provide potentially powerful tools for understanding the mechanism of gravitropism. Various types of mutant screens have been carried out. The most direct approach is to look for seedlings with abnormal gravitropism. Mutagenized seedlings are allowed to grow along the surface of a vertically oriented agar plate, and individuals whose roots do not grow straight downward are selected. Several *Arabidopsis* mutants have been isolated using this screening procedure (Masson 1995). In another approach, mutants initially isolated for their resistance to high auxin concentrations have been shown to be agravitropic as well (Hobbie and Estelle 1994). In two of the mutants, *axr1* and *axr4*, gravitropism is only partly inhibited, whereas gravitropism in *axr2*, *axr3*, *aux1*, and *dwf* is severely impaired. As already noted, the *AUX1* gene has been sequenced and shown to code for a protein similar to an amino acid permease. Taken together, the analyses of these mutants strongly suggest that auxin is required for gravitropism, consistent with the Cholodny–Went model.

Gravity Perception Involves Amyloplasts in Cells of the Starch Sheath and Root Cap

Unlike unilateral light, gravity does not form a gradient between the upper and lower sides of an organ. All parts of the plant experience the gravitational stimulus equally. How do plant cells detect gravity? The only way that gravity can be sensed is through the motion of a falling or sedimenting body. An obvious candidate for intracellular gravity sensors in plants are the large, dense amyloplasts that are present in many plant cells. These specialized amyloplasts are of sufficiently high density relative to the cytosol that they readily sediment to the bottom of the cell (Figure 19.24). Amyloplasts that function as gravity sensors are called **statoliths**, and the specialized gravity-sensing cells in which they occur are called **statocytes**. Whether the statocyte is able to detect the downward motion of the statolith as it passes through the cytoskeleton or whether the stimulus is perceived only when the statolith comes to rest at the bottom of the cell has not yet been determined.

In shoots and coleoptiles evidence suggests that gravity is perceived in the **starch sheath**, a layer of cells that surrounds the vascular tissues of the shoot (Sack 1991). The starch sheath is continuous with the endodermis of the root. Recently, genetic evidence for the role of the starch sheath in the perception of gravity in *Arabidopsis* was obtained by the laboratories of Philip Benfey and Masao Tasaka (Fukaki et al. 1998). As noted in Chapter 16, the *scarecrow* (*scr*) mutant of *Arabidopsis* is missing both the endodermis and starch sheath. As a result, the hypocotyl and inflorescence of the *scr* mutant are agravitropic, whereas the root exhibits a normal gravitropic response. On the basis of the phenotype of the *scarecrow* mutant, we can infer that (1) the starch sheath is

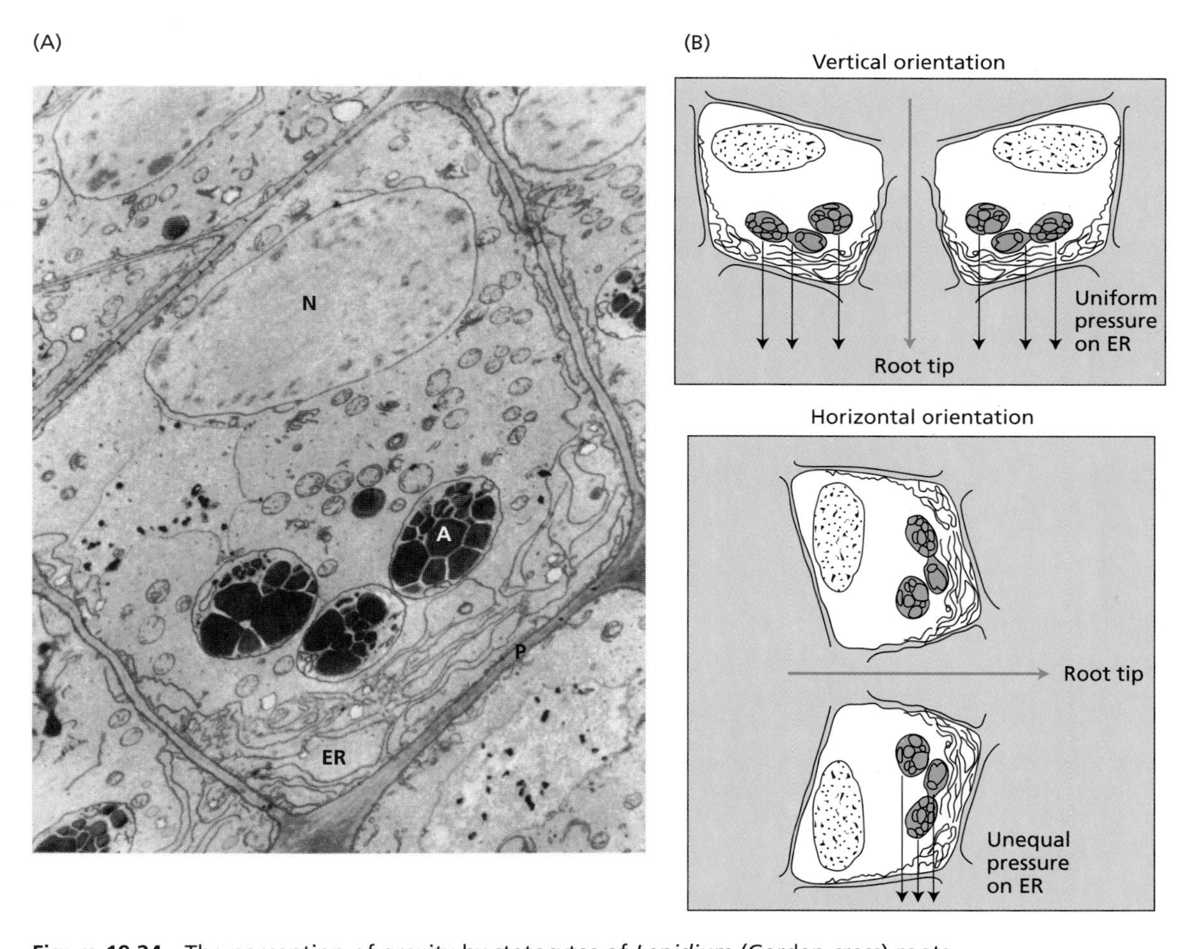

Figure 19.24 The perception of gravity by statocytes of *Lepidium* (Garden cress) roots. (A) A single statocyte found in the columella of the root cap from a primary root of *Lepidium.* The starch-filled amyloplasts (A) rest against the endoplasmic reticulum (ER). (B) These amyloplasts tend to sediment in response to reorientation of the cell and to remain resting against the ER. The green arrows point toward the root tip. When the root is oriented vertically (top), the pressure exerted by the amyloplasts on the ER (solid arrows) is equally distributed. In a horizontal orientation (bottom), the pressure on the ER is unequal on either side of the vertical axis of the root. (From Volkmann and Sievers 1979; reviewed in Sievers et al. 1996.)

required for gravitropism in shoots, and (2) the root endodermis, which does not contain statoliths, is not required for gravitropism in roots.

Gravity perception in primary roots depends on the root cap. Large, graviresponsive amyloplasts are located in the statocytes (see Figure 19.24) in the central cylinder, or columella, of the root cap. Removal of the root cap from otherwise intact roots abolishes root gravitropism without inhibiting growth in most plant species tested. Perhaps contact between the sedimenting amyloplasts and the endoplasmic reticulum near the plasma membrane on the downward side of the cell triggers the response. However, the exact role of the statoliths remains unknown (Sievers et al. 1996).

The **starch–statolith hypothesis** of gravity perception is supported by several lines of data (Masson 1995). Amyloplasts are the only organelles that consistently sediment in the statocytes of different plant species, and the rate of sedimentation correlates well with the time required to perceive the gravitational stimulus. The gravitropic responses of starch-deficient mutants are generally slower than those of wild-type plants. For example, Kiss and colleagues (1996) studied the gravitropism in one starchless *Arabidopsis* mutant and two mutants with intermediate levels of starch. The wild-type roots were more responsive to gravity than the intermediates, and the intermediates were more responsive than the starchless mutant. Nevertheless, starchless mutants exhibit some gravitropism. Thus, while starch is required for a normal gravitropic response, starch-independent gravity perception mechanisms may also exist. For example, nuclei may be dense enough to act as statoliths. It may not even be necessary for a statolith to come to rest at the bottom of the cell. The cytoskeleton may be able to detect a partial vertical displacement of an organelle.

Gravity Can Be Perceived in the Absence of Statoliths

An alternative mechanism of gravity perception that does not involve statoliths has been documented in the giant-celled fresh water alga *Chara*. Despite the absence

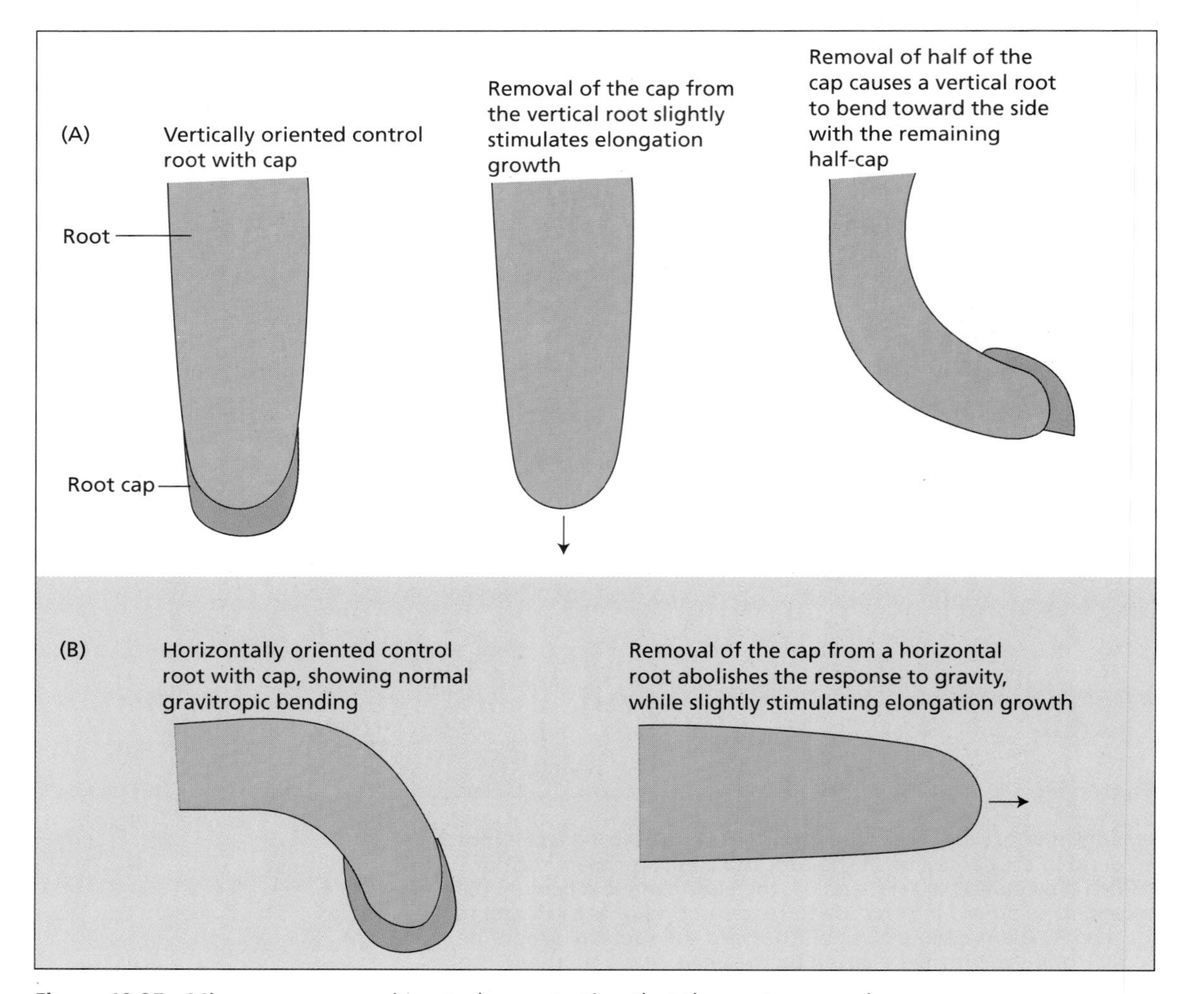

Figure 19.25 Microsurgery experiments demonstrating that the root cap produces an inhibitor that regulates root gravitropism. (After Shaw and Wilkins 1973.)

of any discernible sedimenting statoliths, *Chara* filaments show negative gravitropism. To explain this phenomenon, Randy Wayne and his colleagues at Cornell University have postulated that the entire protoplast acts like a statolith, and that gravity is perceived at the interface between plasma membrane and cell wall (Wayne and Staves 1996).

Barbara Pickard and coworkers at Washington University (Pickard and Ding 1993) have applied this model to the smaller cells of higher plants, and have proposed an alternative to the starch–statolith hypothesis. According to the **plasmalemma central control** (**PCC**) **model**, stretch-activated calcium channels are clustered around attachment centers connecting the cytoskeleton and the cell wall. These channels open in response to tensions at the membrane caused by gravity-induced changes in the distribution of forces exerted by the protoplast, the cytoskeleton, or the cell wall. Something like the PCC model may explain how *Chara* and the starchless mutants of *Arabidopsis* are able to perceive and respond to gravity.

The Root Cap May Play a Role in the Lateral Redistribution of Auxin

In roots gravity is perceived in the root cap, whereas bending occurs some distance back from the tip in the elongation zone. A chemical messenger must therefore be involved in communication between the cap and the elongation zone. Microsurgery experiments in which half of the cap was removed indicated clearly that the cap produces a root growth inhibitor (Figure 19.25). Although root caps contain small amounts of IAA and abscisic acid (ABA) (see Chapter 23), IAA is more inhibitory to root growth than ABA when applied directly to the elongation zone; hence, IAA is generally considered to be the "root cap inhibitor." But what is the relationship between the small amount of IAA in the cap and the greater amounts of IAA transported via the stele from the shoot?

To account for the role of the shoot in supplying auxin to the root, and the role of the cap in perceiving the gravitational stimulus, Hasenstein and Evans (1988) proposed a model in which the root cap is responsible for redirecting the auxin that is transported acropetally within the stele back again (i.e., in the basipetal direction) along the cortex and epidermis of the elongation zone. In a vertically oriented root, basipetal auxin transport is equal on all sides (Figure 19.26A). When the root is oriented horizontally, however, the cap redirects the bulk of the auxin to the lower side, thus inhibiting the growth of the lower side (Figure 19.26B). Consistent with this idea, the transport of [^{3}H]IAA across a horizontally oriented root cap is polar, with a preferential downward movement (Young et al. 1990).

A variety of experiments, mainly from the laboratory of Michael Evans at Ohio State University, have suggested that calcium is required for root gravitropism in maize. Some of these experiments involve EGTA (ethylene glycol-bis(β-aminoethyl ether)-N,N,N′,N′-tetraacetic acid), a compound that can chelate (form a complex with) calcium ions, thus preventing calcium uptake by cells. EGTA inhibits both root gravitropism and the asymmetric distribution of auxin in response to gravity (Young and Evans 1994). Placing a block of agar that contains calcium ions on the side of the cap of a vertically oriented corn root induces the root to grow toward the side with the agar block. As in the case of [^{3}H]IAA, $^{45}Ca^{2+}$ is polarly transported to the lower half of the cap

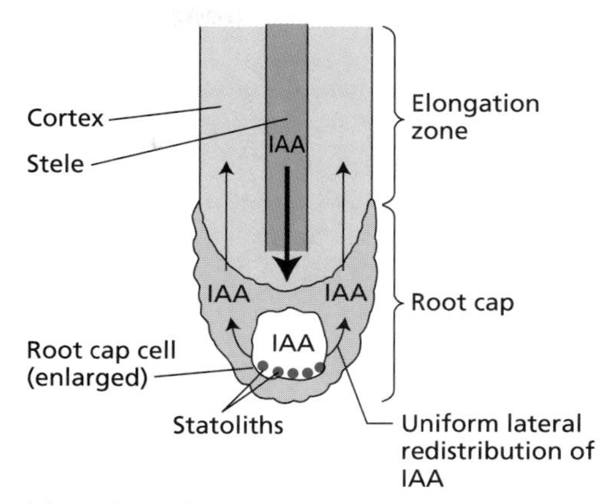

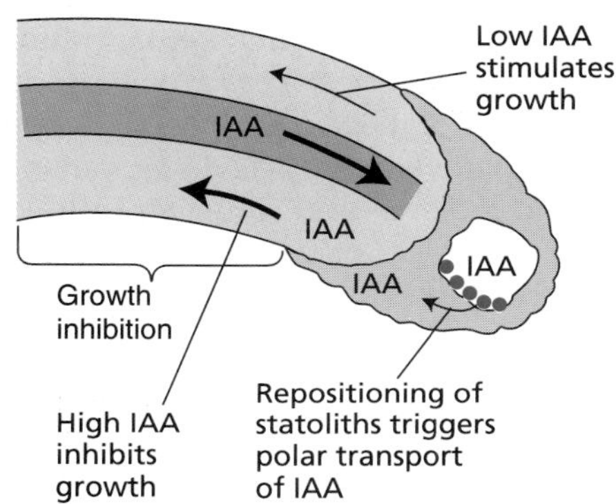

Figure 19.26 Proposed model for the redistribution of auxin during gravitropism in maize roots. (A) IAA is synthesized in the shoot and transported to the root in the stele. When the root is vertical, the statoliths in the cap settle to the basal ends of the cells. Auxin (transported acropetally in the root via the stele) is distributed equally on all sides of the root cap. The IAA is then transported basipetally within the cortex to the elongation zone, where it regulates cell elongation. (B) In a horizontal root the statoliths settle to the side of the cap cells, triggering polar transport of IAA to the lower side of the cap. The majority of the auxin in the cap is then transported basipetally in the cortex on the lower side of the root. The high concentration of auxin on the lower side of the root inhibits growth on that side, while the decreased auxin concentration on the upper side stimulates the upper side to grow. As a result, the root bends downward. (After Hasenstein and Evans 1988.)

of a root stimulated by gravity. Thus in maize roots, gravity induces the polar transport of both IAA and calcium from the upper to the lower half of the maize root cap. However, Simon Gilroy and his colleagues at Penn State University were unable to detect a lateral calcium gradient in horizontal root tips of *Arabidopsis* (Legue et al. 1997). Thus calcium ions may be required for gravitropism, the formation of a lateral calcium gradient in the cap does not appear to be universal.

Other Physiological Effects of Auxin

Although originally discovered in relation to growth, auxin influences nearly every stage of a plant's life cycle from germination to senescence. We assume that the primary mechanism of auxin action is comparable in all cells, involving similar receptors and signal transduction pathways. Since the effect that auxin produces depends on the identity of the target tissue, the response of a tissue to auxin is governed by its developmentally determined genetic program and is further influenced by the presence or absence of other signaling molecules. As we will see in this and subsequent chapters, interaction between two or more hormones is a recurring theme in plant development.

In this section we will examine some additional developmental processes regulated by auxin, including apical dominance, leaf abscission, lateral-root formation, and vascular differentiation.

Auxin Regulates Apical Dominance

In most higher plants, the growing apical bud inhibits the growth of lateral (axillary) buds, a phenomenon called **apical dominance**. Removal of the shoot apex (decapitation) usually results in the growth of one or more of the lateral buds. More than 60 years ago it was found that IAA could substitute for the apical bud in maintaining the inhibition of lateral buds of bean (*Vicia faba*) plants (Thimann and Skoog 1934). This result was soon confirmed for numerous plant species, leading to the hypothesis that the outgrowth of the axillary bud is inhibited by auxin that is transported basipetally from the apical bud. In support of this idea, a ring of the auxin transport inhibitor TIBA in lanolin paste (as a carrier) placed below the shoot apex released the axillary buds from inhibition.

How does auxin from the shoot apex inhibit the growth of lateral buds? Thimann and Skoog originally proposed that auxin from the shoot apex inhibits the growth of the axillary bud directly. According to their model, the optimal auxin concentration for bud growth is low, much lower than the auxin concentration normally found in the stem. The level of auxin normally present in the stem was thought to inhibit the growth of lateral buds.

If the **direct-inhibition model** of apical dominance is correct, the concentration of auxin in the axillary bud should decrease following decapitation of the shoot apex. However, the reverse appears to be true, as was elegantly demonstrated with transgenic plants that contained the genes for bacterial luciferase (*LUXA* and *LUXB*) under the control of an auxin-responsive promoter (Langridge et al. 1989). This reporter gene construct allowed researchers to study the level of auxin in different tissues by monitoring the amount of light emitted by the luciferase-catalyzed reaction. When these transgenic plants were decapitated, the expression of the *LUX* genes, and therefore the light emitted, increased in and around the axillary buds within 12 hours. This experiment demonstrated that after decapitation, the auxin content of the axillary buds increased rather than decreased.

Direct physical measurements of auxin levels in buds have also shown an increase in the auxin of the axillary buds after decapitation (Gocal et al. 1991). The IAA concentration in the axillary bud of *Phaseolus vulgaris* (kidney bean) increased fivefold within 4 hours after decapitation. These and other similar results make it unlikely that auxin from the shoot apex inhibits the axillary bud directly. Other hormones, such as cytokinins and ABA, may be involved.

Direct application of cytokinins to axillary buds stimulates bud growth in many species, overriding the inhibitory effect of the shoot apex. Moreover, measurements of cytokinin levels in axillary buds of Douglas fir (*Pseudotsuga menziesii*) show a very good correlation between endogenous cytokinin levels and bud growth (Pilate et al. 1989). The source of the increase in cytokinin level in the bud has not yet been determined. As will be discussed in Chapter 21, much of the cytokinin of the plant is synthesized in the root and transported to the shoot. Studies with the ^{14}C-labeled cytokinin benzyladenine (BA), have shown that when the labeled compound is applied to roots, more [^{14}C]BA is transported to the shoot apex than to the axillary bud. Decapitation increases the accumulation of [^{14}C]BA by the axillary bud, and application of auxin to the apical stump reduces this accumulation. Thus auxin makes the shoot apex a sink for cytokinin from the root, and this may be one of the factors involved in apical dominance.

Finally, ABA has been found in dormant lateral buds in intact plants. When the shoot apex is removed, the ABA levels in the lateral buds decrease. High levels of IAA in the shoot may help keep ABA levels high in the lateral buds. Removing the apex removes a major source of IAA, which may allow the levels of bud growth inhibitor to fall. Pearce and colleagues (1995) measured ABA and IAA levels in rhizomes (underground stems) of *Elytrigia repens* (Quackgrass), which exhibit strong apical dominance. After the rhizomes were decapitated,

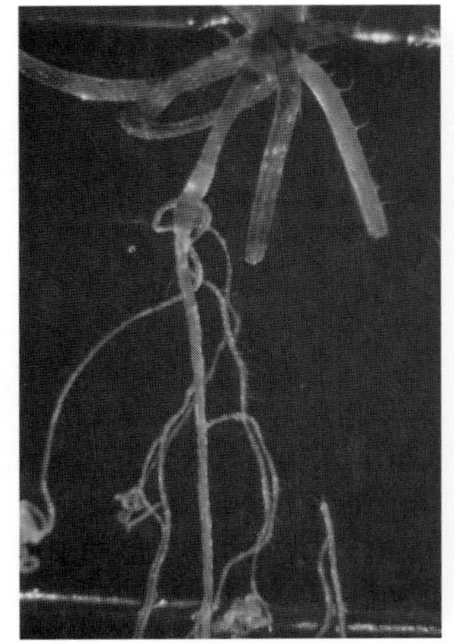

(A) Wild-type *Arabidopsis*

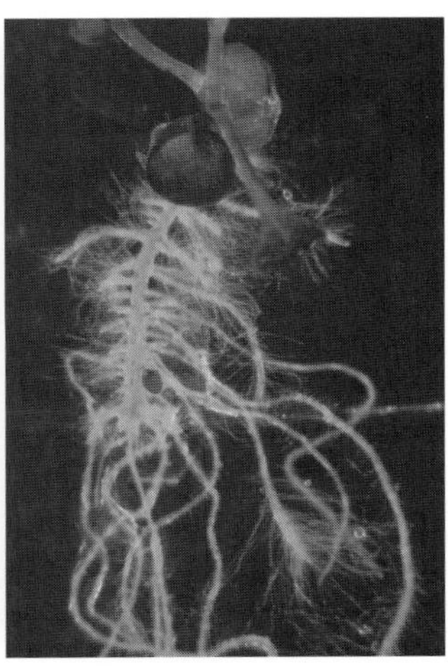

(B) *alf1* mutation, showing extreme proliferation of adventitions and lateral roots

Figure 19.27 Morphology of *Arabidopsis* (A) wild-type and (B) *alf1* seedlings on hormone-free medium. (From Celenza et al. 1995, courtesy of J. Celenza.)

the ABA content of the first axillary bud declined to 20% of control levels within 24 hours, while the IAA content remained constant. The authors concluded that apical dominance in *E. repens* is associated with a reduction in ABA content rather than with a change in IAA content. Further studies are needed in other species to test the generality of this observation.

Auxin Promotes the Formation of Lateral Roots and Adventitious Roots

Although elongation of the primary root is inhibited by auxin concentrations greater than 10^{-8} *M*, initiation of lateral (branch) roots and adventitious roots is stimulated by high auxin levels. Lateral roots are commonly found above the elongation and root hair zone and originate from small groups of cells in the pericycle (see Chapter 16). Auxin stimulates these cells to divide. The dividing cells gradually form into a root apex, and the lateral root grows through the root cortex and epidermis. Adventitious roots can originate in a variety of tissue locations from clusters of mature cells that renew their cell division activity. These dividing cells develop into a root apical meristem in a manner somewhat analogous to the formation of lateral roots.

In horticulture, the stimulatory effect of auxin on the formation of adventitious roots has been very useful for the vegetative propagation of plants by cuttings. When an excised leaf or stem is placed in water or on a moist substrate, adventitious roots often form near the cut surface. The roots form because IAA tends to accumulate immediately above any wound site in shoots or roots as a result of polar auxin transport. The effect can be greatly enhanced if the cut surface of the cutting is dipped in an auxin solution. This relationship is the basis of commercial rooting compounds, which consist mainly of a synthetic auxin mixed with talcum powder.

A series of *Arabidopsis* mutants, named *alf* (*a*berrant *l*ateral root *f*ormation), have provided some insights into the role of auxin in the initiation of lateral roots. The *alf1* mutant exhibits extreme proliferation of adventitious and lateral roots, coupled with a 17-fold increase in endogenous auxin (Figure 19.27). Another mutant, *alf4*, has the opposite phenotype: It is completely devoid of lateral roots. Microscopic analysis of *alf4* roots indicates that lateral-root primordia are absent. The *alf4* phenotype cannot be reversed by application of exogenous IAA. Yet another mutant, *alf3*, is defective in the development of lateral-root primordia into mature lateral roots. The primary root is covered with arrested lateral-root primordia that grow until they protrude through the epidermal cell layer and then stop growing. The arrested growth can be alleviated by application of exogenous IAA.

On the basis of the phenotypes of the *alf* mutants, a model in which IAA is required for at least two steps in the formation of lateral roots has been proposed (Figure 19.28): (1) IAA transported acropetally in the stele is required to initiate cell division in the pericycle; (2) IAA is required to promote cell division and maintain cell viability in the developing lateral root (Celenza et al. 1995).

Auxin Delays the Onset of Leaf Abscission

The shedding of leaves, flowers, and fruits from the living plant is known as **abscission**. These parts abscise in a region called the **abscission zone**, which is located near the base of the petiole of leaves. In most plants, leaf abscission is preceded by the differentiation of a distinct layer of cells, the **abscission layer**, within the abscission zone. During leaf senescence, the walls of the cells in the abscission layer are digested, which causes them to become soft and weak. The leaf eventually breaks off at the abscission layer as a result of stress on the weakened cell walls.

IAA is known to delay the early stages of leaf abscission and to promote the later stages. Auxin levels are high in young leaves, progressively decrease in maturing leaves, and are relatively low in senescing leaves. During the early stages of leaf abscission, application of IAA inhibits leaf drop. During the later stages, however, auxin application hastens the process, probably by inducing the synthesis of ethylene (discussed later in this chapter and in Chapter 22), which promotes leaf abscission. Young leaves are apparently less sensitive to ethylene than older leaves are, and it has been suggested that the high levels of auxin in younger leaves reduce the sensitivity of the abscission zone to ethylene (Reid 1985).

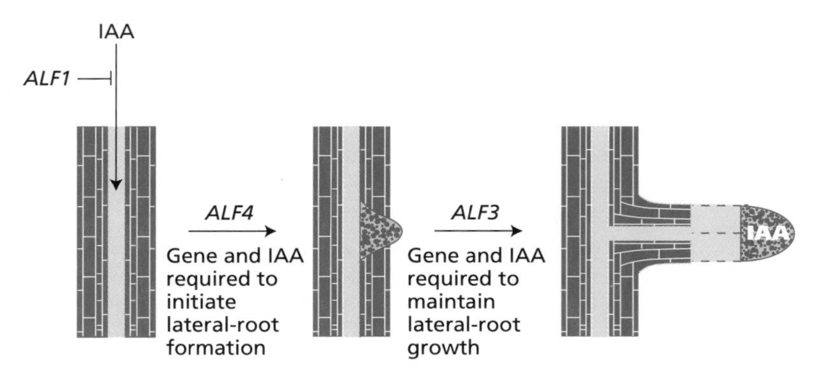

Figure 19.28 A model for the formation of lateral roots, based on the *alf* mutants of *Arabidopsis*. IAA transported acropetally in the vascular cylinder is required to initiate cell division in the pericycle. The *ALF4* gene appears to be required for this process. IAA is also required to maintain the growth of the growing lateral root. The *ALF3* gene is required for lateral-root growth. (After Celenza et al. 1995.)

Auxin Regulates Floral Bud Development

As discussed earlier, treating *Arabidopsis* plants with the polar auxin transport inhibitor NPA causes the loss of gravitropism. If mature plants are treated, NPA causes abnormal floral development, suggesting that polar auxin transport in the inflorescence is required for normal floral development. Recently, an *Arabidopsis* "pin-formed" mutant, *pin1-1*, was isolated whose flowers resemble those of NPA-treated plants. The *pin1-1* mutant exhibits various floral abnormalities—e.g., wide petals, no stamens, pistil-like structures that lack ovules, or in some cases, a complete absence of floral buds. When inflorescence axes were tested for their ability to transport auxin, polar transport in the mutant was 90% inhibited compared to that of the wild type (Okada et al. 1991). Thus the primary function of the *PIN1* gene may be to regulate polar auxin transport in the inflorescence.

Auxin Promotes Fruit Development

Much evidence suggests that auxin is involved in the regulation of fruit development. Auxin is produced in pollen and in the endosperm and the embryo of developing seeds, and the initial stimulus for fruit growth may result from pollination. Successful pollination initiates ovule growth, which is known as **fruit set**. After fertilization, fruit growth may depend on auxin produced in developing seeds. The endosperm may contribute auxin during the first stage of fruit growth, and the developing embryo may take over as the main auxin source during the latter stages.

In some plant species, seedless fruits may be produced naturally or they may be induced by treating the unpollinated flowers with auxin. The production of such seedless fruits is called **parthenocarpy**. In stimulating the formation of parthenocarpic fruits, auxin may act primarily to induce fruit set, which in turn may trigger the endogenous production of auxin by certain fruit tissues to complete the developmental process. However, other plant hormones are also involved in fruit growth. For instance, ethylene is known to influence fruit development, and some of the effects of auxin on fruiting may be mediated through the promotion of ethylene synthesis.

Auxin Induces Vascular Differentiation

New vascular tissues differentiate directly below developing buds and young growing leaves, and removal of the young leaves prevents vascular differentiation (Aloni 1995). Vascular differentiation occurs in a basipetal direction; that is, it is polar. The ability of the shoot apex to stimulate vascular differentiation can be demonstrated also in tissue culture. When a shoot apex is grafted onto a clump of undifferentiated cells, called *callus*, phloem and xylem differentiate underneath the graft (Wetmore and Reir 1963).

As the polarity of vascular differentiation suggests, auxin plays an important role in this process. In woody perennials, auxin produced by growing buds in the spring stimulates activation of the cambium in a basipetal direction; thus, the new round of secondary growth begins at the smallest twigs and progresses downward to the root. The regeneration of vascular tissue following wounding is also controlled by auxin produced by the young leaf directly above the wound site (Figure 19.29). Removal of the leaf prevents the regeneration of vascular tissue, and applied auxin can substitute for the leaf in stimulating regeneration (Thompson and Jacobs 1966).

Another indication that auxin plays a key role in vascular differentiation comes from studies with transgenic plants. When the auxin biosynthesis gene from *Agrobacterium tumefaciens* was overexpressed in petunia plants, the number of xylem tracheary elements increased, although their size decreased (Klee et al. 1987). In con-

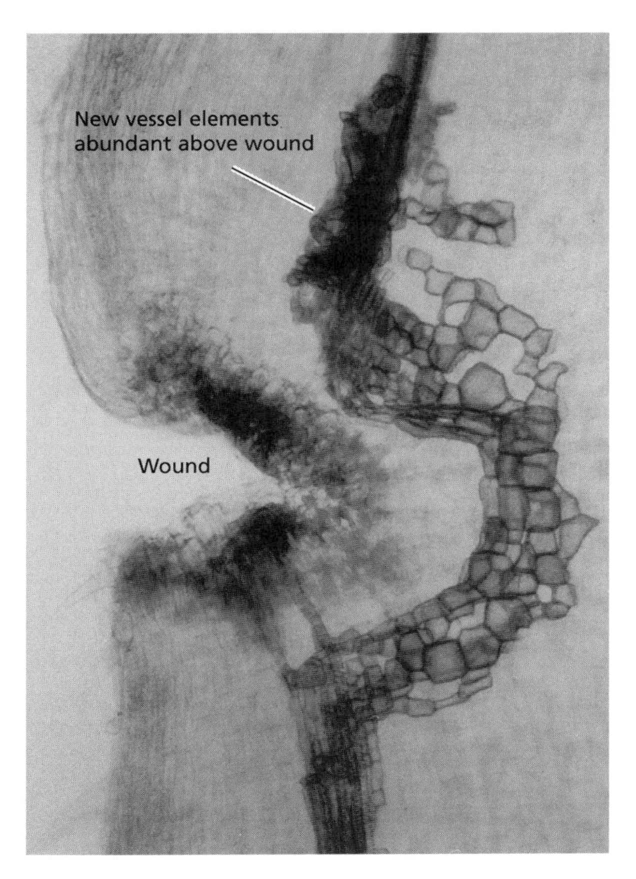

Figure 19.29 IAA-induced xylem regeneration around a wound in cucumber (*Cucumis sativus*) stem tissue. The stem was decapitated, and the leaves and buds above the wound site were removed to deprive it of endogenous auxin. Immediately after wounding, IAA in lanolin paste was applied to the internode above the wound. New vessel elements form above the wound (arrow). Auxin travels down the stem to the wound site via the vascular strand. The polar differentiation of new vascular tissue around the wound site is apparent from the greater abundance of vessel elements at the apical side of the wound. (Courtesy of R. Aloni.)

trast, when the level of free IAA in tobacco plants was decreased by transforming them with a bacterial gene that encodes an IAA conjugating enzyme, the number of vessel elements decreased, and their sizes increased (Romano et al. 1991). Thus the level of free auxin appears to regulate the number of tracheary elements, as well as their size.

Zinnia elegans mesophyll cells provide an elegant system for the study of vessel element differentiation in vitro. As we saw in Chapter 16, *Zinnia* mesophyll cells in culture can differentiate into xylem vessel elements directly without further cell divisions (Fukuda and Komamine 1980). Although auxin is required for tracheary cell differentiation, cytokinins (see Chapter 21) also participate (Fukuda 1997). One possibility is that cytokinins increase the sensitivity of the cells to auxin. Whereas auxin is produced in the shoot and transported downward to the root, cytokinins are produced by the root tips and transported upward into the shoot. Both cambium activation and vascular differentiation are probably regulated by both hormones.

The interactions between auxin and cytokinin in regulating cell division and organogenesis in callus cultures will be discussed in Chapter 21.

Synthetic Auxins Have a Variety of Commercial Uses

Auxins have been used commercially in agriculture and horticulture for more than 50 years. The early commercial uses included rooting of cuttings for plant propagation, promotion of flowering in pineapple, prevention of fruit and leaf drop, induction of parthenocarpic fruit, and thinning of fruit. Today, in addition to these applications, auxins are widely used as herbicides. The chemicals 2,4-D and dicamba (see Figure 19.2) are probably the most widely used synthetic auxins. Synthetic auxins are very effective because they are not metabolized by the plant as quickly as IAA is. Because maize and other monocotyledons can rapidly inactivate these synthetic auxins by conjugation, these auxins are used by farmers for the control of dicot weeds, also called broad-leaved weeds, in commercial cereal fields, and by the home gardener for the control of weeds such as dandelions and daisies in lawns.

The Molecular Mechanism of Auxin Action

Despite the diversity of the effects of auxin on plant development, the primary events are assumed to be similar in all cases, beginning with the binding of auxin to its receptor. Applying Occams's razor—the philosphical principle that the simplest explanation must be chosen—we can envision a single auxin receptor coupled to one or more signal transduction pathways. A more complex situation arises if there is more than one auxin receptor.

Regardless of the number of receptors, the response of each cell to the hormonal signal depends on two main factors: (1) its developmental program—that is, the types of genes that are being expressed at the time of exposure—and (2) the prevailing concentrations of other signaling molecules. For some cells, such as those in the elongation zones of stems and coleoptiles, the response is rapid cell expansion. For other cells, such as future vessel elements, the response involves a relatively slow process of cell differentiation. Although it was initially believed that only the slower auxin responses involved altered patterns of gene expression, it is now recognized that auxin can stimulate the expression of certain genes with a lag time of only 3 minutes. Thus it seems likely that gene expression plays a role even in the rapid effects of auxin.

The ultimate goal of research on the molecular mechanism of hormone action is to reconstruct each step in the chain of events leading from receptor binding to the physiological response. In this last section of the chapter, we'll begin with a description of the current best candidate for the auxin receptor, followed by a discussion of the various signal transduction pathways that have been implicated in auxin action. Next we will turn to auxin-regulated gene expression. Finally, we will contrast the mechanism of auxin action with that of another compound that is also capable of inducing proton extrusion and growth: the fungal phytotoxin fusicoccin (see Figure 19.18).

A Possible Auxin Receptor Has Been Identified

Early efforts to identify auxin receptors relied mainly on in vitro binding studies using radioisotope-labeled auxins (Hertel et al. 1972). At that time, the newly proposed acid growth hypothesis, with its emphasis on the role of proton extrusion, suggested to investigators that the auxin receptor was located on the plasma membrane. Auxin-binding sites were subsequently identified in the endoplasmic reticulum (site I), the plasma membrane (site II), and the tonoplast (site III) (Dohrmann et al. 1978). However, most of the auxin-binding activity was associated with site I, and this is the only binding site on the membrane that has survived the intense scrutiny given it during the past 20 years (Venis and Napier 1997). Besides representing the bulk of the binding activity, the affinity of site I binding for auxin analogs roughly parallels the auxin activity of the analogs (Ray et al. 1977; Napier and Venis 1990). This similarity satisfies the criterion that binding to the receptor should mirror the growth-promoting activity of the analog.

Beginning with the work of M. Löbler and D. Klämbt (1985), numerous laboratories have succeeded in purifying the protein responsible for site I auxin binding in the ER (reviewed by Venis and Napier [1997]). Termed **ABP1** (*a*uxin-*b*inding *p*rotein *1*), the protein consists of a 22 kDa polypeptide that forms a dimer in its native state. Maize has five *ABP1* genes, whereas only a single gene is present in *Arabidopsis*. The *ABP1* gene encodes a hydrophilic glycoprotein containing a signal sequence at its N terminus; the C terminus contains the amino acid sequence that acts as an ER retention signal, lysine-aspartic acid-glutamic acid-leucine (K-D-E-L using the single letter amino acid code). The presence of a signal sequence indicates that ABP1 is synthesized on the ER and translocated into the ER lumen. The presence of the KDEL sequence indicates that, rather than being transported to the Golgi, ABP1 is retained in the ER. This retention is consistent with binding studies showing that the ER is the richest source of ABP1.

The localization of ABP1 in the ER seems to contradict the original assumption that the auxin receptor is located on the plasma membrane. This location was supported by studies using forms of auxin that are unable to cross the membrane and enter the cell. For example, Venis and colleagues (1990) tested the ability of protein conjugates of 5-aminonaphthalene-1-acetic acid and of 5-azidonaphthalene-1-acetic acid to induce elongation in pea epicotyl sections. The impermeant analogs exhibited auxin activity, suggesting that auxin does not have to enter the cell to be active. Moreover, immunocytochemical studies have provided evidence that a small amount of ABP1, perhaps as little as 2% of the total, escapes the ER and is secreted to the surface of maize coleoptile protoplasts (Diekmann et al. 1995). This amount of ABP1 may be sufficient for it to act as an auxin receptor on the cell surface.

ABP1 Participates in the Rapid Auxin-Induced Increase in Membrane Voltage

Direct physiological evidence that ABP1 mediates the rapid effects of auxin has been obtained using isolated protoplasts. Treatment of tobacco mesophyll protoplasts with auxin increases the membrane voltage—that is, causes *hyperpolarization* of the plasma membrane—with a lag time of only 1 to 2 minutes (Ephritikhine et al. 1987). Since antibodies to the plasma membrane H^+-ATPase block the hyperpolarization response to auxin, this response appears to involve a rapid activation of the proton pump. Preincubation of the protoplasts with ABP1 antibodies also inhibited the auxin-induced hyperpolarization (Barbier-Brygoo et al. 1992).

Some ABP1 antibodies can mimic the effect of auxin; that is, they exhibit auxin *agonist* activity. Preincubation of tobacco mesophyll protoplasts with an antibody raised against the putative auxin-binding site of ABP1 exhibited auxin agonist activity in three different assays: hyperpolarization of tobacco mesophyll protoplasts, stimulation of proton extrusion in maize, and modulation of anion channels in tobacco (Venis et al. 1992). Presumably, by specifically binding to the auxin-binding site, the antibodies cause a conformational change similar enough to that induced by auxin to bring about the response.

The MAP Kinase Cascade May Mediate Auxin Efffects

Both auxin and cytokinin are required for the completion of the cell cycle (see Chapter 21). When tobacco cells are deprived of auxin, they arrest at the end of either the G_1 or the G_2 phase and cease dividing; if auxin is added back into the culture medium, the cell cycle resumes (John et al. 1993; Koens et al. 1995). Auxin appears to exert its effect on the cell cycle primarily by stimulating the synthesis of the major cyclin-dependent protein kinase (CDK) Cdc2 (*c*ell *d*ivision *c*ycle 2) (John et al. 1993). (Recall that CDKs, together with their regulatory subunits, the cyclins, regulate the transitions from G_1 to S, and from G_2 to mitosis, during the cell cycle; see Chapter 1.)

A possible clue to the downstream events initiated by auxin was provided by Hirt (1997), who showed that a homolog of the mammalian mitogen-activated protein kinase kinase (MAPKK) and a protein kinase that has the properties of a mitogen-activated protein kinase (MAPK) are activated during auxin-induced cell division in tobacco cells. This finding suggests that a pathway similar to the mammalian MAP kinase cascade (see Chapter 14) may be involved in the auxin signal transduction chain leading to the synthesis of CDK and cell division.

Other Signaling Intermediates Have Been Implicated in Auxin Action

Calcium plays an important role in signal transduction in animals and is thought to be involved in the action of certain plant hormones as well. In animal cells, heterotrimeric G proteins, phospholipase C, and IP_3 (inositol trisphosphate) make up the signal transduction chain leading to increases in cytosolic calcium (see Chapter 14).

The role of calcium in auxin action seems very complex and, at this point in time, very uncertain (Hobbie et al. 1994). Nevertheless, some experimental evidence shows that auxin affects the levels of free calcium in the cell. Using confocal scanning optical microscopy to visualize dyes that exhibit a change in their fluorescence spectrum in the presence of calcium, Gehring and colleagues (1990) studied the response of corn coleoptile cells to the auxin analog 2,4-D. When 2,4-D was applied to the cells, the level of free calcium rose from 280 to 380 n*M* in 4 minutes. The significance of this result remains to be determined. There is as yet no clear evidence that auxin-induced increases in free intracellular calcium involve the participation of G proteins, phospholipase C, or IP_3, nor is there evidence that IAA, directly or indirectly, activates a membrane protein kinase (see Chapter 14).

Changes in cytoplasmic pH can also serve as a second messenger in animals and plants. Two laboratories (Felle 1988 ; Gehring et al. 1990), have reported that auxin induces a decrease in cytosolic pH of about 0.2 units within 4 minutes of application. The cause of this pH drop is not known. Since the cytosolic pH is normally around 7.4, and the pH optimum of the plasma membrane H^+-ATPase is 6.5, a decrease in the cytosolic pH of 0.2 units could cause a marked increase in the activity of the plasma membrane H^+-ATPase. The decrease in cytosolic pH might also account for the auxin-induced increase in free intracellular calcium.

A series of papers from the laboratory of Gunter Scherer has provided evidence for auxin stimulation of phospholipase A_2 (Scherer 1996). Recall that phospholipase A_2 (PLA_2) is the first step in the eicosanoid pathway in animals (see Chapter 14), as well as the jasmonic acid pathway in plants (see Chapter 13). In the jasmonic acid pathway, the fatty acids released from the membrane phospholipids by PLA_2 serve as the substrate for lipoxygenase. The other product of the reaction, *lysophosphatidylcholine,* has detergent properties and can also activate protein kinases.

Scherer and colleagues have shown that auxin increases the activity of PLA_2 by up to 70% within minutes of treatment, and the response to various auxin analogs shows a good correlation with growth-promoting activity (Scherer and André 1989). Lysophosphatidylcholine was also able to activate protein kinases present on plant membranes in vitro and is thus a candidate for a second messenger in the auxin signal transduction pathway. However, thus far there have been no reports of an auxin-induced phosphorylation of membrane proteins, so the significance of the PLA_2 pathway in auxin action remains to be determined.

Could the auxin-stimulation of PLA_2 also play a role in the induction of cell division? In animal cells, the eicosanoid pathway is involved in the synthesis of *prostaglandins,* modified fatty acids that can stimulate cell division and other responses (see Chapter 14). Prostaglandin-stimulated cell division in animal cells has been linked to the activation of a prostaglandin-dependent adenylyl cyclase, the enzyme that synthesizes cyclic AMP (cAMP). Activation of adenylyl cyclase, in turn, causes a transient rise in cAMP, which promotes the onset of S phase of the cell cycle.

Recently, Harry Van Onckelen and his colleagues at the University of Antwerp in Belgium were able to detect a transient rise in cAMP levels during the S and G_1 phases of the cell cycle of tobacco cells (Ehsan et al. 1998). Application of the drug indomethacin, which inhibits prostaglandin synthesis in animal cells, blocked the rise in cAMP and also inhibited cell division. These results suggest three possibilities: (1) plants, which are known to possess the eicosanoid pathway, may also produce prostaglandins; (2) plants may contain a protaglandin-dependent adenylyl cyclase; and (3) cyclic AMP may play a role in plant cell division.

Additional research is needed to ascertain whether prostaglandins are present in plants and whether the activation of PLA_2 by auxin stimulates prostaglandin synthesis via the eicosanoid pathway. As already noted in Chapter 14, there is increasing evidence that cAMP is not only present in plants, but that it serves a signaling function. In support of this hypothesis, genes similar to the adenylyl cyclase gene of yeast have recently been identified in tobacco and *Arabidopsis*. Studies to determine the functions of these genes and their possible role in auxin-induced cell division are currently underway in the laboratory of Jeff Schell at the Max Planck Institute in Cologne.

Auxin May Induce the Ubiquitination of Nuclear Proteins

Mutants of *Arabidopsis* blocked at steps in the auxin response are proving to be useful for elucidating the auxin signal transduction pathway (Hobbie and Estelle 1994; Estelle 1996). Recently, such work has led to the discovery of a pathway, required for auxin action, involving the ubiquitination of nuclear proteins. The first enzyme in the ubiquitin pathway is E1, which binds to and activates ubiquitin using ATP (see Chapter 14). The gene involved in the auxin-resistant *axr1* mutation of *Arabidopsis,* which results in defects in many auxin responses, including gravitropism and gene expression, was found to encode an enzyme related to the *N-terminal half* of E1 (Leyser et al. 1993). This puzzled scientists for many years, since the C-terminal half of the E1-like protein seemed to be missing.

Recently it was discovered that *AXR1* is related to a gene in yeast, *AOS1,* which also lacks the C-terminal half of E1. In yeast, the Aos1p protein forms a heterodimer with another protein, Uba2p, which is homologous to the *C-terminal half* of E1. Based on this, Mark Estelle and his colleagues at Indiana University were able to clone an *Arabidopsis* homolog of the *UBA2* gene, called *ECR1.* The two proteins AXR1 and ECR1 appear to form an E1-like heterodimer similar to that of the yeast Aos1p–Uba2p heterodimer. However, rather than binding to ubiquitin, the **AXR1–ECR1 heterodimer** binds to a family of small, ubiquitin-related proteins called **RUBs** (*r*elated to *ub*iquitin). Binding of AXR1–ECR1 to RUB activates RUB and initiates a pathway similar to the ubiquitination pathway in which RUB is covalently attached to a target protein.

However, unlike ubiquinated proteins, proteins tagged with RUB are usually *activated* rather than marked for destruction. On the basis of biochemical studies with *Arabidopsis,* as well as other studies with yeast, Estelle and colleagues (Ruegger et al. 1998) have proposed a model in which the AXR1–ECR1 heterodimer mediates the transfer of RUB to an E3 enzyme complex of the ubiquitination pathway, thereby activating the E3 complex. According to this model, auxin increases the activity of AXR1–ECR1, either by inducing its synthesis or by activating a preexisting enzyme. Since the AXR1 protein has been localized to the nuclei of dividing and elongating cells, Estelle and colleagues have suggested that this auxin-induced ubiquitination occurs in the nucleus, resulting in the degradation of nuclear proteins by the 26S proteasome (Figure 19.30).

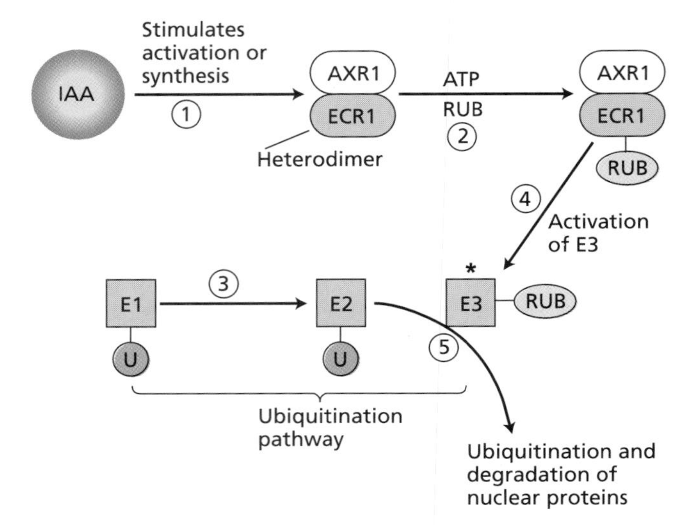

Figure 19.30 Hypothetical model for auxin-induced activation of the ubiquitination pathway. Step 1: IAA induces the synthesis or activation of the AXR1 and ECR1 proteins, which form a heterodimer. Step 2: The heterodimer forms a covalent bond with an RUB protein, using ATP. Step 3: The E1–ubiquitin complex transfers ubiquitin to E2. Step 4: The AXR1–ECR1 heterodimer transfers RUB to E3, thus activating it (asterisk). Step 5: The activated E3 complex transfers ubiquitin (U) from E2 to specific proteins in the nuclei of auxin-responsive cells.

As will be discussed in the sections that follow, repressor proteins that act as negative regulators of auxin-induced gene expression are potential targets of such a ubiquitination pathway.

Auxin Alters Gene Expression

In the early 1960s, it was suggested that auxin acts by inducing the expression of specific genes. However, by the 1970s several laboratories had demonstrated that auxin can stimulate both proton extrusion and growth with a lag time of only 10 to 15 minutes. It was assumed (without direct evidence) that 10 minutes was too rapid for the induction of protein synthesis. It was therefore believed that the initial stimulation of growth by auxin was mediated by direct interaction with membrane components rather than by specific gene expression.

Technical advances in molecular biology of the late 1970s and early 1980s allowed the detection of rapid changes in the levels of translatable mRNA—that is, mRNAs able to direct protein synthesis on ribosomes (Figure 19.31). However, the level of translatable mRNA does not necessarily correspond to the abundance of mRNA, since RNAs may be incompletely processed or slightly degraded. To measure abundance, specific cDNA (complementary DNA) probes have been used in hybridization studies. In particular, the use of "*plus-and-*

minus" screening methods for the identification of auxin-regulated mRNAs (Figure 19.32) has greatly facilitated research.

The plus-and-minus screening method allows the cloning of cDNA sequences that are specifically stimulated by auxin. Once such auxin-specific clones have been isolated, they can be used to prepare radioactive cDNA probes. Hybridization of such radioactive cDNA probes to total mRNA extracted from tissues at various times after auxin treatment makes it possible to determine the kinetics of auxin stimulation of that mRNA. This analysis is typically done on a *Northern blot*—that is, on RNA that has been size-separated by electrophoresis and then blotted onto nitrocellulose paper. The amount and length of the auxin-stimulated mRNA can be estimated from the size and location of the spot that develops on exposure of the blot to X-ray film. This approach has been used to demonstrate that auxin rapidly and specifically alters the abundance of a few RNA species.

Changes in mRNA abundance could be due to an alteration in the rates of transcription or degradation. To distinguish between these two alternatives, nuclear runoff experiments, which measure transcription directly, have been carried out (see Chapter 17) (McClure et al. 1989). Such studies have shown that auxin can stimulate the transcription of specific mRNAs in as little as 2 to 5 minutes. The rapid kinetics of auxin-induced gene expression suggests that gene expression may indeed be involved in the rapid growth response to auxin.

Auxin-Induced Genes Fall into Two Classes: Early and Late

One of the important functions of the signal transduction pathway(s) initiated when auxin binds to its receptor is the activation of a select group of transcription factors. The activated transcription factors enter the nucleus and promote the expression of specific genes. Genes whose expression is stimulated by the activation of *preexisting* transcription factors are called **primary response genes** or **early genes**. This definition implies that all of the proteins required for auxin-induced expression of the early genes are present in the cell at the time of exposure to the hormone; thus, early gene expression cannot be blocked by inhibitors of protein synthesis such as cycloheximide. As a consequence, the time required for the expression of the early genes can be quite short, ranging from a few minutes to several hours (Abel and Theologis 1996).

Studies of animal hormones have shown that primary response genes have three main functions: (1) Some of the early genes encode proteins that regulate the transcription of **secondary response genes**, or **late genes**, that are required for the long-term responses to the hormone. Because late genes require de novo protein synthesis, their expression can be blocked by protein synthesis inhibitors. (2) Other early genes are involved in intercellular communication, or cell–cell signaling. (3) Another group of early genes is involved in stress adaptation.

Abel and Theologis (1996) recently identified five major classes of early auxin-responsive genes: the Aux/IAA gene family, the SAUR gene family, the *GH3* gene family, genes that encode glutathione S-transferase–like (GST-like) proteins, and genes that encode 1-aminocyclopropane-1-carboxylic acid (ACC) synthase, the key enzyme in the ethylene biosynthetic pathway (Chapter 22).

The **Aux/IAA family** genes were the first auxin-inducible genes to be discovered (Ainley et al. 1988). Homologous sequences have been identified in soybean, mung bean, pea, and *Arabidopsis*. The expression of most of the Aux/IAA family of genes is stimulated by auxin within 5 to 60 minutes of hormone addition, and in most cases, induction is insensitive to cycloheximide. All the genes encode small hydrophilic polypeptides, having nuclear localization signals, and ranging from 19 to 36 kDa in molecular mass. Theologis and colleagues at the USDA Plant Gene Expression Center in Albany, California (Abel et al. 1994) determined that Aux/IAA family members have DNA-binding motifs similar to those of bacterial repressors. They also have short half-lives (about 7 minutes), indicating that they are turning over rapidly. These characteristics led Abel and colleagues (1994) to conclude that members of the Aux/IAA gene family encode short-lived transcription factors that function as repressors or activators of the expression of late auxin-inducible genes.

The SAUR gene family, referred to earlier in the chapter in relation to tropisms, was first discovered in soybean hypocotyls in the laboratory of Tom Guilfoyle at the University of Missouri (McClure et al. 1989). Related genes have since been identified in mung bean, pea, and *Arabidopsis*. Auxin stimulates the expression of SAUR genes within 2 to 5 minutes of treatment, and the response is insensitive to cycloheximide. The five SAUR genes of soybean are clustered together, contain no introns, and encode highly similar polypeptides of about 10 kDa of unknown function. Because of the rapidity of the response, expression of SAUR genes has proven to be a convenient probe for the lateral transport of auxin during photo- and gravitropism.

GH3 early-gene family members, also identified in soybean and *Arabidopsis*, are stimulated by auxin within 5 minutes. They encode proteins of about 70 kDa of unknown function.

The Aux/IAA and SAUR gene families are candidates for genes that are required for auxin-induced

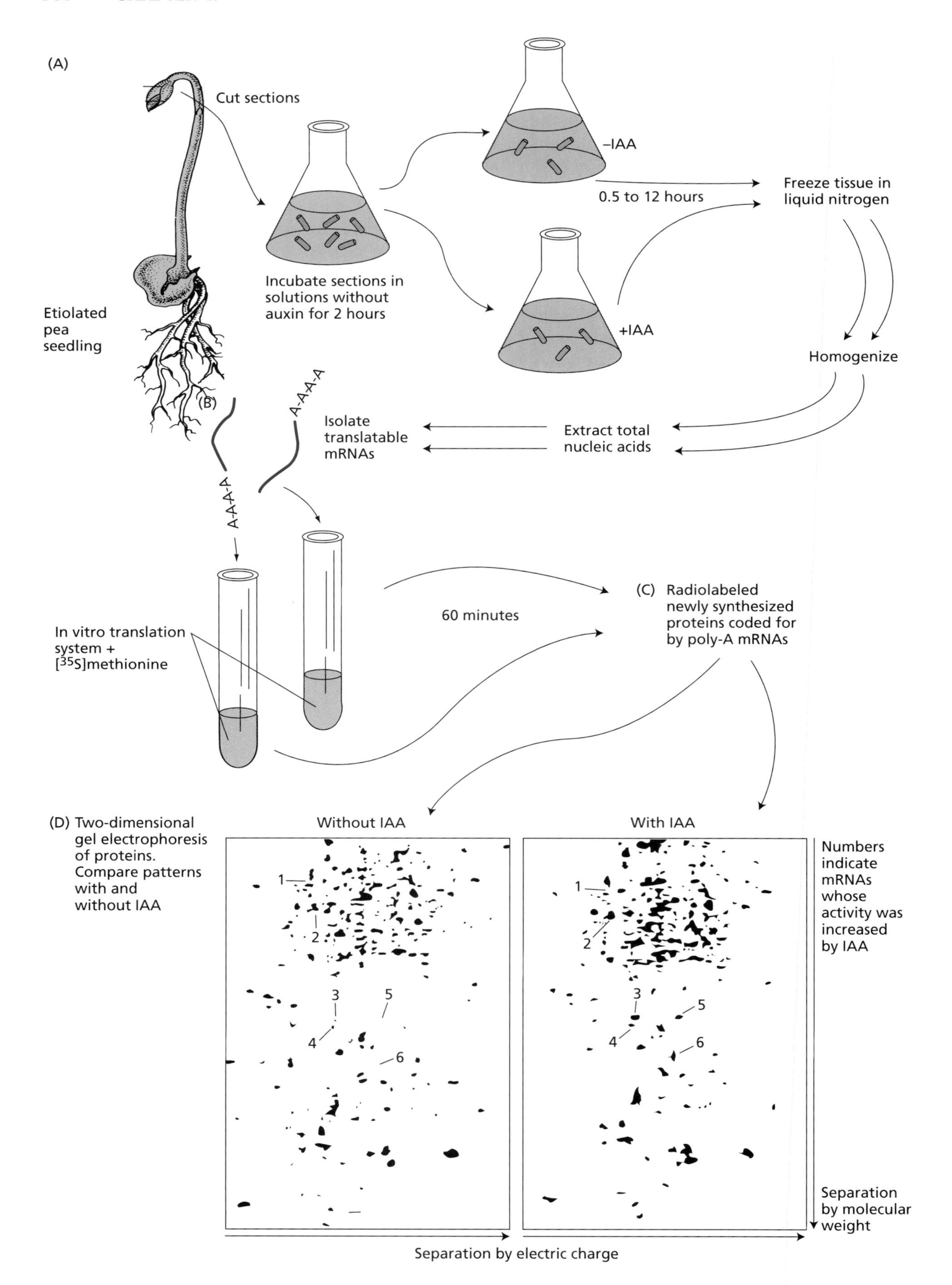
(A)
Cut sections
–IAA
0.5 to 12 hours
Freeze tissue in liquid nitrogen
Incubate sections in solutions without auxin for 2 hours
+IAA
Etiolated pea seedling
Homogenize
(B)
A-A-A-A
Isolate translatable mRNAs
Extract total nucleic acids
A-A-A-A
(C) Radiolabeled newly synthesized proteins coded for by poly-A mRNAs
60 minutes
In vitro translation system + [35S]methionine
(D) Two-dimensional gel electrophoresis of proteins. Compare patterns with and without IAA
Without IAA
With IAA
1
2
3
4
5
6
Numbers indicate mRNAs whose activity was increased by IAA
Separation by molecular weight
Separation by electric charge

◀ **Figure 19.31** Procedure for studying the effects of auxin on translatable messenger RNAs. This simplified diagram, based on the methods of Theologis and Ray (1982) shows the procedure used to detect auxin-induced changes in translatable mRNAs. (A) The tissue sections were first depleted of endogenous auxin by incubation in buffered sucrose, incubated in the presence and absence of IAA, and frozen in liquid nitrogen after various incubation times. (B) Nucleic acids were extracted from the frozen sections, and polyadenylated RNAs (translatable RNAs) were isolated. These mRNAs were translated into proteins in vitro by a cell-free protein synthesis system prepared from wheat germ. The protein synthesis system consists of ribosomes, enzymes, and tRNA, to which are added 20 amino acids, including [^{35}S]methionine, and ATP. (C) After 60 minutes, the in vitro translation was terminated and the newly synthesized polypeptides were separated by two-dimensional gel electrophoresis (the first dimension separated by net electric charge, the second dimension by molecular mass). (D) Since the newly synthesized polypeptides were radioactive, when the gel was exposed to X-ray film the polypeptides showed up as black spots on the developed film, or *autoradiograph*. The results showed that within 2 hours, auxin caused at least six mRNA sequences (numbered 1 through 6) to increase in translational activity.

growth and other developmental phenomena. In contrast, the GST-like genes are stimulated by various environmental factors, including heavy metals and various stress conditions, suggesting that their primary role is in stress adaptation. Likewise, ACC synthase, which is also induced by stress, is the rate-limiting step in ethylene biosynthesis (see Chapter 22). As a gaseous hormone, ethylene probably functions in intercellular communication, especially when the plant is under stress.

To be induced, the promoters of the early auxin genes must contain response elements that bind to the transcription factors that become activated in the presence of auxin. As discussed in the next section, a limited number of these response elements appear to be arranged combinatorily within the promoters of a variety of auxin-induced genes.

The Auxin-Responsive Domain Is a Composite Structure

To identify the **auxin response elements (AuxREs)** within the promoters of the early auxin genes, the promoters were fused to reporter genes, such as luciferase, the bacterial β-glucuronidase, and green fluorescent protein. The AuxREs were then identified by deletion analysis after plants were transformed with the hybrid gene. Using these reporter genes, investigators have identified several specific AuxREs that help confer auxin responsiveness. These AuxREs are usually combined in different ways to form **auxin response domains (AuxRDs)** within the promoter (Abel et al. 1996). For example, the *GH3* gene promoter of soybean is composed of three independently acting AuxRDs that contribute incrementally to the strong auxin inducibility of the promoter. Two distinct AuxRDs of 25 and 32 base pairs, respectively, contain the sequence TGTCTC (Ulmasov et al. 1995). However, when the constructs were expressed in carrot protoplasts, the TGTCTC element in the two AuxRDs was necessary but not sufficient to confer auxin inducibility. A variable sequence upstream of TGTCTC was also required. This upstream sequence caused constitutive expression and no auxin inducibility when part or all of the TGTCTC element was mutated or deleted. Each auxin response domain is thus a composite structure composed of two AuxREs, a variable constitutive element adjacent to a conserved TGTCTC element that confers auxin inducibility (Ulmasov et al. 1995). A minimal model for the activation of auxin early genes would therefore involve an interaction between the transcription factors that bind to these two AuxREs (Figure 19.33).

Early Auxin Genes May Be Under Negative Control of a Short-Lived Repressor

As noted previously, early auxin genes are by definition insensitive to protein synthesis inhibitors such as cycloheximide. Far from being inhibited, the expression of many of the early auxin genes has been found to be *stimulated* by cycloheximide. Cycloheximide stimulation of gene expression is accomplished both by transcriptional activation and by mRNA stabilization (Abel et al. 1996). Transcriptional activation by inhibitors of protein synthesis has usually been interpreted as indicating that the gene is being repressed by a short-lived *repressor protein* or by a regulatory pathway that involves a protein with a high turnover rate.

Although cycloheximide may stimulate the expression of the early auxin genes, it does not cause any of the physiological effects of auxin. In fact, as discussed earlier in the chapter, cycloheximide rapidly inhibits auxin-induced proton extrusion and growth. The reason is that the physiological response depends on the secondary response genes, whose expression would be inhibited by cycloheximide. However, it does suggest a model for auxin action in which the hormone blocks the interaction between the repressor and the activators. Alternatively, auxin might enhance the turnover of the repressor protein. Recall that the gene responsible for the *axr1* auxin-resistant phenotype in *Arabidopsis* encodes a protein thought to be involved in the activation of a ubiquitination pathway in the nucleus (Ruegger et al. 1998). Thus auxin may promote the ubiquitination of the repressor protein, marking it for destruction.

Figure 19.33 shows a model for the auxin-induced expression of the *GH3* early genes that is based on the ubiquitin-mediated turnover of a repressor protein.

Fusicoccin Activates Preexisting H$^+$-ATPases

Our examination of the auxin response would be incomplete without a discussion of the mechanism of **fusic-**

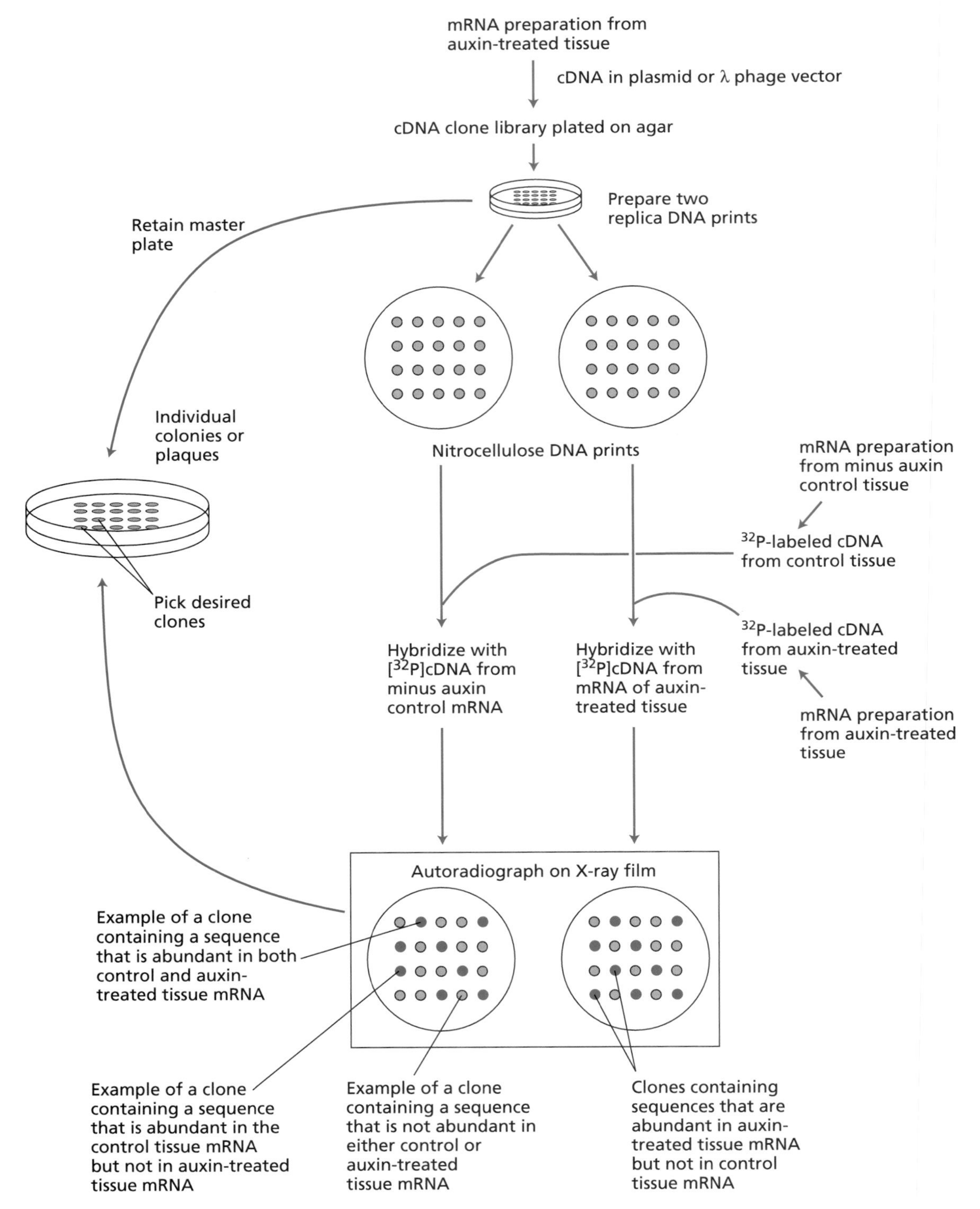

Figure 19.32 The plus-and-minus method of screening for auxin-stimulated mRNAs. Complementary DNA is prepared from mRNA extracted from auxin-treated tissue and cloned into a plasmid or λ phage vector. The cDNA library is plated out on agar, and two nitrocellulose prints are made that contain colonies or plaques in the same positions as on the master plate. The filters are further treated to release and denature the DNA and to bind it to the filters. Radioactively labeled cDNA synthesized from mRNA of control and auxin-treated tissue is hybridized to each of the replica prints. After being washed, the prints are placed under X-ray film. Comparison of the spots that develop on the film makes it possible to identify auxin-specific clones. Such clones can then be isolated from the master plate. Typically, the clones selected initially must be rescreened several times to yield single purified clones.

(A) **Without IAA:** no transcription of *GH3*

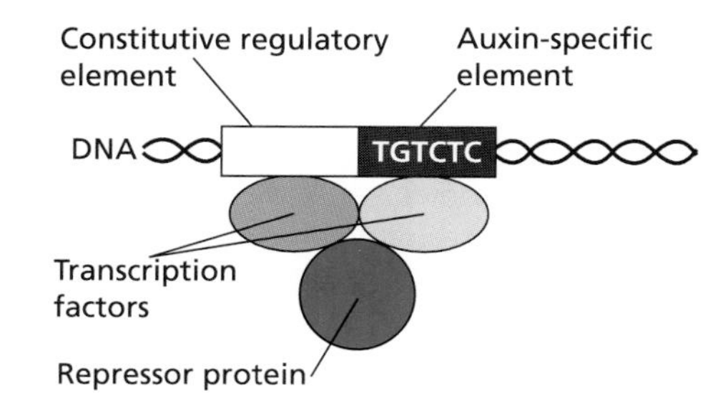

(B) **With IAA:** transcription of *GH3*

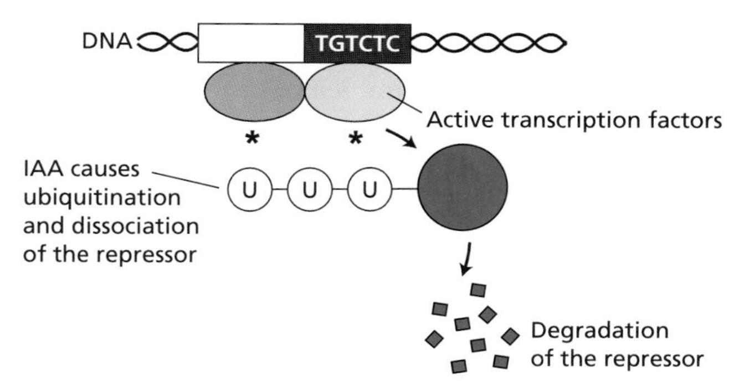

Figure 19.33 A model for transcriptional activation of the *GH3* early gene by auxin. (A) In the absence of IAA, the auxin response domain, consisting of an auxin-specific element, TGTCTC, and an upstream constitutive regulatory element, are bound to their respective transcription factors and a hypothetical repressor protein. (B) IAA causes the ubiquitination and dissociation of the repressor protein from the complex, and the repressor is degraded. As a consequence, the transcription factors become active (asterisks) and initiate transcription. (After Abel et al. 1996.)

occin (**FC**) (see Figure 19.17). This phytotoxin is produced by the fungus *Fusicoccum amygdali,* a parasite of peach and almond trees. Once released into the leaves, the toxin stimulates the H^+-ATPases of the guard cells. At the biochemical level it induces a rapid (10 to 30 seconds) hyperpolarization of the plasma membrane, accompanied by an acidification of the cell wall. This triggers an irreversible opening of the stomata (see Chapter 18), resulting in wilting of the leaves and, eventually, the death of the tree.

FC causes membrane hyperpolarization and proton extrusion in nearly all plant tissues. Treatment of coleoptiles and stem sections with FC leads to a transient growth response, a result that provides support for the acid growth hypothesis for auxin-induced growth. But whereas auxin-induced proton extrusion has a lag time of about 10 minutes and is inhibited by cycloheximide, FC-induced proton extrusion begins after only 1 to 2 minutes, and the response is insensitive to cycloheximide. Moreover, FC-treated cells can acidify down to a much lower extracellular pH than auxin-treated cells can (pH 3 versus pH 4).

Because of these effects, FC has sometimes been referred to as a "super auxin." However, FC acts by a mechanism entirely different from that of auxin. For example, FC does not stimulate the expression of any of the auxin-induced genes, nor can it mimic the effects of IAA on other developmental processes, such as cell division. Thus the effects of FC on plants are much more limited than those of IAA, which is not surprising since FC is a toxin, not an endogenous plant hormone. In the case of proton extrusion, FC is known to activate preexisting H^+-ATPases on the plasma membrane, probably by displacing the autoinhibitory C-terminal domain of the enzyme from the catalytic site (see Chapter 6) (Jahn et al. 1996). In contrast, auxin may promote proton extrusion in part by activation of preexisting H^+-ATPases (perhaps via acidification of the cytosol) and in part by causing the de novo synthesis of the H^+-ATPase.

Fusicoccin Binds to Protein Complexes on the Plasma Membrane

Fusicoccin was first identified as a phytotoxin in the 1960s, but it was not until 1977 that a radioactively labeled version of this toxin was available that could be used to characterize its ligand. Studies with radioactively labeled FC have shown that it binds tightly to a protein on the plasma membrane. Ultimately, these studies led to the purification and identification of the **fusicoccin-binding protein** (**FCBP**).

Initial attempts to purify FCBP after photoaffinity labeling led to conflicting results. Some laboratories identified a homodimer, composed of two 30 kDa subunits. The ability of the purified 30 kDa polypeptides to stimulate the plasma membrane H^+-ATPase in the presence of FC was tested in in vitro proton-pumping experiments. The 30 kDa polypeptide was reconstituted along with purified plasma membrane ATPase into artificial membrane vesicles called *proteoliposomes*. The addition of FC to the reconstituted proteoliposomes caused an increase in their ATP-driven H^+-pumping activity.

Cloning and sequencing of the gene for the 30 kDa polypeptide revealed it to be a member of the so-called **14-3-3 family** of regulatory proteins (Korthout and de Boer 1994; Oecking et al. 1994; Marra et al. 1995). Regulatory proteins of the 14-3-3 family are widespread among plants and animals. Originally identified in mammalian brain tissue (and named after the positions they occupy in chromatographs), 14-3-3 proteins make up a heterogeneous family of soluble proteins ranging from 25 to 32 kDa, which associate to form dimers (Ferl 1996). The precise function of 14-3-3 proteins has not yet been elucidated. However, the numerous functions that have been attributed to them indicate that they play multiple roles in signal transduction pathways.

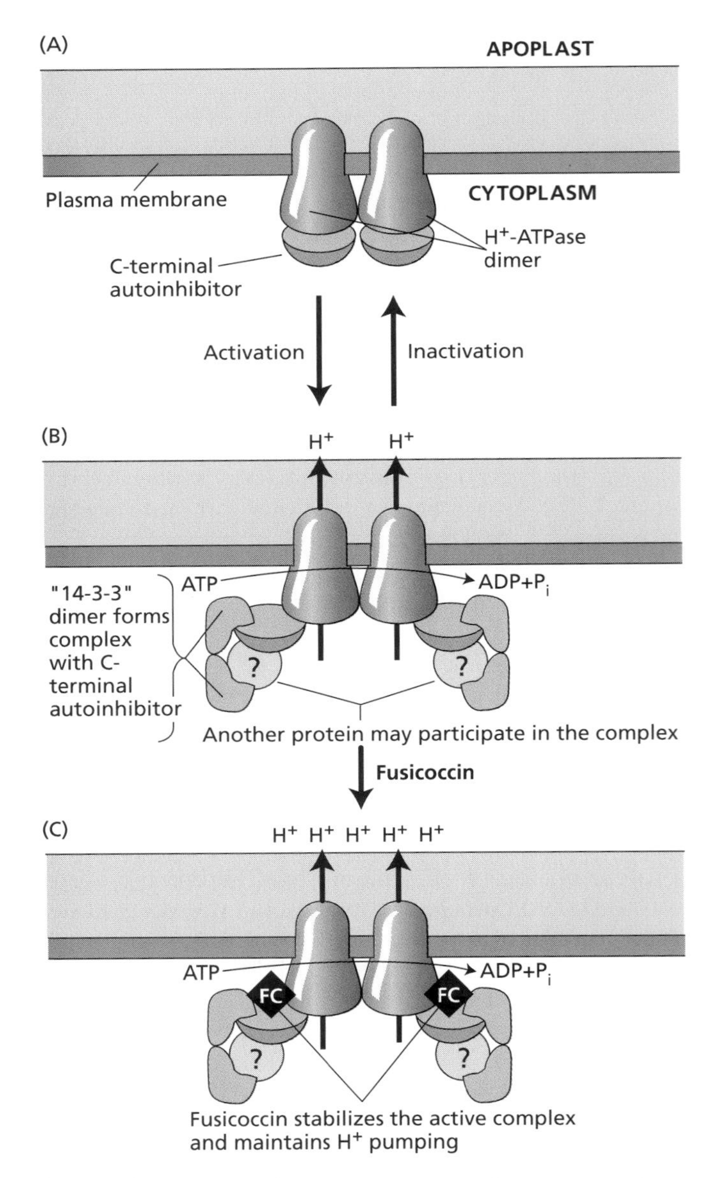

Figure 19.34 Model for the mechanism of fusicoccin activation of H$^+$-ATPases. (A) A pair of plasma membrane H$^+$-ATPases associate to form an active dimer. The activity of the pump is limited by the autoinhibitory effect of the C terminus of each monomer. (B) Dimers of 14-3-3 proteins form a transient complex with the C terminus of the H$^+$-ATPase and possibly another protein. When the autoinhibitory C terminus is bound to the 14-3-3 protein, the activity of the H$^+$-ATPase increases, but the effect is transient because of the instability of the complex. (C) The binding of fusicoccin to the complex stabilizes the complex, locking the H$^+$-ATPase into an active state. (After Oecking et al. 1997.)

Although it was initially believed that the 30 kDa 14-3-3 protein represented the complete FCBP, another laboratory reported that a 90 kDa polypeptide consistently copurified with the two 30 kDa polypeptides (Aducci et al. 1993). More recently it has been demonstrated that the 14-3-3 protein is unable to bind FC by itself. To bind FC, the 14-3-3 protein requires the presence of H$^+$-ATPase (Oecking et al. 1997). This finding confirms the earlier report that a 90 kDa polypeptide (the H$^+$-ATPase) copurifies with the 14-3-3 protein. According to the current model, FC activates the H$^+$-ATPase by stabilizing a transient complex that forms between the autoinhibitory C-terminal domain of the H$^+$-ATPase and the 14-3-3 dimer (Figure 19.34), locking the enzyme into its most active state.

Summary

Auxin was the first hormone to be discovered in plants and is one of an expanding list of chemical signaling agents that regulate plant development. The most common naturally occurring form of auxin is indole-3-acetic acid (IAA). One of the most important roles of auxin in higher plants is the regulation of elongation growth in young stems and coleoptiles. Low levels of auxin are also required for root elongation, although at higher concentrations auxin acts as a root growth inhibitor.

Accurate measurement of the amount of auxin in plant tissues is critical for understanding the role of this hormone in plant physiology. Early coleoptile-based bioassays have been replaced by more accurate techniques, including physicochemical methods and immunoassay.

Regulation of growth in plants may depend in part on the amount of free auxin present in plant cells, tissues, and organs. There are two main pools of auxin in the cell: the cytosol and the chloroplasts. Levels of free auxin can be modulated by several factors, including the synthesis and breakdown of conjugated IAA, IAA metabolism, compartmentation, and polar auxin transport. Several pathways have been implicated in IAA biosynthesis, including tryptophan-dependent and tryptophan-independent pathways. Several degradative pathways for IAA have also been identified.

IAA is synthesized primarily in the apical bud and is transported polarly to the root. Polar transport is thought to occur mainly in the parenchyma cells associated with the vascular tissue. Polar auxin transport can be divided into two main processes: IAA uptake and IAA efflux. In accord with the chemiosmotic model for polar transport, there are two modes of IAA uptake: by a pH-dependent passive transport of the undissociated form, or by an active H$^+$ cotransport mechanism driven by the plasma membrane H$^+$-ATPase. Auxin efflux is thought to occur preferentially at the basal ends of the transporting cells via anion efflux carriers and to be driven by the membrane potential generated by the plasma membrane H$^+$-ATPase. Phytotropins, such as naphthylphthalamic acid (NPA), which interfere with gravitropism, are thought to block polar auxin transport by binding to a protein that regulates the auxin efflux carrier, thereby inhibiting auxin efflux. Auxin can be transported nonpolarly in the phloem.

Auxin-induced cell elongation begins after a lag time of 10 minutes. Auxin promotes elongation growth primarily by increasing the cell wall extensibility. The specific type of wall loosening induced by auxin is termed stress relaxation. Auxin-induced stress relaxation requires continuous metabolic input and is mimicked in part by treatment with acidic buffers. According to the acid growth hypothesis, one of the important actions of auxin is to induce cells to transport protons into the cell wall by stimulating the plasma membrane H^+-ATPase. Two mechanisms have been proposed for auxin-induced proton extrusion: direct activation of the proton pump and enhanced synthesis of the plasma membrane H^+-ATPase. The ability of protons to cause cell wall loosening is mediated by a class of proteins called expansins. Expansins loosen the cell wall by breaking hydrogen bonds between the polysaccharide components of the wall.

In addition to proton extrusion, long-term auxin-induced growth involves the uptake of solutes and the synthesis and deposition of polysaccharides and proteins needed to maintain the acid-induced wall-loosening capacity. Although most studies of auxin-induced growth have been carried out on stem or coleoptile sections, recently, auxin-dependent growth has been demonstrated in intact auxin-deficient mutants of pea.

Promotion of growth in stems and coleoptiles and inhibition of growth in roots are the best-studied physiological effects of auxins. Auxin-promoted differential growth in these organs is responsible for the responses to directional stimuli (i.e., light, gravity) called tropisms. According to the Cholodny–Went model, auxin is transported laterally to the shaded side during phototropism and to the lower side during gravitropism. Statoliths (starch-filled amyloplasts) in the statocytes are involved in the normal percepton of gravity, but they are not absolutely required.

In addition to its roles in growth and tropisms, auxin plays central regulatory roles in apical dominance, lateral-root initiation, leaf abscission, vascular differentiation, floral bud formation, and fruit development. Commercial applications of auxins include rooting compounds and herbicides.

The auxin-binding soluble protein ABP1 is a strong candidate for the auxin receptor. ABP1 is located primarily in the ER lumen, but it is thought to be active on the cell surface.

Studies of the signal transduction pathways involved in auxin action have implicated cyclic AMP and MAP kinases as possible signaling intermediates in auxin-induced cell division. Other possible auxin signaling intermediates include Ca^{2+}, intracellular pH, and lysophosphatidylcholine.

Auxin-induced genes fall into two categories: early and late. Induction of early genes by auxin does not require protein synthesis and is insensitive to protein synthesis inhibitors. The early genes fall into three functional classes: expression of the late genes (secondary response genes), stress adaptation, and intercellular signaling. The auxin response domains of the promoters of the auxin early genes have a composite structure in which an auxin-inducible response element is combined with a constitutive response element. Auxin-induced genes may be negatively regulated by repressor proteins that are degraded via a ubiquitin activation pathway.

The fungal phytotoxin fusicoccin stimulates proton extrusion and growth by activating preexisting plasma membrane H^+-ATPases. Fusicoccin may act by binding and stabilizing a protein complex formed by the plasma membrane H^+-ATPase and a member of the 14-3-3 family of proteins. This complex activates the H^+-ATPase by displacing the autoinhibitory C-terminal region from the catalytic site.

General Reading

*Abel, S., and Theologis, A. (1996) Early genes and auxin action. *Plant Physiol.* 111: 9–17.

*Aducci, P., Ballio, A., Fogliano, V., Fullone, M. R., Marra, M., and Proietti, N. (1993) Purification and photoaffinity labeling of fusicoccin receptors from maize. *Eur. J. Biochem.* 214: 339–345.

Aducci, P., Marra, M., Fogliano, V., and Fullone, M. R. (1995) Fusicoccin receptors: Perception and transduction of the fusicoccin signal. *J. Exp. Bot.* 46: 1463–1478.

*Aloni, R. (1995) The induction of vascular tissue by auxin and cytokinin. In *Plant Hormones and Their Role in Plant Growth Development*, 2nd ed., P. J. Davies, ed., Kluwer, Dordrecht, Netherlands, pp. 531–546.

*Bandurski, R. S., Cohen, J. D., Slovin, J., and Reinecke, D. M. (1995) Auxin biosynthesis and metabolism. In *Plant Hormones and Their Role in Plant Growth Development*, 2nd ed., P. J. Davies, ed., Kluwer, Dordrecht, Netherlands, pp. 39–65.

Brenner, M. L. (1981) Modern methods for plant growth substance analysis. *Annu. Rev. Plant Physiol.* 32: 511–538.

*Caruso, J. L., Pence, V. C., and Leverone, L. A. (1995) Immunoassay methods of plant hormone analysis. In *Plant Hormones and Their Role in Plant Grown Development*, 2nd ed., Davies, P. J. ed., Kluwer, Dordrecht, Netherlands, pp. 433–447.

Cohen, J. D., and Bandurski, R. S. (1982) Chemistry and physiology of the bound auxins. *Annu. Rev. Plant Physiol.* 33: 403–430.

*Cox, D. N., and Muday, G. K. (1994) NPA binding activity is peripheral to the plasma membrane and is associated with the cytoskeleton. *Plant Cell* 6: 1941–1953.

Darwin, C. (1881) *The Power of Movement in Plants*. Appleton-Century-Crofts, New York.

Davies, P. J., ed. (1995) *Plant Hormones and Their Role in Plant Growth Development*, 2nd ed. Kluwer, Dordrecht, Netherlands.

De Boer, B. (1997) Fusicoccin—A key to multiple 14-3-3 locks? *Trends Plant Sci.* 2: 60–66.

Dennison, D. S. (1979) Phototropism. In *Encyclopedia of Plant Physiology*, New Series, Vol. 7, W. Haupt and M. E. Feinleib, eds., Springer, Berlin, pp. 506–566.

*Evans, M. L. (1974) Rapid responses to plant hormones. *Annu. Rev. Plant Physiol.* 25: 195–223.

Evans, M. L. (1985) The action of auxin on plant cell elongation. *Crit. Rev. Plant Sci.* 2: 213–265.

*Firn, R. D. (1994) Phototropism. In *Photomorphogenesis in Plants*, 2nd ed., R. E. Kendrick and G. H. M. Kronenberg, eds., Kluwer, Dordrecht, Netherlands, pp. 659–681.

*Firn, R. D., and Digby, J. (1980) The establishment of tropic curvatures in plants. *Annu. Rev. Plant Physiol.* 31: 131–148.

*Goldsmith, M. H. M. (1977) The polar transport of auxin. *Annu. Rev. Plant Physiol.* 28: 439–478.

*Hobbie, L., and Estelle, M. (1994) Genetic approaches to auxin action. *Plant Cell Environ.* 17: 525–540.

*Hobbie, L., Timpte, C., and Estelle, M. (1994) Molecular genetics of auxin and cytokinin. *Plant Mol. Biol.* 26: 1499–1519.

*Lomax, T. L., Muday, G. K., and Rubery, P. H. (1995) Auxin transport. In *Plant Hormones and Their Role in Plant Growth Development*, 2nd ed., P. J. Davies, ed., Kluwer, Dordrecht, Netherlands, pp. 509–530.

Marre, E. (1979) Fusicoccin: A tool in plant physiology. *Annu. Rev. Plant Physiol.* 30: 273–288.

Millner, P. A., and Causier, B. E. (1996) G-protein coupled receptors in plant cells. *J. Exp. Bot.* 47: 953–992.

Rayle, D. L., and Cleland, R. E. (1992) The acid growth theory of auxin-induced cell elongation is alive and well. *Plant Physiol.* 99: 1271–1274.

Salisbury, F. B., ed. (1996) *Units, Symbols, and Terminology for Plant Physiology*. Oxford University Press, New York.

Sembdner, G., Atzorn, R., and Schneider, G. (1994) Plant hormone conjugation. *Plant Mol. Biol.* 26: 1459–1481.

Skoog, F., ed. (1951) *Plant Growth Substances*. University of Wisconsin Press, Madison.

Takahashi, Y., Ishida, S., and Nagata, T. (1995) Auxin-regulated genes. *Plant Cell Physiol.* 36: 383–390.

Theologis, A. (1986) Rapid gene regulation by auxin. *Annu. Rev. Plant Physiol.* 37: 407–438.

*Thimann, K. V., and Skoog, F. (1934) On the inhibition of bud development and other function of growth substance in *Vicia faba*. *Proc. R. Soc. Lond. [Biol.]* 114: 317–339.

Walden, R., and Lubenow, H. (1996) Genetic dissection of auxin action: More questions than answers? *Trends Plant Sci.* 1: 335–339.

Wareing, P. F., and Phillips, I. D. J. (1981) *Growth and Differentiation in Plants*, 3rd ed. Pergamon, Oxford.

Weaver, R. J. (1972) *Plant Growth Substances in Agriculture*. W. H. Freeman, San Francisco.

Wilkins, M. B. (1979) Growth-control mechanisms in gravitropism. In *Encyclopedia of Plant Physiology*, New Series, Vol. 7, W. Haupt and M. E. Feinleib, eds., Springer, Berlin, pp. 601–626.

* Indicates a reference that is general reading in the field and is also cited in this chapter.

Chapter References

Abel, S., Ballas, N., Wong, L-M., and Theologis, A. (1996) DNA elements responsive to auxin. *Bioessays* 18: 647–654.

Abel, S., Oeller, P. W., and Theologis, A. (1994) Early auxin-induced genes encode short-lived nuclear proteins. *Proc Natl. Acad. Sci. USA* 91: 326–330.

Ainley, W. M., Walker, J. C., Nagao, R., and Key, J. L. (1988) Sequence and characterization of two auxin-regulated genes from soybean. *J. Biol. Chem.* 263: 10658–10666.

Barbier-Brygoo, H., Ephritikhine, G., Maurel, C., and Guern, J. (1992) Perception of the auxin signal at the plasma membrane of tobacco mesophyll protoplasts. *Biochem. Soc. Trans.* 20: 59–63.

Bartel, B. (1997) Auxin biosynthesis. *Annu. Rev. Plant Physiol. Plant Mol. Biol.* 48: 51–66.

Bartel, B., and Fink, G. (1994) Differential regulation of an auxin-producing nitrilase gene family in *Arabidopsis thaliana*. *Proc. Natl. Acad. Sci. USA* 91: 6649–6653.

Bartling, D., Seedorf, M., Schmidt, R. C., and Weiler, E. W. (1994) Molecular characterization of two cloned nitrilases from *Arabidopsis thaliana*: Key enzymes in the biosynthesis of the plant hormone indole-3-acetic acid. *Proc. Natl. Acad. Sci. USA* 91: 6021–6025.

Bennett, M. J., Marchand, A., Green, H. G., May, S. T., Ward, S. P., Millner, P. A., Walker, A. R., Schultz, B., and Feldmann, K. A. (1996) *Arabidopsis* AUX1 gene: A permease-like regulator of root gravitropism. *Science* 273: 948–950.

Bernasconi, P. (1996) Effect of synthetic and natural protein tyrosine kinase inhibitors on auxin efflux in zucchini (*Cucurbita pepo*) hypocotyls. *Physiol. Plantarum* 96: 205–210.

Briggs, W. R. (1963) Mediation of phototropic responses of corn coleoptiles by lateral transport of auxin. *Plant Physiol.* 38: 237–247.

Briggs, W. R. (1998) Signal transduction in phototropism: In pursuit of an elusive photoreceptor. *Plant Biology '98 Final Program*. Annual Meeting of the American Society of Plant Physiologists. Madison, WI, June 27–July 1. Abstract #3002, p. 5.

Cambridge, A. P., and Morris, D. A. (1996) Transfer of exogenous auxin from the phloem to the polar auxin transport pathway in pea (*Pisum sativum* L.). *Planta* 199: 583–588.

Celenza, J. L., Grisafi, P. L., and Fink, G. R. (1995) A pathway for lateral root formation in *Arabidopsis thaliana*. *Genes Dev.* 9: 2131–2142.

Cleland, R. E. (1995) Auxin and cell elongation. In *Plant Hormones and Their Role in Plant Growth Development*, 2nd ed., P. J. Davies, ed., Kluwer, Dordrecht, Netherlands, pp. 214–227.

Cleland, R. E. (1996) Growth substances. In *Units, Symbols and Terminology for Plant Physiology*, F. B. Salisbury, ed., Oxford University Press, New York, pp. 126–128.

Cosgrove, D. J. (1985) Cell wall yield properties of growing tissues: Evaluation by in vivo stress relaxation. *Plant Physiol.* 78: 347–356.

DeLuca, V., Marineau, C., and Brisson, N. (1989) Molecular cloning and analysis of cDNA encoding a plant tryptophan decarboxylase: Comparison with animal dopa decarboxylases. *Proc. Natl. Acad. Sci. USA* 86: 2582–2586.

Diekmann, W., Venis, M. A., and Robinson, D. G. (1995) Auxins induce clustering of the auxin-binding protein at the surface of maize coleoptile protoplasts. *Proc. Natl. Acad. Sci. USA* 92: 3425–3429.

Dohrmann, U., Hertel, R., and Kowalik, H. (1978) Properties of auxin binding sites in different subcellular fractions from maize coleoptiles. *Planta* 140: 97–106.

Edgerton, M. D., Tropsha, A., and Jones, A. M. (1994) Modelling the auxin-binding site of auxin-binding protein 1 of maize. *Phytochemistry* 35: 1111–1123.

Ehsan, H., Reichheld, J.-P., Roef, L., Witters, E., Lardon, F., Van Bockstaele, D., Van Montague, M., Inze, D., and Van Onckelen, H. (1998) Effect of indomethacin on cell cycle dependent cyclic AMP fluxes in tobacco BY-2 cells. *FEBS Letters* 422: 165–169.

Ephritikhine, G., Barbier-Brygoo, H., Muller, J. F., and Guern, J. (1987) Auxin effect on the transmembrane potential difference of wild-type and mutant tobacco protoplasts exhibiting a differential sensitivity of auxin. *Plant Physiol.* 84: 801–804.

Estelle, M. (1996) The ins and outs of auxin. *Curr. Biol.* 6: 1589–1591.

Farrimond, J. A., Elliott, M. C., and Clack, D. W. (1978) Charge separation as a component of the structural requirements for hormone activity. *Nature* 274: 401–402.

Felle, H. (1988) Auxin causes oscillations of cytosolic free calcium and pH in *Zea mays* coleoptiles. *Planta* 174: 495–499.

Ferl, R. J. (1996) 14-3-3 Proteins and signal transduction. *Annu. Rev. Plant Physiol. Plant Mol. Biol.* 47: 49–73.

Frias, I., Caldeira, M. T., Perez-Castineira, J. R., Narvarro-Avino, J. P., Culianez-Macia, F. A., Kuppinger, O., Stransky, H., Pages, M., Hager, A., and Serrano, R. (1996) A major isoform of the maize plasma membrane H^+-ATPase: Characterization and induction by auxin in coleoptiles. *Plant Cell* 8: 1533–1544.

Fukaki, H., Wysocka-Diller, J., Kato, T., Fujisawa, H., Benfey, P. N., and Tasaka, M. (In press) Genetic evidence that the endodermis is essential for shoot gravitropism in *Arabidopsis thaliana*. *Plant J.*

Fukuda, H. (1997) Tracheary element differentiation. *Plant Cell* 9: 1147–1156.

Fukuda, H., and Komanine , A. (1980) Establishment of an experimental system for the tracheary element differentiation from single cells isolated from the mesophyll of *Zinnia elegans*. *Plant Physiol.* 65: 57–60.

Garbers, C., DeLong, A., Deruere, J., Bernasconi, P., and Soll, D. (1996) A mutation in protein phosphatase 2A regulatory subunit affects auxin transport in *Arabidopsis*. *EMBO J.* 15: 2115–2124.

Gehring, C. A., Irving, H. R., and Parish, R. W. (1990) Effects of auxin and abscisic acid on cytosolic calcium and pH in plant cells. *Proc. Natl. Acad. Sci. USA* 87: 9645–9649.

Gocal, G. F. W., Pharis, R. P., Yeung, E. C., and Pearce, D. (1991) Changes after decapitation in concentrations of IAA and abscisic acid in the larger axillary bud of *Phaseolus vulgaris* L. cultivar Tender Green. *Plant Physiol.* (Lancaster, PA.) 95: 344–350.

Guilfoyle, T. J., McClure, B. A., Hagen, G., Brown, D., Gee, M., and Franco, A. (1990) Regulation of plant gene expression by auxins. In *Gene Manipulation in Plant Improvement II*, Gustafson, J. P., ed., Plenum Press: New York, pp. 401–418.

Hager, A., Debus, G., Edel, H. G., Stransky, H., and Serrano, R. (1991) Auxin induces exocytosis and the rapid synthesis of a high turnover pool of plasma-membrane H^+ ATPase. *Planta* 185: 527–537.

Hager, A., Menzel, H., and Krauss, A. (1971) Versuche und Hypothese zur Primärwirkung des Auxins beim Streckungswachstrum. *Planta* 100: 47–75.

Hangarter, R. P., and Good, N. E. (1981) Evidence that IAA conjugates are slow-release sources of free IAA in plant tissues. *Plant Physiol.* 68: 1424–1427.

Hasenstein, K. H., and Evans, M. L. (1988) Effects of cations on hormone transport in primary roots of *Zea mays*. *Plant Physiol.* 86: 890–894.

Hayashi, H., Czaja, I., Lubenow, H., Schell, J., and Walden, R. (1992) Activation of a plant gene by T-DNA tagging auxin-independent growth *in vitro*. *Science* 258: 1350–1353.

Hertel, R., Thomson, K-St., and Russo, V. E. A. (1972) In vitro auxin binding to particulate cell fractions from corn coleoptiles. *Planta* 107: 325–340.

Hirt, H. (1997) Multiple roles of MAP kinases in plant signal transduction. *Trends Plant Sci.* 2: 11–15.

Huala, E., Oeller, P. W., Liscum, E., Han, I.-S., Larsen, E., and Briggs, W. R. (1997) *Arabiodopsis* NPH1: A protein kinase with a putative redox-sensing domain. *Science* 278: 2120–2123.

Iino, M. (1995) Gravitropism and phototropism of maize coleoptile: Evaluation of the Cholodny-Went theory through effects of auxin application and decapitation. *Plant Cell Physiol.* 36: 361–367.

Iino, M., and Briggs, W. R. (1984) Growth distribution during first positive phototropic curvature of maize coleoptiles. *Plant Cell Environ.* 7: 97–104.

Jacobs, M. (1993) The localization of auxin transport carriers using monoclonal antibodies. *What's New in Plant Physiol.* 14: 17–20.

Jacobs, M., and Gilbert, S. F. (1983) Basal localization of the presumptive auxin carrier in pea stem cells. *Science* 220: 1297–1300.

Jacobs, M., and Rubery, P. H. (1988) Naturally occurring auxin transport regulators. *Science* 241: 346–349.

Jahn, T., Johansson, F., Luethen, H., Volkmann, D., and Larsson, C. (1996) Reinvestigation of auxin and fusicoccin stimulation of the plasma-membrane H^+-ATPase activity. *Planta* 199: 359–365.

John, P. C. L., Zhang, K., Dong, C., Diederich, L., and Wightman, F. (1993) P34-cdc2 related proteins in control of cell cycle progression, the switch between division and differentiation in tissue development, and stimulation of division by auxin and cytokinin. *Aust. J. Plant Physiol.* 20: 503–526.

Kaldenhof, R., and Iino, M. (1997) Restoration of phototropic responsiveness in decapitated maize coleoptiles. *Plant Physiol.* 114: 1267–1272.

Katekar, G. F. (1979) Auxins: The nature of the receptor site and molecular requirements for auxin activity. *Phytochemistry* 18: 222–233.

Kiss, J. Z., Wright, J. B., and Caspar, T. (1996) Gravitropism in roots of intermediate-starch mutants of *Arabidopsis*. *Physiol. Plantarum* 97: 237–244.

Klee, H. J., Horsch, R. B., Hinchee, M. A., Hein, M. B., and Hoffman, N. L. (1987) The effects of overproduction of two *Agrobacterium tumefaciens* T-DNA auxin biosynthetic gene products in transgenic petunia plants. *Genes Dev.* 1: 86–96.

Koens, K. B., Nicoloso, F. T., Harteveld, M., Libbenga, K. R., and Kijne, J. W. (1995) Auxin starvation results in G2-arrest in suspension-cultured tobacco cells. *J. Plant Physiol.* 147: 391–396.

Korthout, H. A. A. J., and de Boer, A. H. (1994) A fusicoccin binding protein belongs to the family of 14-3-3 brain protein homologs. *Plant Cell* 6: 1681–1692.

Langridge, W. H. R., Fitzgerald, K. J., Koncz, C., Schell, J., and Szalay, A. A. (1989) Dual promoter of *Agrobacterium tumefaciens* mannopine synthase genes is regulated by plant growth hormones. *Proc. Natl. Acad. Sci. USA* 86: 3219–3223.

Law, D. M., and Davies, P. J. (1990) Comparative indole-3-acetic acid levels in the slender pea and other pea phenotypes. *Plant Physiol.* 93: 1539–1543.

Legue, V., Blancaflor, E., Wymer, C., Perbal, G., Fantin, D., and Gilroy, S. (1997) Cytoplasmic free Ca^{-2+} in *Arabidopsis* roots changes in response to touch but not gravity. *Plant Physiol.* 114: 789–800.

Leyser, H. M. O., Lincoln, C. A., Timpte, C., Lammer, D., Turner, J., and Estelle, M. (1993) *Arabidopsis* auxin-resistance gene AXR1 encodes a protein related to ubiquitin activating enzyme, E1. *Nature* 364: 161–164.

Liscum, E., and Briggs, W. R. (1995) Mutations in the *NPH1* locus of *Arabidopsis* disrupt the perception of phototropic stimuli. *Plant Cell* 7: 473–485.

Löbler, M., and Klämbt, D. (1985) Auxin-binding protein from coleoptile membranes of corn (*Zea mays* L.). *Biol. Chem.* 260: 9848–9853.

Lomax, T. L. (1986) Active auxin uptake by specific plasma membrane carriers. In *Plant Growth Substances*, M. Bopp, ed., Springer Verlag, Berlin, pp. 209–213.

Marra, M., Fullone, M. R., Fogliano, V., Pen, J., Mattei, M., Masi, S., and Aducci, P. (1995) The 30-KD protein present in purified fusicoccin receptor preparations is a 14-3-3 like protein. *Plant Physiol.* 106: 1497–1501.

Masson, P. H. (1995) Root gravitropism. *Bioessays* 17: 119–127.

McClure, B. A., and Guilfoyle, T. (1989) Rapid redistribution of auxin-regulated RNAs during gravitropism. *Science* 243: 91–93.

McClure, B. A., Hagen, G., Brown, C. S., Gee, M. A., and Guilfoyle, T. (1989) Transcription, organization, and sequence of an auxin-regulated gene cluster in soybean. *Plant Cell* 1: 229–239.

Mertens, R., and Weiler, E. W. (1983) Kinetic studies of the redistribution of endogenous growth regulators in gravireacting plant organs. *Planta* 158: 339–348.

Migliaccio, F., and Rayle, D. L. (1984) Sequence of key events in shoot gravitropism. *Plant Physiol.* 75: 78–81.

Morris, D. A., and Johnson, C. F. (1985) Characteristics and mechanism of long distance auxin transport in intact plants. *Acta Univ. Agric. Fac. Agron. (Brno)* 33: 377–383.

Mulkey, T. I., Kuzmanoff, K. M., and Evans, M. L. (1981) Correlations between proton-efflux and growth patterns during geotropism and phototropism in maize and sunflower. *Planta* 152: 239–241.

Napier, R. M., and Venis, M. A. (1990) Monoclonal antibodies detect an auxin-induced conformational change in the maize auxin-binding protein. *Planta* 182: 313–318.

Nonhebel, H. M., Cooney, T. P., and Simpson, R. (1993) The route, control and compartmentation of auxin synthesis. *Aust. J. Plant Physiol.* 20: 527–539.

Normanly, J., Cohen, J. D., and Fink, G. R. (1993) *Arabidopsis thaliana* auxotrophs reveal a tryptophan-independent biosynthetic pathway for indole-3-acetic acid. *Proc. Natl. Acad. Sci. U S A.* 90: 10355–10359.

Normanly, J. P., Slovin, J., and Cohen, J. (1995) Rethinking auxin biosynthesis and metabolism. *Plant Physiol.* 107: 323–329.

Oecking, C., Eckerskorn, C., and Weiler, E. W. (1994) The fusicoccin receptor of plants is a member of the 14-3-3 superfamily of eukaryotic regulatory proteins. *FEBS Lett.* 352: 163–166.

Oecking, C., Piotrowski, M., Hagemeier, J., and Hagemann, K. (1997) Topology and target interaction of the fusicoccin-binding 14-3-3 homologs of *Commelina communis*. *Plant J.* 12: 441–453.

Okada, K., Ueda, J., Komaki, M. K., Bell, C. J., and Shimura, Y. (1991) Requirement of the auxin polar transport system in early stages of *Arabidopsis* floral bud formation. *Plant Cell* 3: 677–684.

Palmer, J. M., Short, T. W., Gallagher, S., and Briggs, W. R. (1993) Blue light-induced phosphorylation of a plasma membrane-associated protein in *Zea mays* L. *Plant Physiol.* 102: 1211–1218.

Palmer, J. M., Warpeha, K. M. F., and Briggs, W. R. (1996) Evidence that zeaxanthin is not the photoreceptor for phototropism in maize coleoptiles. *Plant Physiol.* 110: 1323–1328.

Pearce, D. W., Taylor, J. S., Robertson, J. M., Harker, K. N., and Daly, E. J. (1995) Changes in abscisic acid and indole-3-acetic acid in axillary buds of *Elytrigia repens* released from apical dominance. *Physiol. Plantarum* 94: 110–116.

Peltier, J-B., and Rossignol, M. (1996) Auxin-induced differential sensitivity of the H^+-ATPase in plasma membrane subfractions from tobacco cells. *Biochem. Biophys. Res. Comm.* 219: 492–496.

Pickard, B. G., and Ding, J. P. (1993) The mechanosensory calcium-selective ion channel: Key component of a plasmalemmal control centre? *Aust. J. Plant Physiol.* 20: 439–459.

Pilate, G., Sotta, B., Maldiney, R., Jacques, M., Sossountzov, L., and Miginiac, E. (1989) Abscisic acid, IAA and cytokinin changes in buds of *Pseudotsuga menziesii* during bud quiescence release. *Physiol Plantarum* 76: 100–106.

Poff, K. L., Janoudi, A-K., Rosen, E. S., Orbovic, V., Konjevic, R., Fortin, M-C., and Scott, T. K. (1994) The physiology of tropisms. In *Arabidopsis*, Cold Spring Harbor Monograph Series, No. 27, E. M. Meyerowitz and C. R. Somerville, eds., Cold Spring Harbor Laboratory Press, Plainview, NY, pp. 639–664.

Porter, W. L., and Thimann, K. V. (1965) Molecular requirements for auxin action. Halogenated indoles and indoleacetic acid. *Phytochemistry* 4: 229–243.

Radwanski, E. R., Barczak, A. J., and Last, R. L. (1996) Characterization of tryptophan synthase alpha subunit mutants of *Arabidopsis thaliana*. *Mol. Gen. Genet.* 253: 353–361.

Ray, P. M., Dohrmann, U., and Hertel, R. (1977) Specificity of auxin-binding sites on maize coleoptile membranes as possible receptor sites for auxin action. *Plant Physiol.* 60: 585–591.

Rayle, D. L., and Cleland, R. E. (1970) Enhancement of wall loosening and elongation by acid solutions. *Plant Physiol.* 46: 250–253.

Reid, M. S. (1985) Ethylene and abscission. *HortScience* 20: 45–50.

Romano, C. P., Hein, M. B., and Klee, H. J. (1991) Inactivation of auxin in tobacco transformed with the indoleacetic acid-lysine synthetase gene of *Pseudomonas savastanoi*. *Genes Dev.* 5: 438–446.

Rubery, P. H. (1990) Phytotropins: Receptors and endogenous ligands. *Soc. Exp. Biol. Symp.* 44: 119–146.

Rubery, P. H., and Sheldrake, P. A. (1973) Effect of pH and surface charge on cell uptake of auxin. *Nature* 244: 285–288.

Ruegger, M., Dewey, E., Gray, W. M., Hobbie, L., Turner, J., and Estelle, M. (1998) The TIR1 protein of *Arabidopsis* functions in auxin response and is related to human SKP2 and yeast Grr1p. *Genes Dev.* 12: 198–207.

Ruegger, M., Dewey, E., Hobbie, L., Brown, D., Bernasconi, P., Turner, J., Muday, G., and Estelle, M. (1997) Reduced naphthylphthalamic acid binding in the *tir3* mutant of *Arabidopsis* is associated with a reduction of polar auxin transport and diverse morphological defects. *Plant Cell* 9: 745–757.

Sack, F. D. (1991) Plant gravity sensing. *Int. Rev. Cytol.* 127: 193–252.

Salomon, M., Zacherl, M., Luff, L., and Rüdiger, W. (1997a) Exposure of oat seedlings to blue light results in amplified phosphorylation of the putative photoreceptor for phototropism and in higher sensitivity of the plants to phototropic stimulation. *Plant Physiol.* 115: 493–500.

Salomon, M., Zacherl, M., and Rüdiger, W. (1997b). Asymmetric, blue light-dependent phosphorylation of a 116-kilodalton plasma membrane protein can be correlated with the first- and second-positive phototropic curvature of oat coleoptiles. *Plant Physiol.* 115: 485–491.

Sánchez-Bravo, J., Ortuño, A., Botia, J. M., Acosta, M., and Sabater, F. (1991) Lateral diffusion of polarly transported indoleacetic acid and its role in the growth of *Lupinus albus* L. hypocotyls. *Planta* 185: 391–396.

Scherer, G. F. E. (1996) Phospholipid signaling and lipid-derived second messengers in plants. *Plant Growth Reg.* 18: 125–133.

Scherer, G. F. E., and André, B. (1989) A rapid response to a plant hormone: Auxin stimulates phospholipase A_2 in vivo and in vitro. *Biochem. Biophys. Res. Commun.* 163: 111–117.

Schmidt, R. C., Müller, A., Hain, R., Bartling, D., and Weiler, E. W. (1996) Transgenic tobacco plants expressing *Arabidopsis thaliana* nitrilase II enzyme. *Plant J.* 9: 683–691.

Sievers, A., Buchen, B., and Hodick, D. (1996) Gravity sensing in tip-growing cells. *Trends Plant Sci.* 1: 273–279.

Shaw, S., and Wilkins, M. B. (1973) The source and lateral transport of growth inhibitors in geotropically stimulated roots of *Zea mays* and *Pisum sativum*. *Planta* 109: 11–26.

Sitbon, F., Edlund, A., Gardestrom, P., Olsson, O., and Sandberg, G. (1993) Compartmentation of indole-3-acetic acid metabolism in protoplasts isolated from leaves of wild-type and IAA-overproducing transgenic tobacco plants. *Planta* 191: 274–279.

Sitbon, F., Sundberg, B., Olsson, O., and Sandberg, G. (1991) Free and conjugated indoleacetic acid (IAA) contents in transgenic tobacco plants expressing the iaaM and iaaH IAA biosynthesis genes from *Agrobacterium tumefaciens*. *Plant Physiol.* 95: 480–485.

Sussman, M. R., and Goldsmith, M. H. M. (1981) The action of specific inhibitors of auxin transport on uptake of auxin and binding of N-1-napththylphthalamic acid to a membrane site in maize coleoptiles. *Planta* 152: 13–18.

Szerszen, J. B., Szczyglowski, K., and Bandurski, R. S. (1994) iaglu, a gene from *Zea mays* involved in conjugation of growth hormone indole-3-acetic acid. *Science* 265: 1699–1701.

Taiz, L. 1984. Plant cell expansion: Regulation of cell wall mechanical proterties. *Ann. Rev. Plant Physiol.* 35: 585–657.

Theologis, A., and Ray, P. M. (1982) Early auxin-regulated polyadenylated mRNA sequences in pea stem sections. *Proc. Natl. Acad. Sci. USA* 79: 418–421.

Thimann, K.V., and Mahadevan, S. (1964) Nitrilase. I. Occurrence, preparation and general properties of the enzyme. *Arch. Biochem. Biophys.* 105: 133–141.

Thompson, N. P., and Jacobs, W. P. (1966) Polarity of IAA effect on sieve-tube and xylem regeneration in *Coleus* and tomato stems. *Plant Physiol.* 41: 673–682.

Thomson, K-S., Hertel, R., Muller, S., and Tavares, J. E. (1973) 1-N-Napthylphthalamic acid and 2,3,5-triiodobenzoic acid: In vivo binding to particulate cell fractions and action on auxin transport in corn coleoptiles. *Planta* 109: 337–352.

Tuominen, H., Ostin, A., Sandberg, G., and Sundberg, B. (1994) A novel metabolic pathway for indole-3-acetic acid in apical shoots of *Populus tremula* (L.). *Plant Physiol.* 106: 1511–1520.

Ulmasov, T., Liu, Z-B., Hagen, G., and Guilfoyle, T. J. (1995) Composite structure of auxin response elements. *Plant Cell* 7: 1611–1623.

Van Volkenburgh, E., and Cleland, N. F. (1979) Separation of cell enlargement and division in bean leaves. *Planta* 146: 245–247.

Venis, M. A., and Napier, R. M. (1997) Auxin perception and signal transduction. In *Signal Transduction in Plants*, P. Aducci, ed., Birkhäuser, Basel, Switzerland, pp. 45–63.

Venis, M. A., Napier, R. M., Barbier-Brygoo, H., Maurel, C., Perrot-Rechenmann, C., and Guern, J. (1992) Antibodies to a peptide from the auxin-binding protein have auxin agonist activity. *Proc. Natl. Acad. Sci. USA* 89: 7208–7212.

Venis, M. A., Thomas, E. W. Barbier-Brygoo, H., Ephritikhine, G., and Guern, J. (1990) Impermeant auxin analogues have auxin activity. *Planta* 182: 232–235.

Versel, J-M., and Pilet, P-E. (1986) Distribution of growth and proton efflux in graviresponsive roots of maize (*Zea mays* L.). *Planta* 167: 26–29.

Volkmann, D., and Sievers, A. (1979) Graviperception in multicellular organs. In *Encyclopedia of Plant Physiology*, New Series, Vol. 7, W. Haupt and M. E. Feinleib, eds., Springer, Berlin, pp. 573–600.

Wayne, R., and Staves, M. P. (1996) A down to earth model of gravisensing or Newton's law of gravitation from the apple's perspective. *Physiol. Plantarum* 98: 917–921.

Wetmore, R. H., and Reir, J. P. (1963) Experimental induction of vascular tissues in callus of angiosperms. *Am. J. Bot.* 50: 418–430.

Wright, A. D., Sampson, M. B., Neuffer, M. G., Michalczuk, L. P., Slovin, J., and Cohen, J.(1991) Indole-3-acetic acid biosynthesis in the mutant maize orange pericarp, a tryptophan auxotroph. *Science* 254: 998–1000.

Yang, R. L, Evans, M. L., and Moore, R. (1990) Microsurgical removal of epidermal and cortical cells: Evidence that the gravitropic signal moves through the outer cell layers in primary roots of maize. *Planta* 180: 530–536.

Yang, T., Law, D. M., and Davies, P. J. (1993) Magnitude and kinetics of stem elongation induced by exogenous indole-3-acetic acid in intact light-grown pea seedlings. *Plant Physiol.* 102: 717–724.

Yang, T., Davies, P. J., and Reid, J. B. (1996) Genetic dissection of the relative roles of auxin and gibberellin in the regulation of stem elongation in intact light-grown peas. *Plant Physiol.* 110: 1029–1034.

Young, L. M., and Evans, M. L. (1990) Correlations between gravitropic curvature and auxin movement across gravistimulated roots of *Zea mays*. *Plant Physiol.* 92: 792–796.

Young, L. M., and Evans, M. L. (1994) Calcium-dependent asymmetric movement of ^{3}H-indole-3-acetic acid across gravistimulated isolated root caps of maize. *Plant Growth Reg.* 14: 235–242.

Young, L. M., and Evans, M. L. (1996) Patterns of auxin and abscisic acid movement in the tips of gravistimulated primary roots of maize. *Plant Growth Reg.* 20: 253–258.

Young, L. M., Evans, M. L., and Hertel, R. (1990) Correlations between gravitropic curvature and auxin movement across gravistimulated roots of *Zea mays*. *Plant Physiol.* 92: 792–796.

20 Gibberellins

FOR NEARLY 30 YEARS after the discovery of auxin in 1927 and more than 20 years after its structural elucidation as indole-3-acetic acid, Western plant scientists tried to ascribe the regulation of all developmental phenomena in plants to auxin. However, as we will see in this and subsequent chapters, plant growth and development are regulated by several different types of hormones acting individually and in concert. Only with the discovery of other hormones did the nature of this control become evident.

In the 1950s the second group of hormones, the gibberellins (GAs), was characterized. The gibberellins are a large group of related compounds (more than 110 are known) that, unlike the auxins, are defined by their chemical structure rather than by their biological activity. Gibberellins are most often associated with the promotion of stem growth, and the application of GA to intact plants can induce large increases in plant height. As we will see, however, GAs play important roles in a variety of physiological phenomena. But unlike auxin biosynthesis, GA biosynthesis is under strict developmental control, and numerous GA-deficient mutants have been isolated. Mendel's tall/dwarf alleles in peas are a famous example. Such mutants have been useful in elucidating the complex pathways of GA biosynthesis.

We begin this chapter by describing the discovery, chemical structure, and biosynthesis of the gibberellins, as well as the identification of the active form of the hormone. We then examine the role of GAs in regulating various physiological processes, including seed germination, the mobilization of endosperm storage reserves, shoot growth, flowering, floral development, and fruit set. In recent years, the application of molecular genetic approaches has led to considerable progress in our understanding of the mechanism of GA action at the molecular level. These advances will be

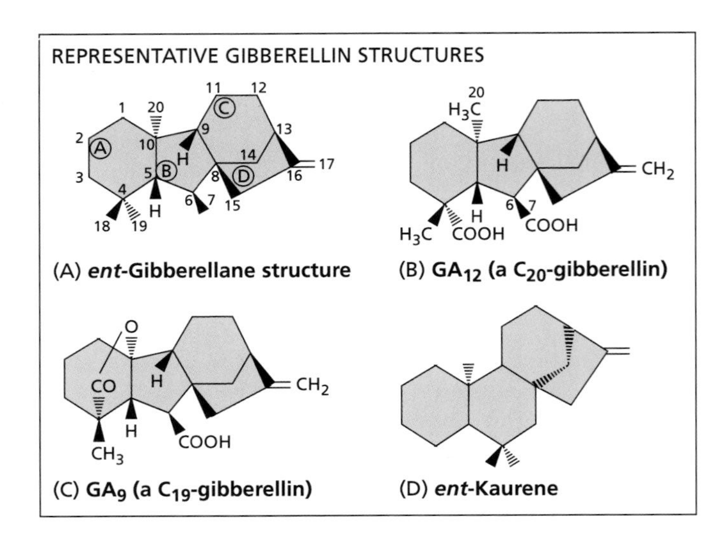

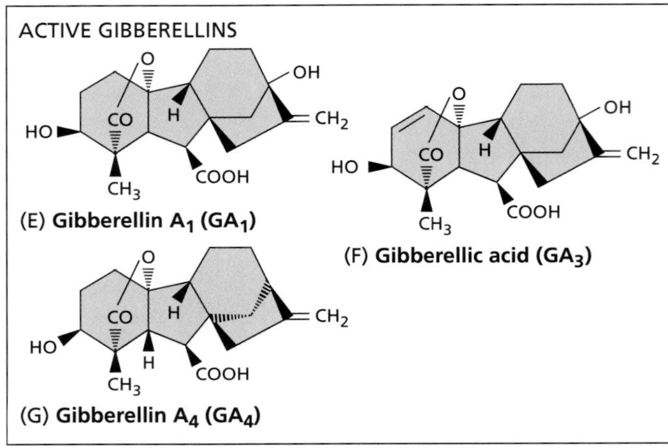

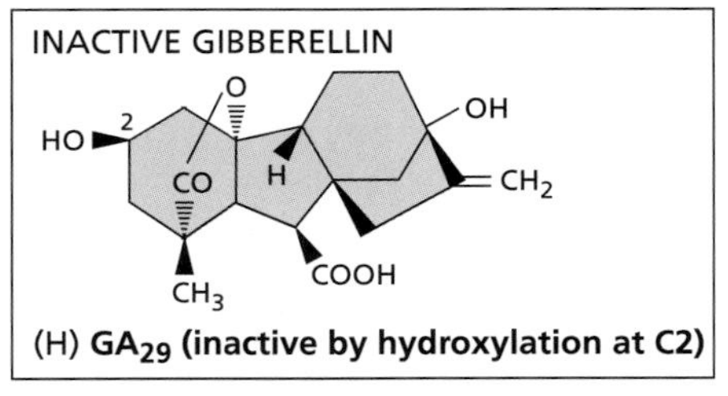

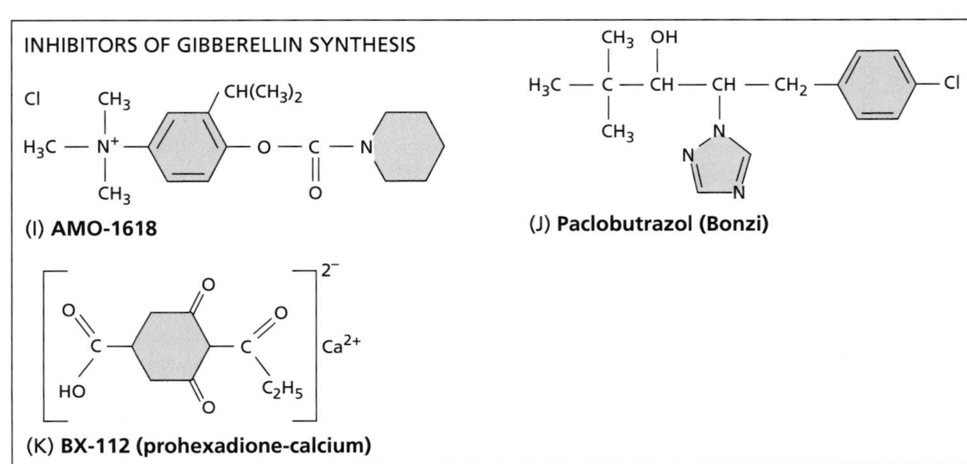

Figure 20.1 Structures of some important gibberellins, their precursors and derivatives, and inhibitors of gibberellin biosynthesis.

discussed at the end of the chapter, particularly in relation to three important experimental systems: internodal elongation in deepwater rice, dwarfism in *Arabidopsis thaliana*, and α-amylase synthesis and secretion by barley aleurone layers.

The Discovery of the Gibberellins

Though gibberellins became known to American and British scientists in the 1950s, they had been discovered much earlier by Japanese scientists. Rice farmers in Asia had long known of a disease that makes the rice plants grow tall but eliminates seed production. In Japan this disease was called the "foolish seedling," or *bakanae*, disease. Plant pathologists investigating the disease found that the tallness of these plants was induced by a chemical secreted by a fungus that had infected the tall plants. This chemical was isolated from filtrates of the cultured fungus and called gibberellin after *Gibberella fujikuroi*, the name of the fungus.

In the 1930s Japanese scientists succeeded in obtaining impure crystals of two fungal growth-active compounds, which they termed *gibberellin A* and *B*, but because of communication barriers and World War II, the information did not reach the West. Not until the mid-1950s did two groups—one at the Imperial Chemical Industries (ICI) research station at Welyn in Britain, the other at the U.S. Department of Agriculture (USDA) in Peoria, Illinois—succeed in elucidating the structure of material that they had purified from fungal culture filtrates, which they named *gibberellic acid* (Figure 20.1F). At about the same time, Nobutaka Takahashi and Saburo Tamura at Tokyo University isolated three gibberellins from the original gibberellin A and named them gibberellin A_1, gibberellin A_2, and gibberellin A_3. Gibberellin A_3 and gibberellic acid proved to be identical.

Figure 20.2 The effect of exogenous GA_1 on normal and dwarf (*d1*) corn. Gibberelin stimulates dramatic stem elongation in the dwarf mutant, but has little or no effect on the tall wild-type plant. (Photo courtesy of B. Phinney.)

It became evident that an entire family of gibberellins exists and that in each fungal culture different gibberellins predominate, though gibberellic acid is always a principal component. As we will see, the structural feature that all gibberellins have in common, and that defines them as a family of molecules, is that they are derived from the *ent*-kaurene ring structure (Figure 20.1D).

As gibberellic acid became available, physiologists began testing it on a wide variety of plants. Spectacular responses were obtained in the elongation growth of dwarf and rosette plants, particularly in genetically dwarf peas (*Pisum sativum*) and dwarf maize (*Zea mays*) (Figure 20.2) and in many rosette plants (Figure 20.3). At the same time, plants that were genetically very tall showed no further response to applied gibberellins. More recently, experiments with dwarf peas and dwarf corn have confirmed that the natural elongation growth of plants is regulated by gibberellins, as we will describe later.

Since applications of gibberellins could increase the height of dwarf plants, it was natural to ask whether plants contain their own gibberellins. Shortly after the discovery of the growth effects of gibberellic acid, Margaret Radley at ICI and Charles West, Bernard Phinney,* and coworkers at the University of California, Los Angeles showed that gibberellin-like substances could be isolated from several species of plants. "Gibberellin-like substance" refers to a compound or an extract that has GA-like biological activity, but whose chemical structure has not yet been defined. Such a response indicates, but does not prove, that the tested substance is a gibberellin.

* Phinney has written a wonderful personal account (Phinney 1983) of the history of the discovery of gibberellins.

Figure 20.3 Cabbage, a long-day plant, remains as a rosette in short days, but can be induced to bolt and flower by applications of gibberellin. In the case illustrated, giant flowering stalks were produced. (© S. Wittwer/Visuals Unlimited.)

In 1958 Jake MacMillan, also working at ICI, was the first to conclusively identify a gibberellin (gibberellin A_1; see Figure 20.1E) from a higher plant (runner bean seeds [*Phaseolus coccineus*]). Because the concentration of gibberellins in immature seeds far exceeds that in vegetative tissue, immature seeds were the tissue of choice for GA extraction. However, since the concentration of gibberellins in plants is very low (a few parts per billion in vegetative tissue and up to 1 part per million in seeds), MacMillan and coworkers had to use kilograms of seeds. A good proportion of the research station staff had to shell truckloads of beans to provide enough material employing the chemical purification and identification methods available at the time. With today's sophisticated methodology we can chemically identify gibberellins isolated from less than 1 g of seed tissue.

As more and more gibberellins from fungal and plant sources were characterized, they were numbered as gibberellin A_X (or GA_X) in the order of their discovery. This scheme was universally adopted for all gibberellins in 1968. However, the number of a gibberellin is simply a cataloging convenience, designed to prevent chaos in the naming of the gibberellins. The system implies no close chemical similarity or metabolic relationship between gibberellins with adjacent numbers.

All gibberellins are based on the *ent*-gibberellane skeleton (see Figure 20.1A). Some gibberellins have the full complement of 20 carbons (C_{20}-GAs) (see Figure 20.1B), while others have only 19 (C_{19}-GAs), having lost one carbon to metabolism. In almost all C_{19}-GAs, the carboxylic acid at carbon 19 bonds to carbon 10 to form a lactone bridge (Figure 20.1C). There are other variations in the basic structure, especially the oxidation state of carbon 20 (in C_{20}-GAs) and the number and position of hydroxyl groups on the molecule. The location of the hydroxyl groups and their stereochemistry (designated by ⦙⦙⦙ or ► for α or β bonds—behind or in front of the formula as viewed on a page, respectively) have a strong bearing on their biological activity. For example, hydroxylation in the β configuration at carbon 2 (see Figure 20.1H) always eliminates biological activity.

Despite the plethora of gibberellins present in plants, genetic analyses have demonstrated that only a few are biologically active as hormones. All the others serve as precursors or represent inactivated forms.

Biosynthesis, Metabolism, and Transport of Gibberellin

Gibberellins constitute a large family of diterpene acids and are synthesized by a branch of the **terpenoid pathway**, which was described in Chapter 13. The elucidation of the GA biosynthetic pathway would not have been possible without the development of sensitive methods of detection. As noted earlier, plants contain a bewildering array of gibberellins, many of which are *biologically inactive*. Bioassays have been important for detecting GA-like activity in partly purified extracts and for assessing the biological activity of known GAs, while physical methods of detection have been essential for identifying the active compounds as authentic gibberellins.

The use of enzyme extracts from seeds and other tissues with high rates of GA biosynthesis has enabled biochemists to identify the various intermediates and branch points in the pathway. More recently, many of the enzymes have been purified, their genes cloned, and their expression patterns in specific tissues determined. Finally, the availability in many species of single-gene mutants that have defects in specific steps of GA biosynthesis has been invaluable for determining the sequence of reactions leading to the biologically active GAs, and for studying the roles of GAs in plant development. In this section we will discuss the biosynthesis of GAs, as well as other factors that regulate the steady-state levels of the biologically active form of the hormone in different plant tissues.

Gibberellin-like Compounds Were First Measured by Bioassays

There are many bioassays for gibberellins (Reeve and Crozier 1975). All of them rely on plant responses thought to be normally regulated by endogenous gibberellins. Three bioassays that have been used widely are the *lettuce hypocotyl elongation bioassay*, the *dwarf rice leaf sheath bioassay* (Figure 20.4), and the *barley aleurone layer α-amylase bioassay* (which we will encounter later, in Figure 20.20). The lettuce hypocotyl and rice leaf sheath bioassays are based on growth responses, while the barley aleurone layer α-amylase bioassay is based on induction of enzyme production and secretion by GA. Bioassays are sensitive to the presence of inhibitory compounds, so some purification of the plant extract is essential. A chromatographic step is also needed to separate the different groups of gibberellins. The sensitivity of the bioassays to the different gibberellins varies, though all the tests are highly responsive to GA_1 and GA_3.

Bioassays cannot provide conclusive information on identity. If the gibberellin content is measured by bioassay without prior chromatographic separation, tall and dwarf plants are found to contain the same amount of "gibberellin-like activity." The reason is that the vast majority of bioassays show no difference between GA_{20} and GA_1 because the bioassay plant has the ability to convert inactive GA_{20} to the active GA_1. Thus, the use of bioassays has declined with the development of highly sensitive physical techniques that allow precise identification and quantification of specific GAs from small amounts of tissue.

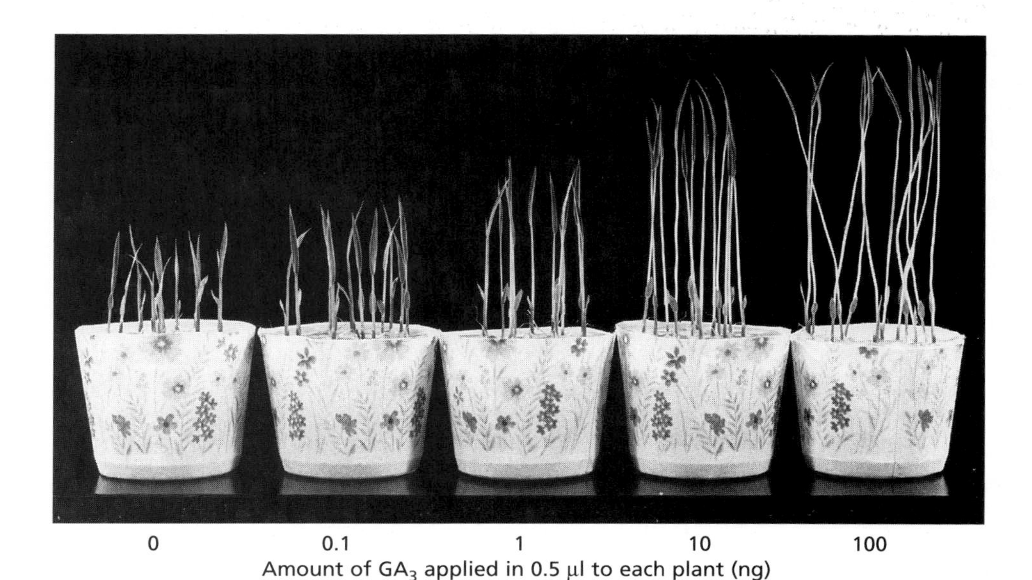

Figure 20.4 Gibberellin causes elongation of the leaf sheath of rice seedlings, and this response is used in the dwarf rice leaf sheath bioassay. Here 4-day-old seedlings were treated with different amounts of GA and allowed to grow for another 5 days. (Courtesy of P. Davies.)

Gibberellins Are Measured by Gas Chromatography Combined with Mass Spectrometry

As with all other plant hormones, the original means of detecting and assaying gibberellins was bioassay, because of its sensitivity and specificity. Measuring gibberellins physically or chemically was extremely difficult, since gibberellins do not absorb ultraviolet light, do not fluoresce, and have no distinguishing chemical characteristics that can form the basis of a specific chemical assay. Thus, except for the highly complex characterization of crystallized gibberellins obtained from vast quantities of plant material, almost all identifications in plants in the past consisted of varying degrees of purification followed by some form of chromatography, often paper, or thin-layer, chromatography (TLC).

Since about 1980, **high-performance liquid chromatography** (**HPLC**) followed by bioassay of the chromatographic fractions has been used. With the increasing availability of **gas chromatography** combined with **mass spectrometry** (GC-MS), or, more recently, **HPLC-MS**, these highly sensitive and selective analytical methods have now become the methods of choice. GC-MS is by far the most widely used method for gibberellin analysis. Data produced by a mass spectrometer are as specific as a fingerprint, and the intensity of the signal can be used to calculate the amount of the substance present. When a gas chromatograph, which can separate up to dozens of compounds in a mixture, is placed in front of the mass spectrometer, the purified compounds are fed one at a time to the mass spectrometer for analysis (Horgan 1995). The more recent development of HPLC-MS enables the direct analysis of compounds in the HPLC effluent, with no chemical modification required for gas chromatography. The instrument removes the solvent just before the mass spectrometer phase of the analysis so that only the dissolved compounds pass directly into the mass spectrometer. This procedure therefore involves far less work, but the instruments are expensive so the method is of limited availability.

Thanks to the pioneering work of J. MacMillan, N. Takahashi, and others, the chemistry of the GAs has been largely worked out. In parallel, studies with mutants blocked at specific steps in the GA biosynthetic pathway have helped elucidate the order in which the reactions occur. However, without the large database of the known GAs, the rapid progress made with the mutants would not have been possible.

Gibberellins Are Synthesized via the Terpenoid Pathway

As discussed in Chapter 13, a terpenoid is a compound made up of five-carbon (isoprene) building blocks:

$$-CH_2-\overset{\overset{\displaystyle CH}{|}}{C}=CH-CH_2-$$

joined head to tail. Gibberellins are tetracyclic diterpenoids made up of four isoprenoid units (Bramley 1997). The biological isoprene unit is isopentenyl pyrophosphate (IPP). Until recently, mevalonic acid was thought to be the immediate precursor of IPP for all terpenoid biosynthesis. However, as discussed in Chapter 13, there now appear to be two terpenoid pathways: a mevalonic acid–dependent pathway and a mevalonic acid–independent pathway (Schwender et al. 1996; Lichtenthaler et al. 1997). The former occurs in the cytosol and is involved primarily in sterol biosynthesis; the latter is localized in the chloroplast and leads to the synthesis of carotenoids and related compounds.

Since the early steps in GA biosynthesis occur in proplastids (see the next section), the IPP used in GA biosynthesis may not be derived from mevalonic acid. In the plastid terpenoid pathway, IPP is synthesized (via an unknown intermediate) from glyceraldehyde-3-phosphate and pyruvate rather than from mevalonate (Schwender et al. 1996; Lichtenthaler et al. 1997). Thus far, the source of the IPP used in GA biosynthesis has not been definitively established.

Regardless of the origin of the IPP, the next few steps are common to both the cytosolic and the plastid pathways: The isoprene units are added successively to produce geranyl pyrophosphate (C_{10}) (the precursor of geraniol, or geranium oil), farnesyl pyrophosphate (C_{15}), and geranylgeranyl pyrophosphate (C_{20}) (see Figure 13.6). At this stage the intermediates are all chains of C_5 units.

Gibberellin Biosynthesis Has Three Stages

In maize, researchers have demonstrated the entire GA biosynthetic pathway in vegetative tissues by feeding them various radioactive intermediates (Kobayashi et al. 1996). On the basis of these and numerous other studies, the GA biosynthetic pathway has been shown to be divided into three stages, each residing in a different cellular compartment (Figure 20.5) (Kende and Zeevaart 1997).

Stage 1: Cyclization reactions. The cyclization reactions that convert geranylgeranyl pyrophosphate (GGPP) to *ent*-kaurene (the prefix "*ent*" refers to the enantiomeric form of kaurene) represents the first step that is specific for the GAs. This conversion is a two-step cyclization process (Figure 20.5A). The two enzymes that catalyze the reactions are localized in the proplastids of meristematic shoot tissues, but they are not present in mature chloroplasts (Sun and Kamiya 1994; Aach et al. 1995; Hedden and Kamiya 1997). Compounds such as AMO-1618, Cycocel, and Phosphon D are specific inhibitors of the first stage of GA biosynthesis.

Stage 2: Oxidations to form GA_{12}-aldehyde. In the second stage of GA biosynthesis, a methyl group is oxidized to a carboxylic acid. Then the B ring contracts from a six- to a five-carbon ring to give GA_{12}-aldehyde, which is the first gibberellin formed in all plants and thus is the precursor of all the other gibberellins (see Figure 20.5B). All the enzymes involved are monooxygenases that utilize cytochrome P450 in their reactions. These P450 monooxygenases are localized on the endoplasmic reticulum, suggesting that the substrate is transported from the plastid to the ER, where it is converted to GA_{12} (Hedden and Kamiya 1997). Paclobutrazol and other inhibitors of P450 monooxygenases specifically inhibit this stage of GA biosynthesis.

Stage 3: Formation of all other gibberellins by GA_{12}-aldehyde. In the first step of the third stage of GA biosynthesis, GA_{12}-aldehyde is oxidized to GA_{12}, the first gibberellin in the pathway (see Figure 20.5C). This reaction can be catalyzed either by a monooxygenase localized on the ER or by a soluble dioxygenase located in the cytosol (Hedden and Kamiya 1997). However, all subsequent steps in the pathway are carried out by a group of soluble dioxygenases in the cytosol. These enzymes require 2-oxoglutarate and molecular oxygen as cosubstrates, and they use Fe^{2+} and ascorbate as cofactors.

The specific steps in the modification of GA_{12} vary from species to species, and between organs of the same species. Two basic chemical changes occur in most plants: (1) hydroxylation at carbon 13 or carbon 3, or both; (2) a successive oxidation at carbon 20 ($CH_2 \rightarrow CH_2OH \rightarrow CHO$), followed by loss of carbon 20 as CO_2. In the pathway that is common in higher plants, the **early-13 hydroxylation pathway**, GA_{12} is first converted to GA_{53} by hydroxylation of carbon 13 (see Figure 20.5C). GA_{53} is then converted to GA_{19} by successive oxidations at carbon 20, followed by the loss of carbon 20 as CO_2 to produce GA_{20}. All of the carbon 20 oxidation steps are carried out by a single, multifunctional enzyme, **GA 20-oxidase** (Xu et al. 1995). GA_{20} is then converted to the biologically active form, GA_1, by the enzyme **3β-hydroxylase**. Finally, 2β-hydroxylation reactions inactivate GA_1 by converting it to GA_8, or they remove GA_{20} from the pathway by converting it to GA_{29}.

Inhibitors of the third stage of the GA biosynthetic pathway interfere with enzymes that utilize 2-oxoglutarate as cosubstrates. Among these, the compound **prohexadione (BX-112)** (see Figure 20.1K), is especially useful because it specifically inhibits gibberellin 3β-hydroxylase, the enzyme that converts inactive GA_{12} to growth-active GA_1.

GA_1 May Be the Only Gibberellin Controlling Stem Growth

When applications of gibberellin to dwarf plants made them grow tall, it was presumed that gibberellins must be the natural regulators of plant stem growth, but proof was lacking. Initial attempts to demonstrate that tall plants have more, or more active, gibberellins than dwarf plants were unsuccessful. In the early 1980s, however, James Reid and coworkers at the University of Tasmania in Australia, in collaboration with Jake MacMillan's group at the University of Bristol in England, finally demonstrated that tall stems do contain more bioactive gibberellin than dwarf stems have and that the level of the endogenous gibberellin mediates the genetic control of tallness (Reid and Howell 1995).

This breakthrough can be attributed to several factors. First, tall and dwarf plants of known genetic makeup were compared. Genes regulating tallness were identified, and the plants used were genetically identical except for the gene that is primarily responsible for tallness. Sec-

(A) **Stage 1: Cyclization reactions**

CH_2OPP CH_2OPP

GGDP → **CPP** → ***ent*-Kaurene**

Location: Proplastids

Enzymes: Cyclases

Inhibitors: Quaternary ammonium and phosphonium compounds; AMO-1618, Cycocel, Phosphon D

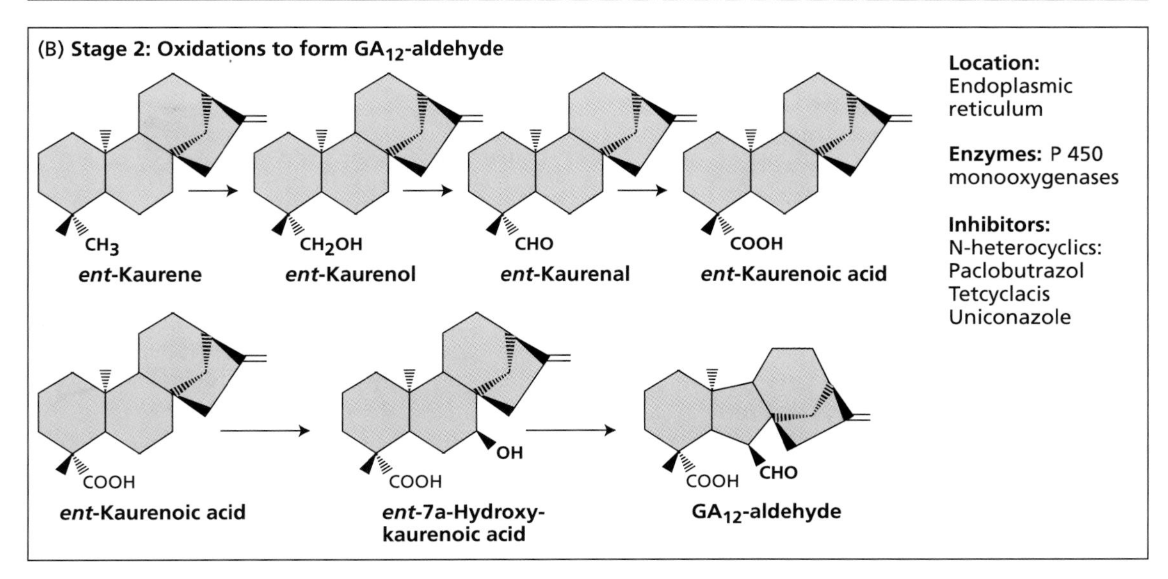

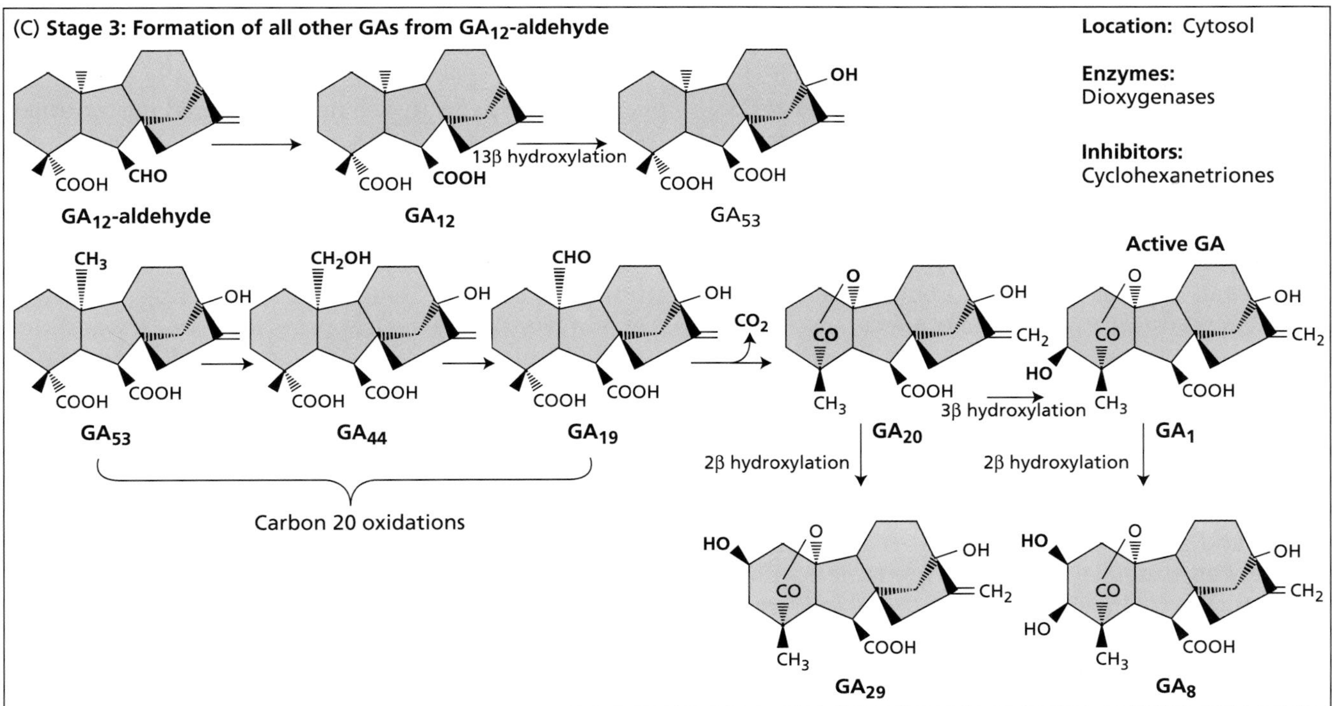

Figure 20.5 The three stages of the gibberellin biosynthetic pathway. (A) In stage 1, geranylgeranyl pyrophosphate (GGPP) is converted to *ent*-kaurene via *ent*-copalyl pyrophosphate (CPP). (B) In stage 2, *ent*-kaurene is converted to GA_{12}-aldehyde. (C) In stage 3, GA_{12}-aldehyde is converted to the first gibberellin, GA_{12}, and other GAs. Shown here is the early-13 hydroxylation pathway from GA_{12} to GA_{53}. This process is followed by a series of oxidations at carbon 20 leading to the production of GA_{20}. Finally, GA_{20} is oxidized to the active gibberellin, GA_1, by a 3β-hydroxylation reaction. Hydroxylation of carbon 2 converts GA_{20} and GA_1 to the inactive forms GA_{29} and GA_8, respectively. (After Kende and Zeevaart 1997.)

ond, gibberellin chemistry and instrumental analysis had advanced so that unequivocal identifications could be made, and custom-tailored gibberellin molecules, labeled with both stable and radioactive isotopes, could be used in sophisticated studies of metabolism. Third, only the expanding internodes where the bioactive GAs are concentrated were compared, rather than whole shoots, including nongrowing mature tissues.

The gibberellins of tall plants containing the *Le* allele* were compared with isogenic dwarf plants (plants with the same genetic makeup except for the genes mentioned) containing the *le* allele. These are the two alleles of the gene that regulates tallness in peas, which was first investigated by Gregor Mendel in the pioneering study in genetics in 1866. Gibberellin extracts of expanding internodes of these plants were separated into the different gibberellins by HPLC, and the resulting fractions were analyzed by GC-MS. The tall plants were found to contain much more GA_1 than the dwarf plants (Potts et al. 1982; Ingram et al. 1983).

As we have seen, the precursor of GA_1 in higher plants is GA_{20} (GA_1 is 3β-OH GA_{20}). If GA_{20} is applied to dwarf (*le*) pea plants, they fail to respond, but they do respond to applied GA_1, so the *Le/le* gene difference is masked. The implication was that the *Le* gene conferred on the plants the ability to convert GA_{20} to GA_1. Researchers proved this relationship by using GA_{20} labeled with both ^3H (for following the GA during purification) and ^{13}C (which can be separated from ^{12}C by GC-MS) (Ingram et al. 1984). The $[^{13}C,^3H]GA_{20}$ was metabolized to GA_1, GA_8 (2β-OH GA_1), and GA_{29} (2β-OH GA_{20}) in *Le* plants, but only GA_{29} and GA_{29}-catabolite were detected in *le* plants. Thus, it was demonstrated conclusively that the *Le* gene, which regulates tallness (or stem length) in peas, does so by causing the synthesis of an enzyme that 3β-hydroxylates GA_{20} to produce GA_1 (Lester et al. 1997). In the absence of this enzyme, little or no GA_1 is produced and the plants are dwarf.

Correlations between GA_1 and tallness. Although the shoots of GA-deficient *le* dwarf peas are much shorter than those of normal plants (3–5 cm in mature dwarf plants versus 15 cm in mature normal plants), the mutation is "leaky" (that is, the mutated gene is partly active) and some endogenous GA_1 remains to cause growth. Different alleles of *le* give rise to peas differing in their height, and the height of the plant has been correlated with the amount of endogenous GA_1 (Figure 20.6). There is also an extreme dwarf mutant of pea that lacks gibberellins altogether. These dwarfs have the allele *na* (the wild-type allele is *Na*), which completely blocks GA biosynthesis between *ent*-kaurene and GA_{12}-aldehyde (Ingram and Reid 1987). As a

* All alleles mentioned in this section are homozygous.

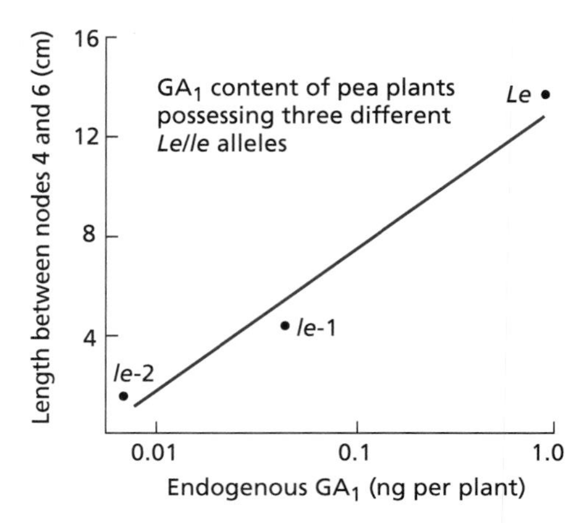

Figure 20.6 Stem elongation corresponds closely to the level of GA_1. Here the GA_1 content in peas with three different alleles at the *Le* locus is plotted against the internode elongation in plants with those alleles. The allele *le*-2 is a more intense dwarfing allele of *Le* than is the regular *le*-1 allele. There is a close correlation between the GA level and internode elongation. (After Ross et al. 1989.)

result, the *nana* mutants, which are almost completely free of gibberellins, achieve a stature of only about 1 cm at maturity (Figure 20.7).

From these observations it has been concluded that GA_1 is the controlling gibberellin in the regulation of tallness in peas (Ingram et al. 1986; Davies 1995). Similar conclusions have been reached for maize, also by the use of genotypes that have blocks in the gibberellin biosynthetic pathway (Phinney 1985), and the control of stem elongation by GA_1 appears to be universal.

Although GA_1 appears to be the primary active GA in stem growth for most species, the possibility still exists that a few other GAs have biological activity in other species or tissues. For example, GA_3, which differs from GA_1 only in having one double bond, is relatively rare in higher plants but appears to be able to substitute for GA_1 in most bioassays. GA_4 (see Figure 20.1G), which is present in both *Arabidopsis* and members of the squash family (Cucurbitaceae), has also been shown to be as active as GA_1 in some bioassays, or even more active, suggesting that GA_4 is an active GA, at least in some species (Kobayashi et al. 1993; Xu et al. 1997).

Gibberellin Intermediates Can Be Transported

Developing seeds and fruits show the highest gibberellin levels. However, there is no evidence that the active gibberellins present in seeds are used for the growth of the seedlings, since the gibberellin level normally decreases to zero in mature seeds (Figure 20.8). On the other hand, mature seeds do contain GA_{12}-aldehyde, the immediate gibberellin precursor, which is converted into growth-active gibberellins during the early stages of germination (Graebe 1986).

Figure 20.7 Phenotypes and genotypes of peas that differ in the gibberellin content of their vegetative tissue. Left to right: Nana (*na*), an ultradwarf containing no detectable GAs; Dwarf (*Na le*), containing GA_{20}; Tall (*Na Le*), containing GA_1; and Slender (*la crys*), an ultratall also containing no detectable GAs but with high auxin and possibly a "switched-on" GA receptor or some subsequent growth-regulating process. (All alleles are homozygous.) (After Davies 1995.)

Work with pea seedlings indicates that GA occurs primarily in the young, actively growing buds, leaves, and upper internodes. These tissues also appear to be the sites of gibberellin synthesis (Coolbaugh 1985; Sherriff et al. 1994). The GAs that are synthesized in the shoot can be transported to the rest of the plant via the phloem. Indeed, the initial steps of gibberellin biosynthesis may occur in one tissue and metabolism to active gibberellins in another. For example, as already noted, mature chloroplasts cannot carry out the stage 1 reactions of GA biosynthesis. Thus the mesophyll cells of mature leaves (which contain mature chloroplasts) are also incapable of carrying out the stage 1 reactions of GA biosynthesis, although they are capable of carrying out stage 3 reactions. This difference suggests that intermediates in the GA biosynthetic pathway can be transported from meristematic shoot tissues to green leaves, where they are converted to active GAs. GAs also have been identified in both root exudates and root extracts, suggesting that roots can synthesize GAs as well. How-

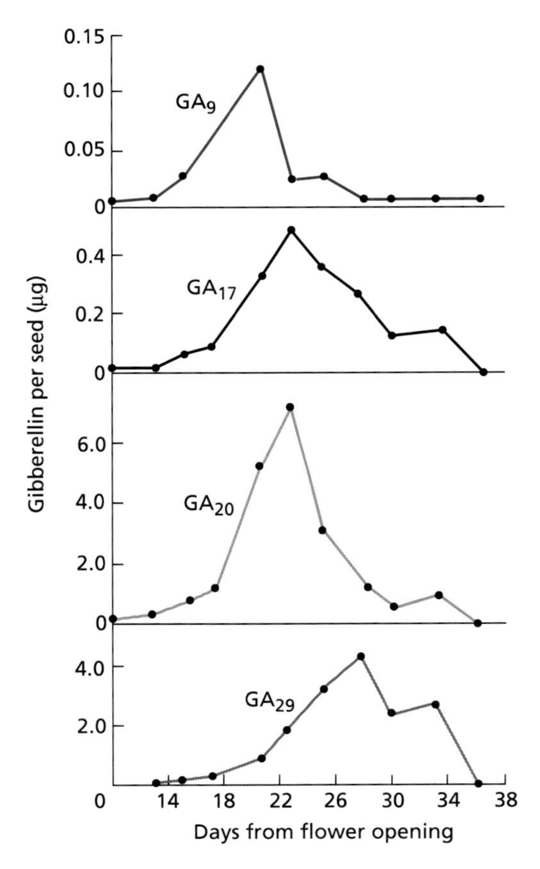

Figure 20.8 The levels of GA_9, GA_{17}, GA_{20}, and GA_{29} in seeds of pea throughout growth and maturation. As seeds near maturation, the amounts of all gibberellins decrease. (Note the differing scales on the ordinates.) (From Frydman et al. 1974.)

Figure 20.9 Spinach plants undergo stem elongation only in long days, remaining in a rosette form in short days. Treatment with the GA biosynthesis inhibitor AMO 1618 prevents stem elongation and maintains the rosette growth habit even under long days. Gibberellic acid can reverse the inhibitory effect of AMO 1618 on stem elongation. As shown in Figure 20.12, long days cause changes in the gibberellin contents of the plant. (Photo courtesy of J. A. D. Zeevaart.)

ever, conclusive evidence that roots synthesize GAs is still lacking.

GA Biosynthesis Is Highly Regulated

Gibberellins play an important role in mediating the effects of environmental stimuli on plant development. Environmental factors such as photoperiod and temperature can alter the levels of active GAs by affecting specific steps in the biosynthetic pathway. In addition, there is now evidence that GA can regulate its own synthesis (feedback regulation).

Photoperiod. When plants that require long days to flower (see Chapter 24) are transferred from short days to long days, alterations in GA metabolism are observed. In spinach (*Spinacia oleracea*), in short days, when the plants maintain a rosette form (Figure 20.9), the level of 13-hydroxylated gibberellins is relatively low. In response to increasing day length, the shoots of spinach plants begin to elongate after about 14 long days. The levels of all the gibberellins of the 13-hydroxylated gibberellin pathway ($GA_{53} \rightarrow GA_{44} \rightarrow GA_{19} \rightarrow GA_{20} \rightarrow GA_1 \rightarrow GA_8$) start to increase after about 4 days (Figure 20.10). Although the level of GA_{20} increases by 16-fold during the first 12 days, the five-fold increase in GA_1 is what induces stem growth.

Researchers have shown the dependence of stem growth on GA_1 by using different inhibitors of gibberellin synthesis and metabolism. The inhibitors AMO-1618 and BX-112 both prevent bolting (see Figure 20.1I and K). The effect of AMO-1618, which blocks gibberellin biosynthesis prior to GA_{12}-aldehyde, can be overcome by applications of GA_{20}, but the effect of BX-112, which blocks the production of GA_1 from GA_{20}, can be overcome only by GA_1 (Figure 20.11). This result demonstrates that the rise in GA_1 is the crucial factor in regulating spinach stem growth.

Northern-blot analysis used to detect the presence of GA 20-oxidase mRNA in spinach tissues showed that mRNA that codes for this enzyme was in the highest amount in shoot tips and elongating stems, and that the level was influenced by photoperiod, being higher under long-day conditions (Wu et al. 1996). The fact that GA 20-oxidase is the enzyme that converts GA_{53} to GA_{20} (see Figure 20.5C) explains why the concentration of GA_{20} was found to be higher in spinach under long-day conditions (Zeevaart and Gage 1993).

Temperature. Cold temperatures are required for the germination of certain seeds (**stratification**) and for

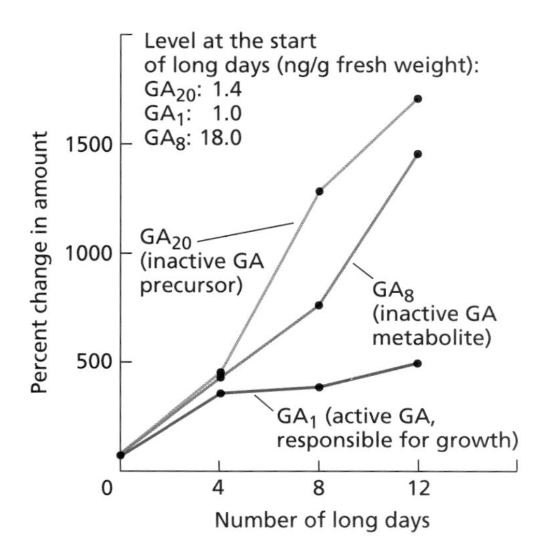

Figure 20.10 Changes in the GA levels in spinach after exposure to an increasing number of long days, but before stem elongation starts at about 14 days. GA_8 is a biologically inactive GA. Although GA_{20} increases 16-fold during this period, the fivefold increase in GA_1 is what causes growth. (After Davies 1995; redrawn from data in Zeevaart et al. 1993.)

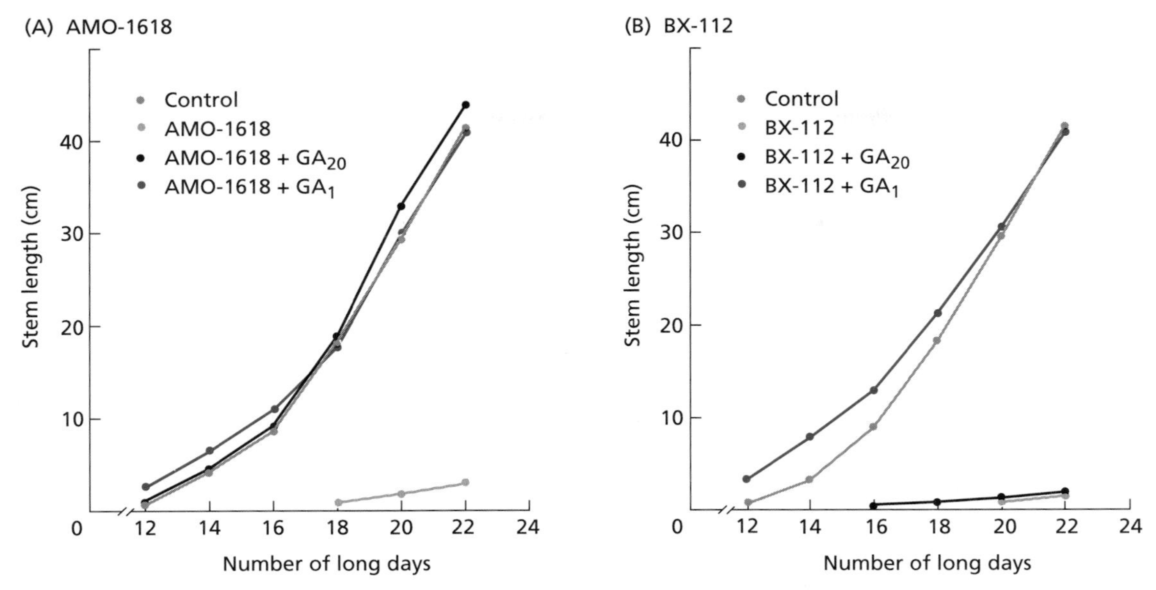

Figure 20.11 The effect of growth retardants (GA biosynthesis and metabolism inhibitors) on stem growth in spinach, induced by transfer to long-day conditions, and the reversal of the effects of the growth retardants by GA_{20} and GA_1. The control lacks inhibitors or added GA. (A) AMO-1618, which blocks GA biosynthesis at the cyclization step, does not inhibit growth in the presence of either GA_{20} or GA_1. (B) In contrast, BX-112, which blocks the conversion of GA_{20} to GA_1, inhibits growth even in the presence of GA_{20}. This result shows that GA_1 is required for growth. (After Zeevaart et al. 1993.)

flowering in certain species (**vernalization**) (see Chapter 24). The best-studied case is the vernalization of *Thlaspi arvense* (field pennycress). A prolonged cold treatment is required for both the stem elongation and the flowering of *T. arvense*, and GAs can substitute for the cold treatment. In the absence of the cold treatment, *ent*-kaurenoic acid accumulates to high levels in the shoot tip, which is also the site of perception of the cold stimulus. After cold treatment and a return to high temperatures, the *ent*-kaurenoic acid is converted to GA_9, the most active GA for stimulating the flowering response. These results are consistent with a cold-induced increase in the activity of *ent*-kaurenoic acid 7β-hydroxylase activity in the shoot tip (reviewed by Hedden and Kamiya [1997]).

Feedback control. Certain GA-insensitive mutants, such as *d8* in maize and *gai* in *Arabidopsis*, have been found to accumulate abnormally high levels of both GA_{20} and GA_1, while the level of GA_{19} is reduced. These findings suggest a possible inverse relationship between the GA response and GA synthesis. In other words, GA appears to repress its own synthesis, perhaps by producing a transcriptional repressor that limits the expression of GA biosynthesis enzymes (Scott 1990).

Because C_{19}-GAs specifically accumulate in the GA-insensitive mutants, GA 20-oxidase may be the primary target of feedback regulation. Researchers recently have confirmed this hypothesis by measuring the transcript levels of three genes for GA 20-oxidase in the *ga1* mutant (deficient in the enzyme, *ent*-copalyl pyrophosphate synthase that converts GGPP to CPP, see Figure 20.5)) of *Arabidopsis*. Because the mutant is blocked at an early stage of GA biosynthesis, feedback inhibition is minimized. Treatment of the mutant with GA_3 substantially reduced the levels of all three GA 20-oxidase transcripts (Phillips et al. 1995). Other enzymes in the pathway may be regulated by feedback inhibition as well.

Gibberellins May Be Conjugated to Sugars

A variety of **gibberellin glycosides** are formed by covalent linkage between GA and a monosaccharide. These GA conjugates are particularly prevalent in some seeds. The conjugating sugar is usually glucose, and it may be attached to the gibberellin via a carboxyl group, forming a gibberellin glycoside, or via a hydroxyl group, forming a gibberellin glycosyl ether. When gibberellins are applied to a plant, a certain proportion usually become glycosylated. Glycosylation may therefore represent another form of inactivation. In some cases, applied glucosides are metabolized back to free GAs, so glucosides may also be a storage form of gibberellins (Schneider and Schmidt 1990).

The various factors that regulate the steady-state level of active gibberellins are summarized in Figure 20.12.

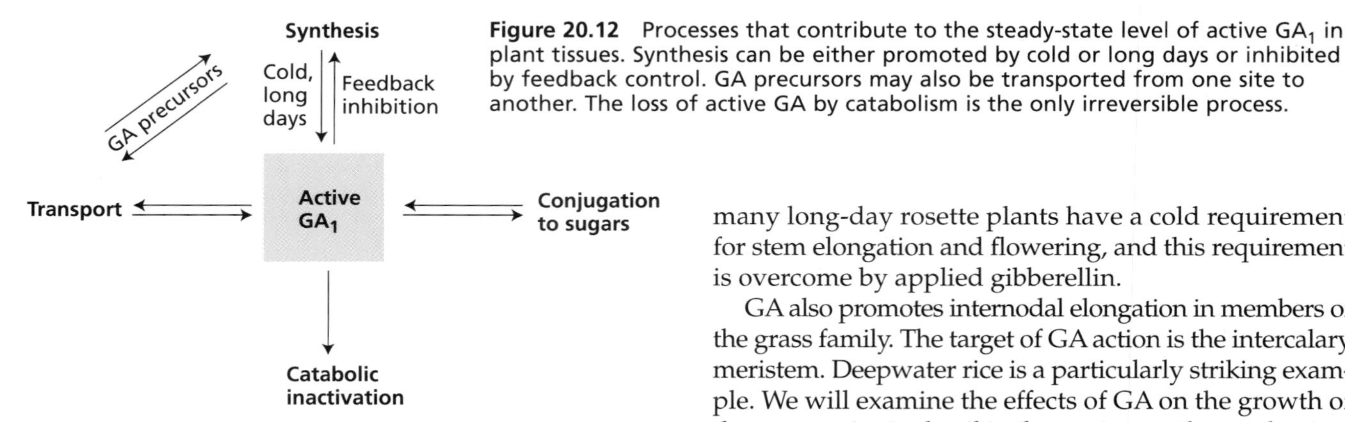

Figure 20.12 Processes that contribute to the steady-state level of active GA_1 in plant tissues. Synthesis can be either promoted by cold or long days or inhibited by feedback control. GA precursors may also be transported from one site to another. The loss of active GA by catabolism is the only irreversible process.

Synthesis of the active GA, GA_1, is promoted by environmental conditions, such as cold and long days. Conversely, GA may also inhibit its own synthesis via feedback. The level of active GA can be reduced by catabolism or by conjugation to sugars. In some cases, active GA can be generated by release from the conjugated form. Finally, the transport of GA (or GA precursors) to or from a tissue can also affect the steady-state level of active GA.

Effects of Gibberellin on Growth and Development

Originally discovered as the cause of a disease of rice that stimulated internode elongation, endogenous gibberellins influence a wide variety of developmental processes. In addition to stem elongation, gibberellins control various aspects of seed germinaton, including the loss of dormancy and the mobilization of endosperm. In reproductive development, GA can affect the transition from the juvenile to the mature stage, as well as floral initiation, sex determinaton, and fruit set. In this section we will review some of these GA-regulated phenomena.

GA Stimulates Stem Growth in Dwarf and Rosette Plants

Applied GA promotes internodal elongation in a wide range of species. However, the most dramatic stimulations are seen in dwarf and rosette species, as well as members of the grass family. Exogenous GA_3 causes such extreme stem elongation in dwarf plants that they resemble the tallest varieties of the same species (see Figure 20.2). Accompanying this effect are a decrease in stem thickness, a decrease in leaf size, and a light green color of the leaves.

Some plants assume a rosette form in short days and bolt and flower only in long days (see Chapter 24). Gibberellin application results in bolting of plants kept in short days, and natural bolting is regulated by gibberellin see Figure 20.3). In addition, as noted earlier, many long-day rosette plants have a cold requirement for stem elongation and flowering, and this requirement is overcome by applied gibberellin.

GA also promotes internodal elongation in members of the grass family. The target of GA action is the intercalary meristem. Deepwater rice is a particularly striking example. We will examine the effects of GA on the growth of deepwater rice in detail in the section on the mechanism of GA-induced stem elongation later in the chapter.

Although stem growth may be dramatically enhanced by GAs, gibberellins have little or no effect on root growth. Thus the signal transduction pathway required for GA-induced growth is probably not expressed in roots.

Gibberellins Regulate the Transition from Juvenile to Adult Phases

Many woody perennials do not flower until they reach a certain stage of maturity; up to that stage they are said to be juvenile (see Chapter 24). The juvenile and mature stages often have different leaf forms (see Figure 24.2). Applied gibberellins can regulate this juvenility in both directions, depending on the species. Thus, in English ivy (*Hedera helix*), GA_3 can cause a reversion from a mature to a juvenile state while many juvenile conifers can be induced to enter the reproductive phase by applications of nonpolar gibberellins such as GA_4 + GA_7 (Metzger 1995). The latter is one instance in which GA_3 is not effective.

Gibberellins Influence Floral Initiation and Sex Determination

As already noted, GA can substitute for the long-day or cold requirement for many plants, especially rosette species, that require either long days or low temperatures to flower (see Chapter 24). GA may thus be a component of the flowering stimulus in some plants, but apparently not in others.

Angiosperm flowers usually consist of four types of floral organs arranged in concentric whorls: sepals, petals, stamens, and pistils (see Chapter 1). Because the male (stamens) and female (pistils) structures are both present, most angiosperm flowers are hermaphroditic, or **perfect**. Certain species, however, produce unisexual, or **imperfect**, flowers. The process whereby unisexual flowers are produced is termed **sex determination**.

In **monoecious** plants, such as cucumber (*Cucumis sativus*) and maize (*Zea mays*), male and female flowers form on the same individual, whereas in **dioecious** plants, such as spinach (*Spinacia*) and hemp (*Cannabis* *sativa*), male and female flowers occur on different individuals. Anatomical studies of a wide range of species have shown that, with some exceptions, unisexual flowers form by the selective abortion of either the stamen or the pistil primordia at an early stage in development.

Sex determination is genetically regulated, but it is also influenced by environmental factors, such as photoperiod and nutritional status, and these environmental effects may be mediated by GA. In maize, for example, the staminate flowers (male) are restricted to the tassel, while the pistillate flowers (female) are contained in the ear. Exposure to short days and cool nights increases the endogenous GA levels in the tassels 100-fold and simultaneously causes feminization of the tassel flowers (Rood et al. 1980). Application of exogenous gibberellic acid to the tassels can also induce pistillate flowers (Hansen et al. 1976).

For studies on genetic regulation, a large collection of maize mutants that have altered patterns of sex determination has been isolated. Mutations in genes that affect either GA biosynthesis or GA signal transduction result in a failure to suppress stamen development in the flowers of the ear. Thus the primary role of GA in sex determination in maize seems to be to suppress stamen development (Irish 1996).

In dicots such as cucumber, hemp, and spinach, GA seems to have the opposite effect. In these species, application of GA promotes the formation of staminate flowers, and inhibitors of GA biosynthesis promote the formation of pistillate flowers. Treatment with ethylene at the bisexual stage, on the other hand, leads to the production of female flowers, suggesting that ethylene suppresses the growth of the male flowers. Thus, GA probably interacts with other hormones in the regulation of sex determination (Metzger 1995).

Gibberellins Promote Fruit Set

Applications of gibberellins can cause *fruit set* (the start of fruit growth following pollination) and growth of some fruit, in cases where auxin may have no effect. Stimulation of fruit set by gibberellin has been observed in apple (*Malus sylvestris*).

Gibberellins Promote Seed Germination

Seed germination may require GAs for one of several possible steps: the activation of vegetative growth of the embryo, the weakening of a growth-constraining endosperm layer surrounding the embryo, and the mobilization of stored food reserves of the endosperm. Some seeds, particularly those of wild plants, require light or cold to induce germination. In such seeds, this **dormancy** (see Chapter 23) can often be overcome by application of gibberellin. Since changes in gibberellin levels are often, but not always, seen in response to chilling of seeds, gibberellins may represent a natural regulator of one or more of the processes involved in germination.

Gibberellin application also stimulates the production of numerous hydrolases, notably α-amylase, by the aleurone layers of germinating cereal grains. This aspect of GA action has led to its use in the brewing industry in the production of malt (discussed in the next section). Since this is the principal system in which GA signal transduction pathways have been analyzed, it will be treated in detail later in the chapter.

Gibberellins Have Commercial Applications

The major uses of gibberellins (GA_3 unless noted otherwise) are in the management of fruit crops, the malting of barley, and the extension of sugarcane, with a resulting increase in sugar yield (Gianfagna 1995). In some crops, a reduction in height is desirable, and this can be accomplished by the use of gibberellin synthesis inhibitors (see Figure 20.1 I–K).

Fruit production. A major use of gibberellins is to increase the stalk length of seedless grapes. Because of the shortness of the individual fruit stalks, bunches of seedless grapes are too compact and the growth of the berries is restricted. Gibberellin stimulates the stalks to grow longer, thereby allowing the grapes to grow larger because they have more room to do so.

A mixture of benzyladenine (a cytokinin) and GA_4 + GA_7 can cause apple fruit to elongate and is used to improve the shape of Delicious-type apples under certain conditions. Although this treatment does not affect yield or taste, it is considered commercially desirable. In citrus fruits, gibberellins delay senescence; thus the fruits can be left on the tree longer to extend the market period.

Malting of barley. Malting is the first step in the brewing process. During malting, barley seeds (*Hordeum vulgare*) are allowed to germinate at temperatures that maximize the production of hydrolytic enzymes by the aleurone layer. Gibberellin is sometimes used to speed up the malting process. The germinated seeds are then heated, dried, and pulverized to produce "malt," consisting mainly of a mixture of amylolytic (starch-degrading) enzymes and partly digested starch. During the subsequent "mashing" step, water is added and the amylases in the malt convert the residual starch, as well as added starch, to the disaccharide maltose, which is converted to glucose by the enzyme maltase. The resulting "wort" is then boiled to stop the reaction. In the final step, yeast converts the glucose in the wort to ethanol by fermentation.

Increasing sugarcane yields. Sugarcane (*Saccharum officinarum*) is one of relatively few plants that store their carbohydrate as sugar (sucrose) instead of starch (the other important sugar-storing crop is sugar beet). Originally from New Guinea, sugarcane is a giant perennial grass that can grow from 4 to 6 m tall. The sucrose is stored in the central vacuoles of the internode parenchyma cells. Spraying the crop with gibberellin can increase the yield of raw cane by up to 20 tons per acre and the sugar yield by 2 tons per acre. This increase is a result of the stimulation of internode elongation during the winter season.

Uses in plant breeding. The long juvenility period in conifers can be detrimental to a breeding program by preventing the reproduction of desirable trees for many years. Spraying with $GA_4 + GA_7$ can considerably reduce the time to seed production by inducing cones to form on very young trees. In addition, the promotion of male flowers in cucurbits and the stimulation of bolting in biennial vegetables such as beet (*Beta vulgaris*) and cabbage (*Brassica oleracea*) are valuable effects of gibberellins that are occasionally used commercially in seed production.

Gibberellin synthesis inhibitors. Bigger is not always better; thus gibberellin biosynthesis inhibitors are used commercially to prevent elongation growth in some plants. In floral crops, short, stocky plants such as lilies, chrysanthemums, and poinsettias are desirable, and restrictions on elongation growth can be achieved by applications of gibberellin synthesis inhibitors such as ancymidol (known commercially as A-Rest) or paclobutrazol (known as Bonzil) (see Figure 20.1J). The same is true of cereal crops grown in cool, damp climates, such as in Europe, where lodging can be a problem. *Lodging,* the bending of stems to the ground in reponse to the weight of moisture collecting on the ripened heads, makes it difficult to harvest the grain with a combine harvester. Shorter internodes reduce the tendency of the plants to lodge, increasing the yield of the crop. Yet another application of GA biosynthesis inhibitors is the restriction of growth in roadside shrub plantings.

Mechanisms of Gibberellin Action: Promoting Stem Growth

Gibberellins are extremely active molecules. In stem elongation, responses to GA_3 can be seen at levels as low as 10^{-10} g (0.1 ng) for lettuce or rice seedlings. Sensitivities to even smaller amounts of GA have been recorded. The methyl ester of GA_{73} has been found in the fern *Lygodium japonicum*, where it induces the formation of antheridia in dark-grown protonemata at a concentration as low as 10^{-14} *M*, which is about 3×10^{-15} g mL^{-1} (3 femtograms mL^{-1}) or one-thirtieth of a part per trillion.

For gibberellin to be effective at such low concentrations, efficient mechanisms for amplifying the hormonal signal must be present in the responding cells. The two GA-regulated phenomena that have been studied in the greatest detail, and about which the most is known, are stem growth and the mobilization of reserve substances in the endosperm. Both of these GA responses involve a similar sequence of events: binding to a receptor, activation of one or more signal transduction pathways, transcription of primary response (early) genes, and transcription of secondary response (late) genes leading to the cellular response. Presumably, some of the steps in the pathway, particularly the early ones, are common to all GA responses. In this section, we focus on stem growth. In the next section, we focus on the mobilization of endosperm reserves.

As described previously, the two cases in which the growth-promoting effects of GA are most evident are dwarf and rosette plants. When dwarf plants are treated with GA they resemble the tallest varieties of the same species (see Figure 20.2). Other examples of GA action include the elongation of hypocotyls and of grass internodes.

A particularly striking example of internode elongation is found in deepwater rice (*Oryza sativa*). In general, rice plants are adapted to conditions of partial submergence. To enable the upper foliage of the plant to stay above water, the internodes elongate as the water level rises. Deepwater rice has the greatest potential for rapid internode elongation. Under field conditions, growth rates of up to 25 cm per day have been measured. Gibberellins have been shown to play a critical role in this extraordinary internodal growth of deepwater rice.

In this section we will discuss the mechanism of stem elongation at the cellular level in relation to the processes of cell division and cell elongation. Then we will examine GA response mutants and the current state of our knowledge of the signal transduction pathways involved in GA-stimulated growth.

GA Stimulates Cell Elongation and Cell Division

The effect of gibberellins applied to intact dwarf plants is so dramatic that it would seem to be a simple task to determine how they act. Unfortunately, this is not the case, because, as we have seen with auxin, so much about plant cell growth is not understood. However, we do know some characteristics of gibberellin-induced stem elongation.

Gibberellin increases both cell elongation and cell division, as evidenced by increases in cell length and cell number in response to applications of GA. For example, internodes of tall peas have more cells than those of dwarf peas, and the cells are longer. Mitosis increases markedly in the subapical region of the meristem of rosette long-day plants after treatment with gibberellin

(Figure 20.13). The dramatic stimulation of internode elongation in deepwater rice is due in part to increased cell division activity in the intercalary meristem. Moreover, only the cells of the intercalary meristem whose division is increased by GA exhibit GA-stimulated cell elongation (Sauter and Kende 1992).

Because GA-induced cell elongation appears to precede GA-induced cell division, we begin our discussion with the role of GA in regulating cell elongation.

Gibberellins Increase Cell Wall Extensibility

The orientation of microfibrils determines the directionality of cell expansion, but it does not influence either the rate or the magnitude of expansion. Normally, the microfibrils in the walls of cells of the stem have a transverse orientation, which favors elongation growth (see Chapter 15). In lettuce hypocotyls, GA was found to promote cell expansion even in the presence of the microtubule poison colchicine, which causes the deposition of randomly oriented cellulose microfibrils in the cell wall (Durnham and Jones 1982). In this case, GA stimulates radial cell expansion instead of cell elongation. This finding indicates that the mechanism of GA-induced cell elongation is independent of the orientation of the cellulose microfibrils.

How does GA increase the *rate* of cell elongation? And how similar is gibberellin-induced growth to auxin-induced growth? The answers to these questions are still forthcoming, although the possibilities have been narrowed somewhat by work during the past decade.

As discussed in Chapter 15, the elongation rate can be influenced by both cell wall extensibility and the osmotically driven rate of water uptake. Which of these parameters is affected by gibberellin? Since the GA stimulation of elongation of lettuce (*Lactuca sativa*) hypocotyls (Métraux 1987) and cucumber (*Cucumis sativus*) hypocotyls (Taylor and Cosgrove 1989) is not associated with an increase in turgor pressure, GA does not appear to act by increasing the rate of water uptake.

In contrast, GA has consistently been observed to cause an increase in both the mechanical extensibility of cell walls and the stress relaxation of the walls of living cells. An analysis of pea genotypes differing in gibberellin content or sensitivity showed that gibberellin decreases the wall yield threshold, which is the minimum force that will cause wall extension (Behringer et al. 1990). Thus, both gibberellin and auxin seem to exert their effects by modifying cell wall properties.

In the case of auxin, cell wall loosening appears to be mediated in part by cell wall acidification (see Chapter 19). However, this does not appear to be the mechanism of gibberellin action. In no case has a gibberellin-stimulated increase in proton extrusion been demonstrated. The typical lag time before gibberellin-stimulated growth begins is longer than for auxin; in deepwater rice it is 40 minutes (Figure 20.14), and in peas it is 2 to 3 hours (see Figure 19.20) (Yang et al. 1996). These longer lag times point to a growth-promoting mechanism different from that of auxin. Consistent with the existence of a separate GA-specific wall-loosening mechanism, the growth responses to applied gibberellin and auxin are additive.

Various suggestions have been made regarding the mechanism of GA-stimulated stem elongation, and all have some experimental support (Metraux 1987), but as yet none provide a clear-cut answer. Recently, a close correlation between GA-stimulated growth and the activity of the enzyme xyloglucan endotransglycosylase (XET) has been observed for many tissues (Potter and Fry 1994; Smith et al. 1996). As discussed in Chapter 15, XET is an enzyme that hydrolyzes xyloglucan internally and transfers one of the cut ends to the free end of an

(A)

Each dot represents a mitotic event

0 h

24 h

48 h

72 h

Distribution of cell division following application of GA

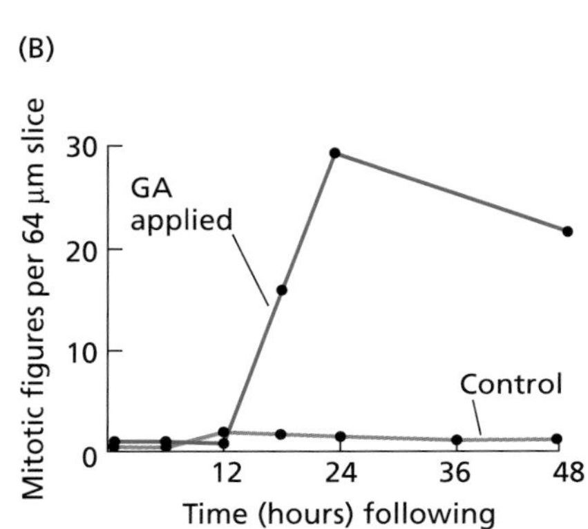

Figure 20.13 Gibberellin applications to rosette plants induce bolting in part by increasing cell division. (A) Longitudinal sections through the axis of *Samolus parviflorus* (brookweed) show an increase in cell division after application of GA. (Each dot represents one mitotic figure in a section 64 μm thick.) (B) The number of such mitotic figures with and without GA in stem apices of *Hyoscyamus niger* (black henbane). (After Sachs 1965.)

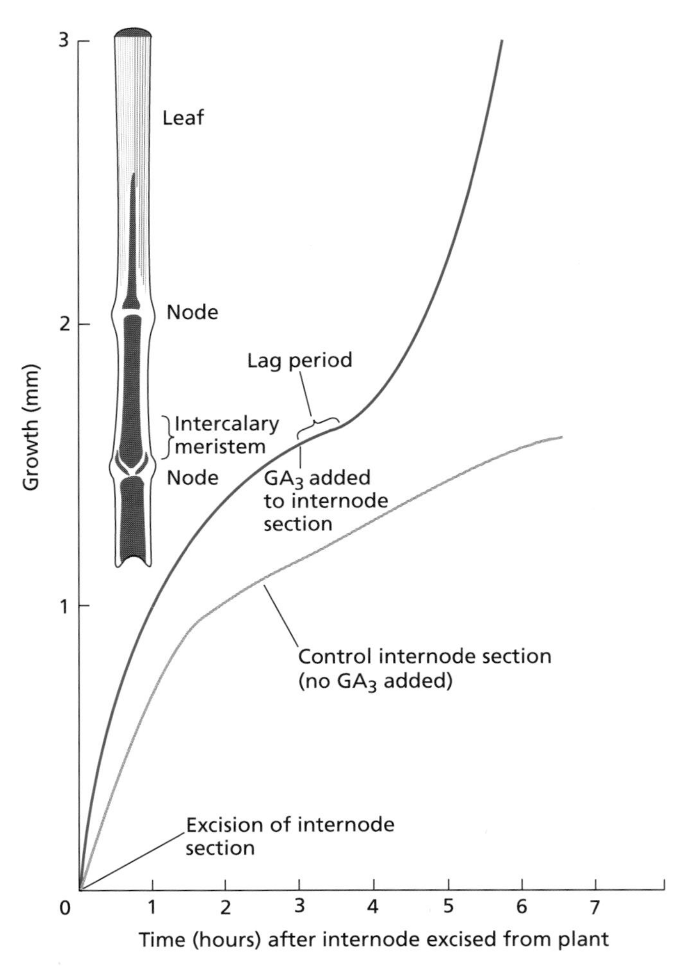

Figure 20.14 Continuous recording of the growth of the upper internode of deepwater rice in the presence or absence of exogenous GA_3. The control internode elongates at a constant rate after an initial growth burst during the first 2 hours after excision of the section. Addition of GA after 3 hours induced a sharp increase in the growth rate after a 40-minute lag period (upper curve). The difference in the initial growth rates of the two treatments is not significant here, but reflects slight variation in experimental materials. The inset shows the internode section of the rice stem used in the experiment. The intercalary meristem just above the node responds to GA. (After Sauter and Kende 1992.)

acceptor xyloglucan molecule. XET thus has the potential to cause molecular rearrangements in the cell wall matrix that could promote wall extension. Auxin-induced growth is not associated with an increase in XET activity. Thus the effect is specific for gibberellins.

Whether or not XET alone can increase wall extensibility or whether it acts in concert with other cell wall–loosening factors, remains to be determined. One possibility is that XET facilitates the penetration of expansins into the cell wall. (Recall that expansins are cell wall proteins that cause cell wall loosening at acid pH by weakening hydrogen bonds between cell wall polysaccharides.) According to this view, GA and auxins may work together to promote cell wall loosening: Auxin induces proton extrusion, while GA stimulates XET activity, which allows expansin proteins to penetrate into the wall, where they become activated by the acidic pH. Consistent with this idea, deepwater rice internodes that were frozen and then thawed have been shown to exhibit acid-stimulated extension (see Chapter 15). Moreover, rice expansins have been purified and shown to increase the mechanical extensibility of isolated cell walls from rice internodes (Cho and Kende 1997). Thus, both expansins and XET may be required for GA-stimulated growth.

GA Regulates the Cell Cycle in Intercalary Meristems

The mechanism of the stimulation of cell division by GA has been investigated most thoroughly in deepwater rice. As already noted, the internodes of deepwater rice dramatically increase their growth rate in response to submergence. The growth response occurs mainly in the intercalary meristem of the youngest internode.

The initial signal is the reduced partial pressure of O_2 resulting from submergence, which induces ethylene synthesis (see Chapter 22) (Raskin and Kende 1984). The ethylene trapped in the submerged tissues, in turn, reduces the level of abscisic acid (see Chapter 23), which acts as an antagonist of GA. The end result is that the tissue becomes more responsive to its endogenous GA (Raskin and Kende 1984; Hoffmann-Benning and Kende 1992). Inasmuch as inhibitors of GA biosynthesis block the stimulatory effect of both submergence and ethylene on growth, and exogenous GA can stimulate growth in the absence of submergence, GA appears to be the hormone directly responsible for growth stimulation.

GA-stimulated growth in deepwater rice can be studied with an excised stem system. Figure 20.14 shows the time course for GA-stimulated growth of such a segment, along with an illustration of the region of stem utilized for the measurement. Control sections exhibited an initial burst of growth that lasted 2 to 3 hours when first excised and placed in the growth-measuring apparatus. After this period, the growth rate decreased to a steady-state rate in the absence of GA. The addition of GA 3 hours into the experiment caused a marked increase in the growth rate after a lag period of about 40 minutes. Cell elongation accounts for about 90% of the length increase during the first 2 hours of GA treatment.

To study the effect of GA on the cell cycle, researchers isolated nuclei from the intercalary meristem, stained the DNA, and quantified the amount of DNA per nucleus by flow cytometry, a technique that allows the absorbance of each stained nucleus in the population to be measured in rapid succession. The amount of DNA in a haploid nucleus is defined as 1C. Postmitotic, pre-

DNA synthesis G_1 nuclei have a 2C amount of DNA, G_2 nuclei have 4C DNA, and S phase nuclei have intermediate DNA contents. At the start of GA treatment, about 83% of the nuclei were in G_1, 10% were in S, and 7% were in G_2 (Figure 20.15). During the first 4 hours of GA treatment, a decrease in the fraction of the nuclei in G_2 was observed. GA caused an initial drop in the fraction of nuclei in S phase and a simultaneous increase in the fraction of nuclei in G_1. Thereafter, the nuclei in G_1 decreased steadily, while the nuclei in the S and G_2 phases continued to increase.

On the basis of the kinetics of growth and the cell cycle, Sauter and Kende (1992) proposed that the first effect of GA is to induce cell elongation in the intercalary meristem. This process is followed by a round of cell division, primarily of cells that have already duplicated their DNA and are in the G_2 phase of the cell division cycle. By 7 hours, the rate of DNA synthesis has increased and the number of cells in G_2 begins to increase, suggesting a general stimulation of cell division. However, the decrease in the population of cells in the G_2 phase after 4 hours of GA treatment suggests that GA specifically regulates the cell cycle at the transition between G_2 and mitosis. In yeast, this is the point at which the cell cycle is regulated in response to cell size, suggesting that the promotion of cell division by GA could be an indirect effect of GA-induced cell elongation.

As we saw in Chapter 1, transitions between the different phases of the cell cycle are regulated by **cyclin-dependent protein kinases** (**CDKs**). Sauter and colleagues (1995) measured the transcript levels of two genes (*CDC2*) encoding cyclin-dependent protein kinases in deepwater rice in the presence or absence of GA. The expression of one of the *CDC2* genes was increased after 1 hour of GA treatment, as was the expression of two corresponding mitotic cyclin genes. These results supported the hypothesis that GA promotes cell division by increasing the level of a specific Cdc2 protein kinase along with the M cyclins required for the entry into mitosis. More recently Sauter (1997) showed that GA also causes an increase in the expression of a protein that has sequence homology to a **Cdc2-activating kinase** (**CAK**). This is the enzyme that activates Cdc2 by phosphorylating it. Thus, GA may stimulate cell division in the intercalary meristem of deepwater rice by several mechanisms.

GA Response Mutants May Be Blocked in Signal Transduction

Single-gene mutants impaired in their response to gibberellin provide potentially valuable tools for identifying genes that encode possible GA receptors or components of GA signal transduction pathways (Hooley 1994). In screenings for such mutants, two main classes of mutations affecting stature have been selected: (1)

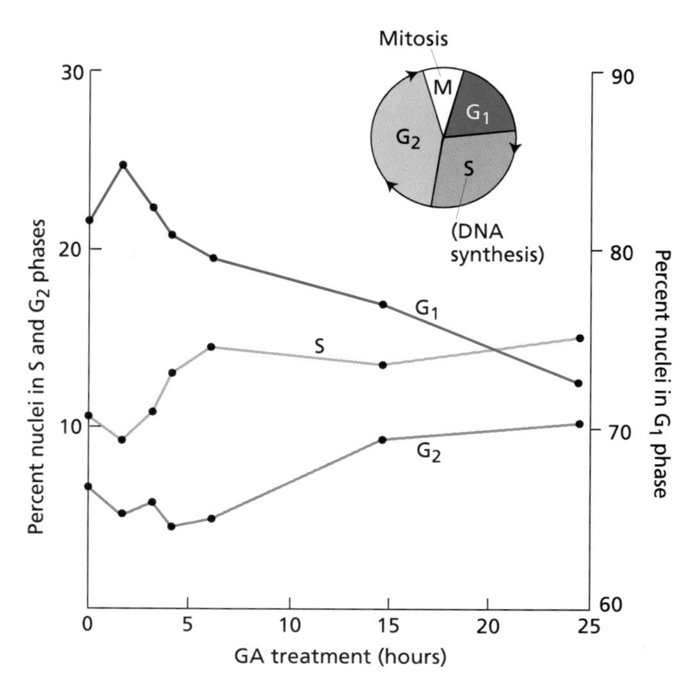

Figure 20.15 Changes in the cell cycle status of nuclei from the intercalary meristems of deepwater rice internodes treated with GA_3. Note that the scale for the G_1 nuclei is on the right-hand side of the graph. (After Sauter and Kende 1992.)

GA-insensitive dwarfs and (2) constitutive GA responders (slender mutants).

Dwarf GA-insensitive mutants. Several dwarf GA-insensitive mutants have been isolated from various species, including the maize *d8* and the *Arabidopsis gai* mutants (Swain and Olszewski 1996). Both *d8* and *gai* resemble GA-deficient mutants, except that they do not respond to exogenous GA. They are therefore considered to be candidates for either receptor or signal transduction pathway mutants. The *Arabidopsis* wild-type *GAI* gene has recently been cloned (Peng et al. 1997). The gene encodes a protein with nuclear localization signals and leucine zipper DNA-binding motifs that are characteristic of transcription factors (see Chapter 14).

The discovery that the *GAI* gene encodes a possible transcription factor brings us a step closer to understanding the GA signal transduction pathway. But how does the GAI protein function in the cell? A genetic analysis by N. P. Harberd and his colleagues at the John Innes Centre in Norwich, England, indicates that GAI acts as a repressor of GA responses, and that GA can prevent this repression (Peng et al. 1997). As illustrated in Figure 20.16, this model suggests that GA, or a signal transduction component of GA, binds to the GAI protein and inactivates it. The mutant *gai* gene was shown to

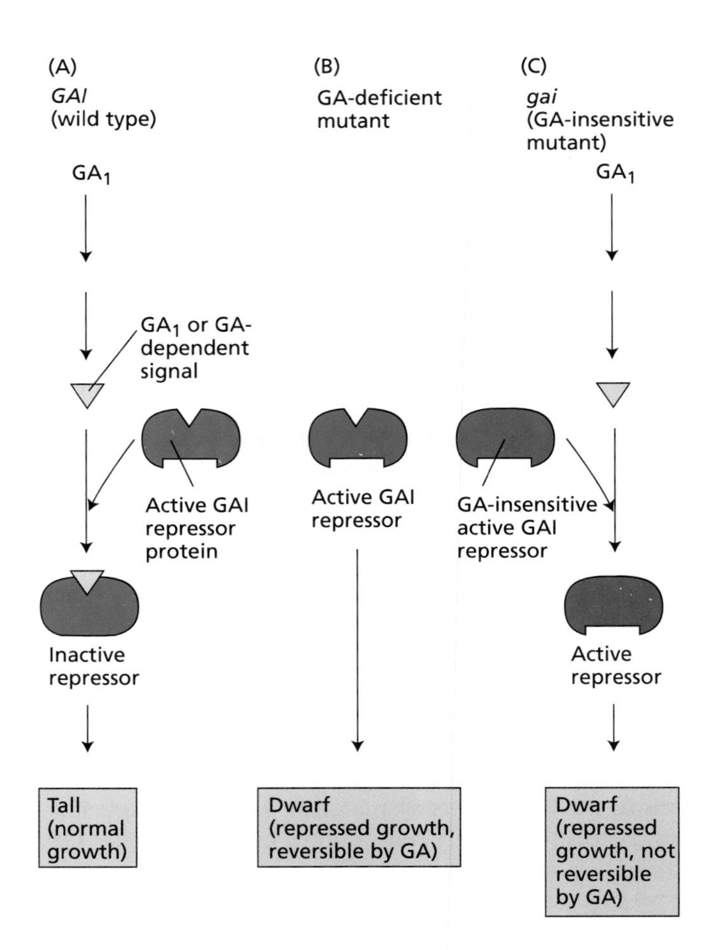

Figure 20.16 Model for the negative regulation of growth by the GAI repressor. (A) In the wild type, the GAI repressor binds to a GA-induced signaling intermediate (or to GA directly) and becomes inactivated. This inactivation of GAI allows normal growth to occur. (B) In a GA-deficient mutant, the lack of GA allows the GAI repressor to remain active, resulting in a dwarf phenotype. This type of dwarfism can be reversed by application of GA. (C) In *gai* mutants, the repressor is unable to bind to the GA-induced signaling intermediate. The result is a type of dwarfism that cannot be reversed by exogenous GA. (After Peng et al. 1997.)

have a deletion of 17 amino acids, which the authors suggest may correspond to the site of binding to GA or a signaling intermediate of GA (Peng et al. 1997). Because of this deletion, the action of the repressor cannot be alleviated by GA, and growth is irreversibly inhibited. This interpretation is consistent with the semidominant nature of the *gai* mutation, since in a heterozygote, the mutant repressor would compete with the wild-type repressor for its binding site on the promoter, but would not be able to respond to GA treatment.

Not all GA-insensitive dwarfs correspond to this pattern. For example, the dwarf phenotype of the *lkb* mutant of pea appears to be caused by deficiencies in both auxin and brassinosteroid, rather than by a GA deficiency or by an inability to respond to GA (Yang et al. 1996; Nomura et al. 1997) (see Box 17.4). These findings indicate that the GA response can also be limited by deficiencies in one or more of the other hormones.

Constitutive GA response mutants. "Slender mutants" resemble wild-type plants that have been treated with GA repeatedly. They exhibit elongated internodes, parthenocarpic fruit growth (in dicots), and poor pollen production. Slender mutants are rare compared to dwarf mutants. One possible cause of the slender phenotype is the overproduction of GA. For example, in the *sln* mutation of peas, a GA deactivation step involving 2β-hydroxylation is blocked in the seed. As a result, the mature seed, which in the wild type contains little or no GA, has abnormally high levels of GA_{20}. The GA_{20} from the seed is then taken up by the germinating seedling and converted to the bioactive GA_1, giving rise to the slender phenotype. However, once the seedling runs out of GA_{20} from the seed, its phenotype returns to normal (Reid et al. 1992, Ross et al. 1995).

If the slender phenotype is *not* due to an overproduction of endogenous GA, the mutant is considered to be a constitutive response mutant. The best characterized of such mutants are *la crys* in pea (representing mutations at two loci: *La* and *Crys*) (Figure 20.7), *pro* in tomato, *sln* in barley, and *spy* in *Arabidopsis*. All of these mutations are recessive and appear to represent loss-of-function mutations of a *negative regulator* or regulators of the GA response pathway. (A loss-of-function mutation of a negative regulator is like a double negative in English grammar: It translates into a positive.) Because the effects of the mutations are **pleiotropic**—that is, they affect processes other than just stem elongation—the steps in the pathway that are being affected may be common to all GA responses.

The *spy* (*spindly*) mutant of *Arabidopsis* was originally isolated by screening of the progeny of chemically mutagenized seeds (designated the *M2 generation*) in the presence of the GA biosynthesis inhibitor paclobutrazol (Jacobsen and Olszewski 1993).This method of screening selects for seedlings that can grow in the absence of GA biosynthesis.

Genetic analysis of the *SPY* gene indicates that the SPY protein acts later than GAI as a negative regulator of GA responses. This was shown by construction of a *spy/gai* double mutant. The double mutant had the same phenotype as the *spy* mutant; in other words, *spy* is completely **epistatic** to *gai*. Because of its pleiotropic effects, the *GAI–SPY* pathway appears to be common to all GA responses (Jacobsen et al. 1996).

The *SPY* gene was cloned and the deduced amino acid sequence suggests that the protein is an *N*-acetyl glucosamine transferase that may add sugar to other proteins involved in GA signaling. Such glycosylation reactions may inhibit the activities of the proteins, either directly or by blocking phosphorylation sites (Taylor 1998).

Mechanisms of Gibberellin Action: Mobilizing Endosperm Reserves

Genetic analyses of gibberellin-regulated growth, such as the studies described in the previous section, have identified some of the genes and their gene products, but not the specific biochemical pathways involved in GA signal transduction. The biochemical and molecular mechanisms, which are probably common to all GA responses, have been studied most extensively in relation to GA-stimulated synthesis and secretion of α-amylase in cereal aleurone layers. In this section we will describe how such studies have shed light on the location of the GA receptor, the transcriptional regulation of the genes for α-amylase and other proteins, and the possible signal transduction pathways involved in the control of α-amylase synthesis and secretion by GA. Since it is widely believed that GA acts through a common pathway or pathways in all of its effects on development, we conclude the chapter with a summary diagram based on information gained from studies of both growth and α-amylase production.

Early Experiments Revealed the Role of GA in Cereal Endosperm Degradation

Cereal grains (or more accurately, caryopses) can be divided into three parts: the diploid embryo, the triploid endosperm, and the fused testa–pericarp (seed coat–fruit wall). The embryo consists of the plant embryo proper, with its specialized absorptive organ, the scutellum. The endosperm is composed of two tissues, the centrally located starchy endosperm and the aleurone layer (Figure 20.17A). The **starchy endosperm**, typically nonliving at maturity, consists of thin-walled cells filled with starch grains. The **aleurone layer** surrounds the starchy endosperm and is cytologically and biochemically distinct from it. Aleurone cells are enclosed in thick primary cell walls and contain large numbers of protein-storing organelles called **protein bodies** (Figure 20.17B), enclosed by a single unit membrane, as well as lipid-storing **oleosomes**, which are surrounded by a half-unit membrane (see Chapter 1).

During germination and early seedling growth, the stored food reserves of the endosperm, chiefly starch and protein, are broken down by a variety of hydrolytic enzymes, and the solubilized sugars, amino acids, and other products are transported to the growing embryo. The two enzymes responsible for starch degradation are α- and β-amylase. α-Amylase hydrolyzes starch chains internally to produce oligosaccharides consisting of α-1,4-linked glucose residues. β-Amylase degrades these oligosaccharides from the ends to produce maltose, a disaccharide. Maltase then converts maltose to glucose.

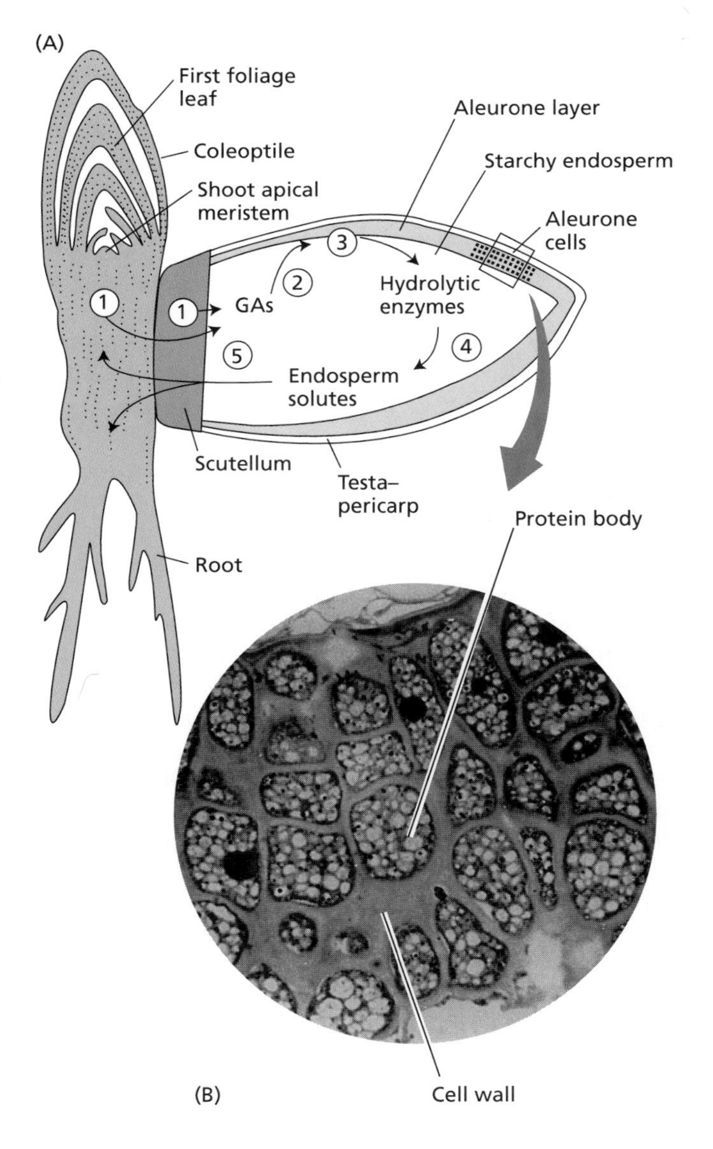

Figure 20.17 The structure of barley seeds and the functions of various tissues during germination. (A) Diagram of the major tissues. The seed is enclosed in a fused seed coat–fruit wall (testa–pericarp). (1) Gibberellins are synthesized by the coleoptile and scutellum of the embryo and released into the starchy endosperm; (2) gibberellins diffuse to the aleurone layer; (3) the aleurone layer is induced to synthesize and secrete α-amylase and other hydrolases into the starchy endosperm; (4) starch and other macromolecules are broken down to small substrate molecules; (5) the endosperm solutes are absorbed by the scutellum and transported to the growing embryo (indicated by arrows). (B) Light micrograph of aleurone cells. Note the presence of thick cell walls and the numerous protein bodies (aleurone grains) within each cell. (From Jones and MacMillan 1984, courtesy of R. Jones.)

α-Amylase is secreted into the starchy endosperm of cereal seeds by both the scutellum and the aleurone layer (see Figure 20.17A). The aleurone layer of graminaceous monocots (e.g., barley, wheat, rice, rye, and oats) is a highly specialized, terminally differentiated secretory tissue whose sole function appears to be the synthesis and release of hydrolytic enzymes. After completing this function, aleurone cells senesce and die.

Experiments carried out in the 1960s confirmed G. Haberlandt's original observation of 1890 that the secretion of starch-degrading enzymes by barley aleurone layers depends on the presence of the embryo. When the embryo was removed to produce a de-embryonated "half-seed," no starch was degraded, even after prolonged hydration of the half-seed on agar. However, when the half-seed was incubated in close proximity to the excised embryo, starch was digested. Thus, the embryo produced a diffusible substance that triggered α-amylase production by the aleurone layer.

It was soon discovered that gibberellic acid could substitute for the embryo in stimulating starch degradation. When half-seeds were incubated in buffered solutions in the presence of gibberellic acid, secretion of α-amylase into the medium was greatly stimulated (relative to the control half-seeds incubated in the absence of gibberellic acid) after an 8-hour lag period. The significance of the GA_3 effect became clear when it was shown that the embryo synthesizes and releases gibberellins (chiefly GA_1) into the endosperm during germination (Lenton et al. 1994). Thus, the cereal embryo efficiently regulates the mobilization of its own food reserves through the secretion of gibberellins, which stimulate the digestive function of the aleurone layer (see Figure 20.17A).

Gibberellin has been found to promote the production and/or secretion of a whole collection of hydrolytic enzymes that are involved in the solubilization of the reserves in the endosperm; principal among these is α-amylase. α-Amylase differs from some of the other GA-regulated hydrolases in that both its synthesis and its secretion are dependent on GA. In contrast, ribonuclease and β-1,3-glucanase are synthesized in aleurone cells in the absence of GA, and GA mainly causes their secretion. This difference in the effects of GA on α-amylase versus ribonuclease and β-1,3-glucanase in aleurone cells indicates that synthesis and secretion are regulated independently of each other. We will return to this point later in the chapter. Because of the predominance of α-amylase production, the majority of research on the stimulation of the production of hydrolases by gibberellin has involved this enzyme.

Since the 1960s, investigators have utilized isolated aleurone layers or even aleurone cell protoplasts rather than half-seeds (as, for example, in Figure 20.18A). The isolated aleurone layer, consisting of a homogeneous population of target cells, provides a unique opportunity to study the molecular aspects of hormone action in the absence of nonresponding cell types.

We begin our discussion of GA-induced α-amylase production with the possible location of the GA receptor. Next we examine the evidence that GA regulates the expression of the gene for α-amylase, including the analysis of promoter regions, and the identification of GA response elements and the transcription factors that bind them. Finally, we discuss what is known about the signal transduction pathways that ultimately lead to expression of α-amylase genes.

The Gibberellin Receptor May Be Located on the Plasma Membrane

Little is known about gibberellin receptors, although advances are beginning to be made. A surface localization of the GA receptor is suggested by studies with GA analogs that are unable to cross the plasma membrane. Such *membrane-impermeant* analogs of GA have been synthesized and shown to be active when added to aleurone protoplast preparations, indicating that entry into the cell is not required for activity (Beale et al. 1992). A more direct approach was taken by Gilroy and Jones (1994), who microinjected GA_3 into barley aleurone protoplasts. When the protoplasts were immersed in GA_3 they produced α-amylase, whereas when the GA_3 was microinjected into the protoplasts it had no effect, suggesting that gibberellin is perceived only on the outer face of the plasma membrane.

Recent evidence from Richard Hooley's laboratory at the University of Bristol suggests that heterotrimeric G proteins may be involved in the early GA signaling events in aleurone cells (Jones et al. 1998). Treatment of wild oat aleurone protoplasts with the peptide Mas7, which stimulates GDP/GTP exchange by heterotrimeric G proteins, induced the synthesis and secretion of α-amylase gene expression. In addition, treatment of aleurone protoplasts with GDP-β-S, a nonhydrolyzable analog of GTP which inhibits GDP/GTP exchange by heterotrimeric proteins, blocked α-amylase induction by GA_1. These findings, together with the evidence presented in the preceding paragraph, lend support to a model for GA action based on the hormone's binding to a plasma membrane receptor followed by interaction of the receptor with a heterotrimeric G protein. However, further work is needed to confirm this hypothesis.

Gibberellic Acid Enhances the Transcription of α-Amylase mRNA

Before molecular biological approaches were developed, there were already indications that gibberellic acid might enhance α-amylase production at the level of transcription (reviewed in Huttly and Phillips 1995 and Jacobsen et al. 1995). First, it was demonstrated by both radioactive and heavy-isotope labeling studies (using ^{14}C-labeled amino acids and $H_2{}^{18}O$) that the stimulation of α-amylase activity by gibberellic acid is due to de novo synthesis of the enzyme (that is, synthesis from amino acids) rather than to activation of preexisting enzyme. Second, GA_3-stimulated α-amylase production is blocked by inhibitors of transcription and translation.

There is now definitive evidence that gibberellin acts primarily by inducing the expression of the gene for α-amylase. (The situation is actually more complicated than this because α-amylase belongs to a multigene family, but for the sake of clarity we will consider these genes as a single group.) First, it was shown that GA_3 enhances the level of translatable mRNA for α-amylase in aleurone layers. Total mRNA was extracted from GA_3-treated and control tissue and translated in vitro in the presence of the radioactively labeled amino acid [^{35}S]methionine. Upon precipitation of the reaction products with α-amylase antibody, the radioactivity in α-amylase was quantified. On the basis of the amount of radioactivity in the immunoprecipitated protein, more α-amylase was synthesized from mRNA extracted from GA-treated tissue than from mRNA of the controls, and the increase in the presumptive α-amylase mRNA preceded the appearance of α-amylase in the medium (Figure 20.18) (Higgins et al. 1976).

The method just described measures only translatable α-amylase mRNA. To obtain a direct measurement of the total amount of α-amylase mRNA, researchers made α-amylase cDNA clones from isolated α-amylase mRNA, which is the principal mRNA made by aleurone cells and thus is relatively easy to obtain, and they used this mRNA to make [^{32}P]cDNA probes. By hybridizing the radioactively labeled cDNA probes to Northern (RNA) blots, researchers showed that the level of α-amylase mRNA is strongly enhanced by gibberellic acid (Chandler et al. 1984).

Two mechanisms can increase the level of α-amylase mRNA: stimulation of transcription or decrease in mRNA turnover. To discriminate between these alternatives, investigators performed a *nuclear runoff* experiment (see Figure 17.17) (Jacobsen and Beach 1985). Isolated nuclei, although incapable of initiating transcription, can complete the transcripts already being synthesized at the time of their isolation if provided the appropriate conditions and substrates. Transcriptionally active nuclei were isolated from aleurone cell protoplasts that had been incubated in the presence or absence of gibberellic acid. The isolated nuclei were then allowed to "run off" their transcripts in the presence of ^{32}P-labeled uridine triphosphate (UTP). This runoff step resulted in the specific labeling of the mRNA sequences being transcribed at the time the nuclei were isolated. When the incorporation of label into α-amylase mRNA was quantified, the results showed that nuclei isolated from GA-treated cells synthesized significantly higher amounts of α-amylase mRNA than did control nuclei, demonstrating that gibberellin promotes the transcription of α-amylase mRNA (Jacobsen and Beach 1985).

The availability of α-amylase mRNA enabled the isolation of genomic clones containing both the structural gene for α-amylase and its upstream promoter sequences. These promoter sequences can be fused to the reporter gene that encodes the enzyme β-glucuronidase (GUS). GUS expression causes the production of a blue color with the artificial substrate called X-gluc, enabling the regulation of a particular gene, in this case α-amylase, to be studied solely by the production of an easily assayed color. The regulation of transcription by gibberellin was proved when such chimeric genes containing α-amylase promoters that were fused to reporter genes were introduced into aleurone protoplasts and the expression of the reporter genes was shown to be regulated by GA_3 (Jacobsen and Close 1991; Jacobsen et al. 1995).

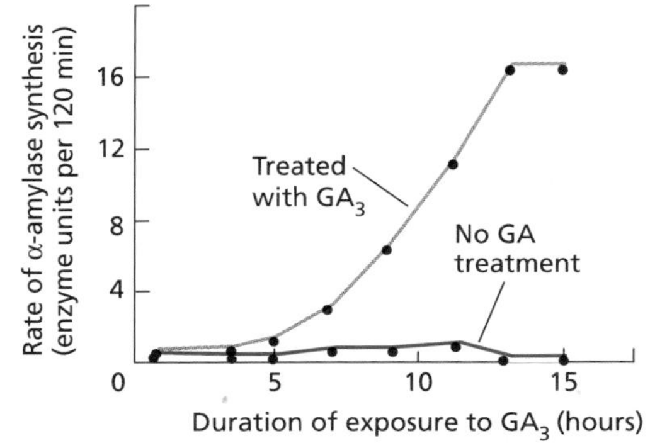

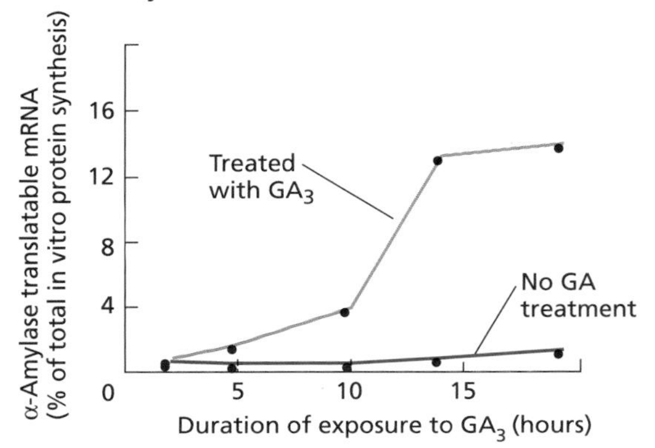

Figure 20.18 Gibberellin effects on enzyme synthesis and mRNA synthesis. (A) Synthesis of α-amylase by isolated barley aleurone layers is evident after 6–8 hours of treatment with GA_3 (10^{-6} *M*). (B) Messenger RNA extracted from aleurone cells and translated in vitro showed an increasing presence of α-amylase mRNA, and the appearance of the mRNA preceded the release of the α-amylase from the aleurone cells by 12 hours. The α-amylase mRNA in this case was measured by the in vitro production of α-amylase as a percentage of the protein produced by the translation of the bulk mRNA. (From Higgins et al. 1976.)

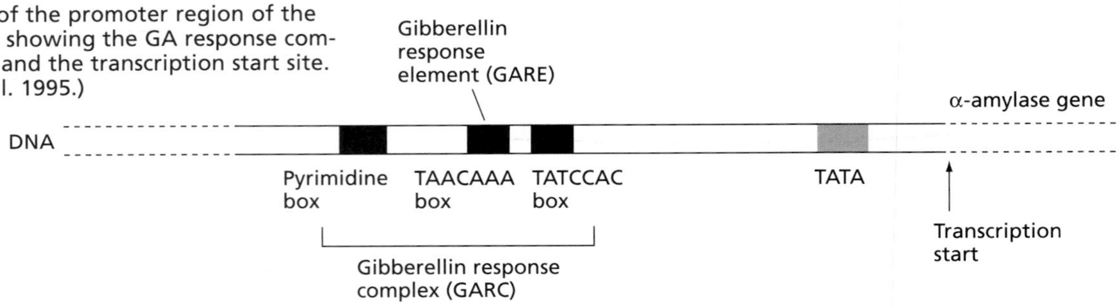

Figure 20.19 Map of the promoter region of the gene for α-amylase, showing the GA response complex, the TATA box, and the transcription start site. (After Jacobsen et al. 1995.)

Gibberellin response element (GARE)
α-amylase gene
DNA
Pyrimidine box
TAACAAA box
TATCCAC box
TATA
Transcription start
Gibberellin response complex (GARC)

Several Promoter Elements Confer GA Responsiveness

The partial deletion of known sequences of bases from α-amylase promoters of several cereals indicates that the sequences conferring GA responsiveness to the α-amylase gene are within 200 to 300 base pairs upstream of the transcription start site (Figure 20.19). The placement, order, and orientation of several sequences appear to be highly conserved in the different cereal species. One particular sequence (TAACAAA) can act alone to induce responsiveness to GA and has been called the **gibberellin response element** (**GARE**) (Skriver et al. 1991; Gubler and Jacobsen 1992). In addition, mutagenesis of another specific sequence, TATCCAC, results in a loss in GA_3-induced expression, but less so than for TAACAAA.

These results indicate that both the TAACAAA and the TATCCAC "boxes" act cooperatively, and a third sequence (C/TCTTTTC/T), referred to as the *pyrimidine box*, may also be required for full gibberellin responsiveness (Jacobsen et al. 1995). Together these three sequences have been referred to as the **gibberellin response complex** (**GARC**). Transcription factors associated with the GARC are assumed to interact with the general transcription factors at the TATA box, although the precise mechanism is unknown.

Transcription Factors Regulate α-Amylase Gene Expression

The promotion of α-amylase gene expression by gibberellin is mediated by specific transcription factors. To demonstrate such DNA-binding proteins in rice, a technique called a *mobility shift assay* was used. First the ends of the upstream regulatory region of the rice α-amylase gene were labeled with [^{32}P]ATP. The labeled DNA was then run on an electrophoretic gel in the presence of proteins extracted from aleurone tissues that had or had not been treated with GA_3. This experiment is based on the fact that if a protein binds to the DNA, the DNA–protein complex is larger than DNA alone and will migrate more slowly on an electrophoretic gel. Such a retardation of α-amylase upstream DNA was found only in the presence of proteins isolated from GA_3-treated aleurone cells (Figure 20.20) (Ou-Lee et al. 1988).

To identify the DNA sequences involved in the protein binding, the DNA containing the upstream

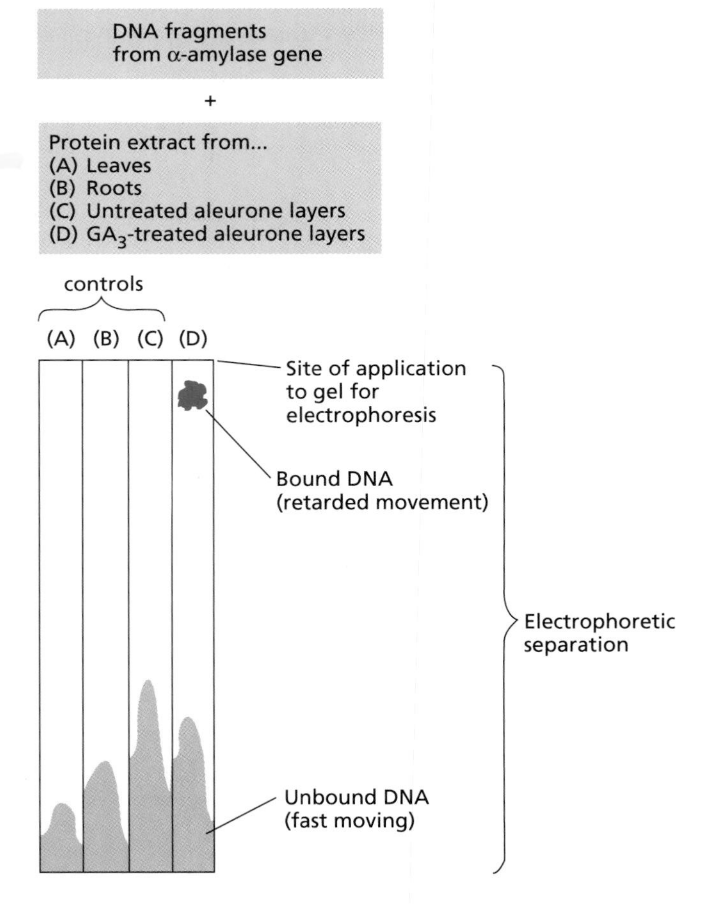

Figure 20.20 Diagram of a mobility shift DNA-binding assay for the detection of gibberellin-induced DNA-binding protein. Labeled DNA fragments from the promoter region of the rice gene for α-amylase were incubated in the presence of protein extracts from GA-treated and control tissues. The DNA–protein mixture was then placed at the top of a polyacrylamide gel and separated by electrophoresis. The fast-moving free DNA is seen as smears at the bottom of each lane; the DNA–protein complex forms a spot near the top of lane D because its mobility has been retarded. (From Ou-Lee et al. 1988.)

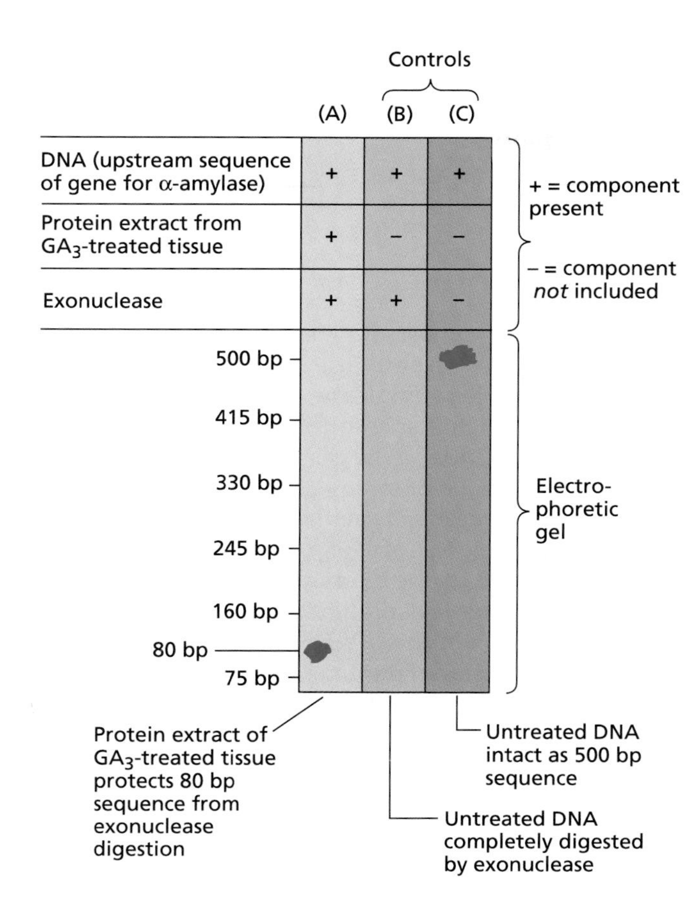

Figure 20.21 Identification of the promoter sequence that binds to the GA-induced DNA-binding protein. To be detected, a DNA fragment from the promoter region of the α-amylase gene, about 500 base pairs (bp) long, was labeled at the 5′ end with ^{32}P. It was then incubated in the presence of an exonuclease that degrades DNA from the 3′ end toward the 5′ end, and separated by gel electrophoresis. In the absence of the exonuclease (lane C), the undegraded 500 bp DNA fragment can be seen at the top of the electrophoretic gel. If exonuclease is added (lane B), the DNA is completely degraded and migrates to the bottom. When the protein extract from GA_3-treated aleurone tissue is added (lane A), the exonuclease digestion is unable to attack an 80 bp fragment of DNA because a GA-induced protein has bound to it. (From Ou-Lee et al. 1988.)

sequence was incubated with an exonuclease (an enzyme that degrades the DNA from the ends only) in the presence of the protein extract from GA_3-treated aleurones. The GA-induced protein was found to protect the region where it bound from digestion by the exonuclease (Figure 20.21). Analysis of the undigested DNA yielded a base sequence to which the GA-induced protein bound. This sequence has a high degree of similarity in both rice and barley and contains the sequences described in the previous section that are essential for gibberellin action. The protected DNA sequence was about 150 base pairs upstream of the transcription start site of the α-amylase gene. From this result we can hypothesize that gibberellin increases either the level or the activity of a protein that switches on the production of α-amylase mRNA by binding to an upstream regulatory sequence of the α-amylase gene.

How does gibberellin promote the synthesis of the DNA-binding protein, and what is the nature of this protein? Analysis of the sequence of one of the barley α-amylase promoters provided a clue to the nature of the DNA-binding protein. The sequence was shown to have two possible MYB-binding sites. MYBs are transcription factors that regulate growth and development (see Chapter 14). This finding enabled Gubler and colleagues (1995) to isolate a cDNA clone from a barley aleurone cDNA library that encodes a MYB protein.

The synthesis of a *MYB* gene transcript is shown to be up-regulated (increased) by GA_3 within 3 hours of gibberellin application, well before the appearance of α-amylase mRNA (Figure 20.22). The inhibitor of translation, cycloheximide, had no effect on the production of *MYB* mRNA, even though it blocked the production of α-amylase mRNA, indicating that the *MYB* gene is a GA primary response gene.

The preceding results suggested that MYB represents one of the transcription factors that binds to the GA response complex of the α-amylase promoter. This hypothesis was confirmed by mobility shift assays in which it was shown that the GA-induced MYB protein

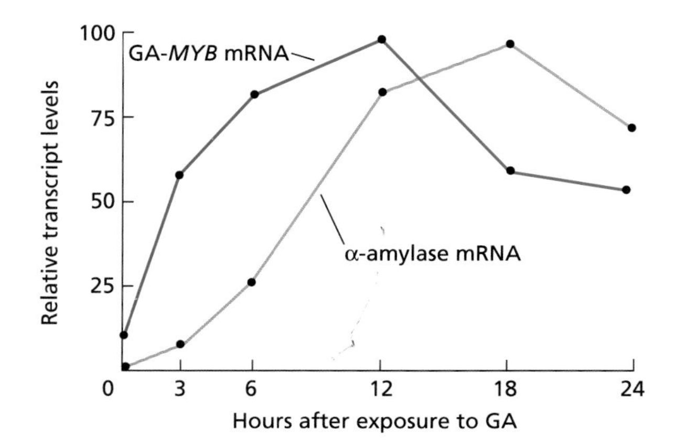

Figure 20.22 Time course for the induction of GA-*MYB* and α-amylase mRNA by gibberellic acid. The production of GA-*MYB* mRNA precedes α-amylase mRNA by about 5 hours. This result is consistent with the role of GA-*MYB* as an early GA response gene that regulates the transcription of the gene for α-amylase. In the absence of GA, the levels of both GA-*MYB* and α-amylase mRNAs are negligible. (After Gubler et al. 1995.)

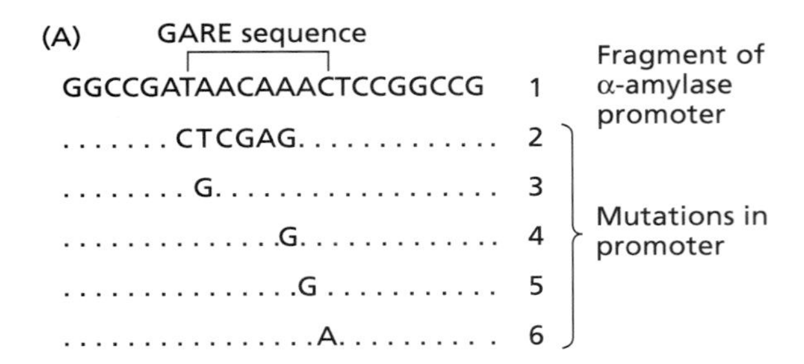

(B)

	Controls	Assay					
Protein	No protein	GA-MYB protein					
	+	+					
Probes	1	1	2	3	4	5	6

Figure 20.23 A mobility shift assay for the binding of various radiolabeled probes to the GA-MYB protein. (A) The probes consist of a 22–base pair fragment of the α-amylase promoter, including the TAACAAA (GARE) sequence (1), and several mutations (2–6) of that sequence. (B) If probe binds to the GA-MYB protein, a higher (retarded) spot is seen on the gel. The TAACAAA sequence of the α-amylase promoter binds to GA-MYB, but most mutations in the sequence inhibit GA-MYB binding. (After Gubler et al. 1995.)

(**GA-MYB**) binds to the gibberellin response element TAACAAA (Figure 20.23).

GA-MYB was also shown to be able to activate the transcription of the gene for α-amylase in a double transformation experiment. Aleurone cells were transformed with two genes: (1) the GA-*MYB* gene fused to a constitutive promoter that caused the aleurone cells to synthesize GA-MYB in the absence of applied GA; (2) the gene that encodes GUS fused to the α-amylase gene promoter. Since the doubly transformed aleurone cells were able to synthesize GUS in the absence of GA, and further addition of GA did not increase GUS activity, the experiment demonstrated that GA-MYB alone can induce the expression of α-amylase (Gubler et al. 1995).

We can conclude that gibberellin acts by causing the synthesis of the transcriptional activation factor GA-MYB, and possibly other transcription factors as well. MYB-type transcription factors may play important roles in regulating gene expression during plant development. As discussed in Chapter 17, a MYB protein has also been implicated in the regulation of the *LHCB* gene by phytochrome. Because its expression is insensitive to cycloheximide, the gene that encodes GA-MYB is an early, or primary response, gene. The protein GA-MYB then binds to the GA response elements in the α-amylase gene promoter, stimulating the synthesis of α-amylase mRNA, which is subsequently translated to the α-amylase enzyme during the germination of cereal grain. Since the expression of α-amylase is inhibited by cycloheximide, the genes for α-amylase and other hydrolases fall into the category of late, or secondary response, genes stimulated by the hormone.

To summarize, GA stimulates the expression of the GA-*MYB* gene, and the MYB protein serves as a transcriptional regulator of the gene for α-amylase. But how does GA cause the *MYB* gene to be expressed? Since protein synthesis is not involved, GA may bring about the activation of one or more *preexisting* transcription factors. The activation of transcription factors is typically mediated by protein phosphorylation events occurring at the end of a signal transduction pathway. In the next and final section of this chapter we describe what is known about the signaling pathways involved in GA-induced α-amylase production.

α-Amylase Synthesis and Secretion Use Two Signal Transduction Pathways

What are the intermediate steps between the binding of GA to its receptor on the cell surface and the activation of the expression of the primary response gene GA-*MYB*? A few pieces of the puzzle have been identified. As we have seen, the genetic approaches applied to the study of GA-stimulated growth led to the identification of the *GAI–SPY* negative regulatory pathway. The proteins GAI and SPY act as repressors of GA responses. Gibberellin acts by deactivating these repressors. Because the aleurone layers of GA-insensitive dwarf wheat are also insensitive to GA (Singh and Paleg 1986; Ross et al. 1997), the same signal transduction pathways that regulate growth appear to regulate GA-induced α-amylase production.

Calcium ions act as second messengers for many hormonal responses in animal cells (see Chapter 14), and they have been implicated in various plant responses to environmental and hormonal stimuli. Barley aleurone protoplasts show a slow rise in cytosolic Ca^{2+} upon incubation with GA (Bethke et al. 1995). This increase begins 1 to 4 hours after exposure to the hormone, and thus precedes the onset of α-amylase synthesis.

By microinjecting calcium-sensitive fluorescent dyes into barley aleurone protoplasts and visualizing the fluorescence by confocal microscopy, researchers showed that the GA-induced increase in calcium ions in aleurone cells is localized in the cytoplasm just below the plasma membrane (Gilroy and Jones 1992). Removal of the extracellular calcium inhibited both the secretion of α-amylase and the increase in calcium, indicating that calcium is being taken up from the external medium. Consistent with a role for calcium in α-amylase production, GA, in the presence of calcium, increases the level of calmodulin in barley aleurone layers by twofold, and the effect begins as early as 2 hours after the start of incubation (Schuurink et al. 1996). Recall that calmodulin binds to calcium ions, and the resulting calcium–calmodulin complex is capable of activating specific enzymes, such as Ca^{2+}–calmodulin-dependent protein kinases (see Chapter 14).

Recent studies by Gilroy (1996) have suggested that GA stimulates the secretion of α-amylase and other hydrolases via a *calcium-dependent pathway*, whereas GA appears to stimulate expression of the α-amylase gene via a *calcium-independent pathway* (Figure 20.24). The operation of different signal transduction pathways for enzyme secretion and gene expression is consistent with the observation that GA stimulates both the secretion and the synthesis of some enzymes (e.g., α-amylase), but only the secretion of other enzymes (e.g., ribonuclease and β-1,3-glucanase).

Cyclic GMP is a possible candidate for a calcium-independent signaling intermediate involved in GA-induced α-amylase gene expression. As discussed in Chapter 17, cGMP has been implicated as a second messenger in phytochrome-regulated gene expression. GA causes a transient rise in cGMP levels in barley aleurone layers after a lag period of only 1 hour (Pensen et al. 1996). An inhibitor of guanylyl cyclase, the enzyme that synthesizes cGMP from GTP, blocks GA-induced α-amylase synthesis and secretion, and the inhibition can be overcome by membrane-permeant analogs of cGMP (Pensen et al. 1996). These findings suggest that cGMP is one of the components of the signal transduction pathway involved in the GA response.

In conclusion, GA signal transduction seems to involve calcium ions as well as cyclic GMP, but the detailed signaling pathways have not been worked out. α-Amylase secretion is regulated by a calcium-dependent pathway, whereas α-amylase gene expression is regulated by a calcium-independent pathway. A few of the genes and some of the biochemical components have now been identified, and they are illustrated in the model shown in Figure 20.24. This model represents a composite of the factors that have been implicated in the GA control of growth and α-amylase gene transcription, and it is based on the assumption that the two responses share a common signal transduction pathway. However, tissue-specific regulatory mechanisms are also possible. Clearly much remains to be learned about how GA works at the molecular level.

Summary

Gibberellins are a family of compounds, now numbering over 110, defined by their structure, some of which are found only in the fungus *Gibberella fujikuroi*. Gibberellins induce dramatic internode elongation in certain types of plants, such as dwarf and rosette species, and grasses. Other physiological effects of GA include changes in juvenility and flower sexuality; and promotion of fruit set, fruit growth, and seed germination. Gibberellins have several commercial applications, mainly in enhancing the size of seedless grapes and in the malting of barley. Gibberellin synthesis inhibitors are used as dwarfing agents.

Gibberellins are identified and quantified by gas chromatography combined with mass spectrometry following separation by high-performance liquid chromatography. Bioassays may be used to give an initial idea of the gibberellins present in a sample. Only certain GAs, notably GA_1 and GA_4, are responsible for the effects in plants; the others are precursors or metabolites.

The gibberellins are terpenoid compounds, made up of isoprene units. The first compound in the isoprenoid pathway committed to gibberellin biosynthesis is *ent*-kaurene, which is converted to GA_{12}-aldehyde, the precursor of all the other gibberellins. GA_{12}-aldehyde, which has 20 carbon atoms, is converted to the other gibberellins by sequential oxidation of carbon 20, followed by the loss of this carbon to give 19-carbon gibberellins. This process is coupled with hydroxylation at one or more positions on the molecule, notably at carbons 13 and/or 3. A subsequent hydroxylation at carbon 2 eliminates biological activity. The genes for GA 20-oxidase, which catalyzes the steps between GA_{53} and GA_{20}, and GA 3β-hydroxylase, which converts GA_{20} into GA_1, have been isolated. Gibberellins may also be glycosylated to give either an inactivated form or a storage form.

Photoperiod is one of the important factors regulating gibberellin metabolism. In spinach, long days promote the conversion of GA_{53} to GA_1 and the increased transcription of the gene for GA 20-oxidase. Gibberellin biosynthesis is also regulated by temperature and by feedback control.

The most pronounced effect of applied gibberellins is stem elongation in dwarf and rosette plants. Gibberellins also cause the elongation of grass internodes, a dramatic example of which is deepwater rice. Gibberellins stimulate stem growth by promoting both cell elongation and cell division. The activity of xyloglucan endotransglycosylase (XET) has been correlated with

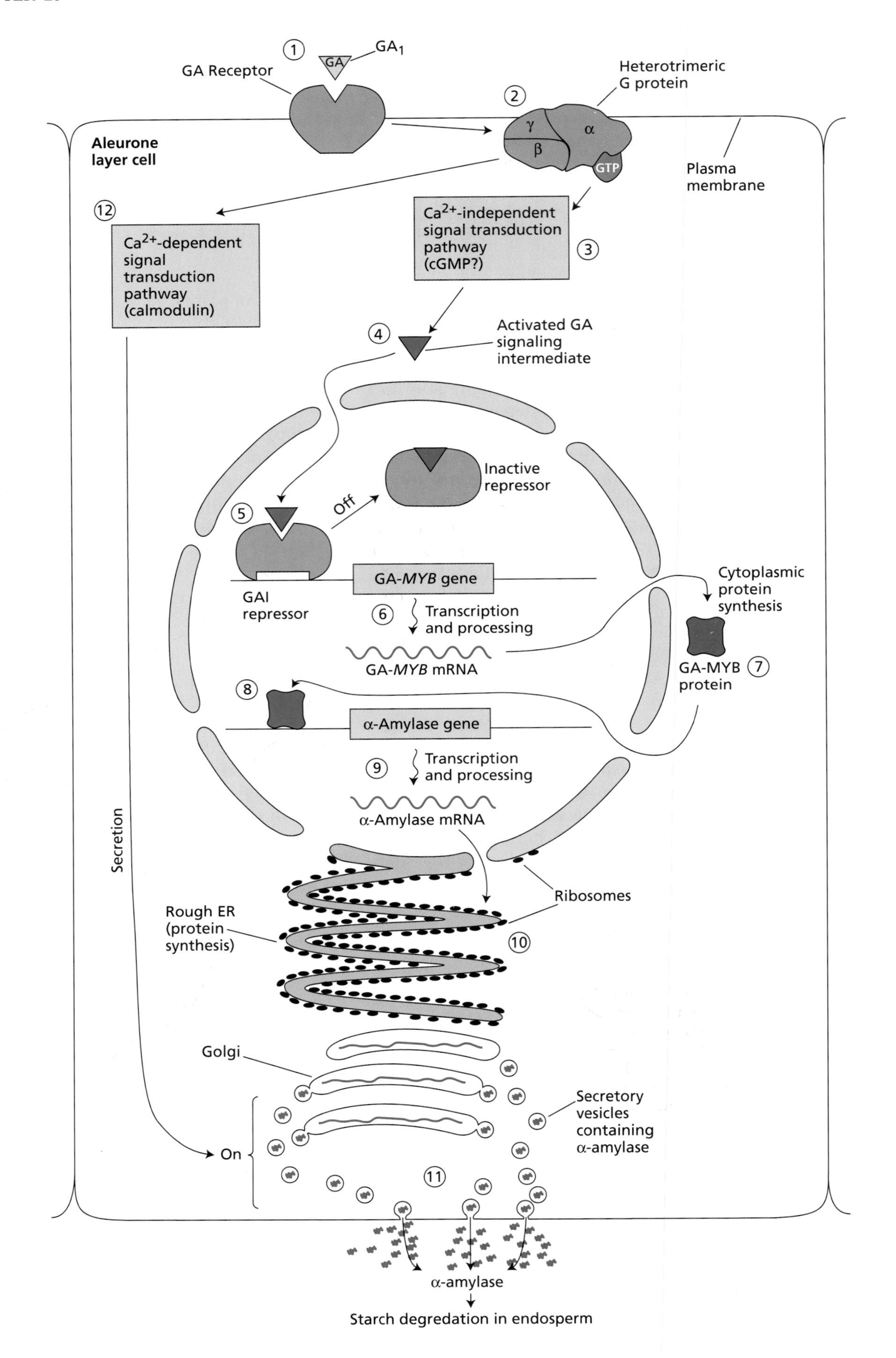

1
GA
GA$_1$
GA Receptor
Heterotrimeric G protein
2
γ
α
β
GTP
Aleurone layer cell
Plasma membrane
12
Ca^{2+}-independent signal transduction pathway (cGMP?)
3
Ca^{2+}-dependent signal transduction pathway (calmodulin)
4
Activated GA signaling intermediate
Inactive repressor
5
Off
GA-*MYB* gene
GAI repressor
6
Transcription and processing
Cytoplasmic protein synthesis
GA-*MYB* mRNA
GA-MYB protein
7
8
α-Amylase gene
9
Transcription and processing
α-Amylase mRNA
Secretion
Ribosomes
Rough ER (protein synthesis)
10
Golgi
Secretory vesicles containing α-amylase
On
11
α-amylase
Starch degradation in endosperm

◀ **Figure 20.24** Composite model for the induction of α-amylase synthesis in barley aleurone layers by gibberellin. (1) GA_1 from the embryo first binds to an unidentified cell surface receptor, initiating two separate signal transduction chains. (2) The cell surface GA-receptor complex interacts with a heterotrimeric G protein, initiating two separate signal transduction chains. (3) A calcium-independent pathway, possibly involving cGMP, results in the activation of a signaling intermediate (4) that binds to the GAI repressor protein (5), thereby inactivating it. This inactivation of the GAI repressor allows the expression of the *MYB* gene (6), as well as other genes repressed by GAI, to proceed through transcription, processing, and translation. (The SPINDLY protein, which may act in concert with GAI to repress gene expression, is omitted for clarity.) The newly synthesized MYB protein (7) then enters the nucleus and binds to the promoter of the gene for α-amylase (8), as well as those of other hydrolytic enzymes, thus activating their transcription (9). After being synthesized on the rough ER (10), the proteins are secreted via the Golgi (11). The secretory pathway may be stimulated by GA via a calcium–calmodulin-dependent signal transduction pathway (12).

GA-induced growth and may interact with expansins to promote cell wall loosening. The stimulation of cell division may be a secondary effect of GA-induced cell expansion. GA-stimulated cell divisions in deepwater rice are believed to be regulated at the transition between the G_2 and the M phases of the cell cycle.

Two types of GA response mutants have been useful in the identification of genes involved in the GA signaling pathway: (1) GA-insensitive dwarfs and (2) constitutive GA responders (slender mutants). Such studies have led to the identification of two repressors of the GA response: GAI and SPY. The *GAI* gene has been cloned and shown to encode a possible transcriptional regulator. Thus, GA appears to induce its effects by derepressing negatively regulated genes.

Gibberellin induction of the enzyme α-amylase in aleurone cells of germinating cereal grains is now well elucidated. The gibberellin from the embryo induces the transcription of the genes for α-amylase mRNA. It appears to do so by causing the production of a regulatory protein (GA-MYB), which binds to the upstream regulatory sequences of the α-amylase gene, thereby switching the gene on. The gibberellin receptor is on the plasma membrane of aleurone cells, with a signal transduction pathway subsequently regulating the production of GA-MYB. Although the details of the signal transduction pathway are unclear, it appears to involve calcium and cyclic GMP as intermediates.

General Reading

Chory, J., and Li, J. (1997) Gibberellins, brassinosteroids and light-regulated development. *Plant Cell Environ.* 20: 801–806.

Cleland, R. E. (1996) Growth substances. In *Units, Symbols, and Terminology for Plant Physiology*, F. B. Salisbury, ed., Oxford University Press, New York, pp. 126–128.

Davies, P. J., ed. (1995) *Plant Hormones: Physiology, Biochemistry and Molecular Biology*. Kluwer, Boston.

Davies, P. J. (1995) The plant hormones: Their nature, occurrence and functions. In *Plant Hormones: Physiology, Biochemistry and Molecular Biology*, P. J. Davies, ed., Kluwer, Boston, pp. 1–12.

*Hooley, R. (1994) Gibberellins: Perception, transduction and responses. *Plant Mol. Biol.* 26: 1529–1555.

Law, D. M., and Davies, P. J. (1990) Comparative indole-3-acetic acid levels in the slender pea and other pea phenotypes. *Plant Physiol.* 93: 1539–1543.

Libbenga, K. R., and Mennes, A. M. (1995) Hormone binding and its role in hormone action. In *Plant Hormones: Physiology, Biochemistry and Molecular Biology*, P. J. Davies, ed., Kluwer, Boston, pp. 272–297.

MacMillan, J. (1997) Biosynthesis of the gibberellin plant hormones. *Nat. Prod. Rep.* 14: 221–243.

*Phinney, B. O. (1983) The history of gibberellins. In *The Biochemistry and Physiology of Gibberellins*, Vol. 1, A. Crozier, ed., Praeger, New York, pp. 19–52.

Sponsel, V. M. (1995) Gibberellin biosynthesis and metabolism. In *Plant Hormones: Physiology, Biochemistry and Molecular Biology*, P. J. Davies, ed., Kluwer, Boston, pp. 66–97.

* Indicates a reference that is general reading in the field and is also cited in this chapter.

Chapter References

Aach, H., Böse, G., and Graebe, J. E. (1995) *ent*-Kaurene biosynthesis in a cell-free system from wheat (*Triticum aestivum* L.) seedlings and the localization of *ent*-kaurene synthetase in plastids of three species. *Planta* 197: 333–342.

Beale, M. H., Ward, J. L., Smith, S. J., and Hooley, R. (1992) A new approach to gibberellin perception in aleurone: Novel, hydrophilic, membrane-impermeant, GA-sulphonic acid derivatives induce α-amylase formation. *Physiol. Plant* 85: A136.

Behringer, F. J., Cosgrove, D. J., Reid, J. B., and Davies, P. J. (1990) Physical basis for altered stem elongation rates in internode length mutants of *Pisum*. *Plant Physiol.* 94: 166–173.

Bethke, P. C., Gilroy, S., and Jones, R. L. (1995) Calcium and plant hormone action. In *Plant Hormones: Physiology, Biochemistry and Molecular Biology*. 2nd ed. P. J. Davies, ed., Kluwer, Dordrecht, Netherlands, pp. 298–317.

Bramley, P. M. (1997) Isoprenoid metabolism. In *Plant Biochemistry*, P. M. Dey and J. B. Harborne, eds., Academic Press, San Diego, pp. 417–437.

Chandler, P. M., Zwar, J. A., Jacobsen, J. V., Higgins, T. J. V., and Inglis, A. S. (1984) The effects of gibberellic acid and abscisic acid on α-amylase mRNA levels in barley aleurone layers. Studies using an α-amylase cDNA clone. *Plant Mol. Biol.* 3: 407–418.

Cho, H-T., and Kende, H. (1997) Expansins in deepwater rice internodes. *Plant Physiol.* 113: 1137–1143.

Coolbaugh, R. C. (1985) Sites of gibberellin biosynthesis in pea seedlings. *Plant Physiol.* 78: 655–657.

Davies, P. J. (1995) The plant hormone concept: Concentration, sensitivity and transport. In *Plant Hormones: Physiology, Biochemistry and Molecular Biology*, P. J. Davies, ed., Kluwer, Boston, pp. 13–38.

Durnham, D. J., and Jones, R. L. (1982) The effects of colchicine and gibberellic acid on growth and microtubules in excised lettuce hypocotyls. *Planta* 154: 204–211.

Frydman, V. M., Gaskin, P., and MacMillan, J. (1974) Qualitative and quantitative analysis of gibberellins through seed maturation in *Pisum sativum* cv. *Planta* 118: 123–132.

Gianfagna, T. J. (1995) Natural and synthetic growth regulators and their use in horticultural and agronomic crops. In *Plant Hormones: Physiology, Biochemistry and Molecular Biology*, P. J. Davies, ed., Kluwer, Boston, pp. 751–773.

Gilroy, S. (1996) Signal transduction in barley aleurone protoplasts is calcium dependent and independent. *Plant Cell* 8: 2193–2209.

Gilroy, S., and Jones, R. L. (1992) Gibberellic acid and abscisic acid coordinately regulate cytoplasmic calcium and secretory activity in barley aleurone protoplasts. *Proc. Natl. Acad. Sci. USA* 89: 3591–3595.

Gilroy, S., and Jones, R. L. (1994) Perception of gibberellin and abscisic acid at the external face of the plasma membrane of barley (*Hordeum vulgare* L.) aleurone protoplasts. *Plant Physiol.* 104: 1185–1192.

Graebe, J. E. (1986) Gibberellin biosynthesis from gibberellin A_{12}-aldehyde. In *Plant Growth Substances 1985*, M. Bopp, ed., Springer, New York, pp. 74–82.

Gubler, F., and Jacobsen, J. V. (1992) Gibberellin-responsive elements in the promoter of a barley high-pI α-amylase gene. *Plant Cell* 4: 1435–1441.

Gubler, F., Kalla, R., Roberts, J. K., and Jacobsen, J. V. (1995) Gibberellin-regulated expression of a MYB gene in barley aleurone cells: Evidence of MYB transactivation of a high-pI α-amylase gene promoter. *Plant Cell* 7: 1879–1891.

Hansen, D. J., Bellman, S. K., and Sacher, R. M. (1976) Gibberellic acid controlled sex expression of corn tassels. *Crop Sci.* 16: 371–374.

Hedden, P., and Kamiya, Y. (1997) Gibberellin biosynthesis: Enzymes, genes and their regulation. *Annu. Rev. Plant Physiol. Plant Mol. Biol.* 48: 431–460.

Higgins, T. J. V., Zwar, J. A., and Jacobsen, J. V. (1976) Gibberellic acid enhances the level of translatable mRNA for α-amylase in barley aleurone layers. *Nature* 260: 166–169.

Hoffmann-Benning, S., and Kende, H. (1992) On the role of abscisic acid and gibberellin in the regulation of growth in rice. *Plant Physiol.* 99: 1156–1161.

Horgan, R. (1995) Instrumental methods of plant hormone analysis. In *Plant Hormones: Physiology, Biochemistry and Molecular Biology*, P. J. Davies, ed., Kluwer, Boston, pp. 415–432.

Huttly, A. K., and Phillips, A. J. (1995) Gibberellin-regulated plant genes. *Physiol. Plantarum* 95: 310–317.

Ingram, T. J., and Reid, J. B. (1987) Internode length in *Pisum*. Gene *na* may block gibberellin synthesis between *ent*-7α-hydroxykaurenoic acid and gibberellin A_{12}-aldehyde. *Plant Physiol.* 83: 1048–1053.

Ingram, T. J., Reid, J. B., and MacMillan, J. (1986) The quantitative relationship between gibberellin A_1 and internode growth in *Pisum sativum* L. *Planta* 168: 414–420.

Ingram, T. J., Reid, J. B., Murfet, I. C., Gaskin, P., Willis, C. L., and MacMillan, J. (1984) Internode length in *Pisum*. The *Le* gene controls the 3β-hydroxylation of gibberellin A_{20} to gibberellin A_1. *Planta* 160: 455–463.

Ingram, T. J., Reid, J. B., Potts W. C., and Murfet, I. C. (1983) Internode length in *Pisum*. IV. The effect of the Le gene on gibberellin metabolism. *Physiol. Plantarum* 59: 607–616.

Irish, E. E. (1996) Regulation of sex determination in maize. *BioEssays* 18: 363–369.

Jacobsen, J. V., and Beach, L. R. (1985) Control of transcription of α-amylase and ribosomal RNA genes in barley aleurone protoplasts by gibberellin and abscisic acid. *Nature* 316: 275–277.

Jacobsen, J. V., and Close, T. J. (1991) Control of transient expression of chimeric genes by gibberellic acid and abscisic acid in protoplasts prepared from mature barley aleurone layers. *Plant Mol. Biol.* 16: 713–724.

Jacobsen, J. V., Gubler, F., and Chandler, P. M. (1995) Gibberellin and abscisic acid in germinating cereals. In *Plant Hormones: Physiology, Biochemistry and Molecular Biology*, P. J. Davies, ed., Kluwer, Boston, pp. 246–271.

Jacobsen, S. E., Binkowski, K. A., and Olszewski, N. E. (1996) SPINDLY, a tetratricopeptide repeat protein involved in gibberellin signal transduction in *Arabidopsis*. *Proc. Natl. Acad. Sci. USA* 93: 9292–9296.

Jacobsen, S. E., and Olszewski, N. E. (1993) Mutations at the *SPINDLY* locus of *Arabidopsis* alter gibberellin signal transduction. *Plant Cell* 5: 887–896.

Jones, H. D., Smith, S. J., Desikan, R., Plakidou-Dymock, S., Lovegrove, A., and Hooley, R. (1998) Heterotrimeric G proteins are implicated in gibberellin induction of α-amylase gene expression in wild oat aleurone layer. *Plant Cell* 10: 245–253.

Jones, R. L., and MacMillan, J. (1984) Gibberellins. In *Advanced Plant Physiology*, M. B. Wilkins, ed., Pitman, London, pp. 21–52.

Kende, H., and Zeevaart, J. A. D. (1997) The five "classical" plant hormones. *Plant Cell* 9: 1197–1210.

Kobayashi, M., Gaskin, P., Spray, C. R., Suzuki, Y., Phinney, B. O., and MacMillan, J. (1993) Metabolism and biological activity of gibberellin A-4 in vegetative shoots of *Zea mays*, *Oryza sativa* and *Arabidopsis thaliana*. *Plant Physiol.* 102: 379–386.

Kobayashi, M., Spray, C. R., Phinney, B. O., Gaskin, P., and MacMillan, J. (1996) Gibberellin metabolism in maize. The stepwise conversion of gibberellin A_{12}-aldehyde to gibberellin A_{20}. *Plant Physiol.* 110: 413–418.

Lang, A. (1957) The effect of gibberellin upon flower formation. *Proc. Natl. Acad. Sci. USA* 43: 709–717.

Lenton, J. R., Appleford, N. E. J., and Croker, S. J. (1994) Gibberellins and α-amylase gene expression in germinating wheat grains. *Plant Growth Reg.* 15: 261–270.

Lester, D. R., Ross, J. J., Davies, P. J., and Reid, J. B. (1997) Mendel's stem length gene (Le) encodes a gibberellin 3-beta hydroxylase. *Plant Cell* 9: 1435–1443.

Lichtenthaler, H. K., Schwender, J., Disch, A., and Rohmer, M. (1997) Biosynthesis of isoprenoids in higher plant chloroplasts proceeds via a mevalonate-independent pathway . *FEBS Lett.* 400: 271–274.

Métraux , J-P. (1987) Gibberellins and plant cell elongation. In *Plant Hormones and Their Role in Plant Growth and Development*, P. J. Davies, ed., Kluwer, Boston, pp. 296–317.

Metzger, J. D. (1995) Hormones and reproductive development. In *Plant Hormones: Physiology, Biochemistry and Molecular Biology*, P. J. Davies, ed., Kluwer, Boston, pp. 617–648.

Nomura, T., Nakayama, M., Reid, J. B., Takeuchi, Y., and Yokota, T. (1997) Blockage of brassinosteroid biosynthesis and sensitivity causes dwarfism in garden pea. *Plant Physiol.* 113: 31–37.

Ou-Lee, T-M., Turgeon, R., and Wu, R. (1988) Interaction of a gibberellin-induced factor with the upstream region of an α-amylase gene in rice aleurone tissue. *Proc. Natl. Acad. Sci. USA* 85: 6366–6369 .

Peng, J., Carol, P., Richards, D. E., King, K. E., Cowling, R. J., Murphy, G. P., and Harberd, N. P. (1997) The *Arabidopsis GAI* gene defines a signalling pathway that negatively regulates gibberellin responses. *Genes Dev.* 11: 1–11.

Penson , S. P., Schuurink, R. C., Fath, A., Gubler, F., Jacobsen, J. V., and Jones, R. L. (1996) cGMP is required for gibberellic acid-induced gene expression in barley aleurone. *Plant Cell* 8: 2325–2333.

Phillips, A. L., Ward, D. A., Uknes, S., Appleford, N. E. J., Lange, T., Huttly, A. K., Gaskin, P., Graebe, J. E., and Hedden, P. (1995) Isolation and expression of three gibberellin 20-oxidase cDNA clones from *Arabidopsis*. *Plant Physiol.* 108: 1049–1057.

Phinney, B. O. (1985) Gibberellin A_1, dwarfism and shoot elongation in higher plants. *Biol. Plant.* 27: 172–179.

Phinney, B. O., and West, C. A. (1959) Gibberellins in the growth of flowering plants. In *Developing Cell Systems and Their Control*, D. Rudnick, ed., Ronald Press, New York, pp. 71–92.

Potter, I., and Fry, S. C. (1994) Changes in xyloglucan endotransglycosylase (XET) activity during hormone-induced growth in lettuce and cucumber hypocotyls and spinach cell suspension cultures. *J. Exp. Bot.* 45: 1703–1710.

Potts, W. C., Reid, J. B., and Murfet, I. C. (1982) Internode length in *Pisum*. I. The effect of the *Le/le* gene difference on endogenous GA-like substances. *Physiol. Plantarum* 55: 323–328.

Raskin, I., and Kende, H. (1984) Role of gibberellin in the growth response of submerged deep water rice. *Plant Physiol.* 76: 947–950.

Reeve, D. R., and Crozier, A. (1975) Gibberellin bioassays. In *Gibberellins and Plant Growth*, H. N. Krishnamoorthy, ed., Wiley Eastern, New Delhi, India, pp. 35–64.

Reid, J. B., and Howell S. H. (1995) Hormone mutants and plant development. In *Plant Hormones: Physiology, Biochemistry and Molecular Biology*, P. J. Davies, ed., Kluwer, Boston, pp. 448–485.

Reid, J. B., Ross, J. J., and Swain, S. M. (1992) Internode length in *Pisum*. A new slender mutant with elevated levels of C_{19} gibberellins. *Planta* 188: 462–467.

Rogler, C. E., and Hackett, W. P. (1975a) Phase change in *Hedera helix:* Induction of the mature to juvenile phase change by GA_3. *Physiol. Plantarum* 34: 141–147.

Rogler, C. E., and Hackett, W. P. (1975b) Phase change in *Hedera helix:* Stabilization of the mature form with abscisic acid and growth retardants. *Physiol. Plantarum* 34: 148–152.

Rood, S. B., Paris, R. P., and Major, D. J. (1980) Changes of endogenous gibberellin-like substances with sex reversal of the apical inflorescence of corn. *Plant Physiol.* 66: 793–796.

Ross, J. J., Murfet, I. C., and Reid, J. B. (1997) Gibberellin mutants. *Physiol. Plantarum* 100: 550–560.

Ross, J. J., Reid, J. B., Gaskin, P., and MacMillan, J. (1989) Internode length in *Pisum.* Estimation of GA_1 levels in genotypes *Le*, *le*, and *led*. *Physiol. Plantarum* 76: 173–176.

Ross, J. J., Reid, J. B., Swain, S. M., Hasan, O., Poole, A. T., Hedden, P., and Willis, C. L. (1995) Genetic regulation of gibberellin deactivation in *Pisum*. *Plant J.* 7: 513–523.

Sachs, R. M. (1965) Stem elongation. *Annu. Rev. Plant Physiol.* 16: 73–96.

Sauter, M. (1997) Differential expression of a CAK (cdc2-activating kinase)-like protein kinase, cyclins and *cdc2* genes from rice during the cell cycle and in response to gibberellin. *Plant J.* 11: 181–190.

Sauter, M., and Kende, H. (1992) Gibberellin-induced growth and regulation of the cell division cycle in deepwater rice. *Planta* 188: 362–368.

Sauter, M., Mekhedov, S. L., and Kende, H. (1995) Gibberellin promotes histone H1 kinase activity and the expression of *cdc2* and cyclin genes during the induction of rapid growth in deepwater rice internodes. *Plant J.* 7: 623–632.

Schneider, G., and Schmidt, J. (1990) Conjugation of gibberellins in *Zea mays* L. In *Plant Growth Substances, 1988*, R. P. Pharis and S. B. Rood, eds., Springer, Heidelberg, pp. 300–306.

Schuurink, R. C., Chan, P. V., and Jones, R. L. (1996) Modulation of calmodulin mRNA and protein levels in barley aleurone. *Plant Physiol.* 111: 371–380.

Schwender, J., Seemann, M., Lichtenthaler, H. K., and Rohmer, M. (1996) Biosynthesis of isoprenoids (carotenoids, sterols, prenyl side-chains of chlorophylls and plastoquinone) via a novel pyruvate/glyceraldehyde 3-phosphate non-mevalonate pathway in the green alga, *Scenedesmus obliquus*. *Biochem. J.* 316: 73–80.

Scott, I. M. (1990) Plant hormone response mutants. *Physiol. Plantarum* 78: 147–152.

Sherriff, L. J., McKay, M. J., Ross, J. J., Reid, J. B., and Willis, C. L. (1994) Decapitation reduces the metabolism of gibberellin A-20 to A-1 in *Pisum sativum* L., decreasing the Le-le difference. *Plant Physiol.* 104: 277–280.

Singh, S. P., and Paleg, L. G. (1986) Low temperature-induced GA_3 sensitivity of wheat. VI. Effect of inhibitors of lipid biosynthesis on α-amylase production by dwarf (*Rht3*) and tall (*rht*) wheat, and on lipid metabolism of tall wheat aleurone tissue. *Aust. J. Plant Physiol.* 13: 409–416.

Skriver, K., Olsen, F. L., Rogers, J. C., and Mundy, J. (1991) *Cis*-acting DNA elements responsive to gibberellin and its antagonist abscisic acid. *Proc. Natl. Acad. Sci. USA* 88: 7266–7270.

Smith, R. C., Matthews, P. R., Schunmann, P. H. D., and Chandler, P. M. (1996) The regulation of leaf elongation and xyloglucan endotransglycosylase by gibberellin in "Himalaya" barley (*Hordeum vulgare* L.). *J. Exp. Bot.* 47: 1395–1404.

Sun, T. P., and Kamiya, Y. (1994) The *Arabidopsis* GA1 locus encodes the cyclase *ent*-kaurene synthetase A of gibberellin biosynthesis. *Plant Cell* 6: 1509–1518.

Swain, S. M., and Olszewski, N. E. (1996) Genetic analysis of gibberellin signal transduction. *Plant Physiol.* 112: 11–17.

Taylor, A., and Cosgrove, D. J. (1989) Gibberellic acid stimulation of cucumber hypocotyl elongation. *Plant Physiol.* 90: 1335–1340.

Taylor, C. (1998) GA signaling: Genes and GTPases. *Plant Cell* 10: 131–133.

Wu, K., Li, L., Gage, D. A., and Zeevaart, J. A. D. (1996) Molecular cloning and photoperiod-regulated expression of gibberellin 20-oxidase from the long-day plant spinach. *Plant Physiol.* 110: 547–554.

Xu, Y-L., Gage, D. A., and Zeevaart, J. A. D. (1997) Gibberellins and stem growth in *Arabidopsis thaliana. Plant Physiol.* 114: 1471–1476.

Xu, Y-L., Li, L., Wu, K., Peeters, A. J. M., Gage, D. A., and Zeevaart, J. A. D. (1995) The GA5 locus of *Arabidopsis thaliana* encodes a multifunctional gibberellin 20-oxidase: Molecular cloning and functional expression. *Proc. Natl. Acad. Sci. USA* 92: 6640–6644.

Yang, T., Davies, P. J., and Reid, J. B. (1996) Genetic dissection of the relative roles of auxin and gibberellin in the regulation of stem elongation in intact light-grown peas. *Plant Physiol.* 110: 1029–1034.

Zeevaart, J. A. D., and Gage, D. A. (1993) *Ent* kaurene biosynthesis is enhanced by long photoperiods in the long-day plants *Spinacia oleracea* L. and *Agrostemma githago* L. *Plant Physiol.* 101: 25–29.

Zeevaart, J. A. D., Gage, D. A., and Talon, M. (1993) Gibberellin A_1 is required for stem elongation in spinach. *Proc. Natl. Acad. Sci. USA* 90: 7401–7405.

21 Cytokinins

THE CYTOKININS WERE DISCOVERED in the course of studies aimed at identifying factors that stimulate plant cells to divide (i.e., undergo cytokinesis). Since their discovery, cytokinins have been shown to have effects on many other physiological and developmental processes as well, including leaf senescence, nutrient mobilization, apical dominance, the formation and activity of shoot apical meristems, floral development, the breaking of bud dormancy, and seed germination. Cytokinins also appear to mediate many aspects of light-regulated development, including chloroplast differentiation, the development of autotrophic metabolism, and leaf and cotyledon expansion.

Although cytokinins regulate many cellular processes, the control of cell division is of considerable significance for plant growth and development and is considered diagnostic for this class of plant growth regulators. For these reasons, we will preface our discussion of cytokinin function with a brief consideration of the roles of cell division in normal development, wounding, gall formation, and tissue culture. Later in the chapter we will examine what is known about the regulation of the plant cell cycle by cytokinins, and about other functions not directly related to cell division: chloroplast differentiation, the repression of leaf senescence, and nutrient mobilization.

Cell Division and Plant Development

Plant cells form as the result of cell divisions in a primary or secondary meristem. Newly formed plant cells typically enlarge and differentiate, but once they assume their function—whether transport, photosynthesis, support, storage, or protection—usually they do not divide again during the life of the plant. In this respect they appear to be similar to animal cells, which are considered to be terminally differentiated. However, many examples could be cited to show that this similarity to the behavior of animal cells is only superficial. Almost every type of plant cell that retains its nucleus at maturity has been shown to be capable of dividing. This property comes into play during such processes as wound healing and leaf abscission.

Differentiated Plant Cells Can Resume Division

Under some circumstances mature, differentiated plant cells may resume cell division in the intact plant. In many species, mature cells of the cortex and/or phloem resume division to form secondary meristems, such as the vascular cambium or the cork cambium. The abscission zone at the base of a leaf petiole is a region where mature parenchyma cells begin to divide again after a period of mitotic inactivity, forming a layer of cells with relatively weak cell walls where abscission can occur (see Chapter 22).

Wounding of plant tissues induces cell divisions at the wound site. Even highly specialized cells, such as phloem fibers and guard cells, may be stimulated by wounding to divide at least once. Wound-induced mitotic activity typically is self-limiting; after a few divisions the derivative cells stop dividing and redifferentiate. However, when the soil-dwelling bacterium *Agrobacterium tumefaciens* invades a wound, it can cause the neoplastic (tumor-forming) disease known as **crown gall**. This phenomenon is dramatic natural evidence of the mitotic potential of mature plant cells (Nester et al. 1984).

Without *Agrobacterium* infection, the wound-induced cell division would subside after a few days and some of the new cells would differentiate as a protective layer of cork cells or vascular tissue. However, *Agrobacterium* changes the character of the cells that divide in response to the wound, making them tumorlike. They do not stop dividing, but continue to divide throughout the life of the plant to produce an unorganized mass of tumorlike tissue called a **gall** (Figure 21.1). We will have more to say about this important disease later in this chapter.

Figure 21.1 Tumors that formed on stems infected with the crown gall bacterium, *Agrobacterium tumefaciens*. Several weeks before this photo was taken the stem of a sunflower plant was wounded with a needle contaminated with a virulent strain of the crown gall bacterium. Some of the stem cells near the wound were transformed with DNA from the *Agrobacterium* Ti plasmid and continued to proliferate after the wound had healed. The bacterium transferred a portion of the Ti plasmid known as the T-DNA into the wounded host cells. The T-DNA subsequently became integrated into the plant's nuclear DNA. The T-DNA contains genes that encode enzymes involved in the synthesis of both auxin and cytokinin. Expression of these genes in a transformed plant cell leads to hormone synthesis, and these hormones in turn stimulate the cells to divide repeatedly to form tumors. (© J. Cunningham/Visuals Unlimited.)

Diffusible Factors May Control Cell Division

The considerations addressed in the previous section suggest that mature plant cells stop dividing because they no longer receive a particular signal, possibly a hormone, that is necessary for the initiation of cell division. The idea that cell division may be initiated by a diffusible factor originated with the Austrian plant physiologist G. Haberlandt, who, around 1913, demonstrated that vascular tissue contains a water-soluble substance or substances that will stimulate the division of wounded potato tuber tissue. The effort to determine the nature of this factor (or factors) led to the discovery of the cytokinins in the 1950s.

Plant Tissues and Organs Can Be Cultured

Biologists have long been intrigued by the possibility that organs, tissues, and cells can be grown in culture on a simple nutrient medium, in the same way that microorganisms are cultured in test tubes or on petri dishes. In the 1930s, Philip White demonstrated that tomato roots can be grown indefinitely in a simple nutrient medium containing only sucrose, mineral salts, and a few vitamins, with no added hormones (White 1934). In contrast to roots, isolated stem tissues exhibit very little growth in culture without added hormones in the medium. Even if auxin is added, only limited growth may occur, and usually this growth is not sustained. Frequently this auxin-induced growth is due to cell enlargement only. The shoots of most plants cannot grow on a simple medium lacking hormones, even if the cultured stem tissue contains apical or lateral meristems, until adventitious roots form. Once the stem tissue has rooted, shoot growth resumes, but now as an integrated, whole plant. These observations indicate that there is a difference in the regulation of cell division in root and shoot meristems. They also suggest that some root-derived factor(s) may regulate growth in the shoot.

Crown gall stem tissue is an exception to these generalizations. After a gall has formed on a plant, heating the plant to 42°C will kill the bacterium that induced gall formation. The plant will survive the heat treatment, and its gall tissue will continue to grow as a bacteria-free tumor (Braun 1958). Tissues removed from these bacteria-free tumors grow on simple, chemically defined culture media that would not support the proliferation of normal stem tissue of the same species. However, these stem-derived tissues are not organized. Instead they grow as a mass of disorganized, relatively undifferentiated cells called **callus tissue**.

Callus tissue sometimes forms naturally in response to wounding, or in graft unions where stems of two different plants are joined. Crown gall tumors are a specific type of callus, whether they are growing attached to the plant or in culture. The finding that crown gall callus tissue can be cultured demonstrated that cells derived from stem tissues are capable of proliferating in culture and that contact with the bacteria may cause the stem cells to produce cell division–stimulating factors.

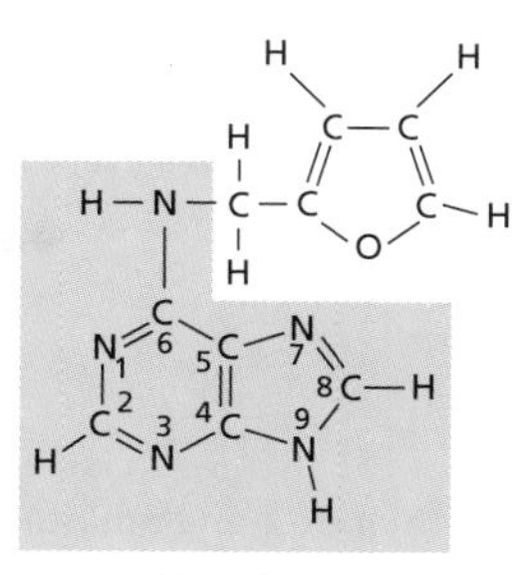

Figure 21.2 Chemical structure of the cytokinin kinetin. Kinetin was first isolated from autoclaved herring sperm DNA, which had been shown to stimulate the proliferation of tobacco pith tissue when it was cultured on a medium that also contained an auxin. Although not found naturally in any plant, kinetin is similar in structure to native cytokinins in that it is an aminopurine derivative with a substitution on the nitrogen at position 6 (N^6). It differs from native cytokinins in the nature of the group found on the 6 nitrogen.

The Discovery, Identification, and Properties of Cytokinins

A great many substances were tested in an effort to initiate and sustain the proliferation of normal stem tissues in culture. Materials ranging from yeast extract to tomato juice were found to have a positive effect, at least with some tissues. However, culture growth was stimulated most dramatically when the liquid endosperm of coconut, also known as coconut milk, was added to the culture medium. White's nutrient medium, supplemented with an auxin and 10 to 20% coconut milk, will support the continued cell division of mature, differentiated cells from a wide variety of tissues and species, leading to the formation of callus tissue (Caplin and Steward 1948).

This finding indicated that coconut milk contains a substance or substances that stimulate mature cells to enter and remain in the cell division cycle. Many attempts were made to determine the chemical nature of the cell division–stimulating substance in coconut milk. Ultimately, coconut milk was shown to contain the cytokinin *zeatin*, but this finding was not obtained until several years after the discovery of the cytokinins (Letham 1974). The first cytokinin to be discovered was the synthetic analog kinetin.

Kinetin Was Discovered as a Breakdown Product of DNA

In the 1940s and 1950s, Folke Skoog at the University of Wisconsin tested many substances for their ability to initiate and sustain the proliferation of cultured tobacco pith tissue. He had observed that the nucleic acid base adenine had a slight promotive effect, so he tested the possibility that nucleic acids would stimulate division in this tissue. Surprisingly, autoclaved herring sperm DNA had a powerful cell division–promoting effect.

After a difficult and time-consuming fractionation of the heat-treated DNA, Skoog and his coworker, Carlos Miller, identified a small molecule that, in the presence of an auxin, would stimulate tobacco pith parenchyma tissue to proliferate in culture. They named the biologically active molecule **kinetin** (Figure 21.2) and demonstrated that it was an adenine (or aminopurine) derivative, 6-furfurylaminopurine (Miller et al. 1955). Kinetin stimulates tobacco pith cells to divide when isolated pith tissue is cultured on a medium that also contains an auxin. No kinetin-induced cell division occurs without auxin in the culture medium.

Kinetin is not a naturally occurring plant growth regulator, and it does not occur as a base in the DNA of any species. It is a by-product of the heat-induced degradation of the DNA, in which the deoxyribose sugar of adenosine is converted to a furfuryl ring and shifted from the 9 position to the 6 position on the adenine ring. The discovery of kinetin was important because it demonstrated that cell division could be induced by a simple chemical substance. Of greater importance, when a synthetic molecule initiates a biological response, often it does so because the synthetic molecule has many of the same properties as naturally occurring molecules that regulate that response in the organism. Thus, the discovery of kinetin suggested that naturally occurring molecules with structures similar to that of kinetin regulate cell division activity within the plant. This hypothesis proved to be correct.

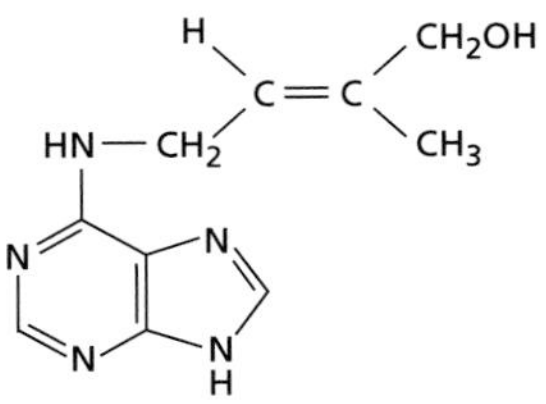

***trans*-Zeatin** 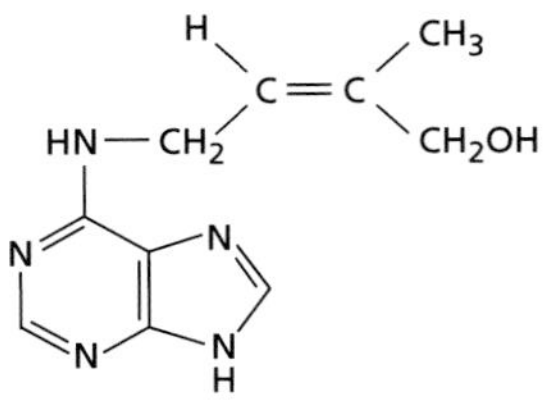 **_cis_-Zeatin**
6-(4-Hydroxy-3-methylbut-2-enylamino)purine

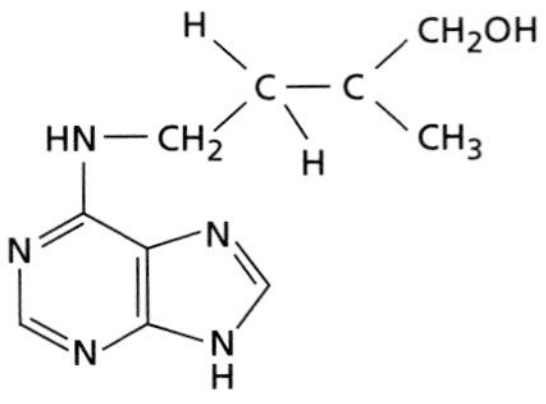

Dihydrozeatin

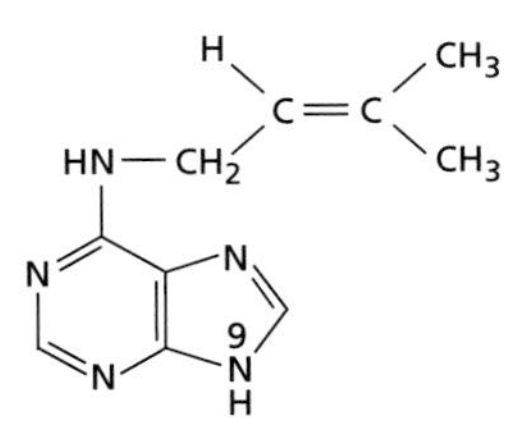

N^6-(Δ^2-Isopentenyl)-adenine (i^{6}Ade)

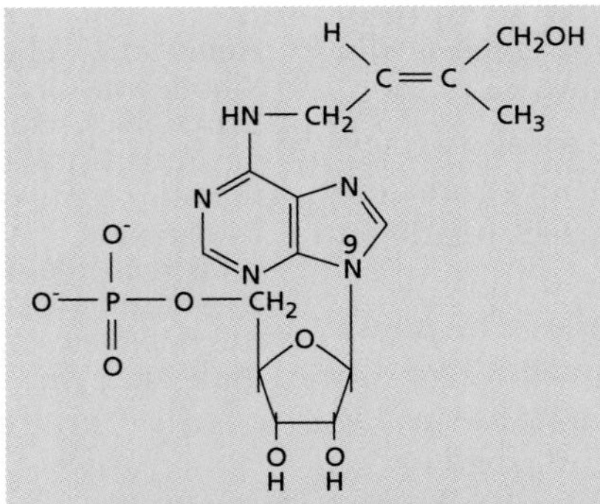

Ribosylzeatin-5′-monophosphate (zeatin ribotide)

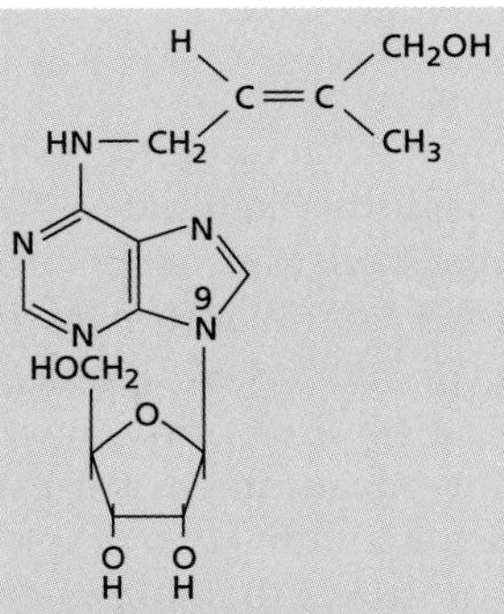

Ribosylzeatin (zeatin riboside)

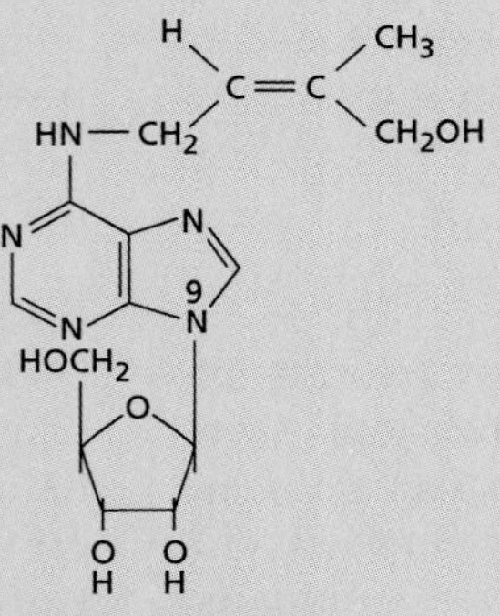

***cis*-Ribosylzeatin**

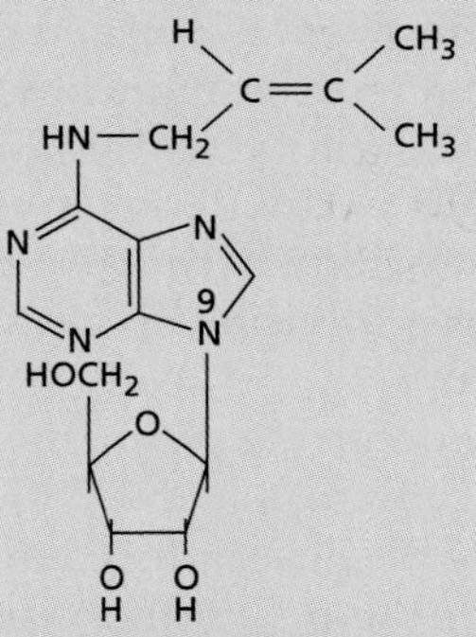

N^6-(Δ^2-Isopentenyl)-adenosine (i^{6}Ado)

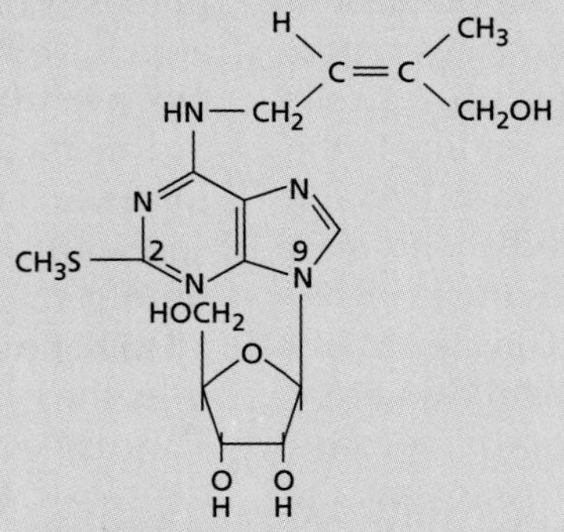

2-Methylthio-*cis*-ribosylzeatin (bacterial)

Figure 21.3 Structures of some naturally occurring cytokinins. Most plants have *trans*-zeatin as the principal free cytokinin, but dihydrozeatin and isopentenyl adenine (i^{6}Ade) are also native plant cytokinins. Free cytokinins also include the ribosides and ribotides of zeatin, dihydrozeatin, and isopentenyladenosine, although these may be active as cytokinins by conversion to the respective bases. Free cytokinins from bacteria include 2-methylthio-*cis*-ribosylzeatin, as well as *cis*- or *trans*-zeatin, and their ribosides and ribotides.

Zeatin Is the Most Abundant Natural Cytokinin

Several years after the discovery of kinetin, Carlos Miller in the United States and D. S. Letham in Australia independently demonstrated that extracts of the immature endosperm of corn (*Zea mays*) contain a substance that has the same biological effect as kinetin. This substance stimulates mature plant cells to divide when added to a culture medium along with an auxin. Letham isolated the molecule responsible for this activity and identified it as *trans*-6-(4-hydroxy-3-methylbut-2-enylamino)purine, which he called **zeatin** (Figure 21.3) (Letham 1973).

The molecular structure of zeatin is similar to that of kinetin. Both molecules are adenine or aminopurine derivatives. Although they have different side chains, in both cases the side chain is attached to the 6 nitrogen of the aminopurine. Because the side chain of zeatin has a double bond, it can exist in either the *cis* or the *trans* configuration. The naturally occurring free zeatin of higher plants has the *trans* configuration, although both the *cis* and the *trans* forms of zeatin are active as cytokinins when given to tissues in bioassays. A zeatin isomerase activity has been demonstrated in some plants, so *cis*-zeatin, when applied to tissues, may be converted to the *trans* form. This conversion may be the reason for the cytokinin activity of *cis*-zeatin.

Since its discovery in immature maize endosperm, zeatin has been found in many plants and in some bacteria. It is the most prevalent cytokinin in higher plants, but other substituted aminopurines that are active as cytokinins have been isolated from many plant and bacterial species. These aminopurines differ from zeatin in the nature of the side chain attached to the 6 nitrogen or in the attachment of a side chain to carbon 2 (see Figure 21.3). In addition, each of these can be present in the plant as a **riboside** (in which a ribose sugar is attached to the 9 nitrogen of the purine ring), a **ribotide** (in which the ribose sugar moiety is esterified with phosphoric acid), or a **glycoside** (in which a sugar molecule is attached to the 7 or 9 nitrogen of the purine ring, or to the oxygen of the zeatin or dihydrozeatin side chain) (see Figure 21.3) (Letham and Palni 1983).

Some Synthetic Compounds Can Mimic or Antagonize Cytokinin Action

Cytokinins are defined as compounds that have biological activities similar to those of *trans*-zeatin. These activities include the ability to "(1) induce cell division in callus cells in the presence of an auxin, (2) promote bud or root formation from callus cultures when in the appropriate molar ratios to auxin, (3) delay senescence of leaves, and (4) promote expansion of dicot cotyledons" (Cleland 1996: 127).

Many chemical compounds have been synthesized and tested for cytokinin activity. Analysis of these compounds provides insight into the structural requirements for activity. Nearly all compounds active as cytokinins are N^6-substituted aminopurines, and all the naturally occurring cytokinins are aminopurine derivatives. The cytokinin **benzylaminopurine** (**BAP**) is an example of a synthetic N^6-substituted aminopurine cytokinin (Figure 21.4A), as is kinetin. The only exceptions to this generalization are certain diphenylurea derivatives that show weak cytokinin activity. These diphenylurea derivatives are not N^6-substituted aminopurines, but they appear to be active as cytokinins by affecting the metabolism of endogenous cytokinins.

In the course of determining the structural requirements for cytokinin activity, investigators found that some molecules act as *cytokinin antagonists* (Figure 21.4B) (Hecht et al. 1971). These molecules are able to block the action of cytokinins, and their effects may be overcome by adding more cytokinin.

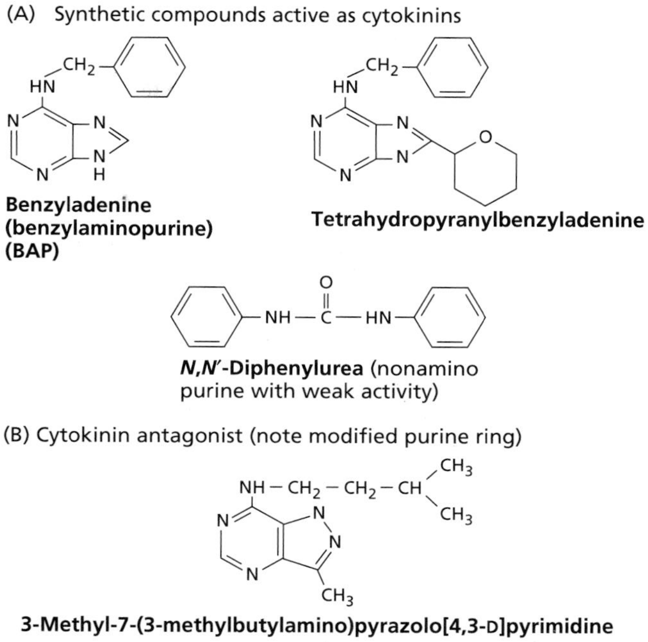

(A) Synthetic compounds active as cytokinins

Benzyladenine (benzylaminopurine) (BAP)

Tetrahydropyranylbenzyladenine

***N,N′*-Diphenylurea** (nonamino purine with weak activity)

(B) Cytokinin antagonist (note modified purine ring)

3-Methyl-7-(3-methylbutylamino)pyrazolo[4,3-D]pyrimidine

Figure 21.4 Structures of some synthetic cytokinins and a cytokinin antagonist. (A) In addition to kinetin, several synthetic compounds are active as cytokinins, such as benzylaminopurine. However, some diphenylurea compounds, which are not aminopurine derivatives, are weak synthetic cytokinins. (B) Modification of the purine ring results in compounds that are cytokinin antagonists. They block the action of cytokinin, possibly by competing with cytokinin for binding to its receptor.

Various Methods Are Used to Detect and Identify Cytokinins

Naturally occurring molecules with cytokinin activity may be detected and identified by a combination of physical methods and bioassays. The introduction of immunoaffinity purification methods combined with liq-

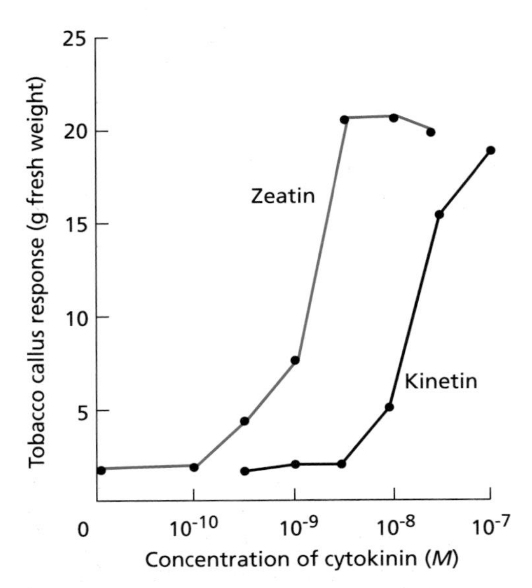

Figure 21.5 Tobacco tissues depend on cytokinin for their growth in culture. Cultured tobacco callus was transferred to fresh medium that contained an auxin and either zeatin or kinetin at the indicated concentrations. The tissues were weighed after growing for 1 month on the various media. The results show that zeatin is more effective than kinetin in supporting tobacco tissue growth. The maximum response was obtained with 5×10^{-8} *M* zeatin, and higher zeatin concentrations either did not stimulate additional growth or were slightly inhibitory. The kinetin concentration had to be at least tenfold higher to achieve the same growth stimulation. (From Leonard et al. 1968.)

uid chromatography and mass spectrometry, along with the use of isotopically labeled cytokinins as internal standards, has resulted in major advances in the measurement of endogenous cytokinins (Redig et al. 1996).

Historically, bioassay was the only method available for the identification of molecules that act as cytokinins. Cell proliferation bioassays using tobacco pith tissue or carrot taproot continue to be important because the ability of these tissues to initiate and sustain cell proliferation is proportional to the concentration of active cytokinin in an extract, provided the extract does not contain growth inhibitors. In addition, some continuously cultured callus tissues of tobacco and soybean that cannot grow in culture without cytokinin have been used for cytokinin bioassays. All of these cytokinin-requiring tissues exhibit a linear increase in growth with increasing cytokinin concentration over a fairly broad range, although high concentrations of cytokinin usually inhibit growth (Figure 21.5).

Immunological methods are very useful for cytokinin identification and quantification. Researchers can produce antibodies against cytokinins by injecting rabbits or mice with cytokinin ribosides conjugated to a protein. Monoclonal antibodies also have been generated that are highly specific for individual cytokinins. These antibodies can be used to quantitate the amount of a cytokinin in a sample by means of a radioimmunoassay (Weiler 1980). For cytokinin isolation, plant extracts are first fractionated, usually by high-performance liquid chromatography (HPLC), and the cytokinins in the fractions are detected and measured by means of a cytokinin radioimmunoassay, similar to the auxin radioimmunoassay described in Chapter 19.

The cytokinin antibodies can also be used to isolate the hormone from extracts by immunoaffinity chromatography (Akiyoshi et al. 1983). Immunological methods hold great promise for the identification and quantification of naturally occurring cytokinins because the antibodies are highly specific and more sensitive than most bioassays (Morris et al. 1991, Nicander et al. 1993). Furthermore, these immunological methods are very rapid.

Cytokinins Occur in Both Free and Bound Forms

Hormonal cytokinins are present as free molecules (not covalently attached to any macromolecule) in plants and certain bacteria. Cytokinins also occur as modified bases in certain transfer RNA molecules of all organisms (Hall et al. 1967).

Free cytokinins have been found in a wide spectrum of angiosperms and probably are universal in this group of plants. They have also been found in algae, diatoms, mosses, ferns, and conifers. Their regulatory role has been demonstrated only in angiosperms, conifers, and mosses, but they may function to regulate the growth, development, and metabolism of all plants. Usually zeatin is the most abundant naturally occurring free cytokinin, but *dihydrozeatin* and *isopentenyl adenine* (i^6Ade) also are commonly found in higher plants and bacteria. Numerous derivatives of these three cytokinins have been identified in plant extracts (see Figure 21.3).

Transfer RNA (tRNA) contains not only the four nucleotides used to construct all other forms of RNA but also some unusual nucleotides in which the base has been modified. Some of these "hypermodified" bases act as cytokinins when the tRNA is hydrolyzed and tested in one of the cytokinin bioassays. Some plant tRNAs contain zeatin as a hypermodified base, although it is present only in the *cis* isomer form, not the active *trans* configuration (Vremarr et al. 1972). However, cytokinins are not confined to plant tRNAs. They are part of certain tRNAs from all organisms, from bacteria to humans.

Some Plant Pathogenic Bacteria Secrete Free Cytokinins

Some bacteria and fungi are intimately associated with higher plants. Many of these microorganisms produce and secrete substantial amounts of cytokinins and/or cause the plant cells to synthesize plant hormones, including cytokinins (Akiyoshi et al. 1987). The cytokinins produced by microorganisms include *trans*-zeatin, i^{6}Ade, *cis*-zeatin, and their ribosides (see Figure 21.3). Infection of plant tissues with these microorganisms can induce the tissues to divide and, in some cases, to form special structures such as mycorrhizae, in which the microorganism can reside in a mutualistic relationship with the plant.

In addition to the crown gall bacterium, *Agrobacterium tumefaciens*, other pathogenic bacteria may stimulate plant cells to divide. For example, *Corynebacterium fascians* is a major cause of the growth abnormality known as witches'-broom. The shoots of plants infected by *C. fascians* resemble an old-fashioned straw broom because the lateral buds, which normally remain dormant, are stimulated to grow by the bacterial cytokinin (Hamilton and Lowe 1972; Nester and Kosuge 1981). Infection with a close relative of the crown gall organism, *Agrobacterium rhizogenes*, causes masses of roots instead of callus tissue to develop from the site of infection. *A. rhizogenes* is able to modify cytokinin metabolism in infected plant tissues through a mechanism that will be described later in this chapter.

Biosynthesis, Metabolism, and Transport of Cytokinins

The side chains of naturally occurring cytokinins are chemically related to rubber, carotenoid pigments, the plant hormones gibberellin and abscisic acid, and some of the plant defense compounds known as phytoalexins. All of these compounds are constructed, at least in part, from isoprene units (see Chapter 13).

$$CH_3-\overset{\overset{\displaystyle CH_3}{|}}{C}=CH-CH_3$$

Isoprene

Isoprene is similar in structure to the side chain of zeatin, of i^{6}Ade , and of other cytokinins (see Figure 21.3). These cytokinin side chains are synthesized from an isoprene derivative. Large molecules of rubber and the carotenoids are constructed by the polymerization of many isoprene units; cytokinins contain just one of these units. The precursor(s) for the formation of these isoprene structures are either mevalonic acid or pyruvate plus 3-phosphoglycerate, depending on which pathway is involved (see Chapter 13). These precursors are converted to the biological isoprene unit Δ^2-isopentenyl pyrophosphate (Δ^2-IPP).

Crown Gall Cells Have Acquired a Gene for Cytokinin Synthesis

Bacteria-free tissues from crown gall tumors proliferate in culture without the addition of any hormones to the culture medium. Crown gall tissues contain substantial amounts of both auxin and free cytokinins. Furthermore, when radioactively labeled adenine is fed to periwinkle (*Vinca rosea*) crown gall tissues, it is incorporated into both zeatin and zeatin riboside, demonstrating that gall tissues contain the cytokinin biosynthesis machinery (Peterson and Miller 1976). Control stem tissue, which has not been transformed by *Agrobacterium*, does not incorporate labeled adenine into cytokinins.

During infection by *Agrobacterium tumefaciens*, plant cells incorporate bacterial DNA into their chromosomes (Binns and Thomashow 1988; Zambryski et al. 1989). The virulent strains of *Agrobacterium* contain a large plasmid known as the **Ti plasmid**. Plasmids are circular pieces of extrachromosomal DNA that are not essential for the life of the bacterium. However, plasmids frequently contain genes that enhance the ability of the bacterium to survive in special environments. A small portion of the Ti plasmid, known as the **T-DNA**, is incorporated into the nuclear DNA of the host plant cell (Figure 21.6) (Chilton et al. 1977). T-DNA carries genes necessary for the biosynthesis of *trans*-zeatin and auxin, as well as a member of a class of unusual nitrogen-containing compounds called *opines* (Figure 21.7). Opines are not synthesized by plants except after crown gall transformation.

The T-DNA gene involved in cytokinin biosynthesis—known as the *ipt* gene—encodes an *i*so*p*entenyl *t*ransferase enzyme that transfers the isopentenyl group from isopentenyl pyrophosphate to AMP to form isopentenyl adenine ribotide (Akiyoshi et al. 1984). It also has been called the *tmr* locus because, when *inactivated* by mutation, it results in "rooty" tumors. Isopentenyl adenine ribotide can be converted to the active cytokinins *trans*-zeatin and dihydrozeatin by endogenous enzymes in plant cells. This conversion route is similar to the pathway for cytokinin synthesis that has been postulated for normal tissue (Figure 21.8).

The T-DNA also contains two genes encoding enzymes that convert tryptophan to the auxin indole-3-acetic acid (IAA). The pathway of tryptophan conversion differs from that in nontransformed cells and involves indoleacetamide as an intermediate. Because their promoters are plant eukaryotic promoters, none of these T-DNA genes are expressed in the bacterium, but are transcribed after they are inserted into the plant chromosomes. Transcription of the genes leads to synthesis of the enzymes they encode, resulting in the production of zeatin, auxin, and an opine. The bacterium can utilize the opine as a nitrogen source, but cells of higher plants cannot. Thus, by transforming the plant cells, the bacterium provides itself with an expanding environment (the gall tissue) in which the host cells are directed to produce a substance (the opine) that only the bacterium can utilize for its nutrition (Bomhoff et al. 1976).

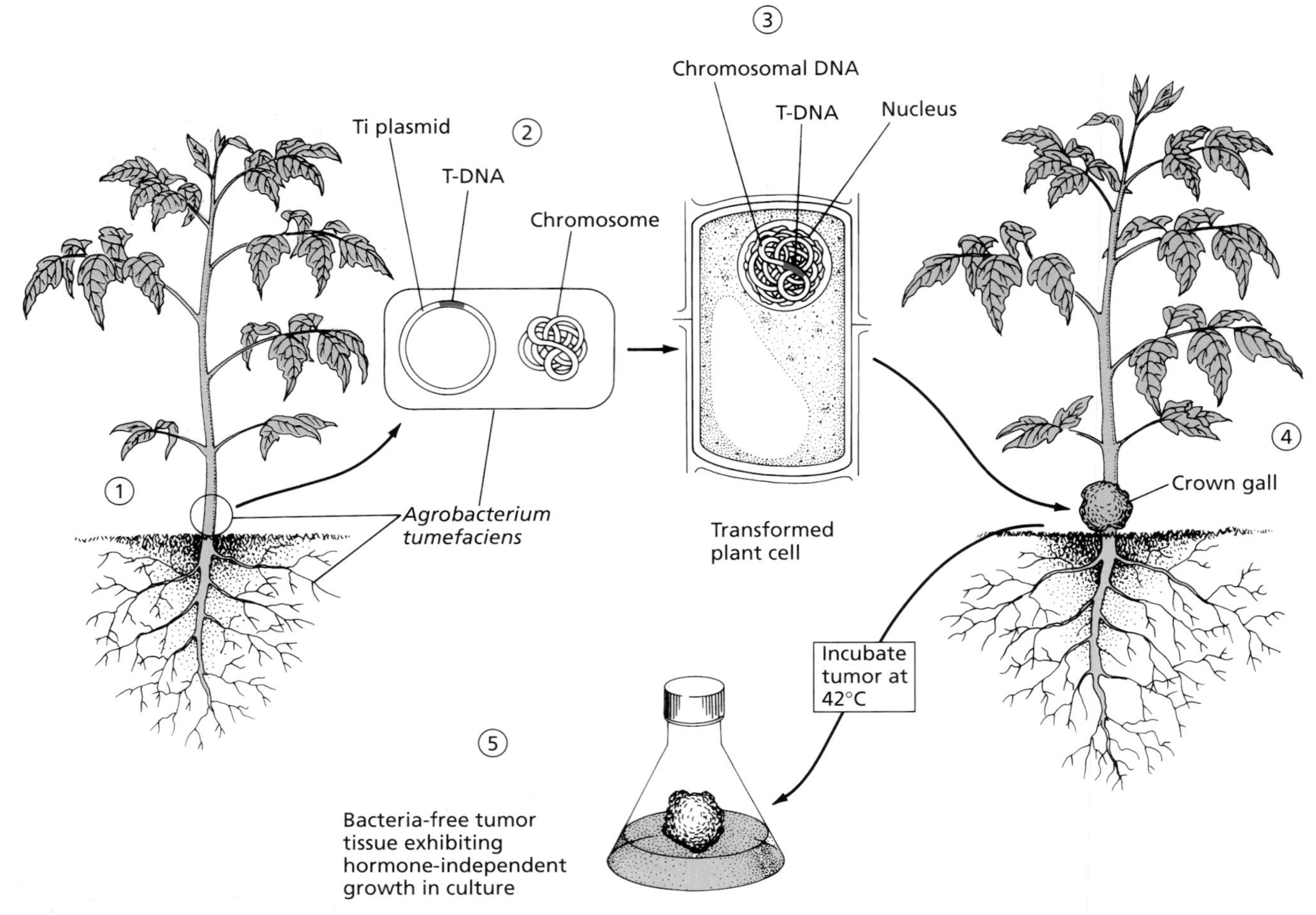

Figure 21.6 Tumor induction by *Agrobacterium tumefaciens*. The tumor is initiated when bacteria enter a lesion, usually near the crown of the plant (the junction of root and stem), and attach themselves to cells (1). A virulent bacterium carries, in addition to its chromosomal DNA, a Ti plasmid (2). The plasmid's T-DNA is introduced into a cell and becomes integrated into the cell's chromosomal DNA (3). Transformed cells proliferate to form a crown gall tumor (4). Tumor tissue can be "cured" of bacteria by incubation at 42°C. The bacteria-free tumor can be cultured indefinitely in the absence of hormones (5). (After Chilton 1983.)

There are important differences between the control of cytokinin biosynthesis in crown gall tissues and that in normal tissues, since the T-DNA genes for cytokinin synthesis are expressed even in cells in which the native plant genes for the hormone are completely shut down. Regulation of the expression of the T-DNA gene that encodes crown gall cytokinin synthase must be very different from that of the native higher-plant gene for the analogous enzyme.

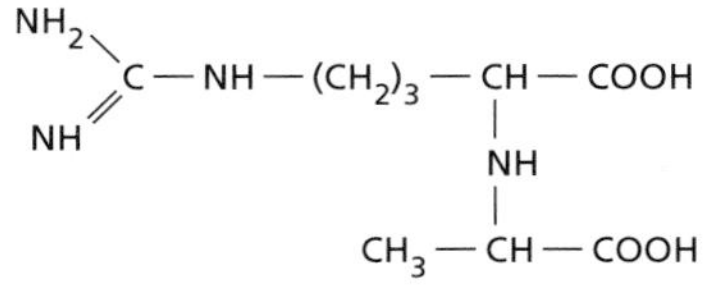

NH_2(NH=)C—NH—$(CH_2)_3$—CH(—COOH)—NH—CH(CH_3)—COOH

Octopine

NH_2(NH=)C—NH—$(CH_2)_3$—CH(—COOH)—NH—CH(COOH—$(CH_2)_2$—)—COOH

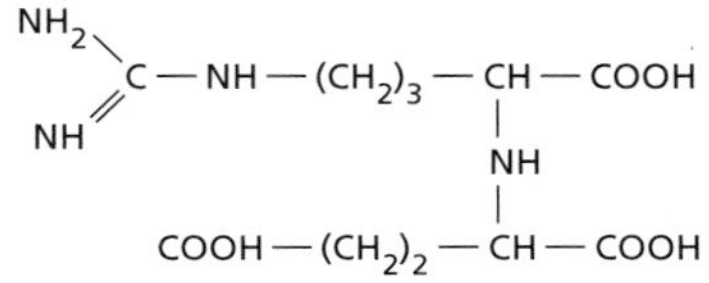

Nopaline

Figure 21.7 The two major opines, octopine and nopaline, are found only in crown gall tumors. The genes required for their synthesis are present in the T-DNA from *Agrobacterium tumefaciens*. The bacteria, but not the plant, can utilize the opines as a nitrogen source.

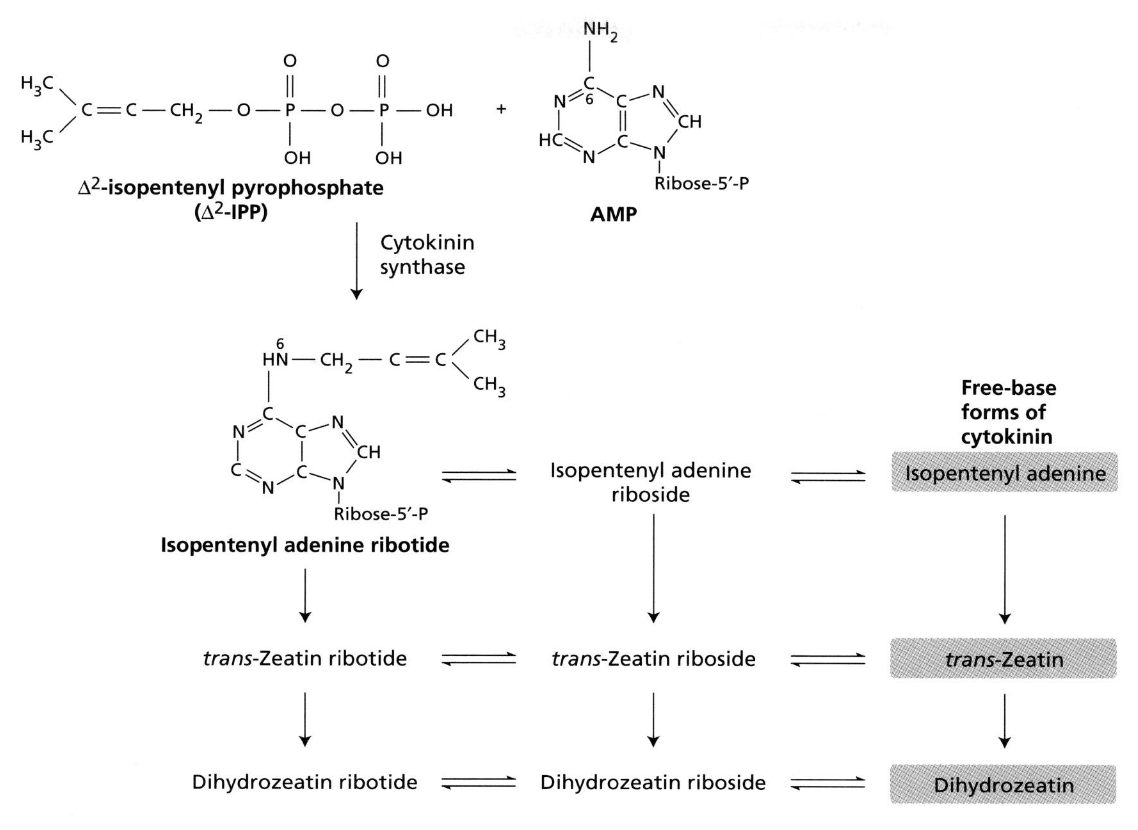

Figure 21.8 Scheme for the biosynthesis of some cytokinins. The first unique enzyme in cytokinin biosynthesis, cytokinin synthase, catalyzes the transfer of an isopentenyl group from isopentenyl pyrophosphate to the 6 nitrogen of adenosine monophosphate (AMP). The product, isopentenyl adenine ribotide, can be readily converted to the cytokinins *trans*-zeatin ribotide and dihydrozeatin ribotide. These, in turn, can be converted to the free-base forms, which are the naturally occurring cytokinins. (After Gan and Amasino 1996, modified from Kaminek 1992.)

Cytokinin Synthase Catalyzes the First Step in Cytokinin Biosynthesis

An enzyme that synthesizes cytokinins—called Δ^2-isopentenyl-pyrophosphate:AMP transferase, or **cytokinin synthase**—has been found in plants (but see Box 21.2). Cytokinin synthase is a type of prenyl transferase, similar to the prenyl transferase enzymes active in the synthesis of other isoprenoid compounds. The prenyl transferase that is active in cytokinin synthesis catalyzes the transfer of the isopentenyl group of Δ^2-IPP to adenosine monophosphate (AMP) (Chen and Melitz 1979). The product of this reaction, the ribotide of i^6Ade, is not a major cytokinin of higher plants. However, it is active as a cytokinin in bioassays and is readily converted to zeatin and other cytokinins (see Figure 21.8).

Although the activity of this enzyme in plant extracts has been described in some detail, it has yet to be purified, and a gene encoding it has not yet been identified or cloned from plants. However, the gene that encodes a functionally related bacterial enzyme, isopentenyl transferase, has been cloned from *Agrobacterium tumefaciens* (to be discussed shortly) (See Box 21.2).

The Cytokinins in tRNA Form after Polymerization

The tRNA cytokinins are synthesized by an entirely different route. Free cytokinins are not involved in the synthesis of the cytokinin-active bases in tRNAs. Instead, tRNAs are made from the four conventional nucleotides during transcription of the genes that encode their sequences. The first product is a tRNA precursor that lacks the hypermodified bases and is somewhat larger than the final product. This precursor is processed, much as mRNA is processed, to yield the functional tRNA molecule. In this processing, specific adenine residues of some of the tRNAs are modified to give the cytokinin.

As with the free cytokinins, Δ^2-isopentenyl groups are transferred to the adenine molecules from Δ^2-isopentenyl pyrophosphate by a prenyl transferase. This enzyme is not the same transferase that is involved in the synthesis of the free cytokinins. The transferase that

BOX 21.1

The Ti Plasmid and Plant Genetic Engineering

THE TI PLASMID (tumor-inducing plasmid) of *Agrobacterium tumefaciens* has been developed as a vehicle for introducing foreign genes into plants (see the figure). When *Agrobacterium* infects plants, a region of the Ti plasmid called the T-DNA is taken up by the plant cell and incorporated into one of its chromosomes. The genes in the T-DNA are referrred to as phyto-oncogenes because they induce neoplastic, or tumor-producing, growth (Matzke and Chilton 1981). To use the Ti plasmid as a vector for introducing new genes into plants, it is necessary to disarm the plasmid so that it does not cause tumors. Researchers accomplished this task by deleting the genes in the T-DNA that encode the enzymes controlling auxin and cytokinin synthesis. In additon, it is necessary to introduce a gene into the T-DNA that will enable the investigator to select the transformed cells. Genes for antibiotic resistance are normally used for this purpose (Fraley et at. 1985; Gleave 1992).

A cloned gene can then be inserted into the T-DNA of the engineered Ti plasmid, and the plasmid can be used to infect cultured cells, leaf discs, or root slices. The infected cells are placed on a culture medium that contains auxin and cytokinin (to induce growth) and the antibiotic. Only the transformed cells can grow in the presence of the antibiotic, because they have received the T-DNA containing not only the foreign gene but also the gene for antibiotic resistance. To obtain a plant containing the foreign gene, it is necessary to regenerate whole plants from the cultured, transformed cells. Fortunately, methods for accomplishing this regeneration have been developed for many plants, although not yet for all important crop species. The regeneration involves adjusting the ratio of cytokinin to auxin to stimulate both shoot and root formation.

Although the transformation of cereal crops with *Agrobacterium* and their regeneration from cultured cells have been difficult, some remarkable successes have resulted from this approach. Investigators have introduced numerous foreign genes into plants such as tobacco, including soybean storage protein genes. Genes for such desirable characteristics as disease resistance, herbicide resistance, and salt tolerance have been transferred to crop plants by these techniques, and commercial crops now are being grown with these genetically engineered strains.

Arabidopsis can also be transformed directly by vacuum infiltration of seeds (Feldmann and Marks 1987) or of whole plants (Bechtold et al. 1993) with *Agrobacterium* carrying engineered or wild-type Ti plasmids. During vacuum infiltration, seeds or whole plants are submerged in liquid medium containing *Agrobacterium* cells, and the air is removed from the intercellular spaces by a vacuum. This procedure enables the *Agrobacterium* cells to penetrate the tissues more deeply than they would otherwise, bypassing tissue culture steps and leading to the rapid generation of transformants. This approach is very useful for molecular genetic studies, such as for characterizing DNA sequences involved in the control of gene expression, or for blocking the expression of specific genes by introducing antisense constructs.

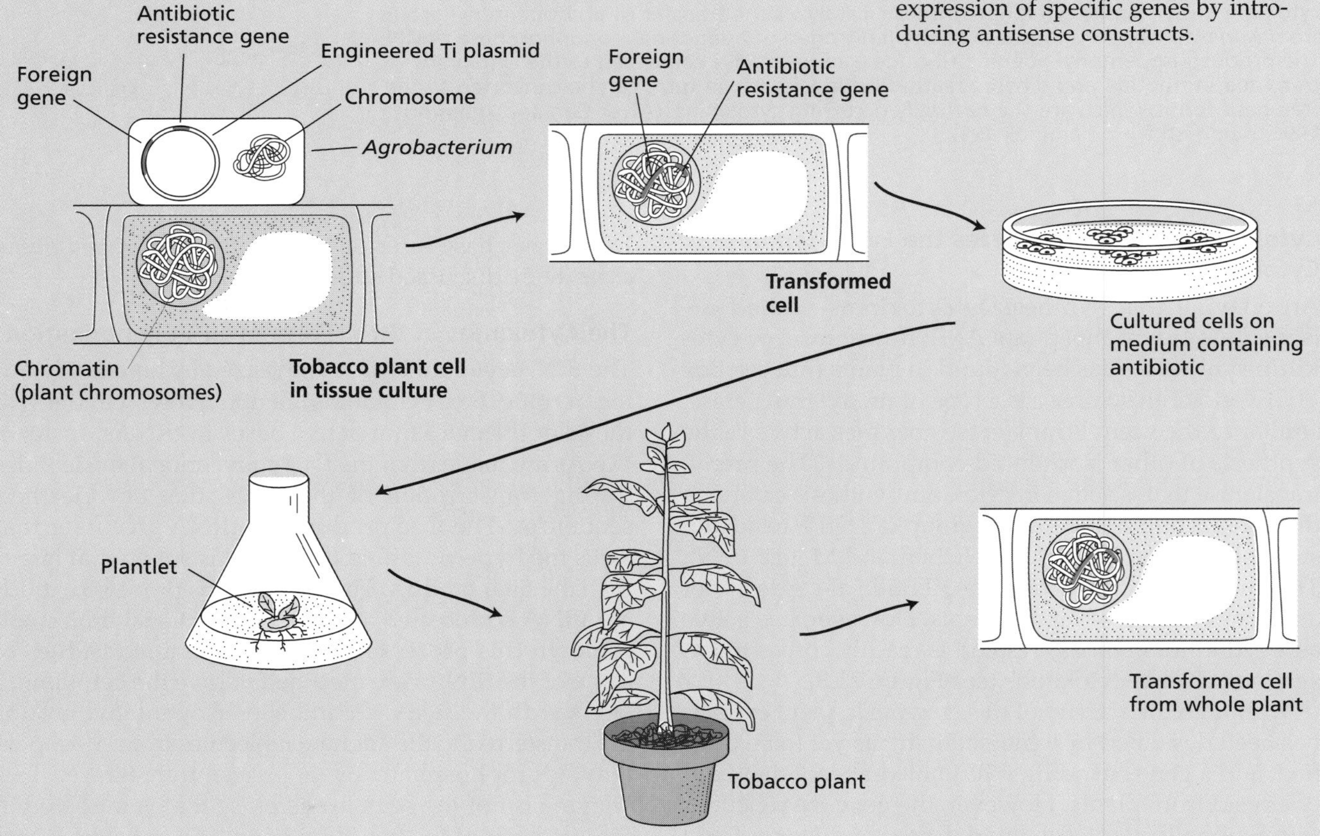

BOX 21.2

Are Symbiotic Bacteria the Source of Plant Hormonal Cytokinins?

PLANTS LIVE IN AN INTIMATE ASSOCIATION with many different microorganisms, most of which we never see and may even have difficulty detecting. Microorganisms not only surround the plant, particularly roots, but they often invade plant tissues, sometimes with beneficial consequences; in other cases the invasion leads to disease. Over the long course of evolution, plants and their specific associated microorganisms have coevolved intricate biochemical and molecular relationships. In Chapter 12 we discussed the importance of symbiotic mycorrhizal fungi in nutrient uptake by roots, as well as the elaborate molecular interrelationship between symbiotic nitrogen-fixing bacteria and the root nodules of legumes. Signaling molecules initiate and maintain these symbiotic or pathogenic relationships. Therefore, it is not surprising that plant-associated microorganisms have evolved the machinery to synthesize plant hormones. What better way to influence the growth and metabolism of the host plant?

We have already noted that the pathogenic bacteria *Corynebacterium fascians*, *Agrobacterium tumefaciens*, and *A. rhizogenes* secrete cytokinins at the site of infection. In addition to such well-known organisms, other less well understood microorganisms make their living on plant surfaces and within the intercellular spaces, such as those in the leaf mesophyll. These surface and endophytic bacteria include various species of *Methylobacterium*, which are present on and within most, if not all, plants (Holland and Polacco 1994).

In addition to utilizing methanol produced during plant metabolism, plant-associated *Methylobacterium* spp. have been shown to produce cytokinins. Recently Holland (1997) speculated that *Methylobacterium* spp. are in a symbiotic relationship with plants in which the bacteria utilize plant by-products that are toxic to the plant for their metabolism. In addition, Holland has put forward the provocative hypothesis that these bacteria serve as the sole source of cytokinins for the plant! Consistent with this idea, there is evidence that even normal developmental processes are impaired when plants are "cured" of their endophytic *methylobacteria*. The hypothesis is further supported by the failure, thus far, to isolate a plant gene for cytokinin synthase. Holland (1997) argues that the cytokinin synthase activity previously measured in plant extracts (Chen and Melitz 1979) is actually derived from bacterial endophytes.

Although highly controversial, the idea that plants might normally depend on endophytic bacteria for cytokinin synthesis needs to be explored. If confirmed, this relationship would represent a remarkable example of the evolution of plant–microbe interdependence. This issue also points out the value of genomic sequencing, for once the sequencing of the *Arabidopsis* genome is complete, it should be possible to identify the gene for cytokinin synthase if it is present.

synthesizes the tRNA cytokinins must be able to recognize a specific base sequence in the tRNA and transfer the isopentenyl group to the adenosine nearest the 3′ end of the anticodon (Figure 21.9). It does not utilize free AMP as a substrate.

The possibility that the free cytokinins are derived from tRNA has been explored extensively. Although the tRNA-bound cytokinins can act as hormonal signals for plant cells if the tRNA is degraded and fed back to the cells, it is unlikely that any significant amount of the free hormonal cytokinin in plants is derived from the turnover of tRNA.

Cytokinins from the Root Are Transported to the Shoot via the Xylem

Root apical meristems are major sites of synthesis of the free cytokinins in whole plants (Torrey 1976). The cytokinins synthesized in roots appear to move through the xylem into the shoot, along with the water and minerals taken up by the roots. This has not been proven to be the pathway for cytokinin movement, but circumstantial evidence suggests that it is.

When the shoot is cut from a rooted plant near the soil line, the xylem sap may continue to flow from the cut stump for some time. This xylem exudate contains cytokinins. If the soil covering the roots is kept moist, the flow of xylem exudate can continue for several days. Since its cytokinin content does not diminish, the cytokinins found in the exudate are likely to be synthesized by the roots. Also, environmental factors that interfere with root function, such as water stress, reduce the cytokinin content of the xylem exudate (Itai and Vaadia 1971).

Cytokinins are not produced exclusively by roots, however. Young maize embryos synthesize cytokinins, as do young developing leaves (Kende 1971). It is possible that cytokinins are synthesized in both root and shoot meristems during their active growth periods, but that the cytokinin production in shoot apical meristems is distributed more locally.

Cytokinins Are Transported in the Xylem as Zeatin Ribosides

The cytokinins in the xylem exudate are mainly in the form of zeatin ribosides. Once they reach the leaves, some portion of these nucleosides is converted to the free base or to glucosides (Noodén and Letham 1993). Cytokinin glucosides may accumulate to high levels in seeds and in leaves, and substantial amounts may be present even in senescing leaves. Although the glucosides are active as cytokinins in bioassays, often they lack hormonal activity after they form within cells, possibly because they are compartmentalized in such a way

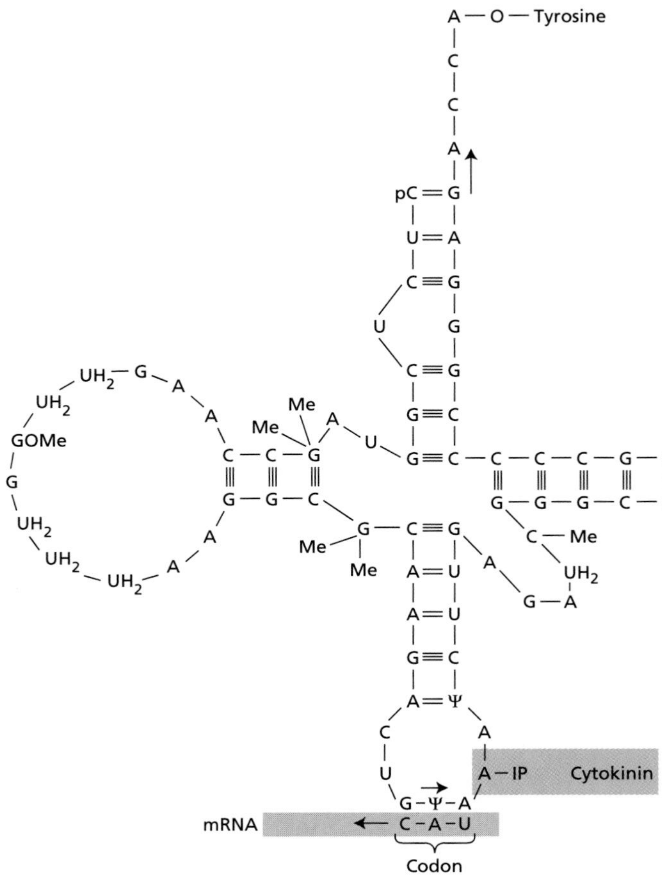

Figure 21.9 Some tRNAs contain a cytokinin. The cytokinin in the examples shown is i[6]Ado (A-IP). Although cytokinins are present in tRNA, tRNA is probably not the source of free cytokinins in plants. (After Hall et al. 1967.)

A—O—Tyrosine

A–IP Cytokinin

mRNA ← C-A-U

Codon

Yeast tRNA for tyrosine

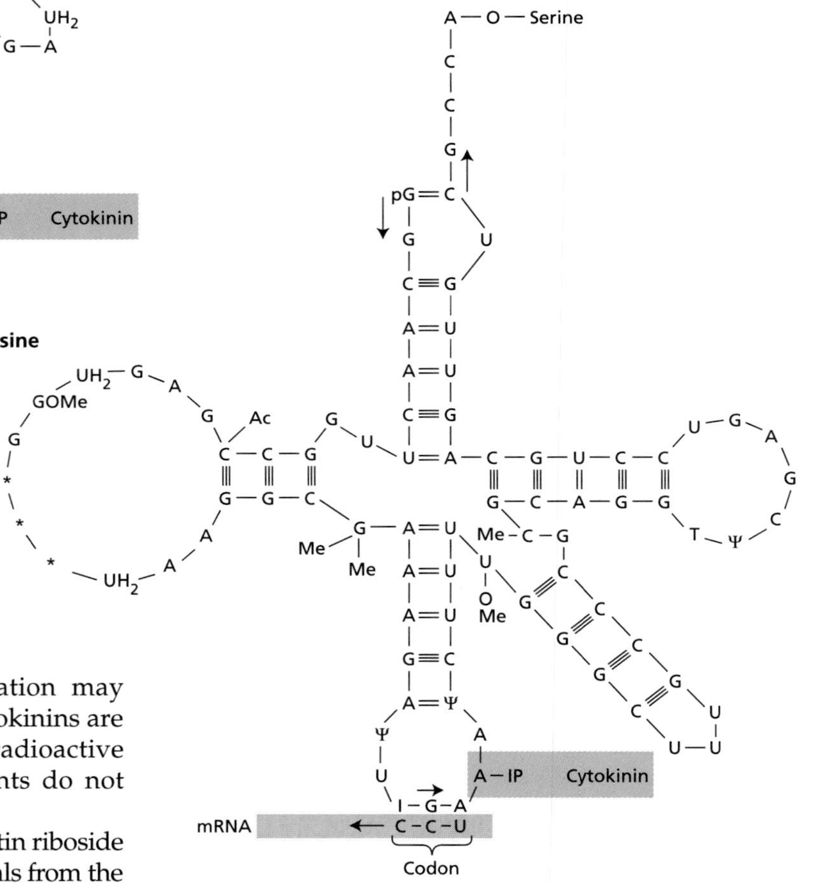

A—O—Serine

A–IP Cytokinin

mRNA ← C-C-U

Codon

Yeast tRNA for serine

that they are unavailable. Compartmentation may explain the conflicting observations that cytokinins are transported readily by the xylem but that radioactive cytokinins applied to leaves in intact plants do not appear to move from the site of application.

Evidence suggests that the transport of zeatin riboside from the root to the shoot is regulated by signals from the shoot. The *rms4* mutant of pea (*Pisum sativum* L.) is characterized by a 40-fold decrease in the concentration of zeatin riboside in the xylem sap of the roots. However, grafting a wild-type shoot onto an *rms4* mutant root increased the zeatin riboside levels in the xylem exudate to wild-type levels. Conversely, grafting an *rms4* mutant shoot onto a wild-type root lowered the concentration of zeatin riboside in the xylem exudate to mutant levels (Beveridge et al. 1997a). These surprising results suggest that a signal from the shoot can regulate cytokinin transport from the root. The identity of this signal has not yet been determined.

We will return to the *rms* mutants later in the chapter when we discuss the role of cytokinins in apical dominance.

Cytokinins Are Rapidly Metabolized by Plant Tissues

Most of the different chemical forms of cytokinins are rapidly interconverted by plant tissues (Letham and Palni 1983). Cytokinin bases, when given to many plant tissues, are converted to their respective nucleotides: zeatin to zeatin ribonucleotide, i^6Ade to i^6Ade ribonucleotide, and so forth. They also may be converted to their glucosides (Brzobohaty et al. 1994). However, glucosides sometimes are not readily converted to free cytokinins.

Cytokinin glucosides have been considered to be a storage form, or metabolically inactive state of these compounds. Nevertheless, glucosidases have been identified that will release the cytokinin base from cytokinin glucoside conjugates. For example, the *rolC* gene of *Agrobacterium rhizogenes* T-DNA encodes a glucosidase that can release free cytokinins* (Estruch et al. 1991). The activity of the *rolC* gene, along with the products of the other *A. rhizogenes* T-DNA genes, *rolA* and *rolB*, is responsible for the abnormal proliferation of roots induced by this bacterium.

A gene encoding another glucosidase that can release cytokinins from sugar conjugates has been cloned from maize, and its expression could play an important role in the germination of maize seeds (Brzobohaty et al. 1993). Dormant seeds often have high levels of cytokinin glucosides, but very low levels of hormonally active, free cytokinins. Levels of free cytokinins increase rapidly, however, as germination is initiated, and this increase in free cytokinins is accompanied by a corresponding decrease in cytokinin glucosides.

Many plant tissues contain the enzyme **cytokinin oxidase**, which converts zeatin, zeatin riboside, and i^6Ade to adenine or its derivatives (Kaminek and Armstrong 1990). This enzyme inactivates the hormone and could be important in regulating or limiting cytokinin effects. The activity of the enzyme is induced by high cytokinin concentrations, but its role in regulating the level of cytokinin in tissues is still unclear.

The Hormonally Active Cytokinin Is the Free Base

The free base is the hormonally active form of cytokinin. Other compounds are active as cytokinins, either because they are readily converted to *trans*-zeatin, dihydrozeatin, or isopentenyl adenine, or because they release these compounds from other molecules, such as cytokinin glucosides. For example, excised radish cotyledons grow when they are cultured in a solution containing the synthetic cytokinin base benzylaminopurine (BAP). The cultured cotyledons readily take up the hormone and convert it to various BAP glucosides, BAP ribonucleoside, and BAP ribonucleotide. When the cotyledons are transferred back to a medium lacking a cytokinin, their growth rate declines, as do the concentrations of BAP, BAP ribonucleoside, and BAP ribonucleotide in the tissues. However, the level of the BAP glucosides remains constant. This finding suggests that the glucosides cannot be the active form of the hormone. Tobacco cells in culture do not grow unless cytokinin ribosides supplied in the culture medium are converted to the free base. Thus, the free base is the most likely candidate for the active form of the hormone.

Cultured Cells Can Acquire the Ability to Synthesize Cytokinins

When cultured normal callus tissues of many species are subcultured repeatedly over a long period, they may become hormone autonomous; that is, they will grow on culture medium lacking an auxin or a cytokinin (Meins et al. 1980). This phenomenon is known as **habituation**. There are several possible explanations for the habituation of plant tissues. First, the tissues may acquire the capacity to synthesize their own cytokinin, and in some cases habituated callus tissues have been shown to contain cytokinins. Alternative explanations are that the rate of cytokinin destruction is reduced or that the tissues become more sensitive to cytokinins, such that lower levels of the hormone are effective.

Frederick Meins and his colleagues in Basel, Switzerland, have shown that different tissues of tobacco plants exhibit different capacities to synthesize cytokinins in tissue culture. Tobacco pith tissue is cytokinin dependent but capable of being habituated. In contrast, tissues cultured from the cortex of the stem are cytokinin autonomous; they can be cultured indefinitely on medium containing auxin as the only hormone supplement. Finally, tobacco leaf tissue has an absolute requirement for cytokinins and cannot be habituated (Meins 1989). These observations indicate that the ability to synthesize cytokinins can be switched on or off during development and that the on and off states can be stably maintained during cell divisions. Such developmental control of cytokinin synthesis may be important in morphogenesis.

The Biological Roles of Cytokinins

Plant hormones rarely, if ever, function alone. Even in cases in which a response can be evoked by application of a single hormone, the tissue may contain additional endogenous hormones that contribute to the response. In some cases we know that a response is evoked by two or more hormones, or that one hormone induces the synthesis or release of another hormone. Unless we can block the synthesis or action of the other endogenous hormones present, we cannot be certain that the response to a single exogenously applied hormone is as simple as it appears to be. Nevertheless, cytokinins can stimulate or inhibit a variety of physiological, metabolic,

* Note that bacterial genes, unlike those of plants, are written in lowercase italics.

biochemical, and developmental processes when they are applied to higher plants, and they probably play an important role in the regulation of these events in the intact plant (Binns 1994).

In this section we will survey some of the diverse effects of cytokinin on plant growth and development, including a discussion of its role in regulating the cell cycle. The discovery of the tumor-inducing Ti-plasmid in the plant pathogenic bacterium, *Agrobacterium tumefaciens*, provided plant scientists with a powerful new tool for introducing foreign genes into plants, and for studying the role of cytokinin in development. In addition to its effects on the cell cycle, cytokinin affects many other processes, including differentiation, apical dominance and senescence.

Cytokinin-Regulated Processes Are Revealed In Plants That Overproduce Cytokinin

The *ipt* gene from the Agrobacterium Ti plasmid, leading to cytokinin biosynthesis, has been introduced into many species of plants. These transgenic plants overproduce cytokinins as a consequence of *ipt* gene expression, resulting in an array of developmental abnormalities that tell us a great deal about the biological role of cytokinins. As discussed above, plant tissues transformed by Agrobacterium carrying a wild type Ti plasmid proliferate as tumors as a result of the overproduction of both auxin and cytokinin. What would happen if all of the other genes in the T-DNA were deleted and plant tissues were transformed with T-DNA containing only a selective antibiotic resistance marker gene and the *ipt* gene? This experiment has been done many times. Instead of tumors, tissues transformed with T-DNA carrying only the *ipt* gene produce shoots. The shoots formed by *ipt*–transformed tissues are difficult to root and, when roots are formed, they tend to be stunted in their growth. As a result, it is difficult to obtain plants from the regenerated shoots expressing the *ipt* gene under the control of its own promoter. To circumvent this problem, a variety of promoters from other plant genes have been used to drive the expression of the *ipt* gene in the transformed tissues. For example, Tom Guilfoyle and his colleagues at the University of Missouri regenerated tobacco plants with an *ipt* gene driven by an auxin-inducible promoter (Li et al. 1992). As a result, cytokinin overproduction was limited to regions of the plant where auxin signaling occurs, such as elongating internodes and expanding leaves. Roots formed, although their growth was reduced. These cytokinin-overproducing plants exhibit several characteristics that point to roles played by cytokinin in plant physiology and development:

1. The shoot apical meristems of cytokinin-overproducing plants produce more leaves.
2. The leaves have higher chlorophyll levels and are much greener.
3. Adventitious shoots may form from unwounded leaf veins and petioles.
4. Leaf senescence is retarded.
5. Apical dominance is greatly reduced.
6. The more extreme cytokinin-overproducing plants also are stunted, with greatly shortened internodes.
7. Rooting of stem cuttings is reduced, as is the root growth rate.

Some of the consequences of cytokinin overproduction could be highly beneficial for agriculture if synthesis of the hormone can be controlled. Since leaf senescence is delayed in the cytokinin-overproducing plants, it should be possible to extend their photosynthetic productivity (which we'll discuss shortly). In addition, tobacco plants transformed with an *ipt* gene under the control of the promoter from a wound-inducible protease inhibitor II gene were more resistant to insect damage. The tobacco hornworm consumed up to 70% fewer tobacco leaves in plants that expressed the protease inhibitor–driven *ipt* gene (Smigocki et al. 1993).

Genetics Can Be Used to Identify Cytokinin-Controlled Processes

Researchers can investigate the processes controlled by hormones by generating mutations in genes whose products are necessary for the synthesis, metabolism or action of the hormone. Whereas it has been possible to isolate numerous mutants that affect the synthesis, metabolism or action of auxin, gibberellin, ethylene, abscisic acid and brassinosteroid, there is a surprising scarcity of mutants that affect the cytokinins.

One approach to the identification of processes regulated by cytokinin is to screen mutagenized plants for developmental abnormalities and then determine whether any of these mutant plants overproduce cytokinins. This approach led to identification of the *Arabidopsis* mutant *amp1* (*a*ltered *m*eristem *p*rogram *1*) (Chaudhury et al. 1993).

The *amp1* mutation is recessive. Plants homozygous for *amp1* have an array of altered phenotypic characteristics when compared to wild-type plants, including an increased ability to form shoots from cultured tissues, a greater number of vegetative leaves initiated by the apical meristem, severely reduced apical dominance with development of a much larger number of inflorescence branches, additional cotyledons and abnormal flowers, and other phenotypic characteristics suggesting cytokinin overproduction (Figure 21.10). Plants homozygous for the *amp1* mutant gene have fourfold higher levels of zeatin and eightfold higher levels of dihydrozeatin.

All of the phenotypic characteristics of the *amp1* mutant can be induced in wild-type plants by treatment

Figure 21.10 The *amp1* mutation results in cytokinin overproduction in *Arabidopsis*. The wild type (A) and the homozygous *amp1* mutant (B) differ in the number of cotyledons, two versus four in A and B respectively, in 7-day-old vegetative plants. The mutant plant has a larger apical meristem, which produces a greater number of leaves in the same time period. The mutant also has much reduced apical dominance, which results in the development of many more inflorescence shoots during flowering (D) than in the wild type (C) (Chaudhury et al. 1993). (Courtesy of A. Chaudhury and S. Craig.)

with high concentrations of cytokinin. That is, treating wild-type seedlings with high concentrations of cytokinin gives a *phenocopy* of the mutant. Presumably the nonmutated form of the *AMP1* gene encodes a protein that participates in regulating either cytokinin biosynthesis or metabolism, but the gene affected by the *amp1* mutation has not yet been identified or further characterized. Nevertheless, this mutation is important for our understanding of the processes regulated by cytokinin in plants. It provides further independent confirmation that this hormone plays a vital part in regulating the formation and activity of the shoot apical meristems.

Cytokinin Overproduction Has Been Implicated in Genetic Tumors

Many species in the genus *Nicotiana* can be crossed to generate interspecific hybrids. More than 300 such interspecific hybrids have been produced; 90% of these hybrids are normal, exhibiting phenotypic characteristics intermediate between those of both parents. The plant used for cigarette tobacco, *Nicotiana tabacum*, for example, is an interspecific hybrid. However, about 10% of these interspecific crosses result in progeny that tend to form spontaneous tumors called **genetic tumors** (Figure 21.11) (Smith 1988).

Genetic tumors are similar morphologically to those induced by *Agrobacterium tumefaciens* (which we discussed at the beginning of this chapter), but genetic tumors form spontaneously in the absence of any external inducing agent. The tumors are composed of masses of rapidly proliferating cells in regions of the plant that ordinarily would contain few dividing cells. Furthermore, the cells divide without differentiating into the cell types normally associated with the tissues giving rise to the tumor. *Nicotiana* hybrids that produce genetic tumors have abnormally high levels of both auxin and cytokinins. Typically, the cytokinin levels in tumor-prone hybrids are five to six times higher than those found in either parent (Ichikawa and Syono 1991).

Figure 21.11 Expression of genetic tumors in a hybrid of *Nicotiana langsdorffii x N. glauca*. (From Smith 1988.)

The tumorous phenotype can be restored in a mutant hybrid that has lost the ability to form tumors by transformation with the *ipt* gene, which encodes an enzyme involved in cytokinin synthesis (see the previous section). When the *ipt* gene is expressed in these transgenic plants, these formerly nontumorous hybrid *Nicotiana* plants produced high levels of cytokinin and developed tumors (Feng et al. 1990).

The Ratio of Auxin to Cytokinin Regulates Morphogenesis in Cultured Tissues

Shortly after the discovery of kinetin, it was observed that cultured tobacco pith segments and tobacco callus tissues produce roots or shoots, depending on the ratio of auxin to cytokinin in the culture medium. High levels of auxin relative to kinetin stimulated the formation of roots, whereas high levels of cytokinin relative to auxin led to the formation of shoots. At intermediate levels the tissue grew as an undifferentiated callus (Figure 21.12) (Skoog and Miller 1965).

Investigators have used molecular genetic methods to investigate the significance of the auxin:cytokinin ratio in regulating morphogenesis in crown gall tissues. They mutated the T-DNA of the *Agrobacterium* Ti plasmid to observe the effects of different T-DNA genes on the growth and development of tumors (Garfinkel et al. 1981). Ti plasmids with mutated T-DNA genes induced the formation of a mass of proliferating shoots or roots instead of the usual tumors. These partly differentiated tumors are known as **teratomas**. When the *tmr* locus of the T-DNA was mutated, the teratomas consisted of an abnormal proliferation of roots. Mutations in the *tms* loci of the T-DNA resulted in an abnormal proliferation of shoots. The *tmr* mutants were said to be "rooty"; the *tms* mutants were dubbed "shooty."

Subsequently the shooty mutations were found to inactivate either of the two genes necessary for the synthesis of the auxin IAA. Similarly, the rooty mutants resulted from mutations in the *ipt* gene required for the synthesis of zeatin (Figure 21.13). Rooty tumors contain high levels of auxin relative to cytokinin, whereas shooty tumors have low levels of auxin and high levels of cytokinin (Akiyoshi et al. 1983). These findings further demonstrate the importance of the auxin:cytokinin ratio in regulating morphogenesis.

Cytokinins Modify Apical Dominance and Promote Lateral Bud Growth

One of the primary determinants of plant form is the degree of apical dominance (see Chapter 19). Plants with strong apical dominance, such as maize, have a single growing axis with few lateral branches. In contrast, many lateral buds initiate growth in shrubby plants. Although apical dominance may be determined primarily by auxin, physiological studies indicate that cytokinins play a role in initiating the growth of lateral buds. For example, direct applications of cytokinins to the axillary buds of many species stimulate cell division activity and growth of the buds. The phenotypes of the cytokinin-overproducing mutants are consistent with this result. Wild-type tobacco shows strong apical dominance during vegetative development, whereas the lateral buds of cytokinin overproducers grow vigorously, developing into shoots that compete with the main shoot. Consequently, cytokinin-overproducing plants tend to be bushy.

Despite the strong correlation between high cytokinin levels and the release from apical dominance already noted, recent work with peas has suggested that hormonelike signals other than cytokinin and auxin may also be involved in regulating the growth of axillary buds (Beveridge et al. 1997b). The *ramosus* mutants of pea (*Pisum sativum*) all exhibit increased branching compared to wild-type plants. Two of the mutants exhibit elevated auxin levels in the shoot, one of the mutants exhibits decreased cytokinin levels in the

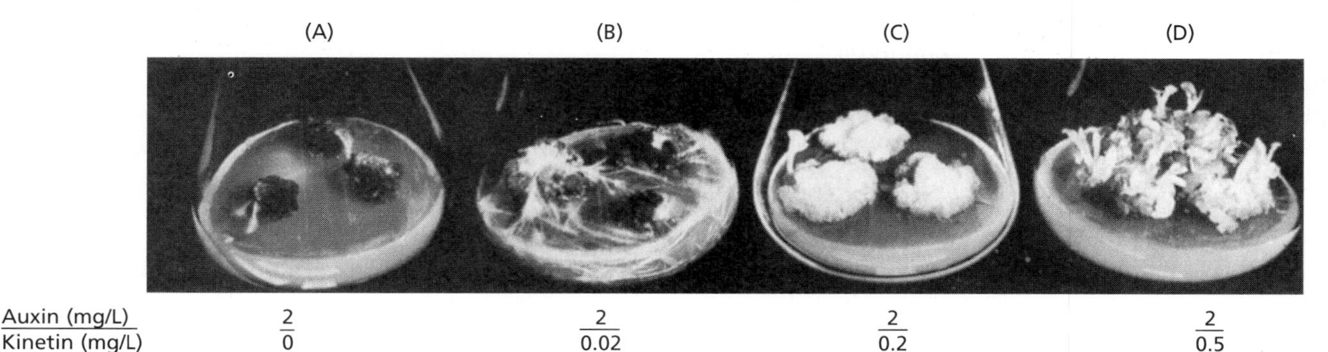

Figure 21.12 The regulation of growth and organ formation in cultured tobacco callus by auxin and cytokinin. These tissues were cultured for approximately 1 month on media that contained an auxin (indoleacetic acid) at a concentration of 11.4 μ*M* and a cytokinin (kinetin) at concentrations between 0 and 50 μ*M*. (A) In the absence of cytokinin, little or no cell division occurred. (B) At a high auxin to cytokinin ratio, roots differentiated. (C) At an intermediate auxin to cytokinin ratio an unorganized callus developed. (D) At a low auxin to cytokinin ratio, masses of buds differentiated. (From Skoog and Miller 1965, courtesy of F. Skoog.)

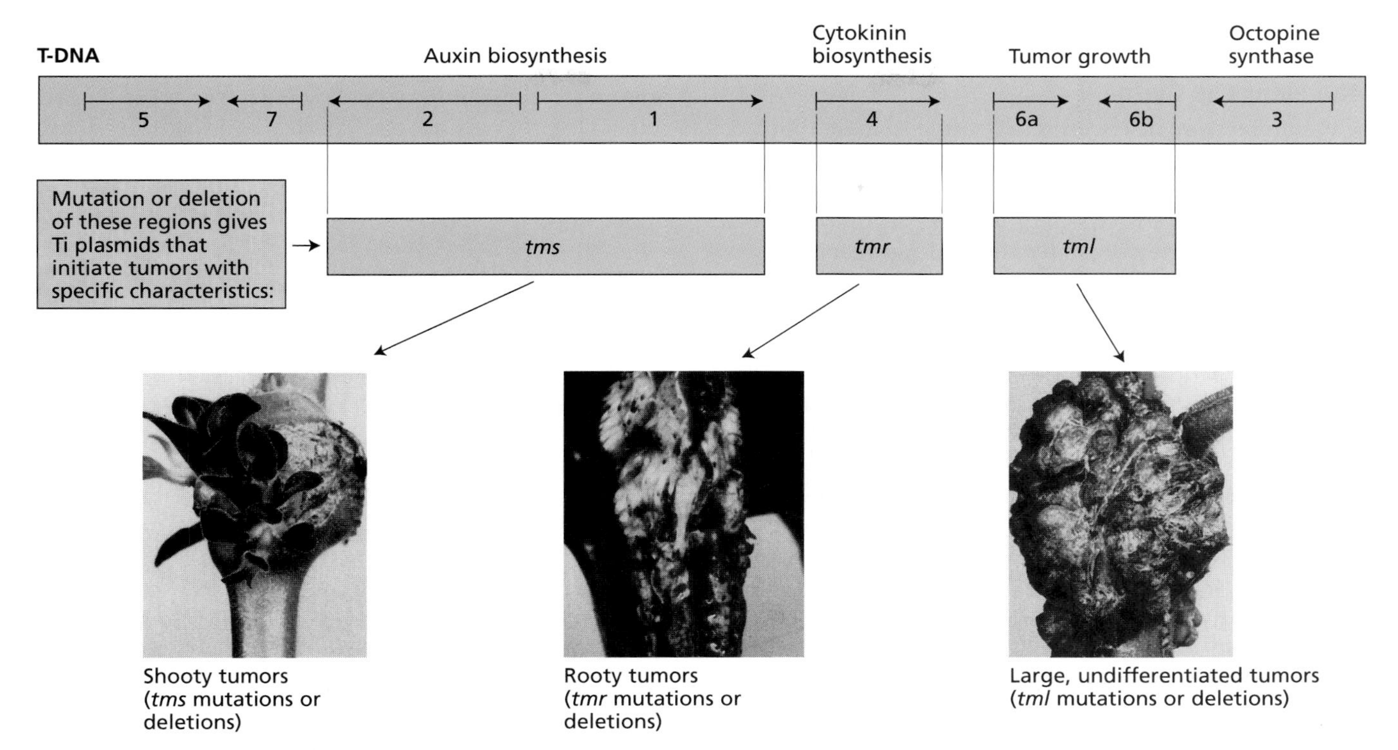

Figure 21.13 Map of the T-DNA from an *Agrobacterium* Ti plasmid, showing the effects of T-DNA mutations on crown gall tumor morphology. The shaded bar represents the T-DNA. The numbered arrows represent the locations of genes in the T-DNA that are expressed in plant cells transformed with the T-DNA. Genes 1 and 2 encode the two enzymes involved in auxin biosynthesis; gene 4 encodes a cytokinin biosynthesis enzyme. Mutations or deletions that inactivate genes 1 and 2 (the *tms* loci) result in a T-DNA that will produce "shooty" tumors. Mutations that inactivate gene 4 (the *tmr* locus) result in a T-DNA that causes "rooty" tumors to form on infected plants. Mutations in gene 6 (the *tml* locus) cause the formation of large, undifferentiated tumors. Gene 3 encodes octopine synthase, the enzyme responsible for the formation of octopine by the transformed cells. (From Morris 1986, courtesy of R. Morris.)

xylem sap from the root, and another mutant exhibits both an increase in shoot auxin levels and a decrease in cytokinin produced in the root. These changes are the opposite of those predicted if cytokinin stimulates, and auxin inhibits, the growth of axillary buds. They suggest that, at least in pea, an additional unknown signal from the root is involved in regulating branching in the shoot.

Cytokinins Induce Bud Formation in a Moss

Thus far we have restricted our discussion of the plant hormones to the angiosperms. However, many of the plant hormones are present and developmentally active in representative species throughout the plant kingdom. The moss *Funaria hygrometrica* is a well-studied example. The germination of moss spores gives rise to a filament of cells called a *protonema* (plural protonemata). The protonema elongates and undergoes cell divisions at the tip, and forms branches some distance back from the tip.

Protonemata pass through two histologically distinct stages, the *chloronema* and the *caulonema* stages. The upright "leafy gametophyte," on which the male and female gametangia form, arises from specific caulonemal cells. The transition from filamentous growth to leafy growth, like the initiation of branch filaments, begins with the formation of a swelling or protuberance near the apical ends of specific cells. An asymmetric cell division follows, creating the **initial cell**. The initial cell then divides mitotically to produce the **bud**, the structure that gives rise to the leafy gametophyte. During normal growth, buds and branches are regularly initiated, usually beginning at the third cell from the tip of the filament.

Light, especially red light, is required for bud formation in *Funaria*. In the dark, buds fail to develop, but the addition of cytokinin to the medium can substitute for the light requirement. Cytokinin not only stimulates normal bud development; it also increases the total number of buds. Even very low levels of cytokinin (picomolar) can stimulate the first step in bud formation: the swelling at the apical end of the specific caulonemal cell. As we will see at the end of the chapter, several lines of evidence suggest that calcium ions act as a second messenger in the cytokinin-induced formation of the initial cell.

Cytokinins and Auxin Help Regulate the Plant Cell Cycle

Cytokinins were discovered in relation to their ability to stimulate cell division in tissues supplied with an optimal level of auxin. Evidence suggests that both auxin and cytokinins participate in the regulation of the cell cycle and that they do so by controlling the activity of cyclin-dependent kinases. As discussed in Chapter 1, **cyclin-dependent protein kinases** (**CDKs**), in concert with their regulatory subunits, the **cyclins**, are enzymes that regulate the eukaryotic cell cycle. The expression of the gene that encodes the major CDK, *CDC2* (*c*ell *d*ivision *c*ycle), is regulated by auxin (see Chapter 19). In pea root tissues, *CDC2* mRNA was induced within 10 minutes after treatment with auxin, and high levels of CDK are induced in tobacco pith when it is cultured on medium containing auxin (John et al. 1993). However, the CDK induced by auxin is enzymatically *inactive*, and high levels of CDK alone are not sufficient to permit cells to divide.

In *Arabidopsis*, cytokinins stimulate the expression of the gene that encodes **δ3 cyclin**, a G_1-type cyclin (Soni et al. 1995). Cultured tissues also require a carbon source, such as sucrose, for cell proliferation in culture, since cultured tissues usually are not photosynthetically active. Sucrose stimulates the expression of another *Arabidopsis* G_1 cyclin gene coding for the protein **δ2**. Thus quiescent tissue, such as tobacco pith, may be induced to divide in culture through the synergistic action of auxin, cytokinins, and a carbon source, leading to the formation of an active CDK–G_1 cyclin complex. CDK activity then triggers the entry of these cells into the S phase of the cell cycle.

The stimulation of δ3 cyclin expression is probably not the only way in which cytokinin regulates the cell cycle. When cultured tobacco cells are specifically deprived of cytokinin, they stop dividing and arrest in the G_2 phase of the cell cycle, rather than at G_1 or at the boundary between G_1 and S, as would be predicted if the only role for cytokinin in regulating the cell cycle were to provide δ3 cyclin. When cytokinin is given to these G_2-arrested cells, they re-enter the cell cycle and initiate mitosis. Although cytokinin-induced mitosis is preceded by increased activity of the CDK–Cdc2 complex, the amount of the enzyme does not change. Instead, its activity is controlled by a change in CDK phosphorylation, apparently by the removal of the phosphate group from the inhibitory phosphorylation site (see Chapter 1). This result suggests that cytokinin controls the activity of a phosphatase whose substrate is the Cdc2 kinase with an inhibitory phosphate group (Zhang et al. 1996). The removal of the inhibitory phosphate group activates the CDK–cyclin complex, permitting it to phosphorylate other proteins to initiate mitosis.

These data suggest the direct involvement of cytokinins in cell cycle regulation. However, these results are only indications of how cytokinins *might* act to control cell cycle progression, and there are many gaps in our understanding. At present, we don't know how cytokinins regulate the transcription of δ3 cyclin or the phosphorylation of CDKs.

Cytokinins Delay Leaf Senescence

Leaves detached from the plant slowly lose chlorophyll, RNA, lipids, and protein, even if they are kept moist and provided with minerals. This programmed aging process leading to death is termed **senescence** (see Chapters 16 and 23). Leaf senescence is more rapid in the dark than in the light. Treating isolated leaves of many species with cytokinins will delay their senescence. Although applied cytokinins do not prevent senescence completely, their effects can be dramatic, particularly when the cytokinin is sprayed directly on the intact plant (Van Staden et al. 1988). If only one leaf is treated, it remains green after other leaves of similar developmental age have yellowed and dropped off the plant. Even a small spot on a leaf will remain green if treated with a cytokinin, after the surrounding tissues on the same leaf begin to senesce.

Although evidence suggests that young leaves can produce cytokinins, mature leaves do not. Mature leaves depend on root-derived cytokinins to postpone their senescence. Senescence is initiated in soybean leaves by seed maturation, a phenomenon known as **monocarpic senescence**, and can be delayed by seed removal. Although the seedpods control the onset of senescence, they do so by controlling the delivery of root-derived cytokinins to the leaves. The cytokinins involved in delaying senescence are primarily zeatin riboside and dihydrozeatin riboside, which are transported into the leaves from the roots through the xylem, along with the transpiration stream (Noodén et al. 1990).

To test the role of cytokinin in regulating the onset of leaf senescence, Richard Amasino and his colleagues at the University of Wisconsin transformed tobacco plants with a chimeric gene in which the promoter of the senescence-specific *Arabidopsis* cysteine protease was used to drive the expression of the *ipt* gene (Gan and Amasino 1995). Since the *ipt* coding sequence in this chimeric gene was transcribed only during leaf senescence, the transformed plants had wild-type levels of cytokinins and developed normally, up to the onset of leaf senescence. As the leaves aged, however, the cysteine protease promoter was activated, triggering the expression of the *ipt* gene within leaf cells just as senescence would have been initiated. The resulting elevated cytokinin levels not only blocked senescence, but also limited further expression of the *ipt* gene, preventing cytokinin overproduction (Figure 21.14). This result

Figure 21.14 Leaf senescence is retarded in a transgenic tobacco plant containing a cytokinin biosynthesis gene, *ipt*. Expression of the *ipt* gene is controlled by the promoter of a cysteine protease gene that is expressed in response to signals that induce senescence. The leaves of the plant expressing the *ipt* gene (left) remain green and photosynthetically active as a result of the autoregulated production of cytokinins in the mature leaves, while those of an age-matched control (right) are in an advanced state of senescence. (From Gan and Amasino 1995, courtesy of R. Amasino.)

strongly suggests that cytokinins are a natural regulator of leaf senescence. However, this straightforward hypothesis to explain the regulation of leaf senescence by cytokinin may be too simplistic. At the present time, little is known about most of the genes expressed during leaf senescence or how their expression is regulated.

Cytokinins Promote Nutrient Mobilization

Cytokinins influence the movement of nutrients into leaves from other parts of the plant, a phenomenon known as cytokinin-induced **nutrient mobilization**. This process is revealed when nutrients (sugars, amino acids, etc.) radiolabeled with ^{14}C or ^{3}H are fed to plants after one leaf or part of a leaf is treated with a cytokinin. Later the whole plant is subjected to autoradiography to reveal the pattern of movement and the sites at which the labeled nutrients accumulate.

Experiments of this nature have demonstrated that nutrients are preferentially transported to, and accumulate in, the cytokinin-treated tissues. It has been postulated that the hormone causes nutrient mobilization by creating a new source–sink relationship. As discussed in Chapter 10, nutrients translocated in the phloem move from a site of production or storage (the source) to a site of utilization (the sink). The metabolism of the treated area may be stimulated by the hormone so that nutrients move toward it. However, it is not necessary for the nutrient itself to be metabolized in the sink cells, since even nonmetabolizable substrate analogs are mobilized by cytokinins (Figure 21.15).

Cytokinins Promote Chloroplast Maturation

Although seeds can germinate in the dark, the morphology of dark-grown seedlings is very different from that of light-grown seedlings. As discussed in Chapter 17, dark-grown seedlings are said to be **etiolated**. The hypocotyl and internodes of etiolated seedlings are more elongated, cotyledons and leaves do not expand, and chloroplasts do not mature. Instead of maturing as chloroplasts, the proplastids of dark-grown seedlings develop into **etioplasts**, in which the inner membranes form a compact, highly regular lattice known as the prolamellar body (see Chapter 1). Etioplasts contain some carotenoids, which is why the etiolated seedlings appear yellow, but they do not synthesize chlorophyll or most of the enzymes and structural proteins required for the formation of the chloroplast thylakoid system and photosynthesis machinery. When seedlings germinate in the light, chloroplasts mature directly from the proplastids present in the embryo, but etioplasts also can mature into chloroplasts when etiolated seedlings are illuminated.

If the etiolated leaves are treated with cytokinin before being illuminated, they form chloroplasts with more extensive grana, and chlorophyll and photosynthetic enzymes are synthesized at a greater rate upon illumination (Lew and Tsuji 1982). These results suggest that cytokinins participate in regulation of the synthesis of photosynthetic pigments and proteins, along with other factors such as light, nutrition, and development.

As we discussed in Chapter 17, the *det* mutants of *Arabidopsis* develop many of the characteristics of light-grown seedlings when they are germinated in the dark. The *det* mutants have short, unhooked hypocotyls and expanded cotyledons, and the apical meristem initiates leaves that subsequently expand without light. Although the *det* mutants do not develop full photosynthetic competence in the dark (recall that the enzymatic conversion of protochlorophyllide *a* to chlorophyllide *a* is directly activated by light; see Chapter 7), there is partial development of the chloroplasts, including the expression of genes encoding many photosynthetic enzymes that are not present in the chloroplasts of dark-grown wild-type seedlings.

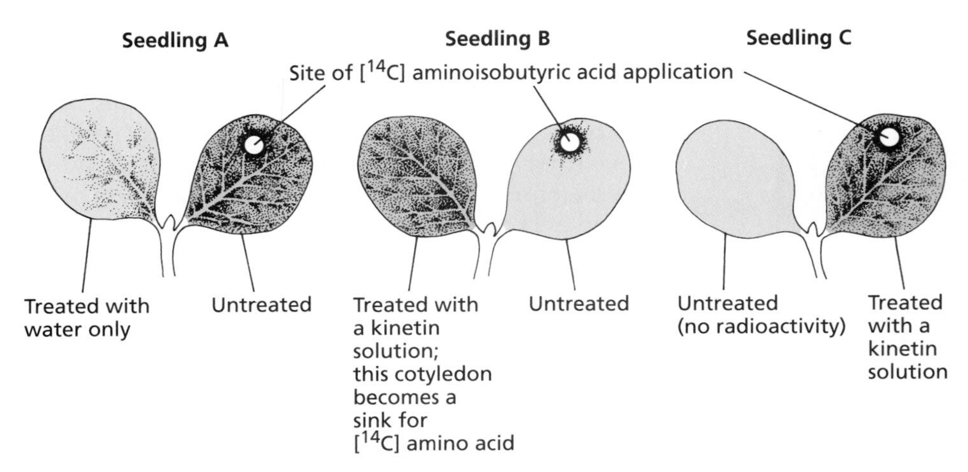

Figure 21.15 The effect of cytokinin on the movement of an amino acid in cucumber seedlings. A radioactively labeled amino acid that cannot be metabolized, such as aminoisobutyric acid, was applied as a discrete spot on the right cotyledon of each of these seedlings. In seedling A, the left cotyledon was sprayed with water as a control. The left cotyledon of seedling B, and the right cotyledon of seedling C, were each sprayed with a solution containing 50 m*M* kinetin. Later, the location of the radioactivity from the labeled amino acid was detected by autoradiography. The dark stippling represents the distribution of the radioactive amino acid as revealed by autoradiography. The results show that kinetin applied to the opposite cotyledon (seedling B) draws the nonmetabolizable amino acid away from the site of its application. The cytokinin-treated cotyledon has become a nutrient sink. However, radioactivity is retained in the cotyledon to which the amino acid was applied when the labeled cotyledon is treated with kinetin (seedling C). (Drawn from data obtained by K. Mothes.)

When wild-type *Arabidopsis* seedlings are germinated in darkness in the presence of cytokinin, they develop many of the characteristics of the *det* mutants: Hypocotyls shorten, cotyledons expand, the apical meristem initiates leaves, and there is partial development of the chloroplasts, including the synthesis of some photosynthetic enzymes. As Figure 21.16 shows, this effect is proportional to the cytokinin dosage given to the seedlings (Chory et al. 1994). Cytokinin is said to produce a *phenocopy* of the *det* mutations.

The ability of exogenous cytokinin to cause de-etiolation of dark-grown seedlings is mimicked by certain mutations that lead to cytokinin overproduction. Dark-grown *Arabidopsis* seedlings carrying the *amp1* mutation have zeatin levels that are approximately 2.5 times higher than those of the wild type and develop the characteristics of *det* mutant seedlings (Chory et al. 1994). Although the *det* mutants do not have elevated levels of cytokinins, they may have altered sensitivity to cytokinins, since *Arabidopsis* tissues from *det1* plants, when cultured in vitro, do not require cytokinin for growth and become green.

Although the functions of the genes altered by the *det* mutants are not completely understood, the *det2* mutation is in a gene required for the synthesis of brassinosteroids (see Chapter 17). Brassinosteroids and cytokinins may have antagonistic effects in regulating light-induced development. These results suggest that cytokinins may be part of the light signal transduction pathway that is responsible for the initiation of normal vegetative development and photosynthetic competence. However, the role for cytokinins in the developmental pathway initiated by light is not clearly established.

Cytokinins Promote Cell Expansion in Leaves and Cotyledons

Cytokinins can promote or inhibit cell enlargement. The promotion of cell enlargement by cytokinins is most clearly demonstrated in the cotyledons of dicots that have leafy cotyledons, such as mustard, cucumber, and sunflower. The cotyledons of these species expand as a result of cell enlargement during seedling growth. Cytokinin treatment promotes additional cell expansion, with no increase in the dry weight of the treated cotyledons.

Leafy cotyledons expand to a much greater extent when the seedlings are grown in the light than in the dark, and cytokinins promote cotyledon growth in both light- and dark-grown seedlings (Figure 21.17). As noted previously, cytokinins also promote the expansion of both leaves and cotyledons in dark-grown *Arabidopsis* seedlings, in which the cytokinin-promoted growth appears to be similar to the auxin-induced promotion of cell elongation in stems. As with auxin-induced growth, cytokinin-stimulated expansion of radish cotyledons is associated with an increase in the mechanical extensibility of the cell walls (Rayle et al. 1982). However,

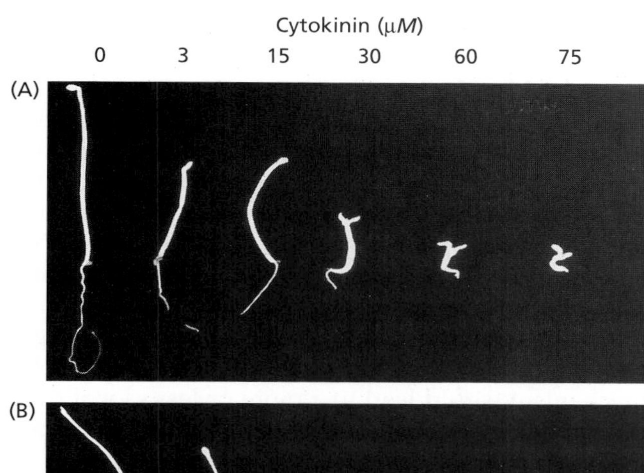

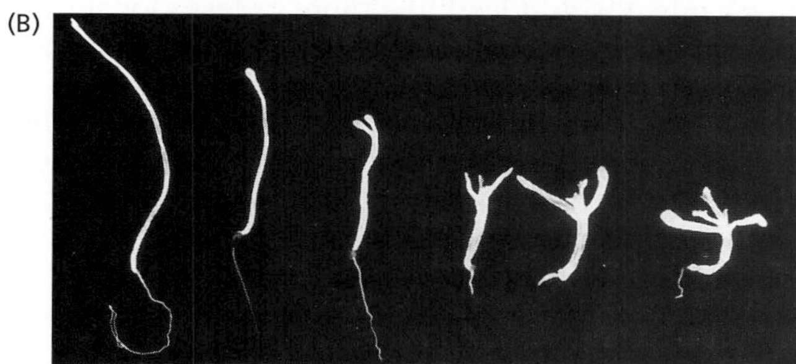

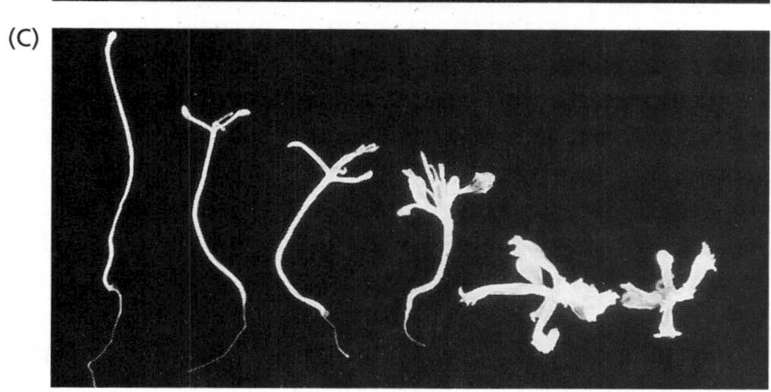

Figure 21.16 The effects of cytokinin on the development of wild-type *Arabidopsis* seedlings grown in darkness. (A–C) The appearance of the seedlings after 1, 2, and 3 weeks, respectively, of growth in the dark with increasing concentrations of cytokinin. The control (no cytokinin) is on the left in each case. The next five seedlings were treated with 3, 15, 30, 60, and 75 μ*M* of cytokinin, respectively. As the cytokinin concentration was increased, the inhibition of hypocotyl elongation became more pronounced, while the cotyledons expanded somewhat and leaves were initiated from the shoot apical meristem. At the higher cytokinin concentrations the seedlings were phenocopies of *det* mutants. Cytokinin treatment also resulted in thylakoid formation in the plastids of dark-grown seedlings (E) as compared to the development of plastids as etioplasts in the untreated, dark-grown wild-type control (D). (From Chory et al. 1994, courtesy of J. Chory, © American Society of Plant Physiologists, reprinted with permission.)

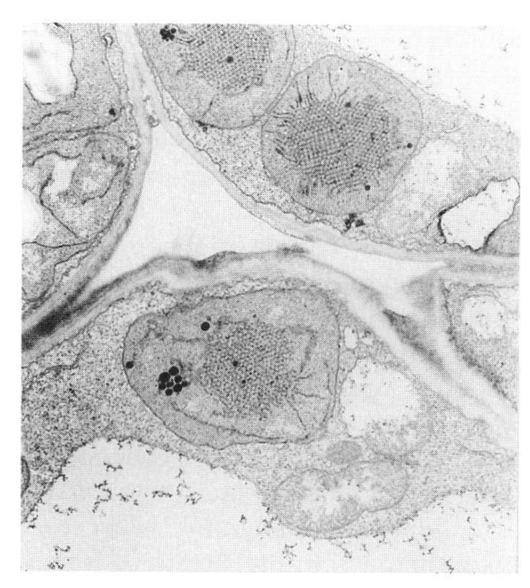

(D) Chloroplasts of dark-grown seedlings without cytokinin

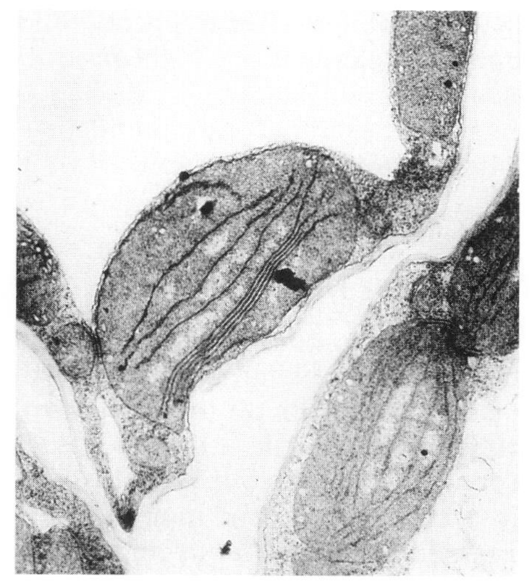

(E) Chloroplasts of dark-grown seedlings with cytokinin, showing thylakoid formation

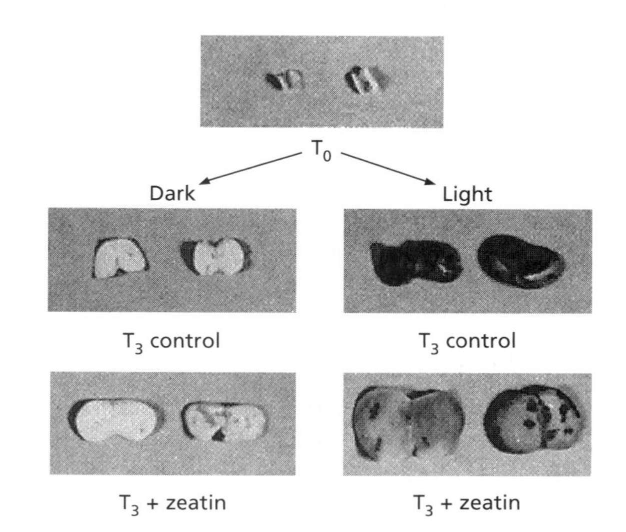

Figure 21.17 The effect of cytokinin on the expansion of radish cotyledons. The upper row shows germinating radish seedlings before the experiment began (T_0). The cotyledons were removed from the seedlings and incubated for 3 days (T_3) in either darkness or in the light in a phosphate buffer that either lacked a cytokinin (control) or contained 2.5 m*M* zeatin. The cotyledons treated with zeatin have expanded to a much greater extent than the controls, both in the light and in the dark. The effects of light and cytokinin are additive. (From Huff and Ross 1975.)

cytokinin-induced wall loosening is not accompanied by proton extrusion (Ross and Rayle 1982). Neither auxin nor gibberellin promotes cell expansion in cotyledons. Thus, cytokinins promote cell expansion in different tissues and by a different mechanism from that of either auxin or gibberellin.

In contrast to its promotive effects on the expansion of cotyledons and leaves, cytokinin usually inhibits cell elongation in stems and roots. Exogenous cytokinin inhibits hypocotyl elongation at concentrations that promote leaf and cotyledon expansion in the dark-grown seedlings. Internode and root elongation are both inhibited in transgenic plants expressing the *ipt* gene and in cytokinin-overproducing mutants, suggesting a role for cytokinin in the regulation of cell elongation. Most studies of the effects of applied cytokinins on stem elongation have used isolated stem segments in which cytokinins antagonize auxin-promoted cell elongation. It is also likely that the inhibition of hypocotyl and internode elongation by cytokinin is due to the production of the gaseous hormone ethylene and this inhibition thus may represent another example of the interdependence of hormonal regulatory pathways (which we'll discuss shortly).

Cellular and Molecular Modes of Cytokinin Action

The diversity of the effects of cytokinin on plant growth and development is consistent with the involvement of signal transduction pathways that have branches leading to specific responses. Although our knowledge of how cytokinin works at the cellular and molecular levels is still quite fragmentary, significant progress has been achieved. In this section we will examine a strong candidate for a cytokinin receptor and several cytokinin-regulated genes. We also consider the evidence that calcium acts as a second messenger in cytokinin action.

A Possible Cytokinin Receptor Has Been Identified

As noted in earlier chapters, plant hormones generally are assumed to interact with specific receptors that reside either on the cell surface or within the cytoplasm. Two candidates for a cytokinin receptor have recently been identified, one of which tends to fit the steroid hormone receptor model (a cytosolic receptor that migrates to the nucleus), while the other fits the membrane receptor model. It is possible, although unlikely, that both of these are cytokinin receptors.

O. N. Kulaeva at the Russian Academy of Sciences in Moscow isolated a 67-kilodalton protein, designated zeatin-binding protein (ZBP) from the cytosol of young barley plants (Kulaeva et al. 1995). ZBP has a high affinity for zeatin, and zeatin binding is highly specific. Furthermore, when ZBP–zeatin was added to an in vitro transcription elongation system prepared from isolated barley nuclei, it resulted in a threefold stimulation of incorporation of labeled nucleotides into RNA. Neither zeatin nor ZBP alone had any effect on RNA synthesis in this system. The stimulation occurred only when zeatin was bound to ZBP. The stimulation involved the activity of both RNA polymerase I and RNA polymerase II, so the transcription of both ribosomal RNA and protein-encoding genes was increased. Apparently, ZBP–zeatin did not affect transcriptional initiation, so this mechanism could not explain how cytokinin regulates the expression of any specific gene. Rather, this mechanism would lead to a more or less global transcriptional enhancement, and the nature of the genes expressed would be determined by other factors.

A molecular genetic approach to identifying a probable cytokinin receptor was taken by Tatsuo Kakimoto at Osaka University in Japan. He generated dominant, gain-of-function mutations in *Arabidopsis* that caused the mutants to be more sensitive than the wild type to endogenous cytokinin levels (Kakimoto 1996). Before presenting Kakimoto's discovery, we will digress for a moment to describe the technique he used, called **activation tagging**.

A **dominant mutation** causes a gene to be expressed at a time or place in which it would be silent in the wild type. Dominant mutations result in new phenotypic characteristics. By identifying the genes affected, one can obtain information about how a process is regulated in a way that is not possible with recessive, loss-of-function mutations. Although dominant mutations can occur spontaneously, they can also be induced experimentally by transformation of plants with a T-DNA element carrying a strong promoter, such as that of the cauliflower mosaic virus (CaMV) 35S promoter. The 35S promoter can then initiate transcription if the T-DNA becomes integrated into a plant chromosome near the transcription start site of a gene. Another advantage of this approach is that the gene is tagged by the T-DNA insertion so that it is relatively easy to clone. Activation tagging has been used to select both tobacco and *Arabidopsis* cell lines that will grow in the absence of auxin or cytokinin (Walden et al. 1994). Plants have been regenerated from several of these mutant cell lines, and some of these resemble cytokinin-overproducing mutants.

Kakimoto screened *Arabidopsis* callus tissue for the ability to form green, growing callus tissue when cultured on medium lacking cytokinin, after the tissue had been transformed with a T-DNA construct containing the 35S promoter. The callus tissue was derived from *Arabidopsis* hypocotyls, which normally require cytokinin for growth, greening, and the formation of shoots in culture. The screen that Kakimoto used was designed to identify genes within the cytokinin signal transduction pathway whose expression could activate cytokinin-induced growth and development. After screening 50,000 callus tissue samples, he obtained six mutant calluses that could grow in the absence of cytokinin. In four of the six mutants, the T-DNA tagged the same gene, which he designated ***CKI1*** (*cytokinin independent 1*). The plants regenerated from these mutant calluses did not form roots or normal flowers and were sterile.

The *CKI1* gene encodes a 125-kilodalton protein whose amino acid sequence is similar to that of the ethylene receptor, encoded by the *ETR1* gene (see Chapter 22) (Chang et al. 1993). The amino acid sequences of both proteins contain a potential histidine kinase domain as well as other similarities to the two-component regulatory systems of bacteria (see Chapter 14). Bacterial two-component regulatory systems mediate a range of responses to environmental stimuli, such as osmoregulation and chemotaxis.

Typically, bacterial two-component systems are composed of two functional elements: a sensor, to which a signal binds, and a response element (usually a histidine kinase). Often these functions reside in separate proteins, but occasionally they are in the same protein, as is the case with the ETR1 protein. They also are integral membrane proteins, with a hydrophobic region that crosses the membrane. The histidine kinase activity is in the carboxy-terminal region of the protein, facing the cytosol. The amino-terminal domain contains the hormone-binding site and resides on the outer face of the plasma membrane, or in the periplasmic space in the case of bacteria. The binding of an effector to the protein induces histidine kinase activity, which results in autophosphorylation of a histidine residue elsewhere on the protein.

The CKI1 protein has not yet been shown to bind cytokinin, but the ETR1 protein is an ethylene receptor and binds ethylene with a high affinity and specificity. The strong homology in the histidine kinase domains of these two proteins argues that the CKI1 protein will be found to bind cytokinin and that it is a cytokinin receptor. As will be described shortly, microinjection studies suggest that the cytokinin receptor is located on the cell surface rather than in the cytosol. Since the CKI1 protein is nearly twice the size of the cytokinin-binding protein found by Kulaeva and is localized in the plasma membrane, it is unlikely that CKI1 and ZBP are the same receptor. Furthermore, CKI1, if it is a cytokinin receptor, is unlikely to directly affect either transcriptional elongation or initiation. By analogy with other two-component receptors, it is more likely to initiate a signal transduction pathway.

Cytokinins Increase the Abundance of Specific mRNAs

Inasmuch as cytokinins initiate or repress numerous developmental pathways, one might expect these hormones to control the synthesis of many proteins, either by regulating the transcription of the genes that encode these proteins, or by having a posttranscriptional effect. To identify genes whose transcription is positively or negatively controlled by cytokinin, researchers have constructed cDNA libraries from cytokinin-treated and cytokinin-starved tissues. The cDNAs whose abundance is affected by cytokinin are then identified by differential or subtractive hybridization. (For a discussion of cDNA libraries and hybridization, see Chapter 19.)

This approach has been taken with many different species and tissues, resulting in the identification of various genes whose transcript abundance is either increased or decreased by cytokinin. For example, 20 cDNA clones corresponding to 20 different mRNAs that increased 2- to 20-fold within 4 hr of cytokinin addition, were isolated from soybean suspension cultures (Crowell et al. 1990). These changes in mRNA levels preceded the greening and cell growth induced by cytokinin in this system. One of these mRNA sequences, *CIM1*, was increased by up to 20-fold with concentrations of BAP (benzylaminopurine) as low as 10^{-8} *M*. When the cytokinin-regulated cDNAs were sequenced, *CIM1* was found to encode a protein with strong homology to a major pollen allergen, and two others encode ribosomal proteins (Crowell 1994). Most of the other cytokinin-induced transcripts showed no similarity to known genes, and their identities remain to be determined. However, since treatment with cycloheximide before cytokinin addition enhances, rather than blocks, the accumulation of these messages (Crowell et al. 1990), they fall under the category of *primary response genes*, as discussed in Chapters 17 and 19.

Nitrate reductase (discussed in Chapter 12) is a well-studied example of an enzyme whose synthesis is induced by cytokinin. Nitrate is the most common form of metabolizable nitrogen available to the plant, but it must be reduced before it can be utilized for the synthesis of nitrogenous compounds. Nitrate reductase, a cytosolic enzyme required for nitrate utilization, reduces nitrate to nitrite ($NO_3^- \rightarrow NO_2^-$). Nitrite then is further reduced to ammonium (NH_4^+) by the chloroplast enzyme nitrite reductase. Nitrate utilization is induced by light in germinating seedlings as they switch from heterotrophic growth (growth that depends on stored food reserves) to autotrophic growth (growth that is fueled by photosynthesis). Light triggers the expression of both the nuclear-encoded nitrate reductase gene and the chloroplastic nitrite reductase gene, as well as additional genes encoding other participants in this important metabolic activity. Cytokinin can stimulate the synthesis of nitrate reductase in etiolated barley leaves, where it at least partly substitutes for light in the induction of the enzyme (Lu et al. 1990). The cytokinin-stimulated increase in activity of the nitrate reductase enzyme is accompanied by an increase in nitrate reductase mRNA. Since inhibitors of both gene transcription and protein synthesis block cytokinin-stimulated nitrate reductase, we can conclude that cytokinin regulates nitrate reductase by controlling transcription of the nitrate reductase gene.

The *zea3* mutant of the tobacco species *Nicotiana plumbaginifolia* provides further indication that cytokinins participate in the induction of nitrate reductase. A mutation in the *ZEA3* gene causes the loss of light-induced nitrate reductase activity (Faure et al. 1994). The *ZEA3* gene was independently identified as a gene that, when mutated, confers resistance to high levels of cytokinin. This finding suggests that the *ZEA3* gene product may be involved in the response to cytokinin, and provides further evidence that part of the cytokinin signal transduction pathway is essential for the induction of nitrate reductase gene expression and the synthesis of the enzyme.

Cytokinins Can Regulate Gene Expression at a Posttranscriptional Step

Some of the effects of cytokinins on mRNA abundance are known to be the result of posttranscriptional regulation. For example, Elaine Tobin and her colleagues at the University of California, Los Angeles, have studied the role of cytokinins in chloroplast maturation in the small aquatic duckweed *Lemna gibba* (Silverthorne and Tobin 1990a). *Lemna* can be grown heterotrophically in the dark for long periods of time if sucrose is added to the culture medium. When green *Lemna* plants are placed in the dark, the nuclear-encoded mRNAs for two important chloroplast proteins, the small subunit of ribulose-1,5-bisphosphate carboxylase (SSU) and the major chlorophyll *a/b*–binding polypeptide of light-harvesting complex II (LHCB), decrease in abundance over the course of 7 days. As a result, the synthesis of SSU and LHCB proteins decreases drastically. However, the decrease in abundance of these mRNAs can be inhibited either by a pulse of dim red light or by the addition of cytokinin to the culture medium. The abundance of *LHCB* mRNA, for example, is about 5-fold greater in the cytokinin-treated plants than in the dark-grown controls. The combination of cytokinin and red light caused a 9.5-fold increase relative to the controls (Silverthorne and Tobin 1990b).

How does cytokinin prevent the decrease in these two important mRNAs in the dark? To determine whether cytokinin treatment influences the *LHCB* mRNA levels by increasing the rate of transcription or by decreasing the rate of mRNA turnover, Flores and Tobin (1988) compared the total abundance of the *LHCB* transcripts with the level of transcription of these genes in nuclei isolated from cytokinin-treated and control *Lemna* using nuclear runoff assays (described in Chapter 17). Briefly, isolated nuclei are allowed to extend their transcripts (initiated in vivo) in the presence of ^{32}P-labeled nucleotides. The labeled RNA is then hybridized to filters containing cloned *LHCB* sequences. The amount of radioactivity bound corresponds to the amount of transcript synthesized.

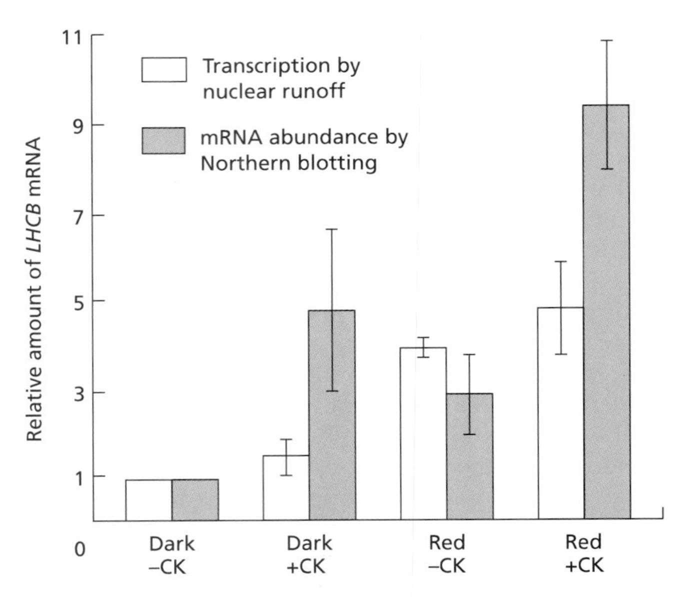

Figure 21.18 Effects of a cytokinin (benzyladenine) and dim red light on the transcription versus total abundance of *LHCB* (chlorophyll *a/b*–binding polypeptide of light-harvesting complex II) mRNA in *Lemna gibba*. Transcription was determined by nuclear runoff experiments with isolated nuclei. Total abundance was measured by hybridization of ^{32}P-labeled cDNA probes to total RNA on Northern blots. In the dark, cytokinin (CK) caused a 5-fold increase in the total mRNA, but only a 1.5-fold increase in transcription. In contrast, red light alone caused a 4-fold enhancement of transcription. The results indicate that red light alters transcription, whereas cytokinin affects a posttranscriptional step. The error bars represent the standard error. (Data from Flores and Tobin 1988.)

As Figure 21.18 shows, cytokinin treatment in the dark caused a 5-fold increase in *LHCB* mRNA abundance, but only a 50% enhancement of transcription. In contrast, a 1-minute exposure to red light in the absence of cytokinin stimulated transcription 4-fold. When cytokinin was given along with red light, a 5-fold increase in the amount of transcript was observed, while transcription was increased by only about 25%. These results suggest that transcription of the gene is mainly under the control of the phytochrome system (see Chapter 17). Cytokinin appears to increase the abundance of *LHCB* mRNA at a posttranscriptional level, possibly by making the mRNA more stable.

Cytokinin-Resistant Mutants May Be Blocked at Steps in the Signaling Pathway

Studies have been initiated to identify genes that are required for cytokinin action by identifying mutations that affect the ability of plants to respond to the hor-

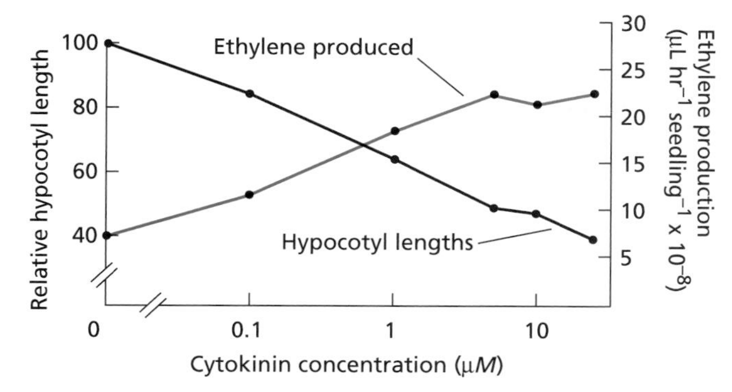

Figure 21.19 Cytokinins stimulate ethylene production by dark-grown *Arabidopsis* seedlings. Ethylene is produced by dark-grown *Arabidopsis* seedlings in increasing amounts as a function of the concentration of cytokinin applied. Ethylene produced during a 24-hour period from the third to the fourth day after germination was measured and plotted as a function of cytokinin concentration. Hypocotyl lengths, also plotted as a function of cytokinin concentration, decreased with increasing cytokinin and ethylene production. The arrows indicate the rate of ethylene production that resulted in a 50% inhibition of hypocotyl elongation. (From Cary et al. 1995, © American Society of Plant Physiologists, reprinted with permission.)

mone. This general approach has yielded important insights into the physiology of the other plant hormones, particularly abscisic acid, gibberellin, and ethylene, but the molecular genetic dissection of the cytokinin signal transduction pathway has not advanced with similar speed. To study this pathway, researchers selected mutants that are defective in their response to cytokinin by screening plants for their ability to develop normally when exposed to cytokinins at levels that cause abnormal growth and morphogenesis in wild-type plants. With this approach, five mutants were selected in *Arabidopsis*, all of which are defective in the same gene, which was designated *CKR1* (*c*ytokinin *r*esistant *1*) (Su and Howell 1992).

Although both root and hypocotyl elongation are inhibited by exogenous cytokinin in dark-grown, wild-type seedlings, neither root nor hypocotyl elongation is inhibited by cytokinin in *ckr1* mutant plants when grown under the same conditions. The *ckr1* mutant plants are also resistant to ethylene, which has a similar ability to inhibit root and hypocotyl elongation in wild-type plants. In fact, ethylene is generated in response to applied cytokinins in a dose-dependent manner in dark-grown seedlings (Figure 21.19). Subsequently, the *CKR1* gene was found to be allelic to the ethylene response gene *EIN2* (see Chapter 22). This result suggests that ethylene is part of the cytokinin signal transduction pathway, or that cytokinin and ethylene share a common signal transduction pathway or components of a signaling pathway. The previously mentioned discovery that a candidate for the cytokinin receptor is structurally similar to the ethylene receptor is a further indication that they are linked to a common response pathway.

Another candidate for a gene that encodes a component of the cytokinin signal transduction pathway in *Arabidopsis* was identified as the *cyr1* (*cy*tokinin *r*esponse *1*) mutant. Like the *ckr1* mutant, the *cyr1* mutant exhibits a tenfold decrease in the sensitivity of roots to cytokinin. However, *cyr1* maps to a different genetic locus and does not affect the sensitivity of the plants to ethylene or other hormones. Except for their response to cytokinins, the *ckr1* mutants are almost identical to wild-type plants. In contrast, the *cyr1* mutants have major developmental abnormalities when they are grown in the absence of exogenous cytokinin (Figure 21.20) (Deik-

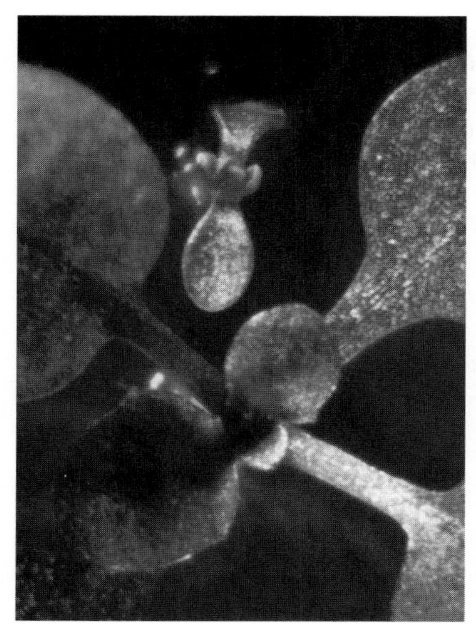
(A)

(B)

Figure 21.20 The effect of the *cyr1* mutation on *Arabidopsis* seedling morphology in the absence of hormone treatment. (A) A *cyr1* mutant (upper plant) and a wild-type sibling (lower plant) after growing for 11 days on nutrient agar. (B) The *cyr1* mutant seedling has produced a single, abnormal flower after 26 days. Two leaves and one sepal were removed to show the inside of the flower. (From Deikman and Ulrich 1995, courtesy of J. Deikman.)

man and Ulrich 1995). The properties of the *cyr1* mutation suggest that the wild-type *CYR1* gene encodes an important component of the signaling pathway for many of the major developmental processes regulated by cytokinin. The cloning and further characterization of this gene that are now under way may provide important insights into the cytokinin signal transduction pathway.

Specific Phenylpropanoid Derivatives Can Also Promote Cell Division

When tobacco leaf tissue is transformed with *Agrobacterium* carrying a Ti plasmid in which the *tmr* locus (*ipt* gene) is inactivated, instead of producing a proliferating tissue with a rooty phenotype, it produces tumors (Black et al. 1994). These tumors grow in vitro as undifferentiated callus masses, without the addition of either an auxin or a cytokinin to the culture medium. Since they acquired fully functional *tms* loci, their ability to synthesize auxins and auxin autonomy is not a surprise. However, since the *tmr* locus encodes the enzyme required for cytokinin synthesis, cells transformed with T-DNA that lack a functional *tmr* gene should require exogenous cytokinin for growth, as in most cases they do. This is not a habituation phenomenon, since callus derived from tobacco leaf tissues does not become habituated. Nor is it the result of an increased sensitivity to cytokinin, or a high level of endogenous cytokinin. In fact, cytokinins were not detected in the *tmr*-inactivated tumors. These tumors do, however, have higher levels of **dehydrodiconiferyl alcohol glucosides (DCGs)** (Figure 21.21), derivatives of the phenylpropanoid pathway (see Chapter 13), which have been shown to induce cell proliferation in the absence of cytokinin (Teutonico et al. 1991).

Dehydrodiconiferyl alcohol glucosides can, in the presence of auxin and the absence of cytokinin, stimulate cell division in tobacco pith tissue, but they cannot substitute for cytokinins in any of the other cytokinin bioassays. Moreover, the level of DCGs increases in cultured tobacco pith tissue after cytokinin treatment. Andrew Binns at the University of Pennsylvania and his collaborator David Lynn at the University of Chicago have proposed that DCGs participate in the cytokinin signal transduction pathway that induces cell division, but not in other cytokinin-regulated processes (Binns 1994). According to this model, tumors that form after transformation with *tmr*-deficient T-DNA are able to bypass the cytokinin requirement by synthesizing a component of the signal transduction pathway that acts downstream of cytokinin. The mechanism by which DCGs might regulate entry into the cell cycle or progression of the cell cycle is not known.

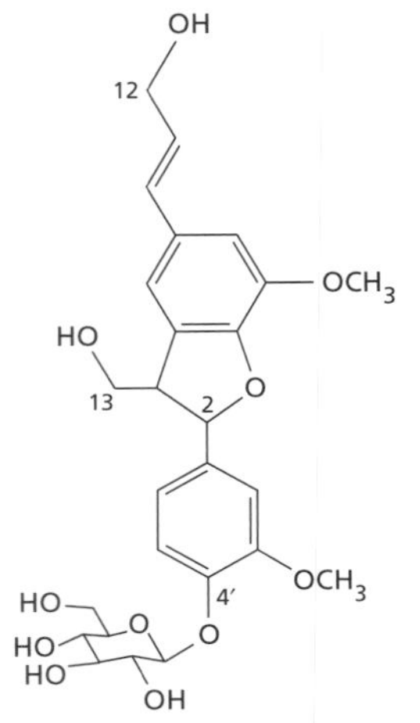

OH
12
OCH_3
HO
13
2
O
OCH_3
4′
HO
HO
HO
O
O
OH

Dehydrodiconiferyl alcohol glucoside

Figure 21.21 Dehydrodiconiferyl alcohol glucoside, a possible intermediate in cytokinin-induced cell division.

Cytokinin Signal Transduction May Involve Calcium as a Second Messenger

As we have seen in earlier chapters, calcium is an important second messenger in many signal transduction pathways in animals, and it has been implicated in plant hormone action, particularly that of gibberellins (see Chapter 20) and of abscisic acid (see Chapter 23). Several studies with the moss *Funaria hygrometrica* have provided evidence for Ca^{2+} as a second messenger in cytokinin-induced bud formation. First, calcium levels have been shown to increase in filament cells that form buds after treatment with cytokinin. Second, cytokinin has been shown to be ineffective in inducing bud formation in calcium-free media. Third, a calcium ionophore* can substitute for cytokinin in inducing bud formation in the presence of external calcium (Saunders and Hepler 1982).

The calcium ionophore A23187 increases the cytosolic calcium concentration by exchanging external Ca^{2+} for internal hydrogen ions. The work of Karen Schumaker and her colleagues at the University of Arizona suggests that calcium ions enter moss cells through voltage-dependent Ca^{2+} channels in the plasma membrane and that the passage of calcium ions through such channels is regulated by cytokinin (Schumaker and Gizinski 1993). More work will be needed to determine which stage of bud formation (initial-cell formation, bud development, or both) requires calcium as an intracellular sec-

* Recall that ionophores are molecules that facilitate the movement of ions across membranes, either alone or in exchange with another ion.

ond messenger. Inasmuch as the formation of initial cells requires only picomolar amounts of cytokinin, whereas bud development requires micromolar levels of the hormone, the two processes may be regulated by different mechanisms.

In many signaling pathways involving calcium, the increases in the concentration of cytosolic Ca^{2+} are rapid and transient. Such calcium "spikes" are characteristic of signals that trigger rapid responses. For example, the induction of stomatal closure by abscisic acid involves such transient increases in cytosolic Ca^{2+} concentrations (see Chapter 23). The effects of cytokinin on calcium fluxes can be measured with fluorescent calcium-sensitive dyes whose fluorescence depends on calcium. Changes in cytosolic Ca^{2+} concentrations, particularly in response to a signal such as cytokinin, can be monitored by measurement of the resulting fluorescence after these dyes have been microinjected into cells.

Using this technique with alfalfa root hairs, Silverman and Bush (1996) showed that within 10 minutes of cytokinin application, the rate of tip growth increases and the plasma membrane hyperpolarizes, and there is a concomitant rapid increase in cytosolic Ca^{2+} concentrations. The fact that microinjection of *cytokinin* into the cells failed to induce either the growth stimulation or the increase in cytosolic Ca^{2+} suggests that the hormonal effects of cytokinin are mediated by binding to a receptor on the cell surface, and that Ca^{2+} acts as an internal, second messenger for cytokinin.

The finding that calcium ions may act as second messengers in cytokinin-induced responses does not provide us with the mechanism of action of the hormone, although it identifies a vital component of the cytokinin signal transduction pathway. In other systems, Ca^{2+} acts through the regulatory protein **calmodulin** (see Chapter 14). Each calmodulin has four high-affinity calcium-binding sites. Calmodulin alone is inactive as a regulator, but the calmodulin–calcium complex can bind to and activate several enzymes, most notably **calmodulin-regulated protein kinases**. In addition to calmodulin-regulated protein kinases, plants have unique protein kinases with an integral calmodulin-like domain, known as **calcium-dependent protein kinases (CDPKs)**, whose activity is regulated by calcium binding (Roberts 1993) (see Chapter 14).

The activity of both calmodulin-regulated protein kinases and CDPKs depends on Ca^{2+}. Plant protein kinases phosphorylate primarily serine or threonine residues in cellular proteins. The phosphorylation of enzymes can either inhibit or stimulate their activity, depending on the residue phosphorylated, as we saw with the cyclin-dependent protein kinases that regulate cell cycle progression (see Chapter 1). Many key regulatory events are mediated by phosphorylation, including the transcription of genes, since the ability of certain transcription factors to bind regulatory elements within promoters is regulated by phosphorylation, as in the case of the bacterial two-component regulatory systems (discussed in Chapter 14). However, a relationship between cytokinin action and transcription factor phosphorylation is largely conjecture at this point.

Summary

Mature plant cells generally do not divide in the intact plant, but they can be stimulated to divide by wounding, by infection with certain bacteria, and by plant hormones, including cytokinins. Cytokinins are N^6-substituted aminopurines that will initiate cell proliferation in mature tobacco pith tissue when it is cultured on a medium that also contains an auxin. The principal cytokinin of higher plants, zeatin, or 6-(4-hydroxy-3-methylbut-*trans*-2-enylamino)purine, is also present in plants as a riboside or ribotide and as glycosides. These forms are also active as cytokinins in bioassays through their enzymatic conversion to the free zeatin base by plant tissue.

Cytokinins may be detected in plant extracts by various bioassays, but in recent years sophisticated chemical methods for cytokinin identification have been developed, including immunochemical methods in which antibodies to cytokinins are used to detect and quantitate cytokinins in plant extracts.

A prenyl transferase enzyme, cytokinin synthase, catalyzes the transfer of the isopentenyl group from Δ^2-isopentenyl pyrophosphate to the 6 nitrogen of adenosine monophosphate. The product of this reaction is readily converted to zeatin and other cytokinins. Cytokinins are synthesized in roots, in developing embryos, and in crown gall tissues. Cytokinins are also synthesized by plant-associated bacteria, and recently it has been proposed that plants derive all their cytokinins from such bacteria. The isolation of a plant gene that encodes cytokinin synthase would resolve this question.

Crown galls originate from plant tissues that have been infected with *Agrobacterium tumefaciens*. The bacterium injects a specific region of its Ti plasmid called T-DNA into wounded plant cells, and the T-DNA is incorporated into the host nuclear genome. The T-DNA contains a gene for cytokinin biosynthesis, as well as genes for auxin biosynthesis. These phyto-oncogenes are expressed in the plant cells, leading to hormone synthesis and unregulated proliferation of the cells to form the gall.

Cytokinins are most abundant in the young, rapidly dividing cells of the shoot and root apical meristems.

They do not appear to be actively transported through living plant tissues. Instead, they are transported passively into the shoot from the root through the xylem, along with water and minerals. However, at least in pea, the shoot can regulate the flow of cytokinin from the root.

Cytokinins participate in the regulation of many plant processes, including cell division, morphogenesis of shoots and roots, chloroplast maturation, cell enlargement, and senescence. Both cytokinin and auxin regulate the plant cell cycle and are needed for cell division. Whereas auxin has been shown to increase the synthesis of Cdc2, cytokinin stimulates the expression of specific G_1 cyclins, and promotes the transition from G_2 to M, apparently by causing the loss of phosphate from the inactivation site of the CDK–cyclin complex.

In addition to cell division, the ratio of auxin to cytokinin determines the differentiation of cultured plant tissues into either roots or buds: High ratios promote roots, low ratios buds. Cytokinins also have been implicated in the release of axillary buds from apical dominance. In the moss *Funaria*, cytokinins greatly increase the number of "buds," the structures that give rise to the leafy gametophyte stage of development.

The mechanism of action of cytokinin is unknown. A possible cytokinin receptor has been identified in *Arabidopsis* by the technique of activation tagging. This transmembrane protein is related to the bacterial two-component system histidine kinases and to the ethylene receptor ETR1. The ethylene response gene *EIN2* was also identified in a screen for cytokinin-resistant mutants, suggesting that ethylene and cytokinin share components of their signal transduction pathways.

Cytokinin has a profound effect on the rate of protein synthesis and on the kinds of proteins made by plant cells. In particular, cytokinins stimulate the synthesis of specific chloroplast proteins that are encoded by nuclear genes and synthesized by cytoplasmic ribosomes. Cytokinins increase the abundance of several specific mRNAs. Some of these appear to be primary response genes, since they are not inhibited by cycloheximide. Cytokinin stimulation of *LHCB* expression in *Lemna* has been shown to be due to a posttranscriptional step, possibly an enhancement of mRNA stability.

In tobacco, dehydrodiconiferyl alcohol glucoside can substitute for cytokinin in promoting cell division, and it may be a component of the signaling pathway involved in cell division. In some systems, such as the growth of root hairs and bud formation in *Funaria*, the initial response to cytokinins may be mediated by an increase in the cytosolic calcium concentration.

General Reading

Alberts, B., Bray, D., Lewis, J., Raff, M., Roberts, K., and Watson, J. D. (1994) *Molecular Biology of the Cell*. Garland, New York.

*Brzobohaty, B., Moore, I., and Palme, K. (1994) Cytokinin metabolism: Implications for regulation of plant growth and development. *Plant Mol. Biol.* 26: 1483–1497.

Bush, D. S. (1995) Calcium regulation in plant cells and its role in signaling. *Annu. Rev. Plant Physiol. Plant Mol. Biol.* 46: 95–122.

*Chilton, M-D. (1983) A vector for introducing new genes into plants. *Sci. Am.* 248: 50–59.

Jacobs, T. W. (1995) Cell cycle control. *Annu. Rev. Plant Physiol. Plant Mol. Biol.* 46: 317–339.

*Kaminek, M. (1992) Progress in cytokinin research. *Trends Biotechnol.* 10: 159–164.

Karin, M. (1994) Signal transduction from the cell surface to the nucleus through the phosphorylation of transcription factors. *Curr. Opin. Cell Biol.* 6: 415–424.

*Letham, D. S., and Palni, L. M. S. (1983) The biosynthesis and metabolism of cytokinin. *Annu. Rev. Plant Physiol.* 34: 163–197.

*Morris, R. O. (1986) Genes specifying auxin and cytokinin biosynthesis in phytopathogens. *Annu. Rev. Plant Physiol.* 37: 509–538.

*Nester, E., Gordon, M. P., Amasino, R., and Yanofsky, M. (1984) Crown gall: A molecular and physiological analysis. *Annu. Rev. Plant Physiol.* 35: 387–413.

* Indicates a reference that is general reading in the field and is also cited in this chapter.

Chapter References

Akiyoshi, D. E., Klee, H., Amasino, R. M., Nester, E. W., and Gordon, M. P. (1984) T-DNA of *Agrobacterium tumefaciens* encodes an enzyme of cytokinin biosynthesis. *Proc. Natl. Acad. Sci. USA* 81: 5994–5998.

Akiyoshi, D. E., Morris, R. O., Hinz, R., Mischke, B. S., Kosuge, T., Garfinkel, D. J., Gordon, M. P., and Nester, E. W. (1983) Cytokinin/auxin balance in crown gall tumors is regulated by specific loci in the T-DNA. *Proc. Natl. Acad. Sci. USA* 80: 407–411.

Akiyoshi, D. E., Regier, D. A., and Gordon, M. P. (1987) Cytokinin production by *Agrobacterium* and *Pseudomonas* spp. *J. Bacteriol.* 169: 4242–4248.

Bechtold, N., Ellis, J., and Pelletier, G. (1993) *In planta Agrobacterium* mediated gene transfer by infiltration of adult *Arabidopsis thaliana* plants. *C. R. Acad. Sci.* 316: 1194–1199.

Beveridge, C. A., Murfet, I. C., Kerhoas, L., Sotta, B., Miginiac, E., and Rameau, C. (1997a) The shoot controls zeatin riboside export from pea roots. Evidence from the branching mutant *rms4*. *Plant J.* 11: 339–345.

Beveridge, C. A., Symons, G. M., Murfet, I. C., Ross, J. J., and Rameau, C. (1997b) The rms1 mutant of pea has elevated indole-3-acetic acid levels and reduced root-sap zeatin riboside content but increased branching controlled by graft-transmissible signal(s). *Plant Physiol.* 115: 1251–1258.

Binns, A. (1994) Cytokinin accumulation and action: Biochemical, genetic, and molecular approaches. *Annu. Rev. Plant Physiol. Plant Mol. Biol.* 45: 173–196.

Binns, A. N., and Thomashow, M. F. (1988) Cell biology of *Agrobacterium* infection and transformation of plants. *Annu. Rev. Microbiol.* 42: 575–606.

Black, R. C., Binns, A. N., Chang, C-F., and Lynn, D. G. (1994) Cell-autonomous cytokinin-independent growth of tobacco cells transformed by *Agrobacterium tumefaciens* strains lacking the cytokinin biosynthesis gene. *Plant Physiol.* 105: 989–998.

Bomhoff, G., Klapwijk, P. M., Kester, H. C. M., and Schilperoort, R. A. (1976) Octopine and nopaline synthesis and breakdown genetically controlled by plasmid of *Agrobacterium tumefaciens*. *Mol. Gen. Genet.* 145: 177–181.

Braun, A. C. (1958) A physiological basis for autonomous growth of the crown-gall tumor cell. *Proc. Natl. Acad. Sci. USA* 44: 344–349.

Brzobohaty, B., Moore, I., Kristoffersen, P., Bako, L., Campos, N., Schell, J., and Palme, K. (1993) Release of active cytokinin by a β-glucosidase localized to the maize root meristem. *Science* 262: 1051–1054.

Caplin, S. M., and Steward, F. C. (1948) Effect of coconut milk on the growth of the explants from carrot root. *Science* 108: 655–657.

Cary, A. J., Liu, W., and Howell, S. H. (1995) Cytokinin action is coupled to ethylene in its effects on the inhibition of root and hypocotyl elongation in *Arabidopsis thaliana* seedlings. *Plant Physiol.* 107: 1075–1082.

Chang, C., Kwok, S. F., Bleecker, A. B., and Meyerowitz, E. M. (1993) *Arabidopsis* ethylene-response gene ETR1: Similarity of product to two-component regulators. *Science* 262: 539–544.

Chaudhury, A. M., Letham, D. S., Craig, S., and Dennis, E. S. (1993) *amp1*—A mutant with high cytokinin levels and altered embryonic pattern, faster vegetative growth, constitutive photomorphogenesis and precocious flowering. *Plant J.* 4: 907–916.

Chen, C-M., and Melitz, D. K. (1979) Cytokinin biosynthesis in a cell-free system from cytokinin-autotrophic tobacco tissue cultures. *FEBS Lett.* 107: 15–20.

Chilton, M-D., Drummond, M. H., Merlo, D. J., Sciaky, D., Montoya, A. L., Gordon, M. P., and Nester, E. W. (1977) Stable incorporation of plasmid DNA into higher plant cells: The molecular basis of crown gall tumorigenesis. *Cell* 11: 263–271.

Chory, J., Reinecke, D., Sim, S., Washburn, T., and Brenner, M. (1994) A role for cytokinins in de-etiolation in *Arabidopsis*. *Det* mutants have an altered response to cytokinins. *Plant Physiol.* 104: 339–347.

Cleland, R. E. (1996) Growth substances. In *Units, Symbols and Terminology for Plant Physiology*, F. B. Salisbury, ed., Oxford University Press. New York, pp. 126–128.

Crowell, D. N. (1994) Cytokinin regulation of a soybean pollen allergen gene. *Plant Mol. Biol.* 25: 829–835.

Crowell, D. N., Kadleck, A. T., John, M. C., and Amasino, R. M. (1990) Cytokinin-induced messenger RNAs in cultured soybean cells. *Proc. Natl. Acad. Sci. USA* 87: 8815–8819.

Deikman, J., and Ulrich, M. (1995) A novel cytokinin-resistant mutant of *Arabidopsis* with abbreviated shoot development. *Planta* 195: 440–449.

Estruch, J. J., Chriqui, D., Grossmann, K., Schell, J., and Spena, A. (1991) The plant oncogene *RolC* is responsible for the release of cytokinins from glucoside conjugates. *EMBO J.* 10: 2889–2895.

Faure, J-D., Jullien, M., and Caboche, M. (1994) *Zea3*: A pleiotropic mutation affecting cotyledon development, cytokinin resistance and carbon-nitrogen metabolism. *Plant J.* 5: 481–491.

Feldmann, K. A., and Marks, M. D. (1987) *Agrobacterium*-mediated transformation of germinating seeds of *Arabidopsis thaliana*: A non-tissue culture approach. *Mol. Gen. Genet.* 208: 1–9.

Feng, X-H., Dube, S. K., Bottino, P. J., and Kung, S. (1990) Restoration of shooty morphology of a nontumorous mutant of *Nicotiana glauca x N. langsdorffii* by cytokinin and the isopentenyltransferase gene. *Plant Mol. Biol.* 15: 407–420.

Flores, S., and Tobin, E. (1988) Cytokinin modulation of LHCP mRNA levels: The involvement of post-transcriptional regulation. *Plant Mol. Biol.* 11: 409–415.

Fraley, R. T., Rogers, S. G., Horsch, R. B., Eichholtz, D. A., Flick, J. S., Fink, C. L., Hoffmann, N. L., and Sanders, P. R. (1985) The SEV system: A new disarmed Ti plasmid vector system for plant transformation. *Bio/Technology* 3: 629–635.

Gan, S., and Amasino, R. M. (1995) Inhibition of leaf senescence by autoregulated production of cytokinin. *Science* 270: 1986–1988.

Gan, S. and Amasino, R.M. (1996) Cytokinins in plant senescence: From spray and pray to clone and play. *Bioessays* 18: 557–565.

Garfinkel, D. J., Simpson, R. B., Ream, L. W., White, F. F., Gordon, M. P., and Nester, E. W. (1981) Genetic analysis of crown gall: Fine structure map of the T-DNA by site-directed mutagenesis. *Cell* 27: 143–153.

Gleave, A. P. (1992) A versatile binary vector system with a T-DNA organisational structure conducive to efficient integration of cloned DNA into the plant genome. *Plant Mol. Biol.* 20: 1203–1207.

Hall, R. H., Csonka, L., David, H., and McLennan, B. (1967) Cytokinins in the soluble RNA of plant tissues. *Science* 156: 69–71.

Hamilton, J. L., and Lowe, R. H. (1972) False broomrape: A physiological disorder caused by growth-regulator imbalance. *Plant Physiol.* 50: 303–304.

Hecht, S. M., Bock, R. M., Schmitz, R. Y., Skoog, F., and Leonard, N. J. (1971) Cytokinins: Development of a potent antagonist. *Proc. Natl. Acad. Sci. USA* 68: 2608–2610.

Holland, M. A. (1997) Occam's razor applied to hormonology. *Plant Physiol.* 115: 865–868.

Holland M. A., and Polacco J. C. (1994) PPFMs and other covert contaminants: Is there more to plant physiology than just plant? *Annu. Rev. Plant Physiol. Plant Mol. Biol.* 45: 197–209.

Huff, A.K., and Ross, C.W. (1975) Promotion of radish cotyledon enlargement and reducing sugar content by zeatin and red light. *Plant Physiol.* 56: 429–433.

Ichikawa, T., and Syono, K. (1991) Tobacco genetic tumors. *Plant Cell Physiol.* 32: 1123–1128.

Itai, C., and Vaadia, Y. (1971) Cytokinin activity in water-stressed shoots. *Plant Physiol.* 47: 87–90.

John, P. C. L., Zhang, K., Dong, C., Diederich, L., and Wightman, F. (1993) P34-cdc2 related proteins in control of cell cycle progression, the switch between division and differentiation in tissue development, and stimulation of division by auxin and cytokinin. *Aust. J. Plant Physiol.* 20: 503–526.

Kakimoto, T. (1996) CKI1, a histidine kinase homolog implicated in cytokinin signal transduction. *Science* 274: 982–985.

Kaminek, M., and Armstrong, D. J. (1990) Genotypic variation in cytokinin oxidase from *Phaseolus* callus cultures. *Plant Physiol.* 93: 1530–1538.

Kende, H. (1971) The cytokinins. *Int. Rev. Cytol.* 31: 301–338.

Kulaeva, O. N., Karavaiko, N. N., Selivankina, S. Y., Zemlyachenko, Y. V., and Shiplova, S. V. (1995) Receptor of *trans*-zeatin involved in transcription activation by cytokinin. *FEBS Lett.* 366: 26–28.

Leonard, N. J., Hecht, S. M., Skoog, F., and Schmitz, R. Y. (1968) Cytokinins: Synthesis of 6-(3-methyl-3-butenylamino)-9-b-D-ribofuranosylpurine (3iPA), and the effect of side-chain unsaturation on the biological activity of isopentenylaminopurines and their ribosides. *Proc. Natl. Acad. Sci. USA* 59: 15–21.

Letham, D. S. (1973) Cytokinins from *Zea mays*. *Phytochemistry* 12: 2445–2455.

Letham, D. S. (1974) Regulators of cell division in plant tissues XX. The cytokinins of coconut milk. *Physiol. Plantarum* 32: 66–70.

Lew, R., and Tsuji, H. (1982) Effects of benzyladenine treatment duration on delta-aminolevulinic acid accumulation in the dark, chlorophyll lag phase abolition , and long-term chlorophyll production in excised cotyledons of dark grown cucumber seedlings. *Plant Physiol.* 69: 663–667.

Li, Y., Hagen, G., and Guilfoyle, T. J. (1992) Altered morphology in transgenic tobacco plants that overproduce cytokinins in specific tissues and organs. *Dev. Biol.* 153: 386–395.

Lu, J., Ertl, J. R., and Chen, C. (1990) Cytokinin enhancement of the light induction of nitrate reductase transcript levels in etiolated barley leaves. *Plant Mol. Biol.* 14: 585–594.

Matzke, A. J. M., and Chilton, M-D. (1981) Site-specific insertion of gene into T-DNA of the *Agrobacterium* tumor-inducing plasmid: An approach to genetic engineering of higher plant cells. *J. Mol. Appl. Genet.* 1: 39–49.

Meins, F., Jr. (1989) A biochemical switch model for cell-heritable variation in cytokinin requirement. In *Molecular Basis of Plant Development*, R. Goldberg, ed., Liss, New York, pp. 13–24.

Meins, F., Jr., Lutz, J., and Binns, A. N. (1980) Variation in the competence of tobacco pith cells for cytokinin-habituation in culture. *Differentiation* 16: 71–75.

Miller, C. O., Skoog, F., Von Saltza, M. H., and Strong, F. (1955) Kinetin, a cell division factor from deoxyribonucleic acid. *J. Am. Chem. Soc.* 77: 1392–1393.

Morris, R. O., Jameson, P. E., Laloue, M., and Morris, J. W. (1991) Rapid identification of cytokinins by an immunological method. *Plant Physiol.* 95: 1156–1161.

Nester, E. W., and Kosuge, T. (1981). Plasmids specifying plant hyperplasias. *Annu. Rev. Microbiol.* 35: 531–565.

Nicander, B., Ståhl, U., Björkman, P-O., and Tillberg, E. (1993) Immunoaffinity co-purification of cytokinins and analysis by high-performance liquid chromatography with ultraviolet-spectrum detection. *Planta* 189: 312–320.

Noodén, L. D., and Letham, D. S. (1993) Cytokinin metabolism and signaling in the soybean plant. *Aust. J. Plant Physiol.* 20: 639–653.

Noodén, L. D., Singh, S., and Letham, D. S. (1990) Correlation of xylem sap cytokinin levels with monocarpic senescence in soybean. *Plant Physiol.* 93: 33–39.

Peterson, J. B., and Miller, C. O. (1976) Cytokinins in *Vinca rosea* L. crown gall tumor tissue as influenced by compounds containing reduced nitrogen. *Plant Physiol.* 57: 393–399.

Rayle, D. L., Ross, C. W., and Robinson, N. (1982) Estimation of osmotic parameters accompanying zeatin-induced growth of detached cucumber cotyledons. *Plant Physiol.* 70: 1634–1636.

Redig, P., Schmulling, T., and Van Onckelen, H. (1996) Analysis of cytokinin metabolism in *ipt* transgenic tobacco by liquid chromatography-tandem mass spectrometry. *Plant Physiol.* 112: 141–148.

Roberts, D. M. (1993) Protein kinases with calmodulin-like domains: Novel targets of calcium signals in plants. *Curr. Opin. Cell Biol.* 5: 242–246.

Ross, C. W., and Rayle, D. L. (1982) Evaluation of H^+ secretion relative to zeatin-induced growth of detached cucumber cotyledons. *Plant Physiol.* 70: 1470–1474.

Saunders, M. J., and Hepler, P. K. (1982) Calcium ionophore A23187 stimulates cytokinin-like mitosis in *Funaria*. *Science* 217: 943–945.

Schumaker, K. S., and Gizinski , M. J. (1993) Cytokinin stimulates dihydropyridine-sensitive calcium uptake in moss protoplasts. *Proc. Natl. Acad. Sci. USA* 90: 10937–10941.

Silverman, F. P., and Bush, D. S. (1996) Membrane transport and cytokinin action in alfalfa. *Mol. Biol. Cell* 7 (Suppl.): 303a (Abstract).

Silverthorne, J., and Tobin, E. M. (1987) Phytochrome regulation of nuclear gene expression. *BioEssays* 7: 18–23.

Silverthorne, J., and Tobin, E. M. (1990) Post-transcriptional regulation of organ-specific expression of individual *rbc*S mRNAs in *Lemna gibba*. *Plant Cell* 2: 1181–1190.

Skoog, F., and Miller, C. O. (1965) Chemical regulation of growth and organ formation in plant tissues cultured *in vitro*. In *Molecular and Cellular Aspects of Development*, E. Bell, ed., Harper and Row, New York, pp. 481–494.

Smigocki, A., Neal, J. W., Jr., McCanna, I., and Douglass, L. (1993) Cytokinin-mediated insect resistance in *Nicotiana* plants transformed with the *ipt* gene. *Plant Mol. Biol.* 23: 325–335.

Smith, H. H. (1988) The inheritance of genetic tumors in *Nicotiana* hybrids. *J. Hered.* 79: 277–284.

Soni, R., Carmichael, J. P., Shah, Z. H., and Murray, J. A. H. (1995) A family of cyclin D homologs from plants differentially controlled by growth regulators and containing the conserved retinoblastoma protein interaction motif. *Plant Cell* 7: 85–103.

Su, W., and Howell, S. H. (1992) A single genetic locus, *Ckr1*, defines *Arabidopsis* mutants in which root growth is resistant to low concentrations of cytokinin. *Plant Physiol.* 99: 1569–1574.

Teutonico, R. A., Dudley, M. E., Orr, J. D., Lynn, D. G., and Binns, A. N. (1991) Activity and accumulation of cell division-promoting phenolics in tobacco tissue cultures. *Plant Physiol.* 97: 288–297.

Torrey, J. G. (1976) Root hormones and plant growth. *Annu. Rev. Plant Physiol.* 27: 435–459.

Van Staden, J., Cook, E., and Noodén, L. D. (1988) Cytokinins and senescence. In *Senescence and Aging in Plants*, L. D. Noodén and A. C. Leopold, eds., Academic Press, San Diego, pp. 281–328.

Vremarr, H. J., Skoog, F., Frihart, C. R., and Leonard, N. J. (1972) Cytokinins in *Pisum* transfer ribonucleic acid. *Plant Physiol.* 49: 848–851.

Walden, R., Fritze, K., Hayashi, H., Miklashevichs, E., Harling, H., and Schell, J. (1994) Activation tagging: A means of isolating genes implicated as playing a role in plant growth and development. *Plant Mol. Biol.* 26: 1521–1528.

Weiler, E. W. (1980) Radioimmunoassays for *trans*-zeatin and related cytokinins. *Planta* 149: 155–162.

White, P. R. (1934) Potentially unlimited growth of excised tomato root tips in a liquid medium. *Plant Physiol.* 9: 585–600.

Zambryski, P., Tempe, J., and Schell, J. (1989) Transfer and function of T-DNA genes from *Agrobacterium* Ti and Ri plasmids in plants. *Cell* 56: 193–201.

Zhang, K., Letham, D. S., and John, P. C. L. (1996) Cytokinin controls the cell cycle at mitosis by stimulating the tyrosine dephosphorylation and activation of p34cdc2-like H1 histone kinase. *Planta* 200: 2–12.

22 Ethylene

DURING THE NINETEENTH CENTURY, when coal gas was used for street illumination, it was observed that trees in the vicinity of streetlamps defoliated more extensively than other trees. Eventually it became apparent that coal gas and air pollutants damage plant tissue, and ethylene was identified as the active component of coal gas by a Russian student in 1901. As a graduate student at the Botanical Institute of St. Petersburg, Dimitry Neljubow observed that dark-grown pea seedlings in the laboratory showed reduced stem elongation, increased lateral growth (swelling), and abnormal, horizontal growth—conditions that were later termed the *triple response.* When the "laboratory air" was removed and the plants were allowed to grow in fresh air, they regained their normal rate of growth. Ethylene, which was present in the laboratory air, was identified as the molecule causing the response.

The first indication that ethylene is a natural product of plant tissues was reported by H. H. Cousins in 1910. Cousins reported that "emanations" from oranges stored in a chamber caused the premature ripening of bananas when these gases were passed through a chamber containing the fruit. However, since we now know that oranges, themselves, synthesize relatively little ethylene compared to other fruits such as apples, it is likely that the oranges used by Cousins were infected with the fungus, *Penicillium,* which produces copious amounts of ethylene. In 1934, R. Gane and others identified ethylene chemically as a natural product of plant metabolism, and because of its effects on the plant it was classified as a hormone.

For 25 years ethylene was not recognized as an important plant hormone, mainly because many physiologists believed that the effects of ethylene could be mediated by auxin, the first plant hormone to be discovered (see Chapter 19). It was thus thought that auxin was the main plant hormone and that ethylene had only an insignificant and indirect physiological role. Work on ethylene was also hampered by the lack of chemical techniques for its quantification. However, after gas chromatography was introduced in ethylene research in 1959, ethylene was rediscovered and its physiological significance as a plant growth regulator was recognized (Burg and Thimann 1959).

In this chapter we will describe the discovery of the ethylene biosynthetic pathway and outline some of the important effects of ethylene on plant growth and development. At the end of the chapter, we will consider how ethylene is believed to act at the cellular and molecular levels.

Recently, the identification of ethylene response mutants of *Arabidopsis* has led to exciting breakthroughs in the identification of components of the ethylene signal transduction pathway. The ethylene receptor is presently the best-characterized plant hormone receptor, and has provided a model for other signaling pathways in plant cells.

Structure, Biosynthesis, and Measurement of Ethylene

Ethylene can be produced by almost all parts of higher plants, although the rate of production depends on the type of tissue and the stage of development. In general, meristematic regions and nodal regions are the most active in ethylene biosynthesis. However, ethylene production also increases during leaf abscission and flower senescence, as well as during fruit ripening. Any type of wounding can induce ethylene biosynthesis, as can physiological stresses such as flooding, chilling, disease, and temperature or drought stress.

Despite its chemical simplicity, the biosynthetic pathway of ethylene eluded plant physiologists for many years. The first breakthrough came in the 1960s with the discovery by Lieberman and Mapson (1964) that the amino acid methionine is the precursor of ethylene. For the next 15 years, scientists tried unsuccessfully to determine the steps in the ethylene biosynthetic pathway from methionine. Finally, in 1979, the second major breakthrough came when Adams and Yang (1979), at the University of California at Davis, discovered that ACC (1-aminocyclopropane-1-carboxylic acid) serves as an intermediate in the conversion of methionine to ethylene. As we will see, the complete pathway turned out to be a cycle, taking its place among the many metabolic cycles that operate in plant cells.

The Properties of Ethylene Are Deceptively Simple

Ethylene is the simplest known olefin (its molecular weight is 28) and is lighter than air under physiological conditions (Figure 22.1A). It is flammable and readily undergoes oxidation. Ethylene can be oxidized to ethylene oxide (Figure 22.1B), and ethylene oxide can be hydrolyzed to ethylene glycol (Figure 22.1C). In most plant tissues, ethylene can be completely oxidized to CO_2 (Figure 22.1D) (Beyer 1979).

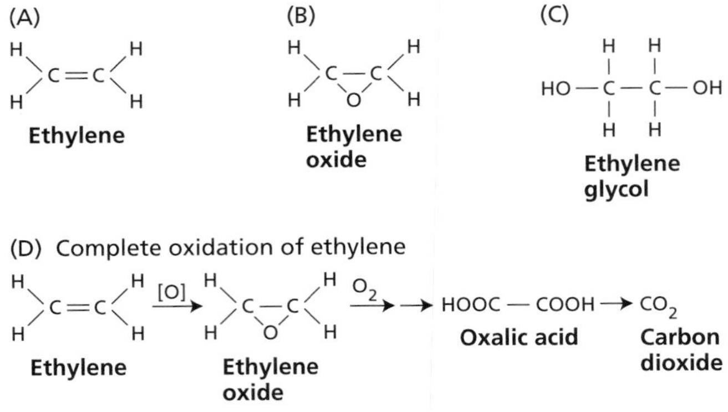

(A) Ethylene (B) Ethylene oxide (C) Ethylene glycol

(D) Complete oxidation of ethylene

Ethylene [O] → Ethylene oxide O_2 → → HOOC — COOH Oxalic acid → CO_2 Carbon dioxide

Figure 22.1 Ethylene (A) can be oxidized to ethylene oxide (B), which can be hydrolyzed to ethylene glycol (C). (D) Alternatively, ethylene can be oxidized completely to CO_2.

Ethylene is released easily from the tissue and diffuses in the gas phase through the intercellular spaces and outside the tissue. At an ethylene concentration of 1 $\mu L\ L^{-1}$ in the gas phase at 25°C, the concentration of ethylene in water is 4.4×10^{-9} *M*. Since ethylene gas is easily lost from the tissue and may affect other tissues or organs, ethylene-trapping systems are used during the storage of fruits, vegetables, and flowers. $KMnO_4$ (potassium permanganate) is an effective absorbent of ethylene and can reduce the concentration of ethylene in apple storage areas from 250 to 10 $\mu L\ L^{-1}$, markedly extending the storage life of the fruit.

Bacteria, Fungi, and Plant Organs Produce Ethylene

Even away from cities and industrial air pollutants, the environment is seldom free of ethylene due to its production by plants and microorganisms. The production of ethylene in plants is highest in senescing tissues and ripening fruits (>1.0 nL g-fresh-weight^{-1} h^{-1}), but all organs of higher plants can synthesize ethylene. Ethylene is biologically active at very low concentrations—less than 1 part per million (1 $\mu L\ L^{-1}$). The internal ethylene concentration in a ripe apple has been reported to be as high as 2500 $\mu L\ L^{-1}$. Young developing leaves produce more ethylene than do fully expanded leaves. In bean (*Phaseolus vulgaris*), young leaves produce 0.4 nL g^{-1} h^{-1}, compared with 0.04 nL g^{-1} h^{-1} for older leaves. With few exceptions, nonsenescent tissues that are wounded or mechanically perturbed will temporarily increase their ethylene production by severalfold within 25 to 30 minutes. Ethylene levels later return to normal.

Gymnosperms and lower plants, including ferns, mosses, liverworts, and certain cyanobacteria, all have shown the ability to produce ethylene. Ethylene production by fungi and bacteria contributes significantly to the ethylene content of soil. In culture, the capacity of microorganisms to synthesize ethylene appears to depend on the nature of the medium on which they grow. Certain strains of the common enteric bacterium *Escherichia coli* and of yeast produce large amounts of ethylene from methionine. However, there is no evidence that healthy mammalian tissues produce ethylene, nor does ethylene appear to be a metabolic product of invertebrates.

Biosynthesis and Catabolism Determine the Physiological Activity of Ethylene

In vivo experiments by M. Lieberman and coworkers at the Agricultural Research Service in Beltsville, Maryland, showed that various plant tissues can convert l-[^{14}C]methionine to [^{14}C]ethylene and that the ethylene is derived from carbons 3 and 4 of methionine (Figure 22.2). Other experiments showed that the CH_3—S group of methionine is recycled in the tissue. Without this recycling, the amount of reduced sulfur present would limit the available methionine and the synthesis of ethylene (Burg and Clagett 1967). Subsequent work showed that *S*-adenosylmethionine (AdoMet), which is synthesized from methionine and ATP, is an intermediate in the ethylene biosynthetic pathway.

Fourteen years after methionine was discovered to be a precursor of ethylene in higher plants, the final step in the pathway became clear. The immediate precursor of ethylene was found to be **1-aminocyclopropane-1-carboxylic acid (ACC)** (Adams and Yang 1979). The role of ACC became evident in experiments in which plants were treated with [^{14}C]methionine. Under anaerobic conditions, ethylene was not produced from the [^{14}C]methionine and labeled ACC accumulated in the tissue, but on exposure to oxygen, ethylene production surged. The labeled ACC was rapidly converted to ethylene by various plant tissues, suggesting that ACC is the immediate precursor of ethylene in higher plants.

When ACC is supplied exogenously to plant tissue that normally produces very little ethylene, ethylene production increases substantially. This observation indicates that the synthesis of ACC is usually the metabolic step that limits ethylene production in plant tissues.

ACC synthase, the enzyme that catalyzes the conversion of AdoMet to ACC (see Figure 22.2), has been characterized in many types of tissues of various plants (Kende 1993). ACC synthase is a cytosolic enzyme with a very short half-life. Its activity is regulated by several environmental and internal factors, such as wounding, drought stress, flooding, and auxin.

Because ACC synthase is present in such low amounts in plant tissues (0.0001% of the total protein of ripe tomato) and is very labile (unstable), it is difficult to purify the enzyme for biochemical analysis. Researchers overcame this obstacle by raising antisera against partly purified ACC synthase from zucchini fruit slices that were induced to produce large amounts of ethylene. The resulting antisera contained a mixture of antibodies, including antibodies to ACC. The investigators then removed the contaminating antibodies by allowing them to bind to proteins extracted from zucchini tissue that lacked the enzyme. The purified ACC synthase antibodies were then used to isolate the gene encoding the enzyme (Sato and Theologis 1989).

Expressing the gene for ACC synthase in the bacterium *E. coli* made it possible to isolate purified ACC synthase in large amounts (Sato et al. 1991). Genes encoding ACC synthase were subsequently cloned from numerous plant species, and the deduced amino acid sequences were found to show a high degree of similarity to the *aminotransferase* superfamily of enzymes. These enzymes catalyze the transfer of amino groups among amino acids and require pyridoxal phosphate as a cofactor. ACC synthase is encoded by members of a divergent multigene family that are differentially regulated by various inducers of ethylene biosynthesis. For example, in tomato, there are at least nine ACC synthase genes, different subsets of which are induced by auxin, wounding, and/or fruit ripening.

The last step in ethylene biosynthesis, the conversion of ACC to ethylene, is catalyzed by the enzyme **ACC oxidase** (see Figure 22.2), which was previously called the "ethylene-forming enzyme." This enzyme is generally not the rate-limiting point in ethylene biosynthesis, although tissues that show high rates of ethylene production, such as ripening fruit and senescing flowers, show increased levels of ACC oxidase activity and mRNA. The gene that encodes ACC oxidase was cloned by a circuitous route. Researchers isolated a set of genes that are highly induced during tomato fruit ripening. They then blocked the expression of these genes by introducing an antisense version of each into transgenic tomatoes. Blocking of one such gene, *pTOM13*, repressed fruit ripening and inhibited ethylene production. Expression of *pTOM13* in *Xenopus* oocytes as well as in yeast conferred the ability to convert ACC to ethylene, indicating that the gene encodes an ACC oxidase enzyme (Spanu et al. 1991). Using this gene, researchers then isolated ACC oxidase clones from numerous plant species.

The deduced amino acid sequences of ACC oxidases revealed that these enzymes belong to the Fe^{2+}/ascorbate oxidase superfamily. This similarity suggested that ACC oxidase might require Fe^{2+} and ascorbate for activity, a requirement that has been confirmed by biochemical analysis of the protein. This requirement for cofactors, along with the low abundance of ACC oxidase, presumably explains why the purification of this enzyme eluded researchers. Like ACC synthase, ACC oxidase is encoded by a multigene family whose members are differentially regulated.

Methionine is found at quite low, nearly constant concentrations in plant tissues, including those that produce large amounts of ethylene, such as ripening fruits. Since methionine is the sole precursor of ethylene in higher plants, tissues with high rates of ethylene production require a continuous supply of methionine. This supply is ensured by methionine recycling via the Yang cycle, as shown in Figure 22.2.

Not all the ACC found in the tissue is converted to ethylene. ACC can also be converted to a nonvolatile conjugated form, *N*-malonyl ACC (see Figure 22.2) (Amrhein et al. 1981; Hoffman et al. 1982), which does not break down and seems to accumulate in the tissue. A second conjugated form of ACC, 1-(γ-L-glutamylamino)cyclopropane-1-carboxylic acid (GACC), has also been identified. The conjugation of ACC is thought to play an important role in the control of ethylene biosynthesis, in a manner analogous to the conjugation of auxin and cytokinin.

Researchers have studied the catabolism of ethylene by supplying $^{14}C_2H_4$ to plant tissues and tracing the radioactive compounds produced. Carbon dioxide, eth-

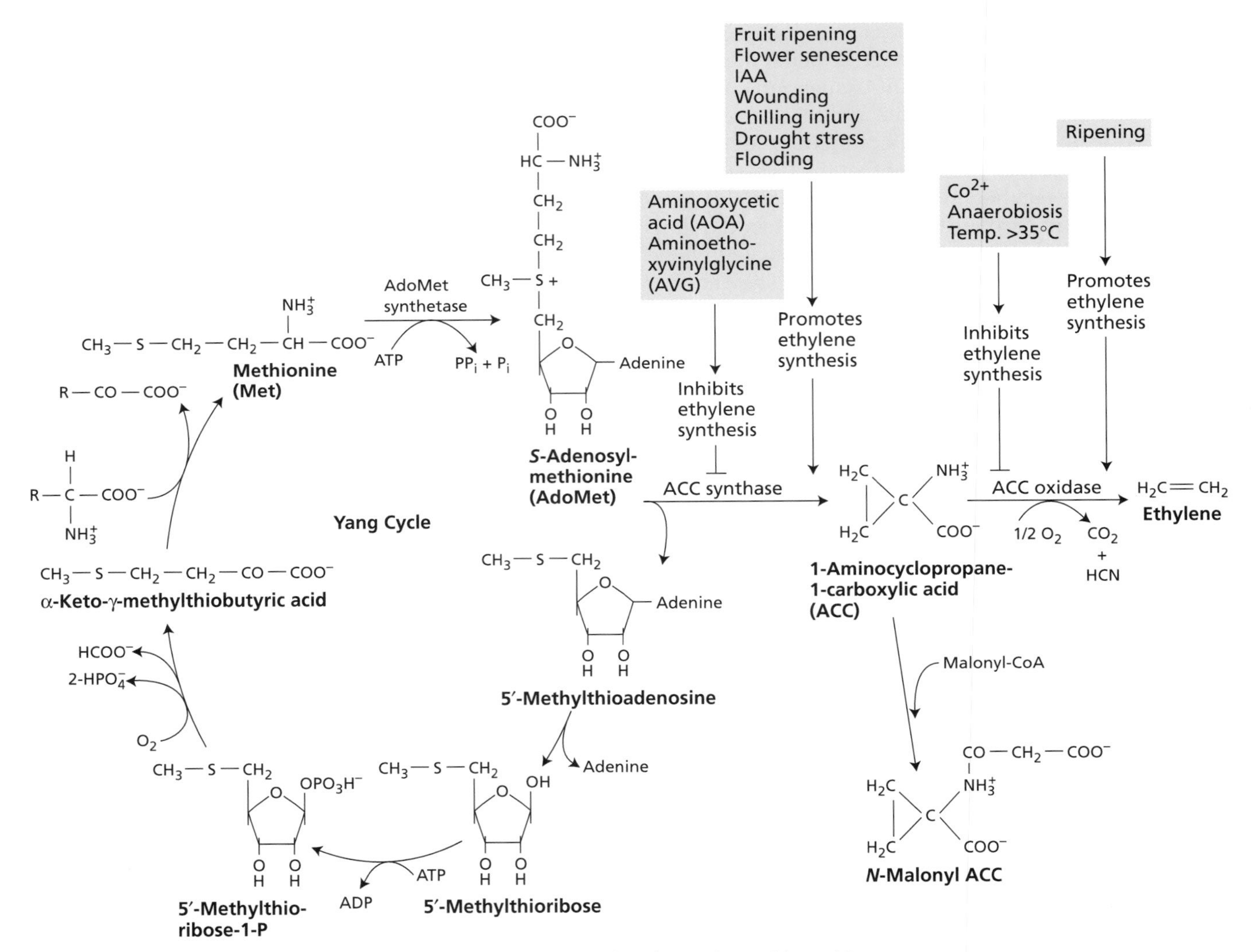

Figure 22.2 Ethylene biosynthetic pathway and the Yang cycle. The amino acid methionine is the precursor of ethylene. The rate-limiting step in the pathway is the conversion of AdoMet to ACC, which is catalyzed by the enzyme ACC synthase. The last step in the pathway, the conversion of ACC to ethylene, requires oxygen and is catalyzed by the enzyme ACC oxidase. The CH_3—S group of methionine is recycled via the Yang cycle and thus conserved for continued synthesis. Besides being converted to ethylene, ACC can be conjugated to *N*-malonyl ACC. (After McKeon et al. 1995.)

ylene oxide, ethylene glycol, and the glucose conjugate of ethylene glycol have been identified as metabolic breakdown products. However, since certain cyclic olefin compounds, such as 1,4-cyclohexadiene, have been shown to block ethylene breakdown without inhibiting ethylene action, ethylene catabolism is not involved in the activity of the hormone (Raskin and Beyer 1989).

Environmental Stresses and Auxins Promote Ethylene Biosynthesis

Ethylene biosynthesis is stimulated by several factors, including developmental state, environmental conditions, other plant hormones, and physical and chemical injury.

BOX 22.1

ACC Synthase Gene Expression and Biotechnology

AS WE HAVE SEEN, ACC synthase is a cytosolic enzyme that catalyzes the first committed step in ethylene biosynthesis in higher plants. It is a key regulatory enzyme that catalyzes the production of the ethylene precursor ACC from AdoMet (*S*-adenosylmethionine). However, this enzyme is difficult to purify because it is labile and present in low abundance in plant tissues. The molecular cloning and functional expression of ACC synthase in *E. coli* and yeasts have facilitated biochemical and structural studies of this enzyme.

ACC synthase cDNAs and genomic sequences have been cloned from numerous plant species. The emerging picture from the study of these genes is that ACC synthase is encoded by members of a divergent multigene family in which each gene is differentially regulated by various environmental and developmental factors during the growth of a plant. Tomato and *Arabidopsis*, for example, have at least nine and five gene members, respectively, which are differentially regulated by inducers such as auxin, fruit ripening, and wounding (Olson et al. 1991; Rottmann et al. 1991; Abel et al. 1995).

The protein synthesis inhibitor cycloheximide is a potent inducer of many ACC synthase genes. Cycloheximide inducibility is a characteristic of *primary response genes* such as the early auxin-inducible genes (see Chapter 19). Two mechanisms have been proposed to explain this inducibility. First, the transcription of these genes may be under the control of a short-lived repressor protein whose synthesis is prevented by the addition of cycloheximide; second, the mRNA transcripts of these genes may be short-lived and are stabilized when cycloheximide prevents the synthesis of a nuclease. The induction of one of the genes for ACC synthase (*ACS2*) in *Arabidopsis* by cycloheximide appears to be due to transcriptional activation (Liang et al. 1996).

A clue to the signal transduction mechanism involved in the regulation of ACC synthase gene expression has come from experiments with lithium (Li^+). Li^+ is known to interfere with the phosphoinositide signaling pathway (see Chapter 14) by inhibiting the activity of inositolphosphate phosphatase. Li^+ is also one of the strongest inducers of ACC synthase activity in plants (Liang et al. 1996). This finding suggests that the regulation of at least some ACC synthase genes may involve Ca^{2+} mobilization mediated by inositol trisphosphate (IP_3) (see Chapter 14).

Because ethylene plays such an important role in fruit ripening, the cloning of the genes for ACC synthase and ACC oxidase has enabled scientists to use biotechnology to reduce ethylene production during fruit ripening, and thus to prolong the storage life of fruit. In the case of tomato, two different strategies have been employed to control ethylene production. The first strategy was to use the Ti plasmid of *Agrobacterium tumefaciens* to transform tomato tissue with antisense DNA for ACC synthase and ACC oxidase. When the genes for the two enzymes are expressed in the reverse direction in the cell, the resulting "antisense transcript" binds to and inactivates the "sense transcript" produced by the wild-type gene, effectively shutting down expression of the wild-type gene (Hamilton et al. 1990; Oeller et al. 1991; Picton et al. 1993). After transformation, intact plants can be regenerated in tissue culture.

In the second strategy for reducing ethylene production, tomato tissue was transformed (using *A. tumefaciens*) with genes encoding enzymes that metabolize ACC or AdoMet (Klee et al. 1991; Good et al. 1993). In this case, the ACC that is produced is rapidly degraded, and no ethylene can be synthesized.

Regardless of the strategy used, all fruits from these genetically engineered plants show significant delays in ripening. Tomato fruits from plants transformed with the ACC synthase antisense DNA fail to ripen unless treated with ethylene (see the figure). Similar approaches are now being carried out in other agronomically important crops, such as melon and banana. In addition, these approaches are being used to retard floral senescence in carnations, thereby reducing postharvest losses. The research on ACC synthase and other ethylene biosynthesis genes thus provides an excellent example of how basic research can lead to improvements in agricultural and horticultural crops, the full benefits of which we are only beginning to realize.

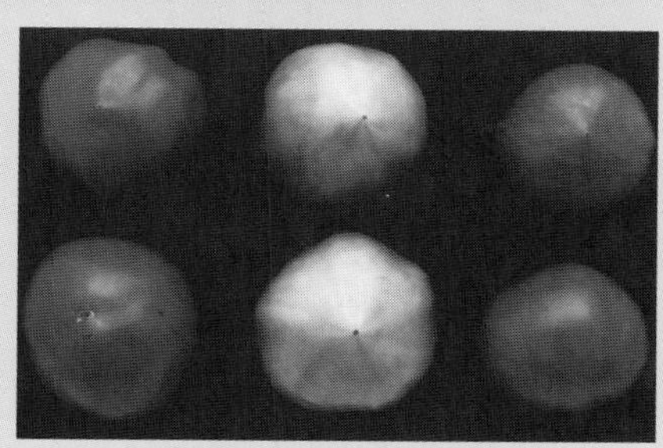

Inhibition of tomato fruit ripening in detached tomato fruits by antisense ACC synthase RNA. Antisense tomato fruits (center) remain firm and green unless treated by ethylene. Ethylene-treated antisense fruits (left) are indistinguishable from naturally ripened fruits (right) in terms of color, flavor, aroma, and texture. (Reprinted with permission from Oeller et al. 1991.)

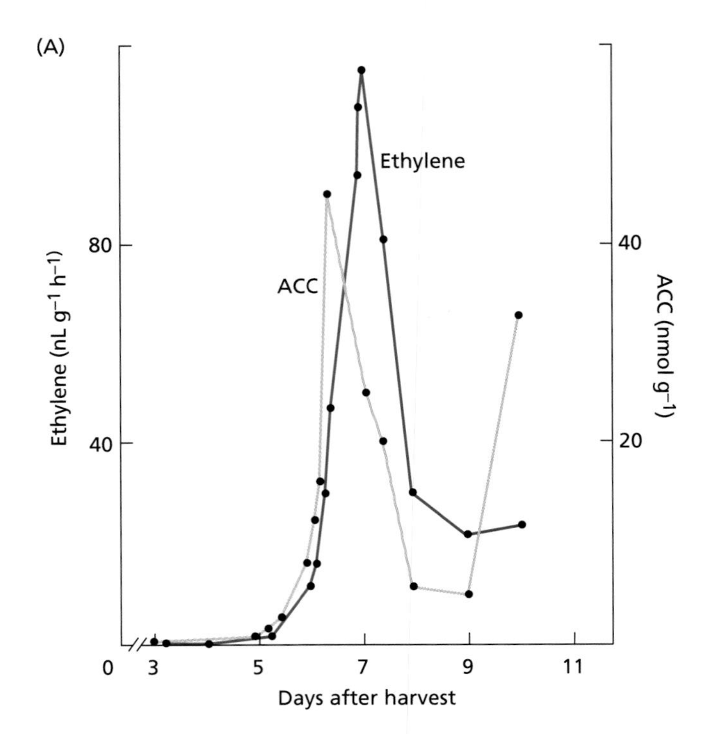

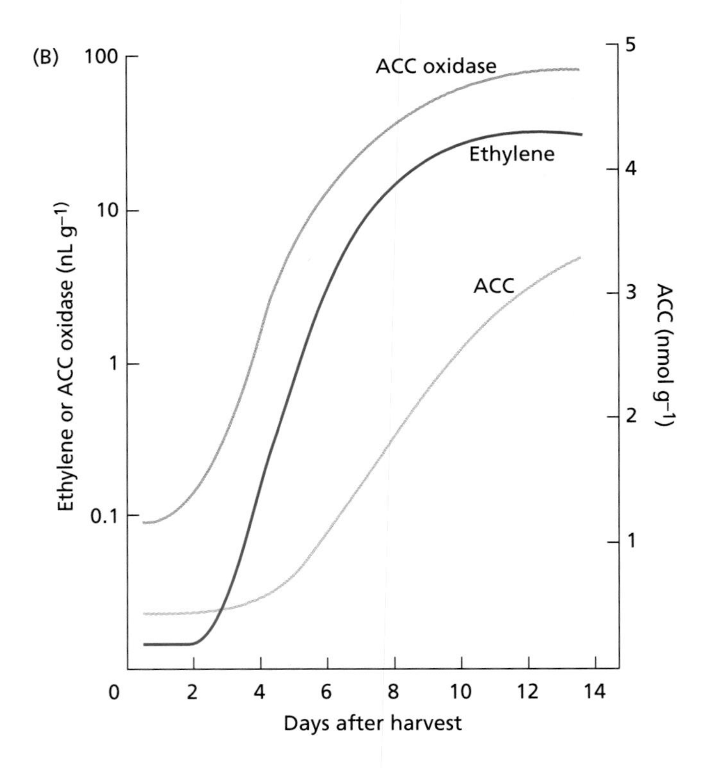

Figure 22.3 Changes in ethylene and ACC contents and ACC oxidase activity during fruit ripening. (A) Simultaneous changes in ethylene and ACC levels in avocado. The avocado fruits do not ripen as long as they are attached to the tree. Changes in ACC and in ethylene contents were monitored after harvesting. (B) Changes in the ACC oxidase activity and ethylene and ACC concentrations of Golden Delicious apples. The data are plotted as a function of days after harvest. Increases in ethylene and ACC concentrations and in ACC oxidase activity are closely correlated with ripening. (A from Hoffman and Yang 1980; B from Yang 1987.)

Fruit ripening. As fruits mature, the rate of ethylene biosynthesis increases. The increase in ethylene biosynthesis is associated with increases in ACC, ACC oxidase activity (Figure 22.3), ACC synthase activity, and the mRNA levels of both ACC synthase and ACC oxidase. However, application of ACC to unripe fruits only slightly enhances ethylene production, indicating that an increase in the activity of ACC oxidase is the critical step in ripening (McKeon et al. 1995).

Stress-induced ethylene production. Ethylene biosynthesis is increased by stress conditions such as drought, flooding, chilling, or mechanical wounding. In all these cases ethylene is produced by the usual biosynthetic pathway, and it has been shown to result at least in part from an increase in transcription of ACC synthase mRNA. This "stress ethylene" is involved in the onset of stress responses such as abscission, senescence, wound healing, and increased disease resistance (see Chapter 25).

Auxin-induced ethylene production. In some instances, auxins and ethylene can cause similar plant responses, such as induction of flowering in pineapple and inhibition of stem elongation (Abeles et al. 1992). These responses might be due to the ability of auxins to promote ethylene synthesis by enhancing the conversion of AdoMet to ACC. These observations suggest that some responses previously attributed to auxin (IAA) are in fact mediated by the ethylene produced in response to auxin.

Inhibitors of protein synthesis block both ACC synthesis and IAA-induced ethylene synthesis, indicating that the synthesis of ACC synthase caused by auxins brings about the marked increase in ethylene production (Imaseki et al. 1982; Yang et al. 1982). The levels of the mRNA that encodes ACC synthase increase in response to application of exogenous IAA, suggesting that increased transcription is at least partly responsible for the increased ethylene production observed in response to auxin (Nakagawa et al. 1991; Liang et al. 1992).

Ethylene Production and Action Can Be Inhibited

Inhibitors of hormone synthesis or action are valuable for the study of the biosynthetic pathways and physiological roles of hormones. Inhibitors are particularly helpful when it is difficult to distinguish between different hormones that have identical effects in plant tissue or when a hormone affects the synthesis or the action of another hormone. Ethylene mimics high concentrations of auxins by inhibiting stem growth and causing epinasty (a downward curvature of leaves). Use of specific inhibitors of ethylene biosynthesis and action made it possible to discriminate between the actions of auxin and ethylene.

These studies showed that ethylene is the primary effector and that auxin acts indirectly by causing a substantial increase in ethylene production. **Aminoethoxyvinylglycine** (**AVG**) and **aminooxyacetic acid** (**AOA**) block the conversion of AdoMet to ACC (see Figure 22.2). AVG and AOA are known to inhibit enzymes that use the cofactor pyridoxal phosphate, consistent with the amino acid sequence homology of ACC synthase to aminotransferases, which use pyridoxal phosphate as a cofactor. The cobalt ion (Co^{2+}) is also an inhibitor of the ethylene biosynthetic pathway, blocking the conversion of ACC to ethylene, the last step in ethylene biosynthesis.

Most of the effects of ethylene can be antagonized by specific ethylene inhibitors. Silver ions (Ag^+) applied as silver nitrate ($AgNO_3$) or as silver thiosulfate [$Ag(S_2O_3)_2^{3-}$] are potent inhibitors of ethylene action. Silver is very specific; the inhibition it causes cannot be induced by any other metal ion. Carbon dioxide at high concentrations (in the range of 5 to 10%) also inhibits many effects of ethylene, such as the induction of fruit ripening, although CO_2 is less efficient than Ag^+. This effect of CO_2 has often been exploited in the storage of fruits, whose ripening is delayed at elevated CO_2 concentrations. The high concentrations of CO_2 required for inhibition make it unlikely that CO_2 acts as an ethylene antagonist under natural conditions.

The volatile compound *trans*-cyclooctene, but not its isomer *cis*-cyclooctene, is the strongest competitive inhibitor of ethylene binding found yet (Sisler et al. 1990); *trans*-cyclooctene is thought to act by competing with ethylene for binding to the receptor.

Ethylene Can Be Measured in Bioassays or by Gas Chromatography

The triple response of etiolated pea seedlings reported by Neljubow in 1901 is still a reliable bioassay for ethylene because of its specificity, sensitivity to low concentrations, and rapidity. When plant tissues are exposed to various concentrations of ethylene (0.1 $\mu L\ L^{-1}$ and higher) in a sealed environment, inhibition of stem elongation, increased lateral growth (swelling), and horizontal growth of the epicotyl are observed (Figure 22.4A). The magnitude of the response is proportional to the ethylene concentration in the sample. Actual concentrations are determined by comparison of the response of the sample with that of tissue exposed to known amounts of ethylene. Other bioassays include epinasty of tomato leaves (Figure 22.4B), flower senescence (Figure 22.4C), root hair formation (Figure 22.4D), and leaf abscission, but these assays are less sensitive than the triple response. All bioassays are diagnostic for the presence of ethylene, but none of them permit quantification over a large range of ethylene concentrations. Such quantification can be accomplished by gas chromatography.

Gas chromatography is the most widely used method for detecting ethylene. As little as 5 parts per billion (ppb) of ethylene can be detected, and the analysis time is only 1 to 5 minutes. Usually, the ethylene produced by a plant tissue is allowed to accumulate in a sealed vial, and a sample is withdrawn with a syringe. The sample is injected into a gas chromatograph column in which the different gases are separated and detected by a flame ionization detector. Quantification of ethylene by this method is very accurate. Recently, a novel method to measure ethylene was developed that uses a laser-driven photoacoustic detector that can detect as little as 50 parts per trillion (ppt) ethylene (Voesenek et al. 1997).

Developmental and Physiological Effects of Ethylene

As we have seen, ethylene was discovered in connection with its effects on seedling growth and fruit ripening. It has since been shown to regulate a wide range of responses in plants, including seed germination, cell expansion, cell differentiation, flowering, senescence, and abscission. In this section we will consider the phenotypic effects of ethylene in more detail.

Ethylene Promotes the Ripening of Some Fruits

In everyday usage, the term "fruit ripening" refers to the changes in fruit that make it ready to eat. Such changes typically include softening due to the enzymatic breakdown of the cell walls, starch hydrolysis, sugar accumulation, and the disappearance of organic acids and phenolic compounds, including tannins. From the perspective of the plant, fruit ripening means that the seeds are ready for dispersal. For seeds whose dispersal depends on animal ingestion, "ripeness" and "edibility" are synonomous. Anthocyanins and carotenoids often accumulate in the epidermis of such fruits, adding to their visibility. However, for seeds that rely on mechanical or other means for dispersal, "fruit ripening" may mean drying followed by splitting. Because of their importance in agriculture, the vast majority of studies on fruit ripening have focused on edible fruits.

For many years, ethylene has been recognized as the hormone that accelerates the ripening of edible fruits. Addition of ethylene to such fruits hastens ripening, and a dramatic increase in ethylene production is closely associated with the initiation of ripening. However, sur-

(A)

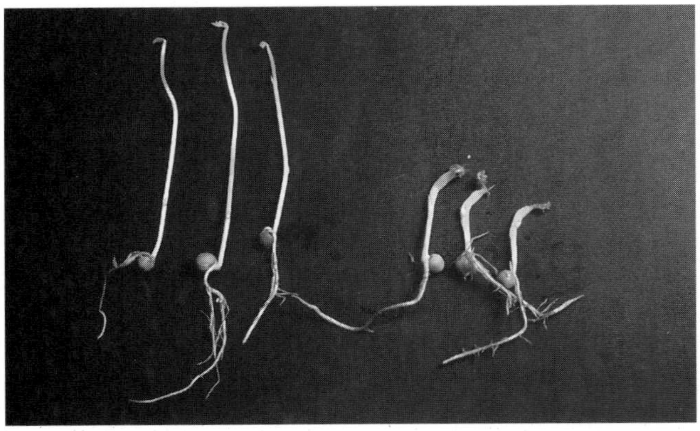

(B)

(C)

(D)

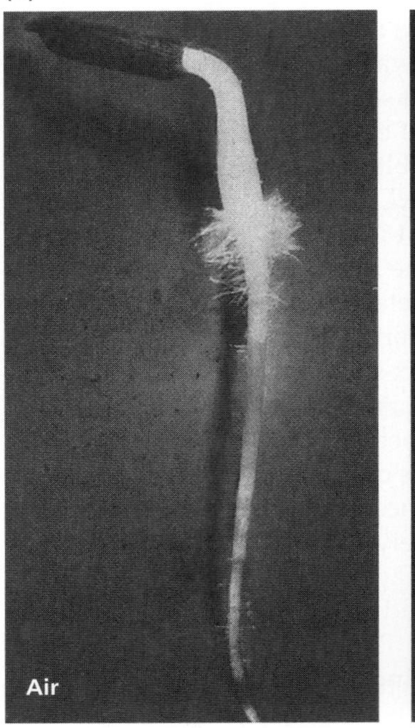

Figure 22.4 Some physiological effects of ethylene on plant tissue in various developmental stages. (A) Triple response of etiolated pea seedlings. Six-day-old pea seedlings were treated with 10 ppm (parts per million) ethylene (right) or left untreated (left). The treated seedlings show a radial swelling, inhibition of elongation of the epicotyl, and horizontal growth (diagravitropism). (B) Epinasty, or downward bending of the tomato leaves (right), is caused by ethylene treatment. Epinasty results when the cells on the upper side of the petiole grow faster than those on the bottom. (C) Inhibition of flower senescence by inhibition of ethylene action. Carnation flowers were held in deionized water for 14 days with (left) or without silver thiosulfate (STS), a potent inhibitor of ethylene action. Blocking ethylene results in a marked inhibition of floral senescence. (D) Promotion of root hair formation by ethylene in lettuce seedlings. Two-day-old seedlings were treated with air (left) or 10 ppm ethylene for 24 hours before the photo was taken. Note the profusion of root hairs on the ethylene-treated seedling. (A and B courtesy of S. Gepstein; C from Reid 1995, courtesy of M. Reid and P. Davies; D from Abeles et al. 1992, courtesy of F. Abeles.)

veys of a wide range of fruits have shown that not all of them respond to ethylene. All fruits that ripen in response to ethylene exhibit a characteristic respiratory rise before the ripening phase called a **climacteric**.* Such fruits also show a spike of ethylene production immediately before the respiratory rise (Figure 22.5). Inasmuch as treatment with ethylene induces the fruit to produce additional ethylene, its action can be described as **autocatalytic**. Apples, bananas, avocados, and tomatoes are examples of climacteric fruits. In contrast, fruits such as citrus fruits and grapes do not exhibit the respiration and ethylene production rise and are called **nonclimacteric** fruits. Other examples of climacteric and nonclimacteric fruits are given in Table 22.1.

When unripe climacteric fruits are treated with ethylene, the onset of the climacteric rise is hastened. When nonclimacteric fruits are treated in the same way, the magnitude of the respiratory rise increases as a function

* Note that the term "climacteric" can be used either as a noun, as in "most fruits exhibit a climacteric during ripening" or as an adjective, as in "a climacteric rise in respiration." The term "nonclimacteric," however, is used only as an adjective.

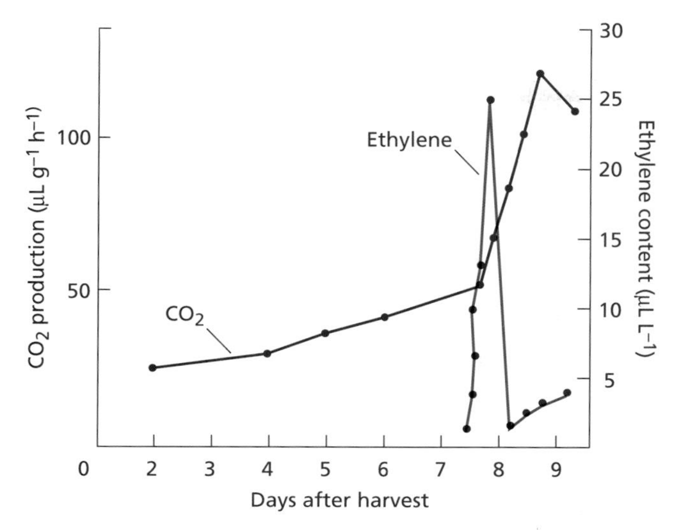

Figure 22.5 Relationship between ethylene production and respiration rate in banana. Ripening is characterized by a climacteric rise in respiration. At a certain point during banana ripening there is a massive increase in CO_2 release, followed by a decrease. A climacteric rise in ethylene production precedes the increase in CO_2 production, suggesting that ethylene is the hormone that triggers the ripening process. (From Burg and Burg 1965.)

of the ethylene concentration, but the treatment does not trigger production of endogenous ethylene and does not accelerate ripening. Elucidation of the role of ethylene in the ripening of climacteric fruits has resulted in many practical applications aimed at either uniform ripening or the delay of ripening.

Although the effects of exogenous ethylene on fruit ripening are straightforward and demonstrable, establishing a causal relation between the level of endogenous ethylene and fruit ripening is more difficult. Inhibitors of ethylene biosynthesis (such as AVG) or of ethylene action (such as CO_2 or Ag^+) have been shown to delay or even prevent ripening. However, the definitive demonstration that ethylene is required for fruit ripening has come from experiments in which ethylene biosynthesis was blocked by expression of an antisense version of either ACC synthase or ACC oxidase in transgenic tomatoes (see Box 22.1). Elimination of ethylene biosynthesis in these transgenic tomatoes completely blocked fruit ripening, and ripening was restored by application of exogenous ethylene. These experiments provided unequivocal proof of the role of ethylene in fruit ripening, and they opened the door to the manipulation of fruit ripening through biotechnology.

Table 22.1
Climacteric and nonclimacteric fruits

Climacteric	Nonclimacteric
Apple	Bell pepper
Avocado	Cherry
Banana	Citrus
Cantaloupe	Grape
Cherimoya	Pineapple
Fig	Snap bean
Mango	Strawberry
Olive	Watermelon
Peach	
Pear	
Persimmon	
Plum	
Tomato	

Leaf Epinasty Results When ACC from the Root Is Transported to the Shoot

The downward curvature of leaves that occurs when the upper (adaxial) side of the petiole grows faster than the lower (abaxial) side is termed **epinasty** (see Figure 22.4B). Ethylene and high concentrations of auxin induce epinasty, and it has now been established that auxin acts indirectly by inducing ethylene production. Flooding (waterlogging) or anaerobic conditions around tomato roots trigger enhanced synthesis of ethylene in the shoot, leading to the epinastic response.

Since these environmental stimuli are sensed by the roots and the response is displayed by the shoot, a "signal" from the roots must be transported to the shoots. This signal is ACC, the immediate precursor of ethylene (Bradford and Yang 1980). ACC levels were found to be significantly higher in the xylem sap after flooding of tomato roots for 1 to 2 days (Figure 22.6). Because water fills the air spaces in waterlogged soil and O_2 diffuses slowly through water, the concentration of oxygen around flooded roots decreases dramatically. The elevated production of ethylene appears to be caused by the accumulation of ACC in the roots under anaerobic conditions, since the conversion of ACC to ethylene requires oxygen (see Figure 22.2). However, the ACC accumulated in the anaerobic roots is transported to shoots via the transpiration stream, where it is readily converted to ethylene.

Ethylene Induces Lateral Cell Expansion

The triple response to ethylene of etiolated pea seedlings has already been described (see Figure 22.4A). At concentrations above 0.1 μL L^{-1}, ethylene changes the growth pattern of seedlings by reducing the rate of elongation and increasing lateral expansion, leading to swelling of the region below the hook. These responses to ethylene are common to growing shoots of most dicots and to coleoptiles and mesocotyls of seedlings of the grass family, such as oat or wheat. In *Arabidopsis*, the triple response consists of inhibition and swelling of the hypocotyl, inhibition of root elongation, and exaggera-

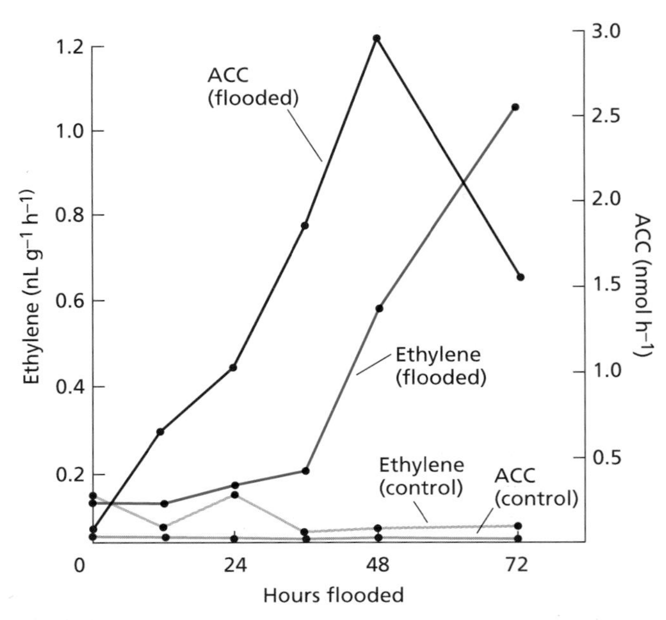

Figure 22.6 Changes in the amounts of ACC in the xylem sap and ethylene production in the petiole following flooding of tomato plants. ACC is synthesized in roots, but it is converted to ethylene very slowly under anaerobic conditions of flooding. ACC is transported via the xylem to the shoot, where it is converted to ethylene. The gaseous ethylene cannot be transported, so it usually affects the tissue near the site of its production. The ethylene precursor ACC is transportable and can produce ethylene far from the site of ACC synthesis. (From Bradford and Yang 1980.)

Figure 22.7 Screen for the *etr1* mutant of *Arabidopsis*. Seedlings were grown for 3 days in the dark in ethylene. Note that all but one of the seedlings are exhibiting the triple response: exaggeration in curvature of the apical hook, inhibition and radial swelling of the hypocotyl, and horizontal growth. The *etr1* mutant is completely insensitive to the hormone and grows like an untreated seedling. (Photograph by K. Stepnitz of the MSU/DOE Plant Research Laboratory.)

tion of the apical hook (Figure 22.7). This response has proven invaluable in the identification of *Arabidopsis* mutations affecting the ethylene signal transduction pathway (discussed later in the chapter).

As described in Chapter 15, the directionality of plant cell expansion is determined by the orientation of the cellulose microfibrils in the cell wall. Transverse microfibrils reinforce the cell wall in the lateral direction, so that turgor pressure is channeled into cell elongation. The orientation of the microfibrils in turn is determined by the orientation of the cortical array of microtubules in the cytoplasm. In typical elongating plant cells, the cortical microtubules are arranged transversely, giving rise to transversely arranged cellulose microfibrils. During the seedling triple response to ethylene, the transverse pattern of microtubule alignment is disrupted, and the microtubules switch over to a longitudinal orientation. This 90° shift in microtubule orientation leads to a parallel shift in cellulose microfibril deposition. The newly deposited wall is reinforced in the longitudinal direction rather than the transverse direction, which promotes lateral expansion instead of elongation (Eisinger 1983).

How do microtubules shift from one orientation to another? To study this phenomenon, Yuan and colleagues (1994) microinjected pea (*Pisum sativum*) epidermal cells with the microtubule protein tubulin, to which a fluorescent dye was covalently attached. The fluorescent "tag" did not interfere with the assembly of microtubules. This procedure permitted them to monitor the assembly of microtubules in living cells using a confocal laser scanning microscope, which can focus in many planes throughout the cell. They found that microtubules do not reorient from the transverse to the longitudinal direction by complete depolymerization of the transverse microtubules followed by repolymerization of a new longitudinal array of microtubules. Instead, increasing numbers of nontransversely aligned microtubules appear in particular locations. Neighboring microtubules then adopt the new alignment, so at one stage different alignments coexist before they adopt a uniformly longitudinal orientation. How ethylene induces this gradual change in microtubule alignment is still a mystery.

The typical horizontal growth that occurs after exposure to ethylene (see Figure 22.4A) may play an important role during germination. When physical barriers in the ground prevent seedling emergence, ethylene production induces horizontal growth, allowing the seedlings to find soil conditions that permit their emergence to the surface.

The Hooks of Dark-Grown Seedlings Are Maintained by Ethylene Production

Etiolated dicotyledonous seedlings are usually characterized by the hook shape of the terminal portion of the shoot apex. This shape facilitates movement of the seedling through the soil and protects the tender apical meristem. Like epinasty, hook closure and opening results from ethylene-induced asymmetric growth. The closed shape of the hook is a consequence of the more rapid elongation of the outer side compared with the inner. When the hook is exposed to light, it opens because the elongation rate of the inner side increases.

Red light induces opening and far-red light prevents opening, indicating that phytochrome is the photoreceptor involved in this process (see Chapter 17). A close interaction between phytochrome and ethylene controls hook opening. As long as ethylene is produced by the hook tissue in the dark, the elongation of the cells on the inner side is inhibited. Red light inhibits ethylene formation, promoting growth on the inner side, thereby causing the hook to open.

Ethylene Breaks Seed and Bud Dormancy in Some Species

When applied to seeds of cereals, ethylene breaks dormancy and initiates germination. In peanuts (*Arachis hypogaea*), ethylene production and seed germination are closely correlated. In addition to its effect on dormancy, ethylene increases the rate of seed germination of several species. Bud dormancy may also be broken by ethylene, and ethylene treatment is sometimes used to promote sprouting in potato tubers and other bulbs.

Ethylene Promotes the Elongation Growth of Submerged Aquatic Species

Although ethylene is closely associated with the inhibition of stem elongation and the promotion of lateral expansion in relation to the triple response of seedlings, it has been shown to promote stem and petiole elongation in various plants that spend at least part of their lives submerged under water. These include the dicots *Ranunculus sceleratus*, *Nymphoides peltata*, and *Callitriche platycarpa*, and the fern *Regnellidium diphyllum*. Another agriculturally important example is deepwater rice (see Chapter 20). In these species, submergence induces rapid internode or petiole elongation, which allows the leaves or upper parts of the shoot to remain above water. Treatment with ethylene mimics the effects of submergence. Growth is stimulated in the submerged plants because ethylene builds up in the tissues. In the absence of O_2, ethylene synthesis is diminished, but the loss of ethylene by diffusion is retarded under water. As a result, the submerged plants are exposed to high ethylene concentrations, which promote the growth of the submerged parts.

As we saw in Chapter 20, in deepwater rice it has been shown that ethylene stimulates internode elongation primarily by increasing the sensitivity of the cells of the intercalary meristem to endogenous gibberellin. Thus the stimulatory effect of ethylene on submerged aquatics in general may be mediated by gibberellins.

Ethylene Induces Roots and Root Hairs

Ethylene is capable of inducing adventitious root formation in leaves, stems, flower stems, and even other roots. This response requires unusually high ethylene concentrations (10 $\mu L\ L^{-1}$). Ethylene has also been shown to act as a positive regulator of root hair formation in several species (see Figure 22.4D). This relationship has been best studied in *Arabidopsis*, in which root hairs normally are located in the epidermal cells that overlie a junction between the underlying cortical cells (Dolan et al. 1994). In ethylene-treated roots, extra hairs form in ectopic (abnormal) locations in the epidermis; that is, cells not overlying a cortical cell junction differentiate into hair cells (Tanimoto et al. 1995). Seedlings grown in the presence of ethylene inhibitors (such as Ag^+) and ethylene-insensitive mutants display a reduction in root hair formation. These observations suggest that ethylene acts as a positive regulator in the differentiation of root hairs.

Ethylene Induces Flowering in the Pineapple Family

Although ethylene inhibits flowering in many species, it induces flowering in pineapple and its relatives, and it is used commercially in pineapple for synchronization of fruit set. Flowering of other species, such as mango, is also initiated by ethylene. On plants that have separate male and female flowers (monoecious species), ethylene may change the sex of developing flowers (see Chapter 24). The promotion of female flower formation in cucumber is one example of this effect (Abeles et al. 1992).

Ethylene Enhances the Rate of Leaf Senescence

As described in Chapter 16, senescence is a genetically programmed developmental process that affects all tissues of the plant. Several lines of physiological evidence support roles for ethylene and cytokinins in the control of leaf senescence. First, exogenous applications of ethylene or ACC (the precursor of ethylene) accelerate leaf senescence (Gepstein and Thimann 1981), while treatment with exogenous cytokinins delays leaf senescence (see Chapter 21). Second, enhanced ethylene production is associated with chlorophyll loss and color fading, which are characteristic features of leaf and flower senescence (see Figure 22.4C); an inverse correlation has been found between cytokinin levels in leaves and the onset of senescence. Third, inhibitors of ethylene synthesis (e.g., AVG or Co^{2+}) and action (e.g., Ag^+ or CO_2)

retard leaf senescence. Taken together, the physiological studies suggest that senescence is regulated by the balance of ethylene and cytokinin. In addition, abscisic acid (ABA) has been implicated in the control of leaf senescence. The role of ABA in senescence will be discussed in Chapter 23.

Senescence in ethylene mutants. Direct evidence for the involvement of ethylene in the regulation of leaf senescence has come from molecular genetic studies on *Arabidopsis.* As will be discussed later in the chapter, several mutants affecting the response to ethylene have been selected and identified. The specific bioassay employed was the triple response assay in which ethylene significantly inhibits seedling hypocotyl elongation and promotes lateral expansion. Ethylene-insensitive mutants, such as *etr1* and *ein2*, were identified by their failure to respond to ethylene. The *etr1* mutant fails to perceive the ethylene signal because of a mutation in the gene that codes for the ethylene receptor protein; the *ein2* mutant is blocked at a later step in the signal transduction pathway.

Consistent with a role for ethylene in leaf senescence, both *etr1* and *ein2* were found to be affected not only during the early stages of germination, but throughout the life cycle, including senescence (Zacarias and Reid 1990; Hensel et al. 1993; Grbić and Bleecker 1995). The ethylene mutants retained their chlorophyll for a longer period of time, as well as other chloroplast components. However, since the total life spans of these mutants were increased by only 30% over that of the wild type, ethylene appears to increase the *rate* of senescence, rather than acting as a developmental switch that initiates the senescence process.

Genetically engineered plants. Another very useful genetic approach that offers direct evidence for the function of specific gene(s) is based on transgenic plants. By using genetic engineering technology, the roles of both ethylene and cytokinins in the regulation of leaf senescence have been confirmed. One way to suppress the expression of a gene is to transform the plant with antisense DNA, which consists of the gene of interest in the reverse orientation with respect to the promoter. When the antisense gene is transcribed, the resulting antisense mRNA is complementary to the sense mRNA and will hybridize to it. Since double-stranded RNA is rapidly degraded in the cell, the effect of the antisense gene is to deplete the cell of the sense mRNA. Thus, transgenic plants expressing antisense versions of genes that encode enzymes involved in the ethylene biosynthetic pathway, such as ACC synthase and ACC oxidase, can synthesize ethylene only at very low levels. Consistent with a role for ethylene in senescence, such antisense mutants have been shown to exhibit delayed leaf senescence as well as fruit ripening in tomato (see Box 22.1).

Ethylene-regulated carpel senescence. In addition to leaf and flower senescence, ethylene appears to regulate the senescence of carpels. For example, in the flowers of the garden pea, which normally self-pollinate, the carpels senesce if pollination is prevented by removal of the stamens. If the carpel is treated with inhibitors of ethylene action, such as silver thiosulfate, senescence of the carpel is delayed. Recently, it has been shown that senescence in pea carpel cells involves nuclear condensation and DNA fragmentation between the nucleosomes, which can be visualized as "DNA ladders" when size-separated by gel electrophoresis (Orzáez and Granell 1997). This type of senescence, first characterized in animal cells, is called **apoptosis** (see Chapter 16).

Ethylene Biosynthesis in the Abscission Zone Is Regulated by Auxin

The shedding of leaves, fruits, flowers, and other plant organs is termed **abscission**. Abscission takes place in specific layers of cells, called **abscission layers**, which become morphologically and biochemically differentiated during organ development. Weakening of the cell walls at the abscission layer depends on cell wall–degrading enzymes such as cellulase and polygalacturonase. Ethylene appears to be the primary regulator of the abscission process, with auxin acting as a suppressor of the ethylene effect. However, supraoptimal auxin concentrations stimulate ethylene production, which has led to the use of auxin analogs as defoliants. For example, 2,4,5-T, the active ingredient in Agent Orange, was widely used as a defoliant during the Vietnam War. Its action is based on its ability to increase ethylene biosynthesis, thereby stimulating leaf abscission.

A model of the hormonal control of leaf abscission describes the process in three distinct sequential phases (Figure 22.8) (Morgan 1984; Reid 1995):

1. *Leaf maintenance phase.* Prior to the perception of any signal (internal or external) that initiates the abscission process, the leaf remains healthy and fully functional in the plant. A gradient of auxin from the leaf to the stem maintains the abscission zone in a nonsensitive state.

2. *Shedding induction phase.* A reduction or reversal in the auxin gradient from the leaf, normally associated with leaf senescence, causes the abscission zone to become sensitive to ethylene. Treatments that enhance leaf senescence may promote abscission by interfering with auxin synthesis and/or transport in the leaf.

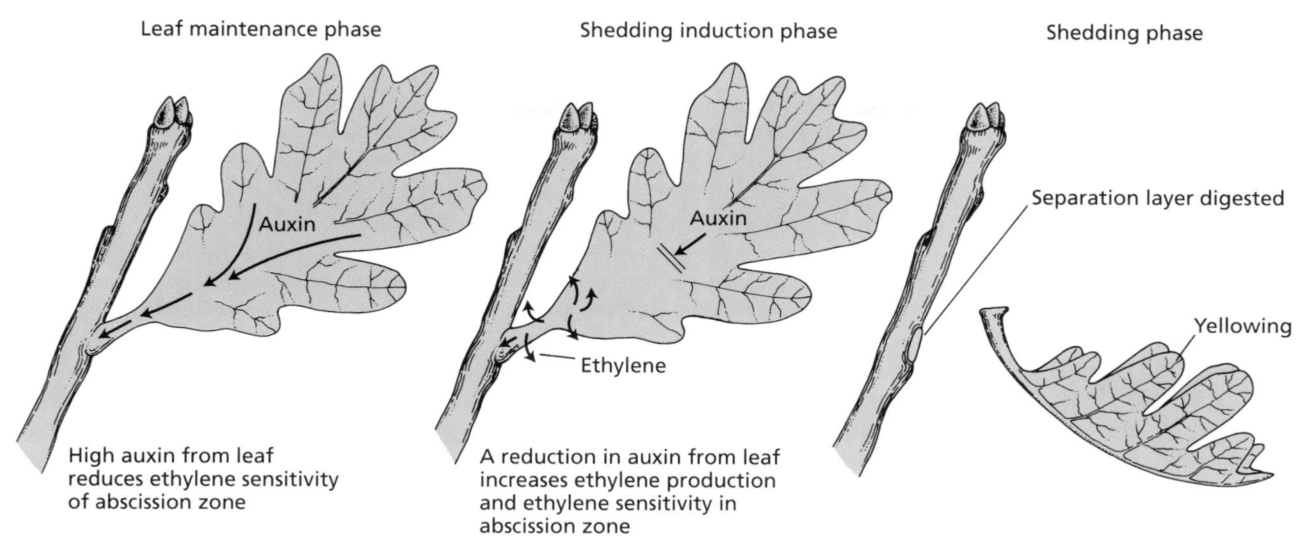

Figure 22.8 Schematic view of roles of auxin and ethylene during leaf abscission. According to this model, auxin prevents leaf shedding during the leaf maintenance phase. In the shedding induction phase, the level of auxin decreases, and the level of ethylene increases. These changes in the hormonal balance increase the sensitivity of the target cells to ethylene, which triggers the events involved in the shedding phase. During the shedding phase, specific enzymes that hydrolyze the cell wall polysaccharides are synthesized. Wall hydrolysis results in cell separation and leaf abscission. (After Morgan 1984.)

3. *Shedding phase*. The sensitized cells of the abscission zone respond to low concentrations of endogenous ethylene by the rapid production and secretion of cellulase and other cell-wall degrading enzymes, resulting in shedding.

During the early phase of leaf maintenance, auxin from the leaf prevents abscission by maintaining the cells of the abscission zone in an ethylene-insensitive state. It has long been known that removal of the leaf blade (the site of auxin production) promotes petiole abscission. Application of exogenous auxin to petioles from which the leaf blade has been removed delays the abscission process. However, application of auxin to the proximal side of the abscission zone (that is, the side closest to the stem) actually *accelerates* the abscission process. This indicates that it is not the absolute amount of auxin at the abscission zone, but rather, the auxin *gradient* that controls the ethylene sensitivity of these cells.

In the shedding induction phase, the amount of auxin from the leaf decreases and the ethylene level rises. Ethylene appears to decrease the activity of auxin by both reducing its synthesis and transport and increasing its destruction. The reduction in the concentration of free auxin increases the response of specific target cells to ethylene. The *target cells*, located in the abscission zone, synthesize cellulase and other polysaccharide-degrading enzymes, and secrete them into the cell wall via secretory vesicles derived from the Golgi. The shedding phase is characterized by induction of genes encoding specific hydrolytic enzymes of cell wall polysaccharides and proteins. The action of these enzymes leads to cell wall loosening, cell separation, and abscission.

Ethylene Has Important Commercial Uses

Since ethylene regulates so many physiological processes in plant development, it is one of the most widely used plant hormones in agriculture. Auxins and ACC can trigger the natural biosynthesis of ethylene and in several cases are used in agricultural practice. Because of its high diffusion rate, ethylene proper is very difficult to apply in the field as a gas, but this limitation can be overcome if an ethylene-releasing compound is used. The most widely used such compound is ethephon, or **2-chloroethylphosphonic acid**, which was discovered in the 1960s and is known by its trade name, Ethrel.

Ethephon is sprayed in aqueous solution and is readily absorbed and transported within the plant. It releases ethylene slowly by a chemical reaction, allowing the hormone to exert its effects:

$$Cl-CH_2-CH_2-\overset{O}{\overset{\|}{P}}(-O)-OH + OH^- \longrightarrow CH_2=CH_2 + H_2PO_4^- + Cl^-$$

2-Chloroethylphosphonic acid (ethephon) → **Ethylene**

BOX 22.2

Abscission and the Dawn of Agriculture

CIVILIZATION AS WE KNOW IT is entirely dependent on an abundant food supply, which can be provided only by agriculture. Before the development of agriculture, people lived in small, nomadic groups whose migration patterns followed the seasonal availability of foods derived from wild plants and animals. The shift from hunting and gathering to cultivation and pastoralism led to the establishment of permanent settlements with ever-expanding populations. This transition, considered to be the most significant (and by some, the most disastrous!) cultural change in human history, was made independently at least seven times at various locations around the world (Smith 1995).

Remains of some of the earliest settlements are found in the region of the Near East called the Fertile Crescent, a broad arching zone of grassland and woodland that begins at the eastern edge of the Mediterranean and curves around some 2000 km eastward to the Zagros Mountains. One of the oldest known agricultural settlements in the world is found here, in the ancient city of Jericho, located 14 miles north of the modern city of Jericho in the Jordan Valley. Here archaeologists have unearthed evidence for a transition from hunting and gathering to agriculture that occurred around 10,000 years ago (Smith 1995).

The first step in the transition from hunting and gathering to agriculture was the domestication of plants and animals. The seven primary domesticates of the Fertile Crescent were barley, emmer wheat, einkorn wheat, sheep, goats, cattle, and pigs. The first of these to be domesticated were the cereals. The Mediterranean region is particularly well stocked with cereals suitable for domestication. Of the world's 56 species of large-seeded grasses, 32 grow wild in the Mediterranean area. Recent DNA analyses have pinpointed the area near the Karacadağ mountains in Turkey as the probable site of origin of einkorn wheat (Heun et al. 1997).

All the evidence suggests that wheat and barley were domesticated over a period of only 2 or 3 centuries, between 10,000 and 9700 B.P. (Smith 1995). Considering the large number of morphological and physiological differences between even the most primitive of the domesticated forms of wheat and barley today and their wild progenitors, it is difficult to envision how so many changes would come about in such a brief period by natural selection alone. Clearly, some type of artificial selection had to be carried out (most likely by women) to accelerate the evolutionary process. Archaeologists and botanists have attempted to reconstruct this selection process.

Of all the anatomical and physiological differences between wild and domesticated cereals, two are regarded as most crucial. The first is the **loss of seed dormancy**—the failure of certain seeds to germinate when planted in moist soil unless given additional environmental signals, such as cold or light (see Chapter 23). The process of plant domestication probably began in small garden plots planted near temporary settlements. It is easy to imagine how the loss of seed dormancy might be selected for by a nomadic group, since seeds that germinated and grew rapidly would be more likely to have been gathered than those whose germination was delayed.

The second critical change that took place during cereal domestication was the development of a **tough rachis**. The **rachis** is the main axis of the inflorescence, or **spike**, of wheat and other cereals, to which the spikelets are attached. **Spikelets** are small inflorescences bearing one or more **florets**, or small flowers, along with a set of miniature bractlike leaves. After pollination, the fruits (**caryopses**) develop within the spikelets. When the caryopses of the wild-type wheat are fully ripened, a series of abscission layers forms that divides the rachis into dispersal units consisting of a single spikelet attached to a short segment of the rachis (see the figure). This **disarticulation** of the rachis begins at the top of the inflorescence and works its way downward. A ripe head of wild wheat is thus easily shattered into dispersal units when touched or blown by the wind and is referred to by archaeologists as a *brittle rachis*.

A brittle rachis is an extremely undesirable trait from a human standpoint. Imagine the plight of a small nomadic group that carefully timed its peregrinations to coincide with the ripening of wild cereals. Because of the extreme fragility of the rachis, it would be crucial to arrive close to the time of ripening, but before disarticulation of the rachis. A slight miscalculation could spell disaster for the group. Arriving too late, they would be greeted by a field of barren stalks yielding only a handful of remnant grains. As a consolation, however, they might find the rare mutant spikes that remained intact because the abscission layers of the rachis failed to develop. This property is referred to by archaeologists as a *tough rachis*. If mutants with the tough-rachis phenotype were then gathered and planted in gardens, their numbers would, over time, increase.

A second scenario for the selection of a tough rachis involves the method of harvesting. The most efficient method for harvesting wheat is to cut

Ethephon hastens fruit ripening of apple and tomato and degreening of citrus, synchronizes flowering and fruit set in pineapple, and accelerates abscission of flowers and fruits. Its application makes it possible to achieve fruit thinning or fruit drop in cotton, cherry, and walnut. It is also used to promote female sex expression in cucumber, to prevent self-pollination and increase yield, and to inhibit terminal growth of some plants in order to promote lateral growth and compact flowering.

Storage facilities developed to inhibit ethylene production and promote preservation of fruits have a controlled atmosphere of low O_2 concentration and low temperature that inhibits ethylene biosynthesis. A relatively high concentration of CO_2 (3 to 5%) prevents ethylene's action as a ripening promoter. Low pressure is used to remove ethylene and oxygen from the storage chambers, reducing the rate of ripening and preventing overripening. Specific inhibitors of ethylene biosynthesis and

BOX 22.2 (continued)

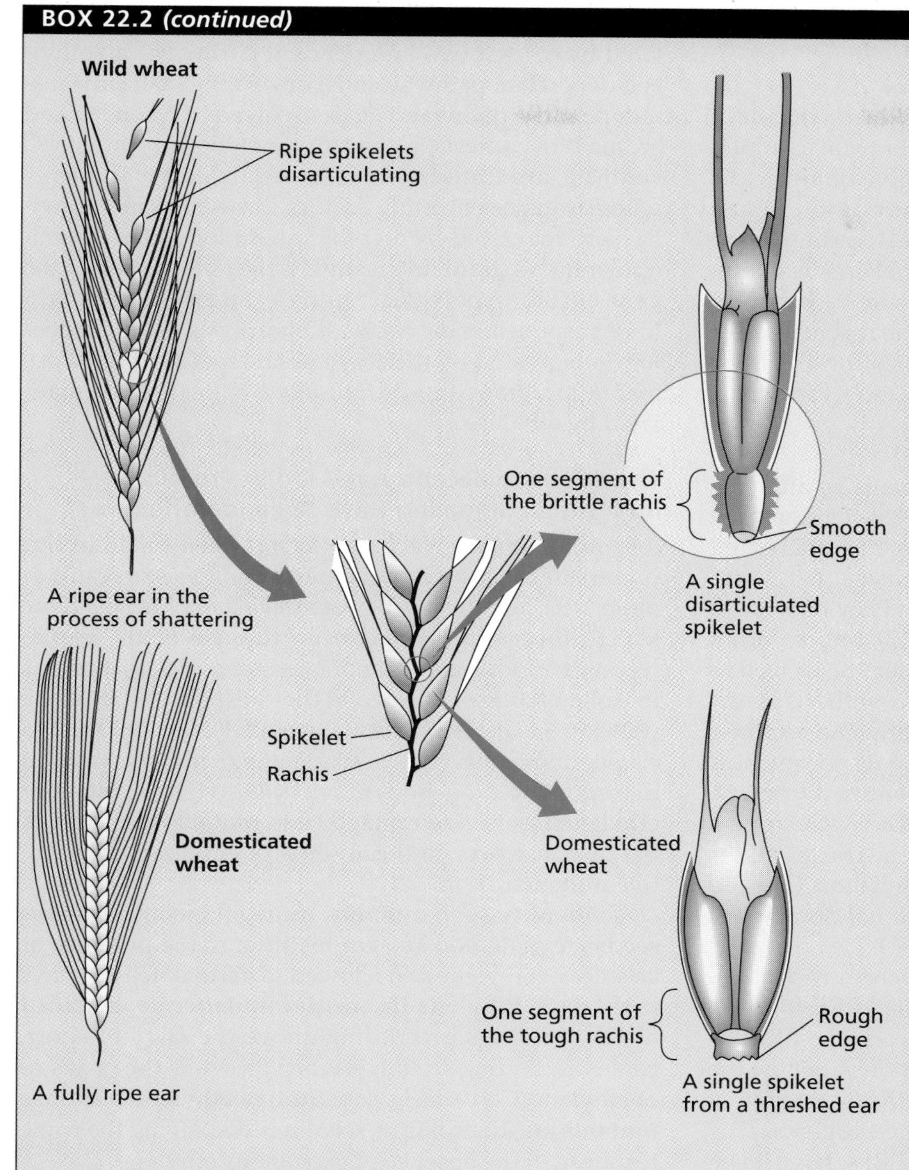

Wild versus domesticated forms of einkorn wheat. The spikelets of the wild variety abscise readily from the ripened head at abscission zones located nearby on the rachis, leaving a smooth abscission scar on the disarticulated spikelet. In contrast, the abscission zones of the domesticated varieties that have tough rachises fail to develop properly, and the spikelets remain on the head until released by threshing. Note that the breakage of the rachis below the spikelet leaves a rough edge, which tends to impede penetration into the ground.

it off at the base of the stalk, and evidence suggests that flint-bladed sickles were used for this purpose. However, harvesting wild wheat by cutting would have to be performed before formation of the abscission layers, or the fragile spikes would shatter, sending their spikelets flying to the ground. This method of harvesting would therefore favor the selection of mutations that delayed the abscission process. Hillman and Davies (1990a, 1990b) have demonstrated in field experiments that over many generations, the population would come to be dominated by the varieties with tough rachises. Once cut, the stalks of these varieties of wheat would then be transported to another location, where the grains would be removed by threshing.

Is the mutation rate in wild cereals rapid enough to account for their domestication in a period of only 200 to 300 years? Sharma and Waines (1980) have determined that two genetic loci are involved in the evolution of the tough-rachis character in einkorn wheat. Even if we assume that a mutation that converts a brittle rachis to a tough rachis occurs in only one out of every million wild-type wheat plants, Hillman and Davies (1990a, 1990b) have estimated that varieties with tough rachises would become the dominant forms in wild wheat populations after only 200 years of human management. Higher rates of mutation would shorten this time considerably, possibly to within 20 years.

Thus the experimental results and calculations are in good agreement with the archaeological evidence. Agriculture in the Fertile Crescent, which ultimately led to the founding of Greece and Rome and all of Western civilization, began with mutations in genes that control abscission.

action are also useful in postharvest preservation. Silver (Ag^+) is used extensively to increase the longevity of cut carnations and several other flowers. The potent inhibitor AVG retards fruit ripening and flower fading, but its commercial use has not yet been approved by regulatory agencies. The strong, offensive odor of *trans*-cyclooctene precludes its use in agriculture.

The near future will see a variety of agriculturally important species that have been engineered by transgenic technology to manipulate the biosynthesis of ethylene or its perception. The inhibition of ripening in tomato by expression of an antisense version of ACC synthase and ACC oxidase has already been mentioned. Another example of this technology is in petunia, in which ethylene biosynthesis has been blocked by transformation of an antisense version of ACC oxidase. Senescence and petal wilting of cut flowers are delayed for weeks in these transgenic plants.

Cellular and Molecular Modes of Ethylene Action

Despite the diversity of ethylene's effects on development, the primary steps in ethylene action are assumed to be similar in all cases: They all involve binding to a receptor, followed by activation of one or more signal transduction pathways (see Chapter 14) leading to the cellular response. Ultimately, ethylene exerts its effects by altering the pattern of gene expression. In recent years, much has been learned about the mechanism of ethylene action through the analysis of ethylene response mutants in *Arabidopsis*.

Ethylene Regulates Gene Expression

One of the primary effects of ethylene is to alter the expression of various target genes. Ethylene increases the mRNA transcript levels of numerous genes, including the genes that encode cellulase, chitinase, β-1,3-glucanase, peroxidase, chalcone synthase (a key enzyme in flavonoid biosynthesis; see Chapter 13), and a pathogenesis-related (PR) protein (see Chapter 13), as well as ripening-related genes and ethylene biosynthesis genes. *Cis*-acting regulatory sequences called **ethylene response elements**, or **EREs**, that confer ethylene responsiveness to a minimum promoter have been identified from the ethylene-regulated PR genes, and a GCCGCC repeat motif was identified, using reporter gene constructs, as both necessary and sufficient for this regulation. (The use of reporter genes to analyze the functional domains of promoters was discussed in Chapter 17.)

Four proteins that bind to ERE sequences were identified in tobacco. These proteins are called **ERE-binding proteins** (**EREBPs**). The genes that encode the EREBPs may represent ethylene primary response genes, the gene products of which may regulate the expression of secondary response genes, such as the PR genes. The DNA-binding domains of these EREBPs have been localized to a 59–amino acid region that shows similarity to the AP2 protein, a predicted nuclear protein that is involved in floral development in *Arabidopsis* (see Chapter 24). The steady-state level of these EREBPs increases dramatically following ethylene treatment.

Ripening-related gene expression has been studied extensively in developing tomato fruits, and several genes have been identified that are highly regulated during ripening (Theologis 1992; Gray et al. 1994). During tomato fruit ripening, the fruit softens as the result of cell wall hydrolysis and changes from green to red as a consequence of chlorophyll loss and the synthesis of the carotenoid pigment lycopene. At the same time, aroma and flavor components are produced.

Analysis of mRNA from tomato fruits from wild-type and transgenic tomato plants genetically engineered to lack ethylene biosynthesis (discussed earlier) has revealed that gene expression during ripening is regulated by at least two independent pathways: an ethylene-dependent pathway and a developmental, ethylene-independent pathway. Genes involved in lycopene and aroma biosynthesis, respiratory metabolism, and ACC synthase are transcriptionally regulated by ethylene, whereas genes encoding ACC oxidase and chlorophyllase are regulated by an ethylene-independent developmental program. Interestingly, the *transcription* of the gene encoding polygalacturonase, an enzyme thought to be responsible for cell wall hydrolysis during ripening, is regulated by the ethylene-independent developmental program, but its *translation* appears to be regulated by ethylene.

The Ethylene Receptor and Other Proteins Involved in Signaling Have Been Identified

Recently, remarkable progress has been made in our understanding of ethylene perception as the result of molecular genetic studies of *Arabidopsis thaliana*. One key to these studies has been the use of the triple-response morphology of etiolated seedlings as a means to isolate mutants affected in their response to ethylene (Bleeker et al. 1988; Guzman and Ecker 1990). Two classes of mutants have been identified: mutants that fail to respond to exogenous ethylene (ethylene-resistant or ethylene-insensitive mutants) and mutants that display the response even in the absence of ethylene (constitutive mutants).

To identify such mutants, mutagenized *Arabidopsis* seeds are plated on an agar medium in the presence or absence of ethylene and allowed to germinate for 3 days in the dark. **Ethylene-insensitive mutants** are identified as tall seedlings protruding above the lawn of short, triple-responding seedlings when grown in the presence of ethylene. Conversely, **constitutive ethylene response mutants** are identified as seedlings displaying the triple response in the absence of exogenous ethylene.

Among the ethylene-insensitive mutants isolated was *etr1* (*e*thylene-*r*esistant *1*) (see Figure 22.7). The *etr1* mutation is a dominant mutation that blocks the response to ethylene (Bleecker et al. 1988). When the *ETR1* gene was cloned and sequenced, the carboxy-terminal half of the protein was shown to have similarity to the bacterial two-component histidine kinases (Chang et al. 1993). As we saw in Chapter 14, two-component regulators are the major route by which bacteria sense and respond to various environmental cues, such as chemosensory stimuli, phosphate availability, and osmolarity. The two components consist of a sensor histidine kinase and a response regulator that often acts as a transcription factor. ETR1 was the first example of a eukaryotic histidine kinase, but others have since been found in yeast, mammals, and plants. Both phytochrome (see Chapter 17) and the putative cytokinin

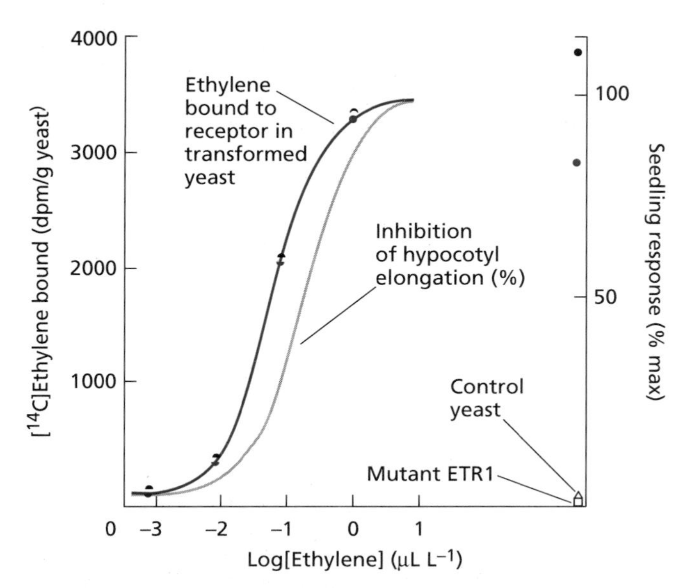

Figure 22.9 Comparison of the response of ethylene binding to ETR1 with the response of seedlings to ethylene. The *ETR1* gene was expressed in yeast, and the ability of the transformed yeast cells to bind radioactive ethylene was tested. The continuous curve (in black) with the open and closed circles represents the amount of ethylene bound to duplicate samples of yeast cells at different ethylene concentrations. The open triangle and square represent ethylene binding by yeast expressing the vector alone or the mutant ETR1 protein. The green line represents the inhibition of hypocotyl elongation by ethylene. The similarity of the two curves strongly suggests that the *ETR1* gene encodes the ethylene receptor. (After Schaller and Bleecker 1995.)

Figure 22.10 Screen for *Arabidopsis* mutants that constitutively display the triple response. Seedlings were grown for 3 days in the dark in air. A single *ctr1* mutant seedling is evident among the taller, wild-type seedlings. (Photo courtesy of J. Kieber.)

receptor (see Chapter 21) are related to the bacterial two-component histidine kinases.

When *ETR1* was expressed in yeast, the yeast cells acquired the capacity to bind radiolabeled ethylene in binding assays. The ethylene-binding properties of the yeast cells expressing ETR1 protein closely match those of the binding sites identified in crude extracts from various plant species. Moreover, the affinity of yeast expressing *ETR1* for ethylene closely parallels the dose–response curve of *Arabidopsis* seedlings to ethylene (Figure 22.9). Thus, the evidence is quite strong that ETR1 is the ethylene receptor.

The amino terminus of ETR1 contains three hydrophobic membrane-spanning domains, which have been shown to be the site of ethylene binding. ETR1 functions as a dimer consisting of two transmembrane proteins linked together by disulfide bonds, similar to the bacterial chemosensory receptors. Thus far, the precise cellular location of ETR1 has not been determined. Since *ETR1* is present as a small gene family in *Arabidopsis*, there may be multiple receptors for ethylene in plants.

The ***ein*** (*e*thylene-*in*sensitive) mutants were also blocked in their ethylene responses. The *EIN2* gene encodes a protein that contains 12 membrane-spanning domains, suggesting that it may act as a channel or pore. The *EIN3* gene encodes a protein with features that suggest it acts as a transcription factor. Like ETR1, EIN3 is encoded by members of a multigene family in *Arabidopsis*.

Among the constitutive mutants identified in the genetic screens was ***ctr1*** (*c*onstitutive *t*riple *r*esponse *1* = triple response in the absence of ethylene), a recessive mutation resulting in a constitutive activation of ethylene responses (Figure 22.10). The fact that the mutation caused an *activation* of the ethylene response suggests that the wild-type protein acts as a *negative regulator* of the response pathway (Kieber et al. 1993). CTR1 shows similarity to the Raf family of serine/threonine protein kinases. Raf is part of a cascade of protein kinases that are involved in the transduction of various external regulatory signals and developmental signaling pathways in organisms ranging from yeast to humans (see Chapter 14).

The order of action of the genes *ETR1, EIN2, EIN3,* and *CTR1* has been determined by analysis of the epistatic relationship between the various mutations—that is, how the mutations interact with each other (Kieber et al. 1993; Roman et al. 1995). Researchers conduct this analysis by constructing a line that is doubly mutant for *ctr1* and an ethylene-insensitive mutation and examining the resultant phenotype. If the double mutant displays a *ctr1* mutant phenotype, it can be inferred that *CTR1* acts downstream of that particular gene (Avery and Wasserman 1992). In this way, the order of action of *ETR1, EIN2,* and *EIN3* were determined relative to *CTR1*.

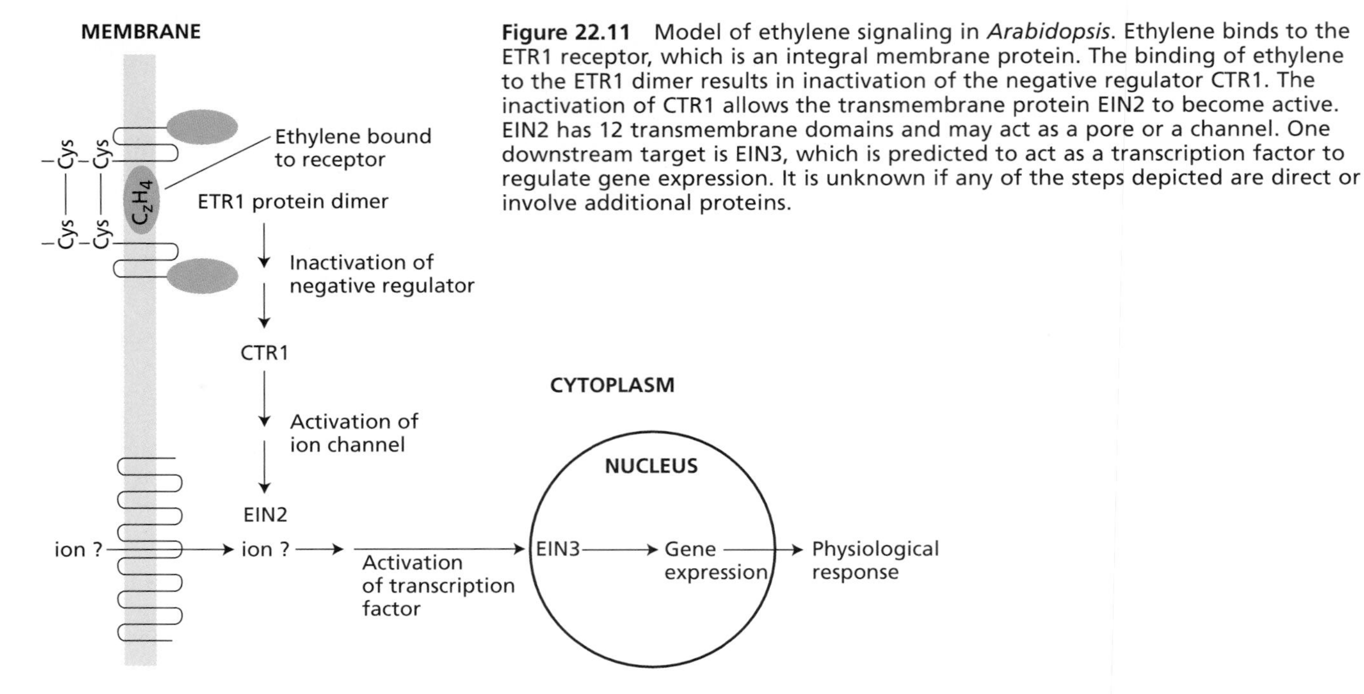

Figure 22.11 Model of ethylene signaling in *Arabidopsis*. Ethylene binds to the ETR1 receptor, which is an integral membrane protein. The binding of ethylene to the ETR1 dimer results in inactivation of the negative regulator CTR1. The inactivation of CTR1 allows the transmembrane protein EIN2 to become active. EIN2 has 12 transmembrane domains and may act as a pore or a channel. One downstream target is EIN3, which is predicted to act as a transcription factor to regulate gene expression. It is unknown if any of the steps depicted are direct or involve additional proteins.

The model in Figure 22.11 summarizes these and other data. This model is still incomplete, since other ethylene response mutations have been identified that act in this pathway. In addition, no data are yet available to determine if any of the interactions depicted are direct or indirect or where in the cells these proteins are located. For the first time, however, we are beginning to glimpse the outline of the molecular basis for the perception and transduction of a plant hormone signal.

Genes that are similar to several of these *Arabidopsis* ethylene signaling genes have been found in other plant species. Most notably, the gene corresponding to the *never-ripe* mutation in tomato (Lanahan et al. 1994) has been shown to have high similarity to *ETR1*. This mutation results in fruits that fail to ripen even when supplied with exogenous ethylene, and it disrupts other ethylene responses. These and other studies suggest that the components of these proteins may represent ubiquitous mediators of ethylene signaling in plants.

Summary

Ethylene is formed in most organs of higher plants. Senescing tissues or ripening fruits produce more ethylene than young or mature tissues do. The precursor of ethylene in vivo is the amino acid methionine, which is converted to *S*-adenosylmethionine (AdoMet), 1-aminocyclopropane-1-carboxylic acid (ACC), and ethylene. The rate-limiting step of this pathway is the conversion of AdoMet to ACC, which is catalyzed by ACC synthase. ACC synthase is encoded by members of a multigene family that are differentially regulated in various plant tissues and in response to various inducers of ethylene biosynthesis.

Ethylene biosynthesis is triggered by various developmental processes, by auxins, and by environmental stresses. In all these cases the level of activity and of mRNA of ACC synthase increases. The physiological effects of ethylene can be blocked by biosynthesis inhibitors or by antagonists. AVG (aminoethoxyvinylglycine) and AOA (aminooxyacetic acid) inhibit the synthesis of ethylene; carbon dioxide, silver ions, and *trans*-cyclooctene inhibit ethylene action. Ethylene can be detected and measured by biological assays and by gas chromatography.

Ethylene regulates fruit ripening and other processes associated with leaf and flower senescence, leaf and fruit abscission, root hair development, seedling growth, and hook opening. Ethylene also regulates the expression of various genes, including ripening-related genes and pathogenesis-related genes.

The ethylene receptor is encoded by the gene *ETR1*. The ETR1 protein is a transmembrane protein that shows sequence similarity to prokaryotic two-component histidine kinases. Downstream signal transduction components include CTR1, a member of the Raf family of protein kinases; EIN2, a channel-like transmembrane protein; and EIN3, a novel transcription factor.

General Reading

*Abeles, F. B., Morgan, P. W., and Saltveit, M. E., Jr. (1992) *Ethylene in Plant Biology*, 2nd ed. Academic Press, San Diego.

*Beyer Jr., E. M. (1979) Effect of silver ion, carbon dioxide, and oxygen on ethylene action and metabolism. *Plant Physiol.* 63: 169–173.

Bleecker, A., and Schaller, G. (1996) The mechanism of ethylene perception. *Plant Physiol.* 111: 653–660.

Borochov, A., and Woodson, W. R. (1989) Physiology and biochemistry of flower petal senescence. *Hort. Rev.* 11: 15–43.

Daum, G., Eisenmann-Tappe, I., Fries, H-W., Troppmair, J., and Rapp, U. (1994) The ins and outs of Raf kinases. *Trends Biochem. Sci.* 19: 474–479.

Ecker, J. (1995) The ethylene signal transduction pathway in plants. *Science* 268: 667–675.

Gray, J., Pictor, S., Shabbeer, J., Schuch, W., and Grierson, D. (1992) Molecular biology of fruit ripening and its manipulation with antisense genes. *Plant Mol. Biol.* 19: 69–87.

Horgan, R. (1995) Instrumental methods of plant hormone analysis. In *Plant Hormones: Physiology, Biochemistry and Molecular Biology*, 2nd ed., P. J. Davies, ed., Kluwer, Dordrecht, Netherlands, pp. 140–157.

*Kende, H. (1993) Ethylene biosynthesis. *Annu. Rev. Plant Physiol. Mol. Biol.* 44: 283–307.

Kieber, J. J. (1997) The ethylene response pathway in *Arabidopsis*. *Annu. Rev. Plant Physiol. Plant Mol. Biol.* 48: 277–296.

Mattoo, A., and Suttle, J., eds. (1991) *The Plant Hormone Ethylene*. CRC Press, Boca Raton, Fla.

Reid, J. R., and Howell, S. H. (1995) The functioning of hormones in plant growth and development. In *Plant Hormones: Physiology, Biochemistry and Molecular Biology*, 2nd ed., P. J. Davies, ed., Kluwer, Dordrecht, Netherlands, pp. 448–485.

*Smith, B. D. (1995) *The Emergence of Agriculture*. Scientific American Library, New York.

Thimann, K. V. (1977) *Hormone Action in the Whole Life of Plants*. University of Massachusetts Press, Amherst.

Zarembinski, T., and Theologis, A. (1994) Ethylene biosynthesis and action: A case of conservation. *Plant Mol. Biol.* 26: 1579–1597.

* Indicates a reference that is general reading in the field and is also cited in this chapter.

Chapter References

Abel, S., Nguyen, M., Chow, W., and Theologis, A. (1995) ACS4, a primary indole-acetic acid-responsive gene encoding 1-aminocyclopropane-1-carboxylate synthase in *Arabidopsis thaliana*. *J. Biol. Chem.* 270: 19093–19099.

Adams, D. O., and Yang, S. F. (1979) Ethylene biosynthesis: Identification of 1-aminocyclopropane-1-carboxylic acid as an intermediate in the conversion of methionine to ethylene. *Proc. Natl. Acad. Sci. USA* 76: 170–174.

Amrhein, N., Schneebeck, D., Skorupka, H., Tophof, S., and Stockigt, J. (1981) Identification of a major metabolite of the ethylene precursor, 1-aminocyclopropane-1-carboxylic acid in higher plants. *Naturwissenschaften* 68: 619–620.

Avery, L., and Wasserman, S. (1992) Ordering gene function: The interpretation of epistasis in regulatory hierarchies. *Trends Genet.* 8: 312–316.

Bleecker, A., Estelle, M., Somerville, C., and Kende, H. (1988) Insensitivity to ethylene conferred by a dominant mutation in *Arabidopsis thaliana*. *Science* 241: 1086–1089.

Bradford, K. J., and Yang, S. F. (1980) Xylem transport of 1–aminocyclopropane–1–carboxylic acid, an ethylene precursor, in waterlogged tomato plants. *Plant Physiol.* 65: 322–326.

Burg, S. P., and Burg, E. A. (1965) Relationship between ethylene production and ripening in bananas. *Bot. Gaz.* 126: 200–204.

Burg, S. P., and Clagett, C. O. (1967) Conversion of methionine to ethylene in vegetative tissue and fruits. *Biochem. Biophys. Res. Commun.* 27: 125–130.

Burg, S. P., and Thimann, K. V. (1959) The physiology of ethylene formation in apples. *Proc. Natl Acad Sci.* (USA) 45: 335–344.

Chang, C., Kwok, S., Bleecker, A., and Meyerowitz, E. (1993) *Arabidopsis* ethylene-response gene *ETR1*: Similarity of product to two-component regulators. *Science* 262: 539–544.

Dolan, L., Duckett, C. M., Grierson, C., Linstead, P., Schneider, K., Lawson, E., Dean, C., Poethig, S., and Roberts, K. (1994) Clonal relationships and cell patterning in the root epidermis of *Arabidopsis*. *Development* 120: 2465–2474.

Eisinger, W. (1983) Regulation of pea internode expansion by ethylene. *Annu. Rev. Plant Physiol.* 34: 225–240.

Gepstein, S., and Thimann, K. V. (1981) The role of ethylene in the senescence of oat leaves. *Plant Physiol.* 68: 349–354.

Good, X., Kellog, J. A., Wagoner, W., Langhoff, D., Matsumara, W., and Bestwick, R. K. 1993. Reduced ethylene synthesis by transgenic tomatoes expressing S-adenosylmethionine hydrolase. *Plant Mol. Biol.* 26: 781–790.

Gray, J. E., Picton, S., Giovannoni, J. J., and Grierson, D. (1994) The use of transgenic and naturally occurring mutants to understand and manipulate tomato fruit ripening. *Plant Cell Environ.* 17: 557–571.

Grbić, V., and Bleecker, A. B. (1995) Ethylene regulates the timing of leaf senescence in *Arabidopsis*. *Plant J.* 8: 595–602.

Guzman, P., and Ecker, J. R. (1990) Exploiting the triple response of *Arabidopsis* to identify ethylene-related mutants. *Plant Cell* 2: 513–523.

Hamilton, A. J., Lycett, G. W., and Grierson, D. (1990) Antisense gene that inhibits synthesis of the hormone ethylene in transgenic plants. *Nature* 346: 284–287.

Hensel, L. L., Grbć, V., Baumgarten, D. A., and Bleecker, A. B. (1993) Developmental and age-related processes that influence the longevity and senescence of photosynthetic tissues in *Arabidopsis*. *Plant Cell* 5: 553–564.

Heun, M., Schäfer-Pregl, R., Klawan, D., Castagna, R., Accerbi, M., Borghi, B., and Salamini, F. (1997) Site of einkorn wheat domestication identified by DNA fingerprinting. *Science* 278: 1312–1314.

Hillman, G. C., and Davies, M. S. (1990a) Domestication rates in wild-type wheats and barley under primitive cultivation. *Biol. J. Linnean Soc.* 39: 39–78.

Hillman, G. C., and Davies, M. S. (1990b) Measured domestication rates in wild wheats and barley under primitive cultivation, and their archaeological implications. *J. World Prehistory* 4: 157–222.

Hoffman, N. E., and Yang, S. F. (1980) Changes of 1-aminocyclopropane-1-carboxylic acid content in ripening fruits in relation to their ethylene production rates. *J. Amer. Soc. Hort. Sci.* 105: 492–495.

Hoffman , N. E., Yang, S. F., and McKeon, T. (1982) Identification of 1-(malonylamino)cyclopropane-1-carboxylic acid as a major conjugate of 1-aminocyclopropane-1-carboxylic acid, an ethylene precursor in higher plants. *Biochem. Biophys. Res. Commun.* 104: 765–770.

Imaseki, H., Yoshii, Y., and Todaka, I. (1982) Regulation of auxin-induced ethylene biosynthesis in plants. In *Plant Growth Substances 1982*, P. F. Wareing, ed., Academic Press, London, pp. 259–268.

Kieber, J. J., Rothenburg, M., Roman, G., Feldmann, K. A., and Ecker, J. R. (1993) CTR1, a negative regulator of the ethylene response pathway in *Arabidopsis*, encodes a member of the Raf family of protein kinases. *Cell* 72: 427–441.

Klee, H. J., Hayford, M. B., Kretzmer, K. A., Barry, G. F., and Kishore, G. M. (1991) Control of ethylene synthesis by expression of a bacterial enzyme in transgenic tomato plants. *Plant Cell* 3: 1187–1193.

Lanahan, M., Yen, H-C., Giovannoni, J., and Klee, H. (1994) The *Never-ripe* mutation blocks ethylene perception in tomato. *Plant Cell* 6: 427–441.

Liang, X., Abel, S., Keller, J., Shen, N., and Theologis, A. (1992) The 1-aminocyclopropane-1-carboxylate synthase gene family of *Arabidopsis thaliana*. *Proc. Natl. Acad. Sci. USA* 89: 11046–11050.

Liang, X., Shen, N. F., and Theologis, A. (1996) Li^+-regulated 1-aminocyclopropane-1-carboxylate synthase gene expression in *Arabidopsis thaliana*. *Plant J.* 10: 1027–1036.

Lieberman, M., and Mapson, L. W. (1964) Genesis and biogenesis of ethylene. *Nature* 204: 343–345.

McKeon, T. A., Fernández-Maculet, J. C., and Yang, S. F. (1995) Biosynthesis and metabolism of ethylene. In *Plant Hormones: Physiology, Biochemistry and Molecular Biology*, 2nd ed., P. J. Davies, ed., Kluwer, Dordrecht, Netherlands, pp. 118–139.

Morgan, P. W. (1984) Is ethylene the natural regulator of abscission? In *Ethylene: Biochemical, Physiological and Applied Aspects*, Y. Fuchs and E. Chalutz, eds., Martinus Nijhoff, The Hague, Netherlands, pp. 231–240.

Nakagawa, J. H., Mori, H., Yamazaki, K., and Imaseki, H. (1991) Cloning of the complementary DNA for auxin-induced 1-aminocyclopropane-1-carboxylate synthase and differential expression of the gene by auxin and wounding. *Plant Cell Physiol.* 32: 1153–1163.

Oeller, P., Min-Wong, L., Taylor, L., Pike, D., and Theologis, A. (1991) Reversible inhibition of tomato fruit senescence by antisense RNA. *Science* 254: 437–439.

Olson, D. C., White, J. A., Edelman, L., Harkins, R. N., and Kende, H. (1991) Differential expression of two genes for 1-aminocyclopropane-1-carboxylate synthase in tomato fruits. *Proc. Natl. Acad. Sci. USA* 88: 5340–5344.

Orzáez, D., and Granell, A. (1997) DNA fragmentation is regulated by ethylene during carpel senescence in *Pisum sativum*. *Plant J.* 11: 137–144.

Picton, S., Barton, S. L., Bouzayen, M., Hamilton, A. J., and Grierson, D. (1993) Altered fruit ripening and leaf senescence in tomatoes expressing an antisense ethylene-forming enzyme transgene. *Plant J.* 3: 469–481.

Raskin, I., and Beyer, E. M., Jr. (1989) Role of ethylene metabolism in *Amaranthus retroflexus*. *Plant Physiol.* 90: 1–5.

Reid, M. S. (1995) Ethylene in plant growth, development and senescence. In *Plant Hormones: Physiology, Biochemistry and Molecular Biology*, 2nd ed., P. J. Davies, ed., Kluwer, Dordrecht, Netherlands, pp. 486–508.

Roman, G., Lubarsky, B., Kieber, J. J., Rothenberg, M., and Ecker, J. R. (1995) Genetic analysis of ethylene signal transduction in *Arabidopsis thaliana*: Five novel mutant loci integrated into a stress response pathway. *Genetics* 139: 1393–1409.

Rottmann, W. H., Peter, G. F., Oeller, P. W., Keller, J. A., Shen, N. F., Nagy, B. P., Taylor, L. P., Campbell, A. D., and Theologis, A. (1991) 1-aminocyclopropane-1-carboxylate synthase in tomato is encoded by a multigene family whose transcription is induced during fruit and floral senescence. *J. Mol. Biol.* 222: 937–961.

Sato, T., and Theologis, A. (1989) Cloning the messenger encoding 1-aminocyclopropane-1-carboxylate synthase, the key enzyme for ethylene biosynthesis in plants. *Proc. Natl. Acad. Sci. USA* 86: 6621–6625.

Sato, T., Oeller, P. W., and Theologis, A. (1991) The 1-aminocyclopropane-1-carboxylate synthase of *Cucurbita*: Purification, properties, expression in *Escherichia coli*, and primary structure determination by DNA sequence analysis. *J. Biol. Chem.* 266: 3752–3759.

Schaller, G., and Bleecker, A. (1995) Ethylene-binding sites generated in yeast expressing the *Arabidopsis ETR1* gene. *Science* 270: 1809–1811.

Sharma, H. C., and Waines, J. G. (1980) Inheritance of tough rachis in crosses of *Triticum monococcum* and *T. boeoticum*. *J. Hered.* 71: 214–216.

Sisler, E., Blankenship, S., and Guest, M. (1990) Competition of cyclooctenes and cyclooctadienes for ethylene binding and activity in plants. *Plant Growth Reg.* 9: 157–164.

Spanu, P., Reinhart, D., and Boller, T. (1991) Analysis and cloning of the ethylene-forming enzyme from tomato by functional expression of its mRNA in *Xenopus laevis* oocytes. *EMBO J.* 10: 2007–2013.

Tanimoto, M., Roberts, K., and Dolan, L. (1995) Ethylene is a positive regulator of root hair development in *Arabidopsis thaliana*. *Plant J.* 8: 943–948.

Theologis, A. (1992) One rotten apple spoils the whole bushel: The role of ethylene in fruit ripening. *Cell* 70: 181–184.

Voesenek, L. A. C. J., Banga, M., Rijnders, J. H. G. M., Visser, E. J. W., Harren, F. J. M.; Brailsford, R. W., Jackson, M. B., and Blom, C. W. P. M. (1997) Laser-driven photoacoustic spectroscopy: What we can do with it in flooding research. *Ann. Bot.* 79: 57–65.

Yang, S. F. (1987) The role of ethylene and ethylene synthesis in fruit ripening. In *Plant Senescence: Its Biochemistry and Physiology*, W. W. Thomson, E. A Nothnagel, and R. C. Huffaker, eds., American Society of Plant Physiologists, Rockville, MD, pp. 156–166.

Yang, S. F., Hoffman, N. E., McKeon, T., Riov, J., Jao, C. H., and Yung, K. H. (1982) Mechanism and regulation of ethylene biosynthesis in fruit ripening. In *Plant Growth Substances*, P. F. Wareing, ed., Academic Press, London, pp. 239–248.

Yuan, M., Shaw, P. J., Warn, R. M., and Lloyd, C. W. (1994) Dynamic reorientation of cortical microtubules, from transverse to longitudinal, in living plant cells. *Proc. Natl. Acad. Sci. USA* 91: 6050–6053.

Zacarias, L., and Reid, M. S. (1990) Role of growth regulators in the senescence of *Arabidopsis thaliana* leaves. *Physiol. Plantarum* 80: 549–554.

23 Abscisic Acid

FOR MANY YEARS, plant physiologists suspected that the phenomena of seed and bud dormancy were caused by inhibitory compounds, and they attempted to extract and isolate such compounds from a variety of plant tissues, especially dormant buds. Early experiments used paper chromatography for the separation of plant extracts, as well as bioassays based on oat coleoptile growth. These early experiments led to the identification of a group of inhibitory compounds that differ from auxin (Bennet-Clark and Kefford 1953). Ten years later, a substance that promotes the abscission of cotton fruits was purified and crystallized and was called *abscisin II* (Ohkuma et al. 1963). At about the same time, a substance that promotes bud dormancy was purified from sycamore leaves and was called *dormin*. When dormin was chemically identified, it was found to be identical to abscisin II, and the compound was renamed **abscisic acid** (**ABA**) (Figure 23.1) because of its supposed involvement in the abscission process.

It is now known that ethylene is the hormone that triggers abscission and that ABA-induced abscission of cotton fruits is due to its ability to stimulate ethylene production. As will be discussed in this chapter, ABA is now recognized as an important plant hormone in its own right. As a growth inhibitor it acts as a negative regulator of growth and stomatal opening, particularly when the plant is under environmental stress. Another important function of the hormone is in regulating seed dormancy. In retrospect, "dormin" would have been a more appropriate name for this hormone, but the name "abscisic acid" is firmly entrenched in the literature.

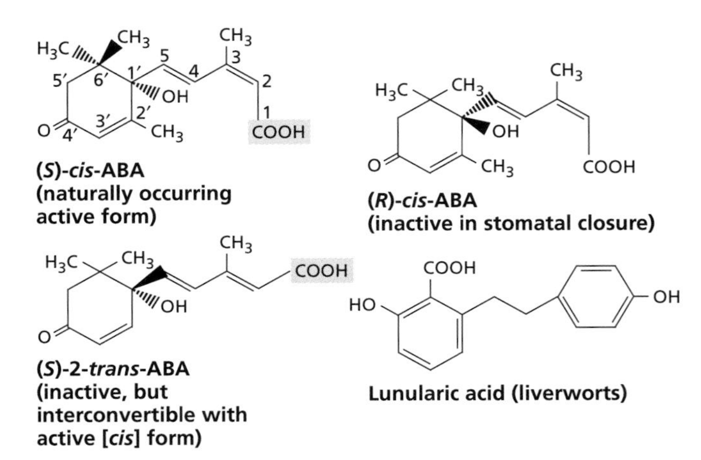

Figure 23.1 The chemical structures of the *S* (counterclockwise array) and *R* (clockwise array) forms of *cis*-ABA, the (*S*)-2-*trans* form of ABA, and lunularic acid. The numbers in the diagram of (*S*)-*cis*-ABA identify the carbon atoms.

Occurrence, Chemical Structure, and Measurement of ABA

Abscisic acid has been found to be a ubiquitous plant hormone in vascular plants. It has been detected in mosses, but appears to be absent in liverworts. Several genera of fungi make ABA as a secondary metabolite (Zeevaart and Creelman 1988). In liverworts, a compound similar to ABA, named **lunularic acid** (see Figure 23.1) appears to play a physiological role similar to that of ABA in higher plants. Within the plant, ABA has been detected in every major organ or living tissue from the root cap to the apical bud (Milborrow 1984). ABA is synthesized in almost all cells that contain chloroplasts or amyloplasts.

The Chemical Structure of ABA Determines Its Physiological Activity

The structure of ABA resembles the terminal portion of some carotenoid molecules (compare Figures 7.5 and 23.1). The 15 carbon atoms of ABA configure an aliphatic ring with one double bond, three methyl groups, and an unsaturated chain that has a terminal carboxyl group. The orientation of the carboxyl group at carbon 2 determines the *cis* and *trans* isomers of ABA (see Figure 23.1). Nearly all the naturally occurring ABA is in the *cis* form, and by convention the name "abscisic acid" refers to that isomer.

ABA also has an asymmetric carbon atom at position 1′ in the ring, resulting in the *S* and *R* (or + and –, respectively) enantiomers. The *S* enantiomer is the natural form; commercially available synthetic ABA is a mixture of approximately equal amounts of the *S* and *R* forms. The *S* enantiomer is the only one that is active in fast responses to ABA, such as stomatal closure (see Chapter 14). In long-term responses, such as changes in protein synthesis, both enantiomers are active. In contrast to the *cis* and *trans* isomers, the *S* and *R* forms cannot be interconverted in the plant tissue.

Studies of the structural requirements for biological activity of ABA have shown that almost any change in the molecule results in loss of activity. The features shown to be essential for biological activity include the carboxyl group, the tertiary hydroxyl group, the 2-*cis*, 4-*trans*-pentadienoic side chain, a 4′-ketone, and a double bond in the cyclohexane ring. Catabolic products of ABA that are present in the tissue and are missing any of these groups are biologically inactive.

ABA Is Assayed by Biological, Physicochemical, and Immunological Methods

Coleoptile growth, the classic bioassay devised for auxins (see Chapter 19), is also used for ABA detection in plant extracts by measurement of coleoptile growth inhibition. This bioassay has adequate sensitivity (minimum detectable level is 10^{-7} *M*) and shows a linear response in the range of 10^{-7} to 10^{-5} *M*, but it has some disadvantages. Plant extracts have other promoters and inhibitors, which decrease the specificity of the bioassay and make preliminary purification steps necessary.

Other available bioassays for ABA include inhibition of germination, inhibition of gibberellic acid–induced α-amylase synthesis in cereal aleurone layers, induction of leaf abscission, and stomatal closure. Stomatal closure is highly specific for ABA because it is affected little by other plant growth regulators. Additional advantages of this bioassay include a fast response of guard cells to ABA, high sensitivity (minimum detectable level 10^{-9} *M*), and a linear response over a wide range of concentrations.

Physical methods of detection are much more reliable than bioassays because of their specificity and suitability for quantitative analysis (Horgan 1995). The most widely used techniques are those based on gas chromatography or high-performance liquid chromatography (HPLC). Gas chromatography allows detection of as little as 10^{-13} g ABA, but it requires several preliminary purification steps, including thin-layer chromatography.

Another way to quantify ABA in plant extracts is by **immunoassay** (Walker-Simmons and Abrams 1991). This method relies on the specific recognition of ABA by antibodies obtained from rabbits or mice injected with the growth regulator. Specific and highly sensitive immunoassays can detect 10^{-13} g of ABA in crude or partly purified extracts, and they are currently available for ABA, auxins, gibberellins, and cytokinins (Weiler 1984).

Biosynthesis, Metabolism, and Transport of ABA

As with the other hormones, the response to ABA depends on its concentration within the tissue, and on the sensitivity of the tissue to the hormone. The processes of biosynthesis, catabolism, compartmentation, and transport all contribute to the concentration of active hormone in the tissue at any given stage of development. The complete biosynthetic pathway of ABA has only recently been elucidated, with the aid of ABA-deficient mutants blocked at specific steps in the pathway.

ABA Is Synthesized from a Xanthophyll Intermediate

The initial steps of ABA biosynthesis take place in chloroplasts and other plastids. Current data support the pathway depicted in Figure 23.2. Several ABA-deficient mutants have been identified with lesions at specific steps of the pathway. These mutants exhibit abnormalities that can be corrected by the application of exogenous ABA. For example, *flacca* (*flc*) and *sitiens* (*sit*) are "wilty mutants" of tomato in which the tendency of the leaves to wilt (due to an inability to close their stomata) can be prevented by the application of exogenous ABA. The *aba* mutant of *Arabidopsis* also exhibits a wilty phenotype. These and other mutants have been useful in elucidating the details of the pathway.

The pathway begins with isopentenyl pyrophosphate (IPP), the biological isoprene unit, and leads to the synthesis of the C_{40} xanthophyll (i.e., oxygenated carotenoid) **violaxanthin** (see Figure 23.2). As discussed in Chapters 13 and 20, IPP, the precursor of all terpenoids, is synthesized via a mevalonic acid–independent pathway in plastids. Support for the role of xanthophylls as intermediates in the ABA biosynthetic pathway has been provided by maize mutants (*vp*) that are blocked in the carotenoid pathway. Such mutants have reduced levels of ABA and also exhibit **vivipary**, the phenomenon whereby seeds germinate in the fruit while still attached to the plant. Vivipary is a feature of many ABA-deficient seeds.

Violaxanthin is converted to **9′-*cis*-neoxanthin**, which is then cleaved to form the C_{15} compound **xanthoxin**, a neutral growth inhibitor that has physiological properties similar to those of ABA. Although xanthophyll is synthesized in chloroplasts, it is not known whether 9′-*cis*-neoxanthin is cleaved within the chloroplast, at the chloroplast membrane, or outside the chloroplast. Finally, xanthoxin is converted to ABA via the intermediate **ABA-aldehyde** in the cytosol (Taylor and Burden 1973 ; Sindhu and Walton 1987).

A direct pathway for ABA biosynthesis has also been characterized; it is thought to occur mainly in phytopathic fungi. Unlike plants, fungi are able to synthesize the C_{15} ABA directly from the C_{15} precursor, farnesyl pyrophosphate (Walton and Li 1995).

ABA Concentrations Can Change Dramatically during Development and in Response to Environmental Stress

Abscisic acid metabolism is particularly interesting because the levels of ABA can fluctuate dramatically in specific tissues during development or in response to changing environmental conditions. In developing seeds, for example, ABA levels can increase 100-fold within a few days, and then decline to vanishingly low levels as maturation proceeds. Under conditions of water stress, ABA in the leaves can increase 50-fold within 4 to 8 hours. Upon rewatering, the ABA level declines to normal in the same amount of time.

Biosynthesis is not the only factor that regulates ABA concentrations in the tissue. As with other plant hormones, the concentration of free ABA in the cytosol is also regulated by degradation, compartmentation, and transport. As we will discuss later in the chapter, cytosolic ABA increases during water stress as a result of synthesis in the leaf, redistribution within the mesophyll cell, and import from the roots, while the decline in ABA concentration after rewatering is a consequence of degradation and export from the leaf, as well as a decrease in the rate of synthesis.

ABA Can Be Inactivated by Oxidation or Conjugation

A major cause of the inactivation of free ABA is oxidation, yielding the unstable intermediate 6-hydroxymethyl ABA, which is rapidly converted to **phaseic acid** (**PA**) and **dihydrophaseic acid** (**DPA**) (see Figure 23.2). PA is usually inactive, or it exhibits greatly reduced activity, in bioassays. However, PA can induce stomatal closure in some species, and it is as active as ABA in inhibiting gibberellic acid–induced α-amylase production in barley aleurone layers (Uknes and Ho 1984). These effects suggest that PA may be able to bind to ABA receptors. In contrast to PA, DPA has no detectable activity in any of the bioassays tested.

Free ABA is also inactivated by covalent conjugation to another molecule, such as a monosaccharide. A common example of an ABA conjugate is **ABA-β-D-glucosyl ester** (**ABA-GE**). Conjugation not only renders ABA inactive as a hormone; it also alters its polarity and cellular distribution. Whereas free ABA is localized in the cytosol, ABA-GE accumulates in vacuoles and thus could potentially serve as a storage form of the hormone. Esterases are present in plant cells that could release free ABA from the conjugated form. However, there is no evidence that ABA-GE hydrolysis contributes to the rapid

Figure 23.2 ABA biosynthesis and metabolism. In higher plants, ABA is synthesized via the terpenoid pathway (see Chapter 13). Isopentenyl pyrophosphate (IPP) serves as the precursor for the biosynthesis of the C_{40} xanthophyll (oxygenated carotenoid) zeaxanthin. Zeaxanthin is then converted by a multistep process to 9′-*cis*-neoxanthin. 9′-*cis*-Neoxanthin is oxidatively cleaved to form the C_{15} molecule xanthoxin, which serves as the immediate precursor of ABA-aldehyde. ABA-aldehyde is then oxidized to form ABA. Some ABA-deficient mutants that have been helpful in elucidating the pathway are shown at the steps at which they are blocked. These include the *vp* mutants of corn, the *aba* mutants of *Arabidopsis*, the *flacca* and *sitiens* mutants of tomato, the *droopy* mutant of potato, and the *nar2a* mutant of barley. The pathways for ABA catabolism include conjugation to form ABA-β-D-glucosyl ester or oxidation to form phaseic acid and then dihydrophaseic acid.

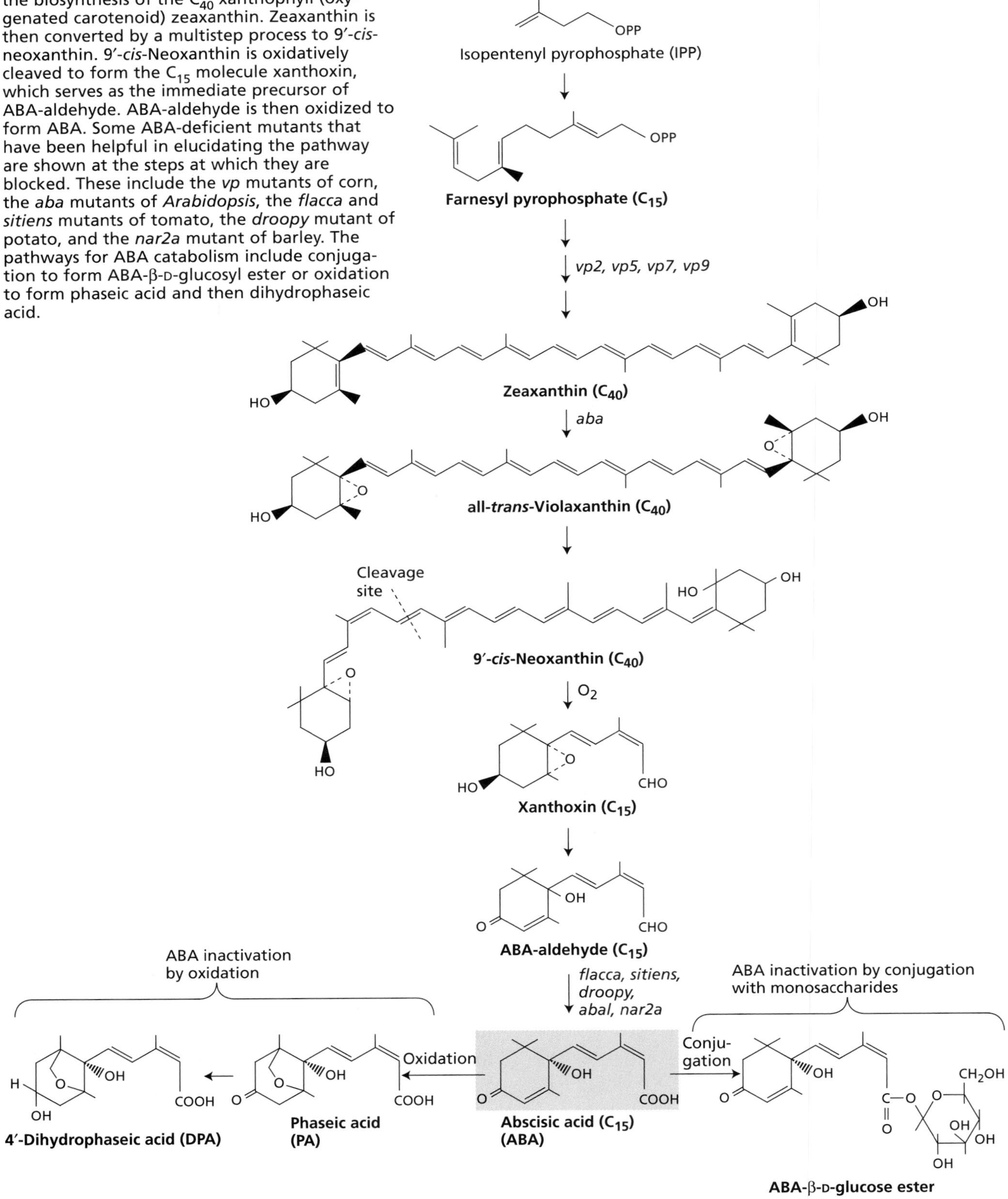

increase in ABA in the leaf during water stress. When plants were subjected to a series of stress and rewatering cycles, the ABA-GE concentration increased steadily, suggesting that the conjugated form is not broken down during water stress (Boyer and Zeevaart 1982).

ABA Is Translocated throughout the Plant in Both Phloem and Xylem

ABA is transported by both the xylem and the phloem, but it is normally much more abundant in the phloem sap. When radioactive ABA is applied to a leaf, it is transported both up the stem and down toward the roots. Most of the radioactive ABA is found in the roots within 24 hours. Destruction of the phloem by a stem girdle prevents ABA accumulation in the roots, indicating that the hormone is transported in the phloem sap.

ABA synthesized in the roots can also be transported to the shoot via the xylem. The concentration of ABA in the xylem sap of well-watered sunflower plants is between 1.0 and 15.0 n*M*, whereas the ABA concentration in water-stressed sunflower plants increases to as much as 3.0 μ*M* (Schurr et al. 1992). As water stress begins, some of the ABA carried by the xylem stream is synthesized in roots that are in direct contact with the drying soil. Because this transport can occur before the low water potential of the soil causes any measurable change in the water status of the leaves, ABA is believed to be a root signal that helps reduce the transpiration rate in the leaves by closing stomata (Davies and Zhang 1991).

Although a concentration of 3.0 μ*M* ABA in the apoplast is sufficient to close stomata, not all of the ABA in the xylem stream reaches the guard cells. Much of the ABA in the transpiration stream is taken up and metabolized by the mesophyll cells. During the early stages of water stress, however, the pH of the xylem sap increases from about 6.3 to about 7.2 (Wilkinson and Davies 1997). Alkalinization of the apoplast favors formation of the dissociated form of abscisic acid, ABA, which does not readily cross membranes (Figure 23.3). Hence, less ABA enters the mesophyll cells and more reaches the guard cells via the transpiration stream. Note that ABA is redistributed in the leaf in this way without any increase in the total ABA level. Wilkinson and Davies (1997) have suggested that this increase in xylem sap pH functions as a root signal that promotes early closure of the stomata.

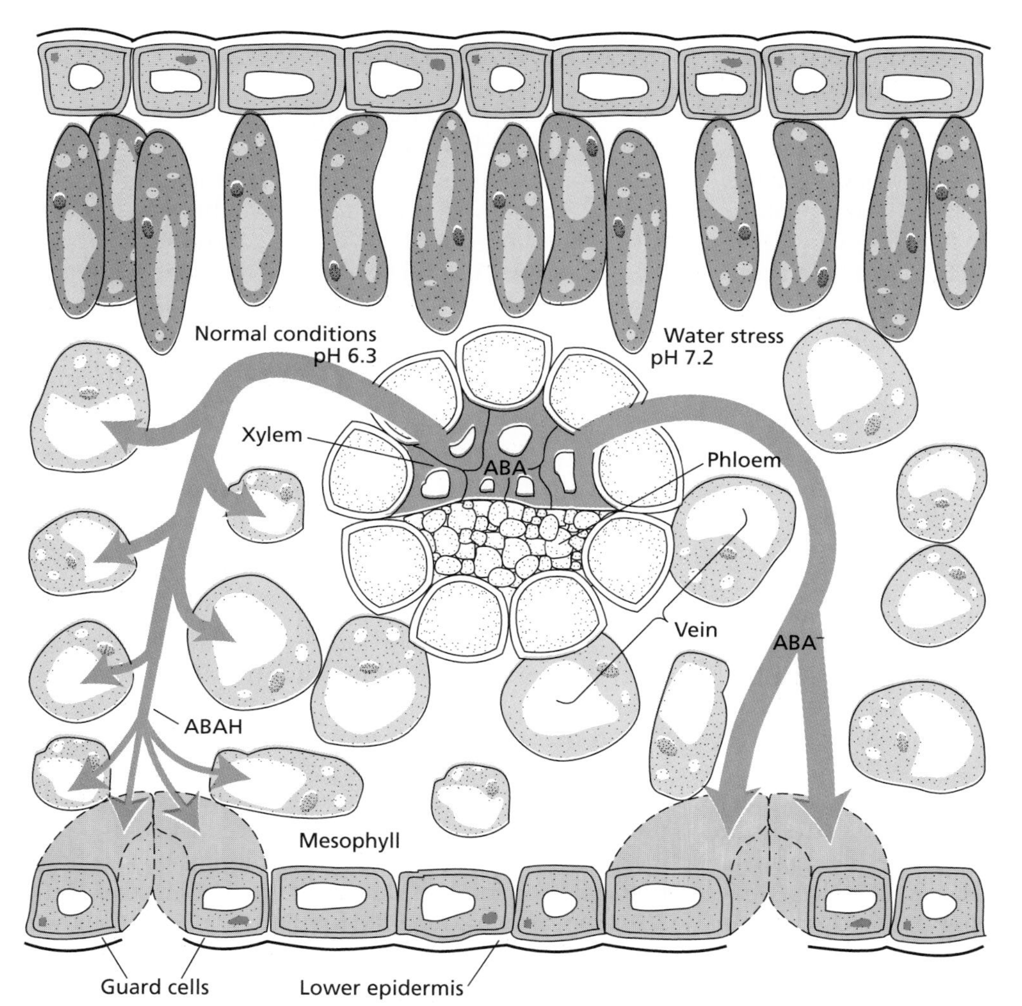

Figure 23.3 Redistribution of ABA in the leaf resulting from alkalinization of the xylem sap during water stress. Under normal conditions the xylem sap is slightly acidic, favoring the uptake of the undissociated form of ABA (ABAH) by the mesophyll cells. During water stress, the xylem sap becomes slightly alkaline, favoring the dissociation of ABAH to ABA^-. As a result, less ABA is taken up by the mesophyll cells and more reaches the guard cells.

Developmental and Physiological Effects of ABA

Abscisic acid plays primary regulatory roles in the initiation and maintenance of seed and bud dormancy and in the plant's response to stress, particularly water stress. In addition, ABA influences many other aspects of plant development by interacting, usually as an antagonist, with auxin, cytokinin, and gibberellin. In this section we will explore the diverse physiological effects of ABA, beginning with its role in seed development.

ABA Levels in Seeds Peak during Embryogenesis

Seed development can be divided into two phases of approximately equal duration. During the first phase, which is characterized by cell divisions, the zygote undergoes embryogenesis and the endosperm tissue proliferates. The second phase begins with the cessation of cell divisions and ends with dehydration and developmental arrest (Bewley and Black 1994). During the second phase, storage compounds accumulate, the embryo becomes tolerant to desiccation, and the seed dehydrates, losing up to 90% of its water.

The mechanisms involved in seed drying, which occurs despite a tremendous decrease in the water potential (equivalent to a $\Psi_{w(cell)}$ of –350 to –50 MPa; see Chapter 3 for a discussion of water potential), are still unknown. As a consequence of dehydration, metabolism comes to a halt and the seed enters a **quiescent** state. Typically, the ABA content of seeds is very low early in embryogenesis, reaches a maximum at about the halfway point, and then gradually falls to low levels as the seed reaches maturity. Thus, there is a broad peak of ABA concentration in the seed corresponding to mid- to late embryogenesis.

The hormonal balance of seeds is complicated by the fact that not all the tissues have the same genotype. The seed coat is derived from maternal tissues (see Chapter 1); the zygote and endosperm are derived from both parents. Genetic studies with ABA-deficient mutants have shown that the zygotic genotype controls the level of ABA in the embryo and endosperm, whereas the maternal genotype determines the amount of ABA in the seed coat (Karssen 1995). Because of the maternal contribution, plants under water stress can accumulate higher levels of ABA in their seed coats as a result of translocation in the phloem from the leaves.

ABA Promotes the Desiccation Tolerance of the Embryo

One of the most important functions of ABA in the developing seed is to promote the synthesis of proteins involved in desiccation tolerance. As will be described in Chapter 25 (on stress physiology), desiccation can severely damage membranes and other cellular constituents. During the mid- to late stages of seed development, specific mRNAs accumulate in embryos at the time of high levels of endogenous ABA. These mRNAs encode so-called **late-embryogenesis-abundant** (**LEA**) proteins thought to be involved in desiccation tolerance. If embryos are isolated at earlier developmental stages and treated with exogenous ABA, they synthesize LEA proteins at an earlier stage than they would normally. Thus the synthesis of LEA proteins is under ABA control.

LEA protein sequences are highly conserved among a wide variety of species. They are related to two other groups of proteins, the **RAB** (*r*esponsive to *AB*A) and **DHN** (*deh*ydri*n*) proteins from *Arabidopsis thaliana*. The LEA, RAB, and DHN proteins are all water soluble, basic, rich in glycine and lysine, and low in hydrophobic residues. As a result they are extremely hydrophilic and stable to boiling. These features led Dure and colleagues (1989) to propose that they function specifically in the protection of membranes and proteins against desiccation damage, possibly by binding water tightly and by preventing crystallization of cellular components through their ability to act as stabilizing "solvents," similar to the effects of sugars. However, a direct relationship between these proteins and desiccation tolerance has not yet been demonstrated.

ABA Promotes the Accumulation of Seed Storage Protein during Embryogenesis

Storage compounds accumulate during late embryogenesis. Because ABA levels are still high at this time, it has been assumed that ABA is the controlling factor in the process. ABA could be affecting the translocation of sugars and amino acids, the synthesis of the reserve materials, or both.

To determine the role of ABA in translocation (the partitioning of assimilates) to the seed, researchers used an *Arabidopsis* double mutant (*aba/abi3-1*) that is impaired in its abilities both to synthesize ABA and to respond to it. When flowers of the double mutant were pollinated with either wild-type or mutant pollen, and the translocation of ^{14}C-labeled photoassimilates to the developing seed (mainly sugars) was measured, no effect of the genotype on the rate of import to the seed was detected (De Bruijn and Vreugdenhil 1993). Thus ABA does not appear to increase translocation to the seed.

In contrast, ABA has been shown to affect the amounts and composition of storage proteins. For example, the accumulation of storage protein by the *aba/abi3-1* double mutant of *Arabidopsis* is strongly reduced and ceases prematurely (Koornneef et al. 1989). Similar results have been obtained for ABA-deficient mutants of maize. Thus, ABA appears to be required for the expression of storage protein genes during embryogenesis.

ABA not only regulates the accumulation of storage proteins during embryogenesis; it can also maintain the mature embryo in a dormant state until the environmental conditions are optimal for growth. Seed dormancy is an important factor in the adaptation of plants to unfavorable environments. As we discuss in the next few sections, plants have evolved a variety of mechanisms, some of them involving ABA, that enable them to maintain their seeds in a dormant state.

Two Types of Seed Dormancy Have Been Recognized

During seed maturation, the embryo enters a quiescent phase in response to desiccation. Seed germination can be defined as the resumption of growth of the embryo of the mature seed; it depends on the same environmental conditions as vegetative growth does. Water and oxygen must be available, the temperature must be suitable, and there must be no inhibitory substances present. However, in many cases a viable (living) seed will not germinate even though all the necessary environmental conditions for growth are satisfied. This phenomenon is termed **seed dormancy.** Seed dormancy introduces a temporal delay in the germination process that provides additional time for seed dispersal over greater geographical distances. It also maximizes seedling survival by preventing germination under unfavorable conditions.

Two types of seed dormancy have been recognized, coat-imposed dormancy and embryo dormancy. **Coat-imposed dormany** is dormancy imposed on the embryo by the seed coat and other enclosing tissues, such as endosperm, pericarp, or extrafloral organs. The embryos of such seeds will germinate readily in the presence of water and oxygen once the seed coat and other surrounding tissues are either removed or damaged. There are five basic mechanisms of coat-imposed dormancy (Bewley and Black 1994):

1. *Prevention of water uptake.* Prevention of water uptake by the seed coat is a common cause of seed dormancy in families found in arid and semiarid regions, especially among legumes, such as clover (*Trifolium* spp.) and alfalfa (*Medicago* spp.). Waxy cuticles, suberized layers, and lignified sclereids all combine to restrict the penetration of water into the seed.
2. *Mechanical constraint.* The first visible sign of germination is typically the radicle breaking through the seed coat. In some cases, however, the seed coat may be too rigid for the radicle to penetrate. Nuts with hard, lignified shells are examples of dormancy caused by mechanical constraint. Such shells must be broken by biotic or environmental forces for the seed to germinate. Even nonlignified tissues, such as the endosperm of lettuce seeds, can suppress expansion of the embryo. For the seeds to germinate, the endosperm cell walls must be weakened by the production of cell wall–degrading enzymes.
3. *Interference with gas exchange.* Some seed coats are considerably less permeable to oxygen than an equivalent thickness of water is—e.g., less by a factor of about 10^4 in seeds of cocklebur (*Xanthium pennsylvanicum*). This lowered permeability to oxygen suggests that the seed coat inhibits germination by limiting the oxygen supply to the embryo. In support of this idea, investigators can break the dormancy of such seeds either by making a small hole in the coat with a pin (without weakening the coat mechanically), or by treating the coat with concentrated oxygen. However, other studies suggest that the oxygen consumption of the embryos from such seeds is considerably less than the amount of oxygen able to penetrate the seed coats under normal aerobic conditions. Thus the role of oxygen impermeability in seed coat dormancy remains unresolved.
4. *Retention of inhibitors.* The seed coat may prevent the escape of inhibitors from the seed. For example, growth inhibitors readily diffuse out of isolated *Xanthium* embryos, but not from intact seeds.
5. *Inhibitor production.* Seed coats and pericarps may contain relatively high concentrations of growth inhibitors that can suppress germination of the embryo. ABA is a common germination inhibitor present in these maternal tissues. In certain cases where repeated washing (leaching) removes dormancy, the effect is thought to be due to the loss of such inhibitory compounds.

The second type of seed dormancy is **embryo dormancy**, which refers to a dormancy that is inherent in the embryo and is not due to any influence of the seed coat or other surrounding tissues. In some cases, embryo dormancy can be relieved by amputation of the cotyledons. Species in which the cotyledons exert an inhibitory effect include European hazel (*Corylus avellana*) and European ash (*Fraxinus excelsior*). A fascinating demonstration of the cotyledon's ability to inhibit growth is found in species (e.g., peach) in which the isolated dormant embryos germinate but grow extremely slowly to form a dwarf plant. If the cotyledons are removed at an early stage of development, however, the plant abruptly shifts to normal growth.

Embryo dormancy is thought to be due to the presence of inhibitors, especially ABA, as well as the absence of growth promoters, such as GA (gibberellic acid). The loss of embryo dormancy is often associated with a sharp drop in the ratio of ABA to GA.

Seed Dormancy May Be Either Primary or Secondary

Seeds that are released from the plant in a dormant state are said to exhibit **primary dormancy**. In contrast, other seeds are nondormant when initially dispersed from the parent plant, but they may be induced to go dormant if the conditions for germination are unfavorable. Such seeds are said to exhibit **induced,** or **secondary, dormancy** (Bewley and Black 1994). For example, seeds of *Avena sativa* (oat) can become dormant in the presence of temperatures higher than the maximum for germination, whereas seeds of *Phacelia dubia* (small-flower scorpionweed) become dormant at temperatures below the minimum for germination. The mechanisms of secondary dormancy are poorly understood.

Environmental Factors Control the Release from Seed Dormancy

Various external factors release the seed from dormancy, and dormant seeds typically respond to more than one condition. Many seeds lose their dormancy when their moisture content is reduced to a certain level by drying. This method of breaking seed dormancy is called **afterripening**, and is usually performed in a special drying oven. On the other hand, if the seed becomes too dry (5% water content or less), the effectiveness of afterripening is diminished.

Another factor that can release seeds from dormancy is low temperature, or **chilling**. Many seeds require a period of cold (0 to 10°C) while in a fully hydrated (imbibed) state in order to germinate. In temperate-zone species, this requirement is of obvious survival value, since such seeds will germinate not in the fall, but only in the following spring. Chilling seeds to break their dormancy is a time-honored practice in horticulture and forestry and traditionally has been referred to as *stratification.* This term is derived from the old agricultural practice of allowing seeds with a chilling requirement to overwinter outdoors in layered mounds of moist soil. Today the seeds are simply stored in a refrigerator.

The third external factor that plays an important role in breaking seed dormancy is **light**. Many seeds have a light requirement for germination, which may involve only a brief exposure, as in the case of lettuce, an intermittent treatment (e.g., succulents of the genus *Kalanchoe*), or even a specific photoperiod involving short or long days. As was discussed in Chapter 17, phytochrome is the main sensor for light-regulated seed germination. Interestingly, all light-requiring seeds exhibit seed coat dormancy, and removal of the outer tissues of the seed allows the embryo to germinate in the absence of light. The effect that light has on the embryo is thus to enable the radicle to penetrate the seed coat. This penetration often involves some enzymatic weakening of the enclosing tissues.

Seed Dormancy Is Controlled by the Ratio of ABA to GA

Mature seeds may be either dormant or nondormant, depending on the species. Nondormant seeds, such as pea, will germinate readily if provided with water only. Dormant seeds, on the other hand, fail to germinate in the presence of water, and instead require some additional treatment or condition. As we have seen, dormancy may arise from the rigidity or impermeability of the seed coat (coat-imposed dormancy) or from the persistence of the state of arrested development of the embryo. Examples of the latter include seeds that require afterripening, chilling, or light to germinate.

ABA mutants have been extremely useful in demonstrating the role of ABA in seed dormancy. *Arabidopsis* seeds exhibit varying degrees of dormancy, depending on the ecotype, which can be overcome with a period of afterripening and/or cold treatment. ABA-deficient (*aba*) mutants of *Arabidopsis* have been shown to be nondormant at maturity. When reciprocal crosses between *aba* and wild-type plants were carried out, the seeds exhibited dormancy only when the embryo itself produced the ABA. Neither maternal ABA, nor ABA applied exogenously to the plant, could induce dormancy in an *aba* embryo.

On the other hand, maternal ABA is required for other aspects of seed development. Thus, the two sources of ABA function in different developmental pathways. Dormancy is also greatly reduced in seeds from the ABA-insensitive mutants *abi1* and *abi3*, even though, for reasons that are unclear, these seeds contain higher ABA concentrations than those of the wild type throughout development. Similar conclusions about the role of ABA in regulating dormancy have been drawn from work on ABA-deficient tomato mutants, indicating that the phenomenon is probably a general one.

Although the role of ABA in initiating and maintaining seed dormancy is well established, other hormones contribute to the overall effect. For example, in most plants the peak of ABA production in the seed coincides with a decline in the levels of IAA and GA.

An elegant demonstration of the importance of the ratio of ABA to GA in seeds is provided by the genetic screen that led to isolation of the first ABA-deficient mutants of *Arabidopsis* (Koornneef et al. 1982). Seeds of a GA-deficient mutant that could not germinate in the absence of exogenous GA were mutagenized and then grown in the greenhouse. The seeds produced by these mutagenized plants were then screened for revertants—that is, seeds that had regained their ability to germinate. Revertants were isolated, and they turned out to be mutants of abscisic acid synthesis. Surprisingly, the revertants germinated because the wild-type ABA:GA ratio had been restored in the GA-deficient

BOX 23.1

The Longevity of Seeds

HOW LONG CAN A SEED REMAIN DORMANT and still remain viable? Seed longevity is of practical importance because of ongoing efforts to preserve plant genetic resources for future agricultural crops by setting up seed gene banks. Spectacular claims of dormant seeds that remained viable for thousands of years have been made, but are considered highly controversial. Extreme examples of ancient seeds that have been reported to retain their viability include submerged lotus seeds found in a 3,000-year-old boat near Tokyo, barley seeds from the 3,000-year-old tomb of King Tutankhamen, and arctic lupine seeds associated with rodent burrows determined to be 14,000 years old. In most cases, however, a scientifically rigorous examination of the data has either disproved or raised serious doubts about the claimed antiquity of the seeds (Bewley and Black 1994).

The most serious criticism is that the dates of the seeds have been inferred from the age of their immediate surroundings or associated artifacts rather than from direct measurements of the seeds themselves. This question raises the possibility that the seeds are much younger than the site at which they were found. In fact, microscopic examination of authenticated ancient cereal grains from various Middle Eastern sites has shown that cells and tissues of these seeds are no longer intact. Such seeds are clearly inviable.

Although the most sensational claims for seed longevity are almost certainly bogus, seeds of *Canna compacta* apparently can live for at least 600 years. Viable *Canna* seeds were obtained from inside a walnut in a tomb in Argentina. The *Canna* seeds had apparently been inserted into the immature seeds of a growing walnut fruit before the hard outer shell formed. Once the shell hardened and the nut dried out, the result was a rattle. Native people strung the rattles together to form a necklace. In this case, the seeds had to be at least as old as the walnut shell, and carbon dating of the shell indicated that it was about 600 years old (Bewley and Black 1994).

Plant collections begun during the late eighteenth century in Europe—for example, at the British Museum in London and at the Museum of Natural History in Paris—have been reliable sources of seeds for viability determinations. *Cassia multijuga* (false sicklepod) seeds, collected in 1776, were still viable after being tested in 1934. In 1879, W. J. Beal initiated the longest-running experiment on seed longevity by burying the seeds of 21 different species in unstoppered bottles in a sandy hilltop near the Michigan Agricultural College in East Lansing. After 100 years, only one species, *Verbascum blattaria* (moth mullein), remained viable.

Why are some seeds able to survive in a dormant state for hundreds of years, whereas others lose viability in less than 5 years? Almost nothing is known about the mechanisms that determine the longevity of seeds. If these mechanisms could be understood, we might someday be able to greatly increase the seed longevity of agriculturally important species and varieties, thereby enhancing our ability to preserve plant genetic resources for generations to come.

---

mutant by blocking of the ABA biosynthetic pathway. This study elegantly illustrates the general principle that the balance of plant hormones is often more critical than are their absolute concentrations in regulating development.

ABA Inhibits Precocious Germination and Vivipary

When immature embryos are removed from their seeds and placed in culture midway through development before the onset of dormancy, they germinate precociously—that is, without passing through the normal dormant stage of development. ABA added to the culture medium inhibits precocious germination. This result suggests that ABA is the natural constraint that keeps developing embryos in their embryogenic state.

Further evidence for the role of ABA in preventing precocious germination has been provided by genetic studies of vivipary. Vivipary is sometimes a natural feature of a plant's life cycle. For example, in mangrove (*Rhizophora mangle*), a tree that grows in swamps, embryos germinate directly on the tree to produce seedlings with long, pointed roots that fall and embed themselves in the mud below. The tendency toward vivipary is a varietal characteristic in grain crops that is favored by wet weather. In maize, several viviparous mutants (*vp* mutants) have been selected in which the embryos germinate directly on the cob while still attached to the plant. Several of these mutants are ABA-deficient (*vp2*, *vp5*, *vp7*); one is ABA-insensitive (*vp1*). Vivipary in the ABA-deficient mutants can be partly prevented by treatment with exogenous ABA (Zeevaart and Creelman 1988 ; Hole et al. 1989).

In contrast with the maize mutants, single-gene mutants of *Arabidopsis* (*aba*, *abi1*, and *abi3-1*) fail to exhibit vivipary, although they are nondormant. The reason may be a lack of moisture, since such seeds will germinate within the fruits (siliques) under conditions of high relative humidity. The double mutant *aba/abi3-1* exhibits vivipary late in development (Koornneef et al. 1989). The seeds of the double mutant also stay green and retain a much higher water content, consistent with the idea that moisture may be a limiting factor in vivipary. When artificially dried, the *aba*/*abi3-1* seeds lose their viability because they have not developed desiccation tolerance.

ABA Accumulates in Dormant Buds

In woody species, dormancy is an important adaptive feature in cold climates. When a tree faces very low temperatures in winter, it protects its meristems with bud scales and temporarily stops bud growth. This response to low temperatures requires a sensory mechanism that detects the environmental changes (sensory signals), and a control system that transduces the sensory signals and triggers the developmental processes leading to bud dormancy. ABA was originally suggested as the dormancy-inducing hormone because it accumulates in dormant buds and decreases after the tissue is exposed to low temperatures. However, later studies showed that the ABA content of buds does not always correlate with the degree of dormancy. As we saw in the case of seed dormancy, this apparent discrepancy could reflect interactions between ABA and other hormones as part of a process in which bud dormancy and growth are regulated by the balance between bud growth inhibitors, such as ABA, and growth-inducing substances, such as cytokinins and gibberellins.

Although much progress has been achieved in elucidating the role of ABA in seed dormancy by the use of ABA-deficient mutants, progress on the role of ABA in bud dormancy, which applies mainly to woody perennials, has lagged because of the lack of a convenient genetic system. This discrepancy illustrates the tremendous contribution that genetics and molecular biology have made to plant physiology, and it underscores the need for extending such approaches to woody species.

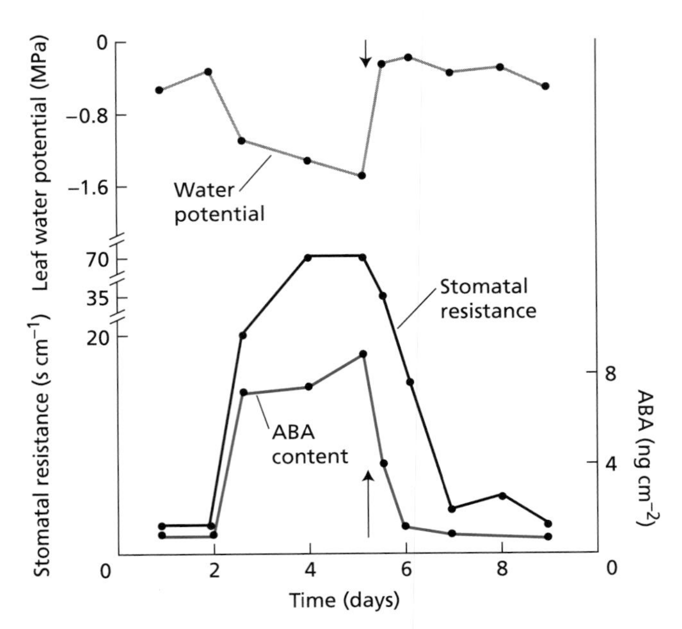

Figure 23.4 Changes in water potential, stomatal resistance (the inverse of stomatal conductance), and ABA content in maize in response to water stress. Watering was interrupted at the beginning of treatment and restarted at the time indicated by the arrows. As the soil dried out, the water potential of the leaf decreased, and the ABA content and stomatal resistance increased. The process was reversed by rewatering. (After Beardsell and Cohen 1975.)

ABA Inhibits GA-Induced Enzymes

ABA inhibits the synthesis of hydrolytic enzymes that are essential for the breakdown of storage reserves in seeds. For example, GA stimulates the aleurone layer of cereal grains to produce α-amylase and other hydrolytic enzymes that catalyze the breakdown of starch and other macromolecules in the starchy endosperm during germination. ABA inhibits this GA-dependent enzyme synthesis by inhibiting the transcription of α-amylase mRNA (see Chapter 20).

ABA Closes Stomata in Response to Water Stress

Elucidation of the roles of ABA in freezing, salt, and water stress (see Chapter 25) led to the characterization of ABA as a stress hormone. As noted earlier, ABA concentrations in leaves can increase up to 50 times under drought conditions, the most dramatic change in concentration reported for any hormone in response to an environmental signal. ABA is very effective in causing stomatal closure, and its accumulation in stressed leaves plays an important role in the reduction of water loss by transpiration under water stress conditions (Figure 23.4). On the other hand, several studies have shown decreases in stomatal apertures before any increases in the total leaf ABA content. This apparent inconsistency is explained by studies showing that the initial stomatal closure is caused by ABA redistribution within the leaf (see Chapter 25).

As noted earlier, stomatal closing can also be caused by ABA produced in the roots and exported to the shoot. Mutants that lack the ability to synthesize ABA show permanent wilting and are called wilty mutants because of their inability to close their stomata. Application of exogenous ABA to such mutants causes stomatal closure and a restoration of turgor pressure (Tal and Imber 1971).

ABA Increases the Hydraulic Conductivity and Ion Flux of Roots

Application of ABA to root tissue stimulates water flow and ion flux, suggesting that ABA regulates turgor not only by decreasing transpiration but also by increasing water influx into roots (Glinka and Reinhold 1971). ABA increases the flow of water by increasing the **hydraulic conductivity** (decreasing the resistance to water movement across the apoplast and membranes) and by enhancing ion uptake, which causes an increase in the water potential gradient between the soil and the root (see Chapters 3 and 4). ABA appears to bring about these effects by modifying membrane properties (Van Steveninck and Van Steveninck 1983). The precise nature of these ABA-induced changes in membrane properties, whether they involve alterations in the lipids or in the protein transporters, has not yet been determined.

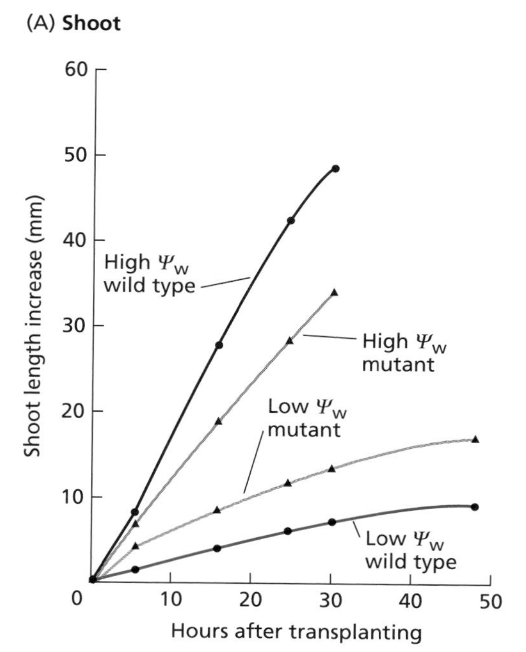

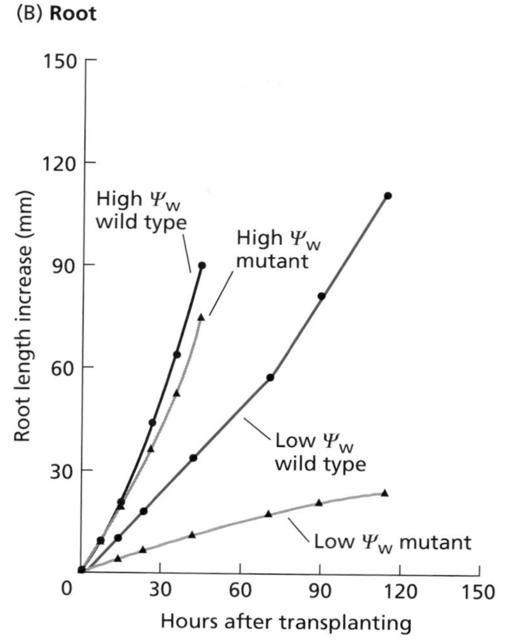

Figure 23.5 Comparison of the growth of the shoots (A) and roots (B) of normal versus ABA-deficient (viviparous) maize plants growing in vermiculite maintained either at high water potential (–0.03 MPa) or at low water potential (–0.3 Mpa in A and –1.6 MPa in B). Water stress (low water potential) depresses the growth of both shoots and roots compared to the controls. However, note that under water stress conditions, the ratio of root growth to shoot growth is much higher when ABA is present (i.e., in the wild type) than when it is absent (in the mutant). (From Saab et al. 1990.)

ABA Promotes Root Growth and Inhibits Shoot Growth at Low Water Potentials

ABA has different effects on the growth of roots and shoots, and the effects are strongly dependent on the water status of the plant. Figure 23.5 compares the growth of shoots and roots of maize seedlings grown either at high water potential or at low water potential. Two types of seedlings were used, wild-type with normal ABA levels and an ABA-deficient, viviparous mutant.

At high water potentials, growth of the shoot is greater in the wild type (plus ABA) than in the mutant (minus ABA). Thus, endogenous ABA seems to promote shoot growth at high water potentials. At low water potentials the opposite occurs: Shoot growth is greater in the mutant (minus ABA) than in the wild type (plus ABA). We can conclude that endogenous ABA acts as an inhibitor of shoot growth, but only at high water potentials. Now let's examine how ABA affects roots. At high water potentials, root growth is slightly greater in the wild type (plus ABA) than in the mutant (minus ABA), similar to growth in shoots. Therefore, at high water potentials (when ABA levels are low), endogenous ABA exerts a slight positive effect on the growth of both roots and shoots. Note, however, that at low water potentials the growth of the roots is much higher in the wild type than in the mutant, although growth is still inhibited relative to root growth at low water potentials.

Summarizing, at low water potentials, when ABA levels are high, the endogenous hormone exerts a strong positive effect on root growth and a slight negative effect on shoot growth. The overall effect is a dramatic increase in the root:shoot ratio at low water potentials, which, along with the effect of ABA on stomatal closure, helps the plant cope with water stress.

ABA Promotes Leaf Senescence Independently of Ethylene

Abscisic acid was originally isolated as an abscission-causing factor. However, it has since become evident that ABA stimulates abscission of organs in only a few species and that the primary hormone causing abscission is ethylene. ABA, on the other hand, is clearly involved in leaf senescence, and through its promotion of senescence it might indirectly increase ethylene formation and stimulate abscission.

Leaf senescence has been studied extensively, and the anatomical, physiological, and biochemical changes that take place during this process have been described (Noodén and Leopold 1988). Leaf segments senesce faster in darkness than in light, and they turn yellow as a result of chlorophyll breakdown. In addition, the breakdown of proteins and nucleic acids is increased by the stimulation of several hydrolases. ABA greatly accelerates the senescence of both leaf segments and attached leaves.

As we saw in Chapter 22, ethylene has also been found to play a role in the senescence of oat leaf segments, but ABA appears to be the initiating agent, whereas ethylene appears to exert its effects at a later stage (Gepstein and Thimann 1981). Although ABA stimulates ethylene production, the effects of ABA on leaf senescence are not mediated by ethylene. Zacarias and Reid (1990) compared the effects of ABA and ethylene on the senescence of leaf discs excised from wild-type *Arabidopsis* and an ethylene-insensitive mutant. Chlorophyll loss from untreated leaf discs of the ethylene-insensitive mutant was much slower than that from the wild type, in the light and in the dark. Ethylene accelerated the chlorophyll loss from the wild-type leaf discs, but it had no effect on the yellowing of the ethylene-insensitive mutant discs. In contrast, treatment with ABA stimulated chlorophyll loss in *both* the wild-type and the mutant discs, whether they were held in the light or in the dark. These results indicate that the promotion of senescence by ABA is not mediated through its stimulation of ethylene production (Zacarias and Reid 1990).

Cellular and Molecular Modes of ABA Action

ABA is involved in long-term developmental processes (e.g., seed maturation) as well as short-term physiological effects (e.g., stomatal closure). Long-term processes inevitably involve changes in the pattern of gene expression, whereas rapid physiological responses frequently involve alterations in the fluxes of ions across membranes. Signal transduction pathways, which amplify the primary signal generated when the hormone binds to its receptor, are required for both the long-term and the short-term effects of ABA. In recent years, much progress has been made in elucidating the steps in these pathways, but many questions remain. In this section we will explore the mechanism of ABA action at the cellular and molecular levels.

ABA Regulates Gene Expression

ABA has been shown to regulate the expression of numerous genes under certain stress conditions, such as heat shock, adaptation to low temperatures, and salt tolerance (see Chapter 25), and during seed maturation (Skriver and Mundy 1990). In a few cases, stimulation of transcription by ABA has been demonstrated directly by the technique of nuclear runoff assay (described in Chapter 17). Gene activation is mediated by DNA-binding proteins. Several **ABA-response elements (ABREs)** have been characterized. Distinct sequence motifs have been identified, and proteins that bind to these sequences have been characterized. One such ABRE-binding protein was cloned and found to have similarity to the basic leucine zipper (bZIP) family of transcription factors (see Chapter 14) (Guiltinan et al. 1990). A few DNA elements have been identified that are involved in transcriptional repression by ABA. The best characterized of these are the **gibberellin response elements (GAREs)** that mediate the gibberellin-inducible, ABA-repressible expression of the barley α-amylase gene (see Chapter 20).

A transcription factor involved in ABA gene activation in maturing seeds has been identified by genetic means. The maize *vp1* (*viviparous-1*) and the *Arabidopsis abi3* (*ABA-insensitive-3*) mutations eliminate ABA responsiveness in developing seeds and encode proteins that are very similar in their primary sequence (McCarty et al. 1991; Giraudat et al. 1992). These proteins have features that suggest they act as transcriptional activators.

Both Extracellular and Intracellular ABA Receptors May Exist

Although ABA has been shown to interact directly with phospholipids, it is widely assumed that the ABA receptor is a protein. To date, however, the protein receptor for ABA has not been identified. Experiments have been performed to determine whether the hormone must enter the cell to be effective, or whether it can act externally by binding to a receptor located on the outer surface of the plasma membrane. The results so far have been contradictory.

Some experiments seem to point to a receptor on the outer surface of the cell. For example, Hartung (1983) showed that ABA induces stomatal closure in *Valerianella locusta* (corn salad plant) as effectively at pH 8.0, at which it is not taken up by guard cells, as at pH 5.0, at which it is taken up readily (recall that ABA dissociates at neutral to alkaline pH). There is also some evidence that radioactively labeled ABA binds to the outer surface of the guard cell plasma membrane (Hornberg and Weiler 1984). Anderson and colleagues (1994) microinjected ABA directly into guard cells of the spiderwort *Commelina* to cytosolic concentrations of 50 to 200 μM, far in excess of the concentration needed to cause closure when applied externally. Stomata with ABA-loaded guard cells showed opening similar to stomata with uninjected guard cells. However, since opening takes two hours for completion, it is possible that the ABA was lost or metabolized.

In similar types of experiments carried out using protoplasts isolated from barley aleurone layers, it has been found that ABA does not inhibit GA-induced α-amylase synthesis when microinjected directly into the cell, suggesting again that at least one of the sites of perception of ABA is on the external face of the plasma membrane (Gilroy and Jones 1994).

Some other experiments, however, support an intracellular location for the ABA receptor. Anderson and colleagues (1994) found that externally applied 10 μM ABA inhibits stomatal closure in *Commelina* by 98% at pH 6.15 and by 57% at pH 8.0, suggesting that ABA must enter the cell to be effective. Schwartz and colleagues (1994), observed that ABA fails to inhibit stomatal *opening* in *Vicia faba* (broad bean) in response to light at pH 8, but as the pH is lowered stomatal opening is increasingly inhibited. The same authors observed that microinjection of ABA into guard cells inhibits inward K^+ currents, an effect sufficient to inhibit stomatal opening. The microinjected ABA also induced stomatal closing.

It could be argued that microinjecting ABA directly into cells might cause artifacts because the cell is being wounded at the same time that it is being given the hormone. To avoid this problem, Allan and colleagues (1994) microinjected guard cells of *Commelina* with a "caged" form of ABA. The caged ABA is inactive as a hormone (Figure 23.6A). However, since the bond between ABA and the "cage" can be broken by a brief exposure to UV light (photolysis), it is possible to control the release of free ABA within the cell. As a control, guard cells can be injected with a nonphotolyzable version of the caged ABA(Figure 23.6B).

After the guard cells were microinjected with either the photolyzable or the nonphotolyzable caged ABA, they were given a 30-minute recovery period in the light, during which the stomata continued to open, showing that they were not damaged by the treatment. The guard cells were then treated for 30 seconds with UV light to release free ABA in the cytoplasm. Upon UV photolysis and the production of ABA, closure was initiated within 2 minutes of UV treatment (Figure 23.6C). Only the stomata whose guard cells were injected with the photolyzable form of caged ABA closed in response to the UV exposure. Neither the uninjected controls nor those injected with the nonphotolyzable conjugate closed, indicating that UV light alone, with or without microinjection, does not cause closure.

The experiment with caged ABA appears to provide the strongest evidence thus far for an intracellular location of the ABA receptor. However, it is difficult to explain some of the other results without invoking an extracellular location as well. Thus, there may be two types of ABA receptors, extracellular and intracellular. The question will be resolved only when the protein or proteins themselves are identified and localized.

ABA Increases Cytosolic Ca^{2+} and pH and Causes Rapid Membrane Depolarization

As discussed in Chapter 18, stomatal closure is driven by a reduction in guard cell turgor pressure caused by a massive long-term efflux of K^+ and anions from the cell. However, the first change detected after exposure of guard cells to ABA is a *transient* membrane depolarization due to the net influx of positive charge. At the same time, transient increases in the cytosolic calcium con-

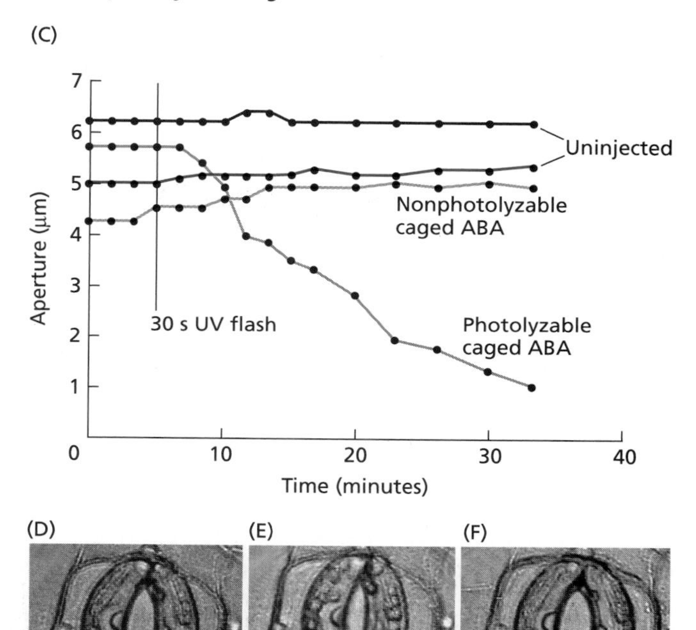

Figure 23.6 Stomatal closure induced by UV photolysis of caged ABA in the guard cell cytoplasm. Single guard cells in stomatal complexes of *Commelina* were microinjected with Calcium Green-1 and caged ABA. (A) The structure of the photolyzable form of caged ABA. (B) The structure of the nonphotolyzable form of caged ABA. (C) The stomatal apertures recorded before and after a 30-second exposure of the cells to UV. (D–F) Light micrographs of the same stomatal complex in which the right-hand guard cell was loaded with the photolyzable caged ABA 10 minutes before UV photolysis (D) and 30 minutes after photolysis (E). Finally, the effect of a 1-hour perfusion with 10 μ*M* fusicoccin is shown in (F) Fusicoccin is a fungal phytotoxin that causes stomatal opening by activating the plasma membrane H^+-ATPase (see Chapter 19). The positive response to fusicoccin indicates that the cells were not damaged by the microinjection and UV treatments. (A–C from Allan et al. 1994; D–F courtesy of A. Allan, from Allan et al. 1994; © American Society of Plant Physiologists, reprinted with permission.)

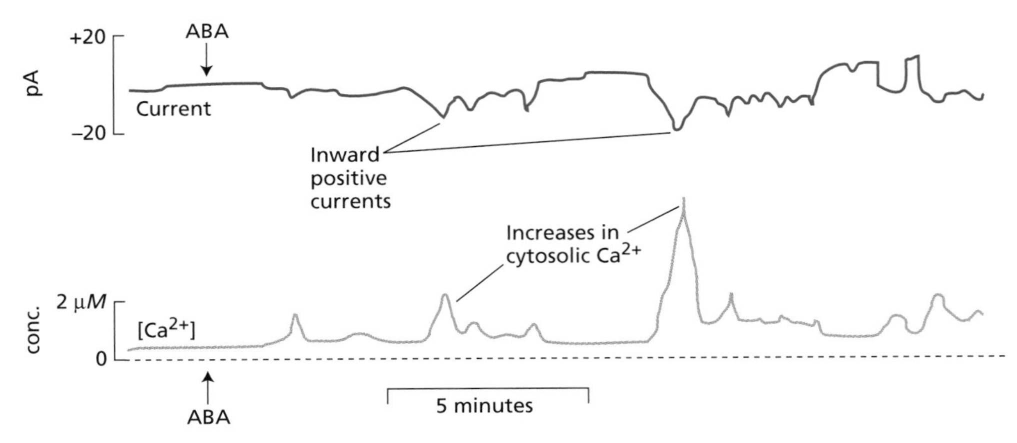

Figure 23.7 Simultaneous measurements of ABA-induced inward positive currents (membrane depolarization events, upper trace) and ABA-induced increases in cytosolic Ca^{2+} concentrations (lower trace) in a guard cell of *Vicia faba* (broad bean). The current was measured by the patch clamp technique; calcium was measured by use of a fluorescent indicator dye. The inward positive currents are represented by the downward peaks of the upper trace; the upward peaks of the lower trace show the transient increases in cytosolic calcium. ABA was added to the system at the arrow in each case. Note that the increases in cytoplasmic Ca^{2+} roughly coincide with the membrane depolarization events. (From Schroeder and Hagiwara 1990.)

centration can be detected (Figure 23.7). Thus it has been suggested that these rapid ABA-induced depolarization events result from activation of calcium channels (Schroeder and Hagiwara 1990).

The rapid, transient depolarizations induced by ABA are insufficient to open the K^+ efflux channels, which require long-term membrane depolarization in order to open. Another effect of ABA is to stimulate the release of calcium into the cytosol from internal compartments, such as the central vacuole. The combination of calcium influx and the release of calcium from internal stores raises the cytosolic calcium concentration from around 50 to 350 n*M* to as high as 1.1 μ*M* (Figure 23.8) (Mansfield and McAinsh 1995). This increase is sufficient to cause stomatal closing, as demonstrated by the following experiment.

Calcium was microinjected into guard cells in a caged form that allowed it to be released into the cytosol in response to a pulse of UV light. This method allowed the investigators careful control of both the concentration and the time of release to the cytosol. At cytosolic concentrations of 600 nM or more, release of

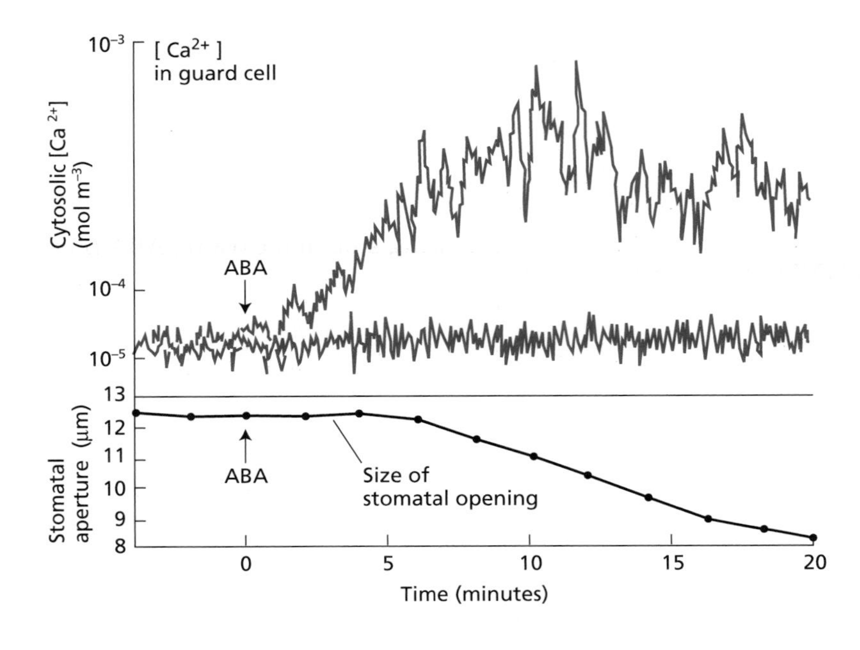

Figure 23.8 Time course of the ABA-induced increase in guard cell cytosolic Ca^{2+} concentration (upper panel) and ABA-induced stomatal aperture (lower panel). (From Mansfield and McAinsh 1995.)

calcium from its cage triggered stomatal closure (Gilroy et al. 1990). This level is well within the concentration range achieved after ABA treatment. Similar microinjection experiments using caged inositol 1,4,5-trisphosphate (IP_3), a second messenger that releases calcium from endoplasmic reticulum and vacuoles (see Chapter 14), led to a rise in cytosolic calcium levels and stomatal closure. These experiments reinforce the hypothesis that an increase in cytosolic calcium, partly derived from intracellular stores, is responsible for stomatal closure.

In addition to increasing the cytosolic calcium concentration, ABA causes an alkalinization of the cytosol from about pH 7.67 to pH 7.94 (Irving et al. 1992). The increase in cytosolic pH has been shown to activate the K^+ efflux channels on the plasma membrane (Blatt and Armstrong 1993) (see Chapter 6). However, activation of K^+ efflux channels alone would not result in K^+ loss because the membrane potential favors K^+ accumulation. For sustained K^+ loss to occur, ABA must induce the long-term depolarization of the membrane. Such long-term depolarizations in response to ABA have been demonstrated (Thiel et al. 1992).

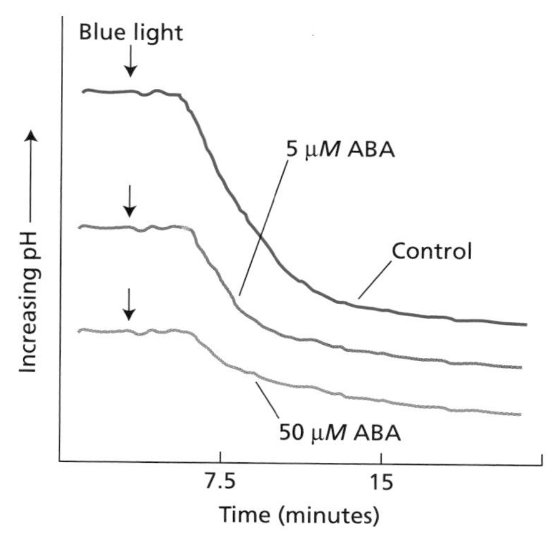

Figure 23.9 ABA inhibition of blue light–stimulated proton pumping by guard cell protoplasts. A suspension of guard cell protoplasts was incubated under red light irradiation, and the pH of the suspension medium was monitored with a pH electrode. The starting pH was the same in all cases (the curves are displaced for ease of viewing). At the time indicated by the arrows, a pulse of blue light was given. Blue light activates the plasma membrane H^+-ATPase, which pumps protons into the external medium and lowers the pH. Addition of 50 μ*M* ABA to the suspension medium inhibits the acidification by 40%. These results suggest that ABA induces changes in the cell that inhibit the H^+-ATPase. Both an increase in the cytosolic Ca^{2+} concentration and an increase in the cytosolic pH would be expected to inhibit the plasma membrane proton pump. (After Shimazaki et al. 1986.)

ABA-Induced Long-Term Depolarization Involves the Opening of Slow Anion Channels

According to a widely accepted model, long-term membrane depolarization is triggered by an ABA-induced transient depolarization of the plasma membrane coupled with an increase in cytosolic calcium (Ward et al. 1995). Both of these conditions are required to open voltage-gated slow (S-type) anion channels on the plasma membrane (see Chapter 6). ABA has been shown to activate slow anion channels in guard cells (Grabov et al. 1997; Pei et al. 1997).

The prolonged opening of these slow anion channels permits large quantities of Cl^- and malate^{2-} ions to escape from the cell, moving down their electrochemical gradient (the outside is positive and has lower Cl^- and malate^{2-} concentrations than the interior). The outward negative ion current that is generated in this way strongly depolarizes the membrane, triggering the voltage-gated K^+ efflux channels to open. In support of this model, inhibitors that block slow anion channels, such as 5-nitro-2,3-phenylpropylaminobenzoic acid (NPPB), also block ABA-induced stomatal closing. However, 4,4′-diisothiocyanatostilbene-2,2′-disulfonic acid (DIDS), which inhibits only the rapid (R-type) anion channels, has no effect on ABA-induced stomatal closing (Ward et al. 1995; Schwartz et al. 1995).

Another factor that may contribute to membrane depolarization is inhibition of the plasma membrane H^+-ATPase. ABA inhibits blue light–stimulated proton pumping by guard cell protoplasts (Figure 23.9), consistent with the idea that a decrease in the activity of the plasma membrane H^+-ATPase helps depolarize the membrane in the presence of ABA. However, ABA does not inhibit the proton pump directly (Goh et al. 1996). In *Vicia faba* (broad bean), at least, the plasma membrane H^+-ATPase of the leaves is strongly inhibited by calcium. A calcium concentration of 0.3 μ*M* blocks half of the enzyme activity, and 1 μ*M* calcium blocks the enzyme completely (Kinoshita et al. 1995). The combination of the increase in cytosolic Ca^{2+} concentration and the alkalinization of the cytosol caused by ABA likely is directly responsible for inhibition of the plasma membrane proton pump.

In addition to causing stomatal closure, ABA prevents light-induced stomatal opening. In this case, ABA acts by inhibiting the inward K^+ channels, which are open when the membrane is hyperpolarized by the proton pump (see Chapters 6 and 18). Inhibition of the inward K^+ channels appears to be mediated by both the ABA-induced increase in cytosolic calcium concentration and the alkalinization of the cytosol. Thus calcium and pH affect guard cell plasma membrane channels in two ways: They inhibit inward K^+ channels, thus preventing stomatal opening, and they activate outward anion channels, thus promoting stomatal closing.

ABA Stimulates Phosphoinositide Metabolism

As already discussed, various lines of evidence suggest that ABA acts by binding to a receptor on the outside of the guard cells, although the results of some experiments are consistent with an intracellular binding site as well. Much evidence supports a role for calcium both in the promotion of stomatal closing and in the inhibition of stomatal opening. Ca^{2+}, inositol 1,4,5-trisphosphate (IP_3), and H^+ are thought to act as secondary messengers that mediate the effects of ABA on the stomata (Figure 23.10).

According to the classic calcium-dependent signal transduction pathway of animal cells, IP_3 is released, along with diacylglycerol (DAG), when phospholipase C is activated by a heterotrimeric GTP-binding protein (G protein) (see Chapter 14). Consistent with this model, ABA has been shown to stimulate phosphoinositide metabolism in *Vicia faba* (broad bean) guard cells (Lee et al. 1996). To determine the effect of ABA on IP_3 synthesis, it was necessary to include Li^+ in the incubation medium as an inhibitor of inositol phosphatase. Under these conditions, ABA caused a 90% increase in the level of IP_3 within 10 seconds of hormone treatment (Lee et al. 1996).

G proteins may mediate the effects of ABA on stomatal movements. For example, in *V. faba* most studies have shown that, at least under certain conditions, G protein activators, such as GTPγS, can inhibit the activity of the inward K^+ channels (Fairley-Grenot and Assmann 1991; Wu and Assmann 1994 ; Kelly et al. 1995). In animal systems, however, G proteins mediate the signals emanating from plasma membrane receptors that bind external ligands, whereas the microinjection studies clearly indicate that ABA can cause stomatal closure from within the plant cytosol. The picture of stomatal regulation that begins to emerge is that of a system capable of responding to multiple signals, possibly involving multiple receptors and overlapping signal transduction pathways.

Protein Kinases and Phosphatases Have Been Implicated in ABA Action

Nearly all biological signaling systems involve protein phosphorylation and dephosphorylation reactions at some step in the pathway (see Chapter 14). Thus we can expect that signal transduction in guard cells, with their multiple sensory inputs, involves protein kinases and phosphatases. Artificially raising the ATP concentration inside guard cells by allowing the cytoplasm to equilibrate with the solution inside a patch pipette (see Chapter 6) strongly activates the slow anion channels.

This activation of the slow anion channels by ATP is abolished by the inclusion of protein kinase inhibitors in the patch pipette solution. Protein kinase inhibitors also block ABA-induced stomatal closing. In contrast,

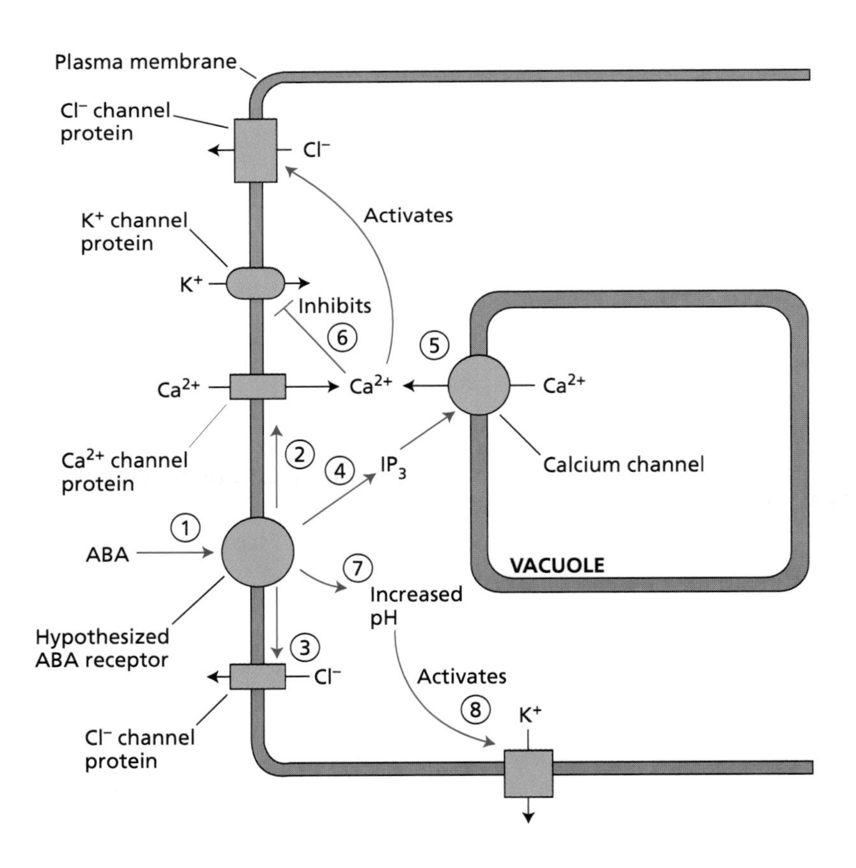

Figure 23.10 Model for ABA action in stomatal guard cells. (1) ABA binds to an as yet uncharacterized receptor, here shown on the plasma membrane. The binding of ABA to the receptor initiates several signaling events. (2) ABA causes the opening of Ca^{2+} channels, and a transient depolarization of the membrane. (3) The transient membrane depolarization promotes the opening of Cl^- channels, which further depolarizes the membrane. (4) ABA increases the level of IP_3. (5) IP_3 opens IP_3-gated Ca^{2+} channels, causing the release of Ca^{2+} from internal stores. (6) The increase in cytosolic Ca^{2+} activates the opening of Cl^- channels and inhibits inward K^+ channels. This net efflux of negative charge results in a large and sustained membrane depolarization. (7) ABA causes an increase in the cytosolic pH. (8) The increase in cytosolic pH activates outward K^+ channels. K^+ leaves the cell in response to the long-term depolarization caused by Cl^- efflux. The turgor of the guard cells thus decreases, and the stomata close. (After Blatt and Thiel 1993.)

lowering the concentration of ATP in the cytosol inactivates the slow anion channels. Presumably, this inactivation is due to the presence of protein phosphatases, which remove phosphate groups that are covalently attached to proteins (Cohen 1989). If so, one would predict that inhibitors of protein phosphatases would maintain the slow anion channels in the phosphorylated (active) state even in the absence of ATP. Indeed, the protein phosphatase inhibitor *okadaic acid* maintains guard cell slow anion channels in the active state even in the absence of ATP, confirming the prediction (Schmidt et al. 1995). These results strongly suggest that protein phosphorylation and dephosphorylation play important roles in the ABA signal transduction pathway in guard cells.

Direct evidence for an **ABA-activated protein kinase (AAPK)** in *Vicia faba* guard cells was provided by Sally Assmann and her colleagues at Pennsylvania State University (Li and Assmann 1996). Treatment of *Vicia faba* guard cell protoplasts with ABA induced the autophosphorylation of a 48 kDa protein in guard cell protoplasts, and the effect was insensitive to the presence of the protein synthesis inhibitor, cycloheximide. This suggests that ABA was directly or indirectly activating the enzyme. Since this same protein was shown to be able to phosphorylate other proteins on serine residues in a Ca^{2+}-independent manner, the enzyme is an autophosphorylating protein kinase which presumably forms part of the Ca^{2+}-independent signal transduction pathway for ABA. (The presence of both Ca^{2+}-dependent and Ca^{2+}-independent pathways for ABA action will be discussed in the next section.)

The analysis of ABA-insensitive mutants has begun to help in the identification of genes coding for components of the signal transduction pathway. The best candidate for an ABA signaling protein identified so far is the *Arabidopsis* protein ABI1. Mutations in the *ABI1* gene result in insensitivity to ABA in both seeds and adult plants. The mutant *abi1* displays phenotypes consistent with a defect in ABA signaling, including reduced seed dormancy, disturbed water relations (due to improper regulation of stomatal aperture), and decreased expression of various ABA-inducible genes. The *ABI1* gene has been cloned and identified as a serine/threonine protein phosphatase (Leung et al. 1994; Meyer et al. 1994). This finding suggests that ABI1 regulates the activity of target proteins by dephosphorylating them.

Armstrong and colleagues (1995) investigated the activity of the channels in the guard cells of tobacco that had been transformed with the *ABI-1* dominant mutant allele of *Arabidopsis*. This mutant gene is able to suppress the activity of the wild-type enzyme in the transformed plants. In tobacco cells the mutant gene renders both the inward and the outward K^+ channels insensitive to ABA. Significantly, the *ABI-1* gene had no effect on either the ABA activation of the slow anion channels or the ABA-induced increase in cytosolic pH, suggesting that the protein phosphatase encoded by *ABI1* specifically regulates K^+ channels.

Experiments with isolated membrane patches have also shown that alkalinization increases the number of outward K^+ channels available for activation (Miedema and Assmann 1996). There is now direct evidence that the *abi1* mutation interferes with the *pH sensitivity* of K^+ channels (Blatt and Grabov 1997). Thus, the channel does not respond to the rise in cytosolic pH that occurs in response to ABA. Taken together, the results suggest that protein dephosphorylation, perhaps of the channels themselves, is required to maintain the pH sensitivity of the outward K^+ channels.

Given that guard cells can respond to multiple environmental signals, such as light, CO_2, and water stress, there may very well be multiple signal transduction pathways leading to stomatal closure. In the next section we address the possibility that ABA regulates stomatal closure by more than one pathway.

The ABA Signaling Pathways Show Redundancy

At least three distinct signaling pathways mediate stomatal closure by ABA: Ca^{2+}, pH, and protein phosphorylation or dephosphorylation. Although an ABA-induced increase in cytosolic calcium concentration has been observed by many investigators and is a key feature of the current model for ABA-induced guard cell closure, such an increase is typically observed in only about a third of the guard cells measured. Moreover, ABA is able to induce stomatal closure even in guard cells that show no increase in cytosolic calcium (Allan et al. 1994). Blatt and Grabov (1997) have now shown that the ABA-induced rise in cytosolic calcium depends on the magnitude of the plasma membrane voltage: Low voltages give no change in cytosolic calcium in response to ABA, yet ABA still induces stomatal closure. These results suggest that ABA can act via one or more calcium-independent pathways.

In addition to calcium, ABA can utilize cytosolic pH as a signaling intermediate. As previously discussed, a rise in cytosolic pH can lead to the activation of outward K^+ channels, and the effect of the *abi1* mutation is to render these K^+ channels insensitive to pH. Moreover, since the *ABI1* gene encodes a protein phosphatase, protein phosphorylation via kinases, as well as dephosphorylation via phosphatases, represents a third level of control over the activity of guard cell channels. Such redundancy in the signal transduction pathways involved in the response to a hormone is wholly consistent with the ability of guard cells to integrate a wide range of hormonal and environmental stimuli that affect stomatal aperture, and such redundancy is probably not unique to guard cells.

A Gene That Encodes a Negative Regulator of the ABA Response Has Been Cloned

Thus far we have focused on components of the signal transduction pathway that amplify the ABA signal and promote its effects within the cell. However, there may also be negative regulators of ABA signal transduction. As we have seen in previous chapters, such negative regulators have been identified as components of the phytochrome, auxin, gibberellin, and ethylene signaling pathways, so it would not be surprising to find negative regulators of ABA responses as well. If a gene that encodes a negative regulator were inactivated by a mutation, the resulting mutant would exhibit an enhanced response to ABA. Cutler and colleagues (1996) set up a screen for such mutants by germinating mutagenized *Arabidopsis* seeds in the presence of a low concentration of ABA.

In their search for mutants, Cutler and colleagues selected 0.3 μ*M* ABA, because at this concentration the hormone had no inhibitory effect on the germination of wild-type seeds. Several mutants were identified whose germination is inhibited by this concentration of ABA. They were termed ***era*** (*e*nhanced *r*esponse to *A*BA) mutants. The *ERA1* gene was cloned, and its protein product was identified as the β subunit of the enzyme farnesyl transferase. (Farnesyl transferases are made up of α and β subunits.)

The function of farnesyl transferases is to catalyze attachment of the isoprenoid intermediate farnesyl pyrophosphate (see Chapter 13) to the carboxy-terminal ends of proteins that contain a specific sequence of amino acids that constitutes a farnesylation signal. Many proteins that have been shown to participate in signal transduction in animals and yeasts, including Ras proteins and the γ subunits of heterotrimeric G proteins, are farnesylated. In animal systems, **farnesylation** anchors the protein to the membrane through hydrophobic interactions between the farnesyl group and the membrane lipids. The identification of ERA1 as part of farnesyl transferase suggests that a protein that normally suppresses the ABA response requires farnesylation and is possibly anchored to the membrane.

Summary

Abscisic acid plays major roles in seed and bud dormancy, as well as responses to water stress. ABA is a 15-carbon terpenoid compound derived from the terminal portion of carotenoids. ABA in tissues can be measured by bioassays based on stomatal closure. Gas chromatography, HPLC, and immunoassays are the most reliable and accurate methods available for measuring ABA levels.

ABA is produced by cleavage of a 40-carbon carotenoid precursor that is synthesized from isopentenyl pyrophosphate via the plastid terpenoid pathway. ABA is inactivated by both oxidative degradation and conjugation.

ABA is synthesized in almost all cells that contain plastids and is transported via both the xylem and the phloem. The level of ABA fluctuates dramatically in response to developmental and environmental changes. During seed maturation, ABA levels reach a peak during mid- to late embryogenesis. ABA is required for the development of desiccation tolerance in the developing embryo, the synthesis of storage proteins, and the acquisition of dormancy. Seed dormancy is controlled by the ratio of ABA to GA, and ABA-deficient embryos may exhibit precocious germination and vivipary.

Although less is known about the role of ABA in buds, ABA is one of the inhibitors that accumulates in dormant buds. It also antagonizes the action of gibberellins by suppressing the synthesis of α-amylase by barley aleurone layers.

During water stress, the ABA level of the leaf can increase 50-fold. In addition to closing stomata, ABA increases the hydraulic conductivity of the root and increases the root:shoot ratio at low water potentials. ABA and an alkalinization of the xylem sap are thought to be two chemical signals that the root sends to the shoot as the soil dries. The increased pH of the xylem sap may allow more of the ABA of the leaf to be translocated to the stomata via the transpiration stream.

ABA exerts both short-term and long-term control over plant development. The long-term effects are mediated by ABA-induced gene expression. ABA stimulates the synthesis of the LEA/RAB/DHN family of proteins during seed development and during water stress. These proteins may protect membranes and other proteins from desiccation damage. ABA-response elements and one of the transcription factors that binds to them have been identified.

There is evidence for both extracellular and intracellular ABA receptors in guard cells. ABA closes stomata by causing long-term depolarization of the guard cell plasma membrane. Depolarization is believed to be caused by an increase in cytosolic Ca^{2+}, as well as alkalinization of the cytosol. The increase in cytosolic calcium is due to a combination of calcium uptake and release of calcium from internal stores. This calcium increase leads to the opening of slow anion channels, which results in membrane depolarization. An increase in IP_3 levels as been demonstrated in ABA-treated guard cells, and some evidence suggests that G proteins participate in the response. Outward K^+ channels open in response to membrane depolarization and to the rise in pH, bringing about massive K^+ efflux. The *ABI1* gene of *Arabidopsis*, which causes insensitivity to ABA when mutated, encodes a protein phosphatase, suggesting that protein dephosphorylation is involved in the ABA response. The *ABI1* gene appears to regulate the pH sensitivity of outward K^+ channels.

In general, the ABA response in guard cells appears to be regulated by more than one signal transduction pathway. This redundancy is consistent with the ability of guard cells to respond to multiple sensory inputs.

General Reading

Assmann, S. M. (1993) Signal transduction in guard cells. *Annu. Rev. Cell Biol.* 9: 345–375.

Avery, L., and Wasserman, S. (1992) Ordering gene function: The interpretation of epistasis in regulatory hierarchies. *Trends Genet.* 8: 312–316.

Bewley, J. D. (1997) Seed germination and dormancy. *Plant Cell* 9: 1055–1066.

*Bewley, J. D., and Black, M. (1994) *Seeds. Physiology of Development and Germination*, 2nd ed. Plenum, New York.

*Blatt, M. R., and Thiel, G. (1993) Hormonal control of ion channel gating. *Annu. Rev. Plant Physiol. Plant Mol. Biol.* 44: 543–567.

Bleecker, A., and Schaller, G. (1996) The mechanism of ethylene perception. *Plant Physiol.* 111: 653–660.

Borochov, A., and Woodson, W. R. (1989) Physiology and biochemistry of flower petal senescence. *Hort. Rev.* 11: 15–43.

*Cohen, P. (1989) The structure and regulation of protein phosphatases. *Annu. Rev. Biochem.* 58: 453–508.

Daum, G., Eisenmann-Tappe, I., Fries, H-W., Troppmair, J., and Rapp, U. (1994) The ins and outs of Raf kinases. *Trends Biochem. Sci.* 19: 474–479.

Davies, P. J., ed. (1995) *Plant Hormones: Physiology, Biochemistry and Molecular Biology*, 2nd ed. Kluwer, Dordrecht, Netherlands.

Davies, W. J., and Jones, H. G., eds. (1991) *Abscisic Acid Physiology and Biochemistry*. BIOS Scientific, Oxford.

*Fairley-Grenot, K., and Assmann, S. M. (1991) Evidence for G-protein regulation of inward K^+ channel current in guard cells of fava bean. *Plant Cell* 3: 1037–1044.

Giraudat, J., Parcy, F., Bertauche, N., Gosti, F., Leung, J., Morris, P-C., Bouvier-Durand, M., and Vartanian, N. (1994) Current advances in abscisic acid action and signaling. *Plant Mol. Biol.* 26: 1557–1577.

*Horgan, R. (1995) Instrumental methods of plant hormone analysis. In *Plant Hormones: Physiology, Biochemistry and Molecular Biology*, 2nd ed., P. J. Davies, ed., Kluwer, Dordrecht, Netherlands, pp. 140–157.

Ingram, J., and Bartels, D. (1996) The molecular basis of dehydration tolerance in plants. *Annu. Rev. Plant Physiol. Mol. Biol.* 47: 377–403.

*Kelly, W. B., Esser, J. E., and Schroeder, J. I. (1995) Effects of cytosolic calcium and limited, possible dual, effects of G protein modulators on guard cell inward potassium channels. *Plant J.* 8: 479–489.

Kigel, J., and Galili, G. (1995) *Seed Development and Germination*. Marcel Dekker, New York.

Krebbs, E. G., and Beavo, J. A. (1979) Phosphorylation-dephosphorylation of enzymes. *Annu. Rev. Biochem.* 48: 923–959.

Lang, G. A., ed. (1996) *Plant Dormancy*. CAB International, Oxford.

McCarty, D. R. (1995) Genetic control and integration of maturation and germination pathways in seed development. *Annu. Rev. Plant Physiol. Plant Mol. Biol.* 46: 71–93.

Rock, C. D., and Quatrano, R. S. (1995) The role of hormones during seed development. In *Plant Hormones: Physiology, Biochemistry and Molecular Biology*, 2nd ed., P. J. Davies, ed., Kluwer, Dordrecht, Netherlands, pp. 140–157.

* Indicates a reference that is general reading in the field and is also cited in this chapter.

Chapter References

Allan, A. C., Fricker, M. D., Ward, J. L., Beale, M. H., and Trewavas, A. J. (1994) Two transduction pathways mediate rapid effects of abscisic acid in *Commelina* guard cells. *Plant Cell* 6: 1319–1328.

Anderson, B. E., Ward, J. M., and Schroeder, J. J. (1994) Evidence for an extracellular reception site for abscisic acid in *Commelina* guard cells. *Plant Physiol.* 104: 1177–1183.

Armstrong, F., Leung, J., Grabov, A., Brearley, J., Giraudat, J., and Blatt, M. R. (1995) Sensitivity to abscisic acid of guard-cell K^+ channels is suppressed by abi1-1, a mutant *Arabidopsis* gene encoding a putative protein phosphatase. *Proc. Natl. Acad. Sci. USA* 92: 9520–9524.

Beardsell, M. F., and Cohen, D. (1975) Relationships between leaf water status, abscisic acid levels, and stomatal resistance in maize and sorghum. *Plant Physiol.* 56:207–212.

Bennet-Clark, T.A. and Kefford, N.P. (1953) Chromatography of the growth substances in plant extracts. *Nature* 171: 645-648.

Blatt, M. R., and Armstrong, F. (1993) Potassium channels of stomatal guard cells: Abscisic acid-evoked control of the outward rectifier mediated by cytoplasmic pH. *Planta* 191: 330–341.

Blatt, M. R., and Grabov, A. (1997) Signal redundancy, gates and integration in the control of ion channels or stomatal movement. *J. Exp Bot.* 48 (Special Issue): 529–537.

Boyer, G. L., and Zeevaart, J. A. D. (1982) Isolation and quantitation of β-D-glucopyranosyl abscisate from leaves of *Xanthium* and spinach. *Plant Physiol.* 70: 227–231.

Cutler, S., Ghassemian, M., Bonetta, D., Cooney, S., and McCourt, P. (1996) A protein farnesyl transferase involved in abscisic acid signal transduction in *Arabidopsis*. *Science* 273: 1239–1241.

Davies, W. J., and Zhang, J. (1991) Root signals and the regulation of growth and development of plants in drying soil. *Annu. Rev. Plant Physiol. Plant Mol. Biol.* 42: 55–76.

De Bruijn, S. M., and Vreugdenhil, D. (1993) Abscisic acid and assimilate partitioning to developing seeds II. Does abscisic acid influence the sink strength of *Arabidopsis* seeds? *Physiol. Plantarum* 88: 583–589.

Dure, L., III, Crouch, M., Harada, J., Ho, T-H. D., Mundy, J., Quatranao, R., Thomas, T., and Sung, Z. R. (1989) Common amino acid sequence domains among the LEA proteins of higher plants. *Plant Mol. Biol.* 12: 475–486.

Gepstein, S., and Thimann, K. V. (1981) The role of ethylene in the senescence of oat leaves. *Plant Physiol.* 68: 349–354.

Gilroy, S., and Jones, R. L. (1994) Perception of gibberellin and abscisic acid at the external face of the plasma membrane of barley (*Hordeum vulgare* L.) aleurone protoplasts. *Plant Physiol.* 104: 1185–1192.

Gilroy, S., Read, N. D., and Trewavas, A. J. (1990) Elevation of cytoplasmic calcium by caged calcium or caged inositol trisphosphate initiates stomatal closure. *Nature* 343: 769–771.

Giraudat, J., Hauge, B. M., Valon, C., Smalle, J., Parcy, F., and Goodman., H. M. (1992) Isolation of the *Arabidopsis AB13* gene by positional cloning. *Plant Cell* 4: 1251–1261.

Glinka, Z., and Reinhold, L. (1971) Abscisic acid raises the permeability of plant cells to water. *Plant Physiol.* 48: 103–105.

Goh, C-H., Kinoshita, T., Oku, T., and Shimazaki, K-I. (1996) Inhibition of blue light-dependent H^+ pumping by abscisic acid in *Vicia* guard-cell protoplasts. *Plant Physiol.* 111: 433–440.

Grabov, A., Leung, J., Giraudat, J., and Blatt, M. (1997) Alteration of anion channel kinetics in wild-type and *abi1-1* transgenic *Nicotiana benthamiana* guard cells by abscisic acid. *Plant J.* 12: 203–213.

Guiltinan, M. J., Marcotte, W. R., and Quatrano, R. S. (1990) A plant leucine zipper protein that recognizes an abscisic acid response element. *Science* 250: 267–271.

Hartung, W. (1983) The site of action of abscisic acid at the guard cell plasmalemma of *Valerianella locusta*. *Plant Cell Environ.* 6: 427–428.

Hole, D. J., Smith, J. D., and Cobb, B. G. (1989) Regulation of embryo dormancy by manipulation of abscisic acid in kernels and associated cob tissue of *Zea mays* L. cultured *in vitro*. *Plant Physiol.* 91: 101–105.

Hornberg, C., and Weiler, E. W. (1984) High-affinity binding sites for abscisic acid on the plasmalemma of *Vicia faba* guard cells. *Nature* 310: 321–324.

Irving, H. R., Gehring, C. A., and Parish, R. W. (1992) Changes in cytosolic pH and calcium of guard cells precede stomatal movements. *Proc. Natl. Acad. Sci. USA* 89: 1790–1794.

Karssen, C. M. (1995) Hormonal regulation of seed development, dormancy, and seed germination studied by genetic control. In *Seed Development and Germination*, J. Kigel and G. Galili, eds., Marcel Dekker, New York, pp. 333–350.

Kinoshita, T., Nishimura, M., and Shimazaki, K-I. (1995) Cytosolic concentration of Ca^{2+} regulates the plasma membrane H^+-ATPase in guard cells of fava bean. *Plant Cell* 7: 1333–1342.

Koornneef, M., Hanhart, C. J., Hilhorst, H. W. M., and Karssen, C. M. (1989) In vivo inhibition of seed development and reserve protein accumulation in recombinants of abscisic acid biosynthesis and responsiveness mutants in *Arabidopsis thaliana*. *Plant Physiol.* 90: 463–469.

Koornneef, M., Jorna, M. L., Brinkhorst-van der Swan, D. L. C., and Karssen, C. M. (1982) The isolation of abscisic acid (ABA) deficient mutants by selection of induced revertants in non-germinating gibberellin sensitive lines of *Arabidopsis thaliana* L. *Heynh. Theor. Appl. Genet.* 61: 385–393.

Lee, Y., Choi, Y. B., Suh, S., Lee, J., Assmann, S. M., Joe, C. O., Kelleher, J. F., and Crain, R. C. (1996) Abscisic acid-induced phosphoinositide turnover in guard cell protoplasts of *Vicia faba*. *Plant Physiol.* 110: 987–996.

Leung, J., Bouvier-Durand, M., Morris, P-C., Guerrier, D., Chefdor, F., and Giraudat, J. (1994) *Arabidopsis* ABA-response gene *ABI1*: Features of a calcium-modulated protein phosphatase. *Science* 264: 1448–1452.

Li, J., and Assmann, S. M. (1996) An abscisic acid-activated and calcium-independent protein kinase from guard cells of Fava Bean. *Plant Cell* 8: 2359–2368.

Mansfield, T. A., and McAinsh, M. R. (1995) Hormones as regulators of water balance. In *Plant Hormones: Physiology, Biochemistry and Molecular Biology*, 2nd ed., P. J. Davies, ed., Kluwer, Dordrecht, Netherlands, pp. 598–616.

McCarty, D., Hattori, T., Carson, C. B., Vasil, V., Lazar, M., and Vasil, I. K. (1991) The viviparous-1 developmental gene of maize encodes a novel transcription activator. *Cell* 66: 895–905.

Meyer, K., Leube, M. P., and Grill, E. (1994) A protein phosphatase 2C involved in ABA signal transduction in *Arabidopsis thaliana*. *Science* 264: 1452–1455.

Miedema, H., and Assmann , S. M. (1996) A membrane-delimited effect of internal pH on the K^+ outward rectifier of *Vicia faba* guard cells. *J. Membr. Biol.* 154: 227–237.

Milborrow, B. V. (1984) Inhibitors. In *Advanced Plant Physiology*, M. B. Wilkins, ed., Pitman, London, pp. 76–110.

Noodén, L. D., and Leopold, A. C., eds. (1988) *Senescence and Aging in Plants*. Academic Press, San Diego.

Okhuma, K., Lyon, J.L., Addicott, F.T., and Smith, O.E. (1963) Abscisin II, an abscission accelerating substance from young cotton fruit. *Science* 142: 1592–1593.

Pei, Z-M., Kuchitsu, K., Ward, J. M., Schwarz, M., and Schroeder, J. I. (1997) Differential abscisic acid regulation of guard cell slow anion channels in *Arabidopsis* wild-type and abi1 and abi2 mutants. *Plant Cell* 9: 409–423.

Saab, I. N., Sharp, R. E., Pritchard, J., and Voetberg, G. A. (1990) Increased endogenous abscisic acid maintains primary root growth of maize seedlings at low water potentials. *Plant Physiol.* 93: 1329–1336.

Schmidt, C., Schelle, I., Liao, Y-J., and Schroeder, J. I. (1995) Strong regulation of slow anion channels and abscisic acid signaling in guard cells by phosphorylation and dephosphorylation events. *Proc. Natl. Acad. Sci. USA* 92: 9535–9539.

Schroeder, J. I., and Hagiwara, S. (1990) Repetitive increases in cystolic Ca^{2+} of guard cells by abscisic acid activation of nonselective Ca^{2+} permeable channels. *Proc. Natl. Acad. Sci. USA* 87: 9305–9309.

Schurr, U., Gollan, T., and Schulze, E-D. (1992) Stomatal response to drying soil in relation to the changes in the xylem sap composition of *Helianthus annuus*. II. Stomatal sensitivity to abscisic acid imported from the xylem sap. *Plant Cell Environ.* 15: 561–567.

Schwartz, A., Wu, W-H., Tucker, E. B., and Assmann, S. M. (1994) Inhibition of inward K^+ channels and stomatal response by abscisic acid: An intracellular locus of phytohormone action. *Proc. Natl. Acad. Sci. USA* 91: 4019–4023.

Schwartz, A., Ilan, N., Schwartz, M., Scheaffer, J., Assmann, S. M. and Schroeder, J. I. (1995) Anion-channel blockers inhibit S-type anion channels and abscisic acid responses in guard cells. *Plant Physiol.* 109: 651–658.

Shimazaki, K., Iino, M., and Zeiger, E. (1986) Blue light–dependent proton extrusion by guard cell protoplasts of *Vicia faba*. *Nature* 319:324–326.

Sindhu, R. K., and Walton, D. C. (1987) Conversion of xanthoxin to abscisic acid by cell–free preparations from bean leaves. *Plant Physiol.* 85:916–921.

Skriver, K., and Mundy, J. (1990) Gene expression in response to abscisic acid and osmotic stress. *Plant Cell* 2: 503–512.

Tal, M., and Imber, D. (1971) Abnormal stomatal behavior and hormonal imbalance in *flacca*, a wilty mutant of tomato. III. Hormonal effects on the water status in the plant. *Plant Physiol.* 47: 849–850.

Taylor, H. F., and Burden, R. S. (1973) Preparation and metabolism of 2-[^{14}C]*cis-trans*-xanthoxin *J. Exp. Bot.* 24: 873–880.

Thiel, G., Macrobbie, E. A. C., and Blatt, M. R. (1992) Membrane transport in stomatal guard cells: The importance of voltage control. *J. Membr. Biol.* 126: 1–18.

Uknes, S. J., and Ho, T. H. D. (1984) Mode of action of abscisic acid in barley aleurone layers: Abscisic acid induces its own conversion to phaseic acid. *Plant Physiol.* 75: 1126–1132.

Van Steveninck, R. F. M., and Van Steveninck, M. E. (1983) Abscisic acid and membrane transport. In *Abscisic Acid*, F. T. Addicott, ed., Praeger, New York, pp. 171–235.

Walker-Simmons, M. K., and Abrams, S. R. (1991) Use of ABA immunoassays. In *Abscisic Acid Physiology and Biochemistry*, W. J. Davies and H. G. Jones, eds., BIOS Scientific, Oxford, pp. 53–61.

Walton, D. C., and Li, Y. (1995) Abscisic acid biosynthesis and metabolism. In *Plant Hormones: Physiology, Biochemistry and Molecular Biology*, 2nd ed., P. J. Davies, ed., Kluwer, Dordrecht, Netherlands, pp. 140–157.

Ward, J. M., Pei, Z-M., and Schroeder, J. I. (1995) Roles of ion channels in initiation of signal transduction in higher plants. *Plant Cell* 7: 833–844.

Weiler, E. W. (1984) Immunoassay of plant growth regulators. *Annu. Rev. Plant Physiol.* 35: 85–95.

Wilkinson, S., and Davies, W. J. (1997) Xylem sap pH increase: A drought signal received at the apoplastic face of the guard cell that involves the suppression of saturable abscisic acid uptake by the epidermal symplast. *Plant Physiol.* 113: 559–573.

Wu, W.-H., and Assmann, S. M. (1994) A membrane-delimited pathway of G-protein regulation of the guard cell inward K^+ channel. *Proc. Natl Acad. Sci. USA* 91: 6310–6314.

Zacarias, L., and Reid, M. S. (1990) Role of growth regulators in the senescence of *Arabidopsis thaliana* leaves. *Physiol. Plantarum* 80: 549–554.

Zeevaart, J. A. D., and Creelman, R. A. (1988) Metabolism and physiology of abscisic acid. *Annu Rev. Plant Physiol. Plant Mol. Biol.* 39: 439–474.

24 The Control of Flowering

MOST PEOPLE LOOK FORWARD to the spring season and the profusion of flowers it brings. Many vacationers carefully time their travels to coincide with specific blooming seasons: *Citrus* along Blossom Trail in southern California, tulips in Holland. In Washington, D.C., and throughout Japan, the cherry blossoms are received with spirited ceremonies. As spring progresses into summer, summer into fall, and fall into winter, wildflowers bloom at their appointed times. In the eastern United States, skunk cabbage flowers poke through the snow of forest floors in late February and early March. Violet and marsh marigold open their flower buds in April. Sometime in May one might see jack-in-the-pulpit, yellow clover, and painted trillium. In June, chicory, evening primrose, and Queen Anne's lace appear. July is the time for pickerelweed, peppermint, and black-eyed Susan. By late July and early August, most of the spring blossoms are gone and in their places are monkey flower, spiked lobelia, and pearly everlasting. Toward the end of August, goldenrod and baneberry have their day in the sun. Finally, in September and October, aster, rose mallow, and gentian ring down the curtain on the blooming season.

Although the strong correlation between flowering and seasons is common knowledge, the phenomenon poses fundamental questions for plant physiologists. How do plants keep track of the seasons of the year and the time of day? Which environmental signals control flowering and how are those signals perceived? Can the environment be manipulated to alter the timing of flowering? How are environmental signals transduced to bring about the developmental changes associated with flowering? And what are the genes that control floral development? This chapter addresses these questions.

In Chapter 16, we discussed the role of the root and shoot apical meristems in vegetative growth and development. The transition to flowering involves major changes in the pattern of morphogenesis and cell differentiation at the shoot apical meristem. Ultimately, this process leads to the production of the floral organs—sepals, petals, stamens, and carpels (see Figure 1.3). Specialized cells in the anther undergo meiosis to produce four haploid microspores that develop into pollen grains. Similarly, a cell within the ovule divides meiotically to produce four haploid megaspores, one of which survives and undergoes three mitotic divisions to produce the cells of the embryo sac (see Figure 1.2). The embryo sac represents the mature female gametophyte; the pollen grain, with its germinating pollen tube, is the mature male gametophyte generation. The two gametophytic structures produce the gametes (egg and sperm cells), which fuse to form the diploid zygote, the first stage of the new sporophyte generation. Clearly, flowers represent a complex array of functionally specialized structures that differ completely from the vegetative plant body in form and cell types. The transition to flowering therefore entails radical changes in cell fate within the shoot apical meristem (Figure 24.1).

In the first part of this chapter we will discuss the transition to flowering in the context of the overall development of the shoot, beginning with the germinating seedling. During vegetative development, the fates of the meristematic cells become altered in ways that cause them to produce new types of structures. This phenomenon is known as *phase change*.

The events occurring in the shoot apex that specifically commit the apical meristem to produce *flowers* are collectively referred to as **floral evocation**. After discussing phase change, we will examine the types of signaling events that lead to floral evocation. The developmental signals that bring about floral evocation include endogenous factors, such as *circadian rhythms* and *hormones*, and external factors, such as *day length* (*photoperiod*) and *temperature*. In the case of photoperiodism, transmissible signals from the leaves, collectively referred to as the *floral stimulus*, are translocated to the shoot apical meristem. The interactions of these endogenous and external factors enable plants to synchronize their reproductive development with the environment.

Finally, at the end of the chapter we address the question of *floral development* itself. Recently, genes have been identified that play crucial roles in the formation of the floral organs. These studies have shed new light on the genetic control of reproductive development.

But before beginning our consideration of the changes that occur at the shoot apex during the transition to flowering, we need to distinguish between the two classes of developmental signals that regulate flowering: internal (autonomous) and external (environmental).

Autonomous Regulation versus Environmental Cues

A plant may flower within a few weeks after germinating, as in annual plants such as groundsel (*Senecio vulgaris*). In contrast, some perennial plants, especially many forest trees, may grow for 20 or more years before they begin to produce flowers. Thus, different species flower at widely different ages, indicating that age, or perhaps size of the plant, is an *internal* factor controlling the switch

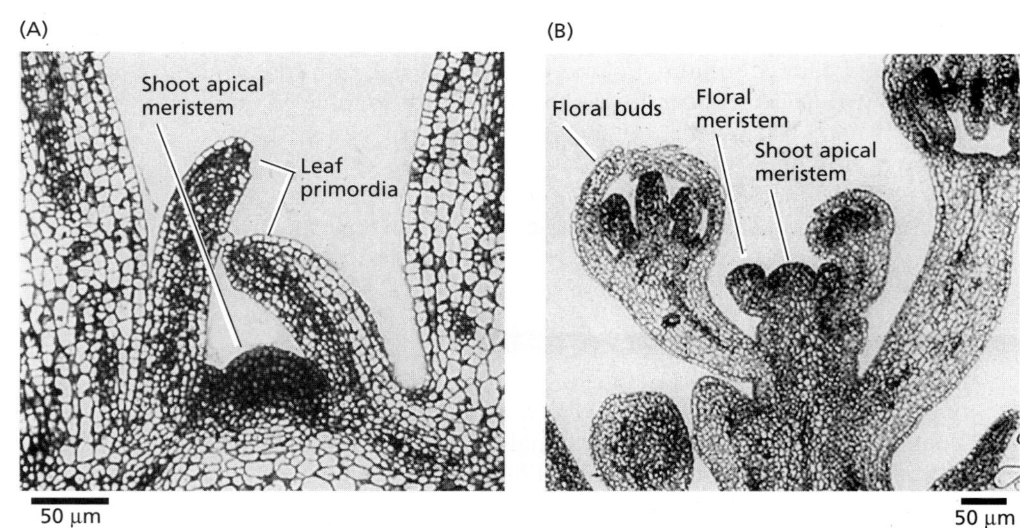

Figure 24.1 Longitudinal sections through a vegetative (A) and a reproductive (B) shoot apical region of *Arabidopsis thaliana*. In *A. thaliana* the shoot apical meristem forms leaf primordia during vegetative growth and floral meristems after the transition to flowering. These floral meristems differentiate into floral buds, which ultimately develop into mature flowers. The vegetative apex in A is presented at approximately twice the magnification of the floral apex in B. (Photographs courtesy of V. Grbić, M. Nelson and assembled and labeled by E. Himelblau.)

to reproductive development. When flowering occurs strictly in response to internal developmental factors and does not depend on any particular environmental conditions, it is referred to as *autonomous regulation*.

Some plants exhibit an absolute requirement for the proper environmental cues in order to flower. This condition is termed an *obligate* or *qualitative* response to an environmental cue. In other plant species, flowering is promoted by certain environmental cues but will eventually occur in the absence of such cues. This is called a *facultative* or *quantitative* response to an environmental cue. The flowering of this latter group of plants thus relies on both environmental and autonomous flowering systems. Photoperiodism and vernalization are two of the most important mechanisms underlying seasonal responses. *Photoperiodism* is a response to the length of day; *vernalization* is an effect on flowering brought about by exposure to cold. Other signals, such as total light radiation and water availability, can also be important external cues.

The evolution of internal (autonomous) and external (environment-sensing) control systems enables plants to carefully regulate the timing of flowering. It is critical for survival that plants flower at the optimal time for reproductive success. For example, in many populations of a particular species, flowering is synchronized. This synchrony favors crossbreeding and allows seeds to be produced in favorable environments, particularly with respect to water and temperature.

The Shoot Apex and Phase Changes

All multicellular organisms pass through a series of more or less defined developmental stages, each with its characteristic features. In humans, infancy, childhood, adolescence, and adulthood represent four general stages of development, and puberty is the dividing line between the nonreproductive and the reproductive phases. The development of higher plants likewise is characterized by developmental phases, but whereas in animals these changes take place throughout the entire organism, in higher plants they occur in a single, dynamic region, the *shoot apical meristem*.

The Shoot Apical Meristem Progresses through Three Developmental Phases

During postembryonic development, the shoot apical meristem passes through three more or less well-defined developmental stages in obligate sequence: the juvenile phase, the adult vegetative phase, and the adult reproductive phase (Poethig 1990). The transition from one phase to another is called **phase change**. The primary distinction between the juvenile and the adult vegetative phases is that the latter has the ability to form reproductive structures: flowers in angiosperms, cones in gymnosperms. As will be discussed later, the reproductive capacity of the shoot apical meristem is referred to as *competence*. However, expression of the reproductive competence of the adult phase (i.e., flowering) often depends on specific environmental and developmental signals. Thus, the absence of flowering itself is not a reliable indicator of juvenility.

The transition from juvenile to adult is frequently accompanied by changes in vegetative characteristics, such as leaf morphology, phyllotaxy (the arrangement of leaves on the stem), thorniness, rooting capacity, and leaf retention in deciduous plants (Figure 24.2). Such changes are most evident in woody perennials, but they

Figure 24.2 Juvenile and adult forms of ivy (*Hedera helix*). The juvenile form (left) has lobed palmate leaves arranged alternately, a climbing growth habit, and no flowers. The adult form (right) has entire ovate leaves arranged in spirals and an upright growth habit. (Photo by L.Taiz.)

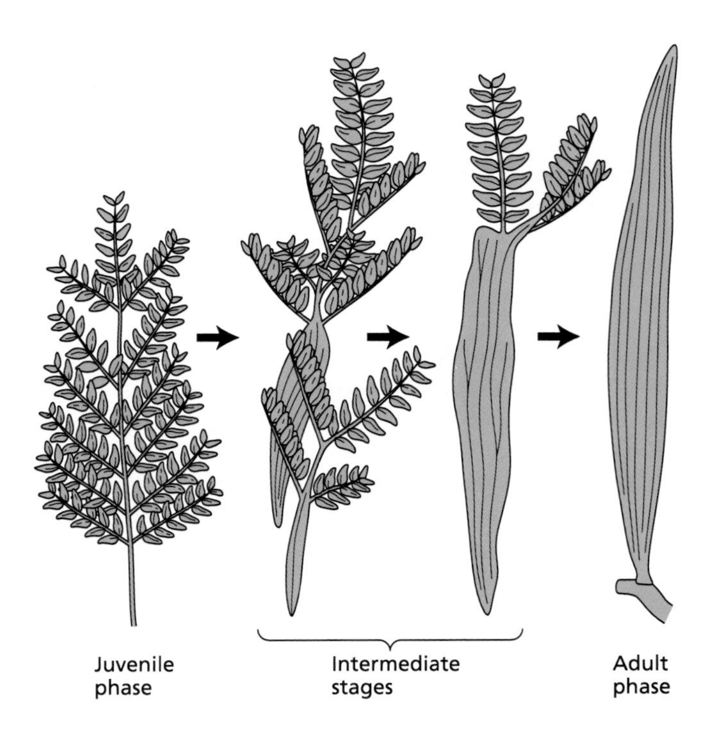

Figure 24.3 Leaves of *Acacia heterophylla*, showing transitions from pinnately compound leaves (juvenile phase) to phyllodes (adult phase). Note that the previous phase is retained at the top of the leaf in the intermediate forms. (After Mueller 1952.)

are apparent in many herbaceous species as well. Unlike the abrupt transition from the adult vegetative phase to the reproductive phase, the transition from juvenile to vegetative adult is usually gradual, involving intermediate forms. A dramatic example is the progressive transformation of juvenile leaves of the leguminous tree, *Acacia heterophylla* into phyllodes, a phenomenon noted by the German poet and natural scientist Goethe (1749–1832). Whereas the juvenile leaves consist of petiole, blade, and leaflets, adult phyllodes are specialized structures consisting of flattened petioles in place of the blade (Figure 24.3). Intermediate structures form also during the transition from aquatic to aerial leaf types in heterophyllous (characterized by different leaf types) aquatic plants such as *Hippuris vulgaris* (common marestail) (Goliber and Feldman 1989). As in the case of *A. heterophylla*, these intermediate forms possess distinct regions with different developmental patterns. Whereas the apex of an intermediate leaf develops according to the earlier developmental program (juvenile or submerged), the base develops according to the new developmental program (adult or aerial) (Poethig 1990).

To account for intermediate forms during the transition from juvenile to adult in maize, Scott Poethig (1990) has proposed a **combinatorial model** (Figure 24.4). According to this model, shoot development can be described as a series of independently regulated, *overlapping* programs (juvenile, adult, and reproductive) that modulate the expression of a common set of developmental processes. In the transition from juvenile to adult leaves, the intermediate forms indicate that different regions of the same leaf can express different developmental programs. Thus the cells at the tip of the leaf remain committed to the juvenile program, while the cells at the base of the leaf become committed to the adult program. The developmental fates of the two sets of cells in the same leaf are quite different.

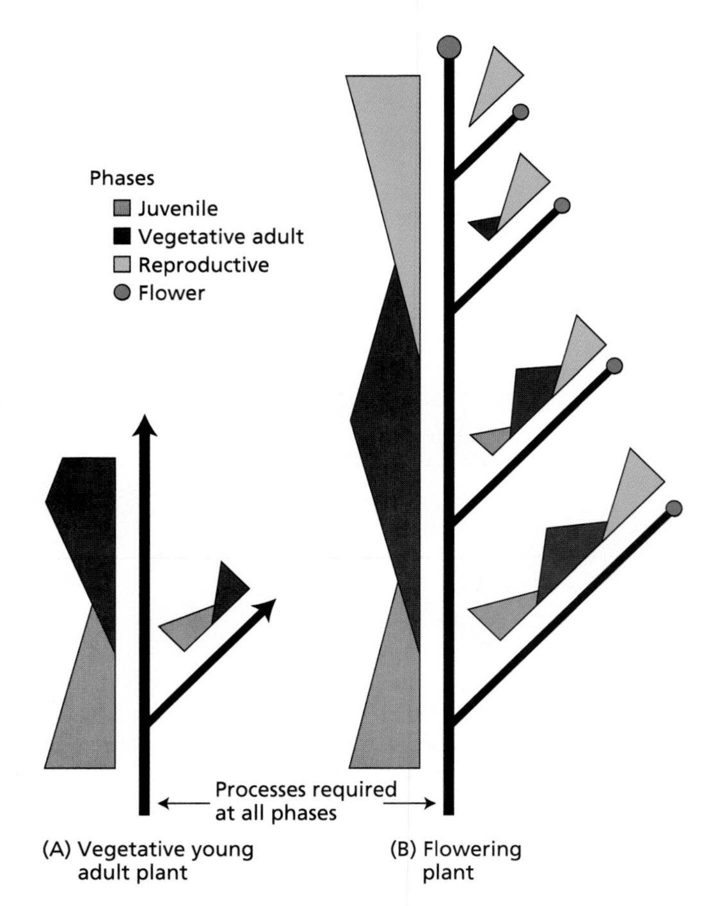

Figure 24.4 Schematic representation of the combinatorial model of shoot development in maize. Overlapping gradients of expression of the juvenile, vegetative adult, and reproductive phases are indicated along the length of the main axis and branches. The continuous black line represents processes that are required during all phases of development. (A) Young plant, including juvenile and adult phases. (B) Mature plant, including juvenile, adult, and reproductive phases. Each of the three phases may be regulated by separated developmental programs, with intermediate phases arising when the programs overlap. (After Poethig 1990.)

Juvenile Tissues Are Produced First and Are Located at the Base of the Shoot

The temporal order of the three developmental phases results in a spatial gradient of juvenility along the shoot axis. Because growth in height is restricted to the apical meristem, the juvenile tissues and organs, which form first, are located at the base of the shoot. In rapidly flowering herbaceous species, the juvenile phase may last only a few days, and few juvenile structures are produced. In contrast, woody species have a more prolonged juvenile phase, in some cases lasting 30 to 40 years (Table 24.1). In these cases, the juvenile structures can account for a significant portion of the mature plant. Once the meristem switches over to the adult phase, only adult vegetative structures are produced, culminating in floral evocation. The adult and reproductive phases are therefore located in the upper and peripheral regions of the shoot.

Attainment of a sufficiently large size appears to be more important than the plant's chronological age in determining the transition to the adult phase. Conditions that retard growth, such as mineral deficiencies, low light, water stress, defoliation, and low temperature tend to prolong the juvenile phase or even cause **rejuvenation** (reversion to juvenility) of adult shoots. In contrast, conditions that promote vigorous growth accelerate the transition to the adult phase (Poethig 1990). When growth is accelerated, exposure to the correct flower-inducing treatment can result in flowering. However, although size seems to be the most important factor, it is not always clear which specific component associated with size is critical. In some *Nicotiana* species, it appears that plants must attain a certain size to transmit a sufficient amount of the floral stimulus to the apex. In addition, the apex may need to undergo a phase transition that renders it more responsive to the floral stimulus (McDaniel et al. 1996).

Table 24.1
Length of juvenile period in some woody plant species

Species	Length of juvenile period
Rose (*Rosa* [hybrid tea])	20–30 days
Grape (*Vitis* spp.)	1 year
Apple (*Malus* spp.)	4–8 years
Citrus spp.	5–8 years
English ivy (*Hedera helix*)	5–10 years
Redwood (*Sequoia sempervirens*)	5–15 years
Sycamore maple (*Acer pseudoplatanus*)	15–20 years
English oak (*Quercus robur*)	25–30 years
European beech (*Fagus sylvatica*)	30–40 years

Source: Clark 1983.

Once the adult phase is attained, it is relatively stable, and is maintained during vegetative propagation or grafting. For example, in mature plants of English ivy (*Hedera helix*), cuttings taken from the basal region develop into juvenile plants, while those from the tip develop into adult plants. Results for grafted plants in which the *scion* (upper graft partner) originated from different parts of an old flowering specimen of silver birch (*Betula verrucosa*) also indicate that the basal part retains the juvenile condition (Longman 1976). When scions were taken from the base of the flowering tree and grafted onto seedling rootstocks, there were no flowers on the grafts within the first 2 years. In contrast, the grafts flowered freely when scions were taken from the top of the flowering tree. In some species, the meristem appears to be competent to flower but does not receive sufficient floral stimulus until the plant becomes large enough. For example, in mango (*Mangifera indica*), juvenile seedlings can be induced to flower when grafted to a mature tree. In other species, however, grafting does not induce flowering.

Phase Changes Can Be Influenced by Nutrients, Hormones, and Other Chemical Signals

The transition at the shoot apex from the juvenile to the adult phase may be affected by transmissible factors from the rest of the plant. In many plants, exposure to low-light conditions prolongs juvenility or causes reversion to juvenility. A major consequence of the low-light regime is a reduction in the supply of carbohydrates to the apex; thus, carbohydrate supply may play a role in the transition between juvenility and maturity. One obvious connection between nutritional status and the adult phase is the effect of nutrition on the size of the apex. For example, in the florist's chrysanthemum (*Chrysanthemum morifolium*), flower primordia are not initiated until a minimum apex size has been reached (Cockshull 1985).

The apex receives a variety of hormonal and other factors from the rest of the plant in addition to carbohydrates and other nutrients. Could any single factor promote the change between juvenility and maturity? Experimental evidence shows that the application of gibberellins causes reproductive structures to form in young, juvenile plants of several conifer families (Pharis and King 1985). For example, although most conifers require several years to attain maturity, male cones have been induced to form in Arizona cypress (*Cupressus arizonica*) plants that were only 2 months old by being sprayed with GAs. The involvement of *endogenous* GAs in the control of reproduction is also indicated by the fact that other treatments that accelerate flowering in pines (e.g., root removal, water stress, and nitrogen starvation) often also result in a buildup of GAs in the plant. On the other hand, while gibberellins promote the attainment of reproductive maturity in conifers and many herbaceous angiosperms, the application of GA_3 causes rejuvenation in *Hedera* and in several other woody angiosperms (Hackett and Srinivasan 1985). The role of gibberellins in the control of maturity is thus complex, varies among species and probably involves interactions with other factors.

Floral Evocation Involves Two Developmental States: Competent and Determined

On the basis of extensive studies on tobacco, Carl McDaniel and his colleagues at the Rensselaer Institute have defined two general categories of developmental states associated with floral evocation: competent and determined (Figure 24.5) (McDaniel et al. 1992). In general terms, a cell or group of cells is said to be **competent** if it can respond in the expected manner when given the appropriate developmental signal. For example, if a vegetative shoot (scion) is grafted onto a flowering stock, and the scion flowers immediately, it is obviously capable of responding to the level of floral stimulus present in the stock and is therefore competent. Failure of the scion to flower would indicate that the shoot apical meristem has not yet attained competence. This is another way of stating that the apical meristem is in the juvenile phase.

Flowering involves several processes that occur in different parts of the plant. In photoperiodic plants (discussed later in the chapter), the *inductive photoperiod* causes the leaves to export a floral stimulus to the rest of the plant that alters the developmental fate of the cells of the shoot apex. Once the meristem becomes committed to the new developmental program (flowering in this case), it is said to be florally determined (see Figure 24.5). The meristem is **determined** if it follows the same developmental program even after being removed from its normal physical and environmental context.

For example, in an experiment to measure the floral determination of axillary buds of a flowering tobacco plant (Figure 24.6A), decapitation of the donor promotes the outgrowth of the axillary bud (McDaniel 1996). After a certain number of leaves (in this case seven) have developed, flowers are produced. However, if the bud is either rooted or grafted onto a stock, it produces many more leaves before it flowers. This result shows that the bud was not florally determined. In contrast, in another experiment (Figure 24.6B), the axillary bud flowered after producing seven leaves in all three situations. This result demonstrates that the axillary bud is florally determined.

Using this approach with growing terminal buds, Singer and McDaniel (1986) showed that in a tobacco plant producing about 41 nodes, the terminal bud became florally determined after initiating 37 leaves. Extensive grafting of shoot tips among tobacco varieties established that the number of nodes a meristem produces before flowering is a function of the strength of the floral stimulus from the leaves and the competence of the meristem to respond to the signal (McDaniel et al. 1996). Thus, juvenility and phase change in tobacco are tightly correlated with, and perhaps explained by, two factors: the strength of the floral stimulus and meristem competence.

In some cases, the **expression** of flowering may be delayed or arrested even after the apex becomes determined, unless it receives a second developmental signal

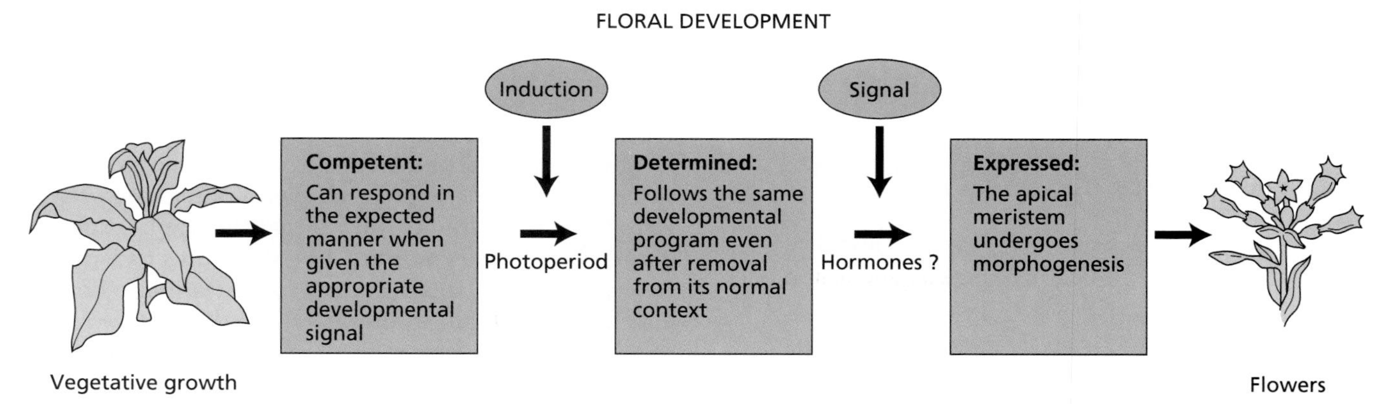

Figure 24.5 A simplified model for floral evocation at the shoot apex in which the cells of the vegetative meristem acquire new developmental fates. To initiate floral development, the cells of the meristem must first become competent. Cells or tissues are competent when they can respond to a developmental signal in the expected way. In this case, a competent vegetative meristem can respond to a floral stimulus (induction) by becoming florally determined (committed to producing a flower). The determined state is subsequently expressed, but expression of the determined state may require an additional developmental signal. (After McDaniel et al. 1992.)

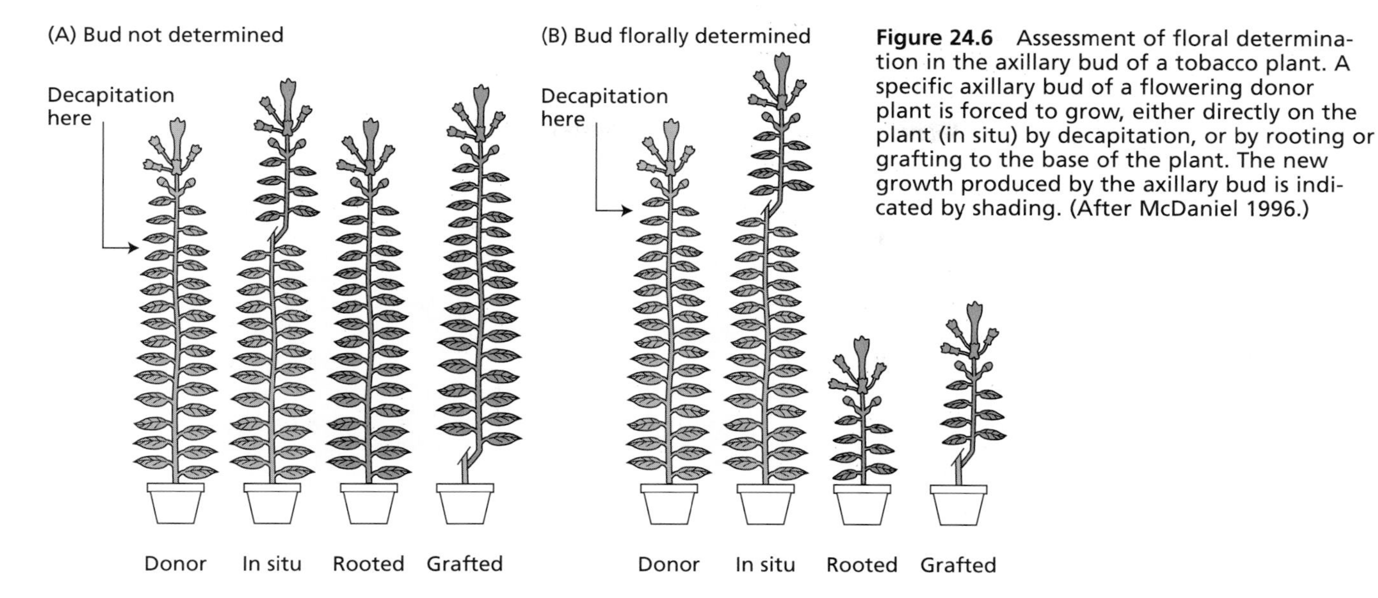

Figure 24.6 Assessment of floral determination in the axillary bud of a tobacco plant. A specific axillary bud of a flowering donor plant is forced to grow, either directly on the plant (in situ) by decapitation, or by rooting or grafting to the base of the plant. The new growth produced by the axillary bud is indicated by shading. (After McDaniel 1996.)

that stimulates expression (see Figure 24.5). For example, intact *Lolium temulentum* (darnel ryegrass) plants become committed to flowering after a single exposure to a long day. If the shoot apical meristem is excised 28 hours after the beginning of the long day and grown in tissue culture, it will produce normal inflorescences in culture, but only if the hormone gibberellic acid (GA) is present in the medium. Since apices cultured from plants grown exclusively on short days never flower, even in the presence of GA, long days are required for determination in *Lolium*, whereas GA is required for *expression* of the determined state (McDaniel and Hartnett 1996).

The *TEOPOD* (*TP*) Genes Regulate Juvenility in Maize

As noted previously, regulation of the transition from juvenile to adult (i.e., the attainment of competence) is influenced by a wide variety of conditions that affect the growth rate, and by transmissible factors such as carbohydrates and gibberellins. Hence it has been difficult to determine exactly what regulates the phase change during normal development. In recent years, plant developmental biologists have begun to take a genetic approach to the problem. The analysis of mutations that affect the timing of phase changes, called **heterochronic mutations**, provides the opportunity to identify specific genes that regulate juvenility. The best-studied example is in maize.

As indicated in Table 24.2, the juvenile and adult phases of corn plants are characterized by differences in vegetative morphology. Scott Poethig and his colleagues at the University of Pennsylvania have been characterizing the heterochronic *teopod* (*tp*) mutants of maize: *tp1*, *tp2* and *tp3* (Poethig 1990; Dudley and Poethig 1991, 1993). These semidominant mutations cause cells and tissues that would normally develop as adult structures to become more juvenile (Figure 24.7). In *tp* plants, leaves in the normally adult part of the shoot are morphologically intermediate between juvenile and adult leaves. Whereas in the wild type, tassels bear male flowers and ears bear female flowers, in *tp* mutants both tas-

Table 24.2
Characteristics of juvenile and adult phases of English ivy (*Hedera helix*) and maize (*Zea mays*)

Traits	Juvenile	Adult
Hedera helix		
Leaf shape	Entire	Lobed
Leaf thickness	230 μm	330 μm
Phyllotaxy[a]	Alternate	Spiral
Plastochron[b]	1 week	2 weeks
Growth habit	Horizontal	Vertical
Anthocyanin	Present	Absent
Aerial roots	Present	Absent
Rooting ability	Good	Poor
Flowers	Absent	Present
Zea mays		
Cuticle thickness	1 μm	3 μm
Epidermal cell shape	Circular	Rectangular
Epicuticular wax	Present	Absent
Aerial roots	Present	Absent
Bulliform cells[c]	Absent	Present
Lateral buds	Tillerlike[d]	Ears or absent

Source: Poethig 1990.
[a] Arrangement of leaves around the stem.
[b] The time period between the successive initiation of leaves.
[c] Large cells on the upper epidermis whose turgor changes control leaf unrolling in members of the grass family.
[d] Branches forming at the base of the stem.

teopod mutants develop juvenile organs and tissues

The wild type develops normal adult organs and tissues

Figure 24.7 Phenotypes of three *teopod* mutants and the wild-type maize plant. From left to right: *tp1*, *tp2*, *tp3*, and wild type. (Photo from Poethig 1990, courtesy of S. Poethig.)

sels and ears possess leaves in addition to flowers. However, *tp* mutations affect neither the rate nor the duration of growth of the plant, nor do they interfere with the timing of tassel formation. The *teopod* mutations thus specifically affect the transition from juvenile to vegetative adult phase without affecting the transition to the reproductive phase.

The possible interaction between the *TEOPOD* genes and the gibberellins has recently been explored (Evans and Poethig 1995). In GA-deficient dwarf mutants of maize, the transitions from juvenile to adult vegetative development and from adult vegetative to reproductive development occur later than in wild-type plants. Thus, endogenous GAs are required for the normal timing of phase changes.

In general, once plants attain the adult phase, they exhibit an increasing tendency to flower with age. For example, in plants controlled by day length, the number of short-day or long-day cycles necessary to achieve flowering is often fewer in older plants (Figure 24.8). These observations suggest that the effect of day length could be to accelerate a flowering process that is already occurring at a slow pace. But before discussing how plants perceive day length, we will lay the foundation by examining how organisms measure time in general. This topic is known as **chronobiology**, or the study of **biological clocks**. The best understood biological clock is the circadian rhythm.

Circadian Rhythms: The Clock Within

Organisms are normally subjected to daily cycles of light and darkness, and both plants and animals often exhibit rhythmic behavior in association with these changes (Brady 1982; Sweeney 1987). Examples of such rhythms include leaf and petal movements (day and night positions), stomatal opening and closing, growth and sporulation patterns in fungi (e.g., *Pilobolus* and *Neurospora*), time of day of pupal emergence (the fruit fly *Drosophila*), and activity cycles in rodents, as well as metabolic processes such as photosynthetic capacity and respiration rate.

When organisms are transferred from daily light–dark cycles to continuous darkness (or continuous dim light), many of these rhythms continue to be expressed, at least for several days. Under such uniform conditions the periodicity of the rhythm is then close to 24 hours, and consequently the term **circadian** (from the Latin for "about one day") **rhythm** is applied. Because they continue in a constant light or dark environment, these circadian rhythms cannot be direct responses to the presence or absence of light but must be based on an endogenous pacemaker that is self-sustaining. This pacemaker is often called an **endogenous oscillator**. The endogenous oscillator is coupled to a process, such as leaf movement or photosynthesis, and maintains the rhythm. For this reason the endogenous oscillator is considered to represent the clock itself, while the physio-

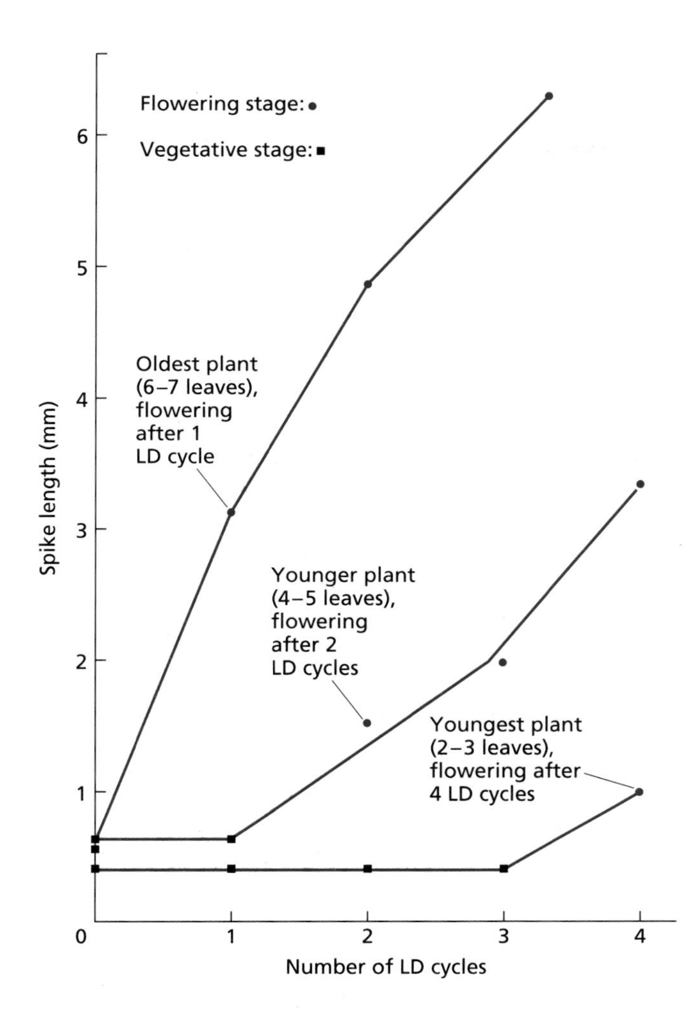

Figure 24.8 Effect of plant age on the number of long-day (LD) inductive cycles required for flowering in the long-day plant *Lolium temulentum* (darnel ryegrass). The older the plant, the fewer cycles needed to produce flowering. An inductive long-day cycle consisted of 8 hours of sunlight followed by 16 hours of low-intensity incandescent light.

logical functions that are being regulated, such as leaf movements or photosynthesis, can be thought of as the *hands* of the clock.

Circadian Rhythms Exhibit Characteristic Features

Rhythms derive from cyclic phenomena, and the pattern repeats over and over. The time between comparable points in the repeating cycle is known as the **period** (Figure 24.9A). The term **phase** is used for any point in the cycle recognizable by its relationship to the rest of the cycle. The most obvious phase points are the maximum (peak) and minimum (trough) positions. The **amplitude** of a biological rhythm is usually considered to be the distance between peak and trough, and it can often vary while the period remains unchanged (see Figure 24.9D).

In constant light or darkness, rhythms depart from an exact 24-hour periodicity. The rhythms then drift in relation to solar time, either gaining or losing time depending on whether the period is shorter or longer than 24 hours. Under natural conditions, the endogenous oscillator is entrained (synchronized) to a true 24-hour periodicity by environmental signals, the most important of which are the light-to-dark transition at dusk and the dark-to-light transition at dawn (Figure 24.9B). Such signals are termed **zeitgebers** (from the German for "time giver"). When the entraining signals are removed—for example, by transfer to continuous darkness—the rhythm is said to be **free-running** and reverts to the circadian period that is characteristic of the particular organism (see Figure 24.9B).

Although the rhythms are innate, they normally require an environmental signal, such as exposure to light or a change in temperature, to initiate their expression. In addition, many rhythms damp out (i.e., the amplitude decreases) when the organism is in a constant environment for some time and then require an environmental zeitgeber, such as a transfer from light to dark or a change in temperature, to restart them (Figure 24.9D). It is important to note that the clock itself does not damp out; only the coupling between the clock (endogenous oscillator) and the hands of the clock (physiological function) is affected. Temperature has little or no effect on the periodicity of the free-running rhythm. This insensitivity to temperature is an essential characteristic; the clock would be of no value if it could not keep accurate time under the fluctuating temperatures experienced in natural conditions.

Phase Shifting Adjusts Circadian Rhythms to Different Day–Night Cycles

In circadian rhythms, the operation of the endogenous oscillator sets a response to occur at a particular time of day. For example, the unicellular alga *Gonyaulax polyedra* generates its maximum "glowing" bioluminescence toward the end of a 12-hour night, and its maximum "flashing" bioluminescence near the middle (see Figure 24.9C). How do such responses remain on time when the daily durations of light and darkness are changing? The answer to this question lies in the fact that the phase of the rhythm can be changed if the whole cycle is moved forward or backward in time without its period being altered (see Figure 24.9E).

As an example, we can take a phase point that occurs 1 hour after the onset of darkness in cycles of 12 hours light and 12 hours dark (see Figure 24.9B). If daylight is extended for 1 hour, this phase point must have drifted forward relative to the dusk zeitgeber, and the appropriate response would be a correcting backward shift (or phase *delay*) such that this phase point still occurs 1 hour

(A)

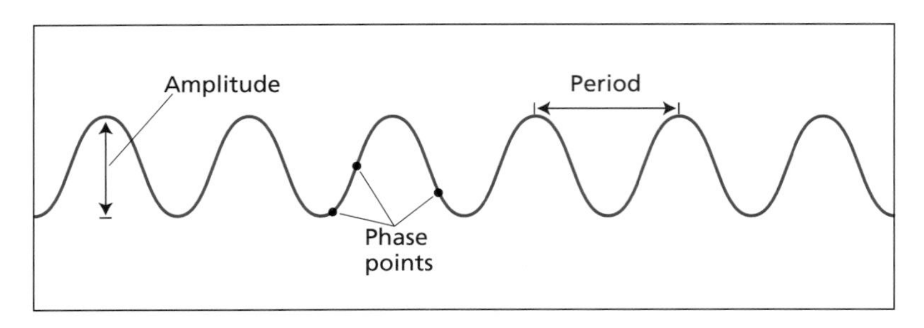

A typical circadian rhythm. The **period** is the time between comparable points in the repeating cycle; the **phase** is any point in the repeating cycle recognizable by its relationship with the rest of the cycle; the **amplitude** is the distance between peak and trough.

(B)

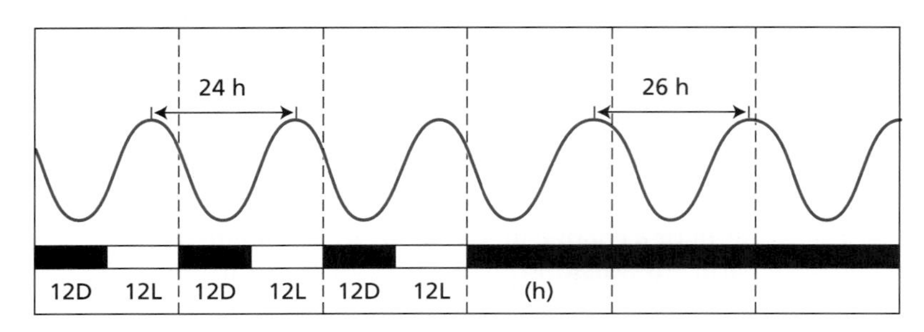

A circadian rhythm entrained to a 24 h light–dark cycle and its reversion to the free-running period (26 h in this example) following transfer to continuous darkness.

(C)

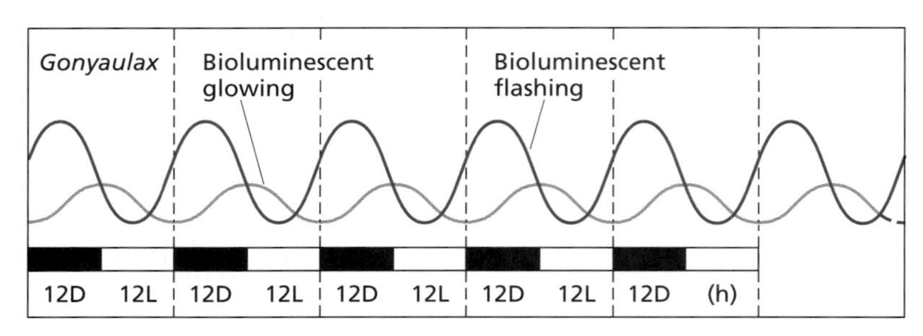

The phase relationship of two different rhythms in the unicellular alga *Gonyaulax polyedra* entrained to cycles of 12 h light and 12 h dark.

(D)

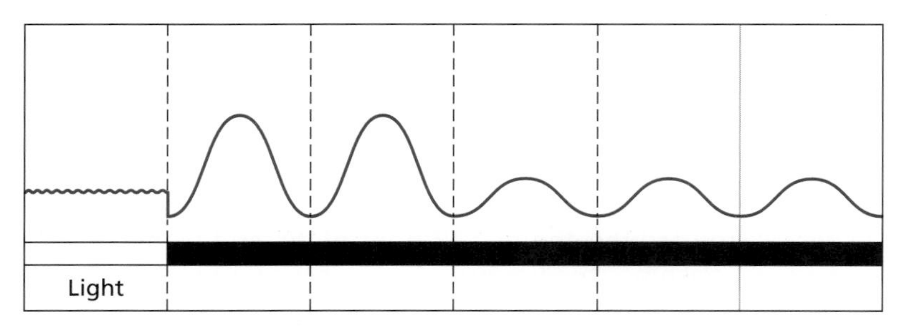

Suspension of a circadian rhythm in continuous bright light and the release or restarting of the rhythm following transfer to darkness.

(E)

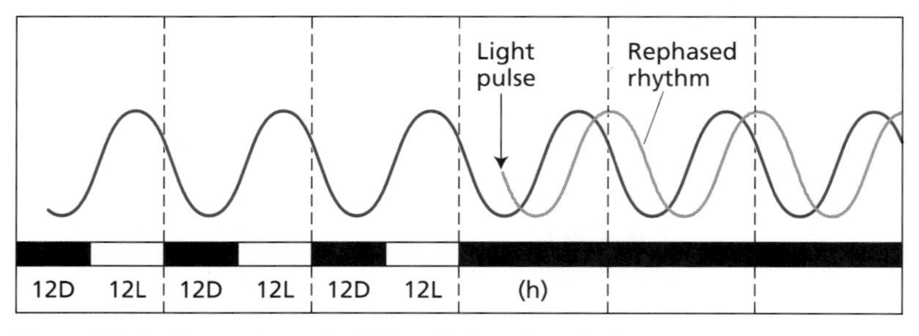

Typical phase-shifting response to a light pulse given shortly after transfer to darkness. The rhythm is rephased (delayed) without its period being changed.

Figure 24.9 Some characteristics of circadian rhythms.

after the end of the light period and the rhythm stays on local time. Similarly, the appropriate answer to light perceived at a phase point that would normally occur 1 hour before dawn would be a phase *advance.* Adaptively, therefore, different *directions* of phase shifting are to be expected in response to light signals given at different phase points in the circadian cycle, and experimentally that is what is observed.

Investigators test the response of the endogenous oscillator usually by placing the organism in continuous darkness and examining the response to a short pulse of light (usually less than 1 hour) given at different phase points in the free-running rhythm. When an organism is entrained to a cycle of 12 hours light and 12 hours dark and then allowed to free-run in darkness, the phase of the rhythm that coincides with the light period of the previous entraining cycle is called the **subjective day** and the phase that coincides with the dark period is called the **subjective night**. If a light pulse is given during the first few hours of the subjective night, the rhythm is delayed; in contrast, a light pulse given toward the end of the subjective night advances the phase of the rhythm (Figure 24.10). As already pointed out, this is precisely the pattern of response that would be expected if the rhythm is to stay on local time. Therefore, these phase-shifting responses enable the rhythm to be entrained to cycles with different durations of light and darkness, and they demonstrate that the rhythm

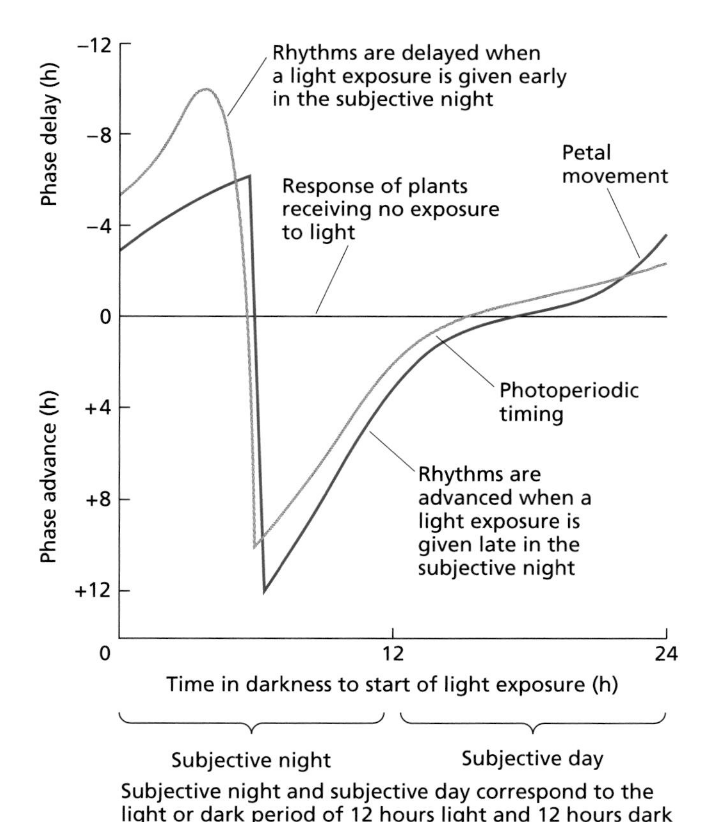

Figure 24.10 Characteristics of the phase-shifting response in circadian rhythms. Researchers obtained the phase-shifting response of the petal movement rhythm in the succulent plant *Kalanchoe blossfeldiana* (darker curve) by giving a 2-hour light treatment at various times after transferring plants from cycles of 12 hours light and 12 hours dark to continuous darkness. The curve shows the delay or advance of petal movement rhythm compared with that of plants receiving no exposure to light. The response of photoperiodic timing in the short-day plant *Chenopodium rubrum* (red goosefoot) (lighter curve) was obtained by a 6-hour light treatment at different times after transfer to darkness. The curve shows the delay or advance of the rhythm of flowering response to a night break. (Data for *Kalanchoe* from Zimmer 1962. Data for *Chenopodium* from King and Cumming 1972.)

will run differently under different natural conditions of day length. It has been demonstrated that different photoperiodic rhythms show a very similar pattern of phase shifting when tested in the same way (see Figure 24.10).

The physiological mechanism whereby a light signal causes phase shifting is not yet known. However, the low levels and specific wavelengths of light that have an effect indicate that the light response must be mediated by a photoreceptor and not by rates of photosynthesis. The role of phytochrome in phase shifting was discussed in Chapter 17. Light reception by phytochrome in the rhythmic leaf movements in *Samanea*, a semitropical leguminous tree, is well characterized (Satter and Galston 1981). Phytochrome is also involved in the phase shifting of photoperiodic rhythms. For example, the circadian rhythm in the flowering response of seedlings of Japanese morning glory (*Pharbitis nil*) can be phase-shifted by exposure to a few seconds of red light (Lumsden et al. 1986). On the other hand, phytochrome does not appear to be the universal photoreceptor for phase shifting, since it is not involved in the circadian rhythm of bioluminescence in *Gonyaulax.*

Circadian Clock Mutants of *Arabidopsis* Have Been Isolated

The isolation of clock mutants has been an important tool for the identification of clock genes in other organisms, such as *Neurospora* and *Drosophila*. Isolating clock mutants in plants requires a convenient assay that allows monitoring of the circadian rhythms of many thousands of individual plants to detect the rare aberrant phenotype. This requirement has proven to be difficult because plants that exhibit externally visible circadian rhythms, such as nyctinastic leaf movements, are not suitable for molecular genetic analysis. In *Arabidopsis*, which is well suited for molecular genetic approaches, nyctinastic leaf movements can be measured, but they are more subtle and are difficult to monitor on a mass scale. However, as noted in Chapter 17, circadian rhythms in the expression of certain genes, such as the gene that encodes the chlorophyll *a*/*b*–binding protein (*CAB* or *LHCB*), have been characterized in *Arabidopsis* (Millar and Kay 1991).

To allow screening for clock mutants in *Arabidopsis*, the promoter region of the *LHCB* gene was fused to the gene that encodes luciferase, an enzyme that emits light in the presence of its substrate, luciferin. This reporter gene construct was then used to transform *Arabidopsis* via the Ti plasmid of *Agrobacterium* as a vector. Using a video system, investigators were able to monitor the temporal and spatial regulation of bioluminescence in individual seedlings in real time (Millar et al. 1992). Using bioluminescence as the circadian phenotype, Millar and colleagues (1995) were able to isolate 21 independent ***toc*** (**t**iming **o**f **C**AB expression) mutants, including both short-period and long-period lines. The *toc* mutants hold great promise for helping us identify the genes that regulate circadian rhythms in plants. As we will discuss later in the chapter, the luciferase imaging technique has also been useful for studying the role of blue light in the control of flowering (see Chapters 17 and 18).

We now return to the question of how changes in day length can influence the transition to flowering, the phenomenon known as photoperiodism.

Photoperiodism: Monitoring Day Length

As we have seen, the circadian clock enables organisms to determine the *time of day* at which a particular molecular or biochemical event occurs. **Photoperiodism**, or the ability of an organism to detect day length, makes it possible for an event to occur at a particular *time of year*, thus allowing for a *seasonal* response. Both circadian rhythms and photoperiodism have the common property of responding to cycles of light and darkness.

Photoperiodic phenomena are found in both animals and plants. In the animal kingdom, day length controls such seasonal activities as hibernation, development of summer or winter coats, and reproductive activity. Plant responses controlled by day length are numerous, including the initiation of flowering, asexual reproduction, the formation of storage organs, and the onset of dormancy (Vince-Prue 1975). Perhaps all plant photoperiodic responses utilize the same photoreceptors, with subsequent specific signal transduction pathways regulating different responses. Since it is clear that monitoring the passage of time is essential to all photoperiodic responses, a timekeeping mechanism must underlie both the time-of-year and the time-of-day responses. The circadian oscillator is thought to provide an autonomous time-measuring mechanism that serves as a reference point for the response to incoming light (or dark) signals from the environment.

Although we now take for granted the ability of plants to detect seasonal change by measuring day length, this was a revolutionary concept when it was first proposed in the 1920s. In this section we will briefly describe the discovery of photoperiodism and the early physiological studies, which showed (1) that night length is more important than day length for the response, (2) that a circadian rhythm is part of the photoperiodic mechanism, (3) that phytochrome is the photoreceptor for many photoperiodic phenomena, and (4) that the blue light receptor, cryptochrome, also appears to be involved.

Plants Can Be Classified by Their Photoperiodic Responses

Many plant species flower during the long days of summer, and for many years plant physiologists believed that the correlation between long days and flowering was a consequence of the accumulation of photosynthetic products synthesized during long days. This hypothesis was shown to be incorrect by the work of Wightman Garner and Henry Allard, conducted in the 1920s at the U.S. Department of Agriculture laboratories in Beltsville, Maryland. They found that a mutant variety of tobacco, Maryland Mammoth, grew profusely to about 5 m in height, but it failed to flower in the prevailing conditions of summer (Figure 24.11). However, the plants flowered in the greenhouse during the winter.

These results ultimately led Garner and Allard to test the effect of artificially providing short days by covering plants grown during the long days of summer with a light-tight tent late in the afternoon. These artificial short days also caused the plants to flower. This requirement for short days was difficult to reconcile with the idea that longer periods of radiation and the resulting increase in photosynthesis promote flowering. Garner and Allard concluded that the length of the day was the determining factor in flowering and were able to confirm this hypothesis in many different species and conditions. This work laid the foundations for the extensive subsequent research on photoperiodic responses.

The classification of plants according to their photoperiodic responses is usually made on the basis of flowering, even though many other aspects of their development may also be affected by day length. The two main photoperiodic response categories are **short-day plants (SDPs)**, in which flowering occurs only in short days (*qualitative* SDP) or is accelerated by short days (*quantitative* SDP), and **long-day plants (LDPs)** in which flowering occurs only in long days (*qualitative* LDP) or is accelerated by long days (*quantitative* LDP). A few plants have more specialized day length requirements. **Intermediate-day plants** flower only between narrow day length limits (for example between 12 and 14 hours in one variety of sugarcane). Plants in another special category, called **ambiphotoperiodic** species,

Figure 24.11 Maryland Mammoth mutant of tobacco (right) compared to wild-type tobacco (left). Both plants were grown during summer in the greenhouse. (Courtesy of R. Amasino and daughter.)

flower in long days or short days but not at intermediate day lengths. The flowering of many species is not regulated by day length; these plants are called **day-neutral plants**.

The essential distinction between long-day and short-day plants is that flowering in LDPs is promoted only when the day length *exceeds* a certain duration in every 24-hour cycle, whereas promotion of flowering in SDPs requires a day length that is *less than* a critical value. The value of the critical day length varies widely between species and only when flowering is examined for a range of day lengths can the correct photoperiodic classification be established (Figure 24.12).

Long-day plants can effectively measure the lengthening days of spring or early summer and delay flowering until the critical day length is reached. Many varieties of wheat (*Triticum aestivum*) behave in this way. SDPs often flower in fall, when the days shorten below the critical day length, as in many varieties of *Chrysanthemum morifolium.* However, day length alone is an ambiguous signal because it cannot distinguish between spring and fall. Plants exhibit several strategies to avoid this ambiguity. One is the coupling of a temperature requirement to a photoperiodic response. Certain plant species, such as winter wheat, do not respond to photoperiod until after vernalization (overwintering) has occurred. (We will discuss vernalization a little later in the chapter.) Another example is found in some short-day varieties of strawberry (*Fragaria x ananassa*), in which flowers are initiated when the critical day length is reached in fall but do not emerge until spring. A different strategy for avoiding seasonal ambiguity would be a mechanism for distinguishing between *lengthening* and *shortening* days. Some animals appear to have this capability, but such a mechanism is not known to operate in plants.

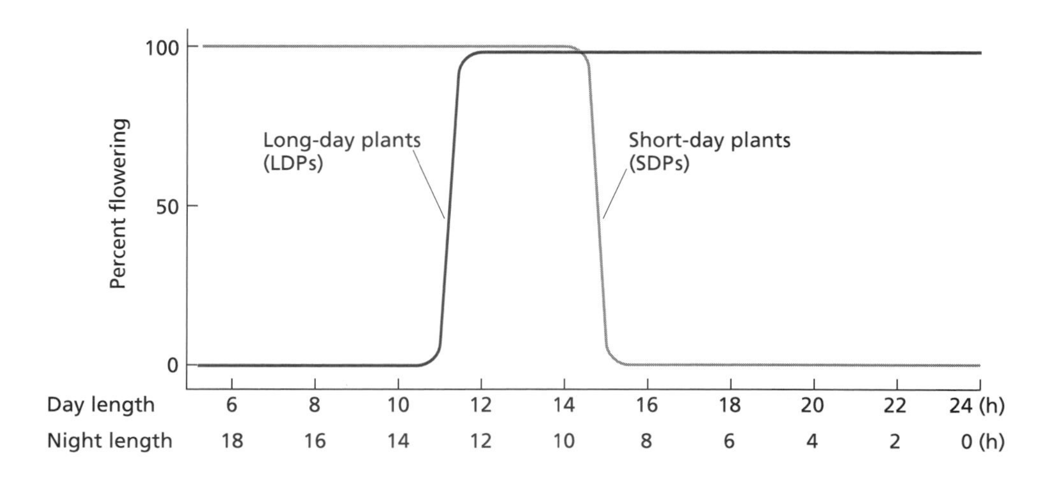

Figure 24.12 The photoperiodic response in long- and short-day plants. Short-day plants flower when the day length is less than (or the night length exceeds) a certain critical duration in the 24-hour cycle. Long-day plants flower when the day length exceeds (or the night length is less than) a certain critical duration in a 24-hour cycle. The critical duration varies between species: In this example, both the SDPs and the LDPs would flower in photoperiods between 12 and 14 h long.

Plants Monitor Day Length by Measuring the Length of the Night

Under natural conditions, day and night lengths configure a 24-hour cycle of light and darkness. A plant thus could perceive a critical day length by measuring the duration of either light or darkness. Much experimental work in the early studies of photoperiodism was devoted to establishing which part of the light–dark cycle is the controlling factor in flowering. Results showed that flowering of SDPs is determined primarily by the duration of darkness (Figure 24.13A). Thus, it was possible to induce flowering in SDPs with light periods longer than the critical value, provided that these were followed by sufficiently long nights (Figure 24.13B). Similarly, SDPs did not flower when short days were followed by short nights. More detailed experiments demonstrated that photoperiodic timekeeping in SDPs is a matter of measuring the duration of darkness (Hamner and Bonner 1938; Hamner 1940). For example, the plants flowered only when the dark period exceeded 8.5 hours in cocklebur (*Xanthium strumarium*) or 10 hours in soybean (*Glycine max*). The duration of darkness was also shown to be important in LDPs (see Figure 24.13B). These plants were found to flower in short days, provided that the

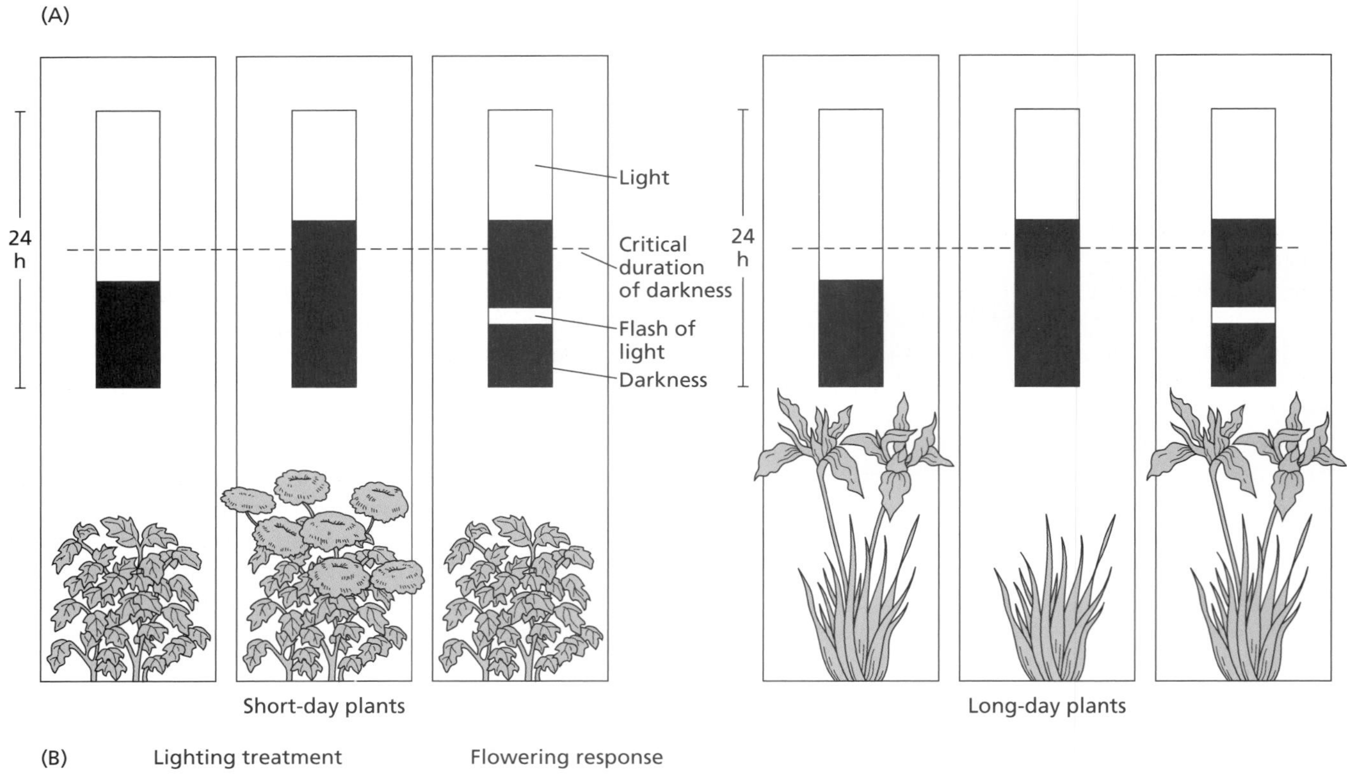

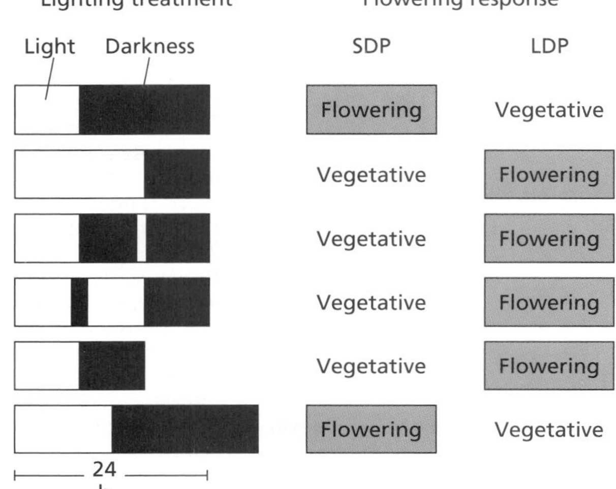

Figure 24.13 The photoperiodic regulation of flowering. (A) Short-day (long-night) plants flower when night length exceeds a critical dark period. Interruption of the dark period by a flash of light (a night break) prevents flowering. Long-day (short-night) plants flower if the night length is shorter than a critical period. In some long-day plants, shortening the night with a flash of light induces flowering. (B) Effects of the duration of the dark period on flowering. Treating short- and long-day plants with different photoperiods clearly shows that the critical variable is the length of the dark period.

accompanying night length was also short; however, long days followed by long nights were ineffective.

A feature that underlines the importance of the dark period is that it can be made ineffective by interruption with a short exposure to light, called a **night break** (see Figure 24.13A). In contrast, interrupting a long day with a brief dark period does not cancel the effect of the long day (see Figure 24.13B). Night break treatments of only a few minutes are effective in *preventing* flowering in many SDPs, including *Xanthium* and *Pharbitis*, but much longer exposures are often required to *promote* flowering in LDPs. In addition, the effect of a night break varies greatly according to the time when it is given. For both LDPs and SDPs, a night break was found to be most effective when given near the middle of a dark period of 16 hours (Figure 24.14). However, when considerably longer dark periods were given, it was established that the time of maximum sensitivity to a night break is related to the time from the beginning of darkness. In *Xanthium*, for example, the maximum effect of a night break still occurred about 8 hours after the beginning of a 24-hour dark period, even though the subsequent 16-hour period of unbroken darkness considerably exceeds the critical night length of 8.5 hours.

The discovery of the night break effect, and its time dependence, had several important consequences. It established the central role of the dark period and provided a valuable probe for studying photoperiodic timekeeping. Because only small amounts of light are needed, it became possible to study the action and identity of the photoreceptor without the interfering effects of photosynthesis and other nonphotoperiodic phenomena. This discovery has also led to the development of commercial methods for regulating the time of flowering in crop plants such as chrysanthemum and poinsettia (*Euphorbia pulcherrima*).

An Endogenous Oscillator Is Involved in Photoperiodic Timekeeping

The decisive effect of night length on flowering indicates that measuring the passage of time in darkness is central to photoperiodic timekeeping. Two different hypotheses have been proposed to explain how plants measure night length. According to the **hourglass hypothesis**, time is measured by a unidirectional series of biochemical reactions that start at the beginning of the dark period. If not interrupted by light, this series of reactions would lead to induction of flowering in SDPs or to inhibition of flowering in LDPs. This type of mechanism can measure only one interval of time and, by analogy with an hourglass device, would be reset each day by the photoperiod.

Some of the characteristics of flowering induction are consistent with the accumulation or decay of biochemical products, as predicted by the hourglass hypothesis (Thomas and Vince-Prue 1984). However, most of the available evidence favors a different mechanism, called the **clock hypothesis** (Bünning 1960). This hypothesis proposes that photoperiodic timekeeping depends on the endogenous circadian oscillator of the type involved in the daily rhythms described earlier in the chapter. In this view, the role of the photoperiod is to set the phase of the rhythm in a way leading to the measurement of time in darkness and of the critical night length.

Measurements of the effect of the night break on flowering may be used to investigate circadian rhythms

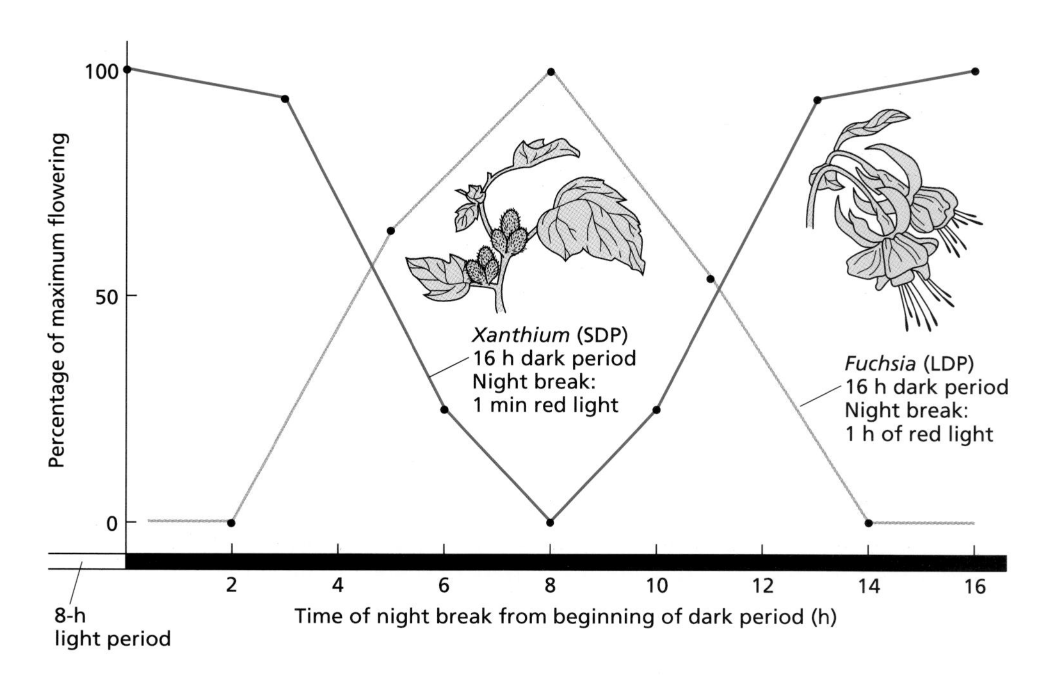

Figure 24.14 The time when a night break is given determines the flowering response. When given during a long dark period, a night break promotes flowering in LDPs and inhibits flowering in SDPs. In both cases, the greatest effect on flowering occurs when the night break is given near the middle of the 16-hour dark period. The LDP *Fuchsia* was given a 1-hour exposure to red light in a 16-hour dark period. *Xanthium* was exposed to red light for 1 minute in a 16-hour dark period. (Data for *Fuchsia* from Vince-Prue 1975. Data for *Xanthium* from Salisbury 1963 and Papenfuss and Salisbury 1967.)

in flowering, and many studies have shown that those rhythms exist. For example, when soybean plants are kept under a long dark cycle after 8 hours of light, and the dark cycle is interrupted at different times by a 4-hour night break, the flowering response to the night break shows a circadian rhythm (Figure 24.15). Many other rhythmic responses of both LDPs and SDPs have been reported (Thomas and Vince-Prue 1984). Further evidence for the presence of a circadian oscillator is the observation that the photoperiodic response can be phase-shifted by a light treatment (see Figure 24.10).

The involvement of a circadian oscillator in photoperiodism poses an important question. How does an oscillation with a 24-hour periodicity measure a critical duration of darkness of, say, 8 to 9 hours, as in the SDP *Xanthium*? Erwin Bünning, working in Tübingen, Germany, proposed in 1936 that the control of flowering by photoperiodism is achieved by an oscillation of phases with different sensitivities to light. Bünning's hypothesis invokes two alternating phases, one requiring 12 hours of darkness, the other 12 hours of light (Bünning 1960).

According to Bünning's hypothesis, light has two distinct roles in the flowering process. Light signals at dawn and dusk, acting as zeitgebers, set the phase of the photoperiodic rhythm. In addition, the rhythm has a light-sensitive phase called the **inducible phase**. When a light signal is received during the inducible phase of the rhythm, the effect is to promote or prevent floral induction. Thus, in nature the phase of the rhythm is initially set by the light-on signal at dawn. The rhythm then continues to run until it reaches a specific phase point at which it is suspended while the plant remains in the light (see Figure 24.9D). At the transition to darkness at dusk, the rhythm is released and runs freely. This entrainment at dusk ensures that the rhythm is released at a specific phase point and that the inducible phase always occurs at a constant real time after transfer to darkness. In SDPs, exposure to light prevents flowering, so flowering would be induced only when dawn occurs after the completion of the inducible phase of the rhythm. In this way, the photoperiod establishes the time of sensitivity to light in the subsequent dark period and sets the conditions for the measurement of the critical night length.

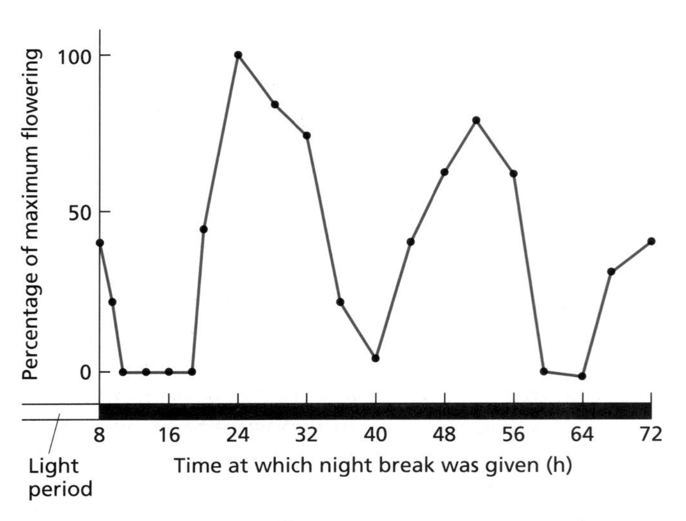

Figure 24.15 Rhythmic flowering in response to night breaks. When plants are given a night break at different times (indicated on the horizontal axis) in a long dark period, the flowering response shows a circadian rhythm. In this experiment, the SDP soybean (*Glycine max*) received cycles of an 8-hour light period followed by a 64-hour dark period. A 4-hour night break was given at various times during the long inductive dark period. (Data from Coulter and Hamner 1964.)

The Leaf Is the Site of Perception of the Photoperiodic Stimulus

One well-established fact about the photoperiodic stimulus is that it is perceived by leaves. Treatment of a single leaf of the SDP *Xanthium* with short photoperiods is sufficient to cause the formation of macroscopically visible flowers, even when the rest of the plant is exposed to long days (Hamner and Bonner 1938). For both LDPs and SDPs, many experiments have confirmed that the photoperiod experienced by the leaves determines the response at the apex. Thus, in response to photoperiod the leaf transmits an unknown signal that regulates the transition to flowering at the shoot apex. The photoperiod-regulated processes that occur in the leaves resulting in the transmission of a floral stimulus to the shoot apex are referred to collectively as **photoperiodic induction**.

Photoperiodic induction can also take place in a leaf that has been separated from the plant. In the SDP *Perilla crispa* (red-leaved perilla), Jan Zeevaart at the Department of Energy Plant Research Lab at Michigan State University has shown that an excised leaf exposed to short days can cause flowering when subsequently grafted to a noninduced plant maintained in long days (Zeevaart and Boyer 1987). Thus, photoperiodic induction appears to depend on events that take place in the leaf. When several photoperiodic cycles are required, these must be given to the same leaf. The cumulative effect of repeated cycles appears to take place in the leaf and not to be a consequence of the accumulation of a metabolite at the apex.

Leaves have also been shown to play an important role in plants in which flowering is under autonomous control. For example, grafting a single leaf of Agate, a day-neutral variety of soybean, caused flowering of the short-day variety Biloxi when the latter was maintained

Table 24.3
Examples of the successful transfer of a flowering induction signal by grafting between plants of different photoperiodic response groups

Donor plants maintained under flower-inducing conditions	Photoperiod type	Vegetative receptor plant induced to flower	Photoperiod type
Helianthus annus	DNP in LD	*H. tuberosus*	SDP in LD
Nicotiana tabacum Delcrest	DNP in SD	*N. sylvestris*	LDP in SD
Nicotiana sylvestris	LDP in LD	*N. tabacum* Maryland Mammoth	SDP in LD
Nicotiana tabacum Maryland Mammoth	SDP in SD	*N. sylvestris*	LDP in SD

Note: Successful transfer of the flowering signal between long-day plants (LDPs), short-day plants (SDPs), and day-neutral plants (DNPs) indicates the existence of a transmissible floral hormone that is effective in plants of different photoperiod response groups.

in noninductive long days (Heinze et al. 1942). Other examples of the transmission of flowering signals from day-neutral plants to LDPs and SDPs are given in Table 24.3. Thus flowering in day-neutral plants may be regulated by the same transmissible signals that control flowering in photoperiod-responsive plants.

The flowering stimulus appears to be transported via the phloem and to be chemical, rather than physical, in nature. For example, grafted leaves cause flowering only when a functional phloem connection is established between the donor and the receptor plants. Treatments that restrict phloem transport, such as girdling (see Chapter 10) or localized heating, prevent movement of the floral signal.

Grafting studies have also demonstrated that in certain situations transmissible *inhibitors* of flowering are produced. For example, when a day-neutral variety of tobacco is grafted to an LDP tobacco species, *Nicotiana sylvestris*, and the graft is maintained in short days, flowering of the day-neutral partner is strongly inhibited. Conversely, when this graft is maintained in long days, flowering of the day-neutral partner is promoted. Thus *N. sylvestris* is capable of transmitting both flowering promoters and flowering inhibitors via a graft junction (Lang et al. 1977). Similar studies in peas have led to the identification of several genetic loci that regulate steps in the biosynthetic pathways of both floral activators and floral inhibitors (see Box 24.1).

It is possible to measure rates of transport of the flowering stimulus by removing a leaf at different times after induction, and comparing the time it takes for the signal to reach two buds located at different distances from the induced leaf. The rationale for this type of measurement is that a threshold amount of the signaling compound has reached the bud when flowering takes place despite the removal of the leaf. Studies using this method (Evans and Wardlaw 1966; King et al. 1968) have shown that the rate of transport of the flowering signal is comparable to, or somewhat slower than, the rate of translocation of sugars in the phloem (see Chapter 10). These rates of transport are consistent with a chemical message and make an electric signal unlikely.

Phytochrome Is the Main Photoreceptor in Photoperiodism

Night break experiments are well suited for studying the nature of the photoreceptors involved in the reception of light signals during the photoperiodic response. The inhibition of flowering in SDPs by night breaks was one of the first physiological processes shown to be under the control of phytochrome (Figure 24.16) (Borthwick et al. 1952). In many SDPs, a night break becomes effective only when the supplied dose of light is sufficient to saturate the photoconversion of Pr (phytochrome that absorbs red light) to Pfr (phytochrome that absorbs far-red light) (see Chapter 17). A subsequent exposure to far-red light, which photoconverts the pigment back to the physiologically inactive Pr form, restores the flowering response (Downs 1956). Red and far-red reversibility has also been demonstrated in some LDPs, in which a night break of red light promoted flowering and a subsequent exposure to far-red light prevented this response.

Action spectra for the inhibition and restoration of the flowering response in SDPs are shown in Figure 24.17. A peak at 660 nm, the absorption maximum of Pr (see Chapter 17), is obtained when dark-grown *Pharbitis* seedlings are used to avoid interference from chlorophyll. In contrast, the spectra for *Xanthium* provide an example of the response in green plants, in which the presence of chlorophyll can cause some discrepancy between the action spectrum and the absorption spec-

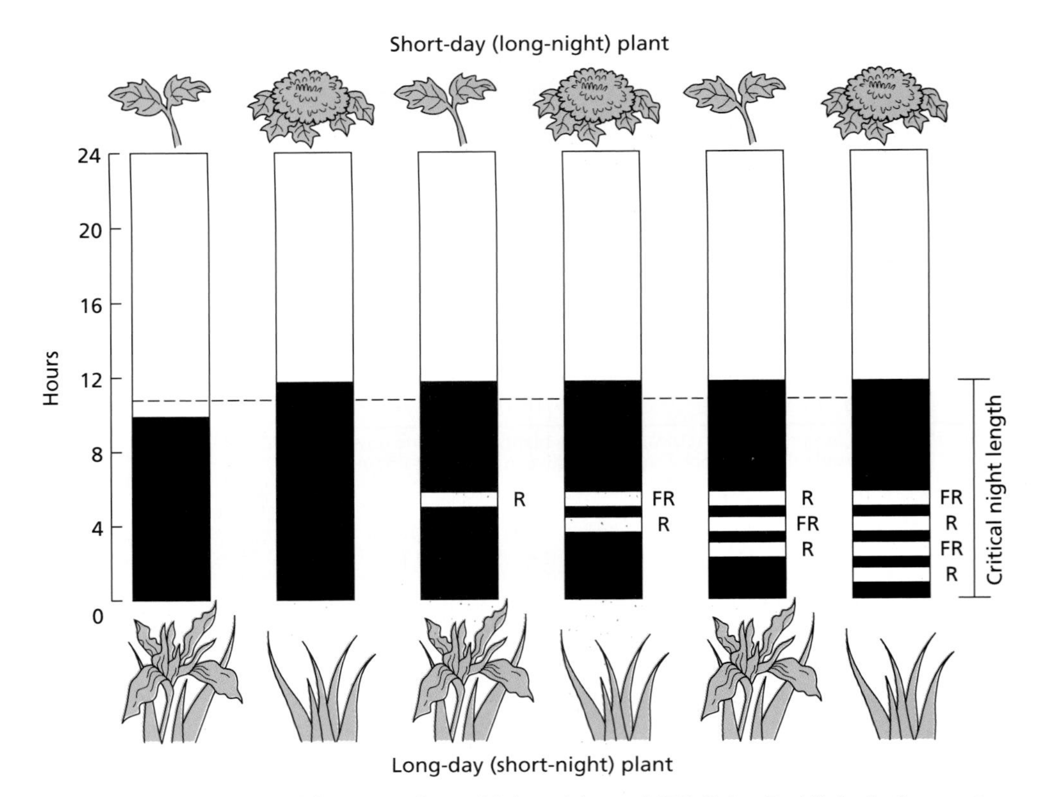

Figure 24.16 Control of flowering by red (R) and far-red (FR) light. Red light is the most effective wavelength in night breaks. A flash of red light during the dark period induces flowering in an LDP, and the effect is reversed by a flash of far-red light. This response indicates the involvement of phytochrome. In SDPs, a flash of red light prevents flowering, and the effect is reversed by a flash of far-red light.

trum of Pr (Vince-Prue and Lumsden 1987). These action spectra and the reversibility between red light and far-red light confirm the role of phytochrome as the photoreceptor that is involved in photoperiod measurement in SDPs. In LDPs the role of phytochrome is more complex, and a blue-light photoreceptor (which we will discuss shortly) also plays a role in controlling flowering.

Far-Red Light Modifies Flowering in Some LDPs

Circadian rhythms have also been found in LDPs. A circadian periodicity in the promotion of flowering by far-

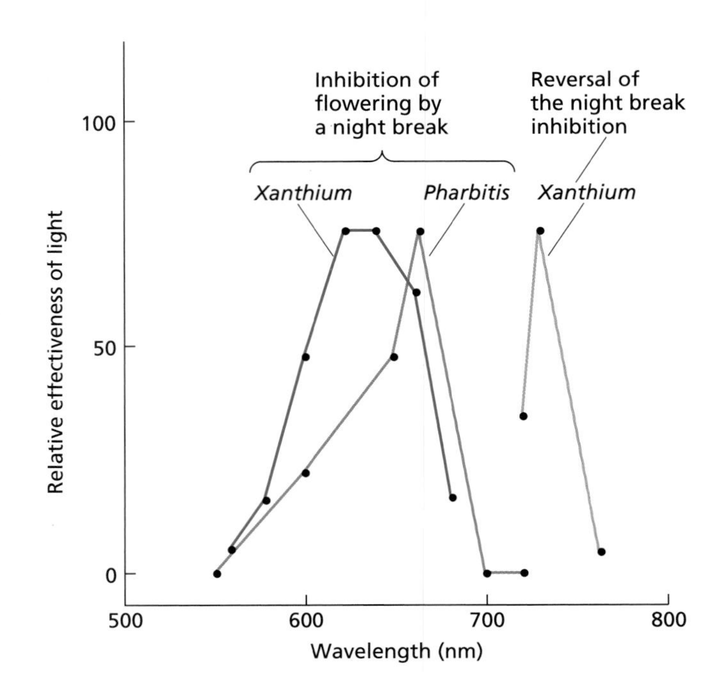

Figure 24.17 Action spectra for the control of flowering by night breaks. Flowering in SDPs is inhibited by a short light treatment (night break) given in an otherwise inductive period. In the SDP *Xanthium strumarium*, red-light night breaks of 620 to 640 nm are the most effective. Reversal of the red-light effect is maximal at 725 nm. In the dark-grown SDP *Pharbitis nil*, which is devoid of chlorophyll, night breaks of 660 nm are the most effective. This maximum coincides with the absorption maximum of phytochrome. (Data for *Xanthium* from Hendricks and Siegelman 1967. Data for *Pharbitis* from Saji et al. 1983.)

BOX 24.1

Genes That Control Flowering Time

AS WE HAVE SEEN, the Maryland Mammoth mutation converts a day-neutral variety of tobacco into an SDP. Analysis of this **flowering-time mutant** by W. Garner and H. Allard led to the discovery that plants can measure photoperiod as an environmental cue for floral induction, and it identified the Maryland Mammoth gene as one of the genes that regulates this process.

The most extensive studies of **flowering-time genes** have been performed in pea (*Pisum sativum*) and *Arabidopsis thaliana* (Weigel 1995; Reid et al. 1996). In these species a range of mutations can alter various aspects of the photoperiodic control of flowering. For example, mutations in certain genes convert these LDPs into day-neutral plants that can flower rapidly. Mutations in other genes result in day-neutral plants that exhibit a long delay in flowering. Because these mutations eliminate the response to photoperiod, the normal (wild-type) products of these genes are thought to be involved in photoperiodism. Other mutations can delay or promote flowering without affecting the photoperiodic response. These mutants are thought to affect genes involved in the *autonomous* regulation of flowering.

Since most biological regulatory systems are under both positive and negative control, it is not surprising to find mutations that promote or delay flowering. Inasmuch as most mutations result in a loss or reduction of gene activity, the wild-type role of genes that delay flowering when mutated is to promote flowering. Similarly, mutations that promote flowering are likely to occur in genes that normally inhibit the process. These genes are promoters and inhibitors of flowering in genetic terms, but the mutant phenotype alone does not usually reveal the flowering mechanism that is affected. For example, a mutation that eliminates the response to photoperiod could either block the production of a flowering hormone (floral stimulus) or, alternatively, interfere with the ability of the target cells in the meristem to receive the hormonal signal (meristem competence).

Arabidopsis plants containing mutations that promote or delay flowering have been isolated. Part A of the figure shows the wild type and a *gigantea* (*gi*) mutant that were grown in inductive photoperiods (long days). The plants are shown at the stage when flower buds have become visible. Flowering in the *gigantea* mutant is delayed; thus this plant has been growing approximately three times as long as the wild type and has formed many more leaves from the primary apical meristem. Because a mutation in the *GIGANTEA* gene delays flowering, the normal role of this gene is to promote flowering. Part B of the figure shows the wild type and an *elf3* mutant plant that were grown in noninductive photoperiods (short days). The mutant phenotype indicates that the normal role of the *ELF3* gene is to repress flowering.

One way to explore the role of these genes is through grafting studies. For example, if the apical bud from a mutant plant with a delayed-flowering phenotype flowers normally when grafted onto a wild-type shoot, the mutant shoot apex is clearly competent to receive and respond to flowering signals. One might then suspect that the mutation affects the production of flowering signals by leaves. If this were the case, a wild-type shoot apex region should exhibit delayed flowering when grafted onto a mutant shoot. Grafting studies of this sort have identified two types of genes in pea: those that control the production of transmissible promoters and inhibitors of flowering, and those that control the sensitivity of the shoot apical meristem to these signals (Reid et al. 1996).

For example, the *GIGAS* gene is thought to be involved in the production of a transmissible flowering promoter. The wild-type pea exhibits a quantitative LDP phenotype. Plants homozygous for the recessive *gigas* mutation exhibit severely delayed flowering. For example, when grown under a 16-hour photoperiod, the wild type flowers after producing 16 leaves, whereas the *gigas* mutant flowers after producing 43 leaves. Delayed flowering in *gigas* plants could be caused either by the presence of inhibitors or by the absence of a floral stimulus.

To test for inhibitors, researchers grafted the shoot of an early flowering day-neutral line onto *gigas* stock. The time of flowering of the scion was unaffected by grafting to the mutant, indicating that the mutant does not produce flowering inhibitors. However, when a *gigas* shoot was grafted onto either the early-flowering or the wild-type stock, flowering was greatly accelerated. Instead of flowering after producing 40 leaves or more, the plants flowered after producing only 14 or 23 leaves, respectively (Beveridge and Murfet 1996). Because

(A From Amasino 1996, courtesy of R. Amasino; B courtesy of K. Hicks and D. R. Meeks-Wagner.)

(A) Long days

(B) Short days

BOX 24.1 *(continued)*

the *gigas* mutant shoot apex flowers readily after grafting, its ability to respond to the flowering stimulus from the leaves is not affected by the mutation. The mutation is therefore thought to affect the production of a floral stimulus.

Another approach to elucidating the floral induction process is to isolate the genes identified by mutations and to study their properties. Once the gene is cloned, the encoded protein can be overexpressed in *E. coli* and purified in sufficient quantities to study its biochemical properties and to produce specific antibodies. The antibodies can be used to determine the tissues in which the protein is produced, as well as its location within the cell. The localization may provide clues to the function of the protein. For example, a protein in the shoot apex may be involved in the reception of flowering signals, whereas a protein present in the leaves may be involved in the production of flowering signals.

The *constans* mutants of *Arabidopsis* are insensitive to day length and flower later than the wild type under long-day conditions. The *CONSTANS* gene of *Arabidopsis* has been cloned and shown to encode a protein that shows similarities to zinc finger transcription factors, suggesting that it acts as a transcriptional regulator (see Chapter 14). The wild-type *CONSTANS* gene is expressed in leaves, and the level of *CONSTANS* mRNA increases during inductive photoperiods (Putterill et al. 1995). This pattern of expression, along with the photoperiod-insensitive mutant phenotype, implicate this gene in the photoperiodic control of flowering.

red light has been observed in barley (*Hordeum vulgare*) (Deitzer et al. 1979) and in darnel ryegrass (*Lolium temulentum*) (Figure 24.18). In both cases, when the plant is exposed to far-red light for 4 to 6 hours, flowering is promoted compared with plants maintained under continuous white or red light. The rhythm continues to run in the light. In SDPs, on the other hand, an essential feature of the circadian timing mechanism seems to be that the rhythm is suspended after a few hours in continuous light and released, or restarted, upon transfer to darkness. The response to far-red light is not the only rhythmic feature in LDPs. Although relatively insensitive to a night break of only a few minutes, many LDPs can be induced to flower with a longer night break, usually of at least 1 hour. A circadian oscillation in the flowering response to such a long night break has been observed in LDPs, showing that a rhythm of responsiveness to light continues to run in darkness.

Thus, circadian rhythms that modify the flowering response in LDPs have been shown to run both in the light (promotion by far-red light) and in the dark (promotion by red or white light). However, we do not yet know how these rhythms affect the photoperiodic response.

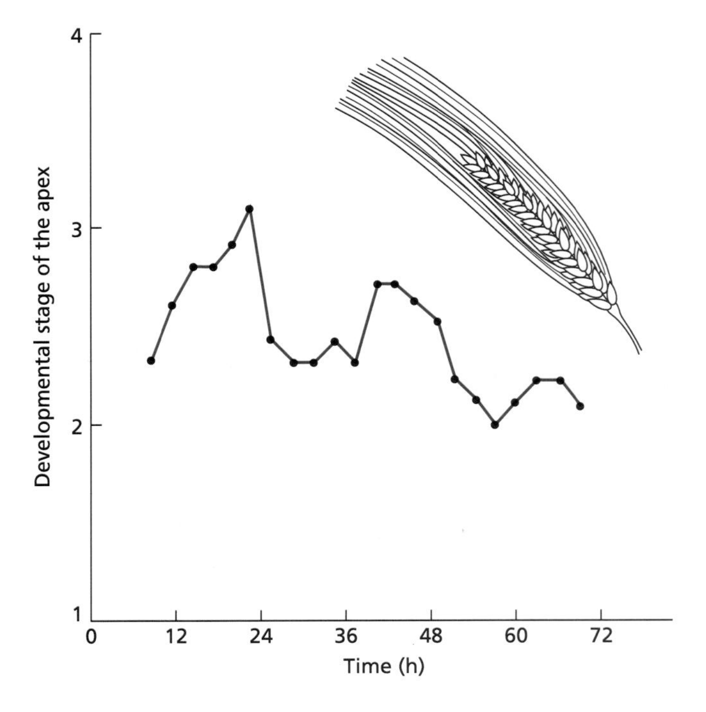

Figure 24.18 Effect of 6 hours of far-red light on floral induction in barley. Far-red light was added at 6 hours at the indicated times during a continuous 72-hour daylight period. Data points in the graph are plotted at the centers of the 6-hour treatments. Developmental stage 1 is an undetermined vegetative apex, stage 2 is a determined stage, and stages 3 and 4 are distinct morphological steps in floral development. (After Deitzer et al. 1979.)

Phytochromes A and B Have Contrasting Effects on Flowering

As noted in Chapter 17, phytochrome is encoded by members of a multigene family, and it is possible that the effects of red and far-red light described in the previous section are mediated by different types of phytochrome (A and B). In LDPs such as *Arabidopsis*, pea (*Pisum sativum*), and sorghum (*Sorghum bicolor*), phytochrome B (phyB) appears to be an *inhibitor* of flowering, since mutations in the gene *PHYB* that eliminate or reduce the amount of the phyB protein cause more rapid flowering (Childs et al. 1992; Reed et al. 1994).

The promotion of flowering by far-red light may result from a decrease in the amount of the Pfr form of phyB, which would have the same effect on the phyB Pfr levels as the mutations that reduce the total amount of phyB. As already discussed, red-light interruptions of a dark period prevent flowering in SDPs; therefore, the Pfr form of phyB may be an *inhibitor* of flowering in SDPs as well. In the LDPs *Arabidopsis* and pea, mutations in the *PHYA* gene interfere with the promotion of flowering by long days (Reed et al. 1994; Weller et al. 1997); thus the Pfr form of phyA may promote flowering in LDPs. The effect of the *phyA* mutation is quite strong in pea (the mutant plants are essentially unable to respond to inductive photoperiods) whereas the *phyA* mutation in *Arabidopsis* has only a slight effect on the ability to respond to inductive photoperiods. The species difference in the effect of *phyA* mutations is probably due to the fact that pea relies entirely on phyA to sense inductive photoperiods, whereas *Arabidopsis* also uses a blue light photoreceptor for this purpose (discussed next).

A Blue-Light Photoreceptor May Play a Role in Photoperiodism

In some LDPs, blue light can promote flowering, suggesting the possible participation of a blue-light photoreceptor in the control of flowering. The role of blue light in flowering and its relationship to circadian rhythms has been investigated by use of the luciferase reporter gene construct described earlier. In continuous white light, the cyclic luminescence has a period of 24.7 hours, but in constant darkness the period lengthens to 30 to 36 hours. Both red and blue light, given individually, shorten the period to 25 hours.

To distinguish between the effects of phytochrome and a blue-light photoreceptor, researchers transformed phytochrome-deficient *hy1* mutants, which are defective in chromophore synthesis and are therefore deficient in *all* phytochromes (see Chapter 17), with the luciferase construct to determine the effect of the mutation on the period length (Millar et al. 1995). Under continuous white light, the *hy1* plants had a period similar to that of the wild type, indicating that little or no phytochrome is required for white light to affect the period. Furthermore, under continuous red light, which would be perceived only by phyB (see Chapter 17), the period of *hy1* was significantly lengthened (i.e., became more like constant darkness), whereas the period was not lengthened by continuous blue light. These results indicate that both phytochrome and a blue-light photoreceptor are involved in period control.

The role of blue light in regulating both circadian rhythmicity and flowering is also supported by studies with an *Arabidopsis* flowering-time mutant: ***elf3*** (**e**arly **f**lowering *3*) (see Box 24.1) (Hicks et al. 1996; Zagotta et al. 1996). The *elf3* mutant flowers earlier than the wild type under a variety of growth conditions and is insensitive to photoperiod; that is, it is day-neutral. Transformation of *elf3* with the luciferase reporter construct revealed a defect in circadian rhythmicity. The mutant displayed no circadian rhythmicity when grown in continuous light. Significantly, the *elf3* mutant also has a long hypocotyl when grown in constant white light, similar to the *hy* mutants, and is less responsive than the wild type to all wavelengths of light, but especially to blue light. This finding suggests that the circadian clock is intact in the mutant, but that the transduction of light signals to the clock is impaired (Hicks et al. 1996; Zagotta et al. 1996). The fact that the *elf3* mutant is more impaired in its response to blue light than in its response to red light further suggests that a blue-light photoreceptor plays a role in the photoperiodic control of flowering.

Confirmation that a blue light photoreceptor is involved in sensing inductive photoperiods in *Arabidopsis* was recently provided by experiments demonstrating that mutations in one of the cryptochrome genes, *CRY2* (see Chapter 18), caused a delay in flowering and an inability to perceive inductive photoperiods (Guo et al. 1998). As discussed in Chapter 18, *CRY1* encodes a possible blue light photoreceptor controlling seedling growth in *Arabidopsis*. Thus, various *CRY* family members have, through evolution, become specialized for different functions in the plant.

Vernalization: Promoting Flowering with Cold Temperature

Vernalization is the process whereby flowering is promoted by a cold treatment given to an imbibed (fully hydrated) seed or to a growing plant. Dry seeds do not respond to the cold treatment. Without the cold treatment, vernalization-requiring plants show delayed flowering or remain vegetative. In many cases these plants grow as rosettes with no elongation of the stem (Figure 24.19).

In this section we will examine some of the characteristics of the cold requirement for flowering, including the range and duration of the inductive temperatures, the sites of perception, the relationship to photoperiodism, and a possible molecular mechanism.

The Shoot Apex Is the Site of Perception of the Vernalization Stimulus

Plants differ considerably in the age at which they become sensitive to vernalization. Winter annuals, such as the winter forms of cereals (which are sown in the fall and flower in the following summer), respond to low temperature very early in their life cycle. They can be vernalized before germination if the seeds have imbibed water and become metabolically active. Other plants, including most biennials (which grow as rosettes during the first season after sowing and flower in the fol-

Figure 24.19 The control of flowering in the SDP *Campanula medium* (Canterbury bell). When grown in continuous long days (LD), the plant grows as a rosette and the stem does not elongate. Eight weeks of short days (SD) followed by long days results in both elongation and flowering. The short days can be substituted by 8 weeks of low temperature. Application of GA_3 results in stem elongation but not in flowering induction. (Photo from Wellensiek 1985.)

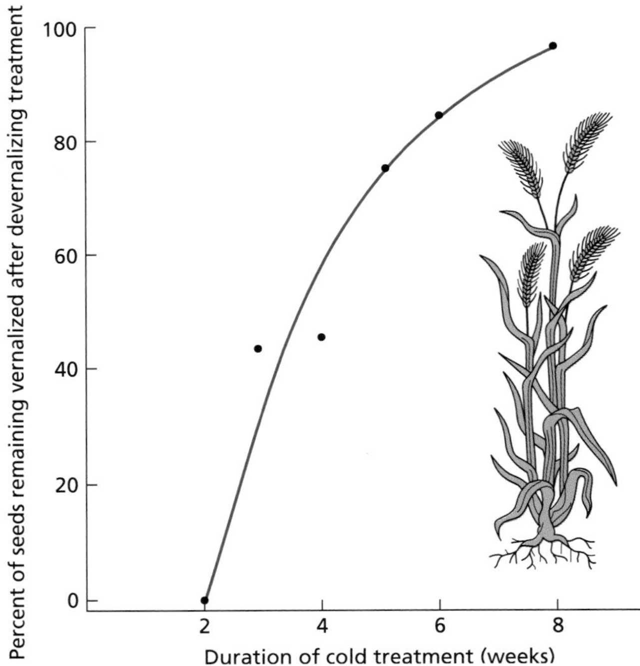

Figure 24.20 The duration of exposure to low temperature affects the stability of the vernalization effect. The longer that winter rye (*Secale cereale*) is exposed to a vernalizing cold treatment, the greater the number of plants that remain vernalized when the cold treatment is followed by a devernalizing treatment. In this experiment, seeds of rye that had imbibed water were exposed to 5°C for different lengths of time, then immediately given a devernalizing treatment of 3 days at 35°C. (Data from Purvis and Gregory 1952.)

lowing summer), must reach a minimal size before they become sensitive to low temperature for vernalization.

The effective temperature range for vernalization is from just below freezing to about 10°C, with a broad optimum usually between about 1 and 7°C (Lang 1965). The effect of cold increases with the duration of the cold treatment until the response is saturated. The response usually requires several weeks of exposure to low temperature, but the precise duration varies widely with species and variety. Vernalization can be lost as a result of exposure to devernalizing conditions, such as high temperature (Figure 24.20), but the longer the exposure to low temperature, the more permanent the vernalization effect.

Vernalization appears to take place primarily in the meristematic zones of the shoot apex (Thomas and Vince-Prue 1984). Localized cooling causes flowering when only the stem apex is chilled, and this effect appears to be largely independent of the temperature experienced by the rest of the plant. Excised shoot tips have been successfully vernalized, and where seed vernalization is possible, fragments of embryos consisting essentially of the shoot tip are sensitive to low temperature.

Photoperiodism and Vernalization May Interact

A vernalization requirement is often linked with a requirement for a particular photoperiod (Napp-Zinn 1984). The most common combination is a requirement for cold treatment *followed* by a requirement for long days—a combination that leads to flowering in early summer at high latitudes. Examples of plants that exhibit this behavior are winter wheat and the biennial form of *Hyoscyamus niger* (henbane). In biennial *H. niger*, prior vernalization renders plants competent to respond to long-day inductive photoperiods for up to 300 days after the vernalization treatment is complete. In this case, vernalization may be necessary for the meristem to become competent to respond to flowering signals produced in long days, because shoot tips of vernalized plants will flower when grafted onto nonvernalized stock plants (Lang 1985).

A different type of behavior is exhibited in *Campanula medium* (Canterbury bell): Short days acting in the leaf can completely substitute for vernalization at the apex (see Figure 24.19). In this case vernalization and photoperiod appear to act as alternative pathways to flowering. Many plant species can substitute short days for vernalization, but the relationship between the two processes is not known. Vernalization of the meristem may cause that meristem to produce leaves capable of making a flowering signal without exposure to inductive photoperiods. Alternatively, vernalization may cause flowers to be initiated at the apex without the product(s) of photoperiodic induction in the leaves.

Vernalization Has Been Correlated with DNA Demethylation

Much remains to be learned about the metabolic processes associated with vernalization. Both sugars and oxygen are required, so it appears unlikely that low temperatures simply arrest some metabolic reactions that inhibit flowering. The sugar and oxygen requirements suggest activation of an aerobic metabolic reaction that is essential for flowering. On the other hand, the rates of most metabolic reactions *decrease* with temperature, so the induction of flowering by low temperatures is likely to be more complex than the simple activation of a metabolic pathway.

The vernalized state of the shoot apex is stable through many rounds of cell division. As discussed in the previous section, *Hyoscyamus niger* remains vernalized for at least 300 days after the end of a cold treatment (Lang 1985). (Investigators determined the stability of the vernalized state by maintaining vernalized plants in short days for various lengths of time before transferring them to inductive long days.) The susceptibility of dividing cells to vernalization and the stability of the vernalized state suggest that cold treatment causes a permanent change in the pattern of gene expression.

It has been proposed that changes in the pattern of DNA methylation can account for altered gene expression induced by vernalization. For example, late-flowering ecotypes of *Arabidopsis* flower early if vernalized. If nonvernalized plants are treated with the DNA-demethylating agent 5-azacytidine, they flower significantly earlier than untreated controls (Burn et al. 1993). *Arabidopsis* lines that do not require vernalization were found not to respond to 5-azacytidine. Moreover, both 5-azacytidine and vernalization caused DNA demethylation in the late-flowering ecotype (Burn et al. 1993). Thus, genes that are needed for early flowering could be blocked by DNA methylation in the late-flowering *Arabidopsis* ecotypes.

Since gibberellins can substitute for the vernalization requirement (as we'll see in the discussion that follows), perhaps a gene that encodes an enzyme in the gibberellin biosynthetic pathway is demethylated during vernalization. Although the demethylation model of vernalization is considered controversial, like all good hypotheses it provides a framework for thinking about the problem and suggests additional experiments in which it can be tested.

Biochemical Signaling Involved in Flowering

In the preceding sections we examined the influence of environmental conditions, such as temperature and day length, versus that of autonomous factors, such as age, on flowering. Although the morphological changes associated with floral evocation occur at the apical meristems of the shoots, the events that result in the morphological changes are triggered by biochemical signals arriving at the apex from other parts of the plant, especially from the leaves. Mutants have been isolated (e.g., *gigas* in pea) that are deficient in the floral stimulus (see Box 24.1).

In this section we will consider the nature of the biochemical signals arriving from the leaves and other parts of the plant in response to photoperiodic stimuli. Such signals may serve either as activators or as inhibitors of flowering. After years of investigation, no single substance has been identified as the universal floral stimulus, although certain hormones, such as gibberellins and ethylene, can induce flowering in some species. Hence, current models of the floral stimulus are based on multiple factors.

Identification of the Hypothetical Florigen Remains Elusive

Production by leaf cells of a biochemical signal that acts on a distant target tissue (the shoot apex) satisfies an important criterion for a hormonal effect. In the 1930s, Mikhail Chailakhyan, working in Russia, postulated the existence of a universal flowering hormone and called it **florigen**. The many attempts to isolate and characterize this hypothetical hormone have been largely unsuccess-

Table 24.4
Examples of the successful transfer of flowering signals by grafting in cold-requiring plants

Flowering donor plant	Receptor plants induced to flower
Hyoscyamus niger, cold-requiring, vernalized *Hyoscyamus niger*, non-cold-requiring annual form *Petunia hybrida*, non-cold-requiring annual	*H. niger*, cold-requiring, nonvernalized
Beta vulgaris, cold-requiring, vernalized *Beta vulgaris*, non-cold-requiring annual form *Beta procumbens*, non-cold-requiring annual	*B. vulgaris*, cold-requiring, nonvernalized

ful. The most common approach has been to make extracts from induced leaf tissue and test for their ability to elicit flowering in noninduced plants. In other experiments, investigators have extracted and analyzed phloem sap from induced plants. In some studies, extracts from one of these sources have induced flowering in test plants, but these results have not been consistently reproduced (Zeevaart and Boyer 1987). Attempts to isolate a specific, graft-transmissible inhibitor of flowering have also been unsuccessful. Thus, despite unequivocal data showing that transmissible factors regulate flowering (see Table 24.3) (Lang et al. 1977; Zeevaart 1984), the substances involved are yet to be characterized.

Table 24.5
Examples of flowering induction by gibberellins in plants with different environmental requirements for flowering

Flowering requirement	Plant	Effect of gibberellins
Long-day plants	*Arabidopsis*	Promotes in SD
	Lolium	Promotes in SD
	Fuchsia	Inhibits in LD
	Anagallis	No effect
Short-day plants	*Zinnia*	Promotes in LD
	Fragaria	Inhibits in SD
	Xanthium	No effect
Dual–day length plants	*Bryophyllum* (LSDP)	Promotes in SD
	Coreopsis (SLDP)	Promotes in SD
	Cestrum (LSDP)	Inhibits
Day-neutral plants	Many conifers	Promotes
	Many woody angiosperms	Inhibits
Plants requiring vernalization	*Daucus*	Promotes
	Oenothera	No effect

Note: LSDP, long-short-day-plants; SLDP, short-long-day plants; LD, long days; SD, short days.

Grafting experiments have also been done to investigate possible signals from vernalized plants (Table 24.4). Cold-requiring, nonvernalized plants have been induced to flower when grafted with a vernalized donor, pointing to a transmissible stimulus. Whether the transmitted message is a specific compound involved in vernalization or is the same substance that induces flowering in plants that do not require vernalization remains to be determined (Thomas and Vince-Prue 1984).

Gibberellins Can Induce Flowering in Some Plants

Among the naturally occurring growth hormones, gibberellins (GAs) (see Chapter 20) can have a strong influence on flowering (Table 24.5). Exogenous gibberellin can substitute for photoperiodic induction when applied to long-day plants that grow as rosettes in short days (Lang 1965; Vince-Prue 1985). In these plants, the flowering response (either to GAs or to long days) is accompanied by elongation of the flowering stem. However, it is important to note that flower formation and stem elongation are independent processes (Zeevaart 1984). In addition, application of GAs can evoke flowering in a few short-day plants in noninductive conditions and can substitute partly or completely for a low-temperature signal in several cold-requiring plants. As previously discussed, cone formation can also be promoted

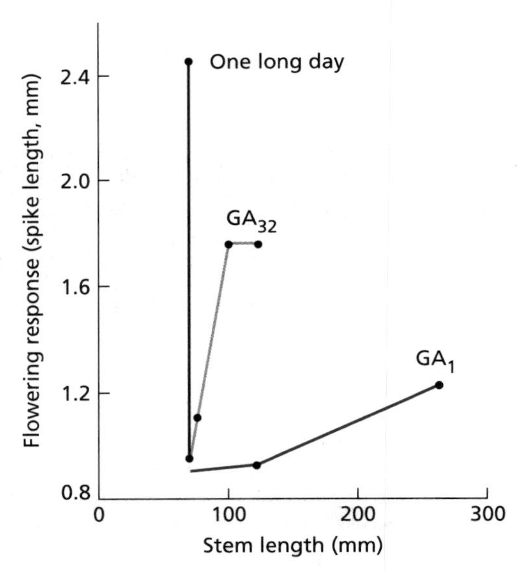

Figure 24.21 The different effects of two different gibberellins on flowering (spike length) and elongation (stem length). In the long-day plant *Lolium temulentum*, GA_{32} strongly promotes flowering, but has only a small effect on stem elongation. In contrast, GA_1 strongly promotes stem growth but has only a small effect on flowering. Gibberellin-treated plants were maintained in short days and, for comparison, the response to a single 24-hour day is shown. This long day strongly promotes flowering but has almost no effect on stem length. (Data from Pharis et al. 1987.)

in juvenile plants of several gymnosperm families by addition of GAs (Pharis and King 1985).

Thus, exogenous GAs can sometimes substitute for the endogenous trigger of age in autonomous flowering and for the primary environmental signals of day length and low temperature. A mutation in GA biosynthesis renders the quantitative LDP *Arabidopsis thaliana* unable to flower in noninductive short days but has little effect on flowering in long days, demonstrating that endogenous GA is required for flowering in specific situations (Wilson et al. 1992).

As discussed in Chapter 20, plants contain many GA-like compounds. Many of these compounds may be precursors to, or inactive metabolites of, the active forms of GA. In some situations different GAs have markedly different effects on flowering and stem elongation, such as in the long-day plant *Lolium temulentum* (darnel ryegrass) (Figure 24.21). These observations suggest that the regulation of flowering may be associated with specific GAs.

Considerable attention has been given to the effects of day length on GA metabolism in the plant (see Chapter 20). For example, in the long-day plant spinach (*Spinacia oleracea*), the levels of gibberellins are relatively low in short days, and the plants maintain a rosette form. After the plants are transferred to long days, the levels of all the gibberellins of the 13-hydroxylated pathway ($GA_{53} \rightarrow GA_{44} \rightarrow GA_{19} \rightarrow GA_{20} \rightarrow GA_1$) increase. However, the fivefold increase in the physiologically active gibberellin, GA_1, is what causes the marked stem elongation that accompanies flowering (see Figure 20.9). Whether other GAs specifically affect flower formation has not yet been determined.

In addition to GAs, other growth hormones can either inhibit or promote flowering (Vince-Prue 1985). One commercially important example is the striking promotion of flowering in pineapple (*Ananas comosus*) by ethylene and ethylene-releasing substances, a response that appears to be restricted to members of the pineapple family (Bromeliaceae).

The Floral Stimulus May Have Several Components

Ultimately, floral primordia form and develop at the shoot apex as a result of biochemical and cellular changes leading to the production of the floral organs. The transition to flowering may involve a complex system of interacting factors that include, among others, carbohydrates, gibberellins, and cytokinins (Bernier 1988). Sometimes changes normally associated with flowering can be caused by treatments that do not themselves bring about floral initiation. For example, one of the earliest events observed at the shoot apex following photoperiodic induction is a transitory increase in the number of cells undergoing mitosis. When cytokinins are applied to the stem apex of the long-day plant white mustard (*Sinapis alba*), they cause an increase in mitotic activity similar to that caused by exposure to a single long day. However, whereas the long day induces flowering, the cytokinin treatment does not. Cytokinin may therefore be a component of the floral stimulus, although it is insufficient by itself to bring about flowering.

(A) $^{+}H_3N—(CH_2)_4—NH_3^{+}$

Putrescine

(B) $^{+}H_3N—(CH_2)_3—\overset{+}{N}H_2—(CH_2)_4—NH_3^{+}$

Spermidine

(C) $^{+}H_3N—(CH_2)_3—\overset{+}{N}H_2—(CH_2)_4—\overset{+}{N}H_2—(CH_2)_3—NH_3^{+}$

Spermine

Figure 24.22 Polyamine structures. (A) Putrescine. (B) Spermidine. (C) Spermine. At neutral pH, polyamines are polycations and can bind to polyanions in the cell, such as DNA, RNA, and phospholipids. The binding of polyamines to nucleic acids and membranes may stabilize these macromolecules. In addition to playing a role in the control of flowering, polyamines have been implicated in the regulation of embryogenesis and senescence.

Other potential components include the **polyamines** (Figure 24.22). For example, photoperiodic induction of the LDP *Sinapis alba* was correlated with a large increase in the concentration of putrescine, the major polyamine, in the phloem exudate from the leaves. Spraying the leaves with an inhibitor of putrescine biosynthesis lowered the level of putrescine in the phloem exudate and also inhibited flowering. This result suggests that putrescine is a component of the floral stimulus in *S. alba* (Havelange et al. 1996).

In addition to receiving signals from the leaves, the shoot apex may accumulate flowering signals from the roots. Classic experiments with cuttings have shown that the initiation and growth of adventitious roots inhibit flowering in the shoot. In many species, root removal promotes flowering. Thus, at least in certain species, the root appears to produce one or more flowering inhibitors. In other cases, the roots may contribute floral activators. For example, both GA and cytokinin are synthesized in the root and transported to the shoot via the xylem. During photoperiodic induction in *Sinapis alba*, the amount of cytokinin present in the xylem exudate from the root increases considerably (Lejeune et al. 1994). This finding indicates that signals emanating from the leaf are transported both to the shoot apex and to the root during photoperiodic induction. In response to the signal from the leaves, the roots produce and export additional chemical signals that modulate the flowering response.

If the floral stimulus does have multiple components, the limiting component may vary among species. Moreover, exposure to a high level of one component, such as a particular gibberellin, may compensate for the lack of a limiting component and induce flowering in its absence. Such ability to compensate could explain the wide range of substances and conditions that can cause flowering. However, many physiological experiments show that the stimulus exported from an induced leaf has some specific properties. Establishing the biochemical basis of flower induction remains a major challenge.

Floral Meristems and Floral Organ Development

Floral meristems usually can be distinguished from vegetative meristems, even in the early stages of reproductive development, by their larger size. The transition from vegetative to reproductive development is marked by an increase in the frequency of cell divisions within the central zone of the shoot apical meristem. In the vegetative meristem, the cells of the central zone complete their division cycles slowly. As reproductive development commences, the increase in the size of the meristem is largely a result of the increased division rate of these central cells. Recently, genetic and molecular studies have identified a network of genes that control floral morphogenesis in *Arabidopsis*, snapdragon (*Antirrhinum*), and other species.

In this section we will focus on floral development in *Arabidopsis*, which has been studied extensively. First we will outline the basic morphological changes that occur during the transition from the vegetative to the reproductive phase. Next we will consider the arrangement of the floral organs in four whorls on the meristem, and the types of genes that govern the normal pattern of floral development. According to the widely accepted ABC model, the specific locations of floral organs in the flower are regulated by the overlapping expression of three types of floral organ identity genes. Although floral development is genetically controlled, it is nevertheless sensitive to environmental and hormonal signals, although perhaps not to the same extent as vegetative development is. We end the chapter by discussing the roles of light and of gibberellin and other hormones in modifying the genetic program for floral morphogenesis.

The Characteristics of Shoot Meristems in *Arabidopsis* Change with Development

During the vegetative phase of growth in the first 25 days after germination, the wild-type *Arabidopsis* vegetative apical meristem produces about 12 phytomeres with very short internodes, resulting in a basal rosette of leaves (Figure 24.23). (Recall from Chapter 16 that a phytomere consists of a leaf, the node to which the leaf is attached, the axillary bud, and the internode below the node.)

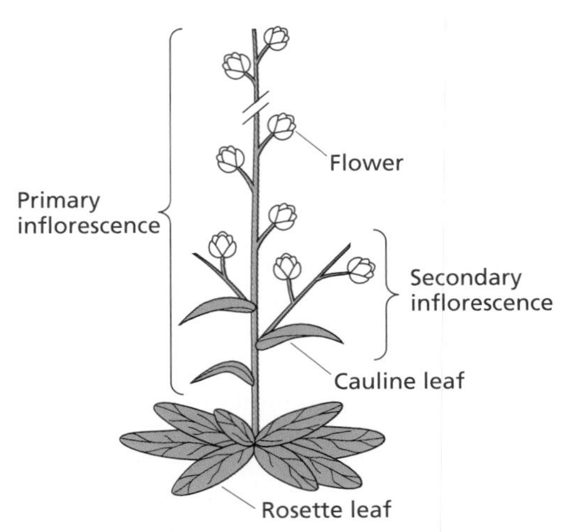

Figure 24.23 The shoot apical meristem in *Arabidopsis* generates different organs at different stages of development, as illustrated in this schematic drawing. Early in development the shoot apical meristem forms a rosette of basal leaves. When the plant makes the transition to flowering, the shoot apical meristem is transformed into a primary inflorescence meristem that directly produces floral buds (see Figure 24.1). Leaf primordia initiated prior to the floral transition become cauline leaves (inflorescence leaves) and secondary inflorescences develop in the axils of the cauline leaves. (From Okamuro et al. 1993, © American Society of Plant Physiologists, reprinted with permission.)

As plants initiate reproductive development, the vegetative meristem is transformed into an indeterminate inflorescence meristem and no longer produces vegetative phytomeres. Instead, the primordia formed on the flanks of the inflorescence meristem develop into flowers as illustrated in Figure 24.1. The leaf primordia that had been initiated before the floral transition of the meristem develop on the elongating flowering stalk and are called the cauline leaves (Hempel and Feldman 1994). Cauline leaves are smaller than rosette leaves and similar in size and structure to the bracts that subtend many angiosperm flowers. However, *Arabidopsis* flowers lack bracts. The lateral buds of the cauline leaves develop into secondary inflorescence meristems, and their activity repeats the pattern of the development of the primary inflorescence meristem.

The Four Different Types of Floral Organs Are Initiated as Separate Whorls

Floral meristems initiate four different types of floral organs: sepals, petals, stamens, and carpels (Coen and Carpenter 1993). Each set of organs is initiated as a **whorl**—that is, arranged in concentric rings around the flanks of the meristem (Figure 24.24). By the time the carpels are initiated, there are no more meristematic cells in the apical dome, and only the floral organ primordia are present. In the wild-type *Arabidopsis* flower, the first (outermost) whorl consists of four sepals, which are green at maturity. The second whorl is composed of four

Figure 24.24 The floral organs are initiated sequentially by the floral meristem. This diagram of a longitudinal section through a developing dicot flower shows that the floral organs are produced as successive whorls (circles), starting with the sepals and progressing inward. The remainder of the apical meristem, enclosed by the carpels, will form a placenta on which the ovules develop. (After Sussex 1989.)

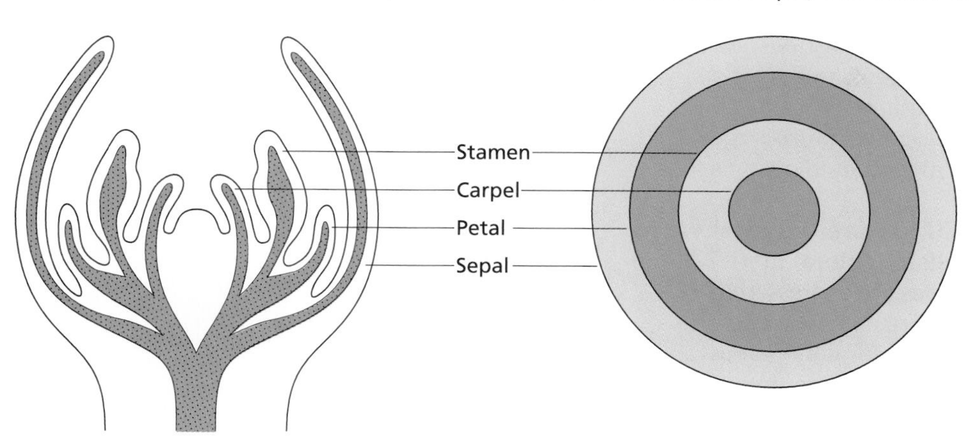

petals, which are white at maturity. The third whorl contains six stamens, two of which are shorter than the other four. Finally, what is considered to be the fourth whorl is a single complex organ, the gynoecium or pistil, which is composed of an ovary with two fused carpels, each containing numerous ovules, and a short style capped with a stigma (Figure 24.25).

Three Types of Genes Regulate Floral Development

Mutations have identified three classes of genes that regulate floral development (Coen 1991). One group of genes directly controls **floral organ identity**. The proteins encoded by these genes are transcription factors that likely control the expression of other genes whose products are involved in the formation and/or function of these organs. Genes of the second group act as spatial regulators of the floral organ identity genes by setting boundaries for their expression. These have been called **cadastral genes** (from the word "cadastre," which is a map or survey showing property boundaries for taxation purposes). The third group of genes is necessary for the initial induction of the organ identity genes. These genes, which are the positive regulators of floral organ identity, are called **meristem identity genes**.

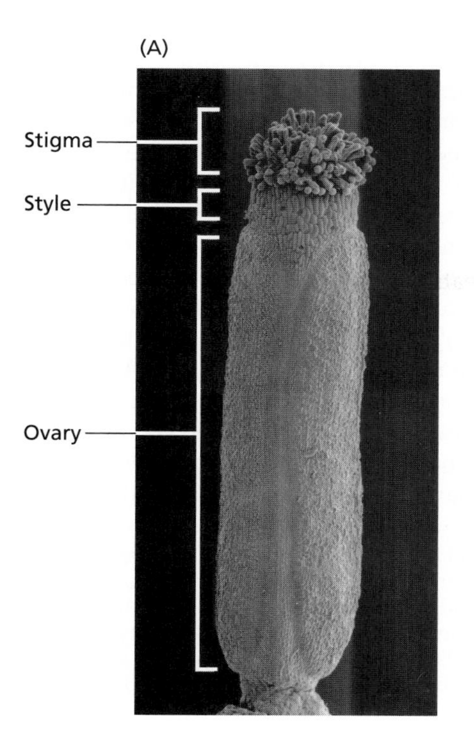

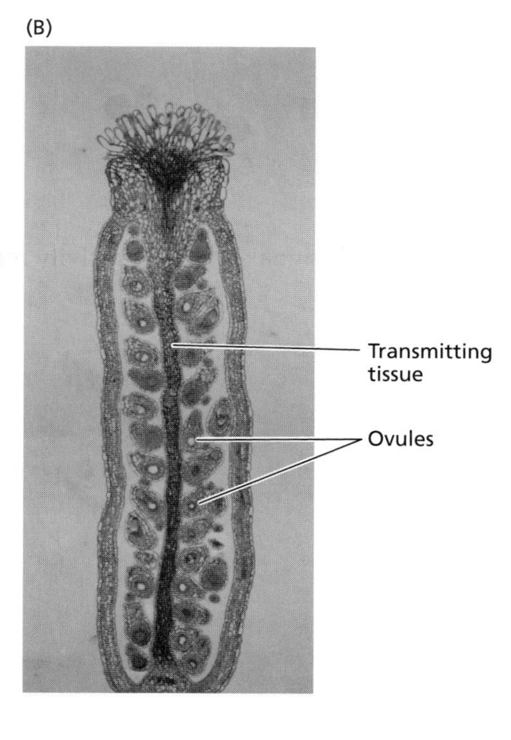

Figure 24.25 The *Arabidopsis* pistil consists of two fused carpels, each containing many ovules. (A) Scanning electron micrograph of a pistil, showing the stigma, a short style, and the ovary. (B) Longitudinal section through the pistil, showing the many ovules attached to the transmitting tissue through which the pollen tubes grow to reach the embryo sac inside the ovules. (From Gasser and Robinson-Beers 1993, courtesy of C. S. Gasser, © American Society of Plant Physiologists, reprinted with permission.)

Meristem Identity Genes Have Been Identified in *Antirrhinum* and *Arabidopsis*

Meristem identity genes must be active for floral meristems to be formed by inflorescence or vegetative meristems. For example, the *floricaula* mutant of *Antirrhinum* (snapdragon) develops an inflorescence but does not produce flowers. Instead of floral meristems in the axils of the bracts, it develops additional inflorescence meristems. The wild-type *FLORICAULA* gene controls the determination step in which floral meristem identity is established.

APETALA1 (*AP1*) and *LEAFY* (*LFY*) are the most important genes involved in establishing floral meristem identity in *Arabidopsis*. These genes are expressed in floral meristems and appear to be part of a genetic pathway that must be activated to establish floral meristem identity. Floral induction triggers the expression first of *LFY* and then of *AP1* (Simon et al. 1996). Recent studies suggest that gibberellin promotes flowering in *Arabidopsis* by activating the expression of the *LEAFY* gene (Blázquez et al. 1998). This suggests that GAs stimulate flowering in some species through a pathway that regulates *LFY* transcription.

In addition to being controlled by positive regulators of floral meristem identity, floral development is subject to negative regulation. An *Arabidopsis* gene designated *EMBRYONIC FLOWER* (*EMF*) has been identified with a mutant phenotype in which plants flower almost immediately after germination (Figure 24.26) (Sung et al. 1992). The shoot apical meristem initiates no vegetative structures in the *emf* mutant, but immediately initiates floral organs as germination proceeds. While the *EMF* gene could regulate the establishment of vegetative meristem identity, it also is possible that it is a negative regulator of the floral meristem identity genes. The flowers also exhibit various abnormalities, suggesting that the *EMF* gene is required for normal inflorescence development (Chen et al. 1997).

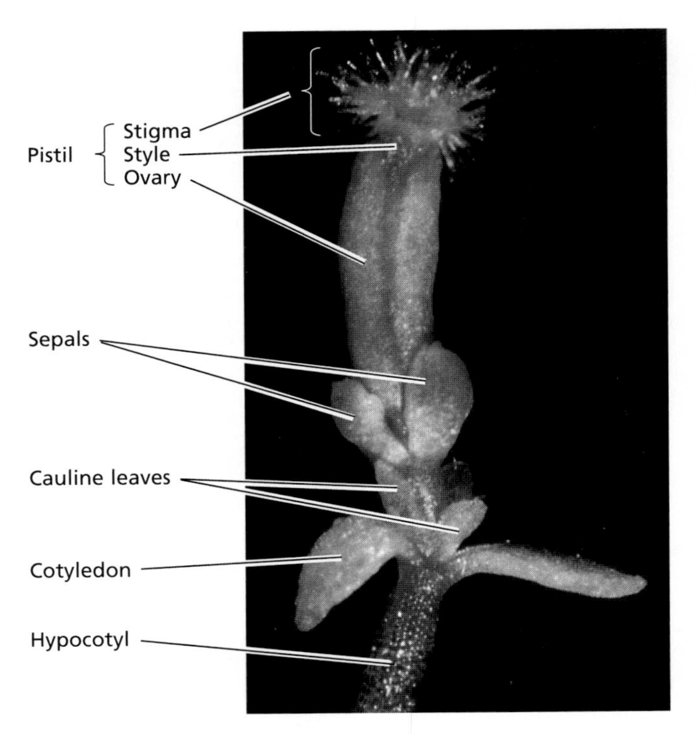

Figure 24.26 The *emf* mutant of *Arabidopsis*. The striking phenotype of this mutant is that the seedling skips the vegetative phase of growth and forms an inflorescence directly. Perhaps because of their early appearance, the flowers are usually abnormal. The hypocotyl is shown at the base with two cotyledons. A short segment of inflorescence stem with a few reduced cauline leaves has formed immediately above the cotyledons. This example of an *emf* mutant flower lacks petals and stamens. A pistil with an enlarged stigma is the most prominent floral structure. (From Sung et al. 1992, courtesy of R. Sung.)

Homeotic Mutations Led to the Identification of Floral Organ Identity Genes

The genes that determine floral organ identity were discovered as **floral homeotic mutants** (see Chapter 14). Recall that in the fruit fly *Drosophila*, homeotic mutations led to the identification of a set of homeotic genes encoding transcription factors that determine the locations at which specific structures develop. Such genes act as major developmental switches that activate the entire genetic program for a particular structure. The expression of homeotic genes thus gives organs their identity.

As we have seen in this chapter, dicot flowers consist of successive whorls of organs that form as a result of the activity of floral meristems: sepals, petals, stamens, and carpels. These organs are produced when and where they are because of the orderly, patterned expression and interactions of a small group of homeotic genes that specify floral organ identity. The floral organ identity genes were identified through homeotic mutations that altered floral organ identity so that some of the floral organs appeared in the wrong place. For example, *Arabidopsis* plants with mutations in the *APETALA2* (*AP2*) gene produce flowers with carpels where sepals should be and stamens where petals normally appear.

The homeotic genes that have been cloned so far encode transcription factors—proteins that control the expression of other genes. Most plant homeotic genes belong to a class of related genes known as **MADS box genes**.* In animals, however, most homeotic genes, which control segment identity, belong to a class of transcription factor–encoding genes known as *homeobox genes*. Plants also have homeobox genes, but at this point

* The acronym "MADS" is derived from the first letter of the first five members of this class of genes: the yeast gene **M***CM-1*, which encodes a transcription factor necessary for mating-type determination, the floral organ identity genes **A***GAMOUS* and **D***EFICIENS*, and the mammalian serum response factor, found to induce the transcription of *fos* proto-oncogenes (see Chapter 14).

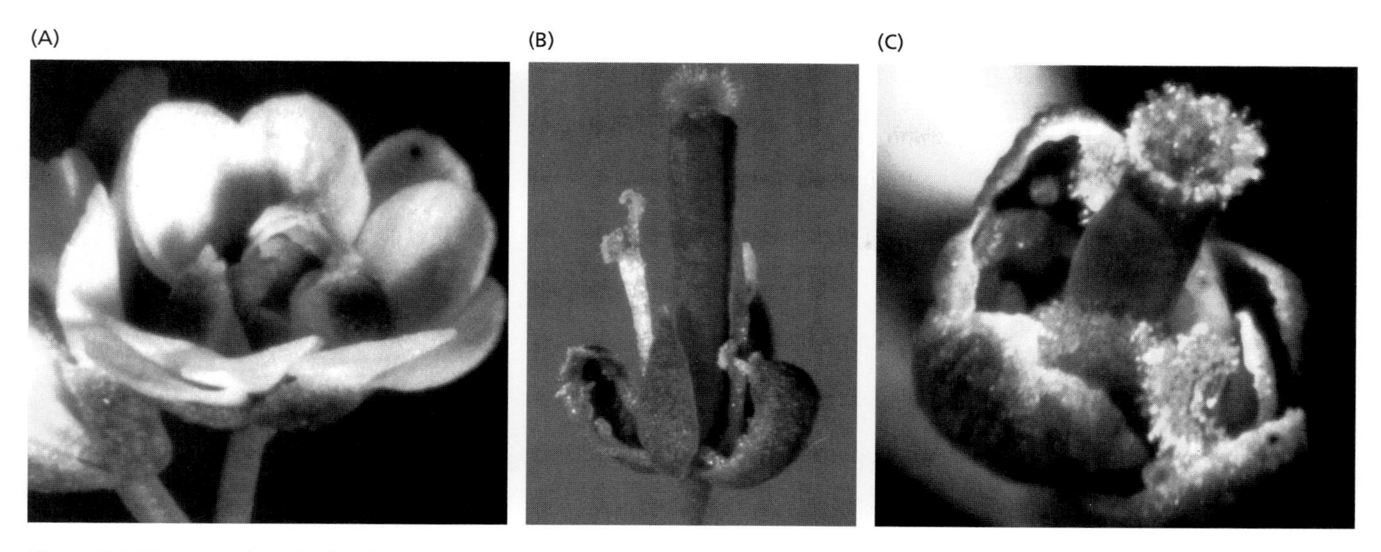

Figure 24.27 Mutations in the floral organ identity genes dramatically alter the structure of the flower. (A) Flowers formed in plants that are homozygous for an *agamous* mutation lack both stamens and petals. (B) Flowers of plants with the *apetala2* mutation lack sepals and petals. (C) Flowers of plants with the *apetala3* mutation lack petals and stamens. (Courtesy of E. Meyerowitz.)

none of these genes are known to have homeotic function. For example, as discussed in Chapter 16, both *KN1* and *STM* genes are homeobox genes. Although they are important in plant development, most plant biologists do not consider them to be homeotic genes.

Many of the genes that determine floral organ identity are MADS box genes, including the *DEFICIENS* gene of snapdragon and the *AGAMOUS, APETALA1,* and *APETALA3* genes of *Arabidopsis*. Both homeobox and MADS box genes encode proteins that function as transcription factors, but they differ in structure. The MADS box genes share a characteristic, conserved nucleotide sequence known as a MADS box, which encodes a protein structure known as the MADS domain. The MADS domain enables these transcription factors to bind to DNA that has a specific nucleotide sequence.

Three Types of Homeotic Genes Control Floral Organ Identity

Five different genes are known to specify floral organ identity in *Arabidopsis*: *APETALA1* (*AP1*), *APETALA2* (*AP2*), *APETALA3* (*AP3*), *PISTILLATA* (*PI*), and *AGAMOUS* (*AG*) (Bowman et al. 1989; Weigel and Meyerowitz 1994). The organ identity genes initially were identified through mutations that alter the identity of the floral organs produced in two adjacent whorls (Figure 24.27). Plants bearing *ap3* mutations, for example, produce sepals instead of petals in the second whorl, and carpels instead of stamens in the third whorl. Since mutations in these genes change floral organ identity without affecting the initiation of flowers, they are homeotic genes (see Box 24.1). These homeotic genes fall into three classes—types A, B, and C—defining three different kinds of activities.

Type A activity controls organ identity in the first and second whorls. Loss of type A activity, encoded by *AP1* and *AP2*, results in the formation of carpels in the first whorl instead of sepals, and of stamens in the second whorl instead of petals. Type B activity, encoded by *AP3* and *PI*, controls organ determination in the second and third whorls. Loss of type B activity results in the formation of sepals in the second whorl instead of petals, and of carpels instead of stamens in the third whorl. The third and fourth whorls are controlled by type C activity, encoded by *AG*. Loss of type C activity results in the formation of petals instead of stamens in the third whorl, and replacement of the fourth whorl by a new flower so that the fourth whorl of the *ag mutant* flower is occupied by sepals.

The role of the organ identity genes in floral development is dramatically illustrated by an experiment in which all three activities are eliminated by loss-of-function mutations. After floral induction, these triple-mutant plants produce floral meristems that develop as pseudoflowers; all the floral organs are replaced with green leaflike structures, although these organs are produced with a whorled phyllotaxy (Figure 24.28). Evolutionary biologists, even the eighteenth-century German philosopher Goethe, have speculated that floral organs are highly modified leaves, but this experiment gives direct support to these ideas.

Figure 24.28 Genetic demonstration that floral organs are modified leaves. (A) The wild-type *Arabidopsis* flower. (B) The flower of the ABC triple mutant (*apetala2*, *pistillata*, and *agamous*) produces only leaves. (From Weigel and Meyerowitz 1994; A courtesy of E. Meyerowitz, B courtesy of J. Bowman.)

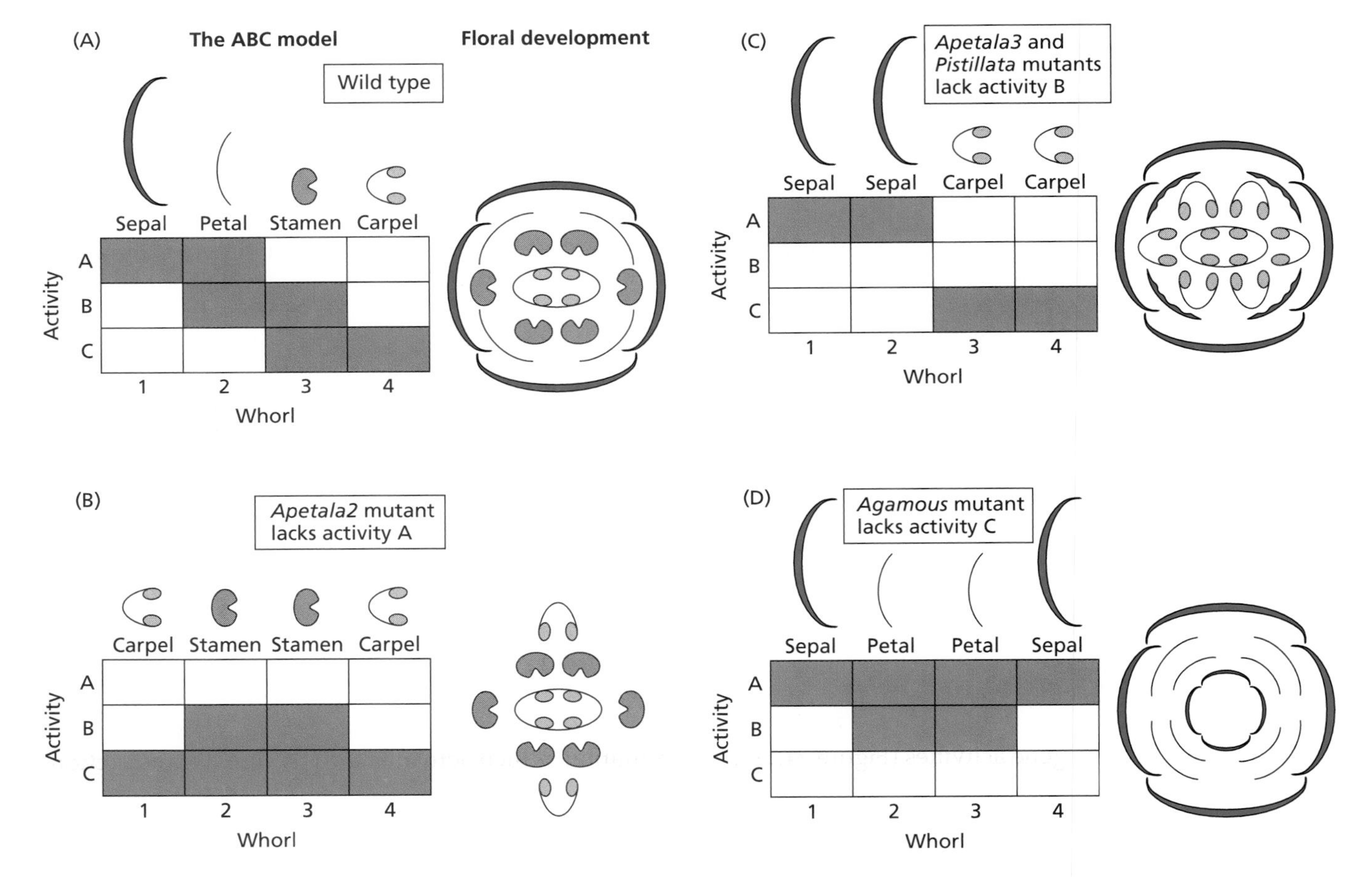

Figure 24.29 The ABC model has been proposed to explain the acquisition of floral organ identity during flower development. The structure of a mature *Arabidopsis* flower is diagrammed (right), showing the arrangement of the floral organs into four distinct whorls. The outermost whorl contains four sepals; whorl 2 consists of four petals. There are six stamens in whorl 3 and two fused carpels in whorl 4. The ABC model (left) proposes that floral organ identity is determined by three different activities: A, B, and C. (A) In the wild type, only activity A occurs in the first whorl of organs, resulting in their differentiation as sepals. As the second whorl of organs is initiated, genes of both type A and type B are expressed, resulting in petals. The combination of activities B and C in the third whorl results in the formation of stamens; activity C alone is expressed in the fourth whorl, specifying carpels. In addition, activity A represses activity C in whorls 1 and 2, while C represses A in whorls 3 and 4. (B) Deletion of the A function results in expansion of the C function throughout the floral meristem, altering the identity of the organs formed as indicated. (C) Loss of B function causes sepals to form instead of petals in the second whorl, and carpels to form instead of stamens in the third whorl. (D) Deletion of the C function results in expansion of the A function throughout the apex, again altering the identity of the organs formed. (From Weigel and Meyerowitz 1994.)

(A)

(B)

Figure 24.30 Two examples of floral reversion. (A) An illustration of a rare case of floral reversion in a rose, probably caused by a mutation in a floral homeotic gene. (B) Floral reversion in *Arabidopsis thaliana*. When *agamous* mutant plants produce flowers under short-day conditions, the floral meristems revert to inflorescence meristems and a new inflorescence grows out from the center of the flower. The response is inhibited by gibberellin. (A from Goethe 1790; B from Okamuro et al. 1996, © National Academy of Sciences, U.S.A.)

Although floral homeotic genes have been identified in several other species, the work of Enrico Coen and Zsusana Schwarz-Sommer leading to the identification of the floral organ identity genes in *Antirrhinum* has been especially important for our general understanding of flowering (Schwarz-Sommer et al. 1990; Coen et al. 1991). Four of the five *Arabidopsis* floral identity genes have close homologs in *Antirrhinum*, in terms of both gene structure and activity.

The ABC Model Explains the Determination of Floral Organ Identity

In 1991, Elliot Meyerowitz and Enrico Coen proposed a model called the **ABC model**, to explain how homeotic genes control organ identity (Coen and Meyerowitz 1991). The ABC model proposes that organ identity in each whorl is determined by a unique combination of the three organ identity gene activities (Figure 24.29A). Activity of type A alone specifies sepals; formation of petals requires both A and B activities. Stamens are formed by a combination of activities B and C, and activity C alone specifies carpels. The model further proposes that activities A and C mutually repress each other (Figure 24.29B and C); that is, both A and C type genes have cadastral function in addition to their function in determining organ identity.

The pattern of organ formation in the wild type and most of the mutant phenotypes are predicted and explained by this model. The model is supported not only by genetic evidence, but also by in situ hybridization studies that demonstrate that most of the organ identity genes are expressed in the areas predicted by their mutant phenotype. The challenge now is to understand not only how the expression of these organ identity genes, which encode transcription factors, alters the pattern of genes expressed in the developing organ, but also how this altered pattern of gene expression results in the development of a specific floral organ.

Both Photoperiod and Hormones Can Influence Floral Development

As we saw in Chapter 20, gibberellins can affect sex determination in flowers. Gibberellin has also been shown to interact with photoperiod in the regulation of meristem identity in *Arabidopsis*. For example, when *agamous* mutants or plants heterozygous for the *leafy* mutation are grown under short days, they undergo a process called floral reversion. In **floral reversion** the floral meristem reverts to an inflorescence meristem, causing an inflorescence shoot to grow directly out of the developing flower (Figure 24.30). Jack Okamuro and his colleagues at the University of California, Santa Cruz, have shown that phytochrome is the photoreceptor involved in the response and that exogenous gibberellin prevents floral reversion under short-day conditions. Floral reversion is also blocked by the *spindly* mutation, which activates the GA signal transduction pathway in the absence of the hormone (Okamuro et al. 1996, 1997) (see Chapter 20). These results indicate that both the GA and the phytochrome signal transduction pathways participate in regulating meristem identity in *Arabidopsis*.

Summary

Flower formation occurs at the shoot apical meristem and is a complex morphological event. The ability to flower (i.e., to make the transition from juvenility to maturity) is attained when the plant has reached a certain age or size. In some plants, the transition to flowering then occurs independently of the environment (autonomously).

Other plants require exposure to appropriate environmental conditions. The most common environmental inputs for flowering are day length and temperature.

The response to day length—photoperiodism—promotes flowering at a particular time of year, and several different categories of response are known. The photoperiodic signal is perceived by the leaf. Exposure to low temperature—vernalization—is required for flowering in some plants, and this requirement is often coupled with a day length requirement. Vernalization occurs at the shoot apical meristem. Photoperiodism and vernalization interact in several ways.

Daily rhythms—circadian rhythms—can locate an event at a particular time of day. Timekeeping in these rhythms is based on an endogenous circadian oscillator. Keeping the rhythm on local time depends on the phase response of the rhythm to environmental signals. The most important signals are dawn and dusk. Short-day plants flower when a critical duration of darkness is exceeded. Long-day plants flower when the length of the dark period is less than a critical value. Light given at certain times in a dark period that is longer than the critical value—a night break—prevents its effect. Light also acts on the circadian oscillator to entrain the photoperiodic rhythm, an effect that is important for timekeeping in the dark. The photoperiodic mechanism shows some variation in short-day and long-day responses, but both appear to involve phytochrome and a circadian oscillator.

When photoperiod-responsive plants are induced to flower by exposure to appropriate day lengths, leaves send a chemical signal to the apex to bring about flowering. This transmissible signal is able to cause flowering in plants of different photoperiodic response groups. In noninductive day lengths, a transmissible inhibitor of flowering may be produced by the leaves. Although physiological experiments, especially grafting, indicate the existence of a transmissible floral stimulus and, in some cases, flowering inhibitors, the chemical identity of these factors is not known. Plant growth hormones, especially the gibberellins, can modify flowering in many plants. The floral stimulus ("florigen") probably consists of a complex of chemical signals, and species may differ as to which particular substance is limiting for flowering.

The rosette plant *Arabidopsis* has been an important model for studies on floral development. The four floral organs (sepals, petals, stamens, and carpels) are initiated as successive whorls. Three classes of genes regulate floral development. The first class contains positive regulators of the floral meristem identity. *APETALA1* (*AP1*) and *LEAFY* (*LFY*) are the most important *Arabidopsis* floral meristem identity genes. Meristem identity genes are positive regulators of another class of genes that determine floral organ identity. There are five known floral organ identity genes in *Arabidopsis*: *APETALA1* (*AP1*), *APETALA2* (*AP2*), *APETALA3* (*AP3*), *PISTILLATA* (*PI*), and *AGAMOUS* (*AG*). Cadastral genes make up the third group. Cadastral genes act as spatial regulators of the floral organ identity genes by setting boundaries for their expression.

The genes that control floral organ identity are homeotic. Most homeotic genes in plants contain the MADS box. Mutations in these genes alter the identity of the floral organs produced in two adjacent whorls. The ABC model seeks to explain how the floral homeotic genes control organ identity through the unique combinations of their products. Type A genes control organ identity in the first and second whorls. Type B activity controls organ determination in the second and third whorls. The third and fourth whorls are controlled by type C activity.

Floral development is also influenced by photoperiod and by hormones. In many dicots, gibberellins promote staminate flowers while ethylene induces pistillate flowers. GA suppresses stamen development in maize. Studies on mutants in floral homeotic genes and meristem identity genes have shown that floral reversion occurs under short-day conditions. Gibberellin suppresses the floral reversion phenotype, indicating that both GA and phytochrome interact with the genes that control floral morphogenesis.

General Reading

Atherton, J. G., ed. (1987) *Manipulation of Flowering*. Butterworths, London.

Bernier, G., Havelange, A., Houssa, C., Petitjeab, A., and Lejeune, P. (1993) Physiological signals that induce flowering. *Plant Cell* 5: 1147–1155.

*Brady, J., ed. (1982) *Biological Timekeeping*. Cambridge University Press, Cambridge.

Bünning, E. (1973) *The Physiological Clock: Circadian Rhythms and Biological Chronometry*, 3rd ed. Springer, New York.

Bünning, E. (1979) Circadian rhythms, light and photoperiodism. A re-evaluation. *Bot. Mag.* 92: 89–103.

Hackett, W. P. (1985) Juvenility, maturation and rejuvenation in woody plants. *Hort. Rev.* 7: 109–155.

Halevy, A. H., ed. (1985) *CRC Handbook of Flowering*. CRC Press, Boca Raton, FL.

Meyerowitz, E. M. (1994) The genetics of flower development. *Sci. Am.* 271: 56–65.

Millar, A. J., and Kay, S. A. (1997) The genetics of phototransduction and circadian rhythms in *Arabidopsis*. *Bioessays* 19: 209–214.

Ross, D., Pharis, P., and Binder, W. D. (1983) Growth regulators and conifers: Their physiology and potential use in forestry. In *Plant Growth Regulating Chemicals*, L. G. Nickell, ed., CRC Press, Boca Raton, FL., pp. 35–78.

Salisbury, F. B., and Ross, C. (1992) *Plant Physiology*, 4th ed. Wadsworth, Belmont, CA.

*Sweeney, B. M. (1987) *Rhythmic Phenomena in Plants.* Academic Press, San Diego.

Vince-Prue, D. (1986) The duration of light and photoperiodic responses. In *Photomorphogenesis in Plants*, R. F. Kendrick and G.

H. M. Kronenberg, eds., Martinus Nijhoff, Dordrecht, Netherlands, pp. 269–305.

Zimmerman, R. H., Hackett, H. P., and Pharis, R. P. (1985) Hormonal aspects of phase change and precocious flowering. In *Encyclopedia of Plant Physiology*, New Series, Vol. 11 (III), R. P. Pharis and D. M. Reid, eds., Springer, Berlin, pp. 79–115.

* Indicates a reference that is general reading in the field and is also cited in this chapter.

Chapter References

Abeles, F. B., Morgan, P. W., and Saltveit, M. E., Jr. (1992) *Ethylene in Plant Biology*, 2nd ed. Academic Press, San Diego.

Bernier, G. (1988) The control of floral evocation and morphogenesis. *Annu. Rev. Plant Physiol. Plant Mol. Biol.* 39: 175–219.

Beveridge, C. A., and Murfet, I. C. (1996) The *gigas* mutant in pea is deficient in the floral stimulus. *Physiol. Plantarum* 96: 637–645.

Blázquez, M. A., Green, R., Nilsson, O., Sussman, M. R., and Weigel, D. (1998) Gibberellins promote flowering of *Arabidopsis* by activating the LEAFY promoter. *Plant Cell* 10: 791–800.

Borthwick, H. A., Hendricks, S. B., and Parker, N. W. (1952) The reaction controlling floral initiation. *Proc. Natl. Acad. Sci. USA* 38: 929–934.

Bowman, J. L., Smyth, D. R., and Meyerowitz, E. M. (1989) Genes directing flower development in *Arabidopsis*. *Plant Cell* 1: 37–52.

Bünning, E. (1960) Biological clocks. *Cold Spring Harbor Symp. Quant. Biol.* 15: 1–9.

Burn, J. E., Bagnall, D. J., Metzger, J. D., Dennis, E. S., and Peacock, W. J. (1993) DNA methylation, vernalization, and the initiation of flowering. *Proc. Natl. Acad. Sci. USA* 90: 287–291.

Chen, L., Cheng, J. C., Castle, L., and Sung, Z. R. (1997) *EMF* genes regulate *Arabidopsis* inflorescence development. *Plant Cell* 9: 2011–2024.

Childs, K. L., Cordonnier-Pratt, M. M., Pratt, L. H., and Morgan, P. W. (1992) Genetic regulation of development in *Sorghum bicolor* VII. ma3 flowering mutant lacks a phytochrome that predominates in green tissue. *Plant Physiol.* 99: 765–770.

Clark, J. R. (1983) Age-related changes in trees. *J. Arboriculture* 9: 201–205.

Cockshull, K. E. (1985) *Chrysanthemum morifolium*. In *Handbook of Flowering*, Vol. 2, A. H. Halevy, ed., CRC Press, Boca Raton, FL., pp. 238–257.

Coen, E. S. (1991) The role of homeotic genes in flower development and evolution. *Annu. Rev. Plant Physiol. Plant Mol. Biol.* 42: 241–279.

Coen, E. S., and Carpenter, R. (1993) The metamorphosis of flowers. *Plant Cell* 5: 1175–1181.

Coen, E. S., Doyle, S., Romero, J. M., Elliott, R., Magrath, R., and Carpenter, R. (1991) Homeotic genes controlling flower development in *Antirrhinum*. *Development* 112 (Suppl. 1): 149–155.

Coen, E. S., and Meyerowitz, E. M. (1991) The war of the whorls: Genetic interactions controlling flower development. *Nature* 353: 31–37.

Coulter, M. W., and Hamner, K. C. (1964) Photoperiodic flowering response of Biloxi soybean in 72 hour cycles. *Plant Physiol.* 39: 848–856.

Deitzer, G. F., Hayes, R., and Jabben, M. (1979) Kinetics and time dependence of the effect of far red light on the photoperiodic induction of flowering in Wintex barley. *Plant Physiol.* 64: 1015–1021.

Downs, R. J. (1956) Photoreversibility of flower initiation. *Plant Physiol.* 31: 279–284.

Dudley, M., and Poethig, R. S. (1991) The effect of a heterochronic mutation, *Teopod2*, on the cell lineage of the maize shoot. *Development* 111: 733–740.

Dudley, M., and Poethig, R. S. (1993) The heterochronic *Teopod1* and *Teopod2* mutations of maize are expressed non-cell-autonomously. *Genetics* 133: 389–399.

Evans, L. T., and Wardlaw, I. F. (1966) Independent translocation of ^{14}C-labelled assimilates and of the floral stimulus in *Lolium temulentum*. *Planta* 68: 310–326.

Evans, M. M. S., and Poethig, R. S. (1995) Gibberellins promote vegetative phase change and reproductive maturity in maize. *Plant Physiol.* 108: 475–487.

Gasser, C. S., and Robinson-Beers, K. (1993) Pistil development. *Plant Cell* 5: 1231–1239.

Goethe, J. W. von (1790) *Goethe's Botany: The Metamorphosis of Plants and Tabler's Ode to Nature*, translated by Agnes Arber. Cronica Botanica, 1946 edition, Waltham, MA.

Goliber, T. E., and Feldman, L. J. (1989) Osmotic stress, endogenous abscisic acid and the control of leaf morphology in *Hippuris vulgaris* L. *Plant Cell Environ.* 12: 163–172.

Guo, H., Yang, H., Mockler, T. C., and Lin, C. (1998) Regulation of flowering time by *Arabidopsis* photoreceptors. *Science* 279: 1360–1363.

Hackett, W. P., and Srinivasan, C. (1985) *Hedera helix* and *Hedera canariensis*. In *Handbook of Flowering*, Vol. 3, A. H. Halevy, ed., CRC Press, Boca Raton, FL., pp. 89–97.

Hamner, K. C. (1940) Interrelation of light and darkness in photoperiodic induction. *Bot. Gaz.* 101: 658–687.

Hamner, K. C., and Bonner, J. (1938) Photoperiodism in relation to hormones as factors in floral initiation and development. *Bot. Gaz.* 100: 388–431.

Havelange, A., Lejeune, P., Bernier, G., Kaur-Sawhney, R., and Galston, A. W. (1996) Putrescine export from leaves in relation to floral transition in *Sinapis alba*. *Physiol. Plantarum* 96: 59–65.

Heinze, P. H., Parker, M. W., and Borthwick, H. A. (1942) Floral initiation in Biloxi soybean as influenced by grafting. *Bot. Gaz.* 103: 518–530.

Hempel, F. D., and Feldman, L. J. (1994) Bidirectional development in *Arabidopsis thaliana*: Acropetal initiation of flowers and basipetal initiation of paraclades. *Planta* 192: 276–286.

Hendricks, S. B., and Siegelman, H. W. (1967) Phytochrome and photoperiodism in plants. *Comp. Biochem.* 27: 211–235.

Hicks, K. A., Millar, A. J., Carre, I. A., Somers, D. E., Straume, M., Meeks-Wagner, D. R., and Kay, S. A. (1996) Conditional circadian dysfunction of the *Arabidopsis* early-flowering 3 mutant. *Science* 274: 790–792.

King, R. W., and Cumming, B. (1972) Rhythms as photoperiodic timers in the control of flowering in *Chenopodium rubrum* L. *Planta* 103: 281–301.

King, R. W., Evans, L. T., and Wardlaw, I. F. (1968) Translocation of the floral stimulus in *Pharbitis nil* in relation to that of assimilates. *Z. Pflanzenphysiol.* 59: 377–388.

Lang, A. (1965) Physiology of flower initiation. In *Encyclopedia of Plant Physiology*, Old Series, Vol. 15, W. Ruhland, ed., Springer, Berlin, pp. 1380–1535.

Lang, A. (1985) *Hyoscyamus niger*. In *CRC Handbook of Flowering*, Vol. 5, A. H. Halevy, ed., CRC Press, Boca Raton, FL., pp. 144–186.

Lang, A., Chailakhyan, M. K., and Frolova, I. A. (1977) Promotion and inhibition of flower formation in a dayneutral plant in grafts with a short-day plant and a long-day plant. *Proc. Natl. Acad. Sci. USA* 74: 2412–2416.

Lejeune, P., Bernier, G., Requier, M-C., and Kinet, J-M. (1994) Cytokinins in phloem and xylem saps of *Sinapis alba* during floral induction. *Physiol. Plantarum* 90: 522–528.

Longman, K. A. (1976) Some experimental approaches to the problem of phase change in forest trees. *Acta Hort.* 56: 81–90.

Lumsden, P. J., Vince-Prue, D., and Furuya, M. (1986) Phase-shifting of the photoperiodic flowering response rhythm in *Pharbitis nil* by redlight pulses. *Physiol. Plantarum* 67: 604–607.

McDaniel, C. N. (1996) Developmental physiology of floral initiation in *Nicotiana tabacum* L. *J. Exp. Bot.* 47: 465–475.

McDaniel, C. N., and Hartnett, L. K. (1996) Flowering as metamorphosis: Two sequential signals regulate floral initiation in *Lolium temulentum*. *Development* 122: 3661–3668.

McDaniel, C. N. , Hartnett, L. K., and Sangrey, K. A. (1996) Regulation of node number in day-neutral *Nicotiana tabacum*: A factor in plant size. *Plant J.* 9: 56–61.

McDaniel, C. N., Singer, S. R., and Smith, S. M. E. (1992) Developmental states associated with the floral transition. *Dev. Biol.* 153: 59–69.

Millar, A. J., Carre, I. A., Strayer, C. A., Chua, N-H., and Kay, S. A. (1995) Circadian clock mutants in *Arabidopsis* identified by luciferase imaging. *Science* 267: 1161–1163.

Millar, A. J., and Kay, S. A. (1991) Circadian control of cab gene transcription and messenger RNA accumulation in *Arabidopsis. Plant Cell* 3: 541–550.

Millar, A. J., Short, S. R., Chua, N-H., and Kay, S. A. (1992) A novel circadian phenotype based on firefly luciferase expression in transgenic plants. *Plant Cell* 4: 1075–1087.

Mueller, B. (1952) *Goethe's Botanical Writings.* University of Hawaii Press, Honolulu.

Napp-Zinn, K. (1984) Light and vernalization. In *Light and the Flowering Process*, D. Vince-Prue, B. Thomas, and K. E. Cockshull, eds., Academic Press, London, pp. 75–88.

Okamuro, J. K., den Boer, B. G. W., and Jofuku, K. D. (1993) Regulation of *Arabidopsis* flower development. *Plant Cell* 5: 1183–1193.

Okamuro, J. K., den Boer, B. G. W., Lotys-Prass, C., Szeto, W., and Jofuku, K. D. (1996) Flowers into shoots: Photo and hormonal control of a meristem identity switch in *Arabidopsis*. *Proc. Natl. Acad. Sci. USA* 93: 13831–13836.

Okamuro, J. K., Szeto, W., Lotys-Prass, C., and Jofuku, K. D. (1997) Photo and hormonal control of meristem identity in the *Arabidopsis* flower mutants *apetala2* and *apetala1*. *Plant Cell* 9: 37–47.

Papenfuss, H. D., and Salisbury, F. B. (1967) Aspects of clock resetting in flowering of *Xanthium*. *Plant Physiol.* 42: 1562–1568.

Pharis, R. P., Evans, L. T., King, R. W., and Mander, L. N. (1987) Gibberellins, endogenous and applied, in relation to flower induction in the long-day plant *Lolium temulentum*. *Plant Physiol.* 84: 1132–1138.

Pharis, R. P., and King, R. W. (1985) Gibberellins and reproductive development in seed plants. *Annu. Rev. Plant Physiol.* 36: 517–568.

Poethig, R. S. (1990) Phase change and the regulation of shoot morphogenesis in plants. *Science* 250: 923–930.

Purvis, O. N., and Gregory, F. G. (1952) Studies in vernalization of cereals. XII. The reversibility by high temperature of the vernalized condition in Petkus winter rye. *Ann. Bot.* 1: 569–592.

Putterill, J., Robson, F., Lee, K., Simon, R., and Coupland, G. (1995) The *CONSTANS* gene of *Arabidopsis* promotes flowering and encodes a protein showing similarities to zinc finger transcription factors. *Cell* 80: 847–857.

Reed, J. W., Nagatani, A., Elich, T. D., Fagan, M., and Chory, J. (1994) Phytochrome A and phytochrome B have overlapping but distinct functions in *Arabidopsis* development. *Plant Physiol.* 104: 1139–1149.

Reid, J. B., Murfet, I. C., Singer, S. R., Weller, J. L., and Taylor, S.A. (1996) Physiological genetics of flowering in *Pisum*. *Sem. Cell Dev. Biol.* 7: 455–463.

Saji, H., Vince-Prue, D., and Furuya, M. (1983) Studies on the photoreceptors for the promotion and inhibition of flowering in dark-grown seedlings of *Pharbitis nil* Choisy. *Plant Cell Physiol.* 67: 1183–1189.

Salisbury, F. B. (1963) Biological timing and hormone synthesis in flowering of *Xanthium*. *Planta* 49: 518–524.

Satter, R. L., and Galston, A. W. (1981) Mechanisms of control of leaf movements. *Annu. Rev. Plant Physiol.* 32: 83–110.

Schwarz-Sommer, Z., Huijser, P., Nacken, W., Saedler, H., and Sommer, H. (1990) Genetic control of flower development: Homeotic genes in *Antirrhinum majus*. *Science* 250: 931–936.

Simon, R., Igeno, M. I., and Coupland, G. (1996) Activation of floral meristem identity genes in *Arabidopsis*. *Nature* 384: 59–62.

Singer, S. R., and McDaniel, C. N. (1986) Floral determination in the terminal and axillary buds of *Nicotiana tabacum* L. *Dev. Biol.* 118: 587–592.

Sung, Z. R., Belachew, A., Shunong, B., and Bertrand-Garcia, R. (1992) EMF, an *Arabidopsis* gene required for vegetative shoot development. *Science* 258: 1645–1647.

Sussex, I. (1989) Developmental programming of the shoot meristem. *Cell* 56: 225–229.

Taylor, S. A., and Murfet, I. C. (1996) Flowering in *Pisum*: Identification of a new *ppd* allele and its physiological action as revealed by grafting. *Physiol. Plantarum* 97: 719–723.

Thomas, B., and Vince-Prue, D. (1984) Juvenility, photoperiodism and vernalization. In *Advanced Plant Physiology*, M. B. Wilkins, ed., Pirman, London, pp. 408–439.

Vince-Prue, D. (1975) *Photoperiodism in Plants.* McGraw-Hill, London.

Vince-Prue, D. (1985) Photoperiod and hormones. In *Encyclopedia of Plant Physiology*, New Series, Vol. 11 (III), R. P. Pharis and D. S. Reid, eds., Springer, Berlin, pp. 308–364.

Vince-Prue, D., and Lumsden, P. I. (1987) Inductive events in the leaves: Time measurement and photoperception in the short-day plant, *Pharbitis nil*. In *Manipulation of Flowering*, J. Atherton, ed., Butterworths, London, pp. 255–269.

Weigel, D. (1995) The genetics of flower development: From floral induction to ovule morphogenesis. *Annu. Rev. Genet.* 29: 19–39.

Weigel, D., and Meyerowitz, E. M. (1994) The ABCs of floral homeotic genes. *Cell* 78: 203–209.

Wellensiek, S. J. (1985) *Campanula medium*. In *Handbook of Flowering*, Vol. 2, A. H. Halevy, ed., CRC Press, Boca Raton, FL., pp. 123–126.

Weller, J. L., Murfet, I. C., and Reid, J. B. (1997) Pea mutants with reduced sensitivity to far-red light define an important role for phytochrome A in day-length detection. *Plant Physiol.* 114: 1225–1236.

Wilson, R. A., Heckman, J. W., and Sommerville, C. R. (1992) Gibberellin is required for flowering in *Arabidopsis thaliana* under short days. *Plant Physiol.* 100: 403–408.

Zagotta, M. T., Hicks, K. A., Jacobs, C. I., Young, J. C., Hangarter, R. P., and Meeks-Wagner, D. R. (1996) The *Arabidopsis* ELF3 gene regulates vegetative photomorphogenesis and the photoperiodic induction of flowering. *Plant J.* 10: 691–702.

Zeevaart, J. A. D. (1984) Photoperiodic induction, the floral stimulus and flower-promoting substances. In *Light and the Flowering Process*, D. Vince-Prue, B. Thomas, and K. E. Cockshull, eds., Academic Press, London, pp. 137–142.

Zeevaart, J. A. D., and Boyer, G. L. (1987) Photoperiodic induction and the floral stimulus in *Perilla*. In *Manipulation of Flowering*, J. G. Atherton, ed., Butterworths, London, pp. 269–277.

Zimmer, R. (1962) Phasenverschiebung und andere Störlichtwirkungen auf die endogen tagesperiodischen Blütenblattbewegungen von *Kalanchoe blossfeldiana* (Phase shift and other light interruption effects on endogenous diurnal petal movements). *Planta* 58: 283–300.

25 Stress Physiology

IT IS USEFUL TO BEGIN A DISCUSSION of stress in plants with some definitions, since the concept is often used imprecisely and the terminology can be confusing. **Stress** is usually defined as an external factor that exerts a disadvantageous influence on the plant. In most cases, stress is measured in relation to plant survival, crop yield, growth (biomass accumulation), or the primary assimilation processes (CO_2 and mineral uptake), which are related to overall growth.

Because stress is defined solely in terms of plant responses, it is sometimes called **strain** in conformance with engineering terminology. The concept of stress is intimately associated with that of **stress tolerance**, which is the plant's fitness to cope with an unfavorable environment. An environment that is stressful for one plant may not be stressful for another. Consider, for example, pea (*Pisum sativum*) and soybean (*Glycine max*), which grow best at about 20°C and 30°C, respectively. As temperature increases, the pea shows signs of heat stress much sooner than the soybean. Thus, the soybean has greater heat stress tolerance. If tolerance increases as a result of exposure to prior stress, the plant is said to be **acclimated** (or **hardened**). Acclimation can be distinguished from *adaptation*, which usually refers to a genetically determined level of resistance acquired by a process of selection over many generations. Unfortunately, the term "adaptation" is also used in the literature to indicate acclimation. Furthermore, we will see later that gene expression plays an important role in acclimation.

Under both natural and agricultural conditions, plants are frequently exposed to stress. Some environmental factors (such as air temperature) can become stressful in just a few minutes; others may take days to weeks (soil water) or even months (some mineral nutrients) to become stressful. It has been estimated that, because of stress resulting from soil conditions that are less than ideal, the yield of field-grown crops in the United States is only 22% of the genetic potential yield (Boyer 1982). In addition, stress plays a major role in determining how soil and climate limit the distribution of plant species. Thus, understanding the physiological processes that underlie adaptation and acclimation, as well as the mechanisms of stress injury, is of immense importance to agriculture and ecology.

In this chapter we examine ways in which plants respond to water deficit, chilling and freezing, heat, salinity, oxygen deficiency in the root zone, and air pollution. We focus mostly on the physiology of stress resistance and the role of gene expression and protein synthesis on plant responses to stress. Although it is convenient to examine each of these factors separately, many are interrelated. For example, water deficit is often associated with salinity in the root zone and/or with heat stress in the leaves. We will also see that plants often display **cross-resistance**, or resistance to one stress induced by acclimation to another. This behavior implies that mechanisms of resistance to several stresses share many common features.

Water Deficit and Drought Resistance

In this section we will examine drought resistance mechanisms, which are divided into several types. First we can distinguish between **desiccation postponement** (the ability to maintain tissue hydration) and **desiccation tolerance** (the ability to function while dehydrated), which are sometimes referred to as drought tolerance at high and low water potentials, respectively. The older literature often uses the term "drought avoidance" rather than "drought tolerance," but these terms are misnomers because drought is a meteorological condition that is tolerated by all plants that survive it and avoided by none. A third category, **drought escape**, comprises plants that complete their life cycles during the wet season, before the onset of drought. These are the only true "drought avoiders."

Among the desiccation postponers are *water savers* and *water spenders*. Water savers use water conservatively, preserving some in the soil for use late in the life cycle; water spenders aggressively consume water, often using prodigious quantities. The mesquite tree (*Prosopis* sp.) is an example of a water spender. This extremely deeply rooted species has ravaged semiarid rangelands in the southwestern United States, and because of its prodigious water use, it has prevented the reestablishment of grasses that have agronomic value.

Drought Resistance Strategies Vary with Climatic or Soil Conditions

Some regions of the world, such as portions of the Sahara Desert of northern Africa, receive an average of 5 mm of rainfall or less per year. Of course, this extreme aridity is atypical of most of the world's land area, particularly arable land that is used for crops or pastures. Nonetheless, most of the world's agriculture is subject to drought problems. Arid and semiarid zones are defined as areas in which plant transpiration totals 50% or less of the transpiration that would occur with unlimited water availability. In these areas water is the major factor limiting plant growth, and water and heat stress are usually alleviated by irrigation whenever possible. But water deficit is not limited to such regions; even in wetter climatic zones the irregular distribution of rainfall leads to periods in which water availability limits growth (Boyer 1982). The year-to-year variability of yields in most rain-fed agricultural areas illustrates the severity of the problem (Table 25.1).

The water-limited productivity of plants depends on the total amount of water available and on the water use efficiency of the plant (see Chapters 4 and 9). A plant that is capable of acquiring more water or that has higher water use efficiency will resist drought better. Some plants possess adaptations, such as the C_4 and CAM modes of metabolism, that allow them to exploit more arid environments. In addition, plants possess acclimation mechanisms that are activated in response to water stress.

When water deficit develops slowly enough to allow changes in developmental processes, water stress has several effects on growth, one of which is a limitation in leaf expansion. Although leaf area is important because photosynthesis is usually proportional to it, rapid leaf expansion can adversely affect water availability. If precipitation occurs only during winter and spring, and summers are dry, accelerated early growth can lead to large leaf areas, rapid early water depletion, and too little residual soil moisture for the plant to complete its life

Table 25.1
Yields of corn and soybean crops in the United States

	Crop yield (percentage of 10-year average)		
Year	Corn	Soybean	
1979	104	106	
1980	87	88	Severe drought
1981	104	100	
1982	108	104	
1983	77	87	Severe drought
1984	101	93	
1985	112	113	
1986	113	110	
1987	114	111	
1988	80	89	Severe drought

Source: U. S. Department of Agriculture 1989.

cycle. In this situation, only plants that retain some water for reproduction late in the season or that complete the life cycle quickly, before the onset of drought (drought escape), will produce seeds for the next generation. Either strategy will allow some reproductive success.

The situation is different if summer rainfall is significant but erratic. In this case, a plant with large leaf area, or one capable of developing large leaf area very quickly, is better suited to take advantage of occasional wet summers. One acclimation strategy in these conditions is a capacity for both vegetative growth and flowering over an extended period. Such plants are said to be **indeterminate** in their growth habit, in contrast to **determinate** plants, which develop preset numbers of leaves and which flower over only very short periods. Sorghum (*Sorghum bicolor*) is more resistant to drought than corn (*Zea mays*) in part because sorghum readily develops secondary shoots that can continue to grow and flower after the main shoot is fully mature. Corn does not form secondary shoots, so if fruiting fails because of drought, the corn plant has no opportunity to recoup the losses if rain occurs later.

In two of the three drought years shown in Table 25.1, the soybean yields were less affected than the corn yields, partly because the indeterminacy of soybean allowed it to take advantage of late-occurring rains to a greater extent than corn could. For maximum reproductive success in an erratic environment, then, a plant must be able to respond rapidly to both stress and relief of stress.

Decreased Leaf Area Is an Early *Adaptive* Response to Water Deficit

Some of the earliest responses to stress appear to be mediated by biophysical events rather than by changes in chemical reactions resulting from dehydration. As the water content of the plant decreases, the cells shrink and the cell walls relax (see Chapter 3). This decrease in cell volume results in lower hydrostatic pressure, or turgor. As water loss progresses and the cells contract further, the solutes in the cells become more concentrated. The plasma membrane becomes thicker and more compressed, as it covers a smaller area than before. Because turgor loss is the earliest significant biophysical effect of water stress, turgor-dependent activities are the most sensitive to water deficits.

Cell expansion is a turgor-driven process and is extremely sensitive to water deficit. As we discussed in Chapter 15, cell expansion is described by the relationship

$$GR = m(\Psi_p - Y) \qquad (25.1)$$

where GR is growth rate, Ψ_p is turgor, Y is the **yield threshold** (the pressure below which the cell wall resists plastic, or nonreversible, deformation), and m is **wall extensibility** (the responsiveness of the wall to pressure). This equation shows that a decrease in turgor causes a decrease in growth rate. Furthermore, not only is growth slowed when stress reduces Ψ_p, but also Ψ_p need decrease only to the value of Y, not to zero, to eliminate expansion—another reason for the high sensitivity of growth to stress. In normal conditions, Y is usually only 0.1 to 0.2 MPa less than Ψ_p, so changes in growth rate take place over a very narrow range of turgor (and of cell water content). The relationships among turgor, cell expansion, and growth under nonlimiting water conditions are discussed in detail in Chapter 19.

Because leaf expansion depends mostly on cell expansion, the principles that underlie the two processes are similar. Inhibition of cell expansion results in a slowing of leaf expansion early in the development of water deficits. The smaller leaf area transpires less water, effectively conserving a limited supply in the soil for use over a longer period. Limitation of leaf area can be considered a first line of defense against drought.

In intact leaves, water stress not only decreases turgor, but also decreases m and increases Y. In unstressed plants, wall extensibility (m) is normally greatest when the cell wall solution is slightly acidic. Stress decreases m in part because it inhibits proton transport across the plasma membrane into the cell wall, raising the cell wall pH. The effects of stress on Y are less well understood, but presumably they involve complex structural changes in the cell wall (see Chapter 15). Cell wall changes are very important because, unlike direct effects of turgor, they are relatively slow to be reversed after relief of stress. Water-deficient plants tend to become rehydrated at night, and as a result substantial leaf growth occurs at night. Nonetheless, because of changes in m and Y, the growth rate is still lower than that of unstressed plants at the same turgor (Figure 25.1).

Water stress limits not only the size of individual leaves, but also the number of leaves on an indeterminate plant, because it decreases both the number and the growth rate of branches. The process of stem growth has been studied less than that of leaf expansion, but it is probably affected by the same forces that limit leaf growth during stress.

Water Deficit Stimulates Leaf Abscission

The total leaf area of a plant does not remain constant after all the leaves have matured. If plants become water stressed after a substantial leaf area has developed, leaves will senesce and eventually fall off (Figure 25.2). This *leaf area adjustment* is an important long-term change that improves the plant's fitness for a water-limited environment. Indeed, many drought-deciduous desert plants drop all their leaves during drought and sprout new ones after a rain. This cycle can occur two or more times in a single season. Abscission during water stress results largely from enhanced synthesis of and responsiveness to the endogenous plant hormone ethylene (see Chapter 22).

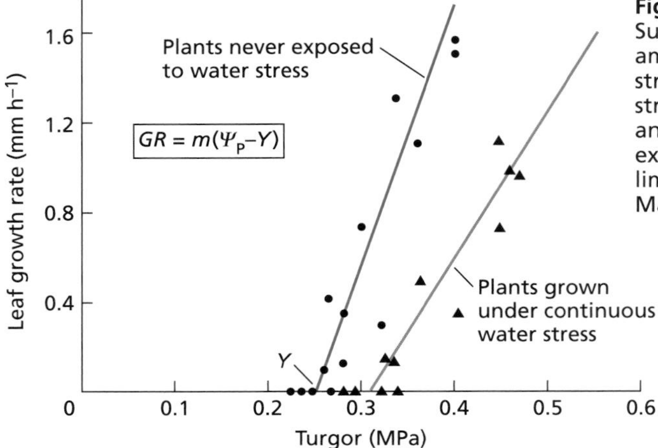

Figure 25.1 Dependence of leaf expansion on leaf turgor. Sunflower (*Helianthus annuus*) plants were grown either with ample water or with limited soil water to produce mild water stress. After rewatering, plants of both treatment groups were stressed by the withholding of water, and leaf growth rates (*GR*) and turgor (Ψ_p) were periodically measured. Both decreased extensibility (*m*) and increased threshold turgor for growth (*Y*) limit the leaf's capacity to grow after exposure to stress. (After Matthews et al. 1984.)

Water Deficit Enhances Root Extension into Deeper, Moist Soil

Mild water deficits also affect the development of the root system. Root–shoot relations appear to be governed by a functional balance between water uptake by the root and photosynthesis by the shoot. Although root–shoot relations depend on complex developmental and nutritional processes, the concept of functional balance can be simply stated: A shoot will grow until it is so large that water uptake by the roots becomes limiting to further growth; conversely, roots will grow until their demand for photosynthate from the shoot equals the supply. This functional balance is shifted if the water supply decreases.

When water uptake is curtailed, leaf expansion is affected very early, but photosynthetic activity is much less affected. Inhibition of leaf expansion reduces the consumption of carbon and energy, and a greater proportion of the plant's assimilates can be distributed to the root system, where they can support further growth. At the same time, the root apices in dry soil lose turgor. All these factors lead to root growth preferentially into the soil zones that remain moist. As water deficits progress, the upper layers of the soil usually dry first. Thus, it is common to see a mainly shallow root system when all soil layers are wetted, and a loss of shallow roots and proliferation of deep roots as water in top layers of the soil is depleted. Deeper root growth into wet soil can be considered a second line of defense against drought.

Enhancement of root growth into moist soil zones during stress depends on the allocation of assimilates to the growing root tips. As alternative sinks for assimilates, fruits usually predominate over roots, and assimilates are directed to the fruits and away from the roots. For this reason, the enhanced water uptake resulting from root growth is less pronounced in reproductive plants than in vegetative plants. Competition for assimilates between roots and fruits is one explanation for the fact that plants are generally more sensitive to water stress during reproduction.

Plant breeders have attempted to increase the capacity for root growth in breeding programs aimed at improved drought resistance of agricultural crops. Most crops show considerable genetic variability in the growth rate of roots, and under the right circumstances breeding for increased root growth can be effective. For this approach to be successful, however, the soil must contain water that would otherwise not be absorbed, which is often the case after a wet winter. On the other hand, if the winter is dry, the cost to the plant of building and maintaining the extra roots is not offset by gains in water uptake, and aboveground yields suffer.

Stomata Close during Water Deficit in Response to Abscisic Acid

The preceding sections focused on changes in plant development during slow, long-term dehydration. When

Figure 25.2 The leaves of young cotton (*Gossypium hirsutum*) plants abscise in response to water stress. The plants at left were watered throughout the experiment; those in the middle and at right were subjected to moderate stress and severe stress, respectively, before being watered again. Only a tuft of leaves at the top of the stem is left on the severely stressed plants. (Photograph courtesy of B. L. McMichael.)

the onset of stress is more rapid or the plant has reached its full leaf area before initiation of stress, other responses protect the plant against immediate desiccation. Under these conditions, the stomata close to reduce evaporation from the existing leaf area. Stomatal closure can be considered a third line of defense against drought.

Uptake and loss of water in guard cells changes their turgor and modulates stomatal opening and closing (see Chapter 4). Because guard cells are exposed to the atmosphere, they can lose water directly by evaporation and so lose turgor, causing the stomata to close by a mechanism called **hydropassive closure**. Hydropassive closure is likely to take place in air of low humidity, when direct water loss from the guard cells is too rapid to be balanced by water movement into the guard cells from adjacent epidermal cells. A second mechanism, called **hydroactive closure**, closes the stomata when the whole leaf or the roots are dehydrated and depends on metabolic processes in the guard cells. The mechanism of hydroactive closure is a reversal of the mechanism of stomatal opening described in Chapters 6 and 18. A reduction in the solute content of the guard cells results in water loss and decreased turgor, causing the stomata to close.

Solute loss from guard cells can be triggered by decreasing water status in the rest of the leaf, and there is much evidence that **abscisic acid** (**ABA**) (see Chapter 23) plays an important role in this process. Abscisic acid is synthesized continuously at a low rate in mesophyll cells and tends to accumulate mostly in the chloroplasts. When the mesophyll becomes mildly dehydrated, two things happen. First, some of the ABA stored in the chloroplasts is released to the apoplast (the cell wall space), making it possible for the transpiration stream to carry some of the ABA to the guard cells (Cornish and Zeevaart 1985). Second, the rate of net synthesis of ABA increases. ABA synthesis increases after closure has begun and appears to enhance or prolong the initial closing effect of the stored ABA. The mechanism of ABA-induced stomatal closure is discussed in Chapter 23.

The redistribution of ABA depends on pH gradients within the leaf, on the weak-acid properties of the ABA molecule, and on the permeability properties of cell membranes (Figure 25.3). In an unstressed photosynthesizing leaf, the pH of the chloroplast stroma is normally higher than that of the cytosol. This pH differential leads to a large accumulation of ABA in chloroplasts. One effect of dehydration is to lower the stroma pH, allowing the release of some ABA. Coupled to this process is an increase in the pH of the apoplast. These concerted changes during water deficit cause the net transfer of ABA from plastid to apoplast (Hartung et al. 1988).

Stomatal responses to leaf dehydration can vary widely both within and across species. The stomata of some dehydration-postponing species, such as cowpea (*Vigna unguiculata*) and cassava (*Manihot esculenta*), are unusually responsive to decreasing water availability, and stomatal conductance and transpiration decrease so much that leaf water potential (Ψ; see Chapters 3 and 4) may remain nearly constant during drought. In cotton (*Gossypium hirsutum*), factors such as nitrogen supply affect ABA accumulation or ABA redistribution or both, and thus greatly alter stomatal responses to water stress (Radin and Hendrix 1988).

Messengers from the root system may affect the stomatal responses to water stress. There are two types of evidence for root messengers. First, stomatal conduc-

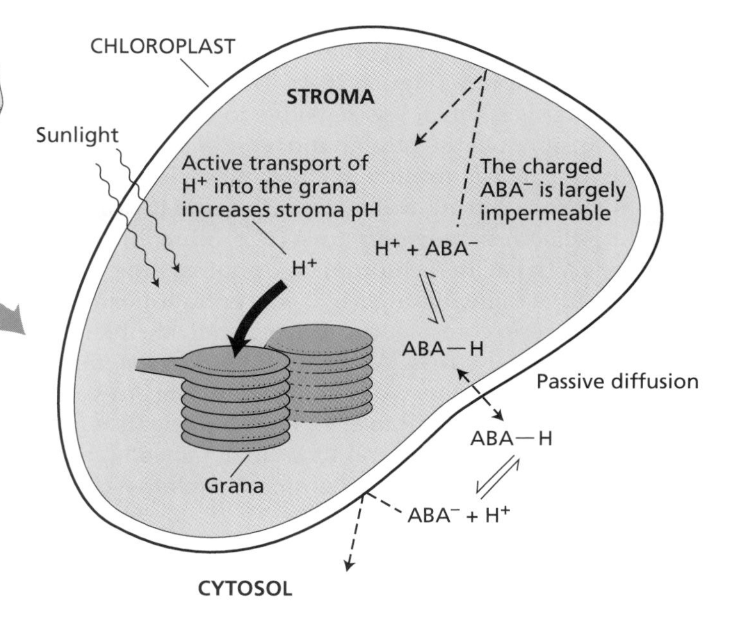

Figure 25.3 Accumulation of ABA by chloroplasts in the light. Dashed lines indicate passive movement of substances by diffusion; the wide arrow indicates movement by active processes. Light stimulates proton uptake into the grana, making the stroma more alkaline. The increased alkalinity causes the weak acid ABA•H to dissociate into H^+ and the ABA^- anion. The concentration of ABA•H in the stroma is lowered below the concentration in the cytosol, and the concentration difference drives the passive diffusion of ABA•H across the chloroplast membrane. At the same time, the concentration of ABA^- in the stroma increases, but the chloroplast membrane is almost impermeable to the anion, which thus remains trapped. This process continues until the ABA•H concentrations in the stroma and the cytosol are equal. But as long as the stroma remains more alkaline, the total ABA concentration (ABA•H + ABA^-) in the stroma greatly exceeds the concentration in the cytosol.

tance is often much more closely related to soil water status than to leaf water status, and the only plant part that can be *directly* affected by soil water status is the root system. In fact, dehydrating only part of the root system can cause stomatal closure even though the well-watered portion of the root system may still deliver ample water to the tops. When corn (*Zea mays*) plants were grown with roots trained into two separate pots and water was withheld from only one of the pots, the stomata partially closed, and the leaf water potential increased, just as in the dehydration postponers already described (Blackman and Davies 1985). These results show that stomata can respond to conditions sensed in the roots.

The second type of evidence for root messengers is that roots are known to produce and export abscisic acid to the xylem sap. When dayflower (*Commelina communis*) plants were grown with their root systems divided between two pots and water was then withheld from one pot, the ABA concentration in roots in the dry pot increased considerably (Zhang et al. 1987). Stomata closed in response to the treatment, despite no change in leaf water potential. The ABA content of guard cells measured in epidermis stripped from the leaf was found to be closely related to the degree of stomatal closure. The ABA from the roots was presumably delivered to the leaf via the transpiration stream. These results imply that stomata respond to two sources of ABA during soil drying: (1) an "early warning system" involving root ABA, indicating that some roots are drying; and (2) ABA translocation within the leaf, resulting from desiccation of the leaves themselves.

Water Deficit Limits Photosynthesis within the Chloroplast

The photosynthetic rate of the leaf (expressed per unit leaf area) is seldom as responsive to mild water stress as leaf expansion is (Figure 25.4). The reason is that photosynthesis is much less sensitive to turgor than is leaf expansion. On the other hand, evidence suggests that the Mg^{2+} concentration in chloroplasts influences photosynthesis during water stress through its role in coupling electron transport to ATP production (see Chapter 7). In isolated chloroplasts, photosynthesis is very sensitive to increasing Mg^{2+} concentration, and a similar process could occur during cell shrinkage induced by water stress. When sunflower (*Helianthus annuus*) plants were grown with different Mg^{2+} levels in the nutrient solution (Rao et al. 1987), the plants with the lower tissue Mg^{2+} concentrations maintained higher photosynthetic rates as leaves became dehydrated (Figure 25.5).

Water stress usually affects both stomatal conductance and photosynthetic activity in the leaf. Upon stomatal closure during early stages of water stress, **water**

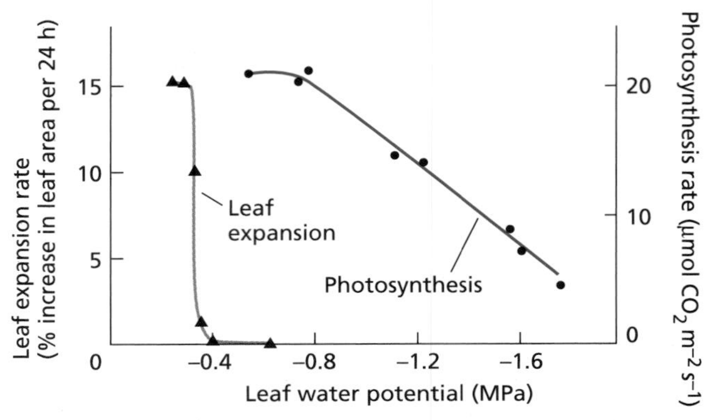

Figure 25.4 Effects of water stress on photosynthesis and leaf expansion of sunflower (*Helianthus annuus*). This species is typical of many plants in the sense that leaf expansion is much more sensitive to dehydration than photosynthetic rate is. (After Boyer 1970.)

use efficiency may increase (more CO_2 is taken up per unit of water transpired) because stomatal closure inhibits transpiration more than it decreases intercellular CO_2 concentrations. Additionally, the dehydration of mesophyll cells inhibits photosynthesis. As stress becomes severe, however, water use efficiency usually decreases, and the inhibition of mesophyll metabolism becomes stronger. We can evaluate the relative effect of the stress on stomatal conductance and photosynthesis by putting stressed leaves into air that contains high concentrations of CO_2. Stomatal limitations to photosynthesis can be overcome by a higher external concentration of CO_2, but any direct effect of water stress on mesophyll metabolism will not be removed by exposure to high CO_2 concentrations.

Does water stress directly affect translocation? Phloem transport is coupled to both photosynthesis, which provides the substrates, and metabolism in the sinks. Water stress decreases photosynthesis and the consumption of assimilates in the expanding leaves. As a consequence, water stress indirectly decreases the amount of photosynthate exported from leaves. Because phloem transport depends on turgor (see Chapter 10), it has been argued that if water potential decreases in the phloem during stress, the movement of assimilates should also be inhibited. However, experiments have shown that translocation is unaffected until late in the stress period, when other processes, such as photosynthesis, have already been strongly inhibited (Figure 25.6). This relative insensitivity of translocation to stress allows plants to mobilize and use reserves where they are needed (for example, in seed growth), even when stress is extremely severe. The ability to continue translocating carbon is a key factor in almost all aspects of plant resistance to drought.

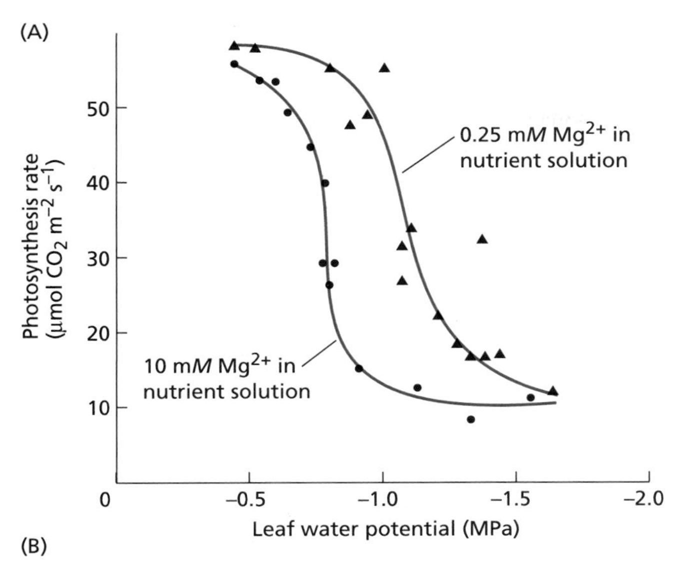

Figure 25.5 (A) Photosynthetic rates (measured at CO_2 and light saturation) in water-stressed sunflower (*Helianthus annuus*). The plants were grown with levels of Mg^{2+} in the nutrient solution chosen to produce different tissue Mg^{2+} concentrations but no change in growth rate, as shown in the photograph (B). As water stress progressed, photosynthesis was affected much earlier in the plants that had higher Mg^{2+} levels (solid curve). (After Rao et al. 1987. Photograph courtesy of J. S. Boyer.)

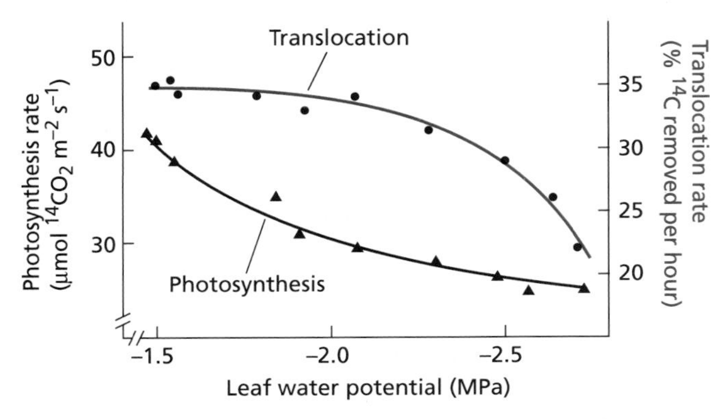

Figure 25.6 Relative effects of water stress on photosynthesis and translocation in sorghum (*Sorghum bicolor*). Plants were exposed to $^{14}CO_2$ for a short interval. The radioactivity fixed in the leaf was taken as a measure of photosynthesis, and the loss of radioactivity after removal of the $^{14}CO_2$ was taken as a measure of the rate of assimilate translocation. Although photosynthesis was affected by lower levels of stress, translocation was unaffected until stress was very severe. (After Sung and Krieg 1979.)

Osmotic Adjustment of Cells Helps Maintain Plant Water Balance

As soil dries, its matric potential (see Chapter 3) becomes more negative. Plants can continue to absorb water only as long as their water potential (Ψ_w) is below (more negative than) that of the source of water. **Osmotic adjustment**, or accumulation of solutes by cells, is a process by which water potential can be decreased without an accompanying decrease in turgor. Recall Equation 3.7 from Chapter 3: $\Psi_w = \Psi_s + \Psi_p$. The change in tissue water potential results simply from changes in solute potential (Ψ_s), the osmotic component of Ψ_w.

Osmotic adjustment should not be confused with the increase in solute concentration that occurs during cell dehydration and shrinkage. Osmotic adjustment is a net increase in solute content per cell that is independent of the volume changes that result from loss of water. The decrease in Ψ_s is typically limited to between 0.2 and 0.8 MPa, except in plants adapted to extremely dry conditions. Most of the adjustment can usually be accounted for by increases in concentration of a variety of common solutes, including sugars, organic acids, and ions (especially K^+).

Enzymes extracted from the cytosol of plant cells have been shown to be severely inhibited by high concentrations of ions. The accumulation of ions during osmotic adjustment appears to occur mainly within the vacuoles, where the ions are kept out of contact with enzymes in the cytosol or subcellular organelles. Because of this compartmentation of ions, other solutes must accumulate in the cytoplasm to maintain water potential equilibrium within the cell. These other solutes, called **compatible solutes** (or compatible osmolytes), are organic compounds that do not interfere with enzyme functions. Proline is a commonly accumulated compatible solute; examples of other compatible solutes are a sugar alcohol, sorbitol, and a quaternary amine, glycine betaine. Synthesis of compatible solutes is also important in the adjustment of plants to increased salinity in the rooting zone, discussed later in this chapter.

Osmotic adjustment develops slowly in response to tissue dehydration. It is therefore important, when studying osmotic adjustment, to allow for a slow onset of stress in order to provide enough time for the completion of substantial shifts in solute synthesis and transport patterns. Over the course of several days, however, other changes (such as growth or photosynthesis) are also taking place. For this reason, it is not clear whether osmotic adjustment is an independent and direct response to water deficit or is a result of another factor, such as decreased growth rate. Nonetheless, leaves that are capable of osmotic adjustment clearly can maintain turgor at lower water potentials than nonadjusted leaves. Maintenance of turgor enables the continuation of cell elongation and facilitates higher stomatal conductances at lower water potentials (Turner and Jones 1980). In this sense, osmotic adjustment is an acclimation that enhances dehydration tolerance.

How much extra water can be acquired by the plant because of osmotic adjustment in the leaf cells? Most of the extractable soil water is held in large air spaces from which it is readily removed by roots (see Chapter 4). As the soil dries, this water is used first, leaving behind the small amount of water that is held more tightly in small pores. Osmotic adjustment enables the plant to extract more of this tightly held water, but the increase in available water is small. Thus, the cost of osmotic adjustment in the leaf is offset by rapidly diminishing returns in terms of water availability to the plant, as can be seen by a comparison of the water relations of adjusting and nonadjusting species.

Figure 25.7 shows such a comparison for pot-grown sugar beet (*Beta vulgaris*), an osmotically adjusting species, and cowpea (*Vigna unguiculata*), a nonadjusting species that instead conserves water during stress by stomatal closure. On any given day after the last watering, the sugar beet leaves maintained a lower water potential than the cowpea leaves, but photosynthesis and transpiration during stress were only slightly greater in the sugar beet. The major difference between the two plants was the leaf water potential. These results show that osmotic adjustment promotes dehydration tolerance but does not have a major effect on productivity (McCree and Richardson 1987).

Osmotic adjustment also occurs in roots, although the process in roots has not been studied so extensively as in leaves. The absolute magnitude of the adjustment is less in roots than in leaves, but as a percentage of the original tissue Ψ_s, it can be greater in roots than in leaves. Again, these changes in many cases only slightly increase water extraction from the previously explored soil. However, osmotic adjustment can occur in the root meristems, enhancing turgor and maintaining root growth, and it is an important component of the

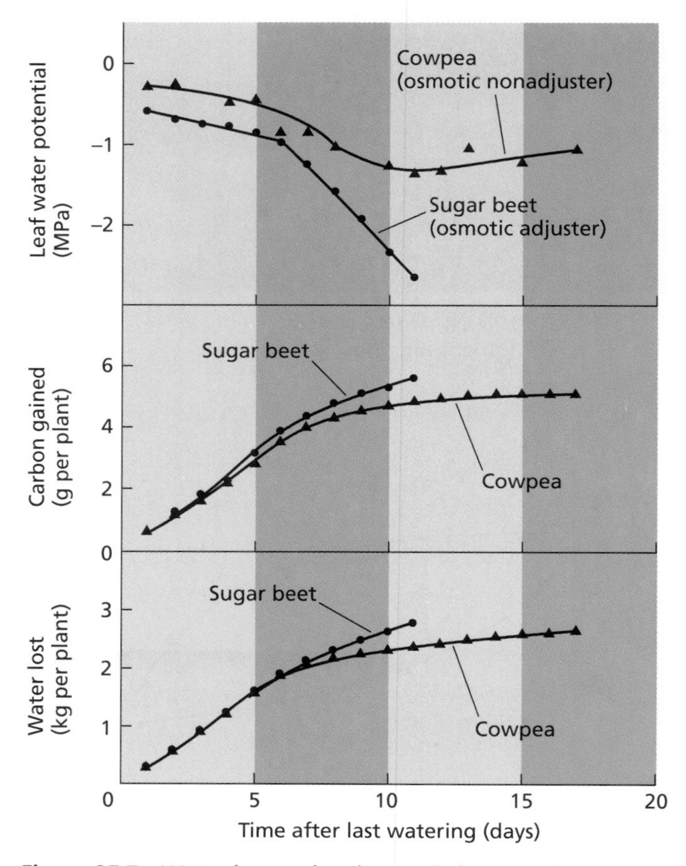

Figure 25.7 Water loss and carbon gain by sugar beet (*Beta vulgaris*), an osmotically adjusting species, and cowpea (*Vigna unguiculata*), a nonadjusting species, grown in pots and subjected to water stress. Although the leaf water potential reached much lower values in sugar beet because of its osmotic adjustment, the total water loss and carbon gain during stress were little affected. (After McCree and Richardson 1987.)

changes in root growth patterns during water depletion in the field (discussed earlier).

Is osmotic adjustment a valuable mechanism of stress acclimation? Do solutes accumulate in one tissue simply because of stress-related inhibition of growth elsewhere? These questions might be difficult to answer. Attempts to increase osmotic adjustment in leaves, either genetically (by breeding and selection) or physiologically (by inducing adjustment with controlled water deficits), have resulted in plants that grow more slowly. Thus, the use of osmotic adjustment to improve agricultural performance is yet to be realized.

Water Deficit Alters Energy Dissipation from Leaves

Recall from Chapter 9 that evaporative heat loss lowers leaf temperature. This cooling effect can be remarkable: In Death Valley, California—one of the hottest places in the world—leaf temperatures were 8°C below air temperatures on plants with access to ample water. In warm, dry climates, leaves from irrigated crop plants can also maintain a temperature difference. In such climates, an experienced farmer can decide whether plants need water simply by touching the leaves, because a rapidly transpiring leaf is distinctly cool to the touch. When water stress limits transpiration, the leaf heats up unless another process offsets the lack of cooling. Because of these interactions, water stress and heat stress are closely interrelated (see the discussion of heat stress later in this chapter).

Maintaining a leaf temperature that is much lower than the air temperature requires evaporation of vast quantities of water. Thus, adaptations that cool leaves by means other than evaporation are very effective in conserving water. When transpiration slows and the leaf temperature becomes warmer than the air temperature, some of the extra energy in the leaf is dissipated as sensible heat loss (see Chapter 9). Many arid-zone plants have very small leaves, which minimize the resistance of the boundary layer to the transfer of heat from the leaf to the air (see Figure 9.13). Because of their low boundary layer resistance, small leaves tend to remain close to air temperature even when transpiration is greatly slowed. In contrast, large leaves have higher boundary layer resistance and dissipate less thermal energy (per unit leaf area) by direct transfer. This limitation can be compensated for by leaf movements that provide additional protection against heating during water stress. Leaves that orient themselves away from the sun are called *paraheliotropic;* leaves that gain energy by orienting themselves normal (perpendicular) to the sunlight are referred to as *diaheliotropic* (see Chapter 9). Figure 25.8 shows the strong effect of water stress on leaf position in soybean.

(A)

(B)

(C)

Figure 25.8 Orientation of leaflets of field-grown soybean (*Glycine max*) plants in the normal, unstressed, position (A); during mild water stress (B); and during severe water stress (C). The large leaf movements induced by mild stress are quite different from wilting, which occurs during severe stress. Note that during mild stress (B), the terminal leaflet has been raised, whereas the two lateral leaflets have been lowered; each is almost vertical. (Courtesy of D. M. Oosterhuis.)

Other factors that can alter the interception of radiation include wilting, which changes the angle of the leaf, and leaf rolling in grasses, which minimizes the profile of tissue exposed to the sun. Absorption of energy can also be decreased by hairs on the leaf surface or by layers of reflective wax outside the cuticle. Leaves of some plants have a gray white appearance because densely packed hairs reflect a large amount of light. This hairiness, or *pubescence*, keeps leaves cooler by reflecting radiation, but it also reflects the visible wavelengths that are active in photosynthesis and thus it decreases carbon assimilation. Because of this problem, attempts to breed pubescence into crops to improve their water use efficiency have been generally unsuccessful.

Water Deficit Increases Resistances to Liquid-Phase Water Flow

When a soil dries, its resistance to the flow of water increases very sharply, particularly near the *permanent wilting point*, usually observed when the soil water potential reaches –1.5 MPa (see Figure 4.3). At the permanent wilting point, water delivery to the roots is too slow to allow overnight rehydration of plants that have wilted during the day. However, the soil is not the only source of increased resistance to flow. In fact, the resistance within the plant has been found to be larger than the resistance within the soil over a wide range of water deficits (Blizzard and Boyer 1980).

Several factors may contribute to the increased plant resistance to water flow during drying. As plant cells lose water, they shrink. When roots shrink, the root surface can move away from the soil particles that hold the water, and the delicate root hairs may be damaged as they are pulled away. Also, as root extension slows during soil drying, the outer layer of the root cortex (the hypodermis) often becomes more extensively covered with **suberin**, a water-impermeable lipid, increasing the resistance to water flow.

Another important factor that increases resistance to water flow is **cavitation**, or the breakage of water columns under tension. As we saw in Chapter 4, transpiration from leaves "pulls" water through the plant by creating a tension on the water column. The cohesive forces that are required to support large tensions are present only in very narrow columns in which the water adheres to the walls. Cavitation begins in most plants at moderate water potentials (–1 to –2 MPa), and the largest vessels cavitate first. Thus, in ring-porous trees such as oak (*Quercus*), the large-diameter vessels that are laid down in the spring function as a low-resistance pathway early in the growing season, when ample water is available. These vessels cease functioning during the summer, leaving the small-diameter vessels produced during the stress period to carry the transpiration stream. This shift has long-lasting consequences: Even if the plant is rewatered, the original low-resistance pathway remains nonfunctional, reducing the efficiency of water flow.

Water Deficit Increases Wax Deposition on the Leaf Surface

A common developmental response to water stress is the production of a thicker cuticle that reduces water loss from the epidermis (cuticular transpiration). A thicker cuticle also decreases CO_2 permeability, but leaf photosynthesis remains unaffected because the epidermal cells underneath the cuticle are nonphotosynthetic. Cuticular transpiration, however, accounts for only 5 to 10% of the total leaf transpiration, so it becomes significant only if stress is extremely severe or if the cuticle has been damaged (e.g., by wind-driven sand).

Water Deficit May Induce Crassulacean Acid Metabolism

Crassulacean acid metabolism (CAM), discussed in Chapters 8 and 9, is a plant adaptation in which stomata open at night and close during the day. The leaf-to-air vapor pressure difference that drives transpiration is much reduced at night, when both leaf and air are cool. As a result, the water use efficiency of CAM plants are among the highest measured in all higher plants. A CAM plant may gain 1 g of dry matter for only 125 g of water used—three to five times greater than the ratio for a typical C_3 plant.

CAM is also characteristic of succulent plants such as cacti. As discussed in Chapter 8, some succulent species display **facultative CAM**, switching to CAM when subjected to water deficits or saline conditions (Hanscom and Ting 1978). This switch in metabolism is a remarkable adaptation to stress, involving accumulation of the enzyme phosphoenolpyruvate (PEP) carboxylase (Figure 25.9), changes in carboxylation and decarboxylation patterns, transport of large quantities of malate into and out of the vacuoles, and reversal of the periodicity of stomatal movements.

Figure 25.9 Increases in the content of phosphoenolpyruvate (PEP) carboxylase in ice plant, *Mesembryanthemum crystallinum*, during the salt-induced shift from C_3 metabolism to CAM. Salt stress was induced by the addition of 500 m*M* NaCl to the irrigation water. The PEP carboxylase protein was revealed in the gels by the use of antibodies and a stain. (After Bohnert et al. 1989.)

Chilling and Freezing

Sensitive species are injured by chilling at temperatures that are too low for normal growth but not low enough for ice to form. Typically, tropical or subtropical species are susceptible to chilling injury. Among crops, maize, *Phaseolus* bean, rice, tomato, cucumber, sweet potato, and cotton are sensitive. *Passiflora*, *Coleus*, and *Gloxinia* are examples of susceptible ornamentals. When plants growing at relatively warm temperatures (25 to 35°C) are cooled to 10 to 15°C, growth is slowed, discolorations or lesions appear on leaves, and the foliage looks as if it has been soaked in water for a long time. If roots are chilled, the plants may wilt.

Species that are considered to be generally sensitive to chilling can show appreciable variation in their response to chilling temperatures. Genetic adaptation to the colder temperatures associated with high altitude improves chill resistance (Figure 25.10). In addition, resistance often increases if plants are first hardened (acclimated) by exposure to cool, but noninjurious, temperatures. Chilling damage thus can be minimized if exposure is slow and gradual. Sudden exposure to temperatures near 0°C, called cold shock, greatly increases the chances of injury. Freezing injury, on the other hand, occurs at temperatures below the freezing point of water. Full induction of tolerance to freezing, as with chilling, requires a period of acclimation at cold temperatures.

In the discussion that follows we will examine how chilling injury alters membrane properties, how ice crystals damage cells and tissues, and how ABA, gene expression, and protein synthesis mediate acclimation to freezing.

Membrane Properties Change in Response to Chilling Injury

Leaves from plants injured by chilling show inhibition of photosynthesis and carbohydrate translocation, slower respiration, inhibition of protein synthesis, and increased degradation of existing proteins. All of these responses probably depend on a common primary mechanism involving loss of membrane function during chilling. For instance, solutes leak from the leaves of chill-sensitive *Passiflora maliformis* (conch apple) floated on water at 0°C but not from those of chill-resistant *Passiflora caerulea* (passion flower). Loss of solutes to the water reflects damage to the plasma membrane and possibly also to the tonoplast. In turn, inhibition of photosynthesis and of respiration reflects injury to chloroplast and mitochondrial membranes.

Why are the membranes of chill-sensitive plants affected by chilling? Plant membranes consist of a lipid bilayer interspersed with proteins and sterols (see Chapter 1). The physical properties of the lipids greatly influence the activities of the integral membrane proteins, including H^+ ATPases, carriers, and channel-forming proteins that regulate the transport of ions and other solutes (see Chapter 6), and of enzymes on which metabolism depends. In chill-sensitive plants, the lipids in the bilayer have a high percentage of saturated fatty acid chains, and this type of membrane tends to solidify into a semicrystalline state at a temperature well above 0°C.* As the membranes become less fluid, their protein components can no longer function normally. The result is inhibition of H^+-ATPase activity, of solute transport into and out of cells, of energy transduction (see Chapters 7 and 11), and of enzyme-dependent metabolism. In addition, chill-sensitive leaves exposed to high photon fluxes and chilling temperatures are photoinhibited (see Chapter 7), causing acute damage to the photosynthetic machinery.

Membrane lipids from chill-resistant plants often have a greater proportion of unsaturated fatty acids than those from chill-sensitive plants (Table 25.2), but does this difference in unsaturated fatty acids confer chill resistance? During acclimation to cool temperatures the activity of desaturase enzymes increases and the proportion of unsaturated lipids rises (Williams et al. 1988; Palta et al. 1993). This modification allows mem-

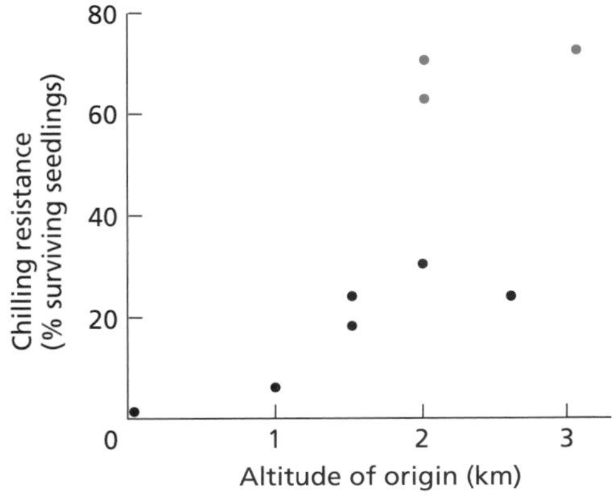

* Saturated fatty acids, which have no double bonds, and *trans*-monounsaturated fatty acids solidify (or melt) at a higher temperature than unsaturated lipids. Butter, which consists of saturated fatty acids, is a solid at room temperature, whereas oil from seeds is more unsaturated (has more double bonds) and remains liquid at much lower temperatures.

Chilling resistance (% surviving seedlings)
80
60
40
20
0
1
2
3
Altitude of origin (km)

Figure 25.10 Survival at low temperature of seedlings of different populations of tomato collected from different altitudes in South America. Seed was collected from wild tomato (*Lycopersicon hirsutum*) and grown in the same greenhouse at 18 to 25°C. All seedlings were then chilled for 7 days at 0°C and then kept for 7 days in a warm growth room, after which the number of survivors was counted. Seedlings from seed collected from high altitudes showed greater resistance to chilling (cold shock) than those from seed collected from lower altitudes. (From Patterson et al. 1978.)

Table 25.2
Fatty acid composition of mitochondria isolated from chill-resistant and chill-sensitive species

	Percent weight of total fatty acid content					
	Chill-resistant species			Chill-sensitive species		
Major fatty acids[a]	**Cauliflower bud**	**Turnip root**	**Pea shoot**	**Bean shoot**	**Sweet potato**	**Maize shoot**
Palmitic (16:0)	21.3	19.0	12.8	24.0	24.9	28.3
Stearic (18:0)	1.9	1.1	2.9	2.2	2.6	1.6
Oleic (18:0)	7.0	12.2	3.1	3.8	0.6	4.6
Linoleic (18:2)	16.4	20.6	61.9	43.6	50.8	54.6
Linolenic (18:3)	49.4	44.9	13.2	24.3	10.6	6.8
Ratio of unsaturated to saturated fatty acids	3.2	3.9	3.8	2.8	1.7	2.1

[a] Shown in parentheses are the number of carbon atoms in the fatty acid chain and the number of double bonds.
Source: After Lyons et al. 1964.

branes to remain fluid by lowering the temperature at which the membrane lipids begin a gradual phase change from fluid to semicrystalline. Thus, desaturation of fatty acids provides some protection against damage from chilling. In these hardened plants, the modified membranes are better able to remain fluid and function at lower temperatures.

The importance of membrane lipids to tolerance of low temperatures has been demonstrated by work with mutant and transgenic strains in which the activity of particular enzymes led to a specific change in membrane lipid composition independent of acclimation to low temperature. The degree of fatty acid saturation of phosphatidylglycerol (PG) is particularly important to the chilling resistance of chloroplasts. In one experiment tobacco was transformed with an acyltransferase gene from either squash (chill sensitive) or *Arabidopsis* (chill tolerant) to modify the fatty acid saturation of PG in the thylakoid membranes of the chloroplasts (Figure 25.11). In transgenic tobacco transformed with the squash gene (see Figure 25.11A), which increased the proportion of high-melting-point (*trans*-monounsaturated and saturated) molecular species of PG, chilling tolerance decreased (88% inhibition of photosynthesis following

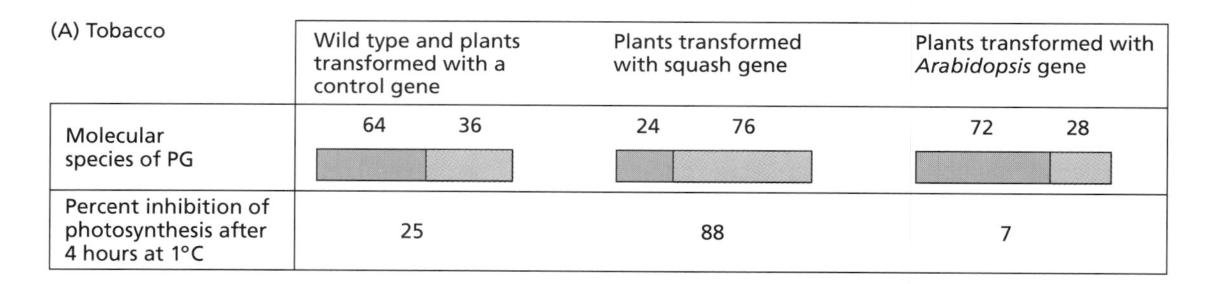

Figure 25.11 How changes in relative levels of *cis*-unsaturated molecular species of phosphatidylglycerol (PG) in thylakoid membranes of (A) tobacco and (B) *Arabidopsis* affect sensitivity to chilling. (After Nishida and Murata 1996.)

4 hours at 1°C). Conversely, when tobacco was transformed with the *Arabidopsis* chill-tolerant gene, the proportion of high-melting-point molecular species of PG decreased, and tolerance of chilling was greater than in wild-type plants.

In *Arabidopsis* (see Figure 25.11B), transformation with a gene from *Escherichia coli* that raised the proportion of high-melting-point molecular species caused the plant to die at 2°C in the light (lethal photoinhibition). Similarly, the *fab1* mutants of *Arabidopsis* have increased levels of saturated fatty acids, particularly 16:0 (see Table 25.2, and the section on lipids in Chapter 11). During a period of 3 to 4 weeks at chilling temperatures, photosynthesis and growth were gradually inhibited, culminating in the destruction of chloroplasts and the total cessation of photosynthesis. At nonchilling temperatures, the mutant grew as well as wild-type controls did (Wu et al. 1997). The genes used to modify the lipid composition of thylakoid membranes in these studies encoded glycerol-3-phosphate acyltransferases (GPATs). These enzymes select fatty acids for incorporation into PG; GPATs from different plant species have different selectivities for saturated and unsaturated fatty acids, leading to the differences in composition illustrated in Figure 25.11.

Freezing Kills Cells by Forming Intracellular Ice Crystals or by Dehydrating the Protoplast

The ability to tolerate freezing temperatures under natural conditions varies greatly among chill-resistant plants. Seeds, other partly dehydrated tissues, and fungal spores can be kept indefinitely at temperatures near absolute zero (0 K, or –273°C), so very low temperatures are not intrinsically harmful. Fully hydrated, vegetative cells can also retain viability if they are cooled very quickly to avoid the formation of large, slowly growing ice crystals that would puncture and destroy subcellular structures within the living protoplast. Ice crystals that form during very rapid freezing are too small to cause mechanical damage to subcellular structures. Rapid warming is required to prevent the growth of small ice crystals into crystals of a damaging size, or to prevent loss of water vapor by sublimation, both of which take place at intermediate temperatures (–100 to –10°C). Under natural conditions, however, cooling of intact, multicellular plant organs is never fast enough to give rise to only small, harmless ice crystals in fully hydrated cells.

When tissue is cooled under natural conditions, ice usually forms first within the intercellular spaces, and in the xylem vessels, along which the ice can quickly propagate. This ice formation is not lethal to hardy plants, and the tissue recovers fully if warmed. However, when plants are exposed to freezing temperatures for an extended period, the growth of extracellular ice crystals results in the movement of liquid water from the protoplast to the extracellular ice, causing excessive dehydration (see Box 25.1).

During rapid freezing, the protoplast, including the vacuole, **supercools**; that is, the cellular water remains liquid even at temperatures several degrees below its theoretical freezing point. Several hundred molecules are needed for an ice crystal to begin forming. The process whereby these hundreds of ice molecules start to form a stable ice crystal is called **ice nucleation**, and it strongly depends on the properties of the involved surfaces. Some large polysaccharides and proteins facilitate ice crystal formation, and are called *ice nucleators*. Some of the ice nucleation proteins made by bacteria appear to facilitate ice nucleation by aligning water molecules along repeated amino acid domains within the protein.

In plant cells, ice crystals begin to grow from endogenous ice nucleators, and the resulting, relatively large intracellular ice crystals cause extensive damage to the cell and are usually lethal.

ABA, Gene Expression, and Protein Synthesis Are Involved in Acclimation to Freezing

In seedlings of alfalfa (*Medicago* sp.), tolerance to freezing at –10°C is greatly improved by previous exposure to cold (4°C) or by treatment with exogenous ABA (see Chapter 23) without exposure to cold. These treatments cause changes in the pattern of newly synthesized proteins that can be resolved on two-dimensional gels. Some of the changes are unique to the particular treatment (cold or ABA), but some of the newly synthesized proteins induced by cold appear to be the same as those induced by ABA or by mild water deficit.

Protein synthesis is necessary for the development of freezing tolerance, and several distinct proteins accumulate during acclimation to cold, as a result of changes in gene expression (Guy 1990). Isolation of the genes for these proteins reveals that several of the proteins that are induced by low temperature share homology with the RAB/LEA/DHN (*r*esponsive to *AB*A, *l*ate *e*mbryo *a*bundant, and *dehydrin*, respectively) protein family. Members of this protein family are synthesized during seed maturation under developmental control, but they are also induced in vegetative cells by water deficit and by exposure to ABA. These proteins are very hydrophilic and have the unusual property of being stable at boiling temperatures. They are thought to protect cells dehydrated by water shortage or freezing, by stabilizing other proteins and membranes.

ABA appears to have a role in inducing freezing tolerance. Winter wheat, rye, spinach, and *Arabidopsis thaliana* are all cold-tolerant species, and when they are hardened by water shortages their freezing tolerance also increases. This tolerance to freezing is increased at nonacclimating temperatures by mild water deficit, or

BOX 25.1

Ice Formation in Higher-Plant Cells

DURING THE INITIAL PHASE of heat loss from tissue, when the temperature drops below the freezing point of the cytosol and the vacuole without ice forming (see the figure), the liquid phase in the cytosol and vacuole is said to be *supercooled*. As the temperature drops further, ice forms in the intercellular spaces, and heat energy (0.33 kJ or 80 calories per gram) is released as a result of the latent heat of fusion of water. At this stage, the temperature of the tissue reflects the balance between heat gain from ice formation and heat loss to the environment. As a result, when ice first forms, the temperature rises rapidly (point B in the figure), and it remains at that level until all the extracellular water is frozen (point C in the figure). When that point is reached, heat release stops, and the temperature begins to fall again. The release of heat energy during ice formation (points B to C in the figure) is the basis for the common practice of spraying crops with water during frost: As long as the water continues to freeze extracellularly, it releases heat that prevents intracellular freezing.

The formation of ice within the intercellular spaces of cells that are sensitive to freezing is not lethal, but extended exposure to freezing temperatures causes water vapor to move through the plasma membrane from the unfrozen protoplast to the cell wall, causing ice crystals to grow within the intercellular spaces. This slow dehydration concentrates solutes within the protoplast, depressing the freezing point by 2 to 3°C. As the temperature continues to drop, a second phase of release of the heat of fusion of water is detectable (points from D to E on the figure). This phase reflects a series of small freezing events: Each "spike" of heat release is thought to represent the freezing of cell protoplasts and coincides with loss of viability. The formation of ice crystals on cell walls or in the protoplasm requires the presence of *ice nucleation points* on which crystals can be initiated and grow.

In some species, acclimation confers an ability to suppress ice nucleation in the protoplast, allowing *deep supercooling* to many degrees below the freezing point without ice formation. However, deep supercooling of intracellular water has a lower limit of about –40°C, the temperature at which ice forms spontaneously. (The homogeneous ice formation temperature of pure water droplets is –38.1°C, but the presence of solutes lowers this temperature.) At or below –40°C, ice crystals form without nucleation points, and intracellular freezing and cell death are unavoidable. Species that tolerate temperatures below –40°C under natural conditions do so not by supercooling, but by tolerating gradual dehydration.

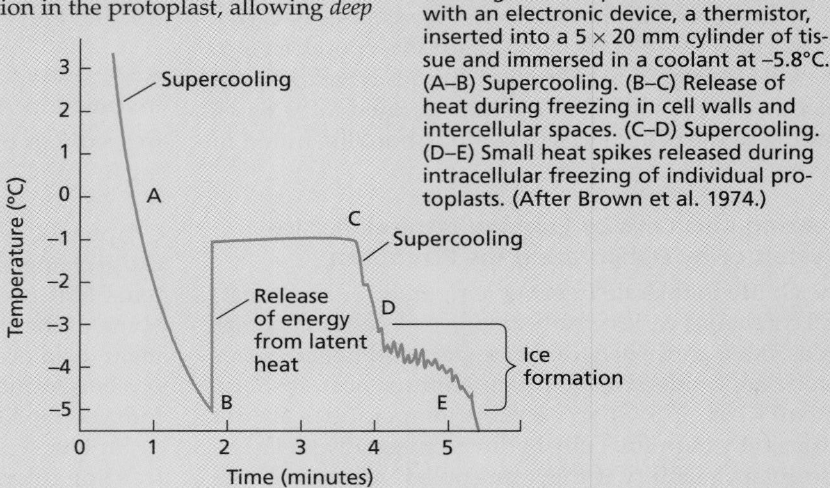

Temperature of parenchyma cells in cucumber (*Cucumis sativus*) fruit during freezing. The temperature was recorded with an electronic device, a thermistor, inserted into a 5 × 20 mm cylinder of tissue and immersed in a coolant at –5.8°C. (A–B) Supercooling. (B–C) Release of heat during freezing in cell walls and intercellular spaces. (C–D) Supercooling. (D–E) Small heat spikes released during intracellular freezing of individual protoplasts. (After Brown et al. 1974.)

at low temperatures, either of which increases endogenous ABA concentrations in leaves. Plants develop freezing tolerance at nonacclimating temperatures when treated with exogenous ABA. Many of the genes or proteins expressed at low temperatures or with water deficit are also inducible by ABA under nonacclimating conditions. All these findings support a role of ABA in tolerance to freezing.

Mutants of *Arabidopsis* that are insensitive to ABA (*abi1*) or ABA deficient (*aba-1*) are unable to undergo low-temperature acclimation to freezing. Only in *aba-1*, however, exposure to ABA restores the ability to develop freezing tolerance (Mantyla et al. 1995). On the other hand, not all the genes induced by low temperature are ABA dependent, and it is not yet clear if their expression is critical for the full development of freezing tolerance. For instance, research on the tolerance of rye crowns to freezing has found that the lethal temperature for 50% of the crowns (LT_{50}) is –2 to –5°C for controls grown at 25°, –8° for ABA-treated crowns, and –28° after acclimation at 2°C. Cell cultures of bromegrass (*Bromus inermis*) show a more dramatic induction of freezing tolerance when treated with ABA: Controls grown at 25°C could survive to –9°C, whereas 7 days of exposure to ABA improved the freezing tolerance to –40°C (Gusta et al. 1996).

Typically, a minimum of several days of exposure to cool temperatures is required for freezing resistance to be induced fully. Potato requires 15 days of exposure to cold. On the other hand, when rewarmed, plants lose their freezing tolerance rapidly, and can become susceptible to freezing once again in 24 hours. The need for cool temperatures to induce acclimation to even lower temperatures, and the rapid loss of acclimation on warming, explain the susceptibility of plants in the southern United States (and similar climatic zones with highly variable winters) to extremes of temperature in the winter months, when air temperature can drop from 20 to 25°C to below 0°C in a few hours.

Limiting Extracellular Ice Formation May Contribute to Freezing Tolerance

Several specialized proteins that have been identified in plants may help limit the growth of ice crystals by a noncolligative mechanism; that is, the effect does not depend on the lowering of the freezing point of water by the presence of solutes. These *antifreeze proteins* are induced by cold temperatures, and they bind to the surfaces of ice crystals and prevent or slow further crystal growth. In rye leaves, antifreeze proteins are localized in the epidermal cells and cells surrounding the intercellular spaces, where they can inhibit the growth of extracellular ice. Plants and animals may use similar mechanisms to limit ice crystals: A cold-inducible gene in *Arabidopsis* has been identified that has homology to a gene that encodes the antifreeze protein in winter flounder.

Sugars and some of the cold-induced proteins are suspected to have cryoprotective effects; they stabilize proteins and membranes during dehydration induced by low temperature. In winter wheat, the greater the sucrose concentration, the greater the freezing tolerance. Sucrose predominates among the soluble sugars associated with freezing tolerance, but in some species raffinose, fructans, sorbitol, or mannitol serves the same function. During cold acclimation of winter cereals, soluble sugars accumulate in the cell walls, where they may help restrict the growth of ice. A cryoprotective glycoprotein has been isolated from leaves of cold-acclimated cabbage (*Brassica oleracea*). In vitro, the protein protects thylakoids isolated from nonacclimated spinach (*Spinacia oleracea*) against damage from freezing and thawing.

Some Woody Plants Can Acclimate to Very Low Temperatures

When in a dormant state, some woody plants are extremely resistant to low temperatures. Resistance is determined in part by previous acclimation to cold, but genetics plays an important role in determining the degree of tolerance to low temperatures. Clones of red osier dogwood (*Cornus stolonifera*) obtained from different climatic regions showed different degrees of resistance to low temperature in the fall and early winter when grown together in the same environment (Weiser 1970). Likewise, native species of *Prunus* (cherry, plum, and other pit fruits) from northern cooler climates in North America are hardier after acclimation than those from milder climates. When the species were tested together in the laboratory, those with a northern geographic distribution showed greater ability to avoid intracellular ice formation, underscoring distinct genetic differences (Burke and Stushnoff 1979).

Under natural conditions, woody species acclimate to cold in two distinct stages (Weiser 1970). In the first stage, hardening is induced in the early autumn by exposure to short days and nonfreezing chilling temperatures, both of which combine to stop growth. A diffusible factor that promotes acclimation (probably ABA) moves in the phloem from leaves to overwintering stems and may be responsible for the changes. During this period, woody species also withdraw water from the xylem vessels, thereby preventing the stem from splitting in response to the expansion of water during later freezing. Cells in the first stage of acclimation can survive temperatures well below 0°C, but they are not fully hardened. In the second stage, direct exposure to freezing is the stimulus; no factor that is known to be translocated within the plant can substitute for this exposure and confer hardening. When fully hardened, the cells can tolerate exposure to temperatures of –50 to –100°C.

Resistance to Freezing Temperatures Involves Supercooling and Slow Dehydration

In many species of the hardwood forests of southeastern Canada and the eastern United States, acclimation to freezing involves the suppression of ice crystal formation at temperatures far below the theoretical freezing point (see Box 25.1). This **deep supercooling** is seen in species such as oak, elm, maple, beech, ash, walnut, hickory, rose, rhododendron, apple, pear, peach, and plum (Burke and Stushnoff 1979). Deep supercooling also takes place in the Rocky Mountains of Colorado in stem and leaf tissue of tree species such as Engelmann spruce (*Picea engelmannii*) and subalpine fir (*Abies lasiocarpa*), but this resistance to freezing is quickly weakened once growth resumes in the spring (Becwar et al. 1981). Stem tissues of subalpine fir, which undergo deep supercooling and remain viable to below –35°C in May, lose their ability to suppress ice formation in June and can be killed then at –10°C.

Cells can supercool only to about –40°C, at which temperature ice forms spontaneously. Spontaneous ice formation sets the low-temperature limit at which many alpine and subarctic species that undergo deep supercooling can survive. It also explains the altitude of the timberline in mountain ranges, which is at or near the –40°C minimum isotherm.

Protoplasts suppress ice nucleation when undergoing deep supercooling. In addition, the cell walls act as barriers both to the growth of ice from the intercellular spaces into the wall, and to the loss of liquid water from the protoplast to the extracellular ice, which is driven by a steep vapor pressure gradient (Wisniewski and Arora 1993). Many flower buds survive the winter by deep supercooling (e.g., grape, blueberry, peach, azalea, and flowering dogwood), and serious economic losses, particularly of peach, can result from the decline in freezing tolerance of the flower buds in the spring. The cells then no longer supercool, and ice crystals that form extracellularly in the bud scales draw water from the apical meristem, killing the floral apex by dehydration.

The floral buds of apple and pear, the vegetative buds of all temperate fruit trees, and the living cells in their bark do not supercool, but they resist dehydration during extracellular ice formation. Resistance to cellular dehydration is highly developed in woody species of northern Canada, Alaska, and northern Europe and Asia that are subject to average annual minima much below –40°C. Ice formation starts at –3 to –5°C in the intercellular spaces, where the crystals continue to grow, fed by the gradual withdrawal of water from the protoplast, which remains unfrozen. Resistance to freezing temperatures depends on the capacity of the extracellular spaces to accommodate the volume of growing ice crystals and on the ability of the protoplast to withstand dehydration.

This restriction of ice crystal formation to extracellular spaces, accompanied by protoplast dehydration, may explain why some woody species that are resistant to freezing are also resistant to water deficit during the growing season. For example, species of willow (*Salix*), white birch (*Betula papyrifera*), quaking aspen (*Populus tremuloides*), pin cherry (*Prunus pensylvanica*), chokecherry (*Prunus virginiana*), and lodgepole pine (*Pinus contorta*) tolerate very low temperatures by limiting the formation of ice crystals to the extracellular spaces. However, acquisition of resistance depends on slow cooling and gradual extracellular ice formation and protoplast dehydration. Sudden exposure to very cold temperatures before full acclimation causes intracellular freezing and cell death.

Some Bacteria That Live on Leaf Surfaces Increase Frost Damage

When leaves are cooled to temperatures in the –3 to –5°C range, the formation of ice crystals on the surface (frost) is accelerated by certain bacteria that naturally inhabit the leaf surface, such as *Pseudomonas syringae* and *Erwinia herbicola*, which act as ice nucleators. When artificially inoculated with cultures of these bacteria, leaves of frost-sensitive species freeze at warmer temperatures than leaves that are bacteria free (Lindow et al. 1982). The surface ice quickly spreads to the intercellular spaces within the leaf, leading to cellular dehydration. Bacterial strains can be genetically modified so that they lose their ice-nucleating characteristics, and such strains have been used commercially in foliar sprays of valuable frost-sensitive crops like strawberry to compete with native bacterial strains and thus minimize the number of potential ice nucleation points.

Freezing of Plant Structures Can Be Viewed by Infrared Thermography

A detector that is sensitive to infrared radiation (8 to 12 μm, see Figure 7.2) and that is connected to a computer can be used to detect and form images from transient outputs of heat (the latent heat of fusion of ice) during the freezing of water in leaves and stems (Wisniewski et al. 1997). Images are color coded for the small temperature changes (less than 1°C) and recorded on video. When the plant tissue is exposed to low temperature, the propagation of ice up the stem or along a leaf petiole can be visualized (Figure 25.12). A droplet of a suspension of ice-nucleating bacteria on a leaf surface can cause freezing to propagate across a leaf like a wave, as heat is momentarily released during the formation of ice.

Heat Stress and Heat Shock

Turning to the other temperature extreme, to what extent are plants able to resist excessively high temperatures? Few higher-plant species survive a steady temperature above 45°C. Nongrowing cells or tissues of higher plants that are dehydrated (e.g., seeds and

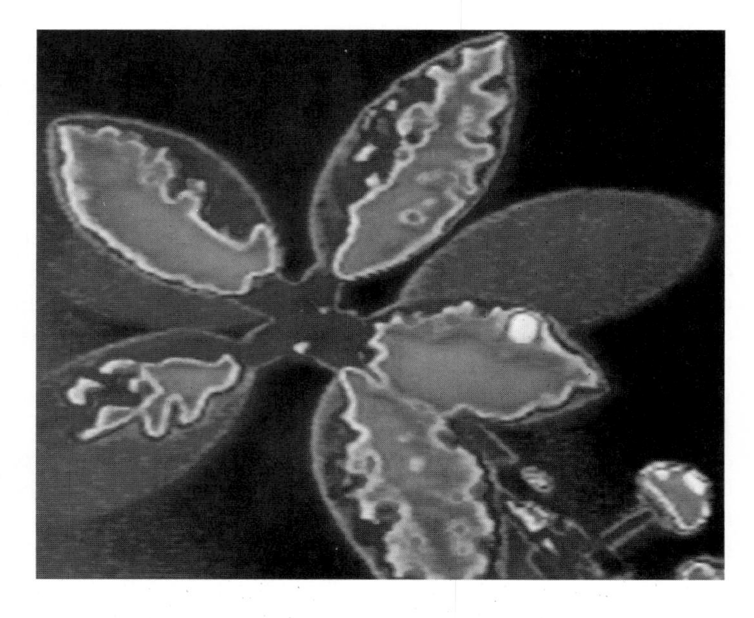

Figure 25.12 The nucleation and propagation of ice in the stem, buds, and leaves of rhododendron (*Rhododendron* sp.). Emission of heat (colored blue and rust) during ice formation was detected by infrared thermography. Ice nucleation was observed to begin in the stem (lower right-hand corner), and then to spread to buds and leaves (blue and rust-colored areas). Even though a drop of a suspension of the ice-nucleating bacteria *Pseudomonas syringae* (yellow spot on the center leaf of the right-hand side) froze, initial nucleation and freezing started in the stem. (From Wisniewski et al. 1997, courtesy of M. Wisniewski, © American Society of Plant Physiologists, reprinted with permission. See frontispiece for a full-color version of this photograph.)

pollen) can survive much higher temperatures than hydrated, vegetative, growing cells can (Table 25.3). Actively growing tissues rarely survive temperatures above 45°C, but dry seeds of some species can endure 120°C and pollen grains of some species can endure 70°C. In general, only single-celled organisms can complete their life cycle at temperatures above 50°C, and only prokaryotes can do it above 60°C. In the following sections we discuss the effect of intense solar radiation on leaf temperature and the ensuing heat stress, the damage inflicted by heat stress to cell membranes and proteins, and the different processes mediating plant adaptations and acclimations to high temperatures.

High Leaf Temperature and Water Deficit Lead to Heat Stress

Some succulent (CAM) higher plants, such as *Opuntia* and *Sempervivum*, are adapted to hot conditions and can tolerate tissue temperatures of 60 to 65°C during intense solar radiation in summer (see Table 25.3). Because CAM plants keep their stomata closed during the day, they cannot cool by transpiration; instead, they depend on re-emission of long-wave (infrared) radiation and loss of heat by conduction and convection to the surrounding air (see Chapter 9). On the other hand, for typical non-irrigated C_3 and C_4 plants that rely on transpirational cooling to lower leaf temperature, leaf temperature can readily rise 4 to 5°C above ambient air temperature in bright sunlight near midday, particularly when soil water deficit causes partial stomatal closure.

Increases in leaf temperature during the day can be pronounced in plants from arid and semiarid regions experiencing drought and high irradiance from sunshine. In general, water deficit is invariably associated with some heat stress, and we can see from Table 25.3 that a rise in temperature may shift leaves from optimal to nearly lethal ranges of temperatures. Heat stress is also a potential danger in greenhouses, where low air speed and high humidity decrease the rate of leaf cooling. A moderate degree of heat stress slows growth of the whole plant. Some irrigated crops have been found to use transpirational cooling as a form of heat resistance. Enhanced transpirational cooling is associated with higher agronomic yields (see Box 25.2).

Table 25.3
Heat-killing temperatures for plants

Plant	Heat-killing temperature (C°)	Time of exposure
Nicotiana rustica (wild tobacco)	49–51	10 min
Cucurbita pepo (squash)	49–51	10 min
Zea mays (corn)	49–51	10 min
Brassica napus (rape)	49–51	10 min
Citrus aurantium (sour orange)	50.5	15–30 min
Opuntia (cactus)	>65	—
sempervivum arachnoideum (succulent)	57–61	—
Potato leaves	42.5	1 hour
Pine and spruce seedlings	54–55	5 min
Medicago seeds (alfalfa)	120	30 min
Grape (ripe fruit)	63	—
Tomato fruit	45	—
Red pine pollen	70	1 hour
Various mosses		
Hydrated	42–51	—
Dehydrated	85–110	—

Source: After Table 11.2 in Levitt 1980.

At High Temperatures, Photosynthesis Is Inhibited before Respiration

Both photosynthesis and respiration are inhibited at high temperatures, but as temperature increases, photosynthetic rates decrease before respiratory rates do (Figure 25.13). The temperature at which the amount of CO_2 fixed by photosynthesis equals the amount of CO_2

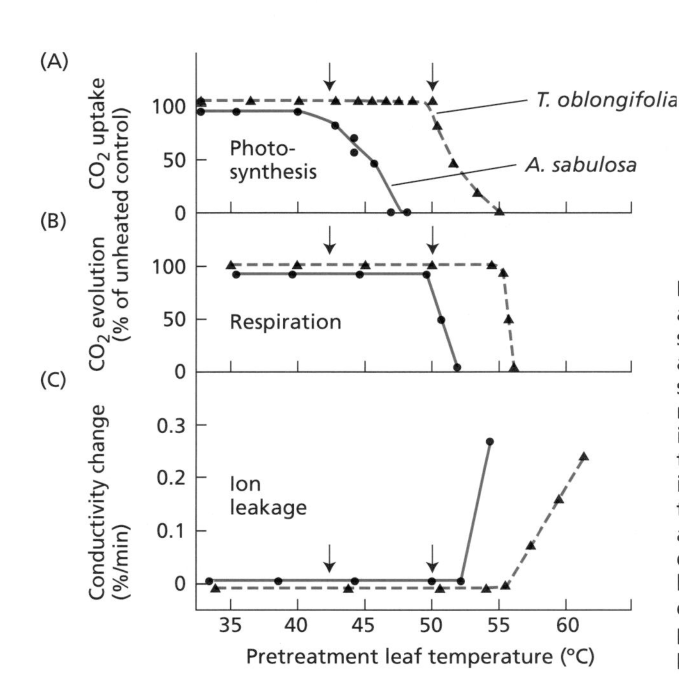

Figure 25.13 Response of *Atriplex sabulosa* (frosted orache) and *Tidestromia oblongifolia* (Arizona honeysweet) to heat stress. Photosynthesis (A) and respiration (B) were measured on attached leaves, and ion leakage (C) was measured with leaf slices submerged in water. Control rates were measured at a noninjurious 30°C. Attached leaves were then exposed to the indicated temperatures for 15 minutes and returned to the initial control conditions before the rates were recorded. Arrows indicate the temperature thresholds for inhibition of photosynthesis in each of the two species. Photosynthesis, respiration, and membrane permeability were all more sensitive to heat damage in *A. sabulosa* than in *T. oblongifolia*. In both species, however, photosynthesis was more sensitive to heat stress than either of the other two processes, and photosynthesis was completely inhibited at temperatures that were noninjurious to respiration. (From Björkman et al. 1980.)

BOX 25.2

Stomatal Conductance Predicts Yields of Irrigated Crops Grown in Hot Environments

IN THE LAST FEW DECADES, the breeding of agricultural crops for higher yields has been very successful. Breeders usually select for high-yielding genotypes empirically, paying no attention to specific plant traits that might be conducive to higher yields. However, comparing old, low-yielding lines of any crop with advanced, high-yielding lines shows clearly that many morphological, physiological, and biochemical traits have been altered by the intense selection pressures for higher yields. These changes indicate that selection for high-yielding genotypes has generated indirect selection pressures on the altered traits. If we exclude effects of genes that regulate the expression of two or more unrelated traits (pleiotropic genes), the study of high-yielding lines might reveal specific traits and genes associated with the higher yield. With this information, breeders could explicitly select for yield-enhancing traits to improve yields further.

Changes in plant traits associated with yield increases can be studied conveniently in a historical series of a crop, constituting successive commercial releases within a breeding program, with each member of the series representing an incremental increase in yield. All the members of the series are grown together under the same conditions, and the traits of interest, such as plant architecture or photosynthetic rates, are measured in parallel with yields. Recent studies of this type focusing on historical series of Pima cotton (*Gossypium barbadense*) and bread wheat (*Triticum aestivum*) have shown a remarkable positive correlation between yield increases and increases in stomatal conductance (Figure 1) (Lu et al. 1998).

Pima cotton is a high-quality fiber cotton grown in hot environments under intensive irrigation. The historical series used in the Pima studies included eight members, released between 1949 and 1996, and encompassing a nearly threefold increase in lint yield. The yield increase attained in each commercial release was accompanied by a corresponding increase in stomatal conductance. As Figure 1A shows, stomatal conductance increased by about 30 mmol m^{-2} s^{-1} for each 100 kg ha^{-1} increase in yield.

Genetic crosses between low- and high-yielding Pima strains have shown that stomatal conductance in Pima cotton has a clear-cut genetic component (Figure 2) (Percy et al. 1996). The studies with stomata of Pima cotton are the first to show a clear-cut genetic regulation of stomatal conductance proper, of the stomatal response to temperature, and of proton pumping rates of isolated guard cells (Lu et al. 1998).

In the absence of explicit selection for higher stomatal conductance in the Pima breeding program, what could be the nature of the indirect selection pressures that caused the increases in conductance paralleling the increases in yield? Higher stomatal conductance increases CO_2 diffusion into the leaf and favors higher photosynthetic rates (see Chapter 9). Higher photosynthetic rates could in turn favor a higher biomass and higher crop yields. Advanced Pima lines show a higher photosynthetic capacity than older, low-yielding lines, but photosynthetic rates measured in the same leaves used to measure stomatal conductance were not positively correlated with yields (Radin et al. 1994). Thus, higher stomatal conductance appears to favor higher yields by a mechanism not directly related to photosynthesis.

Evapotranspiration at the leaf surface lowers leaf temperature (see Chapter 9), and higher stomatal conductance enhances this **leaf cooling**. Optimal daytime temperatures for growth, photosynthesis, and reproduction in Pima cotton are below 30°C, while afternoon air temperature in the hot Pima-growing areas often exceeds 40°C. Evaporative cooling of the leaves thus reduces the gap between optimal growth temperatures and air temperature and provides **heat resistance**. Studies with the Pima historical series showed that, because of leaf cooling, leaf temperatures were several degrees lower than air temperature, and leaf and canopy temperature were lower in the advanced, high-yielding lines than in low-yielding lines.

The same relationship among yields, stomatal conductance, and leaf temperature was found in a historical series of bread wheat grown in the warm Yaqui Valley of northwestern Mexico (see Figure 1B) (Lu et al. 1998). These studies indicate that selection for higher yields in irrigated crops grown at high temperatures imposes indirect selection pressures for high stomatal conductance that lowers leaf temperature and appears to reduce deleterious effects of heat stress on critical flowering and fruiting stages, thus resulting in higher crop yields.

(A) Pima cotton (Arizona)

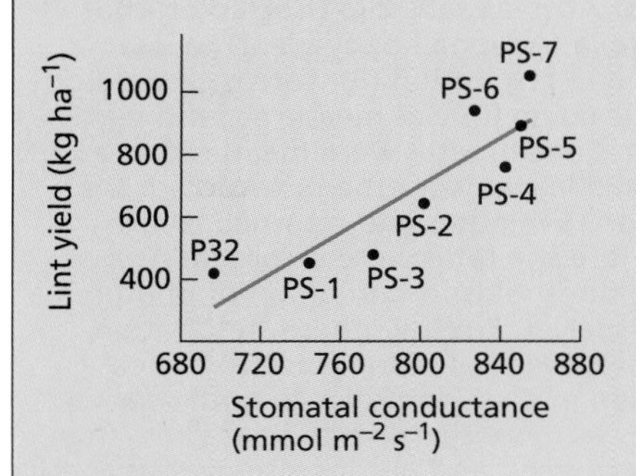

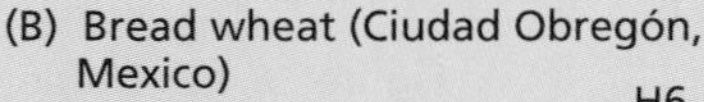

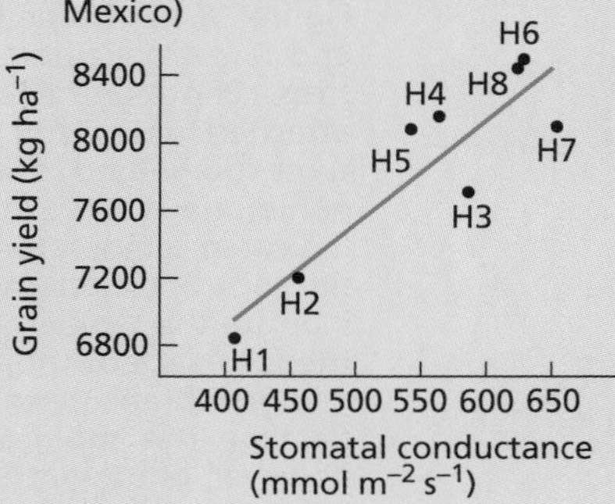

Figure 1 Stomatal conductance has increased in parallel with agronomic yields in irrigated Pima cotton (*Gossypium barbadense*) and bread wheat (*Triticum aestivum*) selected for higher yields at high temperature. (A) The relationship between lint yield and stomatal conductance in a historical series of Pima cotton grown in Arizona. The abbreviations P32 and PS-1 through PS-7 designate successive commercial releases between 1949 and 1996. (B) The relationship between grain yield and stomatal conductance in a historical series of semidwarf bread wheat grown in Ciudad Obregón, Mexico. The abbreviations H1 through H8 designate successive commercial lines released by the International Maize and Wheat Improvement Center between 1962 and 1988. (From Lu et al. 1998.)

BOX 25.2 *(continued)*

If selection for higher yields generates indirect selection pressures for higher stomatal conductance, can breeders explicitly select for higher conductance and thus obtain lines with higher yields? Studies with Pima cotton have shown that selection of high-conductance F_2 progeny from a cross between high- and low-conductance parents produces high-conductance F_4 lines with higher lint yields than low-conductance lines have (Radin et al. 1994). These experiments indicate that high stomatal conductance could be used as a selection trait for high yields in irrigated crops grown at high temperature.

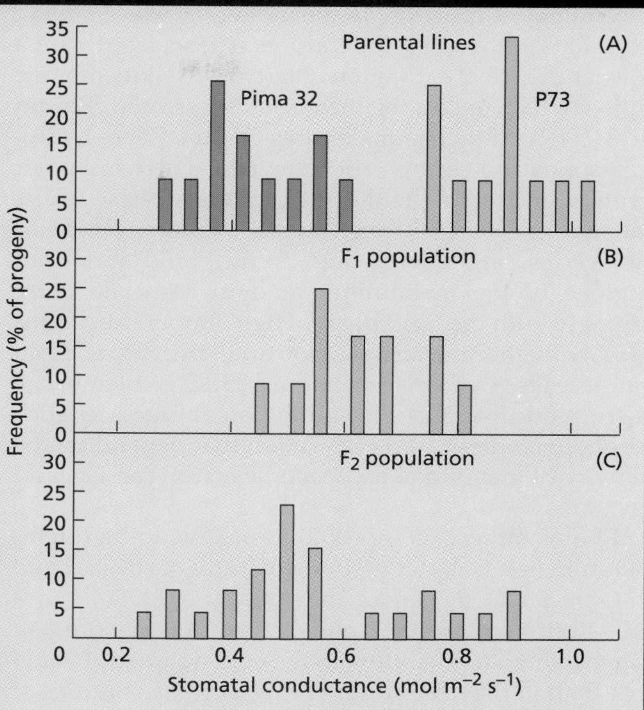

Figure 2 Frequency distribution of stomatal conductance in parental populations of the high-yidling Pima cotton line (P73), the old Pima line (P32), and their F_1 and F_2 progeny. (A) Nonoverlapping distribution of stomatal conductance in the two parental populations. (B) Distribution of stomatal conductance in an F_1 population from a cross between the two parents. (C) Distribution of stomatal conductance in an F_2 population derived from F_1 parents. (After Percy et al. 1996.)

The observed increases in stomatal conductance in crops grown at high temperature also have valuable implications for models developed to predict global climate changes, which are expected to occur because of increases in atmospheric CO_2 (see Chapter 9). Under atmospheric CO_2 concentrations that are twice as high as the present ones, evapotranspiration would decrease over the continents and air temperatures would increase significantly over the tropical land masses, amplifying the changes resulting from atmospheric radiative effects (Sellers et al. 1997). However, these models have yet to take into consideration the effect of higher temperatures on the stomatal control of evapotranspiration. Incorporation of the stomatal response to temperature in climate change models is likely to improve the accuracy of the predictions generated by the models. These studies illustrate both the complex interactions between physical and biological factors in the biosphere, and the advantage of a thorough understanding of the physiological properties of plants for efforts aimed at increasing agronomic yields and at improving models that predict global climate changes.

released by respiration in a given time is called the **temperature compensation point**. Above this temperature, photosynthesis cannot replace the carbon used as a substrate for respiration. As a result, carbohydrate reserves decline, and fruits and vegetables lose sweetness. This imbalance between photosynthesis and respiration is one of the main causes of the deleterious effects of high temperatures. In the same plant, the temperature compensation point is usually lower for shade leaves than for sun leaves that are exposed to light (and heat). Enhanced respiration rates relative to photosynthesis at high temperatures are more detrimental in C_3 plants than in C_4 or CAM plants because the rates of both dark respiration and photorespiration are increased in C_3 plants.

Plants That Are Adapted to Cool Temperatures Acclimate Poorly to High Temperatures

The extent to which plants that are adapted to one temperature range can acclimate to a contrasting temperature range is illustrated by a comparison of the responses of two C_4 species, *Atriplex sabulosa* (frosted orache, family Chenopodiaceae) and *Tidestromia oblongifolia* (Arizona honeysweet, family Amaranthaceae). *A. sabulosa* is native to the cool climate of coastal northern California, and *T. oblongifolia* is native to the very hot climate of Death Valley, California, where it grows in a temperature regime that is lethal for most plant species. When these species were grown in a controlled environment and their growth rates were recorded as a function of temperature, *T. oblongifolia* barely grew at 16°C, while *A. sabulosa* was at 75% of its maximum growth rate. By contrast, the growth rate of *A. sabulosa* began to decline between 25 and 30°C, and growth ceased at 45°C, the temperature at which *T. oblongifolia* growth showed a maximum (Björkman et al. 1980). Clearly, neither species could acclimate to the temperature range of the other.

High Temperature Impairs the Thermal Stability of Membranes and Proteins

The stability of various cellular membranes is important during high-temperature stress, just as it is during chilling and freezing. Excessive fluidity of membrane lipids at high temperatures is correlated with loss of physiological function. In oleander (*Nerium oleander*), acclimation to high temperatures is associated with a greater degree of saturation of fatty acids in membrane lipids, which makes the membranes less fluid (Raison et al. 1982). At high temperatures the strength of hydrogen bonds and electrostatic interactions between polar groups of proteins within the aqueous phase of the membrane decrease. Thus, integral membrane proteins (which associate with both hydrophilic and lipid regions of the membrane) tend to associate more strongly with the lipid phase. High temperatures thus modify membrane composition and structure and can cause leakage of ions (see Figure 25.13C). Membrane disruption also causes the inhibition of processes such as photosynthesis and respiration that depend on the activity of membrane-associated electron carriers and enzymes.

Photosynthesis is especially sensitive to high temperature (see Chapter 9). In their study of *Atriplex* and *Tidestromia*, O. Björkman and colleagues (1980) found that electron transport in photosystem II was more sensitive to high temperature in the cold-adapted *A. sabulosa* than in the heat-adapted *T. oblongifolia*. Studies of enzymes extracted from these plants showed that ribulose-1,5-bisphosphate carboxylase, NADP:glyceraldehyde-3-phosphate dehydrogenase, and phosphoenolpyruvate carboxylase were less stable at high temperatures in *A. sabulosa* than in *T. oblongifolia*. However, the temperatures at which these enzymes began to denature and lose activity were distinctly higher than the temperatures at which photosynthesis began to decline. These results suggest that early stages of heat injury to photosynthesis are more directly related to changes in membrane properties and to uncoupling of the energy transfer mechanisms in chloroplasts than to a general denaturation of proteins.

Several Adaptations Protect Leaves against Excessive Heating

In environments with intense solar radiation and high temperatures, plants avoid excessive heating of their leaves by decreasing their absorption of solar radiation. The anatomical and physiological adaptations that accomplish this task are similar to the adaptations to water stress that decrease water use by energy dissipation, which we discussed earlier in the chapter. In warm, sunny environments in which a transpiring leaf is near its upper limit of temperature tolerance, any further warming arising from decreased evaporation of water or increased energy absorption can damage the leaf.

When energy absorption decreases and the leaf cools, the driving force for evaporation of water also decreases. As a result, both drought resistance and heat resistance depend on the same adaptations: reflective leaf hairs and leaf waxes; leaf rolling and vertical leaf orientation; and growth of small, highly dissected leaves to minimize the boundary layer thickness and thus maximize convective and conductive heat loss (see Chapters 4 and 9). Some desert shrubs—e.g., white brittlebush (*Encelia farinosa*, family Compositae)—have dimorphic leaves to avoid excessive heating: Green, nearly hairless leaves found in the winter are replaced by white, pubescent leaves in the summer.

Increases in Temperature Induce Synthesis of Heat Shock Proteins

The **heat shock proteins** (**HSPs**), which are synthesized in response to high temperatures, help cells withstand heat stress. Heat shock proteins were discovered in the fruit fly (*Drosophila melanogaster*) and have since been identified in other animals, including humans, as well as in plants and microorganisms. When cells or seedlings of soybean are suddenly shifted from 25 to 40°C (just below the lethal temperature), synthesis of the commonly found set of mRNAs and proteins is suppressed, while transcription and translation of a set of 30 to 50 other proteins (HSPs) are enhanced. New HSP transcripts (mRNAs) can be detected 3 to 5 minutes after heat shock (Sachs and Ho 1986). The molecular masses of the HSPs range from 16 to 104 kDa (kilodaltons).

Low-molecular-weight (15 to 30 kDa) HSPs are more abundant in higher plants than in other organisms, and they show little homology with low-molecular-weight HSPs in animals or microorganisms. However, some of the other HSPs are very similar in plants and animals. For instance, structural genes coding for a 70 kDa HSP in maize, *Drosophila*, and humans show a 75% homology. Although plant HSPs were first identified in response to sudden changes in temperature (25 to 40°C) that rarely occur in nature, HSPs are also induced by more gradual rises in temperature that are representative of the natural environment, and they occur in plants under field conditions.

Some HSPs are found in normal, unstressed cells, and some essential cellular proteins are homologous to HSPs but do not increase in response to thermal stress (Vierling 1991). Many HSPs are thus encoded by members of multigene families, only some of which are induced by heat. Different HSPs are localized to the nucleus, mitochondria, chloroplasts, endoplastic reticulum, and cytosol.

Cells or plants that have been induced to synthesize HSPs show improved thermal tolerance and can tolerate exposure to temperatures that were previously lethal. Some of the HSPs are not unique to high-temperature stress; they are also induced by widely different environmental stresses or conditions, including water deficit, ABA treatment, wounding, low temperature, and salinity. Thus cells previously exposed to one stress may gain cross-protection against another stress. Such is the case with tomato fruits, in which heat shock (48 hours at 38°C) has been observed to promote HSP accumulation and to protect cells for 21 days from chilling at 2°C.

The functions of all the different HSPs are not yet known, but HSPs 60, 70, and 90 (60, 70, and 90 kDa, respectively) and others act as molecular chaperones, involving ATP-dependent stabilization and folding of proteins, and the assembly of oligomeric proteins. Some HSPs assist in polypeptide transport across membranes into cellular compartments. HSP 90s are associated with hormone receptors in animal cells and may be required for their activation, but there is no comparable information for plants. Some HSPs temporarily bind and stabilize an enzyme at a particular stage in cell development, later releasing the enzyme to become active. An example is the plastid HSP 70 in the chromoplast of *Narcissus pseudonarcissus* (daffodil, family Amaryllidaceae) flowers, which protects phytoene desaturase, an enzyme associated with carotenoid biosynthesis. The enzyme, nuclear encoded, is imported into the chromoplast early in its development, and there it remains bound and stabilized by the HSP until it is released later to become enzymatically active (Bonk et al. 1996).

Hydrophobic amino acid residues in nonfolded polypeptide chains are protected by HSPs from the aqueous environment that would otherwise lead to hydrophobic interactions and nonnative aggregation of polypeptides. In vitro tests with protein fractions enriched in HSPs 15 to 18 show that they behave as heat-stable proteins to nonspecifically stabilize other proteins that are easily denatured by heat. In vivo, however, the binding of HSPs to particular polypeptides within subcellular compartments is likely to be much more specific. These interactions between polypeptides and HSPs may be especially critical at high temperatures because of the tendency of many proteins to denature at high temperatures.

Conditions that induce thermal tolerance in plants closely match those that induce the accumulation of HSPs, but that correlation alone does not prove that HSPs play an essential role in acclimation to heat stress. Expression of an activated HSP transcription factor induces constitutive synthesis of HSPs and increases the thermotolerance of *Arabidopsis* (Lee et al. 1995). Studies with *Arabidopsis* plants containing an antisense DNA sequence that reduces HSP 70 synthesis showed that the high-temperature extreme at which the plants could survive was reduced by 2°C compared with controls, although the mutant plants grew normally at optimum temperatures (Lee and Schoeffl 1996). Presumably failure to synthesize the entire range of HSPs that are usually induced in the plant would lead to a much more dramatic loss of heat tolerance.

Salinity Stress

Under natural conditions, terrestrial higher plants encounter high concentrations of salts close to the seashore and in estuaries where seawater and fresh water mix or replace each other with the tides. Far inland, natural salt seepage from geologic marine deposits can wash into adjoining areas, rendering them unusable for agriculture. However, a much more extensive problem in agriculture is the accumulation of salts from irrigation water. Evaporation and transpiration remove pure water (as vapor) from the soil, and this water loss concentrates solutes in the soil. When irrigation water contains a high concentration of solutes and when there is no opportunity to flush out accumulated salts to a drainage system, salts can quickly reach levels that are injurious to salt-sensitive species. It is estimated that about one-third of the irrigated land on Earth is affected by salt. In this section, we discuss how plant function is affected by water and soil salinity, and we examine the processes that assist plants in avoiding salinity stress.

Salt Accumulation in Soils Impairs Plant Function and Soil Structure

In discussing the effects of salts in the soil, we distinguish between high concentrations of Na^+ (**sodicity**) and high concentrations of total salts (**salinity**). The two concepts are often related, but in some areas Ca^{2+}, Mg^{2+}, and $SO^{2}_{4}{}^{-}$, as well as NaCl, can contribute substantially to salinity. The high Na^+ concentration of a sodic soil can not only injure plants directly but also degrade the soil structure, decreasing porosity and water permeability. A sodic clay soil known as *caliche* is so hard and impermeable that dynamite is sometimes required to dig through it!

In the field, the salinity of soil water or irrigation water is measured in terms of its electric conductivity or in terms of osmotic potential. Pure water is a very poor conductor of electric current, and the conductivity of a water sample is due to the ions dissolved in it. The higher the salt concentration of water, the greater its electric conductivity and the lower its osmotic potential (higher osmotic pressure) (Table 25.4).

The quality of irrigation water is often poor in semiarid and arid regions. In the United States the salt content of the headwaters of the Colorado River is only 50 mg L^{-1}, but about 2000 km downstream, in southern California, the salt content of the same river reaches about 900 mg L^{-1}—enough to preclude growth of some salt-sensitive crops, such as maize. Water from some wells used for irrigation in Texas may contain as much as 2000 to 3000 mg salt L^{-1}. An annual application of irrigation water totaling 1 m from such wells would add 20 to 30 tonnes of salts per hectare (8 to 12 tons per acre)

Table 25.4
Properties of seawater and of good quality irrigation water

Property	Seawater	Irrigation water
Concentration of ions (m*M*)		
Na^+	457	<2.0
K^+	9.7	<1.0
Ca^{2+}	10	0.5–2.5
Mg^{2+}	56	0.25–1.0
Cl^-	536	<2.0
SO_4^{2-}	28	0.25–2.5
HCO_3^-	2.3	<1.5
Osmotic potential (MPa)	−2.4	−0.039
Total dissolved salts (mg L^{-1} or ppm)	32,000	500

to the soil. These levels of salt are damaging to all but the most resistant crops.

Salinity Depresses Growth and Photosynthesis in Sensitive Species

Plants can be divided into two broad groups on the basis of their response to high concentrations of salts. *Halophytes* are native to saline soils and complete their life cycles in that environment. *Glycophytes* (literally "sweet plants"), or nonhalophytes, are not able to resist salts to the same degree as halophytes. Usually there is a threshold concentration of salt above which glycophytes begin to show signs of growth inhibition, leaf discoloration, and loss of dry weight.

Among crops, maize, onion, citrus, pecan, lettuce, and bean are highly sensitive to salt; cotton and barley are moderately tolerant; and sugar beet and date palms are highly tolerant (Maas and Hoffman 1977; Greenway and Munns 1980). Some species that are highly tolerant of salt, such as *Suaeda maritima* (a salt marsh plant) and *Atriplex nummularia* (a saltbush), show growth stimulation at Cl^- concentrations many times greater than the lethal level for sensitive species (Figure 25.14).

Salt Injury Involves Both Osmotic Effects and Specific Ion Effects

Dissolved solutes in the rooting zone generate a low osmotic potential that lowers the soil water potential. The general water balance of plants is thus affected, because leaves need to develop an even lower water potential to maintain a "downhill" gradient of water potential between the soil and the leaves (see Chapter 4). This effect of dissolved solutes is similar to that of a soil water deficit. Most plants can adjust osmotically when growing in saline soils and in this way prevent loss of turgor, which would slow the extension growth of cells, while generating a lower (more negative) water potential.

Specific ion effects occur when injurious concentrations of Na^+, Cl^-, or SO_4^{2-}accumulate in cells. Under nonsaline conditions, the cytosol of higher-plant cells contains 100 to 200 m*M* K^+ and 1 to 10 m*M* Na^+, an ionic environment in which many enzymes function optimally. An abnormally high ratio of Na^+ to K^+ and high concentrations of total salts inactivate enzymes and inhibit protein synthesis. At a high concentration, Na^+ can displace Ca^{2+} from the plasma membrane of cotton root hairs, resulting in a change in plasma membrane permeability that can be detected as leakage of K^+ from the cells (Cramer et al. 1985).

Photosynthesis is inhibited when high concentrations of Na^+ and/or Cl^- accumulate in chloroplasts. Since photosynthetic electron transport appears relatively insensitive to salts, either carbon metabolism or photophosphorylation may be affected. Enzymes extracted from salt-tolerant species are just as sensitive to the presence of NaCl as enzymes from salt-sensitive glycophytes

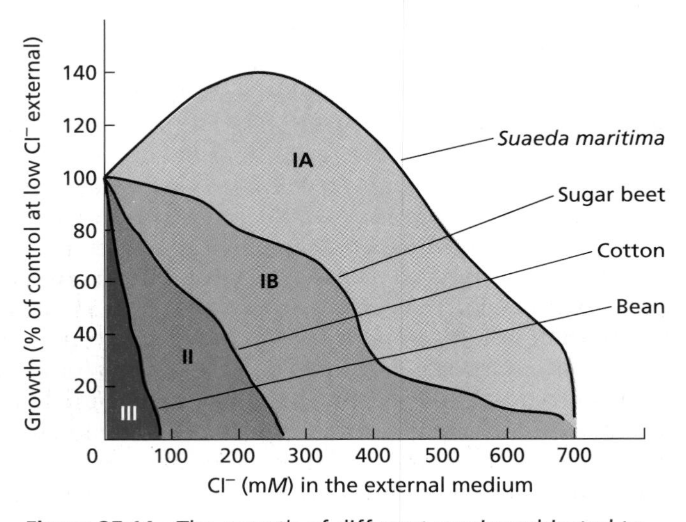

Figure 25.14 The growth of different species subjected to salinity relative to that of unsalinized controls. The curves dividing the regions are based on data for different species. Plants were grown for 1 to 6 months. Representative plant species within the groups are as follows: In Group IA (halophytes), *Suaeda maritima* and *Atriplex nummularia*. These species show growth stimulation with Cl^- levels below 400 m*M*. In Group IB (halophytes), *Spartina* x *townsendii* (Townsend's cordgrass) and sugar beet (*Beta vulgaris*). These plants tolerate salt, but their growth is retarded. Group II (halophytes and nonhalophytes) includes salt-tolerant halophytic grasses that lack salt glands, such as *Festuca rubra* subsp. *littoralis* (red fescue) and *Puccinellia peisonis*, and nonhalophytes, such as cotton (*Gossypium* spp.) and barley (*Hordeum vulgare*). All are inhibited by high salt concentrations. Within this group, tomato (*Lycopersicon esculentum*) is intermediate, and common bean (*Phaseolus vulgaris*) and soybean (*Glycine max*) are sensitive. The species in group III (very salt-sensitive nonhalophytes) are severely inhibited or killed by low salt concentrations. Included are many fruit trees, such as citrus, avocado, and stone fruit. (From Greenway and Munns 1980.)

are. Hence, the resistance of halophytes to salts is not a consequence of a salt-resistant metabolic machinery. Instead, other mechanisms come into play to avoid salt injury, as discussed in the following section.

Plants Use Different Strategies to Avoid Salt Injury

Plants minimize salt injury by excluding salt from meristems, particularly in the shoot, and from leaves that are actively expanding and photosynthesizing. Transport of ions from the root apoplast into the xylem and, subsequently, into the shoot via the transpiration stream is restricted by the requirement that ions move into the symplastic pathway in order to bypass the Casparian strip (see Chapter 4). The extent to which radial ion movement to the xylem must be restricted in order to protect the plant depends on the severity of the stress and the capacity of the specific plant to cope with ions in the shoot. It is generally assumed that halophytes have greater capacity than glycophytes for ion accumulation in cells of the shoots.

Sodium ions can enter roots passively (by moving down an electrochemical-potential gradient; see Chapter 6), so root cells must use energy for active transport of Na^+ back to the outside solution. By contrast, Cl^- is excluded by the low permeability of root plasma membranes to this ion. Movement of Na^+ into leaves is further minimized by absorption of Na^+ from the transpiration stream (xylem sap) during its movement from roots to shoots. Some salt-resistant plants, such as salt cedar (*Tamarix* sp.) and salt bush (*Atriplex* sp.), do not exclude ions at the root, but instead have salt glands at the surface of the leaves. The ions are transported to these glands, where the salt crystallizes and is no longer harmful.

As discussed earlier in relation to water deficit, plant cells can adjust their water potential in response to osmotic stress. Two processes contribute to the decrease in Ψ_s: the accumulation of ions in the vacuole and the synthesis of compatible solutes in the cytosol. The latter compounds include glycine betaine, proline, sorbitol, and sucrose. Specific plant families tend to use one or two of these compounds in preference to others (Wyn Jones and Gorham 1983). The amount of carbon used for the synthesis of these organic solutes can be rather large (about 10% of the plant weight).

In natural vegetation this diversion of carbon to adjust water potential does not affect survival, but in agricultural crops it can reduce yields. Many halophytes exhibit a growth optimum at moderate levels of salinity, and this optimum is correlated with the capacity to accumulate ions in the vacuole, where they can contribute to the cell osmotic potential without damaging the salt-sensitive enzymes. In these leaves, water balance is maintained between the cytoplasm and the vacuoles by the synthesis of organic compounds, such as proline or sucrose, that accumulate in the cytosol.

Oxygen Deficiency

Roots usually obtain sufficient oxygen (O_2) for their aerobic respiration (see Chapter 11) directly from the soil. Gas-filled pores in well-drained, well-structured soil readily permit the diffusion of gaseous O_2 to depths of several meters. Consequently, the O_2 concentration deep in the soil is similar to that in humid air. However, soil can become flooded or waterlogged when it is poorly drained or when rain or irrigation is excessive. Water then fills the pores and blocks the diffusion of O_2 in the gaseous phase. Dissolved oxygen diffuses so slowly in stagnant water that only a few centimeters of soil near the surface remain oxygenated.

When temperatures are low and plants are dormant, oxygen depletion is very slow and the consequences are relatively harmless. When temperatures are higher (greater than 20°C), however, oxygen consumption by plant roots, soil fauna, and soil microorganisms can totally deplete the oxygen from the bulk of the soil water in as little as 24 hours. The growth and survival of many plant species are greatly depressed under such conditions, and crop yields can be severely reduced. Garden pea (*Pisum sativum*) yields can be halved by only 24 hours of flooding. However, specialized natural vegetation (plants of marshes and swamps) and crops such as rice are well adapted to resist oxygen deficiency in the root environment. In the following sections we discuss the damage caused by anaerobiosis to roots and shoots, how wetland vegetation copes with low oxygen tensions, and different acclimations to anoxic stress.

Roots Are Injured in Anaerobic Soil Water

In the absence of O_2, electron transport and oxidative phosphorylation in mitochondria cease, the tricarboxylic acid cycle cannot operate, and ATP can be produced only by fermentation. Thus, when the supply of O_2 is insufficient for aerobic respiration, roots begin to ferment pyruvate (formed in glycolysis; see Chapter 11) to lactate, through the action of lactate dehydrogenase (LDH).

In the root tips of maize, lactate fermentation is transient because the accumulation of lactic acid lowers the cellular pH. As the intracellular pH drops, fermentation switches from the production of lactic acid to ethanol because of the different pH optima of the cytosolic enzymes involved: At acidic pH, LDH is inhibited and pyruvate decarboxylase is activated. The net yield of ATP in fermentation is only 2 moles of ATP per mole of hexose sugar respired (compared with 36 moles of ATP per mole of hexose in aerobic respiration). Thus, injury to root metabolism by O_2 deficiency originates in part from a lack of ATP to drive essential metabolic processes (Drew 1997).

Nuclear magnetic resonance (NMR) spectroscopy was used to measure the intracellular pH of living maize root tips under nondestructive conditions (Roberts et al. 1984, 1992). In healthy cells, the vacuolar contents are more acidic (pH 5.8) than the cytoplasm (pH 7.4). But under conditions of extreme O_2 deficiency, protons gradually leak from the vacuole into the cytoplasm, adding to the acidity generated in the initial burst of lactic acid fermentation. These changes in pH (**cytoplasmic acidosis**) are associated with the onset of cell death. Apparently, active transport of H^+ into the vacuole by tonoplast ATPases is slowed by lack of ATP, and without ATPase activity the normal pH gradient between cytosol and vacuole cannot be maintained, because protons tend to leak across the tonoplast into the cytosol down a steep concentration gradient. Cytoplasmic acidosis irreversibly disrupts metabolism in the cytoplasm of higher-plant cells, as it does in anoxic cells of animals.

When soil is completely depleted of molecular O_2, anaerobic soil microorganisms (anaerobes) derive their energy from the reduction of nitrate (NO_3^-) to nitrite (NO_2^-) or to nitrous oxide (N_2O) and molecular nitrogen (N_2). These gases ($N_2O + N_2$) are lost to the atmosphere in a process called denitrification. As conditions become more reducing, anaerobes reduce Fe^{3+} to Fe^{2+}, and because of its greater solubility Fe^{2+} can rise to toxic concentrations when some soils are anaerobic for many weeks. Other anaerobes may reduce sulfate (SO_4^{2-}) to hydrogen sulfide (H_2S), which is a respiratory poison. When anaerobes have an abundant supply of organic substrate, bacterial metabolites such as acetic acid and butyric acid are released into the soil water, and these acids along with reduced sulfur compounds account for the unpleasant odor of waterlogged soil. All of these substances are toxic to plants at high concentrations (Drew and Lynch 1980).

Root respiration rate and metabolism are modified even before O_2 is completely depleted from the root environment. For the tip of a maize root growing in a well-stirred nutrient solution at 25°C, the **critical oxygen pressure** (**COP**) at which the respiration rate is first slowed by O_2 deficiency is about 0.20 atmosphere (20 kPa, or 20% O_2 by volume), almost the concentration in ambient air. At this oxygen partial pressure, the rate of diffusion of dissolved O_2 from the solution into the tissue and from cell to cell barely keeps pace with the rate of utilization. However, a root tip is metabolically very active, with respiration rates and ATP turnover comparable to those of mammalian tissue. In older zones of the root, where cells are mature and fully vacuolated and the respiration rate is lower, the COP declines to 0.1 to 0.05 atmosphere. When O_2 concentrations are below the COP, the center of the root becomes **anoxic** (completely lacking oxygen) or **hypoxic** (partly deficient in oxygen).

The COP is lower when respiration is slow at cooler temperatures, and it also depends on how bulky the organ is and how tightly the cells are packed. Large, bulky fruits are able to remain fully aerobic because of the large intercellular spaces that readily allow gaseous diffusion. For single cells, an O_2 partial pressure as low as 0.01 atmosphere (1% O_2 in the gaseous phase) can be adequate because diffusion over short distances ensures an adequate O_2 supply to mitochondria. A very low partial pressure of O_2 at the mitochondrion is sufficient to maintain oxidative phosphorylation. The K_m (Michaelis–Menten constant; see Chapter 2) for cytochrome oxidase is 0.1 to 1.0 μM dissolved O_2, a tiny fraction of the concentration of dissolved O_2 in equilibrium with air (277 μM at 20°C). The large difference between the COP values for an organ or tissue and the O_2 requirements of mitochondria is explained by the slow diffusion of dissolved O_2 in aqueous media.

The Failure of O_2-Deficient Roots to Function Injures Shoots

Anoxic or hypoxic roots lack sufficient energy to support physiological processes on which the shoots depend (Jackson and Drew 1984). The failure of the roots of wheat or barley to absorb nutrient ions and transport them to the xylem (and from there to the shoot) leads quickly to a shortage of ions within developing and expanding tissues. Older leaves senesce prematurely because of reallocation of phloem-mobile elements (N, P, K) to younger leaves. The lower permeability of roots to water often leads to a decrease in leaf water potential and wilting, although this decrease is temporary if stomata close, preventing further water loss by transpiration.

Hypoxia also accelerates production of the ethylene precursor, 1-aminocyclopropane-1-carboxylic acid (ACC), in roots (see Chapter 22). In tomato, ACC travels via the xylem sap to the shoot, where, in contact with oxygen, it is converted by ACC oxidase to ethylene. The upper (adaxial) surfaces of the leaf petioles of tomato and sunflower have ethylene-responsive cells that expand more rapidly when ethylene concentrations are high. This expansion results in **epinasty**, the downward growth of the leaves such that they appear to droop. Unlike wilting, epinasty does not involve loss of turgor.

In some species (e.g., pea and tomato), flooding induces stomatal closure apparently without detectable changes in leaf water potential. Oxygen shortage in roots, like water deficit or high concentrations of salts, can stimulate abscisic acid (ABA) production and movement of ABA to leaves. However, stomatal closure under these conditions can be attributed mostly to the additional production of ABA by the older, lower leaves. These leaves do wilt, and they export their ABA to the younger turgid leaves, leading to stomatal closure (Zhang and Zhang 1994).

Submerged Organs Can Acquire O_2 through Specialized Structures

In contrast to flood-sensitive species, wetland vegetation is well adapted to grow for extended periods in water-saturated soil. Even when shoots are partly submerged, they grow vigorously and show no signs of stress. In some wetland species, such as the water lily (*Nymphoides peltata*), submergence traps endogenous ethylene, and the hormone stimulates cell elongation of the petiole, extending it quickly to the water surface so that the leaf is able to reach the air. Internodes of deepwater, or floating, rice respond similarly to trapped ethylene, so the leaves extend above the water surface despite increases in water depth. In the case of pondweed (*Potamogeton pectinatus*), an aquatic monocot, stem extension is insensitive to ethylene; instead it is promoted even under anaerobic conditions, by acidification of the surrounding water caused by the accumulation of respiratory CO_2.

In most wetland plants, and in many plants that acclimate well to wet conditions, the stem and roots develop longitudinally interconnected, gas-filled channels that provide a low-resistance pathway for movement of oxygen and other gases. The gases (air) enter through stomata, or through lenticels on woody stems and roots, and travel by molecular diffusion, or by convection driven by small pressure gradients.

In many wetland plants, exemplified by rice, cells are separated by prominent, gas-filled spaces, which form a tissue called **aerenchyma**, that develop in the roots independently of environmental stimuli. In nonwetland plants, however, including both monocots and dicots, oxygen deficiency induces the formation of aerenchyma in the stem base and newly developing roots (Figure 25.15). In the root tip of maize, hypoxia stimulates the activity of ACC synthase and ACC oxidase, thus causing ACC and ethylene to be produced faster. The ethylene leads to the death and disintegration of cells in the root cortex. The spaces these cells formerly occupied provide the gas-filled voids that facilitate movement of O_2.

Ethylene-signaled cell death is highly selective, for other cells in the root are unaffected. A rise in cytosolic Ca^{2+} concentration is thought to be part of the ethylene signal transduction pathway leading to cell death. Chemicals that elevate cytosolic Ca^{2+} concentration promote cell death under noninducing conditions; conversely, chemicals that lower cytosolic Ca^{2+} concentration block cell death in hypoxic roots that would normally form aerenchyma. It has been suggested (He et al. 1996) that ethylene-dependent cell death in response to hypoxia is an example of programmed cell death.

As the root extends into oxygen-deficient soil, the continuous formation of aerenchyma just behind the tip

(A) Oxygen-rich root (control)

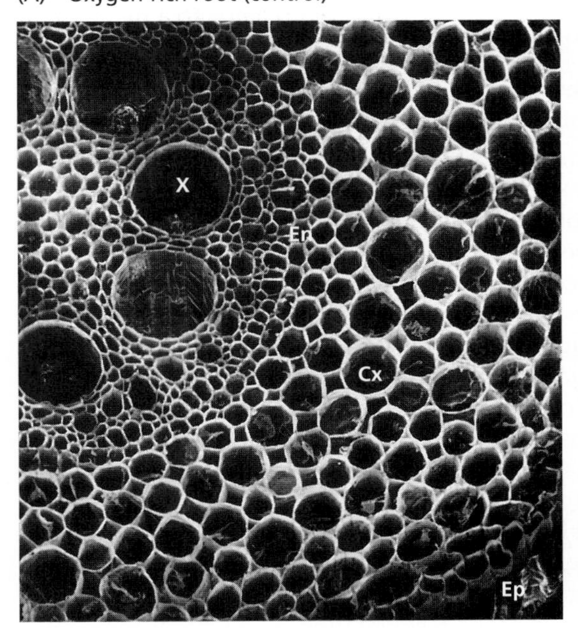

(B) Oxygen-deficient root

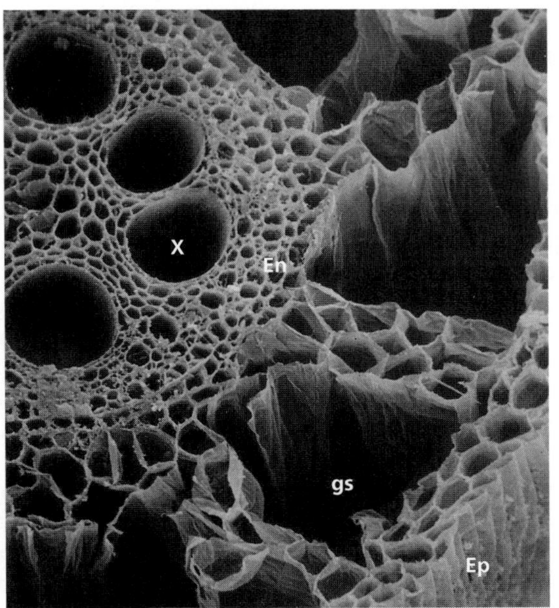

Figure 25.15 Scanning electron micrographs of transverse sections through roots of maize, showing changes in structure with oxygen supply. (150×) (A) Control root, supplied with air, with intact cortical cells. (B) Oxygen-deficient root growing in a nonaerated nutrient solution. Note the prominent gas-filled spaces (gs) in the cortex (Cx), formed by degeneration of cells. The stele (all cells interior to the endodermis, En) and the epidermis (Ep) remain intact. X, xylem. (Courtesy of J. L. Basq and M. C. Drew.)

allows oxygen movement within the root to supply the apical zone. In roots of rice and other typical wetland plants, structural barriers composed of suberized and lignified cells prevent O_2 diffusion outward to the soil. The O_2 thus retained supplies the apical meristem and allows growth to proceed 50 cm or more into anaerobic soil. In contrast, roots of nonwetland species, such as maize, leak O_2, failing to conserve it to the same extent. Thus, in the root apex of these plants, internal O_2 becomes insufficient for aerobic respiration, and this lack of O_2 severely limits the depth to which such roots can extend into anaerobic soil.

Some Plant Tissues Tolerate Anaerobic Conditions

Most tissues of higher plants cannot survive long anaerobically. Root tips of maize, for example, remain viable for only 20 to 24 hours if they are suddenly deprived of O_2. Under anoxia, some ATP is generated slowly by fermentation, but the energy status of cells gradually declines during cytoplasmic acidosis. By contrast, some plants (or parts of them) can tolerate exposure to strictly anaerobic conditions for an extended period (weeks or months). These include the rhizomes (underground horizontal stems) of *Schoenoplectus lacustris* (giant bulrush), *Scirpus maritimus* (salt marsh bulrush), and *Typha angustifolia* (narrow-leaved cattail), which can survive for several months and expand their leaves in an anaerobic atmosphere. The embryo and coleoptile of rice and *Echinochloa crus-galli* var. *oryzicola* (rice grass) can also survive weeks of anoxia.

In nature, the rhizomes overwinter in anaerobic mud at the edges of lakes. In spring, once the leaves have expanded above the mud or water surface, O_2 diffuses down through the aerenchyma into the rhizome. Metabolism then switches from an anaerobic (fermentative) to an aerobic mode, and roots begin to grow using the available oxygen. Likewise, during germination of paddy (wetland) rice and of rice grass, the coleoptile breaks the water surface and becomes a diffusion pathway (a "snorkel") for O_2 to the rest of the plant. Even though rice is a wetland species, its roots are as intolerant of anoxia as are those of maize.

The precise combination of biochemical characteristics that allow some cells to tolerate anoxia for long periods is not fully understood. Root tips of maize and other cereals show a modest degree of acclimation if they first are made hypoxic, whereupon they can survive up to 4 days of anoxia. This acclimation is associated with expression of the genes that encode many of the anaerobic stress proteins (see the next section). After acclimation, the ability to carry out ethanolic fermentation under anoxia (thereby producing ATP to keep some metabolism going) is improved, and it is accompanied by an ability to transport lactate out of the cytosol to the external medium, thus minimizing cytoplasmic acidosis (Drew 1997).

The ability of organs of wetland plants to tolerate chronic anoxia may depend on strategies similar to those just described, but they are clearly employed to greater effect: Critical features appear to be control of cytoplasmic pH, continued generation of ATP by glycolysis and fermentation, and sufficient storage of fuel for anaerobic respiration over extended periods. It has been suggested that synthesis of alanine, succinate, and γ-aminobutyric acid under anoxia consumes protons and minimizes cytoplasmic acidosis. Evidence to this effect has been found in anoxia-tolerant shoots of rice and rice grass, but not in anoxia-sensitive shoots of wheat and barley, or in roots of any species tested so far.

Organs of species that alternate between anaerobic and aerobic metabolism need to deal with the consequences of the entry of O_2 following anoxia. Highly reactive oxygen species are generated during aerobic metabolism (see the section on air pollution), and are normally detoxified by cellular defense mechanisms that involve superoxide dismutase (SOD). This enzyme converts superoxide radicals to hydrogen peroxide, which is then converted to water by peroxidase. In anoxia-tolerant rhizomes of *Iris pseudacorus* (yellow flag), SOD activity increases 13-fold during 28 days of anoxia. This increase is not observed in rhizomes of other *Iris* species that are not anoxia tolerant. In the tolerant species, SOD may be available to cope with the influx of O_2 that occurs when the leaves emerge into the air from water or mud, and so it may assist in resisting **postanoxic stress**.

Acclimation to O_2 Deficit Involves Synthesis of Anaerobic Stress Proteins

When maize roots are made anoxic, protein synthesis ceases except for the continued production of about 20 polypeptides (Sachs and Ho 1986). Most of these **anaerobic stress proteins** have been identified as enzymes of the glycolytic and fermentation pathways. Evidence suggests that intracellular Ca^{2+} is involved in the signal transduction of anoxia. Within minutes of the onset of anoxia, a rise in cytosolic Ca^{2+} concentration acts as a signal for increases in mRNA levels of alcohol dehydrogenase (ADH) and sucrose synthase in maize cells in culture. Chemicals that block a rise in intracellular Ca^{2+} concentration also prevent the expression of the genes for ADH and sucrose synthase from being induced by anoxia, and they greatly enhance the sensitivity of maize seedlings to anoxia (Sachs et al. 1996).

Since ethanolic fermentation is the major metabolic pathway by which ATP is synthesized in anoxic cells of higher plants, and since higher rates of fermentation correlate with improved energy status of cells, these anaerobic stress proteins have been assumed to play a significant role during anoxia. However, among all the enzymes of glycolysis and fermentation in maize roots, only hexokinase and pyruvate decarboxylase have activities low enough to limit the overall rate. The other enzymes, including ADH and sucrose synthase, are present in large excess, even before anoxic induction of their synthesis to higher concentrations, so their role in survival under anoxia is unclear. Further research is needed to resolve these mechanisms and to explain how intracellular Ca^{2+} concentration appears important both in the early survival of cells under anoxia, and in the induction of cell death and aerenchyma formation during prolonged hypoxia.

Air Pollution

The burning of hydrocarbons in motor vehicle engines gives rise to CO_2, CO, SO_2 (sulfur dioxide), NO_x (NO [nitrogen monoxide]) and NO_2^- in varying proportions), and C_2H_4 (ethylene), as well as a variety of other hydrocarbons. Additional SO_2 originates from domestic and industrial burning of fossil fuels. Industrial plants, such as chemical works and metal-smelting plants, release SO_2, H_2S, NO_2, and HF (hydrogen fluoride) into the atmosphere. Tall chimney stacks may be used to carry gases and particles to a high altitude and thus avoid local pollution, but the pollutants return to Earth, sometimes hundreds of kilometers from the original source.

Photochemical smog is the product of chemical reactions driven by sunlight and involving NO_x of urban and industrial origin and volatile organic compounds from either vegetation (*biogenic* hydrocarbons) or human activities (*anthropogenic* hydrocarbons). Ozone (O_3) and peroxyacetylnitrate (PAN) produced in these complex reactions can become injurious to plants and other life forms, depending on concentration and duration of exposure. Hydrogen peroxide, another potentially injurious molecule, can form by the reaction between O_3 and naturally released volatiles (terpenes) from forest trees (see Chapter 9).

The concentrations of polluting gases, or their solutions, to which plants are exposed are thus highly variable, depending on location, wind direction, rainfall, and sunlight. In urban areas, concentrations of SO_2 and NO_x in air are typically 0.02 to 0.5 mL L^{-1}, the upper value being within the range that is inhibitory to plant growth. Relatively long-term experiments at appropriate concentrations of pollutants are necessary to establish the real impact of air pollution on vegetation. The reaction of plants to high concentrations of pollutants in short-term experiments may overwhelm the plant's defense mechanisms and provoke abnormal symptoms.

The responses of plants to polluting gases can also be affected by other ambient conditions, such as light, humidity, temperature, and the supply of water and minerals. Experiments aimed at determining the impact of chronic exposure to low concentrations of gases should allow plants to grow under near-natural conditions. One method is to grow the plants in open-top chambers into which gases are carefully metered, or where plants receiving ambient, polluted air are compared with controls receiving air that has been scrubbed of pollutants.

Polluting Gases and Dust Inhibit Stomatal Movements, Photosynthesis, and Growth

Dust pollution is of localized importance near roads, quarries, cement works, and other industrial areas. Apart from screening out sunlight, dust on leaves blocks stomata and lowers their conductance to CO_2, simultaneously interfering with photosystem II. Polluting gases such as SO_2 and NO_x enter leaves through stomata, following the same diffusion pathway as CO_2. NO_x dissolves in cells and gives rise to nitrite ions (NO_2^-, which are toxic at high concentrations) and nitrate ions (NO_3^-) that enter into nitrogen metabolism as if they had been absorbed through the roots. In some cases, exposure to pollutant gases, particularly SO_2, causes stomatal closure, which protects the leaf against further entry of the pollutant but also curtails photosynthesis. In the cells, SO_2 dissolves to give bisulfite and sulfite ions; sulfite is toxic, but at low concentrations it is metabolized by chloroplasts to sulfate, which is not toxic. At sufficiently low concentrations, bisulfite and sulfite are effectively detoxified by plants, and SO_2 air pollution then provides a sulfur source for the plant.

In urban areas these polluting gases may be present in such high concentrations that they cannot be detoxified rapidly enough to avoid injury. Ozone is presently considered to be the most damaging phytotoxic air pollutant in North America (Heagle 1989; Krupa et al. 1995). It has been estimated that wherever the mean daily O_3 concentration reaches 40, 50, or 60 ppb (parts per billion, or per 10^9), the combined yields of soybean, maize, winter wheat, and cotton would be decreased by 5, 10, and 16%, respectively.

Ozone is highly reactive: It binds to plasma membranes and it alters metabolism. As a result, stomatal apertures are poorly regulated, chloroplast thylakoid membranes are damaged, rubisco is degraded, and photosynthesis is inhibited. Ozone reacts with O_2 and produces reactive oxygen species, including hydrogen peroxide (H_2O_2), superoxide (O_2^-), singlet oxygen ($^1O_2^*$), and the hydroxyl radical (•OH). These denature proteins, damage nucleic acids and thereby give rise to mutations, and cause lipid peroxidation, which breaks down lipids in membranes. Reactive oxygen species form also in the absence of O_3, particularly in electron transport in the mitochondria and chloroplasts, when electrons can be donated to O_2 (see Chapters 7 and 11).

Cells are protected, at least in part, from reactive oxygen species by enzymatic and nonenzymatic defense mechanisms (Bowler et al. 1992 ; Elstner and Osswald 1994). Defense against reactive oxygen species is provided by the scavenging properties of molecules such as ascorbic acid, α-tocopherol, phenolic compounds, and glutathione. Superoxide dismutases (SODs) catalyze the reduction of superoxide to hydrogen peroxide. Hydrogen peroxide is then converted to H_2O by the action of catalases and peroxidases. Of particular importance is the ascorbate-specific peroxidase localized in the chloroplast. Acting in concert, ascorbate peroxidase, dehydroascorbate reductase, and glutathione reductase remove H_2O_2 in a series of reactions called the **Halliwell–Asada pathway**, named after its discoverers (Figure 25.16). Glutathione is a sulfur-containing tripeptide that, in its reduced form, reacts rapidly with dehydroascorbate and becomes oxidized in the process. Glutathione reductase catalyzes the regeneration of reduced glutathione (GSH) from its oxidized form (GSSG) in the following reaction:

$$\text{GSSG} + \text{NADPH} + \text{H}^+ \rightarrow 2\ \text{GSH} + \text{NADP}^+$$

Exposure of plants to reactive oxygen species stimulates the transcription and translation of genes that encode enzymes involved in protection mechanisms. In *Arabidopsis*, exposure for 6 hours per day to low levels of O_3 induces the expression of several genes that encode enzymes associated with protection from reactive oxygen species, including SOD, glutathione S-transferase (which catalyzes detoxification reactions involving glutathione), and phenylalanine ammonia lyase (an important enzyme at the start of the phenylpropanoid pathway leading to the synthesis of flavonoids and other phenolics). In transgenic tobacco transformed with a gene from *Escherichia coli* to give additional glutathione reductase activity in the chloroplast, short-term exposure to high levels of SO_2 is much less damaging than for wild-type tobacco (Aono et al. 1993).

Environmental extremes may either accelerate the production of reactive oxygen species or impair the normal defense mechanisms that protect cells from reactive oxygen species. In water-deficient leaves, for example, greater oxygen photoreduction by photosystems I and II increases superoxide production, and the pool of glutathione, as well as the activity of glutathione reductase, increase, presumably as part of the cell defense mechanism. In contrast, levels of ascorbate, another antioxidant, generally decline with mild water stress. Transgenic plants overexpressing mitochondrion superoxide dismutase (Mn-SOD), the isozyme localized in the mitochondrial matrix, show less water deficit damage and, significantly, improved survival and yield under field conditions (McKersie et al. 1996). In other experiments, transgenic alfalfa overexpressing Mn-SOD , was found to be more tolerant of freezing. Conversely, winter rye, wheat, and barley acclimated at 2°C for several weeks were found to have developed resistance to the herbicides paraquat and acifluorfen, which generate reactive oxygen species. Such investigations support the hypothesis that tolerance of oxidative stress is an important factor in tolerance to a wide range of environmental extremes.

Many deleterious changes in metabolism caused by air pollution precede external symptoms of injury, which appear only at much higher concentrations. For example, when plants are exposed to air containing NO_x, lesions on leaves appear at an NO_x concentration of 5 mL L^{-1}, but photosynthesis starts to be inhibited at a concentration of only 0.1 mL L^{-1}. These low, threshold concentrations refer to the effects of a single pollutant. However, two or more pollutants acting together can have a synergistic effect, producing damage at lower concentrations than if they were acting separately. In addition, vegetation weakened by air pollution can become more susceptible to invasion by pathogens and pests.

Polluting Gases, Dissolved in Rainwater, Fall as "Acid Rain"

Unpolluted rain is slightly acidic, with a pH close to 5.6, because the CO_2 dissolved in it produces the weak acid H_2CO_3. Dissolution of NO_x and SO_2 in water droplets in the atmosphere causes the pH of rain to decrease to 3 to 4, and in southern California polluted droplets in fog can be as acidic as pH 1.7. Dilute acidic solution can remove mineral nutrients from leaves, depending on the age of the leaf and the integrity of the cuticle and surface waxes. The total annual contributions to the soil of acid from acid rain (**wet deposition**) and from particulate matter falling on the soil plus direct absorption from the atmosphere (**dry deposition**) may reach 1.0 to 3.0 kg H^+ per hectare in parts of Europe and the northeastern United States (Schwartz 1989). In soils that lack free calcium carbonate, and therefore are not strongly buffered, such additions of acid can be harmful to plants. Fur-

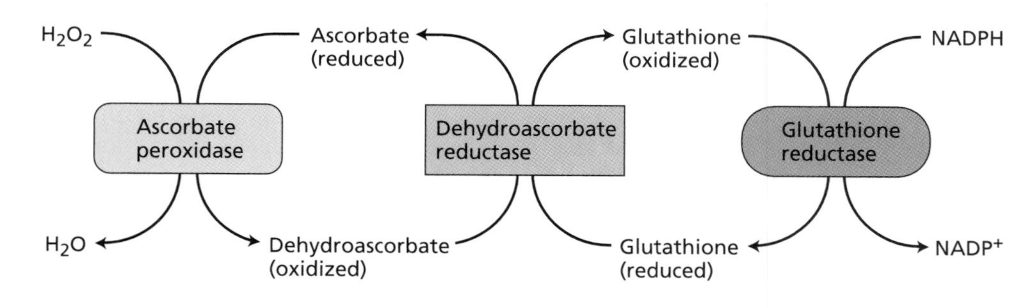

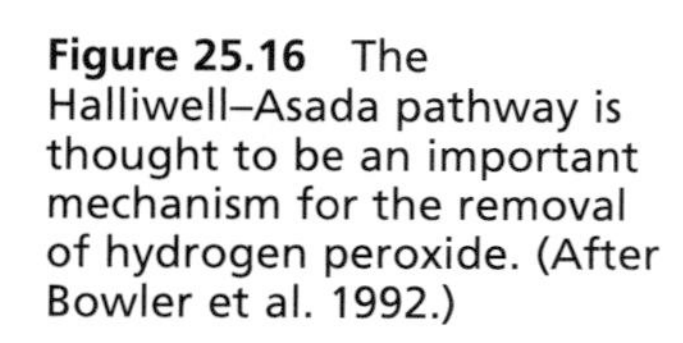
Figure 25.16 The Halliwell–Asada pathway is thought to be an important mechanism for the removal of hydrogen peroxide. (After Bowler et al. 1992.)

thermore, the added acid can result in the release of aluminum ions from soil minerals, causing aluminum toxicity (see Box 5.1). Air pollution is considered to be a major factor in the decline of forests in heavily polluted areas of Europe and North America. There are indications that fast-growing pioneer species are better able to tolerate an acidifying atmosphere than are climax forest trees, possibly because they have a greater potential for assimilation of dissolved NO_x, and more effective acid buffering of the leaf tissue cell sap.

Stress-Induced Gene Expression

Early attempts to identify stress-induced genes in plants began when investigators noticed the appearance of new proteins after exposure to different stresses. As new methods of protein detection and identification became available, many stress-induced proteins were reported. Subsequent development of modern methods for gene isolation led to the discovery of numerous new stress-induced genes. A period of rapid identification and study of these genes followed in the late 1980s and into the 1990s. Most of the genes that were identified by various screening procedures are yet to be associated with specific functions. However, from our understanding of the metabolic and physiological changes that take place during exposure to stress, scientists have also been able to identify genes that appear to play critical roles in stress adaptation. As an example, we next examine the genes induced by osmotic stress. In the following sections we discuss the involvement of gene expression and protein synthesis in stress and in stress tolerance.

A Variety of Genes Are Induced by Osmotic Stress

As noted earlier, the accumulation of compatible solutes in response to osmotic stress requires the activation of the metabolic pathways that biosynthesize these solutes. Several genes that are turned on by salt stress encode enzymes in these pathways, such as Δ^1-pyrroline-5-carboxylate synthase, a key enzyme in the proline biosynthetic pathway, and betaine aldehyde dehydrogenase, an enzyme involved in betaine accumulation.

Several other genes that encode well-known enzymes are induced by osmotic stress. The expression of glyceraldehyde-3-phosphate dehydrogenase increases during osmotic stress, perhaps to allow an increase of carbon flow into organic solutes for osmotic adjustment. In the CAM plant *Mesembryanthemum crystallinum*, several genes that encode enzymes in the CAM pathway, such as PEP carboxylase, pyruvate–orthophosphate dikinase, and NADP malic enzyme, are induced by osmotic or salt stress (Bohnert et al. 1995). Other genes regulated by osmotic stress encode ATPases (Niu et al. 1995) and the water channel proteins, aquaporins (see Chapter 3) (Maggio and Joly 1995).

Several protease genes are also induced by stress, and it has been suggested that these enzymes degrade (remove and recycle) other proteins that are denatured by stress episodes. The protein ubiquitin tags proteins that are targeted for proteolytic degradation. Synthesis of the mRNA for ubiquitin increases in *Arabidopsis* upon desiccation stress. In addition, some heat shock proteins are osmotically induced and may protect or renature proteins inactivated by desiccation.

The sensitivity of cell expansion to osmotic stress (see Figure 25.1), has stimulated studies of various genes that encode proteins involved in the structural composition and integrity of cell walls. Genes coding for enzymes such as S-adenosylmethionine synthase and peroxidases, which may be involved in lignin biosynthesis, have been shown to be controlled by stress.

A large group of genes that are regulated by osmotic stress were discovered by examination of naturally desiccating embryos during seed maturation. As noted earlier, these genes are called *LEA* (*l*ate *e*mbryogenesis *a*bundant) genes, and they are suspected to play a role in cellular membrane protection. The proteins encoded by these genes are strongly hydrophilic, and many are structured as random coils that allow strong binding of water. Their protective function may derive from the ability of such a structure to retain water and prevent crystallization of important cellular proteins and other molecules during desiccation.

Another group of stress-induced genes with unknown function is the *COR* (*co*ld *r*egulated) gene family. *COR* genes are induced by both desiccation and cold, and one of them (*COR15a*) has been shown to protect chloroplasts from freezing injury (Artus et al. 1996).

Specific Promoter Elements Determine Stress-Induced Gene Expression

The multigenic nature of stress tolerance and the simultaneous activation of many genes that are involved in stress tolerance suggest that stress tolerance genes have common regulatory features.

As discussed in Chapter 14, gene transcription is controlled through the interaction of regulatory proteins (transcription factors) with specific regulatory sequences in the promoters of the genes they regulate. Different genes that are induced by the same signal (desiccation or salinity, for example) are controlled by a signaling pathway leading to the activation of these specific transcription factors. The study of the promoters of several stress-induced genes has led to the identification of specific regulatory sequences for genes involved in different stress signals. Most notable among these is the ***RD29* gene**, which contains DNA sequences that make possible activation by osmotic stress, by cold, and by ABA (Yamaguchi-Shinozaki and Shinozaki 1994; Stockinger et al. 1997). A six-nucleotide segment that allows induction by ABA is called an ABRE (ABA-responsive element); a nine-nucleotide segment allowing osmotic induction is called a **dehydration-responsive element (DRE)**. Additional sequences called **coupling elements** may also be required (Shen and Ho 1995).

Ca^{2+} and Protein Kinases Are Involved in Stress Responses

A general scheme for a signal transduction pathway that mediates stress tolerance is shown in Figure 25.17. Little is known about these sensory transduction pathways in higher plants, but recent progress on the understanding of stress tolerance in other organisms should be helpful. These newly discovered environmental signaling systems appear highly conserved, and often involve MAPK (mitogen-activated protein kinase) or ERK (extracellular signal–regulated kinase) proteins (Shinozaki and Yamaguchi-Shinozaki 1997). Genes that encode many protein kinases have been isolated from plants, but their function in stress-induced gene expression remains to be shown.

As indicated earlier, a specific sequence, the ABRE, controls ABA inducibility of many of the genes that respond to osmotic stress. Since both osmotic stress and ABA can increase cytosolic Ca^{2+} levels (see Chapter 23), Ca^{2+}-dependent protein kinases (CDPKs) could be involved in the transduction of stress signals to activate transcription factors that recognize specific elements (e.g., ABRE). It was recently found that activation of specific CDPKs leads to the induced transcription of osmotic stress response genes in plants (Sheen 1996). Calcium also appears to participate in stress signaling through the protein phosphatase **calcineurin**. Calcineurin plays a key role in the induction of genes responsible for osmotic stress tolerance in yeast, and a similar signal system may operate in plants (Mendoza et al. 1996).

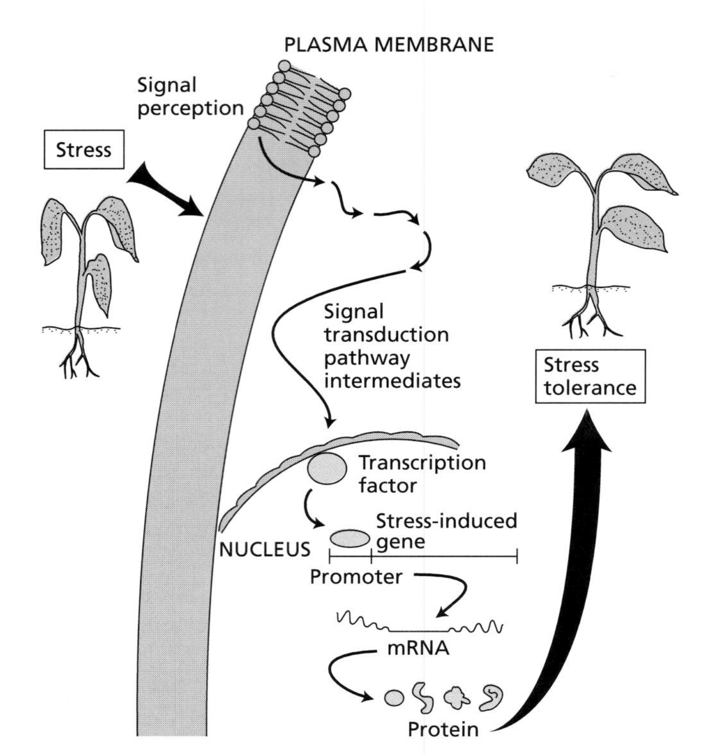

Figure 25.17 Activation of stress response genes by signal perception and transduction. Abiotic stresses such as cold, desiccation, salinity, and heat are believed to activate receptors, probably located in the plasma membrane. These receptors transduce this perception via pathway intermediates (e.g., protein kinases or phosphatases) to transcription factor proteins. Transcription factors in the nucleus activate genes that are important for recovery from stress by binding to specific promoter regions of these responsive genes.

Transgenic Plants Are Valuable Tools to Study Stress Tolerance

The transfer of genes suspected to mediate stress tolerance to a test plant where they are overexpressed or altered in some way has given scientists an important new tool with which to study genes involved in stress tolerance. Results of the first study in which transgenic plants were reported to show altered stress tolerance after transfer of a foreign gene appeared in 1993 (Tarczynski et al. 1993). The transferred gene encoded the enzyme mannitol dehydrogenase, and its expression led to the accumulation of mannitol in the transgenic plants. Transfer of other genes generated transgenic plants that accumulated other solutes, such as trehalose (Holmström et al. 1996), proline (Kishor et al. 1995), and fructan (Pilon-Smits et al. 1995). Interestingly, the amount of accumulated solutes was insufficient to osmotically balance the applied osmotic stress, yet all of the transgenic plants exhibited some increased stress tolerance. These surprising results are not fully understood, but the newly acquired solutes might protect the transgenic plant against free radicals generated in response to stress, rather than function as osmotic-balancing solutes.

Summary

Stress is usually defined as an external factor that exerts a disadvantageous influence on the plant. Under both natural and agricultural conditions, plants are exposed to unfavorable environments that result in some degree of stress. Water deficit, chilling and freezing, heat stress and heat shock, salinity, oxygen deficiency, and air pollution are major stress factors restricting plant growth so that biomass or agronomic yields at the end of the season express only a fraction of the plant's genetic potential.

The capacity of plants to cope with unfavorable environments is known as stress resistance. Plant adaptations that confer stress resistance, such as CAM metabolism, are genetically determined. Acclimation improves resistance as a result of prior exposure of a plant to stress.

Drought resistance mechanisms vary with climate and soil conditions. Indeterminate growth patterns such as that of sorghum and soybean allow these species to take advantage of late-occurring rains; plants with a determinate growth pattern, such as that of corn, lack that form of resistance to water stress. Inhibition of leaf expansion is one of the earliest responses to water stress, occurring when decreases in turgor ensuing from water deficit reduce or eliminate the driving force for cell and leaf expansion. Additional stress resistance mechanisms in response to water stress include leaf abscission, root extension into deeper, wetter soil, and stomatal closure.

Chilling and freezing stress ensue from low temperatures. Chilling injury occurs at temperatures that are too low for normal growth but are above freezing, and it is typical of species of tropical or subtropical origin exposed to temperate climates. Chilling injuries include slow growth, leaf lesions, and wilting. The primary cause of most chilling injuries is the loss of membrane properties ensuing from changes in membrane fluidity. Membrane lipids of chill-resistant plants often have a greater proportion of unsaturated fatty acids than those of chill-sensitive plants. Freezing injury is associated primarily with damage caused by ice crystals formed within cells and organs. Freezing-resistant species have mechanisms that limit the growth of ice crystals to extracellular spaces. Mechanisms that confer the resistance to freezing that is typical of woody plants include dehydration and supercooling.

Heat stress and heat shock are caused by high temperatures. Some CAM species can tolerate temperatures of 60 to 65°C, but most leaves are damaged above 45°C. The temperature of actively transpiring leaves is usually lower than air temperature, but water deficit curtails transpiration and causes overheating and heat stress. Heat stress inhibits photosynthesis and impairs membrane function and protein stability. Adaptations that confer heat resistance include responses that decrease light absorption by the leaves, such as leaf rolling, and a decrease in leaf size that minimizes boundary layer resistance and increases conductive heat loss. Heat shock proteins synthesized at high temperatures act as molecular chaperones that promote stabilization and correct folding of cell proteins.

Salinity stress results from salt accumulation in the soil. Some halophyte species are highly tolerant to salt, but salinity depresses growth and photosynthesis in sensitive species. Salt injury ensues from a decrease in the water potential of the soil that makes soil water less available and from toxicity of specific ions accumulated at injurious concentrations. Plants avoid salt injury by exclusion of excess ions from leaves or by compartmentation of ions in vacuoles.

Oxygen deficiency is typical of flooded or waterlogged soils. Oxygen deficiency depresses growth and survival of many species. On the other hand, plants of marshes and swamps, and crops such as rice, are well adapted to resist oxygen deficiency in the root environment. Most tissues of higher plants cannot survive anaerobically, while some tissues, such as the embryo and coleoptiles from rice, can survive for weeks under anoxic conditions.

Both abscisic acid and ethylene are involved in plant acclimation to water stress, to chilling and heat stress, and to oxygen deficiency. Some of the changes in gene expression during acclimation appear to be signaled by hormones. Some changes in the synthesis of specific mRNAs and stress-related proteins are not unique to a particular stress, and are induced by different factors, such as heat shock, salinity, water deficit, and exposure to ABA in the absence of stress. Some transcriptional and translational changes are restricted to a specific stress signal. Elucidation of the function of stress-activated genes should facilitate breeding for and genetic engineering of stress-resistant plants.

General Reading

*Björkman, O., Badger, M. R., and Armond, P. A. (1980) Response and adaptation of photosynthesis to high temperatures. In *Adaptation of Plants to Water and High Temperature Stress*, N. C. Turner and P. J. Kramer, eds., Wiley, New York, pp. 233–249.

*Boyer, J. S. (1982) Plant productivity and environment. *Science* 218: 443–448.

Cherry, J. H., ed. (1989) *Environmental Stress in Plants: Biochemical and Physiological Mechanisms Associated with Environmental Stress Tolerance in Plants* (NATO ASI Series G, vol. 19). Springer, Berlin.

*Drew, M. C., and Lynch, J. M. (1980) Soil anaerobiosis, microorganisms and root function. *Annu. Rev. Phytopathol.* 18: 37–66.

Fitter, A. H., and Hay, R. K. M. (1987) *Environmental Physiology of Plants*, 2nd ed. Academic Press, London.

*Greenway, H., and Munns, R. (1980) Mechanisms of salt tolerance in nonhalophytes. *Annu. Rev. Plant Physiol.* 31: 149–190.

*Jackson, M. B., and Drew, M. C. (1984) Effects of flooding on growth and metabolism of herbaceous plants. In *Flooding and Plant Growth*, T. T. Kozlowski, ed., Academic Press, New York, pp. 47–128.

*Levitt, J. (1980) *Responses of Plants to Environmental Stresses*, Vol. 1, 2nd ed. Academic Press, New York.

*Sachs, M. M., and Ho, D. T. H. (1986) Alternation gene expression during environmental stress in plants. *Annu. Rev. Plant Physiol.* 37: 363–376.

Schulze, E. D. (1986) Carbon dioxide and water vapor exchange in response to drought in the atmosphere and in the soil. *Annu. Rev. Plant Physiol.* 37: 247–274.

*Schwartz, S. E. (1989) Acid deposition: Unraveling a regional phenomenon. *Science* 243: 753–763.

Steward, F. C., ed. (1986) *Plant Physiology: A Treatise*, Vol. IX: *Water and Solutes in Plants.* Academic Press, New York.

*Turner, N. C., and Jones, M. M. (1980) Turgor maintenance by osmotic adjustment: A review and evaluation. In *Adaptation of Plants to Water and High Temperature Stress*, N. C. Turner and P. J. Kramer, eds., Wiley, New York, pp. 87–103.

Turner, N. C., and Kramer, P. J., eds. (1980) *Adaptation of Plants to Water and High Temperature Stress*. Wiley, New York.

*Wyn Jones, R. G., and Gorham, J. (1983) Osmoregulation. In *Encyclopedia of Plant Physiology*, New Series, Vol. 12C, O. L. Lange, P. S. Nobel, C. B. Osmond, and H. Ziegler, eds., Springer, Berlin, pp. 35–58.

* Indicates a reference that is general reading in the field and is also cited in this chapter.

Chapter References

Aono, M., Kubo, A., Saji, H., Tanaka, K., and Kondo, N. (1993) Enhanced tolerance to photooxidative stress of transgenic *Nicotiana tabacum* with high chloroplastic glutathione reductase activity. *Plant Cell Physiol.* 34: 129–135.

Artus, N. N., Uemura, M., Steponkus, P. L., Gilmour, S. J., Lin, C., and Thomashow, M. F. (1996) Constitutive expression of the cold-regulated *Arabidopsis thaliana* COR15a gene affects both chloroplast and protoplast freezing tolerance. *Proc. Natl. Acad. Sci.* USA 93: 13404–13409.

Becwar, M. R., Rajashekar, C., Bristow, K. J. H., and Burke, M. J. (1981) Deep undercooling of tissue water and winter hardiness limitations in timberline flora. *Plant Physiol.* 68: 111–114.

Blackman, P. G., and Davies, W. J. (1985) Root to shoot communication in maize plants of the effects of soil drying. *J. Exp. Bot.* 36: 39–48.

Blizzard, W. E., and Boyer, J. S. (1980) Comparative resistance of the soil and the plant to water transport. *Plant Physiol.* 66: 809–814.

Bohnert, H. J., Nelson, D. E., and Jensen, R. G. (1995) Adaptations to environmental stresses. *Plant Cell* 7: 1099–1111.

Bohnert, H. J., Ostrem, J. A., and Schmitt, J. M. (1989) Changes in gene expression elicited by salt stress in *Mesembryanthemum crystallinum*. In *Environmental Stress in Plants*, J. H. Cherry, ed., Springer, Berlin, pp. 159–171.

Bonk, M., Tadros, M., Vandekerckhove, J., Al-Babili, S., and Beyer, P. (1996) Purification and characterization of chaperonin 60 and heat-shock protein 70 from chromoplasts of *Narcissus pseudonarcissus*. Involvement of heat-shock protein 70 in a soluble protein complex containing phytoene desaturase. *Plant Physiol.* 111: 931–939.

Bowler, C., Montagu, M. V., and Inze, D. (1992) Superoxide dismutase and stress tolerance. *Annu. Rev. Plant Physiol. Plant Mol. Biol.* 48: 223–250.

Boyer, J. S. (1970) Leaf enlargement and metabolic rates in corn, soybean, and sunflower at various leaf water potentials. *Plant Physiol.* 46: 233–235.

Bray, E. A. (1988) Drought- and ABA-induced changes in polypeptide and mRNA accumulation in tomato leaves. *Plant Physiol.* 88: 1210–1214.

Brown, M. S., Pereira, E. S. B., and Finkle, B. J. (1974) Freezing of nonwoody plant tissues. 2. Cell damage and the fine structure of freezing curves. *Plant Physiol.* 53: 709–711.

Burke, M. J., and Stushnoff, C. (1979) Frost hardiness: A discussion of possible molecular causes of injury with particular reference to deep supercooling of water. In *Stress Physiology in Crop Plants*, H. Mussell and R. C. Staples, eds., Wiley, New York, pp. 197–225.

Cornish, K., and Zeevaart, J. A. D. (1985) Movement of abscisic acid into the apoplast in response to water stress in *Xanthium strumarium* L. *Plant Physiol.* 78: 623–626.

Cramer, G. R., Lauchli, A., and Polito, V. S. (1985) Displacement of Ca^{2+} by Na^+ from the plasmalemma of root cells. A primary response to salt stress? *Plant Physiol.* 79: 207–211.

Drew, M. C. (1997) Oxygen deficiency and root metabolism: Injury and acclimation under hypoxia and anoxia. *Annu. Rev. Plant Physiol. Plant Mol. Biol.* 48: 223–250.

Elstner, E. F., and Osswald, W. (1994) Mechanisms of oxygen activation during plant stress. *Proc. R. Soc. Edinburgh* 102B: 131–154.

Gusta, L. V., Wilen, R. W., and Fu, P. (1996) Low-temperature stress tolerance: The role of abscisic acid, sugars, and heat-stable proteins. *Hort. Sci.* 31: 39–46.

Guy, C. L. (1990) Cold acclimation and freezing stress tolerance: Role of protein metabolism. *Annu. Rev. Plant Physiol. Mol. Biol.* 41: 187–223.

Hanscom, Z., III, and Ting, I. P. (1978) Responses of succulents to plant water stress. *Plant Physiol.* 61: 327–330.

Hartung, W., Radin, J. W., and Hendrix, D. L. (1988) Abscisic acid movement into the apoplastic solution of water-stressed cotton leaves. Role of apoplastic pH. *Plant Physiol.* 86: 908–913.

He, C. J., Morgan, P. W., and Drew, M. C. (1996) Transduction of an ethylene signal is required for cell death and lysis in the root cortex of maize during aerenchyma formation induced by hypoxia. *Plant Physiol.* 112: 463–472.

Heagle, A. S. (1989) Ozone and crop yield. *Annu. Rev. Phytopathol.* 27: 397–423.

Holmström, K. O., Mäntylä, E., Welin, B., Mandal, A., Palva, E. T., Tunnela, O. E., and Londesborough, J. (1996) Drought tolerance in tobacco. *Nature* 379: 683–684.

Kishor, P. B. K., Hong, Z., Miao, G. H., Hu, C. A. A., and Verma, D. P. S. (1995) Overexpression of Δ^1-pyrroline-5-carboxylate synthetase increases proline production and confers osmotolerance in transgenic plants. *Plant Physiol.* 108: 1387–1394.

Krupa, S. V., Gruenhage, L., Jaeger, H. J., Nosal, M., Manning, W. J., Legge, A. H., and Hanewald, K. (1995) Ambient ozone (O_3) and adverse crop response: A unified view of cause and effect. *Environ. Pollut.* 87: 119–126.

Lee, J. H., Huebel, A., and Schoeffl, F. (1995) Depression of the activity of genetically engineered heat shock factor causes constitutive synthesis of heat shock proteins and increased thermotolerance in transgenic *Arabidopsis*. *Plant J.* 8: 603–612.

Lee, J. H., and Schoeffl, F. (1996) An Hsp70 antisense gene affects the expression of HSP70/HSC70, the regulation of HSF, and the acquisition of thermotolerance in transgenic *Arabidopsis thaliana*. *Mol. Gen. Genet.* 252: 11–19.

Lindow, S. E., Arny, D. C., and Upper, C. D. (1982) Bacterial ice nucleation: A factor in frost injury to plants. *Plant Physiol.* 70: 1084–1089.

Lu, Z. M., Percy, R. G., Qualset, C. O., and Zeiger, E. (1998) Stomatal conductance predicts yields in irrigated Pima cotton and bread wheat grown at high temperatures. *J. Exp. Bot.* 49: 453–460.

Lyons, J. M., Wheaton, T. A., and Pratt, H. K. (1964) Relationship between the physical nature of mitochondrial membranes and chilling sensitivity in plants. *Plant Physiol.* 39: 262–268.

Maas, E. V., and Hoffman, G. J. (1977) Crop salt tolerance—Current assessment. *J. Irrig. Drainage Div. Am. Soc. Civ. Eng.* 103: 115–134.

Maggio, A., and Joly, R. J. (1995) Effects of mercuric chloride on the hydraulic conductivity of tomato root systems: Evidence for a channel-mediated water pathway. *Plant Physiol.* 109: 331–335.

Mantyla, E., Lang, V., and Palva, E. T. (1995) Role of abscisic acid in drought-induced freezing tolerance, cold acclimation, and accumulation of LTI78 and RAB18 proteins in *Arabidopsis thaliana*. *Plant Physiol.* 107: 141–148.

Matthews, M. A., Van Volkenburgh, E., and Boyer, J. S. (1984) Acclimation of leaf growth to low water potentials in sunflower. *Plant Cell Environ.* 7: 199–206.

McCree, K. J., and Richardson, S. G. (1987) Stomatal closure vs. osmotic adjustment: A comparison of stress responses. *Crop Sci.* 27: 539–543.

McKersie, B. D., Bowley, S. R., Harjanto, E., and Leprince, O. (1996) Water-deficit tolerance and field performance of transgenic alfalfa overexpressing superoxide dismutase. *Plant Physiol.* 111: 1177–1181.

Mendoza, I., Quintero F. J., Bressan, R. A., Hasegawa, P. M., and Pardo, J. M. (1996) Activated calcineurin confers high tolerance to ion stress and alters the budding pattern and cell morphology of yeast cells. *J. Biol. Chem.* 271: 23061–23067.

Nishida, I., and Murata, N. (1996) Chilling sensitivity in plants and cyanobacteria: The crucial contribution of membrane lipids. *Annu. Rev. Plant Physiol. Plant Mol. Biol.* 47: 541–568.

Niu, X., Bressan, R. A., Hasegawa, P. M., and Pardo, J. M. (1995) Ion homeostasis in NaCl stress environments. *Plant Physiol.* 109: 735–742.

Palta, J. P., Whitaker, B. D., and Weiss, L. S. (1993) Plasma membrane lipids associated with genetic variability in freezing tolerance and cold acclimation of *Solanum* species. *Plant Physiol.* 103: 793–803.

Patterson, B. D., Paull, R., and Smillie, R. M. (1978) Chilling resistance in *Lycopersicon hirsutum* Humb. & Bonpl., a wild tomato with a wide altitudinal distribution. *Aust. J. Plant Physiol.* 5: 609–617.

Percy, R. C., Lu, Z. M., Radin, J. W., Turcotte, E. L., and Zeiger, E. (1996) Inheritance of stomatal conductance in Pima cotton (*Gossypium barbadense*). *Physiol. Plantarum* 96: 389–394.

Pilon-Smits, E. A. H., Ebskamp, M. J. M., Paul, M. J., Jeuken, M. J. W., Weisbeek, P. J., and Smeekens, S. C. M. (1995) Improved performance of transgenic fructan-accumulating tobacco under drought stress. *Plant Physiol.* 107: 125–130.

Radin, J. W., and Hendrix, D. L. (1988) The apoplastic pool of abscisic acid in cotton leaves in relation to stomatal closure. *Planta* 174: 180–186.

Radin, J. W., Lu, Z. M., Percy, R. G., and Zeiger, E. (1994) Genetic variation for stomatal conductance in Pima cotton and its relation to improvements of heat adaptation. *Proc. Natl. Acad. Sci. USA* 91: 7217–7221.

Raison, J. K., Pike, C. S., and Berry, J. A. (1982) Growth temperature-induced alterations in the thermotropic properties of *Nerium oleander* membrane lipids. *Plant Physiol.* 70: 215–218.

Rao, I. M., Sharp, R. E., and Boyer, J. S. (1987) Leaf magnesium alters photosynthetic response to low water potentials in sunflower. *Plant Physiol.* 84: 1214–1219.

Roberts, J. K. M., Callis, J., Jardetzky, O., Walbot, V., and Freeling, M. (1984) Cytoplasmic acidosis as a determinant of flooding intolerance in plants. *Proc. Natl. Acad. Sci. USA* 81: 6029–6033.

Roberts, J. K. M., Hooks, M. A., Miaullis, A. P., Edwards, S., and Webster, C. (1992) Contribution of malate and amino acid metabolism to cytoplasmic pH regulation in hypoxic maize root tips studied using nuclear magnetic resonance spectroscopy. *Plant Physiol.* 98: 480–487.

Sachs, M. M., Subbaiah, C. G., and Saab, I. N. (1996) Anaerobic gene expression and flooding tolerance in maize. *J. Exp. Bot.* 47: 1–15.

Sellers, P. J., Dickinson, R. E., Randall, D. A., Betts, A. K., Hall, F. G., Berry, J. A., Collatz, G. J., Denning, A. S., Mooney, H. A., and Nobre, C. A. (1997) Modeling the exchanges of energy, water, and carbon between continents and the atmosphere. *Science* 275: 502–509.

Sheen, J. (1996) Ca^{2+}-dependent protein kinases and stress signal transduction in plants. *Science* 274: 1900–1902.

Shen, Q., and Ho, D. T. H. (1995) Functional dissection of an abscisic acid (ABA)-inducible gene reveals two independent ABA-responsive complexes each containing a G-box and a novel *cis*-acting element. *Plant Cell* 7: 295–307

Shinozaki, K., and Yamaguchi-Shinozaki, K. (1997) Gene expression and signal transduction in water-stress response. *Plant Physiol.* 115: 327–334.

Stockinger, E. J., Gilmour, S. J., and Thomashow, M. F. (1997) *Arabidopsis thaliana CBF1* encodes an AP2 domain-containing transcriptional activator that binds to the C-repeat-DRE, a cis-acting DNA regulatory element that stimulates transcription in response to low temperature and water deficit. *Proc. Natl. Acad. Sci. USA* 94: 1035–1040.

Sung, F. J. M., and Krieg, D. R. (1979) Relative sensitivity of photosynthetic assimilation and translocation of ^{14}carbon to water stress. *Plant Physiol.* 64: 852–856.

Tarczynski, M. C., Jensen, R. G., and Bohnert, H. J. (1993) Stress protection of transgenic tobacco by production of the osmolyte mannitol. *Science* 259: 508–510.

U.S. Department of Agriculture. (1989) *Agricultural Statistics.* U.S. Government Printing Office, Washington, DC.

Vierling, E. (1991) The roles of heat shock proteins in plants. *Annu. Rev. Plant Physiol. Plant Mol. Biol.* 42: 579–620.

Weiser, C. J. (1970) Cold resistance and injury in woody plants. *Science* 169: 1269–1278.

Williams, J. P., Khan, M. U., Mitchell, K., and Johnson, G. (1988) The effect of temperature on the level and biosynthesis of unsaturated fatty acids in diacylglycerols of *Brassica napus* leaves. *Plant Physiol.* 87: 904–910.

Wisniewski, M., and Arora, R. (1993) Adaptation and response of fruit trees to freezing temperatures. In *Cytology, Histology and Histochemistry of Fruit Tree Diseases*, A. Biggs, ed., CRC Press, Boca Raton, FL, pp. 299–320.

Wisniewski, M., Lindow, S. E., and Ashworth, E. N. (1997) Observations of ice nucleation and propagation in plants using infrared video thermography. *Plant Physiol.* 113: 327–334.

Wu, J., Lightner, J., Warwick, N., and Browse, J. (1997) Low-temperature damage and subsequent recovery of *fab1* mutant *Arabidopsis* exposed to 2°C. *Plant Physiol.* 113: 347–356.

Yamaguchi-Shinozaki, K., and Shinozaki, K. (1994) A novel *cis*-acting element in an *Arabidopsis* gene is involved in responsiveness to drought, low temperature, or high-salt stress. *Plant Cell* 6: 251–264.

Zhang, J., Schurr, U., and Davies, W. J. (1987) Control of stomatal behavior by abscisic acid which apparently originates in the roots. *J. Exp. Bot.* 38: 1174–1181.

Zhang, J., and Zhang, X. (1994) Can early wilting of old leaves account for much of the ABA accumulation in flooded pea plants? *J. Exp. Bot.* 45: 1335–1342.

Author Index

Subject Index

The letters **b**, **f**, or **t** following a page number indicate that the entry is derived from a **box**, a **figure**, or a **table**, respectively.

B

C

F

G

I

N

O

P

Q

R

S

T

About the Book

Editor: Andrew D. Sinauer

Developmental Editor: James Funston

Project Editor: Nan Sinauer

Production Manager: Christopher Small

Electronic Book Production: Janice Holabird, Jefferson Johnson

Illustration Program: Dartmouth Publishing, Inc.

Copy Editor: Stephanie Hiebert

Indexer: Sharon Hughes

Photo Researcher: Suzette Stephens

Book Design: Susan Brown Schmidler

Cover Design: Jean Hammond

Book Manufacturer: Courier Companies, Inc.